UNIT
PROCESSES
OF
EXTRACTIVE
METALLURGY

UNIT PROCESSES OF EXTRACTIVE METALLURGY

ROBERT D. PEHLKE
The University of Michigan
Ann Arbor, Michigan

AMERICAN ELSEVIER PUBLISHING COMPANY INC.

New York London Amsterdam

AMERICAN ELSEVIER PUBLISHING COMPANY, INC.
52 Vanderbilt Avenue, New York, N.Y. 10017

ELSEVIER PUBLISHING COMPANY
335 Jan Van Galenstraat, P.O. Box 211
Amsterdam, The Netherlands

© American Elsevier Publishing Co., Inc., 1973

Library of Congress Cataloging in Publication Data

Pehlke, Robert D
    Unit processes of extractive metallurgy.

    Includes bibliographical references.
    1. Metallurgy. I. Title.
TN665.P422   669   72-87210
ISBN 0-444-00130-1

Manufactured in the United States of America

To my family

CONTENTS

Contents

Chapter 8   *Melting, Pouring, and Solidification*

Chapter 9   *Computer Applications in Extractive Metallurgy*

Contents

Appendix A

Appendix B *Example Problems and Computer-Facilitated Solutions*

Preface

For over a decade the need has existed for a college textbook on extractive metallurgy which reflects developments in the technology and science of this engineering area. In particular, the use of chemical thermodynamics to describe the behavior of metallurgical processes has developed extensively and is now an integral part of the engineering approach to process analysis. During this same time interval, the use of computers has become increasingly important in engineering practice. This text places strong emphasis on the use of both engineering tools as applied to extractive metallurgy.

Most educators in the field of extractive metallurgy have recognized that a unit process treatment of the subject has much more to offer than a descriptive presentation of current practice on a metal-by-metal basis. This text is based on the unit process concept and emphasis is placed on basic principles rather than detailed practice. Nevertheless, some descriptive material has been included to provide a background and engineering context for the student. The instructor can utilize lectures and assigned reading to extend the description of industrial practice while emphasizing fundamental principles as presented in the text. An attempt has been made to extend the unit process concept to consideration of the design of entire extractive processing systems.

The future of metallurgical process engineering undoubtedly will involve, in addition to the advances in metallurgical science, techniques which are adapted from chemical engineering, instrumentation, computing science, economics, operations research, and other allied fields. The metallurgical engineer should be acquainted with these areas, at least to the point of being able to evaluate possible contributions to metallurgical processing operations by specialists from these related fields. Ideally, the trained metallurgist should be prepared to assume a position of leadership in the engineering team which handles the multi-faceted problems of metallurgical process design and operation. This book reflects an effort to expand the teaching emphasis in extractive metallurgy, and includes an introduction to some of the fields which are important in metallurgical processing systems. In addition to an analytical approach to the engineering of metallurgical unit processes, the text includes introductory material on the use of computers in process engineering, operations research procedures which treat system behavior in terms of queueing theory or linear programming models, and mathematical simulation of processing systems.

The objective of providing a textbook on extractive metallurgy for the junior, senior or beginning graduate student placed constraints on the text. These constraints prevented extensive development of each subject area in the broad field encompassed by extractive metallurgy. On the other hand, a detailed development of several subjects is included to emphasize the rigor and thoroughness characteristic of sound engineering.

The text is developed on the basis of three alternative approaches to metal extraction and refining: pyrometallurgy, hydrometallurgy, and electrometallurgy. These production alternatives have their roots in physical chemistry but draw upon many related fields. Furthermore, these three stems of extractive metallurgy are not independent, but often are combined in many extraction and refining systems. Pyrometallurgical unit processes are developed in considerable detail in the first several chapters, followed by hydrometallurgical, and electrometallurgical processes. A chapter on melting, pouring and solidification then follows which involves topics on pyrometallurgy but is focused toward the semi-finished product. The text concludes with a chapter covering the principles and techniques for computer-oriented analysis of extraction and refining systems.

Traditionally the teaching of metallurgical engineering has been problem-oriented. Some engineering problems have specific solutions. Other problems have a range of solutions, some of which are more optimal than others. Both types are encountered in engineering practice, and therefore both have been included in the text. In all cases an effort is made to emphasize the underlying basic principles.

Example problems are a useful supplement in the study of the text material. The text contains a number of such problems as worked examples. Problems and discussion questions are provided with each chapter. The text and problem sets are self-contained. However, the discussion questions, while based on material introduced in the text, suggest avenues for further study.

An appendix of problems with documented computer solutions is included to illustrate the application of this modern technological tool. The problems and approaches to solving them illustrate a variety of numerical methods which can be implemented on the digital computer. The problems have been selected on the basis of their pertinence to engineering practice and hopefully they will have applications beyond the college classroom.

I wish to thank all those who have helped in the preparation of this book. The development of this text has involved several years of teaching a senior course at The University of Michigan entitled Metallurgical Process Design. The students in those classes as well as many interested graduate students contributed immensely to the text material as it developed. Many colleagues and peers, both in industry and in académé, have generously contributed suggestions, comments and encouragement. Special recognition goes to Ralph D. Higgins who reviewed the early manuscripts, to Elaine Galoit who typed the final manuscript, and to Robert E. Marrone who assisted in preparing the computer appendix, to name only a few of the many who contributed. I am grateful to my students for their inspiration and to my family for their help, encouragement, and understanding.

Robert D. Pehlke

Ann Arbor, Michigan

CHAPTER 1

Introduction

Within the broad field of metal sciences and technology, the initial processing steps are extraction and refining. These extractive and purification processes, the subject of this text, are built upon the principles of chemical metallurgy. Chemical metallurgy (as contrasted with physical metallurgy, the study of the solid state and the structure and properties of metallic materials) utilizes physical chemistry, draws upon inorganic chemistry and electrochemistry, and depends upon some areas of organic chemistry to describe metallurgical systems. The unit processes of extractive metallurgy are presented in this text with emphasis on the underlying chemical metallurgical principles.

1-1 Source of Metals

The original source of all metals is the earth's crust, which contains metallic deposits usually in the form of their oxides or sulfides called ores, from which metals can be extracted. In some instances, noble metals, such as gold, silver, and sometimes copper, are found in *native* or free elemental form. Ores are seldom composed of one metallic compound but usually are made up of several, including the mineral of the desired metal as well as other impurity compounds referred to as *gangue*. The important minerals of the most common metals are listed by name and chemical formula in Table 1-1.

Ores are found in a wide variety of locations and at various depths in the crust of the earth. Ores are mined by underground methods or by surface mining techniques, such as open pit or strip mining. Magnesium is obtained primarily from sea water.

The virtually inexhaustible supply of many common metallic elements is reflected by the large amounts of iron, aluminum, silicon, calcium, sodium, and potassium in the earth's crust. In the case of magnesium, a vast supply exists since each cubic mile of sea water contains several thousand tons of magnesium.

A very important metallic element is *iron*, which accounts for the largest part of the metal industry. The primary source of iron is from the minerals magnetite (Fe_3O_4), hematite (Fe_2O_3), and limonite (Fe_2O_3 plus some combined water); the concentration of iron in high-grade, direct smelting ores may range as high as 60%, but ores containing only 25% iron have been smelted. The principal source of high-grade ore within the United States is the Lake Superior district in Upper Michigan and Minnesota. Other important deposits still exist in the Lorraine Basin of France, the English Midlands, the Donetz Basin and the Ural Mountains of Russia, and in northern Sweden. Important deposits which are being mined include those in

1

northern Quebec, Labrador, Brazil, and Venezuela. Until recently, the readily available, iron-bearing minerals were of sufficiently high grade that little attention was paid to methods for concentrating them, but depletion of high-grade deposits has led to development of techniques for concentrating low grade ores, one of which is taconite, a plentiful low-grade iron ore containing approximately 30% iron, in which the iron oxide is finely dispersed throughout a silica matrix. The upgrading of taconite is accomplished by grinding and concentration, followed by agglomeration into pellets for charging to the iron blast furnace.

Copper is found in nature in forms ranging from the pure metal to sulfide, oxide, and carbonate. The pure metal is found in appreciable amounts only in the Upper Peninsula of Michigan, but these "lake copper" deposits are nearly depleted and do not now represent an important source of copper. Copper is most commonly found as a sulfide, either chalcocite, chalcopyrite, or covelite. The action of surface waters on sulfide minerals of copper leads to oxidation, and the upper portions of copper ore deposits are usually high in copper oxides. Copper ores are usually low in copper content, containing less than 5% copper, but ores containing as little as $\frac{1}{2}$ wt % copper can usually be processed at a profit. Shipping costs for these low-grade ores dictate that concentration facilities be located at the mine. The principal sources of copper are the Cordillera of the western United States and South America, the Central African Plateau in northern Rhodesia and Katanga province of the Belgian Congo, and the Canadian Shield. The largest and richest copper mine in the world is at Chuquicamata, Chile, and the second largest mine is in Bingham Canyon, Utah.

Aluminum is one of the most abundant metallic elements and represents approximately 8% of the earth's crust, occurring as the oxide. Although several minerals contain appreciable amounts of aluminum, the only ore utilized commercially is bauxite, a rock containing hydrated oxides of aluminum, principally gibbsite and diaspore. These ores are usually found in tropical or subtropical climates, the most important sources of commercial bauxite being in Surinam, British Guiana, Jamaica, and Indonesia. Other deposits are being mined in southern United States and at several locations in southern Europe.

Lead is found in nature as the sulfide galena, although surface waters have often converted upper portions of lead deposits to the sulfate anglesite or the carbonate cerussite. The ores of lead usually contain, in addition to zinc blende (see below), recoverable amounts of copper, silver, gold, antimony, and bismuth. The largest producers of lead include the United States, Australia, Russia, Canada, and Mexico. Appreciable quantities of lead are also produced in South America, the principal sources being Peru, Bolivia, and Argentina.

The most important *zinc* mineral is the sulfide sphalerite, often referred to as zinc blende. The minerals sphalerite (ZnS) and galena (PbS) nearly always occur together, and a listing of the principal zinc-producing regions would be essentially the same as those for lead. The only exception where zinc deposits do not also contain appreciable amounts of lead is at Franklin Furnace, New Jersey, where the deposits are a mixture of zincite and the iron- and manganese-bearing mineral franklinite.

Tin is a relatively rare metal, accounting for less than 0.001 wt % of the earth's

TABLE 1-1

Minerals of Common Metals

Metal	Mineral	Formula
Iron	Magnetite	Fe_3O_4
	Hematite	Fe_2O_3
	Limonite	$2Fe_2O_3 \cdot 3H_2O$
	Siderite	$FeCO_3$
Copper	Native Copper	Cu
	Chalcocite	Cu_2S
	Covellite	CuS
	Chalcopyrite	$CuFeS_2$
	Cuprite	Cu_2O
Aluminum	Diaspore	$Al_2O_3 \cdot H_2O$
	Gibbsite	$Al_2O_3 \cdot 3H_2O$
	Kaolinite	$Al_2O_3 \cdot 2SiO_2 \cdot 2H_2O$
Lead	Galena	PbS
	Cerussite	$PbCO_3$
	Anglesite	$PbSO_4$
Zinc	Sphalerite	ZnS (Zinc blende)
	Zincite	ZnO
	Franklinite	$(Fe, Zn, Mn) O$
Magnesium	Magnesite	MgO
	Dolomite	$MgCO_3 \cdot CaCO_3$
Tin	Cassiterite	SnO_2
	Stannite	$(Cu_2S \cdot FeS \cdot SnS_2)$
Nickel	Millerite	NiS
	Garnierite	Hydrosilicate of Ni and Mg
	Pentlandite	$(FeNi) S$
Manganese	Pyrolusite	MnO_2
	Rhodochrosite	$MnCO_3$
	Rhodonite	$MnSiO_3$
Chromium	Chromite	$FeCr_2O_4$
Titanium	Ilmenite	$FeO \cdot TiO_2$
	Rutile	TiO_2
Zirconium	Baddeleyite	ZrO_2
	Zircon	$ZrSiO_4$
Vanadium	Patronite	V_2S_5 + sulfur
	Carnotite	$K_2O \cdot 2UO_3 \cdot V_2O_5 \cdot 3H_2O$
	Vanadinite	$3Pb_3(VO_4)_2PbCl_2$
Molybdenum	Molybdenite	MoS_2
	Molybdite	MoO_3
Tungsten	Wolframite	$FeWO_4$
	Scheelite	$CaWO_4$
Silver	Native silver	
	Argentite	Ag_2S
	Cerargyrite	$AgCl$
Gold	Native gold	Au
	Calaverite	$AuTe_2$
	Sylvanite	$(AuAg)Te_2$

TABLE 1-1 (Continued)

Metal	Mineral	Formula
Beryllium	Beryl	$3BeO \cdot Al_2O_3 \cdot 6SiO_2$
Mercury	Cinnabar	HgS
Uranium	Pitchblende	Complex oxide
Cadmium	Greenockite	CdS
Antimony	Stibnite	Sb_2S_3
Cobalt	Cobaltite	CoAsS

crust. The most important mineral of tin is the oxide cassiterite. As a heavy mineral, tin oxide is often concentrated into a placer deposit by the action of water breaking down igneous rock both by physical and chemical means and thus concentrating the heavier minerals. These placer deposits are located principally in Malaya and Indonesia, where the tin is recovered by hydraulic mining techniques, often employing dredges with self-contained equipment for concentrating the ore. Bolivia has substantial tin deposits. The tin is mined there by open cut methods where it occurs in vein deposits as the mineral stannite.

Titanium and *manganese* are relatively common metals, accounting for 0.1–1.0% of the earth's crust. Titanium occurs as ilmenite or rutile, and these deposits are located in the United States, Canada, India, Australia, and Brazil. Manganese occurs primarily as the oxide pyrolusite; the primary producer is the U.S.S.R., accounting for about one-half of the world output. Other important producers of manganese include India, Ghana, South Africa, Brazil, and Morocco.

Other metals such as gold, silver, cadmium, bismuth, arsenic, etc., are the by-products of copper and lead production. Most other metals are found in relatively small, low-grade deposits in the form of minerals listed in Table 1-1.

1-2 Mineral Engineering

Mineral engineering, the first major step in the extraction of metals, involves the separation and concentration of ores. These concentration techniques, which are primarily physical or mechanical separations, are referred to as mineral dressing. It is outside the scope of this text to discuss these concentration techniques in any detail. However, it should be noted that the economics with which the extractive metallurgist must work in optimizing his processes for smelting and refining of metals depend very much on the raw materials with which he is working. These raw materials are the product of the ore-dressing step.

1-3 Extractive Metallurgy

Following ore dressing, the removal of any remaining gangue, decomposition of the metallic mineral to produce a metal, and subsequent refining of the metal are the functions of extractive metallurgy. Extractive metallurgy can be divided into three

TABLE 1-2

1969 World and United States Production of Principal Metals[a]

Metal	World production (thousands of short tons)	U.S.A. production (thousands of short tons)	Percentage of world
Iron			
Pig iron and ferroalloys	459,697	97,593	21
Steel ingots and casting	633,431	141,262	22
Aluminum	10,019	3,793	38
Copper	7,304	1,585	21
Zinc	5,586	1,041	19
Lead	3,671	639	17
Tin	252	<0.5	—
Magnesium	222	100	45

[a] From *Minerals Yearbook*, Vol. I and II, *Metals, Minerals and Fuels*, U.S. Dept. of the Interior, Bureau of Mines; U.S. Government Printing Office, 1971.

principal areas or types of processes:

1. Pyrometallurgy, which employs high temperatures to carry out smelting and refining reactions and melting operations.
2. Hydrometallurgy, which employs liquid solvents, usually aqueous, to separate the desired metal.
3. Electrometallurgy, which employs electrical energy, principally for carrying out electrolysis to extract and refine metals.

The type of process or sequence of processes used depends to a great extent upon the starting material, i.e., the mineral concentrate, and the desired product. Selection of the particular process also depends upon the local economics, including the facilities available, power and fuel costs, and other factors such as the accessibility of water, shipping costs, availability of the market, etc. These considerations, along with the demand for various products and the availability and costs of ores, have determined the makeup of today's world metallurgical industry. A general picture of that industry is provided by Table 1-2, which lists the world and United States production of principal metals for the year 1969.

1-4 Chemical Metallurgy in the Extraction of Metals

Following the mining and concentration of minerals, their extraction is accomplished by application of chemical metallurgy in one of the three areas of extractive metallurgy: pyrometallurgy, hydrometallurgy, or electrometallurgy. In the following chapters these topics are treated in detail. Pyrometallurgy is discussed in terms of the roasting of ores to convert sulfides to the more easily processed oxides. The physical chemistry of roasting and calcining reactions is presented and agglomerating techniques are discussed. The principles of oxidation and reduction reactions are presented in terms of application to smelting of minerals by reduction to the metallic state or to a fused sulfide (matte), and conversion to a semi-refined metal. Subsequent pyrometallurgical refining processes, distillation, liquid phase refining and degassing operations are quantitatively discussed in terms of thermodynamic and kinetic limitations. Hydrometallurgical and electrometallurgical processes are reviewed. Then the pyrometallurgical processes of melting, pouring, casting, and solidification are described. Finally, computer applications in the field of extractive metallurgy are developed, along with the presentation of example programs.

Bibliography

Dennis, W. H., *Extractive Metallurgy*, Philosophical Library, New York, 1965.
Gaudin, A. M., *Flotation*, 2nd ed., McGraw-Hill, New York, 1957.
Gilchrist, J. D., *Extraction Metallurgy*, Pergamon Press, London, 1967.
Hayward, C. R., *An Outline of Metallurgical Practice*, D. VanNostrand Co., New York, 1952.
Newton, J., *Extractive Metallurgy*, John Wiley, New York, 1959.
Pryor, E. J., *Mineral Processing*, Mining Publications, Ltd., London, 1960.
Taggart, A. F., *Elements of Ore Dressing*, John Wiley, New York, 1951.
Taggart, A. F., Handbook of Mineral Dressing, John Wiley, New York, 1945.

Discussion Questions

1. Develop a correlation between the current market price of the more common metals and their level of production or use. What are the major factors in establishing these relationships?

2. What are the current price levels for scrap as contrasted with ingot or primary material for common engineering metals? Discuss the role of secondary metal in each metal market.

3. At what concentrations of mineral are the principal mining operations being worked to recover the major metals?

4. In extracting and refining aluminum, copper, zinc, lead, mercury, nickel, tungsten and magnesium, a key unit process is used. Considering the chemical and physical properties of the metal and its most common mineral, state which process is applicable to each metal and why the particular process is used.

CHAPTER 2

Pyrometallurgy I: Roasting—Agglomeration—Calcination

Minerals recovered from ores are not always in the optimum chemical or physical state for conversion to metals. Oxides are more conveniently reduced to metals than sulfides, or the metal might be more readily leached from the ore if it were present as a sulfate, a chloride, or an oxide. Chemical conversion to the desired species often is an integral segment of the extractive process. Sulfide ores, for example, usually are heated in an oxidizing atmosphere (roasted) to convert them to an oxide or sulfate.

The physical state of an ore may be too fine for charging to a process. Fine ores often are agglomerated by sintering prior to charging to a blast furnace, the principal smelting unit for lead and iron. In the case of iron ore, pelletizing, another very important agglomeration process, recently has achieved commercial adaptation in the iron and steel industry. Pelletizing is a process in which iron oxides fines first are agglomerated into small pellets that are then fired to produce a small oxide briquet suitable for charging to the iron blast furnace.

In the following sections, several commercial techniques for roasting and agglomerating ores are discussed. In addition, the thermodynamics of sulfate roasting is considered, and the thermodynamics and kinetics of calcination are presented.

2-1 Roasting of Sulfide Ores

Many important metals, including copper, lead, and zinc, occur in nature as sulfides. In early civilizations it was discovered that heating sulfide ores in air converted them to a form whereby they could easily be reduced by charcoal to a metallic form. In modern industrial practice, with a considerable amount of fine material being produced in various concentration operations on low-grade ores, several types of roasting equipment have been developed. Roasting is carried out by rabbling over a tier of hearths in a multihearth roaster, by exposing the sulfide ore to oxidizing gases in a shower roaster where the ore falls through hot gases, in fluidized beds, or by exposing the ore in a packed bed to a controlled flow of gas, as on the Dwight–Lloyd sintering machine. The conditions of roasting determine whether or not the sulfide ore is partially roasted to a sulfate or completely roasted to an oxide.

Thermodynamics of Roasting Reactions

A number of reactions take place during the roasting of sulfides. Overall reactions using the roasting of galena (PbS) as an example are given in the following equations. In addition, several other intermediate reactions can occur involving the oxysulfate of lead.

7

Most roasting is carried to completion (so-called "dead" or "sweet" roast) with elimination of most of the sulfur by the overall reactions

$$2PbS + 3O_2 = 2PbO + 2SO_2. \tag{2-1}$$

Similar reactions can be written for zinc, copper, and iron. At lower temperatures, sulfates can be formed:

$$2PbS + 4O_2 = 2PbSO_4, \tag{2-2}$$

$$2PbO + 2SO_2 + O_2 = 2PbSO_4. \tag{2-3}$$

At higher temperatures, the oxide can be reduced by the sulfide to produce the free metal,

$$2PbO + PbS = 3Pb + SO_2. \tag{2-4}$$

Control of temperature and oxygen potential (pressure) can be employed to produce the desired oxidation state. In the case of concentrated zinc ores that later are going to be reduced with carbon, an oxidizing roast with elimination of as much sulfur as possible is desired. In the case of a copper ore which is to be leached, the objective is the formation of as much water-soluble sulfate as possible.

In considering the thermodynamics of roasting, it should be noted that sulfur trioxide is formed in the gas phase according to the reaction,

$$SO_2 + \tfrac{1}{2}O_2 = SO_3. \tag{2-5}$$

This reaction is favored at lower temperatures. Oxide–sulfate equilibria can be considered in the form,

$$MSO_4 = MO + SO_3, \tag{2-6}$$

where M represents a metal ion. These equilibria could also have been treated in a form obtained by combination of Eqs. (2-5) and (2-6), as for example, in Eq. (2-3), where the equilibrium constant* would be expressed as

$$K_3 = \frac{a_{PbSO_4}{}^2}{a_{PbO}{}^2 P_{SO_2}{}^2 P_{O_2}}. \tag{2-7}$$

Standard free energy relationships for several sulfation equilibria of this form are summarized in Table A-I (see Appendix A).

The metal–sulfide equilibria can also be presented as oxygen–sulfur dioxide equilibria with the chemical reaction written:

$$MS + O_2 = M + SO_2. \tag{2-8}$$

The equilibrium constant can be written in the form,

$$K_8 = \frac{a_M P_{SO_2}}{a_{MS} P_{O_2}}. \tag{2-9}$$

---

* An understanding of physical chemistry has been presumed, including thermodynamics, phase equilibria, and elementary kinetics.

Then for pure solids in their standard state at unit activity:

$$\log P_{SO_2} = \log K_8 + \log P_{O_2}. \tag{2-10}$$

The thermodynamics of sulfate systems then can be presented conveniently in terms of predominance area ($\log P_{SO_2}$ versus $\log P_{O_2}$) diagrams. Free energy data for oxide and sulfide systems are required to determine relations of the form shown in Eq. (2-10), or for Eq. (2-7), where for unit activity of the solids

$$\log P_{SO_2} = -\tfrac{1}{2} \log K_3 - \tfrac{1}{2} \log P_{O_2}. \tag{2-11}$$

A predominance area diagram for the nickel–oxygen–sulfur system is presented in Fig. 2-1 for a temperature of 1000°K. The small square in Fig. 2-1 encloses an area where the oxygen and sulfur dioxide pressures range from 10^{-1}–10^{-2} atm. In operations at 1 atm total pressure, gas compositions from 3–10% oxygen and 3–10% sulfur dioxide would yield nickel sulfate as the stable solid phase. At a gas phase

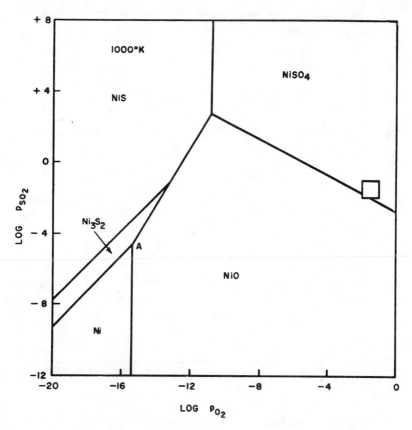

Fig. 2-1. Predominance Area Diagram for the Ni-S-O System at 1000°K

Source: T. R. Ingraham, "Sulphate Stability and Thermodynamic Phase Diagrams with Particular Reference to Roasting" *Applications of Fundamental Thermodnyamics to Metallurgical Processes*, Gordon and Breach Science Publishers Inc., New York, 1967, p. 187.

composition of 1% oxygen and 1% sulfur dioxide, nickel oxide would be stable. At the univariant point A of Fig. 2-1 ($P_{SO_2} = 2.5 \times 10^{-5}$ atm and $P_{O_2} = 5 \times 10^{-16}$ atm), the required conditions are so reducing that it is unlikely that metallic nickel would be formed in commercial roasting operations. The influence of temperature on these equilibria also can be employed advantageously to produce a desired end product by shifting the location of the predominance areas.

Sulfide concentrates of copper and cobalt are sulfated in a fluidized bed operation at 950°K to produce 97% soluble copper and 93.5% soluble cobalt. The roaster gases analyze 8% SO_2 and 4% O_2. Predominance area diagrams for copper and cobalt at 950°K are superimposed in Fig. 2-2. The point representing the commercial operation for roasting copper–cobalt ores is shown to lie well inside an area where $CoSO_4$ and $CuSO_4$ are stable. If a separation of copper and cobalt were desired in the leaching operation, a roasting operation in "area A" of Fig. 2-2 would produce

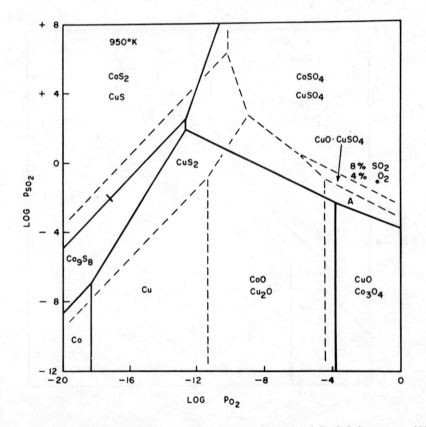

Fig. 2-2. Superimposed Predominance Diagrams for the Co-S-O and Cu-S-O Systems at 950°K

Source: T. R. Ingraham, "Sulphate Stability and Thermodynamic Phase Diagrams with Particular Reference to Roasting" *Applications of Fundamental Thermodynamics to Metallurgical Processes*, Gordon and Breach Science Publishers Inc., New York, 1967, p. 190.

water insoluble cupric oxide and soluble cobaltous sulfate. This separation can be made on a commercial scale.[1]

Example Problem 2-1. A lead roasting operation at 1100°K is proposed with 12% SO_2 and 4% O_2 in the roaster gases at 1 atm total pressure. Is lead sulfate the stable solid phase? With 1% O_2 in the gas, what partial pressure of SO_2 is necessary to form the lead oxysulfate $PbSO_4 \cdot PbO$?

Solution. At 1100°K the equilibrium constant for the reaction

$$SO_3 = SO_2 + \tfrac{1}{2}O_2, \tag{2-a}$$

$$K_a = \frac{P_{SO_2}P_{O_2}^{1/2}}{P_{SO_3}} = 0.19. \tag{2-b}$$

With $P_{SO_2} = 0.12$ and $P_{O_2} = 0.04$, the effective pressure of SO_3 is

$$P_{SO_3} = \frac{(0.12)(0.04)^{1/2}}{0.19} = 0.126 \text{ atm}. \tag{2-c}$$

This pressure then can be utilized for comparison with the SO_3 pressure determined for the equilibrium:

$$2PbSO_4 = PbSO_4 \cdot PbO + SO_3. \tag{2-d}$$

From Table A-I

$$\Delta G^0_d = 74000 + 16.81 \, T \log T - 97.4 \, T \tag{2-e}$$

At 1100°K

$$\Delta G^0_d = 23,060 \text{ cal/mole},$$

$$\log K_d = \log P_{SO_3} = -4.58,$$

$$P_{SO_3} = 2.63 \times 10^{-5} \text{ atm}.$$

This partial pressure of SO_3 is well below that existing in the roaster gases, and, hence, $PbSO_4$ is the thermodynamically stable phase.

If the gases contained 1% oxygen, using Eq. (2-b) the partial pressure of SO_2 in equilibrium with $PbSO_4$ and $PbSO_4 \cdot PbO$ would be

$$P_{SO_2} = \frac{K_a P_{SO_3}}{(P_{O_2})^{1/2}} = \frac{(0.19)(2.63 \times 10^{-5})}{(0.01)^{1/2}} = 4.9 \times 10^{-5} \text{ atm}.$$

Lead has an extremely strong tendency to react with sulfur. This tendency is such that even a trace of SO_2 ($P_{SO_2} = 10^{-4}$ atm) in a blast furnace gas is sufficient to sulfatize PbO to $PbSO_4 \cdot PbO$. Furthermore, the thermodynamic stability of lead sulfates makes it extremely difficult to decompose them to the oxide.

In addition to the complications arising in copper, lead, and zinc processing because of the possibility of forming of oxysulfates, it should be noted that gangue material present in the concentrate charged to the roaster also may react to form complex oxide compounds. This would reduce the activity of the copper, lead or zinc oxide, and correspondingly reduce the equilibrium partial pressure of SO_3.

---

[1] I. F. Theys, *J. Metals* **10**, 46 (1958).

Volatilization During Roasting

The roasting of sulfide ores permits, in addition to the conversion of sulfide to oxide or sulfate, the vaporization of volatile compounds. In particular, antimony and arsenic are found in appreciable concentrations in flue dusts from the roasting of charge materials that contain only traces of these metals. The trioxides of antimony and arsenic, formed by heating the sulfides in air at temperatures as low as 400°C are sublimed in the later stages of roasting. Zinc and cadmium when heated under reducing conditions are vaporized (see Pyrometallurgy II).

A chloridizing roast in which salt (sodium chloride) is mixed with the sulfide ore charged to the roasting operation can result in a conversion of iron, silver, lead, and copper to chlorides that are readily dissolved in water. Such a conversion to chloride is facilitated by the presence of carbon. For example, tin and tungsten are partially separated by roasting in an atmosphere of chlorine and hydrochloric acid. At 600°C the tin and some of the tungsten are volatilized, and at higher temperatures the remainder of the tungsten, and iron and manganese, come off as chlorides. The presence of a reducing agent such as carbon can markedly lower the temperature required for this type of reaction. For example, chlorination of magnesium oxide is carried out at 1000°C in the presence of coke according to the reaction

$$MgO + C + Cl_2(g) = MgCl_2(s) + CO. \tag{2-12}$$

HCl gas also can be used to convert some metallic oxides to chlorides by a reaction of the form

$$MO_2 + 4HCl = MCl_4 + 2H_2O. \tag{2-13}$$

2-2 Industrial Roasting Processes

As mentioned previously, roasting is carried out not only to render the material chemically suitable for a subsequent processing operation, but also to provide a suitable physical state of the material. In the case of copper and other metals that are smelted in reverberatory furnaces, the roasted material (calcine) should be finely divided, whereas in blast furnace operations for materials such as iron and lead, the calcine must be in a reasonably large agglomerated form. Several processes are available for providing these conditions. The important industrial techniques for roasting are mechanical roasting, flash roasting, fluidized bed roasting, and blast or sinter roasting.

Mechanical or hearth roasting is carried out in a vertical multihearth system in which 8–12 hearths are enclosed in a large cylindrical shell 20–25 ft in diameter and 30 ft high. A sectioned view of this type of roaster is shown in Fig. 2-3. The ore enters at the top and drops from hearth to hearth while the sulfide ore particles are roasted as they come in contact with the rising gases. The vessel is lined with fire brick and is equipped with a rotating rabble attached to a central shaft that serves to move, or rabble, the ore across each disc-shaped hearth. The ore is plowed toward an opening on the outer or inner edge of the hearth where, in turn, it falls to the hearth below. The rotating rabble arms are cooled, and openings are provided in the shell for ease of maintenance. This type of equipment can roast from 100–200 tons

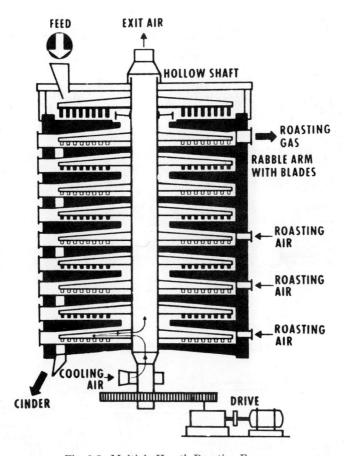

FEED EXIT AIR

HOLLOW SHAFT

ROASTING GAS

RABBLE ARM WITH BLADES

ROASTING AIR

ROASTING AIR

ROASTING AIR

COOLING AIR

CINDER

DRIVE

Fig. 2-3. Multiple Hearth Roasting Furnace

Source: *Lurgi Manual,* Lurgi Gesellschaften Frankfurt (Main) Germany, June 1961, p. 192.

of material per day. Heat can be supplied by combustion of fuel, but usually the process is autogenous, that is the oxidation of the material is sufficient to supply thermal energy to the system.

Flash roasting is a modification of the multihearth roaster. After it was observed that the oxidation occured on the surface of ore exposed to the roasting gases and, in particular during the period of fall from hearth to hearth, flash roasting was developed. This process is carried out in a unit similar to a hearth roaster except that several intermediate hearths are removed. The concentrate which often is wet is dried on the first upper hearth or two and then falls through a combustion chamber. After settling on the lower hearths, the calcine is collected and discharged from the unit. In the case where combustion is not autogenous, auxiliary fuel is burned in the unit to maintain the process temperature. To provide for conversion to sulfate in the case of zinc roasting, additional amounts of SO_2 are often added to control the atmosphere on the lower hearths. A cross-sectional view of a flash roaster for processing zinc concentrates is shown in Fig. 2-4.

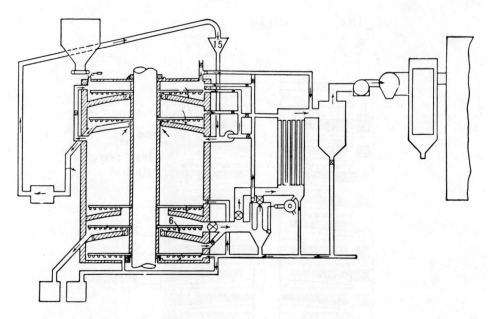

Fig. 2-4. Suspension Roaster for Roasting Zinc Concentrates, Trail, British Columbia

Source: Stimmel, et al., *AIME Trans.*, 1936, Vol. 121, p. 542.

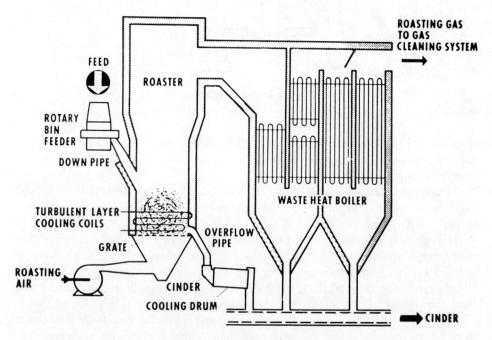

Fig. 2-5. Fluidized Bed Roaster

Source: *Lurgi Manual*, Lurgi Gesellschaften, Frankfurt (Main) Germany, June 1961, p. 195.

Fluid bed roasting has been developed since World War II and is now widely used for roasting iron sulfide, and in several instances the sulfides of copper, cobalt and nickel. A schematic diagram of a fluidized bed system for roasting sulfide concentrates is shown in Fig. 2-5.

The most important variables in the fluidized bed are the particle diameter, or average diameter, and the superficial gas velocity. Figure 2-6 shows the relationship between particle diameter and superficial gas velocity for fluidization at 900°C. It should be noted that because of interaction between the solid particles in a fluidized bed, particles which are fine enough to be carried out of the bed and particles which are coarse enough to sink through the bed may exist in the fluidized state. Consequently, within a reasonable range of particle size, the average diameter may be used to characterize the behavior of the bed.

Temperatures in the fluidized bed can be controlled by proportioning the relative amounts of feed and air. Once initiated, combustion in the bed is self-sustaining and a continuous supply of air and feed is provided to the roaster.

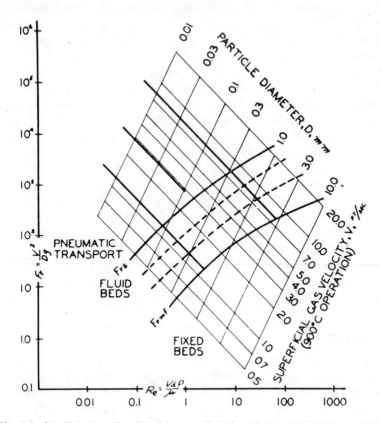

Fig. 2-6. Gas Velocity—Particle Diameter Relationship for Fluidization at 900°C

Source: R. B. Thompson and G. Roesner, *Extractive Metallurgy of Copper, Nickel and Cobalt,* AIME Symposium, New York, Interscience Publishers, Inc., 1961, p. 4.

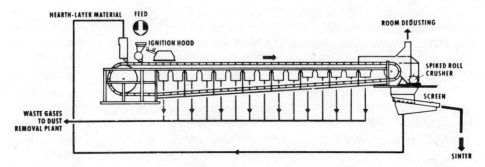

Fig. 2-7. Dwight-Lloyd Sintering Machine

Source: *Lurgi Manual*, Lurgi Gesellschaften, Frankfurt (Main) Germany, June 1961, p. 150.

Sinter-roast offers an advantage over the previously described roasting processes in that agglomeration of the roasted material is accomplished. The blast furnace requires a suitably large particle size, and hence iron and lead sulfide ores are sinter-roasted. This process is usually carried out on a Dwight–Lloyd sintering machine, as shown in Fig. 2-7. Roasting is accompanied by incipient fusion, which produces a porous cinder-like material called sinter.

The Dwight-Lloyd sintering machine, which was developed more than 50 years ago, consists of a series of pallets or grates mounted on an endless track. The concentrate is charged to a depth of about 6–20 in. on the pallets, which move over wind boxes at 2–3 ft/min. Combustion of the bed is initiated on its surface by a burner, and the combustion is maintained and carried through the mass of the charge by the air drawn through the concentrate to the wind box below, which is connected to a suction fan. Relatively high temperatures (900–1200°C) are developed in the material, causing it to fuse into a compact mass. After the sinter has reached the end of the machine it is discharged, cooled, and sized to provide a uniform product. Fines from the sizing operation are returned as charge material.

In sinter-roasting, the sulfur in the ore acts as a fuel. The relatively high temperatures and oxidizing conditions usually provide low sulfur contents, particularly for the roasting of pyrite (FeS_2) or pyrrhotite (FeS). In the case of low sulfur or oxide ores, fuel is added. The latter case is referred to simply as sintering and is used in particular for preparation of charge material to the iron blast furnace.

2-3 Sintering

The requirement for coarse charge material for the blast furnace necessitates agglomeration of fine ores. One method for agglomerating fines is by sintering. Sintering is the process of heating fine materials to an elevated temperature without complete fusion such that the small, solid particles in contact with one another adhere and agglomerate into larger, more useful particles. The predominant mechanisms in the action of sintering are surface diffusion and incipient fusion, and both occur in the commercial sintering of ore.

The sintering of large quantities of material is often necessary in the operation of a metallurgical plant. This process provides an opportunity to use fine material, and often makes a particular process feasible by converting available fine materials to an agglomerated form for use as a charge material. Sintering is sometimes carried out in rotary kilns or by batch processing on sinter pans or hearths. Flow of air through the charge may be by updraft or downdraft methods, but the predominant industrial technique for sintering ore is on a moving hearth, as with the Dwight–Lloyd continuous sintering machine. As originally designed for processing copper ores, the sulfide fines were distributed in a thin layer along a traveling belt made up of grates. The charge was ignited and the sulfur burned out of the ore as air was drawn through the charge by large fans. The fines fused together, forming a strong sinter cake that was desirable for charging to a blast furnace. The basic difference between the processing of sulfide ores and the sintering of oxide-ferrous ores is the self-contained fuel of the sulfide material. In the processing of hematite or magnetite fines, carbon in the form of coal or coke has to be added to provide fuel for the sintering process.

The utilization of the Dwight–Lloyd machine (Fig. 2-7) in the processing of iron ores is essentially the same as for nonferrous sulfide ores. A schematic diagram of an iron ore sintering plant is shown in Fig. 2-8. It is evident that an important part of the sinter plant is the mixing system that blends the fine ores, limestone, coke, plus the fines returned from the sinter strand. The charge mix is loaded onto the moving grates of the sintering machine, where it passes under a burner that ignites the bed. Air is drawn through the burning bed by the suction system below,

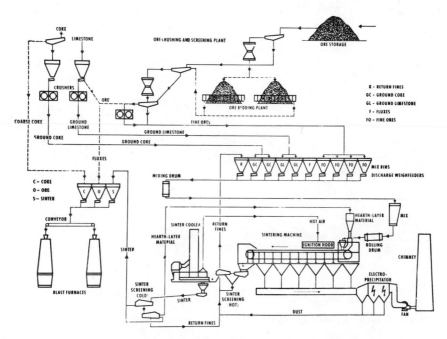

Fig. 2-8. Iron Ore Sintering Plant for Preparation of Self-fluxing Sinter

Source: *Lurgi Manual*, Lurgi Gesellschaften, Frankfurt (Main) Germany, June 1961, p. 151.

and at the end of the strand the sinter drops off the pallets, where it is cooled and screened. The undersize particles are then returned to the sintering process as recycle.

The *fuel requirement* for the sintering of iron oxide ores ranges from 5–8% coal or coke, and is relatively independent of the material to be sintered. The optimum for the fuel requirement varies slightly and depends upon whether or not chemical reactions are involved in the sintering process. The presence of appreciable amounts of limestone or water will require additional fuel, and may depress the maximum temperature achieved. Variation in the carbonate or moisture content of the sinter mix will give a variation in the width of the hot zone that moves down through the ore bed. In normal downdraft sintering, the combustion of the fuel in the sinter mix is initiated in the upper levels of the sinter bed. The hot combustion gases are pulled downward through the bed and preheat the sinter charge.[2] The presence of water in the sinter mix will limit the increase in temperature of the sinter bed until the water is vaporized. The presence of carbonate, such as limestone that is charged to self-fluxing sinter (see Pyrometallurgy III), will result in a broadening of the combusion front that follows the heat front down through the bed.

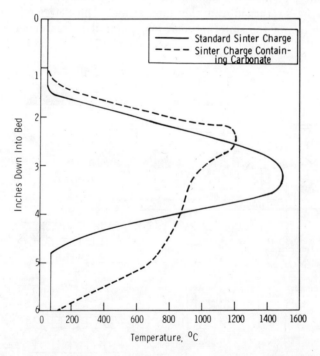

Fig. 2-9. Temperature Distribution for Two Sinter Mixes.

(Combustion is occurring approximately at the midpoint of the bed)

[2] S. Eketorp, *STEELMAKING, The Chipman Conference*, p. 180, MIT Press, Cambridge, Mass., 1965.

Figure 2-9 shows the influence of the presence of a large amount of calcium carbonate in the charge. The peak temperature of the bed is decreased and the width of the hot zone is increased. A well-defined planar front should occur in sintering, as illustrated by the solid line in Fig. 2-9.

The engineering relationships for design of sintering operations are based on flow of air through porous beds. Of particular importance is the bed permeability, which often can be determined by laboratory sintering tests.[3,4] In commercial operations, 1 to $3\frac{1}{2}$ tons of material can be sintered per square foot of hearth area per day on a Dwight–Lloyd sintering machine. A decrease in fuel requirements could be achieved through the use of preheated air supplied from a closed hood over the sinter strand. Up to 40% of the thermal energy required for sintering can be supplied by fuels such as flue gas, natural gas, or fuel oil burned in a hood above the sinter machine. With this "mixed firing process," the mechanical and chemical properties of the sinter can be improved.[5]

Control of the sintering of iron ores by the Dwight–Lloyd process is of particular importance as the productivity of the iron blast furnace increases. The control of material flow and suitable proportioning of raw materials to provide a chemically uniform sinter mix is of prime importance. Maximum utilization of the sinter strand requires control of the "burn through" point such that combustion is completed just as the sinter reaches the discharge end of the strand. The temperature in the wind box often can be used to monitor this "burn through" point.[6]

2-4 Pelletizing

Pelletizing is a recently developed agglomerating process for iron ores that has made possible the economic utilization of low-grade taconite ores. The use of pellets as a blast furnace charge has resulted in marked increases in blast furnace productivity (see Pyrometallurgy III). Taconite is an intimate physical mixture of magnetite or hematite with gangue materials. Successful separation of the iron oxide from the gangue material requires crushing and grinding to extremely fine particle sizes. The agglomeration of this material has been successfully accomplished by balling the ore in various rotating devices such as discs, drums, cones, etc., and once the fine material is balled, it is hardened by heating, generally referred to as *indurating*. A schematic diagram of a pelletizing plant is shown in Fig. 2-10. During induration, several chemical reactions occur, including vaporization of free water, vaporization of combined water, calcination of any limestone present, and in the case of a magnetite (Fe_3O_4) concentrate, the oxidization of the magnetite to hematite (Fe_2O_3). Laboratory tests.are usually employed to determine the optimum time–temperature pattern for induration. The final temperatures during induration reach 2300–2400°F, depending upon the concentrate being pelletized. Shaft furnaces, kilns, and traveling grates with updraft or downdraft techniques are used to

[3] E. W. Voice, S. H. Brooks, and P. K. Gledhill, *J.I.S.I.* **174,** 136 (1953).

[4] E. W. Voice and R. Wild, *J. Metals* **10,** 105 (1958).

[5] *Lurgi Manual*, p. 150, Lurgi Gesellschaften Frankfurt (Main), Germany, June, 1961.

[6] A. Bragard, *J. Metals* **19,** 49–57 (1967).

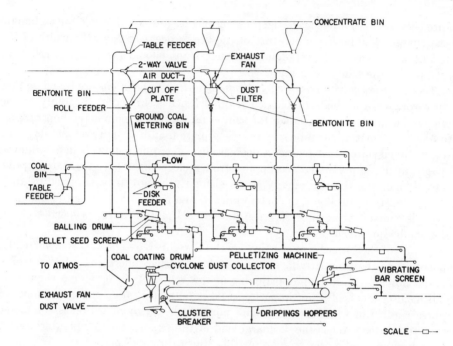

Fig. 2-10. Pelletizing Plant Flow Diagram

Source: M. J. Greaves and A. English, *Agglomeration*, New York, Interscience Publishers, 1962 p. 442.

fire the pellets. Pelletizing machines are designed for producing between 2500–3000 tons of pellets per day and designs are being prepared for machines capable of producing 5000 tons per day.

A typical balling drum circuit, in conjunction with shaft furnaces for indurating the pellets, is shown in Fig. 2-11. With a suitable amount of moisture and a binding agent, usually a clay such as bentonite, the finely divided material is rolled on an inclined drum, disc, etc., causing the ore particles to cling together and roll up into small spheres. The material discharged from the balling unit is sized, and the undersized pellets are recycled. Limestone often is added to provide a fluxing agent during subsequent smelting in the blast furnace.

Other techniques are available for indurating the pellets, including the horizontal grate shown in Fig. 2-12. A recently designed pelletizing process employs a traveling conveyer for drying and initiating induration, a rotary hearth furnace for raising the temperature of the pellets to 1200°F, and an indurating stove or shaft furnace where the reaction from magnetite to hematite is accomplished by a controlled flow of air through the packed bed. The stove includes both a heating and cooling section. The system is shown schematically in Fig. 2-13. Automatic control of the pelletizing plant, utilizing bed-level control, temperature controls, pressure controls, and flow instruments in conjunction with a digital computer, is being installed in several locations.

The rapid development of pelletizing plants on the North American continent stemmed from the need for treating fine size magnetite concentrates produced from taconites of the Massabi range, which could not be sintered successfully because of their extreme fineness. The advantage of this process is such that the bulk of the iron ore produced in North America is now being pelletized. High-grade ores mined underground are also being directly pelletized, and consideration is being given to partial reduction of the pellets during induration, which could lead to even higher productivities for the blast furnace. Fully reduced pellets are being prepared for direct charging to steelmaking furnaces.

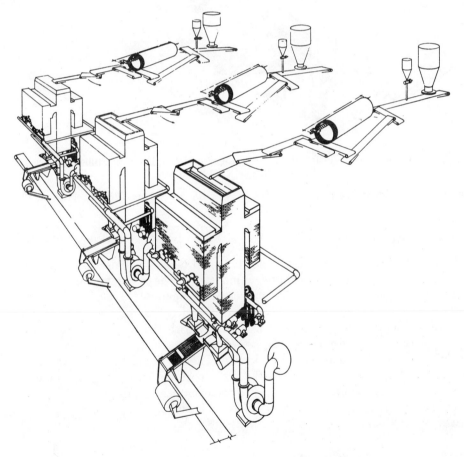

Fig. 2-11. Typical Shaft Furnace Pelletizing Plant

Source: A. English and M. J. Greaves, *Soc. Mining Engineering Transactions*, September 1963, p. 308.

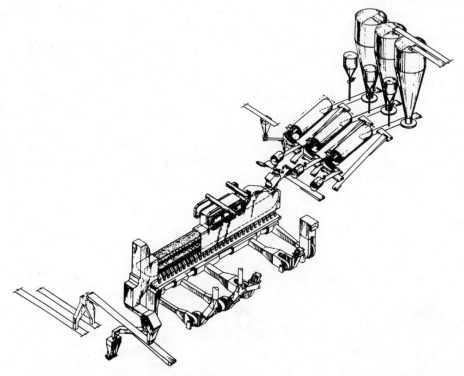

Fig. 2-12. Horizontal Grate Pelletizing Plant of Reserve Mining Co. with Million ton per Year
Capacity, Magnetite Concentrate

Source: A. English and M. J. Greaves, *Soc. Mining Engineers Transactions*, September 1963,
p. 309.

2-5 Calcination

Calcination of natural carbonates yields lime and magnesia which have long been
important materials in the metallurgical industries. The major tonnage outlet for
magnesia is as the refractory "dead-burned grain magnesite." Lime is also a refractory material, but cannot be used in commercial service. Exposure to the atmosphere
after calcination will result in reaction (slaking) with the water vapor in the air,
thus rendering the lime useless. The use of lime as a neutralizing agent for acid
pickling and leaching solutions is well known. The use of lime as a flux for oxygen
steelmaking has markedly increased the lime consumption in the steel industry,
replacing the limestone previously used in the open hearth process. Calcination of
limestone is carried out in many types of kilns, including rotary, shaft, and rotary
hearth.

Thermodynamics of Calcination

The reaction that occurs in calcining a chemically pure limestone may be
expressed as

$$CaCO_3 = CaO + CO_2. \tag{2-14}$$

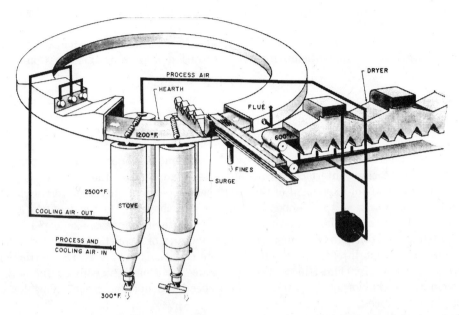

Fig. 2-13. Heat process equipment (in which green balls of magnetite ore are loaded onto a steel, dryer conveyor and dried and preheated at temperatures up to 600°F in air.)

Source: L. W. Smith, D. Beggs, and F. G. Rinker, *J. of Metals*, 1966 Vol. 18, No. 9, p. 1022.

The equilibrium constant for Eq. (2-14) is

$$K_{14} = P_{CO_2}. \qquad (2\text{-}15)$$

The free energy of decomposition may be expressed[7] as a function of temperature as

$$\Delta G^0 = 42{,}490 - 37.7\,T = -4.575\,T \log (P_{CO_2}). \qquad (2\text{-}16)$$

Equations (2-15) and (2-16) can be utilized to calculate the equilibrium pressure of CO_2 in contact with heated limestone.

Kinetics—Mechanism of Calcination

In addition to predicting the temperatures at which decomposition of the limestone can take place, it is necessary as well to consider the kinetics of the reaction. Although several experimental investigations have been made to elucidate the exact manner in which the decomposition takes place, little success has been obtained in defining a mechanism for calcination. It has been found, however, that calcination of limestone takes place at a definite boundary plane, i.e., the boundary of two solid phases, one of lime (CaO) and the other of limestone ($CaCO_3$). This plane progresses from the surface of the piece of limestone being calcined toward

[7] O. Kubaschewski and E. L. Evans, *Metallurgical Thermochemistry*, John Wiley, New York, 1956.

the center at a definite rate and is maintained in about the same shape as the external surface of the stone.

Based upon calcination data, Furnas[8] derived the following equation, which can be used to determine the progressive penetration of the zone of calcination:

$$\log R = 0.003145\ T - 3.3085, \tag{2-17}$$

where

R = rate of advance of calcination interface, cm/hr,

T = temperature of surroundings, °C.

Assuming that the heat absorbed by the limestone is proportional to the surface area of the receding interface, a constant rate of advance of the interface is predicted. If the rate of movement of the interface is dependent only upon temperature and available area and is independent of size and shape of the particle, the time required to calcine is directly proportional to the size of the piece. Conley,[9] Pehlke and Kim,[10] and others have demonstrated the validity of a constant rate of advance of the calcination interface, although the rate equations vary with the particular stone under study. Thus, in operating practice, it is necessary to evaluate the calcination behavior of a material before specifying operating conditions for a calcination process.

Bibliography

Davis, E. W., *Pioneering with Taconite*, Minnesota Historical Society, St. Paul, Minn., 1964.

Greaves, M. J., and A. English, Engineering contributions to new techniques of iron ore agglomeration, *Agglomeration*, pp. 419–453, Interscience Publishers, New York, 1962.

Hopkins, D. W., *Physical Chemistry and Metal Extraction*, The Macmillan Company, New York, 1954.

Kellogg, H. H., A critical review of sulfation equilibria, *Trans. Met. Soc. AIME* **230**, 1622–1634 (1964).

Leva, Max, *Fluidization*, McGraw-Hill, New York, 1959.

Newton, J., *Extractive Metallurgy*, John Wiley, New York, 1959.

Roberts, J. P., Mechanism of sintering, *Metallurgia* **42**, 123 (1950).

Smith, L. W., D. Beggs, and F. G. Rinker, A new process for oxide pellet production, *J. Metals* **18**, 1021–1026 (1966).

Thompson, R. B. and G. Roesner, Fluid bed roasting-principles and practice, *Extractive Metallurgy of Copper, Nickel, and Cobalt*, AIME Symposium, Interscience Publishers, New York, 1961.

Voice, E. W., S. H. Brooks, W. Davies and B. L. Robertson, *Symposium on Sinter*, Special Report No. 53, p. 43, Iron and Steel Institute, London, 1955.

Voice, E. W., and R. Wild, The influence of fundamental factors on the sintering process, *Sintering Symposium*, Australian Inst. Mining and Met., Port Pirie, 1958.

Discussion Questions

1. The sinter-roasting of fine nickel sulfide (-200 mesh) has several inherent disadvantages. Its low melting point favors fusion on oxidation, producing a

---

[8] C. C. Furnas, *Ind. Eng. Chem.* **23**, 534 (1931).

[9] J. E. Conley, Technical Publication Number 1037, *Mining Tech.* **3** (1939).

[10] R. D. Pehlke and Y. U. Kim, Private communication, The University of Michigan, 1965.

relatively dense product which is insufficiently active for certain subsequent processing operations. Suggest a process which will result in a more acceptable product.

2. Discuss the economic factors which determine the optimum length for a sinter strand.

3. Summarize the preliminary processing steps for the principal ores of the common metals.

4. Contrast the advantages and disadvantages of sintering and pelletizing of iron oxide fines.

5. What are the relative merits of the shaft kiln, rotary kiln, and rotary hearth furnace for calcining lime?

6. Diagram the steps involved in concentrating and agglomerating taconite ores.

7. Describe various gas treating systems employed to control the sulfur-bearing gases from roasting operations.

Problems

1. Nickel sulfide roasting is carried out at 1 atm total pressure with gas phase compositions in the ranges 3–10% O_2 and 3–10% SO_2. Over what portion of these composition ranges is nickel sulfate the stable phase?

2. Sulfide concentrates of copper and cobalt are sulfated in a fluidized bed operation at 950°K and near 1 atm pressure to produce 97% soluble copper and 94% soluble cobalt. Roaster gases are reported to analyze 8% SO_2 and 4% O_2. Was gas–solid equilibrium reached in this system?

3. Copper and cobalt are to be separated by roasting at 1 atm and 950°K to produce water soluble cobaltous sulfate and water insoluble cupric oxide. Recommend a control point for roaster gas composition.

4. Roasting of sphalerite is being carried out at 1520°F and at an average roaster gas composition of 10% SO_2 and 4% O_2. The resulting calcine is basic sulfate ($ZnO \cdot 2\ ZnSO_4$). How would you change the gas composition and/or temperature to produce a calcine of:
   - (a) ZnO
   - (b) $ZnSO_4$

5. Evaluate the thermodynamic feasibility of roasting to a sulfate and to an oxide for separation in a subsequent leaching operation for sulfide ores containing:
   - (a) iron and copper
   - (b) iron and zinc
   - (c) zinc and cobalt
   - (d) nickel and copper

   Specify roaster gas compositions if separation can be accomplished.

6. Calculate the maximum pressure of SO_2 at which Pb and PbO can coexist at 1100°K under equilibrium conditions (i.e., without forming the lower sulfate $PbSO_4 \cdot 4PbO$).

7. Consider the fluidized bed roasting of nickel sulfide ore at 2000°F and a rate of 1 ton/hr.

   Given: 136 ft³ (STP) of roaster gas (N_2, O_2, SO_2) per pound of sulfide. The product contains 0.5% S. Heat losses equal 850,000 Btu/hr. Sensible heat of product equals 5 Btu/lb Ni.

   Compute: (a) The heat evolved by oxidation of the nickel sulfide.
            (b) The excess air which is required to fluidize the bed.
            (c) The percentage of thermal energy which is retained in the exiting roaster gas.

8. It is desired to produce a feed for electro-winning of copper from a sulfide ore using a fluidized bed roaster. Specify the required operating temperature of the reactor and the composition of the exiting gas stream.

Ore Component	Composition (wt %)
Copper (as sulfide)	48
Iron (as sulfide)	11
Sulfur (as sulfide)	21
H_2O, free and combined	5
CO_2, as carbonate	3
Inerts, SiO_2, etc.	12

   The oxidizing gas is 10 moles of air/mole of copper supplied to the roaster.

9. One hundred pounds of calcium carbonate is stored in a sealed cannister with a volume of 5 ft³. The cannister is heated to 800°C.

   (a) What is the pressure of CO_2 at equilibrium?
   (b) How much CaO would be formed at equilibrium?

10. Specify operating conditions for a large rotary kiln calcining 1-in. pieces of limestone.

CHAPTER 3

Pyrometallurgy II: Oxidation–Reduction Reactions

3-1 Thermodynamic Stability of Mineral Compounds

Less reactive metals, including gold, silver, copper, and the platinum metals, are sometimes found in nature in the metallic state. Most metals, however, are usually found as mineral compounds such as oxides, sulfides, carbonates, or silicates; the mineral may be a single compound or it may be a physical mixture of a number of compounds. As outlined in Chapter 1, a combination of several separation or concentration processes are employed in the extraction of metals from their ores. When further concentration cannot be carried out by physical means, chemical separations must be employed. The conversion of sulfides to oxides by roasting or conversion of carbonates to oxides by calcination are typical pyrometallurgical processes designed to modify the minerals of useful metals to a more easily separable form. The ease with which a metal can be recovered from its mineral form can be evaluated quantitatively in terms of the thermodynamic stability of the mineral compound; consequently, some attention should be given here to the nature of various mineral compounds.

Thermodynamic Stability of Oxides

The practicality of chemical processing for the extraction of metals can be assessed on the basis of free energy changes for the reactions involved, a negative free energy change indicating that the reaction will tend to occur. The first step in evaluating the free energy change is to determine standard free energy changes. A convenient means for comparing free energy changes occuring in various chemical reactions is to display them graphically. One useful chart, first proposed by Ellingham,[1] is a plot of the standard free energy of formation versus temperature for a group of compounds of a given type, e.g., oxides of various chemical elements. This chart indicates the relative tendency toward oxidation for these elements and shows directly conditions under which reactions occur, thus providing a basis for evaluating the possibility of chemical separation by an oxidation–reduction reaction. The Ellingham diagram showing the standard free energies of formation of oxides as a function of temperature is presented in Fig. 3-1.

The free energy–temperature relationship shown in Fig. 3-1 is nearly linear. The apparent linearity is the result of heat capacities of reactants and products being essentially equal. The slopes of the plots on the Ellingham diagram are

[1] H. J. T. Ellingham, *J. Soc. Chem. Ind.* **63**, 125 (1944).

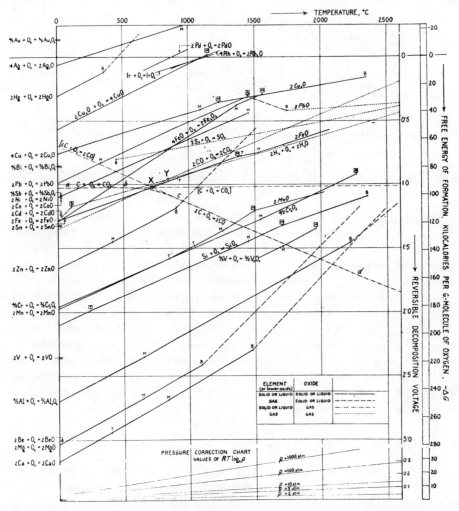

Fig. 3-1. Free Energy of Formation of Oxides

Source: H. J. T. Ellingham, *J. Soc. Chem. Ind.*, 1944, Vol. 63, p. 127, Society of Chemical Industry.

defined by the well-known thermodynamic relationship

$$\left(\frac{\partial \Delta G^0}{\partial T}\right)_P = -\Delta S^0. \tag{3-1}$$

The positive slope for most cases shown in Fig. 3-1 indicates that the entropy change is negative, or that a decrease in entropy occurs for these reactions as written. This is consistent with the fact that a mole of oxygen disappears on reaction, yielding an entropy decrease. The free energy–temperature relationships consist of straight line segments with break points in the lines denoting a phase change. If the entropy change occurs in the reactants, the increase in entropy by a phase change, e.g., solid to liquid or liquid to gas, increases the entropy of the reactants and results in a more negative entropy change for the reaction. In cases where the product shows a phase change, with an accompanying increase in entropy, the slope of the line decreases in a corresponding manner. Standard free energies of formation at high temperatures for several reactions of metallurgical interest are summarized in Table A-II in the form of equations that are linear with temperature.

The standard free energy change for an overall reduction process can be calculated by combining free energy changes for the individual oxidation–reduction reactions involved. Consider the two reactions,

$$2Mg(g) + O_2(g) = 2MgO(s), \qquad (3\text{-}2)$$

$$Si(s) + O_2(g) = SiO_2(s). \qquad (3\text{-}3)$$

The standard free energy change for the reduction of magnesium oxide by silicon can be determined by subtracting Eq. (3-2) from Eq. (3-3) to obtain

$$2MgO(s) + Si(s) = 2Mg(g) + SiO_2(s). \qquad (3\text{-}4)$$

The standard free energy change is given by the difference between the standard free energy changes for Eq. (3-3) and Eq. (3-2), i.e.,

$$\Delta G_4^0 = \Delta G_3^0 - \Delta G_2^0 \qquad (3\text{-}5)$$

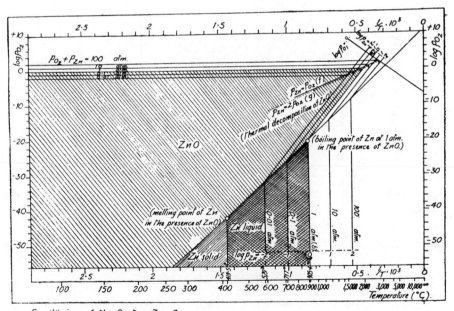

Fig. 3-2. Equilibrium in the System Zn-O

Source: M. J. N. Pourbaix and C. M. Rorive-Boute, *Disc. Faraday Soc.*, 1948, Vol. 4, p. 142, Butterworth and Company (Publ.) Ltd.

This reaction forms the basis for the Pidgeon process[2] for the production of magnesium.

Pourbaix[3] proposed another form of diagram for treating phase equilibria in oxide systems in which he plotted the logarithm of the equilibrium oxygen pressure against reciprocal temperature. An example of a Pourbaix diagram for the zinc–oxygen system is shown in Fig. 3-2. The lines on Fig. 3-2 represent phase equilibria between zinc oxide and zinc in solid, liquid, or gaseous form. The phase equilibria are shifted by pressure, and this is indicated by the vertical lines, which show increasing zinc vapor pressure with increasing temperature. Figure 3-2 shows that heating zinc oxide to high temperatures will result in thermal dissociation, depending upon the partial pressure of oxygen and zinc, according to the reaction,

$$2ZnO(s) = 2Zn(g) + O_2(g). \tag{3-6}$$

If zinc oxide decomposes under high vacuum conditions wherein the pressure of zinc is twice the pressure of oxygen in the atmosphere as shown by curve (g) of Fig. 3-2,

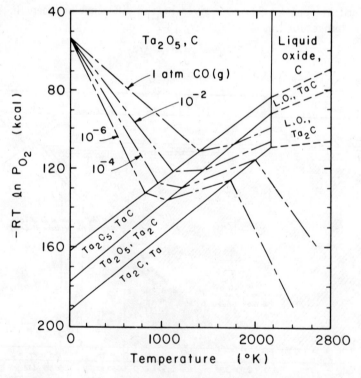

Fig. 3-3. Pourbaix-Ellingham Diagram for the Ta-C-O System

Source: W. L. Worrell and J. Chipman, *Trans. Met. Soc. AIME*, 1964, Vol. 230, p. 1684.

[2] L. M. Pidgeon, *Trans. Canad. Mm. Inst.* **49**, 621 (1946).
[3] M. J. N. Pourbaix, *Bull. Soc. Chim. Belg.* **53**, 145 (1944).

the decomposition pressure or total pressure in the system reaches 0.01 atm at about 1477°C and 1 atm at about 1962°C.[4]

A very useful equilibrium phase diagram can be derived by combining the concepts of the Ellingham and Pourbaix diagrams. This is illustrated in Fig. 3-3, a Pourbaix–Ellingham diagram for the tantalum–carbon–oxygen system. In Fig. 3-3, the near linearity of the Ellingham diagram and the description of phase equilibria of the Pourbaix diagram have been combined by plotting versus temperature the standard free energy of decomposition of the phases in equilibrium to form oxygen gas at 1 atm pressure.

At 1500°K and 1 atm carbon monoxide pressure, tantalum oxide, Ta_2O_5, is in equilibrium with tantalum carbide, TaC. If the temperature is increased to 1800°K, the carbide Ta_2C will form, or if at 1500°K the carbon monoxide pressure were increased to several atmospheres, the tantalum carbide would decompose, forming graphite; a decrease in pressure would result in the formation of the dicarbide Ta_2C.

Example Problem 3-1. Figure 3-3 is a Pourbaix–Ellingham diagram for the Ta–C–O system. Using the data (averaged for the range 298–2000°K) below,[5] calculate the equation for the line representing the gas–solid equilibrium between carbon, Ta_2O_5, and TaC as shown on the diagram. Also calculate the "10^{-4} atm CO(g)" line shown in the tantalum oxide–graphite field of Fig. 3-3.

Compound	$\Delta G_f{}^0$ (cal/mole)	Melting point (°K)
Ta_2O_5	$-480,000 + 96.5\,T$	2145
Ta	—	3273
$Ta_2C(s)$	$-47,000 + 0.5\,T$	3773
$TaC(s)^a$	$-31,000 + 1.6\,T$	3983
$TaC(s)^b$	$-35,600 + 1.6\,T$	—
CO(g)	$-26,760 - 21.0\,T$	—
$CO_2(g)$	$-94,260 - 0.3\,T$	—

[a] Metal rich carbide.
[b] Carbon rich carbide.

Solution. Expressing the reaction for the gas-solid equilibrium in terms of one mole of oxygen:

Reaction	ΔG^0 (cal/g-mole)	
$\tfrac{2}{5}Ta_2O_5(s) = \tfrac{4}{5}Ta(s) + O_2(g)$	$+192,000 - 38.6\,T$	(3-a)
$\tfrac{4}{5}Ta(s) + \tfrac{4}{5}C(s) = \tfrac{4}{5}TaC(s)$	$-28,480 + 1.3\,T$	(3-b)
$\tfrac{2}{5}Ta_2O_5(s) + \tfrac{4}{5}C(s) = \tfrac{4}{5}TaC(s) + O_2(g)$	$+163,520 - 37.3\,T$	(3-c)

[4] M. J. N. Pourbaix and C. M. Rorive-Boute, *Disc. Faraday Soc.* **4,** 139 (1948).
[5] W. L. Worrell and J. Chipman, *Trans. Met. Soc. AIME* **230,** 1682 (1964).

The equilibrium constant for reaction (3-c) can be expressed as

$$K_c = P_{O_2} \qquad (3\text{-d})$$

since the solids are at unit activity. Then the line representing the equilibrium between Ta_2O_5, TaC, and C is

$$-RT \ln P_{O_2} = \Delta G^0 = 163{,}520 - 37.3\,T \text{ cal.} \qquad (3\text{-e})$$

Considering a carbon monoxide potential fixed at 10^{-4} atm in the Ta_2O_5–C field: the equilibrium between carbon, oxygen, and carbon monoxide is

$$2CO(g) = 2C(s) + O_2(g)\,. \qquad (3\text{-f})$$

The equilibrium constant for Eq. (3-f) is

$$K_f = \frac{P_{O_2} a_c^2}{P_{CO}^2} \qquad (3\text{-g})$$

at unit carbon activity and 10^{-4} atm CO pressure.

$$K_f = P_{O_2}(10)^8. \qquad (3\text{-h})$$

Then the line representing 10^{-4} atm of CO pressure is

$$-RT \ln P_{O_2} = 53{,}520 + 42.0\,T + (8)(2.303)(1.987)\,T, \qquad (3\text{-i})$$

$$-RT \ln P_{O_2} = 53{,}520 + 78.6\,T. \qquad (3\text{-j})$$

Carbon–Oxygen Equilibria—The Boudouard Reaction

The Boudouard reaction,

$$2CO(g) = C(s) + CO_2(g), \qquad (3\text{-7})$$

is composed of the reactions for the oxidation of carbon to carbon dioxide and carbon monoxide:

$$C(s) + O_2(g) = CO_2(g), \qquad (3\text{-8})$$

$$C(s) + \tfrac{1}{2}O_2(g) = CO(g). \qquad (3\text{-9})$$

The corresponding standard free energies of formation for carbon monoxide and carbon dioxide can be represented in the temperature range 1500–2000°K by the relationships,

$$\Delta G_8^0 = -94{,}755 + 0.02\,T, \qquad (3\text{-10})$$

$$\Delta G_9^0 = -28{,}200 - 20.16\,T. \qquad (3\text{-11})$$

The corresponding free energy change in the same temperature range for the Boudouard reaction Eq. (3-7) is

$$\Delta G_7^0 = -38{,}355 + 40.34\,T. \qquad (3\text{-12})$$

The Boudouard reaction forms the basis for carbothermic reduction of metals and is extremely important in smelting processes involving carbon as a reducing agent, e.g., the iron or lead blast furnace process or the smelting of tin or zinc. From Fig.

3-1, it can be seen that at 1 atm the standard free energy for the oxidation of carbon to form carbon monoxide and the standard free energy of reaction for the formation of carbon dioxide are the same at a temperature of approximately 710°C. Below this temperature, carbon monoxide is a more active reducing agent, whereas above it, solid graphite is the more active reducing agent. The oxidation of carbon will produce a mixture of carbon monoxide and carbon dioxide, but at low temperatures carbon dioxide will be the predominant component of the gas phase in equilibrium with solid carbon, whereas at higher temperatures carbon monoxide is the predominant gaseous component.

Example Problem 3-2. The standard free energies of formation for CO and CO_2 are equal at 710°C. In a system consisting of only carbon and oxygen with excess solid carbon present, what would the gas phase composition be at 1 atm total pressure? At 10^{-4} atm?

Solution. Since the ΔG^0 for the reactions,

$$C + O_2 = CO_2, \tag{3-a}$$

$$2C + O_2 = 2CO, \tag{3-b}$$

is equal at 710°C, their difference is zero, and $\log K = 0$ for the reaction,

$$2CO = C + CO_2. \tag{3-c}$$

Then at 710°C

$$K = 1 = \frac{P_{CO_2}}{P_{CO}^2} \tag{3-d}$$

but

$$P_{CO} = 1 - P_{CO_2}. \tag{3-e}$$

Substitution of Eq. (3-e) into (3-d) and rearranging,

$$(P_{CO_2})^2 - 3P_{CO_2} + 1 = 0, \qquad P_{CO_2} = 0.382 \text{ atm.} \tag{3-f}$$

Then

$$X_{CO_2} = 0.382,$$

$$X_{CO} = 0.618.$$

At 10^{-4} atm, a similar calculation yields $X_{CO_2} = 0.0001$. The reduction in pressure, following LeChatelier's Principle, drives Eq. (3-c) to the left.

Thermodynamic Stability of Sulfides

The standard free energies of formation of common metallic sulfides are presented on an Ellingham diagram (Fig. 3-4). The position of the curve for the reaction,

$$2O_2 + S_2 = 2SO_2 \tag{3-13}$$

shows that oxygen will displace sulfur from the sulfides of all of the metals except the alkaline earths, i.e., an oxidizing roast can be employed to convert sulfides to oxides, as discussed in the previous chapter. In contrast, the position for the curve for the reaction,

$$2H_2 + S_2 = 2H_2S \qquad (3\text{-}14)$$

shows that hydrogen is not suitable for reducing sulfides, and, similarly, neither is carbon.

The possibility of desulfurizing steels with calcium is clearly shown by the position of the line for the reaction,

$$2Ca + S_2 = 2CaS. \qquad (3\text{-}15)$$

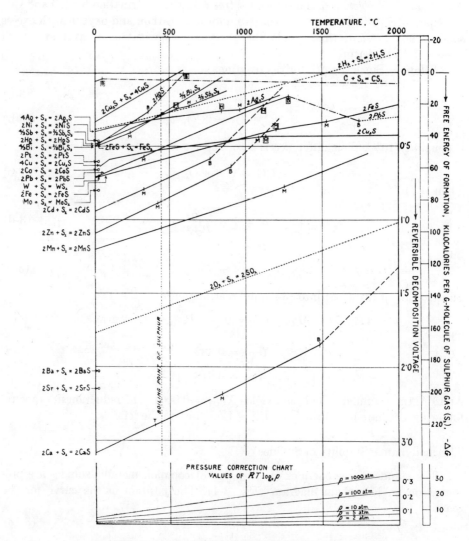

Fig. 3-4. Free Energy of Formation of Sulfides

Source: H. J. T. Ellingham, *J. Soc. Chem. Ind.*, 1944, Vol. 63, p. 128, Society of Chemical Industry.

The ability of manganese to form manganese sulfide in steels during hot rolling is suggested by the position of the line for the formation of manganese sulfide relative to iron sulfide, or the sulfides of other elements commonly present in steels.

Reducibility of Chlorides

The Ellingham diagram giving the standard free energy of formation of chlorides versus temperature is presented in Fig. 3-5. The position of the curve for the formation of carbon tetrachloride clearly shows that carbon is not a suitable reducing agent for chlorides.

The position of the curve for the formation of hydrogen chloride indicates that hydrogen is a potential reducing agent for several of the metallic chlorides. If the curve for the standard free energy of formation for a metallic chloride lies above that for the formation of hydrogen chloride at a given temperature, the metallic chloride will be reduced to the metal by hydrogen, provided the substances are in their standard states. Conversely, a metal whose chloride lies below the hydrogen chloride curve at a given temperature would, under standard conditions, be converted to the chloride by treatment with HCl gas. The hydrogen chloride curve decreases slightly with increasing temperature, whereas the curves for most of the other chlorides increase with temperature, thus providing a more favorable condition for hydrogen reduction at higher temperatures. The hydrogen reduction of chromium chloride is illustrated in the following example problem.

Example Problem 3-3. The reduction of chromium chloride by hydrogen gas has been proposed as a method for producing pure chromium. Calculate the purity of hydrogen gas that must be maintained in a reaction chamber at 800°C to carry out the reduction,

$$CrCl_2(s) + H_2(g) = Cr(s) + 2HCl(g). \qquad (3\text{-}a)$$

Solution. Equation (3-a) can be expressed as the difference between the reactions,

$$H_2(g) + Cl_2(g) = 2HCl(g), \qquad (3\text{-}b)$$

$$Cr(s) + Cl_2(g) = CrCl_2(s). \qquad (3\text{-}c)$$

From Fig. 3-5, the standard free energy change for Eq. (3-a) is

$$\Delta G^0_a = \Delta G^0_b - \Delta G^0_c = -48,500 - (-63,300) = +14,800 \text{ cal}. \qquad (3\text{-}d)$$

Thus it can be seen that the reaction is not spontaneous at standard conditions. At equilibrium at 800°C, with the solid phases at unit activity

$$\Delta G^0_a = -RT \ln \left[\frac{P_{HCl}{}^2 a_{Cr}}{P_{H_2} a_{CrCl_2}} \right] = -RT \ln \left[\frac{P_{HCl}{}^2}{P_{H_2}} \right] = 14,800, \qquad (3\text{-}e)$$

$$\log \left[\frac{P_{HCl}{}^2}{P_{H_2}} \right] = -3.017. \qquad (3\text{-}f)$$

Assuming a total pressure of 1 atm in the system,

$$P_{HCl} + P_{H_2} = 1. \qquad (3\text{-}g)$$

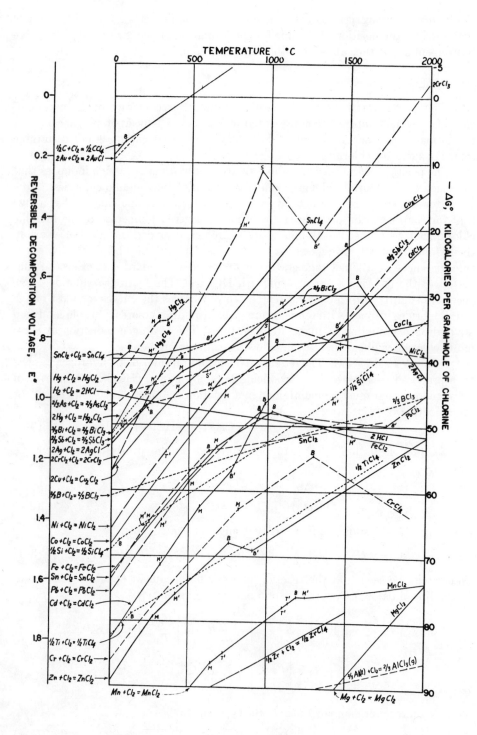

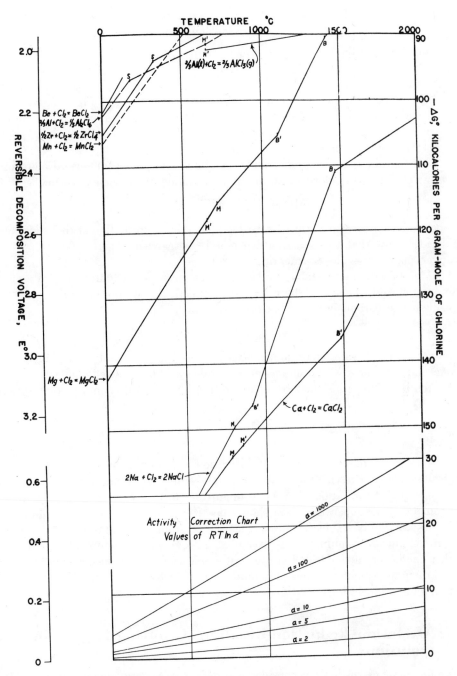

Fig. 3-5. Standard Free Energies of Formation for Metal Chlorides as a Function of Temperature, $\Delta G°$ between $+5,000$ and $-160,000$ cal.

Source: H. H. Kellogg, *Trans. AIME*, 1950, Vol. 188, p. 864.

Then

$$P_{HCl} = 0.031 \text{ atm},$$

$$P_{H_2} = 0.969 \text{ atm}.$$

In order to cause the reaction to occur as written, the concentration of HCl must be maintained below 3.1 vol% (at 1 atm), e.g., by removing HCl and adding dry hydrogen to the system.

Magnesium chloride is the most stable chloride shown in Fig. 3-5, with the exception of the chlorides of sodium and calcium. The high stability of magnesium chloride forms the basis of the Kroll process for the production of titanium sponge by magnesium reduction of titanium tetrachloride.

The chlorination of metallic oxides can be accomplished for several metals by heating the oxide in chlorine gas, to allow a reaction of the general form,

$$MO + Cl_2 = MCl_2 + \tfrac{1}{2} O_2, \tag{3-16}$$

to occur. A comparison of the standard free energies of formation of the oxides and chlorides show that this reaction can readily be carried out at 1000°C for lead, copper, manganese, nickel, zinc, tin, and iron.

Metals with more stable oxides, such as magnesium, chromium, titanium, aluminum, and silicon, can be chlorinated directly in the presence of a reducing agent such as carbon, carbon monoxide, or sulfur. For example,

$$MO + Cl_2 + C = MCl_2 + CO. \tag{3-17}$$

Anhydrous magnesium chloride has been produced commercially from MgO using coke and chlorine gas. Titanium tetrachloride can be produced similarly by the chlorination of ilmenite, the most common titanium mineral (nominal composition, $FeTiO_3$).

Stability of Carbides and Nitrides

The standard free energies of formation for several metallic carbides and nitrides also are presented in Table A-II. A comparison made between standard free energies of formation of carbides or nitrides in temperature ranges typical of pyrometallurgical operations will show that titanium and zirconium carbides and nitrides are highly stable. This precludes the possibility of carbothermic reduction of zirconium or titanium oxides except at extremely high temperatures. On the other hand, iron nitride is one of the least stable metallic nitrides, whereas nitrides of chromium, vanadium, and aluminum are much more stable, and as such are often found as inclusions in steel.

3-2 Oxidation–Reduction Reactions Involving Liquid Metallic Solutions

The foregoing discussion has considered standard free energy changes in which the reactants and products were present in the most commonly defined state, i.e., as pure materials. In practical metallurgical operating situations, reactions are often carried out in the presence of a liquid metallic phase that may have an appreciable solubility for many of the reactants and products present. It is necessary under these circumstances to consider the resulting change in the activities of the various components. In some circumstances an advantage in thermodynamic calculation is realized by changing the standard state.

Thermodynamics of Metallic Solutions

The chemical activity, a, (the tendency of a component of a phase to react), is dependent upon the concentration of the component in that phase, as well as temperature (see Eq. 3-22). In a solution, the chemical activity of a component is related in some fashion to the amount of the component present. As an example, if liquid metal A and liquid metal B are mixed in equal molar proportion, component A might have about one-half the chemical activity of pure liquid A. If the solution were more dilute in A, its activity would be less. A component whose activity is *directly* proportional to its concentration expressed as mole fraction is said to be ideal and to obey *Raoult's law*:

$$a_A = \frac{\text{moles A}}{\text{total moles}} = \frac{n_A}{n} = X_A. \tag{3-18}$$

At $n_A = n$ in Eq. (3-18), $a = 1$. Thus, the pure phase is the reference concentration or the *standard state*. The reference concentration could also have been chosen at some other point. For example, a 1 mol % solution could be used as the standard state and the activities indexed from that point. Then, in an ideal solution, the component activities are equal to molar percentages, i.e., a 1 mol % solution would have an activity equal to 1, whereas a pure solution of a component would have an activity (index) of 100.

Other standard states often are used as well. For a gas, the standard state is taken usually as 1 atm pressure of that particular gas component. An ideal gas with a partial pressure of 1 torr (1 mm of mercury) would have an activity of 1/760. If, however, 1 torr had been chosen as the standard state, the activity would be equal to 1 if the gas component were at a pressure of 1 torr and equal to 760 at 1 standard atm. In metallic solutions, a standard state often is chosen such that the activity at infinite dilution is equal to the concentration in weight percent. This standard state, referred to as the hypothetical 1 wt% solution, often is convenient since compositions in industrial practice generally are expressed in weight percent.

Free Energy and Activity. The free energy of a component, i, is related to the chemical activity of the component by the relationship

$$\bar{G}_i - \bar{G}_i^0 = RT \ln (a_i) = 4.575 \, T \log_{10} (a_i), \tag{3-19}$$

where $\bar{G}_i$ is the free energy of component i within the phase and $\bar{G}_i^0$ is the free energy of the component in its standard state in which $a_i^0 = 1$.

Nonideal Solutions. Most real solutions are not ideal. In order that a solution be ideal, the atomic species of the component must interact with those of the other components in the same manner that they do with their own.

The activity of silicon in molten iron is shown in Fig. 3-6. Silicon does not

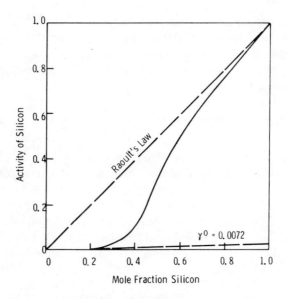

Fig. 3-6. Activity of Silicon in Liquid Iron at 1600°C (2912°F)

Based on *Basic Open Hearth Steelmaking*, 3rd Edition, edited by G. Derge, AIME, New York,
1964, p. 674.

behave ideally. The silicon atoms are more strongly associated with the iron atoms than with other silicon atoms, thus lowering the chemical activity of silicon below the ideal behavior prediction. The strong association between silicon and iron atoms is indicated by the facts that (1) energy is given off (exothermic heat of solution) when silicon is added to iron, and (2) intermetallic compounds of iron and silicon are readily formed in the solid phase.

Figure 3-7 shows the activity of copper in iron. Unlike silicon, copper atoms do not associate as readily with iron atoms as with other copper atoms. Therefore, the chemical activity of copper is increased above that predicted on the basis of ideal behavior. This is also shown by (1) endothermic heat of solution of copper, and (2) the tendency to form two immiscible liquids.

Since few solutions are ideal, it is necessary to account for this when calculating activities from concentrations. This is done by means of an activity coefficient, defined by the ratio of activity to mole fraction:

$$\gamma_i = \frac{a_i}{X_i}. \tag{3-20}$$

Dilute Solution Behavior. In considering copper and silicon solutions with liquid iron, it can be noted from Figs. 3-6 and 3-7 that although solutions of copper and silicon in iron do not behave ideally, their activities in *dilute solution are proportional* to concentration. Such behavior is described by *Henry's law.* The activity coefficient for the Henry's law region is defined by

$$\gamma_i^0 = \lim \, (\gamma_i)_{X_i \to 0} = \lim \left[\frac{a_i}{X_i} \right]_{X_i \to 0}, \tag{3-21}$$

A summary of activity coefficients at infinite dilution for a number of metallic solutions is presented in Table A-III.

From the definition of the activity coefficient in Eq. (3-20) and the defining expression for free energy it can be shown that

$$\frac{d(\ln \gamma_i)}{d(1/T)} = \frac{\Delta H_i}{R},\tag{3-22}$$

where ΔH_i is the partial molar enthalpy of solution for component i. This result suggests that a straight line extrapolation of $\ln \gamma_i$ versus the inverse of the absolute temperature may be used over a short temperature range to obtain values of γ_i at temperatures other than those for which data are available.

For cases where the solutions of interest are dilute, as in most refining operations, it often is convenient to refer activities to the Henry's law behavior of the dilute solution, expressed in weight percent. The *Henrian activity*, h, then is defined

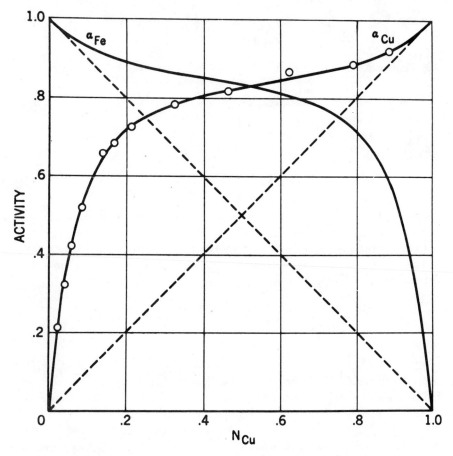

Fig. 3-7. Activities of Copper and Iron in Fe-Cu Alloys at 1550°C. (Standard State: Pure Liquid)

Source: J. P. Morris and G. R. Zellers, *Met. Soc. Trans. AIME*, 206, (1956), p. 1089.

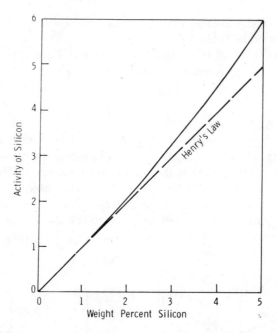

Fig. 3-8. Activity of Silicon in Liquid Iron at 1600°C (Standard State: Infinite Dilution Referred to One Weight Percent Solution)

Based on *Basic Open Hearth Steelmaking*, 3rd Edition, edited by G. Derge, AIME, New York, 1964, p. 674.

by the expression,

$$h_i = \left[\frac{a_i}{(a_i/\mathrm{wt\%}\ i)} \right]_{\mathrm{wt\%}\ i \to 0} \tag{3-23}*$$

and on the limiting slope portion of the activity curve, the concentration in weight percent can be substituted directly for the activity.

At concentrations beyond the Henry's law range, a *Henrian activity coefficient*, f, is used to define the activity relative to the concentration. The Henrian activity coefficient is defined by the relationship,

$$f_i = \left(\frac{h_i}{\mathrm{wt\%}\ i} \right), \tag{3-24}$$

whereby

$$h_i = f_i(\mathrm{wt\%}\ i). \tag{3-25}$$

In accordance with Eq. 3-23 the activity coefficient approaches unity as the concentration of the solute approaches zero. Figures 3-8 and 3-9 show the dilute solution behavior in molten iron for silicon and copper, respectively.

---

* Note: In deriving these dilute solution relationships, the symbol, h, has been used for Henrian activity. However, the reader should be aware that the symbol, a, often is used for activity, regardless of standard state; even in this book, e.g., Fig. 3-9:

Conversion from One Standard State to Another

It is frequently necessary to convert from one standard state to another in order to make use of data which may give ΔG^0 as a function of temperature relative to a reference state which differs from the desired one. For example, the free energy

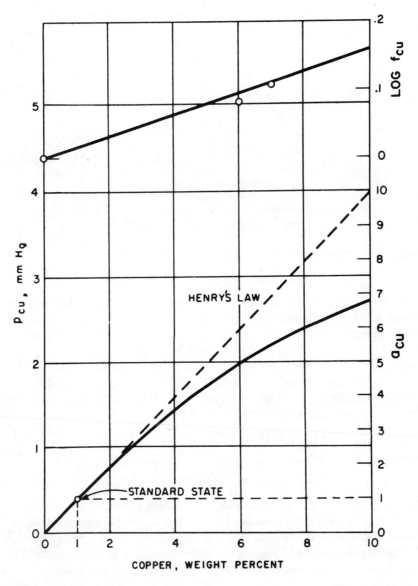

Fig. 3-9. Vapor Pressure and Activity of Copper Dissolved in Liquid Iron at 1540°C
(Standard State: Infinite Dilution Referred to One Weight Percent Solution)

Source: *Basic Open Hearth Steelmaking*, 3rd Edition, edited by G. Derge, AIME, New York, 1964, p. 652.

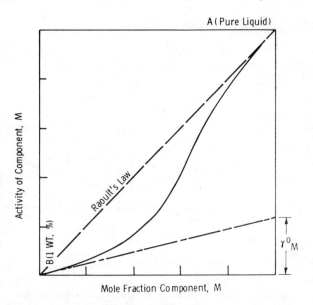

Fig. 3-10. Activity-Composition Diagram Showing Factors Involved in Converting Activities from One Standard State to Another

of formation of a metal oxide may be known for the reaction,

$$M(\text{pure liquid}) + O_2(\text{gas}) = MO_2(\text{solid}), \tag{3-26}$$

but the oxidation of interest may occur from solution in a liquid metal solvent, say liquid iron. The activity of the metal must be expressed in terms of its concentration in solution in liquid iron. Figure 3-10 shows the assumed activity of a given metal, M, in solution in iron as a function of mole fraction, X_M.

If the reaction of interest involves a dilute solution of M in iron such that the activity could be closely represented by Henry's law behavior, a standard state of infinite dilution referred to weight percent (State B of Fig. 3-10) would be more convenient. The reaction of interest then would be*

$$\underline{M}(1 \text{ wt\% solution}) + O_2(\text{gas}) = MO_2(\text{solid}). \tag{3-27}$$

Subtraction of Eq. 3-27 from Eq. 3-26 yields

$$M(\text{pure liquid}) = \underline{M}(1 \text{ wt\% solution}). \tag{3-28}$$

Equation 3-28 represents the conversion of standard states, i.e., from State A to State B on Fig. 3-10. This conversion is usually not given in thermodynamic data references and will be developed, using an iron-base solution as an example. A parallel development could be applied to other solvents, or to other standard states.

The difference in standard free energy changes for the two reactions is

$$\Delta G^0{}_{26} - \Delta G^0{}_{27} = \Delta G_{28} = -RT \ln \frac{K_{26}}{K_{27}} = +RT \ln \frac{a_M}{h_M}, \tag{3-29}$$

---

* The underline notation refers to the element in metallic solution.

where K_{26} and K_{27} are the equilibrium constants for the reactions of Eqs. 3-26 and 3-27, respectively. a_M is the activity of M in solution relative to liquid pure M and h_M is the Henrian activity as defined by Eq. 3-23. The activity of solute M, relative to State A, is a_M. The activity of solute M, relative to State B, is

$$h_M = f_M(\text{wt\% M}). \tag{3-30}$$

The Henrian activity coefficient defines the deviation from Henry's law behavior, and referring to Fig. 3-10

$$f_M = \frac{h_M}{(\text{wt\% M})} = \frac{a_M}{(\gamma_M^0 X_M)}. \tag{3-31}$$

Rearranging,

$$a_M = \frac{h_M \gamma_M^0 X_M}{\text{wt\% M}}. \tag{3-32}$$

The relationship between mole fraction and weight percent is

$$X_M = \frac{\text{wt\% M}}{(AW)_M} \bigg/ \left[\frac{\text{wt\% M}}{(AW)_M} + \frac{100 - \text{wt\% M}}{(AW)_S} \right], \tag{3-33}$$

where $(AW)_M$ is the atomic weight of M and $(AW)_S$ is the atomic weight of the solvent, in this case 55.85 for iron. Application of Eq. 3-33 to a dilute iron solution

$$X_M = \text{wt\% M} \left[\frac{0.5585}{(AW)_M} \right]. \tag{3-34}$$

Noting that the standard state solution follows infinite dilution behavior, Eq. 3-34 may be substituted into 3-32, and Eq. 3-29 yields

$$\Delta G_{28} = RT \ln \left[\frac{0.5585 \gamma_M^0}{(AW)_M} \right]. \tag{3-35}$$

Equation 3-35 represents the free energy change for conversion of standard states according to Eq. 3-28. Thus the standard free energy change for Eq. 3-27 is

$$\Delta G^0_{27} = \Delta G^0_{26} - RT \ln \left[\frac{0.5585 \gamma^0}{(AW)_M} \right] = -RT \ln \left[\frac{a_{MO_2}}{h_M P_{O_2}} \right]. \tag{3-36}$$

The Henrian activity may be taken as the concentration in weight percent, i.e., $h_M \simeq \text{wt\% M}$, in the dilute solution region.† This convenience represents a substantial advantage in most calculations involving dilute solutions.

Free energy changes for conversion of the standard state from the pure component to the infinitely dilute solution referred to 1 wt% are presented in Table A-IV for common alloying elements in liquid iron.

† It should be noted that the extent of the dilute solution region depends upon experimental accuracy.

Solute Interactions in Dilute Solutions

The thermodynamic properties of a solute in dilute solution can be influenced by the presence of other alloying elements. The prediction of these influences ultimately may be accomplished entirely from statistical mechanical considerations. However, at present, liquid solution models that are simple enough for analytical treatment are inadequate for this purpose. As a means for expressing experimental data on dilute solutions, Wagner[6] devised the concept of the *interaction parameter*. The interaction parameter is a coefficient in a Taylor series expansion of the logarithm of the activity coefficient taken at infinite dilution and expressed in terms of the concentration of alloying elements:

$$\ln \gamma_i = \ln \gamma_i^0 + \left[x_i \left(\frac{\partial \ln \gamma_i}{\partial x_i} \right)_{x_{\text{solvent}} \to 1} + x_j \left(\frac{\partial \ln \gamma_i}{\partial x_j} \right)_{x_{\text{solvent}} \to 1} + \cdots \right]$$

$$+ \left[\frac{1}{2} x_i^2 \left(\frac{\partial^2 \ln \gamma_i}{\partial x_i^2} \right)_{x_{\text{solvent}} \to 1} + x_i x_j \left(\frac{\partial^2 \ln \gamma_i}{\partial x_i \partial x_j} \right)_{x_{\text{solvent}} \to 1} + \cdots + \cdots \right].$$

$$(3\text{-}32)$$

If all except the first-order terms are neglected, the expression for the logarithm of the activity coefficient becomes linear with respect to the mole fractions of the solutes present in the dilute alloy.

$$\ln \gamma_i = \ln \gamma_i^0 + x_i \epsilon_i^i + x_j \epsilon_i^j + x_k \epsilon_i^k + \cdots. \qquad (3\text{-}33)$$

The interaction parameters are thus defined as

$$\epsilon_i^i = \left(\frac{\partial \ln \gamma_i}{\partial x_i} \right)_{x_{\text{solvent}} \to 1}, \qquad \begin{array}{l} \text{self-interaction parameter,} \\ \text{i.e., deviation from Henry's law;} \end{array} \qquad (3\text{-}34)$$

$$\epsilon_i^j = \left(\frac{\partial \ln \gamma_i}{\partial x_j} \right)_{x_{\text{solvent}} \to 1}, \qquad \text{ternary interaction parameter.} \qquad (3\text{-}35)$$

The experimental determination of the parameter ϵ_i^i, which represents the influence of alloying element i on its own activity coefficient in dilute solution, can be accomplished from thermodynamic studies of the dilute binary alloy. The parameter ϵ_i^j represents the effect of additions of component j on the activity coefficient of component i when both are very dilute. Thermodynamic data on the dilute ternary alloy are required to compute this parameter.

The inclusion of second-order terms would seem to permit a better representation of the data since a binomial expression could be fitted to the activity data rather than a linear one. Indeed, such a representation can be extended validly to compositions further away from the infinitely dilute solution than can the linear form!

[6] C. Wagner, *Thermodynamics of Alloys*, p. 52, Addison-Wesley, Reading, Mass., 1952.
[7] P. H. Turnock and R. D. Pehlke, *Trans. Met. Soc. AIME* **236,** 1540 (1966).

It can be demonstrated from the Gibbs–Duhem equation that

$$\epsilon_i{}^j = \epsilon_j{}^i. \tag{3-36}$$

This reciprocity relationship is an important advantage of the Wagner representation.

A slight modification of this representation has been proposed[8] that involves the use of weight percent as a concentration variable rather than mole fraction, and also common logarithms rather than natural ones:

$$\log f_i = \log f_i{}^0 + e_i{}^i(\% \, i) + e_i{}^j(\% \, j) + \cdots, \tag{3-37}$$

where

$$f_i = \frac{h_i}{\% \, i} \quad \text{and} \quad e_i{}^j = \left[\frac{\log f_i}{\% \, j}\right]_{\% \, \text{solvent} \to 100}. \tag{3-38}$$

The reciprocal relationship for this parameter is

$$e_j{}^i = e_i{}^j \frac{M_j}{M_i} + 0.434 \times 10^{-2} \left[\frac{M_i - M_j}{M_i}\right], \tag{3-39}$$

where M is the atomic weight of the subscripted specie. The relationship between the two parameters is

$$\epsilon_i{}^j = \frac{(2.303)\,(100)\,M_j}{M_{\text{solvent}}} e_i{}^j + \frac{M_{\text{solvent}} - M_j}{M_{\text{solvent}}}. \tag{3-40}$$

It can be shown that the interaction parameter will vary linearly with reciprocal absolute temperature, providing the partial molar enthalpy is essentially independent of temperature. A summary of interaction parameters in ferrous and non-ferrous systems is presented in Tables A-V and A-VI.

3-3 Industrial Application of Reduction Reactions

The thermodynamics of reduction reactions plays a role in several commercial processes. Examples are presented for reduction by hydrogen, carbon, and metallic elements such as magnesium and silicon.

The *Kroll process*[9] is employed to reduce the tetrachlorides of titanium or zirconium with magnesium metal. The high stability of magnesium chloride (as shown in Fig. 3-5) indicates that the reduction reaction

$$\text{Mg}(1) + \tfrac{1}{2}\text{TiCl}_4(g) = \tfrac{1}{2}\text{Ti}(s) + \text{MgCl}_2(1) \tag{3-41}$$

is favorable for production of titanium. A flow diagram for this process is shown in Figure 3-11. The anhydrous magnesium chloride byproduct is recycled to produce magnesium metal and chlorine gas.

The titanium tetrachloride used in the Kroll Process is prepared by heating titania and carbon in an electric furnace to form titanium carbide which in turn is

[8] J. Chipman, *J. Iron Steel Inst.* **180**, 97 (1955).
[9] W. J. Kroll, *Trans. Electrochem. Soc.* **78**, 35 (1940).

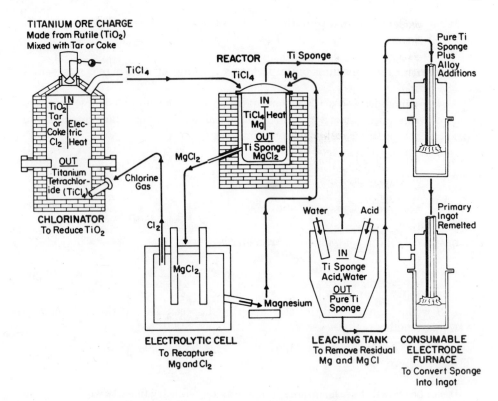

Fig. 3-11. Flow Diagram for Kroll Process

Based on S. C. Williams, *Report on Titanium*, Brundage, Story and Rose, New York, 1965.

reacted with chlorine gas. The titanium tetrachloride is vaporized by the exothermic reaction and condensed as a liquid. Redistillation of the tetrachloride will improve the purity of the titanium produced from it.

The Kroll process is conducted in a heated steel chamber that is maintained at a temperature of about 750°C. Magnesium bars are placed in the bottom of the chamber, which is evacuated and filled with helium; hydrogen flushing at temperature is sometimes used to remove oxygen from the chamber surfaces. The $TiCl_4$ is allowed to drip on the molten magnesium, where it reacts exothermically to form titanium sponge and magnesium chloride.

Sodium reduction of $TiCl_4$ also is used to produce titanium. The chloride processing avoids contamination with carbon, nitrogen, or oxygen which render titanium brittle.

Zirconium sponge is produced in the same manner. $ZrCl_4$ is a solid at ambient temperatures and is vaporized by heating in the upper zone of a closed chamber. The $ZrCl_4$ vapor diffuses to the lower zone where it reacts with liquid magnesium at about 800°C. Zirconium also has been produced by calcium or calcium hydride reduction of zirconia.

Production of Tungsten—The principal minerals of tungsten are wolframite

(FeWO$_4$) and scheelite (CaWO$_4$). By suitable chemical processing, tungsten is separated from these minerals and converted to sodium tungstate, from which the pure oxide is precipitated. The oxide is then reduced to metallic tungsten by reaction with dry hydrogen at a temperature between 1200 and 1400°C.

Example Problem 3-4. In reducing tungsten trioxide with hydrogen at 1400°C, the thermodynamically limiting step is the final reduction of WO$_2$ to metallic tungsten. What is the maximum concentration of water vapor which could be allowed in the reducing atmosphere?

Solution. From Table A-II

$$\text{W(s)} + \tfrac{3}{2}\text{O}_2\text{(g)} = \text{WO}_2\text{(s)}, \qquad \Delta G_a^0 = -134{,}800 + 38.92, \qquad \text{(3-a)}$$

$$\text{H}_2 + \tfrac{1}{2}\text{O}_2 = \text{H}_2\text{O(g)}, \qquad \Delta G_b^0 = -60{,}200 + 13.94\,T. \qquad \text{(3-b)}$$

Combining Eqs. (3-a) and (3-b)

$$\tfrac{1}{2}\text{WO}_2\text{(s)} + \text{H}_2\text{(g)} = \tfrac{1}{2}\text{W(s)} + \text{H}_2\text{O(g)}, \qquad \Delta G_c^0 = 7200 - 5.52\,T. \qquad \text{(3-c)}$$

The equilibrium constant for Eq. (3-c) at 1400°C is

$$K = \frac{a_\text{W}^{1/2} P_{\text{H}_2\text{O}}}{a_{\text{WO}_2}^{1/2} P_{\text{H}_2}} \qquad \text{(3-d)}$$

$$\log K = \frac{-7200}{4.575(1673)} + \frac{5.52}{4.575}$$

$$\log K = -0.941 + 1.207 = 0.266.$$

For this case, the activities of tungsten and tungsten oxide are unity and

$$K = \frac{P_{\text{H}_2\text{O}}}{P_{\text{H}_2}} = 1.84.$$

At 1400°C and any total pressure of hydrogen and water vapor, the reducing gas must be less than about 65% water vapor to effect the conversion to metallic tungsten.

The *Pidgeon process* is employed to produce magnesium from calcined dolomite by the reaction,

$$2\text{MgO(s)} + 2\text{CaO(s)} + \text{Si(s)} = 2\text{Mg(g)} + 2\text{CaO·SiO}_2\text{(s)}. \qquad \text{(3-42)}$$

The magnesium is evolved from the lime as a vapor and the free energy change for the reaction is reduced by operating at a low partial pressure in the reaction vessel. Furthermore, the free energy change for the reaction is lowered because the calcium oxide reacts with the silica to form a silicate.

Example Problem 3-5. What is the minimum temperature at which the Pidgeon process (Eq. 3-42) can be carried out with a partial pressure of magnesium over the solid reactants of 1 mm Hg?

Solution. The standard free energy for Eq. (3-42) can be expressed[10] as

$$\Delta G^0 = 122,400 + 11.74\,T \log T - 100.38\,T. \tag{3-a}$$

The equilibrium constant, assuming unit activity for the solid reactants, is

$$K = \frac{P_{Mg}{}^2 a_{2CaO \cdot SiO_2}{}^2}{a_{MgO}{}^2 a_{CaO}{}^2 a_{Si}} = (P_{Mg}(\text{atm}))^2. \tag{3-b}$$

The equilibrium partial pressure of magnesium, expressed in millimeters of mercury (torr), is given by the relationship

$$\log P_{Mg} = \frac{-\Delta G_a{}^0}{2(4.575)\,T} + 2.88. \tag{3-c}$$

where the latter term converts from atmospheres to millimeters of mercury. A trial and error solution of Eq. (3-c) for T at the specified partial pressure of 1 mm yields a temperature of 1250°C.

Carbothermic Reduction of Magnesia and Alumina

Magnesium oxide can be reduced by carbon at high temperatures. However, as the gases containing carbon monoxide and magnesium vapor cool, the reaction reverses itself to reform the reactants:

$$CO(g) + Mg(g) = MgO(s) + C(s). \tag{3-43}$$

In an effort to prevent the reversion reaction, Hansgirg[11] developed a process in which the magnesium vapor is rapidly chilled, retaining most of it in the metallic state. This "shock cooling" process is accomplished by withdrawing the product gases from a retort furnace and chilling them in a rapidly flowing stream of natural gas or hydrogen. The product usually contains over 50% magnesium metal, with the remainder being MgO and carbon. Although the process was operated in Wales and in California during World War II, both plants were closed following the war because of operating difficulties and failure to achieve projected production output.

Reduction of alumina with carbon will take place to some extent at temperatures in the vicinity of 2000°C. However, the primary reaction products are the carbide and oxy-carbide. The possibility of operating at reduced pressures exists, but thermodynamic calculations indicate that the primary reaction product probably will not be aluminum vapor, but Al_2O. Although considerable research has been carried out on this process, pure metallic aluminum has not been produced commercially by carbothermic reduction.

---

[10] O. Kubaschewski and E. LL. Evans, *Metallurgical Thermochemistry*, p. 375, 3rd ed., Pergamon Press, New York, 1958.

[11] Dungan, *Trans. AIME* **159**, 308 (1944).

Extraction of Zinc

The extraction of zinc is accomplished by a variety of processes with about half of the production by electrolysis[12] (see Zinc Hydrometallurgy and Electro-Winning

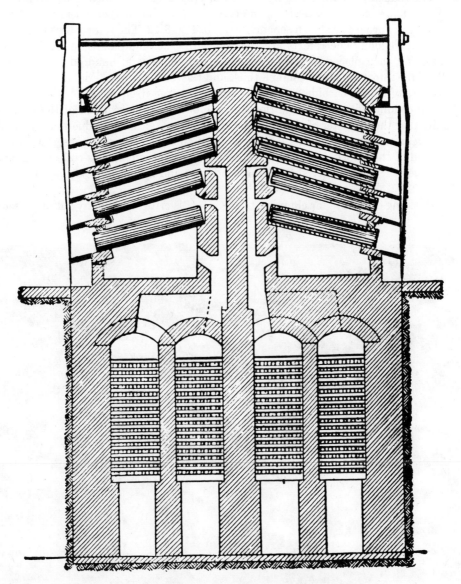

Fig. 3-12. Zinc Retort Furnace (Diagram shows regenerative chambers and gas flow openings. Working floor is shown on upper level)

Source: C. R. Hayward, *Outline of Metallurgical Practice*, 3rd edition, D. Van Nostrand, New York, 1952, p. 226.

[12] S. W. K. Morgan and D. A. Temple, *J. Metals* **19**, 23–29 (1967).

Metals from Leaching Solutions). The remainder is produced primarily by reduction of oxide ores or roasted sulfides with carbon. Lead–zinc ores can be processed advantageously in a modified blast furnace (Imperial Smelting Process).

The reduction of the oxide by carbon usually is carried out in a closed retort. The retorts are of several designs, the classic type being horizontal in a gas-fired regenerative furnace as shown schematically in Fig. 3-12. The retorts are clay tubes about 7–9 ft long. The retorts are charged with zinc oxide and fine coal or coke breeze, and then placed in the furnace which holds several hundred retorts. The walls of the retort are about an inch thick and the refractory material used (usually a mixture of silicon carbide and clay) and the charge itself have a relatively low thermal conductivity. This condition and the endothermic reaction of the process require that the diameter of the retort be small. Since zinc oxide reduction is carried out at temperatures above the boiling point of zinc, a condenser is inserted in the open end of the retort. Zinc vapor condenses to liquid zinc and is collected in the condenser. CO gas is evolved from the condenser and burns in the air as shown in Fig. 3-13.

Roasting of the ore for charging to the retort is critical since any sulfur present

Fig. 3-13. Zinc Retort Process (CO Flame is shown coming from condensers)
Courtesy of American Smelting and Refining Company.

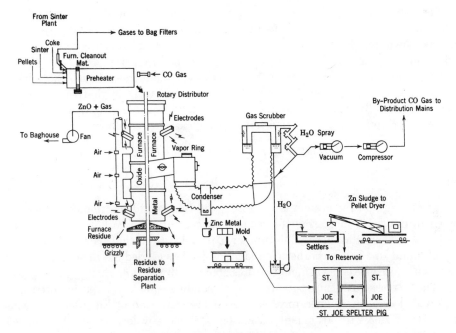

Fig. 3-14. Electrothermic Zinc Process

Source: Weaton, et al. *Trans. AIME, 159,* 1944, p. 155.

as sulfate or unroasted sulfide will not be removed and will retain zinc equal to twice its own weight. If iron oxide is present in the charge, it may be reduced to Fe and react with the sulfur present to release additional zinc. However, slag forming oxides, such as the oxides of iron, will attack the refractory and decrease retort life.

The retorts are charged and discharged every 24–48 hr. Efficient reduction is accomplished with less than 20% reducing agent in the charge. However, the size of the charge is only on the order of 100–150 lb. Consequently, this process is not highly efficient from a material handling viewpoint. St. Joseph Lead Company has developed an electrothermic process which operates continuously and also avoids the use of individual retorts.[13] The flow sheet for this process is presented in Fig. 3-14.

A comparison of the standard free energies of formation of zinc oxide and carbon monoxide in Fig. 3-1 indicates that they are equal at a temperature of about 935°C. The standard free energy change at 935°C is then zero for the reaction,

$$\text{ZnO(s)} + \text{C(s)} = \text{Zn(g)} + \text{CO(g)}. \tag{3-44}$$

Since the standard states for the gaseous components are taken at 1 atm pressure, the equilibrium total pressure with reactants and products in their standard states would be 2 atm. If the total pressure on the system is reduced to 1 atm, and the partial pressures of zinc and carbon monoxide are fixed at 0.5 atm, the equilibrium

[13] C. C. Long, *J. Metals* **17,** 1351, 1352 (1965).

temperature falls just below 900°C, but correspondingly, the boiling point of zinc at 0.5 atm is only 840°C. Any further reduction of the pressure only serves to increase the difference between the two temperatures. Thus, it is necessary that the reduction be carried out with zinc vapor as a reaction product.

Alternatively, the pressure on the system could be increased such that zinc is produced as a liquid phase. However, a complication arises in that the reduction process is not fully represented by Eq. (3-44) but consists of a combination of the Boudouard reaction and a reaction representing gaseous reduction of zinc oxide:

$$ZnO(s) + CO(g) = Zn(g) + CO_2(g), \qquad (3\text{-}45)$$

$$2CO(g) = C(s) + CO_2(g). \qquad (3\text{-}7)$$

As the pressure on the system is increased, the Boudouard reaction, Eq. (3-7), is forced to the right, increasing the relative concentration of carbon dioxide in the gas phase, and shifting the equilibrium for Eq. (3-45) to the left. However, production of liquid zinc in a carbothermic reduction process carried out at several atmospheres pressure is feasible at temperatures above 1040°C.

The increased oxidizing potential (corresponding equilibrium oxygen pressure) of CO_2 at decreasing temperatures can result in a reversion of Eq. (3-45) on condensation of zinc with attendant formation of "blue powder" (partially oxidized zinc). This reaction increases the zinc losses, and several attempts have been made to use a graphite "eliminator,"[14] wherein the Boudouard reaction, Eq. (3-7), is shifted to the left to markedly decrease the concentration of CO_2 in the effluent gases.

Several attempts have been made to extract zinc by reduction in a blast furnace. Because of the large volumes of nitrogen and carbon monoxide present, the difficulties in condensing the metal have prevented development of this process. An alternative has been developed in the Imperial Smelting Process (see Pyrometallurgy III), wherein the zinc is condensed by absorption in liquid lead.

Example Problem 3-6. The reactions in the zinc retort can be represented by Eq. 3-44. The reaction products enter the condenser at 950°C and leave at 450°C. Assuming equilibrium condensation, estimate the efficiency of zinc recovery in the condenser.

Solution. The equilibrium partial pressure of zinc over liquid pure zinc is given by the relationship (Table A-VII),

$$\log P_{mm\ Hg} = -\frac{6670}{T} + 12.00 - 1.126 \log T.$$

At 450°C, $P_{Zn} = 0.36$ mm Hg. The gases escape at one atmosphere so the moles of zinc per mole of CO would be

$$\frac{0.36}{760 - .36} = 4.75 \times 10^{-4}.$$

---

[14] S. W. K. Morgan, Proc. Fourth Empire Min. Met. Cong., Paper HI.2., 1949.

One mole of zinc is released per mole of CO (Eq. 3-44), so the zinc loss is 0.0475%, i.e., the efficiency of recovery under these conditions would be 99.95%.

Another technique for recovering zinc from other oxides is by zinc *fuming*. If a mixture of oxides containing zinc is heated with a reducing agent at a high temperature, the zinc will vaporize and then oxidize in vapor form, creating a finely divided fume. The zinc oxide, *per se*, is not volatile, but vaporization of the metal and subsequent oxidation is the mechanism by which the fume is formed. Zinc oxide has a number of commercial applications, including use in skin medications. Fuming operations are employed to produce both zinc oxide and as a means for recovering zinc from low-grade materials such as slags. Treatment of lead blast furnace slag with a mixture of air and pulverized coal to fume the zinc is a practice carried out at most lead smelters. The zinc fume is usually collected in filters. Specific processes designed for this purpose include the Weatheril process, where zinc fume is produced from calcined material mixed with coal or some other solid reducing agent, and the Waelz process, which employs a rotary kiln for heating a reducing agent and the zinc-bearing material.

Cadmium Production

Cadmium is usually associated with zinc, and being the more volatile constituent, is distilled during the first stages of the carbothermic reduction. The first fraction of zinc to be condensed can be collected and redistilled to recover the cadmium. Thus, cadmium is usually obtained as a by-product of zinc production, e.g., by recovery from lead blast furnace dust and from the residues of electrolytic zinc plants.

Mercury Production

Mercury occurs in nature principally as cinnabar (HgS), and since mercury is not easily oxidized, the roasting of the sulfide mineral results directly in formation of the metal according to the reaction,

$$HgS + O_2 = Hg + SO_2. \tag{3-46}$$

The sulfide may also be decomposed by lime according to the reaction

$$4HgS + 4CaO = 4Hg + 3CaS + CaSO_4. \tag{3-47}$$

Alternatively, iron can be used to reduce the sulfide according to the reaction,

$$HgS + Fe = Hg + FeS. \tag{3-48}$$

These reactions are carried out at temperatures above the boiling point of mercury, 357°C, and means must be provided for condensation of the vapor. Shaft furnaces, small rotary kilns, and multiple hearth furnaces have been used for processing mercury-bearing ores.

The Mond (Carbonyl) Nickel Process

Nickel combines with carbon monoxide at low temperatures to form a volatile nickel carbonyl, $Ni(CO)_4$. The process can be used on ores and roasted calcines that contain nickel in the form of free oxide. The nickel oxide is reduced to nickel by heating at 350–400°C in a reducing atmosphere containing hydrogen and carbon monoxide. If carbon monoxide gas is circulated over the nickel at about 60°C, the nickel is converted to nickel carbonyl according to the reaction,

$$Ni(s) + 4CO(g) = Ni(CO)_4(g). \tag{3-49}$$

The volatile carbonyl gas then can be passed over nickel pellets at about 180°C to reverse the reaction, decomposing the carbonyl and depositing nickel on the pellets. The process is very efficient; on the order of 95% of the nickel is recovered and the product is very pure.

Direct Reduction of Iron Ore

The term "direct reduction" refers to the reduction of oxides to metals at temperatures below the melting point of the metal or any associated impurities. However, in the development of direct reduction processes, some techniques have been developed that do involve some melting of the charge constituents; with regard to iron oxides, the term direct reduction is commonly used to describe any process other than the iron blast furnace. Solid carbonaceous fuels are used for many processes, and in those instances where the reduction is effected by a gaseous reactant such as hydrogen or carbon monoxide, the reaction is often termed "gaseous reduction." One of the advantages of direct reduction processes is that the product

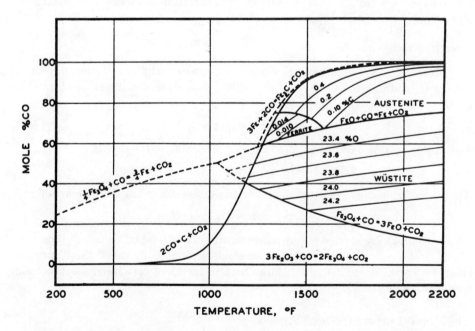

Fig. 3-15. Equilibrium of Iron-oxygen-carbon System at 1 atm

Source: Kun Li, *Proc. Blast Furn. Conf. AIME*, 1960, Vol. 19, p. 162.

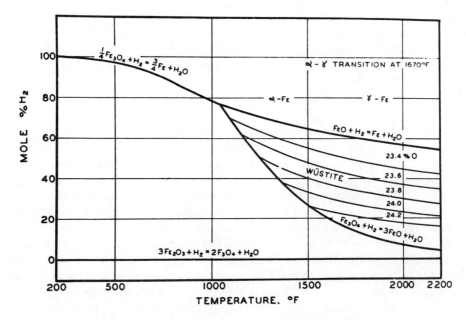

Fig. 3-16. Equilibrium of Iron-oxygen-hydrogen System at 1 atm

Source: Kun Li, *Proc. Blast Furn. Conf. AIME*, 1960, Vol. 19, p. 163.

can be made in relatively small quantities with a capital investment much below that of a blast furnace plant.

Direct reduction processes consume energy. This energy is usually provided by additional fuel, or by electrical energy in the case of the electric furnace smelting processes. The endothermic or exothermic nature of the reduction reactions do not have much effect on the total energy requirements. The net heat effects, however, are of importance since the reduction of iron oxide, e.g., proceeds in stages from hematite (Fe_2O_3) to magnetite (Fe_3O_4) to wüstite (Fe_xO) to iron. Each reaction involves a different heat effect and thus the heat requirement in a process will vary from point to point.

Several materials could be used as reducing agents for iron oxide, but in practice only hydrogen, carbon monoxide, and/or carbon have been used. The equilibrium relationships in the iron–oxygen–carbon system at 1 atm pressure are shown in Fig. 3-15, and the iron–oxygen–hydrogen system is shown in Fig. 3-16. From Figs. 3-15 and 3-16 it can be observed that wüstite (a nonstoichiometric compound usually written as Fe_xO, where x varies between 0.8 and 0.95) is unstable below 1040°F. Thus it is possible to reduce magnetite directly to iron without producing the wüstite phase at temperatures below 1040°F. In the carbon monoxide reduction process, Fig. 3-15, the decomposition of carbon monoxide to carbon and carbon dioxide does not permit magnetite reduction directly to iron at temperatures below 1275°F. Hematite is easily reduced to magnetite at all temperatures with gases of relatively high oxidation potential. As the temperature increases, the ratio of hydrogen to water vapor required to achieve a given degree of reduction decreases,

whereas the effect is reversed with carbon monoxide; here, the higher the temperature, the higher is the ratio of carbon monoxide to carbon dioxide required to
establish a given oxygen potential. In mixtures of hydrogen and carbon monoxide
used as reducing gases, gas phase reactions occur with the formation of methane,
according to the reaction,

$$2CO + 2H_2 = CO_2 + CH_4. \tag{3-50}$$

The kinetics of reduction reactions are made complex by the diverse nature of
iron-bearing minerals. As one might expect, reduction rates are influenced by
temperature, pressure, composition of reducing agent, gas flow rates, the nature of
the iron-bearing material, particle size, state of oxidation, presence of impurities,
etc. The reducing reaction may be controlled by diffusion in the gas phase or in the
solid, or by the rate of chemical reaction. The controlling mechanism is determined
by the specific system and the temperature of the process. The reduction interfaces
move parallel to the outside surface of the iron-bearing material. Porous oxide
minerals usually show a single interface, since the magnetite is fully reduced before a
wüstite phase appears. McKewan[15,16] showed that the rate of reduction per unit area
was constant with time, and in the case of the reduction of iron oxide by hydrogen,
was directly proportional to the partial pressure of hydrogen. In this case the
reaction was controlled at the oxide–metal interface rather than by diffusion through
the reaction product layer.

In view of the foregoing considerations, a rate equation can be derived to fit the
reported data. Consider a sphere of iron oxide of initial radius, r_0, and initial density,
ρ_0. Assume that the rate of formation of a uniform reaction product layer is proportional to the receding surface area of the remaining oxide. If W is the weight of that
part of the original material that has reacted, then

$$\frac{dW}{dt} = kA, \tag{3-51}$$

where k is a proportionality constant having the dimensions $ml^{-2}t^{-1}$ and is a function
of temperature, pressure, and gas composition. From this it follows that

$$r_0\rho_0 f = kt, \tag{3-52}$$

where f is the ratio of the thickness of the reduced layer to the initial radius. Equation (3-51) states that at the oxide metal interface the amount of oxide reacting per
unit area per unit time is constant, or that the reaction interface advances at a
constant rate.

The fractional reduction, R, is defined as the weight of oxygen removed divided
by the total weight of oxygen originally present as iron oxide. The fractional
reduction, R, bears the following geometrical relationship to the fraction thickness, f.

$$f = 1 - (1 - R)^{1/3}. \tag{3-53}$$

[15] W. M. McKewan, *Trans. Met. Soc. AIME* **212**, 791 (1958).
[16] W. M. McKewan, *Trans. Met. Soc. AIME* **218**, 2 (1960).

Substituting this value for f into Eq. (3-52),

$$r_0\rho_0(1 - (1 - R)^{1/3}) = kt. \tag{3-54}$$

The same equation would apply to a cube whose side equals $2r_0$. The reduction rate for any particle whose shape is similar to a sphere or cube can be approximated by Eq. (3-54). Equations also can be derived to fit other geometrical shapes.

The kinetics of reduction reactions vary with temperature. The rate constant which increases with temperature can be treated in terms of the theory of absolute reaction rates where the specific rate constant, k', is defined[17] as

$$k' = \kappa \frac{kT}{h} \epsilon^{-\Delta H^*/RT} \epsilon^{\Delta S^*/R}, \tag{3-55}$$

where

κ = transmission coefficient initially taken to be unity,
k = Boltzmann constant,
h = Planck constant,
ΔH^* = enthalpy of activation,
ΔS^* = entropy of activation.

An alternative approach to the kinetics of reduction of a granular bed of iron ore is to assume that the reduction of iron ore in a gas stream tends to follow a first-order reaction rate according to the expression,

$$-\frac{dc}{dt} = k_r c, \tag{3-56}$$

where

$-\dfrac{dc}{dt}$ = rate at which oxygen is removed from ore, lb O_2/lb of Fe/hr,

c = concentration of oxygen in the material, lb O_2/lb of Fe,
k_r = reduction rate constant, hr^{-1}.

Collecting terms and integrating between c_0 at $t = 0$ and c at time t,

$$\ln\left[\frac{c_0}{c}\right] = k_r t \tag{3-57}$$

and the fraction of oxygen remaining at time t is

$$f_0 = \frac{c}{c_0} e^{-k_r t}. \tag{3-58}$$

As a first approximation it may be assumed that the overall reduction rate of ore particles proceeds according to this type of equation. The actual value of the reaction constant, k_r, could be determined in a small scale reduction bed in which

[17] Ibid., p. 2.

conditions are typical of those in the larger unit that is being designed. It should be noted that with small particles, the mass flow rate at which reactant can be supplied may be a critical design factor.

Several Commercial Processes for Direct Reduction of Iron Ores

One of the earliest commercial processes was the Hoeganaes process, which is used in the United States and has been in use in Southern Sweden for over 50 years. The process involves a relatively pure iron ore charged to a kiln with coal and limestone.

The Wiberg process involves reduction of iron ore with carbon monoxide and hydrogen in a shaft furnace. Recovery of metallic iron averages 85–90%.

The Krupp–Renn process is carried out in a rotary kiln and involves some melting of the slag phase and production of small iron nodules. This process is particularly effective in that it permits separation of the gangue materials and does not require a high purity iron oxide, as do the Hoeganaes and Wiberg processes.

The use of natural gas to directly reduce iron has been put into practice at Monterrey, Mexico, and is called the HyL process. This process requires a high purity ore as a charge material, and its product in turn is charged to a steelmaking furnace. In this operation all gangue present in the ore is concentrated in the sponge iron product. Phosphorus is not eliminated, but about 85% of the sulfur is removed as hydrogen sulfide.

The H-Iron process employs a pressurized fluidized bed operating at 450–550 lb/in.^{2} at a temperature of 1000°F to produce powdered iron. The reducing agent is nearly pure hydrogen gas and the off gas is regenerated and recirculated.

There are other iron ore reduction processes that have received considerable commercial attention. The R-N process uses solid carbonaceous fuels as reducing agents. No fusion takes place in this kiln operation, and the product can be externally beneficiated by magnetic separation to produce a concentrate of 90–95% metallic iron. Electric furnace reduction of iron ores can be carried out in the Strategic-Udy process. Ore, coke, and fluxes are preheated, and the ore is partially reduced in a rotary kiln. This hot charge is moved to an electric furnace where controlled conditions of temperature, reducing potential, and slag composition are used to produce a pig iron or semisteel. The FIOR (fluid-iron-ore reduction) process of Esso Research and the Nu-Iron Process of U. S. Steel Corporation involve fluidization of iron ore in a reducing gas made from natural gas or liquid petroleum components. The product is a briquetted iron consisting of 90–95% metal.

Bibliography

Biswas, A. K. and G. R. Bashforth, *The Physical Chemistry of Metallurgical Processes*, Chapman and Hall, Ltd., London, 1962.

Sims, C. E. (Ed.), *Electric Furnace Steelmaking*, Vol. II, Chap. 16, Interscience Publishers, New York, 1963.

Gilchrist, J. D., *Extraction Metallurgy*, Pergamon Press, London, 1967.

Habashi, F., *Principles of Extractive Metallurgy, Vol. 1 General Principles*, Gordon and Breach, New York, 1969.

Hopkins, D. W., *Physical Chemistry and Metal Extraction*, Chap. 9, The Macmillan Company, New York, 1954.

Kubaschewski, O. and E. LL. Evans, *Metallurgical Thermochemistry*, 3rd ed., p. 75, Pergamon Press, New York, 1954.

Lewis, G. N. and M. Randall, *Thermodynamics*, 2nd ed., McGraw-Hill, New York, 1961.

McGannon, H. E. (Ed.), *The Making, Shaping, and Treating of Steel*, 9th ed., United States Steel Corporation, Pittsburgh, 1971.

Ward, R. G., *An Introduction to the Physical Chemistry of Iron and Steelmaking*, Chap. 3, Edward Arnold Ltd., London, 1962.

Discussion Questions

1. Outline the steps and data required to develop a Pourbaix–Ellingham diagram for the Cr–O–C system. Describe a process which might be employed for carbothermic reduction of chromium oxide.

2. Discuss the advantages of using various standard states.

3. A gas–solid direct reduction process has been proposed to produce a metal "X" from its oxide "XO_2". What information is required to determine the feasibility of the process and required operating conditions? Describe any laboratory experiments which would be required to determine this information.

4. It is desired to recover tin from cassiterite concentrate containing several other oxide minerals (Fe, Ti, Ta, Cb, Pb, and W). Describe a process in terms of equipment and operating conditions which would selectively reduce the tin oxide to an impure metal.

5. Describe procedures which might be used to improve recoveries of metallics in the $CaCl_2$ roasting of low-grade oxide concentrates.

6. Compare the gas cleaning systems required for recovery of particulate material in various reduction processes.

Problems

1. Calculate the equilibrium oxygen pressure over pure calcium oxide in contact with calcium gas at a total pressure of 0.10 mm Hg at 1600°C.
2. Calculate the equilibrium oxygen pressure over tantalum oxide in contact with solid pure tantalum at 1600°C.
3. Calculate the H_2/H_2O ratio required to reduce titanium oxide at 1400°C.
4. Calculate the H_2/H_2O ratio required to reduce tungsten oxide at 1400°C.
5. Calculate the equilibrium CO/CO_2 ratio in contact with graphite at 1600°C at 1 atm total pressure.
6. Air containing 5% (by volume) water vapor is allowed to contact graphite in a closed system. The system comes to equilibrium at 0.5 atm total pressure and 1600°C. What is the composition of the gas phase?
7. Calculate the reduced pressure required to decompose the following nitrides:

    (a) AlN
    (b) BN
    (c) Si_3N_4
    (d) VN

at 1600°C. (Assume presence of pure metal in liquid or solid form.)

8. MoO_2 is to be reduced by dry H_2 at 1400°C to metallic molybdenum. The reaction is to be carried out at 1 atm total pressure.

    (a) What fraction of the dry hydrogen introduced into the reduction chamber will be utilized if we may assume that equilibrium is achieved?
    (b) What would be the effect of pressurizing the system to 10 atm?
    (c) Can dry hydrogen reduce MoO_2 at 1 atm total pressure and 800°C?

9. Specify the minimum operating temperature for carbothermic reduction of molybdenum at 10^{-2} atm.

10. A mixture of carbon monoxide and carbon dioxide at 1 atm pressure is slowly passed through a bed of graphite chips maintained at a temperature of 1000°C in the upper portion of a two zone furnace. The gas is then directly brought into contact with an iron sample at 1500°C in the lower zone of the furnace. If a heating element failure allows the upper zone of the furnace to slowly cool, at what temperature will oxide appear on the iron sample in the lower portion of the furnace?

11. Electromotive force measurements in the cadmium–lead system were taken relative to a liquid pure cadmium electrode at 500°C yielding the following results:

N_{Cd}	Wt % Cd	Activity$_{Cd}$
0.00	0.00	0.00
0.05	2.78	0.15
0.10	5.70	0.27
0.15	8.70	0.375
0.30	18.90	0.57
1.00	100.00	1.00

    (a) Based on the standard state of liquid pure cadmium at 500°C (reference state $N_{Cd} = 1$), compute an activity coefficient for each of the compositions given above.
    (b) Based on the standard state of an infinitely dilute solution at 500°C (reference state: 1 wt% solution), compute an activity coefficient for each of the compositions given above.

12. The equilibrium conditions for the reduction of metal M from its oxides by CO are given below at four different temperatures. The reactions are:

    I. $MO_2 + CO = MO + CO_2$

Temp °C	400	700	900	1100
$\dfrac{CO}{CO + CO_2}$	0.41	0.32	0.26	0.20

II. $MO + CO = M + CO_2$

Temp °C	400	700	900	1100
$\dfrac{CO}{CO + CO_2}$	0.55	0.70	0.80	0.90

(a) Which of these reactions absorbs and which evolves heat? Outline the method by which you would find the numerical value of heat absorbed and evolved at a given temperature for either reaction. Compute a standard free energy change for each equation, expressed as a function of temperature.

(b) Calculate the actual free energy change if 1 mole of MO and 1 mole of CO are mixed and the mixture is allowed to reach equilibrium, the temperature and total pressure being held constant at 700°C and 1 atm, respectively, for the process.

13. The following equilibrium data have been determined for the reaction,

$$NiO(s) + CO(g) \rightarrow Ni(s) + CO_2(g):$$

Temp °C	663	716	754	793	852
$K \times 10^{-3}$	4.535	3.323	2.554	2.037	1.577

(a) Determine ΔH^0 and ΔG^0 at 1000°K

(b) Would an atmosphere of 30% CO_2, 5% CO and 65% N_2 oxidize nickel at 1000°K?

14. Dry air is injected into a hot coke bed and converted to a gaseous fuel. If the temperature is 1400°C and the total pressure is 3 atm:

(a) Calculate the equilibrium gas composition.

(b) Is there any advantage in raising the temperature to 1600°C or in increasing the pressure to 5 atm.

15. The formation of tin oxide from the pure elements is given by the reaction,

$$Sn(l) + O_2(g) = SnO_2(s).$$

What is the standard free energy for the above reaction when the standard state for tin is the infinitely dilute solution in thallium, referred to a 1 wt% solution?

16. Derive the reciprocity relationship, $\epsilon_i{}^j = \epsilon_j{}^i$ (Eq. 3-36).

17. Derive the relationship between $\epsilon_i{}^j$ and $e_i{}^j$ (Eq. 3-40).

18. Show that the interaction parameter will vary linearly with reciprocal absolute temperature, provided that the partial molar enthalpy is essentially independent of temperature.

19. Estimate the concentration of sulfur at 1600°C in the following liquid iron

alloy in equilibrium with an H_2S/H_2 mixture where the partial pressure ratio, P_{H_2S}/P_{H_2} is 1.68×10^{-3}.

Element	Wt %
C	1.01
Al	0.31
P	1.52
Si	3.95
Mn	2.17

20. Estimate the activity of cadmium in the following alloy at 700°C.

Element	Mole fraction
Sn	0.75
Cd	0.10
Pb	0.05
Zn	0.10

21. Two processes have been proposed for the production of titanium at 1 atm pressure from rutile sands.

Process I—Direct carbothermic reduction at 1200°C:

$$TiO_2(S) + 2C(g) = Ti(S) + 2CO(g).$$

Process II—Chlorination under reducing conditions at 1200°C followed by magnesium reduction of $TiCl_4$ at 750°C:

Step 1. $TiO_2(S) + 2C(g) + 2Cl_2(g) = TiCl_4(g) + 2CO(g),$

Step 2. $TiCl_4(g) + 2Mg(l) = Ti(S) + 2MgCl_2(l).$

Evaluate the feasibility of the two processes.

22. Sodium reduction of titanium tetrachloride is used to produce sponge titanium and sodium chloride. Compare the material balance and free energy change for this process with those for the Kroll process.

23. A vacuum reduction system for the production of zinc has been proposed. The system which operates at or below 1 mm Hg consists of two chambers: a high temperature retort (750°C) to which zinc oxide and coke breeze are charged, and a water-cooled condensing chamber where zinc is rapidly condensed as a solid to avoid blue powder formation. Assess the thermodynamic feasibility of this process.

24. Quantitatively assess the effectiveness of a graphite "eliminator" in preventing reversion of the zinc reduction reaction to form blue powder. The eliminator would be a graphite packed bed held at about 1000°C to reduce the

CO_2 content of the gases evolved from the retort. The gases leaving the eliminator then would be cooled, condensing the zinc as a liquid.

25. A zinc retort is charged with ZnO and C at 1000°C and the gases Zn, CO and CO_2 are generated at 1 atm total pressure.

   (a) Compute the theoretical minimum charge ratio of ZnO to C.
   (b) If the process were operated at low pressure, e.g., 10^{-4} atm, how would the charge ratio change?

26. In the "H-Iron" process for direct reduction of iron oxide, dry pure hydrogen is reacted with magnetite at a relatively low temperature and high pressure. The wet hydrogen is recirculated after drying and metallic iron is intermittently removed from the reactor. If the mean temperature in the reactor is 950°F and the pressure is 250 psig, estimate the H_2O content of the exiting gas stream.

27. Direct reduction of magnetite with carbon monoxide at 900°C in the presence of lime has been proposed according to the reaction,

$$Fe_3O_4 + 4CO + 4CaO = 3Fe + 4CaCO_3.$$

Calculate the minimum CO pressure to cause this reaction to go to completion.

28. Derive a relationship of the same general form as Eq. (3-54) for the hydrogen reduction of hematite when the particle is:

   (a) A cube of initial edge length S_0, where $f = (S_0 - S)/S_0$;
   (b) A long cylinder of initial radius r_0, where $f = (r_0 - r)/r_0$;
   (c) A large thin plate of initial thickness X_0, where $f = (X_0 - X)/X_0$.

CHAPTER 4

Pyrometallurgy III: Smelting and Converting

The most extensive industrial applications of oxidation and reduction processes occur in the smelting of metals, i.e., recovering the metallic elements from an oxide or sulfide by heating with fusion while in the presence of a reducing agent. The smelting operation may produce an impure matte (liquid metallic sulfide) or a metal, such as the iron blast furnace which produces liquid pig iron, containing up to 5% carbon as well as appreciable amounts of manganese, silicon, phosphorus, sulfur, etc. Further refining to produce a purer metallic product, such as metallic copper or steel, is accomplished in an oxidizing process referred to as conversion or converting.

4-1 Reverberatory Smelting

Smelting, essentially a melting operation wherein the charge is melted down into two layers consisting of metal or matte, and slag, is often carried out in a reverberatory furnace in which the thermal energy is supplied to the charge primarily by radiation from the furnace roof. These immiscible liquids form two layers on the furnace hearth. In a matting operation, such as that used in copper smelting, the sulfides collect as a molten matte and the slag-forming oxides fuse into a slag layer that floats on the matte. A flow diagram for a reverberatory copper smelting furnace is shown in Fig. 4-1.

In copper smelting, the copper in the charge, in the form of oxide, sulfide, or possibly sulfate, is converted to cuprous sulfide, Cu_2S. The affinity of oxygen for iron and sulfur and of sulfur for copper results in the reverberatory smelting reactions,

$$2\,CuO + 2\,FeS_2 = Cu_2S + 2\,FeS + SO_2, \tag{4-1}$$

$$Cu_2O + FeS = Cu_2S + FeO. \tag{4-2}$$

Some of the iron sulfides present may be oxidized to ferrous oxide according to the reaction,

$$FeS_2 + 5\,Fe_2O_3 = 11\,FeO + 2\,SO_2. \tag{4-3}$$

Magnetite (Fe_3O_4) is usually present in the charge either in the calcines or recycled converter slag (see Copper Converting). Magnetite also can be formed by oxidation of FeO or partial reduction of Fe_2O_3. The magnetite often settles out to form accretions on the bottom or may form a "false bottom" or layer between the slag and matte and hinder matte separation.

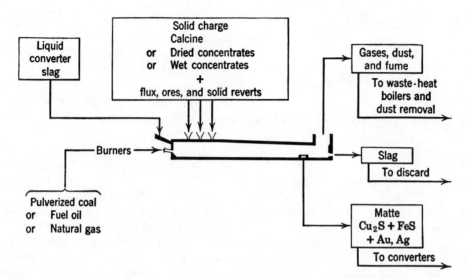

Fig. 4-1. Flow Diagram for Reverberatory Copper Smelting Furnace

Source: J. Newton, *Extractive Metallurgy*, John Wiley and Sons, Inc., 1959, New York, p. 340.

Matte copper content varies between 20 and 70%. The sulfur is in the range of 20–25%. The remainder is iron and minor constituents. The matte can be assumed to be a mixture of Cu_2S and FeS, although the sulfur is usually slightly less than the concentration required to account for stoichiometric formation of these compounds.

The slag is a complex silicate, primarily $2\ FeO \cdot SiO_2$, although substantial quantities of other oxides may be present. The slag contains some sulfur and some copper resulting in a copper loss. The concentration of copper is usually a few tenths of a percent and may be present both as dissolved copper or copper sulfide, and as entrained matte.

A schematic diagram of a reverberatory smelting furnace is shown in Fig. 4-2. The roof is usually arched and is 20–30 ft. wide. The length of the reverberatory furnace may range from 40 ft to as much as 100 ft in large smelting furnaces such as those used in the copper industry. In the past the refractory lining was usually of silica brick, although in modern furnaces, where the charges are now more basic than formerly, the use of magnesite brick has markedly increased. The reverberatory furnace is equipped with charging facilities, often feed hoppers set in the roof, and it is heated by burners, usually located in the end wall of the furnace. The flame extends across the charge, which is heated by radiation from the flame and the hot interior refractories. Natural gas, fuel oil, or pulverized coal is used as fuel.

The metal, or matte, in the case of oxide–sulfide smelting, and slag collect in layers on the furnace hearth. Tap holes and doors are provided in the front wall for additional charging or slag skimming. The hot combustion gases from the furnace often are directed to waste heat boilers to recover their sensible heat.

The charge to the smelting furnace usually consists of calcines from a roasting operation or concentrates directly from a beneficiation process. In addition, molten

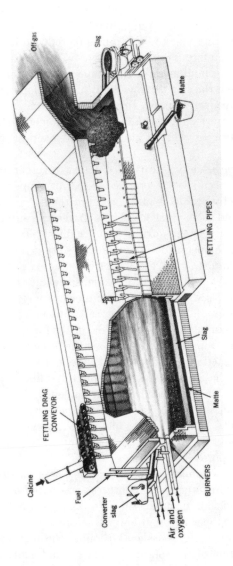

Fig. 4-2. Cutaway View of Reverberatory Smelting Furnace

Source: J. R. Boldt, Jr., *The Winning of Nickel*, D. Van Nostrand Company, Princeton, New Jersey, 1967, p. 242.

slag from converters often is returned as charge to the preceding smelting operation. Flux is added to provide a liquid slag; silica is commonly used in copper
smelting, whereas in smelting operations in which a basic slag is desired lime or
burnt dolomite are common flux materials.

Tin Smelting is carried out in a reverberatory furnace wherein a reducing agent,
such as powdered coal or coke, is charged directly with tin concentrates to produce
metallic tin. One of the primary contaminates of tin is iron, and there is a tendency
for tin to enter the oxide slag and for iron to be reduced into the metallic tin. The
smelting process, therefore, is carried out in two stages, the first with moderate
reducing conditions to minimize the simultaneous reduction of iron, followed by a
second stage in which the slag is remelted at higher temperatures with additions of
reducing agent to recover as much tin as possible. Subsequent purification of the
tin is primarily by fire-refining techniques (see Pyrometallurgy IV) using liquation
and drossing. Electrolytic refining also has been practiced to produce tin of even
higher purity.

Copper Smelting usually is carried out in a reverberatory furnace, although
blast furnaces have been used to produce copper or copper–nickel matte. Calcines
from a roasting operation and liquid slag from the subsequent copper converting
operation are charged to the furnace. These materials contain copper, iron, sulfur,
and oxygen, along with gangue oxides such as silica, alumina, and magnesia. The
mixture is heated to temperatures above 1200°C so that the entire charge is molten.
Two liquid phases result: (1) a copper matte consisting of copper, iron and sulfur,
and (2) a slag that is a solution of iron oxide, silica, alumina, magnesia, and other
oxides but with a low copper content. Any precious metals present are soluble in
the matte and collect in it. The two phases separate under the influence of gravity,
the heavier matte accumulating in the bottom of the furnace. The two liquid phases
are tapped separately from the furnace and are the principal products of the process.
In addition, some reactions yield sulfur dioxide along with other gases from the
burning of the fuel; these gases pass out through the flue.

The composition of the matte produced influences the energy balance for the
subsequent converting operation. The most satisfactory grade of matte is one that
contains from 40–50% copper. Low-grade concentrates that are rich in iron may
yield matte with about 20% copper, whereas some of the rich African ores yield a
matte that is nearly pure Cu_2S.

The processing of *copper–nickel sulfide ores* was carried out until recently
primarily by the Orford process. This process was based on the mutual solubility
of copper, nickel, and sodium sulfides at high temperatures, and the insolubility of
Ni_3S_2 in the sodium–copper matte at lower temperatures. By smelting to produce a
sodium–copper–nickel matte and slowly cooling, the more dense Ni_3S_2 separated
out. The upper portion of the solidified charge would contain most of the copper
with a small amount of nickel whereas the lower portion or "bottoms" would contain
most of the nickel and only a small amount of copper. The copper–sodium matte
was then treated in a copper converter. The nickel sulfide was sintered and then
reduced in a reverberatory smelting operation to impure liquid nickel, which was
cast as anodes for final purification by electrolytic refining.

More recently the separation of copper and nickel sulfides has been carried out directly by reverberatory smelting to produce a copper–nickel matte free of iron sulfide. The matte is slowly cooled in large cakes, which results in the separation of crystals of Cu_2S and Ni_3S_2. The frozen matte is crushed and ground to mechanically break up the copper sulfide and nickel sulfide crystals, which in turn are separated by standard mineral dressing techniques.

4-2 Blast Furnace Smelting

The lead blast furnace, much smaller than the iron blast furnace, is typical of non-ferrous blast furnaces. A cross-sectional view of a lead blast furnace is shown in Fig. 4-3. The walls of the furnace in the bosh (melting zone) are water-cooled steel plates, and the crucible below the bosh zone of the furnace is lined with refractory material. Whereas the iron blast furnace is circular in cross section, the lead blast furnace is rectangular, with tuyeres (nozzles through which the blast is supplied) located along the sides of the furnace but not on the ends. The lead oxide is easily reduced and since it is not necessary to operate the furnace at a high temperature or under strong reducing conditions as in the iron blast furnace, the blast is not preheated. The charge to the lead blast furnace includes lead oxide, usually the product of the Dwight–Lloyd sintering machine, plus about 10% coke and a small amount of scrap iron, which serves to reduce lead sulfide and to minimize the lead content of the slag. Operating the furnace at low temperatures minimizes volatilization of lead, but requires that a fluid slag be provided at a relatively low temperature. Lead blast furnace slags usually contain 30–40% SiO_2, 25–45% iron oxides, with the balance being CaO and ZnO. The lead content of the slags is

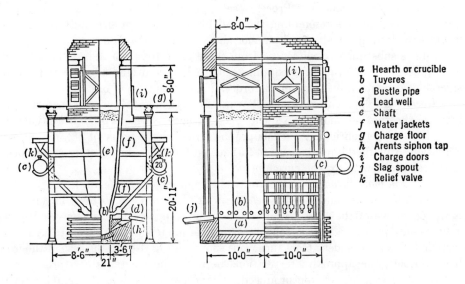

a	Hearth or crucible
b	Tuyeres
c	Bustle pipe
d	Lead well
e	Shaft
f	Water jackets
g	Charge floor
h	Arents siphon tap
i	Charge doors
j	Slag spout
k	Relief valve

Fig. 4-3. Lead Blast Furnace

Source: J. L. Bray, *Non-Ferrous Production Metallurgy*, 2nd Ed., John Wiley and Sons, Inc., New York, 1947, p. 292.

usually about 1%. Zinc is concentrated in the slag, sometimes in concentrations as high as 20 wt%. A separate operation, such as that conducted in a fuming furnace (see Pyrometallurgy II), is usually employed to recover the zinc. The smelted lead is alloyed with small amounts of easily reduced metals such as copper, antimony, tin, arsenic, and any previous metals present. Matte or speiss (arsenides or anti-monides, as contrasted with matte which is sulfide) may form if there is excess sulfur, arsenic, or antimony in the charge. The matte or speiss will collect as a liquid layer between the lead and the slag, and facilities are provided outside the furnace to separate matte or speiss from the slag. If there is an appreciable amount of copper in the charge, it is customary to charge sinter with a sufficient amount of sulfur to form a copper matte and thus minimize the solution of copper in the molten lead.

Nearly all lead is smelted in lead blast furnaces, although it is possible to carry out the smelting operation in a reverberatory furnace. High galena (PbS) ores can be smelted on an oxidizing hearth with coke and a blast of air that oxidizes the galena according to the reaction,

$$2PbS + 3O_2 = 2PbO + 2SO_2. \tag{4-4}$$

The lead oxide reacts with additional galena to produce metallic lead according to the reaction,

$$PbS + 2PbO = 3Pb + SO_2. \tag{4-5}$$

The molten lead collects on the hearth. This procedure is not efficient, particularly with respect to recovery of precious metals, and is now obsolete.

The smelting of lead–zinc ores in the blast furnace for recovery of *both* lead and zinc was considered to be an impossibility for many years. A recent development, the Imperial Smelting Process (see Pyrometallurgy IV), which is carried out in a blast furnace resembling the lead blast furnace in size and appearance, reduces the lead in the standard manner and volatilizes the zinc in the stack gases. The zinc is subsequently condensed from these gases by absorption in liquid lead. This process for treating ores containing both lead and zinc has effected a considerable economic advantage in the processing of this type of ore.

Recent development work[1] has shown that oxygen enrichment of the blast can increase the rate of smelting in a lead blast furnace as well as provide a savings in coke. The oxygen enrichment levels proposed are on the order of 3% for normal operation. A marked improvement in furnace performance during start-up has been noted especially with oxygen enrichment.

The *iron blast furnace* process consists of charging ore, fuel and flux into the top of a shaft furnace and blowing heated air or blast into the bottom. This process is the backbone of the steel industry, producing a high-carbon, silicon- and manganese-containing iron that is converted to steel in steelmaking furnaces. A cross-sectional view of a modern blast furnace is shown in Fig. 4-4. In the case of the iron blast furnace, the approximate relations between the reactants and products is such that for each ton of iron produced, approximately 1.6 tons of ore, 0.5 tons of

[1] R. W. Ruddle, *The Physical Chemistry of Copper Smelting*, Institution of Mining and Metallurgy, London, 1953.

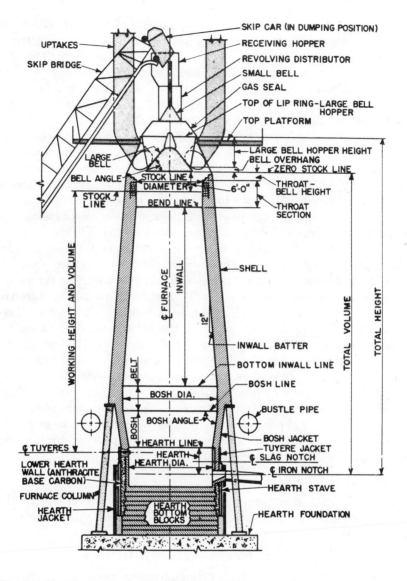

Fig. 4-4. Identification of Principal Dimensions and Components of a Blast Furnace

Source: *The Making, Shaping and Treating of Steel*, edited by Harold E. McGannon, Pittsburgh, United States Steel Corporation, 9th edition, 1971, p. 429.

coke, 0.2 tons of a mixture of limestone and dolomite, and 1.7 tons of air are charged into the furnace. From these materials there is produced 1.0 tons of iron, 0.3 tons of slag, <0.05 tons of flue dust, and 2.7 tons of blast furnace gas. This rough statement relates to the production of pig iron in relatively modern furnaces in north-central United States, accounting for about 75% of the production in the United States.

Iron is supplied to the furnace by ore in the form of the oxide, either hematite (Fe_2O_3) or magnetite (Fe_3O_4). Pure hematite contains 70% iron, but the present ore being mined from the Lake Superior region contains about 50% iron, the difference being represented by gangue, which consists mostly of silica and alumina, and about 12% moisture. Prepared burdens of self-fluxing sinter or pellets (see Pyrometallurgy I) are being used in many furnaces. Iron also is added by charging mill scale, sinter, slag from open hearth furnaces or converters, and scrap. Coke, the fuel for the process is produced by destructive distillation of coal. Coke not only provides the heat for the process, but also is the reducing agent. Limestone and dolomite are added to form a fluid slag and to restrict the amounts of silicon and sulfur entering the pig iron. The blast of heated air supplies oxygen to the process, reacting with the carbon of the coke to form carbon monoxide, the gaseous reducing agent. The blast is sometimes enriched with natural gas or oil which can replace a portion of the coke required. Oxygen gas usually is added with these fuels.

The blast furnace plant consists of several components, the principal one being the blast furnace structure itself. A cross section of a blast furnace plant is shown in Fig. 4-5. The blast is compressed by blowers and passes through stoves, where it is heated to a temperature of 1500–1800°F. The stoves permit regenerative heat transfer, for which fuel is supplied by the blast furnace stack gases which first have been treated to remove dust. In addition, the blast furnace operation requires charging equipment and ladles for the removal and transport of metal and slag. Secondary equipment includes a sinter plant where ore fines and iron oxide from various steelmaking operations, as well as from the dust-collecting equipment of the blast furnace, are sintered into a usable product. Coke ovens are necessary for the production of coke from coal, and often a fully integrated plant will also include production facilities for the manufacture of by-products from coke oven gases. In addition, a fully integrated steel plant will include complete steelmaking facilities and all the auxiliary equipment involved therewith.

Chemistry of the Iron Blast Furnace

The principal reaction taking place in the iron blast furnace process is the reduction of iron oxide by carbon. The actual mechanism for this process is gaseous reduction by carbon monoxide according to the reaction

$$3CO + Fe_2O_3 = 2Fe + 3CO_2. \tag{4-6}$$

In the presence of an excess of carbon at a high temperature, CO_2 is at once reduced to CO:

$$CO_2 + C = 2CO. \tag{4-7}$$

The actual reduction of Fe_2O_3 by CO may take place in three steps: the Fe_2O_3 being successively reduced to Fe_3O_4, FeO, and, finally, Fe (see Pyrometallurgy II). In addition to reducing the oxides of iron, carbon monoxide also reduces the oxides of manganese, silicon, and phosphorous according to the reactions,

$$MnO + CO = \underline{Mn} + CO_2, \tag{4-8}$$

$$SiO_2 + 2CO = \underline{Si} + 2CO_2, \tag{4-9}$$

$$P_2O_5 + 5CO = 2\underline{P} + 5CO_2, \tag{4-10}$$

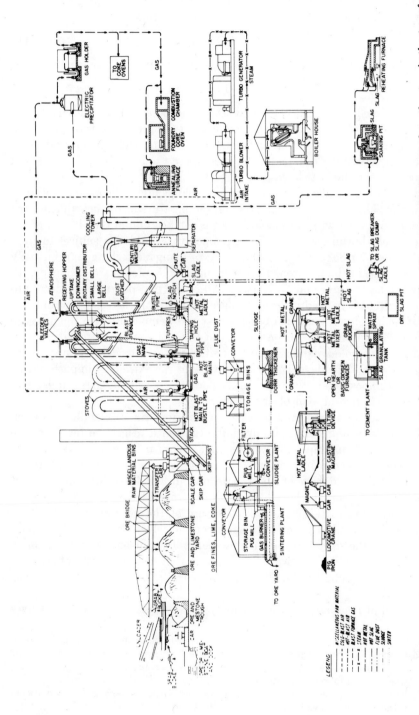

Figure 4-5. Flow Diagram Depicting the Principal Units and Auxiliaries in a Modern Blast Furnace Plant—Illustrated are the steps in the manufacture of pig iron from receipt of raw materials to disposal of pig iron and slag, as well as the methods for utilizing the furnace gases.

Source: *The Making, Shaping and Treating of Steel,* edited by Harold E. McGannon, Pittsburgh, United States Steel Corporation, 9th edition. 1971, p. 425.

where the underline notation refers to the element in metallic solution.

The water vapor in the blast also plays a role in the process:

$$H_2O + C = CO + H_2. \tag{4-11}$$

The hydrogen liberated by the above reaction may react with iron oxide, reducing it:

$$FeO + H_2 = H_2O + Fe. \tag{4-12}$$

The water so formed is again decomposed. It should be noted that relatively few of the reactions involved furnish the heat required for the process; it is the oxidation of carbon and some of the reduction reactions involving carbon monoxide that furnish the heat to dry the raw materials, decompose the limestone, melt the iron and slag, and replace the heat losses.

The distribution of the alloying elements manganese, silicon and sulfur is controlled principally by slag composition and temperature of the hearth since the liquid metal is saturated with carbon and the gas composition in the vicinity of the tuyeres is controlled primarily by the blast temperature and is not subject to great variation. Between 70–80% of the total amount of manganese charged, principally in the ore, is found in the pig iron. Higher manganese contents are obtained with higher hearth temperatures and basic[2] slags. All of the phosphorus is reduced into the iron. Sulfur that is carried into the furnace primarily in the coke, and often in the ore, may be largely retained in a basic slag at high temperatures in the presence of carbon-saturated iron. The silicon content of the pig iron is controlled by temperature and the thermodynamic activity of silica in the slag. High temperatures and acid slags favor high silicon contents. In view of the extremely important role played by slag composition in the blast furnace process, considerable research effort has been expended on studies of blast furnace slags.[3]

In addition to the role played by slag composition, the interactions between dissolved elements in liquid iron also are important with respect to blast furnace metal composition. A principal interaction that occurs in blast furnace metal is between carbon and silicon in solution. The solubility of carbon in liquid iron is greatly reduced by the presence of silicon, as shown in Fig. 4-6. Since the liquid pig iron leaving the blast furnace hearth is essentially saturated with carbon, i.e., at equilibrium with pure carbon at the temperature of the hearth, the activity of silica in the slag (which is a function of silica content, slag composition, and temperature) controls the silicon content of the metal. The silicon content, assuming favorable reaction kinetics, in turn controls the carbon content. The influence of temperature on the solubility of carbon in iron–silicon alloys also is shown in Fig. 4-6.

The control of reactions occuring in iron and steelmaking primarily involves

---

[2] The terms basic and acid follow from the classical scientific definitions. In metallurgical processes, basic refers to oxides such as CaO and MgO which are available for dephosphorization or desulfurization. Basic oxides are contrasted with acid oxides, such as SiO_2 and P_2O_5. These two types of oxides react to form fairly stable compounds such as Ca_2SiO_4 or $Ca_4P_2O_9$.

[3] J. Chipman, *Thermodynamic Properties of Blast Furnace Slags*, Proceedings of International Symposium on the Physical Chemistry of Process Metallurgy, Part I (G. R. St. Pierre, Ed.), Interscience Publishers, New York, 1961.

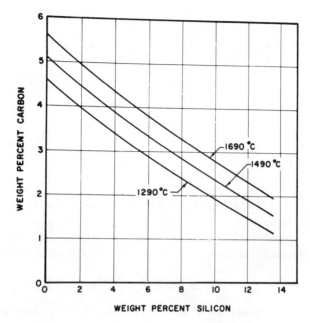

Fig. 4-6. Solubility of Carbon in Iron-Silicon Alloys

Based on "The Solubility of Carbon in Molten Iron and in Iron-Silicon and Iron-Manganese Alloys", J. Chipman, Robert M. Alred, et al., *Transactions ASM*, Vol. 44, 1952, p. 1221.

adjustment of the oxygen potential first to a level to reduce the iron ore to an impure pig iron and then later to increase the oxygen potential to oxidize impurities from that iron in the steelmaking operation. In the iron blast furnace the oxygen potential is extremely low in the hearth zone and continually increases as the reducing gases move up the furnace, reacting with iron oxide and coke according to Eqs. (4-6) and (4-7). The temperature profile of the furnace shows a continually decreasing temperature from about 3000°F in the bosh zone just above the tuyeres down to 400°F at the top of the furnace. The range of oxygen potentials in the furnace is illustrated in the following example problem.

Example Problem 4-1. Compare the oxygen potentials in the blast furnace hearth with those in the blast furnace stack.

Solution. Preheated air is blown into the furnace where the oxygen reacts immediately with hot coke to form carbon monoxide. The pressure is approximately 2 atm and the gas composition would be approximately $\frac{1}{3}$CO and $\frac{2}{3}$N$_2$. Assuming a temperature of 1550°C and equilibrium for the reaction,

$$2CO = 2C(s) + O_2(g), \tag{4-a}$$

$$K_a = \frac{a_C^2 P_{O_2}}{P_{CO}^2} = \frac{P_{O_2}}{(0.67)^2}. \tag{4-b}$$

From Table A-II,

$$-RT \ln K_a = -RT \ln \frac{P_{O_2}}{(0.67)^2} = 56{,}400 + 40.32\,T, \qquad (4\text{-c})$$

$$\log K_a = -\frac{56{,}400}{4.575(1823)} - \frac{40.32}{4.575}, \qquad (4\text{-d})$$

$$\log \frac{P_{O_2}}{(0.67)^2} = -6.75 - 8.81, \qquad (4\text{-e})$$

$$P_{O_2} = 0.67^2 (10)^{-15.56} = 1.24 \times 10^{-16} \text{ atm.} \qquad (4\text{-f})$$

The oxygen potential in the hearth, the lowest oxidation level in the furnace, is on the order of 10^{-16} atm.

Up in the furnace stack, the gases become less reducing. Consider the point at which wüstite is reduced to iron, at a temperature on the order of 750°C according to the reaction,

$$\text{FeO} + \text{CO} = \text{Fe} + \text{CO}_2. \qquad (4\text{-g})$$

The maximum oxygen potential at which this reduction can occur is given by the reaction,

$$2\text{FeO}(s) = 2\text{Fe}(\alpha) + \text{O}_2(g) \qquad (4\text{-h})$$

Approximating the standard free energy change of Eq. (4-h) by extrapolating the linear temperature relationship for delta iron (Table A-II) to lower temperatures, and noting that the solid phases are at unit activity,

$$\log K_g = \log P_{O_2} = -\frac{125{,}200}{4.575(1023)} + \frac{30.36}{4.575}, \qquad P_{O_2} = 10^{-20}. \qquad (4\text{-i})$$

The oxygen potential is then 10^{-20} atm, *but at a temperature of* 750°C. If the oxygen potentials of the gas phases in the hearth and in the stack are compared at the same temperature, e.g., 1550°C, the effective oxygen potential of the gas in the furnace stack would be only 2.5×10^{-8} atm.

Blast Furnace Stoichiometry

A material balance around the blast furnace (or any extractive metallurgical process) not only describes in detail the mass flow of the streams going to and from the system and permits an energy balance to be written based on it, but is also very helpful in performing charge calculations. The material balance for the blast furnace requires a knowledge of the composition and amounts of each stream going to and from the blast furnace system. Often this information is known with limited accuracy and one must, therefore, estimate or make assumptions concerning the disposition of given materials between various streams. The slag volume and theoretical slag composition may be calculated from lime, magnesia, silica, and alumina balances; an iron balance is used to calculate the mass of the metal stream while carbon, oxygen, and hydrogen balances are used to determine the exit gas volume and composition. Several examples of this type of calculation have been presented.[4-6]

Energy Balance

The total enthalpy balance for a blast furnace operation during a given reference period may be written in the form,

$$H_{blast} + H_{burden} + H_{stone} + H_{coke} = H_{hot\ metal} + H_{slag} + H_{top\ gas} + H_{heat\ loss}$$

$$(4\text{-}13)$$

where the enthalpy terms include heats of formation as well as sensible heats. It is necessary to solve the equation for the term $H_{heat\ loss}$. In doing so, the sensible heat of the burden, stone, and coke are taken as zero, since these components may be assumed to enter at 77°F, the usual base temperature for such calculations. The sensible heats of the blast and hot metal may be computed directly. The sensible heat of the top gas may be computed using the reported CO/CO_2 ratio and assuming the ratio $H_2/H_2O = 4$. For normal operation, i.e., without high blast moisture or hydrocarbon injection at the tuyeres, the latter assumption has little or no effect on the overall enthalpy balance. It is usually necessary to assume a temperature for the exit gases, and 400°F is often selected. In calculating the sensible heat of the slag, the stream can be assumed to be a mixture of $2CaO \cdot SiO_2$; $2CaO \cdot Al_2O_3$; SiO_2; $CaO \cdot SiO_2$; $2MgO \cdot SiO_2$; and CaS. Heats of formation for these compounds and their specific heats may be used to compute the sensible heat of the slag. It is a reasonable assumption to neglect the heats of mixing of these slag components. The heat loss term of Eq. (4-13) then can be computed.

The variation in hot metal cost resulting from the use of different blends of available raw materials can be assessed using linear programming techniques (see Computer Applications). This approach was first used by Bailey,[7] who incorporated into his linear program the statistical information developed by Flint.[8] A linear program involving a mass-energy balance, restrictions on the quality requirements for input and output streams, together with the economics of operating and freight costs can be used to optimize the blast furnace burden at different geographic locations with regard to cost and production rate.

Improvements in Blast Furnace Operation

In addition to design improvements, the principal advancements in blast furnace technology during the past two decades have been control of the blast humidity, oxygen enrichment of the blast, natural gas or other fuel injection, pressurized operation, and the increased use of beneficiated materials.

[4] W. K. Lewis, A. H. Radasch, and H. C. Lewis, *Industrial Stoichiometry*, Chemical Engineering Series, Chapter 10, McGraw-Hill, New York, 1954.

[5] H. N. Lander, H. W. Meyer, and F. D. Delve, *AIME Trans.* **221**, 485 (1961).

[6] H. N. Lander, H. W. Meyer, and F. D. Delve, *AIME Blast Furnace Proc.* **19**, 219 (1960).

[7] D. R. Bailey, *AIME Proc. Blast Furnace, Coke Oven and Raw Materials Conf.* **16**, 15 (1956).

[8] R. V. Flint, *AIME Proc. Blast Furnace, Coke Oven and Raw Materials Conf.* **11**, 49–73 (1952).

Almost one-half of the material entering the blast furnace is air. Consequently, relatively small changes in humidity are important with regard to the amount of water carried into the furnace. Indeed, controlled humidity has been proposed as a means of regulating the blast furnace.[9] The endothermic decomposition of the water vapor in the hot zone adjacent to the tuyeres absorbs considerable heat. This additional heat absorbed in the hearth of the furnace increases the amount of fuel required in the form of coke. Some fuel savings can be accomplished by drying the blast stream either by a refrigeration cycle or by passing the air through suitable drying columns.

Oxygen enrichment of the blast as a means of increasing furnace output has been proposed for some time, and the availability of large supplies of low-cost oxygen as the result of developments in air separation processes has made this proposal even more feasible. A 6% oxygen enrichment of the blast should increase production rates from 20–25%. Since the effect of oxygen enrichment would be to decrease the nitrogen content of the blast, this should result in less heat being carried away from the furnaces by the top gas, a decreased volume and velocity of the reacting gas with a corresponding reduction in flue dust, and an increased efficiency of combustion of the coke with higher hearth temperatures.

The amount of oxygen passing through the furnace for a given gas velocity may be increased if the furnace is operated under pressure. Modifications to several existing blast furnaces were made at the time of World War II and showed a 15% increase in production with a decrease in coke rate of about $12\frac{1}{2}\%$ and in flue dust of about 33%. Top pressures of about 1-atm gauge (2 atm absolute pressure) are maintained on some furnaces and attempts to increase this pressure have been made, although there is often considerable mechanical difficulty connected with the operation of a blast furnace under high pressure.

Enrichment of the blast with natural gas also results in a lower coke rate.

The use of pelletized concentrates has markedly increased blast furnace productivity. The first commercial test of taconite pellets was made on the Belle-fonte Furnace of Armco Steel Corporation at Middletown, Ohio, in the summer of 1960. During a 30-day test, the furnace, which was rated at 1500 tons per day using a conventional burden, achieved daily production records of 2700–2800 tons of pig iron.[10]

The beneficiation of materials improves burden properties by concentration and enrichment, and also may improve the physical form of the burden as well. The washing and sizing of the burden is a very important aspect in obtaining optimum furnace performance. Considerable economy also is obtained with regard to transportation of charge materials to the furnace site if they are beneficiated at the mine. In addition, sizing of material prevents channeling of the gases, with resulting improvement in mass and heat transfer in the furnace shaft, and also promotes a more uniform movement of the burden down through the furnace.

[9] A. L. Hodge, AISI Meeting, New York, May 25, 1960; Also *Iron Age*, June 2, 1961.

[10] E. W. Davis, *Pioneering with Taconite*, p. 191, Minnesota Historical Society, St. Paul, Minn., 1961.

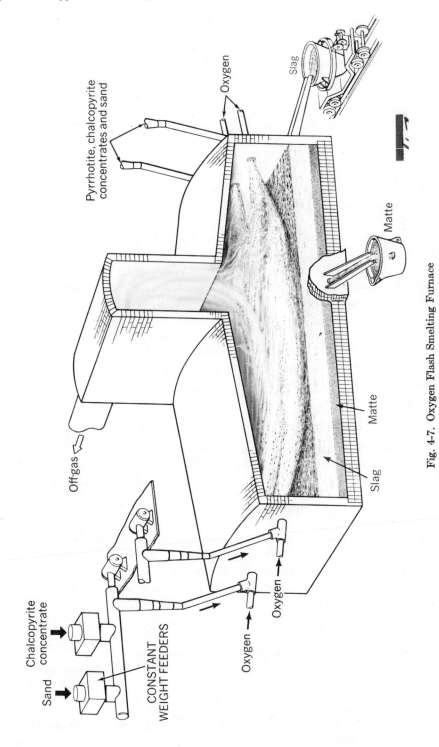

Fig. 4-7. Oxygen Flash Smelting Furnace

Source: J. R. Boldt, Jr., *The Winning of Nickel*, D. Van Nostrand Company, Princeton, New Jersey, 1967, p. 245.

4-3 Autogenous Smelting

Utilization of the heat of combustion of sulfides for smelting to matte and slag has been the objective of considerable development work. The flash roasting of sulfides has been extended into development of an oxygen flash-smelting process. An oxygen flash smelting furnace is shown in Fig. 4-7. Flash smelting takes place without additional fuel when a concentrate is suspended in a stream of oxidizing gases, such as oxygen-enriched air or pure oxygen, wherein the heat of oxidation brings the suspended particles to a temperature that permits melting to occur. Thermal balances for oxidation of sulfides show that it is necessary to preheat the air, enrich it with oxygen, or supply additional fuel to achieve temperatures sufficient to carry out a smelting operation. In flash smelting, a fine, high-grade sulfide ore falls through an oxidizing gas and the slag and matte are collected on a hearth. Mixing of preheated air and dried concentrate in a burner at the top of the reaction shaft spreads a shower of particles over the entire shaft area, permitting smelting to take place as the particles fall through the upwardly rising gases. Control of the furnace is accomplished by altering the preheat temperature of the air, or the fuel rate. The hot gases are cleaned and then cooled by passing them through a waste heat boiler. The concept of flash smelting is to carry out simultaneously the roasting, smelting, and conversion of sulfide ores. The roasting and converting operations are exothermic processes, whereas smelting is endothermic and the heat consuming portion of the process. By combining the three steps, considerable fuel savings can be realized, but it is necessary to carry out this process on relatively low-grade concentrates with sufficient sulfur to serve as fuel.

4-4 Copper Converting

The conversion of copper matte to metallic copper is accomplished by blowing air through the matte to oxidize the iron and sulfur. The conversion takes place in two stages. In the first stage iron sulfide is oxidized to iron oxide according to the reaction,

$$FeS + \tfrac{3}{2}O_2 = FeO + SO_2. \tag{4-14}$$

The iron oxide then combines with the silica flux to form a slag. After the first stage of blowing, the matte consists of essentially pure Cu_2S, which is called "white metal." A very small concentration of Cu_2O, on the order of 1%, is built up in the slag during the first stage of blowing. Copper that is oxidized into the slag tends to be reduced by iron sulfide in the matte until the FeS concentration becomes very small. The oxidation of iron is highly exothermic, and with low grade mattes, i.e., those containing high iron contents, copper scrap is often added to maintain the temperature of the charge being oxidized in the converter.

The second stage of blowing consists of oxidation of the "white metal" according to the reaction,

$$Cu_2S + O_2 = 2Cu + SO_2. \tag{4-15}$$

In operating practice the converter is charged with molten matte and silica flux. The air is turned on and the converter rotated to the position shown sche-

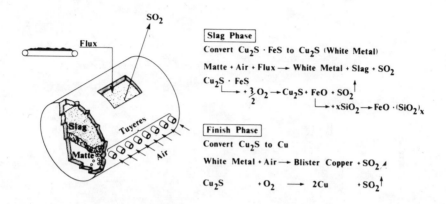

Fig. 4-8. Schematic Illustration of Copper Converter

Source: "Mathematical Model for Copper Converter Control," James H. Foreman, *Quarterly of the Colorado School of Mines*, 1964, Vol. 59, No. 4—Part A, p. 435.

matically in Fig. 4-8. As the iron is oxidized and slag builds up, the slag is run off and more matte and silica are added. This cyclic operation continues until the charge consists of essentially white metal. In the conversion of white metal, the copper sinks to the bottom of the vessel, leaving a layer of Cu_2S above it. Application of the Bessemer converter (see next section) in the copper industry was unsuccessful until the tuyeres were located at the sides of the vessel instead of the bottom. The failure of bottom blowing has been ascribed to separation of the copper-rich and copper sulfide-rich layers in the vessel. Bottom blowing would oxidize the copper and require mixing of the two immiscible liquids to react the copper oxide and copper sulfide according to the reaction,

$$2Cu_2O + Cu_2S = 6Cu + SO_2, \qquad (4\text{-}16)$$

whereas side blowing permits direct exposure of the Cu_2S to the oxidizing air bubbles.

Two general styles of converter have been used, the Great Falls converter, which was a pear-shaped vessel similar to a Bessemer converter (see following section), and the Pierce–Smith converter, which is a longitudinally oriented cylinder. An operating Pierce–Smith converter is shown in Fig. 4-9.

The product of the copper converter is termed "blister copper" because of the formation of gas by combination of dissolved sulfur and oxygen during solidification. The blister copper which is approximately 99% pure, contains the precious

Fig. 4-9a. Charging Matte to Copper Converter at Utah Plant,
Courtesy of Kennecott Copper Corporation

metals present in the matte and is subjected to further refining. Material flow in the smelting and converting of copper is illustrated in Fig. 4-10.

During operation, the tuyeres of the copper converter must be repeatedly opened by punching. One of the mechanisms leading to plugging of the tuyeres is the formation of magnetite in the slag during the first stage of blowing (see Appendix

B, Example Problem 5,). Each converter is equipped with automatic tuyere punching devices.

Overall copper smelter efficiency is highly dependent upon the converter operation. The converter slag, which contains 2–5% copper, is returned to the reverberatory furnace and thus the smelting rates are influenced by the silica and

Fig. 4-9b. Pouring Copper from Converter at Utah Plant,
Courtesy of Kennecott Copper Corporation

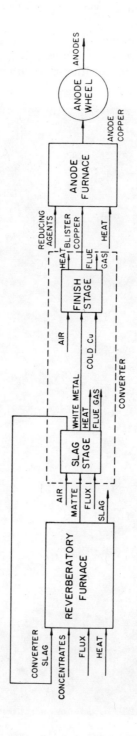

Fig. 4-10. Flow of Materials in Copper Processing

Source: E. H. Dudgeon, L. K. Nenonen, A. G. Balogh, and P. L. Fowler, "Measurements for Control of a Copper Converter," *Continuous Processing and Process Control*, Gordon and Breach Science Publishers Inc., New York, in press.

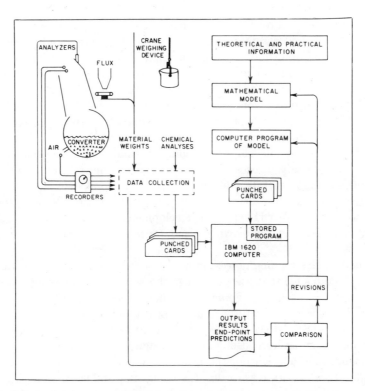

Fig. 4-11. Schematic Diagram of a Simulation Development Procedure for a Copper Converter

Source: J. Foreman, *J. Metals*, 1965, Vol. 17, p. 617.

magnetite concentrations in the returned converter slag. Lining life in both the reverberatory and converter operations is determined by temperature and slag control. Most operations are manually controlled at present, but with suitable instrumentation, computer control could be realized.[11,12] A thermochemical model describing copper converter operation in terms of a mass and energy balance has been described.[13] A schematic diagram showing development of a control model is shown in Fig. 4-11. Application of the concept of oxygen top blowing has been applied to the copper converter, and preliminary experimental work has shown promise for producing blister of high quality. Oxygen-enriched air has also been considered for the conversion of copper matte.

4-5 Steelmaking

From an economic viewpoint, the production of steel is the most important process in extractive metallurgy because of the large tonnages involved. Steel has been

[11] E. H. Dudgeon, L. K. Nenonen, A. G. Balogh, and P. L. Fowler, Measurements for Control of a Copper Converter, presented at 2nd Operating Met. Conf., Philadelphia, December 1966.

[12] J. Foreman, *J. Metals* 17, 616 (1965).

[13] Ibid.

known and used for over 5000 years, although until about 1700 A.D. it was not produced in the liquid state. The reduction of iron ores by carbonaceous fuel directly to a steel composition was practiced in ancient times, but liquid processing was unknown until development of the crucible process, wherein iron sponge or cemented wrought iron, coal, and flux materials were melted in a crucible to produce small quantities of liquid steel. Modern steelmaking processes came into being with the invention of the air-blown converter by Bessemer in 1856. In 1878, the Thomas process, a modified Bessemer process was developed to permit treatment of high phosphorous pig iron. The Siemens–Martin process, commonly referred to as the open hearth process, was developed at about the same time as the Thomas process. The open hearth process utilizes regenerative heat transfer to preheat air used in a burner, and is able to generate sufficient heat to refine solid steel scrap and pig iron in a reverberatory furnace. The availability of high tonnage oxygen at reasonable costs following World War II led to the development of various oxygen steelmaking processes, including the top-blown oxygen process, the Kaldo process, and the rotor process.

Steelmaking is a refining process in which impurity elements are oxidized from pig iron and scrap steel that are charged to the process. Steelmaking processes are classified as acid or basic, depending upon whether the slag is high in silica (acid) or high in lime (basic). The furnace lining in contact with the slag must be of a compatible material; a silica or siliceous material is used in acid processes and a basic material, such as burned dolomite or magnesite, is used in basic processes. Carbon, manganese, and silicon, the principal impurity elements in pig iron, are easily oxidized and separate as oxides, the manganese and silicon oxides going into the slag and carbon being removed primarily as carbon monoxide in the off-gases. Phosphorus also is oxidized but does not separate from the metal unless the slag is basic. Removal of sulfur occurs to some extent by absorption in a basic slag. Thus the basic steelmaking processes are more versatile in the raw materials they can handle, and consequently have become the predominant steelmaking methods.

The pig iron charge to the steelmaking process contains on the order of 4% carbon, 1% manganese, and 1% silicon. The phosphorus and sulfur levels in the pig iron vary, depending upon the steelmaking process, the basic Bessemer using phosphorus levels from 1.5–2%, whereas in other processes it is desirable to have the phosphorus as low as possible. Steel as tapped from the steelmaking furnace generally falls in the composition range 0.04–0.80% carbon, 0.06–0.30 manganese, 0.01–0.05 phosphorus, 0.01–0.05 sulfur, and negligible silicon. The reactions occurring in steelmaking can be represented as*

$$\underline{C} + (FeO) = CO + Fe(l), \qquad (4\text{-}17)$$

$$\underline{Si} + 2(FeO) = (SiO_2) + 2Fe(l), \qquad (4\text{-}18)$$

$$\underline{Mn} + (FeO) = (MnO) + Fe(l), \qquad (4\text{-}19)$$

$$2\underline{P} + 3(CaO) + 5(FeO) = (3CaO \cdot P_2O_5) + 5Fe(l), \qquad (4\text{-}20)$$

$$\underline{S} + CaO = CaS + \underline{O}. \qquad (4\text{-}21)$$

* As noted in Chapter 3, the underline notation refers to the element in metallic solution. The bracket notation refers to the compound in solution in the slag phase.

The equations above are treated in terms of oxidation by liquid iron oxide in the slag. Several alternative mechanisms are considered possible, such as direct oxidation by gaseous oxygen, as illustrated by the reaction,

$$\underline{C} + \tfrac{1}{2}O_2(g) = CO(g) \tag{4-22}$$

or alternatively, the oxygen may dissolve in the liquid iron and react according to the relationship,

$$\underline{C} + \underline{O} = CO. \tag{4-23}$$

The removal of phosphorus from the bath requires a basic oxidizing slag as illustrated by Eq. (4-20). The actual form of the phosphorus in the slag is thought to be tricalcium or tetracalcium phosphate, although the evidence is not entirely clear in this regard. Sulfur is absorbed by a basic slag, although it may or may not form a calcium sulfide molecule as indicated by Eq. (4-21).

Bessemer Converter Processes

The original Bessemer converter employed an acid lining, but this acid process is not used too extensively at present. The process requires a very low phosphorus and low sulfur-bearing pig iron since these elements are not removed in the process. A cross section of a Bessemer converter is shown in Fig. 4-12. Oxidation of the charge is accomplished by blowing air through a wind box on the bottom of the vessel. Elements are eliminated essentially in the same order as the stability of the oxide, i.e., silicon is eliminated first, along with manganese, and finally carbon as shown in Fig. 4-13. The Bessemer process is autogenous, i.e., no additional heat is

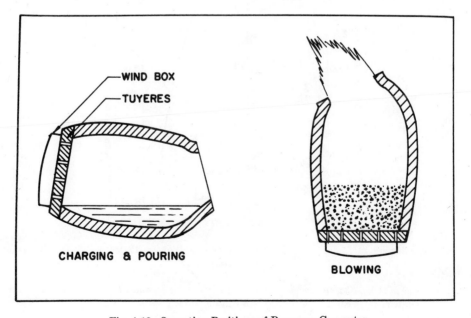

Fig. 4-12. Operating Positions of Bessemer Converter

supplied to the process, and the oxidation of the metalloids, particularly silicon, is sufficient to keep the charge molten and raise the temperature as purification proceeds such that a low carbon steel at a temperature near 2900°F is produced at the end of the process.

The Thomas or Basic Bessemer process, which is used extensively in Europe, is carried out in a basic-lined vessel and utilizes lime or limestone as a charge material to form a basic slag. Pig iron charged to this process contains up to 2 wt% phosphorus, which is oxidized from the bath after most of the carbon has been removed in a so-called "after blow." The oxidation of the phosphorus is exothermic and adds considerable thermal energy to the process.

One of the disadvantages of Bessemer steelmaking is the resulting high phosphorus and nitrogen contents in the steel. The nitrogen content of Bessemer steel is on the order of 0.015 wt%, although lower nitrogen levels can be achieved by careful control of temperature, and by carrying out the final oxidation of the bath with iron ore additions.

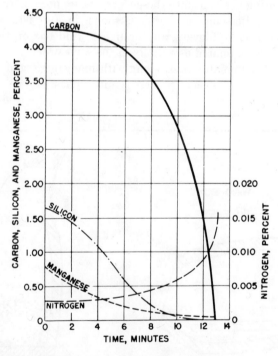

Fig. 4-13. Changes in Metal Bath Composition During Blowing Period of 25 Ton Bessemer Converter

Source: *The Making, Shaping and Treating of Steel*, edited by Harold E. McGannon, Pittsburgh, United States Steel Corporation, 9th edition, 1971, p. 479.

An important aspect of converter steelmaking is control, not only because of the rapidity with which the end point of the process is approached, but because of the necessity for providing a careful thermal- balance for the autogenous process. End point control has been greatly enhanced by the use of photocells to observe the character of the flame and emission from the vessel toward the end of the blow.

Open Hearth Process

Starting at the turn of the century up through the late 1950s, the basic open hearth process was the workhorse of the American industry, accounting for approximately 85% of the steel produced in the United States in 1955. The basic open hearth process (see chapter on Melting, Pouring and Solidification), because of its melting capabilities, can handle a wide range of pig iron to scrap ratios, the optimum operation being 40–60% pig iron with the balance of the metallic charge as steel scrap. Furthermore, in a basic operation a wide range of raw materials can be treated and the phosphorus and sulfur concentrations kept under suitable control. An advantage of the open hearth over the Bessemer process is that the steels produced contain less than 0.005 wt% nitrogen which is low enough to provide good properties in deep-drawing sheet material.

The solid constituents, steel scrap, limestone, and iron ore, are charged to the furnace, and the burner turned on. Burners are built into both ends of the furnace and the direction of the flame is changed from time to time to take advantage of the regenerative heat transfer system for the air, which is located on either side of the furnace at a lower level. As soon as the charge starts to melt down, pig iron is added and the iron oxide reacts with the molten pig iron to produce carbon monoxide in the so-called "ore boil." As the temperature increases, the decomposition of limestone proceeds at the bottom of the furnace, and a "lime boil" occurs, CO_2 being evolved at a relatively rapid rate. As the calcined lime floats up through the charge, phosphorus is removed as well as some sulfur.

The productivity of the open hearth furnace has been limited by both charging time and the extensive melting and refining periods, requiring on the order of 6–8 hr for melting and refining alone. The availability of low-cost oxygen has made it feasible to insert lances in the open hearth as illustrated in Fig. 4-14. Oxygen blowing in the open hearth has decreased the time required from 12–14 hr per heat down to approximately 6 hr. Oxygen blowing, however, creates fume and has necessitated gas cleaning equipment. This limitation on productivity and the added expense of gas cleaning equipment on an older installation has led to a relatively rapid displacement of the open hearth furnace by the oxygen top-blowing process.

Oxygen Steelmaking

The high dissolved nitrogen content in Bessemer steel, which limits its application, has made apparent the necessity for finding a nitrogen-free gas for blowing in the converter. The use of pure oxygen in a bottom blown vessel will result in ex-

Typical Roof Lance For Open Hearth Furnace

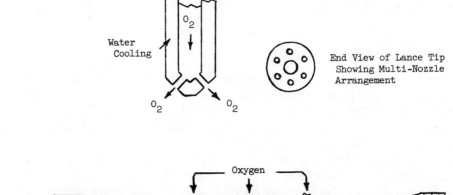

Water Cooling

O_2

O_2 O_2

End View of Lance Tip
Showing Multi-Nozzle
Arrangement

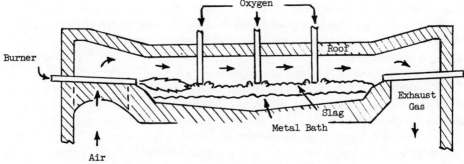

Oxygen

Burner

Roof

Exhaust Gas

Slag

Metal Bath

Air

Fig. 4-14. Oxygen Lancing of an Open Hearth Heat

tremely severe erosion of the bottom lining and tuyere area. An alternative is to blow a Bessemer converter with oxygen and steam to produce low nitrogen steel. This process has received considerable attention in Europe.

In the early 1950s commercial development of the oxygen steelmaking process was brought to fruition at Linz, Austria. Austrian ores are low in phosphorus and could not thermally support the basic Bessemer process, and furthermore a shortage of scrap in Austria following World War II eliminated economic application of the open hearth. These factors, together with availability of low-cost tonnage oxygen, led to development of the top-blown oxygen process, based on previous experience with a small vessel at Gerlafingen, Switzerland. The first large-scale oxygen steelmaking converters were 30-ton vessels. Further development of the process, particularly in the United States, has increased the size of the vessel until most oxygen steelmaking shops employ 250-ton vessels, with vessels as large as 300 tons in operation. The top-blown process is referred to as the BOF, BOP, OP, or L-D process.

Figure 4-15 shows an oxygen steelmaking vessel in which a 250–300 ton heat is being blown. The process is autogenous and uses on the order of 25–35% scrap, the remainder being basic pig iron that is relatively low in sulfur. Burned lime is used as a flux and a sufficient amount is added to produce a lime–silica ratio in the final slag on the order of 3 to 1. Oxygen is blown vertically down into the bath through a Laval nozzle, which produces supersonic flow. With larger vessels the use of a 3-hole lance has improved operation. The flow rate of oxygen gas in standard cubic feet per minute is given by the relationship,

$$Q_{\text{SCFM}} = 17.46 P_d A^*,\tag{4-24}$$

where

P_d = nozzle driving pressure, psia,
A^* = throat area of nozzle, sq in.

This relationship assumes that the oxygen temperature ahead of the nozzle is 60°F. The calculated flow rate can be corrected for oxygen supply temperature by the

Fig. 4-15. A Large Oxygen Steelmaking Furnace in Operation
Courtesy of Bethlehem Steel Corporation, Bethlehem, Pennsylvania.

relationship,

$$Q(\text{actual}) = Q(\text{Eq. 4-24}) \left[\frac{(60 + 460)}{[T_{O_2}(°F) + 460]} \right]^{1/2}. \qquad (4\text{-}25)$$

Copious fume in the form of micron-sized particles of iron oxide is evolved during the oxygen blow. This fume loss amounts to 2–3% of the metallic charge and a large portion of the capital expenditure of a Basic Oxygen Furnace plant is for gas-cleaning equipment. Significant operating variables include the lance height and oxygen flow rate. The required blowing time for a heat, including those approaching 250 tons, is on the order of 20 min. By suitable adjustment of the lance height, some iron can be oxidized early in the blow to dissolve the lime and produce an early basic oxidizing slag to assist in phosphorus removal. Later in the heat, lowering the lance or increasing the oxygen flow rate provides the deep penetration necessary to agitate metal and slag and provide for relatively low iron losses in the slag.

Control of the oxygen steelmaking process requires that the bath temperature at the end of the heat be within ±20°F of a given temperature, usually 2920°F for production of low carbon (below 0.10 wt%) steels. Since the process is autogenous (no external fuel) and the final temperature requires a balanced charge to achieve the desired end-point temperature at the end of the heat, the use of a thermally balanced charge as a control means has met with some success. However, this approach is limited in its performance because of the difficulty in making measurements on the charge materials which are sufficiently accurate, as illustrated in the following example problem.

Example Problem 4-2. Control of the oxygen steelmaking process is based initially on a thermally balanced charge. The calculation of the proper scrap charge (see Appendix B, Example Problem 4) requires accurate knowledge of the charge composition, in particular the hot metal analysis, and the hot metal temperature.

Estimate the influence on end point temperature of a variation of 0.05 in the wt% silicon in the hot metal. (A variation of 0.03–0.05 is the typical standard deviation in silicon analysis of hot metal encountered in operating practice.) Base the calculation on 240,000 lb of hot metal and a heat size of 300,000 lb.

Solution. The assumed value of 0.05 wt% error would yield an error of

$$\frac{0.05}{100} (240{,}000 \text{ lb hot metal}) \frac{1}{28.06} = 4.27 \text{ moles of Si}.$$

The thermal balance for silicon involves

Heat to hot metal temperature	+19700 Btu/lb mole Si
Solution in hot metal	−34150
Heat to slag temperature	−55800
Reaction with oxygen to form SiO_2	+369700
Reaction of SiO_2 with CaO	+53640
Energy available to raise steel bath temperature	353,090

(353090)(4.27) = Btu from error in silicon analysis.
(300000)(0.184)ΔT = change in steel bath energy in Btu = (wt) (Cp) (ΔT).

Equating these energies and solving for ΔT,

$$\Delta T = 27.3°F.$$

Even a deviation of only 0.03 in the wt% silicon would change the predicted turn-down temperature by

$$\frac{0.03}{0.05} (27.3) = 16.3°F.$$

Thus the analysis for silicon in the hot metal is very important in making a proper charge balance. The desired accuracy of the end point temperature, $\pm 20°F$, cannot be reached with present sampling and analysis techniques and instrumentation. (Hot metal temperature is also a very significant variable). Techniques are being developed for "in process" temperature measurements, stack gas analysis, and flow measurements (dynamic control). Temperature can be used as a basis for adjusting the blowing practice and making additions of exothermic or cooling materials. The gas analysis is indicative of proper blowing practice as well as providing a measure of the carbon content, either by determining the carbon lost in the stack from measured CO/CO_2 ratios and stack gas flow rates, or by correlating the CO/CO_2 ratio with bath carbon.

A detailed material and energy balance for the oxygen steelmaking process in the form of a thermochemical model that shows the influence of hot metal silicon on scrap melting is presented as Example Problem 4 in Appendix B.

The processing of high phosphorus irons in the top-blown oxygen steelmaking process is possible by injecting powdered lime into the oxygen stream. This technique is known as the LD-AC process and is used in Europe to process pig iron containing more than 0.5 wt% phosphorus.

Oxygen Converter Gas Recovery Process. The recovery of carbon monoxide evolved from the oxygen steelmaking process is referred to as the O-G Process (oxygen converter gas recovery process). Many oxygen steelmaking vessels have open hoods into which air is ingested to burn the hot evolved converter gases, which are about 90% carbon monoxide during much of the oxygen blowing period. Considerable effort has been carried out, particularly in Japan, to recover the converter gas that averages over 80% carbon monoxide for use as a fuel in the steel plant. The use of a closed hood and special precautions to prevent explosion of the carbon monoxide has resulted in a process which is economic in recovering the thermal energy in the carbon monoxide. In most hoods, a waste heat boiler is employed to recover the sensible heat of the stack gas, but a marked improvement in thermal efficiency is achieved by collecting the carbon monoxide gas for use as a supplementary fuel in the steel plant. Furthermore, the gas cleaning problems are made less difficult because the volume of gas is only approximately $\frac{1}{6}$ as great as when air is ingested in an open hood system. A general view of a gas recovery system is shown in Fig. 4-16.

Kaldo and Rotor Processes. Other oxygen blowing processes have been developed that employ rotating converters. The rotation of the vessel mixes the slag and metal, and also provides for heat transfer from the hot lining to the charge of

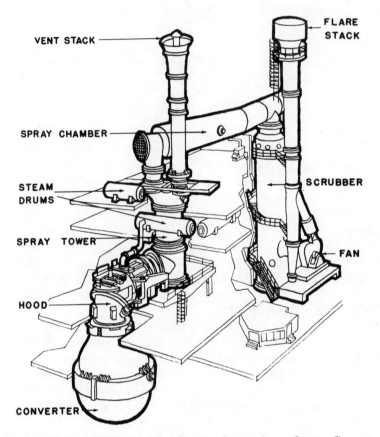

VENT STACK

FLARE STACK

SPRAY CHAMBER

SCRUBBER

STEAM DRUMS

SPRAY TOWER

FAN

HOOD

CONVERTER

Fig. 4-16. Schematic View of a Gas Cleaning System for an Oxygen Converter
Courtesy of Swindell-Dressler, Division of Pullman Inc.

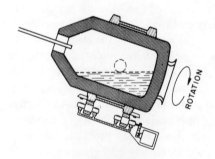

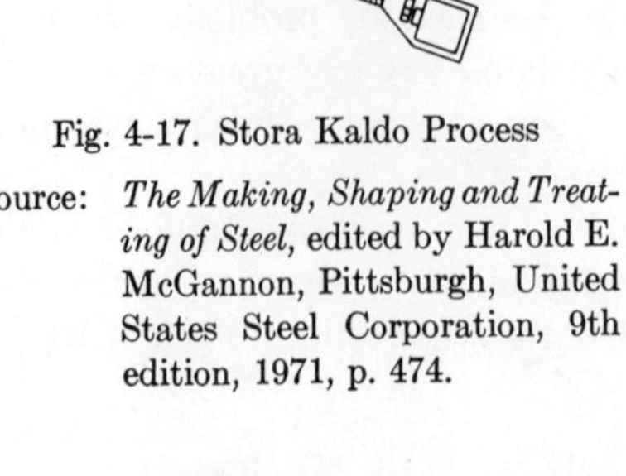

ROTATION

Fig. 4-17. Stora Kaldo Process

Source: *The Making, Shaping and Treating of Steel*, edited by Harold E.
McGannon, Pittsburgh, United
States Steel Corporation, 9th
edition, 1971, p. 474.

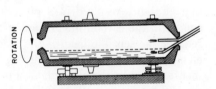

ROTATION

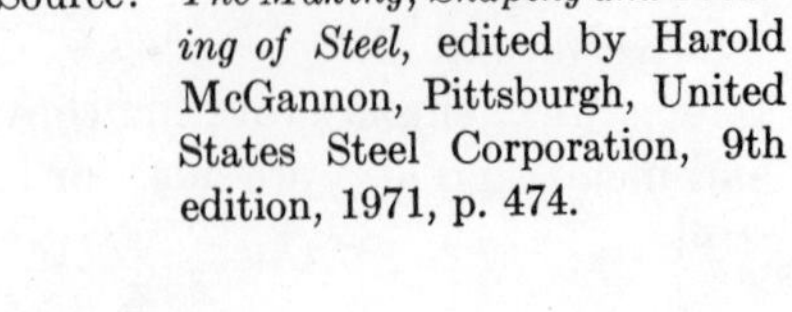

Fig. 4-18. Rotor Process

Source: *The Making, Shaping and Treating of Steel*, edited by Harold
McGannon, Pittsburgh, United
States Steel Corporation, 9th
edition, 1971, p. 474.

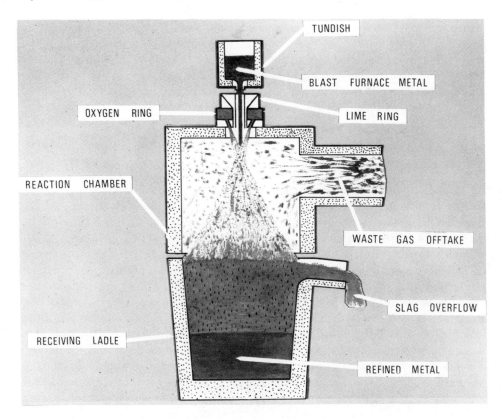

Fig. 4-19. Spray Steelmaking Process

Courtesy of British Iron and Steel Research Association.

the vessel. Schematic diagrams of these two processes are shown in Figs. 4-17 and 4-18. Adjustment of the oxygen lance to provide for limited penetration of the O_2 jet into the bath permits combustion of the carbon monoxide formed by oxidation of carbon in the iron bath to carbon dioxide within the furnace. This release of heat permits the melting of relatively large quantities of steel scrap, which is an economic advantage.

The Kaldo converter is a cylindrical, refractory-lined vessel that rotates on an inclined axis at speeds up to 35 rpm. A single oxygen jet is blown onto the surface of the metal. Oxygen consumption is approximately 2000 ft³ per ton of steel, whereas in the top-blown vessel it averages 1700 to 1800 ft³ per ton of steel produced, depending upon the composition of the charge.

The Rotor vessel is of cylindrical shape and rotates in a horizontal plane at speeds on the order of 1–5 rpm. Oxygen is blown into the vessel through two nozzles, one below and one above the bath surface. This type of oxygen injection provides for combustion of the carbon monoxide with a relatively high thermal efficiency. A two-slag rotor process in which high phosphorus irons are processed has been employed in England.

Spray Steelmaking. The spray steelmaking process offers the possibility of higher processing rates. This method utilizes a stream of oxygen to break up liquid blast furnace iron as it flows from the nozzle of a ladle or tundish (large pouring basin). The stream of oxygen together with powdered lime reacts with the resulting spray of fine droplets removing carbon, silicon, sulfur and phosphorus. The refined metal droplets and slag are collected in a ladle below as shown schematically in Fig. 4-19.

Electric Furnace Steelmaking

Direct arc furnaces ranging in size up to 120 tons are used to produce steel. The electric furnace may be either acid or basic-lined and has some advantage over the open hearth and oxygen converter processes in that very high temperatures can be produced (see Melting, Pouring and Solidification). The high temperature and the flexibility of using either a 100% solid charge, or up to 40 or 50% hot metal, provides considerable versatility for this melting unit. The very high temperatures are particularly useful in producing alloy steels. If large amounts of alloying elements are added to the ladle of steel from an oxygen converter or an open hearth furnace, the steel often is cooled to an extent that good quality ingots cannot be poured, or if the alloying elements are added in the furnace, they may be oxidized into the slag. This is particularly true for elements such as columbium, vanadium, and chromium. Furthermore, the requirement in certain types of stainless steel to provide for very low carbon contents requires that the carbon be removed without excessive chromium losses. This is possible only at high temperatures, as shown in the following problem.

Example Problem 4-3. The decarburization limit of stainless steels is enhanced by increased temperature; hence the electric furnace has been the primary unit for producing stainless steel. To illustrate this aspect of stainless steel production, calculate the minimum carbon content attainable at 1600°C (2912°F) and 1700°C (3092°F) for a steel bath containing 5% Cr, assuming chromite ($FeCr_2O_4$) saturation of the slag.

Solution. The slag–metal reaction limiting decarburization is

$$FeCr_2O_4 + 4\underline{C} = Fe(l) + 2\underline{Cr} + 4CO(g). \tag{4-a}$$

The standard free energy change for Reaction (4-a)[14] is

$$\Delta G^0 = 198{,}200 - 129.2\,T. \tag{4-b}$$

Assuming an effective CO pressure of 1 atm and unit activity of chromite in the slag,

$$\log K_b = \log \frac{(\%Cr)^2(a_{Fe})}{a_C{}^4}. \tag{4-c}$$

---

[14] C. E. Sims, Ed., *Electric Furnace Steelmaking*, Vol. 2, p. 172, Interscience Publishers, New York, 1963.

Rearranging Eq. (4-c) for a steel bath containing 5% chromium,

$$\log (a_C) = \frac{10,830}{T} - 7.06 + \tfrac{1}{2} \log (5) + \tfrac{1}{4} \log (0.95), \qquad (4\text{-d})$$

$$\log (a_C) = \frac{10,830}{T} - 6.716.$$

At 1600°C $\log (a_C) = -0.936$; at 1700°C $\log (a_C) = -1.226$. Neglecting solutes other than chromium, and neglecting the self-interaction of carbon, and assuming that the chromium–carbon interaction parameter is temperature independent,

$$\log (f_C) = (e_C{}^{Cr}) \, \%Cr. \qquad (4\text{-e})$$

From Table A-V,

$$\log f_C = -0.024(5) = -0.12. \qquad (4\text{-f})$$

Then at 1600°C

$$\log (\%C) = \log (a_C) - \log (f_C) = -0.816, \qquad (\%C)_{1600°C} = 0.153.$$

However, at 1700°C, $(\%C)_{1600°C} = 0.078$.

At 1700°C, the limiting carbon content is one-half of that at 1600°C. Oxidation of the bath to achieve carbon levels below these values would only result in oxidation of the chromium into the slag.

Bibliography

Bray, J. L., *Non-Ferrous Metal Production*, John Wiley, Inc., New York.

Dennis, W. H., *Extractive Metallurgy*, Philosophical Library, Inc., New York, 1965.

Sims, C. E. (Ed.), *Electric Furnace Steelmaking*, Vol. I and II, Interscience Publishers, New York, 1963.

Elliott, G. D., *Practical Ironmaking*, The United Steel Companies, Ltd., Sheffield, 1959.

Kirkaldy, J. S. and R. G. Ward (Eds.), *Ferrous Metallurgy*, University of Toronto Press, 1964.

Gilchrist, J. D., *Extraction Metallurgy*, Pergamon Press, London, 1967.

Liddell, D. M. (Ed.), *Handbook of Non-Ferrous Metallurgy*, Vol. I and II, McGraw-Hill, New York, 1926.

Hayward, C. R., *An Outline of Metallurgical Practice*, 3rd ed., D. VanNostrand Company, Inc., New York, 1952.

Hopkins, D. W., *Physical Chemistry and Metal Extraction*, The Macmillan Company, New York, 1954.

McGannon, H. E. (Ed.), *The Making, Shaping and Treating of Steel*, 9th ed., United States Steel Corporation, Pittsburgh, 1971.

Newton, J., *Extractive Metallurgy*, John Wiley, New York, 1959.

Ruddle, R. W., *The Physical Chemistry of Copper Smelting*, Institution of Mining and Metallurgy, London, 1953.

Discussion Questions

1. Discuss the thermal balance involved in the copper reverberatory smelter-converter process. What variations in the material balance can be achieved and how will they influence total production rate?

2. Build-up of magnetite accretions on the floor of copper reverberatory smelting furnaces is a problem which limits productivity and can lead to excessive shut-down for furnace repairs. This magnetite formation is akin in many ways to magnetite formation in the copper converter which causes tuyere blockage. Can you suggest any means by which magnetite formation can be limited or magnetite which has formed can be removed and by a means which does not interfere with normal smelting operations?

3. Based on the following data, describe the smelting of tin.
   Ore—Cassiterite (SnO_2 78.6% Sn)
   Metal—232°C melting point
           2687°C boiling point
           Carbon solubility in liquid metal is limited.

4. Discuss the gas–solid reactions in the iron blast furnace with respect to the specific effects of recently proposed methods for improving furnace operation and efficiency.

5. Briefly discuss the solution loss and carbon deposition reactions in the iron blast furnace with respect to:
   (a) effect of pressure
   (b) effect of temperature
   (c) calcination of limestone
   (d) reduction reactions

6. Briefly discuss the processing conditions which are desirable in order to obtain:
   (a) a high degree of desulfurization in an open hearth bath,
   (b) low phosphorus in iron blast furnace hot metal.

7. Basing the relationships on (1) production rate, (2) capital investment, (3) steel quality (S, P, N) and types, (4) charge materials:
   (a) Compare the Top Blown (BOF), Rotor and Kaldo oxygen processes.
   (b) Compare these oxygen processes as a group with the basic open hearth, electric, and basic Bessemer processes.

8. (a) In the production of low-carbon steel in the open hearth, it is generally not feasible to tap at lower than 0.04% C. Explain.
   (b) Ingot iron sheet averages below 0.02% C. Explain how this is accomplished.
   (c) Explain why it is more difficult to make low-carbon stainless steel.
   (d) Describe briefly a method for making 17% Cr steel containing 0.03% C max.

9. In the electric furnace practice, phosphorous is eliminated early in the oxidizing period before all of the carbon is eliminated. In the basic Bessemer, phosphorous is eliminated at the very end of the blow, after all of the carbon has been removed. Explain.

10. Compare air pollution control systems for fully integrated copper and steel producing plants.

Problems

1. A copper converting operation is carried out in a Pierce–Smith converter at 1100°C. Matte of the composition given below is charged at an average rate of

110 tons/hr. If a constant air blast of 25,000 CFM (100°F and 30 inches Hg) is maintained, what is the efficiency of the oxidation reactions?

Matte Composition: 29% Cu
38% Fe
28% S
5% impurities (gangue, inerts, etc.)

2. The separation of copper in the matte during the smelting of copper ores is highly efficient. Assuming that Cu_2S and FeS behave ideally in the matte, and noting that the activity of iron oxide in a typical reverberatory slag is on the order of 0.5 (relative to liquid pure FeO), justify this industrial experience on a thermodynamic basis.

3. Calculate the partial pressures of $S_2(g)$ and $O_2(g)$ under equilibrium conditions corresponding to converting of copper matte where two phases—matte and copper—coexist and are mutually saturated (assume $a_{Cu} = a_{Cu_2S} = 0.8$). The pressure of SO_2 is about 0.2 atm.

4. The activity of SiO_2 in a slag is 0.0004 based on pure SiO_2 as the standard state. What is the silicon content of carbon-saturated iron in contact with this slag at 1600°C at 1 atm total pressure $(CO + CO_2)$?

5. A roaster calcine charged to a reverberatory smelting furnace contains 27% iron pyrite (FeS) and 32% chalcopyrite. The remainder is gangue. Assuming that FeS and Cu_2S form the matte and that the excess sulfur is removed with the flue gases, what grade of matte will be formed and in what quantity per ton of calcine?

6. Compare the relative quantities of air required for the two stages of the copper converting process for various matte grades (e.g. 20–70% Cu).

7. In preparing a thermochemical model for analysis and control of an iron blast furnace, the energy term "heat loss" is computed from carefully collected operating data. The "heat loss" term which is made up of the heat losses from the furnace and errors in the other terms of the heat balance is subject to variation caused by inaccuracies in the reference period data. Given the following:

2400 TPD (tons per day) furnace
100,000 SCFM (60°F, 14.7 psia) blast rate
1,700,000 BTU/NTHM (net ton hot metal) heat loss
top gas composition: 2.4% H_2
0.6% H_2O
20% CO
18% CO_2
59% N_2

estimate the variation in the "heat loss" term caused by:
(a) an error of 0.1 in top gas CO/CO_2 ratio,
(b) a 20°F error in hot metal temperature.

8. A blast furnace is operated with blast enrichment using natural gas (primarily methane). The wind rate is decreased to keep the gas flow rate after combustion constant. Calculate the increase in tuyere zone temperature for each 1% of natural gas enrichment (present enrichment is based on orig-

inal blast flow rate). State all assumptions. *Hint:* See Appendix B, Example Problem 2.

9. (a) Estimate the activity of sulfur in hot metal containing 0.04% S, 4.0% C, 2% Si, and 1% Mn.
    (b) Calculate the solubility of MnS, as percent S:
        (1) at 1350°C in metal containing 4% C, 1% Si, 1% Mn,
        (2) at 1600°C in metal containing 0.5% C, 0.05% Si, 0.4% Mn.

10. At turndown of a basic oxygen steelmaking heat, the liquid steel contained 0.0005 wt% silicon and 0.06 wt% oxygen at 2920°F. Assuming equilibrium, calculate the activity of silica in the slag phase.

11. At the end of a basic oxygen heat, the FeO content of the slag is approximately 20% which corresponds to a mole fraction of 0.165. Assuming that the activity coefficient of FeO in the slag is unity and that the slag and finished steel are in equilibrium at 1600°C, estimate the carbon content of the metal bath.

12. Compute the required oxygen driving pressure for the following conditions: nozzle—Laval convergent-divergent, 1.825 inches throat diameter, blowing time = 18 min.

	C	Mn	Si	P	Weight (lb)
Pig iron charge	4.1	0.95	1.1	0.25	160,000
Scrap charge	0.18	0.30	0.05	0.03	60,000
Steel produced	0.08	0.25	0.05	0.015	200,000

Note: CO/CO_2 in effluent gas = 9.

13. Recognizing that oxygen cost increases markedly with increasing purity, specify the minimum oxygen purity for a basic oxygen process installation producing low nitrogen (0.004% N maximum) steels for deep drawing sheet.

14. The basic oxygen process is turned down at about 0.05 wt% carbon in the manufacture of low carbon steel. The carbon–oxygen product has been observed to be in equilibrium with 1.5 atm CO pressure. Assuming the slag to be saturated with lime (CaO) and calcium silicate ($2CaO \cdot SiO_2$), *calculate* the equilibrium silicon content of the steel bath.

15. In the basic oxygen process, the correct prediction of end-point temperature and composition is a primary goal for the control system. The end point composition in the production of low carbon steels is relatively easy for the operator to find by observing the "flame drop" which is indicative of a low carbon bath. The end-point temperature, however, depends upon a thermal balance between the enthalpy contents of the various phases, the heat generated by reactions, and heat losses. Computer control has been proposed for this purpose. A limitation placed on any computer "model" for the process is the accuracy of input data. Several variables appear to be of importance including the hot metal temperature and the scrap weight. Estimate the acceptable error in each of these variables which would *individually* cause the normally accepted

limit of error on the pouring temperature ($\pm 20°F$) to be exceeded, i.e., complete the table.

Variable (∅)	Input value	Estimated error in ∅	Change in tap temp. (°F)
Hot metal temp	2500°F	?	20
Scrap weight	75,300 lb	?	20

Base your calculations on the following heat:

	Amount (lb)	Temperature (°F)
Converter input		
Steel scrap	75,300	77
Pig iron	253,000	2500
Ore	2,870	77
Burnt lime	19,100	77
Oxygen gas	286,800 SCF	−60
Converter output		
Steel	300,000	2900
Slag	42,000	3000
Flue gas	—	3000

16. Compute the tons of scrap which can be melted in a BOF under the following conditions:
    (a) Combustion gas is 100% CO at 3000°F.
    (b) Combustion gas is 90% CO_2 and 10% CO at 3500°F.
    Assumptions: heat losses at 2×10^6 Btu/hr,
    pig iron charged 80 tons,
    heat time 45 min,
    flue dust losses are 3% of hot metal charged.
    Data:

	%Si	%Mn	%C	%P	Temperature (°F)
Hot metal	0.8	1.30	4.40	0.13	2500
Blown metal	0.01	0.20	0.05	0.02	3000
Scrap	0.01	0.20	0.05	0.02	77
Oxygen	99.5% oxygen				−60
Burnt lime	Available CaO 92%				77
Mill scale	(0.15 lb/lb burnt lime) 50% FeO				77
Slag	Basicity ratio (%CaO/%SiO₂) = 3.0				3000

17. Preheating of scrap has been investigated in an attempt to develop a means for increasing the scrap melting capability of the basic oxygen converter. Develop a plot of scrap melting capability versus preheat temperature of the scrap for a given converter operation.

18. The recently developed spray steelmaking process can be used to produce steel or as a pretreating process to partially purify liquid pig iron (hot metal) prior to completion of refining by a conventional steelmaking process. Consider a pretreating process carried out with the following specifications:

Hot metal composition—4 wt% carbon; 1.4 wt% silicon.
Required final composition—carbon, 2 wt%; silicon, 0.8 wt%
$$D_{si} = 1.2 \times 10^{-4} \, \text{cm}^2/\text{sec}, \qquad D_c = 6 \times 10^{-4} \, \text{cm}^2/\text{sec}.$$

Estimate the minimum tower height required to accomplish this refining step, assuming that the metal droplets are 2 mm in diameter and spherical in shape, and that the only resistance to mass transfer is the diffusion of silicon and carbon to the surface of the droplet. The atmosphere can be assumed to be highly oxidizing. The diffusion out of a sphere may be described by

$$\left[\frac{c - c_0}{c_s - c_0} \right] = 3.385 \, \frac{\sqrt{Dt}}{r}$$

where

$c =$ the mean concentration of the droplet at time, t,
$c_0 =$ the uniform initial concentration,
$c_s =$ a constant surface concentration,
$r =$ the radius of the sphere,
$D =$ the diffusion coefficient for the species under consideration.

CHAPTER 5

Pyrometallurgy IV: Refining Processes

This chapter considers the refining of liquid metals when the purification step involves the separation of a second phase. Initially, refining by separation of a liquid or solid phase from the liquid phase being treated will be presented. Later sections consider the purification of liquid metals based on vapor–liquid separation by distillation and by degassing techniques.

5-1 Liquid Phase Refining

The use of a limited solubility of either a metallic phase, such as a precipitating solid solution or intermetallic compound, or by precipitation of a nonmetallic phase, such as an oxide, represents a purification or refining step that is used in nearly all of the metallurgical industries. These refining steps include the formation of an insoluble compound, such as in deoxidation or in reactions carried out between a slag and liquid metal phase; the precipitation of a metallic phase, as in liquation; and liquid–solid separations, as employed in purification by zone refining.

One of the principal methods of refining liquid metals is based on the relative insolubility of compounds in liquid metals. The addition of materials that result in the formation of an insoluble compound by reaction with an impurity element permits a refining separation to be carried out on liquid metals.

The calculation of addition requirements to effect a precipitation is based on the solubility product for the separating compound, or more precisely, in terms of the equilibrium constant for the reaction involving formation of the compound from solution by the dissolved reactants.

Deoxidation Equilibria

In view of the fact that oxygen plays a major role in many high-temperature metallurgical process operations, equilibria between dissolved oxygen and other elements in the base material being refined are of prime importance in describing liquid metal refining processes. It should be noted that the formation of insoluble oxides may serve as a refining step in removing either oxygen or the metallic component of the oxide. In addition, the formation of a nitride, sulfide, carbide, or other insoluble compound is also a refining operation that may be described in the same manner.

The reaction equilibrium between the dissolved element oxygen and the oxidation product (which is usually a nonmetallic "slag" or a gas, in the case of

105

carbon, hydrogen, or sulfur), may be expressed by the equation,

$$x\underline{M} + y\underline{O} = M_xO_y, \tag{5-1}$$

where $\underline{M}$ represents the element dissolved in the bulk metal; $\underline{O}$ is the dissolved oxygen, which may be present as oxygen atoms or in combination with bulk metallic atoms (e.g., oxygen dissolved in liquid iron may be considered to be present as dissolved FeO); and M_xO_y is the oxide reaction product. The equilibrium constant for Eq. (5-1) is

$$K_1 = \frac{a_{M_xO_y}}{a_{\underline{M}}{}^x \cdot a_{\underline{O}}{}^y}. \tag{5-2}$$

Experimental determinations of deoxidation constants involve not only the thermodynamics of formation of the oxide product, but the formation of the solutions of oxygen and the reacting element in the bulk metal as well.

Aluminum Deoxidation of Steel. Consider the deoxidation of liquid steel by the addition of aluminum. The reaction is

$$2\underline{Al} + 3\underline{O} = Al_2O_3 \tag{5-3}$$

with equilibrium being specified by the constant,

$$K_3 = \frac{a_{Al_2O_3}}{a_{\underline{Al}}{}^2 \cdot a_{\underline{O}}{}^3}. \tag{5-4}$$

Then since $\Delta G^0 = -RT \ln K$

$$\log K = -\frac{\Delta H^0}{4.575\ T} + \frac{\Delta S^0}{4.575}. \tag{5-5}$$

The aluminum deoxidation relationship given in Eq. (5-3) requires three thermodynamic relationships describing (1) the formation of alumina, (2) the solution of aluminum in iron, and (3) the solution of oxygen in iron. The development of Eq. (5-3) is presented from this standpoint in the following paragraphs.

I. Formation of Al_2O_3 from pure elements.

$$2Al(l) + \tfrac{3}{2}O_2(g) = Al_2O_3(s). \tag{5-6}$$

At high temperatures, i.e., near 1900°K, Elliott and Gleiser[1] give

$$\Delta G^0 = -255,200$$
$$\Delta H^0 = -401,400$$
$$\overline{}$$
$$\Delta S^0 = (\Delta H^0 - \Delta G^0)/T = -77.0$$

Thus the free energy for Eq. (5-6) is given by

$$\Delta G_I{}^0 = -401,400 + 77.0\ T \tag{5-7}$$

II. Solution of aluminum in iron. The activity of aluminum in liquid iron has

[1] J. F. Elliott and M. Gleiser, *Thermochemistry for Steelmaking*, Vol. 1, Addison Wesley, Inc., Reading, Mass., 1960.

been determined by combining data on the activity of aluminum in liquid silver with the results of measurements of the distribution of aluminum between liquid layers of silver and iron. The distribution measurements were reported by Chipman and Floridis,[2] who computed a free energy of solution of aluminum in iron where the activity of aluminum is expressed in wt%. The chemical equation and the thermodynamic relationship is given as

$$\text{Al}(l) = \underline{\text{Al}} \text{ (in liquid iron)}, \qquad (5\text{-}8)$$

$$\Delta G_{II}^0 = -12{,}900 - 7.70 \, T. \qquad (5\text{-}9)$$

III. Solution of oxygen in liquid iron. The solution of oxygen in iron may be considered in terms of the equilibrium between H_2/H_2O gas mixtures and oxygen dissolved in liquid iron, which also has been measured by Floridis and Chipman.[3] That equilibrium may be expressed as

$$H_2(g) + \underline{O} \text{ (in liquid iron)} = H_2O(g), \qquad (5\text{-}10)$$

where

$$K = \frac{P_{H_2O}}{P_{H_2} \cdot a_0}. \qquad (5\text{-}11)$$

It was found that

$$\Delta G^0 = -32{,}200 + 14.63 \, T. \qquad (5\text{-}12)$$

However, the formation of water from the pure elements is given as

$$H_2(g) + \tfrac{1}{2}O_2(g) = H_2O(g) \qquad (5\text{-}13)$$

and

$$\Delta G^0 = -60{,}200 + 13.94 \, T \qquad (5\text{-}14)$$

The difference between Eqs. (5-13) and (5-10) gives

$$\tfrac{1}{2}O_2(g) = \underline{O} \qquad (5\text{-}15)$$

with the free energy change,

$$\Delta G_{III}^0 = -28{,}000 - 0.69 \, T. \qquad (5\text{-}16)$$

IV. Combination of I, II, and III. By combining

$$\text{Al}_2\text{O}_3(s) = 2\text{Al}(l) + \tfrac{3}{2}O_2(g) \qquad \text{(I)}$$
$$-\Delta G_I^0 = 401{,}400 - 77.0 \, T,$$

$$2\text{Al}(l) = 2\underline{\text{Al}} \text{ (in liquid iron)} \qquad \text{(II)}$$
$$\Delta G_{II}^0 = -25{,}800 - 15.40 \, T,$$

$$\tfrac{3}{2}O_2(g) = 3\underline{O} \text{ (in liquid iron)} \qquad \text{(III)}$$
$$\Delta G_{III}^0 = -84{,}000 - 2.07 \, T,$$

[2] John Chipman and T. P. Floridis, *Acta Metallurgica* **3**, 456 (1955).
[3] T. P. Floridis and John Chipman, *Trans. Met. Soc. AIME*, **212**, 549 (1958).

one obtains the result,

$$Al_2O_3(s) = 2\underline{Al} + 3\underline{O} \text{ (both in liquid iron)} \qquad (5\text{-}17)$$

$$\Delta G_{IV}{}^0 = 291,600 - 94.47\ T, \qquad (5\text{-}18)$$

which may be expressed in terms of the deoxidation constant

$$\log K_{IV} = \frac{-63,700}{T} + 20.67. \qquad (5\text{-}19)$$

The deoxidation constant may then be calculated at any temperature in the steel-making range, e.g., K_{IV} at $1600°C = 4.7 \times 10^{-14}$.

Summary of Deoxidation Data. A summary presentation of the free energies of formation of oxide compounds is given in Table A-II. Also included are data for nitrides, sulfides, and carbides, compounds which also may be formed upon removal of an impurity.

In addition to the standard free energies of formation of the insoluble compounds, the standard free energies of formation of the solution containing the reactants also is required. Table A-IV presents the free energy relations for the formation of dilute solutions in liquid iron. The data for other solvent metals is less complete, with only a few values being scattered throughout the literature.

In many cases, an estimation of the deoxidizing power of an element can be made by assuming that an ideal solution is formed between the element and the solvent metal (i.e., the mole fraction of the dissolved element may be substituted for its activity). However, if the free energy of formation of the solution between oxygen and the solvent metal is unknown, the equilibrium oxygen activity may only be specified in terms of an equivalent oxygen partial pressure in the gas phase. This information may be useful to the process engineer if the oxygen potential of the system is determined by other conditions which are known or can be fixed.

The equilibria presented have been derived for the pure reactant phase, e.g., solid pure Al_2O_3. These same basic relationships apply, however, in all cases, and the deoxidizing power of an element may be computed under conditions of equilibrium with complex slag mixtures, including those cases wherein the activity of the oxide formed is reduced below unity resulting in an improvement in deoxidizing power of the dissolved element. The use of silicomanganese as a deoxidizer is an illustration of this case. In this instance, the deoxidation product is an iron–manganese silicate slag in which the activities of the reaction products would be much lower than they would be if either pure silicon or manganese were used. In this manner the deoxidation can be made more complete and thus less deoxidizing material is required for the same degree of deoxidation. Application of vacuum to systems where a gaseous reaction product is formed (e.g., CO) also permits the reaction to reach a greater degree of completion.

These principles apply as well to the formation of nitrides, sulfides, carbides, and other insoluble compounds.

Precipitation Stoichiometry

The total amount of reacting material that must be added to remove an impurity from solution can be divided into two parts. First, a portion of the material added reacts with the impurity, forming an insoluble precipitate that separates from the liquid alloy. A second portion is represented by the added material which remains in solution in sufficient quantity to satisfy the solubility product relationship, e.g., equilibrium between a saturated solution and the precipitate.

Example Problem 5-1. A 100-ton ladle of steel at 1600°C contains 0.03 wt% oxygen. Compute the minimum amount of pure aluminum that must be added to reduce the oxygen content of the steel to 0.001 wt%. Assume Henry's law behavior for both aluminum and oxygen in solution, and that equilibrium conditions are reached.

Solution. The deoxidation constant for aluminum, assuming the conditions given above and that pure alumina is the reaction product, was determined for 1600°C using Eq. (5-19).

$$K = 4.7 \times 10^{-14} = (\%\text{Al})^2(\%\text{O})^3. \tag{5-a}$$

The equilibrium aluminum content is then

$$(\%\text{Al}) = \sqrt{\frac{4.7 \times 10^{-14}}{(0.001)^3}} = 6.85 \times 10^{-3}. \tag{5-b}$$

The residual aluminum in solution in pounds is

$$\text{Al}_{\text{dissolved}} = (100) \times (2000) \times \frac{6.85 \times 10^{-3}}{100} = 13.7 \text{ lb}. \tag{5-c}$$

The aluminum that reacted is given by the pounds of oxygen removed times the weight ratio in the reaction product, or

$$\text{Al}_{\text{reacted}} = \frac{(0.03 - 0.001)}{100} \times (100) \times (2000) \times \frac{(54)}{(48)} = 65.2 \text{ lb}. \tag{5-d}$$

The total aluminum required is given by the sum of the two amounts, or 78.9 lb.

In steel plant practice the deoxidation requirements for a killed steel are often 5 or 6 times this calculated amount. The reasons for the additional aluminum include alloying inefficiency, i.e., all of the aluminum does not go into solution and some floats on the surface of the steel, and absorption from the air during the pouring operation of a considerable amount of oxygen which must be removed by deoxidation. It should be noted also that the deoxidation product is not always pure alumina but often may be an iron–aluminum–oxygen compound, such as hercynite.

Desulfurization of Molten Pig Iron

Liquid pig iron produced in the blast furnace (Pyrometallurgy III) often contains concentrations of sulfur that are too high for production of a low sulfur product in the steelmaking furnace to which the pig iron will be charged. External desulfurization, i.e., the intermediate treatment of molten pig iron with a desulfurizing agent, is sometimes employed to lower the sulfur content of the pig iron. Liquid basic slags and some alkalies, or a solid desulfurizing agent, can be used. The kinetics of the process are limited by the contact of the desulfurizing agent and the liquid metal bath. Bubbling inert gases or shaking the ladle have been employed as mechanical means for increasing process rates.

Strong alkalies such as caustic soda (sodium hydroxide) or soda ash are sometimes used for external desulfurization. Preparation of a synthetic basic slag containing lime and fluxing agents also has been employed. Calcium cyanimide has been used to desulfurize molten pig iron according to the reaction,

$$CaCN_2 + \underline{O} + \underline{S} = CaS + N_2 + CO. \tag{5-20}$$

Injection of the desulfurizing material, using an inert carrier gas such as argon, or nitrogen, has been used to promote contact between the molten bath and the desulfurizing agent. Calcium carbide acts as a desulfurizing agent according to the reaction,

$$CaC_2 + \underline{S} = CaS + 2\underline{C}. \tag{5-21}$$

Powdered lime injected into the bath in a gas stream is a very effective desulfurizing agent. The thermodynamics of this desulfurization technique is illustrated in the following example problem.[4]

Example Problem 5-2. Consider a pig iron containing no silicon and saturated with graphite at 1300°C. Estimate the equilibrium sulfur content when lime is injected in a nitrogen stream into the pig iron bath and desulfurization takes place according to the reaction

$$CaO(s) + \underline{S} + C(gr) = CaS(s) + CO(g), \tag{5-a}$$

$$\Delta G_a{}^0{}_{1300°C} = 27,050 - 27.55\ T\ \text{cal/g-atom}. \tag{5-b}$$

The sulfur activity is referred to the infinitely dilute solution and the activity coefficient of sulfur under these conditions is approximately 5.

Solution.

$$\log K_b = \log \frac{a_{CaS} P_{CO}}{a_{CaO} a_S a_C} = -\frac{27,050}{4.575(1573)} + \frac{27.55}{4.575} = +2.26 \tag{5-c}$$

The activity of the pure solids CaS and CaO can be taken as unity. The activity of carbon in graphite saturated iron is also one. The pressure of carbon monoxide can be assumed to be 1 atm, although the presence of nitrogen would markedly reduce this pressure. It should be noted that a lower effective carbon monoxide pressure would enhance desulfurization. Thus

$$\frac{1}{a_S} = \frac{1}{(\%S)(f_S)} = 182, \tag{5-d}$$

$$\%S = \left(\frac{1}{182}\right)\left(\frac{1}{5}\right) = 0.0011. \tag{5-e}$$

In actual practice, it is more likely that the pig iron would contain an appreciable amount of silicon such that desulfurization would occur according to the

[4] B. Trentini, L. Wahl, and M. Allard, *Trans. Met. Soc. AIME* **209**, 1133 (1957).

reaction,

$$2CaO(s) + 2\underline{S} + \underline{Si} = 2CaS(s) + SiO_2(s). \tag{5-f}$$

The equilibrium sulfur content would be lowered by about a factor of 50, because of formation of dicalcium silicate, and because of an increase in the activity coefficient of sulfur resulting from the presence of silicon.

Parkes Process

In a manner very much parallel to deoxidation, the removal of an alloying element from a liquid metal can be effected by precipitation of an intermetallic compound. This is illustrated by the Parkes process for desilverization of liquid lead with zinc. Zinc readily combines with silver to form a zinc–silver intermetallic compound that is highly insoluble in liquid lead. This process is carried out by adding a sufficient amount of zinc (up to 2 wt%) to lead bullion at about 550°C. The liquid lead is agitated and permitted to cool slightly, and a crust that is rich in silver forms on the bath. The crust contains up to 70% lead along with the silver and zinc.

In operating practice, a two-stage desilverization often is carried out wherein the zinc is added in two batches. The first stage of the refining process produces a crust that is fairly rich in silver, whereas the second crust contains an excess of zinc and is used to remove as much silver from the lead as possible. The zinc-rich crust formed in the second phase of the process is used as a source of zinc for the first stage of the Parkes process on a fresh kettle of lead bullion.

Desilverization of lead bullion with zinc also can be carried out in a continuous manner, as in the Williams continuous process.[5]

Some lead bullions also contain gold. Gold is even stronger in its affinity for zinc than is silver. A two-step refining process can be carried out in which a small addition of zinc is made to the lead bullion to separate the gold, and then a second addition of zinc removes the silver.

To illustrate the thermodynamics involved in carrying out the Parkes process, the following example problem is posed. It should be noted that the compositional relationships may be made more complex in operating practice, in that the crust that forms at the process temperature may actually consist of a ternary alloy of lead, silver, and zinc, rather than the intermetallic silver–zinc compound plus entrained lead.

Example Problem 5-3. A liquid lead bullion initially free of zinc and containing 250 troy ounces of silver per ton is treated at 500°C with zinc to remove the silver. Assume that the reaction product is pure Ag_2Zn_3. Calculate the zinc addition in pounds per ton of lead bullion required to remove 98% of the silver.

$(\gamma_{Zn}^0)_{Pb} = 11$	Atomic Weights
$(\gamma_{Ag}^0)_{Pb} = 2.3$	Ag 107.9
$2Ag(l) + 3Zn(l) = Ag_2Zn_3$	Pb 207.2
$\Delta G^0_{500°C} = -30{,}500$ cal/g-mole	Zn 65.4

[5] G. K. Williams, *Trans. Met. Soc. AIME* **121**, 226 (1936).

Solution. Converting troy ounces to pounds,

$$(250)\frac{(31.1035)}{(28.3495)(16)} = 17.125 \text{ lb Ag/ton bullion},\qquad(5\text{-a})$$

$$\frac{(\text{troy oz/ton})(\text{gm/troy oz})}{(\text{gm/avoir oz})(\text{avoir oz/lb})}.$$

Basis: 1 Ton Lead Bullion.
Initial Composition:

$$\frac{17.125}{107.9} = 0.159 \text{ lb moles Ag/ton bullion},$$

$$\frac{1982.875}{207.2} = 9.60 \text{ lb moles Pb/ton bullion}.$$

Equilibrium Constant at 500°C:

$$\log K = -\frac{\Delta G^0}{2.3RT} = \frac{30,500}{(4.575)(773)} = 8.62,\qquad(5\text{-b})$$

$$K = 4.17 \times 10^8 = \frac{a_{Ag_2Zn_3}}{a_{Ag}{}^2 a_{Zn}{}^3}.\qquad(5\text{-c})$$

Assuming that the reaction product is Ag_2Zn_3 at unit activity,

$$a_{Ag}{}^2 a_{Zn}{}^3 = 2.4 \times 10^{-9},\qquad(5\text{-d})$$

$$X_{Ag}{}^2 \gamma^0{}_{Ag}{}^2 X_{Zn}{}^3 \gamma^0{}_{Zn}{}^3 = 2.4 \times 10^{-9}.\qquad(5\text{-e})$$

The final silver content is

$$(1 - 0.98)(0.159) = 0.00318 \text{ lb moles Ag/ton bullion}.\qquad(5\text{-f})$$

The mole fraction of silver is

$$X_{Ag} = \left(\frac{0.00318}{9.60}\right) = 3.31 \times 10^{-4}.\qquad(5\text{-g})$$

The equilibrium mole fraction of zinc in the bullion in contact with pure Ag_2Zn_3 from Eq. (5-e) is

$$(X_{Zn}) = \sqrt[3]{\frac{(2.4 \times 10^{-9})}{(3.31 \times 10^{-4})^2(2.3)^2(11)^3}},\qquad X_{Zn} = 0.0146\qquad(5\text{-h})$$

The residual zinc content is given to a very good approximation (neglecting the residual silver and assuming that the zinc concentration is small) by the product,

$$(0.0146)(9.6)(65.4) = 9.15 \text{ lb Zn/ton refined bullion}.\qquad(5\text{-i})$$

The zinc reacting with silver is

$$\frac{(0.98)\,(17.125)\,(3)\,(65.4)}{(2)\,(107.9)} = 15.25 \text{ lb zinc.} \qquad (5\text{-}j)$$

The zinc required to carry out the refining to 98% of completion is given by the sum of the zinc remaining in solution and the zinc reacting with silver $(9.15 + 15.25) = 24.4$ lb/ton bullion. In practice, the actual zinc requirement is slightly greater than predicted by the foregoing analysis. Failure to achieve equilibrium and a reduction in the activity of the zinc–silver compound by the presence of lead in the crust could both contribute to the differences. The residual zinc content of the lead bullion is on the order of 0.6 wt% or about 12 lb/ton of bullion, and the total addition often approaches 40 lb/ton.

Kinetics of Nucleation of a Separating Phase

The principal barrier to reaching equilibrium often manifests itself in the rate of nucleation of the separating phase. This is particularly true for crystallization processes, where large degrees of supercooling often have been noted. The surface energy between the precipitate and the solution must be considered, particularly for solids, in addition to the difference in free energies of the bulk phases. For homogeneous nucleation of spherical particles, the total free energy change may be expressed as

$$\Delta G = \frac{4\pi r^3}{3}\,\Delta G_v + 4\pi r^2 \gamma, \qquad (5\text{-}22)$$

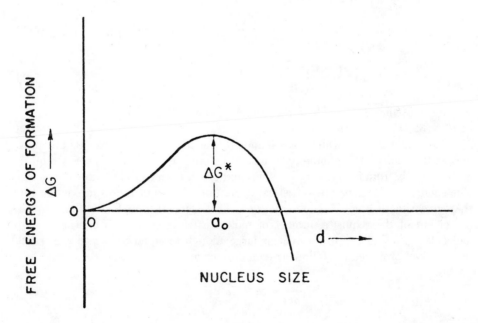

Fig. 5-1. Free Energy of Nucleus Formation as a Function of Nucleus Size

where

ΔG_v = bulk free energy change,

γ = surface energy of a unit area of interface between parent liquid and crystal nucleus,

r = radius of a spherical particle.

It should be noted that these terms vary with the specific geometry of the nucleus. The geometrical factors introduce a size effect into the problem, and, since ΔG_v is negative for a supercooled melt, these factors determine the critical size for a nucleus. When the nuclei are small, the positive term predominates, and the curve shown in Fig. 5-1 rises to a maximum as the size increases. A nucleus is stable if the size is greater than r^*, the critical nucleus size, since the free energy decreases if the nucleus grows. The free energy change for the formation of a nucleus of critical size is called the "work of nucleus formation." The nucleus of critical size, r^*, is in unstable equilibrium with the parent phase since any change in its size results in a decrease in free energy, and a tendency either to grow or disappear.

The work of formation is determined by r^*, which may be evaluated by maximizing ΔG.

$$\frac{\partial(\Delta G)}{\partial r} = 4\pi r^2 \Delta G_v + 8\pi r\gamma = 0, \tag{5-23}$$

$$r^* = -\frac{2\gamma}{\Delta G_v}. \tag{5-24}$$

Substitution in Eq. (5-22) gives

$$\Delta G^* = \Delta G_{(a=a^*)} = \tfrac{4}{3}\pi \left(\frac{2\gamma}{\Delta G_v}\right)^3 \Delta G_v + 4\pi \left(\frac{2\gamma}{\Delta G_v}\right)^2 \gamma, \tag{5-25}$$

$$\Delta G^* = \frac{\beta\gamma^3}{(\Delta G_v)^2}, \tag{5-26}$$

where the coefficient $\beta = \tfrac{16}{3}\pi$ for a sphere. If the nucleus were a cube, β would be equal to 32, where r is the cube edge.

The value of ΔG^*, which corresponds to the critical size, represents an energy barrier that must be overcome by a nucleus before it is stable. The energy required can come only from momentary local fluctuations of both concentration and energy. The energy fluctuations are statistical in nature and are of the usual kinetic type that give rise to homogeneous reactions. The concentration fluctuation requires movement of the requisite number of molecules close enough to one another by molecular diffusion to form a nucleus large enough to equal or exceed the critical size. Becker[6] proposed the following equation for nucleation rate,

$$\frac{dN}{dt} = ce^{-Q/kT}e^{-A(T)/kT}, \tag{5-27}$$

[6] R. Becker, *Ann. Phys.* **27**, 128 (1938).

where

$$\frac{dN}{dt} = \text{nucleation rate, number/unit volume–unit time,}$$

Q = activation energy for diffusion,
$A(T)$ = work required to form a nucleus,
T = absolute temperature,
c = a constant,
k = Boltzmann constant.

The work of nucleation, $A(T)$, increases markedly with decrease in supersaturation and is infinite at the saturation point. The term $e^{-Q/kT}$ decreases with decreasing temperature. The plot of N versus T thus has a pronounced maximum that corresponds to a definite supersaturation. However, if enough time is allowed, homogeneous nucleation will, in theory, eventually occur at any supersaturation. The presence of foreign particles or of seed crystals can increase appreciably the nucleation rate by innoculation. The alteration of surface energies caused, e.g., by the presence of trace amounts of surface active elements also can drastically affect the rate of nucleation. These effects occur because of the change in the surface energy term of the nucleation rate equation.

It may be concluded from the foregoing discussion that heterogeneous nucleation caused by the presence of seed crystals or innoculating agents that promote nucleation will occur prior to homogeneous nucleation. This is desirable in phase separation processes since a rapid nucleation of the precipitating phase promotes the kinetics of phase separation, and prevents nucleation of the precipitate from becoming rate-limiting. In practice, it is often necessary to "seed" the melt with foreign particles to promote heterogeneous nucleation (see Bayer Process, nickel reduction from aqueous solution, Chapter 6, Hydrometallurgy). Agitation in the form of vigorous stirring or vibrating techniques also has been used to induce nucleation.

Growth and Flotation of a Separating Phase

The limitation on the effectiveness of a phase separation may reside in the rate of growth of the precipitate and its subsequent separation from the bulk liquid. The rate of growth of the inclusion will be limited by the process of diffusion of reactants from the bulk liquid to the surface of the nucleated precipitate. Simultaneous with the diffusion process established at the surface of the particle is the relative motion of the particle in the bulk liquid.

The precipitating phase usually will have a different density than the bulk liquid and consequently will tend to rise (or sink). If the particle size is fixed, the terminal velocity of the particle can be estimated from Stokes law concepts. However, because the particle can grow during its residence in the liquid metal, its velocity will continuously increase. Nevertheless, Stokes law is applicable because of the relatively low Reynolds numbers encountered in situations of this nature (see Degassing). The additional possibility of collision and coalesence of small particles also exists. For the specific situation of deoxidation of liquid steel, theoretical[7] and experimental[8] studies have been carried out on the rate of growth and separation of the deoxidation product.

Liquid Phase Separation

Liquid phase separation in metallic systems often is described by the term "liquation." Liquation refers to the separation of a liquid phase both by liquid immiscibility and by melting point difference. The recovery from steel of lead, solders, or low melting alloys by liquation consists of heating a steel scrap charge to a temperature above the melting point of the low melting alloys, but below the melting point of steel, and permitting the low melting alloy to drain from the scrap. Liquation carried out by slowly cooling a liquid metal to permit a second phase, either liquid or solid, to separate is a common technique in the refining of nonferrous metals.

Processing Operations—Lead–Silver System. The recovery of silver from lead may be accomplished by cooling the silver-containing lead bullion (Pattinson Process). Consider a lead–silver alloy containing 0.9 wt% silver (260 troy ounces per ton). As shown in the phase diagram, Fig. 5-2, high purity lead will precipitate from lead–silver solutions at a temperature slightly above 320°C and will continue to do so until the eutectic point is reached at 304°C and a liquid phase composition of 2.5 wt% silver. At this point, the liquid which has been enriched in silver can be separated; hence, crystals of alpha phase (pure or nearly pure lead) separate, contaminated only by mechanical attachment or "wetting" by the high silver alloy, which is still liquid. This process is no longer used industrially but has been replaced by the Parkes Process in treating lead bullion for recovery of silver.

Lead–Copper System. A first step in the refining of lead is removal of copper by kettle drossing. The copper–lead phase diagram is shown in Fig. 5-3. This system exhibits a low melting point eutectic in which 99.8% lead as a liquid is in equilibrium with solid pure copper. Although lead bullion may contain less than 1% copper, most of this copper can be removed effectively by slowly cooling the lead from 450–350°C. A copper-rich dross appears on the surface of the molten lead. Commercial practice of the "drossing" of liquid lead is shown in Fig. 5-4.

Zinc Recovery by Liquid Phase Separation. An example of the commercial application of liquation in a continuous process involving separation of two liquid phases is the recovery of zinc from the Imperial Smelting Process, a combined zinc-lead blast furnace. A diagram of the process is shown in Fig. 5-5. The zinc recovery system and operating conditions are given in Fig. 5-6. Ores that are high in zinc are smelted in a lead blast furnace, producing a stack gas containing up to 6% zinc. The stack gases are passed through a splash-condenser containing lead, which absorbs the zinc from the vapor phase. Under typical conditions, the gas entering the condensers contains 5.9% zinc at a temperature of 1000°C, which represents over 90% of the zinc charged. The lead enters the condenser at 450°C and leaves at about 560°C. It is saturated with zinc on entering, 2.15 wt%. During

[7] E. T. Turkdogan, *J.I.S.I.* **204**, 914–919 (1966).

[8] G. C. Duderstadt and R. D. Weller, *J. Metals*, **18**, No. 6, 714–718 (1966).

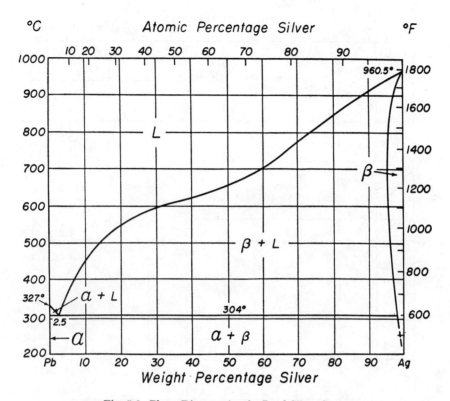

Fig. 5-2. Phase Diagram for the Lead-Silver System

Source: *Metals Handbook*, edited by Taylor Lyman, American Society for Metals, Metals Park, Ohio, 1948, p. 1153.

the condensation, the lead is heated and dissolves an additional 0.25% zinc. The lead is then circulated through a cooling launder. The concentration of zinc, 2.4%, is far below the saturation point on exiting from the condenser, i.e., 4.4% at 560°C. However, on cooling, at 470°C essentially pure liquid zinc begins to precipitate out from the lead, and the additional 0.25% absorbed in the condensers separates out. The upper liquid layer of zinc is then separated by means of a weir or baffle, and the underflow of lead-rich alloy is recirculated.

Engineering Design. The major considerations of a phase separation which concern the engineer are: (1) the yield of a given product, (2) the purity of the product, (3) the energy requirements, (4) the rate of production.

Process Yield. The yield of the product may be estimated by constructing a material balance for the system, assuming that equilibrium conditions prevail. This would represent the maximum yield, a quantity that is given in pounds by

$$Y = W_0 \frac{c_2^I - c_2^0}{c_2^I - c_2^{II}}, \tag{5-28}$$

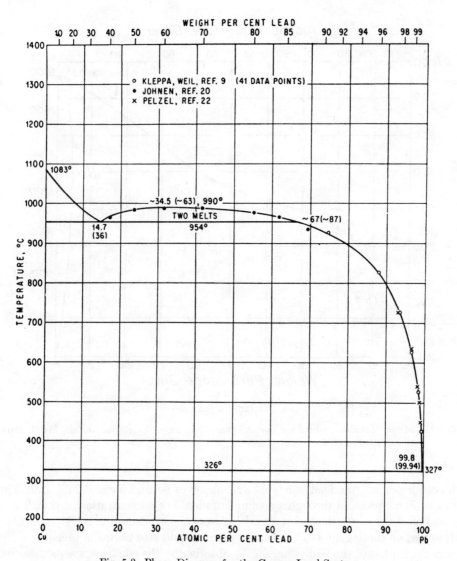

Fig. 5-3. Phase Diagram for the Copper-Lead System

Source: Hansen, *Constitution of Binary Alloys*, McGraw-Hill Book Company, New York, 1961,
          p. 610.

where

   W_0 = initial weight of the charge, lb,
   c_2^{I} = equilibrium concentration, in wt% of component 2 in the parent phase
          at the lower temperature.
   c_2^{0} = concentration, in wt% of component 2 in the parent phase initially.
   c_2^{II} = equilibrium concentration, in wt% of component 2 in the precipitated
          phase at the lower temperature.

Fig. 5-4. Refining of Liquid Lead by Drossing

Courtesy of American Smelting and Refining Company.

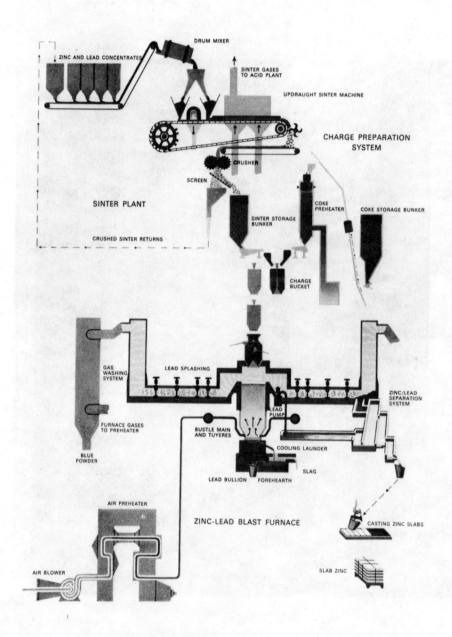

Fig. 5-5. Imperial Smelting Process

Courtesy of Imperial Smelting Processes Limited, Avonmouth, Bristol, England.

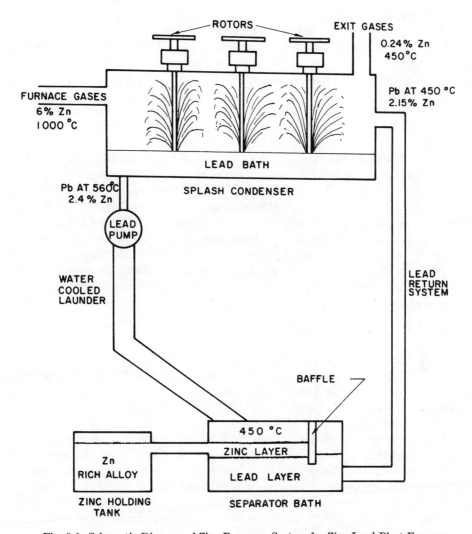

Fig. 5-6. Schematic Diagram of Zinc Recovery System for Zinc-Lead Blast Furnace

Equation (5-28) is an expression of the "lever-rule" as used in the study of binary phase equilibria. Mole fractions may be substituted for weight percentages and N_0, the total number of moles of initial solution, may be substituted for W_0. The yield then is expressed in moles, and the weight may be computed from the molecular weights of the solute and solvent and the concentrations in the product.

When the process involves a highly volatile solvent, or reactions occur between the lining of the process container and the material being processed, a correction for the losses must be made to the yield relation given above.

Purity of Product. For the equilibrium case, the purity of the product is given directly by the phase diagram. In practice, the adherence of the parent liquid to the precipitated crystals, or the entrainment of one phase within the other often decreases the purity of the recovered material. The purity usually can be improved by repeating the separation on the product, at a correspondingly greater cost.

Thermal Requirements. The energy requirements for a phase separation can be calculated from a heat balance on the process. The cooling energy represents heat which must be removed from the system, and is given by the algebraic sum of the sensible heat change in the solution, plus the heat evolved during the separation of phases (a quantity which may be either negative or positive, and often is approximated by the heat of solution for dilute solutions whose composition does not vary greatly during the operation), minus the heat losses to the surroundings, and any other cooling effects, e.g., vaporization of the solvent.

Process Kinetics. The rate at which a phase separation can be carried out may be limited either by the kinetics of the process, by the capacities of the equipment, or both may play a role. The kinetics of nucleation, introduced in a previous section, may be applied to the case where the rate-limiting step is the nucleation of the second phase. The growth of the separating phase also may be rate-limiting.

The ability of the equipment to extract heat rapidly from the solution may be a primary factor in the rate of the process. The removal of the material from the equipment, or the charging methods, i.e., material handling, also may limit the operation. A consideration of these factors is of major importance in the engineering design of phase separation processes.

Refining by Oxidation

Application of slightly oxidizing (or sulfurizing) conditions to an impure metal can result in the removal of the impurity as an oxide (or sulfide). These operations are illustrated best by the practices of "softening" lead bullion. The removal of copper, antimony, arsenic, and tin result in softer or more ductile lead, hence use of the term "softening."

The initial removal of copper from lead bullion by drossing was described in a previous section. Complete removal of the copper is accomplished by the addition of sulfur. About 2 lb of sulfur per ton of lead is stirred into the bath and a second dross, made up primarily of cuprous sulfide, floats out of the bath and can be skimmed from the surface. This treatment can reduce the copper content of the bullion to less than 0.005 wt%. Some removal of arsenic, antimony, and iron also is accomplished in this process.

Antimony, arsenic, and tin oxidize more readily than lead. Oxidizing conditions, imposed by agitation of molten lead with compressed air in a small reverberatory furnace, produce a lead-bearing slag containing the oxides of these impurities.

Example Problem 5-4—"Softening" of Lead–Removal of Tin. Lead bullion containing tin is oxidized by bubbling compressed air through the liquid metal at 800°K. Assuming that the reaction products are pure $PbO(s)$ and pure $SnO_2(s)$, what is the minimum tin composition that can be reached by this "softening" process?

$$Sn(l) + O_2(g) = SnO_2(s) \quad \Delta G^0_{800°K} = -139{,}550 \text{ cal/mole} \quad \text{Atomic weights}$$
$$Pb(l) + \tfrac{1}{2}O_2 = PbO(s) \quad \Delta G^0_{800°K} = -52{,}500 \text{ cal/mole} \qquad Pb \quad 207.2$$
$$(\gamma^0_{Sn})_{Pb} \text{ at } 800°K = 2.3 \qquad\qquad\qquad Sn \quad 118.7$$

Solution. For the combined reactions,

$$Sn(l) + 2PbO(s) = 2Pb(l) + SnO_2$$

$$\Delta G^0 = -34{,}550 \text{ cal/mole.}$$

The equilibrium constant at 800°K is

$$\log K = \frac{34{,}550}{(4.575)(800)} = 9.44,$$

$$K = 2.76 \times 10^9 = \frac{a_{SnO_2} a_{Pb}^2}{a_{Sn} a_{PbO}^2}.$$

The bath is nearly pure lead, hence, $a_{Pb} = 1$. The respective oxides are assumed to be present as pure solids and at unit activity.

$$K = 2.76 \times 10^9 = \frac{1}{(X_{Sn}) \gamma^0_{Sn}},$$

$$X_{Sn} = \frac{1}{(2.76 \times 10^9)(2.3)} = 1.57 \times 10^{-10},$$

$$\%Sn = \frac{(100)(1.57 \times 10^{-10})(118.7)}{(207.2)} = 9 \times 10^{-9}.$$

The oxidation of tin from lead is a very effective refining process. In plant operations, tin is removed to the chemical "trace" level. The formation of compounds or solid solutions of the oxides would tend to lower the activity of the tin oxide and make the process even more effective. Agitation of the metal to promote reaction with the entire bath such that the tin is removed with a minimum lead loss is a primary objective in commercial practice.

Alternative oxidizing procedures for purifying lead can be carried out with a reduced loss of lead by using oxidizing agents such as sodium nitrate in fluxes of sodium hydroxide or sodium chloride to dissolve the alkaline stannates, antimonates, and arsenates.

The fire refining of crude tin is carried out using liquation and drossing techniques to achieve high purity tin. Liquation can be used to remove iron, arsenic, and copper from tin, while oxidation with air or steam removes iron, aluminum, zinc, and nickel. The copper can be removed with sulfur additions, and formation of intermetallic compounds with aluminum additions removes antimony, arsenic, copper, and nickel. Further refining for the removal of lead and small remaining amounts of antimony and arsenic can be accomplished by electrolytic refining.

The use of chlorine gas to remove zinc from liquid lead as a chloride is the basis of the Betterton process. This means of zinc removal for purification of lead is illustrated in the following example problem.[9,10]

[9] H. H. Kellogg, *Trans. Met. Soc. AIME* **188**, 862 (1950).
[10] R. D. Pehlke and K. Okajima, *Trans. Met. Soc. AIME* **239**, 1351 (1967).

Example Problem 5-5. After the Parkes process in which an appreciable zinc concentration remains in the lead, chlorine gas is bubbled through lead bullion at 390°C. Calculate the limiting zinc concentration, assuming that the chlorides form an ideal solution, and that the mole fraction of zinc chloride in the salt formed is 0.983.

$$Pb(l) + Cl_2(g) = PbCl_2(l), \quad \Delta G^0_{390°C} = -61,550 \text{ cal/g-mole},$$

$$Zn(l) + Cl_2(g) = ZnCl_2(l), \quad \Delta G^0_{390°C} = -76,640 \text{ cal/g-mole},$$

$$\log (\gamma^0_{Zn})_{Pb} = \frac{1215}{T°_K} - 0.418.$$

Solution. At 390°C, $\log \gamma^0_{Zn} = 1.417$; $\gamma^0_{Zn} = 26.1$

$$Zn(l) + PbCl_2(l) = Pb(l) + ZnCl_2(l) \quad \Delta G^0 = -15,090 \text{ cal/mole.} \quad (5\text{-}a)$$

The equilibrium constant is given by

$$\log K_a = + \frac{15,090}{(4.575)(663)} = +4.97, \quad (5\text{-}b)$$

$$K_a = 93,500 = \frac{a_{Pb}a_{ZnCl_2}}{a_{Zn}a_{PbCl_2}}. \quad (5\text{-}c)$$

The activities of the chlorides are assumed to be given by the mole fractions

$$a_{ZnCl_2} = X_{ZnCl_2} = 0.983$$

and

$$a_{PbCl_2} = X_{PbCl_2} = 0.017.$$

The activity of the lead is unity, since the chloride is in contact with the purified bullion. The activity of zinc in equilibrium is given by

$$a_{Zn} = X_{Zn}\gamma^0_{Zn} = \frac{(1)(0.983)}{(0.017)(93,500)} = 6.18 \times 10^{-4}, \quad (5\text{-}d)$$

$$X_{Zn} = \frac{6.18 \times 10^{-4}}{26.1} = 2.37 \times 10^{-5}. \quad (5\text{-}e)$$

The theoretical wt% zinc remaining in the lead is then 0.00075. In actual practice, the residual zinc is usually below 0.005 wt% for these conditions. The kinetics of reaction of chlorine gas (or the salt phase) with the zinc remaining in very dilute solution in the lead limits the attainment of the calculated theoretical limit.

Zone Refining

The advent of the semiconductor industry with its requirement for ultrahigh purity materials led to the development of zone refining. In zone refining, a small molten zone moves through a long charge of alloy or impure material. The differences in impurity concentration between liquid and crystallizing solid lead to segregation of impurities into the moving liquid zone and a corresponding purifica-

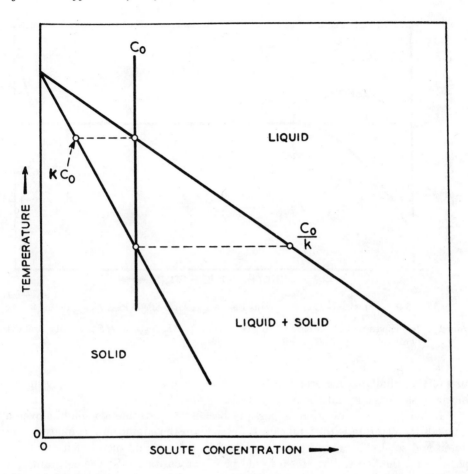

Fig. 5-7. Portion of Constitutional Diagram for a Solute Which Lowers the Freezing Point of the Solvent

Source: "Principles of Zone-Melting," W. G. Pfann, *Transactions AIME*, (Metals Branch), 1952, Vol. 194, p. 748.

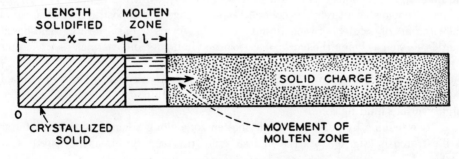

Fig. 5-8. Solidification by zone-melting

Source: "Principles of Zone-Melting," W. G. Pfann, *Transactions AIME*, (Metals Branch), 1952, Vol. 194, p. 748.

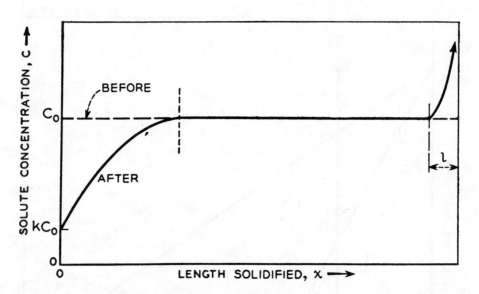

Fig. 5-9. Approximate Concentrations Before and After Single-Pass Zone-Melting

Source: "Principles of Zone-Melting," W. G. Pfann, *Transactions AIME*, (Metals Branch), 1952, Vol. 194, p. 748.

tion of the solidifying material.[11,12] Zone refining is used to purify not only metals but various inorganic and organic materials.

Refining by solidification is possible because impure metals solidify over a range of temperature with significant differences in the equilibrium composition of liquid and solid. This is illustrated in Fig. 5-7. For a liquid metal of initial composition C_0, the first solid to form would be of composition kC_0. The constant k of Fig. 5-7 is defined by the ratio of the equilibrium concentration of solute in the solid at the liquid–solid interface to the concentration of the liquid. In the case illustrated by Fig. 5-7, the distribution coefficient, k, is less than unity, and the liquid would be enriched in solute during solidification. In dilute solutions, the solidus and liquidus concentrations can be approximated closely by straight lines. Hence, the distribution coefficient remains constant over a range of compositions. The unequal distribution of solute between liquid and solid forms a basis for purification of metals, as well as being the cause of segregation in ingots and castings (see Chapter 8).

In a classic paper[13] W. G. Pfann first described zone melting and developed mathematical relationships for the redistribution of solute in a zone-melted material under various limiting conditions.

Zone refining by passing a single molten zone along a length of solid material is illustrated in Fig. 5-8. As the molten zone traverses the first zone length, l, a

[11] W. G. Pfann, *Zone Melting*, John Wiley, New York, 1958.

[12] H. Schildknecht, *Zone Melting*, Verlag Chemie, Academic Press, New York, 1966.

[13] W. G. Pfann, *Trans. Met. Soc. AIME* **194**, 747 (1952).

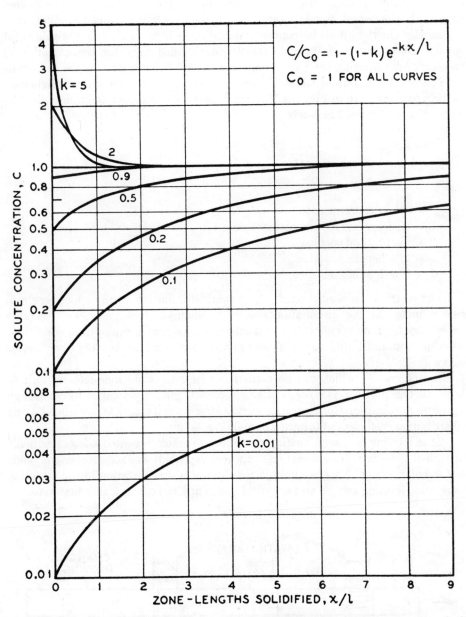

Fig. 5-10. Curves for Zone-Melting, Showing Solute Concentration in Solid as a Function of x/1, the Zone-Lengths Solidified, Calculated from Eq. 5-29, for Various k's

Source: "Principles of Zone-Melting," W. G. Pfann, *Transactions AIME*, (Metals Branch), 1952, Vol. 194, p. 749.

solid with solute concentration kC_0 freezes out as shown in Fig. 5-7. The molten zone is thus enriched in solute, and as the zone traverses along the ingot, it is continually enriched until its concentration reaches a value C_0/k. From this point on, the concentration of the zone remains unchanged since the concentration of unmelted solid entering the zone is the same as that freezing out. At the end of the pass, there is a build-up of solute. The concentration distribution after a single zone melting pass is shown in Fig. 5-9. The mathematical expression describing the concentration gradient of Fig. 5-9 is

$$\frac{C}{C_0} = 1 - (1 - k)e^{-kx/l}, \qquad (5\text{-}29)$$

where

C = solute concentration,
C_0 = initial solute concentration,
k = distribution coefficient,
x = total length solidified,
l = length of molten zone.

The distribution coefficient, k, has a marked influence on the effectiveness of zone refining. As the distribution coefficient changes from unity, i.e., becomes either very large or very small, the purification, assuming equilibrium conditions between solid and liquid and complete mixing in the liquid, is more effective, as shown in Fig. 5-10.

Several passes, which can be carried out in a multiple zone-refining unit as shown in Fig. 5-11, or as a series of single passes, can be employed to reach high degrees of purification. Concentration curves after several passes for a system having a distribution coefficient of 0.1 are shown in Fig. 5-12.

Zone leveling is a zone melting procedure in which repeated passes are made through the rod in reverse directions or where repeated-pass zone melting is made on a circular charge. This results in a very uniform concentration in the material. Thus, zone leveling can be an important technique in homogenizing materials.

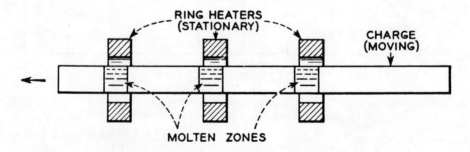

Fig. 5-11. Multiple-Zone-Refining, Using a Straight Charge

Source: "Principles of Zone-Melting," W. G. Pfann, *Transactions AIME*, (Metals Branch),
          1952, Vol. 194, p. 750.

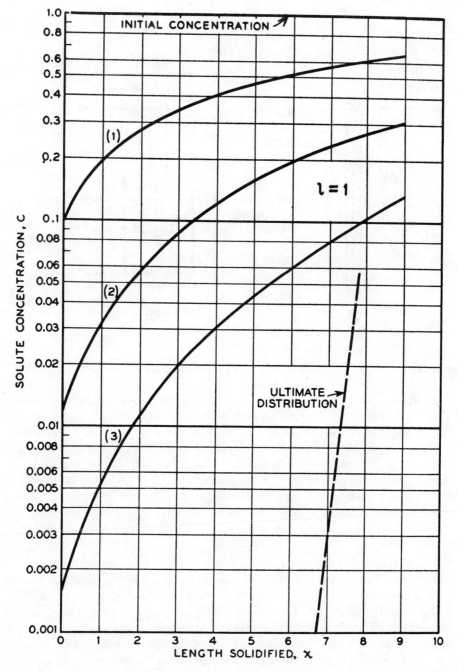

Fig. 5-12. Concentration Curves, Showing Concentration of Solute in Solid as a Function of *x*, the Length Solidified, After 1, 2, and 3 Successive Single-Zone Passes From Left to Right, Through an Initially Uniform Charge Ten Units in Length, for Unit Zone-Length.

Source: "Principles of Zone-Melting," W. G. Pfann, *Transaction AIME*, (Metals Branch), 1952, Vol. 194, p. 751.

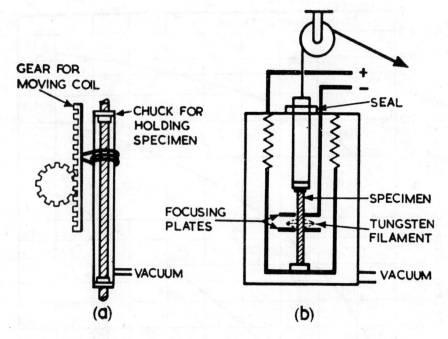

Fig. 5-13. Floating-zone Techniques

Source: W. H. Dennis, *Extractive Metallurgy*, Philosophical Library, New York, 1965, p. 321.

Zone refining is primarily employed in the semiconductor industry for refining of germanium and silicon. However, nearly all metals can be refined, including those which are highly reactive. To avoid reaction with refractory materials, the floating zone technique, Fig. 5-13, has been developed, in which surface tension of the metal is sufficient to contain the zone in a vertical rod, free from contact with any supporting material. Zone refining is usually employed in refining small batches of metals, but developments in this field point toward refining of larger masses of material and a more general use of this highly effective purification technique.

Secondary Metal Refining

A large segment of the metallurgical process industry is engaged in up-grading secondary and scrap materials to produce a prime metallurgical product. The recycle of scrap in the iron and steel industry as well as in most nonferrous industries is well known. Remelting and fire refining, using many of the techniques described in the previous paragraphs, are employed in up-grading secondary metallic material. Units for remelting copper, aluminum, and other nonferrous materials to make specific commercial alloys are an important part of the metallurgical industry.

5-2 Distillation Processes

Distillation is the separation of the constituents of a liquid mixture by partial vaporization, followed by separate recovery of vapor and residue. Distillation processes are an important aspect of process metallurgical engineering. They play a principal role in the extractive metallurgy of zinc and mercury and have been used to remove volatile constituents from other metals, including metals with rather high melting points, such as iron. The impurity element arsenic is present in many iron ores, but no suitable method has been developed to lower its content in the liquid iron produced from these ores. Fractional distillation, however, has been shown to be a possible method for the removal of this element from liquid iron.[14]

The engineering design of a distillation process requires a knowledge of the partial pressures of the vapor constituents in equilibrium with the condensed phase. Relatively few directly measured data are available on the vapor pressures of elements in liquid metallic solutions. However, the vapor pressures of pure elements are known to a relatively high degree of accuracy. The equations giving the vapor pressures of several metals are presented in Appendix A as Table A-VII.

If the vapor pressure of the pure element is known, the partial pressure of that element in equilibrium with a metallic solution can be computed, provided the activity coefficient is known, using the relation,

$$P_a = P_a^0 \gamma_a x_a, \tag{5-30}$$

where

P_a^0 = vapor pressure of pure component a at the temperature under consideration,

γ_a = activity coefficient of component a in solution,

x_a = mole fraction,

For dilute solutions, the activity coefficient of the solvent may be taken as one; also, if the solution is known to have ideal behavior, the activity coefficient of the solute is one. Otherwise, specific experimental data are required to determine the activity coefficient of the solute. The vapor pressures of several alloying elements over dilute solutions in liquid iron at 1600°C are presented in Table 5-1.

Clausius–Clapeyron Equation

From a combined statement of the first and second laws of thermodynamics, one may derive for a vapor–liquid equilibrium the relationship,

$$(S_v - S_1) \, dT = (V_v - V_1) \, dP \tag{5-31}$$

or

$$\frac{dP}{dT} = \frac{\Delta S_v}{\Delta V_v} = \frac{\Delta H_v}{T \Delta V_v}. \tag{5-32}$$

The relation is known as the Clausius–Clapeyron equation. Its integration requires an experimental knowledge of ΔH_v and ΔV_v. If the volume of the liquid is small in comparison to that of the vapor, then $\Delta V_v \cong V_v$, and if the vapor behaves ideally, then $V_v = RT/P$, assumptions that are reasonable for metals at elevated tem-

---

[14] M. Olette, "Vacuum Distillation of Minor Elements from Liquid Ferrous Alloys," *Proc. Physical Chemistry of Process Met. Conf., Pittsburgh,* 1959, Vol. 2. G. K. St. Pierre (ed.), pp. 1065–1087, Interscience, New York, 1961.

TABLE 5-1

Thermodynamic Data for Some Metallic Elements

Elements	Atomic wt. A_y	Vapor pressure for pure elements at 1600°C^a mm Hg, P^0_y	Activity coefficient in dilute Fe solutions, γ^0 at 1600°C	Source of data for γ^0	Vapor pressure over dilute Fe solutions at 1600°C^a mm Hg			Evaporation coefficient $\gamma = \dfrac{\gamma^0(P^0_y)}{P_{Fe}}\sqrt{\dfrac{M_{Fe}}{M_y}}$
					0.050%	0.2%	1%	
Pb	207.21	340	1	Assuming ideality	0.046	(two liquids)		(3000)
Mn	54.94	42	1.3	Sanbongi and Masa-yasu	0.028	0.11	0.55	900
Al	26.98	1.9	0.031	Chipman and Floridis	0.00006	0.00024	0.0012	1.4
Cu	63.54	1.0	8.0	Morris and Zellars, Koros and Chipman	0.0035	0.014	0.070	125
Sn	118.70	0.8	1	Assuming ideality	0.00019	0.00076	0.0038	9.1
Si	28.09	0.42	0.0072	Chipman et al.	0.000003	0.000012	0.00006	0.07
Cr	52.01	0.19	1	Assuming ideality	0.0010	0.004	0.020	3.3
Fe	55.85	0.060	1			0.060		1
Ni	58.71	0.043	1	Assuming ideality	0.00002	0.0008	0.00041	0.7
Co	58.94	0.031	1	Assuming ideality	0.000015	0.00006	0.0003	0.5

^{a} Approximate values.

Source: Michel Olette, Vacuum distillation of minor elements from liquid ferrous alloys, *Physical Chemistry of Process Metallurgy, Part* **2**, Vol. **8**, p. 1069, 1961, copyright AIME.

peratures. Substitution in Eq. (5-32) gives

$$\frac{dP}{P} = \frac{\Delta H_v}{R}\left[\frac{dT}{T^2}\right].$$ (5-33)

By further assuming that ΔH_v is independent of temperature and pressure, i.e., ΔH_v, the enthalpy change accompanying the transformation from the liquid to the vapor state at equilibrium is essentially equal to that accompanying the change from liquid to vapor at the boiling point, ΔH_v^0, Eq. (5-33) can be integrated to the form,

$$\ln\frac{P_2}{P_1} = -\frac{\Delta H_v^0}{R}\left[\frac{1}{T} - \frac{1}{T_1}\right].$$ (5-34)

This relationship may be used to extrapolate vapor pressure data to any desired temperature.

Eq. (5-33) may be integrated to a general form,

$$\log P = -\frac{A}{T} + B,$$ (5-35)

where A and B are constants whose values depend upon the units in which the vapor pressure, P, and the absolute temperature, T, are expressed. This relationship permits the vapor pressures of materials to be tabulated in the form of two constants. Since ΔH_v^0 is not entirely independent of temperature, an improved fit for the vapor pressure relationship can be made by adding additional terms, e.g., $C \log T$ as given in Table A-VII.

Relative Volatility

The term volatility is used to compare the vapor pressure of one pure substance with another. However, since the vapor pressure is a function of the presence of other components, the vapor pressure of a pure component does not in itself define the ease of component separation from solutions. The partial pressure ratio of two components over a solution is given by the relationship,

$$\text{Partial pressure ratio} = \frac{\gamma_a x_a P_a^0}{\gamma_b x_b P_b^0}.$$ (5-36)

The criterion for successful distillation is that the molar (or mass) ratio of a component to be separated must be greater in the vapor phase than in the condensed phase. From Eq. (5-36), the coefficient of the molar ratio in solution may be termed α where

$$\alpha = \frac{\gamma_a P_a^0}{\gamma_b P_b^0}.$$ (5-37)

The relative volatility, α, is thus a direct measure of the ease of component separation by a distillation process. If α is unity, no separation is possible, since the concentration of the two components in the vapor will be the same as in the condensed phase. Substances that can be readily separated show either large or small values of α.

The relative volatilities of certain elements in liquid iron solutions are included in Table 5-1. The computed values of α indicate that lead, manganese, copper, tin, and chromium may be distilled from liquid iron but that it would be impossible to remove silicon, nickel, or cobalt from liquid iron; that is, the relative volatility of iron is greater than that of the alloying element in these solutions. The calculations also indicate that aluminum has a relative volatility of one and thus cannot be separated.

The relative volatility of two components changes with temperature. Viewed from the standpoint of ease of separation alone, the optimum temperature range in which to conduct the distillation is that in which the value of α is maximum.

Equation (5-36) may be used to derive the following relationship between the

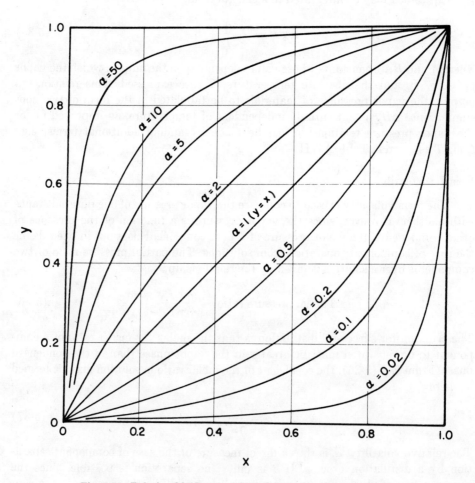

Fig. 5-14. Relationship Between X and Y for Various Values of α

concentration of the volatile component in the vapor phase, y_a, and in the liquid phase, x_a:

$$y_a = \frac{\alpha x_a}{1 + x_a(\alpha - 1)}. \tag{5-38}$$

Figure 5-14 shows a plot of y, the vapor phase mole fraction, versus x, the liquid phase mole fraction for a given component at various constant values of α.

Example Problem 5-6. Calculate the equilibrium partial pressures of the three components cadmium, bismuth, and lead over a ternary bismuth-rich solution containing 20 mol% lead and 20 mol% cadmium at 500°C. Compare the volatilities of lead and cadmium in this solution. The activity coefficients are: $\gamma_{Bi} = 0.9$, $\gamma_{Cd} = 1.2$, $\gamma_{Pb} = 0.9$.

Solution. From Table A-VII the vapor pressures of the three metals can be calculated. For pure cadmium,

$$\log P^0_{Cd}(\text{mm Hg}) = -\frac{5819}{773} + 12.287 - 1.257\,(\log 773) \tag{5-a}$$

$$= -7.525 + 12.287 - (1.257)\,(2.888)$$
$$= 1.127$$
$$P^0_{Cd} = 13.4 \text{ mm Hg}.$$

The partial pressure of cadmium over the alloy is given by the relationship,

$$P_{Cd} = X_{Cd}\gamma_{Cd}P^0_{Cd} = (0.2)\,(1.2)\,(13.4) = 3.22 \text{ mm Hg} \tag{5-b}$$

Similarly for lead and bismuth

$$P_{Pb} = X_{Pb}\gamma_{Pb}P^0_{Pb} = (0.2)\,(0.9)\,(1.64 \times 10^{-5}) = 2.95 \times 10^{-6} \text{ mm Hg}, \tag{5-c}$$

$$P_{Bi} = X_{Bi}\gamma_{Bi}P^0_{Bi} = (0.6)\,(0.9)\,(1.82 \times 10^{-5}) = 9.83 \times 10^{-5} \text{ mm Hg}. \tag{5-d}$$

The relative volatility of cadmium and lead in this alloy is given by Eq. (5-37):

$$\alpha = \frac{(1.2)\,(13.4)}{(0.9)\,(1.64 \times 10^{-5})} = 1.09 \times 10^6. \tag{5-e}$$

A similar calculation for bismuth yields a correspondingly large ratio, thus indicating that cadmium can be separated from these elements by distillation. The extent of purification depends, of course, on the concentration levels as indicated by Eq. (5-36).

Equilibrium Distillation

Simple batch distillation is the principal method of distillation used in metallurgical processes. In this technique increments of vapor upon formation are removed from the remaining liquid solution. If the operation is not conducted under high vacuum conditions, and under the assumption that the vapor coming off the solution is in equilibrium with the bulk liquid, the following mathematical analysis may be applied, as first done by Rayleigh.[15]

[15] Rayleigh, *Philosophical Magazine* **4**, 521 (1902).

The mass balance of such a process may be written in terms of a differential equation, the Rayleigh equation, derived as follows: assume a solution of two components, the total moles of which are L. Let the mole fraction of the more volatile component be x in the liquid and y in the vapor phase. Let dL moles be vaporized. The liquid will lose and the vapor will gain a differential quantity of the more volatile component. By a material balance,

$$(L - dL)(x - dx) + (y + dy)\, dL = Lx. \tag{5-39}$$

Neglecting differentials of second order and rearranging,

$$\frac{dL}{L} = \frac{dx}{y - x}. \tag{5-40}$$

Integrating Eq. (5-40) between limits

$$\ln \frac{L_1}{L_2} = \int_{x_2}^{x_1} \frac{dx}{y - x}, \tag{5-41}$$

where

L_1 = moles of original charge,
L_2 = moles of residual charge after $L_1 - L_2$ have been distilled off,
x_1 = the mole fraction of more volatile component in the original charge L_1,
x_2 = mole fraction of more volatile component in the residual charge L_2.

Consistent weight units may be substituted for the molar units in Eq. (5-41)).

If experimental data for the relationship between y and x are available, the right side of Eq. (5-41) may be integrated graphically. If a mathematical relationship exists between y and x, the integration may be carried out analytically, as in the following cases. For example, during a simple batch distillation at constant pressure, the temperature may rise as the residual liquid becomes poorer in the more volatile component. If α does not vary with temperature, Eq. (5-38) may be substituted in Eq. (5-41) with the integrated result

$$\ln \frac{L_1}{L_2} = \frac{1}{\alpha - 1} \left[\ln \left[\frac{x_1}{x_2} \right] + \alpha \ln \left[\frac{1 - x_2}{1 - x_1} \right] \right]. \tag{5-42}$$

If Henry's law applies to the solute that is being removed from the solution by distillation, the relationship $y = kx$ may be substituted with the result,

$$\ln \frac{L_1}{L_2} = \frac{1}{k - 1} \ln \left[\frac{x_1}{x_2} \right]. \tag{5-43}$$

It is important to note that Henry's law is an isothermal relationship, and in order to assure valid use of Eqs. (5-42) and (5-43), one should be aware of any temperature influences on the relationships involved.

The application also may be made to multicomponent mixtures. In the event that only one of the components of the complex solution is highly volatile, the solution may be treated as a pseudo binary. However, if two or more components have a volatility of the same order, one may compute the concentrations. The mathematics are more involved, but the principles remain the same.

Molecular Distillation

In most metallurgical distillation operations, particularly those carried out under high vacuum conditions, the rate of distillation is controlled by the rate of evaporation of atoms from the surface of the melt, and therefore it is not valid to assume that equilibrium is maintained between the concentrations in the vapor and in the liquid. The process conducted under these conditions is termed molecular distillation, and under conditions of complete nonreturn condensation, the quantitative rate of distillation is given by the Langmuir equation.

In 1913 Langmuir[16] using the kinetic theory of gases derived a relationship for the rate of evaporation of single component systems in terms of the vapor pressure. Under isothermal conditions there is a maximum rate at which a volatile substance will evaporate from an exposed surface. From the kinetic theory of gases, the rate of collision of vapor molecules with the surface can be calculated from the pressure of the vapor. Assuming that the vapor molecules obeyed a Maxwell–Boltzmann velocity distribution, and by making use of the ideal gas law, Langmuir showed that the mass of vapor molecules, striking a unit area of surface per unit time is given by the relation,

$$\mu = P \sqrt{\frac{M}{2\pi RT}} \tag{5-44}$$

where

P = pressure of vapor,
R = gas constant,
M = molecular or atomic weight of the vapor species.

The number of molecules returning to the surface can be assumed to be the same as the number striking it, i.e., all of the molecules striking the surface are held there. Under equilibrium conditions between vapor and liquid, this rate will be the same as the maximum rate of evaporation. Therefore,

$$w_0 = P_0 \sqrt{\frac{M}{2\pi RT}}, \tag{5-45}$$

where

w_0 = rate of evaporation, g/cm² per second,
P_0 = saturation pressure, mm Hg at solution temperature,
T = absolute temperature, °K,
M = molecular or atomic weight.

Evaporation can occur at a rate of w_0 only in a perfect vacuum and then only when the rate of evaporation is so small that the mean-free-path of the vapor

---

[16] I. Langmuir, *Phys. Rev.* **2,** 329 (1913).

molecules exceeds the distance between the evaporating and condensing surfaces. At appreciable rates of evaporation, the vapor molecules will collide with each other. Some will rebound to the surface, so the actual rate of evaporation will be the difference between w_0 and the rate of return to the surface w_1. The net rate of evaporation will then be

$$w = w_0 - w_1. \tag{5-46}$$

The two rates w_0 and w_1 are related to the pressures P_0 and P_1. Therefore, the net rate of evaporation is given by

$$w = (P_0 - P_1) \sqrt{\frac{M}{2\pi RT}}, \tag{5-47}$$

where P_1 is the partial pressure of the vapor at the evaporating surface.

The ratio between P_1 and P_0 is termed ϕ, the degree of saturation of the vapor; it varies from 0 to 1 as the evaporating conditions vary from molecular to equilibrium vaporization. The degree of saturation is related in the following way to the relation between w, the observed rate, and w_0, the maximum rate of evaporation.

$$\phi = \frac{P_1}{P_0} = 1 - \frac{w}{w_0}. \tag{5-48}$$

For the case of multicomponent systems, the relation given in Eq. (5-47) is modified and is referred to as the Langmuir–Knudsen equation:

$$W_A = (P^0_A - P_A') \gamma_A X_A \sqrt{\frac{M_A}{2\pi RT}}. \tag{5-49}$$

where

W_A = rate of evaporation of component A from a multicomponent solution,
P^0_A = equilibrium partial pressure of pure component A,
P'_A = effective partial pressure of component A in the vapor phase,
γ_A = activity coefficient of component A in the liquid phase,
X_A = mole fraction of component A in the liquid phase,
M_A = atomic or molecular weight of component A.

Example Problem 5-7. Using the Langmuir–Knudsen equation, estimate the concentration of manganese in a 1% Mn steel after holding in a high vacuum at 1600°C for 10 min. Assume the maximum rate of molecular distillation. The melt is inductively stirred, is 70 mm deep, and is contained in a 40-mm ID crucible.

Solution. From Table 5-1

$$P^0{}_{Mn} = 42 \text{ mm Hg}, \quad \gamma^0{}_{Mn} = 1.3, \quad M_{Mn} = 54.94.$$

Neglecting iron evaporation, the maximum evaporation rate for manganese is

$$W = 0.0583 \, P^0{}_{Mn} \gamma_{Mn} X_{Mn} \sqrt{\frac{M}{T}}. \tag{5-a}$$

Since the atomic weights of manganese and iron are nearly the same, the mole fraction can be converted directly to wt%.

$$\frac{d(\%\mathrm{Mn})}{dt} = \frac{(0.0583)\,S}{V\rho}\, P^0{}_{\mathrm{Mn}}\gamma_{\mathrm{Mn}} \sqrt{\frac{M}{T}}\,(\%\mathrm{Mn}). \tag{5-b}$$

Rearranging and integrating,

$$\log_{10}\left[\frac{(\%\mathrm{Mn})_{t=0}}{(\%\mathrm{Mn})_t}\right] = \frac{0.0583\,S}{2.3\,V\rho}\, P^0{}_{\mathrm{Mn}}\gamma_{\mathrm{Mn}} \sqrt{\frac{M}{T}}\,t, \tag{5-c}$$

where

S = surface area of the melt, cm^2
V = volume of the melt, cm^3
ρ = density of the melt, 7.1 gm/cm^3
t = time, sec.

$$\log\left[\frac{1}{(\%\mathrm{Mn})_t}\right] = \frac{(0.0583)}{(2.3)\,(7.1)\,(7)}\,(42)\,(1.3)\,\sqrt{\frac{54.94}{1873}}\,(600) = 2.87,$$

$(\%\mathrm{Mn})_{t=600 \text{ sec}} = 0.00135\%.$

The Langmuir-Knudsen equation describes distillation phenomena with good accuracy under conditions where the vapor pressure of the component is reflected by the bulk concentration, i.e., the rate of mass transport is controlled by vaporization at the surface. In commercial operations, attempts to achieve high distillation rates often create situations that place the Langmuir–Knudsen equation in error. One cause of the discrepancy between the Langmuir–Knudsen theory and observed distillation rates has been ascribed to depletion of the surface which occurs if liquid diffusion is the rate-limiting step which controls mass transfer to the surface. Furthermore, the surface may be covered with oxides that restrict mass transfer to the vapor, or the distillation rate may be limited by the rate of heat transfer to the surface which provides the necessary heat of evaporation.

In commercial operations the rate of distillation also may be controlled by the rate of transfer of vapor away from the surface. This rate of transfer is not as easy to calculate as the rate of evaporation. Factors which must be considered are (1) pressure gradient of vapor, (2) effective pressure of permanent gas in the system, and (3) dimensions of still and condensing system.

Finally, it should be noted that the vapor pressure is difficult to treat analytically in a gas stream whose molecules are moving predominately in one direction. In general, thermodynamic quantities are defined for systems at equilibrium and cannot be used rigorously under other conditions. Despite these limitations and the fact that conditions which lead to deviations from the Langmuir–Knudsen theory are difficult to describe analytically, these factors may become important in commercial operations, and must be considered in analyzing or modeling a distillation process.

Vacuum Dezincing of Desilverized Lead Bullion

The first commercial vacuum dezincing process for the purification of lead was developed by the St. Joseph Lead Company.[17] This process is used to remove zinc remaining in the lead after the Parkes desilverizing process.

The lead bullion is refined in batches at a temperature of 450–500°C. The surface exposed to vacuum is free of dross, and, since distillation is more rapid than diffusion of zinc through the molten lead, vigorous agitation must be employed to prevent surface depletion of zinc and also to prevent a temperature drop at the surface caused by radiation and energy consumed as latent heat of vaporization of zinc. In the St. Joseph batch operation, agitation is produced by a mechanical stirrer. In the continuous process used by the Broken Hill Associated Smelters, in Australia, agitation is produced by circulating a thin stream of lead in turbulent flow. The zinc is condensed as a solid from these operations.

The amount of lead distilling over with the zinc increases with temperature because the relative difference in vapor pressures becomes less at higher temperatures. Consequently, the best separation of zinc from lead is obtained at the lowest temperatures. However, the zinc recovered from the lead bullion is recycled for desilverizing, so that contamination of lead in this instance is of little practical consequence. Operating practice shows that the lead in the condensate is less than 10%, which represents less than 0.05% of the lead treated. This is not a great circulating load for the process. It is also important that oxygen be avoided. A vacuum of 200 μ or less is adequate to provide a distillate that is shiny in appearance. The rate of distillation in these processes has been quantitatively treated and compared with the Langmuir equation.[18] The actual operating times are on the order of 5 hr, or approximately 10 times the theoretical treatment time based on the Langmuir equation. This discrepancy may result from surface depletion of zinc or from inefficient condensation, which permits the zinc vapor to condense and fall back into the lead bath. In the continuous process used in Australia, a horizontal flow of vapor from the liquid lead surface to the condenser prevents precondensed zinc droplets from falling back into the bath, with the result that the distillation rates more closely approach the theoretical limit.

Observations on Liquid Iron Alloys

Under conditions of molecular distillation, the influence of the molecular weights of the components on the rate of evaporation must be considered. Thus, for distillations from iron baths the relative volatility coefficient, α, as defined in Eq. (5-37), may be modified to the form,

$$\alpha' = \frac{\gamma_A}{\gamma_{Fe}} \left[\frac{P^0{}_A}{P^0{}_{Fe}} \right] \sqrt{\frac{M_{Fe}}{M_A}}, \qquad (5\text{-}50)$$

[17] W. T. Isbell, *Trans. Met. Soc. AIME* **182**, 186 (1949).
[18] T. R. A. Davey, *J. Metals* **5**, 991 (1953).

where

M_{Fe} = atomic weight of iron,
M_A = atomic weight of solute.

In a dilute solution, iron (the solvent) obeys Raoult's law and hence, γ_{Fe} is unity. Also, the solute A at low concentrations obeys Henry's law, and γ_A is nearly constant and equal to γ^0_A. (It would be equal to one if the solution were ideal). Under such conditions, α' becomes

$$\alpha' = \gamma^0_A \left[\frac{P^0_A}{P^0_{Fe}}\right] \sqrt{\frac{55.85}{M_A}} . \tag{5-51}$$

A comparison is made in Table 5-1 between the computed values of α' and those measured experimentally on liquid iron solutions. Although there is some discrepancy, it should be noted that the comparative evaporation coefficients remain in the same order. The unusual results for silicon (and also for aluminum, where it was noted that in some cases aluminum was highly volatile and in other cases did not distill at all) are explained in terms of the formation of a highly volatile suboxide.[19] This effect also has been noted by Floridis,[20] who showed that the vacuum distillation of sulfur from liquid iron was greatly enhanced by the presence of silicon, presumably because of the formation of a volatile silicon sulfide, which assists in removing both elements from solution.

5-3 Degassing Operations

An important problem in producing a suitable metal product is that of gas control. Hydrogen is a particularly troublesome gas—causing bleeding ingots, embrittlement, low ductility, and the presence of blow holes. Nitrogen also may have a desirable or undesirable effect on the properties of a metal, particularly steel, depending upon the metal's composition, subsequent treatment, and intended use. Oxygen, as a principal refining agent, plays a role in determining the final composition and properties of metals. In addition to these simple gases, the metallurgist must also cope with complex gases, those which contain more than one type of atom in the molecule, principally CO, CO_2, SO_2, and H_2O. In the design of processes to control these gases in metals, data are required on the solubility, rates of solution, and chemical reactions involved in formation or decomposition in the metal, particularly in the liquid state. Much of this information is presently available, and an attempt will be made in this section to apply that data to the design of degassing operations as refining processes.

Fundamentals of Gas–Metal Reactions

Diatomic Gases. The solubility of a gas in a metal is a function of the pressure of that gas in equilibrium with the liquid metal. The solubility of a diatomic gas,

[19] M. Olette, Vacuum Distillation of Minor Elements from Liquid Ferrous Alloys, *Proc. Physical Chemistry of Process Met. Conf.*, *Pittsburgh*, 1959, Vol. 2. G. K. St. Pierre (ed.), pp. 1065–1087, Interscience, New York, 1961.

[20] T. P. Floridis, *Trans. Met. Soc. AIME* **215,** 870 (1959).

e.g., nitrogen, is proportional to the square root of the pressure of nitrogen, which is in equilibrium with the melt. This relationship has come to be known as Sieverts' law,[21] and follows directly from consideration of the equilibrium of the reaction,

$$\tfrac{1}{2}N_2 = \underline{N},\qquad\qquad (5\text{-}52)$$

where the equilibrium constant may be expressed as

$$K_{52} = \frac{a_N}{\sqrt{P_{N_2}}} = \frac{f_N(\%N)}{\sqrt{P_{N_2}}}.\qquad\qquad (5\text{-}53)$$

Since the solubility of nitrogen is directly proportional to the square root of the nitrogen pressure as shown in Fig. 5-15 for nitrogen dissolving in liquid iron, the activity coefficient of nitrogen in the liquid pure metal is unity and independent of concentration, i.e., nitrogen pressure, up to several atmospheres pressure. From this behavior, it follows that a diatomic gas, such as nitrogen, dissolves monatomically in a liquid metal at high temperatures. Eq. (5-53) can be rewritten in Sieverts' law form as

$$\%N = K_{52}\sqrt{P_{N_2}}.\qquad\qquad (5\text{-}54)$$

Similar relationships can be rewritten for hydrogen and other diatomic gases.

The solubility of a gas in an alloy is also a function of the concentrations of solutes in the alloy. As an example, the influence of alloying elements on the solubility of nitrogen in liquid iron is shown in Fig. 5-16. Since the equilibrium constant, K_{52} of Eq. (5-53), remains fixed, a change in solubility of nitrogen occasioned by the presence of an alloying element may be represented by a corresponding change in the activity coefficient, f_N. Customarily the infinitely dilute solution of nitrogen is taken as the reference state,

$$\lim \left[\frac{a_N}{\%N} \right]_{\%N\to 0} = f_N = 1.\qquad\qquad (5\text{-}55)$$

Since nitrogen solution follows Sieverts' law in Fe–N solutions, f_N remains unity and the self-interaction parameter for nitrogen, e_N^N, is zero. As a result of the reference state taken, the activity coefficient of nitrogen in an Fe-j alloy is obtained by the following equation:

$$f_N = \left[\frac{\%N(\text{pure Fe})}{\%N(\text{alloy})} \right]_{P_{N_2},T}.\qquad\qquad (5\text{-}56)$$

A plot of the logarithm of the activity coefficient of nitrogen in liquid iron as a function of added third elements is shown in Fig. 5-17. The slope of a curve of Fig. 5-17, taken as the alloying element approaches zero concentration, represents the interaction parameter e_N^j.

A summary of the standard free energy changes for the solution of various gases in liquid metal solvents is presented in Table 5-2. Additional interaction data for solutes in liquid iron is presented in Table A-V of Appendix A.

---

[21] A. Sieverts, *Z. Phys. Chem.* **60**, 129 (1907).

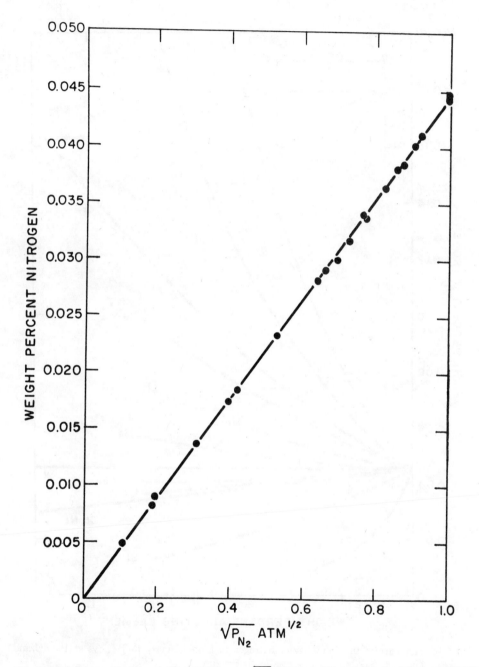

Fig. 5-15. Adherence to Sieverts' Law, $\%N \propto \sqrt{P_{N_2}}$, Nitrogen Dissolved in Pure Iron at 1606°C

Source: "Solubility of Nitrogen in Liquid Iron Alloys," R. D. Pehlke and J. F. Elliott, *Trans. Met. Soc. AIME*, 1960, Vol. 218, p. 1091.

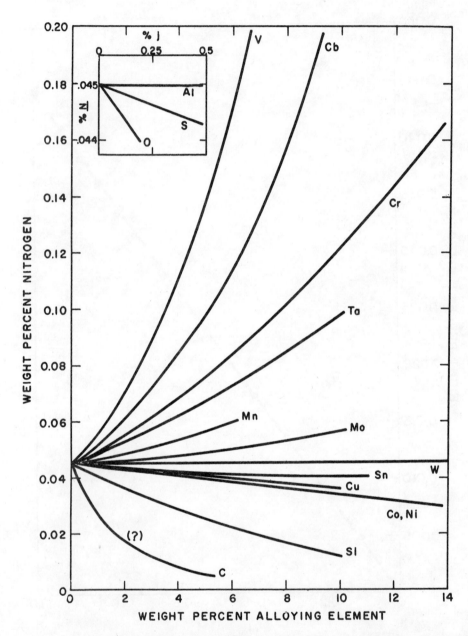

Fig. 5-16. Summary of the Solubility of Nitrogen in Liquid Binary Fe-j Alloys at 1600°C and
1-atm N_2 Pressure

Source: "Solubility of Nitrogen in Liquid Iron Alloys," R. D. Pehlke and J. F. Elliott, *Trans.
Met. Soc. AIME*, 1960, Vol. 218, p. 1099.

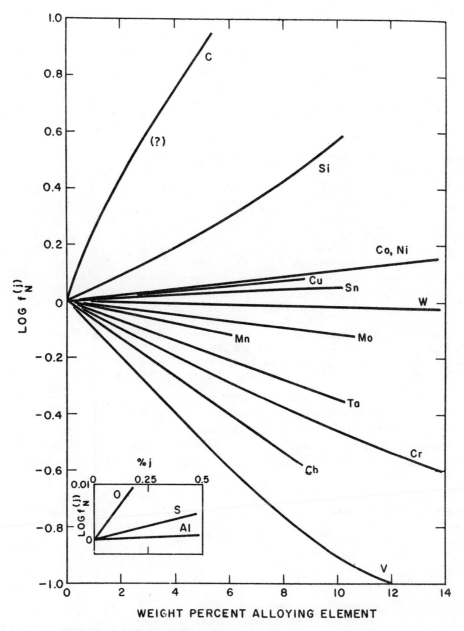

Fig. 5-17. Effect of Alloying Elements on the Activity Coefficient of Nitrogen in Binary Fe-j Alloys at 1600°C

Source: "Solubility of Nitrogen in Liquid Iron Alloys", R. D. Pehlke and J. F. Elliott, *Trans. Met. Soc. AIME*, 1960, Vol. 218, p. 1100.

TABLE 5-2

Standard Free Energy of Solution of Gas Forming Solutes in Various Liquid Metal Solvents

(Standard state in solvent is hypothetical 1 wt% solution)

Solvent	Element	$\Delta G°$ cal/g-atom	Reference
Aluminum	Hydrogen $\frac{1}{2}$ H$_2$ = $\underline{H}$	$11,650 + 6.55\ T$	1
Cobalt	Hydrogen $\frac{1}{2}$H$_2$ = $\underline{H}$	$9,800 + 7.03\ T$	2
	Nitrogen $\frac{1}{2}$ N$_2$ = $\underline{N}$	$10,000 + 5.24\ T$	3
	Oxygen $\frac{1}{2}$ O$_2$ = $\underline{O}$	$-14,700 - 4.91\ T$	4
Copper	Hydrogen $\frac{1}{2}$ H$_2$ = $\underline{H}$	$10,400 - 10.8\ T$	5
	Oxygen $\frac{1}{2}$O$_2$ = $\underline{O}$	$-18,150$	6
Iron	Carbon C(g) = $\underline{C}$	$5100 - 10.0\ T$	See Table A-VI
	Hydrogen $\frac{1}{2}$ H$_2$ = $\underline{H}$	$8715 + 7.28\ T$	7
	Nitrogen $\frac{1}{2}$ N$_2$ = $\underline{N}$	$860 + 5.71\ T$	See Table A-VI
	Oxygen $\frac{1}{2}$ O$_2$ = $\underline{O}$	$-28,000 - 0.69\ T$	See Table A-VI
	Sulfur $\frac{1}{2}$ S$_2$ = $\underline{S}$	$-31,520 + 5.27\ T$	See Table A-VI
Nickel	Hydrogen $\frac{1}{2}$ H$_2$ = $\underline{H}$	$4800 - 9.9\ T$	5
	Oxygen $\frac{1}{2}$ O$_2$ = $\underline{O}$	$-11,920 - 2.28\ T$	4

1. W. R. Opie and N. J. Grant, *Trans. AIME* **188**, 1237–1241 (1950).
2. F. E. Woolley and R. D. Pehlke, *Trans. AIME* **233**, 1454–1461 (1965).
3. R. G. Blossey and R. D. Pehlke, *Trans. AIME* **236**, 28–31 (1966).
4. E. S. Tankins, N. A. Gokcen, and G. R. Belton, *Trans. AIME* **230**, 820–826 (1964).
5. M. Weinstein and J. F. Elliott, *Trans. AIME* **227**, 285–286 (1963).
6. G. R. Belton and E. S. Tankins, *Trans. AIME* **233**, 1892–1898 (1965).
7. M. Weinstein and J. F. Elliott, *Trans. AIME* **227**, 382–393 (1963).

Compound Gases. Equilibrium constants may be written for the reactions between gases dissolved in metals and other constituents in solution. For example, the equilibrium partial pressure of CO over liquid steel is a function of the carbon and oxygen activities in the liquid metal. This may be seen from the expression,

$$\underline{C} + \underline{O} = CO, \tag{5-57}$$

where the equilibrium constant is written as

$$K_{57} = \frac{P_{CO}}{a_C \cdot a_O}. \tag{5-58}$$

This expression may be rewritten in the form,

$$P_{CO} = K_{57} \cdot a_C \cdot a_O, \tag{5-59}$$

which indicates that the partial pressure of CO in equilibrium with the steel is determined by the equilibrium constant and the product of the activities of carbon and oxygen in solution.

Equation (5-59) implies a hyperbolic relationship between the concentrations of carbon and oxygen in liquid iron at a given pressure of CO. In Fig. 5-18 this form of relationship is shown for equilibrium at 1 atm carbon monoxide pressure. Figure

5-18 also shows average values from electric and open hearth furnaces where the steel bath compositions are slightly above equilibrium with 1 atm pressure of carbon monoxide. (See Appendix B, Example Problem 9.) Since the activity coefficients for oxygen or carbon can be influenced by the concentrations of the other element, the relationship is not perfectly hyperbolic. The presence of other solutes will also slightly influence the shape of the equilibrium curve while the location of the equilibrium curve of Fig. 5-18 is determined by the CO pressure.

As another case, consider the dissolution of a compound which contains an atomic species of a gaseous component. For example, the solubility product of aluminum nitride is given by the reaction

$$AlN = \underline{Al} + \underline{N},\tag{5-60}$$

where the activity of the nitride is taken as unity and the equilibrium constant is expressed as

$$K_{60} = \frac{a_{Al} \cdot a_N}{a_{AlN}} = a_{Al} \cdot a_N.\tag{5-61}$$

Combining this expression with Eq. (5-53), one can compute the partial pressure

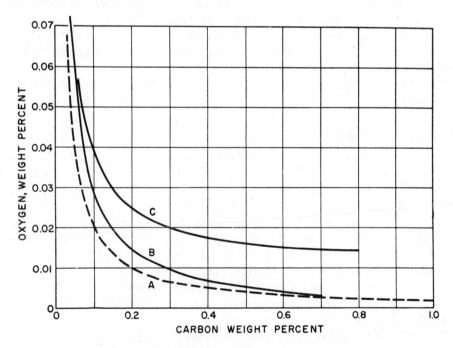

Fig. 5-18. Carbon-Oxygen Equilibrium in Steel Furnaces

Source: *Electric Furnace Steelmaking, Vol. II:* Theory and Fundamentals, edited by Clarence E. Sims, Interscience Publishers, New York 1963, p. 126, copyright AIME
A. Equilibrium at 1600°C and 1 atmosphere CO pressure
B. Electric Furnaces, Marsh J., *Electric Furnace Proc. AIME*, 1944, Vol. 16, p. 135.
C. Open Hearth, Fetters K., and J. Chipman, *Trans. AIME*, 1940, vol. 140, p. 170.

of nitrogen in equilibrium with a melt containing a given wt% aluminum, and in contact with aluminum nitride, provided the activity coefficients for the solution are known. The reaction would be expressed as

$$AlN = \tfrac{1}{2}N_2 + \underline{Al},\qquad\qquad (5\text{-}62)$$

where the equilibrium relationship is given by

$$K_{62} = \frac{K_{60}}{K_{52}} = \sqrt{(P_{N_2})}f_{Al}(\%Al).\qquad\qquad (5\text{-}63)$$

The activity coefficient of aluminum, f_{Al}, is a function of the composition of the melt and is determined by the concentrations of dissolved nitrogen and aluminum and any other alloying elements that may be present. This activity coefficient can be estimated using interaction parameters.

During degassing of steel melts, dissolved carbon can react with the refractories. An example of this is given by the following reaction:

$$MgO(s) + \underline{C} = Mg(g) + CO(g).\qquad\qquad (5\text{-}64)$$

Experimental support for this phenomenon has been developed by Bennett et al.,[22] who showed that carbon was lost by reaction with a magnesia refractory under vacuum. Similar reactions can take place with silica and alumina. The refractories used for degassing should be as resistant as possible to this type of attack. The greater thermodynamic stability of alumina under vacuum degassing conditions (note that aluminum reaction product would not vaporize as magnesium does) supports the practical experience that high alumina refractories are preferable for this application.

Inert Flush Degassing

Inert flush degassing is a process for removing a dissolved gas from a liquid metal by bubbling an inert or insoluble gas through the liquid metal. The solubility of the dissolved gas is a function of the partial pressure of that component in the gas phase with which the metal is in contact. Exposure of a melt to an atmosphere which is very dilute in the dissolved gas will provide a driving force for elimination of the dissolved gas. The mechanism for elimination of a dissolved gas from the liquid metal involves diffusion to the surface of the melt, reaction at the surface, and subsequent removal from the melt. Consequently, any process which provides a dilute atmosphere and which also shortens the diffusion path, or provides a greater surface area, should provide a more rapid means for removing the dissolved gas. The blowing of an inert flush gas through a liquid metal as a fine dispersion of bubbles is such a process. A schematic diagram of an argon degassing operation is shown in Fig. 5-19.

An Equilibrium Process. The process engineer is interested in the rate at which inert flush degassing will remove a dissolved gas. If it is assumed that the flush gas is in equilibrium with the melt when it leaves the melt surface, i.e., that the partial

[22] G. H. J. Bennett et al., *J.I.S.I.* **195**, 174–180 (1960).

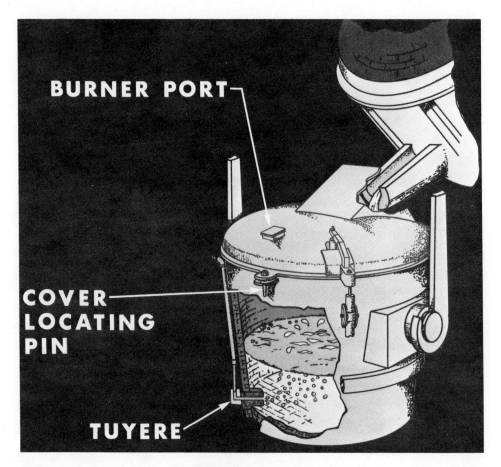

Fig. 5-19. Argon Degassing Vessel

Source: "Argon Degassing Practice and Results," R. J. Choulet, R. L. W. Holmes, and L. R. Chrzan, *J. of Metals*, 1966, *18*, p. 73.

pressure of dissolved gas in the bubble leaving the melt is equal to the equilibrium partial pressure for the instantaneous concentration of dissolved gas in the melt, the following differential relationship applies

$$-dN_G = (dN_F + dN_G) \left[\frac{P_G}{(P_G + P_F)} \right], \tag{5-65}$$

where

N_G = cubic feet of dissolved gas,
N_F = cubic feet of flushing gas,
P_G and P_F = respective partial pressures in the existing gas bubble.

If the total pressure on the system is 1 atm, then $(P_G + P_F)$ is equal to 1, and Eq. (5-65) may be rewritten as

$$dN_F = -dN_G \left[\frac{1 + P_G}{P_G} \right]. \tag{5-66}$$

Converting from standard cubic feet of dissolved gas to wt%

$$dN_F = - \left[\frac{(379)(2000)(W)}{(100)(M)} \right] \left[\frac{1 + P_G}{P_G} \right] dC_G, \tag{5-67}$$

where

W = tons of metal,
C_G = concentration of dissolved gas,
M = molecular weight of the dissolved gas.

For the case where the dissolved gas is diatomic, the equilibrium pressure of dissolved gas, P_G, is related to the concentration of dissolved gas by Sieverts' law,

$$C_G = K_G (P_G)^{1/2}, \tag{5-68}$$

where the concentration is expressed in wt% and the pressure in atmospheres. [The Sieverts' law constant is given in the units $\%/(atm)^{1/2}$.] The partial pressure of dissolved gas is then

$$P_G = \left[\frac{C_G}{K_G} \right]^2. \tag{5-69}$$

Equation (5-67) may be integrated in the form,

$$\int_0^{V_F} dN_F = - \left[\frac{7580\,W}{M} \right] \int_{C_0}^{C} \left(dC_G + K_G^2 \frac{dC_G}{C_G^2} \right). \tag{5-70}$$

The volume of flush gas in standard cubic feet, V_F, required to reduce the dissolved gas content in wt% from C_0 to C is

$$V_F = \left[\frac{7580\,W}{M} \right] \left[\frac{(C_0 - C)(K_G^2 + C\,C_0)}{C\,C_0} \right]. \tag{5-71}$$

In most cases of inert flush degassing, the concentration of dissolved gas in the bubble leaving the melt is very much less than that of the purge gas. Hence, Eq. (5-65) can be rewritten in the form,

$$-dN_G = dN_F \left[\frac{P_G}{P_F} \right] \tag{5-72}$$

and when $P_F = 1$ atm, then

$$dN_F = - \left[\frac{7580\,W}{M} \right] \left[\frac{dC_G}{P_G} \right], \tag{5-73}$$

which integrates in a simpler form,

$$\int_0^{V_F} dN_F = \left[\frac{7580\,W}{M}\right] K_G{}^2 \int_{C_0}^{C} -\left[\frac{dC_G}{C_G{}^2}\right], \tag{5-74}$$

$$V_F = \left[\frac{7580\,W}{M}\right] K_G{}^2 \left[\frac{1}{C} - \frac{1}{C_0}\right]. \tag{5-75}$$

Equations of slightly different form result when other types of gases are considered. Deoxidation of liquid iron can be carried out by inert flush degassing to purge the melt of carbon monoxide. If the exiting gas is assumed to leave at 1 atm with a very small CO content and the carbon content is low, but large relative to the oxygen content, such that it remains nearly constant; and the process is assumed to be carried out at equilibrium, then

$$dN_F = -\left[\frac{7580\,W}{M}\right]\left[\frac{1}{K_{57}\%\mathrm{C}}\right]\left[\frac{dC_0}{\%\mathrm{O}}\right], \tag{5-76}$$

$$V_F = \left[\frac{7580\,W}{M}\right]\left[\frac{1}{(K_{57})\,(\%\mathrm{C})}\right] \ln\left[\frac{(\%\mathrm{O})\,\mathrm{initial}}{(\%\mathrm{O})\,\mathrm{final}}\right]. \tag{5-77}$$

In cases where the carbon content is of the same order as the oxygen content, the concentrations of both components will decrease during degassing and hence the carbon content cannot be assumed to remain constant. The equilibrium degassing relationships become more complex under these circumstances.

The foregoing equations, describing an equilibrium process, define the minimum volume of flushing gas required for a given degree of degassing. In actual practice, the exiting bubble does not reach equilibrium with the melt, and a larger volume of flushing gas is required than predicted by equilibrium calculations. Geller,[23] in his original work on this subject, corrected the equilibrium calculations with an empirical efficiency factor. An alternative approach is to consider the mechanism by which dissolved gas is transferred to the rising gas bubbles.

Boundary Layer Model. The transfer of dissolved gas from the liquid melt to the inert flush bubble is controlled by diffusion of the dissolved species through a boundary layer surrounding the bubble. Since the diffusion coefficient is finite and the residence times for bubbles are relatively small, it is quite conceivable that the exiting bubble would not have sufficient time to reach equilibrium. If it is assumed that the melt remains homogeneous throughout the flush process, and that the concentration of dissolved gas in the flush bubble leaving the melt is quite small, i.e., that the concentration gradient through the boundary layer film surrounding the bubble is essentially given by the concentration of the bulk metal divided by the boundary layer thickness, a mass transfer model may be derived. It should be noted that this selection of the concentration gradient through the boundary layer film assumes that the concentration in the boundary layer at the bubble surface is zero. Under these conditions, a linear concentration gradient, approximated as

[23] W. Geller, *Z. Metallkunde* **35,** 213 (1934).

$\Delta C/\Delta x$, is equal to $(C - O)/\delta$. The kinetics of inert flush degassing are described by the relation,

$$-\frac{dC}{dt} = \left[\frac{DA_B}{\delta V_M}\right] F\, t_R\, C, \qquad (5\text{-}78)$$

where

$\quad D$ = diffusion coefficient of atomic species of the gas in the metal,
$\quad A_B$ = average surface area of a gas bubble,
$\quad \delta$ = boundary layer thickness,
$\quad V_M$ = volume of liquid metal phase,
$\quad F$ = flow rate of flush gas,
$\quad t_R$ = residence time of average gas bubble,
$\quad C$ = instantaneous concentration of dissolved gas in metal.

The expression integrates to the form,

$$\log\left[\frac{C_0}{C}\right] = k't, \qquad (5\text{-}79)$$

where k' is a constant which may be evaluated from the factors given on the right side of Eq. (5-78). In general, one can estimate a diffusion coefficient and a boundary layer thickness and determine the remaining quantities from the geometry of the degassing equipment. It should be noted that t_R, the residence time, is a constant which represents the time required for a gas bubble to leave the source which is injecting the flush gas and rise to the surface of the melt.

The boundary layer model, which expresses the concentration of dissolved gas in the melt in the general form of Eq. (5-79), is in excellent agreement with experimental results as shown in Fig. 5-20.

Calculation of Residence Time. The residence time may be computed by assuming that the bubble is a sphere and then writing a force balance about the bubble. The forces on the bubble are the buoyant force, its weight, and the friction force caused by viscous drag. Thus,

$$F = ma = wg - mg - F_d, \qquad (5\text{-}80)$$

where

$\quad g$ = acceleration due to gravity,
$\quad w$ = mass of fluid displaced by the bubble, i.e., having the same volume,
$\quad m$ = mass of the bubble.

Hence,

$\quad wg$ = buoyant force on the bubble,
$\quad mg$ = gravitational force,
$\quad F_d$ = resisting force caused by friction effects in the viscous fluid.

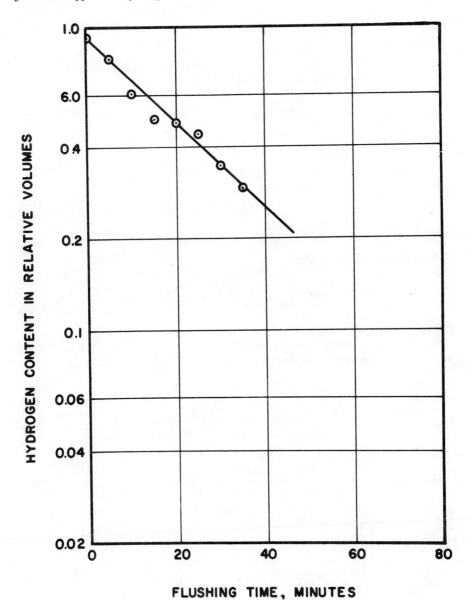

Fig. 5-20. Hydrogen Content Versus Flushing Time with Argon

Based on *Electric Furnace Steel Proceedings*, C. E. Sims, "Flushing Molten Steel With Neutral Gases," 1949, Vol 7, p. 311.

Combining the first two terms of Eq. (5-80),

$$\frac{dv}{dt} = \left[\frac{\rho_L - \rho_B}{\rho_B}\right] g - \left[\frac{3f_D \rho_L v^2}{4 d \rho_B}\right], \tag{5-81}$$

where

d = diameter of bubble,

ρ_B = density of bubble,

v = velocity of bubble in vertical direction,

ρ_L = density of liquid metal,

f_D = drag coefficient, which is given as a function of Reynolds number in Table 5-3.

If a bubble is assumed to reach its terminal velocity instantaneously and to rise with no rapid changes in acceleration, the velocity at any point is given as

$$v = -\frac{dh}{dt} = \frac{4(\rho_L - \rho_B)gd}{3\rho_L f_D}, \qquad (5\text{-}82)$$

where h is instantaneous depth in the melt. Rearranging Eq. (5-82), the residence time, t_R, is given by the integral,

$$t_R = \int_H^0 \frac{-dh}{4(\rho_L - \rho_B)gd/3\rho_L f_d}, \qquad (5\text{-}83)$$

where H is the depth at which the flush gas is discharged.

Equation (5-83) may be evaluated in either of two ways. First, an average value can be taken for the density of the gas in the bubbles as they are released,

TABLE 5-3

Drag Coefficient as a Function of Reynolds Number for Air Bubbles
Rising at Their Terminal Velocity (Average of Various Liquids)

f_d	R_e
10^7	2×10^{-6}
10^6	2×10^{-5}
10^5	1.8×10^{-4}
10^4	1.75×10^{-3}
10^3	2×10^{-2}
10^2	1.8×10^{-1}
10	2.0
1	35
0.2	200
0.18	300
0.2	350
0.3	400
0.9	900
2.5	2000
2.5	10,000

Source: W. L. Haberman and R. K. Morton, *An Experimental Investigation of the Drag and Shape of Air Bubbles Rising in Various Liquids*, Report No. 802, Naval Ship Research and Development Center, September 1953, p. 25.

and an average value estimated for the bubble diameter based on the size of the discharge orifices and the flow rate. Then the average velocity of the bubbles upward through the melt can be computed using a trial and error solution involving the drag coefficient and the Reynolds number. Alternatively, a more rigorous approach may be taken. The density of the gas can be expressed as a function of atmospheric pressure and depth in the melt, and perhaps temperature as well, since the bubble is heated during its passage through the liquid metal. The diameter of the bubble can be expressed in terms of the density and mass of gaseous material in the bubble. These expressions may be substituted in the integral of Eq. (5-83), which then can be evaluated graphically. Instead, using the expressions for bubble diameter and density, the velocity may be determined at several depths in the melt using Eq. (5-82). The average of these velocities would give a reasonable value for the velocity up through the melt, at least a more reliable value than one calculated using the first procedure outlined. In the event that the velocity remains essentially constant, the assumption of negligible acceleration is valid, and the residence time is given by the expression

$$t_R = \frac{H}{v}. \tag{5-84}$$

In analyzing the removal of a dissolved gas by inert flushing on a small scale, e.g., the degassing of aluminum in small ladles prior to casting, one must also consider losses of hydrogen from the surface. Two paths exist for removal of hydrogen: (1) the diffusion through the surface to the atmosphere above, and (2) the diffusion into the inert flush bubbles. This problem has been considered in detail and an analysis has been presented in the literature for this particular case.[24]

Vacuum Degassing

An operation that is capable of producing liquid metals containing extremely low concentrations of dissolved gases is that of vacuum degassing. In this process, the liquid metal is held in a closed container under extremely low pressures. The mechanism by which the dissolved gases are removed is essentially the same as that previously described, namely, diffusion through a boundary layer at the surface and evaporation from that surface into the vapor space above. Since the partial pressure of dissolved gases above the surface of the melt is extremely low in the vacuum and remains so throughout the operation, the process is a very effective way of removing dissolved gases.

Assuming the rate-controlling step to be diffusion of atomic species of the gaseous component through a boundary layer in the metal, the kinetics of vacuum degassing processes may be described by the equation,

$$\frac{dC}{dt} = -\frac{DA}{\delta V} C, \tag{5-85}$$

[24] R. D. Pehlke and A. L. Bement, Jr., *Trans. AIME* **214**, 1237–1242 (1962).

where

A = area of the gas–metal interface (generally taken as the cross-sectional area of the container)

V = volume of melt

C = concentration of dissolved gas.

Since the concentration at the gas–metal interface is assumed to be in equilibrium with the gas phase, in this case a vacuum, the concentration gradient is given by the ratio C/δ. If the process is carried out at reduced pressures rather than under vacuum, the gradient is given by $(C - C_e)/\delta$, where C_e is the concentration of dissolved gas in equilibrium with its reduced partial pressure in the process atmosphere (Fig. 5-21).

Eq. (5-85) can be integrated to

$$\log \left(\frac{C_0}{C} \right) = \frac{DA}{2.3\,\delta V}\, t. \tag{5-86}$$

It should be noted that according to this model, as long as the reaction taking place at the surface of the melt is much more rapid than the rate of diffusion through the liquid boundary layer, the degassing rate is independent of the molecular form of the gas species formed in the vapor space above the melt.

Figure 5-22 shows the concentration of hydrogen in a melt of steel as a function of time under vacuum.[25] The slope of the curve, k', i.e., the reaction rate constant for degassing, is given by the coefficient of Eq. (5-86).

$$k' = \frac{DA}{2.3\,\delta V}. \tag{5-87}$$

Assuming that D equals 6×10^{-3} cm^2/min, δ equals 0.003 cm, and noting that V/A was approximately 50 cm, the predicted rate constant is

$$k' = \frac{6 \times 10^{-3}}{(2.3)(0.003)(50)} = 0.017 \text{ min}^{-1}, \tag{5-88}$$

which is in approximate agreement with the average slope of the curve in Fig. 5-22, about 0.02 min^{-1}.

The desulfurization of a liquid 80% Ni–20% Co alloy under a vacuum of 10^{-5} atm is shown as a function of time in Fig. 5-23. The mechanism of desulfurization in this case was shown to be by evolution of SO_2 gas. However, the rate limiting step in the presence of dissolved oxygen was shown to be diffusion of sulfur. In the case under consideration, the melt was 10 cm deep, and if a diffusion coefficient for sulfur of 3×10^{-5} cm^2/sec and a boundary layer thickness of 0.003 cm are selected, the estimated rate constant is

$$k' = \frac{3 \times 10^{-5}}{(2.3)(0.003)(10)} = 4.3 \times 10^{-4} \text{ sec}^{-1}. \tag{5-89}$$

[25] A. M. Aksoy, *Thermodynamics and kinetics in vacuum induction melting, Vacuum Metallurgy* (R. S. Bunshah, Ed.), Reinhold Publishing Corp., New York, 1958.

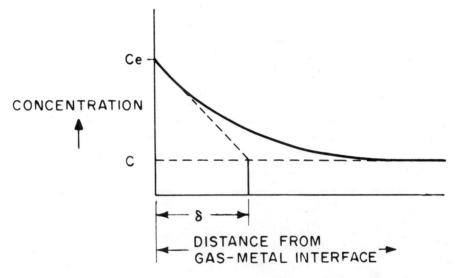

Fig. 5-21. Diagrammatic Representation of Static Boundary Layer Model

An average observed value of 2.1×10^{-4} sec^{-1} is in reasonable agreement with the predicted result.

These examples serve to illustrate the general validity of the assumption that diffusion is the rate controlling step in the kinetics of vacuum degassing processes at high temperatures. However, under some circumstances, the rate limiting step may be the reaction at the metal surface, and other kinetic models must be used.

An alternative model can be developed for the case of the surface degassing of a liquid metal where diffusion is rate controlling and the presence of a boundary layer is unlikely because of the marked difference in viscosity of the two fluids, gas, and metal. Here the surface degassing can be described in terms of semiinfinite diffusion into a continuously renewed surface layer.[26,27] The melt is assumed to be made up of rigid quanta of fluid that appear on the surface, absorb or desorb gas, and then are stirred into the melt. This concept is particularly applicable where induction stirring is present, but also can be applied to cases where the rate of surface renewal can be specified.

The rate of degassing of a melt under these conditions can be described using relationships which assume that the liquid metal is moving without shear gradients, i.e., as a "rigid body" at the gas–metal interface. The rigid body motion, referred to as streamline flow, is assumed to be maintained until the melt adjacent to the gas–metal surface is no longer in contact with the gas phase. The streamline flow layer is assumed to be free of convective currents and to be of sufficient thickness such that the mathematical assumption of diffusion into a semiinfinite medium is valid for the absorption of atoms into the streamline flow layer. Under these condi-

[26] P. V. Danckwerts, *Ind. Eng. Chem.* **43**, 1460 (1951).
[27] E. S. Machlin, *Trans. AIME* **218**, 314 (1960).

tions, transport into the metal is normal to the direction of flow as illustrated in Fig. 5-24. After the surface element has been in contact with the atmosphere, as shown in Fig. 5-24 for position A, the surface concentration is C_e, the concentration in equilibrium with the gas phase, and a sharp concentration gradient exists to C, the concentration of the bulk phase. This gradient extends down into the streamline flow layer but does not extend below this volume, which is treated as a rigid body. At a later time, the volume element has reached position B, shown in Fig. 5-24. The concentration gradient is less steep and goes further into the element, but still extends from C_e at the surface down to C, the bulk concentration of the

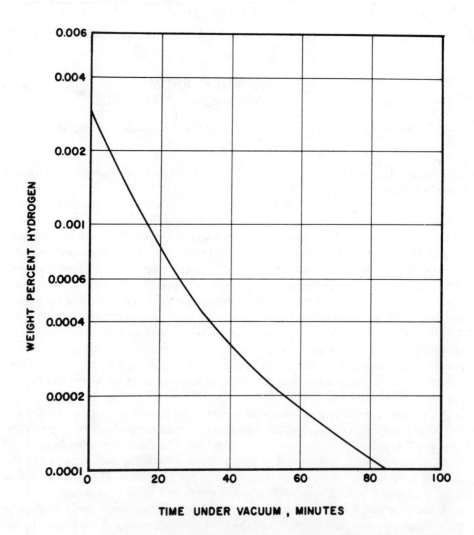

TIME UNDER VACUUM , MINUTES

Fig. 5-22. Vacuum Degassing of Liquid Iron in Magnesia Crucible

Based on R. F. Bunshah, *Vacuum Metallurgy*, Reinhold Book Corporation, a subsidiary of Chapman-Reinhold, Inc., New York, 1958, p. 68.

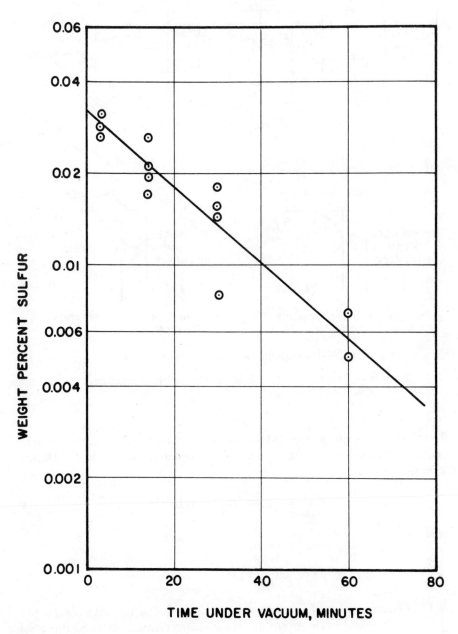

Fig. 5-23. Desulfurization of an 80% Ni–20% Co Alloy at 1520°C by Vacuum Treatment

Based on "Kinetics of Vacuum Induction Refining-Theory," E. S. Machlin, *Trans. Met. Soc. AIME*, 1960, Vol. 218, p. 315.

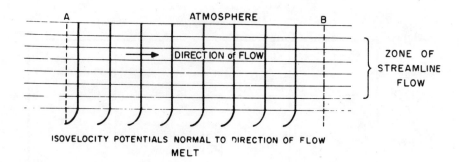

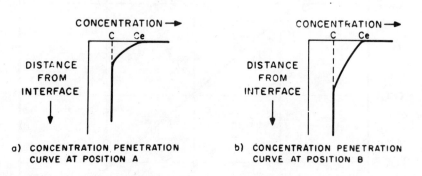

a) CONCENTRATION PENETRATION   b) CONCENTRATION PENETRATION
   CURVE AT POSITION A   CURVE AT POSITION B

Fig. 5-24. Diagrammatic Representation of Streamline Flow Model

Based on "Kinetics of Vacuum Induction Refining-Theory," E. S. Machlin, *Trans. Met. Soc. AIME*, 1960, Vol. 218, p. 315.

melt at the time the surface element was created. Under these boundary conditions, the equation for diffusion into a semiinfinite medium, as derived by Darken,[28] may be applied:

$$\frac{dM}{dA} = 1.1284 \ (C_e - C)(Dt')^{1/2}, \tag{5-90}$$

where

$\dfrac{dM}{dA}$ = change in solute per unit area of gas–metal surface,

C_e = concentration of solute in the metal in equilibrium with the gas phase (expressed in consistent units, e.g., cubic centimeters at STP per cubic centimeter of liquid metal),

C = concentration of solute in the bulk phase of metal at any time, t, in the same units as C_e.

[28] L. S. Darken and R. W. Gurry, *Physical Chemistry of Metals*, p. 445, McGraw-Hill, New York, 1952.

D = diffusion coefficient of solute in the liquid metal,

t' = time interval during which the solute is transferred from the surface element.

Converting variables

$$\frac{dc}{dt} = \frac{dM}{dA} \times \frac{dA}{dt} \times \frac{1}{V},$$ (5-91)

where

$\dfrac{dc}{dt}$ = rate of change of solute concentration in liquid metal,

$\dfrac{dA}{dt}$ = rate of creation of gas–metal interface,

V = volume of melt.

Considering the case where inductive stirring exists and the flow of metal is upward at the center and out toward the walls, the life of the element, t', may be expressed as x/v, the ratio of the distance traveled (approximately one-half the crucible radius) to the flow velocity. For cylindrical geometry, the volume of the melt is equal to $\pi r^2 h$, where r is the radius of the melt and h is the depth. The rate of creation of surface dA/dt is also the rate of disappearance of melt surface, which may be expressed by the relation,

$$\frac{dA}{dt} = 2\pi rv.$$ (5-92)

If Eqs. (5-90), (5-91), and (5-92) are combined

$$\frac{dC}{dt} = \left[\frac{2.56\,Dv}{rh^2}\right]^{1/2} (C_e - C),$$ (5-93)

which integrates to

$$\log\left[\frac{C_e - C_0}{C_e - C}\right] = \frac{1}{2.3}\left[\frac{2.56\,Dv}{rh^2}\right]^{1/2} (t - t_0).$$ (5-94)

Equation (5-94) thus takes a form similar to that derived for the boundary layer diffusion model, excepting that the adjusting parameter, δ, is absent. However, there is an uncertainty as to the true velocity of streamline flow. If the surface elements are to move as rigid bodies, the velocity of streamline flow must be smaller at the periphery of the melt than near the center, where the surface is being generated. Furthermore, the distance the element travels has been approximated by one-half the radius, which must be considered only as a reasonable guess. Thus the adjustable parameter of the boundary layer model has been replaced with other parameters which are not well defined. It should be noted, however, that the streamline flow model more closely describes the physical system which exists in the process of degassing an inductively stirred melt.

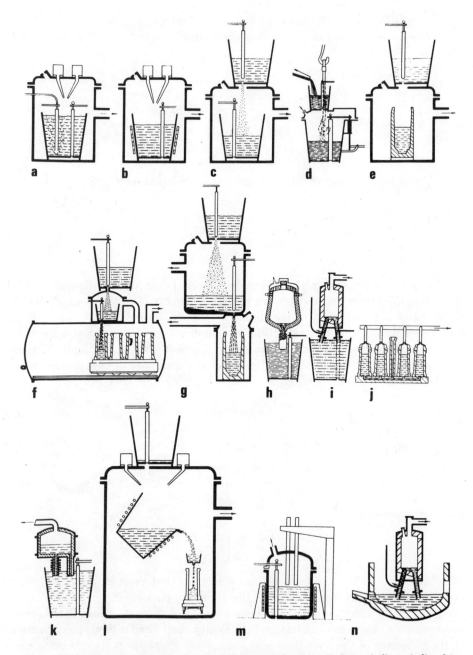

Fig. 5-25. Methods of Degassing—*a* purging gas; *b* induction stirring; *c* ladle to ladle; *d* tap degassing; *e* ladle to mold; *f* ladle to multiple mold; *g* two-stage treatment; *h* vacuum lift; *i* circulation by gas lift; *j* ingot degassing; *k* circulation by induction; *l* combination process; *m* arc heating; *n* treatment in the furnace.

Source: "Vacuum Degassing 1955–1965 . . .", J. H. Flux, *J. Iron and Steel Inst.*, 1965, Vol. 203, p. 1196.

Application of Eq. (5-94) to one of the preceding examples, the desulfurization of a nickel–cobalt alloy, is appropriate since the stirring conditions and melt geometry were reasonably well defined.[29] In that experiment, the desulfurization was carried out in a 6.4 cm dia crucible with a melt height of 10 cm, and under inductive stirring conditions where the velocity on the surface was between 1 and 10 cm/sec. A range for the rate coefficient may be defined using Eq. (5–94).

Since

$$k' = \frac{1}{2.3} \left[\frac{2.56 \, Dv}{rh^2} \right]^{1/2}$$

then it follows that

$$\frac{1}{2.3} \left[\frac{(2.56)(3 \times 10^{-5})(1)}{(3.2)(10)^2} \right]^{1/2} < k' < \frac{1}{2.3} \left[\frac{(2.56)(3 \times 10^{-5})(10)}{(3.2)(10)^2} \right]^{1/2}$$

$$2.1 \times 10^{-4} < k' < 6.7 \times 10^{-4} \, \text{sec}^{-1}. \tag{5-95}$$

This result compares favorably with the observed value of $2.4 \times 10^{-4} \, \text{sec}^{-1}$ (Fig. 5-24).

Other Degassing Processes

Several other degassing processes have been proposed which utilize vacuum, or a combination of vacuum and inert flushing to produce a larger gas–metal surface area and, consequently, an increased rate of degassing. Several of these processes are shown in Fig. 5-25. An increased degassing rate is particularly important in the treating of liquid metals, since the metal temperature decreases during the operation thus limiting the time available for processing.

The Dortmund–Hörder degassing process[30,31] is used to remove dissolved gases and control the extent of the carbon–oxygen reaction in liquid steel. The process utilizes a vacuum chamber into which low carbon steel is drawn up by vacuum from a ladle and then released to flow by gravity back to the ladle. Only a small fraction of the liquid metal in the ladle can be drawn up into the vacuum chamber, and to obtain extensive degassing, several cycles of the process are required. A schematic description of the process is shown in Fig. 5-26 and also in Fig. 5-25-h.

The question facing the process designer who is considering use of the Dortmund–Hörder process is, "How many cycles of the degassing operation may be required to lower the gas content of the metal in the ladle to a desired level?" In the face of deficient thermodynamic and kinetic data, the assumption that a given reaction goes to completion within the allotted process time often is used as a basis for design purposes. This approach to the Dortmund–Hörder process results in the assumption that the metal drawn up into the vacuum chamber is completely degassed, and, upon being returned to the ladle, is completely mixed with the ladle material. For this case, the material balance for the degassing operation is given

[29] E. S. Machlin, *Trans. AIME* **218**, 314 (1960).
[30] F. Harders, H. Knuppel, and K. Brotzmann, *J. Metals* **12**, 398 (1960).
[31] P. J. Wooding and W. Sieckman, *Metal Progress* **77**, 1–116 (1960).

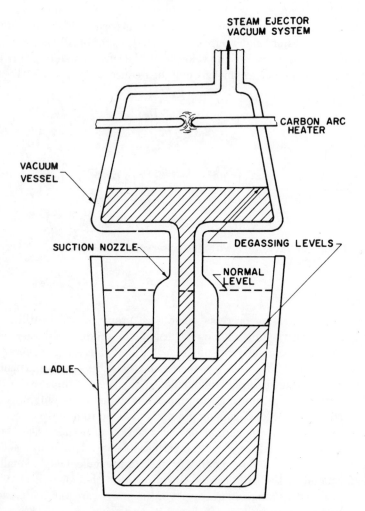

Fig. 5-26. Schematic View of Dortmund-Hörder Hüttenunion Vacuum Treating Process

by the relation,

$$C = C_0(1 - X)^n, \tag{5-96}$$

where

n = number of cycles,
C = concentration of gas in the metal in the ladle after the nth cycle,
C_0 = initial gas concentration in the metal,
X = fraction of ladle material which is drawn into the vacuum chamber during each cycle.

The results of actual data taken from the operation of the process are compared with Eq. (5-96) in Fig. 5-27. The agreement between the two curves shows the assumption to be quite reasonable.

A process has been developed in Germany called the Ruhrstahl–Heraus Continuous Vacuum Process. This process, shown in Fig. 5-25-i, is used in ladle degassing. The ladle of molten steel is placed in a pit, and the vacuum apparatus positioned above it by an overhead crane. Tubes are lowered into the melt while the vacuum chamber is being pumped out by a multiple stage pumping system. Argon is introduced into one of the tubes, and causes the liquid steel to move up in this tube. The metal then splashes into the vacuum chamber where it is degassed. The time required to degas 100 tons of liquid steel is 15–20 min, and the temperature drop during this period is only about 30°C. If necessary, it is possible to heat the liquid steel by auxiliary equipment during the vacuum treatment.[32,33]

Several other vacuum degassing techniques include vacuum degassing in the

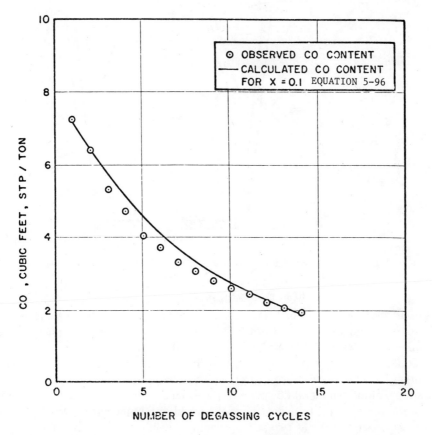

Fig. 5-27. Removal of CO from Liquid Steel by DHHU Vacuum Treating Process

Based on P. J. Wooding and W. Sieckman, *Metal Progress*, 77, (1960), p. 121.

[32] H. W. Balster, *Elect. Furnace Steel Proc. AIME* **16**, 192 (1958).
[33] F. W. Starratt, *J. Metals* **10**, 465 (1958).

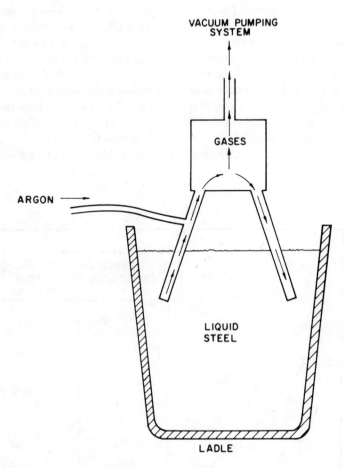

Fig. 5-28. Schematic Diagram of Vacuum-Flow Steel Degassing Process

ladle, a process devised by A. Finkl and Sons Co., Chicago.[34] Vacuum stream degassing is another method of bringing more of the surface of the melt in contact with the vacuum. In this process a bottom-pouring ladle of steel is placed over a vacuum chamber and the liquid metal is permitted to flow into the vacuum chamber. The reduced pressure causes the gases to diffuse from the metal, which is received in another ladle below, from which in turn it can be poured into an ingot mold. The thin stream of molten metal flowing into the evacuated chamber allows an excellent chance for the vacuum to degas the melt quickly and efficiently. Results by this method have been as low as two parts per million residual hydrogen.[35] This vacuum degassing technique is shown in Fig. 5-28.

Another effective method of controlling dissolved gases (particularly for oxygen

[34] C. W. Finkl, *Metal Progress* **76,** 111 (1959).
[35] A. Tix, *Elect. Furnace Steel Proc. AIME* **13,** p. 70 (1955).

and nitrogen) is that of adding a scavenging element, such as titanium or zirconium. These elements form stable nitrides or oxides which are insoluble in liquid metals. The principles involved here are the same as those outlined in previous sections of this chapter. Inert flush degassing may also be accomplished by placing a volatile solid in the liquid melt, which then forms purging gases that are evolved in the bulk of the metal.

The effectiveness of the methods outlined here and continued development effort in perfecting degassing techniques will result in a rapid increase in the use of degassing processes as a refining operation in the metallurgical industries.

Temperature Drop of the Melt During Degassing

The temperature drop of a melt during degassing may be estimated from a consideration of the radiative and convective heat losses from the metal surface during the degassing operation. Absorption of heat into the refractory lining of the ladle or other degassing equipment also can represent a significant heat loss, although preheating of the refractories can diminish this latter source. The heat effect caused by the removal of the dissolved gas from solution (and in the case of inert flush degassing, the sensible heat lost to the flush gases) is negligible compared with the principal source of heat loss, direct radiation from exposed metal surfaces. This is true especially for steels. The loss of heat by direct radiation is enhanced further by the agitation and stirring occasioned by gas evolution. Indeed, the bubbling of argon through liquid steel in the ladle often is used to cool heats prior to tapping. The agitation and breaking of the insulating slag cover results in direct radiation from the metal surface, which can reduce the temperature by 10–15°F within 1 min or less. A balance between the various sources of heat loss and any energy input such as by induction or arc heating (see Fig. 5-25-b, -h, -m) permits an evaluation to be made of the time available for degassing, or the required superheat of the metal.

Bibliography

Bradshaw, A. V. and F. D. Richardson, Thermodynamic and kinetic aspects of vacuum degassing, *Vacuum Degassing of Steel*, Iron and Steel Institute, London, Special Report No. 92, 1965.

Butts, A., *Metallurgical Problems*, Chap. 22 and 23, McGraw-Hill, New York, 1943.

Dennis, W. H., *Extractive Metallurgy*, Philosophical Library, Inc., New York, 1965.

Electric Furnace Steelmaking, Chap. 14 and 16, AIME, New York, 1964.

Flux, H., Vacuum degassing 1955–1965: A survey of processes and plant for bulk steel production, *J. Iron Steel Inst.* **203,** 1194 (1965).

Gilchrist, J. D., *Extraction Metallurgy*, Pergamon Press, London, 1967.

Hayward, C. R., *An Outline of Metallurgical Practice*, 3rd ed., D. Van Nostrand Company, Inc., New York, 1952.

Hopkins, D. W., *Physical Chemistry and Metal Extraction*, The Macmillan Company, New York, 1954.

Newton, J., *Extractive Metallurgy*, John Wiley, New York, 1959.

Pfann, W. G., *Zone Melting*, John Wiley, New York, 1958.

G. R. St Pierre, Ed., *Physical Chemistry of Process Metallurgy*, Vol. 1 and 2, Interscience Publishers, New York, 1961.

Discussion Questions

1. Silico-manganese, an alloy containing silicon and manganese, is made in several grades including a 3% carbon, 15% silicon, 65% manganese, balance iron grade. What are the advantages and disadvantages of using this material versus either aluminum or silicon as a deoxidizer?

2. Briefly describe economically and technically feasible liquid phase techniques for:
    (a) removing cadmium from lead at cadmium concentrations below 1 wt%,
    (b) removing bismuth from lead,
    (c) removing oxygen from steel,
    (d) removing hydrogen from aluminum.

3. A secondary metals company has purchased large quantities of scrap bearings at a very low cost. The bearings consist of cylinders of an alloy (80% Cu, 10% Pb, 10% Zn) mounted in steel sleeves. Outline an integrated process for recovering the various metals.

4. Discuss the potential success and feasibility of processes for external desulfurization of iron blast furnace hot metal using magnesium, calcium carbide, or lime.

5. The impurities in gold bullion are sometimes removed by bubbling chlorine gas through the impure liquid gold. What information is required to thermodynamically describe this process?

6. Although thermodynamics shows distillation processes to be very favorable for refining in many metallurgical systems, they are not often used. Why?

7. Present a more general statement of Sieverts' Law which is modified to include the solution behavior of compound gases such as CO_2 or SO_2.

Problems

1. Compute the minimum theoretical weight of ferro-silicon (75% Si) required to deoxidize 150 tons of basic oxygen blown steel with a turndown carbon content of 0.05 and temperature of 2900°F to an oxygen content of 0.01 wt%. Make the same calculation for pure vanadium and chromium.

2. In a recent experiment, liquid pure iron was held in a silica crucible at 1600°C under an argon atmosphere. After several hours, the melt was quenched in an argon blast. Chemical analysis revealed that the metal was of the following composition:

Silicon	0.0120 wt%
Oxygen	0.0550 wt%
Iron	Remainder

    (a) Is this experimental result in agreement with tabulated data?
    (b) Is silicon a suitable deoxidizer for steels and how does it compare with aluminum?

3. Calculate the equilibrium activity of nickel oxide in a slag in contact with a liquid iron alloy containing 4 wt% nickel and 0.05 wt% oxygen at 1600°C. Specify the reference state for nickel oxide.

4. Can a low carbon steel be desulfurized to concentrations of 0.01% S or below by treatment with magnesium at 1600°C?

5. Chemical analyses for silicon in low carbon steels (below 0.10% C) produced in the BOF are usually reported as "nil." Explain.

6. The following quotation appears in *Trudy Leningradskogo Politekhn. Instituta, Im. M. l. Kalinina* No. 125, 1963, pp. 143–148:

   "It has often been established that when pig iron is transferred in ladles from the blast furnace to the mixer or casting ladle, the sulfur content is reduced, the more so when the initial sulfur and manganese contents of the pig iron are high and when it is frequently shaken up during transit by rail, at switches and rail joints. This is explained by the associated acceleration of the coalescence of the manganous sulfide particles. Moreover, a reduction in the sulfur content leads to a displacement of the state of equilibrium of the reaction in favor of the formation of manganous sulfide as the pig iron cools upon removal from the furnace and during its stay in the ladles. The high melting point of manganous sulfide (1620°C–2950°F) and its low solubility in pig iron result in its separation in the form of solid particles. There is no direct experimental confirmation of such mechanism."

   The article further reports that a pig iron of the composition 0.6% Si, 0.2% Mn, 4.0% C and 0.143% S showed a final composition of 0.03% S and 0.06% S after adding sufficient manganese to raise the Mn content to 2% and cooling over a period of 17 min to temperatures of 1200 and 1350°C, respectively. Evaluate the thermodynamic feasibility of precipitation of manganese sulfide from liquid pig under these two circumstances.

7. Desulfurization of an alloy was accomplished by contacting the metal in the liquid state with a sulfur-free basic slag. The metal bath was 7.5 cm deep and the slag-metal interfacial area was 20 cm². The sulfur concentration of the metal was observed to be:

Time (min)	% S
0	0.03
10	0.02
20	0.013
30	0.0086
40	0.006
50	0.005
60	0.0045

   (a) Explain the concentration-time behavior.

   (b) Assuming sulfur diffusion in the metal is rate limiting, estimate the diffusion coefficient of sulfur in the liquid alloy.

8. A steel containing 0.40 wt% carbon (AISI 1040) is held in an alumina crucible under an atmosphere of CO at a pressure of 760 mm Hg. Estimate the equilibrium aluminum content of the metal. Neglect interaction between solutes.

9. A steel company proposes to deoxidize low carbon heats with vanadium. If the grade specification calls for oxygen below 0.01 and vanadium below 2 wt%, is this a feasible process?

10. Phosphorous often is used to deoxidize liquid copper. The solubility product for this reaction at 1200°C is

$$\log\,(\%P)\,(\%O)^{2.5} = -11.07.$$

What concentration of phosphorous will limit the oxygen concentration to 0.0005% at 1200°C? What will be the volume percent of phosphorous rich phase in the solidified copper (if any)?

11. Starting with a value of 62,360 mm Hg-cm³/°K g-mole for R, the gas constant, make a conversion of units to develop the value of R used in Eq. 5-44.

12. What fraction of a 3 wt% Mn steel would have to be evaporated at 1600°C under equilibrium conditions to reduce the manganese concentration to 0.5 wt%.

13. In the vacuum dezincing of lead bullion at 600°C, estimate the maximum evaporation rates for bullion containing 0.5 and 0.05% zinc.

14. Estimate the lead contamination of the zinc condensate for a lead bullion being vacuum treated at zinc concentrations of 0.5 and 0.05 wt%.

15. Calculate the thermal requirement (exclusive of heat losses) to remove 0.5 wt% zinc from lead bullion by distillation. The lead bath is at 600°C and the zinc is condensed to a solid at 200°C.

16. Calculate the relative volatility at 700°C for a lead alloy dilute in cadmium.

17. The solubility of N_2 in liquid Fe has been reported as 0.0447 wt% at 1535°C and 1 atm pressure of nitrogen, and 0.0368 wt% at 1760°C and 0.639 atm of nitrogen pressure. Express this information as a ΔG^0 equation.

18. The solubility of nitrogen at 1 atm pressure in liquid pure iron at 1600°C is 36 CC (STP)/100 g Fe. The total dissolved gas $(H_2 + N_2)$ under a 20% H_2–80% N_2 mixture at 1 atm total pressure in 100 g of pure liquid iron at 1600°C is 45 CC (STP). Compute the total amount of dissolved gas CC(STP) in 100 g of the following alloy at 1600°C under a gas mixture of 5% H_2–1% N_2–94% A at 1 atm total pressure.

Element	Wt%
Cb	2
Co	3
Ni	4
Si	2
Fe	Balance

19. Steel, tapped from a steelmaking furnace at 2920°F, contains 0.003% nitrogen. Will the steel tend to pick up or lose nitrogen from the atmosphere?

20. The solubility product $(\%C \times \%O)$ is approximately 0.0022 at 2850°F for medium and low carbon steels. Calculate the equilibrium oxygen content in a steel at 2850°F having the following carbon content:
    (a) 0.60% C
    (b) 0.40% C

(c)   0.10% C

(d)   0.06% C

(e)   0.02% C

In actual steelmaking practices, a heat entering the ladle often may have either more or less oxygen than the solubility product would indicate. List a number of conditions which can lead to this.

21.   Calculate the partial pressure of SO_2 over a bath of low carbon steel containing 0.05% S and 0.10% O at 1600°C. Comment on the feasibility of removing S by vacuum treatment.

22.   What is the total volume of dissolved gas in 100 lb of liquid iron containing 4% silicon in equilibrium with a gas at 1 atm pressure consisting of 50% H_2 and 50% N_2?

23.   Calculate the interaction parameter, e_H^{Si}, based on the measurement: 20.4 cc H_2 (STP)/100 gm in a 3% Si–Co alloy at 1600°C and 1 atm H_2 pressure.

24.   Calculate the degassing time required to remove 95% of the dissolved hydrogen from liquid copper held in a zirconia crucible at 1420°C. The pressure of the system is maintained at 1 mm Hg. The Sieverts Law constant for hydrogen in copper at 1420°C is 1.097 cc/100 g-(atm)$^{1/2}$.

   Operating Conditions:

   weight of copper = 50 lb

   D_H in liquid copper at 1420°C = 10^{-3} cm^2/sec

   δ = 0.003 cm

   crucible diameter = 6 in.

   ρ, density of liquid copper = 7.68 gm/cc

   initial hydrogen content = 0.8 cc/100 gm

25.   An induction melting unit is stationed in a vacuum chamber. Five hundred pounds of cobalt alloy is melted and held under vacuum at 1600°C. The chamber then is pressurized to 800 mm Hg with hydrogen gas. The melt is maintained at 1600°C under the hydrogen atmosphere for 5 min. Then the chamber is pumped out and maintained at 5 mm Hg for 3 min. The alloy is immediately cast into a water-cooled copper mold. Estimate the final hydrogen content of the alloy given:

   alloy = 10% Mo, 10% Ta, balance Co

   ρ_{alloy} at 1600°C = 10 gm/cm^3

   I.D. of crucible = 20 cm

   D_H in liquid cobalt = 2 × 10^{-3} cm^2/sec

   δ = 0.003 cm

26.   A dilute liquid iron alloy is melted in a vacuum induction furnace. The alloy contains 5 ppm (by weight) of hydrogen. Estimate the degassing time required to reach 0.5 ppm of hydrogen given:

   D_H = 2.5 × 10^{-4} cm$_2$/sec

   δ = 10^{-3} cm

   weight of melt = 12 kg

   ρ_{melt} = 7.16 gm/cc

   area of bath surface = 75 cm^2

   pressure in vacuum chamber = 48.5 μ

27. Special heats of OFHC copper are prepared by treating liquid copper with hydrogen at 1 atm in an inductively heated crucible under a reducing slag. This treatment removes dissolved oxygen. The copper is then held under a vacuum to remove the dissolved hydrogen. Estimate the time required to decrease the dissolved hydrogen content to 0.5 cc H_2/100 gm (STP).

   Data:
   $D_H = 5 \times 10^{-4}$ cm²/sec
   $\rho_{Cu} = 7.76$ gm/cc
   melt—10 in. diameter $\times$ 15 in. deep
   metal temperature—1300°C
   vacuum system—1.0 mm Hg

   Assume well stirred melt with rate limiting transport being diffusion across a boundary layer 0.01 cm thick in the metal phase.

28. Evaluate the thermodynamic feasibility of the following refining processes.
    (a) Desulfurization of steel with manganese.
        Process conditions: maximum residual sulfur—0.03 wt%
                              maximum residual manganese—0.5 wt%
                              temperature—1600°C
    (b) Distillation of aluminum from silver.
        Process conditions: temperature—1000°C where $\gamma^0_{Al} = 0.11$
                              maximum residual aluminum—1 mol%
    (c) Phase separation of zinc from lead.
        Process conditions: minimum temperature—350°C
                              maximum residual zinc—1 wt%
    (d) Vacuum degassing of copper.
        Process conditions: minimum chamber pressure—0.1 mm Hg
                              maximum residual hydrogen—0.1 cc(STP)/100 g
                              temperature—1400°C

29. Estimate the volume of hydrogen which would have to be blown through a low-carbon steel at 1600°C to reduce the sulfur content from 0.04 to 0.02 wt%.

$$H_2(g) + \underline{S} = H_2S(g) \quad K_{1600°C} = 5 \times 10^{-3},$$

where K can be expressed as $(\%)^{-1}$.

30. A 100-ton ladle of liquid steel at 1600°C contains 0.04% C and 0.05% O. Argon is bubbled through the bath at a low rate in finely dispersed bubbles such that the gas mixture leaving the melt at 1 atm pressure is in equilibrium with the bath at all times.
    (a) How much argon (ft³ STP) must be blown into the bath to reduce the carbon content to 0.01%?
    (b) What will the oxygen content be when the carbon content just reaches 0.01%?

31. A 100-ton heat of copper is fire-refined in a fuel-fired reverberatory furnace; the furnace atmosphere contains 20% (by volume) water vapor. At tap, the oxygen content of the copper is 0.03% and the temperature is 1227°C. The heat is tapped into a holding ladle, during which operation the temperature

falls to 1177°C and thereafter may be assumed constant. Nitrogen is bubbled through the copper to lower the hydrogen content before casting. What is the hydrogen content after 200 ft³ STP of N_2 have been used?

$$H_2O\,(g) \;=\; 2\underline{H} + \underline{O} \qquad \log K = \frac{-10640}{T} - 3.09.$$

Activities of H and O are in weight percent based on the infinitely dilute solution. The gas mixture leaving the copper may be assumed to have reached equilibrium with respect to its dissolved constituents. Any assumptions which are made to facilitate solutions should be stated.

32. An inert flush degassing operation supplies 100 ft³ (STP)/min of argon at 100°F to a 100-ton ladle of steel at 3000°F to reduce the hydrogen content from 7.5 to 4 ppm in 30 min. *Estimate* in °F the difference in temperature decrease of the melt caused by inert flushing as compared to vacuum degassing for the same period and hydrogen removal.

$$C_p \text{ (Fe liquid)} = 10.00 \text{ Btu/\# mole-°F}$$
$$C_p \text{ (A)} \qquad\;\; = 4.97 \text{ Btu/\# mole-°F}$$

CHAPTER 6

Hydrometallurgy

Hydrometallurgy is that phase of metallurgy which involves the extraction and recovery of metals using aqueous or organic solutions. The unit processes of hydrometallurgy include the leaching of ores (particularly low-grade ores) for recovery of metals, the separation of the leaching solution from the spent ore, and also the recovery of the dissolved metal from the solution. A typical example is the recovery of copper from a sulfide ore (usually a low-grade ore containing less than 1.5 wt% copper that has been partially oxidized to copper sulfate) by extraction with a solvent such as water, subsequent washing of the ore with fresh water to recover all of the copper solution, and precipitation of copper from the dilute copper sulfate solution by "cementation," i.e., displacement of the copper ions in solution by iron on the surface of iron scrap.

Hydrometallurgical extraction always involves three steps:

1. The solvent or reacting solution must come in contact with the solid to be treated in order to transfer the soluble constituent to the solvent.
2. The solution is separated from the residual solid and the solid is washed. Liquid always adheres to the solids, which must be washed to prevent either the loss of solution or contamination of the solids, depending upon which is the desired material.
3. The desired solute is recovered from the solution, or in the case where the residual solid is the desired material, a process also must be provided for recovery of the solvent. The techniques employed to treat the enriched solution include evaporation, distillation, electrolysis, ion exchange, solvent extraction using an insoluble liquid, or precipitation, the latter being one of the most common in the metallurgical industries.

Further examples of hydrometallurgical operations include the extraction of gold and silver by cyanidation and precipitation of these elements from the cyanide solution with zinc. Leaching with acid is employed to dissolve zinc oxide and copper oxide as sulfates although copper oxide also can be dissolved in alkaline solutions of ammonia. Caustic soda is used to dissolve alumina, and uranium oxide can be dissolved with acid or carbonate. The recovery of these solutes can be accomplished by several methods. Copper is recovered from acid solutions by cementation with iron or by electrolysis. Uranium and vanadium are recovered from acid or carbonate solutions by solvent extraction or ion exchange, and metallic ions can be precipitated from aqueous solutions as insoluble sulfides.

In recent years interest in hydrometallurgical operations has been intensified, largely a result of the necessity for turning to lower grade raw materials that must be concentrated for processing. Many of the new developments involve leaching and precipitation at elevated temperatures and pressures. The advantages in operating under these conditions are greatly increased rates of reaction, favorable displacement of thermodynamic equilibria, and the possibility of using certain gaseous or highly volatile reagents such as oxygen, hydrogen, or ammonia. Among the new processes which have resulted from the application of high-temperature and high-pressure techniques are direct leaching of nickel, copper, and cobalt sulfides with ammonia; carbonate leaching of uranium ores; leaching of certain sulfide ores with water; precipitation of metallic nickel, cobalt and copper from aqueous solutions of their salts by hydrogen reduction; and precipitation of uranium and vanadium oxides from aqueous carbonate solution by hydrogen reduction. Although there is a demand for advances relating to the chemistry of some of the older hydrometallurgical processes, undoubtedly the employment of higher temperatures and pressures in conjunction with new extraction techniques will bring the greatest advances to the field.

6-1 Hydrometallurgical Processes

Leaching

In preparation for leaching, the ore is usually reduced to a small size by crushing and grinding; in some cases, chemical treatment may be necessary to render the ore to its optimum state for recovery. Thus, zinc sulfide and cobalt sulfide, e.g., are insoluble in most solvents, but with an oxidizing roast they can be converted to water soluble sulfates, and roasting the ore with salt can convert the metallic material to a water-soluble chloride. The material selected for a leaching process is usually a low-grade ore, but in some instances may be concentrates, calcines (sulfide materials converted to oxides by roasting), sulfides, alloys, etc.

The leaching of low-grade ore is a large scale operation and requires large volumes of solvent and great quantities of ore. The solvent must be inexpensive and capable of being cheaply regenerated. If the ore is concentrated, the solvent or reactant costs are not too significant because of the high metallic yield of the ore, and more expensive reactants (such as concentrated acids) often are used. Furthermore, concentrated acid leaching of concentrates results in rapid solution rates and requires a smaller installation.

Ores are prepared to two general conditions, sands or slimes. Sands refer to ores that are coarse enough to permit circulation of the solvent through the void spaces between the ore particles. Metallics in ores of this type are recovered by percolating the leaching solution through the ore. Very fine ore products which are not suitable for a "percolating" leach are called slimes; leaching of slimes is carried out by agitating the slime and solvent together in a tank.

As mentioned earlier, sulfide ores must be roasted by heating in contact with an oxidizing gas to convert the sulfide to an oxide or a sulfate, thus transforming the metal into a soluble form. The process volatilizes certain soluble impurities and also renders the ore porous enough so the metallic compounds are readily attacked by the solvent. Roasting is a pyrometallurgical operation (see Pyrometallurgy I) and is relatively expensive; hence, roasting is limited in its application to high-grade, concentrated ore. Low-grade sulfide ores can be oxidized to oxide by natural weathering—the slow oxidation of sulfides by the action of the atmosphere, accomplished by outdoor exposure of crushed ore. This process is slow and inefficient, and because it requires the storage of a large amount of metallic material for a long time, the technique is expensive. Of course, the advantages in weathering are that as a process, it is inexpensive, and the technique has been used to advantage with certain copper ores that are thus converted into water soluble sulfates.

Solvents are liquids which from an engineering point of view should be inexpensive, capable of regeneration, and able to dissolve rapid'y the desired minerals, thus separating them from the gangue material. Several solvents are used in leaching operations. The most important solvent for sulfate minerals is water; for example, rain water or natural mine waters have been used to recover copper sulfate from ore. Another solvent is dilute sulfuric acid, used to dissolve copper and zinc oxides according to the reactions,

$$CuO + H_2SO_4 \rightarrow CuSO_4 + H_2O, \tag{6-1}$$

$$ZnO + H_2SO_4 \rightarrow ZnSO_4 + H_2O. \tag{6-2}$$

Hydrochloric and nitric acids also have been utilized as solvents.

Alkaline solutions such as ammonium hydroxide with oxygen or ammonium carbonate plus ammonium hydroxide are used to leach native copper ores and metallic carbonates. The advantage of using an alkaline leaching solution with a carbonate gangue is that the alkaline solution does not attack the carbonate, whereas the use of an acid leach would consume a considerable quantity of acid by extraneous reaction with the carbonate. When copper is leached with an alkaline solution containing ammonia, the copper can be precipitated as CuO by heating, which also drives the ammonia and carbon dioxide out of solution and allows recovery of these gases. Sodium hydroxide is used in the Bayer process for the purification of bauxite. Cyanide solutions, such as NaCN, are used to dissolve gold and silver according to the reactions,

$$4Au + 8NaCN + O_2 + 2H_2O \rightarrow 4NaAu(CN)_2 + 4NaOH, \tag{6-3}$$

$$4Ag + 8NaCN + O_2 + 2H_2O \rightarrow 4NaAg(CN)_2 + 4NaOH. \tag{6-4}$$

Sulfite and chloride solutions are used for recovery of metallic chlorides.

Leaching Techniques

Copper ores often are leached *in situ* by circulating water through underground copper deposits. Exposure of the ore body to water and oxygen oxidizes the sulfide to a sulfate, which dissolves in the water percolating through the ore. The water is collected then and the dissolved copper recovered, usually by precipitation (cementation) on iron scrap. The "cemented" copper is loosely adherent to the scrap and can be flushed off the scrap into a collecting tank.

Heap leaching also depends upon oxidation of sulfide minerals by exposure to water and air. Ore is placed above ground in large heaps which are exposed to the atmosphere and water, either natural rain water or water sprays. The solution percolates through the ore, which usually contains iron, and by one postulated mechanism, the pyrite is oxidized according to the reaction,

$$2FeS_2 + H_2O + 7\tfrac{1}{2} O_2 = Fe_2(SO_4)_3 + H_2SO_4. \qquad (6\text{-}5)$$

The ferric sulfate then reacts with the copper converting it to a soluble sulfate as

$$Cu_2S + Fe_2(SO_4)_3 = CuSO_4 + 2FeSO_4 + CuS. \qquad (6\text{-}6)$$

An extraction carried out in this manner may require several years and the loss of metal in solution which seeps into the ground represents an inherent inefficiency.

Leaching is more commonly carried out in a tank, either by *percolation* through a sand or by *agitation*, either mechanically or with compressed air, with a slime.

A continuous method of leaching and separation of the enriched solution from the gangue is *counter–current decantation* (C.C.D.). This process was used originally in the cyanide extraction of silver and gold but now is used in several other hydro-metallurgical applications. The technique employs a series of thickening tanks arranged to permit counter–current flow of solids and liquid solution as illustrated in Fig. 6-1. The underflow or sludge is pumped to each succeeding tank at a higher elevation, while the overflow solution is permitted to move in the opposite direction. The pulp or gangue material is stripped of the desired metal as it moves toward the discharge, while the liquid solution moving counter–current is enriched.

Pressure leaching is carried out in a autoclave at high pressures and at higher temperatures than are feasible in open tanks. The advantages of this technique are that gaseous reagents, such as oxygen, can be used and the high pressure enhances the solubility of the gas in the leaching solution. Oxygen is necessary for the solution of gold by a cyanide solution (Eq. 6-3) and the rate of reaction between the cyanide solution and the gold is increased by increased oxygen pressure. Sulfides can be directly dissolved in acid or alkaline solutions at higher temperatures and higher

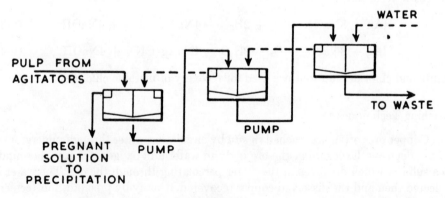

Fig. 6-1. Counter-Current Decantation

Source: W. H. Dennis, *Extractive Metallurgy*, Philosophical Library, Inc., New York, 1965, p. 245.

oxygen pressures. Cobalt and cobalt–nickel sulfides, as well as certain uranium ores, are treated by pressure leaching.

The chemistry of high pressure leaching of sulfides is very complex. Nickel and cobalt sulfides are leached in sulfuric acid and air under high pressure (100 psig) and high temperature (250°F) conditions. The sulfides are oxidized to sulfates which may be described by the simplified overall reactions:

$$NiS + 2O_2 \rightarrow NiSO_4, \qquad (6\text{-}7)$$

$$CoS + 2O_2 \rightarrow CoSO_4. \qquad (6\text{-}8)$$

Liquid–solid separation techniques are required to remove the leaching solution from the residual gangue. Several methods are employed including decanting, thickening, and filtering.

Recovery of Metal from Solution

Several techniques are available for recovering metals from leaching solutions. The most important industrial process is electrolytic deposition (see Electro-metallurgy), but other methods include gaseous reduction, precipitation, solvent extraction, and ion exchange. The development of commercial processes for gaseous reduction of metallic ions from aqueous solutions has been a major metallurgical advance in recent years. This technique, although known for many decades, only recently has achieved a significant industrial status. The thermodynamic aspects of the process are reasonably well understood, and are developed in some detail in the following paragraphs.

Thermodynamics of Reduction of Metals from Aqueous Solution. The use of chemical displacement reactions to precipitate metals from ionic solutions is a time-worn practice in the metallurgical art. The classical example of such a reduction is the "cementation" process in which copper is displaced from aqueous solution by reaction with metallic iron according to the ionic reaction,

$$Cu^{++} + Fe = Cu + Fe^{++}. \qquad (6\text{-}9)$$

Recently developed hydrometallurgical processes employing reducing gases at high pressure to precipitate metals from aqueous solutions are based on the same concepts. These gaseous reduction processes are carried out at elevated temperatures (over 100°C) and at pressures ranging up to several hundred psi. Although the principal reducing agent in these processes is gaseous hydrogen, other reducing gases, such as carbon monoxide or hydrogen sulfide, can be used; the technique has been applied primarily in the production of cobalt and nickel, but with these other gases, however, the formation of undesirable by-products and high cost are disadvantages. Many other metals can be recovered by this process. A generalized representation of the reaction for hydrogen reduction of a metallic ion in aqueous solution is

$$M^{n+} + \frac{n}{2} H_2 = M + nH^+, \qquad (6\text{-}10)$$

where M is the metallic element and n is the valence of the metallic ion in aqueous solution.

The feasibility of these reduction processes can be tested on the basis of metal–metal ion potentials in aqueous solutions.

Electrode Potentials. The equilibrium relation for a displacement reaction in aqueous solution can be developed from standard free energies of reaction, but it is more convenient to use electrode potentials as a measure of the free energy change, since the tabulated data appear in this form, and in fact, for the most part are based on electrode potential measurements.

Standard electrode potentials have been determined for aqueous electrolytes at ambient temperature. The standard potential for several metal–metal ion electrodes and their corresponding reactions can be defined from experimental measurements of the potential of such electrodes relative to a *standard* cell whose potential is defined by the hydrogen electrode (zero potential at 1 atm hydrogen gas and 25°C). A summary of the standard electrode potentials for several half-cell reactions is presented in Table A-VIII of Appendix A.

Cell potentials depend not only upon the temperature, the materials making up the electrodes, and the overall cell reaction, but also upon the activities of the participating ions in solution. Utilizing the definition of net electrical work, which is equivalent to free energy,

$$\Delta G = -n\mathbf{F}\epsilon, \tag{6-11}$$

where

 n = number of equivalents transferred,
 $\mathbf{F}$ = Faraday's constant, 23066 cal/V-g equiv,
 ϵ = emf of call.

And considering the reaction isotherm for a reaction

$$aA + bB + \cdots = cC + dD + \cdots, \tag{6-12}$$

where

$$\Delta G = \Delta G^0 + RT \ln \left[\frac{a_C{}^c a_D{}^d \cdots}{a_A{}^a a_B{}^b \cdots} \right]. \tag{6-13}$$

Then

$$-n\mathbf{F}\epsilon = -n\mathbf{F}\epsilon^0 + RT \ln \left[\frac{a_C{}^c a_D \cdots}{a_A{}^a a_B{}^b \cdots} \right] \tag{6-14}$$

and

$$\epsilon = \epsilon^0 - \frac{RT}{n\mathbf{F}} \ln \left[\frac{a_C{}^c a_D{}^d \cdots}{a_A{}^a a_B{}^b \cdots} \right]. \tag{6-15}$$

The Hydrogen Electrode. The hydrogen electrode is defined by the half-cell reaction,

$$H_2 \rightarrow 2H^+ + 2e^-. \tag{6-16}$$

The potential of the hydrogen electrode is defined by the concentration of hydrogen ions in solution and the hydrogen pressure. The potential of a hydrogen electrode, according to Eq. (6-15), is given by the expression,

$$\epsilon_{H_2} = \epsilon^0_{H_2} - \frac{RT}{2F} \ln (a_{H^+})^2 + \frac{RT}{2F} \ln (P_{H_2}). \tag{6-17}$$

[The pressure of hydrogen in atmospheres is commonly substituted for fugacity. At 100 atm hydrogen pressure, the fugacity is 106.1 atm, which changes the last term of Eq. (6-17) by only 1%.] The activity of the hydrogen ions in solution is defined in terms of the pH scale by the relationship,

$$pH = -\log_{10} (a_{H^+}) = -\frac{\ln (a_{H^+})}{2.303}. \tag{6-18}$$

Since the standard potential for the hydrogen electrode, $\epsilon^0_{H_2}$, is zero by definition, Eq. (6-17) at 25°C reduces to

$$\epsilon_{H_2} = 0.0592 \, pH + 0.0296 \log_{10} (P_{H_2}). \tag{6-19}$$

Metal–Metal Ion Electrode Potentials. The electrode potential for a metallic ion in aqueous solution is determined by the standard electrode potential for the metal–metal ion half-cell and the activity of the ion in solution. The cell potential for a half-cell reaction,

$$M = M^{n+} + ne^-, \tag{6-20}$$

is given by the relationship,

$$\epsilon_M = \epsilon_M^0 - \frac{RT}{nF} \ln (a_{M^{n+}}). \tag{6-21}$$

The activity of the metal ion in solution is related to the concentration according to the relationship,

$$a_{M^{n+}} = f_{M^{n+}}(M^{n+}). \tag{6-22}$$

The standard state is chosen so that the activity coefficient, $f_{M^{n+}}$, approaches unity at infinite dilution. Thus, by definition

$$f_{M^{n+}} = \left[\frac{a_{M^{n+}}}{(M^{n+})} \right]_{(M^{n+}) \to 0} \to 1. \tag{6-23}$$

The concentration in Eq. (6-22) is expressed in terms of molality (moles per 1000 g of solvent). A one molal solution which is common in hydrometallurgical processes, may *not* result in a cell potential, ϵ_M, which is equal to the standard cell potential, ϵ^0_M, since the activity coefficient *could* be much less than unity. Although the concentration would be unity, the activity would not be, and the last term in Eq. (6-21) would alter the cell potential. The activity coefficients for aqueous sulfates and chlorides, common metallurgical processing solutions, are usually less than one, even in relatively dilute solutions, as illustrated in Table 6-1.

Example Problem 6-1. Compute the half-cell potential at 25°C for the reaction,

$$Ni = Ni^{++} + 2e^-, \tag{6-a}$$

in a 0.1 molal solution of nickel sulfate.

TABLE 6-1

Activity Coefficients in Some Metal Salt Solutions

Salt	Activity Coefficients in solutions of molality									
	0.001	0.002	0.005	0.01	0.02	0.05	0.1	0.2	0.5	1.0
MgSO$_4$	—	—	—	0.40	0.32	0.22	0.18	0.13	0.088	0.064
NiSO$_4$	—	—	—	—	—	—	0.18	0.13	0.075	0.051
CuSO$_4$	0.74	—	0.53	0.41	0.31	0.21	0.16	0.11	0.068	0.047
ZnSO$_4$	0.70	0.61	0.48	0.39	– –	—	0.15	0.11	0.065	0.045
MgCl$_2$	—	—	—	—	—	—	0.56	0.53	0.52	0.62
FeCl$_2$	0.89	0.86	0.80	0.75	0.70	0.62	0.58	0.55	0.59	0.67
CuCl$_2$	0.89	0.85	0.78	0.72	0.66	0.58	0.52	0.47	0.42	0.43
ZnCl$_2$	0.88	0.84	0.77	0.71	0.64	0.56	0.50	0.45	0.38	0.33

Source: B. Meddings and V. N. Mackiw, Gaseous reduction of metals from aqueous solutions, *Applications of Fundamental Thermodynamics to Metallurgical Processes*, p. 110, Gordon and Breach Science Publishers Inc., New York, 1967.

Solution. Using Eq. (6-21) and the cell potential from Table A-VIII

$$\epsilon_{Ni} = 0.236 - \frac{0.0592}{2} \log_{10}(a_{Ni^{++}}). \tag{6-b}$$

From Table 6-1, $f_{Ni^{++}} = 0.18$, and Eq. (6-b) yields

$$\epsilon_{Ni} = 0.236 - 0.0296 \log_{10}[(0.18)(0.1)] \tag{6-c}$$
$$\epsilon_{Ni} = 0.236 + 0.052 = 0.288 \text{ V}$$

Hydrogen Reduction of Metal Ions. The generalized hydrogen reduction reaction, Eq. (6-10), may be considered to be made up of two half-cell reactions, as given by Eqs. (6-16) and (6-20). The corresponding cell potentials are given by Eqs. (6-17) and (6-21). If the cell potential for the oxidation of metal, Eq. (6-20), is greater than for the oxidation of hydrogen, Eq. (6-16), then the metal will oxidize and go into solution and hydrogen gas will be liberated. If, however, the cell potential for the reaction of Eq. (6-16) is the larger, the metal will precipitate from solution and hydrogen will be oxidized.

It is especially important to note that the cell potentials are independent of the number of electrons taking part in the oxidation or reduction reaction. This is evident from rewriting Eq. (6-11) in the form,

$$\epsilon = -\frac{\Delta G}{n\mathbf{F}}. \tag{6-24}$$

The total free energy change is dependent upon the number of gram atoms, moles, or equivalents taking place in a reaction but the corresponding emf is independent of the form in which the equation is written. Thus, Eqs. (6-10) and (6-20) can be combined to yield

$$\frac{n}{2}\,H_2 = nH^+ + ne^-. \tag{6-25}$$

As stated above, at constant temperature the potential of the hydrogen electrode is dependent only on the pH and the hydrogen pressure. Thus, for a given condition, the cell potential for Eq. (6-16) and for Eq. (6-25) are the same, regardless of the value assigned to n in Eq. (6-25). This is evident from Eq. (6-17).

The thermodynamic feasibility of a reduction by hydrogen can be determined by comparing cell potentials. The cell potential for reduction of a pure metal is dependent upon the concentration of the metallic ion in solution, and the cell potential for the hydrogen reaction is dependent upon pH and hydrogen pressure. Figure 6-2 presents a comparison of the hydrogen potentials at 25°C for 1 and 100 atm on a pH scale with the potentials for reduction of various metallic ions at two levels of concentration. Activities rather than concentrations have been used for calculation of the metal–metal ion potentials, yielding smaller slopes. Since the values of pH, hydrogen pressure, and metal ion concentration cover the range normally encountered in industrial operations, copper can always be reduced by hydrogen whereas zinc cannot. Reduction of zinc, iron, and cadmium are considered in the following example problem.

Example Problem 6-2. Compare the hydrogen pressures required to precipitate Zn^{++}, Fe^{++}, and Cd^{++} at 25°C in solutions of pH 6 and 10. The potentials for precipitation can be taken as 0.83, 0.50, and 0.47 V for Zn^{++}, Fe^{++}, and Cd^{++}, respectively.

Solution. At pH 6, 25°C, and 1 atm hydrogen pressure, the hydrogen potential, from Eq. (6-19) is

$$\epsilon_{H_2} = \epsilon^0{}_{H_2} + 0.0592(\text{pH}) + 0.0296\,\log_{10}(P_{H_2}), \tag{6-a}$$

$$\epsilon_{H_2} = +0.0592(6) = 0.36\,\text{V}. \tag{6-b}$$

The potentials for reduction of Zn^{++}, Fe^{++}, and Cd^{++} are thus 0.47, 0.14, and 0.11 V above the hydrogen potentials, respectively. The emf of the hydrogen half-cell can be increased by increasing the hydrogen pressure by an amount calculable from the last term of Eq. (6-19)

$$\Delta\epsilon = 0.0296\,\log_{10}(P_{H_2}). \tag{6-c}$$

The corresponding pressures at pH 6 ($\epsilon^0{}_{H_2} = 0.36$) and at pH 10 ($\epsilon^0{}_{H_2} = 0.592$ V) are

Equilibrium Hydrogen Pressure in Atmospheres

	pH = 6	pH = 10
Zn^{++}	8×10^{15}	1.1×10^8
Fe^{++}	5.4×10^4	1^a
Cd^{++}	5.2×10^4	1

^{a} Above a pH of 8.5, ferrous ions are unstable with respect to hydrolysis. The equilibrium solubility product for ferrous hydroxide is on the order of 10^{-14} and hence the ferrous ion concentration would be about 10^{-10} mole/liter and the hydrogen pressure would have to be increased to about 10^6 atm to precipitate metallic iron under these circumstances.

At pH 10, $\epsilon_{Zn^{++}}$ is still much greater than ϵ_{H_2}, and it is unlikely that zinc will ever be recovered by hydrogen reduction. However, cadmium could be precipitated at less than 1 atm hydrogen pressure.

It should be noted that as the reduction reaction is carried out, there is a limiting concentration of metallic ion at which the cell potential for the reduction of the metal ion and the oxidation of hydrogen become equal. For 1 atm hydrogen pressure and 25°C this potential is

$$\epsilon_{H_2} = 0.0592 \text{ pH} = \epsilon_M^0 - \frac{0.0592}{n} \log_{10}(a_{M^{n+}}). \qquad (6\text{-}26)$$

At low ion concentrations, the activity of the metal ion can be replaced by concentration. Eq. (6-26) then becomes

$$-\log_{10}(C_{M^{n+}}) = (n)(\text{pH}) - \frac{(n)(\epsilon_M^0)}{0.0592}. \qquad (6\text{-}27)$$

For a concentration of 10^{-2} mole/liter of metallic ion and at 1 atm hydrogen pressure and 25°C, the equilibrium pH values calculated by Eq. (6-27) are tabulated in Table 6-2. As reduction of the metal ion continues, hydrogen ions are released in the solution and the pH decreases. It is evident from Table 6-2 that precautions must be taken to assure that the pH is maintained above the desired level. This is particularly true for metals with positive ϵ^0 values.

It is apparent from Fig. 6-2 that reduction by hydrogen is favored at higher pH values. However, in alkaline solutions, the solubility of many metal cations is limited particularly for the transition metals. In commercial practice the metal ions are "complexed" by adding ammonia, which raises the pH and forms a soluble amine. The addition of ammonia to solutions containing cobalt ions, e.g., would cause formation of a series of amines with increasing cobalt-to-ammonia ratios. The formation of the first amine is given by the equation,

$$Co^{++} + NH_3 = Co(NH_3)^{++}. \qquad (6\text{-}28)$$

TABLE 6-2

Equilibrium pH Values at $C_{M^{n+}} = 10^{-2}$ Molar

Ion	E^0	pH
Zn^{++}	0.762	13.9
Fe^{++}	0.440	8.5
Cd^{++}	0.403	7.8
Co^{++}	0.267	5.5
Ni^{++}	0.241	5.1
Cu^{++}	-0.337	-4.7
Ag^{++}	-0.799	-11.5

Source: B. Meddings and V. N. Mackiw, Gaseous reduction of metals from aqueous solutions, *Applications of Fundamental Thermodynamics to Metallurgical Processes*, p. 114, Gordon and Breach Science Publishers Inc., New York, 1967.

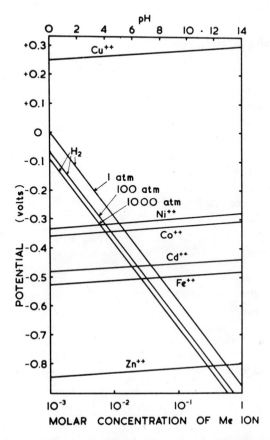

Fig. 6-2. Variation of Hydrogen and Metal Electrode Potentials with Ion Concentration in Sulfate Solutions (corrected for solute activity interactions)

Source: D. J. I. Evans, "Advances in Extractive Metallurgy", Symposium Volume, p. 881, American-Elsevier, England, 1968.

The distribution of types of amines in the solution determines the reduction potential for the cobalt ion. The greatest driving force occurs near the ratio $NH_3/Co^{++} = 2$, and it is this concentration which is usually employed in industrial processing of both cobalt and nickel.[1] The addition of ammonia and formation of an alkaline solution for the high pressure reduction of metals from aqueous solutions has a further practical advantage in that the extremely corrosive conditions existing in acid solutions under hydrogen pressure are avoided.

Kinetics of Gaseous Reduction in Aqueous Solutions. The rate of reduction of a metallic ion from aqueous solution can be limited by a number of factors; nucleation of solid particles; area of the nucleation surface; the thermodynamic factors of hydrogen pressure, pH, or metal ion concentration; etc. The rate limiting step in

---

[1] T. M. Kaneko and M. E. Wadsworth, *J. Phys. Chem.* **60,** 457 (1956).

the conversion varies with the system and the stage of the recovery process. The significant kinetic factors for a few industrial hydrometallurgical processes are outlined in the following paragraphs.

Nucleation of a precipitate is often a rate limiting step in the reduction process. The reduction of nickel is a heterogeneous process and requires the presence of nuclei. Nickel powder usually is suspended in the solution and vigorously agitated. Ferrous sulfate also can be used. The rate of reduction of cobalt also is nucleation dependent.[2] The reduction rate of nickel has been shown to be proportional to the surface area of the nickel "seed" particles.[3]

The pressure of hydrogen is also important in determining the rate of reaction. The thermodynamic equilibrium depends upon the hydrogen pressure; for example, the rate of reduction of nickel from an ammoniacal solution varies with the first power of the hydrogen pressure,[4] such that

$$-\frac{d(\mathrm{Ni}^{++})}{dt} = kA\,(P_{\mathrm{H}_2}), \tag{6-29}$$

where

k = temperature dependent rate constant,
A = area of nucleating surface,
P_{H_2} = pressure of hydrogen in atm.

In a closed reduction vessel, the pressure decreases with time in direct proportion to the nickel reduced, so that

$$-\frac{dP_{\mathrm{H}_2}}{dt} = k'A\,(P_{\mathrm{H}_2}), \tag{6-30}$$

where k' is a constant combining the rate constant k and the volume of the system, volume of solution, etc. Eq. (6-30) can be integrated to

$$\log_{10}\left[\frac{(P_{\mathrm{H}_2})_{t=0}}{(P_{\mathrm{H}_2})_t}\right] = \frac{k'At}{2.303}. \tag{6-31}$$

Catalysts, such as anthraquinone in the case of nickel reduction, have been used to increase the rate of reaction.

The rate of reduction is *temperature dependent* and Arrhenius plots have been used to determine activation energies. In the case of nickel reduction the activation energy was 10,200 cal/mole.[5]

To achieve success in industrial practice, the hydrogen pressure, pH, and metallic ion concentration are well above those predicted by equilibrium calculations. The actual equilibria are complicated by complexing reactions in ammoniacal solutions. Agitation and the presence of nuclei, particularly in the cases

---

[2] J. Halpern and A. G. Webster, *J. Phys. Chem.* **61**, 1245 (1957).
[3] V. N. Mackiw, W.C. Lin, and W. Kunda, *Trans. AIME* **209**, 786 (1957).
[4] Ibid.
[5] Ibid.

of nickel and cobalt reduction, are especially important in assuring a reasonably rapid reaction and the precipitation of the desired sand-like deposit.

Other Recovery Methods. The precipitation of a metal ion from a solution by formation of an insoluble compound is a well-known technique in chemical analysis. It can be applied to the recovery of metals from leaching solutions or for the separation of elements if the leaching solution contains more than one element, as in the sulfuric acid–oxygen leaching of nickel–cobalt–iron sulfides. By using ammonia to raise the pH to approximately 5 and oxidizing the solution with air at atmospheric pressure, the ferrous ions are oxidized to ferric ions and precipitated as hydrated ferric oxide. This separation process is described by the reaction,

$$2FeSO_4 + 4NH_4OH + \tfrac{1}{2} O_2 = Fe_2O_3 + 2(NH_4)_2SO_4 + 2H_2O. \qquad (6\text{-}32)$$

Another example of this technique is in the precipitation of gold or silver from cyanide solutions by using zinc dust. The cyanide solution is agitated with zinc dust under conditions which are relatively oxygen-free. The zinc precipitates the gold (or silver) according to the reaction,

$$Zn + 2NaAu(CN)_2 = Na_2Zn(CN)_4 + 2Au. \qquad (6\text{-}33)$$

The long-standing technique for recovery of copper from dilute solutions has been by cementation. These dilute solutions of copper usually contain large quantities of impurities that would make electrolytic refining difficult. Precipitation of the copper is accomplished by bringing the solution in contact with low-cost scrap iron. The copper precipitates as a finely divided, loosely adherent sponge according to the reaction,

$$CuSO_4 + Fe = FeSO_4 + Cu. \qquad (6\text{-}34)$$

Any adherent copper is easily removed from the iron by blasting with high pressure water. The cemented copper is not very pure (50–75% copper). In theory, it requires less than 1 lb of iron to precipitate a pound of copper, but because of reaction with the acid solution and reduction of ferric ions, the consumption of iron ranges from 1.5–2 lb per pound of copper precipitated.

The efficiency of copper recovery by cementation depends upon the surface area of iron exposed to the solution, extent of agitation of the solution, and length of contact. A standard gravity launder provides 90% recovery of copper, but a recently designed cone precipitator (see Fig. 6-3) improved cementation efficiencies such that over 93% of the copper is recovered. The use of shredded scrap iron in the cone precipitator with its turbulent action zone where kinetic factors are favorable results in the improvement. The automatic recovery of the cemented copper is also a marked advantage in this process.

Example Problem 6-3. In the cementation of copper, the copper concentration in the leaching solution can be reduced to very low levels. Assume a dilute sulfuric acid solution in which the activity coefficient of the ferrous ion is 0.2. The solution leaves the cementation launder at 25°C and contains 0.6 g/liter of iron. Estimate the residual equilibrium copper content of the exit stream.

Solution. The displacement reaction is given by Eq. (6-9).

$$Cu^{++} + Fe = Cu + Fe^{++}. \qquad (6\text{-}a)$$

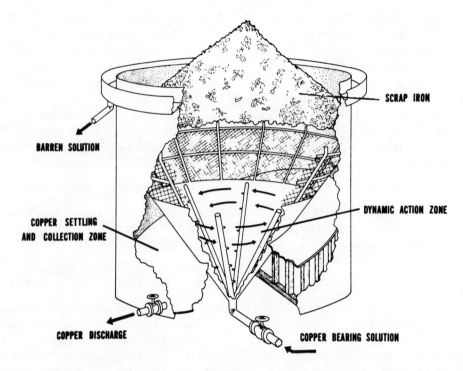

Fig. 6-3. Sketch of Cone Precipitator Showing Solution Inflow and Copper Precipitate Discharge

Source: "Cone-Type Precipitators for Improved Copper Recovery," H. R. Spedden, E. E. Malouf, and J. D. Prater, *J. of Metals*, 1966, Vol. 18, p. 1139.

The half-cell potential for iron is

$$\epsilon = \epsilon^0 - \frac{4.575\,T}{(2)\,(23066)} \log_{10}\,(a_{Fe^{++}}). \tag{6-b}$$

From Table A-VIII, $\epsilon^0_{Fe} = 0.441$, and

$$\epsilon = 0.441 - 0.0296 \log_{10}\,(0.2)\,\frac{0.6}{55.85}, \tag{6-c}$$

$$\epsilon = 0.441 + 0.0795 = 0.521 \text{ V.} \tag{6-d}$$

At equilibrium, the copper half-cell must have the same potential, and

$$\epsilon = \epsilon^0_{Cu} - 0.0296 \log_{10}\,(a_{Cu^{++}}). \tag{6-e}$$

From Table A-VIII, $\epsilon^0_{Cu} = -0.3441$. Assuming that the activity coefficient of copper is unity in the dilute solution and substituting concentration for activity,

$$\log_{10}\,(C_{Cu^{++}}) = -\frac{(0.521 + 0.3441)}{0.0296} = 29.2. \tag{6-f}$$

The molality is 6.3×10^{-30} which corresponds to 4×10^{-28} g/liter.

The cementation process is highly favored thermodynamically, and in practice is a very efficient way to recover copper from dilute aqueous solutions. The extent of recovery in practice is limited by the effectiveness of contact between the copper solution and the scrap iron, and the spent solution usually contains more than 0.06 g of Cu/liter.

Solvent extraction has been used for many years in the chemical and oil industries, and has become increasingly important in the metals field, particularly with the advent of uranium ore processing. This method also is referred to as liquid–liquid extraction. The solvents used with aqueous solutions are usually organics, since it is necessary that the extracting solution be relatively insoluble in the leaching solution and also be capable of a clean separation, that is, it is important

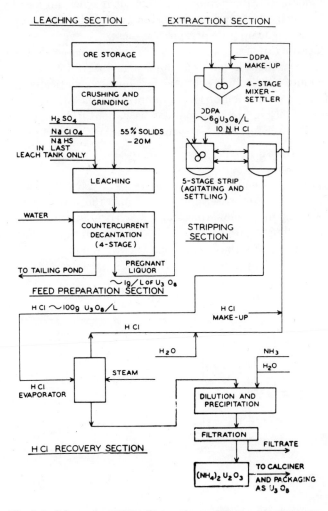

Fig. 6-4. Scheme for DDPA Extraction and Recovery of Uranium

Source: W. D. Wilkinson, *Uranium Metallurgy*, Vol. 1, "Uranium Process Metallurgy", Interscience Publishers; John Wiley and Sons, Inc., New York, 1962, p. 141.

that the extracting solution does not form an emulsion. The extractant must be highly selective in its ability to react with or absorb the desired material.

In a system for processing uranium ores, di(2-ethyhexyl) phosphoric acid (EHPA) is dissolved in kerosene and contacted with the leaching solution containing uranium. The phosphoric acid forms an extractable complex with uranium, and the extraction is carried out with various types of mixing units, such as stirred tanks, columns, etc. A typical acid-leach system employing a similar phosphoric acid compound as an extractant for uranium is diagrammed in Fig. 6-4.

Ion exchange is another technique for recovering metals from leaching solutions, and this method also has received considerable impetus from the processing of uranium ores. The same technique is used in water softeners, in which a zeolite (a sodium aluminate–silicate) removes calcium or magnesium ions by exchanging them with sodium ions. Ion exchangers used in most hydrometallurgical operations are synthetic resins made up from organic polymers.

In the processing of uranium ores, an anionic exchanger is used because the uranium is present in the solution as a complex anion. An ion exchange process for recovery of uranium from sulfate solutions is shown schematically in Fig. 6-5. This permits an effective separation of the uranium since the other metals in solution, aluminum, cobalt, nickel, etc., are present as cations. The active group on the surface of the resin exchanges with the complex ion according to the reaction,

$$UO_2(SO_4)_3^{-4} + 4RX = R_4UO_2(SO_4)_3 + 4X^-, \qquad (6\text{-}35)$$

where R indicates the resin and X^- is a negative ion. Adsorption takes place rapidly until the surface of the resin is fully reacted and the leaching solution passing through the resin begins to show traces of dissolved uranium. When the resin has fully reacted, it is brought into contact with a strong solution of chloride or nitrate ions (X^- ions), which puts the uranyl ions back into solution and recharges the

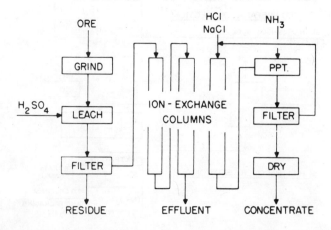

Fig. 6-5. Uranium Ore Processing—Ion-Exchange for Sulfate Solutions

Source: W. D. Wilkinson, *Uranium Metallurgy, Vol. I: Uranium Process Metallurgy*, Interscience Publishers; John Wiley and Sons, Inc., New York, 1962, p. 111.

surface of the resin. This solution (eluate) is treated subsequently for recovery of the uranium.

Ion exchange is a process requiring several steps: absorbing metal ions from solution; removal of the retained ions by replacement reactions using a solution containing the proper ion (anion in the case of uranium processing), and, finally, recovering the metal from the eluate. Ion exchange is relatively expensive and has been restricted to separation processes like uranium extraction and the processing of rare-earth metals.

6-2 Industrial Hydrometallurgical Operations

In the following paragraphs, a brief description is given of some of the more important hydrometallurgical processes used in the metallurgical industries. The examples cited include the recovery of copper, gold and silver, uranium, zinc, nickel, cobalt, and the purification of alumina for production of metallic aluminum.

Hydrometallurgy of Copper

As mentioned previously, the leaching of copper as a recovery process is usually practiced on low-grade ores. Dilute sulfuric acid is used to leach ores containing copper oxide, whereas water will leach copper sulfate from partially roasted, or weathered ores.

The presence of iron in sulfide minerals as ferric salts enhances the solution of copper. Copper is usually recovered by cementation or electrolysis.

Gold and Silver Hydrometallurgy

The recovery of gold is classically described by stories of the old prospector and his panning operations. The placer deposits of gold where the waste material has been washed away leaving the gold behind permit recovery of the gold by panning. Large operations in Alaska use hydraulic mining, in which water is used to wash the gold bearing dirt into sluiceways where gravity separation is carried out.

Gold also is recovered by amalgamation. The solubility of gold in mercury is quite high, over 15 wt% at 100°C, and finely divided gold readily dissolves when brought into contact with a fresh mercury surface. By washing the gold-bearing material in a finely divided state over a copper surface coated with mercury, the gold is alloyed readily with the mercury for later recovery by distillation. Various types of milling operations are carried out in the presence of mercury to recover gold by amalgamation.

The chemical recovery of gold in a hydrometallurgical manner was formerly accomplished primarily by chlorination. However, in 1888, the McArthur–Forrest process for cyanidation was introduced, and since then this method has been a primary process for production of gold and other precious metals. Dissolution of gold in dilute cyanide solution according to Eq. (6-3) shows that oxygen is an important factor, and gold will not dissolve in cyanide solutions *unless* oxygen is present, i.e., as dissolved oxygen gas or possibly as an oxidizing agent. Sodium cyanide is used because it is less expensive than potassium cyanide, the original reagent material.

The solutions used contain approximately 0.1% sodium cyanide. One difficulty arising in cyanidation is the presence of cyanicides, a term given to materials (other than the precious metals) that react with and consume the cyanide solution. Acids or salts which react with strong bases often are present and react with cyanide. Also, most copper minerals react with sodium cyanide to form complex compounds. Zinc carbonate and iron also consume the cyanide. Cyanicides must be removed from the solution and this often can be accomplished after acid leaching by addition

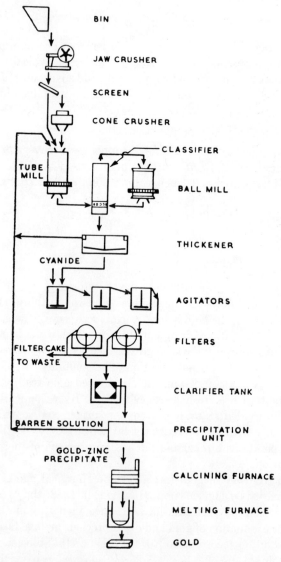

Fig. 6-6. Schematic Diagram Outline the All-sliming Gold-recovery Process

Source: W. H. Dennis, *Extractive Metallurgy*, Philosophical Library, Inc., New York, 1965, p. 251.

of lime, which neutralizes the free acid and precipitates some of the interfering metallic ions such as iron. A schematic diagram outlining a gold recovery process in which an ore slime is treated is shown in Fig. 6-6.

Gold is recovered from the leaching solution by precipitation with zinc [Eq. (6-33)]. Several metals could be used to displace the gold in an ionic reaction, and, in fact, sodium and potassium would be most desirable because they would regenerate the solution. Aluminum sometimes is used to produce sodium cyanide and a gold precipitate, but wide adoption of this method has not taken place because of the interference of calcium oxide, often used to supply alkalinity to the solution, which precipitates along with the gold. The reaction with aluminum is

$$2NaAu(CN)_2 + 4NaOH + 2Al = 4NaCN + 2Au + Na_2Al_2O_4 + 2H_2. \quad (6\text{-}36)$$

Charcoal also could be used. Nevertheless, in industrial practice because of its low cost and suitability, zinc is used most widely for this purpose.

The gold precipitate is treated with sulfuric acid to dissolve the zinc, and the residual material then is fused with borax, sand, and soda to produce metallic gold in a pyrometallurgical operation. The gold thus produced is about 90% pure and usually contains silver. Further refining is carried out by electrolysis.

Uranium Hydrometallurgy

Uranium minerals usually are extracted by alkaline or acid leaching. In the acid process sulfuric acid is used, whereas in alkali leaching sodium carbonate and bicarbonate solutions are used. In acid leaching the uranium is dissolved according to the ionic equation,

$$2UO_2^{+2} + 5SO_4^{-2} = UO_2(SO_4)_3^{-4} + UO_2(SO_4)_2^{-2}. \quad (6\text{-}37)$$

In alkaline leaching the corresponding reaction would be

$$UO_2^{+2} + 3CO_3^{-2} = UO_2(CO_3)_3^{-4}. \quad (6\text{-}38)$$

Alkali leaching is much slower but the kinetics can be enhanced by treating a finer material and using heated leaching solutions.

Acid leaching usually is employed and is carried out in the presence of an oxidizing agent to convert the uranium to U_3O_8. The concentration of uranium in the leaching solutions is relatively low, corresponding to about 1 g of uranium oxide/liter. The dissolved uranium is present in complex anions that can be advantageously treated with ion exchange or solvent extraction processes. Direct precipitation of the uranium is possible but is not desirable because of the diluteness of the solution and the fact that the resulting precipitate would be contaminated by the metallic impurities present. *Anionic* removal of the uranium using ion exchange, although expensive, is very effective in purifying the uranium oxide.

Uranium as a molten metal is extremely reactive but can be produced by electrolysis of fused salts, the uranium being present as KUF_5. Calcium hydride also has been used to decompose uranium oxide, producing calcium oxide, hydrogen, and uranium powder upon reaction at about 1000°C in a vacuum or inert atmosphere.

Zinc Hydrometallurgy

Zinc usually is found as a sulfide and such ores are roasted to convert the sulfide to an oxide which in turn is leached with sulfuric acid according to Eq. (6-2). The solution usually contains other impurities, such as copper, cadmium, etc. These metals are precipitated by displacement reactions upon agitating the solution with zinc dust. The zinc sulfate solution then is electrolyzed to produce metallic zinc. In

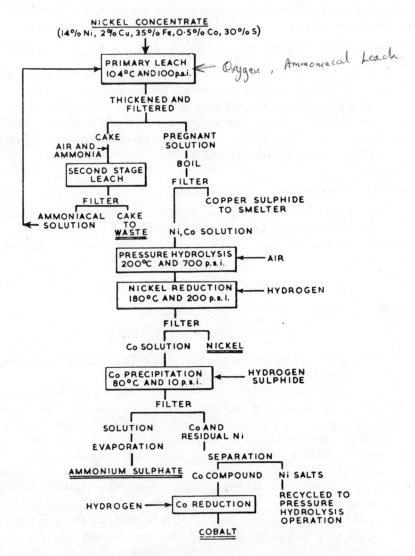

Fig. 6-7. Pressure Leach Separation of Copper, Nickel and Cobalt

Source: W. H. Dennis, *Extractive Metallurgy*, Philosophical Library, Inc., New York, 1965, p. 270.

the electrolytic process, the sulfuric acid is regenerated and returned to the leaching operation.

Hydrometallurgy of Nickel and Cobalt

Sulfide minerals including nickel–cobalt–copper ores are dissolved readily in ammonia or sodium hydroxide solutions, or in sulfuric acid, under oxygen pressure at slightly elevated temperatures. This process for the recovery of nickel, cobalt, and copper has been practiced in Canada since the early 1950s. A flow sheet for the process is given in Fig. 6-7.

The finely ground concentrate is given a preliminary leach in an ammoniacal solution in an autoclave at about 100°C and 100 psi. Most of the copper and nickel are dissolved in this leach. After thickening and filtering, the pulp from the first leach is given a second leach under similar conditions. The iron is removed from solution as ferric hydroxide, which is filtered off with the residual ore. The "pregnant liquor" from the primary leach contains nickel, copper, and cobalt as amines and various ammonium salts. On heating this solution, ammonia is driven off and the copper is precipitated as a sulfide and filtered off. The solution which contains nickel and a small amount of cobalt is treated in an autoclave at 200°C under a high oxidizing pressure to decompose the complex ions containing sulfur. The solution then is treated in a two-step process at temperatures up to 200°C at high hydrogen pressures. Nickel is first precipitated. Precipitation is enhanced by addition of metallic nickel powder or by introduction of a suitable catalyst such as ferrous sulfate. The resulting solution contains cobalt and traces of nickel. By suitable adjustment of the temperature and pressure conditions, selective precipitation can be carried out in two stages on this solution. Overall recoveries are very high, ranging up to 90% or higher for nickel, about 90% for copper, and as high as 75% for the cobalt. This direct production of the metals eliminates electrolysis and the expensive pyrometallurgical processes that would have to be carried out to accomplish the same purification.

Hydrometallurgy of Aluminum—Bayer Process

The commercial production of high purity aluminum (99+% or pure) by electrolysis in the Hall cell is made possible through the use of high purity alumina. The alumina for use in the Hall process is produced from bauxite ores using the Bayer process, a chemical process producing sodium aluminate that is soluble in alkali solutions. The aluminate is decomposed to produce aluminum hydroxide, which then is calcined to alumina. A schematic flow chart for the Bayer process is shown in Fig. 6-8. The steps in this process can be outlined as follows:

1. The bauxite is dried, calcined, and ground.
2. The bauxite powder is digested in a concentrated sodium hydroxide solution. To facilitate this digestion the reaction,

$$Al_2O_3 + 2NaOH = 2NaAlO_2 + H_2O,$$

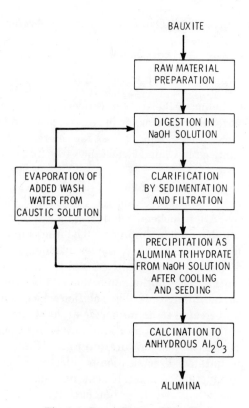

Fig. 6-8. Bayer Process Flow Sheet

is carried out in autoclaves at 50 to 70 lb psi pressure and a temperature of 160–170°C.

3. The solution and any residue is transferred to settling tanks where undissolved ferric oxide, silica, and sodium–aluminum silicates settle out.

4. The sodium aluminate solution is cooled to 25–35°C, precipitating the Al(OH)₃ and regenerating NaOH. Freshly precipitated aluminum hydroxide sometimes is stirred into the solution to "seed" the reaction. The sodium hydroxide is recirculated, concentrated by evaporation and new sodium hydroxide is added to compensate for losses.

5. The aluminum hydroxide is calcined at about 1100°C to produce alumina.

Bibliography

Brown, G. C., et al., *Unit Operations*, John Wiley and Sons, New York, 1950.

Burkin, A. R., *The Chemistry of Hydrometallurgical Processes*, E. and F. N. Spon, Ltd., London, 1966.

Burkin, A. R. and F. D. Richardson, The production of metal powders from aqueous solutions, *Powder Metallurgy* **10**, 33–57 (1967).

Butts, A., *Metallurgical Problems*, McGraw-Hill, New York, 1943.

Dennis, W. H., *Extractive Metallurgy*, Philosophical Library, Inc., New York, 1965.

Extraction and Refining of the Rarer Metals, Institution of Mining and Metallurgy, London, 1957.

Foust, A. S., et al., *Principles of Unit Operations*, John Wiley, New York, 1959.

Hayward, C. R., *An Outline of Metallurgical Practice*, D. Van Nostrand Co., Inc., New York, 1952.

Meddings, B., and V. N. Mackiw, The gaseous reduction of metals from aqueous solutions, *Applications of Fundamental Thermodynamics to Metallurgical Processes* (G. R. Fitterer, Ed.) pp. 105–151, Gordon and Breach Science Publishers, New York, 1967.

Newton, J., *Extractive Metallurgy*, John Wiley, New York, 1959.

John H. Perry, Ed., *Chemical Engineers' Handbook*, 4th ed., McGraw-Hill, New York, 1963.

Unit Processes in Hydrometallurgy, AIME Met. Soc. Conf., Vol. 24, Gordon and Breach Science Publishers, Ltd., London, 1964.

VanArsdale, G. D., *Hydrometallurgy of Base Metals*, McGraw-Hill, New York, 1953.

Discussion Questions

1. Uranium ore is leached with acid under an oxygen pressure of 10–15 psi according to the reaction,

$$U_3O_8 + \tfrac{1}{2} O_2 + 3H_2SO_4 = 3UO_2SO_4 + 3H_2O.$$

   What influence do the following variables have on the rate of pressure leaching?
   (a) mesh size of ore
   (b) temperature
   (c) oxygen pressure

2. Describe briefly the influences of temperature, bath agitation, solute concentrations, and particle geometry on the kinetics of leaching.

3. Bacterial oxidation offers considerable potential as a mechanism which can enhance leaching of various ores. Discuss the mechanism and outline possible application of leaching reactions involving microorganisms.

4. Hydrometallurgical processes involving the gaseous reduction of metals from solution have been successfully operated on a commercial scale since 1955. However, production in the first decade or more has been limited to nickel, cobalt and copper. Discuss the reasons for this and project the potential for application of this processing operation to other metals.

5. The extension of hydrogen reduction of metals from aqueous solution to production of coated metallic powders and intimate mixtures of finely subdivided materials has been proposed. Discuss the technology and possible market for these processes.

6. Discuss Fig. 6-2 in terms of separating copper, nickel and cobalt. What is the possibility of developing a process for hydrogen reduction of zinc from aqueous solution?

7. Nearly all of the alumina used for metal production is prepared by the Bayer process. A recent investigation of the rate of solution of Gibbsite, the most important aluminum mineral in natural bauxites, in sodium hydroxide has shown that the reaction is reaction rate limited. In deriving a kinetic model, what variables should be included?

8. Discuss the economic and tecnological advantages and disadvantages of regeneration, neutralization, and disposal as alternative treatments for acid leaching solutions.

Problems

1. The rate of leaching of a copper ore with sulfuric acid has been measured under the following conditions:

Rate (Mg Cu^{++}/cm²-min)	Temperature (°C)	Acid concentration (Molarity)
1.09	40	0.685
1.86	49.5	0.685
0.554	40.9	0.103
1.37	50.1	0.353
1.83	41	1.377

Derive an equation or equations describing the rate of leaching as a function of temperature and acid concentration.

2. Copper, cadmium, and silver are recovered from a leaching solution by precipitation with excess zinc dust. Assuming that the zinc ion concentration is 114 g/liter and that its activity coefficient at this concentration level is close to unity, estimate the residual ionic concentrations of the recovered metals.

3. Under certain circumstances, cementation of copper on nickel powder can be carried out economically, e.g., the cementation of copper with nickel from a nickel-rich electrolyte and recovery of nickel by conventional means. If during precipitation of copper with nickel powder in a sulfate solution the nickel ion concentration is 15 g/liter, estimate the residual copper ion concentration.

4. A copper ore containing 0.63% CuO and 0.39% Cu_2S is leached with a sulfuric acid-ferric sulfate solution. Upward percolation is carried out in a 9000 T (ore) vat. The system is on an 8-day leach—1-day wash cycle with the overflow composition averaging 23.4 g Cu/liter, 12.5 g H_2SO_4/liter and 2.2 g of Fe^{++} ion/liter. Calculate the composition of the original leaching solution if 90% of the oxide and 45% of the sulfide is recovered.

5. The rate of cementation of copper with iron follows first-order kinetics with no appreciable back reaction. Derive a relation describing this behavior. What additional variables are likely to influence the cementation rate and how could they be measured?

6. (a) What is the ionic activity of nickel under equilibrium conditions following precipitation of metallic nickel from a solution of pH = 5.5 in contact with 1 atm of hydrogen gas at 25°C?
    (b) Is this activity increased or decreased with
        (1) a decrease in pH of the solution.
        (2) an increase in hydrogen pressure.
    (c) In part (a), what would be the corresponding activity of ferrous ion?

7. Separation of nickel and cobalt in acid solution depends on the pH and hydrogen pressure. At low concentrations, calculate the equilibrium ratio of nickel to cobalt in solution. What can be done to increase the ratio?

8. What is the limiting equilibrium residual cobalt ion activity for precipitation from aqueous solution (pH = 6.6) with hydrogen at 50 atm?
9. What is the electrode potential for nickel–nickel ion (0.005 molal) in a sulfate solution at 25°C? At what pH would hydrogen at 10 atm be in equilibrium with the solution and pure nickel?

CHAPTER 7

Electrometallurgy

7-1 Principles of Electrolysis

The use of electrical energy to win metals from ores, to recover metals from leaching solutions, and to refine metals establishes electrochemistry as an extremely important part of extractive metallurgy. Electrochemical engineering is employed in the separation of materials in electrolyzing solutions (as in the copper and zinc industries), in electro-refining of metals (as in the production of electrolytic copper, nickel, lead, etc.), and in the electrolysis of fused salt electrolytes (for the production of aluminum, magnesium, and alkali metals). Additional extensions of this science are found in electric furnace operations and in the conversion of chemical to electrical energy.

Half-Cell Reactions

The relation between the chemical and electrical nature of the reactions taking place during electrolysis may be represented by electrochemical equations. Consider, for example, two strips of pure copper immersed in a copper sulfate solution. If an external circuit is set up so electrons flow between the two copper pieces, one will be the cathode in an electrolytic cell and the other will be the anode. Under the influence of the electrical potential between the two electrodes, copper ions in solution will migrate to the more negatively charged electrode, the cathode in this instance. At the cathode,[1] electrons are released which combine with the positively charged copper ions in the electrolyte to form atoms of copper deposited on the cathode. This half-cell reaction is represented by the relationship,

$$Cu^{++} + 2e^- = Cu \text{ (at cathode).} \qquad (7\text{-}1)$$

At the anode, to which the negatively charged sulfate ions are migrating, the reverse reaction occurs:

$$Cu \text{ (at anode)} = Cu^{++} + 2e^-. \qquad (7\text{-}2)$$

The copper ions combine with the sulfate ions in solution to form copper sulfate which in turn dissolves in the electrolyte and ionizes. Combining reactions (7-1) and (7-2), one obtains the overall cell reaction,

$$Cu \text{ (at anode)} = Cu \text{ (at cathode).} \qquad (7\text{-}3)$$

---

[1] As a general principle of electrochemistry, the cathode is the electrode where electrons enter the cell, and the anode is the electrode where electrons leave the cell.

This cell reaction represents the process occurring in electrolytic refining of copper. The theoretical voltage required to activate this cell reaction is infinitesimal, but in actual practice a finite voltage must be applied to the cell to force the reaction, as well as to overcome cell inefficiencies and losses that are described later.

Faraday's Laws

The research of Michael Faraday (1791–1867) led to quantitative relationships describing the amount of electricity passing between electrodes in an electrolyte and the amount of material separated from or deposited on the electrodes. These relationships are defined by two laws which state that

1. The quantity of material released at an electrode is directly proportional to the quantity of electricity passed through the cell.
2. The same quantity of electricity will release the same number of equivalents of a substance.

The electrochemical equivalent of a substance is the number of grams of that substance released by 1 C (A-sec) of electricity. Thus, Faraday's second law says that chemical equivalents (atomic weight/valence) are proportional to electrochemical equivalents. The proportionality factor is Faraday's constant, given by the ratio of the gram equivalent weight to the electrochemical equivalent. This ratio, as determined for silver, is

$$\frac{107.88 \text{ g/g equiv}}{0.0011180 \text{ g/C}} = 96,496 \text{ C/g-equiv.}$$

TABLE 7-1

Electrochemical Equivalent of Metals

Metal	Atomic weight	Valency	Chemical equivalent (atomic wt/ valency)	Amount deposited (g/Ah)	Electro-chemical equivalent (mg/C)
Aluminum	26.98	3	8.99	0.3358	0.0930
Antimony	121.76	3	40.58	1.4949	0.2523
Cobalt	58.94	2	29.97	1.1000	0.3054
Copper	63.54	2	32.77	1.1858	0.3290
Gold	197.00	3	65.67	2.4524	2.041
Hydrogen	1.008	1	1.008	0.0376	0.0104
Iron	55.85	3	18.62	0.6946	0.2894
Lead	207.21	2	103.6	3.8654	0.5368
Nickel	58.71	2	29.35	1.0946	0.0340
Silver	107.88	1	107.88	4.0258	1.1180
Tin	118.70	2	59.35	2.2144	0.6150
Zinc	65.38	2	32.69	1.2192	0.3385

Source: Dennis, W. H., *Extractive Metallurgy*, p. 275, Philosophical Library, New York, 1965.

The quantity of electricity required to release 1 g equivalent weight of any substance is 1 $\mathbf{F}$ or 96,496 C (23066 cal/V-g equiv). Table 7-1 summarizes these electrochemical quantities for several common metals.

Faraday's second law may be expressed in differential form,

$$W = \frac{M\,i\,t}{n\mathbf{F}}\,; \qquad \frac{dW}{M} = dm = \frac{i}{n\mathbf{F}}\,dt, \tag{7-4}$$

where

dW = weight of metal deposited, grams in time interval dt,
M = atomic weight of metal deposited,
dm = number of gram moles of metal deposited in time interval dt,
n = valence of metallic ion in electrolyte,
i = current, amp,
t = time, sec,
$\mathbf{F}$ = Faraday's constant, 96,496 abs C/equiv.

Thus the rate of deposition or flux of material proportional to the electrical current is

$$\frac{dm}{dt} = \frac{i}{n\mathbf{F}}\,. \tag{7-5}$$

Integration of Eq. (7-4) yields an expression for the maximum theoretical quantity of material that can be liberated by a given amount of current, i.e., at 100% current efficiency.

Reversible Cell Potential

Cell potentials depend not only upon the materials making up the electrodes and the overall cell reaction, but also upon the activities of the participating ions in solution and the temperature. Utilizing the definition of net electrical work which is equivalent to free energy,

$$\Delta G = -n\mathbf{F}\epsilon, \tag{7-6}$$

where

n = number of equiv transferred/g-atom,
$\mathbf{F}$ = Faraday's constant,
ϵ = cell emf;

and considering the reaction isotherm for a reaction,

$$aA + bB + \cdots = cC + dD + \cdots, \tag{7-7}$$

where

$$\Delta G = \Delta G^0 + RT \ln \left[\frac{a_C{}^c a_D{}^d \cdots}{a_A{}^a a_B{}^b \cdots} \right], \tag{7-8}$$

then

$$-n\mathbf{F}\epsilon = -n\mathbf{F}\epsilon^0 + RT \ln \left[\frac{a_C{}^c a_D{}^d \cdots}{a_A{}^a a_B{}^b \cdots} \right] \tag{7-9}$$

which yields the Nernst equation,

$$\epsilon = \epsilon^0 - \frac{RT}{n\mathbf{F}} \ln \left[\frac{a_C{}^c a_D{}^d \cdots}{a_A{}^a a_B{}^b \cdots} \right]. \tag{7-10}$$

The standard free energy change is constant at constant temperature, and hence ϵ^0 is constant at constant temperature.

Example Problem 7-1. Calculate the reversible cell potential for the electrolysis of sodium and chlorine (1 atm pressure) from a molten mixture of sodium and calcium chlorides at 600°C. The sodium chloride is present in the salt solution at an activity of 0.4. The standard free energy of formation of liquid pure NaCl at 600°C is −82,830 cal/mole. The liquid pure chloride freezes at 800°C, but the data on "supercooled" liquid can be used to represent the thermodynamics of NaCl in the liquid electrolyte.

Solution. This situation is representative of conditions in the Downs cell during production of sodium and chlorine. Using Eq. (7-10) applied to the cell reaction,

$$\text{NaCl (in solution)} = \text{Na(l)} + \tfrac{1}{2}\text{Cl}_2(\text{g}), \qquad \Delta G^0 = 82,830 \text{ cal/g mole},$$

$$\epsilon = \epsilon^0 - \frac{RT}{n\mathbf{F}} \ln \left[\frac{a_{\text{Na}} \text{Cl}_2{}^{1/2}}{a_{\text{NaCl}}} \right].$$

The sodium and chlorine are released at unit activity, hence

$$\epsilon = -\frac{\Delta G^0}{n\mathbf{F}} + \frac{RT}{n\mathbf{F}} \ln (a_{\text{NaCl}})$$

$$= -\frac{(82,830)}{(1)(23066)} + \frac{(4.575)(873)}{(1)(23066)} \log_{10} (0.4)$$

$$= -3.59 - 0.069 = -3.66 \text{ V}.$$

The reversible cell potential would be −3.59 V if the NaCl were at unit activity. Since the activity of the NaCl is reduced in the salt solution, the reversible cell potential is more negative by 69 mV. However, the actual cell voltage is approximately 6 V.

Standard electrode potentials have been determined for aqueous electrolytes at ambient temperature. By experimentally measuring the potential of metal–metal ion electrodes relative to a standard cell whose potential is defined by the hydrogen electrode (zero potential at 1 atm hydrogen gas and 25°C, the standard potential for several metal–metal ion electrodes and for other cell reactions can be defined. A summary of the standard electrode potentials for several half-cell reactions is presented in Appendix A, Table A-VIII.

Cell Voltages

Cell voltages are comprised of several components: (1) the *IR* drop to overcome the resistance of the electrolyte, bus bars, connections, etc., (2) the reversible cell potential, (3) polarization, and (4) over-voltage. The total cell voltage is the sum of these four potentials.

Ohmic Resistance. From Ohm's law, $E = IR$, the contribution of ohmic resistance of the solution to the cell voltage may be computed. R, the resistance of the electrolyte in question, may be calculated from the equation,

$$R = \rho \frac{L}{A}, \tag{7-11}$$

where

L = average length of flow path,
A = cross-sectional area,
ρ = resistivity of electrolyte.

Reversible Cell Potential. The reversible cell potential can be calculated from Eq. (7-10).

Polarization. A polarization potential arises from a concentration difference between ions in the neighborhood of the electrode and those in the bulk of the electrolyte solution. The concentration of ions in the bulk phase can be measured directly, while the concentration at an electrode surface often can be estimated, thus permitting a calculation of the concentration polarization potential. The basis for the relationships by which the polarization potential may be computed are shown in the following manner.

Consider, e.g., a cell of the type,

$$Cu(a_{cu} = 1) \mid Cu^{++}(a_1) \mid Cu^{++}(a_2) \mid Cu(a_{cu} = 1), \tag{7-12}$$

consisting of two pure copper electrodes immersed in solutions of copper-containing electrolyte in which the activities of the copper ions are equal to a_1 and a_2, respectively. The electrode reaction on the left side is

$$Cu(pure) = Cu^{++}(a_1) + 2e^-, \tag{7-13}$$

with ϵ given by

$$\epsilon_{13} = \epsilon^0{}_{Cu} - \frac{RT}{2\mathbf{F}} \ln (a_1). \tag{7-14}$$

Again, for the reduction on the right side

$$Cu^{++}(a_2) + 2e^- = Cu(pure), \tag{7-15}$$

$$\epsilon_{15} = -\epsilon^0{}_{Cu} - \frac{RT}{2\mathbf{F}} \ln \left(\frac{1}{a_2}\right). \tag{7-16}$$

The overall cell reaction may be represented as

$$Cu^{++}(a_2) = Cu^{++}(a_1) \tag{7-17}$$

and the cell emf is

$$\epsilon_{17} = (\epsilon_{13} + \epsilon_{15}) = -\frac{RT}{2\mathbf{F}} \ln \left(\frac{a_1}{a_2}\right). \tag{7-18}$$

Since a_1 (concentration near the anode) is greater than a_2 (concentration near the cathode), the sign of ϵ_{17} will be negative. This decrease in cell potential indicates that any applied potential would have to increase, i.e. the polarization potential opposes the applied potential. The decrease in cell emf reflects the fact that at the anode the copper ions are being dissolved in a solution whose concentration is greater than the bulk electrolyte liquid and at the cathode copper ions are being removed from solution at a concentration level which is below that of the bulk electrolyte liquid.

A practical example of such a concentration cell may exist in an electrolytic processing operation if the solution near the anode builds up in metallic ion concentration or if there is a depletion of ions in the solution at the cathode. The case of a build-up of ionic concentration at the anode and a simultaneous depletion at the cathode is illustrated in Fig. 7-1. The concentration difference between the solution near the electrodes creates a potential difference that must be overcome and acts to increase the total cell voltage.

From Eq. (7-18), it is evident that the emf of this concentration cell arises from a transfer of copper ions from the solution where the ionic activity is a_1 to the solution where the ionic activity is a_2. Furthermore, emf depends only on the ratio of the activities of copper ion at the electrode surfaces and not on the absolute value of the ionic activity. In the final cell emf, Eq. (7-18), ϵ^0 does not appear. This is true of all concentration cells. It may be concluded therefore, that for concentration cells, ϵ^0 is 0, and the emf equation takes on the simplified form,

$$\epsilon = -\frac{RT}{n\mathbf{F}} \ln \left(\frac{a_1}{a_2}\right). \tag{7-19}$$

Since in cells of the type under discussion, a_2 and a_1 refer to the activities of the

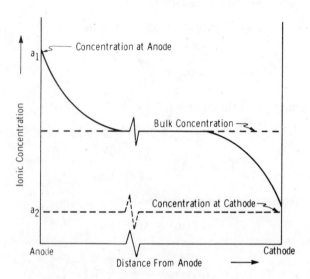

Fig. 7-1. Concentration Polarization in Electrolytic Cell.

metallic ion in solution, we may assume that for dilute solutions, these are essentially equal to or are proportional in the same manner to the concentrations, c_1 and c_2. Thus, Eq. (7-19) may be written approximately as

$$\epsilon \simeq -\frac{RT}{n\mathbf{F}} \ln \left(\frac{c_1}{c_2}\right). \tag{7-20}$$

The equation in this form may be used to calculate the concentration polarization potential of a cell where c_1 refers to the surface concentration of the metallic ion in solution at the anode and c_2 refers to the surface concentration at the cathode. Since we are interested in the ratio of the two concentrations, any consistent units may be used to express them.

Over-Voltage. In general, it is not possible to compute accurately the cell voltage contribution arising from gas over-voltage. The over-voltage is related to the potential required for evolution of gas on an electrode. The over-voltage has been found to increase with increasing current density and to decrease with increasing temperature.

The over-voltage is also a function of the nature of the electrode. The hydrogen over-voltage in ambient temperature, aqueous electrolyte cells for refining lead, mercury, zinc, and tin is high. These metals are poor catalysts for the reaction,

$$H_2 = 2H^+ + 2e^-. \tag{7-21}$$

This over-voltage, however, is quite low for platinum, iron, and silver—metals which are good catalysts for the above reaction.

Other factors influencing the over-voltage are roughness of surface and electrode impurities. A rough surface has a low over-voltage, presumably because of the ability of the surface to act as a nucleating agent for the formation and evolution of the gas. And the hydrogen over-voltage tends toward the over-voltage of those impurities which are present in the electrodes. The source of impurities may lie not only in the electrodes themselves, but in traces of metal ions in the electrolyte which may plate out at the cathode.

7-2 Electrolysis of Fused Salts

An important process for the production of reactive metals is the electrolysis of fused salts. Such processes are carried out at relatively high temperatures, and consequently, differ in some aspects from electrolytic processes carried out in aqueous solutions, even though the principles involved are essentially the same for both. The engineering approach to processes used for electrolytic purification, or for the recovery of metals from leaching solutions are parallel to those for fused salt electrolysis and are described in the following sections.

Since highly electropositive metals tend to react at high temperatures with most materials, common pyrometallurgical processes cannot be used for smelting them. A straightforward method for producing these metals is first to purify a salt of a reactive metal and then electrolyze the salt for recovery of the metallic element. This technique is used for the production of aluminum, magnesium, beryllium, cerium, lithium, sodium, potassium, and calcium. It also could be used to produce less reactive materials, although usually without economic advantage.

The electrolysis of a fused salt bath requires a container for the bath equipped with a device for collecting the metal produced, and a cathode and an anode to deliver the electric power for decomposition. A direct current is passed through the molten bath. The bath may be the pure salt of the metal or the salt of a more reactive metal into which a compound containing the desired metallic element is dissolved. Examples of these two cases would be the production of sodium from pure liquid sodium chloride, and the production of aluminum from a fused cryolite bath (AlF_3–$3NaF$) in which a small amount of alumina is dissolved, respectively.

The salt or oxide of the element to be produced must be pure. If less electrochemically active impurities were present in the electrolyte, they would be reduced and migrate to the cathode, thus contaminating the element being produced. Furthermore, displacement reactions could occur at the cathode, causing transfer of impurities from the electrolyte into the reduced metal. Consequently, one of the critical steps in fused salt electrolysis is purification of the electrolyte. For example, the Bayer process for producing pure alumina from bauxite ores has permitted the electrolytic production of high purity aluminum on a commercial basis.

Some of the engineering aspects and operating characteristics of fused salt electrolytic cells are presented in the following examples. Descriptions of commercial production cells are given for aluminum, magnesium, and sodium.

Aluminum

Aluminum is produced by electrolysis of alumina from a bath of fused cryolite. This process was developed independently in 1886 by Hall in the United States and Heroult in France. Cryolite melts just below 1000°C, and because of a depressed freezing point in the dilute solution region, is able to dissolve from 4–20 wt% alumina while remaining liquid at temperatures above 980°C. Beyond a concentration of 20 wt% alumina, the melting point rises rapidly. Aluminum production cells use commercial cryolite, which consists of aluminum fluoride and sodium fluoride in a 3–1 molar ratio. Potassium or calcium fluorides also can be substituted for the sodium fluoride. Since these fluoride solutions are highly corrosive, successful development of the electrolysis technique depended upon coping with this material's corrosive nature. The problem was solved by using an iron vessel lined with carbon as a cathode and a carbon electrode as an anode, a design still in use today.

The commercial production of high purity aluminum (99+% or pure) requires high purity alumina. The alumina for use in the Hall process is produced from bauxite ores using the Bayer process (see Hydrometallurgy).

Hall (Heroult) Process. A cross-sectional view of an aluminum reduction cell is shown in Fig. 7-2. This cell consists of a steel shell with a thin lining of insulating brick. Inside the refractory lining is a carbon lining several inches thick which serves as the cathode. The anodes are carbon and project through a crust of electrolyte frozen at the surface. Modern cells use the continuous Soderberg electrode, a single anode in which a mixture of carbon and pitch is continuously fed from above through a rectangular metal box. The heat of the cell bakes the mixture and forms a continuous graphite anode.

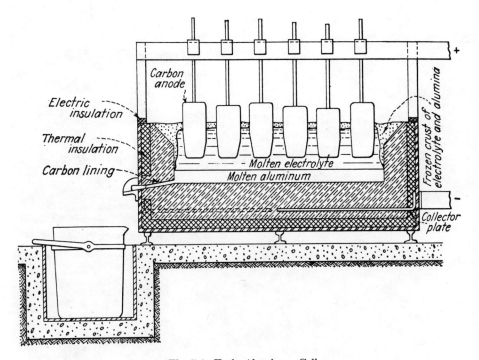

Fig. 7-2. Early Aluminum Cell

Source: Electrochemical Engineering, C. L. Mantell, McGraw-Hill Book Company, Inc., New York, 1960, p. 379.

The alumina concentration in the electrolyte ranges from $2\frac{1}{2}$–20 wt%. The cell temperature is in the vicinity of 900–1000°C. The cell operating voltage is 6–$7\frac{1}{2}$ V, and the current density ranges up to 1000 A/ft² of cathode. At 100% current efficiency this current density would produce $\frac{3}{4}$ lb metal/hr-ft² of cathode; the actual current efficiency in the Hall cell is 80–90%. Anode consumption ranges 0.6–0.8 lb of carbon/lb of alumina, and the alumina requirements are approximately 2 lb of alumina per pound of aluminum. The total amperage for an aluminum cell varies with the size and cell design but may range up to 50,000 A or higher. An operating aluminum cell is shown in Fig. 7-3.

Cell Potential. The reversible cell potential can be calculated using the principles outlined in the previous chapter. At the cathode, pure aluminum (99+%) is deposited as a liquid. At the anode, pure oxygen gas is liberated, which immediately reacts with the carbon anode material giving off carbon monoxide and some carbon dioxide gas. These reactions may be summarized as

$$Al_2O_3(s) = 2Al(l) + -O_2(g), \tag{7-22}$$

$$\Delta G^0 = +401,400 - 77.0\ T\ \text{cal/mole}; \tag{7-23}$$

$$C(gr) + \tfrac{1}{2}O_2(g) = CO(g), \tag{7-24}$$

$$\Delta G^0 = -28,200 - 20.16\ T\ \text{cal/mole}. \tag{7-25}$$

Fig. 7-3. Electrolytic Cells for Aluminum Production at Badin Smelting Works
Courtesy of Aluminum Company of America

A craneman at Alcoa's Badin (N.C.) Smelting Works is manipulating a large crucible
into position to tap one of the electrolytic cells containing molten aluminum. Assisting
him is a pot serviceman. One spout of the crucible is lowered into the pot, then, through
vacuum, molten metal is drawn into the crucible from the pot bottom.

At cell operating temperatures the CO/CO_2 mixture in equilibrium with graphite is
nearly pure CO (actual CO/CO_2 ratios are on the order of 0.5–1). In calculating the
theoretical minimum voltage at which aluminum will be deposited at the cathode in
the Hall process, it may be assumed that the anode gas is essentially pure carbon
monoxide and that the salt bath is saturated with alumina. Calculating the free
energy of the sum of the reactions given in Eqs. (7-22) and (7-24),

$$Al_2O_3(s) + 3C(gr) = 2Al(l) + 3CO(g), \qquad (7\text{-}26)$$

$$\Delta G^0 = 316{,}800 - 137.48\ T\ \text{cal/mole.} \qquad (7\text{-}27)$$

The decomposition potential, ϵ, (in this case, ϵ^0, since reactants and products are in their standard states) is related to the free energy of reaction by the relation,

$$\Delta G^0 = -n\mathbf{F}\epsilon^0. \tag{7-28}$$

For the above case, at 900°C, the result is

$$\epsilon^0 = -1.13 \text{ V}. \tag{7-29}$$

There is still considerable speculation about how much effect the oxidation of the carbon anode has on the reaction voltage of the aluminum cell. The point in question is the influence of the energy of formation of the carbon monoxide and carbon dioxide formed at the anode.

The reversible cell potential at 900°C, (assuming that the carbon–oxygen reaction at the anode does *not* influence the cell potential) is -2.24 V for the decomposition of alumina into liquid aluminum and gaseous oxygen. However a typical operating cell potential is 6 V or more. This additional voltage is due to over-voltage at the anode and ohmic drop through the bus bars, cathode lining, electrolyte, and anode. The over-voltage at the anode is a function of current density, while the resistance of the electrolyte varies with alumina concentration. In addition the decomposition voltage will increase as the alumina activity decreases, i.e., as the concentration of alumina decreases from the saturation value. Optimum operation of the reduction cell requires that changes be made in operating conditions as the alumina concentration changes.

Another interesting aspect of Hall cell operation is the "anode effect," a sharp increase in voltage across the cell which raises the cell potential from about 6 V to something over 30 V. The "anode effect" occurs when the electrolyte does not wet the anode, creating a gas-filled gap between the anode and the electrolyte. The voltage increase is localized in this region and the cell requires stirring or the addition of alumina to correct the condition.

Magnesium

Approximately 85% of the world's magnesium is produced electrolytically and the remaining 15% is produced by thermal reduction processes. Early production of magnesium by electrolysis was carried out from a fused salt bath in a manner similar to the Hall process for producing aluminum. This process is not used at the present time, and magnesium is produced electrolytically from anhydrous $MgCl_2$ or from $MgCl_2$ containing about 25% water.

The Dow Chemical Company, primary producer of magnesium in the United States, developed its first magnesium processing plant at Midland, Michigan, using brine from wells. Electrolysis of the solution of mixed salts magnesium chloride, sodium chloride, and calcium chloride yields magnesium metal and chlorine gas. A typical electrolyte composition is 25% $MgCl_2$, 15% $CaCl_2$, and 60% NaCl. A schematic diagram of the Dow Magnesium Cell is shown in Fig. 7-4. The temperature in the cell is maintained between 700 and 750°C by controlled firing of natural gas. The graphite anodes are inserted from the top. The cathodes in the Dow Cell are coiled iron liners spaced very closely and symmetrically around each

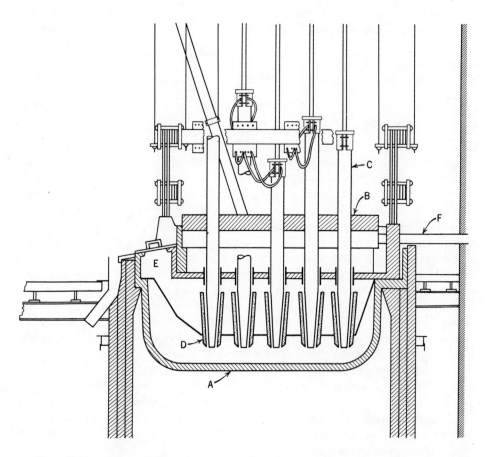

Fig. 7-4. Diagram of Dow Magnesium Cell—The steel container (A) is equipped with ceramic cover (B) through which pass graphite anodes (C). The magnesium is deposited on the cathodes (D) and is diverted as it rises into collection sump (E). The chlorine is withdrawn through vent (F).

Courtesy of Dow Chemical Company

graphite anode. The cell voltage ranges from 5–8 V, and typical Dow Cell currents may range up to 100,000 A. The power requirement is approximately 10 kWhr/lb of magnesium produced.

The magnesium is lighter than the salt electrolyte and after being released on the cathode shell, it floats to the surface where it is trapped in inverted troughs. Magnesium is recovered as a liquid and in subsequent handling it is protected from atmospheric oxidation with a chloride flux or by molten sulfur. Chlorine gas is evolved at the anodes and is collected for use in other chemical operations, principally in the production of hydrochloric acid.

The present primary source of magnesium is from seawater. At Freeport, Texas, the Dow Chemical Company, using calcined oyster shells as a source of calcium hydroxide, precipitates magnesium from seawater as magnesium hydroxide. The magnesium hydroxide is treated with HCl and the $MgCl_2$ slurry is dried to

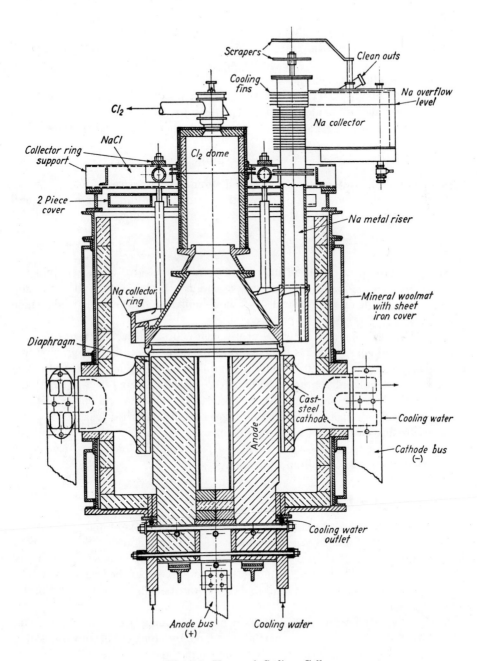

Fig. 7-5. Knapsack Sodium Cell

Source: *Electrochemical Engineering*, C. L. Mantell, McGraw-Hill Book Company, Inc., New York, 1960, p. 413.

$MgCl_2 \cdot 1.5\ H_2O$. It is not necessary to dry the $MgCl_2 \cdot 1.5\ H_2O$ further, and this material is added directly to the chloride electrolyte of the magnesium cell. Chlorine gas, HCl, and water vapor are liberated at the anodes. The water vapor driven out of the electrolyte attacks the graphite electrodes, whereas in an anhydrous magnesium chloride cell operation, the carbon anodes are permanent since they are not attacked by the liberated chlorine gas.

The magnesium industry in Norway employs a chlorination process to produce $MgCl_2$ using magnesium oxide [$Mg(OH)_2$ is precipitated from sea water and dried to MgO] and carbon, which are heated in an atmosphere of chlorine gas according to the reaction,

$$2MgO + C + 2Cl_2 = 2MgCl_2 + CO_2. \tag{7-30}$$

The magnesium chloride is treated in a cell similar in principle to that shown in Fig. 7-4, except that the cell is of refractory construction and utilizes anode and cathode compartments to keep the liquid magnesium and chlorine gas separated.

Sodium

Metallic sodium is produced in a Castner cell by electrolysis of sodium hydroxide or in a Downs cell by electrolysis of sodium chloride dissolved in calcium chloride. The design of the two cells is fairly similar. A Downs cell is shown in Fig. 7-5.

The Downs cell operates at temperatures near 600°C. The cell current furnishes sufficient thermal energy for the cell to maintain an electrolyte of approximately 60% calcium chloride and 40% sodium chloride in the liquid state. Sodium is collected at wire gauze diaphragms and the chlorine is collected in a dome at the top of the cell. Current efficiencies are in the area of 75–80%, and cell amperages may range as high as 50,000 A. The advantages of sodium chloride electrolysis over sodium hydroxide electrolysis lie primarily in use of a cheaper raw material and production of chlorine gas as a by-product. Approximately 5 kWhr of electricity are expended per pound of sodium recovered.

7-3 Electrolytic Refining

The electrolysis of solutions containing metallic ions often is employed to refine metals or to recover them from the products of leaching operations. Solutions containing metallic ions are electrolyzed to deposit the element in metallic form. By taking advantage of the difference in electrode potentials between various metals, it is often possible to separate several metallic elements from the same solution by electrolytic means.

Electrolytic refining usually is carried out by passing a current through an ionic solution between an anode partially refined by some other process and a cathode. The anode is electrochemically dissolved in the electrolyte to form ions of the desired element, which are transported through the electrolyte to the cathode where they are deposited as atoms, thus effecting the refining of the anode material. A design engineer dealing with electrolytic processes is interested in the power and voltage requirements, plating rates, effects of impurities, size of the operation (i.e., the number of tanks, the volume of solution, number of cathodes and anodes), and other factors that influence the economics of the operation.

Electrolytic refining is used throughout the metallurgical industry, but it is the principal refining method used for copper, nickel, and much of the lead produced. This type of refining also permits recovery of precious metals and other impurities as by-products. Electrolytic refining is practiced to some extent for purification of other metals such as iron, cobalt, tin, silver, and gold.

Copper Refining

Electrolytic refining yields the very pure copper used by the electrical industry. In addition, gold, silver, and the platinum metals are recovered from the crude copper supplied to the electrolytic cell. The primary source of selenium is from the anode slimes of copper refining cells, while nickel, arsenic, antimony, tellurium and other metals also are recovered as impurities. Based on average analyses, 1 million lb of crude copper may contain on the order of 400 oz of gold, and 25,000 oz of silver, as well as a few ounces of palladium and platinum. Concentrations are quite low, but because of the large volume of copper handled, this source of precious metals is extremely significant. About 75% of the total world production of silver comes as a by-product of copper and lead refining. A flow sheet for a copper refinery is presented in Fig. 7-6.

Anode Preparation. Blister copper, the product of a converter or a blast furnace, is usually porous and brittle if directly cast into anodes. If further refining to improve physical properties is necessary, the blister copper is remelted in large reverberatory furnaces and subjected to oxidizing conditions to remove various impurities, particularly the residual iron and sulfur. The oxygen often is supplied from air blown through steel pipes into the furnace below the surface of the copper. The oxygen oxidizes some of the copper and also reacts with iron and sulfur, with the iron and copper oxides combining with silica present to form a slag. The charge then is subjected to a reducing stage in which the oxygen dissolved in the copper is removed by reaction with a reducing agent, carbon or a hydrocarbon gas. At this point the copper is cast into an anode weighing 500–700 lb and measuring about 3 ft square; the anodes contain less then 1% total impurities. Figure 7-7 is a picture of copper anodes, while a Walker casting machine for continuous production of anodes is shown in Fig. 7-8.

Refining Systems. Several basic electrical hook-ups, each with its own advantages, are possible in electrolytic process equipment. The two principal systems, parallel (multiple) and series, are shown in Fig. 7-9. A parallel system is one in which the electrode plates within each tank are connected in parallel so that the voltage drop across the entire tank is the same as that between any pair of plates in the tank. The advantage of this system is that there is a smaller voltage drop than in the series connections and, therefore, fewer losses due to short circuiting.

A series connection is one in which the electrodes within the tanks are connected in series. Consequently, the total voltage drop across the cell is the sum of the voltage drops between each electrode. In the series system, each plate acts as an anode on one side and a cathode on the other, whereas in the parallel system, each

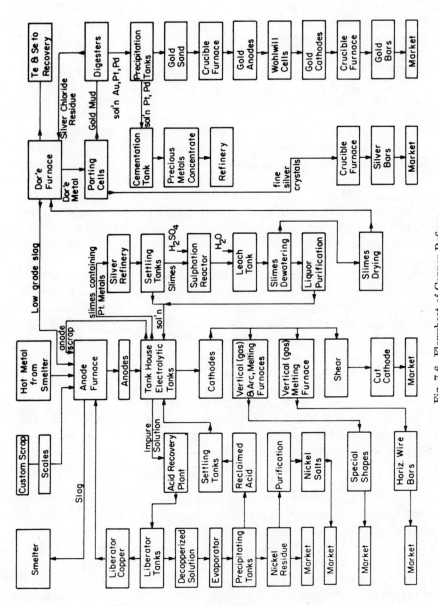

Fig. 7-6. Flowsheet of Copper Refinery

Courtesy of International Nickel Company of Canada

Fig. 7-7. A Stack of Copper Anodes at Anoconda

Source: C. Hayward, *Outline of Metallurgical Practice*, D. Van Nostrand Company, Inc.,
      New York, 1952, p. 95.

electrode behaves wholly as either a cathode or an anode, the reaction taking place
on both surfaces. Thus, less space is required for the series arrangement, and there
is no necessity for preparing special starting sheets. Furthermore, the tanks themselves are usually connected in series with the total number in the circuit being
determined by the supply voltage and power of the generator.

Although the pure series system is used in a few large refineries, the system
that is most commonly employed in metallurgical plants is the multiple system. In
this system, the plates within each tank are connected in parallel, but the tanks
themselves are connected in series. In the multiple system, cathode starting sheets
are required. These sheets are approximately $\frac{1}{16}$ in thick and the same shape as
the anodes. The cathode starting sheets are prepared in special stripper cells where
a copper deposit is built up on oiled copper blanks and subsequently stripped off
for starting the cathode in a refining cell. Hard-rolled copper blanks often are used
as cathode starting sheets.

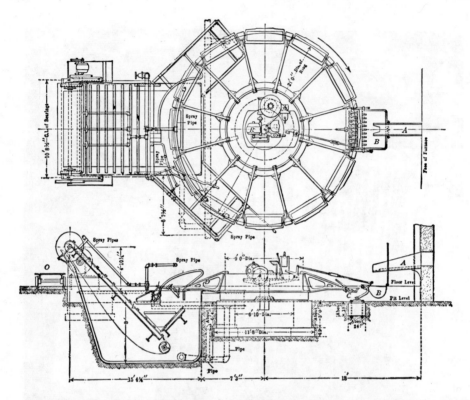

Fig. 7-8. Walker Casting Machine for Copper—Note the furnace spout *A* delivering metal to tipping ladle *B* which fills the molds *C*. The wheel is rotated in a clockwise direction by a motor controlled by an operator. When solid the copper passes under a water spray. The molds are automatically dumped at the left and the copper falls into a water bosh, from which it is withdrawn by endless metallic belt *L* and discharged to car *O*. The inverted empty mold is sprayed with clay, bone ash or special wash, righted and again passes under the ladle. Various details are shown in the figure.

Source: C. Hayward, *Outline of Metallurgical Practice*, D. Van Nostrand Company, Inc., New York, 1952, p. 93.

Tank House Practice. A typical tank house is shown in Fig. 7-10. The tanks are usually constructed of concrete and are lined with antimonial lead (3–6% antimony), but considerable research is being carried out on plastic and fiberglass linings for tanks.

The tanks contain from 20–40 cathodes and anodes in most instances. The electrolyte is a copper sulfate solution (40–50 g/liter) containing 100–200 g/liter of sulfuric acid. The electrolyte is circulated to permit temperature control and purification and also to minimize concentration polarization at the surfaces of the electrodes. Purification of the electrolyte is required since soluble impurities such as iron, zinc, nickel, and arsenic tend to accumulate in the solution, and there may be some build-up of copper concentration. Electrolysis often is used in a separate system to purify the electrolyte.

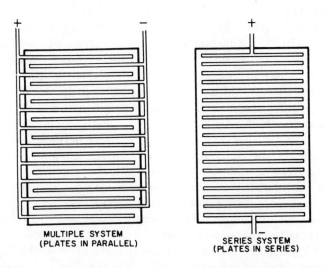

Fig. 7-9. Electrical Connections for Electrolytic Refining Cells

Fig. 7-10. Tank Room in an Electrolytic Copper Refinery

Source: J. Newton, *Extractive Metallurgy*, John Wiley and Sons, Inc., 1959, New York, p. 459.

In a typical large refinery operation, the electrolyte is circulated at a rate of approximately 4–5 gal/min. The tanks are 10–12 ft long, 3–4 ft wide, and about 4 ft deep. The electrolyte is maintained at a temperature of 45–60°C, since the resistance of the solution to passage of current accounts for over one-half of the total cell resistance, and this resistance decreases with increasing temperature. Glue and other additives are added to the electrolyte to improve the physical characteristics of the cathode deposit. Other chemical additives may be put into the electrolyte to depress the deposition of impurity ions.

The current density on the cathode usually ranges between 20 and 30 A/ft². Low current densities offer the advantages of a better grade of cathode copper, more efficient recovery of precious metals, and a lower power cost of electrolysis. On the other hand, a larger plant then is required for a given copper production; however, an economic optimum exists for all these cost factors.

The voltage per tank varies with the current density and the resistivity of the solution. The theoretical cell voltage is zero, but the usual plant voltage is on the order of 0.2–0.4 V per tank. The cathodes are changed every 10–15 days, and the anode life is approximately twice that of the cathode. The anodes are removed from the tank when approximately 10–15% of the original weight remains; this material is returned to the anode furnace for remelting. The tanks are periodically drained and the slimes washed out, filtered, and processed for the recovery of selenium, tellurium, and precious metals.

Example Problem 7-2. How much copper is plated out in a copper refining tank operating at 20 A/ft² (current density at cathode). The cell is arranged in a parallel circuit and contains 38 anodes and 39 cathodes. The cathodes are 36 in. square and current efficiency for the tank is 90%.

Solution. On a per day basis for 1 ft² of operating surface

$$\text{Copper Deposited} = (0.90)(20)(24 \times 3600)\frac{(63.6/2)}{(96500)}$$

$$= (\text{current efficiency})\left(\frac{A}{ft^2}\right)(\text{time, sec})\left(\frac{g/\text{equiv wt}}{A\ \text{sec/equiv wt}}\right)$$

$$= 512\ \text{g Cu/ft}^2\ \text{day}$$

With this cell arrangement the area is

$$(38)(2)(3 \times 3) = 684\ \text{ft}^2$$

$$\frac{(512)(684)}{453.6} = 770\ \text{lb Cu/day}.$$

Example Problem 7-3. Copper is electrolytically refined at 60°C using an electrolyte containing 42 g/liter of copper. The voltage drop from anode to cathode is 0.03 V. At what silver ion concentration in the electrolyte could silver be plated out at the cathode?

Solution. Considering the half-cell reactions, from Table A-VIII,

$$\text{Ag(s)} = \text{Ag}^+ + e^- \qquad \epsilon^0 = -0.7990\ \text{V},$$

$$\text{Cu(s)} = \text{Cu}^{++} + 2e^- \qquad \epsilon^0 = -0.3411\ \text{V}.$$

The half-cell potential for the oxidation of copper is

$$\epsilon = \epsilon^0 - \frac{RT}{n\mathbf{F}} \ln \left[\frac{a_{Cu^{++}}}{a_{Cu}} \right].$$

The concentration of copper ion is $(42/63.6) = 0.66$ g-atom/liter. Substituting concentrations for activities and noting that the metals are at unit activity,

$$\epsilon_{half-cell} = -0.3441 - \frac{(1.987)\,(333)\,(2.303)}{(2)\,(23066)} \log_{10} (0.66)$$

$$= -0.3441 + 0.0060$$

$$= -0.3381.$$

The half-cell potential for the reduction of silver is

$$\epsilon = -\left(\epsilon^0 - RT \ln \left[\frac{a_{Ag^+}}{a_{Ag}} \right] \right).$$

The overall cell potential is

$$\epsilon_{cell} = \epsilon_{anode} + \epsilon_{cathode},$$

$$-0.03 = -0.3381 + 0.7990 + \frac{RT}{n\mathbf{F}} \ln (a_{Ag^+}).$$

Substituting concentration for activity,

$$-0.4909 = \frac{(1.987)\,(333)\,(2.303)}{(1)\,(23066)} \log (C_{Ag^+}),$$

$$\log (C_{Ag^+}) = -\frac{(0.4909)\,(23066)}{(1.987)\,(333)\,(2.303)} = -7.43,$$

$$C_{Ag^+} = 3.7 \times 10^{-8} \text{ g-atom/liter.}$$

This low concentration of silver is not reached because of the presence of a small amount of chloride ion in the electrolyte that precipitates any electrolyzed silver as AgCl. The silver also is found in the anode mud as Ag_2Se.

Treatment of Insolubles. The "anode mud" or slime accumulating in a copper refining cell may contain slightly less than 1% gold, from 5–25% silver, and from 15–40% copper. The remainder consists of selenium (often 5–20%) and small concentrations of tellurium, arsenic, antimony, lead, and traces of nickel and other metallics. A typical slime analysis is given in Table 7-2.

Special treatment of the "anode mud" is carried out, particularly for the recovery of silver, gold, platinum, palladium, selenium, and tellurium. The raw slime is washed, filtered, and roasted in an oxidizing atmosphere at 700°F. This converts the copper to copper oxide which is leached out with sulfuric acid. The slime then is filtered, and melted and refined in a small "Dore" reverberatory furnace. During refining in the Dore furnace, the lead is oxidized to a slag, and antimony and some

TABLE 7-2

Analysis of Blister Copper, Anodes, and Cathodes

Analysis	Blister	Anode	Cathode
Copper	99.0140%	99.4220%	99.9800%
Oxygen	0.6910%	0.1540%	
Sulfur	0.0185%	0.0026%	
Arsenic	0.0464%	0.0435%	0.0010%
Antimony	0.0350%	0.0317%	0.0010%
Lead	0.0039%	0.0042%	Nil
Selenium	0.0184%	0.0176%	0.0003%
Tellurium	0.0122%	0.0122%	0.0002%
Nickel	0.0140%	0.0146%	0.0010%
Silver	63.31 oz/ton	62.44 oz/ton	0.30 oz/ton
Gold	0.265 oz/ton	0.265 oz/ton	

Log of Tank House (One Tank)

Total weight of anodes	14,065 lb
Weight of anode scrap	1,678 lb
Weight corroded from anodes	12,387 lb
Total electrolysis, 31 days, 17 hr, 55 min	
Weight of deposited copper	11,837 lb
Average amp during run	6,700
Current efficiency, %	91.3
Slime	51 lb 13 oz
Pounds of slime per ton of copper refined	8.37

Slime analysis

Copper	18.80%	
Silver	51.008%	14,877.73 oz/ton
Gold	0.2171%	63.32 oz/ton
Arsenic	3.90%	
Antimony	8.04%	
Lead	1.06%	
Selenium	4.37%	
Tellurium	2.99%	
Nickel	0.05%	

Source: C. L. Mantell, *Electrochemical Engineering*, p. 165, McGraw-Hill, New York, 1960.

selenium and tellurium are volatilized and recovered as flue dust. Selenium and tellurium are removed primarily by alkaline fluxes. The product, Dore metal, often contains over 98% silver, approximately 1% gold (sometimes as high as 10%) and a small amount of other metallic impurities. The crude silver is cast into anodes and electrolyzed to remove silver in Thum or Moebius electrolytic cells. The electrolyte in these cells is a solution of silver nitrate. These two refining systems, shown in Fig. 7-11, employ canvas diaphragms to retain the gold slime and prevent mixing with the cathode deposit. The silver deposit is very pure, often 99.99% silver. The Wohlwill process employs a hot acidic gold chloride solution for electrolyzing gold anodes which are prepared from the slime of the Thum or Moebius cells. Since platinum and palladium are soluble in the Wohlwill process electrolyte, they are permitted to accumulate and then are precipitated with ammonium chloride.

Refining of Copper Cathodes. The cathodes produced in the electrolytic refining of copper are of extremely high purity but contain some sulfur. This impurity is removed by remelting the copper, usually in an electric furnace.

Oxygen content also is important since it controls the "set" (shape of cast surface) of the copper. The oxygen concentration in the cathodes is very low, of the order of 10 ppm or less. Usually no care is taken to keep the oxygen concentration low, and often it is necessary to leave 0.03–0.05 wt% oxygen in the finished copper to produce a casting with a level surface, especially if bars for wire drawing are cast from the remelted cathodes. Alternatively, very low oxygen contents are often desired.

Copper, free from oxygen, is used in electrical products. The very low concen-

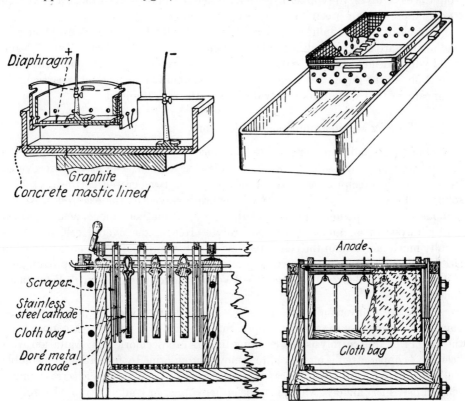

Fig. 7-11. Thum and Moebius Cells for Separating Silver from Other Noble Metals

Source: *Electrochemical Engineering*, C. L. Mantell, McGraw-Hill Book Company, Inc., New York, 1960, pp. 168–9.

trations of oxygen can be reached by removing oxygen from the molten copper in several ways: (1) by using carbon and casting under reducing conditions, (2) by using a deoxidizing agent other than carbon, usually phosphorus, or (3) by extruding directly an electrolytically deposited brittle cathode. The removal of oxygen with carbon or reducing gases, and casting under reducing conditions produces oxygen-free, high-conductivity copper.

Nickel and Lead Refining

Nickel is electrolytically refined in a nickel sulfate solution containing about 50 g/liter nickel, 20 g/liter boric acid and 50 g/liter sodium chloride. The anodes are approximately 95% nickel and contain 2-$2\frac{1}{2}$ wt% copper together with small amounts of iron and sulfur. The current density ranges from 12–15 A/ft². Precautions are taken to avoid contamination of the cathode with the principal impurities of the nickel anode, copper, and iron. As with silver recovery, each cathode is enclosed in a canvas diaphragm, and the electrolyte is circulated to minimize iron and copper contamination of the cathode. Then the copper and iron are removed from the electrolyte, the copper by "cementing out" on an iron surface, and the iron by oxidation with air precipitating it as a hydrate.

The electrolytic refining of lead is carried out by the Betts process, in which cast lead anodes are electrolyzed in a solution containing lead fluosilicate ($PbSiF_6$) and free hydrofluosilicic acid (H_2SiF_6). The process produces a very pure lead, and practically all the impurities accumulate as anode mud. This method is particularly significant in that it permits removal of bismuth from the lead.

Electro-Winning Metals from Leaching Solutions

Many low-grade ores are treated for recovery of metallics by leaching the mineral with an acidic solution which then is electrolyzed. The metal is recovered by electrolytic precipitation and the electrolyte is regenerated as fresh leaching solution. Two distinctly different reactions are involved at the electrodes. At the cathode, the metallic ion is reduced to the metal. At an inert anode, acid is formed by hydrolysis of an anion in the electrolyte. Hence, the overall cell voltage is usually much larger than the cell potentials of a few tenths of a volt encountered in electrolytic refining. In electro-winning of copper, the cell voltage is approximately 2 V, whereas it is 3–3.7 V for zinc, 2.5–3 V for antimony, and about 5 V for manganese.

There are many copper ores found in the world that are very low in metallic value, containing less than 0.5% copper. Sulfuric acid is used to leach copper from the ore. Prior to electrolysis the leaching liquors are dechloridized. Agitation with "cement" copper often is employed to remove the chlorine by formation of insoluble $CuCl$. This treatment also reduces the ferric ions to ferrous ions. Electro-winning cells employ inert anodes, which may be either copper or lead alloys for electrolysis in sulfate solutions.

The control of electrolyte composition in electro-winning is particularly important. This may be emphasized by considering the deposition of zinc from sulfate solutions. The feed to the electrolytic cells may contain about 100 g of recoverable zinc/liter, and the cell would correspondingly discharge over 100 g of sulfuric acid/liter of electrolyte. The presence of antimony in the electrolyte is particularly troublesome, since a concentration of as low as 1 mg/liter will result in a poor cathode deposit. Arsenic and cobalt cause similar problems. Other electrolyte reactions, such as oxidation of ferrous or reduction of ferric ions, often cause fluctuations of current efficiency and may cause redissolution of the deposit. The presence of iron in zinc ores requires a high acid concentration in the leaching solution to dissolve the zinc ferrite ($ZnFe_2O_4$). The use of concentrated acid results in the solution of more iron, which precipitates arsenic and antimony. Copper, cadmium, and cobalt often are removed from zinc leaching solutions by agitating the electrolyte with zinc dust at a temperature of about 80°C. These metals are precipitated by the zinc, so that the purified solution is relatively free of these elements. The starting cathodes for zinc recovery are made of aluminum sheet with an aluminum conductor bar welded to them, and the cathodes are stripped approximately once a day. Current densities range from 50–100 A/ft².

High purity (99.98%) manganese is produced by electro-deposition. The potential of manganese in the electromotive force series makes it difficult to deposit electrolytically, and requires considerable care in controlling electrolyte composition and temperature. Preliminary concentration of manganese ores usually is made by gravity methods. The ores are leached with sulfuric acid, and the leaching solution is purified by depositing copper on iron. Iron and arsenic are removed by oxidizing with air and adding MnO_2. Copper and nickel are removed by addition of potassium or sodium ethyl xanthate. Electrolysis is carried out with lead anodes and stainless steel cathodes separated by canvas diaphragms. Current density is on the order of 20 A/ft². Current efficiency is only 50–60% and the electrical energy required ranges from 3.5–5.5 kWhr/lb manganese. This is 2–3 times the requirement for electro-deposition of copper.

Electro-winning is a very effective and efficient means for recovering metals from low-grade ores. The electrolytic refining of copper, lead, and nickel is carried out primarily for recovery of precious metals. In the case of copper, electrolytic refining is the only practical method for removing gold and silver. However, in the case of lead, the Parkes process represents an alternative approach.

Bibliography

Blum, W. and G. Hogaboom, *Principles of Electroplating and Electroforming*, McGraw-Hill, New York, 1930.

Butts, A., *Metallurgical Problems*, McGraw-Hill, New York, 1943.

Bray, J. L., *Non-Ferrous Production Metallurgy*, 2nd ed. John Wiley, New York, 1947.

Davies, C. W., *Principles of Electrolysis*, Royal Institute of Chemistry Monographs for Teachers, No. 1, 1962.

Dennis, W. H., *Extractive Metallurgy*, Philosophical Library, Inc., New York, 1965.

Queneau, P., Ed., *Extractive Metallurgy of Copper, Nickel, and Cobalt*, AIME Symposium, Interscience Publishers, New York, 1961.

Hayward, C. R., *An Outline of Metallurgical Practice*, D. VanNostrand Company, Inc., New York, 1952.

Lidell, D. M. (ed.), *Handbook of Non-Ferrous Metallurgy*, Vol. I, II McGraw-Hill, New York, 1926.

Mantell, C. L., *Electrochemical Engineering*, McGraw-Hill, New York, 1960.

Newton, J., *Extractive Metallurgy*, John Wiley, New York, 1959.
Pearson, T. G., *The Chemical Background of the Aluminum Industry*, Royal Inst. Chem. London, Monograph No. 3, 1955.
Prutton, C. F. and S. H. Maron, *Fundamental Principles of Physical Chemistry*, Chap. 17 and 18, The MacMillan Co., New York, 1951.

Discussion Questions

1. What are the reasons for electrorefining nickel? What is the theoretical voltage per cell? Account for the differences between the theoretical voltage and the practical operating voltage. If a nickel ion of the type (Ni complex)$^+$ could be electrodeposited, would there be any advantage?
2. Discuss the electrowinning of metals from aqueous chloride solution as contrasted with electrolysis of sulfate solution.
3. Discuss possible improvement of the Hall cell by development of inert anodes. What are the material requirements? What materials might be used? How would cell operation be affected?
4. Describe a process which would permit continuous metal reduction from a fused salt bath.
5. Consider the electrolysis of magnesium chloride.
    (a) Why is magnesium never electrolyzed from pure $MgCl_2$?
    (b) Why are high concentrations of KCl to be avoided in the salt bath?
    (c) What happens to impurities such as Al, Cu, Ni, and Zn in the bath?
6. The feed analysis to a magnesium reduction cell based on a feed of 5200 lb and the production of 1000 lb of magnesium is:

Salt	Feed Analysis	Impurities lb Added	Sludge lb Removed
NaCl	1.0%	52	152
CaCl$_2$	0.75%	39	38
MgCl$_2$	75.0%		63
MgO	3.0%	156	65

What is happening in the cell in addition to reduction of magnesium? What measures should be taken to provide for continuous production?
7. What pollution control techniques are available for disposal of aqueous electrolyte solutions?

Problems

1. Determine the standard potential for a cell composed of a hydrogen electrode and a copper–copper chloride electrode.
2. Nickel is electrolytically refined at 40°C using an electrolyte containing 50 g/liter of nickel ion. The impure anodes contain 2–2½ wt% copper which can plate out at the cathode. At what level of copper ion must the electrolyte be

maintained to avoid copper contamination of the cathode?

3. Quantitatively define a procedure for electrolytically separating copper, nickel, and iron from an anode which is essentially an impure copper containing about 1% iron and 5% nickel.

4. Starting sheets are used in a lead electrolytic refining process with a submerged cross section of 32 × 27 in. The potential drop between anodes and cathodes is maintained at 0.4 V. The average current density is 18.5 A/ft².

    (a) At a current efficiency of 85%, calculate the time required to deposit 4200 lb of lead per starting sheet?

    (b) Compute the energy required per pound of refined lead.

5. A copper refining cell is operating with an electrolyte which contains 0.05 mole/liter of stannous ion. At what concentration of cupric ion will the tin start to precipitate in preference to copper?

6. The overall reaction describing electrowinning of copper from a sulfate solution is:

$$CuSO_4 + H_2O = Cu + H_2SO_4 + \tfrac{1}{2}O_2.$$

The anode reaction has a standard electrode potential of +1.23 V. Calculate the reversible cell potential if the pH is 0.4 and the concentration of copper ion is 40 g/liter. In a commercial operation of this type, the cell voltage is on the order of 2 V. Explain the difference, if any, between the theoretical and the operating voltages.

7. Compute the minimum theoretical voltage for electrolytically reducing magnesium in the cell described below.

Cell:	Steel shell construction
	Natural gas fired at 700°C
	Cathode—steel shell
	Anode—carbon rods
Electrolyte:	25% $MgCl_2$
	60% NaCl
	15% $CaCl_2$
Product:	Liquid magnesium floating on bath
	Gaseous chlorine evolved at anode

Assume the electrolyte behaves ideally.

8. Compute the reversible cell potential required to reduce cadmium from a liquid chloride melt containing NaCl, KCl, and $CdCl_2$ at 500°C if the cadmium chloride activity is 0.25.

9. An aluminum reduction cell may be regarded as operating so that the gas evolved at the carbon anodes is in equilibrium with graphite at the operating temperature.

    (a) Calculate anode carbon consumption per pound of aluminum produced when the operating temperature is 800°C.

    (b) At an operating temperature of 1000°C, the anode gas is essentially pure CO. The cell bath is saturated with Al_2O_3. What is the theoretical minimum voltage at which aluminum will be deposited at the cathode?

CHAPTER 8

Melting, Pouring, and Solidification

8-1 Melting Practice

The advantages of a molten phase make it desirable to produce and refine metals in the liquid state. Subsequent to this, the metal is poured into a mold of desired shape and permitted to solidify in either a casting or an ingot form.

Thermal Requirements for Melting

The minimum power required to raise a metal from room temperature to some temperature above its melting point is equivalent to the enthalpy increase that the metal experiences in changing from the lower to the higher temperature. However, the actual power requirements are much greater than this, and may be determined from an overall heat balance for the melting process. In writing such a heat balance, several factors must be considered: (1) the energy input, (2) the sensible heat of the metal before and after the operation, (3) the sensible heat increase of the melting unit itself, and (4) the heat losses, which include not only radiation and convection from the exposed surfaces of the melting unit, but also any heat removed by the cooling water that often is used to protect critical areas of the melting unit. The heat balance is written most conveniently for a given period of time. The calculation of the required power input then depends upon the type of unit used.

Melting Units

Two types of electrical melting units are schematically represented in Fig. 8-1. The induction melting unit, Fig. 8-1A, is quite efficient since the thermal energy is generated directly in the metal. The charge is placed in a crucible surrounded by a coil which is supplied with a rapidly alternating current. The rise and fall of alternating current flowing through the coil causes a varying magnetic inductive force; this force produces electrical eddy currents in the metal which are dissipated in the form of heat. The power taken up from the coil by the metal is thus dependent upon the induction coil, its size and number of turns, crucible geometry and frequency of the electrical input, and also upon the electrical properties of the metal being melted.

Metals such as copper, brass, or aluminum, which are not melted easily by induction because of their relatively low resistivity, may be placed in a container made of a material such as carbon, which can be heated by the induction coil. Or, if a reaction would take place between the metal and the carbon crucible, it is possible to surround a refractory oxide crucible with an inductively heated cylinder. The cylinder, usually a high melting point metal, is called a susceptor. A material

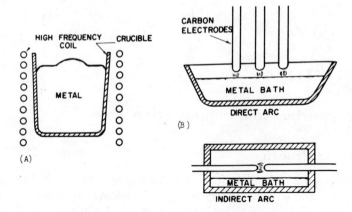

Fig. 8-1. Electrical Melting Units

(A) Induction Melting
(B) Arc Melting

that often is used under protective atmospheres for such applications is molybdenum.

Induction melting may be done either in air or, in the case of metals that are easily oxidized or that are desired in a high state of purity, the melting may take place under a protective atmosphere or vacuum. Vacuum induction melting often is considered to be more of a refining operation, and a great deal of research effort has been expended in demonstrating that vacuum-melted materials usually are superior to those that are air-melted.

Another type of melting unit that utilizes electric power is shown in Fig. 8-1B. This is known as arc melting and may be either of the direct or indirect type, depending upon whether the arc passes between electrode and metal or between two electrodes in the system. The large melting units used in electric furnace operations for steelmaking or electro-smelting generally use three-phase direct-arc systems that have melting capacities ranging up to 400 tons. In the case of indirect arc furnaces the electrodes often are placed along the centerline of the furnace. This allows the furnace to be rotated to provide a cooling of the refractory surfaces by the liquid metal itself, thus avoiding excessive attack of the refractory surfaces.

Another form of arc melting which is used to melt and refine reactive metals with high melting points is consumable-electrode arc melting. A schematic diagram of a consumable-electrode arc melting furnace is shown in Fig. 8-2.

Metals such as titanium and zirconium commonly are melted and cast in this type of unit. Titanium or zirconium sponge is pressed into rods to form an electrode. The electrode diameter is slightly smaller than the diameter of the finished ingot.

In this process, an arc is struck between the electrode and a small amount of metal or "starter button" placed in the bottom of a water-cooled copper mold. A small molten pool is maintained at the top of the ingot, and metal from the consumable electrode melts and flows down into the pool. As metal is added to the mold, the electrode distance is maintained constant, and the ingot builds up. The metal is in contact only with itself and the cold walls of the water-cooled copper mold. An inert atmosphere or vacuum is maintained in the melting chamber, preventing contamination and providing some refining by removal of volatile impurities.

A modification of this process is nonconsumable-electrode arc melting in which the electrode is carbon or tungsten, and sponge of the metal to be cast is continuously fed into the arc.

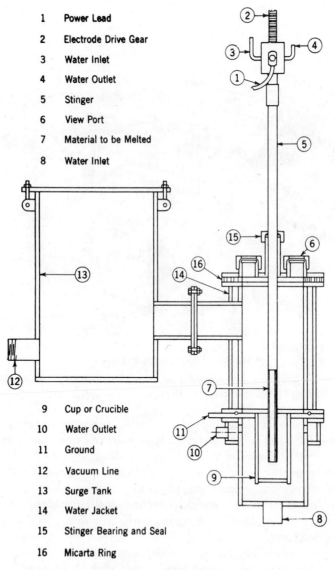

1	Power Lead
2	Electrode Drive Gear
3	Water Inlet
4	Water Outlet
5	Stinger
6	View Port
7	Material to be Melted
8	Water Inlet

9	Cup or Crucible
10	Water Outlet
11	Ground
12	Vacuum Line
13	Surge Tank
14	Water Jacket
15	Stinger Bearing and Seal
16	Micarta Ring

Fig. 8-2. Arc Melting Ingot Furnace Model R-1

Courtesy of Oregon Metallurgical Corporation.

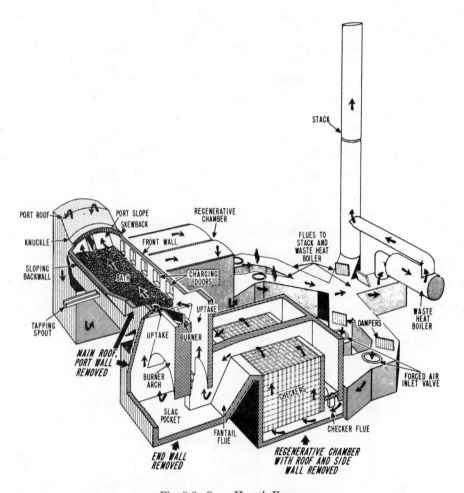

Fig. 8-3. Open Hearth Furnace

Source: *The Making, Shaping and Treating of Steel*, edited by Harold E. McGannon, Pittsburgh, United States Steel Corporation, 9th edition, 1971, p. 508.

A third type of electrical melting unit, one utilizing resistance heating elements, is sometimes used to melt metals, but usually is restricted to low-melting metals or small laboratory melts. Several different types of resistance melting units are in use, but the energy generally is supplied by radiation from a heating element. For operation at relatively low temperature, oxidation-resistant alloys (usually nickel–chromium base) often are used as heating elements; for slightly higher temperatures, nonmetallic elements, such as silicon carbide may be employed; and for operation at very high temperatures (approaching 4000°F) carbon resistors may be used, provided they are protected by an inert atmosphere.

In contrast to the use of electric power, heat may be supplied to a melting furnace by burning a low-cost fuel such as natural gas, fuel oil, or coke. Such a furnace may be of the shaft, retort, or reverberatory type. However, the apparent economic advantage of using these types of fuels often is offset by the deleterious effects of melt contamination by sulfur in the fuel, e.g., as in the case of steel or by the costs of gas cleaning equipment to prevent air pollution.

Nevertheless, a large portion of the steelmaking in the United States still is carried out in open-hearth furnaces, a gas or oil fired reverberatory type furnace shown in Fig. 8-3 (see also Fig. 4-1 for reverberatory type furnace). Heat transfer in the open hearth furnace includes all of the common heat flow mechanisms— conduction, convective heat transfer in liquids and gases, and radiation from the gas flame and the hot interior refractories. In an effort to obtain a high flame temperature, the open hearth employs a regenerative heat transfer system, as seen in Fig. 8-4. By periodically reversing the direction of firing in the furnace, the hot combustion gases are passed out of the furnace through a set of brick checkers that recover heat from the expended combustion gases. The incoming air is passed through the previously heated checker work at the opposite end of the furnace before being fed to the burner. This technique for preheating the combustion air in the regenerative type of furnace provides fuel economy.

The cupola melting unit, a coke-fueled shaft furnace, is shown in Fig. 8-5. This furnace is used primarily in foundries for the melting of cast iron, although it is sometimes used to prepare a modified pig iron for use in steelmaking operations. Modern furnaces often employ a water-cooled shell and an air blast that is preheated before being blown into the furnace to react with the coke. The reaction chemistry of the melting zone of the cupola is shown in Fig. 8-6.

The choice of a melting unit depends upon economics and the required metallurgical and temperature control. The replacement of Bessemer steelmaking facilities by open hearths at the turn of the century, and the present replacement of open hearths by basic oxygen steelmaking furnaces are examples of improved melting techniques which provide economy of operation and improved product quality. In the iron foundry industry, at the present time the electric furnace is replacing in part or supplementing the melting capacity normally carried by the cupola furnace.

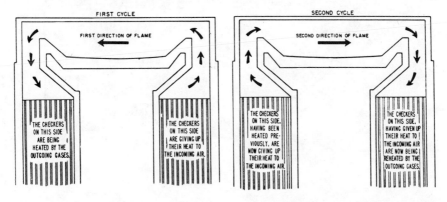

Fig. 8-4. Preheating Combustion Gases in Regenerative Furnaces

Source: *The Making, Shaping and Treating of Steel,* edited by Harold E. McGannon, Pittsburgh, United States Steel Corporation, 9th edition, 1971, p. 505.

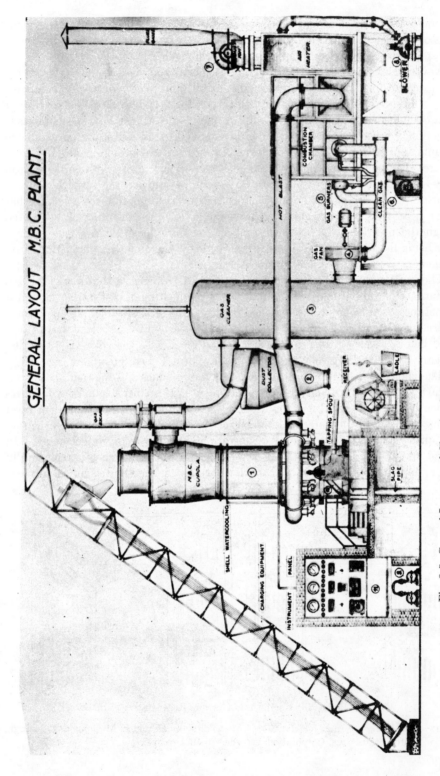

Fig. 8-5. General Layout and Charging System for Metallurgical Blast Cupola Plant

Source: *The Cupola and Its Operation,* 3rd edition, The American Foundrymen's Society, 1965, p. 30.

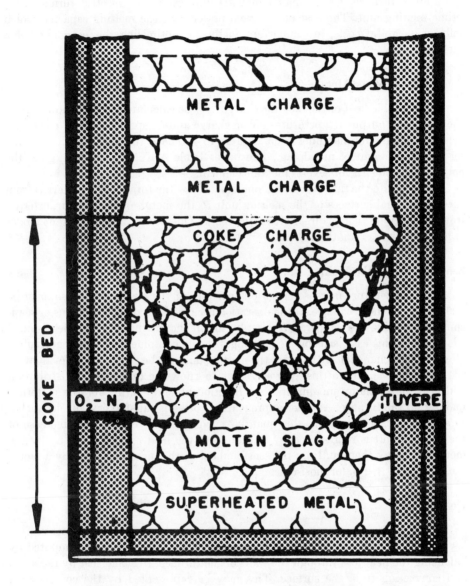

Fig. 8-6. Cross-section of Cupola Showing Reaction Areas

Source: *The Cupola and Its Operation*, 3rd edition, The American Foundrymen's Society, 1965, p. 26.

The greater metallurgical control, as well as greater production flexibility for the electric arc furnace, has increased industry interest in the electric furnace as a prime melting unit. The cost of electrical power, and the melting capacity which influences capital costs, are important factors in optimizing the economics of a melting facility.

Sensible Heat of the Metal

The sensible heat of the metallic charge may be calculated from a knowledge of the input and output temperatures of the charge and its specific heat. A more convenient method for handling such data is to tabulate the enthalpy increments for a given amount of metal, e.g., 1 lb mole. Table A-IX, Appendix A, gives the enthalpy increments for several metallic elements.

To compute the rate of heating, one must know the total heat transferred from the hot internal surfaces of the melting unit to the metal, or the heat transferred from the burning fuel or direct arc to the metal. One of the most important modes of heat transfer is radiation.

Sensible Heat of the Melting Unit

The temperature increase of various portions of the melting unit also must be included in the heat balance. If the operation is carried out on a continuous or semi-continuous basis, the temperature distribution in the melting unit can be assumed constant. In this case, sensible heat changes in the equipment itself are negligible. However, the change in enthalpy of the melting unit can be considered in evaluating the heat balance. To do this, it is necessary to know the temperature distribution and also the enthalpy increments as a function of temperature for the materials making up the furnace. Another approach that often is satisfactory, but less accurate, is to assign a factor to this quantity such that the sensible heat increase of the melting unit is represented by a fraction of the heat energy taken up by the metal. One may then multiply the sensible heat of the metallic charge by a factor ranging from 1.1–1.5, for most cases.

Heat Losses

The principal heat losses in melting operations are the radiation and convection losses from the exposed surfaces of the furnace. These losses may be computed by balancing the heat loss through the walls of the furnace with that lost by radiation and convection from the surface. This may be represented by the conservation relation,

$$q_{cond} = q_{conv} + q_r. \tag{8-1}$$

A trial-and-error approach to this problem is necessary where one solves the following equation for the interface temperature T_0:

$$q = \frac{kAm}{\Delta x}(T_i - T_0) = h_c A_0(T_0 - T_s) + \sigma e_0 A_0(T_0^4 - T_s^4), \tag{8-2}$$

where

T_i = temperature of internal surface of furnace,

T_0 = interface temperature between outer surface of the melting unit and surroundings,

T_s = temperature of surroundings,

k = thermal conductivity of furnace lining,

Am = mean area of lining,

Δx = thickness of lining,

σ = Stefan–Boltzmann constant, $.173 \times 10^{-8}$ Btu/ft²-hr-⁰R⁴,

e_0 = emissivity of the outer surface and surroundings, assumed equal,

A_0 = outer area of melting unit,

h_c = convective coefficient from surface of unit (which may be taken on the average as being $0.3 \, [T_i - T_s]^{0.25}$ for plane surfaces).

An example problem utilizing computer techniques for solution is presented in Appendix B (Problem 6).

In addition to heat losses from the exterior surfaces of a melting unit, heat may be lost to cooling water that is used to maintain low temperatures in certain critical areas of the furnace. In the case where the metal surface is exposed directly to the surroundings, one may compute the heat losses on the basis of radiation from that surface.

Mass and Energy Balance on Melting Unit

A material and energy balance on the cupola melting furnace is presented below to illustrate the engineering approach and point out the factors which must be considered in preparing a mass and heat balance on a melting unit.

The mass balance is the first step in providing a quantitative description of the melting operation. The streams entering the cupola are the blast, metallic charge, coke, and flux. The streams leaving the process are the stack gas, liquid metal, and slag. In order to reduce the material balance to its simplest form, and for ease in treating the energy balance, the materials are broken down into element or compound form for each stream. Material flow in the cupola furnace is illustrated schematically in Fig. 8-7. A material balance for a small cupola (approximately one order of magnitude smaller than the modern hot-blast, water-cooled cupola used in large foundry operations) is presented in Table 8-1.

The energy balance for the system can be calculated using thermochemical data for the compounds entering and leaving the process. The total enthalpy balance may be written as

$$H_{blast} + H_{charge} + H_{flux} + H_{coke} = H_{top\ gas} + H_{metal} + H_{slag} + H_{losses} \quad (8\text{-}3)$$

Each of the enthalpy terms in the foregoing equation can be calculated from a reference temperature, and are based on two terms—the energy of reaction to form the compounds in the stream, and the sensible heat in the stream relative to the selected base temperature.

TABLE 8-1

Reference Period Data[a] on Cupola Furnace

Chemical analyses, (Wt %)

Charge Material	Molten metal (wt. lb/net ton)	Si	S	Mn	P	C	Fe
Hematite pig	596.85	2.940	0.00	0.940	0.09	4.00	92.03
Low P pig	198.95	2.610	0.00	0.800	0.24	3.17	93.18
Return scrap	795.80	2.000	0.00	0.730	0.14	3.30	93.83
Steel scrap	397.90	0.040	0.00	0.600	0.05	0.10	99.21
Ferromanganese	7.16	0.000	0.00	50.000	0.00	0.00	50.00

	Molten metal	SiO_2	Al_2O_3	CaO	MgO	P	Fe	S	Fixed C	Vols	Ash	H_2O
Coke	265	4.51	2.09	0.15	0.06	0.016	0.98	1.18	91.25	0.95	7.80	3.00

Flux	Molten metal	SiO_2	Al_2O_3	CaO	MgO	P	S	CaF_2
Limestone	87	1.81	0.12	51.80	1.86	0.000	0.000	0.000
Lining loss	55	84.10	15.40	0.05	0.15	0.000	0.000	0.000

	SiO₂	Al₂O₃	CaO	MgO	S	Mn	C	Fe	CaF₂	Temp, °F	
Slag	48.30	10.96	33.90	0.80	0.000	2.47	0.00	2.22	0.00	2642	136

	Si	S	Mn	P	C	Fe	Temp, °F	
Metal	1.79	0.000	0.79	0.130	3.33	93.96	2642	—

	O₂	N₂	H₂O	CO	CO₂	H₂	Temp, °F	Blast Rate, scfm	
Blast	20.658	77.714	1.627	—	—	—	77	3293	—
Top gas	—	70.62	0.30	15.04	12.86	1.18	442	—	

[a] Based on mass balance presented in Report of Sub-committee TS 52, Developments in the melting of metals for foundries, *Institute of British Foundrymen* **54**, 103 (1961).
Source: R. D. Pehlke, Thermochemical Model for Computer Prediction of Cupola Performance, *Modern Castings* **44**, 583 (1963).

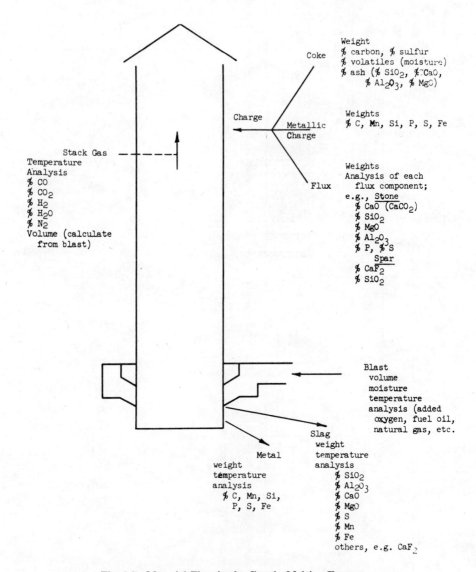

Fig. 8-7. Material Flow in the Cupola Melting Furnace

Source: Thermochemical Model for Computer Prediction of Cupola Performance," R. D.
          Pehlke, *Modern Castings*, 1963, Vol. 44, p. 582.

A summary of the enthalpies of the streams presented in the material balance of Table 8-1 are shown in Table 8-2. The weights of materials are given on a pound basis and the enthalpy terms are expressed in BTU/lb of product. The energy balance is used to calculate the heat loss per net ton of metal produced by the furnace.

TABLE 8-2

Energy Balance for Reference Period of Operation[a]

	H, Btu/ lb Product	lb Prod-uct/Ton Metal	H, Btu/ Ton Metal	Reaction	H,/Btu lb Product	lb Prod-uct/Ton Metal	H,/Btu Ton Metal
Blast (77 F)							
$H_2 + \frac{1}{2}O_2 \rightarrow H_2O$	−5872	18.67	−109,630				
		H-Blast =	−109,630				
				Stack gas (442°F)			
				$C(77°F) + \frac{1}{2}O_2(77°F) \rightarrow CO(442°F)$	−1605	295	−473,475
				$C(77°F) + O_2(77°F) \rightarrow CO_2(442°F)$	−3763	397	−1,493,910
				$H_2(77°F) \rightarrow H_2(442°F)$	+1271	1.68	+2,135
				$H_2(77°F) + \frac{1}{2}O_2(77°F) \rightarrow H_2O(442°F)$	−5608	3.63	−20,360
				$N_2(77°F) \rightarrow N_2(442°F)$	+91	1385	+126,035
					H-Stack gas =		−1,859,575
Flux (77 F)							
$CaO + C + O_2 \rightarrow CaCO_3$	−2467	80.18	−197,800				
$MgO + C + O_2 \rightarrow MgCO_3$	−2607	3.55	−9,250				
		H-Flux =	−207,050				
				Slag (2642°F)			
				$Slag\ (77°F) \rightarrow Slag\ (2642°F)$	+822	136	+111,790
				$CaO + SiO_2 \rightarrow CaO\cdot SiO_2(77°F)$	−330	82.8	−27,320
				$2MgO + SiO_2 \rightarrow 2MgO\cdot SiO_2(77°F)$	−194	3.22	−624
				$CaO + \frac{1}{2}S_2(g) \rightarrow CaS + \frac{1}{2}O_2(77°F)$	+915	6.81	+6,230
					H-Slag =		+90,076
Charge (77 F)							
loss by oxidation				**Metal (2642°F)**			
$Si(77\ F) + O_2(77\ F) \rightarrow SiO_2(77\ F)$	−6156	−6.45	+39,700	$Fe(77°F) \rightarrow Fe(2642°F)$	+549	1879	+1,031,571
$Mn(77\ F) + \frac{1}{2}O_2(77\ F) \rightarrow MnO(77\ F)$	−2335	−2.70	+6,300	$Si(77°F) \rightarrow \underline{Si}(2642°F)$	−587	35.7	−20,956
$Fe(77\ F) + \frac{1}{2}O_2(77\ F) \rightarrow FeO(77\ F)$	−1598	−3.29	+5,260	$P(77°F) \rightarrow \underline{P}(2642°F)$	−2606	2.6	−6,776
				$Mn(77°F) \rightarrow \underline{Mn}(2642°F)$	+582	15.8	+9,116
$C(77\ F) \rightarrow \underline{C}(77\ F)$	+1750	56.83	+99,450	$C(77°F) \rightarrow \underline{C}(2642°F)$	+1987	66.5	+132,135
		H-Charge =	+150,710		H-Metal =		+1,145,090

Heat Loss (by difference) = 458,439 Btu/Net Ton Metal

[a] Based on Table 8-1 with a reference temperature of 77°F.

Source: R. D. Pehlke, Thermochemical model for computer prediction of cupola performance, *Modern Castings*, **44**, 583 (1963).

8-2 Pouring Operations

Liquid metals are transported in metallurgical plants by ladles, cylindrically shaped vessels consisting of a steel shell lined with refractories. The type of refractory in most common use is fire clay for pig or cast iron, and magnesia for steel. Ladles vary in capacity from 10 lb to 300 ton. Smaller ladles usually are lip-poured, but in the case of large heats of steel, the metal usually is bottom-poured to allow non-metallic inclusions to float out of the metal. (See Fig. B-8, of Example Problem 7, Appendix B.) With this method of pouring, a stopper assembly acts as a valve to control the flow of metal through a nozzle at the bottom of the ladle.

One of the most important factors affecting the surface of metal cast into an ingot mold, and the subsequent product made from that ingot, is the rate of rise of metal in the mold, which depends on the rate of flow of the pouring stream and the size of the mold. The pouring rate is determined by nozzle size, the height of metal in the ladle, and the erosion of the stopper and nozzle. For example, in the case of steel, the optimum pouring rate is determined by the type of product, but is also dependent upon mold design and temperature of pour. In general, a slow pour minimizes ingot cracking but increases the tendency toward rippled ingot surfaces. The rapid rise of molten steel in the mold prevents a rippled ingot surface. However, a rapidly poured ingot will result in a thin shell of steel frozen at the mold surface that shrinks away from the mold wall, thus losing its support against the pressure caused by the liquid iron contained in it. Since the shell is thin and hot and consequently very weak, it cannot stand the pressure of the liquid metal; as a result it ruptures, permitting fresh liquid to break through to the mold surface. This produces a fold in the ingot surface that may open on cooling or subsequent forming operations.

An intermediate pouring vessel is often used in lieu of casting the metal directly from the ladle to the mold. Such vessels are called pouring baskets, pouring boxes, or tundishes. This method usually involves pouring from the teeming (pouring) ladle into a vessel that has a nozzle leading into the ingot mold. Tundish or basket pouring is more expensive than conventional top pouring, but has the advantages of a greater opportunity for separation of nonmetallic inclusions from the metal and less splashing in the mold.

Some of the disadvantages of top-pouring mentioned above may be eliminated through the use of a bottom-pour mold assembly. The metal is cast into a runner system and finally emerges at the bottom of the mold through a special outlet. In bottom pouring, the metal rises steadily in the mold with very little agitation. Bottom pouring is much more expensive than top pouring, but is occasionally used to ensure optimum surface quality.

Rate of Pouring

Since the rate of pouring greatly influences the quality of a cast ingot, much attention is given to the control of this factor. By considering the flow of metal through the nozzle of a bottom pour ladle in terms of the generalized flow equation, the rate at which metal is entering the mold at any time during the pouring operation can be estimated.[1,2] Since the nozzle is eroded by the liquid stream, it is necessary to have some relationship which indicates the size of the nozzle as a function of the amount of metal which has passed through it. Such correlations often may be obtained from plant data. If estimates for the entrance orifice coefficient and the friction factor also are available, the rate of pouring may be computed.

The gradual enlargement of the nozzle due to erosion has an effect that is desirable, i.e., the rate of pouring does not decline as rapidly with the ferrostatic head in the ladle as it would if the nozzle opening remained constant. While the head of metal has been decreasing, the diameter of the orifice has been increasing, and the rate of pouring can remain nearly constant or even increase slightly. Since the rate of pouring has a significant influence on the structure of the ingot, this is an important aspect in the design of the pouring operation. Knowing the behavior of several materials that are available for nozzles, the process engineer can select the material that provides a relatively constant pouring rate throughout the casting operation. Application of the generalized flow equation to the estimation of teeming rates from a bottom pour ladle is illustrated by Example Problem 7, Appendix B.

Temperature Drop

Metal temperature during the pouring of steel into ingot molds is an important process variable. The furnace tapping temperature is controlled to establish the desired teeming temperature, but is not sufficient of itself to define the temperature during pouring into ingot molds. Several factors control the temperature drop that occurs between tapping and teeming. The principal factors are:

1. Radiation and convection losses from the steel surface, or slag surface, during the time the metal is in the ladle.
2. Heat losses from the tapping stream.
3. Heat losses into the brick lining of the ladle. These heat losses may be estimated in several ways. One assumption that can be made is that the brick lining acts as a semi-infinite heat sink and that heat continually diffuses into the refractory lining of the ladle. During both tapping of the furnace and teeming into ingot molds, the steel is in contact with the refractory near the bottom of the ladle for a longer period than it is in contact with the upper surfaces of the ladle. One way of treating the heat absorption process is to divide the lining into zones, the more zones the more precise the calculation, and to assign different times for contact with these zones during the filling and emptying of the vessel.
4. Thermal effects of deoxidation, including the energy required to heat the deoxidizing material up to temperature, and the heat of reaction.

Samways and Dancy[3] have reported on the temperature drop of liquid steel

---

[1] J. G. Henzel, Jr., and J. Keverian, *Modern Castings* **50,** 69 (1966).

[2] R. D. Pehlke, Digital computer simulation of a bottom pouring ladle, *Computers in the Mineral Industries*, pp. 227–236 Stanford University School of Earth Sciences, Palo Alto, Calif., 1964.

[3] N. L. Samways and T. E. Dancy, *J. Metals* **12,** 331 (1960).

between tapping and teeming and have compared statistically derived relationships describing plant practice with theoretically derived values.

8-3 Casting and Solidification

After pouring into molds, the liquid metal is permitted to solidify; the solidified castings are called ingots. The eventual use of the ingots may be in any of several mechanical forming operations, or may even include remelting and recasting to a desired shape. The mechanical forming operations used in the metallurgical industries are extensive subjects in themselves; discussion in this section consequently will be limited to the casting of ingot shapes for use in further forming operations and will be concerned chiefly with rates of solidification, ingot structure, and some consideration of special casting processes.

Rate of Solidification

There are two fundamental approaches to the mathematical treatment of solidification rates. One method involves development of rigorous equations describing the solidification process and the solution of these equations by analytical techniques. The equations, however, are usually complex and their solution is difficult and sometimes impossible to obtain for even the simplest of cases. Alternatively, approximate relationships can be developed that are representative of a given practical situation. With the assistance of high-speed computers, such approximations can be made to approach very closely actual conditions. Numerical approximations are much more versatile and do not require the limiting assumptions often necessary to develop a soluble analytical expression. As an example, the use of finite difference approximations to simulate heat flow in a slab is presented in Appendix B, Example Problem 8.

In order to indicate the nature of the analytical method, the following semi-quantitative approach to solidification is developed. The unidirectional unsteady state heat flow in a homogeneous solid is given by the relationship,

$$\frac{\partial T}{\partial t} = \left[\frac{k}{C_p \rho} \right] \frac{\partial^2 T}{\partial x^2} , \tag{8-4}$$

where

T = temperature in the heat-conducting medium,
t = time,
k = thermal conductivity,
C_p = specific heat,
ρ = density,
x = distance.

The quantity $k/C_p\rho$ is termed the thermal diffusivity and usually is represented by the symbol α.

The solution for Eq. (8-4) depends upon the boundary conditions chosen. Where the heat-conducting medium is taken as semi-infinite, at an initial uniform temperature T_0, and at time $t = 0$, has its boundary temperature instantaneously raised to and maintained at T_i, the ensuing temperature distribution through the medium is given by the expression,

$$T_x = T_0 + (T_i - T_0) \operatorname{erfc}\left[\frac{x}{2\sqrt{\alpha t}}\right], \tag{8-5}$$

where

T_x = temperature at a distance x in heat-conducting medium,
α = thermal diffusivity as defined above,

$$\operatorname{erfc}\left[\frac{x}{2\sqrt{\alpha t}}\right] = \left[1 - \text{error function of } \frac{x}{2\sqrt{\alpha t}}\right] \text{ as defined in Table A-X,}$$

Appendix A.

Ruddle,[4,5] and Feild[6] have developed a parabolic rate equation for solidification of an ingot using Eq. (8-5) in the following manner. Assuming that the mold into which liquid metal is cast meets the boundary conditions for Eq. (8-5), the temperature distribution in the mold is given by Eq. (8-5) where T_x is the temperature in the mold material at any distance, x, from the mold–metal interface. The rate of heat transfer across the mold–metal interface is given by the relationship,

$$\frac{q}{A} = -k\left[\frac{\partial T}{\partial x}\right]_{x=0}, \tag{8-6}$$

where

$\dfrac{q}{A}$ = heat transfer rate per unit area,

k = thermal conductivity of mold material,

$\left[\dfrac{\partial T}{\partial x}\right]_{x=0}$ = thermal gradient in mold material at mold-metal interface.

Differentiating Eq. (8-5) with respect to x and evaluating the differential at $x = 0$ yields

$$\left[\frac{\partial T_x}{\partial x}\right]_{x=0} = \frac{T_i - T_0}{\sqrt{\pi \alpha t}}. \tag{8-7}$$

Then

$$\frac{q}{A} = \frac{1}{A}\left[\frac{\partial Q}{\partial t}\right] = -\frac{k(T_i - T_0)}{\sqrt{\pi \alpha t}}, \tag{8-8}$$

---

[4] R. W. Ruddle, *The Solidification of Castings*, Institute of Metals Monograph and Report Series, 7, Institute of Metals, London, 1957.
[5] R. W. Ruddle, *J. Inst. Metals* **77**, 1 (1950).
[6] A. L. Feild, *Amer. Soc. Steel Treating Trans.* **11**, p 264 (1927).

where Q = total quantity of heat transferred; thus $q = \dfrac{\partial Q}{\partial t}$. Integrating Eq. (8-8)

$$\frac{Q}{A} = \frac{2k(T_i - T_0)}{\sqrt{\pi\alpha}} \sqrt{t}. \tag{8-9}$$

Under these conditions, the heat removed is proportional to the square root of time. Assuming that the freezing thickness depends upon removing a given quantity of heat per unit area

$$d = \beta'\sqrt{t}, \tag{8-10}$$

where

d = thickness of frozen metal,
β' = solidification constant,
t = freezing time.

For an infinite plate, the total freezing distance is one-half the plate thickness, which is also the volume to surface area ratio. Hence the total freezing time is

$$t_f = \gamma'\left[\frac{V}{A}\right]^2 \tag{8-11}$$

where

t_f = total freezing time,
γ' = empirical constant,
$\dfrac{V}{A}$ = volume to surface area ratio.

Equations (8-10) and (8-11) can be applied to the solidification behavior of the simple shapes encountered in ingot casting, although they are not strictly rigorous since the boundary conditions assumed in deriving these relationships are not fully valid. Nevertheless, over a wide range of industrial conditions, the solidification thickness is observed to be linear with the square root of time, particularly during the early stages of solidification. Thus, if the empirical constants are evaluated from previous operating data, Eqs. (8-10) and (8-11) are useful for interpolating solidification thicknesses and freezing times under similar conditions.

Ingot Structure

Important factors that affect ingot structure are metal composition and gas evolution. In addition, pouring conditions, such as temperature of the liquid metal, the nature, temperature and geometry of the mold, and the rate of pouring, influence the rate of heat removal and, hence, the ingot structure.

Effect of Rate of Heat Removal on Ingot Structure. A cast ingot will generally freeze in three zones. Near the surface of the ingot the structure consists of very fine, randomly oriented grains. This region is called the chill zone and its structure is the result of rapid solidification with high nucleation rates during the early stages of freezing. Underneath this region is a second zone where the crystals are thin and elongated, being oriented perpendicular to the mold wall. This zone is called the columnar zone and is the result of relatively rapid heat removal, giving rise to directional solidification and grain growth. Underneath the columnar zone, the ingot has a structure consisting of randomly oriented crystals that are much coarser than in the chill zone. The increase in grain size occurs because of a lower rate of heat removal in the final stage of solidification. All ingots that do not evolve gas during solidification will show these three structures, with the extent and nature of the zones being a function of pouring conditions, composition, and average rate of solidification. The rate of solidification varies widely, being nearly infinite at the mold surface and decreasing to a very small value near the center of the ingot. It is the combination of the variation in cooling rate during ingot solidification, together with metal composition and gas evolution, which markedly influences ingot structure.

Influence of Metal Composition. Since most liquid metals involved in metallurgical operations are not pure, solidification occurs over a temperature range, so that the first liquid to solidify has a different composition than that which is last to solidify. The distribution of solutes between liquid and solid is described by a distribution coefficient, k (see Zone Refining, Chap. 5). The separation of solute between liquid and solid is the cause of segregation.

Most alloying elements lower the melting point of a liquid pure metal, and consequently upon freezing the first crystals of solid more closely approach the pure major component than the composition of the bulk liquid. However, as the first crystals of solid form, the adjacent liquid is left slightly less pure than the bulk liquid. The solidification of a metal containing a solute then can be visualized as a stepwise crystallization process in which the material separating out is more pure in the major component, and, because of the rejection of solute on solidification, is separated from the bulk liquid by a film of liquid metal that is less pure than the bulk liquid.

In the columnar zone of solidification, less pure liquid is trapped between the long columnar crystals. Since this liquid is enriched in solute, it freezes in the interstices between the columnar grains at a lower temperature than the crystallized solid, resulting in the phenomenon described as micro-segregation. Contrasted with microsegregation is macrosegregation, a term used to describe relatively large-scale segregation that may be visible on a polished section to the naked eye and is exemplified by differences in concentration existing in different areas of an ingot or casting.

As the solidification front in a metallic solution advances, solute is rejected to the liquid phase, e.g., where k is less than one. In order for solidification to take place, a temperature gradient must exist across the liquid–solid interface such that the interface is at a temperature below that where solid and liquid are in equilibrium. In addition, there is a temperature gradient in the liquid phase. Depending upon the temperature gradient and the liquidus temperature for the solute-enriched liquid existing ahead of the freezing interface, the temperature in the liquid phase ahead of the solidification front may drop below the liquidus temperature. This condition is referred to as constitutional supercooling[7] and can lead to the formation

[7] B. Chalmers, *Trans. Met. Soc. AIME* **200,** p 519 (1954).

of solid crystals in the bulk liquid ahead of the normal solidification front. The formation of crystals in the supercooled liquid depends upon nucleation conditions, and it has been shown that in order for homogeneous nucleation to occur, extensive supercooling is required.[8] In actual operating practice, only a few degrees of under-cooling are required to initiate solidification. This is ascribed to the presence of small nucleating particles, e.g., the products of deoxidation, etc. Hence, inclusions enhance the nucleation of solidification.

In the case of a liquid metal containing some solute, freezing may proceed in any of three ways: (1) At high rates of solidification, the liquid solidifies quite rapidly, so that the solidification process is not distinguishable from that of a liquid pure metal and no segregation occurs; (2) At lower rates of solidification, the boundary layer of pure liquid that builds up in front of the solidification front may delay solidification slightly; or (3) At low rates of solidification, the bulk liquid beyond the impure film may reach its liquidus temperature before the impure liquid which exists at the solidification front. In this latter circumstance, free crystals will form in the liquid and settle out to the bottom of the ingot. As the solidification rate changes during the freezing of an ingot, all three types of freezing can take place in sequential steps. In the absence of gas evolution, the segregation char-acteristics of an ingot are determined by its composition, its rate of cooling and the tendency of any alloying elements to segregate between liquid and solid.

These segregation characteristics are illustrated by the freezing of a killed steel ingot. Figure 8-8 illustrates the mechanism of solidification of a killed steel ingot. Figure 8-8A shows the initial solidified chill zone made up of small, randomly oriented crystallites. Certain crystals will have an orientation such that the pre-ferred- direction of more rapid growth extends out into the liquid. These crystals (Marked "a" in Fig. 8-8A) become elongated and produce the columnar zone, and in trapping less pure liquid between the columnar crystals, as shown in Fig. 8-8B, create microsegregation in this area. As the freezing rate decreases and the tem-perature gradient becomes less steep through the liquid and solid, the temperature beyond the less pure liquid film at the solidification front drops below the liquidus temperature of the unaltered liquid metal near the center of the ingot. Free crystals then form in this supercooled region and drop to the bottom of the ingot. These falling crystals and the corresponding rise in liquid level along the solidification front lead to macrosegregation, as shown in Fig. 8-9. The conical zone of negative segregation in the lower portion of the ingot is caused by the settling of free crystals that are purer and more dense than the bulk liquid. As the free crystals settle, they cause an upward movement of less pure liquid along the side of the ingot near the solidification front, thus leading to positive segregation in this region. Similarly, the V-shaped zones of positive segregation in the upper central portion of the ingot are caused by the movement of less pure liquid down into the ingot cavity by shrinkage during the final stages of solidification.[9,10]

[8] J. H. Holomon and D. Turnbull, *Progress in Metal Physics*, Vol. 4, p 333, Pergamon Press, London, 1953.

[9] G. Derge, Ed., *Basic Open Hearth Steelmaking*, Chap. 10 and 11, AIME, New York, 1964.

[10] J. K. McCauley, and F. C. Langenberg, *J. Metals*, *17*, 318 (1965).

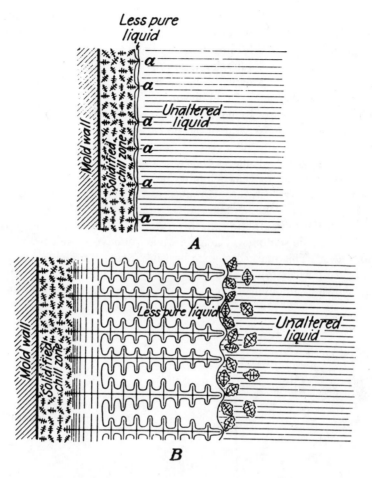

Fig. 8-8. Schematic Drawing Illustrating the Mechanism of Solidification in a Killed Ingot

Source: *Basic Open Hearth Steelmaking*, edited by G. Derge, 1964, AIME, New York, Chapter 11, p. 440.

Influence of Shrinkage. The overall structure of cast ingots also is influenced by the change in specific volume of the metal with temperature. The change in volume of an ingot on casting occurs because of liquid shrinkage as the metal cools; shrinkage upon solidification, either at the melting point or over a small temperature range in the case of an alloy; and the thermal contraction of the solid material as it cools to room temperature. The volume contraction accompanying the transition from liquid to solid at the melting point depends upon the specific metal, but may range from 3–7%. The shrinkage in the liquid state depends upon the superheat and can be as high as 2%, whereas the solid state shrinkage for most metals averages about 3%.

These volume changes, particularly the shrinkage occurring during freezing, can lead to the formation of a large depression in the top of the ingot commonly referred to as *pipe*. The exposed surface of a pipe will oxidize and will not weld during subsequent forming operations, necessitating that the upper portion of the ingot be cut off or *cropped*, leading to a decrease in ingot yield. One means for allowing for shrinkage on solidification is to provide an auxiliary liquid reservoir or *hot top* such that the top of the ingot remains liquid while solidification is promoted from the bottom of the mold. Other means for avoiding shrinkage pipe include controlled evolution of gases by control of the oxygen content in copper ingots to provide the proper set or by control of the carbon and oxygen contents in steel (such as in production of semikilled or rimming ingots), to counteract the contraction in volume by a controlled formation of gas within the ingot.

Effect of Dissolved Gases. The solubility of gases in both liquid and solid metals usually increases with temperature. However, at the melting point where liquid and solid exist in equilibrium, the solubility in the liquid is usually several times that of the solubility in the solid. If the concentration of the dissolved gas in the liquid metal exceeds the solid solubility, the metal thus will have a tendency to evolve gas upon solidification. This marked change in solubility at the melting point is illustrated in Fig. 8-10, where the solubilities of hydrogen and nitrogen at 1 atm pressure in iron are shown as functions of temperature. At the melting point, an approxi-

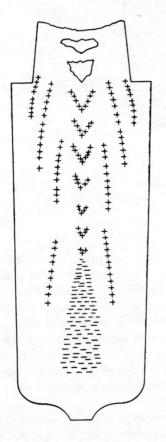

+zones of positive segregation,
−zones of negative segregation

Fig. 8-9. Segregation Pattern in Hot-topped
Killed Steel Ingot

Source: *Basic Open Hearth Steelmaking*, edited
by G. Derge, 1964, AIME, New York,
Chapter 11, p. 441.

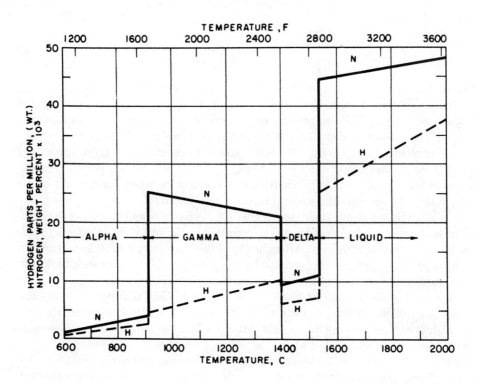

Fig. 8-10. Effect of Temperature on the Solubility of Hydrogen and Nitrogen at 1 atm Pressure in Iron

Source: *Basic Open Hearth Steelmaking*, edited by G. Derge, 1964, AIME, New York, Chapter 16, p. 646.

mately fourfold increase in the solubility of these gases exists in the liquid as compared with the solid.

The evolution of gas from an ingot during solidification may change not only its physical nature through formation of blowholes, but also may affect the segregation tendencies since the presence of a less dense gas phase results in fluid flow in the ingot during solidification. However, the presence of gas in an ingot is often an advantage since, as pointed out earlier, gas evolution can be used to compensate for shrinkage and to reduce or eliminate the ingot pipe, thus giving a higher yield.

In the casting of steel, the carbon–oxygen equilibrium and the influence of temperature on it are of primary importance in the control of ingot structure by gas evolution. Carbon and oxygen react to form carbon monoxide whenever the activity product exceeds the equilibrium constant for the solution of the gas in liquid iron. The reaction may be written

$$\underline{C} + \underline{O} = CO, \tag{8-12}$$

where

$$K = \frac{P_{CO}}{(a_C)(a_O)}. \tag{8-13}$$

The equilibrium constant is thus proportional to the pressure of CO, which in turn is a linear function of depth in the ingot mold. At atmospheric pressure near the top of the ingot, the equilibrium product of the carbon and oxygen activities would be one-half the value at a depth of 5 ft where the pressure is twice as great.

In practice, deoxidation in the ladle is used to control the oxygen content of liquid steel. As freezing of the ingot proceeds, carbon and oxygen become concentrated in the liquid, eventually resulting in gas evolution when the product of the oxygen and carbon activities exceeds the equilibrium value.

In a semikilled ingot, the concentration of oxygen is adjusted with deoxidizers in the ladle and also during the pouring operation in the mold, such that gas evolution compensates for shrinkage. After the top of the ingot freezes over, gas formation in the interior of the ingot is controlled by the pressure. As the steel shrinks during solidification, the pressure is relieved and additional gas is evolved.

Rimmed steels are characterized by rapid gas evolution that begins during the solidification of the chill zone at the mold wall. With the first solidification, marked gas evolution begins at the liquid–solid interface and the rising bubbles cause the liquid metal to move upward along the sides of the ingot and downward in the center. The motion of the liquid tends to limit columnar grain growth and keeps the interior liquid at a nearly uniform temperature. Just inside the rim zone of a rimmed ingot, a series of primary blowholes is found. The mechanism of formation of these blowholes has been described by Hultgren and Phragmen.[11] Gas is evolved from the solute-enriched liquid at the liquid–solid interface. As the rate of solidification decreases, the gas bubbles are not swept away but attach themselves to the solid and grow with the solidifying metal, resulting in a primary blowhole. After the top of the ingot freezes over, enough impure liquid accumulates at the solid–liquid interface to start gas formation in much the same manner as it occurs in semikilled ingots. This results in an interior series of spherical gas pockets called secondary blowholes.

The upward motion of the liquid along the solidifying wall sweeps away impure liquid and mixes it with the unsolidified body of the ingot. The eventual result is a large positive segregation in the upper central portion of the ingot, the last part to solidify.

The presence of other gases (especially hydrogen) may cause blowholes. In the case of copper, the evolution of hydrogen and water vapor or SO_2 may also influence ingot structure, as can hydrogen in aluminum and many other alloys. The control of gases in liquid metals is discussed in Chapter 5.

---

[11] A. Hultgren and G. Phragmen, *Met. Soc. Trans. AIME* **135**, 133 (1939).

Example Problem 8-1. If a 3000 lb deoxidized copper cake solidifies slowly from a liquid containing 4 ppm (by weight) of dissolved hydrogen, how much hydrogen could be evolved?

Solution. At the freezing point the solubility of hydrogen in copper at 1 atm hydrogen pressure is about 1.3 ppm in the solid and 5 ppm in the liquid phase. The hydrogen evolved, neglecting any transfer between metal and the surrounding atmosphere and assuming that the freezing solid is saturated at 1 atm pressure, is $4 - 1.3 = 2.7$ ppm. Converting to cubic feet at the freezing point,

$$\text{Volume of cake} = \frac{(3000)\ \text{lb}}{(680)\ \text{lb/ft}^3} = 4.42\ \text{ft}^3,$$

$$\text{Volume of H}_2 = \frac{(2.7 \times 10^{-6})\ (3000)\ (359)\ (1356)}{(2)\ (273)}$$

$$\frac{(\text{ppm})\ (\text{lb})\ (\text{ft}^3/\text{lb mole})\ (°\text{K}_{actual})}{(\text{lb/lb mole})\ (°\text{K}_{reference})} = \text{ft}^3$$

$$= 7.20\ \text{ft}^3,$$

$$\text{Relative volume} = \frac{7.2}{4.42} = 1.63.$$

In view of the limited solid solubilities of gases in metals, it is not surprising that the production of a gas-free casting is difficult.

In the case of rimming steel ingots, the relative volume of gas evolved at casting temperatures may be 30 or higher.[12] This large volume of evolved gas is utilized to promote better surface quality and higher yield.

Defects in Ingots. In addition to cavities caused by shrinkage or gas evolution, several other ingot defects have important influences on subsequent product quality.

Nonmetallic inclusions, depending upon their size, quantity, and distribution, can be detrimental to the properties of a metal in its service application. Nonmetallic inclusions originate from two primary sources and are designated accordingly as either indigenous or exogenous. Indigenous inclusions are the result of precipitation reactions within the metal itself, e.g., in the processing of steel, the entrapment of oxides formed during deoxidation or the precipitation of sulfides such as manganese sulfide during solidification. Exogenous inclusions are foreign particles trapped in the metal, such as particles of refractory eroded from linings, etc.

Another common defect in ingots is the formation of a plane of weakness along the diagonal of the cross section of the ingot. Since columnar grains form at right angles to the walls of the mold, their line of intersection represents a plane of weakness that can fail during mechanical working of the ingotism. This defect can be minimized by using rounded ingot mold corners and by promoting higher initial rates of solidification.

[12] A. Hayes and J. Chipman, *Met. Soc. Trans. AIME*, **135,** 85 (1939).

8-4 Continuous Casting

Continuous casting processes have received a great deal of attention in recent years and have come into wide use in the casting of light alloy slabs and ingots; application is now being made in the processing of copper and steel. The purpose of continuous casting is to bypass conventional ingot casting and cast to a form that is directly rollable on finishing mills. Furthermore, there is good indication that use of this process should result in improvement in yield, in surface condition, and in internal quality of product.

Continuous casting involves the following sequence of operations:

1. Delivery of liquid metal to the casting strand.
2. Flow of metal through a distributor into the casting mold.
3. Formation of the cast section in a water-cooled mold.
4. Withdrawal of the casting from the mold.
5. Further heat removal from the casting, e.g., water spray beneath the mold.
6. Cutting and removal of the cast bars.

Several arrangements are now in commercial use to accomplish these objectives. The principal types of continuous casting machines are vertical, vertical with bending, and curved mold, as shown in Fig. 8-11A, -B, and -C. A curved mold continuous casting machine is shown in greater detail in Fig. 8-12.

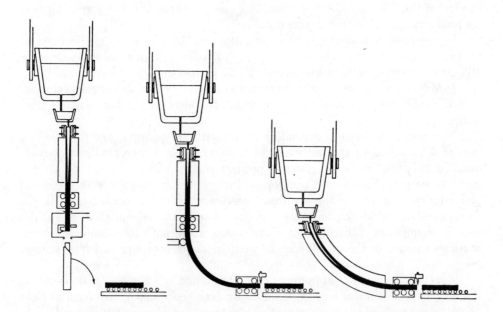

Fig. 8-11. Vertical, Bent-strand, and Curved-mould Continuous Casting Machines

Source: R. S. Wagstaff and G. E. Stock, "A Decade of Development in the Continuous Casting of Special Steels," Special Report No. 89, *Continuous Casting of Steel*, Iron and Steel Institute, London, 1965, p. 122.

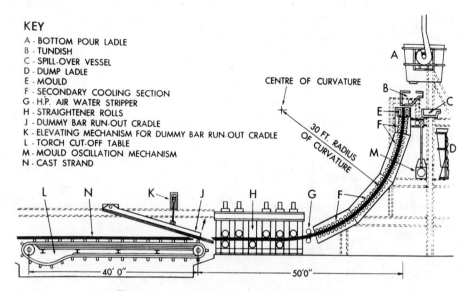

KEY

A - BOTTOM POUR LADLE
B - TUNDISH
C - SPILL-OVER VESSEL
D - DUMP LADLE
E - MOULD
F - SECONDARY COOLING SECTION
G - H.P. AIR WATER STRIPPER
H - STRAIGHTENER ROLLS
J - DUMMY BAR RUN-OUT CRADLE
K - ELEVATING MECHANISM FOR DUMMY BAR RUN-OUT CRADLE
L - TORCH CUT-OFF TABLE
M - MOULD OSCILLATION MECHANISM
N - CAST STRAND

CENTRE OF CURVATURE

30 FT. RADIUS OF CURVATURE

Fig. 8-12. Curved Mold Continuous Casting Machine

Source: *Iron and Steel Engineer*, July 1965, p. 147.

Most of the original continuous casting machines for steel were vertical machines. The vertical machine with bending and the curved mold machine, although more complicated in their construction, were developed to minimize the height of the machine and allow installation in present plants without modification of the crane height.

A primary factor influencing the casting rate is the rate of heat removal. The limiting casting rate is one which makes it just possible to remove the heat required for complete solidification prior to the time that the cast section is cut. Another more important limitation on casting rate involves the need to develop a solid shell within the water-cooled mold that can contain the liquid metal as the section leaves the mold. The governing factor is the thickness of the shell emerging from the mold, and this is determined largely by the rate at which heat is extracted in the mold. On this basis, then, the critical part of the continuous casting process is the removal of heat in the water-cooled mold. It is on this phase of the operation that the following analytical discussion will focus.

One of the earliest studies of solidification during continuous casting was made by Roth[13] who derived an expression showing that the freezing rate is proportional to the horizontal thickness already solidified, and predicting a parabolic-shaped freezing front. Ruddle[14] has reviewed this and other studies of the solidification of continuously cast metals.

[13] W. Roth, *Aluminum* **25**, 283 (1943).

[14] R. W. Ruddle, *The Solidification of Castings*, Chap. 7, Institute of Metals Monograph and Report Series, 7, Institute of Metals, London, 1957.

Savage and Pritchard[15] investigated the rate of heat transfer in continuous casting of steel by measuring the heat transferred to a water stream passing through the cooled copper mold. Up to about 40 secs after casting, their results could be represented by the relation,

$$\frac{1}{A}\left[\frac{\partial Q}{\partial t}\right] = 64 - 8\sqrt{t}, \text{ cal/cm}^2 - \text{sec.} \tag{8-14}$$

This equation may be integrated to give the average rate of heat transfer over a given period

$$\frac{1}{A}\left[\frac{\partial Q}{\partial t_{\text{av}}}\right] = \frac{1}{t}\int_0^t \left[\frac{\partial Q}{\partial t}\right] dt = 64 - \frac{16}{3}\sqrt{t}. \tag{8-15}$$

In the case of continuous casting with a withdrawal rate of v cm/sec and a mold length of L cm, t may be replaced by L/v so that Eq. (8-15) becomes

$$\frac{1}{A}\left[\frac{\partial Q}{\partial t_{\text{av}}}\right] = 64 - \frac{16}{3}\sqrt{\frac{L}{v}}. \tag{8-16}$$

The prediction of this equation, e.g., that the average rate of heat transfer in the mold increases with the casting velocity, has been confirmed by experiment,[16] where it also was shown that the rate of freezing is proportional to the square root of t. The rate of freezing was found to be of the same order as that occurring in an ordinary chill casting during the first stage of cooling, about 0.65 in./(min)$^{1/2}$. A large increase in the rate of freezing takes place during the second phase where the cooling is intensified by a water spray beneath the mold. In this stage the rate is equal to about 8.44 in./(min)$^{1/2}$, a value much greater than that encountered in chill casting. For continuous casting, a plot of thickness frozen against the square root of t thus consists of two intersecting straight lines.

An investigation was carried out on the continuous casting of low carbon steel in which the thickness of the shell emerging from the water-cooled mold was estimated by measuring the outer slab temperature.[17] Using these temperatures, and assuming a constant temperature gradient across a uniform solidified shell, the thickness of the frozen outer skin was estimated at the time the casting emerged from a 24 × 6½-in. mold. The thickness of the emerging shell was found to decrease with increasing casting rate in qualitative agreement with the results of Savage and Pritchard.[18]

8-5 Other Casting Procedures

Several other casting processes of the continuous or semicontinuous type are either in use or under study in the basic metals industry.

[15] J. Savage and W. H. Pritchard, *J. Iron and Steel Inst.* **178,** 269 (1954).
[16] D. M. Lewis and J. Savage, *Metallurgical Rev.* **1,** 65 (1956).
[17] F. D. Jaicks, L. E. Kraay, and M. Tennenbaum, *AIME TRANS.*, 209, 1057 (1957).
[18] Savage and Pritchard, *op. cit.*

In the Hazelett process, liquid metal is poured between two water-cooled rolls or bands to produce a continuous strip, usually $\frac{1}{2}$ to 2 in. in thickness. This process has been used to produce aluminum and attempts have been made to apply it to other nonferrous as well as ferrous materials.

The pressure pouring process developed by Griffin Wheel Company for casting steel freight car wheels has been adopted for casting steel billets and slabs. In this process, molten metal is forced from a closed ladle up through a refractory tube into a mold by compressed air. The molds are made up of an assembly of graphite blocks, and hence, are of adjustable size—lending flexibility to the process. The surface quality of the semifinished mill shapes cast using this method has been very good. The process is being used to cast high alloy and stainless steel slabs.

Bibliography

Basic Open Hearth Steelmaking, Derge, G. Ed., AIME, New York, 1964.

Boichenko, M. C., *Continuous Casting of Steel*, Butterworths, London. 1961.

Butts, A., *Metallurgical Problems*, McGraw-Hill, New York, 1943.

The Making, Shaping and Treating of Steel, McGannon, H. E. (ed.), 9th Ed. United States Steel Corporation, Pittsburgh, 1971.

Continuous Casting of Steel, Special Report No. 89, Iron and Steel Institute, London, 1964.

McBride, D. L. and T. E. Dancy, Eds., Continuous Casting, New York, Interscience Publishers, 1964.

Electric Furnace Steelmaking. AIME, New York, 1962.

Elliott, J. F., and M. Gleiser, *Thermochemistry for Steelmaking*, Addison-Wesley, Reading, Mass. 1960.

Holomon, J. H. and D. Turnbull, *Progress in Metal Physics*, Vol. 4, Pergamon Press, London, 1953.

Jakob, M., *Heat Transfer*, John Wiley, New York, 1949.

Kelley, K. K., *Contributions to Data on Theoretical Metallurgy*, Vol. 10, U. S. Bureau of Mines Bulletin 476, 1949.

Korotkov, K. P., H. P. Mayorov, A. A. S. K. Koortsov, and A. D. Akimendo, *The Continuous Casting of Steel in Commercial Use*, Pergamon Press, New York, 1960.

McAdams, W. H., *Heat Transmission*, 3rd ed., McGraw-Hill, New York, 1954.

Ruddle, R. W., *The Solidification of Castings*, Institute of Metals Monograph and Report Series, 7, Institute of Metals, London, 1957.

Trinks, W., *Industrial Furnaces*, Vol. 1, John Wiley, New York, 1956.

Veynik, A. I., *Theory of Special Casting Methods*, ASME, New York, 1962.

Discussion Questions

1. Discuss the factors to be evaluated in choosing a melting system for a large cast iron foundry.
2. What factors control the rate of scrap melting in a steelmaking operation using the basic open hearth process?
3. Discuss deoxidation practice on steel heats with respect to the advantages and disadvantages of carrying out the practice in the furnace, ladle, and ingot mold.
4. A steel mill has been encountering difficulties in its rimming steel ingots. The gas blowholes in the upper half of the ingot are too numerous and large giving rise to excessive croppage losses. It has been suggested that the steel be poured

under vacuum to remove these gases thus permitting the casting of a sounder ingot. Would you support this suggestion? Why?

5. Sketch a simplified flow diagram of a fully integrated steel mill. The diagram should include the size and number of operating units (typical of today's operation) for producing 3.5 million tons of sheet, strip and bar products, of which 0.5 million tons are high carbon, alloy, and stainless steels.

6. In the manufacture of steel sheets for deep drawing, the product of the hot strip mill is cold-reduced to slightly greater than final gauge, box annealed at 1400°F and temper-rolled to reduce its thickness approximately 3%. The susceptibility of rimming steel, thus treated, to strain aging is known to be due mainly to the presence of nitrogen in the steel.

    (a) Comment on the relative suitability of open hearth, basic Bessemer, and oxygen blown steels for this application.

    (b) A fully killed aluminum-deoxidized steel is not subject to strain aging. Explain.

    (c) It is claimed that the addition of vanadium to the mold does not interfere with the rimming action and produces a nonaging material. Discuss the physico-chemical background for this claim.

    (d) What other elements might be considered for use in producing rimmed or killed ingots for nonaging sheet?

Problems

1. The production of a fine grain in low carbon rimming steels for deep drawing applications has been related to the formation of fine precipitates of carbides, oxides or nitrides in the solid which prevent grain coarsening during annealing. A slight excess of aluminum will ensure a fine grain, but other elements are often more desirable. Vanadium and titanium have both been proposed as additives to ensure a fine grain. One objection raised is that even if the addition is made after aluminum deoxidation, the residual oxygen (an average of about 0.03 wt%) will react with the additive and eliminate the rimming action. If it is necessary to retain 0.02 wt% of either V or Ti in the steel, which, if either, would you recommend as a ladle additive? Justify your answer with calculations. (Teeming Temperature—1550°C.)

2. It has been observed in vacuum melting of iron in magnesia crucibles that the oxygen content often increases markedly, sometimes approaching the saturation point. Demonstrate by calculation that decomposition of the magnesia refractory producing magnesium gas at a low pressure and oxygen in solution in liquid iron is a mechanism by which this phenomenon can occur.

3. Estimate the length of time liquid pig iron can be held in a torpedo ladle. Data:

    Blast furnace tap temperature—2700°F

    Pig iron liquidus—2100°F

    C_p of liquid pig iron—0.18 Btu/lb °F

    Metal weight—150 tons

    Ladle shell emissivity—0.8

Ladle shell convection coefficient—0.3 Btu/hr °F ft²

Ladle shell area—800 ft³

Average ladle shell temperature—550°F

Weight of ladle refractory—30 tons

C_p of refractory—0.24 Btu/lb °F

4. A low carbon steel heat is bottom-poured into the tundish of a continuous casting machine. *Estimate the maximum* amount of nitrogen that could be absorbed by the pouring stream during the transfer to the tundish.

$D_N = 7.5 \times 10^{-4}$ cm²/sec

$\delta = 0.003$ cm

d, diameter of stream = 3 in. (7.6 cm)

h, pouring height = 30 in. (76 cm)

v, average velocity of stream during the pouring operation, 10 ft/sec (305 cm/sec)

5. A 100-g sample of liquid iron–carbon alloy at its melting point is brought into equilibrium with CO gas at a partial pressure of 0.5 atm and then is allowed to solidify under a total pressure of 1 atm. The alloy contains 0.20% C. Estimate how much metal will solidify before bubbles can form. At this stage the concentration of carbon in the solid is 0.14 times that in the liquid. Assume that oxygen is insoluble in solid iron.

6. A plant produces a full range of steels, including some by continuous casting, and also has finishing facilities for most products. One of the principal products of the mill is low carbon killed steel for sheet and plate applications. Two steels are used for this purpose, a silicon and an aluminum killed grade. The specifications are:

	C	Mn	P	Si	Al	S
Al killed	0.03/0.10	0.25/0.40 max	0.035 max	—	0.020/0.030	0.04 max
Si killed	0.03/0.10	0.25/0.40 max	0.035 max	0.12 max	—	0.04 max

Explain the limits on aluminum and silicon. Estimate the residual oxygen in each case as the metal is teemed into ingot molds.

7. It is proposed to treat liquid "rimming" steel with 0.25 wt% vanadium. This practice would provide a deep drawing sheet material which is free from "stretcher strains" by virtue of tying up the dissolved nitrogen as vanadium nitride in the steel. However, in order to maintain surface quality of the steel the rimming action must not be impaired. It has been found in a given plant practice that at least 0.02 wt% oxygen is necessary to provide a satisfactory rim in low carbon steels. Can the vanadium addition be made without degrading surface quality?

8. A large plate is cast and slowly solidified. Assume the following conditions:

    (1) The metal is a dilute solution of solute B in A.

    (2) The segregation coefficient is a constant, k.

    (3)   Equilibrium is obtained at the solidifying phase interface.

    (4)   No diffusion occurs in the solid.

    (5)   The liquid composition is homogeneous.

Derive an expression relating the concentration of solute B in the liquid or at the solidifying interface to the initial concentration of solute, the fraction solidified, and the segregation coefficient.

9. Low carbon steel is being continuously cast as a 10 × 52-in. slab. Using the result of Savage and Pritchard (Eq. 8-16), estimate the solidified thickness as the steel leaves the mold. The steel is cast 70°F above the liquidus. The mold is 30 in. long (27 in. are in contact with the solidifying steel). State your assumptions regarding heat transfer within the slab, particularly for the liquid core.

10. During continuous casting, cooling of the frozen outer shell results in a contraction which increases along the strand as further decreases in temperature take place. In the mold where the initial skin formation takes place, it is important that the frozen skin be supported by the water-cooled copper mold walls. Consequently, continuous casting molds are usually built with a slight taper to prevent the pressure of the ferrostatic head from stretching and tearing the frozen skin. Assume that the frozen thickness of a 10 × 48-in. low carbon steel slab is 0.6 inches as it leaves the mold and that the surface temperature is 400°F below the solidus temperature. The mold is 30 in. long and the liquid meniscus is 3 in. below the top of the mold. The thermal coefficient of expansion of the steel is 1.2×10^{-5} in./in. °F. Estimate the taper required on the wide and narrow faces to compensate for 90% of the contraction in the mold during an average cast.

11. Estimate the local power costs to melt 200 lb of steel in an induction melting furnace. Compare the cost per pound with a 10 ton induction melting unit. Specify power supply, design of melting unit, charge, tapping temperature, etc.

CHAPTER 9

Computer Applications in Extractive Metallurgy

One of the major developments recently made in the scientific world has been the perfection of the electronic computer in both its analog and digital form. The utility of electronic computers in the engineering of metallurgical systems is now well established, and computers undoubtedly will take preeminence as a tool of the metallurgical process engineer. Computers are used presently in research and in operating practice for a variety of applications involving metallurgical calculations and data analysis. Computer applications involving simulation and control of metallurgical processes are rapidly expanding; furthermore, the storage and retrieval of metallurgical information is a subject of increasing import.

The purpose of the present chapter is to outline several computer applications that are particularly germane to extractive metallurgy, and to emphasize the potential applications and the increasing importance of this approach to the engineering of metallurgical processes. In the following chapter, a brief description is presented of digital and analog computers and their potential applications in process engineering. The techniques of operations research in the form of queueing theory and linear programming are introduced, along with a brief outline of metallurgical system simulation. Appendix B contains ten documented computer programs illustrating applications in extractive metallurgy.

9-1 General Nature of Digital and Analog Computers

There are two basic types of large-scale calculating machines in use today, the analog, or continuous variable machine, and the digital, or discrete variable machine. An analog computer is a physical system, mechanical, electrical or optical which is designed in such a way that the variables of the system satisfy the same mathematical laws as do the variables of the problem of interest. The physical process of computation is replaced by a measurement of the physical quantities corresponding to the values of the unknown variables. Digital computers, in contrast, perform their operations by counting, in a manner similar to a desk calculator.

The Digital Computer

The large-scale digital calculator will perform only the simplest types of operations. First of all, it will read, i.e., assimilate information which is supplied to it on punched cards, punched paper tape, magnetic tape, or by other techniques. Secondly, it will remember what it has read. The memory cells of most digital computers consist of magnetic cores, mercury delay lines, cathode-ray tubes, magnetic drums, banks of relays, or electrostatic storage tubes. A relay computer remembers information by having certain relays in an energized condition while others are in the unenergized condition. Information is stored in a similar yes–no manner by other basic electrical components. Thirdly, the digital machine will perform arithmetical operations. Most computers will add, subtract, multiply, and divide. By combination of these basic operations, subroutines may be developed to take roots, raise numbers to powers, or evaluate other commonly used functions. Fourth, the large-scale digital computer may exercise choice. It is able to follow a set of logical expressions, choosing between one set of operations and another on the basis of relative magnitudes of quantities stored within its memory. Finally, the computer can write. Based on instructions given to it, the computer may record on punched cards, tape, or other output devices, the values of quantities stored within its memory.

Figure 9-1 illustrates the performance of a digital computer. The computer reads instructions and data which are presented to it, and then performs the operations upon the data according to the instructions. In performing these operations, the computer may use any of the five facilities which it possesses. These facilities are indicated in the square of Fig. 9-1 which shows the computer as a black box.[1] The output of the computer is then the numerical results called for by the original machine instructions.

A simple example may serve to illustrate the way in which a digital computer may be used. Suppose one wishes to evaluate the polynomial,

$$R = a_0 + a_1x + a_2x^2 + a_3x^3 + \cdots + a_nx^n, \tag{9-1}$$

at each integer value from 0 to 100.

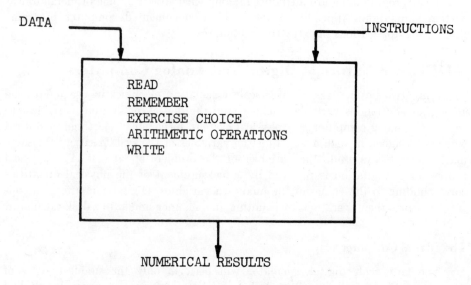

Fig. 9-1. Digital Computer Operation

---

[1] A detailed examination of the internal workings of a digital computer as well as a listing of references on the subject has been prepared: H. H. Goode and R. E. Machol, *System Engineering*, McGraw-Hill, New York, 1957.

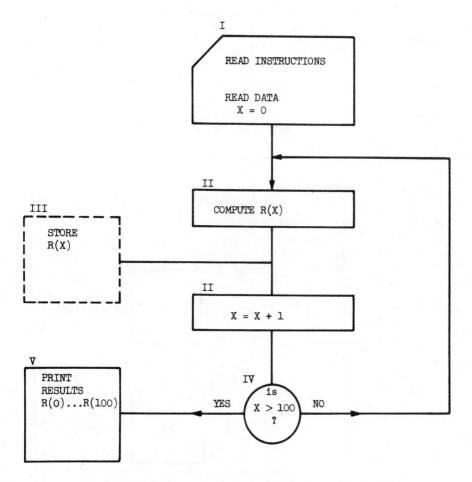

Fig. 9-2. Flow Diagram for Successive Evaluation of Polynomial

The overall organization of the computation is shown by a flow diagram in Fig. 9-2. The instructions and initial data are read into the machine in Input Box I. The machine has been instructed to compute the result, R, for the present value of x, which has been initialized at zero, Box II. The machine then stores the result, $R(x)$, in its memory, Box III. The value of x is then incremented by one and the computer is asked to exercise choice, Box IV. If the value of x exceeds 100, the limit of the range for calculation, the computer proceeds to print the results, Box V; if not, the loop containing instructions for computation and storage, II and III, is repeated, and the test at IV again imposed. When the value of x exceeds 100, the computer will call the values of $R(0) \ldots R(100)$ from memory and write them as Output V.

The flow diagram includes all of the operations which the digital computer is capable of performing: I Read, II Compute, III Remember, IV Exercise Choice, and V Write.

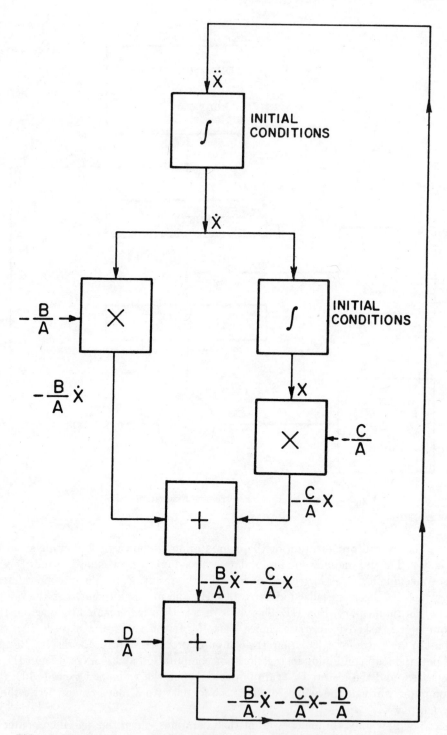

Fig. 9-3. Flow Diagram for Solution of Differential Equation with Analog Computer

The Analog Computer

A particular form of the analog computer that is finding wide application in engineering is the electronic analog computer, or electronic differential analyzer. This device uses operational amplifiers to add, integrate, change sign, and multiply by constants greater than one. Linear potentiometers are used to multiply by constants less than one. Servo-driven potentiometers are used for general multiplication. Amplifier-servo combinations are used to divide one variable by another, and the device often includes special generators to provide certain types of functions. These components are connected to provide an electrical circuit with the same mathematical description as the problem to be solved. The circuit forms a complete loop, one variable being solved for in terms of the remaining variables and then fed back into itself.

Consider an elementary example, a second-order, linear, ordinary differential equation of the form,

$$A \frac{d^2x}{dt^2} + B \frac{dx}{dt} + Cx + D = 0. \tag{9-2}$$

Using dot notation and solving for the highest derivative,

$$\ddot{x} = -\frac{B}{A}\dot{x} - \frac{C}{A}x - \frac{D}{A}. \tag{9-3}$$

A flow diagram may be drawn indicating the solution to Eq. (9-3), as shown in Fig. 9-3. If $\ddot{x}$ is known, $\dot{x}$ and x may be found by integration. By performing the appropriate additions and multiplications indicated by the right side of Eq. (9-3), $\ddot{x}$ is obtained and fed back into the input. The circuit behavior then corresponds to Eqs. (9-2) and (9-3). This type of feedback loop is basic to solutions of equations on analog computers. The initial conditions place constraints on the voltages in the loop in order to satisfy the conditions of the problem, and the results are obtained by output voltages in the circuit.

9-2 Applications of Electronic Computers

The applications of analog and digital computers have, in general, been in keeping with the characteristics of the two types of machines.[2] The analog machine is best suited to performing dynamic simulation of processes, particularly systems that are described in terms of differential relationships with respect to time. Consequently, applications of analog computers in process design have been primarily in process simulation[3] and in the solution of problems involving rate equations.[4–6]

[2] T. J. Williams, Systems Engineering for the Process Industries, Fourth E. P. Schoch Lecture at the University of Texas, Austin, Texas, October 16 and 17, 1959.

[3] H. W. Meyer, N. F. Simcic, W. H. Ceckler and H. N. Lander, *Blast Furnace, Coke Oven, and Raw Materials Proc.* AIME **19**, 580–93 (1960).

[4] R. M. Howe, *Application of Difference Technique to Heat Flow Problems Using the Electronic Analyzer*, Engineering Research Institute, University of Michigan, Ann Arbor, May, 1950.

[5] K. L. Chien, J. A. Hrones and J. B. Reswick, *Trans. ASME*, **74**, 175 (1952).

[6] A. Acrivos and N. R. Amundson, *Ind. Eng. Chem.* **45**, 467 (1953).

The digital computer is best suited to algebraic or logistic problems, as well as to data reduction and processing. Applications include the solution of linear programming models, statistical analysis, and process simulations that are too large or complex for the analog machine.[7] Several applications of a small electronic digital computer to problems in metallurgical research have been explored.[8]

Comparison of Analog and Digital Machines

The question as to whether analog or digital techniques should be employed to obtain a given problem solution often arises. The decision is not an easy one, since the theoretical advantages of one over the other may be essentially lost in practice. The decision should be based primarily on effectiveness and cost, but flexibility for use in other potential applications, ease of procurement, and considerations relative to the implementation of the machine should also be weighed.

In general, the analog machine is less expensive. But whereas both computers can solve any soluble mathematical problem, the digital machine is more suitable for problems which involve logic or massive sets of data. Other writers have discussed in some detail the relative desirability of analog or digital machines for use in solving particular classes of mathematical problems.[9,10] Table 9-1 presents a comparison of the two types of computers for use in process simulation, design, and analysis.

Extractive Metallurgy and System Engineering

The use of computers in metallurgical engineering applications is rapidly increasing, particularly in areas of extractive metallurgy, along with other engineering disciplines, accounting, production control, etc. These interdisciplinary activities are usually staffed by personnel with a variety of backgrounds and directed on a mission, often computer oriented, which requires a contribution from each of these areas. The metallurgist, as a viable and contributing member of such a team, should be prepared to consider his engineering problems in terms of the approach of other disciplines. In the following sections of this chapter, computer-oriented applications of system analysis techniques are introduced. The system optimization approaches considered are queueing theory, linear programming and simulation.

[7] A. L. Hodge, Predicting Effects of Oxygen, Moisture and Fuel Additions on Blast Furnace Operation with Electronic Computers, presented at the General Meeting of American Iron and Steel Institute, New York, May 25, 1960.

[8] R. J. Leary, R. W. Smith, Jr., and B. J. Mitchel, Applications of a Small Electronic Digital Computer to Pyrometallurgical Research, *Bureau of Mines Information Circular 7959*, U. S. Department of the Interior, 1960.

[9] M. Rubinoff, *Proc. IRE* **41**, 1254 (1953).

[10] J. Forrester and A. Vance, Institute of Radio Engineers Meeting, New York, 1951.

TABLE 9-1

Comparison of Analog and Digital Computers[a]

The advantages of analog computers relative to digital computers are:

Analog computer	Digital computer
Simulates behavior of any system by action of easily manipulated and measured variables	Performs arithmetic operations with numbers
Simulation is continuous, permits inclusion of concepts such as distance, velocity, acceleration	Operates discontinuously; can only approximate higher order effects
Results presented as family of graphs of variation of dependent variable (same data obtained with recorder connected to process)	Results presented as tables of numbers, although plotting subroutines are available for most systems
Speed of problem solving is direct function of actual speed of physical system, is independent of size or complexity of system	Speed is direct function of problem complexity and size, relatively independent of operating speed of process
Programmed so that parameter magnitudes are entered as setting on variable potentiometers; parameter values can be changed at will	Flexibility not present without special programming precautions

The disadvantages of the analog relative to the digital computer are:

Analog computer	Digital computer
Size of computer (number of computing components) determines size of problem which can be solved; simplifying assumptions often made to reduce complexity	Problem size reflected in computing time; no limit to size of problem if time is available
Specifically designed for solution of ordinary differential equations; other problems, e.g., solution of simultaneous algebraic equations, treated by trial and error. Less complex partial differential equations solved if converted to ordinary differential equations and one variable assumed constant; family of solutions for different values of variable results	Useful in nearly all types of problems, Complex partial differential equations more amenable to solution by digital equipment

[a] T. J. Williams, *Ind. Eng. Chem.* **50,** 1632 (1958).

The relative advantages of the two types of computers for various applications are:

Application	Analog computers	Digital computers
Equipment and plant design	Show great promise for working out relative sizes of plant equipment by solving dynamic equations of plant performance	Superior for details of mechanical design, e.g., stresses, sizing of members, costs, which are trial and error arithmetical problems
Instrumentation and data reduction	Small, special purpose analogs, important in converting and plotting single variable and directly correlated multiple variable data in form of graphs	
Process control simulations (research)	Best tool in existence for studying resulting phenomenon; limitation on system complexity imposed by computer size	
Plant control applications	Definite promise as final control elements for units of highly automated plants	Flexibility and accuracy of digital computers better for control of whole plants
Research in basic processes	Well suited to simulation of reaction kinetics involved in process development	Better for determination of kinetic parameters by statistical data reduction where necessary
Research in unit operations	Well suited for study of transient state such as heat transfer, distillation	Better suited for steady-state studies which are usually arithmetical or statistical
Plant operations and management		Better here because statistics and arithmetical operations predominate

9-3 Material Handling—A Queueing Theory Approach

The handling of materials between unit processes in a system plays an extremely important role in determining the overall efficiency of the operation. As a prime example of a computer application in metallurgy, a discussion will follow of a queueing theory approach to material handling. Although it is not an exhaustive treatment of the subject[11] it should lay a good foundation for the metallurgical

---

[11]A sound approach to plant design engineering (T. C. Vilbrands and C. E. Dryden, *Chemical Engineering Plant Design*, McGraw-Hill, New York, 1959.) and a discussion of some of the details involved in material handling [Session on materials handling, *Electric Furnace Steel Proceedings* **7**, 5 (1949)] are available in the literature.

engineer in outlining some of the methods and procedures that are available to him when considering the design of systems that involve the transport of materials.

Queueing Theory

Queueing theory is used to describe the mathematical approaches to problems arising whenever delays occur or priorities must be arranged regarding the sequence of a given operation. The mathematical models and solutions of the problems are the same whether the individuals or things are waiting or whether a gate is moving around to the waiting persons or objects. Applications of queueing theory have been made to a large number of problems in which waiting time is involved, including the landing of aircraft, the loading and unloading of ore ships, the design of automobile parking, waiting and traffic facilities, the passage of travelers through customs, pedestrian movements, and servicing machine failures.[12,13]

In the development of a queueing model one must provide for (1) gate or service points, (2) an input process, (3) some queue discipline, and (4) some service mechanism. The model then describes the situation in which customers arrive at a gate or service point, are serviced, and then leave the service area. The arrival of customers is the input process, and may be described in terms of the distribution of arrivals at the service point. As the product of the input process arrives at the service point and joins a queue, it is termed a customer. The behavior of the customer is described in terms of a queue discipline that determines how the customer reacts to the exsiting condition of the queue. If the queue is too long, the customer may leave the service area and be lost to the system. In the case of a multiple channel servicing system, the customer may join the queue of shortest length. Or the customer may be required to remain in the queue, regardless of the apparent waiting time. The service mechanism is a description of the service provided in terms of the service time involved, which may be a constant or may decrease slightly with increase in queue length, or even be statistically distributed in some manner.

The elements of interest in an analysis of the queueing problem are: (1) waiting times for the customers, (2) the number of customers in the queue, (3) the ratio of waiting time to service time.

The purpose in seeking a model which describes the queueing problem is to provide for the servicing of as many customers with as few facilities as possible. Generally, one optimizes the economics of the situation by seeking the minimum cost involved for some balance of waiting time between servicing unit and customer.

Single Station Queueing Problem

Let us consider the problem of determining the probability of a given queue length and the expected queue length for the case of a single station for which both input and output are assumed to be random. (Specifically, a single channel queue system, having a Poisson input and a negative exponential service time distribution

[12] Marshalling and Queueing, *Operational Res. Quart.* **3**, No. 1 (1952).
[13] D. G. Kendall, *J. Roy. Stat. Soc. B* **13**, 151 (1951).

in the steady state). A queue discipline is assumed such that each arriving unit takes the last position in the queue and that the units which make up the line are serviced in order of appearance in the line, i.e., first in, first out. It may be shown[14] that the probability of a waiting line of length n units P_n is given by

$$P_n = \left(\frac{\lambda}{\mu}\right)^n \left(\frac{1 - \lambda}{\mu}\right), \tag{9-4}$$

if λ/μ is less than one, where λ is the mean arrival rate, and μ the mean service rate; (if λ/μ were greater than one, the line would grow infinitely long). The ratio λ/μ is sometimes called "traffic intensity" and is the average number of arrivals per unit of service.

Then the mean length of a waiting line is given by the expression,

$$\bar{n} = \frac{\lambda/\mu}{[1 - (\lambda/\mu)]} \tag{9-5}$$

if λ/μ is less than one.

The expected waiting time of arrivals at a single station with random input can be formulated as follows:

Let $\bar{t}_w$ = expected waiting time and $\bar{t}_s$ = expected time spent in service; then $\bar{t}_w + \bar{t}_s$ = total expected time consumed in both waiting and service. When the mean arrival rate is λ,

$$\bar{n} = \lambda(\bar{t}_w + \bar{t}_s) \tag{9-6}$$

from which

$$t_w = \frac{\bar{n}}{\lambda} - \bar{t}_s. \tag{9-7}$$

Since

$$t_s = \frac{1}{\mu}. \tag{9-8}$$

By substitution in Eq. (9-5)

$$\bar{t}_w = \frac{1}{(\mu - \lambda)} - \frac{1}{\mu} \tag{9-9}$$

describes the expected waiting time at a single station.

In order that the real life situation be more closely approached by the queueing model, specific distributions for the probability of arrival often have been substituted in the model. In several cases data taken under actual operating conditions for a sufficiently large sample have been shown to be approximated very closely by exponential or Poisson distributions.

Example of Single Station Queueing Problem

Consider the case of a heat treating section consisting of a single furnace that provides service for the many areas of a manufacturing plant. The job requirements

[14] C. W. Churchman, R. L. Ackoff, and E. L. Arnoff, *Introduction to Operations Research*, John Wiley, New York, 1957.

vary in a random manner, as do the arrivals of parts for heat treating. The section is operated as a closed system, and parts arriving remain until they are treated with each job handled in the order in which it arrives. This hypothetical situation corresponds very closely with the queueing model derived above.

If the mean arrival rate λ is 5 jobs per day and the mean service rate μ is 10 jobs per day, the traffic intensity $\lambda/\mu = \frac{5}{10} = \frac{1}{2}$. The probability of a waiting line of given length, n, is

$$P_n = (\tfrac{1}{2})^n (1 - \tfrac{1}{2}) \tag{9-10}$$

and the average number of jobs waiting is

$$\bar{n} = \frac{\tfrac{1}{2}}{(1 - \tfrac{1}{2})} = 1. \tag{9-11}$$

If the traffic intensity λ/μ increases, the average queue length increases, approaching infinity, as the service time approaches the arrival time, i.e. $(1 - \lambda/\mu) \rightarrow 0$.

The analysis above is important to the management of the manufacturing firm in determining the optimum heat treating facilities for handling a given distribution of job requests. In the simplified case just stated, an increase in the arrival rate to a level approaching 10 jobs per day would require that either the service times be decreased or that a multiple station arrangement be substituted, i.e., that the capacity of the facilities be increased.

Multistation Queueing Problem

Consider the case in which customers arrive at a service station and receive service from several units, S, under the assumptions made for the single station model. The probability of having to wait in line, which is the sum of all probabilities that all service facilities are being used or that S or more customers are in line, is given by the expression,

$$W = \sum_{n=S}^{\infty} P_n = \left(\frac{\lambda}{\mu}\right)^S \frac{P_0}{S!(1 - \lambda/\mu S)}, \tag{9-12}$$

where W is the probability that all service facilities are being used and P_0 is the probability that zero customers are being served. The average waiting time for the case of multichannel servicing facilities is

$$\bar{t}_w = \frac{P_0}{\mu S(S!)[1 - (\lambda/\mu S)]^2} \left(\frac{\lambda}{\mu}\right)^S \tag{9-13}$$

and it may also be shown that the probability of no customers being served, P_0, is given by the expression,

$$P_0 = 1 \left/ \left\{ \left[\sum_{n=0}^{S-1} \frac{(\lambda/\mu)^n}{n!} \right] + \left[\frac{(\lambda/\mu)^S}{\{S![1 - (\lambda/\mu S)]\}} \right] \right\} \right. \tag{9-14}$$

Example of Multistation Model

The crane facilities provided in an open-hearth shop for tapping, teeming, and other pouring floor service requirements may be optimized by application of the multistation queueing model described above.

While strictly speaking, the behavior of the handling system in a steelmaking plant does not follow the assumption of random input and output (in fact in most plants a large staff of people are engaged in preventing it from becoming so!), the model permits some examination of the problem and represents a situation more extreme than normal. It may be assumed that the facilities, one, two, three, or more cranes, are the service gates which provide service to the customers (the furnaces), and do so on a basis that permits any free crane to handle the service requirements of a furnace, the furnaces queueing until service may be provided. This last assumption also is not accurate, since the cranes usually operate on the same track and are not free to move around each other to change their relative position. However, the assumption may be justified on the basis that a crane that has received a ladle of metal may transfer that ladle to the next crane and then proceed to provide service to a furnace further down in the shop.

Example Problem 9-1. Consider an open hearth shop with twelve operating 200-ton furnaces which average 10 hr/heat. The service times for tapping, pouring, and removing the slag ladles average 1.0 hr/tap. Assuming a cost of \$300/hr for the furnaces and \$50/hr for the cranes, determine the optimum number of cranes.

Solution. The average time between arrivals is 0.833 hr, since in a 10-hr period, an average of 12 heats would require servicing. In a 24-hr period, an average of 28.8 arrivals would occur, requiring a total of 28.8 hr of service. This service requirement dictates that at least two cranes are necessary.

Evaluating the expected waiting times for the furnaces for two cranes may be done as follows. From Eq. (9-14), where the traffic intensity $\lambda/\mu = 1/0.833/1/1.0 = 1.2$ heats per average service time

$$P_0 = 1 \bigg/ \left\{ \left[\sum_{n=0}^{1} \frac{(1.2)^n}{n!} \right] + \left[\frac{(1.2)^2}{2![1 - (1.2/2)]} \right] \right\} \tag{9-a}$$

$$= 1 \bigg/ \left[(1 + 1.2) + \frac{(1.2)^2}{0.8} \right] = 0.250. \tag{9-b}$$

$$\bar{l}_w = \frac{0.250}{(1)(2)(2)[1 - (1.2/2)]^2} (1.2)^2 = 0.561. \tag{9-c}$$

One may thus prepare a table, Table 9-2, indicating the waiting times for the furnaces in average service time units.

The idle hours of the furnaces and cranes are given in Table 9-3. The total waiting time for the furnaces is given by the average waiting time per heat times the number of heats per day, 28.8. The total waiting time of the cranes is given by the available hours, 24 times the number of cranes, minus the required hours of service, 28.8.

TABLE 9-2

Summary of Calculations for Multistation Queueing Example

Arrivals per service time	service time	No. of cranes, S	P_0	Average furnace waiting time	
				Av. Ser. Time units	Hours
1.2	1.0	2	0.250	0.561	0.561
1.2	1.0	3	0.294	0.078	0.078
1.2	1.0	4	0.300	0.013	0.013

The total cost of idle time may be computed as shown in Table 9-3. The optimum service facilities are indicated by the minimum cost, and in this case are shown to be 3 cranes, a result in keeping with practice in the industry.

Other Approaches to Queueing Problems

The analytical methods presented above are relatively straightforward and the examples used to illustrate them have been greatly simplified. Only a restricted number of cases can be validly treated by these methods. There are, however, techniques for determining the characteristics of a queueing system when the arrival and service distributions are not conveniently expressed mathematically. These methods are described in the literature and listed under several headings, which include Monte Carlo techniques, Theory of Games, and Stochastic Processes.

The problems involved in analyzing queueing situations which arise in actual integrated plant operations are extremely complex and often prohibitively difficult. However, advances are being rapidly made and new applications are being generated at a high rate. Consideration of these concepts is inherent to the system approach in

TABLE 9-3

Cost Analysis for Service Facilities in Open Hearth Shop

No. of cranes	Idle hr cranes	Cost of cranes	Idle hr furnaces	Cost of furnaces	Total cost per 24 hr
2	19.2	$ 960	16.2	$4860	$5820
3	43.2	$2160	2.25	$ 675	$2835
4	67.2	$3360	0.38	$ 114	$3474

metallurgical process design, and greater emphasis will undoubtedly be placed on the queueing theory approach to material transportation problems.[15-23]

9-4 Linear Programming

Linear programming is a mathematical procedure used for solving a general class of optimization problems that involve a combination of a number of interacting factors to produce a maximum or minimum result. In order to apply linear programming, the following criteria must be met: (1) There is some function that is to be made maximum or minimum; (2) There are a variety of solutions, each of which is subject to a set of well-defined restrictions that may be expressed as equalities or inequalities; and (3) Among the variables of the problem, the relationship between any two may be represented by a straight line or linear relationship at least to an acceptable degree of approximation.

The use of the linear programming model may be illustrated by the following example.

Example Problem 9-2. Consider a metallurgical firm that wishes to produce an alloy steel of the following specifications: vanadium—0.8% minimum; chromium—1.8% minimum; manganese—0.6% minimum. In addition, the firm possesses a supply of two alloying materials of the following compositions:

	Alloy 1	Alloy 2
%V	40%	10%
%Cr	40%	30%
%Mn	10%	30%
%Fe	Balance	Balance

where Alloy 1 costs \$4/lb and Alloy 2 costs \$2/lb.

Solution. On the basis of the production of 100 lb of the desired alloy steel, let x_1 equal the number of pounds of Alloy 1 used and x_2 equal the number of pounds of Alloy 2 used, both as additives to molten low carbon steel scrap of negligible cost.

A combination of x_1 and x_2 will produce a solution to the problem if and only if that combination satisfies the specifications for the alloy steel. This may be expressed as

$$0.4x_1 + 0.1x_2 \geq 0.8 \quad \text{Vanadium balance,} \tag{9-a}$$

[15] J. F. McCloskey and F. N. Trefethen, Eds., *Operations Research for Management*, The Johns Hopkins Press, Baltimore, 1954.

[16] P. M. Morse, *Queues, Inventories, and Maintenance*, New York, John Wiley, New York, 1958.

[17] W. Feller, *An Introduction to Probability Theory and its Application*, John Wiley, New York, 1950.

[18] M. D. J. Brisby and R. T. Eddison, *J. Iron and Steel Inst.* **172**, 171 (1952).

[19] D. P. Gayer, *J. Operations Res. Soc. Amer.* **2**, No. 2 (1954).

[20] L. C. Eddie, *J. Operations Res. Soc. Amer.* **2**, 107 (1954).

[21] T. Fry, *Probability and Its Engineering Uses*, 2nd ed., D. Van Nostrand Co., Inc., Princeton, N.J., 1965.

[22] M. Sasieni, A. Yaspan, and L. Friedman, *Operations Research*, John Wiley, New York, 1959.

[23] A. Charnes, W. Cooper, and A. E. Henderson, *An Introduction to Linear Programming*, John Wiley, New York, 1953.

$$0.4x_1 + 0.3x_2 \geq 1.8 \quad \text{Chromium balance,} \qquad \text{(9-b)}$$

$$0.1x_1 + 0.3x_2 \geq 0.6 \quad \text{Manganese balance,} \qquad \text{(9-c)}$$

where

$$x_1 \geq 0 \qquad \text{(9-d)}$$

and

$$x_2 \geq 0 \qquad \text{(9-e)}$$

It is possible to show graphically the feasible solutions which satisfy the set of inequalities presented above. The shaded region of Fig. 9-4 contains all the allowable solutions.

The solution to the problem is found in the optimum combination of alloying additions which minimize the cost function,

$$C = 4x_1 + 2x_2. \qquad \text{(9-f)}$$

The cost function may be plotted as a family of parallel lines whose parameter is total cost. The family of parallel lines may be superimposed on the graph as shown in Fig. 9-4 from which it may be concluded that the optimum alloying addition will occur at the vertex of the line-bounded region of feasible additions. Since the firm seeks the lowest cost, the optimum alloying addition occurs at the point $(0.75, 5)$ indicating that the optimum addition consists of

$$\left. \begin{array}{l} \tfrac{3}{4} \text{ lb of Alloy 1} \\[2mm] 5 \text{ lb of Alloy 2} \end{array} \right\} \text{ per 100 lb of steel}$$

at a cost of $(\tfrac{3}{4})4 + (5)2 = \$13/100$ lb of steel.

As the number of variables in a linear programming problem increases, visualization of the region of feasible solutions is more difficult. In three dimensions, the region of feasible solutions is represented by a polyhedron bounded by planes corresponding to the problem restrictions. As in the two-dimensional case, the optimum solution will occur at a vertex of the polyhedron, and the cost function will define a family of planes. As the number of variables increases, the geometry becomes more and more complex. As a result, it is necessary to turn to an algebraic approach that may be applied to any number of variables and is most suitably carried out on a digital computer. The simplex technique is the most general linear programming model,[24-31] and permits a directed exploration of the extreme points of the problem. A documented computer program for executing the simplex tech-

[24] G. B. Dantzig, *Activity Analysis of Production and Allocation* (T. C. Koopmans, Ed.), Cowles Commission Monograph, Vol. 13, Chap. 21, John Wiley, New York, 1951.

[25] Churchman, Ackoff, and Arnoff, *op. cit.*

[26] *Operations Research for Management, op. cit.*

[27] A. Charnes, W. Cooper, and A. E. Henderson, *An Introduction to Linear Programming*, John Wiley, New York, 1953.

[28] An-Min Chung, *Linear Programming*, Charles E. Merrill Books, Inc., Columbus, Ohio, 1963.

[29] S. Gass, *Linear Programming*, McGraw-Hill, New York, 1964.

[30] M. Simonnard, *Linear Programming*, Prentice-Hall, Inc., Englewood Cliffs, N. J., 1960.

[31] G. B. Dantzig, *Linear Programming and Extensions*, Princeton University Press, Princeton, N. J., 1963.

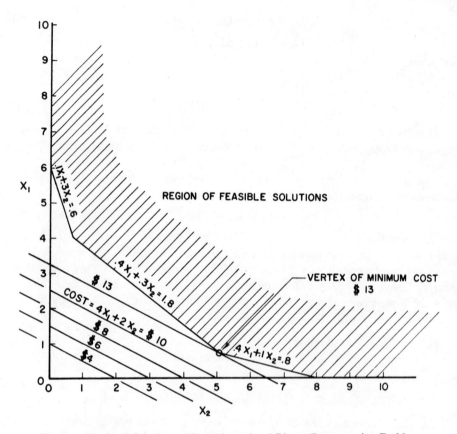

Fig. 9-4. Graphical Solution of Two-Dimensional Linear Programming Problem

nique has been presented.[32,33] (See Appendix B, Example Problem 10). In addition, most computing centers have linear programming models available as subroutines.

In certain special cases, the linear programming problem can be solved by the "transportation technique." Problems of this type also can be solved by the simplex method, but are solved more directly by the transportation technique. The general transportation problem involves a number of origins from which products are shipped to a given number of destinations, each requiring a given number of units of product. The cost of shipping is to be minimized. Techniques for treating this problem have been presented.[34]

Applications of Linear Programming

Linear programming has found considerable application in the field of business planning and operation. A few problems that have been solved by linear program-

[32] Simonnard, *op. cit.*

[33] B. Carnahan, Private Communication, University of Michigan, 1967.

[34] A. Charnes and W. W. Cooper, Appendix, *Management Sci.* **1,** No 1 (1954).

ming techniques include the blending of aviation gases,[35] contract awards,[36] allocation of manufactured products,[37,38] optimum treatment of a natural resource,[39] and the long-range planning of coke oven replacements.[40]

The use of linear programming for control of ·technical processes has been developing rapidly in all areas of engineering, particularly for the chemical industry.[41–43] Fabian[44] has proposed the use of linear programming to determine the least cost rate of input of materials into an integrated steel mill. Although Fabian's treatment of the problem is limited, it clearly illustrates the potential of the method and represents an excellent approach to a relatively difficult problem. Linear programming has also been used to predict the minimum materials cost for melting stainless steels.[45,46]

It should be noted that the foregoing discussion is by no means complete. Optimization techniques extend beyond linear programming and include dynamic programming.[47–50] Dynamic programming has been applied to the optimization of stainless steel melting practice.[51] Other systematic techniques for determining optima have been reviewed.[52] The use of these optimization techniques, particularly in conjunction with large-scale digital computers, should offer an excellent means for utilizing the vast amount of data that is being produced in various areas of the process metallurgical industry. By combining these data with available operating and economic information in the form of optimization models, many problems— both economic and technical—that face the metallurgical process engineer may be greatly clarified and suitable solutions may be worked out using these techniques.[53,54]

[35] A. Charnes, W. W. Cooper, and B. Mellon, Blending Aviation Gasolines—A Study in Programming Interdependent Activities, Project SCOOP, Symposium of Linear Inequalities and Programming, Headquarters, U. S. Air Force, Washington, 1952.

[36] L. Goldstein, *The Theory of Gains and Economic Behavior*, Princeton University Press, Princeton, N.J., 1947.

[37] E. L. Arnoff, An application of linear programming, *Preceedings of the Conference on Operations Research in Production and Inventory Control*, Case Institute of Technology, Cleveland, Ohio, 1954.

[38] W. W. Cooper, A. Charnes and D. Farr, *Linear Programming Models for Scheduling Manufactured Products*, Carnegie Institute of Technology, Pittsburgh, September 1, 1952.

[39] R. Klimpel and M. Klein, *Symposium for the Use of Computers and Operations Research in the Mineral Industries*, University Park, Penn., April, 1966.

[40] T. D. Truan and D. A. Porco, *Management Sci.* 4, 337 (1958).

[41] W. B. White, S. M. Johnson, and G. B. Dantzig, *J. Chem. Phys.* 28, 751 (1958).

[42] P. Stillson, *Ind. Eng. Chem.* 48, 402 (1956).

[43] J. M. Henderson, *Rev. Econ. Stat.* 37, 336 (1955).

[44] T. Fabian, *Management Sci.* 4, 415 (1958).

[45] D. C. Hilty, R. W. Taylor, and R. H. Gillespie, *J. Metals* 11, 458 (1959).

[46] IBM: Data Processing Application, Paper No. B 20-0147-0.

[47] R. Bellman, *Dynamic Programming*, Princeton University Press, Princeton, N. J., 1957.

[48] G. Nemhauser, *Introduction to Dynamic Programming*, John Wiley, New York, 1966.

[49] D. F. Rudd and C. C. Watson, *Strategy of Process Engineering*, John Wiley, New York, 1968.

[50] R. Aris, *Discrete Dynamic Programming*, Blaisdell Publishing Company, New York, 1964.

[51] E. Calanog and G. H. Geiger, *J. Metals* 19, No. 7, 96 (1967).

[52] B. Carnahan, H. A. Luther, and J. O. Wilkes, *Applied Numerical Methods*, John Wiley, New York; 1969.

[53] J. J. Taborek, *Chem. Eng. Prog.* 56, 37 (1960).

[54] C. Berg, *Chem. Eng. Prog.* 56, 42 (1960).

9-5 Process Simulation

One of the principal applications of computers to process engineering is in the area of simulation.[55-62] A detailed computer simulation of a process to determine the effect of all process parameters on yields and on by-product formation requires the derivation of a complete reaction model. An accurate simulation of a process permits optimum processing conditions to be specified and serves as a basis for control of a process or combination of processes.[63]

The derivation of an accurate dynamic model is often not an easy task, and one is faced with several alternatives.

1. The dynamics of the process may be described in terms of a series of differential equations that represent the mass and energy transfer processes which occur in the reactor.

2. The process dynamics may be described in terms of the operating results of similar reactors that are already in commercial use. A detailed description of the change in chemical concentrations, temperatures, flow rates, etc., as functions of time may be reduced to descriptive functions by numerical analysis with the assistance of a digital computer, or by trial and error techniques using an analog computer.

3. The system may be described in terms of the average operating characteristics for the particular type of reactor, based on the observations made on commercial-sized units. The particular reactor may present a characteristic concentration-time relationship, or may involve temperature dependences that are easily observable. In this case, the average operating results may be applied to the system which is in the design stage. This model is not entirely a dynamic one, but it is certainly highly descriptive of the process and may be used to some advantage.

4. In the case of metallurgical reactors in which the mechanisms of the processes being carried out are not completely understood, one may neglect the process dynamics and select a fixed process time for a batch unit or a suitable flow rate for a continuous unit on the basis of either previous experience or good engineering judgment. Although this final choice is far

[55] T. J. Williams, *Systems Engineering for the Process Industries*, The Fourth E. P. Schoch Lecture at the University of Texas, Austin, Texas, October 16 and 17, 1959.

[56] D. P. Campbell, *Process Dynamics*, John Wiley, New York, 1958.

[57] D. P. Eckman, *Automatic Process Control*, Chap. 2, John Wiley, New York, 1958.

[58] *Project Cyclone Symposium I on Reac Techniques*, Port Washington, N. Y., March 15 and 16, 1951, SDC, USN.

[59] *Project Cyclone Symposium II on Simulation and Computing Techniques*, Port Washington N. Y., April 28–May 2, 1952, SDC, and BUAER, USN.

[60] *Symposium III on Simulation and Computing Techniques*, Port Washington, N. Y., October 12–14, 1953, BAUER and NADC, USN.

[61] *Proc. Natl. Simulation Conference*, Sponsored by Institute of Radio Engineers, Dallas, Texas, January 19–21, 1956.

[62] H. H. Goode, *Proc. IRE* **39,** 1501 (1951).

[63] A. B. Chelyustkin, *The Application of Computing Technique to Automatic Control Systems in Metallurgical Plant*, Pergamon Press, New York, 1964.

from a satisfactory one, the reduction of the process model to a mass and energy balance does permit the evaluation of many of the process variables and furnishes a basis on which the dynamics of the process may be either estimated or evaluated after construction of the pilot plant or commercial-sized unit.

The set of simultaneous equations describing the process system includes one for each separate chemical species formed or utilized in the chemical reactions comprising the process. The mass and energy balances for each chemical species are interrelated in the form,

$$\frac{dQ}{dt} = -V \frac{dC}{dt} \Delta H, \qquad (9\text{-}15)$$

where

V = factor that converts concentration to weight units,

C = concentration of product,

ΔH = change in thermal energy during the reaction for each weight unit formed,

$\dfrac{dQ}{dt}$ = rate of heat gain to the complete reactor system.

Heat transfer equations may then be set up to include:

1. The energy manifested as a temperature change in the reaction mass.
2. The energy representing that which appears as a temperature change in the reaction vessel.
3. The energy transferred to or from the surroundings.

Having expressed the process variables in such a set of simultaneous differential equations, the dynamics of the system may be included in the process model in the form of a differential equation involving the changes in concentration with respect to time as functions of process geometry and chemical driving forces.

Consider, for example, the reaction which takes place when a reactive gas is contacted with a liquid metal. The differential equation describing the rate of solution of gas is

$$\frac{dc}{dt} = k' \frac{A}{V} (c_e - c), \qquad (9\text{-}16)$$

where

$\dfrac{dc}{dt}$ = rate of change of concentration, wt%/sec,

k' = rate constant, cm/sec,

A = area of gas–metal interface, cm^2,

V = volume of melt, cm^3,

c_e = equilibrium concentration of dissolved gas, wt%,

c = instantaneous concentration of dissolved gas, wt%.

If the particular gas under consideration is nitrogen, and the melt is liquid pure iron, the mass and heat balances may be interrelated by the relationship,

$$\frac{dQ}{dt} = - \left[\frac{dc}{dt}\right] \left[\frac{W}{100}\right] \left[\frac{860}{14}\right], \qquad (9\text{-}17)$$

where

$\dfrac{dQ}{dt}$ = energy transferred to the system, cal/sec,

W = weight of melt, g,

$\dfrac{860}{14}$ = heat of solution, cal/g.

The dynamics of the process may then be expressed in terms of Eqs. (9-16) and (9-17). These equations, when coupled with the mass and energy balances, may then form a process model suitable for simulation. The dynamics of this simple process are very well understood, and the computer simulation of such a process would yield highly accurate results. This particular model would correspond to the dynamic process Model 1 indicated above.

Process Model 2 would involve the observation of the behavior of the system with respect to mass changes, that is, gas going into solution in the metal and temperature changes along with the variation in those quantities entering into the mass and energy balances. Such data then could be reduced on a computer to a dynamic model that would yield a highly accurate simulation of the process, accuracy, of course, depending upon the reliability and extensiveness of the data taken.

Process Model 3 would involve the averaging of data observed on many systems involving gas–metal reactions, and the application of such generalized gas–metal behavior to the system under consideration. One might assume, for example, that the gas content of the metal varies linearly with time and that the average time of solution is given by a certain number of minutes. Such a model, of course, is not highly accurate, but does permit a reasonable prediction of the behavior of such a gas–metal equilibration process.

Process Model 4, which ignores the variations in process variables with respect to time and considers only the overall result, also can be applied to this particular process. If the equilibrium state of the system is known, and one has reason to believe that sufficient time will be allowed to approach this equilibrium condition, the mass and energy balances of the system may be written on this basis.

The example system selected above was a highly simplified one. In the general case, there are a large number of concentration variables as well as many thermal and geometrical parameters involved. In most metallurgical process operations, one is faced with the relatively difficult task of not only evaluating each of these variables, but also attempting to write a descriptive model that involves their interactions as well. Much of this data has become available in the last few years and considerable work is presently being expended in this direction. It is thus expected that Approach 1 will become more and more important as metallurgical process operations are better understood. In the present situation, the availability of high-speed computing equipment as well as the installation of better instrumentation and control systems in metallurgical plants should permit very advantageous use to be made of Approach 2. The development of data processing techniques and the success with which simulation of processes has been used, particularly in the chemical industries, should be an indication to the practicing metallurgist that considerable clarification may be brought to bear on the operation of metallurgical processing plants.

In attempting to write descriptive models for the purpose of process simulation, one should bear in mind some of the basic elements of process dynamics. It may be noted that the dynamic behavior of a chemical process depends upon the kinetics of the reactions involved, the manner in which the environment variables change with time, and the geometry of the system. To relate these factors quantitatively requires that differential equations be formed to express the interrelation that exists between physical variables. As indicated above, these data generally are not available, or if available, the resulting equations are not solved easily by means of arithmetical methods because they are nonlinear. However, it often is unnecessary to make a precise study of the dynamics of the process in order to approximate sufficiently the reactor dynamics to a degree that may be useful in obtaining engineering results.

The rates of metallurgical reactions may range from very slow to very rapid. The relative speed of the reaction with respect to the residence time in the system determines whether the chemical reaction rates or the material handling steps have the dominant role in the process kinetics. When mixing lags are small and reaction rates are slow, the dynamic behavior of the system may be determined largely by the chemical kinetics.

However, when reaction rates are high and material handling lags large, the situation reverses. Instantaneous conversion of charge material into product can be assumed and the kinetics of the process treated on the basis of the material-handling dynamics of the product emerging from the system. If material handling is the rate-controlling step, static material balance equations may be used to describe the conversion of raw material into product in an "ideal system." These equations can be modified then by the appropriate functions to describe mixing and transportation of material.

In addition to the above concepts, a number of assumptions often are made to facilitate the formulation of a descriptive process mode. Batch and continuous stirred systems generally are considered to be perfectly mixed, i.e., there are no temperature or concentration gradients in the reacting materials. The heat transfer coefficients throughout the system are considered to be constant at the ambient operating temperatures of the system. The use of these assumptions, of course, involves their justification in each application for which they are used. One must often rely on engineering judgment, particularly where previous experience in the area is not available.

Bibliography

An-Min Chung, *Linear Programming*, Charles E. Merrill Books, Inc., Columbus, Ohio, 1963.
Bellman, R. *Dynamic Programming*, Princeton University Press, Princeton, N.J., 1957.

Campbell, D. P., Process Dynamics, John Wiley, New York, 1958.

Carnahan, B., H. A. Luther, and J. O. Wilkes, *Applied Numerical Methods*, John Wiley, New York, 1969.

Charnes, A., W. Cooper, and A. E. Henderson, *An Introduction to Linear Programming*, John Wiley, New York, 1953.

Chelyustkin, A. B., *The Application of Computing Technique to Automatic Control Systems in Metallurgical Plant*, Pergamon Press, New York, 1964.

Churchman, C. W., R. L. Ackoff, and E. L. Arnoff, *Introduction to Operations Research*, John Wiley, New York, 1957.

Pehlke, R. D. and M. J. Sinnott, Eds., *Computer Applications in Metallurgical Engineering*, American Society for Metals, Metals Park, Ohio, 1964.

Dantzig, G. B., *Linear Programming and Extensions*, Princeton University Press, Princeton, N. J., 1963.

Eckman, D. P., *Automatic Process Control*, John Wiley, New York, 1958.

Electronic Computers in Engineering Education, First Annual Report of a Ford Founcation Project in the College of Engineering, The University of Michigan, August 26, 1960.

Fry, T., *Probability and Its Engineering Uses*, 2nd ed., D. Van Nostrand Co., Inc., Princeton, N.J., 1965.

Gass, S., *Linear Programming*, McGraw-Hill, New York, 1964.

Goode, H. H. and R. E. Machol, *System Engineering*, McGraw-Hill, New York, 1957.

Morse, P. M., *Queues, Inventories, and Maintenance*, John Wiley, New York, 1958.

Nemhauser, *Introduction to Dynamic Programming*, John Wiley and Sons, Inc., New York, 1966.

McCloskey, J. F. and F. N. Trefethen, Eds., *Operations Research for Management*, The Johns Hopkins Press, Baltimore, 1954.

Rudd, D. F. and Watson, C. C. *Strategy in Process Engineering*, John Wiley and Sons, Inc., New York, 1968.

Sasieni, M., A. Yaspan, and L. Friedman, *Operations Research*, John Wiley, New York, 1959.

Simonnard, M., *Linear Programming*, Prentice-Hall, Inc., Englewood Cliffs, N. J., 1960.

Vilbrands, T. C. and C. E. Dryden, *Chemical Engineering Plant Design*, McGraw-Hill, New York, 1959.

Discussion Questions

1. Describe a general systematic approach to the development of a computer control system for an extractive metallurgical process. Include in addition to a philosophy, instrumentation, mathematical model building with special emphasis on the key links to kinetic descriptions of processes, open-loop versus closed loop control, integration of control system and process, and optimization of the individual process with the entire plant.

2. Diagram and set up a flow chart describing the movement of all materials in and out of a three vessel basic oxygen steelmaking plant.

3. Discuss computer control of the iron blast furnace. What variables should be monitored? What changes in operation can be effected to provide process control? Outline a means for developing the interrelationship between the monitored process variables and the control parameters.

4. Prepare an algorithm for a computation procedure which would simulate the heat flow and temperature loss for steel between tapping from the steelmaking furnace and teeming into a continuous casting strand or ingot molds.

5. Hybrid (analog-digital) computers have been developed to take advantage of combining the unique characteristics of both types of computers. Discuss possible applications of hybrid computers in simulation and control of metallurgical processes.

6. Sintering plants for ore preparation represent a highly important portion of the process of steel production. A high rate of production and the need for presetting the mixture of material to be sintered with continuous regulation requires that sintering be done with a high degree of automation. Consider the feed to a sinter bed to consist of the following:
    - (1) ore
    - (2) mill scale
    - (3) flue dust—from blast furnace and oxygen converter
    - (4) sinter fines—recycled
    - (5) coke fines

Draw a flow diagram for a process which would automatically control the proportions of these constituents. Indicate all instrumentation necessary to effect this control.

Problems

1. Available at plants A and B of American Steel Company are 5000 tons of steel for shipment to fill an order for 5000 tons at plants 1, 2, and 3 of United Auto Company. The distribution of material is:

American Steel Company	United Auto Company
Supply	Demand
Plant A—3000 tons	Plant 1—1000 tons
Plant B—2000 tons	Plant 2—1500 tons
	Plant 3—2500 tons

The cost of transportation is proportional to mileage. How should the steel be shipped?

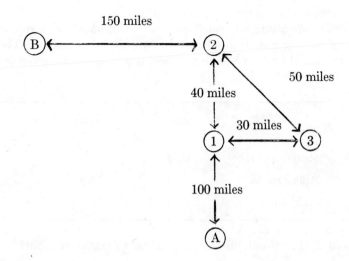

2. The Bright Metal Company operates a general heat treating plant. A long-term contract has been written with two neighboring steel companies for the batch annealing of hot-band coils. The contract calls for the treating of 1500 coils/month in a random supply. The contract includes a penalty clause for delays in processing. The processing equipment to handle this contract, as well as the required storage and handling equipment, are to be purchased. The batch annealing units will handle 3 coils and the complete annealing cycle requires 28 hr. The annealing furnaces cost $15/hr of actual operation including labor. The capital and installation costs can be amortized at $6/day/unit. The cost of storing coils, including the delay penalty, will be $20/day/coil. How many annealing units should Bright Metal Company purchase to supply the requirements of this contract?

3. A secondary brass and copper refiner produces an alloy which is guaranteed to contain a minimum of 90% copper and 5–8% nickel with the remainder other metallics. The refining company receives an order for 1000 lb of alloy at $0.45/lb. Two scrap materials as well as pure copper and nickel are available. The analyses and costs for the materials are:

	Scrap I	Scrap II	Copper	Nickel
Cu	95	85	100	—
Ni	3	1	—	100
Other	2	14	—	—
Cost ($/lb)	0.30	0.10	0.60	0.90

What is the optimum charge make-up?

4. A "mini-mill" operating one small electric arc furnace and a rolling mill to produce reinforcing rods has entered the low alloy bar market. The facility requirement and the relative profit for these two products are given below.

	Alloy Bar	Re-Rod
Melting hr/ton	3	5
Rolling hr/ton	4.5	3.5
Profit $/ton	15	12

How should the "mini-mill" utilize its facilities to maximize profit?

5. Evaluate and plot the relationship between log (P_{CO}/P_{CO_2}) versus temperature for the equilibrium $C(gr) + CO_2(g) = 2CO(g)$ with total pressure as a parameter. Compare this result with the equilibrium $\underline{Si} + 2CO_2(g) = SiO_2(s) + 2CO(g)$ on the same plot with the activities of silicon and silica as parameters.

6. Write and test a computer program which will permit calculation of the volume of pure argon gas which must be bubbled through a ladle of liquid steel to reduce the hydrogen concentration from some initial value, c_0, to a final value, c. Assume the purging gas leaves the melt at 1 atm pressure and with a hydrogen gas pressure in equilibrium with various fractions of the residual hydrogen content of the liquid metal. Apply the results of this computer program to the following practical problem.

A steel company using 50-ton ladles has experienced hydrogen problems in the following three areas.

Steel purpose	Hydrogen analysis PPM	
	Ladle	Maximum tolerable
Rolling stock	14–15	6
Light forging blanks	7–8	4
Heavy forging blanks	4	3

Inert flush degassing in the ladle has been recommended in these three cases to solve the gas–metal problem. The equipment available will supply 250 ft³/min of purified argon, and the purging operation is limited to less than 30 min. Are you in accord with these recommendations?

7. Write and test a computer program which will provide design calculations for the mold of a continuous casting process for wide steel slabs. Calculate the thickness of solid skin building up in the mold, the rate of heat removal, and the temperatures of the mold and slab surfaces. Extend calculations to a skin thickness of 2 in. Assume that a one-dimensional analysis applied to the center of the wide face is adequate for design purposes. State all assumptions. Prepare a recommendation for the length of mold to be used in the installation for various casting speeds. What additional information is necessary to finalize your design?

8. The following solidification profile has been determined for a continuous casting installation:

Time (min)	Thickness (in.)
0.25	0.25
0.50	0.50
0.75	0.70
1.00	0.75
1.25	0.80
1.50	1.00
1.75	1.06
2.00	1.13
2.25	1.30
2.50	1.35
2.75	1.45
3.00	1.50
3.25	1.58
3.50	1.65
3.75	1.74
4.00	1.85
4.25	1.93
4.50	2.00
4.75	2.08
5.00	2.15

Evaluate the constants in a parabolic rate law describing the solidification process. With what quantitative accuracy can this parabolic law be applied to the given installation?

9. Current output from hot strip rolling mill facilities are to be increased. The heating and subsequent temperature control of slabs is a major factor in mill productivity. Slabs are delivered from a blooming mill to pusher-type slab heating facilities. Usually, the slabs have cooled after a substantial storage period in the slab yard. They are heated in gas-fired furnaces which are held at 2400°F. Experience has shown that the center of the slab must be at a temperature of 1700°F or higher for adequate sheet quality. The minimum heating time for 10″ × 60″ slabs must be known to determine whether another furnace will be required to optimize the mill.

    Furthermore, delays on the rolling mill have resulted in edge cracks on the hot band. These cracks are caused when a slab is taken out or the furnace, and then is delayed on the table allowing the corners to reach temperatures below 1900°F. If a uniform temperature of 2250°F is assumed for the edge portion (e.g. outer 12″ × 10″ thick), how extensive a delay could be tolerated before the slab must be returned to the furnace?

Appendix A

Summary of Free Energy Equations[a]

$$\Delta G^0 = a + bT \log T + cT$$

Reaction	a	b	c	Range (°K)	Accuracy
$Ag_2SO_4(l) = 2\ Ag(c) + SO_2 + O_2$	96,640	30.49	−162.44	930 to 1234	500
$\frac{1}{3}Al_2(SO_4)_3 = \frac{1}{3}Al_2O_3(\alpha) + SO_3$	47,090	9.25	−73.51	600 to 1100	500
$\frac{1}{3}Al_2(SO_4)_3 = \frac{1}{3}Al_2O_3(\gamma) + SO_3$	48,670	—	−42.52	700 to 1100	2000
$BeSO_4(\gamma) = BeO + SO_3$	50,180	23.03	−114.14	900 to 1100	500
$CaSO_4 = CaO + SO_2 + \frac{1}{2}O_2$	136,830	41.61	−206.73	1100 to 1638	700
$\frac{3}{2}CdSO_4 = \frac{1}{2}(CdSO_4 \cdot 2\ CdO) + SO_3$	38,700	—	−22.0	1100 to 1273	2000
$\frac{1}{3}Ce_2(SO_4)_3 = \frac{2}{3}CeO_2 + SO_2 + \frac{1}{3}O_2$	63,000	—	−50.97	900 to 1200	500
$CoSO_4 = CoO + SO_3$	69,230	27.63	−137.18	890 to 1250	500
$CoSO_4 = \frac{1}{3}Co_3O_4 + SO_2 + \frac{1}{3}O_2$	77,250	32.24	−160.62	1000 to 1100	500
$2\ CuSO_4 = CuO \cdot CuSO_4 + SO_3$	51,780	5.16	−60.60	700 to 1100	500
$CuO \cdot CuSO_4 = 2\ CuO + SO_3$	52,030	5.16	−57.59	800 to 1200	500
$\frac{1}{2}Er_2(SO_4)_3 = \frac{1}{2}(Er_2O_3 \cdot SO_3) + SO_3$	57,810	—	−38.56	1000 to 1300	500
$\frac{1}{3}Fe_2(SO_4)_3 = \frac{1}{3}Fe_2O_3 + SO_3$	48,590	8.151	−71.03	700 to 1000	500
$\frac{1}{2}La_2(SO_4)_3 = \frac{1}{2}(La_2O_3 \cdot SO_3) + SO_3$	60,480	—	−38.82	1000 to 1300	500
$\frac{1}{2}Lu_2(SO_4)_3 = \frac{1}{2}(Lu_2O_3 \cdot SO_3) + SO_3$	57,840	—	−37.71	1000 to 1300	500
$MgSO_4 = MgO + SO_3$	78,560	16.58	−100.69	1000 to 1428	1000
$MnSO_4 = \frac{1}{3}Mn_3O_4 + SO_2 + \frac{1}{3}O_2$	75,890	14.90	−106.69	700 to 1100	500
$MnSO_4(c,l) = \frac{1}{3}Mn_3O_4(?) + SO_2 + \frac{1}{3}O_2$	63,510	—	−47.25	1100 to 1300	?
$Na_2SO_4(c,I) = Na_2O + SO_3$	137,480	14.90	−83.76	600 to 1157	3000
$\frac{1}{2}Nd_2(SO_4)_3 = \frac{1}{2}(Nd_2O_3 \cdot SO_3) + SO_3$	56,160	—	−38.03	1000 to 1300	500
$NiSO_4 = NiO + SO_3$	59,290	—	−47.52	900 to 1200	500
$2PbSO_4 = PbSO_4 \cdot PbO + SO_3$	74,000	16.81	−97.40	800 to 1139	500
$3(PbSO_4 \cdot PbO) = 2(PbSO_4 \cdot 2PbO) + SO_3$	87,920	16.81	−104.08	889 to 1223	500
$\frac{5}{3}(PbSO_4 \cdot PbO) = \frac{2}{3}(PbSO_4 \cdot 4PbO) + SO_3$	80,570	16.81	−95.81	800 to 889	500
$\frac{5}{2}(PbSO_4 \cdot 2PbO) = \frac{3}{2}(PbSO_4 \cdot 4PbO) + SO_3$	71,380	16.81	−85.47	889 to 1168	500
$PbSO_4 \cdot 4PbO = 5PbO + SO_3$	72,870	—	−30.70	800 to 1108	500
$\frac{1}{2}Pr_2(SO_4)_3 = \frac{1}{2}(Pr_2O_3 \cdot SO_3) + SO_3$	56,600	—	−37.99	1000 to 1300	500
$\frac{1}{3}Sc_2(SO_4)_3 = \frac{1}{3}Sc_2O_3 + SO_3$	60,980	—	−44.82	1000 to 1300	500
$\frac{1}{2}Sm_2(SO_4)_3 = \frac{1}{2}(Sm_2O_3 \cdot SO_3) + SO_3$	56,190	—	−38.82	1000 to 1300	500
$\frac{1}{2}Th(SO_4)_2 = \frac{1}{2}ThO_2 + SO_3$	70,290	25.33	−133.60	700 to 1100	500
$2\ VOSO_4 = V_2O_5 + SO_2 + SO_3$	85,470	18.42	−150.83	700 to 900	500
$\frac{1}{2}Y_2(SO_4)_3 = \frac{1}{2}(Y_2O_3 \cdot SO_3) + SO_3$	60,070	—	−39.25	1000 to 1300	500
$\frac{1}{2}Yb_2(SO_4)_3 = \frac{1}{2}(Yb_2O_3 \cdot SO_3) + SO_3$	57,150	—	−37.48	1000 to 1300	500

287

TABLE A-I (Continued)

Summary of Free Energy Equations[a]

$$(\Delta G^0 = a + bT \log T + cT)$$

Reaction	a	b	c	Range(°K)	Accuracy
$3\,ZnSO_4(\alpha) = ZnO \cdot 2\,ZnSO_4 + SO_3$	53,730	—	−45.21	800 to 1007	500
$3\,ZnSO_4(\beta) = ZnO \cdot 2\,ZnSO_4 + SO_3$	39,280	—	−30.87	1007 to 1200	500
$\frac{1}{2}(ZnO \cdot 2\,ZnSO_4) = \frac{3}{2}ZnO + SO_3$	57,190	7.60	−65.63	800 to 1200	500

[a] The standard state for all sulfates and oxides is the normally stable crystal form unless otherwise indicated.

Note: $SO_3 = SO_2 + \frac{1}{2}O_2$; $\log K = 8.8557 - \dfrac{5465.5}{T} - 1.21572 \log T$.

Source: H. H. Kellogg, *Trans. Met. Soc. AIME* **230,** 1624 (1964). copyright AIME.

TABLE A-II

Standard Free Energies of Formation of Compounds at Steelmaking Temperatures[a,b,c]

$$(\Delta G^0 = A + BT)$$

Reaction	Temp (°K)	(A)	(B)
Elements Dissolved in Liquid Iron			
$CO_2(g) + C\,(gr) = 2CO\,(g)$	1700–2000	+38,400	40.40
$CO_2\,(g) + \underline{C}\,(\%) = 2CO\,(g)$	1700–2000	+33,300	−30.40
$CO\,(g) + \underline{O}\,(\%) = CO_2\,(g)$	1700–2000	−38,700	+20.9
$\underline{C}\,(\%) + \underline{O}\,(\%) = CO\,(g)$	1700–2000	−5,350	−9.48
$Cr_3O_4\,(s) = 3\underline{Cr}\,(\%) + 4\underline{O}\,(\%)$	1700–2000	+244,800	−109.6
$H_2O\,(g) = 2\underline{H}\,(ppm) + \underline{O}\,(\%)$	1700–2000	+47,480	−35.87
$Fe\,(l) + \frac{1}{2}O_2\,(g) = FeO\,(l)$	1700–2000	−56,900	+11.82
$MnO\,(l\,in\,FeO) = \underline{Mn}\,(\%) + \underline{O}\,(\%)$	1700–2000	+58,400	−25.98
$\underline{O}\,(\%) + H_2\,(g) = H_2O\,(g)$	1700–2000	−32,200	+14.63
$SiO_2\,(s) = \underline{Si} + 2\underline{O}$	1700–2000	+142,000	−55.0
$\underline{S}\,(\%) + H_2\,(g) = H_2S\,(g)$	1700–2000	+9,980	+6.46
$S\,(g) + 2\underline{O}\,(\%) = SO_2\,(g)$	1700–2000	+1,780	+13.40
$V_2O_3\,(s) = 2\underline{V}\,(\%) + 3\underline{O}\,(\%)$	(1600 C)	−49,300	
Carbides (from graphite)			
$4Al\,(l) + 3C\,(s) = Al_4C_3\,(s)$	1500–2000	−41,550	+5.10
$4B\,(s) + C\,(s) = \text{``}B_4C\text{''}\,(s)^c$	1500–2000	−16,700	+4.48
$Ca\,(l) + 2C\,(s) = CaC_2\,(\beta)$	1500–1765	−14,900	−5.87
$Ca\,(g) + 2C\,(s) = CaC_2\,(\beta)$	1765–2000	−50,900	+14.58
$23Cr\,(s) + 6C\,(s) = \text{``}Cr_{23}C_6\text{''}\,(s)$	1500–2000	−98,300	−9.21
$7Cr\,(s) + 3C\,(s) = \text{``}Cr_7C_3\text{''}\,(s)$	1500–2000	−45,100	−4.43
$3Cr\,(s) + 2C\,(s) = \text{``}Cr_3C_2\text{''}\,(s)$	1500–2000	−21,500	−4.11
$2Co\,(s) + C\,(s) = Co_2C\,(s)$	1500–1766	+3,950	−2.09
$2Co\,(l) + C\,(s) = Co_2C\,(s)$	1766–2000	−3,300	+2.00

TABLE A-II (Continued)

Standard Free Energies of Formation of Compounds at Steelmaking Temperatures[a,b,c]

$$(\Delta G^0 = A + BT)$$

Reaction	Temp (°K)	(A)	(B)
Carbides (from graphite)—*Continued*			
$2C\ (s) + H_2\ (g) = C_2H_2\ (g)$	1500–2000	+52,000	−11.80
$C\ (s) + 2H_2\ (g) = CH_4\ (g)$	1500–2000	−22,100	+26.45
$3Fe\ (\gamma) + C\ (s) = Fe_3C\ (s)$	1500–1665	+2,900	−2.73
$Mg\ (g) + 2C\ (s) = MgC_2\ (s)$	1500–2000	−11,900	+24.45
$7Mn\ (l) + 3C\ (s) = Mn_7C_3\ (s)$	1516–2000	−15,800	−13.91
$2Mo\ (s) + C\ (s) = \text{``}Mo_2C\text{''}\ (s)$	1500–2000	+15,800	−26.00
$Mo\ (s) + C\ (s) = MoC\ (s)$	1500–2000	+9,700	−14.01
$3Ni\ (s) + C\ (s) = Ni_3C\ (s)$	1500–1726	+9,200	−2.37
$Si\ (s) + C\ (s) = SiC\ (s)$	1500–1686	−12,800	+1.68
$Si\ (l) + C\ (s) = SiC\ (s)$	1686–2000	−24,900	+8.85
$2Ta\ (s) + C\ (s) = \text{``}Ta_2C\text{''}\ (s)$	1500–2000	−34,000	+1.32
$Ta\ (s) + C\ (s) = \text{``}TaC\text{''}\ (s)$	1500–2000	−38,500	+1.32
$Ti\ (\beta) + C\ (s) = \text{``}TiC\text{''}\ (s)$	1500–1940	−44,400	+3.41
Nitrides			
$Al\ (l) + \tfrac{1}{2}N_2\ (g) = AlN\ (s)$	1500–2000	−77,700	+26.80
$B\ (s) + \tfrac{1}{2}N_2\ (g) = BN\ (s)$	1500–2000	−60,600	+21.80
$B\ (s) + \tfrac{1}{2}N_2\ (g) = BN\ (g)$	1500–2000	+155,500	−26.85
$3Ca\ (l) + N_2\ (g) = Ca_3N_2\ (l)$	1500–1765	−112,000	+56.62
$3Ca\ (g) + N_2\ (g) = Ca_3N_2\ (l)$	1765–2000	−219,500	+117.55
$2Cr\ (s) + \tfrac{1}{2}N_2\ (g) = \text{``}Cr_2N\text{''}\ (s)$	1500–2000	−23,500	+11.25
$Cr\ (s) + \tfrac{1}{2}N_2\ (g) = \text{``}CrN\text{''}\ (s)$	1500–2000	−24,500	+16.05
$4Fe\ (\gamma) + \tfrac{1}{2}N_2\ (g) = \text{``}Fe_4N\text{''}\ (s)$	1500–1665	−2,000	+10.39
$3Mg\ (g) + N_2\ (g) = Mg_3N_2\ (\gamma)$	1500–2000	−207,500	+121.25
$2Mo\ (s) + \tfrac{1}{2}N_2\ (g) = \text{``}Mo_2N\text{''}\ (s)$	1500–2000	−17,100	+13.83
$3Si\ (s) + 2N_2\ (g) = Si_3N_4\ (s)$	1500–1686	−173,000	+75.48
$3Si\ (l) + 2N_2\ (g) = Si_3N_4\ (s)$	1686–2000	−209,000	+96.78
$Ta\ (s) + \tfrac{1}{2}N_2\ (g) = TaN\ (s)$	1500–2000	−56,150	+18.87
$Ti\ (\beta) + \tfrac{1}{2}N_2\ (g) = \text{``}TiN\text{''}\ (s)$	1500–1940	−80,000	+22.21
$V\ (s) + \tfrac{1}{2}N_2\ (g) = \text{``}VN\text{''}\ (s)$	1500–2000	−40,300	+19.15
$Zr\ (\beta) + \tfrac{1}{2}N_2\ (g) = \text{``}ZrN\text{''}\ (s)$	1500–2000	−86,300	+21.67
Oxides			
$2Al\ (l) + \tfrac{1}{2}O_2\ (g) = Al_2O\ (g)$	1500–2000	−47,000	−13.06
$Al\ (l) + \tfrac{1}{2}O_2\ (g) = AlO\ (g)$	1500–2000	+3,500	−13.31
$2Al\ (l) + \tfrac{3}{2}O_2\ (g) = Al_2O_3\ (s)$	1500–2000	−401,500	+76.91
$Ba\ (l) + \tfrac{1}{2}O_2\ (g) = BaO\ (s)$	1500–1910	−132,000	+22.10
$Ba\ (g) + \tfrac{1}{2}O_2\ (g) = BaO\ (s)$	1910–2000	−167,250	+40.60
$Be\ (l) + \tfrac{1}{2}O_2\ (g) = BeO\ (s)$	1555–2000	−143,750	+23.30
$B\ (s) + \tfrac{1}{2}O_2\ (g) = BO\ (g)$	1500–2000	−16,500	−19.43
$2B\ (s) + \tfrac{3}{2}O_2\ (g) = B_2O_3\ (s) \text{ or } (gl)$	1500–2000	−295,850	+51.92

TABLE A-II (Continued)

Standard Free Energies of Formation of Compounds at Steelmaking Temperatures[a,b,c]

$$(\Delta G^0 = A + BT)$$

Reaction	Temp (°K)	(A)	(B)
Oxides—Continued			
Ca (l) + $\frac{1}{2}$O$_2$ (g) = CaO (s)	1500–1765	$-152,850$	$+25.78$
Ca (g) + $\frac{1}{2}$O$_2$ (g) = CaO (s)	1765–2000	$-187,900$	$+45.70$
C (s) + $\frac{1}{2}$O$_2$ (g) = CO (g)	1500–2000	$-28,200$	-20.16
C (s) + O$_2$ (g) = CO$_2$ (g)	1500–2000	$-94,755$	$+0.02$
2Cb (s) + 2O$_2$ (g) = Cb$_2$O$_4$ (s)	1500–2000	$-374,500$	$+80.56$
2Cb (s) + $\frac{5}{2}$O$_2$ (g) = Cb$_2$O$_5$ (s)	1500–1785	$-446,200$	$+97.56$
2Cb (s) + $\frac{5}{2}$O$_2$ (g) = Cb$_2$O$_5$ (l)	1785–2000	$-417,500$	$+81.42$
2Ce (l) + $\frac{3}{2}$O$_2$ (g) = Ce$_2$O$_3$ (s)	1500–2000	$-436,500$	$+80.46$
Ce (l) + O$_2$ (g) = CeO$_2$ (s)	1500–2000	$-246,000$	$+51.20$
2Cr (s) + $\frac{3}{2}$O$_2$ (g) = Cr$_2$O$_3$ (β)	1500–2000	$-270,550$	$+61.35$
Co(γ) + $\frac{1}{2}$O$_2$ (g) = CoO (s)	1500–1766	$-56,900$	$+17.50$
Co (l) + $\frac{1}{2}$O$_2$ (g) = CoO (s)	1766–2000	$-60,500$	$+19.53$
3Co (γ) + 2O$_2$ (g) = Co$_3$O$_4$ (s)	1500–1766	$-208,900$	$+83.31$
2Cu (l) + $\frac{1}{2}$O$_2$ (g) = Cu$_2$O (l)	1502–2000	$-34,950$	$+14.40$
Cu (l) + $\frac{1}{2}$O$_2$ (g) = CuO (s)	1500–1720	$-40,000$	$+23.06$
Cu (l) + $\frac{1}{2}$O$_2$ (g) = CuO (l)	1720–2000	$-36,800$	$+21.26$
H$_2$ (g) + $\frac{1}{2}$O$_2$ (g) = H$_2$O (g)	1500–2000	$-60,200$	$+13.94$
Fe (γ) + $\frac{1}{2}$O$_2$ (g) = "FeO" (s)	1500–1650	$-62,600$	$+15.18$
Fe (δ) + $\frac{1}{2}$O$_2$ (g) = "FeO" (l)	1665–1809	$-54,850$	$+10.47$
Fe (l) + $\frac{1}{2}$O$_2$ (g) = "FeO" (l)	1809–2000	$-56,900$	$+11.82$
2Fe (γ) + $\frac{3}{2}$O$_2$ (g) = Fe$_2$O$_3$ (γ)	1500–1665	$-191,300$	$+57.60$
2Fe (δ) + $\frac{3}{2}$O$_2$ (g) = Fe$_2$O$_3$ (γ)	1665–1809	$-190,950$	$+57.46$
3Fe (γ) + 2O$_2$ (g) = Fe$_3$O$_4$ (β)	1500–1665	$-259,450$	$+70.84$
3Fe (δ) + 2O$_2$ (g) = Fe$_3$O$_4$ (β)	1665–1809	$-259,550$	$+70.91$
Mg (g) + $\frac{1}{2}$O$_2$ (g) = MgO (s)	1500–2000	$-174,750$	$+49.09$
Mn (l) + $\frac{1}{2}$O$_2$ (g) = MnO (s)	1516–2000	$-97,550$	$+21.22$
3Mn (l) + 2O$_2$ (g) = Mn$_3$O$_4$ (β)	1516–1833	$-339,050$	$+89.06$
Mo (s) + O$_2$ (g) = MoO$_2$ (s)	1500–2000	$-130,800$	$+34.17$
Mo (s) + $\frac{3}{2}$O$_2$ (g) = MoO$_3$ (g)	1553–2000	$-113,800$	$+12.92$
Ni (s) + $\frac{1}{2}$O$_2$ (g) = NiO (s)	1500–1726	$-55,000$	$+20.01$
Ni (l) + $\frac{1}{2}$O$_2$ (g) = NiO (s)	1726–2000	$-60,350$	$+22.61$
$\frac{1}{2}$P$_2$ (g) + $\frac{1}{2}$O$_2$ (g) = PO (g)	1500–2000	$-27,000$	$+2.28$
2P$_2$ (g) + 5O$_2$ (g) = P$_4$O$_{10}$ (g)	1500–2000	$-750,700$	$+230.6$
Si (l) + $\frac{1}{2}$O$_2$ (g) = SiO (g)	1686–2000	$-37,100$	-11.3
Si (l) + O$_2$ (g) = SiO$_2$ (l)	1883–2000	$-223,800$	$+46.08$
Si (l) + O$_2$ (g) = SiO$_2$ $(\beta$-crist.$)$	1686–1986	$-226,500$	$+47.50$
2Ta (s) + $\frac{5}{2}$O$_2$ (g) = Ta$_2$O$_5$ (s)	1500–2000	$-475,450$	$+94.08$
Sn (l) + $\frac{1}{2}$O$_2$ (g) = SnO (s)	1500–2000	$-67,600$	$+24.64$
Sn (l) + $\frac{1}{2}$O$_2$ (g) = SnO (g)	1500–2000	$-6,800$	-10.39
Sn (l) + O$_2$ (g) = SnO$_2$ (s)	1500–1898	$-135,500$	$+47.15$
Sn (l) + O$_2$ (g) = SnO$_2$ (l)	1898–2000	$-122,650$	$+40.33$
Ti (s) + $\frac{1}{2}$O$_2$ (g) = TiO (β)	1500–1940	$-120,100$	$+19.83$
Ti (s) + $\frac{1}{2}$O$_2$ (g) = TiO (g)	1500–1940	$+6,300$	-18.22
2Ti (s) + $\frac{3}{2}$O$_2$ (g) = Ti$_2$O$_3$ (β)	1500–1940	$-354,000$	$+58.36$

TABLE A-II (Continued)

Standard Free Energies of Formation of Compounds at Steelmaking Temperatures[a,b,c]

$$(\Delta G^0 = A + BT)$$

Reaction	Temp (°K)	(A)	(B)
Oxides—Continued			
$3Ti\ (s) + \frac{5}{2}O_2\ (g) = Ti_3O_5\ (\beta)$	1500–1940	−577,500	+97.69
$Ti\ (s) + O_2\ (g) = TiO_2\ (s)$	1500–1940	−223,500	+41.55
$W\ (s) + O_2\ (g) = WO_2\ (s)$	1500–2000	−134,800	+38.92
$W\ (s) + \frac{3}{2}O_2\ (g) = WO_3\ (s)$	1500–1743	−194,550	+54.47
$W\ (s) + \frac{3}{2}O_2\ (g) = WO_3\ (l)$	1743–2000	−177,750	+44.85
$V\ (s) + \frac{1}{2}O_2\ (g) = VO\ (s)$	1500–2000	−96,000	+17.78
$V\ (s) = \frac{1}{2}O_2\ (g) = VO\ (g)$	1500–2000	+47,000	−17.78
$2V\ (s) + \frac{3}{2}O_2\ (g) = V_2O_3\ (s)$	1500–2000	−287,000	+54.00
$2V\ (s) + 2O_2\ (g) = V_2O_4\ (s)$	1500–1818	−331,000	+70.80
$2V\ (s) + 2O_2\ (g) = V_2O_4\ (l)$	1818–2000	−301,750	+54.49
$2V\ (s) + \frac{5}{2}O_2\ (g) = V_2O_5\ (l)$	1500–2000	−346,500	+75.83
$Zr\ (\beta) + O_2\ (g) = ZrO_2\ (\beta)$	1500–2000	−258,000	+42.50
Sulfides			
$Ca\ (l) + \frac{1}{2}S_2\ (g) = CaS\ (s)$	1500–1765	−132,060	+25.91
$Ca\ (g) + \frac{1}{2}S_2\ (g) = CaS\ (s)$	1765–2000	−167,900	+46.21
$C\ (s) + \frac{1}{2}S_2\ (g) = CS\ (g)$	1500–2000	+59,000	−22.84
$C\ (s) + S_2\ (g) = CS_2\ (g)$	1500–2000	−3,250	−1.61
$C\ (s) + \frac{1}{2}O_2\ (g) + \frac{1}{2}S_2\ (g) = COS\ (g)$	1500–2000	−49,700	−2.22
$H_2\ (g) + \frac{1}{2}S_2\ (g) = H_2S\ (g)$	1500–2000	−21,540	+11.73
$Fe\ (\gamma) + \frac{1}{2}S_2\ (g) = $ "FeS" (l)	1500–1665	−27,130	+6.32
$Fe\ (\delta) + \frac{1}{2}S_2\ (g) = $ "FeS" (l)	1665–1809	−26,700	+6.06
$Fe\ (l) + \frac{1}{2}S_2\ (g) = $ "FeS" (l)	1809–2000	−29,970	+7.90
$Mg\ (g) + \frac{1}{2}S_2\ (g) = MgS\ (s)$	1500–2000	−132,540	+47.24
$Mn\ (l) + \frac{1}{2}S_2\ (g) = MnS\ (s)$	1516–1803	−69,250	+19.18
$Mn\ (l) + \frac{1}{2}S_2\ (g) = MnS\ (l)$	1803–2000	−63,100	+15.77
$Mo\ (s) + S_2\ (g) = MoS_2\ (s)$	1500–2000	−86,650	+46.36
$Pt\ (s) + \frac{1}{2}S_2\ (g) = $ "PtS" (s)	1500–1700	−33,050	+21.78
$\frac{1}{2}S_2\ (g) + O_2\ (g) = SO_2\ (g)$	1500–2000	−86,130	+17.27
$\frac{1}{2}S_2\ (g) + \frac{1}{2}O_2\ (g) = SO\ (g)$	1500–2000	−15,360	−1.24

[a] The reference state of an element of compound is the form that is stable normally at 1 atm pressure and the temperature being considered, *i.e.*, the pure stable solid, the pure stable liquid, and the equilibrium gas at 1 atm pressure. For elements dissolved in iron, the reference state is the hypothetical 1 per cent solution.

[b] See Elliott, J. F., and M. Gleiser: *Thermochemistry for Steelmaking, I*, Addison-Wesley (1960).

[c] Quotation marks indicate that compound is not necessarily of the stoichiometric composition.

Source: *Electric Furnace Steelmaking* (Clarence E. Sims, Ed.), Vol. 2, Chap. 16, pp. 134–137, Interscience Publishers, New York, 1963.

TABLE A-III

Activity Coefficients at Infinite Dilution in Liquid Metals

Solvent	Solute	γ^0	Temp, °C
Aluminum	Ag	0.38	700
		0.47	900
		0.53	1000
Antimony	Cd	0.25	500
	Pb	0.72	500
	Sn	0.40	905
	Zn	0.30	550
Bismuth	Au	0.78	700
	Cd	1.0	500
	Hg	0.39	321
	Mg	5.5×10^{-3}	700
	Pb	0.44	475
		0.46	500
		0.48	527
		0.53	665
	Sn	1.12	330
	Tl	0.04	270
		0.14	480
	Zn	5.0	352
		3.5	600
	U	1.6×10^{-5}	450
		2.5×10^{-5}	507
		4.6×10^{-5}	540
		7.8×10^{-5}	570
Cadmium	Bi	0.85	500
	Hg	0.10	327
		0.063	350
	Na	0.01	350
		0.01	395
	Pb	6.3	432
		5.5	480
		5.1	500
		4.7	544
		4.3	572
	Sb	0.27	500
	Sn	2.2	431
		2.1	483
		1.9	500
		1.9	544
		1.8	585
	U	6.58	467
		8.15	507
	Zn	6.9	435
		6.3	466
		5.1	540
		2.8	682

TABLE A-III (Continued)

Activity Coefficients at Infinite Dilution in Liquid Metals

Solvent	Solute	γ^0	Temp, °C
Copper	Zn	0.14	604
		0.17	653
		0.21	727
		0.24	802
Gold	Bi	0.90	700
	Pb	0.142	600
	Sn	1.0×10^{-5}	600
	Tl	0.49	700
Indium	Zn	5.15	352
Iron	Al	0.063	1600
	B	0.022	1600
	Cb	0.2	1600
		0.27	1600
	Co	1	1600
	Cr	1	1600
	Cu	8.6	1600
		8.0	1600
		8.8	1600
	Mn	1	1600
	Mo	1	1600
	Ni	0.47	1510
		0.66	1600
		0.66	1600
	Si	0.0072	1600
	Ti	0.011	1627
		0.048	1600
	V	0.12	1600
	W	1	1600
	Zr	0.0076	1600
Lead	Ag	2.3	1000
		1.5	1085
	Au	0.25	600
	Bi	0.52	475
		0.488	500
		0.48	527
		0.43	665
	Cd	3.7	432
		3.5	480
		3.22	500
		3.0	544
		2.9	572
	Mg	0.066	560
		0.064	560
	Mn	28.8	750
		20.5	800
		15.2	850

TABLE A-III (Continued)

Activity Coefficients at Infinite Dilution in Liquid Metals

Solvent	Solute	γ^0	Temp, °C
		11.4	900
		9.0	950
		7.7	980
		7.2	1000
	Na	0.0022	400
	Sb	0.80	500
	Sn	2.3	500
	Tl	0.74	450
	Zn	11.0	653
Mercury	Ba	1.0×10^{-55}	25
	Ce	1.0×10^{-55}	25
		2.2×10^{-29}	281
		3.0×10^{-27}	301
		1.8×10^{-25}	327
		6.9×10^{-23}	378
	K	6.3×10^{-22}	0
		4.0×10^{-20}	25
		1.3×10^{-18}	50
		2.5×10^{-24}	200
		1.6×10^{-22}	250
		4.0×10^{-21}	310
		1.0×10^{-19}	390
	Na	1.0×10^{-20}	25
		4.7×10^{-14}	335.5
		5.6×10^{-6}	350
		1.5×10^{-13}	375
		1.2×10^{-7}	400
	Tl	0.12	20
		0.43	325
Nickel	Fe	0.26	1510
		0.39	1600
		0.40	1600
Silver	Al	0.022	700
		0.028	800
		0.039	900
		0.11	1000
	Cd	0.18	827
	Pb	2.0	1000
Sodium	Hg	1.6×10^{-3}	350
Thalium	Au	0.31	700
	Bi	0.15	480
	Hg	0.80	325
	Sn	2.8	352
		2.4	414
		2.1	478

TABLE A-III (Continued)

Activity Coefficients at Infinite Dilution in Liquid Metals

Solvent	Solute	γ^0	Temp, °C
Tin	Au	0.006	600
	Cd	2.0	431
		1.9	483
		1.80	500
		1.8	544
		1.7	585
	Pb	2.33	500
	Sb	0.46	632
	Tl	2.9	352
		2.4	414
		1.8	478
	Zn	2.67	352
		2.2	431
		2.2	437
		2.0	466
		1.7	539
		1.6	570
Zinc	Cd	3.7	435
		3.5	466
		3.1	540
		3.5	682
	Sb	0.3	550
	Sn	5.8	431
		5.3	466
		4.0	539
		3.5	570

Source: J. M. Dealy and R. D. Pehlke, *Trans. Met. Soc. AIME* **227**, 1031, 1032 (1963).

TABLE A-IV

Standard Free Energy of Solution of Various Elements in Liquid Iron

$(M \text{ (pure)} = \underline{M}(1\% \text{ solution, hypothetical})$

Element, state	$\gamma°_{1873}$ [a]	$\Delta G°$	Basis
Aluminum (l)	0.063	$-10,300 - 7.71T$	—
Carbon (graphite)	—	$5,100 - 10.00T$	—
Chromium (s)	1	$5,000 - 11.31T$	(Ideal soln.)
Cobalt (l)	1	$- 9.31T$	(Ideal soln.)
Copper (l)	8.5	$8,000 - 9.40T$	—
Hydrogen $\frac{1}{2}H_2$ (g) $= \underline{H}$ (ppm)	—	$7,640 - 10.62T$	—
Manganese (l)	1	$- 9.11T$	(Ideal soln.)
Molybdenum (s)	1	$5,800 - 13.3T$	(Ideal soln.)
Nickel (l)	0.66	$-5,000 - 7.42T$	—
Nitrogen $\frac{1}{2}N_2$ (g) $= \underline{N}$ (%)	—	$860 + 5.71T$	—
Oxygen $\frac{1}{2}O_2$ (g) $= \underline{O}$ (%)	—	$-28,000 - 0.69T$	—
FeO (l) $= \underline{O}$ (%) $+$ Fe (l)	—	$+28,900 - 12.51T$	—
Phosphorus, $\frac{1}{2}P_2$ (g) $= \underline{P}$ (%)	—	$-29,200 - 4.6T$	—
Silicon (l)	0.0072[d]	$-28,500 - 6.09T$	—
Sulfur $\frac{1}{2}S_2$ (g) $= \underline{S}$ (%)	—	$-31,520 + 5.27T$	—
Titanium (s)	0.011	$-13,100 - 10.7T$[b,c]	—
Tungsten (s)	(1)	$8,000 - 13.4T$[c]	(Ideal soln.)
Vanadium (s)	0.12	$-3,700 - 10.9T$[b,c]	—
Zirconium (s)	0.0076[d]	$-12,800 - 12.0T$[b]	—

[a] $\gamma°$ referred to pure liquid element.

[b] Calculated assuming regular solution behavior.

[c] Estimated on basis of heat and entropy of melting at 1600 C.

[d] Altered to agree with Table A-III.

Source: *Electric Furnace Steelmaking* (Clarence E. Sims, Ed.), Vol. 2, Chap. 16, p. 133, Interscience Publishers, New York, 1963

TABLE A-V

Interaction Coefficients $e^j_i \times 10^{2}$[a] for Elements Dissolved in Liquid Iron at 1600°C

Dissolved element	Al	As	Au	B	C	Cb	Co	Cr	Cu	H	Mn	Mo	N	Ni	O	P	Pt	S	Sb	Se	Si	Sn	Ta	Ti	V	W
Al	4.8				11					*34*			0.5	0.5	−160			4.9			6.0				−3.8	−3.3
As													*9.6*													
Au																										
B										*53*					−4.5											
C	4.8				22	−6.0	1.2	−2.4	1.6	*72*		−0.9	*11.1*	1.2	*−9.7*			9			*10*					
Cb					*−46*	8.3				*−21*			*−44.4*		*82*						*280*					
Co					*6*					*11*			*4.7*		*2.6*			*−2.4*								
Cr					−10					*−11*			*−16.6*		−13										−3.8	
Cu					*8.5*				−2.1	*3.1*			4.1		−3.8			*−4.3*			~0					
H	1.3			5	6	−0.23	0.18	−0.22	0.05	*0*	−0.14			0				0.8			2.7	0.53	−2.0	−8		
Mn					*−7*								−7.8		0	1.1					0					
Mo										*−7.7*			−7.8		2.1	2.1										
N	0.3	1.8			13	−6.7	1.1	−4.5	0.9	*0*	−2.0	−1.1	0	1.0	*2.1*	5.1		1.3	0.9	0	4.7	0.7	−3.4	−63	−10	−0.2
Ni			−0.4		*5.9*	−14	0.7	−4.1	−0.9		0		*4.2*	−0.0				0			*1.0*					
O	−94				−13				−0.9			0.35	5.7	0.6	−20	5.0	0.4	*−9.1*			−14	0				0.8
P															*13.5*	7.0										
Pt										*34*			*11.3*		*5.5*						*9.5*					
S	5.8				*24*			−2.2	−1.2	*26*	−2.5		3.0	0	−18	4.5		−2.8			6.6					
Sb													*7.8*													
Se													0													
Si	6.3				24					*76*	0		9.3	0.5	−25	8.6		5.7			32					
Sn					28					*62*			5.9		0									−19	−27	
Ta													−44		−56											
Ti										*−360*			−210													
V					*−16*					*−385*			−36		*−86.4*											
W					*−4.7*								−2.6		9.7											

[a] $e^{(j)}_i = [\partial \log f_i/\partial(\%j)] = 0.2425/M_j$. $e^{(j)}_i = M_i/M_j \cdot e^{(i)}_j$; f = activity coefficient; $\%_{ij} = wt\%$ of (j); M = atomic weight (of i or j). Plain numbers are obtained from values reported in the literature. Italicized numbers are obtained by calculation from the reciprocal relationship noted above. *Author's Note:* (These reciprocal relationships are only approximate– See Chapter 3, Equations 3-39 and 3-40).

Source: J. F. Elliot and M. Gleiser, *Thermochemistry for Steelmaking, II*, p. 566–567, Addison Wesley Publishing Co., Inc., Reading, Mass. 1963.

TABLE A-VI

Nonferrous Interaction Parameters

Solvent	i	j	ϵ^j_i	Temperature (°C)	Basis
Aluminum	Ag	Ag	-3.1	1000	
	Cu	H	see ϵ^{Cu}_H	700–1000	*
	H	Cu	39.0	700	
	H	Cu	16.6	800	
	H	Cu	20.1	900	
	H	Cu	4.3	1000	
	H	H	0	700–1000	
	H	Si	11.5	700	
	H	Si	6.2	800	
	H	Si	4.2	900	
	H	Si	1.8	1000	
	Si	H	see ϵ^{Si}_H	700–1000	*
Antimony	Cd	Cd	1.5	500	
	Cd	Pb	2.8	500	
	Pb	Cd	2.8	500	*
	Pb	Pb	0.59	500	
	Sn	Sn	2.2	905	
Bismuth	Au	Au	2.1	700	
	Cd	Cd	-1.22	500	
	Cd	Pb	1.62	500	
	Hg	Hg	-0.33	321	
	Mg	Mg	0.85	700	
	Pb	Cd	1.62	500	*
	Pb	Pb	1.44	500	
	Pb	Pb	0.75	665	
	Pb	Pb	0.91	475	
	Sn	Sn	-0.056	330	
	Tl	Tl	3.22	270	
	Tl	Tl	2.14	480	
	Zn	Zn	-3.3	600	
Cadmium	Bi	Bi	-6.5	500	
	Bi	Pb	-3.2	500	
	Hg	Hg	3.36	327	
	Hg	Hg	2.5	350	
	Hg	Na	-6	350	
	Na	Hg	-6	350	*
	Na	Na	17.8	350	
	Na	Na	15.8	395	
	Pb	Bi	-3.2	500	*
	Pb	Pb	-4.56	500	
	Pb	Sb	0	500	
	Pb	Sn	2.86	500	
	Sb	Pb	0	500	*
	Sb	Sb	-6.5	500	
	Sn	Pb	2.86	500	*

TABLE A-VI (Continued)

Nonferrous Interaction Parameters

Solvent	i	j	ϵ^j_i	Temperature (°C)	Basis
	Sn	Sn	−1.5	500	
	Zn	Zn	−1.78	682	
Cobalt	Cr	N	−19.5	1600	*
	Fe	H	0.53	1600	*
	H	Fe	0.53	1600	
	H	Ni	0.09	1600	
	N	Cr	−19.5	1600	
	N	Si	80	1600	
	N	V	−72	1600	
	Ni	H	0.09	1600	*
	Si	N	80	1600	*
	S	Cu	−4(±2)	1500	**
	V	N	−72	1600	*
Copper	H	Sn	5 ± .3	1100–1300	
	S	Au	6.9(±1)	1115	**
	S	Co	−4.8(±0.8)	1300	**
	S	Fe	−7.4(±0.5)	1300	**
	S	Ni	−6.6(±1)	1300	**
	S	Pt	9.2(±0.2)	1200	**
	S	Si	6.9(±2)	1200	**
	Sn	H	5 ± .3	1100–1300	*
	Zn	Zn	0.38	802	
	Zn	Zn	0.72	727	
	Zn	Zn	1.185	653	
	Zn	Zn	1.40	604	
Gold	Bi	Bi	0.1	700	
	Pb	Pb	5	600	
	Sn	Sn	38.9	600	
	Tl	Tl	0.1	700	
Lead	Ag	Ag	−0.6	1085	
	Ag	Ag	−0.92	1000	
	Au	Au	3.2	600	
	Bi	Bi	2.6	500	
	Bi	Cd	−0.86	500	
	Cd	Bi	−0.86	500	*
	Cd	Cd	−2.6	500	
	Cd	Sb	−1.6	500	
	Cd	Sn	1.35	500	
	Hg	Na	−5.25	400	*
	Mg	Mg	0.6	833	
	Mn	Mn	0	750–1000	
	Na	Hg	−5.25	400	
	Na	Na	3.6	400	
	Sb	Cd	−1.6	500	*

TABLE A-VI (Continued)

Nonferrous Interaction Parameters

Solvent	i	j	ϵ^j_i	Temperature (°C)	Basis
	Sb	Sb	0.16	500	
	Sn	Cd	1.35	500	*
	Sn	Sn	−1.17	500	
	Tl	Tl	0.1	450	
	Zn	Zn	−5.3	653	
Mercury	Ba	Ba	127	298	
	Cd	Na	6.0	350	*
	Ce	Ce	129.9	25	
			66.0	281	
			61.0	301	
			57.0	327	
			51.0	378	
	K	K	73	0	
			65.3	25	
			61.6	50	
			67	?	
			54.4	200	
			50.3	250	
			47.0	310	
			43.8	390	
	K	Na	46.5	?	*
	K	Tl	30.8	?	
	Li	Li	26.5	25	
	Li	Tl	17.5	?	
	Na	Cd	6.0	350	
	Na	K	46.5	?	
	Na	Na	46.1	25	
			30.7	335.5	
			29.6	375	
			23.4	350	
			36.0	?	
			43.0	400	
	Na	Pb	6	400	
	Na	Tl	22.1	?	
	Pb	Na	6	400	*
	Tl	K	30.8	?	*
	Tl	Li	17.5	?	*
	Tl	Na	22.1	?	*
	Tl	Tl	6.0	325	
	Tl	Tl	13.1	?	
Nickel	Co	H	2.58	1600	*
	Cr	N	−23.6	1600	*
	Fe	H	0.009	1600	*
	Fe	N	−3.25	1600	*
	Fe	O	132	1594	*

TABLE A-VI (Continued)

Nonferrous Interaction Parameters

Solvent	i	j	ϵ^j_i	Temperature (°C)	Basis
	H	Co	2.58	1600	
	H	Fe	0.009	1600	
	N	Cr	−23.6	1600	
	N	Fe	−3.25	1600	
	O	Fe	132	1594	
	S	Cu	0(±1)	1600	**
Silver	Al	Al	6.4	700	
			5.6	800	
			4.6	900	
			4.22	1000	
	Cd	Cd	1.47	827	
	Pb	Pb	−0.4	1000	
Sodium	Cd	Hg	4.2	350	*
	Hg	Cd	4.2	350	
	Hg	Hg	2.0	350	
Thallium	Au	Au	2.6	700	
	Hg	Hg	0.091	325	
	Sn	Sn	−1.7	352	
	Sn	Sn	−0.8	414	
	Sn	Sn	0	478	
Tin	Au	Au	5.8	600	
	Cd	Cd	−1.05	500	
	Cd	Cd	−2.36	700	
	Cd	Pb	0	500	
	Cd	Zn	−1.65	700	*
	Cu	H	see $\epsilon^{Cu}{}_H$	1000–1300	*
	H	Cu	−9.65	1000	
	H	Cu	−8.05	1100	
	H	Cu	−5.64	1200	
	H	Cu	−4.45	1300	
	Pb	Cd	0	500	*
	Pb	Pb	0.65	500	
	Tl	Tl	−3.2	352	
	Tl	Tl	−1.78	414	
	Tl	Tl	−0.475	478	
	Zn	Cd	−1.65	700	
	Zn	Zn	−1.0	437	
	Zn	Zn	−0.6	684	
Zinc	Cd	Cd	−3.33	682	

* Calculated from reciprocity relationship—See Chapter 3, Equation 3-36.
** Values using data where x_s not approaching zero.

Source: J. M. Dealy and R. D. Pehlke, *Trans. Met. Soc. AIME 227*, 91, 92 (1963)

TABLE A-VII

Vapor Pressures of Liquid Metals

$$\log p = -A/T + B + C \log T \text{ (mm Hg)}$$

Element	A	B	C	Temp. Range (°K)
Ag	14,240	12.23	−1.055	1234–2400
Al	16,450	12.36	−1.023	1200–2800
Au	19,280	12.38	−1.01	1336–3240
Bi	10,400	12.35	−1.26	544–1833
Ca	9,600	12.55	−1.21	1123–1765
Cd	5,819	12.287	−1.257	594–1050
Cr	20,400	16.23	−1.82	2171–2938
Cu	17,650	13.39	−1.273	1356–2870
Hg	3,308	10.373	−0.8	298–630
Li	8,415	11.34	−1.0	453–1603
Mg	7,550	12.79	−1.41	923–1390
Mn	13,900	17.27	−2.52	1520–2309
Na	6,540	11.33	−1.718	400–1200
Ni	18,000	8.17	—	1728–3000
Pb	10,130	11.16	−0.985	600–2030
Pt	28,900	16.05	−1.77	2042–4105
Se	17,180	8.30	—	1505–1748
Si	20,900	10.84	−0.565	1690–3000
Sn	15,500	11.50	−1.0	500–2960
Ti	22,100	9.135	—	2000–3300
U	23,380	8.58	—	1630–1970
Zn	6,670	12.00	−1.126	692–1000

Source: C. J. Smithells, *Metals Reference Book*, 3rd ed., Vol. 2, pp. 655, 656, Butterworth, Inc., Washington, 1962.

TABLE A-VIII

Standard Electrode Potentials at 25°C for Oxidation Reactions

Electrode	Electrode Reaction	ϵ^0 (volts)
Li | Li$^+$	Li(s) = Li$^+$ + θ	+3.024
K | K$^+$	K(s) = K$^+$ + θ	+2.9241
Ca | Ca^{++}	Ca(s) = Ca^{++} + 2θ	+2.763
Na | Na$^+$	Na(s) = Na$^+$ + θ	+2.7146
Zn | Zn^{++}	Zn(s) = Zn^{++} + 2θ	+0.7611
Fe | Fe^{++}	Fe(s) = Fe^{++} + 2θ	+0.441
Cd | Cd^{++}	Cd(s) = Cd^{++} + 2θ	+0.4021
Pb | PbSO$_4$(s), SO$_4^{--}$	Pb(s) + SO$_4^{--}$ = PbSO$_4$(s) + 2θ	+0.3546
Tl | Tl$^+$	Tl(s) = Tl$^+$ + θ	+0.3385
Ni | Ni^{++}	Ni(s) = Ni^{++} + 2θ	+0.236
Ag | AgI(s), I$^-$	Ag(s) + I$^-$ = AgI(s) + θ	+0.1522
Sn | Sn^{++}	Sn(s) = Sn^{++} + 2θ	+0.1406
Pb | Pb^{++}	Pb(s) = Pb^{++} + 2θ	+0.1264
H$_2$ | H$^+$	H$_2$(g, 1 atm) = 2 H$^+$ + 2θ	±0.0000
Ag | AgBr(s), Br$^-$	Ag(s) + Br$^-$ = AgBr(s) + θ	−0.0711
Hg | Hg$_2$Br$_2$(s), Br$^-$	2 Hg(l) + 2 Br$^-$ = Hg$_2$Br$_2$(s) + 2θ	−0.1385
Pt | Sn^{++}, Sn^{++++}	Sn^{++} = Sn^{++++} + 2θ	−0.14
Ag | AgCl(s), Cl$^-$ |	Ag(s) + Cl$^-$ = AgCl(s) + θ	−0.2225
Hg | Hg$_2$Cl$_2$(s), Cl$^-$ |	2 Hg(l) + 2 Cl$^-$ = Hg$_2$Cl$_2$ + 2θ	−0.2680
Cu | Cu^{++}	Cu(s) = Cu^{++} + 2θ	−0.3441
I$_2$ | I$^-$	2 I$^-$ = I$_2$(s) + 2θ	−0.5362
Hg | Hg$_2$SO$_4$(s), SO$_4^{--}$	2 Hg(l) + SO$_4^{--}$ = Hg$_2$SO$_4$(s) + 2θ	−0.6141
Pt | Fe^{++}, Fe^{+++}	Fe^{++} = Fe^{+++} + θ	−0.7477
Ag | Ag$^+$	Ag(s) = Ag$^+$ + θ	−0.7990
Br$_2$ | Br$^-$	2 Br$^-$ = Br$_2$(l) + 2θ	−1.0659
Pt | Tl$^+$, Tl^{+++}	Tl$^+$ = Tl^{+++} + 2θ	−1.211
Cl$_2$ | Cl$^-$	2 Cl$^-$ = Cl$_2$(g, 1 atm) + 2θ	−1.3583
Pt | Ce^{+++}, Ce^{++++}	Ce^{+++} = Ce^{++++} + θ	−1.55
Pt | Co^{++}, Co^{+++}	Co^{++} = Co^{+++} + θ	−1.817

Source: C. F. Prutton and S. H. Maron, *Fundamental Principles of Physical Chemistry*, p. 531, The MacMillan Company, New York, 1951.

TABLE A-IX

Heat Content of Several Metals

Temperature °K	$H_T - H^\circ_{298.15}$ cal/gm atom					
	Aluminum	Copper	Iron[a]	Lead	Magnesium	Zinc
298	0	0	0(α)	0	0	0
400	600	600	642	656	620	630
500	1230	1215	1318	1324	1256	1270
600	1890	1845	2044	2014	1920	1940
700	2580	2480	2832	3884(liq)	2610	4400(liq)
800	3310	3130	3703	4605	3330	5150
900	4060	3800	4682	5318	4095	5900
1000	7330(liq)	4490	5819	6024	7010(liq)	6650
1100	8030	5190	7135	6723	7810	7400
1200	8730	5895	8347(γ)	7415	8640	
1300	9430	6615	9203	8100	9490	
1400	10130	10480(liq)	10059	8780		
1500	10830	11230	10915	9450		
1600	11530	11980	11771	10110		
1700	12230	12730	12939(δ)	10760		
1800	12930	13460	13914	11410		
1900	13630	14230	18658(liq)	12050		
2000	14330	14980	19714	12680		
2100	15030	15730	20774			
2200	15730	16480	21838			
2300	16430	17230	22906			
2400	17130	17980	23978			
2500	17830	18730	25054			
2600	18530	19480	26134			
2700	19230	20230	27218			
2800		20980	28306			
2900			29398			
3000			30494			
Melting point (°K)	932	1356	1809	606.6	923	692.7
ΔH_f cal/g-atom	2550	3120	3700	1141	2140	1765

[a] Phase changes: Curie pt 1042°K; $\alpha \rightarrow \gamma$ 1184°K 215 cal/g-atom; $\gamma \rightarrow \delta$ 1665°K 270 cal/g-atom.

Source: J. F. Elliott and M. Gleiser, *Thermochemistry for Steelmaking*, *I*, pp. 16, 48, 58, 61, 64, 118, Addison-Wesley Publishing Company, Inc., Reading, Mass., 1960.

Appendix A–X: The Error Function

Diffusion of energy or mass often is described by differential equations that are solved by an integral of the form,

$$\operatorname{erf} x = \frac{2}{\sqrt{\pi}} \int_0^x e^{-\alpha^2}\, d\alpha,$$

which is referred to as the error function. The factor, $\dfrac{2}{\sqrt{\pi}}$, normalizes the function such that it approaches 1 as x approaches ∞. The error function takes on values between 0 and 1 as x ranges from 0 to ∞, i.e., erf $(0) = 0$, erf $(\infty) = 1$. Differentiating the error function

$$\frac{d(\operatorname{erf} x)}{dx} = \frac{2}{\sqrt{\pi}}\, e^{-x^2}.$$

Then at $x = 0$

$$\left(\frac{d(\operatorname{erf} x)}{dx}\right)_{x=0} = \frac{2}{\sqrt{\pi}}\,.$$

For small values of x, the first two terms of a Taylor series expansion of the function

TABLE A-X

Values of the Error Function[a]

x	erf x	x	erf x
0.000	0.00000	0.80	0.74210
0.05	0.05637	0.85	0.77067
0.10	0.11246	0.90	0.79691
0.15	0.16800	0.95	0.82089
0.20	0.22270	1.00	0.84270
0.25	0.27633	1.10	0.88021
0.30	0.32863	1.20	0.91031
0.35	0.37938	1.30	0.93401
0.40	0.42839	1.40	0.95229
0.45	0.47548	1.50	0.96611
0.50	0.52050	1.60	0.97635
0.55	0.56332	1.70	0.98379
0.60	0.60386	1.80	0.98909
0.65	0.64203	1.90	0.99279
0.70	0.67780	2.00	0.995322
0.75	0.71116	2.50	0.999593
		3.00	0.999978

[a] The values above are taken from the National Bureau of Standards tabulation, Applied Mathematics Series 41, *Tables of the Error Function and Its Derivatives*, 1954, 302 pages.

about zero yield

$$\mathrm{erf}\ x = \mathrm{erf}\ (0) + x\left(\frac{d(\mathrm{erf}\ x)}{dx}\right)_{x=0} = \frac{2}{\sqrt{\pi}}\, x,$$

which is valid for $x \ll 1$.

Other special properties of the error function and application in diffusion problems are presented elsewhere.[1-3]

[1] R. M. Barrer, *Diffusion in and through Solids*, University Press, Cambridge, 1941.
[2] J. Crank, *The Mathematics of Diffusion*, Clarendon Press, Oxford, 1955.
[3] W. Jost, *Diffusion*, Academic Press, New York, 1952.

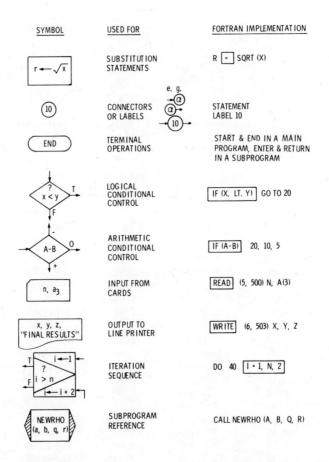

Flow Charting Convention used for Example Problems in Appendix B.

APPENDIX B Example Problems and Computer-facilitated Solutions

This appendix consists of ten example problems and their computer-facilitated solutions. The solutions are documented and include a flow diagram or other descriptive material of the solution algorithm, as well as a listing of the computer program and output. The programs were prepared in FORTRAN IV and were executed on the IBM 360/67 duplex computer at The University of Michigan, using the *Michigan Terminal System*, a multiprocessing facility. FORTRAN IV was chosen as the programming language to permit the most extensive direct use of these example programs.

A logic flowchart is presented for each computer program. The flowcharting convention used for the example problems in this appendix was designed to provide a system by which the *logic* of each computer program could be represented clearly, and concisely without sacrificing generality. Although the example problems have been solved using the FORTRAN language, the logic flow diagrams contained herein are language-independent.

The symbols and their interpretations follow. It should be noted that the larger example problems have only a system flow chart, i.e. each detailed step is not listed separately. A symbol table is provided for each program. Minor alterations to a few symbols shown below occur in some flowcharts.

Example Problem No. 1: Thermochemical Data Calculation— Enthalpy Increments

Problem Statement

This problem requires that a computer program be written to integrate a specific heat equation over the temperature range 298.16 to 2500°K. Up to nine phase changes may be involved at various temperatures. The enthalpy increments are to be printed at 100°K intervals as well as for the primary phase at each transformation temperature and for the transformed phase at the same temperature.

The input data include in addition to the name of the material the coefficients, A, B, and C, of a specific heat equation of the form:

$$C_p = A + BT + C/T^2,$$

for each phase; the temperature T of transformation; the first temperature X above T which is an integral multiple of 100; and H, the enthalpy of the transformation.

Symbol Table—Problem 1

Algebraic Symbol	Definition	FORTRAN Symbol
A_j	Constant A for phase j	$A(J)$
B_j	Constant B for phase j	$B(J)$
C_j	Constant C for phase j	$C(J)$
T_j	Transition temperature between the j-1st and jth phases	$T(J)$
X_j	Temperature just above T_j which is an integral multiple of 100	$X(J)$
n	Number of phase changes ($N < 10$)	N
H_j	Enthalpy change at phase transition temperature T_j	$H(J)$
j	Counter on phases	J
i	Counter on 100° intervals	I
T_L	Lower temperature limit for integration	TL
ΔT	Temperature interval (in this case 100°)	$DELT$
...	Locations for storage of material names	MA, TE, RI, AL, and S
$(H_T - H_{298.16}°)$	Enthalpy at T above reference	$E(J)$

A flow diagram, tested program in the FORTRAN language and printed output for a few elements are to be presented.

Method of Solution—Problem 1

In regions where there is no phase change the increase in enthalpy between two temperatures is

$$\Delta H = \int_{T_1}^{T_2} \left(A_j + B_j T + \frac{C_j}{T^2} \right) dT. \tag{a}$$

This may be integrated and $(T_2 - T_1)$ factored out to give

$$\Delta H = \left[A_j + \frac{B_j(T_2 + T_1)}{2} + \frac{C_j}{T_1 T_2} \right] (T_2 - T_1). \tag{b}$$

It is this form of the integral that is used in the computations.

At a phase transition point the increase in enthalpy will simply be that stipulated by input data. Since enthalpies are to be printed every 100°K and also at the beginning and end of each phase change, the program must be alert to the occurrence of a transition at any given temperature between 298.16°K and 2500°K. The method by which this is done is to input not only the transition temperature, but also the next higher temperature above each transition that is an integral multiple of 100. Use of this integral multiple temperature X_j, is demonstrated in the following program.

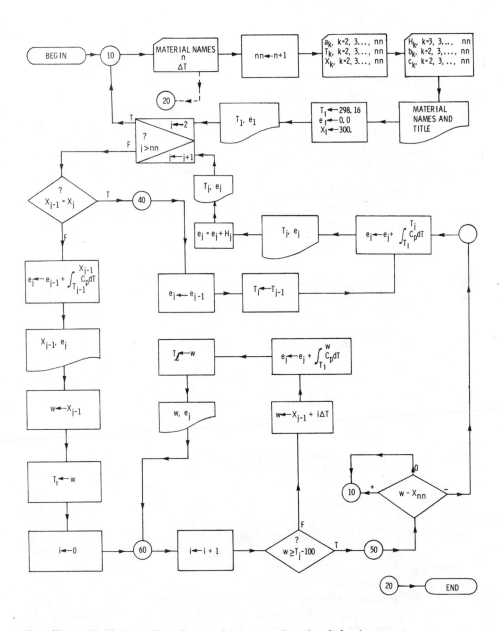

Flow Chart—Problem 1.   Flow diagram for computation of enthalpy increment.

FORTRAN Listing—Problem 1

```
        REAL MA,I
        DIMENSION A(10),B(10),C(10),T(10),X(10),H(10),E(10)
  10    READ (5,500,END=20) MA,TE,RI,AL,S,N,DELT
        NN=N+1
        READ (5,510) (A(K),K=2,NN)
        READ (5,510) (T(K),K=2,NN)
        READ (5,510) (X(K),K=2,NN)
        READ (5,510) (H(K),K=2,NN)
        READ (5,520) (B(K),K=2,NN)
        READ (5,520) (C(K),K=2,NN)
        WRITE (6,600) MA,TE,RI,AL,S
        T(1)=298.16
        E(1)=0.
        X(1)=300.
        WRITE (6,610) T(1),E(1)
        DO 30 J=2,NN
        IF (X(J-1).EQ.X(J)) GO TO 40
        E(J)=E(J-1)+(A(J)+(B(J)*(X(J-1)+T(J-1))/2.)+C(J)/(X(J-1)*T(J-1)))*
       1(X(J-1)-T(J-1))
        WRITE (6,610) X(J-1),E(J)
        TEMP=X(J-1)
        TL=TEMP
        I=0.
  60    I=I+1.
        IF (TEMP.GE.(T(J)-100.)) GO TO 50
        TEMP=X(J-1)+I*DELT
        E(J)=E(J)+(A(J)+(B(J)*(TEMP+TL)/2.)+C(J)/(TEMP*TL))*(TEMP-TL)
        TL=TEMP
        WRITE (6,610) TEMP, E(J)
        GO TO 60
  50    IF (TEMP-X(NN)) 61,10,10
  40    E(J)=E(J-1)
        TL=T(J-1)
  61    E(J)=E(J)+(A(J)+(B(J)*(T(J)+TL)/2.)+C(J)/(T(J)*TL))*(T(J)-TL)
        WRITE (6,610) T(J),E(J)
        E(J)=E(J)+H(J)
  30    WRITE (6,610) T(J),E(J)
        GO TO 10
  20    CALL SYSTEM
 500    FORMAT (5A4,10X,I5,F5.0)
 510    FORMAT (8F9.2)
 520    FORMAT (8E9.2)
 600    FORMAT (35H1HIGH TEMPERATURE HEAT CONTENT FOR 5A4/1H-,12X,16HT DEG
       1REES KELVIN,9X,16HH(T) - H(298.16)//1H-)
 610    FORMAT (F20.0,F30.1)
        END
```

Consider a metal which has two transitions within a 100° interval, say one at 627°K which will be denoted T_1 and the other at 685°K which will be denoted T_2. A comparison is first made between x_{j-1}, which for $j = 1$ is $X_0 = 300$ and X_j which is $X_1 = 700$. The subscript j indicates the stable phase. Since 300 does not equal 700, the procedure is to calculate the increase in enthalpy between 298.16 and 300 using A_1, B_1, and C_1. The temperature T is then incremented by 100 and the enthalpy increase between 300° and 400° computed and added to the value at 300. This process is repeated until T at the lower limit of the integral is equal to or greater than $T_1 - 100$ or 527°. Thus, the last calculation will be between 500° and 600°, since 600° is greater than 527°, and the procedure will stop short of that. The next calculation is to integrate between 600° and T_1 which is 627°, using still A_1, B_1, and C_1, for so far we are dealing with only the first phase. At 627°, ΔH of transi-

Input (Data)—Problem 1

```
IRON                                5  100
          337     1040      485   1030    1000
       103300   117900   167400  180300  260000
       110000   120000   170000  190000  250000
        41000    21000    11000  370000     000
       7.10E-3  0.00E+0  3.00E-3  0.00E+0  0.00E+0
       0.43E+5  0.00E+0  0.00E+0  0.00E+0  0.00E+0
MANGANESE                           6  100
          570      833     1070   1130    1130    626
       100000   137400   141000  151700  236800  260000
       110000   140000   150000  160000  240000  250000
        53500    54500    43000  350000  5370000    000
       3.38E-3  0.66E-3  0.00E+0  0.00E+0  0.00E+0  0.00E+0
       0.37E+5  0.00E+0  0.00E+0  0.00E+0  0.00E+0  0.00E+0
CARBON                              1  100
          410
       260000
       250000
          000
       1.02E-3
       2.10E+5
SILICON                             2  100
          579      613
       168600   260000
       170000   250000
      1210000      000
       0.56E-3  0.00E+0
       1.09E+5  0.00E+0
```

tion is added to the enthalpy to give the enthalpy of the transformed phase at 627°.

In considering phase 2, a comparison is made between $X_1 = 700°$ and X_2. Since the second transition is in the same 100° interval as the first, X_2 also is 700°. Because these two are equal, the program transfers to a calculation of the integral between 627° and 685°. Then the ΔH at 685° is added to give the enthalpy of the transformed or third phase at 685°. Continuing with the third phase the program determines the integral between 685° and 700° first, and then proceeds again by 100° intervals until a new transition temperature is reached.

Discussion of Results—Problem 1

Examination of the computed enthalpies for iron shows how the calculation advanced by 100° intervals up to 1000°K. At this point the program proceeded to 1033°, the first transition temperature. There it added a ΔH of 410, and then went on to 1100° etc.

The computed enthalpies (H) are in calories per g-mole.

Output (Results)—Problem 1

HIGH TEMPERATURE HEAT CONTENT FOR IRON

T DEGREES KELVIN	H(T) - H(298.16)
298.	0.0
300.	11.0
400.	632.3
500.	1310.3
600.	2052.2
700.	2860.9

Output (Results)—Problem 1

HIGH TEMPERATURE HEAT CONTENT FOR IRON

T DEGREES KELVIN	H(T) - H(298.16)
800.	3738.1
900.	4684.5
1000.	5700.8
1033.	6051.6
1033.	6461.6
1100.	7158.4
1179.	7980.0
1179.	8190.0
1200.	8366.7
1300.	9226.7
1400.	10116.7
1500.	11036.7
1600.	11986.7
1674.	12709.0
1674.	12819.0
1700.	13086.8
1800.	14116.8
1803.	14147.7
1803.	17847.7
1900.	18817.7
2000.	19817.7
2100.	20817.7
2200.	21817.7
2300.	22817.7
2400.	23817.7
2500.	24817.7

HIGH TEMPERATURE HEAT CONTENT FOR MANGANESE

T DEGREES KELVIN	H(T) - H(298.16)
298.	0.0
300.	13.1
400.	732.2
500.	1472.8
600.	2241.1
700.	3039.6
800.	3869.7
900.	4732.1
1000.	5627.3
1000.	6162.3
1100.	7064.6
1200.	7973.5
1300.	8889.0
1374.	9570.8
1374.	10115.7
1400.	10393.9
1410.	10500.9
1410.	10930.9
1500.	11947.9
1517.	12140.0
1517.	15640.0
1600.	16577.9
1700.	17707.9
1800.	18837.9
1900.	19967.9
2000.	21097.9
2100.	22227.9
2200.	23357.9
2300.	24487.9
2368.	25256.3
2368.	78956.3
2400.	79156.6
2500.	79782.5

HIGH TEMPERATURE HEAT CONTENT FOR CARBON

T DEGREES KELVIN	H(T) - H(298.16)
298.	0.0
300.	12.4
400.	633.1
500.	1194.0
600.	1730.1
700.	2256.4
800.	2780.4
900.	3306.3
1000.	3836.5
1100.	4372.7
1200.	4915.9
1300.	5466.9
1400.	6026.1
1500.	6594.0
1600.	7170.9
1700.	7756.9
1800.	8352.2
1900.	8957.1
2000.	9571.5
2100.	10195.6
2200.	10829.4
2300.	11473.1
2400.	12126.6
2500.	12790.0

HIGH TEMPERATURE HEAT CONTENT FOR SILICON

T DEGREES KELVIN	H(T) - H(298.16)
298.	0.0
300.	13.2
400.	702.6
500.	1361.3
600.	2007.5
700.	2648.8
800.	3289.3
900.	3931.0
1000.	4575.3
1100.	5223.0
1200.	5874.7
1300.	6530.7
1400.	7191.3
1500.	7856.7
1600.	8527.0
1686.	9107.5
1686.	21207.5
1700.	21293.4
1800.	21906.3
1900.	22519.3
2000.	23132.3
2100.	23745.3
2200.	24358.3
2300.	24971.3
2400.	25584.3
2500.	26197.3

Example Problem No. 2: Calculation of Adiabatic Flame Temperatures at the Tuyeres of a Coke-Fueled Shaft Furnace

Problem Statement

One of the physical limitations placed on the coke-fueled shaft furnace (such as the cupola or iron-blast furnace) is the necessity for a specific temperature distribution to exist in the melting zone just above the tuyeres. A number of possibilities for coke reduction and increased production from the shaft furnace exist through the addition of blast injectants such as oxygen, natural gas, and fuel oil. In order to estimate the changes in blast composition and temperature required to maintain the desired temperature distribution about the tuyere zone, the concept of adiabatic flame temperature has been adopted.

The adiabatic flame temperature is assumed to relate to the combustion of the blast with the downwardly moving coke in the zone just in front of and above the tuyeres. If the furnace has been operating satisfactorily on a given burden and with a given blast composition and temperature, the adiabatic flame temperature for these conditions can be used as a required condition for smooth furnace operation.

Write and test a FORTRAN program to calculate the adiabatic flame temperature at the tuyeres of a coke-fueled shaft furnace (Problem 2A). Assume a temperature of 2800°F for the coke. Base the calculation on known values of blast temperature and composition (including moisture) and a mass balance based on the combustion gases nitrogen, hydrogen, and carbon monoxide. Changes in temperature of the metal and slag components as they move down through this zone can be assumed to be independent of the blast volume per unit of coke. Consider only oxygen enrichment and natural gas (methane) injection.

This calculation can be used to specify a balanced blast to maintain a desired flame temperature as blast operating conditions change. If the necessary flame temperature has been determined for a given furnace practice, the appropriate adjustments in composition and temperature of the blast to maintain this flame temperature also can be computed. One could, for example, enrich the blast in oxygen and compensate for the increased flame temperature by adding moisture or natural gas. Alternatively, one could increase the blast temperature, thereby increasing the flame temperature, and compensate for this by adding moisture or natural gas.

Write and test another program for calculating the required oxygen enrichment to maintain a specified flame temperature, given the other blast variables (Problem 2B).

Method of Solution—Problem 2

The carbon–oxygen reaction provides the thermal energy for operation of the coke-fueled shaft furnace. Oxygen may react with carbon to produce carbon monoxide or carbon dioxide. At the flame temperatures involved in the tuyere zone of the cupola or blast furnace, the reaction product is essentially carbon monoxide. Nitrogen, the major constituent in air, acts as an inert material and passes through the furnace unchanged.

An important blast variable is the moisture content of the air. In the presence of graphite at high temperature, water vapor is decomposed into hydrogen and oxygen, with the oxygen immediately reacting to form carbon monoxide. This overall reaction,

$$H_2O + C = H_2 + CO, \tag{a}$$

is endothermic, and absorbs approximately 34,000 BTU per lb. mole. The flame temperature is thus highly dependent upon the moisture content of the blast, and decreases with increasing moisture content.

The enrichment of the blast with natural gas also will decrease the flame temperature. This is caused by the absorption of energy in decomposing methane (and other alkanes) into hydrogen and carbon, which will eventually react to form carbon monoxide according to the reactions:

$$CH_4 = C + 2H_2, \qquad C + \tfrac{1}{2}O_2 = CO. \tag{b}$$

The chemistry of tuyere zone reactions is illustrated in Fig. B-1. The reactants

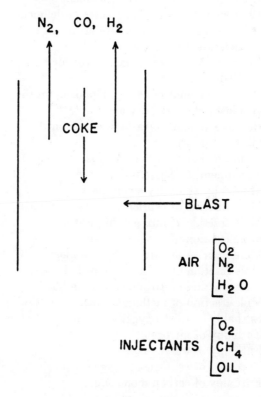

Fig. B-1.   Components in Tuyere zone

are air, containing moisture, and in some cases added injectants such as oxygen or a fuel such as natural gas, oil, or powdered coal. The blast, sometimes preheated, reacts in the zone in front of the tuyeres with the coke, which has been preheated by the hot gases rising up the furnace stack. As discussed above, the products of combustion are hydrogen and carbon monoxide, which are essentially free of any moisture and carbon dioxide. The nitrogen passes through the furnace unchanged.

A mass and energy balance for combustion can be carried out taking into account the sensible heat of the reactants and the heats of reaction (including vaporization or decomposition reactions) and equating these energies to the sensible heat of the products. Usually this is accomplished by trial and error evaluation of the temperature of the reaction products.

Symbol Table—Problem 2A

Algebraic Symbol	Definition	FORTRAN Symbol
· · ·	Grains of moisture per ft³ in blast	$WIQH2O$
· · ·	Per cent oxygen enrichment	$WIQ2$
· · ·	Per cent carbon monoxide enrichment	$WICO$
· · ·	Per cent carbon dioxide enrichment	$WICO2$
· · ·	Per cent natural gas (methane) enrichment	$WICH4$
· · ·	Relative amount of moisture in the air blast, ft³ moisture/ft³ dry air	$RH2O$
· · ·	Relative amount of oxygen in the air blast, ft³ O_2/ft³ dry air	$RQ2$
· · ·	Relative amount of carbon monoxide in the air blast, ft³ CO/ft³ dry air	RCO
· · ·	Relative amount of carbon dioxide in the air blast, ft³ CO_2/ft³ dry air	$RCO2$
· · ·	Relative amount of natural gas (methane) enrichment, ft³ CH_4/ft³ dry air	$RCH4$
X_{air}	Mole fraction of dry air in blast	AR
$X_{O_2(Q)}$	Mole fraction of (enriched) oxygen in blast	$XWIQ2$
X_{N_2}	Mole fraction of nitrogen in blast	$XWIN2$
X_{H_2O}	Mole fraction of water in blast	$XWIH2O$
X_{O_2}	Mole fraction of (air) oxygen in blast	$XWIO2$
X_{CO}	Mole fraction of carbon monoxide in blast	$XWICO$
X_{CO_2}	Mole fraction of carbon dioxide in blast	$XWICO2$
X_{CH_4}	Mole fraction of methane in blast	$XWICH4$
H_{O_2}	Enthalpy of (air) oxygen	$HWO2$
H_{N_2}	Enthalpy of nitrogen	$HWN2$
H_{H_2O}	Enthalpy of water	$HWH2O$
$H_{O_2(q)}$	Enthalpy of (enriched) oxygen	$HWQ2$
H_{CO}	Enthalpy of carbon monoxide	$HWCO$
H_{CO_2}	Enthalpy of carbon dioxide	$HWCO2$
H_{CO_4}	Enthalpy of methane	$HWCH4$

Symbol Table—Problem 2A—(Continued)

Algebraic Symbol	Definition	FORTRAN Symbol
H_C	Enthalpy of coke	HC
H_{BL}	Enthalpy of blast	HBL
$H_{CO(CG)}$	Enthalpy of carbon monoxide in combustion gases	HTGCO
$H_{H_2(CG)}$	Enthalpy of hydrogen in combustion gases	HTGH2
$H_{N_2(CG)}$	Enthalpy of nitrogen in combustion gases	HTGN2
H_{CG}	Enthalpy of combustion products (total)	HTG
$N_{CO(CG)}$	Number of moles of carbon monoxide in combustion gases	NTGCO
$N_{H_2(CG)}$	Number of moles of hydrogen in combustion gases	NTGH2
$N_{N_2(CG)}$	Number of moles of nitrogen in combustion gases	NTGN2
$N_{C(tuyere)}$	Moles of carbon burned in reaction per mole of blast in tuyere zone	NBC
T_{blast}	Temperature of the blast	TA
$T_{O_2(Q)}$	Temperature of oxygen added	TQ2
T_C	Temperature of coke	TC
T_{CO}	Temperature of carbon monoxide	TCO
T_{CO_2}	Temperature of carbon dioxide	TCO2
T_{CH_4}	Temperature of methane	TCH4
T_f	Temperature of flame	TFL
ΔT	Current increment in flame temperature	TDELT
n	Iteration counter	N
H_{err_1}	$(H_{products} - H_{reactants})$ for the last iteration	ODIF
H_{err_2}	$(H_{products} - H_{reactants})$ for the current iteration	NDIF

Convergence by Half-Interval Method

The half-interval method is a technique for determining the root of an equation. In the present case, the sensible heat of the reaction products, expressed in terms of the adiabatic flame temperature, are set equal to the sensible heat of the reactants plus the energy released by reaction. The appropriate value of the flame temperature is the root of the energy balance equation. To determine this value, a flame temperature is assumed and the sensible heat of the products calculated. If this energy is less than the sensible heat of the reactants plus the energy of reaction, the flame temperature is incremented by a given amount and the process repeated. This technique is illustrated with a specific example in Fig. B-2.

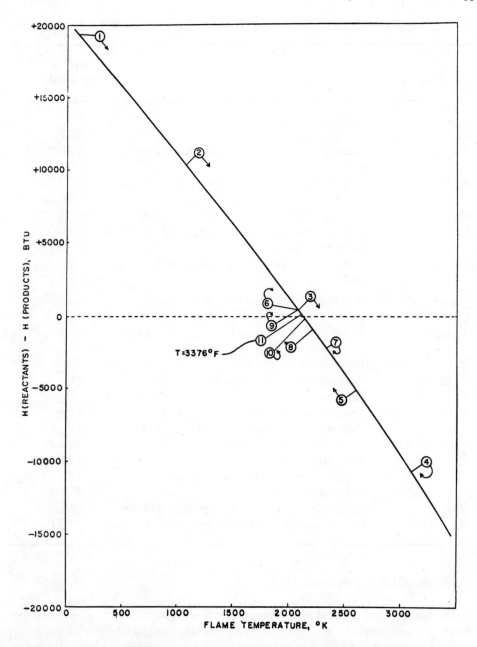

Fig. B-2.   Flame temperature calculation by half-interval method.

The first point shown in Fig. B-2 is located at a temperature of 100°K, the first assumed value for the flame temperature. This value is obviously too low, and the difference between total energy for the reactants (the sensible heat and energy of reaction) and the products is quite large. The flame temperature then has been incremented by 1000°K to point 2. This assumed temperature also is too low, as is

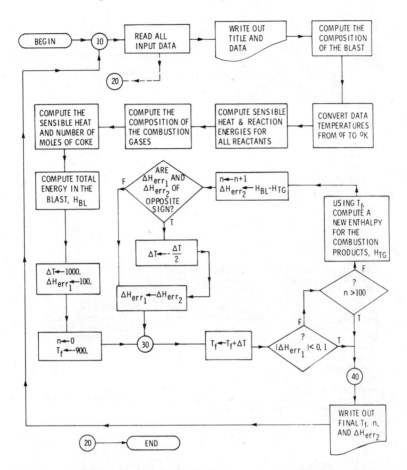

Flow Chart—Problem 2A.   Flow diagram for computation of flame temperature.

point 3. However, at point 4 the assumed flame temperature is too high, as evidenced by the fact that the sign of the difference between energy of reactants and products has changed from positive to negative. The increment to be added to the flame temperature has been decreased by a factor of one half and the sign changed. The next assumed flame temperature is at point 5. This assumed flame temperature is still too high, and an additional increment, in this case negative, has been added to the flame temperature, returning the assumed value to point 3 (also point 6). This value was too low, and the increment to the flame temperature is decreased by a factor of one half again and the sign changed, yielding point 7. This process continues until convergence is achieved at a flame temperature where the total enthalpy of the reactants is within an acceptable limit of difference from that of the products at point 11.

    This routine procedure could be shortened somewhat by a person making a hand calculation. However, by cutting the interval in half each time the convergence point is passed, the maximum number of trials is well defined, and convergence is achieved rapidly. If the calculation were made by hand, the direction of imbalance in the total enthalpies for reactants and products readily would be observable.

FORTRAN Listing—Problem 2A

```
          REAL NTGCO,NTGH2,NTGN2,NBC,NDIF
   10     READ (5,500,END=20) WIGH2O,WIQ2,WICO,WICO2,WCH4,TA,TQ2,TCO,TCO2,TC
          1H4,TC
          WRITE (6,600) TA,WCH4,TC,WICO,TCH4,WICO2,TCO,WIGH2O,TCO2,WIQ2,TQ2
          RH2O=379.*WIGH2O/(7000.*18.)
          RQ2=WIQ2/100./(1.-RH2O/(1.+RH2O))
          RCO=WICO/100./(1.-RH2O/(1.+RH2O))
          RCO2=WICO2/100./(1.-RH2O/(1.+RH2O))
          RCH4=WCH4/100./(1.-RH2O/(1.+RH2O))
          AR=1./(1.+RH2O+RQ2+RCO+RCO2+RCH4)
          XWIO2=.21*AR
          XWIN2=.79*AR
           XWIH2O=RH2O*AR
          XWIQ2=RQ2*AR
          XWICO=RCO*AR
          XWICO2=RCO2*AR
          XWICH4=RCH4*AR
          TA=(TA+460.)/1.8
          TQ2=(TQ2+460.)/1.8
          TCO=(TCO+460.)/1.8
          TCO2=(TCO2+460.)/1.8
          TCH4=(TCH4+460.)/1.8
          HWO2=7.16*TA+.5E-3*TA*TA+.4E5/TA-2313.
          HWN2=6.83*TA+.45E-3*TA*TA+.12E5/TA-2117.
          HWH2O=7.3*TA+1.23E-3*TA*TA-2286.-57800.
          HWQ2=7.16*TQ2+5E-3*TQ2*TQ2+.4E5/TQ2-2313.
          HWCO=6.79*TCO+.49E-3*TCO*TCO+.11E5/TCO-2105.-26620.
          HWCO2=10.57*TCO2+1.05E-3*TCO2*TCO2+2.06E5/TCO2-3936.-94030.
          HWCH4=5.65*TCH4+5.72E-3*TCH4*TCH4+.46E5/TCH4-2347.-17889.
          NTGCO=2.*XWIO2+XWIH2O+2.*XWIO2+XWICO+2.*XWICO2+XWICH4
          NTGH2=XWIH2O+2.*XWICH4
          NTGN2=XWIN2
          NBC=NTGCO-XWICO-XWICO2-XWICH4
          TC=(TC+460.)/1.8
          HC=4.03*TC+.57E-3*TC*TC+2.04E5/TC-1936.
          HBL=HWO2*XWIO2+HWN2*XWIN2+HWH2O*XWIH2O+HWQ2*XWIQ2+HWCO*XWICO+HWCO2
          1 *XWICO2+HWCH4*XWICH4+NBC*HC
          TDELT=1000.
          ODIF=100.
          N=0
          TFL=-900.
   30     TFL=TFL+TDELT
          IF (ABS(ODIF).LT..1.OR.N.GT.100) GO TO 40
          N=N+1
          HTGCO=6.79*TFL+.49E-3*TFL*TFL+.11E5/TFL-2105.-26620.
          HTGH2=6.52*TFL+.39E-3*TFL*TFL-.12E5/TFL-1938.
          HTGN2=6.83*TFL+.45E-3*TFL*TFL+.12E5/TFL-2117.
          HTG=NTGCO*HTGCO+NTGH2*HTGH2+NTGN2*HTGN2
          NDIF=HBL-HTG
          IF (ODIF*NDIF.LT.0.) TDELT=-TDELT/2.
          ODIF=NDIF
          GO TO 30
   40     TFL=TFL-TDELT
          TA=TA*1.8-460.
          TQ=TQ*1.8-460.
          TCO=TCO*1.8-460.
          TCO2=TCO2*1.8-460.
          TCH4=TCH4*1.8-460.
          TFL=TFL*1.8-460.
          WRITE (6,610) TFL,NDIF,N
          WRITE (6,611) HBL,HWH2O,ODIF,HC,HWN2,TC,HTG,HWO2,XWICH4,HTGCO,HWQ2
          1,XWICO,HTGH2,NBC,XWICO2,HTGN2,NDIF,XWIH2O,HWCH4,NTGCO,XWIN2,HWCO,N
          2TGH2,XWIO2,HWCO2,NTGN2,XWIQ2
          GO TO 10
   20     CALL SYSTEM
  500     FORMAT (5F6.2,6F5.0)
  600     FORMAT (30H1FLAME TEMPERATURE CALCULATION/5H0DATA/1H0,1X,3HTA=F6.0
          1,11X,5HWCH4=F6.2/2X,3HTC=F6.0,11X,5HWICO=F6.2/2X,5HTCH4=F6.0,9X,6H
          2WICO2=F6.2/2X,4HTCO=F6.0,10X,7HWICH2O=F6.2/2X,5HTCO2=F6.0,9X,5HWIQ
          32=F6.2/2X,4HTQ2=F6.0)
```

```
610    FORMAT (8H-RESULTS/32HOFLAME TEMPERATURE, DEGREES F. =F6.0/   58HO
    1ENTHALPY DIFFERENCE BETWEEN REACTANTS AND PRODUCTS, CAL =F8.0/41HO
    2ITERATIONS REQUIRED TO ATTAIN SOLUTION =I3)
611    FORMAT (1H-,1X,4HHBL=F8.0,8X,6HHWH2O=F8.0,6X,5HODIF=F6.0/2X,3HHC=F
    18.0,9X,5HHWN2=F8.0,7X,3HTC=F6.0/2X,4HHTG=F8.0,8X,5HHWO2=F8.0,7X,7H
    2XWICH4=F5.2/2X,6HHTGCO=F8.0,6X,5HHWO2=F9.0,6X,6HXWICO=F5.2/2X,6HHT
    3GH2=F8.0,6X,4HNBC=F6.1,10X,7HXWICO2=F5.2/2X,6HHTGN2=F8.0,6X,5HNDIF
    4=F6.1,9X,7HXWIH2O=F5.2/2X,6HHHWCH4=F8.0,6X,6HNTGCO=F6.1,8X,6HXWIN2=
    5F5.2/2X,5HHWCO=F8.0,7X,6HNTGH2=F6.1,8X,6HXWIO2=F5.2/2X,6HHWCO2=F8.
    60,6X,6HNTGN2=F6.1,8X,6HXWIQ2=F5.0)
    END
```

Input (Data)—Problem 2A

```
        300     00      00     00    00 1000 1000    77    77    77 2800
```

Output (Results)—Problem 2A

```
FLAME TEMPERATURE CALCULATION
DATA

    TA= 1000.              WCH4=   0.0
    TC= 2800.              WICO=   0.0
    TCH4=    77.           WICO2=  0.0
    TCO=    77.            WICH2O= 3.00
    TCO2=   77.            WIQ2=   0.0
    TQ2= 1000.

RESULTS

FLAME TEMPERATURE, DEGREES F. = 3380.

ENTHALPY DIFFERENCE BETWEEN REACTANTS AND PRODUCTS, CAL =          0.

ITERATIONS REQUIRED TO ATTAIN SOLUTION = 28

    HBL=    6375.        HWH2O= -53356.      ODIF=     0.
    HC=     7345.        HWN2=    3734.      TC= 1811.
    HTG=    6375.        HWO2=    3873.      XWICH4= 0.0
    HTGCO= -12006.       HWQ2=    6833.      XWICO= 0.0
    HTGH2=  13739.       NBC=     0.4        XWICO2= 0.0
    HTGN2=  14506.       NDIF=    0.0        XWIH2O= 0.01
    HWCH4= -17887.       NTGCO=   0.4        XWIN2= 0.78
    HWCO= -26619.        NTGH2=   0.0        XWIO2= 0.21
    HWCO2= -94029.       NTGN2=   0.8        XWIQ2= 0.
```

However, in machine computation, and particularly for use of the technique in other applications, a direct and simple way to indicate that the convergence point has been overshot is to multiply the difference between the calculated quantities, in this case the enthalpies of the reactants and products, by the difference on the previous iteration. If the convergence point has been passed, the sign of this product will be negative (a negative quantity times a positive quantity) and the direction of search is reversed and the increment is decreased to one half of its former magnitude. If, however, the product is positive (the product of two positive or of two negative quantities), the procedure is continued in the same direction and the magnitude of the increment is maintained at the same level.

It is necessary, of course, to set some criterion for convergence or the iterative procedure would continue indefinitely. Two criteria can be adoped, and either should prove successful. One is to set an acceptable limiting maximum increment for the flame temperature, whereas the second (and the one used in this particular case) is to set the acceptable maximum increment on the difference between the enthalpy values for the reactants and products.

Problem 2B—Calculation of Oxygen Enrichment Level

In precisely the same manner outlined above, a blast variable (such as per cent oxygen enrichment, per cent natural gas, moisture content, blast temperature, etc.) can be calculated to provide a given adiabatic flame temperature. The iterative technique employed involves incremental changes in the particular blast variable of interest, until the enthalpy balance for reactants and products converges. A flow chart showing how this calculation is carried out for enrichment of the blast with oxygen is presented below. Parallel programs can be developed for other blast variables. Results of calculations are discussed in later paragraphs.

Symbol Table—Problem 2B

The symbols used in the program are the same as those for the flame temperature program. *QDELT* is used to indicate the current increment in oxygen enrichment.

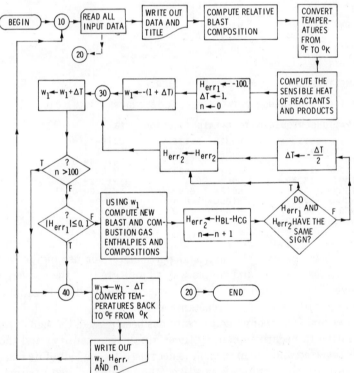

Flow Chart—Problem 2B.   Flow diagram for computation of oxygen enrichment of blast.

FORTRAN Listing—Problem 2B

```
      REAL NTGCO,NTGH2,NTGN2,NBC,NDIF
   10 READ (5,500,END=20) WIGH2O,WICO,WICO2,WCH4,TA,TQ2,TCO,TCO2,TCH4,TC
     1,TFL
      WRITE (6,600) TA,TQ2,TC,WCH4,TCH4,WICO,TCO,WICO2,TCO2,WIGH2O,TFL
      RH2O=379.*WIGH2O/(7000.*18.)
      RCO=WICO/100./(1.-RH2O/(1.+RH2O))
      RCO2=WICO2/100./(1.-RH2O/(1.+RH2O))
      RCH4=WCH4/100./(1.-RH2O/(1.+RH2O))
```

```
          TA=(TA+460.)/1.8
          TQ2=(TQ2+460.)/1.8
          TCO=(TCO+460.)/1.8
          TCO2=(TCO2+460.)/1.8
          TCH4=(TCH4+460.)/1.8
          TFL=(TFL+460.)/1.8
          HWO2=7.16*TA+.5E-3*TA*TA+.4E5/TA-2313.
          HWN2=6.83*TA+.45E-3*TA*TA+.12E5/TA-2117.
          HWH2O=7.3*TA+1.23E-3*TA*TA-2286-57800.
          HWO2=7.16*TQ2+.50E-3*TQ2*TQ2+.4E5/TQ2-2313.
          HWCO=6.79*TCO+.49E-3*TCO*TCO+.11E5/TCO-2105.-26620.
          HWCO2=10.57*TCO2+1.05E-3*TCO2*TCO2+2.06E5/TCO2-3936.-94030.
          HWCH4=5.65*TCH4+5.72E-3*TCH4*TCH4+.46E5/TCH4-2347.-17889.
          TC=(TC+460.)/1.8
          HC=4.03*TC+.57E-3*TC*TC+2.04E5/TC-1936.
          HTGCO=6.79*TFL+.49E-3*TFL*TFL+.11E5/TFL-2105.-26620.
          HTGH2=6.52*TFL+.39E-3*TFL*TFL-.12E5/TFL-1938.
          HTGN2=6.83*TFL+.45E-3*TFL*TFL+.12E5/TFL-2117.
          ODIF=-100.
          QDELT=1.
          N=0
          WIQ2=-1.-QDELT
   30     WIQ2=WIQ2+QDELT
          IF (ABS(ODIF).LT..1.OR.N.GT.100) GO TO 40
          N=N+1
          RQ2=WIQ2/100./(1.-RH2O/(1.+RH2O))
          AR=1./(1.+RH2O+RQ2+RCO+RCO2+RCH4)
          XWIO2=.21*AR
          XWIN2=.79*AR
           XWIH2O=RH2O*AR
          XWIQ2=RQ2*AR
          XWICO=RCO*AR
          XWICO2=RCO2*AR
          XWICH4=RCH4*AR
          NTGCO=2.*XWIO2+XWIH2O+2.*XWIO2+XWICO+2.*XWICO2+XWICH4
          NTGH2=XWIH2O+2.*XWICH4
          NTGN2=XWIN2
          NBC=NTGCO-XWICO-XWICO2-XWICH4
          HBL=HWO2*XWIO2+HWN2*XWIN2+HWH2O*XWIH2O+HWO2*XWIQ2+HWCO*XWICO+HWCO2
         1 *XWICO2+HWCH4*XWICH4+NBC*HC
          HTG=NTGCO*HTGCO+NTGH2*HTGH2+NTGN2*HTGN2
          NDIF=HBL-HTG
          IF (ODIF*NDIF.LT.0.) QDELT=-QDELT/2.
          ODIF=NDIF
          GO TO 30
   40     WIQ2=WIQ2-QDELT
          TA=TA*1.8-460.
          TQ=TQ*1.8-460.
          TCO=TCO*1.8-460.
          TCO2=TCO2*1.8-460.
          TCH4=TCH4*1.8-460.
          TFL=TFL*1.8-460.
          WRITE (6,610) WIQ2,NDIF,N
          WRITE (6,611) HBL,HWH2O,ODIF,HC,HWN2,TC,HTG,HWO2,XWICH4,HTGCO,HWO2
         1,XWICO,HTGH2,NBC,XWICO2,HTGN2,NDIF,XWIH2O,HWCH4,NTGCO,XWIN2,HWCO,N
         2TGH2,XWIO2,HWCO2,NTGN2,XWIQ2
        . GO TO 10
   20     CALL SYSTEM
  500     FORMAT (4F5.2,7F6.0)
```

```
600   FORMAT (30H1OXYGEN ENRICHMENT CALCULATION/5HODATA/1H0,1X,3HTA=F6.0
     1,11X,4HTQ2=F6.0/2X,3HTC=F6.0,11X,5HWCH4=F6.2/2X,5HTCH4=F6.0,9X,5HW
     2ICO=F6.2/2X,4HTCO=F6.0,10X,6HWICO2=F6.2/2X,5HTCO2=F6.0,9X,7HWICH2O
     3=F6.2/2X,4HTFL=F6.0)
610   FORMAT (8H-RESULTS/26H0OXYGEN ENRICHMENT LEVEL =F5.2/58H0ENTHALPY
     1DIFFERENCE BETWEEN REACTANTS AND PRODUCTS, CAL =F4.0/41H0ITERATION
     2S REQUIRED TO ATTAIN SOLUTION =I3)
611   FORMAT (1H-,1X,4HHBL=F8.0,8X,6HHWH2O=F8.0,6X,5HODIF=F6.0/2X,3HHC=F
     18.0,9X,5HHWN2=F8.0,7X,3HTC=F6.0/2X,4HHTG=F8.0,8X,5HHWO2=F8.0,7X,7H
     2XWICH4=F5.2/2X,6HHTGCO=F8.0,6X,5HHWQ2=F9.0,6X,6HXWICO=F5.2/2X,6HHT
     3GH2=F8.0,6X,4HNBC=F6.1,10X,7HXWICO2=F5.2/2X,6HHTGN2=F8.0,6X,5HNDIF
     4=F6.1,9X,7HXWIH2O=F5.2/2X,6HHHWCH4=F8.0,6X,6HNTGCO=F6.1,8X,6HXWIN2=
     5F5.2/2X,5HHHWCO=F8.0,7X,6HNTGH2=F6.1,8X,6HXWIO2=F5.2/2X,6HHHWCO2=F8.
     60,6X,6HNTGN2=F6.1,8X,6HXWIQ2=F5.0)
      END
```

Input (Data)—Problem 2B

```
300   00    00    800   1000   1000      77      77      77   2800   3380
```

Output (Results)—Problem 2B

```
OXYGEN ENRICHMENT CALCULATION

DATA

   TA= 1000.              TQ2= 1000.
   TC= 2800.              WCH4=   8.00
   TCH4=    77.           WICO=   0.0
   TCO=     77.           WICO2=  0.0
   TCO2=    77.           WICH2O= 3.00
   TFL= 3380.


RESULTS

OXYGEN ENRICHMENT LEVEL = 6.27

ENTHALPY DIFFERENCE BETWEEN REACTANTS AND PRODUCTS, CAL = -0.

ITERATIONS REQUIRED TO ATTAIN SOLUTION = 26


   HBL=   5346.          HWH2O= -53356.        ODIF=   -0.
   HC=    7345.          HWN2=    3734.        TC= 1811.
   HTG=   5346.          HWO2=    3873.        XWICH4= 0.07
   HTGCO= -12004.        HWQ2=    3873.        XWICO= 0.0
   HTGH2=  13741.        NBC=     0.5          XWICO2= 0.0
   HTGN2=  14507.        NDIF=   -0.0          XWIH2O= 0.01
   HWCH4= -17887.        NTGCO=   0.6          XWIN2= 0.69
   HWCO=  -26619.        NTGH2=   0.1          XWIO2= 0.18
   HWCO2= -94029.        NTGN2=   0.7          XWIQ2=  0.
```

Discussion of Results—Problem 2

Calculation of the adiabatic flame temperature for air preheated to temperatures ranging up to 1200°F and containing 5.5 grains of moisture per cubic foot are

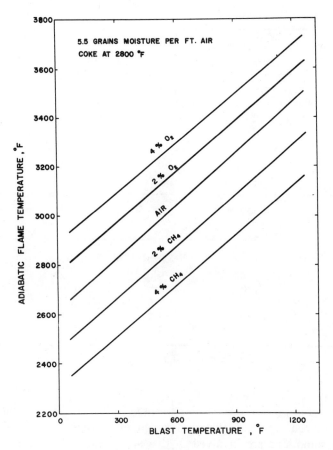

Fig. B-3. Calculated adiabatic flame temperature in tuyere zone for various levels of blast enrichment.

presented in Fig. B-3. These calculations have been extended to include various levels of enrichment of oxygen and natural gas. The injection of cold natural gas at the tuyeres decreases the temperature in the melting zone because of the decomposition of methane to carbon and hydrogen. The ensuing combustion in the presence of hot coke does not proceed beyond carbon monoxide and hydrogen, and the overall result is that less energy is developed in the melting zone with injection. Consequently, natural gas injection should be accompanied by oxygen enrichment or an increased blast temperature to maintain a given flame temperature in the melting zone. The enrichment with oxygen shows an increase in flame temperature since the oxygen reacts with hot coke to form carbon monoxide, releasing energy in the combustion zone.

The most useful part of this approach to developing guidelines for operating practice is that adjustments in blast variables can be made to maintain a constant flame temperature at the tuyeres and provide smooth furnace operation for a given charge and blowing practice. Several examples of simultaneous enrichment with

TABLE B-1[a]

Blast temp (°F)	Moisture (gr/ft³)	%CH₄	%O₂	Flame temp (°F) (calculated)

Blast temp (°F)	Moisture (gr/ft³)	$\%CH_4$	$\%O_2$	Flame temp (°F) (calculated)
Calculated Adiabatic Flame Temperatures				
1000	3	0	0	3380
1000	5	0	0	3332
1000	3	3	0	3222
1800	3	0	0	3957
1800	5	0	0	3910
1800	3	3	0	3766

Flame temp (°F)	Moisture (gr/ft³)	$\%CH_4$	$\%O_2$ (calculated)	Blast temp (°F)
Calculated Oxygen Enrichment				
3380	3.0	8.0	6.27	1000
3766	3.0	3.0	0.01	1800
3957	3.0	3.0	3.09	1800
3957	5.0	3.0	3.81	1800
3957	3.0	8.0	8.24	1800

[a] Table based on 2800°F coke, 77°F CH_4, with O_2 enrichment at blast temperature.

natural gas and oxygen injection are presented in Table B-1. The blast variables have been adjusted to provide the same flame temperature.

Several assumptions are involved in this approach, and place a limitation on the accuracy and potential usefulness of this technique. It should be recognized that the adiabatic flame temperature is an idealized concept, and that actual flame temperatures would be lower than those calculated on the basis of adiabatic conditions. This limitation should not prove to be particularly serious, however, since the actual combustion temperatures should be proportional or directly related to the theoretical flame temperatures. The application of this approach should be limited to blast volume rates within the same range, since heat transfer and temperature distribution undoubtedly are related to gas flowing through this zone of the furnace. Another assumption which goes to the very heart of the analysis is whether or not the reactions taking place in this particular zone are accurately represented by the calculated equilibria. There has been some indication that carbon dioxide may be formed at low blast temperatures, such as found in the cupola, but that this carbon dioxide is reduced to carbon monoxide within a very short distance from the tuyere. Consequently, the kinetics of combustion are a major factor in the validity of this equilibrium assumption, and conditions may vary markedly with the level of enrichment in the blast. Indications from operating practice are that the adiabatic flame temperature concept is quite useful in predicting a required blast condition to provide smooth furnace operation, particularly when combined with operating experience on various sizes of furnaces, tuyere designs, blast conditions, and other related operating factors.[1-3]

[1] J. A. Cordier, Injection of different materials in the blast, *Blast Furnace, Coke Oven, and Raw Materials Proceedings, AIME* **19,** 238 (1960).

[2] H. N. Lander, H. W. Meyer, and F. D. Delve, Prediction of blast furnace performance from operating and thermal data, *Blast Furnace, Coke Oven, and Raw Materials Proceedings, AIME* **19,** 219 (1960).

[3] R. D. Pehlke, Computer prediction of cupola performance utilizing oxygen and natural gas in the blast, *AFS Trans.* **72,** 806 (1964).

Example Problem No. 3: Cooling of Pig Iron in a Transfer Ladle

Problem Statement

The pig iron produced in a blast furnace is transferred to steelmaking facilities in a ladle mounted on a railroad car. Several types of ladles are used, the most prominent one being the Pugh-type which is a cigar-shaped tilting ladle. The pig iron is tapped from the blast furnace into the ladle and then transferred to the steelmaking facilities which may be some distance from the blast furnace. It is of interest to know the time which one can allow the metal to remain in the ladle before solidification begins.

Write and test a FORTRAN program which will calculate the temperature drop of liquid pig iron held in a closed transport ladle of the Pugh-type (see Fig. B-4). Using the test data, extend the computations to either the initial solidification temperature or to an elapsed time of 24 hr.

Suggested Approach.   Assume the following:

1. The temperature of the liquid metal is uniform and equal to the internal wall temperature of the refractory lining of the ladle car.
2. The heat losses may be computed by determining the rate of loss over short intervals of time and approximating the heat loss as the rate-time product.
3. The thermal resistance of the thin steel shell which surrounds the ladle may be neglected.

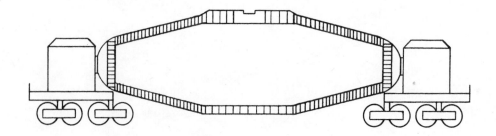

Fig. B-4.   150 Ton transfer ladle car.

Method of Solution—Problem 3

Heat is transferred to the surroundings by conduction through the refractory and then by radiation and convection to the ambient surroundings at temperature T_a. The conduction heat transfer during a unit time is

$$q_c = \frac{k \cdot A_{lm}}{l} (T_0 - T_s), \tag{a}$$

where the cylindrical wall is being considered; the log mean area for the heat conduction calculation is given by the relationship,

$$A_{lm} = \frac{A_0 - A_i}{\ln(A_0/A_i)}.$$ (b)

The radiation loss during a unit time period of one hour is

$$q_{rad} = 0.173 \times 10^{-8} \, (\epsilon) \, (T_s{}^4 - T_a{}^4) \, A_0.$$ (c)

The convection loss is

$$q_{conv} = h \, A_0 (T_s - T_a).$$ (d)

The convection coefficient h is a function of the temperature difference to the one-quarter power, $(T_s - T_a)^{0.25}$. Therefore, one may write

$$q_{conv} = h_c \, A_0 (T_s - T_a)^{1.25}.$$ (e)

The combined convection and radiation loss is

$$q_r = q_{rad} + q_{conv}.$$ (f)

The heat loss from the metal results in a decrease in its temperature. This decrease in energy is

$$\Delta H = 2000 \, w_m \, C_p F (T_0 - T_n).$$ (g)

By energy balance

$$q_c = q_r.$$ (h)

If either of these is multiplied by a short time interval, Δt, a quantity of heat is computed which determines the decrease in metal temperature during the interval. Thus,

$$\Delta t \cdot q_c = 2000 \, w \cdot C_p \cdot F (T_0 - T_n) = CPM (T_0 - T_n),$$ (i)

where CPM is the combination of factors given.

At the start of any time interval the metal temperature T_0 and the ambient temperature T_a are known. The surface temperature T_s is not known, so an initial estimate is made. This allows q_c and q_r to be calculated. If T_s has been estimated correctly, q_c is equal to q_r, and the temperature of the metal at the end of the time interval can be calculated as

$$T_n = T_0 - \frac{\Delta t \cdot q_c}{CPM}.$$ (j)

If T_s has not been determined within a small limit of error, it is necessary to correct the assumed temperature. The magnitude of the change to be made in T_s depends upon how different q_c and q_r turn out to be for a given T_s. In this solution it was

Symbol Table—Problem 3

Algebraic Symbol	Definition	FORTRAN Symbol
T_m	Temperature of charged metal	TM
T_a	Temperature of surroundings	TA
T_{sol}	Temperature of initial solidification, °F	$TLOW$
T_0	Temperature of metal at time t, °R	TO
T_s	Temperature of outside surface, °R	TS
T_{s1}	Most recent value of T_s during iteration, °R	TI
T_n	Metal temperature after latest iteration, °R	TN
T_x	T_n in °F	TX
T_y	T_{s1} in °F	TY
ΔT	Tolerable error in surface temperature, °R	EPS
r	Radiation from surroundings to the ladle	R
q_c	Conductive heat transfer, BTU/hr	QC
q_r	Radiative and convective heat transfer, BTU/hr	QR
h	Convective heat transfer coefficient	—
h_c	Coefficient for h	HC
ϵ	Emissivity of steel shell	E
C_p	Specific heat of metal, BTU/lb°F	CP
CPM	Gross heat capacity for ladle and metal, BTU/lb °F	CPM
k	Thermal conductivity of refractory	K
w_m	Weight of metal charged	W
F	Factor to account for specific heat of ladle	F
A_0	Area of external surface	AO
A_i	Area of internal surface	AI
A_{lm}	Log mean area for heat transfer through the refractory	AM
l	Thickness of refractory	L
t	Elapsed time from charging of ladle, hours	P
Δt	Time interval, hours	I
j	Number of time intervals	J
i_{max}	Maximum time of holding, hours	$IMAX$
n_{max}	Maximum number of iterations to determine surface temperature	$NMAX$
n	Number of iterations required for temperature convergence	N

decided that if

$$| T_{s_1} - T_s | \geq \Delta T, \tag{k}$$

where ΔT is 0.1°R, the surface temperature would be changed to

$$T_s = T_s + (100) \cdot \left(\frac{q_c - q_r}{q_r} \right). \tag{l}$$

It is clear that a higher order approximation might be made in the heat transfer equations by putting in the average metal or average surface temperatures during the time interval. Thus, for conduction, one might write

$$q_c = \frac{k \cdot A_{lm}}{l} \left(\frac{T_0 + T_n}{2} - \frac{T_s + T_{s_1}}{2} \right). \tag{m}$$

The anticipated change in temperature, however, does not merit this refinement, so it is assumed for the heat transfer calculations that T_0 and T_s remain constant during each time step.

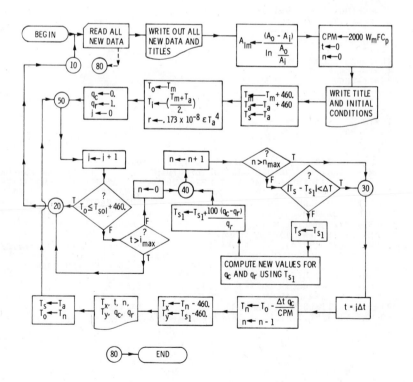

Flow Chart—Problem 3.   Flow diagram for computation of cooling of pig iron.

FORTRAN Listing—Problem 3

```
      REAL K,L,I,J,IMAX
10    READ (5,500,END=80) TM,TA,AI,AO,K,L,HC,E,W,CP,F,EPS,I,IMAX,NMAX,TL
      1OW
      WRITE (6,600) TM,TA,AI,AO,K,L,HC,E,W,CP
      WRITE (6,601) F,EPS,I,IMAX,NMAX,TLOW
      AM=(AO-AI)/ALOG(AO/AI)
      CPM=W*2000.*CP*F
      P=0.
      N=0
      WRITE (6,610) TM,P,N
      TM=TM+460.
      TA=TA+460.
      R=.173E-8*E*TA**4
      TS=TA
      TO=TM
      TI=(TM+TA)/2.
      QC=0.
      QR=1.
      J=0.
50    J=J+1.
      IF (TO.LE.(TLOW+460.).OR.P.GE.IMAX) GO TO 20
      N=0
40    N=N+1
      IF (N.GT.NMAX.OR.ABS(TS-TI).LT.EPS) GO TO 30
      TS=TI
      QR=AO*(HC*(TI-TA)**1.25+.173E-8*E*TI**4-R)
      QC=K*AM*(TO-TI)/L
      TI=TI+100.*(QC-QR)/QR
      GO TO 40
30    P=I*J
      TN=TO-QC*I/CPM
      TX=TN-460.
      TY=TI-460.
      N=N-1
      WRITE (6,620) TX,P,N,TY,QC,QR
      TS=TA
      TO=TN
      GO TO 50
20    GO TO 10
80    CALL SYSTEM
500   FORMAT (4F5.0,4F5.2,F4.0,4F4.2,F4.0,I3,F5.0)
600   FORMAT (48H1TEMPERATURE DROP OF LIQUID METAL IN A LADLE CAR//5H-DA
      1TA//1H-,1X,27HINITIAL METAL TEMPERATURE =F6.0/2X,17HAIR TEMPERATUR
      2E =F4.0/2X,20HINSIDE AREA, SQ FT =F5.0/2X,21HOUTSIDE AREA, SQ FT =
      3F5.0/2X,22HTHERMAL CONDUCTIVITY =F4.1/2X,26HREFRACTORY THICKNESS,
      4FT =F4.1/2X,24HCONVECTION COEFFICIENT =F3.1/2X,18HSHELL EMISSIVITY
      5 =F3.1/2X,20HMETAL WEIGHT, TONS =F5.0/2X,36HSPECIFIC HEAT OF THE M
      6ETAL, BTU/LB =F3.1)
601   FORMAT (2X,13HHEAT FACTOR =F4.1/2X,23HTEMPERATURE DEVIATION =F4.2/
      12X,20HTIME INCREMENT, HR =F3.1/2X,30HMAXIMUM NUMBER OF INCREMENTS
      2=F4.0/2X,26HMAXIMUM NUMBER OF TRIALS =I4/2X,16HFREEZING POINT =F6.
      30/40H-METAL TEMPERATURE AS A FUNCTION OF TIME)
610   FORMAT (6H-METAL,5X,4HTIME,4X,10HITERATIONS,4X,7HSURFACE,4X,10HCON
      1DUCTION,4X,8HRAD-CONV/1X,6HTEMP F,5X,2HHR,19X,6HTEMP F,4X,11HLOSS
      3BTU/HR,3X,11HLOSS BTU/HR/1H0,F6.1,4X,F4.1,8X,I2)
620   FORMAT (1H9,F6.1,4X,F4.1,8X,I2,8X,F5.1,7X,F8.0,5X,F8.0)
      END
```

Input (Data)—Problem 3

```
2700   80   530   800   240   150   30   80 150   20 140   10   50   48100 2100
```

Output (Results)—Problem 3

TEMPERATURE DROP OF LIQUID METAL IN A LADLE CAR

DATA

    INITIAL METAL TEMPERATURE = 2700.
    AIR TEMPERATURE = 80.
    INSIDE AREA, SQ FT = 530.
    OUTSIDE AREA, SQ FT = 300.
    THERMAL CONDUCTIVITY = 2.4
    REFRACTORY THICKNESS, FT = 1.5
    CONVECTION COEFFICIENT =0.3
    SHELL EMISSIVITY =0.8
    METAL WEIGHT, TONS = 150.
    SPECIFIC HEAT OF THE METAL, BTU/LB =0.2
    HEAT FACTOR = 1.4
    TEMPERATURE DEVIATION =0.10
    TIME INCREMENT, HR =0.5
    MAXIMUM NUMBER OF INCREMENTS = 48.
    MAXIMUM NUMBER OF TRIALS = 100
    FREEZING POINT = 2100.

METAL TEMPERATURE AS A FUNCTION OF TIME

METAL TEMP F	TIME HR	ITERATIONS	SURFACE TEMP F	CONDUCTION LOSS BTU/HR	RAD-CONV LOSS BTU/HR
2700.0	0.0	0			
2687.1	0.5	24	637.8	2163620.	2165121.
2674.3	1.0	5	636.2	2151737.	2153777.
2661.6	1.5	5	634.6	2139988.	2142052.
2648.9	2.0	5	633.0	2128315.	2130358.
2636.3	2.5	5	631.4	2116711.	2118731.
2623.8	3.0	5	629.8	2105176.	2107169.
2611.3	3.5	5	628.2	2093709.	2095679.
2598.9	4.0	5	626.6	2082310.	2084257.
2586.6	4.5	5	625.0	2070978.	2072902.
2574.3	5.0	5	623.4	2059713.	2061616.
2562.1	5.5	5	621.8	2048515.	2050392.
2550.0	6.0	5	620.2	2037383.	2039239.
2537.9	6.5	5	618.7	2026316.	2028153.
2526.0	7.0	5	617.1	2015316.	2017127.
2514.0	7.5	5	615.5	2004379.	2006170.
2502.2	8.0	5	613.9	1993508.	1995274.
2490.4	8.5	5	612.4	1982700.	1984447.
2478.6	9.0	5	610.8	1971956.	1973679.
2466.9	9.5	5	609.3	1961276.	1962982.
2455.3	10.0	5	607.7	1950658.	1952341.
2443.8	10.5	5	606.2	1940103.	1941763.
2432.3	11.0	5	604.6	1929610.	1931252.
2420.9	11.5	5	603.1	1919179.	1920799.
2409.5	12.0	5	601.5	1908809.	1910410.
2398.2	12.5	5	600.0	1898500.	1900081.
2387.0	13.0	5	598.5	1888252.	1889814.
2375.8	13.5	5	596.9	1878064.	1879607.
2364.7	14.0	5	595.4	1867936.	1869460.
2353.6	14.5	5	593.9	1857868.	1859373.
2342.6	15.0	5	592.4	1847858.	1849346.
2331.7	15.5	5	590.8	1837908.	1839375.
2320.8	16.0	5	589.3	1828016.	1829464.
2310.0	16.5	5	587.8	1818182.	1819612.
2299.2	17.0	5	586.3	1808406.	1809816.
2288.5	17.5	5	584.8	1798686.	1800080.
2277.8	18.0	5	583.3	1789024.	1790399.
2267.3	18.5	5	581.8	1779418.	1780776.

2256.7	19.0	5	580.3	1769869.	1771208.
2246.2	19.5	5	578.9	1760376.	1761700.
2235.8	20.0	5	577.4	1750938.	1752246.
2225.5	20.5	5	575.9	1741556.	1742849.
2215.1	21.0	5	574.4	1732228.	1733501.
2204.9	21.5	5	572.9	1722955.	1724212.
2194.7	22.0	5	571.5	1713736.	1714976.
2184.5	22.5	5	570.0	1704571.	1705795.
2174.4	23.0	5	568.5	1695460.	1696668.
2164.4	23.5	5	567.1	1686402.	1687594.
2154.4	24.0	5	565.6	1677396.	1678573.
2144.5	24.5	5	564.2	1668444.	1669606.
2134.6	25.0	5	562.7	1659543.	1660691.
2124.8	25.5	5	561.3	1650695.	1651826.
2115.0	26.0	5	559.8	1641898.	1643015.
2105.3	26.5	5	558.4	1633152.	1634254.
2095.6	27.0	5	557.0	1624458.	1625544.

Discussion of Results—Problem 3

The computed result indicates that liquid pig iron can be held in a transfer ladle for periods of about 24 hours prior to the start of solidification. The exact time is dependent upon the initial hot metal temperature and the condition of the ladle lining.

Example Problem No. 4: Predicting the Scrap Requirement for the Oxygen Steelmaking Process

Problem Statement

The computer analysis of oxygen steel converting has become operational at most steel companies. The computer programs consider many process variables which include hot metal analysis, temperature, oxygen flow rate, process geometries, and several others which influence the control of the converting process. Most commercial computer programs are highly empirical, and although of limited accuracy, have produced marked improvement in control of the operation.

In an effort to present the magnitude of some of the thermochemical interactions which play a role in design and operation of oxygen converters, a problem was selected for computer analysis which requires a mass and thermal balance for the oxygen steelmaking process. The relationship selected for investigation is the influence of silicon content of the hot metal on the required scrap addition. Computation of the mass balance involves an iterative calculation. The thermal balance requires lengthy calculations of the heat effect of each reaction, a task suitable for the digital computer.

Write and test a FORTRAN program which will permit a calculation of the scrap requirement for an oxygen steelmaking converter. Use the program to determine the influence of silicon content of the hot metal on scrap consumption for production of a 250 ton heat.

The following thermodynamic data may be used.

1. Heats of reaction at 77°F ΔH (Btu/lb-mole)

$P_2(g) + 3\,CaO + \tfrac{5}{2}\,O_2(g) \rightarrow Ca_3P_2O_8$	$-1,016,000$
$2\,Fe + \tfrac{3}{2}\,O_2(g) \rightarrow Fe_2O_3$	$-354,000$
$Fe + \tfrac{1}{2}\,O_2 \rightarrow FeO$	$-114,800$
$Si + O_2(g) \rightarrow SiO_2$	$-369,700$
$Mn + \tfrac{1}{2}\,O_2(g) \rightarrow MnO$	$-165,600$
$2\,CaO + SiO_2 \rightarrow 2\,CaO \cdot SiO_2$	$-53,640$
$FeO + SiO_2 \rightarrow FeSiO_3$	$-8,100$
$MnO + SiO_2 \rightarrow MnSiO_3$	$-10,600$
$C(gr) + \tfrac{1}{2}\,O_2(g) \rightarrow CO(g)$	$-47,600$
$C(gr) + O_2(g) \rightarrow CO_2(g)$	$-169,000$

2. Heats of solution (Btu/lb-mole)   T in °F

$Si_{77°F} \rightarrow \underline{Si}_T$	$\Delta H = 6.7\,T - 34,150$
$Mn_{77°F} \rightarrow \underline{Mn}_T$	$\Delta H = 11.0\,T + 2,900$
$\tfrac{1}{2}\,P_2(g)_{77°F} \rightarrow \underline{P}_T$	$\Delta H = 4.45\,T - 92,500$
$C(gr)_{77°F} \rightarrow \underline{C}_T$	$\Delta H = 5.23\,T + 10,020$

3. Sensible heats (Btu/lb mole)—Reference temperature 77°F (T in °F)

Iron–Base metal (liquid)	$\Delta H = 0.184\,T + 63$ Btu/lb.
Slag (liquid)	$\Delta H = 875 + (T - 2820) \times 0.3$ Btu/lb.
CO	$\Delta H = 23,200 + (T - 3000) \times 8.5$ Btu/lb.-mole
CO_2	$\Delta H = 37,400 + (T - 3000) \times 14.4$ Btu/lb.-mole
N_2	$\Delta H = 23,000 + (T - 3000) \times 8.6$ Btu/lb.-mole

Method of Solution—Problem 4

The diagramming of a sequence to solve this problem involves a relatively complete presentation of the steelmaking process. The oxygen converting process for producing steel is presented in Chapter 4. The detailed examination of the process in terms of its mass and energy requirements is presented here, and one of the principal process variables is explored in detail.

The solution requires that a mass balance be completed which satisfies the production rate to which the heat losses are tied. Upon completion of the mass balance, excess thermal energy is calculated to determine the amount of scrap which is required to achieve the desired final process temperature. This scrap plus the refined steel produced by oxidizing the impurities from the blast furnace hot metal charged to the converter must satisfy the specified production tonnage.

The mass balance involves an iterative computation sequence because the burnt lime added depends upon both the silicon which is oxidized from the hot metal and the basicity ratio of the slag. Mill scale, added to supply additional oxygen to the process, is defined in terms of the lime added. The thread which runs through the several input and output streams of the converting process is the requirement that the charged iron, a constituent of all of the condensed phase streams (i.e., all except the oxygen lanced into the converter and the combustion gases exiting from the process) must be equal to the iron removed. The mass balance then is defined in terms of an iron balance.

Symbol Table—Problem 4

FORTRAN Symbol	Definition and/or Purpose
A	Percent available base in the charged lime
BLMTID	Holds the 14-character identification for the blown metal
BLOTIM	Time of oxygen blow in minutes
BTIME	$BTIME(1)$ has the number of whole minutes for the oxygen blow; $BTIME(2)$ has the number of seconds
BR	Basicity ratio expressed as moles CaO/mole SiO_2
CORFAC	Correction factor applied to hot metal and scrap weight after each energy balance iteration
CSCPF	Weight—% carbon in the scrap whose weight is fixed
DANAL	Holds the long version of the date and time of analysis
DELT	Maximum allowable error for the ore balance and the slag FeO balance
EPS	Maximum allowable error for the blown metal production balance
EXIT	Called when there is no more data; A subprogram which terminates execution of the calling program
FDLOSS	Flue dust loss, as a percentage of the charged hot metal
FESCPF	Weight—% iron in the scrap whose weight is fixed

Symbol Table—Problem 4—(Continued)

FORTRAN Symbol	Definition and/or Purpose
FSCRAP	Basic material matrix for the two possible fixed scraps; 2 rows by 7 columns, where the first row is for the first fixed scrap and the second row is for the second fixed scrap (if used). The column definitions are the same as those for *HM*
FSC17	Used to store the weight in pounds of the first fixed scrap
FSC27	Used to store the weight in pounds of the second fixed scrap
FV23RS	Ratio of iron valence 2 (FeO) to iron valence 3 (Fe_2O_3) in the slag
HCG	Energy content of the combustion gases, BTU
HHM	Energy content of the charged hot metal, BTU
HL	Total of all energy losses per heat, BTU
HM	Basic material vector for the charged hot metal. $HM(1)$ is wt-% Si, $HM(2)$ is wt-% Mn, $HM(3)$ is wt-% C, $HM(4)$ is wt-% P, $HM(5)$ is wt-% S, $HM(6)$ is temperature in degrees Fahrenheit, and $HM(7)$ is weight in pounds.
HO2	Energy content of the total oxygen blown, BTU
HS	Energy content of the final slag, BTU
HSCRAP	Energy available for melting scrap, BTU
HST	Energy content of the final blown metal minus the scrap, BTU
HTMLID	Holds the 14-character identification for the hot metal charged
I, J, K	Iteration variables, used as needed
IHL	Energy loss in BTU's per heat (same as *HL*) used for output
MNSCPF	Weight-% manganese in the scrap whose weight is fixed
NBMC	Number of moles of carbon in the blown metal
NBMMN	Number of moles of manganese in the blown metal
NBMP	Number of moles of phosphorus in the blown metal
NBMSI	Number of moles of silicon in the blown metal
NCCG	Number of moles of carbon in the combustion gases
NCGC	Number of moles of carbon in the total charge
NCGMN	Number of moles of manganese in the total charge
NCGP	Number of moles of phosphorus in the total charge
NCGSI	Number of moles of silicon in the total charge
NFE2SI	Number of moles of $2FeO \cdot SiO_2$ in the slag, zero in basic slag
NMNSI	Number of moles of $MnO \cdot SiO_2$ in the slag, zero in basic slag
NOXYO	Number of moles of O required for the charge
NSCAO	Number of moles of lime in the slag
NSCRPS	Number of scraps to be used
NSFEO	Number of moles of FeO in the slag
NSFEOX	Number of moles of iron oxides (FeO and Fe_2O_3) in the slag
NSF2O3	Number of moles of Fe_2O_3 in the slag
NSMNO	Number of moles of MnO in the slag
NSP2O5	Number of moles of P_2O_5 in the slag

Symbol Table—Problem 4—(Continued)

FORTRAN Symbol	Definition and/or Purpose
NSSIOT	Number of moles of SiO_2 (total) in the slag
NSSIO2	Number of moles of SiO_2 in the slag which came from oxidation of charged materials
N2CS	Number of moles of $2CaO \cdot SiO_2$ in the slag
ONE	The integer 1, used in output
ORE	Pounds of ore charged
OREFRC	Weight ratio of ore charge to lime charge
OXYFLO	Oxygen flow rate in standard cubic feet per minute
PCSIOR	Weight percent SiO_2 in the iron ore
PORFEO	Weight percent FeO in the iron ore
POXYO2	Purity of oxygen blown, in volume percent
PSC	Weight percent scrap in the total charge
PSCPF	Weight—% phosphorus in the scrap whose weight is fixed
QNSFEO	Temporary location for comparison with *NSFEO* to check for slag composition convergence
RCO2	Ratio of CO_2/CO in the combustion gases
RSTM	Used for storage of the results of the call to the *TIME* subprogram
SCFOX	Standard cubic feet of oxygen (as O_2) required for the blow
SCRAP	Basic material vector for the variable (first) scrap charge; the column definitions are the same as those for *HM*
SCRPID	Array for storage of the identifications of the three scraps; 3 rows by 2 columns; each row corresponds one of three possible scraps and the two columns hold the *ID*
SISCPF	Weight—% silicon in the scrap whose weight is fixed
SSCPF	Weight—% sulfur in the scrap whose weight is fixed
STEEL	Basic material vector for the blown metal to be produced; the column definitions are the same as those for *HM* (see program symbol *HM*)
TCG	Temperature of the combustion gases, degrees Fahrenheit
THMCG	Tons of hot metal charged
TIME	A subprogram called to obtain the current date and time of analysis
TLIME	Tons of lime charged
TORE	Tons of ore charged
TO2	Temperature at the nozzle of the incoming oxygen jet, degrees Fahrenheit
TS	Temperature of the slag, degrees Fahrenheit
TSC	Temperature of the total scrap charge
TSCP	Tons of scrap charged
TSCPF	Temperature of the scrap whose weight is fixed, degrees Fahrenheit
TVSCP	Tons of the variable (first) scrap charged
WBM	Weight of blown metal, pounds
WFENET	Weight of iron in the charge which goes into the final blown metal

Symbol Table—Problem 4—(Continued)

FORTRAN Symbol	Definition and/or Purpose
WFESCF	Weight percent iron in the scrap whose weight is fixed
WFESCP	Pounds of iron in the total scrap charge
WHMFE	Pounds of iron in the hot metal, corrected for flue dust loss
WTHM	Pounds of hot metal charged, identical to $HM(7)$ (see program symbol HM)
WTLIME	Pounds of lime charged
WTORE	Pounds of ore charged
WTSCP	Pounds of total scrap charged
WTSCPF	Pounds of fixed scrap charged
WTSCP1	Pounds of the variable (first) scrap, identical to $SCRAP(7)$ (see program symbol $SCRAP$)
WTSLAG	Pounds of slag
XSFEO	Mole fraction of iron oxides in the slag, expressed as FeO

The method of calculation is as follows.

1. Read the input data and calculate the combined weights and components in any fixed scrap charged.

2. Calculate the moles of silicon, manganese, carbon, and phosphorous in the hot metal which is assumed to be charged, after correcting for flue-dust losses.

3. Initialize the weight of ore added, and the moles of iron in the slag stream at zero.

4. Determine the value for the net weight of iron from the hot metal and scrap by adding the weight of iron in the ore and subtracting the weight of iron in the slag.

5. Calculate the weight of blown metal produced from the hot metal charged on the basis of an iron balance.

6. Calculate the moles of silicon, manganese, carbon, and phosphorous which remain in the blown metal, subtracting them from the results of steps 1 and 2 to determine the moles which are oxidized.

7. Calculate the weight of burnt lime added to provide the correct basicity ratio in the slag, based on silica as the only acidic slag component (P_2O_5 could be included).

8. Calculate the weight of ore added on the basis of the result of step 7, and calculate the weight of slag produced.

9. Compare the weight of ore calculated with that used in Step 4. If they agree within a limit of error continue on; otherwise, return to step 4, substituting the result of step 8. Repeat for FeO in slag.

10. Calculate the oxygen gas required and determine blowing time.

11. Calculate the composition of the slag in terms of the compounds which form, assuming a preferred order to the reactions.

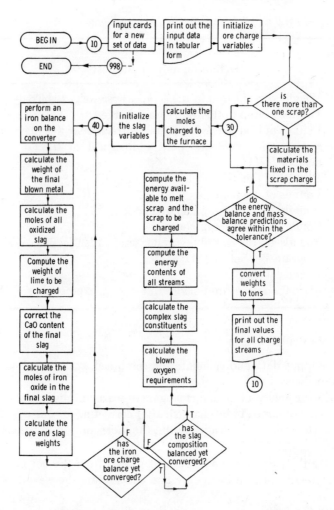

Flow Chart—Problem 4.   Flow diagram for computation of BOF charge.

FORTRAN Listing—Problem 4

```
      IMPLICIT REAL(N)
      REAL*8 HTMLID(2),SCRPID(3,2),BLMTID(2)
      REAL MNSCPF
      INTEGER NSCRPS,ONE/1/,BTIME(2)
      LOGICAL*1 DANAL(20)
      DIMENSION HM(7),SCRAP(7),FSCRAP(2,7),STEEL(7),RSTM(5)
      EQUIVALENCE(DANAL(1),RSTM(1))
C
C   READ THE DATA AND PRINT IT OUT IN TABULAR FORM
C
  10  READ(5,500,END=998) HTMLID,HM,BLMTID,STEEL,NSCRPS,SCRPID(1,1),
     1SCRPID(1,2),SCRAP,((SCRPID(K+1,I),I=1,2),(FSCRAP(K,J),J=1,7),K=1,2
     2),POXYO2,A,PORFEO,BR,RCO2,TO2,TS,TCG,HL,FDLOSS,DELT,EPS,OREFRC,
     3XSFEO,FV23RS,PCSIOR,OXYFLO
      CALL TIME(6,0,RSTM)
      IHL=HL
      PRINT 501,RSTM(1),RSTM(2),DANAL(13),DANAL(14),RSTM(3),RSTM(5),
     1HTMLID,HM,SCRPID(1,1),SCRPID(1,2),ONE,SCRAP
      IF(NSCRPS-1)10,14,11
```

```
11      DO 13 I=2,NSCRPS
13      PRINT 502,SCRPID(I,1),SCRPID(I,2),I,(FSCRAP(I-1,J),J=1,7)
14      PRINT 503,BLMTID,STEEL,POXYO2,TO2,OXYFLO,FDLOSS,TCG,RCO2,IHL,BR,
       1TS,XSFEO,FV23RS,PORFEO,PCSIOR,OREFRC,A,DELT,EPS
C
C    INITIALIZE THE FIXED SCRAP AND ITERATION VARIABLES
C
        WTORE=0.
15      WTSCPF=0.
        WFESCF=0.
        WTSCP=SCRAP(7)+FSCRAP(1,7)+FSCRAP(2,7)
C
C    IF THERE IS MORE THAN ONE SCRAP, CALCULATE THE MATERIALS FIXED IN THE CHARGE
C
        IF(NSCRPS-1)10,24,16
16      FSC17=FSCRAP(1,7)
        FSC27=FSCRAP(2,7)
        WTSCPF=FSC17+FSC27
        SISCPF=(FSCRAP(1,1)*FSC17+FSCRAP(2,1)*FSC27)/WTSCPF
        MNSCPF=(FSCRAP(1,2)*FSC17+FSCRAP(2,2)*FSC27)/WTSCPF
        CSCPF=(FSCRAP(1,3)*FSC17+FSCRAP(2,3)*FSC27)/WTSCPF
        PSCPF=(FSCRAP(1,4)*FSC17+FSCRAP(2,4)*FSC27)/WTSCPF
        SSCPF=(FSCRAP(1,5)*FSC17+FSCRAP(2,5)*FSC27)/WTSCPF
        IF(NSCRPS.EQ.3)GOTO21
        TSCPF=FSCRAP(1,6)
        FESCPF=100.-(FSCRAP(1,1)+FSCRAP(1,2)+FSCRAP(1,3)+FSCRAP(1,4)+FSCR
       1AP(1,5))
        GO TO 23
21      TSCPF=(FSCRAP(1,6)*FSC17+FSCRAP(2,6)*FSC27)/WTSCPF
        FESCPF=100.-(SISCPF+MNSCPF+CSCPF+PSCPF+SSCPF)
23      WFESCF=(WTSCPF*FESCPF)/100.
        GO TO 25
24      TSCPF=SCRAP(6)
C
C    STORE THE COMMONLY USED WEIGHTS IN SINGLE VARIABLE LOCATIONS FOR REFERENCE
C
25      WTHM=HM(7)
30      WTSCP1=SCRAP(7)
C
C    CALCULATE MOLES OF PRIMARY CONSITITUENTS CHARGED TO THE FURNACE
C
        NCGSI=(WTHM*HM(1)+WTSCP1*SCRAP(1)+SISCPF*WTSCPF)/2806.
        NCGMN=(WTHM*HM(2)+WTSCP1*SCRAP(2)+MNSCPF*WTSCPF)/5494.
        NCGC=(WTHM*HM(3)+WTSCP1*SCRAP(3)+CSCPF*WTSCPF)/1201.
        NCGP=(WTHM*HM(4)+WTSCP1*SCRAP(4)+PSCPF*WTSCPF)/3098.
        WFESCP=WFESCF+WTSCP1*(100.-(SCRAP(1)+SCRAP(2)+SCRAP(3)+SCRAP(4)+
       1SCRAP(5)))/100.
        WHMFE=(100.-FDLOSS)/100.*WTHM*(100.-(HM(1)+HM(2)+HM(3)+HM(4)+HM(5
       1)))/100.
C
C    INITIALIZE THE ORE CHARGE VARIABLES
C
        NSF2O3=0.
        NSFEO=0.
C
C    BEGIN THE INNER ITERATIVE MASS BALANCE
C
40      ORE=WTORE
        WFENET=WHMFE+WFESCP+WTORE*PORFEO*55.85/7185.+WTORE*(100.-PORFEO-P
       1CSIOR)*111.7/15970.-NSFEO*55.85-NSF2O3*111.7
C
C    CALCULATE THE WEIGHT OF BLOWN METAL FROM THE IRON CHARGED AND CALCULATE
C    THE MOLES OF THE PRIMARY OTHER CONSTITUENTS IN THE  FINAL BLOWN METAL
C
        WBM=100.*WFENET/(100.-(STEEL(1)+STEEL(2)+STEEL(3)+STEEL(4)+STEEL(
       15)))
        NBMSI=WBM*STEEL(1)/2806.
        NBMMN=WBM*STEEL(2)/5494.
        NBMC=WBM*STEEL(3)/1201.
        NBMP=WBM*STEEL(4)/3098.
C
C    CALCULATE THE MOLES OF OXIDIZED CONSTITUENTS IN THE SLAG
C
```

```
      NSSIO2=NCGSI-NBMSI
      NSMNO=NCGMN-NBMMN
      NSP2O5=(NCGP-NBMP)/2.
      NCCG=NCGC-NBMC
C
C     COMPUTE THE WEIGHTS OF LIME AND ORE TO BE ADDED
C
      NSCAO=(NSSIO2+PCSIOR*WTORE/6006.)*BR
      WTLIME=NSCAO*5608./A
C
C     COMPUTE THE CORRECTION OF THE SLAG CAO FOR AVAILABILITY OF LIME
C
      NSCAO=NSCAO+(WTLIME*(100.-A)/100.)*(BR/(BR+1.))/56.08
C
C     COMPUTE THE MOLES OF FEO AND FE2O3 IN THE SLAG, THE REQUIRED
C     CHARGE OF ORE, AND THE TOTAL WEIGHT OF THE FINAL SLAG
C
      NSSIOT=NSCAO/BR
      NSFEOX=(NSSIOT+NSMNO+NSP2O5+NSCAO)*XSFEO/(1.-XSFEO)
      QNSFEO=NSFEO
      NSFEO=NSFEOX*FV23RS/(FV23RS+1.)
      NSF2O3=NSFEOX-NSFEO
      WTORE=OREFRC*WTLIME
      WTSLAG=NSSIOT*60.06+NSMNO*70.94+NSP2O5*141.96+NSFEO*71.85+
     1NSF2O3*159.7+NSCAO*56.08
C
C     IF THE ORE CHARGE ITERATIVE CALCULATION HAS CONVERGED, PROCEED; IF
C     NOT, CONTINUE THE ITERATION BY GOING BACK TO THE IRON BALANCE
C
C     IF THE ORE CHARGE ITERATIVE CALCULATION HAS CONVERGED, PROCEED; IF
C     NOT, CONTINUE THE ITERATION BY GOING BACK TO THE IRON BALANCE
C
      IF(ABS(ORE-WTORE).LT.DELT)GO TO 45
      GO TO 40
C
C     IF THE SLAG COMPOSITION ITERATIVE CALCULATION HAS CONVERGED, PROCEED:
C     IF NOT, CONTINUE THE ITERATION BY GOING BACK TO THE IRON BALANCE
C
   45 IF(ABS(QNSFEO-NSFEO).LT.DELT/71.85)GO TO 50
      GO TO 40
C
C     CALCULATION OF BLOWN OXYGEN REQUIREMENT
C
   50 NOXYO=NSSIO2*2.+NSMNO+NSP2O5*5.+NSFEO+NCCG*(1.+2.*RCO2)/(1.+RCO2)
     1-WTORE*PORFEO/7185.-WTORE*(100.-PORFEO-PCSIOR)*3./15970.+3.*NSF2O3
     2+WHMFE*FDLOSS/(100.-FDLOSS)/55.85
      SCFOX=NOXYO/2.*37900./POXYO2
      BLOTIM=SCFOX/OXYFLO
C
C     CALCULATION OF SLAG CONSTITUENTS
C
      N2CS=NSSIOT
      IF(NSCAO.GT.2.*NSSIOT+3.*NSP2O5)GO TO 80
      N2CS=(NSCAO-3.*NSP2O5)/2.
      NMNSI=NSSIOT-N2CS
      IF(NMNSI.GT.NSMNO) GO TO 70
      NFE2SI=0.
      GO TO 90
   70 NFE2SI=NSSIOT-N2CS-NSMNO
      NMNSI=NSMNO
      GO TO 90
   80 NMNSI=0.
      NFE2SI=0.
C
C     BEGIN ENERGY BALANCE ON ENTIRE OXYGEN CONVERTER SYSTEM
C
   90 HHM=(.184*HM(6)+63.)*WHMFE+(6.7*HM(6)-34150.)*NSSIO2+(11.*HM(6)+29
     100.)*NSMNO+(4.45*HM(6)-92500.)*NSP2O5*2.+(5.23*HM(6)+10020.)*NCCG
      HS=(875.+(TS-2820.)*.3)*WTSLAG-1016000.*NSP2O5-114800.*(NSFEO+WHMF
     1E*FDLOSS/100./55.85-WTORE*PORFEO/7185.)-369700.*NSSIO2-165600.*NSM
     2NO-53640.*N2CS-10600.*NMNSI-8100.*NFE2SI-353160.*(NSF2O3-WTORE*(10
     30.-PORFEO-PCSIOR)/15970.)
      HCG=(23200.+(TCG-3000.)*8.5-47600.)*NCCG/(1.+RCO2)+(37400.+(TCG-30
     100.)*14.4-169000.*NCCG*RCO2/(1.+RCO2)
```

```
          HST=(.184*STEEL(6)+63.)*(WBM-WTSCP)
          HO2=7.*(NOXYO/2.)*(TO2-77.)
         HSCRAP=HHM-HST-HS-HCG-HL+HO2
          TSC=(SCRAP(7)*SCRAP(6)+WTSCPF*TSCPF)/(SCRAP(7)+WTSCPF)
          WTSCP=HSCRAP/(.184*(STEEL(6)-TSC)+63.+14.2)
          SCRAP(7)=WTSCP-WTSCPF
C
C    IF THE RESULTS OF THE ITERATIVE ENERGY BALANCE AGREE WITH THE TARGET
C    BLOWN METAL PRODUCTION, CALCULATE THE FINAL RESULTS; IF NOT, GO BACK
C    TO THE INITIAL STAGES OF THE CHARGE CALCULATION
C
          IF(ABS(WBM-STEEL(7)).LT.EPS)GOTO100
          CORFAC=STEEL(7)/WBM
          WTHM=WTHM*CORFAC
          HM(7)=WTHM
          SCRAP(7)=(SCRAP(7)+WTSCPF)*CORFAC-WTSCPF
          GO TO 30
C
C    CONVERT WEIGHTS TO TONS AND PRINT OUT THE RESULTS FOR THIS DATA SET
C
  100    TVSCP=SCRAP(7)/2000.
         TSCP=WTSCP/2000.
          THMCG=HM(7)/2000.
         TORE=WTORE/2000.
         TLIME=WTLIME/2000.
          BTIME(1)=BLOTIM
        BTIME(2)=INT(60.*(BLOTIM-BTIME(1)))
         PSC=100.*WTSCP/(WTSCP+WTHM)
         PRINT 504,PSC,TSCP,TVSCP,THMCG,TORE,TLIME,BTIME
         GO TO 10
  500 0FORMAT(2(A8,A6,7F8.0/),I2/3(A8,A6,7F8.0/),10F7.0/9F7.0)
  501 0FORMAT('1CHARGE CALCULATION FOR A BASIC OXYGEN STEELMAKING CONTROL
     2 SYSTEM,'/         /'-THIS ANALYSIS IS BEING RUN AT ',2A4,2X,2A1,1X,
     3A3,2H, ,A4/'- BASIC PROCESS         COMPOSITIONS, IN WEIGHT PERCENT
     4 TEMPER-  WEIGHT' / ' MATERIALS MATRIX:',37X,'ATURE,   IN LBS' /
     5 22X,2HSI,5X,2HMN,6X,1HC,6X,1HP,6X, 'S   DEG. F  (*=EST.)'/ ' HOT
     6METAL' / 2X,A8,A6,F10.3,4F7.3,F7.0,F9.0,2H */ 'OSCRAP' / 2X,A8,A6,
     72H (,I1,1H),F6.3,4F7.3,F7.0,F9.0,2H *)
  502 0FORMAT(2X,A8,A6,2H (,I1,1H),F6.3,4F7.3,F7.0,F9.0)
  503 0FORMAT('0BLOWN METAL' / 2X,A8,A6,F10.3,4F7.3,F7.0,F9.0 / '-% OXYGE
     1N IN BLAST   =',F6.2,' %    OXYGEN TEMPERATURE    =',F6.0,' DEG. F
     2.' / ' FLUE DUST LOSS, AS',15X,'BLOWN O2 FLOW RATE    =',F7.0,' SCF
     3M' / ' PERCENT OF HOT METAL=',F6.2,' %    COMBUSTION GAS TEMP  =',
     44F7.0,' DEG. F.' / 'CO2/CO, COMBUST GAS =',F6.3,6X,'ENERGY LOSS EST
     5IMATE =',I8,' BTU/HT' / ' SLAG MOLE FRACTION',15X,'SLAG BASICITY R
     6ATIO  =',F5.1,' 'OF IRON OXIDES,',18X,'TEMPERATURE OF SLAG  =',F6.
     70,' DEG. F.' / 'EXPRESSED AS FEO    =',F5.2,7X,'SLAG FEO/FE2O3 RA
     8TIO =',F5.2,' ' PERCENT FEO IN ORE  =',F6.2,' %   PERCENT SIO2 IN
     9 ORE =',F5.2,' %' / ' ORE-TO-LIME RATIO   =',F6.2,6X,'PERCENT BA
     1SE IN LIME =',F6.2,' %' / ' MAX ERROR, CHARGE WT=',F5.1,' LB  MAX
     2 ERROR, METAL OUT =',F6.0,' POUNDS ' / '0')
  504 0FORMAT('-'/'-RESULTS FOLLOW:'/'0PERCENT SCRAP CHARGED   =',F6.2,5X
     1,'TONS SCRAP CHARGED      =',F7.2 / '0TONS VARIABLE SCRAP (1) =',
     2F6.2,5X,'TONS HOT METAL CHARGED =',F7.2 / '0TONS IRON ORE CHARGED
     3 =',F6.3,5X,'TONS OF LIME CHARGED  =',F7.3 / '002 BLOW TIME (MIN
     4:SEC)  =',I3,1H:,I2 / '-')
  998    CALL EXIT
         END
```

Input (Data)—Problem 4

```
BOTTLE 701632   0.8      0.72     4.5      0.229    0.1     2445.    400000.BOF DATA 1
TARGET STEEL    0.0      0.11     0.052    0.013    0.005   2920.    500000.BOF DATA 2
1
YARD SCRAP      0.0      0.45     0.15     0.015    0.015   77.      130000.BOD DATA 4


99.8   92.    0.    3.     0.111   -200.   2920.   3000. 6000000 2.
5.0    500.   0.0   0.128  4.0     4.0     22000.
```

Output (Results)—Problem 4

```
    CHARGE CALCULATION FOR A BASIC OXYGEN STEELMAKING CONTROL SYSTEM,

    THIS ANALYSIS IS BEING RUN AT
      BASIC PROCESS        COMPOSITIONS, IN WEIGHT PERCENT    TEMPER-  WEIGHT
      MATERIALS MATRIX:                                       ATURE,   IN LBS
                           SI     MN     C      P      S      DEG. F  (*=EST.)
      HOT METAL
        BOTTLE 701632      0.800  0.720  4.500  0.229  0.100  2445.   400000. *

      SCRAP
        YARD SCRAP    (1) 0.0     0.450  0.150  0.015  0.015   77.    130000. *

      BLOWN METAL
        TARGET STEEL      0.0     0.110  0.052  0.013  0.005  2920.   500000.
      % OXYGEN IN BLAST  = 99.80 %    OXYGEN TEMPERATURE   = -200.  DEG. F.
      FLUE DUST LOSS, AS             BLOWN O2 FLOW RATE    = 22000. SCFM
      PERCENT OF HOT METAL= 2.00 %   COMBUSTION GAS TEMP   = 3000.  DEG. F.
      CO2/CO, COMBUST GAS = 0.111    ENERGY LOSS ESTIMATE = 6000000 BTU/HT
      SLAG MOLE FRACTION             SLAG BASICITY RATIO   = 3.0
      OF IRON OXIDES,                TEMPERATURE OF SLAG   = 2920.  DEG. F.
      EXPRESSED AS FEO   = 0.13      SLAG FEO/FE2O3 RATIO = 4.00
      PERCENT FEO IN ORE = 0.0  %    PERCENT SIO2 IN ORE  = 4.00  %
      ORE-TO-LIME RATIO  = 0.0       PERCENT BASE IN LIME = 92.00 %
      MAX ERROR, CHARGE WT= 5.0 LB   MAX ERROR, METAL OUT = 500. POUNDS

      RESULTS FOLLOW:

      PERCENT SCRAP CHARGED    = 31.45      TONS SCRAP CHARGED       =  84.20

      TONS VARIABLE SCRAP (1) = 84.20      TONS HOT METAL CHARGED = 183.52

      TONS IRON ORE CHARGED    = 0.0        TONS OF LIME CHARGED    =   9.568

      O2 BLOW TIME (MIN:SEC)   = 17:49
```

12. Compute the enthalpy of each of the streams, the hot metal charge, the slag, combustion gases, and blown metal produced. The inlet enthalpy of the hot metal (including the oxidizing reactions which occur) minus the enthalpy of the slag, combustion gases, blown metal (minus scrap) and heat lost from the vessel will yield the excess energy which is available to melt scrap.

13. Calculate the scrap which can be melted, and the metal produced which is the scrap plus the blown metal (minus the scrap).

14. Compare the metal produced with that specified for the heat. If they do not agree, correct the assumed hot metal and scrap charge proportionally and return to step 2.

Discussion of Results—Problem 4

As the silicon content of the hot metal increases, the excess heat available for melting scrap increases resulting in a higher scrap percentage in the charge as shown in Figure B-5.

The structure of the computer program is such that other variables may be investigated, including the influences of changing the temperatures of the process. One commercial proposal for increasing the ability of the process to handle scrap involves scrap preheating, which decreases the energy required to accomplish melting. This program can be used to study the effect of preheating to various temperature levels. Other compositional variables also can be studied.

Example Problem No. 5: Digital Computer Analysis of Heat Flow and Temperature Distribution Around a Copper Converter Tuyere

Problem Statement

Write and test a FORTRAN program which will perform the relaxation calculations outlined by Krivsky and Schuhmann[1] for determining the heat flow and temperature distribution around a copper converter tuyere.

The cross section of a copper converter through a tuyere center line is shown schematically in Fig. B-6. A relaxation technique was employed to obtain a solution. This involves dividing a solid into volume elements represented by grid points, specifying boundary conditions, and assuming a steady-state temperature distribution which dictates that the heat flow into each grid point shall be zero. The heat flow is evaluated in a two-dimensional matrix with radial symmetry about the center line of the tuyere, through which the blast air is supplied. The equations for heat flow into each type of grid point, including the interior points, interior or exterior surface points, tuyere surface points, and the points at the tuyere source and tuyere mouth have been derived.[1] These heat flow equations are solved in terms of the steady state temperatures of each grid point, where the calculation of the steady state temperature at each grid point is a function of the temperatures of the surrounding grid points or boundary temperatures. It should be noted that in Fig. 5

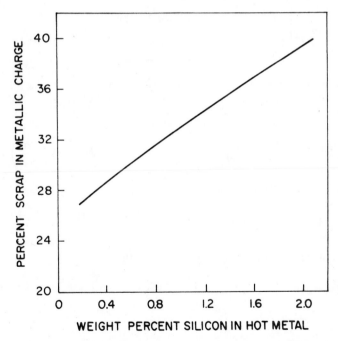

Fig. B-5. Effect of silicon content of hot metal on percent scrap in metallic charge to BOF.

[1] Krivsky, W. A., and R. Schuhmann, Jr., Heat flow and temperature distribution around a copper converter tuyere, *AIME Trans.* **215,** 82 (1959).

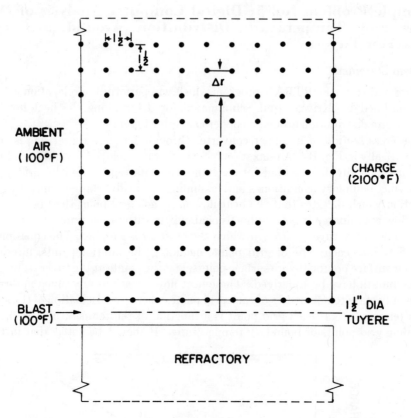

Fig. B-6.   Idealized cross section of single $1\frac{1}{2}$ in. tuyere in 15 in. wall, showing grid of points for temperature distribution calculation.[1]

of reference 1, the equation which appeared in print as

$$\frac{\Sigma Q}{\pi r k} = 2B_1T_1 + \left(1 + \frac{\Delta r}{4r}\right)(T_2 + T_4 + 2T_3) - \left(2B_1 + 4 + \frac{\Delta r}{r}\right)T_0 = 0$$

should read

$$\frac{\Sigma Q}{\pi r k} = 2B_1T_1 + \left(1 + \frac{\Delta r}{4r}\right)(T_2 + T_4) + 2\left(1 + \frac{\Delta r}{2r}\right)T_3 - \left(2B_1 + 4 + \frac{3\Delta r}{2r}\right)T_0 = 0$$

and in Fig. 6 of reference 1, the equation which appeared as

$$Q_{4 \to 0} = h_4\pi\left(r + \frac{\Delta r}{4}\right)(T_4 - T_0)$$

should read

$$Q_{4 \to 0} = h_4\Delta r\pi\left(r + \frac{\Delta r}{4}\right)(T_4 - T_0).$$

Method of Solution—Problem 5

1. Initialize the temperature distribution using an estimate of the temperature at each grid point.
2. Make a point by point calculation of the steady state temperature by systematically moving through the grid, utilizing the surrounding temperatures and heat flow equations at each grid point.
3. After calculating the temperature at each grid point, make a comparison of the calculated temperature with the temperature which existed on the previous pass through the grid. When there are no temperature adjustments which are greater than a specified temperature increment ΔT_{err} for any of the points in the grid from one pass to the next, the calculation is considered complete and the temperature distribution is printed out.

Symbol Table—Problem 5

Algebraic Symbol	Definition	FORTRAN Symbol
$T_{i,j}$	Temperature array (grid)	T
$d_{\max}$	Maximum number of grid points up from the tuyere/refractory interface	$DIST$
$w_{\max}$	Maximum number of grid points across the tuyere	$THICK$
s	Grid spaces, per foot	$SCALE$
r_x	Tuyere radius	$RADIUS$
T_n	Current new temperature at a point	TN
T_m	Melt temperature	$TMELT$
T_{bl}	Blast temperature	$TBLAST$
T_{air}	Air temperature	$TAIR$
n	Iteration counter	N
$n_{\max}$	Maximum number of iterations	$NMAX$
—	Heat transfer coefficient for outside surface	$HAIR$
—	Heat transfer coefficient for blast/tuyere interface	$HBLAST$
—	Heat transfer coefficient for melt/refractory interface	$HMELT$
k	Thermal conductivity of refractory	$KCOND$
ΔT_{err}	Maximum error for each point at convergence	ERR
c	Check digit to test convergence	C
r	Radial distance from tuyere centerline	R
a	Iteration counter for distance *across* the refractory	A

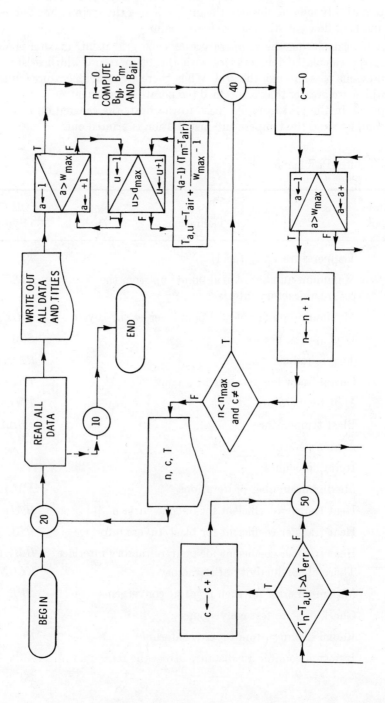

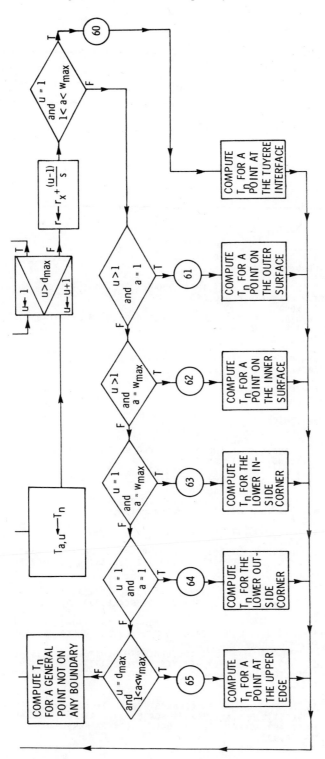

Flow Chart—Problem 5.   Flow diagram for computation of temperature distribution around copper converter tuyere.

Symbol Table—Problem 5 —(Continued)

Algebraic Symbol	Definition	FORTRAN Symbol
u	Iteration counter for distance *up* from the tuyere	U
B_{air}	Biot number for outside surface	$BAIR$
B_{bl}	Biot number for blast	$BBLAST$
B_m	Biot number for melt	$BMELT$

FORTRAN Listing—Problem 5

```
       INTEGER A,U,THICK,DIST,C
       REAL KCOND
       DIMENSION T(50,50)
  20   READ (5,500,END=10) DIST,THICK,NMAX,SCALE,RADIUS,TMELT,TBLAST,TAIR
      1,HBLAST,HMELT,HAIR,KCOND,ERR
       WRITE (6,600)
       WRITE (6,610) DIST,THICK,NMAX,SCALE,RADIUS,TMELT,TBLAST,TAIR,HBLAS
      1T,HMELT,HAIR,KCOND,ERR
       DO 30 A=1,THICK
       DO 30 U=1,DIST
  30   T(A,U)=TAIR+(TMELT-TAIR)*(A-L.)/(THICK-1.)
       BBLAST=HBLAST/(KCOND*SCALE)
       BMELT=HMELT/(KCOND*SCALE)
       BAIR=HAIR/(KCOND*SCALE)
       N=0
  40   C=0
       DO 50 A=1,THICK
       DO 50 U=1,DIST
       R=RADIUS+( U-1.)/SCALE
       IF (U.EQ.1.AND.A.GT.1.AND.A.LT.THICK) GO TO 60
       IF (U.GT.1.AND.A.EQ.1) GO TO 61
       IF (U.GT.1.AND.A.EQ.THICK) GO TO 62
       IF (U.EQ.1.AND.A.EQ.THICK) GO TO 63
       IF (U.EQ.1.AND.A.EQ.1) GO TO 64
       IF (U.EQ.DIST.AND.A.GT.1.AND.A.LT.THICK) GO TO 65
       TN=(T(A-1,U)+T(A+1,U)+T(A,U+1)+T(A,U-1)+((T(A,U+1)-T(A,U-1))/(2.*R
      1*SCALE)))/4.
  52   IF (ABS(TN-T(A,U)).GT.ERR) C=C+1
       GO TO 50
  60   TN=((2.*BBLAST*TBLAST)+(1.+(1./(4.*R*SCALE)))*(T(A-1,U)+T(A+1,U))
      2+2.*(1.+1./(2.*R*SCALE))*T(A,U+1))/(2.*BBLAST+4.+(3./(2.*R*SCALE))
      3)
       GO TO 52
  61                     TN=(T(1,U-1)+T(1,U+1)+(2.*T(2,U))+(2.*BAIR*
      1TAIR)+((T(1,U+1)-T(1,U-1)*(1./(2.*R*SCALE))))/(4.+2.*BAIR)
       GO TO 52
  62                     TN=(T(A,U-1)+T(A,U+1)+(2.*T(A=1,U))+(2.
      1*BMELT*TMELT)+(T(A,U+1)-T(A,U-1))*(1./(2.*R*SCALE)))/(4.+2.*BMELT)
       GO TO 52
  63                     TN=((BBLAST*TBLAST)+(T(A-1,1)*(1.+(1./(
      14.*R*SCALE))))+(T(A,2)*(1.+(1./(2.*R*SCALE))))+(TMELT*BMELT*(1.+(1
      2./(4.*R*SCALE)))))/(2.+BBLAST+(1./(4.*R*SCALE))+1./(2.*R*SCALE)
      3+(BMELT*(1.+(1./(4.*R*SCALE)))))
        GO TO 52
  64                     TN=((BBLAST*TBLAST)+(T(2,1)*(1.+(1./(4.*R*S
      1CALE))))+(T(1,2)*(1.+(1./(2.*R*SCALE))))+(TAIR*BAIR*(SCALE+(1./(4.
      3*R)))))/(2.+BBLAST+(1./(4.*R*SCALE))+(1./(2.*R*SCALE))+(BAIR*(SCAL
      4E+(1./(4.*R)))))
        GO TO 52
```

```
65   TN=T(1,U)+ (T(THICK,U)-T(1,U))*(A-1.)/(THICK-1.)
     GO TO 52
50   T(A,U)=TN
     N=N+1
     IF (N.LT.NMAX.AND.C.NE.O) GO TO 40
     WRITE (6,620) N,C
```

Input (Data)—Problem 5

```
11   11   100 8000   0625210000 10000 10000 20000 50000   600   200  50
11   11   100 8000   0625210000 10000 10000  1000 50000   600   200  50
```

Output (Results)—Problem 5

TEMPERATURE DISTRIBUTION AROUND A COPPER CONVERTER TUYERE

DATA

```
DISTANCE FROM TUYERE, SPACES = 11
REFRACTORY THICKNESS, SPACES = 11
MAXIMUM NUMBER OF ITERATIONS = 100
SCALE, SPACES/FT =  8.0
TUYERE RADIUS, FT =0.0625
MELT TEMPERATURE, DEGREES F. = 2100.
BLAST TEMPERATURE, DEGREES F. = 100.
AIR TEMPERATURE, DEGREES F. = 100.
HBLAST, BTU/HR/SQ FT/DEGREE F. = 200.
HMELT, BTU/HR/SQ FT/DEGREE F. = 500.
HAIR, BTU/HR/SQ FT/DEGREE F. =  6.
THERMAL CONDUCTIVITY OF REFRACTORY,
        BTU-FT/HR/SQ FT/DEGREE F. =  2.
ALLOWABLE ERROR, DEGREES F. =.50
```

RESULTS

```
N = 37
C =  0
```

TEMPERATURE DISTRIBUTION

```
285.7   463.1   640.6   818.1   995.5 1173.0 1350.4 1527.9 1705.4 1882.8 2060.3

373.5   507.1   662.7   827.7   997.9 1171.6 1347.9 1526.9 1708.8 1895.5 2093.1

406.1   532.8   677.3   833.2   997.2 1167.2 1342.4 1522.6 1707.7 1898.0 2093.7

417.7   543.2   681.8   831.3   990.1 1156.9 1331.2 1512.6 1700.9 1895.2 2093.6

418.0   541.7   675.7   819.9   973.9 1137.6 1311.0 1494.2 1686.8 1887.6 2093.4

409.8   529.6   658.1   796.5   945.3 1105.2 1277.4 1462.8 1661.8 1873.4 2093.0

393.1   506.2   627.3   757.9   899.4 1053.7 1223.0 1410.4 1618.4 1847.5 2092.2

366.3   469.0   579.1   698.3   828.7  973.0 1135.5 1322.1 1540.5 1797.9 2090.6

325.0   412.1   505.5   607.0   719.1  845.3  991.4 1167.7 1391.6 1690.7 2087.2

258.0   319.8   386.0   458.1   538.2  629.6  738.4  877.0 1074.8 1408.0 2074.5

117.6   130.1   139.4   149.4   160.5  173.1  188.2  207.7  238.3  343.1 1659.7
```

TEMPERATURE DISTRIBUTION AROUND A COPPER CONVERTER TUYERE

DATA

DISTANCE FROM TUYERE, SPACES = 11
REFRACTORY THICKNESS, SPACES = 11
MAXIMUM NUMBER OF ITERATIONS = 100
SCALE, SPACES/FT = 8.0
TUYERE RADIUS, FT =0.0625
MELT TEMPERATURE, DEGREES F. = 2100.
BLAST TEMPERATURE, DEGREES F. = 100.
AIR TEMPERATURE, DEGREES F. = 100.
HBLAST, BTU/HR/SQ FT/DEGREE F. = 10.
HMELT, BTU/HR/SQ FT/DEGREE F. = 500.
HAIR, BTU/HR/SQ FT/DEGREE F. = 6.
THERMAL CONDUCTIVITY OF REFRACTORY,
        BTU-FT/HR/SQ FT/DEGREE F. = 2.
ALLOWABLE ERROR, DEGREES F. =.50

RESULTS

N = 52
C = 0

TEMPERATURE DISTRIBUTION

287.5	464.7	642.0	819.3	996.6	1173.9	1351.2	1528.4	1705.7	1883.0 2060.3
379.0	514.3	671.1	837.1	1007.7	1181.1	1356.7	1534.3	1714.2	1898.4 2093.2
416.3	546.6	693.9	852.0	1017.1	1187.0	1360.9	1538.3	1719.2	1904.2 2093.9
433.9	565.1	708.4	861.7	1022.6	1189.7	1362.1	1539.1	1720.5	1905.8 2094.0
441.5	573.6	714.8	864.8	1022.7	1187.5	1358.6	1535.6	1718.0	1904.6 2093.9
442.5	574.1	713.4	860.6	1015.7	1178.5	1348.7	1526.1	1710.4	1900.4 2093.8
437.5	567.0	703.5	847.7	999.7	1160.1	1329.3	1507.6	1695.4	1891.7 2093.5
425.6	551.1	683.6	823.4	971.3	1128.1	1295.2	1474.4	1667.4	1874.7 2093.0
403.2	522.7	649.4	782.8	923.8	1074.3	1236.6	1414.6	1613.5	1839.3 2091.9
356.3	472.0	591.2	713.7	842.2	979.7	1130.0	1299.4	1499.6	1752.8 2089.0
205.7	366.9	471.5	567.9	667.0	772.9	889.5	1023.6	1192.9	1455.7 2056.1

Discussion of Results—Problem 5

The corner temperature at the tuyere mouth is particularly critical in copper converter operation. A low temperature at this point permits precipitation of magnetite from the slag, which blocks the tuyere and requires that the operation be stopped and the tuyere punched out. Tuyere punching is presently one of the limiting steps in copper production rates and adds to the expense and difficulty of the operation. The results of the calculation show that a more highly insulating refractory at the tuyere wall, which in effect decreases the rate of heat transfer to the incoming air, would increase the theoretical temperature at the tuyere mouth and prevent the precipitation of magnetite from the slag, thus eliminating the necessity of punching the tuyeres.

The accuracy of the result which is obtained depends upon the number of grid points selected. The solution of Krivsky and Schuhmann was carried out by hand

TABLE B-2

Computed Corner Temperatures for Various Grid Spacings
(Allowable temperature change at grid point, 0.5°F)

Tuyere heat transfer coefficient (Btu/hr–ft²–°F)	Matrix size	Grid spacing (in.)	Iterations	Corner temperature (°F)
10	4 × 4	5	11	2076
10	6 × 6	3	21	2068
10	11 × 11	1.5	52	2056
10	16 × 16	1	78	2049
200	4 × 4	5	9	1830
200	6 × 6	3	17	1754
200	11 × 11	1.5	37	1660
200	16 × 16	1	52	1614

calculation and involved first selecting a five inch grid spacing (4 × 4 matrix). Then, utilizing this calculated temperature distribution, they estimated a temperature distribution for the temperature calculation at the tuyere mouth based on a finer grid spacing in the lower right-hand corner of the grid shown in Figure B-6.

The calculated result is influenced markedly by the grid spacing, as shown by the results presented in Table B-2. The calculations carried out for presentation of the example problem were not extended to extremely fine grid sizes because of time and cost requirements on the computer. One of the major factors in the time requirement (in addition to the number of matrix points) is the accuracy of the initial temperature distribution. In the present problem a linear distribution was assumed between the melt and the outside or ambient air temperature. This assumption would be true for temperature independent thermal properties in the case of a perfectly insulated tuyere, but is far from precise in the case considered in the example problem.

Example Problem No. 6: Heat Loss Through a Furnace Roof

Problem Statement

Write and test a FORTRAN program for evaluating the heat loss through the roof of a reverberatory furnace.
Suggested conditions for the furnace operation are:
Furnace temperature—2800°F
Temperature of surroundings—70°F
Roof size—50 × 22, i.e. 1100 ft²
Refractory thickness—1 ft.
Convective heat transfer coefficient—.38 $\Delta T^{.25}$
Thermal conductivity of refractory—.27 − 5 × 10⁻⁴ t_{avg}
Emissivity of refractory—0.85
Method of Solution—Problem 6

Under steady state operation, the temperature distribution through the refractory roof of a reverberatory furnace is constant. The mechanism of heat loss consists of conduction through the refractory roof and convection and radiation from the outer surface of the roof. Since the temperature distribution in a roof is constant, the heat flux is constant at all points and the rate of energy transmission through the roof by conduction must be equal to the heat loss from the outer surface of the roof by convection and radiation.

Thus,

$$q_{conduction} = q_{convection} + q_{radiation}. \tag{a}$$

The thermal loss by conduction through the refractory is given by the expression

$$q_{conduction} = k_{av}A_r \frac{\Delta T}{\Delta X} = k_{av}Ar(T_f - T_r)/w_r \tag{b}$$

The thermal loss from the outer surface of the roof by radiation is

$$q_{radiation} = (0.173)(\epsilon)\left[\frac{(T_r + 460)^4}{100} - \frac{(T_s + 460)^4}{100}\right](Ar) \tag{c}$$

and by convection is

$$q_{convection} = C'(T_r - T_s)^{1.25}(A_r), \tag{d}$$

$$h_c = C'(T_r - T_s)^{0.25}. \tag{e}$$

The steady state condition is defined by a given surface temperature for the refractory, T_r, which satisfies Eq. a. Since T_r appears in the equations in a non-linear manner, this problem is best solved by an iterative procedure.

Symbol Table—Problem 6

Algebraic Symbol	Definition	FORTRAN Symbol
T_r	Exterior surface temperature of roof	TREFR
T_f	Temperature of furnace	TFCE
T_s	Temperature of surroundings	TSURR

T_{old}	Previously computed value of roof temperature	*TOLD*
ΔT	Initial temperature increment in iteration	*INC*
ΔT_{err}	Acceptable temperature difference for convergence	*EPS*
ϵ	Emissivity of refractory	*EMIS*
C'	Convection coefficient for horizontal surface facing up	*CPRIME*
a	Intercept in equation for average thermal conductivity	*TCONDA*
b	Slope in equation for average thermal conductivity	*TCONDB*
Ar	Area of furnace roof	*AREA*
w_r	Thickness of refractory roof	*THICK*
q_c	Conductive energy loss	*QC*
q_r	Convective and radiative energy loss	*QR*

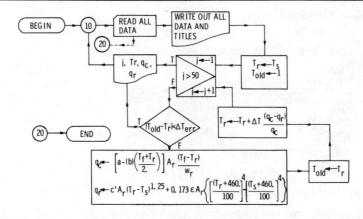

Flow Chart—Problem 6.   Flow diagram for computation of heat loss from furnace roof.

FORTRAN Listing—Problem 6

```
C     HEAT LOSS THROUGH FURNACE ROOF
      REAL INC
   10 READ (5,500,END=20) TFCE,TSURR,CPRIME,TCONDA,TCONDB,EMIS,AREA,THIC
     1K,EPS,INC
      WRITE (6,600) AREA,TCONDA,CPRIME,TCONDB,EMIS,TFCE,EPS,THICK,INC,TS
     1URR
      TREFR=TSURR
      TOLD=1.
      DO 30 J=1,50
      IF (ABS(TOLD-TREFR).LT.EPS) GO TO 40
      QC=(TCONDA-TCONDB*(TFCE+TREFR)/2.)*AREA*(TFCE-TREFR)/THICK
      QR=CPRIME*AREA*(TREFR-TSURR)**1.25+.173*EMIS*(((TREFR+460.)/100.)*
     1*4-((TSURR+460.)/100.)**4)*AREA
      TOLD=TREFR
   30 TREFR=TREFR+INC*(QC-QR)/QC
   40 WRITE (6,610) J,TREFR,QC,QR
      GO TO 10
```

```
 20   CALL SYSTEM
500   FORMAT (F5.0,F4.0,F4.2,F4.1,F6.4,F4.2,F5.0,F4.2,F3.1,F4.0)
600   FORMAT('1HEAT LOSS THROUGH FURNACE ROOF'/'0DATA'/'0AREA =',F5.0,13
     1X,'TCONDA =',F4.1/' CPRIME =',F4.2,12X,'TCONDB =',F6.4/' EMIS ',
     2F4.2,15X,'TFCE =',F5.0/' EPS =',F3.1,16X,'THICK =',F4.2/' INC =',
     3F4.0,15X,'TSURR =',F4.0)
610   FORMAT ('0RESULTS'/'0ITERATIONS REQUIRED TO ATTAIN SOLUTION =',I3/
     1'0EXTERIOR SURFACE TEMPERATURE OF ROOF, DEGREES F. =',F6.1/'0THERM
     2AL LOSS BY CONDUCTION, BTU/HR =',F8.0/'0THERMAL LOSS BY CONVECTION
     3 AND RADIATION, BTU/HR =',F8.0)
      END
```

Input (Data)—Problem 6

```
          2800  70  38 2.7     5  85 1100 100  1 100
```

Output (Results)—Problem 6

```
HEAT LOSS THROUGH FURNACE ROOF

DATA

AREA =1100.              TCONDA = 2.7
CPRIME =0.38             TCONDB =0.0005
EMIS 0.85               TFCE =2800.
EPS =0.1                THICK =1.00
INC =100.               TSURR = 70.

RESULTS

ITERATIONS REQUIRED TO ATTAIN SOLUTION = 23

EXTERIOR SURFACE TEMPERATURE OF ROOF, DEGREES F. = 706.1

THERMAL LOSS BY CONDUCTION, BTU/HR =4200313.

THERMAL LOSS BY CONVECTION AND RADIATION, BTU/HR =4196966.
```

Discussion of Results—Problem 6

A computer calculation of the steady state temperature was achieved using a "sledge-hammer" approach. This technique involved selection of an initial value for T_r and incrementing it depending upon how far away the estimated value is from the correct value, and also whether it is too high or too low. One such correction might be given by the expression

$$T_r = T_r + \Delta T * (QC - QR)/QC,$$

where $QC = q_c$ and $QR = q_{convection} + q_{radiation}$

Classical trial and error solutions also can be achieved graphically. An initial series of surface temperatures are assumed in this approach. At each selected value of surface temperature, the heat loss by conduction through the refractory and the heat loss by convection and radiation from the roof surface is calculated. The two heat losses are plotted against surface temperature yielding the curves shown in Fig. B-7. The steady state temperature is that temperature at which the same heat flux is calculated for the two consecutive transfer processes, i.e., the surface temperature corresponding to the intersection of the two curves in Fig. B-7.

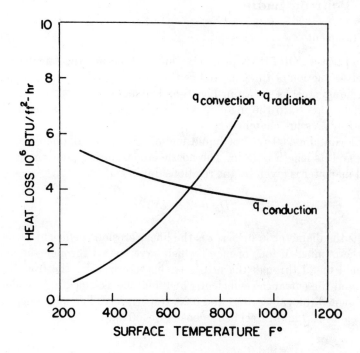

Fig. B-7.   Heat loss by radiation and convection from a furnace roof.

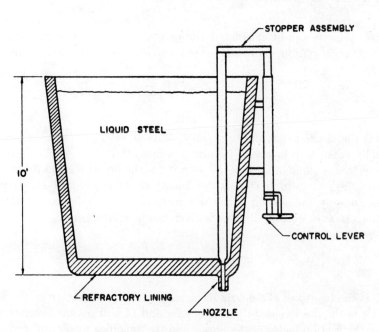

Fig. B-8.   200 Ton bottom-pour ladle.

Example Problem No. 7: Digital Computer Simulation of a Bottom-Pouring Ladle

Problem Statement

Write and test a FORTRAN program which will permit the calculation of pouring rates as a function of time, as well as the total pouring time, for a ladle which is teeming liquid steel as shown in Fig. B-8. The metal flows out of a nozzle in the bottom of the ladle into the ingot mold. The ladle also may be used to vacuum pour into a low pressure chamber.

The ladle contains 200 tons of liquid metal. The initial diameter of the nozzle is 2 inches and its length is 18 in. The nozzle erodes during pouring, however, and the eroded diameter is given by the relationship,

$$D_N = 2 + \epsilon N, \tag{a}$$

where D_N is the diameter in inches, ϵ is the linear erosion coefficient, 0.006 in./ton, and N is the number of tons of metal which have passed through the nozzle. The flow of liquid metal through the nozzle is characterized by the friction factor of the nozzle and the entrance coefficient which may be assumed independent of flow rate and assigned the values, 0.04 and 0.20, respectively. The initial depth of metal in the ladle is 10 ft. Liquid steel has a density of 450 lb/ft³.

Method of Solution—Problem 7

The flow velocity can be calculated at any stage of the pouring operation by the generalized flow equation, expressed for any two locations in the system as

$$\frac{g}{g_c}(z_2 - z_1) + \frac{(V_2^2 - V_1^2)}{2g_c} + \frac{(P_2 - P_1)}{\rho} = -\bar{l}w - w, \tag{b}$$

where g is the acceleration due to gravity, 32.17 ft/sec², g_c is a conversion factor, numerically equal to g but with the units lb-mass ft/lb-force sec², z is the height above a reference plane in ft, V is the velocity of the liquid steel in ft/sec, P is the pressure of the system in lb/ft², ρ is the density of the liquid in lb-mass/ft³, $-w$ is the work done by the system which is zero in the case under consideration, and $-\bar{l}w$ is the lost work in ft lb$_f$/lb$_m$ and is given by the relationship,

$$\bar{l}w = \frac{f_N V_N L}{2g_c D_N} + \frac{\phi_N V_N^2}{2g_c}, \tag{c}$$

where L is the length of the nozzle in inches, D_N is the diameter of the nozzle in inches, V_N is the fluid velocity in the nozzle, and f_N and ϕ_N are dimensionless coefficients, the friction factor of the nozzle and the entrance coefficient, respectively.

For purposes of illustrating the approach to programming the problem for the computer, the flow velocity and pouring rate at the onset of the operation are calculated below.

Selecting point 1 at the upper surface of the metal in the ladle and point 2 just above the nozzle exit the pressure drop $(P_1 - P_2)$ is zero. V_1, the velocity of the surface of the metal in the ladle, is negligibly small in comparison to the exiting velocity from the nozzle. Then,

$$-\frac{g}{g_c}\left[10\left(\frac{200 - N}{200}\right) + L\right] + \frac{V_2^2}{2g_c} = \frac{-V_2^2}{2g_c}\left(f_N\frac{L}{D_N} + \phi_N\right) \tag{d}$$

$$V_2 = \left[(2g_c)\frac{10[(200 - N)/200] + L}{(1 + f_N L/D_N + \phi_N)}\right]^{1/2}. \tag{e}$$

V_2, the exiting flow velocity in ft/sec can be converted to a flow rate by the relationship,

$$\dot{N} \text{ (flow rate in tons per minute)} = \frac{V_2(60)D_N^2\pi\rho}{(2000)(4)} \tag{f}$$

Then

$$\dot{N} = \left\{(2g_c)\frac{10[(200 - N)/200] + L}{(1 + f_N L/D_N + \phi_N)}\right\}^{1/2} \cdot \frac{60\,D_N^2\pi\rho}{(2000)(4)}, \tag{g}$$

where $D_N = 2 + 0.006\,N$.

Evaluating the last two expressions at the start of pouring:

$$D_N = 2 + 0.006 \times 0 = 2 \text{ in..} \tag{h}$$

$$\dot{N} = \left\{(2)(32.2)\frac{10[(200 - 0)/200] + \frac{18}{12}}{1 + (0.04)\frac{18}{2} + 0.2}\right\}^{1/2} \cdot \frac{(60)(2)^2\pi\,450}{(2000)(144)4}$$

$$= \left[(64.4)\frac{11.5}{2.56}\right]^{1/2}\frac{240\pi(450)}{(2000)(576)} = 6.41 \text{ tons/min.} \tag{i}$$

In addition to the stream velocity and the pouring rate, a principal variable in the design of the operation is the time required for pouring. The total pouring time may be estimated by breaking the operation down into short intervals in which the average pouring rate and the number of tons poured are determined. The total time is given by the expression,

$$t = \sum_1^m \frac{(\text{Tons poured in interval } m)}{(\text{Average rate})}. \tag{j}$$

The Basic Approach Involves the Steps:

1. Calculate the stream velocity at the nozzle and the flow rate at the initial conditions.
2. Assume the flow rate remains constant over a short interval of time, and calculate the tons of steel poured during the interval.

Symbol Table—Problem 7

Algebraic Symbol	Definition	FORTRAN Symbol
P_1	Ladle pressure, atm absolute	PL
P_a	Mold pressure, atm absolute	PA
ρ	Metal density, lb/ft³	RHO
w_m	Weight of metal, tons	WT
n	Tons of metal poured during this increment	N
w	Tons of metal poured (total)	W
r	Pouring rate, tons/min	R
r_{old}	Previous estimate for r	RL
h_m	Height of metal, feet	HI
l	Length of nozzle, inches	L
d_0	Initial diameter of nozzle, inches	DI
d	Current diameter of nozzle, inches	D
d_{avg}	Average diameter over current time interval, inches	DAVG
i_{max}	Maximum number of iterations	IMAX
j_{max}	Maximum number of intervals	JMAX
f_n	Nozzle friction factor	F
f_0	Nozzle entrance loss factor	ENT
ϵ	Erosion coefficient, inches/ton	EROS
v	Liquid velocity, ft/sec	V
v_{old}	Previous estimate for v	VL
Δt	Time increment, minutes	INC
t	Current elapsed time, minutes	T
$\Delta \dot{m}_{err}$	Maximum allowable error in rate, tons/min	RERR
$\Delta \dot{V}_{err}$	Maximum allowable error in velocity, ft/min	VERR
ℓ	Number of iterations	J

3. Calculate a new metal depth in the ladle, and a new nozzle diameter based on the tons poured during the last interval.
4. Return to 1 and perform the calculation using the Results of 3 until the entire 200 tons of metal has been poured from the ladle.

One should be aware of the fact that more accurate results could be obtained by using shorter time intervals. In the solution presented below, an average flow rate for each time interval is used to calculate the amount of metal poured during each interval.

1. Calculate the stream velocity at the nozzle exit and the flow rate at the initial conditions.
2. Calculate the number of tons poured during the next interval by multiplying the last calculated pouring rate by the length of the time interval.
3. Calculate an average diameter for the interval based on the diameter at the end of the last interval and the predicted eroded diameter at the end of the present interval.
4. Calculate an average pouring rate based on the average diameter calculated in 3 and the last calculated flow velocity.
5. Return to 2 until the pouring rate calculated in 4 agrees with the previous one to within a given error tolerance.
6. Calculate an average stream velocity based on the average diameter calculated in 3.
7. Return to 2 until the stream velocity calculated in 6 compares with the previous one within a given limit of error.
8. Calculate the total time, the tons poured at the end of the interval, and the eroded diameter. Print these results.
9. Return to 2 to evaluate the next interval. Keep doing so until the entire heat of metal has been poured from the ladle.

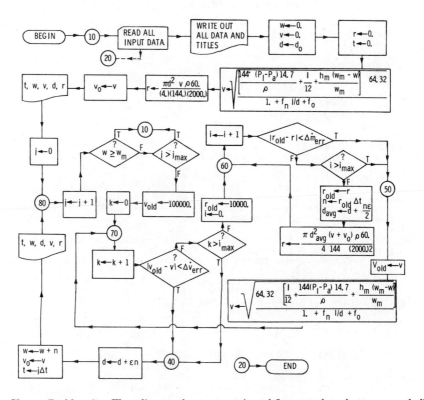

Flow Chart—Problem 7.   Flow diagram for computation of flow rate from bottom-pour ladle.

FORTRAN Listing—Problem 7

```
          REAL L,INC,N,JJ
    10    READ (5,500,END=20) DI,L,HI,WT,RHO,F,ENT,EROS,PL,PA,INC,RERR,VERR,
          1JMAX,IMAX
          WRITE (6,600) DI,L,HI,WT,RHO,F,ENT,EROS,PL,PA
          WRITE (6,601) INC,RERR,VERR,JMAX,IMAX
          W=0.
          V=0.
          D=DI
          R=0.
          T=0.
          V=SQRT(((PL-PA)*14.7*144./RHO+L/12.+HI*(WT-W)/WT)*64.32/(1.+F*L/D+
          1ENT))
          R=D**2*3.1416/(4.*144.)*V*RHO*60./2000.
          VO=V
          WRITE (6,610) T,W,D,V,R
          J=0
    80    J=J+1
          IF (W.GE.WT.OR.J.GT.JMAX) GO TO 30
          VL=100000.
          K=0
    70    K=K+1
          IF (ABS(VL-V).LT.VERR.OR.K.GT.IMAX) GO TO 40
          RL=10000.
          I=0
    60    I=I+1
          IF (ABS(RL-R).LT.RERR.OR.I.GT.IMAX) GO TO 50
          RL=R
          N=RL*INC
          DAVG=D+EROS*N/2.
          R=DAVG**2*3.1416*(V+VO)*RHO*60./(4.*144.*2000.*2.)
          GO TO 60
    50    VL=V
          V=SQRT(((PL-PA)*14.7*144./RHO+L/12.+HI*(WT-W)/WT)*64.32/(1.+F*L/DA
          1VG+ENT))
          GO TO 70
    40    D=D+EROS*N
          VO=V
          W=W+N
          T=J*INC
          WRITE (6,610) T,W,D,V,R
          GO TO 80
    30    GO TO 10
    20    CALL SYSTEM
   500    FORMAT (8F9.3/5F9.3,2I4)
   600    FORMAT (21H1BOTTOM POURING LADLE/1H-,1X,26HINITIAL DIAMETER, INCHE
          1S =F5.1/2X,26HLENGTH OF NOZZLE, INCHES =F5.1/2X,31HINITIAL HEIGHT
          2OF METAL, FEET =F5.1/2X,23HWEIGHT OF METAL, TONS =F5.1/2X,38HDENSI
          3TY OF METAL IN LB PER CUBIC FT. =F5.1/2X,38HFRICTION FACTOR OF NOZ
          4ZLE, FT PER FT =F5.3/2X,29HNOZZLE ENTRANCE LOSS FACTOR =F5.2/2X,37
          5HEROSION COEFFICIENT, INCHES PER TON =F6.3/2X,38HLADLE PRESSURE, A
          6TMOSPHERES ABSOLUTE =F5.1/2X,37HMOLD PRESSURE, ATMOSPHERES ABSOLUT
          7E =F5.1)
   601    FORMAT (2X,27HTIME INCREMENT IN MINUTES =F5.1/2X,29HERROR IN RATE,
          1 TONS PER MIN =F7.3/2X,41HERROR IN STREAM VELOCITY, FT PER MINUTE
          2=F7.3/2X,29HMAXIMUM NUMBER OF INTERVALS =I5/2X,30HMAXIMUM NUMBER O
          3F ITERATIONS =I5/1H-,1X,4HTIME,4X,6HPOURED,3X,3HDIA,5X,15HSTREAM V
          4ELOCITY,3X,12HPOURING RATE/2X,3HMIN,6X,4HTONS,3X,6HINCHES,5X,10HFT
          5 PER SEC,6X,12HTONS PER MIN/1H0,//)
   610    FORMAT (F6.2,F9.2,F8.2,F13.2,F17.2)
          END
```

Input (Data)—Problem 7

```
    2000      18000      10000     200000     450000        040       200        006
    1000       1000       1000        010        010 100 100
    2000      18000      10000     200000     450000        040       200        006
    1000        000       1000        010        010 100 100
```

Output (Results)—Problem 7
Results—Standard Pouring Practice

```
BOTTOM POURING LADLE

    INITIAL DIAMETER, INCHES =  2.0
    LENGTH OF NOZZLE, INCHES = 18.0
    INITIAL HEIGHT OF METAL, FEET = 10.0
    WEIGHT OF METAL, TONS =200.0
    DENSITY OF METAL IN LB PER CUBIC FT. =450.0
    FRICTION FACTOR OF NOZZLE, FT PER FT =0.040
    NOZZLE ENTRANCE LOSS FACTOR = 0.20
    EROSION COEFFICIENT, INCHES PER TON = 0.006
    LADLE PRESSURE, ATMOSPHERES ABSOLUTE =  1.0
    MOLD PRESSURE, ATMOSPHERES ABSOLUTE =  1.0
    TIME INCREMENT IN MINUTES =  1.0
    ERROR IN RATE, TONS PER MIN =  0.010
    ERROR IN STREAM VELOCITY, FT PER MINUTE =  0.010
    MAXIMUM NUMBER OF INTERVALS =  100
    MAXIMUM NUMBER OF ITERATIONS =  100
```

TIME MIN	POURED TONS	DIA INCHES	STREAM VELOCITY FT PER SEC	POURING RATE TONS PER MIN
0.0	0.0	2.00	21.78	6.41
1.00	6.54	2.04	21.80	6.54
2.00	13.31	2.08	21.54	6.77
3.00	20.26	2.12	21.25	6.95
4.00	27.40	2.16	20.96	7.14
5.00	34.72	2.21	20.64	7.32
6.00	42.22	2.25	20.31	7.50
7.00	49.91	2.30	19.96	7.68
8.00	57.76	2.35	19.59	7.86
9.00	65.79	2.39	19.19	8.02
10.00	73.97	2.44	18.78	8.18
11.00	82.30	2.49	18.34	8.33
12.00	90.77	2.54	17.88	8.46
13.00	99.35	2.60	17.39	8.58
14.00	108.03	2.65	16.88	8.68
15.00	116.78	2.70	16.34	8.75
16.00	125.57	2.75	15.77	8.79
17.00	134.38	2.81	15.17	8.80
18.00	143.16	2.86	14.54	8.78
19.00	151.87	2.91	13.88	8.71
20.00	160.47	2.96	13.19	8.60
21.00	168.91	3.01	12.46	8.43
22.00	177.12	3.06	11.69	8.21
23.00	185.04	3.11	10.89	7.92
24.00	192.62	3.16	10.06	7.57
25.00	199.77	3.20	9.18	7.15
26.00	206.43	3.24	8.26	6.65

Results—Vacuum Pouring Practice

```
BOTTOM POURING LADLE

    INITIAL DIAMETER, INCHES =  2.0
    LENGTH OF NOZZLE, INCHES = 18.0
    INITIAL HEIGHT OF METAL, FEET = 10.0
    WEIGHT OF METAL, TONS =200.0
    DENSITY OF METAL IN LB PER CUBIC FT. =450.0
    FRICTION FACTOR OF NOZZLE, FT PER FT =0.040
    NOZZLE ENTRANCE LOSS FACTOR = 0.20
    EROSION COEFFICIENT, INCHES PER TON = 0.006
    LADLE PRESSURE, ATMOSPHERES ABSOLUTE =  1.0
    MOLD PRESSURE, ATMOSPHERES ABSOLUTE =  0.0
    TIME INCREMENT IN MINUTES =  1.0
    ERROR IN RATE, TONS PER MIN =  0.010
    ERROR IN STREAM VELOCITY, FT PER MINUTE =  0.010
    MAXIMUM NUMBER OF INTERVALS =  100
    MAXIMUM NUMBER OF ITERATIONS =  100
```

TIME MIN	POURED TONS	DIA INCHES	STREAM VELOCITY FT PER SEC	POURING RATE TONS PER MIN
0.0	0.0	2.00	25.85	7.61
1.00	7.79	2.05	25.88	7.80
2.00	15.93	2.10	25.64	8.14
3.00	24.38	2.15	25.37	8.45
4.00	33.15	2.20	25.09	8.77
5.00	42.25	2.25	24.78	9.10
6.00	51.69	2.31	24.46	9.44
7.00	61.47	2.37	24.10	9.78
8.00	71.61	2.43	23.73	10.14
9.00	82.11	2.49	23.32	10.49
10.00	92.96	2.56	22.89	10.85
11.00	104.17	2.62	22.43	11.20
12.00	115.72	2.69	21.93	11.55
13.00	127.61	2.77	21.40	11.89
14.00	139.82	2.84	20.82	12.21
15.00	152.33	2.91	20.21	12.50
16.00	165.10	2.99	19.56	12.76
17.00	178.08	3.07	18.86	12.98
18.00	191.24	3.15	18.11	13.15
19.00	204.49	3.23	17.31	13.25

Discussion of Results—Problem 7

The results printed out by the computer program include the pouring time at intervals of one minute, the total number of tons of steel poured up to that time, the current diameter of the pouring nozzle in inches, the stream velocity in feet per second, and pouring rate in tons per minute as averaged for the time increment. Results are presented for a standard pouring practice and for vacuum pouring into an evacuated chamber, a practice which is increasing in importance in steel processing.

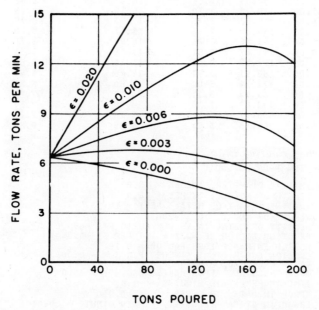

Fig. B-9.   Flow rates from bottom-pour ladle for various nozzle materials with erosion coefficient, ϵ. (Standard pouring conditions.)

The nozzle gradually enlarges during pouring, an effect which is desirable since the rate of pouring does not decline continuously with the ferrostatic head in the ladle as it would if the nozzle diameter remained constant. While the head of metal decreases, the diameter of the orifice increases, and the pouring rate can remain nearly constant or even increase slightly. The rate of pouring is an important variable in ingot quality, and a reasonably constant pouring rate is desired throughout the entire pouring operation. Knowing the erosion coefficients for various nozzle materials, a process engineer can select the optimum nozzle material for a given pouring operation, based on the above analysis. The pouring practice as given in the problem statement is presented in Fig. B-9.

The pouring rate continually increases throughout the vacuum pouring operation. The result indicates that a more stable material (lower erosion coefficient) would better suit those operating conditions, such that the pouring rate would undergo a minimum change during the pouring operation.

Example Problem No. 8: Unsteady State Heat Flow—Cooling of a Slab

Problem Statement

Heat conduction in one dimension in a solid semi-infinite plate can be represented by the second order partial differential equation,

$$\left(\frac{\partial \theta}{\partial t}\right)_x = \frac{k}{C_p\rho}\left(\frac{\partial^2\theta}{\partial x^2}\right),$$ (a)

where k is the thermal conductivity of the solid, C_p is its specific heat, ρ is its density, x is distance, θ is temperature, and t is time. The quantity $k/C_p\rho$ is the thermal diffusivity, α.

The solution of Eq. (a) can be developed by finite difference techniques which involve a network of grid points in terms of the independent variables. The grid spacings are small increments of the independent variables, e.g., in Eq. (a), Δt and Δx.

Write and test a FORTRAN program which will calculate the temperature distribution after thirty minutes in a large slab (semi-infinite in length $\times$ 12 inches thick) which is removed from a preheating furnace at a uniform temperature of 1600°F and set on its edge in the slab yard. Assume constant thermal properties.

Method of Solution—Problem 8

Finite Difference Approximation

The derivatives in the conduction equation with respect to distance may be

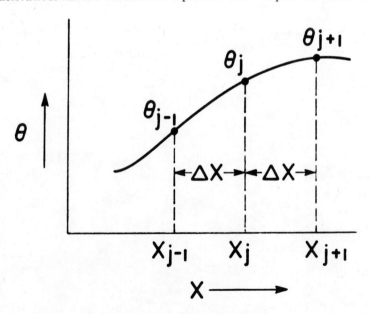

Fig. B-10.   Temperature distribution in solid.

represented by numerical approximations which are computed from a temperature distribution for a network of space points within the conducting medium.

Consider a temperature distribution as given in Fig. B-10.

The variations of temperature, θ, with distance, x, can be expressed by a Taylor Series expansion of the form,

$$\theta_{j+1} = \theta_j + \frac{(\Delta x)^1}{1!} \frac{\partial \theta_j}{\partial x} + \frac{(\Delta x)^2}{2!} \frac{\partial^2 \theta_j}{\partial x^2} + \frac{(\Delta x)^3}{3!} \frac{\partial^3 \theta_j}{\partial x^3} + \cdots, \qquad (b)$$

$$\theta_{j-1} = \theta_j - \Delta x \frac{\partial \theta_j}{\partial x} + \frac{(\Delta x)^2}{2!} \frac{\partial^2 \theta_j}{\partial x^2} - \frac{(\Delta x)^3}{3!} \frac{\partial^3 \theta_j}{\partial x^3} + \cdots, \qquad (c)$$

where θ_{j+1} and θ_{j-1} are points on either side of θ_j, Δx units away. Subtracting Eq. (c) from Eq. (b)

$$\theta_{j+1} - \theta_{j-1} = 2\Delta x \frac{\partial \theta_j}{\partial x} + \frac{2(\Delta x)^3}{3!} \frac{\partial^3 \theta_j}{\partial x^3} + \cdots. \qquad (d)$$

Neglecting terms of third order or higher, and solving for the first approximation to the first derivative:

$$\frac{\partial \theta_j}{\partial x} = \frac{\theta_{j+1} - \theta_{j-1}}{2\Delta x}. \qquad (e)$$

Adding Eqs. (b) and (c), the second derivative can be solved for in a similar manner.

$$\theta_{j+1} + \theta_{j-1} = 2\theta_j + (\Delta x)^2 \frac{\partial^2 \theta_j}{\partial x^2} + \frac{2(\Delta x)^4}{4!} \frac{\partial^4 \theta_j}{\partial x^4} + \cdots. \qquad (f)$$

Then, neglecting terms of 4th order or higher, rearrangement yields:

$$\frac{\partial^2 \theta_j}{\partial x^2} = \frac{\theta_{j+1} + \theta_{j-1} - 2\theta_j}{(\Delta x)^2}. \qquad (g)$$

The derivative with respect to time is first order and is given by the forward difference relationship:

$$\frac{\partial \theta}{\partial t} = \frac{(\theta_j)_{t+\Delta t} - (\theta_j)_t}{\Delta t}, \qquad (h)$$

where $(\theta_j)_{t+\Delta t}$ is the value of θ_j after a time interval Δt and $(\theta_j)_t$ is the temperature at time, t.

Explicit Form of the Difference Equation

Substituting Eqs. (g) and (h) into Eq. (a) yields

$$(\theta_j)_{t+\Delta t} \doteq (\theta_j)_t + \alpha \frac{\Delta t}{(\Delta x)^2} \left[(\theta_{j+1})_t + (\theta_{j-1})_t - 2(\theta_j)_t \right]. \qquad (i)$$

It is often more convenient to work with dimensionless variables, and if the thermal properties of the conducting medium can be assumed constant, the variables t and x can be replaced by

$$X = \frac{x}{L}, \tag{j}$$

and

$$\tau = \frac{\alpha t}{L^2}, \tag{k}$$

where X is a dimensionless space variable ranging from 0 to 1, and defined by the ratio of x, the variable distance to L the total distance; τ is a dimensionless time variable which is related to real time by the thermal diffusivity, α, and total distance, L. Equation (i) then becomes

$$(\theta_j)_{\tau+\Delta\tau} = (\theta_j)_\tau + \frac{\Delta\tau}{(\Delta X)^2} [\theta_{j+1} + \theta_{j-1} - 2\theta_j]_\tau. \tag{l}$$

If $\Delta\tau/(\Delta X)^2 = \lambda$, then

$$(\theta_j)_{\tau+\Delta\tau} = [\lambda(\theta_{j+1} + \theta_{j-1}) + (1 - 2\lambda)\theta_j]_\tau. \tag{m}$$

The temperature at point j at dimensionless time, $\tau + \Delta\tau$, is defined in terms of the temperature at points j and neighboring points, $j + 1$ and $j - 1$, at dimensionless time, τ. An iterative procedure, advancing $\Delta\tau$ units and calculating temperatures at each space point, X_j, yields the time-dependent temperature distribution.

Stability of the Explicit Form

It can be shown that a sufficient condition for stability of the explicit form is that $0 < \lambda \le \frac{1}{2}$.[1]

Boundary Conditions

The temperature at any point, X_j, has been defined in terms of temperatures at neighboring points. The solution also requires that boundary conditions be defined at the end points of the space interval.

Heat transfer from the surface of an object can be by the transport processes convection, radiation, conduction, or some combination of these. For radiation and/or convection the heat transfer can be defined by a heat transfer coefficient, h, where continuity at the boundary (in this case, a left-hand boundary where X increases to the right) requires

$$- q = k\left(\frac{\partial \theta}{\partial x}\right)_{x=0} = h(\theta_{x=0} - \theta_s), \tag{n}$$

where θ_s is the temperature of the surroundings.

Several conditions for heat flow may be considered.

[1] B. Carnahan, H. A. Luther, and J. O. Wilkes, *Applied Numerical Methods*, John Wiley, New York, 1969.

Case A—h very large, relative to k. If the heat transfer coefficient is very large, $\theta_{x=0} \approx \theta_s$, and the surface of the conducting medium can be assumed to be maintained at θ_s. If this is the case and the conducting medium is initially at a temperature θ_0, it is often helpful to define a dimensionless temperature T where

$$T = \frac{\theta_j - \theta_0}{\theta_s - \theta_0}. \tag{o}$$

Equation (a) then is simplified to

$$\frac{\partial T}{\partial \tau} = \frac{\partial^2 T}{\partial X^2}. \tag{p}$$

Equation (p), applied to an infinite parallel sided slab with the boundary conditions,

$$T = 0 \text{ at } \tau = 0 \quad \text{for} \quad 0 \le X \le 1,$$

$$T = 1 \text{ at } X = 0 \quad \text{and} \quad X = 1 \text{ for } \tau > 0,$$

has been solved by finite difference approximation using the computer, and the calculated results compared with the analytical solution.[2]

Case B—k very large, relative to h. In this case, all the resistance to heat flow is at the surface of the conducting medium, and all interior temperatures are equal (Newton's Law of Cooling). Then

$$\text{Heat Loss} = -V\rho C_p \left(\frac{d\theta}{dt}\right) = hA(\theta - \theta_s), \tag{q}$$

where V and A are the volume and surface area of the conducting medium, respectively, and θ is the uniform temperature of the body at any time t. Equation (q) integrates to

$$\log_{10}\left[\frac{\theta - \theta_s}{\theta_{t=0} - \theta_s}\right] = -\frac{hA}{2.3\, V\rho C_p} t. \tag{r}$$

Case C—General values of h and k. For values of h and k which are intermediate between Cases A and B, the heat transfer at the surface is given by Eq. (n). Rearranging at $x = 0$:

$$\left(\frac{\partial \theta}{\partial x}\right)_{x=0} = \frac{h}{k}(\theta_j - \theta_s). \tag{s}$$

Substituting Eq. (s) in Eq. (b), dropping terms of higher order, and solving for the second-order partial differential:

$$\left(\frac{\partial^2 \theta}{\partial x^2}\right) = \frac{2}{(\Delta x)^2}\left\{\theta_{j+1} + \frac{h(\Delta x)}{k}\theta_s - \theta_j\left(1 + \frac{h(\Delta x)}{k}\right)\right\}. \tag{t}$$

Substituting Eqs. (h) and (t) into Eq. (a) yields

$$\frac{(\theta_j)_{t+\Delta t} - (\theta_j)_t}{\Delta t} = \frac{2\alpha}{(\Delta x)^2}\left\{\theta_{j+1} + \frac{h(\Delta x)}{k}\theta_s - \theta_j\left(1 + \frac{h(\Delta x)}{k}\right)\right\}_t. \tag{u}$$

[2] Ibid.

Converting to dimensionless time and space, variables, Eq. (u) becomes

$$(\theta_j)_{\tau+\Delta\tau} = \left\{2\lambda\left[\theta_{j+1} + \left(\frac{h(\Delta X)L}{k}\right)\theta_s\right] + \theta_j\left[1 - 2\lambda\left(1 + \frac{h(\Delta X)L}{k}\right)\right]\right\}_\tau. \quad (v)$$

This expression corresponds to the left hand boundary of a semi-infinite plate, with j increasing to the right. Equation (c) could be used to develop the finite difference relationship for the right hand boundary, an insulated surface as assumed from the problem statement.

Central Point Approximation

In the case of a symmetrical body, such as a semi-infinite plate where the boundary conditions usually are the same on each side, and where the initial temperature distribution is symmetrical, only one-half of the body need be considered. The central point can be approximated by the finite difference relation for interior points, Eq. (m), except that the $j + $1st point, which is outside the range of the space variable can be approximated by the $j - $1st point, such that

$$(\theta_j)_{\tau+\Delta\tau} = [\theta_j + 2\lambda(\theta_{j-1} - \theta_j)]_\tau. \quad (w)$$

Symbol Table—Problem 8

Algebraic Symbol	Definition	FORTRAN Symbol
n	Number of grid spaces	N
t_{max}	Maximum time of simulation, minutes	$TMAX$
Δt	Time Increment	$DELT$
T_j	Temperature array for the last time step	T
T_{nj}	Temperature array for the current time step	TN
T_0	Initial uniform temperature of slab	$TINIT$
T_s	Temperature of surroundings	$TSURR$
τ	Dimensionless Time	TAU
t	Real time	$TIME$
k	Thermal Conductivity, BTU-ft/(hr-ft²-°F)	K
C_p	Specific Heat, BTU/°F-lb.	CP
ρ	Density, 490 lb./ft³	RHO
α	Thermal Diffusivity	$ALPH$
ϵ	Emissivity of slab surface	$EMIS$
h	Linear Heat Transfer coefficient combining radiation and convection terms	H
j	Subscript for grid points	J
λ	Combination of terms	$LAMBDA$
Δx	Grid Spacing	$DELX$
L	Half width of slab, ft.	L
F	Number of time steps between printouts of temperature distribution	$FREQ$
c	Counter for output frequency	$COUNT$
q_1	Convective and radiative flux term	$Q1$

Convergence of Explicit Method

Verification of convergence can be made by using ever decreasing increments of τ and X, until the solution does not change, within some acceptable tolerance limit.

Implicit Form of the Difference Equations

In an implicit method, the finite difference approximations are written in terms of a series of simultaneous equations involving the temperature terms at the next time step. The computational procedure is slightly more complex, but offers the advantage of not being limited to small time increments as in the explicit method where $0 < \Delta\tau/(\Delta X)^2 \leq \frac{1}{2}$. In many problems the advantage of a larger time step far outweighs the added programming complexity. A detailed development of some implicit methods and several example problems using this form have been presented.[3]

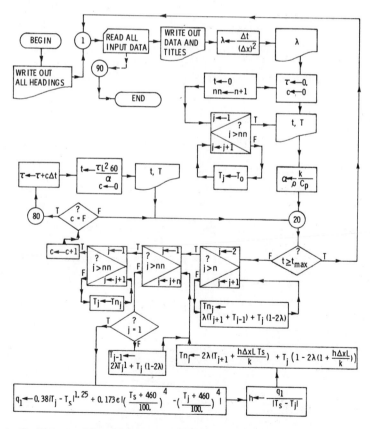

Flow Chart—Problem 8.   Flow diagram for computation of cooling of slab.

[3] Ibid

FORTRAN Listing—Problem 8

```
          INTEGER R
          REAL LAMBDA,L,K
          DIMENSION T(100),TN(100)
          WRITE (6,600)
   1      READ (5,500,END=90) N,TMAX,DELT,DELX,L,FREQ,TINIT,TSURR,K,CP,RHO,E
         1MIS
          WRITE (6,610) N,TINIT,TMAX,TSURR,DELT,K,DELX,CP,L,RHO,FREQ,EMIS
          LAMBDA=DELT/DELX**2
          WRITE (6,620) LAMBDA
          TAU=0.
          COUNT=0.
          TIME=0.
          NN=N+1
          DO 10 J=1, NN
   10     T(J)=TINIT
          WRITE (6,630)
          WRITE (6,640) TIME, (T(J), J=1, NN)
          ALPH=K/(CP*RHO)
   20     IF (TIME.GE.TMAX) GO TO 30
          DO 40 J=2,N
   40     TN(J)=LAMBDA*(T(J+1)+T(J-1))+T(J)*(1.-2.*LAMBDA)
          DO 50 J=1, NN, N
          IF (J.EQ.1) GO TO 60
          TN(J)=2.*LAMBDA*T(J-1)+T(J)*(1.-2.*LAMBDA)
          GO TO 50
   60     Q1=.38*ABS(T(J)-TSURR )**1.25+.173*EMIS*ABS(((TSURR+460.)/100.)**4
         1-((T(J)+460.)/100.)**4)
          H=Q1/ABS(TSURR-T(J))
          TN(J)=2.*LAMBDA*(T(J+1)+H*DELX*L/K*TSURR)+T(J)*(1.-2.*LAMBDA*(1.+H
         1*DELX*L/K))
   50     CONTINUE
          DO 70 J=1, NN
   70     T(J)=TN(J)
          COUNT=COUNT+1.
          IF (COUNT.EQ.FREQ) GO TO 80
          GO TO 20
   80     TAU=TAU+COUNT*DELT
          TIME=TAU*L**2*60./ALPH
          COUNT=0.
          WRITE (6,640) TIME, (T(J), J=1, NN)
          GO TO 20
   30     GO TO 1
   90     CALL SYSTEM
  500     FORMAT (I3,F4.1,F6.4,3F4.1,F6.1,F6.2,F4.1,F4.2,F5.1,F3.1)
  600     FORMAT (30H1COOLING OF SEMI-INFINITE SLAB)
  610     FORMAT (1H-,4X,2HN=I3,15X,6HTINIT=F7.1/5X,5HTMAX=F5.1,10X,
         16HTSURR=F6.1/5X,5HDELT=F7.4,8X,2HK=F5.1/5X,5HDELX=F4.1,
         211X,3HCP=F5.2/5X,2HL=F4.1,14X,4HRHO=F6.1/5X,
         35HFREQ=F5.1,10X,5HEMIS=F4.1)

  620     FORMAT (1H-,5X,7HLAMBDA=F6.3)
  630     FORMAT (1H-,1X,5HTIME,,20X,24HTEMPERATURE DISTRIBUTION/2X,3HMIN///
         1)
  640     FORMAT (1X,F5.2,11F7.1)
          END
```

Input (Data)—Problem 8

```
          10 30      25    1    5 100 16000 10000 250   11 4900   8
```

Output (Results)—Problem 8

```
COOLING OF SEMI-INFINITE SLAB


     N= 10              TINIT= 1600.0
     TMAX= 30.0         TSURR= 100.0
     DELT= 0.0025       K= 25.0
     DELX= 0.1          CP= 0.11
     L= 0.5             RHO= 490.0
     FREQ= 10.0         EMIS= 0.8


     LAMBDA= 0.250
```

```
TIME,                   TEMPERATURE DISTRIBUTION
MIN
0.0    1600.0 1600.0 1600.0 1600.0 1600.0 1600.0 1600.0 1600.0 1600.0 1600.0 1600.0
0.81   1511.5 1551.9 1577.3 1590.9 1597.0 1599.2 1599.8 1600.0 1600.0 1600.0 1600.0
1.62   1480.3 1520.6 1550.7 1571.6 1584.8 1592.1 1596.6 1598.6 1599.5 1599.8 1599.9
2.43   1458.2 1497.9 1529.5 1553.6 1570.9 1582.6 1590.2 1594.7 1597.3 1598.5 1598.9
3.23   1440.7 1479.6 1511.8 1537.6 1557.4 1571.9 1582.2 1589.0 1593.3 1595.5 1596.3
4.04   1426.0 1464.1 1496.6 1523.3 1544.7 1561.2 1573.4 1582.1 1587.7 1590.9 1592.0
4.85   1413.2 1450.7 1483.1 1510.4 1532.8 1550.7 1564.3 1574.3 1581.1 1585.0 1586.2
5.66   1401.9 1438.7 1470.9 1498.6 1521.6 1540.4 1555.1 1566.0 1573.6 1578.0 1579.5
6.47   1391.6 1427.8 1459.8 1487.5 1511.0 1530.4 1545.7 1557.4 1565.5 1570.3 1571.9
7.28   1382.1 1417.7 1449.4 1477.1 1500.8 1520.6 1536.4 1548.5 1557.0 1562.1 1563.7
8.08   1373.2 1408.3 1439.7 1467.3 1491.0 1511.0 1527.1 1539.5 1548.3 1553.5 1555.2
8.89   1364.8 1399.3 1430.4 1457.8 1481.5 1501.5 1517.8 1530.4 1539.3 1544.7 1546.4
9.70   1356.8 1390.8 1421.5 1448.7 1472.3 1492.3 1508.6 1521.3 1530.3 1535.7 1537.5
10.51  1349.1 1382.6 1412.9 1439.9 1463.3 1483.2 1499.5 1512.2 1521.2 1526.7 1528.5
11.32  1341.7 1374.7 1404.6 1431.3 1454.5 1474.3 1490.5 1503.1 1512.2 1517.6 1519.4
12.13  1334.4 1367.0 1396.5 1422.9 1445.9 1465.5 1481.6 1494.2 1503.2 1508.6 1510.4
12.94  1327.4 1359.5 1388.6 1414.6 1437.4 1456.8 1472.8 1485.3 1494.2 1499.6 1501.4
13.74  1320.5 1352.1 1380.9 1406.6 1429.1 1448.3 1464.1 1476.5 1485.3 1490.7 1492.4
14.55  1313.7 1344.9 1373.3 1398.7 1420.9 1439.9 1455.5 1467.8 1476.5 1481.8 1483.6
15.36  1307.0 1337.8 1365.8 1390.9 1412.8 1431.6 1447.0 1459.2 1467.8 1473.1 1474.8
16.17  1300.5 1330.9 1358.5 1383.2 1404.9 1423.4 1438.7 1450.6 1459.2 1464.4 1466.1
16.98  1294.0 1324.0 1351.3 1375.7 1397.1 1415.3 1430.4 1442.3 1450.7 1455.8 1457.5
17.79  1287.7 1317.2 1344.1 1368.2 1389.3 1407.4 1422.3 1434.0 1442.3 1447.4 1449.1
18.60  1281.4 1310.6 1337.1 1360.9 1381.7 1399.5 1414.3 1425.8 1434.0 1439.0 1440.7
19.40  1275.2 1304.0 1330.2 1353.6 1374.2 1391.8 1406.3 1417.7 1425.9 1430.8 1432.4
20.21  1269.0 1297.5 1323.3 1346.5 1366.8 1384.2 1398.5 1409.7 1417.8 1422.6 1424.3
21.02  1263.0 1291.1 1316.6 1339.4 1359.5 1376.6 1390.8 1401.9 1409.8 1414.6 1416.2
21.83  1257.0 1284.7 1309.9 1332.5 1352.3 1369.2 1383.1 1394.1 1401.9 1406.7 1408.2
22.64  1251.1 1278.4 1303.3 1325.6 1345.1 1361.8 1375.6 1386.4 1394.2 1398.8 1400.4
23.45  1245.2 1272.3 1296.8 1318.8 1338.1 1354.6 1368.2 1378.8 1386.5 1391.1 1392.6
24.25  1239.4 1266.1 1290.4 1312.1 1331.1 1347.4 1360.9 1371.4 1378.9 1383.5 1385.0
25.06  1233.7 1260.1 1284.0 1305.5 1324.3 1340.3 1353.6 1364.0 1371.5 1375.9 1377.4
25.87  1228.0 1254.1 1277.7 1298.9 1317.5 1333.4 1346.5 1356.7 1364.1 1368.5 1370.0
26.68  1222.4 1248.2 1271.5 1292.4 1310.8 1326.5 1339.4 1349.5 1356.8 1361.2 1362.6
27.49  1216.9 1242.3 1265.4 1286.1 1304.2 1319.6 1332.4 1342.4 1349.6 1353.9 1355.4
28.30  1211.4 1236.5 1259.3 1279.7 1297.6 1312.9 1325.5 1335.4 1342.5 1346.8 1348.2
29.11  1206.0 1230.8 1253.3 1273.5 1291.2 1306.3 1318.7 1328.5 1335.5 1339.7 1341.1
29.91  1200.6 1225.1 1247.4 1267.3 1284.8 1299.7 1312.0 1321.6 1328.5 1332.7 1334.1
30.72  1195.3 1219.5 1241.5 1261.2 1278.4 1293.2 1305.3 1314.9 1321.7 1325.8 1327.2
```

Discussion of Results—Problem 8

The results present the temperature profile developing across the half-thickness of the slab. They show that the rate of temperature decrease at the surface of the slab decreases with time. The program also can be used to investigate heating rates.

Example Problem No. 9: Linear Regression Analysis—BOF Steel Chemistry

Problem Statement

The oxygen content of an oxygen-blown heat of steel is very important in that it determines the quantity of deoxidizing elements which must be added to control the oxygen content of the steel. The oxygen content of the liquid steel bath is proportional to the reciprocal of the carbon content according to the reaction,

$$\underline{C} + \underline{O} = CO(g), \tag{a}$$

where the equilibrium constant for reaction (a) is

$$K_a = \frac{P_{CO}}{a_C a_O}. \tag{b}$$

At equilibrium with one atmosphere carbon monoxide pressure and at 1600°C, and neglecting the small influence of solute elements on the oxygen and carbon activity

$$(a_O) = \frac{P_{CO}}{(K_a)(a_c)} = \%O = \frac{0.00203}{(\%C)}. \tag{c}$$

At the end of a BOF heat, the bath oxygen content will be greater than for equilibrium with one atmosphere carbon monoxide pressure. Nevertheless, it is expected that the oxygen content will be related to the carbon content, perhaps by an expression of the form

$$\%O = A + B/(\%C), \tag{d}$$

where A is the intercept and B the slope for a postulated linear relationship for percent oxygen versus the reciprocal of percent carbon.

Prepare a computer program which will fit a linear relationship of the form given in Eq. (d) to N data sets of the form $[(\%O), (\%C)], \ldots, [(\%O)_i, (\%C)_i], \ldots$ $[(\%O)_n, (\%C)_n]$. The least-square criterion should of the form given in equation (d) to N data sets of the form $[(\%O_i), (\%C_i)]$, $i = 1, 2, \ldots, N - 1, N$. A least squares method can be used as the criterion for "best" fit of the data. The program also should provide a measure of the "goodness of fit." To provide generality for the program (which might be used for fitting a linear relationship to any set of data), the regression analysis should be written as a subprogram for a general program which reads data, modifies it (e.g. takes the reciprocal, squares, etc.), calls on the subprogram, and prints both data and results.

Method of Solution—Problem 9

The purpose of regression analysis is to fit a curve to a set of data points. This curve will best represent the data according to some criterion. If the criterion adopted is that the sum of the squares of the point deviations from the regression curve is a minimum, the technique is called a "least squares" regression. If the regression curve is first order, the technique is called "linear regression."

Assume a relationship of the form

$$Y = A + BX \tag{e}$$

To minimize the sum of the squares of the deviations of the measured values of the independent variable from the values predicted by equation (e),

$$\frac{\partial \left(\sum\limits_{i=1}^{N} (Y_i - A - BX_i)^2 \right)}{\partial B} = 0 \tag{f}$$

and

$$\frac{\partial \left(\sum\limits_{i=1}^{N} (Y_i - A - BX_i)^2 \right)}{\partial A} = 0 \tag{g}$$

Taking the indicated derivatives and setting them equal to zero we have:

$$-2 \sum_{i=1}^{N} (Y_i - A - BX_i) = 0 \tag{h}$$

$$-2 \sum_{i=1}^{N} X_i (Y_i - A - BX_i) = 0 \tag{i}$$

Dividing each of these equations by -2, and realizing that $\sum\limits_{i=1}^{N} A = NA$

$$NA + B \sum_{i=1}^{N} X_i = \sum_{i=1}^{N} Y_i \tag{j}$$

$$A \sum_{i=1}^{N} X_i + B \sum_{i=1}^{N} (X_i)^2 = \sum_{i=1}^{N} X_i Y_i \tag{k}$$

Solving equations (j) and (k), usually called the *normal equations*, simultaneously for the coefficients A and B

$$B = \frac{N \sum\limits_{i=1}^{N} X_i Y_i - \sum\limits_{i=1}^{N} X_i \sum\limits_{i=1}^{N} Y_i}{N \sum\limits_{i=1}^{N} (X_i)^2 - \left(\sum\limits_{i=1}^{N} X_i \right)^2} \tag{l}$$

$$A = \frac{\sum\limits_{i=1}^{N} Y_i - B \sum\limits_{i=1}^{N} X_i}{N} \tag{m}$$

The average squared error, known as the *variance* of the data points about the regression line, is estimated from the data as

$$s_{y,x}^2 = \frac{\sum_{i=1}^{N} Y_i^2 - A \sum_{i=1}^{N} Y_i - B \sum_{i=1}^{N} X_i Y_i}{N - 2} \tag{n}$$

A measurement of how well we can predict a value of Y, given a value of X, is a *product moment correlation coefficient*, usually denoted r. It ranges in value from -1 to 1, where a value of 1 represents perfect correlation, 0 represents no correlation, and -1 represents perfect inverse correlation. The value of r is estimated from data by:

$$r = \frac{N \sum_{i=1}^{N} X_i Y_i - \sum_{i=1}^{N} Y_i \sum_{i=1}^{N} X_i}{\sqrt{N \sum_{i=1}^{N} X_i^2 - (\sum_{i=1}^{N} X_i)^2} \sqrt{N \sum_{i=1}^{N} Y_i^2 - (\sum_{i=1}^{N} Y_i)^2}} \tag{o}$$

For statistical reasons, linear regression results usually are presented with the regression constants, the standard deviation (square root of the variance), and product moment correlation coefficient.

Two cautions on the use of linear regression must be observed. First, the correlation coefficient is a measure of how much uncertainty in the prediction of a value of Y is removed by using the regression equation. Values of $r > 0.8$ usually indicate a strong deterministic association, i.e. a chosen form of equation appears to be the correct one. In other fields, such as the social sciences, $r > 0.3$ often is considered a significant correlation because stochastic (probabilistic) processes continually affect each measurement in unknown ways. This example problem is deterministic because we have sound thermodynamic knowledge on which to base our chosen form of the regression equation.

Second, this type of analysis is rigorous only under the assumption that the absolute errors of measurement are the same over the entire range of the data. Here this is not the case, since the X values are reciprocals of the %C analysis. The "standard deviation" as computed is actually the *minimum* standard deviation at a value of X equal to the average of all the data values of X. A final determination of the appropriateness of a computed regression line always must be based on a visual comparison of the computed regression line with the original data points.

Computer Solution

The computer program to solve the problem of correlating BOF turndown oxygen and carbon analyses consists of a main program which handles the data and a subroutine which performs the actual regression calculation.

Symbol Table—Problem 9 Main Program

Algebraic Symbol	Definition	FORTRAN Symbol
x	Measured values of the independent variable	X
y	Measured values of the dependent variable	Y
n	Number of data points	N
inv	Logical check variable to test for inversion of X values	$INVERT$
a	Best estimate of the intercept of the regression line	A
b	Best estimate of the regression line slope	B
s	Best estimate of the standard deviation of the regression line	$STDDEV$
r	Best estimate of the correlation coefficient	R

Symbol Table—Problem 9 Subroutine LINREG

Algebraic Symbol	Definition	FORTRAN Symbol
n, x, y, a b, r, s	Defined the same as for the calling program, above	$N, X, Y, A,$ $B, R, STDDEV$
sx	Sum of x_i values	SX
sy	Sum of y_i values	SY
sx^2	Sum of x_i^2 values	$SX2$
sy^2	Sum of y_i^2 values	$SY2$
sxy	Sum of $x_i y_i$ products	SXY
xi	Temporary storage for current x_i	XI
yi	Temporary storage for current y_i	YI
$s_2 y_x$	Common term (variance)	$S2YX$

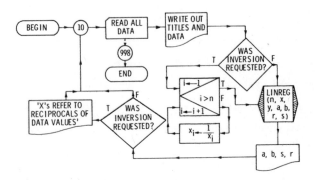

Flow Chart—Problem 9.   Flow diagram of main program for linear regression analysis.

FORTRAN Listing—Problem 9

FORTRAN Listing—Main Program

```
        LOGICAL INVERT
        DIMENSION X(200),Y(200)
10      READ(5,500,END=998) N,INVERT,(X(I),I=1,N)
        READ(5,501) (Y(I),I=1,N)
        WRITE(6,502) N,(X(I),Y(I),I=1,N)
        IF(.NOT.INVERT) GO TO 30
        DO 20 I=1,N
20      X(I)=1./X(I)
30      CALL LINREG(N,X,Y,A,B,R,STDDEV)
        WRITE(6,503) A,B,STDDEV,R
        IF(INVERT) WRITE(6,504)
        GO TO 10
500     FORMAT(I5,L2/(10F8.0))
501     FORMAT(10F8.0)
502     FORMAT('1LEAST SQUARES LINEAR REGRESSION'/'-THE',I3,' DATA POINTS
       1 ARE:'/ '0   X        Y'/(F7.3,F9.4))
503     FORMAT('-RESULTS:    Y =',F9.5,'  +  (',F8.5,') * X'/
       1'0ESTIMATE OF STANDARD DEVIATION =',F9.5/'0ESTIMATE OF CORRELATION
       2 COEFFICIENT =',F7.3)
504     FORMAT('0WHERE THE X''S ABOVE REFER TO THE'/' RECIPROCAL OF THE D
       1ATA VALUES FOR X.')
998     CALL EXIT
        END
```

Main Program

The main program consists of input and output statements which handle the
data and results. N is the number of data points, stored in X and Y as number
pairs. The X's are turndown carbon contents and the Y's are turndown oxygen
contents. The program provides a logical variable INVERT which may be set to
a value of .TRUE. or .FALSE. by the data and which provides for possible inversion
of the X values prior to regression. After calling on the regression subprogram,
the main program prints out the resulting equation and two statistical measures
for "goodness of fit," namely the standard deviation and the correlation coefficient.

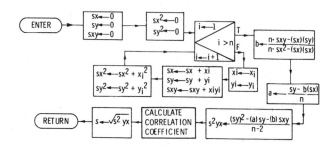

Flow Chart—Problem 9.   Flow diagram of subroutine LINREG for linear regression analysis.

FORTRAN Listing—Subroutine LINREG

```
      SUBROUTINE LINREG(N,X,Y,A,B,R,STDDEV)
      DIMENSION X(N),Y(N)
      SX=0.
      SY=0.
      SX2=0.
      SY2=0.
      SXY=0.
      DO 20 I=1,N
      XI=X(I)
      YI=Y(I)
      SX=SX+XI
      SY=SY+YI
      SX2=SX2+XI*XI
      SY2=SY2+YI*YI
   20 SXY=SXY+XI*YI
      B=(N*SXY-SX*SY)/(N*SX2-SX*SX)
      A=(SY-B*SX)/N
      S2YX=(SY2-A*SY-B*SXY)/(N-2)
      R=(N*SXY-SX*SY)/(SQRT(N*SX2-SX*SX)*SQRT(N*SY2-SY*SY))
      STDDEV=SQRT(S2YX)
      RETURN
      END
```

Subroutine LINREG

The subroutine LINREG is a subprogram which operates on a given set of N values stored in the vectors X and Y and computes the regression constants A and B as well as the standard deviation and correlation coefficient.

Input (Data)—Problem 9

```
 30 T
0.06      0.04      0.05      0.066     0.07      0.055     0.04      0.15      0.06      0.066
0.042     0.04      0.03      0.04      0.24      0.03      0.06      0.05      0.045     0.06
0.05      0.06      0.06      0.035     0.05      0.07      0.12      0.18      0.055     0.2
0.068     0.083     0.095     0.074     0.069     0.075     0.108     0.037     0.0714    0.0698
0.1070    0.082     0.114     0.095     0.019     0.136     0.065     0.084     0.105     0.0675
0.078     0.0686    0.0728    0.1140    0.0835    0.0560    0.036     0.025     0.0626    0.024
```

Output (Results)—Problem 9

LEAST SQUARES LINEAR REGRESSION

THE 30 DATA POINTS ARE:

X	Y
0.060	0.0680
0.040	0.0830
0.050	0.0950
0.066	0.0740
0.070	0.0690
0.055	0.0750
0.040	0.1080
0.150	0.0370
0.060	0.0714
0.066	0.0698
0.042	0.1070
0.040	0.0820
0.030	0.1140
0.040	0.0950
0.240	0.0190
0.030	0.1360
0.060	0.0650
0.050	0.0840
0.045	0.1050
0.060	0.0675
0.050	0.0780
0.060	0.0686
0.060	0.0728
0.035	0.1140
0.050	0.0835
0.070	0.0560
0.120	0.0360
0.180	0.0250
0.055	0.0626
0.200	0.0240

RESULTS: Y = 0.00983 + (0.00357) * X

ESTIMATE OF STANDARD DEVIATION = 0.00854

ESTIMATE OF CORRELATION COEFFICIENT = 0.954

WHERE THE X'S ABOVE REFER TO THE
RECIPROCAL OF THE DATA VALUES FOR X.

Discussion of Results—Problem 9

The turndown oxygen content of the BOF heats used in the regression analysis can be predicted by the relationship,

$$\%O = 0.0093 + \frac{0.000357}{(\%C)}, \tag{o}$$

with a standard deviation of 0.00854% O. The correlation coefficient of 0.994 indicates that a good fit is obtained with this linear relationship.

The data are plotted in Fig. B-11 together with the regression line and the equilibrium relationship for 1 atm CO pressure at 1600°C. The data in the vicinity of 0.06% carbon, a typical turndown analysis for low carbon steel heats, indicate average oxygen concentrations corresponding to an equilibrium CO pressure of about 1.8 atm.

Example Problem No. 10: Linear Programming—Optimization of Cupola Charge

Problem Statement

A problem which often arises in melting operations is the calculation of a charge to meet specific composition limits. This calculation is relatively easy if one is concerned only with the composition and is not seeking the absolute minimum cost of meeting the charge requirements. Rule of thumb techniques can usually produce a charge that is low in cost, but there is no guarantee that the estimated mix is the most economical. Linear Programming techniques can produce the most economical solution which lies within the required composition limits. The following problem illustrates the application of linear programming to optimize the charge to a cupola melting furnace producing a specified cast iron composition.

Available Materials

Material	Compositions (w/o)[a]				Cost ($)/Ton
	C	Si	Mn	Cr	
Pig iron	4.2	2.25	0.8	—	58
Silvery pig	—	15.0	4.5	10.0	120
Ferrosilicon #1	—	45.0	—	—	128
Ferrosilicon #2	—	42.0	—	—	120
Master alloy #1	—	18.0	60.0	—	200
Master alloy #2	—	30.0	9.0	20.0	260
Master alloy #3	—	25.0	33.0	8.0	238
Silicon carbide	15.0	30.0	—	—	160
Steel scrap #1	0.4	—	0.9	—	39
Steel scrap #2	0.1	—	0.3	—	30
Steel scrap #3	0.1	—	0.3	—	33

[a] Balance Fe.

Composition Limits and Other Constraints

> $3.2 \leq$ w/o C ≤ 3.5
> $2.7 \leq$ w/o Si ≤ 3.0
> $1.4 \leq$ w/o Mn ≤ 1.6
> $0.3 \leq$ w/o Cr ≤ 0.45
> SiC ≤ 0.01 ton
> Steel scrap #1 ≤ 0.3 ton
> Steel scrap #2 ≤ 0.3 ton
> Steel scrap #3 ≤ 0.3 ton
> Total steel scrap ≤ 0.3 ton
> Total weight of charge is to be one ton.

Write a computer program which will minimize the cost per ton of charge while maintaining the above constraints.

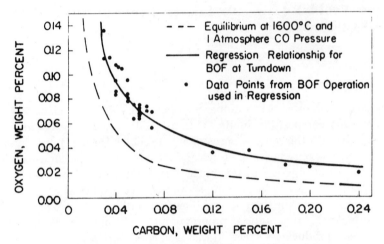

Fig. B-11.   Carbon-Oxygen relationship in BOF steelmaking.

Method of Solution—Problem 10

This problem can be solved using a Simplex routine. A computer program could consist of a Main Program which calls on a Subroutine to carry out matrix manipulations according to the Simplex technique. A computer program using this approach is presented, as adopted from Carnahan et al.[1] and including modifications suggested by Simonnard.[2]

The following notation is used to represent the mass of each material in the charge, expressed in tons:

$$
\begin{aligned}
X(1) &= \text{Pig iron,} \\
X(2) &= \text{Silvery pig,} \\
X(3) &= \text{Ferrosilicon \#1,} \\
X(4) &= \text{Ferrosilicon \#2,} \\
X(5) &= \text{Master alloy \#1,} \\
X(6) &= \text{Master alloy \#2,} \\
X(7) &= \text{Master alloy \#3,} \\
X(8) &= \text{Silicon carbide,} \\
X(9) &= \text{Steel scrap \#1,} \\
X(10) &= \text{Steel scrap \#2,} \\
X(11) &= \text{Steel scrap \#3.}
\end{aligned}
$$

Converting the real variables to those of the program and setting up the constraints, one obtains:

Carbon balance.

$$4.2X(1) + 15X(8) + 0.4X(9) + 0.1X(10) + 0.1X(11) \geq 3.2.$$

$$4.2X(1) + 15X(8) + 0.4X(9) + 0.1X(10) + 0.1X(11) \leq 3.5.$$

[1] B. Carnahan, private communication, University of Michigan, 1967.
[2] M. Simonnard, *Linear Programming*, Prentice-Hall, Inc., Englewood Cliffs, N.J., 1960.

Silicon balance.

$$2.25X(1) + 15X(2) + 45X(3) + 42X(4) + 18X(5) + 30X(6)$$
$$+ 25X(7) + 30X(8) \geq 2.7.$$

$$2.25X(1) + 15X(2) + 45X(3) + 42X(4) + 18X(5) + 30X(6)$$
$$+25X(7) + 30X(8) \leq 3.0.$$

Manganese balance.

$$0.8X(1) + 4.5X(2) + 60X(5) + 9X(6) + 33X(7) + 0.9X(9)$$
$$+ 0.3X(10) + 0.3X(11) \geq 1.4.$$

$$0.8X(1) + 4.5X(2) + 60X(5) + 9X(6) + 33X(7) + 0.9X(9)$$
$$+ 0.3X(10) + 0.3X(11) \leq 1.6.$$

Chromium balance.

$$10X(2) + 20X(6) + 8X(7) \geq 0.3.$$

$$10X(2) + 20X(6) + 8X(7) \leq 0.45.$$

Other constraints.

$$X(8) \leq 0.01,$$
$$X(9) \leq 0.3,$$
$$X(10) \leq 0.3,$$
$$X(11) \leq 0.3,$$
$$X(9) + X(10) + X(11) \leq 0.3,$$
$$X(1) + X(2) + \ldots + X(11) = 1.0.$$

This approach is designed to handle only the "$\leq$" and "$=$" type constraints, so a slack variable must be subtracted from each of the "$\geq$" type constraints. The method automatically adds a slack variable to the "$\leq$" type constraints and an artificial variable to the "$=$" type constraints. This results in the following constraints:

$$4.2X(1) + 15X(8) + 0.4X(9) + 0.1X(10) + 0.1X(11) - X(12)$$
$$+ X(16) = 3.2.$$

$$4.2X(1) + 15X(8) + 0.4X(9) + 0.1X(10) + 0.1X(11) + X(17) = 3.5.$$

$$2.25X(1) + 15X(2) + 45X(3) + 42X(4) + 18X(5) + 30X(6)$$
$$+ 25X(7) + 30X(8) - X(13) + X(18) = 2.7.$$

$$2.25X(1) + 15X(2) + 45X(3) + 42X(4) + 18X(5) + 30X(6)$$
$$+ 25X(7) + 30X(8) + X(19) = 3.0.$$

$$0.8X(1) + 4.5X(2) + 60X(5) + 9X(6) + 33X(7) + 0.9X(9)$$
$$+ 0.3X(10) + 0.3X(11) - X(14) + X(20) = 1.4.$$

$$0.8X(1) + 4.5X(2) + 60X(5) + 9X(6) + 33X(7) + 0.9X(9)$$
$$+ 0.3X(10) + 0.3X(11) + X(21) = 1.6.$$

$$10X(2) + 20X(6) + 8X(7) - X(15) + X(22) = 0.3$$

$$10X(2) + 20X(6) + 8X(7) + X(23) = 0.45,$$

$$X(8) + X(24) = 0.01,$$

$$X(9) + X(25) = 0.3,$$

$$X(10) + X(26) = 0.3,$$

$$X(11) + X(27) = 0.3,$$

$$X(9) + X(10) + X(11) + X(28) = 0.3,$$

$$X(1) + X(2) + \cdots + X(11) + X(29) = 1.0,$$

where $X(1)$, $X(2)$, ..., $X(29) \geq 0$; $X(16)$, $X(18)$, $X(20)$, $X(22)$, and $X(29)$ are artificial variables; $X(12)$, ..., $X(15)$, $X(17)$, $X(19)$, $X(21)$, $X(23)$, ..., $X(28)$ are slack variables.

The objective function to be minimized is

$$P = 58X(1) + 120X(2) + 128X(3) + 120X(4) + 200X(5) + 260X(6)$$

$$+238X(7) + 160X(8) + 39X(9) + 30X(10) + 33X(11)$$

$$+ M(X(16) + X(18) + X(20) + X(22) + X(29)).$$

where M is some arbitrarily large value so as to force the artificial variables out of the solution. The cost coefficients of the slack variables are 0, as it costs nothing to not use part of a stockpile.

Symbol Table—Problem 10

Algebraic Symbol	Definition	FORTRAN Symbol
n	Number of variables in the original problem	N
m	Number of constraints	M
a	The $m \times n$ matrix of constraint coefficients	A
b	The m-element vector which contains the right-hand side of the constraint equations	B
c	The $n + m$ vector of cost coefficients	C
x	The $n + m$ vector of solutions	X
row_x	$n + m_1$ used for the maximum length of c	$ROWX$
col_t	$n + m + 1$, used by the subprogram for the number of columns in the full (augmented) tableau	$COLT$
m_1	$m + 1$, used by the subprogram for the number of rows in the full (augmented) tableau	$M1$
pt	A logical variable used to determine if all intermediate results are to be printed	PT
min	A logical variable used to determine if this is a minimization or maximization problem	MIN

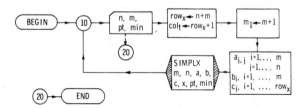

Flow Chart—Problem 10.   Flow diagram for computation of linear programming optimization.

The program requires as data the X coefficients, the composition limits, the other constraining limits, and the cost coefficients.

FORTRAN Listing—Problem 10

FORTRAN Listing—Main Program

```
       DIMENSION A(30,80),B(30),C(30),X(30)
       COMMON M1,COLT,ROWX
       INTEGER ROWX,COLT
       LOGICAL PT,MIN
   10  READ (5,500,END=20) N,M,PT,MIN
       ROWX=N+M
       COLT=ROWX+1
       M1=M+1
       READ (5,510) ((A(I,J),J=1,M),I=1,M),(B(I),I=1,M),(C(I),I=1,ROWX)
       CALL SIMPLX(M,N,A,B,C,X,PT,MIN)
       GO TO 10
   20  CALL SYSTEM
  500  FORMAT (2I4,2L3)
  510   FORMAT ((12F6.2))
  999  FORMAT (1H0)
       END
```

FORTRAN Listing—Subroutine SIMPLX

```
       SUBROUTINE SIMPLX(M,N,A,B,C,X,PT,MIN)
       REAL LIMIT
       INTEGER ROWS,COLT,ROWX,BASIS,BAS
       DIMENSION BASIS(30),A(30,80),B(30),C(30),X(30),TABLO(30,80)
       COMMON M1,COLT,ROWX
       LOGICAL PT,MIN
       IF (M.GT.30.OR.ROWX.GT.80) WRITE (6,600)
       DO 10 I=1,M1
       DO 10 J=1,COLT
   10  TABLO(I,J)=0.
       DO 20 I=1,M
       DO 30 J=1,N
   30  TABLO(I,J)=A(I,J)
       TABLO(I,N+I)=1.
       TABLO(I,COLT)=B(I)
       TABLO(M1,COLT)=TABLO(M1,COLT)+B(I)*C(N+I)
   20  BASIS(I)=N+I
       DO 40 J=1,N
       TABLO(M1,J)=-C(J)
       DO 40 I=1,M
```

```
40        TABLO(M1,J)=TABLO(M1,J)+TABLO(I,J)*C(N+I)
          WRITE (6,610)      ((TABLO(I,J),J=1,COLT),I=1,M1)
          IT=0
50        ZMINC=TABLO(M1,1)
          K=1
          DO 60 J=2,ROWX
          IF (TABLO(M1,J).LE.ZMINC) GO TO 60
          ZMINC=TABLO(M1,J)
          K=J
60        CONTINUE
          IF (ZMINC) 70,70,80
70        DO 90 I=1,ROWX
90        X(I)=0.
          DO 100 I=1,M
          BAS=BASIS(I)
100       X(BAS)=TABLO(I,COLT)
          TAB=TABLO(M1,COLT)
          IF (.NOT.MIN) TAB=-TAB
          WRITE (6,620) IT,(BASIS(I),I=1,M)
          WRITE (6,690) TAB
          WRITE (6,640) (I,X(I),I=1,ROWX)
          WRITE (6,650)      ((TABLO(I,J),J=1,COLT),I=1,M1)
          RETURN
80        DO 110 I=1,M
          J=I
          IF (TABLO(I,K).GT.0.) GO TO 120
110       CONTINUE
120       IF (J-M) 130,130,140
140       WRITE (6,680)
          GO TO 70
130       IF (TABLO(J,K).LE.0.) GO TO 140
          THET=TABLO(J,COLT)/TABLO(J,K)
          ROWS=J
          II=J+1
          DO 150 I=II,M
          IF (TABLO(I,K)) 150,150,160
160       THETI=TABLO(I,COLT)/TABLO(I,K)
          IF (THETI-THET) 170,150,150
170       THET=THETI
          ROWS=I
150       CONTINUE
          BASIS(ROWS)=K
          PIV=TABLO(ROWS,K)
          DO 180 J=1,COLT
180       TABLO(ROWS,J)=TABLO(ROWS,J)/PIV
          DO 190 I=1,M1
          LIMIT=TABLO(I,K)
          IF (LIMIT.EQ.0..OR.I.EQ.ROWS) GO TO 190
          DO 200 J=1,COLT
200       TABLO(I,J)=TABLO(I,J)-TABLO(ROWS,J)*LIMIT
190       CONTINUE
          IT=IT+1
          IF (.NOT.PT) GO TO 50
          DO 210 J=1,ROWX
210       X(J)=0.
          DO 220 I=1,M
          BAS=BASIS(I)
220       X(BAS)=TABLO(I,COLT)
          TAB=TABLO(M1,COLT)
          IF (.NOT.MIN) TAB=-TAB
          WRITE (6,630) IT, (BASIS(I),I=1,M)
          WRITE (6,691) TAB
          WRITE (6,660) (I,X(I),I=1,ROWX)
          WRITE (6,670)      ((TABLO(I,J),J=1,COLT),I=1,M1)
          GO TO 50
600       FORMAT ('1THE PROBLEM IS TOO LARGE.')
610       FORMAT ('1INITIAL TABLEAU....'/('0'/1X,10F10.3/5X,10F10.3/9X,11F1
         10.3))
620       FORMAT ('1THE RESULTS AND FINAL TABLEAU ARE:'/'0ITERATIONS REQUIRE
         1D TO ATTAIN SOLUTION ='I3/'0THE INDICES OF THE CURRENT BASIS VECTO
         2R ARE:'I6/(45X,I6))
```

```
630    FORMAT ('1THE CURRENT RESULTS AND TABLEAU ARE:'/'0ITERATIONS UP TO
       1 THIS POINT ='I3/'0THE INDICES OF THE CURRENT BASIS VECTOR ARE:'
       2I6/(45X,I6))
640    FORMAT ('0THE FINAL MIXTURE IS:  X('I2,') ='F8.3/(24X,'X('I2,') ='
       1F8.3/))
650    FORMAT ('0FINAL TABLEAU....'/(/'0'/1X,10F10.3/5X,10F10.3/9X,11F10.
       13))
660    FORMAT ('0THE CURRENT MIXTURE IS:  X('I2,') ='F8.3/(26X,'X('I2,')
       1='F8.3/))
670    FORMAT ('0CURRENT TABLEAU....'/(/'0'/1X,10F10.3/5X,10F10.3/9X,11F1
       10.3))
680    FORMAT ('1THE OBJECTIVE FUNCTION IS UNBOUNDED.  COMPUTATION HAS BE
       1EN TERMINATED.')
690    FORMAT (/' THE OPTIMUM SOLUTION IS:'F15.3)
691    FORMAT (/' THE CURRENT "OPTIMUM" SOLUTION IS:'F15.3)
       END
```

Input (Data)—Problem 10

```
 15   14   T   T
4.2    0.    0.    0.    0.    0.    0.    15.    .4    .1    .1   -1.
0.     0.    0.    4.2   0.    0.    0.    0.     0.    0.    15.   .4
.1     .1    0.    0.    0.    0.    2.25  15.    45.   42.   18.   30.
25.    30.   0.    0.    0.    0.   -1.    0.     0.    2.25  15.   45.
42.    18.   30.   25.   30.   0.    0.    0.     0.    0.    0.    0.
.8     4.5   0.    0.    60.   9.    33.   0.     .9    .3    .3    0.
0.    -1.    0.    .8    4.5   0.    0.    60.    9.    33.   0.    .9
.3     .3    0.    0.    0.    0.    0.    10.    0.    0.    0.    20.
8.     0.    0.    0.    0.    0.    0.    0.    -1.    0.    10.   0.
 0.    0.    20.   8.    0.    0.    0.    0.     0.    0.    0.    0.
0.     0.    0.    0.    0.    0.    0.    1.     0.    0.    0.    0.
0.     0.    0.    0.    0.    0.    0.    0.     0.    0.    0.    1.
0.     0.    0.    0.    0.    0.    0.    0.     0.    0.    0.    0.
0.     0.    0.    1.    0.    0.    0.    0.     0.    0.    0.    0.
0.     0.    0.    0.    0.    0.    0.    1.     0.    0.    0.    0.
0.     0.    0.    0.    0.    0.    0.    0.     1.    1.    1.    0.
0.     0.    0.    1.    1.    1.    1.    1.     1.    1.    1.    1.
1.     1.    0.    0.    0.    0.    3.2   3.5    2.7   3.    1.4   1.6
.3     .45   .01   .3    .3    .3    .3    1.     58.   120.  128.  120.
200.   260.  238.  160.  39.   30.   33.   0.     0.    0.    0.    1000.
0.     1000. 0.    1000. 0.    1000. 0.    0.     0.    0.    0.    0.
1000.
```

Output (Results)—Problem 10

```
THE RESULTS AND FINAL TABLEAU ARE:

ITERATIONS REQUIRED TO ATTAIN SOLUTION = 10

THE INDICES OF THE CURRENT BASIS VECTOR ARE:      1
                                                 17
                                                  4
                                                 19
                                                  5
                                                 21
                                                  2
                                                 23
                                                  8
                                                 25
                                                 26
                                                 27
                                                 28
                                                 10

THE OPTIMUM SOLUTION IS:          56.240

THE FINAL MIXTURE IS:   X( 1) =   0.721
                        X( 2) =   0.030
                        X( 3) =   0.0
                        X( 4) =   0.003
                        X( 5) =   0.010
                        X( 6) =   0.0
                        X( 7) =   0.0
                        X( 8) =   0.010
                        X( 9) =   0.0
                        X(10) =   0.225
                        X(11) =   0.0
                        X(12) =   0.0
                        X(13) =   0.0
                        X(14) =   0.0
                        X(15) =   0.0
                        X(16) =   0.0
                        X(17) =   0.300
                        X(18) =   0.0
                        X(19) =   0.300
                        X(20) =   0.0
                        X(21) =   0.200
                        X(22) =   0.0
                        X(23) =   0.150
                        X(24) =   0.0
                        X(25) =   0.300
                        X(26) =   0.075
                        X(27) =   0.300
                        X(28) =   0.075
                        X(29) =   0.0
```

Discussion of Results—Problem 10

M, the cost of the artificial variables, was chosen as \$1000 for this problem. All of these variables were driven from the final tableau, i.e. took on zero values. As the calculation approached an optimum, in this case, minimum cost, the total cost of raw materials to produce a ton of cast iron decreased.

The computer output is not shown, as the matrix is very large (30×15). Converted to the original variables, the solution is:

Material	w/o
Pig iron	= 72.1
Silvery pig	= 3.0
Ferrosilicon #1	= 0
Ferrosilicon #2	= 0.3
Master alloy #1	= 1.0
Master alloy #2	= 0
Master alloy #3	= 0
Silicon carbide	= 1.0
Steel scrap #1	= 0
Steel scrap #2	= 22.5
Steel scrap #3	= 0

Cost per ton of charge = $56.24.

This program computes the optimum charge to a melting unit on the basis of cost of raw materials. Electric arc and induction melting operations can be monitored similarly. The use of continually up-dated cost information in conjunction with an optimization program of this type can be a valuable operator guide in charge control of melting furnaces.

INDEX

ROBERT DONALD PEHLKE is a Fulbright Scholar who has authored over 80 publications. Currently a professor of metallurgical engineering at The University of Michigan in Ann Arbor and a Registered Professional Engineer, he has been employed in metallurgical research at both the General Motors Research Laboratory and the Ford Scientific Laboratory, and he has been a consultant for numerous firms. Active in the field, Dr. Pehlke lectures throughout the U.S.A. and is a member of AIME, ASM, ASEE, the Iron and Steel Institute in both London and Japan, Verein Deutscher Eisenhüttenleute, Sigma Xi, and Tau Beta Pi. Dr. Pehlke received his B.S.E. in Metallurgical Engineering at The University of Michigan and his S.M. and Sc.D. in Metallurgy at the Massachusetts Institute of Technology.

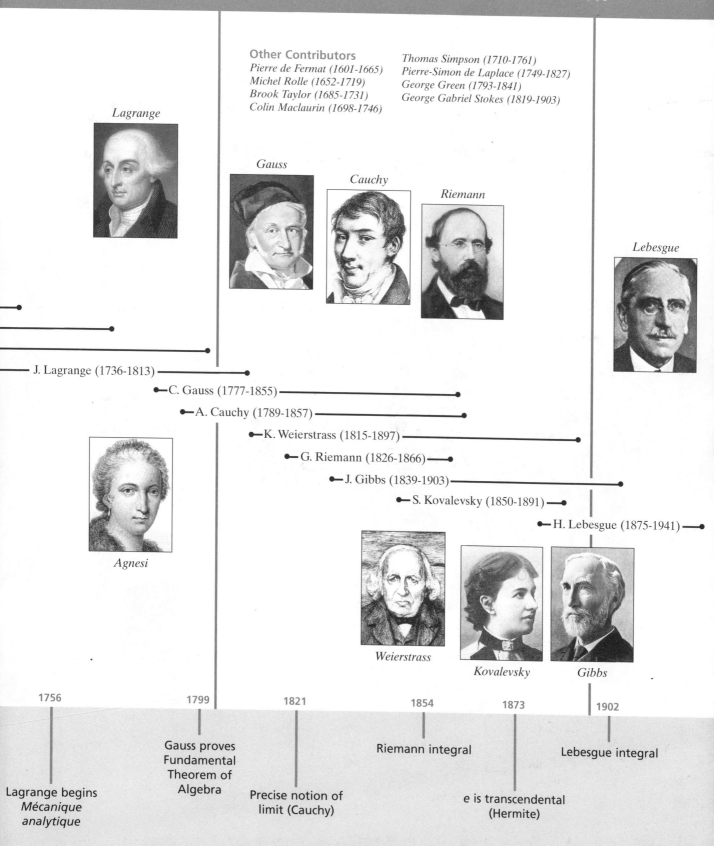

1800

1900

Lagrange

Other Contributors
Pierre de Fermat (1601-1665)
Michel Rolle (1652-1719)
Brook Taylor (1685-1731)
Colin Maclaurin (1698-1746)

Thomas Simpson (1710-1761)
Pierre-Simon de Laplace (1749-1827)
George Green (1793-1841)
George Gabriel Stokes (1819-1903)

Gauss

Cauchy

Riemann

Lebesgue

J. Lagrange (1736-1813)

C. Gauss (1777-1855)

A. Cauchy (1789-1857)

K. Weierstrass (1815-1897)

G. Riemann (1826-1866)

J. Gibbs (1839-1903)

S. Kovalevsky (1850-1891)

H. Lebesgue (1875-1941)

Agnesi

Weierstrass

Kovalevsky

Gibbs

1756

1799

1821

1854

1873

1902

Gauss proves
Fundamental
Theorem of
Algebra

Riemann integral

Lebesgue integral

Lagrange begins
*Mécanique
analytique*

Precise notion of
limit (Cauchy)

e is transcendental
(Hermite)

Calculus
FOR BIOLOGY & MEDICINE

CLAUDIA NEUHAUSER

Prepared Exclusively for
the University of California, Davis Mathematics Department

Taken from:
Calculus for Biology and Medicine, Second Edition
by Claudia Neuhauser

Taken from:

Calculus for Biology and Medicine, Second Edition
by Claudia Neuhaser
Copyright © 2004, 2000 by Pearson Education, Inc.
Published by Prentice-Hall, Inc.
Upper Saddle River, New Jersey 07458

This special edition published in cooperation with Pearson Custom Publishing.

Printed in the United States of America

10 9 8 7 6 5 4

ISBN 0-536-97558-2

2005360526

KK

Please visit our web site at *www.pearsoncustom.com*

PEARSON CUSTOM PUBLISHING
75 Arlington Street, Suite 300, Boston, MA 02116
A Pearson Education Company

CONTENTS

Preface xiii

6 INTEGRATION 335

PREFACE

When the first edition of this book appeared three years ago, I immediately started thinking about the second edition. Topics absent from the first edition were needed in a calculus book for life science majors: difference equations, extrema for functions of two variables, optimization under constraints, and expanded probability theory. I also wanted to add more biological examples, in particular in the first half of the book, and add more problems (the number of problems in many sections doubled or tripled compared with the first edition).

Despite these changes, the goals of the first edition remain: To model and analyze phenomena in the life sciences using calculus. In a traditional calculus course, biology students rarely see why the material is relevant to their training. This text is written exclusively for students in the biological and medical sciences. It makes an effort to show them from the beginning how calculus can help to understand phenomena in nature.

This text differs from traditional calculus texts. **First**, it is written in a life science context; concepts are motivated with biological examples to emphasize that calculus is an important tool in the life sciences. The second edition has many more biological examples than the first edition, particularly in the first half of the book. **Second**, difference equations are now extensively treated in the book. They are introduced in Chapter 2, where they are accessible to calculus students without a knowledge of calculus and provide an easier entrance to population models than differential equations. They are picked up again in Chapters 5 and 10, where they receive a more formal treatment using calculus. **Third**, differential equations, one of the most important modeling tools in the life sciences, are introduced early, immediately after the formal definition of derivatives in Chapter 4. Two chapters deal exclusively with differential equations and systems of differential equations; both chapters contain numerous up-to-date applications. **Fourth**, biological applications of differentiation and integration are integrated throughout the text. **Fifth**, multivariable calculus is taught in the first year, recognizing that most students in the life sciences will not take the second year of calculus and that multivariable calculus is needed to analyze systems of difference and differential equations, which students encounter later in their science courses. The chapter on multivariable calculus now has a treatment of extrema and Lagrange multipliers.

This text does not teach modeling; the objective is to teach calculus. Modeling is an art that should be taught in a separate course. However, throughout the text, students encounter mathematical models for biological phenomena. This will facilitate the transition to actual modeling and allows them to see how calculus provides useful tools for the life sciences.

Examples   Each topic is motivated with biological examples. This is followed by a thorough discussion outside of the life science context to enable students to become familiar with both the meaning and the mechanics of the topic. Finally, biological examples are given to teach students how to use the material in a life science context.

Examples in the text are completely worked out; steps in calculations are frequently explained in words.

Problems   Calculus cannot be learned by watching someone do it. This is recognized by providing the students with both drill and word problems. Word problems are an integral part of teaching calculus in a life science context. The word problems are up to date; they are adapted from either standard biology texts or original research. Many new problems have been added to the second edition. Since this text is written for college freshmen, the examples were chosen so that no formal training in biology is needed.

Technology   The book takes advantage of graphing calculators. This allows students to develop a much better visual understanding of the concepts in calculus. Beyond this, no special software is required.

CHAPTER SUMMARY

Chapter 1   Basic tools from algebra and trigonometry are summarized in Section 1.1. Section 1.2 contains the basic functions used in this text, including exponential and logarithmic functions. Their graphical properties and their biological relevance are emphasized. Section 1.3 covers log-log and semilog plots; these are graphical tools that are frequently used in the life sciences. In addition, a section on translating verbal descriptions of biological phenomena into graphs will provide students with much needed skills when they read biological literature.

Chapter 2   This chapter was added to the second edition. It covers difference equations (or discrete time models) and sequences. This provides a more natural way to explain the need for limits. Classical models of population growth round up this chapter; this gives students a first glimpse at the excitement of using models to understand biological phenomena.

Chapter 3   Limits and continuity are key concepts for understanding the conceptual parts of calculus. Visual intuition is emphasized before the theory is discussed. The formal definition of limits is now at the end of the chapter and can be omitted.

Chapter 4   The geometric definition of a derivative as the slope of a tangent line is given before the formal treatment. After the formal definition of the derivative, differential equations are introduced as models for biological phenomena. Differentiation rules are discussed. These sections give students time to acquaint themselves with the basic rules of differentiation before applications are discussed. Related rates and error propagation, in addition to differential equations, are the main applications.

Chapter 5   This chapter presents biological and more traditional applications of differentiation. Many of the applications are consequences of the mean value theorem. Many of the word problems are adapted from either biology textbooks or original research articles; this puts the traditional applications (such as extrema, monotonicity, and concavity) in a biological context. A section on analyzing difference equations is added.

Chapter 6   Integration is motivated geometrically. The fundamental theorem of calculus and its consequences are discussed in depth. Both biological and traditional applications of integration are provided before integration techniques are covered.

Chapter 7   This chapter contains integration techniques. However, only the most important techniques are covered. Tables of integrals are used to integrate more complicated integrals. The use of computer software is not integrated in the book, though their usefulness in evaluating integrals is acknowledged. This chapter also contains a section on Taylor polynomials.

Chapter 8   This chapter provides an introduction to differential equations. The treatment is not complete, but it will equip students with both analytical and graphical skills to analyze differential equations. Eigenvalues are introduced early to facilitate the analytical treatment of systems of differential equations in Chapter 11. Many of the differential equations discussed in the text are important models in biology. Although this text is not a modeling text, students will see how differential equations are used to model biological phenomena and will be able to interpret differential equations. Chapter 8 contains a large number of up-to-date applications of differential equations in biology.

Chapter 9   Matrix algebra is an indispensable tool for every life scientist. The material in this chapter covers the most basic concepts; it is tailored to Chapters 10 and 11, where matrix algebra is frequently used. Special emphasis is given to the treatment of eigenvalues and eigenvectors because of their importance in analyzing systems of differential equations.

Chapter 10   This is an introduction to multidimensional calculus. The treatment is brief and tailored to Chapter 11, where systems of differential equations are discussed. The main topics are partial derivatives and linearization of vector-valued functions. The discussion of gradient and diffusion is not needed for Chapter 11. A section on extrema and Lagrange multipliers is added, which is also not required for Chapter 11. If difference equations were covered early in a course, the final section in this chapter will provide an introduction to systems of difference equations with many biological examples.

Chapter 11   This material is most relevant for students in the life sciences. Both graphical and analytical tools are developed to enable students to analyze systems of differential equations. The material is divided into linear and nonlinear systems. Understanding the stability of linear systems in terms of vector fields, eigenvectors, and eigenvalues helps students to master the more difficult analysis of nonlinear systems. Theory is explained before applications are given—this allows students to become familiar with the mechanics before delving into applications. An extensive problem set allows students to experience the power of this modeling tool in a biological context.

Chapter 12   This chapter contains some basic probabilistic and statistical tools. It is greatly expanded compared with the first edition; in particular, Chapter 8 of the first edition is now incorporated into Chapter 12. It cannot replace a full-semester course in probability and statistics but allows students to see early some of the concepts needed in population genetics and experimental design.

HOW TO USE THIS BOOK

This book contains more material that can be covered in one year. This is to allow for more flexibility in the choice of material. Sections that are labeled optional can be omitted; their material is not needed in subsequent sections.

The material can be arranged so that the course can be taught as a one-semester, two-quarter, two-semester, four-quarter, or three-semester course. Chapters 1–4 must be covered in that order before any of the other sections are covered. In addition to Chapters 1–4, the following sections can be chosen:

One semester 5.1, 5.2, 5.3, 5.4, 5.5, 5.6, 5.8, 6.1, 6.2, 6.3 (without 6.3.4 and 6.3.5)

One semester 5.1, 5.2, 5.3, 5.4, 5.5, 5.6, 5.8, 6.1, 6.2, 8.2 (without solving any of the differential equations)

Two quarters 5.1, 5.2, 5.3, 5.4, 5.5, 5.6, 5.8, 6.1, 6.2, 6.3 (without 6.3.4 and 6.3.5), Chapter 7, Chapter 8

Two semesters 5.1, 5.2, 5.3, 5.4, 5.5, 5.6, 5.8, 6.1, 6.2, 6.3, Chapters 7, 8, and 9, 10.1, 10.2, 10.3, 10.4, 10.7, 11.1, 11.2, 11.3, 11.4 (two of the subsections)

Four quarters or three semesters All sections that are not labeled optional; optional sections should be chosen as time permits

One semester — probability emphasis Chapter 1, only Section 2.2.1 and 2.2.2 in Chapter 2, Chapter 3 (except 3.6), Chapter 4, 5.1, 5.2, 5.3, 5.4, 5.8, 6.1, 6.2, 12.1, 12.2, 12.3, 12.4, 12.5, 12.6

ACKNOWLEDGMENTS

This book would not have been possible without the help of numerous people. The book greatly benefited from critical reviews of an earlier draft.

Andrea Brose
University of California, Los Angeles

Shandelle Henson
Andrews University

Yang Kuang
Arizona State University

En-Bing Lin
University of Toledo

Jennifer Mueller
Colorado State University

Bruce Peckham
University of Minnesota, Duluth

Scott Rimbey
University of South Florida

Roberto Schonmann
University of California, Los Angeles

Christopher Sogge
Johns Hopkins University

Glenn Tesler
University of California, San Diego

Paul Van Steenberghe
University of Maine

Nathaniel Whitaker
University of Massachusetts, Amherst

Sarah Witherspoon
Amherst College

Thanks to all my students who took this course over the years at the University of Minnesota and the University of California–Davis for their constructive criticism and enthusiasm. I owe special thanks to George Lobell, my former mathematics editor at Prentice Hall, who made this book possible in the first place; Eric Frank, my current math editor; Dawn Murrin, associate editor and Jeanne Audino, the math production editor all deserve a big thank you for getting the second edition done in a timely manner. Finally, I wish to thank my husband, Maury Bramson, who again patiently put up with my long hours of working on the second edition of the book.

Claudia Neuhauser
cneuhaus@cbs.umn.edu
Minneapolis, Saint Paul, Minnesota

PREVIEW AND REVIEW

1

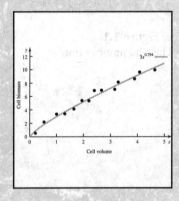

Isaac Newton (1642–1727) and Gottfried Wilhelm Leibniz (1646–1716) are typically credited with the invention of calculus. They were not the sole inventors but rather the first to develop it systematically.

Calculus has two parts, differential and integral calculus. Historically, differential calculus was concerned with finding tangent lines to curves and extrema (i.e., maxima and minima) of curves. Integral calculus, on the other hand, has its roots in attempts to determine the areas of regions bounded by curves, or the volumes of solids. The two parts of calculus are closely related: The basic operation of one can be considered the inverse of the other. This result is known as the fundamental theorem of calculus and goes back to Newton and Leibniz. They were the first to understand the meaning of this inverse relationship and to put this relationship to use in solving difficult problems.

Finding tangents, extrema, and areas is a basic geometric problem, and it is maybe somewhat surprising that the solution of this problem led to the development of methods that are useful in a wide range of scientific problems. The main reason for this is that the slope of a tangent line at a given point is related to how quickly the function changes at this point. Knowing how quickly a function changes at a point opens up the possibility of a dynamic description of biology, such as population growth, the speed at which a chemical reaction proceeds, the firing rate of neurons, and the speed at which an invasive species invades new habitat. For this reason, calculus has been one of the most powerful tools in the mathematical formulation of scientific concepts. Applications of calculus are not restricted to biology. In fact, physics was the driving force in developing calculus originally. In this text, however, we will be primarily concerned with how calculus is used in biology.

In addition to developing the theory of differential and integral calculus, we will consider many examples in which calculus is used to describe or model situations in the biological sciences. The use of quantitative reasoning is becoming increasingly more important in biology—for instance, when modeling interactions between species in a community, describing neuron activities, explaining genetic diversity in populations, and predicting the impact of global warming on vegetation. Calculus (Chapters 1–11) and probability and statistics (Chapter 12) are among the most important quantitative tools of a biologist.

1.1 PRELIMINARIES

We provide a brief review of some of the concepts and techniques from precalculus that are frequently used in calculus. (The problems at the end of this section will help you to reacquaint yourself with this material.)

1.1.1 The Real Numbers

The **real numbers** can most easily be visualized on the **real number line** (see Figure 1.1).

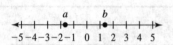

▲ **Figure 1.1**
The real number line

Numbers are ordered on the real number line so that if $a < b$, then a is to the left of b. Sets (collections) of real numbers are typically denoted by capital letters A, B, or alike. To describe the set A, we write

$$A = \{x : \text{condition}\}$$

where "condition" tells us which numbers are in the set A. The most important sets in calculus are **intervals**. We use the following notations: If $a < b$, then

the **open** interval $(a, b) = \{x : a < x < b\}$

and

the **closed** interval $[a, b] = \{x : a \leq x \leq b\}$

We also use **half-open** intervals

$$[a, b) = \{x : a \leq x < b\} \qquad \text{and} \qquad (a, b] = \{x : a < x \leq b\}$$

Unbounded intervals are sets of the form $\{x : x > a\}$. Here are the possible cases:

$$[a, \infty) = \{x : x \geq a\}$$
$$(-\infty, a] = \{x : x \leq a\}$$
$$(a, \infty) = \{x : x > a\}$$
$$(-\infty, a) = \{x : x < a\}$$

The symbols "∞" and "$-\infty$" mean "plus infinity" and "minus infinity"; these are *not* real numbers but are merely used for notational convenience. The real number line, denoted by $\mathbf{R}$, does not have endpoints, and we can write $\mathbf{R}$ in the following equivalent forms:

$$\mathbf{R} = \{x : -\infty < x < \infty\} = (-\infty, \infty)$$

The location of the number 0 on the real number line is called the **origin**, and we can measure the distance of the number x to the origin. For instance, -5 is 5 units to the left of the origin. There is a convenient notation for measuring distances from the origin on the real number line, namely, the absolute value of a real number.

> Definition The **absolute value** of a real number a, denoted by $|a|$, is
> $$|a| = \begin{cases} a & \text{if } a \geq 0 \\ -a & \text{if } a < 0 \end{cases}$$

For example, $|-7| = -(-7) = 7$. We can use absolute values to find the distance between any two numbers x_1 and x_2; namely,

$$\text{distance between } x_1 \text{ and } x_2 = |x_1 - x_2|$$

Note that $|x_1 - x_2| = |x_2 - x_1|$. To find the distance between -2 and 4, we compute $|-2 - 4| = |-6| = 6$, or $|4 - (-2)| = |4 + 2| = 6$.

We will frequently need to solve equations containing absolute values, for which the following property is useful. Assume $b \geq 0$.

---

1. For $a \geq 0$, $|a| = b$ is equivalent to $a = b$.

2. For $a < 0$, $|a| = b$ is equivalent to $-a = b$.

▶ **Example 1** Solve $|x - 4| = 2$.

Solution

If $x - 4 \geq 0$, then $x - 4 = 2$ and thus $x = 6$. If $x - 4 < 0$, then $-(x - 4) = 2$ and thus $x = 2$. The solutions are therefore $x = 6$ and $x = 2$. This can be illustrated graphically (see Figure 1.2). The points of intersection of $y = |x - 4|$ and $y = 2$ are at $x = 6$ and $x = 2$. Solving $|x - 4| = 2$ can also be interpreted as finding the two numbers that have distance 2 from 4. ◀

If we need to solve an equation of the form $|a| = |b|$, then we write this as either $a = b$ or $a = -b$, which is illustrated in the next example.

▶ **Example 2** Solve $\left|\frac{3}{2}x - 1\right| = \left|\frac{1}{2}x + 1\right|$.

Solution

Either

$$\frac{3}{2}x - 1 = \frac{1}{2}x + 1$$
$$x = 2$$

or

$$\frac{3}{2}x - 1 = -\left(\frac{1}{2}x + 1\right)$$
$$\frac{3}{2}x - 1 = -\frac{1}{2}x - 1$$
$$2x = 0$$
$$x = 0$$

A graphical solution of this example is shown in Figure 1.3. ◀

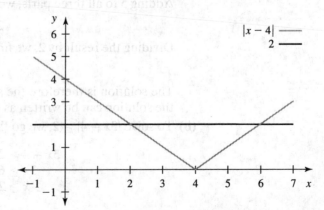

▲ **Figure 1.2**

The graphs of $y = |x - 4|$ and $y = 2$. The points of intersection are at $x = 6$ and $x = 2$

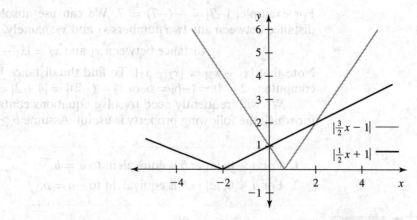

▲ **Figure 1.3**
The graphs of $y = |\frac{3}{2}x - 1|$ and $y = |\frac{1}{2}x + 1|$. The points of intersection are at $x = 0$ and $x = 2$

Returning to Example 1, where we found the two points whose distance from 4 was equal to 2, we can also try to find those points whose distance from 4 is less than (or greater than) 2. This amounts to solving inequalities with absolute values. Looking back at Figure 1.2, we see that the set of x values whose distance from 4 is less than 2 (i.e., $|x - 4| < 2$) is the interval $(2, 6)$. Similarly, the set of x values whose distance from 4 is greater than 2 (i.e., $|x - 4| > 2$) is the union of the two intervals $(-\infty, 2) \cup (6, \infty)$.

In general, to solve absolute value inequalities, the following two properties are useful. Assume $b > 0$.

---

1. $|a| < b$ is equivalent to $-b < a < b$.

2. $|a| > b$ is equivalent to $a > b$ or $a < -b$.

▶ **Example 3** (a) Solve $|2x - 5| < 3$. (b) Solve $|3x - 4| \geq 2$.

Solution

(a) We rewrite $|2x - 5| < 3$ as

$$-3 < 2x - 5 < 3$$

Adding 5 to all three parts, we obtain

$$2 < 2x < 8$$

Dividing the result by 2, we find

$$1 < x < 4$$

The solution is therefore the set $\{x : 1 < x < 4\}$. In interval notation, the solution can be written as the open interval $(1, 4)$.

(b) To solve $|3x - 4| \geq 2$, we go through the following steps:

$$
\begin{aligned}
3x - 4 &\geq 2 & & & 3x - 4 &\leq -2 \\
3x &\geq 6 & &\text{or} & 3x &\leq 2 \\
x &\geq 2 & & & x &\leq \frac{2}{3}
\end{aligned}
$$

The solution is the set $\{x : x \geq 2 \text{ or } x \leq \frac{2}{3}\}$, or, in interval notation, $(-\infty, \frac{2}{3}] \cup [2, \infty)$. ◀

1.1.2 Lines in the Plane

We will frequently encounter situations where the relationship between quantities can be described by a **linear equation**. For example, when a weight is attached to a helical spring made of some elastic material (and the weight is not too heavy), the relationship between the length y of the spring and the weight x is

$$y = y_0 + kx \tag{1.1}$$

Here, y_0 denotes the length of the spring when no weight is attached to it and k is a positive constant. Equation (1.1) is an example of a linear equation, and we say that x and y satisfy a linear equation.

The **standard** form of a linear equation is given by

$$Ax + By + C = 0$$

where A, B, and C are constants, A and B are not both equal to 0, and x and y are the two variables. In algebra, you learned that the graph of a linear equation is a straight line.

If you know two points (x_1, y_1) and (x_2, y_2) on a straight line, then the **slope** of the line is

$$m = \frac{y_2 - y_1}{x_2 - x_1}$$

This is illustrated in Figure 1.4. Two points (or one point and the slope) are sufficient to determine the equation of a straight line.

If you are given one point and the slope, the **point-slope** form of a straight line can be used to write down its equation. It is given by

$$y - y_0 = m(x - x_0)$$

where m is the slope and (x_0, y_0) is a point on the line. If you are given two points, first compute the slope, then use one of the points and the slope to find the equation of the straight line in point-slope form.

Lastly, the most frequently used form of a linear equation is the **slope-intercept** form

$$y = mx + b$$

where m is the slope and b is the y-intercept, which is the point of intersection of the line with the y-axis; it has coordinates $(0, b)$.

We summarize these three forms of linear equations in the following box.

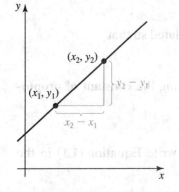

▲ **Figure 1.4**
The slope of a straight line

$$
\begin{array}{ll}
Ax + By + C = 0 & \text{(Standard Form)} \\
y - y_0 = m(x - x_0) & \text{(Point-Slope Form)} \\
y = mx + b & \text{(Slope-Intercept Form)}
\end{array}
$$

▶ **Example 4** Determine the equation of the line passing through $(-2, 1)$ and $(3, -\frac{1}{2})$ in slope-intercept form.

Solution

The slope of the line is

$$m = \frac{y_2 - y_1}{x_2 - x_1} = \frac{-\frac{1}{2} - 1}{3 - (-2)} = \frac{-\frac{3}{2}}{5} = -\frac{3}{10}$$

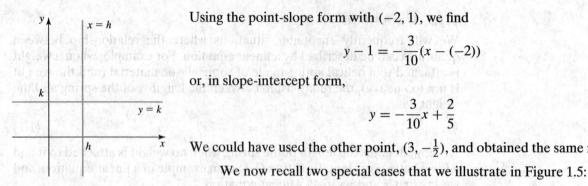

▲ **Figure 1.5**
The horizontal line $y = k$ and
the vertical line $x = h$

Using the point-slope form with $(-2, 1)$, we find

$$y - 1 = -\frac{3}{10}(x - (-2))$$

or, in slope-intercept form,

$$y = -\frac{3}{10}x + \frac{2}{5}$$

We could have used the other point, $(3, -\frac{1}{2})$, and obtained the same result. ◀

We now recall two special cases that we illustrate in Figure 1.5:

$y = k$	horizontal line (slope 0)
$x = h$	vertical line (slope undefined)

In the next example, we show how to determine the slope and the y-intercept of a given straight line.

▶ **Example 5**   Determine the slope and the y-intercept of the line $3y - 2x + 9 = 0$.

Solution

We solve for y : $3y = 2x - 9$ and thus $y = \frac{2}{3}x - 3$. We can now read off the slope $m = \frac{2}{3}$ and the y-intercept $b = -3$. ◀

When two quantities x and y are linearly related so that

$$y = mx$$

we say that y is **proportional** to x, with m denoting the **constant of proportionality**, and we write

$$y \propto x$$

The symbol $\propto$ is read "proportional to." If we write Equation (1.1) in the form

$$y - y_0 = kx$$

then the change in length $y - y_0$ is proportional to the attached weight with constant of proportionality k, and we can write

$$y - y_0 \propto x$$

There are two more properties of straight lines we wish to mention. When two lines l_1 and l_2 in the plane have no points in common, they are called **parallel**, denoted by $l_1 \parallel l_2$. The following criterion is useful for deciding whether two lines are parallel: Two noncoincident lines l_1 and l_2 are parallel ($l_1 \parallel l_2$) if and only if their slopes are identical. For two noncoincident, nonvertical lines l_1 and l_2 with slopes m_1 and m_2, respectively, the criterion becomes

$$l_1 \parallel l_2 \quad \text{if and only if} \quad m_1 = m_2$$

Two lines l_1 and l_2 are called **perpendicular** ($l_1 \perp l_2$) if their intersection forms an angle of $90°$. The following criterion is useful for deciding whether two

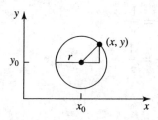

▲ Figure 1.6
Circle with radius r centered at (x_0, y_0)

lines are perpendicular: Two nonvertical lines are perpendicular if and only if their slopes are negative reciprocals. That is, if l_1 and l_2 are nonvertical lines with slopes m_1 and m_2, then

$$l_1 \perp l_2 \quad \text{if and only if} \quad m_1 m_2 = -1$$

We will prove this result in Problem 54 at the end of this section.

1.1.3 Equation of the Circle

A **circle** is the set of all points at a given distance, called the **radius**, from a given point, called the **center**. If r is the distance from (x_0, y_0) to (x, y) (see Figure 1.6), then using the Pythagorean theorem,

$$r^2 = (x - x_0)^2 + (y - y_0)^2$$

If $r = 1$ and $(x_0, y_0) = (0, 0)$, the circle is called the **unit circle**.

▶ **Example 6**   Find the equation of the circle with center $(2, 3)$ and passing through $(5, 7)$.

Solution

Using the Pythagorean theorem, we can compute the distance in the plane between $(2, 3)$ and $(5, 7)$, namely,

$$\sqrt{(5 - 2)^2 + (7 - 3)^2} = \sqrt{9 + 16} = 5$$

Thus the circle has radius 5 and center $(2, 3)$, and its equation is given by

$$25 = (x - 2)^2 + (y - 3)^2 \qquad \blacktriangleleft$$

1.1.4 Trigonometry

We will need a few results from trigonometry. Recall that angles are measured in either degrees or radians, and that a complete revolution on a unit circle corresponds to $360°$ or 2π. For reasons that will become clear, the radian measure is preferred in calculus. To convert between degree and radian measure, we use the formula

$$\frac{\theta \text{ measured in degrees}}{360°} = \frac{\theta \text{ measured in radians}}{2\pi}$$

For instance, to convert $23°$ into radian measure, we compute

$$\theta = 23° \frac{2\pi}{360°} = 0.401$$

To convert $\frac{\pi}{6}$ into degrees, we compute

$$\theta = \frac{\pi}{6} \frac{360°}{2\pi} = 30°$$

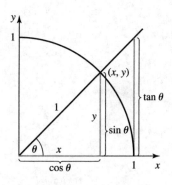

▲ Figure 1.7
The trigonometric functions on a unit circle

There are six trigonometric functions that you should be familiar with: sine, cosine, tangent, and secant (the other two, cotangent and cosecant, are rarely used). They are defined on a unit circle (see Figure 1.7) and abbreviated as sin, cos, tan, sec, cot, and csc. Recall that a positive angle is measured counterclockwise from the positive x-axis, whereas a negative angle is measured clockwise.

$$\sin\theta = \frac{y}{1} = y \qquad \csc\theta = \frac{1}{\sin\theta} = \frac{1}{y}$$

$$\cos\theta = \frac{x}{1} = x \qquad \sec\theta = \frac{1}{\cos\theta} = \frac{1}{x}$$

$$\tan\theta = \frac{y}{x} \qquad \cot\theta = \frac{1}{\tan\theta} = \frac{x}{y}$$

The following trigonometric identities are used frequently. Since $\tan\theta = y/x$ with $y = \sin\theta$ and $x = \cos\theta$, it follows that

$$\tan\theta = \frac{\sin\theta}{\cos\theta}$$

Applying the Pythagorean theorem to the triangle in Figure 1.7, we find [using the notation $\sin^2\theta = (\sin\theta)^2$]

$$\sin^2\theta + \cos^2\theta = 1$$

This identity is worth memorizing. With some algebra, we can derive another useful identity. Namely, if we divide the identity by $\cos^2\theta$, we find

$$\frac{\sin^2\theta}{\cos^2\theta} + 1 = \frac{1}{\cos^2\theta}$$

Using $\tan\theta = \sin\theta/\cos\theta$ and $\sec\theta = 1/\cos\theta$, this can be written as

$$\tan^2\theta + 1 = \sec^2\theta$$

In the next example, we will solve a trigonometric equation.

▶ **Example 7** Solve

$$2\sin\theta\cos\theta = \cos\theta \quad \text{on } [0, 2\pi)$$

Solution

Do not be tempted to cancel $\cos\theta$ on each side; this would cause you to lose solutions. Instead, bring $\cos\theta$ to the left side and factor $\cos\theta$. We obtain

$$\cos\theta(2\sin\theta - 1) = 0$$

that is,

$$\cos\theta = 0 \quad \text{or} \quad 2\sin\theta - 1 = 0$$

Solving $\cos\theta = 0$, we find

$$\theta = \frac{\pi}{2} \quad \text{or} \quad \theta = \frac{3\pi}{2}$$

Solving $2\sin\theta - 1 = 0$, we get

$$\sin\theta = \frac{1}{2}$$

which yields

$$\theta = \frac{\pi}{6} \quad \text{or} \quad \theta = \frac{5\pi}{6}$$

The solution set is therefore $\{\frac{\pi}{6}, \frac{\pi}{2}, \frac{5\pi}{6}, \frac{3\pi}{2}\}$. ◀

The following two identities follow from Figure 1.8 when we compare the two angles θ and $-\theta$. (A positive angle is measured counterclockwise

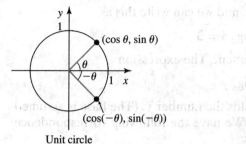

Figure 1.8
Using the unit circle to define trigonometric identities

TABLE 1.1 SOME EXACT TRIGONOMETRIC VALUES					
Angle θ	0	$\frac{\pi}{6}$	$\frac{\pi}{4}$	$\frac{\pi}{3}$	$\frac{\pi}{2}$
	(0°)	(30°)	(45°)	(60°)	(90°)
$\sin\theta$	$\frac{1}{2}\sqrt{0}$	$\frac{1}{2}\sqrt{1}$	$\frac{1}{2}\sqrt{2}$	$\frac{1}{2}\sqrt{3}$	$\frac{1}{2}\sqrt{4}$
$\cos\theta$	$\frac{1}{2}\sqrt{4}$	$\frac{1}{2}\sqrt{3}$	$\frac{1}{2}\sqrt{2}$	$\frac{1}{2}\sqrt{1}$	$\frac{1}{2}\sqrt{0}$

from the positive x-axis, whereas a negative angle is measured clockwise.)

$$\sin(-\theta) = -\sin\theta \quad \text{and} \quad \cos(-\theta) = \cos\theta$$

You should memorize certain exact trigonometric values; these are collected in Table 1.1. Of course, $\frac{1}{2}\sqrt{0} = 0$, $\frac{1}{2}\sqrt{1} = \frac{1}{2}$, and $\frac{1}{2}\sqrt{4} = 1$, and you should memorize the simplified values. Writing it as in Table 1.1 will make it easier to re-create the table in case you forget the exact values. Using $\tan\theta = \sin\theta/\cos\theta$, you immediately get the values for $\tan\theta$.

1.1.5 Exponentials and Logarithms

Exponentials and logarithms are particularly important in biological contexts. An exponential is an expression of the form

$$a^r$$

where a is called the **base** and r the **exponent**. Unless r is an integer or a rational number of the form p/q, where p is an integer and q is an odd integer, we will assume that a is positive. We summarize some of the properties of an exponential.

$$a^r a^s = a^{r+s}$$

$$\frac{a^r}{a^s} = a^{r-s}$$

$$a^{-r} = \frac{1}{a^r}$$

$$(ab)^r = a^r b^r$$

$$\left(\frac{a}{b}\right)^r = \frac{a^r}{b^r}$$

$$(a^r)^s = a^{rs}$$

▶ **Example 8** Evaluate the following exponential expressions:

(a)
$$3^2 3^{5/2} = 3^{2+5/2} = 3^{9/2}$$

(b)
$$\frac{2^{-4} 2^3}{2^2} = \frac{2^{-1}}{2^2} = 2^{-1-2} = 2^{-3} = \frac{1}{2^3} = \frac{1}{8}$$

(c)
$$\frac{a^k a^{3k}}{a^{5k}} = a^{k+3k-5k} = a^{-k} = \frac{1}{a^k}$$

Logarithms allow us to solve equations of the form

$$2^x = 8$$

The solution of this equation is $x = 3$, and we can write this as

$$x = \log_2 8 = 3$$

In other words, a logarithm is an exponent. The expression

$$\log_a y$$

is the exponent on the base a that yields the number y. (The base is assumed to be positive and different from 1.) We have the following correspondence between logarithms and exponentials:

$$x = \log_a y \quad \text{is equivalent to} \quad y = a^x$$

▶ Example 9   Which real number x satisfies

(a) $\log_3 x = -2$?

(b) $\log_{1/2} 8 = x$?

Solution

(a) We write this in the equivalent form:

$$x = 3^{-2}$$

Hence,

$$x = \frac{1}{3^2} = \frac{1}{9}$$

(b) We write this in the equivalent form:

$$\left(\frac{1}{2}\right)^x = 8$$

$$2^{-x} = 2^3$$

$$2^x = 2^{-3}$$

Setting the exponents equal to each other, we find $x = -3$. Note that in order to compare exponents, their bases must be the same.   ◀

Some important properties for logarithms are as follows:

$$\log_a(xy) = \log_a x + \log_a y$$

$$\log_a\left(\frac{x}{y}\right) = \log_a x - \log_a y$$

$$\log_a x^r = r \log_a x$$

The most important logarithm is the **natural logarithm**, which has the number e as its base. The number e is an irrational number whose value is approximately 2.7182818. The natural logarithm is written $\ln x$; that is, $\log_e x = \ln x$.

▶ Example 10   Simplify the following expressions.

(a) $\log_3(9x^2) = \log_3 9 + \log_3 x^2 = 2 + 2\log_3 x$

(b) $\log_5 \frac{x^2+3}{5x} = \log_5(x^2 + 3) - \log_5 5 - \log_5 x = \log_5(x^2 + 3) - 1 - \log_5 x$

(Note that $\log_5(x^2 + 3)$ cannot be simplified any further.)

(c) $-\ln\frac{1}{2} = \ln(\frac{1}{2})^{-1} = \ln 2$

(d) $\ln\frac{3x^2}{\sqrt{y}} = \ln 3 + \ln x^2 - \ln\sqrt{y} = \ln 3 + 2\ln x - \frac{1}{2}\ln y$

(In the last step, we used the fact that $\sqrt{y} = y^{1/2}$.)

In algebra, you learned how to solve equations of the form $e^{2x} = 3$ or $\ln(x + 1) = 5$. We will need to do this frequently. The key to solving such equations are the two identities

$$\log_a a^x = x \quad \text{and} \quad a^{\log_a x} = x$$

The following example illustrates how to use these identities.

▶ **Example 11** Solve for x.

(a) $e^{2x} = 3$

(b) $\ln(x + 1) = 5$

(c) $5^{2x-1} = 2^x$

Solution

(a) To solve $e^{2x} = 3$, we take logarithms to base e on both sides,

$$\ln e^{2x} = \ln 3$$

But $\ln e^{2x} = 2x$; hence,

$$2x = \ln 3 \quad \text{or} \quad x = \frac{1}{2}\ln 3$$

(b) To solve $\ln(x + 1) = 5$, we write the equation in exponential form,

$$e^{\ln(x+1)} = e^5$$

which simplifies to

$$x + 1 = e^5 \quad \text{or} \quad x = e^5 - 1$$

(c) To solve $5^{2x-1} = 2^x$ for x, we observe that the two bases are different. We therefore cannot compare the exponents directly. Instead, we take logarithms on both sides. Any positive base (different from 1) for the logarithm would work and we choose base e since it is the most commonly used base in calculus. This yields

$$\ln 5^{2x-1} = \ln 2^x$$

or, after simplifying,

$$(2x - 1)\ln 5 = x\ln 2$$

Solving for x, we find

$$2x\ln 5 - x\ln 2 = \ln 5$$

$$x(2\ln 5 - \ln 2) = \ln 5$$

Hence,

$$x = \frac{\ln 5}{2\ln 5 - \ln 2}$$

◀

1.1.6 Complex Numbers and Quadratic Equations

The square of a real number is always nonnegative. However, there are situations where we need to take a square root of a negative number. Since this cannot be a real number, we introduce a new symbol, which we denote by i, that will allow us to deal with this case. We set

$$i^2 = -1$$

The symbol i is called the **imaginary unit**. Instead of writing $\sqrt{-17}$, for instance, we can now write $i\sqrt{17}$.

The symbol i allows us to introduce a new number system, the set of **complex numbers**.

A **complex number** is of the form

$$z = a + bi$$

where a and b are real numbers. The real number a is the **real part** of $a + bi$, and the real number b is the **imaginary part**.

For instance, $-3 + 7i$ has real part -3 and imaginary part 7; $2 - 5i$ has real part 2 and imaginary part -5. Since $a + 0i = a$, we see that all real numbers are contained in the set of complex numbers. Complex numbers of the form bi are called **purely imaginary numbers**.

Two complex numbers are equal if their respective real and imaginary parts agree; that is,

$$a + bi = c + di \quad \text{if and only if} \quad a = c \quad \text{and} \quad b = d$$

To add two complex numbers, we use the following rule:

$$(a + bi) + (c + di) = (a + c) + (b + d)i$$

This says that real and imaginary parts are added separately. To calculate the product of two complex numbers, we proceed as follows:

$$(a + bi)(c + di) = ac + adi + bci + bdi^2$$
$$= ac + (ad + bc)i - bd$$
$$= (ac - bd) + (ad + bc)i$$

There is no need to memorize this since we can always compute it using the distributive law.

▶ **Example 12** Find (a) $(2 + 3i) - (5 - 6i)$, (b) $(5 - 3i)(1 + 2i)$.

Solution

(a) $(2 + 3i) - (5 - 6i) = 2 + 3i - 5 + 6i = -3 + 9i$, (b) $(5 - 3i)(1 + 2i) = 5 + 10i - 3i - 6i^2 = 5 + 7i - (6)(-1) = 11 + 7i$. ◀

If $z = a + bi$ is a complex number, its **conjugate**, denoted by $\bar{z}$, is defined as

$$\bar{z} = a - bi$$

For complex numbers z and w, it can be shown (see Problems 113–115) that

$$\overline{(\bar{z})} = z$$
$$\overline{z + w} = \bar{z} + \bar{w}$$
$$\overline{zw} = \bar{z}\bar{w}$$

Furthermore, if we multiply a complex number by its conjugate, we find

$$z\bar{z} = (a + bi)(a - bi) = a^2 - abi + abi - b^2 i^2$$
$$= a^2 + b^2$$

That is,

$$z\bar{z} = a^2 + b^2$$

▶ **Example 13**   Let $z = 3 + 2i$. (a) Find $\bar{z}$. (b) Compute $z\bar{z}$.

Solution

(a) $\bar{z} = 3 - 2i$. (b) $z\bar{z} = (3 + 2i)(3 - 2i) = 9 - 4i^2 = 9 + 4 = 13.$ ◀

We will encounter complex numbers primarily when solving quadratic equations. Recall that to solve, for $a \neq 0$,

$$ax^2 + bx + c = 0$$

we use the quadratic formula

$$x_{1,2} = \frac{-b \pm \sqrt{b^2 - 4ac}}{2a}$$

where $x_{1,2}$ refers to the two solutions x_1 ("+" sign) and x_2 ("−" sign).

▶ **Example 14**   Solve

$$x^2 + 4x + 5 = 0$$

Solution

Using the quadratic formula, we find

$$x_{1,2} = \frac{-4 \pm \sqrt{4^2 - (4)(1)(5)}}{(2)(1)}$$
$$= \frac{-4 \pm \sqrt{16 - 20}}{2} = \frac{-4 \pm \sqrt{-4}}{2}$$

If we only allow solutions in the real number system, we would conclude that $x^2 + 4x + 5 = 0$ has no solutions. But if we allow solutions in the complex number system, we find

$$x_{1,2} = \frac{-4 \pm \sqrt{4i^2}}{2} = \frac{-4 \pm 2i}{2} = \frac{2(-2 \pm i)}{2} = -2 \pm i$$

That is, $x_1 = -2 + i$ and $x_2 = -2 - i$.   ◀

The term $b^2 - 4ac$ under the square root in the quadratic formula is called the **discriminant**. If the discriminant is nonnegative, the two solutions of the corresponding quadratic equation are real. (When the discriminant is equal to 0, the two solutions are identical.) If the discriminant is negative, the two solutions are complex conjugates of each other.

▶ **Example 15** Without solving

$$2x^2 - 3x + 7 = 0$$

what can you say about the solution?

Solution

We compute the discriminant

$$b^2 - 4ac = (-3)^2 - (4)(2)(7) = 9 - 56 = -47 < 0$$

Since the discriminant is negative, the equation $2x^2 - 3x + 7 = 0$ has two complex solutions, which are conjugates of each other. ◀

1.1.7 Problems

(1.1.1)

1. Find the two numbers that have distance 3 from -1 by (a) measuring the distances on the real number line and (b) solving an appropriate equation involving an absolute value.

2. Find all pairwise distances between the numbers $-5, 2$, and 7 by (a) measuring the distances on the real number line and (b) computing the distances using absolute values.

3. Solve the following equations.

(a) $|2x - 4| = 6$

(b) $|x - 3| = 2$

(c) $|2x + 3| = 5$

(d) $|7 - 3x| = -2$

4. Solve the following equations.

(a) $|2x + 4| = |5x - 2|$

(b) $|5 - 3u| = |3 + 2u|$

(c) $|4 + \frac{t}{2}| = |\frac{3}{2}t - 2|$

(d) $|2s - 3| = |7 - s|$

5. Solve the following inequalities.

(a) $|5x - 2| \le 4$

(b) $|1 - 3x| > 8$

(c) $|7x + 4| \ge 3$

(d) $|6 - 5x| < 7$

6. Solve the following inequalities.

(a) $|2x + 3| < 6$

(b) $|3 - 4x| \ge 2$

(c) $|x + 5| \le 1$

(d) $|7 - 2x| < 0$

(1.1.2)

In Problems 7–42, determine the equation in standard form for the line that satisfies the stated requirements.

7. The line passing through $(2, 4)$ with slope $-\frac{1}{3}$

8. The line passing through $(1, -2)$ with slope 2

9. The line passing through $(0, -2)$ with slope -3

10. The line passing through $(-3, 5)$ with slope $1/2$

11. The line passing through $(-2, -3)$ and $(1, 4)$

12. The line passing through $(-1, 4)$ and $(2, -\frac{1}{2})$

13. The line passing through $(0, 4)$ and $(3, 0)$

14. The line passing through $(1, -1)$ and $(4, 5)$

15. The horizontal line through $(3, \frac{3}{2})$

16. The horizontal line through $(0, -1)$

17. The vertical line through $(-1, \frac{7}{2})$

18. The vertical line through $(2, -3)$

19. The line with slope 3 and y-intercept $(0, 2)$

20. The line with slope -1 and y-intercept $(0, -3)$

21. The line with slope $1/2$ and y-intercept $(0, 2)$

22. The line with slope $-1/3$ and y-intercept $(0, -1)$

23. The line with slope -2 and x-intercept $(1, 0)$

24. The line with slope 1 and x-intercept $(-2, 0)$

25. The line with slope $-1/4$ and x-intercept $(3, 0)$

26. The line with slope $1/5$ and x-intercept $(-1/2, 0)$

27. The line passing through $(2, -3)$ and parallel to
$$x + 2y - 4 = 0$$

28. The line passing through $(1, 2)$ and parallel to
$$x - 3y - 6 = 0$$

29. The line passing through $(-1, -1)$ and parallel to the line passing through $(0, 1)$ and $(3, 0)$

30. The line passing through $(2, -3)$ and parallel to the line passing through $(0, -1)$ and $(2, 1)$

31. The line passing through $(1, 4)$ and perpendicular to
$$2y - 5x + 7 = 0$$

32. The line passing through $(-1, -1)$ and perpendicular to
$$x - y + 3 = 0$$

33. The line passing through $(5, -1)$ and perpendicular to the line passing through $(-2, 1)$ and $(1, -2)$

34. The line passing through $(4, -1)$ and perpendicular to the line passing through $(-2, 0)$ and $(1, 1)$

35. The line passing through $(4, 2)$ and parallel to the horizontal line passing through $(1, -2)$

36. The line passing through $(-1, 5)$ and parallel to the horizontal line passing through $(2, -1)$

37. The line passing through $(-1, 1)$ and parallel to the vertical line passing through $(2, -1)$

38. The line passing through $(3, 1)$ and parallel to the vertical line passing through $(-1, -2)$

39. The line passing through $(1, -3)$ and perpendicular to the horizontal line passing through $(-1, -1)$

40. The line passing through $(4, 2)$ and perpendicular to the horizontal line passing through $(3, 1)$

41. The line passing through $(7, 3)$ and perpendicular to the vertical line passing through $(-2, 4)$

42. The line passing through $(-2, 5)$ and perpendicular to the vertical line passing through $(1, 4)$

43. To convert a length measured in feet to a length measured in centimeters, we use the fact that a length measured in feet is proportional to a length measured in centimeters and that 1 ft corresponds to 30.5 cm. If x denotes the length measured in ft and y denotes the length measured in cm, then
$$y = 30.5x$$

(a) Explain how to use this relationship.

(b) Use this relationship to convert the following measurements into cm.

(i) 6 ft **(ii)** 3 ft 2 in **(iii)** 1 ft 7 in

(c) Use this relationship to convert the following measurements into ft.

(i) 173 cm **(ii)** 75 cm **(iii)** 48 cm

44. (a) To convert the weight of an object from kilograms (kg) to pounds (lb), you use the fact that a weight measured in kilograms is proportional to a weight measured in pounds, and that 1 kg corresponds to 2.20 lb. Find an equation that relates weight measured in kilograms to weight measured in pounds.

(b) Use your answer in (a) to convert the following measurements.

(i) 63 lb **(ii)** 150 lb **(iii)** 2.5 kg **(iv)** 140 kg

45. Assume that the distance a car travels is proportional to the time it takes to cover the distance. Find the equation that relates distance and time if it takes the car 15 min to travel 10 mi. What is the constant of proportionality if distance is measured in miles and time is measured in hours?

46. Assume that the number of seeds a plant produces is proportional to its above-ground biomass. Find the equation that relates number of seeds and above-ground biomass if a plant that weighs 217 g has 17 seeds.

47. Experimental study plots are often squares of length 1 m. If 1 ft corresponds to 0.305 m, compute the area of a square plot of length 1 m in ft^2.

48. Large areas are often measured in hectares (ha) or in acres. If 1 ha $= 10{,}000$ m^2 and 1 acre $= 4046.86$ m^2, how many acres are one hectare?

49. To convert the volume of a liquid measured in ounces to a volume measured in liters, we use the fact that 1 liter equals 33.81 ounces. Denote by x the volume measured in ounces and by y the volume measured in liters. Assume a linear relationship between these two units of measurements.

(a) Find the equation relating x and y.

(b) A typical soda can contains 12 ounces of liquid. How many liters is this?

50. To convert a distance measured in miles to a distance measured in kilometers, we use that 1 mile equals 1.609 kilometers. Denote by x the distance measured in miles and by y the distance measured in kilometers. Assume a linear relationship between these two units of measurements.

(a) Find an equation relating x and y.

(b) The distance between Minneapolis and Madison is 261 miles. How many kilometers is this?

51. Car speed in many countries is measured in kilometers per hour. In the U.S., car speed is measured in miles per hour. To convert between these units, use that 1 mile equals 1.609 kilometers.

(a) The speed limit on many U.S. highways is 55 miles per hour. Convert this into kilometers per hour.

(b) The recommended speed limit on German highways is 130 kilometers per hour. Convert this into miles per hour.

To measure temperature, three scales are commonly used: Fahrenheit, Celsius, and Kelvin. These scales are linearly related. We discuss these scales in Problems 52 and 53.

52. (a) The Celsius scale is devised so that $0°$C is the freezing point of water (at one atmosphere of pressure), and $100°$C is the boiling point of water (at one atmosphere of pressure). If you are more familiar with the Fahrenheit scale, then you know that water freezes at $32°$F and boils at $212°$F. Find a linear equation that relates temperature measured in Celsius and temperature measured in Fahrenheit.

(b) The normal body temperature in humans ranges from $97.6°$F to $99.6°$F. Convert this temperature range into degrees Celsius.

53. (a) The Kelvin (K) scale is an **absolute** scale of temperature. The zero point of the scale (0 K) denotes **absolute zero**, the coldest possible temperature; that is, no body can have a temperature below 0 K. It has been determined experimentally that 0 K corresponds to $-273.15°$C. If 1 K denotes the same temperature difference as $1°$C, find an equation that relates the Kelvin and Celsius scales.

(b) Pure nitrogen and pure oxygen can be produced cheaply by first liquefying purified air and then allowing the temperature of the liquid air to rise slowly. Since nitrogen and oxygen have different boiling points, they are distilled at different temperatures. The boiling point of nitrogen is 77.4 K; of oxygen, 90.2 K. Convert each of the boiling point temperatures into Celsius. If you solved Problem 52(a), also convert the boiling point temperatures into Fahrenheit. Using the techniques described for distilling nitrogen and oxygen, which gets distilled first?

54. By carrying out the following steps, we will show that if two nonvertical lines l_1 and l_2 with slopes m_1 and m_2 are perpendicular, then $m_1 m_2 = -1$. Assume that $m_1 < 0$ and $m_2 > 0$.

(a) Use a graph to show that if θ_1 and θ_2 are the respective angles of inclination of the lines l_1 and l_2, then $\theta_1 = \theta_2 + \frac{\pi}{2}$. (The angle of inclination of a line is the angle $\theta \in [0, \pi)$ between the line and the positively directed x-axis.)

(b) Show that $m_1 = \tan\theta_1$ and $m_2 = \tan\theta_2$. [Use the fact that $\tan(\pi - x) = -\tan x$.]

(c) Use the fact that $\tan(\frac{\pi}{2} - x) = \cot x$ and $\cot(-x) = -\cot x$ to show that $m_1 = -\cot\theta_2$. Deduce from this the truth of our claim.

(1.1.3)

55. Find the equation of a circle with center $(-1, 4)$ and radius 3.

56. Find the equation of a circle with center $(2, 3)$ and radius 4.

57. (a) Find the equation of a circle with center $(2, 5)$ and radius 3.

(b) Where does the circle intersect the y-axis?

(c) Does the circle intersect the x-axis? Explain.

58. (a) Find possible radii of a circle centered at $(3, 6)$ so that the circle intersects only one axis.

(b) Find possible radii of a circle centered at $(3, 6)$ so that the circle intersects both axes.

59. Find the center and the radius of the circle given by the equation

$$(x - 2)^2 + y^2 = 16$$

60. Find the center and the radius of the circle given by the equation

$$(x + 1)^2 + (y - 3)^2 = 9$$

61. Find the center and the radius of the circle given by the equation

$$0 = x^2 + y^2 - 4x + 2y - 11$$

(To do this, you must complete the squares.)

62. Find the center and the radius of the circle given by the equation

$$x^2 + y^2 + 2x - 4y + 1 = 0$$

(To do this, you must complete the squares.)

(1.1.4)

63. (a) Convert $75°$ to radian measure.

(b) Convert $\frac{17}{12}\pi$ to degree measure.

64. (a) Convert $-15°$ to radian measure.

(b) Convert $\frac{3}{4}\pi$ to degree measure.

65. Evaluate the following expressions without using a calculator.

(a) $\sin(-\frac{5\pi}{4})$

(b) $\cos(\frac{5\pi}{6})$

(c) $\tan(\frac{\pi}{3})$

66. Evaluate the following expressions without using your calculator.

(a) $\sin(\frac{3\pi}{4})$

(b) $\cos(-\frac{13\pi}{6})$

(c) $\tan(\frac{4\pi}{3})$

67. (a) Find the values of $\alpha \in [0, 2\pi)$ that satisfy

$$\sin\alpha = -\frac{1}{2}\sqrt{3}$$

(b) Find the values of $\alpha \in [0, 2\pi)$ that satisfy

$$\tan\alpha = \sqrt{3}$$

68. (a) Find the values of $\alpha \in [0, 2\pi)$ that satisfy

$$\cos\alpha = -\frac{1}{2}\sqrt{2}$$

(b) Find the values of $\alpha \in [0, 2\pi)$ that satisfy

$$\sec\alpha = 2$$

69. Show that the identity

$$1 + \tan^2\theta = \sec^2\theta$$

follows from

$$\sin^2\theta + \cos^2\theta = 1$$

70. Show that the identity

$$1 + \cot^2\theta = \csc^2\theta$$

follows from

$$\sin^2\theta + \cos^2\theta = 1$$

71. Solve $2\cos\theta\sin\theta = \sin\theta$ on $[0, 2\pi)$.

72. Solve $\sec^2 x = \sqrt{3}\tan x + 1$ on $[0, \pi)$.

(1.1.5)

73. Evaluate the following exponential expressions.

(a) $4^3 4^{-2/3}$

(b) $\frac{3^2 3^{1/2}}{3^{-1/2}}$

(c) $\frac{5^k 5^{2k-1}}{5^{1-k}}$

74. Evaluate the following exponential expressions.

(a) $(2^4 2^{-3/2})^2$

(b) $\left(\frac{6^{5/2} 6^{2/3}}{6^{1/3}}\right)^3$

(c) $\left(\frac{3^{-2k+3}}{3^{4+k}}\right)^3$

75. Which real number x satisfies

(a) $\log_4 x = -2$?

(b) $\log_{1/3} x = -3$?

(c) $\log_{10} x = -2$?

76. Which real number x satisfies

(a) $\log_{1/2} x = -4$?

(b) $\log_{1/4} x = 2$?

(c) $\log_5 x = 3$?

77. Which real number x satisfies

(a) $\log_{1/2} 32 = x$?

(b) $\log_{1/3} 81 = x$?

(c) $\log_{10} 0.001 = x$?

78. Which real number x satisfies

(a) $\log_4 64 = x$?

(b) $\log_{1/5} 625 = x$?

(c) $\log_{10} 10{,}000 = x$?

79. Simplify the following expressions.

(a) $-\ln\frac{1}{3}$

(b) $\log_4(x^2 - 4)$

(c) $\log_2 4^{3x-1}$

80. Simplify the following expressions.

(a) $-\ln\frac{1}{5}$

(b) $\ln\frac{x^2-y^2}{\sqrt{x}}$

(c) $\log_3 3^{2x+1}$

81. Solve for x.

(a) $e^{3x-1} = 2$

(b) $e^{-2x} = 10$

(c) $e^{x^2-1} = 10$

82. Solve for x.

(a) $3^x = 81$

(b) $9^{2x+1} = 27$

(c) $10^{5x} = 1000$

83. Solve for x.

(a) $\ln(x - 3) = 5$

(b) $\ln(x + 2) + \ln(x - 2) = 1$

(c) $\log_3 x^2 - \log_3 2x = 2$

84. Solve for x.

(a) $\ln(2x - 3) = 0$

(b) $\log_2(1 - x) = 3$

(c) $\ln x^3 - 2\ln x = 1$

(1.1.6)

In Problems 85–92, simplify and write each expression in the standard form $a + bi$.

85. $(3 - 2i) - (-2 + 5i)$

86. $(7 + i) - 4$

87. $(4 - 2i) + (9 + 4i)$

88. $(6 - 4i) + (2 + 5i)$

89. $3(5 + 3i)$

90. $(2 - 3i)(5 + 2i)$

91. $(6 - i)(6 + i)$

92. $(-4 - 3i)(4 + 2i)$

In Problems 93–98, let $z = 3 - 2i$, $u = -4 + 3i$, $v = 3 + 5i$, and $w = 1 - i$. Compute the following expressions.

93. $\bar{z}$

94. $\overline{z + u}$

95. $\overline{z + v}$

96. $\overline{v - w}$

97. $\overline{vw}$

98. $\overline{uz}$

99. If $z = a + bi$, find $z + \bar{z}$ and $z - \bar{z}$.

100. If $z = a + bi$, find $\bar{z}$. Use your answer to compute $\overline{(\bar{z})}$ and compare your answer with z.

In Problems 101–106, solve each quadratic equation in the complex number system.

101. $2x^2 - 3x + 2 = 0$

102. $3x^2 - 2x + 1 = 0$

103. $-x^2 + x + 2 = 0$

104. $-2x^2 + x + 3 = 0$

105. $4x^2 - 3x + 1 = 0$

106. $-2x^2 + 4x - 3 = 0$

In Problems 107–112, first determine whether the solutions of each quadratic equation are real or complex without solving them. Then solve them.

107. $3x^2 - 4x - 7 = 0$

108. $3x^2 - 4x + 7 = 0$

109. $-x^2 + 2x - 1 = 0$

110. $4x^2 - x + 1 = 0$

111. $3x^2 - 5x + 6 = 0$

112. $-x^2 + 7x - 2 = 0$

113. Show $\overline{(\bar{z})} = z$.

114. Show $\overline{z + w} = \bar{z} + \bar{w}$.

115. Show $\overline{zw} = \bar{z}\,\bar{w}$.

1.2 ELEMENTARY FUNCTIONS

1.2.1 What Is a Function?

Scientific investigations are often concerned with studying relationships between quantities, such as how seedling density depends on the distance to the adult plant or how tree ring width depends on the amount of light available to the tree. To describe such relationships mathematically, the concept of a function is useful.

It seems that the word function (or, more precisely, its Latin equivalent *functio*, which means "execution") was introduced by Leibniz in 1694 in order to describe curves. Later, Euler (1707–1783) used it to describe any equation involving variables and constants. The modern definition is much broader and emphasizes the basic idea of expressing relationships between any two sets.

> **Definition** A **function** f is a rule that assigns each element x in the set A exactly one element y in the set B. The element y is called the **image** (or **value**) of x under f and is denoted by $f(x)$ (read: "f of x"). The set A is called the **domain** of f, the set B is called the **codomain** of f, and the set $f(A) = \{y : y = f(x) \text{ for some } x \in A\}$ is called the **range** of f.

To define a function, we use the notation

$$f : \quad A \to B$$
$$x \to f(x)$$

Here, A and B are subsets of the set of real numbers. Frequently, we simply write $y = f(x)$ and call x the **independent** variable and y the **dependent** variable. We can illustrate functions graphically in the x-y plane. In Figure 1.9, we see the graph of $y = f(x)$ with domain A, codomain B, and range $f(A)$.

Of course, $f(x)$ must be specified; for example, $f(x) = x^2$. Note that $f(A) \subset B$, but not every element in the codomain B must be in $f(A)$. For instance, let

$$f : \quad \mathbf{R} \to \mathbf{R}$$
$$x \to x^2$$

The domain of f is $\mathbf{R}$, but the range of f is only $[0, \infty)$ because the square of a real number is nonnegative; that is, $f(\mathbf{R}) = [0, \infty) \neq \mathbf{R}$. Also, the domain of a function need not be the largest possible set on which we can define the function, such as $\mathbf{R}$ in the preceding example. For instance, we could have defined f on a smaller set, such as $[0, 1]$, calling the new function g,

$$g : \quad [0, 1] \to \mathbf{R}$$
$$x \to x^2$$

Though the same rule is used for f and g, the two functions are not the same because their respective domains are different.

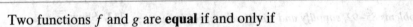

> Two functions f and g are **equal** if and only if
> 1. f and g are defined on the same domain, and
> 2. $f(x) = g(x)$ for all x in the domain.

▶ **Example 1** Let

$$f_1 : \quad [0, 1] \to \mathbf{R}$$
$$x \to x^2$$

$$f_2 : \quad [0, 1] \to \mathbf{R}$$
$$x \to \sqrt{x^4}$$

and

$$f_3 : \quad \mathbf{R} \to \mathbf{R}$$
$$x \to x^2$$

Determine which of the functions are equal.

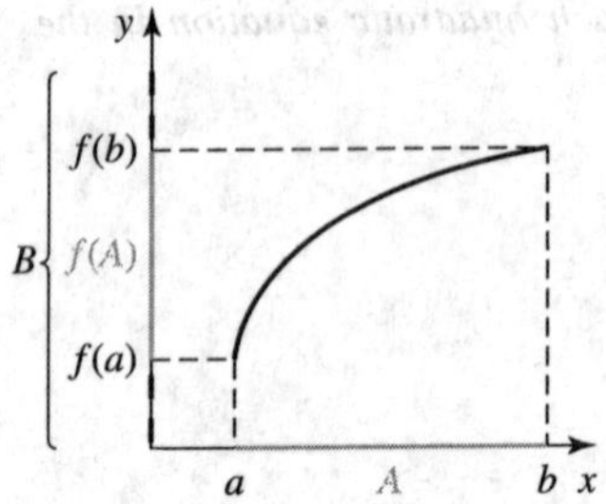

▲ Figure 1.9
A function $f(x)$ with domain A, codomain B, and range $f(A)$

Solution

Because f_1 and f_2 are defined on the same domain, and $f_1(x) = f_2(x) = x^2$ for all $x \in [0, 1]$, it follows that f_1 and f_2 are equal.

Neither f_1 nor f_2 is equal to f_3 because the domain of f_3 is different from the domains of f_1 and f_2.   ◀

The choices of domains for the functions that we have considered may look somewhat arbitrary (and they are arbitrary in the examples we have seen so far). In applications, however, there is often a natural choice of domain. For instance, if we look at a certain plant response (such as total biomass or the ratio of above to below biomass) as a function of nitrogen concentration in the soil, then, given that nitrogen concentration cannot be negative, the domain for this function could be the set of non-negative real numbers. To look at another example, suppose we define a function that depends on the fraction of a population infected with a certain virus; then a natural choice for the domain of this function would be the interval $[0, 1]$ because a fraction of a population must be a number between 0 and 1.

In the definition of a function, we stated that it is a rule that assigns to each element $x \in A$ *exactly* one element $y \in B$. When we graph $y = f(x)$ in the x-y plane, there is a simple test to decide whether or not $f(x)$ is a function: Each vertical line intersects the graph of $y = f(x)$ at most once. Figure 1.10 shows the graph of a function: Each vertical line intersects the graph of $y = f(x)$ at most once. The graph of $y = f(x)$ in Figure 1.11 is not a function since there are x-values that are assigned to more than one y-value, as illustrated by the vertical line that intersects the graph more than once.

Sometimes functions show certain symmetries. For example, looking at Figure 1.12, $f(x) = x$ is symmetric about the origin; that is, $f(x) = -f(-x)$. Looking at Figure 1.13, $g(x) = x^2$ is symmetric about the y-axis; that is, $g(x) = g(-x)$. In the first case, we say that f is odd; in the second case, that g is even. To check whether a function is even or odd, we use the following definition.

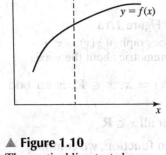

▲ **Figure 1.10**
The vertical line test shows that the graph of $y = f(x)$ is a function

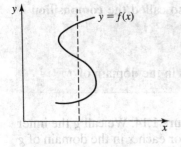

▲ **Figure 1.11**
The vertical line test shows that the graph of $y = f(x)$ is not a function

---

A function $f : A \to B$ is called

1. **even** if $f(x) = f(-x)$ for all $x \in A$, and
2. **odd** if $f(x) = -f(-x)$ for all $x \in A$.

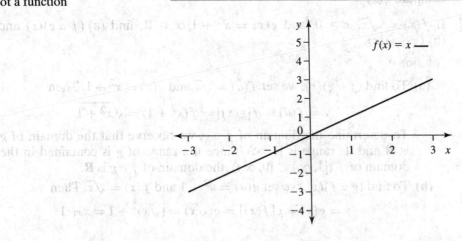

◀ **Figure 1.12**
The graph of $f(x) = x$ is symmetric about the origin

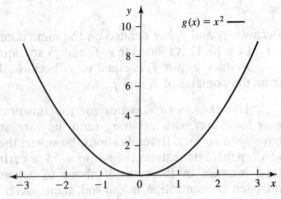

◀ Figure 1.13
The graph of $g(x) = x^2$ is
symmetric about the y-axis

Using this criterion, we can check that $f(x) = x$, $x \in \mathbf{R}$, is an odd function, namely,

$$-f(-x) = -(-x) = x = f(x) \quad \text{for all } x \in \mathbf{R}$$

Likewise, to check that $g(x) = x^2$, $x \in \mathbf{R}$, is an even function, we compute

$$g(-x) = (-x)^2 = x^2 = g(x) \quad \text{for all } x \in \mathbf{R}$$

We will now look at the case where one quantity is given as a function of another quantity, which in turn can be written as a function of yet another quantity. To illustrate this, suppose that $y = f(u)$ and $u = g(x)$. We can express f as a function of x by substituting $g(x)$ for u; that is, $y = f[g(x)]$. Functions that are defined in such a way are called composite functions.

> **Definition** The **composite function** $f \circ g$ (also called the **composition** of f and g) is defined as
>
> $$(f \circ g)(x) = f[g(x)]$$
>
> for each x in the domain of g for which $g(x)$ is in the domain of f.

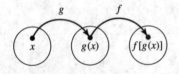

▲ Figure 1.14
The composition of functions

The composition of functions is illustrated in Figure 1.14. We call g the inner function and f the outer function. The phrase "for each x in the domain of g for which $g(x)$ is in the domain of f" is best explained using Figure 1.14. In order to compute $f(u)$, u needs to be in the domain of f. But since $u = g(x)$, we really require that $g(x)$ is in the domain of f for the values of x we use to compute $g(x)$.

▶ **Example 2**   If $f(x) = \sqrt{x}$, $x \geq 0$, and $g(x) = x^2 + 1$, $x \in \mathbf{R}$, find (a) $(f \circ g)(x)$ and (b) $(g \circ f)(x)$.

Solution

(a) To find $(f \circ g)(x)$, we set $f(u) = \sqrt{u}$ and $g(x) = x^2 + 1$. Then

$$y = f(u) = f[g(x)] = f(x^2 + 1) = \sqrt{x^2 + 1}$$

To determine the domain of $f \circ g$, we observe that the domain of g is $\mathbf{R}$ and its range is $[1, \infty)$. Since the range of g is contained in the domain of f ($[1, \infty) \subset [0, \infty)$), the domain of $f \circ g$ is $\mathbf{R}$.

(b) To find $(g \circ f)(x)$, we set $g(u) = u^2 + 1$ and $f(x) = \sqrt{x}$. Then

$$y = g(u) = g[f(x)] = g(\sqrt{x}) = (\sqrt{x})^2 + 1 = x + 1$$

To determine the domain of $g \circ f$, we observe that the domain of f is $[0, \infty)$ and its range is $[0, \infty)$. The range of f is contained in the domain of g ($[0, \infty) \subset \mathbf{R}$), so the domain of $g \circ f$ is $[0, \infty)$. ◄

In the last example, you should observe that $f \circ g$ is different from $g \circ f$, which implies that the order in which you compose functions is important. The notation $f \circ g$ means that you apply g first and then f. In addition, you should pay attention to the domains of composite functions. In the next example, the domain is harder to find.

▶ **Example 3**   If $f(x) = 2x^2$, $x \geq 2$, and $g(x) = \sqrt{x}$, $x \geq 0$, find $(f \circ g)(x)$ together with its domain.

Solution

We compute

$$(f \circ g)(x) = f[g(x)] = f(\sqrt{x}) = 2(\sqrt{x})^2 = 2x$$

This part was not difficult. However, finding the domain of $f \circ g$ is more complicated. The domain of the inner function g is the interval $[0, \infty)$; hence, the range of g is the interval $[0, \infty)$. The domain of f is only $[2, \infty)$, which means that the range of g is *not* contained in the domain of f. We therefore need to restrict the domain of g to ensure that its range is contained in the domain of f. We can only choose values of x so that $g(x) \in [2, \infty)$. Since $g(x) = \sqrt{x}$, this means we need to restrict x to $[4, \infty)$. For every $x \in [4, \infty)$, $g(x) \in [2, \infty)$, which is equal to the domain of f. Therefore,

$$(f \circ g)(x) = 2x, \quad x \geq 4$$

(see Figure 1.15). ◄

In the following subsections we introduce the basic functions that are used throughout the remainder of this book.

1.2.2 Polynomial Functions

Polynomial functions are the simplest elementary functions.

> **Definition**   A **polynomial** function is a function of the form
>
> $$f(x) = a_0 + a_1 x + a_2 x^2 + \cdots + a_n x^n$$
>
> where n is a nonnegative integer and $a_0, a_1, \ldots, a_n$ are (real-valued) constants with $a_n \neq 0$. The coefficient a_n is called the **leading coefficient** and n is called the **degree** of the polynomial function. The largest possible domain of f is $\mathbf{R}$.

We have already encountered polynomials, namely, the constant function $f(x) = c$, the linear function $f(x) = mx + b$, and the quadratic function $f(x) = ax^2$. The constant, nonzero function has degree 0, the linear function has degree 1, and the quadratic function has degree 2. Other examples are $f(x) = 4x^3 - 3x + 1$, $x \in \mathbf{R}$, which is a polynomial of degree 3, and $f(x) = 2 - x^7$, $x \in \mathbf{R}$, which is a polynomial of degree 7. In Figure 1.16, we display $y = x^n$ for $n = 2$ and 3. Looking at Figure 1.16, we see that $y = x^n$ is an even function (i.e., symmetric about the y-axis) when $n = 2$ and an odd function (i.e., symmetric about the origin) when $n = 3$. This holds in general,

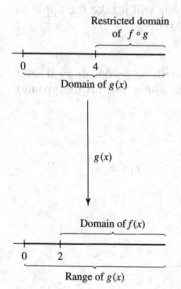

Restricted domain of $f \circ g$

0   4

Domain of $g(x)$

$g(x)$

Domain of $f(x)$

0   2

Range of $g(x)$

▲ **Figure 1.15**
Finding the domain of a composite function: The domain of $g(x)$ must be restricted in Example 3

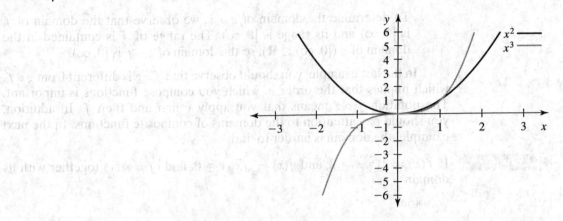

▲ **Figure 1.16**
The graphs of $y = x^n$ for $n = 2$ and 3

namely, $y = x^n$ is an even function when n is even and an odd function when n is odd; this can be shown algebraically using the criterion in Section 1.2.1 (see Problem 28 at the end of this section).

Polynomials arise naturally in many situations. We present two examples.

 **Example 4**   Suppose that at time 0 an apple begins to drop from a tree that is 64 ft tall. Ignoring air resistance, we can show that at time t (measured in seconds) the apple is at height $h(t)$ (measured in feet), with

$$h(t) = 64 - 16t^2$$

We assume that the height of the ground level is equal to 0. Explain that $h(t)$ is a polynomial and determine its degree. How long will it take the apple to hit the ground? Find an appropriate domain for $h(t)$.

Solution

The function $h(t)$ is a polynomial of degree 2 with $a_0 = 64$, $a_1 = 0$, and $a_2 = -16$. Its graph is shown in Figure 1.17. The apple will hit the ground

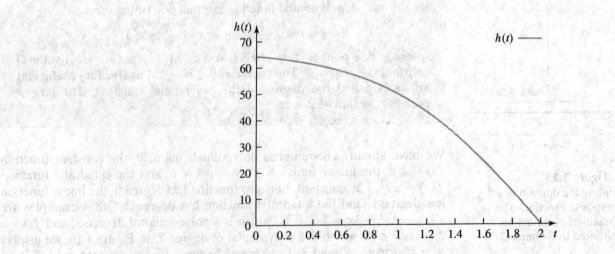

▲ **Figure 1.17**
The graph of $h(t) = 64 - t^2$ for $0 \le t \le 2$ of Example 4

when $h(t) = 0$. That is, we must solve

$$0 = 64 - 16t^2$$
$$t^2 = \frac{64}{16} = 4$$
$$t = 2 \quad (\text{or } t = -2)$$

Since $t = -2 < 0$ and the apple begins to drop at time $t = 0$, we can ignore this solution. We find that it takes the apple 2 seconds to hit the ground (ignoring air resistance). Note that because $h(t) \geq 0$ [where $h(t)$ is the height above the ground and the height of the ground level is equal to 0], the range is $[0, 64]$. Because $t \geq 0$, the domain of $h(t)$ is the interval $[0, 2]$. ◀

▶ Example 5   **(A Chemical Reaction)** Consider the reaction rate of the chemical reaction

$$A + B \longrightarrow AB$$

in which the molecular reactants A and B form the molecular product AB. The rate at which this reaction proceeds depends on how often A and B molecules collide. The **law of mass action** states that the rate at which this reaction proceeds is proportional to the product of the respective concentrations of the reactants. Here, *concentration* means the number of molecules per fixed volume. If we denote the reaction rate by R and the concentration of A and B by [A] and [B], respectively, then the law of mass action says that

$$R \propto [A] \cdot [B]$$

Introducing the proportionality factor k, we obtain

$$R = k[A] \cdot [B]$$

Note that $k > 0$ because [A], [B], and R are positive. We assume now that the reaction occurs in a closed vessel; that is, we add specific amounts of A and B to the vessel at the beginning of the reaction and then let the reaction proceed without further additions.

We can express the concentrations of the reactants A and B during the reaction in terms of their initial concentrations a and b and the concentration of the molecular product [AB]. If $x = [AB]$, then

$$[A] = a - x \quad \text{for } 0 \leq x \leq a \quad \text{and} \quad [B] = b - x \quad \text{for } 0 \leq x \leq b$$

The concentration of AB cannot exceed either of the concentrations of A or B. (Suppose five A molecules and seven B molecules are allowed to react; a maximum of five AB molecules can result, at which point all of the A molecules are used up and the reaction ceases. The two B molecules left over have no A molecules to react with.) Therefore, we get

$$R(x) = k(a - x)(b - x) \quad \text{for } 0 \leq x \leq a \quad \text{and } 0 \leq x \leq b$$

The condition "$0 \leq x \leq a$ and $0 \leq x \leq b$" can be written as $0 \leq x \leq \min(a, b)$, where $\min(a, b)$ denotes the minimum of a and b. To see that $R(x)$ is indeed a polynomial function, we expand the expression for $R(x)$.

$$R(x) = k(ab - ax - bx + x^2)$$
$$= kx^2 - k(a + b)x + kab$$

for $0 \leq x \leq \min(a, b)$. We see that $R(x)$ is a polynomial of degree 2.

A graph of $R(x)$, $0 \leq x \leq a$, is shown in Figure 1.18 for the case $a \leq b$ (we chose $k = 2$, $a = 2$, and $b = 5$ in the figure). Notice that when $x = 0$

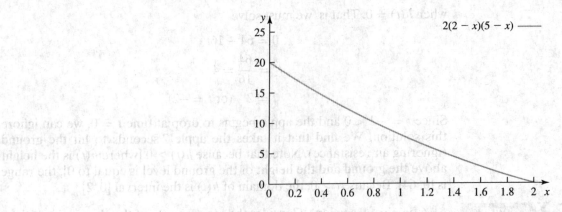

▲ **Figure 1.18**
The graph of $R(x) = 2(2 - x)(5 - x)$ for $0 \leq x \leq 2$

(that is, when no AB molecules have yet formed), the rate at which the reaction proceeds is at a maximum. As more and more AB molecules form and, consequently, the concentrations of the reactants decline, the reaction rate decreases. This should also be intuitively clear: As fewer and fewer A and B molecules are in the vessel, it becomes less and less likely that they will collide to form the molecular product AB. When $x = a = \min(a, b)$, the reaction rate $R(a) = 0$. This is the point at which all A molecules are exhausted and the reaction necessarily ceases.

1.2.3 Rational Functions

Rational functions are built from polynomial functions.

> **Definition** A **rational** function is the quotient of two polynomial functions $p(x)$ and $q(x)$:
> $$f(x) = \frac{p(x)}{q(x)} \quad \text{for } q(x) \neq 0$$

Since division by 0 is not allowed, we must exclude those values of x for which $q(x) = 0$. Here are a couple of examples of rational functions together with their largest possible domains:

$$y = \frac{1}{x}, \quad x \neq 0$$

$$y = \frac{x^2 + 2x - 1}{x - 3}, \quad x \neq 3$$

An important example of a rational function is the hyperbola together with its largest possible domain:

$$y = \frac{1}{x}, \quad x \neq 0$$

Its graph is shown in Figure 1.19.

Throughout this text we will encounter populations whose sizes change with time. The change in population size is described by the **growth rate**. Roughly speaking, the growth rate tells you how much a population changes during a small time interval. (This is analogous to the velocity of a car:

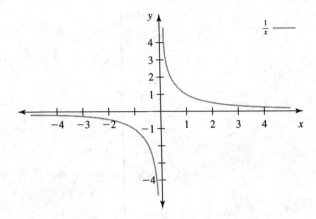

▲ **Figure 1.19**
The graph of $y = \frac{1}{x}$ for $x \neq 0$

Velocity is also a rate; it tells you how much the position changes in a small time interval. We will give a precise definition of rates in Section 4.1.) The **per capita** growth rate is the growth rate divided by the population size. The per capita growth rate is also called the **specific** growth rate. The following example introduces a function that is frequently used to describe growth rates.

▶ **Example 6**   **(Monod Growth Function)** The following function is frequently used to describe the per capita growth rate of organisms when the growth rate depends on the concentration of some nutrient and becomes saturated for large enough nutrient concentrations. We denote the concentration of the nutrient by N; the per capita growth rate $r(N)$ is given by the Monod growth function

$$r(N) = \frac{aN}{k + N}, \quad N \geq 0$$

where a and k are positive constants. The graph of $r(N)$ is shown in Figure 1.20; it is a piece of a hyperbola. It shows a decelerating rise approaching the saturation level a, which is the maximal specific growth rate. When $N = k$, then $r(N) = a/2$; for this reason, k is called the half-saturation constant. The growth rate increases with nutrient concentration N; however, doubling the nutrient concentration has a much bigger effect on the growth rate for small values of N than when N is already large. This type of function is also used in biochemistry to describe enzymatic reactions; it is then called the Michaelis-Menten function.

1.2.4 Power Functions

Definition A **power** function is of the form
$$f(x) = x^r$$
where r is a real number.

Examples of power functions, with their largest possible domains, are

$$y = x^{1/3}, \quad x \in \mathbf{R}$$
$$y = x^{5/2}, \quad x \geq 0$$

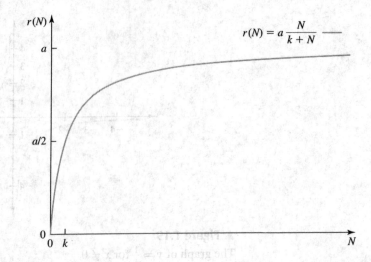

▲ **Figure 1.20**

The graph of the Monod function $r(N) = a\frac{N}{k+N}$ for $N \geq 0$

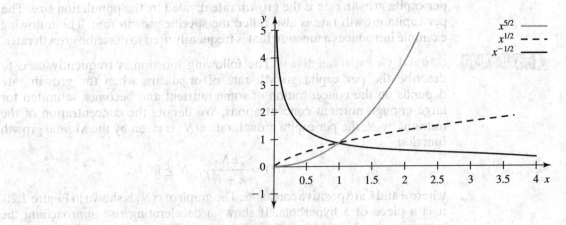

▲ **Figure 1.21**

Some power functions with rational exponents

$$y = x^{1/2}, \quad x \geq 0$$
$$y = x^{-1/2}, \quad x > 0$$

Polynomials of the form $y = x^n$, $n = 1, 2, \ldots$ are a special case of power functions. Since power functions may involve even roots, such as in $y = x^{3/2} = (\sqrt{x})^3$, we frequently need to restrict their domain.

In Figure 1.21, we compare some power functions for $x > 0$, namely $y = x^{5/2}$, $y = x^{1/2}$, and $y = x^{-1/2}$. You should pay close attention to how the exponent determines the ranking according to size for x between 0 and 1 and for $x > 1$. We find that $x^{5/2} < x^{1/2} < x^{-1/2}$ for $0 < x < 1$, but $x^{5/2} > x^{1/2} > x^{-1/2}$ for $x > 1$.

 Example 7   Power functions are frequently found in "scaling relations" between biological variables (for instance, organ sizes). These are relations of the form

$$y \propto x^r$$

where r is a nonzero real number. That is, y is proportional to some power of x. Recall that we can write this as an equation if we introduce the proportionality

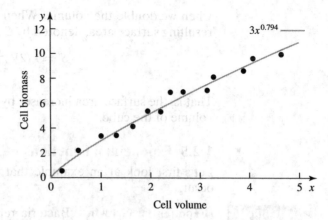

▲ **Figure 1.22**
Some data points and the fitted curve of Example 7 (*Note*: the "data points" are not real data)

factor k, namely,

$$y = kx^r$$

Finding such relationships is the objective of **allometry**. In a study of 45 species of unicellular algae, a relationship between cell volume and cell biomass was sought. It was found [see, for instance, Niklas (1994)] that

$$\text{cell biomass} \propto (\text{cell volume})^{0.794}$$

Most of these scaling relations are to be interpreted in a statistical sense; they are obtained by fitting a curve to data points. The data points are typically scattered about the fitted curve given by the scaling relation (see Figure 1.22).

The following example relates the volume and the surface area of a cube. This is not to be understood in a statistical sense because it is an exact relationship resulting from geometric considerations.

▶ **Example 8** Suppose that we wish to know the scaling relation between the surface area S and the volume V of a cube. For each of these quantities, we know their respective scaling relation with the length L of the cube:

$$S \propto L^2 \quad \text{or} \quad S = k_1 L^2$$
$$V \propto L^3 \quad \text{or} \quad V = k_2 L^3$$

Here, k_1 and k_2 denote the constants of proportionality. (We label them with different subscripts to indicate that they might be different.) To express S in terms of V, we must first solve L in terms of V and then substitute L in the equation for S. Because $L = (V/k_2)^{1/3}$,

$$S = k_1 \left[\left(\frac{V}{k_2} \right)^{1/3} \right]^2 = \frac{k_1}{k_2^{2/3}} V^{2/3}$$

Introducing the constant of proportionality $k = k_1 / k_2^{2/3}$, we find

$$S = kV^{2/3} \quad \text{or simply} \quad S \propto V^{2/3}$$

In words, the surface area of a cube scales with the volume in proportion to $V^{2/3}$. We can now ask, for instance, by what factor the surface area increases when we double the volume. When we double the volume, we find that the resulting surface area, denoted by S', is

$$S' = k(2V)^{2/3} = 2^{2/3} \underbrace{kV^{2/3}}_{S}$$

That is, the surface area increases by a factor of $2^{2/3} \approx 1.587$ if we double the volume of the cube.

1.2.5 Exponential Functions

Let's first look at an example that illustrates where exponential functions occur.

▶ **Example 9**

(Exponential Growth) Bacteria reproduce asexually by cellular fission, in which the parent cell splits into two daughter cells after duplication of the genetic material. This division may happen as often as every 20 minutes; under ideal conditions, a bacterial colony can double in size in that time.

Let us measure time such that one unit of time corresponds to the doubling time of the colony. If we denote the size of the population at time t by $N(t)$, then the size of the population at time $t + 1$ is twice the population size at time t. We write this as

$$N(t + 1) = 2N(t) \tag{1.2}$$

The function

$$N(t) = 2^t, \quad t = 0, 1, 2, \ldots$$

satisfies Equation (1.2):

$$N(t + 1) = 2^{t+1} = 2 \cdot 2^t = 2N(t)$$

The function $N(t) = 2^t$, $t = 0, 1, 2, \ldots$ is an exponential function because the variable t is in the exponent. We call the number 2 the base of the exponential function $N(t) = 2^t$. We chose $t = 0, 1, 2, \ldots$ as the domain for the function $N(t)$ because we think of $t = 0$ as the time when we start measuring the population size and we measure the population size each unit of time.

When $t = 0$, we find that $N(0) = 1$; that is, there is just one individual in the population at time $t = 0$. If at time $t = 0$ forty individuals were present in the population, we would write $N(0) = 40$ and

$$N(t) = 40 \cdot 2^t, \quad t = 0, 1, 2, \ldots \tag{1.3}$$

You can check that $N(t)$ in (1.3) also satisfies (1.2).

It is often desirable not to specify the initial number of individuals in the equation describing $N(t)$. This has the advantage that the equation for $N(t)$ then describes a more general situation, in the sense that we can use the same equation for different initial population sizes. We often denote the population size at time 0 by N_0 (read: "N sub 0") instead of $N(0)$. The equation for $N(t)$ is then

$$N(t) = N_0 2^t, \quad t = 0, 1, 2, \ldots$$

We can check that $N(0) = N_0 2^0 = N_0$ and that $N(t+1) = N_0 2^{t+1} = 2(N_0 2^t) = 2N(t)$. We will discuss this example in more detail in Chapter 2.

The function $f(t) = 2^t$ can be defined for all $t \in \mathbf{R}$; its graph is shown in Figure 1.23.

Here is the definition of an exponential function.

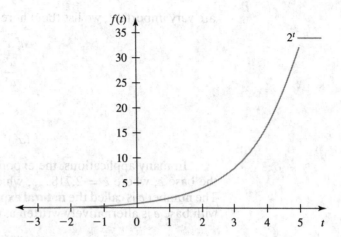

▲ **Figure 1.23**
The function $f(t) = 2^t, t \in \mathbf{R}$

> **Definition**   The function f is an **exponential** function with base a if
>
> $$f(x) = a^x$$
>
> where a is a positive constant other than 1. The largest possible domain of f is **R**.

When $a = 1$, then $f(x) = 1$ for all values of x, a case that will occur in biological examples but is excluded from the definition since it is simply the constant function.

The basic shape of the exponential function $f(x) = a^x$ depends on the base a; two examples are shown in Figure 1.24. As x increases, the graph of $f(x) = 2^x$ shows a rapid increase, whereas the graph of $f(x) = (1/3)^x$ shows a rapid decrease toward 0. We find the rapid increase whenever $a > 1$ and the rapid decrease whenever $0 < a < 1$. Therefore, we say that we have exponential growth when $a > 1$ and exponential decay when $0 < a < 1$.

Recall that $a^0 = 1$ and $a^{1/k} = \sqrt[k]{a}$, where k is a positive integer. In Subsection 1.1.5, we summarized the properties of exponentials. Since they

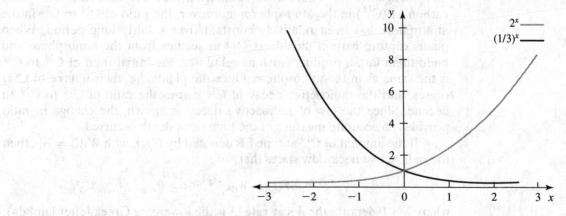

▲ **Figure 1.24**
Exponential growth and exponential decay

are very important, we list them here once more.

$$a^r a^s = a^{r+s}$$

$$\frac{a^r}{a^s} = a^{r-s}$$

$$a^{-r} = \frac{1}{a^r}$$

$$(a^r)^s = a^{rs}$$

In many applications, the exponential function is expressed in terms of the base e, where $e = 2.718\ldots$, which we encountered in Subsection 1.1.5. The number e is called the **natural exponential base**. The exponential function with base e is alternatively written as $\exp(x)$. That is,

$$\exp(x) = e^x$$

The advantage of this alternative can be seen when we try to write something like $e^{x^2/\sqrt{x^3+1}}$; $\exp(x^2/\sqrt{x^3 + 1})$ is easier to read. More generally, if $g(x)$ is a function in x, then we can write, equivalently,

$$\exp[g(x)] \qquad \text{or} \qquad e^{g(x)}$$

Bases 2 and 10 are also frequently used; in calculus, however, e will turn out to be the most common base.

The next two examples provide an important application of exponential functions.

▶ **Example 10** (Radioactive Decay) Radioactive isotopes such as carbon 14 are used to determine the absolute age of fossils or minerals, establishing an absolute chronology of the geological time scale. This technique was discovered in the early years of the twentieth century and is based on the property of certain atoms to transform spontaneously by giving off protons, neutrons, or electrons. This phenomenon is called radioactive decay; it occurs at a constant rate, and the rate is independent of environmental conditions. The method was used, for instance, to trace the successive emergence of the Hawaiian islands, from the oldest, Kauai, to the youngest, Hawaii (which is about 100,000 years old).

Carbon 14 is formed high in the atmosphere. It is radioactive and decays into nitrogen (N^{14}). There is an equilibrium between carbon 12 (C^{12}) and carbon 14 (C^{14}) in the atmosphere; moreover, the ratio of C^{14} to C^{12} in the atmosphere has been relatively constant over a fairly long period. When plants capture carbon dioxide (CO_2) molecules from the atmosphere and build them into a product (such as cellulose), the initial ratio of C^{14} to C^{12} is the same as in the atmosphere. Once the plants die, their uptake of CO_2 ceases, and the radioactive decay of C^{14} causes the ratio of C^{14} to C^{12} to decline. Since the law of radioactive decay is known, the change in ratio provides an accurate measure of the time since death occurred.

If the amount of C^{14} at time t is denoted by $W(t)$, with $W(0) = W_0$, then the radioactive decay law states that

$$W(t) = W_0 e^{-\lambda t}, \quad t \geq 0$$

where $\lambda > 0$ denotes the **decay rate** (λ is the lowercase Greek letter lambda). The function $W(t) = W_0 e^{-\lambda t}$ is another example of an exponential function. Its graph is shown in Figure 1.25.

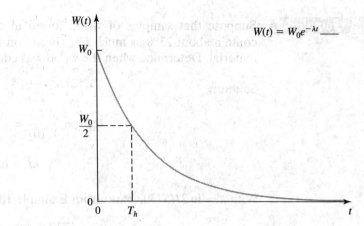

$$W(t) = W_0 e^{-\lambda t}$$

▲ **Figure 1.25**
The function $W(t) = W_0 e^{-\lambda t}$

Frequently, the decay rate is expressed in terms of the **half-life** of the material, which is the length of time that it takes for half of the material to decay. If we denote this time by T_h, then (see Figure 1.25)

$$W(T_h) = \frac{1}{2} W_0 = W_0 e^{-\lambda T_h}$$

from which we obtain

$$\frac{1}{2} = e^{-\lambda T_h}$$

$$2 = e^{\lambda T_h}$$

Recall from algebra (or Subsection 1.1.5) that to solve for the exponent λT_h we must take logarithms on both sides. Since the exponent has base e, we use natural logarithms and find

$$\ln 2 = \lambda T_h$$

Solving for T_h or λ yields

$$T_h = \frac{\ln 2}{\lambda} \quad \text{or} \quad \lambda = \frac{\ln 2}{T_h}$$

It is known that the half-life of C^{14} is 5730 years. Hence,

$$\lambda = \frac{\ln 2}{5730 \text{ years}}$$

Note that the unit "years" appears in the denominator. It is important to carry the units along. When we compute λt in the exponent of $e^{-\lambda t}$ and use this expression for λ, we need to measure t in units of years in order for the units to cancel properly: Say $t = 2000$ years, then

$$\lambda t = \frac{\ln 2}{5730 \text{ years}} \, 2000 \text{ years} = \frac{(\ln 2)(2000)}{5730} \approx 0.2419$$

and we see that "years" appears in both the numerator and the denominator and can thus be canceled.

An application of the C^{14} dating method is shown in the next example.

▶ Example 11   Suppose that samples of wood found in an archeological excavation site contain about 23% as much C^{14} (based on their C^{12} content) as living plant material. Determine when the wood was cut.

Solution

$$0.23 = \frac{W(t)}{W(0)} = e^{-\lambda t}$$

$$\lambda t = \ln \frac{1}{0.23}$$

With $\lambda = \ln 2/(5730 \text{ years})$ from Example 10,

$$t = \frac{5730 \text{ years}}{\ln 2} \ln \frac{1}{0.23}$$

Using a calculator to compute this result, we find that the wood was cut about 12,150 years ago.   ◀

1.2.6 Inverse Functions

Before we can introduce logarithmic functions, we need the concept of inverse functions. Roughly speaking, the inverse of a function f reverses the effect of f. That is, if f maps x into $y = f(x)$, the inverse function, denoted by f^{-1} (read "f inverse"), takes y and maps it back into x (see Figure 1.26). Not every function can have an inverse; since an inverse function is a function itself, we require that every value y in the range of f be mapped into exactly one value x. In other words, we need that whenever $x_1 \neq x_2$, then $f(x_1) \neq f(x_2)$ or, equivalently, that $f(x_1) = f(x_2)$ implies $x_1 = x_2$. (Recall the definition of a function, where we required that a function assigns each element in the domain to *exactly one* element in the range.)

Functions that have the property "$x_1 \neq x_2$ implies $f(x_1) \neq f(x_2)$" [or, equivalently, "$f(x_1) = f(x_2)$ implies $x_1 = x_2$"] are called **one to one**. If you know what the graph of a particular function looks like over its domain, then it is easy to determine whether or not the function is one to one. Namely, if no horizontal line intersects the graph of the function f more than once, then f is one to one. This is called the horizontal line test. We illustrate how to use this test in Figures 1.27 and 1.28.

Consider $y = x^3$ and $y = x^2$, for $x \in \mathbf{R}$. The function $y = x^3$, $x \in \mathbf{R}$, has an inverse function because $x_1^3 \neq x_2^3$ whenever $x_1 \neq x_2$ (see Figure 1.27). The function $y = x^2$, $x \in \mathbf{R}$, does not have an inverse function because $x_1 \neq x_2$ does not imply $x_1^2 \neq x_2^2$ (or, equivalently, $x_1^2 = x_2^2$ does not imply $x_1 = x_2$) (see Figure 1.28). The equation $x_1^2 = x_2^2$ only implies $|x_1| = |x_2|$. Since x_1 and x_2 can be positive or negative, we cannot simply drop the absolute value signs.

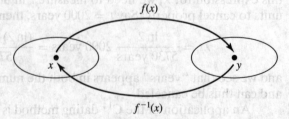

◀ **Figure 1.26**
The function $y = f(x)$ and its inverse function

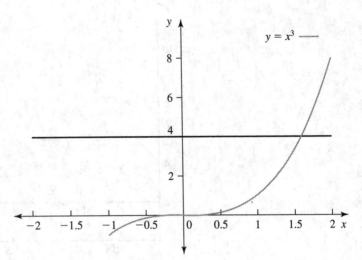

▲ **Figure 1.27**
Horizontal line test is successful

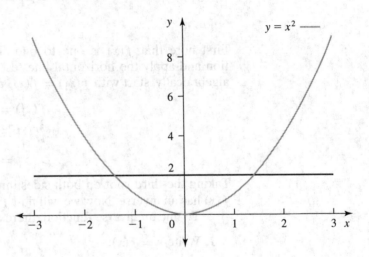

▲ **Figure 1.28**
Horizontal line test is unsuccessful

For instance, both -2 and 2 are mapped into 4, and we find that $f(-2) = f(2)$ but $-2 \neq 2$ (note that $|-2| = |2|$). To invert this function, we would have to map 4 into -2 and 2, but then it would no longer be a function by our definition. By restricting the domain of $y = x^2$ to, say, $x \geq 0$, we can define an inverse function of $y = x^2$, $x \geq 0$.

Here is the formal definition of an inverse function.

> **Definition** Let $f : A \to B$ be a one-to-one function with range $f(A)$. The **inverse** function f^{-1} has domain $f(A)$ and range A and is defined by
>
> $$f^{-1}(y) = x \quad \text{if and only if} \quad y = f(x)$$
>
> for all $y \in f(A)$.

▶ Example 12   Find the inverse function of $f(x) = x^3 + 1$, $x \geq 0$.

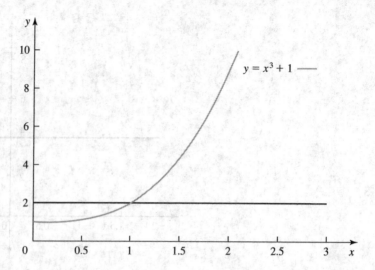

▲ **Figure 1.29**
The graph of $f(x) = x^3 + 1$ in Example 12. The horizontal line test is successful

Solution

First note that $f(x)$ is one to one. To see this quickly, graph the function and apply the horizontal line test (see Figure 1.29). To demonstrate it algebraically, start with $f(x_1) = f(x_2)$ and show that this implies $x_1 = x_2$:

$$f(x_1) = f(x_2)$$
$$x_1^3 + 1 = x_2^3 + 1$$
$$x_1^3 = x_2^3$$

Taking the third root on both sides implies that $x_1 = x_2$, which tells us that $f(x)$ has an inverse. Now we will find f^{-1}.

To find an inverse function, there are three steps:

1. Write $y = f(x)$:

$$y = x^3 + 1$$

2. Solve for x:

$$x^3 = y - 1$$
$$x = \sqrt[3]{y - 1}$$

The range of f is $[1, \infty)$, and the range of f becomes the domain of f^{-1}, so we obtain

$$f^{-1}(y) = \sqrt[3]{y - 1}, \quad y \geq 1$$

Typically, we write functions in terms of x. To do this, we need to interchange x and y in $x = f^{-1}(y)$. This is the third step.

3. Interchange x and y:

$$y = f^{-1}(x) = \sqrt[3]{x - 1}, x \geq 1$$

Note that switching x and y in step 3 corresponds to reflecting the graph of $y = f(x)$ about the line $y = x$. The graphs of f and f^{-1} are shown in

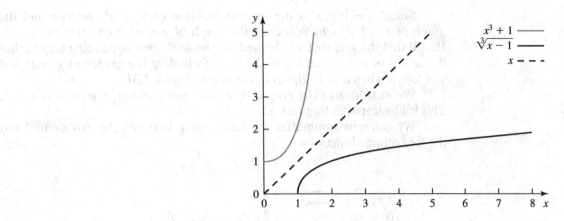

▲ Figure 1.30
Inverse functions

Figure 1.30. Look at the graphs carefully to see that the two graphs are indeed related to each other by reflecting either one about the line $y = x$.   ◀

As mentioned in the beginning of this subsection, the inverse of a function f reverses the effect of f. If we first apply f to x and then f^{-1} to $f(x)$, we obtain the original value x. Likewise, if we first apply f^{-1} to x and then f to $f^{-1}(x)$, we obtain the original value x. That is, if $f : A \to B$ has an inverse f^{-1}, then

$$f^{-1}[f(x)] = x \quad \text{for all } x \in A$$
$$f[f^{-1}(x)] = x \quad \text{for all } x \in f(A)$$

A note of warning: The superscript in f^{-1} does *not* mean the reciprocal of f (i.e., $1/f$). This difference is further explained in Problem 74 at the end of this section.

1.2.7 Logarithmic Functions

Recall from algebra (or Subsection 1.1.5) that to solve the equation

$$e^x = 3$$

for x, you must take logarithms on both sides:

$$x = \ln 3$$

In other words, the natural logarithm undoes the operation of raising e to the x power—it is the inverse of the exponential function. (Consequently, the exponential function is the inverse of the logarithmic function.)

We will now define the inverse of the exponential function $f(x) = a^x$, $x \in \mathbf{R}$. The base a can be any positive number, except 1.

> **Definition**   The inverse of $f(x) = a^x$ is called the **logarithm to base a** and is written $f^{-1}(x) = \log_a x$.

The domain of $f(x) = a^x$ is the set of all real numbers, and its range is the set of all positive numbers. Since the range of f is the domain of f^{-1}, we find that the domain of $f^{-1}(x) = \log_a x$ is the set of positive numbers.

Since $y = \log_a x$ is the inverse function of $y = a^x$, we can find the graph of $y = \log_a x$ by reflecting the graph of $y = a^x$ about the line $y = x$. Recall that the graph of $y = a^x$ had two basic shapes, depending on whether $0 < a < 1$ or $a > 1$ (see Figure 1.24). Let's look at the graphs of $y = a^x$ and $y = \log_a x$ when $a > 1$. This is illustrated in Figure 1.31.

We next look at the graphs of $y = a^x$ and $y = \log_a x$ when $0 < a < 1$. This is illustrated in Figure 1.32.

We can now summarize the relationship between the exponential and the logarithmic functions.

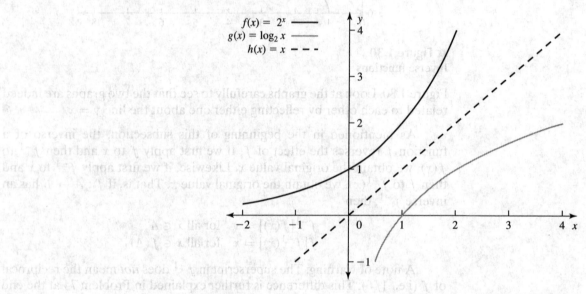

▲ **Figure 1.31**
The graph of $y = a^x$ and the graph of $y = \log_a x$ for $a > 1$. (Here, $a = 2$)

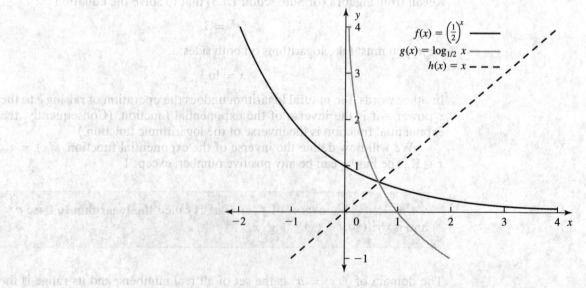

▲ **Figure 1.32**
The graph of $y = a^x$ and the graph of $y = \log_a x$ for $0 < a < 1$. (Here, $a = 1/2$)

1. $a^{\log_a x} = x$ for $x > 0$
2. $\log_a a^x = x$ for $x \in \mathbf{R}$

It is important to remember that the logarithm is only defined for positive numbers; that is, $y = \log_a x$ is only defined for $x > 0$. The logarithm satisfies the following properties:

$$\log_a(st) = \log_a s + \log_a t$$

$$\log_a\left(\frac{s}{t}\right) = \log_a s - \log_a t$$

$$\log_a s^r = r \log_a s$$

The inverse of the exponential function with the natural base e is denoted by $\ln x$ and is called the natural logarithm of x. The graphs of $y = e^x$ and $y = \ln x$ are shown in Figure 1.33. Note that both e^x and $\ln x$ are increasing functions. However, whereas e^x climbs very quickly for large values of x, $\ln x$ increases very slowly for large values of x. Looking at both graphs, we can see that each can be obtained as the reflection of the other about the line $y = x$.

The logarithm to base 10 is frequently written as $\log x$ (i.e., the base of 10 in $\log_{10} x$ is omitted).

▶ **Example 13**   Simplify the following expressions: (a) $\log_2[8(x-2)]$, (b) $\log_3 9^x$, (c) $\ln e^{3x^2+1}$.

Solution

(a) We simplify

$$\log_2[8(x - 2)] = \log_2 8 + \log_2(x - 2) = 3 + \log_2(x - 2)$$

No further simplification is possible.

(b) Simplifying yields

$$\log_3 9^x = x \log_3 9 = x \log_3 3^2 = 2x$$

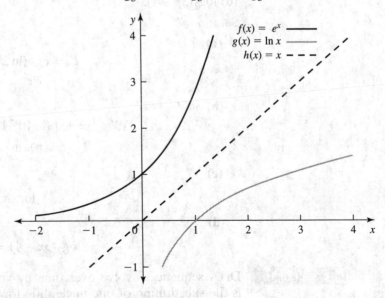

▲ **Figure 1.33**
The graphs of $y = e^x$ and $y = \ln x$.

The fact that $\log_3 9 = 2$ can be seen in two ways: We can write $9 = 3^2$ and say that $\log_3$ undoes raising 3 to the second power (as we did previously), or we can say that $\log_3 9$ denotes the exponent to which we must raise 3 in order to get 9.

(c) We use the fact that $\ln x$ and e^x are inverse functions and find that

$$\ln e^{3x^2+1} = 3x^2 + 1 \qquad \blacktriangleleft$$

Any exponential function with base a can be written as an exponential function with base e. Likewise, any logarithm to base a can be written in terms of the natural logarithm. The following two identities show how.

$$a^x = \exp[x \ln a]$$

$$\log_a x = \frac{\ln x}{\ln a}$$

The first identity follows from the fact that exp and ln are inversely related, which implies that $a^x = \exp[\ln a^x]$ and the fact that $\ln a^x = x \ln a$. To understand the second identity, note that

$$y = \log_a x \quad \text{means} \quad a^y = x$$

Taking logarithms to base e on both sides of $a^y = x$, we get

$$\ln a^y = \ln x$$

or

$$y \ln a = \ln x$$

Hence,

$$y = \frac{\ln x}{\ln a}$$

▶ **Example 14**   Write the following expressions in terms of base e: (a) 2^x, (b) 10^{x^2+1}, (c) $\log_3 x$, (d) $\log_2(3x - 1)$.

Solution

(a)

$$2^x = \exp(\ln 2^x) = \exp(x \ln 2)$$
$$= e^{x \ln 2}$$

(b)

$$10^{x^2+1} = \exp(\ln 10^{x^2+1}) = \exp[(x^2 + 1) \ln 10]$$
$$= e^{(x^2+1) \ln 10}$$

(c)

$$\log_3 x = \frac{\ln x}{\ln 3}$$

(d)

$$\log_2(3x - 1) = \frac{\ln(3x - 1)}{\ln 2} \qquad \blacktriangleleft$$

▶ **Example 15**   DNA sequences evolve over time by various processes. One such process is the substitution of one nucleotide for another. The simplest substitution scheme is that of Jukes and Cantor (1969), which assumes that substitutions are equally likely among the four nucleotide types. When comparing two

DNA sequences that have a common origin, it is possible to estimate the number of substitutions per site. Since more than one substitution can occur per site, the number of observed substitutions may be smaller than the number of actual substitutions, particularly when the time of divergence is large. Mathematical models are used to correct for this difference. The proportion p of observed nucleotide differences between two sequences that share a common ancestor can be used to find an estimate for the actual number K of substitutions per site since the time of divergence. Using the substitution scheme of Jukes and Cantor, K and p are related by

$$K = -\frac{3}{4} \ln \left(1 - \frac{4}{3} p \right)$$

provided p is not too large. Assume that two sequences of length 150 nucleotides differ from each other by 23 nucleotides. Find K.

Solution

The variable p denotes the proportion of observed nucleotide differences, which is $23/150 \approx 0.1533$ in our example. We thus find

$$K = -\frac{3}{4} \ln \left(1 - \frac{4}{3} \frac{23}{150} \right) \approx 0.1715 \qquad \blacktriangleleft$$

1.2.8 Trigonometric Functions

The trigonometric functions are examples of periodic functions.

> **Definition** A function $f(x)$ is **periodic** if there is a positive constant a such that
>
> $$f(x + a) = f(x)$$
>
> for all x in the domain of f. If a is the smallest number with this property, we call it the **period** of $f(x)$.

We begin with the sine and cosine functions. In Subsection 1.1.4, we recalled the definition of sine and cosine in a unit circle. There, $\sin \theta$ and $\cos \theta$ represented trigonometric functions of angles and θ was either measured in degrees or radians. Now we define the trigonometric functions as functions of *real* numbers. For instance, we define $f(x) = \sin x$ for $x \in \mathbf{R}$. The value of $\sin x$ is then *by definition* the sine of an angle of x radians (similarly for all the other trigonometric functions).

The graphs of the sine and cosine functions are shown in Figures 1.34 and 1.35, respectively.

The sine function $y = \sin x$ is defined for all $x \in \mathbf{R}$. Its range is $-1 \le y \le 1$. Likewise, the cosine function $y = \cos x$ is defined for all $x \in \mathbf{R}$ with range $-1 \le y \le 1$. Both functions are periodic with period 2π. That is, $\sin(x + 2\pi) = \sin x$ and $\cos(x + 2\pi) = \cos x$. [We also have $\sin(x + 4\pi) = \sin x$, $\sin(x + 6\pi) = \sin x, \ldots$, and $\cos(x + 4\pi) = \cos x$, $\cos(x + 6\pi) = \cos x, \ldots$, but by convention, we use the smallest possible value to specify the period.] We see from Figures 1.34 and 1.35 that the graph of the cosine function can be obtained by shifting the graph of the sine function a distance of $\pi/2$ units to the left. (We will discuss horizontal shifts of graphs in more detail in the next section.)

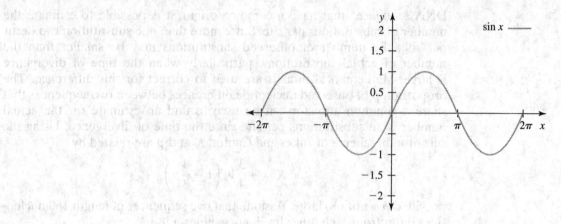

▲ Figure 1.34
The graph of $y = \sin x$

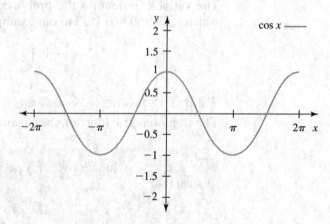

▲ Figure 1.35
The graph of $y = \cos x$

To define the tangent function, $y = \tan x$, recall that

$$\tan x = \frac{\sin x}{\cos x}$$

Because $\cos x = 0$ for values of x that are odd integer multiples of $\pi/2$, the domain of $\tan x$ consists of all real numbers with the exception of odd integer multiples of $\pi/2$. The range of $y = \tan x$ is $-\infty < y < \infty$. The graph of $y = \tan x$ is shown in Figure 1.36, from which we see that $\tan x$ is periodic with period π.

The graphs of the remaining three trigonometric functions are shown in Figures 1.37–1.39. The domain of the secant function $y = \sec x$ consists of all real numbers with the exception of odd integer multiples of $\pi/2$; the range is $|y| \geq 1$. The domain of the cosecant function $y = \csc x$ consists of all real numbers with the exception of integer multiples of π; the range is $|y| \geq 1$. The domain of the cotangent function $y = \cot x$ consists of all real numbers with the exception of integer multiples of π; the range is $-\infty < y < \infty$.

Since the sine and cosine functions are of particular importance, we now describe them in more detail. Consider the function

$$f(x) = a \sin(kx) \quad \text{for } x \in \mathbf{R}$$

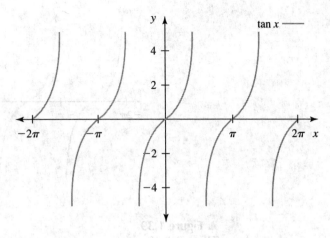

▲ **Figure 1.36**
The graph of $y = \tan x$

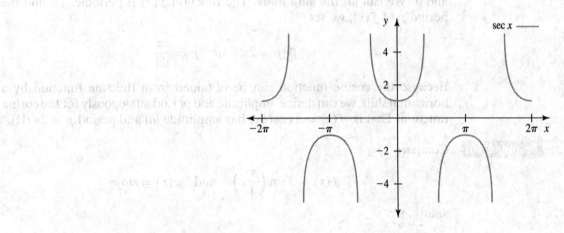

▲ **Figure 1.37**
The graph of $y = \sec x$

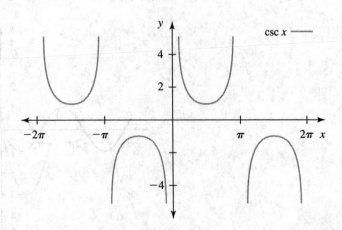

▲ **Figure 1.38**
The graph of $y = \csc x$

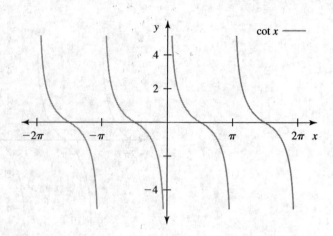

▲ **Figure 1.39**
The graph of $y = \cot x$

where a is a real number and $k \neq 0$. Now, $f(x)$ takes on values between $-a$ and a. We call $|a|$ the **amplitude**. The function $f(x)$ is periodic. To find the period p of $f(x)$, we set

$$|k|p = 2\pi \quad \text{or} \quad p = \frac{2\pi}{|k|}$$

Because the cosine function can be obtained from the sine function by a horizontal shift, we can define amplitude and period analogously for the cosine function. That is, $f(x) = a\cos(kx)$ has amplitude $|a|$ and period $p = 2\pi/|k|$.

▶ **Example 16**   Compare

$$f(x) = 3\sin\left(\frac{\pi}{4}x\right) \quad \text{and} \quad g(x) = \sin x$$

Solution

The amplitude of $f(x)$ is 3, whereas the amplitude of $g(x)$ is 1. The period p of $f(x)$ satisfies $\frac{\pi}{4}p = 2\pi$ or $p = 8$, whereas the period of $g(x)$ is 2π. Graphs of $f(x)$ and $g(x)$ are shown in Figure 1.40.   ◀

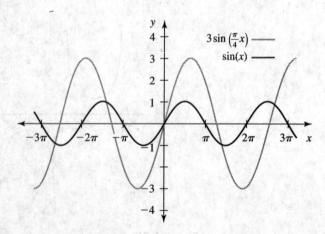

▲ **Figure 1.40**
The graphs of $y = 3\sin(\frac{\pi}{4}x)$ and $g(x) = \sin x$ in Example 16

Remark. A number is called **algebraic** if it is the solution of a polynomial equation with rational coefficients. For instance, $\sqrt{2}$ is algebraic as it satisfies the equation $x^2 - 2 = 0$. Numbers that are not algebraic are called **transcendental**. For instance, π and e are transcendental.

A similar distinction is also made for functions. We call a function $y = f(x)$ algebraic if it is the solution of an equation of the form

$$P_n(x)y^n + \cdots + P_1(x)y + P_0(x) = 0$$

in which the coefficients are polynomial functions in x with rational coefficients. For instance, the function $y = 1/(1 + x)$ is algebraic as it satisfies the equation $(x + 1)y - 1 = 0$. Here, $P_1(x) = x + 1$ and $P_0(x) = -1$. Other examples of algebraic functions include polynomials and rational functions with rational coefficients.

Functions that are not algebraic are called transcendental. All the trigonometric, exponential, and logarithmic functions that we introduced in this section are examples of transcendental functions.

1.2.9 Problems

(1.2.1)

In Problems 1–4, state the range for the given functions. Graph each function.

1. $f(x) = x^2, x \in \mathbf{R}$
2. $f(x) = x^2, x \in [0, 1]$
3. $f(x) = x^2, -1 < x \leq 0$
4. $f(x) = x^2, -\frac{1}{2} < x < \frac{1}{2}$
5. (a) Show that for $x \neq 1$,
$$\frac{x^2 - 1}{x - 1} = x + 1$$
   (b) Are the functions
$$f(x) = \frac{x^2 - 1}{x - 1}, \quad x \neq 1$$
   and
$$g(x) = x + 1, \quad x \in \mathbf{R}$$
   equal?
6. (a) Show that
$$2|x - 1| = \begin{cases} 2(x - 1) & \text{for } x \geq 1 \\ 2(1 - x) & \text{for } x \leq 1 \end{cases}$$
   (b) Are the functions
$$f(x) = \begin{cases} 2 - 2x & \text{for } 0 \leq x \leq 1 \\ 2x - 2 & \text{for } 1 \leq x \leq 2 \end{cases}$$
   and
$$g(x) = 2|x - 1|, x \in [0, 2]$$
   equal?

In Problems 7–12, sketch the graph of each function, and decide in each case whether the function is (i) even, (ii) odd, or (iii) does not show any obvious symmetry. Then check your answers using the criteria in Subsection 1.2.1.

7. $f(x) = 2x$
8. $f(x) = 3x^2$
9. $f(x) = |3x|$
10. $f(x) = 2x + 1$
11. $f(x) = -|x|$
12. $f(x) = 3x^3$
13. Suppose that
$$f(x) = x^2, \quad x \in \mathbf{R}$$
   and
$$g(x) = 3 + x, \quad x \in \mathbf{R}$$
   (a) Show that
$$(f \circ g)(x) = (3 + x)^2, \quad x \in \mathbf{R}$$
   (b) Show that
$$(g \circ f)(x) = 3 + x^2, \quad x \in \mathbf{R}$$
14. Suppose that
$$f(x) = x^3, \quad x \in \mathbf{R}$$
   and
$$g(x) = 1 - x, \quad x \in \mathbf{R}$$
   (a) Show that
$$(f \circ g)(x) = (1 - x)^3, \quad x \in \mathbf{R}$$
   (b) Show that
$$(g \circ f)(x) = 1 - x^3, \quad x \in \mathbf{R}$$
15. Suppose that
$$f(x) = 1 - x^2, \quad x \in \mathbf{R}$$
   and
$$g(x) = 2x, \quad x \geq 0$$
   (a) Find
$$(f \circ g)(x)$$
   together with its domain.
   (b) Find
$$(g \circ f)(x)$$
   together with its domain.
16. Suppose that
$$f(x) = \frac{1}{x + 1}, \quad x \neq -1$$

and
$$g(x) = 2x^2, \quad x \in \mathbf{R}$$

(a) Find $(f \circ g)(x)$.

(b) Find $(g \circ f)(x)$.

In both **(a)** and **(b)**, find the domain.

17. Suppose that
$$f(x) = 3x^2, \quad x \geq 3$$
and
$$g(x) = \sqrt{x}, \quad x \geq 0$$
Find $(f \circ g)(x)$ together with its domain.

18. Suppose that
$$f(x) = x^4, \quad x \geq 3$$
and
$$g(x) = \sqrt{x+1}, \quad x \geq 3$$
Find $(f \circ g)(x)$ together with its domain.

19. Suppose that $f(x) = x^2$, $x \geq 0$, and $g(x) = \sqrt{x}$, $x \geq 0$. Typically, $f \circ g \neq g \circ f$, but here is an example where the order of composition does not matter. Show that $f \circ g = g \circ f$.

20. Suppose that $f(x) = x^4$, $x \geq 0$. Find $g(x)$ so that $f \circ g = g \circ f$.

(1.2.2)

21. Use a graphing calculator to graph $f(x) = x^2$, $x \geq 0$, and $g(x) = x^4$, $x \geq 0$, together. For which values of x is $f(x) > g(x)$, and for which is $f(x) < g(x)$?

22. Use a graphing calculator to graph $f(x) = x^3$, $x \geq 0$, and $g(x) = x^5$, $x \geq 0$, together. When is $f(x) > g(x)$, and when is $f(x) < g(x)$?

23. Graph $y = x^n$, $x \geq 0$, for $n = 1, 2, 3$, and 4 in one coordinate system. Where do the curves intersect?

24. (a) Graph $f(x) = x$, $x \geq 0$, and $g(x) = x^2$, $x \geq 0$, together, in one coordinate system.

(b) For which values of x is $f(x) \geq g(x)$, and for which values of x is $f(x) \leq g(x)$?

25. (a) Graph $f(x) = x^2$ and $g(x) = x^3$ for $x \geq 0$ together, in one coordinate system.

(b) Show algebraically that
$$x^2 \geq x^3$$
for $0 \leq x \leq 1$.

(c) Show algebraically that
$$x^2 \leq x^3$$
for $x \geq 1$.

26. Show algebraically that if $n \geq m$,
$$x^n \leq x^m \quad \text{for } 0 \leq x \leq 1$$
and
$$x^n \geq x^m \quad \text{for } x \geq 1$$

27. (a) Show that $y = x^2$, $x \in \mathbf{R}$, is an even function.

(b) Show that $y = x^3$, $x \in \mathbf{R}$, is an odd function.

28. Show that

(a) $y = x^n$, $x \in \mathbf{R}$, is an even function when n is an even integer.

(b) $y = x^n$, $x \in \mathbf{R}$, is an odd function when n is an odd integer.

29. In Example 5 of this section, we considered the chemical reaction
$$A + B \rightarrow AB$$

Assume that initially only A and B are in the reaction vessel and that the initial concentrations are $a = [A] = 3$ and $b = [B] = 4$.

(a) We found that the reaction rate $R(x)$, where x is the concentration of AB, is given by
$$R(x) = k(a - x)(b - x)$$
where a is the initial concentration of A, b is the initial concentration of B, and k is the constant of proportionality. Suppose that the reaction rate $R(x)$ is equal to 9 when the concentration of AB is $x = 1$. Use this to find the reaction rate $R(x)$.

(b) Determine the appropriate domain of $R(x)$, and use a graphing calculator to sketch the graph of $R(x)$.

30. An autocatalytic reaction uses its resulting product for the formation of a new product, as in the reaction
$$A + X \rightarrow X$$
If we assume that this reaction occurs in a closed vessel, then the reaction rate is given by
$$R(x) = kx(a - x)$$
for $0 \leq x \leq a$, where a is the initial concentration of A and x is the concentration of X.

(a) Show that $R(x)$ is a polynomial and determine its degree.

(b) Graph $R(x)$ for $k = 2$ and $a = 6$. Find the value of x at which the reaction rate is maximal.

31. Suppose that a beetle walks up a tree along a straight line at a constant speed of 1 meter per hour. What distance will the beetle have covered after 1 hour, 2 hours, and 3 hours? Write an equation that expresses the distance (in meters) as a function of the time (in hours), and show that this function is a polynomial of degree 1.

32. Suppose that a fungal disease originates in the middle of an orchard, initially affecting only one tree. The disease spreads out radially at a constant speed of 10 feet per day. What area will be affected after 2 days, 4 days, and 8 days? Write an equation that expresses the affected area as a function of time measured in days, and show that this function is a polynomial of degree 2.

(1.2.3)

In Problems 33–36, for each function, find the largest possible domain and determine the range.

33. $f(x) = \frac{1}{1-x}$

34. $f(x) = \frac{2x}{(x-2)(x+3)}$

35. $f(x) = \frac{x-2}{x^2-9}$

36. $f(x) = \frac{1}{x^2+1}$

37. Compare $y = \frac{1}{x}$ and $y = \frac{1}{x^2}$ for $x > 0$ by graphing the two functions. Where do the curves intersect? Which function is greater for small values of x? for large values of x?

38. Let n and m be two positive integers with $m \leq n$. Answer the following questions about $y = x^{-n}$ and $y = x^{-m}$ for $x > 0$. Where do the curves intersect? Which function is greater for small values of x? for large values of x?

39. Let

$$f(x) = \frac{1}{x+1}, \quad x > -1$$

(a) Use a graphing calculator to graph $f(x)$.

(b) Based on the graph in (a), determine the range of $f(x)$.

(c) For which values of x is $f(x) = 2$?

(d) Based on the graph in (a), determine how many solutions $f(x) = a$ has, where a is in the range of $f(x)$.

40. Let

$$f(x) = \frac{2x}{3+x}, \quad x \geq 0$$

(a) Use a graphing calculator to graph $f(x)$.

(b) Find the range of $f(x)$.

(c) For which values of x is $f(x) = 1$?

(d) Based on the graph in (a), explain in words why for any value a in the range of $f(x)$, you can find exactly one value $x \geq 0$ so that $f(x) = a$. Determine x by solving $f(x) = a$.

41. Let

$$f(x) = \frac{3x}{1+x}, \quad x \geq 0$$

(a) Use a graphing calculator to graph $f(x)$.

(b) Find the range of $f(x)$.

(c) For which values of x is $f(x) = 2$?

(d) Based on the graph in (a), explain in words why for any value a in the range of $f(x)$, you can find exactly one value $x \geq 0$ so that $f(x) = a$. Determine x by solving $f(x) = a$.

In Problems 42–44, we discuss the Monod growth function, which was introduced in Example 6 of this section.

42. Use a graphing calculator to investigate the Monod growth function

$$r(N) = \frac{aN}{k+N}, \quad N \geq 0$$

where a and k are positive constants.

(a) Graph $r(N)$ for (i) $a = 5$ and $k = 1$, (ii) $a = 5$ and $k = 3$, and (iii) $a = 8$ and $k = 1$. Place all three graphs in one coordinate system.

(b) Based on the graphs in (a), describe in words what happens when you change a.

(c) Based on the graphs in (a), describe in words what happens when you change k.

43. The Monod growth function $r(N)$ describes growth as a function of nutrient concentration N. Assume

$$r(N) = 5\frac{N}{1+N}, \quad N \geq 0$$

Find the percentage increase when the nutrient concentration is doubled from $N = 0.1$ to $N = 0.2$. Compare this to your findings when you double the nutrient concentration from $N = 10$ to $N = 20$. This is an example of *diminishing return*.

44. The Monod growth function $r(N)$ describes growth as a function of nutrient concentration N. Assume that

$$r(N) = a\frac{N}{k+N}, \quad N \geq 0$$

where a and k are positive constants.

(a) What happens to $r(N)$ as N increases? Use this to explain why a is called the saturation level.

(b) Show that k is the half-saturation constant; that is, show that if $N = k$, then $r(N) = a/2$.

45. Let

$$f(x) = \frac{x^2}{4+x^2}, \quad x \geq 0$$

(a) Use a graphing calculator to graph $f(x)$.

(b) Based on your graph in (a), find the range of $f(x)$.

(c) What happens to $f(x)$ as x gets larger?

46. The following function is used in biochemistry to model reaction rates as a function of the concentration of some reactants.

$$f(x) = \frac{x^n}{b^n + x^n}, \quad x \geq 0$$

where n is a positive integer and b is a positive real number.

(a) Use a graphing calculator to graph $f(x)$ for $n = 1, 2$, and 3 in one coordinate system when $b = 2$.

(b) Where do the three graphs in (a) intersect?

(c) What happens to $f(x)$ as x gets larger?

(d) For an arbitrary positive value of b, show that $f(b) = 1/2$. Based on this and your answer in (c), explain why b is called the half-saturation constant.

(1.2.4)

In Problems 47–50, use a graphing calculator to sketch the graphs of the functions.

47. $y = x^{3/2}, x \geq 0$

48. $y = x^{1/3}, x \geq 0$

49. $y = x^{-1/4}, x > 0$

50. $y = 2x^{-7/8}, x > 0$

51. (a) Graph $y = x^{-1/2}, x > 0$, and $y = x^{1/2}, x \geq 0$, together in one coordinate system.

(b) Show algebraically that

$$x^{-1/2} \geq x^{1/2}$$

for $0 < x \leq 1$.

(c) Show algebraically that

$$x^{-1/2} \leq x^{1/2}$$

for $x \geq 1$.

52. (a) Graph $y = x^{5/2}, x \geq 0$, and $y = x^{1/2}, x \geq 0$, together in one coordinate system.

(b) Show algebraically that

$$x^{5/2} \leq x^{1/2}$$

for $0 \leq x \leq 1$. (*Hint:* Show that $x^{1/2}/x^{-1/2} = x \leq 1$ for $0 < x \leq 1$.)

(c) Show algebraically that

$$x^{5/2} \geq x^{1/2}$$

for $x \geq 1$.

In Problems 53–56, sketch each scaling relation (Niklas, 1994).

53. In a sample based on 46 species, leaf area was found to be proportional to (stem diameter)$^{1.84}$. Based on your graph, as stem diameter increases, does leaf area increase of decrease?

54. In a sample based on 28 species, the volume fraction of spongy mesophyll was found to be proportional to (leaf thickness)$^{-0.49}$. (The spongy mesophyll is part of the internal tissue of a leaf blade.) Based on your graph, as leaf thickness increases, does the volume fraction of spongy mesophyll increase or decrease?

55. In a sample of 60 species of trees, wood density was found to be proportional to (breaking strength)$^{0.82}$. Based on your graph, when does breaking strength increase: as wood density increases or decreases?

56. Suppose that a cube of length L and volume V has mass M and that $M = 0.35V$. How does the length of the cube depend on its mass?

(1.2.5)

57. Assume that a population size at time t is $N(t)$ and that
$$N(t) = 2^t, \quad t \geq 0$$

(a) Find the population size for $t = 0, 1, 2, 3,$ and 4.

(b) Graph $N(t)$ for $t \geq 0$.

58. Assume that a population size at time t is $N(t)$ and that
$$N(t) = 40 \cdot 2^t, \quad t \geq 0$$

(a) Find the population size at time $t = 0$.

(b) Show that
$$N(t) = 40e^{t \ln 2}, \quad t \geq 0$$

(c) How long will it take until the population size reaches 1000? [*Hint:* Find t so that $N(t) = 1000$.]

59. The half-life of C^{14} is 5730 years. If a sample of C^{14} has a mass of 20 micrograms at time $t = 0$, how much is left after 2000 years?

60. The half-life of C^{14} is 5730 years. If a sample of C^{14} has a mass of 20 micrograms at time 0, how long will it take until (a) 10 micrograms, and (b) 5 micrograms are left?

61. After 7 days, a particular radioactive substance decays to half of its original amount. Find the decay rate of this substance.

62. After 5 days, a particular radioactive substance decays to 37% of its original amount. Find the half-life of this substance.

63. Polonium 210 (Po^{210}) has a half-life of 140 days.

(a) If a sample of Po^{210} has a mass of 300 micrograms, find a formula for the mass after t days.

(b) How long would it take this sample to decay to 20% of its original amount?

(c) Sketch the graph of the amount of mass left after t days.

64. The half-life of C^{14} is 5730 years. Suppose that wood found at an archeological excavation site contains about 35% as much C^{14} (in relation to C^{12}) as living plant material. Determine when the wood was cut.

65. The half-life of C^{14} is 5730 years. Suppose that wood found at an archeological excavation site is 15,000 years old. How much C^{14} (based on C^{12} content) does the wood contain relative to living plant material?

66. The age of rocks of volcanic origin can be estimated using isotopes of argon 40 (Ar^{40}) and potassium 40 (K^{40}). K^{40} decays into Ar^{40} over time. If a mineral that contains potassium is buried under the right circumstances, argon forms and is trapped. Since argon is driven off when the mineral is heated to very high temperatures, rocks of volcanic origin do not contain argon when they are formed. The amount of argon found in such rocks can therefore be used to determine the age of the rock. Assume that a sample of volcanic rock contains 0.00047% K^{40}. The sample also contains 0.000079% Ar^{40}. How old is the rock? (The decay rate of K^{40} to Ar^{40} is 5.335×10^{-10}/yr.)

67. (*Adapted from Moss, 1980*) Hall (1964) investigated the change in population size of the zooplankton species *Daphnia galeata mendota* in Base Line Lake, Michigan. The population size at time t, $N(t)$, was modeled by the equation
$$N(t) = N_0 e^{rt}$$
where N_0 denotes the population size at time 0. The constant r is called the **intrinsic rate of growth**.

(a) Plot $N(t)$ as a function of t if $N_0 = 100$ and $r = 2$. Compare your graph to the graph of $N(t)$ when $N_0 = 100$ and $r = 3$. Which population grows faster?

(b) The constant r is an important quantity since it describes how quickly the population changes. Suppose that you determine the size of the population at the beginning and at the end of a time period of length 1, and you find that in the beginning there were 200 individuals and after one unit of time there were 250 individuals. Determine r. [*Hint:* Consider the ratio $N(t + 1)/N(t)$.]

68. Fish are indeterminate growers: that is, they grow throughout their lifetime. The growth can be described by the von Bertalanffy function
$$L(x) = L_\infty(1 - e^{-kx})$$
for $x \geq 0$, where $L(x)$ is the length at age x and k and L_∞ are positive constants.

(a) Use a graphing calculator to graph $L(x)$ for $L_\infty = 20$, for (i) $k = 1$, and (ii) $k = 0.1$.

(b) For $k = 1$, find x so that the length is 90% of L_∞. Repeat for 99% of L_∞. Can the fish ever attain length L_∞? Interpret the meaning of L_∞.

(c) Compare the graphs obtained in (a). Which growth curve reaches 90% of L_∞ faster? Can you explain what happens to the curve of $L(x)$ when you vary k (for fixed L_∞)?

(1.2.6)

69. Which of the following functions is one to one? (Use the horizontal line test.)

(a) $f(x) = x^2, x \geq 0$

(b) $f(x) = x^2, x \in \mathbb{R}$

(c) $f(x) = \frac{1}{x}, x > 0$

(d) $f(x) = e^x, x \in \mathbb{R}$

(e) $f(x) = \frac{1}{x^2}, x \neq 0$

(f) $f(x) = \frac{1}{x^2}, x > 0$

70. **(a)** Show that $f(x) = x^3 - 1, x \in \mathbf{R}$, is one to one, and find its inverse together with its domain.

(b) Graph $f(x)$ and $f^{-1}(x)$ in one coordinate system, together with the line $y = x$, and convince yourself that the graph of $f^{-1}(x)$ can be obtained by reflecting the graph of $f(x)$ about the line $y = x$.

71. **(a)** Show that $f(x) = x^2 + 1, x \geq 0$, is one to one, and find its inverse together with its domain.

(b) Graph $f(x)$ and $f^{-1}(x)$ in one coordinate system, together with the line $y = x$, and convince yourself that the graph of $f^{-1}(x)$ can be obtained by reflecting the graph of $f(x)$ about the line $y = x$.

72. **(a)** Show that $f(x) = \sqrt{x}, x \geq 0$, is one to one, and find its inverse together with its domain.

(b) Graph $f(x)$ and $f^{-1}(x)$ in one coordinate system, together with the line $y = x$, and convince yourself that the graph of $f^{-1}(x)$ can be obtained by reflecting the graph of $f(x)$ about the line $y = x$.

73. **(a)** Show that $f(x) = 1/x^3, x > 0$, is one to one, and find its inverse together with its domain.

(b) Graph $f(x)$ and $f^{-1}(x)$ in one coordinate system, together with the line $y = x$, and convince yourself that the graph of $f^{-1}(x)$ can be obtained by reflecting the graph of $f(x)$ about the line $y = x$.

74. The reciprocal of a function $f(x)$ can be written as either $1/f(x)$ or $[f(x)]^{-1}$. The point of this problem is to make clear that a reciprocal of a function has nothing to do with the inverse of a function. As an example, let $f(x) = 2x + 1$, $x \in \mathbf{R}$. Find both $[f(x)]^{-1}$ and $f^{-1}(x)$, and compare the two functions. Graph all three functions together.

(1.2.7)

75. Find the inverse of $f(x) = 3^x, x \in \mathbf{R}$, together with its domain and graph both functions in the same coordinate system.

76. Find the inverse of $f(x) = 5^x, x \in \mathbf{R}$, together with its domain and graph both functions in the same coordinate system.

77. Find the inverse of $f(x) = (\frac{1}{4})^x, x \in \mathbf{R}$, together with its domain and graph both functions in the same coordinate system.

78. Find the inverse of $f(x) = (\frac{1}{3})^x, x \in \mathbf{R}$, together with its domain and graph both functions in the same coordinate system.

79. Find the inverse of $f(x) = 2^x, x \geq 0$, together with its domain and graph both functions in the same coordinate system.

80. Find the inverse of $f(x) = (\frac{1}{2})^x, x \geq 0$, together with its domain and graph both functions in the same coordinate system.

81. Simplify the following expressions.

(a) $2^{5\log_2 x}$

(b) $3^{4\log_3 x}$

(c) $5^{5\log_{1/5} x}$

(d) $4^{-2\log_2 x}$

(e) $2^{3\log_{1/2} x}$

(f) $4^{-\log_{1/2} x}$

82. Simplify the following expressions.

(a) $\log_4 16^x$

(b) $\log_2 16^x$

(c) $\log_3 27^x$

(d) $\log_{1/2} 4^x$

(e) $\log_{1/2} 8^{-x}$

(f) $\log_3 9^{-x}$

83. Simplify the following expressions.

(a) $\ln x^2 + \ln x^3$

(b) $\ln x^4 - \ln x^{-2}$

(c) $\ln(x^2 - 1) - \ln(x + 1)$

(d) $\ln x^{-1} + \ln x^{-3}$

84. Simplify the following expressions.

(a) $e^{3\ln x}$

(b) $e^{-\ln(x^2+1)}$

(c) $e^{-2\ln(1/x)}$

(d) $e^{-2\ln x}$

85. Write the following expressions in terms of base e and simplify.

(a) 3^x

(b) 4^{x^2-1}

(c) 2^{-x-1}

(d) 3^{-4x+1}

86. Write the following expressions in terms of base e.

(a) $\log_2(x^2 - 1)$

(b) $\log_3(5x + 1)$

(c) $\log(x + 2)$

(d) $\log_2(2x^2 - 1)$

87. Show that the function $y = (1/2)^x$ can be written in the form $y = e^{-\mu x}$, where μ is a positive constant. Determine μ.

88. Show that if $0 < a < 1$, the function $y = a^x$ can be written in the form $y = e^{-\mu x}$, where μ is a positive constant. Write μ in terms of a.

89. Assume that two DNA sequences of common origin, each of length 300 nucleotides, differ from each other by 47 nucleotides. Use the Jukes and Cantor correction of Example 15 to find an estimate for the number of substitutions per site, K.

90. A community measure that takes both species abundance and species richness into account is the Shannon diversity index, H. To calculate it, the proportion p_i of species i in the community is used. Assume that the community consists of S species. Then

$$H = -(p_1 \ln p_1 + p_2 \ln p_2 + \cdots + p_S \ln p_S)$$

(a) Assume that $S = 5$ and that all species are equally abundant; that is, $p_1 = p_2 = \cdots = p_5$. Compute H.

(b) Assume that $S = 10$ and that all species are equally abundant; that is, $p_1 = p_2 = \cdots = p_{10}$. Compute H.

(c) A measure of equitability (or evenness) of the species distribution can be measured by dividing the diversity index H by $\ln S$. Compute $H / \ln S$ for $S = 5$ and $S = 10$.

(d) Show that, in general, if there are N species and all species are equally abundant, then

$$\frac{H}{\ln S} = 1$$

(1.2.8)

In Problems 91–96, use a graphing calculator to compare the following pairs of functions. Describe what you see.

91. $y = \sin x$, and $y = 2 \sin x$

92. $y = \sin x$, and $y = \sin(2x)$

93. $y = \cos x$, and $y = 2 \cos x$

94. $y = \cos x$, and $y = \cos(2x)$

95. $y = \tan x$, and $y = 2 \tan x$

96. $y = \tan x$, and $y = \tan(2x)$

97. Let
$$f(x) = 3 \sin(4x), \quad x \in \mathbf{R}$$
Find the amplitude and the period of $f(x)$.

98. Let
$$f(x) = -2 \sin\left(\frac{x}{2}\right), \quad x \in \mathbf{R}$$

Find the amplitude and the period of $f(x)$.

99. Let
$$f(x) = 4 \sin(2\pi x), \quad x \in \mathbf{R}$$
Find the amplitude and the period of $f(x)$.

100. Let
$$f(x) = -\frac{3}{2} \sin\left(\frac{\pi}{3}x\right), \quad x \in \mathbf{R}$$
Find the amplitude and the period of $f(x)$.

101. Let
$$f(x) = 4 \cos\left(\frac{x}{4}\right), \quad x \in \mathbf{R}$$
Find the amplitude and the period of $f(x)$.

102. Let
$$f(x) = 7 \cos(2x), \quad x \in \mathbf{R}$$
Find the amplitude and the period of $f(x)$.

103. Let
$$f(x) = -3 \cos\left(\frac{\pi x}{5}\right), \quad x \in \mathbf{R}$$
Find the amplitude and the period of $f(x)$.

104. Let
$$f(x) = -\frac{2}{3} \cos\left(\frac{3x}{\pi}\right), \quad x \in \mathbf{R}$$
Find the amplitude and the period of $f(x)$.

1.3 GRAPHING

We introduced the functions most important to our study in the preceding section. You must be able to graph the following functions without a calculator: $y = c, x, x^2, x^3, 1/x, e^x, \ln x, \sin x, \cos x, \sec x,$ and $\tan x$. This will help you to sketch functions quickly and to come up with an analytical description of a function based on a graph. In this section, you will learn how to obtain new functions from these basic functions and how to graph them. In addition, we will introduce important transformations that are often used to display data graphically.

1.3.1 Graphing and Basic Transformations of Functions

In this subsection, we will recall some basic transformations: vertical and horizontal translations, reflections about $x = 0$ and $y = 0$, and stretching and compressing.

Definition The graph of

$$y = f(x) + a$$

is a **vertical translation** of $y = f(x)$. If $a > 0$, the graph of $y = f(x)$ is shifted up a units; if $a < 0$, it is shifted down $|a|$ units.

This is illustrated in Figure 1.41, where we display $y = x^2$, $y = x^2 + 2$, and $y = x^2 - 2$.

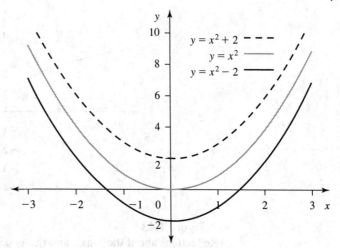

▲ **Figure 1.41**
The graphs of $y = x^2$, $y = x^2 + 2$ and $y = x^2 - 2$

Definition   The graph of

$$y = f(x - c)$$

is a **horizontal translation** of $y = f(x)$. If $c > 0$, the graph of $y = f(x)$ is shifted c units to the right; if $c < 0$, it is shifted $|c|$ units to the left.

This is illustrated in Figure 1.42, where we display $y = x^2$, $y = (x - 3)^2$, and $y = (x + 3)^2$. Note that $y = (x + 3)^2$ is shifted to the left since $y = (x - (-3))^2$ and therefore $c = -3 < 0$.

Reflections about the x-axis ($y = 0$) and the y-axis ($x = 0$) are illustrated in Figure 1.43. We graph $y = \sqrt{x}$, its reflection about the x-axis, $y = -\sqrt{x}$, and its reflection about the y-axis, $y = \sqrt{-x}$.

Multiplying a function by a factor between 0 and 1 compresses its graph; multiplying a function by a factor greater than 1 stretches its graph. This is illustrated in Figure 1.44, where we graph $y = x^2$ and $y = \frac{1}{2}x^2$, and in Figure 1.45, where we graph $y = x^2$ and $y = 2x^2$.

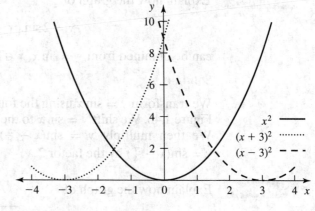

▲ **Figure 1.42**
The graphs of $y = x^2$, $y = (x - 3)^2$ and $y = (x + 3)^2$

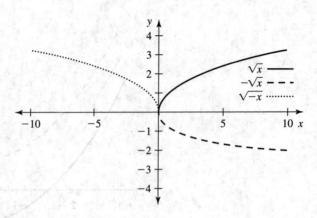

▲ **Figure 1.43**
Reflections about the x-axis and the y-axis

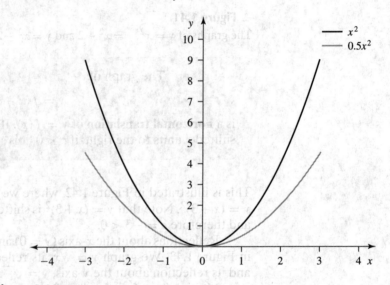

▲ **Figure 1.44**
The graphs of $y = x^2$ and $y = \frac{1}{2}x^2$

We illustrate these transformations in the following two examples.

▶ **Example 1** Explain how the graph of

$$y = 2\sin\left(x - \frac{\pi}{4}\right) \quad \text{for } x \in \mathbf{R}$$

can be obtained from $y = \sin x$, $x \in \mathbf{R}$.

Solution

We transform $y = \sin x$ using the following two steps, which are illustrated in Figure 1.46. We shift $y = \sin x$ to the right $\frac{\pi}{4}$ units. This yields $y = \sin(x - \frac{\pi}{4})$. We then multiply $y = \sin(x - \frac{\pi}{4})$ by 2. This corresponds to stretching $y = \sin(x - \frac{\pi}{4})$ by the factor 2. ◀

▶ **Example 2** Explain how the graph of

$$y = -\sqrt{x - 3} - 1, \quad x \geq 3$$

can be obtained from $y = \sqrt{x}$, $x \geq 0$.

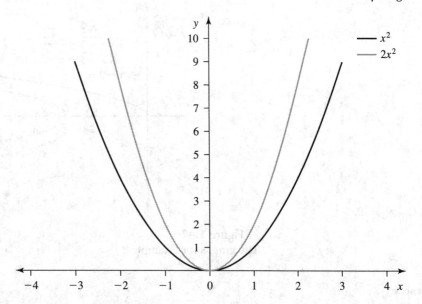

▲ **Figure 1.45**
The graphs of $y = x^2$ and $y = 2x^2$

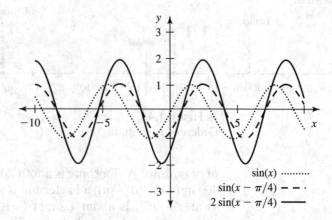

▲ **Figure 1.46**
The graphs of $y = \sin x$, $y = \sin(x - \frac{\pi}{4})$, and $y = 2\sin(x - \frac{\pi}{4})$

Solution

We transform $y = \sqrt{x}$ using the following three steps, which are illustrated in Figure 1.47. We shift $y = \sqrt{x}$ three units to the right and obtain $y = \sqrt{x - 3}$. We reflect $y = \sqrt{x - 3}$ about the x-axis, which yields $y = -\sqrt{x - 3}$. We shift $y = -\sqrt{x - 3}$ down one unit. This is the graph of $y = -\sqrt{x - 3} - 1$. ◀

1.3.2 The Logarithmic Scale

We often encounter sizes that vary over a wide range. Lengths in the metric system are measured in meters (m). (A meter is a bit longer than a yard: 1 meter is equal to 1.0936 yards.) A longer metric unit that is commonly used is a kilometer (km), which is 1000 m. Shorter commonly used metric units are a millimeter (mm), which is 1/1000 of a meter; a micrometer (μm), which is 1/1,000,000 (one millionth) of a meter; and a nanometer (nm), which is 1/1,000,000,000 (one billionth) of a meter. Here are some examples of lengths

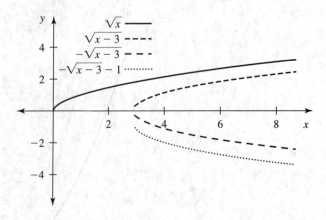

▲ **Figure 1.47**
The graphs for Example 2

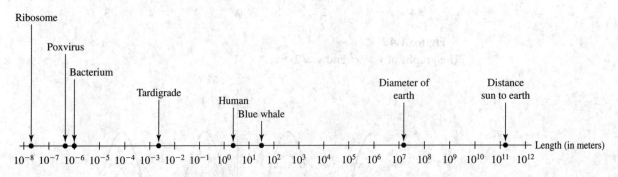

▲ **Figure 1.48**
Orders of magnitude

of organisms: A ribosome is about 20 nm ($=2 \times 10^{-8}$ m), a poxvirus is about 400 nm (4×10^{-7} m), a bacterium is about 1 μm ($=10^{-6}$ m), a tardigrade (or "water bear") is about 1.2 mm ($=1.2 \times 10^{-3}$ m), an adult human is about 1.8 m, a blue whale is between 25 and 35 m, the diameter of the earth is 12,755 km ($\approx 1.3 \times 10^{7}$ m), and the average distance from the sun to the earth is about 150 million km ($=1.5 \times 10^{11}$ m). These sizes are conveniently illustrated using a **logarithmic scale**. This is a scale where multiples of ten are equally distant (Figure 1.48).

When we take logarithms to base 10 of the quantities displayed in Figure 1.48, we find that this transformed scale looks like the arithmetic scale we are familiar with (Figure 1.49). The numbers on the logarithmic scale in Figure 1.49 correspond to exponents.

Let's look at the two number lines in more detail. The origin of the number line in Figure 1.49 corresponds to the number 1 in Figure 1.48 since $\log 1 = 0$. If we go to the left of 1 on the line in Figure 1.48, we get smaller and smaller numbers, but they are *all* positive. (Since $\log x$ is only defined for $x > 0$, we cannot logarithmically transform negative numbers.) Going to the left of 1 in Figure 1.48 corresponds to going to the left of 0 in Figure 1.49. The negative numbers on the number line in Figure 1.49 correspond to negative exponents; for instance, the -8 in Figure 1.49 means $\log x = -8$ or $x = 10^{-8}$. A similar interpretation holds when we go to the right of 1 in Figure 1.48 (or to the right of 0 in Figure 1.49). A logarithmic scale is typically

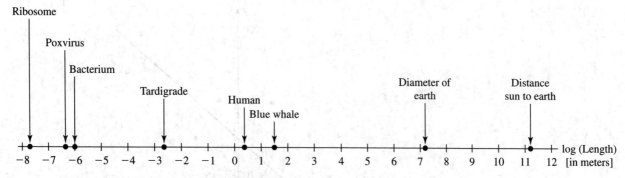

▲ Figure 1.49
Orders of magnitude

based on logarithms to base 10 since this makes conversion between the two representations in Figures 1.48 and 1.49 easier.

The lengths in the preceding examples differed by many factors of 10. Instead of saying that quantities differ by many factors of 10, we will say that they differ by many **orders of magnitude**: If two quantities differ by a factor of 10, they differ by one order of magnitude (likewise, if they differ by a factor of 100, they differ by two orders of magnitude, and so on). Orders of magnitude are approximate comparisons: A 1.8-m-tall human and a 25-m-long blue whale differ by about one order of magnitude.

▶ Example 3 Display the numbers 0.003, 0.1, 0.5, 6, 200, and 4000 on a logarithmic scale.

Solution

To display the numbers, we need to take logarithms first:

x	0.003	0.1	0.5	6	200	4000
$\log x$	−2.5229	−1	−0.3010	0.7782	2.3010	3.6021

Since $\log 0.003 = -2.5229$, we find this number 2.5229 units to the left of 0 on the logarithmic scale. Similarly, since $\log 0.1 = -1$, this number is one unit to the left of 0, and $\log 200 = 2.3010$ is 2.3010 units to the right of 0 (Figure 1.50).

In the biological literature, x and not $\log x$ is used for labeling logarithmic number lines. The locations of the numbers are the same; only the labeling changes. That is, 0.003 would be −2.5229 units to the left of the origin of the line (which is now at 1). The line in Figure 1.50 would then look like the line in Figure 1.51. ◀

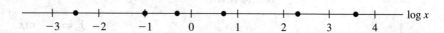

▲ Figure 1.50
Example 3

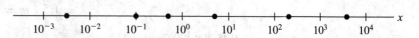

▲ Figure 1.51
Example 3

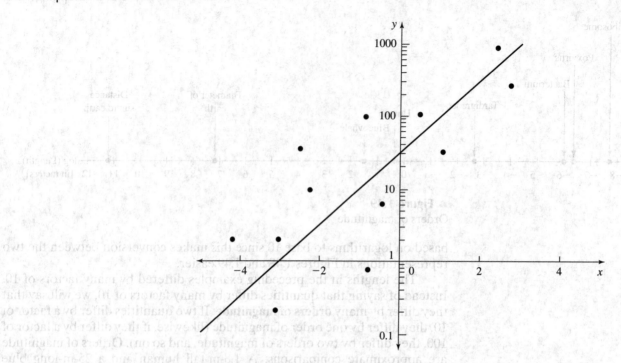

A straight line when the vertical axis is logarithmically transformed

1.3.3 Transformations into Linear Functions

When you look through a biology textbook, you very likely find graphs like the ones in Figures 1.52 and 1.53. In either graph, you see a straight line (with data points scattered about it). In Figure 1.52, the vertical axis is logarithmically transformed and the horizontal axis is on a linear scale; in Figure 1.53, both axes are logarithmically transformed. Why do we display data like this, and what do these graphs mean?

The first question is quick to answer: Straight lines (or linear relationships) are easy to recognize visually. If transforming data results in data points lying along a straight line, we should do the transformation. We will see that this will allow us obtain a functional relationship between quantities. Now on to the second question: What do these graphs mean?

Exponential Functions   Let's look at Figure 1.52 and redraw just the straight line using log y (instead of y) on the vertical axis and x on the horizontal axis (Figure 1.54). Set $Y = \log y$ and forget for a moment where the graph came from. We see a linear relationship between Y and x. A linear relationship is of the form

$$Y = c + mx$$

where c is the Y-intercept and m is the slope. We can read these two quantities off of the graph in Figure 1.54.

$$c = 1.5 \quad m = 0.5$$

That is, we have

$$Y = 1.5 + 0.5x$$

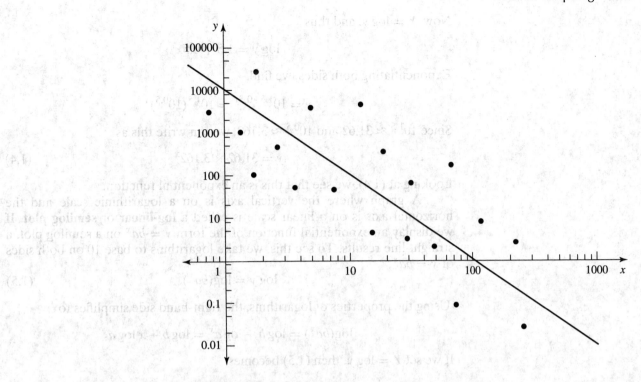

▲ **Figure 1.53**
A straight line when both axes are logarithmically transformed

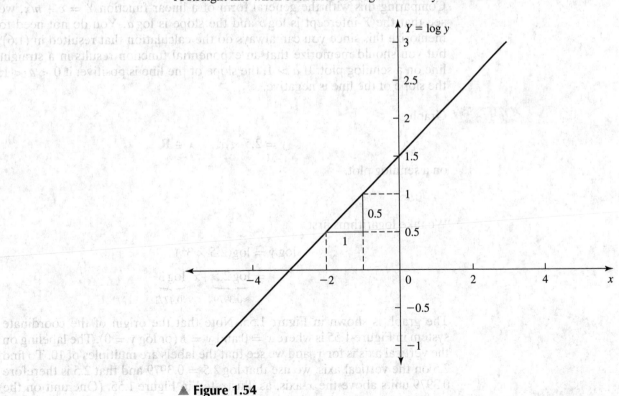

▲ **Figure 1.54**
Now the vertical axis is labeled log(*y*)

Now, $Y = \log y$, and thus

$$\log y = 1.5 + 0.5x$$

Exponentiating both sides, we find

$$y = 10^{1.5+0.5x} = 10^{1.5}(10^{0.5})^x$$

Since $10^{1.5} \approx 31.62$ and $10^{0.5} \approx 3.162$, we can write this as

$$y = 31.62 \times 3.162^x \tag{1.4}$$

Looking at (1.4), we see that this is an exponential function.

A graph where the vertical axis is on a logarithmic scale and the horizontal axis is on a linear scale is called a **log-linear** or **semilog plot**. If we display an exponential function of the form $y = ba^x$ on a semilog plot, a straight line results. To see this, we take logarithms to base 10 on both sides of $y = ba^x$,

$$\log y = \log(ba^x) \tag{1.5}$$

Using the properties of logarithms, the right-hand side simplifies to

$$\log(ba^x) = \log b + \log a^x = \log b + x \log a$$

If we set $Y = \log y$, then (1.5) becomes

$$Y = \log b + (\log a)x \tag{1.6}$$

Comparing this with the general form of a linear function $Y = c + mx$, we see that the Y-intercept is $\log b$ and the slope is $\log a$. You do not need to memorize this since you can always do the calculation that resulted in (1.6), but you should memorize that an exponential function results in a straight line on a semilog plot. If $a > 1$, the slope of the line is positive; if $0 < a < 1$, the slope of the line is negative.

▶ **Example 4**   Graph

$$y = 2.5 \times 3^x, \quad x \in \mathbf{R}$$

on a semilog plot.

Solution

We take logarithms first

$$\log y = \log(2.5 \times 3^x)$$
$$= \underbrace{\log 2.5}_{\approx 0.3979} + x \underbrace{\log 3}_{\approx 0.4771}$$

The graph is shown in Figure 1.55. Note that the origin of the coordinate system in Figure 1.55 is where $x = 0$ and $y = 1$ (or $\log y = 0$). The labeling on the vertical axis is for y, and we see that the labels are multiples of 10. To find 2.5 on the vertical axis, we use that $\log 2.5 = 0.3979$ and that 2.5 is therefore 0.3979 units above the x-axis, as illustrated in Figure 1.55. (One unit on the vertical axis corresponds to a factor of 10.)   ◀

▶ **Example 5**   Find the functional relationship between x and y based on the graph in Figure 1.56.

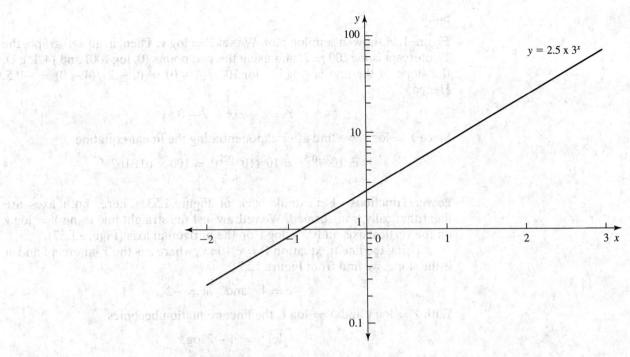

▲ **Figure 1.55**
The graph of 2.5×3^x on a semilog plot

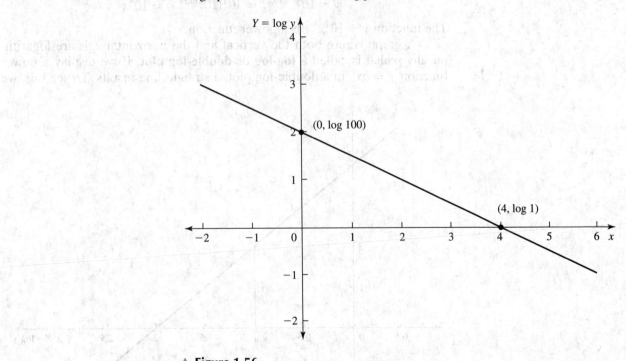

▲ **Figure 1.56**
The graph for Example 5. The line goes through the points (0,2) and (4,0)

Solution

Figure 1.56 shows a semilog plot. We set $Y = \log y$. Then, in an x-Y graph, the Y-intercept is $\log 100 = 2$, and using the two points $(0, \log 100)$ and $(4, \log 1)$, the slope of the line is $(\log 1 - \log 100)/(4 - 0) = (0 - 2)/(4 - 0) = -0.5$. Hence,

$$Y = c + mx = 2 - 0.5x$$

Since $Y = \log y$, we find after exponentiating the linear equation

$$y = 10^{2-0.5x} = 10^2(10^{-0.5})^x = 100 \times (0.3162)^x \qquad \blacktriangleleft$$

Power Functions Let's look back at Figure 1.53. There, both axes are logarithmically transformed. We redraw just the straight line using $Y = \log y$ on the vertical axis and $X = \log x$ on the horizontal axis (Figure 1.57).

Using the linear equation $Y = c + mX$, where c is the Y-intercept and m is the slope, we find from Figure 1.57

$$c = 4 \quad \text{and} \quad m = -2$$

With $Y = \log y$ and $X = \log x$, the linear equation becomes

$$\log y = 4 - 2\log x$$

Exponentiating both sides, we get

$$y = 10^{4-2\log x} = 10^4(10^{\log x^{-2}}) = 10^4 x^{-2}$$

The function $y = 10^4 x^{-2}$ is a power function.

A graph where both the vertical and the horizontal axis are logarithmically scaled is called a **log-log** or **double-log plot**. If we display a power function $y = bx^r$ in a double-log plot, a straight line results. To see this, we

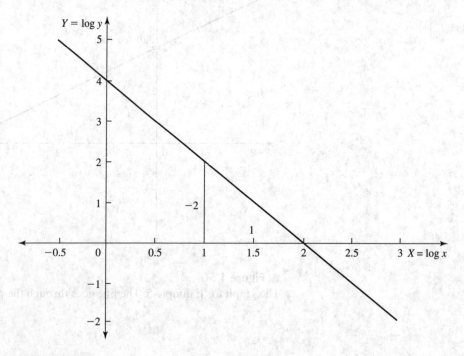

▲ **Figure 1.57**
Now the vertical axis is labeled $\log(y)$ and the horizontal axis is labeled $\log(x)$

take logarithms to base 10 on both sides of $y = bx^r$:

$$\log y = \log(bx^r) \tag{1.7}$$

Using the properties of logarithms on the right-hand side of (1.7), we get

$$\log(bx^r) = \log b + \log x^r = \log b + r \log x$$

If we set $Y = \log y$ and $X = \log x$, then (1.7) becomes

$$Y = \log b + rX$$

Comparing this with the general form of a linear function, $Y = c + mX$, we see that the Y-intercept is $\log b$ and the slope is r. If $r > 0$, the slope is positive. If $r < 0$, the slope is negative.

▶ **Example 6**   Graph

$$y = 100x^{-2/3}, \quad x > 0$$

on a double-log plot.

Solution

We take logarithms first:

$$\log y = \log(100x^{-2/3}) = \log 100 - \frac{2}{3}\log x$$

We set $Y = \log y$ and $X = \log x$. Then with $\log 100 = 2$, we find

$$Y = 2 - \frac{2}{3}X$$

This is the equation of a straight line with X-intercept 3 and Y-intercept 2 (and thus slope $-2/3$). We graph this function in Figure 1.58, where we have X and Y on the two axes. If we use x and y on the two axes (Figure 1.59), the labels change: The y-intercept is now at 100 (corresponding to $\log 100 = 2$) and the x-intercept is at 1000 (corresponding to $\log 1000 = 3$). Note that the origin in Figure 1.58 is at $X = 0$ and $Y = 0$; the origin in Figure 1.59 is at $x = 1$ and $y = 1$. ◀

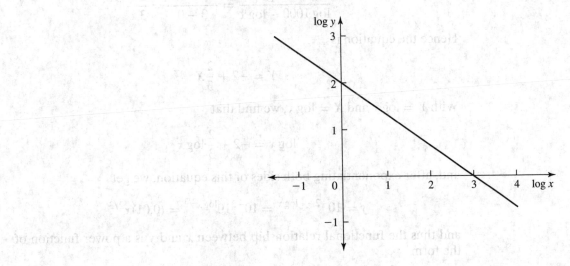

▲ **Figure 1.58**
The graph of $y = 100x^{-2/3}$ on a double-log plot where the axes are labeled $X = \log x$ and $Y = \log y$

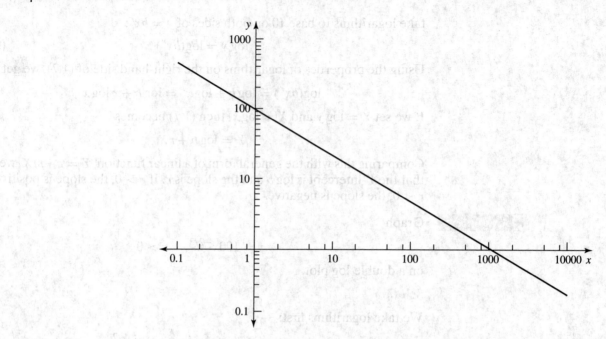

▲ **Figure 1.59**
The graph of $y = 100x^{-2/3}$ on a double-log plot where the axes are labeled x and y

▶ **Example 7** Find the functional relationship between x and y based on the graph in Figure 1.60.

Solution

If $Y = \log y$ and $X = \log x$, then in an X-Y graph, the Y-intercept is at $\log 0.01 = -2$ and, using the two points $(\log 1, \log 0.01)$ and $(\log 1000, \log 1)$, the slope is

$$\frac{\log 1 - \log 0.01}{\log 1000 - \log 1} = \frac{0 - (-2)}{3 - 0} = \frac{2}{3}$$

Hence the equation is

$$Y = -2 + \frac{2}{3}X$$

With $Y = \log y$ and $X = \log x$, we find that

$$\log y = -2 + \frac{2}{3}\log x$$

and, after exponentiating both sides of this equation, we get

$$y = 10^{-2 + \frac{2}{3}\log x} = 10^{-2}10^{\log x^{2/3}} = (0.01)x^{2/3}$$

and thus the functional relationship between x and y is a power function of the form

$$y = (0.01)x^{2/3}$$

◀

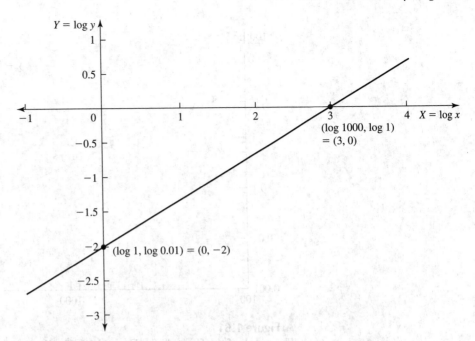

▲ **Figure 1.60**
The graph of the function for Example 7: The two points on the log-log graph used for finding the relationship are $(\log 1, \log 0.01)$ and $(\log 1000, \log 1)$

Applications

▶ Example 8

When growing plants at sufficiently high initial densities, we often observe that the number of plants decreases as the size of the plants grows. This is called self-thinning. When the per plant dry weight of the above-ground biomass is plotted as a function of the density of survivors on a log-log plot, we frequently find that the data lie along a straight line with slope $-3/2$. Assume that, for a particular plant, such a relationship holds for plant densities between 10^2 and 10^4 plants per square meter and that, at a density of 100 plants per square meter, the dry weight per plant is about 10 grams. Find the functional relationship between dry weight and plant density, and graph this function in a log-log plot.

Solution

Since the relationship between density (x) and dry weight (y) follows a straight line with slope $-3/2$ on a log-log plot, we set

$$\log y = C - \frac{3}{2}\log x \quad \text{for } 10^2 \leq x \leq 10^4$$

where C is a constant. To find C, we use the fact that when $x = 100$, $y = 10$. Therefore,

$$\log 10 = C - \frac{3}{2}\log 100$$

or

$$1 = C - \frac{3}{2}\cdot 2, \quad \text{which implies} \quad C = 4$$

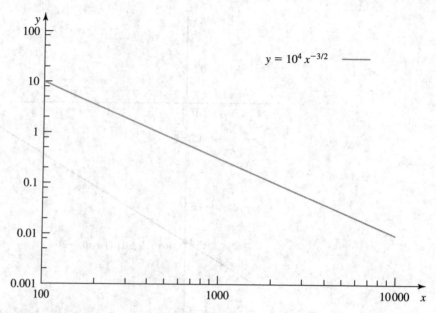

▲ **Figure 1.61**
The graph of the function for Example 8: The line has slope $-3/2$ and goes through the point $(\log 100, \log 10)$ on a double-log plot

Hence,

$$\log y = 4 - \frac{3}{2} \log x$$

Exponentiating both sides (and remembering that log is the logarithm to base 10), we find that

$$y = 10^4 x^{-3/2} \quad \text{for } 10^2 \le x \le 10^4$$

The graph of this function on a log-log scale is shown in Figure 1.61. ◀

▶ **Example 9** Polonium 210 (Po^{210}) is a radioactive material. To determine the half-life of Po^{210} experimentally, we measure the amount of radioactive material left after time t for various values of t. When we plot the data on a semilog plot, we find that we can fit a straight line to the curve. The slope of the straight line is -0.0022/day. Find the half-life of Po^{210}.

Solution

Radioactive decay follows the equation

$$W(t) = W(0)e^{-\lambda t} \quad \text{for } t \ge 0$$

where $W(t)$ is the amount of radioactive material left after time t. If we log transform this equation, we obtain

$$\log W(t) = \log W(0) - \lambda t \log e$$

Note that we use logarithms to base 10. If we plot $W(t)$ as a function of t on a semilog plot, we obtain a straight line with slope $-\lambda \log e$. Matching this with the number given in the example, we obtain

$$\lambda \log e = 0.0022/\text{day}$$

Solving for λ yields

$$\lambda = \frac{1}{\log e} 0.0022/\text{day}$$

To find the half-life T_h, we use the formula (see Subsection 1.2.5)

$$T_h = \frac{\ln 2}{\lambda} = \frac{\ln 2}{0.0022}(\log e) \text{ days}$$
$$\approx 136.8 \text{ days}$$

◀

Note that in the preceding example, we used logarithms to base 10 to do the log transformation. The radioactive law was given in terms of the natural exponent e. The slope therefore contained the factor $\log e \approx 0.4343$.

▶ Example 10   Light intensity in lakes decreases with depth. Denote by $I(z)$ the light intensity at depth z, with $z = 0$ representing the surface. The percentage surface radiation at depth z, denoted by PSR(z), is computed as

$$\text{PSR}(z) = 100 \frac{I(z)}{I(0)}$$

When we graph the percentage surface radiation as a function of depth on a semilog plot, a straight line results. An example of such a curve is given in Figure 1.62, where the coordinate system is rotated clockwise by 90° so that

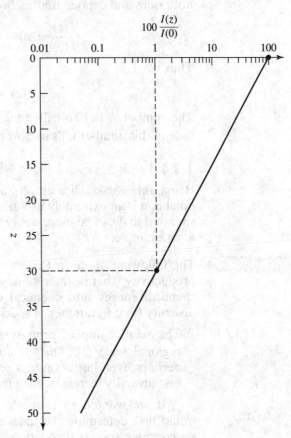

▲ **Figure 1.62**
The graph of percentage surface radiation as a function of depth

the depth axis points downward. Derive an equation for $I(z)$ based on the graph.

Solution

We see that the dependent variable, $100I(z)/I(0)$, is logarithmically transformed, whereas the independent variable, z, is on a linear scale. The graph is a straight line. We thus find

$$\log 100 \frac{I(z)}{I(0)} = c + mz$$

where c is the intercept on the percentage surface radiation axis and m is the slope. We see that

$$c = \log 100$$

and, using the points $(0, 100)$ and $(30, 1)$, we get

$$m = \frac{\log 100 - \log 1}{0 - 30} = -\frac{2}{30} = -\frac{1}{15}$$

Thus,

$$\log 100 \frac{I(z)}{I(0)} = \log 100 - \frac{1}{15}z$$

The left-hand side simplifies to $\log 100 + \log \frac{I(z)}{I(0)}$. After canceling $\log 100$ on both sides and exponentiating both sides, we find

$$\frac{I(z)}{I(0)} = 10^{-(1/15)z} = \exp[\ln 10^{-(1/15)z}]$$

Thus

$$I(z) = I(0)e^{-(\frac{1}{15}\ln 10)z}$$

The number $\frac{1}{15}\ln 10$ is called the vertical attenuation coefficient. The magnitude of this number tells us how quickly light is absorbed in a lake. ◀

1.3.4 From a Verbal Description to a Graph (Optional)

Being able to sketch a graph based on a verbal explanation of some phenomenon is an extremely useful skill since a graph can summarize a complex situation that can be more easily communicated and remembered. Let's look at an example.

▶ **Example 11**   The following quote in Rosenzweig and Abramsky (1993) relates primary productivity (that is, the rate at which autotrophs convert light or inorganic chemical energy into chemical energy of organic compounds) to species diversity (that is, number of species):

> The relationship of primary productivity and species diversity on a regional scale (10^6 km^2) is not simple. But within such regions, and perhaps even larger ones, a pattern is emerging: as productivity rises, first diversity increases, then it declines.

If we wanted to translate this verbal description into a graph, we would first determine the independent and the dependent variable. Here, we consider species diversity as a function of primary productivity; hence primary productivity is the independent variable and species diversity is the dependent variable. We will therefore use a coordinate system whose

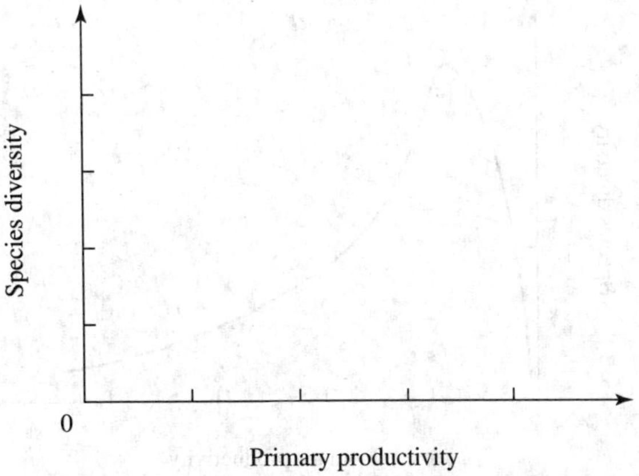

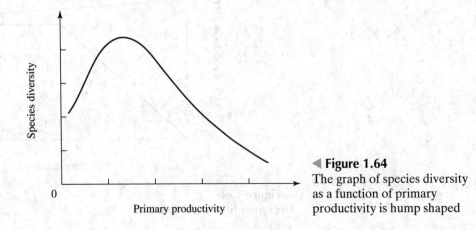

◀ **Figure 1.63**
The coordinate system for species diversity as a function of primary productivity can be restricted to the first quadrant

horizontal axis denotes primary productivity and whose vertical axis denotes species diversity. Since both primary productivity and species diversity are nonnegative, we only need to draw the first quadrant (Figure 1.63).

Going back to the quote, we see that as productivity increases, diversity first increases, then decreases. The graph in Figure 1.64 illustrates this behavior.

The exact shape of the curve cannot be inferred from the quote and will depend on the system studied. For instance, the graph in Figure 1.65 resembles the curve from a study in the Costa Rican forests; as productivity increases, the curve shows an initial increase followed by a decrease in species diversity. The shape of the curve in Figure 1.65 is quantitatively different from the graph in Figure 1.64, but both have the same qualitative features of an initial increase followed by a decrease.

As another example, we will look at the *functional response* of a predator to its prey density.

▶ **Example 12**   The functional response of a predator to its prey density relates the number of prey consumed per predator (dependent variable) to the prey density (independent variable). Holling (1959) introduced three basic types. Type 1 describes a response where the number of prey eaten per predator as a function of prey density rises linearly to a plateau. The type 2 functional response increases at a decelerating rate and eventually levels off. The type 3 functional response is S-shaped or sigmoidal and also eventually levels off. Now let's translate these three ways into graphs. All graphs will be placed

◀ **Figure 1.64**
The graph of species diversity as a function of primary productivity is hump shaped

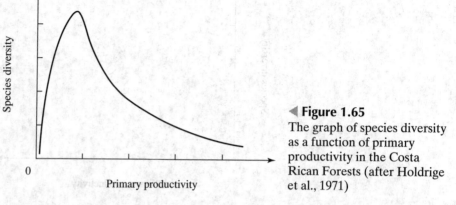

◄ Figure 1.65
The graph of species diversity as a function of primary productivity in the Costa Rican Forests (after Holdrige et al., 1971)

in coordinate systems, where we plot prey density on the horizontal axis and the number of prey eaten per predator on the vertical axis. Since both prey density and number of prey eaten per predator are nonnegative variables, we only need to draw the first quadrant.

Even though this was not mentioned, we will assume that when the prey density is equal to zero, the number of prey eaten per predator will also be zero, and once the prey density is positive, the number of prey eaten per predator will be positive. This means that the three functional response curves all go through the origin.

The type 1 functional response first increases linearly, then reaches a plateau. Increasing linearly results in a straight line. Reaching a plateau means that the value stays constant once the plateau is reached (Figure 1.66).

The type 2 functional response is described as a function that increases at a decelerating rate. This means that the function will increase less quickly as prey density increases (Figure 1.67). In contrast to the type 1 functional response, the type 2 function will continue to increase and approach the plateau but not actually reach the plateau at a finite value of prey density.

The type 3 functional response is described as sigmoidal. Sigmoidal curves are characterized by an initial accelerating increase followed by an increase at a decelerating rate (Figure 1.68). Similar to the type 2 functional response curve, the type 3 functional response curve approaches a plateau as

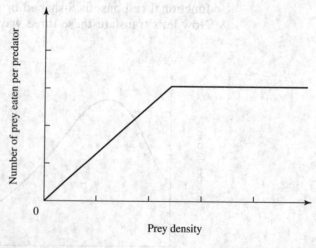

▲ Figure 1.66
The type 1 functional response

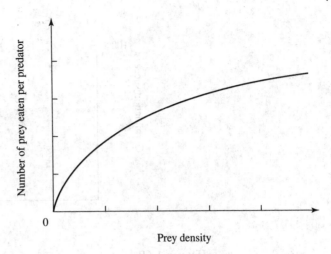

▲ **Figure 1.67**
The type 2 functional response

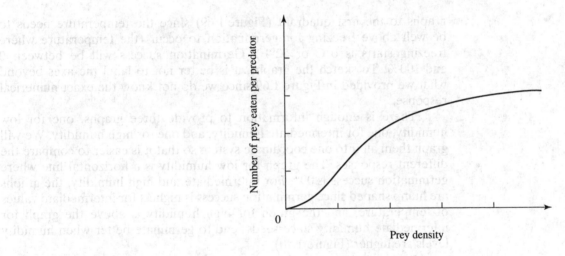

▲ **Figure 1.68**
The type 3 functional response

prey density increases but will not reach the plateau at a finite value of prey density.

For each example discussed so far, the functional relationship depended on just one variable, like number of prey eaten per predator as a function of prey density. Very often, a response depends on more than one independent variable. The following example has a response depending on two independent variables and shows how to draw a graph for this more complex relationship.

▶ Example 13   Germination success of seeds depends on both temperature and humidity. When humidity is too low, seeds tend not to germinate at all, regardless of temperature; germination success is highest for intermediate values of temperature; finally, seeds tend to germinate better when humidity levels are higher.

One way to translate this information into a graph is to graph germination success as a function of temperature for different levels of humidity. If we measure temperature in Fahrenheit or Celsius, we can restrict the

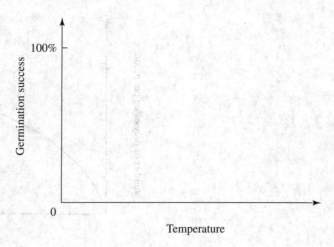

▲ **Figure 1.69**
The coordinate system for germination success as a function of temperature can be restricted to the first quadrant. Germination success will be between 0 and 100%

graphs to the first quadrant (Figure 1.69) since the temperature needs to be well above freezing for germination to occur (the temperature where freezing starts is 0°C or 32°F). Germination success will be between 0 and 100%. To sketch the graphs, it is better not to label the axes beyond what we provided in Figure 1.69 since we do not know the exact numerical response.

There is enough information to provide three graphs, one for low humidity, one for intermediate humidity, and one for high humidity. We will graph them all into one coordinate system so that it is easier to compare the different responses. The graph for low humidity is a horizontal line where germination success is 0%. For intermediate and high humidity, the graphs are hump shaped since germination success is highest for intermediate values of temperature, and the graph for high humidity is above the graph for intermediate humidity since seeds tend to germinate better when humidity levels are higher (Figure 1.70).

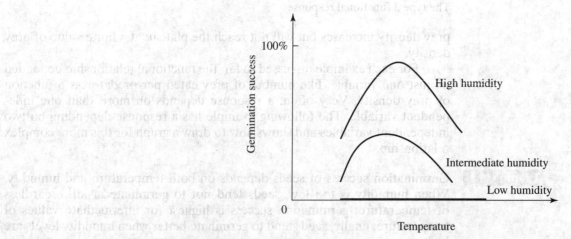

▲ **Figure 1.70**
Germination success as a function of temperature for three humidity levels (low, intermediate, high)

1.3.5 Problems

(1.3.1)

In Problems 1–22, sketch the graph of each function. Do not use a graphing calculator. (Assume the largest possible domain.)

1. $y = x^2 + 1$
2. $y = -(x - 2)^2 + 1$
3. $y = x^3 - 2$
4. $y = -x^4 + 1$
5. $y = -2x^2 - 3$
6. $y = -(3 - x)^2$
7. $y = 3 + 1/x$
8. $y = 1 - 1/x$
9. $y = 1/(x - 1)$
10. $y = 1 + 1/(x + 2)^2$
11. $y = \exp(x - 2)$
12. $y = \exp(-x)$
13. $y = e^{-(x+3)}$
14. $y = 3e^{2x+1}$
15. $y = \ln(x + 1)$
16. $y = \ln(x - 3)$
17. $y = -\ln(x - 1) + 1$
18. $y = -\ln(1 - x)$
19. $y = 2\sin(x + \pi/4)$
20. $y = 0.2\cos(-x)$
21. $y = -\sin(\pi x/2)$
22. $y = -2\cos(\pi x/4)$

23. Explain how the following functions can be obtained from $y = x^2$ by basic transformations.
    (a) $y = x^2 - 2$
    (b) $y = (x - 1)^2 + 1$
    (c) $y = -2(x + 2)^2$

24. Explain how the following functions can be obtained from $y = x^3$ by basic transformations.
    (a) $y = x^3 + 1$
    (b) $y = (x + 1)^3 - 1$
    (c) $y = -3(x - 2)^3$

25. Explain how the following functions can be obtained from $y = 1/x$ by basic transformations.
    (a) $\quad y = 1 - \dfrac{1}{x}$
    (b) $\quad y = -\dfrac{1}{x - 1}$
    (c) $\quad y = \dfrac{x}{x + 1}$

26. Explain how the following functions can be obtained from $y = 1/x^2$ by basic transformations.
    (a) $\quad y = \dfrac{1}{x^2} + 1$
    (b) $\quad y = -\dfrac{1}{(x + 1)^2}$

(c) $\quad y = -\dfrac{1}{x^2} - 2$

27. Explain how the following functions can be obtained from $y = e^x$ by basic transformations.
    (a) $\quad\quad y = 2e^x - 1$
    (b) $\quad\quad y = -e^{-x}$
    (c) $\quad\quad y = e^{x-2} + 1$

28. Explain how the following functions can be obtained from $y = e^x$ by basic transformations.
    (a) $\quad\quad y = e^{-x} - 1$
    (b) $\quad\quad y = -e^x + 1$
    (c) $\quad\quad y = -e^{x-3} - 2$

29. Explain how the following functions can be obtained from $y = \ln x$ by basic transformations.
    (a) $\quad\quad y = \ln(x - 1)$
    (b) $\quad\quad y = -\ln x + 1$
    (c) $\quad\quad y = \ln(x + 3) - 1$

30. Explain how the following functions can be obtained from $y = \ln x$ by basic transformations.
    (a) $\quad\quad y = \ln(1 - x)$
    (b) $\quad\quad y = \ln(2 + x) - 1$
    (c) $\quad\quad y = -\ln(2 - x) + 1$

31. Explain how the following functions can be obtained from $y = \sin x$ by basic transformations.
    (a) $\quad\quad y = 1 - \sin x$
    (b) $\quad\quad y = \sin(x - \frac{\pi}{4})$
    (c) $\quad\quad y = -\sin(x + \frac{\pi}{3})$

32. Explain how the following functions can be obtained from $y = \cos x$ by basic transformations.
    (a) $\quad\quad y = 1 + 2\cos x$
    (b) $\quad\quad y = -\cos(x + \frac{\pi}{4})$
    (c) $\quad\quad y = -\cos\left(\frac{\pi}{2} - x\right)$

(1.3.2)

33. Find the following numbers on a number line that is on a logarithmic scale (base 10): 0.0002, 0.02, 1, 5, 50, 100, 1000, 8000, and 20000.

34. Find the following numbers on a number line that is on a logarithmic scale (base 10): 0.03, 0.7, 1, 2, 5, 10, 17, 100, 150, and 2000.

35. (a) Find the following numbers on a number line that is on a logarithmic scale (base 10): 10^2, 10^{-3}, 10^{-4}, 10^{-7}, and 10^{-10}.
    (b) Can you find 0 on a number line that is on a logarithmic scale?
    (c) Can you find negative numbers on a number line that is on a logarithmic scale?

36. Find the following numbers on a number line that is on a logarithmic scale (base 10):
    (a) $10^{-3}, 2 \times 10^{-3}, 5 \times 10^{-3}$
    (b) $10^{-1}, 2 \times 10^{-1}, 5 \times 10^{-1}$

(c) $10^2, 2 \times 10^2, 5 \times 10^2$

(d) Using (a)–(c), how many units (on a logarithmic scale) is 2×10^{-3} from 10^{-3} (2×10^{-1} from 10^{-1} and 2×10^2 from 10^2)?

(e) Using (a)–(c), how many units (on a logarithmic scale) is 5×10^{-3} from 10^{-3} (5×10^{-1} from 10^{-1} and 5×10^2 from 10^2)?

In Problems 37–42, insert an appropriate number in the blank space.

37. The longest known species of worms is the earth worm *Microchaetus rappi* of South Africa; in 1937, a 6.7-m-long specimen was collected from the Transvaal. The shortest worm is *Chaetogaster annandalei*, which measures less than 0.51 mm in length. *M. rappi* is ____ order(s) of magnitude longer than *C. annandalei*.

38. Both the La Plata river dolphin (*Pontoporia blainvillei*) and the sperm whale (*Physeter macrocephalus*) belong to the suborder Odontoceti (individuals of which have teeth). A La Plata river dolphin weighs between 30 and 50 kg, whereas a sperm whale weighs between 35,000 and 40,000 kg. A sperm whale is ____ order(s) of magnitude heavier than a La Plata river dolphin.

39. Compare a ball of radius 1 cm to a ball of radius 10 cm. The radius of the larger ball is ____ order(s) of magnitude bigger than the radius of the smaller ball. The volume of the larger ball is ____ order(s) of magnitude bigger than the volume of the smaller ball.

40. Compare a square with side length 1 m to a square with side length 100 m. The area of the larger square is ____ order(s) of magnitude larger than the area of the smaller square.

41. The diameter of a typical bacterium is about 0.5 to 1 μm. An exception is the bacterium *Eplopiscium fishelsoni*, which is about 600 μm long and 80 μm wide. The volume of *E. fishelsoni* is about ____ order(s) of magnitude larger than that of a typical bacterium. (*Hint:* Approximate the shape of a typical bacterium by a sphere and the shape of *E. fishelsoni* by a cylinder.)

42. The length of a typical bacterial cell is about one tenth that of a small eukaryotic cell. Consequently, the cell volume of a bacterium is about ____ order(s) of magnitude smaller than that of a small eukaryotic cell. (*Hint:* Approximate the shapes of both types of cells by spheres.)

(1.3.3)

In Problems 43–46, when $\log y$ is graphed as a function of x, a straight line results. Graph straight lines, each given by two points, on a log-linear plot, and determine the functional relationship. (The original x-y coordinates are given.)

43. $(x_1, y_1) = (0, 5), (x_2, y_2) = (3, 1)$

44. $(x_1, y_1) = (-1, 4), (x_2, y_2) = (2, 8)$

45. $(x_1, y_1) = (-2, 3), (x_2, y_2) = (1, 1)$

46. $(x_1, y_1) = (1, 4), (x_2, y_2) = (6, 1)$

In Problems 47–54, use a logarithmic transformation to find a linear relationship between the given quantities and graph the resulting linear relationship on a log-linear plot.

47. $y = 3 \times 10^{-2x}$

48. $y = 4 \times 10^{5x}$

49. $y = 2e^{-1.2x}$

50. $y = 7e^{3x}$

51. $y = 5 \times 2^{4x}$

52. $y = 6 \times 2^{-0.9x}$

53. $y = 4 \times 3^{2x}$

54. $y = 5^{-6x}$

In Problems 55–58, when $\log y$ is graphed as a function of $\log x$, a straight line results. Graph straight lines, each given by two points, on a log-log plot, and determine the functional relationship. (The original x-y coordinates are given.)

55. $(x_1, y_1) = (1, 2), (x_2, y_2) = (5, 1)$

56. $(x_1, y_1) = (3, 5), (x_2, y_2) = (1, 5)$

57. $(x_1, y_1) = (4, 2), (x_2, y_2) = (8, 8)$

58. $(x_1, y_1) = (2, 5), (x_2, y_2) = (5, 2)$

In Problems 59–66, use a logarithmic transformation to find a linear relationship between the given quantities and graph the resulting linear relationship on a log-log plot.

59. $y = 2x^5$

60. $y = 3x^2$

61. $y = x^6$

62. $y = 5x^3$

63. $y = x^{-2}$

64. $y = 6x^{-1}$

65. $y = 4x^{-3}$

66. $y = 7x^{-5}$

In Problems 67–72, use a logarithmic transformation to find a linear relationship between the given quantities, and determine whether a log-log or log-linear plot should be used to graph the resulting linear relationship.

67. $f(x) = 3x^{1.7}$

68. $g(s) = 1.8e^{-0.2s}$

69. $N(t) = 130 \times 2^{1.2t}$

70. $I(u) = 4.8u^{-0.89}$

71. $R(t) = 3.6t^{1.2}$

72. $L(c) = 1.7 \times 10^{2.3c}$

73. The following table is based on a functional relationship between x and y, which is either an exponential or a power function. Use an appropriate logarithmic transformation and a graph to decide whether the table comes from a power function or an exponential function and find the functional relationship between x and y.

x	y
1	1.8
2	2.07
4	2.38
10	2.85
20	3.28

74. The following table is based on a functional relationship between x and y, which is either an exponential or a power function. Use an appropriate logarithmic transformation and a graph to decide whether the table comes from a power function or an exponential function and find the functional relationship between x and y.

x	y
0.5	7.81
1	3.4
1.5	2.09
2	1.48
2.5	1.13

75. The following table is based on a functional relationship between x and y, which is either an exponential or a power function. Use an appropriate logarithmic transformation and a graph to decide whether the table comes from a power function or an exponential function and find the functional relationship between x and y.

x	y
−1	0.398
−0.5	1.26
0	4
0.5	12.68
1	40.18

76. The following table is based on a functional relationship between x and y, which is either an exponential or a power function. Use an appropriate logarithmic transformation and a graph to decide whether the table comes from a power function or an exponential function and find the functional relationship between x and y.

x	y
0	3
0.5	2.20
1	1.61
1.5	1.18
2	0.862

77. The following table is based on a functional relationship between x and y, which is either an exponential or a power function. Use an appropriate logarithmic transformation and a graph to decide whether the table comes from a power function or an exponential function and find the functional relationship between x and y.

x	y
0.1	0.045
0.5	1.33
1	5.7
1.5	13.36
2	24.44

78. The following table is based on a functional relationship between x and y, which is either an exponential or a power function. Use an appropriate logarithmic transformation and a graph to decide whether the table comes from a power function or an exponential function and find the functional relationship between x and y.

x	y
0.1	1.72
0.5	1.41
1	1.11
1.5	0.872
2	0.685

So far, we have always used base 10 for a logarithmic transformation. The reason for this is that our number system is based on base 10 and it is therefore easy to logarithmically transform numbers of the form ... , 0.01, 0.1, 1, 10, 100, 1000, ... *when we use base 10. In Problems 79–82, use the indicated base to logarithmically transform each exponential relationship so that a linear relationship results. Graph each relationship in a coordinate system whose axes are accordingly transformed so that a straight line results (using the indicated base).*

79. $y = 2^x$; base 2
80. $y = 3^x$; base 3
81. $y = 2^{-x}$; base 2
82. $y = 3^{-x}$; base 3
83. Suppose that $N(t)$ denotes a population size at time t and satisfies the equation

$$N(t) = 2e^{3t} \quad \text{for } t \geq 0$$

(a) If you graph $N(t)$ as a function of t on a semilog plot, a straight line results. Explain why.
(b) Graph $N(t)$ as a function of t on a semilog plot, and determine the slope of the resulting straight line.
84. Suppose that you follow a population over time. When you plot your data on a semilog plot, a straight line with slope 0.03 results. Furthermore, assume that the population size at time 0 was 20. If $N(t)$ denotes the population size at time t, what function best describes the population size at time t?
85. (*Species-area curves*) Many studies have shown that the number of species on an island increases with the area of the island. Frequently, the functional relationship between the number of species (S) and the area (A) is approximated by

$S = CA^z$, where z is a constant that depends on the particular species and habitat in the study. (Actual values of z range from about 0.2 to 0.35.) Suppose that the best fit to your data points on a log-log scale is a straight line. Is your model $S = CA^z$ an appropriate description of your data? If yes, how would you find z?

86. (*Michaelis-Menten equation*) Enzymes serve as catalysts in many chemical reactions in living systems. The simplest such reactions transform a single substrate into a product with the help of an enzyme. The Michaelis-Menten equation describes the initial velocity of such enzymatically controlled reactions. The equation gives the relationship between the initial velocity of the reaction (v_0) and the concentration of the substrate (s_0):

$$v_0 = \frac{v_{max} s_0}{s_0 + K_m}$$

where v_{max} is the maximum velocity at which the product may be formed and K_m is the Michaelis-Menten constant. Note that this equation has the same form as the Monod growth function.

(a) Show that the Michaelis-Menten equation can be written in the form

$$\frac{1}{v_0} = \frac{K_m}{v_{max}} \frac{1}{s_0} + \frac{1}{v_{max}}$$

This is known as the Lineweaver-Burk equation and shows that there is a linear relationship between $1/v_0$ and $1/s_0$.

(b) Sketch the graph of the Lineweaver-Burk equation. Use a coordinate system where $1/s_0$ is on the horizontal axis and $1/v_0$ is on the vertical axis. Show that the resulting graph is a line that intersects the horizontal axis at $-1/K_m$ and the vertical axis at $1/v_{max}$.

(c) To determine K_m and v_{max}, we measure the initial velocity of the reaction, denoted by v_0, as a function of the concentration of the substrate, denoted by s_0, and fit a straight line through the points on a coordinate system where the horizontal axis is $1/s_0$ and the vertical axis is $1/v_0$. Explain how to determine K_m and v_{max} from the graph.

(Note that this is an example where a nonlogarithmic transformation is used to obtain a linear relationship. Since the reciprocals of both quantities of interest are used, the resulting plot is called a double-reciprocal plot.)

87. (*Continuation of Problem 86*) Estimating v_{max} and K_m from the Lineweaver-Burk graph as described in Problem 86 is not always satisfactory. A different transformation typically yields better estimates (Dowd and Riggs, 1965). Show that the Michaelis-Menten equation can be written as

$$\frac{v_0}{s_0} = \frac{v_{max}}{K_m} - \frac{1}{K_m} v_0$$

and explain why this transformation results in a straight line when you graph v_0 on the horizontal axis and $\frac{v_0}{s_0}$ on the vertical axis. Explain how you can estimate v_{max} and K_m from this graph.

88. (*Adapted from Reiss, 1989*) In a case study where the maximal rates of oxygen consumption (in ml/s) for nine species of wild African mammals (Taylor et al.,

1980) were plotted against body mass (in kg) on a log-log plot, it was found that the data points fell on a straight line with slope approximately equal to 0.8 and vertical axis intercept of approximately 0.105. Find an equation that relates maximal oxygen consumption and body mass.

89. (*Adapted from Benton and Harper, 1997*) In vertebrates, embryos and juveniles have large heads relative to overall body size. As the animal grows older, proportions change; for instance, the ratio of skull length to body length diminishes. That this is not only the case for living vertebrates but also for fossil vertebrates is shown by the following example.

Ichthyosaurs are a group of marine reptiles that appeared in the early Triassic and died out well before the end of the Cretaceous.[1] They were fish shaped and comparable in size to dolphins. In a study of 20 fossil skeletons, the following allometric relationship between skull length (S) (measured in cm) and backbone length (B) (measured in cm) was found.

$$S = 1.162 B^{0.93}$$

(a) Choose suitable transformations of S and B so that the resulting relationship is linear. Plot the transformed relationship, and find the slope and the y-intercept.

(b) Explain why the allometric equation confirms that juveniles had relatively large heads. (*Hint:* Compute the ratio of S to B for a number of different values of B—say, 10 cm, 100 cm, 500 cm—and compare.)

90. Light intensity in lakes decreases exponentially with depth. If $I(z)$ denotes the light intensity at depth z, with $z = 0$ representing the surface, then

$$I(z) = I(0)e^{-\alpha z}, \quad z \geq 0$$

where α is a positive constant called the vertical attenuation coefficient. Figure 1.71 shows the percentage surface radiation, defined as $100 I(z)/I(0)$, as a function of depth in different lakes.

(a) Based on the graph, estimate α for each lake.

(b) Reproduce a graph like the one in Figure 1.71 for Lake Constance (Germany) in May ($\alpha = 0.768\,\text{m}^{-1}$) and December ($\alpha = 0.219\,\text{m}^{-1}$) (data from Tilzer et al., 1982).

(c) Explain why the graphs are straight lines.

91. Absorption of light in a uniform water column follows an exponential law—that is, the intensity $I(z)$ at depth z is

$$I(z) = I(0)e^{-\alpha z}$$

where $I(0)$ is the intensity at the surface (i.e., when $z = 0$) and α is the *vertical attenuation coefficient*. (We assume here that α is constant. In reality, α depends on the wave length of the light penetrating the surface.)

---

[1]The Triassic is a geologic period that began about 248 million years ago and ended about 213 million years ago; the Cretaceous began about 144 million years ago and ended 65 million years ago.

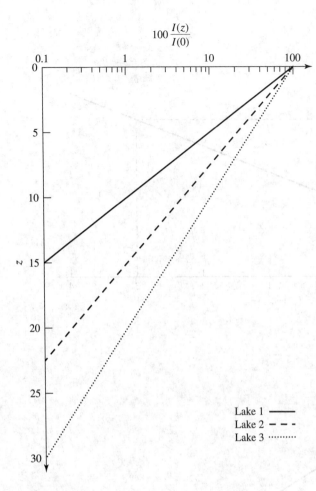

$$100\,\frac{I(z)}{I(0)}$$

Lake 1 ——
Lake 2 – – –
Lake 3 ·········

▲ **Figure 1.71**
Light intensity as a function of depth for Problem 90

(a) Suppose that 10% of the light is absorbed in the uppermost meter. Find α. What are the units of α?

(b) What percentage of the remaining intensity at 1 m is absorbed in the second meter? What percentage of the remaining intensity at 2 m is absorbed in the third meter?

(c) What percentage of the initial intensity remains at 1 m, at 2 m, and at 3 m?

(d) Plot the light intensity as a percentage of the surface intensity, on both a linear plot and a log-linear plot.

(e) Relate the slope of the curve on the log-linear plot to the attenuation coefficient α.

(f) The level at which 1% of the surface intensity remains is of biological significance. Approximately, it is the level where algal growth ceases. The zone above this level is called the euphotic zone. Express the depth of the euphotic zone as a function of α.

(g) Compare a very clear lake to a milky glacier stream. Is the attenuation coefficient α for the clear lake greater or smaller than the attenuation coefficient α for the milky stream?

92. When plants are grown at high densities, we often observe that the number of plants decreases as plant weights increase (due to plant growth). If we plot the logarithm of the total above ground dry weight biomass per plant, $\log w$, against the logarithm of the density of survivors, $\log d$ (base 10), a straight line with slope $-3/2$ results. Find the equation that relates w and d, assuming that $w = 1$ g when $d = 10^3$ m^{-2}.

In Problems 93–98, find each functional relationship based on the given graph.

93. Figure 1.72
94. Figure 1.73
95. Figure 1.74
96. Figure 1.75
97. Figure 1.76 (*Hint:* This relationship is different from the ones considered so far. The x-axis is logarithmically transformed but the y-axis is linear.)
98. Figure 1.77 (*Hint:* This relationship is different from the ones considered so far. The x-axis is logarithmically transformed but the y-axis is linear.)
99. The free energy ΔG in transporting an uncharged solute across a membrane from concentration c_1 to one of concentration c_2 follows the equation

$$\Delta G = 2.303\,RT\,\log\frac{c_2}{c_1}$$

where $R = 1.99$ kcal K^{-1} kmol^{-1} is the universal gas constant and T is temperature measured in Kelvin (K). Plot ΔG as a function of the concentration ratio c_2/c_1 when $T = 298$ K (25° C). Use a coordinate system where the vertical axis is on a linear scale and the horizontal axis is on a logarithmic scale.

100. (*Logistic transformation*) Suppose that

$$f(x) = \frac{1}{1 + e^{-(b+mx)}} \qquad (1.8)$$

A function of the form (1.8) is called a **logistic** function. The logistic function was introduced by the Dutch mathematical biologist Verhulst around 1840 to describe the growth of populations with limited food resources. Show that

$$\ln\frac{f(x)}{1 - f(x)} = b + mx \qquad (1.9)$$

This transformation is called the logistic transformation. It is a standard transformation for linearizing functions of the form (1.8).

101. Not every study of species richness as a function of productivity produces a hump-shaped curve. Owen (1988) studied rodent assemblages in Texas. He found that the number of species was a decreasing function of productivity. Sketch a graph that would describe this situation.

102. Species diversity in a community may be controlled by disturbance frequency. The intermediate disturbance hypothesis states that species diversity is greatest at intermediate disturbance levels. Sketch a graph of species diversity as a function of disturbance level that illustrates this hypothesis.

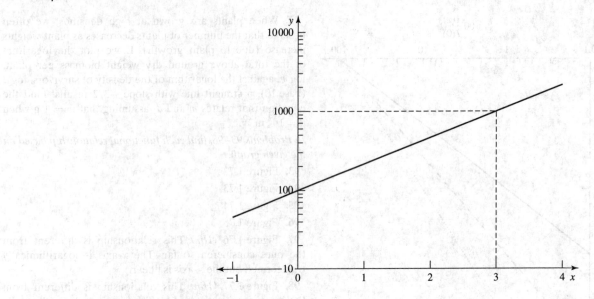

▲ Figure 1.72
Graph for problem 93

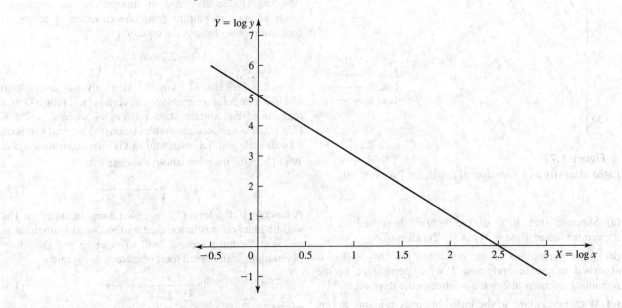

▲ Figure 1.73
Graph for problem 94

103. Preston (1962) investigated the dependence of number of bird species on island area in the West Indian islands. He found that the number of bird species increased at a decelerating rate as island area increased. Sketch this relationship.

104. Phytoplankton convert carbon dioxide to organic compounds during photosynthesis. This process requires sunlight. It has been observed that the rate of photosynthesis is a function of light intensity: At low intensity, the rate of photosynthesis increases approximately linearly with light intensity; it saturates at intermediate levels, and decreases slightly at high intensity. Sketch a graph of the rate of photosynthesis as a function of light intensity.

105. Brown lemming densities in the tundra areas of North America and Eurasia show cyclic behavior: Every three to four years, lemming densities build up very rapidly and typically crash the following year. Sketch a graph that describes this situation.

106. Nitrogen productivity can be defined as the amount of dry matter produced per unit of nitrogen and unit of time. Experimental studies suggest that nitrogen productivity increases as a function of light intensity at a decelerating rate. Sketch a graph of nitrogen productivity as a function of light intensity.

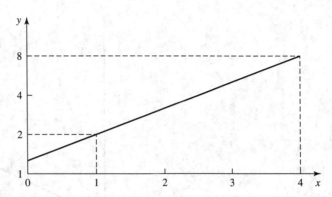

▲ **Figure 1.74**
Graph for problem 95

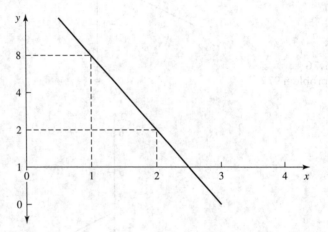

▲ **Figure 1.75**
Graph for problem 96

107. A study of Borchert (1994) investigated the relationship between stem water storage and wood density in a number of tree species in Costa Rica. The study showed that water storage is inversely related to wood density; that is, higher wood density corresponds to lower water content. Sketch a graph of water content as a function of wood density that illustrates this situation.

108. Species richness can be a hump-shaped function of productivity. In the same coordinate system, sketch two hump-shaped graphs of species richness as a function of productivity, one where the maximum occurs at low productivity and one where the maximum occurs at high productivity.

109. The size distribution of zooplankton in a lake is typically a hump-shaped curve; that is, if the frequency (in %) of zooplankton is plotted against the body length of zooplankton, a curve that first increases and then decreases results. Brooks and Dodson (1965) studied the effects of introducing a planktivorous fish in a lake. They found that the composition of zooplankton after the introduction shifted to smaller zooplankton. In the same coordinate system, sketch the size distribution of zooplankton before and after introduction of the planktivorous fish.

110. *Daphnia* is a genus of zooplankton that comprises a number of species. The body growth rate of *Daphnia* depends on food concentration. A minimum food concentration is required for growth: Below this level the growth rate is negative; above it, it is positive. In a study by Gliwicz (1990), it was found that growth rate is an increasing function of food concentration and that the minimum food concentration required for growth decreases with increasing size of the animal. Sketch two graphs in the same coordinate system, one for a large and one for a small *Daphnia* species, that illustrates this situation.

111. Grant (1982) investigated egg weight as a function of adult body weight among 10 species of Darwin's finches. He found that the relationship between the logarithm of the average egg size and the logarithm of the average body size is linear and that smaller species lay smaller eggs and larger species lay larger eggs. Graph this relationship.

112. Grant et al. (1985) investigated the relationship between mean wing length and mean weight among males of populations of six ground finch species. They found a positive and nearly linear relationship between these two quantities. Graph this relationship.

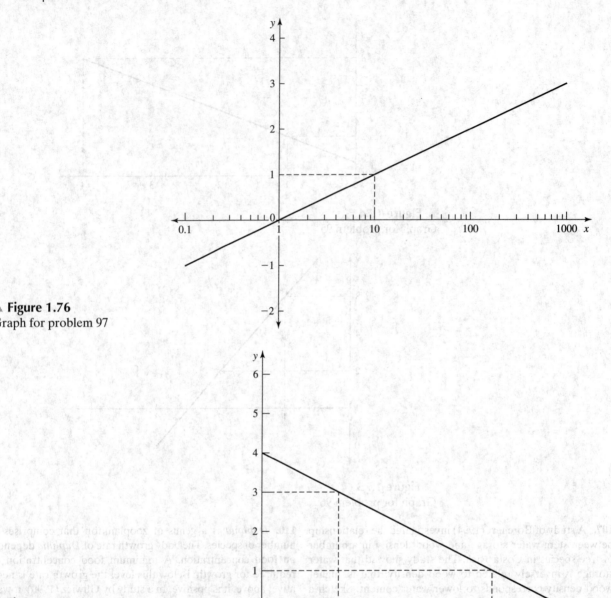

▲ **Figure 1.76**
Graph for problem 97

▲ **Figure 1.77**
Graph for problem 98

113. Bohlen et al. (2001) investigated stream nitrate concentration along an elevation gradient at the Hubbard Brook Experimental Forest in New Hampshire. They found that the nitrate concentration in stream water declined with decreasing elevation. Sketch stream nitrate concentration as a function of elevation.

114. In Example 13, we discussed germination success as a function of temperature for varying levels of humidity. We can also consider germination success as a function of humidity for various levels of temperature. Sketch the following graphs of germination success as a function of humidity: one for low temperature, one for intermediate temperature, and one for high temperature.

115. Boulinier et al. (2001) studied the dynamics of forest bird communities. They found that the mean local extinction

rate of area sensitive species declined with mean forest patch size, whereas the mean extinction rate of non-area-sensitive species did not depend on mean forest size. In the same coordinate system, graph the mean extinction rate as a function of mean forest patch size for (a) an area-sensitive species and (b) a non-area-sensitive species.

116. Dalling et al. (2001) compared net photosynthetic rates of two pioneer trees, *Alseis blackiana* and *Miconia argenta*, as a function of gap size in Barro Colorado Island. They found that net photosynthetic rates (measured on a per unit basis) increased with gap size for both trees and that the photosynthetic rate for *Miconia argenta* was higher than for *Alseis blackiana*. In the same coordinate system, graph the net photosynthetic rates as functions of gap size for both tree species.

1.4 Key Terms

Chapter 1 Review: Topics *Discuss the following definitions and concepts:*

1. Real numbers
2. Intervals: open, closed, half-open
3. Absolute value
4. Proportional
5. Lines: standard form, point-slope form, slope-intercept form
6. Parallel and perpendicular lines
7. Circle: radius, center, equation of circle, unit circle
8. Angle: radians, degrees
9. Trigonometric identities
10. Complex numbers: real part, imaginary part
11. Function: domain, codomain, range, image
12. Symmetry of functions: even, odd
13. Composition of functions
14. Polynomial
15. Degree of a polynomial
16. Chemical reaction: law of mass action
17. Rational function
18. Growth rate
19. Specific growth rate and per capita growth rate
20. Monod growth function
21. Power function
22. Allometry and scaling relations
23. Exponential function
24. Exponential growth
25. Natural exponential base
26. Radioactive decay
27. Half-life
28. Inverse function, one to one
29. Logarithmic function
30. Relationship between exponential and logarithmic functions
31. Periodic function
32. Trigonometric function
33. Amplitude, period
34. Translation: horizontal, vertical
35. Logarithmic scale
36. Order of magnitude
37. Logarithmic transformation
38. Log-log plot
39. Semilog plot

1.5 Review Problems

1. Suppose that the number of bacteria in a petri dish is given by

$$B(t) = 10{,}000e^{0.1t}$$

where t is measured in hours.

(a) How many bacteria are present at $t = 0, 1, 2, 3$, and 4?

(b) Find the time t when the number of bacteria reaches 100,000.

2. Suppose that a pathogen is introduced into a population of bacteria at time 0. The number of bacteria then declines as

$$B(t) = 25{,}000e^{-2t}$$

where t is measured in hours.

(a) How many bacteria are left after 3 hours?

(b) How long will it be until only 1% of the initial number of bacteria are left?

3. Consider the chemical reaction

$$A + 2B \rightarrow AB_2$$

Assume that the reaction occurs in a closed vessel and that the initial concentrations of A and B are $a = [A]$ and $b = [B]$, respectively.

(a) Explain why the reaction rate $R(x)$ is given by

$$R(x) = k(a - x)(b - 2x)^2$$

where $x = [AB_2]$.

(b) Show that $R(x)$ is a polynomial and determine its degree.

(c) Graph $R(x)$ for the relevant values of x when $a = 5$, $b = 6$, and $k = 0.3$.

4. Euclid, a Greek mathematician who lived around 300 B.C., wrote the *Elements*, which is by far the most important mathematical text of that period. It is a systematic exposition of most of the mathematical knowledge of that time, arranged in 13 volumes. In Book III, Euclid discusses the construction of a tangent to a circle at a point P on the circle. To phrase the construction in modern terminology, we draw a straight line through the point P that is perpendicular to the line through the center of the circle and the point P on the circle.

(a) Use this geometric construction to find the equation of the line that is tangent to the unit circle at the point $(\frac{1}{2}\sqrt{3}, \frac{1}{2})$.

(b) Determine the angle θ between the positive x-axis and the tangent line found in (a). What is the relationship between the angle θ and the slope of the tangent line found in (a)?

5. To compare logarithmic and exponential growth, we consider the following two hypothetical plants that are of the same genus but exhibit rather different growth rates. Both plants produce a single leaf whose length continues to increase as long as the plant is alive. One plant is called *Growthus logarithmiensis*, the other one is called *Growthus exponentialis*. The length L (measured in feet) of the leaf of *G. logarithmiensis* at age t (measured in years) is given by

$$L(t) = \ln(t + 1), \quad t \geq 0$$

The length E (measured in feet) of the leaf of *G. exponentialis* at age t (measured in years), is given by

$$E(t) = e^t - 1, \quad t \geq 0$$

(a) Find the length of each leaf after 1, 10, 100, and 1000 years.

(b) How long would it take for the leaf of *G. exponentialis* to reach a length of 233,810 mi, the average distance from the earth to the moon? (Note that 1 mi = 5280 ft.) How long would the leaf of *G. logarithmiensis* then be?

(c) How many years would it take the leaf of *G. logarithmiensis* to reach a length of 233,810 mi? Compare this to the length of time since life appeared on earth, about 3,500 million years. If *G. logarithmiensis* had appeared 3,500 million years ago, and if there was a plant of this species that had actually survived throughout the entire period, how long would its leaf be today?

(d) Plants only started to conquer land in the late Ordovician around 450 million years ago.[2] If both *G. exponentialis* and *G. logarithmiensis* had appeared then, and there was a plant of each species that had actually survived throughout the entire period, how long would their respective leaves be today?

6. Charles Darwin asserts in Chapter 3 of *The Origin of Species* (Darwin, 1859) that a "struggle for existence inevitably follows from the high rate at which all organic beings tend to increase. ... Although some species may be now increasing, more or less rapidly, in numbers, all cannot do so, for the world would not hold them." To illustrate this point, he continues as follows:

> There is no exception to the rule that every organic being naturally increases at so high a rate, that, if not destroyed, the earth would soon be covered by the progeny of a single pair. Even slow-breeding man has doubled in twenty-five years, and at this rate, in a few thousand years, there would literally not be standing room for his progeny.

Starting with a single pair, compute the world's population after 1000 years and after 2000 years under Darwin's assumption that the world's population doubles every 25 years, and find the resulting population densities (number of people per square foot). To answer the last part, you need to know that the earth's diameter is about 7900 mi, the surface of a sphere is $4\pi r^2$, where r is the radius of a sphere, and the continents make up about 29% of the earth's surface. (Note that 1 mi = 5280 ft.)

7. Assume that a population grows $q\%$ each year. How many years will it take the population to double in size? Give the functional relationship between the doubling time T and the annual percentage increase q. Produce a table that shows the doubling time T as a function of q for $q = 1, 2, \ldots, 10$, and graph T as a function of q. What happens to T as q gets closer to 0?

8. (*Beverton-Holt recruitment curve*) Many organisms show density-dependent mortality. The following is a simple mathematical model that incorporates this effect. Denote by N_b the density of parents and by N_a the density of *surviving* offspring.

(a) Suppose that without density-dependent mortality the number of surviving offspring per parent is equal to R. Show that if we plot N_b/N_a versus N_b, then this results in a horizontal line with y-intercept $1/R$. That is,

$$\frac{N_b}{N_a} = \frac{1}{R}$$

or

$$N_a = R \cdot N_b$$

The constant R is called the **net reproductive rate**.

(b) To include density-dependent mortality, we assume that N_b/N_a is an increasing function of N_b. The simplest way to do this is to assume that the graph of N_b/N_a versus N_b is a straight line, with y-intercept $1/R$ that goes through the point $(K, 1)$. Show that this implies that

$$N_a = \frac{RN_b}{1 + \frac{(R-1)N_b}{K}}$$

This relationship is called the Beverton-Holt recruitment curve.

(c) Explain in words why the model described by the Beverton-Holt recruitment curve behaves for small initial densities N_b like the model for density-independent mortality described in (a).

(d) Show that if $N_b = K$, then $N_a = K$. Furthermore, show that $N_b < K$ implies $N_b < N_a < K$ and that $N_b > K$ implies $K < N_a < N_b$. Explain in words what this means. (Note that K is called the carrying capacity.)

(e) Plot N_a as a function of N_b for $R = 2$ and $K = 20$. What happens for large values of N_b? Explain in words what this means.

9. (*Adapted from Moss, 1980*) Oglesby (1977) investigated the relationship between annual fish yield (Y) and summer phytoplankton chlorophyll concentration (C). Fish yield was measured in grams dry weight per square meter and per year, and the chlorophyll concentration was measured in micrograms per liter. Data from 19 lakes, mostly in the northern hemisphere, yielded the following relationship:

$$\log_{10} Y = 1.17 \log_{10} C - 1.92 \qquad (1.10)$$

(a) Plot $\log_{10} Y$ as a function of $\log_{10} C$.

(b) Find the relationship between Y and C; that is, write Y as a function of C. Explain the advantage of the log-log transformation resulting in (1.10) versus writing Y as a function of C. [*Hint:* Try to plot Y as a function of C and compare with your answer in (a).]

(c) Find the predicted yield (Y_p) as a function of the current yield (Y_c) if the current summer phytoplankton chlorophyll concentration were to double.

(d) By what percentage would the summer phytoplankton chlorophyll concentration need to increase to obtain a 10% increase in fish yield?

10. (*Adapted from Moss, 1980*) To trace the history of a lake, a sample of mud from a core is taken and dated. One such

---

[2] The Ordovician lasted from about 505 million years ago to about 438 million years ago.

dating method uses radioactive isotopes. The C^{14} method is effective for sediments that are younger than 60,000 years. The $C^{14} : C^{12}$ ratio has been essentially constant in the atmosphere over a long period of time, and living organisms take up carbon in this ratio. Upon death, the uptake of carbon ceases and C^{14} decays, which changes the $C^{14} : C^{12}$ ratio according to

$$\left(\frac{C^{14}}{C^{12}}\right)_t = \left(\frac{C^{14}}{C^{12}}\right)_{initial} e^{-\lambda t}$$

where t is the time since death.

(a) If the $C^{14} : C^{12}$ ratio in the atmosphere is 10^{-12} and the half-life of C^{14} is 5730 years, find an expression for t, the age of the material being dated, as a function of the $C^{14} : C^{12}$ ratio in the material being dated.

(b) Use your answer in (a) to find the age of a mud sample from a core for which the $C^{14} : C^{12}$ ratio is 1.61×10^{-13}.

11. (*Adapted from Futuyama, 1995, and Dott and Batten, 1976*) Corals deposit a single layer of lime each day. In addition, seasonal fluctuation in the thickness of the layers allows for grouping these layers into years. In modern corals, we can therefore count 365 layers per year. J. Wells, a paleontologist, counted such growth layers on fossil corals. To his astonishment, he found that Devonian[3] corals that lived about 380 million years ago had about 400 daily layers per year.

(a) Today, the earth rotates about its axis every 24 hours and revolves around the sun every $365\frac{1}{4}$ days. Astronomers have determined that the earth's rotation has slowed down in recent centuries at the rate of about 2 seconds every 100,000 years. That is, 100,000 years ago, a day was 2 seconds shorter than today. Extrapolate this back to the Devonian, and determine the length of a day and the length of a year back when Wells's corals lived. (*Hint:* The number of hours per year remains constant. Why?)

(b) Find a linear equation that relates geologic time (in million of years) to the number of hours per day at a given time.

(c) Algal stromatolites also show daily layers. A sample of some fossil stromatolites showed 400 to 420 daily layers per year. Use your answer in (b) to date the stromatolites.

12. The height in feet (y) of a certain tree as a function of age in years (x) can be approximated by

$$y = 132e^{-20/x}$$

(a) Use a graphing calculator to plot the graph of this function. Describe in words how the tree grows, paying particular attention to questions like the following: Does the tree grow equally fast over time? What happens when the tree is young? What happens when the tree is old?

(b) How many years will it take for the tree to reach 100 ft in height?

(c) Can the tree ever reach a height of 200 ft? Is there a final height, that is, a maximum height that the tree will eventually reach?

---

[3] The Devonian period lasted from about 408 million years ago to about 360 million years ago.

13. (*Model for aging*) The probability that an individual lives beyond age t is called the *survivorship function* and is denoted by $S(t)$. The Weibull model is a popular model in reliability theory and in studies of biological aging. Its survivorship function is described by two parameters, λ and β, and is given by

$$S(t) = \exp[-(\lambda t)^\beta]$$

Mortality data from a *Drosophila melanogaster* population in Dr. Jim Curtsinger's lab at the University of Minnesota were collected and fitted to this model separately for males and females (Pletcher, 1998). The following parameter values were obtained (t was measured in days):

Sex	λ	β
Males	0.019	3.41
Females	0.022	3.24

(a) Use a graphing calculator to sketch the survivorship function for both the female and male populations.

(b) For each population, find the value of t for which the probability of living beyond that age is $1/2$.

(c) If you had a male and a female of this species, which would you expect to live longer?

14. Carbon has two stable isotopes, C^{12} and C^{13}. Organic material contains both stable isotopes but the ratio $[C^{13}] : [C^{12}]$ in organic material is smaller than that in inorganic material, reflecting the fact that light carbon (C^{12}) is preferentially taken up by plants during photosynthesis. This is called isotopic fractionation and is measured as

$$\delta^{13}C = \left[\frac{([C^{13}] : [C^{12}])_{sample}}{([C^{13}] : [C^{12}])_{standard}} - 1 \right]$$

The standard is taken from the isotope ratio in the carbon of belemnite shells found in the Cretaceous Pedee formation of South Carolina. Based on the preceding information, explain why the following quotation from Krauskopf and Bird (1995) makes sense:

The low [negative] values of $\delta^{13}C$ in the hydrocarbons of petroleum are one of the important bits of evidence for ascribing the origin of petroleum to the alteration of organic material rather than to condensation of primeval gases from the Earth's interior.

15. The speed of an enzymatic reaction is frequently described by the Michaelis-Menten equation

$$v = \frac{ax}{k + x}$$

where v is the velocity of the reaction, x is the concentration of the substrate, a is the maximum reaction velocity, and k is the substrate concentration at which the velocity is half of the maximum velocity. This curve describes how the reaction velocity depends on the substrate concentration.

(a) Show that when $x = k$, the velocity of the reaction is half the maximum velocity.

(b) Show that an 81-fold change in substrate concentration is needed to change the velocity from 10% to 90% of the maximum velocity, regardless of the value of k.

16. Atmospheric pollutants can cause acidification of lakes (acid rain). This can be a serious problem for lake organisms; for instance, in fish the ability of hemoglobin to transport oxygen decreases with decreasing pH levels of the water. Experiments with the zooplankton *Daphnia magna* showed a negligible decline in survivorship at pH = 6 but a marked decline in survivorship at pH = 3.5, resulting in no survivors after just eight hours. Illustrate graphically the percentage survivorship as a function of time for pH = 6 and pH = 3.5.

17. The pH level of lakes controls the concentrations of the harmless ammonium ions (NH_4^+) and the toxic ammonia (NH_3). For pH levels below 8, concentrations of NH_4^+ ions are little affected by changes in the pH value, but they decline over many orders of magnitude as pH levels increase beyond pH = 8. On the other hand, NH_3 concentrations are negligible at low pH, increase over many orders of magnitude as the pH level increases, and reach a high plateau at about pH = 10 (after which levels of NH_3 are little affected by changes in pH levels). Illustrate this graphically.

18. Egg development times of the zooplankton *Daphnia longispina* depend on temperature. It takes only about 3 days at 20°C but almost 20 days at 5°C. When graphed on a log-log plot, egg development (in days) as a function of temperature (in °C) is a straight line.

(a) Sketch a graph of egg development time as a function of temperature on a log-log plot.

(b) Use the data to find the function that relates egg development time and temperature for *D. longispina*.

(c) Use your answer in (b) to predict egg development time of *D. longispina* at 10°C.

(d) Suppose you measured egg development times in hours and temperature in Fahrenheit. Would you still find a straight line on a log-log plot?

19. Organisms consume resources. The rate of resource consumption, denoted by v, depends on resource concentration, denoted by S. The *Blackman* model of resource consumption assumes a linear relationship between resource consumption rate and resource concentration: Below a threshold concentration (S_k) the consumption rate increases linearly with $S = 0$ when $v = 0$; when $S = S_k$, the consumption rate v reaches its

maximum value v_{max}; for $S > S_k$, the resource consumption rate stays at the maximum value v_{max}. A function like this with a sharp transition cannot be described analytically by just one expression; it needs to be defined piecewise:

$$v = \begin{cases} g(S) & \text{for } 0 \le S \le S_k \\ v_{max} & \text{for } S > S_k \end{cases}$$

Find $g(S)$ and graph the resource consumption rate v as a function of resource concentration S.

20. Light intensity in lakes decreases exponentially with depth. If $I(z)$ denotes the light intensity at depth z, with $z = 0$ representing the surface, then

$$I(z) = I(0)e^{-\alpha z}, \quad z \ge 0$$

where α is a positive constant called the vertical attenuation coefficient. The attenuation coefficient depends on the wavelength of the light and on the amount of dissolved matter and particles in the water. In the following, we assume that the water is pure.

(a) About 65% of red light (720 nm) is absorbed in the first meter. Find α.

(b) About 5% of blue light (475 nm) is absorbed in the first meter. Find α.

(c) Explain in words why a diver would not see red hues a few meters below the surface of a lake.

21. Light intensity in lakes decreases with depth according to

$$I(z) = I(0)e^{-\alpha z}, \quad z \ge 0$$

where $I(z)$ denotes the light intensity at depth z, with $z = 0$ representing the surface, and α is a positive constant, denoting the vertical attenuation coefficient. The depth where light intensity is about 1% of the surface light intensity is important for photosynthetic activity of phytoplankton: Below this level, photosynthesis is insufficient to compensate for respiratory losses. The 1% level is called the *compensation level*. An often used and relatively reliable method for determining the compensation level is the *Secchi disk* method. A Secchi disk is a white disk with radius 10 cm. The disk depth is the depth where the disk disappears from the viewer. Twice this depth approximately coincides with the compensation level.

(a) Find α for a lake with Secchi disk depth of 9 m.

(b) Find the Secchi disk depth for a lake with $\alpha = 0.473$ m^{-1}.

Discrete Time Models, Sequences, and Difference Equations

CHAPTER 2

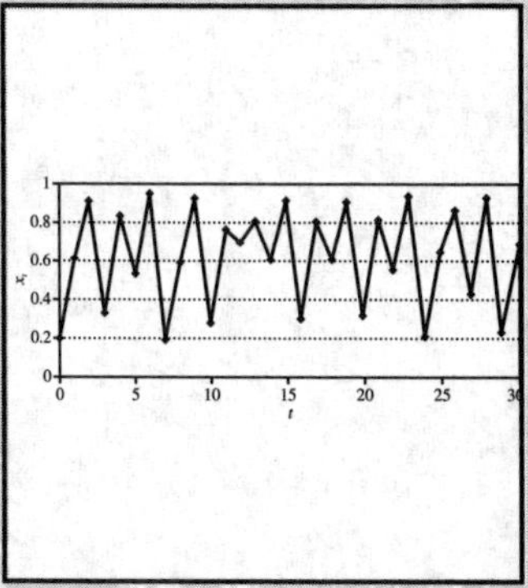

In this chapter, we discuss models for populations that reproduce at discrete times. They are given by functions whose domains are subsets of the set of nonnegative integers $\mathbf{N} = \{0, 1, 2, \ldots\}$. We have seen one such function in Example 9 of Section 1.2. These functions are used extensively in biology to describe, for instance, the population size of a plant that reproduces once a year and then dies (an annual plant). The functions are of the form

$$f: \mathbf{N} \to \mathbf{R}$$
$$n \to f(n)$$

When the independent variable denotes time, we will frequently use t instead of n. As in Chapter 1, tables and graphs are useful tools to illustrate these functions.

Example 1 Let

$$f: \mathbf{N} \to \mathbf{R}$$
$$n \to f(n) = \frac{1}{n+1}$$

Produce a table for $n = 0, 1, 2 \ldots, 5$ and graph the function.

Solution

Here is the table:

n	0	1	2	3	4	5
$\frac{1}{n+1}$	1	$\frac{1}{2}$	$\frac{1}{3}$	$\frac{1}{4}$	$\frac{1}{5}$	$\frac{1}{6}$

The graph of this function consists of discrete points (Figure 2.1).◀

2.1 EXPONENTIAL GROWTH AND DECAY

2.1.1 Modeling Population Growth in Discrete Time

Imagine that we observe bacteria that divide every 20 minutes and that at the start of the experiment, there was one bacterium. How will the number of bacteria change over time? We call the time when we started the observation, time 0. At time 0, there is one bacterium. After 20 minutes, the bacterium splits in two, so there

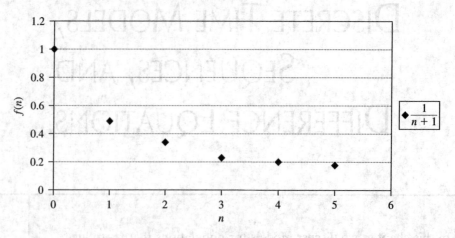

▲ **Figure 2.1**
The graph of the function $f(n) = \frac{1}{n+1}$ in Example 1

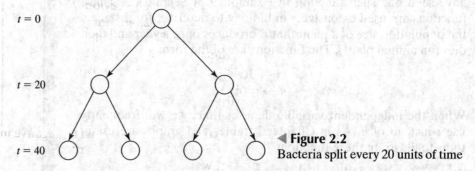

◀ **Figure 2.2**
Bacteria split every 20 units of time

are two bacteria at time 20. Twenty minutes later, each of the bacteria splits again, resulting in four bacteria at time 40, and so on (Figure 2.2).

We can produce a table that describes the growth of this population.

Time (min)	0	20	40	60	80	100	120
Population size	1	2	4	8	16	32	64

We can simplify the description of the growth of this population if we measure time in more convenient units. We say that one unit of time equals 20 minutes. Two units of time then correspond to 40 minutes, three units of time to 60 minutes, and so on. We reproduce the table of population growth with these new units.

Time	0	1	2	3	4	5	6
Population size	1	2	4	8	16	32	64

These new time units make it easier to write a general formula for the population size at time t. Namely, denoting by $N(t)$ the population size at time t, where t is now measured in these new units (one unit is equal to 20 minutes), we guess from the second table that

$$N(t) = 2^t, \quad t = 0, 1, 2, \ldots \tag{2.1}$$

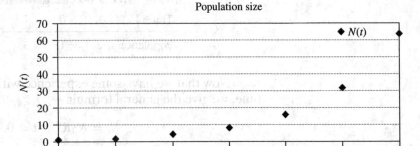

Population size

▲ **Figure 2.3**
The graph of $N(t) = 2^t$ for $t = 0, 1, 2, \ldots, 6$

We saw this function in Section 1.2. Equation (2.1) allows us to determine the population size at any discrete time t directly without first calculating the population sizes at all previous time steps. For instance, at time $t = 5$, we find $N(5) = 2^5 = 32$, as shown in the second table, or, at time $t = 10$, $N(10) = 2^{10} = 1024$. The graph of $N(t) = 2^t$ is shown in Figure 2.3.

The function $N(t) = 2^t$, $t = 0, 1, 2, \ldots$, is an exponential function, and we call this type of population growth **exponential growth**. The base 2 reflects the fact that the population size doubles every unit of time; we will see this more clearly later.

Instead of $N(t)$, we will often write N_t. The subscript notation is only used for functions $N(t)$ where t is a nonnegative integer. So, instead of writing $N(t) = 2^t$, $t = 0, 1, 2, \ldots$, we can write $N_t = 2^t$, $t = 0, 1, 2, \ldots$.

So far, we assumed that $N(0) = N_0 = 1$. Let's see what N_t looks like if $N_0 = 100$. Regardless of N_0, the population size doubles every unit of time. We obtain the following table, where time is again measured in units of 20 minutes.

Time	0	1	2	3	4	5	6
Population size	100	200	400	800	1600	3200	6400

We can guess the general form of N_t with $N_0 = 100$ from the table:

$$N_t = 100 \cdot 2^t, \quad t = 0, 1, 2, \ldots$$

We see that the initial population size $N_0 = 100$ appears as a multiplicative factor in front of the term 2^t. If we do not want to specify a numerical value for the population size at time 0, N_0, we can write

$$N_t = N_0 2^t, \quad t = 0, 1, 2, \ldots$$

We already mentioned that the base 2 indicates that the population size doubles every unit of time. Replacing 2 by another number, we can describe other populations. For instance,

$$N_t = 3^t, \quad t = 0, 1, 2, \ldots$$

describes a population with $N_0 = 1$ that triples in size every unit of time.

Time	0	1	2	3	4
Population size	1	3	9	27	81

Now that we have some experience with exponential growth in discrete time, we give the general formula

$$N_t = N_0 R^t, \quad t = 0, 1, 2, \ldots \tag{2.2}$$

The parameter R is a positive constant and called the **growth constant**. The constant N_0 is nonnegative and denotes the population size at time 0. The assumptions $R > 0$ and $N_0 \geq 0$ are made for biological reasons: Negative values for R or N_0 would result in negative population sizes, and $R = 0$ would be uninteresting.

The function $N_t = N_0 R^t$, $t = 0, 1, 2, \ldots$, is an exponential function. We discussed exponential functions in the previous chapter. There, we looked at $f(x) = a^x$, $x \in \mathbf{R}$. To make the comparison easier, we choose $N_0 = 1$ in (2.2) and restrict the function $f(x) = a^x$ to $x \geq 0$. If we choose the same values for R and a, then the two functions N_t and $f(x)$ use the same rule to compute function values. The difference is in the domain: N_t is only defined for nonnegative integers, whereas $f(x)$ is defined for all nonnegative real numbers. The two functions agree where they are both defined. This can be seen when we graph N_t and $f(x)$ in the same coordinate system for $R = a$ (Figure 2.4).

In Chapter 1, we learned how $f(x) = a^x$, $x \in \mathbf{R}$, behaves for different values of a. We can use this now to describe the behavior of $N_t = N_0 R^t$, $t = 0, 1, 2, \ldots$. In Figure 2.5, we show the function $f(x) = a^x$, $x \geq 0$, for different values of a. Superimposed are the graphs of $N_t = N_0 R^t$, $t = 0, 1, 2, \ldots$, for $R = a$ and $N_0 = 1$.

We see that when $R > 1$, the population size N_t increases indefinitely; when $R = 1$, the population size N_t stays the same for all $t = 0, 1, 2, \ldots$; when $0 < R < 1$, the population size N_t declines and approaches 0 as t increases. The behavior is the same for other positive initial population sizes ($N_0 > 0$). When $N_0 = 0$, then $N_t = 0$ for all $t = 1, 2, 3, \ldots$, implying that the population size does not change over time. A population size that does not change over time is called a **fixed point** or an **equilibrium**. We will discuss fixed points after introducing other ways to represent the function N_t.

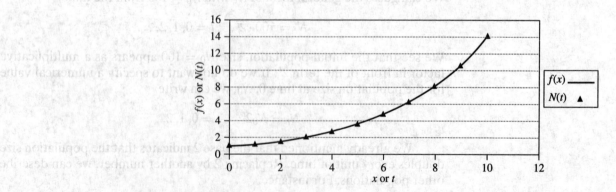

▲ **Figure 2.4**
The graphs of $f(x) = a^x$, $0 \leq x \leq 10$, and $N(t) = R^t$, $t = 0, 1, 2, \ldots, 10$ when $a = R = 1.3$.

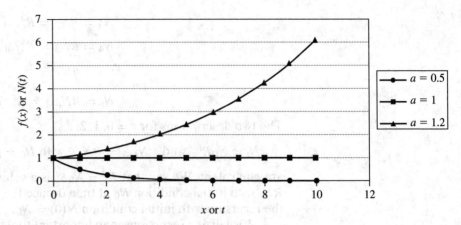

▲ **Figure 2.5**
The graphs of $f(x) = a^x$, $0 \leq x \leq 10$, and $N(t) = R^t$, $t = 0, 1, 2, \ldots, 10$ for three different values of $a = R$: $a = R = 0.5$, $a = R = 1$, and $a = R = 1.2$

2.1.2 Recursions

When we constructed the tables for the bacterial population size with $R = 2$ at consecutive time steps, we simply doubled the population size from time step to time step. In other words, we computed the population size at time $t + 1$ based on the population size at time t using the equation

$$N_{t+1} = 2N_t \tag{2.3}$$

Equation (2.3) is called a **recursion** since it is a rule that is applied repeatedly to go from one time step to the next. We say that equation (2.3) defines the population size **recursively**.

   If we want to use equation (2.3) to find the population size say at time $t = 4$, we need to know the population size at some earlier time, say at time $t = 0$. Let's assume $N_0 = 1$. Then applying the recursion (2.3) repeatedly, we find

$$N_1 = 2N_0 = 2$$
$$N_2 = 2N_1 = 4$$
$$N_3 = 2N_2 = 8$$
$$N_4 = 2N_3 = 16$$

We thus have two equivalent ways to describe this population. For $t = 0, 1, 2, \ldots,$

$$N_t = 2^t \quad \text{is equivalent to} \quad N_{t+1} = 2N_t \quad \text{with } N_0 = 1$$

The recursion for a general value of R is

$$N_{t+1} = RN_t \quad \text{with } N_0 = \text{population size at time 0} \tag{2.4}$$

Applying (2.4) repeatedly, we find

$$N_1 = RN_0$$
$$N_2 = RN_1 = R^2 N_0$$

$$N_3 = RN_2 = R^3 N_0$$
$$N_4 = RN_3 = R^4 N_0$$
$$\vdots$$
$$N_t = RN_{t-1} = R^t N_0$$

The two descriptions for $t = 0, 1, 2, \ldots,$

$$N_t = N_0 R^t \quad \text{and} \quad N_{t+1} = RN_t \quad \text{with } N_0 = \text{population size at time 0}$$

are equivalent. We say that $N_t = N_0 R^t$ is a **solution** of the recursion $N_{t+1} = RN_t$ with initial condition N_0 at time 0 since the function $N_t = N_0 R^t$ satisfies the recursion with initial condition $N(0) = N_0$.

Visualizing recursions is an important tool for understanding them. One way to do this is to plot N_t on the horizontal axis and N_{t+1} on the vertical axis. The exponential growth recursion

$$N_{t+1} = RN_t \tag{2.5}$$

is then a straight line through the origin with slope R (Figure 2.6). Since $N_t \geq 0$ for biological reasons, we restrict the graph to the first quadrant.

What does this graph tell us? For any current population size N_t, it allows us to find the population size in the next time step, namely N_{t+1}. For instance, if we choose $R = 2$ and $N_0 = 1$, then successive population sizes are $1, 2, 4, 8, 16, 32, \ldots$. For this choice of N_0, we will never see a population size of, say, 5 or 10. Thus, for a specific choice of N_0, only a selected number of points on the graph $N_{t+1} = RN_t$ will be realized (Figure 2.7). A different choice of initial condition would yield a different set of points.

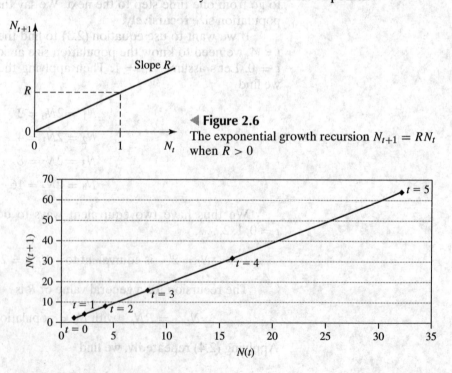

◄ **Figure 2.6**
The exponential growth recursion $N_{t+1} = RN_t$ when $R > 0$

▲ **Figure 2.7**
Successive population sizes on the graph of the exponential growth recursion when $R = 2$ for $t = 0, 1, 2, \ldots, 5$

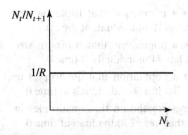

▲ Figure 2.8
The graph of the parent to offspring ratio $\frac{N_t}{N_{t+1}}$ as a function of N_t when $N_t > 0$

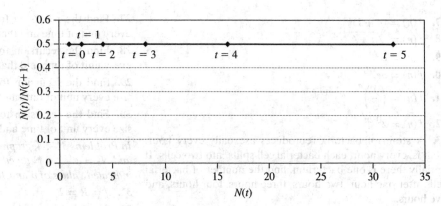

▲ Figure 2.9
The graph of the parent to offspring ratio $\frac{N_t}{N_{t+1}}$ as a function of N_t when $N_t = 1$ and $R = 2$.

We also see from Figure 2.7 that unless we label the points according to the corresponding t-value, we would not be able to tell at what time a point (N_t, N_{t+1}) was realized. We say that time is *implicit* in this graph. Compare this to Figure 2.3, where we graphed N_t as a function of t for the same values of R and N_0; in Figure 2.3, time is *explicit*.

The hallmark of exponential growth is that the ratio of successive population sizes, N_t/N_{t+1}, is constant. When $N_t > 0$ (and hence $N_{t+1} > 0$), it follows from $N_{t+1} = RN_t$ that

$$\frac{N_t}{N_{t+1}} = \frac{1}{R}$$

If the population consists of annual plants, we can interpret the ratio N_t/N_{t+1} as the parent-offspring ratio. If this ratio is constant, parents produce the same number of offspring regardless of the current population density. Such growth is called **density independent**.

When $R > 1$, then $1/R$, the parent-offspring ratio, is less than 1, implying that the number of offspring exceeds the number of parents. Density-independent growth with $R > 1$ results in an ever-increasing population size. This eventually becomes biologically unrealistic since any population will sooner or later experience food or habitat limitations that will limit its growth. We will discuss models that include such limitations in Section 2.3.

The density independence in exponential growth is reflected in a graph of N_t/N_{t+1} as a function of N_t, which is a horizontal line at level $1/R$ (Figure 2.8).

As before, only a selected number of points are realized on the graph of N_t/N_{t+1} as a function of N_t and time is implicit in the graph. See Figure 2.9 when $R = 2$ and $N_0 = 1$.

2.1.3 Problems

In Problems 1–12, produce a table for $n = 0, 1, 2, \ldots, 5$, and graph the function.

1. $f(n) = \dfrac{1}{n+2}$

2. $f(n) = \dfrac{1}{1+n^2}$

3. $f(n) = \dfrac{1}{(1+n)^2}$

4. $f(n) = \dfrac{1}{\sqrt{n+1}}$

5. $f(n) = n^2 - 1$

6. $f(n) = n^3 + 1$

7. $f(n) = (n+1)^2$

8. $f(n) = \sqrt{n+4}$

9. $f(n) = e^{\sqrt{n}}$

10. $f(n) = 3e^{-0.1n}$

11. $f(n) = \left(\dfrac{1}{3}\right)^n$

12. $f(n) = 2^{0.2n}$

13. A strain of bacteria reproduces asexually every hour. That is, every hour, each bacterial cell splits into two cells. If initially there is one bacterium, find the number of bacterial cells after one hour, two hours, three hours, four hours, and five hours.

14. A strain of bacteria reproduces asexually every 30 minutes. That is, every 30 minutes, each bacterial cell splits into two cells. If initially there is one bacterium, find the number of bacterial cells after one hour, two hours, three hours, four hours, and five hours.

15. A strain of bacteria reproduces asexually every 23 minutes. That is, every 23 minutes, each bacterial cell splits into two cells. If initially there is one bacterium, how long will it take until there are 128 bacteria?

16. A strain of bacteria reproduces asexually every 42 minutes. That is, every 42 minutes, each bacterial cell splits into two cells. If initially there is one bacterium, how long will it take until there are 512 bacteria?

17. A strain of bacteria reproduces asexually every 10 minutes. That is, every 10 minutes, each bacterial cell splits into two cells. If initially there are three bacteria, how long will it take until there are 96 bacteria?

18. A strain of bacteria reproduces asexually every 50 minutes. That is, every 50 minutes, each bacterial cell splits into two cells. If initially there are 10 bacteria, how long will it take until there are 640 bacteria?

19. Find the exponential growth equation for a population that doubles in size every unit of time and that has forty individuals at time 0.

20. Find the exponential growth equation for a population that doubles in size every unit of time and that has 53 individuals at time 0.

21. Find the exponential growth equation for a population that triples in size every unit of time and that has 20 individuals at time 0.

22. Find the exponential growth equation for a population that triples in size every unit of time and that has 72 individuals at time 0.

23. Find the exponential growth equation for a population that quadruples in size every unit of time and that has five individuals at time 0.

24. Find the exponential growth equation for a population that quadruples in size every unit of time and that has 17 individuals at time 0.

25. Find the recursion for a population that doubles in size every unit of time and that has 20 individuals at time 0.

26. Find the recursion for a population that doubles in size every unit of time and that has 37 individuals at time 0.

27. Find the recursion for a population that triples in size every unit of time and that has 10 individuals at time 0.

28. Find the recursion for a population that triples in size every unit of time and that has 84 individuals at time 0.

29. Find the recursion for a population that quadruples in size every unit of time and that has 30 individuals at time 0.

30. Find the recursion for a population that quadruples in size every unit of time and that has 62 individuals at time 0.

In Problems 31–34, graph the functions $f(x) = a^x$, $x \in [0, \infty)$ and $N_t = R^t$, $t \in \mathbf{N}$ together in one coordinate system for the indicated values of a and R.

31. $a = R = 2$

32. $a = R = 3$

33. $a = R = 1/2$

34. $a = R = 1/3$

In Problems 35–46, find the population sizes for $t = 0, 1, 2, \ldots, 5$ for each recursion.

35. $N_{t+1} = 2N_t$ with $N_0 = 3$

36. $N_{t+1} = 2N_t$ with $N_0 = 5$

37. $N_{t+1} = 3N_t$ with $N_0 = 2$

38. $N_{t+1} = 3N_t$ with $N_0 = 7$

39. $N_{t+1} = 5N_t$ with $N_0 = 1$

40. $N_{t+1} = 7N_t$ with $N_0 = 4$

41. $N_{t+1} = \frac{1}{2}N_t$ with $N_0 = 1024$

42. $N_{t+1} = \frac{1}{2}N_t$ with $N_0 = 4096$

43. $N_{t+1} = \frac{1}{3}N_t$ with $N_0 = 729$

44. $N_{t+1} = \frac{1}{3}N_t$ with $N_0 = 3645$

45. $N_{t+1} = \frac{1}{5}N_t$ with $N_0 = 31250$

46. $N_{t+1} = \frac{1}{4}N_t$ with $N_0 = 8192$

In Problems 47–58, write N_t as a function of t for each recursion.

47. $N_{t+1} = 2N_t$ with $N_0 = 15$

48. $N_{t+1} = 2N_t$ with $N_0 = 7$

49. $N_{t+1} = 3N_t$ with $N_0 = 12$

50. $N_{t+1} = 3N_t$ with $N_0 = 3$

51. $N_{t+1} = 4N_t$ with $N_0 = 24$

52. $N_{t+1} = 5N_t$ with $N_0 = 17$

53. $N_{t+1} = \frac{1}{2}N_t$ with $N_0 = 5000$

54. $N_{t+1} = \frac{1}{2}N_t$ with $N_0 = 2300$

55. $N_{t+1} = \frac{1}{3}N_t$ with $N_0 = 8000$

56. $N_{t+1} = \frac{1}{3}N_t$ with $N_0 = 3500$

57. $N_{t+1} = \frac{1}{5}N_t$ with $N_0 = 1200$

58. $N_{t+1} = \frac{1}{7}N_t$ with $N_0 = 6400$

In Problems 59–66, graph the line $N_{t+1} = RN_t$ in the N_t-N_{t+1} plane for the indicated value of R and locate the points (N_t, N_{t+1}), $t = 0, 1,$ and 2 for the given value of N_0.

59. $R = 2, N_0 = 2$

60. $R = 2, N_0 = 3$

61. $R = 3, N_0 = 1$

62. $R = 4, N_0 = 2$
63. $R = \frac{1}{2}, N_0 = 16$
64. $R = \frac{1}{2}, N_0 = 64$
65. $R = \frac{1}{3}, N_0 = 81$
66. $R = \frac{1}{4}, N_0 = 16$

In Problems 67–74, graph the line $\frac{N_t}{N_{t+1}} = \frac{1}{R}$ in the N_t-$\frac{N_t}{N_{t+1}}$ plane for the indicated value of R and locate the points $(N_t, \frac{N_t}{N_{t+1}})$, t = 0, 1, and 2 for the given value of N_0.

67. $R = 2, N_0 = 2$
68. $R = 2, N_0 = 4$
69. $R = 3, N_0 = 2$
70. $R = 4, N_0 = 1$
71. $R = \frac{1}{2}, N_0 = 16$
72. $R = \frac{1}{2}, N_0 = 128$
73. $R = \frac{1}{3}, N_0 = 27$
74. $R = \frac{1}{4}, N_0 = 64$

75. A bird population lives in a habitat where the number of nesting sites is a limiting factor in population growth. In which of the following cases would you expect that the population growth of this bird population over the next few generations could be reasonably well approximated by exponential growth?

(a) All nesting sites are occupied.

(b) The bird population just invaded the habitat and the population size is still much smaller than the available nesting sites.

(c) In the previous year, a hurricane killed more than 90% of the birds in this habitat.

76. Pollen records show that the number of Scotch pine (*Pinus sylvestris*) grew exponentially for about 500 years after colonization of the Norfolk region of Great Britain about 9500 years ago. Can you find a possible explanation for this?

77. Exponential growth generally occurs when population growth is density independent. List conditions under which a population might stop growing exponentially.

2.2 SEQUENCES

2.2.1 What Are Sequences?

Before we explore other discrete time population models, we need to develop further the theory of functions with domain **N**. A function

$$f : \mathbf{N} \to \mathbf{R}$$
$$n \to f(n)$$

is called a **sequence**. We will also use the notation $a_n = f(n)$, and write $\{a_n\}$ if we mean the entire sequence. We can list the values of the sequence $\{a_n\}$ in the order of increasing n:

$$a_0, a_1, a_2, \ldots$$

▶ Example 1   The sequence

$$a_n = (-1)^n, \quad n = 0, 1, 2, \ldots$$

takes on values

$$1, -1, 1, -1, 1, \ldots \qquad \blacktriangleleft$$

▶ Example 2   The sequence

$$a_n = n^2, \quad n = 0, 1, 2, \ldots$$

takes on values

$$0, 1, 4, 9, 16, \ldots \qquad \blacktriangleleft$$

When we see a sequence and recognize a pattern, we can often write an expression for a_n.

▶ **Example 3**   Find a_n for the sequence

$$1, \frac{1}{2}, \frac{1}{3}, \frac{1}{4}, \frac{1}{5}, \ldots$$

Solution

Looking at the sequence, we can guess the next terms, namely, $\frac{1}{6}, \frac{1}{7}, \frac{1}{8}$, and so on. We thus find

$$a_n = \frac{1}{n+1}, \quad n = 0, 1, 2, \ldots$$

We do not need to start a sequence at $n = 0$. If we start the sequence at $n = 1$, we would write

$$a_n = \frac{1}{n}, \quad n = 1, 2, 3, \ldots$$

In either case, it is important to include the domain of the sequence.   ◀

▶ **Example 4**   Find a_n for the sequence

$$1, -\frac{1}{4}, \frac{1}{9}, -\frac{1}{16}, \frac{1}{25}, \ldots$$

Solution

This sequence has alternating signs: The first term is positive, the second is negative, the third is positive, and so on. This indicates that we need a factor $(-1)^n$, $n = 0, 1, 2, \ldots$. The numerators are all equal to 1, and the denominators are successive squares of integers, starting with the integer 1. We can thus write

$$a_0 = (-1)^0 \frac{1}{(1)^2} = 1$$

$$a_1 = (-1)^1 \frac{1}{(2)^2} = -\frac{1}{4}$$

$$a_2 = (-1)^2 \frac{1}{(3)^2} = \frac{1}{9}$$

and so on. This suggests that

$$a_n = (-1)^n \frac{1}{(n+1)^2}, \quad n = 0, 1, 2, \ldots$$

If we wanted to start the sequence at $n = 1$, we could write

$$a_n = (-1)^{n-1} \frac{1}{n^2}, \quad n = 1, 2, 3, \ldots$$

or

$$a_n = (-1)^{n+1} \frac{1}{n^2}, \quad n = 1, 2, 3, \ldots$$

Look carefully at the exponent of (-1). Any of the terms $(-1)^n$, $(-1)^{n-1}$, or $(-1)^{n+1}$ produces alternating signs. Since the first term in the sequence $\{a_n\}$ is positive, we need to use $(-1)^n$ if we start with $n = 0$, and either $(-1)^{n-1}$ or $(-1)^{n+1}$ if we start with $n = 1$.   ◀

The exponential growth model we considered in the previous section is an example of a sequence. We gave two descriptions, one explicit and the other recursive. These two descriptions can be used for sequences in general. An explicit description is of the form

$$a_n = f(n), \quad n = 0, 1, 2, \ldots$$

where $f(n)$ is a function of n.

A recursive description is of the form

$$a_{n+1} = f(a_n), \quad n = 0, 1, 2, \ldots$$

where $f(a_n)$ is a function of a_n. (The functions f in the two descriptions are different in general.) If, as is shown here, the value of a_{n+1} only depends on the value one time step back, namely a_n, then the recursion is called a **first-order recursion**. Later in the chapter we will see an example of a second-order recursion, where the value of a_{n+1} depends on the values a_n and a_{n-1}; that is, on the values one and two time steps back. To determine the value of successive members of a sequence given in recursive form, we need to specify an initial value, namely a_0, if we start the sequence at $n = 0$ (or a_1 if we start the sequence at $n = 1$).

Using the notation of this section, the exponential growth of the previous section is given explicitly by

$$a_n = a_0 R^n, \quad n = 0, 1, 2, \ldots$$

and recursively by

$$a_{n+1} = R a_n, \quad n = 0, 1, 2, \ldots$$

2.2.2 Limits

When studying populations over time, we are often interested in their **long-term behavior**. Specifically, if N_t is the population size at time t, $t = 0, 1, 2, \ldots$, we want to know how N_t behaves as t becomes larger and larger, or, in mathematical terms, as t tends to infinity. Using the notation of this section, we want to know the behavior of a_n as n tends to infinity. When we let n tend to infinity, we say that "we take the limit as n goes to infinity of the sequence a_n" and use the short-hand notation

$$\lim_{n \to \infty} a_n \quad \text{or} \quad \lim_{n \to \infty} a_n$$

read as "limit of a_n as n tends to infinity." The first notation is used in displayed equations, the second when the expression appears in text. Let's first discuss limits informally to get an idea of what can happen.

▶ **Example 5**   Let

$$a_n = \frac{1}{n+1}, \quad n = 0, 1, 2, \ldots$$

Find $\lim_{n \to \infty} a_n$.

Solution

Plugging successive values of n into a_n, we find that a_n is the sequence

$$1, \frac{1}{2}, \frac{1}{3}, \frac{1}{4}, \frac{1}{5}, \ldots$$

and guess that the terms will approach 0 as n tends to infinity. This is indeed the case, and we can show that

$$\lim_{n \to \infty} \frac{1}{n+1} = 0$$

Since the limiting value is a unique number, we say that the limit exists. ◀

You should note that plugging in successive values of n into a_n is only a heuristic way of determining how a_n behaves as $n \to \infty$.

▶ **Example 6** Let

$$a_n = (-1)^n, \quad n = 0, 1, 2, \ldots$$

Find $\lim_{n \to \infty} a_n$.

Solution

The sequence is of the form

$$1, -1, 1, -1, 1, \ldots$$

and we see that its terms alternate between 1 and −1. There is thus no single number we could assign as the limit of a_n as $n \to \infty$. We say that the limit does not exist. ◀

▶ **Example 7** Let

$$a_n = 2^n, \quad n = 0, 1, 2, \ldots$$

Find $\lim_{n \to \infty} a_n$.

Solution

Successive terms of a_n,

$$1, 2, 4, 8, 16, 32, \ldots$$

indicate that the terms continue to grow. Thus a_n goes to infinity as $n \to \infty$, and we can write $\lim_{n \to \infty} a_n = \infty$. Since infinity ($\infty$) is not a real number, we say that the limit does not exist. ◀

Let's look at one more example of a limit that exists before we give a formal definition.

▶ **Example 8** Find

$$\lim_{n \to \infty} \frac{n+1}{n}$$

Solution

Computing successive terms starting with $n = 1$, we find

$$2, \frac{3}{2}, \frac{4}{3}, \frac{5}{4}, \frac{6}{5}, \ldots$$

We see that the terms get closer and closer to 1, and indeed we can show that

$$\lim_{n \to \infty} \frac{n+1}{n} = 1$$ ◀

The way we solved the first four examples is unsatisfying: We guessed the limiting values. How do we know that our guesses are correct? There is a formal definition of limits that can be used to compute limits. However, except in the simplest cases, the formal definition is quite cumbersome to use. Fortunately, there are limit laws that build on simple limits (which can be computed using the formal definition of limits). We will first discuss the formal definition (as an optional topic) and then introduce the limit laws.

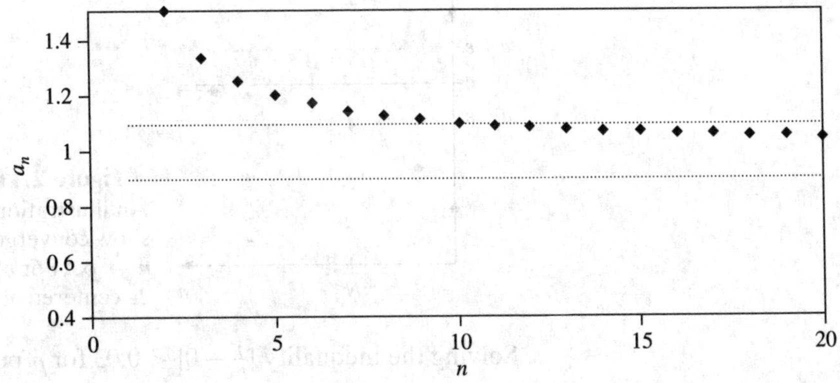

▲ Figure 2.10
Convergence of the sequence $a_n = \frac{n+1}{n}$ to $a = 1$

Formal Definition of Limits (Optional) Example 8 will motivate the formal definition of limits. When we guessed the limit in Example 8, we realized that successive terms approached 1. This means that no matter how small an interval about 1 we choose, all points must lie in this interval for all sufficiently large values of n. Graphically, this means that the points of the graph of a_n must lie between the two dashed lines in Figure 2.10 for all large enough values of n, no matter how close the two dashed lines are to the horizontal line at height 1. Translating this into a formal statement for the general case, we arrive at the following definition.

> **Formal Definition of Limits** The sequence $\{a_n\}$ has **limit** a, written as $\lim_{n \to \infty} a_n = a$, if for every $\epsilon > 0$ there exists an integer N so that
>
> $$|a_n - a| < \epsilon \quad \text{whenever } n > N$$
>
> If the limit exists, the sequence is called **convergent** and we say that a_n **converges** to a as n tends to infinity. If the sequence has no limit, it is called **divergent**.

The value of N will typically depend on ϵ: The smaller ϵ, the larger N. Since it is important to understand what it means when a sequence is convergent, we illustrate this in Figure 2.11. The horizontal dashed lines are at heights $a + \epsilon$ and $a - \epsilon$, respectively. They form a strip of width 2ϵ centered at the horizontal line at height a. Points a_n within this strip satisfy the inequality $|a_n - a| < \epsilon$. For a sequence to be convergent, we require that *all* points a_n lie in this strip for *all* n sufficiently large (namely, larger than some N).

▶ **Example 9** Show that

$$\lim_{n \to \infty} \frac{1}{n} = 0$$

Solution

Before we do this for an arbitrary ϵ, let's try to find N for a particular choice of ϵ, say $\epsilon = 0.03$. We need to find an integer N so that

$$\left| \frac{1}{n} - 0 \right| < 0.03 \quad \text{whenever } n > N$$

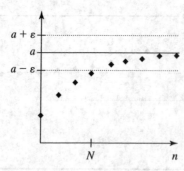

◀ **Figure 2.11**
An illustration of the formal definition of limits to show convergence of the sequence a_n to a as $n \to \infty$: For all $n > N$, a_n lies in the strip of width 2ϵ centered at a

Solving the inequality $|\frac{1}{n} - 0| < 0.03$ for n positive, we find

$$\left|\frac{1}{n}\right| < 0.03 \quad \text{or} \quad n > \frac{1}{0.03} \approx 33.33$$

The smallest value for N we can choose is $N = 33$, which is the largest integer less than or equal to $1/0.03$. Successive values for $n > 33$ give us confidence that we are on the right track but don't prove that our choice is correct:

$$a_{34} = \frac{1}{34} \approx 0.0294, \quad a_{35} = \frac{1}{35} \approx 0.0286, \text{ and so on}$$

To see that our choice for N works, we need to show that $n > 33$ implies $|1/n| < 0.03$. Now, $n > 33$ implies $1/n \leq 1/34 \approx 0.0294$. Since $n > 0$, we have that

$$\left|\frac{1}{n} - 0\right| < 0.03 \quad \text{whenever } n > 33$$

To show that $a_n = \frac{1}{n}$ converges to 0, we need to do the same calculation for an arbitrary ϵ. That is, we need to show that for every $\epsilon > 0$ we can find N so that

$$\left|\frac{1}{n} - 0\right| < \epsilon \quad \text{whenever } n > N$$

To find a candidate for N, we solve the inequality $|\frac{1}{n}| < \epsilon$. Since $\frac{1}{n} > 0$, we can drop the absolute value signs and find

$$\frac{1}{n} < \epsilon \quad \text{or} \quad n > \frac{1}{\epsilon}$$

Let's choose N to be the largest integer less than or equal to $1/\epsilon$. If $n > N$, then $n \geq N + 1$, which is equivalent to $1/n \leq 1/(N+1)$. Since N is the largest integer less than or equal to $1/\epsilon$, it follows that $1/n \leq 1/(N+1) < \epsilon \leq 1/N$ for $n > N$. This, together with $n > 0$, shows that if N is the largest integer less than or equal to $1/\epsilon$, then

$$\left|\frac{1}{n} - 0\right| < \epsilon \quad \text{whenever } n > N \qquad ◀$$

Limit Laws The formal definition of limits is cumbersome when we want to compute limits in specific examples. Fortunately, there are limit laws that facilitate the computation of limits.

LIMIT LAWS

If $\lim_{n \to \infty} a_n$ and $\lim_{n \to \infty} b_n$ exist and c is a constant, then

$$\lim_{n \to \infty} (a_n + b_n) = \lim_{n \to \infty} a_n + \lim_{n \to \infty} b_n$$

$$\lim_{n \to \infty} (c a_n) = c \lim_{n \to \infty} a_n$$

$$\lim_{n \to \infty} (a_n b_n) = (\lim_{n \to \infty} a_n)(\lim_{n \to \infty} b_n)$$

$$\lim_{n \to \infty} \frac{a_n}{b_n} = \frac{\lim_{n \to \infty} a_n}{\lim_{n \to \infty} b_n}, \text{ provided } \lim_{n \to \infty} b_n \ne 0$$

Although for using the limit laws we do not need to know the formal definition of limits, we will need to know in the following two examples that

$$\lim_{n \to \infty} \frac{1}{n} = 0 \tag{2.6}$$

which was proved in Example 9 (using the formal definition of limits).

▶ **Example 10**   Find

$$\lim_{n \to \infty} \frac{n + 1}{n}$$

Solution

We break $\frac{n+1}{n}$ into a sum of two terms, namely $1 + \frac{1}{n}$. Since $\lim_{n \to \infty} 1$ and $\lim_{n \to \infty} \frac{1}{n}$ exist (the former is equal to 1 and the latter equal to 0 according to (2.6)), we find

$$\lim_{n \to \infty} \frac{n + 1}{n} = \lim_{n \to \infty} \left(1 + \frac{1}{n}\right) = \lim_{n \to \infty} (1) + \lim_{n \to \infty} \frac{1}{n} = 1 + 0 = 1 \qquad ◀$$

▶ **Example 11**   Find

$$\lim_{n \to \infty} \frac{4n^2 - 1}{n^2}$$

Solution

We rewrite a_n:

$$a_n = \frac{4n^2 - 1}{n^2} = 4 - \frac{1}{n^2} = 4 - \frac{1}{n} \cdot \frac{1}{n}$$

Since $\lim_{n \to \infty} 4$ and $\lim_{n \to \infty} \frac{1}{n}$ exist, we find

$$\lim_{n \to \infty} \frac{4n^2 - 1}{n^2} = \lim_{n \to \infty} \left(4 - \frac{1}{n} \cdot \frac{1}{n}\right) = \lim_{n \to \infty} 4 - \left(\lim_{n \to \infty} \frac{1}{n}\right)\left(\lim_{n \to \infty} \frac{1}{n}\right)$$

$$= 4 - 0 \cdot 0 = 4 \qquad ◀$$

▶ **Example 12**   Without proof, we will state the long-term behavior of exponential growth. For $R > 0$, exponential growth is given by

$$a_n = a_0 R^n, \quad n = 0, 1, 2, \ldots$$

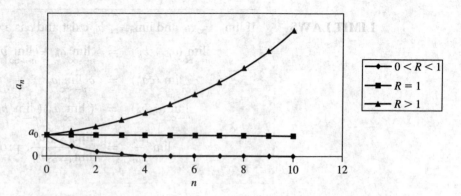

▲ **Figure 2.12**
Exponential growth in Example 12 for three different values of R

Figure 2.12 indicates that

$$\lim_{n \to \infty} a_n = \begin{cases} 0 & \text{if } 0 < R < 1 \\ a_0 & \text{if } R = 1 \\ \infty & \text{if } R > 1 \end{cases}$$

This can also be shown rigorously using the formal definition of limits. ◄

2.2.3 Recursions

We learned in the previous subsection how to find $\lim_{n \to \infty} a_n$ when a_n is given explicitly as a function of n. But how can we find such limits when a_n is defined recursively?

When we define a first-order sequence $\{a_n\}$ recursively, we express a_{n+1} in terms of a_n and specify a value for a_0. We can then compute successive values of a_n, which might allow us to guess the limit if it exists. In some cases, we can find a solution of the recursion, as in the following example.

▶ **Example 13**   Compute a_n for $n = 1, 2, \ldots, 5$ when

$$a_{n+1} = \frac{1}{4}a_n + \frac{3}{4} \quad \text{with } a_0 = 2 \tag{2.7}$$

Find a solution of the recursion and then take a guess for the limiting behavior of the sequence.

Solution

By repeatedly applying the recursion, we find

$$a_1 = \frac{1}{4}a_0 + \frac{3}{4} = \frac{1}{4} \cdot 2 + \frac{3}{4} = \frac{5}{4} = 1.25$$

$$a_2 = \frac{1}{4}a_1 + \frac{3}{4} = \frac{1}{4} \cdot \frac{5}{4} + \frac{3}{4} = \frac{17}{16} = 1.0625$$

$$a_3 = \frac{1}{4}a_2 + \frac{3}{4} = \frac{1}{4} \cdot \frac{17}{16} + \frac{3}{4} = \frac{65}{64} \approx 1.0156$$

$$a_4 = \frac{1}{4}a_3 + \frac{3}{4} = \frac{1}{4} \cdot \frac{65}{64} + \frac{3}{4} = \frac{257}{256} \approx 1.0039$$

$$a_5 = \frac{1}{4}a_4 + \frac{3}{4} = \frac{1}{4} \cdot \frac{257}{256} + \frac{3}{4} = \frac{1025}{1024} \approx 1.0010$$

There seems to be a pattern—namely, the denominators are powers of 4 and the numerators are just 1 larger than the denominators. We therefore set

$$a_n = \frac{4^n + 1}{4^n} \tag{2.8}$$

and check whether this is indeed a solution to the recursion. First, we need to check the initial condition: $a_0 = \frac{4^0 + 1}{4^0} = \frac{2}{1} = 2$, which agrees with the given initial condition. Next we need to check whether a_n satisfies the recursion. We write

$$a_{n+1} = \frac{4^{n+1} + 1}{4^{n+1}} = 1 + \frac{1}{4 \cdot 4^n} = 1 + \frac{1}{4} \frac{1}{4^n}$$

Now, $a_n = \frac{4^n + 1}{4^n}$ implies that $a_n = 1 + \frac{1}{4^n}$ or $\frac{1}{4^n} = a_n - 1$. Using this and simplifying then yields

$$a_{n+1} = 1 + \frac{1}{4} \frac{1}{4^n} = 1 + \frac{1}{4}(a_n - 1) = \frac{1}{4} a_n + \frac{3}{4}$$

which is the given recursion and thus proves that (2.8) is a solution of (2.7). We can now find the limit using (2.8):

$$\lim_{n \to \infty} a_n = \lim_{n \to \infty} \frac{4^n + 1}{4^n} = \lim_{n \to \infty} \left(1 + \frac{1}{4^n}\right) = 1$$

since $\lim_{n \to \infty} \frac{1}{4^n} = \lim_{n \to \infty} \left(\frac{1}{4}\right)^n = 0$ according to Example 12. ◀

Finding an explicit expression for a_n as in Example 13 is not a feasible strategy in general since solving recursions can be very difficult or even impossible. How then can we say anything about the limiting behavior of a recursively defined sequence?

The following procedure will allow us to identify *candidates* for limits. These are the fixed points we mentioned in Section 2.1. A **fixed point** is a point such that, if a_0 is equal to the fixed point, then all successive values of a_n are also equal to the fixed point. (This is the definition of a fixed point.) In mathematical terms, if we call the fixed point a, then if $a_0 = a$, we have $a_1 = a$, $a_2 = a$, and so on.

Now, if $a_{n+1} = f(a_n)$, then this means that if $a_0 = a$ and a is a fixed point, then $a_1 = f(a_0) = f(a) = a$, $a_2 = f(a_1) = f(a) = a$, and so on. That is, a fixed point satisfies the equation

$$a = f(a) \tag{2.9}$$

We will use (2.9) to find fixed points.

In Example 13, we had the recursion $a_{n+1} = \frac{1}{4} a_n + \frac{3}{4}$. Fixed points for the recursion thus satisfy

$$a = \frac{1}{4} a + \frac{3}{4}$$

Solving this for a, we find $a = 1$. It turns out that in Example 13, the fixed point is also the limiting point. This will not always be the case: A fixed point is only a candidate for a limit; a sequence does not have to converge to a given fixed point (unless a_0 is already equal to the fixed point). We illustrate this in the next example.

▶ Example 14   Let

$$a_{n+1} = \frac{3}{a_n}$$

Find the fixed points of this recursion and investigate the limiting behavior of a_n when a_0 is not equal to a fixed point.

Solution

To find the fixed points, we need to solve

$$a = \frac{3}{a}$$

This is equivalent to $a^2 = 3$, and hence $a = \sqrt{3}$ or $-\sqrt{3}$. These are the two fixed points. If $a_0 = \sqrt{3}$, then $a_1 = \sqrt{3}$, $a_2 = \sqrt{3}$, and so on. (Likewise, if $a_0 = -\sqrt{3}$, then $a_1 = -\sqrt{3}$, $a_2 = -\sqrt{3}$, and so on.)

Let's start with a value that is not equal to one of the fixed points, say $a_0 = 2$. Using the recursion, we find

$$a_1 = \frac{3}{a_0} = \frac{3}{2}$$

$$a_2 = \frac{3}{a_1} = \frac{3}{\frac{3}{2}} = 3 \cdot \frac{2}{3} = 2$$

$$a_3 = \frac{3}{a_2} = \frac{3}{2}$$

$$a_4 = \frac{3}{a_3} = \frac{3}{\frac{3}{2}} = 3 \cdot \frac{2}{3} = 2$$

and so on. That is, successive terms alternate between 2 and 3/2. Let's try another initial value, say $a_0 = -3$. Then

$$a_1 = \frac{3}{a_0} = \frac{3}{-3} = -1$$

$$a_2 = \frac{3}{a_1} = \frac{3}{-1} = -3$$

$$a_3 = \frac{3}{a_2} = \frac{3}{-3} = -1$$

$$a_4 = \frac{3}{a_3} = \frac{3}{-1} = -3$$

and so on. Successive terms alternate between -3 and -1. Alternating between two values, one of which is the initial value, happens with any initial value that is not one of the fixed points. Namely,

$$a_1 = \frac{3}{a_0} \quad \text{and} \quad a_2 = \frac{3}{a_1} = \frac{3}{\frac{3}{a_0}} = a_0$$

and hence a_3 is the same as a_1, a_4 is the same as a_2 and hence a_0, and so on. ◀

This last example illustrates that fixed points are only candidates for limits and that, depending on the initial condition, the sequence $\{a_n\}$ may or may not converge. If we know, however, that a sequence $\{a_n\}$ converges, the limit of the sequence must be one of the fixed points.

▶ Example 15   Assume that $\lim_{n \to \infty} a_n$ exists for

$$a_{n+1} = \sqrt{3a_n} \quad \text{with } a_0 = 2$$

Find $\lim_{n \to \infty} a_n$.

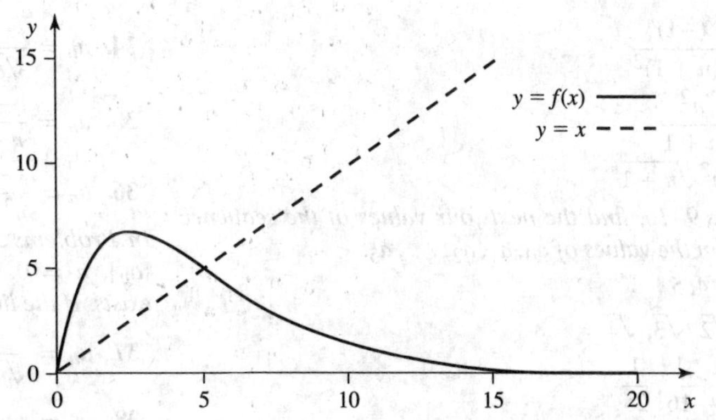

▲ Figure 2.13
A graphical way to find fixed points. (See text for explanation)

Solution

Since the problem tells us that the limit exists, we don't have to worry about existence. The problem that remains is to identify the limit. To do this, we compute the fixed points. We solve

$$a = \sqrt{3a}$$

which has two solutions, namely $a = 0$ and $a = 3$. When $a_0 = 2$, then $a_n > 2$ for all $n = 1, 2, 3, \ldots$, so we can exclude $a = 0$ as the limiting value. This leaves only one possibility, and we conclude

$$\lim_{n \to \infty} a_n = 3$$

Using a calculator, we can find successive values of a_n, which we collect in the following table (accurate to two decimals). The tabulated values suggest that the limit is indeed 3.

n	0	1	2	3	4	5	6	7
a_n	2	2.45	2.71	2.85	2.92	2.96	2.98	2.99

There is a graphical method to find fixed points, which we will mention briefly here and explore in more detail in Section 5.6. If the recursion is of the form $a_{n+1} = f(a_n)$, then a fixed point satisfies $a = f(a)$. This suggests that if we graph $y = f(x)$ and $y = x$ in the same coordinate system, then fixed points are located where the two graphs intersect, as shown in Figure 2.13.

We will return to the relationship between fixed points and limits in Section 5.6, where we will learn methods that allow us to determine whether a sequence converges to a particular fixed point.

2.2.4 Problems

(2.2.1)

In Problems 1–8, determine the values of the sequence $\{a_n\}$ for $n = 0, 1, 2, \ldots, 5$.

1. $a_n = n$

2. $a_n = 3n^2$

3. $a_n = \dfrac{1}{n + 2}$

4. $a_n = \dfrac{1}{\sqrt{n + 1}}$

5. $a_n = (-1)^n n$

6. $a_n = \dfrac{(-1)^n}{(n+1)^2}$

7. $a_n = \dfrac{n^2}{n+1}$

8. $a_n = n^3 \sqrt{n+1}$

In Problems 9–16, find the next four values of the sequence $\{a_n\}$ *based on the values of* $a_0, a_1, a_2, \ldots, a_5$.

9. $1, 2, 3, 4, 5$

10. $0, 1, \sqrt{2}, \sqrt{3}, \sqrt{4}$

11. $1, \dfrac{1}{4}, \dfrac{1}{9}, \dfrac{1}{16}, \dfrac{1}{25}$

12. $-1, \dfrac{1}{4}, -\dfrac{1}{9}, \dfrac{1}{16}, -\dfrac{1}{25}$

13. $\dfrac{1}{2}, \dfrac{2}{3}, \dfrac{3}{4}, \dfrac{4}{5}, \dfrac{5}{6}$

14. $\dfrac{1}{5}, \dfrac{4}{10}, \dfrac{9}{17}, \dfrac{16}{26}, \dfrac{25}{37}$

15. $\sqrt{1+e}, \sqrt{2+e^2}, \sqrt{3+e^3}, \sqrt{4+e^4}, \sqrt{5+e^5}$

16. $\sin\dfrac{\pi}{2}, -\sin\dfrac{\pi}{4}, \sin\dfrac{\pi}{6}, -\sin\dfrac{\pi}{8}, \sin\dfrac{\pi}{10}$

In Problems 17–28, find an expression for a_n *based on the values of* $a_0, a_1, a_2, \ldots$.

17. $0, 1, 2, 3, 4, \ldots$

18. $0, 2, 4, 6, 8, \ldots$

19. $1, 2, 4, 8, 16, \ldots$

20. $1, 3, 5, 7, 9, \ldots$

21. $1, \dfrac{1}{3}, \dfrac{1}{9}, \dfrac{1}{27}, \dfrac{1}{81}, \ldots$

22. $\dfrac{1}{3}, \dfrac{2}{5}, \dfrac{3}{7}, \dfrac{4}{9}, \dfrac{5}{11}, \ldots$

23. $-1, 2, -3, 4, -5, \ldots$

24. $2, -4, 6, -8, 10, \ldots$

25. $-\dfrac{1}{2}, \dfrac{1}{3}, -\dfrac{1}{4}, \dfrac{1}{5}, -\dfrac{1}{6}, \ldots$

26. $\dfrac{1}{2}, -\dfrac{1}{8}, \dfrac{1}{18}, -\dfrac{1}{32}, \dfrac{1}{50}, \ldots$

27. $\sin(\pi), \sin(2\pi), \sin(3\pi), \sin(4\pi), \sin(5\pi), \ldots$

28. $-\cos\dfrac{\pi}{2}, \cos\dfrac{\pi}{4}, -\cos\dfrac{\pi}{6}, \cos\dfrac{\pi}{8}, -\cos\dfrac{\pi}{10}, \ldots$

(2.2.2)

In Problems 29–36, write the first five terms of the sequence $\{a_n\}$, $n = 0, 1, 2, 3, \ldots$, *and find* $\lim_{n\to\infty} a_n$.

29. $a_n = \dfrac{1}{n+2}$

30. $a_n = \dfrac{2}{n+1}$

31. $a_n = \dfrac{n}{n+1}$

32. $a_n = \dfrac{2n}{n+2}$

33. $a_n = \dfrac{1}{n^2+1}$

34. $a_n = \dfrac{1}{\sqrt{n+1}}$

35. $a_n = \dfrac{(-1)^n}{n+1}$

36. $a_n = \dfrac{(-1)^n}{n^3+3}$

In Problems 37–48, write the first five terms of the sequence $\{a_n\}$, $n = 0, 1, 2, 3, \ldots$, *and determine whether* $\lim_{n\to\infty} a_n$ *exists. If the limit exists, find it.*

37. $a_n = \dfrac{1}{\sqrt{n+1}}$

38. $a_n = \dfrac{1}{\sqrt{n+3}}$

39. $a_n = \dfrac{1}{n+1}$

40. $a_n = \dfrac{n}{n+1}$

41. $a_n = \dfrac{n^2}{n+1}$

42. $a_n = \dfrac{n^3}{n+1}$

43. $a_n = \sqrt{n}$

44. $a_n = n^2$

45. $a_n = 2^n$

46. $a_n = \left(\dfrac{1}{2}\right)^n$

47. $a_n = 3^n$

48. $a_n = \left(\dfrac{1}{3}\right)^n$

Formal Definition of Limits: In Problems 49–64, $\lim_{n\to\infty} a_n = a$. *Find the limit* a *and determine* N *so that* $|a_n - a| < \epsilon$ *for all* $n > N$ *for the given value of* ϵ.

49. $a_n = \dfrac{1}{n}, \epsilon = 0.01$

50. $a_n = \dfrac{1}{n}, \epsilon = 0.02$

51. $a_n = \dfrac{1}{n^2}, \epsilon = 0.01$

52. $a_n = \dfrac{1}{n^2}, \epsilon = 0.001$

53. $a_n = \dfrac{1}{\sqrt{n}}, \epsilon = 0.1$

54. $a_n = \dfrac{1}{\sqrt{n}}, \epsilon = 0.05$

55. $a_n = \dfrac{(-1)^n}{n}, \epsilon = 0.01$

56. $a_n = \dfrac{(-1)^n}{n}, \epsilon = .001$

57. $a_n = \dfrac{(-1)^n}{n^2}, \epsilon = 0.001$

58. $a_n = \dfrac{(-1)^n}{n^2}, \epsilon = .0001$

59. $a_n = \dfrac{n}{n+1}, \epsilon = 0.01$

60. $a_n = \dfrac{n}{n+1}, \epsilon = .05$

61. $a_n = \dfrac{n+1}{n}, \epsilon = 0.01$

62. $a_n = \dfrac{n+1}{n}, \epsilon = .05$

63. $a_n = \dfrac{n^2}{n^2+1}, \epsilon = 0.01$

64. $a_n = \dfrac{n^2}{n^2+1}, \epsilon = .001$

Formal Definition of Limits: In Problems 65–70, use the formal definition of limits to show that $\lim_{n\to\infty} a_n = a$; *that is, find N so that for every* $\epsilon > 0$, *there exists N so that* $|a_n - a| < \epsilon$ *whenever* $n > N$.

65. $\lim\limits_{n\to\infty} \dfrac{1}{n} = 0$

66. $\lim\limits_{n\to\infty} \dfrac{1}{n+1} = 0$

67. $\lim\limits_{n\to\infty} \dfrac{1}{n^2} = 0$

68. $\lim\limits_{n\to\infty} \dfrac{1}{n^2+1} = 0$

69. $\lim\limits_{n\to\infty} \dfrac{n+1}{n} = 1$

70. $\lim\limits_{n\to\infty} \dfrac{n}{n+1} = 1$

In Problems 71–82, use the limit laws to determine $\lim_{n\to\infty} a_n = a$.

71. $\lim\limits_{n\to\infty} \left(\dfrac{1}{n} + \dfrac{1}{n^2} \right)$

72. $\lim\limits_{n\to\infty} \left(\dfrac{2}{n} - \dfrac{1}{n^2+1} \right)$

73. $\lim\limits_{n\to\infty} \left(\dfrac{n+1}{n} \right)$

74. $\lim\limits_{n\to\infty} \left(\dfrac{2n-3}{n} \right)$

75. $\lim\limits_{n\to\infty} \left(\dfrac{n^2+1}{n^2} \right)$

76. $\lim\limits_{n\to\infty} \left(\dfrac{3n^2-5}{n^2} \right)$

77. $\lim\limits_{n\to\infty} \left(\dfrac{n+1}{n^2-1} \right)$

78. $\lim\limits_{n\to\infty} \left(\dfrac{n+2}{n^2-4} \right)$

79. $\lim\limits_{n\to\infty} \left[\left(\dfrac{1}{3} \right)^n + \left(\dfrac{1}{2} \right)^n \right]$

80. $\lim\limits_{n\to\infty} \left(3^{-n} - 4^{-n} \right)$

81. $\lim\limits_{n\to\infty} \dfrac{n+2^{-n}}{n}$

82. $\lim\limits_{n\to\infty} \dfrac{n+3^{-n}}{n}$

(2.2.3)

In Problems 83–92, the sequence $\{a_n\}$ *is recursively defined. Compute* a_n *for* $n = 1, 2, \dots, 5$.

83. $a_{n+1} = 2a_n, a_0 = 1$

84. $a_{n+1} = 2a_n, a_0 = 3$

85. $a_{n+1} = 3a_n - 2, a_0 = 1$

86. $a_{n+1} = 3a_n - 2, a_0 = 2$

87. $a_{n+1} = 4 - 2a_n, a_0 = 5$

88. $a_{n+1} = 4 - 2a_n, a_0 = \dfrac{4}{3}$

89. $a_{n+1} = \dfrac{a_n}{1+a_n}, a_0 = 1$

90. $a_{n+1} = \dfrac{a_n}{a_n+3}, a_0 = 2$

91. $a_{n+1} = a_n + \dfrac{1}{a_n}, a_0 = 1$

92. $a_{n+1} = 5a_n - \dfrac{5}{a_n}, a_0 = 2$

In Problems 93–102, the sequence $\{a_n\}$ *is recursively defined. Find all fixed points of* $\{a_n\}$.

93. $a_{n+1} = \dfrac{1}{2}a_n + 2$

94. $a_{n+1} = \dfrac{1}{3}a_n + \dfrac{4}{3}$

95. $a_{n+1} = \dfrac{2}{5}a_n - \dfrac{9}{5}$

96. $a_{n+1} = -\dfrac{1}{3}a_n + \dfrac{1}{4}$

97. $a_{n+1} = \dfrac{4}{a_n}$

98. $a_{n+1} = \dfrac{7}{a_n}$

99. $a_{n+1} = \dfrac{2}{a_n+2}$

100. $a_{n+1} = \dfrac{3}{a_n-2}$

101. $a_{n+1} = \sqrt{5a_n}$

102. $a_{n+1} = \sqrt{7a_n}$

In Problems 103–110, assume that $\lim_{n\to\infty} a_n$ *exists. Find all fixed points of* $\{a_n\}$ *and use a table or other reasoning to guess which fixed point is the limiting value for the given initial condition.*

103. $a_{n+1} = \dfrac{1}{2}(a_n + 5), a_0 = 1$

104. $a_{n+1} = \dfrac{1}{3}\left(a_n + \dfrac{1}{9} \right), a_0 = 1$

105. $a_{n+1} = \sqrt{2a_n}, a_0 = 1$

106. $a_{n+1} = \sqrt{2a_n}, a_0 = 0$

107. $a_{n+1} = 2a_n(1 - a_n), a_0 = 0.1$

108. $a_{n+1} = 2a_n(1 - a_n), a_0 = 0$

109. $a_{n+1} = \dfrac{1}{2}\left(a_n + \dfrac{4}{a_n} \right), a_0 = 1$

110. $a_{n+1} = \dfrac{1}{2}\left(a_n + \dfrac{9}{a_n} \right), a_0 = -1$

2.3 MORE POPULATION MODELS

An important biological application of sequences are models of seasonally breeding populations with non-overlapping generations where the population size at one generation only depends on the population size of the previous generation. The exponential growth model of Section 2.1 fits into this category. We denote the population size at time t by $N(t)$ or N_t, $t = 0, 1, 2, \ldots$. To model how the population size at generation $t + 1$ is related to the population size at generation t, we write

$$N_{t+1} = f(N_t) \tag{2.10}$$

where the function f describes the density dependence of the population dynamics.

As explained in Section 2.2, a recursion of the form (2.10) is called a first-order recursion since to obtain the population size at time $t + 1$, only the population size at the previous time step t needs to be known. A recursion is also called a **difference equation** or an **iterated map**. [The name *difference equation* comes from writing the dynamics in the form $N_{t+1} - N_t = g(N_t)$. The name *iterated map* refers to the recursive definition.]

When we study population models, we are frequently interested in the long-term behavior of the population, such as, will the population size reach a constant value, will it oscillate predictably, or will it fluctuate widely without any recognizable patterns? We will explore this in the following examples, where we will see that discrete time population models show very rich and complex behavior.

2.3.1 Restricted Population Growth: The Beverton-Holt Recruitment Curve

In Section 2.1, we discussed exponential growth defined by the recursion

$$N_{t+1} = RN_t \quad \text{with } N_0 = \text{population size at time } 0$$

When $R > 1$, the population size will grow indefinitely, provided $N_0 > 0$. We can understand why this happens if we look at the parent to offspring ratio for $N_t > 0$, N_t/N_{t+1}, which is $1/R$, a constant, in this case. This means that regardless of the current population density, the number of offspring per parent is a constant. This is called density-independent growth. Such growth is biologically unrealistic. As the size of the population increases, individuals will start to compete with each other for resources, such as food or nesting sites. This will reduce population growth. We call population growth that depends on population density density-dependent growth.

To find a model that incorporates a reduction in growth when the population size gets large, we start with the ratio of successive population sizes in the exponential growth model and assume that N_0 is positive so that all successive population sizes are positive:

$$\frac{N_t}{N_{t+1}} = \frac{1}{R} \tag{2.11}$$

When we can consider the ratio as a function of the current population size N_t, then the ratio N_t/N_{t+1} is a constant function of the current population size N_t. This corresponds to a horizontal line in a graph where N_t is on the horizontal axis and the ratio N_t/N_{t+1} is on the vertical axis. Note that as long as the parent to offspring ratio N_t/N_{t+1} is less than 1, the population size increases since there are fewer parents than offspring. Once the ratio is equal

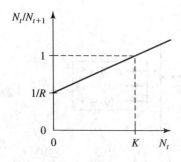

▲ **Figure 2.14**
Density dependent growth:
The parent to offspring ratio
increases as a function of
current population size. (Note
that this is only defined for
$N_t > 0$)

to 1, the population size stays the same from one time step to the next. When this ratio is greater than 1, the population size decreases.

To model the reduction in growth when the population size gets larger, we drop the assumption that the ratio N_t/N_{t+1} is constant and instead assume that the parent to offspring ratio N_t/N_{t+1} is an increasing function of the population size N_t. That is, we replace the constant $1/R$ in (2.11) by a function that increases with N_t. The simplest such function is linear. Graphically, this is a straight line with positive slope. To compare the model with density dependence to the exponential growth model (2.11), we assume that the two models agree when the population sizes are very small. We can achieve this by assuming that the line corresponding to density-dependent growth goes through the point $(0, 1/R)$. To make sure that the population grows at low densities, we assume that $R > 1$. The population density where the parent to offspring ratio is equal to 1 is of particular importance since it corresponds to the population size that does not change from one generation to the next. We call this population size the **carrying capacity** and denote it by K, where K is a positive constant. We thus require that the line corresponding to density-dependent growth connects the points $(0, 1/R)$ and $(K, 1)$ in a graph where N_t is on the horizontal axis and the ratio N_t/N_{t+1} is on the vertical axis (Figure 2.14).

The straight line in Figure 2.14 has slope $(1 - 1/R)/K$ and vertical axis intercept $1/R$, which yields the equation

$$\frac{N_t}{N_{t+1}} = \frac{1}{R} + \frac{1 - \frac{1}{R}}{K} N_t$$

We solve this for N_{t+1} to obtain a recursion. Multiply both sides by N_{t+1},

$$N_t = N_{t+1}\left(\frac{1}{R} + \frac{1 - \frac{1}{R}}{K} N_t\right)$$

then isolate N_{t+1},

$$N_{t+1} = \frac{N_t}{\frac{1}{R} + \frac{1 - \frac{1}{R}}{K} N_t}$$

and simplify the right-hand expression by multiplying numerator and denominator by R:

$$N_{t+1} = \frac{R N_t}{1 + \frac{R-1}{K} N_t} \qquad (2.12)$$

This recursion is known as the **Beverton-Holt recruitment curve**.

Using results from Section 2.2, we can compute the fixed points of (2.12). Solving

$$N = \frac{RN}{1 + \frac{R-1}{K} N}$$

for N, we immediately find $N = 0$. If $N \neq 0$, we divide both sides by N,

$$1 = \frac{R}{1 + \frac{R-1}{K} N}$$

and solve for N,

$$1 + \frac{R-1}{K} N = R \quad \text{or} \quad \frac{R-1}{K} N = R - 1$$

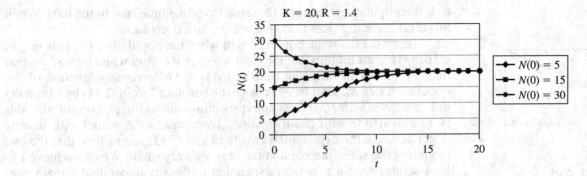

▲ Figure 2.15
The population sizes N_t when $K = 20$ and $R = 1.4$ in the Beverton-Holt recruitment model for different initial population densities

from which we obtain

$$N = \frac{R-1}{\frac{R-1}{K}} = (R-1)\frac{K}{R-1} = K$$

We thus have two fixed points when $R > 1$: the fixed point $N = 0$, which we call **trivial** since it corresponds to the absence of the population, and the fixed point $N = K$, which we call **nontrivial** since it corresponds to a positive population size.

In Figure 2.15, we set $K = 20$ and plot N_t as a function of t for three different initial population sizes. For clarity, we include the lines that connect successive population sizes. We see from Figure 2.15 that if $N_0 > 0$, then N_t will approach K eventually (if $N_0 = K$, then $N_t = K$ for all $t = 1, 2, 3, \ldots$ since K is a fixed point). This is the reason for calling K the carrying capacity. Based on Figure 2.15, we conclude that when $K > 0$, $R > 1$, and $N_0 > 0$,

$$\lim_{t \to \infty} N_t = K$$

At this point, we need to rely on graphs and tables to say anything about the long-term behavior of the population. This is a serious limitation since it restricts our investigations to specific parameter values and consequently our conclusions only hold for the parameter values we checked. Obviously, we cannot explore all possible parameter values this way. It turns out that this example has the same qualitative behavior for all R and K provided $R > 1$ and $K > 0$. We will learn analytical methods in Section 5.6 that will allow us to make general statements (like the one in the previous sentence) about the behavior of discrete time population models that will not depend on tables and graphs. In the next subsection, we will see an example where the behavior depends strongly on the choice of parameters.

2.3.2 The Discrete Logistic Equation

The most popular, discrete time, single-species model is the discrete logistic equation. Its recursion is given by

$$N_{t+1} = N_t \left[1 + R\left(1 - \frac{N_t}{K}\right) \right] \tag{2.13}$$

where R and K are positive constants. R is called the **growth parameter** and K is called the **carrying capacity**. The following analysis will explain this terminology. This model of population growth exhibits very complicated dynamics, which are described in an influential review paper by Robert May (1976).

Before we analyze the model, we will rewrite it in what is called the *canonical form*. The advantage of this is that the resulting recursion will be of a simpler algebraic form. The following algebraic steps are not obvious but will lead to the canonical form of the discrete logistic equation. We write

$$N_{t+1} = N_t\left[1 + R\left(1 - \frac{N_t}{K}\right)\right]$$

$$= N_t\left[1 + R - \frac{R}{K}N_t\right]$$

$$= N_t(1+R)\left[1 - \frac{R}{K(1+R)}N_t\right]$$

Now, dividing by $1 + R$ yields

$$\frac{1}{1+R}N_{t+1} = N_t\left[1 - \frac{R}{K(1+R)}N_t\right]$$

Let's multiply both sides by R/K (you'll see in a moment why).

$$\frac{R}{K(1+R)}N_{t+1} = \frac{R}{K}N_t\left[1 - \frac{R}{K(1+R)}N_t\right] \tag{2.14}$$

If we define the new variable

$$x_t = \frac{R}{K(1+R)}N_t \tag{2.15}$$

then

$$\frac{R}{K(1+R)}N_{t+1} = x_{t+1} \quad \text{and} \quad \frac{R}{K}N_t = (1+R)x_t$$

and (2.14) becomes

$$x_{t+1} = (1+R)x_t(1-x_t)$$

At this point, customarily, a new parameter is introduced, namely $r = 1 + R$. Note that $r > 1$ since $R > 0$. We arrive at the canonical form of the logistic recursion

$$x_{t+1} = rx_t(1-x_t) \tag{2.16}$$

The advantage of this is threefold: (1) The recursion (2.16) looks simpler than the original recursion (2.13); (2) instead of two parameters (R and K) there is just one (r); and (3) the quantity x_t is *dimensionless*. The last point needs some explanation. The original variable N_t has units (or dimension) "number of individuals"; the parameter K has the same units. By dividing N_t by K in (2.15), the units cancel and we say that the quantity x_t is dimensionless. [The parameter R does not have a dimension, so multiplying N_t/K by $R/(1+R)$ does not introduce any additional units.] A dimensionless variable has the advantage that it has the same numerical value regardless of what the units of measurement are in the original variable (see Problems 31–34). The process of making a quantity dimensionless is called **nondimensionalization**.

Let's go back to the discrete logistic equation in its canonical form (2.16) and see what its behavior is. The function $f(x) = rx(1-x)$ is an upside-down parabola since $r > 1$ (Figure 2.16). We see from Figure 2.16 that if x is outside of the interval $(0, 1)$, $f(x)$ is nonpositive. Since $x_t = \frac{R}{K(1+R)}N_t$ [see (2.15)] and we want N_t to be positive (it is a population size), we require x_t to be positive.

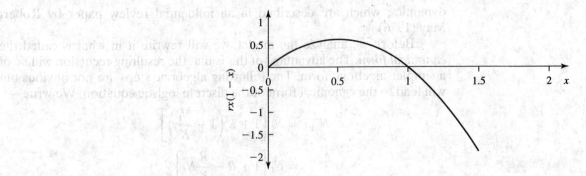

▲ **Figure 2.16**
A graph of the discrete logistic equation in its canonical form (here $r = 2.5$)

This means that we need to ensure that $x_{t+1} = f(x_t)$ stays within the interval $(0, 1)$. The maximum value of $f(x)$ occurs at $x = 1/2$ and $f(1/2) = r/4$, and so we require $r/4 < 1$ or $r < 4$. To summarize, if $1 < r < 4$, then x_t stays within the interval $(0, 1)$ for all $t = 1, 2, 3, \ldots$ provided $x_0 \in (0, 1)$. We will therefore assume in the following that $1 < r < 4$ and $x_0 \in (0, 1)$.

We first compute fixed points of (2.16). We need to solve

$$x = rx(1 - x)$$

This immediately yields the trivial solution $x = 0$. If $x \neq 0$, we can divide both sides by x and find

$$1 = r(1 - x) \quad \text{or} \quad x = 1 - \frac{1}{r}$$

(see Figure 2.17). Provided $r > 1$, both fixed points are in $[0, 1)$.

We return to the original variable N_t for a moment to see what $x = 0$ and $x = 1 - 1/r$ mean in terms of N. Since $x = \frac{R}{K(1+R)} N$, the fixed point $x = 0$ corresponds to the fixed point $N = 0$, which justifies calling $x = 0$ a trivial equilibrium. When $x = 1 - 1/r$, then using $r = 1 + R$,

$$N = \frac{K(1+R)}{R} x = \frac{K(1+R)}{R} \left(1 - \frac{1}{1+R}\right)$$

$$= \frac{K(1+R)}{R} \frac{1+R-1}{1+R} = K$$

so $N = K$ is the other fixed point.

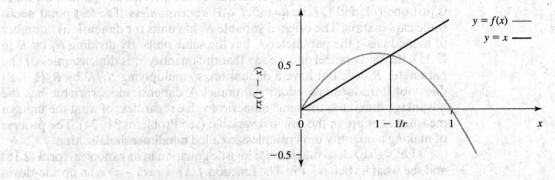

▲ **Figure 2.17**
A graphical illustration of the fixed points of the discrete logistic equation in its canonical form. The fixed points are where the parabola and the line $y = x$ intersect

As pointed out earlier, the long-term behavior of the discrete logistic equation is very complicated. We will go through the different cases by simply listing them. Later, in Section 5.6, we will be able, at least to some extent, to understand why this equation has such complicated behavior.

When $1 < r < 3$ and $x_0 \in (0, 1)$, then x_t converges to the fixed point $1 - 1/r$ (Figure 2.18). Increasing r to a value between 3 and $3.449\ldots$, x_t settles into a cycle of period 2 (Figure 2.19). This means that for large enough times, x_t will oscillate back and forth between a larger and a smaller value. For r between $3.449\ldots$ and $3.544\ldots$, the period doubles: A cycle of period 4 appears for large enough times. The population size now oscillates between the same four values (Figure 2.20). Increasing r continues to double the period: A cycle of period 8 is born when $r = 3.544\ldots$, a cycle of period 16 is born when $r = 3.564\ldots$, a cycle of period 32 is born when $r = 3.567\ldots$. This period doubling continues to occur until r reaches a value of about 3.57 when the population pattern becomes **chaotic** (Figure 2.21). The population dynamics seem to be random, though the rules are entirely deterministic! There is no regular pattern we can discern; x_t no longer oscillates between the same values; the dynamics are **aperiodic**. Furthermore, starting from ever so slightly different initial conditions quickly produces very different trajectories (Figure 2.22). This sensitivity to initial conditions is characteristic of chaotic behavior.

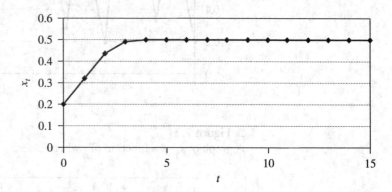

▲ **Figure 2.18**
A graph of x_t as a function of t when $r = 2$

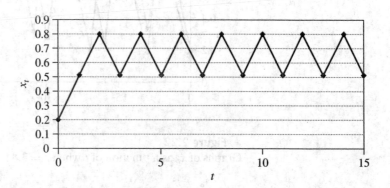

▲ **Figure 2.19**
A graph of x_t as a function of t when $r = 3.2$

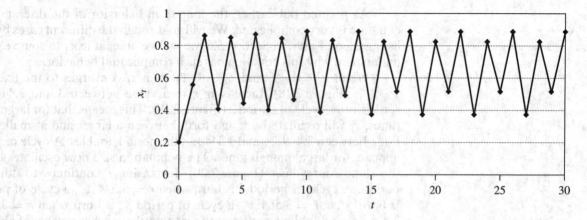

▲ **Figure 2.20**
A graph of x_t as a function of t when $r = 3.52$

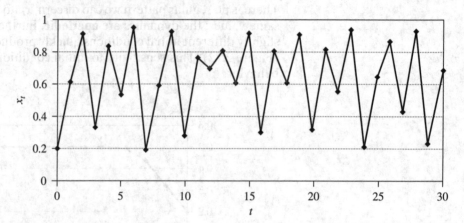

▲ **Figure 2.21**
A graph of x_t as a function of t when $r = 3.8$

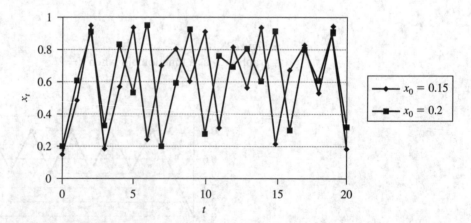

▲ **Figure 2.22**
Graphs of x_t as a function of t when $r = 3.8$ for two different initial values of x_0

To obtain biologically sensible results, we needed to restrict both r and x_0. The reason for this was that if $x_t > 1$, then x_{t+1} is negative. This can be easily remedied by slightly changing the dynamics. We discuss such a model in the next subsection.

2.3.3 Ricker's Curve

The discrete logistic map has the biologically unrealistic feature that unless one restricts the initial population size and the growth parameter, negative population sizes can occur. The reason for this is that the function $f(x) = rx(1 - x)$ takes on negative values for $x > 1$ and so if $x_t > 1$, then $x_{t+1} < 0$. It is not difficult to avoid this problem. One example of an iterated map that has the same (desirable) properties as the logistic map but does not admit negative population sizes (provided the population size at time 0 is positive) is **Ricker's curve**. The recursion, called the **Ricker logistic equation**, is given by

$$N_{t+1} = N_t \exp\left[R\left(1 - \frac{N_t}{K}\right)\right] \tag{2.17}$$

where R and K are positive parameters. As in the discrete logistic model, R is the growth parameter and K is the carrying capacity. The graph of Ricker's curve (Figure 2.23), $f(N_t) = N_t \exp\left[R\left(1 - \frac{N_t}{K}\right)\right]$, is positive for all $N_t > 0$, thus avoiding the problem of negative population sizes.

Fixed points of (2.17) satisfy

$$N = N \exp\left[R\left(1 - \frac{N}{K}\right)\right] \tag{2.18}$$

We find the trivial fixed point $N = 0$. If $N \neq 0$, we can divide both sides of (2.18) by N,

$$1 = \exp\left[R\left(1 - \frac{N}{K}\right)\right]$$

which is satisfied if $R(1 - N/K) = 0$ or $N = K$. The parameter K has the same meaning as in the discrete logistic equation, namely it is the carrying capacity.

The Ricker logistic equation shows the same complex dynamics as the discrete logistic map [convergence to the fixed point for small positive values of R (Figure 2.24), periodic behavior with period doubling as R increases,

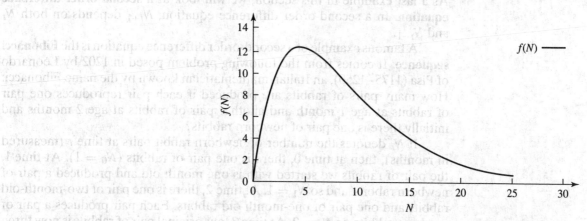

▲ **Figure 2.23**
Ricker's curve when $R = 2.8$ and $K = 9$

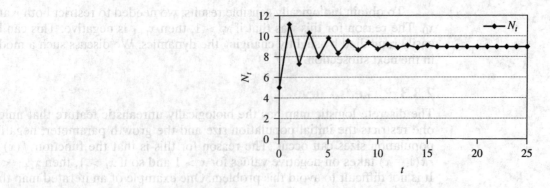

▲ **Figure 2.24**
The population size N_t as a function of t when $R = 1.8$ and $K = 9$

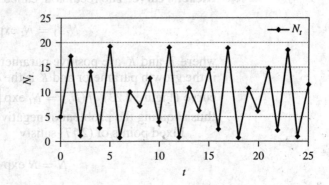

▲ **Figure 2.25**
The population size N_t as a function of t when $R = 2.8$ and $K = 9$

chaotic behavior for larger values of R (Figure 2.25)]. The values of R where the behavior changes are different than in the discrete logistic equation. For instance, onset of chaos in the discrete logistic equation occurs for $R = 2.570\ldots$, whereas onset of chaos in the Ricker logistic equation occurs for $R = 2.692\ldots$.

2.3.4 Fibonacci Sequences

As a last example in this section, we will look at a **second-order difference equation**. In a second-order difference equation, N_{t+1} depends on both N_t and N_{t-1}.

A famous example of a second-order difference equation is the **Fibonacci sequence**. It comes from the following problem posed in 1202 by Leonardo of Pisa (1175–1250), an Italian mathematician known by the name Fibonacci: How many pairs of rabbits are produced if each pair reproduces one pair of rabbits at age 1 month and another pair of rabbits at age 2 months and initially there is one pair of newborn rabbits?

If N_t denotes the number of newborn rabbit pairs at time t (measured in months), then at time 0, there is one pair of rabbits ($N_0 = 1$). At time 1, the pair of rabbits we started with is one month old and produced a pair of newborn rabbits, and so $N_1 = 1$. At time 2, there is one pair of two-month-old rabbits and one pair of one-month-old rabbits. Each pair produces a pair of newborn rabbits, so $N_2 = 2$. At time 3, our original pair of rabbits is now three months old and will stop reproducing; there is then one pair of two-month-old rabbits and two pairs of one-month-old rabbits. Since each pair of one-month and two-month-old rabbits produces a pair of newborn rabbits, at time $t = 3$, there will be $2 + 1 = 3$ newborn rabbits. More generally, to find the number of pairs of newborn rabbits, we need to add up the number of pairs of one-month-old rabbits and two-month-old rabbits. The one-month-old rabbits at time $t + 1$ were newborn rabbits at time t; the two-month-old rabbits were newborns at time $t - 1$. So the number of pairs of newborn rabbits at time $t + 1$ is

$$N_{t+1} = N_t + N_{t-1}, t = 1, 2, 3, \ldots \quad \text{with } N_0 = 1 \text{ and } N_1 = 1$$

Note that we need to specify N_t for $t = 0$ and $t = 1$ in order to be able to use the recursion. Using the recursion, we find the sequence

$$1, 1, 2, 3, 5, 8, 13, \ldots$$

We see that the number of newborn pairs of rabbits will go to infinity as t tends to infinity, so N_t will not converge to a finite limit. It turns out, however, that the ratio N_{t+1}/N_t converges. We can find a candidate for the limiting value as follows. Start with the recursion

$$N_{t+1} = N_t + N_{t-1}$$

and divide both sides by N_t, yielding

$$\frac{N_{t+1}}{N_t} = 1 + \frac{N_{t-1}}{N_t} \tag{2.19}$$

If we now assume that

$$\lim_{t \to \infty} \frac{N_{t+1}}{N_t} = \lambda$$

which also implies that

$$\lim_{t \to \infty} \frac{N_t}{N_{t-1}} = \lambda$$

(λ is the lowercase Greek letter lambda), then

$$\lim_{t \to \infty} \frac{N_{t-1}}{N_t} = \lim_{t \to \infty} \frac{1}{\frac{N_t}{N_{t-1}}} = \frac{1}{\lim_{t \to \infty} \frac{N_t}{N_{t-1}}} = \frac{1}{\lambda}$$

Taking the limit $t \to \infty$ in (2.19), we find

$$\lambda = 1 + \frac{1}{\lambda}$$

which is $\lambda^2 = \lambda + 1$ after multiplying both sides by λ. We thus need to solve

$$\lambda^2 - \lambda - 1 = 0$$

The formula for solving quadratic equations yields

$$\lambda_{1,2} = \frac{1 \pm \sqrt{1 + 4}}{2} = \frac{1 \pm \sqrt{5}}{2}$$

One solution is positive; the other is negative. Only the positive solution is relevant when $N_0 = N_1 = 1$ since then $N_{t+1}/N_t > 0$ for all $t = 0, 1, 2, \ldots$. The ratio

$$\frac{1 + \sqrt{5}}{2} \approx 1.61803$$

is the limit of N_{t+1}/N_t as $t \to \infty$ and is called the **golden mean**.

A rectangle whose sides bear this ratio is called a **golden rectangle**; it is thought to be the visually most pleasing proportion a rectangle can have. This was known to the ancient Greeks, who used it to scale the dimensions of their buildings (for example, the Parthenon). Ratios of successive Fibonacci numbers can be found in nature as well—for instance, the florets on plants, such as the sunflower, run in spirals and the ratio of the number of spirals running in opposite directions are often successive Fibonacci numbers.

2.3.5 Problems

(2.3.1)

In Problems 1–6, assume that the population growth is described by the Beverton-Holt recruitment curve with growth parameter R and carrying capacity K. For the given values of R and K, graph N_t/N_{t+1} as a function of N_t and find the recursion for the Beverton-Holt recruitment curve.

1. $R = 2, K = 15$
2. $R = 2, K = 50$
3. $R = 1.5, K = 40$
4. $R = 3, K = 120$
5. $R = 2.5, K = 90$
6. $R = 2, K = 150$

In Problems 7–12, assume that the population growth is described by the Beverton-Holt recruitment curve with growth parameter R and carrying capacity K. Find R and K.

7. $N_{t+1} = \dfrac{2N_t}{1 + N_t/20}$

8. $N_{t+1} = \dfrac{3N_t}{1 + 2N_t/40}$

9. $N_{t+1} = \dfrac{1.5N_t}{1 + 0.5N_t/30}$

10. $N_{t+1} = \dfrac{2N_t}{1 + N_t/200}$

11. $N_{t+1} = \dfrac{4N_t}{1 + N_t/150}$

12. $N_{t+1} = \dfrac{5N_t}{1 + N_t/20}$

In Problems 13–18, assume that the population growth is described by the Beverton-Holt recruitment curve with growth parameter R and carrying capacity K. Find all fixed points.

13. $N_{t+1} = \dfrac{4N_t}{1 + N_t/30}$

14. $N_{t+1} = \dfrac{3N_t}{1 + N_t/60}$

15. $N_{t+1} = \dfrac{2N_t}{1 + N_t/30}$

16. $N_{t+1} = \dfrac{2N_t}{1 + N_t/100}$

17. $N_{t+1} = \dfrac{3N_t}{1 + N_t/30}$

18. $N_{t+1} = \dfrac{5N_t}{1 + N_t/120}$

In Problems 19–24, assume that the population growth is described by the Beverton-Holt recruitment curve with growth parameter R and carrying capacity K. Find the population sizes for $t = 1, 2, \ldots, 5$ and $\lim_{t\to\infty} N_t$ for the given initial value N_0.

19. $R = 2, K = 10, N_0 = 2$
20. $R = 2, K = 20, N_0 = 5$
21. $R = 3, K = 15, N_0 = 1$
22. $R = 3, K = 30, N_0 = 0$
23. $R = 4, K = 40, N_0 = 3$
24. $R = 4, K = 20, N_0 = 10$

(2.3.2)

In Problems 25–30, assume the discrete logistic equation is used with parameters R and K. Write the equation in the canonical form $x_{t+1} = rx_t(1 - x_t)$ and determine r and x_t in terms of R, K, and N_t.

25. $R = 1, K = 10$
26. $R = 1, K = 20$
27. $R = 2, K = 15$
28. $R = 2, K = 20$
29. $R = 2.5, K = 30$
30. $R = 2.5, K = 50$

In Problems 31–34, we will investigate the advantage of dimensionless variables.

31. (a) Let N_t denote the population size at time t and let K denote the carrying capacity. Both quantities are measured in units of number of individuals. Show that $x_t = N_t/K$ is dimensionless.

(b) Let M_t denote the population size at time t and let L denote the carrying capacity. Assume that M_t and L are measured in units of 1000 individuals. Show that $y_t = M_t/L$ is dimensionless.

(c) How are N_t and M_t related? How are K and L related?

(d) Use (c) to find M_t and L if there are 20,000 individuals at time t and the carrying capacity is 5000.

(e) For the population size and the carrying capacity in (d), show that $x_t = y_t$.

32. To quantify the spatial structure of a plant population, it might be convenient to introduce a characteristic length scale. This length scale might be characterized by the average dispersal distance of the plant under study. Assume that the characteristic length scale is denoted by L. Denote by x the distance of seeds from their source. Define $z = x/L$. Find z if $x = 100$ cm and $L = 50$ cm, and show that z has the same value if x and L are measured in units of meters instead.

33. Suppose a bacterium divides every 20 minutes, which we call the characteristic time scale and denote by T. Let t be the time elapsed since the beginning of an experiment that involves this bacterium. Define $z = t/T$. Find z if $t = 120$ minutes and show that z has the same value if t and T are measured in units of hours instead.

34. The time to the most recent common ancestor of a pair of individuals from a randomly mating population depends on population size. Let t denote the time measured in units of generations to the most recent common ancestor and let T be equal to N generations, where N is the population size of the randomly mating population. Define $z = t/T$. Show that z is dimensionless and that the value of z does not change regardless of whether t and T are measured in units of generations or in units of, say, years. (Assume that one generation is equal to n years.)

In Problems 35–46, we will investigate the behavior of the discrete logistic equation

$$x_{t+1} = rx_t(1 - x_t)$$

Compute x_t for $t = 0, 1, 2, \ldots, 20$ for the given values of r and x_0 and graph x_t as a function of t.

35. $r = 2, x_0 = 0.2$
36. $r = 2, x_0 = 0.1$
37. $r = 2, x_0 = 0.9$
38. $r = 2, x_0 = 0$
39. $r = 3.1, x_0 = 0.5$
40. $r = 3.1, x_0 = 0.1$
41. $r = 3.1, x_0 = 0.9$
42. $r = 3.1, x_0 = 0$
43. $r = 3.8, x_0 = 0.5$
44. $r = 3.8, x_0 = 0.1$
45. $r = 3.8, x_0 = 0.9$
46. $r = 3.8, x_0 = 0$

(2.3.3)

In Problems 47–50, graph the Ricker's curve

$$N_{t+1} = N_t \exp\left[R \left(1 - \frac{N_t}{K} \right) \right]$$

in the N_t-N_{t+1} plane for the given values of R and K. Find the points of intersection of this graph with the line $N_{t+1} = N_t$.

47. $R = 2, K = 10$
48. $R = 3, K = 15$
49. $R = 2.5, K = 12$
50. $R = 4, K = 20$

In Problems 51–54, we investigate the behavior of the Ricker's curve

$$N_{t+1} = N_t \exp\left[R \left(1 - \frac{N_t}{K} \right) \right]$$

Compute N_t for $t = 1, 2, \ldots, 20$ for the given values of R, K, and N_0 and graph N_t as a function of t.

51. (a) $R = 1, K = 20, N_0 = 5$
(b) $R = 1, K = 20, N_0 = 10$
(c) $R = 1, K = 20, N_0 = 20$
(d) $R = 1, K = 20, N_0 = 0$
52. (a) $R = 1.8, K = 20, N_0 = 5$
(b) $R = 1.8, K = 20, N_0 = 10$
(c) $R = 1.8, K = 20, N_0 = 20$
(d) $R = 1.8, K = 20, N_0 = 0$
53. (a) $R = 2.1, K = 20, N_0 = 5$
(b) $R = 2.1, K = 20, N_0 = 10$
(c) $R = 2.1, K = 20, N_0 = 20$
(d) $R = 2.1, K = 20, N_0 = 0$
54. (a) $R = 2.8, K = 20, N_0 = 5$
(b) $R = 2.8, K = 20, N_0 = 10$
(c) $R = 2.8, K = 20, N_0 = 20$
(d) $R = 2.8, K = 20, N_0 = 0$

(2.3.4)

55. Compute N_t and N_t/N_{t-1} for $t = 2, 3, 4, \ldots, 20$ when

$$N_{t+1} = N_t + N_{t-1}$$

with $N_0 = 1$ and $N_1 = 1$.

56. Compute N_t and N_t/N_{t-1} for $t = 2, 3, 4, \ldots, 20$ when

$$N_{t+1} = N_t + 2N_{t-1}$$

with $N_0 = 1$ and $N_1 = 1$.

57. In the text, an interpretation of the Fibonacci recursion

$$N_{t+1} = N_t + N_{t-1}$$

is given. Use a similar example to give an interpretation of the recursion

$$N_{t+1} = N_t + 2N_{t-1}$$

58. In the text, an interpretation of the Fibonacci recursion

$$N_{t+1} = N_t + N_{t-1}$$

is given. Use a similar example to give an interpretation of the recursion

$$N_{t+1} = 2N_t + N_{t-1}$$

2.4 Key Terms

Chapter 2 Review: Topics *Discuss the following definitions and concepts:*

1. Exponential growth
2. Growth constant
3. Fixed point
4. Equilibrium
5. Recursion
6. Solution
7. Density independence
8. Sequence
9. First-order recursion
10. Limit
11. Long-term behavior
12. Convergence, divergence

13. Limit laws
14. Difference equation
15. Beverton-Holt recruitment curve
16. Density dependence
17. Carrying capacity
18. Growth parameter
19. Discrete logistic equation
20. Nondimensionalization
21. Periodic behavior
22. Chaos
23. Ricker's curve
24. Fibonacci sequence
25. Golden mean

2.5 Review Problems

In Problems 1–10, find the limits.

1. $\lim\limits_{n\to\infty} 2^{-n}$

2. $\lim\limits_{n\to\infty} 3^n$

3. $\lim\limits_{n\to\infty} 40(1 - 4^{-n})$

4. $\lim\limits_{n\to\infty} \dfrac{2}{1 + 2^{-n}}$

5. $\lim\limits_{n\to\infty} a^n$ when $a > 1$

6. $\lim\limits_{n\to\infty} a^n$ when $0 < a < 1$

7. $\lim\limits_{n\to\infty} \dfrac{n(n+1)}{n^2 - 1}$

8. $\lim\limits_{n\to\infty} \dfrac{n^2 + n - 6}{n - 2}$

9. $\lim\limits_{n\to\infty} \dfrac{\sqrt{n}}{n + 1}$

10. $\lim\limits_{n\to\infty} \dfrac{n + 1}{\sqrt{n}}$

In Problems 11–14, write a_n explicitly as a function of n based on the first five terms of the sequence a_n, $n = 0, 1, 2, \ldots$.

11. $\dfrac{1}{2}, \dfrac{3}{4}, \dfrac{5}{6}, \dfrac{7}{8}, \dfrac{9}{10}$

12. $\dfrac{2}{2}, \dfrac{6}{4}, \dfrac{12}{8}, \dfrac{20}{16}, \dfrac{30}{32}$

13. $\dfrac{1}{2}, \dfrac{2}{5}, \dfrac{3}{10}, \dfrac{4}{17}, \dfrac{5}{26}$

14. $0, \dfrac{1}{3}, \dfrac{2}{4}, \dfrac{3}{5}, \dfrac{4}{6}$

15. The Beverton-Holt recruitment curve is given by the recursion

$$N_{t+1} = \frac{RN_t}{1 + \frac{R-1}{K}N_t}$$

where $R > 1$ and $K > 0$. When $N_0 > 0$, then $\lim_{t\to\infty} N_t = K$ for all values of $R > 0$. To investigate how R affects the limiting behavior of N_t, find N_t for $t = 1, 2, 3, \ldots, 10$ for $K = 100$ and $N_0 = 20$ when (a) $R = 2$, (b) $R = 5$, and (c) $R = 10$, and plot N_t as a function of t for the three choices of R in one coordinate system.

16. (*Temporally varying environment*) The recursion

$$N_{t+1} = R_t N_t$$

describes growth in a temporally varying environment if we interpret R_t as the growth parameter in generation t. A population was followed over ten years and the population sizes were recorded each year. Use the data provided to find R_t for $t = 0, 1, 2, \ldots, 9$.

t	N_t	t	N_t
0	10	6	95.1
1	15.5	7	103.2
2	15.6	8	165.0
3	10.8	9	418.7
4	15.6	10	15.7
5	32.2		

17. (*Temporally varying environment*) The recursion

$$N_{t+1} = R_t N_t$$

describes growth in a temporally varying environment if we interpret R_t as the growth parameter in generation t. A population was followed over 20 years and the population sizes were recorded every year. The table provides the population size data and the inferred values of R_t for each of the twenty years. The values of N_t indicate that the population heads toward extinction. The long-term behavior of the geometric mean of the growth parameter, denoted by $\hat{R}_t$ (read "R sub t hat"), is defined as

$$\hat{R}_t = \left(R_0 R_1 \cdots R_{t-1}\right)^{1/t}$$

and determines whether the population will go extinct. Namely, if

$$\lim_{t\to\infty} \hat{R}_t < 1$$

then the population will go extinct. Compute $\hat{R}_t$ for $t = 1, 2, \ldots, 20$.

t	N_t	R_t	t	N_t	R_t
0	10.0	2.78	11	0.45	0.88
1	27.8	0.29	12	0.40	2.69
2	8.10	0.43	13	1.06	0.36
3	3.49	0.25	14	0.38	0.08
4	0.87	2.90	15	0.03	2.34
5	2.52	1.67	16	0.07	2.13
6	4.21	1.17	17	0.15	2.20
7	4.94	0.69	18	0.34	2.80
8	3.39	1.45	19	0.94	0.29
9	4.92	1.13	20	0.28	1.22
10	5.56	0.08			

18. (*Temporally varying environment*) The recursion

$$N_{t+1} = R_t N_t$$

describes growth in a temporally varying environment if we interpret R_t as the growth parameter in generation t.

(a) Show that

$$N_t = (R_{t-1} R_{t-2} \cdots R_1 R_0) N_0$$

(b) The quantity $\hat{R}_t$ (read "R sub t hat"), defined as

$$\hat{R}_t = (R_{t-1} R_{t-2} \cdots R_1 R_0)^{1/t}$$

is called the **geometric mean**. Show that

$$N_t = (\hat{R}_t)^t N_0$$

(c) The **arithmetic mean** of a sequence of numbers $x_0, x_1, \ldots, x_{n-1}$ is defined as

$$\bar{x}_n = \frac{x_0 + x_1 + \cdots + x_{n-1}}{n}$$

Set $r_t = \ln R_t$ and show that

$$\bar{r}_t = \frac{\ln R_{t-1} + \ln R_{t-2} + \cdots + \ln R_0}{t}$$

(d) Use (c) to show that

$$N_t = N_0 e^{\bar{r}_t t}$$

19. (*Harvesting model*) Let N_t denote the population size at time t and assume that

$$N_{t+1} = (1 - c) N_t \exp\left[R\left(1 - \frac{(1-c)N_t}{K} \right) \right]$$

where R and K are positive constants and c is the fraction harvested. Find N_t for $t = 1, 2, \ldots, 20$ when $R = 1$, $K = 100$, and $N_0 = 50$ for (a) $c = 0.1$, (b) $c = 0.5$, and (c) $c = 0.9$.

20. (*Harvesting model*) Let N_t denote the population size at time t and assume that

$$N_{t+1} = (1 - c) N_t \exp\left[R\left(1 - \frac{(1-c)N_t}{K} \right) \right]$$

where R and K are positive constants and c is the fraction harvested. Find N_t for $t = 1, 2, \ldots, 20$ when $R = 3$, $K = 100$, and $N_0 = 50$ for (a) $c = 0.1$, (b) $c = 0.5$, and (c) $c = 0.9$.

CHAPTER

3

LIMITS AND CONTINUITY

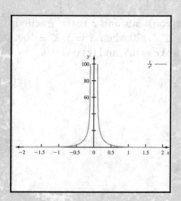

3.1 LIMITS

In Chapter 2, we discussed limits of the form $\lim_{n\to\infty} a_n$, where n took on integer values. In this chapter, we will consider limits of the form

$$\lim_{x\to c} f(x) \tag{3.1}$$

where x is now a continuously varying real variable that tends to a fixed value c (c may be finite or infinite). Let's look at an example that will motivate the need for limits of the form (3.1).

Population growth in populations with discrete breeding seasons (as in Chapter 2) can be described by the change in population size from generation to generation. In populations that breed continuously, there is no such natural time scale as generations. Instead, we will look at how the population size changes over small time intervals. We denote by $N(t)$ the population size at time t, where t is now varying continuously over the interval $[0, \infty)$. We will investigate how the population size changes during the time interval $[t, t + h]$, where $h > 0$. The absolute change during this time interval, denoted by ΔN, is

$$\Delta N = N(t + h) - N(t)$$

(The symbol Δ indicates that we take a difference.) To obtain the change relative to the length of the time interval $[t, t+h]$, we divide ΔN by the length of the time interval, denoted by Δt, which is $(t + h) - t = h$. We find

$$\frac{\Delta N}{\Delta t} = \frac{N(t + h) - N(t)}{h}$$

This ratio is called the **average growth rate**.

We see from Figure 3.1 that $\Delta N/\Delta t$ is the slope of the secant line connecting the points $(t, N(t))$ and $(t + h, N(t + h))$. The average growth rate $\Delta N/\Delta t$ depends on the length of the time interval Δt. This is illustrated in Figure 3.2, where we see that the slopes of the two secant lines (lines 1 and 2) are different. But we also see that as we choose smaller and smaller time intervals, the secant lines converge to the tangent line at the point $(t, N(t))$ of the graph of $N(t)$ (line 3).

The slope of the tangent line is called the **instantaneous growth rate** and is a convenient way to describe the growth of a continuously breeding population. To obtain this quantity, we need

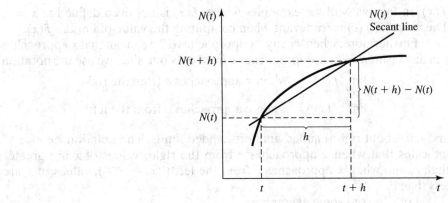

◀ **Figure 3.1**
The slope of the secant line is the average growth rate

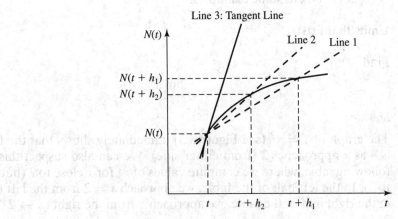

◀ **Figure 3.2**
The slope of the secant line converges to the slope of the tangent line as the length of the interval $[t, t + h]$ shrinks to 0

to take a limit, namely, we need to shrink the length of the time interval $[t, t + h]$ to 0 by letting h tend to 0. We express this as

$$\lim_{h \to 0} \frac{N(t + h) - N(t)}{h} \tag{3.2}$$

In (3.2), we take a limit of a quantity where a continuously varying variable, namely h, approaches some fixed value, namely 0. This is a limit of the form (3.1).

3.1.1 An Informal Discussion of Limits

Informal Definition The "**limit of $f(x)$, as x approaches c, is equal to L**" means that $f(x)$ can be made arbitrarily close to L whenever x is sufficiently close to c (but not equal to c). We denote this by

$$\lim_{x \to c} f(x) = L$$

or $f(x) \to L$ as $x \to c$.

If $\lim_{x \to c} f(x) = L$ and L is a finite number, we say that the limit **exists** and that $f(x)$ **converges** to L. If the limit does not exist, we say that $f(x)$ **diverges** as x tends to c.

Note that we say that we choose x close to c but not equal to c. That is, when finding the limit of $f(x)$ as x approaches c, we do not simply plug c into

$f(x)$. In fact, we will see examples where $f(x)$ is not even defined at $x = c$. The value of $f(c)$ is irrelevant when computing the value of $\lim_{x \to c} f(x)$.

Furthermore, when we say "x approaches c," we mean that x approaches c in any fashion. When x approaches c from only one side, we use the notation

$$\lim_{x \to c^+} f(x) \qquad \text{when } x \text{ approaches } c \text{ from the right}$$

$$\lim_{x \to c^-} f(x) \qquad \text{when } x \text{ approaches } c \text{ from the left}$$

and talk about right-handed and left-handed limits. The notation "$x \to c^+$" indicates that when x approaches c from the right, values of x are greater than c, and when x approaches c from the left ("$x \to c^-$"), values of x are less than c.

Let's look at some examples.

Limits that Exist

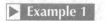 Find

$$\lim_{x \to 2} x^2$$

Solution

The graph of $y = x^2$ (see Figure 3.3) immediately shows that the limit of x^2 is 4 as x approaches 2 (from either side). We can also suspect this from the following table, where we compute values of x^2 for x close to 2 (but not equal to 2). In the left half of the table, we approach $x = 2$ from the left ($x \to 2^-$); in the right half of the table, we approach x from the right ($x \to 2^+$).

x	x^2	x	x^2
1.9	3.61	2.1	4.41
1.99	3.9601	2.01	4.0401
1.999	3.996001	2.001	4.004001
1.9999	3.99960001	2.0001	4.00040001

We find that

$$\lim_{x \to 2} x^2 = 4$$

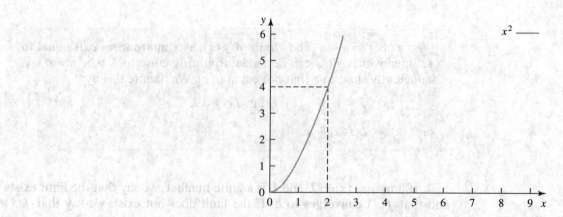

▲ **Figure 3.3**
As x approaches 2, $f(x) = x^2$ approaches 4

Since this limit is a finite number, we say that the limit exists and that x^2 converges to 4 as x tends to 2. The fact that $f(x) = x^2$ at $x = 2$ is 4 as well is a nice property that will be introduced and named later. Not all functions are like that. ◀

▶ **Example 2** Find

$$\lim_{x \to 3} \frac{x^2 - 9}{x - 3}$$

Solution

We define $f(x) = \frac{x^2-9}{x-3}$, $x \neq 3$. Since the denominator of $f(x)$ is equal to 0 when $x = 3$, we exclude $x = 3$ from the domain. When $x \neq 3$, we can simplify the expression, namely

$$f(x) = \frac{x^2 - 9}{x - 3} = \frac{(x-3)(x+3)}{x-3} = x + 3 \quad \text{for } x \neq 3$$

We were able to cancel the term $x - 3$ since we assumed $x \neq 3$ and $x - 3 \neq 0$ for $x \neq 3$. (If we allowed $x = 3$, then canceling $x - 3$ would mean dividing by 0.) The graph of $f(x)$ is a straight line with one point deleted at $x = 3$ (see Figure 3.4). Taking the limit, we find

$$\lim_{x \to 3} \frac{x^2 - 9}{x - 3} = \lim_{x \to 3}(x + 3)$$

Now, either using the graph of $y = x + 3$ for $x \neq 3$ or a table, we suspect that

$$\lim_{x \to 3}(x + 3) = 6$$

(One goal of this chapter is to learn how to show this without looking at graphs or relying on tables.) We conclude that $\lim_{x \to 3} f(x)$ exists and that $f(x)$ converges to 6 as x tends to 3. Note that $f(3)$ does not even exist. ◀

One-Sided Limits To compute one-sided limits, we use the notation

$$\lim_{x \to c^+} f(x) \quad \text{when } x \text{ approaches } c \text{ from the right}$$

$$\lim_{x \to c^-} f(x) \quad \text{when } x \text{ approaches } c \text{ from the left}$$

that was introduced previously.

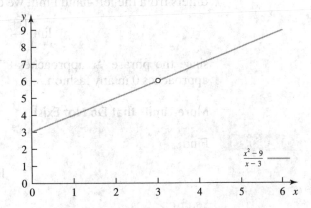

▲ **Figure 3.4**

The graph of $f(x) = \frac{x^2-9}{x-3}$ is a straight line with the point $(3, 6)$ removed

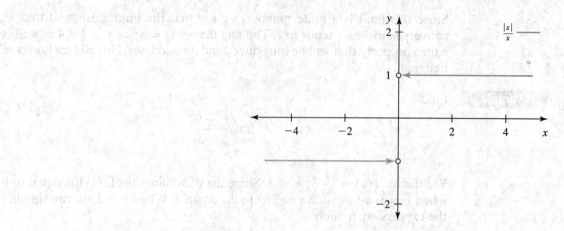

▲ **Figure 3.5**

The graph of $\frac{|x|}{x}$ in Example 3: The function is not defined at $x = 0$

▶ **Example 3** Find

$$\lim_{x \to 0^+} \frac{|x|}{x} \quad \text{and} \quad \lim_{x \to 0^-} \frac{|x|}{x}$$

Solution

We set $f(x) = \dfrac{|x|}{x}$, $x \ne 0$. Since $|x| = x$ for $x \ge 0$ and $|x| = -x$ for $x \le 0$, we find

$$f(x) = \frac{|x|}{x} = \begin{cases} +1 & \text{for } x > 0 \\ -1 & \text{for } x < 0 \end{cases}$$

The graph of $f(x)$ is shown in Figure 3.5. We see that $f(x)$ converges to 1 as x tends to 0 from the right and that $f(x)$ converges to -1 as x tends to 0 from the left. We can write this as

$$\lim_{x \to 0^+} \frac{|x|}{x} = 1 \quad \text{and} \quad \lim_{x \to 0^-} \frac{|x|}{x} = -1$$

and observe that the one-sided limits exist. ◀

In Example 3, we computed one-sided limits. Since the right-hand limit differs from the left-hand limit, we conclude that

$$\lim_{x \to 0} \frac{|x|}{x} \quad \text{does not exist}$$

since the phrase "x approaches 0" (or, in symbols, $\lim_{x \to 0}$) means that x approaches 0 in any fashion.

More Limits that Do Not Exist

▶ **Example 4** Find

$$\lim_{x \to 0} \frac{1}{x^2}$$

Solution

A graph of $f(x) = 1/x^2$, $x \ne 0$, reveals that $f(x)$ increases without bound as $x \to 0$ (see Figure 3.6). We can also suspect this when we plug in values close

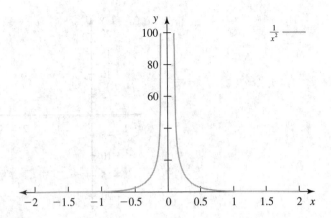

▲ Figure 3.6

The graph of $f(x) = \frac{1}{x^2}$ in Example 4: The function grows without bound as x tends to 0

to 0. By choosing values sufficiently close to 0, we can get arbitrarily large values of $f(x)$.

x	-0.1	-0.01	-0.001	0.001	0.01	0.1
$f(x)$	100	10,000	10^6	10^6	10,000	100

This indicates that the limit does not exist. ◄

When $\lim_{x \to c} f(x)$ does not exist, we say that $f(x)$ diverges as x tends to c. The divergence in Example 4 was such that the function grew without bound. This is an important case, and we define it in the following box.

$$\lim_{x \to c} f(x) = +\infty \quad \text{if } f(x) \text{ increases without bound as } x \to c$$

$$\lim_{x \to c} f(x) = -\infty \quad \text{if } f(x) \text{ decreases without bound as } x \to c$$

Similar definitions can be given for one-sided limits, which we will need in the next example. You should note that when we write $\lim_{x \to c} f(x) = +\infty$ (or $-\infty$), we say that $f(x)$ diverges as $x \to c$. In particular, this means that $\lim_{x \to c} f(x)$ does *not* exist. (The symbols $+\infty$ and $-\infty$ do not refer to real numbers.) Nevertheless, we write $\lim_{x \to c} f(x) = +\infty$ (or $-\infty$) if $f(x)$ increases (or decreases) without bound as $x \to c$ since it is useful to know when a function does that.

▶ **Example 5** Find

$$\lim_{x \to 3} \frac{1}{x - 3}$$

Solution

The graph of $f(x) = 1/(x - 3)$, $x \neq 3$, in Figure 3.7 reveals that

$$\lim_{x \to 3^+} \frac{1}{x - 3} = +\infty \quad \text{and} \quad \lim_{x \to 3^-} \frac{1}{x - 3} = -\infty$$

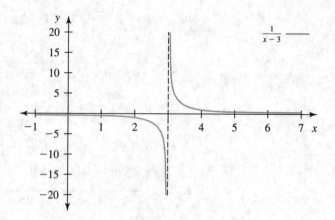

▲ **Figure 3.7**

The graph of $f(x) = \frac{1}{x-3}$ in Example 5 grows without bound as x approaches 3 from the right and decreases without bound as x approaches 3 from the left

We arrive at the same conclusion when we compute values of $f(x)$ for x close to 3. If x is slightly larger than 3, we see that $f(x)$ is positive and increases without bound as x approaches 3 from the right. Likewise, if x is slightly smaller than 3, $f(x)$ is negative and decreases without bound as x approaches 3 from the left. We conclude that $f(x)$ diverges as x approaches 3. ◀

The next example shows that a function can diverge without having one-sided limits or going to $+\infty$ or $-\infty$.

▶ **Example 6** Find

$$\lim_{x \to 0} \sin \frac{\pi}{x}$$

Solution

Simply using a calculator and plugging in values to find limits can yield wrong answers if one does not exercise proper caution. If we produced a table of values of $f(x) = \sin \frac{\pi}{x}$ for $x = 0.1, 0.01, 0.001, \ldots$, we would find $\sin \frac{\pi}{0.1} = 0$, $\sin \frac{\pi}{0.01} = 0$, $\sin \frac{\pi}{0.001} = 0$, and so on. (Note that we measure angles in radians.) This might prompt us to conclude that the limit of the function is 0. But let's look at its graph, which is shown in Figure 3.8. The graph does not support our calculator-based conclusion.

What we find instead is that the values of $f(x)$ oscillate infinitely often between -1 and $+1$ as $x \to 0$. We can see why as follows: As $x \to 0^+$, the argument in the sine function goes to infinity (likewise, as $x \to 0^-$, the argument goes to negative infinity).

$$\lim_{x \to 0^+} \frac{\pi}{x} = \infty \quad \text{and} \quad \lim_{x \to 0^-} \frac{\pi}{x} = -\infty$$

As the argument of the sine function goes to $+\infty$ or $-\infty$, the function values oscillate between -1 and $+1$. Therefore, $\sin \frac{\pi}{x}$ continues to oscillate between -1 and $+1$ as $x \to 0$. ◀

The behavior exhibited in Example 6 is called **divergence by oscillation**.

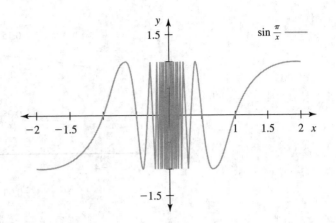

▲ Figure 3.8
The graph of $f(x) = \sin \frac{\pi}{x}$ in Example 6

Pitfalls The following is an interesting example that shows other limitations of using a calculator when computing limits.

▶ **Example 7** Find

$$\lim_{x \to 0} \frac{\sqrt{x^2 + 16} - 4}{x^2}$$

Solution

The graph of $f(x) = \frac{\sqrt{x^2+16}-4}{x^2}$, $x \neq 0$, in Figure 3.9 indicates that the limit exists and, based on the graph, we conjecture that it is equal to 0.125. If, instead, we use a calculator to produce a table for values of $f(x)$ close to 0, something strange seems to happen.

x	0.01	0.001	0.0001	0.00001	0.000001	0.0000001
$f(x)$	0.1249998	0.125	0.125	0.125	0.1	0

As we get closer to 0, we first find that $f(x)$ gets closer to 0.125, but when we get very close to 0, $f(x)$ seems to drop to 0. What is going on? First, before you worry too much, $\lim_{x \to 0} f(x) = 0.125$. In the next section, we will learn how to compute this limit without resorting to the (somewhat dubious) help of the calculator. The strange behavior of the calculated values happens since when x is very small, the difference in the numerator is so close to 0 that the calculator can no longer accurately determine its value. The calculator can only accurately compute a certain number of digits, which is good enough for most cases. Here, however, we need greater accuracy. The same strange thing happens when you try to graph this function on a graphing calculator. When the x range of the viewing window is too small, the graph is no longer accurate. (Try, for instance, $-0.00001 \le x \le 0.00001$ and $-0.03 \le y \le 0.15$ as the range for the viewing window.)   ◀

At the end of this chapter, we will discuss how limits are formally defined. The formal definition is conceptually similar to the one we used to define limits of the form $\lim_{n \to \infty} a_n$, but we will not use it to compute limits. As in Chapter 2, there are limit laws that will allow us to compute limits much more easily.

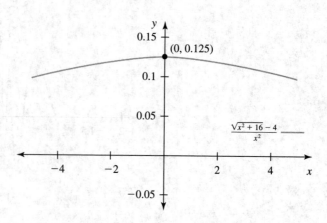

▲ Figure 3.9
The graph of $f(x)$ in Example 7: as x tends to 0, the function approaches 0.125

3.1.2 Limit Laws

We encountered limit laws in Chapter 2. Analogous laws also hold for limits of the type $\lim_{x \to c} f(x)$.

LIMIT LAWS

Suppose that a is a constant and that

$$\lim_{x \to c} f(x) \quad \text{and} \quad \lim_{x \to c} g(x)$$

exist. Then the following rules hold:

1. $\lim_{x \to c} af(x) = a \lim_{x \to c} f(x)$

2. $\lim_{x \to c} [f(x) + g(x)] = \lim_{x \to c} f(x) + \lim_{x \to c} g(x)$

3. $\lim_{x \to c} [f(x) \cdot g(x)] = \lim_{x \to c} f(x) \cdot \lim_{x \to c} g(x)$

4. $\lim_{x \to c} \dfrac{f(x)}{g(x)} = \dfrac{\lim_{x \to c} f(x)}{\lim_{x \to c} g(x)}$ provided $\lim_{x \to c} g(x) \neq 0$

You are probably easily convinced that

$$\lim_{x \to c} x = c \tag{3.3}$$

We will use the formal definition for limits to show this in Section 3.6. For now, we accept (3.3) as a fact. Starting from (3.3), we can use the limit laws to compute limits of polynomials and rational functions.

▶ **Example 8**   Find

$$\lim_{x \to 2} [x^3 + 4x - 1]$$

Solution

Using Rules 1 and 2, this becomes

$$\lim_{x \to 2} x^3 + 4 \lim_{x \to 2} x - \lim_{x \to 2} 1$$

provided the individual limits exist. For the first term, we use Rule 3,

$$\lim_{x \to 2} x^3 = (\lim_{x \to 2} x)(\lim_{x \to 2} x)(\lim_{x \to 2} x)$$

provided $\lim_{x \to 2} x$ exists. Using (3.3), it follows that $\lim_{x \to 2} = 2$ and we find that

$$(\lim_{x \to 2} x)(\lim_{x \to 2} x)(\lim_{x \to 2} x) = (2)(2)(2) = 8$$

To compute the second term, we use (3.3) again, $\lim_{x \to 2} x = 2$. For the last term, we find $\lim_{x \to 2} 1 = 1$. Now that we showed hat the individual limits exist, we can use Rules 1 and 2 to evaluate

$$\lim_{x \to 2}[x^3 + 4x - 1] = \lim_{x \to 2} x^3 + 4 \lim_{x \to 2} x - \lim_{x \to 2} 1 = 8 + (4)(2) - 1 = 15 \quad \blacktriangleleft$$

▶ **Example 9** Find

$$\lim_{x \to 4} \frac{x^2 + 1}{x - 3}$$

Solution

Using Rule 4, we find

$$\lim_{x \to 4} \frac{x^2 + 1}{x - 3} = \frac{\lim_{x \to 4}(x^2 + 1)}{\lim_{x \to 4}(x - 3)}$$

provided the limits in the numerator and denominator exist and the limit in the denominator is not equal to 0. Using Rules 2 and 3 in the numerator, we find

$$\lim_{x \to 4}(x^2 + 1) = \left(\lim_{x \to 4} x^2\right) + \left(\lim_{x \to 4} 1\right) = (4)(4) + 1 = 17$$

Breaking up the limit of the sum in the numerator into a sum of limits is only justified once we show that the individual limits exist. Using Rules 1 and 2 in the denominator, we find

$$\lim_{x \to 4}(x - 3) = \lim_{x \to 4} x - \lim_{x \to 4} 3 = 4 - 3 = 1$$

Again using the limit laws is only justified once we demonstrate that the individual limits exist. Since the limits in both the denominator and the numerator exist and the limit in the denominator is not equal to 0, we obtain

$$\lim_{x \to 4} \frac{x^2 + 1}{x - 3} = \frac{17}{1} = 17 \quad \blacktriangleleft$$

The computations in Examples 8 and 9 look somewhat awkward and it appears that what we have done is plug 2 into the expression $x^3 + 4x - 1$ in Example 8, and 4 into the expression $\frac{x^2+1}{x-3}$ in Example 9, even though we emphasized in the informal definition of limits that we are not allowed to simply plug c into $f(x)$ when computing $\lim_{x \to c} f(x)$. But in essence, we did the following calculation in Example 8,

$$\lim_{x \to 2}[x^3 + 4x - 1] = 2^3 + (4)(2) - 1 = 15$$

and the following calculation in Example 9:

$$\lim_{x \to 4} \frac{x^2 + 1}{x - 3} = \frac{17}{1} = 17$$

Even though we made a point that we cannot simply substitute the value c into $f(x)$ when we take the limit $x \to c$ of $f(x)$, the limit laws and (3.3) (which we will prove in Section 3.6) show that we can do just that when we take a limit of a polynomial or a rational function. Let's summarize this and then look at two more examples that show how to compute limits of polynomials or rational functions using these results.

If $f(x)$ is a polynomial, then

$$\lim_{x \to c} f(x) = f(c)$$

If $f(x)$ is a rational function, that is,

$$f(x) = \frac{p(x)}{q(x)}$$

where $p(x)$ and $q(x)$ are polynomials and $q(c) \neq 0$, then

$$\lim_{x \to c} f(x) = \lim_{x \to c} \frac{p(x)}{q(x)} = \frac{p(c)}{q(c)} = f(c)$$

▶ **Example 10**   Find

$$\lim_{x \to 3} [x^2 - 2x + 1]$$

Solution

Since $f(x) = x^2 - 2x + 1$ is a polynomial, we find

$$\lim_{x \to 3} [x^2 - 2x + 1] = 9 - 6 + 1 = 4$$ ◀

▶ **Example 11**   Find

$$\lim_{x \to -1} \frac{2x^3 - x + 5}{x^2 + 3x + 1}$$

Solution

Note that

$$f(x) = \frac{2x^3 - x + 5}{x^2 + 3x + 1}$$

is a rational function that is defined for $x = -1$ (the denominator is not equal to 0 when we substitute $x = -1$). We find

$$\lim_{x \to -1} \frac{2x^3 - x + 5}{x^2 + 3x + 1} = \frac{2(-1)^3 - (-1) + 5}{(-1)^2 + 3(-1) + 1} = \frac{4}{-1} = -4$$ ◀

When you use the limit laws for finding limits of the form

$$\lim_{x \to c} [f(x) + g(x)] \quad \text{or} \quad \lim_{x \to c} [f(x) \cdot g(x)] \quad \text{or} \quad \lim_{x \to c} \frac{f(x)}{g(x)}$$

you need to check first that both $\lim_{x \to c} f(x)$ and $\lim_{x \to c} g(x)$ exist and, in the case of $\lim_{x \to c} \frac{f(x)}{g(x)}$, that $\lim_{x \to c} g(x) \neq 0$. We illustrate in the next two examples the importance of checking the assumptions in the limit laws before applying them.

▶ **Example 12**   Find

$$\lim_{x \to 0} \frac{\frac{1}{x}}{\frac{1}{x} + 1}$$

Solution

We observe that neither

$$\lim_{x \to 0} \frac{1}{x} \quad \text{nor} \quad \lim_{x \to 0} \left(\frac{1}{x} + 1 \right)$$

exist. So we cannot use Rule 4 right away. Multiplying both numerator and denominator by x, however, will help:

$$\lim_{x \to 0} \frac{\frac{1}{x}}{\frac{1}{x} + 1} = \lim_{x \to 0} \frac{1}{1 + x}$$

Now, we have a rational function on the right-hand side and we can plug in 0 since the denominator $1 + x$ will be different from 0.

$$\lim_{x \to 0} \frac{1}{1 + x} = \frac{1}{1 + 0} = 1 \qquad \blacktriangleleft$$

▶ Example 13   Find

$$\lim_{x \to 4} \frac{x^2 - 16}{x - 4}$$

Solution

The function $f(x) = \frac{x^2 - 16}{x - 4}$ is a rational function, but since $\lim_{x \to 4}(x - 4) = 0$, we cannot use Rule 4. Instead, we need to simplify $f(x)$ first.

$$\lim_{x \to 4} \frac{x^2 - 16}{x - 4} = \lim_{x \to 4} \frac{(x - 4)(x + 4)}{x - 4}$$

Since $x \neq 4$, we can cancel $x - 4$ in the numerator and denominator, which yields

$$\lim_{x \to 4}(x + 4) = 8$$

where we used that $x + 4$ is a polynomial when computing the limit. $\blacktriangleleft$

3.1.3 Problems

(3.1.1)

In Problems 1–32, evaluate each limit using a table or a graph.

1. $\lim\limits_{x \to 2} (x^2 - 3x + 1)$

2. $\lim\limits_{x \to 2} \dfrac{x^2 + 3}{x + 2}$

3. $\lim\limits_{x \to -1} \dfrac{x}{1 + x^2}$

4. $\lim\limits_{s \to 2} s(s^2 - 4)$

5. $\lim\limits_{x \to \pi} 3 \cos \dfrac{x}{4}$

6. $\lim\limits_{t \to \pi/12} \sin(4t)$

7. $\lim\limits_{x \to \pi/2} 2 \sec \dfrac{x}{3}$

8. $\lim\limits_{x \to \pi/2} \tan \dfrac{x - \pi/2}{2}$

9. $\lim\limits_{x \to -1} e^{-x^2/2}$

10. $\lim\limits_{x \to 0} \dfrac{e^x + 1}{2x + 3}$

11. $\lim\limits_{x \to 0} \ln(x + 1)$

12. $\lim\limits_{t \to e} \ln t^2$

13. $\lim\limits_{x \to 3} \dfrac{x^2 - 16}{x - 4}$

14. $\lim\limits_{x \to 2} \dfrac{x^2 - 4}{x + 2}$

15. $\lim\limits_{x \to \infty} \dfrac{2}{x}$

16. $\lim\limits_{x \to -\infty} \dfrac{3}{x}$

17. $\lim\limits_{x \to \infty} \dfrac{x}{1 + x}$

18. $\lim\limits_{x \to \infty} \dfrac{x^2}{2 - x^2}$

19. $\lim\limits_{x \to -\infty} e^x$

20. $\lim\limits_{x \to \infty} e^{-x}$

21. $\lim\limits_{x \to 4^-} \dfrac{2}{x - 4}$

22. $\lim\limits_{x \to 3^+} \dfrac{1}{x-3}$

23. $\lim\limits_{x \to 1^-} \dfrac{2}{1-x}$

24. $\lim\limits_{x \to 2^+} \dfrac{3}{2-x}$

25. $\lim\limits_{x \to 1^-} \dfrac{1}{1-x^2}$

26. $\lim\limits_{x \to 2^+} \dfrac{2}{x^2-4}$

27. $\lim\limits_{x \to 1} \dfrac{1}{(x-1)^2}$

28. $\lim\limits_{x \to 0} \dfrac{1-x^2}{x^2}$

29. $\lim\limits_{x \to 0} \dfrac{\sqrt{x^2+9}-3}{x^2}$

30. $\lim\limits_{x \to 0} \dfrac{\sqrt{x^2+4}-2}{x}$

31. $\lim\limits_{x \to 0} \dfrac{1-\sqrt{1-x^2}}{x^2}$

32. $\lim\limits_{x \to 0} \dfrac{\sqrt{2-x}-\sqrt{2}}{2x}$

33. Use a table and a graph to find out what happens to

$$f(x) = \frac{2}{x^2}$$

as $x \to \infty$. What happens when $x \to -\infty$? What happens as $x \to 0$?

34. Use a table and a graph to find out what happens to

$$f(x) = \frac{2x}{x-1}$$

as $x \to \infty$. What happens when $x \to -\infty$? What happens as $x \to 1$?

35. Use a graphing calculator to investigate

$$\lim\limits_{x \to 1} \sin \frac{1}{x-1}$$

36. Use a graphing calculator to investigate

$$\lim\limits_{x \to 0} \cos \frac{1}{x}$$

(3.1.2)

In Problems 37–54, use the limit laws to evaluate each limit.

37. $\lim\limits_{x \to -1} (x^3 + 7x - 1)$

38. $\lim\limits_{x \to 2} (3x^4 - 2x + 1)$

39. $\lim\limits_{x \to -5} (4 + 2x^2)$

40. $\lim\limits_{x \to 1} (8x^3 - 2x + 3)$

41. $\lim\limits_{x \to 3} (2x^2 - \frac{1}{x})$

42. $\lim\limits_{x \to -2} \left(\dfrac{x^2}{2} - \dfrac{2}{x^2} \right)$

43. $\lim\limits_{x \to -3} \dfrac{x^3 - 20}{x+1}$

44. $\lim\limits_{x \to 1} \dfrac{x^3 - 1}{x+2}$

45. $\lim\limits_{x \to 3} \dfrac{3x^2 + 1}{2x - 3}$

46. $\lim\limits_{x \to -2} \dfrac{1+x}{1-x}$

47. $\lim\limits_{x \to 1} \dfrac{1-x^2}{1-x}$

48. $\lim\limits_{u \to 2} \dfrac{4 - u^2}{2 - u}$

49. $\lim\limits_{x \to 3} \dfrac{x^2 - 2x - 3}{x - 3}$

50. $\lim\limits_{x \to 1} \dfrac{(x-1)^2}{x^2 - 1}$

51. $\lim\limits_{x \to 3} \dfrac{3-x}{x^2 - 9}$

52. $\lim\limits_{x \to -4} \dfrac{x+4}{16 - x^2}$

53. $\lim\limits_{x \to -2} \dfrac{2x^2 + 3x - 2}{x + 2}$

54. $\lim\limits_{x \to 1/2} \dfrac{1 - x - 2x^2}{1 - 2x}$

3.2 CONTINUITY

3.2.1 What is Continuity?

Let's look at the following two functions:

$$f(x) = \begin{cases} \dfrac{x^2 - 9}{x - 3} & \text{if } x \neq 3 \\ 6 & \text{if } x = 3 \end{cases}$$

and

$$g(x) = \begin{cases} \dfrac{x^2 - 9}{x - 3} & \text{if } x \neq 3 \\ 7 & \text{if } x = 3 \end{cases}$$

We are interested in how these functions behave for x close to 3. Both functions are defined for all $x \in \mathbf{R}$ and are the same for $x \neq 3$. Furthermore, as we saw in Example 2 of Section 3.1,

$$\lim_{x \to 3} f(x) = \lim_{x \to 3} g(x) = \lim_{x \to 3} \frac{x^2 - 9}{x - 3} = 6 \tag{3.4}$$

But the two functions differ at $x = 3$: $f(3) = 6$ and $g(3) = 7$. Comparing this with (3.4), we see that

$$\lim_{x \to 3} f(x) = f(3) \quad \text{but} \quad \lim_{x \to 3} g(x) \neq g(3)$$

This difference can also be seen graphically (Figures 3.10 and 3.11): The graph of $f(x)$ can be drawn without lifting the pencil; whereas when graphing $g(x)$, we need to lift the pencil at $x = 3$ since $\lim_{x \to 3} g(x) \neq g(3)$. We say that the function $f(x)$ is **continuous** at $x = 3$, whereas $g(x)$ is **discontinuous** at $x = 3$. Here is the definition of continuity at a point.

> **Definition** A function f is said to be **continuous** at $x = c$ if
>
> $$\lim_{x \to c} f(x) = f(c)$$

In order to check whether a function is continuous at $x = c$, we therefore need to check the following three conditions:

1. $f(x)$ is defined at $x = c$.
2. $\lim_{x \to c} f(x)$ exists.
3. $\lim_{x \to c} f(x)$ is equal to $f(c)$.

If any of these three conditions fails, the function is **discontinuous** at $x = c$.

▶ **Example 1**   Show that $f(x) = 2x - 3$, $x \in \mathbf{R}$, is continuous at $x = 1$.

Solution

We must check all three conditions.

1. $f(x)$ is defined at $x = 1$ since $f(1) = 2 \cdot 1 - 3 = -1$.
2. We use that $\lim_{x \to c} x = c$ to conclude that $\lim_{x \to 1} f(x)$ exists.

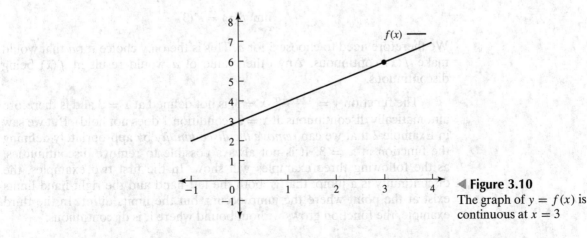

◀ **Figure 3.10**
The graph of $y = f(x)$ is continuous at $x = 3$

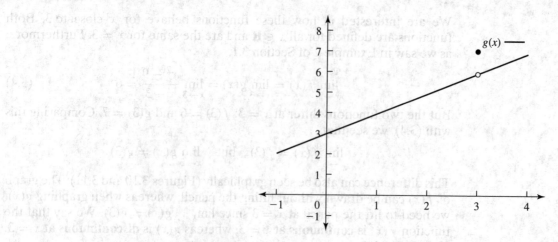

▲ **Figure 3.11**
The graph of $y = f(x)$ is discontinuous at $x = 3$

3. Using the limit laws, we find that $\lim_{x \to 1} f(x) = -1$. This is the same as $f(1)$.

   Since all three conditions are satisfied, $f(x) = 2x - 3$ is continuous at $x = 1$.   ◀

▶ **Example 2**   Let

$$f(x) = \begin{cases} \dfrac{x^2 - x - 6}{x - 3} & \text{if } x \neq 3 \\ a & \text{if } x = 3 \end{cases}$$

and find a so that $f(x)$ is continuous at $x = 3$.

Solution

To compute

$$\lim_{x \to 3} \frac{x^2 - x - 6}{x - 3}$$

we factor the numerator, $x^2 - x - 6 = (x - 3)(x + 2)$. Hence, since $x \neq 3$,

$$\lim_{x \to 3} \frac{x^2 - x - 6}{x - 3} = \lim_{x \to 3} \frac{(x - 3)(x + 2)}{x - 3} = \lim_{x \to 3} (x + 2) = 5$$

To ensure that $f(x)$ is continuous at $x = 3$, we require that

$$\lim_{x \to 3} f(x) = f(3)$$

We therefore need to choose 5 for a. This is the only choice for a that would make $f(x)$ continuous. Any other value of a would result in $f(x)$ being discontinuous.   ◀

The function $y = \frac{x^2 - x - 6}{x - 3}$, $x \neq 3$, is not defined at $x = 3$ and is therefore automatically discontinuous at $x = 3$ (condition 1 does not hold). But we saw in Example 2 that we can *remove the discontinuity* by appropriately defining the function at $x = 3$. It is not always possible to remove discontinuities, as the following three examples will show. In the first two examples, the discontinuity is a jump; that is, both the left-hand and the right-hand limits exist at the point where the jump occurs, but the limits differ. In the third example, the function grows without bound where it is discontinuous.

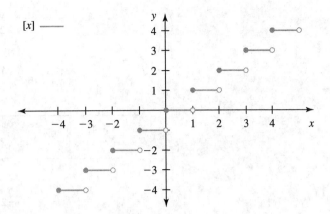

▲ **Figure 3.12**
The floor function $f(x) = \lfloor x \rfloor$

▶ **Example 3**   The floor function

$$f(x) = \lfloor x \rfloor = \text{the largest integer less than or equal to } x$$

is graphed in Figure 3.12. The closed circles in the figure correspond to endpoints that are contained in the graph of the function, whereas the open circles correspond to endpoints that are not contained in the graph of the function. To explain this function, we compute a few values: $f(2.1) = 2$, $f(2) = 2$, $f(1.9999) = 1$. The function jumps whenever x is an integer. Let k be an integer; then $f(k) = k$ and

$$\lim_{x \to k^+} f(x) = k, \quad \lim_{x \to k^-} f(x) = k - 1$$

That is, only when x approaches an integer from the right is the limit equal to the value of the function. The function is therefore discontinuous at integer values and the discontinuity cannot be removed. If c is not an integer, then $f(x)$ is continuous at $x = c$.   ◀

Example 3 motivates the definition of one-sided continuity.

> **Definition** A function f is said to be continuous from the right at $x = c$ if
>
> $$\lim_{x \to c^+} f(x) = f(c)$$
>
> and continuous from the left at $x = c$ if
>
> $$\lim_{x \to c^-} f(x) = f(c)$$

The function $f(x) = \lfloor x \rfloor$, $x \in \mathbf{R}$, of Example 3, is therefore continuous from the right but not from the left. In the next example, the discontinuity is again a jump, but this time we do not even have one-sided continuity.

▶ **Example 4**   Show that

$$f(x) = \begin{cases} \dfrac{|x|}{x} & \text{if } x \neq 0 \\ 0 & \text{if } x = 0 \end{cases}$$

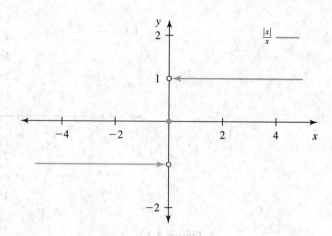

▲ **Figure 3.13**
The function $f(x) = \frac{|x|}{x}$ is discontinuous at $x = 0$

is discontinuous at $x = 0$ and that the discontinuity at $x = 0$ cannot be removed.

Solution

The graph of $f(x)$ is shown in Figure 3.13. We can write

$$f(x) = \begin{cases} 1 & \text{for } x > 0 \\ 0 & \text{for } x = 0 \\ -1 & \text{for } x < 0 \end{cases}$$

since $|x| = x$ for $x > 0$ and $|x| = -x$ for $x < 0$. We therefore get

$$\lim_{x \to 0^+} f(x) = 1 \quad \text{and} \quad \lim_{x \to 0^-} f(x) = -1$$

The one-sided limits exist but they are not equal [which implies that $\lim_{x \to 0} f(x)$ does not exist]. When we graph the function, a jump occurs at $x = 0$ (see Figure 3.13). This function does not exhibit even one-sided continuity because $f(x)$ is neither 1 nor -1 at $x = 0$. There is no way that we could assign a value to $f(0)$ so that the function would be continuous at $x = 0$. ◀

▶ Example 5 Show that

$$f(x) = \frac{1}{x^2}, \quad x \neq 0$$

is discontinuous at $x = 0$ and that the discontinuity at $x = 0$ cannot be removed.

Solution

The graph of $f(x)$ is shown in Figure 3.14. To show that $f(x) = \frac{1}{x^2}$ is discontinuous at $x = 0$ is easy since $f(x)$ is not defined at $x = 0$ and hence condition 1 does not hold. We already know what happens at $x = 0$; we showed in Example 4 of Section 3.1 that

$$\lim_{x \to 0} \frac{1}{x^2} = \infty \quad \text{(limit does not exist)}$$

Because ∞ is not a real number, we cannot assign a value to $f(0)$ so that $f(x)$ would be continuous at $x = 0$. ◀

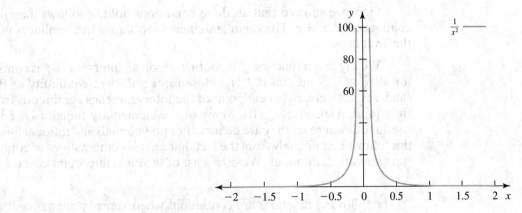

▲ **Figure 3.14**
The function $f(x) = \frac{1}{x^2}$ is discontinuous at $x = 0$

3.2.2 Combinations of Continuous Functions

Using the limit laws, we find that the following hold for combinations of continuous functions.

> Suppose that a is a constant and the functions f and g are continuous at $x = c$. Then the following functions are continuous at $x = c$:
>
> **1.** $a \cdot f$
> **2.** $f + g$
> **3.** $f \cdot g$
> **4.** $\frac{f}{g}$ provided $g(c) \neq 0$

Proof We will prove only the second statement. We must show that conditions 1–3 of the previous subsection hold.

1. Note that $[f + g](x) = f(x) + g(x)$. Therefore, $f + g$ is defined at $x = c$ and $[f + g](c) = f(c) + g(c)$.

2. We assumed that f and g are continuous at $x = c$. This means, in particular, that

$$\lim_{x \to c} f(x) \quad \text{and} \quad \lim_{x \to c} g(x)$$

both exist. That is, the hypothesis in the limit laws holds, and we can apply Rule 2 for limits and find

$$\lim_{x \to c} [f + g](x) = \lim_{x \to c} [f(x) + g(x)] = \lim_{x \to c} f(x) + \lim_{x \to c} g(x) \quad (3.5)$$

That is, $\lim_{x \to c} [f + g](x)$ exists, and condition 2 holds.

3. Since f and g are continuous at $x = c$,

$$\lim_{x \to c} f(x) = f(c) \quad \text{and} \quad \lim_{x \to c} g(x) = g(c) \quad (3.6)$$

Therefore, combining (3.5) and (3.6),

$$\lim_{x \to c} [f + g](x) = \lim_{x \to c} f(x) + \lim_{x \to c} g(x) = f(c) + g(c)$$

which is equal to $[f + g](c)$ and hence condition 3 holds.

Since we showed that all three conditions hold, it follows that $f + g$ is continuous at $x = c$. The other statements are shown in a similar way, using the limit laws. ■

We say that a function f is continuous on an interval I if f is continuous for all $x \in I$. Note that if I is a closed interval, then continuity at the left (and, respectively, right) endpoint of the interval means continuous from the right (and, respectively, left). Many of the elementary functions are indeed continuous wherever they are defined. For polynomials and rational functions, this follows immediately from the fact that certain combinations of continuous functions are continuous. We give a list of the most important cases.

> The following functions are continuous wherever they are defined:
> 1. polynomial functions
> 2. rational functions
> 3. power functions
> 4. trigonometric functions
> 5. exponential functions of the form a^x, $a > 0$ and $a \neq 1$
> 6. logarithmic functions of the form $\log_a x$, $a > 0$ and $a \neq 1$

The phrase "wherever they are defined" is crucial. It helps us to identify points where a function might be discontinuous. For instance, the power function $1/x^2$ is only defined for $x \neq 0$, and the logarithmic function $\log_a x$ is only defined for $x > 0$. We will illustrate these six cases in the following example, paying particular attention to the phrase "wherever they are defined."

▶ **Example 6**   For which values of $x \in \mathbf{R}$ are the following functions continuous?

(a) $f(x) = 2x^3 - 3x + 1$ **(b)** $f(x) = \dfrac{x^2 + x + 1}{x - 2}$ **(c)** $f(x) = x^{1/4}$

(d) $f(x) = 3 \sin x$ **(e)** $f(x) = \tan x$ **(f)** $f(x) = 3^x$

(g) $2 \ln(x + 1)$

Solution

(a) $f(x)$ is a polynomial and is defined for all $x \in \mathbf{R}$; it is therefore continuous for all $x \in \mathbf{R}$.

(b) $f(x)$ is a rational function. It is defined for all $x \neq 2$; it is therefore continuous for all $x \neq 2$.

(c) $f(x) = x^{1/4} = \sqrt[4]{x}$ is a power function that is defined for $x \geq 0$. It is therefore continuous for $x \geq 0$.

(d) $f(x)$ is a trigonometric function. Because $\sin x$ is defined for all $x \in \mathbf{R}$, $3 \sin x$ is continuous for all $x \in \mathbf{R}$.

(e) $f(x)$ is a trigonometric function. The tangent function is defined for all $x \neq \frac{\pi}{2} + k\pi$, where k is an integer. It is therefore continuous for all $x \neq \frac{\pi}{2} + k\pi$, where k is an integer.

(f) $f(x)$ is an exponential function. $f(x) = 3^x$ is defined for all $x \in \mathbf{R}$ and is therefore continuous for all $x \in \mathbf{R}$.

(g) $f(x)$ is a logarithmic function. $f(x) = 2 \ln(x + 1)$ is defined as long as $x + 1 > 0$ or $x > -1$. It is therefore continuous for all $x > -1$. ◀

The following result is useful in determining whether a composition of functions is continuous.

> **Theorem** If $g(x)$ is continuous at $x = c$ with $g(c) = L$ and $f(x)$ is continuous at $x = L$, then $(f \circ g)(x)$ is continuous at $x = c$. In particular,
>
> $$\lim_{x \to c}(f \circ g)(x) = \lim_{x \to c} f[g(x)] = f[\lim_{x \to c} g(x)] = f[g(c)] = f(L)$$

To explain this theorem, we recall what it means to compute $(f \circ g)(c) = f[g(c)]$. When we compute $f[g(c)]$, we take the value c, compute $g(c)$, and then take the result $g(c)$ and plug it into the function f to obtain $f[g(c)]$. If at each step the functions are continuous, the resulting function will be continuous.

▶ **Example 7**   Determine where the following functions are continuous.

(a) $h(x) = e^{-x^2}$

(b) $h(x) = \sin \frac{\pi}{x}$

(c) $h(x) = \frac{1}{1+2x^{1/3}}$

Solution

(a) Set $g(x) = -x^2$ and $f(x) = e^x$; then $h(x) = (f \circ g)(x)$. Since $g(x)$ is a polynomial, it is continuous for all $x \in \mathbf{R}$, and the range of $g(x)$ is $(-\infty, 0]$. $f(x)$ is continuous for all values in the range of $g(x)$ [in fact, $f(x)$ is continuous for all $x \in \mathbf{R}$]. It therefore follows that $h(x)$ is continuous for all $x \in \mathbf{R}$.

(b) Set $g(x) = \frac{\pi}{x}$ and $f(x) = \sin x$. $g(x)$ is continuous for all $x \neq 0$. The range of $g(x)$ is the set of all real numbers excluding 0. $f(x)$ is continuous for all x in the range of $g(x)$. Hence, $h(x)$ is continuous for all $x \neq 0$. Recall that we showed in Example 6 of Section 3.1 that

$$\lim_{x \to 0} \sin \frac{\pi}{x}$$

does not exist. That is, $h(x)$ is discontinuous at $x = 0$.

(c) Set $g(x) = x^{1/3}$ and $f(x) = \frac{1}{1+2x}$. Then $h(x) = (f \circ g)(x)$. $g(x)$ is continuous for all $x \in \mathbf{R}$ since $g(x) = x^{1/3} = \sqrt[3]{x}$ and 3 is an odd integer. The range of $g(x)$ is $(-\infty, \infty)$. $f(x)$ is continuous for all real x different from $-1/2$. Since $g(-\frac{1}{8}) = -\frac{1}{2}$, $h(x)$ is continuous for all real x different from $-1/8$. Another way to see that we need to exclude $-\frac{1}{8}$ from the domain of $h(x)$ is by directly looking at the denominator of $h(x)$. Namely, $1 + 2x^{1/3} = 0$ when $x = -\frac{1}{8}$.   ◀

When we compute $\lim_{x \to c} f(x)$ and we know that $f(x)$ is continuous at $x = c$, then $\lim_{x \to c} f(x) = f(c)$. The next three examples illustrate this.

▶ **Example 8**   Find

$$\lim_{x \to 3} \sin \left(\pi \frac{x^2 - 1}{4} \right)$$

Solution

The function $f(x) = \sin(\pi \frac{x^2-1}{4})$ is continuous at $x = 3$. Hence,

$$\lim_{x \to 3} \sin\left(\pi \frac{x^2-1}{4}\right) = \sin\left(\pi \frac{9-1}{4}\right) = \sin(2\pi) = 0 \qquad \blacktriangleleft$$

▶ **Example 9**   Find

$$\lim_{x \to 1} \sqrt{2x^3 - 1}$$

Solution

The function $f(x) = \sqrt{2x^3 - 1}$ is continuous at $x = 1$. Hence,

$$\lim_{x \to 1} \sqrt{2x^3 - 1} = \sqrt{(2)(1)^3 - 1} = \sqrt{1} = 1 \qquad \blacktriangleleft$$

▶ **Example 10**   Find

$$\lim_{x \to 0} e^{x-1}$$

Solution

The function $f(x) = e^{x-1}$ is continuous at $x = 0$. Hence,

$$\lim_{x \to 0} e^{x-1} = e^{0-1} = e^{-1} \qquad \blacktriangleleft$$

We will conclude this section by calculating the limit in Example 7 of Section 3.1:

▶ **Example 11**   Find

$$\lim_{x \to 0} \frac{\sqrt{x^2 + 16} - 4}{x^2}$$

Solution

We cannot apply Rule 4 of Section 3.1 since $f(x) = (\sqrt{x^2 + 16} - 4)/x^2$ is not defined for $x = 0$. (If we plug in 0, we get the expression $0/0$.) We use a trick that will allow us to find the limit: namely, we rationalize the numerator. For $x \neq 0$, we find

$$\frac{\sqrt{x^2 + 16} - 4}{x^2} = \frac{(\sqrt{x^2 + 16} - 4)}{x^2} \frac{(\sqrt{x^2 + 16} + 4)}{(\sqrt{x^2 + 16} + 4)}$$

$$= \frac{x^2 + 16 - 16}{x^2(\sqrt{x^2 + 16} + 4)} = \frac{x^2}{x^2(\sqrt{x^2 + 16} + 4)}$$

$$= \frac{1}{\sqrt{x^2 + 16} + 4}$$

Note that we are allowed to divide by x^2 in the last step since we assume that $x \neq 0$. We can now apply Rule 4 to $1/(\sqrt{x^2 + 16} + 4)$ and find

$$\lim_{x \to 0} \frac{\sqrt{x^2 + 16} - 4}{x^2} = \lim_{x \to 0} \frac{1}{\sqrt{x^2 + 16} + 4} = \frac{1}{8} = 0.125$$

as we saw in Example 7 of Section 3.1. In Chapter 5, we will learn another method for finding the limit of expressions of the form $0/0$. $\qquad \blacktriangleleft$

3.2.3 Problems

(3.2.1)

In Problems 1–4, show that each function is continuous at the given value.

1. $f(x) = x^3 - 2x + 1, c = 2$

2. $f(x) = \sqrt{x^2 + 1}, c = -1$

3. $f(x) = \sin(2x), c = \dfrac{\pi}{4}$

4. $f(x) = e^{-x}, c = 1$

5. Show that
$$f(x) = \begin{cases} \frac{x^2 - x - 2}{x - 2} & \text{if } x \neq 2 \\ 3 & \text{if } x = 2 \end{cases}$$
is continuous at $x = 2$.

6. Show that
$$f(x) = \begin{cases} \frac{2x^2 + x - 6}{x + 2} & \text{if } x \neq -2 \\ -7 & \text{if } x = -2 \end{cases}$$
is continuous at $x = -2$.

7. Let
$$f(x) = \begin{cases} \frac{x^2 - 9}{x - 3} & \text{if } x \neq 3 \\ a & \text{if } x = 3 \end{cases}$$
Which value must you assign to a so that $f(x)$ is continuous at $x = 3$?

8. Let
$$f(x) = \begin{cases} \frac{x^2 + x - 2}{x - 1} & \text{if } x \neq 1 \\ a & \text{if } x = 1 \end{cases}$$
Which value must you assign to a so that $f(x)$ is continuous at $x = 1$?

9. Show that
$$f(x) = \begin{cases} \frac{1}{x - 3} & \text{for } x \neq 3 \\ 0 & \text{for } x = 3 \end{cases}$$
is discontinuous at $x = 3$.

10. Show that
$$f(x) = \begin{cases} \frac{1}{x^2 - 1} & \text{for } x \neq -1, 1 \\ 0 & \text{for } x = -1 \text{ or } 1 \end{cases}$$
is discontinuous at $x = -1$ and $x = 1$.

11. Show that
$$f(x) = \begin{cases} \frac{x^2 - 3x + 2}{x - 2} & \text{if } x \neq 1 \\ 1 & \text{if } x = 1 \end{cases}$$
is discontinuous at $x = 1$.

12. Show that
$$f(x) = \begin{cases} x^2 - 1 & \text{for } x \leq 0 \\ x & \text{for } x > 0 \end{cases}$$
is discontinuous at $x = 0$.

13. Show that the floor function $f(x) = \lfloor x \rfloor$ is continuous at $x = 5/2$ but discontinuous at $x = 3$.

14. Show that the floor function $f(x) = \lfloor x \rfloor$ is continuous from the right at $x = 2$.

15. (a) Show that
$$f(x) = \sqrt{x - 1}, \quad x \geq 1$$
is continuous from the right at $x = 1$.
(b) Graph $f(x)$.
(c) Does it make sense to look at continuity from the left at $x = 1$?

16. (a) Show that
$$f(x) = \sqrt{x^2 - 4}, \quad |x| \geq 2$$
is continuous from the right at $x = 2$ and continuous from the left at $x = -2$.
(b) Graph $f(x)$.
(c) Does it make sense to look at continuity from the left at $x = 2$ and at continuity from the right at $x = -2$?

(3.2.2)

In Problems 17–26, find the values of $x \in \mathbf{R}$ for which the given functions are continuous.

17. $f(x) = 3x^4 - x^2 + 4$

18. $f(x) = \sqrt{x^2 - 1}$

19. $f(x) = \dfrac{x^2 + 1}{x - 1}$

20. $f(x) = \cos(2x)$

21. $f(x) = e^{-|x|}$

22. $f(x) = \ln(x - 2)$

23. $f(x) = \ln \dfrac{x}{x + 1}$

24. $f(x) = \exp[-\sqrt{x - 1}]$

25. $f(x) = \tan(2\pi x)$

26. $f(x) = \sin\left(\dfrac{2x}{3 + x}\right)$

27. Let
$$f(x) = \begin{cases} x^2 + 2 & \text{for } x \leq 0 \\ x + c & \text{for } x > 0 \end{cases}$$
(a) Graph $f(x)$ when $c = 1$, and determine whether $f(x)$ is continuous for this choice of c.
(b) How must you choose c so that $f(x)$ is continuous for all $x \in (-\infty, \infty)$?

28. Let
$$f(x) = \begin{cases} \dfrac{1}{x} & \text{for } x \geq 1 \\ 2x + c & \text{for } x < 1 \end{cases}$$
(a) Graph $f(x)$ when $c = 0$, and determine whether $f(x)$ is continuous for this choice of c.
(b) How must you choose c so that $f(x)$ is continuous for all $x \in (-\infty, \infty)$?

In Problems 29–48, find the limits.

29. $\displaystyle \lim_{x \to \pi/3} \sin\left(\dfrac{x}{2}\right)$

30. $\lim\limits_{x \to -\pi/2} \cos(2x)$

31. $\lim\limits_{x \to \pi/2} \dfrac{\cos^2 x}{1 - \sin^2 x}$

32. $\lim\limits_{x \to -\pi/2} \dfrac{1 + \tan^2 x}{\sec^2 x}$

33. $\lim\limits_{x \to 1} \sqrt{1 + 8x^4}$

34. $\lim\limits_{x \to -2} \sqrt{6 + x}$

35. $\lim\limits_{x \to -1} \sqrt{x^2 + 2x + 2}$

36. $\lim\limits_{x \to 1} \sqrt{x^3 + 4x - 1}$

37. $\lim\limits_{x \to 0} e^{-x^2/2}$

38. $\lim\limits_{x \to 0} e^{3x+1}$

39. $\lim\limits_{x \to 2} e^{x^2-4}$

40. $\lim\limits_{x \to -1} e^{x^2/2-1}$

41. $\lim\limits_{x \to 0} \dfrac{e^{2x} - 1}{e^x - 1}$

42. $\lim\limits_{x \to 0} \dfrac{e^{-x} - e^x}{e^{-x} + 1}$

43. $\lim\limits_{x \to -2} \dfrac{1}{\sqrt{5x^2 - 4}}$

44. $\lim\limits_{x \to 1} \dfrac{1}{\sqrt{3 - 2x^2}}$

45. $\lim\limits_{x \to 0} \dfrac{\sqrt{x^2 + 9} - 3}{x^2}$

46. $\lim\limits_{x \to 0} \dfrac{5 - \sqrt{25 + x^2}}{2x^2}$

47. $\lim\limits_{x \to 0} \ln(1 - x)$

48. $\lim\limits_{x \to 1} \ln(e^x + 1)$

3.3 LIMITS AT INFINITY

The limit laws discussed in Subsection 3.1.2 also hold as x tends to ∞ (or $-\infty$).

▶ **Example 1**   Find

$$\lim_{x \to \infty} \frac{x}{x + 1}$$

Solution

We set $f(x) = x$ and $g(x) = x + 1$. Obviously, neither $\lim_{x \to \infty} f(x)$ nor $\lim_{x \to \infty} g(x)$ exists. Thus, we cannot use Rule 4 from Section 3.1. But we can divide both numerator and denominator by x. We find

$$\lim_{x \to \infty} \frac{x}{x + 1} = \lim_{x \to \infty} \frac{1}{1 + \frac{1}{x}}$$

Since $\lim_{x \to \infty} 1 = 1$ and $\lim_{x \to \infty}(1 + \frac{1}{x}) = 1$, both limits exist. Furthermore, $\lim_{x \to \infty}(1 + \frac{1}{x}) \neq 0$. We can now apply Rule 4 of Section 3.1 after having done the algebraic manipulation and find

$$\lim_{x \to \infty} \frac{x}{x + 1} = \lim_{x \to \infty} \frac{1}{1 + \frac{1}{x}} = \frac{\lim_{x \to \infty} 1}{\lim_{x \to \infty} \left(1 + \frac{1}{x}\right)} = \frac{1}{1} = 1 \qquad ◀$$

In Example 1, we computed the limit of a rational function as x tended to infinity. Rational functions are ratios of polynomials. To find out how the limit of a rational function behaves as x tends to infinity, we will first compare the relative growth of functions of the form $y = x^n$: If $n > m$, then x^n dominates x^m for large x, in the sense that

$$\lim_{x \to \infty} \frac{x^n}{x^m} = \infty \quad \text{and} \quad \lim_{x \to \infty} \frac{x^m}{x^n} = 0$$

This follows immediately if we simplify the fractions,

$$\frac{x^n}{x^m} = x^{n-m} \quad \text{with } n - m > 0$$

and

$$\frac{x^m}{x^n} = \frac{1}{x^{n-m}} \quad \text{with } n - m > 0$$

This limiting behavior is important when we compute limits of rational functions as $x \to \infty$. We compare the following three limits:

(a) $\displaystyle\lim_{x \to \infty} \frac{x^2 + 2x - 1}{x^3 - 3x + 1}$

(b) $\displaystyle\lim_{x \to \infty} \frac{2x^3 - 4x + 7}{3x^3 + 7x^2 - 1}$

(c) $\displaystyle\lim_{x \to \infty} \frac{x^4 + 2x - 5}{x^2 - x + 2}$

To determine whether the numerator or the denominator dominates, we look at the leading term of the polynomials in the numerator and the denominator. (The leading term is the term with the largest exponent.) The leading term of a polynomial tells us how quickly a polynomial increases as x increases.

(a) The leading term in the numerator is x^2, and the leading term in the denominator is x^3. As $x \to \infty$, the denominator grows much faster than the numerator. We therefore expect the limit to be equal to 0. We can show this by dividing both numerator and denominator by the higher of the two powers, namely x^3. We find

$$\lim_{x \to \infty} \frac{x^2 + 2x - 1}{x^3 - 3x + 1} = \lim_{x \to \infty} \frac{\frac{1}{x} + \frac{2}{x^2} - \frac{1}{x^3}}{1 - \frac{3}{x^2} + \frac{1}{x^3}}$$

Since $\lim_{x \to \infty}(\frac{1}{x} + \frac{2}{x^2} - \frac{1}{x^3})$ exists (it is equal to 0), and $\lim_{x \to \infty}(1 - \frac{3}{x^2} + \frac{1}{x^3})$ exists and is not equal to 0 (it is equal to 1), we can apply Rule 4 and find

$$\lim_{x \to \infty} \frac{\frac{1}{x} + \frac{2}{x^2} - \frac{1}{x^3}}{1 - \frac{3}{x^2} + \frac{1}{x^3}} = \frac{\lim_{x \to \infty}\left(\frac{1}{x} + \frac{2}{x^2} - \frac{1}{x^3}\right)}{\lim_{x \to \infty}\left(1 - \frac{3}{x^2} + \frac{1}{x^3}\right)} = \frac{0}{1} = 0$$

(b) The leading term in both the numerator and the denominator is x^3. We divide both numerator and denominator by x^3 and obtain

$$\lim_{x \to \infty} \frac{2x^3 - 4x + 7}{3x^3 + 7x^2 - 1} = \lim_{x \to \infty} \frac{2 - \frac{4}{x^2} + \frac{7}{x^3}}{3 + \frac{7}{x} - \frac{1}{x^3}} = \frac{2}{3}$$

In the last step, we used the facts that the limits in both the numerator and the denominator exist and that the limit in the denominator is not equal to 0. Applying Rule 4 yields the limiting value. Note that the limiting value is equal to the ratio of the coefficients of the leading terms in the numerator and the denominator.

(c) The leading term in the numerator is x^4 and the leading term in the denominator is x^2. Since the leading term in the numerator grows much more quickly than the leading term in the denominator, we expect the limit to be undefined. This is indeed the case and can be seen if we divide the numerator by the denominator. We find

$$\lim_{x \to \infty} \frac{x^4 + 2x - 5}{x^2 - x + 2} = \lim_{x \to \infty} \left(x^2 + x - 1 - \frac{x + 3}{x^2 - x + 2}\right) \quad \text{does not exist}$$

It is often useful to determine whether the limit tends to $+\infty$ or $-\infty$. Since $x^2 + x - 1$ tends to $+\infty$ as $x \to \infty$ and the ratio $\frac{x+3}{x^2-x+2}$ tends to 0 as $x \to \infty$, the limit of $\frac{x^4+2x-5}{x^2-x+2}$ tends to $+\infty$ as $x \to +\infty$.

Let's summarize our findings. If $f(x)$ is a rational function of the form $f(x) = p(x)/q(x)$, where $p(x)$ is a polynomial of degree $\deg(p)$ and $q(x)$ is a polynomial of $\deg(q)$, we find that

$$\lim_{x\to\infty} f(x) = \lim_{x\to\infty} \frac{p(x)}{q(x)} = \begin{cases} 0 & \text{if } \deg(p) < \deg(q) \\ L \neq 0 & \text{if } \deg(p) = \deg(q) \\ \text{does not exist} & \text{if } \deg(p) > \deg(q) \end{cases}$$

Here L is a real number that is the ratio of the coefficients of the leading terms in the numerator and denominator. The same behavior holds as $x \to -\infty$.

▶ **Example 2**   Compute

(a) $\lim\limits_{x\to-\infty} \dfrac{1 - x + 2x^2}{3x - 5x^2}$

(b) $\lim\limits_{x\to\infty} \dfrac{1 - x^3}{1 + x^5}$

(c) $\lim\limits_{x\to\infty} \dfrac{2 - x^2}{1 + 2x}$

(d) $\lim\limits_{x\to-\infty} \dfrac{4 + 3x^2}{1 - 7x}$

Solution

(a) Since the degree of the numerator is equal to the degree of the denominator, we find

$$\lim_{x\to-\infty} \frac{1 - x + 2x^2}{3x - 5x^2} = \frac{2}{-5} = -\frac{2}{5}$$

(b) Since the degree of the numerator is less than the degree of the denominator, we find

$$\lim_{x\to\infty} \frac{1 - x^3}{1 + x^5} = 0$$

(c) Since the degree of the numerator is greater than the degree of the denominator, the limit does not exist. When x is very large, then the expression $\frac{2-x^2}{1+2x}$ behaves like $\frac{-x^2}{2x} = -\frac{x}{2}$, which tends to $-\infty$ as $x \to \infty$. Hence,

$$\lim_{x\to\infty} \frac{2 - x^2}{1 + 2x} = -\infty \qquad \text{(limit does not exist)}$$

(d) The degree of the numerator is greater than the degree of the denominator. We find

$$\lim_{x\to-\infty} \frac{4 + 3x^2}{1 - 7x} = \infty \qquad \text{(limit does not exist)}$$

since $\frac{4+3x^2}{1-7x}$ behaves like $\frac{3x^2}{-7x} = -\frac{3}{7}x$ for x large, which tends to $+\infty$ as $x \to -\infty$. ◀

Rational functions are not the only functions that involve limits as $x \to \infty$ (or $x \to -\infty$). Many important applications in biology involve exponential functions. We will use the following result repeatedly; it is one of the most important limits.

$$\lim_{x \to \infty} e^{-x} = 0$$

The graph of $f(x) = e^{-x}$ is given in Figure 3.15. Be certain to familiarize yourself with the basic shape of the function $f(x) = e^{-x}$ and its behavior as $x \to \infty$.

▶ Example 3 **Logistic Growth** The logistic curve describes the density of a population over time, where the rate of growth depends on the population size. We will discuss this function in more detail in coming chapters. It suffices here to say that the per capita rate of growth decreases with increasing population size. If $N(t)$ denotes the size of the population at time t, then the logistic curve is given by

$$N(t) = \frac{K}{1 + \left(\frac{K}{N(0)} - 1\right)e^{-rt}} \quad \text{for } t \geq 0$$

The parameters K and r are positive numbers that describe the population dynamics. You can check that $N(0)$ on the right-hand side is indeed the population size at time 0 [evaluate $N(t)$ at $t = 0$] and we assume that $N(0)$ is positive. The graph of $N(t)$ is shown in Figure 3.16. We will interpret K now; the interpretation of r must wait until the next chapter.

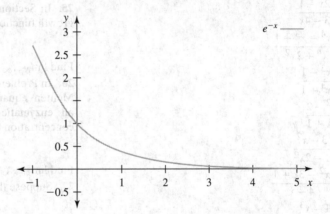

▲ **Figure 3.15**
The graph of $f(x) = e^{-x}$

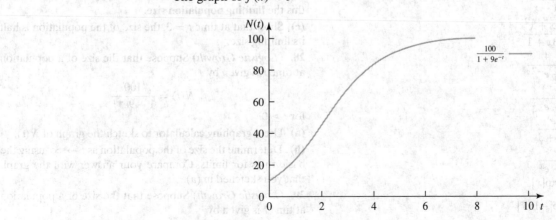

▲ **Figure 3.16**
The graph of the logistic curve with $K = 100$, $N_0 = 10$ and $r = 1$

If we are interested in the long-term behavior of the population evolving according to the logistic growth curve, we need to investigate what happens to $N(t)$ as $t \to \infty$. We find that

$$\lim_{t \to \infty} \frac{K}{1 + \left(\frac{K}{N(0)} - 1\right) e^{-rt}} = K$$

since $\lim_{t \to \infty} e^{-rt} = 0$ for $r > 0$. That is, as $t \to \infty$, the population size approaches K. This value is called the **carrying capacity** of the population. You will encounter logistic growth repeatedly in this text; it is one of the most fundamental equations for describing population growth.   ◀

3.3.1 Problems

Evaluate the limits in Problems 1–24.

1. $\lim_{x \to \infty} \dfrac{2x^2 - 3x + 5}{x^4 - 2x + 1}$

2. $\lim_{x \to \infty} \dfrac{x^2 + 3}{5x^2 - 2x + 1}$

3. $\lim_{x \to \infty} \dfrac{x^3 + 3}{x - 2}$

4. $\lim_{x \to -\infty} \dfrac{2x - 1}{3 - 4x}$

5. $\lim_{x \to \infty} \dfrac{3x^4 - x^3 + 1}{x^4 + 2x^2}$

6. $\lim_{x \to \infty} \dfrac{5x^3 - 1}{4x^4 + 1}$

7. $\lim_{x \to \infty} \dfrac{x^2 - 2}{2x + 1}$

8. $\lim_{x \to -\infty} \dfrac{3 - x^2}{1 - 2x^2}$

9. $\lim_{x \to -\infty} \dfrac{x^2 - 3x + 1}{4 - x}$

10. $\lim_{x \to -\infty} \dfrac{1 - x^3}{2 + x}$

11. $\lim_{x \to -\infty} \dfrac{2 + x^2}{1 - x}$

12. $\lim_{x \to -\infty} \dfrac{2x + x^2}{3x + 1}$

13. $\lim_{x \to \infty} \dfrac{4}{1 + e^{-2x}}$

14. $\lim_{x \to \infty} \dfrac{e^{-x}}{1 - e^{-x}}$

15. $\lim_{x \to \infty} \dfrac{2e^x}{e^x + 3}$

16. $\lim_{x \to \infty} \dfrac{e^x}{2 - e^x}$

17. $\lim_{x \to -\infty} \exp[x]$

18. $\lim_{x \to \infty} \exp[-\ln x]$

19. $\lim_{x \to \infty} e^{-x} \sin x$

20. $\lim_{x \to \infty} e^{-x} \cos x$

21. $\lim_{x \to \infty} \dfrac{3}{2 + e^{-x}}$

22. $\lim_{x \to -\infty} \dfrac{4}{1 + e^{-x}}$

23. $\lim_{x \to -\infty} \dfrac{e^x}{1 + x}$

24. $\lim_{x \to \infty} \dfrac{2}{e^x(1 + x)}$

25. In Section 1.2.3, Example 6, we introduced the Monod growth function

$$r(N) = a \frac{N}{k + N}, \quad N \geq 0$$

Find $\lim_{N \to \infty} r(N)$.

26. In Problem 86 of Section 1.3, we discussed the Michaelis-Menten equation, which describes the initial velocity of an enzymatic reaction (v_0) as a function of substrate concentration (s_0). The equation was given by

$$v_0 = \frac{v_{max} s_0}{s_0 + K_m}$$

Find $\lim_{s_0 \to \infty} v_0$.

27. Suppose the size of a population at time t is given by

$$N(t) = \frac{500t}{3 + t}, \quad t \geq 0$$

(a) Use a graphing calculator to sketch the graph of $N(t)$.

(b) Determine the size of the population as $t \to \infty$. We call this the **limiting population size**.

(c) Show that at time $t = 3$, the size of the population is half its limiting size.

28. (*Logistic Growth*) Suppose that the size of a population at time t is given by

$$N(t) = \frac{100}{1 + 9e^{-t}}$$

for $t \geq 0$.

(a) Use a graphing calculator to sketch the graph of $N(t)$.

(b) Determine the size of the population as $t \to \infty$, using the basic rules for limits. Compare your answer with the graph that you sketched in (a).

29. (*Logistic Growth*) Suppose that the size of a population at time t is given by

$$N(t) = \frac{50}{1 + 3e^{-t}}$$

for $t \geq 0$.

(a) Use a graphing calculator to sketch the graph of $N(t)$.

(b) Determine the size of the population as $t \to \infty$, using the basic rules for limits. Compare your answer with the graph that you sketched in (a).

30. **(a)** Use a graphing calculator to sketch the graph of

$$f(x) = e^{ax} \sin x, \quad x \geq 0$$

for $a = -0.1, -0.01, 0, 0.01$, and 0.1.

(b) Which part of the function $f(x)$ produces the oscillations that you see in the graphs sketched in (a)?

(c) Describe in words the effect that the value of a has on the shape of the graph of $f(x)$.

(d) Graph $f(x) = e^{ax} \sin x$, $g(x) = -e^{ax}$, and $h(x) = e^{ax}$ together in one coordinate system for (i) $a = 0.1$ and (ii) $a = -0.1$. [Use separate coordinate systems for (i) and (ii).] Explain what you see in each case. Show that

$$-e^{ax} \leq e^{ax} \sin x \leq e^{ax}$$

Use this to determine the values of a for which

$$\lim_{x \to \infty} f(x)$$

exists, and find the limiting value.

3.4 THE SANDWICH THEOREM AND SOME TRIGONOMETRIC LIMITS

The following theorem is both useful and intuitive. We will not prove it.

> **Sandwich Theorem** If $f(x) \leq g(x) \leq h(x)$ for all x in an open interval that contains c (except possibly at c) and
>
> $$\lim_{x \to c} f(x) = \lim_{x \to c} h(x) = L$$
>
> then
>
> $$\lim_{x \to c} g(x) = L$$

The theorem is called the sandwich theorem because we "sandwich" the function $g(x)$ between the two functions $f(x)$ and $h(x)$. Since $f(x)$ and $h(x)$ converge to the same value as $x \to c$, $g(x)$ also must converge to that value as $x \to c$, because it is squeezed in between $f(x)$ and $h(x)$. The sandwich theorem also applies to one-sided limits. We demonstrate how to use the sandwich theorem in the next example.

▶ **Example 1**   Show that

$$\lim_{x \to 0} x \sin \frac{1}{x} = 0$$

Solution

First, note that we cannot use Rule 3, which says that the limit of a product is equal to the product of the limits because it requires that the limits of both factors exist. The limit of $\sin(1/x)$ as $x \to 0$ does not exist. It diverges by oscillations (see Example 6 of Section 3.1 for a similar limit). However, we know that

$$-1 \leq \sin \frac{1}{x} \leq 1$$

for all $x \neq 0$. To go from this set of inequalities to one that involves $x \sin \frac{1}{x}$, we need to multiply all three parts by x. Since multiplying an inequality by x reverses inequality signs when $x < 0$, we need to split the discussion into two cases, one involving $x > 0$, the other $x < 0$.

Multiplying all three parts by $x > 0$, we find

$$-x \leq x \sin \frac{1}{x} \leq x$$

Since $\lim_{x \to 0^+}(-x) = \lim_{x \to 0^+} x = 0$, we can apply the sandwich theorem and find

$$\lim_{x \to 0^+} x \sin \frac{1}{x} = 0$$

We can repeat the same steps when we multiply by $x < 0$, except we now need to reverse the inequality signs. That is, for $x < 0$,

$$-x \geq x \sin \frac{1}{x} \geq x$$

Since $\lim_{x \to 0^-}(-x) = \lim_{x \to 0^-} x = 0$, we can again apply the sandwich theorem and find

$$\lim_{x \to 0^-} x \sin \frac{1}{x} = 0$$

Combining the two results, we find

$$\lim_{x \to 0} x \sin \frac{1}{x} = 0$$

This limit is illustrated in Figure 3.17. ◀

The sandwich theorem can also be used for limits at infinity.

▶ **Example 2**   Compute

$$\lim_{x \to \infty} e^{-x} \sin x$$

Solution

Since

$$-1 \leq \sin x \leq 1$$

and $e^{-x} > 0$, we find (see Figure 3.18 for $0 \leq x \leq 8$ and Figure 3.19 for $6 \leq x \leq 15$; note different scales on the y-axes of the two graphs)

$$-e^{-x} \leq e^{-x} \sin x \leq e^{-x}$$

Now,

$$\lim_{x \to \infty} (-e^{-x}) = 0 \quad \text{and} \quad \lim_{x \to \infty} e^{-x} = 0$$

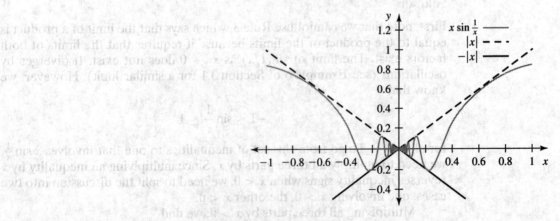

▲ **Figure 3.17**
The Sandwich Theorem illustrated on $\lim_{x \to 0} x \sin(1/x)$

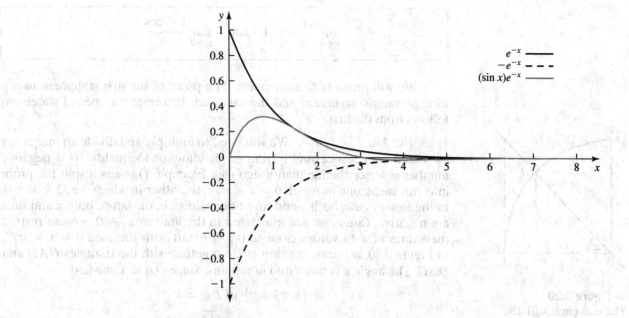

▲ **Figure 3.18**

The graphs of $y = e^{-x}$, $y = -e^{-x}$, and $y = (\sin x)e^{-x}$ from $x = 0$ to $x = 8$

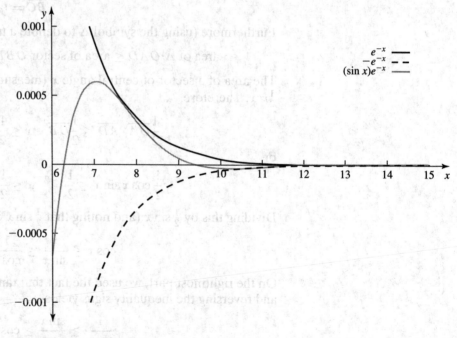

▲ **Figure 3.19**

The graphs of $y = e^{-x}$, $y = -e^{-x}$, and $y = (\sin x)e^{-x}$ from $x = 6$ to $x = 15$

Hence,

$$\lim_{x \to \infty} e^{-x} \sin x = 0 \qquad \blacktriangleleft$$

There are two trigonometric limits that are important for developing the differential calculus for trigonometric functions; you should memorize them. Note that the angle x is measured in radians.

$$\lim_{x \to 0} \frac{\sin x}{x} = 1 \quad \text{and} \quad \lim_{x \to 0} \frac{1 - \cos x}{x} = 0$$

We will prove both statements. The proof of the first statement uses a nice geometric argument and the sandwich theorem; the second statement follows from the first.

Proof that $\lim_{x \to 0} \frac{\sin x}{x} = 1$ We will need to multiply and divide an inequality by x and $\sin x$. Since multiplying or dividing an inequality by a negative number reverses the inequality sign (see Example 1), we will split the proof into two cases, one in which $0 < x < \pi/2$, the other in which $-\pi/2 < x < 0$. In the former case, both x and $\sin x$ are positive; in the latter, both x and $\sin x$ are negative. (Since we are interested in the limit as $x \to 0$, we can restrict the values of x to values close to 0.) We start with the case $0 < x < \pi/2$. In Figure 3.20, we draw the unit circle together with the triangles OAD and OBC. The angle x is measured in radians. Since $\overline{OB} = 1$, we find

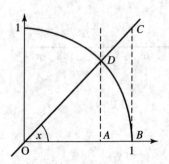

▲ **Figure 3.20**
The unit circle with the triangles OAD and OBC

$$\text{arc length of } BD = x$$

$$\overline{OA} = \cos x$$

$$\overline{AD} = \sin x$$

$$\overline{BC} = \tan x$$

Furthermore (using the symbol $\triangle$ to denote a triangle),

$$\text{area of } \triangle OAD \le \text{area of sector } OBD \le \text{area of } \triangle OBC$$

The area of a sector of central angle x (measured in radians) and radius r is $\frac{1}{2}r^2 x$. Therefore,

$$\frac{1}{2}\overline{OA} \cdot \overline{AD} \le \frac{1}{2}\overline{OB}^2 \cdot x \le \frac{1}{2}\overline{OB} \cdot \overline{BC}$$

or

$$\frac{1}{2}\cos x \sin x \le \frac{1}{2} \cdot 1^2 \cdot x \le \frac{1}{2} \cdot 1 \cdot \tan x$$

Dividing this by $\frac{1}{2}\sin x$ (and noting that $\frac{1}{2}\sin x > 0$ for $0 < x < \pi/2$) yields

$$\cos x \le \frac{x}{\sin x} \le \frac{1}{\cos x}$$

On the rightmost part, we used the fact that $\tan x = \frac{\sin x}{\cos x}$. Taking reciprocals and reversing the inequality signs yields

$$\frac{1}{\cos x} \ge \frac{\sin x}{x} \ge \cos x$$

We can now take the limit as $x \to 0^+$. (Remember, we assumed that $0 < x < \pi/2$, so we can only approach 0 from the right.) Note that

$$\lim_{x \to 0^+} \cos x = 1 \quad \text{and} \quad \lim_{x \to 0^+} \frac{1}{\cos x} = \frac{1}{\lim_{x \to 0^+} \cos x} = 1$$

We now apply the sandwich theorem and find

$$\lim_{x \to 0^+} \frac{\sin x}{x} = 1$$

We only showed that $\lim_{x\to 0^+} \frac{\sin x}{x} = 1$, but a similar argument can be carried out when $-\frac{\pi}{2} < x < 0$. In this case, $\lim_{x\to 0^-} \frac{\sin x}{x} = 1$. The left-hand and the right-hand limits are the same and we conclude that

$$\lim_{x\to 0} \frac{\sin x}{x} = 1$$ ■

Proof that $\lim_{x\to 0} \frac{1-\cos x}{x} = 0$ Multiplying both numerator and denominator of $f(x) = (1 - \cos x)/x$ by $1 + \cos x$, we can reduce the second statement to the first statement.

$$\lim_{x\to 0} \frac{1 - \cos x}{x} = \lim_{x\to 0} \frac{1 - \cos x}{x} \frac{1 + \cos x}{1 + \cos x}$$

$$= \lim_{x\to 0} \frac{1 - \cos^2 x}{x(1 + \cos x)}$$

Using the identity $\sin^2 x + \cos^2 x = 1$, we can write this as

$$\lim_{x\to 0} \frac{\sin^2 x}{x(1 + \cos x)}$$

Rewriting this now as

$$\lim_{x\to 0} \frac{\sin x}{x} \frac{\sin x}{1 + \cos x}$$

we can determine the limit. Note that $\lim_{x\to 0} \frac{\sin x}{x}$ exists by the first statement, and $\lim_{x\to 0} \frac{\sin x}{1+\cos x}$ exists because $1 + \cos x \neq 0$ for x close to 0. Applying Rule 3 of the limit laws, it follows that the limit of the product is the product of the limits. We therefore find that

$$\lim_{x\to 0} \frac{\sin x}{x} \frac{\sin x}{1 + \cos x} = \lim_{x\to 0} \frac{\sin x}{x} \lim_{x\to 0} \frac{\sin x}{1 + \cos x} = 1 \cdot 0 = 0$$ ■

▶ **Example 3** Find the following limits.

(a) $\lim_{x\to 0} \dfrac{\sin 3x}{5x}$

(b) $\lim_{x\to 0} \dfrac{\sin^2 x}{x^2}$

(c) $\lim_{x\to 0} \dfrac{\sec x - 1}{x \sec x}$

Solution

(a) We cannot apply the first trigonometric limit directly. The trick is to substitute $z = 3x$ and observe that $z \to 0$ as $x \to 0$. Then

$$\lim_{x\to 0} \frac{\sin 3x}{5x} = \lim_{z\to 0} \frac{\sin z}{5z/3} = \frac{3}{5} \lim_{z\to 0} \frac{\sin z}{z} = \frac{3}{5}$$

(b) We find

$$\lim_{x\to 0} \frac{\sin^2 x}{x^2} = \lim_{x\to 0} \left(\frac{\sin x}{x}\right)^2 = \left(\lim_{x\to 0} \frac{\sin x}{x}\right)^2 = 1$$

Here, we used the fact that the limit of a product is the product of the limits, provided that the individual limits exist.

(c) We first write $\sec x = 1/\cos x$. Then we multiply both numerator and denominator by $\cos x$. This yields

$$\lim_{x\to 0} \frac{\sec x - 1}{x \sec x} = \lim_{x\to 0} \frac{\frac{1}{\cos x} - 1}{\frac{x}{\cos x}}$$

$$= \lim_{x\to 0} \frac{\left(\frac{1}{\cos x} - 1\right)\cos x}{\frac{x}{\cos x}\cos x} = \lim_{x\to 0} \frac{1 - \cos x}{x} = 0$$

◀

3.4.1 Problems

1. Let

$$f(x) = x^2 \cos\frac{1}{x}, \quad x \neq 0$$

(a) Use a graphing calculator to sketch the graph of $y = f(x)$.

(b) Show that for $x \neq 0$,

$$-x^2 \leq x^2 \cos\frac{1}{x} \leq x^2$$

holds.

(c) Use your result in (b) and the sandwich theorem to show that

$$\lim_{x\to 0} x^2 \cos\frac{1}{x} = 0$$

2. Let

$$f(x) = x^2 \sin\frac{1}{x}, \quad x \neq 0$$

(a) Use a graphing calculator to sketch the graph of $y = f(x)$.

(b) Show that for $x \neq 0$,

$$-x^2 \leq x^2 \sin\frac{1}{x} \leq x^2$$

holds.

(c) Use your result in (b) and the sandwich theorem to show that

$$\lim_{x\to 0} x^2 \sin\frac{1}{x} = 0$$

3. Let

$$f(x) = \frac{\ln x}{x}, \quad x > 0$$

(a) Use a graphing calculator to graph $y = f(x)$.

(b) Use a graphing calculator to investigate the values of x for which

$$\frac{1}{x} \leq \frac{\ln x}{x} \leq \frac{1}{\sqrt{x}}$$

holds.

(c) Use your result in (b) to explain why the following is true:

$$\lim_{x\to\infty} \frac{\ln x}{x} = 0$$

4. Let

$$f(x) = \frac{\sin x}{x}, \quad x > 0$$

(a) Use a graphing calculator to graph $y = f(x)$.

(b) Explain why you cannot use the basic rules for finding limits to compute

$$\lim_{x\to\infty} \frac{\sin x}{x}$$

(c) Show that for $x > 0$,

$$-\frac{1}{x} \leq \frac{\sin x}{x} \leq \frac{1}{x}$$

holds, and use the sandwich theorem to compute

$$\lim_{x\to\infty} \frac{\sin x}{x}$$

In Problems 5–18, evaluate the trigonometric limits.

5. $\displaystyle\lim_{x\to 0} \frac{\sin(2x)}{2x}$

6. $\displaystyle\lim_{x\to 0} \frac{\sin(2x)}{3x}$

7. $\displaystyle\lim_{x\to 0} \frac{\sin(5x)}{x}$

8. $\displaystyle\lim_{x\to 0} \frac{\sin x}{-x}$

9. $\displaystyle\lim_{x\to 0} \frac{\sin(\pi x)}{x}$

10. $\displaystyle\lim_{x\to 0} \frac{\sin(-\pi x/2)}{2x}$

11. $\displaystyle\lim_{x\to 0} \frac{\sin(\pi x)}{\sqrt{x}}$

12. $\displaystyle\lim_{x\to 0} \frac{\sin^2 x}{x}$

13. $\displaystyle\lim_{x\to 0} \frac{\sin x \cos x}{x(1-x)}$

14. $\displaystyle\lim_{x\to 0} \frac{1 - \cos^2 x}{x^2}$

15. $\displaystyle\lim_{x\to 0} \frac{1 - \cos x}{2x}$

16. $\displaystyle\lim_{x\to 0} \frac{1 - \cos(2x)}{3x}$

17. $\displaystyle\lim_{x\to 0} \frac{\sin x(1 - \cos x)}{x^2}$

18. $\displaystyle\lim_{x\to 0} \frac{\csc x - \cot x}{x \csc x}$

3.5 PROPERTIES OF CONTINUOUS FUNCTIONS

3.5.1 The Intermediate Value Theorem

As you hike up a mountain, the temperature decreases with increasing elevation. Suppose the temperature at the bottom of the mountain is 70°F and the temperature at the top of the mountain is 40°F. How do you know that at some time during your hike you must have crossed a point where the temperature was exactly 50°F? Your answer will probably be something like the following: "To go from 70°F to 40°F, I must have passed through 50°F, since 50°F is between 40°F and 70°F and the temperature changed continuously as I hiked up the mountain." This represents the content of the intermediate value theorem.

> **The Intermediate Value Theorem** Suppose that f is continuous on the closed interval $[a, b]$. If L is any real number with $f(a) < L < f(b)$ or $f(b) < L < f(a)$, then there exists at least one number c on the open interval (a, b) such that $f(c) = L$.

We will not prove this theorem, but Figure 3.21 should convince you that this result is true. In Figure 3.21, f is continuous and defined on the closed interval $[a, b]$ with $f(a) < L < f(b)$. Therefore, the graph of $f(x)$ must intersect the line $y = L$ at least once in the open interval (a, b).

▶ **Example 1** Let

$$f(x) = 3 + \sin x \quad \text{for } 0 \le x \le \frac{3\pi}{2}$$

Show that there exists at least one point c in $(0, 3\pi/2)$ such that $f(c) = 5/2$.

Solution

The graph of $f(x)$ is shown in Figure 3.22. First, note that $f(x)$ is defined on a closed interval and is continuous on $[0, 3\pi/2]$. Furthermore, we find that

$$f(0) = 3 + \sin 0 = 3 + 0 = 3$$

$$f\left(\frac{3\pi}{2}\right) = 3 + \sin\frac{3\pi}{2} = 3 + (-1) = 2$$

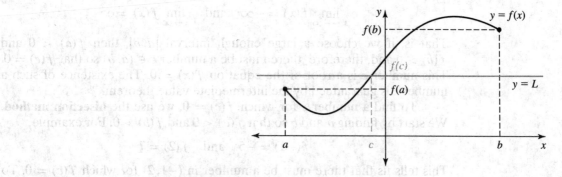

▲ **Figure 3.21**
The intermediate value theorem

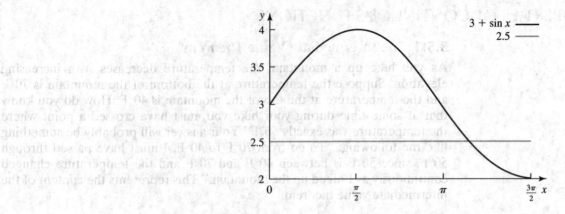

▲ **Figure 3.22**

The intermediate value theorem for $f(x) = 3 + \sin x$, $0 \leq x \leq 3\pi/2$ and $L = 2.5$

Given that

$$2 < \frac{5}{2} < 3$$

we conclude from the intermediate value theorem that there exists a number c such that $f(c) = 5/2$. The intermediate value theorem does not tell us where c is, or whether there is more than one such number.   ◀

When applying the intermediate value theorem, it is important to check that f is continuous. Discontinuous functions can easily miss values; for example, the floor function in Example 3 of Section 3.2 misses all numbers that are not integers.

As mentioned in the previous example, the intermediate value theorem only gives us the existence of a number c, it does not tell us how many such points there are or where they are located.

You might wonder how such a result can be of any use. One important application is that the theorem can be used to find approximate roots (or solutions) of equations of the form $f(x) = 0$. We show how in the following example.

▶ **Example 2**   Find a root of the equation $x^5 - 7x^2 + 3 = 0$.

Solution

Let $f(x) = x^5 - 7x^2 + 3 = 0$. Because $f(x)$ is a polynomial, it is continuous for all $x \in \mathbf{R}$. Furthermore,

$$\lim_{x \to -\infty} f(x) = -\infty \quad \text{and} \quad \lim_{x \to \infty} f(x) = \infty$$

That is, if we choose a large enough interval $[a, b]$, then $f(a) < 0$ and $f(b) > 0$ and, therefore, there must be a number $c \in (a, b)$ so that $f(c) = 0$. This number c is a root of the equation $f(x) = 0$. The existence of such a number c is guaranteed by the intermediate value theorem.

To find a number c for which $f(c) = 0$, we use the **bisection method**. We start by finding a and b so that $f(a) < 0$ and $f(b) > 0$. For example,

$$f(-1) = -5 \quad \text{and} \quad f(2) = 7$$

This tells us that there must be a number in $(-1, 2)$ for which $f(c) = 0$. To locate this root with more precision, we take the midpoint of $(-1, 2)$, which is 0.5, and evaluate the function at $x = 0.5$. [The midpoint of the interval (a, b)

is $(a + b)/2$.] Now, $f(0.5) \approx 1.28$ (rounded to two decimals). We thus have

$$f(-1) = -5 \quad f(0.5) \approx 1.28 \quad f(2) = 7$$

Using the intermediate value theorem again, we can now guarantee a root in $(-1, 0.5)$, since $f(-1) < 0$ and $f(0.5) > 0$. Bisecting the new interval and computing the respective values of $f(x)$, we find

$$f(-1) = -5 \quad f(-0.25) \approx 2.562 \quad f(0.5) \approx 1.28$$

Using the intermediate value theorem again, we can guarantee a root in $(-1, -0.25)$, since $f(-1) < 0$ and $f(-0.25) > 0$. Repeating this procedure of bisecting and selecting a new (smaller) interval will eventually produce an interval that is small enough so that we can locate the root to any desired accuracy. The first several steps are summarized in Table 3.1.

After nine steps, we find that there exists a root in

$$(-0.6484375, -0.642578125)$$

The length of this interval is 0.005859375, which gives us a solution that is accurate to two decimal places. If we are satisfied with this precision, we can stop here and choose, for instance, the midpoint of the last interval as an approximate value for a root of the equation $x^5 - 7x^2 + 3 = 0$. The midpoint is

$$\frac{-0.642578125 + (-0.6484375)}{2} = -0.6455078125$$

$$\approx -0.646$$

after rounding to three decimals.

Note that the length of the interval decreases by a factor of $1/2$ each step. That is, after nine steps, the length of the interval is $(1/2)^9$ of the length of the original interval. In our case, the length of the original interval was 3 and hence the length of the interval after nine steps is

$$3 \cdot \left(\frac{1}{2}\right)^9 = \frac{3}{512} = 0.005859375$$

as we saw. The bisection method is fairly slow when we need high accuracy. For instance, to reduce the length of the interval to 10^{-6}, we would need at least 22 steps since

$$3 \cdot \left(\frac{1}{2}\right)^{21} > 10^{-6} > 3 \cdot \left(\frac{1}{2}\right)^{22}$$

In Section 5.7, we will learn a faster method.

TABLE 3.1 BISECTION METHOD

a	$\frac{a+b}{2}$	b	$f(a)$	$f\left(\frac{a+b}{2}\right)$	$f(b)$
-1	0.5	2	-5	1.28	7
-1	-0.25	0.5	-5	2.562	1.28
-1	-0.625	-0.25	-5	0.170	2.562
-1	-0.8125	-0.625	-5	-1.975	0.170
-0.8125	-0.71875	-0.625	-1.975	-0.808	0.170
-0.71875	-0.671875	-0.625	-0.808	-0.297	0.170
-0.671875	-0.6484375	-0.625	-0.297	-0.0579	0.170
-0.6484375	-0.63671875	-0.625	-0.0579	0.0575	0.170
-0.6484375	-0.642578125	-0.63671875	-0.0579	9.9×10^{-5}	0.0575
-0.6484375		-0.642578125			

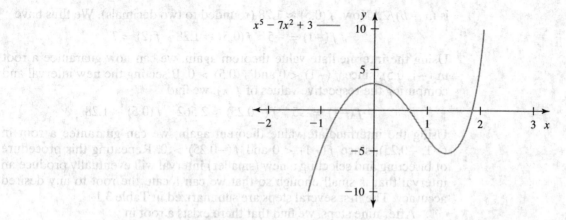

▲ **Figure 3.23**
The graph of $f(x) = x^5 - 7x^2 + 3$

Figure 3.23 shows the graph of $f(x) = x^5 - 7x^2 + 3$. We see that the graph intersects the x-axis three times. We found an approximation of the leftmost root of the equation $x^5 - 7x^2 + 3$. If we had used another starting interval, say $[1, 2]$, we would have located an approximation of the rightmost root of the equation.   ◄

3.5.2 A Final Remark on Continuous Functions

Many functions in biology are in fact discontinuous. For example, if we measure the size of a population over time, then this quantity takes on discrete values only (namely, nonnegative integers) and therefore changes discontinuously. However, if the population size is sufficiently large, an increase or decrease by one changes the population size so slightly that it might be justified to approximate the population size by a continuous function. For example, if we measure the number of bacteria in a petri dish in millions, then the number 2.1 would correspond to 2,100,000 bacteria. An increase by one results in 2,100,001 bacteria, or if we measure the size in millions, in 2.100001, an increase of 10^{-6}.

3.5.3 Problems

(3.5.1), (3.5.2)

1. Let
$$f(x) = x^2 - 1, \quad 0 \le x \le 2$$
(a) Graph $y = f(x)$ for $0 \le x \le 2$.

(b) Show that
$$f(0) < 0 < f(2)$$
and use the intermediate value theorem to conclude that there exists a number $c \in (0, 2)$ so that $f(c) = 0$.

2. Let
$$f(x) = x^3 - 2x + 3, \quad -3 \le x \le -1$$
(a) Graph $y = f(x)$ for $-3 \le x \le -1$.

(b) Use the intermediate value theorem to conclude that
$$x^3 - 2x + 3 = 0$$
has a solution in $(-3, -1)$.

3. Let
$$f(x) = \sqrt{x^2 + 2}, \quad 1 \le x \le 2$$
(a) Graph $y = f(x)$ for $1 \le x \le 2$.

(b) Use the intermediate value theorem to conclude that
$$\sqrt{x^2 + 2} = 2$$
has a solution in $(1, 2)$.

4. Let
$$f(x) = \sin x - x, \quad -1 \le x \le 1$$
(a) Graph $y = f(x)$ for $-1 \le x \le 1$.

(b) Use the intermediate value theorem to conclude that
$$\sin x = x$$
has a solution in $(-1, 1)$.

5. Use the intermediate value theorem to show that
$$e^{-x} = x$$
has a solution in $(0, 1)$.

6. Use the intermediate value theorem to show that

$$\cos x = x$$

has a solution in $(0, 1)$.

7. Use the bisection method to find a solution of

$$e^{-x} = x$$

that is accurate to two decimal places.

8. Use the bisection method to find a solution of

$$\cos x = x$$

that is accurate to two decimal places.

9. (a) Use the bisection method to find a solution of $3x^3 - 4x^2 - x + 2 = 0$ that is accurate to two decimal places.

(b) Graph the function $f(x) = 3x^3 - 4x^2 - x + 2$.

(c) Which solution did you locate in (a)? Is it possible in this case to find the other solution using the bisection method together with the intermediate value theorem?

10. In Example 2, how many steps are required to guarantee that the approximate root is within 0.0001 of the true value of the root?

11. Suppose that the number of individuals of a population at time t is given by

$$N(t) = \frac{54t}{13 + t}, \quad t \geq 0 \tag{3.7}$$

(a) Use a calculator to confirm that $N(10)$ is approximately 23.47826. Considering that the number of individuals in a population is an integer, how should you report your answer?

(b) Now suppose that $N(t)$ is given by the same function (3.7), but that the size of the population is measured in millions. How should you report the population size at time $t = 10$? Make some reasonable assumptions about the accuracy of a measurement for the size of such a large population.

(c) Discuss the use of continuous functions in both (a) and (b).

12. Explain why a polynomial of degree 3 has at least one root.

13. Explain why $y = x^2 - 4$ has at least two roots.

14. Explain why a polynomial of odd degree has at least one root.

3.6 A FORMAL DEFINITION OF LIMITS (OPTIONAL)

The ancient Greeks used limiting procedures to compute areas, such as the area of a circle, by using the "method of exhaustion." There, a region was covered (or "exhausted") as closely as possible by triangles. Adding the areas of the triangles then yielded an approximation of the area of the region of interest. Newton and Leibniz, the inventors of calculus, were very much aware of the importance of taking limits in their development of calculus; however, they did not give a rigorous definition of the procedure. The French mathematician Augustine-Louis Cauchy (1789–1857) was the first to develop a rigorous definition of limits; the definition we will use goes back to the German mathematician Karl Weierstrass (1815–1897).

Before we write the formal definition, let's return to the informal definition. There we stated that $\lim_{x \to c} f(x) = L$ means that the value of $f(x)$ can be made arbitrarily close to L whenever x is sufficiently close to c. But just how close is sufficient? Take Example 1 from Section 3.1: Suppose we wish to show that

$$\lim_{x \to 2} x^2 = 4$$

without using continuity of $y = x^2$, which itself was based on $\lim_{x \to x} x = c$ [Equation (3.3)]. What would we have to do? We would need to show that x^2 can be made arbitrarily close to 4 for all values of x sufficiently close to 2 but not equal to 2. (In the following, we will always exclude $x = 2$ from the discussion since the value of x^2 at $x = 2$ is irrelevant for finding the limit.) Suppose we wish to make x^2 within 0.01 of 4, that is, we want $|x^2 - 4| < 0.01$. Does this hold for all x sufficiently close to 2 but not equal to 2? We begin with

$$|x^2 - 4| < 0.01$$

This is equivalent to

$$-0.01 < x^2 - 4 < 0.01$$

$$3.99 < x^2 < 4.01$$

$$\sqrt{3.99} < |x| < \sqrt{4.01}$$

Now $\sqrt{3.99} = 1.997498\ldots$, and $\sqrt{4.01} = 2.002498\ldots$. We therefore find that values of $x \neq 2$ in the interval $(1.998, 2.002)$ satisfy $|x^2 - 4| < 0.01$. (We chose a somewhat smaller interval than indicated, to get an interval that is symmetric about 2.) That is, for all values of x within 0.002 of 2 but not equal to 2 (i.e., $0 < |x - 2| < 0.002$), x^2 is within the prescribed precision, namely within 0.01 of 4.

You might think about this example in the following way: Suppose that you wish to stake out a square of area 4 m². Each side of your square is 2 m long. You bring along a stick which you cut to a length of 2 m. We can then ask how accurately do we need to cut the stick so that the area will be within a prescribed precision. Our prescribed precision was 0.01, and we found that if we cut the stick within 0.002 of 2 m, we would be able to obtain the prescribed precision.

There is nothing special about 0.01; we could have chosen any other degree of precision and would have found a corresponding interval of x values. We translate this into a formal definition for limits (see Figure 3.24).

> **Definition** The statement
>
> $$\lim_{x \to c} f(x) = L$$
>
> means that for every $\epsilon > 0$ there exists a number $\delta > 0$ so that
>
> $$|f(x) - L| < \epsilon \qquad \text{whenever} \quad 0 < |x - c| < \delta$$

Note that, as in the informal definition of limits, we exclude the value $x = c$ from the statement. (This is done in the inequality $0 < |x - c|$.) To apply this definition, we first need to guess the limiting value L. We then choose an $\epsilon > 0$, the prescribed precision, and try to find a $\delta > 0$ so that $f(x)$ is

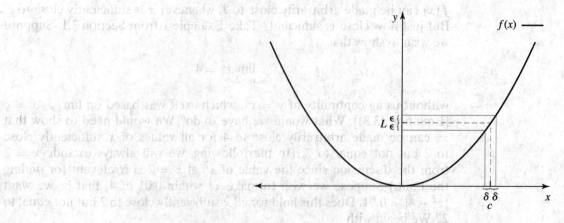

▲ **Figure 3.24**
The ϵ-δ definition of limits

within ϵ of L whenever x is within δ of c but not equal to c. [In our example, $f(x) = x^2, c = 2, L = 4, \epsilon = 0.01$, and $\delta = 0.002$.]

▶ **Example 1**   Show that

$$\lim_{x \to 1} (2x - 3) = -1$$

Solution

We let $f(x) = 2x - 3$. Our guess for the limiting value is $L = -1$. Then

$$|f(x) - L| = |2x - 3 - (-1)|$$
$$= |2x - 2|$$
$$= 2|x - 1|$$

We now choose $\epsilon > 0$ (ϵ is arbitrary, and we do not specify it because our statement needs to hold for all $\epsilon > 0$). Our goal is to find $\delta > 0$ so that $2|x - 1| < \epsilon$ whenever x is within δ of 1 but not equal to 1; that is, $0 < |x - 1| < \delta$. The value of δ will typically depend on our choice of ϵ. Since $|x - 1| < \delta$ implies $2|x - 1| < 2\delta$, this suggests that we should try $2\delta = \epsilon$. If we choose $\delta = \epsilon/2$, then indeed

$$|f(x) - L| = 2|x - 1| < 2\delta = 2\frac{\epsilon}{2} = \epsilon$$

This means that for every $\epsilon > 0$, we can find a number $\delta > 0$ (namely $\delta = \epsilon/2$), so that

$$|f(x) - (-1)| < \epsilon \quad \text{whenever} \quad 0 < |x - 1| < \delta$$

But this is exactly the definition of

$$\lim_{x \to 1} (2x - 3) = -1 \qquad \blacktriangleleft$$

▶ **Example 2**   We promised in Section 3.2 that we would show

$$\lim_{x \to c} x = c$$

Solution

Let $f(x) = x$. We need to show that for every $\epsilon > 0$ there corresponds a number $\delta > 0$ so that

$$|x - c| < \epsilon \quad \text{whenever} \quad 0 < |x - c| < \delta \qquad (3.8)$$

This immediately suggests that we should choose $\delta = \epsilon$, and indeed if $\delta = \epsilon$, then (3.8) holds. $\qquad \blacktriangleleft$

Let's look at an example where $f(x)$ is not linear.

▶ **Example 3**   Use the formal definition of limits to show that

$$\lim_{x \to 4} \sqrt{x} = 2$$

Solution

We assume throughout that $x > 0$. We need to show that for every $\epsilon > 0$ there corresponds a number $\delta > 0$ so that

$$|\sqrt{x} - 2| < \epsilon \quad \text{whenever} \quad 0 < |x - 4| < \delta \qquad (3.9)$$

Now $|\sqrt{x} - 2| < \epsilon$ is equivalent to

$$-\epsilon < \sqrt{x} - 2 < \epsilon$$
$$2 - \epsilon < \sqrt{x} < 2 + \epsilon$$

To square all sides, we need to make sure that $2 - \epsilon \geq 0$ and $2 + \epsilon \geq 0$. Since $\epsilon > 0$ by assumption, $2 - \epsilon \geq 0$ requires us to assume $0 < \epsilon \leq 2$. (The expression $2 + \epsilon$ is positive as long as $\epsilon > 0$.) With $0 < \epsilon \leq 2$, squaring yields

$$(2 - \epsilon)^2 < x < (2 + \epsilon)^2$$

$$4 - 4\epsilon + \epsilon^2 < x < 4 + 4\epsilon + \epsilon^2$$

Since $4 - 4\epsilon + \epsilon^2 > 4 - 4\epsilon - \epsilon^2$, we have

$$4 - (4\epsilon + \epsilon^2) < x < 4 + (4\epsilon + \epsilon^2)$$

or

$$|x - 4| < 4\epsilon + \epsilon^2$$

This suggests that if we set $\delta = 4\epsilon + \epsilon^2$ when $0 < \epsilon \leq 2$, then $0 < |x - 4| < \delta$ implies $|\sqrt{x} - 2| < \epsilon$. This is indeed the case, namely if $0 < |x - 4| < 4\epsilon + \epsilon^2$, then $-4\epsilon - \epsilon^2 < x - 4 < 4\epsilon + \epsilon^2$, or $4 - 4\epsilon - \epsilon^2 < x < 4 + 4\epsilon + \epsilon^2$. This shows that $(2 - \epsilon)^2 < x < (2 + \epsilon)^2$. Taking square roots (and observing that $x > 0$), $2 - \epsilon < \sqrt{x} < 2 + \epsilon$ or $-\epsilon < \sqrt{x} - 2 < \epsilon$, which is the same as $|\sqrt{x} - 2| < \epsilon$.

When $\epsilon > 2$, we can choose $\delta = 12$ (or any other smaller value). The value $\delta = 12$ comes from plugging $\epsilon = 2$ into $\delta = 4\epsilon + \epsilon^2$. With $\delta = 12$, if $0 < |x - 4| < 12$, then $-12 < x - 4 < 12$ or $-8 < x < 16$. Since $x > 0$, we have $0 < x < 16$, or after taking square roots, $0 < \sqrt{x} < 4$, which is the same as $-2 < \sqrt{x} - 2 < 2$ or $|\sqrt{x} - 2| < 2 < \epsilon$. ◀

We can also use the definition to show that a limit does not exist.

▶ **Example 4**   Show that

$$\lim_{x \to 0} \frac{|x|}{x}$$

does not exist.

Solution   This is trickier (see Figure 3.25). Set $f(x) = |x|/x$, $x \neq 0$. The approach is as follows: We assume the limit exists and then try to find a contradiction.

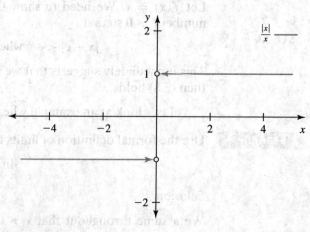

▲ **Figure 3.25**
The graph of $f(x) = \frac{|x|}{x}$ in Example 4: The limit of $\frac{|x|}{x}$ as x tends to 0 does not exist

Suppose there exists an L so that

$$\lim_{x \to 0} \frac{|x|}{x} = L$$

If we look at Figure 3.25, we see that if, for instance, we choose $L = 1$, then we cannot get close to L when x is less than 0. Similarly, we see that for any value of L, either the distance to $+1$ exceeds 1, or the distance to -1 exceeds 1. That is, regardless of the value of L, if $\epsilon < 1$, we will not be able to find a value of δ so that if $0 < |x| < \delta$, then $|f(x) - L| < \epsilon$, since $f(x)$ takes on both the values $+1$ and -1 for $0 < |x| < \delta$. Therefore, $\lim_{x \to 0} \frac{|x|}{x}$ does not exist. ◀

In the previous section, we considered an example where $\lim_{x \to c} f(x) = \infty$. This can be made precise as well.

> **Definition** The statement
>
> $$\lim_{x \to c} f(x) = \infty$$
>
> means that for every $M > 0$ there exists $\delta > 0$ so that
>
> $$f(x) > M \qquad \text{whenever} \quad 0 < |x - c| < \delta$$

Similar definitions hold for the case when $f(x)$ decreases without bound as $x \to c$ and for one-sided limits. We will not give definitions for all possible cases but rather illustrate how one would use such a definition.

▶ **Example 5**   Show that

$$\lim_{x \to 0} \frac{1}{x^2} = \infty$$

Solution

The graph of $f(x) = 1/x^2$, $x \neq 0$, is shown in Figure 3.26. We fix $M > 0$ (again, M is arbitrary, because our solution must hold for all $M > 0$). We need to find $\delta > 0$ so that $f(x) > M$ whenever $0 < |x| < \delta$ (note that $c = 0$). We start with the inequality $f(x) > M$ and try to determine how to choose δ.

$$\frac{1}{x^2} > M \quad \text{is the same as} \quad x^2 < \frac{1}{M}$$

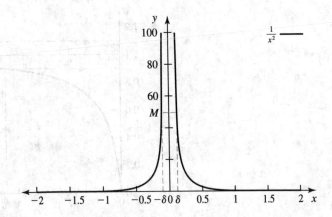

▲ **Figure 3.26**
The function $f(x) = \frac{1}{x^2}$ in Example 5: The limit of $\frac{1}{x^2}$ as x tends to 0 does not exist

Taking square roots on both sides, we find

$$|x| < \frac{1}{\sqrt{M}}$$

This suggests that we should choose $\delta = 1/\sqrt{M}$. Let's try this value: Given $M > 0$, we choose $\delta = 1/\sqrt{M}$. If $0 < |x| < \delta$, this implies

$$x^2 < \delta^2 \quad \text{or} \quad \frac{1}{x^2} > \frac{1}{\delta^2} = M$$

That is, $1/x^2 > M$ whenever $0 < |x| < \delta = 1/\sqrt{M}$. ◀

There is also a formal definition when $x \to \infty$ (and a similar one for $x \to -\infty$). This is analogous to the definition in Chapter 2.

> **Definition** The statement
>
> $$\lim_{x \to \infty} f(x) = L$$
>
> means that for every $\epsilon > 0$ there exists an $x_0 > 0$ so that
>
> $$|f(x) - L| < \epsilon \quad \text{whenever} \quad x > x_0$$

Note that in the definition x_0 is a real number.

▶ **Example 6** Show that

$$\lim_{x \to \infty} \frac{x}{0.5 + x} = 1$$

Solution

This limit is illustrated in Figure 3.27. You can see that $f(x) = x/(0.5 + x)$, $x \geq 0$, is in the strip of width 2ϵ centered at the limiting value $L = 1$ for all values of x which are greater than x_0. (We assume $\epsilon < 1$ since when $\epsilon \geq 1$, the choice $x_0 = 1$ works.) We now determine x_0 when $\epsilon < 1$. To do this, we try to solve

$$\left| \frac{x}{0.5 + x} - 1 \right| < \epsilon$$

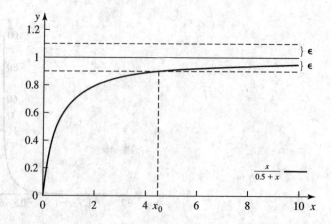

▲ **Figure 3.27**
The function $f(x) = \frac{x}{0.5+x}$ in Example 6: The limit of $f(x)$ as x tends to infinity is 1

for $\epsilon > 0$. This is equivalent to

$$-\epsilon < \frac{x}{0.5 + x} - 1 < \epsilon$$

or, after adding 1 to all three parts,

$$1 - \epsilon < \frac{x}{0.5 + x} < 1 + \epsilon$$

Since $\frac{x}{0.5+x} < 1$ for $x > 0$, the right-hand side of the inequality always holds. We therefore need only consider

$$1 - \epsilon < \frac{x}{0.5 + x}$$

Since we are interested in the behavior of $f(x)$ as $x \to \infty$, we need only look at large values of x. Multiplying by $0.5 + x$ (and noticing that we can assume $0.5 + x > 0$, because we let $x \to \infty$), we find

$$(1 - \epsilon)(0.5 + x) < x$$

Solving for x yields

$$(1 - \epsilon)(0.5) < x - x(1 - \epsilon)$$
$$(1 - \epsilon)(0.5) < \epsilon x$$
$$\frac{1 - \epsilon}{2\epsilon} < x$$

For instance, if $\epsilon = 0.1$ (as in Figure 3.27), then

$$x > \frac{0.9}{0.2} = 4.5$$

That is, we would set $x_0 = 4.5$ and conclude that for $x > 4.5$, $|f(x) - 1| < 0.1$. More generally, we find that for every $0 < \epsilon < 1$, there exists

$$x_0 = \frac{1 - \epsilon}{2\epsilon}$$

so that

$$|f(x) - 1| < \epsilon \quad \text{whenever} \quad x > x_0$$

3.6.1 Problems

1. Find the values of x such that
$$|2x - 1| < 0.01$$

2. Find the values of x such that
$$|x^2 - 9| < 0.01$$

3. Find the values of x such that
$$|3x^2 + 1| < 0.1$$

4. Find the values of x such that
$$|2\sqrt{x} - 5| < 0.1$$

5. Let
$$f(x) = 2x - 1, \quad x \in \mathbb{R}$$
(a) Graph $y = f(x)$ for $-3 \le x \le 5$.

(b) For which values of x is $y = f(x)$ within 0.1 of 3? (*Hint:* Find values for x so that $|(2x - 1) - 3| < 0.1$.)

(c) Illustrate your result in (b) on the graph that you obtained in (a).

6. Let
$$f(x) = \sqrt{x}, \quad x \ge 0$$
(a) Graph $y = f(x)$ for $0 \le x \le 6$.

(b) For which values of x is $y = f(x)$ within 0.2 of 1? (*Hint:* Find values for x so that $|\sqrt{x} - 1| < 0.2$.)

(c) Illustrate your result in (b) on the graph that you obtained in (a).

7. Let
$$f(x) = \frac{1}{x}, \quad x > 0$$
(a) Graph $y = f(x)$ for $0 < x \le 4$.

(b) For which values of x is $y = f(x)$ greater than 4?

(c) Illustrate your result in (b) on the graph that you obtained in (a).

8. Let
$$f(x) = e^{-x}, \quad x \geq 0$$
(a) Graph $y = f(x)$ for $0 \leq x \leq 6$.

(b) For which values of x is $y = f(x)$ less than 0.1?

(c) Illustrate your result in (b) on the graph that you obtained in (a).

In Problems 9–22, use the formal definition of limits to prove each statement.

9. $\lim\limits_{x \to 2} (2x - 1) = 3$

10. $\lim\limits_{x \to 0} x^2 = 0$

11. $\lim\limits_{x \to 9} \sqrt{x} = 3$

12. $\lim\limits_{x \to 1} \dfrac{1}{x} = 1$

13. $\lim\limits_{x \to 0} \dfrac{4}{x^2} = \infty$

14. $\lim\limits_{x \to 0} \dfrac{-2}{x^2} = -\infty$

15. $\lim\limits_{x \to 0} \dfrac{1}{x^4} = \infty$

16. $\lim\limits_{x \to 3} \dfrac{1}{(x-3)^2} = \infty$

17. $\lim\limits_{x \to \infty} \dfrac{3}{x^2} = 0$

18. $\lim\limits_{x \to \infty} e^{-x} = 0$

19. $\lim\limits_{x \to \infty} \dfrac{x}{x+1} = 1$

20. $\lim\limits_{x \to -\infty} \dfrac{x}{x+1} = 1$

21. $\lim\limits_{x \to c} (mx + b) = mc + b$, where m and b are constants.

22. $\lim\limits_{x \to c} \sqrt{x} = \sqrt{c}$

3.7 Key Terms

Chapter 3 Review: Topics *Discuss the following definitions and concepts:*

1. Limit of $f(x)$ as x approaches c

2. One-sided limits

3. Infinite limits

4. Divergence by oscillations

5. Convergence

6. Divergence

7. Limit laws

8. Continuity

9. One-sided continuity

10. Continuous function

11. Removable discontinuity

12. Sandwich theorem

13. Trigonometric limits

14. Intermediate value theorem

15. Bisection method

16. ϵ-δ definition of limits

3.8 Review Problems

In Problems 1–4, determine where each function is continuous. Investigate the behavior as $x \to \pm\infty$. Use a graphing calculator to sketch the corresponding graphs.

1. $f(x) = e^{-|x|}$

2. $f(x) = \begin{cases} \frac{\sin x}{x} & \text{if } x \neq 0 \\ 1 & \text{if } x = 0 \end{cases}$

3. $f(x) = \dfrac{2}{e^x + e^{-x}}$

4. $f(x) = \dfrac{1}{\sqrt{x^2 - 1}}$

5. Sketch the graph of a function that is discontinuous from the left and continuous from the right at $x = 1$.

6. Sketch the graph of a function $f(x)$ that is continuous on $[0, 2]$ except at $x = 1$, where $f(1) = 4$, $\lim_{x \to 1^-} f(x) = 2$, and $\lim_{x \to 1^+} f(x) = 3$.

7. Sketch the graph of a continuous function on $[0, \infty)$ with $f(0) = 0$ and $\lim_{x \to \infty} f(x) = 1$.

8. Sketch the graph of a continuous function on $(-\infty, \infty)$ with $f(0) = 1$, $f(x) \geq 0$ for all $x \in \mathbf{R}$, and $\lim_{x \to \pm\infty} f(x) = 0$.

9. Show that the floor function
$$f(x) = \lfloor x \rfloor$$
is continuous from the right but discontinuous from the left at $x = -2$.

10. Suppose $f(x)$ is continuous on the interval $[1, 3]$. If $f(1) = 0$ and $f(3) = 2$, explain why there must be a number $c \in (1, 3)$ such that $f(c) = 1$.

11. Assume that the size of a population at time t is
$$N(t) = \dfrac{at}{k + t}, \quad t \geq 0$$
where a and k are positive constants. Suppose that the limiting population size is
$$\lim\limits_{t \to \infty} N(t) = 1.24 \times 10^6$$
and that at time $t = 5$, the population size is half the limiting population size. Use the preceding information to determine the constants a and k.

12. Suppose that
$$N(t) = 10 + 2e^{-0.3t} \sin t, \quad t \geq 0$$

describes the size of a population (in millions) at time t (measured in weeks).

(a) Use a graphing calculator to sketch the graph of $N(t)$, and describe in words what you see.

(b) Give lower and upper bounds on the size of the population; that is, find N_1 and N_2 so that, for all $t \geq 0$,

$$N_1 \leq N(t) \leq N_2$$

(c) Find $\lim_{t \to \infty} N(t)$. Interpret this expression.

13. Suppose that an organism reacts to a stimulus only when the stimulus exceeds a certain threshold. Assume that the stimulus is a function of time t, and that it is given by

$$s(t) = \sin(\pi t), \quad t \geq 0$$

The organism reacts to the stimulus and shows a certain reaction when $s(t) \geq 1/2$. Define a function $g(t)$ so that $g(t) = 0$ when the organism shows no reaction at time t, and $g(t) = 1$ when the organism shows the reaction.

(a) Plot $s(t)$ and $g(t)$ in the same coordinate system.

(b) Is $s(t)$ continuous? Is $g(t)$ continuous?

14. The following function describes the height of a tree as a function of age:

$$f(x) = 132e^{-20/x}, \quad x \geq 0$$

Find $\lim_{x \to \infty} f(x)$.

15. There are a number of mathematical models that describe predator-prey interactions. Typically, they share the feature that the number of prey eaten per predator increases with the prey density. In the simplest version, the number of encounters with prey per predator is proportional to the product of the total number of prey and the time period over which the predators search for prey. That is, if we let N be the number of prey, P be the number of predators, T be the time period available for searching, and N_e be the number of prey encounters, then

$$\frac{N_e}{P} = aTN \tag{3.10}$$

where a is a positive constant. The quantity N_e/P is the number of prey encountered per predator.

(a) Set $f(N) = aTN$, and sketch the graph of $f(N)$ when $a = 0.1$ and $T = 2$ for $N \geq 0$.

(b) Predators usually spend some time eating the prey that they find. Therefore, not all of the time T can be used for searching. The actual searching time is reduced by the per-prey handling time T_h and can be written as

$$T - T_h \frac{N_e}{P}$$

Show that if $T - T_h \frac{N_e}{P}$ is substituted for T in (3.10), then

$$\frac{N_e}{P} = \frac{aTN}{1 + aT_hN} \tag{3.11}$$

Define

$$g(N) = \frac{aTN}{1 + aT_hN}$$

and graph $g(N)$ for $N \geq 0$ when $a = 0.1$, $T = 2$, and $T_h = 0.1$.

(c) Show that (3.11) reduces to (3.10) when $T_h = 0$.

(d) Find

$$\lim_{N \to \infty} \frac{N_e}{P}$$

in the cases when $T_h = 0$ and when $T_h > 0$. Explain the difference between the two cases in words.

16. Duarte and Agustí (1998) investigated the CO_2 balance of aquatic ecosystems. They related the community respiration rates (R) to the gross primary production rates (P) of aquatic ecosystems. (Both quantities were measured in the same units.) They made the following statement:

> Our results confirm the generality of earlier reports that the relation between community respiration rate and gross production is not linear. Community respiration is scaled as the approximate two-thirds power of gross production.

(a) Use the preceding quote to explain why

$$R = aP^b$$

can be used to describe the relationship between the community respiration rates (R) and the gross primary production rates (P). What value would you assign to b based on their quote?

(b) Suppose that you obtained data for gross production and respiration rates of a number of freshwater lakes. How would you display your data graphically to quickly convince an audience that the exponent b in the power equation relating R and P is indeed approximately 2/3? (*Hint:* Use an appropriate log transformation.)

(c) The ratio R/P for an ecosystem is of importance in assessing the global CO_2 budget. If respiration exceeds production (i.e., $R > P$), then the ecosystem acts as a carbon dioxide source, whereas if production exceeds respiration (i.e., $P > R$), then the ecosystem acts as a carbon dioxide sink. Assume now that the exponent in the power equation relating R and P is 2/3. Show that the ratio R/P as a function of P is continuous for $P > 0$. Furthermore, show that

$$\lim_{P \to 0+} \frac{R}{P} = \infty$$

and

$$\lim_{P \to \infty} \frac{R}{P} = 0$$

Use a graphing calculator to sketch the graph of the ratio R/P as a function of P for $P > 0$. (Experiment with the graphing calculator to see how the value of a affects the graph.)

(d) Use your results in (c) and the intermediate value theorem to conclude that there exists a value P^* so that the ratio R/P at P^* is equal to 1. Based on your graph in (c), is there more than one such value P^*?

(e) Use your results in (d) to identify production rates P where the ratio $R/P > 1$, that is, where respiration exceeds production.

(f) Use your results in (a)–(e) to explain the following quote from Duarte and Agustí:

> Unproductive aquatic ecosystems ... tend to be heterotrophic ($R > P$), and act as carbon dioxide sources.

17. Hyperbolic functions are used in the sciences. We take a look at the following three examples: the hyperbolic sine, sinh x; the hyperbolic cosine, cosh x; and the hyperbolic tangent, tanh x. They are defined as

$$\sinh x = \frac{e^x - e^{-x}}{2}, \quad x \in \mathbf{R}$$

$$\cosh x = \frac{e^x + e^{-x}}{2}, \quad x \in \mathbf{R}$$

$$\tanh x = \frac{e^x - e^{-x}}{e^x + e^{-x}}, \quad x \in \mathbf{R}$$

(a) Show that these three hyperbolic functions are continuous for all $x \in \mathbf{R}$. Use a graphing calculator to sketch the graphs of all three functions.

(b) Find

$$\lim_{x \to \infty} \sinh x$$

$$\lim_{x \to -\infty} \sinh x$$

$$\lim_{x \to \infty} \cosh x$$

$$\lim_{x \to -\infty} \cosh x$$

$$\lim_{x \to \infty} \tanh x$$

$$\lim_{x \to -\infty} \tanh x$$

(c) Show that the two identities

$$\cosh^2 x - \sinh^2 x = 1$$

and

$$\tanh x = \frac{\sinh x}{\cosh x}$$

are valid.

(d) Show that sinh x and tanh x are odd functions and that cosh x is even.

(*Note*: It can be shown that if a flexible cable is suspended between two points at equal heights, the shape of the resulting curve is given by the hyperbolic cosine function. This curve is called a catenary.)

DIFFERENTIATION

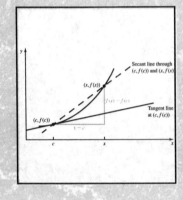

Differential calculus allows us to solve two of the basic problems that we mentioned in Chapter 1, namely, constructing a tangent line to a curve (Figure 4.1) and finding maxima and minima of a curve (Figure 4.2). The solutions to these two problems, by themselves, cannot explain the impact calculus has had on the sciences. Calculus is one of the most important analytical tools for investigating dynamical problems. Applications of differential calculus in the life sciences include simple growth models, interactions between organisms, invasions of organisms, the working of neurons, enzymatic reactions, harvesting models in fishery, epidemiological modeling, change of gene frequencies under random mating, evolutionary strategies, and many others.

Growth models will be of particular interest to us. Let's revisit the example at the beginning of Chapter 3, where we looked at a population whose size at time t is given by $N(t)$. The average growth rate during the time interval $[t, t+h]$ is equal to

$$[\text{average growth rate}] = \frac{[\text{change in population size}]}{[\text{length of time interval}]} = \frac{\Delta N}{\Delta t}$$

where

$$\Delta N = N(t+h) - N(t) \quad \text{and} \quad \Delta t = (t+h) - t = h$$

Thus

$$\frac{\Delta N}{\Delta t} = \frac{N(t+h) - N(t)}{h}$$

The instantaneous rate of growth is defined as the limit of $\Delta N / \Delta t$ as $\Delta t \to 0$ (or $h \to 0$):

$$\lim_{\Delta t \to 0} \frac{\Delta N}{\Delta t} = \lim_{h \to 0} \frac{N(t+h) - N(t)}{h}$$

provided this limit exists.

We are interested in the geometric interpretation of this limit when it exists. When we draw a straight line through the points $(t, N(t))$ and $(t+h, N(t+h))$, we obtain the **secant line**. The slope of this line is given by the quantity $\Delta N / \Delta t$ (Figure 4.3). In the limit, as $\Delta t \to 0$, the secant line converges to the line that touches the graph at the point $(t, N(t))$. This line is called the **tangent line** at the point $(t, N(t))$ (Figure 4.4). The limit of $\Delta N / \Delta t$ as the length

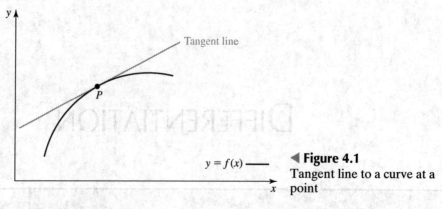

◀ **Figure 4.1**
Tangent line to a curve at a point

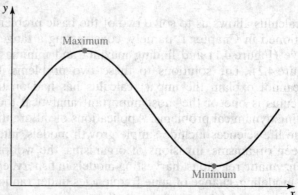

▲ **Figure 4.2**
Maxima and minima of a curve

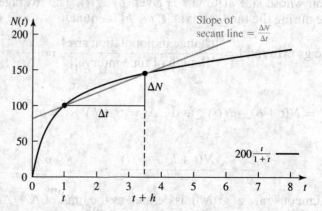

▲ **Figure 4.3**
The average growth rate $\frac{\Delta N}{\Delta t}$ is equal to the slope of the secant line

of the time interval $[t, t+h]$ goes to 0 (i.e., $\Delta t \to 0$ or $h \to 0$) will therefore be equal to the slope of the tangent line at $(t, N(t))$. We denote the limiting value of $\Delta N/\Delta t$, as $\Delta t \to 0$, by $N'(t)$ (read: N prime of t) and call this quantity the **derivative** of $N(t)$. That is,

$$N'(t) = \lim_{\Delta t \to 0} \frac{\Delta N}{\Delta t} = \lim_{h \to 0} \frac{N(t+h) - N(t)}{h}$$

provided this limit exists.

Finding derivatives is the topic of this chapter.

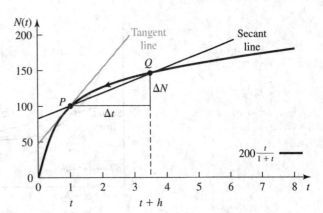

▲ Figure 4.4
The instantaneous growth rate is the limit $\Delta N/\Delta t$ as $\Delta t \to 0$. Geometrically, the point Q moves towards the point P on the graph of $N(t)$, and the secant line through P and Q becomes the tangent line at P. The instantaneous growth rate is then equal to the slope of the tangent line

4.1 FORMAL DEFINITION OF THE DERIVATIVE

> **Definition** The derivative of a function f at x, denoted by $f'(x)$, is
> $$f'(x) = \lim_{h \to 0} \frac{f(x+h) - f(x)}{h}$$
> provided that the limit exists.

If the limit exists, then we say f is differentiable at x. The phrase "provided that the limit exists" is crucial; if we take an arbitrary function f, the limit may not exist. In fact, we saw many examples in the previous chapter where limits did not exist. The geometric interpretation will help us to understand when the limit exists and under which conditions we cannot expect the limit to exist. Notice that $\lim_{h \to 0}$ is a two-sided limit (that is, we approach 0 from both the negative and the positive side). The quotient

$$\frac{f(x+h) - f(x)}{h}$$

is called the **difference quotient**, and we denote it by $\frac{\Delta f}{\Delta x}$ (see Figure 4.5).

We say that f is differentiable on (a, b) if it is differentiable at every $x \in (a, b)$. (Since the limit in the definition is two-sided, we exclude the endpoints of the interval. At endpoints, only one-sided limits can be computed, which yield one-sided derivatives.)

If we want to compute the derivative at $x = c$, we can also write

$$f'(c) = \lim_{x \to c} \frac{f(x) - f(c)}{x - c}$$

which emphasizes that the point $(x, f(x))$ converges to the point $(c, f(c))$ as we take the limit $x \to c$ (Figure 4.6). This will be important when we discuss the geometric interpretation of the derivative in the next subsection.

There is more than one way to write the derivative of a function $y = f(x)$. The following expressions are equivalent:

$$y' = \frac{dy}{dx} = f'(x) = \frac{df}{dx} = \frac{d}{dx}f(x)$$

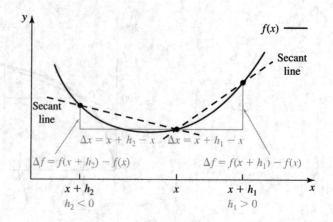

▲ Figure 4.5

The difference quotient $\frac{f(x+h)-f(x)}{h}$ when $h = h_1 > 0$ and $h = h_2 < 0$

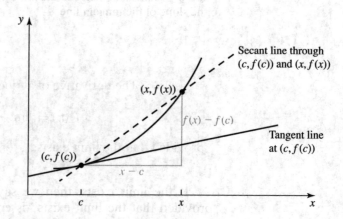

▲ Figure 4.6

The derivative $f'(c) = \lim_{x \to c} \frac{f(x)-f(c)}{x-c}$ is the slope of the tangent line at $(c, f(c))$

The notation $\frac{df}{dx}$ goes back to Leibniz and is called **Leibniz notation** (Leibniz had a real talent for finding good notation). It should remind you that we take the limit of $\Delta f/\Delta x$ as Δx approaches 0.

If we wish to emphasize that we evaluate the derivative of $f(x)$ at $x = c$, we write

$$f'(c) = \frac{df}{dx}\Big|_{x=c}$$

Newton used different notation to denote the derivative of a function. He wrote $\dot{y}$ (read: y dot) for the derivative of y. This notation is still common in physics when derivatives are taken with respect to a variable that denotes time. We will use either Leibniz notation or the notation "$f'(x)$."

4.1.1 Geometric Interpretation and Using the Definition

Let's look at $f(x) = x^2$, $x \in \mathbf{R}$ (refer to Figures 4.7 and 4.8 as we go along). To compute the derivative of f at, say, $x = 1$ using the definition, we first

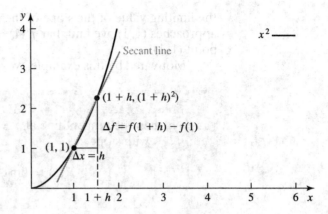

▲ **Figure 4.7**
The slope of the secant line through $(1, 1)$ and $(1 + h, (1+h)^2)$ is $\Delta f / \Delta x$

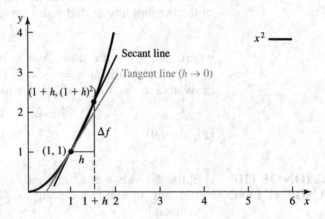

▲ **Figure 4.8**
Taking the limit $h \to 0$, the secant line converges to the tangent line at $(1, 1)$

compute the difference quotient at $x = 1$. For $h \neq 0$,

$$\frac{\Delta f}{\Delta x} = \frac{f(1 + h) - f(1)}{h} = \frac{(1+h)^2 - 1^2}{h}$$

$$= \frac{1 + 2h + h^2 - 1}{h} = \frac{2h + h^2}{h} = \frac{h(2 + h)}{h}$$

$$= 2 + h$$

The difference quotient $\Delta f / \Delta x$ is the slope of the secant line through the points $(1, 1)$ and $(1 + h, (1 + h)^2)$ (Figure 4.7).

To find the derivative $f'(1)$, we need to take the limit as $h \to 0$ (Figure 4.8):

$$f'(1) = \lim_{h \to 0} \frac{f(1 + h) - f(1)}{h} = \lim_{h \to 0} (2 + h) = 2$$

Taking the limit as $h \to 0$ means that the point $(1 + h, (1 + h)^2)$ approaches the point $(1, 1)$. (The limit as $h \to 0$ is a two-sided limit; in Figure 4.8, we only drew one point $(1 + h, (1 + h)^2)$ for some $h > 0$.) As $h \to 0$, the secant lines through the points $(1, 1)$ and $(1+h, (1+h)^2)$ converge to the line that touches the graph at $(1, 1)$. The limiting line is called the tangent line. Since $f'(1)$ is the limiting value of the slope of the secant line as the point $(1 + h, (1 + h)^2)$ approaches $(1, 1)$, we find that $f'(1) = 2$ is the slope of the tangent line at the point $(1, 1)$.

Motivated by this example, we define the tangent line formally.

> **Definition of the Tangent Line** If the derivative of a function f exists at $x = c$, then the tangent line at $x = c$ is the line going through the point $(c, f(c))$ with slope
>
> $$f'(c) = \lim_{h \to 0} \frac{f(c + h) - f(c)}{h}$$

Knowing the derivative at a point (which is the slope of the tangent line at that point) and the coordinates of that point allows us to find the equation of the tangent line at that point using the point-slope form of a straight line,

$$y - y_0 = m(x - x_0)$$

where (x_0, y_0) is the point and m is its slope. Going back to the function $y = x^2$, the point at $c = x_0 = 1$ has coordinates $(x_0, y_0) = (1, 1)$ and its derivative at $c = 1$ is $m = 2$. The equation of the tangent line is then given by

$$y - 1 = 2(x - 1) \quad \text{or} \quad y = 2x - 1$$

(Figure 4.9).

EQUATION OF THE TANGENT LINE

> If the derivative of a function f exists at $x = c$, then $f'(c)$ is the slope of the tangent line at the point $(c, f(c))$. The equation of the tangent line is given by
>
> $$y - f(c) = f'(c)(x - c)$$

The geometric interpretation will help us to compute derivatives in the first two examples.

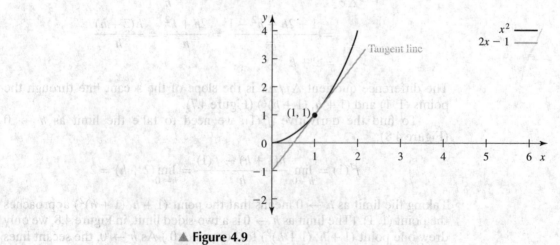

▲ **Figure 4.9**
The slope of $f(x) = x^2$ at $(1, 1)$ is $m = 2$. The equation of the tangent line at $(1, 1)$ is $y = 2x - 1$

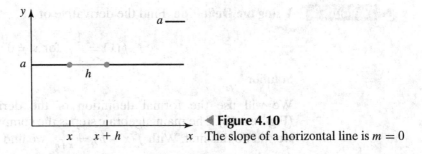

◀ **Figure 4.10**
The slope of a horizontal line is $m = 0$

▶ **Example 1**   **The Derivative of a Constant Function**   The graph of $f(x) = a$ is a horizontal line that intersects the y-axis at $(0, a)$ (Figure 4.10). Since the graph is a straight line, the tangent line at x coincides with the graph of $f(x)$ and, therefore, the slope of the tangent line at x is equal to the slope of the straight line. The slope of a horizontal line is 0; we therefore expect $f'(x) = 0$. Using the formal definition with $f(x) = a$ and $f(x + h) = a$, we find

$$f'(x) = \lim_{h \to 0} \frac{f(x + h) - f(x)}{h} = \lim_{h \to 0} \frac{a - a}{h} = \lim_{h \to 0} \frac{0}{h} = \lim_{h \to 0} 0 = 0$$

Here again, it is important to remember that when we take the limit as $h \to 0$, h approaches 0 (from both sides) but is not equal to 0. Since $h \neq 0$, the expression $0/h = 0$. This was used in the second-to-last step.   ◀

▶ **Example 2**   **The Derivative of a Linear Function**   The graph of $f(x) = mx + b$ is a straight line with slope m and y-intercept b (Figure 4.11). The derivative of $f(x)$ is the slope of the tangent line at x. Since the graph is a straight line, the tangent line at x coincides with the graph of $f(x)$ and, therefore, the slope of the tangent line at x is equal to the slope of the straight line. We therefore expect $f'(x) = m$. Using the formal definition, we can confirm this:

$$f'(x) = \lim_{h \to 0} \frac{f(x + h) - f(x)}{h} = \lim_{h \to 0} \frac{m(x + h) + b - (mx + b)}{h}$$

$$= \lim_{h \to 0} \frac{mx + mh + b - mx - b}{h} = \lim_{h \to 0} \frac{mh}{h} = \lim_{h \to 0} m = m$$

In the second-to-last step we were able to cancel h since $h \neq 0$.

We find the following: If $f(x) = mx + b$, then $f'(x) = m$. This includes the special case of a constant function where $m = 0$ (Example 1).   ◀

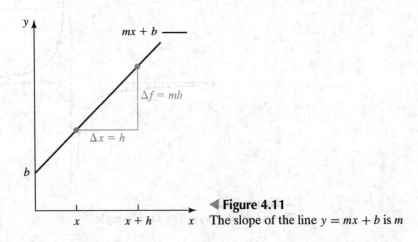

◀ **Figure 4.11**
The slope of the line $y = mx + b$ is m

▶ **Example 3**   **Using the Definition** Find the derivative of

$$f(x) = \frac{1}{x} \quad \text{for } x \neq 0$$

Solution

We will use the formal definition of the derivative to compute $f'(x)$ (Figure 4.12). The main algebraic step is the computation of $f(x+h) - f(x)$; we will do this first. With $f(x+h) = \frac{1}{x+h}$, we find

$$f(x+h) - f(x) = \frac{1}{x+h} - \frac{1}{x}$$

$$= \frac{x - (x+h)}{x(x+h)} = \frac{-h}{x(x+h)}$$

To compute $f'(x)$, we need to divide this by h and take the limit $h \to 0$:

$$f'(x) = \lim_{h \to 0} \frac{f(x+h) - f(x)}{h} = \lim_{h \to 0} \frac{\frac{-h}{x(x+h)}}{h}$$

$$= \lim_{h \to 0} \left(-\frac{h}{x(x+h)} \frac{1}{h} \right) = \lim_{h \to 0} \left(-\frac{1}{x(x+h)} \right) = -\frac{1}{x^2}$$

That is, if $f(x) = \frac{1}{x}$, $x \neq 0$, then

$$f'(x) = -\frac{1}{x^2}, \quad x \neq 0 \qquad \blacktriangleleft$$

Looking back at the first three examples, we see that in order to compute $f'(x)$ using the formal definition of the derivative, we evaluate the limit

$$\lim_{h \to 0} \frac{f(x+h) - f(x)}{h}$$

Since both $\lim_{h \to 0}[f(x+h) - f(x)]$ and $\lim_{h \to 0} h$ are equal to 0, we cannot simply evaluate the limits in the numerator and the denominator separately, since this would result in the undefined expression 0/0. It is important to simplify the difference quotient before we take the limit.

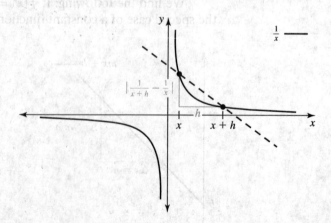

▲ **Figure 4.12**
The graph of $f(x) = 1/x$ for Example 3

4.1.2 The Derivative as an Instantaneous Rate of Change: A First Look at Differential Equations

Velocity   Suppose that you ride your bike on a straight road. Your position (in miles) at time t (in hours) is given by (Figure 4.13)

$$s(t) = -t^3 + 6t^2 \quad \text{for } 0 \leq t \leq 6$$

You might ask what the average velocity during the time interval (say) $[2, 4]$ is, which is defined as the net change in position during the interval divided by the length of the interval. To compute the average velocity, find the position at time 2 and at time 4, and take the difference of these two quantities. Then divide this difference by the time that it took you to travel this distance. At time $t = 2$, $s(2) = -8 + 24 = 16$, and at time $t = 4$, $s(4) = -64 + 96 = 32$. Hence, the average velocity is

$$\frac{s(4) - s(2)}{4 - 2} = \frac{32 - 16}{4 - 2} = 8 \text{ mph}$$

We recognize this ratio as the difference quotient

$$\frac{\Delta s}{\Delta t} = \frac{s(t + h) - s(t)}{h}$$

and call the difference quotient $\Delta s/\Delta t$ the **average velocity**, which is an average rate of change.

The **instantaneous velocity** at time t is defined as the limit of $\frac{\Delta s}{\Delta t}$ as $\Delta t \to 0$:

$$\lim_{\Delta t \to 0} \frac{\Delta s}{\Delta t} = \lim_{h \to 0} \frac{s(t + h) - s(t)}{h}$$

provided the limit exists. This is the derivative of $s(t)$ at time t, which we denote by

$$\frac{ds}{dt} = \lim_{\Delta t \to 0} \frac{\Delta s}{\Delta t}$$

This is an instantaneous rate of change. Instantaneous velocity (or simply velocity) is then the slope of the tangent line at a given point of the position function $s(t)$ provided the derivative at this point exists.

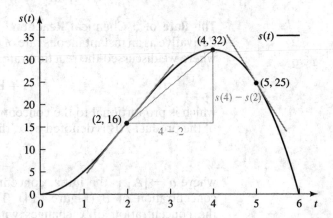

▲ **Figure 4.13**

The average velocity $\frac{s(4)-s(2)}{4-2}$ is the slope of the secant line through $(2, 16)$ and $(4, 32)$. The velocity at time t is the slope of the tangent line t at t: At $t = 2$, the velocity is positive; at $t = 5$, the velocity is negative

Let's look at two points on the graph of $s(t)$, namely, $(2, 16)$ and $(5, 25)$. We find that the slope of the tangent line is positive at $(2, 16)$ and negative at $(5, 25)$. The velocity is therefore positive at time $t = 2$ and negative at time $t = 5$. At $t = 2$, we move away from our starting point, whereas at $t = 5$, we move toward our starting point. At these two times, we move in opposite directions.

There is a difference between *velocity* and *speed*. If you had a speedometer on your bike, it would tell you the speed and not the velocity. Speed is the absolute value of velocity; it ignores direction.

Interpreting the derivative as an instantaneous rate of change will turn out to be extremely important to us. In fact, when you encounter derivatives in your science courses, this will be the interpretation most often used. This interpretation will allow us to describe a quantity in terms of how quickly it changes with respect to another quantity. To illustrate this point, we revisit two previous examples and introduce one new application.

Population Growth At the beginning of this chapter, we described the growth of a population at time t by the continuous function $N(t)$. If the derivative of $N(t)$ exists at time t, we can define the instantaneous growth rate of the population by

$$N'(t) = \frac{dN}{dt} = [\text{instantaneous population growth rate at time } t]$$

We are frequently interested in the **instantaneous per capita growth rate**. This is the growth rate per individual, and it can be obtained by dividing the instantaneous growth rate of the population by the population size at that time. That is,

$$\frac{1}{N(t)} \frac{dN}{dt} = [\text{instantaneous per capita growth rate at time } t]$$

In biology textbooks (and in this book), the dependence on t is often not explicitly spelled out and we write

$$\frac{1}{N} \frac{dN}{dt} \quad \text{instead of} \quad \frac{1}{N(t)} \frac{dN}{dt}$$

The Rate of a Chemical Reaction Another illustration of the use of the derivative as an instantaneous rate of change is Example 5 of Subsection 1.2.2, where we discussed the reaction rate of the irreversible chemical reaction

$$A + B \to AB$$

which is proportional to the concentrations of A and B. If the concentration of the product AB is denoted by x, then the reaction rate is equal to

$$k(a - x)(b - x)$$

where $a = [A]$ is the initial concentration of A, and $b = [B]$ is the initial concentration of B (Figure 4.14). The reaction rate tells us how quickly the concentration of x changes with time as the reaction proceeds. The concentration x is thus a function of time t, $x = x(t)$. The reaction rate is an instantaneous rate of change, namely,

$$\lim_{\Delta t \to 0} \frac{x(t + \Delta t) - x(t)}{\Delta t}$$

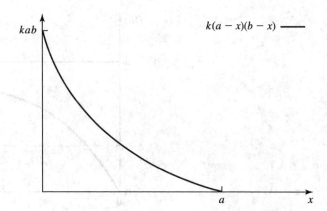

▲ **Figure 4.14**
The reaction rate for $a \leq b$

We can identify the limit $\Delta t \to 0$ as the derivative of the function $x(t)$ with respect to t and therefore write

$$\frac{dx}{dt} = k(a - x)(b - x) \tag{4.1}$$

Equation (4.1) is an example of a **differential equation**, that is, an equation that contains the derivative of a function. We will discuss such equations extensively in later chapters.

From this point on, when we say "rate of change," we will always mean "instantaneous rate of change." When we are interested in the average rate of change, we will always state this explicitly.

The rate of change in the chemical reaction is described by a differential equation. It is sometimes possible to solve such differential equations, that is, to give explicitly a function whose derivative satisfies the given equation. We will discuss this in detail later. More often, it is not possible (or not necessary) to explicitly find a solution. Without solving the differential equation, we can still obtain useful information about its behavior. We illustrate this in the next application.

Tilman's Model for Resource Competition   Dr. David Tilman of the University of Minnesota developed a theoretical framework to describe the outcome of competition for limiting resources (Tilman, 1982). This applies, for instance, to the grassland habitat at Cedar Creek Natural History Area, Minnesota, where Tilman conducted many experiments to test the predictions of this theory.

For this grassland habitat, nitrogen is a limiting resource; that is, adding nitrogen to the soil will result in an increase in biomass. We will discuss the case where one species competes for a single limiting resource. We assume that the rate of biomass change has two components: rate of growth and rate of loss. We write

[rate of biomass change] = [rate of growth] − [rate of loss]

We denote the biomass of the plant population at time t by $B(t)$ and assume that the rate of growth depends on a single resource whose concentration is denoted by R. We will write an equation for the **specific rate of change** of biomass, which is defined as the change of biomass per unit of biomass, that is, $\frac{1}{B}\frac{dB}{dt}$. We assume that the per unit biomass rate of loss is constant and

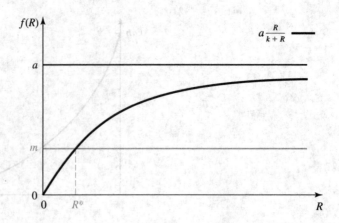

▲ Figure 4.15
Growth balances loss when $R = R^\star$

denote this quantity by m. A simple model for how the biomass changes over time is then

$$\frac{1}{B}\frac{dB}{dt} = f(R) - m \qquad (4.2)$$

where the function $f(R)$ describes the specific growth rate as a function of resource concentration. A common choice for $f(R)$ is the Monod growth function (or Michaelis-Menten equation) that we considered in Example 6 of Subsection 1.2.3:

$$f(R) = a\frac{R}{k + R} \qquad (4.3)$$

where a and k are positive constants. Let's graph both $f(R)$ and m together in Figure 4.15. This yields the following observations. When $0 < m < a$, the graphs of the functions $y = f(R)$ and $y = m$ intersect at $R = R^\star$ (read: "R star"). At this resource level, $f(R) = m$, and thus the specific rate of change $\frac{1}{B}\frac{dB}{dt}$ is equal to 0. That is, growth balances loss, and the biomass of the species no longer changes. We say that the biomass is at **equilibrium**. If the resource level R was held at a value less than $R^\star$, then $f(R) - m < 0$, and the specific rate of growth would be negative, that is, biomass would decrease. If $R > R^\star$, then $f(R) - m > 0$, and biomass would increase.

We can compute $R^\star$ in the case when $f(R)$ is given by (4.3). Since $R^\star$ satisfies $f(R^\star) = m$, we find

$$a\frac{R^\star}{k + R^\star} = m \quad \text{or} \quad R^\star = \frac{mk}{a - m}$$

4.1.3 Differentiability and Continuity

Using the geometric interpretation, we can find situations in which $f'(c)$ does not exist at one or more values of c.

▶ **Example 4**   **A Function with a "Corner."** Let

$$f(x) = |x| = \begin{cases} x & \text{for } x \geq 0 \\ -x & \text{for } x < 0 \end{cases}$$

The graph of $f(x)$ is shown in Figure 4.16. Looking at the graph, we realize that there is no tangent line at $x = 0$ and therefore we do not expect that $f'(0)$ exists. We can define the slope of the secant line when we approach 0 from the right and also when we approach 0 from the left; however, the

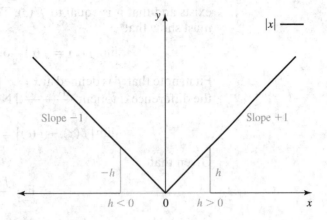

▲ Figure 4.16
f is not differentiable at $x = 0$

slopes converge to different values in the limit. The former is $+1$, the latter is -1. In this example, we can read off the slopes from the graph. But we can also find the slopes formally by taking appropriate limits. When $h > 0$, $f(h) = |h| = h$ and

$$\lim_{h \to 0+} \frac{f(0+h) - f(0)}{h} = \frac{h - 0}{h} = 1$$

When $h < 0$, $f(h) = |h| = -h$ and

$$\lim_{h \to 0-} \frac{f(0+h) - f(0)}{h} = \frac{-h - 0}{h} = -1$$

Since $1 \neq -1$, it follows that

$$\lim_{h \to 0} \frac{f(0+h) - f(0)}{h}$$

and thus $f'(0)$ do not exist.

At all other points, the derivative exists. We can find the derivative by simply looking at the graph. We see that

$$f'(x) = \begin{cases} +1 & \text{for} \quad x > 0 \\ -1 & \text{for} \quad x < 0 \end{cases}$$ ◀

Example 4 shows that continuity alone is not enough for a function to be differentiable: The function $f(x) = |x|$ is continuous at all values of x but it is not differentiable at $x = 0$. To draw the graph of a continuous function that is not differentiable at a point, put in a "corner" at that point (Figure 4.17).

However, if a function is differentiable, it is also continuous. This means that continuity is a necessary but not a sufficient condition for differentiability. This result is important enough that we will formulate it as a theorem and prove it.

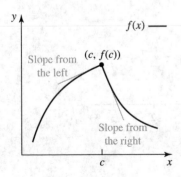

▲ Figure 4.17
$f(x)$ is continuous at $x = c$ but not differentiable at $x = c$: The derivatives from the left and the right are not equal

Theorem If f is differentiable at $x = c$, it is also continuous at $x = c$.

Proof Since f is differentiable at $x = c$, we know that the limit

$$\lim_{x \to c} \frac{f(x) - f(c)}{x - c} \qquad (4.4)$$

exists and that it is equal to $f'(c)$. To show that f is continuous at $x = c$, we must show that

$$\lim_{x \to c} f(x) = f(c) \quad \text{or} \quad \lim_{x \to c} [f(x) - f(c)] = 0$$

First, note that f is defined at $x = c$. [Otherwise, we could not have computed the difference quotient $\frac{f(x)-f(c)}{x-c}$.] Now,

$$\lim_{x \to c} [f(x) - f(c)] = \lim_{x \to c} \frac{f(x) - f(c)}{x - c}(x - c)$$

Given that

$$\lim_{x \to c} \frac{f(x) - f(c)}{x - c}$$

exists and is equal to $f'(c)$ [this is Equation (4.4)] and that

$$\lim_{x \to c} (x - c)$$

exists (it is equal to 0), we can apply the product rule for limits and find

$$\lim_{x \to c} \frac{f(x) - f(c)}{x - c}(x - c) = \lim_{x \to c} \frac{f(x) - f(c)}{x - c} \lim_{x \to c} (x - c) = f'(c) \cdot 0 = 0$$

This shows that

$$\lim_{x \to c} [f(x) - f(c)] = 0$$

and consequently that f is continuous at $x = c$. ■

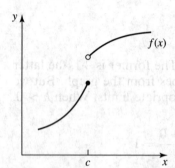

▲ Figure 4.18
The function $y = f(x)$ is not differentiable at $x = c$

It follows from the theorem that if a function f is not continuous at $x = c$, then f is not differentiable at $x = c$. The function $y = f(x)$ in Figure 4.18 is discontinuous at $x = c$; we cannot draw a tangent line at $x = c$.

Functions can have vertical tangent lines. But since the slope of a vertical line is not defined, the function would not be differentiable at any point where the tangent line is vertical. This is illustrated in the following example.

▶ **Example 5** **Vertical Tangent Line** Show that

$$f(x) = x^{1/3}$$

is not differentiable at $x = 0$.

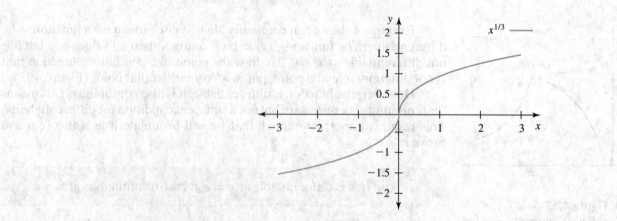

▲ Figure 4.19
The function $f(x) = x^{1/3}$ has a vertical tangent line at $x = 0$. It is therefore not differentiable at $x = 0$

Solution

We see from the graph of $f(x)$ in Figure 4.19 that $f(x)$ is continuous at $x = 0$. Using the formal definition, we find

$$f'(0) = \lim_{h \to 0} \frac{f(h) - f(0)}{h} = \lim_{h \to 0} \frac{h^{1/3} - 0}{h}$$

$$= \lim_{h \to 0} \frac{1}{h^{2/3}} = \infty \quad \text{does not exist}$$

Since the limit does not exist, $f(x)$ is not differentiable at $x = 0$. We see from the graph that the tangent line at $x = 0$ is vertical. ◀

4.1.4 Problems

(4.1.1)

In Problems 1–8, find the derivative at the indicated point from the graph of each function.

1. $f(x) = 2; x = 1$
2. $f(x) = -3x; x = 2$
3. $f(x) = 2x - 5; x = -1$
4. $f(x) = -5x + 1; x = 0$
5. $f(x) = x^2; x = 0$
6. $f(x) = (x - 1)^2; x = 1$
7. $f(x) = \cos x; x = 0$
8. $f(x) = \sin x; x = \dfrac{\pi}{2}$

In Problems 9–12, compute $f(c + h) - f(c)$ at the indicated point.

9. $f(x) = -2x + 1; c = 2$
10. $f(x) = 3x^2; c = 1$
11. $f(x) = \sqrt{x}; c = 4$
12. $f(x) = \dfrac{1}{x}; c = -2$

13. (a) Use the formal definition of the derivative to find the derivative of $y = 5x^2$ at $x = -1$.

(b) Show that the point $(-1, 5)$ is on the graph of $y = 5x^2$, and find the equation of the tangent line at the point $(-1, 5)$.

(c) Graph $y = 5x^2$ and the tangent line at the point $(-1, 5)$ in the same coordinate system.

14. (a) Use the formal definition to find the derivative of $y = -2x^2$ at $x = 1$.

(b) Show that the point $(1, -2)$ is on the graph of $y = -2x^2$, and find the equation of the tangent line at the point $(1, -2)$.

(c) Graph $y = -2x^2$ and the tangent line at the point $(1, -2)$ in the same coordinate system.

15. (a) Use the formal definition to find the derivative of $y = 1 - x^3$ at $x = 2$.

(b) Show that the point $(2, -7)$ is on the graph of $y = 1 - x^3$, and find the equation of the normal line at the point $(2, -7)$.

(c) Graph $y = 1 - x^3$ and the tangent line at the point $(2, -7)$ in the same coordinate system.

16. (a) Use the formal definition to find the derivative of $y = \dfrac{1}{x}$ at $x = 2$.

(b) Show that the point $(2, \frac{1}{2})$ is on the graph of $y = \frac{1}{x}$, and find the equation of the normal line at the point $(2, \frac{1}{2})$.

(c) Graph $y = \frac{1}{x}$ and the tangent line at the point $(2, \frac{1}{2})$ in the same coordinate system.

17. Use the formal definition to find the derivative of

$$y = \sqrt{x}$$

for $x > 0$.

18. Use the formal definition to find the derivative of

$$f(x) = \frac{1}{x + 1}$$

for $x \neq -1$.

19. Find the equation of the tangent line to the curve $y = 2x^2$ at the point $(1, 2)$.

20. Find the equation of the tangent line to the curve $y = 3/x$ at the point $(3, 1)$.

21. Find the equation of the tangent line to the curve $y = \sqrt{x}$ at the point $(4, 2)$.

22. Find the equation of the tangent line to the curve $y = x^2 - 3x + 1$ at the point $(2, -1)$.

23. Find the equation of the normal line to the curve $y = -3x^2$ at the point $(-1, -3)$.

24. Find the equation of the normal line to the curve $y = 4/x$ at the point $(-1, -4)$.

25. Find the equation of the normal line to the curve $y = 2x^2 - 1$ at the point $(1, 1)$.

26. Find the equation of the normal line to the curve $y = \sqrt{x - 1}$ at the point $(5, 2)$.

27. The following limit represents the derivative of a function f at the point $(a, f(a))$. Find $f(x)$.

$$\lim_{h \to 0} \frac{2(a + h)^2 - 2a^2}{h}$$

28. The following limit represents the derivative of a function f at the point $(a, f(a))$. Find $f(x)$.

$$\lim_{h \to 0} \frac{4(a + h)^3 - 4a^3}{h}$$

29. The following limit represents the derivative of a function f at the point $(a, f(a))$. Find f and a.

$$\lim_{h \to 0} \frac{\dfrac{1}{(2 + h)^2 + 1} - \dfrac{1}{5}}{h}$$

30. The following limit represents the derivative of a function f at the point $(a, f(a))$. Find f and a.

$$\lim_{h \to 0} \frac{\sin\left(\frac{\pi}{6} + h\right) - \sin\frac{\pi}{6}}{h}$$

(4.1.2)

31. A car moves along a straight road. Its location at time t is given by

$$s(t) = 20t^2, \quad 0 \le t \le 2$$

where t is measured in hours and $s(t)$ is measured in kilometers.

(a) Graph $s(t)$ for $0 \le t \le 2$.

(b) Find the average velocity of the car between $t = 0$ and $t = 2$. Illustrate the average velocity on the graph of $s(t)$.

(c) Use calculus to find the instantaneous velocity of the car at $t = 1$. Illustrate the instantaneous velocity on the graph of $s(t)$.

32. A train moves along a straight line. Its location at time t is given by

$$s(t) = \frac{100}{t}, \quad 1 \le t \le 5$$

where t is measured in hours and $s(t)$ is measured in kilometers.

(a) Graph $s(t)$ for $1 \le t \le 5$.

(b) Find the average velocity of the train between $t = 1$ and $t = 5$. Where on the graph of $s(t)$ can you find the average velocity?

(c) Use calculus to find the instantaneous velocity of the train at $t = 2$. Where on the graph of $s(t)$ can you find the instantaneous velocity? What is the speed of the train at $t = 2$?

33. If $s(t)$ denotes the position of an object that moves along a straight line, then $\Delta s / \Delta t$ is the average rate of change of $s(t)$, called the average velocity, and $v(t) = ds/dt$ is the instantaneous rate of change of $s(t)$, called the (instantaneous) velocity. The speed of the object is the absolute value of the velocity, $|v(t)|$.

Suppose now that a car moves along a straight road. The location at time t is given by

$$s(t) = \frac{160}{3}t^2, \quad 0 \le t \le 1$$

where t is measured in hours and $s(t)$ is measured in kilometers.

(a) Where is the car at $t = 3/4$, and where is it at $t = 1$?

(b) Find the average velocity of the car between $t = 3/4$ and $t = 1$.

(c) Find the velocity and the speed of the car at $t = 3/4$.

34. Suppose a particle moves along a straight line. The position at time t is given by

$$s(t) = 3t - t^2, \quad t \ge 0$$

where t is measured in seconds and $s(t)$ is measured in meters.

(a) Graph $s(t)$ for $t \ge 0$.

(b) Use the graph in (a) to answer the following questions.

(i) Where is the particle at time 0?

(ii) Is there another time at which the particle visits the location where it was at time 0?

(iii) How far to the right on the straight line does the particle travel?

(iv) How far to the left on the straight line does the particle travel?

(v) Where is the velocity positive? where negative? equal to 0?

(c) Find the velocity of the particle.

(d) When is the velocity of the particle equal to 1 m/s?

35. In Subsection 4.1.2, we considered Tilman's resource model. Denote by $B(t)$ the biomass at time t, and assume that

$$\frac{1}{B}\frac{dB}{dt} = f(R) - m$$

where R denotes the resource level,

$$f(R) = 200\frac{R}{5 + R}$$

and $m = 40$. Use the graphical approach to find the value $R^\star$ at which $\frac{1}{B}\frac{dB}{dt} = 0$. Then compute $R^\star$ by solving $\frac{1}{B}\frac{dB}{dt} = 0$.

36. Assume that $N(t)$ denotes the size of a population at time t, and that $N(t)$ satisfies the differential equation

$$\frac{dN}{dt} = rN$$

where r is a constant.

(a) Find the per capita growth rate.

(b) Assume that $r > 0$ and that $N(0) = 20$. Is the population size at time 1 greater than 20 or less than 20? Explain your answer.

37. Assume that $N(t)$ denotes the size of a population at time t, and that $N(t)$ satisfies the differential equation

$$\frac{dN}{dt} = 3N\left(1 - \frac{N}{20}\right)$$

Let $f(N) = 3N(1 - \frac{N}{20})$ for $N \ge 0$. Graph $f(N)$ as a function of N and identify all equilibria, that is, all points where $\frac{dN}{dt} = 0$.

38. Assume that a species lives in a habitat that consists of many islands close to a mainland. The species occupies both the mainland and the islands, but, while it is present on the mainland at all times, it frequently goes extinct on the islands. Islands can be recolonized by migrants from the mainland. The following model keeps track of the fraction of islands occupied. Denote by $p(t)$ the fraction of islands occupied at time t. Assume that each island experiences a constant extinction risk and that vacant islands (fraction $1 - p$) are colonized from the mainland at a constant rate. Then

$$\frac{dp}{dt} = c(1 - p) - ep$$

where c and e are positive constants.

(a) The gain from colonization is $f(p) = c(1 - p)$ and the loss from extinction is $g(p) = ep$. Graph $f(p)$ and $g(p)$ for $0 \le p \le 1$ in the same coordinate system. Explain why the two graphs intersect whenever e and c are both positive. Compute the point of intersection and interpret its biological meaning.

(b) The parameter c measures how quickly a vacant island becomes colonized from the mainland. The closer the islands, the larger the value of c. Use your graph in (a) to explain what happens to the point of intersection of the two lines in (a) as c increases. Interpret your result in biological terms.

39. Consider the chemical reaction

$$A + B \longrightarrow AB$$

If $x(t)$ denotes the concentration of AB at time t, then

$$\frac{dx}{dt} = k(a - x)(b - x)$$

where k is a positive constant, and a and b denote the concentrations of A and B at time 0. Assume $k = 3, a = 7$ and $b = 4$. For what values of x is $dx/dt = 0$? How can you interpret this?

40. Consider the autocatalytic reaction

$$A + X \longrightarrow X$$

which was introduced in Problem 30 of Subsection 1.2.9. Find a differential equation that describes the rate of change of the concentration of the product X.

41. Suppose that the rate of change of the size of a population is given by

$$\frac{dN}{dt} = rN\left(1 - \frac{N}{K}\right)$$

where $N = N(t)$ denotes the size of the population at time t, and r and K are positive constants. Find the equilibrium size of the population, that is, the size where the rate of change is equal to 0. Use your answer to explain why K is called the carrying capacity.

42. (*Biotic Diversity*) (Adapted from Valentine 1985.) Walker and Valentine (1984) suggested a model for species diversity that assumes that species extinction rates are independent of diversity, but speciation rates are regulated by competition. They denoted the number of species at time t by $N(t)$, the speciation rate by b, and the extinction rate by a. They used the following model:

$$\frac{dN}{dt} = N\left[b\left(1 - \frac{N}{K}\right) - a\right]$$

where K denotes the number of "niches," or potential places for species in the ecosystem.

(a) Find possible equilibria under the condition $a < b$.

(b) Use your result in (a) to explain the following statement by Valentine (1985):

In this situation, ecosystems are never "full," with all potential niches occupied by species so long as the extinction rate is above zero.

(c) What happens when $a \geq b$?

(4.1.3)

43. Which of the following statements is true?

(A) If $f(x)$ is continuous, then $f(x)$ is differentiable.

(B) If $f(x)$ is differentiable, then $f(x)$ is continuous.

44. Explain the relationship between continuity and differentiability.

45. Sketch the graph of a function that is continuous at all points in its domain and differentiable in the domain except at one point.

In Problems 46–59, graph each function and, based on the graph, guess where the function is not differentiable. (Assume the largest possible domain.)

46. $y = |x - 2|$

47. $y = -|x + 5|$

48. $y = 2 - |x - 1|$

49. $y = |x + 2| - 1$

50. $y = \dfrac{1}{1 + x}$

51. $y = \dfrac{1}{x - 3}$

52. $y = \dfrac{3 - x}{2 + x}$

53. $y = \dfrac{x - 1}{x + 1}$

54. $y = |x^2 - 3|$

55. $y = |2x^2 - 1|$

56. $f(x) = \begin{cases} x & \text{for } x \leq 0 \\ x + 1 & \text{for } x > 0 \end{cases}$

57. $f(x) = \begin{cases} 2x & \text{for } x \leq 1 \\ x + 2 & \text{for } x > 1 \end{cases}$

58. $f(x) = \begin{cases} x^2 & \text{for } x \leq -1 \\ 2 - x^2 & \text{for } x > -1 \end{cases}$

59. $f(x) = \begin{cases} x^2 + 2 & \text{for } x \leq 0 \\ e^{-x} & \text{for } x > 0 \end{cases}$

60. Suppose the function $f(x)$ is piecewise defined, that is, $f(x) = f_1(x)$ for $x \leq a$ and $f(x) = f_2(x)$ for $x > a$. Assume that $f_1(x)$ is continuous and differentiable for $x < a$ and that $f_2(x)$ is continuous and differentiable for $x > a$. Sketch graphs of $f(x)$ for the following three cases.

(a) $f(x)$ is continuous and differentiable at $x = a$.

(b) $f(x)$ is continuous but not differentiable at $x = a$.

(c) $f(x)$ is neither continuous nor differentiable at $x = a$.

4.2 THE POWER RULE, THE BASIC RULES OF DIFFERENTIATION, AND THE DERIVATIVES OF POLYNOMIALS

In this section, we will begin a systematic treatment of the computation of derivatives. Knowing how to differentiate is fundamental to your understanding of the rest of the course. Although computer software is now available to compute derivatives of many functions (such as $y = cx^n$ or $y = e^{\sin x}$), it is nonetheless important that you master the techniques of differentiation.

The power rule is the simplest of the differentiation rules. It allows us to compute the derivative of a function of the form $y = x^n$, where n is a positive integer.

POWER RULE

Let n be a positive integer; then

$$\frac{d}{dx}(x^n) = nx^{n-1}$$

We found the rule for the constant function $f(x) = a$ in the previous section.

If $f(x)$ is the constant function, namely $f(x) = a$, then

$$\frac{d}{dx}f(x) = 0$$

We prove the power rule first for $n = 2$, that is, for $f(x) = x^2$ (Figure 4.20). In Subsection 4.1.1, we computed the derivative of $y = x^2$ at $x = 1$. Instead of $x = 1$, we can compute the difference quotient $\frac{\Delta f}{\Delta x}$ at an arbitrary x,

$$\frac{\Delta f}{\Delta x} = \frac{f(x+h) - f(x)}{h} = \frac{(x+h)^2 - x^2}{h}$$

We use the expansion $(x+h)^2 = x^2 + 2xh + h^2$ and find

$$\frac{\Delta f}{\Delta x} = \frac{x^2 + 2xh + h^2 - x^2}{h} = \frac{2xh + h^2}{h} = 2x + h \qquad (4.5)$$

after canceling h in both the numerator and the denominator. To find the derivative, we need to let $h \to 0$.

$$f'(x) = \lim_{\Delta x \to 0} \frac{\Delta f}{\Delta x} = \lim_{h \to 0}(2x + h) = 2x$$

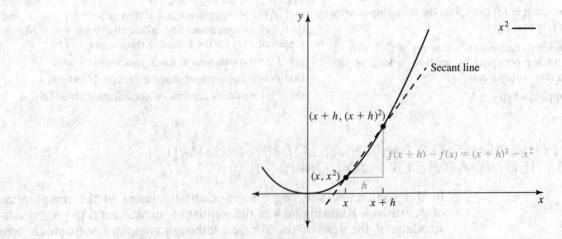

▲ **Figure 4.20**

The slope of the secant line through (x, x^2) and $(x+h, (x+h)^2)$ is $\frac{(x+h)^2 - x^2}{h}$

This proves the power rule for $n = 2$. The proof of the power rule for other positive integers of n is conceptually not any different from the case $n = 2$. But it gets algebraically much more involved. For general n, we need the expansion for $(x + h)^n$, which is provided by the **binomial theorem**, which we will not prove.

Binomial Theorem If n is a positive integer, then

$$(x + y)^n = x^n + nx^{n-1}y + \frac{n(n - 1)}{2 \cdot 1}x^{n-2}y^2$$

$$+ \frac{n(n - 1)(n - 2)}{3 \cdot 2 \cdot 1}x^{n-3}y^3$$

$$+ \cdots + \frac{n(n - 1) \cdots (n - k + 1)}{k(k - 1) \cdots 2 \cdot 1}x^{n-k}y^k$$

$$+ \cdots + nxy^{n-1} + y^n$$

The expansion of $(x + y)^n$ is thus a sum of terms of the form

$$C_{n,k}x^{n-k}y^k, \quad k = 0, 1, \ldots, n$$

where $C_{n,k}$ is a coefficient that depends on n and k. The exact form of the coefficients $C_{n,k}$ will not be important in the proof of the power rule, except for the two terms $C_{n,0} = 1$ and $C_{n,1} = n$, which are the coefficients for x^n and $x^{n-1}y$, respectively.

Proof of the Power Rule We use the expansion provided by the binomial theorem to compute the difference in the numerator of the difference quotient.

$$\Delta f = f(x + h) - f(x) = (x + h)^n - x^n$$

$$= (C_{n,0}x^n + C_{n,1}x^{n-1}h + C_{n,2}x^{n-2}h^2 + C_{n,3}x^{n-3}h^3$$

$$+ \cdots + C_{n,n-1}xh^{n-1} + C_{n,n}h^n) - x^n$$

Since $C_{n,0} = 1$, the terms x^n cancel. We can then factor h from the remaining terms and find

$$f(x + h) - f(x) = h\left[C_{n,1}x^{n-1} + C_{n,2}x^{n-2}h + C_{n,3}x^{n-3}h^2 + \cdots + C_{n,n}h^{n-1}\right]$$

When we divide by h and let $h \to 0$, we see that

$$f'(x) = \lim_{h \to 0} \frac{f(x + h) - f(x)}{h}$$

$$= \lim_{h \to 0}\left[C_{n,1}x^{n-1} + C_{n,2}x^{n-2}h + C_{n,3}x^{n-3}h^2 + \cdots + C_{n,n}h^{n-1}\right]$$

All terms except for the first have h as a factor and thus tend to 0 as $h \to 0$. (The first term does not depend on h.) We find that

$$f'(x) = C_{n,1}x^{n-1}$$

With $C_{n,1} = n$, this is then

$$f'(x) = nx^{n-1}$$

which proves the power rule. ■

▶ **Example 1**   We apply the power rule to various examples and take the opportunity to practice the different notations.

(a) If $f(x) = x^6$, then $f'(x) = 6x^5$.

(b) If $f(x) = x^{300}$, then $f'(x) = 300x^{299}$.

(c) If $g(t) = t^5$, then $\frac{d}{dt} g(t) = 5t^4$.

(d) If $z = s^3$, then $\frac{dz}{ds} = 3s^2$.

(e) If $x = y^4$, then $\frac{dx}{dy} = 4y^3$.   ◀

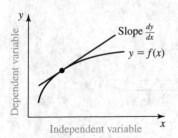

Slope $\frac{dy}{dx}$

$y = f(x)$

Independent variable

▲ **Figure 4.21**
If $y = f(x)$, then x is the independent variable and y is the dependent variable

Example 1 illustrates the importance of knowing how the variables depend on each other (Figure 4.21). If $y = f(x)$, we call x the **independent variable** and y the **dependent variable** because y depends on the variable x. For instance, in (a), y is a function of x, thus, x is the independent and y is the dependent variable; whereas in (e), x is a function of y, thus, y is now the independent and x the dependent variable. The Leibniz notation $\frac{dy}{dx}$ emphasizes this dependence. When we write $\frac{dy}{dx}$, we consider y as a function of x (that is, y is the dependent and x is the independent variable) and differentiate y with respect to x.

Since polynomials and rational functions are built from power functions of the form $y = x^n$, $n = 0, 1, 2, \ldots$, by the basic operations of addition, subtraction, multiplication, and division, we need basic differentiation rules for such operations. We begin with the following rules.

> **Theorem**   Suppose a is a constant and $f(x)$ and $g(x)$ are differentiable at x. Then the following hold:
>
> **1.** $\dfrac{d}{dx}[af(x)] = a\dfrac{d}{dx} f(x)$
>
> **2.** $\dfrac{d}{dx}[f(x) + g(x)] = \dfrac{d}{dx} f(x) + \dfrac{d}{dx} g(x)$

Rule 1 says that a constant factor can be pulled out; Rule 2 says that the derivative of a sum of two functions is equal to the sum of the derivatives of the functions. Since $f(x) - g(x) = f(x) + (-1)g(x)$, the derivative of a difference of functions is the difference of the derivatives:

$$\frac{d}{dx}[f(x) - g(x)] = \frac{d}{dx}[f(x) + (-1)g(x)] = \frac{d}{dx} f(x) + \frac{d}{dx}[(-1)g(x)]$$

Using Rule 1 on the rightmost term, we find that $\frac{d}{dx}[(-1)g(x)] = (-1)\frac{d}{dx} g(x)$. Therefore,

$$\frac{d}{dx}[f(x) - g(x)] = \frac{d}{dx} f(x) - \frac{d}{dx} g(x)$$

Rules 1 and 2 allow us to differentiate polynomials, as illustrated in the following three examples.

▶ **Example 2**   Differentiate $y = 2x^4 - 3x^3 + x - 7$.

Solution

$$\frac{d}{dx}(2x^4 - 3x^3 + x - 7) = \frac{d}{dx}(2x^4) - \frac{d}{dx}(3x^3) + \frac{d}{dx}x - \frac{d}{dx}7$$

$$= 2\frac{d}{dx}x^4 - 3\frac{d}{dx}x^3 + \frac{d}{dx}x - \frac{d}{dx}7$$

$$= 2(4x^3) - 3(3x^2) + 1 - 0 = 8x^3 - 9x^2 + 1 \quad ◀$$

▶ **Example 3**

(a) $\frac{d}{dx}(-5x^7 + 2x^3 - 10) = -35x^6 + 6x^2$

(b) $\frac{d}{dt}(t^3 - 8t^2 + 3t) = 3t^2 - 16t + 3$

(c) Suppose that n is a positive integer and a is a constant. Then $\frac{d}{ds}(as^n) = ans^{n-1}$.

(d) $\frac{d}{dN}(\ln 5 + N \ln 7) = \ln 7$

(e) $\frac{d}{dr}(r^2 \sin \frac{\pi}{4} - r^3 \cos \frac{\pi}{12} + \sin \frac{\pi}{6}) = 2r \sin \frac{\pi}{4} - 3r^2 \cos \frac{\pi}{12}$ ◀

In the previous section we related the derivative to the slope of the tangent line; the following example uses this interpretation.

▶ **Example 4**   If $f(x) = 2x^3 - 3x + 1$, find the tangent and the normal line at $(-1, 2)$.

Solution

The slope of the tangent line at $(-1, 2)$ is $f'(-1)$. We calculate this as follows:

$$f'(x) = 6x^2 - 3$$

Evaluating $f'(x)$ at $x = -1$, we get

$$f'(-1) = 6(-1)^2 - 3 = 3$$

Therefore, the equation of the tangent line at $(-1, 2)$ is

$$y - 2 = 3(x - (-1)) \quad \text{or} \quad y = 3x + 5$$

To find the equation of the normal line, recall that the normal line is perpendicular to the tangent line; hence, the slope of the normal line, m, is given by

$$m = -\frac{1}{f'(-1)} = -\frac{1}{3}$$

The normal line goes through the point $(-1, 2)$ as well. The equation of the normal line is therefore

$$y - 2 = -\frac{1}{3}(x - (-1)) \quad \text{or} \quad y = -\frac{1}{3}x + \frac{5}{3}$$

The graph of $f(x)$, including the tangent and the normal line at $(-1, 2)$, is shown in Figure 4.22. ◀

Look again at the last example: When we computed $f'(-1)$, we *first* computed $f'(x)$; the *second* step was to evaluate $f'(x)$ at $x = -1$. It makes no sense to plug -1 into $f(x)$ and then differentiate the result. Since $f(-1) = 2$ is a constant, the derivative would be 0, which is obviously not $f'(-1)$. The notation $f'(-1)$ means that we evaluate the function $f'(x)$ at $x = -1$.

4.2.1 Problems

Differentiate the functions given in Problems 1–22, with respect to the independent variable.

1. $f(x) = 2x^3 - 3x + 1$

2. $f(x) = 3x^2 - 4x^4$

3. $f(x) = -2x^5 + 7x - 4$

4. $f(x) = -3x^4 + 6x^2 - 2$

5. $f(x) = 3 - 4x - 5x^2$

6. $f(x) = 8x^4 + 2x^2 - 1$

7. $g(s) = 5s^7 + 2s^3 - 5s$

8. $g(s) = 3 - 4s^2 - 4s^3$

9. $h(t) = \frac{3}{6}t^4 + 4t$

10. $h(t) = \frac{1}{2}t^2 - 3t + 2$

11. $f(x) = x^2 \sin \frac{\pi}{3} + \tan \frac{\pi}{4}$

12. $f(x) = 2x^3 \cos \frac{\pi}{4} + \cos \frac{\pi}{4}$

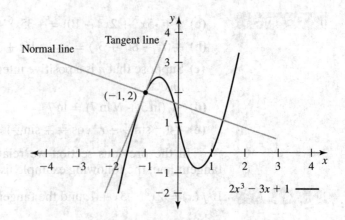

▲ **Figure 4.22**
The graph of $f(x) = 2x^3 - 3x + 1$, together with the tangent and normal lines at $(-1, 2)$

13. $f(x) = -3x^4 \tan \dfrac{\pi}{6} - \cot \dfrac{\pi}{6}$

14. $f(x) = x^2 \sec \dfrac{\pi}{6} + 3x \sec \dfrac{\pi}{4}$

15. $f(t) = t^3 e^{-2} + t$

16. $f(x) = \dfrac{1}{2}x^2 e^3 - x^4$

17. $f(s) = s^3 e^3 + 3e$

18. $f(x) = \dfrac{x}{\pi} + \pi x + \pi$

19. $f(x) = 20x^3 - 4x^6 + 9x^8$

20. $f(x) = \dfrac{x^3}{15} - \dfrac{x^4}{20} + \dfrac{2}{15}$

21. $f(x) = 2^3 x^3 - \dfrac{1}{2^3} + \dfrac{x}{2^3}$

22. $f(x) = \pi x e^2 - \dfrac{x^2 \pi}{e}$

23. Differentiate
$$f(x) = ax^3$$
with respect to x. Assume that a is a constant.

24. Differentiate
$$f(x) = x^3 + a$$
with respect to x. Assume that a is a constant.

25. Differentiate
$$f(x) = ax^2 - 2a$$
with respect to x. Assume that a is a constant.

26. Differentiate
$$f(x) = a^2 x^4 - 2ax^2$$
with respect to x. Assume that a is a constant.

27. Differentiate
$$h(s) = rs^2 - r$$
with respect to s. Assume that r is a constant.

28. Differentiate
$$f(r) = rs^2 - r$$
with respect to r. Assume that s is a constant.

29. Differentiate
$$f(x) = rs^2 x^3 - rx + s$$
with respect to x. Assume that r and s are constants.

30. Differentiate
$$f(x) = \dfrac{r + x}{rs^2} - rsx + (r + s)x - rs$$
with respect to x. Assume that r and s are nonzero constants.

31. Differentiate
$$f(N) = (b - 1)N^4 - \dfrac{N^2}{b}$$
with respect to N. Assume that b is a nonzero constant.

32. Differentiate
$$f(N) = \dfrac{bN^2 + N}{K + b}$$
with respect to N. Assume that b and K are positive constants.

33. Differentiate
$$g(t) = a^3 t - at^3$$
with respect to t. Assume that a is a constant.

34. Differentiate
$$h(s) = a^4 s^2 - as^4 + \dfrac{s^2}{a^4}$$
with respect to s. Assume that a is a positive constant.

35. Differentiate
$$V(t) = V_0(1 + \gamma t)$$
with respect to t. Assume that V_0 and γ are positive constants.

36. Differentiate
$$p(T) = \dfrac{NkT}{V}$$
with respect to T. Assume that N, k, and V are positive constants.

37. Differentiate
$$g(N) = N\left(1 - \dfrac{N}{K}\right)$$
with respect to N. Assume that K is a positive constant.

38. Differentiate
$$g(N) = rN\left(1 - \dfrac{N}{K}\right)$$
with respect to N. Assume that K and r are positive constants.

39. Differentiate

$$g(N) = rN^2 \left(1 - \frac{N}{K}\right)$$

with respect to N. Assume that K and r are positive constants.

40. Differentiate

$$g(N) = rN(a - N)\left(1 - \frac{N}{K}\right)$$

with respect to N. Assume that r, a, and K are positive constants.

41. Differentiate

$$R(T) = \frac{2\pi^5}{15} \frac{k^4}{c^2 h^3} T^4$$

with respect to T. Assume that k, c, and h are positive constants.

In Problems 42–48, find the tangent line, in standard form, to $y = f(x)$ at the indicated point.

42. $y = 3x^2 - 4x + 7$, at $x = 2$

43. $y = 7x^3 + 2x - 1$, at $x = -3$

44. $y = -2x^3 - 3x + 1$, at $x = 1$

45. $y = \frac{1}{\sqrt{2}} x^2 - \sqrt{2}$, at $x = 4$

46. $y = 3\pi x^5 - \frac{\pi}{2} x^3$, at $x = -1$

47. $y = 2x^4 - 3x$, at $x = 2$

48. $y = -3x^3 - 2x^2$, at $x = 0$

In Problems 49–54, find the normal line, in standard form, to $y = f(x)$ at the indicated point.

49. $y = 4x^3 - 3x^3$, at $x = -1$

50. $y = 1 - 3x^2$, at $x = -2$

51. $y = \sqrt{3}x^4 - 2\sqrt{3}x^2$, at $x = -\sqrt{3}$

52. $y = -e^2 x^2 - ex$, at $x = 0$

53. $y = \frac{x^3}{\sqrt{3}} - \sqrt{3}x^3$, at $x = 1$

54. $y = 1 - \pi x^2$, at $x = -1$

55. Find the tangent line to

$$f(x) = ax^2$$

at $x = 1$. Assume that a is a positive constant.

56. Find the tangent line to

$$f(x) = ax^3 - 2ax$$

at $x = -1$. Assume that a is a positive constant.

57. Find the tangent line to

$$f(x) = \frac{ax^2}{a^2 + 2}$$

at $x = 2$. Assume that a is a positive constant.

58. Find the tangent line to

$$f(x) = \frac{x^2}{a + 1}$$

at $x = a$. Assume that a is a positive constant.

59. Find the normal line to

$$f(x) = ax^3$$

at $x = -1$. Assume that a is a positive constant.

60. Find the normal line to

$$f(x) = ax^2 - 3ax$$

at $x = 2$. Assume that a is a positive constant.

61. Find the normal line to

$$f(x) = \frac{ax^2}{a + 1}$$

at $x = 2$. Assume that a is a positive constant.

62. Find the normal line to

$$f(x) = \frac{x^3}{a + 1}$$

at $x = 2a$. Assume that a is a positive constant.

In Problems 63–70, find the coordinates of all the points of the graph of $y = f(x)$ that have horizontal tangents.

63. $f(x) = x^2$

64. $f(x) = 2 - x^2$

65. $f(x) = 3x - x^2$

66. $f(x) = 4x + 2x^2$

67. $f(x) = 3x^3 - x^2$

68. $f(x) = -4x^4 + x^3$

69. $f(x) = \frac{1}{2}x^4 - \frac{7}{3}x^3 - 2x^2$

70. $f(x) = 3x^5 - \frac{3}{2}x^4$

71. Find a point on the curve

$$y = 4 - x^2$$

whose tangent line is parallel to the line $y = 2$. Is there more than one such point? If so, find all other points with this property.

72. Find a point on the curve

$$y = 4 - x^2$$

whose tangent line is parallel to the line $y = -3$. Is there more than one such point? If so, find all other points with this property.

73. Find a point on the curve

$$y = 4 - x^2$$

whose tangent line is parallel to the line $y = x$. Is there more than one such point? If so, find all other points with this property.

74. Find a point on the curve

$$y = 4 - x^2$$

whose tangent line is parallel to the line $y = -x$. Is there more than one such point? If so, find all other points with this property.

75. Find a point on the curve

$$y = x^3 + 2x + 2$$

whose tangent line is parallel to the line $3x - y = 2$. Is there more than one such point? If so, find all other points with this property.

76. Find a point on the curve

$$y = 2x^3 - 4x + 1$$

whose tangent line is parallel to the line $y - 2x = 1$. Is there more than one such point? If so, find all other points with this property.

77. Show that the tangent line to the curve

$$y = x^2$$

at the point $(1, 1)$ passes through the point $(0, -1)$.

78. Find all tangent lines to the curve
$$y = x^2$$
that pass through the point $(0, -1)$.

79. Find all tangent lines to the curve
$$y = x^2$$
that pass through the point $(0, -a^2)$ where a is a positive number.

80. How many tangent lines to the curve
$$y = x^2 + 2x$$
pass through the point $(-\frac{1}{2}, -3)$?

81. Suppose that $P(x)$ is a polynomial of degree 4. Is $P'(x)$ a polynomial as well? If yes, what is its degree?

82. Suppose that $P(x)$ is a polynomial of degree k. Is $P'(x)$ a polynomial as well? If yes, what is its degree?

4.3 THE PRODUCT AND QUOTIENT RULES, AND THE DERIVATIVES OF RATIONAL AND POWER FUNCTIONS

4.3.1 The Product Rule

The derivative of a sum of differentiable functions is the sum of the derivatives of the functions; the rule for products is not so simple. This can be seen from the following example. Consider $y = x^5 = (x^3)(x^2)$. We know that

$$\frac{d}{dx}x^5 = 5x^4$$

$$\frac{d}{dx}x^3 = 3x^2$$

$$\frac{d}{dx}x^2 = 2x$$

This shows that

$$\frac{d}{dx}x^5 \quad \text{is not equal to} \quad \left(\frac{d}{dx}x^3\right)\left(\frac{d}{dx}x^2\right)$$

(Leibniz first thought that the multiplication rule was as simple as that, but he quickly realized his mistake and found the correct formula for differentiating products of functions.)

THE PRODUCT RULE

If $h(x) = f(x)g(x)$ and both $f(x)$ and $g(x)$ are differentiable at x, then
$$h'(x) = f'(x)g(x) + f(x)g'(x)$$
If we set $u = f(x)$ and $v = g(x)$, then
$$(uv)' = u'v + uv'$$

Proof Since $h(x) = f(x)g(x)$ is a product of two functions, we can visualize $h(x)$ as the area of a rectangle with sides $f(x)$ and $g(x)$. We need $h(x + \Delta x)$ to compute the derivative; this is given by

$$h(x + \Delta x) = f(x + \Delta x)g(x + \Delta x)$$

To compute $h'(x)$, we must compute $h(x + \Delta x) - h(x)$ (Figure 4.23). We find

$$h(x + \Delta x) - h(x) = \text{area of I} + \text{area of II}$$
$$= [f(x + \Delta x) - f(x)]g(x)$$
$$\quad + [g(x + \Delta x) - g(x)]f(x + \Delta x)$$

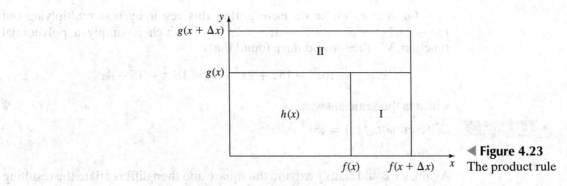

◀ **Figure 4.23**
The product rule

Dividing this result by Δx and taking the limit $\Delta x \to 0$, we obtain

$$h'(x) = \lim_{\Delta x \to 0} \frac{h(x + \Delta x) - h(x)}{\Delta x}$$

$$= \lim_{\Delta x \to 0} \frac{[f(x + \Delta x) - f(x)]g(x) + [g(x + \Delta x) - g(x)]f(x + \Delta x)}{\Delta x}$$

$$= \lim_{\Delta x \to 0} \left[\frac{f(x + \Delta x) - f(x)}{\Delta x} g(x) + \frac{g(x + \Delta x) - g(x)}{\Delta x} f(x + \Delta x) \right]$$

Now, we need the assumption that $f'(x)$ and $g'(x)$ exist and that $f(x)$ is continuous at x [which follows from the fact that $f(x)$ is differentiable at x]. This allows us to use the basic rules for limits, and we write the last expression as

$$\left(\lim_{\Delta x \to 0} \frac{f(x + \Delta x) - f(x)}{\Delta x} \right) g(x) + \left(\lim_{\Delta x \to 0} \frac{g(x + \Delta x) - g(x)}{\Delta x} \right)$$

$$\times \left(\lim_{\Delta x \to 0} f(x + \Delta x) \right)$$

The limits of the difference quotients are the respective derivatives. Using the fact that $f(x)$ is continuous at x, we find $\lim_{\Delta x \to 0} f(x + \Delta x) = f(x)$. Therefore,

$$h'(x) = f'(x)g(x) + g'(x)f(x)$$

as claimed. ■

▶ **Example 1** Differentiate $f(x) = (3x + 1)(2x^2 - 5)$.

Solution

We write $u = 3x + 1$ and $v = 2x^2 - 5$. The product rule says that $(uv)' = u'v + uv'$. That is, we need the derivatives of both u and v:

$$u = 3x + 1 \qquad\qquad v = 2x^2 - 5$$
$$u' = 3 \qquad\qquad v' = 4x$$

Then

$$(uv)' = u'v + uv'$$

$$= 3(2x^2 - 5) + (3x + 1)(4x)$$

$$= 6x^2 - 15 + 12x^2 + 4x = 18x^2 + 4x - 15$$

Of course, we could have gotten this result by first multiplying out $(3x + 1)(2x^2 - 5) = 6x^3 - 15x + 2x^2 - 5$, which is simply a polynomial function. We then would have found that

$$\frac{d}{dx}(6x^3 - 15x + 2x^2 - 5) = 18x^2 - 15 + 4x$$

which is the same answer. ◄

▶ **Example 2** Differentiate $f(x) = (3x^3 - 2x)^2$.

Solution

Again, we could simply expand the square and then differentiate the resulting polynomial—but we can also use the product rule. To do so, we write $u = v = 3x^3 - 2x$. Then $f(x) = uv$ and $(uv)' = u'v + uv'$. Since $u = v$, it follows that $u' = v'$ and the formula simplifies to $(uv)' = (u^2)' = u'u + uu' = 2uu'$. Since $u' = 9x^2 - 2$, we find that

$$f'(x) = 2(3x^3 - 2x)(9x^2 - 2)$$ ◄

▶ **Example 3** In many population models, the population growth rate depends only on the current population size. We can express this by

$$\frac{dN}{dt} = f(N)$$

where $N(t)$ denotes the size of the population at time t and $f(N)$ is the population growth rate, which depends only on the current population size $N = N(t)$. The per capita growth rate $\frac{1}{N}\frac{dN}{dt}$ is then also just a function of N:

$$\frac{1}{N}\frac{dN}{dt} = g(N)$$

with

$$f(N) = Ng(N)$$

Assume that $g(N)$ is differentiable and that $\lim_{N\to 0^+} g(N)$ and $\lim_{N\to 0^+} g'(N)$ exist. Show that

$$g(0) = \lim_{N\to 0^+} \frac{d}{dN} f(N)$$

Solution

We compute the derivative of the population growth rate $f(N) = Ng(N)$ using the product rule

$$\frac{d}{dN}(Ng(N)) = g(N) + Ng'(N)$$

and then take the limit $N \to 0^+$:

$$\lim_{N\to 0^+} \frac{d}{dN}(Ng(N)) = \lim_{N\to 0^+} [g(N) + Ng'(N)] = g(0)$$

(Note that we can only take one-sided limits here since $N \geq 0$ for biological reasons.) ◄

4.3.2 The Quotient Rule

The quotient rule will allow us to compute the derivative of a quotient of two functions. In particular, it will allow us to compute the derivative of a rational function, because it is the quotient of two polynomial functions.

THE QUOTIENT RULE

> If $h(x) = \frac{f(x)}{g(x)}$, $g(x) \neq 0$ and both $f'(x)$ and $g'(x)$ exist, then
>
> $$h'(x) = \frac{f'(x)g(x) - f(x)g'(x)}{[g(x)]^2}$$
>
> In short, with $u = f(x)$ and $v = g(x)$,
>
> $$\left(\frac{u}{v}\right)' = \frac{u'v - uv'}{v^2}$$

We could prove the quotient rule much as we did the product rule, by using the formal definition of derivatives, but that would not be very exciting. Instead, we will give a different proof of the quotient rule in the next subsection.

Carefully note the exact form of the product and the quotient rules. In the product rule, we add $f'g$ and fg', whereas in the quotient rule, we subtract fg' from $f'g$. As mentioned, we can use the quotient rule to find the derivative of rational functions. We illustrate this in the following two examples.

▶ **Example 4**   Differentiate $y = \frac{x^3 - 3x + 2}{x^2 + 1}$. (This function is defined for all $x \in \mathbf{R}$, since $x^2 + 1 \neq 0$.)

Solution

We set $u = x^3 - 3x + 2$ and $v = x^2 + 1$. Both u and v are polynomials, which we know how to differentiate. We find

$$u = x^3 - 3x + 2 \qquad\qquad v = x^2 + 1$$
$$u' = 3x^2 - 3 \qquad\qquad v' = 2x$$

Using the quotient rule, we can compute y':

$$y' = \frac{u'v - uv'}{v^2} = \frac{(3x^2 - 3)(x^2 + 1) - (x^3 - 3x + 2)2x}{(x^2 + 1)^2}$$

$$= \frac{3x^4 + 3x^2 - 3x^2 - 3 - 2x^4 + 6x^2 - 4x}{(x^2 + 1)^2}$$

$$= \frac{x^4 + 6x^2 - 4x - 3}{(x^2 + 1)^2}$$

◀

▶ **Example 5**   Differentiate the Monod growth function

$$f(R) = \frac{aR}{k + R}, \qquad R \geq 0$$

where a and k are positive constants.

Solution

Since a and k are positive constants, $f(R)$ is defined for all $R \geq 0$. We write $u = aR$ and $v = k + R$ and obtain

$$u = aR \qquad\qquad v = k + R$$
$$u' = a \qquad\qquad v' = 1$$

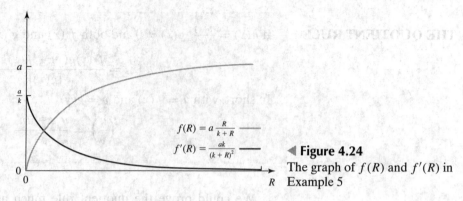

◀ Figure 4.24
The graph of $f(R)$ and $f'(R)$ in Example 5

Hence,

$$\frac{d}{dR}f(R) = \frac{u'v - uv'}{v^2} = \frac{a(k+R) - aR \cdot 1}{(k+R)^2} = \frac{ak}{(k+R)^2}$$

In Figure 4.24, we graph both $f(R)$ and $f'(R)$. We see that the slope of the tangent line at $(R, f(R))$ is positive for all $R \geq 0$. We can also conclude this from the graph of $f'(R)$ since it is positive for all $R \geq 0$. Furthermore, we see that $f(R)$ becomes less steep as R increases, which is reflected in the fact that $f'(R)$ becomes smaller as R increases. ◀

The quotient rule allows us to extend the power rule to the case where the exponent is a negative integer.

POWER RULE (NEGATIVE INTEGER EXPONENTS)

> If $f(x) = x^{-n}$, where n is a positive integer, then
> $$f'(x) = -nx^{-n-1}$$

Note that the power rule for negative integer exponents works the same way as the power rule for positive integer exponents: We write the exponent of the original function in front and decrease the exponent by 1. We will now prove the power rule for negative integer exponents.

Proof We write $f(x) = \frac{1}{x^n}$, and set $u = 1$ and $v = x^n$. Then

$$u = 1 \qquad\qquad v = x^n$$
$$u' = 0 \qquad\qquad v' = nx^{n-1}$$

and, therefore,

$$f'(x) = \frac{u'v - uv'}{v^2} = \frac{0 \cdot x^n - 1 \cdot nx^{n-1}}{(x^n)^2} = -\frac{nx^{n-1}}{x^{2n}} = -nx^{-n-1} \qquad ■$$

▶ **Example 6**   **(a)** If $y = \frac{1}{x}$, then

$$y' = \frac{d}{dx}(x^{-1}) = (-1)x^{-1-1} = -\frac{1}{x^2}$$

(b) If $g(x) = \frac{3}{x^4}$, then

$$g'(x) = \frac{d}{dx}(3x^{-4}) = 3\frac{d}{dx}x^{-4} = 3(-4)x^{-4-1}$$

$$= -12x^{-5} = -\frac{12}{x^5} \qquad ◀$$

There is a general form of the power rule, in which the exponent can be any real number. In the next section, we will give the proof for the case when the exponent is rational, and we will prove the general case in Section 4.7.

POWER RULE (GENERAL FORM)

> Let $f(x) = x^r$, where r is any real number. Then
> $$f'(x) = rx^{r-1}$$

▶ Example 7

(a) If $y = \sqrt{x}$, then

$$y' = \frac{d}{dx}\left(x^{1/2}\right) = \frac{1}{2}x^{\frac{1}{2}-1} = \frac{1}{2}x^{-1/2} = \frac{1}{2\sqrt{x}}$$

(b) If $y = \sqrt[5]{x}$, then

$$y' = \frac{d}{dx}\left(x^{1/5}\right) = \frac{1}{5}x^{(1/5)-1} = \frac{1}{5}x^{-4/5} = \frac{1}{5x^{4/5}}$$

(c) If $g(t) = \frac{1}{\sqrt[3]{t}}$, then

$$g'(t) = \frac{d}{dt}\left(t^{-1/3}\right) = \left(-\frac{1}{3}\right)t^{(-1/3)-1} = \left(-\frac{1}{3}\right)t^{-4/3} = -\frac{1}{3t^{4/3}}$$

(d) If $h(s) = s^{\pi}$, then $h'(s) = \pi s^{\pi-1}$. ◀

The function $f(x) = \sqrt{x}$, $x \geq 0$, appears quite frequently. It is therefore worthwhile to memorize its derivative, which is only defined for $x > 0$:

$$\frac{d}{dx}\sqrt{x} = \frac{1}{2\sqrt{x}}$$

▶ Example 8 **Combining the Rules** Differentiate $f(x) = \sqrt{x}(x^2 - 1)$.

Solution 1

We can consider this as a product of two functions. Let

$$u = \sqrt{x} \qquad v = x^2 - 1$$

$$u' = \frac{1}{2\sqrt{x}} \qquad v' = 2x$$

Hence,

$$f'(x) = u'v + uv' = \frac{1}{2\sqrt{x}}(x^2 - 1) + \sqrt{x}(2x)$$

$$= \frac{x^2 - 1 + \sqrt{x}(2x)2\sqrt{x}}{2\sqrt{x}} = \frac{x^2 - 1 + 4x^2}{2\sqrt{x}} = \frac{5x^2 - 1}{2\sqrt{x}}$$

Solution 2

Since $f(x) = \sqrt{x}(x^2 - 1) = x^{5/2} - x^{1/2}$, we can also use the general version of the power rule. We find that

$$f'(x) = \frac{5}{2}x^{(5/2)-1} - \frac{1}{2}x^{(1/2)-1} = \frac{5}{2}x^{3/2} - \frac{1}{2\sqrt{x}} = \frac{5x^{3/2}\sqrt{x} - 1}{2\sqrt{x}} = \frac{5x^2 - 1}{2\sqrt{x}}$$ ◀

▶ **Example 9** **A Function that Contains a Constant** Differentiate $h(t) = (at)^{1/3}(a+1) - a$, where a is a positive constant.

Solution

Since $h(t)$ is a function of t, we need to differentiate with respect to t, keeping in mind that a is a constant. Rewriting $h(t)$ will make this easier:

$$h(t) = a^{1/3}(a+1)t^{1/3} - a$$

The factor $a^{1/3}(a+1)$ in front of $t^{1/3}$ is a constant. Thus

$$h'(t) = a^{1/3}(a+1)\frac{1}{3}t^{-2/3} - 0 = \frac{a^{1/3}(a+1)}{3t^{2/3}}$$

◀

▶ **Example 10** **Differentiating a Function that Is Not Specified** Suppose $f(2) = 3$ and $f'(2) = 1/4$. Find

$$\frac{d}{dx}[xf(x)]$$

at $x = 2$.

Solution

Since $xf(x)$ is a product, we can use the product rule

$$\frac{d}{dx}[xf(x)] = f(x) + xf'(x)$$

Hence

$$\frac{d}{dx}[xf(x)]|_{x=2} = f(2) + 2f'(2) = 3 + \frac{1}{2} = \frac{7}{2}$$

◀

▶ **Example 11** **Differentiating a Function that Is Not Specified** Suppose that $f(x)$ is differentiable. Find an expression for the derivative of

$$y = \frac{f(x)}{x^2}$$

Solution

We set

$$u = f(x) \qquad\qquad v = x^2$$
$$u' = f'(x) \qquad\qquad v' = 2x$$

and use the quotient rule. We find that

$$y' = \frac{f'(x)x^2 - f(x)2x}{x^4} = \frac{xf'(x) - 2f(x)}{x^3}$$

◀

4.3.3 Problems

(4.3.1)

In Problems 1–16, differentiate with respect to the independent variable.

1. $f(x) = (2x - 1)(2 - x^2)$

2. $f(x) = (3x + 2x^2)(5x^3 - 2)$

3. $f(x) = (x^3 + 17)(3x - 14x^2)$

4. $f(x) = (2x^4 - 3x^2 + 1)(2x - 5x^3)$

5. $f(x) = \left(\frac{1}{2}x^2 - 1\right)(2x + 3x^2)$

6. $f(x) = 2(3x^2 - 2x^3)(1 - 5x^2)$

7. $f(x) = \frac{(x-1)(x+1)}{5}$

8. $f(x) = 3(x^2 + 2)(4x^2 - 5x^4) - 3$

9. $f(x) = (3x - 1)^2$

10. $f(x) = (4 - 2x^2)^2$

11. $f(x) = 3(1 - 2x)^2$

12. $f(x) = \dfrac{(2x^2 - 3x + 1)^2}{4} + 2$

13. $g(s) = (2s^2 - 5s)^2$

14. $h(t) = 4(3t^2 - 1)(2t + 1)$

15. $g(t) = 3(2t^2 - 5t^4)^2$

16. $h(s) = (4 - 3s^2 + 4s^3)^2$

In Problems 17–20, find the tangent line, in slope-intercept form, of $y = f(x)$ at the specified point.

17. $f(x) = (3x^2 - 2)(x - 1)$, at $x = 1$

18. $f(x) = (1 - 2x)(1 + 2x)$, at $x = 2$

19. $f(x) = 4(2x^4 + 3x)(4 - 2x^2)$, at $x = -1$

20. $f(x) = (3x^3 - 3)(2 - 2x^2)$, at $x = 0$

In Problems 21–24, find the normal line, in slope-intercept form, of $y = f(x)$ at the specified point.

21. $f(x) = (1 - x)(2 - x^2)$, at $x = 2$

22. $f(x) = (2x + 1)(3x^2 - 1)$, at $x = 1$

23. $f(x) = 5(1 - 2x)(x + 1) - 3$, at $x = 0$

24. $f(x) = \dfrac{(2 - x)(3 - x)}{4}$, at $x = -1$

In Problems 25–28, apply the product rule repeatedly to find the derivative of $y = f(x)$.

25. $f(x) = (2x - 1)(3x + 4)(1 - x)$

26. $f(x) = (x - 3)(2 - 3x)(5 - x)$

27. $f(x) = (x - 3)(2x^2 + 1)(1 - x^2)$

28. $f(x) = (2x + 1)(4 - x^2)(1 + x^2)$

29. Differentiate
$$f(x) = a(x - 1)(2x - 1)$$
with respect to x. Assume that a is a positive constant.

30. Differentiate
$$f(x) = (a - x)(a + x)$$
with respect to x. Assume that a is a positive constant.

31. Differentiate
$$f(x) = 2a(x^2 - a)^2 + a$$
with respect to x. Assume that a is a positive constant.

32. Differentiate
$$f(x) = \frac{3(x - 1)^2}{2 + a}$$
with respect to x. Assume that a is a positive constant.

33. Differentiate
$$g(t) = (at + 1)^2$$
with respect to t. Assume that a is a positive constant.

34. Differentiate
$$h(t) = \sqrt{a}(t - a) + a$$
with respect to t. Assume that a is a positive constant.

35. Suppose that $f(2) = -4$, $g(2) = 3$, $f'(2) = 1$, and $g'(2) = -2$. Find
$$(fg)'(2)$$

36. Suppose that $f(2) = -4$, $g(2) = 3$, $f'(2) = 1$, and $g'(2) = -2$. Find
$$(f^2 + g^2)'(2)$$

In Problems 37–40, assume that $f(x)$ is differentiable. Find an expression for the derivative of y.

37. $y = 2xf(x)$

38. $y = 3x^2 f(x)$

39. $y = -5x^3 f(x) - 2x$

40. $y = \dfrac{xf(x)}{2}$

In Problems 41–44, assume that $f(x)$ and $g(x)$ are differentiable at x. Find an expression for the derivative of y.

41. $y = 3f(x)g(x)$

42. $y = [f(x) - 3]g(x)$

43. $y = [f(x) + 2g(x)]g(x)$

44. $y = [-2f(x) - 3g(x)]g(x) + \dfrac{2g(x)}{3}$

45. Let $B(t)$ denote the biomass at time t with specific growth rate $g(B)$. Show that the specific growth rate at $B = 0$ is given by the slope of the tangent line on the graph of the growth rate at $B = 0$.

46. Let $N(t)$ denote the size of a population at time t. Differentiate
$$f(N) = rN\left(1 - \frac{N}{K}\right)$$
with respect to N, where r and K are positive constants.

47. Let $N(t)$ denote the size of a population at time t. Differentiate
$$f(N) = r\left(aN - N^2\right)\left(1 - \frac{N}{K}\right)$$
with respect to N, where r, K, and a are positive constants.

48. Consider the chemical reaction
$$A + B \longrightarrow AB$$
If x denotes the concentration of AB at time t, then the reaction rate $R(x)$ is given by
$$R(x) = k(a - x)(b - x)$$
where k, a, and b are positive constants. Differentiate $R(x)$.

(4.3.2)

In Problems 49–70, differentiate with respect to the independent variable.

49. $f(x) = \dfrac{2x + 1}{x + 1}$

50. $f(x) = \dfrac{3x^2 - 2x + 1}{2x + 1}$

51. $f(x) = \dfrac{1 - 2x^2}{1 - x}$

52. $f(x) = \dfrac{3x^3 + 2x - 1}{5x^2 - 2x + 1}$

53. $f(x) = \dfrac{3 - x^3}{1 - x}$

54. $f(x) = \dfrac{1 + 2x^2 - 4x^4}{3x^3 - 5x^5}$

55. $h(t) = \dfrac{t^2 - 3t + 1}{t + 1}$

56. $h(t) = \dfrac{3 - t^2}{(t+1)^2}$

57. $f(s) = \dfrac{4 - 2s^2}{1 - s}$

58. $f(s) = \dfrac{2s^3 - 4s^2 + 5s - 7}{(s^2 - 3)^2}$

59. $f(x) = \sqrt{x}(x - 1)$

60. $f(x) = \sqrt{x}(x^4 - 5x^2)$

61. $f(x) = \sqrt{3x}(x^2 - 1)$

62. $f(x) = \dfrac{\sqrt{5x}(1 + x^2)}{\sqrt{2}}$

63. $f(x) = x^3 - \dfrac{1}{x^3}$

64. $f(x) = x^5 - \dfrac{1}{x^5}$

65. $f(x) = 2x^2 - \dfrac{3x - 1}{x^3}$

66. $f(x) = -x^3 + \dfrac{2x^2 - 3}{4x^4}$

67. $g(s) = \dfrac{s^{1/3} - 1}{s^{2/3} - 1}$

68. $g(s) = \dfrac{s^{1/7} - s^{2/7}}{s^{3/7} + s^{4/7}}$

69. $f(x) = (1 - 2x)\left(\sqrt{2x} + \dfrac{2}{\sqrt{x}}\right)$

70. $f(x) = (x^3 - 3x^2 + 2)\left(\sqrt{x} + \dfrac{1}{\sqrt{x}} - 1\right)$

In Problems 71–74, find the tangent line, in slope-intercept form, of $y = f(x)$ at the specified point.

71. $f(x) = \dfrac{x^2 + 3}{x^3 + 5}$, at $x = -2$

72. $f(x) = \dfrac{3}{x} - \dfrac{4}{\sqrt{x}} + \dfrac{2}{x^2}$, at $x = 1$

73. $f(x) = \dfrac{2x - 5}{x^3}$, at $x = 2$

74. $f(x) = \sqrt{x}(x^3 - 1)$, at $x = 1$

75. Differentiate
$$f(x) = \dfrac{ax}{3 + x}$$
with respect to x. Assume that a is a positive constant.

76. Differentiate
$$f(x) = \dfrac{ax}{k + x}$$
with respect to x. Assume that a and k are positive constants.

77. Differentiate
$$f(x) = \dfrac{ax^2}{4 + x^2}$$
with respect to x. Assume that a is a positive constant.

78. Differentiate
$$f(x) = \dfrac{ax^2}{k^2 + x^2}$$
with respect to x. Assume that a and k are positive constants.

79. Differentiate
$$f(R) = \dfrac{R^2}{k + R}$$
with respect to R. Assume that k is a positive constant.

80. Differentiate
$$h(t) = \sqrt{at}(1 - a) + a$$
with respect to t. Assume that a is a positive constant.

81. Differentiate
$$h(t) = \sqrt{at}(t - a) + at$$
with respect to t. Assume that a is a positive constant.

82. Suppose that $f(2) = -4$, $g(2) = 3$, $f'(2) = 1$, and $g'(2) = -2$. Find
$$\left(\dfrac{1}{f}\right)'(2)$$

83. Suppose that $f(2) = -4$, $g(2) = 3$, $f'(2) = 1$, and $g'(2) = -2$. Find
$$\left(\dfrac{f}{2g}\right)'(2)$$

In Problems 84–87, assume that $f(x)$ is differentiable. Find an expression for the derivative of y.

84. $y = \dfrac{f(x)}{x^2 + 1}$

85. $y = \dfrac{x^2 + 4f(x)}{f(x)}$

86. $y = [f(x)]^2 - \dfrac{x}{f(x)}$

87. $y = \dfrac{f(x)}{f(x) + x}$

In Problems 88–91, assume that $f(x)$ and $g(x)$ are differentiable at x. Find an expression for the derivative of y.

88. $y = \dfrac{2f(x) + 1}{3g(x)}$

89. $y = \dfrac{f(x)}{[g(x)]^2}$

90. $y = \dfrac{x^2}{f(x) - g(x)}$

91. $y = \sqrt{x}\, f(x) g(x)$

92. Assume that $f(x)$ is a differentiable function. Find the derivative of the reciprocal function $g(x) = 1/f(x)$ at those points x where $f(x) \neq 0$.

93. Find the tangent line to the hyperbola $yx = c$, c a positive constant, at the point (x_1, y_1) with $x_1 > 0$. Show that the tangent line intersects the x-axis at a point that does not depend on c.

94. (*Adapted from Roff, 1992*) The males in the frog species *Eleutherodactylus coqui* (found in Puerto Rico) take care of their brood. While they protect the eggs, they cannot find other mates, and therefore cannot increase their number of offspring. On the other hand, if they do not spend enough time with their brood, then the offspring might not survive. Simple mathematical models are used to give the proportion of hatching offspring per unit time, $w(t)$, as a function of the probability of hatching if time t is spent brooding, $f(t)$, and the cost C associated with the time spent searching for other mates. Given the following relationship
$$w(t) = \dfrac{f(t)}{C + t}$$
find the derivative of $w(t)$.

4.4 THE CHAIN RULE AND HIGHER DERIVATIVES

4.4.1 The Chain Rule

In Section 1.2, we defined the composition of functions. To find the derivative of composite functions, we need the **chain rule** (the proof of the chain rule is at the end of this section).

CHAIN RULE
> If g is differentiable at x and f is differentiable at $y = g(x)$, then the composite function $(f \circ g)(x) = f[g(x)]$ is differentiable at x, and the derivative is given by
> $$(f \circ g)'(x) = f'[g(x)]g'(x)$$

This formula looks complicated. Let's take a moment to see what we need to do to find the derivative of the composite function $(f \circ g)(x)$. The function g is the inner function; the function f is the outer function. The expression $f'[g(x)]g'(x)$ thus means that we need to find the derivative of the outer function evaluated at $g(x)$ and the derivative of the inner function evaluated at x, and then multiply the two together.

▶ **Example 1**   **A Polynomial**   Find the derivative of

$$h(x) = (3x^2 - 1)^2$$

Solution

The inner function is $g(x) = 3x^2 - 1$; the outer function is $f(u) = u^2$. Then

$$g'(x) = 6x \quad \text{and} \quad f'(u) = 2u$$

Evaluating $f'(u)$ at $u = g(x)$ yields

$$f'[g(x)] = 2g(x) = 2(3x^2 - 1)$$

Thus,

$$h'(x) = (f \circ g)'(x) = f'[g(x)]g'(x)$$
$$= 2(3x^2 - 1)6x = 12x(3x^2 - 1) \qquad \blacktriangleleft$$

The derivative of $f \circ g$ can be written in Leibniz notation. If we set $u = g(x)$, then

$$\frac{d}{dx}[(f \circ g)(x)] = \frac{df}{du}\frac{du}{dx}$$

This form of the chain rule emphasizes that in order to differentiate $f \circ g$, we multiply the derivative of the outer function and the derivative of the inner function, the former evaluated at u, the latter at x.

▶ **Example 2**   **A Polynomial**   Find the derivative of

$$h(x) = (2x + 1)^3$$

Solution

If we set $u = g(x) = 2x + 1$ and $f(u) = u^3$, then $h(x) = (f \circ g)(x)$. We need to find both $f'[g(x)]$ and $g'(x)$ to compute $h'(x)$. Now,

$$g'(x) = 2 \quad \text{and} \quad f'(u) = 3u^2$$

Hence, since $f'[g(x)] = 3(g(x))^2 = 3(2x+1)^2$,

$$h'(x) = f'[g(x)]g'(x) = 3(2x+1)^2 \cdot 2$$

$$= 6(2x+1)^2$$

If we use Leibniz notation, this becomes

$$h'(x) = \frac{df}{du}\frac{du}{dx} = 3u^2 \cdot 2 = 3(2x+1)^2 \cdot 2$$

$$= 6(2x+1)^2 \qquad \blacktriangleleft$$

▶ **Example 3** **A Radical** Find the derivative of $h(x) = \sqrt{x^2+1}$.

Solution

If we set $u = g(x) = x^2 + 1$ and $f(u) = \sqrt{u}$, then $h(x) = (f \circ g)(x)$. We find that

$$g'(x) = 2x \quad \text{and} \quad f'(u) = \frac{1}{2\sqrt{u}}$$

We need to evaluate f' at $g(x)$, that is,

$$f'[g(x)] = \frac{1}{2\sqrt{g(x)}} = \frac{1}{2\sqrt{x^2+1}}$$

Therefore,

$$h'(x) = f'[g(x)]g'(x) = \frac{1}{2\sqrt{x^2+1}}2x = \frac{x}{\sqrt{x^2+1}} \qquad \blacktriangleleft$$

▶ **Example 4** **A Radical** Find the derivative of

$$h(x) = \sqrt[7]{2x^2+3x}$$

Solution

We write

$$h(x) = (2x^2 + 3x)^{1/7}$$

The inner function is $u = g(x) = 2x^2 + 3x$ and the outer function is $f(u) = u^{1/7}$. Thus, we find

$$h'(x) = \frac{df}{du}\frac{du}{dx} = \frac{1}{7}u^{1/7-1}(4x+3)$$

$$= \frac{1}{7}(2x^2+3x)^{-6/7}(4x+3)$$

$$= \frac{4x+3}{7(2x^2+3x)^{6/7}} \qquad \blacktriangleleft$$

▶ **Example 5** **A Rational Function** Find the derivative of $h(x) = \left(\frac{x}{x+1}\right)^2$.

Solution

If we set $u = g(x) = \frac{x}{x+1}$ and $f(u) = u^2$, then $h(x) = (f \circ g)(x)$. We use the quotient rule to compute the derivative of $g(x)$:

$$g'(x) = \frac{1 \cdot (x+1) - x \cdot 1}{(x+1)^2} = \frac{1}{(x+1)^2}$$

Since $f'(u) = 2u$, we find

$$h'(x) = f'[g(x)]g'(x) = 2\frac{x}{x+1}\frac{1}{(x+1)^2} = \frac{2x}{(x+1)^3} \qquad \blacktriangleleft$$

The Proof of the Quotient Rule We can use the chain rule to prove the quotient rule. Assume that $g(x) \neq 0$ for all x in the domain of g. If we define $h(x) = \frac{1}{x}$, then

$$(h \circ g)(x) = h[g(x)] = \frac{1}{g(x)}$$

We used the formal definition of the derivative in Example 3 in Section 4.1 to show that $h'(x) = -\frac{1}{x^2}$. This, together with the chain rule, yields

$$(h \circ g)'(x) = -\frac{1}{[g(x)]^2} g'(x) \quad \text{or} \quad \left(\frac{1}{g}\right)' = -\frac{g'}{g^2}$$

Since $\frac{f}{g} = f\frac{1}{g}$, we can use the product rule to find the derivative of $\frac{f}{g}$.

$$\left(\frac{f}{g}\right)' = f'\frac{1}{g} + f\left(\frac{1}{g}\right)' = f'\frac{1}{g} + f\left(-\frac{g'}{g^2}\right)$$

$$= \frac{f'g - fg'}{g^2}$$

Note that we did not use the power rule for negative integer exponents (Subsection 4.3.2) to compute $h'(x)$, but instead used the formal definition of derivatives to compute the derivative of $1/x$. Using the power rule for negative integer exponents would have been circular reasoning; we used the quotient rule to prove the power rule for negative integer exponents, and so we cannot use the power rule for negative integer exponents to prove the quotient rule. ■

▶ **Example 6** **A Function with Parameters** Find the derivative of

$$h(x) = (ax^2 - 2)^n$$

where $a > 0$ and n is a positive integer.

Solution

If we set $u = g(x) = ax^2 - 2$ and $f(u) = u^n$, then $h(x) = (f \circ g)(x)$. Since

$$g'(x) = 2ax \quad \text{and} \quad f'(u) = nu^{n-1}$$

we find

$$h'(x) = f'[g(x)]g'(x) = n(ax^2 - 2)^{n-1}2ax$$

$$= 2anx(ax^2 - 2)^{n-1}$$

Looking at $h'(x) = n(ax^2 - 2)^{n-1} \cdot 2ax$, we see that we first differentiated the outer function f, which yielded $n(ax^2 - 2)^{n-1}$ using the power rule, and then multiplied the result by the derivative of the inner function g, which is $2ax$. ◀

▶ **Example 7** **Differentiating a Function that Is Not Specified** Suppose $f(x)$ is differentiable. Find

$$\frac{d}{dx} \frac{1}{\sqrt{f(x)}}$$

Solution

We set

$$h(x) = \frac{1}{\sqrt{f(x)}} = [f(x)]^{-1/2}$$

Now, $u = f(x)$ is the inner function and $h(u) = u^{-1/2}$ is the outer function; hence

$$\frac{d}{dx}h(x) = \frac{dh}{du}\frac{du}{dx} = -\frac{1}{2}u^{-3/2}f'(x)$$

$$= -\frac{1}{2u^{3/2}}f'(x) = -\frac{f'(x)}{2[f(x)]^{3/2}} \quad \blacktriangleleft$$

▶ **Example 8** **Generalized Power Rule** Suppose $f(x)$ is differentiable and r is a real number. Find

$$\frac{d}{dx}[f(x)]^r$$

Solution

Using the general form of the power rule and the chain rule, we find

$$\frac{d}{dx}[f(x)]^r = r[f(x)]^{r-1}f'(x) \quad \blacktriangleleft$$

▶ **Example 9** **Differentiating a Function that Is Not Specified** Suppose that $f'(x) = 3x - 1$. Find

$$\frac{d}{dx}f(x^2) \quad \text{at } x = 3$$

Solution

The inner function is $u = x^2$, the outer function is $f(u)$, and we find

$$\frac{d}{dx}f(x^2) = 2xf'(x^2)$$

If we plug $x = 3$ into $f'(x^2)$, we find $f'(3^2) = f'(9) = (3)(9) - 1 = 26$. Thus

$$\frac{d}{dx}f(x^2)|_{x=3} = (2)(3)f'(9) = (6)(26) = 156 \quad \blacktriangleleft$$

The chain rule can be applied repeatedly, as shown in the next two examples.

▶ **Example 10** **Nested Chain Rule** Find the derivative of

$$h(x) = \left(\sqrt{x^2+1}+1\right)^2$$

Solution

If we set $h(x) = (f \circ g)(x)$, then $g(x) = \sqrt{x^2+1}+1$ and $f(u) = u^2$. We see that $g(x)$ is itself a composition of two functions, with inner function $v = x^2 + 1$ and outer function $\sqrt{v} + 1$. To differentiate $h(x)$, we proceed stepwise,

$$h'(x) = \frac{d}{dx}\left(\sqrt{x^2+1}+1\right)^2 = 2\left(\sqrt{x^2+1}+1\right)\frac{d}{dx}\left(\sqrt{x^2+1}+1\right)$$

Since

$$\frac{d}{dx}\left(\sqrt{x^2+1}+1\right) = \frac{2x}{2\sqrt{x^2+1}} = \frac{x}{\sqrt{x^2+1}}$$

where we used the chain rule to differentiate $\sqrt{x^2+1}$, we find

$$h'(x) = 2\left(\sqrt{x^2+1}+1\right)\frac{x}{\sqrt{x^2+1}} \quad \blacktriangleleft$$

▶ **Example 11** **Nested Chain Rule** Find the derivative of

$$h(x) = \left(2x^3 - \sqrt{3x^4 - 2}\right)^3$$

Solution

As in the previous example, we proceed stepwise:

$$h'(x) = 3\left(2x^3 - \sqrt{3x^4 - 2}\right)^2 \frac{d}{dx}\left(2x^3 - \sqrt{3x^4 - 2}\right)$$

$$= 3\left(2x^3 - \sqrt{3x^4 - 2}\right)^2 \left(6x^2 - \frac{12x^3}{2\sqrt{3x^4 - 2}}\right)$$

$$= 18x^2\left(2x^3 - \sqrt{3x^4 - 2}\right)^2 \left(1 - \frac{x}{\sqrt{3x^4 - 2}}\right) \qquad ◀$$

We conclude this subsection with the proof of the chain rule. The first part of the proof follows along the lines of the argument we sketched out in the beginning of this section, but the second part is much more technical and deals with the problem that Δu could be zero.

Proof of the Chain Rule We will use the definition of the derivative to prove the chain rule. Formally,

$$(f \circ g)'(x) = \lim_{x \to c} \frac{f[g(x)] - f[g(c)]}{x - c}$$

We need to show that the right-hand side is equal to $f'[g(c)]g'(c)$. As long as $g(x) \neq g(c)$, we can write

$$\lim_{x \to c} \frac{f[g(x)] - f[g(c)]}{x - c} = \lim_{x \to c} \frac{\frac{f[g(x)] - f[g(c)]}{g(x) - g(c)}[g(x) - g(c)]}{x - c}$$

$$= \lim_{x \to c} \frac{f[g(x)] - f[g(c)]}{g(x) - g(c)} \frac{g(x) - g(c)}{x - c}$$

Since

$$\lim_{x \to c} \frac{f[g(x)] - f[g(c)]}{g(x) - g(c)} = f'[g(c)] \quad \text{and} \quad \lim_{x \to c} \frac{g(x) - g(c)}{x - c} = g'(c),$$

these limits exist, and we can use the fact that the limit of a product is the product of the limits. We find

$$\lim_{x \to c} \frac{f[g(x)] - f[g(c)]}{x - c} = \lim_{x \to c} \frac{f[g(x)] - f[g(c)]}{g(x) - g(c)} \lim_{x \to c} \frac{g(x) - g(c)}{x - c}$$

$$= f'[g(c)]g'(c)$$

In the preceding calculation, we needed to assume that $g(x) - g(c) \neq 0$. Of course, when we take the limit $x \to c$, there might be x values with $g(x) = g(c)$, and we must deal with this possibility.

We set $y = g(x)$ and $d = g(c)$. The expression

$$f^\star(y) \equiv \frac{f(y) - f(d)}{y - d}$$

is only defined for $y \neq d$. Since

$$\lim_{y \to d} \frac{f(y) - f(d)}{y - d} = f'(d)$$

a natural extension of $f^\star[g(x)]$ that makes $f^\star[g(x)]$ a continuous function is, therefore,

$$f^\star[g(x)] = \begin{cases} \dfrac{f[g(x)] - f[g(c)]}{g(x) - g(c)} & \text{for } g(x) \neq g(c) \\[2ex] f'[g(c)] & \text{for } g(x) = g(c) \end{cases}$$

This means that for all x,

$$f[g(x)] - f[g(c)] = f^\star[g(x)][g(x) - g(c)]$$

With this equivalence, we can repeat our calculations and find

$$\lim_{x \to c} \frac{f[g(x)] - f[g(c)]}{x - c} = \lim_{x \to c} \frac{f^\star[g(x)][g(x) - g(c)]}{x - c}$$

$$\lim_{x \to c} f^\star[g(x)] \lim_{x \to c} \frac{g(x) - g(c)}{x - c} = f'[g(c)] \cdot g'(c)$$

Note that in the last step we used the fact that $f^\star[g(x)]$ is continuous at $x = c$. ∎

4.4.2 Implicit Functions and Implicit Differentiation

So far, we have considered only functions of the form $y = f(x)$, which define y *explicitly* as a function of x. It is also possible to define y *implicitly* as a function of x, as in the following equation:

$$y^5 x^2 - yx + 2y^2 = \sqrt{x}$$

Here, y is given as a function of x, that is, y is the dependent variable. There is no obvious way to solve for y. There is a very useful technique, based on the chain rule, which will allow us to find dy/dx for implicitly defined functions. This technique is called **implicit differentiation**. We explain the procedure in the following example.

▶ **Example 12**   Find $\frac{dy}{dx}$ if $x^2 + y^2 = 1$.

Solution

We differentiate both sides of the equation $x^2 + y^2 = 1$ with respect to x. We need to remember that y is a function of x.

$$\frac{d}{dx}\left(x^2 + y^2\right) = \frac{d}{dx}(1)$$

Since the derivative of a sum is the sum of the derivatives, we find

$$\frac{d}{dx}(x^2) + \frac{d}{dx}(y^2) = \frac{d}{dx}(1)$$

Starting with the left-hand side and using the power rule, we find $\frac{d}{dx}(x^2) = 2x$. To differentiate y^2 with respect to x, we must apply the chain rule. We find $\frac{d}{dx}(y^2) = 2y\frac{dy}{dx}$. On the right-hand side, we obtain $\frac{d}{dx}(1) = 0$. We therefore have

$$2x + 2y\frac{dy}{dx} = 0$$

We can now solve for $\frac{dy}{dx}$ and find

$$\frac{dy}{dx} = -\frac{2x}{2y} = -\frac{x}{y}$$

Since $x^2 + y^2 = 1$ is the equation for the unit circle centered at the origin (Figure 4.25), we can convince ourselves that this is indeed the correct

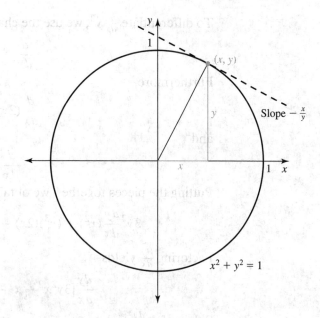

▲ **Figure 4.25**

The slope of the tangent line at the unit circle $x^2 + y^2 = 1$ at (x, y) is $m = -\frac{x}{y}$

derivative by using a geometric argument. The line that connects $(0, 0)$ and (x, y) has slope y/x and is perpendicular to the tangent line at (x, y). Since the slopes of perpendicular lines are negative reciprocals of each other, the slope of the tangent line at (x, y) must be $-x/y$.

We could have solved $x^2 + y^2 = 1$ for y and then differentiated with respect to x; this would have yielded the same answer but would have been more complicated.   ◀

We summarize the steps to find dy/dx when an equation defines y implicitly as a differentiable function of x.

Step 1. Differentiate both sides of the equation with respect to x, keeping in mind that y is a function of x.

Step 2. Solve the resulting equation for dy/dx.

Note that differentiating terms involving y typically requires the chain rule. Here is another example. This time, we can use neither a geometric argument nor easily solve for y.

▶ **Example 13**   Find $\frac{dy}{dx}$ when $y^3 x^2 - yx + 2y^2 = x$.

Solution

We differentiate both sides of the equation with respect to x:

$$\frac{d}{dx}(y^3 x^2) - \frac{d}{dx}(yx) + \frac{d}{dx}(2y^2) = \frac{d}{dx}(x)$$

To differentiate $y^3 x^2$ and yx with respect to x, we use the product rule:

$$\frac{d}{dx}(y^3 x^2) = \left(\frac{d}{dx}y^3\right)x^2 + y^3\left(\frac{d}{dx}x^2\right) = \left(\frac{d}{dx}y^3\right)x^2 + (y^3)(2x)$$

$$\frac{d}{dx}(yx) = \left(\frac{d}{dx}y\right)(x) + y\left(\frac{d}{dx}x\right) = \left(\frac{dy}{dx}\right)x + y$$

To differentiate $\frac{d}{dx}y^3$, we use the chain rule and find

$$\frac{d}{dx}y^3 = 3y^2\frac{dy}{dx}$$

Furthermore,

$$\frac{d}{dx}(2y^2) = 4y\frac{dy}{dx}$$

and

$$\frac{d}{dx}(x) = 1$$

Putting the pieces together, we obtain

$$3y^2\frac{dy}{dx}(x^2) + (y^3)(2x) - \left[\left(\frac{dy}{dx}\right)x + y\right] + 4y\frac{dy}{dx} = 1$$

Factoring $\frac{dy}{dx}$ yields

$$\frac{dy}{dx}[3y^2x^2 - x + 4y] + 2xy^3 - y = 1$$

Solving for $\frac{dy}{dx}$ yields

$$\frac{dy}{dx} = \frac{y + 1 - 2xy^3}{3y^2x^2 - x + 4y}$$

◄

The next example prepares us for the power rule for rational exponents.

▶ **Example 14** Find $\frac{dy}{dx}$ when $y^2 = x^3$. Assume $x > 0$.

Solution

We differentiate both sides with respect to x:

$$\frac{d}{dx}(y^2) = \frac{d}{dx}(x^3)$$

$$2y\frac{dy}{dx} = 3x^2$$

Therefore,

$$\frac{dy}{dx} = \frac{3}{2}\frac{x^2}{y}$$

Since $y = x^{3/2}$, we find

$$\frac{dy}{dx} = \frac{3}{2}\frac{x^2}{y} = \frac{3}{2}\frac{x^2}{x^{3/2}} = \frac{3}{2}x^{1/2}$$

This is the answer we expect from the general version of the power rule:

$$\frac{dy}{dx} = \frac{d}{dx}x^{3/2} = \frac{3}{2}x^{1/2}$$

◄

Power Rule for Rational Exponents We can generalize this example to functions of the form $y = x^r$, where r is a rational number. This will provide a proof of the generalized form of the power rule when the exponent is a rational number, something we promised in the previous section. We write $r = p/q$, where p and q are integers. (If q is even, we require x positive.) Then

$$y = x^{p/q} \iff y^q = x^p$$

Differentiating both sides of $y^q = x^p$ with respect to x, we find

$$qy^{q-1}\frac{dy}{dx} = px^{p-1}$$

Hence

$$\frac{dy}{dx} = \frac{p}{q}\frac{x^{p-1}}{y^{q-1}} = \frac{p}{q}\frac{x^{p-1}}{(x^{p/q})^{q-1}}$$

$$= \frac{p}{q}\frac{x^{p-1}}{x^{p-p/q}} = \frac{p}{q}x^{p-1-p+p/q}$$

$$= \frac{p}{q}x^{p/q-1} = rx^{r-1}$$

We summarize this result:

If r is a rational number, then

$$\frac{d}{dx}(x^r) = rx^{r-1}$$

4.4.3 Related Rates

An important application of implicit differentiation is related rates problems. We begin with a motivating example.

We consider the case of a parcel of air rising quickly in the atmosphere. As a consequence, it expands without exchanging heat with the surrounding air. Laws of physics tell us that the volume (V) and the temperature (T) of the parcel of air are related via

$$TV^{\gamma-1} = C$$

where γ is approximately 1.4 for sufficiently dry air (γ is the small Greek letter gamma) and C is a constant. The temperature is measured in Kelvin,[1] a scale that is chosen so that the temperature is always positive (absolute temperature scale). Since rising air expands, the volume of the parcel of air increases with time; we express this mathematically as $dV/dt > 0$, where t denotes time.

To determine how the temperature of the air parcel changes as it rises, we implicitly differentiate $TV^{\gamma-1} = C$ with respect to t:

$$\frac{dT}{dt}V^{\gamma-1} + T(\gamma-1)V^{\gamma-2}\frac{dV}{dt} = 0$$

or

$$\frac{dT}{dt} = -\frac{T(\gamma-1)V^{\gamma-2}\frac{dV}{dt}}{V^{\gamma-1}}$$

$$= -T(\gamma-1)\frac{1}{V}\frac{dV}{dt}$$

If we use $\gamma = 1.4$, then

$$\frac{dT}{dt} = -T(0.4)\frac{1}{V}\frac{dV}{dt}$$

implying that if air expands (i.e., $dV/dt > 0$), then temperature decreases (i.e., $dT/dt < 0$) since both T and V are positive: The temperature of a parcel

---

[1] To compare the Celsius and the Kelvin scales, note that a temperature difference of 1°C is equal to a temperature difference of 1 K, and that 0°C = 273.15 K and 100°C = 373.15 K.

of air decreases as the parcel rises, and the temperature of a falling air parcel increases. This can be observed close to high mountains.

In a typical related rates problem, one quantity is expressed in terms of another quantity, and both quantities change with time. We usually know how one of the quantities changes with time and are interested in finding out how the other quantity changes. For instance, suppose that y is a function of x, and both y and x depend on time. If we know how x changes with time (i.e., if we know dx/dt), then we might want to know how y changes with time (i.e., dy/dt). We illustrate this in the following example.

▶ **Example 15**   Find $\frac{dy}{dt}$ when $x^2 + y^3 = 1$ and $\frac{dx}{dt} = 2$ for $x = \sqrt{7/8}$.

Solution

In this example, both x and y are functions of t. Implicit differentiation with respect to t yields

$$\frac{d}{dt}(x^2 + y^3) = \frac{d}{dt}(1)$$

and hence

$$2x\frac{dx}{dt} + 3y^2\frac{dy}{dt} = 0$$

Solving for $\frac{dy}{dt}$ yields

$$\frac{dy}{dt} = -\frac{2}{3}\frac{x}{y^2}\frac{dx}{dt}$$

When $x = \sqrt{7/8}$,

$$y^3 = 1 - x^2 = 1 - \frac{7}{8} = \frac{1}{8}$$

and hence $y = 1/2$. Therefore,

$$\frac{dy}{dt} = -\frac{2}{3}\frac{\sqrt{7/8}}{1/4} \cdot 2 = -\frac{16}{3}\sqrt{\frac{7}{8}} = -\frac{4}{3}\sqrt{14}$$ ◀

We present two more applications of related rates.

▶ **Example 16**   A spherical balloon is being filled with air. When the radius $r = 6$ cm, the radius is increasing at a rate of 2 cm/s. How fast is the volume changing at this time?

Solution

The volume V of a sphere of radius r is given by

$$V = \frac{4}{3}\pi r^3 \qquad (4.6)$$

(see Figure 4.26). Note that V is a function of r. Since r is increasing at a certain rate, we think of r as a function of time t, that is, $r = r(t)$. Since the volume V depends on r, it changes with time t as well. We therefore consider V also as a function of time t. Differentiating both sides of (4.6) with respect to t, we find

$$\frac{dV}{dt} = \frac{4}{3}\pi 3r^2\frac{dr}{dt}$$

$$= 4\pi r^2\frac{dr}{dt}$$

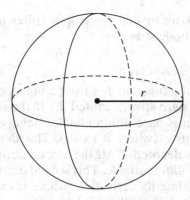

◀ **Figure 4.26**
The volume of a sphere with radius r is $V = \frac{4}{3}\pi r^3$

When $r = 6$ cm and $dr/dt = 2$ cm/s, then

$$\frac{dV}{dt} = 4\pi 6^2 \text{cm}^2 2 \frac{\text{cm}}{\text{s}} = 288\pi \frac{\text{cm}^3}{\text{s}}$$

Note that the unit of dV/dt is cm^3/s, which is what you should expect because the unit of the volume is cm^3 and time is measured in seconds.   ◀

▶ **Example 17**   **(Adapted from Benton and Harper, 1997)** Ichthyosaurs are a group of marine reptiles that were fishshaped and comparable in size to dolphins. They became extinct during the Cretaceous.[2] Based on a study of 20 fossil skeletons, it was found that the skull length (in cm) and backbone length (in cm) of an individual were related through the allometric equation (we introduced allometric equations in Example 7 of Section 1.2)

$$[\text{skull length}] = 1.162[\text{backbone length}]^{0.933}$$

How is the growth rate of the backbone related to the growth rate of the skull?

Solution

Let x denote the age of the ichthyosaur, and set

$$S = S(x) = \text{skull length at age } x$$

$$B = B(x) = \text{backbone length at age } x$$

so that

$$S(x) = (1.162)[B(x)]^{0.933}$$

We are interested in the relationship between dS/dx and dB/dx, the growth rates of the skull and the backbone, respectively. Differentiating the equation for $S(x)$ with respect to x, we find

$$\frac{dS}{dx} = (1.162)(0.933)[B(x)]^{0.933-1}\frac{dB}{dx}$$

Rearranging terms on the right-hand side, we write this as

$$\frac{dS}{dx} = \underbrace{(1.162)[B(x)]^{0.933}}_{S(x)}(0.933)\frac{1}{B(x)}\frac{dB}{dx}$$

Hence

$$\frac{1}{S(x)}\frac{dS}{dx} = 0.933\frac{1}{B(x)}\frac{dB}{dx}$$

This equation relates the relative growth rates $\frac{1}{S}\frac{dS}{dx}$ and $\frac{1}{B}\frac{dB}{dx}$. The factor 0.933 is less than 1, which indicates that skulls grow less quickly than backbones.

[2]The Cretaceous period began about 144 million years ago and ended about 65 million years ago.

This is familiar to us: Juvenile vertebrates often have larger heads than adults, relative to their body sizes. ◀

4.4.4 Higher Derivatives

The derivative of a function f is itself a function. We refer to this derivative also as the **first derivative**, denoted f'. If the first derivative exists, we say that the function is once differentiable. Being a function, we can therefore define its derivative (where it exists). This derivative is called the **second derivative** and is denoted f''. If the second derivative exists, we say that the function is twice differentiable. This second derivative is again a function and hence we can define its derivative (where it exists). The result is the **third derivative**, denoted f'''. If the third derivative exists, we say that the function is three times differentiable. We can continue in this manner; from the fourth derivative on, we denote the derivatives by $f^{(4)}$, $f^{(5)}$, and so on. If the nth derivative exists, we say that the function is n times differentiable.

Polynomials are functions that can be differentiated as many times as desired. The reason for this is that the first derivative of a polynomial of degree n is a polynomial of degree $n - 1$. Since the derivative is a polynomial as well, we can find its derivative, and so on. Eventually, the derivative will be equal to 0. This is illustrated in the next example.

▶ **Example 18** Find the nth derivative of $f(x) = x^5$ for $n = 1, 2, \ldots$.

Solution

Differentiating $f(x)$, we find the first derivative to be

$$f'(x) = 5x^4$$

Differentiating $f'(x)$, we find the second derivative to be

$$f''(x) = 5(4x^3) = 20x^3$$

Differentiating $f''(x)$, we find the third derivative:

$$f'''(x) = 20(3x^2) = 60x^2$$

Differentiating $f'''(x)$, we find the fourth derivative:

$$f^{(4)}(x) = 60(2x) = 120x$$

Differentiating $f^{(4)}(x)$, we find the fifth derivative:

$$f^{(5)}(x) = 120$$

Differentiating $f^{(5)}(x)$, we find the sixth derivative:

$$f^{(6)}(x) = 0$$

All higher-order derivatives, that is, $f^{(7)}$, $f^{(8)}$ ..., are equal to 0 as well. ◀

We can write higher-order derivatives in Leibniz notation: The nth derivative of $f(x)$ is denoted by

$$\frac{d^n f}{dx^n}$$

▶ **Example 19** Find the second derivative of $f(x) = \sqrt{x}$, $x \geq 0$.

Solution

First, we must find the first derivative,

$$\frac{d}{dx}\sqrt{x} = \frac{d}{dx}x^{1/2} = \frac{1}{2}x^{-1/2} \quad \text{for } x > 0$$

To find the second derivative, we differentiate the first derivative

$$\frac{d^2}{dx^2}\sqrt{x} = \frac{d}{dx}\left(\frac{d}{dx}\sqrt{x}\right) = \frac{d}{dx}\left(\frac{1}{2}x^{-1/2}\right) = \frac{1}{2}\left(-\frac{1}{2}\right)x^{(-1/2)-1}$$

$$= -\frac{1}{4}x^{-3/2}$$

When functions are implicitly defined, we can use the technique of implicit differentiation to find higher derivatives.

▶ **Example 20**   Find $\frac{d^2y}{dx^2}$ when $x^2 + y^2 = 1$.

Solution

We found

$$\frac{dy}{dx} = -\frac{x}{y}$$

in Example 12. Differentiating both sides of this equation with respect to x, we find

$$\frac{d}{dx}\left[\frac{dy}{dx}\right] = \frac{d}{dx}\left[-\frac{x}{y}\right]$$

The left-hand side can be written as

$$\frac{d^2y}{dx^2}$$

On the right-hand side, we use the quotient rule. Hence,

$$\frac{d^2y}{dx^2} = -\frac{1 \cdot y - x\frac{dy}{dx}}{y^2}$$

Substituting $-\frac{x}{y}$ for $\frac{dy}{dx}$, we obtain

$$\frac{d^2y}{dx^2} = -\frac{y - x(-\frac{x}{y})}{y^2}$$

$$= -\frac{y + \frac{x^2}{y}}{y^2} = -\frac{y^2 + x^2}{y^3}$$

Since $x^2 + y^2 = 1$, we can simplify this further and obtain

$$\frac{d^2y}{dx^2} = -\frac{1}{y^3}$$

We introduced the velocity of an object that moves on a straight line as the derivative of its position. The derivative of the velocity is the **acceleration**. If $s(t)$ denotes the position of the object moving on a straight line, $v(t)$ its velocity, and $a(t)$ its acceleration, then the three quantities are related as follows:

$$v(t) = \frac{ds}{dt} \quad \text{and} \quad a(t) = \frac{dv}{dt} = \frac{d^2s}{dt^2}$$

▶ **Example 21**   **Acceleration** Assume the position of a car moving along a straight line is given by

$$s(t) = 3t^3 - 2t + 1$$

Find its velocity and acceleration.

Solution

To find the velocity, we need to differentiate the position

$$v(t) = \frac{ds}{dt} = 9t^2 - 2$$

To find its acceleration, we differentiate the velocity

$$a(t) = \frac{dv}{dt} = \frac{d^2s}{dt^2} = 18t$$

◀

▶ **Example 22**   Neglecting air resistance, the distance (in meters) an object falls when dropped from rest from a height is

$$s(t) = \frac{1}{2}gt^2$$

where $g = 9.81$ m/s^2 is the earth's gravitational constant and t is the amount of time (in seconds) elapsed since the object was released.

(a) Find its velocity and acceleration.

(b) If the height is 30 m, how long will it take until the object hits the ground and what is its velocity at the time of impact?

Solution

(a) The velocity is

$$v(t) = \frac{ds}{dt} = gt$$

and the acceleration is

$$a(t) = \frac{dv}{dt} = g$$

Note that the acceleration is constant.

(b) To find the time it takes the object to hit the ground, we set $s(t) = 30$ m and solve for t:

$$30 \text{ m} = \frac{1}{2}(9.81)\frac{\text{m}}{\text{s}^2}t^2$$

This yields

$$t^2 = \frac{60}{9.81}\text{s}^2 \quad \text{or} \quad t = \sqrt{\frac{60}{9.81}}\text{ s} \approx 2.47 \text{ s}$$

(We need only consider the positive solution.) The velocity at the time of impact is

$$v(t) = gt = (9.81)\frac{\text{m}}{\text{s}^2}\sqrt{\frac{60}{9.81}}\text{ s} \approx 24.3 \frac{\text{m}}{\text{s}}$$

◀

4.4.5 Problems

(4.4.1)

In Problems 1–28, differentiate the functions with respect to the independent variable.

1. $f(x) = (x - 2)^2$

2. $f(x) = (2x - 1)^3$

3. $f(x) = (1 - 3x^2)^4$

4. $f(x) = (2x^2 - 3x)^2$

5. $f(x) = \sqrt{x^2 + 3}$

6. $f(x) = \sqrt{2x - 5}$

7. $f(x) = \sqrt{3 - x^3}$

8. $f(x) = \sqrt{5x + 3x^3}$

9. $f(x) = \dfrac{1}{(x^3 - 2)^4}$

10. $f(x) = \dfrac{2}{(1 - 2x^2)^3}$

11. $f(x) = \dfrac{3x - 1}{\sqrt{2x^2 - 1}}$

12. $f(x) = \dfrac{(1 - 2x^2)^2}{(3 - x^2)^3}$

13. $f(x) = \dfrac{\sqrt{2x - 1}}{(x - 1)^2}$

14. $f(x) = \dfrac{\sqrt{x^2 - 1}}{1 + \sqrt{x^2 + 1}}$

15. $f(s) = \sqrt{s + \sqrt{s}}$

16. $g(t) = \sqrt{t^2 + \sqrt{t + 1}}$

17. $g(t) = \left(\dfrac{t}{t - 3}\right)^3$

18. $h(s) = \left(\dfrac{2s^2}{s + 1}\right)^4$

19. $f(r) = (r^2 - r)^3(r + 3r^3)^{-4}$

20. $h(s) = \dfrac{2(3 - s)^2}{s^2 + (7s - 1)^2}$

21. $h(x) = \sqrt[5]{3 - x^4}$

22. $h(x) = \sqrt[3]{1 - 2x}$

23. $f(x) = \sqrt[7]{x^2 - 2x + 1}$

24. $f(x) = \sqrt[4]{2 - 4x^2}$

25. $g(s) = (3s^7 - 7s)^{\pi}$

26. $h(t) = (t^4 - 4t)^e$

27. $h(t) = \left(3t + \dfrac{3}{t}\right)^{2/5}$

28. $h(t) = \left(4t^4 + \dfrac{4}{t^4}\right)^{1/4}$

29. Differentiate
$$f(x) = (ax + 1)^3$$
with respect to x. Assume that a is a positive constant.

30. Differentiate
$$f(x) = \sqrt{ax^2 - 2}$$
with respect to x. Assume that a is a positive constant.

31. Differentiate
$$g(N) = \dfrac{bN}{(k + N)^2}$$
with respect to N. Assume that b and k are positive constants.

32. Differentiate
$$g(N) = \dfrac{N}{(k + bN)^2}$$
with respect to N. Assume that b and k are positive constants.

33. Differentiate
$$g(T) = a(T_0 - T)^3 - b$$

with respect to T. Assume that a, b, and T_0 are positive constants.

34. Suppose that $f'(x) = 2x + 1$. Find the following.

(a) $$\dfrac{d}{dx} f(x^2)$$
at $x = -1$.

(b) $$\dfrac{d}{dx} f(\sqrt{x})$$
at $x = 4$.

35. Suppose that $f'(x) = \frac{1}{x}$. Find the following.

(a) $$\dfrac{d}{dx} f(x^2 + 3)$$

(b) $$\dfrac{d}{dx} f(\sqrt{x} - 1)$$

In Problems 36–39, assume that $f(x)$ and $g(x)$ are differentiable.

36. Find
$$\dfrac{d}{dx} \sqrt{f(x) + g(x)}$$

37. Find
$$\dfrac{d}{dx} \left(\dfrac{f(x)}{g(x)} + 1\right)^2$$

38. Find
$$\dfrac{d}{dx} f\left[\dfrac{1}{g(x)}\right]$$

39. Find
$$\dfrac{d}{dx} \dfrac{[f(x)]^2}{g(2x) + 2x}$$

In Problems 40–46, find $\frac{dy}{dx}$ by applying the chain rule repeatedly.

40. $y = (\sqrt{1 - x^2} + 2)^2$

41. $y = (\sqrt{x^3 - 3x} + 3x)^3$

42. $y = \left(1 + (x - 1)^2\right)^2$

43. $y = \left(1 + (3x^2 - 1)^3\right)^2$

44. $y = \left(\dfrac{x}{2(x^2 - 1)^2 - 1}\right)^2$

45. $y = \left(\dfrac{2x + 1}{3(x^3 - 1)^3 - 1}\right)^3$

46. $y = \left(\dfrac{(2x + 1)^2 - x}{(3x^3 + 1)^3 - x}\right)^2$

(4.4.2)

In Problems 47–54, find $\frac{dy}{dx}$ by implicit differentiation.

47. $x^2 + y^2 = 4$

48. $y = x^2 + yx$

49. $x^{3/4} + y^{3/4} = 1$

50. $xy - y^3 = 1$

51. $\sqrt{xy} = x^2 + 1$

52. $\dfrac{1}{xy} - y^2 = 2$

53. $\dfrac{x}{y} = \dfrac{y}{x}$

54. $\dfrac{x}{xy+1} = 2xy$

In Problems 55–57, find the lines that are (a) tangential and (b) normal to each curve at the given point.

55. $x^2 + y^2 = 25$, $(4, -3)$ (circle)

56. $\dfrac{x^2}{4} + \dfrac{y^2}{9} = 1$, $\left(1, \frac{3}{2}\sqrt{3}\right)$ (ellipse)

57. $\dfrac{x^2}{25} - \dfrac{y^2}{9} = 1$, $\left(\frac{25}{3}, 4\right)$ (hyperbola)

58. (*Lemniscate*)

(a) The curve with equation $y^2 = x^2 - x^4$ is shaped like the numeral eight. Find $\frac{dy}{dx}$ at $\left(\frac{1}{2}, \frac{1}{4}\sqrt{3}\right)$.

(b) Use a graphing calculator to graph the curve in (a). If the graphing calculator cannot graph implicit functions, graph the upper and the lower half of the curve separately, that is, graph

$$y_1 = \sqrt{x^2 - x^4}$$
$$y_2 = -\sqrt{x^2 - x^4}$$

Choose the viewing rectangle $-2 \le x \le 2, -1 \le y \le 1$.

59. (*Astroid*)

(a) Consider the curve with equation $x^{2/3} + y^{2/3} = 4$. Find $\frac{dy}{dx}$ at $(-1, 3\sqrt{3})$.

(b) Use a graphing calculator to graph the curve in (a). If the graphing calculator cannot graph implicit functions, graph the upper and the lower half of the curve separately. To get the left half of the graph, make sure that your calculator evaluates $x^{2/3}$ in the order $(x^2)^{1/3}$. Choose the viewing rectangle $-10 \le x \le 10, -10 \le y \le 10$.

60. (*Kampyle of Eudoxus*)

(a) Consider the curve with equation $y^2 = 10x^4 - x^2$. Find $\frac{dy}{dx}$ at $(1, 3)$.

(b) Use a graphing calculator to graph the curve in (a). If the graphing calculator cannot graph implicit functions, graph the upper and the lower half of the curve separately. Choose the viewing rectangle $-3 \le x \le 3, -10 \le y \le 10$.

(4.4.3)

61. Assume that x and y are differentiable functions of t. Find $\frac{dy}{dt}$ when $x^2 + y^2 = 1$, $\frac{dx}{dt} = 2$ for $x = \frac{1}{2}$, and $y > 0$.

62. Assume that x and y are differentiable functions of t. Find $\frac{dy}{dt}$ when $y^2 = x^2 - x^4$, $\frac{dx}{dt} = 1$ for $x = \frac{1}{2}$, and $y > 0$.

63. Assume that x and y are differentiable functions of t. Find $\frac{dy}{dt}$ when $x^2 y = 1$ and $\frac{dx}{dt} = 3$ for $x = 2$.

64. Assume that u and v are differentiable functions of t. Find $\frac{du}{dt}$ when $u^2 + v^3 = 12$, $\frac{dv}{dt} = 2$ for $v = 2$, and $u > 0$.

65. Assume that the side length x and the volume $V = x^3$ of a cube are differentiable functions of t. Express dV/dt in terms of dx/dt.

66. Assume that the radius r and the area $A = \pi r^2$ of a circle are differentiable functions of t. Express dA/dt in terms of dr/dt.

67. Assume that the radius r and the surface area $S = 4\pi r^2$ of a sphere are differentiable functions of t. Express dS/dt in terms of dr/dt.

68. Assume that the radius r and the volume $V = \frac{4}{3}\pi r^3$ of a sphere are differentiable functions of t. Express dV/dt in terms of dr/dt.

69. Suppose that water is stored in a cylindrical tank of radius 5 m. If the height of the water in the tank is h, then the volume of the water is $V = \pi r^2 h = (25 \text{ m}^2)\pi h = 25\pi h \text{ m}^2$. If we drain the water at a rate of 250 liters per minute, what is the rate at which the water level inside the tank drops? (Note that one cubic meter contains 1000 liters.)

70. Suppose that we pump water into an inverted right-circular conical tank at the rate of 5 cubic feet per minute (i.e., the tank stands point down). The tank has a height of 6 ft and the radius on top is 3 ft. What is the rate at which water level is rising when the water is 2 ft deep? (Note that the volume of a right-circular cone of radius r and height h is $V = \frac{1}{3}\pi r^2 h$.)

71. Two people start biking from the same point. One bikes east at 15 mph, the other south at 18 mph. What is the rate at which the distance between the two people is changing after 20 minutes, and after 40 minutes?

72. Allometric equations describe the scaling relationship between two measurements, such as skull length versus body length. In vertebrates, we typically find

[skull length] $\propto$ [body length]^a

for some $a \in (0, 1)$. Express the growth rate of the skull length in terms of the growth rate of the body length.

(4.4.4)

In Problems 73–82, find the first and the second derivative of each function.

73. $f(x) = x^3 - 3x^2 + 1$

74. $f(x) = (x^2 - 3)^2$

75. $g(x) = \dfrac{x-1}{x+1}$

76. $h(s) = \dfrac{1}{s^2+2}$

77. $g(t) = \sqrt{3t^3 + 2t}$

78. $f(x) = \dfrac{1}{x^2} + x - 1$

79. $f(s) = \sqrt{s^{3/2} - 1}$

80. $f(x) = \dfrac{x}{x+1}$

81. $g(t) = t^{-5/2} - t^{1/2}$

82. $f(x) = x^3 - \dfrac{1}{x^3}$

83. Find the first ten derivatives of $y = x^5$.

84. Find $f^{(n)}(x)$ and $f^{(n+1)}(x)$ of $f(x) = x^n$.

85. Find a second-degree polynomial $p(x) = ax^2 + bx + c$ with $p(0) = 3$, $p'(0) = 2$, and $p''(0) = 6$.

86. The position at time t of a particle that moves along a straight line is given by the function $s(t)$. The first derivative of $s(t)$ is called the velocity, denoted by $v(t)$; that is, the velocity is the rate of change of the position. The rate of change of the velocity is called **acceleration**, denoted by $a(t)$, that is,

$$\frac{d}{dt}v(t) = a(t)$$

Given that $v(t) = s'(t)$, it follows that

$$\frac{d^2}{dt^2} s(t) = a(t)$$

Find the velocity and the acceleration at time $t = 1$ for the following position functions.

(a) $s(t) = t^2 - 3t$

(b) $s(t) = \sqrt{t^2 + 1}$

(c) $s(t) = t^4 - 2t$

87. Neglecting air resistance, the height h (in meters) of an object thrown vertically from the ground with initial velocity v_0 is given by

$$h(t) = v_0 t - \frac{1}{2} g t^2$$

where $g = 9.81$ m/s^2 is the earth's gravitational constant, and t is the time (in seconds) elapsed since the object was released.

(a) Find the velocity and the acceleration of the object.

(b) Find the time when the velocity is equal to 0. In which direction is the object traveling right before this time, in which direction right after this time?

4.5 DERIVATIVES OF TRIGONOMETRIC FUNCTIONS

We will need the trigonometric limits from Section 3.4 to compute the derivatives of sine and cosine. Note that all angles are measured in radians.

Theorem The functions $\sin x$ and $\cos x$ are differentiable for all x, and

$$\frac{d}{dx} \sin x = \cos x \quad \text{and} \quad \frac{d}{dx} \cos x = -\sin x$$

A sketch of the derivatives of each of the trigonometric functions based on the geometric interpretation of a derivative as the slope of the tangent line confirms these rules (see Figures 4.27 and 4.28).

Proof We prove the first formula; the similar proof of the second formula is discussed in Problem 61. We need the trigonometric identity

$$\sin(\alpha + \beta) = \sin \alpha \cos \beta + \cos \alpha \sin \beta$$

Using the formal definition of derivatives, we find

$$\frac{d}{dx} \sin x = \lim_{h \to 0} \frac{\sin(x + h) - \sin x}{h}$$

$$= \lim_{h \to 0} \frac{\sin x \cos h + \cos x \sin h - \sin x}{h}$$

$$= \lim_{h \to 0} \left[\sin x \frac{\cos h - 1}{h} + \cos x \frac{\sin h}{h} \right]$$

We showed in Section 3.4 that

$$\lim_{h \to 0} \frac{\cos h - 1}{h} = 0 \quad \text{and} \quad \lim_{h \to 0} \frac{\sin h}{h} = 1$$

We can therefore apply the basic rules for limits and find

$$\frac{d}{dx} \sin x = \sin x \lim_{h \to 0} \frac{\cos h - 1}{h} + \cos x \lim_{h \to 0} \frac{\sin h}{h}$$

$$= (\sin x)(0) + (\cos x)(1) = \cos x \qquad ■$$

▶ **Example 1** Find the derivative of $f(x) = -4 \sin x + \cos \frac{\pi}{6}$.

Solution

$$f'(x) = \frac{d}{dx} \left(-4 \sin x + \cos \frac{\pi}{6} \right)$$

$$= -4 \frac{d}{dx} \sin x + \frac{d}{dx} \cos \frac{\pi}{6}$$

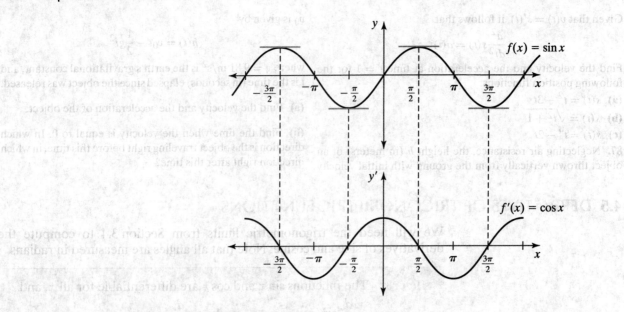

▲ Figure 4.27
The function $f(x) = \sin x$ and its derivative $f'(x) = \cos x$. The derivative $f'(x) = 0$ where $f(x)$ has a horizontal tangent line

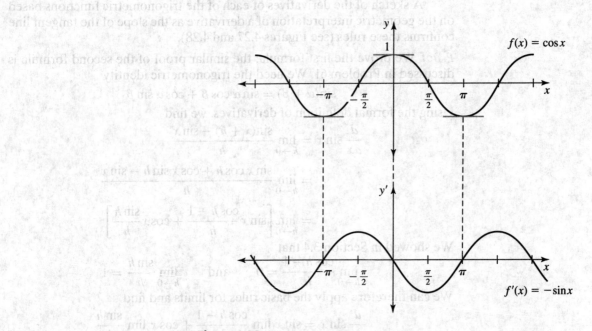

▲ Figure 4.28
The function $f(x) = \cos x$ and its derivative $f'(x) = -\sin x$. The derivative $f'(x) = 0$ where $f(x)$ has a horizontal tangent line

The first term is a trigonometric function and the second term is a constant (namely, $\frac{1}{2}\sqrt{3}$). Differentiating each term, we find

$$= -4(\cos x) + 0$$

$$= -4 \cos x$$

◀

▶ **Example 2**   Find the derivative of $y = \cos(x^2 + 1)$.

Solution

We set $u = g(x) = x^2 + 1$ and $f(u) = \cos u$; then $y = f[g(x)]$. Using the chain rule, we find

$$y' = \frac{df}{du}\frac{du}{dx} = \frac{d}{du}(\cos u)\frac{d}{dx}(x^2 + 1) = (-\sin u)(2x)$$

$$= -[\sin(x^2 + 1)]2x = -2x\sin(x^2 + 1) \quad \blacktriangleleft$$

▶ **Example 3**   Find the derivative of $y = x^2 \sin(3x) - \cos(5x)$.

Solution

We will use the product rule for the first term; in addition, we will need the chain rule for both $\sin(3x)$ and $\cos(5x)$.

$$y' = \frac{d}{dx}[x^2\sin(3x) - \cos(5x)]$$

$$= \frac{d}{dx}[x^2\sin(3x)] - \frac{d}{dx}\cos(5x)$$

$$= \left(\frac{d}{dx}x^2\right)\sin(3x) + x^2\frac{d}{dx}\sin(3x) - \frac{d}{dx}\cos(5x)$$

$$= 2x\sin(3x) + x^2 3\cos(3x) - 5(-\sin(5x))$$

$$= 2x\sin(3x) + 3x^2\cos(3x) + 5\sin(5x) \quad \blacktriangleleft$$

The derivatives of the other trigonometric functions can be found using the following identities:

$$\tan x = \frac{\sin x}{\cos x} \qquad \cot x = \frac{\cos x}{\sin x}$$

$$\sec x = \frac{1}{\cos x} \qquad \csc x = \frac{1}{\sin x}$$

For instance, to find the derivative of the tangent, we use the quotient rule:

$$\frac{d}{dx}\tan x = \frac{d}{dx}\frac{\sin x}{\cos x}$$

$$= \frac{\left(\frac{d}{dx}\sin x\right)\cos x - \sin x\left(\frac{d}{dx}\cos x\right)}{\cos^2 x}$$

$$= \frac{(\cos x)(\cos x) - (\sin x)(-\sin x)}{\cos^2 x}$$

$$= \frac{\cos^2 x + \sin^2 x}{\cos^2 x} = \frac{1}{\cos^2 x} = \sec^2 x$$

In the penultimate step, we used the identity $\cos^2 x + \sin^2 x = 1$.

The other derivatives can be found in a similar fashion, as explained in Problems 62–64.

We summarize the derivatives of all trigonometric functions in the following box.

$$\frac{d}{dx}\sin x = \cos x \qquad \frac{d}{dx}\cos x = -\sin x$$

$$\frac{d}{dx}\tan x = \sec^2 x \qquad \frac{d}{dx}\cot x = -\csc^2 x$$

$$\frac{d}{dx}\sec x = \sec x \tan x \qquad \frac{d}{dx}\csc x = -\csc x \cot x$$

▶ **Example 4** Compare the derivatives of (a) $\tan x^2$ and (b) $\tan^2 x$.

Solution

(a) If $y = \tan x^2 = \tan(x^2)$, then using the chain rule, we find

$$\frac{dy}{dx} = \frac{d}{dx}\tan(x^2) = (\sec^2(x^2))(2x) = 2x\sec^2(x^2)$$

(b) If $y = \tan^2 x = (\tan x)^2$, then using the chain rule, we find

$$\frac{dy}{dx} = \frac{d}{dx}(\tan x)^2 = 2(\tan x)\frac{d}{dx}\tan x = 2\tan x \sec^2 x$$

The two derivatives are clearly different, and you should look again at $\tan x^2$ and $\tan^2 x$ to make sure that you understand which is the inner and which the outer function. ◀

▶ **Example 5** *Repeated Application of the Chain Rule* Find the derivative of $f(x) = \sec\sqrt{x^2+1}$.

Solution

This is a composite function; the inner function is $\sqrt{x^2+1}$, and the outer function is $\sec x$. Applying the chain rule once, we find

$$\frac{df}{dx} = \frac{d}{dx}\sec\sqrt{x^2+1} = \sec\sqrt{x^2+1}\tan\sqrt{x^2+1}\frac{d}{dx}\sqrt{x^2+1}$$

To evaluate $\frac{d}{dx}\sqrt{x^2+1}$, we need to apply the chain rule a second time:

$$\frac{d}{dx}\sqrt{x^2+1} = \frac{1}{2\sqrt{x^2+1}}2x = \frac{x}{\sqrt{x^2+1}}$$

Combining the two steps, the result is

$$\frac{df}{dx} = \left(\sec\sqrt{x^2+1}\right)\left(\tan\sqrt{x^2+1}\right)\frac{x}{\sqrt{x^2+1}}$$

The function $f(x)$ can be thought of as a composition of three functions. The innermost function is $u = g(x) = x^2+1$, the middle is $v = h(u) = \sqrt{u}$, and the outermost function is $f(v) = \sec v$. When we computed the derivative, we applied the chain rule twice in the form

$$\frac{df}{dx} = \frac{df}{dv}\frac{dv}{du}\frac{du}{dx}$$

◀

4.5.1 Problems

In Problems 1–58 find the derivative with respect to the independent variable.

1. $f(x) = 2\sin x - \cos x$

2. $f(x) = 3\cos x - 2\sec x$

3. $f(x) = 3\sin x + 5\cos x - 2\sec x$

4. $f(x) = -\sin x + \cos x$

5. $f(x) = \tan x - \cot x$

6. $f(x) = \sec x - \csc x$

7. $f(x) = \sin(3x)$

8. $f(x) = \cos(-5x)$

9. $f(x) = 2\sin(3x + 1)$

10. $f(x) = -3\cos(1 - 2x)$

11. $f(x) = \tan(4x)$

12. $f(x) = \tan(2 - 3x)$

13. $f(x) = 2\sec(1 + 2x)$

14. $f(x) = -3\sec(3 - 5x)$

15. $f(x) = 3\sin(x^2)$

16. $f(x) = 2\sin(x^3 - 3x)$

17. $f(x) = \sec(x^2 - 3)$

18. $f(x) = \cos^2(x^2 - 1)$

19. $f(x) = 3\sin^2 x$

20. $f(x) = -\sin^2(2x - 1)$

21. $f(x) = 4\cos(x^2)$

22. $f(x) = -5\cos(2 - x^3)$

23. $f(x) = 4\cos^2 x$

24. $f(x) = -3\cos^2(3x^2 - 1)$

25. $f(x) = 2\tan(1 - x^2)$

26. $f(x) = -\tan(3x^3 - 4x)$

27. $f(x) = -2\tan^3(3x - 1)$

28. $f(x) = \sqrt{\sin x}$

29. $f(x) = \sqrt{\sin(2x^2 - 1)}$

30. $g(s) = (\cos^2 s - 3s^2)^2$

31. $g(s) = \sqrt{\cos s}$

32. $g(t) = \dfrac{\sin(3t)}{\cos(5t)}$

33. $g(t) = \dfrac{\sin(2t) + 1}{\cos(6t) - 1}$

34. $f(x) = \dfrac{\cot(2x)}{\tan(4x)}$

35. $f(x) = \dfrac{\sec(x^2 - 1)}{\csc(x^2 + 1)}$

36. $f(x) = \sin x \cos x$

37. $f(x) = \sin(2x - 1)\cos(3x + 1)$

38. $f(x) = \tan x \cot x$

39. $f(x) = \tan(3x^2 - 1)\cot(3x^2 + 1)$

40. $f(x) = \sec x \cos x$

41. $f(x) = \sin x \sec x$

42. $f(x) = \dfrac{1}{\sin x + \cos x}$

43. $f(x) = \dfrac{1}{\sec x + \tan x}$

44. $g(x) = \dfrac{1}{\sin(3x)}$

45. $g(x) = \dfrac{1}{\sin(3x^2 - 1)}$

46. $g(x) = \dfrac{1}{\csc(5x)}$

47. $g(x) = \dfrac{1}{\csc(1 - 5x^2)}$

48. $h(x) = \cot(3x)\csc(3x)$

49. $h(x) = \dfrac{3}{\tan(2x) - x}$

50. $g(t) = \dfrac{1}{\sqrt{\sin t^2}}$

51. $h(s) = \sin^2 s + \cos^2 s$

52. $f(x) = (2x^3 - x)\cos(1 - x^2)$

53. $f(x) = \dfrac{\sin(2x)}{1 + x^2}$

54. $f(x) = \dfrac{1 + \cos(3x)}{2x^3 - x}$

55. $f(x) = \tan\dfrac{1}{x}$

56. $f(x) = \sec\dfrac{1}{1 + x}$

57. $f(x) = \dfrac{\sec x^2}{\sec^2 x}$

58. $f(x) = \dfrac{\csc(3 - x^2)}{1 - x^2}$

59. Find the points on the curve $y = \sin\left(\frac{\pi}{3}x\right)$ that have a horizontal tangent.

60. Find the points on the curve $y = \cos^2 x$ that have a horizontal tangent.

61. Use the identity
$$\cos(\alpha + \beta) = \cos\alpha\cos\beta - \sin\alpha\sin\beta$$
and the definition of the derivative to show that
$$\frac{d}{dx}\cos x = -\sin x$$

62. Use the quotient rule to show that
$$\frac{d}{dx}\cot x = -\csc^2 x$$
(*Hint:* Write $\cot x = \frac{\cos x}{\sin x}$.)

63. Use the chain rule to show that
$$\frac{d}{dx}\sec x = \sec x \tan x$$
[*Hint:* Write $\sec x = (\cos x)^{-1}$.]

64. Use the chain rule to show that
$$\frac{d}{dx}\csc x = -\csc x \cot x$$
[*Hint:* Write $\csc x = (\sin x)^{-1}$.]

65. Find the derivative of
$$f(x) = \sin \sqrt{x^2 + 1}$$
66. Find the derivative of
$$f(x) = \cos \sqrt{x^2 + 1}$$
67. Find the derivative of
$$f(x) = \sin \sqrt{3x^3 + 3x}$$
68. Find the derivative of
$$f(x) = \cos \sqrt{1 - 4x^4}$$
69. Find the derivative of
$$f(x) = \sin^2(x^2 - 1)$$
70. Find the derivative of
$$f(x) = \cos^2(2x^2 + 3)$$
71. Find the derivative of
$$f(x) = \tan^3(3x^3 - 3)$$
72. Find the derivative of
$$f(x) = \sec^2(2x^2 - 2)$$

73. Suppose that the concentration of nitrogen in a lake shows periodic behavior. That is, if we denote by $c(t)$ the concentration of nitrogen at time t, then we assume
$$c(t) = 2 + \sin\left(\frac{\pi}{2}t\right)$$

(a) Find
$$\frac{dc}{dt}$$

(b) Use a graphing calculator to graph both $c(t)$ and $\frac{dc}{dt}$ in the same coordinate system.

(c) By inspecting the graph in (b), answer the following questions:

(i) When $c(t)$ reaches a maximum, what is the value of dc/dt?

(ii) When dc/dt is positive, is $c(t)$ increasing or decreasing?

(iii) What can you say about $c(t)$ when $dc/dt = 0$?

4.6 DERIVATIVES OF EXPONENTIAL FUNCTIONS

Pierre de Fermat (1601–1665) devised a method of finding the tangent line at a given point (x, y) on a curve by constructing the **subtangent**. The subtangent is defined as the line segment between the point where the tangent line intersects the x-axis and the point $(x, 0)$ (see Figure 4.29).

Fermat's procedure essentially amounted to finding the slope of the tangent line by considering the secant line through (x, y) and a nearby point of the graph of $y = f(x)$. After computing the slope of the secant line, he set the two points he used for computing the slope of the secant line equal to each other, thus obtaining the slope of the tangent line. This sounds very much like the definition of the derivative that we use today and, in fact, it is the same idea. Fermat did not, however, develop and formalize a general framework for the differential calculus; this was done by Leibniz and Newton.

Using the definition of derivatives, we can relate the subtangent to the slope of the tangent at the corresponding point of the graph of $y = f(x)$. Suppose the tangent line at (x, y), where $y = f(x)$, intersects the x-axis at $(p(x), 0)$; the location of $p(x)$ depends on x. We set $c(x) = x - p(x)$. This is the equation of the subtangent. We see from Figure 4.29 that the slope of the

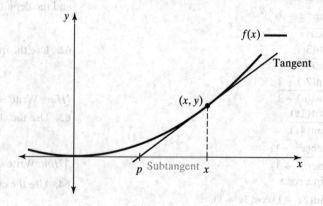

▲ **Figure 4.29**
The subtangent problem

tangent line at (x, y) is given by $y/c(x)$. Since the slope of the tangent line at (x, y) is the derivative of the function of the curve evaluated at x [i.e., $f'(x)$], we find

$$\frac{dy}{dx} = \frac{y}{c(x)}$$

A natural problem (which was posed to Descartes by Debaune in 1639) is to find a curve whose subtangent is a given constant. That is, we wish to find the function $y = f(x)$ that satisfies

$$\frac{dy}{dx} = \frac{y}{c}$$

where c is a constant other than 0. (This problem was solved by Leibniz in 1684, when he published his differential calculus for the first time.) In words, we are looking for a function $y = f(x)$ whose derivative is proportional to the function itself. As we will see in the following, exponential functions are the solutions to this problem.

Recall from Section 1.2 that the function f is an exponential function with base a if

$$f(x) = a^x, \quad x \in \mathbf{R}$$

where a is a positive constant other than 1 (see Figure 4.30). We can use the formal definition of the derivative to compute $f'(x)$, namely,

$$f'(x) = \frac{d}{dx}a^x = \lim_{h \to 0} \frac{f(x+h) - f(x)}{h} = \lim_{h \to 0} \frac{a^{x+h} - a^x}{h}$$

$$= \lim_{h \to 0} \frac{a^x(a^h - 1)}{h} = a^x \lim_{h \to 0} \frac{a^h - 1}{h}$$

In the final step we were able to write the term a^x in front of the limit since a^x does not depend on h. Thus, we are left with investigating

$$\lim_{h \to 0} \frac{a^h - 1}{h}$$

We first note that the limit does not depend on x. If we assume that this limit exists, then it follows that it is equal to $f'(0)$. To see why, use the formal definition of the derivative to compute $f'(0)$. (It can be shown that this limit exists, but this is beyond the scope of this course.)

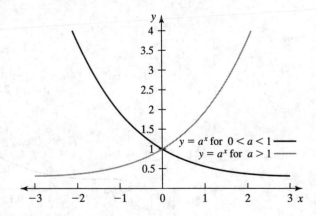

▲ **Figure 4.30**
The function $y = a^x$

It follows from the preceding calculation that if $f'(0)$ exists, then $f'(x)$ exists and

$$f'(x) = a^x f'(0)$$

This shows that the exponential function is a function whose derivative is proportional to the function itself, provided $f'(0)$ exists. [The constant of proportionality is $f'(0)$.] That is, exponential functions solve the subtangent problems just mentioned. We single out the case where the value of the base a is such that

$$\lim_{h \to 0} \frac{a^h - 1}{h}$$

is equal to 1. We denote this base by e. The number e is thus defined by

$$\lim_{h \to 0} \frac{e^h - 1}{h} = 1 \tag{4.7}$$

We find for the derivative of $f(x) = e^x$,

$$\frac{d}{dx} e^x = e^x \tag{4.8}$$

A graph of $f(x) = e^x$ is shown in Figure 4.31. The domain of this function is $\mathbf{R}$ and its range is the open interval $(0, \infty)$ (in particular, $e^x > 0$ for all $x \in \mathbf{R}$). Denoting by e the base of the exponential function for which (4.7) and (4.8) hold is no accident. It is indeed the natural exponential base that we introduced in Section 1.2. Though we cannot prove this here, the following table should convince you. With $e = 2.71828\ldots$, we find

h	0.1	0.01	0.001	0.0001
$\frac{e^h - 1}{h}$	1.0517	1.0050	1.00050	1.000050

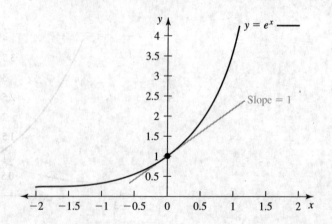

▲ **Figure 4.31**
The function $y = e^x$. The slope of the tangent line at $x = 0$ is $m = 1$

Recall also that there is an alternative notation for e^x, nam

the identity

$$a^x = \exp[\ln a^x]$$

and the fact that $\ln a^x = x \ln a$, we can find the derivative of a^x with

of the chain rule, namely,

$$\frac{d}{dx} a^x = \frac{d}{dx} \exp[\ln a^x] = \frac{d}{dx} \exp[x \ln a]$$

$$= \exp[x \ln a] \ln a = (\ln a) a^x$$

That is, we have

$$\frac{d}{dx} a^x = (\ln a) a^x \tag{4.9}$$

which allows us to obtain the following identity:

$$\lim_{h \to 0} \frac{a^h - 1}{h} = \ln a \tag{4.10}$$

▶ **Example 1**   Find the derivative of $f(x) = e^{-x^2/2}$.

Solution

We use the chain rule:

$$f'(x) = e^{-x^2/2} \left(-\frac{2x}{2} \right) = -x e^{-x^2/2} \qquad \blacktriangleleft$$

▶ **Example 2**   Find the derivative of $f(x) = 3^{\sqrt{x}}$.

Solution

We can use (4.9) and the chain rule to find

$$\frac{d}{dx} 3^{\sqrt{x}} = (\ln 3) 3^{\sqrt{x}} \frac{1}{2\sqrt{x}}$$

However, since every exponential function can be written in terms of the base e, and the differentiation rule for e^x is particularly simple ($\frac{d}{dx} e^x = e^x$), it is often easier to rewrite the exponential function in terms of e and then to differentiate. That is, we write

$$3^{\sqrt{x}} = \exp[\ln 3^{\sqrt{x}}] = \exp[\sqrt{x} \ln 3]$$

Then, using the chain rule, we find

$$\frac{d}{dx} \exp[\sqrt{x} \ln 3] = \frac{\ln 3}{2\sqrt{x}} \exp[\sqrt{x} \ln 3] = \frac{\ln 3}{2\sqrt{x}} 3^{\sqrt{x}} \qquad \blacktriangleleft$$

Since we must frequently differentiate functions of the form $y = e^{g(x)}$, we state this as a separate rule. Using the chain rule, we have

$$\frac{d}{dx} e^{g(x)} = g'(x) e^{g(x)} \tag{4.11}$$

Find the derivative of $f(x) = \exp[\sin \sqrt{x}]$.

Solution

Ve set $g(x) = \sin \sqrt{x}$. To differentiate $g(x)$, we must apply the chain rule,

$$\frac{d}{dx} g(x) = (\cos \sqrt{x}) \frac{1}{2\sqrt{x}}$$

:an now differentiate $f(x)$ using Equation (4.11),

$$\frac{d}{dx} f(x) = (\cos \sqrt{x}) \frac{1}{2\sqrt{x}} \exp[\sin \sqrt{x}]$$

◀

Here is an example that shows how (4.10) is used.

▶ **Example 4**   Find

$$\lim_{h \to 0} \frac{3^{2h} - 1}{h}$$

Solution

We make the substitution $l = 2h$, and note that $l \to 0$ as $h \to 0$. Then

$$\lim_{h \to 0} \frac{3^{2h} - 1}{h} = \lim_{l \to 0} \frac{3^l - 1}{l/2}$$

$$= 2 \lim_{l \to 0} \frac{3^l - 1}{l} = 2 \ln 3$$

◀

The exponential function with base e appears in many scientific problems; the following example involves radioactive decay.

▶ **Example 5**   **(Radioactive Decay)** Find the derivative of the radioactive decay function, which describes the amount of material left after t units of time (see Example 10 in Subsection 1.2.5):

$$W(t) = W_0 e^{-\lambda t}, \quad t \geq 0$$

where W_0 is the amount of material at time 0 and λ is called the radioactive decay rate. Show that $W(t)$ satisfies the differential equation

$$\frac{dW}{dt} = -\lambda W(t)$$

Solution

We use the chain rule to find the derivative of $W(t)$,

$$\frac{d}{dt} W(t) = \underbrace{W_0 e^{-\lambda t}}_{W(t)} (-\lambda)$$

that is,

$$\frac{dW}{dt} = -\lambda W(t)$$

In words, the rate of decay is proportional to the amount of material left. This should remind you of the subtangent problem; there, we wanted to find a function whose derivative is proportional to the function itself. This is exactly the situation that we have in this example: The derivative of $W(t)$ is proportional to $W(t)$.

◀

▶ **Example 6** **Exponential Growth** Find the per capita growth rate of a population whose population size at time t, $N(t)$, follows exponential growth

$$N(t) = N(0)e^{rt}$$

where $N(0)$ is the population size at time 0 and r is a constant.

Solution

We first find the derivative of $N(t)$.

$$\frac{dN}{dt} = N(0)re^{rt}$$

Since $N(0)e^{rt} = N(t)$, we can write

$$\frac{dN}{dt} = rN(t)$$

The per capita growth rate of this exponentially growing population is then

$$\frac{1}{N}\frac{dN}{dt} = r \qquad \blacktriangleleft$$

4.6.1 Problems

Differentiate the functions in Problems 1–52 with respect to the independent variable.

1. $f(x) = e^{2x}$

2. $f(x) = e^{-4x}$

3. $f(x) = 4e^{1-3x}$

4. $f(x) = 3e^{2-5x}$

5. $f(x) = e^{-2x^2+3x-1}$

6. $f(x) = e^{4x^2-2x+1}$

7. $f(x) = e^{7x^3-\sqrt{x}+3}$

8. $f(x) = e^{-x^2-\sqrt{2x+1}}$

9. $f(x) = xe^x$

10. $f(x) = 2xe^{3x}$

11. $f(x) = x^2e^{-x}$

12. $f(x) = (3x^2-1)e^{1-x^2}$

13. $f(x) = \frac{1+e^x}{1+x^2}$

14. $f(x) = \frac{x-e^{-x}}{1+xe^{-x}}$

15. $f(x) = \frac{e^x+e^{-x}}{2+e^x}$

16. $f(x) = \frac{x}{e^x+e^{-x}}$

17. $f(x) = e^{\sin x}$

18. $f(x) = e^{\cos x}$

19. $f(x) = e^{\sin(x^2-1)}$

20. $f(x) = e^{\cos(1-2x^3)}$

21. $f(x) = \sin(e^x)$

22. $f(x) = \cos(e^x)$

23. $f(x) = \sin(e^{2x}+x)$

24. $f(x) = \cos(3x - e^{x^2-1})$

25. $f(x) = \exp[x - \sin x]$

26. $f(x) = \exp[x^2 - 2\cos x]$

27. $g(s) = \exp[\sec s^2]$

28. $g(s) = \exp[\tan s^3]$

29. $f(x) = e^{x\sin x}$

30. $f(x) = e^{1-x\cos x}$

31. $f(x) = -3e^{x^2+\tan x}$

32. $f(x) = 2e^{-x}\sec(3x)$

33. $f(x) = 2^x$

34. $f(x) = 3^x$

35. $f(x) = 2^{x+1}$

36. $f(x) = 3^{x-1}$

37. $f(x) = 5^{2x-1}$

38. $f(x) = 3^{1-3x}$

39. $f(x) = 2^{x^2+1}$

40. $f(x) = 3^{x^3-1}$

41. $h(t) = 2^{t^2-1}$

42. $h(t) = 4^{2t^3-t}$

43. $f(x) = 2^{\sqrt{x}}$

44. $f(x) = 3^{\sqrt{x+1}}$

45. $f(x) = 2^{\sqrt{x^2-1}}$

46. $f(x) = 4^{\sqrt{1-2x^3}}$

47. $h(t) = 5^{\sqrt{t}}$

48. $h(t) = 6^{\sqrt{6t^6-6}}$

49. $g(x) = 2^{2\cos x}$

50. $g(r) = 2^{-3\sin r}$

51. $g(r) = 3^{r^{1/5}}$

52. $g(r) = 4^{r^{1/4}}$

Compute the limits in Problems 53–56.

53. $\lim\limits_{h\to 0} \dfrac{e^{2h}-1}{h}$

54. $\lim_{h \to 0} \dfrac{e^{5h} - 1}{3h}$

55. $\lim_{h \to 0} \dfrac{e^{h} - 1}{\sqrt{h}}$

56. $\lim_{h \to 0} \dfrac{2^{h} - 1}{h}$

57. Find the length of the subtangent at the point $(1, 2)$ for the curve $y = 2^x$.

58. Find the length of the subtangent at the point $(2, e^4)$ for the curve $y = \exp[x^2]$.

59. Suppose that the population size at time t is
$$N(t) = e^{2t}, t \geq 0$$

(a) What is the population size at time 0?

(b) Show that
$$\frac{dN}{dt} = 2N$$

60. Suppose that the population size at time t is
$$N(t) = N_0 e^{rt}, t \geq 0$$
where N_0 is a positive constant and r is a real number.

(a) What is the population size at time 0?

(b) Show that
$$\frac{dN}{dt} = rN$$

61. Suppose that a bacterial colony grows in such a way that at time t, the population size is
$$N(t) = N(0)2^t$$
where $N(0)$ is the population size at time 0. Find the rate of growth; that is, find dN/dt and express your solution in terms of $N(t)$. Show that the growth rate of the population is proportional to the population size.

62. Suppose that a bacterial colony grows in such a way that at time t, the population size is
$$N(t) = N(0)2^t$$
where $N(0)$ is the population size at time 0. Find the per capita growth rate.

63. **(a)** Find the derivative of the logistic growth curve (see Example 3 in Section 3.3)
$$N(t) = \frac{K}{1 + \left(\frac{K}{N(0)} - 1\right)e^{-rt}}$$
where r and K are positive constants and $N(0)$ is the population size at time 0.

(b) Show that $N(t)$ satisfies the equation
$$\frac{dN}{dt} = rN\left(1 - \frac{N}{K}\right)$$
[Hint: Use the function $N(t)$ given in (a) for the right-hand side, and simplify until you obtain the derivative of $N(t)$ that you computed in (a).]

(c) Plot the per capita rate of growth, $\frac{1}{N}\frac{dN}{dt}$, as a function of N, and note that it decreases with increasing population size.

64. The following model is used in the fisheries literature to describe the recruitment of fish as a function of the size of the parent stock. If we denote the number of recruits by R and

the size of the parent stock by P, then
$$R(P) = \alpha P e^{-\beta P}, \quad P \geq 0$$
where α and β are positive constants.

(a) Sketch the graph of the function $R(P)$ when $\beta = 1$ and $\alpha = 2$.

(b) Differentiate $R(P)$ with respect to P.

(c) Find all the points on the curve that have a horizontal tangent.

65. The growth of fish can be described by the von Bertalanffy growth function
$$L(x) = L_\infty - (L_\infty - L_0)e^{-kx}$$
where x denotes the age of the fish and k, L_∞, and L_0 are positive constants.

(a) Set $L_0 = 1$ and $L_\infty = 10$. Graph $L(x)$ for $k = 1.0$ and $k = 0.1$.

(b) Interpret L_∞ and L_0.

(c) If you compare the graphs for $k = 0.1$ and $k = 1.0$, in which do fish reach $L = 5$ more quickly?

(d) Show that
$$\frac{d}{dx}L(x) = k(L_\infty - L(x))$$
that is, $dL/dx \propto L_\infty - L$. What does this say about how the rate of growth changes with age?

(e) The constant k is the proportionality constant in (d). What does the value of k tell you about how quickly a fish grows?

66. Suppose $W(t)$ denotes the amount of a radioactive material left after time t (measured in days). Assume that the radioactive decay rate of the material is 0.2/day. Find the differential equation for the radioactive decay function $W(t)$.

67. Suppose $W(t)$ denotes the amount of a radioactive material left after time t (measured in days). Assume that the radioactive decay rate of the material is 4/day. Find the differential equation for the radioactive decay function $W(t)$.

68. Suppose $W(t)$ denotes the amount of a radioactive material left after time t (measured in days). Assume that the half-life of the material is 3 days. Find the differential equation for the radioactive decay function $W(t)$.

69. Suppose $W(t)$ denotes the amount of a radioactive material left after time t (measured in days). Assume that the half-life of the material is 5 days. Find the differential equation for the radioactive decay function $W(t)$.

70. Suppose $W(t)$ denotes the amount of a radioactive material left after time t. Assume that $W(0) = 15$ and that
$$\frac{dW}{dt} = -2W(t)$$

(a) How much material is left at time $t = 2$?

(b) What is the half-life of this material?

71. Suppose $W(t)$ denotes the amount of a radioactive material left after time t. Assume that $W(0) = 6$ and that
$$\frac{dW}{dt} = -3W(t)$$

(a) How much material is left at time $t = 4$?

(b) What is the half-life of this material?

72. Suppose $W(t)$ denotes the amount of a radioactive material left after time t. Assume that $W(0) = 10$ and $W(1) = 8$.

(a) Find the differential equation that describes this situation.

(b) How much material is left at time $t = 5$?

(c) What is the half-life of this material?

73. Suppose $W(t)$ denotes the amount of a radioactive material left after time t. Assume that $W(0) = 5$ and $W(1) = 2$.

(a) Find the differential equation that describes this situation.

(b) How much material is left at time $t = 3$?

(c) What is the half-life of this material?

4.7 DERIVATIVES OF INVERSE AND LOGARITHMIC FUNCTIONS

Recall that the logarithmic function is the inverse of the exponential function: To find the derivative of the logarithmic function, we must therefore learn how to compute the derivative of an inverse function.

4.7.1 Derivatives of Inverse Functions

We begin with an example (Figure 4.32). Let $f(x) = x^2, x \geq 0$. We computed the inverse function of f in Subsection 1.2.6. First note that $f(x) = x^2, x \geq 0$, is one to one (use the horizontal line test); hence, we can define its inverse. We repeat the steps to find an inverse function. [Recall that we obtain the graph of the inverse function by reflecting $y = f(x)$ about the line $y = x$.]

1. Write $y = f(x)$:

$$y = x^2$$

2. Solve for x:

$$x = \sqrt{y}$$

3. Interchange x and y:

$$y = \sqrt{x}$$

Since the range of $f(x)$, which is the interval $[0, \infty)$, becomes the domain for the inverse function, we find

$$f^{-1}(x) = \sqrt{x} \quad \text{for } x \geq 0$$

We already know the derivative of $\sqrt{x}$, namely, $1/(2\sqrt{x})$. But we will try to find the derivative in a different way that we can generalize to get a formula

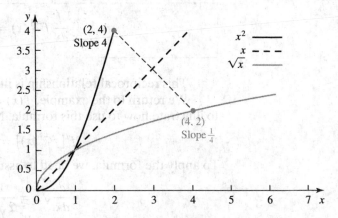

▲ **Figure 4.32**

The function $y = x^2, x \geq 0$, and its inverse function $y = \sqrt{x}, x \geq 0$

for finding derivatives of inverse functions. Let $g(x) = f^{-1}(x)$. Then

$$(f \circ g)(x) = f[g(x)] = (\sqrt{x})^2 = x, \quad x \geq 0$$

Therefore, the derivatives of $(\sqrt{x})^2$ and x must be equal. Applying the chain rule, we find

$$\frac{d}{dx}(\sqrt{x})^2 = 2\sqrt{x}\frac{d}{dx}\sqrt{x}$$

Since $\frac{d}{dx}x = 1$, we obtain

$$2\sqrt{x}\frac{d}{dx}\sqrt{x} = 1$$

or, for $x > 0$,

$$\frac{d}{dx}\sqrt{x} = \frac{1}{2\sqrt{x}}$$

To prepare for how the derivatives of a function and its inverse function are related geometrically, look at Figure 4.32, where the slope at the point $(2, 4)$ of $f(x) = x^2$ is $m = 4$, and the slope at the point $(4, 2)$ of $f^{-1}(x) = \sqrt{x}$ is $m = 1/4$. We will find this reciprocal relationship of slopes at related points also in the general case.

The steps that led us to the derivative of $\sqrt{x}$ can be used to find a general formula for the derivative of inverse functions. If $g(x)$ is the inverse function of $f(x)$, then $f[g(x)] = x$. Following the same steps, for the general case $f[g(x)] = x$, we find

$$\frac{d}{dx}f[g(x)] = f'[g(x)]g'(x)$$

Since $\frac{d}{dx}x = 1$, we obtain

$$f'[g(x)]g'(x) = 1$$

If $f'[g(x)] \neq 0$, we can divide by $f'[g(x)]$ and find

$$g'(x) = \frac{1}{f'[g(x)]}$$

Since $g(x) = f^{-1}(x)$ and $g'(x) = \frac{d}{dx}g(x) = \frac{d}{dx}f^{-1}(x)$, we obtain

$$\frac{d}{dx}f^{-1}(x) = \frac{1}{f'[f^{-1}(x)]} \tag{4.12}$$

This reciprocal relationship is illustrated in Figure 4.33.

We return to the example, $f(x) = x^2, x \geq 0$, where $f^{-1}(x) = \sqrt{x}, x \geq 0$, to illustrate how to use this formula. Now, $f'(x) = 2x$. We need to evaluate

$$f'[f^{-1}(x)] = f'[\sqrt{x}] = 2\sqrt{x}$$

To apply the formula, we need to assume that $f'[f^{-1}(x)] \neq 0$. Then

$$\frac{d}{dx}\sqrt{x} = \frac{1}{2\sqrt{x}} \quad \text{for } x > 0$$

Looking at the graphs of $y = x^2$ and $y = \sqrt{x}, x \geq 0$, it is easy to see why $f^{-1}(x)$ is not differentiable at $x = 0$. [Recall that we obtain the graph of the

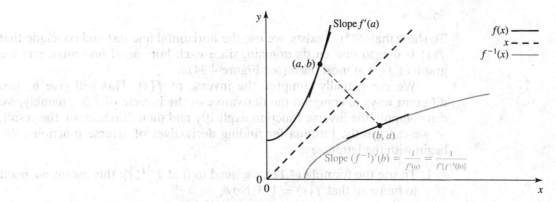

▲ Figure 4.33

The graphs of $y = f(x)$ and its inverse function $y = f^{-1}(x)$ have reciprocal slopes at the points (a, b) and (b, a)

inverse function by reflecting $y = f(x)$ about the line $y = x$.] If we draw the tangent line at $x = 0$ on the curve $y = x^2$, we find that the tangent line is horizontal, that is, its slope is 0. Reflecting a horizontal line about $y = x$ results in a vertical line for which the slope is not defined (Figure 4.32).

The formula for finding derivatives of inverse functions takes on a particularly easy to remember form when we use Leibniz notation. To see this, note that (without interchanging x and y)

$$y = f(x) \quad \Leftrightarrow \quad x = f^{-1}(y)$$

and hence

$$\frac{dx}{dy} = \frac{1}{\dfrac{dy}{dx}}$$

This formula again emphasizes the reciprocal relationship. We illustrate the formula using the example

$$y = x^2 \quad \Leftrightarrow \quad x = \sqrt{y}$$

for $x > 0$. Since $\frac{dy}{dx} = 2x$, we find

$$\frac{dx}{dy} = \frac{1}{\dfrac{dy}{dx}} = \frac{1}{2x} = \frac{1}{2\sqrt{y}}$$

That is,

$$\frac{d}{dy}\sqrt{y} = \frac{1}{2\sqrt{y}}$$

The answer is now in terms of y since we did not interchange x and y when we computed the inverse function. If we do so, we again find

$$\frac{d}{dx}\sqrt{x} = \frac{1}{2\sqrt{x}}$$

▶ **Example 1**   Let

$$f(x) = \frac{x}{1 + x} \quad \text{for } x \geq 0$$

Find $\frac{d}{dx} f^{-1}\left(\frac{1}{3}\right)$.

Solution

To show that $f^{-1}(x)$ exists, we use the horizontal line test and conclude that $f(x)$ is one to one on its domain, since each horizontal line intersects the graph of $f(x)$ at most once (see Figure 4.34).

We can actually compute the inverse of $f(x)$. This will give us two different ways to compute the derivative of the inverse of $f(x)$; namely, we can compute the inverse function explicitly and then differentiate the result, or we can use the formula for finding derivatives of inverse functions. We begin with the latter way.

1. To use the formula (4.12), we need to find $f^{-1}(\frac{1}{3})$; this means we need to find x so that $f(x) = 1/3$. Now,

$$\frac{x}{1+x} = \frac{1}{3} \quad \text{implies} \quad 2x = 1 \quad \text{or} \quad x = \frac{1}{2}$$

Therefore, $f^{-1}(\frac{1}{3}) = \frac{1}{2}$. Formula (4.12) becomes

$$\frac{d}{dx} f^{-1}\left(\frac{1}{3}\right) = \frac{1}{f'[f^{-1}(\frac{1}{3})]} = \frac{1}{f'(\frac{1}{2})}$$

We use the quotient rule to find the derivative of $f(x)$.

$$f'(x) = \frac{(1)(1+x) - (x)(1)}{(1+x)^2} = \frac{1}{(1+x)^2}$$

At $x = 1/2$, $f'(\frac{1}{2}) = \frac{1}{(1+1/2)^2} = \frac{4}{9}$. Therefore,

$$\frac{d}{dx} f^{-1}\left(\frac{1}{3}\right) = \frac{1}{\frac{4}{9}} = \frac{9}{4}$$

2. We can compute the inverse function: Set $y = \frac{x}{1+x}$. Solving for x yields

$$x = \frac{y}{1-y}$$

Interchanging x and y, we find

$$y = \frac{x}{1-x}$$

Since the domain of $f(x)$ is $[0, \infty)$, the range of f is $[0, 1)$. The range of f becomes the domain of the inverse; therefore, the inverse function is

$$f^{-1}(x) = \frac{x}{1-x} \quad \text{for } 0 \le x < 1$$

We use the quotient rule to find $\frac{d}{dx} f^{-1}(x)$:

$$\frac{d}{dx} f^{-1}(x) = \frac{(1)(1-x) - (x)(-1)}{(1-x)^2} = \frac{1}{(1-x)^2}$$

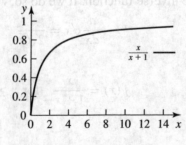

◀ **Figure 4.34**

The graph of $f(x) = \frac{x}{x+1}$ for $x \ge 0$

and therefore

$$\frac{d}{dx} f^{-1}\left(\frac{1}{3}\right) = \frac{1}{(1 - 1/3)^2} = \frac{9}{4}$$

which agrees with the answer in (1).   ◀

The inverse cannot always be computed explicitly.

▶ **Example 2**   Let

$$f(x) = 2x + e^x \quad \text{for } x \in \mathbf{R}$$

Find $\frac{d}{dx} f^{-1}(1)$.

Solution

In this case, it is not possible to solve $y = 2x + e^x$ for x. Therefore, we must use (4.12) if we wish to compute the derivative of the inverse function at a particular point. Equation (4.12) becomes

$$\frac{d}{dx} f^{-1}(1) = \frac{1}{f'[f^{-1}(1)]}$$

when $x = 1$. We need to find $f'(x)$:

$$f'(x) = 2 + e^x$$

Since $f(0) = 1$, it follows that $f^{-1}(1) = 0$ and hence

$$\frac{1}{f'[f^{-1}(1)]} = \frac{1}{f'(0)} = \frac{1}{2 + 1} = \frac{1}{3}$$
◀

The next example, in which we again need to use (4.12), involves finding the derivative of the inverse of a trigonometric function.

▶ **Example 3**   Let $f(x) = \tan x$, $-\frac{\pi}{2} < x < \frac{\pi}{2}$. Find $\frac{d}{dx} f^{-1}(1)$.

Solution

Since $f(\frac{\pi}{4}) = \tan \frac{\pi}{4} = 1$, it follows that $f^{-1}(1) = \frac{\pi}{4}$. Recall that $f'(x) = \sec^2 x$. We therefore have

$$\frac{d}{dx} f^{-1}(1) = \frac{1}{f'[f^{-1}(1)]} = \frac{1}{f'(\frac{\pi}{4})} = \frac{1}{\sec^2(\frac{\pi}{4})}$$

$$= \cos^2\left(\frac{\pi}{4}\right) = \left(\frac{1}{2}\sqrt{2}\right)^2 = \frac{1}{2}$$
◀

If we define $f(x) = \tan x$ on the domain $(-\frac{\pi}{2}, \frac{\pi}{2})$, then $f(x)$ is one to one, as can be seen from Figure 4.35 (use the horizontal line test). The range of $f(x)$ is $(-\infty, \infty)$. We cannot use algebra to solve $y = \tan x$ for x; instead, the inverse of the tangent function gets its own name. It is called $y = \arctan x$ (or $y = \tan^{-1} x$) and its domain is $(-\infty, \infty)$. In the next example, we will find the derivative of $y = \arctan x$, which will turn out to have a surprisingly simple form.

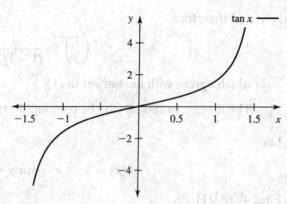

◀ **Figure 4.35**
The function $f(x) = \tan x$, $(-\frac{\pi}{2}, \frac{\pi}{2})$, is one to one on its domain

▶ **Example 4**   Let $f(x) = \tan x$, $-\frac{\pi}{2} < x < \frac{\pi}{2}$. Find $\frac{d}{dx} f^{-1}(x)$.

Solution

As mentioned previously, $f^{-1}(x)$ exists since $f(x)$ is one to one on its domain. Recall that

$$\frac{d}{dx} \tan x = \sec^2 x$$

The inverse of the tangent function is denoted by $\tan^{-1} x$ or $\arctan x$. (Note that $\tan^{-1} x$ is different from $\frac{1}{\tan x}$. The superscript "−1" refers to the function being an inverse function.) We set $y = \arctan x$ (and hence $x = \tan y$). Then

$$\frac{dy}{dx} = \frac{d}{dx} \arctan x = \frac{1}{\frac{dx}{dy}} = \frac{1}{\frac{d}{dy} \tan y} = \frac{1}{\sec^2 y} = \cos^2 y$$

We can simplify this expression using the following trick, which will allow us to write $\cos^2 y$ in terms of x. Since $x = \tan y$, we find

$$x^2 = \tan^2 y = \frac{\sin^2 y}{\cos^2 y} = \frac{1 - \cos^2 y}{\cos^2 y} = \frac{1}{\cos^2 y} - 1$$

In the second step, we used the identity $\tan y = \frac{\sin y}{\cos y}$, and in the third step, we used the identity $\sin^2 y + \cos^2 y = 1$. Solving the preceding expression for $\cos^2 y$, we get

$$\cos^2 y = \frac{1}{1 + x^2}$$

Therefore,

$$\frac{d}{dx} \arctan x = \cos^2 y = \frac{1}{1 + x^2}$$ ◀

The result in the preceding example is important, and we summarize it in the following box.

$$\frac{d}{dx} \arctan x = \frac{d}{dx} \tan^{-1} x = \frac{1}{1 + x^2}$$

The derivative of the inverse sine function, $y = \arcsin x$, is discussed in Problem 22 of this section. The derivatives of the remaining inverse trigonometric functions are listed in the table of derivatives on the inside back cover of this book.

4.7.2 The Derivative of the Logarithmic Function

We introduced the logarithmic function to base a, $\log_a x$, as the inverse function of the exponential function a^x (Figures 4.36 and 4.37). We can therefore use the formula for derivatives of inverse functions to find the derivative of $y = \log_a x$. Since

$$\log_a x = \frac{\ln x}{\ln a}$$

and $\ln a$ is a constant, it is enough to find the derivative of $\ln x$ (Figure 4.38). We set $f(x) = e^x$; then $f'(x) = e^x$ and $f^{-1}(x) = \ln x$. Therefore,

$$\frac{d}{dx} \ln x = \frac{d}{dx}[f^{-1}(x)] = \frac{1}{f'[f^{-1}(x)]} = \frac{1}{\exp[\ln x]} = \frac{1}{x}$$

We summarize this in the following box.

$$\frac{d}{dx} \ln x = \frac{1}{x}$$

$$\frac{d}{dx} \log_a x = \frac{1}{(\ln a)x}$$

▶ **Example 5** Find the derivative of $y = \ln(3x)$.

Solution

We use the chain rule with $u = g(x) = 3x$ and $f(u) = \ln u$:

$$\frac{dy}{dx} = \frac{dy}{du}\frac{du}{dx} = \frac{1}{u}3 = \frac{3}{3x} = \frac{1}{x}$$

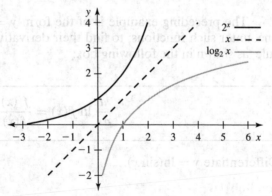

◀ **Figure 4.36**
The function $y = \log_2 x$ as the inverse function of $y = 2^x$

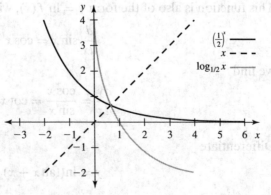

◀ **Figure 4.37**
The function $y = \log_{1/2} x$ as the inverse function of $y = \left(\frac{1}{2}\right)^x$

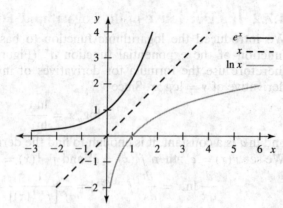

◄ Figure 4.38
The natural logarithm $y = \ln x$ as the inverse function of the natural exponential function $y = e^x$

If you are surprised that the factor 3 disappeared, note that

$$y = \ln(3x) = \ln 3 + \ln x$$

Since $\ln 3$ is a constant, its derivative is 0. Hence,

$$\frac{d}{dx}(\ln 3 + \ln x) = \frac{d}{dx}\ln 3 + \frac{d}{dx}\ln x = 0 + \frac{1}{x} = \frac{1}{x} \qquad ◄$$

▶ **Example 6** Find the derivative of $y = \ln x^2$, $x \neq 0$.

Solution

We can use the chain rule with $u = g(x) = x^2$ and $f(u) = \ln u$. We find

$$y' = \frac{df}{du}\frac{du}{dx} = \frac{1}{u}2x = \frac{1}{x^2}2x = \frac{2}{x} \quad \text{for } x \neq 0 \qquad ◄$$

The preceding example is of the form $y = \ln f(x)$. We will frequently encounter such functions; to find their derivatives, we need to use the chain rule, as shown in the following box:

$$\frac{d}{dx}\ln f(x) = \frac{f'(x)}{f(x)}$$

▶ **Example 7** Differentiate $y = \ln(\sin x)$.

Solution

This function is also of the form $y = \ln f(x)$, with $f(x) = \sin x$. Since

$$\frac{d}{dx}\sin x = \cos x$$

we find

$$\frac{dy}{dx} = \frac{\cos x}{\sin x} = \cot x \qquad ◄$$

▶ **Example 8** Differentiate

$$y = \ln(\tan x + x)$$

Solution

This function is of the form $y = \ln f(x)$ with $f(x) = \tan x + x$. Thus

$$y' = \frac{\sec^2 x + 1}{\tan x + x}$$ ◀

▶ **Example 9** Differentiate

$$y = \log(2x^3 - 1)$$

Solution

This function is of the form $y = \log f(x)$ with $f(x) = 2x^3 - 1$. The logarithm is to base 10. Thus

$$y' = \frac{1}{\ln 10} \frac{6x^2}{2x^3 - 1}$$ ◀

4.7.3 Logarithmic Differentiation

In 1695, Leibniz introduced logarithmic differentiation, following a suggestion of Johann Bernoulli to find derivatives of functions of the form $y = [f(x)]^x$. Bernoulli generalized this method and published his results two years later. The basic idea is to take logarithms on both sides and then to differentiate using the technique of implicit differentiation.

▶ **Example 10** Find $\frac{dy}{dx}$ when $y = x^x$.

Solution

We take logarithms on both sides of the equation $y = x^x$.

$$\ln y = \ln x^x$$

Applying properties of the logarithm, we can simplify the right-hand side as $\ln x^x = x \ln x$. We can now differentiate both sides with respect to x. Since y is a function of x, we need to use the chain rule to differentiate $\ln y$ (as we learned in the section on implicit differentiation):

$$\frac{d}{dx}[\ln y] = \frac{d}{dx}[x \ln x]$$

$$\frac{1}{y}\frac{dy}{dx} = 1 \cdot \ln x + x\frac{1}{x}$$

$$\frac{dy}{dx} = y[\ln x + 1]$$

$$\frac{dy}{dx} = (\ln x + 1)x^x$$ ◀

If the function $y = x^x$ looks strange, write it as

$$y = x^x = \exp[\ln x^x] = \exp[x \ln x]$$

that is, $y = e^{x \ln x}$. We can differentiate this function without using logarithmic differentiation; namely,

$$\frac{dy}{dx} = e^{x \ln x}\frac{d}{dx}(x \ln x)$$

$$= e^{x \ln x}\left(1 \cdot \ln x + x\frac{1}{x}\right)$$

$$= e^{x \ln x}(\ln x + 1)$$

Either approach will give you the correct answer.

▶ Example 11   Find the derivative of $y = (\sin x)^x$.

Solution

We take logarithms on both sides of the equation and simplify:

$$\ln y = x \ln(\sin x)$$

Differentiating with respect to x yields

$$\frac{d}{dx}[\ln y] = \frac{d}{dx}[x \ln(\sin x)]$$

$$\frac{1}{y}\frac{dy}{dx} = 1 \cdot \ln(\sin x) + x\frac{d}{dx}[\ln(\sin x)]$$

$$= \ln(\sin x) + x\frac{\cos x}{\sin x}$$

$$= \ln(\sin x) + x \cot x$$

Hence, after multiplying by y and substituting $(\sin x)^x$ for y,

$$\frac{dy}{dx} = [\ln(\sin x) + x \cot x](\sin x)^x \qquad ◀$$

The following example should convince you that logarithmic differentiation can simplify finding the derivatives of complicated expressions.

▶ Example 12   Differentiate

$$y = \frac{e^x x^{3/2}\sqrt{1+x}}{(x^2+3)^4(3x-2)^3}$$

Solution

Without logarithmic differentiation, this would be rather difficult. Taking logarithms on both sides, however, we see that the expression simplifies. It is very important that we apply the properties of the logarithm before differentiating, as this will simplify the expressions that we must differentiate.

$$\ln y = \ln \frac{e^x x^{3/2}\sqrt{1+x}}{(x^2+3)^4(3x-2)^3}$$

$$= \ln e^x + \ln x^{3/2} + \ln\sqrt{1+x} - \ln(x^2+3)^4 - \ln(3x-2)^3$$

$$= x + \frac{3}{2}\ln x + \frac{1}{2}\ln(1+x) - 4\ln(x^2+3) - 3\ln(3x-2)$$

This no longer looks so daunting, and we can differentiate both sides:

$$\frac{d}{dx}[\ln y] = \frac{d}{dx}\left[x + \frac{3}{2}\ln x + \frac{1}{2}\ln(1+x) - 4\ln(x^2+3) - 3\ln(3x-2)\right]$$

$$\frac{1}{y}\frac{dy}{dx} = 1 + \frac{3}{2}\frac{1}{x} + \frac{1}{2}\frac{1}{1+x} - 4\frac{2x}{x^2+3} - 3\frac{3}{3x-2}$$

Finally, solving for dy/dx yields

$$\frac{dy}{dx} = \left(1 + \frac{3}{2x} + \frac{1}{2(1+x)} - \frac{8x}{x^2+3} - \frac{9}{3x-2}\right) \frac{e^x x^{3/2}\sqrt{1+x}}{(x^2+3)^4(3x-2)^3}$$ ◀

We can also use this method to prove the general power rule (as promised in Section 4.3).

POWER RULE (GENERAL FORM)

> Let $f(x) = x^r$, where r is any real number. Then
>
> $$\frac{d}{dx}(x^r) = rx^{r-1}$$

Proof We set $y = x^r$ and use logarithmic differentiation to find

$$\frac{d}{dx}[\ln y] = \frac{d}{dx}[\ln x^r]$$

$$\frac{1}{y}\frac{dy}{dx} = \frac{d}{dx}[r\ln x]$$

$$\frac{1}{y}\frac{dy}{dx} = r\frac{1}{x}$$

Solving for dy/dx yields

$$\frac{dy}{dx} = r\frac{1}{x}y$$

$$= r\frac{1}{x}x^r$$

$$= rx^{r-1}$$ ■

4.7.4 Problems

(4.7.1)

In Problems 1–6, find the inverse of each function, and differentiate each inverse in two ways: (i) Differentiate the inverse function directly, and (ii) use (4.12) to find the derivative of the inverse.

1. $f(x) = \sqrt{2x}, x \geq 0$

2. $f(x) = \sqrt{x-1}, x \geq 1$

3. $f(x) = 3x^2, x \geq 0$

4. $f(x) = x^2 - 2, x \geq 0$

5. $f(x) = 3 - 2x^3, x \geq 0$

6. $f(x) = \dfrac{2x^2 - 1}{x^2 - 1}, x > 1$

In Problems 7–23, use (4.12) to find the derivative of the inverse at the indicated point.

7. Let

$$f(x) = 2x^2 - 2, \quad x \geq 0$$

Find $\frac{d}{dx}f^{-1}(0)$. [Note that $f(1) = 0$.]

8. Let

$$f(x) = -x^2 + 3, \quad x \geq 0$$

Find $\frac{d}{dx}f^{-1}(-1)$. [Note that $f(2) = -1$.]

9. Let

$$f(x) = \sqrt{x+1}, \quad x \geq 0$$

Find $\frac{d}{dx}f^{-1}(2)$. [Note that $f(3) = 2$.]

10. Let

$$f(x) = \sqrt{5+x^2}, \quad x \geq 0$$

Find $\frac{d}{dx}f^{-1}(3)$. [Note that $f(2) = 3$.]

11. Let

$$f(x) = x + e^x, \quad x \in \mathbf{R}$$

Find $\frac{d}{dx}f^{-1}(1)$. [Note that $f(0) = 1$.]

12. Let

$$f(x) = x + e^x, \quad x > 0$$

Find $\frac{d}{dx}f^{-1}(1+e)$. [Note that $f(1) = 1+e$.]

13. Let
$$f(x) = x - \sin x, \quad x \in \mathbf{R}$$
Find $\frac{d}{dx} f^{-1}(\pi)$. [Note that $f(\pi) = \pi$.]

14. Let
$$f(x) = x - \cos x, \quad x \in \mathbf{R}$$
Find $\frac{d}{dx} f^{-1}(-1)$. [Note that $f(0) = -1$.]

15. Let
$$f(x) = x^2 + \tan x, \quad x \in \left(-\frac{\pi}{2}, \frac{\pi}{2}\right)$$
Find $\frac{d}{dx} f^{-1}(0)$. [Note that $f(0) = 0$.]

16. Let
$$f(x) = x^2 + \tan x, \quad x \in \left(-\frac{\pi}{2}, \frac{\pi}{2}\right)$$
Find $\frac{d}{dx} f^{-1}(\frac{\pi^2}{16} + 1)$. [Note that $f(\frac{\pi}{4}) = \frac{\pi^2}{16} + 1$.]

17. Let $f(x) = \sin x, 0 < x < \pi/2$. Find $\frac{d}{dx} f^{-1}(1/2)$.

18. Let $f(x) = \tan x, 0 < x < \pi/2$. Find $\frac{d}{dx} f^{-1}(\sqrt{3})$.

19. Let $f(x) = x^5 + x + 1, -1 < x < 1$. Find $\frac{d}{dx} f^{-1}(1)$.

20. Let $f(x) = e^{-x^2} + x$. Find $\frac{d}{dx} f^{-1}(1)$.

21. Let $f(x) = e^{-x^2/2} + 2x$. Find $\frac{d}{dx} f^{-1}(1)$.

22. Denote by $y = \arcsin x, -1 \le x \le 1$, the inverse of $y = \sin x, -\frac{\pi}{2} \le x \le \frac{\pi}{2}$. Show that
$$\frac{d}{dx} \arcsin x = \frac{1}{\sqrt{1 - x^2}}, \quad -1 < x < 1$$

(4.7.2)

In Problems 23–60, differentiate the functions with respect to the independent variable. (Note that log *denotes the logarithm to base 10.)*

23. $f(x) = \ln(x + 1)$
24. $f(x) = \ln(3x)$
25. $f(x) = \ln(1 - 2x)$
26. $f(x) = \ln(3 + 4x)$
27. $f(x) = \ln x^2$
28. $f(x) = \ln(1 - x^2)$
29. $f(x) = \ln(x^3)$
30. $f(x) = \ln(x^3 - 1)$
31. $f(x) = (\ln x)^2$
32. $f(x) = (\ln x)^3$
33. $f(x) = (\ln x^2)^2$
34. $f(x) = (\ln(1 - x^2))^3$
35. $f(x) = \ln \sqrt{x^2 + 1}$
36. $f(x) = \ln \sqrt{2x^2 - x}$
37. $f(x) = \ln \frac{x}{x + 1}$
38. $f(x) = \ln \frac{2x}{1 + x^2}$
39. $f(x) = \ln \frac{1 - x}{1 + 2x}$
40. $f(x) = \ln \frac{x^2 - 1}{x^3 - 1}$
41. $f(x) = \exp[x - \ln x]$
42. $g(s) = \exp[s^2 + \ln s]$
43. $f(x) = \ln(\sin x)$
44. $f(x) = \ln(\cos(1 - x))$
45. $f(x) = \ln(\tan x^2))$
46. $g(s) = \ln(\sin^2(3s))$

47. $f(x) = x \ln x$
48. $f(x) = x^2 \ln x^2$
49. $f(x) = \frac{\ln x}{x}$
50. $h(t) = \dfrac{\ln t}{1 + t^2}$
51. $h(t) = \sin(\ln(3t))$
52. $h(s) = \ln(\ln s)$
53. $f(x) = \ln |x^2 - 3|$
54. $f(x) = \log(2x - 1)$
55. $f(x) = \log(1 - x^2)$
56. $f(x) = \log(3x^2 - 5x)$
57. $f(x) = \log(x^3 - 3x)$
58. $f(x) = \log(\sqrt[3]{\tan x^2})$
59. $f(u) = \log_3(3 + u^4)$
60. $g(s) = \log_5(3^s - 2)$

61. Let $f(x) = \ln x$. We know that $f'(x) = \frac{1}{x}$. We will use this fact and the definition of derivatives to show that
$$\lim_{n \to \infty} \left(1 + \frac{1}{n}\right)^n = e$$

(a) Use the definition of the derivative to show that
$$f'(1) = \lim_{h \to 0} \frac{\ln(1 + h)}{h}$$

(b) Show that (a) implies that
$$\ln[\lim_{h \to 0}(1 + h)^{1/h}] = 1$$

(c) Set $h = \frac{1}{n}$ in (b) and let $n \to \infty$. Show that this implies
$$\lim_{n \to \infty} \left(1 + \frac{1}{n}\right)^n = e$$

62. Assume that $f(x)$ is differentiable with respect to x. Show that
$$\frac{d}{dx} \ln\left[\frac{f(x)}{x}\right] = \frac{f'(x)}{f(x)} - \frac{1}{x}$$

(4.7.3)

In Problems 63–74, use logarithmic differentiation to find the first derivative of the given functions.

63. $f(x) = x^x$
64. $f(x) = x^{2x}$
65. $f(x) = (\ln x)^x$
66. $f(x) = (\ln x)^{3x}$
67. $f(x) = x^{\ln x}$
68. $f(x) = x^{2 \ln x}$
69. $f(x) = x^{1/x}$
70. $f(x) = x^{3/x}$
71. $y = x^{x^x}$
72. $y = (x^x)^x$
73. $y = x^{\cos x}$
74. $y = (\cos x)^x$
75. Differentiate
$$y = \frac{e^{2x}(9x - 2)^3}{\sqrt[4]{(x^2 + 1)(3x^3 - 7)}}$$

76. Differentiate
$$y = \frac{e^{x-1} \sin^2 x}{(x^2 + 5)^{2x}}$$

4.8 APPROXIMATION AND LOCAL LINEARITY

Suppose we want to find an approximation for $\ln(1.05)$ without using a calculator. The method to solve this problem will be useful in many other applications. Let's look at the graph of $f(x) = \ln x$ (Figure 4.39a). We know $\ln 1 = 0$ and see that 1.05 is quite close to 1. So close in fact that the curve connecting $(1, 0)$ to $(1.05, \ln 1.05)$ is close to a straight line (Figure 4.39b). This suggests that we should approximate the curve by a straight line. But not just any straight line: We choose the tangent line to the graph of $f(x) = \ln x$ at $x = 1$ (Figure 4.39a). We can find the equation of the tangent line without a calculator. Namely, the slope of $f(x) = \ln x$ at $x = 1$ is $f'(1) = \frac{1}{x}|_{x=1} = 1$. This, together with the point $(1, 0)$, allows us to find the tangent line at $x = 1$:

$$L(x) = f(1) + f'(1)(x - 1)$$

$$= 0 + (1)(x - 1) = x - 1$$

We call $L(x)$ the **tangent line approximation** or **the linearization** of $f(x)$ at $x = 1$. If we evaluate $L(x)$ at $x = 1.05$, we find $L(1.05) = 1.05 - 1 = 0.05$, which is a good approximation to $\ln 1.05 = 0.048790\ldots$. (Here we used the calculator to see how close the approximation is to the exact value.)

The Tangent Line Approximation

> Assume that $y = f(x)$ is differentiable at $x = a$; then
>
> $$L(x) = f(a) + f'(a)(x - a)$$
>
> is the **tangent line approximation** or the **linearization** of f at $x = a$.

Geometrically, the linearization of f at $x = a$, $L(x) = f(a) + f'(a)(x-a)$, is the equation of the tangent line to the graph of $f(x)$ at the point $(a, f(a))$ (see Figure 4.40).

If $|x - a|$ is sufficiently small, then $f(x)$ can be linearly approximated by $L(x)$, that is,

$$f(x) \approx f(a) + f'(a)(x - a)$$

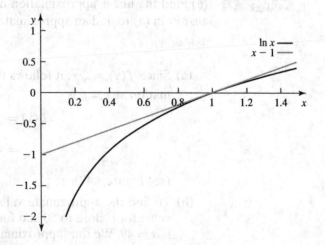

▲ **Figure 4.39a**
The tangent line approximation for $\ln x$ at $x = 1$ to approximate $\ln(1.05)$

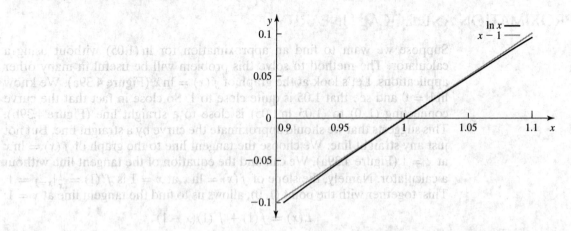

▲ **Figure 4.39b**
The tangent line approximation for $\ln x$ at $x = 1$ to approximate $\ln(1.05)$. When x is close to 1, the tangent line and the graph of $y = \ln x$ are close.

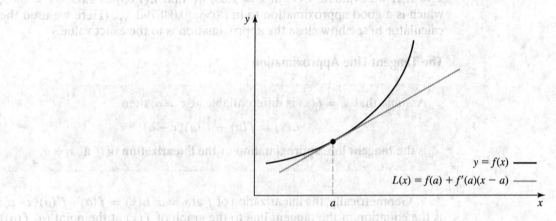

▲ **Figure 4.40**
The tangent line approximation of $y = f(x)$ at $x = a$

This is illustrated in Figure 4.41.

▶ **Example 1** (a) Find the linear approximation of $f(x) = \sqrt{x}$ at $x = a$, and (b) use your answer in (a) to find an approximate value of $\sqrt{50}$.

Solution

(a) Since $f(x) = \sqrt{x}$, it follows that $f'(x) = \frac{1}{2\sqrt{x}}$, and the linear approximation at $x = a$ is

$$L(x) = f(a) + f'(a)(x - a)$$

$$= \sqrt{a} + \frac{1}{2\sqrt{a}}(x - a)$$

(see Figure 4.42).

(b) To find the approximate value of $f(50) = \sqrt{50}$, we need to choose a value for a close to 50 and for which we know $\sqrt{a}$ exactly. Our choice is $a = 49$. We thus approximate $f(50)$ by $L(50)$ with $a = 49$ and find

$$\sqrt{50} \approx \sqrt{49} + \frac{50 - 49}{2\sqrt{49}} = 7 + \frac{1}{14} \approx 7.0714$$

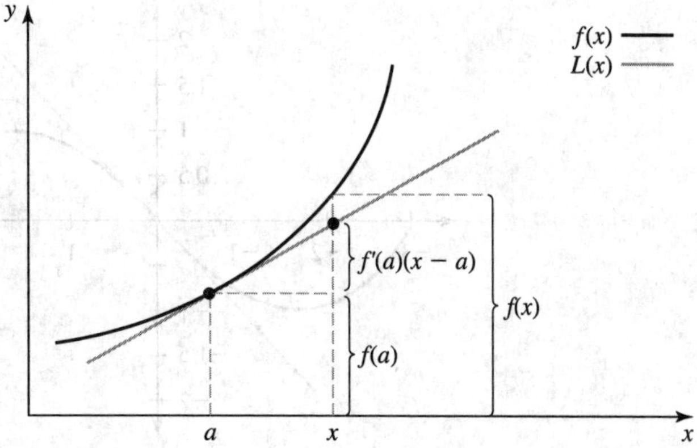

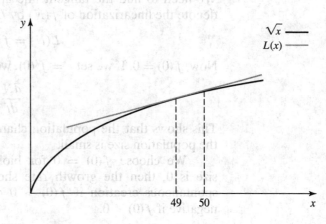

▲ **Figure 4.41**
The linearization of f at $x = a$ can be used to approximate $f(x)$ for x close to a

▲ **Figure 4.42**
The linear approximation of $f(x) = \sqrt{x}$ at $x = 49$, is the line $y = L(x)$

Using a calculator to compute $\sqrt{50} = 7.0711\ldots$, we see that the error in the linear approximation is quite small. ◀

▶ **Example 2**   Find the linear approximation of $f(x) = \sin x$ at $x = 0$.

Solution
Since $f'(x) = \cos x$, it follows that

$$L(x) = f(0) + f'(0)(x - 0)$$
$$= \sin 0 + (\cos 0)x = x$$

(Figure 4.43). That is, for small values of x we can approximate $\sin x$ by x. This is often used in physics. (Note that x is measured in radians.) ◀

▶ **Example 3**   Let $N(t)$ be the size of a population at time t and assume that its growth rate dN/dt is given by

$$\frac{dN}{dt} = f(N)$$

where $f(N)$ is a differentiable function with $f(0) = 0$. Find the linearization of the growth rate at $N = 0$.

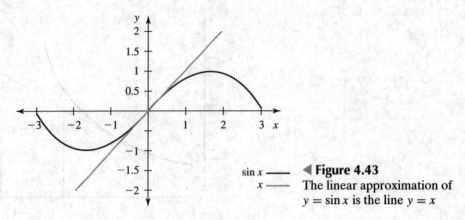

sin x ——   ◀ **Figure 4.43**
x ——   The linear approximation of
y = sin x is the line y = x

Solution

We need to find the tangent line approximation of $f(N)$ at $N = 0$. If we denote the linearization of $f(N)$ by $L(N)$, we find

$$L(N) = f(0) + f'(0)N$$

Now, $f(0) = 0$. If we set $r = f'(0)$, we find that for N close to 0,

$$\frac{dN}{dt} \approx rN$$

This shows that the population changes approximately exponentially when the population size is small.

We choose $f(0) = 0$ for biological reasons: When the population size is 0, then the growth rate should be 0, otherwise, we would have spontaneous creation if $f(0) > 0$ or the population size would become negative if $f(0) < 0$.   ◀

▶ **Example 4**   Let $N(t)$ be the size of a bacterial population at time t (measured in millions), and assume that the per capita growth rate is equal to 2%. We can express this as a differential equation, namely,

$$\frac{1}{N}\frac{dN}{dt} = 0.02$$

Suppose we know that at time $t = 10$, the size of the population is 250,000,000, that is, $N(10) = 250$ (since we measure the population size in millions). Use a linear approximation to predict the approximate population size at time $t = 10.2$.

Solution

To predict the population size at time 10.2, we use the linearization of $N(t)$ at $t = 10$, namely,

$$L(t) = N(10) + N'(10)(t - 10)$$

To evaluate $L(t)$ at $t = 10.2$, we need to find $N'(10)$.

Using the differential equation, we find

$$N'(t) = (0.02)N(t)$$

When $t = 10$, we obtain

$$N'(10) = (0.02)N(10) = (0.02)(250) = 5$$

Hence,

$$L(10.2) = N(10) + N'(10)(10.2 - 10)$$
$$= 250 + (5)(0.2) = 251$$

Thus, we predict that the population size at time 10.2 is approximately 251,000,000. Note that this approximation is only good if the time at which we want to predict the population size is very close to the time at which we know the population size. ◀

Error Propagation   Linear approximations are used in problems of **error propagation**. Suppose that you wish to determine the surface area of a spherical cell. Since the surface area S of a sphere with radius r is given by

$$S = 4\pi r^2$$

it suffices to measure the radius r of the cell. If your measurement of the radius is accurate within 2%, how does this affect the accuracy of the surface area?

First we must discuss what it means for a measurement to be accurate within a certain percentage. Suppose that x_0 is the true value of an observation, and x is the measured value. Then $|\Delta x| = |x - x_0|$ is the **absolute error** or tolerance in measurement. The **relative error** is defined as $|\Delta x / x_0|$, and the **percentage error** as $100|\Delta x / x_0|$.

Returning to our example, let's find the error when computing the surface area. We start with the absolute error of the surface area,

$$|\Delta S| = |S(r_0 + \Delta r) - S(r_0)|$$

where r_0 is the true radius and $|\Delta r|$ is the absolute error in the measurement of the radius. We approximate $S(r_0 + \Delta r) - S(r_0)$ by its linear approximation $S'(r_0)\Delta r$, that is,

$$\Delta S \approx S'(r_0)\Delta r$$

Since $S'(r) = 8\pi r$, the percentage error in the measurement of the surface area is

$$100 \left| \frac{\Delta S}{S(r_0)} \right| \approx 100 \left| \frac{S'(r_0)\Delta r}{S(r_0)} \right|$$

$$= 100 \left| \frac{8\pi r_0}{4\pi r_0^2} \Delta r \right| = 2\left(100 \left| \frac{\Delta r}{r_0} \right| \right) = 4$$

since $100|\Delta r / r_0| = 2$. That is, the surface area is (approximately) accurate within 4% if the radius is accurate within 2%. Where the doubling of the percentage error comes from can be seen in the following example.

▶ **Example 5**   Suppose that you wish to determine $f(x)$ from a measurement of x. If $f(x)$ is given by a power function, namely $f(x) = cx^s$, then an error in measurement of x is propagated as follows:

$$\Delta f \approx f'(x)\Delta x = csx^{s-1}\Delta x$$

The percentage error $100\left|\frac{\Delta f}{f}\right|$ is related to the percentage error $100\left|\frac{\Delta x}{x}\right|$ as follows:

$$100\left|\frac{\Delta f}{f}\right| \approx 100\left|\frac{f'(x)\Delta x}{f}\right| = 100\left|\frac{csx^{s-1}\Delta x}{cx^s}\right|$$

$$= |s|\left(100\left|\frac{\Delta x}{x}\right|\right)$$

In our previous example, $s = 2$; hence the percentage error in the surface area measurement is twice the percentage error in the radius measurement. ◀

▶ **Example 6**   Suppose that you wish to estimate total leaf area of a tree in a certain plot. Experimental data in your study plot (Niklas, 1994) indicate that

$$[\text{leaf area}] \propto [\text{stem diameter}]^{1.84}$$

Instead of trying to measure total leaf area directly, you measure the stem diameter and then use the scaling relationship to estimate total leaf area. How accurately must you measure stem diameter if you want to estimate leaf area within an error of 10%?

Solution

We denote leaf area by A and stem diameter by d. Then

$$A(d) = cd^{1.84}$$

where c is the constant of proportionality. An error in measurement of d is propagated as

$$\Delta A \approx A'(d)\Delta d = c(1.84)d^{0.84}\Delta d$$

The percentage error $100\left|\frac{\Delta A}{A}\right|$ is related to the percentage error $100\left|\frac{\Delta d}{d}\right|$ as

$$100\left|\frac{\Delta A}{A}\right| \approx 100\left|\frac{A'(d)\Delta d}{A}\right|$$

$$= 100\left|\frac{c(1.84)d^{0.84}}{cd^{1.84}}\Delta d\right|$$

$$= (1.84)\left(100\left|\frac{\Delta d}{d}\right|\right)$$

We require $100\left|\frac{\Delta A}{A}\right| = 10$. Hence

$$10 = (1.84)\left(100\left|\frac{\Delta d}{d}\right|\right)$$

or

$$100\left|\frac{\Delta d}{d}\right| = \frac{10}{1.84} = 5.4$$

That is, we must measure stem diameter within an error of 5.4%.

Using the result of Example 5, we could have found this immediately, since

$$A(d) = cd^{1.84}$$

we get $s = 1.84$, where s is the exponent defined in Example 5. Using

$$100 \left| \frac{\Delta A}{A} \right| = |s| \left(100 \left| \frac{\Delta d}{d} \right| \right)$$

we obtain

$$100 \left| \frac{\Delta d}{d} \right| = \frac{1}{|s|} \left(100 \left| \frac{\Delta A}{A} \right| \right) = \frac{10}{1.84} = 5.4$$

as before.   ◄

▶ **Example 7**   Suppose that you wish to determine the percentage error of $f(x)$ from a measurement of x, where $f(x) = \ln x$, $x = 10$ and the percentage error for $x = 2\%$. Find the percentage error of $f(x)$.

Solution

The function $f(x)$ is not a power function, so there is no simple rule. We find

$$100 \frac{\Delta f}{f} \approx 100 \frac{f'(x) \Delta x}{f(x)}$$

Since we know $100 \left| \frac{\Delta x}{x} \right|$, we multiply and divide the right-hand side by x and rearrange terms.

$$100 \frac{f'(x) \Delta x}{f(x)} = 100 \frac{\Delta x}{x} \frac{x f'(x)}{f(x)}$$

Since $f'(x) = 1/x$, at $x = 10$, we find

$$100 \left| \frac{\Delta x}{x} \right| \left| \frac{x f'(x)}{f(x)} \right| \Big|_{x=10} = 2 \frac{(10)(1/10)}{\ln 10} = \frac{2}{\ln 10} \approx 0.869$$

Thus, the percentage error of f is approximately 0.9%.   ◄

4.8.1 Problems

In Problems 1–10, use the formula
$$f(x) \approx f(a) + f'(a)(x - a)$$
to approximate the value of the given function, and compare your result with the value from a calculator.

1. $\sqrt{65}$; let $f(x) = \sqrt{x}$, $a = 64$, and $x = 65$

2. $\sqrt{35}$; let $f(x) = \sqrt{x}$, $a = 36$, and $x = 35$

3. $\sqrt[3]{124}$

4. $(3.9)^3$

5. $(0.99)^{25}$

6. $\sin(0.01)$

7. $\sin\left(\frac{\pi}{2} + 0.02\right)$

8. $\cos\left(\frac{\pi}{4} - 0.01\right)$

9. $\ln(1.01)$

10. $\frac{1}{0.99}$

In Problems 11–30, approximate $f(x)$ at a by the linear approximation
$$L(x) = f(a) + f'(a)(x - a)$$

11. $f(x) = \dfrac{1}{1 + x}$ at $a = 0$

12. $f(x) = \dfrac{1}{1 - x}$ at $a = 0$

13. $f(x) = \dfrac{1}{1 + x}$ at $a = 1$

14. $f(x) = \dfrac{1}{1 - x}$ at $a = 2$

15. $f(x) = \dfrac{1}{(1 + x)^2}$ at $a = 0$

16. $f(x) = \dfrac{1}{(1 - x)^2}$ at $a = 0$

17. $f(x) = \ln(1 + x)$ at $a = 0$

18. $f(x) = \ln(1 + 2x)$ at $a = 0$

19. $f(x) = \ln x$ at $a = e$

20. $f(x) = \ln x$ at $a = e^2$

21. $f(x) = e^x$ at $a = 0$

22. $f(x) = e^{2x}$ at $a = 0$

23. $f(x) = e^{-x}$ at $a = 0$

24. $f(x) = e^{-3x}$ at $a = 0$

25. $f(x) = e^{x-1}$ at $a = 1$

26. $f(x) = e^{2x+1}$ at $a = -1/2$

27. $f(x) = (1 + x)^{-n}$ at $a = 0$. (Assume that n is a positive integer.)

28. $f(x) = (1 - x)^{-n}$ at $a = 0$. (Assume that n is a positive integer.)

29. $f(x) = \tan x$ at $a = 0$

30. $f(x) = \cos x$ at $a = 0$

31. Suppose that the per capita growth rate of a population is 3%, that is, if $N(t)$ denotes the population size at time t, then
$$\frac{1}{N}\frac{dN}{dt} = 0.03$$
Suppose that the population size at time $t = 4$ is equal to 100. Use a linear approximation to compute the population size at time $t = 4.1$.

32. Suppose that the per capita growth rate of a population is 2%, that is, if $N(t)$ denotes the population size at time t, then
$$\frac{1}{N}\frac{dN}{dt} = 0.02$$
Suppose that the population size at time $t = 2$ is equal to 50. Use a linear approximation to compute the population size at time $t = 2.1$.

33. Suppose that the specific growth rate of a plant is 1%, that is, if $B(t)$ denotes the biomass at time t, then
$$\frac{1}{B(t)}\frac{dB}{dt} = 0.01$$
Suppose that the biomass at time $t = 1$ is equal to 5 grams. Use a linear approximation to compute the biomass at time $t = 1.1$.

34. Suppose that a certain plant is grown along a gradient ranging from nitrogen poor to nitrogen-rich soil. Experimental data show that the average mass per plant grown in a soil with total nitrogen content of 1000 mg nitrogen per kg of soil is 2.7 g and the rate of change of the average mass per plant at this nitrogen level is 1.05×10^{-3} g per mg change in total nitrogen per kg soil. Use a linear approximation to predict the average mass per plant grown in a soil with total nitrogen content of 1100 mg nitrogen per kg of soil.

In Problems 35–40, a measurement error in x affects the accuracy of the value f(x). In each case determine an interval of the form
$$[f(x) - \Delta f, f(x) + \Delta f]$$
that reflects the measurement error Δx. In the problems, given is $f(x)$ and $x = $ true value of $x \pm |\Delta x|$.

35. $f(x) = 2x, x = 1 \pm 0.1$

36. $f(x) = 1 - 3x, x = -2 \pm 0.3$

37. $f(x) = 3x^2, x = 2 \pm 0.1$

38. $f(x) = \sqrt{x}, x = 10 \pm 0.5$

39. $f(x) = e^x, x = 2 \pm 0.2$

40. $f(x) = \sin x, x = -1 \pm 0.05$

In Problems 41–44, assume that the measurement of x is accurate within 2%. Determine the error Δf in the calculation of f and the percentage error $100\frac{\Delta f}{f}$ in each case. Given is $f(x)$ and the true value of x.

41. $f(x) = 4x^3, x = 1.5$

42. $f(x) = x^{1/4}, x = 10$

43. $f(x) = \ln x, x = 20$

44. $f(x) = \dfrac{1}{1 + x}, x = 4$

45. The volume V of a spherical cell of radius r is given by
$$V(r) = \frac{4}{3}\pi r^3$$
If you can determine the radius within an accuracy of 3%, how accurate is your calculation of the volume?

46. The speed v of blood flowing along the central axis of an artery of radius R is given by Poiseuille's law
$$v(R) = cR^2$$
where c is a constant. If you can determine the radius of the artery within an accuracy of 5%, how accurate is your calculation of the speed?

47. Suppose that you are studying reproduction in moss. A scaling relation has been found (Niklas, 1994) between the number of moss spores (N) and the capsule length (L), namely,
$$N \propto L^{2.11}$$
This relationship is not very accurate but it turns out that it suffices for your purpose. To estimate the number of moss spores you measure capsule length. If you wish to estimate the number of moss spores within an error of 5%, how accurately must you measure capsule length?

48. (*Tilman's resource model*) Suppose that the rate of growth of a plant in a certain habitat depends on a single resource: for instance, nitrogen. Assume that the growth rate $f(R)$ depends on the resource level R as
$$f(R) = a\frac{R}{k + R}$$
where a and k are constants. Express the percentage error of the growth rate, $100\frac{\Delta f}{f}$, as a function of the percentage error of the resource level, $100\frac{\Delta R}{R}$.

49. The reaction rate $R(x)$ of the irreversible reaction
$$A + B \rightarrow AB$$
is a function of the concentration x of the product AB and is given by
$$R(x) = k(a - x)(b - x)$$
where k is a constant, a is the concentration of A in the beginning of the reaction, and b is the concentration of B in the beginning of the reaction. Express the percentage error of the reaction rate, $100\frac{\Delta R}{R}$, as a function of the percentage error of the concentration x, $100\frac{\Delta x}{x}$.

4.9 Key Terms

Chapter 4 Review: Topics *Discuss the following definitions and concepts:*

1. Derivative, formal definition
2. Difference quotient
3. Secant line and tangent line
4. Instantaneous rate of change
5. Average rate of change
6. Differential equation
7. Differentiability and continuity
8. Power rule
9. Basic rules of differentiation
10. Product rule
11. Quotient rule
12. Chain rule
13. Implicit function
14. Implicit differentiation
15. Related rates
16. Higher derivatives
17. Derivatives of trigonometric functions
18. Derivatives of exponential functions
19. Derivatives of inverse and logarithmic functions
20. Logarithmic differentiation
21. Tangent line approximation
22. Error propagation
23. Absolute error, relative error, percentage error

4.10 Review Problems

In Problems 1–8, differentiate with respect to the independent variable.

1. $f(x) = -3x^4 + \dfrac{2}{\sqrt{x}} + 1$

2. $g(x) = \dfrac{1}{\sqrt{x^3 + 4}}$

3. $h(t) = \left(\dfrac{1-t}{1+t}\right)^{1/3}$

4. $f(x) = (x^2 + 1)e^{1-x}$

5. $f(x) = e^{2x} \sin\left(\dfrac{\pi}{2}x\right)$

6. $g(s) = \dfrac{\sin(3s + 1)}{\cos(3s)}$

7. $f(x) = 2\dfrac{\ln(x+1)}{\ln x^2}$

8. $g(x) = e^{-x} \ln(x + 1)$

In Problems 9–12, find the first and second derivatives of the given functions.

9. $f(x) = e^{-x^2/2}$

10. $g(x) = \tan(x^2 + 1)$

11. $h(x) = \dfrac{x}{x + 1}$

12. $f(x) = \dfrac{e^{-x}}{e^{-x} + 1}$

In Problems 13–16, find dy/dx.

13. $x^2 y - y^2 x = \sin x$

14. $e^{x^2 + y^2} = 2x$

15. $\ln(x - y) = 2x$

16. $\tan(x - y) = x^2$

In Problems 17–19, find dy/dx and d^2y/dx^2.

17. $x^2 + y^2 = 16$

18. $x = \tan y$

19. $e^y = \ln x$

20. Assume that x is a function of t. Find $\dfrac{dy}{dt}$ when $y = \cos x$ and $\dfrac{dx}{dt} = \sqrt{3}$ for $x = \dfrac{\pi}{3}$.

21. A flock of birds passes directly overhead, flying horizontally at an altitude of 100 feet and at a speed of 6 feet per second. How quickly is the distance between you and the birds increasing when the distance is 320 feet? (You are on the ground and are not moving.)

22. Find the derivative of

$$y = \ln|\cos x|$$

23. Suppose that $f(x)$ is differentiable. Find an expression for the derivative of each of the following functions.

(a) $y = e^{f(x)}$

(b) $y = \ln f(x)$

(c) $y = [f(x)]^2$

24. Find the tangent line and the normal line to $y = \ln(x + 1)$ at $x = 1$.

25. Suppose that

$$f(x) = \dfrac{x^2}{1 + x^2}, \quad x \geq 0$$

(a) Use a graphing calculator to graph $f(x)$ for $x \geq 0$. Note that the graph is "S" shaped.

(b) Find a line through the origin that touches the graph of $f(x)$ at some point $[c, f(c)]$ with $c > 0$. This is the tangent line at $(c, f(c))$ that goes through the origin. Graph the tangent line in the same coordinate system that you used in (a).

In Problems 26–29, find an equation for the tangent line to the curve at the specified point.

26. $y = (\sin x)^{\cos x}$ at $x = \dfrac{\pi}{2}$

27. $y = e^{-x^2} \cos x$ at $x = \dfrac{\pi}{3}$

28. $x^2 + y = e^y$ at $x = \sqrt{e - 1}$

29. $x \ln y = y \ln x$ at $x = 1$

30. In Review Problem 17 of Chapter 3, we introduced the hyperbolic functions

$$\sinh x = \frac{e^x - e^{-x}}{2}, \quad x \in \mathbf{R}$$

$$\cosh x = \frac{e^x + e^{-x}}{2}, \quad x \in \mathbf{R}$$

$$\tanh x = \frac{e^x - e^{-x}}{e^x + e^{-x}}, \quad x \in \mathbf{R}$$

(a) Show that

$$\frac{d}{dx}\sinh x = \cosh x$$

and

$$\frac{d}{dx}\cosh x = \sinh x$$

(b) Use the facts that

$$\tanh x = \frac{\sinh x}{\cosh x}$$

and

$$\cosh^2 x - \sinh^2 x = 1$$

together with your results in (a) to show that

$$\frac{d}{dx}\tanh x = \frac{1}{\cosh^2 x}$$

31. Find a second-degree polynomial

$$p(x) = ax^2 + bx + c$$

with $p(-1) = 6$, $p'(1) = 8$, and $p''(0) = 4$.

32. Use the geometric interpretation of the derivative to find the equations of the tangent lines to the curve

$$x^2 + y^2 = 1$$

at the following points.

(a) $(1, 0)$

(b) $\left(\frac{1}{2}, \frac{1}{2}\sqrt{3}\right)$

(c) $\left(-\frac{1}{2}\sqrt{2}, -\frac{1}{2}\sqrt{2}\right)$

(d) $(0, -1)$

33. Geradedorf and Straightville are connected by a very straight but rather hilly road. Biking from Geradedorf to Straightville, your position at time t (measured in hours) is given by the function

$$s(t) = 3\pi t + 3(1 - \cos(\pi t))$$

for $0 \le t \le 5.5$, where $s(t)$ is measured in miles.

(a) Use a graphing calculator to convince yourself that you didn't backtrack during your trip. How can you check this? Assuming that your trip takes 5.5 hours, find the distance between Geradedorf and Straightville.

(b) Find the velocity $v(t)$ and the acceleration $a(t)$.

(c) Use a graphing calculator to graph $s(t)$, $v(t)$, and $a(t)$. In (a), you used the function $s(t)$ to conclude that you didn't backtrack during your trip. Can you use any of the other two functions to answer the question of backtracking? Explain your answer.

(d) Assuming that you slow down going uphill and speed up going downhill, how many peaks and valleys does this road have?

34. Suppose your position at time t on a straight road is given by

$$s(t) = \cos(\pi t)$$

for $0 \le t \le 2$, where t is measured in hours.

(a) What is your position at the beginning and end of your trip?

(b) Use a graphing calculator to help describe your trip in words.

(c) What is the total distance you traveled?

(d) Determine your velocity and your acceleration during the trip. When is your velocity equal to 0? Relate this to your position, and explain what it means.

35. In the following very simple population model, the growth rate at time t depends on the number of individuals at time $t - T$, where T is a positive constant. (That is, the model incorporates a time delay in the birth rate.) This is useful, for instance, if one wishes to take into account the fact that individuals must mature before reproducing.

Denote by $N(t)$ the size of the population at time t, and assume that

$$\frac{dN}{dt} = \frac{\pi}{2T}(K - N(t - T)) \tag{4.13}$$

where K and T are positive constants.

(a) Show that

$$N(t) = K + A\cos\frac{\pi t}{2T}$$

is a solution of (4.13).

(b) Graph $N(t)$ for $K = 100$, $A = 50$, and $T = 1$.

(c) Explain in words how the size of the population changes over time.

36. We denote by $W(t)$ the amount of a radioactive material left at time t if the initial amount present was $W(0) = W_0$.

(a) Show that

$$W(t) = W_0 e^{-\lambda t}$$

solves the differential equation

$$\frac{dW}{dt} = -\lambda W(t)$$

(b) Show that if you graph $W(t)$ on semilog paper, then the result is a straight line.

(c) Use your result in (b) to explain why

$$\frac{d\ln W(t)}{dt} = \text{constant}$$

Determine the constant, and relate it to the graph in (b).

(d) Show that

$$\frac{d\ln W(t)}{dt} = \text{constant}$$

implies that

$$\frac{dW}{dt} \propto W(t)$$

37. In Example 17 of Subsection 4.4.3, we introduced an allometric relationship between skull length (in cm) and backbone length (in cm) of Ichthyosaurs, a group of extinct marine reptiles. The relationship is

$$S = (1.162)B^{0.933}$$

where S and B denote skull length and backbone length, respectively. Suppose that you found only the skull of an individual and that based on the skull length you wish to estimate the backbone length of this specimen. How accurately must you measure skull length if you want to estimate backbone length within an error of 10%?

APPLICATIONS
OF DIFFERENTIATION

5.1 EXTREMA AND THE MEAN VALUE THEOREM

The primary focus of this chapter is how calculus can help us to understand the behavior of functions. Points where the function is smallest or largest, called extrema, are of particular importance. This section defines extrema, gives conditions that guarantee extrema (extreme value theorem), and provides a characterization of extrema (Fermat's theorem). Fermat's theorem will be crucial in establishing the mean value theorem, a result that can be understood graphically. The mean value theorem has far reaching consequences that are felt in the subsequent sections where we learn methods to characterize the behavior of functions.

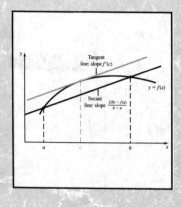

5.1.1 The Extreme Value Theorem

Suppose that you measure the depth of a creek along a transect between two points A and B (see Figure 5.1). Looking at the profile of the creek, you see that there is a location of maximum depth and a location of minimum depth. The existence of such locations is the content of the extreme value theorem. To formalize this result, we must introduce some terminology.

> **Definition** Let f be a function defined on the set D that contains the number c. Then
> f has a **global** (or **absolute**) **maximum** at $x = c$ if
>
> $$f(c) \geq f(x) \quad \text{for all } x \in D$$
>
> and
> f has a **global** (or **absolute**) **minimum** at $x = c$ if
>
> $$f(c) \leq f(x) \quad \text{for all } x \in D$$

The following result gives conditions under which global maxima and global minima, collectively called **global** (or **absolute**) **extrema**, exist.

▲ Figure 5.1
A transect of a creek between
the points A and B

The Extreme Value Theorem If f is continuous on a closed interval $[a, b]$, $-\infty < a < b < \infty$, then f has a global maximum and a global minimum on $[a, b]$.

The proof of this theorem is beyond the scope of this text and will be omitted. However, the result is quite intuitive, and we illustrate it in Figures 5.2 and 5.3. Figure 5.2 shows that the function may attain its extreme values at the endpoints of the interval $[a, b]$, whereas in Figure 5.3 the extreme values are attained in the interior of the interval $[a, b]$. The function must be continuous and defined on a closed interval in order to conclude that it has global maxima and global minima. But note that the extreme value theorem only tells us that global extrema exist, not where they are. Furthermore, they need not be unique, meaning that a function can have more than one global maximum or global minimum.

▶ **Example 1** (Optimal Strategy) Suppose a plant has two reproductive strategies; one is asexual by clonal reproduction, the other is sexual by seed production. The plant's fitness depends on how it allocates its resources to the two strategies. Suppose that a plant allocates a fixed amount of resources to reproduction, a fraction p of which is allocated to clonal reproduction ($0 \leq p \leq 1$) and a fraction $1 - p$ to sexual reproduction. Denote by $f(p)$ the plant's fitness as a function of p. Assuming that $f(p)$ is a continuous function, why is there a strategy of resource allocation (called an optimal strategy) that maximizes the plant's fitness?

Solution

Since $f(p)$ is continuous on the closed interval $[0, 1]$, $f(p)$ has a global maximum (and a global minimum) on $[0, 1]$ according to the extreme value

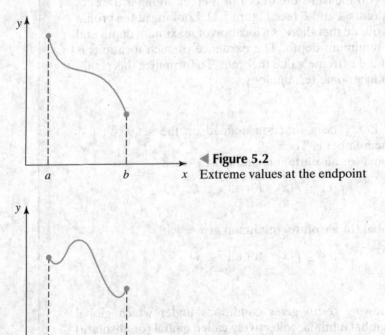

◀ Figure 5.2
Extreme values at the endpoint

◀ Figure 5.3
Extreme values in the interior

theorem. The global maximum represents the optimal strategy. Note that the extreme value theorem only guarantees the existence of an optimal strategy but does not tell us which strategy is optimal. Furthermore, there could be more than one global maximum, meaning that there could be more than one optimal strategy of resource allocation.   ◀

Figure 5.2 illustrates the importance of one of the assumptions in the extreme value theorem, namely that the interval is *closed*. If the interval from *a* to *b* in Figure 5.2 did not include the endpoints, we would have neither a global maximum nor a global minimum.

The following two examples illustrate that the theorem cannot be used if either *f* is discontinuous or the interval is not closed.

 Example 2   Let

$$f(x) = \begin{cases} 2x & \text{if } 0 \le x < 2 \\ 3 - x & \text{if } 2 \le x \le 3 \end{cases}$$

Note that $f(x)$ is defined on a *closed* interval, namely [0, 3]. However, f is discontinuous at $x = 2$, as can be seen in Figure 5.4. The graph of $f(x)$ shows that there is no value $c \in [0, 3]$ where $f(c)$ attains a global maximum. Do not be tempted to say that there should be a global maximum close to $x = 2$: For any candidate of a global maximum you might come up with, you will be able to find a point whose y-coordinate exceeds the y-coordinate of your previous candidate. Try it! The reason for this is that

$$\lim_{x \to 2^-} f(x) = 4$$

but the function f takes on value 1 at $x = 2$. We conclude that the function does not have a global maximum. It does, however, have global minima, namely at $x = 0$ and $x = 3$, where $f(x)$ takes on the value 0. (This is a function that has more than one global minimum.)   ◀

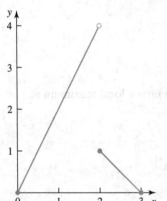

▲ **Figure 5.4**
The graph of $f(x)$ in Example 2

▶ **Example 3**   Let

$$f(x) = x \quad \text{for } 0 < x < 1$$

Note that $f(x)$ is continuous on its domain (0, 1), but it is *not* defined on a closed interval (Figure 5.5). The function $f(x)$ attains neither a global maximum nor a global minimum. Although

$$\lim_{x \to 0^+} f(x) = 0 \quad \text{and} \quad \lim_{x \to 1^-} f(x) = 1$$

and $0 < f(x) < 1$ for all $x \in (0, 1)$, there is no number c in the open interval (0, 1) where $f(c) = 0$ or $f(c) = 1$.   ◀

▲ **Figure 5.5**
The graph of $f(x)$ in Example 3

5.1.2 Local Extrema

We will now discuss local (or relative) extrema, which are points where a graph is higher or lower than all *nearby* points. This will allow us to identify the peaks and the valleys of the graph of a function (see Figure 5.6). The graph of the function in Figure 5.6 has three peaks, at $x = a, c$, and e, and two valleys, at $x = b$ and d. A peak (or local maximum) has the property that the graph is lower nearby; a valley (or local minimum) has the property that the graph is higher nearby.

The formal definition follows (see Figures 5.7 and 5.8).

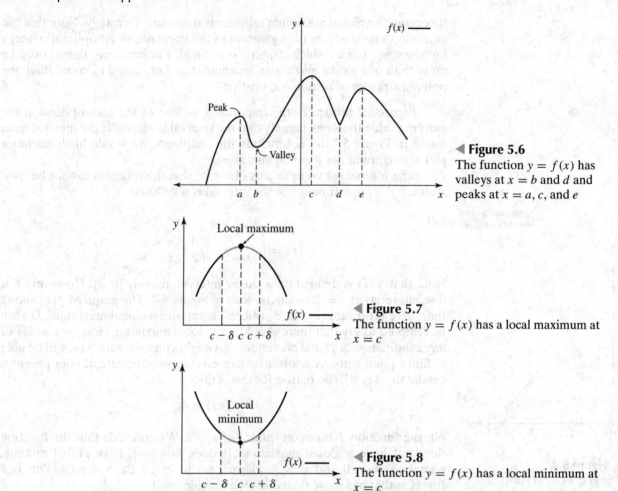

◀ **Figure 5.6**
The function $y = f(x)$ has valleys at $x = b$ and d and peaks at $x = a, c$, and e

◀ **Figure 5.7**
The function $y = f(x)$ has a local maximum at $x = c$

◀ **Figure 5.8**
The function $y = f(x)$ has a local minimum at $x = c$

> **Definition** A function f defined on a set D has a **local** (or **relative**) **maximum** at a point c if there exists a $\delta > 0$ so that
>
> $$f(c) \geq f(x) \qquad \text{for all } x \in (c - \delta, c + \delta) \cap D$$
>
> A function f defined on a set D has a **local** (or **relative**) **minimum** at a point c if there exists a $\delta > 0$ so that
>
> $$f(c) \leq f(x) \qquad \text{for all } x \in (c - \delta, c + \delta) \cap D$$

Local maxima and local minima are collectively called local (or relative) extrema. If D is an interval and c is in the interior of D (that is, not a boundary point), then the definition simplifies: The function f has a local maximum at c if there exists an open interval I so that $f(c) \geq f(x)$ for all $x \in I$. Likewise, the function f has a local minimum at c if there exists an open interval I so that $f(c) \leq f(x)$ for all $x \in I$. In the definition, we wrote $(c - \delta, c + \delta) \cap D$. If c is an interior point of D, δ can be chosen small enough so that $(c - \delta, c + \delta)$ is contained in D, and we can set $I = (c - \delta, c + \delta)$. Intersecting the interval $(c - \delta, c + \delta)$ with D becomes important when c is a boundary point, as we will see in Example 4.

We examine local and global extrema in the following two examples; the discussion is based on looking at the graphs of functions. In the first example, we consider a function that is defined on a closed interval; this allows us to compute the value of the function at both endpoints of its domain. In the second example, we consider a function that is defined on a half-open interval; thus, the value of the function can be computed at one endpoint of its domain but not at the other endpoint.

▶ **Example 4**   Let

$$f(x) = (x - 1)^2(x + 2) \quad \text{for } -2 \le x \le 3$$

(a) Use the graph of $f(x)$ to find all local extrema.

(b) Find the global extrema.

Solution

(a) The graph of $f(x)$ is illustrated in Figure 5.9. The function f is defined on the closed interval $[-2, 3]$. We begin with local extrema that occur at interior points of the domain $D = [-2, 3]$; looking at Figure 5.9, we see that a local maximum occurs at $x = -1$, as there are no greater values of f nearby. That is, we can find a small interval I about $x = -1$ so that $f(-1) \ge f(x)$ for all $x \in I$. For instance, we can choose $\delta = 0.1$ in the preceding definition and obtain $I = (-1.1, -0.9)$ (Figure 5.10).

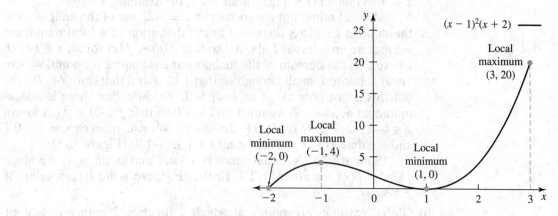

▲ **Figure 5.9**

The graph of $f(x) = (x - 1)^2(x + 2)$ for $-2 \le x \le 3$ in Example 4

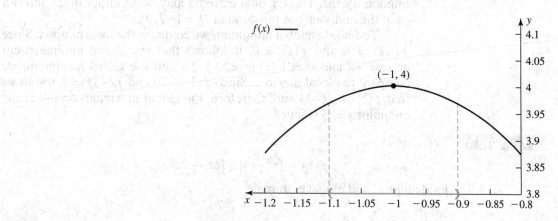

▲ **Figure 5.10**

The graph of $f(x) = (x - 1)^2(x + 2)$ near $x = -1$. The point $(-1, 4)$ is a local maximum: $f(-1) \ge f(x)$ for all x nearby in the domain of f

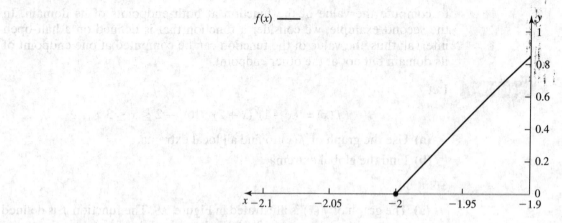

$f(x)$ ——

▲ **Figure 5.11**
The graph of $f(x) = (x - 1)^2(x + 2)$ near $x = -2$. The point $(-2, 0)$ is a local minimum: $f(-2) \leq f(x)$ for all x nearby in the domain of f

There is a local minimum at $x = 1$ since there are no smaller values of f nearby. This time, we need to find a small interval about $x = 1$ so that $f(1) \leq f(x)$ for all $x \in I$; for example, $I = (0.9, 1.1)$.

A local minimum also occurs at $x = -2$, one of the endpoints of the domain of f. As discussed in the definition of a local minimum, we require an interval I about c so that $f(c) \leq f(x)$ for all $x \in I \cap D$, where D is the domain of the function. If c is an interior point, we can always choose I small enough so that $I \subset D$ and therefore $I \cap D = I$, but this is not possible at an endpoint. To show that there is a local minimum at $x = -2$, we must find $\delta > 0$ so that $f(-2) \leq f(x)$ for all $x \in (-2 - \delta, -2 + \delta) \cap D = [-2, -2 + \delta)$. We can again choose $\delta = 0.1$ and see that $f(-2) \leq f(x)$ for all $x \in [-2, -1.9)$ (Figure 5.11).

Similarly, we see that there is a local maximum at $x = 3$ since $f(3) \geq f(x)$ for all $x \in (2.9, 3]$; that is, there is no larger value of f nearby.

(b) Global extrema are points at which a function is either largest or smallest. Since f is defined on a closed interval, we know from the extreme value theorem that both a global maximum and a global minimum exist. These global extrema may occur either in the interior or at the endpoints of the domain $D = [-2, 3]$.

To find the global minimum, we compare the local minima. Since $f(-2) = 0$ and $f(1) = 0$, it follows that the global minima occur at $x = -2$ and $x = 1$ (Figure 5.9). To find the global maximum, we compare the local maxima. Since $f(3) = 20$ and $f(-1) = 4$, it follows that $f(3) > f(-1)$ and, therefore, the global maximum occurs at the endpoint $x = 3$ (Figure 5.9). ◄

▶ **Example 5** Let

$$f(x) = |x^2 - 4| \quad \text{for} \ -2.5 \leq x < 3$$

Find all local and global extrema.

Solution

The graph of $f(x)$, illustrated in Figure 5.12, reveals that local minima occur at $x = -2$ and $x = 2$, and local maxima occur at $x = -2.5$ and $x = 0$. Note that $f(x)$ is not defined at $x = 3$ and, therefore, $x = 3$ cannot be a local maximum.

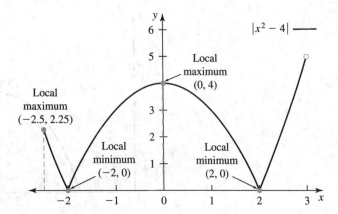

▲ Figure 5.12

The graph of $f(x) = |x^2 - 4|$ for $-2.5 \le x < 3$ in Example 5

To find the global extrema, however, we need to look at the function values close to the boundary $x = 3$ as well. Candidates for global extrema are all the local extrema, which then must be compared to the value of the function near the boundary $x = 3$. We discuss the global maximum first. Since

$$f(-2.5) = 2.25 \quad f(0) = 4 \quad \lim_{x \to 3^-} f(x) = 5$$

the function is largest near the point $x = 3$. But since $f(x)$ is not defined at $x = 3$, there exists no global maximum for the function. (This does not contradict the extreme value theorem since the function is not defined on a closed interval, which is an assumption in the theorem.) To find the global minimum, we need only compare $f(-2)$ and $f(2)$. We find that $f(-2) = 0$ and $f(2) = 0$; therefore, global minima occur at $x = -2$ and $x = 2$. ◀

Looking at Figures 5.9 and 5.12, we see that if the function f is differentiable at an interior point where f has a local extremum, then there is a horizontal tangent line at that point. This is known as Fermat's theorem (see Figure 5.13).

> **Fermat's Theorem** If f has a local extremum at an interior point c and $f'(c)$ exists, then $f'(c) = 0$.

Proof We prove Fermat's theorem for the case where the local extremum is a maximum; the proof where the local extremum is a minimum is similar. We need to show that $f'(c) = 0$. To do so, we use the formal definition of the derivative to compute $f'(c)$, namely,

$$f'(c) = \lim_{x \to c} \frac{f(x) - f(c)}{x - c}$$

To compute the limit, we will compute the left-hand limit ($x \to c^-$) and the right-hand limit ($x \to c^+$). We begin with the following observation (Figure 5.14): Suppose that f has a local maximum at an interior point c. Then there exists a $\delta > 0$ so that

$$f(x) \le f(c) \quad \text{for all } x \in (c - \delta, c + \delta)$$

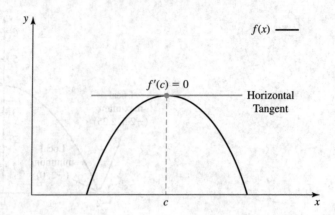

▲ **Figure 5.13**
Fermat's theorem

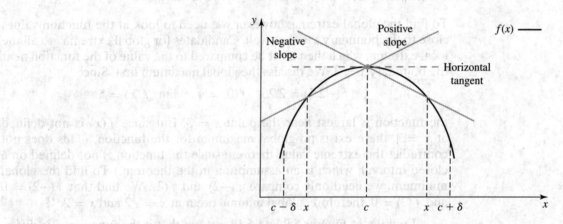

▲ **Figure 5.14**
An illustration of the proof of Fermat's theorem: For $x < c$, the slope of the secant line is positive, for $x > c$, the slope of the secant line is negative. In the limit $x \to c$, the secant lines converge to the horizontal tangent line

Since $f(x) - f(c) \leq 0$ and $x - c < 0$ if $x < c$, we find for the left-hand limit that

$$\lim_{x \to c^-} \frac{f(x) - f(c)}{x - c} \geq 0 \tag{5.1}$$

and since $f(x) - f(c) \leq 0$ and $x - c > 0$ if $x > c$, we find for the right-hand limit that

$$\lim_{x \to c^+} \frac{f(x) - f(c)}{x - c} \leq 0 \tag{5.2}$$

Now, since f is differentiable at c,

$$f'(c) = \lim_{x \to c^-} \frac{f(x) - f(c)}{x - c} = \lim_{x \to c^+} \frac{f(x) - f(c)}{x - c}$$

This, together with Equation (5.1), shows that $f'(c) \geq 0$, and together with Equation (5.2), that $f'(c) \leq 0$. Now, if you have a number that is simultaneously nonnegative and nonpositive, the number must be 0. Therefore, $f'(c) = 0$. ■

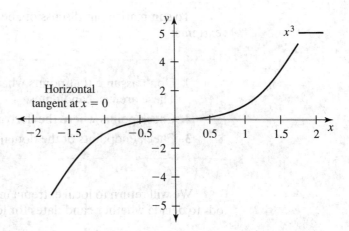

▲ Figure 5.15
The graph of $y = x^3$ has a horizontal tangent at $x = 0$ but $(0, 0)$ is not an extremum

▶ **Example 6** Explain why $y = \tan x$ does not have a local extremum at $x = 0$.

Solution

$y = \tan x$ is differentiable at $x = 0$ with

$$\frac{d}{dx}\tan x = \sec^2 x$$

and hence, the derivative of $y = \tan x$ at $x = 0$ is equal to 1. Since the derivative is not equal to 0, Fermat's theorem (more precisely, its contrapositive) implies that $x = 0$ is not a local extremum. ◀

Caution **1.** The condition that $f'(c) = 0$ is a necessary but not sufficient condition for local extrema at interior points where $f'(c)$ exists. In particular, the fact that f is differentiable at c with $f'(c) = 0$ tells us nothing about whether f has a local extremum at $x = c$. For instance, $f(x) = x^3$, $x \in \mathbf{R}$, is differentiable at $x = 0$ and $f'(0) = 0$, but there is no local extremum at $x = 0$. The graph of $y = x^3$ is shown in Figure 5.15. Though there is a horizontal tangent at $x = 0$, there is no local extremum at $x = 0$. Fermat's theorem does tell you, however, that if $x = c$ is an interior point with $f'(c) \neq 0$, then $x = c$ cannot be a local extremum (Example 6). Interior points with horizontal tangents are *candidates* for local extrema.

2. The function f may not be differentiable at a local extremum. For instance, in Example 5, the function $f(x)$ is not differentiable at $x = -2$ and $x = 2$, but both points turned out to be local extrema. This means that in order to identify candidates for local extrema, it will not be enough to simply look at points with horizontal tangents. You also must look at points where the function $f(x)$ is not differentiable.

3. Local extrema may occur at endpoints of the domain. Since Fermat's theorem says nothing about what happens at endpoints, you will have to look at endpoints separately.

To summarize our discussion, here are some guidelines for finding local extrema:

> **1.** Don't assume that points where $f'(x) = 0$ give you local extrema; These are just candidates.
> **2.** Check points where the derivative is not defined.
> **3.** Check endpoints of the domain.

We will return to local extrema in Section 5.3, where we will learn methods to decide whether candidates for local extrema are indeed local extrema.

5.1.3 The Mean Value Theorem

The mean value theorem (MVT) is a very important yet easily understood result in calculus. Its consequences are far reaching, and we will use it in every section in this chapter to derive important results that will help us to analyze functions.

To see the content of the mean value theorem, consider

$$f(x) = x^2 \quad \text{for } 0 \le x \le 1$$

The secant line connecting the endpoints of the graph of $f(x)$, $(0,0)$ and $(1,1)$, has slope

$$m = \frac{f(1) - f(0)}{1 - 0} = \frac{1 - 0}{1 - 0} = 1$$

The graph of $f(x)$ and the secant line are shown in Figure 5.16. Note that $f(x)$ is differentiable in $(0, 1)$; that is, you can draw a tangent line at every point of the graph in the open interval $(0, 1)$. If you look at the graph of $f(x)$, you see that there exists a number $c \in (0, 1)$ such that the slope of the tangent line at $(c, f(c))$ is the same as the slope of the secant line through the points $(0, 0)$ and $(1, 1)$. That is, we claim that there exists a number $c \in (0, 1)$ such that

$$\frac{f(1) - f(0)}{1 - 0} = f'(c)$$

The existence of such a value c is the content of the mean value theorem.

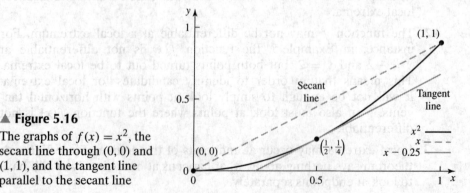

▲ **Figure 5.16**
The graphs of $f(x) = x^2$, the secant line through $(0, 0)$ and $(1, 1)$, and the tangent line parallel to the secant line

We can compute the value of c in this example. Since $f'(x) = 2x$ and the slope of the secant line is $m = 1$, we must solve

$$1 = 2c \quad \text{or} \quad c = \frac{1}{2}$$

Using the point-slope form $[y - y_0 = m(x - x_0)]$, we can find the equation of the tangent line at $(\frac{1}{2}, f(\frac{1}{2})) = (\frac{1}{2}, \frac{1}{4})$, namely

$$y - \frac{1}{4} = 1\left(x - \frac{1}{2}\right), \quad \text{or} \quad y = x - \frac{1}{4}$$

(this tangent line is also shown in Figure 5.16).

> **The Mean Value Theorem (MVT)** If f is continuous on the closed interval $[a, b]$ and differentiable on the open interval (a, b), then there exists at least one number $c \in (a, b)$ such that
>
> $$\frac{f(b) - f(a)}{b - a} = f'(c)$$

The fraction on the left-hand side of the equation in the theorem is the slope of the secant line connecting the points $(a, f(a))$ and $(b, f(b))$, and the quantity on the right-hand side is the slope of the tangent line at $(c, f(c))$ (see Figure 5.17).

Geometrically, the MVT is indeed easily understood: It states that there exists a point on the graph between $(a, f(a))$ and $(b, f(b))$ where the tangent line at this point is parallel to the secant line through $(a, f(a))$ and $(b, f(b))$. [We denoted this point by $(c, f(c))$.] The MVT is an "existence" result; it tells us neither how many such points there are nor where they are in the interval (a, b).

Going back to the example $f(x) = x^2$, $0 \le x \le 1$, we see that $f(x)$ satisfies the assumptions of the MVT, namely, $f(x)$ is continuous on the closed interval $[0, 1]$ and differentiable on the open interval $(0, 1)$. The MVT then guarantees the existence of at least one number $c \in (0, 1)$ such that the slope of the secant line through the points $(0, 0)$ and $(1, 1)$ is equal to the slope of the tangent line at $(c, f(c))$.

At this point, you might be wondering how such a seemingly simple theorem can be so important. In the following sections, you will encounter this theorem mostly in proofs of other important results that will allow us to understand properties of functions using calculus. But here is an application that gives physical meaning to the theorem.

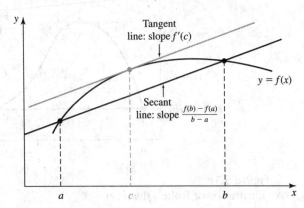

▲ Figure 5.17

The mean value guarantees the existence of a number $c \in (a, b)$ such that the tangent line at $(c, f(c))$ has the same slope as the secant line through $(a, f(a))$ and $(b, f(b))$

▶ Example 7   A car moves in a straight line. At time t (measured in seconds) its position (measured in meters) is

$$s(t) = \frac{1}{25}t^3, \quad 0 \le t \le 10$$

Show that there is a time $t \in (0, 10)$ where the velocity is equal to the average velocity between $t = 0$ and $t = 10$.

Solution

The average velocity between $t = 0$ and $t = 10$ is

$$\frac{s(10) - s(0)}{10 - 0} = \frac{\frac{1}{25} \cdot 1000 \text{ m}}{10 \text{ s}} = 4 \, \frac{\text{m}}{\text{s}}$$

This is the slope of the secant line connecting the points $(0, 0)$ and $(10, 40)$. Since $s(t)$ is continuous on $[0, 10]$ and differentiable on $(0, 10)$, the MVT tells us that there must exist a number $c \in (0, 10)$ such that $s'(c) = 4 \, \frac{\text{m}}{\text{s}}$. Now, $s'(t)$ is the (instantaneous) velocity. So, at some point during this short trip, the speedometer must have read $4 \, \frac{\text{m}}{\text{s}}$.   ◀

The rest of this section is devoted to the proof of the MVT. The MVT is typically proved by first showing a special case of the theorem, called Rolle's theorem.

> **Rolle's Theorem** If f is continuous on the closed interval $[a, b]$, differentiable on the open interval (a, b), and $f(a) = f(b)$, then there exists a number $c \in (a, b)$ such that $f'(c) = 0$.

Figure 5.18 illustrates Rolle's theorem. The function in the graph is defined on the closed interval $[a, b]$ and takes on the same values at the two endpoints of $[a, b]$ [namely, $f(a) = f(b)$]. Thus the secant line connecting the two endpoints is a horizontal line. We see that there is a point in (a, b) with a horizontal tangent line.

Before we prove Rolle's theorem, we check why it is a special case of the MVT. If we compare the assumptions in the two theorems, we find

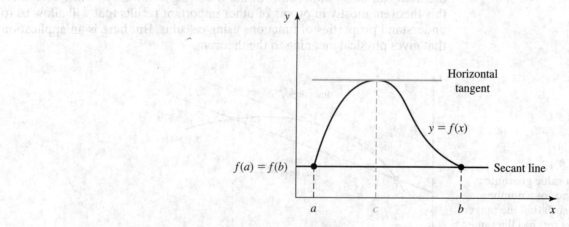

▲ **Figure 5.18**
An illustration of Rolle's theorem

that Rolle's theorem has an additional requirement, $f(a) = f(b)$; that is, the function values must agree at the endpoints of the interval on which f is defined. If we apply the MVT to such a function, it says that there exists a number $c \in (a, b)$ such that

$$f'(c) = \frac{f(b) - f(a)}{b - a}$$

But since $f(b) = f(a)$, it follows that the expression on the right-hand side is equal to 0, that is, $f'(c) = 0$, which is the conclusion of Rolle's theorem.

Proof of Rolle's Theorem If f is the constant function, then $f'(x) = 0$ for all $x \in (a, b)$ and the result is true. We assume in the following that f is not constant. Since $f(x)$ is continuous on the closed interval $[a, b]$, it follows from the extreme value theorem that the function has a global maximum and a global minimum in the closed interval $[a, b]$. To see that the function must have a global extremum inside the open interval (a, b), we observe that if f is not constant, then there exists $x_0 \in (a, b)$ so that either $f(x_0) > f(a) = f(b)$ or $f(x_0) < f(a) = f(b)$. This global extremum is also a local extremum. Suppose that this local extremum is at $c \in (a, b)$; it follows from Fermat's theorem that $f'(c) = 0$. (In Figure 5.18, the global minima occur at the endpoints of the interval $[a, b]$, but the global maximum occurs in the open interval (a, b), and that's where the horizontal tangent is.) ■

The mean value theorem follows from Rolle's theorem and can be thought of as a "tilted" version of Rolle's theorem. (The secant and tangent lines in the MVT are no longer necessarily horizontal [Figure 5.17], as in Rolle's theorem [Figure 5.18], but "tilted;" they are still parallel, though.)

Proof of the MVT We define the following function:

$$F(x) = f(x) - \frac{f(b) - f(a)}{b - a}(x - a)$$

The function F is continuous on $[a, b]$ and differentiable on (a, b). Furthermore,

$$F(a) = f(a) - \frac{f(b) - f(a)}{b - a}(a - a) = f(a)$$

$$F(b) = f(b) - \frac{f(b) - f(a)}{b - a}(b - a) = f(a)$$

and, therefore, $F(a) = F(b)$. We can apply Rolle's theorem to the function $F(x)$: There exists $c \in (a, b)$ with $F'(c) = 0$. Since

$$F'(x) = f'(x) - \frac{f(b) - f(a)}{b - a}$$

it follows that for this value of c,

$$0 = F'(c) = f'(c) - \frac{f(b) - f(a)}{b - a}$$

and hence

$$f'(c) = \frac{f(b) - f(a)}{b - a}$$

■

We discuss two consequences of the mean value theorem.

Corollary 1 If f is continuous on the closed interval $[a, b]$ and differentiable on the open interval (a, b), such that

$$m \leq f'(x) \leq M \quad \text{for all } x \in (a, b)$$

then

$$m(b - a) \leq f(b) - f(a) \leq M(b - a)$$

This corollary is useful in obtaining information about a function based on its derivative.

▶ **Example 8** Denote the population size at time t by $N(t)$ and assume that $N(t)$ is continuous on the interval $[0, 10]$ and differentiable on the interval $(0, 10)$ with $N(0) = 100$ and $|dN/dt| \leq 3$ for all $t \in (0, 10)$. What can you say about $N(10)$?

Solution

Since $|dN/dt| \leq 3$ implies $-3 \leq dN/dt \leq 3$, we can set $m = -3$ and $M = 3$ in Corollary 1. With $a = 0$ and $b = 10$, Corollary 1 yields the following estimate:

$$(-3)(10 - 0) \leq N(10) - N(0) \leq (3)(10 - 0)$$

Simplifying and solving for $N(10)$ gives

$$-30 + N(0) \leq N(10) \leq 30 + N(0)$$

Since $N(0) = 100$, we find

$$70 \leq N(10) \leq 130$$

That is, the population size at time $t = 10$ is bounded between 70 and 130. ◀

▶ **Example 9** Show that

$$|\sin b - \sin a| \leq |b - a|$$

Solution

If $a = b$, then $|\sin a - \sin a| \leq |a - a|$ is a true statement. We therefore assume that $a < b$ (the case $a > b$ is similar). Let $f(x) = \sin x$, $a \leq x \leq b$. $f(x)$ is continuous on $[a, b]$ and differentiable on (a, b). Since $f'(x) = \cos x$, it follows that

$$-1 \leq f'(x) \leq 1$$

for all $x \in (a, b)$. Applying Corollary 1, with $m = -1$ and $M = 1$, to $f(x) = \sin x$, $a < x < b$, we find

$$-(b - a) \leq \sin b - \sin a \leq (b - a)$$

which is the same as

$$|\sin b - \sin a| \leq |b - a| \quad ◀$$

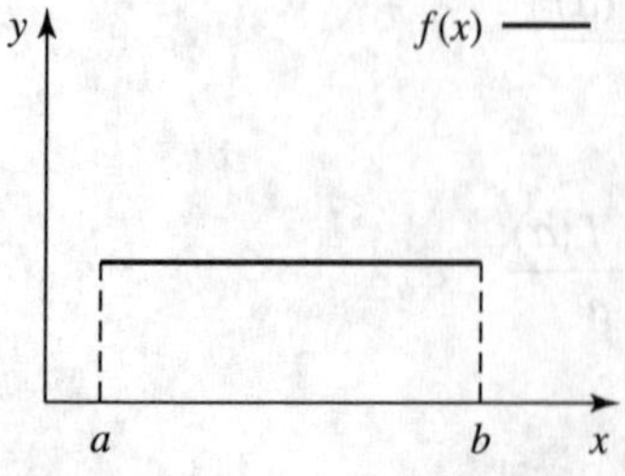

The next corollary is important, and we will see it again in Section 5.8.

Corollary 2 If f is continuous on the closed interval $[a, b]$ and differentiable on the open interval (a, b), with $f'(x) = 0$ for all $x \in (a, b)$, then f is constant on $[a, b]$.

▲ **Figure 5.19**
An illustration of Corollary 2

Figure 5.19 explains why the result should be true—each point on the graph has a horizontal tangent, so the function must be constant.

▶ **Example 10** Assume that f is continuous on $[-1, 1]$ and differentiable on $(-1, 1)$, with $f(0) = 2$ and $f'(x) = 0$ for all $x \in (-1, 1)$. Find $f(x)$.

Solution

Corollary 2 tells us that $f(x)$ is a constant. Since we know that $f(0) = 2$, this implies that $f(x) = 2$ for all $x \in [-1, 1]$. ◀

Proof of Corollary 2 Let $x_1, x_2 \in (a, b)$, $x_1 < x_2$. Then f satisfies the assumptions of the MVT on the closed interval $[x_1, x_2]$. Therefore, there exists a number $c \in (x_1, x_2)$, such that

$$\frac{f(x_2) - f(x_1)}{x_2 - x_1} = f'(c)$$

Since $f'(c) = 0$, it follows that $f(x_2) = f(x_1)$. Since x_1, x_2 are arbitrary numbers from the interval (a, b), we conclude that f is constant. ■

▶ **Example 11** Show that

$$\sin^2 x + \cos^2 x = 1 \quad \text{for all } x \in [0, 2\pi]$$

Solution

This identity can be shown without calculus. But let's see what we get if we use Corollary 2. We define $f(x) = \sin^2 x + \cos^2 x$, $0 \le x \le 2\pi$. Then $f(x)$ is continuous on $[0, 2\pi]$ and differentiable on $(0, 2\pi)$ with

$$f'(x) = 2 \sin x \cos x - 2 \cos x \sin x = 0$$

Using Corollary 2 now, we conclude that $f(x)$ is equal to a constant on $[0, 2\pi]$. To find the constant, we only need to evaluate $f(x)$ at one point in the interval, say, $x = 0$. We find

$$f(0) = \sin^2 0 + \cos^2 0 = 1$$

This proves the identity. ◀

5.1.4 Problems

(5.1.1)

In Problems 1–8, each function is continuous and defined on a closed interval. It therefore satisfies the assumptions of the extreme value theorem. With the help of a graphing calculator, graph each function and locate its global extrema. (Note that a function may assume a global extremum at more than one point.)

1. $f(x) = 2x - 1, 0 \le x \le 1$
2. $f(x) = -x^2 + 1, -1 \le x \le 1$
3. $f(x) = \sin x, 0 \le x \le 2\pi$
4. $f(x) = \cos x, 0 \le x \le 2\pi$
5. $f(x) = |x|, -1 \le x \le 1$
6. $f(x) = (x - 1)^2(x + 2), -2 \le x \le 2$
7. $f(x) = e^{-x}, 0 \le x \le 2$
8. $f(x) = \ln(x + 1), 0 \le x \le 2$

9. Sketch the graph of a function that is continuous on the closed interval $[0, 3]$ and has a global maximum at the left endpoint and a global minimum at the right endpoint.

10. Sketch the graph of a function that is continuous on the closed interval $[-2, 1]$ and has a global maximum and a global minimum in the interior of the domain of the function.

11. Sketch the graph of a function that is continuous on the open interval $(0, 2)$ and has neither a global maximum nor a global minimum in its domain.

12. Sketch the graph of a function that is continuous on the closed interval $[1, 4]$ except at $x = 2$ and has neither a global maximum nor a global minimum in its domain.

(5.1.2)

In Problems 13–18, use a graphing calculator to determine all local and global extrema of the functions on their respective domains.

13. $f(x) = 2 - x, x \in [-1, 3)$
14. $f(x) = 5 + 2x, x \in (-2, 1)$
15. $f(x) = x^2 - 2, x \in [-1, 1]$
16. $f(x) = x^2 - 4x + 2, x \in [0, 3]$
17. $f(x) = -x^2 + 1, x \in [-2, 1]$
18. $f(x) = x(x - 1), x \in [0, 1]$

19. Show that $f(x) = x^2$ has a local minimum at $x = 0$, and compute $f'(0)$.

20. Show that $f(x) = (x - 1)^2$ has a local minimum at $x = 1$, and compute $f'(1)$.

21. Show that $f(x) = -x^2$ has a local maximum at $x = 0$, and compute $f'(0)$.

22. Show that $f(x) = -(x - 2)^2$ has a local maximum at $x = 2$, and compute $f'(2)$.

23. Suppose that $f(x) = x^3$. Show that $f'(0) = 0$ but $x = 0$ is not a local extremum.

24. Suppose that $f(x) = x^5$. Show that $f'(0) = 0$ but $x = 0$ is not a local extremum.

25. Suppose that $f(x) = (x + 2)^3$. Show that $f'(-2) = 0$ but $x = -2$ is not a local extremum.

26. Suppose that $f(x) = -(x - 1)^5$. Show that $f'(1) = 0$ but $x = 1$ is not a local extremum.

27. Show that $f(x) = |x|$ has a local minimum at $x = 0$ but $f(x)$ is not differentiable at $x = 0$.

28. Show that $f(x) = |x - 1|$ has a local minimum at $x = 1$ but $f(x)$ is not differentiable at $x = 1$.

29. Show that $f(x) = |x^2 - 1|$ has local minima at $x = 1$ and $x = -1$ but $f(x)$ is not differentiable at $x = 1$ or $x = -1$.

30. Show that $f(x) = -|x^2 - 4|$ has local maxima at $x = 2$ and $x = -2$ but $f(x)$ is not differentiable at $x = 2$ or $x = -2$.

31. Graph
$$f(x) = \left|1 - |x|\right|, \quad -1 \le x \le 2$$
and determine all local and global extrema on $[-1, 2]$.

32. Graph
$$f(x) = -\left||x| - 2\right|, \quad -3 \le x \le 3$$
and determine all local and global extrema on $[-3, 3]$.

33. Suppose the size of a population at time t is $N(t)$, and its growth rate is given by the logistic growth function
$$\frac{dN}{dt} = rN\left(1 - \frac{N}{K}\right), \quad t \ge 0$$
where r and K are positive constants.

(a) Graph the growth rate $\frac{dN}{dt}$ as a function of N for $r = 2$ and $K = 100$, and find the population size for which the growth rate is maximal.

(b) Show that $f(N) = rN(1 - N/K)$, $N \ge 0$, is differentiable for $N > 0$, and compute $f'(N)$.

(c) Show that $f'(N) = 0$ for the value of N that you determined in (a) when $r = 2$ and $K = 100$.

34. Suppose that the size of a population at time t is $N(t)$ and its growth rate is given by the logistic growth function
$$\frac{dN}{dt} = rN\left(1 - \frac{N}{K}\right), \quad t \ge 0$$
where r and K are positive constants. The per capita growth rate is defined by
$$g(N) = \frac{1}{N}\frac{dN}{dt}$$

(a) Show that
$$g(N) = r\left(1 - \frac{N}{K}\right)$$

(b) Graph $g(N)$ as a function of N for $N \ge 0$ when $r = 2$ and $K = 100$, and find the population size for which the per capita growth rate is maximal.

(5.1.3)

35. Suppose $f(x) = x^2$, $x \in [0, 2]$.

(a) Find the slope of the secant line connecting the points $(0, 0)$ and $(2, 4)$.

(b) Find a number $c \in (0, 2)$ so that $f'(c)$ is equal to the slope of the secant line you computed in (a) and explain why such a number must exist in $(0, 2)$.

36. Suppose $f(x) = 1/x$, $x \in [1, 2]$.

(a) Find the slope of the secant line connecting the points $(1, 1)$ and $(2, 1/2)$.

(b) Find a number $c \in (1, 2)$ so that $f'(c)$ is equal to the slope of the secant line you computed in (a) and explain why such a number must exist in $(1, 2)$.

37. Suppose that $f(x) = x^2$, $x \in [-1, 1]$. Use the MVT to show that there exists a point $c \in [-1, 1]$ with a horizontal tangent. Find c.

38. Suppose that $f(x) = x^2 - x - 2$, $x \in [-1, 2]$. Use the MVT to show that there exists a point $c \in [-1, 2]$ with a horizontal tangent. Find c.

39. Let $f(x) = x(1 - x)$. Use the MVT to find an interval that contains a number c so that $f'(c) = 0$.

40. Let $f(x) = 1/(1 + x^2)$. Use the MVT to find an interval that contains a number c so that $f'(c) = 0$.

41. Suppose that $f(x) = -x^2 + 2$. Explain why there exists a point c in the interval $(-1, 2)$ such that $f'(c) = -1$.

42. Suppose that $f(x) = x^3$. Explain why there exists a point c in the interval $(-1, 1)$ such that $f'(c) = 1$.

43. Sketch the graph of a function $f(x)$ that is continuous on the closed interval $[0, 1]$ and differentiable on the open interval $(0, 1)$, such that there exists exactly one point on the graph, $(c, f(c))$, at which the slope of the tangent line is equal to the slope of the secant line connecting the points $(0, f(0))$ and $(1, f(1))$. Why can you be sure that there is such a point?

44. Sketch the graph of a function $f(x)$ that is continuous on the closed interval $[0, 1]$ and differentiable on the open interval $(0, 1)$, such that there exist exactly two points on the graph, $(c_1, f(c_1))$ and $(c_2, f(c_2))$, at which the slope of the tangent lines is equal to the slope of the secant line connecting the points $(0, f(0))$ and $(1, f(1))$. Why can you be sure that there is at least one such point?

45. Suppose that $f(x) = x^2$, $x \in [a, b]$.

(a) Compute the slope of the secant line through the points $(a, f(a))$ and $(b, f(b))$.

(b) Find the point $c \in (a, b)$ such that the slope of the tangent line to the graph of f at $(c, f(c))$ is equal to the slope of the secant line determined in (a). How do you know that such a point exists? Show that c is the midpoint of the interval (a, b), that is, $c = (a + b)/2$.

46. Assume that f is continuous on $[a, b]$ and differentiable on (a, b). Show that if $f(a) < f(b)$, then f' is positive at some point between a and b.

47. Assume that f is continuous on $[a, b]$ and differentiable on (a, b). Furthermore, assume that $f(a) = f(b) = 0$ but f is not constant on $[a, b]$. Explain why there must be a point $c_1 \in (a, b)$ with $f'(c_1) > 0$, and a point $c_2 \in (a, b)$ with $f'(c_2) < 0$.

48. A car moves in a straight line. At time t (measured in seconds) its position (measured in meters) is

$$s(t) = \frac{1}{10}t^2, 0 \le t \le 10$$

(a) Find its average velocity between $t = 0$ and $t = 10$.

(b) Find its instantaneous velocity for $t \in (0, 10)$.

(c) At what time is the instantaneous velocity of the car equal to its average velocity?

49. A car moves in a straight line. At time t (measured in seconds) its position (measured in meters) is

$$s(t) = \frac{1}{100}t^3, 0 \le t \le 5$$

(a) Find its average velocity between $t = 0$ and $t = 5$.

(b) Find its instantaneous velocity for $t \in (0, 5)$.

(c) At what time is the instantaneous velocity of the car equal to its average velocity?

50. Denote the population size at time t by $N(t)$ and assume that $N(0) = 50$ and $|dN/dt| \le 2$ for all $t \in [0, 5]$. What can you say about $N(5)$?

51. Denote the biomass at time t by $B(t)$ and assume that $B(0) = 3$ and $|dB/dt| \le 1$ for all $t \in [0, 3]$. What can you say about $B(3)$?

52. Suppose that f is differentiable for all $x \in \mathbf{R}$ and, furthermore, that f satisfies $f(0) = 0$ and $1 \le f'(x) \le 2$ for all $x > 0$.

(a) Use Corollary 1 of the MVT to show that

$$x \le f(x) \le 2x$$

for all $x \ge 0$.

(b) Use your result in (a) to explain why $f(1)$ cannot be equal to 3.

(c) Find an upper and a lower bound for the value of $f(1)$.

53. Suppose that f is differentiable for all $x \in \mathbf{R}$ with $f(2) = 3$ and $f'(x) = 0$ for all $x \in \mathbf{R}$. Find $f(x)$.

54. Suppose that $f(x) = e^{-|x|}$, $x \in [-2, 2]$.

(a) Show that $f(-2) = f(2)$.

(b) Compute $f'(x)$, where defined.

(c) Show that there is no number $c \in (-2, 2)$ such that $f'(c) = 0$.

(d) Explain why your results in (a) and (c) do not contradict Rolle's theorem.

(e) Use a graphing calculator to sketch the graph of $f(x)$.

55. Use Corollary 2 of the MVT to show that if $f(x)$ is differentiable for all $x \in \mathbf{R}$ and satisfies

$$|f(x) - f(y)| \le |x - y|^2 \tag{5.3}$$

for all $x, y \in \mathbf{R}$, then $f(x)$ is constant. [*Hint:* Show that (5.3) implies that

$$\lim_{x \to y} \frac{f(x) - f(y)}{x - y} = 0 \tag{5.4}$$

and use the definition of the derivative to interpret the left-hand side of (5.4).]

56. We have seen that

$$f(x) = f_0 e^{rx}$$

satisfies the differential equation

$$\frac{df}{dx} = rf(x)$$

with $f(0) = f_0$. This exercise will show that it is in fact the only solution. Suppose that r is a constant and f is a differentiable function, where

$$\frac{df}{dx} = rf(x) \tag{5.5}$$

for all $x \in \mathbf{R}$ and where $f(0) = f_0$. The following steps will show that $f(x) = f_0 e^{rx}$, $x \in \mathbf{R}$, is the only solution to (5.5).

(a) Define the function

$$F(x) = f(x)e^{-rx}, \quad x \in \mathbf{R}$$

Use the product rule to show that

$$F'(x) = e^{-rx}[f'(x) - rf(x)]$$

(b) Use (a) and (5.5) to show that $F'(x) = 0$ for all $x \in \mathbf{R}$.

(c) Use Corollary 2 to show that $F(x)$ is a constant, and hence $F(x) = F(0) = f_0$.

(d) Show that (c) implies that

$$f_0 = f(x)e^{-rx}$$

and therefore

$$f(x) = f_0 e^{rx}$$

5.2 MONOTONICITY AND CONCAVITY

Fish are indeterminate growers; they increase in body size throughout their life. However, as they become older, they grow proportionately more slowly. Their growth is often described mathematically by the von Bertalanffy equation, which fits a large number of both freshwater and marine fishes. This equation is given by

$$L(x) = L_\infty - (L_\infty - L_0)e^{-Kx}$$

where $L(x)$ denotes the length at age x, L_0 the length at age 0, and L_∞ the asymptotic maximum attainable length. We assume $L_\infty > L_0$. K is related to how quickly the fish grows. Figure 5.20 shows examples for two different

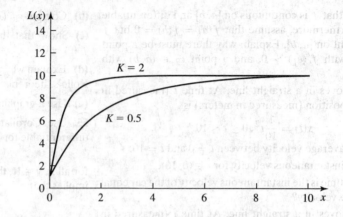

▲ **Figure 5.20**
The function $L(x)$ for $L_0 = 1$ and $L_\infty = 10$ with $K = 0.5$ and $K = 2$

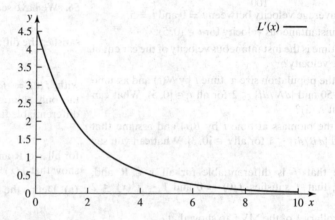

▲ **Figure 5.21**
The graph of $L'(x)$ with $L_0 = 1$, $L_\infty = 10$ and $K = 0.5$

values of K; L_∞ and L_0 are the same in both cases. We see from the graphs that for larger K, the asymptotic length L_∞ is approached more quickly.

The fact that fish increase their body size throughout their life can be expressed mathematically using the first derivative of the function $L(x)$. Looking at the graph of $L(x)$, we see that $L(x)$ is an increasing function of x: The tangent line at any point of the graph has a positive slope, or, equivalently, $L'(x) > 0$. We can compute $L'(x)$, namely,

$$L'(x) = K(L_\infty - L_0)e^{-Kx}$$

Since $L_\infty > L_0$ (by assumption) and $e^{-Kx} > 0$ (this holds for all x, regardless of K), we see that indeed $L'(x) > 0$. The graph of $L'(x)$ is shown in Figure 5.21.

The graph of $L'(x)$ shows that $L'(x)$ is a decreasing function of x; though fish increase their body size throughout their life, they do so at a rate that decreases with age. Mathematically, this can be expressed using the second derivative of $L(x)$—the derivative of the first derivative. The tangent line at any point on the graph of $L'(x)$ has a negative slope; that is, the derivative of $L'(x)$ is negative: $L''(x) < 0$. The fact that the rate of growth decreases with age can also be seen directly from the graph of $L(x)$: It bends downward. The second derivative thus tells us something about which way the graph of $L(x)$ bends.

This section discusses the important concepts of monotonicity, whether a function is decreasing or increasing, and concavity, whether a function bends upwards or downwards.

5.2.1 Monotonicity

We saw in the motivating example that the first derivative tells us something about whether a function increases or decreases. Not every function is differentiable, however, so we phrase the definitions of increasing and decreasing in terms of the function f alone (see Figures 5.22 and 5.23).

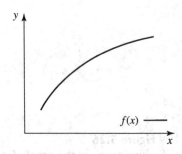

▲ **Figure 5.22**
An increasing function

> **Definition** A function f defined on an interval I is called **(strictly) increasing** on I if
>
> $$f(x_1) < f(x_2) \quad \text{whenever } x_1 < x_2 \text{ in } I$$
>
> It is called **(strictly) decreasing** on I if
>
> $$f(x_1) > f(x_2) \quad \text{whenever } x_1 < x_2 \text{ in } I$$

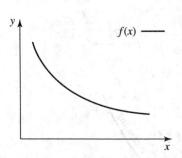

▲ **Figure 5.23**
A decreasing function

An increasing or decreasing function is called **monotonic**. The word *strictly* in the preceding definition refers to having a strict inequality ($f(x_1) < f(x_2)$ and $f(x_1) > f(x_2)$). We will frequently drop *strictly*. If instead of the strict inequality $f(x_1) < f(x_2)$ we have the inequality $f(x_1) \le f(x_2)$ whenever $x_1 < x_2$ in I, we call f nondecreasing. If $f(x_1) \ge f(x_2)$ whenever $x_1 < x_2$ in I, then f is called nonincreasing (see Figures 5.24 and 5.25).

When the function f is differentiable, there is a useful test to determine whether a function is increasing or decreasing. This criterion is a consequence of the mean value theorem.

FIRST DERIVATIVE TEST FOR MONOTONICITY

> Suppose f is continuous on $[a, b]$ and differentiable on (a, b).
>
> **(a)** If $f'(x) > 0$ for all $x \in (a, b)$, then f is increasing on $[a, b]$.
> **(b)** If $f'(x) < 0$ for all $x \in (a, b)$, then f is decreasing on $[a, b]$.

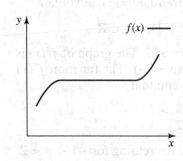

▲ **Figure 5.24**
A nondecreasing function may have regions where the function is constant

Proof (See Figure 5.26.) We choose two numbers x_1 and x_2 in $[a, b]$, $x_1 < x_2$. Then f is continuous on $[x_1, x_2]$ and differentiable on (x_1, x_2). We can therefore apply the mean value theorem to f defined on $[x_1, x_2]$: There exists a number $c \in (x_1, x_2)$ such that

$$\frac{f(x_2) - f(x_1)}{x_2 - x_1} = f'(c)$$

In part (a) of the theorem, we assume $f'(x) > 0$ for all $x \in (a, b)$. Since $c \in (x_1, x_2) \subset (a, b)$, it follows that $f'(c) > 0$. Since, in addition, $x_2 > x_1$, it follows that

$$f(x_2) - f(x_1) = f'(c)(x_2 - x_1) > 0$$

which implies that $f(x_2) > f(x_1)$. Since x_1 and x_2 are arbitrary numbers in $[a, b]$ satisfying $x_1 < x_2$, it follows that f is increasing. The proof of part (b) is similar and relegated to Problem 24. ■

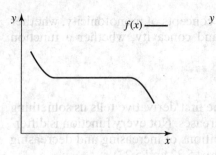

▲ Figure 5.25
A nonincreasing function may have regions where the function is constant

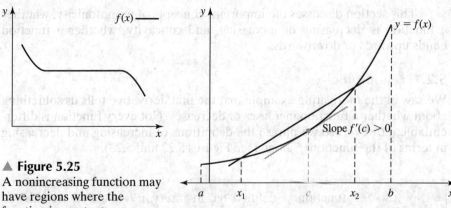

Slope $f'(c) > 0$

◀ Figure 5.26
An illustration of the proof of "$f'(x) > 0$ for all $x \in (a, b)$ implies $f(x)$ increasing on $[a, b]$"

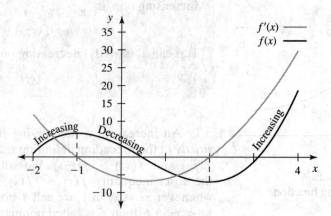

▲ Figure 5.27
The graph of $f(x) = x^3 - \frac{3}{2}x^2 - 6x + 3$ and $f'(x) = 3x^2 - 3x - 6$

▶ **Example 1** Determine where the function

$$f(x) = x^3 - \frac{3}{2}x^2 - 6x + 3, \quad x \in \mathbf{R}$$

is increasing and where it is decreasing.

Solution

Since $f(x)$ is continuous and differentiable for all $x \in \mathbf{R}$, we can use the first derivative test for monotonic functions. We differentiate $f(x)$ and obtain

$$f'(x) = 3x^2 - 3x - 6 = 3(x - 2)(x + 1), \quad x \in \mathbf{R}$$

The graphs of $f(x)$ and $f'(x)$ are shown in Figure 5.27. The graph of $f'(x)$ is a parabola that intersects the x-axis at $x = 2$ and $x = -1$. The function $f'(x)$ therefore changes sign at $x = -1$ and $x = 2$. We find that

$$f'(x) \begin{cases} > 0 & \text{if } x < -1 \text{ or } x > 2 \\ < 0 & \text{if } -1 < x < 2 \end{cases}$$

Thus, $f(x)$ is increasing for $x < -1$ or $x > 2$ and decreasing for $-1 < x < 2$. This is also confirmed when looking at the graph of $f(x)$ in Figure 5.27. ◀

▶ **Example 2** Parasitoids are insects whose larvae develop inside other host insects. This eventually kills the host. An example is the parasitoid *Macrocentrus grandii*, a wasp, which parasitizes *Ostrinia nubilis*, the European corn borer.

To understand host-parasitoid interactions, a large number of models have been developed. The function

$$f(N) = \left(1 + \frac{a\beta P}{k(\beta + aN)}\right)^{-k}$$

is one example that describes the likelihood of a host escaping parasitism as a function of host density N, where a, β, and k are positive parameters and P is the density of the parasitoid. What is the effect of an increase in host density on the likelihood of escaping parasitism based on $f(N)$?

Solution

To find the effect of an increase in host density, we compute the derivative of $f(N)$:

$$\frac{df}{dN} = (-k)\left[1 + \frac{a\beta P}{k(\beta + aN)}\right]^{-k-1} \frac{-a^2\beta P}{k(\beta + aN)^2}$$

$$= \left[1 + \frac{a\beta P}{k(\beta + aN)}\right]^{-k-1} \frac{a^2\beta P}{(\beta + aN)^2}$$

Both factors in the final expression are positive; hence $df/dN > 0$. In words, if the density of the parasitoid is fixed and the host density increases, a host is more likely to escape parasitism in cases where the interaction is described by $f(N)$. ◀

5.2.2 Concavity

We saw in the motivating example at the beginning of this section that the second derivative tells us something about whether a function bends upward or downward. We arrived at this conclusion by checking whether the first derivative was increasing or decreasing.

A function is called **concave up** if it bends upward, and **concave down** if it bends downward. Before stating a precise definition of concavity for differentiable functions, we give two examples, which are illustrated in Figure 5.28.

First, look at the graph of the differentiable function $y = x^2$: It bends upward, so we call it concave up. Bending upward means that the slopes of the tangent lines are increasing as x increases. We can check this by computing

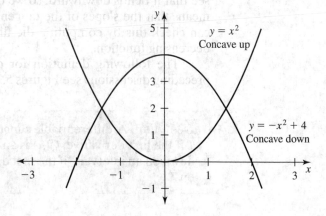

▲ **Figure 5.28**
The graphs of $f(x) = x^2$ and $g(x) = -x^2 + 4$

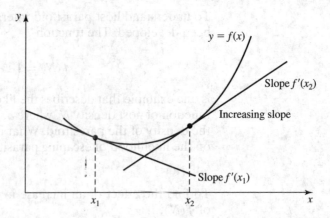

▲ **Figure 5.29**
A function is concave up if its derivative is increasing

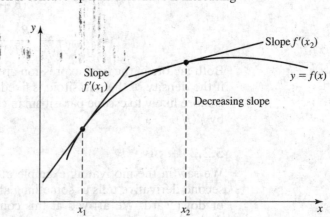

▲ **Figure 5.30**
A function is concave down if its derivative is decreasing

the slope of the tangent line at x, which is given by the first derivative, $y' = 2x$. Since $y' = 2x$ is an increasing function, the slopes of the tangent lines are increasing as x increases.

Looking at the graph of the differentiable function $y = -x^2 + 4$, we see that it bends downward, so we call it concave down. Bending downward means that the slopes of the tangent lines are decreasing as x increases. We can check this by computing the first derivative of y, which is $y' = -2x$, a decreasing function.

The following definition for differentiable functions is based on the preceding discussion (see Figures 5.29 and 5.30).

> **Definition** A differentiable function $f(x)$ is **concave up** on an interval I if the first derivative $f'(x)$ is an increasing function on I. It is **concave down** on an interval I if the first derivative $f'(x)$ is a decreasing function on I.

Note that the definition assumes that $f(x)$ is differentiable. There is a more general definition that does not require differentiability (after all, not all functions are differentiable). The more general definition is more difficult to use. The definition given here suffices for our purposes: It has the added advantage that it provides a criterion we can use to determine whether a twice-differentiable function is concave up or concave down.

SECOND DERIVATIVE TEST FOR CONCAVITY

> Suppose that f is twice differentiable on an open interval I.
>
> **(a)** If $f''(x) > 0$ for all $x \in I$, then f is concave up on I.
>
> **(b)** If $f''(x) < 0$ for all $x \in I$, then f is concave down on I.

Proof Since f is twice differentiable, we can apply the first derivative criterion to the function $f'(x)$. The proof of part (a) proceeds, then, as follows. If $f''(x) > 0$ on I, then $f'(x)$ is an increasing function on I. Using the definition of *concave up*, it follows that f is concave up on I. The proof of part (b) is similar and relegated to Problem 25. ■

You can use the function $y = x^2$ to remember which functions are concave up. The "u" in concave "up" should remind you of the U-shaped form of the graph of $y = x^2$. You can also use the function $y = x^2$ to remember the second derivative criterion. You already know that the graph of $y = x^2$ is concave up, and you can easily compute the second derivative of $y = x^2$, namely, $y'' = 2 > 0$.

▶ **Example 3**   Determine where the function

$$f(x) = x^3 - \frac{3}{2}x^2 - 6x + 3, \quad x \in \mathbf{R}$$

is concave up and where it is concave down.

Solution

This is the same function as in Example 1 (redrawn in Figure 5.31). Since $f(x)$ is a polynomial, it is twice differentiable. In Example 1, we found $f'(x) = 3x^2 - 3x - 6$; differentiating $f'(x)$, we get the second derivative of f,

$$f''(x) = 6x - 3$$

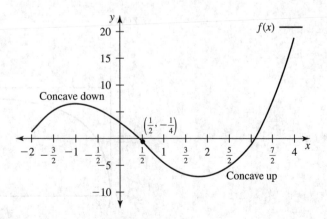

▲ **Figure 5.31**
The graph of $f(x) = x^3 - \frac{3}{2}x^2 - 6x + 3$

We find

$$f''(x) \begin{cases} > 0 & \text{if } x > \dfrac{1}{2} \\[2mm] < 0 & \text{if } x < \dfrac{1}{2} \end{cases}$$

Thus, $f(x)$ is concave up for $x > 1/2$ and concave down for $x < 1/2$. A look at Figure 5.31 confirms this result. ◀

A very common mistake is to associate monotonicity and concavity. One has nothing to do with the other. For instance, an increasing function can bend downward or upward (this will be discussed in Problem 21).

There are many biological examples of increasing functions that have a decreasing derivative and are therefore concave down.

▶ **Example 4** The response of crop yield Y to soil nitrogen level N can often be described by a function of the form

$$Y(N) = Y_{\max} \frac{N}{K + N}, \quad N \geq 0$$

where $Y_{\max}$ is the maximum attainable yield and K is a positive constant. The graph of $Y(N)$ is shown in Figure 5.32. It is obvious from the graph that $Y(N)$ is an increasing function of N. The graph bends downward and hence is concave down. Before continuing, we will check this using the results from this section. Using the quotient rule, we find

$$Y'(N) = Y_{\max} \frac{K + N - N}{(K + N)^2} = Y_{\max} \frac{K}{(K + N)^2}$$

and using the chain rule, we find

$$Y''(N) = \frac{d}{dN} \left(Y_{\max} K (K + N)^{-2} \right)$$

$$= Y_{\max} K (-2)(K + N)^{-3}(1) = -Y_{\max} \frac{2K}{(K + N)^3}$$

Since $Y_{\max}$ and K are positive constants and $N \geq 0$, it follows that

$$Y'(N) > 0$$

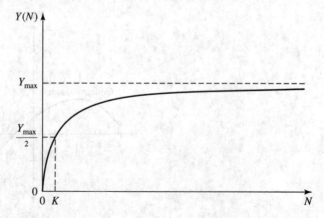

▲ **Figure 5.32**
The graph of $Y(N)$ in Example 4

which implies that $Y(N)$ is an increasing function. Furthermore,

$$Y''(N) < 0$$

which implies that $Y(N)$ is concave down. That is, $Y(N)$ is an increasing function but the rate of increase is decreasing. We say that Y is *increasing at a decelerating rate*. What does this mean? As we increase fertilizer levels, the yield will increase but at a proportionally lesser rate. This is called *diminishing return*. To be concrete, we choose values for Y_{max} and K:

$$Y(N) = 50\frac{N}{5+N}, \quad N \ge 0$$

The graph of this function is shown in Figure 5.33.

Suppose that initially $N = 5$. If we increase N by 5, that is, from 5 to 10, then $Y(N)$ changes from $Y(5) = 25$ to $Y(10) = 33.3$, an increase of 8.3. If we increase N by double the amount, namely 10, that is, from 5 to 15, then $Y(15) = 37.5$ and the increase in yield is only 12.5, less than twice 8.3. This can also be understood by comparing successive increments. If we start with $N = 5$ and increase by 5 to $N = 10$, then the change in Y (the Y increment) is 8.3. Increasing N by the same amount starting at 10, the Y increment changes

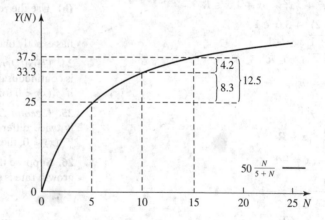

▲ **Figure 5.33**
The graph of $Y(N)$ in Example 4 for $Y_{max} = 50$ and $K = 5$

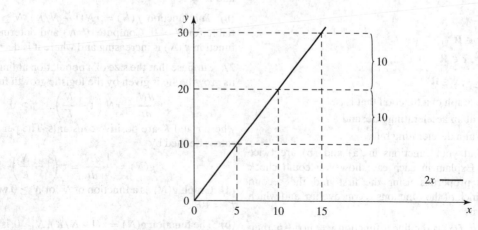

▲ **Figure 5.34**
The graph of a linear function: Increases are proportional

by $Y(15) - Y(10) = 4.2$. Changing N by equal increments has less of an effect for larger values of N, thus we say that the return is diminishing.

You should compare this to a linear function, say $f(x) = 2x$, which is neither concave up nor concave down (see Figure 5.34). If we increase x from 5 to 10, $f(x)$ changes from 10 to 20, that is, $f(x)$ increases by 10. If we increase x from 10 to 15, $f(x)$ changes from 20 to 30, again an increase of 10. That is, for linear functions, the increase is proportional. ◄

5.2.3 Problems

In Problems 1–20, determine where each function is increasing, decreasing, concave up, and concave down. With the help of a graphing calculator, sketch the graph of each function and label the intervals where it is increasing, decreasing, concave up, and concave down. Make sure that your graphs and your calculations agree.

1. $y = 3x - x^2, x \in \mathbf{R}$
2. $y = x^2 + 5x, x \in \mathbf{R}$
3. $y = x^2 + x - 4, x \in \mathbf{R}$
4. $y = x^3, x \in \mathbf{R}$
5. $y = -\frac{2}{3}x^3 + \frac{7}{2}x^2 - 3x + 4, x \in \mathbf{R}$
6. $y = (x - 2)^3 + 3, x \in \mathbf{R}$
7. $y = \sqrt{x + 1}, x \geq -1$
8. $y = \sqrt[3]{2x - 3}, x \in \mathbf{R}$
9. $y = \frac{1}{x}, x \neq 0$
10. $y = \frac{1}{x^2}, x \neq 0$
11. $(x^2 + 1)^{1/3}, x \in \mathbf{R}$
12. $y = \frac{1}{1 + x}, x \neq -1$
13. $y = \frac{1}{(1 + x)^2}, x \neq -1$
14. $y = \frac{x}{x + 1}, x \geq 0$
15. $y = \sin x, 0 \leq x \leq 2\pi$
16. $y = \tan x, -\frac{\pi}{2} < x < \frac{\pi}{2}$
17. $y = e^x, x \in \mathbf{R}$
18. $y = e^{-x}, x \in \mathbf{R}$
19. $y = e^{-x^2/2}, x \in \mathbf{R}$
20. $y = \frac{1}{1 + e^{-x}}, x \in \mathbf{R}$

21. Sketch the graph of a function that is

(a) increasing at an accelerating rate; and

(b) increasing at a decelerating rate.

(c) Assume that your functions in (a) and (b) are twice differentiable. Explain in each case how you could check the respective properties using the first and the second derivatives. Which of the functions is concave up, and which is concave down?

22. Show that if $f(x)$ is the linear function $y = mx + b$, then increases in $f(x)$ are proportional to increases in x. That is, if we increase x by Δx, then $f(x)$ increases by the same amount

Δy regardless of the value of x. Compute Δy as a function of Δx. Illustrate this graphically.

23. We frequently must solve equations of the form $f(x) = 0$. When f is a continuous function on $[a, b]$ and $f(a)$ and $f(b)$ have opposite signs, then the intermediate value theorem guarantees that there exists at least one solution to the equation $f(x) = 0$ in $[a, b]$.

(a) Explain in words why there exists exactly one solution in (a, b) if, in addition, f is differentiable in (a, b) and $f'(x)$ is either strictly positive or strictly negative throughout (a, b).

(b) Use the result in (a) to show that

$$x^3 - 4x + 1 = 0$$

has exactly one solution in $[-1, 1]$.

24. (*First Derivative Test for Monotonicity*) Suppose that f is continuous on $[a, b]$ and differentiable on (a, b). Show that if $f'(x) < 0$ for all $x \in (a, b)$, then f is decreasing on $[a, b]$.

25. (*Second Derivative Test for Concavity*) Suppose that f is twice differentiable on an open interval I. Show that if $f''(x) < 0$, then f is concave down.

26. Suppose the size of a population at time t is $N(t)$, and its growth rate is given by the logistic growth function

$$\frac{dN}{dt} = rN\left(1 - \frac{N}{K}\right), \quad t \geq 0$$

where r and K are positive constants.

(a) Graph the growth rate $\frac{dN}{dt}$ as a function of N for $r = 3$ and $K = 10$.

(b) The function $f(N) = rN(1 - N/K), N \geq 0$, is differentiable for $N > 0$. Compute $f'(N)$ and determine where the function $f(N)$ is increasing and where it is decreasing.

27. Suppose that the size of a population at time t is $N(t)$ and its growth rate is given by the logistic growth function

$$\frac{dN}{dt} = rN\left(1 - \frac{N}{K}\right), \quad t \geq 0$$

where r and K are positive constants. The per capita growth rate is defined by

$$g(N) = \frac{1}{N}\frac{dN}{dt} = r\left(1 - \frac{N}{K}\right)$$

(a) Graph $g(N)$ as a function of N for $N \geq 0$ when $r = 3$ and $K = 10$.

(b) The function $g(N) = r(1 - N/K), N \geq 0$, is differentiable for $N > 0$. Compute $g'(N)$ and determine where the function $g(N)$ is increasing and where it is decreasing.

28. The growth rate of a plant depends on the amount of resources available. A simple and frequently used model for resource dependent growth is the Monod model, where the growth rate is equal to

$$f(R) = \frac{aR}{k+R}, \quad R \geq 0$$

Here, R denotes the resource level and a and k are positive constants. When is the growth rate increasing? when is it decreasing?

29. Suppose that the growth rate of a population is given by

$$f(N) = N\left(1 - \left(\frac{N}{K}\right)^{\theta}\right)$$

where N is the size of the population, K is a positive constant denoting the carrying capacity, and θ is a parameter greater than 1. Find $f'(N)$ and determine where the growth rate is increasing and where it is decreasing.

30. Spruce budworms are a major pest that defoliate balsam fir. They are preyed upon by birds. A model for the per capita predation rate is given by

$$f(N) = \frac{aN}{k^2 + N^2}$$

where N denotes the density of spruce budworm and a and k are positive constants. Find $f'(N)$ and determine where the predation rate is increasing and where it is decreasing.

31. Parasitoids are insects that lay their eggs in, on or close to other (host) insects. Parasitoid larvae then devour the host insect. This results in the death of the host insect. The likelihood of escaping parasitism may depend on parasitoid density. One such model sets the probability of escaping parasitism equal to

$$f(P) = e^{-aP}$$

where P is the parasitoid density and a is a positive constant. Determine whether the probability of escaping parasitism increases or decreases with parasitoid density.

32. Parasitoids are insects that lay their eggs in, on, or close to other (host) insects. Parasitoid larvae then devour the host insect. This results in the death of the host insect. The likelihood of escaping parasitism may depend on parasitoid density. One such model sets the probability of escaping parasitism equal to

$$f(P) = \left(1 + \frac{aP}{k}\right)^{-k}$$

where P is the parasitoid density and a and k are positive constants. Determine whether the probability of escaping parasitism increases or decreases with parasitoid density.

33. Suppose that the height in feet (y) of a tree as a function of tree age in years (x) is given by

$$y = 117e^{-10/x}, \quad x > 0$$

(a) Show that the height of the tree increases with age. What is the maximum attainable height?

(b) Where is the graph of height versus age concave up, and where is it concave down?

(c) Use a graphing calculator to sketch the graph of the height versus age.

(d) Use a graphing calculator to verify that the rate of growth is greatest at the point where the graph in (c) changes concavity.

34. Plants employ two basic reproductive strategies: *polycarpy*, in which reproduction occurs repeatedly during the life time of the organism, and *monocarpy*, in which reproduction occurs only once during the life time of the organism. (Bamboo, for instance, is a monocarpic plant.) The following quote taken from Iwasa et al. (1995):

> The optimal strategy is polycarpy (repeated reproduction) if reproductive success increases with the investment at a decreasing rate, while monocarpy ("big bang" reproduction) or intermittent reproduction if the reproductive success increases at an increasing rate.

(a) Sketch the graph of reproductive success as a function of reproductive investment for the cases of (i) polycarpy and (ii) monocarpy.

(b) Given that the second derivative describes whether a curve bends upward or downward, explain the quote in terms of the second derivative of the reproductive success function.

35. Assume that the following formula (Iwasa et al., 1995) expresses the relationship between the number of flowers on a plant, F, and the average number of pollinator visits, $X(F)$:

$$X(F) = cF^{\gamma}$$

where c is a positive constant. Find the range of values for the parameter γ such that the average number of pollinator visits to a plant increases with the number of flowers F but the rate of increase decreases with F. Explain your answer in terms of appropriate derivatives of the function $X(F)$.

36. Assume that the dependence of the average number of pollinator visits to a plant, X, on the number of flowers, F, is given by

$$X(F) = cF^{\gamma}$$

where γ is a positive constant less than 1 and c is a positive constant (Iwasa et al., 1995). How does the average number of pollen grains exported per flower, $E(F)$, change with the number of flowers on the plant, F, if $E(F)$ is proportional to

$$1 - \exp\left[-k\frac{X(F)}{F}\right]$$

where k is a positive constant?

37. Denote the size of a population by $N(t)$ and assume that $N(t)$ satisfies

$$\frac{dN}{dt} = Ne^{-aN} - N^2$$

where a is a positive constant.

(a) Show that the nontrivial equilibrium $N^{\star}$ satisfies

$$e^{-aN^{\star}} = N^{\star}$$

(b) Assume now that the nontrivial equilibrium $N^{\star}$ is a function of the parameter a. Use implicit differentiation to show that $N^{\star}$ is a decreasing function of a.

38. Denote the size of a population by $N(t)$ and assume that $N(t)$ satisfies

$$\frac{dN}{dt} = N\left(1 - \frac{N}{K}\right) - N\ln N$$

where K is a positive constant.

(a) Show that if $K > 1$, then there exists a nontrivial equilibrium $N^* > 0$, which satisfies

$$1 - \frac{N^*}{K} = \ln N^*$$

(b) Assume now that the nontrivial equilibrium N^* is a function of the parameter K. Use implicit differentiation to show that N^* is an increasing function of K.

39. (*A simple model of intraspecific competition*) (Adapted from Bellows, 1981) Suppose that a study plot contains N annual plants, each of which produces S seeds that are sown within the same plot. The number of surviving plants in the following year is given by

$$A(N) = \frac{NS}{1 + (aN)^b} \tag{5.6}$$

for some positive constants a and b. This mathematical model incorporates density-dependent mortality; namely, the greater the number of plants in the plot, the lower the number of surviving offspring per plant, which is given by $A(N)/N$ and called the *net reproductive rate*.

(a) Use calculus to show that $A(N)/N$ is a decreasing function of N.

(b) The following quantity, called the *k-value*, can be used to quantify the effects of intraspecific competition (i.e., competition between individuals of the same species):

$$k = \log[\text{initial density}]$$
$$- \log[\text{final density}]$$

where log denotes the logarithm to base 10. The initial density is the product of the number of plants (N) and the number of seeds each plant produces (S). The final density is given by (5.6). Show that for the given model,

$$k = \log[NS] - \log\left[\frac{NS}{1 + (aN)^b}\right]$$
$$= \log[1 + (aN)^b]$$

We typically plot k versus $\log N$; the slope of the resulting curve is then used to quantify the effects of competition.

(i) Show that

$$\frac{d \log N}{dN} = \frac{1}{N \ln 10}$$

where ln denotes the natural logarithm.

(ii) Show that

$$\frac{dk}{d \log N} = (\ln 10) N \frac{dk}{dN}$$
$$= \frac{b}{1 + (aN)^{-b}}$$

(iii) Find

$$\lim_{N \to \infty} \frac{dk}{d \log N}$$

(iv) Show that if

$$\frac{dk}{d \log N} < 1$$

then $A(N)$ is increasing, whereas if

$$\frac{dk}{d \log N} > 1$$

then $A(N)$ is decreasing. [*Hint*: Compute $A'(N)$.] Explain in words what this would mean with respect to varying

the initial density of seeds and observing the number of surviving plants in the following year. (*Note*: The first case is called *undercompensation* and the second case is called *overcompensation*.)

(v) The case

$$\frac{dk}{d \log N} = 1$$

is referred to as *exact compensation*. Suppose that you plot k versus $\log N$, and observe that over a certain range of values of N the slope of the resulting curve is equal to 1. Explain what this means.

40. (*Adapted from Reiss, 1989*) Suppose that the rate at which body weight W changes with age x is

$$\frac{dW}{dx} \propto W^a \tag{5.7}$$

where a is some species-specific positive constant.

(a) The relative growth rate (percentage weight gained per unit of time) is defined as

$$\frac{1}{W}\frac{dW}{dx}$$

What is the relationship between the relative growth rate and body weight? For which values of a is the relative growth rate increasing, and for which values is it decreasing?

(b) As fish grow larger, their weight increases each day but the relative growth rate is decreasing. If the rate of growth is described by (5.7), what values of a can you exclude, based on your results in (a)? Explain in words how the percentage weight increase (relative to the current body weight) differs for juvenile fish and for adult fish.

41. (*Allometric equations*) Allometric equations describe the scaling relationship between two measurements, such as tree height versus tree diameter or skull length versus backbone length. These equations are often of the form

$$Y = bX^a \tag{5.8}$$

where b is some positive constant and a is a constant that can be positive, negative or zero.

(a) Assume that X and Y are body measurements (and therefore positive) and their relationship is described by an allometric equation of the form (5.8). For what values of a is Y an increasing function of X, but such that the ratio Y/X decreases with increasing X? Is Y as a function of X concave up or concave down in this case?

(b) In vertebrates, we typically find

$$[\text{skull length}] \propto [\text{body length}]^a$$

for some $a \in (0, 1)$. Use your answer in (a) to explain what this means for skull length versus body length in juveniles versus adults; that is, at which developmental stage do vertebrates have larger skulls relative to their body length?

42. The pH value of a solution measures the concentration of hydrogen ions, denoted by $[H^+]$. It is defined as

$$\text{pH} = -\log[H^+]$$

Use calculus to decide whether the pH value of a solution increases or decreases as the concentration of H^+ increases.

43. The differential equation

$$\frac{dy}{dx} = k\frac{y}{x}$$

describes allometric growth, where k is a positive constant. Assume that x and y are both positive variables and that $y = f(x)$ is twice differentiable. Use implicit differentiation to determine for which values of k the function $y = f(x)$ is concave up.

44. Let $N(t)$ denote the population size at time t and assume that $N(t)$ is twice differentiable and satisfies the differential equation

$$\frac{dN}{dt} = rN$$

where r is a real number. Differentiate the differential equation with respect to t to determine whether $N(t)$ is concave up or down.

5.3 EXTREMA, INFLECTION POINTS, AND GRAPHING

5.3.1 Extrema

If f is a continuous function on the closed interval $[a, b]$, then f has a global maximum and a global minimum in $[a, b]$. This is the content of the extreme value theorem, which is an existence result: It only tells us that global extrema exist under certain conditions, but it does not tell us how to find them.

Our strategy for finding global extrema in the case where f is a continuous function defined on a closed interval will be, first, to identify all local extrema of the function, and then to select the global extrema from the set of the local extrema. If f is a continuous function defined on an open interval or half-open interval, the existence of global extrema is no longer guaranteed, and we must compare the local extrema with the behavior of the function near the open boundaries of the domain (see Example 5 in Section 5.1). In particular, if $f(x)$ is defined on **R**, we need to investigate the behavior of $f(x)$ as $x \to \pm\infty$. For if the function $f(x)$ goes to $+\infty$ (respectively, $-\infty$) as $x \to +\infty$ or $-\infty$, it cannot have a global maximum (respectively, a global minimum). We discuss this in Example 1 of this section.

Local extrema can be found in a very systematic way, using a straightforward recipe to identify candidates for local extrema. We showed in Section 5.1 that if f has a local extremum at an interior point c and $f'(c)$ exists, then $f'(c) = 0$ (Fermat's theorem). That is, points where f is differentiable and where the first derivative is equal to 0 are certainly candidates for local extrema in the interior of the domain. Of course, these are only candidates, as explained in Section 5.1 (recall that $y = x^3$ has a horizontal tangent at $x = 0$, but $y = x^3$ does not have a local extremum at $x = 0$). In addition to points where the first derivative is equal to 0, we must check all points where the function is not differentiable (for instance, $y = |x|$ has a local minimum at $x = 0$ though it is not differentiable at 0). Points where the first derivative is equal to 0 or does not exist are called **critical points**. In addition to the critical points, we must always check the endpoints of the interval on which f is defined (provided that there are such endpoints).

There are no other points where local extrema can occur. We are thus equipped with a systematic way of searching for *candidates* for local extrema:

1. Find all numbers c where $f'(c) = 0$.
2. Find all numbers c where $f'(c)$ does not exist.
3. Find the endpoints of the domain of f.

We illustrate this procedure in the following example. We wish to find all local and global extrema of the function

$$f(x) = |x^2 - 4|, \quad -3 \le x < 2.5$$

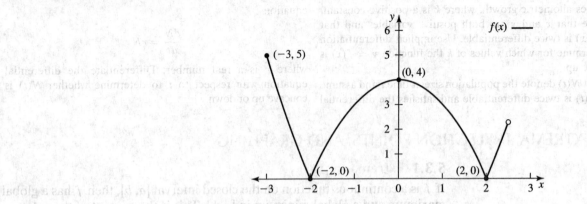

▲ **Figure 5.35**
The graph of $f(x) = |x^2 - 4|$ for $-3 \le x < 2.5$.

We know from Example 5 of Subsection 5.1.2 what the graph of the function looks like. We graph it again (Figure 5.35), which will make it easier to understand the procedure for finding relative extrema. But note that very often we do not know what a graph looks like, and we find relative extrema to gain a better understanding of the graph!

We will first rewrite $f(x)$ as a piecewise-defined function in order to get rid of the absolute value sign

$$f(x) = \begin{cases} x^2 - 4 & \text{for } -3 \le x \le -2 \quad \text{or} \quad 2 \le x < 2.5 \\ -x^2 + 4 & \text{for } -2 \le x \le 2 \end{cases}$$

This piecewise-defined function is differentiable on the open intervals $(-3, -2)$, $(-2, 2)$, and $(2, 2.5)$. We find

$$f'(x) = \begin{cases} 2x & \text{for } -3 < x < -2 \quad \text{or} \quad 2 < x < 2.5 \\ -2x & \text{for } -2 < x < 2 \end{cases}$$

Since $f'(x) = 0$ for $x = 0$ and $0 \in (-2, 2)$, $(0, f(0))$ is a critical point and is our first candidate for a local extremum. There are no other points where $f'(x) = 0$.

The second step is to identify interior points where the function is not differentiable. Since the function is differentiable in the open intervals $(-3, -2)$, $(-2, 2)$, and $(2, 2.5)$, we must look at the points where the function is pieced together, namely at $x = -2$ and at $x = 2$. We find

$$\lim_{x \to -2^-} f'(x) = -4 \quad \text{and} \quad \lim_{x \to -2^+} f'(x) = 4$$

and

$$\lim_{x \to 2^-} f'(x) = -4 \quad \text{and} \quad \lim_{x \to 2^+} f'(x) = 4$$

This shows that the function is not differentiable at $x = -2$ and $x = 2$. Therefore, there are critical points at $x = -2$ and $x = 2$, and these are also candidates for local extrema.

The third step is to identify endpoints of the domain. Since f is defined on $[-3, 2.5)$, there is an endpoint at $x = -3$. The fourth candidate is thus at $x = -3$. The interval $[-3, 2.5)$ is open at $x = 2.5$ and hence 2.5 is not in the domain of the function. The number 2.5 is therefore not a candidate for an extremum.

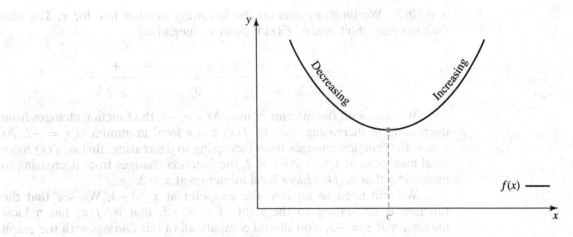

▲ **Figure 5.36**
The function $y = f(x)$ has a local minimum at $x = c$

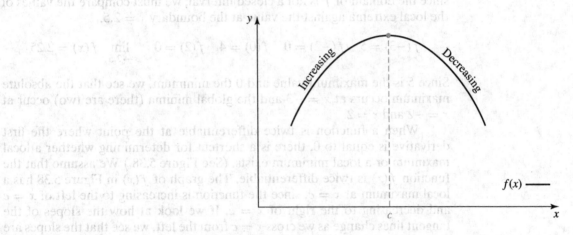

▲ **Figure 5.37**
The function $y = f(x)$ has a local maximum at $x = c$

Our systematic procedure has provided us with four candidates for local extrema at $x = -3, -2, 0,$ and 2. In each case, we must decide whether it is in fact a local extremum and, if so, whether it is a local maximum or minimum. The following observation, although rather obvious, is the key (see Figures 5.36 and 5.37):

> A continuous function has a local minimum at c if the function is decreasing to the left of c and increasing to the right of c. A function has a local maximum at c if the function is increasing to the left of c and decreasing to the right of c.

If the function is differentiable, as in our example, we can use the first derivative test to identify regions where the function is increasing and where it is decreasing.

Since $f'(x) = 2x$ for $-3 < x < -2$ and $2 < x < 2.5$, we see that $f'(x) > 0$ for $2 < x < 2.5$ and $f'(x) < 0$ for $-3 < x < -2$. Since $f'(x) = -2x$ for $-2 < x < 2$, we find that $f'(x) > 0$ for $x \in (-2, 0)$ and $f'(x) < 0$ for

$x \in (0, 2)$. We illustrate this on the following number line for x. The plus (minus) signs show where $f'(x)$ is positive (negative).

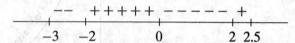

We start with the interior points. At $x = -2$, the function changes from decreasing to increasing, that is, $f(x)$ has a local minimum at $x = -2$. At $x = 0$, the function changes from increasing to decreasing, that is, $f(x)$ has a local maximum at $x = 0$. At $x = 2$, the function changes from decreasing to increasing, that is, $f(x)$ has a local minimum at $x = 2$.

We still need to analyze the endpoint at $x = -3$. We see that the function is decreasing to the right of $x = -3$, that is, $f(x)$ has a local maximum at $x = -3$. You should compare all of our findings with the graph of $f(x)$.

The last step is to select the global extrema from the local extrema. But since the domain of f is not a closed interval, we must compare the values of the local extrema against the value at the boundary $x = 2.5$.

$$f(-3) = 5 \quad f(-2) = 0 \quad f(0) = 4 \quad f(2) = 0 \quad \lim_{x \to 2.5^-} f(x) = 2.25$$

Since 5 is the maximum value and 0 the minimum, we see that the absolute maximum occurs at $x = -3$ and the global minima (there are two) occur at $x = -2$ and $x = 2$.

When a function is twice differentiable at the point where the first derivative is equal to 0, there is a shortcut for determining whether a local maximum or a local minimum exists. (See Figure 5.38.) We assume that the function $f(x)$ is twice differentiable. The graph of $f(x)$ in Figure 5.38 has a local maximum at $x = c$, since the function is increasing to the left of $x = c$ and decreasing to the right of $x = c$. If we look at how the slopes of the tangent lines change as we cross $x = c$ from the left, we see that the slopes are decreasing; that is, $f''(c) < 0$. In other words, the function is concave down at $x = c$ (which is immediately apparent when you look at the graph, but remember that typically you don't have the graph in front of you). There is an analogous result where f has a local minimum at $x = c$ (see Figure 5.39). This yields the following test.

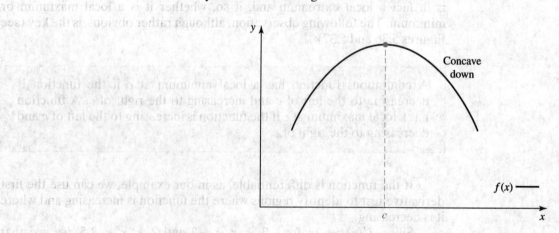

▲ **Figure 5.38**
The function $y = f(x)$ has a local maximum at $x = c$

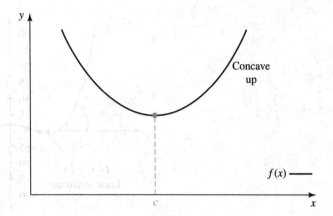

▲ **Figure 5.39**
The function $y = f(x)$ has a local minimum at $x = c$

THE SECOND DERIVATIVE TEST FOR LOCAL EXTREMA

Suppose that f is twice differentiable on an open interval containing c.

If $f'(c) = 0$ and $f''(c) < 0$, then f has a local maximum at $x = c$.

If $f'(c) = 0$ and $f''(c) > 0$, then f has a local minimum at $x = c$.

Note that finding the point c where $f'(c) = 0$ gave us a candidate for local extrema; if the second derivative $f''(c) \neq 0$, it nails it down: Not only does this tell us whether there is a local extremum, it identifies it. This test is easy to apply, as we only have to check the sign of the second derivative at $x = c$; we do not have to check the behavior of the function in a neighborhood of $x = c$. The second derivative test will not always work. For instance, if $f(x) = x^4$, then $f'(x) = 4x^3$ and $f''(x) = 12x^2$. Based on the graph of $y = f(x)$, we know that $f(x)$ has a local minimum at $x = 0$. We find $f'(0) = 0$ and $f''(0) = 0$. Thus the theorem cannot be used to draw any conclusions about $y = f(x)$ at $x = 0$. We look at two examples where this test can be applied.

▶ **Example 1**   Find all local and global extrema of

$$f(x) = \frac{3}{2}x^4 - 2x^3 - 6x^2 + 2, \quad x \in \mathbf{R}$$

Solution

[See Figure 5.40 for a graph of $f(x)$.] Since $f(x)$ is twice differentiable for all $x \in \mathbf{R}$, we begin by finding the first two derivatives of f. The first derivative is

$$f'(x) = 6x^3 - 6x^2 - 12x = 6x(x - 2)(x + 1)$$

Factoring $f'(x)$ will make it easier to find its zeros. The second derivative is

$$f''(x) = 18x^2 - 12x - 12$$

Since $f'(x)$ exists for all $x \in \mathbf{R}$ and there are no endpoints of the domain, the only candidates for local extrema are points where $f'(x) = 0$:

$$6x(x - 2)(x + 1) = 0$$

We find $x = 0$, $x = 2$, and $x = -1$. Since $f''(x)$ exists, we can use the second derivative test to determine whether these candidates are local extrema and

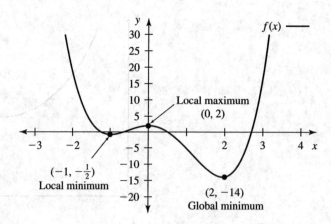

▲ **Figure 5.40**
The graph of $f(x) = \frac{3}{2}x^4 - 2x^3 - 6x^2 + 2$ in Example 1

of what type. We need to evaluate the second derivative for each candidate.

$$f''(0) = -12 < 0 \Rightarrow \text{local maximum at } x = 0$$

$$f''(2) = 36 > 0 \Rightarrow \text{local minimum at } x = 2$$

$$f''(-1) = 18 > 0 \Rightarrow \text{local minimum at } x = -1$$

The function $f(x)$ is defined on **R**. As mentioned at the beginning of this section, in order to find global extrema, we must check the local extrema and compare their values against each other and against the function values as $x \to \infty$ and $x \to -\infty$.

$$f(0) = 2 \quad f(2) = -14 \quad f(-1) = -\frac{1}{2}$$

and

$$\lim_{x \to \infty} f(x) = \infty \quad \lim_{x \to -\infty} f(x) = \infty$$

Though the local maximum is 2 at $x = 0$, the point $(0, 2)$ is not a global maximum. Since the function goes to ∞ as $|x| \to \infty$, it certainly exceeds the value 2. In fact, there is no global maximum; there is, however, a global minimum at $x = 2$.   ◀

▶ **Example 2**   Find all local and global extrema of

$$f(x) = x(1 - x)^{2/3}, \quad x \in \mathbf{R}$$

Solution

[A graph of $f(x)$ is shown in Figure 5.41.] We differentiate $f(x)$ using the product rule:

$$f'(x) = (1 - x)^{2/3} + x\frac{2}{3}(1 - x)^{-1/3}(-1)$$

$$= (1 - x)^{2/3} - \frac{2x}{3(1 - x)^{1/3}} \quad \text{for } x \neq 1$$

$$f''(x) = \frac{2}{3}(1 - x)^{-1/3}(-1) - \frac{2}{3}\left[(1 - x)^{-1/3} + x\left(-\frac{1}{3}\right)(1 - x)^{-4/3}(-1)\right]$$

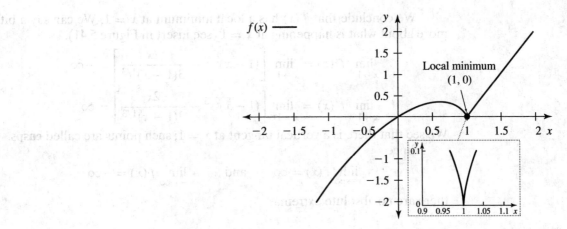

▲ **Figure 5.41**
The graph of $f(x) = x(1-x)^{2/3}$.

$$= -\frac{2}{3(1-x)^{1/3}} - \frac{2}{3(1-x)^{1/3}} - \frac{2x}{9(1-x)^{4/3}}$$

$$= -\frac{4}{3(1-x)^{1/3}} - \frac{2x}{9(1-x)^{4/3}} \quad \text{for } x \neq 1$$

We first find points where $f'(x) = 0$. If $x \neq 1$, then $f'(x) = 0$ when

$$(1-x)^{2/3} = \frac{2x}{3(1-x)^{1/3}}$$

$$3(1-x) = 2x$$

$$3 = 5x$$

$$x = \frac{3}{5}$$

We check the second derivative at $x = \frac{3}{5}$:

$$f''\left(\frac{3}{5}\right) = -\frac{4}{3\left(\frac{2}{5}\right)^{1/3}} - \frac{(2)\left(\frac{3}{5}\right)}{9\left(\frac{2}{5}\right)^{4/3}} < 0$$

That is, $f(x)$ has a local maximum at $x = \frac{3}{5}$.

The first derivative is not defined at $x = 1$. We therefore must investigate the function in the neighborhood of $x = 1$. Note that $f(x)$ is continuous at $x = 1$. We investigate where the function is increasing and where decreasing. The only x-values at which the derivative of the function can change sign are $x = \frac{3}{5}$ and $x = 1$. To determine whether $f'(x)$ is positive or negative in each of these intervals, we simply need to evaluate $f'(x)$ at one value in each interval. For instance, $f'(0) = 1 > 0$; that is, $f'(x) > 0$ for $x < 3/5$. A similar computation can be carried out for the other subintervals. We find that $f'(x)$ is positive to the left of $\frac{3}{5}$, negative between $\frac{3}{5}$ and 1, and positive for $x > 1$. This is summarized in the following number line.

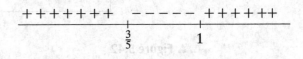

We conclude that $f(x)$ has a local minimum at $x = 1$. We can say a bit more about what is happening at $x = 1$ (see insert in Figure 5.41).

$$\lim_{x \to 1^-} f'(x) = \lim_{x \to 1^-} \left[(1-x)^{2/3} - \frac{2x}{3(1-x)^{1/3}} \right] = -\infty$$

$$\lim_{x \to 1^+} f'(x) = \lim_{x \to 1^+} \left[(1-x)^{2/3} - \frac{2x}{3(1-x)^{1/3}} \right] = \infty$$

We see that there is a vertical tangent at $x = 1$; such points are called **cusps**. Since

$$\lim_{x \to \infty} f(x) = \infty \qquad \text{and} \qquad \lim_{x \to -\infty} f(x) = -\infty$$

there are no absolute extrema.   ◀

5.3.2 Inflection Points

We begin with a verbal definition of **inflection points** and then give a procedure for locating such points (see Figure 5.42).

> Inflection points are points where the concavity of a function changes—that is, where the function changes from concave up to concave down or from concave down to concave up.

If the function is twice differentiable, there is an algebraic condition for finding candidates for inflection points. Recall that if a function f is twice differentiable, it is concave up if $f'' > 0$ and concave down if $f'' < 0$. At an inflection point, f'' must therefore change sign; that is, the second derivative must be 0 at an inflection point.

> If $f(x)$ is twice differentiable and has an inflection point at $x = c$, then $f''(c) = 0$.

Note that $f''(c) = 0$ is a necessary but not sufficient condition for an inflection point of a twice-differentiable function. For instance, $f(x) = x^4$ has

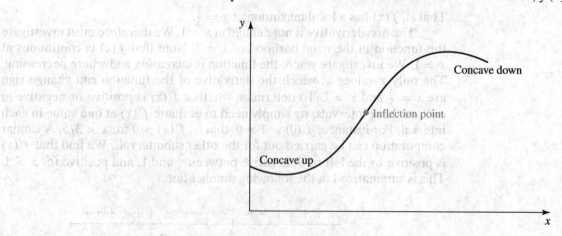

▲ **Figure 5.42**
Inflection point

$f''(0) = 0$ but $f(x)$ does not have an inflection point at $x = 0$. The function $f(x) = x^4$ is concave up and has a local minimum at $x = 0$ (see Figure 5.43). (We encountered a similar situation in Fermat's theorem, where we had a necessary but not sufficient condition for local extrema.) We can therefore use this test only for finding candidates for inflection points. To determine whether a candidate is an inflection point, we must check whether the second derivative changes sign.

▶ **Example 3**   Show that the function

$$f(x) = \frac{1}{2}x^3 - \frac{3}{2}x^2 + 2x + 1, \quad x \in \mathbf{R}$$

has an inflection point at $x = 1$.

Solution

[The graph of $f(x)$ is shown in Figure 5.44.] We compute the first two derivatives:

$$f'(x) = \frac{3}{2}x^2 - 3x + 2$$

$$f''(x) = 3x - 3$$

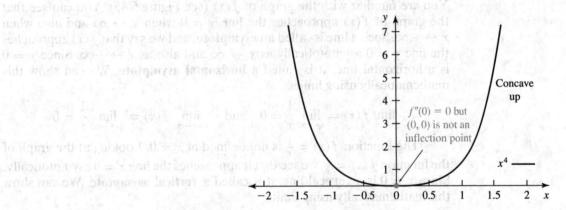

▲ **Figure 5.43**
The function $f(x) = x^4$ has $f''(0) = 0$ but no inflection point at $x = 0$

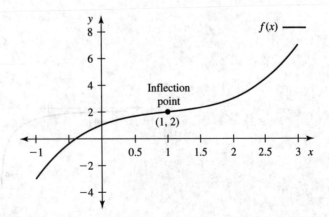

▲ **Figure 5.44**
The graph of $f(x) = \frac{1}{2}x^3 - \frac{3}{2}x^2 + 2x + 1$ has an inflection point at $x = 1$

Now, $f''(x) = 0$ if $x = 1$. Therefore, $x = 1$ is a candidate for an inflection point. Since $f''(x)$ is positive for $x > 1$ and negative for $x < 1$, $f''(x)$ changes sign at $x = 1$. We therefore conclude that $f(x)$ has an inflection point at $x = 1$. ◀

5.3.3 Graphing and Asymptotes

Using the first and the second derivatives of a twice differentiable function, we can obtain a fair amount of information about the function. We can determine intervals where the function is increasing, decreasing, concave up, and concave down. We can identify local and global extrema and find inflection points. To graph the function, we also need to know how the function behaves in the neighborhood of points where either the function or its derivative is not defined, and we need to know how the function behaves at the endpoints of its domain (or, if the function is defined for all $x \in \mathbf{R}$, how the function behaves for $x \to \pm\infty$).

We will need limits again—this time, to determine the behavior of a function at points where it is not defined and when $x \to \pm\infty$. We illustrate this in the following example. Consider

$$f(x) = \frac{1}{x}, \quad x \neq 0$$

You are familiar with the graph of $f(x)$ (see Figure 5.45). You can see that the graph of $f(x)$ approaches the line $y = 0$ when $x \to \infty$ and also when $x \to -\infty$. Such a line is called an **asymptote**, and we say that $f(x)$ approaches the line $y = 0$ asymptotically as $x \to \infty$ and also as $x \to -\infty$. Since $y = 0$ is a horizontal line, it is called a **horizontal asymptote**. We can show this mathematically using limits:

$$\lim_{x \to \infty} f(x) = \lim_{x \to \infty} \frac{1}{x} = 0 \quad \text{and} \quad \lim_{x \to -\infty} f(x) = \lim_{x \to -\infty} \frac{1}{x} = 0$$

The function $f(x) = \frac{1}{x}$ is not defined at $x = 0$. Looking at the graph of the function $f(x) = \frac{1}{x}$, we see that it approaches the line $x = 0$ asymptotically. Since $x = 0$ is a vertical line, it is called a **vertical asymptote**. We can show this mathematically using limits:

$$\lim_{x \to 0^-} f(x) = \lim_{x \to 0^-} \frac{1}{x} = -\infty \quad \text{and} \quad \lim_{x \to 0^+} f(x) = \lim_{x \to 0^+} \frac{1}{x} = +\infty$$

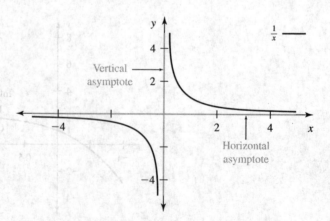

▲ **Figure 5.45**

The graph of $f(x) = \frac{1}{x}$ with horizontal asymptote $y = 0$ and vertical asymptote $x = 0$

Definition A line $y = b$ is a **horizontal** asymptote if either

$$\lim_{x \to -\infty} f(x) = b \quad \text{or} \quad \lim_{x \to \infty} f(x) = b$$

A line $x = c$ is a **vertical** asymptote if

$$\lim_{x \to c^+} f(x) = +\infty \quad \text{or} \quad \lim_{x \to c^+} f(x) = -\infty$$

or

$$\lim_{x \to c^-} f(x) = +\infty \quad \text{or} \quad \lim_{x \to c^-} f(x) = -\infty$$

In addition to horizontal and vertical asymptotes, there are **oblique** asymptotes (see Figure 5.46). These are straight lines that are neither horizontal nor vertical and such that the graph of the function approaches this line as either $x \to +\infty$ or $x \to -\infty$. Mathematically, we express this in the following way. If

$$\lim_{x \to +\infty} [f(x) - (mx + b)] = 0 \quad \text{or} \quad \lim_{x \to -\infty} [f(x) - (mx + b)] = 0$$

then the line $y = mx + b$ is an oblique asymptote. The simplest case of an oblique asymptote occurs with a rational function in which the degree of the numerator is one higher than the degree of the denominator.

▶ **Example 4**   **Oblique Asymptote** Let

$$f(x) = \frac{x^2 - 3}{x - 2}, \quad x \neq 2$$

To determine whether $f(x)$ has an oblique asymptote, we use long division. We find

$$\frac{x^2 - 3}{x - 2} = x + 2 + \frac{1}{x - 2}$$

We see that $f(x)$ is the sum of a linear term, namely $x + 2$, and a remainder term, namely $\frac{1}{x-2}$, which goes to 0 as $x \to \pm\infty$. To check that $y = x + 2$ is

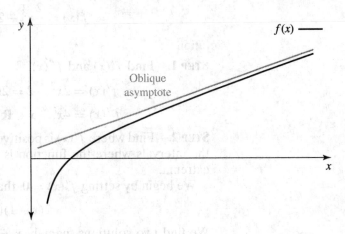

▲ **Figure 5.46**
The function $y = f(x)$ has an oblique asymptote

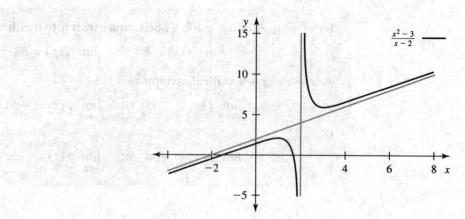

$\dfrac{x^2-3}{x-2}$

▲ **Figure 5.47**

The graph of $f(x) = \dfrac{x^2-3}{x-2}$ with oblique asymptote $g(x) = x+2$ (and a vertical asymptote at $x = 2$)

indeed an oblique asymptote, we carry out the following computation:

$$\lim_{x\to\pm\infty}[f(x)-(x+2)] = \lim_{x\to\pm\infty}\left[\frac{x^2-3}{x-2}-(x+2)\right]$$

$$= \lim_{x\to\pm\infty}\left[x+2+\frac{1}{x-2}-(x+2)\right]$$

$$= \lim_{x\to\pm\infty}\frac{1}{x-2} = 0$$

Hence, $y = x + 2$ is an oblique asymptote. The graph of $f(x) = \dfrac{x^2-3}{x-2}$ together with its oblique asymptote is shown in Figure 5.47. [The graph of $f(x)$ also has a vertical asymptote at $x = 2$.] ◀

We can now combine the results that we have obtained so far to produce the graph of a given function. We illustrate the steps leading to the graph in the following two examples.

▶ **Example 5**   Sketch the graph of the function

$$f(x) = \frac{2}{3}x^3 - 2x + 1, \quad x \in \mathbf{R}$$

Solution

STEP 1. Find $f'(x)$ and $f''(x)$.

$$f'(x) = 2x^2 - 2 = 2(x-1)(x+1), \quad x \in \mathbf{R}$$

$$f''(x) = 4x, \quad x \in \mathbf{R}$$

STEP 2. Find where $f'(x)$ is positive, negative, zero, or undefined. Identify the intervals where the function is increasing and decreasing. Find local extrema.

We begin by setting $f'(x) = 0$, that is,

$$2(x-1)(x+1) = 0$$

We find two solutions, namely $x = 1$ and $x = -1$. We illustrate on the following number line where $f'(x)$ is positive and where negative. Since $f'(x)$ is a polynomial, it is differentiable for all $x \in \mathbf{R}$; it can only change sign at $x = 1$ and at $x = -1$. This breaks the number line into three intervals, $x < -1$, $-1 < x < 1$, and $x > 1$. To determine whether $f'(x)$ is positive or negative in each of these intervals, we simply need to evaluate $f'(x)$ at one value in each interval. For instance, $f'(-2) = 2(-3)(-1) = 6 > 0$, that is, $f'(x) > 0$ for $x < -1$. A similar computation can be carried out for the other subintervals.

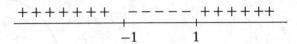

That is, $f(x)$ is increasing for $x < -1$ and $x > 1$ and decreasing for $-1 < x < 1$. This also shows that the function has a local maximum at $x = -1$, namely $(-1, \frac{7}{3})$, and a local minimum at $x = 1$, namely $(1, -\frac{1}{3})$. (Alternatively, we could have used the second derivative test.)

Step 3. Find where $f''(x)$ is positive, negative, zero, or undefined. Find inflection points.

We set $f''(x) = 0$, that is,

$$4x = 0$$

We find one solution, namely $x = 0$. Since $f''(x)$ exists for all $x \in \mathbf{R}$, it can only change sign at $x = 0$. Evaluating $f''(x)$ at a value to the left of 0, say at $x = -1$, shows that $f''(x) < 0$ for $x < 0$. Similarly, evaluating $f''(x)$ at a value to the right of 0, say at $x = 1$, shows that $f''(x) > 0$ for $x > 0$. This is illustrated on the following number line.

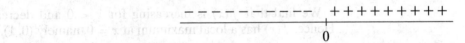

Since $f''(x)$ changes sign at $x = 0$, we conclude that the function has an inflection point at $x = 0$, namely $(0, 1)$. The function is concave down for $x < 0$ and concave up for $x > 0$.

Step 4. Determine the behavior at endpoints of the domain.

The domain is $(-\infty, +\infty)$. We therefore need to check the behavior of $f(x)$ as $x \to +\infty$ and as $x \to -\infty$. We find

$$\lim_{x \to -\infty} f(x) = -\infty \quad \text{and} \quad \lim_{x \to +\infty} f(x) = +\infty$$

Combining the results of these four steps allows us to sketch the graph of the function, as illustrated in Figure 5.48. You should label all extrema and inflection points. ◀

▶ **Example 6** Sketch the graph of the function

$$f(x) = e^{-x^2/2}, \quad x \in \mathbf{R}$$

Solution

Step 1.

$$f'(x) = -xe^{-x^2/2}, \quad x \in \mathbf{R}$$

and

$$f''(x) = (-1)e^{-x^2/2} + (-x)(-x)e^{-x^2/2}$$
$$= e^{-x^2/2}(x^2 - 1), \quad x \in \mathbf{R}$$

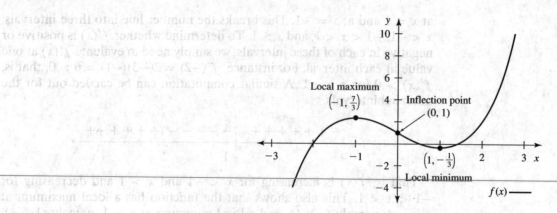

▲ **Figure 5.48**

The graph of $f(x) = \frac{2}{3}x^3 - 2x + 1$

STEP 2. Since $f'(x)$ is defined for all $x \in \mathbf{R}$, we only need to identify those points where $f'(x) = 0$.

$$f'(x) = 0 \quad \text{for } x = 0$$

The sign of $f'(x)$ is illustrated on the following number line for x.

We find that $f(x)$ is increasing for $x < 0$ and decreasing for $x > 0$. Hence, $f(x)$ has a local maximum at $x = 0$, namely $(0, 1)$.

STEP 3. $f''(x) = 0 \quad \text{for } x = 1 \text{ and } x = -1$

The sign of $f''(x)$ is illustrated on the following number line for x.

$$+ + + + + + \quad - - - - - \quad + + + + +$$
$$\overline{\qquad\qquad -1 \qquad\qquad 1 \qquad\qquad}$$

We find that $f(x)$ is concave up for $x < -1$ and $x > 1$, and it is concave down for $-1 < x < 1$. There are two inflection points, one at $x = -1$,

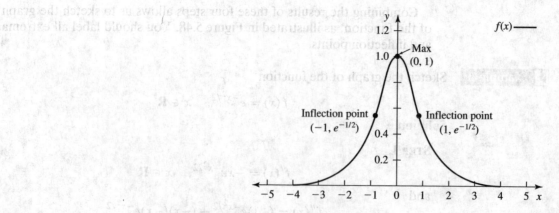

▲ **Figure 5.49**

The graph of $f(x) = e^{-x^2/2}$

namely $(-1, e^{-1/2})$, and the other at $x = 1$, namely $(1, e^{-1/2})$. There are no other inflection points, since $f''(x)$ is defined for all $x \in \mathbf{R}$.

STEP 4.

$$\lim_{x \to -\infty} f(x) = 0 \quad \text{and} \quad \lim_{x \to +\infty} f(x) = 0$$

This shows that $y = 0$ is a horizontal asymptote.
The graph of $f(x)$ is shown in Figure 5.49. ◀

5.3.4 Problems

(5.3.1)

Find the local maxima and minima of each of the functions in Problems 1–16. Determine whether each function has absolute maxima and minima and find their coordinates. Find the intervals where each function is increasing and decreasing.

1. $y = 4 - x^2, -2 \le x \le 3$
2. $y = \sqrt{x}, 0 \le x \le 4$
3. $y = \sin x, 0 \le x \le 2\pi$
4. $y = \ln x, x > 0$
5. $y = e^{-x}, x \ge 0$
6. $y = |16 - x^2|, -5 \le x \le 8$
7. $y = (x - 1)^3 + 1, x \in \mathbf{R}$
8. $y = \tan x, -\dfrac{\pi}{2} < x < \dfrac{\pi}{2}$
9. $y = \cos x, 0 \le x \le \pi$
10. $y = \ln(x^2 + 1) - x, x \in \mathbf{R}$
11. $y = e^{-|x|}, x \in \mathbf{R}$
12. $y = e^{-x^2/4}, x \in \mathbf{R}$
13. $y = \frac{1}{3}x^3 + \frac{1}{2}x^2 - 6x + 2, x \in \mathbf{R}$
14. $y = x^2(1 - x), x \in \mathbf{R}$
15. $y = \sin(\pi x^2), -1 \le x \le 1$
16. $y = \sqrt{1 + x^2}, x \in \mathbf{R}$

17. [This problem illustrates the fact that $f'(c) = 0$ is not a sufficient condition for a local extremum of a differentiable function.] Show that the function $f(x) = x^3$ has a horizontal tangent at $x = 0$; that is, $f'(0) = 0$ but $f'(x)$ does not change sign at $x = 0$ and, hence, $f(x)$ does not have a local extremum at $x = 0$.

18. Suppose that $f(x)$ is twice differentiable on $\mathbf{R}$ with $f(x) > 0$ for $x \in \mathbf{R}$. Show that if $f(x)$ has a local maximum at $x = c$, then $g(x) = \ln f(x)$ also has a local maximum at $x = c$.

(5.3.2)

In Problems 19–24, determine all inflection points.

19. $f(x) = x^3 - 2, x \in \mathbf{R}$
20. $f(x) = \cos x, 0 \le x \le \pi$
21. $f(x) = e^{-x^2}, x \ge 0$
22. $f(x) = xe^{-x}, x \ge 0$
23. $f(x) = \tan x, -\dfrac{\pi}{2} < x < \dfrac{\pi}{2}$
24. $f(x) = \ln x + \frac{1}{x}, x > 0$
25. [This problem illustrates that $f''(c) = 0$ is not a sufficient condition for an inflection point of a twice differentiable function.] Show that the function $f(x) = x^4$ has $f''(0) = 0$, but that $f''(x)$ does not change sign at $x = 0$ and, hence, $f(x)$ does not have an inflection point at $x = 0$.

26. (*Logistic equation*) Suppose that the size of a population at time t is denoted by $N(t)$ and satisfies

$$N(t) = \frac{100}{1 + 3e^{-2t}}$$

for $t \ge 0$.

(a) Show that $N(0) = 25$.

(b) Show that $N(t)$ is strictly increasing.

(c) Show that

$$\lim_{t \to \infty} N(t) = 100$$

(d) Show that $N(t)$ has an inflection point when $N(t) = 50$, that is, when the size of the population is at half its limiting value.

(e) Use your results in (a)–(d) to sketch the graph of $N(t)$.

(5.3.3)

Find the local maxima and minima of the functions in Problems 27–34. Determine whether the functions have absolute maxima and minima, and find their coordinates. Find inflection points. Find the intervals where the function is increasing, decreasing, concave up, and concave down. Sketch the graph of each function.

27. $y = \frac{2}{3}x^3 - 2x^2 - 6x + 2$ for $-2 \le x \le 5$
28. $y = x^4 - 2x^2, x \in \mathbf{R}$
29. $y = |x^2 - 9|, -4 \le x \le 5$
30. $y = \sqrt{|x|}, x \in \mathbf{R}$
31. $y = x + \cos x, x \in \mathbf{R}$
32. $y = \tan x - x, x \in \left(-\dfrac{\pi}{2}, \dfrac{\pi}{2}\right)$
33. $y = \dfrac{x^2 - 1}{x^2 + 1}, x \in \mathbf{R}$
34. $y = \ln(x^2 + 1), x \in \mathbf{R}$
35. Let

$$f(x) = \frac{x}{x - 1}, \quad x \ne 1$$

(a) Show that

$$\lim_{x \to -\infty} f(x) = 1$$

and

$$\lim_{x \to +\infty} f(x) = 1$$

That is, $y = 1$ is a horizontal asymptote of the curve $y = \frac{x}{x-1}$.

(b) Show that

$$\lim_{x \to 1^-} f(x) = -\infty$$

and

$$\lim_{x \to 1^+} f(x) = +\infty$$

That is, $x = 1$ is a vertical asymptote of the curve $y = \frac{x}{x-1}$.

(c) Determine where $f(x)$ is increasing and where it is decreasing. Does $f(x)$ have local extrema?

(d) Determine where $f(x)$ is concave up and where it is concave down. Does $f(x)$ have inflection points?

(e) Sketch the graph of $f(x)$ together with its asymptotes.

36. Let

$$f(x) = -\frac{2}{x^2 - 1}, \quad x \neq -1, 1$$

(a) Show that

$$\lim_{x \to +\infty} f(x) = 0$$

and

$$\lim_{x \to -\infty} f(x) = 0$$

That is, $y = 0$ is a horizontal asymptote of $f(x)$.

(b) Show that

$$\lim_{x \to -1^-} f(x) = -\infty$$

and

$$\lim_{x \to -1^+} f(x) = +\infty$$

and that

$$\lim_{x \to 1^-} f(x) = +\infty$$

and

$$\lim_{x \to 1^+} f(x) = -\infty$$

That is, $x = -1$ and $x = 1$ are vertical asymptotes of $f(x)$.

(c) Determine where $f(x)$ is increasing and where it is decreasing. Does $f(x)$ have local extrema?

(d) Determine where $f(x)$ is concave up and where it is concave down. Does $f(x)$ have inflection points?

(e) Sketch the graph of $f(x)$ together with its asymptotes.

37. Let

$$f(x) = \frac{2x^2 - 5}{x + 2}, \quad x \neq -2$$

(a) Show that $x = -2$ is a vertical asymptote.

(b) Determine where $f(x)$ is increasing and where it is decreasing. Does $f(x)$ have local extrema?

(c) Determine where $f(x)$ is concave up and where it is concave down. Does $f(x)$ have inflection points?

(d) Since the degree of the numerator is one higher than the degree of the denominator, $f(x)$ has an oblique asymptote. Find it.

(e) Sketch the graph of $f(x)$ together with its asymptotes.

38. Let

$$f(x) = \frac{\sin x}{x}, \quad x \neq 0$$

(a) Show that $y = 0$ is a horizontal asymptote.

(b) Since $f(x)$ is not defined at $x = 0$, does this mean that $f(x)$ has a vertical asymptote at $x = 0$? Find $\lim_{x \to 0^+} f(x)$ and $\lim_{x \to 0^-} f(x)$.

(c) Use a graphing calculator to sketch the graph of $f(x)$.

39. Let

$$f(x) = \frac{x^2}{1 + x^2}, \quad x \in \mathbf{R}$$

(a) Determine where $f(x)$ is increasing and decreasing.

(b) Where is the function concave up and where is it concave down? Find all inflection points of $f(x)$.

(c) Find $\lim_{x \to \pm\infty} f(x)$ and decide whether $f(x)$ has a horizontal asymptote.

(d) Sketch the graph of $f(x)$ together with its asymptotes and inflection points (if they exist).

40. Let

$$f(x) = \frac{x^k}{1 + x^k}, \quad x \geq 0$$

where k is a positive integer greater than 1.

(a) Determine where $f(x)$ is increasing and decreasing.

(b) Where is the function concave up and where is it concave down? Find all inflection points of $f(x)$.

(c) Find $\lim_{x \to \infty} f(x)$ and decide whether $f(x)$ has a horizontal asymptote.

(d) Sketch the graph of $f(x)$ together with its asymptotes and inflection points (if they exist).

41. Let

$$f(x) = \frac{x}{a + x}, \quad x \geq 0$$

where a is a positive constant.

(a) Determine where $f(x)$ is increasing and decreasing.

(b) Where is the function concave up and where is it concave down? Find all inflection points of $f(x)$.

(c) Find $\lim_{x \to \infty} f(x)$ and decide whether $f(x)$ has a horizontal asymptote.

(d) Sketch the graph of $f(x)$ together with its asymptotes and inflection points (if they exist).

42. Let

$$f(x) = \frac{x^2}{a^2 + x^2}, \quad x \geq 0$$

where a is a positive constant.

(a) Determine where $f(x)$ is increasing and decreasing.

(b) Where is the function concave up and where is it concave down? Find all inflection points of $f(x)$.

(c) Find $\lim_{x \to \infty} f(x)$ and decide whether $f(x)$ has a horizontal asymptote.

(d) Sketch the graph of $f(x)$ together with its asymptotes and inflection points (if they exist).

43. Suppose that the growth rate of a population is given by

$$f(N) = N\left(1 - \left(\frac{N}{K}\right)^\theta\right) \quad N \geq 0$$

where N is the size of the population, K is a positive constant denoting the carrying capacity, and θ is a parameter greater than 1. Find the population size for which the growth rate is maximal.

44. Spruce budworms are a major pest that defoliate balsam fir. They are preyed upon by birds. A model for the per capita predation rate is given by

$$f(N) = \frac{aN}{k^2 + N^2}$$

where N denotes the density of spruce budworm and a and k are positive constants. For which density of spruce budworms is the per capita predation rate maximal?

5.4 OPTIMIZATION

There are many situations in which we wish to maximize or minimize certain quantities. For instance, in a chemical reaction, you might wish to know under which conditions the reaction rate is maximized. In an agricultural setting, you might be interested in finding the amount of fertilizer that would maximize yield. Optimization problems arise in the study of the evolution of life histories, involving questions such as when an organism should begin reproduction in order to maximize the number of surviving offspring. In each case, we are interested in finding global extrema.

▶ **Example 1** **Chemical Reaction** We consider the chemical reaction

$$A + B \to AB$$

In Example 5 of Subsection 1.2.2, we found that the reaction rate is given by the function

$$R(x) = k(a - x)(b - x), \quad 0 \le x \le \min(a, b)$$

where x is the concentration of the product AB and $\min(a, b)$ denotes the minimum of the two values of a and b. The constants a and b are the concentrations of the reactants A and B at the beginning of the reaction. To be concrete, we choose $k = 2$, $a = 2$, and $b = 5$. Then

$$R(x) = 2(2 - x)(5 - x) \quad \text{for } 0 \le x \le 2$$

(see Figure 5.50).

We are interested in finding the concentration x that maximizes the reaction rate: This is the absolute maximum of $R(x)$. Since $R(x)$ is differentiable on $(0, 2)$, we can find all local extrema in $(0, 2)$ by investigating the first derivative. To compute the first derivative of $R(x)$, we multiply $R(x)$ out. This yields

$$R(x) = 20 - 14x + 2x^2 \quad \text{for } 0 \le x \le 2$$

Differentiating with respect to x yields

$$R'(x) = -14 + 4x \quad \text{for } 0 < x < 2$$

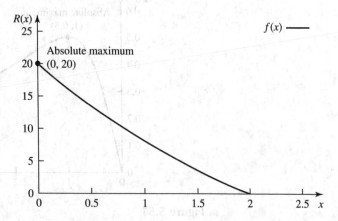

▲ **Figure 5.50**
The chemical reaction rate $R(x)$ in Example 1. The graph of $R(x) = 2(2 - x)(5 - x)$, $0 \le x \le 2$, has an absolute maximum at $(0, 20)$

To find candidates for local extrema, we set $R'(x) = 0$:

$$-14 + 4x = 0 \quad \text{or} \quad x = \frac{7}{2}$$

Since $\frac{7}{2} \notin (0, 2)$, there are no points in the interval $(0, 2)$ with horizontal tangents. Given that

$$R'(x) = -14 + 4x < 0 \quad \text{for } x \in (0, 2)$$

we conclude that $R(x)$ is decreasing in $(0, 2)$. The absolute maximum is therefore attained at the left endpoint of the interval $[0, 2]$, namely at $x = 0$. Thus, the reaction rate is maximal when the concentration of the product AB is equal to 0. You should compare this result with the graph of $R(x)$ in Figure 5.50. Since the reaction rate is proportional to the product of the concentrations of A and B, and A and B react to form the product AB, their concentrations decrease during the reaction and hence the reaction rate should be highest at the beginning of the reaction, when the concentrations of A and B are highest. ◀

▶ **Example 2**   **Crop Yield** Let $Y(N)$ be the yield of an agricultural crop as a function of nitrogen level N in the soil. A model that is used for this is

$$Y(N) = \frac{N}{1 + N^2} \quad \text{for } N \geq 0$$

(where N is measured in appropriate units). Find the nitrogen level that maximizes yield.

Solution

The function $Y(N)$, shown in Figure 5.51, is differentiable for $N > 0$. We find

$$Y'(N) = \frac{(1 + N^2) - N \cdot 2N}{(1 + N^2)^2} = \frac{1 - N^2}{(1 + N^2)^2}$$

Setting $Y'(N) = 0$, we obtain candidates for local extrema:

$$Y'(N) = 0 \quad \text{if } 1 - N^2 = 0 \text{ or } N = \pm 1$$

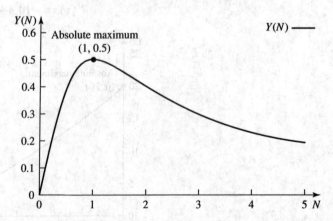

▲ **Figure 5.51**
Crop yield $Y(N)$ in Example 2. The graph of $Y(N) = \frac{N}{1+N^2}$, $N \geq 0$, has an absolute maximum at $\left(1, \frac{1}{2}\right)$

Since $N = -1$ is not in the domain of $Y(N)$, we can discard this candidate. The other candidate $N = 1$ is in the domain, and we see that

$$Y'(N) \begin{cases} > 0 & \text{for } 0 < N < 1 \\ < 0 & \text{for } N > 1 \end{cases}$$

Since $Y(N)$ changes from increasing to decreasing at $N = 1$, $N = 1$ is a local maximum. To find the global maximum, we still need to check the endpoints of the domain. We find

$$Y(0) = 0 \quad \text{and} \quad \lim_{N \to \infty} Y(N) = \lim_{N \to \infty} \frac{N}{1 + N^2} = 0$$

Since $Y(1) = \frac{1}{2}$, we conclude that $N = 1$ is the global maximum. That is, $N = 1$ is the nitrogen level that maximizes yield. ◀

▶ **Example 3** **Maximizing Area** A field biologist wants to enclose a rectangular study plot. She has 1600 ft of fencing. Using this fencing, what are the dimensions of the study plot that will have the largest area?

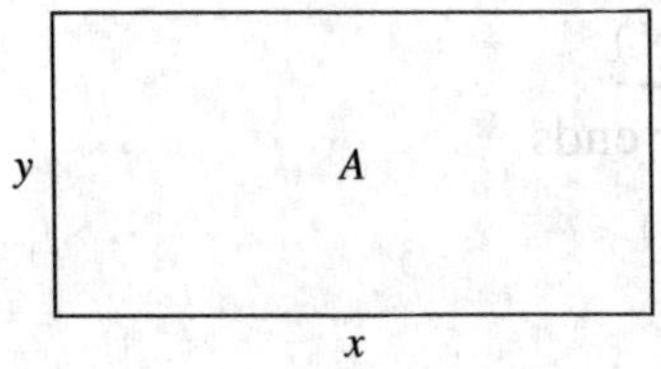

Solution

Figure 5.52 illustrates the situation. The area A of this study plot is given by

$$A = xy \tag{5.9}$$

and the perimeter of the study plot is given by

$$1600 = 2x + 2y \tag{5.10}$$

Solving equation (5.10) for y yields

$$y = 800 - x$$

We can substitute this for y in equation (5.9) and obtain

$$A(x) = x(800 - x) = 800x - x^2 \quad \text{for } 0 \le x \le 800$$

It is important to state the domain of the function. Clearly, the smallest value of x is 0, in which case the enclosed area is also 0 since $A(0) = 0$. The largest possible value for x is 800; this will also produce a rectangle where one side has length 0; the corresponding area is $A(800) = 0$.

We wish to maximize the enclosed area $A(x)$. The function $A(x)$ is differentiable for $x \in (0, 800)$:

$$A'(x) = 800 - 2x \quad \text{for } 0 < x < 800$$
$$A''(x) = -2 \quad \text{for } 0 < x < 800$$

To find candidates for local extrema, we set $A'(x) = 0$.

$$800 - 2x = 0 \quad \text{or} \quad x = 400$$

Since $A''(400) < 0$, $x = 400$ is a local maximum. To find the global maximum, we need to check the function $A(x)$ at the endpoints of the interval $[0, 800]$.

$$A(0) = A(800) = 0$$

Since $A(400) = 400^2 = 160,000$, the area is maximized for $x = 400$, which implies that the study plot is a square. This is true in general: For a rectangle with fixed perimeter, the maximum area occurs when the rectangle is a square (see Problem 2 in this section). ◀

▲ **Figure 5.52**
The rectangular study plot in Example 3

▶ **Example 4**   **Minimizing Material** Aluminum soda cans are shaped like a right-circular cylinder and hold about 12 ounces of liquid. Production of aluminum requires a lot of energy, so it is desirable to design soda cans that use the least amount of material. What dimensions would such an optimal soda can have?

Solution

We approximate the soda can by a right-circular cylinder in which the cylinder wall and both ends are made of aluminum. We denote the height of the cylinder by h and the radius by r (see Figure 5.53).

If we measure h and r in centimeters, we must convert the 12 ounces into a volume measured in cubic centimeters (cm^3). For this, we need to know that 12 ounces are about 0.355 liters, and 1 liter equals 1000 cm^3.

We are now ready to set up the problem. The can with the least amount of material is the cylinder whose surface area is minimal for a given volume. Using formulas from geometry, we find that the surface area A of a right circular cylinder with top and bottom closed is given by

$$A = \underbrace{2\pi rh}_{\text{cylinder wall}} + \underbrace{2(\pi r^2)}_{\text{cylinder ends}}$$

The volume V of a right cylinder is given by

$$V = \pi r^2 h$$

We therefore need to minimize A when the volume $V = 12$ ounces $= 355$ cm^3. Solving $\pi r^2 h = 355$ for h yields

$$h = \frac{355}{\pi r^2}$$

Substituting this for h in the formula for A, we find

$$A(r) = 2\pi r \frac{355}{\pi r^2} + 2\pi r^2 = \frac{710}{r} + 2\pi r^2 \quad \text{for } r > 0$$

The graph of $A(r)$ is shown in Figure 5.54.

To find the global minimum, differentiate $A = A(r)$ and set the derivative equal to 0.

$$A'(r) = -\frac{710}{r^2} + 4\pi r = 0$$

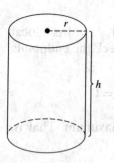

▲ **Figure 5.53**
A right circular cylinder with height h and radius r

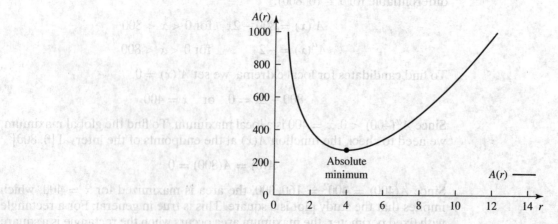

▲ **Figure 5.54**
The surface area A of a right cylinder with given volume as a function of radius r

Solving for r yields

$$4\pi r = \frac{710}{r^2} \quad \text{or} \quad r^3 = \frac{710}{4\pi} = \frac{355}{2\pi}$$

We thus find that

$$r = \left(\frac{355}{2\pi}\right)^{1/3} = 3.84 \text{ cm}$$

To check whether this is indeed a minimum, we compute the second derivative of $A(r)$:

$$A''(r) = 2\frac{710}{r^3} + 4\pi > 0 \quad \text{for } r > 0$$

Since $A''(r) > 0$, $r = \left(\frac{355}{2\pi}\right)^{1/3}$ is a local minimum. To determine whether this value of r is the global minimum, we need to compute the surface area at the boundaries of the domain. Given that

$$\lim_{r \to 0+} A(r) = \infty \quad \text{and} \quad \lim_{r \to \infty} A(r) = \infty$$

it follows that the global minimum is achieved at $r = \left(\frac{355}{2\pi}\right)^{1/3}$.

To find h, we use

$$h = \frac{355}{\pi r^2} = \frac{355}{\pi \left(\frac{355}{2\pi}\right)^{2/3}} = \frac{\frac{355}{\pi}}{\left(\frac{355}{2\pi}\right)^{2/3}}$$

$$= 2\left(\frac{355}{2\pi}\right)^{1/3} = 2r = 7.67 \text{ cm}$$

We thus find that the can that uses the least amount of material is one where the height of the can is equal to its diameter.

A real soda can has $h = 12.5$ cm and $r = 3.1$ cm. In our computation, we assumed that the material used to manufacture the can is of equal thickness. The top of a real soda can, however, is of thicker material, which could explain why soda cans don't have the dimensions we computed in this example. ◄

The previous four examples illustrate the basic types of optimization problems. In the first two examples, the functions that we wished to optimize were given; in the third and fourth examples, we needed to set up equations for the functions that we wished to optimize. Once you obtain a function that you wish to optimize, you must use the results from the previous sections in this chapter to find the global extremum. It is important to state the domain of the function, since global extrema may be found at the endpoints of the domain.

Life histories of organisms are thought to evolve to some optimal state within given constraints. Theoretical models can help to find such optimal states. As an example, we will look at clutch size. Suppose an organism can produce more than one offspring at a time. What is the optimal clutch size? Clutch size is determined by the amount of resources the parents can provide to each offspring. If resources are limited (as they usually are), the more offspring per clutch, the less is available to each individual offspring. On the other hand, if the number of offspring is too small, then the chances of having any offspring that survive to reproductive age might be quite small for reasons other than insufficient resources. This trade-off between not enough resources per offspring if too many offspring and the chance of losing the entire clutch if too few offspring suggests that an intermediate number of offspring might be optimal. We discuss a simple model that addresses this trade-off in the following example.

▶ Example 5

(Adapted from Roff, 1992) Lloyd (1987) proposed the following model to determine optimal clutch size. If clutch size is equal to N and the total amount of resources allocated is R, then the amount of resources allocated to each offspring is $x = R/N$. The survival chance for an individual, denoted by $f(x)$, is related to the investment x per individual. Lloyd proposed an S-shaped or sigmoidal curve; that is, survival chances are very low when the investment is low and the curve shows a saturation effect for large investments. With $N = N(x) = R/x$, the success of a clutch, or its **fitness**, can be measured as

$$w(x) = [\text{number of offspring}] \times [\text{survival probability of offspring}]$$

$$= N(x)f(x) = \frac{R}{x}f(x) \quad \text{for } x > 0$$

where R is a positive constant. We wish to find the value of x that maximizes the fitness $w(x)$. If we assume that $f(x)$ is differentiable for $x > 0$, then differentiating $w(x)$ with respect to x yields

$$\frac{dw}{dx} = R\frac{d}{dx}\left(\frac{f(x)}{x}\right)$$

$$= R\frac{f'(x)x - f(x)}{x^2} = \frac{R}{x}\left(f'(x) - \frac{f(x)}{x}\right) \quad \text{for } x > 0$$

Setting $w'(x) = 0$ yields

$$f'(x) = \frac{f(x)}{x} \tag{5.11}$$

We denote the solution of equation (5.11) by $\hat{x}$ (read: "x hat"). If $f(x)$ is twice differentiable at $\hat{x}$, we can use the second derivative test to test whether $\hat{x}$ is a local maximum or minimum. We find

$$\frac{d^2w}{dx^2} = -\frac{R}{x^2}\left(f'(x) - \frac{f(x)}{x}\right) + \frac{R}{x}\left(f''(x) - \frac{d}{dx}\frac{f(x)}{x}\right)$$

Since $f'(\hat{x}) = f(\hat{x})/\hat{x}$, the first term on the right-hand side is equal to 0 when $x = \hat{x}$. Furthermore, because of (5.11), $\frac{d}{dx}[f(x)/x] = \frac{xf'(x)-f(x)}{x^2} = 0$ when $x = \hat{x}$. Hence,

$$\left.\frac{d^2w}{dx^2}\right|_{x=\hat{x}} = \frac{R}{\hat{x}}f''(\hat{x})$$

If we choose a function $f(x)$ that is concave down at $\hat{x}$, then $f''(\hat{x}) < 0$ and $w(x)$ has a local maximum at $\hat{x}$.

A common choice for $f(x)$ is

$$f(x) = \frac{x^2}{k^2 + x^2} \quad \text{for } x \geq 0$$

where k is a positive constant. The graph of $f(x)$ is shown in Figure 5.55; the curve is sigmoidal. Differentiating $f(x)$ with respect to x yields

$$f'(x) = \frac{2x(k^2 + x^2) - x^2(2x)}{(k^2 + x^2)^2} = \frac{2k^2x}{(k^2 + x^2)^2}$$

Since

$$\frac{dw}{dx} = 0 \quad \text{for } f'(x) = \frac{f(x)}{x}$$

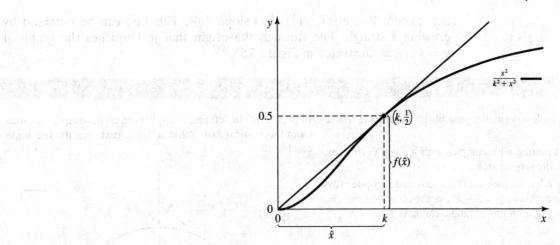

▲ Figure 5.55

The graph of $f(x) = \frac{x^2}{k^2+x^2}$ together with the tangent line at $(k, 1/2)$

we find

$$\frac{2k^2x}{(k^2+x^2)^2} = \frac{1}{x}\frac{x^2}{k^2+x^2}$$

which yields

$$2k^2 = k^2 + x^2 \quad \text{or} \quad k^2 = x^2$$

Since k is a positive constant and $x \geq 0$, we can discard the solution $x = -k$ and find $\hat{x} = k$. To see whether $\hat{x} = k$ is a local maximum for the function $w(x)$, we evaluate $w''(k)$; we need to find $f''(x)$ first.

$$f''(x) = \frac{2k^2(k^2+x^2)^2 - (2k^2x)2(k^2+x^2)(2x)}{(k^2+x^2)^4}$$

$$= \frac{2k^2(k^2+x^2) - 8k^2x^2}{(k^2+x^2)^3} = \frac{2k^4 - 6k^2x^2}{(k^2+x^2)^3}$$

Since

$$\frac{d^2w}{dx^2}\bigg|_{x=k} = \frac{R}{k}f''(k) = \frac{R}{k}\frac{-4k^4}{(2k^2)^3} < 0$$

we conclude that there is a local maximum at $\hat{x} = k$. To see whether it is a global maximum, we compare $w(k)$ with $w(0)$ and $\lim_{x\to\infty} w(x)$. Since

$$w(x) = \frac{R}{x}f(x) = \frac{R}{x}\frac{x^2}{k^2+x^2} = R\frac{x}{k^2+x^2}$$

we find

$$w(0) = 0 \quad w(k) = \frac{R}{2k} \quad \lim_{x\to\infty} w(x) = 0$$

Hence, $\hat{x} = k$ is the absolute maximum; for our choice of $f(x) = \frac{x^2}{k^2+x^2}$, the optimal clutch size N_{opt} satisfies $N_{\text{opt}} = R/k$. [Other choices of $f(x)$ would give a different result.]

There is a geometric way of finding $\hat{x}$. Namely, since

$$f'(\hat{x}) = \frac{f(\hat{x})}{\hat{x}}$$

the tangent line at $(\hat{x}, f(\hat{x}))$ has slope $\frac{f(\hat{x})}{\hat{x}}$. This line can be obtained by drawing a straight line through the origin that just touches the graph of $y = f(x)$, as illustrated in Figure 5.55. ◀

5.4.1 Problems

1. Find the smallest perimeter possible for a rectangle whose area is 25 in.^{2}.

2. Show that among all rectangles with a given perimeter, the square has the largest area.

3. A rectangle has its base on the x-axis and its upper two vertices on the parabola $y = 3 - x^2$, as shown in Figure 5.56. What is the largest area the rectangle can have?

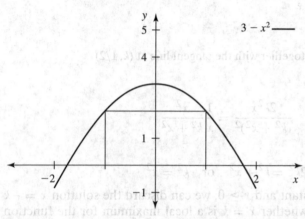

▲ **Figure 5.56**

The graph of $y = 3 - x^2$ together with the inscribed rectangle in Problem 3

4. A rectangular study area is to be enclosed by a fence and divided into two equal parts, with a fence running along the division parallel to one of the sides. If the total area is 384 ft^2, find the dimensions of the study area that will minimize the total length of the fence. How much fence will be required?

5. A rectangular field is bounded on one side by a river and on the other three sides by a fence. Find the dimensions of the field that will maximize the enclosed area if the fence has total length 320 ft.

6. Find the largest possible area of a right triangle whose hypotenuse is 4 cm long.

7. Suppose that a and b are the side lengths in a right triangle whose hypotenuse is 5 cm long. What is the smallest perimeter possible?

8. Suppose that a and b are the side lengths in a right triangle whose hypotenuse is 10 cm long. Show that the area of the triangle is largest when $a = b$.

9. A rectangle has its base on the x-axis, its lower left corner at $(0, 0)$, and its upper right corner on the curve $y = 1/x$. What is the smallest perimeter the rectangle can have?

10. A rectangle has its base on the x-axis and its upper left and right corners on the curve $y = \sqrt{4 - x^2}$, as shown in

Figure 5.57. The left and the right corners have equal distance from the vertical axis. What is the largest area the rectangle can have?

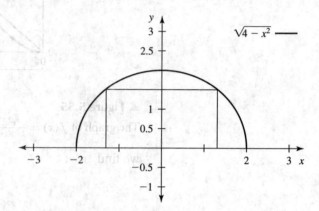

▲ **Figure 5.57**

The graph of $y = (4 - x^2)^{1/2}$ together with the inscribed rectangle in Problem 10

11. Denote by (x, y) a point on the straight line $y = 4 - 3x$ (see Figure 5.58).

(a) Show that the distance of (x, y) to the origin is given by

$$f(x) = \sqrt{x^2 + (4 - 3x)^2}$$

(b) Give the coordinates of the point on the line $y = 4 - 3x$ closest to the origin. [*Hint:* Find x so that the distance you computed in (a) is minimized.]

(c) Show that the *square* of the distance between the point (x, y) on the line and the origin is given by

$$g(x) = [f(x)]^2 = x^2 + (4 - 3x)^2$$

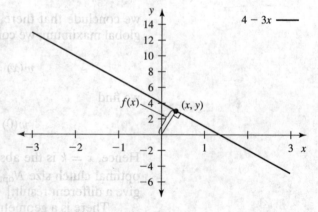

▲ **Figure 5.58**

The graph of $y = 4 - 3x$ in Problem 11

and find the minimum of $g(x)$. Show that this agrees with your answer in (b).

12. How close does the curve $y = 1/x$ come to the origin? (*Hint:* Find the point on the curve that minimizes the *square* of the distance between the origin and the point on the curve. If you use the square of the distance instead of the distance, you avoid dealing with square roots.)

13. Show that if $f(x)$ is a positive differentiable function that has a local minimum at $x = c$, then $g(x) = [f(x)]^2$ has a local minimum at $x = c$ as well.

14. Show that if $f(x)$ is a differentiable function with $f(x) < 0$ for all $x \in \mathbf{R}$ and with a local maximum at $x = c$, then $g(x) = [f(x)]^2$ has a local minimum at $x = c$.

15. Find the dimensions of a right-circular cylindrical can (bottom and top closed) that has volume 1 liter and that minimizes the amount of material used. (*Note:* One liter corresponds to 1000 cm^3.)

16. Find the dimensions of a right-circular cylinder that is open on the top and closed on the bottom, so that the can holds 1 liter and uses the least amount of material.

17. A circular sector with radius r and angle θ has area A. Find r and θ so that the perimeter is smallest when (a) $A = 2$ and (b) $A = 10$. (*Note:* $A = \frac{1}{2}r^2\theta$, and the length of the arc $s = r\theta$, when θ is measured in radians; see Figure 5.59.)

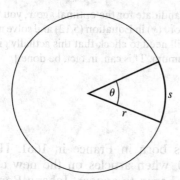

▲ Figure 5.59
The circular sector in Problems 17 and 18

18. A circular sector with radius r and angle θ has area A. Find r and θ so that the perimeter is smallest for a given area A. (*Note:* $A = \frac{1}{2}r^2\theta$, and the length of the arc $s = r\theta$, when θ is measured in radians; see Figure 5.59.)

19. Repeat Example 4 under the assumption that the top of the can is made out of aluminum that is three times as thick as the aluminum used for the wall and the bottom.

20. (*Classical model of viability selection*) We consider a population of diploid organisms (that is, each individual carries two copies of each chromosome). Genes reside on chromosomes, and we call the location of a gene on a chromosome a **locus**. Different versions of the same gene are called **alleles**. We consider the case of one locus with two possible alleles, A_1 and A_2. Since the individuals are diploid, the following types, called **genotypes**, may occur: A_1A_1, A_1A_2, and A_2A_2 (where A_1A_2 and A_2A_1 are considered to be equivalent). If two parents mate and produce an offspring,

the offspring receives one gene from each parent. If mating is random, then we can imagine all genes being put in one big gene pool from which we choose two genes at random. If we assume that the frequency of A_1 in the population is p and the frequency of A_2 is $q = 1 - p$, then the combination A_1A_1 is picked with probability p^2, the combination A_1A_2 with probability $2pq$ (the factor 2 appears since A_1 can come from either the father or the mother), and the combination A_2A_2 with probability q^2.

We assume that the survival chances of offspring depend on their genotypes. We define the quantities w_{11}, w_{12}, and w_{22} to describe the differential survival chances of the types A_1A_1, A_1A_2, and A_2A_2. The ratio of $A_1A_1 : A_1A_2 : A_2A_2$ among adults is given by

$$p^2 w_{11} : 2pq w_{12} : q^2 w_{22}$$

The average fitness of this population is defined as

$$\overline{w} = p^2 w_{11} + 2pq w_{12} + q^2 w_{22}$$

We will investigate this function. Since $q = 1 - p$, $\overline{w}$ is a function of p only:

$$\overline{w}(p) = p^2 w_{11} + 2p(1 - p)w_{12}$$
$$+ (1 - p)^2 w_{22}$$

for $0 \le p \le 1$. We consider the following three cases:

(i) Directional selection:

$$w_{11} > w_{12} > w_{22}$$

(ii) Overdominance:

$$w_{12} > w_{11}, w_{22}$$

(iii) Underdominance:

$$w_{12} < w_{11}, w_{22}$$

(a) Show that

$$\overline{w}(p) = p^2(w_{11} - 2w_{12} + w_{22})$$
$$+ 2p(w_{12} - w_{22}) + w_{22}$$

and graph $\overline{w}(p)$ for each of the three cases, where we choose the parameters as follows:

(i) $w_{11} = 1$, $w_{12} = 0.7$, $w_{22} = 0.3$
(ii) $w_{11} = 0.7$, $w_{12} = 1$, $w_{22} = 0.3$
(iii) $w_{11} = 1$, $w_{12} = 0.3$, $w_{22} = 0.7$

(b) Show that

$$\frac{d\overline{w}}{dp} = 2p(w_{11} - 2w_{12} + w_{22})$$
$$+ 2(w_{12} - w_{22})$$

(c) Find the global maximum of $\overline{w}(p)$ in each of the three cases considered in (a). (Note that the global maximum may occur at the boundary of the domain of $\overline{w}$.)

(d) We can show that under a certain mating scheme the gene frequencies change until $\overline{w}$ reaches its global maximum. Assume that this is the case, and state for each of the three cases considered in (a) what the equilibrium frequency will be.

21. (*Continuation of Problem 94 in Subsection 4.3.3*) We discussed the properties of hatching offspring per unit time, $w(t)$, in the species *Eleutherodactylus coqui*. The function $w(t)$ was given by

$$w(t) = \frac{f(t)}{C + t}$$

where $f(t)$ is the proportion of offspring that survive if t is the time spent brooding, and C is the cost associated with the time spent searching for other mates.

We assume now that $f(t)$, $t \geq 0$, is concave down with $f(0) = 0$ and $0 \leq f \leq 1$. The optimal brooding time is defined as the time that maximizes $w(t)$.

(a) Show that the optimal brooding time can be obtained by finding the point on the curve $f(t)$ where the line through $(-C, 0)$ is tangential to the curve $f(t)$.

(b) Use the procedure in (a) to find the optimal brooding time for $f(t) = \frac{t}{1+t}$ and $C = 2$. Determine the equation of the line through $(-2, 0)$ that is tangential to the curve $f(t) = \frac{t}{1+t}$, and graph both $f(t)$ and the tangent together.

22. (*Optimal age of reproduction*) (from Roff, 1992) Semelparous organisms breed only once during their lifetime. Examples of this type of reproduction can be found in Pacific salmon and bamboo. The per capita rate of increase, r, can be thought of as a measure of reproductive fitness. The greater r, the more offspring an individual produces. The intrinsic rate of increase is typically a function of age, x. Models for age-structured populations of semelparous organisms predict that the intrinsic rate of increase as a function of x is given by

$$r(x) = \frac{\ln[l(x)m(x)]}{x}$$

where $l(x)$ is the probability of surviving to age x and $m(x)$ is the number of female births at age x. The optimal age of reproduction is the age x that maximizes $r(x)$.

(a) Find the optimal age of reproduction for

$$l(x) = e^{-ax}$$

and

$$m(x) = bx^c$$

where a, b and c are positive constants.

(b) Use a graphing calculator to sketch the graph of $r(x)$ when $a = 0.1$, $b = 4$, and $c = 0.9$.

23. (*Optimal age at first reproduction*) (From Lloyd, 1987) Iteroparous organisms breed more than once during their lifetime. We consider a model where the intrinsic rate of increase, r, depends on the age of first reproduction, denoted by x, and satisfies the equation

$$\frac{e^{-x(r(x)+L)}(1 - e^{-kx})^3 c}{1 - e^{-(r(x)+L)}} = 1 \qquad (5.12)$$

where k, L, and c are positive constants describing the life history of the organism. The optimal age of first reproduction is the age x for which $r(x)$ is maximized. Since we cannot separate $r(x)$ in the preceding equation, we must use implicit differentiation to find a candidate for the optimal age of reproduction.

(a) Find an equation for $\frac{dr}{dx}$. [*Hint:* Take logarithms on both sides of (5.12) before differentiating with respect to x.]

(b) Set $\frac{dr}{dx} = 0$ and show that this gives

$$r(x) = \frac{3ke^{-kx}}{1 - e^{-kx}} - L$$

[To find the candidate for the optimal age x, you would need to substitute for $r(x)$ in Equation (5.12) and solve numerically. You would still need to check that this actually gives you the absolute maximum. This can, in fact, be done.]

5.5 L'HOSPITAL'S RULE

Guillaume François l'Hospital was born in France in 1661. He became interested in calculus around 1690, when articles on the new calculus by Leibniz and the Bernoulli brothers began to appear. Johann Bernoulli was in Paris in 1691, and l'Hospital asked Bernoulli to teach him some calculus. Bernoulli left Paris a year later but continued to provide l'Hospital with new material on calculus. Bernoulli received a monthly salary for his service and also agreed that he would not give anyone else access to the material. Once l'Hospital thought he understood the material well enough, he decided to write a book on the subject, which was published under his name and met with great success. Bernoulli was not particularly happy about this, as his contributions were hardly acknowledged in the book; l'Hospital perhaps felt that because he had paid for the course material, he had a right to publish it.

Today, l'Hospital is most famous for his treatment of the limits of fractions for which both the numerator and the denominator tend to 0 in the limit. The rule that bears his name was discovered by Johann Bernoulli but published in l'Hospital's book. The rule also works when both the numerator and the denominator tend to infinity. We have encountered such examples before; for instance,

$$\lim_{x \to 3} \frac{x^2 - 9}{x - 3} = \lim_{x \to 3} (x + 3) = 6$$

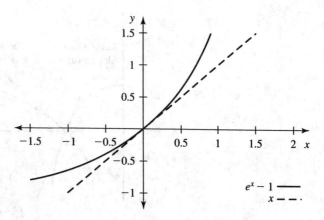

▲ **Figure 5.60**
The graphs of $y = e^x - 1$ and $y = x$

or

$$\lim_{x \to \infty} \frac{kx}{1 + x} = \lim_{x \to \infty} \frac{k}{\frac{1}{x} + 1} = k$$

In both examples we were able to find the limit by algebraic manipulations. This is not always possible, as in the following example:

$$\lim_{x \to 0} \frac{e^x - 1}{x}$$

Both numerator and denominator tend to 0 as $x \to 0$ (see Figure 5.60). There is no way of algebraically simplifying the ratio. Instead, we linearize both numerator and denominator at $a = 0$. The linearization serves as an approximation. Recall from Section 4.8 that the linear approximation of a function $f(x)$ at $x = a$ is defined as

$$L(x) = f(a) + f'(a)(x - a)$$

If $f(x) = e^x - 1$, then $f'(x) = e^x$, $f(0) = 0$, and $f'(0) = 1$. That is, the linear approximation of the numerator at $a = 0$ is

$$L(x) = 0 + (1)(x - 0) = x$$

The denominator is already a linear function, namely, $g(x) = x$. We use the linearization to approximate $\frac{e^x - 1}{x}$. This yields the ratio $\frac{x}{x} = 1$. We might then expect that the limiting value of $\frac{e^x - 1}{x}$ as $x \to 0$ is 1. This can indeed be shown.

To see clearly what we have just done, we look at the general case

$$\lim_{x \to a} \frac{f(x)}{g(x)}$$

and assume that both

$$\lim_{x \to a} f(x) = 0 \quad \text{and} \quad \lim_{x \to a} g(x) = 0$$

(see Figure 5.61). Using a linear approximation as before, we find that, for x close to a,

$$\frac{f(x)}{g(x)} \approx \frac{f(a) + f'(a)(x - a)}{g(a) + g'(a)(x - a)}$$

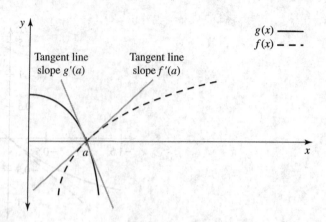

▲ **Figure 5.61**
L'Hospital's rule

Since $f(a) = g(a) = 0$ and $x \neq a$, the right-hand side is equal to

$$\frac{f'(a)(x - a)}{g'(a)(x - a)} = \frac{f'(a)}{g'(a)}$$

provided that $\frac{f'(a)}{g'(a)}$ is defined. We therefore hope that something like

$$\lim_{x \to a} \frac{f(x)}{g(x)} = \frac{f'(a)}{g'(a)}$$

holds when $\frac{f(a)}{g(a)}$ is of the form $\frac{0}{0}$ and $\frac{f'(a)}{g'(a)}$ is defined. In fact, something like this does hold. It is called l'Hospital's rule. This rule is more general than what we just did. The proof of this result would require a generalized version of the mean value theorem, which is beyond the scope of this book.

L'HOSPITAL'S RULE

Suppose that f and g are differentiable functions and that

$$\lim_{x \to a} f(x) = \lim_{x \to a} g(x) = 0$$

or

$$\lim_{x \to a} f(x) = \lim_{x \to a} g(x) = \infty$$

If

$$\lim_{x \to a} \frac{f'(x)}{g'(x)} = L$$

then

$$\lim_{x \to a} \frac{f(x)}{g(x)} = L$$

L'Hospital's rule works for $a = +\infty$ or $-\infty$ as well. It also applies to one-sided limits.

Using l'Hospital's rule, we can redo the first three examples. In each case, the ratio $\frac{f(x)}{g(x)}$ is an **indeterminate expression**; when a is substituted for x, it is either of the form $\frac{0}{0}$ or $\frac{\infty}{\infty}$. We apply l'Hospital's rule in each case:

We differentiate both numerator and denominator and then take the limit of $\frac{f'(x)}{g'(x)}$ as $x \to a$. If this limit exists, it is equal to the limit of $\frac{f(x)}{g(x)}$ as $x \to a$.

$$\lim_{x \to 3} \frac{x^2 - 9}{x - 3} = \lim_{x \to 3} \frac{2x}{1} = 6$$

$$\lim_{x \to \infty} \frac{kx}{1 + x} = \lim_{x \to \infty} \frac{k}{1} = k$$

$$\lim_{x \to 0} \frac{e^x - 1}{x} = \lim_{x \to 0} \frac{e^x}{1} = 1$$

As we just said, if the limit $\frac{f'(x)}{g'(x)}$ exists as $x \to a$, then we can conclude that the limit of $\frac{f(x)}{g(x)}$ exists as $x \to a$ and the two limits are equal. Thus, each of the first equalities in the three examples are true *because* the second equality holds. This should be kept in mind in the following examples. We look at a few more examples; in each case, you might wish to use a graphing calculator to graph the function. The first three examples are straightforward applications of l'Hospital's rule.

▶ **Example 1** **Indeterminate Expression 0/0** Evaluate

$$\lim_{x \to 2} \frac{x^6 - 64}{x^2 - 4}$$

Solution

This limit is of the form $\frac{0}{0}$, since $2^6 - 64 = 0$ and $2^2 - 4 = 0$. Applying l'Hospital's rule yields

$$\lim_{x \to 2} \frac{x^6 - 64}{x^2 - 4} = \lim_{x \to 2} \frac{6x^5}{2x} = \frac{(6)(2^5)}{(2)(2)} = (6)(2^3) = 48 \qquad ◀$$

▶ **Example 2** **Indeterminate Expression 0/0** Evaluate

$$\lim_{x \to 0} \frac{1 - \cos^2 x}{\sin x}$$

Solution

This limit is of the form $\frac{0}{0}$, since $1 - \cos^2 0 = 0$ and $\sin 0 = 0$. Applying l'Hospital's rule yields

$$\lim_{x \to 0} \frac{1 - \cos^2 x}{\sin x} = \lim_{x \to 0} \frac{2 \cos x \sin x}{\cos x} = \lim_{x \to 0} 2 \sin x = 0 \qquad ◀$$

▶ **Example 3** **Indeterminate Expression ∞/∞** Evaluate

$$\lim_{x \to \infty} \frac{\ln x}{x}$$

Solution

This limit is of the form $\frac{\infty}{\infty}$. We can apply l'Hospital's rule:

$$\lim_{x \to \infty} \frac{\frac{1}{x}}{1} = \lim_{x \to \infty} \frac{1}{x} = 0 \qquad ◀$$

▶ **Example 4**   **Indeterminate Expression** ∞/∞—**One-Sided Limit** Evaluate

$$\lim_{x \to (\pi/2)^-} \frac{\tan x}{1 + \tan x}$$

Solution

This is of the form $\frac{\infty}{\infty}$. We apply l'Hospital's rule and find

$$\lim_{x \to (\pi/2)^-} \frac{\tan x}{1 + \tan x} = \lim_{x \to (\pi/2)^-} \frac{\sec^2 x}{\sec^2 x} = \lim_{x \to (\pi/2)^-} 1 = 1 \qquad \blacktriangleleft$$

▶ **Example 5**   **Applying L'Hospital's Rule More Than Once** Evaluate

$$\lim_{x \to \infty} \frac{x^3 - 3x + 1}{3x^3 - 2x^2}$$

Solution

This limit is of the form $\frac{\infty}{\infty}$. Applying l'Hospital's rule yields

$$\lim_{x \to \infty} \frac{x^3 - 3x + 1}{3x^3 - 2x^2} = \lim_{x \to \infty} \frac{3x^2 - 3}{9x^2 - 4x}$$

which is still of the form $\frac{\infty}{\infty}$. Applying l'Hospital's rule again, we find

$$\lim_{x \to \infty} \frac{3x^2 - 3}{9x^2 - 4x} = \lim_{x \to \infty} \frac{6x}{18x - 4}$$

Since this is still of the form $\frac{\infty}{\infty}$, we can apply l'Hospital's rule yet again. We find

$$\lim_{x \to \infty} \frac{6x}{18x - 4} = \lim_{x \to \infty} \frac{6}{18} = \frac{1}{3}$$

In this example, we could have found the answer without using l'Hospital's rule, by dividing both numerator and denominator by x^3. Namely,

$$\lim_{x \to \infty} \frac{x^3 - 3x + 1}{3x^3 - 2x^2} = \lim_{x \to \infty} \frac{x^3 \left(1 - \frac{3}{x^2} + \frac{1}{x^3}\right)}{x^3 \left(3 - \frac{2}{x}\right)}$$

$$= \lim_{x \to \infty} \frac{1 - \frac{3}{x^2} + \frac{1}{x^3}}{3 - \frac{2}{x}} = \frac{1}{3} \qquad \blacktriangleleft$$

When you apply l'Hospital's rule, it can certainly happen that the limit is infinite, as in the following example.

▶ **Example 6**   **Infinite Limit** Evaluate

$$\lim_{x \to \infty} \frac{e^x}{x}$$

Solution

This is of the form $\frac{\infty}{\infty}$. We apply l'Hospital's rule.

$$\lim_{x \to \infty} \frac{e^x}{x} = \lim_{x \to \infty} \frac{e^x}{1} = \infty \qquad \text{(limit does not exist)} \qquad \blacktriangleleft$$

L'Hospital's rule can sometimes be applied to limits of the form

$$\lim_{x \to a} f(x)g(x)$$

where

$$\lim_{x \to a} f(x) = 0 \quad \text{and} \quad \lim_{x \to a} g(x) = \infty$$

since we can write the limit in the form

$$\lim_{x \to a} f(x)g(x) = \lim_{x \to a} \frac{f(x)}{\frac{1}{g(x)}} = \lim_{x \to a} \frac{g(x)}{\frac{1}{f(x)}}$$

which is again of form either $\frac{0}{0}$ or $\frac{\infty}{\infty}$.

▶ **Example 7** **Indeterminate Expression** $0 \cdot \infty$ Evaluate

$$\lim_{x \to 0^+} x \ln x$$

Solution

This is of the form $(0)(-\infty)$. We apply l'Hospital's rule after rewriting it in the form $\frac{\infty}{\infty}$.

$$\lim_{x \to 0^+} x \ln x = \lim_{x \to 0^+} \frac{\ln x}{\frac{1}{x}} = \lim_{x \to 0^+} \frac{\frac{1}{x}}{-\frac{1}{x^2}}$$

$$= \lim_{x \to 0^+} \frac{1}{x}\left(-\frac{x^2}{1}\right) = \lim_{x \to 0^+} (-x) = 0$$

We could have written the limit in the form

$$\lim_{x \to 0^+} \frac{x}{\frac{1}{\ln x}}$$

and then applied l'Hospital's rule. In this case,

$$\lim_{x \to 0^+} \frac{x}{\frac{1}{\ln x}} = \lim_{x \to 0^+} \frac{1}{(-1)(\ln x)\frac{1}{x}} = \lim_{x \to 0^+} \frac{x}{-\ln x} = 0$$

You will probably agree that the first way was easier, because it is more difficult to differentiate $1/\ln x$ than $\ln x$. Before you apply l'Hospital's rule to expressions of the form $0 \cdot \infty$, you should always determine which way will be easier to evaluate. ◀

▶ **Example 8** **Indeterminate Expression** $0 \cdot \infty$ Evaluate

$$\lim_{x \to 0^+} x \cot x$$

Solution

This is of the form $0 \cdot \infty$. We have two choices: We can evaluate

$$\lim_{x \to 0^+} x \cot x = \lim_{x \to 0^+} \frac{x}{\frac{1}{\cot x}} = \lim_{x \to 0^+} \frac{x}{\tan x}$$

where we use the fact that $\cot x = \frac{1}{\tan x}$, or

$$\lim_{x \to 0^+} x \cot x = \lim_{x \to 0^+} \frac{\cot x}{\frac{1}{x}}$$

It certainly seems easier to apply l'Hospital's rule to the first form.

$$\lim_{x \to 0^+} x \cot x = \lim_{x \to 0^+} \frac{x}{\tan x} = \lim_{x \to 0^+} \frac{1}{\sec^2 x}$$

$$= \lim_{x \to 0^+} \cos^2 x = 1$$

In the penultimate step, we used the fact that $\sec x = \frac{1}{\cos x}$. ◀

Limits of the form $\infty - \infty$ sometimes can be evaluated using l'Hospital's rule, if we can algebraically transform such limits into the form $\frac{0}{0}$ or $\frac{\infty}{\infty}$.

▶ **Example 9**   **Indeterminate Expression $\infty - \infty$** Evaluate

$$\lim_{x \to \left(\frac{\pi}{2}\right)^-} (\tan x - \sec x)$$

Solution

This limit is of the form $\infty - \infty$. Note that

$$\tan x = \frac{\sin x}{\cos x} \quad \text{and} \quad \sec x = \frac{1}{\cos x}$$

Using these two identities, we can write the limit as

$$\lim_{x \to \left(\frac{\pi}{2}\right)^-} (\tan x - \sec x) = \lim_{x \to \left(\frac{\pi}{2}\right)^-} \left(\frac{\sin x}{\cos x} - \frac{1}{\cos x}\right)$$

$$= \lim_{x \to \left(\frac{\pi}{2}\right)^-} \frac{\sin x - 1}{\cos x}$$

This is now of the form $\frac{0}{0}$, and we can apply l'Hospital's rule to find

$$\lim_{x \to \left(\frac{\pi}{2}\right)^-} \frac{\sin x - 1}{\cos x} = \lim_{x \to \left(\frac{\pi}{2}\right)^-} \frac{\cos x}{-\sin x} = \frac{0}{-1} = 0$$

◀

▶ **Example 10**   **Indeterminate Expression $\infty - \infty$** Evaluate

$$\lim_{x \to \infty} \left(x - \sqrt{x^2 + x}\right)$$

Solution

This is of the form $\infty - \infty$. We need to get a product or a ratio: We can factor x and find

$$\lim_{x \to \infty} \left(x - \sqrt{x^2 + x}\right) = \lim_{x \to \infty} x \left(1 - \sqrt{1 + \frac{1}{x}}\right)$$

This is of the form $\infty \cdot 0$. We can transform it to the form $\frac{0}{0}$ and then apply l'Hospital's rule.

$$\lim_{x \to \infty} x \left(1 - \sqrt{1 + \frac{1}{x}}\right) = \lim_{x \to \infty} \frac{1 - \sqrt{1 + \frac{1}{x}}}{\frac{1}{x}}$$

$$= \lim_{x \to \infty} \frac{-\frac{1}{2}\left(1 + \frac{1}{x}\right)^{-1/2}\left(-\frac{1}{x^2}\right)}{-\frac{1}{x^2}}$$

$$= \lim_{x \to \infty} \frac{-1}{2\sqrt{1 + \frac{1}{x}}} = -\frac{1}{2}$$

◀

Finally, we will consider expressions of the form

$$\lim_{x \to a} [f(x)]^{g(x)}$$

when they are of the types 0^0, ∞^0, or 1^∞. The key to solving such limits is to rewrite them as

$$\lim_{x \to a} [f(x)]^{g(x)} = \lim_{x \to a} \exp \left\{ \ln [f(x)]^{g(x)} \right\}$$

$$= \lim_{x \to a} \exp [g(x) \cdot \ln f(x)]$$

$$= \exp \left[\lim_{x \to a} (g(x) \cdot \ln f(x)) \right]$$

The last step, where we interchanged lim and exp, uses the fact that the exponential function is continuous. Rewriting the limit in this way transforms

$$0^0 \quad \text{into} \quad \exp[0 \cdot (-\infty)]$$

$$\infty^0 \quad \text{into} \quad \exp[0 \cdot (\infty)]$$

$$1^\infty \quad \text{into} \quad \exp[\infty \cdot \ln 1] = \exp[\infty \cdot 0]$$

Since we know how to deal with limits of the form $0 \cdot \infty$, we are in good shape again. We present a couple of examples.

▶ **Example 11**   **Indeterminate Expression 0^0** Evaluate

$$\lim_{x \to 0^+} x^x$$

Solution

This is of the form 0^0; we rewrite the limit first.

$$\lim_{x \to 0^+} x^x = \lim_{x \to 0^+} \exp[\ln x^x] = \lim_{x \to 0^+} \exp[x \ln x]$$

$$= \exp[\lim_{x \to 0^+} (x \ln x)]$$

We evaluated this limit in Example 7 and found

$$\lim_{x \to 0^+} (x \ln x) = 0$$

Hence,

$$\lim_{x \to 0^+} x^x = \exp[\lim_{x \to 0^+} (x \ln x)] = \exp[0] = 1$$

(see Figure 5.62). ◀

▶ **Example 12**   **Indeterminate Expression 1^∞** Evaluate

$$\lim_{x \to (\pi/4)^-} (\tan x)^{\tan(2x)}$$

Solution

Since $\tan \frac{\pi}{4} = 1$ and $\tan \left(2\frac{\pi}{4} \right) = \infty$, this is of the form 1^∞. We rewrite the limit as

$$\lim_{x \to (\pi/4)^-} (\tan x)^{\tan(2x)} = \lim_{x \to (\pi/4)^-} \exp[\tan(2x) \cdot \ln \tan x]$$

$$= \exp \left[\lim_{x \to (\pi/4)^-} (\tan(2x) \cdot \ln \tan x) \right]$$

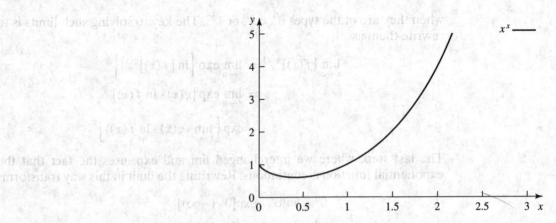

▲ Figure 5.62
The graph of $y = x^x$.

The limit is now of the form $\infty \cdot 0$ (since $\ln \tan \frac{\pi}{4} = \ln 1 = 0$). We evaluate the limit by writing it in the form $\frac{0}{0}$ and then applying l'Hospital's rule.

$$\lim_{x \to (\pi/4)^-} (\tan(2x) \cdot \ln \tan x) = \lim_{x \to (\pi/4)^-} \frac{\ln \tan x}{\frac{1}{\tan(2x)}}$$

$$= \lim_{x \to (\pi/4)^-} \frac{\ln \tan x}{\cot(2x)}$$

Since

$$\frac{d}{dx} \ln \tan x = \frac{\sec^2 x}{\tan x} = \frac{\cos x}{\cos^2 x \sin x} = \frac{1}{\sin x \cos x}$$

and

$$\frac{d}{dx} \cot(2x) = -(\csc^2(2x)) \cdot 2 = \frac{-2}{\sin^2(2x)}$$

we find

$$\lim_{x \to (\pi/4)^-} \frac{\ln \tan x}{\cot(2x)} = \lim_{x \to (\pi/4)^-} \frac{\frac{1}{\sin x \cos x}}{\frac{-2}{\sin^2(2x)}} = \lim_{x \to (\pi/4)^-} \frac{\sin^2(2x)}{-2 \sin x \cos x}$$

$$= \frac{1}{(-2) \left(\frac{1}{2}\sqrt{2}\right) \left(\frac{1}{2}\sqrt{2}\right)} = -1$$

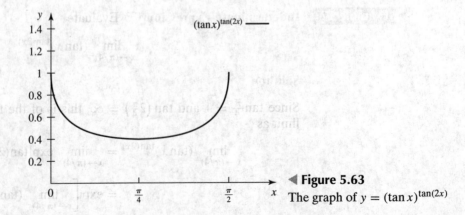

◀ Figure 5.63
The graph of $y = (\tan x)^{\tan(2x)}$

Therefore,

$$\lim_{x \to (\pi/4)^-} (\tan x)^{\tan(2x)} = \exp\left[\lim_{x \to (\pi/4)^-} (\tan(2x) \ln \tan x)\right]$$

$$= \exp[-1] = e^{-1}$$

The graph of $f(x) = (\tan x)^{\tan(2x)}$ is shown in Figure 5.63. ◀

5.5.1 Problems

Use l'Hospital's rule to find the limits in Problems 1–40.

1. $\lim\limits_{x \to 4} \dfrac{x^2 - 16}{x - 4}$

2. $\lim\limits_{x \to 1} \dfrac{x - 1}{x^2 - 1}$

3. $\lim\limits_{x \to -2} \dfrac{2x^2 + x - 6}{x + 2}$

4. $\lim\limits_{x \to -3} \dfrac{x + 3}{x^2 + 2x - 3}$

5. $\lim\limits_{x \to 0} \dfrac{\sqrt{2x + 4} - 2}{x}$

6. $\lim\limits_{x \to 0} \dfrac{3 - \sqrt{2x + 9}}{2x}$

7. $\lim\limits_{x \to 0} \dfrac{\sin x}{x \cos x}$

8. $\lim\limits_{x \to 0} \dfrac{x \sin x}{1 - \cos x}$

9. $\lim\limits_{x \to 0} \dfrac{1 - \cos x}{x \tan x}$

10. $\lim\limits_{x \to \pi/2} \dfrac{\sin\left(\frac{\pi}{2} - x\right)}{\cos x}$

11. $\lim\limits_{x \to 0^+} \dfrac{\sqrt{x}}{\ln(x + 1)}$

12. $\lim\limits_{x \to \infty} \dfrac{\ln x}{\sqrt{x}}$

13. $\lim\limits_{x \to 0} \dfrac{2^x - 1}{3^x - 1}$

14. $\lim\limits_{x \to 0} \dfrac{2^{-x} - 1}{5^x - 1}$

15. $\lim\limits_{x \to 0} \dfrac{e^x - 1 - x}{x^2}$

16. $\lim\limits_{x \to 0} \dfrac{e^x - 1 - x - \frac{x^2}{2}}{x^3}$

17. $\lim\limits_{x \to \infty} \dfrac{(\ln x)^2}{x^2}$

18. $\lim\limits_{x \to \infty} \dfrac{x^4}{e^x}$

19. $\lim\limits_{x \to (\pi/2)^-} \dfrac{\tan x}{\sec^2 x}$

20. $\lim\limits_{x \to 0} \dfrac{e^x - 1}{\sin x}$

21. $\lim\limits_{x \to \infty} xe^{-x}$

22. $\lim\limits_{x \to \infty} x^2 e^{-x}$

23. $\lim\limits_{x \to 0^+} \sqrt{x} \ln x$

24. $\lim\limits_{x \to 0^+} x^2 \ln x$

25. $\lim\limits_{x \to (\pi/2)^-} \left(\dfrac{\pi}{2} - x\right) \sec x$

26. $\lim\limits_{x \to 1^-} (1 - x) \tan\left(\dfrac{\pi}{2} x\right)$

27. $\lim\limits_{x \to \infty} \sqrt{x} \sin \dfrac{1}{x}$

28. $\lim\limits_{x \to \infty} x^2 \sin \dfrac{1}{x^2}$

29. $\lim\limits_{x \to 0^+} (\cot x - \csc x)$

30. $\lim\limits_{x \to \infty} \left(x - \sqrt{x^2 - 1}\right)$

31. $\lim\limits_{x \to 0^+} \left(\dfrac{1}{\sin x} - \dfrac{1}{x}\right)$

32. $\lim\limits_{x \to 0^+} \left(\dfrac{1}{\sin^2 x} - \dfrac{1}{x}\right)$

33. $\lim\limits_{x \to 0^+} x^{2x}$

34. $\lim\limits_{x \to 0^+} x^{\sin x}$

35. $\lim\limits_{x \to \infty} x^{1/x}$

36. $\lim\limits_{x \to \infty} (1 + e^x)^{1/x}$

37. $\lim\limits_{x \to \infty} \left(1 + \dfrac{3}{x}\right)^x$

38. $\lim\limits_{x \to \infty} \left(1 + \dfrac{3}{x^2}\right)^x$

39. $\lim\limits_{x \to \infty} \left(\dfrac{x}{1 + x}\right)^x$

40. $\lim\limits_{x \to 0^+} (\cos(2x))^{3/x}$

Find the limits in Problems 41–50. Be sure to check whether l'Hospital's rule can be applied before you evaluate the limit.

41. $\lim\limits_{x \to 0} xe^x$

42. $\lim\limits_{x \to 0^+} \dfrac{e^x}{x}$

43. $\lim\limits_{x \to (\pi/2)^-} (\tan x + \sec x)$

44. $\lim\limits_{x \to (\pi/2)^-} \dfrac{\tan x}{1 + \sec x}$

45. $\lim\limits_{x \to 1} \dfrac{x^2 + 5}{x + 1}$

46. $\lim\limits_{x\to 0} \dfrac{1-\cos x}{\sec x}$

47. $\lim\limits_{x\to -\infty} xe^x$

48. $\lim\limits_{x\to 0^+} \left(\dfrac{1}{x} - \dfrac{1}{\sqrt{x}}\right)$

49. $\lim\limits_{x\to 0^+} x^{3x}$

50. $\lim\limits_{x\to \infty} \left(\dfrac{x+1}{x+2}\right)^x$

51. Use l'Hospital's rule to find

$$\lim_{x\to 0} \frac{a^x - 1}{b^x - 1}$$

where $a, b > 0$.

52. Use l'Hospital's rule to find

$$\lim_{x\to \infty} \left(1 + \frac{c}{x}\right)^x$$

where c is a constant.

53. For $p > 0$, determine the values of p for which the following limit is either 1 or ∞ or a constant that is neither 1 nor ∞:

$$\lim_{x\to \infty} \left(1 + \frac{c}{x^p}\right)^x$$

54. Show that

$$\lim_{x\to \infty} x^p e^{-x} = 0$$

for any positive number p. Graph $f(x) = x^p e^{-x}$, $x > 0$, for $p = 1/2, 1$, and 2. Since $f(x) = x^p e^{-x} = x^p/e^x$, this shows that the exponential function grows faster than any power of x as $x \to \infty$.

55. Show that

$$\lim_{x\to \infty} \frac{\ln x}{x^p} = 0$$

for any number $p > 0$. This shows that the logarithmic function grows more slowly than any positive power of x as $x \to \infty$.

56. When l'Hospital introduced indeterminate limits in his textbook, his *first* example was

$$\lim_{x\to a} \frac{\sqrt{2a^3 x - x^4} - a\sqrt[3]{a^2 x}}{a - \sqrt[4]{ax^3}}$$

where a is a positive constant. (This example was communicated to him by Bernoulli.) Show that this limit is equal to $(16/9)a$.

57. The height in feet (y) of a tree as a function of tree age in years (x) is given by

$$y = 121 e^{-17/x} \quad \text{for } x > 0$$

(a) Determine the rate of growth when $x \to 0^+$ and the limit of the height as $x \to \infty$.

(b) Find the age at which the growth rate is maximal.

(c) Show that the height of the tree is an increasing function of age. At what age is the height increasing at an accelerating rate, and at what age at a decelerating rate?

(d) Sketch the graph of both the height and the rate of growth as functions of age.

5.6 DIFFERENCE EQUATIONS: STABILITY (OPTIONAL)

In Chapter 2, we introduced difference equations and saw that first-order difference equations can be described by recursions of the form

$$x_{t+1} = f(x_t), \quad t = 0, 1, 2, \ldots \tag{5.13}$$

where $f(x)$ is a function. There, we were only able to analyze difference equations numerically (except for exponential growth, which we were able to solve). We saw that fixed points (or equilibria) played a special role. A fixed point $x^\star$ of (5.13) satisfies the equation

$$x^\star = f(x^\star) \tag{5.14}$$

and has the property that if $x_0 = x^\star$, then $x_t = x^\star$ for $t = 1, 2, 3, \ldots$. We also saw in a number of applications that under certain conditions, x_t converged to the fixed point as $t \to \infty$ even if $x_0 \neq x^\star$. However, back then, we were not able to predict when this behavior would occur.

In this section, we will return to fixed points and use calculus to come up with a condition that allows us to predict when convergence to a fixed point occurs. We start with the simplest example, exponential growth.

5.6.1 Exponential Growth

Exponential growth in discrete time is given by the recursion

$$N_{t+1} = RN_t, \quad t = 0, 1, 2, \ldots \tag{5.15}$$

where N_t is the population size at time t and $R > 0$ is the growth parameter. We assume throughout that $N_0 \geq 0$, which implies $N_t \geq 0$.

The fixed point of (5.15) can be found by solving $N = RN$. The only solution of this is $N^\star = 0$ unless $R = 1$. If $R = 1$, then the population size never changes regardless of N_0. A consequence of being a fixed point is that if $N_0 = N^\star$, then $N_t = N^\star$ for $t = 1, 2, 3, \ldots$. That is, with $N^\star = 0$, if $N_0 = 0$, then $N_t = 0$ for $t = 1, 2, 3, \ldots$. But what happens if we start with a value that is different from 0? In Chapter 2, we found that $N_t = N_0 R^t$ is a solution of (5.15) with initial condition N_0. Using this fact, we concluded that if $N_0 > 0$ and $0 < R < 1$, then $N_t \to 0$ as $t \to \infty$; whereas if $N_0 > 0$ and $R > 1$, then $N_t \to \infty$ as $t \to \infty$. If $R = 1$, then $N_t = N_0$ for $t = 1, 2, 3, \ldots$.

We can interpret the behavior of N_t as follows: If $0 < R < 1$ and $N_0 > 0$, then N_t will return to the equilibrium $N^\star = 0$; if $R \geq 1$ and $N_0 > 0$, then N_t will not return to the equilibrium $N^\star = 0$; if $R = 1$, N_t will stay at N_0; if $R > 1$, N_t will go to infinity. We say that $N^\star = 0$ is **stable** if $0 < R < 1$, and **unstable** if $R > 1$. The case $R = 1$ is called **neutral** since no matter what the value of N_0 is, $N_t = N_0$ for $t = 1, 2, 3, \ldots$.

Cobwebbing   There is a graphical approach to determining whether a fixed point is stable or unstable. The fixed points of (5.15) are found graphically where the graphs of $N_{t+1} = RN_t$ and $N_{t+1} = N_t$ intersect, as already pointed out in Chapter 2. We see (Figure 5.64) that the two graphs only intersect where $N_t = 0$ when $R \neq 1$, confirming what we found earlier.

We can use the two graphs in Figure 5.65 to follow successive population sizes ($R > 1$ in Figure 5.65). Start at N_0 on the horizontal axis. Since $N_1 = RN_0$, we find N_1 on the vertical axis as shown by the solid horizontal and vertical line segments in Figure 5.65. Using the line $N_{t+1} = N_t$, we can locate N_1 on the horizontal axis as shown in Figure 5.65 by the dotted horizontal and vertical line segments. Using the line $N_{t+1} = RN_t$ again, we can find N_2 on the vertical axis as shown in Figure 5.65 by the broken horizontal and vertical line

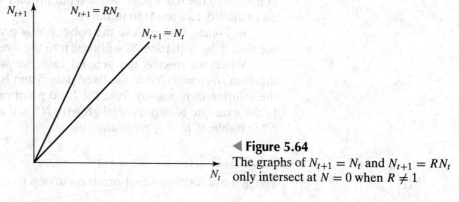

◀ **Figure 5.64**
The graphs of $N_{t+1} = N_t$ and $N_{t+1} = RN_t$ only intersect at $N = 0$ when $R \neq 1$

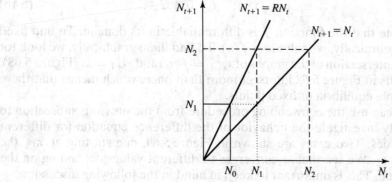

◀ **Figure 5.65**
The graphs of $N_{t+1} = N_t$ and $N_{t+1} = RN_t$ can be used to determine successive population sizes

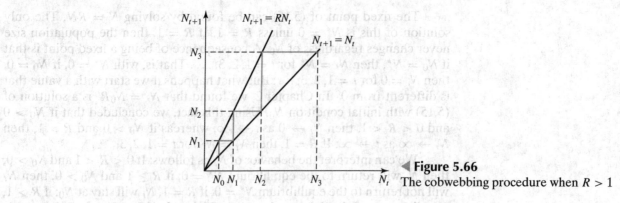

◄ Figure 5.66
The cobwebbing procedure when $R > 1$

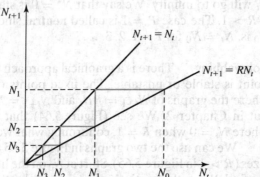

◄ Figure 5.67
The cobwebbing procedure when $0 < R < 1$

segments. Using the line $N_{t+1} = N_t$ again, we can locate N_2 on the horizontal axis and then repeat the preceding steps to find N_3 on the vertical axis, and so on (Figure 5.66). This procedure is called **cobwebbing**.

In Figures 5.64–5.66, $R > 1$, and we see that if $N_0 > 0$, then N_t will not converge to the fixed point $N^\star = 0$ but instead move away from 0 (and in fact go to infinity as t tends to infinity).

In Figure 5.67, we use the cobwebbing procedure when $0 < R < 1$. We see that if $N_0 > 0$, then N_t will return to the fixed point $N^\star = 0$.

When we discuss the general case, we will find that the slope of the function $N_{t+1} = f(N_t)$ at the fixed point [that is, $f'(N^\star)$] determines whether the solution moves away from the fixed point or converges to the fixed point. In the example of exponential growth, $N^\star = 0$ and $f'(0) = R$. For $0 < R < 1$, $N^\star$ is stable; if $R > 1$, $N^\star$ is unstable.

5.6.2 Stability: General Case

The general form of a first-order recursion is

$$x_{t+1} = f(x_t), \quad t = 0, 1, 2, \ldots \tag{5.16}$$

We assume that the function f is differentiable in its domain. To find fixed points algebraically, we solve $x = f(x)$. To find them graphically, we look for points of intersection of the graphs of $x_{t+1} = f(x_t)$ and $x_{t+1} = x_t$ (Figure 5.68). The graphs in Figure 5.68 intersect more than once, which means that there are multiple equilibria or fixed points.

We can use the cobwebbing procedure from the previous subsection to graphically investigate the behavior of the difference equation for different initial values. Two cases are shown in Figure 5.69, one starting at $x_{0,1}$ the other at $x_{0,2}$. We see that x_t converges to different values depending on the initial value. This is important to keep in mind in the following discussion.

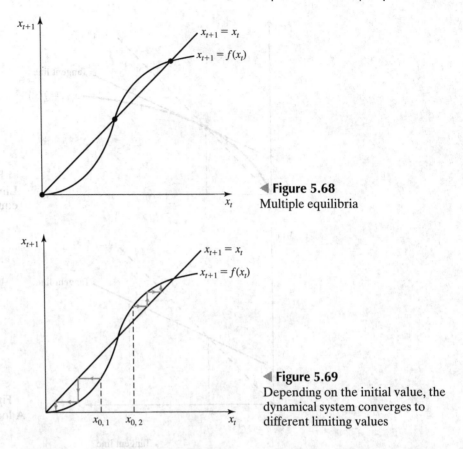

◀ **Figure 5.68**
Multiple equilibria

◀ **Figure 5.69**
Depending on the initial value, the
dynamical system converges to
different limiting values

Stability   To determine **stability** of an equilibrium—that is, whether it is
stable or unstable—we will proceed as in the previous subsection, namely, we
will start at a value that is different from the equilibrium and check whether
the solution will return to the equilibrium. There is one important difference:
We will *not* allow just any initial value that is different from the equilibrium,
but only initial values that are "close" to the equilibrium. We think of starting
at a different value as a **perturbation** of the equilibrium, and since the initial
value is close to the equilibrium, we call it a **small perturbation**. The reason
for only looking at small perturbations is that if there are multiple equilibria
and if we start too far away from the equilibrium of interest, we might end up
at a different equilibrium not because the equilibrium of interest is unstable
but simply because we are drawn to another equilibrium (as in Figure 5.69).

   If we are only concerned with small perturbations, we can approximate
the function $f(x)$ by its tangent line approximation at the equilibrium $x^\star$
(Figure 5.70). We will therefore first look at graphs where we replace $f(x)$ by
its tangent line approximation at $x^\star$.

   There are four different cases that can be divided according to whether
the slope of the tangent line at $x^\star$ is between 0 and 1 (Figure 5.71a), greater
than 1 (Figure 5.71b), between -1 and 0 (Figure 5.72a), and less than -1
(Figure 5.72b).

   We see that when the slope of the tangent line is between -1 and 1,
x_t converges to the equilibrium (Figures 5.71a and 5.72a). The difference
between Figures 5.71a and 5.72a is that in Figure 5.72a, the solution x_t
approaches the equilibrium in a spiral (thus exhibiting oscillatory behavior),
whereas in Figure 5.71a it approaches it in one direction (thus exhibiting non-
oscillatory behavior). Looking at Figure 5.71b, where the slope is greater than

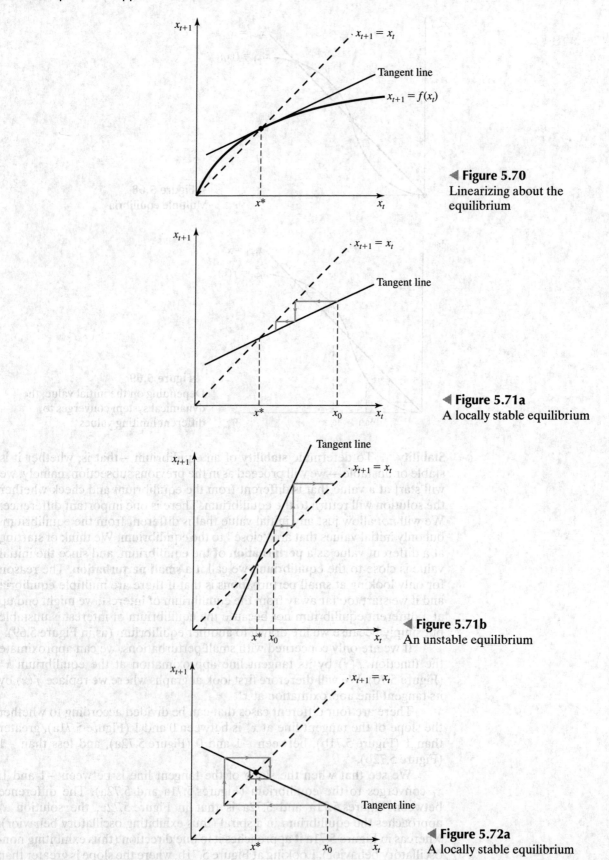

◄ Figure 5.70
Linearizing about the equilibrium

◄ Figure 5.71a
A locally stable equilibrium

◄ Figure 5.71b
An unstable equilibrium

◄ Figure 5.72a
A locally stable equilibrium

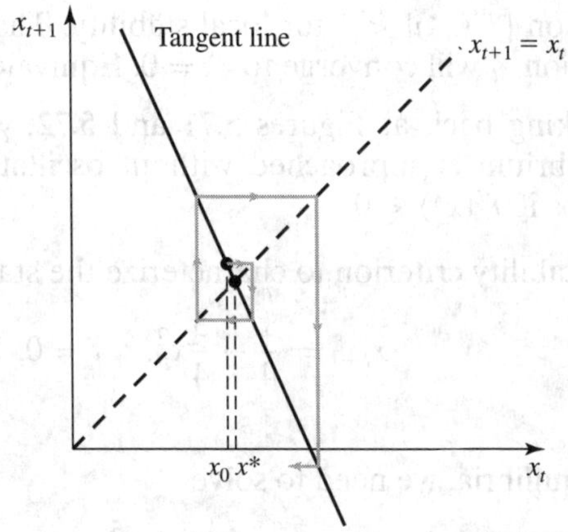

◀ **Figure 5.72b**
An unstable equilibrium

1, and Figure 5.72b, where the slope is less than -1, we see that the solution x_t does not return to the equilibrium. In Figure 5.72b, the solution moves away from the equilibrium in a spiral (thus exhibiting oscillatory behavior), whereas in Figure 5.71b, it moves away in one direction (thus exhibiting non-oscillatory behavior). We call the equilibria in Figures 5.71a and 5.71b **locally stable**, and in Figures 5.71b and 5.72b **unstable**. Note that we added the word *locally* to *stable* to emphasize that this is a local property since we only consider perturbations close to the equilibrium.

Since the slope of the tangent line approximation of $f(x)$ at $x^\star$ is given by $f'(x^\star)$, we are led to the following criterion, which we will prove using calculus.

CRITERION

> An equilibrium $x^\star$ of $x_{t+1} = f(x_t)$ is locally stable if
> $$|f'(x^\star)| < 1$$

Proof In Figures 5.71 and 5.72, we looked at the linearization of $f(x)$ about the equilibrium $x^\star$ and investigated how a small perturbation affects the future of the solution. Translating this into equations, we denote a small perturbation at time t by z_t and write

$$x_t = x^\star + z_t$$

Then

$$x_{t+1} = f(x_t) = f(x^\star + z_t)$$

Recall that the linear approximation for $f(x)$ at $x = a$ is $L(x) = f(a) + f'(a)(x - a)$. With $x = x^\star + z_t$ and $a = x^\star$, the linear approximation for $f(x^\star + z_t)$ at $x^\star$ is

$$L(x^\star + z_t) = f(x^\star) + f'(x^\star)z_t$$

We can approximate $x_{t+1} = x^\star + z_{t+1}$ by

$$x^\star + z_{t+1} \approx f(x^\star) + f'(x^\star)z_t$$

Since $f(x^\star) = x^\star$ ($x^\star$ is an equilibrium), we find

$$z_{t+1} \approx f'(x^\star)z_t \tag{5.17}$$

This last expression should remind you of exponential growth $y_{t+1} = Ry_t$, where we can identify y_t with z_t and R with $f'(x^\star)$. Since the solution of $y_{t+1} = Ry_t$ is equal to $y_t = y_0 R^t$ and $R^t \to 0$ as $t \to \infty$ for $|R| < 1$, we obtain the criterion $|f'(x^\star)| < 1$ for local stability. That is, if $|f'(x^\star)| < 1$, then the perturbation z_t will converge to $z^\star = 0$. Equivalently, $x_t \to x^\star$ as $t \to \infty$. ■

Looking back at Figures 5.71 and 5.72, we can see, in addition, that the equilibrium is approached without oscillations if $f'(x^\star) > 0$ and with oscillations if $f'(x^\star) < 0$.

▶ **Example 1** Use the stability criterion to characterize the stability of the equilibria of

$$x_{t+1} = \frac{1}{4} - \frac{5}{4}x_t^2, \quad t = 0, 1, 2, \ldots$$

Solution

To find equilibria, we need to solve

$$x = \frac{1}{4} - \frac{5}{4}x^2$$

$$\frac{5}{4}x^2 + x - \frac{1}{4} = 0$$

$$5x^2 + 4x - 1 = 0$$

The left-hand side can be factored into $(5x - 1)(x + 1)$, and we find that

$$(5x - 1)(x + 1) = 0 \quad \text{if} \quad x = \frac{1}{5} \quad \text{or} \quad x = -1$$

To determine stability, we need to evaluate the derivative of $f(x) = \frac{1}{4} - \frac{5}{4}x^2$ at the equilibria. Now,

$$f'(x) = -\frac{5}{2}x$$

and so if $x = \frac{1}{5}$, then $|f'(\frac{1}{5})| = |-\frac{1}{2}| = \frac{1}{2} < 1$ and if $x = -1$, then $|f'(-1)| = |\frac{5}{2}| = \frac{5}{2} > 1$. Thus $x = \frac{1}{5}$ is locally stable and $x = -1$ is unstable.

We can say a bit more, namely, since $f'(\frac{1}{5}) = -\frac{1}{2} < 0$, the equilibrium $x^\star = 1/5$ is approached with oscillations. ◀

▶ **Example 2** Use the stability criterion to characterize the stability of the equilibria of

$$x_{t+1} = \frac{x_t}{0.1 + x_t}, \quad t = 0, 1, 2, \ldots$$

Solution

To find equilibria, we need to solve

$$x = \frac{x}{0.1 + x}$$

This immediately yields $x = 0$ as a solution. If $x \neq 0$, then after dividing by x, we have

$$1 = \frac{1}{0.1 + x} \quad \text{or} \quad 0.1 + x = 1 \quad \text{or} \quad x = 0.9$$

With $f(x) = \frac{x}{0.1+x}$, we find

$$f'(x) = \frac{0.1 + x - x}{(0.1 + x)^2} = \frac{0.1}{(0.1 + x)^2}$$

Since $f'(0) = \frac{1}{0.1} = 10 > 1$, we conclude that $x^\star = 0$ is unstable. Since $f'(0.9) = 0.1 \in (0, 1)$, we conclude that $x^\star = 0.9$ is stable and is approached without oscillations. ◀

5.6.3 Examples

In the following examples, we will revisit three of the growth models we discussed in Chapter 2. There, we analyzed these models by simulations and you simply had to believe for which parameters a nontrivial locally stable solution existed. We are now in the position to determine stability analytically using the criterion from the previous subsection.

▶ **Example 3** **(Beverton-Holt Recruitment Curve)** Denote by N_t the size of a population at time t, $t = 0, 1, 2, \ldots$. Find all equilibria and determine their stability for the Beverton-Holt recruitment curve

$$N_{t+1} = \frac{RN_t}{1 + \frac{R-1}{K}N_t}$$

where we assume that the parameters R and K satisfy $R > 1$ and $K > 0$.

Solution

To find equilibria, we set

$$N = \frac{RN}{1 + \frac{R-1}{K}N}$$

and solve for N. This gives immediately the trivial solution $N = 0$ and, after division by $N \neq 0$,

$$1 = \frac{R}{1 + \frac{R-1}{K}N}$$

Solving this last expression for N yields the nontrivial solution

$$1 + \frac{R-1}{K}N = R \quad \text{or} \quad N = K$$

To determine stability of the two equilibria, we need to differentiate

$$f(N) = \frac{RN}{1 + \frac{R-1}{K}N}$$

We find (using the quotient rule)

$$f'(N) = \frac{R\left(1 + \frac{R-1}{K}N\right) - RN\frac{R-1}{K}}{\left(1 + \frac{R-1}{K}N\right)^2}$$

$$= \frac{R}{\left(1 + \frac{R-1}{K}N\right)^2}$$

To determine the stability of the trivial equilibrium $N^\star = 0$, we compute

$$f'(0) = R > 1$$

(since we assumed $R > 1$). Thus $N^\star = 0$ is unstable. The stability of the nontrivial equilibrium $N^\star = K$ can be determined by computing

$$f'(K) = \frac{R}{\left(1 + \frac{R-1}{K}K\right)^2} = \frac{1}{R}$$

and hence $|f'(K)| < 1$ since $R > 1$. Thus, $N^\star = K$ is locally stable when $R > 1$, as we found in Chapter 2. Since $f'(K) > 0$, the equilibrium is approached without oscillations. ◀

▶ **Example 4** (**Logistic Growth**) Denote by N_t the size of a population at time t, $t = 0, 1, 2, \ldots$. Find all equilibria and determine their stability for the discrete logistic growth

$$N_{t+1} = N_t \left[1 + R \left(1 - \frac{N_t}{K} \right) \right]$$

where we assume that the parameters R and K are both positive.

Solution

To find equilibria, we set

$$N = N \left[1 + R \left(1 - \frac{N}{K} \right) \right]$$

This yields the trivial solution $N = 0$ and the nontrivial solution $N = K$.
To determine stability, we need to differentiate

$$f(N) = N \left[1 + R \left(1 - \frac{N}{K} \right) \right]$$

We find (using the product rule)

$$f'(N) = 1 + R \left(1 - \frac{N}{K} \right) + N \left(-\frac{R}{K} \right)$$

$$= 1 + R - \frac{2NR}{K}$$

Since $f'(0) = 1 + R > 1$, we conclude that $N^\star = 0$ is unstable. Now,

$$f'(K) = 1 + R - 2R = 1 - R$$

Since $|f'(K)| = |1 - R| < 1$ if $-1 < 1 - R < 1$ or $2 > R > 0$, we conclude that $N^\star = K$ is locally stable if $0 < R < 2$, as we saw in Chapter 2. We can say a bit more now: If $0 < R < 1$, then $N^\star = K$ is approached without oscillations since $f'(K) > 0$; if $1 < R < 2$, it is approached with oscillations since $f'(K) < 0$. ◀

▶ **Example 5** (**Ricker's Curve**) Denote by N_t the size of a population at time t, $t = 0, 1, 2, \ldots$. Find all equilibria and determine their stability for the Ricker's curve

$$N_{t+1} = N_t \exp \left[R \left(1 - \frac{N_t}{K} \right) \right]$$

where we assume that the parameter R is positive.

Solution

To find equilibria, we set

$$N = N \exp \left[R \left(1 - \frac{N}{K} \right) \right]$$

This gives the trivial equilibrium $N = 0$ and

$$1 = \exp \left[R \left(1 - \frac{N}{K} \right) \right]$$

Solving this for N yields

$$R \left(1 - \frac{N}{K} \right) = 0 \qquad \text{or} \qquad N = K$$

To determine stability, we need to differentiate

$$f(N) = N \exp\left[R\left(1 - \frac{N}{K}\right)\right]$$

Using the product rule and the chain rule, we find

$$f'(N) = \exp\left[R\left(1 - \frac{N}{K}\right)\right] + N \exp\left[R\left(1 - \frac{N}{K}\right)\right]\left(-\frac{R}{K}\right)$$

$$= \exp\left[R\left(1 - \frac{N}{K}\right)\right]\left(1 - \frac{NR}{K}\right)$$

Now,

$$f'(0) = e^R > 1$$

for $R > 0$ and so $N^\star = 0$ is unstable. Since

$$f'(K) = 1 - R$$

and $|f'(K)| = |1 - R| < 1$ if $-1 < 1 - R < 1$ or $0 < R < 2$, we conclude that $N^\star = K$ is locally stable if $0 < R < 2$. We can say a bit more now: If $0 < R < 1$, then $N^\star = K$ is approached without oscillations since $f'(K) > 0$; if $1 < R < 2$, it is approached with oscillations since $f'(K) < 0$.   ◀

5.6.4 Problems

(5.6.1)

1. Assume a discrete time population whose size at generation $t + 1$ is related to the size of the population at generation t by

$$N_{t+1} = (1.03)N_t, \quad t = 0, 1, 2, \ldots$$

(a) If $N_0 = 10$, how large will the population be at generation $t = 5$?

(b) How many generations will it take for the population size to reach double the size at generation 0?

2. Suppose a discrete time population evolves according to

$$N_{t+1} = (0.9)N_t, \quad t = 0, 1, 2, \ldots$$

(a) If $N_0 = 50$, how large will the population be at generation $t = 6$?

(b) After how many generations will the size of the population be one-quarter of its original size?

(c) What will happen to the population in the long run, that is, as $t \to \infty$?

3. Assume the discrete time population model

$$N_{t+1} = bN_t, \quad t = 0, 1, 2, \ldots$$

Assume that the population increases by 2% each generation.

(a) Determine b.

(b) Find the size of the population at generation 10 when $N_0 = 20$.

(c) After how many generations will the population size have doubled?

4. Assume the discrete time population model

$$N_{t+1} = bN_t, \quad t = 0, 1, 2, \ldots$$

Assume that the population decreases by 3% each generation.

(a) Determine b.

(b) Find the size of the population at generation 10 when $N_0 = 50$.

(c) How long will it take until the population is one-half its original size?

5. Assume the discrete time population model

$$N_{t+1} = bN_t, \quad t = 0, 1, 2, \ldots$$

Assume that the population increases by x% each generation.

(a) Determine b.

(b) After how many generations will the population size have doubled? Compute the doubling time for $x = 0.1, 0.5, 1, 2, 5,$ and 10.

6. (a) Find all equilibria of

$$N_{t+1} = 1.3N_t, \quad t = 0, 1, 2, \ldots$$

(b) Use cobwebbing to determine the stability of the equilibria you found in (a).

7. (a) Find all equilibria of

$$N_{t+1} = 0.9N_t, \quad t = 0, 1, 2, \ldots$$

(b) Use cobwebbing to determine the stability of the equilibria you found in (a).

8. (a) Find all equilibria of

$$N_{t+1} = N_t, \quad t = 0, 1, 2, \ldots$$

(b) How will the population size N_t change over time starting at time 0 with N_0?

(5.6.2)

9. Use the stability criterion to characterize the stability of the equilibria of

$$x_{t+1} = \frac{2}{3} - \frac{2}{3}x_t^2, \quad t = 0, 1, 2, \ldots$$

10. Use the stability criterion to characterize the stability of the equilibria of

$$x_{t+1} = \frac{3}{5}x_t^2 - \frac{2}{5}, \quad t = 0, 1, 2, \ldots$$

11. Use the stability criterion to characterize the stability of the equilibria of

$$x_{t+1} = \frac{x_t}{0.5 + x_t}, \quad t = 0, 1, 2, \ldots$$

12. Use the stability criterion to characterize the stability of the equilibria of

$$x_{t+1} = \frac{x_t}{0.3 + x_t}, \quad t = 0, 1, 2, \ldots$$

13. (a) Use the stability criterion to characterize the stability of the equilibria of

$$x_{t+1} = \frac{5x_t^2}{4 + x_t^2}, \quad t = 0, 1, 2, \ldots$$

(b) Use cobwebbing to decide to which value x_t converges as $t \to \infty$ if (i) $x_0 = 0.5$ and (ii) $x_0 = 2$.

14. (a) Use the stability criterion to characterize the stability of the equilibria of

$$x_{t+1} = \frac{10x_t^2}{9 + x_t^2}, \quad t = 0, 1, 2, \ldots$$

(b) Use cobwebbing to decide to which value x_t converges as $t \to \infty$ if (i) $x_0 = 0.5$ and (ii) $x_0 = 3$.

(5.6.3)

15. The Ricker curve is given by

$$R(P) = \alpha P e^{-\beta P}$$

for $P \geq 0$, where P denotes the size of the parental stock and $R(P)$ the number of recruits. The parameters α and β are positive constants.

(a) Show that $R(0) = 0$ and $R(P) > 0$ for $P > 0$.

(b) Find

$$\lim_{P \to \infty} R(P)$$

(c) For what size of the parental stock is the number of recruits maximal?

(d) Does $R(P)$ have inflection points? If so, find them.

(e) Sketch the graph of $f(x)$ when $\alpha = 2$ and $\beta = 1/2$.

16. Suppose that the size of a fish population at generation t is given by

$$N_{t+1} = 1.5N_t e^{-0.001N_t}$$

for $t = 0, 1, 2, \ldots$.

(a) Assume that $N_0 = 100$. Find the size of the fish population at generation t for $t = 1, 2, \ldots, 20$.

(b) Assume that $N_0 = 800$. Find the size of the fish population at generation t for $t = 1, 2, \ldots, 20$.

(c) Determine all fixed points. Based on your computations in (a) and (b), can you make a guess as to what will happen to the population in the long run when starting from $N_0 = 100$ and $N_0 = 800$, respectively?

(d) Use the cobwebbing method to illustrate your answer in (a).

(e) Explain why the dynamical system converges to the nontrivial fixed point.

17. Suppose that the size of a fish population at generation t is given by

$$N_{t+1} = 10N_t e^{-0.01N_t}$$

for $t = 0, 1, 2, \ldots$.

(a) Assume that $N_0 = 100$. Find the size of the fish population at generation t for $t = 1, 2, \ldots, 20$.

(b) Show that if $N_0 = 100 \ln 10$, then $N_t = 100 \ln 10$ for $t = 1, 2, 3, \ldots$; that is, $N^* = 100 \ln 10$ is a nontrivial fixed point or equilibrium. How would you find N^*? Are there any other equilibria?

(c) Based on your computations in (a), can you make a prediction about the long-term behavior of the fish population when $N_0 = 100$? How does this compare to your answer in (b)?

(d) Use the cobwebbing method to illustrate your answer in (c).

In Problems 18–20, consider the following discrete time dynamical system, which is called the discrete logistic model and which models the size of a population over time:

$$N_{t+1} = rN_t \left(1 - \frac{N_t}{100}\right)$$

for $t = 0, 1, 2, \ldots$.

18. (a) Find all equilibria when $r = 0.5$.

(b) Investigate the system when $N_0 = 10$ and describe what you see.

19. (a) Find all equilibria when $r = 2$.

(b) Investigate the system when $N_0 = 10$ and describe what you see.

20. (a) Find all equilibria when $r = 3.1$.

(b) Investigate the system when $N_0 = 10$ and describe what you see.

5.7 NUMERICAL METHODS: THE NEWTON-RAPHSON METHOD (OPTIONAL)

Numerical methods are very important in the sciences, as we frequently encounter situations where exact solutions are impossible. In Section 3.5, we encountered one method, the bisection method, to solve equations of the form $f(x) = 0$. Here is another method that is often much more efficient than the bisection method.

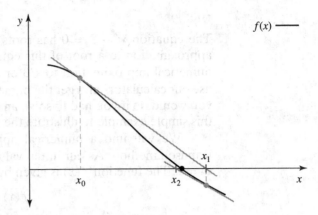

▲ **Figure 5.73**
The graph of $y = f(x)$ and the first two iterations in the Newton-Raphson method

The Newton-Raphson method allows us to find solutions to equations of the form

$$f(x) = 0$$

The idea behind it can be best explained graphically (see Figure 5.73).

Suppose that we wish to find the roots of $f(x) = 0$. We begin by choosing a value x_0 as our initial guess; then we replace the graph of $f(x)$ by its tangent line $y = L(x)$ at our initial guess x_0. Since the slope of the tangent line $y = L(x)$ is equal to $f'(x_0)$, and $(x_0, f(x_0))$ is a point on the tangent line, we can use the point-slope form to find the equation of the tangent line. It is given by

$$y - f(x_0) = f'(x_0)(x - x_0)$$

This line will intersect the x-axis at some point $x = x_1$, provided $f'(x_0) \neq 0$. To find x_1, we set $x = x_1$ and $y = 0$ and solve for x_1:

$$0 - f(x_0) = f'(x_0)(x_1 - x_0) \quad \text{or} \quad x_1 = x_0 - \frac{f(x_0)}{f'(x_0)}$$

We use x_1 as our next guess and repeat the procedure, which will result in x_2, and so on. The value of x_{n+1} is then given by

$$x_{n+1} = x_n - \frac{f(x_n)}{f'(x_n)} \quad \text{for } n = 0, 1, 2, \ldots \tag{5.18}$$

This produces a sequence of numbers $x_1, x_2, \ldots$. If these values converge to the root of $f(x) = 0$, denoted by r, as $n \to \infty$,

$$\lim_{n \to \infty} x_n = r$$

then computing $x_1, x_2, \ldots$ provides a numerical way to approximate r. The method will not always converge, but before we discuss situations in which the method fails, we give two examples where the method works.

▶ **Example 1** Use the Newton-Raphson method to find a numerical approximation to a solution of the equation

$$x^2 - 3 = 0$$

Solution

The equation $x^2 - 3 = 0$ has roots $r = \sqrt{3}$ and $-\sqrt{3}$. Finding a numerical approximation to a root of this equation is therefore the same as finding a numerical approximation to $\sqrt{3}$ or $-\sqrt{3}$. You might ask why we don't just use our calculators and get the numerical value of $\sqrt{3}$, and in fact this is what you would do if you had to solve an equation like $x^2 - 3 = 0$, but we will use this simple example to illustrate the method.

We will find a numerical approximation to $\sqrt{3}$ using the Newton-Raphson method. As our initial value, we choose a number close to $\sqrt{3}$, say, $x_0 = 2$. The function $f(x)$ is given by

$$f(x) = x^2 - 3$$

and its derivative is given by

$$f'(x) = 2x$$

Using (5.18), we find

$$x_{n+1} = x_n - \frac{x_n^2 - 3}{2x_n}$$

$$= x_n - \frac{x_n}{2} + \frac{3}{2x_n} = \frac{x_n}{2} + \frac{3}{2x_n} \quad \text{for } n = 0, 1, 2, \ldots$$

(See Figure 5.74 for the first step in the approximation.)

The following table shows the results of the procedure. We can use a calculator to compare our numerical approximation to an accurate numerical approximation of $\sqrt{3}$, namely, a calculator yields the approximation 1.732050080757 for $\sqrt{3}$.

n	x_n	$x_{n+1} = \frac{x_n}{2} + \frac{3}{2x_n}$	$\left\lvert \sqrt{3} - x_{n+1} \right\rvert$
0	2	1.75	0.0179
1	1.75	1.7321429	9.2×10^{-5}
2	1.7321429	1.7320508	2.45×10^{-9}

With the starting value $x_0 = 2$, after three steps we obtain the approximation 1.732050, which is correct to six decimal places. As you can see, this

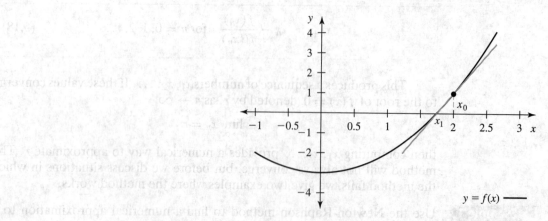

▲ **Figure 5.74**
The graph of $f(x) = x^2 - 3$ and the first step in the Newton-Raphson method

method can converge very quickly. In fact, it is used in many calculators to calculate roots.   ◄

In the next example, we cannot simply solve for x by algebraic manipulations. To find a solution of the given equation, we must resort to a numerical method.

▶ **Example 2**   Solve the equation

$$e^x + 1 + x = 0$$

Solution

It turns out that this can only be done numerically. We will use the Newton-Raphson method to determine the root of $e^x + 1 + x = 0$. Since $e^x + 1 + x$ is positive for $x = 0$ and negative for $x = -2$, we conclude that there is a root of $e^x + 1 + x = 0$ in the interval $(-2, 0)$. We therefore choose a starting value in this interval, say $x_0 = -1$. We find

$$f(x) = e^x + x + 1$$

$$f'(x) = e^x + 1$$

Therefore,

$$x_{n+1} = x_n - \frac{f(x_n)}{f'(x_n)}$$

$$= x_n - \frac{e^{x_n} + x_n + 1}{e^{x_n} + 1}$$

$$= x_n - 1 - \frac{x_n}{e^{x_n} + 1} \quad \text{for } n = 0, 1, 2, \ldots$$

The first step in the approximation is shown in Figure 5.75.

We summarize the first few steps of the iteration in the following table, where $x_0 = -1$.

n	x_n	$x_{n+1} = x_n - 1 - \frac{x_n}{e^{x_n} + 1}$
0	-1	-1.26894142
1	-1.26894142	-1.27845462
2	-1.27845462	-1.27846454
3	-1.27846454	-1.27846454

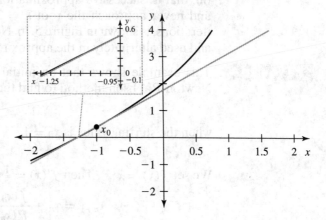

▲ **Figure 5.75**
The graph of $f(x) = e^x + x + 1$ and the first step in the approximations of the Newton-Raphson method

Note that x_2 and x_3 agree to four decimal places and that x_3 and x_4 agree to eight decimal places. We therefore suspect that $x = -1.278464$ is a root of $e^x + x + 1 = 0$, accurate to six decimal places. This can be confirmed using a graphing calculator, for instance. ◀

It is not our goal here to give a complete description of the Newton-Raphson method (when it works, how quickly it converges, etc.). Instead, we give a few examples that illustrate some of the problems one can encounter when using this method.

▶ **Example 3**   This is an example of a situation where the method does not work at all. Let

$$f(x) = \begin{cases} \sqrt{x - 2.5} & \text{for } x \geq 2.5 \\ -\sqrt{2.5 - x} & \text{for } x \leq 2.5 \end{cases}$$

We wish to solve the equation $f(x) = 0$. We see immediately that $x = 2.5$ is a solution. Suppose we wish to use the Newton-Raphson method. Given that

$$f'(x) = \begin{cases} \dfrac{1}{2\sqrt{x - 2.5}} & \text{for } x > 2.5 \\ \dfrac{1}{2\sqrt{2.5 - x}} & \text{for } x < 2.5 \end{cases}$$

we find

$$x_{n+1} = x_n - \frac{f(x_n)}{f'(x_n)}$$

$$= \begin{cases} x_n - \dfrac{\sqrt{x_n - 2.5}}{\frac{1}{2\sqrt{x_n - 2.5}}} & \text{for } x_n > 2.5 \\ x_n - \dfrac{-\sqrt{2.5 - x_n}}{\frac{1}{2\sqrt{2.5 - x_n}}} & \text{for } x_n < 2.5 \end{cases}$$

$$= \begin{cases} x_n - 2(x_n - 2.5) = -x_n + 5 & \text{for } x_n > 2.5 \\ x_n + 2(2.5 - x_n) = -x_n + 5 & \text{for } x_n < 2.5 \end{cases}$$

$$= -x_n + 5 \quad \text{for } x_n \neq 2.5$$

If $x_0 = 2.5 + h$, then $x_1 = 2.5 - h$, $x_2 = 2.5 + h$, $x_3 = 2.5 - h$, and so on; that is, successive approximations oscillate between $2.5 + h$ and $2.5 - h$, and never approach the root $r = 2.5$. A graph of $f(x)$, together with the iterations, is shown in Figure 5.76. Note that the two tangent lines are parallel and used alternately in the approximation. ◀

▶ **Example 4**   This example shows that an initial approximation can get worse. Use the Newton-Raphson method to find the root of

$$x^{1/3} = 0$$

when the starting value is $x_0 = 1$.

Solution

We set $f(x) = x^{1/3}$. Then $f'(x) = \frac{1}{3}x^{-2/3}$, and

$$x_{n+1} = x_n - \frac{f(x_n)}{f'(x_n)}$$

$$= x_n - \frac{x_n^{1/3}}{\frac{1}{3}x_n^{-2/3}} = x_n - 3x_n = -2x_n$$

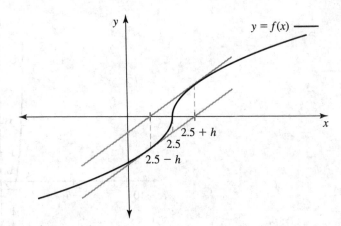

▲ Figure 5.76
The graph of $f(x)$ in Example 3 where the Newton-Raphson method does not converge

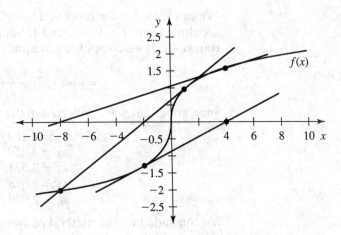

▲ Figure 5.77
The graph of $f(x)$ in Example 4 where the Newton-Raphson method does not converge

n	x_n	$x_{n+1} = -2x_n$
0	1	-2
1	-2	4
2	4	-8
3	-8	16

The following table shows successive values of x_n.

This does not converge to the root $r = 0$. This is graphically illustrated in Figure 5.77.

Graphing calculators have an option in the graphing menu that allows you to find roots of equations $f(x) = 0$ if you graph $f(x)$ and put the cursor close to the root. For this example, the calculator cannot produce an answer.

◀

▶ **Example 5** This example shows that one does not necessarily converge to the closest root. We wish to find a root of the equation

$$x^4 - x^2 = 0$$

If we set $f(x) = x^4 - x^2$, then $f'(x) = 4x^3 - 2x$ and

$$x_{n+1} = x_n - \frac{f(x_n)}{f'(x_n)}$$

$$= x_n - \frac{x_n^4 - x_n^2}{4x_n^3 - 2x_n} = x_n - \frac{x_n^3 - x_n}{4x_n^2 - 2}$$

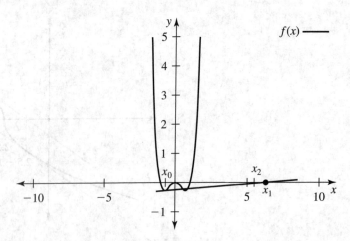

▲ Figure 5.78
The graph of $f(x)$ in Example 5 together with the first two approximations

We can solve for the roots of $x^4 - x^2 = 0$. Since $x^4 - x^2 = x^2(x^2 - 1)$, there are three roots: $r = 0, -1$, and 1. Suppose that we set $x_0 = -0.7$. The closest root is -1. If we compute x_1, we find

$$x_1 = (-0.7) - \frac{(-0.7)^3 - (-0.7)}{4(-0.7)^2 - 2} = 8.225$$

Successive values are collected in the following list.

$$\begin{aligned} x_2 &= 6.184 & x_7 &= 1.613 \\ x_3 &= 4.659 & x_8 &= 1.306 \\ x_4 &= 3.521 & x_9 &= 1.115 \\ x_5 &= 2.678 & x_{10} &= 1.024 \\ x_6 &= 2.059 & x_{11} &= 1.001 \end{aligned}$$

We conclude that the method converges to the root $r = 1$. This is illustrated in Figure 5.78. ◀

5.7.1 Problems

1. Use the Newton-Raphson method to find a numerical approximation to the solution of
$$x^2 - 7 = 0$$
that is correct to six decimal places.

2. Use the Newton-Raphson method to find a numerical approximation to the solution of
$$e^{-x} = x$$
that is correct to six decimal places.

3. Use the Newton-Raphson method to find a numerical approximation to the solution of
$$x^2 + \ln x = 0$$
that is correct to six decimal places.

4. The equation
$$x^2 - 5 = 0$$
has two solutions. Use the Newton-Raphson method to approximate the two solutions.

5. Use the Newton-Raphson method to solve the equation
$$\sin x = \frac{1}{2}x$$
in the interval $(0, \pi)$.

6. Let
$$f(x) = \begin{cases} \sqrt{x-1} & \text{for } x \geq 1 \\ -\sqrt{1-x} & \text{for } x \leq 1 \end{cases}$$

(a) Show that if you use the Newton-Raphson method to solve $f(x) = 0$, the following holds: If $x_0 = 1 + h$, then $x_1 = 1 - h$, and if $x_0 = 1 - h$, then $x_1 = 1 + h$.

(b) Does the Newton-Raphson method converge? Use a graph to explain what happens.

7. In Example 4, we discussed the case of finding the root of $x^{1/3} = 0$.

(a) Given x_0, find a formula for $|x_n|$.

(b) Find
$$\lim_{n \to \infty} |x_n|$$

(c) Graph $f(x) = x^{1/3}$ and illustrate what happens when you apply the Newton-Raphson method.

8. In Example 5, we considered the equation

$$x^4 - x^2 = 0$$

(a) What happens if you choose

$$x_0 = -\frac{1}{2}\sqrt{2}$$

in the Newton-Raphson method? Give a graphical illustration.

(b) Repeat the procedure in (a) for $x_0 = -0.71$, and compare with the result we obtained in Example 5 when $x_0 = -0.70$. Give a graphical illustration, and explain it in words. What happens when $x_0 = -0.6$? (This is an example where small changes in the initial value can drastically change the outcome.)

9. Use the Newton-Raphson method to find a numerical approximation to the solution of

$$x^2 - 16 = 0$$

when your initial guess is (a) $x_0 = 3$ and (b) $x_0 = 4$.

10. Suppose that you wish to solve

$$f(x) = 0$$

numerically using the Newton-Raphson method. It just so happens that your initial guess x_0 satisfies $f(x_0) = 0$. What happens to subsequent iterations? Give a graphical illustration of your results. [Assume that $f'(x_0) \neq 0$.]

5.8 ANTIDERIVATIVES

Throughout this and the previous chapter, we have repeatedly encountered differential equations. Occasionally, we showed that a certain function would solve a given differential equation. In this section, we will discuss a particular type of differential equation and address two important general questions: First, given a differential equation, how can we find solutions? Second, given a solution, how do we know if it is the only one?

We will consider differential equations of the form

$$\frac{dy}{dx} = f(x)$$

that is, where the rate of change of y with respect to x depends *only* on x. Our goal is to find functions y that satisfy $y' = f(x)$. We will see that if we can find one function with $y' = f(x)$, then there is a whole family of functions with this property, all related by vertical translations. If we want to pick out one of these functions, we need to specify an **initial condition**. This is a point (x_0, y_0) on the graph of the function. Such a function is called a **solution** of the **initial value problem**

$$\frac{dy}{dx} = f(x) \quad \text{with } y = y_0 \text{ when } x = x_0 \tag{5.19}$$

Let's look at an example before we begin a systematic treatment of the solution of differential equations of the form (5.19). Consider a population whose size at time t is denoted by $N(t)$, and assume that the growth rate is given by

$$\frac{dN}{dt} = \frac{1}{2\sqrt{t}} \quad \text{for } t > 0 \tag{5.20}$$

and that the size of the population at time 0 is $N(0) = 20$ [that is, the initial condition is $N(0) = 20$]. Then

$$N(t) = \sqrt{t} + 20 \quad \text{for } t \geq 0 \tag{5.21}$$

is a solution of the differential equation (5.20) that satisfies the initial condition $N(0) = 20$. This is easy to check: First, note that $N(0) = \sqrt{0} + 20 = 20$. Second, differentiating $N(t)$, we find

$$\frac{dN}{dt} = \frac{d}{dt}(\sqrt{t} + 20) = \frac{1}{2\sqrt{t}} \quad \text{for } t > 0$$

That is, $N(t) = \sqrt{t} + 20$ satisfies (5.20) with $N(0) = 20$.

This example shows that if we have a solution of a differential equation, we can check that the solution indeed satisfies the differential equation by differentiating the solution. This suggests that, in order to find solutions, we reverse the process of differentiation. This leads us to what is called an antiderivative, which is defined as follows.

> **Definition** A function F is called an **antiderivative** of f on an interval I if $F'(x) = f(x)$ for all $x \in I$.

How can we find antiderivatives? Let

$$f(x) = 3x^2 \quad \text{for } x \in \mathbf{R}$$

To find the antiderivative of $f(x) = 3x^2$, we need to find a function whose derivative is $3x^2$. We can guess an answer, namely

$$F(x) = x^3 \quad \text{for } x \in \mathbf{R}$$

satisfies $F'(x) = 3x^2$. But this is not the only answer. Take $F(x) = x^3 + 4$. Then $F'(x) = 3x^2$; hence, $x^3 + 4$ is also an antiderivative of $3x^2$. In fact, $F(x) = x^3 + C$, $x \in \mathbf{R}$, where C is any constant, is an antiderivative of $3x^2$. [We will soon show that there are no other functions $F(x)$ such that $F'(x) = 3x^2$.] The function $f(x)$ and some of its antiderivatives are shown in Figure 5.79. All antiderivatives are related through vertical shifts since they all have the same derivative.

Though we will learn rules that will allow us to compute antiderivatives, this process is typically much more difficult than finding derivatives, and sometimes it takes ingenuity to come up with the correct answer; in addition, there are even cases where it is impossible to find an expression for an antiderivative.

We begin with two corollaries of the mean value theorem that will help us in finding antiderivatives. The first of these is Corollary 2 from Section 5.1, which we recall here.

> **Corollary 2** If f is continuous on the closed interval $[a, b]$ and differentiable on the open interval (a, b), with $f'(x) = 0$ for all $x \in (a, b)$, then f is constant on $[a, b]$.

Note that Corollary 2 is the converse of the rule that states that $f'(x) = 0$ when $f(x) = c$, where c is a constant. It tells us that *all* antiderivatives of a function that is identically 0 are constant functions.

The next corollary tells us that functions with identical derivatives differ only by a constant; that is, to find all antiderivatives of a given function, we need only find one.

> **Corollary 3** If $F(x)$ and $G(x)$ are antiderivatives of the continuous function $f(x)$ on an interval I, then there exists a constant C so that
>
> $$G(x) = F(x) + C \quad \text{for all } x \in I$$

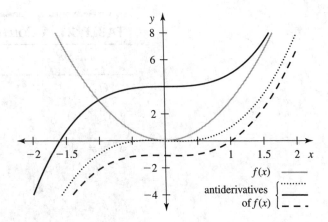

▲ **Figure 5.79**
The function $f(x) = 3x^2$ and some of its antiderivatives

Proof Since $F(x)$ and $G(x)$ are both antiderivatives of $f(x)$, it follows that $F'(x) = f(x)$ and $G'(x) = f(x)$; that is, $F'(x) = G'(x)$ or $F'(x) - G'(x) = 0$. Since

$$0 = F'(x) - G'(x) = \frac{d}{dx}[F(x) - G(x)]$$

it follows from Corollary 2 applied to the function $F - G$ that $F(x) - G(x) = C$, where C is a constant. ■

▶ **Example 1** Find general antiderivatives for the given functions. Assume that all functions are defined for $x \in \mathbf{R}$.

(a) $f(x) = 3x^2$ (b) $f(x) = \cos x$ (c) $f(x) = e^x$

Solution

(a) If $F(x) = x^3$, then $F'(x) = 3x^2$, that is, $F(x) = x^3$ is a particular antiderivative of $3x^2$. Using Corollary 3, we find the general antiderivative simply by adding a constant; that is, the general antiderivative of $f(x) = 3x^2$ is the function $G(x) = x^3 + C$, where C is a constant.

(b) If $F(x) = \sin x$, then $F'(x) = \cos x$. Hence, the general antiderivative of $f(x) = \cos x$ is the function $G(x) = \sin x + C$, where C is a constant.

(c) If $F(x) = e^x$, then $F'(x) = e^x$ and the general antiderivative of $f(x) = e^x$ is the function $G(x) = e^x + C$, where C is a constant. ◀

▶ **Example 2** Find general antiderivatives for the given functions. (Assume the largest possible domain.)

(a) $f(x) = 3x^5$ (b) $f(x) = x^2 + 2x - 1$
(c) $f(x) = e^{2x}$ (d) $f(x) = \sec^2(3x)$

Solution

(a) Since $\frac{d}{dx}(\frac{1}{2}x^6) = 3x^5$, $F(x) = \frac{1}{2}x^6 + C$ is the general antiderivative of $f(x) = 3x^5$.

(b) Since $\frac{d}{dx}(\frac{1}{3}x^3 + x^2 - x) = x^2 + 2x - 1$, $F(x) = \frac{1}{3}x^3 + x^2 - x + C$ is the general antiderivative of $f(x) = x^2 + 2x - 1$.

TABLE 5.1 A COLLECTION OF ANTIDERIVATIVES

Function	Particular Antiderivative
$kf(x)$	$kF(x)$
$f(x) + g(x)$	$F(x) + G(x)$
$x^n, n \neq -1$	$\dfrac{1}{n+1}x^{n+1}$
$\dfrac{1}{x}$	$\ln|x|$
e^{ax}	$\dfrac{1}{a}e^{ax}$
$\sin(ax)$	$-\dfrac{1}{a}\cos(ax)$
$\cos(ax)$	$\dfrac{1}{a}\sin(ax)$
$\sec^2(ax)$	$\dfrac{1}{a}\tan(ax)$

(c) Since $\frac{d}{dx}(\frac{1}{2}e^{2x}) = \frac{1}{2}e^{2x}(2) = e^{2x}$, $F(x) = \frac{1}{2}e^{2x} + C$ is the general antiderivative of $f(x) = e^{2x}$.

(d) Since $\frac{d}{dx}(\frac{1}{3}\tan(3x)) = \frac{1}{3}(\sec^2(3x))(3) = \sec^2(3x)$, $F(x) = \frac{1}{3}\tan(3x) + C$ is the general antiderivative of $f(x) = \sec^2(3x)$. ◀

Table 5.1 summarizes some of the rules for finding antiderivatives. We denote functions by $f(x)$ or $g(x)$, and their particular antiderivatives by $F(x)$ or $G(x)$. The general antiderivative is then obtained simply by adding a constant. The quantities a and k denote nonzero constants.

We can now return to our initial question, namely, how to solve differential equations of the form (5.19).

▶ **Example 3** Find the general solution of

$$\frac{dy}{dx} = \frac{3}{x^2} - 2x^2, \quad x \neq 0$$

Solution

Finding the general solution of this differential equation means finding the antiderivative of the function $f(x) = \frac{3}{x^2} - 2x^2$. Using Table 5.1, we obtain as a particular antiderivative

$$\frac{3}{-1}x^{-1} - \frac{2}{3}x^3 = -\frac{3}{x} - \frac{2}{3}x^3$$

That is, the general solution is

$$y = -\frac{3}{x} - \frac{2}{3}x^3 + C, \quad x \neq 0$$ ◀

In Example 3, we found the general solution of the given differential equation. Often, we wish to select a particular solution; for instance, we may

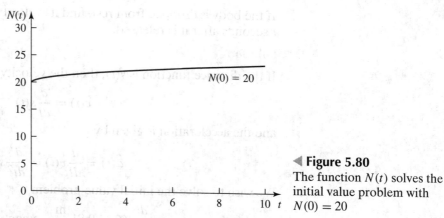

◀ **Figure 5.80**
The function $N(t)$ solves the
initial value problem with
$N(0) = 20$

know that the solution has to pass through a particular point (x_0, y_0). Such a
problem is called an initial value problem, as explained in the beginning of
this section. We consider the example again now.

▶ **Example 4**   Solve the initial value problem

$$\frac{dN}{dt} = \frac{1}{2\sqrt{t}} \quad \text{for } t > 0 \quad \text{with } N(0) = 20$$

Solution

The general antiderivative of $f(t) = \frac{1}{2\sqrt{t}}$ is $F(t) = \sqrt{t} + C$. Since $N(0) = 20$,
we find

$$N(0) = \sqrt{0} + C = 20 \quad \text{or} \quad C = 20$$

That is, the function

$$N(t) = \sqrt{t} + 20, \quad t \geq 0$$

solves the initial value problem, and it is the only solution (see Figure 5.80).

◀

▶ **Example 5**   Solve the initial value problem

$$\frac{dy}{dx} = -2x^2 + 3 \quad \text{for } x \in \mathbf{R} \text{ and } y = 10 \text{ when } x = 3$$

Solution

The general antiderivative of $f(x) = -2x^2 + 3$ is $F(x) = -\frac{2}{3}x^3 + 3x + C$.
Since

$$F(3) = -\frac{2}{3}3^3 + (3)(3) + C = -9 + C = 10$$

it follows that $C = 19$. That is,

$$y = -\frac{2}{3}x^3 + 3x + 19, \quad x \in \mathbf{R}$$

solves the initial value problem and it is the only solution.   ◀

▶ **Example 6**   An object that falls freely in a vacuum, close to the surface of the earth,
experiences a constant acceleration of

$$g = 9.81 \, \frac{m}{s^2}$$

If the body is dropped from rest, find its velocity and the distance it travels in t seconds after it is released.

Solution

If the distance function is $s(t)$, then the velocity $v(t)$ is given by

$$v(t) = \frac{d}{dt} s(t)$$

and the acceleration is given by

$$a(t) = \frac{d}{dt} v(t) = \frac{d^2}{dt^2} s(t)$$

We wish to solve the initial value problem

$$\frac{d}{dt} v(t) = 9.81 \, \frac{m}{s^2} \quad \text{when } v(0) = 0$$

A general solution is

$$v(t) = \left(9.81 \, \frac{m}{s^2} \right) t + C$$

Since $v(0) = 0$, it follows that $C = 0$. Hence,

$$v(t) = \left(9.81 \, \frac{m}{s^2} \right) t, \quad t \geq 0$$

To find the distance traveled, note that $s(0) = 0$ and

$$\frac{d}{dt} s(t) = v(t) = \left(9.81 \, \frac{m}{s^2} \right) t, \quad t \geq 0$$

A general solution is

$$s(t) = \frac{1}{2} \left(9.81 \, \frac{m}{s^2} \right) t^2 + C$$

Since $s(0) = 0$, it follows that $C = 0$. Hence,

$$s(t) = \frac{1}{2} \left(9.81 \, \frac{m}{s^2} \right) t^2, \quad t \geq 0$$

Note that if t is measured in seconds, the unit of $v(t)$ is $\frac{m}{s}$ and the unit of $s(t)$ is m. ◀

5.8.1 Problems

In Problems 1–36, find the general antiderivative for the given function.

1. $f(x) = 2x^2$

2. $f(x) = 1 - 3x^2$

3. $f(x) = x^2 + 2x - 1$

4. $f(x) = 2x - 4x^3$

5. $f(x) = x^4 - 3x^2 + 1$

6. $f(x) = 2x^2 + x - 5$

7. $f(x) = 4x^3 - 2x + 3$

8. $f(x) = x - 2x^2 - 3x^3 - 4x^4$

9. $f(x) = 1 + \dfrac{1}{x}$

10. $f(x) = x^2 - \dfrac{2}{x}$

11. $f(x) = 1 - \dfrac{1}{x^2}$

12. $f(x) = x^3 - \dfrac{1}{x^3}$

13. $f(x) = \dfrac{1}{1+x}$

14. $f(x) = \dfrac{x}{1+x}$

15. $f(x) = 5x^4 + \dfrac{5}{x^4}$

16. $f(x) = x^7 + \dfrac{1}{x^7}$

17. $f(x) = \dfrac{1}{2+x}$

18. $f(x) = \dfrac{1}{3+x}$

19. $f(x) = e^{-3x}$

20. $f(x) = e^{x/2} + e^{-x/2}$

21. $f(x) = 2e^{2x}$

22. $f(x) = -3e^{-4x}$

23. $f(x) = \dfrac{1}{e^{2x}}$

24. $f(x) = \dfrac{3}{e^{-x}}$

25. $f(x) = \sin(2x)$

26. $f(x) = \cos(3x)$

27. $f(x) = \sin\left(\dfrac{x}{3}\right)$

28. $f(x) = \cos\left(\dfrac{x}{5}\right)$

29. $f(x) = 2\sin\left(\dfrac{\pi}{2}x\right) - 3\cos\left(\dfrac{\pi}{2}x\right)$

30. $f(x) = -3\sin\left(\dfrac{\pi}{3}x\right) + 4\cos\left(-\dfrac{\pi}{4}x\right)$

31. $f(x) = \sec^2(2x)$

32. $f(x) = \sec^2(-4x)$

33. $f(x) = \sec^2\left(\dfrac{x}{3}\right)$

34. $f(x) = \sec^2\left(\dfrac{x}{5}\right)$

35. $f(x) = \dfrac{\sec x + \cos x}{\cos x}$

36. $f(x) = \sin^2 x + \cos^2 x$

In Problems 37–48, find the general solution of the differential equation.

37. $\dfrac{dy}{dx} = \dfrac{2}{x} - x, \; x > 0$

38. $\dfrac{dy}{dx} = \dfrac{2}{x^2} - x^2, \; x > 0$

39. $\dfrac{dy}{dx} = x(1 + x), \; x > 0$

40. $\dfrac{dy}{dx} = e^{-2x}, \; x > 0$

41. $\dfrac{dy}{dt} = t(1 - t), \; t \geq 0$

42. $\dfrac{dy}{dt} = t^2(1 - t^2), \; t \geq 0$

43. $\dfrac{dy}{dt} = e^{-t/2}, \; t \geq 0$

44. $\dfrac{dy}{dt} = 1 - e^{3t}, \; t \geq 0$

45. $\dfrac{dy}{ds} = \sin(\pi s), \; 0 \leq s \leq 1$

46. $\dfrac{dy}{ds} = \cos(2\pi s), \; 0 \leq s \leq 1$

47. $\dfrac{dy}{dx} = \sec^2\left(\dfrac{x}{2}\right), \; -1 < x < 1$

48. $\dfrac{dy}{dx} = 1 + \sec^2\left(\dfrac{x}{4}\right), \; -1 < x < 1$

In Problems 49–62, solve the initial value problem.

49. $\dfrac{dy}{dx} = 3x^2$, for $x \geq 0$ with $y = 1$ when $x = 0$

50. $\dfrac{dy}{dx} = \dfrac{x^2}{3}$, for $x \geq 0$ with $y = 2$ when $x = 0$

51. $\dfrac{dy}{dx} = 2\sqrt{x}$, for $x \geq 0$ with $y = 2$ when $x = 1$

52. $\dfrac{dy}{dx} = \dfrac{1}{2\sqrt{x}}$, for $x \geq 1$ with $y = 3$ when $x = 4$

53. $\dfrac{dN}{dt} = \dfrac{1}{t}$, for $t \geq 1$ with $N(1) = 10$

54. $\dfrac{dN}{dt} = \dfrac{t}{t+1}$, for $t \geq 0$ with $N(0) = 5$

55. $\dfrac{dW}{dt} = e^t$, for $t \geq 0$ with $W(0) = 1$

56. $\dfrac{dW}{dt} = e^{2t}$, for $t \geq 0$ with $W(0) = 5$

57. $\dfrac{dW}{dt} = e^{-3t}$, for $t \geq 0$ with $W(0) = 2/3$

58. $\dfrac{dW}{dt} = e^{-4t}$, for $t \geq 0$ with $W(0) = 3$

59. $\dfrac{dT}{dt} = \sin(\pi t)$, for $t \geq 0$ with $T(0) = 3$

60. $\dfrac{dT}{dt} = \cos(\pi t)$, for $t \geq 0$ with $T(0) = 3$

61. $\dfrac{dy}{dx} = \dfrac{e^{-x} + e^x}{2}$, for $x \geq 0$ with $y = 0$ when $x = 0$

62. $\dfrac{dN}{dt} = t^{-1/3}$, for $t > 0$ with $N(0) = 60$

63. Suppose that the length of a certain organism at age x is given by $L(x)$, which satisfies the differential equation

$$\frac{dL}{dx} = e^{-0.1x}, \quad x \geq 0$$

Find $L(x)$ if the limiting length L_∞ is given by

$$L_\infty = \lim_{x \to \infty} L(x) = 25$$

How big is the organism at age $x = 0$?

64. Fish are indeterminate growers; that is, their length $L(x)$ increases with age x throughout their lifetime. If we plot the growth rate dL/dx versus age x on semilog paper, a straight line with negative slope results. Set up a differential equation that relates growth rate and age. Solve this equation under the assumption that $L(0) = 5$, $L(1) = 10$, and

$$\lim_{x \to \infty} L(x) = 20$$

Graph the solution $L(x)$ as a function of x.

65. An object is dropped from a height of 100 ft. The acceleration is 32 ft/s^2. When will the object hit the ground, and what will its speed be at impact?

66. Suppose that the growth rate of a population at time t undergoes seasonal fluctuations in size according to

$$\frac{dN}{dt} = 3\sin(2\pi t)$$

where t is measured in years and $N(t)$ denotes the size of the population at time t. If $N(0) = 10$ (measured in thousands), find an expression for $N(t)$. How are the seasonal fluctuations in the growth rate reflected in the population size?

67. Suppose that the amount of water contained in a plant at time t is denoted by $V(t)$. Due to evaporation, $V(t)$ changes over time. Suppose that the volume change at time t, measured over a 24-hour period, is proportional to $t(24 - t)$ measured in grams per hour. To offset the water loss, you water the plant at a constant rate of 4 grams of water per hour.

(a) Explain why

$$\frac{dV}{dt} = -at(24 - t) + 4$$

$0 \leq t \leq 24$, for some positive constant a, describes this situation.

(b) Determine the constant a for which the net water loss over a 24-hour period is equal to 0.

5.9 Key Terms

Chapter 5 Review: Topics *Discuss the following definitions and concepts.*

1. Global or absolute extrema
2. Local or relative extrema: local minimum and local maximum
3. The extreme value theorem
4. Fermat's theorem
5. Mean value theorem
6. Rolle's theorem
7. Increasing and decreasing function
8. Monotonicity and the first derivative
9. Concavity: concave up and concave down
10. Concavity and the second derivative
11. Diminishing return
12. Candidates for local extrema
13. Monotonicity and local extrema
14. The second derivative test for local extrema
15. Inflection points
16. Inflection points and the second derivative
17. Asymptotes: horizontal, vertical, and oblique
18. Using calculus to graph functions
19. L'Hospital's rule
20. Dynamical systems: cobwebbing
21. Stability of equilibria
22. Newton-Raphson method for finding roots
23. Antiderivative

5.10 Review Problems

1. Suppose that

$$f(x) = xe^{-x}, \quad x \geq 0$$

(a) Show that $f(0) = 0$, $f(x) > 0$ for $x > 0$, and

$$\lim_{x \to \infty} f(x) = 0$$

(b) Find local and absolute extrema.

(c) Find inflection points.

(d) Use the foregoing information to graph $f(x)$.

2. Suppose that

$$f(x) = x \ln x, \quad x > 0$$

(a) Define $f(x)$ at $x = 0$ so that $f(x)$ is continuous for all $x \geq 0$.

(b) Find extrema and inflection points.

(c) Graph $f(x)$.

3. In Review Problem 17 of Chapter 3, we introduced the hyperbolic functions

$$\sinh x = \frac{e^x - e^{-x}}{2}, \quad x \in \mathbf{R}$$

$$\cosh x = \frac{e^x + e^{-x}}{2}, \quad x \in \mathbf{R}$$

$$\tanh x = \frac{e^x - e^{-x}}{e^x + e^{-x}}, \quad x \in \mathbf{R}$$

(a) Show that $f(x) = \tanh x$, $x \in \mathbf{R}$, is a strictly increasing function on $\mathbf{R}$. Evaluate

$$\lim_{x \to -\infty} \tanh x$$

and

$$\lim_{x \to \infty} \tanh x$$

(b) Use your results in (a) to explain why $f(x) = \tanh x$, $x \in \mathbf{R}$, is invertible, and show that its inverse function $f^{-1}(x) = \tanh^{-1} x$ is given by

$$f^{-1}(x) = \frac{1}{2} \ln \frac{1+x}{1-x}$$

What is the domain of $f^{-1}(x)$?

(c) Show that

$$\frac{d}{dx} f^{-1}(x) = \frac{1}{1 - x^2}$$

(d) Use your result in (c) and the facts that

$$\tanh x = \frac{\sinh x}{\cosh x}$$

and

$$\cosh^2 x - \sinh^2 x = 1$$

to show that

$$\frac{d}{dx} \tanh x = \frac{1}{\cosh^2 x}$$

4. Let

$$f(x) = \frac{x}{1 + e^{-x}}, \quad x \in \mathbf{R}$$

(a) Show that $y = 0$ is a horizontal asymptote as $x \to -\infty$.

(b) Show that $y = x$ is an oblique asymptote as $x \to +\infty$.

(c) Show that

$$f'(x) = \frac{1 + e^{-x}(1 + x)}{(1 + e^{-x})^2}$$

(d) Use your result in (c) to show that $f(x)$ has exactly one local extrema at $x = c$, where c satisfies the equation
$$1 + c + e^c = 0$$
[*Hint:* Use your result in (c) to show that $f'(x) = 0$ if and only if $1 + e^{-x}(1 + x) = 0$. Let $g(x) = 1 + e^{-x}(1 + x)$. Show that $g(x)$ is strictly increasing for $x < 0$ and that $g(0) > 0$ and $g(-2) < 0$. This implies that $g(x) = 0$ has exactly one solution in $(-2, 0)$. Since $g(-2) < 0$ and $g(x)$ is strictly increasing for $x < 0$, there are no solutions of $g(x) = 0$ for $x < -2$. Furthermore, $g(x) > 0$ for $x > 0$; hence there are no solutions of $g(x) = 0$ for $x > 0$.]

(e) The equation $1 + c + e^c = 0$ can only be solved numerically for c. With the help of a calculator, find a numerical approximation for c. [*Hint:* From (d) you know that $c \in (-2, 0)$.]

(f) Show that $f(x) < 0$ for $x < 0$. [This implies that for $x < 0$ the graph of $f(x)$ is below the horizontal asymptote $y = 0$.]

(g) Show that $x - f(x) > 0$ for $x > 0$. [This implies that for $x > 0$, the graph of $f(x)$ is below the oblique asymptote $y = x$.]

(h) Use your results in (a)–(g) and the fact that $f(0) = 0$ and $f'(0) = \frac{1}{2}$ to sketch the graph of $f(x)$.

5. (*Recruitment model*) The **Ricker curve** describes the relationship between the size of the parental stock of some fish and the number of recruits. If we denote the size of the parental stock by P and the number of recruits by R, then the Ricker curve is given by
$$R(P) = \alpha P e^{-\beta P} \quad \text{for } P \geq 0$$
where α and β are positive constants. [Note that $R(0) = 0$, that is, without parents there are no offspring. Furthermore, $R(P) > 0$ when $P > 0$.]

We are interested in the size of the parental stock (P) that maximizes the number of recruits ($R(P)$). Since $R(P)$ is differentiable, we can use the first derivative of $R(P)$ to solve this problem.

(a) Use the product rule to show that for $P > 0$
$$R'(P) = \alpha e^{-\beta P}(1 - \beta P)$$
$$R''(P) = -\alpha\beta e^{-\beta P}(2 - \beta P)$$

(b) Show that $R'(P) = 0$ if $P = 1/\beta$ and that $R''(1/\beta) < 0$. This shows that $R(P)$ has a *local* maximum at $P = \frac{1}{\beta}$. Show that $R(1/\beta) = \frac{\alpha}{\beta}e^{-1} > 0$.

(c) To find the global maximum, you need to check $R(0)$ and $\lim_{P \to \infty} R(P)$. Show that
$$R(0) = 0 \quad \text{and} \quad \lim_{P \to \infty} R(P) = 0$$
and that this implies that there is a global maximum at $P = 1/\beta$.

(d) Show that $R(P)$ has an inflection point at $P = 2/\beta$.

(e) Sketch the graph of $R(P)$ for $\alpha = 2$ and $\beta = 1$.

6. The *Gompertz growth curve* is sometimes used to study the growth of populations. Its properties are quite similar to the properties of the logistic growth curve. The Gompertz growth curve is given by
$$N(t) = K \exp[-ae^{-bt}]$$
for $t \geq 0$, where K and b are positive constants.

(a) Show that $N(0) = Ke^{-a}$ and hence
$$a = \ln \frac{K}{N_0}$$
if $N_0 = N(0)$.

(b) Show that $y = K$ is a horizontal asymptote and that $N(t) < K$ if $N_0 < K$, $N(t) = K$ if $N_0 = K$, and $N(t) > K$ if $N_0 > K$.

(c) Show that
$$\frac{dN}{dt} = bN(\ln K - \ln N)$$
and
$$\frac{d^2N}{dt^2} = b\frac{dN}{dt}[\ln K - \ln N - 1]$$

(d) Use your results in (b) and (c) to show that $N(t)$ is strictly increasing if $N_0 < K$ and strictly decreasing if $N_0 > K$.

(e) When does $N(t)$, $t \geq 0$, have an inflection point? Discuss its concavity.

(f) Graph $N(t)$ when $K = 100$ and $b = 1$, if (i) $N_0 = 20$, (ii) $N_0 = 70$, and (iii) $N_0 = 150$, and compare your graphs with your answers in (b)–(e).

7. The Monod growth curve is given by
$$f(x) = \frac{cx}{k + x}$$
for $x \geq 0$, where c and k are positive constants. It can be used to describe the specific growth rate of a species, dependent on a resource level x.

(a) Show that $y = c$ is a horizontal asymptote for $x \to \infty$. The constant c is called the *saturation value*.

(b) Show that $f(x)$, $x \geq 0$, is strictly increasing and concave down. Explain why this implies that the saturation value is equal to the maximal specific growth rate.

(c) Show that if $x = k$, then $f(x)$ is equal to half the saturation value. (For this reason, the constant k is called the *half-saturation constant*.)

(d) Sketch a graph of $f(x)$ for $k = 2$ and $c = 5$, clearly marking the saturation value and the half-saturation constant. Compare this graph to one where $k = 3$ and $c = 5$.

(e) Without graphing the three curves, explain how you can use the saturation value and the half-saturation constant to decide quickly that
$$\frac{10x}{3 + x} > \frac{10x}{5 + x} > \frac{8x}{5 + x}$$
for $x \geq 0$.

8. The logistic growth curve is given by
$$N(t) = \frac{K}{1 + \left(\frac{K}{N_0} - 1\right)e^{-rt}}$$
for $t \geq 0$, where K, N_0, and r are positive constants and $N(t)$ denotes the population size at time t.

(a) Show that $N(0) = N_0$ and that $y = K$ is a horizontal asymptote as $t \to \infty$.

(b) Show that $N(t) < K$ if $N_0 < K$, $N(t) = K$ if $N_0 = K$, and $N(t) > K$ if $N_0 > K$.

(c) Show that
$$\frac{dN}{dt} = rN\left(1 - \frac{N}{K}\right)$$

and

$$\frac{d^2N}{dt^2} = r\frac{dN}{dt}\left(1 - \frac{2N}{K}\right)$$

(d) Use your results in (b) and (c) to show that $N(t)$ is strictly increasing if $N_0 < K$ and strictly decreasing if $N_0 > K$.

(e) Show that if $N_0 < K/2$, then $N(t)$, $t \geq 0$, has exactly one inflection point $(t^\star, N(t^\star))$, with $t^\star > 0$ and

$$N(t^\star) = \frac{K}{2}$$

(i.e., half the carrying capacity). What happens if $K/2 < N_0 < K$? What if $N_0 > K$? Where is the function $N(t)$, $t \geq 0$, concave up, and where concave down?

(f) Sketch the graphs of $N(t)$ for $t \geq 0$ when

(i) $K = 100$, $N_0 = 10$, $r = 1$

(ii) $K = 100$, $N_0 = 70$, $r = 1$

(iii) $K = 100$, $N_0 = 150$, $r = 1$

together with their respective horizontal asymptotes. Clearly mark the inflection point if it exists.

9. A population is said to be in Hardy-Weinberg equilibrium, with respect to a single gene with two alleles A and a, if the three genotypes AA, Aa, and aa have frequencies $p_{AA} = \theta^2$, $p_{Aa} = 2\theta(1-\theta)$, and $p_{aa} = (1-\theta)^2$ for some $\theta \in [0, 1]$. Suppose that we take a random sample of size n from a population. We can show that the probability of observing n_1 individuals of type AA, n_2 individuals of type Aa, and n_3 individuals of type aa is given by

$$\frac{n!}{n_1! n_2! n_3!} p_{AA}^{n_1} p_{Aa}^{n_2} p_{aa}^{n_3}$$

where $n! = n(n-1)(n-2)\cdots 3 \cdot 2 \cdot 1$ (read: "n factorial"). Here $n_1 + n_2 + n_3 = n$. This probability depends on θ. The following method, called the *maximum likelihood method*, can be used to estimate θ. The principle is simple: We find the value of θ that maximizes the probability of the observed data. Since the coefficient

$$\frac{n!}{n_1! n_2! n_3!}$$

does not depend on θ, we need only maximize

$$L(\theta) = p_{AA}^{n_1} p_{Aa}^{n_2} p_{aa}^{n_3}$$

(a) Suppose $n_1 = 8$, $n_2 = 6$, and $n_3 = 3$. Compute $L(\theta)$.

(b) Show that if $L(\theta)$ is maximal for $\theta = \hat{\theta}$ (read: "theta hat"), then $\ln L(\theta)$ is also maximal for $\theta = \hat{\theta}$.

(c) Use your result in (b) to find the value $\hat{\theta}$ that maximizes $L(\theta)$ for the data given in (a). The number $\hat{\theta}$ is the maximum likelihood estimate.

10. Suppose the volume of a cell is increasing at a constant rate of 10^{-12} cm^3/s.

(a) If $V(t)$ denotes the cell volume at time t, set up an initial value problem that describes this situation if the initial volume is 10^{-10} cm^3.

(b) Solve the initial value problem given in (a) and determine the volume of the cell after 10 seconds.

11. Suppose the concentration $c(t)$ of a drug in the bloodstream at time t satisfies

$$\frac{dc}{dt} = -0.1e^{-0.3t}$$

for $t \geq 0$.

(a) Solve the differential equation, under the assumption that there will eventually be no trace of the drug in the blood.

(b) How long does it take until the concentration reaches half its initial value?

12. Sterner (1997) investigated the effect of food quality on zooplankton dynamics. In his model, zooplankton may be limited by either carbon (C) or phosphorus (P). He argued that when food quantity is low, demand for carbon increases relative to demand for phosphorus to satisfy basic metabolic requirements and that there should be a curve separating C- and P-limited growth when food quantity, C_F (measured in amount of carbon per liter), is graphed as a function of the C : P ratio of the food, $f = C_F : P_F$. He derived the following equation for the curve separating the two regions:

$$C_F = \frac{m}{a_C g - \frac{C_Z a_P g}{P_Z f}}$$

where m is equal to respiration rate, g is the ingestion rate, and a_C (a_P) is the assimilation rate of carbon (phosphorus). C_Z and P_Z are the carbon and the phosphorus content of the zooplankton.

(a) The graph of $C_F(f)$ has a horizontal asymptote. Show that the horizontal asymptote occurs when $m = a_C g C_F$, which occurs when the rate at which carbon is assimilated equals metabolic rate.

(b) The graph of $C_F(f)$ has a vertical asymptote. Let $f = C_F : P_F$ (the C : P ratio of the food). Show that the vertical asymptote occurs at

$$\frac{C_F}{P_F} = \frac{C_Z}{P_Z}\frac{a_P}{a_C}$$

(c) Sketch a graph of $C_F(f)$ as a function of f.

(d) The graph of $C_F(f)$ separates C-limited (below the curve) from P-limited (above the curve) growth. Explain why this graph indicates that when food quantity is low, the demand for carbon relative to phosphorus increases.

13. Neglecting air resistance, the height (in meters) of an object thrown vertically from the ground with initial velocity v_0 is given by

$$h(t) = v_0 t - \frac{1}{2}gt^2$$

where $g = 9.81$ m/s^2 is the earth's gravitational constant, and t is the time (in seconds) elapsed since the object was released

(a) Find the time at which the object reaches its maximum height.

(b) Find the maximum height.

(c) Find the velocity of the object at the time it reaches its maximum height.

(d) At what time $t > 0$ will the object reach the initial height again?

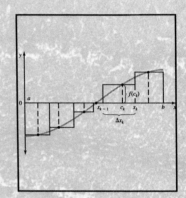

INTEGRATION

6.1 THE DEFINITE INTEGRAL

Computing the area of a region bounded by curves is an ancient problem that was solved in certain cases by Greek mathematicians, foremost Archimedes (about 287–212 B.C.), more than 2000 years ago. The Greeks used a method called exhaustion, which goes back to the Greek mathematician Eudoxus (c. 408–355 B.C.). The basic idea is to divide an area into very small regions consisting mostly of rectilinear figures of known area (such as triangles), so that the total area of the rectilinear figures is close to the area of the region of interest.

Interest in this problem resurfaced in the seventeenth century, when many new curves had been defined and an attempt was made to determine the areas of regions bounded by such curves. The first curves that were considered were of the form $y = x^n$, where n is a positive integer. Fermat wrote in a letter to Roberval on September 22, 1636, that he had succeeded in computing the area under the curve $y = x^n$. He noted that his method was different from the method employed by the Greeks, notably Archimedes. Whereas Archimedes used triangles to exhaust the area bounded by curved lines, Fermat used rectangles. This sounds like a minor difference, but it enabled Fermat to compute areas bounded by other curves that previously could not have been computed. He found that the area under the curve $y = x^n$ inscribed in a rectangle of width b and height b^n is $1/(n+1)$ times the area of the rectangle, that is,

$$\frac{1}{n+1}b \cdot b^n = \frac{1}{n+1}b^{n+1}$$

(see Figure 6.1). (This problem was independently solved around 1640 by Cavalieri, Pascal, Roberval, and Torricelli as well.)

Augustin-Louis Cauchy (1789–1857) was the first to define areas based on the limit of the sum of areas of approximating rectangles. The definition we will use goes back to Georg Bernhard Riemann (1826–1866). His definition is more general than Cauchy's and allows for a larger class of functions to be used as boundary curves for areas.

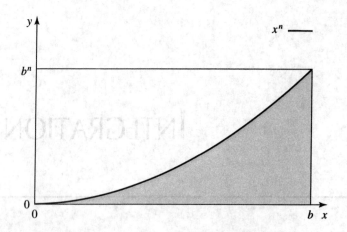

▲ **Figure 6.1**
The area of the shaded region under the curve $y = x^n$ inscribed in the rectangle is $1/(n+1)$ times the area of the rectangle

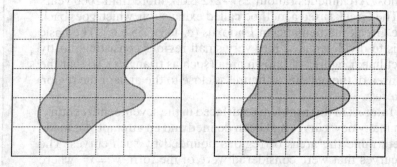

▲ **Figure 6.2**
The outline of a lake with superimposed grid: The finer the grid, the more accurately the area of the lake can be determined

6.1.1 The Area Problem

We wish to find the surface area of the lake shown in Figure 6.2; to do so, we overlay a grid and count the number of squares that have a nonempty intersection with the lake. This will approximate the area of the lake. The finer the grid, the closer our approximation will be to the true area of the lake.

Dividing a region into smaller regions of known area is the basic principle we will employ in this section to find the area of a region bounded by curves of continuous functions.

▶ **Example 1** We will try to find the area of the region below the parabola $f(x) = x^2$ and above the x-axis between 0 and 1 (see Figure 6.3). To do this, we divide the interval $[0, 1]$ into n subintervals of equal length and approximate the area of interest by a sum of the areas of rectangles whose bases are the subintervals and whose heights are the values of the function at the left endpoints of these subintervals. This is illustrated in Figure 6.4 with $n = 5$.

In Figure 6.4, the base of each rectangle has width $1/5 = 0.2$. The height of the first rectangle is $f(0) = 0$, the height of the second rectangle is $f(0.2) = (0.2)^2$, the height of the third rectangle is $f(0.4) = (0.4)^2$, and so on. The area of a rectangle is width times height; adding up the areas of the

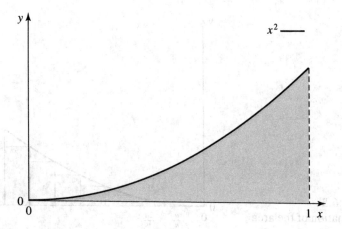

▲ **Figure 6.3**
The region under the curve $y = x^2$ in Example 1

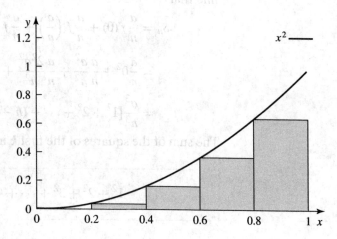

▲ **Figure 6.4**
Approximation of the area under the curve $y = x^2$ from $x = 0$ to $x = 1$ by five rectangles

approximating rectangles in Figure 6.4 yields

$$(0.2)(0)^2 + (0.2)(0.2)^2 + (0.2)(0.4)^2 + (0.2)(0.6)^2 + (0.2)(0.8)^2$$
$$= (0.2)\left[0^2 + (0.2)^2 + (0.4)^2 + (0.6)^2 + (0.8)^2\right] = 0.24$$

Thus, an approximation of the area between 0 and 1 is 0.24 when we use five subintervals.

We turn now to the general case (illustrated in Figure 6.5), where the interval is $[0, a]$ and the number of subintervals is n. Since the interval $[0, a]$ has length a and the number of subintervals is n, each subinterval has length a/n. The left endpoints of successive subintervals are therefore $0, a/n, 2a/n, 3a/n, \ldots, (n - 1)a/n$. The heights of the successive rectangles are then $f(0) = 0$, $f(a/n) = (a/n)^2$, $f(2a/n) = (2a/n)^2, \ldots, f((n-1)a/n) = ((n-1)a/n)^2$. We denote the sum of the areas of the n rectangles by S_n, where S stands for sum and the subscript n for the number of subintervals. We

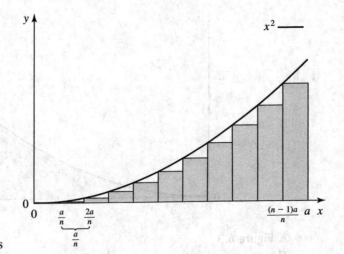

▲ **Figure 6.5**
Approximation of the area under the curve $y = x^2$ from $x = 0$ to $x = a$ by n rectangles

find that

$$S_n = \frac{a}{n}f(0) + \frac{a}{n}f\left(\frac{a}{n}\right) + \frac{a}{n}f\left(\frac{2a}{n}\right) + \cdots + \frac{a}{n}f\left(\frac{(n-1)a}{n}\right)$$

$$= \frac{a}{n}0^2 + \frac{a}{n}\frac{a^2}{n^2} + \frac{a}{n}\frac{2^2a^2}{n^2} + \cdots + \frac{a}{n}\frac{(n-1)^2a^2}{n^2}$$

$$= \frac{a^3}{n^3}[1^2 + 2^2 + \cdots + (n-1)^2]$$

The sum of the squares of the first k integers can be computed:

$$1^2 + 2^2 + 3^2 + \cdots + k^2 = \frac{k(k+1)(2k+1)}{6} \qquad (6.1)$$

(For a proof of this formula, see Problem 31.) Using this formula for $k = n - 1$, we find

$$S_n = \frac{a^3}{n^3}\frac{(n-1)(n-1+1)(2(n-1)+1)}{6}$$

$$= \frac{a^3}{n^3}\frac{(n-1)n(2n-1)}{6} = \frac{a^3}{6}\frac{n-1}{n}\frac{n}{n}\frac{2n-1}{n}$$

$$= \frac{a^3}{6}\left(1 - \frac{1}{n}\right) \cdot 1 \cdot \left(2 - \frac{1}{n}\right)$$

The finer the subdivision of $[0, a]$ (that is, the larger n), the more accurate the approximation, as illustrated in Figure 6.6, where we see that the area of the region below the parabola and above the x-axis between 0 and a is more accurately approximated when we use a larger number of rectangles. Choosing finer and finer subdivisions means that we let n go to infinity. We find

$$\lim_{n\to\infty} S_n = \lim_{n\to\infty} \frac{a^3}{6}\left(1 - \frac{1}{n}\right) \cdot 1 \cdot \left(2 - \frac{1}{n}\right) = \frac{a^3}{6}(1)(1)(2) = \frac{a^3}{3}$$

That is, the area under the parabola $y = x^2$ from 0 to a is equal to $a^3/3$. ◄

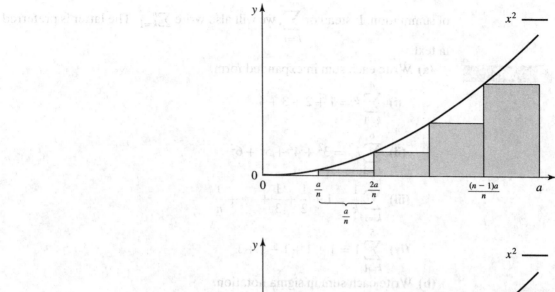

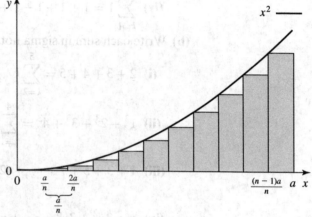

▲ **Figure 6.6**
Increasing the number of approximating rectangles improves the accuracy of the approximation

We see from Example 1 that computing areas entails summing a large number of terms. Before we continue our discussion of the computation of areas, we will therefore spend some time on sums of the type we encountered in Example 1.

It will be convenient to have a shorthand notation for sums that involve a large number of terms.

Definition (Sigma Notation for Finite Sums) Let $a_1, a_2, \ldots, a_n$ be real numbers and n be a positive integer. Then

$$\sum_{k=1}^{n} a_k = a_1 + a_2 + \cdots + a_n$$

The letter Σ is the capital Greek letter sigma, and the symbol $\sum_{k=1}^{n}$ means that we sum from $k = 1$ to n, where k is called the index of summation, the number 1 is the lower limit of summation, and the number n is the upper limit of summation. Instead of $\sum\limits_{k=1}^{n}$, we will also write $\sum_{k=1}^{n}$. The latter is preferred in text.

▶ Example 2

(a) Write each sum in expanded form.

(i) $\sum\limits_{k=1}^{4} k = 1 + 2 + 3 + 4$

(ii) $\sum\limits_{k=3}^{6} k^2 = 3^2 + 4^2 + 5^2 + 6^2$

(iii) $\sum\limits_{k=1}^{n} \dfrac{1}{k} = 1 + \dfrac{1}{2} + \dfrac{1}{3} + \cdots + \dfrac{1}{n}$

(iv) $\sum\limits_{k=1}^{5} 1 = 1 + 1 + 1 + 1 + 1$

(b) Write each sum in sigma notation.

(i) $2 + 3 + 4 + 5 = \sum\limits_{k=2}^{5} k$

(ii) $1^3 + 2^3 + 3^3 + 4^3 = \sum\limits_{k=1}^{4} k^3$

(iii) $1 + 3 + 5 + 7 + \cdots + (2n + 1) = \sum\limits_{k=0}^{n} (2k + 1)$

(iv) $x + 2x^2 + 3x^3 + \cdots + nx^n = \sum\limits_{k=1}^{n} kx^k$ ◀

Occasionally, we will need a formula to sum the first n integers. The next example shows how this is done.

▶ Example 3   Show that

$$S_n = \sum_{k=1}^{n} k = 1 + 2 + 3 + \cdots + n = \frac{n(n+1)}{2}$$

Solution

The following trick will enable us to compute the sum S_n. We write the sum in the usual order and in reverse order.

$$S_n = 1 + 2 + 3 + \cdots + n$$
$$S_n = n + (n-1) + (n-2) + \cdots + 2 + 1$$

Adding vertically, we find

$$2S_n = (1 + n) + (2 + n - 1) + (3 + n - 2) + \cdots + (n + 1)$$
$$= (n + 1) + (n + 1) + (n + 1) + \cdots + (n + 1)$$

There are n terms (each $n + 1$) on the right-hand side. Hence

$$2S_n = n(n + 1) \quad \text{or} \quad S_n = \frac{n(n+1)}{2}$$

This method was used by Karl Friedrich Gauss (1777–1855) when he was ten years old: One day, to keep the students busy, his teacher asked them to add all the numbers from 1 to 100. To his teacher's astonishment, Karl Friedrich quickly gave the correct answer, 5050. To find the answer, he did not add the numbers in their numerical order but rather added $1 + 100$, $2 + 99$, $3 + 98, \ldots, 50 + 51$ (just as we did in the preceding derivation). Each term is equal to 101 and there are 50 such terms; hence the answer is $50 \cdot 101 = 5050$. (Gauss went on to become one of the greatest mathematicians in history, contributing to geometry, number theory, astronomy, and other areas.) ◀

The following rules are useful when evaluating finite sums.

ALGEBRAIC RULES

1. Constant value rule: $\displaystyle\sum_{k=1}^{n} 1 = n$

2. Constant multiple rule: $\displaystyle\sum_{k=1}^{n} c \cdot a_k = c \sum_{k=1}^{n} a_k$, where c is a constant that does not depend on k

3. Sum rule: $\displaystyle\sum_{k=1}^{n} (a_k + b_k) = \sum_{k=1}^{n} a_k + \sum_{k=1}^{n} b_k$

▶ **Example 4**   Use the algebraic rules to simplify the following sums.

(a) $\displaystyle\sum_{k=2}^{4} (3 + k) = \sum_{k=2}^{4} 3 + \sum_{k=2}^{4} k = (3 + 3 + 3) + (2 + 3 + 4) = 18$

(b) $\displaystyle\sum_{k=1}^{n} (k^2 - 2k) = \sum_{k=1}^{n} k^2 - 2 \sum_{k=1}^{n} k$

$$= \frac{n(n+1)(2n+1)}{6} - 2\frac{n(n+1)}{2} = \frac{n(n+1)(2n-5)}{6}$$

[We used (6.1) to evaluate the first sum and Example 3 to evaluate the second sum.] ◀

6.1.2 Riemann Integrals

We will now develop a more systematic solution to the area problem. Though our approach will be very similar to that in the previous subsection, we will look at a more general situation. We allow the function whose graph makes up the boundary of the region of interest to take on negative values as well. Furthermore, the rectangles that we use for approximating the area may vary in width, and the points that are chosen to compute the heights of the rectangles may be anywhere in their respective subintervals (which form the bases of the rectangles).

Let f be a continuous function on the interval $[a, b]$ (see Figure 6.7). We partition $[a, b]$ into n subintervals by choosing $n - 1$ numbers $x_1, x_2, \ldots, x_{n-1}$ in (a, b), such that

$$a = x_0 < x_1 < x_2 < \cdots < x_{n-1} < x_n = b$$

The n subintervals

$$[x_0, x_1], [x_1, x_2], \ldots, [x_{n-1}, x_n]$$

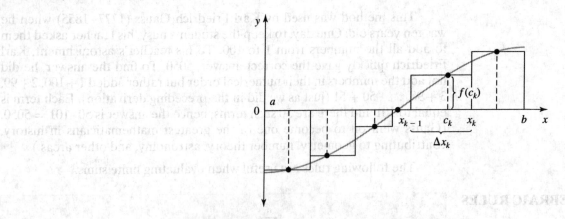

▲ **Figure 6.7**
An illustration of a Riemann sum

form a **partition** of $[a, b]$, which we denote by $P = [x_0, x_1, x_2, \ldots, x_n]$. The partition P depends on the number of subintervals n and on the choice of points $x_0, x_1, \ldots, x_n$. For notational convenience, however, we will simply call a partition P. The length of the kth subinterval $[x_{k-1}, x_k]$ is denoted by Δx_k. The length of the longest subinterval is called the **norm** of P and denoted by $\|P\|$ (read "norm of P"):

$$\|P\| = \max\{\Delta x_1, \Delta x_2, \ldots, \Delta x_n\}$$

where $\max\{\Delta x_1, \Delta x_2, \ldots, \Delta x_n\}$ denotes the largest element of the set $\{\Delta x_1, \Delta x_2, \ldots, \Delta x_n\}$. In each subinterval $[x_{k-1}, x_k]$, we choose a point c_k and construct a rectangle with base Δx_k and height $|f(c_k)|$, as shown in Figure 6.7. If $f(c_k)$ is positive, then $f(c_k)\Delta x_k$ is the area of the rectangle. If $f(c_k)$ is negative, then $f(c_k)\Delta x_k$ is the negative of the rectangle's area. The sum of these products is denoted by S_P. That is,

$$S_P = \sum_{k=1}^{n} f(c_k)\,\Delta x_k$$

The value of the sum depends on the choice of the partition P (hence the subscript P on S) and the choice of the points $c_k \in [x_{k-1}, x_k]$ and is called a **Riemann sum** for f on $[a, b]$.

▶ **Example 5**   Find the Riemann sum for $f(x) = x^2$ on $[0, 1]$ using five equal subintervals with (a) left endpoints, (b) midpoints, (c) right endpoints.

Solution

We partition $[0, 1]$ into five equal subintervals, each of length 0.2: $[0, 0.2]$, $[0.2, 0.4]$, $[0.4, 0.6]$, $[0.6, 0.8]$, and $[0.8, 1.0]$. The Riemann sum is given by

$$S_P = \sum_{k=1}^{5} f(c_k)\,\Delta x_k$$

where $\Delta x_k = 0.2$ in all three cases (a)–(c).

(a) We use left endpoints; thus $c_1 = 0$, $c_2 = 0.2$, $c_3 = 0.4$, $c_4 = 0.6$, and $c_5 = 0.8$. We find

$$S_P = (0.2)[0^2 + (0.2)^2 + (0.4)^2 + (0.6)^2 + (0.8)^2] = 0.24$$

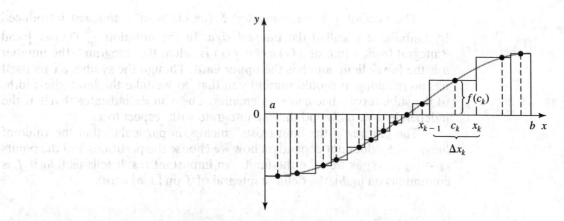

▲ **Figure 6.8**
Definite integral

(b) We use midpoints; thus $c_1 = 0.1$, $c_2 = 0.3$, $c_3 = 0.5$, $c_4 = 0.7$, and $c_5 = 0.9$. We find

$$S_P = (0.2)[(0.1)^2 + (0.3)^2 + (0.5)^2 + (0.7)^2 + (0.9)^2] = 0.33$$

(c) We use right endpoints; thus $c_1 = 0.2$, $c_2 = 0.4$, $c_3 = 0.6$, $c_4 = 0.8$, and $c_5 = 1.0$. We find

$$S_P = (0.2)[(0.2)^2 + (0.4)^2 + (0.6)^2 + (0.8)^2 + (1.0)^2] = 0.44$$

Comparing S_P in (a)–(c) shows that the Riemann sum depends on the choice of the points $c_k \in [x_{k-1}, x_k]$. ◄

To obtain a better approximation, we need to choose finer and finer partitions of $[a, b]$ so that the rectangles fill out the region between the curve and the x-axis more and more accurately (see Figure 6.8). A finer partition means that both the number of subintervals becomes larger and the length of the longest subinterval becomes smaller so that the norm of the partition P becomes smaller. One way to do this is to choose a sequence of partitions $P = P(n) = [x_0, x_1, x_2, \ldots, x_n]$, $n = 1, 2, 3, \ldots$, in such a way that $\|P(n)\| > \|P(n+1)\|$ for $n = 1, 2, 3, \ldots$. In fact, we will take the limit $\|P(n)\| \to 0$ as $n \to \infty$. For notational convenience, we will omit n and simply write $\|P\| \to 0$. But keep in mind that this means that, simultaneously, the number of subintervals goes to infinity and the length of the longest subinterval goes to 0. The limit of S_P as $\|P\| \to 0$ (if it exists) is called the definite integral of f from a to b.

> **Definition (Definite Integral)** Let $P = [x_0, x_1, x_2, \ldots, x_n]$, $n = 1, 2, \ldots$, be a sequence of partitions of $[a, b]$ with $\|P\| \to 0$. Set $\Delta x_k = x_k - x_{k-1}$ and $c_k \in [x_{k-1}, x_k]$. The **definite integral** of f from a to b is
>
> $$\int_a^b f(x)\, dx = \lim_{\|P\| \to 0} \sum_{k=1}^{n} f(c_k)\, \Delta x_k$$
>
> if the limit exists, in which case f is said to be (Riemann) **integrable** on the interval $[a, b]$.

The symbol $\int$ is an elongated S (as in "sum") and was introduced by Leibniz. It is called the **integral sign**. In the notation $\int_a^b f(x)\,dx$ [read "integral from a to b of $f(x)\,dx$"]; $f(x)$ is called the **integrand**, the number a is the **lower limit**, and b is the **upper limit**. Though the symbol dx by itself has no meaning, it should remind you that, as we take the limit, the widths of the subintervals become ever smaller. The x in dx indicates that x is the independent variable and that we integrate with respect to x.

The phrase "if the limit exists" means, in particular, that the value of $\lim_{\|P\|\to 0} S_P$ does not depend on how we choose the partitions and the points $c_k \in [x_{k-1}, x_k]$ as we take the limit. An important result tells us that if f is continuous on $[a, b]$, the definite integral of f on $[a, b]$ exists.

> **Theorem** All continuous functions are Riemann integrable; that is, if $f(x)$ is continuous on $[a, b]$, then
> $$\int_a^b f(x)\,dx$$
> exists.

The class of functions that are Riemann integrable is quite a bit larger than the set of continuous functions. For instance, functions that are both bounded (for which there exists an $M < \infty$ such that $|f(x)| < M$ for all x over which we wish to integrate) and piecewise continuous (continuous except for a finite number of discontinuities) are integrable (see Figure 6.9). We will primarily be concerned with continuous functions in this text; knowing that continuous functions are Riemann integrable will therefore suffice for the most part.

Note that $\int_a^b f(x)\,dx$ is a number that does not depend on x. We could have written $\int_a^b f(u)\,du$ (or any other letter in place of x) and meant the same thing.

▶ **Example 6**   Express the definite integral
$$\int_3^7 (x^2 - 1)\,dx$$
as a limit of Riemann sums.

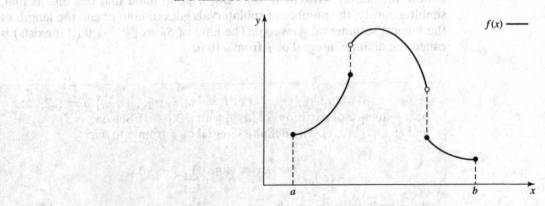

▲ **Figure 6.9**
The function $y = f(x)$ is piecewise continuous and bounded on $[a, b]$

Solution

$$\int_3^7 (x^2 - 1)\, dx = \lim_{\|P\| \to 0} \sum_{k=1}^n (c_k^2 - 1)\, \Delta x_k$$

where $x_0 = 3 < x_1 < x_2 < \cdots < x_n = 7$, $n = 1, 2, \ldots$, is a sequence of partitions of $[3, 7]$, $c_k \in [x_{k-1}, x_k]$, $\Delta x_k = x_k - x_{k-1}$, and the limit $\|P\| \to 0$ means that the norm of the partition tends to 0 (and, simultaneously, the number of subintervals goes to infinity). ◄

▶ **Example 7**   Express the limit

$$\lim_{\|P\| \to 0} \sum_{k=1}^n \sqrt{c_k - 1}\, \Delta x_k$$

as a definite integral, where $P = [x_0, x_1, \ldots, x_n]$, $n = 1, 2, \ldots$, is a sequence of partitions of $[2, 4]$ into n subintervals, $\Delta x_k = x_k - x_{k-1}$, and $c_k \in [x_{k-1}, x_k]$.

Solution

$$\lim_{\|P\| \to 0} \sum_{k=1}^n \sqrt{c_k - 1}\, \Delta x_k = \int_2^4 \sqrt{x - 1}\, dx \qquad ◄$$

▶ **Example 8**   Evaluate

$$\int_0^2 x^2\, dx$$

Solution

We evaluated the Riemann sum and its limit for $y = x^2$ from 0 to a in Example 1 and found

$$\lim_{n \to \infty} S_n = \int_0^a x^2\, dx = \frac{a^3}{3}$$

With $a = 2$, we therefore have

$$\int_0^2 x^2\, dx = \frac{8}{3} \qquad ◄$$

Geometric Interpretation of Definite Integrals   In Example 1, we computed the area of the region below the parabola $y = x^2$ and above the x-axis between 0 and a by approximating the region with n rectangles of equal width and then taking the limit as $n \to \infty$. More generally, we can now define the **area** of a region A above the x-axis, as shown in Figure 6.10, as the limiting value (if it exists) of the Riemann sum of approximating rectangles. (Note that an area is always a positive number.) This definition allows us to interpret the definite integral of a nonnegative function as an area.

If $f(x) \leq 0$ on $[a, b]$, then the definite integral $\int_a^b f(x)\, dx$ is less than or equal to 0, and the value is the negative of the area of the region above the graph of f and below the x-axis between a and b (see Figure 6.11). We refer to this as a "signed area." (A signed area may be either positive or negative.)

In general, a definite integral can thus be interpreted as a difference of areas, as illustrated in Figure 6.12. If A_+ denotes the total area of the region above the x-axis and below the graph of f (where $f \geq 0$), and A_- denotes the total area of the region below the x-axis and above the graph of f (where $f \leq 0$), then

$$\int_a^b f(x)\, dx = A_+ - A_-$$

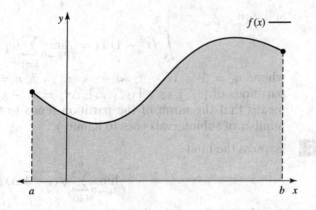

▲ **Figure 6.10**
The area of a region under the curve of a positive function is given by the definite integral $\int_a^b f(x)\,dx$

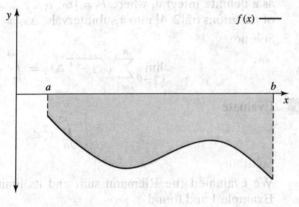

▲ **Figure 6.11**
The area of a region under the curve of a negative function is given by $-\int_a^b f(x)\,dx$

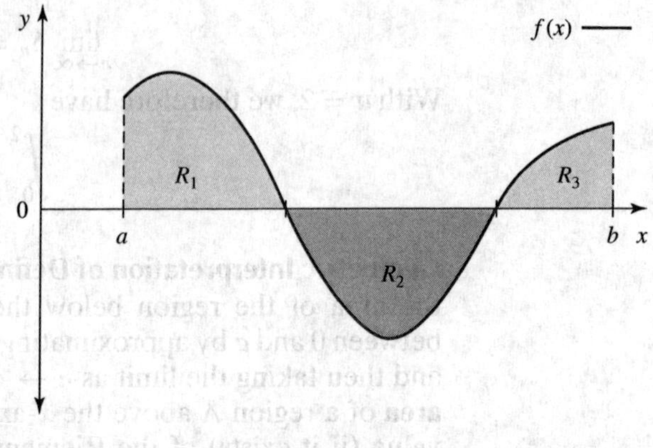

▲ **Figure 6.12**
A_+ is the combined area of R_1 and R_3, A_- is the area of R_2. Then $\int_a^b f(x)\,dx = A_+ - A_-$

---

1. If f is integrable on $[a, b]$ and $f(x) \geq 0$ on $[a, b]$, then

$$\int_a^b f(x)\,dx = \left[\begin{array}{l} \text{the area of the region between the} \\ \text{graph of } f \text{ and the } x\text{-axis from } a \text{ to } b \end{array} \right]$$

2. If f is integrable on $[a, b]$, then

$$\int_a^b f(x)\,dx = [\text{area above } x\text{-axis}] - [\text{area below } x\text{-axis}]$$

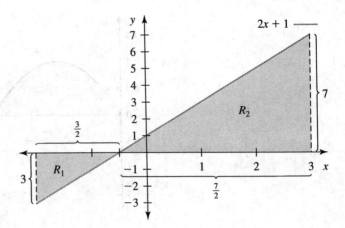

▲ Figure 6.13
The area of R_1 is A_-; the area of R_2 is A_+

▶ **Example 9** Find the value of

$$\int_{-2}^{3} (2x + 1)\, dx$$

by interpreting it as the signed area of an appropriately chosen region.

Solution

We graph $y = 2x + 1$ between -2 and 3 (see Figure 6.13). The line intersects the x-axis at $x = -1/2$. The area of the region to the left of $-1/2$ between the graph of $y = 2x + 1$ and the x-axis is denoted by A_-; the area of the region to the right of $-1/2$ between the graph of $y = 2x + 1$ and the x-axis is denoted by A_+. Both regions are triangles whose areas can be computed using a formula from geometry.

$$A_- = \frac{1}{2} \cdot \frac{3}{2} \cdot 3 = \frac{9}{4}$$

$$A_+ = \frac{1}{2} \cdot \frac{7}{2} \cdot 7 = \frac{49}{4}$$

Therefore,

$$\int_{-2}^{3} (2x + 1)\, dx = A_+ - A_- = \frac{49}{4} - \frac{9}{4} = \frac{40}{4} = 10$$ ◀

▶ **Example 10** Find the value of

$$\int_{0}^{2\pi} \sin x\, dx$$

by interpreting it as the signed area of an appropriately chosen region.

Solution

We graph $y = \sin x$ from 0 to 2π (see Figure 6.14). The function $f(x) = \sin x$ is symmetric about $x = \pi$. It follows from this symmetry that the area of the region below the graph of f and above the x-axis between 0 and π (denoted

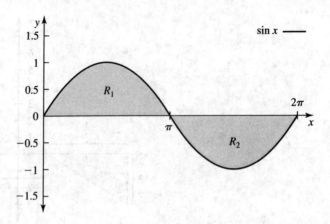

▲ **Figure 6.14**
The graph of $f(x) = \sin x$, $0 \le x \le 2\pi$, in Example 10. The area of R_1 is A_+; the area of R_2 is A_-

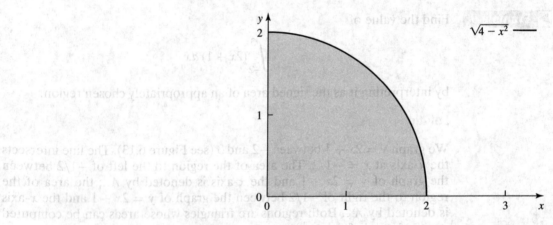

▲ **Figure 6.15**
The graph of $f(x) = \sqrt{4 - x^2}$ is a quarter circle. The area of the shaded region is equal to $\int_0^2 \sqrt{4 - x^2}\, dx$

by A_+) is the same as the area of the region above the graph of f and below the x-axis between π and 2π (denoted by A_-). Therefore, $A_+ = A_-$ and

$$\int_0^{2\pi} \sin x\, dx = A_+ - A_- = 0$$ ◀

▶ **Example 11**   Find the value of

$$\int_0^2 \sqrt{4 - x^2}\, dx$$

by interpreting it as the signed area of an appropriately chosen region.

Solution

The graph of $y = \sqrt{4 - x^2}$, $0 \le x \le 2$, is the quarter-circle with center at $(0, 0)$ and radius 2 in the first quadrant (see Figure 6.15). Since the area of a circle with radius 2 is $\pi(2)^2 = 4\pi$, the area of a quarter-circle is $4\pi/4 = \pi$. Hence,

$$\int_0^2 \sqrt{4 - x^2}\, dx = \pi$$ ◀

6.1.3 Properties of the Riemann Integral

In this subsection, we will collect important properties that will help us to evaluate definite integrals.

PROPERTIES

> Assume that f is integrable over $[a, b]$. We set
>
> **1.** $\displaystyle\int_a^a f(x)\,dx = 0$ and
>
> **2.** $\displaystyle\int_b^a f(x)\,dx = -\int_a^b f(x)\,dx$

Both properties make sense geometrically. The first integral says that the signed area between a and a is equal to 0; given that the width of the area is equal to 0, we expect the area to be 0 as well. The second property gives an orientation to the integral; for instance, if $f(x)$ is nonnegative on $[a, b]$, then $\int_a^b f(x)\,dx$ is nonnegative and can be interpreted as the area of the region between the graph of $f(x)$ and the x-axis from a to b. If we reverse the direction of the integration—that is, compute $\int_b^a f(x)\,dx$—we want the integral to be negative.

The next three properties follow immediately from the definition of the definite integral as the limit of a sum of areas of approximating rectangles.

PROPERTIES

> Assume that f and g are integrable over $[a, b]$.
>
> **3.** If k is a constant, then
>
> $$\int_a^b k f(x)\,dx = k \int_a^b f(x)\,dx$$
>
> **4.** $$\int_a^b [f(x) + g(x)]\,dx = \int_a^b f(x)\,dx + \int_a^b g(x)\,dx$$
>
> **5.** If f is integrable over an interval containing the three numbers a, b, and c, then
>
> $$\int_a^b f(x)\,dx = \int_a^c f(x)\,dx + \int_c^b f(x)\,dx$$

We prove Property (4) to illustrate how the definition of the definite integral can be used to prove its properties.

Proof of (4) We choose a sequence of partitions $P = [x_0, x_1, \ldots, x_n]$ of $[a, b]$ into n subintervals, $n = 1, 2, \ldots$, with $\|P\| \to 0$, $\Delta x_k = x_k - x_{k-1}$, and $c_k \in [x_{k-1}, x_k]$, and use the definition of definite integrals.

$$\int_a^b [f(x) + g(x)]\,dx = \lim_{\|P\|\to 0} \sum_{k=1}^{n} [f(c_k) + g(c_k)]\,\Delta x_k$$

$$= \lim_{\|P\|\to 0} \sum_{k=1}^{n} [f(c_k)\,\Delta x_k + g(c_k)\,\Delta x_k]$$

Applying the sum rule for finite sums, we find

$$= \lim_{\|P\| \to 0} \left[\sum_{k=1}^{n} f(c_k) \, \Delta x_k + \sum_{k=1}^{n} g(c_k) \, \Delta x_k \right]$$

Since f and g are integrable, the individual limits exist, and we find

$$= \lim_{\|P\| \to 0} \sum_{k=1}^{n} f(c_k) \, \Delta x_k + \lim_{\|P\| \to 0} \sum_{k=1}^{n} g(c_k) \, \Delta x_k$$

Using the definition again, we obtain

$$= \int_{a}^{b} f(x) \, dx + \int_{a}^{b} g(x) \, dx \qquad ■$$

Property (5) is an addition property. Rather than proving the statement, we give two special cases (illustrated in Figures 6.16 and 6.17). In the first case (Figure 6.16), $a < c < b$ and $f(x) \geq 0$ for $x \in [a, b]$. The definite integral $\int_{a}^{b} f(x) \, dx$ can then be interpreted as the area between the graph of $f(x)$ and the x-axis from a to b. We see from Figure 6.16 that this area is composed of two areas: the area between the graph of $f(x)$ and the x-axis from a to c and the area between the graph of $f(x)$ and the x-axis from c to b. We can

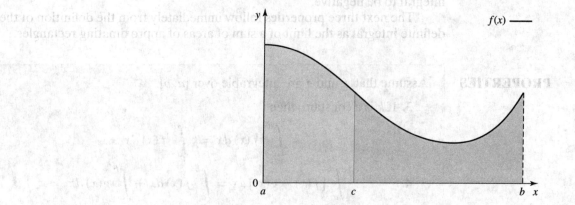

▲ Figure 6.16
Property (5) when $a < c < b$

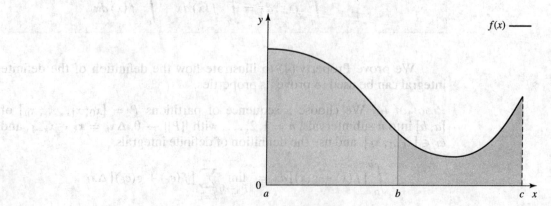

▲ Figure 6.17
Property (5) when $a < b < c$

express this relationship mathematically as

$$\int_a^b f(x)\, dx = \int_a^c f(x)\, dx + \int_c^b f(x)\, dx$$

which is Property (5) in this special case.

In the second case we wish to discuss, $a < b < c$ and $f(x) \geq 0$ for $x \in [a, c]$ (Figure 6.17). From Figure 6.17, we see that

$$\int_a^b f(x)\, dx = \int_a^c f(x)\, dx - \int_b^c f(x)\, dx$$

But because of Property (2),

$$\int_b^c f(x)\, dx = -\int_c^b f(x)\, dx$$

and, therefore,

$$\int_a^b f(x)\, dx = \int_a^c f(x)\, dx + \int_c^b f(x)\, dx$$

as stated in Property (5).

Property (5) is much more general: The function f need not be positive as in Figures 6.16 and 6.17 (it merely needs to be integrable), and the numbers a, b, and c can be arranged in any order on the number line (not just $a < c < b$, as in Figure 6.16, or $a < b < c$, as in Figure 6.17). The next example shows how to use this property.

▶ **Example 12** Given that $\int_0^a x^2 = a^3/3$, evaluate

$$\int_1^4 (3x^2 + 2)\, dx$$

Solution

$$\int_1^4 (3x^2 + 2)\, dx = 3\int_1^4 x^2\, dx + \int_1^4 2\, dx$$

To evaluate $\int_1^4 x^2\, dx$, use the addition property (5), and write

$$\int_1^4 x^2\, dx = \int_1^0 x^2\, dx + \int_0^4 x^2\, dx$$

Since

$$\int_1^0 x^2\, dx = -\int_0^1 x^2\, dx$$

it follows that

$$\int_1^4 x^2\, dx = -\int_0^1 x^2\, dx + \int_0^4 x^2\, dx$$

which can be evaluated using $\int_0^a x^2 = a^3/3$. To evaluate $\int_1^4 2\, dx$, we note that $y = 2$ is a horizontal line that intersects the y-axis at $y = 2$. The region under

$y = 2$ from 1 to 4 is therefore a rectangle with base $4 - 1 = 3$ and height 2. Therefore,

$$\int_1^4 (3x^2 + 2)\, dx = 3\left[\int_0^4 x^2\, dx - \int_0^1 x^2\, dx\right] + \int_1^4 2\, dx$$

$$= 3\left(\frac{4^3}{3} - \frac{1^3}{3}\right) + (2)(3)$$

$$= 64 - 1 + 6 = 69 \qquad \blacktriangleleft$$

The next three properties are called order properties. They allow us either to compare definite integrals or say something about how big or small a particular definite integral can be. We will first state the properties and then explain what they mean geometrically.

PROPERTIES

Assume that f and g are integrable over $[a, b]$.

6. If $f(x) \geq 0$ on $[a, b]$, then $\displaystyle\int_a^b f(x)\, dx \geq 0$.

7. If $f(x) \leq g(x)$ on $[a, b]$, then $\displaystyle\int_a^b f(x)\, dx \leq \int_a^b g(x)\, dx$.

8. If $m \leq f(x) \leq M$ on $[a, b]$, then

$$m(b - a) \leq \int_a^b f(x)\, dx \leq M(b - a)$$

Property (6), illustrated in Figure 6.18, says that if f is nonnegative over the interval $[a, b]$, then the definite integral over this interval is also nonnegative. We can understand this from the geometric interpretation: If $f(x) \geq 0$ for $x \in [a, b]$, then $\int_a^b f(x)\, dx$ is the area between the curve and the x-axis between a and b. But an area must be a nonnegative number.

Property (7) is explained in Figure 6.19 when both f and g are positive functions on $[a, b]$. We use the fact that in this case, the definite integral can be interpreted as an area. Looking at Figure 6.19, we see that the function

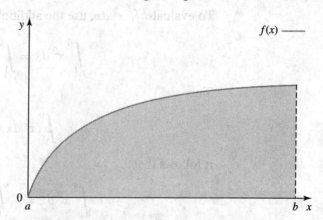

▲ **Figure 6.18**
An illustration of Property (6)

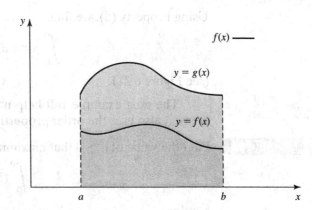

▲ Figure 6.19
An illustration of Property (7)

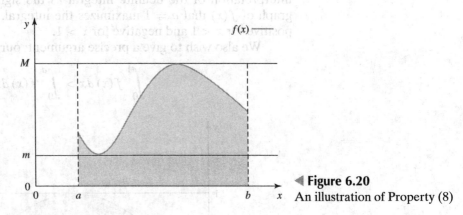

◀ Figure 6.20
An illustration of Property (8)

f has a smaller area than g. Property (7) holds without the assumption that both f and g are positive, and we can draw an analogous figure for the general case as well. The definite integral then needs to be interpreted as a signed area.

Property (8) is explained in Figure 6.20 for $f(x) \geq 0$ in $[a, b]$. We see that the rectangle with height m is contained in the area between the graph of f and the x-axis, which in turn is contained in the rectangle with height M. Since $m(b - a)$ is the area of the small rectangle, $\int_a^b f(x)\, dx$ is the area between the graph of f and the x-axis for nonnegative f, and $M(b - a)$ is the area of the big rectangle, the inequalities in (8) follow. Note that the statement does not require that f be nonnegative; you can draw an analogous figure when f is negative on parts or all of $[a, b]$.

The next example illustrates how these order properties are used.

▶ **Example 13**   Show that

$$0 \leq \int_0^{\pi} \sin x \, dx \leq \pi$$

Solution

Note that $0 \leq \sin x \leq 1$ for $x \in [0, \pi]$. Using Property (6), we find

$$\int_0^{\pi} \sin x \, dx \geq 0$$

Using Property (8), we find

$$\int_0^\pi \sin x \, dx \le (1)(\pi) = \pi$$

(see Figure 6.21).   ◄

The next example will help us to deepen our understanding of signed areas; it also uses the order properties.

▶ Example 14   Find the value of $a \ge 0$ that maximizes

$$\int_0^a (1 - x^2) \, dx$$

Solution

We graph the integrand $f(x) = 1 - x^2$ for $x \ge 0$ in Figure 6.22. Using the interpretation of the definite integral as the signed area, we see from the graph of $f(x)$ that $a = 1$ maximizes the integral, since the graph of $f(x)$ is positive for $x < 1$ and negative for $x > 1$.

We also wish to give a precise argument; our goal is to show that

$$\int_0^1 f(x) \, dx > \int_0^a f(x) \, dx$$

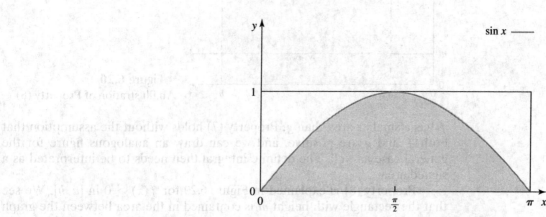

▲ Figure 6.21
An illustration of the integral in Example 13. The shaded area is non-negative and less than the area of the rectangle with base length π and height 1

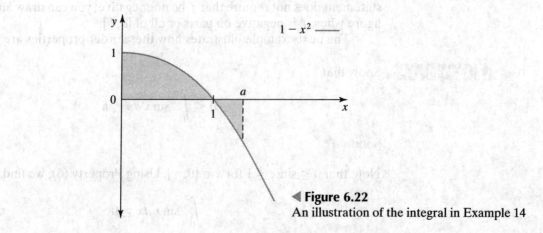

◄ Figure 6.22
An illustration of the integral in Example 14

for all $a \geq 0$, provided that $a \neq 1$. This would then imply that $a = 1$ maximizes the integral

$$\int_0^a (1 - x^2)\,dx$$

First, note that $f(x)$ is continuous for $x \geq 0$ and that

$$f(x) \begin{cases} > 0 & \text{for } 0 \leq x < 1 \\ < 0 & \text{for } x > 1 \end{cases}$$

which implies that for $0 \leq a < 1$,

$$\int_a^1 f(x)\,dx > 0$$

and, therefore, for $0 \leq a < 1$,

$$\int_0^1 f(x)\,dx = \int_0^a f(x)\,dx + \underbrace{\int_a^1 f(x)\,dx}_{>0} > \int_0^a f(x)\,dx \qquad (6.2)$$

On the other hand, for $a > 1$,

$$\int_1^a f(x)\,dx < 0$$

and, therefore, for $a > 1$,

$$\int_0^a f(x)\,dx = \int_0^1 f(x)\,dx + \underbrace{\int_1^a f(x)\,dx}_{<0} < \int_0^1 f(x)\,dx \qquad (6.3)$$

Combining (6.2) and (6.3) shows that

$$\int_0^a f(x)\,dx < \int_0^1 f(x)\,dx$$

for all $a \geq 0$ and $a \neq 1$. Hence, $a = 1$ maximizes the integral $\int_0^a (1 - x^2)\,dx$. ◀

6.1.4 Problems

(6.1.1)

1. Approximate the area under the parabola $y = x^2$ from 0 to 1, using four equal subintervals with left endpoints.

2. Approximate the area under the parabola $y = x^2$ from 0 to 1, using five equal subintervals with midpoints.

3. Approximate the area under the parabola $y = x^2$ from 0 to 1, using four equal subintervals with right endpoints.

4. Approximate the area under the parabola $y = 1 - x^2$ from 0 to 1, using five equal subintervals with (a) left endpoints, and (b) right endpoints.

In Problems 5–14, write each sum in expanded form.

5. $\displaystyle\sum_{k=1}^{4} \sqrt{k}$

6. $\displaystyle\sum_{k=3}^{6} k^2$

7. $\displaystyle\sum_{k=0}^{5} 2^k$

8. $\displaystyle\sum_{k=1}^{3} \frac{k}{k+1}$

9. $\displaystyle\sum_{k=0}^{4} x^k$

10. $\displaystyle\sum_{k=0}^{4} k^x$

11. $\sum_{k=0}^{3} (-1)^k$

12. $\sum_{k=1}^{n} f(c_k) \Delta x_k$

13. $\sum_{k=1}^{n} \left(\frac{k}{n}\right)^2 \frac{1}{n}$

14. $\sum_{k=1}^{n} \cos\left(k\frac{\pi}{n}\right) \frac{\pi}{n}$

In Problems 15–22, write each sum in sigma notation.

15. $1 + 2 + 3 + 4 + 5 + 6$

16. $\frac{1}{\sqrt{1}} + \frac{1}{\sqrt{2}} + \frac{1}{\sqrt{3}} + \frac{1}{\sqrt{4}}$

17. $\ln 2 + \ln 3 + \ln 4 + \ln 5$

18. $\frac{3}{5} + \frac{4}{6} + \frac{5}{7} + \frac{6}{8} + \frac{7}{9}$

19. $2 + 4 + 6 + 8 + \cdots + 2n$

20. $\frac{1}{1} + \frac{1}{2} + \frac{1}{4} + \frac{1}{8} + \frac{1}{16} + \cdots + \frac{1}{2^n}$

21. $1 + q + q^2 + q^3 + q^4 + \cdots + q^{n-1}$

22. $1 - a + a^2 - a^3 + a^4 - a^5 + \cdots + (-1)^n a^n$

In Problems 23–30, use the algebraic rules for sums to evaluate each sum. Recall that

$$\sum_{k=1}^{n} k = \frac{n(n+1)}{2}$$

and

$$\sum_{k=1}^{n} k^2 = \frac{n(n+1)(2n+1)}{6}$$

23. $\sum_{k=1}^{20} (3k + 2)$

24. $\sum_{k=1}^{10} (2 - k^2)$

25. $\sum_{k=0}^{6} k(k + 1)$

26. $\sum_{k=1}^{n} 3k$

27. $\sum_{k=1}^{n} 2k^2$

28. $\sum_{k=1}^{n} (k + 2)(k - 2)$

29. $\sum_{k=1}^{10} (-1)^k$

30. $\sum_{k=1}^{11} (-1)^k$

31. The following steps will show that

$$\sum_{k=1}^{n} k^2 = \frac{n(n+1)(2n+1)}{6}$$

(a) Show that

$$\sum_{k=1}^{n} [(1+k)^3 - k^3]$$

$$= (2^3 - 1^3) + (3^3 - 2^3) + (4^3 - 3^3)$$

$$+ \cdots + [(1+n)^3 - n^3]$$

$$= (1+n)^3 - 1^3$$

(Sums that "collapse" like this due to cancellation of terms are called telescoping or collapsing sums.)

(b) Use the algebraic rules for sums and Example 3 to show that

$$\sum_{k=1}^{n} [(1+k)^3 - k^3]$$

$$= 3\sum_{k=1}^{n} k^2 + 3\frac{n(n+1)}{2} + n$$

(c) In (a) and (b), we found two expressions for the sum

$$\sum_{k=1}^{n} [(1+k)^3 - k^3]$$

They are therefore equal; that is,

$$(1+n)^3 - 1^3$$

$$= 3\sum_{k=1}^{n} k^2 + 3\frac{n(n+1)}{2} + n$$

Solve this equation for $\sum_{k=1}^{n} k^2$, and show that

$$\sum_{k=1}^{n} k^2 = \frac{n(n+1)(2n+1)}{6}$$

(6.1.2)

32. Approximate

$$\int_{-1}^{1} (1 - x^2)\, dx$$

using five equal subintervals and left endpoints.

33. Approximate

$$\int_{-1}^{1} (1 - x^2)\, dx$$

using five equal subintervals and midpoints.

34. Approximate

$$\int_{-1}^{1} (1 - x^2)\, dx$$

using five equal subintervals and right endpoints.

35. Approximate

$$\int_{-2}^{2} (2 + x^2)\, dx$$

using four equal subintervals, with left endpoints.

36. Approximate

$$\int_{-1}^{2} e^{-x}\, dx$$

using three equal subintervals, with midpoints.

37. Approximate

$$\int_0^{3\pi/2} \sin x \, dx$$

using three equal subintervals, with right endpoints.

38. (a) Assume that $a > 0$. Evaluate $\int_0^a x \, dx$ using the fact that the region bounded by $y = x$ and the x-axis between 0 to a is a triangle (see Figure 6.23).

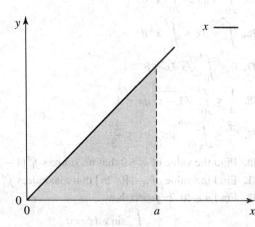

▲ **Figure 6.23**
The region for Problem 38

(b) Assume that $a > 0$. Evaluate $\int_0^a x \, dx$ by approximating the region bounded by $y = x$ and the x-axis from 0 to a with rectangles. Use equal subintervals and take right endpoints. (*Hint:* Use the result in Example 3 to evaluate the sum of the areas of the rectangles.)

39. Assume that $0 < a < b < \infty$. Use a geometric argument to show that

$$\int_a^b x \, dx = \frac{b^2 - a^2}{2}$$

40. Assume that $0 < a < b < \infty$. Use a geometric argument and Example 1 to show that

$$\int_a^b x^2 \, dx = \frac{b^3 - a^3}{3}$$

Express the limits in Problems 41–47 as definite integrals. Note that $P = [x_0, x_1, \ldots, x_n]$ is a partition of the indicated interval, $c_k \in [x_{k-1}, x_k]$, and $\Delta x_k = x_k - x_{k-1}$.

41. $\displaystyle \lim_{\|P\| \to 0} \sum_{k=1}^{n} c_k^2 \Delta x_k$, where P is a partition of $[0, 1]$

42. $\displaystyle \lim_{\|P\| \to 0} \sum_{k=1}^{n} \sqrt{c_k} \Delta x_k$, where P is a partition of $[1, 4]$

43. $\displaystyle \lim_{\|P\| \to 0} \sum_{k=1}^{n} (2c_k - 1) \Delta x_k$, where P is a partition of $[-3, 2]$

44. $\displaystyle \lim_{\|P\| \to 0} \sum_{k=1}^{n} \frac{1}{c_k + 1} \Delta x_k$, where P is a partition of $[1, 2]$

45. $\displaystyle \lim_{\|P\| \to 0} \sum_{k=1}^{n} (c_k - 1)(c_k + 2) \Delta x_k$, where P is a partition of $[-3, 3]$

46. $\displaystyle \lim_{\|P\| \to 0} \sum_{k=1}^{n} (\sin c_k) \Delta x_k$, where P is a partition of $[0, \pi]$

47. $\displaystyle \lim_{\|P\| \to 0} \sum_{k=1}^{n} e^{c_k} \Delta x_k$, where P is a partition of $[-5, 2]$

In Problems 48–53, express the definite integrals as limits of Riemann sums.

48. $\displaystyle \int_{-2}^{-1} \frac{x}{1 - x} \, dx$

49. $\displaystyle \int_2^6 \sqrt{x + 1} \, dx$

50. $\displaystyle \int_1^3 e^{-x} \, dx$

51. $\displaystyle \int_1^e \ln x \, dx$

52. $\displaystyle \int_0^\pi \cos(2x) \, dx$

53. $\displaystyle \int_0^5 g(x) \, dx$, where $g(x)$ is a continuous function on $[0, 5]$

In Problems 54–60, use a graph to interpret the definite integral in terms of areas. Do not compute the integrals.

54. $\displaystyle \int_{-2}^3 (x - 3) \, dx$

55. $\displaystyle \int_{-1}^2 (x^2 - 1) \, dx$

56. $\displaystyle \int_{-2}^2 \frac{1}{2} x^3 \, dx$

57. $\displaystyle \int_0^5 e^{-x} \, dx$

58. $\displaystyle \int_{-\pi}^\pi \cos x \, dx$

59. $\displaystyle \int_{1/2}^4 \ln x \, dx$

60. $\displaystyle \int_{-3}^2 \left(1 - \frac{1}{2} x\right) dx$

In Problems 61–67, find the value of each integral by interpreting it as the (signed) area under the graph of an appropriately chosen function, and using an area formula from geometry.

61. $\displaystyle \int_{-2}^3 |x| \, dx$

62. $\displaystyle \int_{-3}^3 \sqrt{9 - x^2} \, dx$

63. $\displaystyle \int_2^5 \left(\frac{1}{2} x - 4\right) dx$

64. $\displaystyle \int_{1/2}^1 \sqrt{1 - x^2} \, dx$

65. $\int_{-2}^{2} (\sqrt{4-x^2} - 2)\, dx$

66. $\int_{0}^{1} \sqrt{2-x^2}\, dx$

67. $\int_{-3}^{0} (4 - \sqrt{9-x^2})\, dx$

(6.1.3)

68. Given that

$$\int_{0}^{a} x^2\, dx = \frac{1}{3}a^3$$

evaluate the following.

(a) $\int_{0}^{2} \frac{1}{2}x^2\, dx$

(b) $\int_{-3}^{0} 2x^2\, dx$

(c) $\int_{1}^{3} \frac{1}{3}x^2\, dx$

(d) $\int_{1}^{1} 3x^2\, dx$

(e) $\int_{-2}^{3} \frac{3}{2}x^2\, dx$

(f) $\int_{2}^{4} (x-2)^2\, dx$

69. Find $\int_{2}^{2} \cos(3x^2)\, dx$.

70. Find $\int_{-3}^{-3} e^{-x^2/2}\, dx$.

71. Find $\int_{-2}^{2} (x+1)\, dx$.

72. Find $\int_{0}^{-5} (1-x)\, dx$.

73. Find $\int_{-1}^{1} \tan x\, dx$.

74. Explain geometrically why

$$\int_{1}^{2} x^2\, dx = \int_{0}^{2} x^2\, dx - \int_{0}^{1} x^2\, dx \qquad (6.4)$$

and show that (6.4) can be written as

$$\int_{1}^{2} x^2\, dx = \int_{1}^{0} x^2\, dx + \int_{0}^{2} x^2\, dx \qquad (6.5)$$

Relate (6.5) to addition property (5).

In Problems 75–79, verify each inequality without evaluating the integrals.

75. $\int_{0}^{1} x\, dx \geq \int_{0}^{1} x^2\, dx$

76. $\int_{1}^{2} x\, dx \leq \int_{1}^{2} x^2\, dx$

77. $0 \leq \int_{0}^{4} \sqrt{x}\, dx \leq 8$

78. $\frac{1}{2} \leq \int_{0}^{1} \sqrt{1-x^2}\, dx \leq 1$

79. $\frac{\pi}{3} \leq \int_{\pi/6}^{5\pi/6} \sin x\, dx \leq \frac{2\pi}{3}$

80. Find the value of $a \geq 0$ that maximizes $\int_{0}^{a} (4-x^2)\, dx$.

81. Find the value of $a \in [0, 2\pi]$ that maximizes $\int_{0}^{a} \cos x\, dx$.

82. Find $a \in (0, 2\pi]$ such that

$$\int_{0}^{a} \sin x\, dx = 0$$

83. Find $a > 1$ such that

$$\int_{1}^{a} (x-2)^3\, dx = 0$$

84. Find $a > 0$ such that

$$\int_{-a}^{a} (1-|x|)\, dx = 0$$

85. To determine age-specific mortality, a group of individuals, all born at the same time, is followed over time. If $N(t)$ denotes the number still alive at time t, then $N(t)/N(0)$ is the fraction surviving at time t. The quantity $r(t)$, called the hazard rate function, measures the rate at which individuals die at time t, namely, $r(t)\, dt$ is the probability that an individual that is alive at time t dies during the infinitesimal time interval $(t, t + dt)$. The cumulative hazard during the time interval $[0, t]$, $\int_{0}^{t} r(s)\, ds$, can be estimated as $-\ln \frac{N(t)}{N(0)}$. Show that the cumulative hazard during the time interval $[t, t + 1]$, $\int_{t}^{t+1} r(s)\, ds$, can be estimated as $-\ln \frac{N(t+1)}{N(t)}$.

6.2 THE FUNDAMENTAL THEOREM OF CALCULUS

In Section 6.1, we used the definition of definite integrals to compute $\int_{0}^{a} x^2\, dx$. This required the summation of a large number of terms, which was facilitated by the explicit summation formula for $\sum_{k=1}^{n} k^2$. Fermat and others were able to carry out similar calculations for the area under curves of the form $y = x^r$, where r was a rational number different from -1. The solution to the case $r = -1$ was found by the Belgian mathematician Gregory of St. Vincent (1584–1667) and published in 1647. At that time, it seemed that methods specific to a given function needed to be developed to compute the area under the curve of that function. This would not have been practical.

Fortunately, it turns out that the area problem is related to the tangent problem. This is not at all obvious; among the first to notice this relationship were Isaac Barrow (1630–1677) and James Gregory (1638–1675). Each presented this relationship in geometrical terms, without realizing the importance of their discovery.

Both Newton and Leibniz are to be credited with systematically developing the connection between the tangent and area problems, which ultimately resulted in a method for computing areas and solving problems that can be translated into area problems. The result is known as the fundamental theorem of calculus, which says that the tangent and area problems are inversely related.

The fundamental theorem of calculus has two parts: The first part links antiderivatives and integrals, and the second part provides a method for computing definite integrals.

6.2.1 The Fundamental Theorem of Calculus (Part I)

Let $f(x)$ be a continuous function on $[a, b]$, and let

$$F(x) = \int_a^x f(u)\, du$$

Geometrically, $F(x) = \int_a^x f(u)\, du$ represents the signed area between the graph of $f(u)$ and the horizontal axis between a and x (see Figure 6.24). Note that the independent variable x appears as the upper limit of integration. We can now ask how the signed area $F(x)$ changes as x varies. To answer this question, we compute $\frac{dF}{dx}$ using the definition of the derivative. That is,

$$\frac{d}{dx} F(x) = \lim_{h \to 0} \frac{F(x+h) - F(x)}{h}$$

$$= \lim_{h \to 0} \frac{1}{h} \left[\int_a^{x+h} f(u)\, du - \int_a^x f(u)\, du \right] \tag{6.6}$$

$$= \lim_{h \to 0} \frac{1}{h} \int_x^{x+h} f(u)\, du$$

[In the last step, we used property (5) of Subsection 6.1.3.] To evaluate

$$\lim_{h \to 0} \frac{1}{h} \int_x^{x+h} f(u)\, du$$

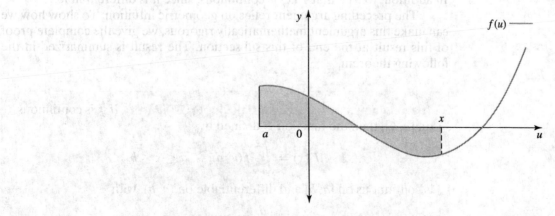

▲ **Figure 6.24**
The shaded signed area is $F(x)$

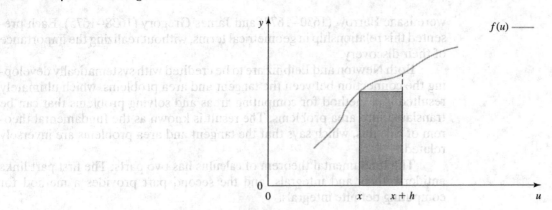

▲ Figure 6.25
The approximate rectangle in the fundamental theorem of calculus

we will resort to the geometric interpretation of definite integrals. The following argument is illustrated in Figure 6.25. Note that

$$\int_x^{x+h} f(u)\, du$$

is the signed area of the region bounded by the graph of $f(u)$ and the horizontal axis between x and $x + h$. If h is small, then this area is closely approximated by the area of the inscribed rectangle with height $|f(x)|$. The signed area of this rectangle is $f(x)h$. Hence,

$$\int_x^{x+h} f(u)\, du \approx f(x)h$$

If we divide both sides by h and then let h tend to 0, then

$$\lim_{h \to 0} \frac{1}{h} \int_x^{x+h} f(u)\, du = f(x) \tag{6.7}$$

(as will be shown mathematically rigorously at the end of this subsection). Combining (6.7) and (6.7), we arrive at the remarkable result

$$\frac{d}{dx} F(x) = f(x)$$

In addition, we see that $F(x)$ is continuous, since it is differentiable.

The preceding argument relies on geometric intuition. To show how we can make this argument mathematically rigorous, we give the complete proof of this result at the end of this subsection. The result is summarized in the following theorem.

The Fundamental Theorem of Calculus (FTC) (Part I) If f is continuous on $[a, b]$, then the function F defined by

$$F(x) = \int_a^x f(u)\, du, \quad a \le x \le b$$

is continuous on $[a, b]$ and differentiable on (a, b), with

$$\frac{d}{dx} F(x) = f(x)$$

Simply stated, the FTC (part I) says that if we first integrate $f(x)$ and then differentiate the result, we get $f(x)$ again. In this sense, it shows that integration and differentiation are inverse operations.

We begin with an example that can be immediately solved using FTC.

▶ **Example 1**   Compute

$$\frac{d}{dx} \int_0^x (\sin u - e^{-u}) \, du$$

for $x > 0$.

Solution

First, note that $f(x) = \sin x - e^{-x}$ is continuous for $x \geq 0$. If we set $F(x) = \int_0^x (\sin u - e^{-u}) \, du$ and apply the FTC, then

$$\frac{d}{dx} F(x) = \frac{d}{dx} \int_0^x (\sin u - e^{-u}) \, du = \sin x - e^{-x}$$

◀

▶ **Example 2**   Compute

$$\frac{d}{dx} \int_3^x \frac{1}{1 + u^2} \, du$$

for $x > 3$.

Solution

First, note that $f(x) = \frac{1}{1+x^2}$ is continuous for $x \geq 3$. If we set $F(x) = \int_3^x \frac{1}{1+u^2} \, du$ and apply the FTC, then

$$\frac{d}{dx} F(x) = \frac{d}{dx} \int_3^x \frac{1}{1 + u^2} \, du = \frac{1}{1 + x^2}$$

◀

The remainder of this subsection can be omitted.

Leibniz's Rule (Optional)   Combining the chain rule and the FTC (part I), we can differentiate integrals with respect to x when the upper and/or lower limits of integration are functions of x.

In the first example the upper limit of integration is a function of x.

▶ **Example 3**   Compute

$$\frac{d}{dx} \int_0^{x^2} (u^3 - 2) \, du, \quad x > 0$$

Solution

Note that $f(u) = u^3 - 2$ is continuous for all $u \in \mathbf{R}$. We set $F(v) = \int_0^v (u^3 - 2) \, du$, $v > 0$. Then, for $x > 0$,

$$F(x^2) = \int_0^{x^2} (u^3 - 2) \, du$$

We wish to compute $\frac{d}{dx} F(x^2)$. To do so, we need to apply the chain rule. We set $v(x) = x^2$. Then

$$\frac{d}{dx} F(x^2) = \frac{dF(v)}{dv} \frac{dv}{dx}$$

To evaluate $\frac{d}{dv}F(v) = \frac{d}{dv}\int_0^v (u^3 - 2)\,du$, we use the FTC:

$$\frac{d}{dv}\int_0^v (u^3 - 2)\,du = v^3 - 2$$

Since $\frac{dv}{dx} = \frac{d}{dx}(x^2) = 2x$, we therefore find that

$$\frac{d}{dx}F(x^2) = (v^3 - 2)2x = [(x^2)^3 - 2]2x$$

$$= (x^6 - 2)2x \qquad \blacktriangleleft$$

Thus far, we have dealt only with the case where the upper limit of integration depends on x. The following example shows what we must do when the lower limit of integration depends on x.

▶ **Example 4**   Compute

$$\frac{d}{dx}\int_{\sin x}^1 u^2\,du$$

Solution

Note that $f(u) = u^2$ is continuous for all $u \in \mathbf{R}$. We use the fact that

$$\int_{\sin x}^1 u^2\,du = -\int_1^{\sin x} u^2\,du$$

The upper limit now depends on x, but we introduced a minus sign. Hence

$$\frac{d}{dx}\int_{\sin x}^1 u^2\,du = -\frac{d}{dx}\int_1^{\sin x} u^2\,du$$

$$= -(\sin x)^2 \cos x \qquad \blacktriangleleft$$

The preceding example makes an important point: We need to be careful about whether the upper or the lower limit of integration depends on x. In the following example, we show what we must do when both limits of integration depend on x.

▶ **Example 5**   For $x \in \mathbf{R}$, compute

$$\frac{d}{dx}\int_{x^2}^{x^3} e^u\,du$$

Solution

Note that $f(u) = e^u$ is continuous for all $u \in \mathbf{R}$. The given integral is therefore defined for all $x \in \mathbf{R}$, and we can split it into two integrals at any $a \in \mathbf{R}$. We choose $a = 0$, which yields

$$\int_{x^2}^{x^3} e^u\,du = \int_{x^2}^0 e^u\,du + \int_0^{x^3} e^u\,du = -\int_0^{x^2} e^u\,du + \int_0^{x^3} e^u\,du$$

The right-hand side is now written in a form that we know how to differentiate, and we find that

$$\frac{d}{dx}\int_{x^2}^{x^3} e^u\,du = -\frac{d}{dx}\int_0^{x^2} e^u\,du + \frac{d}{dx}\int_0^{x^3} e^u\,du$$

$$= -\left[e^{x^2}\frac{d}{dx}x^2\right] + \left[e^{x^3}\frac{d}{dx}x^3\right]$$

$$= -e^{x^2}2x + e^{x^3}3x^2 \qquad \blacktriangleleft$$

The preceding example illustrates the most general case that we can encounter, namely, when both limits of integration depend on x. We summarize this case in the following box, known as Leibniz's rule.

LEIBNIZ'S RULE

If $g(x)$ and $h(x)$ are differentiable functions and $f(u)$ is continuous for u between $g(x)$ and $h(x)$, then

$$\frac{d}{dx} \int_{g(x)}^{h(x)} f(u)\, du = f[h(x)]h'(x) - f[g(x)]g'(x)$$

We can check that Examples 3–5 can be solved using this formula; for instance, in Example 4, we have $f(u) = u^2$, $g(x) = \sin x$, and $h(x) = 1$. Then $g'(x) = \cos x$ and $h'(x) = 0$. We therefore find that

$$f[h(x)]h'(x) - f[g(x)]g'(x) = 0 - (\sin x)^2 \cos x$$

which is the answer we found in Example 4.

Proof of the Fundamental Theorem of Calculus (Part I) (Optional) At the beginning of this subsection, we found that if

$$F(x) = \int_a^x f(u)\, du$$

then

$$\frac{d}{dx} F(x) = \lim_{h \to 0} \frac{1}{h} \int_x^{x+h} f(u)\, du \tag{6.8}$$

We will now give a mathematically rigorous argument that will show that

$$\frac{d}{dx} F(x) = f(x)$$

We begin with the observation that the continuous function $f(u)$ defined on the closed interval $[x, x + h]$ attains an absolute minimum and an absolute maximum on $[x, x + h]$, according to the extreme value theorem. That is, there exist m and M such that m is the minimum of f on $[x, x + h]$ and M is the maximum of f on $[x, x + h]$, which implies that

$$m \le f(u) \le M \quad \text{for all } u \in [x, x + h] \tag{6.9}$$

Of course, m and M depend on both x and h. Applying property (8) from Subsection 6.1.3 to (6.9), we find

$$\int_x^{x+h} m\, du \le \int_x^{x+h} f(u)\, du \le \int_x^{x+h} M\, du$$

and hence

$$mh \le \int_x^{x+h} f(u)\, du \le Mh$$

Dividing by h, we obtain

$$m \le \frac{1}{h} \int_x^{x+h} f(u)\, du \le M \tag{6.10}$$

We set

$$I = \frac{1}{h} \int_x^{x+h} f(u)\, du$$

then (6.10) reads $m \leq I \leq M$. That is, I is a number between m and M. We compare this with (6.9), which says that $f(u)$ also lies between m, the minimum of f on $[x, x+h]$, and M, the maximum of f on $[x, x+h]$, for all $u \in [x, x+h]$. The intermediate value theorem applied to $f(u)$ tells us that any value between m and M is attained by $f(u)$ for some number in the interval $[x, x+h]$. Specifically, there must exist a number $c_h \in [x, x+h]$ such that $f(c_h) = I$, since I lies between m and M; that is,

$$f(c_h) = \frac{1}{h} \int_x^{x+h} f(u)\, du \tag{6.11}$$

Since $x \leq c_h \leq x + h$, it follows that

$$\lim_{h \to 0} c_h = x$$

Since f is continuous,

$$\lim_{h \to 0} f(c_h) = f\left(\lim_{h \to 0} c_h\right) = f(x) \tag{6.12}$$

Combining (6.8), (6.11), and (6.12), we find the result

$$\frac{d}{dx} F(x) = f(x) \qquad ■$$

6.2.2 Antiderivatives and Indefinite Integrals

The first part of the fundamental theorem of calculus tells us that if

$$F(x) = \int_a^x f(u)\, du$$

then $F'(x) = f(x)$ [provided $f(x)$ is continuous over the range of integration]. This says that $F(x)$ is an antiderivative of $f(x)$ (we introduced antiderivatives in Section 5.8). Now, if we let

$$F(x) = \int_a^x f(u)\, du \quad \text{and} \quad G(x) = \int_b^x f(u)\, du$$

where a and b are two numbers, then both integrals have the same derivative, namely $F'(x) = G'(x) = f(x)$ [provided $f(x)$ is continuous over the range of integration]. That is, both $F(x)$ and $G(x)$ are antiderivatives of $f(x)$. We saw in Section 5.8 that antiderivatives of a given function differ only by a constant; we can check this for $F(x)$ and $G(x)$, namely,

$$F(x) = \int_a^x f(u)\, du = \int_a^b f(u)\, du + \int_b^x f(u)\, du = C + G(x)$$

where C is a constant, denoting the number $\int_a^b f(u)\, du$.

The general antiderivative of a function $f(x)$ is $F(x) + C$, where $F'(x) = f(x)$ and C is a constant. We therefore see that $C + \int_a^x f(u)\, du$ is the general antiderivative of $f(x)$. We will use the notation $\int f(x)\, dx$ to denote both the general antiderivative of $f(x)$ and the function $C + \int_a^x f(u)\, du$, that is,

$$\int f(x)\, dx = C + \int_a^x f(u)\, du \tag{6.13}$$

We call $\int f(x)\, dx$ an **indefinite integral**. Thus, the first part of the FTC says that indefinite integrals and antiderivatives are the same.

When we write $\int_a^x f(u)\, du$, we use a letter other than x in the integrand because x already appears as the upper limit of integration. However, in the symbolic notation $\int f(x)\, dx$ we write x. This is to be interpreted as in

(6.13). The notation $\int f(x)\,dx$ is a convenient shorthand for $C + \int_a^x f(u)\,du$. The choice of value of a for the lower limit of integration on the right side of (6.13) is not important, because different indefinite integrals of the same function $f(x)$ differ only by an additive constant that can be absorbed in the constant C.

Examples 6–8 show how to compute indefinite integrals.

▶ **Example 6** Compute $\int x^4\,dx$.

Solution

We need to find a function $F(x)$ such that $F'(x) = x^4$. The solution is

$$\int x^4\,dx = \frac{1}{5}x^5 + C$$

where C is a constant. We check that indeed

$$\frac{d}{dx}\left(\frac{1}{5}x^5 + C\right) = x^4$$

◀

▶ **Example 7** Compute $\int (e^x + \sin x)\,dx$.

Solution

We need to find an antiderivative of $f(x) = e^x + \sin x$. Since

$$\frac{d}{dx}(e^x - \cos x) = e^x - (-\sin x) = e^x + \sin x$$

we find that

$$\int (e^x + \sin x)\,dx = e^x - \cos x + C$$

◀

When we compute the indefinite integral $\int f(x)\,dx$, we want to know the general antiderivative of $f(x)$; this is why we added the constant C in the previous two examples.

▶ **Example 8** Show that

$$\int \frac{1}{x}\,dx = \ln|x| + C \quad \text{for } x \neq 0$$

Solution

Since the absolute value of x appears on the right-hand side, we split our discussion into two parts, according to whether $x \geq 0$ or $x \leq 0$. Recall that

$$|x| = \begin{cases} x & \text{for } x \geq 0 \\ -x & \text{for } x \leq 0 \end{cases}$$

Since $\ln x$ is not defined at $x = 0$, we consider the two cases $x > 0$ and $x < 0$.

(i) $x > 0$: Since $\ln|x| = \ln x$ when $x > 0$, we find

$$\frac{d}{dx}\ln x = \frac{1}{x}$$

and hence

$$\int \frac{1}{x}\,dx = \ln x + C \quad \text{for } x > 0$$

(ii) $x < 0$: Since $\ln|x| = \ln(-x)$ when $x < 0$, we find

$$\frac{d}{dx}\ln(-x) = \frac{1}{-x}(-1) = \frac{1}{x}$$

and hence

$$\int \frac{1}{x}\,dx = \ln(-x) + C \quad \text{for } x < 0$$

Combining (i) and (ii), we obtain

$$\int \frac{1}{x}\,dx = \ln|x| + C \quad \text{for } x \neq 0 \qquad \blacktriangleleft$$

When we introduced logarithmic and exponential functions in Sections 4.6 and 4.7, we had to resort to the calculator to convince ourselves that e^x and $\ln x$ were indeed the functions we knew from precalculus. To give a mathematically rigorous definition of these functions, we typically start by *defining* $\ln x$ as $\int_1^x \frac{1}{u}\,du$ and then derives the algebraic rules for $\ln x$ from this integral representation. The exponential function e^x is defined as the inverse function of $\ln x$, and the number e is defined so that $\ln e = 1$. This definition is then consistent with the definition in Section 4.6.

We have seen that in order to evaluate indefinite integrals, we must find antiderivatives. We provide a list of indefinite integrals in Table 6.1 (this is a slightly expanded form of the table of antiderivatives from Section 5.8). Examples 9–10 show how to use this list to compute indefinite integrals.

▶ **Example 9** Evaluate

$$\int \frac{1}{\sin^2 x - 1}\,dx$$

Solution

We first work on the integrand. Using the fact that $\sin^2 x + \cos^2 x = 1$, we find

$$\frac{1}{\sin^2 x - 1} = -\frac{1}{\cos^2 x} = -\sec^2 x$$

TABLE 6.1 A COLLECTION OF INDEFINITE INTEGRALS

$$\int x^n\,dx = \frac{x^{n+1}}{n+1} + C \quad (n \neq -1) \qquad \int \frac{1}{x}\,dx = \ln|x| + C$$

$$\int e^x\,dx = e^x + C \qquad\qquad \int a^x\,dx = \frac{a^x}{\ln a} + C$$

$$\int \cos x\,dx = \sin x + C \qquad\qquad \int \sin x\,dx = -\cos x + C$$

$$\int \sec^2 x\,dx = \tan x + C \qquad\qquad \int \csc^2 x\,dx = -\cot x + C$$

$$\int \sec x \tan x\,dx = \sec x + C \qquad\qquad \int \csc x \cot x\,dx = -\csc x + C$$

$$\int \tan x\,dx = \ln|\sec x| + C \qquad\qquad \int \cot x\,dx = -\ln|\csc x| + C$$

$$\int \frac{1}{1+x^2}\,dx = \tan^{-1} x + C \qquad\qquad \int \frac{1}{\sqrt{1-x^2}}\,dx = \sin^{-1} x + C$$

Hence,

$$\int \frac{1}{\sin^2 x - 1}\, dx = -\int \sec^2 x\, dx = -\tan x + C \qquad \blacktriangleleft$$

▶ **Example 10** Evaluate

$$\int \frac{x^2}{x^2 + 1}\, dx$$

Solution

We will first rewrite the integrand. Either by using long division, or by the following algebraic manipulation, we find that

$$\frac{x^2}{x^2 + 1} = \frac{x^2 + 1 - 1}{x^2 + 1} = \frac{x^2 + 1}{x^2 + 1} - \frac{1}{x^2 + 1} = 1 - \frac{1}{x^2 + 1}$$

Hence, using the table of indefinite integrals,

$$\int \frac{x^2}{x^2 + 1}\, dx = \int \left(1 - \frac{1}{x^2 + 1}\right) dx = x - \tan^{-1} x + C \qquad \blacktriangleleft$$

6.2.3 The Fundamental Theorem of Calculus (Part II)

The first part of the fundamental theorem of calculus only allows us to compute integrals of the form $\int_a^x f(u)\, du$ up to an additive constant; for instance,

$$F(x) = \int_1^x u^2\, du = \frac{1}{3}x^3 + C$$

To evaluate the definite integral $F(2) = \int_1^2 u^2\, du$, which represents the area under the graph of $f(x) = x^2$ between $x = 1$ and $x = 2$, we would need to know the value of the constant. This value is provided by the second part of the fundamental theorem of calculus, which allows us to evaluate definite integrals. The following calculation shows us how this constant is determined.

 We saw in the last subsection that if we set

$$G(x) = \int_a^x f(u)\, du$$

then $G(x)$ is an antiderivative of $f(x)$. Furthermore, if $F(x)$ is another antiderivative of $f(x)$, then $G(x)$ and $F(x)$ differ only by an additive constant. That is,

$$G(x) = F(x) + C$$

where C is a constant. Now,

$$G(a) = \int_a^a f(u)\, du = 0$$

Hence,

$$0 = G(a) = F(a) + C$$

which implies that $C = -F(a)$ and, therefore, $G(x) = F(x) - F(a)$, or, using the integral representation of $G(x)$,

$$\int_a^x f(u)\, du = F(x) - F(a)$$

If we set $x = b$, then

$$\int_a^b f(u)\, du = F(b) - F(a)$$

This formula allows us to evaluate definite integrals and is the content of the second part of the fundamental theorem of calculus.

> **The Fundamental Theorem of Calculus (Part II)** Assume that f is continuous on $[a, b]$; then
> $$\int_a^b f(x)\, dx = F(b) - F(a)$$
> where $F(x)$ is an antiderivative of $f(x)$, that is, $F'(x) = f(x)$.

So how do we use this theorem? To compute the definite integral $\int_a^b f(x)\, dx$, when f is continuous on $[a, b]$, we first need to find an antiderivative $F(x)$ of $f(x)$ (any antiderivative will do) and then compute $F(b) - F(a)$. This number is then equal to $\int_a^b f(x)\, dx$. The table of indefinite integrals in the previous subsection will help us to find such antiderivatives. Note that an indefinite integral is a function, whereas a definite integral is simply a number.

Using the FTC (Part II) to Evaluate Definite Integrals

▶ **Example 11**   Evaluate $\int_{-1}^2 (x^2 - 3x)\, dx$.

Solution

Note that $f(x) = x^2 - 3x$ is continuous on $[-1, 2]$. We need to find an antiderivative of $f(x) = x^2 - 3x$; for instance, $F(x) = \frac{1}{3}x^3 - \frac{3}{2}x^2$ is an antiderivative of $f(x)$ since $F'(x) = f(x)$. We then must evaluate $F(2) - F(-1)$:

$$F(2) = \frac{1}{3}2^3 - \frac{3}{2}2^2 = \frac{8}{3} - 6 = -\frac{10}{3}$$

$$F(-1) = \frac{1}{3}(-1)^3 - \frac{3}{2}(-1)^2 = -\frac{1}{3} - \frac{3}{2} = -\frac{11}{6}$$

We find $F(2) - F(-1) = -\frac{10}{3} - (-\frac{11}{6}) = -\frac{9}{6} = -\frac{3}{2}$ and, therefore,

$$\int_{-1}^2 (x^2 - 3x)\, dx = F(2) - F(-1) = -\frac{3}{2}$$

In the preceding calculation, we chose the simplest antiderivative, that is, the one where the constant C is equal to 0. We could have chosen any $C \neq 0$, and the answer would have been the same. Let's see why. The general antiderivative of $f(x)$ is $G(x) = \frac{1}{3}x^3 - \frac{3}{2}x^2 + C$. We can write this as $G(x) = F(x) + C$, where $F(x)$ is the antiderivative we used previously. Then, using $G(x)$ now to evaluate the integral, we find

$$\int_{-1}^2 (x^2 - 3)\, dx = G(2) - G(-1)$$

$$= [F(2) + C] - [F(-1) + C] = F(2) - F(-1)$$

which is the same answer as before since the constant C cancels out. We thus see that we can use the simplest antiderivative (the one where $C = 0$), and we will do so from now on. ◀

▶ **Example 12**   Evaluate $\int_0^\pi \sin x \, dx$.

Solution

Note that $\sin x$ is continuous on $[0, \pi]$. Since $F(x) = -\cos x$ is an antiderivative of $\sin x$,

$$\int_0^\pi \sin x \, dx = F(\pi) - F(0) = -\cos \pi - (-\cos 0) = -(-1) + 1 = 2 \quad ◀$$

▶ **Example 13**   Evaluate

$$\int_{-5}^{-1} \frac{1}{x} \, dx$$

Solution

Note that $\frac{1}{x}$ is continuous on $[-5, -1]$. Now $\ln |x|$ is an antiderivative of $\frac{1}{x}$. We use this antiderivative to evaluate the integral and find

$$\int_{-5}^{-1} \frac{1}{x} \, dx = (\ln |-1|) - (\ln |-5|) = -\ln 5$$

since $\ln |-1| = \ln 1 = 0$ and $|-5| = 5$. ◀

We introduce additional notation. If $F(x)$ is an antiderivative of $f(x)$, we write

$$\int_a^b f(x) \, dx = F(x)\Big]_a^b = F(b) - F(a)$$

For instance,

$$\int_{-5}^{-1} \frac{1}{x} \, dx = \ln |x|\Big]_{-5}^{-1} = \ln |-1| - \ln |-5|$$

The notation $F(x)\big]_a^b$ indicates that we evaluate the antiderivative $F(x)$ at b and a, respectively, and compute the difference $F(b) - F(a)$.

▶ **Example 14**   Evaluate

$$\int_0^3 2x e^{x^2} \, dx$$

Solution

Observe that $2x e^{x^2}$ is continuous on $[0, 3]$ and that $F(x) = e^{x^2}$ is an antiderivative of $f(x) = 2x e^{x^2}$ since, applying the chain rule, we find

$$F'(x) = e^{x^2} \left(\frac{d}{dx} x^2 \right) = e^{x^2} 2x$$

Therefore,

$$\int_0^3 2x e^{x^2} \, dx = e^{x^2}\Big]_0^3 = e^9 - e^0 = e^9 - 1 \quad ◀$$

▶ Example 15   Evaluate

$$\int_1^4 \frac{2x^2 - 3x + \sqrt{x}}{\sqrt{x}} \, dx$$

Solution

The integrand is continuous on [1, 4]. We first simplify the integrand:

$$f(x) = \frac{2x^2 - 3x + \sqrt{x}}{\sqrt{x}} = 2x^{3/2} - 3\sqrt{x} + 1$$

An antiderivative of $f(x)$ is, therefore,

$$F(x) = 2 \cdot \frac{2}{5}x^{5/2} - 3 \cdot \frac{2}{3}x^{3/2} + x = \frac{4}{5}x^{5/2} - 2x^{3/2} + x$$

which can be checked by differentiating $F(x)$. We can now evaluate the integral:

$$\int_1^4 \frac{2x^2 - 3x + \sqrt{x}}{\sqrt{x}} \, dx = \frac{4}{5}x^{5/2} - 2x^{3/2} + x \Big]_1^4$$

$$= \left(\frac{4}{5} \cdot 4^{5/2} - 2 \cdot 4^{3/2} + 4 \right) - \left(\frac{4}{5} \cdot 1^{5/2} - 2 \cdot 1^{3/2} + 1 \right)$$

$$= \left(\frac{4}{5} \cdot 32 - (2)(8) + 4 \right) - \left(\frac{4}{5} - 2 + 1 \right)$$

$$= \frac{68}{5} - \left(-\frac{1}{5} \right) = \frac{69}{5}$$ ◀

Finding an Integrand

▶ Example 16   Suppose that

$$\int_0^x f(t) \, dt = \cos(2x) + a$$

where a is a constant. Find $f(x)$ and a.

Solution

We solve this problem in two steps. First, we use the FTC, part I, to conclude that

$$\frac{d}{dx} \int_0^x f(t) \, dt = f(x)$$

Hence,

$$f(x) = \frac{d}{dx} [\cos(2x) + a] = -2\sin(2x)$$

In the second step, we use the FTC, part II, to determine a. Namely,

$$\int_0^x (-2\sin(2t)) \, dt = \cos(2t)]_0^x$$

$$= \cos(2x) - \cos(0) = \cos(2x) - 1$$

We conclude that $a = -1$.   ◀

You might wonder why we always check that the integrand is continuous on the interval between the lower and the upper limit of integration.

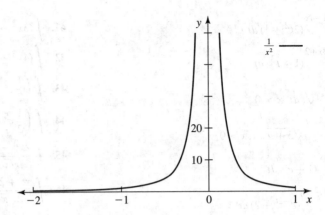

▲ **Figure 6.26**

The graph of $y = \frac{1}{x^2}$ between $x = -2$ and $x = 1$. The function is discontinuous at $x = 0$

The following is an example of what can go wrong when the integrand is discontinuous. Try to evaluate

$$\int_{-2}^{1} \frac{1}{x^2} \, dx$$

An antiderivative of $f(x) = 1/x^2$ is $F(x) = -\frac{1}{x}$. We find $F(1) = -1$ and $F(-2) = \frac{1}{2}$. When we compute $F(1) - F(-2)$, we get $-\frac{3}{2}$. This is obviously not equal to $\int_{-2}^{1} \frac{1}{x^2} \, dx$ since $f(x) = \frac{1}{x^2}$ is positive on $[-2, 1]$ (see Figure 6.26) and, therefore, the integral of $f(x)$ between -2 and 1 should not be negative. The function $f(x)$ is discontinuous at $x = 0$ (it has a vertical asymptote at $x = 0$). The second part of the fundamental theorem of calculus therefore cannot be applied. We will learn how to deal with such discontinuities (see Section 7.4). In any case, before you evaluate an integral, always check whether the integrand is continuous between the limits of integration.

6.2.4 Problems

(6.2.1)

In Problems 1–14, find $\frac{dy}{dx}$.

1. $y = \displaystyle\int_{0}^{x} u^2 \, du$

2. $y = \displaystyle\int_{0}^{x} (1 - u^3) \, du$

3. $y = \displaystyle\int_{0}^{x} (4u - 3) \, du$

4. $y = \displaystyle\int_{0}^{x} (1 + u^4) \, du$

5. $y = \displaystyle\int_{0}^{x} \sqrt{2 + u} \, du, \; x > 0$

6. $y = \displaystyle\int_{0}^{x} \sqrt{1 + u^2} \, du, \; x > 0$

7. $y = \displaystyle\int_{0}^{x} \sqrt{1 + \sin^2 u} \, du, \; x > 0$

8. $y = \displaystyle\int_{0}^{x} \sqrt{\tan^2 u + 2} \, du, \; x > 0$

9. $y = \displaystyle\int_{3}^{x} u e^{4u} \, du$

10. $y = \displaystyle\int_{1}^{x} u e^{-u^2} \, du$

11. $y = \displaystyle\int_{-2}^{x} \frac{1}{u + 3} \, du, \; x > -2$

12. $y = \displaystyle\int_{-2}^{x} \frac{1}{1 + u^2} \, du$

13. $y = \displaystyle\int_{\pi/2}^{x} \sin(u^2 + 1) \, du$

14. $y = \displaystyle\int_{\pi/4}^{x} \cos^2(u - 3) \, du$

In Problems 15–38, use Leibniz's rule to find $\frac{dy}{dx}$.

15. $y = \displaystyle\int_{0}^{3x} (1 + t) \, dt$

16. $y = \displaystyle\int_{0}^{2x-1} (t^2 - 1) \, dt$

17. $y = \int_0^{1-4x} (2t^2 + 1)\, dt$

18. $y = \int_0^{3x+2} (1 + t^3)\, dt$

19. $y = \int_4^{x^2} \sqrt{t}\, dt,\ x > 0$

20. $y = \int_2^{x^2-2} \sqrt{3 + u}\, du,\ x > 0$

21. $y = \int_0^{3x} (1 + e^t)\, dt$

22. $y = \int_0^{2x^2-1} (e^{-2t} + 2)\, dt$

23. $y = \int_1^{3x^2+x} (1 + te^t)\, dt$

24. $y = \int_2^{\ln x} e^{-t}\, dt,\ x > 0$

25. $y = \int_x^3 (1 + t)\, dt$

26. $y = \int_x^5 (1 + e^t)\, dt$

27. $y = \int_{2x}^3 (1 + \sin t)\, dt$

28. $y = \int_{2x^2}^6 (1 + t^2)\, dt$

29. $y = \int_x^5 \frac{1}{u^2}\, du,\ x > 0$

30. $y = \int_{x^2}^3 \frac{1}{1+t}\, dt,\ x > 0$

31. $y = \int_{x^2}^1 \sec t\, dt,\ -1 < x < 1$

32. $y = \int_{1+x^2}^2 \tan t\, dt$

33. $y = \int_x^{2x} (1 + t^2)\, dt$

34. $y = \int_{-x}^x \tan u\, du,\ 0 < x < \frac{\pi}{4}$

35. $y = \int_{x^2}^{x^3} \ln t\, dt,\ x > 0$

36. $y = \int_{x^3}^{x^4} \ln(1 + t)\, dt,\ x > 0$

37. $y = \int_{2-x^2}^{x+x^3} \sin t\, dt$

38. $y = \int_{1+x^2}^{x^3-2x} \cos t\, dt$

(6.2.2)

In Problems 39–96, compute the indefinite integrals.

39. $\int (1 - x^2)\, dx$

40. $\int (x^3 - 4)\, dx$

41. $\int (3x^2 - 2x)\, dx$

42. $\int (4x^3 + 5x^2)\, dx$

43. $\int \left(\frac{1}{2}x^2 + 3x - \frac{1}{3} \right) dx$

44. $\int \left(\frac{1}{2}x^5 + 2x^3 - 1 \right) dx$

45. $\int \frac{2x^2 - x}{\sqrt{x}}\, dx$

46. $\int \frac{x^2 + 2x}{2\sqrt{x}}\, dx$

47. $\int x^2 \sqrt{x}\, dx$

48. $\int (1 + x^3)\sqrt{x}\, dx$

49. $\int (x^{7/2} + x^{2/7})\, dx$

50. $\int (x^{3/5} + x^{5/3})\, dx$

51. $\int \left(\sqrt{x} + \frac{1}{\sqrt{x}} \right) dx$

52. $\int \left(2\sqrt{x} + \frac{1}{2\sqrt{x}} \right) dx$

53. $\int (x - 1)(x + 1)\, dx$

54. $\int (x - 1)^2\, dx$

55. $\int (x - 2)(3 - x)\, dx$

56. $\int (2x + 3)^2\, dx$

57. $\int e^{2x}\, dx$

58. $\int 2e^{3x}\, dx$

59. $\int 3e^{-x}\, dx$

60. $\int 2e^{-x/3}\, dx$

61. $\int xe^{-x^2/2}\, dx$

62. $\int e^x (1 - e^{-x})\, dx$

63. $\int \sin(2x)\, dx$

64. $\int \sin(1 - x)\, dx$

65. $\int \cos(3x)\, dx$

66. $\int \cos(2 - 4x)\, dx$

67. $\int \sec^2(3x)\, dx$

68. $\int \csc^2(2x)\, dx$

69. $\displaystyle\int \frac{\sin x}{1 - \sin^2 x}\, dx$

70. $\displaystyle\int \frac{\cos x}{1 - \cos^2 x}\, dx$

71. $\displaystyle\int \tan(2x)\, dx$

72. $\displaystyle\int \cot(3x)\, dx$

73. $\displaystyle\int (\sec^2 x + \tan x)\, dx$

74. $\displaystyle\int (\cot x - \csc^2 x)\, dx$

75. $\displaystyle\int \frac{4}{1 + x^2}\, dx$

76. $\displaystyle\int \left(1 - \frac{x^2}{1 + x^2}\right) dx$

77. $\displaystyle\int \frac{1}{\sqrt{1 - x^2}}\, dx$

78. $\displaystyle\int \frac{5}{\sqrt{1 - x^2}}\, dx$

79. $\displaystyle\int \frac{1}{x + 1}\, dx$

80. $\displaystyle\int \frac{1}{x - 1}\, dx$

81. $\displaystyle\int \frac{x - 1}{x}\, dx$

82. $\displaystyle\int \frac{2x - 3}{x}\, dx$

83. $\displaystyle\int \frac{x + 3}{x^2 - 9}\, dx$

84. $\displaystyle\int \frac{x + 4}{x^2 - 16}\, dx$

85. $\displaystyle\int \frac{3 - x}{x^2 - 9}\, dx$

86. $\displaystyle\int \frac{4 - x}{x^2 - 16}\, dx$

87. $\displaystyle\int \frac{5x^2}{x^2 + 1}\, dx$

88. $\displaystyle\int \frac{2x^2}{1 + x^2}\, dx$

89. $\displaystyle\int 3^x\, dx$

90. $\displaystyle\int 2^x\, dx$

91. $\displaystyle\int 2^{-x}\, dx$

92. $\displaystyle\int 4^{-x}\, dx$

93. $\displaystyle\int (x^2 + 2^x)\, dx$

94. $\displaystyle\int (x^{-3} + 3^{-x})\, dx$

95. $\displaystyle\int (\sqrt{x} + \sqrt{e^x})\, dx$

96. $\displaystyle\int \left(\frac{1}{\sqrt{x}} + \frac{1}{\sqrt{e^x}}\right) dx$

(6.2.3)

In Problems 97–122, evaluate the definite integrals.

97. $\displaystyle\int_2^4 (3 - 2x)\, dx$

98. $\displaystyle\int_{-1}^3 (2x^2 - 1)\, dx$

99. $\displaystyle\int_0^1 (x^2 - \sqrt{x})\, dx$

100. $\displaystyle\int_1^2 x^{5/2}\, dx$

101. $\displaystyle\int_1^8 x^{-2/3}\, dx$

102. $\displaystyle\int_4^9 \frac{1 + \sqrt{x}}{\sqrt{x}}\, dx$

103. $\displaystyle\int_0^2 (2t - 1)(t + 3)\, dt$

104. $\displaystyle\int_{-1}^2 (4 - 3t)^2\, dt$

105. $\displaystyle\int_0^{\pi/4} \sin(2x)\, dx$

106. $\displaystyle\int_{-\pi/3}^{\pi/3} 2\cos\left(\frac{x}{2}\right) dx$

107. $\displaystyle\int_0^{\pi/8} \sec^2(2x)\, dx$

108. $\displaystyle\int_{-\pi/4}^{\pi/4} \tan x\, dx$

109. $\displaystyle\int_0^1 \frac{1}{1 + x^2}\, dx$

110. $\displaystyle\int_{-\sqrt{3}}^{-1} \frac{4}{1 + x^2}\, dx$

111. $\displaystyle\int_0^{1/2} \frac{1}{\sqrt{1 - x^2}}\, dx$

112. $\displaystyle\int_{-1}^1 \frac{3}{\sqrt{1 - x^2}}\, dx$

113. $\displaystyle\int_0^{\pi/6} \tan(2x)\, dx$

114. $\displaystyle\int_{\pi/20}^{\pi/15} \sec(5x)\tan(5x)\, dx$

115. $\displaystyle\int_{-1}^0 e^{3x}\, dx$

116. $\displaystyle\int_0^2 2te^{t^2}\, dt$

117. $\displaystyle\int_{-1}^1 |x|\, dx$

118. $\displaystyle\int_{-1}^1 e^{-|s|}\, ds$

119. $\displaystyle\int_1^e \frac{1}{x}\, dx$

120. $\int_2^3 \dfrac{1}{z+1}\,dz$

121. $\int_{-2}^{-1} \dfrac{1}{1-u}\,du$

122. $\int_2^3 \dfrac{2}{t-1}\,dt$

123. Use l'Hospital's rule to compute
$$\lim_{x\to 0} \frac{1}{x^2} \int_0^x \sin t\,dt$$

124. Use l'Hospital's rule to compute
$$\lim_{h\to 0} \frac{1}{h} \int_0^h e^x\,dx$$

125. Suppose that
$$\int_0^x f(t)\,dt = 2x^2$$
Find $f(x)$.

126. Suppose that
$$\int_0^x f(t)\,dt = \frac{1}{2}\tan(2x)$$
Find $f(x)$.

6.3 APPLICATIONS OF INTEGRATION

In this section, we will discuss a number of applications of integrals. In the first application, we will revisit the interpretation of integrals as areas; the second interprets integrals as cumulative (or net) change; the third will allow us to compute averages using integrals; and, finally, we will use integrals to compute volumes. In each application, you will see that integrals can be interpreted as "sums of many small increments."

6.3.1 Areas

The first application is already familiar to us. If f is a nonnegative, continuous function on $[a, b]$, then

$$A = \int_a^b f(x)\,dx$$

represents the area of the region bounded by the graph of $f(x)$ between a and b, the vertical lines $x = a$ and $x = b$, and the x-axis between a and b. In all of the preceding examples, one of the boundaries of the region whose area we wanted to know has been the x-axis. We will now discuss how to find the geometric area between two arbitrary curves. We emphasize that we want to compute geometric areas; that is, the areas we compute in this subsection will always be positive.

Suppose that $f(x)$ and $g(x)$ are continuous functions on $[a, b]$. We wish to find the area between the graphs of f and g. We start with a simple example (see Figure 6.27). We assume for the moment that both f and

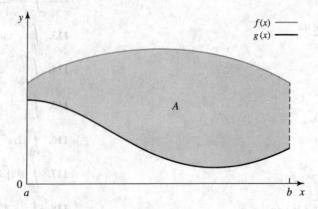

▲ **Figure 6.27**
Computing the area between the two curves

g are nonnegative on $[a, b]$ and that $f(x) \geq g(x)$ on $[a, b]$. We see from Figure 6.27 that

$$A = \begin{bmatrix} \text{area between} \\ f \text{ and } x\text{-axis} \end{bmatrix} - \begin{bmatrix} \text{area between} \\ g \text{ and } x\text{-axis} \end{bmatrix}$$

$$= \int_a^b f(x)\,dx - \int_a^b g(x)\,dx$$

Using Property (4) of Subsection 6.1.3, we can write this as

$$A = \int_a^b [f(x) - g(x)]\,dx$$

We obtained this formula under the assumption that both f and g are nonnegative on $[a, b]$; we now show that this assumption is not necessary. To do so, we derive a formula for the area between two curves from Riemann sums.

We assume that f and g are continuous on $[a, b]$ and that $f(x) \geq g(x)$ for all $x \in [a, b]$. We approximate the area between the two curves by rectangles: We divide $[a, b]$ into n equal subintervals, each of length Δx; that is, we set $a = x_0 < x_1 < x_2 < \cdots < x_n = b$ with $\Delta x = x_k - x_{k-1} = (b - a)/n$. The kth subinterval is thus between x_{k-1} and x_k. We choose left endpoints to compute the heights of the approximating rectangles. From Figure 6.28, we see that the height of the kth rectangle is equal to $f(x_{k-1}) - g(x_{k-1})$. We therefore find

$$A = \lim_{n \to \infty} \sum_{k=1}^{n} [f(x_{k-1}) - g(x_{k-1})]\,\Delta x$$

$$= \int_a^b [f(x) - g(x)]\,dx$$

> If f and g are continuous on $[a, b]$, with $f(x) \geq g(x)$ for all $x \in [a, b]$, then the area of the region between the curves $y = f(x)$ and $y = g(x)$ from a to b is equal to
>
> $$\text{Area} = \int_a^b [f(x) - g(x)]\,dx$$

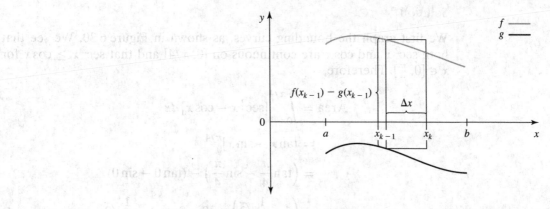

▲ **Figure 6.28**
The kth rectangle between x_{k-1} and x_k

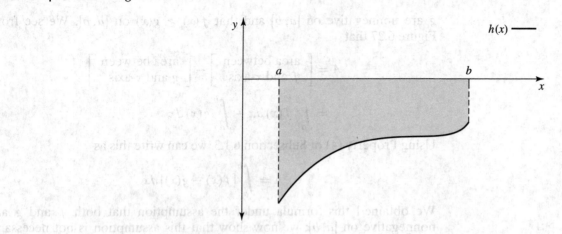

▲ Figure 6.29

The definite integral $\int_a^b h(x)\,dx$ represents a signed area; it is negative here

Before looking at a number of examples, we point out once more that this area formula always yields a nonnegative number since it computes the geometric area. To contrast this with the concept of a signed area, let us consider a function $h(x)$ on $[a, b]$, with $h(x) \le 0$ for all $x \in [a, b]$ (see Figure 6.29). The definite integral $\int_a^b h(x)\,dx$ represents a signed area and is negative in this case; more precisely, $\int_a^b h(x)\,dx$ is the negative of the geometric area of the region between the x-axis and the graph of $h(x)$ from $x = a$ to $x = b$. That this is consistent with our definition of area can be seen as follows. The region of interest is bounded by the two curves $y = 0$ and $y = h(x)$. Since $h(x) \le 0$ for $x \in [a, b]$, the area formula yields

$$\text{Area} = \int_a^b [0 - h(x)]\,dx = -\int_a^b h(x)\,dx$$

which is a positive number.

When you compute the area of the region between two curves, you should always graph the bounding curves. This will show you how to set up the appropriate integral(s).

▶ **Example 1** Find the area between the curves $y = \sec^2 x$ and $y = \cos x$, from $x = 0$ to $x = \pi/4$.

Solution

We first graph the bounding curves, as shown in Figure 6.30. We see that both $\sec^2 x$ and $\cos x$ are continuous on $[0, \pi/4]$ and that $\sec^2 x \ge \cos x$ for $x \in [0, \frac{\pi}{4}]$. Therefore,

$$\begin{aligned}
\text{Area} &= \int_0^{\pi/4} [\sec^2 x - \cos x]\,dx \\
&= \tan x - \sin x\Big]_0^{\pi/4} \\
&= \left(\tan\frac{\pi}{4} - \sin\frac{\pi}{4}\right) - (\tan 0 - \sin 0) \\
&= \left(1 - \frac{1}{2}\sqrt{2}\right) - (0 - 0) = 1 - \frac{1}{2}\sqrt{2}
\end{aligned}$$

◀

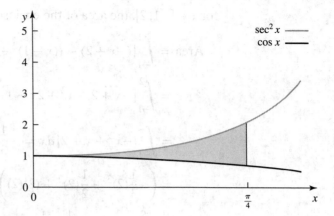

▲ **Figure 6.30**
The region for Example 1

▶ Example 2 Find the area of the region enclosed by $y = (x-1)^2 - 1$ and $y = -x + 2$.

Solution

The bounding curves are graphed in Figure 6.31. To find the points where the two curves intersect, we solve

$$(x-1)^2 - 1 = -x + 2$$
$$x^2 - 2x + 1 - 1 = -x + 2$$
$$x^2 - x - 2 = 0$$
$$(x+1)(x-2) = 0$$

Therefore,

$$x = -1 \quad \text{and} \quad x = 2$$

are the x-coordinates of the points of intersection. Note that both $y = (x-1)^2 - 1$ and $y = -x + 2$ are continuous on $[-1, 2]$. Since $-x + 2 \geq (x-1)^2 - 1$

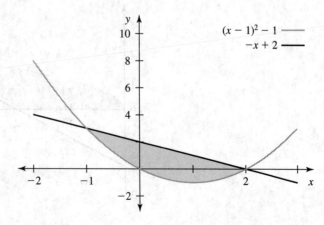

▲ **Figure 6.31**
The region for Example 2

for $x \in [-1, 2]$, the area of the enclosed region is

$$\text{Area} = \int_{-1}^{2} [(-x + 2) - ((x - 1)^2 - 1)] \, dx$$

$$= \int_{-1}^{2} [-x + 2 - x^2 + 2x - 1 + 1] \, dx$$

$$= \int_{-1}^{2} [-x^2 + x + 2] \, dx = -\frac{1}{3}x^3 + \frac{1}{2}x^2 + 2x \Big]_{-1}^{2}$$

$$= \left(-\frac{1}{3}(2)^3 + \frac{1}{2}(2)^2 + (2)(2)\right) - \left(-\frac{1}{3}(-1)^3 + \frac{1}{2}(-1)^2 + (2)(-1)\right)$$

$$= -\frac{8}{3} + 2 + 4 - \frac{1}{3} - \frac{1}{2} + 2 = \frac{9}{2}$$

◀

▶ **Example 3** Find the area of the region bounded by $y = \sqrt{x}$, $y = x - 2$, and the x-axis.

Solution

We first graph the bounding curves in Figure 6.32. We see that $y = \sqrt{x}$ and $y = x - 2$ intersect. To find the point of intersection, we solve

$$x - 2 = \sqrt{x}$$

Squaring both sides, we find

$$(x - 2)^2 = (\sqrt{x})^2 \quad \text{and thus} \quad x^2 - 4x + 4 = x$$

or

$$x^2 - 5x + 4 = 0$$

We can factor the equation,

$$(x - 4)(x - 1) = 0$$

which yields the solutions $x = 4$ and $x = 1$. Since squaring an equation can introduce extraneous solutions, we need to check whether the solutions satisfy $x - 2 = \sqrt{x}$. When $x = 4$, we find $4 - 2 = \sqrt{4}$, which is a true statement;

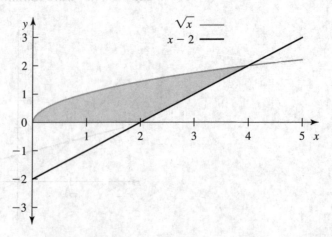

▲ **Figure 6.32**
The region for Example 3

when $x = 1$, we find $1 - 2 = \sqrt{1}$, which is a false statement. Hence, $x = 4$ is the only solution.

The graph of $y = x - 2$ intersects the x-axis at $x = 2$. To compute the area, we need to split the integral into two parts, since the lower bounding curve is composed of two parts; namely, the x-axis from $x = 0$ to $x = 2$ and the line $y = x - 2$ from $x = 2$ to $x = 4$. We see from the graph that all bounding curves are continuous on their respective intervals. We find

$$
\text{Area} = \int_0^2 \sqrt{x}\,dx + \int_2^4 [\sqrt{x} - (x - 2)]\,dx
$$

$$
= \left[\frac{2}{3} x^{3/2} \right]_0^2 + \left[\frac{2}{3} x^{3/2} - \frac{1}{2} x^2 + 2x \right]_2^4
$$

$$
= \frac{2}{3} \cdot 2^{3/2} + \frac{2}{3} \cdot 4^{3/2} - \frac{1}{2} \cdot 16 + 8 - \frac{2}{3} \cdot 2^{3/2} + 2 - 4 = \frac{10}{3} \quad \blacktriangleleft
$$

In the preceding example, we needed to split the integral into two parts since the lower boundary of the area was composed of two different curves that determined the heights of the approximating rectangles in the Riemann integral. Recall that the rectangles are obtained by partitioning the x-axis. It is sometimes more convenient to partition the y-axis. We illustrate this in Figure 6.33 for the region of Example 3, where we partition the interval $[0, 2]$ on the y-axis into n equal subintervals, each of length Δy. We need to express the boundary curves as functions of y. In the case of Example 3, the right boundary curve is $x = f(y) = y + 2$ and the left boundary curve is $x = g(y) = y^2$.

We set $y_0 = 0 < y_1 < y_2 < \cdots < y_n = 2$, where $y_k - y_{k-1} = \Delta y$ for $k = 1, 2, \ldots, n$. As shown in Figure 6.33, the kth rectangle has width $y_k - y_{k-1}$ and height $f(y_k) - g(y_k)$. Therefore, the area of the kth rectangle is

$$
[f(y_k) - g(y_k)]\,\Delta y
$$

If we sum this from $k = 1$ to $k = n$ and let $\Delta y \to 0$, we find

$$
\lim_{\Delta y \to 0} \sum_{k=1}^n [f(y_k) - g(y_k)]\,\Delta y = \int_0^2 [f(y) - g(y)]\,dy
$$

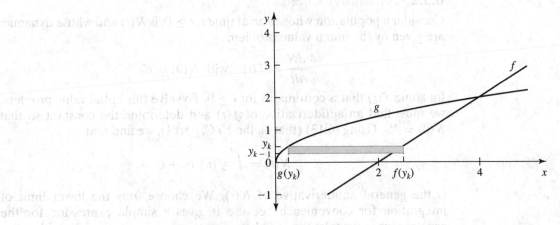

▲ Figure 6.33
The area between $f(y)$ and $g(y)$ in Example 3 together with an approximating rectangle

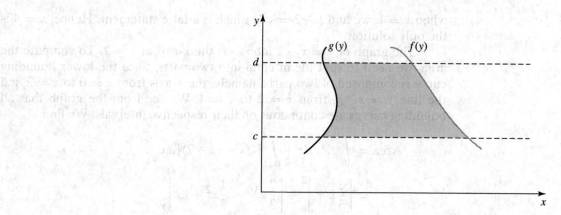

▲ Figure 6.34
The area between $f(y)$ and $g(y)$

Since each rectangle is bounded by the same two curves, there is no need to split the integral. Using $f(y) = y+2$ and $g(y) = y^2$, we find, for the total area,

$$\text{Area} = \int_0^2 (y + 2 - y^2)\, dy$$

$$= \frac{1}{2}y^2 + 2y - \frac{1}{3}y^3 \Big]_0^2 = 2 + 4 - \frac{8}{3} = \frac{10}{3}$$

which is the same as the result in Example 3.

To summarize the general case, as illustrated in Figure 6.34, suppose that a region is bounded by $x = f(y)$ and $x = g(y)$, with $g(y) \le f(y)$ for $c \le y \le d$; that is, $f(y)$ is to the right of $g(y)$ for all $y \in [c, d]$. Then the area of the shaded region is given by the following formula.

$$\text{Area} = \int_c^d [f(y) - g(y)]\, dy$$

6.3.2 Cumulative Change

Consider a population whose size at time t, $t \ge 0$, is $N(t)$ and whose dynamics are given by the initial value problem

$$\frac{dN}{dt} = f(t) \quad \text{with } N(0) = N_0$$

for some $f(t)$ that is continuous for $t \ge 0$. To solve this initial value problem, we must find an antiderivative of $f(t)$ and determine the constant so that $N(0) = N_0$. Using (6.13) (that is, the FTC, part I), we find that

$$N(t) = \int_0^t f(u)\, du + C$$

is the general antiderivative of $N(t)$. We choose 0 as the lower limit of integration for convenience because it gives a simple expression for the constant when we take the initial condition into account. Namely, with

$$N(0) = \int_0^0 f(u)\, du + C = C$$

it follows that $C = N(0) = N_0$. Therefore,

$$N(t) = N_0 + \int_0^t f(u)\, du$$

or

$$N(t) - N_0 = \int_0^t f(u)\, du \tag{6.14}$$

Since $f(u) = dN/du$ and $N(0) = N_0$, we can write

$$N(t) - N(0) = \int_0^t \frac{dN}{du}\, du$$

which allows us to interpret the definite integral $\int_0^t \frac{dN}{du}\, du$ as the **net** or **cumulative change** in population size between times 0 and t, since it is a "sum" of instantaneous changes accumulated over time. That is,

$$\begin{bmatrix} \text{cumulative} \\ \text{change in } [0,\, t] \end{bmatrix} = \int_0^t \begin{bmatrix} \text{instantaneous rate of} \\ \text{change at time } u \end{bmatrix} du$$

We present another example in which we can interpret an integral as the cumulative change of a quantity. Recall that velocity is the instantaneous rate of change of distance. That is, if a particle moves along a straight line, and we denote by $s(t)$ the location of the particle at time t, with $s(0) = s_0$, and by $v(t)$ the velocity at time t, then $s(t)$ and $v(t)$ are related via

$$\frac{ds}{dt} = v(t) \quad \text{for } t > 0 \text{ with } s(0) = s_0$$

To solve this initial value problem, we must find an antiderivative of $v(t)$ that satisfies $s(0) = s_0$. That is,

$$s(t) = \int_0^t v(u)\, du + C$$

with

$$s_0 = s(0) = \int_0^0 v(u)\, du + C = 0 + C$$

which implies that $C = s_0$. Hence,

$$s(t) = s_0 + \int_0^t v(u)\, du$$

or, with $s(0) = s_0$,

$$s(t) - s(0) = \int_0^t v(u)\, du$$

Again, the cumulative change in distance, $s(t) - s(0)$, can be represented as a "sum" of instantaneous changes.

6.3.3 Average Values

The concentration in g/m³ of soil nitrogen was measured every meter along a transect in moist tundra and yielded the following data.

Distance from origin [m]	1	2	3	4	5
Concentration [g/m³]	589.3	602.7	618.5	667.2	641.2
Distance from origin [m]	6	7	8	9	10
Concentration [g/m³]	658.3	672.8	661.2	652.3	669.8

If we denote the concentration at distance x from the origin by $c(x)$, then the average concentration, denoted by $\bar{c}$ (read "c bar"), is the arithmetic average

$$\bar{c} = \frac{1}{10} \sum_{k=1}^{10} c(k) = 643.3 \text{ g/m}^3$$

More generally, to find the average concentration between two points a and b along a transect, we measure the concentration at equal distances. To formulate this mathematically, we divide $[a, b]$ into n subintervals of equal lengths $\Delta x = \frac{b-a}{n}$ and measure the concentration at, say, the right endpoint of each subinterval. If the concentration at location x_k is denoted by $c(x_k)$, then the average concentration $\bar{c}$ is

$$\bar{c} = \frac{1}{n} \sum_{k=1}^{n} c(x_k)$$

Since $\Delta x = \frac{b-a}{n}$, we can write $n = \frac{b-a}{\Delta x}$. Hence,

$$\bar{c} = \frac{1}{b-a} \sum_{k=1}^{n} c(x_k) \, \Delta x$$

If we let the number of subintervals grow ($n \to \infty$), then the length of each subinterval goes to 0 ($\Delta x \to 0$) and

$$\bar{c} = \frac{1}{b-a} \lim_{n \to \infty} \sum_{k=1}^{n} c(x_k) \, \Delta x$$

$$= \frac{1}{b-a} \int_a^b c(x) \, dx$$

That is, the average concentration can be expressed as an integral over $c(x)$ between a and b, divided by the length of the interval $[a, b]$.

Assume that $f(x)$ is a continuous function on $[a, b]$. The average value of f on the interval $[a, b]$ is

$$f_{\text{avg}} = \frac{1}{b-a} \int_a^b f(x) \, dx$$

▶ Example 4   Find the average value of $f(x) = 4 - x^2$ on the interval $[-2, 2]$.

Solution

We use the formula for computing average values. Note that $f(x) = 4 - x^2$ is continuous on $[-2, 2]$. Then the average value of f on the interval $[-2, 2]$ is

$$f_{avg} = \frac{1}{2 - (-2)} \int_{-2}^{2} (4 - x^2)\, dx = \frac{1}{4}\left[4x - \frac{1}{3}x^3\right]_{-2}^{2}$$

$$= \frac{1}{4}\left[8 - \frac{8}{3} + 8 - \frac{8}{3}\right] = \frac{1}{4} \cdot \frac{32}{3} = \frac{8}{3}$$ ◀

The following theorem says a bit more about the value of f_{avg}.

> **The Mean Value Theorem for Definite Integrals**   Assume that $f(x)$ is a continuous function on $[a, b]$. Then there exists a number $c \in [a, b]$ such that
>
> $$f(c)(b - a) = \int_{a}^{b} f(x)\, dx$$

That is, when we compute the average value of a function that is continuous on $[a, b]$, there exists a number c such that $f(c) = f_{avg}$. This should be fairly obvious when you look at the graph of a function f and at f_{avg}. For simplicity, let's assume that $f(x) \geq 0$. Since $\int_{a}^{b} f(x)\, dx$ is then equal to the area between the graph of $f(x)$ and the x-axis, and $f_{avg}(b - a)$ is equal to the area of the rectangle with height f_{avg} and width $b - a$, and since the two areas are equal, the horizontal line $y = f_{avg}$ must intersect the graph of $f(x)$ at some point in the interval $[a, b]$ (see Figure 6.35). The x-coordinate of this point of intersection is then the value c in the mean value theorem for definite integrals. (Note that there could be more than one such number.) A similar argument can be made when we do not assume that $f(x)$ is positive. In this case, "area" is replaced by "signed area." The proof of this theorem is short, and we supply it for completeness at the end of this subsection.

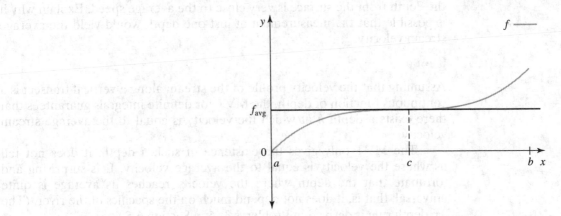

▲ **Figure 6.35**
An illustration of the average value of a function: $\int_{a}^{b} f(x)\, dx = f_{avg}(b - a)$

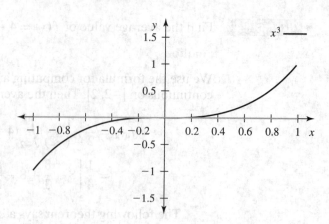

▲ **Figure 6.36**
The graph of $y = x^3$, $-1 \le x \le 1$. The average value is 0

▶ **Example 5** Find the average value of $f(x) = x^3$ on the interval $[-1, 1]$, and determine $x \in [-1, 1]$ such that $f(x)$ equals the average value.

Solution

The function $f(x) = x^3$ is continuous on $[-1, 1]$. Then

$$f_{avg} = \frac{1}{2} \int_{-1}^{1} x^3 \, dx = \frac{1}{2} \left[\frac{1}{4} x^4 \right]_{-1}^{1} = \frac{1}{8}(1^4 - (-1)^4) = 0$$

The graph of $y = x^3$ (Figure 6.36) is symmetric about the origin, which explains why the average value is 0. Since $x^3 = 0$ for $x = 0$, the function $f(x)$ takes on its average value at $x = 0$. ◀

▶ **Example 6** The speed of water in a channel varies considerably with depth. Due to friction, the velocity reaches zero at the bottom and along the sides of the channel; the velocity is greatest near the surface of the water. The average velocity of a stream is of interest in characterizing rivers. One way to obtain this average value would be to measure stream velocity at various depths along a vertical transect and then average these values. In practice, however, a much simpler method is employed; namely, the speed measured at 60% of the depth from the surface is very close to the average speed. Explain why it is possible that the measurement at just one depth would yield the average stream velocity.

Solution

Assuming that the velocity profile of the stream along a vertical transect is a continuous function of depth, the MVT for definite integrals guarantees that there exists a depth h at which the velocity is equal to the average stream velocity.

The MVT only gives the existence of such a depth; it does not tell us where the velocity is equal to the average velocity. It is surprising and fortunate that the depth where the velocity reaches its average is quite universal; that is, it does not depend much on the specifics of the river. (The 0.6 depth rule is derived in Problems 3–5 in Section 6.5.) ◀

Proof of the Mean Value Theorem for Definite Integrals. Since $f(x)$ is continuous on $[a, b]$, we can apply the extreme value theorem to conclude that f attains an absolute maximum and an absolute minimum in $[a, b]$. If we denote the absolute maximum by M and the absolute minimum by m, then

$$m \leq f(x) \leq M \quad \text{for all } x \in [a, b]$$

and f takes on both m and M for some values in $[a, b]$. We therefore find

$$m(b - a) \leq \int_a^b f(x)\,dx \leq M(b - a)$$

or

$$m \leq \frac{1}{b - a} \int_a^b f(x)\,dx \leq M$$

We set $I = \frac{1}{b-a} \int_a^b f(x)\,dx$; then $m \leq I \leq M$.

Using the facts that $f(x)$ takes on all values between m and M in the interval $[a, b]$ (this follows from the intermediate value theorem) and that I is a number between m and M, it follows (from the intermediate value theorem) that there must be a number $c \in [a, b]$ such that $f(c) = I$; that is,

$$f(c) = \frac{1}{b - a} \int_a^b f(x)\,dx \qquad ■$$

6.3.4 The Volume of a Solid (Optional)

For certain regular solids, such as a circular cylinder, we know formulas from geometry for computing the volume. To compute the volume of a less regularly shaped solid, we will use an approach that is very similar to that for computing areas of irregularly shaped regions; there, we approximated the area by rectangles whose areas were easy to compute using simple formulas from geometry.

We begin with the volume of a right cylinder; the volume is the base area times the height. The base can be an arbitrarily shaped region (see Figure 6.37, for example).

If we denote the base area by A and the height of the cylinder by h, then the volume of the cylinder is

$$V = Ah$$

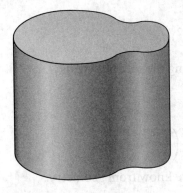

▲ **Figure 6.37**
A right cylinder with an irregularly shaped base

As an example, consider the circular cylinder whose base is a disk. If the disk has radius r and the cylinder has height h, then the volume of the circular cylinder is $\pi r^2 h$. We will use cylinders to approximate volumes of more complicated solids.

Suppose that we wish to compute the volume of the solid in Figure 6.38. We can slice the solid into small slabs by cutting it perpendicular to the x-axis at points $x_0 = a < x_1 < x_2 < \cdots < x_n = b$ that partition the interval $[a, b]$ and then slice the solid into planes. The intersection of such a plane and the solid is called a cross section. We denote the area of the cross section at x_k by $A(x_k)$. By cutting the solid along these planes, we obtain slices, just like when cutting bread. We will approximate the volume of a slice between x_{k-1} and x_k by the volume of a cylinder with base area equal to that of the slice at x_k and height $\Delta x_k = x_k - x_{k-1}$. The volume of the slice between x_{k-1} and x_k is then approximately

$$A(x_k)\,\Delta x_k$$

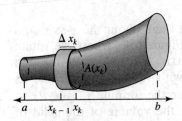

▲ **Figure 6.38**
The volume of an irregularly shaped solid using the disk method

Adding the volumes of all the slices gives us an approximation for the total volume of the solid, that is,

$$V \approx \sum_{k=1}^{n} A(x_k)\, \Delta x_k$$

By making the partition of $[a, b]$ finer, we can improve the approximation.

Recall that we used $\|P\|$ (norm of P) as a measure for how fine the partition is, that is, $\|P\| = \max_{k=1,2,\dots,n} \Delta x_k$. This suggests that we should define the volume of the solid as the limit of our approximation as $\|P\| \to 0$. We summarize this in the following.

> **Definition** The **volume of a solid** of integrable cross-sectional area $A(x)$ between a and b is
>
> $$\int_a^b A(x)\, dx$$

▶ **Example 7** Find the volume of the sphere of radius r centered at the origin.

Solution

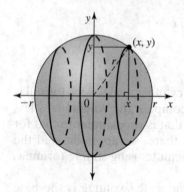

▲ **Figure 6.39**
The volume of a sphere

The cross section at x is perpendicular to the x-axis (see Figure 6.39). It is a disk of radius $y = \sqrt{r^2 - x^2}$ whose area is

$$A(x) = \pi y^2 = \pi(r^2 - x^2)$$

Since the solid is between $-r$ and r, we find

$$\text{Volume} = \int_{-r}^{r} \pi(r^2 - x^2)\, dx$$

The integrand is continuous on $[-r, r]$. Evaluating the integral yields

$$= \pi \left[r^2 x - \frac{1}{3} x^3 \right]_{-r}^{r}$$

$$= \pi \left[\left(r^3 - \frac{1}{3} r^3 \right) - \left(-r^3 + \frac{1}{3} r^3 \right) \right] = \pi \left(\frac{2}{3} r^3 + \frac{2}{3} r^3 \right) = \frac{4}{3} \pi r^3$$

This agrees with the formula for a sphere that we know from geometry. ◀

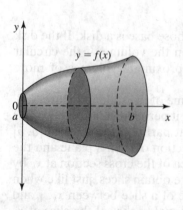

▲ **Figure 6.40**
The solid of rotation when rotating $f(x)$ about the x-axis

The cross sections of the sphere in the last example were of a particularly simple form, namely disks. We can think of a sphere as a **solid of revolution**, that is, a solid that is obtained by revolving a curve about the x-axis (or the y-axis). In this case, we rotate the curve $y = \sqrt{r^2 - x^2}$, $-r \le x \le r$, about the x-axis, which creates circular cross sections.

We can use other curves $y = f(x)$, rotate them about the x-axis, and obtain solids in this way. This is illustrated in Figure 6.40, where we rotate the graph of $y = f(x)$, $a \le x \le b$, about the x-axis. A cross section through x perpendicular to the x-axis is then a disk with radius $f(x)$; hence, its cross-sectional area is $A(x) = \pi[f(x)]^2$. If we use the formula $\int_a^b A(x)\, dx$ to compute the volume of the solid, we find that the **volume** of the solid of revolution is

$$V = \int_a^b \pi[f(x)]^2\, dx \tag{6.15}$$

Computing volumes using (6.15) is called the **disk method**.

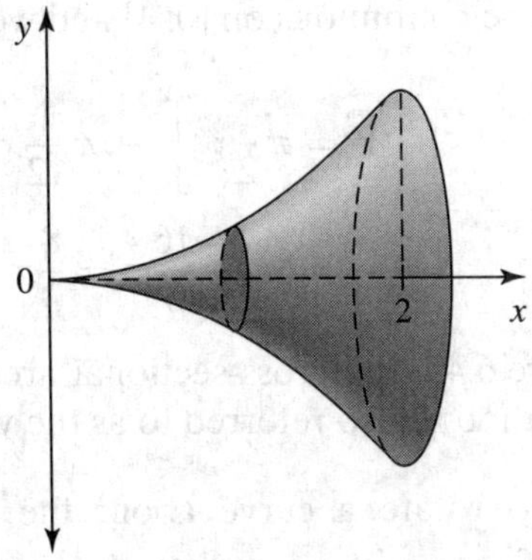

◀ **Figure 6.41**
The solid of rotation for Example 8

▶ Example 8   Compute the volume of the solid obtained by rotating $y = x^2$, $0 \le x \le 2$, about the x-axis.

Solution

We illustrate this in Figure 6.41. When we rotate the graph about the x-axis, we find that the cross-section at x is a disk with radius $y = f(x) = x^2$. The cross-sectional area at x, $A(x)$, is then $\pi(x^2)^2 = \pi x^4$, which is integrable on $[0, 2]$. Thus the volume is

$$V = \int_0^2 \pi [f(x)]^2 \, dx = \int_0^2 \pi x^4 \, dx$$

$$= \frac{\pi}{5} x^5 \Big]_0^2 = \frac{32}{5} \pi$$

◀

▶ Example 9   Rotate the area bounded by the curves $y = \sqrt{x}$ and $y = x/2$ about the x-axis, and compute the volume of the solid of rotation.

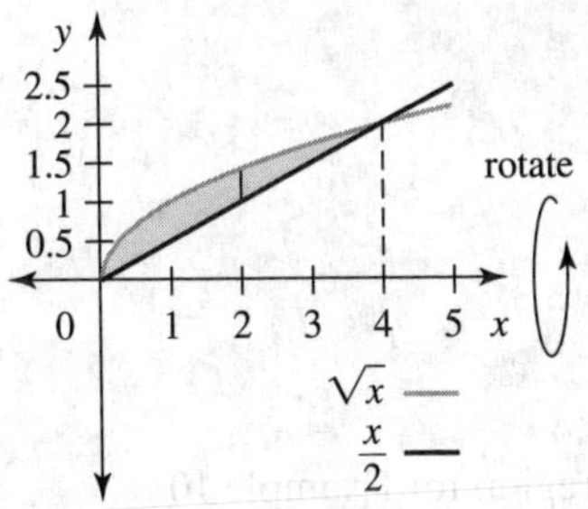

Solution

▲ **Figure 6.42**
The plane region for
Example 9

The curves $y = \sqrt{x}$ and $y = x/2$ are graphed in Figure 6.42, together with a vertical bar to indicate the cross-section. We see from the graph that the curves intersect at $x = 0$ and $x = 4$. To find the points of intersection algebraically, we need to equate $\sqrt{x}$ and $x/2$ and solve for x:

$$\sqrt{x} = \frac{x}{2}$$

This yields immediately $x = 0$. If $x > 0$, we can divide by $\sqrt{x}$ and find

$$1 = \frac{\sqrt{x}}{2} \quad \text{or} \quad 2 = \sqrt{x}$$

Squaring this yields $x = 4$. Thus, the curves intersect at $x = 0$ and $x = 4$. We can compute the volume of this solid of rotation by first rotating $y = \sqrt{x}$, $0 \le x \le 4$, about the x-axis, computing the volume of this solid, and then subtracting the volume of the solid obtained by rotating $y = x/2$, $0 \le x \le 4$, yielding

$$V = \int_0^4 \pi (\sqrt{x})^2 \, dx - \int_0^4 \pi \left(\frac{1}{2} x\right)^2 \, dx$$

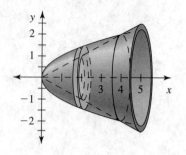

▲ Figure 6.43
The solid of rotation for
Example 9

Both integrands are continuous on $[0, 4]$ and we find

$$V = \pi \frac{1}{2}x^2 \Big]_0^4 - \pi \frac{1}{12}x^3 \Big]_0^4$$

$$= 8\pi - \frac{16}{3}\pi = \frac{8}{3}\pi$$

Looking at Figure 6.43, the cross-sectional area is that of a washer. In this case, the disk method is also referred to as the washer method.   ◀

We can also rotate a curve about the y-axis, as illustrated in the following example.

▶ Example 10   Rotate the region bounded by $y = 2$, $x = 0$, $y = 0$, and $y = \ln x$ about the y-axis, and compute the volume of the resulting solid.

Solution

When we rotate about the y-axis, the cross sections are perpendicular to the y-axis. At $y = \ln x$, the radius of the cross-sectional disk is x (see Figure 6.44). The solid is shown in Figure 6.45. Since we "sum" the slices along the y-axis, we must integrate with respect to y; since $y = \ln x$, we find $x = e^y$. Therefore, the cross-sectional area at y is $A(y) = \pi(e^y)^2$, which is integrable on $[0, 2]$, and the volume is

$$V = \int_0^2 \pi(e^y)^2 \, dy = \int_0^2 \pi e^{2y} \, dy$$

$$= \pi \frac{1}{2}e^{2y} \Big]_0^2 = \frac{\pi}{2}(e^4 - 1)$$ ◀

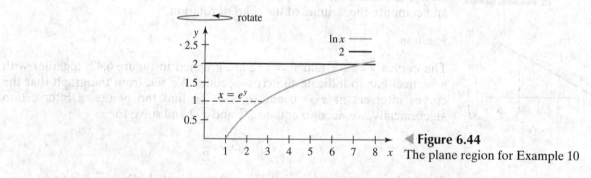

◀ Figure 6.44
The plane region for Example 10

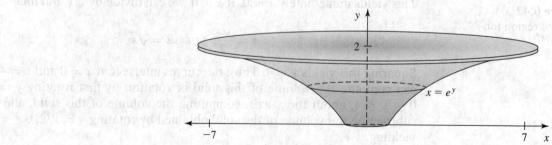

▲ Figure 6.45
The solid of rotation for Example 10

6.3.5 Rectification of Curves (Optional)

In this subsection, we will study how to compute the lengths of curves in the plane. This is yet another example in which integrals appear as we add up a large number of small increments. The first curve whose length was determined was the semicubical parabola $y^2 = x^3$; this was done by William Neile (1637–1670) in 1657. Interestingly enough, only about 20 years earlier, Descartes asserted that there was no rigorous way to determine the exact length of a curve. It turns out that the exact formula is in essence an infinitesimal version of the Pythagorean theorem.

How would we rectify a curve (that is, determine its length) with a ruler? We could approximate the curve by short line segments, measure the length of each segment and add up these measurements, as illustrated in Figure 6.46. By choosing smaller line segments, our approximation would improve. This is precisely the method that we will employ to find an exact formula.

To find the exact formula, assume that the curve whose length we want to find is given by a function $y = f(x), a \le x \le b$, which has a continuous first derivative on (a, b). We partition the interval $[a, b]$ into subintervals using the partition $P = [x_0, x_1, x_2, \ldots, x_n]$, where $a = x_0 < x_1 < x_2 < \cdots < x_n = b$, and approximate the curve by a polygon that consists of the straight line segments connecting neighboring points on the curve, as shown in Figure 6.46. A typical line segment connecting the points P_{k-1} and P_k is shown in Figure 6.47. Using the Pythagorean theorem, we can find its length. We set

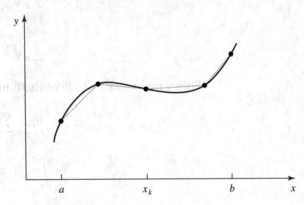

▲ **Figure 6.46**
The length of a curve in the plane

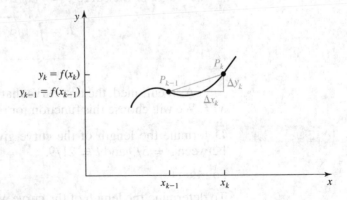

▲ **Figure 6.47**
A typical line segment

$\Delta x_k = x_k - x_{k-1}$ and $\Delta y_k = y_k - y_{k-1}$. Then the length of the line segment is given by

$$\sqrt{(\Delta x_k)^2 + (\Delta y_k)^2}$$

The length of the polygon using the partition P is then

$$L_P = \sum_{k=1}^{n} \sqrt{(\Delta x_k)^2 + (\Delta y_k)^2} \qquad (6.16)$$

By choosing finer and finer partitions, the length of the polygon will become a better and better approximation of the length of the corresponding curve. However, before we can take this limit, we need to work on the sum.

The difference Δy_k is equal to $f(x_k) - f(x_{k-1})$. The mean value theorem guarantees that there is a number $c_k \in [x_{k-1}, x_k]$ such that

$$f'(c_k) = \frac{f(x_k) - f(x_{k-1})}{x_k - x_{k-1}}$$

Since $\Delta x_k = x_k - x_{k-1}$, we find

$$\Delta y_k = f'(c_k)(x_k - x_{k-1}) = f'(c_k)\,\Delta x_k \qquad (6.17)$$

Replacing Δy_k in (6.16) by (6.17), we find for the length of the polygon

$$L_P = \sum_{k=1}^{n} \sqrt{(\Delta x_k)^2 + [f'(c_k)\,\Delta x_k]^2}$$

$$= \sum_{k=1}^{n} \sqrt{1 + [f'(c_k)]^2}\,\Delta x_k$$

This form allows us to take the limit as $\|P\| \to 0$. We find

$$L = \lim_{\|P\| \to 0} \sum_{k=1}^{n} \sqrt{1 + [f'(c_k)]^2}\,\Delta x_k$$

$$= \int_a^b \sqrt{1 + [f'(x)]^2}\,dx$$

If $f(x)$ is differentiable on (a, b) and $f'(x)$ is continuous on $[a, b]$, then the length of the curve $y = f(x)$ from a to b is given by

$$L = \int_a^b \sqrt{1 + [f'(x)]^2}\,dx$$

As mentioned, the first curve that was rectified was given by $y = f(x) = x^{3/2}$. We will choose this function for our first example.

▶ Example 11 ⬛ Determine the length of the curve given by the graph of $y = f(x) = x^{3/2}$ between $a = 5/9$ and $b = 21/9$.

Solution

To determine the length of the curve, we need to find $f'(x)$ first.

$$f'(x) = \frac{3}{2}x^{1/2}$$

Then

$$L = \int_{5/9}^{21/9} \sqrt{1 + \left[\frac{3}{2}x^{1/2}\right]^2}\, dx = \int_{5/9}^{21/9} \sqrt{1 + \frac{9}{4}x}\, dx$$

An antiderivative of $\sqrt{1 + \frac{9}{4}x}$ is $\frac{4}{9} \cdot \frac{2}{3}(1 + \frac{9}{4}x)^{3/2}$, as can be checked by differentiating $\frac{4}{9} \cdot \frac{2}{3}(1 + \frac{9}{4}x)^{3/2}$ with respect to x. Thus the length is

$$L = \left[\frac{4}{9} \cdot \frac{2}{3}\left(1 + \frac{9}{4}x\right)^{3/2}\right]_{5/9}^{21/9} = \frac{8}{27}\left[\left(1 + \frac{9}{4} \cdot \frac{21}{9}\right)^{3/2} - \left(1 + \frac{9}{4} \cdot \frac{5}{9}\right)^{3/2}\right]$$

$$= \frac{8}{27}\left[\left(\frac{5}{2}\right)^3 - \left(\frac{3}{2}\right)^3\right] = \frac{8}{27}\left(\frac{125}{8} - \frac{27}{8}\right) = \frac{98}{27}$$

The length of the curve is therefore 98/27. ◀

Before we show another example, we discuss the formula in more detail: namely, using

$$\frac{dy}{dx} = f'(x)$$

we can write for the length

$$L = \int_a^b \sqrt{1 + \left(\frac{dy}{dx}\right)^2}\, dx$$

Treating dy and dx like regular numbers, we can rewrite this as

$$L = \int_a^b \sqrt{(dx)^2 + (dy)^2}$$

We call the expression $\sqrt{(dx)^2 + (dy)^2}$ the **arc length differential** and denote it by ds. We can think of ds as a typical infinitesimal line segment. The Pythagorean theorem in this infinitesimal form then reads $(ds)^2 = (dx)^2 + (dy)^2$. "Adding up" these line segments (that is, computing $\int_a^b ds$) then yields the length of the curve.

▶ **Example 12** Determine the length of

$$f(x) = \frac{1}{4}x^2 - \frac{1}{2}\ln x \quad \text{from } x = 1 \text{ to } x = e$$

Solution

Differentiating $f(x)$, we find

$$f'(x) = \frac{x}{2} - \frac{1}{2x}$$

The length of the curve is then given by

$$L = \int_1^e \sqrt{1 + \left(\frac{x}{2} - \frac{1}{2x}\right)^2}\, dx = \int_1^e \sqrt{1 + \left(\frac{x^2}{4} - \frac{1}{2} + \frac{1}{4x^2}\right)}\, dx$$

$$= \int_1^e \sqrt{\frac{x^2}{4} + \frac{1}{2} + \frac{1}{4x^2}}\, dx$$

We notice that the expression under the square root is a perfect square, namely,

$$\frac{x^2}{4} + \frac{1}{2} + \frac{1}{4x^2} = \left(\frac{x}{2} + \frac{1}{2x}\right)^2$$

Hence, the integral for the length simplifies to

$$L = \int_1^e \sqrt{\left(\frac{x}{2} + \frac{1}{2x}\right)^2}\, dx$$

$$= \int_1^e \left(\frac{x}{2} + \frac{1}{2x}\right)\, dx = \frac{1}{4}x^2 + \frac{1}{2}\ln|x|\Big]_1^e$$

$$= \frac{1}{4}e^2 + \frac{1}{2} - \frac{1}{4} = \frac{1}{4}(e^2 + 1) \qquad \blacktriangleleft$$

Due to the somewhat complicated form of the integrand in the computation of the length, we quickly run into problems when we actually try to compute the integral. The integrand rarely simplifies enough for easy computation, such as in Example 12, where the integrand turned out to be a perfect square.

Even seemingly simple looking functions, like $y = 1/x$, quickly turn into complicated integrals when we compute the length of the curve.

▶ **Example 13**   Set up but do not evaluate the length of the curve of the hyperbola $f(x) = \frac{1}{x}$ between $a = 1$ and $b = 2$.

Solution

To determine the length of the curve, we need to find $f'(x)$ first.

$$f'(x) = -\frac{1}{x^2}$$

Then the length of the curve is given by the integral

$$L = \int_1^2 \sqrt{1 + \left(-\frac{1}{x^2}\right)^2}\, dx = \int_1^2 \sqrt{1 + \frac{1}{x^4}}\, dx \qquad \blacktriangleleft$$

The antiderivative of the integrand in Example 13 is quite complicated, and we will not be able to find it with the techniques available in this text. In Section 7.5, we will learn numerical methods to evaluate integrals, which can be used to evaluate the integral in Example 13. There are also computer software packages that can numerically evaluate such integrals. If we did this for the integral in Example 13, we would find that the length L is approximately 1.13.

6.3.6 Problems

(6.3.1)

Find the areas of the regions bounded by the lines and curves in Problems 1–12.

1. $y = e^x$, $y = -x$, $x = 0$, $x = 2$
2. $y = \sin x$, $y = 1$, $x = 0$, $x = \frac{\pi}{2}$
3. $y = x^2 - 1$, $y = x + 1$
4. $y = x^2$, $y = 2 - x^4$
5. $y = x^2 + 1$, $y = 4x - 2$ (in the first quadrant)
6. $y = x^2$, $y = 2 - x$, $y = 0$ (in the first quadrant)
7. $y = x^2$, $y = \frac{1}{x}$, $y = 4$ (in the first quadrant)
8. $y = \sin x$, $y = \cos x$ from $x = 0$ to $x = \frac{\pi}{2}$
9. $y = \frac{1}{x}$, $y = 1$ from $x = \frac{1}{2}$ to $x = \frac{3}{2}$

10. $y = x^2$, $y = (x-2)^2$, $y = 0$ from $x = 0$ to $x = 2$

11. $y = x^2$, $y = x^3$ from $x = 0$ to $x = 2$

12. $y = e^{-x}$, $y = x + 1$ from $x = -1$ to $x = 1$

In Problems 13–16, find the areas of the regions bounded by the lines and curves by expressing x as a function of y and integrating with respect to y.

13. $y = x^2$, $y = (x-2)^2$, $y = 0$ from $x = 0$ to $x = 2$

14. $y = x$, $yx = 1$, $y = \frac{1}{2}$ (in the first quadrant)

15. $x = (y-1)^2 + 3$, $x = 1 - (y-1)^2$ from $y = 0$ to $y = 2$ (in the first quadrant)

16. $x = (y-1)^2 - 1$, $x = (y-1)^2 + 1$ from $y = 0$ to $y = 2$

(6.3.2)

17. Consider a population whose size at time t is $N(t)$ and whose dynamics are given by the initial value problem

$$\frac{dN}{dt} = e^{-t}$$

with $N(0) = 100$.

(a) Find $N(t)$ by solving the initial value problem.

(b) Compute the cumulative change in population size between $t = 0$ and $t = 5$.

(c) Express the cumulative change in population size between time 0 and time t as an integral. Provide a geometric interpretation of this quantity.

18. Suppose that a change in biomass $B(t)$ at time t during the interval $[0, 12]$ follows the equation

$$\frac{d}{dt} B(t) = \cos\left(\frac{\pi}{6}t\right)$$

for $0 \le t \le 12$.

(a) Graph $\frac{dB}{dt}$ as a function of t.

(b) Suppose that $B(0) = B_0$. Express the cumulative change of biomass during the time interval $[0, t]$ as an integral. Give a geometric interpretation. What is the value of the biomass at the end of the interval $[0, 12]$ compared to the value at time 0? How are these two quantities related to the cumulative change of the biomass during the interval $[0, 12]$?

19. A particle moves along the x-axis with velocity

$$v(t) = -(t-2)^2 + 1$$

for $0 \le t \le 5$. Assume that the particle is at the origin at time 0.

(a) Graph $v(t)$ as a function of t.

(b) Use the graph of $v(t)$ to determine when the particle moves to the left and when it moves to the right.

(c) Find the location $s(t)$ of the particle at time t for $0 \le t \le 5$. Give a geometric interpretation of $s(t)$ in terms of the graph of $v(t)$.

(d) Graph $s(t)$ and find the leftmost and rightmost positions.

20. Recall that the acceleration $a(t)$ of a particle moving along a straight line is the instantaneous rate of change of the velocity $v(t)$, that is,

$$a(t) = \frac{d}{dt} v(t)$$

Assume that $a(t) = 32$ ft/s^2. Express the cumulative change in velocity during the time interval $[0, t]$ as a definite integral, and compute the integral.

21. If $\frac{dl}{dt}$ represents the growth rate of an organism at time t (measured in months), explain what

$$\int_2^7 \frac{dl}{dt}\, dt$$

represents.

22. If $\frac{dw}{dx}$ represents the rate of change of the weight of an organism of age x, explain what

$$\int_3^5 \frac{dw}{dx}\, dx$$

means.

23. If $\frac{dB}{dt}$ represents the rate of change of biomass at time t, explain what

$$\int_1^6 \frac{dB}{dt}\, dt$$

means.

24. Let $N(t)$ denote the size of a population at time t, and assume that

$$\frac{dN}{dt} = f(t)$$

Express the cumulative change of the population size in the interval $[2, 4]$ as an integral.

(6.3.3)

25. Let $f(x) = x^2 - 2$. Compute the average value of $f(x)$ over the interval $[0, 2]$.

26. Let $g(t) = e^{-t}$. Compute the average value of $g(t)$ over the interval $[-1, 1]$.

27. Suppose that the temperature T in a growing chamber (measured in Fahrenheit) varies over a 24-hour period according to

$$T(t) = 68 + \sin\left(\frac{\pi}{12}t\right)$$

for $0 \le t \le 24$.

(a) Graph the temperature T as a function of time t.

(b) Find the average temperature and explain your answer graphically.

28. Suppose that the concentration of soil nitrogen (measured in g m^{-3}) along a transect in moist tundra yields data points that follow a straight line with equation

$$y = 673.8 - 34.7x$$

for $0 \le x \le 10$, where x is the distance to the beginning of the transect. What is the average concentration of soil nitrogen along this transect?

29. Let $f(x) = \tan x$. Give a geometric argument to explain why the average value of $f(x)$ over $[-1, 1]$ is equal to 0.

30. Suppose that you drive from St. Paul to Duluth and you average 50 mph. Explain why there must be a time during your trip at which your speed is exactly 50 mph.

31. Let $f(x) = 2x$, $0 \le x \le 2$. Use a geometric argument to find the average value of f over the interval $[0, 2]$, and find x such that $f(x)$ is equal to this average value.

32. A particle moves along the x-axis with velocity

$$v(t) = -(t-3)^2 + 5$$

for $0 \le t \le 6$.

(a) Graph $v(t)$ as a function of t for $0 \le t \le 6$.

(b) Find the average velocity of this particle during the time interval $[0, 6]$.

(c) Find a time $t^* \in [0, 6]$ such that the velocity at time t^* is equal to the average velocity during the time interval $[0, 6]$. Is it clear that such a point exists? Is there more than one such point in this case? Use your graph in (a) to explain how you would find t^* graphically.

(6.3.4)

33. Find the volume of a right-circular cone with base radius r and height h.

34. Find the volume of a pyramid with square base of side length a and height h.

In Problems 35–40, find the volumes of the solids obtained by rotating the region bounded by the given curves about the x-axis. In each case, sketch the region and a typical disk element.

35. $y = 4 - x^2, y = 0, x = 0$ (in the first quadrant)
36. $y = \sqrt{x}, y = 0, x = 4$
37. $y = \sqrt{\sin x}, 0 \le x \le \pi, y = 0$
38. $y = e^x, y = 0, x = 0, x = \ln 2$
39. $y = \sec x, -\dfrac{\pi}{3} \le x \le \dfrac{\pi}{3}, y = 0$
40. $y = \sqrt{4 - x^2}, -2 \le x \le 2, y = 0$

In Problems 41–46, find the volumes of the solids obtained by rotating the region bounded by the given curves about the x-axis. In each case, sketch the region together with a typical disk element.

41. $y = x^2, y = x, 0 \le x \le 1$
42. $y = 1 - x^2, y = 1 + x^2, 0 \le x \le 1$
43. $y = e^x, y = e^{-x}, 0 \le x \le 2$
44. $y = \sqrt{1 - x^2}, y = 1, x = 1$ (in the first quadrant)
45. $y = \sqrt{\cos x}, y = 1, x = \dfrac{\pi}{2}$
46. $y = \dfrac{1}{x}, x = 0, y = 1, y = 2$ (in the first quadrant)

In Problems 47–52, find the volumes of the solids obtained by rotating the region bounded by the given curves about the y-axis. In each case, sketch the region together with a typical disk element.

47. $y = \sqrt{x}, y = 2, x = 0$
48. $y = x^2, y = 4, x = 0$ (in the first quadrant)
49. $y = \ln(x + 1), y = \ln 3, x = 0$
50. $y = \sqrt{x}, y = x, 0 \le x \le 1$
51. $y = x^2, y = \sqrt{x}, 0 \le x \le 1$
52. $y = \dfrac{1}{x}, x = 0, y = \frac{1}{2}, y = 1$

(6.3.5)

53. Find the length of the straight line
$$y = 2x$$
from $x = 0$ to $x = 2$ by the following methods.

(a) using planar geometry

(b) using the integral formula for the lengths of curves, derived in Subsection 6.3.5

54. Find the length of the straight line
$$y = mx$$
from $x = 0$ to $x = a$, where m and a are positive constants, by the following methods.

(a) using planar geometry

(b) using the integral formula for the lengths of curves, derived in Subsection 6.3.5

55. Find the length of the curve
$$y^2 = x^3$$
from $x = 1$ to $x = 4$.

56. Find the length of the curve
$$2y^2 = 3x^3$$
from $x = 0$ to $x = 1$.

57. Find the length of the curve
$$y = \frac{x^3}{6} + \frac{1}{2x}$$
from $x = 1$ to $x = 3$.

58. Find the length of the curve
$$y = \frac{x^4}{4} + \frac{1}{8x^2}$$
from $x = 2$ to $x = 4$.

In Problems 59–62, set up but do not evaluate the integrals for the lengths of the following curves.

59. $y = x^2, -1 \le x \le 1$
60. $y = \sin x, 0 \le x \le \dfrac{\pi}{2}$
61. $y = e^{-x}, 0 \le x \le 1$
62. $y = \ln x, 1 \le x \le e$

63. Find the length of the quarter-circle
$$y = \sqrt{1 - x^2}$$
for $0 \le x \le 1$ by the following methods.

(a) using a formula from geometry

(b) using the integral formula from Subsection 6.3.5

64. A cable that hangs between two poles at $x = -M$ and $x = M$ takes the shape of a catenary, with equation
$$y = \frac{1}{2a}(e^{ax} + e^{-ax})$$
where a is a positive constant. Compute the length of the cable when $a = 1$ and $M = \ln 2$.

65. Show that if
$$f(x) = \frac{e^x + e^{-x}}{2}$$
then the length of the curve of $f(x)$ between $x = 0$ and $x = a$ for any $a > 0$ is given by $f'(a)$.

6.4 Key Terms

Chapter 6 Review: Topics *Discuss the following definitions and concepts.*

1. Area

2. Summation notation

3. Algebraic rules for sums

4. A partition of an interval and the norm of a partition

5. Riemann sum

6. Definite integral

7. Riemann integrable

8. Geometric interpretation of definite integrals

9. The constant multiple and sum rule for integrals
10. The definite integral over a union of intervals
11. Comparison rules for definite integrals
12. The fundamental theorem of calculus, part I
13. Leibniz's rule
14. Antiderivatives
15. The fundamental theorem of calculus, part II
16. Evaluating definite integrals using the FTC, part II

17. Computing the area between curves using definite integrals
18. Cumulative change and definite integrals
19. The mean value theorem for definite integrals
20. The volume of a solid and definite integrals
21. Rectification of curves
22. Length of a curve
23. Arc length differential

6.5 Review Problems

1. (Discharge of a river)
When studying the flow of water in an open channel, such as a river in its bed, the amount of water passing through a cross section per second, the discharge (Q), is of interest. The following formula is used to compute the discharge:

$$Q = \int_0^B \overline{v}(b)h(b)\,db \qquad (6.18)$$

where b is the distance from one bank of the river to the point where the depth of the river $h(b)$ and the average velocity $\overline{v}(b)$ of the vertical velocity profile of the river at b were measured. The total width of the cross section is B (see Figure 6.48).

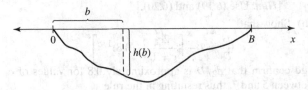

▲ **Figure 6.48**
The river for Problem 1

To evaluate the integral in (6.18), we would need to know $\overline{v}(b)$ and $h(b)$ at every location b along the cross section. In practice, the cross section is divided into a finite number of subintervals, and measurements of $\overline{v}$ and h are taken at the right endpoints, say, of each subinterval. The following table contains an example of such measurements. The location 0 corresponds to the left bank and the location $B = 16$ to the right bank of the river. The units of the location and of h are meters, and of $\overline{v}$, meter per second.

Location	h	$\overline{v}$
0	0	0
1	0.28	0.172
3	0.76	0.213
5	1.34	0.230
7	1.57	0.256
9	1.42	0.241
11	1.21	0.206
13	0.83	0.187
15	0.42	0.116
16	0	0

Approximate the integral in (6.18) by a Riemann sum, using the locations in the table, and find the approximate discharge using the data from the table.

2. Suppose that you grow plants in several study plots and wish to measure the response of total biomass to the respective treatment in each plot. One way to measure this would be to determine the average specific growth rate of the biomass for each plot over the course of the growing season.

We denote by $B(t)$ the biomass in a given plot at time t. Then the specific growth rate of the biomass at time t is given by

$$\frac{1}{B(t)}\frac{dB}{dt}$$

(a) Explain why

$$\frac{1}{t}\int_0^t \frac{1}{B(s)}\frac{dB(s)}{ds}\,ds$$

is a way to express the average specific growth rate over the time interval $[0, t]$.

(b) Use the chain rule to show that

$$\frac{1}{B(t)}\frac{dB}{dt} = \frac{d}{dt}(\ln B(t))$$

(c) Use the results in (a) and (b) to show that the average specific growth rate of $B(s)$ over the interval $[0, t]$ is given by

$$\frac{1}{t}\int_0^t \frac{d}{ds}(\ln(B(s)))\,ds = \frac{1}{t}\ln\frac{B(t)}{B(0)}$$

provided that $B(s) > 0$ for $s \in [0, t]$.

(d) Explain the measurements that you would need to take if you wanted to determine the average specific growth rate of biomass in a given plot over the time interval $[0, t]$.

Problems 3–6 discuss stream velocity profiles and provide a justification for the two measurement methods described in the following.

(*Adapted from Herschy, 1995*) The speed of water in a channel varies considerably with depth. Due to friction, the velocity reaches zero at the bottom and along the sides of the channel. The velocity is greatest near the surface of the stream. To find the average velocity for the vertical velocity profile, two methods are frequently employed in practice:

(a) The 0.6 depth method: The speed is measured at 0.6 of the depth from the surface, and this value is taken as the average speed.

(b) The 0.2 and 0.8 depth method: The speed is measured at 0.2 and 0.8 of the depth from the surface, and the average of the two readings is taken as the average speed.

The theoretical velocity distribution of water flowing in an open channel is given approximately by

$$v(d) = \left(\frac{D-d}{a}\right)^{1/c} \tag{6.19}$$

where $v(d)$ is the velocity at depth d below the water surface, c is a constant varying from 5 for coarse beds to 7 for smooth beds, D is the total depth of the channel, and a is a constant that is equal to the distance above the bottom of the channel at which the velocity has unit value.

3. (a) Sketch the graph of $v(d)$ as a function of d for $D = 3$ m and $a = 1$ m for (i) $c = 5$ and (ii) $c = 7$.

(b) Show that the velocity is equal to 0 at the bottom ($d = D$) and is maximal at the surface ($d = 0$).

4. (a) Show by integration that the average velocity in the vertical profile ($\bar{v}$) is given by

$$\bar{v} = \frac{c}{c+1}\left(\frac{D}{a}\right)^{1/c} \tag{6.20}$$

(b) What fraction of the maximum speed is the average speed $\bar{v}$?

(c) If you knew that the maximum speed occurred at the surface of the river [as predicted in the approximate formula for $v(d)$], how could you find $\bar{v}$? (In practice, the maximum speed may occur quite a bit below the surface due to friction between the water on the surface and the atmosphere. Therefore, the speed at the surface would not provide an accurate measurement of the maximum speed.)

5. Explain why the depth d_1, at which $v = \bar{v}$, is given by the equation

$$\bar{v} = \left(\frac{D-d_1}{a}\right)^{1/c} \tag{6.21}$$

We can find d_1 by equating (6.20) and (6.21). Show that

$$\frac{d_1}{D} = 1 - \left(\frac{c}{c+1}\right)^c$$

and that d_1/D is approximately 0.6 for values of c between 5 and 7, thus resulting in the rule

$$\bar{v} \approx v_{0.6}$$

where $v_{0.6}$ is the velocity at depth $0.6D$. (*Hint:* Graph $1 - (c/(c+1))^c$ as a function of c for $c \in [5, 7]$, and investigate the range of this function.)

6. We denote by $v_{0.2}$ the velocity at depth $0.2D$. We will now find the depth d_2 so that

$$\bar{v} = \frac{1}{2}(v_{0.2} + v_{d_2})$$

(a) Show that d_2 satisfies

$$\frac{1}{2}\left[\left(\frac{D - 0.2D}{a}\right)^{1/c} + \left(\frac{D - d_2}{a}\right)^{1/c}\right]$$

$$= \frac{c}{c+1}\left(\frac{D}{a}\right)^{1/c}$$

[*Hint:* Use (6.19) and (6.20).]

(b) Show that

$$\frac{d_2}{D} = 1 - \left[\frac{2c}{c+1} - (0.8)^{1/c}\right]^c$$

and confirm that d_2/D is approximately 0.8 for values of c between 5 and 7, thus resulting in the rule

$$\bar{v} \approx \frac{1}{2}(v_{0.2} + v_{0.8})$$

INTEGRATION TECHNIQUES AND COMPUTATIONAL METHODS

In the first two sections of this chapter, we will learn two important integration techniques that are essentially differentiation rules applied backward. (Because of the connection between differentiation and integration, it should not come as a surprise that some integration techniques are closely related to differentiation rules.) The first technique, called the substitution rule, is the chain rule backward; the second, called integration by parts, is the product rule backward. The third section is devoted to these integration techniques and an additional technique called partial fractions. The fourth section deals with improper integrals, which are integrals where the integrand goes to infinity somewhere over the integration interval or the integration interval is unbounded. Finally, we devote a section to numerical integration, another to the use of tables to evaluate more complicated integrals, and a last section to the approximation of functions by polynomials.

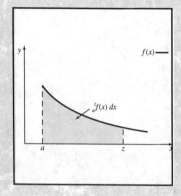

7.1 THE SUBSTITUTION RULE

7.1.1 Indefinite Integrals

The substitution rule is the chain rule in integral form. We therefore begin by recalling the chain rule. Suppose that we wish to differentiate

$$f(x) = \sin(3x^2 + 1)$$

This is clearly a situation in which we need to use the chain rule. We set

$$f(u) = \sin u \quad \text{and} \quad u = 3x^2 + 1$$

and find that $f'(u) = \cos u$. To differentiate the inner function $u = 3x^2 + 1$, we use Leibniz notation:

$$\frac{du}{dx} = 6x$$

or, if we treat du and dx like any other variables,

$$du = 6x \, dx$$

This latter form will be particularly convenient when we reverse the chain rule. But let's first use the chain rule. We obtain

$$\frac{d}{dx}\sin(3x^2 + 1) = \frac{df}{du}\frac{du}{dx} = (\cos u)(6x) = \cos(3x^2 + 1) \cdot 6x$$

Reversing this, we get

$$\int \underbrace{\cos(3x^2 + 1)}_{\cos u} \cdot \underbrace{6x \, dx}_{du} = \int \cos u \, du = \underbrace{\sin u}_{\sin(3x^2+1)} + C = \sin(3x^2 + 1) + C$$

In the first step, we substituted u for $3x^2 + 1$ and used $du = 6x \, dx$. This substitution simplified the integrand. At the end, we substitute back $3x^2 + 1$ for u to get the final answer in terms of x.

To see the general principle behind this, we write $u = g(x)$ [and hence $du = g'(x) \, dx$]. Our integral in the preceding is of the form

$$\int f[g(x)]g'(x) \, dx$$

If we denote by $F(x)$ an antiderivative of $f(x)$ [that is, $F'(x) = f(x)$], then, using the chain rule to differentiate $F[g(x)]$, we find

$$\frac{d}{dx} F[g(x)] = F'[g(x)]g'(x) = f[g(x)]g'(x)$$

which shows that $F[g(x)]$ is an antiderivative of $f[g(x)]g'(x)$. We can therefore write

$$\int f[g(x)]g'(x) \, dx = F[g(x)] + C \tag{7.1}$$

If we set $u = g(x)$, then the right-hand side of (7.1) can be written as $F(u) + C$. But since $F(u)$ is an antiderivative of $f(u)$, it also has the representation

$$\int f(u) \, du = F(u) + C \tag{7.2}$$

which shows that the left-hand side of (7.1) is the same as the left-hand side of (7.2). Equating the left-hand sides of (7.1) and (7.2) results in the substitution rule.

SUBSTITUTION RULE FOR INDEFINITE INTEGRALS

If $u = g(x)$, then

$$\int f[g(x)]g'(x) \, dx = \int f(u) \, du$$

We present a number of examples that will illustrate how to use this rule, and when the rule can be successfully applied. The first two examples are straightforward.

▶ **Example 1**   **Using Substitution** Evaluate

$$\int (2x + 1)e^{x^2 + x} \, dx$$

Solution

The expression $2x + 1$ is the derivative of $x^2 + x$, which is the inner function of $e^{x^2 + x}$. This suggests the following substitution:

$$u = x^2 + x \quad \text{with} \quad \frac{du}{dx} = 2x + 1 \quad \text{or } du = (2x + 1) \, dx$$

Hence,

$$\int \underbrace{e^{x^2+x}}_{e^u} \underbrace{(2x+1)\,dx}_{du} = \int e^u\,du = e^u + C = e^{x^2+x} + C$$

In the last step, we substituted $x^2 + x$ back for u since we want the final result in terms of x. ◀

▶ Example 2 **Using Substitution** Evaluate

$$\int \frac{1}{x \ln x}\,dx$$

Solution

We see that $1/x$ is the derivative of $\ln x$. We try

$$u = \ln x \quad \text{with} \quad \frac{du}{dx} = \frac{1}{x} \quad \text{or} \quad du = \frac{1}{x}\,dx$$

Then

$$\int \underbrace{\frac{1}{\ln x}}_{\frac{1}{u}} \underbrace{\frac{1}{x}\,dx}_{du} = \int \frac{1}{u}\,du = \ln|u| + C = \ln|\ln x| + C$$ ◀

Examples 1 and 2 illustrate types of integrals that are frequently encountered, and we display them in the following for ease of reference.

$$\int g'(x)e^{g(x)}\,dx = e^{g(x)} + C$$

$$\int \frac{g'(x)}{g(x)}\,dx = \ln|g(x)| + C$$

In the previous two examples, the derivative of the inner function appeared exactly in the integrand. This will not always be the case, as shown in the following example, where the derivative of the inner function appears in the integrand only up to a multiplicative constant.

▶ Example 3 **Multiplicative Constant** Evaluate

$$\int 4x\sqrt{x^2 + 1}\,dx$$

Solution

If we set $u = x^2 + 1$, then

$$\frac{du}{dx} = 2x \quad \text{or} \quad \frac{du}{2} = x\,dx$$

The integrand contains the derivative of the inner function up to a multiplicative constant. But this is good enough:

$$\int 4x\sqrt{x^2 + 1}\,dx = \int 4\underbrace{\sqrt{x^2 + 1}}_{\sqrt{u}}\underbrace{x\,dx}_{\frac{du}{2}} = \int 4\sqrt{u}\,\frac{du}{2}$$

$$= \int 2\sqrt{u}\,du = 2 \cdot \frac{2}{3}u^{3/2} + C = \frac{4}{3}(x^2 + 1)^{3/2} + C$$ ◀

▶ **Example 4** **Rewriting the Integrand** Evaluate

$$\int \tan x \, dx$$

Solution

The trick here is to rewrite $\tan x = \frac{\sin x}{\cos x}$ and realize that the derivative of the function in the denominator is the function in the numerator (up to a minus sign). The integral is therefore of the type $\int [g'(x)/g(x)] \, dx$, which we discussed before. We use the substitution

$$u = \cos x \quad \text{with} \quad \frac{du}{dx} = -\sin x \quad \text{or} \quad -du = \sin x \, dx$$

Then

$$\int \tan x \, dx = \int \frac{\sin x}{\cos x} \, dx = \int \underbrace{\frac{1}{\cos x}}_{\frac{1}{u}} \underbrace{\sin x \, dx}_{-du}$$

$$= -\int \frac{1}{u} \, du = -\ln |u| + C = -\ln |\cos x| + C$$

In Table 6.1 of Section 6.2, we listed $\int \tan x \, dx = \ln |\sec x| + C$. This is the same result that we have just obtained, since $-\ln |\cos x| = \ln |\cos x|^{-1} = \ln |\sec x|$. ◀

In Problem 59, you will evaluate $\int \cot x \, dx$, where the same trick as in Example 4 is used. That is, both $\int \tan x \, dx$ and $\int \cot x \, dx$ are special cases of $\int [g'(x)/g(x)] \, dx$. We collect both integrals in the following.

$$\int \tan x \, dx = -\ln |\cos x| + C \qquad (7.3)$$

$$\int \cot x \, dx = \ln |\sin x| + C \qquad (7.4)$$

It is not always obvious that substitution will be successful, as in the following example.

▶ **Example 5** **Substitution and Square Roots** Evaluate

$$\int x\sqrt{2x - 1} \, dx$$

Solution

Obviously, x is not the derivative of $2x - 1$, so this does not seem to fit our scheme. But watch what we do: Set

$$u = 2x - 1 \quad \text{with} \quad \frac{du}{dx} = 2 \quad \text{or} \quad dx = \frac{du}{2}$$

Since $u = 2x - 1$, we find $x = \frac{1}{2}(u + 1)$. Making all the substitutions, we find

$$\int x\sqrt{2x - 1}\,dx = \int \frac{1}{2}(u + 1)\sqrt{u}\frac{du}{2} = \frac{1}{4}\int (u^{3/2} + u^{1/2})\,du$$

$$= \frac{1}{4}\left(\frac{2}{5}u^{5/2} + \frac{2}{3}u^{3/2}\right) + C$$

$$= \frac{1}{10}(2x - 1)^{5/2} + \frac{1}{6}(2x - 1)^{3/2} + C \qquad ◀$$

Functions in the integrand are not always explicitly given, as in the following example.

▶ **Example 6** Assume that $g(x)$ is a differentiable function whose derivative $g'(x)$ is continuous. Evaluate

$$\int g'(x)\cos[g(x)]\,dx$$

Solution

If we set

$$u = g(x) \quad \text{with} \quad \frac{du}{dx} = g'(x) \quad \text{or } du = g'(x)\,dx$$

then

$$\int g'(x)\cos[g(x)]\,dx = \int \cos u\,du = \sin u + C$$

$$= \sin[g(x)] + C \qquad ◀$$

7.1.2 Definite Integrals

When we evaluate a definite integral, the FTC (part II) says that we must find an antiderivative of the integrand and then evaluate the antiderivative at the limits of integration. When we use the substitution $u = g(x)$ to find an antiderivative of an integrand, the antiderivative will be given in terms of u at first. To complete the calculation, we can proceed in either of two ways: (1) We can substitute $g(x)$ for u in the antiderivative and then evaluate the antiderivative at the limits of integration in terms of x, or (2) we can leave the antiderivative in terms of u and change the limits of integration according to $u = g(x)$.

We illustrate these two ways in the following example, where we wish to evaluate

$$\int_0^4 2x\sqrt{x^2 + 1}\,dx$$

Recall that when we compute definite integrals, we need to check whether the integrand is continuous over the integration interval. This is the case here.

First Way   To find an antiderivative of $f(x) = 2x\sqrt{x^2 + 1}$, we choose the substitution

$$u = x^2 + 1 \quad \text{with} \quad \frac{du}{dx} = 2x \quad \text{or } du = 2x\,dx$$

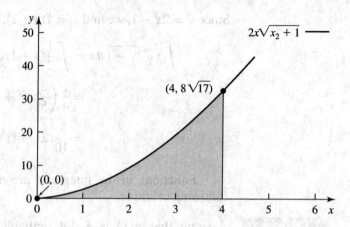

▲ Figure 7.1

The region corresponding to $\int_0^4 2x\sqrt{x^2+1}\,dx$ before substitution has an area of $\frac{2}{3}[(17)^{3/2}-1]$

Then

$$\int 2x\sqrt{x^2+1}\,dx = \int \sqrt{u}\,du = \frac{2}{3}u^{3/2} + C = \frac{2}{3}(x^2+1)^{3/2} + C$$

and $F(x) = \frac{2}{3}(x^2+1)^{3/2}$ is an antiderivative of $2x\sqrt{x^2+1}$. Using the FTC, part II, we can now compute the definite integral:

$$\int_0^4 2x\sqrt{x^2+1}\,dx = F(4) - F(0)$$

$$= \frac{2}{3}(17)^{3/2} - \frac{2}{3}(1)^{3/2} = \frac{2}{3}[(17)^{3/2}-1]$$

The region corresponding to the definite integral before substitution is shown in Figure 7.1.

Second Way   We change the limits of integration along with the substitution. That is, we set

$$u = x^2 + 1 \quad \text{with } \frac{du}{dx} = 2x \text{ or } du = 2x\,dx$$

as before, and note that

$$\text{if } x = 0 \quad \text{then } u = 1$$
$$\text{if } x = 4 \quad \text{then } u = 17$$

Hence

$$\int_0^4 2x\sqrt{x^2+1}\,dx = \int_1^{17} \sqrt{u}\,du = \frac{2}{3}u^{3/2}\bigg]_1^{17} = \frac{2}{3}[(17)^{3/2}-1]$$

After substitution, the integrand is $\sqrt{u}$, and the limits of integration are $u = 1$ and $u = 17$. The region corresponding to the definite integral after substitution is shown in Figure 7.2.

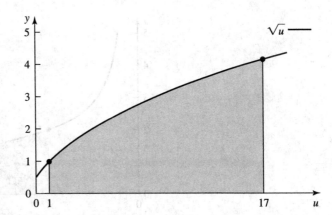

▲ Figure 7.2

The definite integral $\int_0^4 2x\sqrt{x^2+1}\,dx$ becomes $\int_1^{17}\sqrt{u}\,du$ after the substitution $u = x^2 + 1$. The region corresponding to $\int_1^{17}\sqrt{u}\,du$ has an area of $\frac{2}{3}[(17)^{3/2}-1]$

The second way is often more convenient, and we summarize the procedure in the following.

SUBSTITUTION RULE FOR DEFINITE INTEGRALS

If $u = g(x)$, then

$$\int_a^b f[g(x)]g'(x)\,dx = \int_{g(a)}^{g(b)} f(u)\,du$$

▶ **Example 7** **Definite Integral** Compute

$$\int_1^2 \frac{3x^2+1}{x^3+x}\,dx$$

Solution

(The region corresponding to the definite integral is shown in Figure 7.3.) The integrand is continuous on $[1, 2]$ and is of the form $\frac{g'(x)}{g(x)}$. We set

$$u = x^3 + x \quad \text{with} \quad \frac{du}{dx} = 3x^2 + 1 \quad \text{or } du = (3x^2+1)\,dx$$

and change the limits of integration:

$$\text{if } x = 1 \quad \text{then } u = 2$$
$$\text{if } x = 2 \quad \text{then } u = 10$$

Therefore,

$$\int_1^2 \frac{3x^2+1}{x^3+x}\,dx = \int_2^{10} \frac{1}{u}\,du = \ln|u|]_2^{10} = \ln 10 - \ln 2 = \ln\frac{10}{2} = \ln 5 \quad ◀$$

▶ **Example 8** **Substitution Function is Decreasing** Compute

$$\int_{1/2}^1 \frac{1}{x^2}e^{1/x}\,dx$$

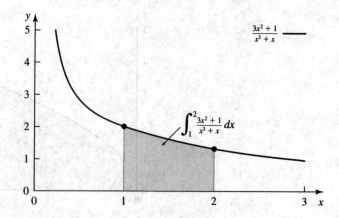

▲ **Figure 7.3**
The region corresponding to the definite integral in Example 7

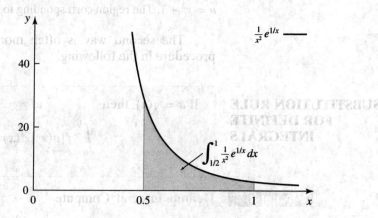

▲ **Figure 7.4**
The region corresponding to the definite integral in Example 8

Solution

(The region corresponding to the definite integral is shown in Figure 7.4.) The integrand is continuous on $[1/2, 1]$. Since $-1/x^2$ is the derivative of $1/x$, we set

$$u = \frac{1}{x} \quad \text{with} \quad \frac{du}{dx} = -\frac{1}{x^2} \quad \text{or} \quad -du = \frac{1}{x^2}\, dx$$

and change the limits of integration:

$$\text{if } x = \frac{1}{2} \quad \text{then } u = 2$$
$$\text{if } x = 1 \quad \text{then } u = 1$$

Therefore,

$$\int_{1/2}^{1} \frac{1}{x^2} e^{1/x}\, dx = -\int_{2}^{1} e^u\, du = \int_{1}^{2} e^u\, du = e^u\big]_1^2 = e^2 - e$$

Note that since $\frac{1}{x}$ is a decreasing function for $x > 0$, the lower limit is greater than the upper limit of integration after the substitution. When we reversed the order of integration in the second step, we removed the negative sign.

However, you need not reverse the order of integration. Instead, you can compute directly:

$$-\int_2^1 e^u \, du = - \, e^u]_2^1 = -(e^1 - e^2) = e^2 - e$$

◀

▶ **Example 9**   **Trigonometric Substitution** Compute

$$\int_0^{\pi/6} \cos x \, e^{\sin x} \, dx$$

Solution

(The region corresponding to the definite integral is shown in Figure 7.5.) The integrand is continuous on the interval $[0, \pi/6]$ and is of the form $g'(x)e^{g(x)}$, which suggests the substitution

$$u = \sin x \quad \text{with} \quad \frac{du}{dx} = \cos x \quad \text{or} \quad du = \cos x \, dx$$

Now, change the limits of integration:

$$\text{if } x = 0 \quad \text{then } u = \sin 0 = 0$$
$$\text{if } x = \frac{\pi}{6} \quad \text{then } u = \sin \frac{\pi}{6} = \frac{1}{2}$$

Therefore,

$$\int_0^{\pi/6} \cos x \, e^{\sin x} \, dx = \int_0^{1/2} e^u \, du = e^u]_0^{1/2} = e^{1/2} - 1$$

◀

▶ **Example 10**   **Rational Function** Compute

$$\int_4^9 \frac{2}{x - 3} \, dx$$

Solution

(The region corresponding to the definite integral is shown in Figure 7.6.) The integrand is continuous on the interval $[4, 9]$. We set

$$u = x - 3 \quad \text{with} \quad \frac{du}{dx} = 1 \quad \text{or} \quad du = dx$$

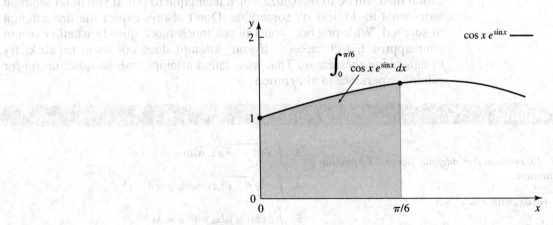

▲ Figure 7.5
The region corresponding to the definite integral in Example 9

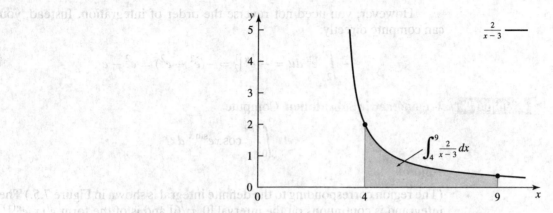

▲ **Figure 7.6**
The region corresponding to the definite integral in Example 10

and change the limits of integration:

$$\text{if } x = 4 \quad \text{then } u = 1$$
$$\text{if } x = 9 \quad \text{then } u = 6$$

Therefore,

$$\int_4^9 \frac{2}{x-3}\, dx = \int_1^6 \frac{2}{u}\, du = 2\ln|u|\Big]_1^6 = 2(\ln 6 - \ln 1) = 2\ln 6 \quad \triangleleft$$

We can easily spend a great deal of time on integration techniques. The problems can get very involved: To solve them all, we would need a big bag full of tricks. There are excellent software programs (such as *Mathematica* and *Matlab*) that can integrate symbolically. These programs do not render integration techniques useless; in fact, they use them. Understanding the basic techniques conceptually and being able to apply them in simple situations makes these software packages less of a "black box." Nevertheless, their availability has made it less important to acquire a large number of tricks.

So far, we have only learned one technique: substitution. Unless you can immediately recognize an antiderivative, substitution is the only method you can try at this point.

As we proceed, you will learn other techniques. An additional complication then will be to recognize which technique to use. If you don't see right away what to do, just try something. Don't always expect the first attempt to succeed. With practice, you will see much more quickly whether or not your approach will succeed. If your attempt does not seem to work, try to determine the reason. That way, failed attempts can be quite useful for gaining experience in integration.

7.1.3 Problems

(7.1.1)

In Problems 1–16, evaluate the indefinite integral by making the given substitution.

1. $\displaystyle\int 2x\sqrt{x^2+3}\, dx$, with $u = x^2 + 3$

2. $\displaystyle\int 3x^2\sqrt{x^3+1}\, dx$, with $u = x^3 + 1$

3. $\displaystyle\int x\sqrt{1-x^2}\, dx$, with $u = 1 - x^2$

4. $\displaystyle\int x^3\sqrt{4+x^4}\, dx$, with $u = 4 + x^4$

5. $\displaystyle\int 5\cos(3x)\, dx$, with $u = 3x$

6. $\displaystyle\int 5\sin(1-2x)\, dx$, with $u = 1 - 2x$

7. $\int 7x^2 \sin(4x^3)\,dx$, with $u = 4x^3$

8. $\int x \cos(x^2 - 1)\,dx$, with $u = x^2 - 1$

9. $\int e^{2x+3}\,dx$, with $u = 2x + 3$

10. $\int 3e^{1-x}\,dx$, with $u = 1 - x$

11. $\int xe^{-x^2/2}\,dx$, with $u = -x^2/2$

12. $\int xe^{1-3x^2}\,dx$, with $u = 1 - 3x^2$

13. $\int \dfrac{x+2}{x^2+4x}\,dx$, with $u = x^2 + 4x$

14. $\int \dfrac{2x}{3-x^2}\,dx$, with $u = 3 - x^2$

15. $\int \dfrac{3}{x+4}\,dx$, with $u = x + 4$

16. $\int \dfrac{1}{5-x}\,dx$, with $u = 5 - x$

In Problems 17–36, use substitution to evaluate the indefinite integrals.

17. $\int \sqrt{x+3}\,dx$

18. $\int \sqrt{4-x}\,dx$

19. $\int (4x-3)\sqrt{2x^2-3x+2}\,dx$

20. $\int (x^2-2x)\sqrt{x^3-3x^2+3}\,dx$

21. $\int \dfrac{4x-1}{1+x-2x^2}\,dx$

22. $\int \dfrac{x^2-1}{x^3-3x+1}\,dx$

23. $\int \dfrac{x}{1+2x^2}\,dx$

24. $\int \dfrac{x^3-1}{x^4-4x}\,dx$

25. $\int 2xe^{x^2}\,dx$

26. $\int \cos x\, e^{\sin x}\,dx$

27. $\int \dfrac{1}{x}e^{1+\ln x}\,dx$

28. $\int \sec^2 x\, e^{\tan x}\,dx$

29. $\int \sin\left(\dfrac{3\pi}{2}x + \dfrac{\pi}{4}\right)dx$

30. $\int \cos(2x-1)\,dx$

31. $\int \tan x \sec^2 x\,dx$

32. $\int \sin^3 x \cos x\,dx$

33. $\int \dfrac{(\ln x)^2}{x}\,dx$

34. $\int \dfrac{dx}{(x-3)\ln(x-3)}$

35. $\int x\sqrt{5+x}\,dx$

36. $\int \sqrt{1+\ln x}\,\dfrac{\ln x}{x}\,dx$

In Problems 37–42, a, b, and c are constants, and g(x) is a continuous function whose derivative g′(x) is also continuous. Use substitution to evaluate the indefinite integrals.

37. $\int \dfrac{2ax+b}{ax^2+bx+c}\,dx$

38. $\int \dfrac{1}{ax+b}\,dx$

39. $\int g'(x)[g(x)]^n\,dx$

40. $\int g'(x)\sin[g(x)]\,dx$

41. $\int g'(x)e^{-g(x)}\,dx$

42. $\int \dfrac{g'(x)}{[g(x)]^2+1}\,dx$

(7.1.2)

In Problems 43–58, use substitution to evaluate the definite integrals.

43. $\int_0^3 x\sqrt{x^2+1}\,dx$

44. $\int_1^2 x^2\sqrt{x^3+2}\,dx$

45. $\int_2^3 \dfrac{2x+3}{(x^2+3x)^3}\,dx$

46. $\int_0^2 \dfrac{2x}{\sqrt{4x^2+3}}\,dx$

47. $\int_2^5 (x-2)e^{-(x-2)^2/2}\,dx$

48. $\int_{\ln 4}^{\ln 7} \dfrac{e^x}{(e^x-3)^2}\,dx$

49. $\int_0^{\pi/3} \sin x \cos x\,dx$

50. $\int_{-\pi/6}^{\pi/6} \sin^2 x \cos x\,dx$

51. $\int_0^{\pi/4} \tan x \sec^2 x\,dx$

52. $\int_0^{\pi/3} \dfrac{\sin x}{\cos^2 x}\,dx$

53. $\int_5^9 \dfrac{x}{x-3}\,dx$

54. $\int_0^2 \dfrac{x}{x+2}\,dx$

55. $\int_e^{e^2} \dfrac{dx}{x(\ln x)^2}$

56. $\int_1^2 \dfrac{x\,dx}{(x^2+1)\ln(x^2+1)}$

57. $\displaystyle\int_1^9 \frac{1}{\sqrt{x}} e^{-\sqrt{x}}\, dx$

58. $\displaystyle\int_0^2 x\sqrt{4-x^2}\, dx$

59. Use the fact that

$$\cot x = \frac{\cos x}{\sin x}$$

to evaluate

$$\int \cot x \, dx$$

7.2 INTEGRATION BY PARTS

As mentioned at the beginning of this chapter, integration by parts is the product rule in integral form. Let $u = u(x)$ and $v = v(x)$ be differentiable functions, then differentiating with respect to x

$$(uv)' = u'v + uv'$$

or, after rearranging,

$$uv' = (uv)' - u'v$$

Integrating both sides with respect to x, we find

$$\int uv' \, dx = \int (uv)' \, dx - \int u'v \, dx$$

Since uv is an antiderivative of $(uv)'$,

$$\int (uv)' \, dx = uv + C$$

We therefore find

$$\int uv' \, dx = uv - \int u'v \, dx$$

(Note that the constant C can be absorbed into the indefinite integral on the right-hand side.) Since $u' = du/dx$ and $v' = dv/dx$, we can write this in short form:

$$\int u \, dv = uv - \int v \, du$$

We summarize this result in the following.

INTEGRATION BY PARTS RULE

If $u(x)$ and $v(x)$ are differentiable functions, then

$$\int u(x)v'(x) \, dx = u(x)v(x) - \int u'(x)v(x) \, dx$$

or, in short form,

$$\int u \, dv = uv - \int v \, du$$

You are probably wondering how this will help, given that we traded one integral for another one. Here is a first example.

▶ **Example 1** **Integration by Parts** Evaluate

$$\int x \sin x \, dx$$

Solution

The integrand $x \sin x$ is a product of two functions, one of which will be designated as u, the other as v'. Since applying the integration by parts rule will result in another integral of the form $\int u'v\,dx$, we must choose u and v' so that $u'v$ is of a simpler form. This suggests the following choices:

$$u = x \quad \text{and} \quad v' = \sin x$$

Since $v = -\cos x$ and $u' = 1$, the integral $\int u'v\,dx$ is then of the form $-\int \cos x\,dx$, which is indeed simpler. We obtain

$$\int x \sin x\,dx = (-\cos x)(x) - \int (-\cos x)(1)\,dx$$

$$= -x \cos x + \int \cos x\,dx$$

$$= -x \cos x + \sin x + C$$

If we had chosen $v' = x$ and $u = \sin x$, then we would have had $v = \frac{1}{2}x^2$ and $u' = \cos x$. The integral $\int u'v\,dx$ would have been of the form $\int \frac{1}{2}x^2 \cos x\,dx$, which is even more complicated than $\int x \sin x\,dx$.

If we use the short form $\int u\,dv = uv - \int v\,du$, we would write

$$u = x \quad \text{and} \quad dv = \sin x\,dx$$

Then

$$du = dx \quad \text{and} \quad v = -\cos x$$

and

$$\int \underbrace{x}_{u}\, \underbrace{\sin x\,dx}_{dv} = \underbrace{x}_{u}\,\underbrace{(-\cos x)}_{v} - \int \underbrace{(-\cos x)}_{v}\,\underbrace{dx}_{du}$$

$$= -x \cos x + \sin x + C \qquad \blacktriangleleft$$

▶ **Example 2** **Integration by Parts** Evaluate

$$\int x \ln x\,dx$$

Solution

Since we do not know an antiderivative of $\ln x$, we try

$$u = \ln x \quad \text{and} \quad v' = x$$

Then

$$u' = \frac{1}{x} \quad \text{and} \quad v = \frac{1}{2}x^2$$

and

$$\int x \ln x\,dx = \frac{1}{2}x^2 \ln x - \int \frac{1}{2}x^2 \cdot \frac{1}{x}\,dx$$

$$= \frac{1}{2}x^2 \ln x - \int \frac{1}{2}x\,dx = \frac{1}{2}x^2 \ln x - \frac{1}{4}x^2 + C \qquad \blacktriangleleft$$

Before we present a few more useful tricks, we show how to evaluate definite integrals with this method.

▶ **Example 3** **Definite Integral** Compute

$$\int_0^1 xe^{-x}\,dx$$

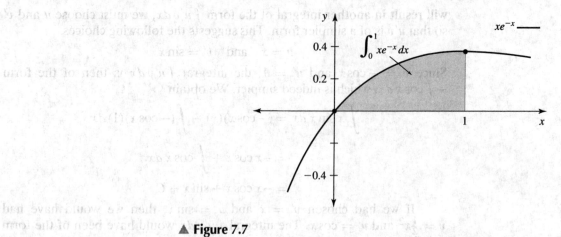

▲ **Figure 7.7**
The region corresponding to the definite integral in Example 3

Solution

The region representing the definite integral is shown in Figure 7.7. The integrand is continuous on $[0, 1]$. We set

$$u = x \quad \text{and} \quad dv = e^{-x}\, dx$$

Then

$$du = dx \quad \text{and} \quad v = -e^{-x}$$

Therefore,

$$\int_0^1 xe^{-x}\, dx = -xe^{-x}\Big]_0^1 - \int_0^1 (-e^{-x})\, dx$$

$$= -1e^{-1} - (-0e^{-0}) + \int_0^1 e^{-x}\, dx$$

$$= -e^{-1} + \left[-e^{-x}\right]_0^1 = -e^{-1} + (-e^{-1} - (-e^{-0}))$$

$$= -e^{-1} - e^{-1} + 1 = 1 - 2e^{-1}$$ ◀

In the next two examples, we demonstrate a trick that is sometimes useful when integrating by parts. The technique is called "multiplying by 1."

▶ **Example 4** **Multiplying by 1** Evaluate

$$\int \ln x\, dx$$

Solution

The integrand $\ln x$ is not a product of two functions, but we can write it as $(1)(\ln x)$ and set

$$u = \ln x \quad \text{and} \quad v' = 1$$

Then

$$u' = \frac{1}{x} \quad \text{and} \quad v = x$$

We find

$$\int \ln x \, dx = \int (1)(\ln x) \, dx = x \ln x - \int x \frac{1}{x} \, dx$$

$$= x \ln x - \int 1 \, dx = x \ln x - x + C$$

Our choices for u and v' might surprise you, as we said that our goal was to make the integral look simpler, which often means that we try to reduce the power of functions of the form x^n. In this case, however, integrating 1 and differentiating $\ln x$ yielded a simpler integral. In fact, if we had chosen $u' = \ln x$ and $v = 1$, we would not have been able to carry out the integration by parts since we would have needed the antiderivative of $\ln x$ to compute uv and $\int uv' \, dx$.

If you prefer the short form notation, you don't need to multiply by 1 since

$$u = \ln x \quad \text{and} \quad dv = dx$$

with

$$du = \frac{1}{x} dx \quad \text{and} \quad v = x$$

produces immediately

$$\int \ln x \, dx = x \ln x - \int x \frac{1}{x} \, dx = x \ln x - \int dx$$

$$= x \ln x - x + C$$

▶ **Example 5** *Multiplying by 1* Evaluate

$$\int \tan^{-1} x \, dx$$

Solution

We write $\tan^{-1} x = (1)(\tan^{-1} x)$, and

$$u = \tan^{-1} x \quad \text{and} \quad v' = 1$$

Then

$$u' = \frac{1}{x^2 + 1} \quad \text{and} \quad v = x$$

We find

$$\int \tan^{-1} x \, dx = x \tan^{-1} x - \int \frac{x}{x^2 + 1} \, dx$$

We need to use substitution to evaluate the integral on the right-hand side. With

$$w = x^2 + 1 \quad \text{and} \quad \frac{dw}{dx} = 2x$$

we obtain

$$\int \frac{x}{x^2 + 1} \, dx = \frac{1}{2} \int \frac{dw}{w} = \frac{1}{2} \ln |w| + C_1 = \frac{1}{2} \ln(x^2 + 1) + C_1$$

where C_1 is the constant of integration. Hence,

$$\int \tan^{-1} x \, dx = x \tan^{-1} x - \frac{1}{2} \ln(x^2 + 1) - C_1$$

We write the final answer as

$$\int \tan^{-1} x \, dx = x \tan^{-1} x - \frac{1}{2} \ln(x^2 + 1) + C$$

where C is the constant of integration with $C = -C_1$. Replacing $-C_1$ by C is done for purely aesthetical reasons. ◀

▶ **Example 6** *Using Integration by Parts Repeatedly* Compute

$$\int_0^1 x^2 e^x \, dx$$

Solution

When you evaluate a definite integral using integration by parts, it is often easier to evaluate the indefinite integral first and then to use the FTC, part II, to compute the definite integral. To evaluate $\int x^2 e^x \, dx$, we set

$$u = x^2 \quad \text{and} \quad v' = e^x$$

Then

$$u' = 2x \quad \text{and} \quad v = e^x$$

Therefore

$$\int x^2 e^x \, dx = x^2 e^x - \int 2x e^x \, dx \tag{7.5}$$

To evaluate the integral $\int x e^x \, dx$, we must use integration by parts for a second time. We set

$$u = x \quad \text{and} \quad v' = e^x$$

Then

$$u' = 1 \quad \text{and} \quad v = e^x$$

Therefore,

$$\int x e^x \, dx = x e^x - \int e^x \, dx = x e^x - e^x + C \tag{7.6}$$

Combining (7.5) and (7.6), we find

$$\int x^2 e^x \, dx = x^2 e^x - 2[x e^x - e^x + C]$$

$$= x^2 e^x - 2x e^x + 2e^x - 2C$$

After evaluating the indefinite integral, we can now compute the definite integral. Note that the integrand is continuous on $[0, 1]$. We set $F(x) = x^2 e^x - 2x e^x + 2e^x$. Then

$$\int_0^1 x^2 e^x \, dx = F(1) - F(0)$$

$$= (e - 2e + 2e) - (0 - 0 + 2) = e - 2$$ ◀

▶ **Example 7** *Using Integration by Parts Repeatedly* Evaluate

$$\int e^x \cos x \, dx$$

Solution

You can check that it does not matter which of the functions you call u and which v'. We set

$$u = \cos x \quad \text{and} \quad v' = e^x$$

Then

$$u' = -\sin x \quad \text{and} \quad v = e^x$$

Therefore,

$$\int e^x \cos x \, dx = e^x \cos x + \int e^x \sin x \, dx \tag{7.7}$$

We apply the integration by parts rule a second time. This time, the choice matters. We need to set

$$u = \sin x \quad \text{and} \quad v' = e^x$$

Then

$$u' = \cos x \quad \text{and} \quad v = e^x$$

Therefore,

$$\int e^x \sin x \, dx = e^x \sin x - \int e^x \cos x \, dx \tag{7.8}$$

Combining (7.7) and (7.8) yields

$$\int e^x \cos x \, dx = e^x \cos x + e^x \sin x - \int e^x \cos x \, dx$$

We see that the integral $\int e^x \cos x \, dx$ appears on both sides. Rearranging the equation, we obtain

$$2 \int e^x \cos x \, dx = e^x \cos x + e^x \sin x + C_1$$

or

$$\int e^x \cos x \, dx = \frac{1}{2} e^x (\cos x + \sin x) + C$$

with $C = C_1/2$. [Note that we introduced the constant C_1 (and C) in the final answer.]

We said that the choices for u and v' in the second application of the integration by parts matters. If we had designated $u = e^x$ and $v' = \sin x$, then $u' = e^x$ and $v = -\cos x$, yielding

$$\int e^x \sin x \, dx = -e^x \cos x + \int e^x \cos x \, dx$$

and thus, combining this with (7.7),

$$\int e^x \cos x \, dx = e^x \cos x - e^x \cos x + \int e^x \cos x \, dx$$

which is a correct but useless statement. ◀

We conclude this section with a piece of practical advice: In integrals of the form $\int P(x) \sin(ax) \, dx$, $\int P(x) \cos(ax) \, dx$, and $\int P(x) e^{ax} \, dx$, where $P(x)$ is a polynomial and a is a constant, the polynomial $P(x)$ should be considered as u and the expressions $\sin(ax)$, $\cos(ax)$, and e^{ax} as v'. If an integral contains functions $\ln x$, $\tan^{-1} x$, or $\sin^{-1} x$, then these functions are usually treated as u. After practicing the problems, you can confirm this advice.

7.2.1 Problems

In Problems 1–30, use integration by parts to evaluate the integrals.

1. $\int x \cos x \, dx$

2. $\int 3x \cos x \, dx$

3. $\int 2x \cos(3x) \, dx$

4. $\int 3x \cos(1 - x) \, dx$

5. $\int 2x \sin x \, dx$

6. $\int x \sin(2x) \, dx$

7. $\int xe^x \, dx$

8. $\int 2xe^{-x} \, dx$

9. $\int x^2 e^x \, dx$

10. $\int 2x^2 e^{-x} \, dx$

11. $\int x \ln x \, dx$

12. $\int x^2 \ln x \, dx$

13. $\int x \ln(3x) \, dx$

14. $\int x^2 \ln x^2 \, dx$

15. $\int x \sec^2 x \, dx$

16. $\int x \csc^2 x \, dx$

17. $\int_0^{\pi/3} x \sin x \, dx$

18. $\int_0^{\pi/4} x \cos x \, dx$

19. $\int_1^2 \ln x \, dx$

20. $\int_1^e \ln x^2 \, dx$

21. $\int_1^4 \ln \sqrt{x} \, dx$

22. $\int_1^4 \sqrt{x} \ln \sqrt{x} \, dx$

23. $\int_0^1 xe^{-x} \, dx$

24. $\int_0^3 x^2 e^{-x} \, dx$

25. $\int_0^{\pi/3} e^x \sin x \, dx$

26. $\int_0^{\pi/6} e^x \cos x \, dx$

27. $\int e^{-3x} \cos\left(\frac{\pi}{2}x\right) dx$

28. $\int e^{-2x} \sin\left(\frac{x}{2}\right) dx$

29. $\int \sin(\ln x) \, dx$

30. $\int \cos(\ln x) \, dx$

31. Evaluating the integral
$$\int \cos^2 x \, dx$$
requires two steps.

First, write
$$\cos^2 x = (\cos x)(\cos x)$$
and apply integration by parts to show that
$$\int \cos^2 x \, dx = \sin x \cos x$$
$$+ \int \sin^2 x \, dx$$

Then, use $\sin^2 x + \cos^2 x = 1$ to replace $\sin^2 x$ in the integral on the right-hand side, and complete the integration of $\int \cos^2 x \, dx$.

32. Evaluating the integral
$$\int \sin^2 x \, dx$$
requires two steps.

First, write
$$\sin^2 x = (\sin x)(\sin x)$$
and apply integration by parts to show that
$$\int \sin^2 x \, dx = -\sin x \cos x$$
$$+ \int \cos^2 x \, dx$$

Then, use $\sin^2 x + \cos^2 x = 1$ to replace $\cos^2 x$ in the integral on the right-hand side, and complete the integration of $\int \sin^2 x \, dx$.

33. Evaluating the integral
$$\int \arcsin x \, dx$$
requires two steps.

(a) Write
$$\arcsin x = 1 \cdot \arcsin x$$
and apply integration by parts once to show that
$$\int \arcsin x \, dx$$
$$= x \arcsin x - \int \frac{x}{\sqrt{1 - x^2}} \, dx$$

(b) Use substitution to compute
$$\int \frac{x}{\sqrt{1 - x^2}} \, dx \qquad (7.9)$$
and combine your result in (a) and (7.9) to complete the computation of $\int \arcsin x \, dx$.

34. Evaluating the integral
$$\int \arccos x \, dx$$
requires two steps.

(a) Write
$$\arccos x = 1 \cdot \arccos x$$

and apply integration by parts once to show that

$$\int \arccos x \, dx$$

$$= x \arccos x + \int \frac{x}{\sqrt{1-x^2}} \, dx$$

(b) Use substitution to compute

$$\int \frac{x}{\sqrt{1-x^2}} \, dx \qquad\qquad (7.10)$$

and combine your result in (a) and (7.10) to complete the computation of $\int \arccos x \, dx$.

35. (a) Use integration by parts to show that for $x > 0$

$$\int \frac{1}{x} \ln x \, dx = (\ln x)^2 - \int \frac{1}{x} \ln x \, dx$$

(b) Use your result in (a) to evaluate

$$\int \frac{1}{x} \ln x \, dx$$

36. (a) Use integration by parts to show that

$$\int x^n e^x \, dx = x^n e^x$$

$$- n \int x^{n-1} e^x \, dx$$

Such formulas are called **reduction formulas** since they reduce the exponent of x by 1 each time they are applied.

(b) Apply the reduction formula in (a) repeatedly to compute

$$\int x^3 e^x \, dx$$

37. (a) Use integration by parts to verify the validity of the reduction formula

$$\int x^n e^{ax} \, dx = \frac{1}{a} x^n e^{ax}$$

$$- \frac{n}{a} \int x^{n-1} e^{ax} \, dx$$

where a is a constant not equal to 0.

(b) Apply the reduction formula in (a) to compute

$$\int x^2 e^{-3x} \, dx$$

38. (a) Use integration by parts to verify the validity of the reduction formula

$$\int (\ln x)^n \, dx = x(\ln x)^n$$

$$- n \int (\ln x)^{n-1} \, dx$$

(b) Apply the reduction formula in (a) repeatedly to compute

$$\int (\ln x)^3 \, dx$$

In Problems 39–48, first make an appropriate substitution, and then use integration by parts to evaluate the indefinite integrals.

39. $\int \cos \sqrt{x} \, dx$

40. $\int \sin \sqrt{x} \, dx$

41. $\int x^3 e^{-x^2/2} \, dx$

42. $\int x^5 e^{x^2} \, dx$

43. $\int \sin x \cos x e^{\sin x} \, dx$

44. $\int \sin x \cos^3 x e^{1-\sin^2 x} \, dx$

45. $\int e^{\sqrt{x}} \, dx$

46. $\int e^{\sqrt{x+1}} \, dx$

47. $\int \ln(\sqrt{x}+1) \, dx$

48. $\int x^3 \ln(x^2+1) \, dx$

7.3 PRACTICING INTEGRATION AND PARTIAL FRACTIONS

In the preceding two sections, we learned the two main integration techniques, substitution and integration by parts. One of the major difficulties in integration is deciding which rule to use. This takes a lot of practice. This section is devoted to practicing integration; in addition, we will learn how to integrate some rational functions.

7.3.1 Practicing Integration

▶ **Example 1** Find

$$\int \tan x \sec^2 x e^{\tan x} \, dx$$

Solution

Since $\frac{d}{dx} \tan x = \sec^2 x$, we try the substitution $w = \tan x$. Then $dw = \sec^2 x \, dx$ and

$$\int \tan x \sec^2 x e^{\tan x} \, dx = \int w e^w \, dw$$

To continue, we need to use integration by parts, with

$$u = w \quad \text{and} \quad v' = e^w$$

Then

$$u' = 1 \quad \text{and} \quad v = e^w$$

and

$$\int w e^w \, dw = w e^w - \int e^w \, dw = w e^w - e^w + C$$

With $w = \tan x$, we therefore find

$$\int \tan x \sec^2 x e^{\tan x} \, dx = e^{\tan x}(\tan x - 1) + C \qquad \blacktriangleleft$$

Frequently, we must perform algebraic manipulations of the integrand before we can integrate.

▶ Example 2   Find

$$\int_0^{\sqrt{3}} \frac{1}{9 + x^2} \, dx$$

Solution

The region corresponding to the definite integral is shown in Figure 7.8. The integrand is continuous on $[0, \sqrt{3}]$. The integrand should remind you of the function $\frac{1}{1+u^2}$, whose antiderivative is $\tan^{-1} u$. However, we have a 9 in the denominator. To get a 1 there, we factor 9 in the denominator and find

$$\frac{1}{9 + x^2} = \frac{1}{9(1 + \frac{x^2}{9})} = \frac{1}{9(1 + (\frac{x}{3})^2)}$$

The last expression now suggests that we should try the substitution

$$u = \frac{x}{3} \quad \text{with} \quad dx = 3 \, du$$

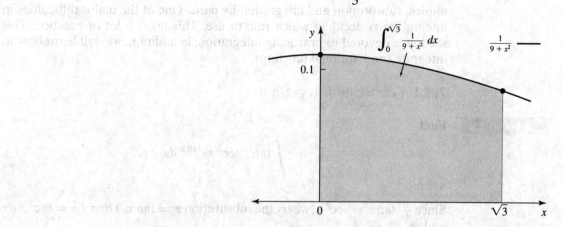

▲ **Figure 7.8**
The region corresponding to the definite integral in Example 2

Since we wish to evaluate a definite integral, we must change the limits of integration as well. We find that $x = 0$ corresponds to $u = 0$ and $x = \sqrt{3}$ corresponds to $u = \frac{1}{3}\sqrt{3}$. We end up with

$$\int_0^{\sqrt{3}} \frac{1}{9+x^2}\,dx = \frac{1}{9}\int_0^{\sqrt{3}} \frac{1}{1+(\frac{x}{3})^2}\,dx = \frac{1}{9}\int_0^{\frac{1}{3}\sqrt{3}} \frac{3}{1+u^2}\,du$$

$$= \frac{1}{3}\int_0^{\frac{1}{3}\sqrt{3}} \frac{1}{1+u^2}\,du = \frac{1}{3}\tan^{-1}u\Big]_0^{\frac{1}{3}\sqrt{3}}$$

$$= \frac{1}{3}\left[\tan^{-1}\left(\frac{1}{3}\sqrt{3}\right) - \tan^{-1}0\right] = \frac{1}{3}\left(\frac{\pi}{6} - 0\right) = \frac{\pi}{18} \quad \blacktriangleleft$$

The next example shows that simplifying the integrand first can help greatly.

▶ **Example 3** Find

$$\int x^{1/2}\ln(x^{1/2}e^x)\,dx$$

Solution

Before we try any of our techniques, let's simplify the logarithm first. We find

$$\ln(x^{1/2}e^x) = \ln x^{1/2} + \ln e^x = \frac{1}{2}\ln x + x$$

The integral now becomes

$$\int x^{1/2}\ln(x^{1/2}e^x)\,dx = \int x^{1/2}\left(\frac{1}{2}\ln x + x\right)\,dx$$

$$= \frac{1}{2}\int x^{1/2}\ln x\,dx + \int x^{3/2}\,dx$$

We can evaluate the first integral using integration by parts, with

$$u = \ln x \quad \text{and} \quad v' = x^{1/2}$$

Then

$$u' = \frac{1}{x} \quad \text{and} \quad v = \frac{2}{3}x^{3/2}$$

We find for the first integral

$$\int x^{1/2}\ln x\,dx = \frac{2}{3}x^{3/2}\ln x - \int \frac{1}{x}\cdot\frac{2}{3}x^{3/2}\,dx$$

$$= \frac{2}{3}x^{3/2}\ln x - \frac{2}{3}\int x^{1/2}\,dx$$

$$= \frac{2}{3}x^{3/2}\ln x - \frac{2}{3}\cdot\frac{2}{3}x^{3/2} + C$$

$$= \frac{2}{3}x^{3/2}\left(\ln x - \frac{2}{3}\right) + C$$

The other integral is straightforward:

$$\int x^{3/2}\,dx = \frac{2}{5}x^{5/2} + C$$

Combining our results, we find

$$\int x^{1/2} \ln(x^{1/2} e^x) \, dx = \frac{1}{2} \int x^{1/2} \ln x \, dx + \int x^{3/2} dx$$

$$= \frac{1}{3} x^{3/2} \left(\ln x - \frac{2}{3} \right) + \frac{2}{5} x^{5/2} + C$$

Note that we used the same symbol C to denote the integration constants. We could have called them C_1 and C_2 and then combined them into $C = C_1 + C_2$. But since they simply stand for arbitrary constants, we need not keep track of how they are related and can simply denote them all by the same symbol. However, we should keep in mind that they are not all the same. ◀

7.3.2 Rational Functions and Partial Fractions

A rational function f is the quotient of two polynomials. That is,

$$f(x) = \frac{P(x)}{Q(x)} \tag{7.11}$$

where $P(x)$ and $Q(x)$ are polynomials. To integrate rational functions, we use an algebraic technique, called the method of **partial fractions**, to write $f(x)$ as a sum of a polynomial and simpler rational functions. Such a sum is called a **partial fraction decomposition**. These simpler rational functions, which can be integrated with the methods we have learned, are of the form

$$\frac{A}{(ax + b)^n} \quad \text{or} \quad \frac{Bx + C}{(ax^2 + bx + c)^n} \tag{7.12}$$

where $A, B, C, a, b,$ and c are constants, and n is a positive integer. In this form, the quadratic polynomial $ax^2 + bx + c$ can no longer be factored into a product of two linear functions with real coefficients. Such polynomials are called **irreducible**. (In other words, $ax^2 + bx + c$ is irreducible if and only if $ax^2 + bx + c = 0$ has no real roots.)

If the degree of $P(x)$ in (7.11) is greater than or equal to the degree of $Q(x)$, then the first step in the partial fraction decomposition is to use long division to write $f(x)$ as a sum of a polynomial and a rational function, where the rational function is such that the degree of the polynomial in the numerator is less than the degree of the polynomial in the denominator (such rational functions are called **proper**). We illustrate this step in the following two examples.

▶ **Example 4** **Long Division before Integration** Find

$$\int \frac{x}{x + 2} \, dx$$

Solution

The degree of the numerator is equal to the degree of the denominator; using long division or writing the integrand in the form

$$\frac{x}{x + 2} = \frac{x + 2 - 2}{x + 2} = 1 - \frac{2}{x + 2}$$

results in a polynomial of degree 0 and a proper rational function. We can integrate the integrand in this new form:

$$\int \frac{x}{x + 2} \, dx = \int \left(1 - \frac{2}{x + 2} \right) dx = x - 2 \ln |x + 2| + C \qquad ◀$$

▶ Example 5   **Long Division before Integration** Find

$$\int \frac{3x^3 - 7x^2 + 17x - 3}{x^2 - 2x + 5} \, dx$$

Solution

Since the degree of the numerator is higher than the degree of the denominator, we use long division to simplify the integrand.

$$
\begin{array}{r}
3x \quad -1 \\
x^2 \ -2x \ +5\overline{)\ 3x^3 \ -7x^2 \ +17x \ -3} \\
3x^3 \ -6x^2 \ +15x \\
\hline
-x^2 \ +2x \ -3 \\
-x^2 \ +2x \ -5 \\
\hline
2
\end{array}
$$

That is,

$$\frac{3x^3 - 7x^2 + 17x - 3}{x^2 - 2x + 5} = 3x - 1 + \frac{2}{x^2 - 2x + 5}$$

Now, $x^2 - 2x + 5 = 0$ does not have real solutions; this can be checked using the quadratic formula. This means that it is irreducible. Instead, we will complete the square. That is,

$$x^2 - 2x + 5 = (x^2 - 2x + 1) + 4 = (x - 1)^2 + 4$$

The integral we wish to evaluate is therefore

$$\int \left(3x - 1 + \frac{2}{(x-1)^2 + 4} \right) dx = \int (3x - 1) \, dx + 2 \int \frac{1}{(x-1)^2 + 4} \, dx$$

$$(7.13)$$

The first integral is easy; we find

$$\int (3x - 1) \, dx = \frac{3}{2}x^2 - x + C$$

To evaluate the second integral, we use a similar trick as in Example 2 of this section. Namely, we factor 4 in the denominator,

$$\int \frac{1}{(x-1)^2 + 4} \, dx = \frac{1}{4} \int \frac{1}{1 + \left(\frac{x-1}{2}\right)^2} \, dx$$

Setting

$$u = \frac{x-1}{2} \quad \text{with} \quad dx = 2 \, du$$

yields

$$\frac{1}{4} \int \frac{1}{1 + \left(\frac{x-1}{2}\right)^2} \, dx = \frac{1}{4} \int \frac{2 \, du}{1 + u^2}$$

$$= \frac{1}{2} \tan^{-1} u + C = \frac{1}{2} \tan^{-1}\left(\frac{x-1}{2}\right) + C$$

Putting the pieces together (and remembering that there was a factor 2 in front of the second integral in (7.13), we find

$$\int \frac{3x^3 - 7x^2 + 17x - 3}{x^2 - 2x + 5} \, dx = \frac{3}{2}x^2 - x + \tan^{-1}\left(\frac{x-1}{2}\right) + C$$

Assume now that the rational function is proper. Unless the integrand is already of one of the types in (7.12), we need to decompose it further. A result in algebra tells us that every polynomial can be written as a product of linear and irreducible quadratic factors. This is the key to the partial fraction decomposition. We factor the denominator into a product of linear and irreducible quadratic factors. The linear factors then correspond to a sum of rational functions given as follows:

If the linear factor $ax + b$ is contained n times in the factorization of the denominator of a proper rational function, then the partial fraction decomposition contains terms of the form

$$\frac{A_1}{ax + b} + \frac{A_2}{(ax + b)^2} + \cdots + \frac{A_n}{(ax + b)^n}$$

where $A_1, A_2, \ldots, A_n$ are constants.

▶ Example 6   **Distinct Linear Factors** Find

$$\int \frac{1}{x(x - 1)}\, dx$$

Solution

The integrand is a rational function whose denominator is a product of two distinct linear functions. We claim that the integrand can be written in the form

$$\frac{1}{x(x - 1)} = \frac{A}{x} + \frac{B}{x - 1} \tag{7.14}$$

where A and B are constants that we need to determine. To find A and B, we write the right-hand side of (7.14) with a common denominator. That is,

$$\frac{A}{x} + \frac{B}{x - 1} = \frac{A(x - 1) + Bx}{x(x - 1)} = \frac{(A + B)x - A}{x(x - 1)}$$

Since this must be equal to $\frac{1}{x(x-1)}$, we conclude that

$$(A + B)x - A = 1$$

Therefore,

$$A + B = 0 \quad \text{and} \quad -A = 1$$

This yields $A = -1$ and $B = -A = 1$. We thus find

$$\frac{1}{x(x - 1)} = -\frac{1}{x} + \frac{1}{x - 1}$$

The integrand can now be written as a sum of two rational functions, which can be integrated immediately:

$$\int \frac{1}{x(x - 1)}\, dx = \int \left[\frac{1}{x - 1} - \frac{1}{x} \right] dx = \int \frac{1}{x - 1}\, dx - \int \frac{1}{x}\, dx$$

$$= \ln |x - 1| - \ln |x| + C = \ln \left| \frac{x - 1}{x} \right| + C \quad ◀$$

▶ **Example 7**   **Repeated Linear Factors** Evaluate

$$\int \frac{dx}{x^2(x+1)}$$

Solution

The integrand is a proper rational function whose denominator is a product of three linear functions, namely, x, x (again), and $x+1$. We write the integrand in the form

$$\frac{1}{x^2(x+1)} = \frac{A}{x} + \frac{B}{x^2} + \frac{C}{x+1} \tag{7.15}$$

where A, B, and C are constants. As in the previous example, we find A, B, and C by writing the right-hand side of (7.15) with a common denominator. This yields

$$\frac{A}{x} + \frac{B}{x^2} + \frac{C}{x+1} = \frac{Ax(x+1) + B(x+1) + Cx^2}{x^2(x+1)}$$

$$= \frac{Ax^2 + Ax + Bx + B + Cx^2}{x^2(x+1)}$$

$$= \frac{(A+C)x^2 + (A+B)x + B}{x^2(x+1)}$$

Comparing the last expression with the left-hand side of (7.15), we conclude that

$$A + C = 0, \quad A + B = 0, \quad \text{and } B = 1$$

This implies $A = -1$ and $C = 1$. Therefore,

$$\frac{1}{x^2(x+1)} = -\frac{1}{x} + \frac{1}{x^2} + \frac{1}{x+1}$$

and

$$\int \frac{1}{x^2(x+1)} \, dx = \int \left(-\frac{1}{x} + \frac{1}{x^2} + \frac{1}{x+1} \right) dx$$

$$= -\ln|x| - \frac{1}{x} + \ln|x+1| + C \qquad ◀$$

Irreducible quadratic factors in the denominator of a proper rational functions are dealt with in the partial fraction decomposition as follows.

If the irreducible quadratic factor $ax^2 + bx + c$ is contained n times in the factorization of the denominator of a proper rational function, then the partial fraction decomposition contains terms of the form

$$\frac{B_1x + C_1}{ax^2 + bx + c} + \frac{B_2x + C_2}{(ax^2 + bx + c)^2} + \cdots + \frac{B_nx + C_n}{(ax^2 + bx + c)^n}$$

▶ **Example 8**   **Distinct Irreducible Quadratic Factors** Evaluate

$$\int \frac{2x^3 - x^2 + 2x - 2}{(x^2 + 2)(x^2 + 1)} \, dx$$

Solution

The rational function in the integrand is proper. The denominator is already factored, each factor is an irreducible quadratic polynomial, and the two factors are distinct. We can therefore write the integrand as

$$\frac{2x^3 - x^2 + 2x - 2}{(x^2 + 2)(x^2 + 1)} = \frac{Ax + B}{x^2 + 2} + \frac{Cx + D}{x^2 + 1}$$

$$= \frac{(Ax + B)(x^2 + 1) + (Cx + D)(x^2 + 2)}{(x^2 + 2)(x^2 + 1)}$$

$$= \frac{Ax^3 + Ax + Bx^2 + B + Cx^3 + 2Cx + Dx^2 + 2D}{(x^2 + 2)(x^2 + 1)}$$

$$= \frac{(A + C)x^3 + (B + D)x^2 + (A + 2C)x + B + 2D}{(x^2 + 2)(x^2 + 1)}$$

Comparing the last expression with the integrand, we find

$$A + C = 2, \quad B + D = -1, \quad A + 2C = 2, \quad \text{and } B + 2D = -2$$

which yields $C = 0$ (write $A + 2C = 2$ as $A + C + C = 2$ and use $A + C = 2$) and $D = -1$ (write $B + 2D = -2$ as $B + D + D = -2$ and use $B + D = -1$). Then $A = 2$ and $B = 0$. Therefore,

$$\frac{2x^3 - x^2 + 2x - 2}{(x^2 + 2)(x^2 + 1)} = \frac{2x}{x^2 + 2} - \frac{1}{x^2 + 1}$$

and

$$\int \frac{2x^3 - x^2 + 2x - 2}{(x^2 + 2)(x^2 + 1)} \, dx = \int \frac{2x}{x^2 + 2} \, dx - \int \frac{1}{x^2 + 1} \, dx$$

$$= \int \frac{du}{u} - \tan^{-1} x + C = \ln |u| - \tan^{-1} x + C$$

$$= \ln |x^2 + 2| - \tan^{-1} x + C$$

where we used the substitution $u = x^2 + 2$. ◀

▶ **Example 9** **Repeated Irreducible Quadratic Factors** Evaluate

$$\int \frac{x^2 + x + 1}{(x^2 + 1)^2} \, dx$$

Solution

The rational function $\frac{x^2+x+1}{(x^2+1)^2}$ is proper since the numerator is a polynomial of degree 2 and the denominator is of degree 4 [because $(x^2+1)^2 = x^4+2x^2+1$]. The denominator contains the irreducible quadratic factor $x^2 + 1$ twice. We can therefore write the integrand as

$$\frac{x^2 + x + 1}{(x^2 + 1)^2} = \frac{Ax + B}{x^2 + 1} + \frac{Cx + D}{(x^2 + 1)^2} = \frac{(Ax + B)(x^2 + 1) + Cx + D}{(x^2 + 1)^2}$$

$$= \frac{Ax^3 + Ax + Bx^2 + B + Cx + D}{(x^2 + 1)^2}$$

$$= \frac{Ax^3 + Bx^2 + (A + C)x + (B + D)}{(x^2 + 1)^2}$$

Comparing the last expression with the integrand, we conclude

$$A = 0, \quad B = 1, \quad A + C = 1, \quad \text{and } B + D = 1$$

which implies $C = 1$ and $D = 0$. Therefore,

$$\frac{x^2 + x + 1}{(x^2 + 1)^2} = \frac{1}{x^2 + 1} + \frac{x}{(x^2 + 1)^2}$$

and

$$\int \frac{x^2 + x + 1}{(x^2 + 1)^2} \, dx = \int \frac{1}{x^2 + 1} \, dx + \int \frac{x}{(x^2 + 1)^2} \, dx$$

The first integral on the right-hand side is $\tan^{-1} x + C$. To evaluate the second integral on the right-hand side, we use substitution: $u = x^2 + 1$ with $du/2 = x \, dx$. This yields

$$\int \frac{x}{(x^2 + 1)^2} \, dx = \frac{1}{2} \int \frac{du}{u^2} = -\frac{1}{2u} + C = -\frac{1}{2(x^2 + 1)} + C$$

Combining these two results, we find

$$\int \frac{x^2 + x + 1}{(x^2 + 1)^2} \, dx = \tan^{-1} x - \frac{1}{2(x^2 + 1)} + C \qquad ◀$$

We conclude this section by providing a summary of the two most important cases, namely, when the integrand is a rational function for which the denominator is a polynomial of degree 2 and either a product of two not necessarily distinct linear factors or an irreducible quadratic polynomial.

The first step is to make sure that the degree of the numerator is less than the degree of the denominator. If not, use long division to simplify the integrand.

We will now assume that the degree of the numerator is strictly less than the degree of the denominator (that is, the integrand is a proper rational function). We write the rational function $f(x)$ as

$$f(x) = \frac{P(x)}{Q(x)}$$

with $Q(x) = ax^2 + bx + c$, $a \neq 0$, and $P(x) = rx + s$. Either $Q(x)$ can be factored into two linear factors, or it is irreducible (that is, does not have real roots).

Case 1a: *$Q(x)$ is a product of two distinct linear factors.* In this case, we write

$$Q(x) = a(x - x_1)(x - x_2)$$

where x_1 and x_2 are the two distinct roots of $Q(x)$. We then use the method of partial fractions to simplify the rational function:

$$\frac{P(x)}{Q(x)} = \frac{rx + s}{ax^2 + bx + c} = \frac{1}{a} \left[\frac{A}{x - x_1} + \frac{B}{x - x_2} \right]$$

The constants A and B must be determined, as in Example 6.

Case 1b: *Q(x) is a product of two identical linear factors.* In this case, we write

$$Q(x) = a(x - x_1)^2$$

where x_1 is the root of $Q(x)$. We then use the method of partial fractions to simplify the rational function:

$$\frac{P(x)}{Q(x)} = \frac{rx + s}{ax^2 + bx + c} = \frac{1}{a}\left[\frac{A}{x - x_1} + \frac{B}{(x - x_1)^2}\right]$$

The constants A and B must be determined, as in Example 7.

Case 2: *Q(x) is an irreducible quadratic polynomial.* In this case,

$$Q(x) = ax^2 + bx + c \quad \text{with } b^2 - 4ac < 0$$

and we must complete the square, as in Example 5. This then leads to integrals of the form

$$\int \frac{dx}{x^2 + 1} \quad \text{or} \quad \int \frac{x}{x^2 + 1}\, dx$$

The first integral is $\tan^{-1} x + C$, whereas the second integral can be evaluated using substitution.

7.3.3 Problems

(7.3.1)

In Problems 1–12, use either substitution or integration by parts to evaluate each integral.

1. $\int xe^{-2x}\, dx$

2. $\int xe^{-2x^2}\, dx$

3. $\int \frac{1}{\tan x}\, dx$

4. $\int \frac{1}{\csc x \sec x}\, dx$

5. $\int 2x \sin(x^2)\, dx$

6. $\int 2x^2 \sin x\, dx$

7. $\int \frac{1}{16 + x^2}\, dx$

8. $\int \frac{1}{x^2 + 5}\, dx$

9. $\int \frac{x}{x + 3}\, dx$

10. $\int \frac{1}{x^2 + 3}\, dx$

11. $\int \frac{x}{x^2 + 3}\, dx$

12. $\int \frac{x + 2}{x^2 + 2}\, dx$

13. The integral

$$\int \ln x\, dx$$

can be evaluated in two ways.

(a) Write $\ln x = 1 \cdot \ln x$ and use integration by parts to evaluate the integral.

(b) Use the substitution $u = \ln x$ and integration by parts to evaluate the integral.

14. Use an appropriate substitution followed by integration by parts to evaluate

$$\int x^3 e^{-x^2/2}\, dx$$

15. Use an appropriate substitution to evaluate

$$\int x(x - 2)^{1/4}\, dx$$

16. Use an appropriate substitution after simplifying the integrand to evaluate

$$\int \frac{\sin^2 x - \cos^2 x}{(\sin x - \cos x)^2}\, dx$$

In Problems 17–22, evaluate each definite integral.

17. $\int_1^4 e^{\sqrt{x}}\, dx$

18. $\int_1^2 \ln(x^2 e^x)\, dx$

19. $\int_{-1}^0 \frac{2}{1 + x^2}\, dx$

20. $\int_1^2 x^2 \ln x\, dx$

21. $\int_0^{\pi/4} e^x \sin x\, dx$

22. $\int_0^{\pi/6} (1 + \tan^2 x)\, dx$

(7.3.2)

23. (a) Find A and B so that

$$\frac{1}{x(x-2)} = \frac{A}{x} + \frac{B}{x-2}$$

(b) Use the partial fraction decomposition in (a) to evaluate

$$\int \frac{1}{x(x-2)}\, dx$$

24. (a) Find A and B so that

$$\frac{1}{(x-1)(x+2)} = \frac{A}{x-1} + \frac{B}{x+2}$$

(b) Use the partial fraction decomposition in (a) to evaluate

$$\int \frac{1}{(x-1)(x+2)}\, dx$$

25. Complete the square in the denominator to evaluate

$$\int \frac{1}{x^2 - 4x + 13}\, dx$$

26. Complete the square in the denominator to evaluate

$$\int \frac{1}{x^2 + 2x + 5}\, dx$$

In Problems 27–36, evaluate each integral.

27. $\displaystyle\int \frac{1}{(x-3)(x+2)}\, dx$

28. $\displaystyle\int \frac{2x-1}{(x+4)(x+1)}\, dx$

29. $\displaystyle\int \frac{1}{x^2-9}\, dx$

30. $\displaystyle\int \frac{1}{x^2+9}\, dx$

31. $\displaystyle\int \frac{1}{x^2-x-2}\, dx$

32. $\displaystyle\int \frac{1}{x^2-x+2}\, dx$

33. $\displaystyle\int \frac{x^2+1}{x^2+3x+2}\, dx$

34. $\displaystyle\int \frac{x^3+1}{x^2+3}\, dx$

35. $\displaystyle\int \frac{x^2+4}{x^2-4}\, dx$

36. $\displaystyle\int \frac{x^4+3}{x^2-4x+3}\, dx$

In Problems 37–44, evaluate each definite integral.

37. $\displaystyle\int_3^5 \frac{x-1}{x}\, dx$

38. $\displaystyle\int_3^5 \frac{x}{x-1}\, dx$

39. $\displaystyle\int_0^1 \frac{x}{x^2+1}\, dx$

40. $\displaystyle\int_1^2 \frac{x^2+1}{x}\, dx$

41. $\displaystyle\int_2^3 \frac{1}{1-x}\, dx$

42. $\displaystyle\int_2^3 \frac{1}{1-x^2}\, dx$

43. $\displaystyle\int_0^1 \tan^{-1} x\, dx$

44. $\displaystyle\int_0^1 x \tan^{-1} x\, dx$

In Problems 45–52, evaluate each integral.

45. $\displaystyle\int \frac{1}{(x+1)^2 x}\, dx$

46. $\displaystyle\int \frac{1}{x^2(x-1)^2}\, dx$

47. $\displaystyle\int \frac{4}{(1-x)(1+x)^2}\, dx$

48. $\displaystyle\int \frac{2x^2+2x-1}{x^3(x-3)}\, dx$

49. $\displaystyle\int \frac{1}{(x^2-9)^2}\, dx$

50. $\displaystyle\int \frac{1}{(x^2-x-2)^2}\, dx$

51. $\displaystyle\int \frac{1}{x^2(x^2+1)}\, dx$

52. $\displaystyle\int \frac{1}{(x+1)^2(x^2+1)}\, dx$

7.4 IMPROPER INTEGRALS

In this section we discuss definite integrals of two types with the following characteristics:

1. One or both limits of integration are infinite; that is, the integration interval is unbounded; or

2. The integrand becomes infinite at one or more points of the interval of integration.

We call such integrals **improper integrals**.

7.4.1 Type 1: Unbounded Intervals

Suppose that we wanted to compute the area of the unbounded region below the graph of $f(x) = e^{-x}$ and above the x-axis for $x \geq 0$ (see Figure 7.9). How

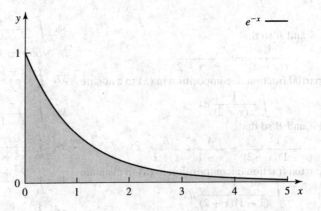

▲ Figure 7.9

The unbounded region between the graph of $y = e^{-x}$ and the x-axis for $x \geq 0$

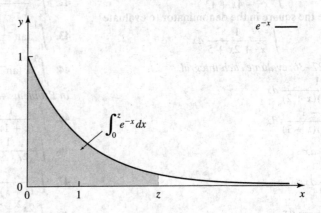

▲ Figure 7.10

The region between 0 and z

would we proceed? We know how to find the area of a region bounded by the graph of a continuous function [here: $f(x) = e^{-x}$] and the x-axis between 0 and z, namely,

$$A(z) = \int_0^z e^{-x} dx = -e^{-x}\Big]_0^z = 1 - e^{-z}$$

This is the shaded area in Figure 7.10. If we now let z tend to infinity, we may regard the limiting value (if it exists) as the area of the unbounded region below the graph of $f(x) = e^{-x}$ and above the x-axis for $x \geq 0$ (see Figure 7.9):

$$A = \lim_{z \to \infty} A(z) = \lim_{z \to \infty} (1 - e^{-z}) = 1$$

We write

$$\int_0^\infty e^{-x} dx = 1$$

Therefore, for functions that are continuous on unbounded intervals (see Figures 7.11 and 7.12), we define

$$\int_a^\infty f(x) dx = \lim_{z \to \infty} \int_a^z f(x) dx$$

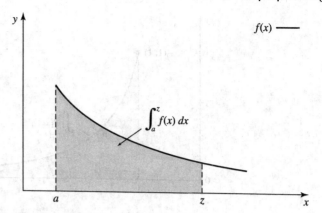

▲ **Figure 7.11**

The definition of the improper integral $\int_a^{\infty} f(x)\,dx$ as the limit of $\int_a^z f(x)\,dx$ as $z \to \infty$

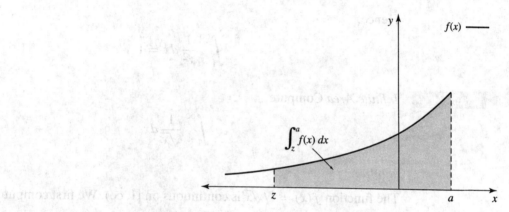

▲ **Figure 7.12**

The definition of the improper integral $\int_{-\infty}^a f(x)\,dx$ as the limit of $\int_z^a f(x)\,dx$ as $z \to -\infty$

and

$$\int_{-\infty}^a f(x)\,dx = \lim_{z \to -\infty} \int_z^a f(x)\,dx$$

You might be surprised that the area of an unbounded region can be finite. This need not be the case and only happens if the graph of $f(x)$ approaches the x-axis sufficiently fast. We illustrate this in the next two examples.

▶ **Example 1**   *Finite Area* Compute

$$\int_1^{\infty} \frac{1}{x^2}\,dx$$

Solution

The function $y = 1/x^2$ is continuous on $[1, \infty)$. We first compute

$$A(z) = \int_1^z \frac{1}{x^2}\,dx = -\frac{1}{x}\Big]_1^z = 1 - \frac{1}{z}$$

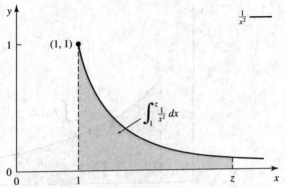

◀ **Figure 7.13**
The region corresponding to $A(z)$ in Example 1

(see Figure 7.13) and then let $z \to \infty$. We find

$$\lim_{z \to \infty} A(z) = \lim_{z \to \infty} \left(1 - \frac{1}{z}\right) = 1$$

Hence,

$$\int_1^\infty \frac{1}{x^2}\, dx = 1$$

▶ **Example 2**   *Infinite Area* Compute

$$\int_1^\infty \frac{1}{\sqrt{x}}\, dx$$

Solution

The function $f(x) = 1/\sqrt{x}$ is continuous on $[1, \infty)$. We first compute

$$A(z) = \int_1^z \frac{1}{\sqrt{x}}\, dx = 2\sqrt{x}\,]_1^z = 2(\sqrt{z} - 1)$$

(see Figure 7.14) and then let $z \to \infty$. We find

$$\lim_{z \to \infty} A(z) = \lim_{z \to \infty} 2(\sqrt{z} - 1) = \infty$$

Hence,

$$\int_1^\infty \frac{1}{\sqrt{x}}\, dx$$

does not exist.

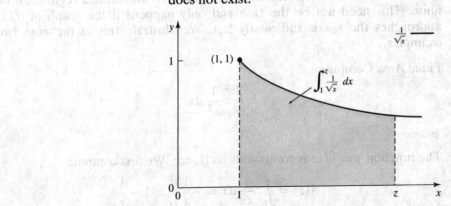

◀ **Figure 7.14**
The region corresponding to $A(z)$ in Example 2

Looking back at Examples 1 and 2, we see that, in both cases, the respective integrands approached the x-axis as $x \to \infty$, that is, both

$$\lim_{x \to \infty} \frac{1}{x^2} = 0 \quad \text{and} \quad \lim_{x \to \infty} \frac{1}{\sqrt{x}} = 0$$

However, $\frac{1}{x^2}$ approaches the x-axis much faster than $\frac{1}{\sqrt{x}}$, as can be seen from the graphs in Figure 7.15. The exponent of x in the denominator determines how fast the function approaches the x-axis. The area between the graph and the x-axis from $x = 1$ to infinity is only finite if the graph approaches the x-axis fast enough. Indeed, if we tried to compute

$$\int_1^\infty \frac{1}{x^p} \, dx$$

for $0 < p < \infty$, we would find that

$$\int_1^\infty \frac{1}{x^p} \, dx = \begin{cases} \frac{1}{p-1} & \text{for } p > 1 \\ \infty & \text{for } 0 < p \le 1 \end{cases}$$

(Note that $y = 1/x^p$ is continuous on $[1, \infty)$.) That is, for $p > 1$, the function $\frac{1}{x^p}$ approaches the x-axis fast enough as $x \to \infty$ for the area under the graph to be finite. (We investigate this further in Problem 33.)

We will use the following terminology to indicate whether an improper integral is finite or infinite.

Let $f(x)$ be continuous on the interval $[a, \infty)$. If

$$\lim_{z \to \infty} \int_a^z f(x) \, dx$$

exists and has a finite value, we say that the improper integral

$$\int_a^\infty f(x) \, dx$$

converges and define

$$\int_a^\infty f(x) \, dx = \lim_{z \to \infty} \int_a^z f(x) \, dx$$

Otherwise, we say that the improper integral **diverges**.

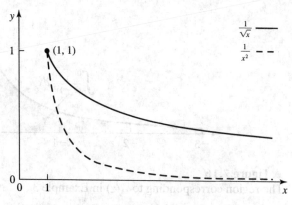

◀ **Figure 7.15**

The function $y = \frac{1}{x^2}$ approaches the x-axis much faster than the function $y = \frac{1}{\sqrt{x}}$

Analogous definitions can be given when the lower limit of integration is infinite.

▶ **Example 3**   **Infinite Lower Limit**   Show that the improper integral

$$\int_{-\infty}^{0} \frac{1}{(x-1)^2}\, dx$$

converges.

Solution

Note that $y = 1/(x-1)^2$ is continuous on $(-\infty, 0]$. To show that the integral converges, we compute its value. We need to find

$$A(z) = \int_{z}^{0} \frac{1}{(x-1)^2}\, dx \quad \text{for } z < 0$$

and then let $z \to -\infty$ (see Figure 7.16). We find

$$\int_{z}^{0} (x-1)^{-2}\, dx = -(x-1)^{-1}\Big]_{z}^{0}$$

$$= -\frac{1}{x-1}\Big]_{z}^{0} = -\frac{1}{-1} + \frac{1}{z-1} = 1 + \frac{1}{z-1}$$

and

$$\lim_{z \to -\infty} \left(1 + \frac{1}{z-1}\right) = 1$$

Therefore,

$$\int_{-\infty}^{0} \frac{1}{(x-1)^2}\, dx = 1$$ ◀

We next discuss the case when both limits of integration are infinite.

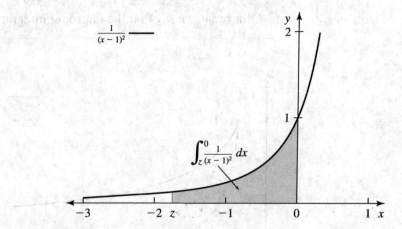

▲ **Figure 7.16**
The region corresponding to $A(z)$ in Example 3

Assume that $f(x)$ is continuous on $(-\infty, \infty)$. Then

$$\int_{-\infty}^{\infty} f(x)\,dx = \int_{-\infty}^{a} f(x)\,dx + \int_{a}^{\infty} f(x)\,dx \qquad (7.16)$$

where a is a real number. If *both* improper integrals on the right-hand side of (7.16) are convergent, then the value of the improper integral on the left-hand side of (7.16) is the sum of the two limiting values on the right-hand side.

Suppose that we wish to compute

$$\int_{-\infty}^{\infty} x^3\,dx$$

We choose a value $a \in (-\infty, \infty)$; for instance, $a = 0$. Then

$$\int_{-\infty}^{\infty} x^3\,dx = \int_{-\infty}^{0} x^3\,dx + \int_{0}^{\infty} x^3\,dx$$

Looking at Figure 7.17, you will probably agree that both improper integrals on the right-hand side are divergent. We check this for the second one:

$$\int_{0}^{\infty} x^3\,dx = \lim_{z \to \infty} \int_{0}^{z} x^3\,dx = \frac{1}{4}x^4 \Big]_{0}^{z} = \frac{1}{4} \lim_{z \to \infty} (z^4 - 0)$$

which does not exist. Hence,

$$\int_{-\infty}^{\infty} x^3\,dx$$

is divergent.

It is important to realize that the definition of $\int_{-\infty}^{\infty} f(x)\,dx$ is different from

$$\lim_{b \to \infty} \int_{-b}^{b} f(x)\,dx$$

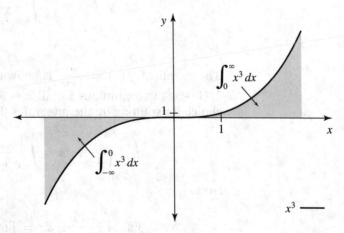

▲ **Figure 7.17**
The integral $\int_{-\infty}^{\infty} x^3\,dx$ is divergent

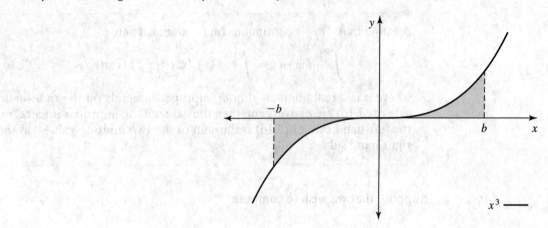

▲ **Figure 7.18**

Because of symmetry, $\int_{-b}^{b} x^3\, dx = 0$

We illustrate this using $f(x) = x^3$ again. We find, for any $b > 0$, that

$$\int_{-b}^{b} x^3\, dx = \frac{1}{4}x^4 \Big]_{-b}^{b} = \frac{1}{4}(b^4 - (-b)^4) = 0$$

(see Figure 7.18). Therefore,

$$\lim_{b \to \infty} \int_{-b}^{b} x^3\, dx = 0$$

This limit is not the same as $\int_{-\infty}^{\infty} x^3\, dx$.

Looking at (7.16), we see that in order to evaluate $\int_{-\infty}^{\infty} f(x)\, dx$, we need to split up the integral at some $a \in \mathbf{R}$. There are often natural choices for a. We illustrate this in the next example.

▶ **Example 4** *Infinite Upper and Lower Limit* Compute

$$\int_{-\infty}^{\infty} \frac{1}{1 + x^2}\, dx$$

Solution

The graph of $f(x) = \frac{1}{1+x^2}$ is shown in Figure 7.19. The function $f(x) = 1/(1 + x^2)$ is continuous for all $x \in \mathbf{R}$. It is symmetric about $x = 0$; a good choice for splitting up the integral is therefore $a = 0$. We write

$$\int_{-\infty}^{\infty} \frac{1}{1 + x^2}\, dx = \int_{-\infty}^{0} \frac{1}{1 + x^2}\, dx + \int_{0}^{\infty} \frac{1}{1 + x^2}\, dx$$

Now,

$$\lim_{z \to \infty} \int_{0}^{z} \frac{1}{1 + x^2}\, dx = \lim_{z \to \infty} \left[\tan^{-1} x \right]_{0}^{z}$$

$$= \lim_{z \to \infty} (\tan^{-1} z - \tan^{-1} 0) = \frac{\pi}{2}$$

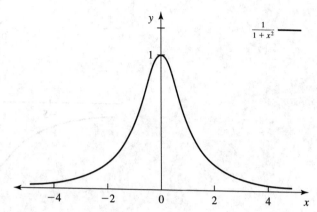

▲ **Figure 7.19**
The graph of $f(x) = \frac{1}{1+x^2}$ in Example 4

and

$$\lim_{z \to -\infty} \int_z^0 \frac{1}{1+x^2}\, dx = \lim_{z \to -\infty} \left[\tan^{-1} x\right]_z^0$$

$$= \lim_{z \to -\infty} (\tan^{-1} 0 - \tan^{-1} z) = \frac{\pi}{2}$$

That $\int_{-\infty}^0 \frac{1}{1+x^2}\, dx = \frac{\pi}{2}$ is expected because of symmetry. The area of the region to the left of the y-axis is equal to the area of the region to the right of the y-axis. Putting things together, we find

$$\int_{-\infty}^{\infty} \frac{1}{1+x^2}\, dx = \frac{\pi}{2} + \frac{\pi}{2} = \pi$$

◀

▶ **Example 5** **Infinite Upper and Lower Limit** Compute

$$\int_{-\infty}^{\infty} \frac{x}{1+x^2}\, dx$$

Solution

The graph of $f(x) = \frac{x}{1+x^2}$ is shown in Figure 7.20. The function $f(x) = x/(1+x^2)$ is continuous for all $x \in \mathbf{R}$. Because of the symmetry about the origin, you might be tempted to say that the signed area to the left of 0 is the negative of the area to the right of 0 and, therefore, the value of the improper integral should be 0. But this is wrong! We choose $a = 0$, and write

$$\int_{-\infty}^{\infty} \frac{x}{1+x^2}\, dx = \int_{-\infty}^{0} \frac{x}{1+x^2}\, dx + \int_{0}^{\infty} \frac{x}{1+x^2}\, dx$$

We begin by computing

$$\int_0^z \frac{x}{1+x^2}\, dx$$

Using the substitution $u = 1 + x^2$ and $du = 2x\, dx$, we find

$$\int_0^z \frac{x}{1+x^2}\, dx = \int_1^{1+z^2} \frac{1}{2u}\, du = \frac{1}{2} \ln |u| \Big]_1^{1+z^2}$$

$$= \frac{1}{2}[\ln(1+z^2) - \ln 1] = \frac{1}{2} \ln(1+z^2)$$

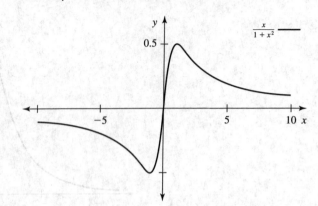

▲ **Figure 7.20**
The graph of $f(x) = \frac{x}{1+x^2}$ in Example 5

Taking the limit $z \to \infty$, we find

$$\int_0^\infty \frac{x}{1+x^2}\,dx = \lim_{z\to\infty} \frac{1}{2}\ln(1+z^2) = \infty$$

Since one of the integrals is already divergent, we conclude that

$$\int_{-\infty}^\infty \frac{x}{1+x^2}\,dx$$

is divergent and therefore cannot be equal to 0. This example has an important take-home message: Before we can use symmetry to compute an improper integral, we need to make sure that the integral exists. ◀

7.4.2 Type 2: Unbounded Integrand

So far, when we computed a definite integral, we made sure that the integrand was continuous over the integration interval. We will now explain what to do when the integrand becomes infinite at one or more points of the integration interval. Suppose we wish to integrate

$$\int_0^1 \frac{dx}{\sqrt{x}}$$

The graph of $f(x) = \frac{1}{\sqrt{x}}$ is shown in Figure 7.21. We see immediately that $f(x)$ is continuous on $(0, 1]$, undefined at $x = 0$, and that

$$\lim_{x\to 0^+} \frac{1}{\sqrt{x}} = \infty$$

Let's choose a number $c \in (0, 1)$ and compute

$$\int_c^1 \frac{dx}{\sqrt{x}} = 2\sqrt{x}\,\Big]_c^1 = 2(1 - \sqrt{c})$$

(see Figure 7.22). If we now let $c \to 0^+$, we may regard the limiting value (if it exists) as the definite integral $\int_0^1 \frac{1}{\sqrt{x}}\,dx$. That is,

$$\int_0^1 \frac{dx}{\sqrt{x}} = \lim_{c\to 0^+} \int_c^1 \frac{dx}{\sqrt{x}} = \lim_{c\to 0^+} 2(1 - \sqrt{c}) = 2$$

If f is continuous on $(a, b]$ and $\lim_{x\to a^+} f(x) = \pm\infty$ (see Figure 7.23), we define

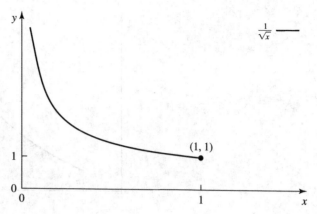

▲ **Figure 7.21**
The graph of $y = \frac{1}{\sqrt{x}}$

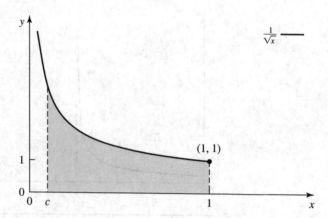

▲ **Figure 7.22**
The area of the shaded region is $\int_c^1 \frac{1}{\sqrt{x}} dx = 2(1 - \sqrt{c})$

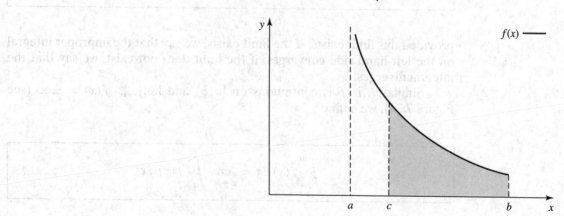

▲ **Figure 7.23**
The improper integral $\int_a^b f(x)\, dx$ is defined as the limit of $\int_c^b f(x)\, dx$ as $c \to a^+$

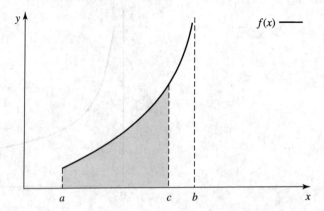

▲ Figure 7.24
The improper integral $\int_a^b f(x)\, dx$ is defined as the limit of $\int_a^c f(x)\, dx$ as $c \to b^-$

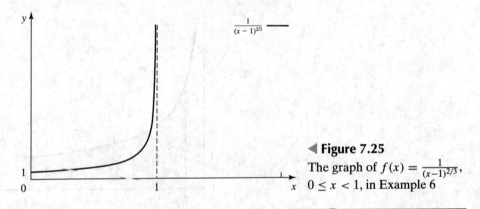

◄ Figure 7.25
The graph of $f(x) = \frac{1}{(x-1)^{2/3}}$,
$0 \le x < 1$, in Example 6

$$\int_a^b f(x)\, dx = \lim_{c \to a^+} \int_c^b f(x)\, dx$$

provided this limit exists. If the limit exists, we say that the improper integral on the left-hand side **converges**; if the limit does not exist, we say that the integral **diverges**.

Similarly, if f is continuous on $[a, b)$ and $\lim_{x \to b^-} f(x) = \pm\infty$ (see Figure 7.24), we define

$$\int_a^b f(x)\, dx = \lim_{c \to b^-} \int_a^c f(x)\, dx$$

provided this limit exists.

▶ **Example 6** **Integrand Undefined at Right Endpoint** Compute

$$\int_0^1 \frac{dx}{(x-1)^{2/3}}$$

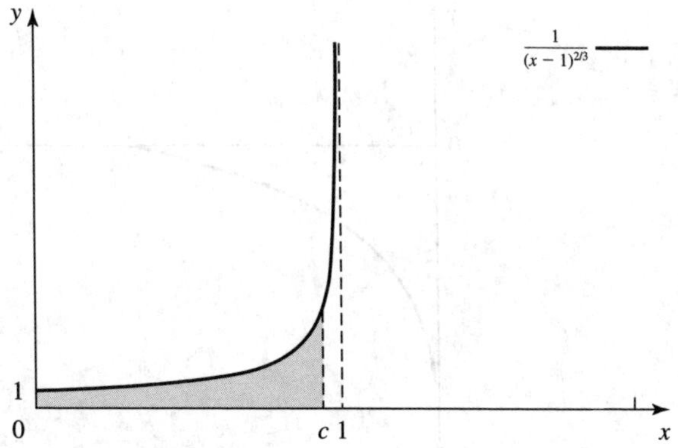

◀ **Figure 7.26**
The area of the shaded region
is $\int_0^c \frac{1}{(x-1)^{2/3}}\, dx$.

Solution

The graph of $f(x) = \frac{1}{(x-1)^{2/3}}$ is shown in Figure 7.25. We see immediately that $f(x)$ is continuous on $[0, 1)$, undefined at $x = 1$, and that

$$\lim_{x \to 1^-} f(x) = \infty$$

To compute the integral, we choose a number $c \in (0, 1)$ and compute

$$\int_0^c \frac{dx}{(x-1)^{2/3}}$$

(see Figure 7.26). Letting $c \to 1^-$ will then produce the desired integral. That is,

$$\int_0^1 \frac{dx}{(x-1)^{2/3}} = \lim_{c \to 1^-} \int_0^c \frac{dx}{(x-1)^{2/3}}$$

We first compute the indefinite integral

$$\int \frac{dx}{(x-1)^{2/3}} = 3(x-1)^{1/3} + C$$

If we set $F(x) = 3(x-1)^{1/3}$, then

$$\lim_{c \to 1^-} \int_0^c \frac{dx}{(x-1)^{2/3}} = \lim_{c \to 1^-} [F(c) - F(0)]$$

$$= \lim_{c \to 1^-} \left[3(c-1)^{1/3} - 3(-1)^{1/3} \right] = 3 \qquad (7.17)$$

We therefore find

$$\int_0^1 \frac{dx}{(x-1)^{2/3}} = 3$$

◀

▶ **Example 7**   **Integrand Undefined at Left Endpoint** Compute

$$\int_0^1 \ln x \, dx$$

Solution

The graph of $f(x) = \ln x, 0 < x \le 1$, is shown in Figure 7.27. We immediately see that $f(x)$ is continuous on $(0, 1]$, not defined at $x = 0$, and that

$$\lim_{x \to 0^+} f(x) = -\infty$$

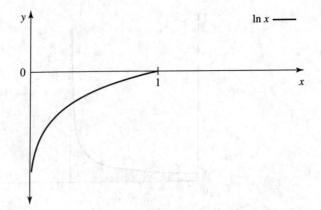

▲ **Figure 7.27**
The graph of $f(x) = \ln x$, $0 < x \leq 1$, in Example 7

To determine the definite integral, we need to compute

$$\lim_{c \to 0^+} \int_c^1 \ln x \, dx$$

Since $F(x) = x \ln x - x$ is an antiderivative of $f(x) = \ln x$, we find

$$\lim_{c \to 0^+} \int_c^1 \ln x \, dx = \lim_{c \to 0^+} [F(1) - F(c)]$$

$$= \lim_{c \to 0^+} [1 \ln 1 - 1 - c \ln c + c]$$

We need to find $\lim_{c \to 0^+} c \ln c$. The limit is of the form $0 \cdot \infty$. L'Hospital's rule yields

$$\lim_{c \to 0^+} c \ln c = \lim_{c \to 0^+} \frac{\ln c}{\frac{1}{c}} = \lim_{c \to 0^+} \frac{\frac{1}{c}}{-\frac{1}{c^2}}$$

$$= \lim_{c \to 0^+} \left(-\frac{1}{c} \cdot \frac{c^2}{1} \right) = -\lim_{c \to 0^+} c = 0$$

Together with $\lim_{c \to 0^+} c = 0$, we therefore find

$$\int_0^1 \ln x \, dx = \lim_{c \to 0^+} \int_c^1 \ln x \, dx = -1$$

◀

▶ **Example 8**   **Integrand Discontinuous in Interval** Compute

$$\int_{-1}^1 \frac{1}{x^2} \, dx$$

Solution

The function $f(x) = \frac{1}{x^2}$ is not defined at $x = 0$. In fact, it has a vertical asymptote at $x = 0$, since

$$\lim_{x \to 0^-} \frac{1}{x^2} = \infty \quad \text{and} \quad \lim_{x \to 0^+} \frac{1}{x^2} = \infty$$

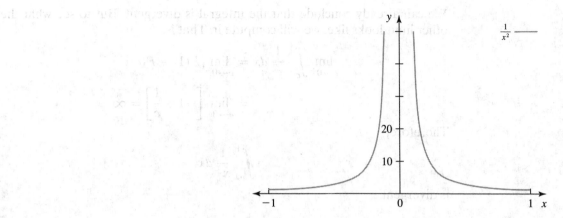

▲ Figure 7.28
The graph of $f(x) = \frac{1}{x^2}$

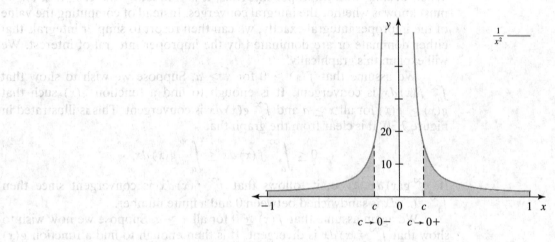

▲ Figure 7.29
The improper integral $\int_{-1}^{1} \frac{1}{x^2}\, dx$

The graph of $f(x) = \frac{1}{x^2}$, $x \neq 0$, is shown in Figure 7.28. We see that $f(x) = 1/x^2$ is continuous except at $x = 0$. To deal with this discontinuity, we split the integral at $x = 0$. We write

$$\int_{-1}^{1} \frac{1}{x^2}\, dx = \lim_{c \to 0^-} \int_{-1}^{c} \frac{1}{x^2}\, dx + \lim_{c \to 0^+} \int_{c}^{1} \frac{1}{x^2}\, dx$$

(see Figure 7.29). The function

$$F(x) = -\frac{1}{x}$$

is an antiderivative of $\frac{1}{x^2}$. Therefore,

$$\lim_{c \to 0^-} \int_{-1}^{c} \frac{1}{x^2}\, dx = \lim_{c \to 0^-} \left[F(c) - F(-1) \right]$$

$$= \lim_{c \to 0^-} \left[-\frac{1}{c} - 1 \right] = \infty$$

We can already conclude that the integral is divergent. But to see what the other limit looks like, we will compute it. That is,

$$\lim_{c \to 0^+} \int_c^1 \frac{1}{x^2}\,dx = \lim_{c \to 0^+} [F(1) - F(c)]$$

$$= \lim_{c \to 0^+} \left[-1 + \frac{1}{c}\right] = \infty$$

Therefore,

$$\int_{-1}^1 \frac{1}{x^2}\,dx$$

is divergent. ◀

7.4.3 A Comparison Result for Improper Integrals

There are many cases where it is impossible to evaluate an integral exactly. When dealing with improper integrals, it is frequently the case that all we must know is whether the integral converges. Instead of computing the value of the improper integral exactly, we can then resort to simpler integrals that either dominate or are dominated by the improper integral of interest. We will explain this graphically.

We assume that $f(x) \geq 0$ for $x \geq a$. Suppose we wish to show that $\int_a^\infty f(x)\,dx$ is convergent. It is enough to find a function $g(x)$ such that $g(x) \geq f(x)$ for all $x \geq a$ and $\int_a^\infty g(x)\,dx$ is convergent. This is illustrated in Figure 7.30. It is clear from the graph that

$$0 \leq \int_a^\infty f(x)\,dx \leq \int_a^\infty g(x)\,dx$$

If $\int_a^\infty g(x)\,dx < \infty$, it follows that $\int_a^\infty f(x)\,dx$ is convergent since then $\int_a^\infty f(x)\,dx$ is sandwiched between 0 and a finite number.

We again assume that $f(x) \geq 0$ for all $x \geq a$. Suppose we now wish to show that $\int_a^\infty f(x)\,dx$ is divergent. It is then enough to find a function $g(x)$ such that $0 \leq g(x) \leq f(x)$ for all $x \geq a$ and $\int_a^\infty g(x)\,dx$ is divergent. This is illustrated in Figure 7.31. It is clear from the graph that

$$\int_a^\infty f(x)\,dx \geq \int_a^\infty g(x)\,dx \geq 0$$

If $\int_a^\infty g(x)\,dx$ is divergent, it follows that $\int_a^\infty f(x)\,dx$ is divergent.

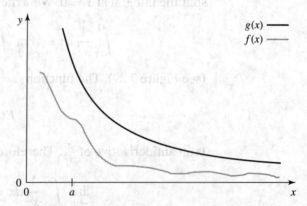

▲ **Figure 7.30**
The function $g(x)$ lies above the function $f(x)$

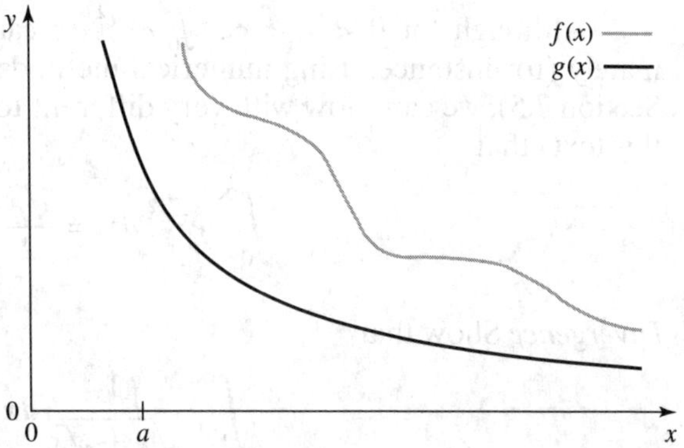

◀ **Figure 7.31**
The graph of $g(x)$ is below
the graph of $f(x)$

You can see from the preceding discussion that in one case we selected a function that dominated $f(x)$, whereas in the other case we selected a function that was dominated by $f(x)$. This indicates that you must first guess whether the integral is likely to converge, before you find a comparison function. (With practice, you get better at guessing whether an integral converges or diverges.) Sketching the functions involved can help to convince yourself that you are making the comparison in the right direction. Your comparison function, of course, should be simple enough so that you can integrate it without any problems. We present two examples that illustrate both cases.

▶ Example 9   *Convergence* Show that

$$\int_0^\infty e^{-x^2}\, dx$$

is convergent.

Solution

The function $f(x) = e^{-x^2}$ is continuous and positive for $x \in [0, \infty)$. We cannot compute the antiderivative of $f(x) = e^{-x^2}$ with any of the techniques we learn in this text. In fact, there is no simple way to express the value of $\int_0^z e^{-x^2}\, dx$ for $z > 0$ (it can be expressed as a sum of infinitely many terms). But we can still determine whether the integral is convergent. We write $\int_0^\infty e^{-x^2}\, dx$ as a sum of two integrals and show that each one is finite:

$$\int_0^\infty e^{-x^2}\, dx = \int_0^1 e^{-x^2}\, dx + \int_1^\infty e^{-x^2}\, dx$$

Since $0 < e^{-x^2} \le 1$, it follows that

$$0 < \int_0^1 e^{-x^2}\, dx \le \int_0^1 dx = 1 < \infty$$

To show that $\int_1^\infty e^{-x^2}\, dx$ is convergent, we use that e^{-x} is a decreasing function and that if $x \ge 1$, then $x \le x^2$. It follows that

$$0 \le e^{-x^2} \le e^{-x} \quad \text{for } x \ge 1$$

Therefore,

$$0 \le \int_1^\infty e^{-x^2}\, dx \le \int_1^\infty e^{-x}\, dx = \lim_{c \to \infty} \left[-e^{-x} \right]_1^c = e^{-1} < \infty$$

Since both integrals are convergent, it follows that $\int_0^\infty e^{-x^2}\, dx$ is convergent.

Although for $0 < z < \infty$, $\int_0^z e^{-x^2}\, dx$ can only be computed approximately (for instance, using numerical methods of the sort we will discuss in Section 7.5), we can show with very different tools (which we do not cover in this text) that

$$\int_0^\infty e^{-x^2}\, dx = \frac{\sqrt{\pi}}{2}$$

◀

▶ Example 10   *Divergence* Show that

$$\int_1^\infty \frac{1}{\sqrt{x + \sqrt{x}}}\, dx$$

is divergent.

Solution

The function $f(x) = 1/\sqrt{x + \sqrt{x}}$ is continuous on $[1, \infty)$. The integrand looks rather complicated. But since $x + \sqrt{x} \le x + x$ for $x \ge 1$, it follows that

$$\frac{1}{\sqrt{x + \sqrt{x}}} \ge \frac{1}{\sqrt{2x}} \quad \text{for } x \ge 1$$

Hence,

$$\int_1^\infty \frac{1}{\sqrt{x + \sqrt{x}}}\, dx \ge \int_1^\infty \frac{1}{\sqrt{2x}}\, dx = \frac{1}{\sqrt{2}} \int_1^\infty \frac{1}{\sqrt{x}}\, dx = \infty$$

as shown in Example 2. Therefore,

$$\int_1^\infty \frac{dx}{\sqrt{x + \sqrt{x}}}$$

is divergent.

◀

7.4.4 Problems

(7.4.1), (7.4.2)

The integrals in Problems 1–16 are all improper and all converge. Explain in each case why the integral is improper, and evaluate each integral.

1. $\int_0^\infty 3e^{-3x}\, dx$

2. $\int_0^\infty x e^{-x}\, dx$

3. $\int_0^\infty \frac{2}{1 + x^2}\, dx$

4. $\int_e^\infty \frac{dx}{x(\ln x)^2}$

5. $\int_1^\infty \frac{1}{x^{3/2}}\, dx$

6. $\int_{-\infty}^{-1} \frac{1}{1 + x^2}\, dx$

7. $\int_{-\infty}^\infty e^{-|x|}\, dx$

8. $\int_{-\infty}^\infty x e^{-x^2/2}\, dx$

9. $\int_{-\infty}^\infty \frac{x}{(1 + x^2)^2}\, dx$

10. $\int_{-\infty}^\infty x^3 e^{-x^4}\, dx$

11. $\int_0^9 \frac{dx}{\sqrt{9 - x}}$

12. $\int_1^e \frac{dx}{x\sqrt{\ln x}}$

13. $\int_0^{\pi/2} \frac{\cos x}{\sqrt{\sin x}}\, dx$

14. $\int_{-2}^0 \frac{dx}{(x + 1)^{1/3}}$

15. $\int_{-1}^1 \ln|x|\, dx$

16. $\int_0^2 \frac{dx}{(x - 1)^{2/5}}$

In Problems 17–28, determine whether each integral is convergent. If the integral is convergent, compute its value.

17. $\int_1^\infty \frac{1}{x^3}\, dx$

18. $\displaystyle\int_1^\infty \frac{1}{x^{1/3}}\,dx$

19. $\displaystyle\int_0^4 \frac{1}{x^4}\,dx$

20. $\displaystyle\int_0^4 \frac{1}{x^{1/4}}\,dx$

21. $\displaystyle\int_0^2 \frac{1}{(x-1)^{1/3}}\,dx$

22. $\displaystyle\int_0^2 \frac{1}{(x-1)^4}\,dx$

23. $\displaystyle\int_0^\infty \frac{1}{\sqrt{x+1}}\,dx$

24. $\displaystyle\int_{-1}^0 \frac{1}{\sqrt{x+1}}\,dx$

25. $\displaystyle\int_e^\infty \frac{dx}{x\ln x}$

26. $\displaystyle\int_1^e \frac{dx}{x\ln x}$

27. $\displaystyle\int_{-2}^2 \frac{2x\,dx}{(x^2-1)^{1/3}}$

28. $\displaystyle\int_{-\infty}^1 \frac{3}{1+x^2}\,dx$

29. Determine whether
$$\int_{-\infty}^\infty \frac{1}{x^2-1}\,dx$$
is convergent.

  Hint: Use the partial fraction decomposition
$$\frac{1}{x^2-1} = \frac{1}{2}\left(\frac{1}{x-1} - \frac{1}{x+1}\right).$$

30. Though we cannot compute the antiderivative of $f(x) = e^{-x^2/2}$, it is known that
$$\int_{-\infty}^\infty e^{-x^2/2}\,dx = \sqrt{2\pi}.$$
Use this fact to show that
$$\int_{-\infty}^\infty x^2 e^{-x^2/2}\,dx = \sqrt{2\pi}.$$
Hint: Write the integrand as
$$x \cdot (xe^{-x^2/2})$$
and use integration by parts.

31. Determine the constant c so that
$$\int_0^\infty ce^{-3x}\,dx = 1.$$

32. Determine the constant c so that
$$\int_{-\infty}^\infty \frac{c}{1+x^2}\,dx = 1.$$

33. In this problem, we investigate the integral
$$\int_1^\infty \frac{1}{x^p}\,dx$$
for $0 < p < \infty$.

(a) For $z > 1$, set
$$A(z) = \int_1^z \frac{1}{x^p}\,dx$$

and show that
$$A(z) = \frac{1}{1-p}(z^{-p+1}-1)$$
for $p \neq 1$ and
$$A(z) = \ln z$$
for $p = 1$.

(b) Use your results in (a) to show that for $0 < p \le 1$
$$\lim_{z\to\infty} A(z) = \infty.$$

(c) Use your results in (a) to show that for $p > 1$
$$\lim_{z\to\infty} A(z) = \frac{1}{p-1}.$$

34. In this problem, we investigate the integral
$$\int_0^1 \frac{1}{x^p}\,dx$$
for $0 < p < \infty$.

(a) Compute
$$\int \frac{1}{x^p}\,dx$$
for $0 < p < \infty$.
(*Hint:* Treat the case where $p = 1$ separately.)

(b) Use your result in (a) to compute
$$\int_c^1 \frac{1}{x^p}\,dx$$
for $0 < c < 1$.

(c) Use your result in (b) to show that
$$\int_0^1 \frac{1}{x^p}\,dx = \frac{1}{1-p}$$
for $0 < p < 1$.

(d) Show that
$$\int_0^1 \frac{1}{x^p}\,dx$$
is divergent for $p \ge 1$.
(7.4.3)

35. (a) Show that
$$0 \le e^{-x^2} \le e^{-x}$$
for $x \ge 1$.

(b) Use your result in (a) to show that
$$\int_1^\infty e^{-x^2}\,dx$$
is convergent.

36. (a) Show that
$$0 \le \frac{1}{\sqrt{1+x^4}} \le \frac{1}{x^2}$$
for $x > 0$.

(b) Use your result in (a) to show that
$$\int_1^\infty \frac{1}{\sqrt{1+x^4}}\,dx$$
is convergent.

37. (a) Show that
$$\frac{1}{\sqrt{1+x^2}} \ge \frac{1}{2x} > 0$$
for $x \ge 1$.

(b) Use your result in (a) to show that

$$\int_1^\infty \frac{1}{\sqrt{1+x^2}}\, dx$$

is divergent.

38. (a) Show that

$$\frac{1}{\sqrt{x+\ln x}} \geq \frac{1}{\sqrt{2x}} > 0$$

for $x \geq 1$.

(b) Use your result in (a) to show that

$$\int_1^\infty \frac{1}{\sqrt{x+\ln x}}\, dx$$

is divergent.

In Problems 39–42, find a comparison function for each integrand and determine whether the integral is convergent.

39. $\displaystyle\int_{-\infty}^\infty e^{-x^2/2}\, dx$

40. $\displaystyle\int_1^\infty \frac{1}{\sqrt{1+x^6}}\, dx$

41. $\displaystyle\int_1^\infty \frac{1}{\sqrt{1+x}}\, dx$

42. $\displaystyle\int_{-\infty}^\infty \frac{1}{e^x + e^{-x}}\, dx$

43. (a) Show that

$$\lim_{x\to\infty} \frac{\ln x}{\sqrt{x}} = 0$$

(b) Use your result in (a) to show that

$$2\ln x \leq \sqrt{x} \qquad (7.18)$$

for sufficiently large x. Use a graphing calculator to determine just how large x must be for (7.18) to hold.

(c) Use your result in (b) to show that

$$\int_0^\infty e^{-\sqrt{x}}\, dx \qquad (7.19)$$

converges. Use a graphing calculator to sketch the function $f(x) = e^{-\sqrt{x}}$ together with its comparison function(s), and use your graph to explain how you showed that the integral in (7.19) is convergent.

44. (a) Show that

$$\lim_{x\to\infty} \frac{\ln x}{x} = 0$$

(b) Use your result in (a) to show that for any $c > 0$,

$$cx \geq \ln x$$

for sufficiently large x.

(c) Use your result in (b) to show that for any $p > 0$,

$$x^p e^{-x} \leq e^{-x/2}$$

provided that x is sufficiently large.

(d) Use your result in (c) to show that for any $p > 0$,

$$\int_0^\infty x^p e^{-x}\, dx$$

is convergent.

7.5 NUMERICAL INTEGRATION

There are integrals that are impossible to evaluate exactly, such as

$$\int_0^4 e^{-x^2}\, dx$$

In such situations, numerical approximations are needed.

One way to numerically approximate an integral should be obvious from our initial approach to the area problem—namely, approximating areas by rectangles, that is, the Riemann sum approximation. Recall that for f continuous,

$$\int_a^b f(x)\, dx = \lim_{\|P\|\to 0} \sum_{k=1}^n f(c_k)\, \Delta x_k$$

where $P = [x_0, x_1, \ldots, x_n]$, $n = 1, 2, \ldots$, is a sequence of partitions of $[a, b]$ with $x_0 = a$ and $x_n = b$ and $\|P\| \to 0$ as $n \to \infty$. The number c_k is in $[x_{k-1}, x_k]$, and $\Delta x_k = x_k - x_{k-1}$ for $1 \leq k \leq n$.

In the following, we will assume that we partition the interval $[a, b]$ into n equal subintervals; that is, each subinterval is of length

$$\Delta x = \frac{b-a}{n}$$

We assume that the function f is continuous on $[a, b]$. We will discuss two methods, the midpoint rule and the trapezoidal rule.

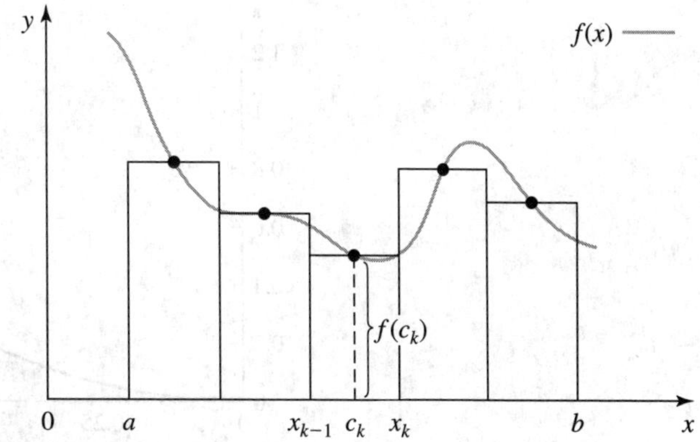

▲ **Figure 7.32**
The midpoint rule

7.5.1 The Midpoint Rule

This is the Riemann sum approximation, where we choose the midpoint of each subinterval for the point c_k. The midpoint of the interval $[x_{k-1}, x_k]$ is

$$c_k = \frac{x_{k-1} + x_k}{2}$$

The rule is defined as follows (also see Figure 7.32).

MIDPOINT RULE

Suppose that $f(x)$ is continuous on $[a, b]$ and that $[x_0, x_1, \ldots, x_n]$ is a partition of $[a, b]$ into n subintervals of equal lengths. We approximate

$$\int_a^b f(x)\, dx$$

by

$$M_n = \frac{b - a}{n} \sum_{k=1}^n f(c_k)$$

where $c_k = \frac{x_{k-1}+x_k}{2}$ is the midpoint of $[x_{k-1}, x_k]$.

In the following example, we choose an integral that we can evaluate exactly, so that we can see how close the approximation is.

▶ **Example 1**   *Midpoint Rule* Use the midpoint rule with $n = 4$ to approximate

$$\int_0^1 x^2\, dx.$$

Solution

The function $f(x) = x^2$ is continuous on $[0, 1]$. For $n = 4$, we find $\Delta x = \frac{b-a}{4} = \frac{1}{4}$ and we obtain four subintervals, each of length $\frac{1}{4}$ (see Figure 7.33).

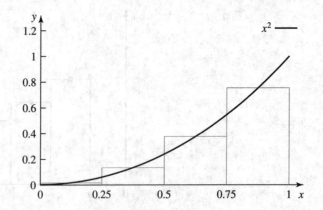

▲ **Figure 7.33**

The midpoint rule for $\int_0^1 x^2 \, dx$ with $n = 4$

Subinterval $[x_{k-1}, x_k]$	Midpoint c_k	$f(c_k)$	$f(c_k)\,\Delta x$
$\left[0, \dfrac{1}{4}\right]$	$\dfrac{1}{8}$	$\dfrac{1}{64}$	$\dfrac{1}{64}\dfrac{1}{4}$
$\left[\dfrac{1}{4}, \dfrac{1}{2}\right]$	$\dfrac{3}{8}$	$\dfrac{9}{64}$	$\dfrac{9}{64}\dfrac{1}{4}$
$\left[\dfrac{1}{2}, \dfrac{3}{4}\right]$	$\dfrac{5}{8}$	$\dfrac{25}{64}$	$\dfrac{25}{64}\dfrac{1}{4}$
$\left[\dfrac{3}{4}, 1\right]$	$\dfrac{7}{8}$	$\dfrac{49}{64}$	$\dfrac{49}{64}\dfrac{1}{4}$

We find the approximation

$$M_4 = \frac{b-a}{4} \sum_{k=1}^{4} f(c_k) = \frac{1}{4}\left(\frac{1}{64} + \frac{9}{64} + \frac{25}{64} + \frac{49}{64}\right) = \frac{1}{4}\frac{84}{64} = \frac{21}{64} \approx 0.3281$$

We know that $\int_0^1 x^2 \, dx = \frac{1}{3}$. Hence, the error is

$$\left| \int_0^1 x^2 \, dx - M_4 \right| \approx 0.0052$$

Larger values of n improve the approximation. ◀

Instead of memorizing the formula for the midpoint rule, it is easier to keep a picture in mind. We illustrate this in the following example.

▶ **Example 2** *Midpoint Rule* Use the midpoint rule with $n = 5$ to approximate

$$\int_1^2 \frac{1}{x} \, dx$$

Solution

The graph of $f(x) = \frac{1}{x}$ together with the five approximating rectangles is shown in Figure 7.34. We see that $f(x) = 1/x$ is continuous on $[1, 2]$.

With $n = 5$, the partition of $[1, 2]$ is given by $P = [1, 1.2, 1.4, 1.6, 1.8, 2]$ and the midpoints are $1.1, 1.3, 1.5, 1.7,$ and 1.9. Since the width of each

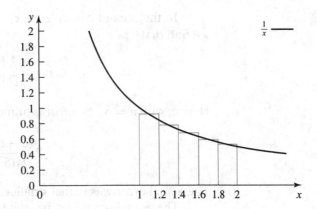

▲ Figure 7.34
The midpoint rule for Example 2

rectangle is 0.2 and $f(x) = \frac{1}{x}$, the area of the first rectangle is $(0.2)\frac{1}{1.1}$, the area of the second rectangle is $(0.2)\frac{1}{1.3}$, and so on. We thus find

$$M_5 = (0.2)\left[\frac{1}{1.1} + \frac{1}{1.3} + \frac{1}{1.5} + \frac{1}{1.7} + \frac{1}{1.9}\right] = 0.6919$$

Note that we factored 0.2, the width of each rectangle, since it is a common factor of the areas of the five rectangles.

We know that

$$\int_1^2 \frac{1}{x}\,dx = \ln x\big]_1^2 = \ln 2 - \ln 1 = \ln 2$$

Hence the error in the approximation is

$$\left|\int_1^2 \frac{1}{x}\,dx - M_5\right| = |\ln 2 - 0.6919| = 0.0012 \qquad ◀$$

We typically use an approximation when we cannot evaluate the integral exactly. Thus, we cannot use the exact value to determine how close the approximation is. Fortunately, there are results that allow us to obtain upper bounds for the error.

ERROR BOUND FOR THE MIDPOINT RULE

Suppose that $|f''(x)| \le K$ for all $x \in [a, b]$. Then the error in the midpoint rule is at most

$$\left|\int_a^b f(x)\,dx - M_n\right| \le K\frac{(b-a)^3}{24n^2}$$

Let's check this bound for our two examples. In the first example, $f(x) = x^2$, and therefore $f''(x) = 2$. Hence with $n = 4$, the error is at most

$$2\frac{(1-0)^3}{24(4^2)} \approx 0.0052$$

This is in fact the error that we obtained.

In the second example, $f(x) = \frac{1}{x}$. Since $f'(x) = -\frac{1}{x^2}$ and $f''(x) = \frac{2}{x^3}$, we find that

$$|f''(x)| = \left|\frac{2}{x^3}\right| \le 2 \quad \text{for } 1 \le x \le 2$$

Hence, with $n = 5$, the error is at most

$$2\frac{(2-1)^3}{24(5^2)} = 0.0033$$

The actual error was in fact smaller, only 0.0012.

The actual error can be quite a bit smaller than the theoretical error bound, which is the worst-case scenario, but it will never be larger. The advantage of such an error bound is that it allows us to find the number of subintervals required to obtain a certain accuracy. For instance, if we want to numerically approximate $\int_0^1 x^2\,dx$ so that the error was at most 10^{-4}, then we must choose n so that

$$K\frac{(b-a)^3}{24n^2} \le 10^{-4}$$

$$2\frac{1}{24n^2} \le 10^{-4}$$

$$\frac{1}{12}10^4 \le n^2$$

$$28.9 \le n$$

That is, $n = 29$ would suffice to produce an error of at most 10^{-4}.

Finding a value for K in the estimate is not always easy. A graph of $f''(x)$ over the interval of interest can facilitate finding a bound on the second derivative. We need not find the best possible bound. For instance, if we wanted to integrate $f(x) = e^x$ over the interval $[1, 2]$, we would need to find a bound on $f''(x) = e^x$ over the interval $[1, 2]$. Since $|e^x| \le e^2$ over the interval $[1, 2]$, we could use $K = 9$, for instance (see Figure 7.35). This is not the best possible bound but a number that we can find without using a calculator. (The best possible bound would be $K = e^2$.)

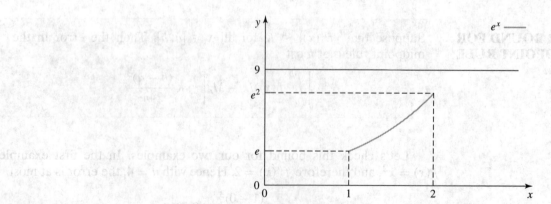

▲ **Figure 7.35**
An upper bound on $|e^x|$ over $[1, 2]$ is 9

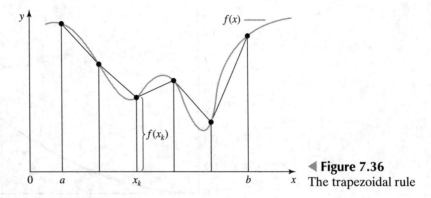

◀ **Figure 7.36**
The trapezoidal rule

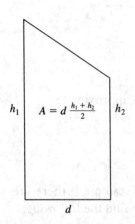

▲ **Figure 7.37**
The area of a trapezoid

7.5.2 The Trapezoidal Rule

Instead of rectangles, we use trapezoids to approximate integrals in this method, as illustrated in Figure 7.36. We assume again that f is a continuous function on $[a, b]$ and divide $[a, b]$ into n equal subintervals. But this time we approximate the function $f(x)$ by a polygon $P(x)$. To obtain the polygon $P(x)$, we connect the points $(x_k, f(x_k))$, $k = 0, 1, 2, \ldots, n$, by straight lines, as shown in Figure 7.36. The integral $\int_a^b P(x)\,dx$ is then the approximation to $\int_a^b f(x)\,dx$.

We see from Figure 7.36 that this amounts to adding up (signed) areas of trapezoids. Recall from planar geometry that the area of the trapezoid in Figure 7.37 is

$$A = d\frac{h_1 + h_2}{2}$$

The width of each trapezoid in Figure 7.36 is $d = \frac{b-a}{n}$. Adding up the areas of the trapezoids in Figure 7.36 then yields

$$T_n = \frac{b-a}{n}\left[\frac{f(x_0) + f(x_1)}{2} + \frac{f(x_1) + f(x_2)}{2} + \frac{f(x_2) + f(x_3)}{2}\right.$$
$$\left. + \cdots + \frac{f(x_{n-2}) + f(x_{n-1})}{2} + \frac{f(x_{n-1}) + f(x_n)}{2}\right]$$
$$= \frac{b-a}{n}\left[\frac{f(x_0)}{2} + f(x_1) + f(x_2) + \cdots + f(x_{n-1}) + \frac{f(x_n)}{2}\right]$$

TRAPEZOIDAL RULE
Suppose that $f(x)$ is continuous on $[a, b]$ and that $P = [x_0, x_1, x_2, \ldots, x_n]$ is a partition of $[a, b]$ into n subintervals of equal length. We approximate

$$\int_a^b f(x)\,dx$$

by

$$T_n = \frac{b-a}{n}\left[\frac{f(x_0)}{2} + f(x_1) + f(x_2) + \cdots + f(x_{n-1}) + \frac{f(x_n)}{2}\right]$$

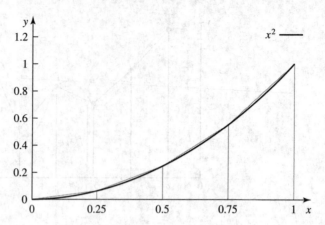

▲ **Figure 7.38**

The trapezoidal rule for $\int_0^1 x^2\, dx$ with $n = 4$

▶ **Example 3** *Trapezoidal Rule* Use the trapezoidal rule with $n = 4$ to approximate

$$\int_0^1 x^2\, dx$$

Solution

The function $f(x) = x^2$ is continuous on $[0, 1]$. As in Example 1, there are four subintervals, each of length $\frac{1}{4}$ (see Figure 7.38). We find the following: The approximation is

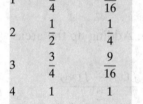

k	x_k	$f(x_k)$
0	0	0
1	$\dfrac{1}{4}$	$\dfrac{1}{16}$
2	$\dfrac{1}{2}$	$\dfrac{1}{4}$
3	$\dfrac{3}{4}$	$\dfrac{9}{16}$
4	1	1

$$T_4 = \frac{1}{4}\left[\frac{0}{2} + \frac{1}{16} + \frac{1}{4} + \frac{9}{16} + \frac{1}{2}\right] = 0.34375$$

Since we know that $\int_0^1 x^2\, dx = \frac{1}{3}$, we can compute the error:

$$\left|\int_0^1 x^2\, dx - T_4\right| = 0.0104 \qquad \blacktriangleleft$$

▶ **Example 4** *Trapezoidal Rule* Use the trapezoidal rule with $n = 5$ to approximate

$$\int_1^2 \frac{1}{x}\, dx$$

Solution

This is illustrated in Figure 7.39. The function $1/x$ is continuous on $[1, 2]$. With $n = 5$, the partition of $[1, 2]$ is given by $P = [1.0, 1.2, 1.4, 1.6, 1.8, 2.0]$. The base of each trapezoid has length 0.2. Hence

$$T_5 = (0.2)\left[\frac{1}{2}\cdot\frac{1}{1.0} + \frac{1}{1.2} + \frac{1}{1.4} + \frac{1}{1.6} + \frac{1}{1.8} + \frac{1}{2}\cdot\frac{1}{2.0}\right] = 0.69563$$

Since we know from Example 2 that $\int_1^2 \frac{1}{x}\, dx = \ln 2$, we can compute the error:

$$\left|\int_1^2 \frac{1}{x}\, dx - T_5\right| = 0.00249 \qquad \blacktriangleleft$$

There is also a theoretical error bound for the trapezoidal rule.

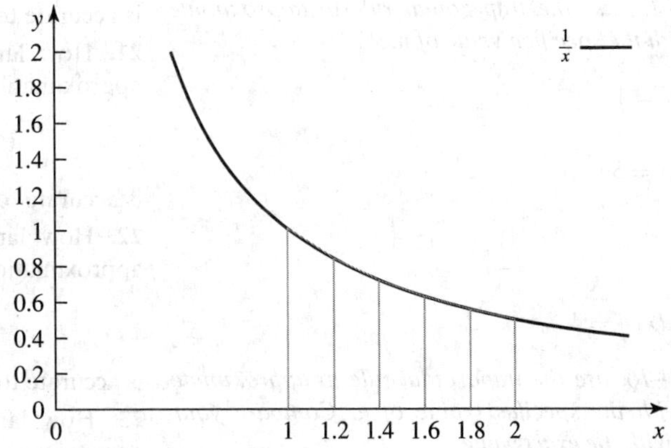

▲ Figure 7.39

The trapezoidal rule for $\int_1^2 \frac{1}{x} dx$ with $n = 5$

ERROR BOUND FOR THE TRAPEZOIDAL RULE

Suppose that $|f''(x)| \leq K$ for all $x \in [a, b]$. Then the error in the trapezoidal rule is at most

$$\left| \int_a^b f(x)\, dx - T_n \right| \leq K \frac{(b-a)^3}{12n^2}$$

In Example 3, since $f(x) = x^2$, we find $f''(x) = 2$ and hence $K = 2$. The error is therefore bounded by

$$2\frac{1}{12(4^2)} = 0.0104$$

which is the same as the actual error.

In Example 4, since $f(x) = 1/x$, we find $|f''(x)| = 2/x^3 \leq 2$ for $1 \leq x \leq 2$ (as in Example 2). Hence, with $n = 5$, the error bound is at most

$$2\frac{(2-1)^3}{12(5^2)} = 0.0067$$

The actual error was in fact smaller, only 0.00249. As in the midpoint rule, the theoretical error can be quite a bit larger than the actual error.

7.5.3 Problems

(7.5.1), (7.5.2)

In Problems 1–4, use the midpoint rule to approximate each integral with the specified value of n.

1. $\int_1^2 x^2\, dx, n = 4$

2. $\int_{-1}^0 x^3\, dx, n = 5$

3. $\int_0^1 e^{-x}\, dx, n = 3$

4. $\int_0^{\pi/2} \sin x\, dx, n = 4$

In Problems 5–8, use the midpoint rule to approximate each integral with the specified value of n. Compare your approximation with the exact value.

5. $\int_2^4 \frac{1}{x}\, dx, n = 4$

6. $\int_{-1}^1 (e^x - 1)\, dx, n = 4$

7. $\int_0^4 \sqrt{x}\, dx, n = 4$

8. $\int_1^3 \frac{1}{\sqrt{x}}\, dx, n = 5$

In Problems 9–12, use the trapezoidal rule to approximate each integral with the specified value of n.

9. $\int_1^2 x^2 \, dx, n = 4$

10. $\int_{-1}^0 x^3 \, dx, n = 5$

11. $\int_0^1 e^{-x} \, dx, n = 3$

12. $\int_0^{\pi/2} \sin x \, dx, n = 4$

In Problems 13–16, use the trapezoidal rule to approximate each integral with the specified value of n. Compare your approximation with the exact value.

13. $\int_1^3 x^3 \, dx, n = 5$

14. $\int_{-1}^1 (1 - e^{-x}) \, dx, n = 4$

15. $\int_0^2 \sqrt{x} \, dx, n = 4$

16. $\int_1^2 \frac{1}{x} \, dx, n = 5$

17. How large should n be so that the midpoint rule approximation of

$$\int_0^2 x^2 \, dx$$

is accurate to within 10^{-4}?

In Problems 18–24, use the theoretical error bound to determine how large n should be. [Hint: In each case, find the second derivative of the integrand, graph it, and use a graphing calculator to find an upper bound on $|f''(x)|$.]

18. How large should n be so that the midpoint rule approximation of

$$\int_1^2 \frac{1}{x} \, dx$$

is accurate to within 10^{-3}?

19. How large should n be so that the midpoint rule approximation of

$$\int_0^2 e^{-x^2/2} \, dx$$

is accurate to within 10^{-4}?

20. How large should n be so that the midpoint rule approximation of

$$\int_2^8 \frac{1}{\ln t} \, dt$$

is accurate to within 10^{-3}?

21. How large should n be so that the trapezoidal rule approximation of

$$\int_0^1 e^{-x} \, dx$$

is accurate to within 10^{-5}?

22. How large should n be so that the trapezoidal rule approximation of

$$\int_0^2 \sin x \, dx$$

is accurate to within 10^{-4}?

23. How large should n be so that the trapezoidal rule approximation of

$$\int_1^2 \frac{e^t}{t} \, dt$$

is accurate to within 10^{-4}?

24. How large should n be so that the trapezoidal rule approximation of

$$\int_1^2 \frac{\cos x}{x} \, dx$$

is accurate to within 10^{-3}?

25. (a) Show graphically that for $n = 5$, the trapezoidal rule overestimates and the midpoint rule underestimates

$$\int_0^1 x^3 \, dx$$

In each case, compute the approximate value of the integral and compare it to the exact value.

(b) The result in (a) has to do with the fact that $y = x^3$ is concave up on $[0, 1]$. To generalize this, we assume that the function $f(x)$ is continuous, nonnegative, and concave up on the interval $[a, b]$. Denote by M_n the midpoint rule and by T_n the trapezoidal rule approximation of $\int_a^b f(x) \, dx$. Explain in words why

$$M_n \leq \int_a^b f(x) \, dx \leq T_n$$

(c) If we assume that $f(x)$ is continuous, nonnegative, and concave *down* on $[a, b]$, then

$$M_n \geq \int_a^b f(x) \, dx \geq T_n$$

Explain why this is so. Use this result to give an upper and a lower bound on

$$\int_0^1 \sqrt{x} \, dx$$

when $n = 4$ in the approximation.

7.6 TABLES OF INTEGRALS

Tables of indefinite integrals are useful aids when evaluating integrals. Often, it is still necessary to bring the integrand of interest into a form that is listed in the table; and there are many integrals that simply cannot be evaluated exactly and must be evaluated numerically. We will give a very brief list of indefinite integrals and explain how to use such tables.

I. Basic Functions

1. $\displaystyle\int x^n dx = \frac{1}{n+1}x^{n+1} + C, n \neq -1$

2. $\displaystyle\int \frac{1}{x} dx = \ln|x| + C$

3. $\displaystyle\int e^x dx = e^x + C$

4. $\displaystyle\int a^x dx = \frac{a^x}{\ln a} + C$ with $a > 0, a \neq 1$

5. $\displaystyle\int \ln x \, dx = x\ln x - x + C$

6. $\displaystyle\int \sin x \, dx = -\cos x + C$

7. $\displaystyle\int \cos x \, dx = \sin x + C$

8. $\displaystyle\int \tan x \, dx = -\ln|\cos x| + C$

II. Rational Functions

9. $\displaystyle\int \frac{1}{ax+b} dx = \frac{1}{a}\ln|ax+b| + C$

10. $\displaystyle\int \frac{x}{ax+b} dx = \frac{x}{a} - \frac{b}{a^2}\ln|ax+b| + C$

11. $\displaystyle\int \frac{x}{(ax+b)^2} dx = \frac{b}{a^2(ax+b)} + \frac{1}{a^2}\ln|ax+b| + C$

12. $\displaystyle\int \frac{x}{ax^2+bx+c} dx = \frac{1}{2a}\ln|ax^2+bx+c| - \frac{b}{2a}\int \frac{1}{ax^2+bx+c} dx$

13. $\displaystyle\int \frac{1}{a^2+x^2} dx = \frac{1}{a}\arctan\frac{x}{a} + C$

14. $\displaystyle\int \frac{1}{a^2-x^2} dx = \frac{1}{2a}\ln\left|\frac{x+a}{x-a}\right| + C$

III. Integrands Involving $\sqrt{a^2+x^2}$, $\sqrt{a^2-x^2}$, or $\sqrt{x^2-a^2}$

15. $\displaystyle\int \frac{1}{\sqrt{a^2-x^2}} dx = \arcsin\frac{x}{a} + C$

16. $\displaystyle\int \frac{1}{\sqrt{x^2\pm a^2}} dx = \ln|x + \sqrt{x^2\pm a^2}| + C$

17. $\displaystyle\int \sqrt{a^2\pm x^2}\, dx = \frac{1}{2}\left(x\sqrt{a^2\pm x^2} + a^2\int \frac{1}{\sqrt{a^2\pm x^2}} dx\right)$

18. $\displaystyle\int \sqrt{x^2-a^2}\, dx = \frac{1}{2}\left(x\sqrt{x^2-a^2} - a^2\int \frac{1}{\sqrt{x^2-a^2}} dx\right)$

IV.   Integrands Involving Trigonometric Functions

19. $\int \sin(ax)\,dx = -\dfrac{1}{a}\cos(ax) + C$

20. $\int \sin^2(ax)\,dx = \dfrac{1}{2}x - \dfrac{1}{4a}\sin(2ax) + C$

21. $\int \sin(ax)\sin(bx)\,dx = \dfrac{\sin(a-b)x}{2(a-b)} - \dfrac{\sin(a+b)x}{2(a+b)} + C,\ \text{for } a^2 \neq b^2$

22. $\int \cos(ax)\,dx = \dfrac{1}{a}\sin(ax) + C$

23. $\int \cos^2(ax)\,dx = \dfrac{1}{2}x + \dfrac{1}{4a}\sin(2ax) + C$

24. $\int \cos(ax)\cos(bx)\,dx = \dfrac{\sin(a-b)x}{2(a-b)} + \dfrac{\sin(a+b)x}{2(a+b)} + C,\ \text{for } a^2 \neq b^2$

25. $\int \sin(ax)\cos(ax)\,dx = \dfrac{1}{2a}\sin^2(ax) + C$

26. $\int \sin(ax)\cos(bx)\,dx = -\dfrac{\cos(a+b)x}{2(a+b)} - \dfrac{\cos(a-b)x}{2(a-b)} + C,\ \text{for } a^2 \neq b^2$

V.   Integrands Involving Exponential Functions

27. $\int e^{ax}\,dx = \dfrac{1}{a}e^{ax} + C$

28. $\int x e^{ax}\,dx = \dfrac{e^{ax}}{a^2}(ax - 1) + C$

29. $\int x^n e^{ax}\,dx = \dfrac{1}{a}x^n e^{ax} - \dfrac{n}{a}\int x^{n-1} e^{ax}\,dx$

30. $\int e^{ax}\sin(bx)\,dx = \dfrac{e^{ax}}{a^2 + b^2}(a\sin(bx) - b\cos(bx)) + C$

31. $\int e^{ax}\cos(bx)\,dx = \dfrac{e^{ax}}{a^2 + b^2}(a\cos(bx) + b\sin(bx)) + C$

VI.   Integrands Involving Logarithmic Functions

32. $\int \ln x\,dx = x\ln x - x + C$

33. $\int (\ln x)^2\,dx = x(\ln x)^2 - 2x\ln x + 2x + C$

34. $\int x^m \ln x\,dx = x^{m+1}\left[\dfrac{\ln x}{m+1} - \dfrac{1}{(m+1)^2}\right] + C,\ m \neq -1$

35. $\int \dfrac{\ln x}{x}\,dx = \dfrac{(\ln x)^2}{2} + C$

36. $\int \dfrac{1}{x\ln x}\,dx = \ln(\ln x) + C$

37. $\int \sin(\ln x)\,dx = \dfrac{x}{2}(\sin(\ln x) - \cos(\ln x)) + C$

38. $\int \cos(\ln x)\,dx = \dfrac{x}{2}(\sin(\ln x) + \cos(\ln x)) + C$

We will now illustrate how to use the table. We begin with examples that fit one of the listed integrals exactly.

▶ **Example 1**   *Square Root* Find

$$\int \sqrt{3 - x^2}\, dx$$

Solution

The integrand involves $\sqrt{a^2 - x^2}$ and is of the form III.17 with $a^2 = 3$. Hence,

$$\int \sqrt{3 - x^2}\, dx = \frac{1}{2}\left(x\sqrt{3 - x^2} + 3\int \frac{1}{\sqrt{3 - x^2}}\, dx \right)$$

To evaluate $\int \frac{1}{\sqrt{3-x^2}}\, dx$, we use III.15 with $a^2 = 3$ and find

$$\int \frac{1}{\sqrt{3 - x^2}}\, dx = \arcsin \frac{x}{\sqrt{3}} + C$$

Hence,

$$\int \sqrt{3 - x^2}\, dx = \frac{1}{2}\left(x\sqrt{3 - x^2} + 3 \arcsin \frac{x}{\sqrt{3}} \right) + C \qquad ◀$$

▶ **Example 2**   *Trigonometric Function* Find

$$\int \sin(3x) \cos(4x)\, dx$$

Solution

The integrand involves trigonometric functions, and we can find it in IV.26 with $a = 3$ and $b = 4$. Hence,

$$\int \sin(3x) \cos(4x)\, dx = -\frac{\cos(7x)}{14} - \frac{\cos(-x)}{(2)(-1)} + C$$

Since $\cos(-x) = \cos x$, this simplifies to

$$\int \sin(3x) \cos(4x)\, dx = -\frac{\cos(7x)}{14} + \frac{\cos x}{2} + C \qquad ◀$$

▶ **Example 3**   *Exponential Function* Find

$$\int x^2 e^{3x}\, dx$$

Solution

This is of the form V.29 with $n = 2$ and $a = 3$. Hence,

$$\int x^2 e^{3x}\, dx = \frac{1}{3}x^2 e^{3x} - \frac{2}{3}\int x e^{3x}\, dx$$

We now use V.28 to continue the evaluation of the integral and find

$$\int x e^{3x}\, dx = \frac{e^{3x}}{9}(3x - 1) + C$$

Hence,

$$\int x^2 e^{3x}\, dx = \frac{1}{3}x^2 e^{3x} - \frac{2}{3}\left[\frac{e^{3x}}{9}(3x - 1) \right] + C \qquad ◀$$

Thus far, each of our examples matched exactly one of the integrals in our table. Often, this will not be the case, and the integrand must be manipulated until it matches one of the integrals in the table. These manipulations include expansions, long division, completion of the square, and substitution. We give a few examples to illustrate this.

▶ **Example 4**   *Exponential Function* Find

$$\int e^{2x} \sin(3x - 1)\, dx$$

Solution

This looks similar to V.30. If we use the substitution

$$u = 3x - 1 \quad \text{with } dx = \frac{1}{3}\, du \quad \text{and } 2x = \frac{2}{3}(u + 1)$$

then the integrand can be transformed so that it matches V.30 exactly:

$$\int e^{2x} \sin(3x - 1)\, dx = \int e^{2(u+1)/3}(\sin u)\frac{1}{3}\, du$$

$$= \frac{e^{2/3}}{3} \int e^{2u/3} \sin u\, du$$

$$= \frac{e^{2/3}}{3} \frac{e^{2u/3}}{\frac{4}{9} + 1} \left[\frac{2}{3} \sin u - \cos u \right] + C$$

$$= \frac{e^{2/3}}{3 \cdot \frac{13}{9}} e^{2(3x-1)/3} \left[\frac{2}{3} \sin(3x - 1) - \cos(3x - 1) \right] + C$$

$$= \frac{3}{13} e^{2x} \left[\frac{2}{3} \sin(3x - 1) - \cos(3x - 1) \right] + C \qquad ◀$$

▶ **Example 5**   *Rational Function* Find

$$\int \frac{x^2}{9 + x^2}\, dx$$

Solution

The integrand is a rational function; we can use long division to simplify it:

$$\frac{x^2}{9 + x^2} = 1 - \frac{9}{9 + x^2}$$

Then, using II.13 with $a = 3$, we find

$$\int \frac{x^2}{9 + x^2}\, dx = \int dx - 9 \int \frac{1}{9 + x^2}\, dx$$

$$= x - 9 \left(\frac{1}{3} \arctan \frac{x}{3} \right) + C$$

$$= x - 3 \arctan \frac{x}{3} + C \qquad ◀$$

▶ **Example 6**   *Rational Function* Find

$$\int \frac{1}{x^2 - 2x - 3}\, dx$$

Solution

The first step is to complete the square in the denominator:

$$\frac{1}{x^2 - 2x - 3} = \frac{1}{(x^2 - 2x + 1) - 1 - 3}$$

$$= \frac{1}{(x - 1)^2 - 4}$$

Hence, using the substitution $u = x - 1$ with $du = dx$, we find

$$\int \frac{dx}{(x - 1)^2 - 4} = \int \frac{du}{u^2 - 4} = -\int \frac{du}{4 - u^2}$$

which is of the form II.14 with $a = 2$. Therefore,

$$\int \frac{1}{x^2 - 2x - 3}\,dx = -\int \frac{du}{4 - u^2} = -\frac{1}{4}\ln\left|\frac{u + 2}{u - 2}\right| + C$$

$$= -\frac{1}{4}\ln\left|\frac{x + 1}{x - 3}\right| + C$$

◄

7.6.1 Problems

In Problems 1–8, use the table to compute each integral.

1. $\displaystyle\int \frac{x}{2x - 3}\,dx$

2. $\displaystyle\int \frac{dx}{16 + x^2}$

3. $\displaystyle\int \sqrt{x^2 - 16}\,dx$

4. $\displaystyle\int \sin(2x)\cos(2x)\,dx$

5. $\displaystyle\int_0^1 x^3 e^{-x}\,dx$

6. $\displaystyle\int_0^{\pi/4} e^{-x}\cos(2x)\,dx$

7. $\displaystyle\int_1^e x^2 \ln x\,dx$

8. $\displaystyle\int_e^{e^2} \frac{dx}{x \ln x}$

In Problems 9–22, use the table to compute each integral, after manipulating the integrand in a suitable way.

9. $\displaystyle\int_0^{\pi/6} e^x \cos\left(x - \frac{\pi}{6}\right) dx$

10. $\displaystyle\int_1^2 x \ln(2x - 1)\,dx$

11. $\displaystyle\int (x^2 - 1)e^{-x/2}\,dx$

12. $\displaystyle\int (x + 1)^2 e^{-2x}\,dx$

13. $\displaystyle\int \cos^2(5x - 3)\,dx$

14. $\displaystyle\int \frac{x^2}{4x^2 + 4x + 1}\,dx$

15. $\displaystyle\int \sqrt{9 + 4x^2}\,dx$

16. $\displaystyle\int \frac{1}{\sqrt{16 - 9x^2}}\,dx$

17. $\displaystyle\int e^{2x+1}\sin\left(\frac{\pi}{2}x\right) dx$

18. $\displaystyle\int (x - 1)^2 e^{2x}\,dx$

19. $\displaystyle\int_2^4 \frac{1}{x \ln\sqrt{x}}\,dx$

20. $\displaystyle\int_1^e (x + 2)^2 \ln x\,dx$

21. $\displaystyle\int \cos(\ln(3x))\,dx$

22. $\displaystyle\int \frac{3}{x^2 - 4x + 8}\,dx$

7.7 THE TAYLOR APPROXIMATION

In many ways, polynomials are the easiest functions to work with. In this section, we will learn how to approximate functions by polynomials. We will see that the approximation typically improves when we use higher-degree polynomials.

7.7.1 Taylor Polynomials

In Section 4.8, we discussed how to linearize a function about a given point. This led to the linear or tangent approximation. We found the following.

> The linear approximation of $f(x)$ at $x = a$:
>
> $$L(x) = f(a) + f'(a)(x - a)$$

As an example, we look at

$$f(x) = e^x$$

and approximate this function by its linearization at $x = 0$. We find

$$L(x) = f(0) + f'(0)x = 1 + x \qquad (7.20)$$

since $f'(x) = e^x$ and $f(0) = f'(0) = 1$. To see how close the approximation is, we graph both $f(x)$ and $L(x)$ in the same coordinate system (the result is shown in Figure 7.40). The approximation is quite good as long as x is close to 0. Figure 7.40 suggests that it gets gradually worse as we move away from 0. In the approximation, we only required that $f(x)$ and $L(x)$ have in common $f(0) = L(0)$ and $f'(0) = L'(0)$.

To improve the approximation, we may wish to use an approximating function whose higher-order derivatives also agree with those of $f(x)$ at $x = 0$. The function $L(x)$ is a polynomial of degree 1. To improve the approximation, we will continue to work with polynomials but require that the function and its first n derivatives at $x = 0$ agree with those of the polynomial. To be able to match up the first n derivatives, the polynomial must be of degree n. (If the degree of the polynomial is $k < n$, then all derivatives of degree $k + 1$ or higher are equal to 0.) A polynomial of degree n can be written as

$$P_n(x) = a_0 + a_1 x + a_2 x^2 + \cdots + a_n x^n \qquad (7.21)$$

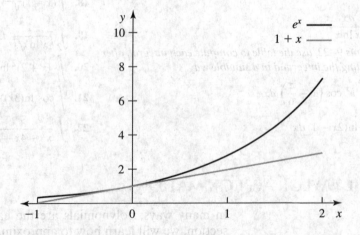

▲ **Figure 7.40**
The graph of $y = e^x$ and its linear approximation at 0

If we want to approximate $f(x)$ at $x = 0$, then we require

$$f(0) = P_n(0)$$
$$f'(0) = P'_n(0)$$
$$f''(0) = P''_n(0)$$
$$\vdots$$
$$f^{(n)}(0) = P_n^{(n)}(0)$$

(7.22)

Now

$$P_n(0) = a_0 + a_1 x + \cdots + a_n x^n \big|_{x=0} = a_0$$

$$P'_n(0) = a_1 + 2a_2 x + 3a_3 x^2 + \cdots + n a_n x^{n-1} \big|_{x=0} = a_1$$

$$P''_n(0) = 2a_2 + (3)(2)a_3 x + \cdots + n(n-1)a_n x^{n-2} \big|_{x=0} = 2a_2$$

$$P'''_n(0) = (3)(2)a_3 + (4)(3)(2)a_4 x + \cdots + n(n-1)(n-2)a_n x^{n-3} \big|_{x=0}$$
$$= (3)(2)a_3$$

$$\vdots$$

$$P_n^{(n)}(0) = n(n-1)(n-2) \cdots (3)(2)(1)a_n |_{x=0}$$
$$= n(n-1)(n-2) \cdots (3)(2)(1)a_n$$

We introduce the notation

$$k! = k(k-1)(k-2) \cdots (3)(2)(1)$$

where $k!$ is read "k factorial." Solving these equations for $a_k, k = 0, 1, 2, \ldots, n$, and using $f^{(k)}(0) = P^{(k)}(0), k = 0, 1, 2, \ldots, n$, we find

$$a_0 = P_n(0) = f(0)$$

$$a_1 = P'_n(0) = f'(0)$$

$$a_2 = \frac{1}{2} P''_n(0) = \frac{1}{2!} f''(0)$$

$$a_3 = \frac{1}{3 \cdot 2} P'''_n(0) = \frac{1}{3!} f'''(0)$$

$$\vdots$$

$$a_n = \frac{1}{n(n-1) \cdots 2 \cdot 1} P_n^{(n)}(0) = \frac{1}{n!} f^{(n)}(0)$$

(7.23)

A polynomial of degree n of the form (7.21), whose coefficients satisfy (7.24), is called a **Taylor polynomial** of degree n. We summarize this definition in the following.

Definition The Taylor polynomial of degree n about $x = 0$ for the function $f(x)$ is given by

$$P_n(x) = f(0) + f'(0)x + \frac{f''(0)}{2!}x^2 + \frac{f'''(0)}{3!}x^3$$

$$+ \frac{f^{(4)}(0)}{4!}x^4 + \cdots + \frac{f^{(n)}(0)}{n!}x^n$$

▶ **Example 1** Compute the Taylor polynomial of degree 3 about $x = 0$ for the function $f(x) = e^x$.

Solution

To find the Taylor polynomial of degree 3, we need the first three derivatives of $f(x)$ at $x = 0$. We find

$$f(x) = e^x \quad \text{so } f(0) = 1$$
$$f'(x) = e^x \quad \text{so } f'(0) = 1$$
$$f''(x) = e^x \quad \text{so } f''(0) = 1$$
$$f'''(x) = e^x \quad \text{so } f'''(0) = 1$$

Therefore,

$$P_3(x) = 1 + x + \frac{x^2}{2!} + \frac{x^3}{3!} = 1 + x + \frac{x^2}{2} + \frac{x^3}{6}$$

since $2! = (2)(1) = 2$ and $3! = (3)(2)(1) = 6$.

Our claim was that this polynomial would provide a better approximation to e^x than the linearization $1 + x$. We check this by evaluating e^x, $L(x)$, and $P_3(x)$ at a few values. The results are summarized in the following table.

x	e^x	$1 + x$	$1 + x + \dfrac{x^2}{2} + \dfrac{x^3}{6}$
-1	0.36788	0	0.3333
-0.1	0.90484	0.9	0.9048
0	1	1	1.0000
0.1	1.1052	1.1	1.1052
1	2.7183	2	2.6667

We see from the table that the third-degree Taylor polynomial provides a better approximation. Indeed, for x sufficiently close to 0, the values of $f(x)$ and $P_3(x)$ are very close. For instance,

$$f(0.1) = 1.105170918 \quad \text{and} \quad P_3(0.1) = 1.105166667$$

That is, their first five digits are identical. The error of approximation is

$$|f(0.1) - P_3(0.1)| = 4.25 \times 10^{-6}$$

which is quite small.

In Figure 7.41, we display the graphs of $f(x)$ and the Taylor polynomials $P_1(x)$, $P_2(x)$, and $P_3(x)$. We see from the graphs that the approximation is only good as long as x is close to 0. We also see that increasing the degree of the Taylor polynomial improves the approximation. ◀

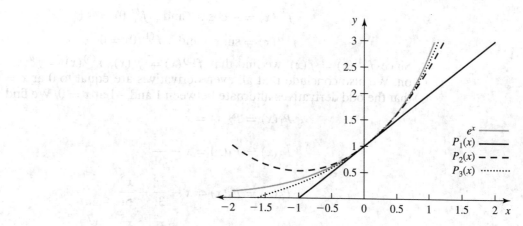

▲ **Figure 7.41**
The graph of $y = e^x$ and the first three Taylor polynomials

When we look at Example 1, we find that the successive Taylor polynomials for $f(x) = e^x$ about $x = 0$ are

$$P_0(x) = 1$$

$$P_1(x) = 1 + x$$

$$P_2(x) = 1 + x + \frac{x^2}{2!}$$

$$P_3(x) = 1 + x + \frac{x^2}{2!} + \frac{x^3}{3!}$$

The first thing we notice is that $P_1(x)$ is the linear approximation $L(x)$ that we found in (7.20).

The next thing we notice is that there is a pattern, and we might be tempted to guess the form of $P_n(x)$ for an arbitrary n. Our guess would be

$$P_n(x) = 1 + x + \frac{x^2}{2!} + \frac{x^3}{3!} + \cdots + \frac{x^n}{n!}$$

and this is indeed the case.

You might wonder why we bother to find an approximation for the function $f(x) = e^x$. To compute $f(1) = e$, for instance, it seems a lot easier to simply use a calculator. However, since $f(x) = e^x$ is not an algebraic function, the values of $f(x)$ cannot be found exactly using only the basic algebraic operations. Taylor polynomials are one way to evaluate such functions on a computer.

We now give additional functions for which we can find the Taylor polynomial of degree n.

▶ **Example 2** Compute the Taylor polynomial of degree n about $x = 0$ for the function $f(x) = \sin x$.

Solution

We begin by computing successive derivatives of $f(x) = \sin x$ at $x = 0$.

$$f(x) = \sin x \quad \text{and} \quad f(0) = 0$$

$$f'(x) = \cos x \quad \text{and} \quad f'(0) = 1$$

$$f''(x) = -\sin x \quad \text{and} \quad f''(0) = 0$$

$$f'''(x) = -\cos x \quad \text{and} \quad f'''(0) = -1$$

$$f^{(4)}(x) = \sin x \quad \text{and} \quad f^{(4)}(0) = 0$$

Since $f^{(4)}(x) = f(x)$, we find that $f^{(5)}(x) = f'(x)$, $f^{(6)}(x) = f''(x)$, and so on. We also conclude that all even derivatives are equal to 0 at $x = 0$, and that the odd derivatives alternate between 1 and -1 at $x = 0$. We find

$$P_1(x) = P_2(x) = x$$

$$P_3(x) = P_4(x) = x - \frac{x^3}{3!}$$

$$P_5(x) = P_6(x) = x - \frac{x^3}{3!} + \frac{x^5}{5!}$$

$$P_7(x) = P_8(x) = x - \frac{x^3}{3!} + \frac{x^5}{5!} - \frac{x^7}{7!}$$

and so on. To find the Taylor polynomial of degree n, we must find out how to write the last term. Note that the sign in front of successive terms alternates between plus and minus. To account for this alternating sign, we introduce the factor

$$(-1)^n = \begin{cases} 1 & \text{if } n \text{ is even} \\ -1 & \text{if } n \text{ is odd} \end{cases}$$

An odd number can be written as $2n + 1$ for an integer n. For a term of the form $\pm\frac{x^k}{k!}$ with k odd, we write

$$(-1)^n \frac{x^{2n+1}}{(2n+1)!} \tag{7.24}$$

where n is an integer. Inserting successive values of n into (7.24), we find the following (see table).

We see from the table that the term (7.24) produces successive terms in the Taylor polynomial for $f(x) = \sin x$. The Taylor polynomial of degree $2n + 1$ is

$$P_{2n+1}(x) = x - \frac{x^3}{3!} + \frac{x^5}{5!} - \frac{x^7}{7!} + \cdots + (-1)^n \frac{x^{2n+1}}{(2n+1)!} \quad \blacktriangleleft$$

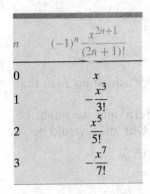

n	$(-1)^n \dfrac{x^{2n+1}}{(2n+1)!}$
0	x
1	$-\dfrac{x^3}{3!}$
2	$\dfrac{x^5}{5!}$
3	$-\dfrac{x^7}{7!}$

▶ **Example 3** Compute the Taylor polynomial of degree n about $x = 0$ for the function $f(x) = \frac{1}{1-x}, x \neq 1$.

Solution

We begin by computing successive derivatives of $f(x) = \frac{1}{1-x}$ at $x = 0$.

$$f(x) = \frac{1}{1-x} \qquad \text{so } f(0) = 1$$

$$f'(x) = \frac{1}{(1-x)^2} \qquad \text{so } f'(0) = 1$$

$$f''(x) = \frac{2}{(1-x)^3} \qquad \text{so } f''(0) = 2 = 2!$$

$$f'''(x) = \frac{(2)(3)}{(1-x)^4} \qquad \text{so } f'''(0) = (2)(3) = 3!$$

$$f^{(4)}(x) = \frac{(2)(3)(4)}{(1-x)^5} \qquad \text{so } f^{(4)}(0) = (2)(3)(4) = 4!$$

and so on. Continuing in this way, we find

$$f^{(k)}(x) = \frac{(2)(3)(4)\cdots(k)}{(1-x)^{k+1}} \quad \text{so } f^{(k)}(0) = k!$$

We obtain for the Taylor polynomial of degree n about $x = 0$:

$$P_n(x) = 1 + x + \frac{2!}{2!}x^2 + \frac{3!}{3!}x^3 + \frac{4!}{4!}x^4 + \frac{5!}{5!}x^5 + \cdots + \frac{n!}{n!}x^n$$

$$= 1 + x + x^2 + x^3 + \cdots + x^n$$

◀

Taylor approximations are widely used in biology. Here is an example that is already familiar to us.

▶ **Example 4**   Denote by $N(t)$ the size of a population at time t. A general model that describes the dynamics of this population is given by

$$\frac{dN}{dt} = f(N) \quad \text{with } f(0) = 0$$

Find the linear and the quadratic approximation of $f(N)$ about $N = 0$.

Solution

The linear approximation of $f(N)$ about $N = 0$ is the Taylor polynomial of degree 1:

$$P_1(N) = \underbrace{f(0)}_{=0} + f'(0)N$$

If we set $r = f'(0)$, then the first-order approximation of this growth model is

$$\frac{dN}{dt} = rN$$

which is the equation that describes exponential growth.

The quadratic approximation of $f(N)$ about $N = 0$ is the Taylor polynomial of degree 2:

$$P_2(N) = \underbrace{f(0)}_{=0} + f'(0)N + \frac{f''(0)}{2}N^2$$

Factoring $f'(0)N$ yields

$$P_2(N) = f'(0)N\left[1 + \frac{f''(0)}{2f'(0)}N\right]$$

If we set $r = f'(0)$ and $K = -\frac{2f'(0)}{f''(0)}$, then the second-order approximation of the growth model is

$$\frac{dN}{dt} = rN\left(1 - \frac{N}{K}\right)$$

which is the equation that describes logistic growth if K and r are positive. In either approximation, $r = f'(0)$ is the intrinsic rate of growth. ◀

7.7.2 The Taylor Polynomial about $x = a$

Thus far, we have only considered Taylor polynomials about $x = 0$. Because Taylor polynomials typically are good approximations only close to the point of approximation, it is useful to have approximations about points other than $x = 0$. We have already done this for linear approximations. For instance, the tangent line approximation of $f(x)$ at $x = a$ is

$$L(x) = f(a) + f'(a)(x - a) \qquad (7.25)$$

Note that $L(a) = f(a)$ and $L'(a) = f'(a)$. That is, the linear approximation and the original function, together with their first derivatives, agree at $x = a$. If we want to approximate $f(x)$ at $x = a$ by a polynomial of degree n, we then require that the polynomial and the original function, together with their first n derivatives, agree at $x = a$. This leads us to a polynomial of the form

$$P_n(x) = c_0 + c_1(x - a) + c_2(x - a)^2 + \cdots + c_n(x - a)^n \qquad (7.26)$$

Comparing (7.26) and (7.25), we conclude that $c_0 = f(a)$ and $c_1 = f'(a)$. To find the remaining coefficients, we proceed as in the case $a = 0$. That is, we differentiate $f(x)$ and $P_n(x)$ and require that their first n derivatives agree at $x = a$. We then arrive at the following formula.

> The Taylor polynomial of degree n about $x = a$ for the function $f(x)$ is given by
>
> $$P_n(x) = f(a) + f'(a)(x - a) + \frac{f''(a)}{2!}(x - a)^2$$
> $$+ \frac{f'''(a)}{3!}(x - a)^3 + \cdots + \frac{f^{(n)}(a)}{n!}(x - a)^n$$

▶ **Example 5** Find the Taylor polynomial of degree 3 for

$$f(x) = \ln x$$

at $x = 1$.

Solution

We need to evaluate $f(x)$ and its first three derivatives at $x = 1$. We find

$$f(x) = \ln x \quad \text{so } f(1) = 0$$

$$f'(x) = \frac{1}{x} \quad \text{so } f'(1) = 1$$

$$f''(x) = -\frac{1}{x^2} \quad \text{so } f''(1) = -1$$

$$f'''(x) = \frac{2}{x^3} \quad \text{so } f'''(1) = 2$$

Using the definition of the Taylor polynomial, we get

$$P_3(x) = 0 + (1)(x - 1) + \frac{(-1)}{2!}(x - 1)^2 + \frac{2}{3!}(x - 1)^3$$

$$= (x - 1) - \frac{1}{2}(x - 1)^2 + \frac{1}{3}(x - 1)^3$$

Figure 7.42 shows $f(x)$, the linear approximation $P_1(x) = x - 1$, and $P_3(x)$. We see that the approximation is good when x is close to 1 and that the approximation $P_3(x)$ is better than the linear approximation. ◀

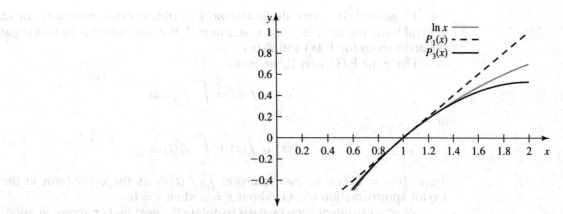

▲ **Figure 7.42**
The graph of $y = \ln x$, the linear approximation, and the Taylor polynomial of degree 3

7.7.3 How Accurate Is the Approximation? (Optional)

We saw in Example 1 that the approximation improved when the degree of the polynomial was higher. We will now investigate how accurate the Taylor approximation is. We can assess the accuracy of the approximation directly for the function in Example 3.

In Example 3, we showed that the Taylor polynomial of degree n about $x = 0$ for $f(x) = \frac{1}{1-x}$, $x \neq 1$, is

$$P_n(x) = 1 + x + x^2 + x^3 + \cdots + x^{n-1} + x^n \tag{7.27}$$

There is a nice trick to find an expression for the sum (7.27). Note that

$$x P_n(x) = x + x^2 + x^3 + \cdots + x^n + x^{n+1} \tag{7.28}$$

Subtracting (7.28) from (7.27), we find

$$P_n(x) - x P_n(x) = 1 + x + x^2 + \cdots + x^n - x - x^2 - \cdots - x^n - x^{n+1}$$

$$= 1 - x^{n+1}$$

that is,

$$(1 - x) P_n(x) = 1 - x^{n+1}$$

or

$$P_n(x) = \frac{1 - x^{n+1}}{1 - x} = \frac{1}{1 - x} - \frac{x^{n+1}}{1 - x}$$

provided $x \neq 1$. We therefore conclude that

$$|f(x) - P_n(x)| = \left| \frac{x^{n+1}}{1 - x} \right|$$

We can interpret the term $x^{n+1}/(1 - x)$ as the error of approximation. Since

$$\lim_{n \to \infty} \left| \frac{x^{n+1}}{1 - x} \right| = \begin{cases} \infty & \text{if } |x| > 1 \\ 0 & \text{if } |x| < 1 \end{cases}$$

we conclude that the error of approximation can be made small only when $|x| < 1$. For $|x| > 1$, the error of approximation increases with increasing n. [When $x = 1$, the function $f(x)$ is not defined.]

In general, it is not straightforward to obtain error estimates. In its general form, the error is given as an integral. We motivate this by looking at the error terms for $P_0(x)$ and $P_1(x)$.

Using the FTC, part II, we find

$$f(x) - f(a) = \int_a^x f'(t)\,dt$$

or

$$f(x) = f(a) + \int_a^x f'(t)\,dt$$

Since $f(a) = P_0(x)$, we can interpret $\int_a^x f'(t)\,dt$ as the error term in the Taylor approximation of $f(x)$ about $x = a$ when $n = 0$.

We can use integration by parts to obtain the next higher approximation:

$$\int_a^x f'(t)\,dt = \int_a^x 1 \cdot f'(t)\,dt$$

We set $u' = 1$ with $u = -(x - t)$, and $v = f'(t)$ with $v' = f''(t)$. [Writing $u = -(x - t)$ turns out to be a more convenient antiderivative of $u' = 1$ than $u = t$, as you will see shortly.] We find

$$\int_a^x f'(t)\,dt = -(x - t)f'(t)\Big]_a^x + \int_a^x (x - t)f''(t)\,dt$$

$$= (x - a)f'(a) + \int_a^x (x - t)f''(t)\,dt$$

That is,

$$f(x) = f(a) + f'(a)(x - a) + \int_a^x (x - t)f''(t)\,dt$$

The expression $f(a) + f'(a)(x - a)$ is the linear approximation $P_1(x)$; the integral can then be considered as the error term.

Continuing in this way, we find the general formula.

TAYLOR'S FORMULA

Suppose that $f : I \to \mathbf{R}$, where I is an interval, $a \in I$, and f and its first $n + 1$ derivatives are continuous at $a \in I$. Then, for $x, a \in I$,

$$f(x) = f(a) + \frac{f'(a)}{1!}(x - a) + \frac{f''(a)}{2!}(x - a)^2$$

$$+ \cdots + \frac{f^{(n)}(a)}{n!}(x - a)^n + R_{n+1}(x)$$

where

$$R_{n+1}(x) = \frac{1}{n!}\int_a^x (x - t)^n f^{(n+1)}(t)\,dt$$

We will now examine the error term in Taylor's formula more closely. The error term is given in integral form, and it is often difficult (or impossible) to evaluate the integral. We will first look at the case $n = 0$; that is, we approximate $f(x)$ by the constant function $f(a)$. The error term is then $R_{n+1}(x)$ when $n = 0$; that is,

$$R_1(x) = \int_a^x f'(t)\,dt$$

Using the mean value theorem for integrals, we can find a value c in the interval between a and x, such that

$$\int_a^x f'(t)\, dt = f'(c)(x - a)$$

That is, we find

$$R_1(x) = f'(c)(x - a)$$

for some number c between a and x. Though we don't know the value of c, this form is quite useful. Namely, if we set

$$K = \left[\begin{array}{c} \text{largest value of } |f'(t)| \\ \text{for } t \text{ between } a \text{ and } x \end{array}\right]$$

then

$$|R_1(x)| \le K|x - a|$$

Before we give the corresponding results for $R_{n+1}(x)$, we look at one example that illustrates how to find K.

▶ **Example 6**   Estimate the error in the approximation of $f(x) = e^x$ by $P_0(x)$ about $x = 0$ on the interval $[0, 1]$.

Solution

Since $f(0) = 1$, we find

$$P_0(x) = 1$$

and

$$f(x) = 1 + R_1(x)$$

with

$$R_1(x) = \int_0^x f'(t)\, dt = xf'(c)$$

for some c between 0 and x. Since $f'(t) = e^t$, the largest value of $|f'(t)|$ in the interval $[0, 1]$ occurs when $t = 1$, namely, $|f'(1)| = e$. Since we want to find an approximation of $f(x) = e^x$ for $x \in [0, 1]$, we should not use e in our error estimate, as e is one of the values we want to estimate. Instead, we use $|f'(t)| \le 3$ for $t \in [0, 1]$. We thus have

$$|R_1(x)| \le 3x \quad \text{for } x \in [0, 1] \qquad \blacktriangleleft$$

The error term for general n can be dealt with in a similar fashion, so that we find the following.

There exists a c between a and x such that the error term in Taylor's formula is of the form

$$R_{n+1}(x) = \frac{f^{(n+1)}(c)}{(n + 1)!}(x - a)^{n+1}$$

As in the case where $n = 0$, this form of the error term is quite useful. Though we don't know what c is, we can try to estimate $f^{(n+1)}(c)$ between a and x as before, when $n = 0$. Let

$$K = \left[\begin{array}{c} \text{largest value of } |f^{(n+1)}(t)| \\ \text{for } t \text{ between } a \text{ and } x \end{array} \right]$$

Then

$$|R_{n+1}(x)| \leq \frac{K|x-a|^{n+1}}{(n+1)!}$$

We will use this in the following example to determine in advance what degree of Taylor polynomial will allow us to achieve a given accuracy.

▶ **Example 7** Suppose that $f(x) = e^x$. What degree of Taylor polynomial about $x = 0$ will allow us to approximate $f(1)$ so that the error is less than 10^{-5}?

Solution

In Example 1, we found that for any $n \geq 1$,

$$f^{(n+1)}(t) = e^t$$

We need to find out how large $f^{(n+1)}(t)$ can get for $t \in [0, 1]$. We find

$$|f^{(n+1)}(t)| = e^t \leq e \quad \text{for } 0 \leq t \leq 1$$

As in Example 6, instead of using e as a bound, we use a slightly larger value, namely 3. Therefore,

$$|R_{n+1}(1)| \leq \frac{3|1|^{n+1}}{(n+1)!} = \frac{3}{(n+1)!} \tag{7.29}$$

We want the error to be less than 10^{-5}; that is, we want

$$|R_{n+1}(1)| < 10^{-5}$$

Inserting different values of n shows that

$$\frac{3}{8!} = 7.44 \times 10^{-5} \quad \text{and} \quad \frac{3}{9!} = 8.27 \times 10^{-6}$$

That is, when $n = 8$,

$$|R_{n+1}(1)| = |R_9(1)| \leq 8.27 \times 10^{-6} < 10^{-5}$$

Because the estimate of the error is greater than 10^{-5} when $n = 7$, we conclude that a polynomial of degree 8 would certainly give us the desired accuracy, whereas a polynomial of degree 7 might not. We can easily check this; we find

$$1 + 1 + \frac{1}{2!} + \frac{1}{3!} + \frac{1}{4!} + \cdots + \frac{1}{7!} = 2.71825396825$$

$$1 + 1 + \frac{1}{2!} + \frac{1}{3!} + \frac{1}{4!} + \cdots + \frac{1}{8!} = 2.71827876984$$

Comparing this to $e = 2.71828182846\ldots$, we see that the error is equal to 2.79×10^{-5} when $n = 7$ and 3.06×10^{-6} when $n = 8$. The error that we computed using (7.29) is a worst-case scenario; that is, the true error can be (and typically is) smaller than the error bound. ◀

We have already seen one example where a Taylor polynomial was useful only for values close to the point at which we approximated the function, regardless of n, the degree of the polynomial. There are situations in which the error in the approximation cannot be made small for *any* value close to the point of approximation, regardless of n, the degree of the polynomial. One such example is the continuous function

$$f(x) = \begin{cases} e^{-1/x} & \text{for } x > 0 \\ 0 & \text{for } x \le 0 \end{cases}$$

which is used, for instance, to describe the height of a tree as a function of age. We can show that $f^{(k)}(0) = 0$ for *all* $k \ge 1$, which implies that a Taylor polynomial of degree n about $x = 0$ is

$$P_n(x) = 0$$

for all n. This clearly shows that it will not help to increase n; the approximation just will not improve.

When we use Taylor polynomials to approximate functions, it is important to know for which values of x the approximation can be made arbitrarily close by choosing n large.

In the following, we list a few of the most important functions, together with their Taylor polynomials about $x = 0$ and the range of x values where the approximation can be made arbitrarily close by choosing n large enough.

$$e^x = 1 + x + \frac{x^2}{2!} + \frac{x^3}{3!} + \cdots + \frac{x^n}{n!} + R_{n+1}(x), \quad -\infty < x < \infty$$

$$\sin x = x - \frac{x^3}{3!} + \frac{x^5}{5!} - \frac{x^7}{7!} + \frac{x^9}{9!} - \cdots + (-1)^n \frac{x^{2n+1}}{(2n+1)!}$$

$$+ R_{2n+2}(x), \quad -\infty < x < \infty$$

$$\cos x = 1 - \frac{x^2}{2!} + \frac{x^4}{4!} - \frac{x^6}{6!} + \frac{x^8}{8!} - \cdots + (-1)^n \frac{x^{2n}}{(2n)!}$$

$$+ R_{2n+1}(x), \quad -\infty < x < \infty$$

$$\ln(1 + x) = x - \frac{x^2}{2} + \frac{x^3}{3} - \frac{x^4}{4} + \frac{x^5}{5} - \cdots + (-1)^{n+1} \frac{x^n}{n}$$

$$+ R_{n+1}(x), \quad -1 < x \le 1$$

$$\frac{1}{1-x} = 1 + x + x^2 + x^3 + x^4 + \cdots + x^n + R_{n+1}(x), \quad -1 < x < 1$$

7.7.4 Problems

(7.7.1)

In Problems 1–5, find the linear approximation of $f(x)$ at $x = 0$.

1. $f(x) = e^{2x}$
2. $f(x) = \sin x$
3. $f(x) = \dfrac{1}{1-x}$
4. $f(x) = x^2$
5. $f(x) = \ln(2 + x^2)$

In Problems 6–10, compute the Taylor polynomial of degree n about $a = 0$ for the indicated functions.

6. $f(x) = \dfrac{1}{1+x}, n = 4$
7. $f(x) = \cos x, n = 5$
8. $f(x) = e^{3x}, n = 3$
9. $f(x) = x^5, n = 6$
10. $f(x) = \sqrt{1+x}, n = 3$

In Problems 11–16, compute the Taylor polynomial of degree n about $a = 0$ for the indicated functions, and compare the value of the functions at the indicated point to the value of the corresponding Taylor polynomial.

11. $f(x) = \sqrt{2+x}, n = 3, x = 0.1$
12. $f(x) = \dfrac{1}{1-x}, n = 3, x = 0.1$
13. $f(x) = \sin x, n = 5, x = 1$
14. $f(x) = e^{-x}, n = 5, x = 0.3$
15. $f(x) = \tan x, n = 2, x = 0.1$
16. $f(x) = \ln(1 + x), n = 4, x = 0.2$
17. **(a)** Find the Taylor polynomial of degree 3 about $a = 0$ for $f(x) = \sin x$.

(b) Use your result in (a) to give an intuitive explanation why

$$\lim_{x \to 0} \frac{\sin x}{x} = 1$$

18. (a) Find the Taylor polynomial of degree 2 about $a = 0$ for $f(x) = \cos x$.

(b) Use your result in (a) to give an intuitive explanation why

$$\lim_{x \to 0} \frac{\cos x - 1}{x} = 0$$

(7.7.2)

In Problems 19–23, compute the Taylor polynomial of degree n about a, and compare the value of the approximation to the value of the function at the given point x.

19. $f(x) = \sqrt{x}, a = 1, n = 3; x = 2$

20. $f(x) = \ln x, a = 1, n = 3; x = 2$

21. $f(x) = \cos x, a = \frac{\pi}{6}, n = 3; x = \frac{\pi}{7}$

22. $f(x) = x^{1/3}, a = -1, n = 3; x = -0.9$

23. $f(x) = e^x, a = 2, n = 3; x = 2.1$

24. Show that

$$T^4 \approx T_a^4 + 4T_a^3(T - T_a)$$

for T close to T_a.

25. Show that

$$rN\left(1 - \frac{N}{K}\right) \approx rN$$

for N close to 0 (where r and K are positive constants).

26. (a) Show that

$$f(R) = \frac{aR}{k + R} \approx \frac{a}{k}R$$

for R close to 0 (where a and k are positive constants).

(b) Show that

$$f(R) = \frac{aR}{k + R} \approx \frac{a}{2} + \frac{a}{4k}(R - k)$$

for R close to k (where a and k are positive constants).

(7.7.3)

In Problems 27–30, use the following form of the error term

$$R_{n+1}(x) = \frac{f^{(n+1)}(c)}{(n+1)!}x^{n+1}$$

where c is between 0 and x, to determine in advance the degree of Taylor polynomial at $a = 0$ that would achieve the indicated accuracy in the interval $[0, x]$. (Do not compute the Taylor polynomial.)

27. $f(x) = e^x, x = 2,$ error $< 10^{-3}$

28. $f(x) = \sin x, x = 1,$ error $< 10^{-2}$

29. $f(x) = 1/(1 + x), x = 0.2,$ error $< 10^{-2}$

30. $f(x) = \ln(1 + x), x = 0.1,$ error $< 10^{-3}$

31. Let $f(x) = e^{-1/x}$ for $x > 0$ and $f(x) = 0$ for $x = 0$. Compute a Taylor polynomial of degree 2 at $x = 0$, and determine how large the error is.

32. We can show that the Taylor polynomial for $f(x) = (1 + x)^\alpha$ about $x = 0$, with α a positive constant, converges for $x \in (-1, 1)$. Show that

$$(1 + x)^\alpha = 1 + \alpha x + \frac{\alpha(\alpha - 1)}{2!}x^2$$

$$+ \frac{\alpha(\alpha - 1)(\alpha - 2)}{3!}x^3 + \cdots + R_{n+1}(x)$$

33. We can show that the Taylor polynomial for $f(x) = \tan^{-1} x$ about $x = 0$ converges for $|x| \le 1$.

(a) Show that the following is true:

$$\tan^{-1} x = x - \frac{x^3}{3} + \frac{x^5}{5} - \frac{x^7}{7} + \cdots + R_{n+1}(x)$$

(b) Explain why the following holds:

$$\frac{\pi}{4} = 1 - \frac{1}{3} + \frac{1}{5} - \frac{1}{7} + \cdots$$

(This series converges very slowly, as you would see if you used it to approximate π.)

7.8 Key Terms

Chapter 7 Review: Topics Discuss the following definitions and concepts.

1. The substitution rule for indefinite integrals
2. The substitution rule for definite integrals
3. Integration by parts
4. The "multiplying by 1" trick
5. Partial fraction decomposition
6. Partial fraction method
7. Proper rational function
8. Irreducible quadratic factor
9. Improper integral
10. Integration when the interval is unbounded
11. Integration when the integrand is discontinuous
12. Convergence and divergence of improper integrals
13. Comparison results for improper integrals
14. Numerical integration: midpoint and trapezoidal rule
15. Error bounds for the midpoint and the trapezoidal rule
16. Using tables of integrals for integration
17. Linear approximation
18. Taylor polynomial of degree n
19. Taylor's formula

7.9 Review Problems

In Problems 1–30, evaluate the given indefinite integrals.

1. $\displaystyle\int x^2(1 - x^3)^2 \, dx$

2. $\displaystyle\int \frac{\cos x}{1 + \sin^2 x} \, dx$

3. $\displaystyle\int 4xe^{-x^2} \, dx$

4. $\displaystyle\int \frac{x \ln(1 + x^2)}{1 + x^2} \, dx$

5. $\int (1 + \sqrt{x})^{1/3} \, dx$

6. $\int x\sqrt{3x + 1} \, dx$

7. $\int x \sec^2(3x^2) \, dx$

8. $\int \tan x \sec^2 x \, dx$

9. $\int x \ln x \, dx$

10. $\int x^3 \ln x^2 \, dx$

11. $\int \sec^2 x \ln(\tan x) \, dx$

12. $\int \sqrt{x} \ln \sqrt{x} \, dx$

13. $\int \dfrac{1}{4 + x^2} \, dx$

14. $\int \dfrac{1}{4 - x^2} \, dx$

15. $\int \tan x \, dx$

16. $\int \tan^{-1} x \, dx$

17. $\int e^{2x} \sin x \, dx$

18. $\int x \sin x \, dx$

19. $\int \sqrt{e^x} \, dx$

20. $\int \ln \sqrt{x} \, dx$

21. $\int \sin^2 x \, dx$

22. $\int \sin x \cos x \, e^{\sin x} \, dx$

23. $\int \dfrac{1}{x(x - 1)} \, dx$

24. $\int \dfrac{1}{(x + 1)(x - 2)} \, dx$

25. $\int \dfrac{x}{x + 5} \, dx$

26. $\int \dfrac{x}{x^2 + 5} \, dx$

27. $\int \dfrac{1}{x + 5} \, dx$

28. $\int \dfrac{1}{x^2 + 5} \, dx$

29. $\int \dfrac{(x + 1)^2}{x - 1} \, dx$

30. $\int \dfrac{2x + 1}{\sqrt{1 - x^2}} \, dx$

In Problems 31–50, evaluate the given definite integrals.

31. $\int_1^3 \dfrac{x^2 + 1}{x} \, dx$

32. $\int_0^{\pi/2} x \sin x \, dx$

33. $\int_0^1 x e^{-x^2/2} \, dx$

34. $\int_1^2 \ln x \, dx$

35. $\int_0^2 \dfrac{1}{4 + x^2} \, dx$

36. $\int_0^{1/2} \dfrac{2}{\sqrt{1 - x^2}} \, dx$

37. $\int_2^6 \dfrac{1}{\sqrt{x - 2}} \, dx$

38. $\int_0^2 \dfrac{1}{x - 2} \, dx$

39. $\int_0^\infty \dfrac{1}{9 + x^2} \, dx$

40. $\int_0^\infty \dfrac{1}{x^2 + 3} \, dx$

41. $\int_0^\infty \dfrac{1}{x + 3} \, dx$

42. $\int_0^\infty \dfrac{1}{(x + 3)^2} \, dx$

43. $\int_0^1 \dfrac{1}{x^2} \, dx$

44. $\int_1^\infty \dfrac{1}{x^2} \, dx$

45. $\int_0^1 \dfrac{1}{\sqrt{x}} \, dx$

46. $\int_1^\infty \dfrac{1}{\sqrt{x}} \, dx$

47. $\int_0^1 x \ln x \, dx$

48. $\int_0^1 x 2^x \, dx$

49. $\int_0^{\pi/4} e^{\cos x} \sin x \, dx$

50. $\int_0^{\pi/4} x \sin(2x) \, dx$

In Problems 51–54, use (a) the midpoint rule, and (b) the trapezoidal rule to approximate each integral with the specified value of n.

51. $\int_0^2 (x^2 - 1) \, dx, n = 4$

52. $\int_{-1}^1 (x^3 - 1) \, dx, n = 4$

53. $\int_0^1 e^{-x} \, dx, n = 5$

54. $\int_0^{\pi/4} \sin(4x) \, dx, n = 4$

In Problems 55–58, find the Taylor polynomial of degree n about $x = a$ for each function.

55. $f(x) = \sin(2x), a = 0, n = 3$

56. $f(x) = e^{-x^2/2}, a = 0, n = 3$

57. $f(x) = \ln x, a = 1, n = 3$

58. $f(x) = \dfrac{1}{x-3}, a = 4, n = 4$

59. (*Cost of gene substitution*)

(*Adapted from Roughgarden, 1996*) Suppose that an advantageous mutation arises in a population. Initially, the gene carrying this mutation is at a low frequency. As the gene carrying the mutation spreads through the population, the average fitness of the population increases. We denote by $f_{avg}(t)$ the average fitness of the population at time t, by $f_{avg}(0)$ the average fitness of the population at time 0 (when the mutation arose), and by K the final value of the average fitness after the mutation has spread through the population. Haldane (1957) suggested measuring the *cost of evolution* (now known as the *cost of gene substitution*) by the cumulative difference between the current and the final fitness; that is,

$$\int_0^\infty (K - f_{avg}(t))\, dt$$

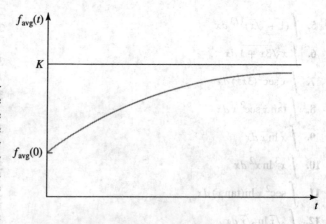

▲ **Figure 7.43**
The cost of gene substitution. See Problem 59

In Figure 7.43, shade the region whose area is equal to the cost of gene substitution.

DIFFERENTIAL EQUATIONS

In the simplest model of population growth, the growth rate at any time is proportional to the population size at that time. How would we express such a situation mathematically? Let $N(t)$ denote the size of a population at time t, $t \geq 0$. Then

$$\frac{dN}{dt} = rN(t), \quad t \geq 0 \tag{8.1}$$

(We introduced this equation in Section 4.6.) Equation (8.1) contains the derivative of a function and such equations are therefore called **differential equations**. Modeling biological situations frequently leads to differential equations.

Differential equations can contain any order derivative; for example,

$$\frac{d^2y}{dx^2} + \frac{dy}{dx} = xy$$

is a differential equation that contains the first and second derivative of the function $y = y(x)$. If a differential equation, such as (8.1), contains only the first derivative, it is called a **first-order differential equation**.

Throughout this chapter, we will restrict ourselves to first order differential equations of the form

$$\frac{dy}{dx} = f(x)g(y) \tag{8.2}$$

The right-hand side of (8.2) is the product of two functions; one depends only on x, the other only on y. Such equations are called **separable differential equations** (the reason for this name will become clear shortly). This type of differential equation includes two special cases, namely,

$$\frac{dy}{dx} = f(x) \tag{8.3}$$

and

$$\frac{dy}{dx} = g(y) \tag{8.4}$$

We discussed differential equations of the form (8.3) in Section 5.8. Differential equations of the form (8.4) include (8.1) and are frequently used in biological models.

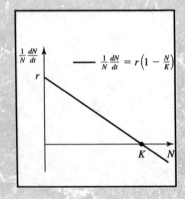

8.1 SOLVING DIFFERENTIAL EQUATIONS

Let's return to the growth model in (8.1), and, to be concrete, let's choose $r = 2$. This results in the equation

$$\frac{dN}{dt} = 2N(t), \quad t \geq 0 \tag{8.5}$$

We are interested in finding a function $N(t)$ that satisfies (8.5). Such a function is called a **solution** of the differential equation. We claim that with $N_0 = N(0)$,

$$N(t) = N_0 e^{2t}, \quad t \geq 0$$

is a solution of (8.5). To check this, we differentiate $N(t)$:

$$\frac{dN}{dt} = 2 \underbrace{N_0 e^{2t}}_{N(t)} = 2N(t)$$

At any point $(t, N(t))$ of the graph of $N(t)$, the slope is equal to $2N(t)$ (see Figure 8.1).

A first-order differential equation tells us what the derivative of a function is. In order to find a solution, we must therefore integrate. (Since it is not always possible to integrate a function, it is not always possible to write the solution of a differential equation in explicit form.)

We begin with a general method for solving differential equations of the form (8.2)

$$\frac{dy}{dx} = f(x)g(y) \tag{8.6}$$

We divide both sides of (8.6) by $g(y)$ [assuming $g(y) \neq 0$]:

$$\frac{1}{g(y)}\frac{dy}{dx} = f(x) \tag{8.7}$$

Now, if $y = u(x)$ is a solution of (8.7), then $u(x)$ satisfies

$$\frac{1}{g[u(x)]}u'(x) = f(x)$$

and, therefore,

$$\int \frac{1}{g[u(x)]}u'(x)\,dx = \int f(x)\,dx$$

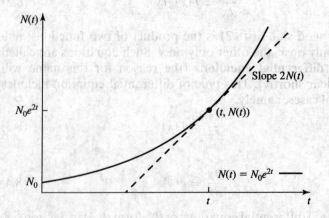

▲ **Figure 8.1**
The solution $N(t) = N_0 e^{2t}$: The slope at $(t, N(t))$ is $2N(t)$

But, since $g[u(x)] = g(y)$ and $u'(x)\,dx = dy$, this is the same as

$$\int \frac{1}{g(y)}\,dy = \int f(x)\,dx \qquad (8.8)$$

This suggests the following procedure of solving separable differential equations. Namely, we separate the variables x and y so that one side of the equation depends only on y and the other side only on x. To do so, we treat dy and dx as if they were regular numbers: Writing (8.6) formally as

$$\frac{dy}{g(y)} = f(x)\,dx$$

and then integrating both sides

$$\int \frac{dy}{g(y)} = \int f(x)\,dx$$

yields equation (8.8). The method of separating the variables x and y works since the right-hand side of (8.6) is of the special form $f(x)g(y)$, which gave this type of differential equation its name. A note of caution: We simply divided (8.6) by $g(y)$. We must be careful, since $g(y)$ might be 0 for some values of y, in which case we cannot divide. We will address this problem in Subsection 8.1.2. In the next two subsections, we discuss how to solve differential equations of the form (8.3) and (8.4); we give an example of the general type (8.2) in Subsection 8.1.3.

8.1.1 Pure Time Differential Equations

In many applications, the independent variable represents time. If the rate of change of a function depends only on time, we call the resulting differential equation a **pure time differential equation**. Such a differential equation is of the form

$$\frac{dy}{dx} = f(x), \ x \in I \qquad (8.9)$$

where I is an interval and x represents time. We discussed such equations in Section 5.8, where we found that the solution is of the form

$$y(x) = \int_{x_0}^{x} f(u)\,du + C \qquad (8.10)$$

The constant C comes from finding the general antiderivative of $f(x)$; the number x_0 is in the interval I. To determine C, we must phrase the problem as an initial value problem (see Section 5.8); if we assume $y(x_0) = y_0$, then plugging x_0 into (8.10) yields $y(x_0) = C$ and thus $C = y_0$. The solution can then be written as

$$y(x) = y_0 + \int_{x_0}^{x} f(u)\,du$$

To solve (8.9) formally, we separate variables; we write the differential equation (8.9) in the form

$$dy = f(x)\,dx$$

and then integrate both sides:

$$\int dy = \int f(x)\,dx$$

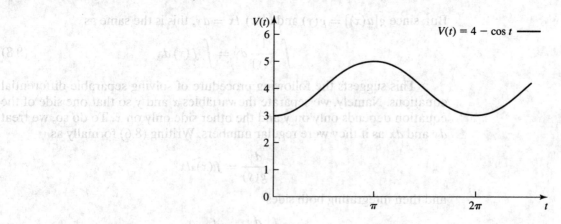

▲ **Figure 8.2**
The solution $V(t) = 4 - \cos t$ in Example 1

or

$$y(x) = \int f(x)\,dx$$

which is the same as (8.10).

▶ **Example 1**   Suppose that the volume $V(t)$ of a cell at time t changes according to

$$\frac{dV}{dt} = \sin t \quad \text{with } V(0) = 3$$

Find $V(t)$.

Solution

Since

$$V(t) = V(0) + \int_0^t \sin u\,du$$

we find

$$V(t) = 3 + [-\cos u]_0^t$$
$$= 3 + (-\cos t + \cos 0)$$
$$= 4 - \cos t$$

since $\cos 0 = 1$ (see Figure 8.2).   ◀

If we changed the initial condition in Example 1 to some other value, the graph of the new solution would be obtained from the old solution by shifting the solution vertically so that it would satisfy the new initial condition (see Figure 8.3). (This was discussed in Section 5.8.)

8.1.2 Autonomous Differential Equations

Many of the differential equations that model biological situations, such as (8.1), are of the form

$$\frac{dy}{dx} = g(y) \tag{8.11}$$

where the right-hand side of (8.11) does not explicitly depend on x. These equations are called **autonomous differential equations**.

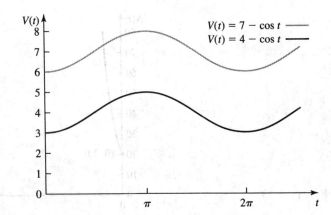

▲ **Figure 8.3**
The curve $V(t) = 7 - \cos t$ is obtained from $V(t) = 4 - \cos t$ by a vertical shift. The function $V(t) = 7 - \cos t$ solves the differential equation in Example 1 with $V(0) = 6$

To interpret the biological meaning of autonomous, let's return to the growth model

$$\frac{dN}{dt} = 2N(t) \tag{8.12}$$

We will see below that the general solution of (8.12) is

$$N(t) = Ce^{2t} \tag{8.13}$$

where C is a constant that can be determined if the population size at one time is known. Assume we conduct an experiment, where we follow a population over time, which satisfies (8.12) with $N(0) = 20$. Using (8.13), we find $N(0) = C = 20$. Then the size of the population at time t is given by

$$N(t) = 20e^{2t}$$

If we repeat the experiment at time $t = 10$ (say) with the exact same initial population size, then, everything else being equal, the population evolves in exactly the same way as the one starting at $t = 0$. Using (8.13) now with $N(10) = 20$, we find $N(10) = Ce^{20} = 20$ or $C = 20e^{-20}$. The size of the population is then given by

$$N(t) = 20e^{-20}e^{2t} = 20e^{2(t-10)}$$

The graph of this solution can be obtained from the previous graph, where $N(0) = 20$, by shifting the old graph 10 units to the right to the new starting point $(10, 20)$ (see Figure 8.4).

This means that a population starting with $N = 20$ follows the same trajectory, regardless of when we start the experiment. This makes sense biologically: If the growth conditions do not depend explicitly on time, the experiment should yield the same outcome regardless of when the experiment is performed. If the growth conditions for the population do change over time, we would not be able to use an autonomous differential equation to describe the growth of the population; we would need to explicitly include the time dependence in the dynamics.

Formally, we can solve (8.11) by separation of variables; namely, dividing both sides of (8.11) by $g(y)$ and multiplying both sides of (8.11) by dx, we obtain

$$\frac{dy}{g(y)} = dx$$

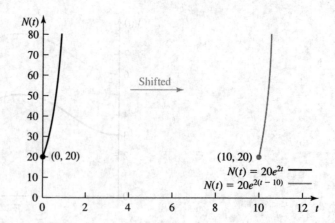

▲ **Figure 8.4**

The graph of the solution $N(t) = 20e^{2t}$ is shifted to the new starting point $(10, 20)$

Integrating both sides gives

$$\int \frac{dy}{g(y)} = \int dx$$

We will discuss two cases: $g(y) = k(y - a)$, and $g(y) = k(y - a)(y - b)$. The growth model (8.1) is an example of the first case; the logistic equation, which we saw in Section 3.3 for the first time, is an example of the second case.

Case 1: $g(y) = k(y - a)$ We wish to solve

$$\frac{dy}{dx} = k(y - a) \tag{8.14}$$

where k and a are constants. We assume $k \neq 0$. Separating variables yields

$$\frac{dy}{y - a} = k\, dx \tag{8.15}$$

where we need to assume $y \neq a$ in order to be able to divide by $y - a$. It is somewhat arbitrary whether we leave k on the right-hand side or move it to the left-hand side. Leaving it on the right-hand side is more convenient. Integrating both sides of (8.15) results in

$$\int \frac{dy}{y - a} = \int k\, dx$$

or

$$\ln |y - a| = kx + C_1$$

Exponentiating both sides yields

$$|y - a| = e^{kx + C_1}$$

or

$$|y - a| = e^{C_1} e^{kx}$$

Removing the absolute value signs, we find

$$y - a = \pm e^{C_1} e^{kx}$$

Renaming the constant by setting $C = \pm e^{C_1}$, we can write the solution as

$$y = Ce^{kx} + a \tag{8.16}$$

(Renaming the constant only serves the purpose to get the equation in a more readable form.) As in the case of a pure time differential equation, integration introduces a constant. If we know a point (x_0, y_0) of the solution, then C can be determined. We will refer to such a point as an **initial condition**.

To obtain (8.15), we divided by $y - a$. We are only allowed to do this as long as $y \neq a$. If $y = a$, then $dy/dx = 0$ and the constant function $y = a$ is a solution of (8.14). We *lost* this solution when we divided (8.14) by $y - a$. Note that the constant C in (8.16) is different from 0 since $C = \pm e^{C_1}$ and $e^{C_1} \neq 0$ for any real number C_1. But we can combine the constant solution $y = a$ and the solution in (8.16) by allowing C to be equal to 0 in (8.16).

Before we turn to biological applications, we give an example where we see how to solve such a differential equation and how to determine the value of C.

▶ **Example 2**   Solve

$$\frac{dy}{dx} = 2 - 3y \quad \text{where } y_0 = 1 \text{ when } x_0 = 1$$

Solution

Instead of trying to identify the constants C, k, and a in equation (8.16), it is easier to solve the equation directly. We separate variables and then integrate, which results in

$$\int \frac{dy}{2 - 3y} = \int dx$$

Since an antiderivative of $\frac{1}{2-3y}$ is $-\frac{1}{3} \ln |2 - 3y|$, we find

$$-\frac{1}{3} \ln |2 - 3y| = x + C_1$$

Solving for y yields

$$\ln |2 - 3y| = -3x - 3C_1$$
$$|2 - 3y| = e^{-3x - 3C_1}$$
$$2 - 3y = \pm e^{-3C_1} e^{-3x}$$

Setting $C = \pm e^{-3C_1}$, we find

$$2 - 3y = Ce^{-3x}$$

To determine C, we use the initial condition $y_0 = 1$ when $x_0 = 1$. That is,

$$2 - 3 = Ce^{-3} \quad \text{or} \quad C = -e^3$$

Hence,

$$2 - 3y = -e^3 e^{-3x}$$

or

$$y = \frac{2}{3} + \frac{1}{3} e^{3-3x}$$

(see Figure 8.5).   ◀

We will now turn to two biological applications that are covered by Case 1.

▶ **Example 3**   **(Exponential Population Growth)** This model is given by (8.1) and was introduced in Section 4.6. We denote by $N(t)$ the population size at time $t \geq 0$

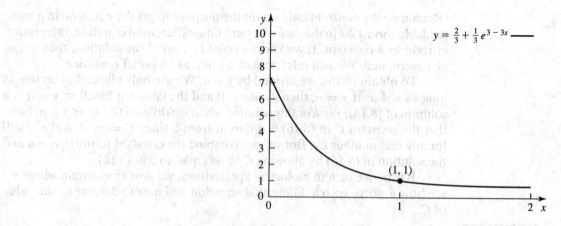

▲ **Figure 8.5**
The solution to Example 2

and assume $N(0) = N_0 > 0$. The change in population size is described by the initial value problem

$$\frac{dN}{dt} = rN \quad \text{where } N(0) = N_0 \tag{8.17}$$

The parameter r is called the intrinsic rate of growth and is the per capita rate of growth since

$$r = \frac{1}{N}\frac{dN}{dt}$$

When $r > 0$, this represents a growing population. When $r < 0$, the size of the population decreases. (Note that the per capita growth rate r is independent of the population size.)

We can either solve (8.17) directly as in Example 2, or use (8.14). Let's use (8.14). If we compare (8.17) with (8.14), we find $k = r$ and $a = 0$. Hence, using (8.16), we find the solution

$$N(t) = Ce^{rt}$$

Since $N(0) = N_0 = C$, we write the solution as

$$N(t) = N_0 e^{rt} \tag{8.18}$$

Equation (8.18) shows that the population grows exponentially when $r > 0$. When $r < 0$, the population size decreases exponentially fast. When $r = 0$, the population size stays constant.

We show solution curves of $N(t) = N_0 e^{rt}$ for $r > 0$, $r = 0$, and $r < 0$ in Figure 8.6. Exponential growth (or decay) is one of the most important growth equations in biology. You should therefore memorize both the differential equation (8.17) and its solution (8.18), together with the graphs in Figure 8.6, and know that (8.17) describes a situation where the per capita growth rate (or intrinsic rate of growth) is a constant.

When $r > 0$, the population size grows without bound ($\lim_{t \to \infty} N(t) = \infty$). This can be found when individuals are not limited by food availability or by competition. If we start a bacterial colony on a nutrient-rich substrate by inoculating the substrate with a few bacteria, then the bacteria, initially, can grow and divide unrestricted. Subsequently, when the substrate becomes more crowded and the food source depleted, the growth will be restricted, and a different differential equation will be required to describe

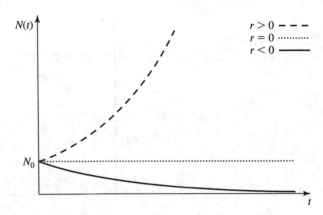

▲ Figure 8.6
Solution curves for $dN/dt = rN$

this situation. (We will discuss this in Case 2.) Exponential decay in a population can be seen when the death rate exceeds the birth rate (for instance, in cases of starvation). ◀

The type of growth in Example 3 is referred to as Malthusian growth, named after Thomas Malthus (1766–1834), a British clergyman and economist. Malthus wrote about the consequences of unrestricted growth on the welfare of humans. He observed that exponential growth of the human population would lead to starvation, because food supply would remain constant.

Recall from Section 4.6 that when $r < 0$, (8.17) has the same form as the differential equation that describes radioactive decay. $N(t)$ would then denote the amount of radioactive material left at time t. We will revisit this application in Problem 20.

▶ **Example 4** (Restricted Growth: von Bertalanffy Equation) This example describes the simplest form of restricted growth, and can be used to describe the growth of fish. We denote by $L(t)$ the length of the fish at age t and assume $L(0) = L_0$. Then

$$\frac{dL}{dt} = k(A - L) \tag{8.19}$$

where A is a positive constant that we will interpret in the following. We assume $L_0 < A$ and explain this restriction subsequently. The constant k is also positive; the equation says that the growth rate dL/dt is proportional to $A - L$, so k should be interpreted as a constant of proportionality. We see that the growth rate dL/dt is positive and decreases linearly with length as long as $L < A$, and that the growth stops (that is, $dL/dt = 0$) when $L = A$. To solve (8.19), we separate variables and integrate. This yields

$$\int \frac{dL}{A - L} = \int k\, dt$$

Hence,

$$-\ln |A - L| = kt + C_1$$

After multiplying this equation by -1 and exponentiating,

$$|A - L| = e^{-C_1} e^{-kt}$$

or

$$A - L = Ce^{-kt}$$

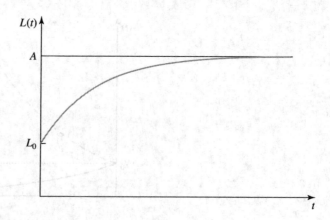

▲ **Figure 8.7**
The graph of the von Bertalanffy equation

with $C = \pm e^{-C_1}$. Since $L(0) = L_0$, we find

$$A - L_0 = C$$

The solution is then given by

$$A - L(t) = (A - L_0)e^{-kt}$$

or

$$L(t) = A\left[1 - \left(1 - \frac{L_0}{A}\right)e^{-kt}\right] \tag{8.20}$$

(see Figure 8.7).

This is the von Bertalanffy equation that we encountered previously. Since

$$\lim_{t \to \infty} L(t) = A$$

we see that the parameter A denotes the **asymptotic length** of the fish. Mathematically, there are no restrictions on L_0; biologically, however, we require that $0 < L_0 < A$. Otherwise, the growth rate would be negative, meaning that the fish shrinks in size. Note that A is an asymptotic length that is never reached, since there is no finite age T with $L(T) = A$ if $L(0) < A$.

Returning to (8.19), we can now interpret this differential equation: The growth rate is proportional to the difference in the current length and the asymptotic length, with k representing the constant of proportionality. Since the difference between the current length and the asymptotic length is decreasing over time, this also shows that the growth rate decreases over time, implying that juveniles grow at a faster rate than adults. Moreover, the growth rate is always positive. This means that fish grow throughout their lives, which is indeed the case. ◀

Case 2: $g(y) = k(y - a)(y - b)$ We now turn to differential equations of the form

$$\frac{dy}{dx} = k(y - a)(y - b) \tag{8.21}$$

where $k, a,$ and b are constant. We assume $k \neq 0$. Separating variables and integrating both sides yields

$$\int \frac{dy}{(y - a)(y - b)} = \int k\,dx \tag{8.22}$$

provided $y \neq a$ and $y \neq b$. When $a = b$, we must find an antiderivative of $\frac{1}{(y-a)^2}$, which is $-\frac{1}{y-a}$. In this case, we find

$$-\frac{1}{y-a} = kx + C$$

or

$$y = a - \frac{1}{kx + C}$$

The constant C can then be determined from the initial condition. When $y = a$, then $dy/dx = 0$ and consequently $y = $ constant (namely, $y = a$).

To find the solution when $a \neq b$, we must use the partial fraction method that we introduced in Section 7.3. Let's first do an example.

▶ **Example 5** Solve

$$\frac{dy}{dx} = 2(y-1)(y+2) \quad \text{with } y_0 = 2 \text{ when } x_0 = 0$$

Solution

Separation of variables yields

$$\int \frac{dy}{(y-1)(y+2)} = \int 2\, dx$$

We use the partial fraction method to integrate the left-hand side.

$$\frac{1}{(y-1)(y+2)} = \frac{A}{y-1} + \frac{B}{y+2}$$

$$= \frac{A(y+2) + B(y-1)}{(y-1)(y+2)}$$

$$= \frac{(A+B)y + 2A - B}{(y-1)(y+2)}$$

Comparing the last term to the integrand, we find

$$A + B = 0 \quad \text{and} \quad 2A - B = 1$$

which yields

$$A = -B \quad \text{and} \quad 2A - B = 3A = 1$$

and thus

$$A = \frac{1}{3} \quad \text{and} \quad B = -\frac{1}{3}$$

Using the partial fraction decomposition, we must integrate

$$\frac{1}{3}\int \left(\frac{1}{y-1} - \frac{1}{y+2} \right) dy = \int 2\, dx$$

which yields

$$\frac{1}{3}[\ln|y-1| - \ln|y+2|] = 2x + C_1$$

Simplifying this results in

$$\ln\left|\frac{y-1}{y+2}\right| = 6x + 3C_1$$

$$\left|\frac{y-1}{y+2}\right| = e^{3C_1}e^{6x}$$

$$\frac{y-1}{y+2} = \pm e^{3C_1}e^{6x}$$

$$\frac{y-1}{y+2} = Ce^{6x}$$

Using the initial condition $y_0 = 2$ when $x_0 = 0$, we find

$$\frac{1}{4} = C$$

The solution is therefore

$$\frac{y-1}{y+2} = \frac{1}{4}e^{6x}$$

If we want the solution in the form $y = f(x)$, we must solve for y. We find

$$y - 1 = (y+2)\frac{1}{4}e^{6x}$$

$$y\left(1 - \frac{1}{4}e^{6x}\right) = \frac{1}{2}e^{6x} + 1$$

$$y = \frac{\frac{1}{2}e^{6x} + 1}{1 - \frac{1}{4}e^{6x}}$$

$$y = \frac{2e^{6x} + 4}{4 - e^{6x}}$$

(see Figure 8.8).   ◀

We return now to (8.22) when $a \neq b$ and $y \neq a$ or b. We use the partial fraction method to simplify the integral on the left-hand side of (8.22). We decompose the integrand on the left-hand side of (8.22) into a sum of

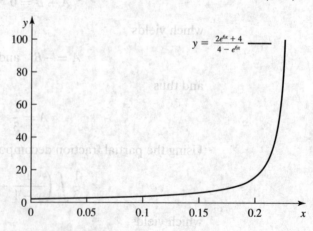

▲ **Figure 8.8**
The solution for Example 5

simpler rational functions that we know how to integrate; namely, we write the integrand in the form

$$\frac{1}{(y-a)(y-b)} = \frac{A}{y-a} + \frac{B}{y-b} \tag{8.23}$$

where A and B are constants that we must find. We do the following algebraic manipulation on the right-hand side of (8.23):

$$\frac{A}{y-a} + \frac{B}{y-b} = \frac{A(y-b) + B(y-a)}{(y-a)(y-b)}$$

$$= \frac{y(A+B) - (Ab + Ba)}{(y-a)(y-b)} \tag{8.24}$$

Comparing the last expression in (8.24) to the left-hand side of (8.23), we conclude that the constants A and B must satisfy

$$A + B = 0 \quad \text{and} \quad Ab + Ba = -1$$

Substituting $B = -A$ from the first equation into the second, we find

$$Ab - Aa = -1 \quad \text{or} \quad A(b-a) = -1 \quad \text{or} \quad A = \frac{1}{a-b}$$

and, therefore,

$$B = -\frac{1}{a-b}$$

That is,

$$\frac{1}{(y-a)(y-b)} = \frac{1}{a-b}\left[\frac{1}{y-a} - \frac{1}{y-b}\right] \tag{8.25}$$

Equation (8.25) allows us to evaluate

$$\int \frac{dy}{(y-a)(y-b)}$$

which is the left-hand side of (8.22). Namely,

$$\int \frac{dy}{(y-a)(y-b)} = \frac{1}{a-b}\int \left(\frac{1}{y-a} - \frac{1}{y-b}\right) dy$$

$$= \frac{1}{a-b}\left[\ln|y-a| - \ln|y-b|\right] + C_1$$

Integrating the right-hand side of (8.22) as well and combining the constants of integration into a new constant, C_2, we find

$$\frac{1}{a-b}\left[\ln|y-a| - \ln|y-b|\right] = kx + C_2$$

or

$$\ln\left|\frac{y-a}{y-b}\right| = k(a-b)x + C_2(a-b)$$

Exponentiating this yields

$$\left|\frac{y-a}{y-b}\right| = e^{C_2(a-b)}e^{k(a-b)x}$$

or

$$\frac{y-a}{y-b} = \pm e^{C_2(a-b)}e^{k(a-b)x}$$

Defining $C = \pm e^{C_2(a-b)}$, we find as the solution of (8.21)

$$\frac{y-a}{y-b} = Ce^{k(a-b)x}$$

We can solve this for y, namely,

$$y = a + (y-b)Ce^{k(a-b)x}$$

$$y(1 - Ce^{k(a-b)x}) = a - bCe^{k(a-b)x}$$

or

$$y = \frac{a - bCe^{k(a-b)x}}{1 - Ce^{k(a-b)x}} \tag{8.26}$$

The constant C can then be determined from the initial condition. When $y = a$ or b, then $dy/dx = 0$ and, consequently, $y = $ constant (namely, $y = a$ or b).

The following example of density-dependent growth is one of the most important applications of this case. In Example 3, we considered unrestricted (or density-*in*dependent) growth. It has the unrealistic feature that if the intrinsic rate of growth is positive, the size of a population grows without bound. Frequently, the per capita growth rate decreases as the population size increases. The growth rate thus depends on the population density. The simplest such model for which the growth rate is density dependent is the logistic equation. In the logistic equation, the size of the population cannot grow without bound.

The logistic equation was originally developed around 1835 by Pierre-François Verhulst, who used the term *logistic* for the equation. His work was completely forgotten until 1920, when Raymond Pearl and Lowell J. Reed published a series of papers on population growth. They used the same equation as Verhulst. After discovering Verhulst's work, Pearl and Reed adopted the name logistic for the equation. Pearl and Reed (1920) used the logistic equation to predict future growth of the U.S. population based on census data from 1790 to 1920. Their equation would have predicted about 185 million people in the United State for the year 2000, which is a gross underestimate of the actual population size (over 260 million people). Though the equation does not seem to fit actual populations very well, it is a useful model for analyzing growth under limiting resources.

▶ **Example 6** **(The Logistic Equation)** The logistic equation describes the change in size of a population for which per capita growth is density dependent. If we denote by $N(t)$ the population size at time t, then the change in growth is given by the initial value problem:

$$\frac{dN}{dt} = rN\left(1 - \frac{N}{K}\right) \quad \text{with } N(0) = N_0 \tag{8.27}$$

where r and K are positive constants. This is the simplest way of incorporating density dependence in the per capita growth rate, namely, it decreases linearly with population size (see Figure 8.9):

$$\frac{1}{N}\frac{dN}{dt} = r\left(1 - \frac{N}{K}\right)$$

We can interpret the parameter r as follows: The per capita growth rate is equal to r when $N = 0$. Therefore, one way to measure r is to grow the organism at a very low density, that is, when N is much smaller than K, so that

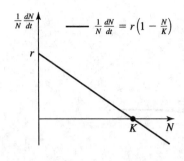

▲ Figure 8.9
The per capita growth rate in the logistic equation is a linearly decreasing function of population size

the per capita growth rate is close to r. (In Problem 42, we discuss a different way to find r.)

The quantity K is called the **carrying capacity**. Looking at Figure 8.9, we see that the per capita growth rate is 0 when the population size is at the carrying capacity. Since the per capita growth rate is positive below K and negative above K, the size of the population will increase below K and decrease above K. The number K thus determines the population size that can be supported by the environment. To show this mathematically, we need to solve (8.27). We will then see from the solution of (8.27) that starting from any positive initial population size the size of the population will eventually reach K [that is, $\lim_{t\to\infty} N(t) = K$ if $N(0) > 0$].

Let's solve (8.27). We write the right-hand side of (8.27) as

$$g(N) = -\frac{r}{K}(N-0)(N-K) \tag{8.28}$$

Comparing (8.28) with the right-hand side of (8.21), we find

$$k = -\frac{r}{K} \quad a = 0 \quad b = K \tag{8.29}$$

The solution of (8.27) is given in (8.26). Using (8.29), the solution of (8.27) is therefore

$$N(t) = \frac{0 - KCe^{-(r/K)(0-K)t}}{1 - Ce^{-(r/K)(0-K)t}}$$

or

$$N(t) = \frac{-CKe^{rt}}{1 - Ce^{rt}} = \frac{CK}{C - e^{-rt}}$$

$$= \frac{K}{1 - e^{-rt}/C} \tag{8.30}$$

Since $N(0) = N_0$, we find

$$N_0 = \frac{CK}{C - 1}$$

Solving this for C,

$$N_0(C - 1) = CK \quad \text{or} \quad C(N_0 - K) = N_0$$

thus

$$C = \frac{N_0}{N_0 - K} \tag{8.31}$$

Substituting (8.31) into (8.30), we find

$$N(t) = \frac{K}{1 - \frac{N_0 - K}{N_0}e^{-rt}}$$

or

$$N(t) = \frac{K}{1 + \left(\frac{K}{N_0} - 1\right)e^{-rt}} \tag{8.32}$$

It follows from (8.32) that

$$\lim_{t\to\infty} N(t) = K$$

We see from Figure 8.10, where we plot (8.32) for $N_0 > K$ and $N_0 < K$, that the solution $N(t)$ approaches K, the carrying capacity, as $t \to \infty$. When $N_0 > K$, then $N(t)$ approaches K from above; when $0 < N_0 < K$, then $N(t)$

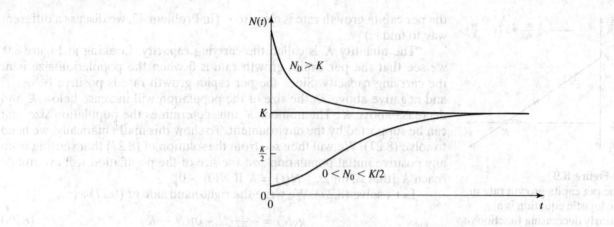

▲ Figure 8.10
Solution curves for different initial values N_0

approaches K from below. To use separation of variables to solve (8.27), we need to exclude $N = 0$ and $N = K$. When $N(0) = 0$, then $dN/dt = 0$ and, consequently, $N(t) = \text{constant} = N(0) = 0$ for all $t \geq 0$. When $N(0) = K$, then $dN/dt = 0$, and, consequently, $N(t) = \text{constant} = K$ for all $t \geq 0$. [The constant solution $N(t) = K$ is contained in (8.32) if we choose $N_0 = K$.]

The constant solutions K and 0 are called equilibria. The constant solution $N(t) = 0$ is not a very interesting one. We call it the trivial equilibrium. If $N(0) = 0$, then nothing happens, that is, $N(t)$ stays equal to 0 for all later times. This makes sense; if there aren't any individuals to begin with, then there won't be any later on.

We can show that if $0 < N_0 < \frac{K}{2}$, then the solution curve is S-shaped, as seen in Figure 8.10. This is characteristic of populations that show this type of density-dependent growth. At low densities, the growth is almost like unrestricted growth. At higher densities, the growth is restricted and the curve bends around and eventually levels off at the carrying capacity. If the population size is initially greater than the carrying capacity, the population size decreases and becomes asymptotically (that is, when $t \to \infty$) equal to the carrying capacity K. ◀

8.1.3 Allometric Growth

Solving differential equations of the form (8.2) gets complicated quickly. Before we show the role such differential equations play in biology, we will give an example in which we solve a differential equation of this type.

▶ Example 7 Solve

$$\frac{dy}{dx} = \frac{y+1}{x} \quad \text{with } y_0 = 0 \quad \text{when } x_0 = 1$$

Solution

To solve this differential equation, we separate variables first and then integrate. We find

$$\int \frac{dy}{y+1} = \int \frac{dx}{x}$$

Carrying out the integration on both sides, we obtain

$$\ln|y+1| = \ln|x| + C_1 \tag{8.33}$$

Solving for y, we find

$$|y + 1| = e^{C_1}|x|$$

$$y + 1 = \pm e^{C_1} x$$

Setting $C = \pm e^{C_1}$, we obtain

$$y = Cx - 1$$

Using the initial condition $y_0 = 0$ when $x_0 = 1$ allows us to determine C:

$$0 = C - 1 \quad \text{or} \quad C = 1$$

The solution is therefore

$$y = x - 1 \qquad \blacktriangleleft$$

We will now turn to a biological application, namely allometric growth. We have seen a number of allometric relationships throughout the earlier chapters of this book. These are typically relationships between sizes of parts of an organism (for instance, skull length and body length, or leaf area and stem diameter). We denote by $L_1(t)$ and $L_2(t)$ the sizes of two organs of an individual of age t. We say that L_1 and L_2 are related through an allometric law if their specific growth rates are proportional, that is, if

$$\frac{1}{L_1}\frac{dL_1}{dt} = k\frac{1}{L_2}\frac{dL_2}{dt} \tag{8.34}$$

If the constant k in (8.34) is equal to 1, then the growth is called isometric; otherwise it is called allometric. Cancelling dt on both sides of (8.34) and integrating, we find

$$\int \frac{dL_1}{L_1} = k \int \frac{dL_2}{L_2}$$

or

$$\ln |L_1| = k \ln |L_2| + C_1 \tag{8.35}$$

Solving for L_1, we obtain

$$L_1 = CL_2^k \tag{8.36}$$

where $C = \pm e^{C_1}$. (Since L_1 and L_2 are typically positive, the constant C will typically be positive.)

▶ Example 8　In a study of 45 species of unicellular algae, the relationship between cell volume (V) and cell biomass (B) was found to be

$$B \propto V^{0.794}$$

Find a differential equation that relates the relative growth rates of cell biomass and volume.

Solution

The relationship between cell biomass and volume can be expressed as

$$B(V) = CV^{0.794} \tag{8.37}$$

where C is the constant of proportionality. Comparing (8.37) with (8.36), we see that $k = 0.794$. It therefore follows from (8.34) that

$$\frac{1}{B}\frac{dB}{dt} = (0.794)\frac{1}{V}\frac{dV}{dt} \tag{8.38}$$

We can also get (8.38) by differentiating (8.37) with respect to V, that is,

$$\frac{dB}{dV} = C(0.794)V^{0.794-1}$$

Equation (8.37) allows us to eliminate C. Solving (8.37) for C, we find $C = BV^{-0.794}$ and, therefore,

$$\frac{dB}{dV} = BV^{-0.794}(0.794)V^{0.794-1}$$

$$= (0.794)BV^{-1}$$

Rearranging terms yields

$$\frac{dB}{B} = (0.794)\frac{dV}{V}$$

Dividing both sides by dt, we get

$$\frac{1}{B}\frac{dB}{dt} = (0.794)\frac{1}{V}\frac{dV}{dt}$$

which is the same as (8.38). ◀

▶ **Example 9** (Homeostasis) The nutrient content of a consumer (for instance, the percent nitrogen of the consumer's biomass) can range from reflecting the nutrient content of its food to being constant. The former is referred to as *absence of homeostasis*, the latter as *strict homeostasis*. A model for homeostatic regulation is provided in Sterner and Elser (2002). The model relates consumer's nutrient content (denoted by y) to its food's nutrient content (denoted by x) as follows:

$$\frac{dy}{dx} = \frac{1}{\theta}\frac{y}{x} \tag{8.39}$$

where $\theta \geq 1$ is a constant. Solve the differential equation and relate θ to absence of homeostasis and strict homeostasis.

Solution

We can solve (8.39) using separation of variables

$$\int \frac{dy}{y} = \frac{1}{\theta}\int \frac{dx}{x}$$

Integrating and simplifying yields

$$\ln|y| = \frac{1}{\theta}\ln|x| + C_1$$

$$|y| = e^{(1/\theta)\ln|x|+C_1}$$

$$|y| = |x|^{1/\theta}e^{C_1}$$

$$y = \pm e^{C_1}x^{1/\theta}$$

Since x and y are positive (they denote nutrient contents), we find

$$y = Cx^{1/\theta}$$

where C is a positive constant.

Absence of homeostasis refers to when the consumer reflects the food's nutrient content. This occurs when $y = Cx$ and thus when $\theta = 1$. *Strict homeostasis* means that the nutrient content of the consumer is independent of the nutrient content of the food, that is, $y = C$; this occurs in the limit $\theta \to \infty$. ◀

8.1.4 Problems

(8.1.1)

In Problems 1–8, solve each pure time differential equation.

1. $\frac{dy}{dx} = x + \sin x$, where $y_0 = 0$ for $x_0 = 0$

2. $\frac{dy}{dx} = e^{-x}$, where $y_0 = 10$ for $x_0 = 0$

3. $\frac{dy}{dx} = \frac{1}{x}$, where $y_0 = 0$ when $x_0 = 1$

4. $\frac{dy}{dx} = \frac{1}{1 + x^2}$, where $y_0 = 1$ when $x_0 = 0$

5. $\frac{dx}{dt} = \frac{1}{1 - t}$, where $x(0) = 2$

6. $\frac{dx}{dt} = \cos(t - 3)$, where $x(3) = 1$

7. $\frac{ds}{dt} = \sqrt{3t + 1}$, where $s(0) = 1$

8. $\frac{dh}{dt} = 5 - 16t^2$, where $h(3) = -11$

9. Suppose that the volume $V(t)$ of a cell at time t changes according to

$$\frac{dV}{dt} = \cos t \quad \text{with } V(0) = 5$$

Find $V(t)$.

10. Suppose that the amount of phosphorus in a lake at time t, denoted by $P(t)$, follows the equation

$$\frac{dP}{dt} = 3t + 1 \quad \text{with } P(0) = 0$$

Find the amount of phosphorus at time $t = 10$.

(8.1.2)

In Problems 11–16, solve the given autonomous differential equations.

11. $\frac{dy}{dx} = 3y$, where $y_0 = 2$ for $x_0 = 0$

12. $\frac{dy}{dx} = 2(1 - y)$, where $y_0 = 2$ for $x_0 = 0$

13. $\frac{dx}{dt} = -2x$, where $x(1) = 5$

14. $\frac{dx}{dt} = 1 - 3x$, where $x(-1) = -2$

15. $\frac{dh}{ds} = 2h + 1$, where $h(0) = 4$

16. $\frac{dN}{dt} = 5 - \frac{1}{2}N$, where $N(2) = 3$

17. Suppose that a population, whose size at time t is denoted by $N(t)$, grows according to

$$\frac{dN}{dt} = 0.3N(t) \quad \text{with } N(0) = 20$$

Solve this differential equation, and find the size of the population at time $t = 5$.

18. Suppose that you follow the size of a population over time. When you plot the size of the population versus time on a semilog plot (that is, the horizontal axis, representing time,

is on a linear scale; the vertical axis, representing the size of the population, is on a logarithmic scale), you find that your data fit a straight line that intercepts the vertical axis at 1 (log scale) and has slope -0.43. Find a differential equation that relates the growth rate of the population at time t to the size of the population at time t.

19. Suppose that a population, whose size at time t is denoted by $N(t)$, grows according to

$$\frac{1}{N}\frac{dN}{dt} = r \quad (8.40)$$

where r is a constant.

(a) Solve (8.40).

(b) Transform your solution in (a) appropriately so that the resulting graph is a straight line. How can you determine the constant r from your graph?

(c) Suppose now that you followed a population over time that evolved according to (8.40). Describe how you would determine r from your data.

20. Assume that $W(t)$ denotes the amount of radioactive material at time t. Radioactive decay is then described by the differential equation

$$\frac{dW}{dt} = -\lambda W(t) \quad \text{with } W(0) = W_0 \quad (8.41)$$

where λ is a positive constant, called the decay constant.

(a) Solve (8.41).

(b) Assume that $W(0) = 123\,\text{gr}$ and $W(5) = 20\,\text{gr}$ and that time is measured in minutes. Find the decay constant λ and determine the half-life of the radioactive substance.

21. Suppose that a population, whose size at time t is $N(t)$, grows according to

$$\frac{dN}{dt} = \frac{1}{100}N^2 \quad \text{with } N(0) = 10 \quad (8.42)$$

(a) Solve (8.42).

(b) Graph $N(t)$ as a function of t for $0 \leq t < 10$. What happens as $t \to 10$? Explain in words what this means.

22. Denote by $L(t)$ the length of a fish at time t and assume that the fish grows according to the von Bertalanffy equation

$$\frac{dL}{dt} = k(34 - L(t)) \quad \text{with } L(0) = 2 \quad (8.43)$$

(a) Solve (8.43).

(b) Use your solution in (a) to determine k under the assumption that $L(4) = 10$. Sketch the graph of $L(t)$ for this value of k.

(c) Find the length of the fish when $t = 10$.

(d) Find the asymptotic length of the fish, that is, find $\lim_{t \to \infty} L(t)$.

23. Denote by $L(t)$ the length of a certain fish at time t and assume that this fish grows according to the von Bertalanffy equation

$$\frac{dL}{dt} = k(L_\infty - L(t)) \quad \text{with } L(0) = 1 \quad (8.44)$$

where k and L_∞ are positive constants. A study showed that the asymptotic length is equal to 123 in. and that it takes this fish 27 months to reach half its asymptotic length.

(a) Use this information to determine the constants k and L_∞ in (8.44).

[*Hint:* Solve (8.44).]

(b) Determine the length of the fish after 10 months.

(c) How long will it take until the fish reaches 90% of its asymptotic length?

24. Let $N(t)$ denote the size of a population at time t. Assume that the population exhibits exponential growth.

(a) If you plot $\log N(t)$ versus t, what do you get?

(b) Find a differential equation that describes the growth of this population and sketch possible solution curves.

25. Use the partial fraction method to solve
$$\frac{dy}{dx} = y(1+y)$$
where $y_0 = 2$ for $x_0 = 0$.

26. Use the partial fraction method to solve
$$\frac{dy}{dx} = y(1-y)$$
where $y_0 = 2$ for $x_0 = 0$.

27. Use the partial fraction method to solve
$$\frac{dy}{dx} = y(y-5)$$
where $y_0 = 1$ for $x_0 = 0$.

28. Use the partial fraction method to solve
$$\frac{dy}{dx} = (y-1)(y-2)$$
where $y_0 = 0$ for $x_0 = 0$.

29. Use the partial fraction method to solve
$$\frac{dy}{dx} = 2y(3-y)$$
where $y_0 = 5$ for $x_0 = 1$.

30. Use the partial fraction method to solve
$$\frac{dy}{dt} = \frac{1}{2}y^2 - 2y$$
where $y_0 = -3$ for $t_0 = 0$.

In Problems 31–34, solve the given differential equations.

31. $\dfrac{dy}{dx} = y(1+y)$

32. $\dfrac{dy}{dx} = (1+y)^2$

33. $\dfrac{dy}{dx} = (1+y)^3$

34. $\dfrac{dy}{dx} = (3-y)(2+y)$

35. (a) Use the partial fraction method to show that
$$\int \frac{du}{u^2 - a^2} = \frac{1}{2a} \ln \left| \frac{u-a}{u+a} \right| + C$$

(b) Use your result in (a) to find a solution of
$$\frac{dy}{dx} = y^2 - 4$$
that passes through (i) $(0,0)$, (ii) $(0,2)$, and (iii) $(0,4)$.

36. Find a solution of
$$\frac{dy}{dx} = y^2 + 4$$
that passes through $(0,2)$.

37. Suppose that the size of a population at time t is denoted by $N(t)$ and that $N(t)$ satisfies the differential equation
$$\frac{dN}{dt} = 0.34N \left(1 - \frac{N}{200}\right) \quad \text{with } N(0) = 50$$
Solve this differential equation, and determine the size of the population in the long run; that is, find $\lim_{t\to\infty} N(t)$.

38. Assume that the size of a population, denoted by $N(t)$, evolves according to the logistic equation. Find the intrinsic rate of growth if the carrying capacity is 100, $N(0) = 10$, and $N(1) = 20$.

39. Suppose that $N(t)$ denotes the size of a population at time t and that
$$\frac{dN}{dt} = 1.5N \left(1 - \frac{N}{50}\right)$$

(a) Solve this differential equation when $N(0) = 10$.

(b) Solve this differential equation when $N(0) = 90$.

(c) Graph your solutions in (a) and (b) in the same coordinate system.

(d) Find $\lim_{t\to\infty} N(t)$ for your solutions in (a) and (b).

40. Suppose that the size of a population, denoted by $N(t)$, satisfies
$$\frac{dN}{dt} = 0.7N \left(1 - \frac{N}{35}\right) \tag{8.45}$$

(a) Determine all equilibria by solving $dN/dt = 0$.

(b) Solve (8.45) for (i) $N(0) = 10$, (ii) $N(0) = 35$, (iii) $N(0) = 50$, and (iv) $N(0) = 0$. Find $\lim_{t\to\infty} N(t)$ for each of the four initial conditions.

(c) Compare your answer in (a) to the limiting values you found in (b).

41. Let $N(t)$ denote the size of a population at time t. Assume that the population evolves according to the logistic equation. Furthermore, assume that the intrinsic growth rate is 5 and that the carrying capacity is 30.

(a) Find a differential equation that describes the growth of this population.

(b) Without solving the differential equation in (a), sketch solution curves of $N(t)$ as a function of t when (i) $N(0) = 10$, (ii) $N(0) = 20$, and (iii) $N(0) = 40$.

42. Logistic growth is described by the differential equation
$$\frac{dN}{dt} = rN \left(1 - \frac{N}{K}\right)$$
The solution of this differential equation with initial condition $N(0) = N_0$ is given by
$$N(t) = \frac{K}{1 + \left(\frac{K}{N_0} - 1\right)e^{-rt}} \tag{8.46}$$

(a) Show that
$$r = \frac{1}{t} \ln \left(\frac{K - N_0}{N_0}\right) + \frac{1}{t} \ln \left(\frac{N(t)}{K - N(t)}\right) \tag{8.47}$$
by solving (8.46) for r.

(b) The expression for r in (8.47) can be used to estimate r. Suppose we follow a population that grows according to the logistic equation, and find that $N(0) = 10$, $N(5) = 22$, $N(100) = 30$, and $N(200) = 30$. Estimate r.

43. (*Selection at a single locus*) We consider one locus with two alleles, A_1 and A_2, in a randomly mating diploid population. That is, each individual in the population is either of type $A_1 A_1$, $A_1 A_2$, or $A_2 A_2$. We denote by $p(t)$ the frequency of the A_1 allele and by $q(t)$ the frequency of the A_2 allele in the population a time t. Note that $p(t) + q(t) = 1$. We denote the fitness of the $A_i A_j$ type by w_{ij} and assume that $w_{11} = 1$, $w_{12} = 1 - s/2$, and $w_{22} = 1 - s$, where s is a nonnegative constant less than or equal to 1. That is, the fitness of the heterozygote $A_1 A_2$ is halfway between the fitness of the two homozygotes, and the type $A_1 A_1$ is the fittest. If s is small, we can show that approximately

$$\frac{dp}{dt} = \frac{1}{2}sp(1-p) \quad \text{with } p(0) = p_0 \qquad (8.48)$$

(a) Use separation of variables and the partial fraction method to find the solution of (8.48).

(b) Suppose $p_0 = 0.1$ and $s = 0.01$; how long will take until $p(t) = 0.5$?

(c) Find $\lim_{t \to \infty} p(t)$. Explain in words what this limit means. (8.1.3)

In Problems 44–52, solve each differential equation with the given initial condition.

44. $\dfrac{dy}{dx} = 2\dfrac{y}{x}$ with $y_0 = 1$ if $x_0 = 1$

45. $\dfrac{dy}{dx} = \dfrac{x+1}{y}$ with $y_0 = 2$ if $x_0 = 0$

46. $\dfrac{dy}{dx} = \dfrac{y}{x+1}$ with $y_0 = 2$ if $x_0 = 0$

47. $\dfrac{dy}{dx} = (y+1)e^{-x}$ with $y_0 = 2$ if $x_0 = 0$

48. $\dfrac{dy}{dx} = x^2 y^2$ with $y_0 = 1$ if $x_0 = 1$

49. $\dfrac{dy}{dx} = \dfrac{y+1}{x-1}$ with $y_0 = 5$ if $x_0 = 2$

50. $\dfrac{du}{dt} = \dfrac{\sin t}{u^2 + 1}$ with $u_0 = 3$ if $t_0 = 0$

51. $\dfrac{dr}{dt} = re^{-t}$ with $r_0 = 1$ if $t_0 = 0$

52. $\dfrac{dx}{dy} = \dfrac{1}{2}\dfrac{x}{y}$ with $x_0 = 2$ if $y_0 = 3$

53. (*Adapted from Reiss, 1989*) In a case study by Taylor et al. (1980), where the maximal rate of oxygen consumption (in ml s^{-1}) for nine species of wild African mammals was plotted against body mass (in kg) on a log-log plot, it was found that the data points fall on a straight line with slope approximately equal to 0.8. Find a differential equation that relates maximal oxygen consumption and body mass.

54. Consider the following differential equation, which is of importance in population genetics:

$$a(x)g(x) - \frac{1}{2}\frac{d}{dx}[b(x)g(x)] = 0$$

where $b(x) > 0$.

(a) Define $y = b(x)g(x)$, and show that y satisfies

$$\frac{a(x)}{b(x)}y - \frac{1}{2}\frac{dy}{dx} = 0 \qquad (8.49)$$

(b) Separate variables in (8.49), and show that if $y > 0$, then

$$y = C \exp\left[2\int \frac{a(x)}{b(x)}\,dx\right]$$

55. When phosphorus content in *Daphnia* was plotted against phosphorus content of its algal food on a log-log plot, a straight line with slope $1/7.7$ resulted (see Sterner and Elser, 2002; data from DeMott et al., 1998). Find a differential equation that relates phosphorus content of *Daphnia* to the phosphorus content of its algal food.

8.2 EQUILIBRIA AND THEIR STABILITY

In Subsection 8.1.2, we learned how to solve autonomous differential equations and graphed their solutions as functions of the independent variable for given initial conditions. For instance, logistic growth

$$\frac{dN}{dt} = rN\left(1 - \frac{N}{K}\right) \qquad (8.50)$$

with initial condition $N(0) = N_0$ has the solution given in (8.32) and graphed in Figure 8.10 for different initial values.

The solution of a differential equation can inform us about the long-term behavior, as we saw in the case of logistic growth. Namely, if $N_0 > 0$, then $N(t) \to K$, the carrying capacity, as $t \to \infty$, and if $N_0 = 0$, then $N(t) = 0$ for all $t > 0$. Also, if $N_0 = K$, then $N(t) = K$ for all $t > 0$. What is so special about $N_0 = K$ or $N_0 = 0$? We see from Equation (8.50) that if $N = K$ or $N = 0$, then $dN/dt = 0$, implying that $N(t)$ is constant.

Constant solutions form a very special class of solutions of autonomous differential equations. They are called **point equilibria** or simply equilibria.

The constant solutions $N = K$ and $N = 0$ are point equilibria of the logistic equation.

In this section, we will consider autonomous differential equations of the form

$$\frac{dy}{dx} = g(y) \qquad (8.51)$$

where we will typically think of x as time. We will learn how to find point equilibria, and we will discuss what they can tell us about the long-term behavior of the solution $y = y(x)$; that is, the behavior of $y(x)$ as $x \to \infty$. If we can solve (8.51), we can study the solution directly to obtain information about the long-term behavior. But what should we do if we cannot solve (8.51)?

Candidates for describing the long-term behavior are the constant solutions or equilibria $y = \hat{y}$ (read "y hat") that satisfy $g(\hat{y}) = 0$. (Such solutions, of course, need not exist.)

If $\hat{y}$ satisfies

$$g(\hat{y}) = 0$$

then $\hat{y}$ is an equilibrium of

$$\frac{dy}{dx} = g(y)$$

Let's look at equilibria in more detail before we discuss specific examples. The basic property of equilibria is that if, initially (say, at $x = 0$), $y(0) = \hat{y}$ and $\hat{y}$ is an equilibrium, then $y(x) = \hat{y}$ for all $x > 0$.

A physical analogue of equilibria is provided in Figure 8.11. On the left side, a ball rests on top of a hill; on the right side, a ball rests at the bottom of a valley. In either case, the ball is in equilibrium because it does not move.

Of great interest is the **stability** of equilibria. What we mean by this is best explained using our example of a ball on a hill versus a ball in the valley. If we perturb the ball by a small amount—that is, if we move it out of its equilibrium slightly—the ball on the left side will roll down the hill and not return to the top, whereas the ball on the right side will return to the bottom of the valley. We call the situation on the left side unstable and the situation on the right side stable.

The analogue of stability for equilibria of differential equations is as follows. Suppose that $\hat{y}$ is an equilibrium of $\frac{dy}{dx} = g(y)$, that is, $g(\hat{y}) = 0$. We say $\hat{y}$ is **locally stable** if the solution returns to $\hat{y}$ after a small perturbation; this means that we look at what happens to the solution when we start close to the equilibrium interpreting this as having moved the solution away from the equilibrium by a small amount—this is called a small perturbation. If the solution does not return after a small perturbation, we say $\hat{y}$ is **unstable**. These concepts will be developed in the following subsection, in which we will discuss a graphical and an analytical method for analyzing stability of equilibria. This is then followed by a number of applications.

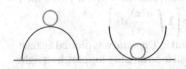

▲ **Figure 8.11**
Stability illustrated with a ball on a hill and in a valley

8.2.1 A First Look at Stability

Graphical Approach   Suppose that $g(y)$ is of the form given in Figure 8.12. To find the equilibria of (8.51), we set $g(y) = 0$. Graphically, this means that if we graph $g(y)$ (that is, the *derivative* of y with respect to x) as a function of y, then the equilibria are the points of intersection of $g(y)$ with the horizontal

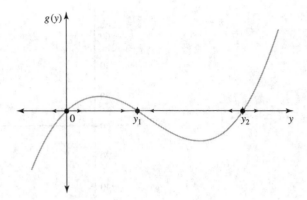

▲ **Figure 8.12**
The function $g(y)$. The arrows close to the equilibria indicate the type of stability

axis, which is the y-axis in this case since y is the independent variable. (Look at the labels of the axes in Figure 8.12 to see what is graphed in Figure 8.12.) We see in Figure 8.12 that for our choice of $g(y)$, the equilibria are at $y = 0$, y_1, and y_2.

Why does this work? Remember, we are discussing autonomous differential equations. This means that the derivative of y (dy/dx) is a function of y; it does *not* depend explicitly on x. This allows us to graph the derivative of y as a function of y. Since $dy/dx = g(y)$, this means graphing $g(y)$ as a function of y. We can then use the graph of $g(y)$ to say something about the fate of a solution based on its current value. Namely, if the current value y is such that $g(y) > 0$ (that is, $dy/dx > 0$), then y will increase as a function of x. If y is such that $g(y) < 0$ (that is, $dy/dx < 0$), then y will decrease as a function of x. The points y where $g(y) = 0$ are the points where y will not change as a function of x [since $g(y) = dy/dx = 0$]. These are the (point) equilibria.

Equilibria are characterized by the property that a system in an equilibrium state stays there for all later times (unless some external force disturbs the system). This does *not* imply that the system will necessarily reach a particular equilibrium when starting from some initial value that is different from the equilibrium, nor does it imply that it will return to the equilibrium after a small perturbation.

Whether or not the system will return to an equilibrium after a small perturbation depends on the local stability of the equilibrium. By this we mean the following. Suppose that the system is in equilibrium, which we denote by $\hat{y}$. We apply a small perturbation to the system, so that after the perturbation the new state of the system is

$$y = \hat{y} + z \tag{8.52}$$

where z is small and may be positive or negative. We explain what can happen using the function $g(y)$ in Figure 8.12.

Suppose that $\hat{y} = 0$ and we subject this equilibrium to a small perturbation. If the new value $y = \hat{y} + z = z > 0$ for z small, then $dy/dx > 0$, that is, y will increase (see arrow). If the new value $y = \hat{y} + z = z < 0$ for z small, then $dy/dx < 0$ and y will decrease (see arrow). In either case, the system will not return to 0. We say that $\hat{y} = 0$ is unstable. Drawing arrows on the horizontal axis helps determine stability.

We now turn to the equilibrium $\hat{y} = y_1$. We perturb the equilibrium to $y = y_1 + z$ for z small. Looking at Figure 8.12, if $z > 0$, then $dy/dx < 0$

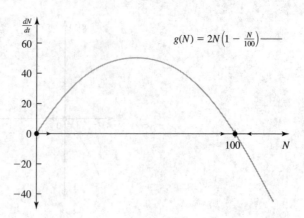

▲ Figure 8.13
The graph of dN/dt versus N in Example 1

and hence y decreases. If $z < 0$, then $dy/dx > 0$ and y increases. The system will therefore return to the equilibrium y_1 after a small perturbation. We say that y_1 is locally stable. The attribute *locally* refers to the fact that the system will return to y_1 if the perturbation is sufficiently small. It does not say anything about what happens when the perturbation is large. For instance, if the perturbation z is large and the new value is less than 0, then $dy/dx < 0$ and the system will not return to y_1.

Analyzing the equilibrium y_2 in the same way shows that it is an unstable equilibrium, just like the equilibrium $\hat{y} = 0$.

The preceding discussion illustrates that it is not necessarily the case that the system will reach an equilibrium value. The only equilibrium in Figure 8.12 that can be reached is y_1: If initially $y(0) \in (0, y_2)$, then $y(x)$ will approach y_1. If $y(0) < 0$, $y(x) \to -\infty$, and if $y > y_2$, then $y(x) \to \infty$.

The preceding discussion also illustrates that we can only subject an equilibrium to a small perturbation if we want to learn something about its stability: If we perturb y_1 too much so that the value after the perturbation is either less than 0 or greater than y_2, the solution will not return to the equilibrium value y_1.

▶ **Example 1** Let $N(t)$ denote the size of a population at time t that evolves according to the logistic equation

$$\frac{dN}{dt} = 2N\left(1 - \frac{N}{100}\right) \quad \text{for } N \geq 0$$

Find the equilibria and analyze their stability.

Solution

To find the equilibria, we set $\frac{dN}{dt} = 0$. That is,

$$2N\left(1 - \frac{N}{100}\right) = 0$$

We see that either

$$N_1 = 0 \quad \text{or} \quad N_2 = 100$$

To analyze the stability, we draw the graph of dN/dt versus N in Figure 8.13. Note that $N \geq 0$, since N represents the size of a population. To perturb the trivial equilibrium $N_1 = 0$, we therefore choose only values that are slightly bigger than 0. We see from the graph that such small perturbations result in

$dN/dt > 0$, and hence N increases (see arrow). This implies that $N_1 = 0$ is an unstable equilibrium.

To perturb the nontrivial equilibrium $N_2 = 100$, we can either increase or decrease the population size a bit. If we decrease it, then $dN/dt > 0$ and the population size will increase. If we increase it, then $dN/dt < 0$ and the population size will decrease. That is, if the system is subjected to a small perturbation about the nontrivial equilibrium $N_2 = 100$, the population size will return to the equilibrium value of $N_2 = 100$, as indicated by the arrows. This implies that $N_2 = 100$ is a locally stable equilibrium. ◀

Analytical Approach   We assume that $\hat{y}$ is an equilibrium of

$$\frac{dy}{dx} = g(y)$$

[thus $\hat{y}$ satisfies $g(\hat{y}) = 0$]. We consider a small perturbation about the equilibrium $\hat{y}$, which we express as

$$y = \hat{y} + z$$

where z is small and may be either positive or negative. Then

$$\frac{dy}{dx} = \frac{d}{dx}(\hat{y} + z) = \frac{dz}{dx}$$

since $d\hat{y}/dx = 0$ ($\hat{y}$ is a constant). We find

$$\frac{dz}{dx} = g(\hat{y} + z)$$

If z is sufficiently small, we can approximate $g(\hat{y} + z)$ by its linear approximation. The linear approximation of $g(y)$ about $\hat{y}$ is given by

$$L(y) = g(\hat{y}) + (y - \hat{y})g'(\hat{y})$$

Since $g(\hat{y}) = 0$,

$$L(y) = (y - \hat{y})g'(\hat{y})$$

Therefore, the linear approximation of $g(\hat{y} + z)$ is given by

$$L(\hat{y} + z) = (\hat{y} + z - \hat{y})g'(\hat{y}) = zg'(\hat{y})$$

If we set

$$\lambda = g'(\hat{y})$$

then

$$\frac{dz}{dx} = \lambda z$$

is the first-order approximation of the perturbation. It has the solution

$$z(x) = z(0)e^{\lambda x} \tag{8.53}$$

which has the property

$$\lim_{x \to \infty} z(x) = 0 \quad \text{if } \lambda < 0$$

Since $y(x) = \hat{y} + z(x)$, it follows that if $\lambda < 0$, the system returns to the equilibrium $\hat{y}$ after a small perturbation $z(0)$. This means that $\hat{y}$ is locally stable if $\lambda < 0$. On the other hand, if $\lambda > 0$, then $z(x)$ does not go to 0 as $x \to \infty$, implying that the system will not return to the equilibrium $\hat{y}$ after a small perturbation, and $\hat{y}$ is called unstable. The value $\lambda = g'(\hat{y})$ is called an **eigenvalue**; it is the slope of the tangent line of $g(y)$ at $\hat{y}$.

STABILITY CRITERION

Consider the differential equation

$$\frac{dy}{dx} = g(y)$$

where $g(y)$ is a differentiable function. Assume that $\hat{y}$ is an equilibrium, that is, $g(\hat{y}) = 0$. Then

$\hat{y}$ *is locally stable if* $g'(\hat{y}) < 0$

$\hat{y}$ *is unstable if* $g'(\hat{y}) > 0$

When the eigenvalue $\lambda = 0$, the first-order approximation $\frac{dz}{dx} = \lambda z$ does not allow us to draw any conclusions about the behavior of $z(x)$ since higher-order terms then become important.

We wish to tie this analytical approach in with the more informal graphical analysis in the beginning of this section. Looking at Figure 8.12, we find that the slope of the tangent line at $y = 0$ is positive, that is, $g'(0) > 0$; similarly, we find $g'(y_1) < 0$ and $g'(y_2) > 0$. Hence, the equilibria 0 and y_2 are unstable and the equilibrium y_1 is locally stable, as found in the graphical analysis.

Let's try the analytical approach out on Example 1. There

$$g(N) = 2N\left(1 - \frac{N}{100}\right)$$

To differentiate $g(N)$, we multiply out,

$$g(N) = 2N - \frac{N^2}{50}$$

Differentiating $g(N)$ yields

$$g'(N) = 2 - \frac{2N}{50}$$

If $N = 0$, then $g'(0) = 2 > 0$; thus $N = 0$ is an unstable equilibrium. If $N = 100$, then $g'(100) = 2 - 200/50 = -2 < 0$; thus $N = 100$ is a locally stable equilibrium.

The analytical approach is more powerful than the graphical approach: In addition to determining whether an equilibrium is locally stable or unstable, it allows us to say something about how quickly a solution returns to an equilibrium after a small perturbation. This follows from (8.53). There, we found that the perturbation has the approximate solution

$$z(x) = z(0)e^{\lambda x}$$

If $\lambda > 0$, then the larger λ, the faster the solution moves away from the equilibrium. If $\lambda < 0$, then the more negative λ, the faster the solution will return to the equilibrium after a small perturbation (see Figure 8.14).

The preceding derivation for the analytical stability criterion was based on linearizing $g(y)$ about the equilibrium $\hat{y}$. Since the linearization is only close for values close to $\hat{y}$ [unless $g(y)$ is linear], the perturbations about the equilibrium must be small, and hence the stability analysis is always local, that is, within close vicinity of the equilibrium. When $g(y)$ is linear, the analysis is exact; in particular, we can compute exactly how quickly a solution returns to a locally stable equilibrium (or moves away from an unstable equilibrium) after a perturbation.

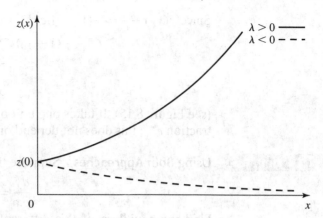

▲ **Figure 8.14**
The graph of $z(x) = z(0)e^{\lambda x}$ for $\lambda > 0$ and $\lambda < 0$

▶ **Example 2** $g(y)$ *is linear* Show that the differential equation

$$\frac{dy}{dx} = 1 - y \tag{8.54}$$

has a locally stable equilibrium at $\hat{y} = 1$, and determine how quickly a solution starting at $y(0) = y_0 \neq 1$ will reach $\hat{y} = 1$.

Solution

Since $g(y) = 0$ for $y = 1$, $\hat{y} = 1$ is an equilibrium. Differentiating $g(y) = 1 - y$, we find $g'(y) = -1$, which is negative regardless of y. Therefore, $\hat{y} = 1$ is a locally stable equilibrium.

We can solve (8.54) exactly using separation of variables:

$$\int \frac{dy}{1 - y} = \int dx$$

$$-\ln|1 - y| = x + C_1$$

$$1 - y = \pm e^{-C_1} e^{-x}$$

$$1 - y = Ce^{-x}$$

where we set $\pm e^{-C_1} = C$. Solving for y yields

$$y(x) = 1 - Ce^{-x}$$

If we set $y(0) = y_0$, then $y_0 = 1 - C$, and the solution is

$$y(x) = 1 - (1 - y_0)e^{-x}$$

We see that for any initial value y_0,

$$\lim_{x \to \infty} y(x) = 1$$

and it takes an infinite amount of time to reach the equilibrium $\hat{y} = 1$ if $y_0 \neq 1$.

Instead of asking how long it takes until the equilibrium is reached (which yielded the uninformative answer: an infinite amount of time), we compute the time it takes to reduce the initial deviation from the equilibrium, $y_0 - 1$, to a fraction e^{-1}; that is, we wish to determine the number x_R such that

$$\underbrace{y(x_R) - 1}_{\text{deviation from } x = x_R} = \underbrace{e^{-1}(y_0 - 1)}_{\text{initial deviation}}$$

Since $y(x_R) - 1 = -(1 - y_0)e^{-x_R}$, we find

$$-(1 - y_0)e^{-x_R} = e^{-1}(y_0 - 1)$$
$$e^{-x_R} = e^{-1}$$
$$x_R = 1$$

(see Figure 8.15). It takes one unit of time to reduce the initial deviation to a fraction e^{-1}; this does not depend on how large the initial deviation is. ◀

▶ **Example 3** **Using Both Approaches** Suppose that

$$\frac{dy}{dx} = y(4 - y)$$

Find the equilibria of this differential equation and discuss their stability, using both the graphical and the analytical approach.

Solution

To find the equilibria, we set $dy/dx = 0$. That is,

$$y(4 - y) = 0$$

which yields

$$y_1 = 0 \quad \text{and} \quad y_2 = 4$$

If we set $g(y) = y(4 - y)$ and graph $g(y)$ (see Figure 8.16), we see that $y_1 = 0$ is unstable, since if we perturb $y_1 = 0$ to $y = z > 0$ where z is small, then $g(y) = dy/dx > 0$, and if we perturb $y_1 = 0$ to $y = z < 0$ where z is small, then $g(y) = dy/dx < 0$. That is, in either case, y will not return to 0. On the other hand, y_2 is a locally stable equilibrium. A small perturbation to the right of $y = y_2$ results in $dy/dx < 0$, whereas a small perturbation to the left results in $dy/dx > 0$. In either case, the solution y will return to $y_2 = 4$.

If we use the analytical approach, we need to compute the eigenvalues. The eigenvalue associated with $y_1 = 0$ is

$$\lambda_1 = g'(0) = 4 - 2y|_{y=0} = 4 > 0$$

which implies that y_1 is unstable. The eigenvalue associated with $y_2 = 4$ is

$$\lambda_2 = g'(4) = 4 - 2y|_{y=4} = -4 < 0$$

which implies that $y_2 = 4$ is locally stable. ◀

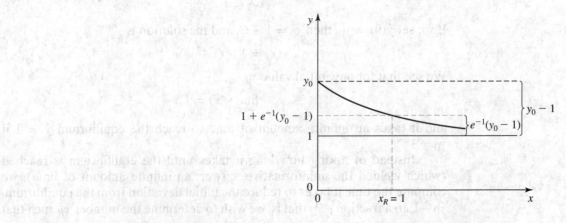

▲ **Figure 8.15**
An illustration of the time x_R in Example 2

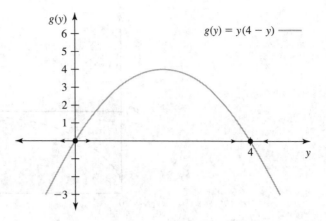

▲ Figure 8.16
The graph of $g(y)$ in Example 3

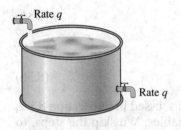

▲ Figure 8.17
Input and output rates are the same: The water in the tank remains at the same level

8.2.2 Single Compartment or Pool

This example is adapted from DeAngelis (1992). Compartment models are frequently used to model the flow of matter (nutrients, energy, and so forth). The simplest such model consists of one compartment, for instance, a fixed volume V of water (such as a tank or lake) containing a solute (such as phosphorus). Assume that water enters the compartment at a constant rate q and leaves the compartment at the same rate (see Figure 8.17). (Having the same input and output rate keeps the volume of the pool constant.) We will investigate the effects of different input concentrations of the solution on the concentration of the solution in the pool.

We denote by $C(t)$ the concentration of the solution in the compartment at time t. Then the total mass of the solute is $C(t)V$, where V is the volume of the compartment. For instance, if the concentration of the solution is 2 grams per liter and the volume of the compartment is 10 liters, then the total mass of the solute in the compartment is 2 g liters^{-1} times 10 liters, which is equal to 20 g.

If C_I is the concentration of the incoming solution and q is the rate at which water enters, then qC_I, the **input loading**, is the rate at which mass enters. For instance, if the concentration of the incoming solution is 5 g liters^{-1} and the rate at which the solution enters is 0.1 liter s^{-1}, then the input loading—that is, the rate at which mass enters—is 5 g liters^{-1} times 0.1 liter s^{-1}, which is equal to 0.5 g s^{-1}.

If we assume that the solution in the compartment is well mixed, so that the outflowing solution has the same concentration as the solution in the compartment—namely $C(t)$ at time t—then $qC(t)$ is the rate at which mass leaves the compartment at time t.

These different processes can be schematically illustrated in a flow diagram, as shown in Figure 8.18. Because mass is conserved in the system, we can use the **conservation of mass law** to derive an equation that describes the flow of matter in this system:

▲ Figure 8.18
Flow diagram for the single compartment model

$$\begin{bmatrix} \text{rate of change} \\ \text{of mass of} \\ \text{solute in pool} \end{bmatrix} = \begin{bmatrix} \text{rate at which} \\ \text{mass enters} \end{bmatrix} - \begin{bmatrix} \text{rate at which} \\ \text{mass leaves} \end{bmatrix}$$

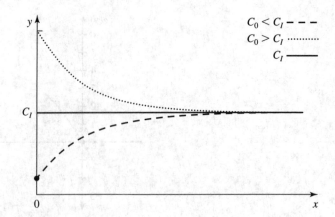

▲ **Figure 8.19**
The solution curves for the single compartment model for different values of C_0

or, writing C for $C(t)$ (and being careful not to confuse C with a constant),

$$\frac{d}{dt}(CV) = qC_I - qC \tag{8.55}$$

Since V is constant, we can write (8.55) as

$$\frac{dC}{dt} = \frac{q}{V}(C_I - C) \tag{8.56}$$

This is a linear differential equation of the type discussed in Subsection 8.1.2, Case 1. It can be solved using separation of variables. We skip the steps, to concentrate on the discussion of the system. If $C(0) = C_0$, then the solution of the differential equation is

$$C(t) = C_I\left[1 - \left(1 - \frac{C_0}{C_I}\right)e^{-(q/V)t}\right] \tag{8.57}$$

Solution curves for different values of C_0 are shown in Figure 8.19. We conclude from (8.56) that C_I is the only equilibrium. Looking at the solution $C(t)$ in (8.57), we see that

$$\lim_{t \to \infty} C(t) = C_I$$

regardless of the initial concentration C_0 in the compartment. This shows that C_I is **globally stable**. Global stability implies that no matter how much we perturb the equilibrium, the system will return to the equilibrium.

We can also obtain this result using the eigenvalue method. We write

$$C(t) = C_I + z(t)$$

Then, as before,

$$\frac{dC(t)}{dt} = \frac{d}{dt}[C_I + z(t)] = \frac{dz(t)}{dt}$$

that is,

$$\frac{dz}{dt} = -\frac{q}{V}z(t) \tag{8.58}$$

since $C_I - C(t) = -z(t)$. Note that (8.58) is exact, that is, we need not use a linear approximation for the right-hand side of (8.56), since it is already linear. We see from (8.58) that $-q/V$ is the eigenvalue associated with the equilibrium C_I. Since q and V are both positive, the eigenvalue is negative, and it follows that C_I is locally stable. We can obtain more information here, since (8.58) is exact with solution

$$z(t) = z(0)e^{-(q/V)t} \tag{8.59}$$

which shows that for *any* perturbation $z(0)$, $z(t) \to 0$ as $t \to \infty$. That is, the system returns to the equilibrium C_I [that is, $C(t) \to C_I$ as $t \to \infty$] regardless of how much we perturb the system. The reason that the eigenvalue method allows us to show global stability lies in the fact that the differential equation (8.55) is *linear*. In other cases, the eigenvalue method only allows us to obtain local stability, because we must first linearize and, consequently, the linearized differential equation is only an approximation.

Equation (8.59) shows that the system returns to the equilibrium C_I exponentially fast and that the eigenvalue $-(q/V)$ determines the time scale, that is, how quickly the system reaches the equilibrium. The larger q/V, the faster the system recovers from a perturbation. In this context, we can define a return time to equilibrium (just as in Example 2), denoted by T_R. By convention, the return time to equilibrium, T_R, is defined as the amount of time it takes to reduce the initial difference $C_0 - C_I$ to a fraction e^{-1}, that is,

$$C(T_R) - C_I = e^{-1}(C_0 - C_I) \tag{8.60}$$

Using (8.57), we see that

$$C(T_R) = C_I \left[1 - \left(1 - \frac{C_0}{C_I} \right) e^{-(q/V)T_R} \right]$$

which we can rewrite as

$$C(T_R) - C_I = (C_0 - C_I)e^{-(q/V)T_R} \tag{8.61}$$

Equating (8.60) and (8.61), we find

$$e^{-1}(C_0 - C_I) = (C_0 - C_I)e^{-(q/V)T_R}$$

which yields

$$1 = \frac{q}{V}T_R \quad \text{or} \quad T_R = \frac{V}{q}$$

That is, the return time increases with the volume V and decreases with the flow rate q. This should be intuitively clear; the larger the volume and the smaller the input rate, the longer it takes to return to equilibrium. We can show that T_R can also be interpreted as the mean residence time of a molecule of the solute, that is, T_R is the average time a molecule of the solute spends in the compartment before leaving, when the system is in equilibrium.

8.2.3 The Levins Model

The ecological importance of spatial structure to the maintenance of populations was pointed out by Andrewartha and Birch (1954) based on studies of insect populations. They observed that although local populations become frequently extinct, their patches become subsequently recolonized by migrants from other patches occupied by this population, thus allowing the population to persist globally. Fifteen years later, in 1969, Richard Levins introduced the concept of metapopulations (Levins, 1969). This was a major theoretical advance; this concept provided a theoretical framework for studying spatially structured populations.

A metapopulation is a collection of subpopulations. Each subpopulation occupies a patch, and different patches are linked via migration of individuals

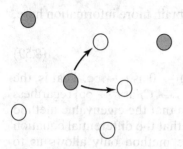

▲ Figure 8.20
A schematic description of a metapopulation model. The shaded patches are occupied; arrows indicate migration events

between patches (see Figure 8.20). In this setting, we only keep track of what proportion of patches are occupied by subpopulations. Subpopulations go extinct at a constant rate, denoted by m (m stands for mortality). Vacant patches can be colonized at a rate that is proportional to the fraction of occupied patches; the constant of proportionality is denoted by c (c stands for colonization rate). If we denote by $p(t)$ the fraction of patches that are occupied at time t, then writing $p = p(t)$,

$$\frac{dp}{dt} = cp(1 - p) - mp \tag{8.62}$$

The first term on the right hand-side describes the colonization process. Note that an increase in the fraction of occupied patches occurs only if a vacant patch becomes occupied; hence the product $p(1 - p)$ in the first term on the right-hand side. The minus sign in front of m shows that an extinction event decreases the fraction of occupied patches.

We will not solve (8.62), but focus on its equilibria. We set

$$cp(1 - p) - mp = 0$$

After factoring cp, we obtain

$$cp \left(1 - \frac{m}{c} - p\right) = 0$$

which has the two solutions:

$$p_1 = 0 \quad \text{and} \quad p_2 = 1 - \frac{m}{c}$$

We call the solution $p_1 = 0$ a trivial solution because it corresponds to the situation in which all patches are vacant. Since individuals are not created spontaneously, a vacant patch can be recolonized only through migration from other occupied patches. Therefore, once a metapopulation is extinct, it stays extinct. The other equilibrium $p_2 = 1 - m/c$ is only relevant when $p_2 \in (0, 1]$, because p represents a fraction that is a number between 0 and 1. Since m and c are both positive, it follows immediately that $p_2 < 1$ for all choices of m and c. To see when $p_2 > 0$, we check

$$1 - \frac{m}{c} > 0$$

which holds when

$$m < c$$

That is, the nontrivial equilibrium $p_2 = 1 - m/c$ is in $(0, 1]$ if the extinction rate m is less than the colonization rate c. If $m \geq c$, then there is only one equilibrium in $[0, 1]$, namely $p_1 = 0$. We illustrate this in Figures 8.21 and 8.22; looking at the graphs, we can analyze the stability of the equilibria.

Case 1: $m \geq c$. There is only the trivial equilibrium $p_1 = 0$. For any $p \in (0, 1]$, we see that $dp/dt < 0$; hence the fraction of occupied patches declines. The equilibrium is locally and globally stable.

Case 2: $m < c$. There are two equilibria, namely 0 and $1 - m/c$. The trivial equilibrium $p_1 = 0$ is now unstable, since if we perturb $p_1 = 0$ to some value in $(0, 1 - m/c)$, then $dp/dt > 0$, which implies that $p(t)$ increases. The system will therefore not return to 0.

The other equilibrium, $p_2 = 1 - m/c$, is locally stable. After a small perturbation of this equilibrium, to the right of p_2, $dp/dt < 0$, and to the left of p_2, $dp/dt > 0$; therefore, the system will return to p_2.

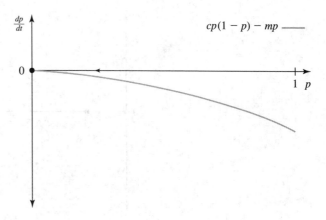

▲ Figure 8.21
The case $m \geq c$

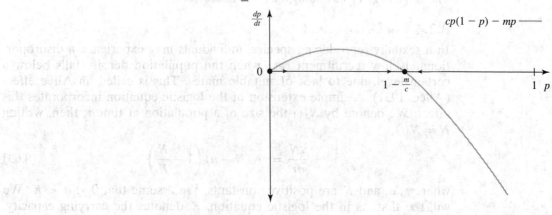

▲ Figure 8.22
The case $m < c$

We can also use the eigenvalue approach to analyze the stability of the equilibria. In addition, this will allow us to obtain information on how quickly the system returns to the stable equilibrium. We set

$$g(p) = cp(1 - p) - mp$$

To linearize this function about the equilibrium values, we must find

$$g'(p) = c - 2cp - m$$

Now, if $p_1 = 0$, then

$$g'(0) = c - m$$

whereas, if $p_2 = 1 - m/c$, then

$$g'\left(1 - \frac{m}{c}\right) = c - 2c\left(1 - \frac{m}{c}\right) - m = c - 2c + 2m - m = m - c$$

From this we see that $c - m$ is the eigenvalue corresponding to $p_1 = 0$ and $m - c$ is the eigenvalue corresponding to $p_2 = 1 - m/c$. We find that

if $c - m < 0$ then $p_1 = 0$ is locally stable

if $m - c < 0$ then $p_1 = 0$ is unstable and $p_2 = 1 - \frac{m}{c}$ is locally stable

To summarize our results, if $m > c$, then $p_1 = 0$ is the only equilibrium in [0, 1] and $p_1 = 0$ is locally stable. (In fact, the graphical analysis showed that $p_1 = 0$ is globally stable.) If $m < c$, then there are two equilibria in [0, 1]. The equilibrium $p_1 = 0$ is now unstable and the equilibrium $p_2 = 1 - m/c$ is locally stable.

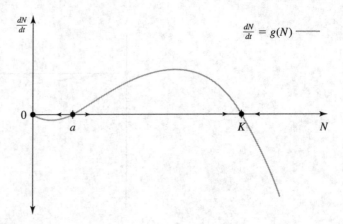

▲ Figure 8.23
The graph of $g(N)$ illustrating the Allee effect

8.2.4 The Allee Effect

In a sexually reproducing species, individuals may experience a disproportionately low recruitment rate when the population density falls below a certain level, due to lack of suitable mates. This is called an Allee effect (Allee, 1931). A simple extension of the logistic equation incorporates this effect. We denote by $N(t)$ the size of a population at time t; then, writing $N = N(t)$,

$$\frac{dN}{dt} = rN(N - a)\left(1 - \frac{N}{K}\right) \tag{8.63}$$

where r, a, and K are positive constants. We assume that $0 < a < K$. We will see that, as in the logistic equation, K denotes the carrying capacity. The constant a is a threshold population size, below which the recruitment rate is negative, meaning that the population will shrink and ultimately go to extinction.

The equilibria of (8.63) are given by $\hat{N} = 0, a$, and K. We set

$$g(N) = rN(N - a)\left(1 - \frac{N}{K}\right) = r\left(N^2 + \frac{a}{K}N^2 - \frac{N^3}{K} - aN\right)$$

A graph of $g(N)$ is shown in Figure 8.23. Differentiating $g(N)$ yields

$$g'(N) = r\left(2N + \frac{2a}{K}N - \frac{3N^2}{K} - a\right) = \frac{r}{K}(2NK + 2aN - 3N^2 - aK)$$

We can compute the eigenvalue $g'(\hat{N})$ associated with the equilibrium $\hat{N}$.

if $\hat{N} = 0$ then $g'(0) = \frac{r}{K}(-aK) < 0$

if $\hat{N} = a$ then $g'(a) = \frac{r}{K}a(K - a) > 0$

if $\hat{N} = K$ then $g'(K) = \frac{r}{K}K(a - K) < 0$

As we continue, you should compare the results from the eigenvalue method with the graph of $g(N)$.

Since $g'(0) < 0$, it follows that $\hat{N} = 0$ is locally stable. Likewise, since $g'(K) < 0$, it follows that $\hat{N} = K$ is locally stable. The equilibrium $\hat{N} = a$ is unstable since $g'(a) > 0$. This is also evident from Figure 8.23. This is an example where both stable equilibria are locally but not globally stable.

We see from Figure 8.23 that if $0 \leq N(0) < a$, then $N(t) \to 0$ as $t \to \infty$. If $a < N(0) \leq K$ or $N(0) \geq K$, then $N(t) \to K$ as $t \to \infty$. To interpret our results, we see that if the initial population $N(0)$ is too small [that is, $N(0) < a$], then the population goes extinct. If the initial population is large enough [that is, $N(0) > a$], the population persists. That is, the parameter a is a threshold level. The recruitment rate is only large enough when the population size exceeds this level.

8.2.5 Problems

(8.2.1)

1. Suppose that
$$\frac{dy}{dx} = y(2 - y)$$

(a) Find the equilibria of this differential equation.

(b) Graph dy/dx as a function of y, and use your graph to discuss the stability of the equilibria.

(c) Compute the eigenvalues associated with each equilibrium and discuss the stability of the equilibria.

2. Suppose that
$$\frac{dy}{dx} = (2 - y)(3 - y)$$

(a) Find the equilibria of this differential equation.

(b) Graph dy/dx as a function of y, and use your graph to discuss the stability of the equilibria.

(c) Compute the eigenvalues associated with each equilibrium and discuss the stability of the equilibria.

3. Suppose that
$$\frac{dy}{dx} = y(y - 1)(y - 2)$$

(a) Find the equilibria of this differential equation.

(b) Graph dy/dx as a function of y, and use your graph to discuss the stability of the equilibria.

(c) Compute the eigenvalues associated with each equilibrium and discuss the stability of the equilibria.

4. Suppose that
$$\frac{dy}{dx} = y(1 - y)(y - 2)$$

(a) Find the equilibria of this differential equation.

(b) Graph dy/dx as a function of y, and use your graph to discuss the stability of the equilibria.

(c) Compute the eigenvalues associated with each equilibrium and discuss the stability of the equilibria.

5. (*Logistic equation*) Assume that the size of a population evolves according to the logistic equation with intrinsic rate of growth $r = 1.5$. Assume that the carrying capacity $K = 100$.

(a) Find the differential equation that describes the rate of growth of this population.

(b) Find all equilibria, and discuss the stability of the equilibria using the graphical approach.

(c) Find the eigenvalues associated with the equilibria, and use the eigenvalues to determine stability of the equilibria. Compare your answers with your results in (b).

6. (*A simple model of predation*) Suppose that $N(t)$ denotes the size of a population at time t. The population evolves according to the logistic equation but, in addition, predation reduces the size of the population so that the rate of change is given by
$$\frac{dN}{dt} = N\left(1 - \frac{N}{50}\right) - \frac{9N}{5 + N} \qquad (8.64)$$
The first term on the right-hand side describes the logistic growth; the second term describes the effect of predation.

(a) Set
$$g(N) = N\left(1 - \frac{N}{50}\right) - \frac{9N}{5 + N}$$
and graph $g(N)$.

(b) Find all equilibria of (8.64).

(c) Use your graph in (a) to determine the stability of the equilibria found in (b).

(d) Use the method of eigenvalues to determine the stability of the equilibria found in (b).

7. (*Logistic equation*) Assume that the size of a population evolves according to the logistic equation with intrinsic rate of growth $r = 2$. Assume that $N(0) = 10$.

(a) Determine the carrying capacity K if the population grows fastest when the population size is 1000. (*Hint:* Show that the graph of dN/dt as a function of N has a maximum at $K/2$.)

(b) If $N(0) = 10$, how long will it take the population size to reach 1000?

(c) Find $\lim_{t \to \infty} N(t)$.

8. (*Logistic equation*) The logistic curve $N(t)$ is an S-shaped curve that satisfies
$$\frac{dN}{dt} = rN\left(1 - \frac{N}{K}\right) \quad \text{with } N(0) = N_0 \qquad (8.65)$$
when $N_0 < K$.

(a) Use the differential equation (8.65) to show that the inflection point of the logistic curve is at exactly half the saturation value of the curve. (*Hint:* do not solve (8.65); instead differentiate the right-hand side of (8.65) with respect to t.)

(b) The solution $N(t)$ of (8.65) can be defined for all $t \in \mathbf{R}$. Show that $N(t)$ is symmetric about the inflection point and that $N(0) = N_0$. That is, first use the solution of (8.65) that is given in (8.32), and find the time t_0 so that $N(t_0) = K/2$, that is, the inflection point is at $t = t_0$. Compute $N(t_0 + h)$ and $N(t_0 - h)$ for $h > 0$, and show that

$$N(t_0 + h) - \frac{K}{2} = \frac{K}{2} - N(t_0 - h)$$

Use a sketch of the graph of $N(t)$ to explain why this shows that $N(t)$ is symmetric about the inflection point $(t_0, N(t_0))$.

9. Suppose that a fish population evolves according to the logistic equation, and that a fixed number of fish per unit time are removed. That is,

$$\frac{dN}{dt} = rN\left(1 - \frac{N}{K}\right) - H$$

Assume in the following that $r = 2$ and $K = 1000$.

(a) Find possible equilibria, and discuss their stability when $H = 100$.

(b) What is the maximal harvesting rate that maintains a positive population size?

10. Suppose that a fish population evolves according to a logistic equation and that fish are harvested at a rate proportional to the population size. If $N(t)$ denotes the population size at time t, then

$$\frac{dN}{dt} = rN\left(1 - \frac{N}{K}\right) - hN$$

Assume that $r = 2$ and $K = 1000$.

(a) Find possible equilibria and discuss their stability when $h = 0.1$, using the graphical approach, and find the maximal harvesting rate that maintains a positive population size.

(b) Show that if $h < r = 2$, then there is a nontrivial equilibrium. Find the equilibrium.

(c) Analyze the stability of the equilibrium found in (b) using (i) the eigenvalue approach and (ii) the graphical approach.

(8.2.2)

11. Assume the single compartment model defined in Subsection 8.2.2: If $C(t)$ is the concentration of the solute at time t, then dC/dt is given by (8.56), that is,

$$\frac{dC}{dt} = \frac{q}{V}(C_I - C)$$

where q, V, and C_I are defined as in Subsection 8.2.2. Use graphical approach to discuss the stability of the equilibrium $\hat{C} = C_I$.

12. Assume the single-compartment model defined in Subsection 8.2.2; that is, denote by $C(t)$ the concentration of the solute at time t and assume that

$$\frac{dC}{dt} = 3(20 - C(t)) \quad \text{for } t \geq 0 \qquad (8.66)$$

(a) Solve (8.66) when $C(0) = 5$.

(b) Find $\lim_{t\to\infty} C(t)$.

(c) Use your answer in (a) to determine t so that $C(t) = 10$.

13. Assume the single-compartment model defined in Subsection 8.2.2; that is, denote by $C(t)$ the concentration of the solution at time t and assume that the concentration of the incoming solution is $3\,\text{g liters}^{-1}$ and the rate at which mass enters is $0.2\,\text{liter s}^{-1}$. Furthermore, assume that the volume of the compartment $V = 400$ liters.

(a) Find the differential equation for the rate of change of the concentration at time t.

(b) Solve the differential equation in (a) when $C(0) = 0$ and find $\lim_{t\to\infty} C(t)$.

(c) Find all equilibria of the differential equation and discuss their stability.

14. Suppose that a tank holds 1000 liters of water, and 2 kg of salt is poured into the tank.

(a) Compute the concentration of salt in g liter^{-1}.

(b) Assume now that you want to reduce the salt concentration. One method would be to remove a certain amount of the salt water from the tank and then replace it by pure water. How much salt water do you have to replace by pure water to obtain a salt concentration of $1\,\text{g liter}^{-1}$?

(c) Another method to reduce the salt concentration would be to hook up an overflow pipe and pump pure water into the tank. This way, the salt concentration would be gradually reduced. Assume that you have two pumps, one that pumps water at a rate of $1\,\text{liter s}^{-1}$, the other at a rate of $2\,\text{liters s}^{-1}$. For each pump, find out how long it would take to reduce the salt concentration from the original concentration to $1\,\text{g liter}^{-1}$, and how much pure water is needed in each case. (Note that the rate at which water enters the tank is equal to the rate at which water leaves the tank.) Compare the amount of water needed using the pumps with the amount of water needed in part (b).

15. Assume the single compartment model we introduced in Subsection 8.2.2. Denote by $C(t)$ the concentration at time t and assume

$$\frac{dC}{dt} = 0.37(254 - C(t)) \quad \text{for } t \geq 0$$

(a) Find the equilibrium concentration.

(b) Assume that the concentration is suddenly increased from the equilibrium concentration to 400. Find the return time to equilibrium, denoted by T_R, which is the amount of time until the initial difference is reduced to a fraction e^{-1}.

(c) Repeat (b) for the case when the concentration is suddenly increased from the equilibrium concentration to 800.

(d) Are the values for T_R computed in (b) and (c) different?

16. Assume the compartment model as in Subsection 8.2.2. Suppose that the equilibrium concentration is C_I, and the initial concentration is C_0. Express the time it takes until the initial deviation $C_0 - C_I$ is reduced to a fraction p in terms of T_R.

17. Assume the compartment model as in Subsection 8.2.2. Suppose that the equilibrium concentration is C_I. The time T_R has an integral representation which can be generalized to systems with more than one compartment. Show that

$$T_R = \int_0^\infty \frac{C(t) - C_I}{C(0) - C_I}\,dt$$

[*Hint:* Use (8.57) to show that

$$\frac{C(t) - C_I}{C(0) - C_I} = e^{-(q/V)t}$$

and integrate both sides with respect to t from 0 to ∞.]

18. Use the compartment model defined in Subsection 8.2.2 to investigate how the size of a lake influences nutrient dynamics in the lake after a perturbation. Mary Lake and Elizabeth Lake are two fictitious lakes in the North Woods that are used as experimental lakes to study nutrient dynamics. Mary Lake has a volume of $6.8 \times 10^3 \, m^3$, Elizabeth Lake has twice this volume, namely, $13.6 \times 10^3 \, m^3$. Both lakes have the same inflow/outflow rate $q = 170$ liters s^{-1}. Because both lakes share the same drainage area, the concentration C_I of the incoming solute is the same for both lakes, namely, $C_I = 0.7$ mg liter^{-1}. Assume that in the beginning of the experiment both lakes are in equilibrium, that is, the concentration of the solution in both lakes is 0.7 mg liter^{-1}. Your experiment consists of increasing the concentration of the solution by 10% in each lake at time 0 and then watching how the concentration of the solution in each lake changes with time. Assume the single compartment model to make predictions on how the concentration of the solution will evolve. (Note that $1 \, m^3$ of water corresponds to 1000 liters of water.)

(a) Find the initial concentration C_0 of the solution in each lake at time 0 (that is, immediately after the 10% increase in solution concentration).

(b) Use Equation (8.57) to determine how the concentration of the solution changes over time in each lake. Graph your results.

(c) Which lake returns to equilibrium faster? Compute the return time to equilibrium, T_R, for each lake, and explain how this is related to the eigenvalues corresponding to the equilibrium concentration C_I for each lake.

19. Use the compartment model defined in Subsection 8.2.2 to investigate the effect of an increase in the input concentration C_I on the nutrient concentration in a lake. Suppose a lake in a pristine environment has an equilibrium phosphorus concentration of 0.3 mg^{-1}. The volume V of the lake is $12.3 \times 10^6 \, m^3$ and the inflow/outflow rate q is equal to 220 liters s^{-1}. Conversion of land in the drainage area of the lake to agricultural use has increased the input concentration from 0.3 mg liter^{-1} to 1.1 mg liter^{-1}. Assume that this increase happened instantaneously. Compute the return time to the new equilibrium, denoted by T_R, in days, and find the nutrient concentration in the lake T_R units of time after the change in input concentration. (Note that $1 \, m^3$ of water corresponds to about 1000 liters of water.)

(8.2.3)

20. (*Levins model*) Denote by $p = p(t)$ the fraction of occupied patches in a metapopulation model, and assume that

$$\frac{dp}{dt} = 2p(1-p) - p \quad \text{for } t \geq 0 \qquad (8.67)$$

(a) Set $g(p) = 2p(1-p) - p$. Graph $g(p)$ for $p \in [0,1]$.

(b) Find all equilibria of (8.67) that are in $[0,1]$, and determine their stability using your graph in (a).

(c) Use the eigenvalue approach to analyze the stability of the equilibria that you found in (b).

21. (*Levins model*) Denote by $p = p(t)$ the fraction of occupied patches in a metapopulation model, and assume that

$$\frac{dp}{dt} = 0.5p(1-p) - 1.5p \quad \text{for } t \geq 0 \qquad (8.68)$$

(a) Set $g(p) = 0.5p(1-p) - 1.5p$. Graph $g(p)$ for $p \in [0,1]$.

(b) Find all equilibria of (8.68) that are in $[0,1]$, and determine their stability using your graph in (a).

(c) Use the eigenvalue approach to analyze the stability of the equilibria that you found in (b).

22. (*A metapopulation model with density-dependent extinction*) Denote by $p = p(t)$ the fraction of occupied patches in a metapopulation model, and assume that

$$\frac{dp}{dt} = cp(1-p) - p^2 \quad \text{for } t \geq 0 \qquad (8.69)$$

where $c > 0$. The term p^2 describes density-dependent extinction of patches, namely the per patch extinction rate is p and a fraction p of patches are occupied, resulting in an extinction rate of p^2. The colonization of vacant patches is the same as in Levins model.

(a) Set $g(p) = cp(1-p) - p^2$ and sketch the graph of $g(p)$.

(b) Find all equilibria of (8.69) in $[0,1]$, and determine their stability.

(c) Is there a nontrivial equilibrium when $c > 0$? Contrast your findings with the corresponding results in Levins model.

23. (*Habitat destruction*) In Subsection 8.2.3, we introduced Levins model. To study the effects of habitat destruction on a single species, we modify Equation (8.62) in the following way. We assume that a fraction D of patches is permanently destroyed. Consequently, only patches that are vacant and undestroyed can be successfully colonized. These patches have frequency $1 - p(t) - D$ if $p(t)$ denotes the fraction of occupied patches at time t. Then

$$\frac{dp}{dt} = cp(1-p-D) - mp \qquad (8.70)$$

(a) Explain in words the meaning of the different terms in (8.70).

(b) Show that there are two possible equilibria, the trivial equilibrium $p_1 = 0$ and the nontrivial equilibrium $p_2 = 1 - D - \frac{m}{c}$. Sketch the graph of p_2 as a function of D.

(c) Assume that $m < c$, such that the nontrivial equilibrium is stable when $D = 0$. Find a condition for D such that the nontrivial equilibrium is between 0 and 1, and investigate the stability of both the nontrivial equilibrium and the trivial equilibrium under this condition.

(d) Assume the condition that you derived in (c), that is, the nontrivial equilibrium is between 0 and 1. Show that when the system is in equilibrium, the fraction of patches that are vacant and undestroyed, that is, the sites that are *available* for colonization, is independent of D. Show that the **effective colonization rate** in equilibrium, that is, c times the fraction of available patches, is equal to the extinction rate. This shows that the effective birth rate of new colonies balances their extinction rate at equilibrium.

(8.2.4)

24. (*Allee effect*) Denote by $N(t)$ the size of a population at time t, and assume that

$$\frac{dN}{dt} = 2N(N - 10)\left(1 - \frac{N}{100}\right) \quad \text{for } t \geq 0 \qquad (8.71)$$

(a) Find all equilibria of (8.71).

(b) Use the eigenvalue approach to determine the stability of the equilibria found in (a).

(c) Set

$$g(N) = 2N(N - 10)\left(1 - \frac{N}{100}\right)$$

for $N \geq 0$ and graph $g(N)$. Identify the equilibria of (8.71) on your graph, and use your graph to determine the stability of the equilibria. Compare your results with your findings in (b). Use your graph to give a graphical interpretation of the eigenvalues associated with the equilibria.

25. (*Allee effect*) Denote by $N(t)$ the size of a population at time t, and assume that

$$\frac{dN}{dt} = 0.3N(N - 17)\left(1 - \frac{N}{200}\right) \quad \text{for } t \geq 0 \qquad (8.72)$$

(a) Find all equilibria of (8.72).

(b) Use the eigenvalue approach to determine the stability of the equilibria found in (a).

(c) Set

$$g(N) = 0.3N(N - 17)\left(1 - \frac{N}{200}\right)$$

for $N \geq 0$ and graph $g(N)$. Identify the equilibria of (8.72) on your graph, and use your graph to determine the stability of the equilibria. Compare your results with your findings in (b). Use your graph to give a graphical interpretation of the eigenvalues associated with the equilibria.

8.3 SYSTEMS OF AUTONOMOUS EQUATIONS (OPTIONAL)

In the preceding two sections, we discussed models that could be described by a single differential equation. If we wish to describe models in which several quantities interact, such as a competition model in which various species interact, more than one differential equation is needed. We call this then a system of differential equations. We will restrict ourselves again to autonomous systems, that is, systems whose dynamics do not depend explicitly on the independent variable (which is typically time).

This section is a preview of Chapter 11, where we will discuss systems of differential equations in detail. A thorough analysis of systems of differential equations requires a fair amount of theory, which we will develop in Chapters 9 and 10. Since we are not yet equipped with the right tools to analyze systems of differential equations, this section will be rather informal. As with movie previews, you will not know the full story after you finished this sections, but reading it will (hopefully) convince you that systems of differential equations provide a rich tool for modeling biological systems.

8.3.1 A Simple Epidemic Model

We begin our discussion of systems of autonomous differential equations with a classical model of an infectious disease, the Kermack-McKendrick model (1927; 1932; 1933). We consider a population of fixed size N that, at time t, can be divided into three classes: the susceptibles, $S(t)$, who can get infected; the infectives, $I(t)$, who are infected and can transmit the disease; and the removed class, $R(t)$, who are immune to the disease. The flow among these classes can be described by

$$S \longrightarrow I \longrightarrow R$$

We assume that the infection spreads according to the mass action law that we encountered in the discussion of chemical reactions. Each susceptible becomes infected at a rate that is proportional to the number of infectives I. Each infected individual recovers at a constant rate. A gain in the class of infectives is a simultaneous loss in the class of susceptibles. Likewise, a gain in the class of recovered individuals is a loss in the class of infectives. We can therefore describe the dynamics by

$$\frac{dS}{dt} = -bSI \tag{8.73}$$

$$\frac{dI}{dt} = bSI - aI \tag{8.74}$$

$$\frac{dR}{dt} = aI \tag{8.75}$$

Note that

$$\frac{dS}{dt} + \frac{dI}{dt} + \frac{dR}{dt} = -bSI + bSI - aI + aI = 0$$

Since

$$\frac{dS}{dt} + \frac{dI}{dt} + \frac{dR}{dt} = \frac{d}{dt}(S + I + R)$$

it follows that $S(t) + I(t) + R(t)$ is a constant that we can identify as the population size N. To analyze the system, we assume that at time 0

$$S(0) > 0 \qquad I(0) > 0 \qquad R(0) = 0$$

A question of interest is whether the infection will spread. We say that the infection spreads if

$$I(t) > I(0) \quad \text{for some } t > 0$$

Equation (8.74) allows us to answer this question. If

$$\frac{dI}{dt} = I(bS - a) > 0 \quad \text{at } t = 0$$

then $I(t)$ increases at the beginning and hence the infection can spread. This condition can be written as

$$\frac{bS(0)}{a} > 1$$

The value $bS(0)/a$ is called the **basic reproductive rate** of the infection and is typically denoted by R_0 [not to be confused with the number of recovered individuals at time 0, $R(0)$]. The quantity R_0 is of great importance in epidemiology, because R_0 tells us whether an infection can spread. It is the key to understanding why vaccination programs work. It explains why it is not necessary to vaccinate every one against an infectious disease: As long as the number of susceptibles is reduced below a certain threshold, the infection will not spread. The theoretical threshold, based on the Kermack-McKendrick model, is a/b. In practice, there are additional factors that influence whether or not an infection will spread (such as spatial proximity of infected individuals). But the basic conclusion is the same: As long as the number of susceptibles is below a certain threshold, the infection will not spread.

To find out how the infection progresses over time, we divide (8.73) by (8.75). For $I > 0$, we obtain

$$\frac{dS/dt}{dR/dt} = \frac{dS}{dR} = -\frac{bSI}{aI} = -\frac{b}{a}S$$

That is,

$$\frac{dS}{dR} = -\frac{b}{a}S$$

Separating variables and integrating yields

$$\int \frac{dS}{S} = -\frac{b}{a} \int dR$$

Since $S(t) > 0$, we find

$$\ln S(t) = -\frac{b}{a}R(t) + C$$

With $R(0) = 0$, we obtain

$$\ln S(0) = C$$

Hence,

$$S(t) = S(0)e^{-(b/a)R(t)} \qquad (8.76)$$

Since $R(t)$ is nondecreasing (no one can leave the removed class), it follows from (8.76) that $S(t)$ is nonincreasing (see Figure 8.24).

Letting $t \to \infty$, we find

$$S(\infty) = \lim_{t \to \infty} S(t) = \lim_{t \to \infty} S(0)e^{-(b/a)R(t)}$$
$$= S(0)e^{-(b/a)R(\infty)}$$

Since $R(\infty) \le N$, it follows that

$$S(\infty) \ge S(0)e^{-(b/a)N} > 0$$

That is, not everyone becomes infected. As long as the number of susceptibles is greater than a/b, $dI/dt > 0$, and the number of infected individuals increases. Since the populations size is constant, the infection "uses up" susceptibles, and there will be a time when the number of susceptibles falls below a/b (see Figure 8.25). From then on, dI/dt is negative, and the number of infected individuals begins to decline. The infection will eventually cease, that is,

$$\lim_{t \to \infty} I(t) = 0$$

Since not everyone becomes infected ($S(\infty) > 0$), when the infection finally comes to an end, it is because it runs out of infectives, not because of a lack of susceptibles.

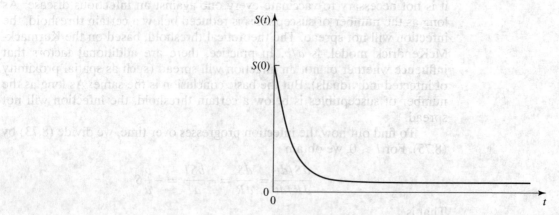

▲ **Figure 8.24**
The solution $S(t)$ of the Kermack-McKendrick model

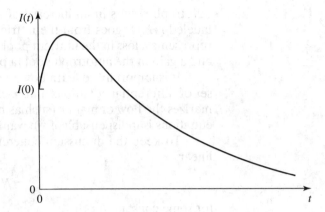

▲ Figure 8.25
The solution $I(t)$ of the Kermack-McKendrick model

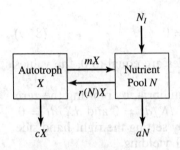

▲ Figure 8.26
A schematic description of the interaction between autotrophs and the nutrient pool

8.3.2 A Compartment Model

In Subsection 8.2.2, we introduced a single-compartment model that led to an autonomous differential equation with one dependent variable. In this subsection, we introduce a model with two compartments that describes the interaction of an autotroph[1] and its nutrient pool. (The following discussion is partially adapted from DeAngelis, 1992.) A schematic description of the interactions can be found in Figure 8.26.

We assume that the nutrient pool has an external source and denote the input rate of nutrients by N_I. That is, N_I denotes the total mass per unit time that flows into the system. Nutrients may be washed out, and we assume that the output rate is proportional to the total mass of nutrients in the compartment, N, with proportionality constant a.

The autotroph feeds on the nutrients in the nutrient pool and its growth rate is proportional to the quantity of autotroph biomass, X. The specific rate of growth, $r(N)$, depends on the amount of available nutrients.

Autotrophs can leave the autotroph compartment in two ways. At rate c, they get completely lost (for instance, through harvesting or by being washed out of their habitat). At rate m, the biomass of the autotroph gets recycled back into the nutrient pool (for instance, after the death of an organism).

This system is then described by the following set of differential equations.

$$\frac{dN}{dt} = N_I - aN - r(N)X + mX \tag{8.77}$$

$$\frac{dX}{dt} = r(N)X - (m + c)X \tag{8.78}$$

You should compare the set of differential equations to the schematic description of the model in Figure 8.26. In particular, you should pay attention to the direction of the arrows. For instance, the term $r(N)X$ shows up in both equations: Because this term corresponds to the nutrient uptake by the autotrophs, it appears as a loss to the nutrient pool, which is indicated by the minus sign in front of the term in equation (8.77). The same term appears with a plus sign in equation (8.78), because the uptake of nutrients by the

---

[1] An autotroph is an organism that can manufacture organic compounds entirely from inorganic components. Examples include most chlorophyll-containing plants and blue-green algae.

autotroph results in an increase of autotroph biomass. Note that the arrow labeled $r(N)X$ goes from the nutrient pool to the autotroph pool. Hence it represents a loss in the nutrient pool [a minus sign in (8.77) in front of $r(N)X$] and a gain in the autotroph pool [a plus sign in (8.78) in front of $r(N)X$].

It is important to learn how to go from the schematic description to the set of differential equations and back. A schematic description quickly summarizes the flow of matter (such as nutrients), whereas the set of differential equations is indispensable if we want to analyze the system.

To keep the discussion concrete, we assume that the function $r(N)$ is linear,

$$r(N) = bN$$

for some constant $b > 0$.

As in Section 8.2, we can introduce the concept of equilibria. An equilibrium for the system given by (8.77) and (8.78) is characterized by simultaneously requiring that

$$\frac{dN}{dt} = 0 \quad \text{and} \quad \frac{dX}{dt} = 0 \tag{8.79}$$

since when the rates of change of both quantities are equal to 0, the values for N and X no longer change.

There is a graphical method for finding equilibria, by plotting the **zero isoclines**. These curves are obtained by setting $dN/dt = 0$ and $dX/dt = 0$. The curves for which $dX/dt = 0$ are obtained by setting the right-hand side of (8.78) equal to 0, that is, $X(bN - (m + c)) = 0$, yielding

$$N = \frac{m + c}{b} \quad \text{or} \quad X = 0$$

The curve for which $dN/dt = 0$, is obtained by setting the right-hand side of (8.77) equal to 0, yielding

$$X = \frac{N_I - aN}{bN - m}$$

We plot the three curves in the N-X plane, as illustrated in Figure 8.27.

The zero isocline for N intersects the horizontal line $X = 0$ at $(N_1, 0)$, where N_1 satisfies

$$0 = \frac{N_I - aN_1}{bN_1 - m}$$

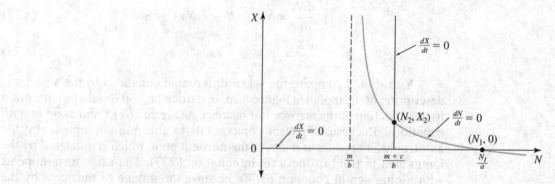

▲ **Figure 8.27**
The zero-isoclines in the N-X plane

and thus $N_1 = N_I/a$. We call this a trivial equilibrium since it corresponds to the case in which there is no autotroph in the system (the system reduces to the single-compartment model we discussed in Section 8.2).

Depending on where the vertical zero-isocline for X is located, there can be another equilibrium (N_1, N_2) for which both N_2 and X_2 are positive. Namely, looking at Figure 8.27, when

$$\frac{m}{b} < \frac{m+c}{b} < \frac{N_I}{a} \tag{8.80}$$

the two zero isoclines $N = (m+c)/b$ and $X = (N_I - aN)/(bN - m)$ intersect in the first quadrant. We call this equilibrium a nontrivial equilibrium.

Since $(m+c)/b > m/b$ when $c > 0$, the vertical zero isocline $dX/dt = 0$ is always to the right of the vertical asymptote $N = m/b$. The first inequality in (8.80) therefore always holds. If we solve the second inequality

$$\frac{m+c}{b} < \frac{N_I}{a}$$

for b, we find

$$b > \frac{a}{N_I}(m+c)$$

That is, the growth parameter b must exceed a certain threshold in order for the autotroph to survive. This should be intuitively clear, since m and c are the rates at which the autotroph pool is depleted. This depletion must be balanced by an increase in biomass of the autotroph.

Suppose that the ratio m/b is fixed. Then the smaller c (the rate at which autotrophs get lost), the further to the left is the vertical zero isocline $dX/dt = 0$ and, consequently, the larger the equilibrium value X_2. This should also be intuitively clear, since losing fewer autotrophs should result in a higher equilibrium value.

To find the nontrivial equilibrium (N_2, X_2), we need to solve the following system of equations:

$$0 = N_I - aN_2 - bN_2X_2 + mX_2 \tag{8.81}$$

$$0 = bN_2X_2 - (m+c)X_2 \tag{8.82}$$

where we now assume that $X_2 \neq 0$. Equation (8.82) yields

$$N_2 = \frac{m+c}{b}$$

Using this in (8.81), we find

$$0 = N_I - a\frac{m+c}{b} - (m+c)X_2 + mX_2$$

or

$$X_2 = \frac{1}{c}\left(N_I - a\frac{m+c}{b}\right)$$

It can be shown that if the nontrivial equilibrium exists, it is locally stable, that is, the system will return to this equilibrium after a small perturbation. In Chapter 11, we will learn two methods for analyzing the stability of a system of differential equations, a graphical and an analytical one; these are extensions of the methods we developed in Section 8.1.

8.3.3 A Hierarchical Competition Model

We now extend the metapopulation model (Levins model), introduced in Subsection 8.2.3, to a multispecies setting. We recall Levins model, but use a somewhat different interpretation. In Levins model, the dynamics of subpopulations was described. We will now view it as an occupancy model of sites for single individuals; that is, we consider the habitat as divided into patches that are now so small that at most one individual occupies a patch. Using this interpretation, we denote by $p(t)$ the fraction of sites that are occupied by a single individual at time t. Then

$$\frac{dp}{dt} = cp(1 - p) - mp$$

where m is the extinction rate (the death rate of an individual) and c is the colonization rate (the rate at which individuals send out propagules). Tilman (1994) extended this model to a system in which species are ranked according to their competitiveness. The fraction of sites that are occupied by species i at time t is denoted by $p_i(t)$. We assume that species 1 is the best competitor, species 2 the next best, and so on. A superior competitor can invade a site that is occupied by an inferior competitor. The inferior competitor is displaced upon invasion of the superior competitor. The dynamics are then given by

$$\frac{dp_1}{dt} = c_1 p_1(1 - p_1) - m_1 p_1$$

$$\frac{dp_2}{dt} = c_2 p_2(1 - p_1 - p_2) - m_2 p_2 - c_1 p_1 p_2$$

$$\frac{dp_3}{dt} = c_3 p_3(1 - p_1 - p_2 - p_3) - m_3 p_3 - c_1 p_1 p_3 - c_2 p_2 p_3$$

$$\vdots$$

$$\frac{dp_i}{dt} = c_i \left(1 - \sum_{j=1}^{i} p_j\right) - m_i p_i - \sum_{j=1}^{i-1} c_j p_j p_i$$

$$\vdots$$

The first term describes the colonization by species i of sites that are either occupied by an inferior competitor or vacant; the second term describes the extinction of sites that are occupied by species i; and the remainder of the term(s) describe the competitive displacement by superior competitors.

To simplify the discussion, we assume that there are only two species and that $m_1 = m_2 = 1$. The equations are then given by

$$\frac{dp_1}{dt} = c_1 p_1(1 - p_1) - p_1$$

$$\frac{dp_2}{dt} = c_2 p_2(1 - p_1 - p_2) - p_2 - c_1 p_1 p_2$$

The hierarchical structure of the model makes it easy to find possible equilibria. Setting $dp_1/dt = 0$, we find

$$0 = \hat{p}_1[c_1(1 - \hat{p}_1) - 1]$$

which, aside from the trivial equilibrium 0, gives

$$\hat{p}_1 = 1 - \frac{1}{c_1}$$

The equation for species 1 is identical to Levins model in Subsection 8.2.3, and we can use the results from that subsection. If $c_1 > 1$, $\hat{p}_1 \in (0, 1)$, and the nontrivial equilibrium is locally stable. Because we are interested in coexistence of species, we assume that $c_1 > 1$ in the following, so that species 1 can survive.

Setting $dp_2/dt = 0$ with $\hat{p}_1 = 1 - 1/c_1$ allows us to find $\hat{p}_2$:

$$0 = \hat{p}_2[c_2(1 - \hat{p}_1 - \hat{p}_2) - 1 - c_1\hat{p}_1]$$

which, aside from the trivial equilibrium 0, yields

$$\hat{p}_2 = (1 - \hat{p}_1) - \frac{1}{c_2} - \frac{c_1}{c_2}\hat{p}_1$$

$$= \frac{1}{c_1} - \frac{1}{c_2} - \frac{c_1}{c_2} + \frac{1}{c_2}$$

$$= \frac{1}{c_1} - \frac{c_1}{c_2}$$

Coexistence of the two species means that both $\hat{p}_1$ and $\hat{p}_2$ are positive and that their sum $\hat{p}_1 + \hat{p}_2$, which denotes the total fraction of occupied sites is less than 1. Now, the sum of the two nontrivial equilibria $\hat{p}_1 + \hat{p}_2 = 1 - c_1/c_2$ is automatically less than 1. Furthermore, since we assumed $c_1 > 1$, we have $\hat{p}_1 > 0$. Therefore, we only need to find out when $\hat{p}_2 > 0$, that is, when

$$\frac{1}{c_1} - \frac{c_1}{c_2} > 0$$

To satisfy this inequality, we need

$$c_2 > c_1^2$$

So far, we only know that there is a nontrivial equilibrium if $c_1 > 1$ and $c_2 > c_1^2$, but this does not tell us anything about stability. Even though we cannot yet analyze stability directly, we can take an approach that is very common in the ecological literature. We determine whether species 2 can invade the *monoculture equilibrium* of species 1, that is, the positive equilibrium of species 1 in the absence of species 2. Why does this help us? First, note that species 1 is unaffected by the presence of species 2: As long as species 1 can survive in the absence of species 2, it can also survive in its presence. Therefore, we need only worry about species 2. If species 2 can invade the monoculture equilibrium of species 1, then this says that if species 2 is at a low density, it will be able to increase its density, and it will therefore be able to coexist with species 1. The invasion criterion is then

$$\left.\frac{dp_2}{dt}\right|_{p_1=\hat{p}_1} > 0 \quad \text{when } p_2 \text{ is small}$$

This works as follows:

$$\frac{dp_2}{dt} = p_2[c_2(1 - p_1 - p_2) - 1 - c_1p_1]$$

We assume that $p_1 = \hat{p}_1 = 1 - 1/c_1$ and that p_2 is very small. If p_2 is small, then $1 - \hat{p}_1 - p_2 \approx 1 - \hat{p}_1$. Hence,

$$\frac{dp_2}{dt} \approx p_2\left[c_2\frac{1}{c_1} - 1 - c_1 + 1\right]$$

$$= p_2\left[\frac{c_2}{c_1} - c_1\right] > 0$$

if

$$\frac{c_2}{c_1} - c_1 > 0 \quad \text{or} \quad c_2 > c_1^2$$

Since $dp_2/dt > 0$ when species 1 is in equilibrium and species 2 has a low abundance, this means that species 2 can invade. We conclude that species 1 and 2 can coexist when $c_2 > c_1^2$.

This mechanism of coexistence is referred to as the **competition-colonization trade-off**. That is, the weaker competitor (species 2) can compensate for inferior competitiveness by being a superior colonizer ($c_2 > c_1^2$).

8.3.4 Problems

(8.3.1)

In Problems 1–4, we will investigate the classical Kermack-McKendrick model for the spread of an infectious disease in a population of fixed size N. (This model was introduced in Subsection 8.3.1, and you should refer to this subsection when working out the problems.) If S(t) denotes the number of susceptibles at time t, I(t) the number of infectives at time t, and R(t) the number of immune individuals at time t, then

$$\frac{dS}{dt} = -bSI$$
$$\frac{dI}{dt} = bSI - aI$$

and $R(t) = N - S(t) - I(t)$.

1. Determine in each of the following cases whether or not the disease can spread. (*Hint:* Compute R_0.)

(a) $S(0) = 1000, a = 200, b = 0.3$

(b) $S(0) = 1000, a = 200, b = 0.1$

2. Assume that $a = 100$ and $b = 0.2$. The **critical number** of susceptibles $S_c(0)$ at time 0 for the spread of a disease that is introduced into the population at time 0, is defined as the minimum number of susceptibles for which the disease can spread. Find $S_c(0)$.

3. Suppose that $a = 100, b = 0.01$, and $N = 10,000$. Can the disease spread if at time 0 there is one infected individual?

4. (a) Show that when $I > 0$,

$$\frac{dI}{dS} = \frac{a}{b}\frac{1}{S} - 1 \qquad (8.83)$$

and solve (8.83) when $R(0) = 0, I(0) = I_0$, and $S(0) = S_0$.

(b) Since $I(t)$ gives the number of infectives at time t and $dI/dt = bSI - aI$, if $S(0) > a/b$, then $dI/dt > 0$ at time $t = 0$. Since $\lim_{t \to \infty} I(t) = 0$, there is a time $t > 0$ at which $I(t)$ is maximal. Show that the number of susceptibles when $I(t)$ is maximal is given by $S = a/b$.

(c) In part (a), you expressed $I(t)$ as a function of $S(t)$. Use your result in (b) to show that the maximal number of infectives, I_{max}, is given by

$$I_{max} = N - \frac{a}{b} + \frac{a}{b} \ln\left(\frac{a/b}{S_0}\right)$$

(d) Use your result in (c) to show that I_{max} is a decreasing function of the parameter a/b for $a/b < S_0$ (that is, in the case in which the infection can spread). Use this to explain how a and b determine the severity (as measured by I_{max}) of the disease. Does this make sense?

(8.3.2)

5. Assume the compartment model of Subsection 8.3.2, with $a = 5, b = 0.02, m = 1$, and $c = 1$.

(a) Find the system of differential equations that corresponds to these values.

(b) Determine which values of N_I result in a nontrivial equilibrium, and find the equilibrium values for both the autotroph and the nutrient pool.

6. Assume the compartment model of Subsection 8.2.3, with $a = 1, b = 0.01, m = 2, c = 1$, and $N_I = 500$.

(a) Find the system of differential equations that corresponds to these values.

(b) Plot the zero isoclines corresponding to this system.

(c) Use your graph in (b) to determine whether this system has a nontrivial equilibrium.

7. Assume the compartment model of Subsection 8.3.2, with $a = 1, b = 0.01, m = 2, c = 1$ and $N_I = 200$.

(a) Find the system of differential equations that corresponds to these values.

(b) Plot the zero isoclines corresponding to this system.

(c) Use your graph in (b) to determine whether this system has a nontrivial equilibrium.

(8.3.3)

8. Assume the hierarchical competition model with two species, introduced in Subsection 8.3.3. Specifically, assume

$$\frac{dp_1}{dt} = 2p_1(1 - p_1) - p_1$$
$$\frac{dp_2}{dt} = 5p_2(1 - p_1 - p_2) - p_2 - 2p_1p_2$$

(a) Find all equilibria.

(b) Determine whether species 2 can invade a monoculture of species 1. (Assume that species 1 is in equilibrium.)

9. Assume the hierarchical competition model with two species, introduced in Subsection 8.3.3. Specifically, assume

$$\frac{dp_1}{dt} = 2p_1(1 - p_1) - p_1$$
$$\frac{dp_2}{dt} = 3p_2(1 - p_1 - p_2) - p_2 - 2p_1p_2$$

(a) Find all equilibria.

(b) Determine whether species 2 can invade a monoculture equilibrium of species 1.

10. Assume the hierarchical competition model with two species, introduced in Subsection 8.3.3. Specifically, assume

$$\frac{dp_1}{dt} = 2p_1(1 - p_1) - p_1$$
$$\frac{dp_2}{dt} = 6p_2(1 - p_1 - p_2) - p_2 - 2p_1p_2$$

(a) Use the zero-isocline approach to find all equilibria graphically.

(b) Determine the numerical values of all equilibria.

11. Assume the hierarchical competition model with two species, introduced in Subsection 8.3.3. Specifically, assume

$$\frac{dp_1}{dt} = 3p_1(1 - p_1) - p_1$$
$$\frac{dp_2}{dt} = 5p_2(1 - p_1 - p_2) - p_2 - 3p_1p_2$$

(a) Use the zero-isocline approach to find all equilibria graphically.

(b) Determine the numerical values of all equilibria.

12. (*Adapted from Crawley, 1997*) Denote by V plant biomass, and by N herbivore number. The plant-herbivore interaction is modeled as

$$\frac{dV}{dt} = aV(1 - \frac{V}{K}) - bVN$$
$$\frac{dN}{dt} = cVN - dN$$

(a) Suppose the herbivore number is equal to 0. What differential equation describes the dynamics of the plant biomass? Can you explain the resulting equation? Determine the plant biomass equilibrium in the absence of herbivores.

(b) Now assume that herbivores are present. Describe the effect of herbivores on plant biomass, that is, explain the term $-bVN$ in the first equation. Describe the dynamics of the herbivores, that is, how their population size increases and what contributes to decreases in population size.

(c) Determine the equilibria by solving

$$\frac{dV}{dt} = 0 \quad \text{and} \quad \frac{dN}{dt} = 0$$

Also determine the equilibria also graphically. Explain why this model implies that "plant abundance is determined solely by attributes of the herbivore," as stated in Crawley (1997).

8.4 Key Terms

Chapter 8 Review: Topics *Discuss the following definitions and concepts.*

1. Differential equation
2. Separable differential equation
3. Solution of a differential equation
4. Pure time differential equation
5. Autonomous differential equation
6. Exponential growth
7. Von Bertalanffy equation
8. Logistic equation
9. Allometric growth
10. Equilibrium
11. Stability
12. Eigenvalue
13. Single-compartment model
14. Levins model
15. Allee effect
16. Kermack-McKendrick model
17. Zero isocline
18. Hierarchical competition model

8.5 Review Problems

1. (*Newton's law of cooling*) Suppose that an object has temperature T and is brought into a room that is kept at a constant temperature T_a. Newton's law of cooling states that the rate of temperature change of the object is proportional to the difference between the temperature of the object and the surrounding medium.

(a) Denote by $T(t)$ the temperature at time t, and explain why

$$\frac{dT}{dt} = k(T - T_a)$$

is the differential equation that expresses Newton's law of cooling.

(b) Suppose that it takes the object 20 min to cool from 30°C to 28°C in a room whose temperature is 21°C. How long will it take the object to cool to 25°C, if it is at 30°C when it is brought into the room? [*Hint:* Solve the differential equation in (a) with the initial condition $T(0) = 30°C$

and with $T_a = 21°C$. Use $T(20) = 28°C$ to determine the constant k.]

2. (*Adapted from Cain et al., 1995*) In this problem we discuss a model for clonal growth in the white clover *Trifolium repens*. *T. repens* is a widespread perennial clonal plant species. It spreads through stolon growth (a stolon is a horizontal stem). By mapping the shape of a clone over time, Cain et al. estimated stolon elongation and die-back rates as follows. Denote by $S(t)$ the stolon length of the clone at time t. Cain et al. observed that the change in stolon length was proportional to stolon length, that is,

$$\frac{dS}{dt} \propto S$$

Introducing the proportionality constant r, called the net growth rate, we find

$$\frac{dS}{dt} = rS \tag{8.84}$$

(a) Suppose that S_f and S_0 are the final and the initial stolon lengths, respectively, and T denotes the time period of observation. Use (8.84) to show that r, the net growth rate, can be estimated from

$$r = \frac{1}{T} \ln \frac{S_f}{S_0}$$

[*Hint:* Solve the differential equation (8.84) with initial condition $S(0) = S_0$, and use the fact that $S(T) = S_f$.]

(b) The net growth rate r is the difference between the stolon elongation rate b and the stolon die-back rate m, that is,

$$r = b - m$$

Let B be the total amount of stolon elongation and D be the total amount of stolon die-back over the observation period of length T. Show that

$$B = \int_0^T bS(t)\, dt = \frac{bS_0}{r}(e^{rT} - 1)$$

$$D = \int_0^T mS(t)\, dt = \frac{mS_0}{r}(e^{rT} - 1)$$

(c) Show that $B - D = S_f - S_0$, and rearrange the equations for B and D in (b) so that you could estimate b and m from r, B, and D; that is, show that

$$b = \frac{rB}{S_f - S_0} = \frac{rB}{B - D}$$

$$m = \frac{rD}{S_f - S_0} = \frac{rD}{B - D}$$

(d) Explain how B and r can be estimated if S_f, S_0 and D are known from field measurements. Use your result in (c) to explain how you would then find estimates for b and m.

3. (Diversification of life) (*Adapted from Benton, 1997, and Walker, 1985*) Several models have been proposed to explain the diversification of life during geological periods. According to Benton (1997),

> The diversification of marine families in the past 600 million years (Myr) appears to have followed two or three logistic curves, with equilibrium levels that lasted for up to 200 Myr. In contrast, continental organisms clearly show an exponential pattern of diversification, and although it is not clear whether the empirical diversification patterns are real or are artifacts of a poor fossil record, the latter explanation seems unlikely.

In the following, we will investigate three models for diversification. They are analogous to models for population growth; however, the quantities involved have a different interpretation. We denote by $N(t)$ the diversification function, which counts the number of taxa as a function of time, and by r the intrinsic rate of diversification.

(a) (*Exponential model*) This model is described by

$$\frac{dN}{dt} = r_e N \tag{8.85}$$

Solve (8.85) with the initial condition $N(0)$ at time 0, and show that r_e can be estimated from

$$r_e = \frac{1}{t} \ln \left[\frac{N(t)}{N(0)} \right] \tag{8.86}$$

[*Hint:* To find (8.86), solve for r in the solution of (8.85).]

(b) (*Logistic growth*) This model is described by

$$\frac{dN}{dt} = r_l N \left(1 - \frac{N}{K} \right) \tag{8.87}$$

where K is the equilibrium value. Solve (8.87) with the initial condition $N(0)$ at time 0, and show that r_l can be estimated from

$$r_l = \frac{1}{t} \ln \left[\frac{K - N(0)}{N(0)} \right] + \frac{1}{t} \ln \left[\frac{N(t)}{K - N(t)} \right] \tag{8.88}$$

for $N(t) < K$.

(c) Assume that $N(0) = 1$ and $N(10) = 1000$. Estimate r_e and r_l for both $K = 1001$ and $K = 10,000$.

(d) Use your answer in (c) to explain the following quote from Stanley (1979):

> There must be a general tendency for calculated values of [r] to represent underestimates of exponential rates, because some radiation will have followed distinctly sigmoid paths during the interval evaluated .... .

(e) Explain why the exponential model is a good approximation to the logistic model when N/K is small compared to 1.

4. (A simple model for photosynthesis of individual leaves) (*Adapted from Horn, 1971*) Photosynthesis is a complex mechanism; the following model is a very simplified caricature. Suppose that a leaf contains a number of traps that can capture light. If a trap captures light, it becomes energized. The energy in the trap can then be used to produce sugar, which causes the energized trap to become unenergized. The number of traps that can become energized is proportional to the number of unenergized traps and the light intensity. Denote by T the total number of traps (unenergized and energized) in a leaf, by I the light intensity, and by x the number of energized traps. Then the following differential equation describes how the number of energized traps changes over time:

$$\frac{dx}{dt} = k_1(T - x)I - k_2 x$$

where k_1 and k_2 are positive constants. Find all equilibria and study their stability using the eigenvalue approach.

5. (Gompertz growth model) This model is sometimes used to study the growth of a population for which the per capita growth rate is density dependent. Denote by $N(t)$ the size of a population at time t; then for $N \geq 0$,

$$\frac{dN}{dt} = kN(\ln K - \ln N) \quad \text{with } N(0) = N_0 \tag{8.89}$$

(a) Show that

$$N(t) = K \exp \left[- \left(\ln \frac{K}{N_0} \right) e^{-kt} \right]$$

is a solution of (8.89). To do this, differentiate $N(t)$ with respect to t and show that it can be written in the form

(8.89). Don't forget to show that $N(0) = N_0$. Use a graphing calculator to sketch the graph of $N(t)$ for $N_0 = 100$, $k = 2$, and $K = 1000$. The function $N(t)$ is called the Gompertz growth curve.

(b) Use l'Hospital's rule to show that
$$\lim_{N \to 0} N \ln N = 0$$
and use this to show that $\lim_{N \to 0} dN/dt = 0$. Are there any other values of N where $dN/dt = 0$?

(c) Sketch the graph of dN/dt as a function of N for $k = 2$ and $K = 1000$. Find the equilibria and discuss their stability using your graph. Explain the meaning of K.

6. (*Island biogeography*) Preston (1962) and MacArthur and Wilson (1963) investigated the effect of area on species diversity in oceanic islands. It is assumed that species can immigrate to an island from a species pool of size P, and that species on the island can go extinct. We denote the immigration rate by $I(S)$ and the extinction rate by $E(S)$, where S is the number of species on the island. Then the change in species diversity over time is
$$\frac{dS}{dt} = I(S) - E(S) \tag{8.90}$$
The simplest functional forms for $I(S)$ and $E(S)$, for a fixed island, are
$$I(S) = c\left(1 - \frac{S}{P}\right) \tag{8.91}$$
$$E(S) = m\frac{S}{P} \tag{8.92}$$
where c, m, and P are positive constants.

(a) Find the equilibrium species diversity $\hat{S}$ of (8.90) with $I(S)$ and $E(S)$ given in (8.91) and (8.92).

(b) It is reasonable to assume that the extinction rate is a decreasing function of island size. That is, if A denotes the area of the island, we assume that m is a function of the island size A with $dm/dA < 0$. Furthermore, we assume that the immigration rate I does not depend on the size of the island. Use these assumptions to investigate how the equilibrium species diversity changes with island size.

(c) Assume that $S(0) = S_0$. Solve (8.90) with $I(S)$ and $E(S)$ as given in (8.91) and (8.92), respectively.

(d) Assume that $S_0 = 0$. That is, the island is initially void of species. The time constant T for the system is defined as
$$S(T) = (1 - e^{-1})\hat{S}$$
Show that under the assumption $S_0 = 0$,
$$T = \frac{P}{c + m}$$

(e) Use the assumptions in (b) and your answer in (d) to investigate the effect of island size on the time constant T; that is, determine whether $T(A)$ is an increasing or decreasing function of A.

7. (*Chemostat*) A chemostat is an apparatus for growing bacteria in a medium in which all nutrients but one are available in excess. One nutrient, whose concentration can be controlled, is held at a concentration that limits the growth of bacteria. The growth chamber of the chemostat is continually flushed by adding nutrients dissolved in liquid at a constant rate and

allowing the liquid in the growth chamber, which contains bacteria, to leave the growth chamber at the same rate. If X denotes the number of bacteria in the growth chamber, then the growth dynamics of the bacteria are given by
$$\frac{dX}{dt} = r(N)X - qX \tag{8.93}$$
where $r(N)$ is the growth rate depending on the nutrient concentration N and q is the input and output flow rate. The equation for the nutrient flow is given by
$$\frac{dN}{dt} = qN_0 - qN - r(N)X \tag{8.94}$$
Note that Equation (8.93) is (8.78) with $m = 0$, $N_I = qN_0$, and $a = e = q$, and (8.94) is (8.77) with $m = 0$.

(a) Explain in words the meaning of the terms in (8.93) and (8.94).

(b) Assume that $r(N)$ is given by the Monod growth function
$$r(N) = b\frac{N}{k + N}$$
where k and b are positive constants. Draw the zero isoclines in the N-X plane and explain how to find equilibria $(\hat{N}, \hat{X})$ graphically.

(c) Show that a nontrivial equilibrium (an equilibrium for which $\hat{N}$ and $\hat{X}$ are both positive) satisfies
$$r(\hat{N}) - q = 0 \tag{8.95}$$
$$qN_0 - q\hat{N} - r(\hat{N})\hat{X} = 0 \tag{8.96}$$
Show that (8.95) has a positive solution $\hat{N}$ if $q < b$, and find an expression for $\hat{N}$. Use this and (8.96) to find $\hat{X}$.

(d) Assume that $q < b$. Use your results in (c) to show that $\hat{X} > 0$ if $\hat{N} < N_0$ and $\hat{N} < N_0$ if $q < bN_0/(k + N_0)$. Furthermore, show that $\hat{N}$ is an increasing function of q for $q < b$.

(e) Use your results in (d) to explain why the following is true. As we increase the flow rate q from 0 to $bN_0/(k + N_0)$, the nutrient concentration $\hat{N}$ increases until it reaches the value N_0 and the number of bacteria decreases to 0.

8. (*Adapted from Nee and May, 1992, and Tilman, 1994*) In Subsection 8.3.3, we introduced a hierarchical competition model. We will use this model to investigate the effects of habitat destruction on coexistence. We assume that a fraction D of the sites is permanently destroyed. Furthermore, we restrict our discussion to two species and assume that species 1 is the superior and species 2 the inferior competitor. In the case in which both species have the same mortality ($m_1 = m_2$), which we set equal to 1, the dynamics are described by
$$\frac{dp_1}{dt} = c_1 p_1(1 - p_1 - D) - p_1 \tag{8.97}$$
$$\frac{dp_2}{dt} = c_2 p_2(1 - p_1 - p_2 - D) - p_2 - c_1 p_1 p_2 \tag{8.98}$$
where $p_i, i = 1, 2$, is the fraction of sites occupied by species i.

(a) Explain in words the meanings of the different terms in (8.97) and (8.98).

(b) Show that
$$\hat{p}_1 = 1 - \frac{1}{c_1} - D$$

is an equilibrium for species 1 that is in $(0, 1)$ and is stable if $D < 1 - 1/c_1$ and $c_1 > 1$.

(c) Assume that $c_1 > 1$ and $D < 1 - 1/c_1$. Show that species 2 can invade the nontrivial equilibrium of species 1 [computed in (b)] if

$$c_2 > c_1^2(1 - D)$$

(d) Assume that $c_1 = 2$ and $c_2 = 5$. Then species 1 can survive as long as $D < 1/2$. Show that the fraction of sites that are occupied by species 1 is then

$$\hat{p}_1 = \begin{cases} \dfrac{1}{2} - D & \text{for } 0 \le D \le \dfrac{1}{2} \\ 0 & \text{for } \dfrac{1}{2} \le D \le 1 \end{cases}$$

Show that

$$\hat{p}_2 = \frac{1}{10} + \frac{2}{5}D \quad \text{for} \quad 0 \le D \le \frac{1}{2}$$

For $D > 1/2$, species 1 can no longer persist. Explain why the dynamics for species 2 reduce in this case to

$$\frac{dp_2}{dt} = 5p_2(1 - p_2 - D) - p_2$$

Show that the nontrivial equilibrium is of the form

$$\hat{p}_2 = 1 - \frac{1}{5} - D \quad \text{for } \frac{1}{2} \le D \le 1 - \frac{1}{5}$$

Plot $\hat{p}_1$ and $\hat{p}_2$ as functions of D in the same coordinate system. What happens for $D > 1 - 1/5$? Use the plot to explain in words how each species is affected by habitat destruction.

(e) Repeat (d) for $c_1 = 2$ and $c_2 = 3$.

LINEAR ALGEBRA AND ANALYTIC GEOMETRY

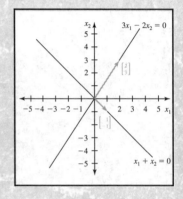

9.1 LINEAR SYSTEMS

Two different species of insects are reared together in a laboratory cage. They are supplied with two different types of food each day. Each individual of species 1 consumes 5 units of food A and 3 units of food B, whereas each individual of species 2 consumes 2 units of food A and 4 units of food B, on average, per day. Each day, a lab technician supplies 900 units of food A and 960 units of food B. How many of each species are reared together?

To solve such a problem, we will set up a system of equations. If

$$x = \text{number of individuals of species 1}$$
$$y = \text{number of individuals of species 2}$$

then the following two equations must be satisfied:

$$\text{food A:} \quad 5x + 2y = 900$$
$$\text{food B:} \quad 3x + 4y = 960$$

We refer to these two equations as a system of two linear equations in two variables. This section is devoted to finding solutions to such systems.

9.1.1 Graphical Solution

In this subsection, we restrict ourselves to systems of two linear equations in two variables. Recall that the standard form of a linear equation in two variables is

$$Ax + By = C$$

where A, B, and C are constants, A and B are not both equal to 0, and x and y are the two variables; its graph is a straight line (see Figure 9.1). Any point (x, y) on this straight line satisfies (or solves) the equation $Ax + By = C$. We can extend this to more than one equation, in which case we talk about a **system** of linear equations. In the case of two linear equations in two variables, the system is of the form

$$Ax + By = C$$
$$Dx + Ey = F \tag{9.1}$$

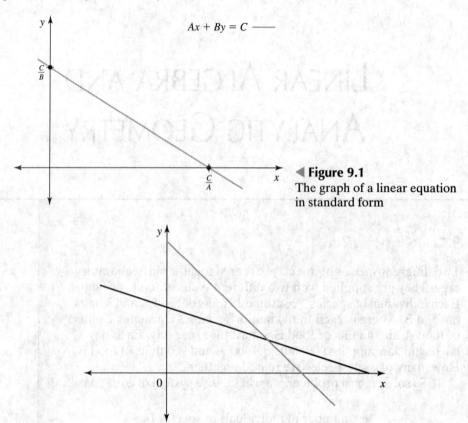

◀ **Figure 9.1**
The graph of a linear equation in standard form

▲ **Figure 9.2**
The two lines have exactly one point of intersection

where A, B, C, D, E, and F are constants, and x and y are the two variables. (We require that A and B [D and E, respectively] are not both equal to 0.) When we say that we "solve" (9.1) for x and y, we mean that we find an ordered pair (x, y) that satisfies *each* equation of the system (9.1). Because each equation in (9.1) describes a straight line, we are therefore asking for the point of intersection of these two lines. The following three cases are possible:

1. The two lines have exactly one point of intersection. In this case, the system (9.1) has exactly one solution, as illustrated in Figure 9.2.

2. The two lines are parallel and do not intersect. In this case, the system (9.1) has no solution, as illustrated in Figure 9.3.

3. The two lines are parallel and intersect (that is, they are identical). In this case, the system (9.1) has infinitely many solutions, namely each point on the line, as illustrated in Figure 9.4.

Exactly One Solution

▶ **Example 1**   Find the solution of

$$2x + 3y = 6$$
$$2x + \ y = 4$$

(9.2)

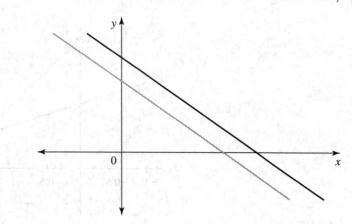

▲ **Figure 9.3**
The two lines are parallel but do not intersect

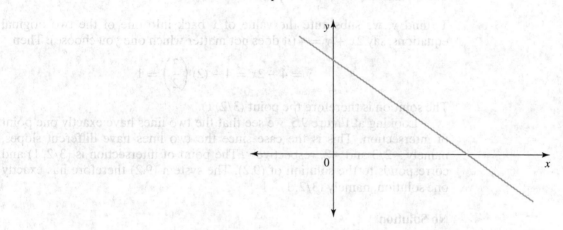

▲ **Figure 9.4**
The two lines are identical

Solution

The line corresponding to $2x + 3y = 6$ has y-intercept $(0, 2)$ and x-intercept $(3, 0)$; the line corresponding to $2x + y = 4$ has y-intercept $(0, 4)$ and x-intercept $(2, 0)$ (see Figure 9.5). To find the solution of the linear system (9.2), we need to find the point of intersection of the two lines. Solving each equation for y produces the new set of equations

$$y = 2 - \frac{2}{3}x$$
$$y = 4 - 2x$$

Setting the right-hand sides equal to each other, we find

$$2 - \frac{2}{3}x = 4 - 2x$$

$$\frac{4}{3}x = 2$$

$$x = \frac{3}{2}$$

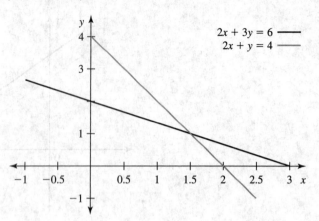

▲ **Figure 9.5**
The two lines have exactly one point of intersection

To find y, we substitute the value of x back into one of the two original equations, say $2x + y = 4$ (it does not matter which one you choose). Then

$$y = 4 - 2x = 4 - (2)\left(\frac{3}{2}\right) = 1$$

The solution is therefore the point $(3/2, 1)$.

Looking at Figure 9.5, we see that the two lines have exactly one point of intersection. This is the case since the two lines have different slopes, namely $-2/3$ and -2, respectively. The point of intersection is $(3/2, 1)$ and corresponds to the solution of (9.2). The system (9.2) therefore has exactly one solution, namely $(3/2, 1)$. ◀

No Solution

▶ **Example 2**   Solve

$$\begin{array}{r} 2x + y = 4 \\ 4x + 2y = 6 \end{array} \qquad (9.3)$$

Solution

The line corresponding to $2x + y = 4$ has y-intercept $(0, 4)$ and x-intercept $(2, 0)$; the line corresponding to $4x + 2y = 6$ has y-intercept $(0, 3)$ and x-intercept $(3/2, 0)$ (see Figure 9.6). Since

$$2x + y = 4 \Longleftrightarrow y = 4 - 2x$$

$$4x + 2y = 6 \Longleftrightarrow y = 3 - 2x$$

we immediately see that both lines have the same slope, namely -2, but different y-intercepts, namely 4 and 3, respectively. This implies that the two lines are parallel and do not intersect.

Let's see what happens when we solve the system (9.3). We equate the two equations $y = 4 - 2x$ and $y = 3 - 2x$:

$$4 - 2x = 3 - 2x$$

$$4 = 3$$

which is obviously wrong; that is, there is no point (x, y) that satisfies both equations in (9.3) simultaneously.

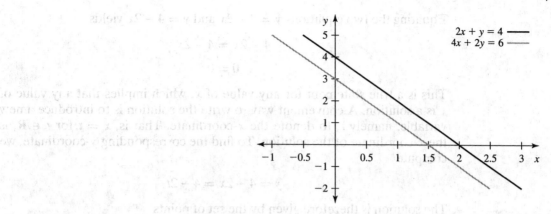

▲ **Figure 9.6**
The two lines are parallel but do not intersect

Looking at Figure 9.6, we see that the two lines are parallel. Since parallel lines that are not identical do not intersect, we see from the graph that (9.3) has no solution. In this case, we write the solution as Ø, the symbol for the empty set. ◄

Infinitely Many Solutions

▶ **Example 3**   Solve

$$2x + y = 4$$
$$4x + 2y = 8$$
(9.4)

Solution

If we divide the second equation by 2, we find that both equations are identical, namely $2x + y = 4$. That is, both equations describe the same line with x-intercept $(2, 0)$ and y-intercept $(0, 4)$, as shown in Figure 9.7. Every point (x, y) on this line is therefore a solution of (9.4). To find the solution algebraically, we use the same procedure as in Examples 1 and 2. We first solve each equation for y. This produces

$$2x + y = 4 \Longleftrightarrow y = 4 - 2x$$
$$4x + 2y = 8 \Longleftrightarrow y = 4 - 2x$$

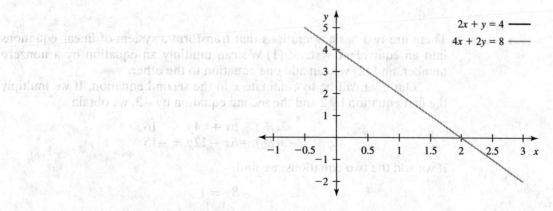

▲ **Figure 9.7**
The two lines are identical

Equating the two equations $y = 4 - 2x$ and $y = 4 - 2x$ yields

$$4 - 2x = 4 - 2x$$

$$0 = 0$$

This is a true statement for any value of x, which implies that any value of x is a solution. A convenient way to write the solution is to introduce a new variable, namely t, to denote the x-coordinate. That is, $x = t$ for $t \in \mathbf{R}$, is the x-coordinate of the solution. To find the corresponding y-coordinate, we compute

$$y = 4 - 2x = 4 - 2t$$

The solution is therefore given by the set of points

$$\{(t, 4 - 2t) : t \in \mathbf{R}\}$$

This shows that the system (9.4) has infinitely many solutions, as expected from the graphical considerations. [Figure 9.7 shows that the two lines representing (9.4) are identical and hence only one line is visible.] ◀

In Example 3, we introduced a new variable, namely t, to describe the set of solutions. We call t a **dummy variable**; it stands for any real number. Introducing a dummy variable is a convenient way to describe the set of solutions when there are infinitely many.

A Solution Method   The graphical and algebraic way of solving systems of linear equations we have employed so far only works for systems in two variables. To solve systems of m linear equations in n variables, we will need to develop a method that will work for any size system. The basic strategy will be to transform the system of linear equations into a new system of equations that has the same solutions as the original. The new system is called an **equivalent system**. It will be of a simpler form, so that we can solve for the unknown variables one by one and thus arrive at a solution. We illustrate this in the following example. We label all equations with labels of the form (R_i); R_i stands for "ith row." This will allow us to keep track of our computations.

▶ **Example 4**   Solve

$$3x + 2y = 8 \quad (R_1)$$
$$2x + 4y = 5 \quad (R_2)$$

Solution

There are two basic operations that transform a system of linear equations into an equivalent system: (1) We can multiply an equation by a nonzero number, and (2) we can add one equation to the other.

Our goal will be to eliminate x in the second equation. If we multiply the first equation by 2 and the second equation by -3, we obtain

$$2(R_1) \quad 6x + 4y = 16$$
$$-3(R_2) \quad -6x - 12y = -15$$

If we add the two equations, we find

$$-8y = 1$$

We thus eliminated x from the second equation. We can replace the original system of equations by a new (equivalent) system by leaving the first equation unchanged and replacing the second equation by $-8y = 1$. We obtain the equivalent system of equations [labeled (R_3) and (R_4)]:

$$(R_1) \quad 3x + 2y = 8 \quad (R_3)$$
$$2(R_1) - 3(R_2) \quad -8y = 1 \quad (R_4)$$

We can now successively solve the system. It follows from equation (R_4) that

$$y = -\frac{1}{8}$$

Substituting this value into equation (R_3) and solving for x, we find

$$3x + (2)\left(-\frac{1}{8}\right) = 8$$

$$3x = 8 + \frac{1}{4}$$

$$x = \frac{11}{4}$$

The solution is therefore $(11/4, -1/8)$. ◀

When we look at equations (R_3) and (R_4), the left-hand side has the shape of a triangle

$$*x + *y = *$$
$$*y = *$$

We therefore call a system that is written in this form **upper triangular**. In the preceding example, we reduced the system of linear equations to upper triangular form and then solved it by back-substitution. As we will see in the next subsection, this method can be generalized to larger systems of equations.

9.1.2 Solving Systems of Linear Equations

In this subsection, we will extend the solution method of Example 4 to systems of m equations in n variables, which we write in the form

$$
\begin{aligned}
a_{11}x_1 + a_{12}x_2 + \cdots + a_{1n}x_n &= b_1 \\
a_{21}x_1 + a_{22}x_2 + \cdots + a_{2n}x_n &= b_2 \\
&\cdots\cdots\cdots\cdots\cdots\cdots\cdots\cdots \\
a_{m1}x_1 + a_{m2}x_2 + \cdots + a_{mn}x_n &= b_m
\end{aligned}
\tag{9.5}
$$

The variables are now $x_1, x_2, \ldots, x_n$. The coefficients a_{ij} on the left-hand side have two subscripts. The first subscript indicates the equation and the second subscript to which x_j it belongs. For instance, you would find a_{21} in the second equation as the coefficient of x_1; or a_{43} would be in the fourth equation as the coefficient of x_3. Using double subscripts is a convenient way of labeling the coefficients. The subscripts on the numbers b_i on the right-hand side of (9.5) indicate the equation.

We will transform this system into an equivalent system in upper triangular form. (Recall that *equivalent* means that the new system has the same solutions as the old system.) To do so we will use the following three basic operations:

1. Multiplying an equation by a nonzero constant
2. Adding one equation to another
3. Rearranging the order of the equations

This method is also called **Gaussian elimination**.

As in the case of a linear system with two equations in two variables, the general system (9.5) may have

1. Exactly one solution
2. No solutions
3. Infinitely many solutions

When a system has no solutions, we say that the system is **inconsistent**.

Exactly One Solution

▶ **Example 5**   Solve

$$3x + 5y - z = 10 \quad (R_1)$$
$$2x - y + 3z = 9 \quad (R_2)$$
$$4x + 2y - 3z = -1 \quad (R_3)$$

Solution

Our goal is to reduce this system to upper triangular form. The first step is to eliminate the elements below $3x$ in the first column; that is, we wish to eliminate $2x$ from the second equation and $4x$ from the third equation. We leave the first equation unchanged. Multiplying the first equation by 2 and the second equation by -3 and then adding the two equations yields a new equation that replaces the second equation. In the new equation the term involving x has been eliminated. That is, we add

$$2(R_1) \quad 6x + 10y - 2z = 20$$
$$-3(R_2) \quad -6x + 3y - 9z = -27$$

which yields

$$13y - 11z = -7$$

We transform the third equation by multiplying the second equation by 2 and the third equation by -1 and then adding the two equations. In the resulting equation, the term involving x has been eliminated. That is, we add

$$2(R_2) \quad 4x - 2y + 6z = 18$$
$$-(R_3) \quad -4x - 2y + 3z = 1$$

which yields

$$-4y + 9z = 19$$

These two steps transform the original set of equations into the following equivalent set of equations, labeled (R_4)–(R_6).

$$(R_1) \quad 3x + 5y - z = 10 \quad (R_4)$$
$$2(R_1) - 3(R_2) \quad 13y - 11z = -7 \quad (R_5)$$
$$2(R_2) - (R_3) \quad -4y + 9z = 19 \quad (R_6)$$

Our next step is to eliminate y from (R_6). To do this, we multiply (R_5) by 4 and (R_6) by 13 and then add the two equations:

$$4(R_5) \quad 52y - 44z = -28$$
$$13(R_6) \quad -52y + 117z = 247$$

which yields

$$73z = 219$$

We leave the first two equations unchanged. The new (equivalent) system of equations is thus

$$\begin{array}{llll}
(R_4) & 3x + 5y - z = 10 & (R_7) & \\
(R_5) & 13y - 11z = -7 & (R_8) & (9.6)\\
4(R_5) + 13(R_6) & 73z = 219 & (R_9) &
\end{array}$$

The system of equations is now in upper triangular form, as can be seen from its shape. We use back-substitution to find the solution. Solving equation (R_9) for z yields

$$z = \frac{219}{73} = 3$$

Solving (R_8) for y and substituting the value of z yields

$$y = \frac{1}{13}(-7 + 11z) = \frac{1}{13}(-7 + (11)(3)) = \frac{26}{13} = 2$$

Solving (R_7) for x and substituting the values of y and z, we find

$$x = \frac{1}{3}(10 - 5y + z)$$

$$= \frac{1}{3}(10 - (5)(2) + 3) = 1$$

Hence the solution is $x = 1$, $y = 2$, and $z = 3$. ◀

A comment on the various steps in Example 5: We eliminated the terms involving x in (R_2) and (R_3); typically, we have a choice in how to do this. We used equations (R_1) and (R_2) to replace (R_2), and used (R_2) and (R_3) to replace (R_3). We could have used (R_2) and (R_3) to replace (R_2). In this case, (R_2) would have to be replaced by $(R_3) - 2(R_2)$, resulting in $4y - 9z = -19$. Then we could not have used (R_2) and (R_3) again to replace (R_3). If we used (R_2) and (R_3) again, we would replace (R_3) by $(R_3) - 2(R_2)$, which is $4y - 9z = -19$, and we would replace both (R_2) and (R_3) by the same equation, thus effectively losing one equation. To replace (R_3), we would then have had to choose (R_1) and (R_3). Namely, (R_3) would have to be replaced by $4(R_1) - 3(R_3)$, which is $14y + 5z = 43$.

A further comment about eliminating y from (R_5) and (R_6) is in place: When we replaced (R_6), we had to use (R_5) and (R_6). If we had used (R_6) and (R_4), say, to eliminate y, we would have introduced the variable x into the resulting equation and then the system would not have been reduced to upper triangular form. Namely, to eliminate y using (R_6) and (R_4), we would need to compute $(R_6) + 2(R_4)$, which results in the equation $4x + z = 8$.

No Solution

▶ Example 6 Solve

$$\begin{array}{ll}
2x - y + z = 3 & (R_1)\\
4x - 4y + 3z = 2 & (R_2)\\
2x - 3y + 2z = 1 & (R_3)
\end{array}$$

Solution

As in Example 5, we try to reduce the system to upper triangular form. We begin by eliminating terms involving x from the second and third equation. We leave the first equation unchanged. We replace the second equation by the result of subtracting the second equation from the first, after multiplying the first equation by 2. We replace the third equation by the result of subtracting the third equation from the first. We obtain

$$(R_1) \quad 2x - y + z = 3 \quad (R_4)$$
$$2(R_1) - (R_2) \quad \qquad 2y - z = 4 \quad (R_5)$$
$$(R_1) - (R_3) \quad \qquad 2y - z = 2 \quad (R_6)$$

To obtain (R_5), we also could have computed $(R_2) - 2(R_1)$; we would have found $-2y + z = -4$. Similarly, for (R_6), we could have computed $(R_3) - (R_1)$. As long as we do not use the same pair of equations twice, we have some freedom in how we reduce the system to upper triangular form.

Now, let's continue with our calculations. To eliminate terms involving y from Equation (R_6), we replace (R_6) by the difference between (R_5) and (R_6). We keep the first two equations. Then

$$(R_4) \quad 2x - y + z = 3$$
$$(R_5) \quad \qquad 2y - z = 4$$
$$(R_5) - (R_6) \quad \qquad 0 = 2$$

The last equation, $0 = 2$, is a wrong statement, which means that this system does not have a solution. We write the solution as Ø, the symbol denoting the empty set. ◄

Infinitely Many Solutions

▶ **Example 7** Solve

$$x - 3y + z = 4 \quad (R_1)$$
$$x - 2y + 3z = 6 \quad (R_2)$$
$$2x - 6y + 2z = 8 \quad (R_3)$$

Solution

We proceed as in the preceding examples. The first step is to eliminate terms involving x from the second and third equation. We leave the first equation unchanged. We replace the second equation by the difference between the first and the second equation. We replace the third equation by the difference between the first and the third equation after dividing the third equation by 2. We find

$$(R_1) \qquad \qquad x - 3y + z = 4 \quad (R_4)$$
$$(R_2) - (R_1) \qquad \qquad y + 2z = 2 \quad (R_5)$$
$$(R_1) - \frac{1}{2}(R_3) \qquad \qquad 0z = 0 \quad (R_6)$$

The third equation, $0z = 0$, is a correct statement. It means that we can substitute any number for z and still obtain a correct result. We introduce the dummy variable t, and set

$$z = t \quad \text{for } t \in \mathbf{R}$$

If we solve (R_5) for y and substitute $z = t$ into the resulting equation, we find

$$y = 2 - 2z = 2 - 2t$$

If we solve (R_4) for x and substitute $y = 2 - 2t$ and $z = t$ into the resulting equation, we find

$$x = 4 + 3y - z = 4 + 3(2 - 2t) - t$$
$$= 4 + 6 - 6t - t = 10 - 7t$$

The solution is therefore the set

$$\{(x, y, z) : x = 10 - 7t, y = 2 - 2t, z = t, \text{ for } t \in \mathbf{R}\} \quad ◄$$

A Shorthand Notation When we transform a system of linear equations, we make changes only to the coefficients of the variables. It is therefore convenient to introduce notation that will simply keep track of all the coefficients. This motivates the following definition.

Definition A **matrix** is a rectangular array of numbers

$$A = \begin{bmatrix} a_{11} & a_{12} & \cdots & a_{1n} \\ a_{21} & a_{22} & \cdots & a_{2n} \\ \cdots\cdots\cdots\cdots\cdots \\ a_{m1} & a_{m2} & \cdots & a_{mn} \end{bmatrix}$$

The elements a_{ij} of the matrix A are called **entries**. If the matrix has m rows and n columns, it is called an $m \times n$ matrix.

We can think of the matrix A in the definition as the **coefficient matrix** of the linear system we introduced in (9.5).

$$\begin{aligned} a_{11}x_1 + a_{12}x_2 + \cdots + a_{1n}x_n &= b_1 \\ a_{21}x_1 + a_{22}x_2 + \cdots + a_{2n}x_n &= b_2 \\ &\cdots\cdots\cdots\cdots\cdots\cdots \\ a_{m1}x_1 + a_{m2}x_2 + \cdots + a_{mn}x_n &= b_m \end{aligned} \qquad (9.7)$$

The entries a_{ij} of the matrix have two subscripts: The entry a_{ij} is located in the ith row and the jth column. You can think of this numbering system as the street address of the entries, using streets and avenues. Suppose that avenues go north-south and streets go east-west; finding the corner of 2nd Street and 3rd Avenue would be like finding the element of A that is in the second row and third column, which is a_{23} in the matrix.

If a matrix has the same number of rows as columns, it is called a **square matrix**. An $m \times 1$ matrix is called a **column vector** and a $1 \times n$ matrix is called a **row vector**. Examples follow of a 3×3 square matrix, a 3×1 column vector, and a 1×4 row vector. The shape of each of the matrices explains their names.

$$\begin{bmatrix} 1 & 3 & 0 \\ 0 & 1 & -1 \\ 5 & 4 & 3 \end{bmatrix}, \qquad \begin{bmatrix} 2 \\ 7 \\ 4 \end{bmatrix}, \qquad \begin{bmatrix} 1 & 3 & 0 & 5 \end{bmatrix}$$

If A is a square matrix, then the **diagonal line** of A consists of the elements $a_{11}, a_{22}, \ldots, a_{nn}$. In the preceding 3×3 square matrix, the diagonal line thus consists of the elements 1, 1, and 3.

To solve systems of linear equations using matrices, we introduce the **augmented matrix**. The augmented matrix is the coefficient matrix of the linear system (9.7), augmented by an additional column representing the right-hand side of (9.7). The augmented matrix representing the linear system (9.7) is therefore

$$\begin{bmatrix} a_{11} & a_{12} & \cdots & a_{1n} & b_1 \\ a_{21} & a_{22} & \cdots & a_{2n} & b_2 \\ & & \cdots\cdots & & \cdots \\ a_{m1} & a_{m2} & \cdots & a_{mn} & b_m \end{bmatrix}$$

So how can we use augmented matrices when solving systems of linear equations?

▶ **Example 8** Find the augmented matrix for the system given in Example 5, and solve the system using the augmented matrix.

Solution

The augmented matrix is given by

$$\left[\begin{array}{ccc|c} 3 & 5 & -1 & 10 \\ 2 & -1 & 3 & 9 \\ 4 & 2 & -3 & -1 \end{array}\right] \begin{array}{l} (R_1) \\ (R_2) \\ (R_3) \end{array}$$

Our goal is to transform the augmented matrix into a simpler form, also called upper triangular form, in which all the entries below the diagonal line are zero. We use the same basic transformations as before, and you should compare the following to the steps in Example 5. We indicate on the left side of the matrix the transformation that we used, and relabel the rows on the right side of the matrix.

$$\begin{array}{c} (R_1) \\ 2(R_1) - 3(R_2) \\ 2(R_2) - (R_3) \end{array} \left[\begin{array}{ccc|c} 3 & 5 & -1 & 10 \\ 0 & 13 & -11 & -7 \\ 0 & -4 & 9 & 19 \end{array}\right] \begin{array}{l} (R_4) \\ (R_5) \\ (R_6) \end{array}$$

That is, we copied row 1 which we now call row 4; to get the second row, we multiplied the first row by 2 and the second row by 3 and then took the difference $(2(R_1) - 3(R_2))$, calling the resulting row (R_5), and so on.

The next step is to eliminate the -4 in (R_6). We find

$$\begin{array}{c} (R_4) \\ (R_5) \\ 4(R_5) + 13(R_6) \end{array} \left[\begin{array}{ccc|c} 3 & 5 & -1 & 10 \\ 0 & 13 & -11 & -7 \\ 0 & 0 & 73 & 219 \end{array}\right]$$

This is now the augmented matrix for the system of linear equations in (9.6), and we can proceed as in Example 5 to solve the system by back-substitution. ◀

So far, all of the systems that we have considered have had the same number of equations as variables. This need not be the case. When a system has fewer equations than variables, we say that it is **underdetermined**. Though underdetermined systems can be inconsistent (have no solutions), they frequently have infinitely many solutions. When a system has more equations than variables, we say that it is **overdetermined**. Overdetermined systems are frequently inconsistent.

In Example 9, we will look at an underdetermined system, and in Example 10, an overdetermined system.

▶ **Example 9** Solve the following underdetermined system.

$$\begin{array}{l} 2x + 2y - z = 1 \quad (R_1) \\ 2x - y + z = 2 \quad (R_2) \end{array}$$

Solution

The system has fewer equations than variables and is therefore called underdetermined. We use the augmented matrix to solve the system.

$$\left[\begin{array}{ccc|c} 2 & 2 & -1 & 1 \\ 2 & -1 & 1 & 2 \end{array}\right] \begin{array}{l} (R_1) \\ (R_2) \end{array}$$

Transforming the augmented matrix into upper triangular form, we find

$$\begin{array}{c} (R_1) \\ (R_1) - (R_2) \end{array} \left[\begin{array}{ccc|c} 2 & 2 & -1 & 1 \\ 0 & 3 & -2 & -1 \end{array}\right]$$

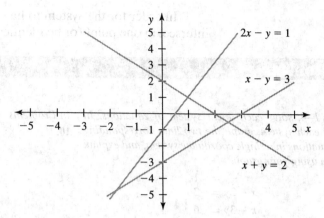

▲ Figure 9.8
The straight lines in Example 10 do not intersect in one point

Translating this back into a system of equations, we find

$$2x + 2y - z = 1$$
$$3y - 2z = -1$$

We see that

$$y = -\frac{1}{3} + \frac{2}{3}z$$

and

$$2x = 1 - 2y + z = 1 - 2\left(-\frac{1}{3} + \frac{2}{3}z\right) + z = \frac{5}{3} - \frac{1}{3}z$$

or

$$x = \frac{5}{6} - \frac{1}{6}z$$

We use a dummy variable again, and set $z = t$, $t \in \mathbf{R}$; then $x = \frac{5}{6} - \frac{1}{6}t$ and $y = -\frac{1}{3} + \frac{2}{3}t$. The solution can then be written as

$$\left\{(x, y, z) : x = \frac{5}{6} - \frac{1}{6}t, y = -\frac{1}{3} + \frac{2}{3}t, z = t, t \in \mathbf{R}\right\}$$ ◀

▶ **Example 10** Solve the following overdetermined system.

$$2x - y = 1 \quad (R_1)$$
$$x + y = 2 \quad (R_2)$$
$$x - y = 3 \quad (R_3)$$

Solution

The system has more equations than variables and is therefore called overdetermined. To solve it, we write

$$(R_1) \quad 2x - y = 1 \quad (R_4)$$
$$2(R_2) - (R_1) \quad 3y = 3 \quad (R_5)$$
$$(R_2) - (R_3) \quad 2y = -1 \quad (R_6)$$

It follows from (R_6) that $y = -\frac{1}{2}$, and from (R_5) that $y = 1$. Since $1 \neq -\frac{1}{2}$, there cannot be a solution. The system is inconsistent, and the solution set is the empty set. In this example, we can illustrate that this system is inconsistent if we graph the three straight lines represented by (R_1), (R_2), and (R_3), as in Figure 9.8.

In order for the system to have solutions, the three lines would need to intersect in one point (or be identical). They don't, so there is no solution. ◀

9.1.3 Problems

(9.1.1)

In Problems 1–4, solve each linear system of equations. In addition, for each system, graph the two lines corresponding to the two equations in a single coordinate system, and explain your solution using your graph.

1.
$$x - y = 1$$
$$x - 2y = -2$$

2.
$$x + 3y = 6$$
$$x - 4y = -4$$

3.
$$x - 3y = 6$$
$$y = 3 + \tfrac{1}{3}x$$

4.
$$2x + y = \tfrac{1}{3}$$
$$6x + 3y = 1$$

5. Determine c such that
$$2x - 3y = 5$$
$$4x - 6y = c$$
has (a) infinitely many solutions and (b) no solutions. (c) Is it possible to choose a number for c so that the system has exactly one solution? Explain your answer.

6. Show that the solution of
$$a_{11}x_1 + a_{12}x_2 = b_1$$
$$a_{21}x_1 + a_{22}x_2 = b_2$$
is given by
$$x_1 = \frac{a_{22}b_1 - a_{12}b_2}{a_{11}a_{22} - a_{21}a_{12}}$$
and
$$x_2 = \frac{-a_{21}b_1 + a_{11}b_2}{a_{11}a_{22} - a_{21}a_{12}}.$$

7. Zach wants to buy fish and plants for his aquarium. Each fish costs \$2.30; each plant costs \$1.70. He buys a total of 11 items and spends a total of \$21.70. Set up a system of linear equations that will allow you to determine how many fish and how many plants Zach bought, and solve the system.

8. Laboratory mice are fed with a mixture of two foods that contain two essential nutrients. Food 1 contains 3 units of nutrient A and 2 units of nutrient B per ounce; food 2 contains 4 units of nutrient A and 5 units of nutrient B per ounce.

(a) In what proportion should you mix the food if the mice require the nutrients A and B in equal amounts?

(b) Assume now that the mice require the nutrients A and B in the ratio 1 : 2. Is it possible to satisfy their dietary needs with the two foods available?

9. Show that if
$$a_{11}a_{22} - a_{21}a_{12} \neq 0$$
then the system
$$a_{11}x_1 + a_{12}x_2 = 0$$
$$a_{21}x_1 + a_{22}x_2 = 0$$
has exactly one solution, namely $x_1 = 0$ and $x_2 = 0$.

(9.1.2)

In Problems 10–14, solve each system of linear equations.

10.
$$2x - 3y + z = -1$$
$$x + y - 2z = -3$$
$$3x - 2y + z = 2$$

11.
$$5x - y + 2z = 6$$
$$x + 2y - z = -1$$
$$3x + 2y - 2z = 1$$

12.
$$x + 4y - 3z = -13$$
$$2x - 3y + 5z = 18$$
$$3x + y - 2z = 1$$

13.
$$-2x + 4y - z = -1$$
$$x + 7y + 2z = -4$$
$$3x - 2y + 3z = -3$$

14.
$$2x - y + 3z = 3$$
$$2x + y + 4z = 4$$
$$2x - 3y + 2z = 2$$

In Problems 15–18, find the augmented matrix and use it to solve the system of linear equations.

15.
$$-x - 2y + 3z = -9$$
$$2x + y - z = 5$$
$$4x - 3y + 5z = -9$$

16.
$$3x - 2y + z = 4$$
$$4x + y - 2z = -12$$
$$2x - 3y + z = 7$$

17.
$$y + x = 3$$
$$z - y = -1$$
$$x + z = 2$$

18.
$$2x - z = 1$$
$$y + 3z = x - 1$$
$$x + z = y - 3$$

In Problems 19–22, determine whether each system is overdetermined or underdetermined; then solve each system.

19.
$$x - 2y + z = 3$$
$$2x - 3y + z = 8$$

20.
$$x - y = 2$$
$$x + y + z = 3$$

21.
$$2x - y = 3$$
$$x - y = 4$$
$$3x - y = 1$$

22.
$$4y - 3z = 6$$
$$2y + z = 1$$
$$y + z = 0$$

23. SplendidLawn sells three types of lawn fertilizer, SL 24-4-8, SL 21-7-12, and SL 17-0-0. The three numbers refer to the percentages of nitrogen, phosphate, and potassium, in that order, of the contents. (For instance, 100 g of SL 24-4-8 contains 24 g of nitrogen.) Suppose that each year your lawn requires 500 g of nitrogen, 100 g of phosphate, and 180 g of potassium per 1000 square feet. How much of each of the three types of fertilizer should you apply per 1000 square feet per year?

24. Three different species of insects are reared together in a laboratory cage. They are supplied with two different types of food each day. Each individual of species 1 consumes 3 units of food A and 5 units of food B; each individual of species 2 consumes 2 units of food A and 3 units of food B; and individual of species 3 consumes 1 unit of food A and 2

units of food B. Each day, 500 units of food A and 900 units of food B are supplied. How many individuals of each species can be reared together? Is there more than one solution? What happens if we add a third type of food, called C, and each individual of species 1 consumes 2 units of food C, each individual of species 2 consumes 4 units of food C, and each individual of species 3 consumes 1 unit of food C, and 550 units of food C are supplied?

9.2 MATRICES

We introduced matrices in the previous section; in this section, we will learn various matrix operations.

9.2.1 Basic Matrix Operations

We recall the definition of a matrix. An $m \times n$ matrix A is a rectangular array of numbers with m rows and n columns. We write this as

$$A = \begin{bmatrix} a_{11} & a_{12} & \cdots & a_{1n} \\ a_{21} & a_{22} & \cdots & a_{2n} \\ \cdots\cdots\cdots\cdots\cdots\cdots \\ a_{m1} & a_{m2} & \cdots & a_{mn} \end{bmatrix} = [a_{ij}]_{\substack{1 \le i \le m \\ 1 \le j \le n}}$$

We will also use the shorthand notation $A = [a_{ij}]$ if the size of the matrix is clear. We list a few simple definitions that do not need much explanation.

> **Definition** Suppose that $A = [a_{ij}]$ and $B = [b_{ij}]$ are two $m \times n$ matrices. Then
>
> $$A = B$$
>
> if and only if, for all $1 \le i \le m$ and $1 \le j \le n$,
>
> $$a_{ij} = b_{ij}$$

This definition says that we can compare matrices of the same size, and they are equal if all their corresponding entries are equal. The next definition shows how to add matrices.

> **Definition** Suppose that $A = [a_{ij}]$ and $B = [b_{ij}]$ are two $m \times n$ matrices. Then
>
> $$C = A + B$$
>
> is an $m \times n$ matrix with entries
>
> $$c_{ij} = a_{ij} + b_{ij} \quad \text{for } 1 \le i \le m, 1 \le j \le n$$

Note that addition is only defined for matrices of equal sizes. Matrix addition satisfies the following two properties:

1. $A + B = B + A$
2. $(A + B) + C = A + (B + C)$

The matrix with all its entries equal to zero is called the **zero matrix** and denoted by **0**. The following holds:

$$A + 0 = A$$

We can multiply matrices by numbers. Such numbers are called **scalars**.

> **Definition** Suppose that $A = [a_{ij}]$ is an $m \times n$ matrix and c is a scalar. Then cA is an $m \times n$ matrix with entries ca_{ij} for $1 \le i \le m$ and $1 \le j \le n$.

▶ **Example 1** Find $A + 2B - 3C$, if

$$A = \begin{bmatrix} 2 & 3 \\ 1 & 0 \end{bmatrix}, \quad B = \begin{bmatrix} 0 & 1 \\ -1 & -3 \end{bmatrix}, \quad C = \begin{bmatrix} 1 & 0 \\ 0 & 3 \end{bmatrix}$$

Solution

$$A + 2B - 3C = \begin{bmatrix} 2 & 3 \\ 1 & 0 \end{bmatrix} + 2 \begin{bmatrix} 0 & 1 \\ -1 & -3 \end{bmatrix} - 3 \begin{bmatrix} 1 & 0 \\ 0 & 3 \end{bmatrix}$$

$$= \begin{bmatrix} 2 & 3 \\ 1 & 0 \end{bmatrix} + \begin{bmatrix} 0 & 2 \\ -2 & -6 \end{bmatrix} + \begin{bmatrix} -3 & 0 \\ 0 & -9 \end{bmatrix}$$

$$= \begin{bmatrix} 2+0-3 & 3+2+0 \\ 1-2+0 & 0-6-9 \end{bmatrix} = \begin{bmatrix} -1 & 5 \\ -1 & -15 \end{bmatrix} \quad ◀$$

▶ **Example 2** Let

$$A = \begin{bmatrix} 1 & 3 \\ 0 & 4 \end{bmatrix}, \quad B = \begin{bmatrix} 5 & 3 \\ -2 & 1 \end{bmatrix}, \quad C = \begin{bmatrix} -1 & -3 \\ 0 & -4 \end{bmatrix}$$

Show that $A + B = B - C$.

Solution

One way is to simply compute $A + B$ and $B - C$ and then compare the results. We find

$$A + B = \begin{bmatrix} 1 & 3 \\ 0 & 4 \end{bmatrix} + \begin{bmatrix} 5 & 3 \\ -2 & 1 \end{bmatrix} = \begin{bmatrix} 6 & 6 \\ -2 & 5 \end{bmatrix}$$

and

$$B - C = \begin{bmatrix} 5 & 3 \\ -2 & 1 \end{bmatrix} - \begin{bmatrix} -1 & -3 \\ 0 & -4 \end{bmatrix} = \begin{bmatrix} 6 & 6 \\ -2 & 5 \end{bmatrix}$$

Comparing $A + B$ with $B - C$, we find that they are equal since all their entries are the same.

Another way to reach this conclusion is to subtract B from both sides of the assertion $A + B = B - C$, which results in $A = -C$. Adding C to both sides, we find $A + C = 0$. That is, to show the assertion $A + B = B - C$, we can also show that $A + C = 0$. Computing $A + C$, we obtain

$$A + C = \begin{bmatrix} 1 & 3 \\ 0 & 4 \end{bmatrix} + \begin{bmatrix} -1 & -3 \\ 0 & -4 \end{bmatrix} = \begin{bmatrix} 0 & 0 \\ 0 & 0 \end{bmatrix} = 0$$

which shows that $A + C = 0$ and implies that $A + B = B - C$ is true. ◀

The operation that interchanges rows and columns is called **transposition**.

Definition Suppose that $A = [a_{ij}]$ is an $m \times n$ matrix. The **transpose** of A, denoted by A', is an $n \times m$ matrix with entries

$$a'_{ij} = a_{ji}$$

To understand how this operation works, we look at the following example.

▶ **Example 3** Transpose the following matrices.

$$A = \begin{bmatrix} 1 & 2 & 3 \\ 4 & 5 & 6 \end{bmatrix}, \quad B = \begin{bmatrix} 1 \\ 2 \end{bmatrix}, \quad C = \begin{bmatrix} 3 & 4 \end{bmatrix}$$

Solution

To find the transpose, we need to interchange rows and columns. Since A is a 2×3 matrix, its transpose A' is a 3×2 matrix, namely

$$A' = \begin{bmatrix} 1 & 4 \\ 2 & 5 \\ 3 & 6 \end{bmatrix}$$

To check this with our definition, note that

$$a'_{11} = a_{11} = 1, \quad a'_{12} = a_{21} = 4, \quad a'_{21} = a_{12} = 2, \ldots$$

The matrix B is a 2×1 matrix, which is a column vector. Its transpose is a 1×2 matrix, which is a row vector. We find

$$B' = \begin{bmatrix} 1 & 2 \end{bmatrix}$$

Similarly, C is a 1×2 row vector; its transpose is then a 2×1 column vector. We find

$$C' = \begin{bmatrix} 3 \\ 4 \end{bmatrix}$$
◀

In the preceding example, we saw that the transpose of a row vector is a column vector, and vice versa. When we need to write a column vector in text, such as $X = \begin{bmatrix} 1 \\ 2 \\ 3 \end{bmatrix}$, we can instead write $X = \begin{bmatrix} 1 & 2 & 3 \end{bmatrix}'$, which, written as the transpose of a row vector, is really a column vector. A large column vector written as the transpose of a row vector is more legible in text.

9.2.2 Matrix Multiplication

The multiplication of two matrices is more complicated. We give the definition first.

Definition Suppose that $A = [a_{ij}]$ is an $m \times l$ matrix and $B = [b_{ij}]$ is an $l \times n$ matrix. Then

$$C = AB$$

is an $m \times n$ matrix with

$$c_{ij} = a_{i1}b_{1j} + a_{i2}b_{2j} + \cdots + a_{il}b_{lj} = \sum_{k=1}^{l} a_{ik}b_{kj}$$

for $1 \leq i \leq m$ and $1 \leq j \leq n$.

To explain this in words, note that c_{ij} is the entry in C that is located in the ith row and the jth column. To obtain it, we multiply the entries of the ith row of A with the entries of the jth column of B as indicated in the definition. For the product AB to be defined, the number of columns in A must equal the number of rows in B. The definition looks rather formidable; let's see how it actually works.

▶ Example 4 (Matrix multiplication) Compute AB when

$$A = \begin{bmatrix} 1 & 2 & 3 \\ -1 & 0 & 4 \end{bmatrix} \quad \text{and} \quad B = \begin{bmatrix} 1 & 2 & 3 & -3 \\ 0 & -1 & 4 & 0 \\ -1 & 0 & -2 & 1 \end{bmatrix}$$

Solution

First, note that A is a 2×3 matrix and B is a 3×4 matrix. That is, A has 3 columns and B has 3 rows. Therefore, the product AB is defined and AB is a 2×4 matrix. We write the product as

$$C = AB = \begin{bmatrix} 1 & 2 & 3 \\ -1 & 0 & 4 \end{bmatrix} \begin{bmatrix} 1 & 2 & 3 & -3 \\ 0 & -1 & 4 & 0 \\ -1 & 0 & -2 & 1 \end{bmatrix}$$

For instance, to find c_{11}, we multiply the first row of A with the first column of B as follows:

$$c_{11} = \begin{bmatrix} 1 & 2 & 3 \end{bmatrix} \begin{bmatrix} 1 \\ 0 \\ -1 \end{bmatrix} = (1)(1) + (2)(0) + (3)(-1) = -2$$

You can see from this calculation that the number of columns of A must be equal to the number of rows of B. Otherwise, we would run out of numbers when multiplying the corresponding entries.

To find c_{23}, we multiply the second row of A with the third column of B:

$$c_{23} = \begin{bmatrix} -1 & 0 & 4 \end{bmatrix} \begin{bmatrix} 3 \\ 4 \\ -2 \end{bmatrix} = (-1)(3) + (0)(4) + (4)(-2) = -11$$

The other entries are obtained in a similar way, and we find

$$C = \begin{bmatrix} -2 & 0 & 5 & 0 \\ -5 & -2 & -11 & 7 \end{bmatrix} \quad ◀$$

When multiplying matrices, the order is important. For instance, BA is not defined for the matrices in Example 4. From this, it should be clear that, typically, $AB \neq BA$. Even if both AB and BA exist, the resulting matrices are usually not the same. They can differ in both the number of rows and columns, and in the actual entries. The next two examples illustrate this important point.

▶ Example 5 (Order is important) Suppose that

$$A = \begin{bmatrix} 2 & 1 & -1 \end{bmatrix} \quad \text{and} \quad B = \begin{bmatrix} 1 \\ -1 \\ 0 \end{bmatrix}$$

Show that both AB and BA are defined but $AB \neq BA$.

Solution

Note that A is a 1×3 matrix and B is a 3×1 matrix. The product AB is defined since the number of columns of A is the same as the number of rows of B (namely 3); the product AB is a 1×1 matrix. When we carry out the matrix multiplication, we find

$$AB = \begin{bmatrix} 2 & 1 & -1 \end{bmatrix} \begin{bmatrix} 1 \\ -1 \\ 0 \end{bmatrix} = [2 - 1 + 0] = [1] = 1$$

Note that a 1×1 matrix is simply a number.

The product BA is also defined, since the number of columns of B is the same as the number of rows of A (namely 1); the product BA is a 3×3 matrix. When we carry out the matrix multiplication, we find

$$BA = \begin{bmatrix} 1 \\ -1 \\ 0 \end{bmatrix} \begin{bmatrix} 2 & 1 & -1 \end{bmatrix} = \begin{bmatrix} 2 & 1 & -1 \\ -2 & -1 & 1 \\ 0 & 0 & 0 \end{bmatrix}$$

When we compare AB and BA, we immediately see that $AB \neq BA$ since the two matrices are not even of the same size. ◄

▶ **Example 6** **(Order is important)** Suppose that

$$A = \begin{bmatrix} 2 & -1 \\ 4 & -2 \end{bmatrix} \quad \text{and} \quad B = \begin{bmatrix} 1 & -1 \\ 2 & -2 \end{bmatrix}$$

Show that both AB and BA are defined but $AB \neq BA$.

Solution

Both A and B are 2×2 matrices. We conclude that both AB and BA are 2×2 matrices as well. We find

$$AB = \begin{bmatrix} 0 & 0 \\ 0 & 0 \end{bmatrix} \quad \text{and} \quad BA = \begin{bmatrix} -2 & 1 \\ -4 & 2 \end{bmatrix}$$

Since the entries in AB are not the same as in BA, it follows that $AB \neq BA$. ◄

The product AB in Example 6 resulted in a matrix with all entries equal to 0. Earlier, we called this matrix the zero matrix and denoted it by **0**. This example shows that a product of two matrices can be the zero matrix without either matrix being the zero matrix. This is an important fact and different from multiplying real numbers. When multiplying two real numbers, we know that if the product is equal to 0, at least one of the two factors must have been equal to 0. This does no longer hold for matrix multiplication: If A and B are two matrices whose product AB is defined, then the product AB can be equal to the zero matrix **0** without either A or B being equal to the zero matrix **0**.

We list the following properties of matrix multiplication. We assume that all matrices are of appropriate sizes so that all matrix multiplications are defined.

1. $(A + B)C = AC + BC$
2. $A(B + C) = AB + AC$

3. $(AB)C = A(BC)$

4. $A0 = 0A = 0$

(We will practice applying these properties in the problem section.)

Next, we note that if A is a square matrix (that is, if it has the same number of rows as columns), we can define powers of A; namely, if k is a positive integer, then

$$A^k = A^{k-1}A = AA^{k-1} = \underbrace{AA\ldots A}_{k \text{ factors}}$$

For instance, $A^2 = AA$, $A^3 = AAA$, and so on.

▶ **Example 7**   **(Powers of matrices)** Find A^3 if

$$A = \begin{bmatrix} 1 & 2 \\ 3 & 4 \end{bmatrix}$$

Solution

To find A^3, we first need to compute A^2. We obtain

$$A^2 = AA = \begin{bmatrix} 1 & 2 \\ 3 & 4 \end{bmatrix}\begin{bmatrix} 1 & 2 \\ 3 & 4 \end{bmatrix} = \begin{bmatrix} 7 & 10 \\ 15 & 22 \end{bmatrix}$$

To compute A^3 now, we compute either A^2A or AA^2. Both will yield A^3. This is a case in which the order of multiplication does not matter, since $A^3 = (AA)A = A(AA) = AAA$. We do it both ways.

$$A^3 = A^2A = \begin{bmatrix} 7 & 10 \\ 15 & 22 \end{bmatrix}\begin{bmatrix} 1 & 2 \\ 3 & 4 \end{bmatrix} = \begin{bmatrix} 37 & 54 \\ 81 & 118 \end{bmatrix}$$

and

$$A^3 = AA^2 = \begin{bmatrix} 1 & 2 \\ 3 & 4 \end{bmatrix}\begin{bmatrix} 7 & 10 \\ 15 & 22 \end{bmatrix} = \begin{bmatrix} 37 & 54 \\ 81 & 118 \end{bmatrix}$$ ◀

An important square matrix is the **identity matrix**, denoted by I_n. The identity matrix is an $n \times n$ matrix with 1s on its diagonal line and 0s elsewhere, that is,

$$I_n = \begin{bmatrix} 1 & 0 & 0 & \cdots & 0 \\ 0 & 1 & 0 & \cdots & 0 \\ \cdots\cdots\cdots\cdots\cdots \\ 0 & 0 & 0 & \cdots & 1 \end{bmatrix}$$

For instance,

$$I_1 = [1], \qquad I_2 = \begin{bmatrix} 1 & 0 \\ 0 & 1 \end{bmatrix}, \qquad I_3 = \begin{bmatrix} 1 & 0 & 0 \\ 0 & 1 & 0 \\ 0 & 0 & 1 \end{bmatrix}$$

(We frequently write I instead of I_n if the size of I_n is clear from the context.) The identity matrix serves the same role as the number 1 in the multiplication of real numbers: If A is an $m \times n$ matrix, then

$$AI_n = I_mA = A$$

It follows that $I^k = I$ for any positive integer k.

If a and x are numbers, then you know that $ax = x$ can be written as $ax - x = 0$. To factor x on the left-hand side, we transform the left-hand side as follows: $ax - 1x = (a - 1)x$. To factor x, we needed to introduce the factor 1 in front of x. We obtain $(a - 1)x = 0$. There is a similar procedure for matrices, but we must be more careful, since the order in which we multiply matrices is important.

▶ **Example 8**   **(Matrix equation)** Let A be a 2×2 matrix and X be a 2×1 matrix. Show that $AX = X$ can be written as $(A - I_2)X = \mathbf{0}$.

Solution

We find that

$$AX = X \qquad \text{is equivalent to} \qquad AX - X = \mathbf{0}$$

To factor X, we multiply X from the left by I_2. The matrix I_2 will then allow us to factor X just as in the case of real numbers where we needed to introduce the factor 1 to factor x in $ax - x = 0$. This yields

$$AX - I_2X = \mathbf{0}$$

We can now factor X and find

$$(A - I_2)X = \mathbf{0}$$

An important observation: We multiply X from the left by $A - I_2$. We are not allowed to reverse the order [the product $X(A - I_2)$ is not even defined]. ◀

There is a close connection between systems of linear equations and matrices. The system of linear equations

$$
\begin{aligned}
a_{11}x_1 + a_{12}x_2 + \cdots + a_{1n}x_n &= b_1 \\
a_{21}x_1 + a_{22}x_2 + \cdots + a_{2n}x_n &= b_2 \\
&\cdots\cdots\cdots \\
a_{m1}x_1 + a_{m2}x_2 + \cdots + a_{mn}x_n &= b_m
\end{aligned}
$$

can be written in matrix form

$$AX = B$$

with

$$
A = \begin{bmatrix} a_{11} & \cdots & a_{1n} \\ a_{21} & \cdots & a_{2n} \\ \cdots\cdots\cdots \\ a_{m1} & \cdots & a_{mn} \end{bmatrix}, \quad X = \begin{bmatrix} x_1 \\ x_2 \\ \vdots \\ x_n \end{bmatrix}, \quad B = \begin{bmatrix} b_1 \\ b_2 \\ \vdots \\ b_m \end{bmatrix}
$$

▶ **Example 9**   **(Matrix representation of linear systems)** Write

$$
\begin{aligned}
2x_1 + 3x_2 + 4x_3 &= 1 \\
-x_1 + 5x_2 - 6x_3 &= 7
\end{aligned}
$$

in matrix form.

Solution

When we write this system in matrix form, we find

$$
\begin{bmatrix} 2 & 3 & 4 \\ -1 & 5 & -6 \end{bmatrix} \begin{bmatrix} x_1 \\ x_2 \\ x_3 \end{bmatrix} = \begin{bmatrix} 1 \\ 7 \end{bmatrix}
$$

We see that A is a 2×3 matrix and X is a 3×1 matrix. We therefore expect (based on the rules of multiplication) that B is a 2×1 matrix, which it is indeed. ◀

We will use matrix representation to solve systems of linear equations in the next subsection.

9.2.3 Inverse Matrices

In this subsection, we will learn how to solve systems of n linear equations in n unknowns when they are written in matrix form,

$$AX = B \tag{9.8}$$

where A is an $n \times n$ square matrix and X and B are $n \times 1$ column vectors. We start with a simple example in which $n = 1$. When we solve

$$5x = 10$$

for x, we simply divide both sides by 5 or, equivalently, multiply both sides by $\frac{1}{5} = 5^{-1}$, and obtain

$$5^{-1} \cdot 5x = 5^{-1} \cdot 10$$

Since $5^{-1} \cdot 5 = 1$ and $5^{-1} \cdot 10 = 2$, we find that $x = 2$. To solve (9.8), we therefore need an operation that is analogous to division or multiplication by the "reciprocal" of A. We will define a matrix A^{-1} that will serve this function. It is called the **inverse matrix** of A. This matrix will allow us to write the solution of (9.8) as

$$A^{-1}AX = A^{-1}B$$

(As with finding reciprocals, it will not always be possible to find inverse matrices.) In order for the inverse matrix to have the same effect as multiplying a number by its reciprocal, it should have the property $A^{-1}A = I$, where I is the identity matrix. The solution would then be of the form

$$X = A^{-1}B$$

> **Definition** Suppose that $A = [a_{ij}]$ is an $n \times n$ square matrix. If there exists an $n \times n$ square matrix B such that
>
> $$AB = BA = I_n$$
>
> then B is called the inverse matrix of A and denoted by A^{-1}.

If A has an inverse matrix, A is called **invertible** or **nonsingular**; if A does not have an inverse, then A is called **singular**. If A is invertible, its inverse matrix is unique; that is, if B and C are both inverse matrices of A, then $B = C$. (In other words, if you and your friend compute the inverse of A, you both should get the exact same answer.) To see why, assume that B and C are both inverse matrices of A. Using $BA = I_n$ and $AC = I_n$, we find that

$$B = BI_n = B(AC) = (BA)C = I_nC = C$$

The following hold for inverse matrices.

1. If A is an invertible $n \times n$ matrix, then

$$(A^{-1})^{-1} = A$$

2. If A and B are invertible $n \times n$ matrices, then

$$(AB)^{-1} = B^{-1}A^{-1}$$

The first statement says that if we apply the inverse operation twice, we get the original matrix back. This is familiar from dealing with real numbers: Take the number 2; its inverse is $2^{-1} = 1/2$. If we take the inverse of $1/2$, we get 2 again. To see why this is true for matrices, assume that A is an $n \times n$ matrix with inverse matrix A^{-1}. Then, by definition,

$$AA^{-1} = A^{-1}A = I_n$$

But this equation also says that A is the inverse matrix of A^{-1}, that is, $A = (A^{-1})^{-1}$.

To see why the second statement is true, we must check that

$$(AB)(B^{-1}A^{-1}) = (B^{-1}A^{-1})(AB) = I_n$$

We compute

$$(AB)(B^{-1}A^{-1}) = A(BB^{-1})A^{-1} = AI_nA^{-1} = AA^{-1} = I_n$$

and

$$(B^{-1}A^{-1})(AB) = B^{-1}(A^{-1}A)B = B^{-1}I_nB = B^{-1}B = I_n$$

In the next example, we will see how we can check whether two matrices are inverses of each other. The matrices will be of size 3×3. It is somewhat cumbersome to find the inverse of a matrix of size 3×3 or larger (unless you use a calculator or computer software). Here, we start out with the two matrices, and simply check whether they are inverse matrices of each other.

▶ Example 10   (**Checking whether matrices are inverses of each other**) Show that the inverse of

$$A = \begin{bmatrix} 3 & 5 & -1 \\ 2 & -1 & 3 \\ 4 & 2 & -3 \end{bmatrix} \quad \text{is} \quad B = \begin{bmatrix} -\dfrac{3}{73} & \dfrac{13}{73} & \dfrac{14}{73} \\ \dfrac{18}{73} & -\dfrac{5}{73} & -\dfrac{11}{73} \\ \dfrac{8}{73} & \dfrac{14}{73} & -\dfrac{13}{73} \end{bmatrix}$$

Solution

Carrying out the matrix multiplications AB and BA, we see that

$$AB = I_3 \quad \text{and} \quad BA = I_3$$

Therefore, $B = A^{-1}$.   ◀

In this book, we will only need to invert 2×2 matrices. In the next example, we will show one way to do this.

▶ Example 11   (**Finding an inverse**) Find the inverse of

$$A = \begin{bmatrix} 2 & 5 \\ 1 & 3 \end{bmatrix}$$

Solution

We need to find a matrix

$$B = \begin{bmatrix} b_{11} & b_{12} \\ b_{21} & b_{22} \end{bmatrix}$$

such that $AB = I_2$. We will then check that $BA = I_2$ and, hence, that B is the inverse of A. We must solve

$$\begin{bmatrix} 2 & 5 \\ 1 & 3 \end{bmatrix} \begin{bmatrix} b_{11} & b_{12} \\ b_{21} & b_{22} \end{bmatrix} = \begin{bmatrix} 1 & 0 \\ 0 & 1 \end{bmatrix}$$

or

$$\begin{array}{ll} 2b_{11} + 5b_{21} = 1 & \qquad 2b_{12} + 5b_{22} = 0 \\ b_{11} + 3b_{21} = 0 \quad \text{and} & \qquad b_{12} + 3b_{22} = 1 \end{array} \qquad (9.9)$$

We solve both equations by reducing them to upper triangular form. Leaving the first equation of either set of equations and eliminating b_{11} from the first set of equations and b_{12} from the second set of equations by multiplying the second equation by (-2) and adding the first and second equation, we find

$$\begin{array}{ll} 2b_{11} + 5b_{21} = 1 & \qquad 2b_{12} + 5b_{22} = 0 \\ \quad\quad -b_{21} = 1 \quad \text{and} & \qquad \quad\quad -b_{22} = -2 \end{array} \qquad (9.10)$$

The left set of equations has the solution

$$b_{21} = -1 \quad \text{and} \quad b_{11} = \frac{1}{2}(1 - 5b_{21}) = \frac{1}{2}(1 + 5) = 3$$

The right set of equations has the solution

$$b_{22} = 2 \quad \text{and} \quad b_{12} = \frac{1}{2}(-5b_{22}) = -5$$

Hence

$$B = \begin{bmatrix} 3 & -5 \\ -1 & 2 \end{bmatrix}$$

You can verify that both $AB = I_2$ and $BA = I_2$, hence $B = A^{-1}$.   ◀

In the optional Subsection 9.2.4, we will present a method that can be employed for larger matrices. It is essentially the same method as in Example 11 but uses augmented matrices and is computationally quite intensive. Graphing calculators or computer software are quite useful for inverting larger matrices.

At the end of Subsection 9.2.2, we mentioned the connection between systems of linear equations and matrix multiplication. A system of linear equations can be written as

$$AX = B \qquad (9.11)$$

If we consider a linear system of n equations in n unknowns, then A is an $n \times n$ matrix. If A is invertible, then multiplying (9.11) by A^{-1} from the left, we find

$$A^{-1}AX = A^{-1}B$$

Since $A^{-1}A = I_n$ and $I_n X = X$, it follows that

$$X = A^{-1}B$$

We will use this to repeat Example 5 of Section 9.1.

▶ **Example 12**   **(Using inverse matrices to solve linear systems)** Solve

$$3x + 5y - z = 10$$
$$2x - y + 3z = 9$$
$$4x + 2y - 3z = -1$$

Solution

The coefficient matrix of A is

$$A = \begin{bmatrix} 3 & 5 & -1 \\ 2 & -1 & 3 \\ 4 & 2 & -3 \end{bmatrix}$$

which we encountered in Example 10, where we saw that

$$A^{-1} = \begin{bmatrix} -\dfrac{3}{73} & \dfrac{13}{73} & \dfrac{14}{73} \\ \dfrac{18}{73} & -\dfrac{5}{73} & -\dfrac{11}{73} \\ \dfrac{8}{73} & \dfrac{14}{73} & -\dfrac{13}{73} \end{bmatrix}$$

We set

$$B = \begin{bmatrix} 10 \\ 9 \\ -1 \end{bmatrix}$$

and compute $A^{-1}B$. We find

$$A^{-1} \begin{bmatrix} 10 \\ 9 \\ -1 \end{bmatrix} = \begin{bmatrix} 1 \\ 2 \\ 3 \end{bmatrix}$$

that is, $x = 1$, $y = 2$, and $z = 3$, as we found in Example 5 of Section 9.1. ◀

We now have two methods for solving systems of linear equations when the number of equations is equal to the number of variables. We can either reduce the system to upper triangular form and then use back-substitution, or we can write the system in matrix form $AX = B$, find A^{-1}, and then compute $X = A^{-1}B$. Of course, the second method only works when A^{-1} exists. If A^{-1} does not exist, the system has either no solution or infinitely many solutions. But when A^{-1} exists, then $AX = B$ has exactly one solution.

There is a simple and very useful criterion for checking whether a 2×2 matrix is invertible. Deriving this criterion will also provide us with a formula for the inverse of an invertible 2×2 matrix. We set

$$A = \begin{bmatrix} a_{11} & a_{12} \\ a_{21} & a_{22} \end{bmatrix} \quad \text{and} \quad B = \begin{bmatrix} b_{11} & b_{12} \\ b_{21} & b_{22} \end{bmatrix}$$

We wish to write the entries of B in terms of the entries of A, such that $AB = BA = I$ or $B = A^{-1}$. From

$$\begin{bmatrix} a_{11} & a_{12} \\ a_{21} & a_{22} \end{bmatrix} \begin{bmatrix} b_{11} & b_{12} \\ b_{21} & b_{22} \end{bmatrix} = \begin{bmatrix} 1 & 0 \\ 0 & 1 \end{bmatrix}$$

we obtain the following set of equations:

$$\begin{aligned} a_{11}b_{11} + a_{12}b_{21} &= 1 \\ a_{21}b_{11} + a_{22}b_{21} &= 0 \end{aligned} \tag{9.12}$$

$$\begin{aligned} a_{11}b_{12} + a_{12}b_{22} &= 0 \\ a_{21}b_{12} + a_{22}b_{22} &= 1 \end{aligned} \tag{9.13}$$

We solve the system of linear equations (9.12) for b_{11} and b_{21}:

$$\begin{aligned} a_{11}b_{11} + a_{12}b_{21} &= 1 \quad (R_1) \\ a_{21}b_{11} + a_{22}b_{21} &= 0 \quad (R_2) \end{aligned}$$

If we compute $a_{11}(R_2) - a_{21}(R_1)$, we find

$$a_{11}a_{22}b_{21} - a_{12}a_{21}b_{21} = -a_{21}$$

Solving this for b_{21}, we first factor b_{21} on the left-hand side. This yields

$$(a_{11}a_{22} - a_{12}a_{21})b_{21} = -a_{21}$$

If $a_{11}a_{22} - a_{12}a_{21} \neq 0$, we can isolate b_{21}. We find

$$b_{21} = -\frac{a_{21}}{a_{11}a_{22} - a_{12}a_{21}}$$

Substituting b_{21} into (R_1) and solving for b_{11}, we find, for $a_{11} \neq 0$,

$$b_{11} = \frac{1}{a_{11}}[1 - a_{12}b_{21}]$$

$$= \frac{1}{a_{11}}\left[1 - \frac{-a_{12}a_{21}}{a_{11}a_{22} - a_{12}a_{21}}\right]$$

$$= \frac{1}{a_{11}}\frac{a_{11}a_{22}}{a_{11}a_{22} - a_{12}a_{21}}$$

$$= \frac{a_{22}}{a_{11}a_{22} - a_{12}a_{21}}$$

(When $a_{11} = 0$, the same final result holds.) When we solve (9.13), we find

$$b_{12} = -\frac{a_{12}}{a_{11}a_{22} - a_{12}a_{21}}$$

$$b_{22} = \frac{a_{11}}{a_{11}a_{22} - a_{12}a_{21}}$$

Our calculations yield two important results.

> If
>
> $$A = \begin{bmatrix} a_{11} & a_{12} \\ a_{21} & a_{22} \end{bmatrix}$$
>
> and $a_{11}a_{22} - a_{12}a_{21} \neq 0$, then A is invertible, and
>
> $$A^{-1} = \frac{1}{a_{11}a_{22} - a_{12}a_{21}}\begin{bmatrix} a_{22} & -a_{12} \\ -a_{21} & a_{11} \end{bmatrix}$$

The expression $a_{11}a_{22} - a_{12}a_{21}$ is called the **determinant** of A and is denoted by det A.

> **Definition** If
>
> $$A = \begin{bmatrix} a_{11} & a_{12} \\ a_{21} & a_{22} \end{bmatrix}$$
>
> then the determinant of A is defined as
>
> $$\det A = a_{11}a_{22} - a_{12}a_{21}$$

Looking back now at the formula for A^{-1}, where A is a 2×2 matrix whose determinant is nonzero, to find the inverse of A, we divide A by the determinant of A, switch the diagonal elements of A, and change the sign of the off-diagonal elements. If the determinant is equal to 0, then the inverse of A does not exist.

The determinant can be defined for any $n \times n$ matrix. We will not give the general formula, which is computationally complicated for $n \geq 3$. Graphing calculators or more sophisticated computer software can compute determinants. But we mention the following result, which allows us to determine whether or not an $n \times n$ matrix has an inverse.

> **Theorem** Suppose that A is an $n \times n$ matrix. A is nonsingular if and only if $\det A \neq 0$.

▶ **Example 13** (**Using the determinant to find inverse matrices**) Determine which of the following matrices is invertible. In each case, compute the inverse if it exists.

(a) $A = \begin{bmatrix} 3 & 5 \\ 2 & 4 \end{bmatrix}$ **(b)** $B = \begin{bmatrix} 2 & 6 \\ 1 & 3 \end{bmatrix}$

Solution

(a) To check whether A is invertible, we compute the determinant of A.

$$\det A = (3)(4) - (2)(5) = 2 \neq 0$$

Hence, A is invertible and we find

$$A^{-1} = \frac{1}{2} \begin{bmatrix} 4 & -5 \\ -2 & 3 \end{bmatrix} = \begin{bmatrix} 2 & -\dfrac{5}{2} \\ -1 & \dfrac{3}{2} \end{bmatrix}$$

(b) Since

$$\det B = (2)(3) - (1)(6) = 0$$

B is not invertible. ◀

We mentioned that if A is invertible, then

$$AX = B$$

has exactly one solution, namely $X = A^{-1}B$. Of particular importance is the case when $B = \mathbf{0}$. That is, assume that A is a 2×2 matrix and $B = \begin{bmatrix} 0 \\ 0 \end{bmatrix}$;

then $X = \begin{bmatrix} 0 \\ 0 \end{bmatrix}$ is a solution of

$$AX = \begin{bmatrix} 0 \\ 0 \end{bmatrix}$$

We call the solution $\begin{bmatrix} 0 \\ 0 \end{bmatrix}$ a **trivial** solution. It is the only solution of $AX = \mathbf{0}$

when A is invertible. If $AX = \mathbf{0}$ has a solution $X \neq \begin{bmatrix} 0 \\ 0 \end{bmatrix}$, then X is called a

nontrivial solution. In order to get a nontrivial solution for

$$AX = \mathbf{0},$$

A must be singular. We formulate this as a theorem in the more general case when A is an $n \times n$ matrix; we will need this result repeatedly.

Theorem Suppose that A is an $n \times n$ matrix, and X and $\mathbf{0}$ are $n \times 1$ matrices. The equation

$$AX = \mathbf{0}$$

has a nontrivial solution if and only if A is singular.

▶ **Example 14** **(Nontrivial solutions)** Let

$$A = \begin{bmatrix} a & 6 \\ 3 & 2 \end{bmatrix}$$

Determine a so that $AX = \mathbf{0}$ has a nontrivial solution, and find a nontrivial solution.

Solution

To determine when $AX = \mathbf{0}$ has a nontrivial solution, we must find conditions for which A is singular or, equivalently, for which $\det A = 0$.

Since $\det A = 2a - 18$, it follows that A is singular if $2a - 18 = 0$ or $a = 9$. Therefore, if $a = 9$, $AX = \mathbf{0}$ has a nontrivial solution. To compute a nontrivial solution, we must solve

$$\begin{bmatrix} 9 & 6 \\ 3 & 2 \end{bmatrix} \begin{bmatrix} x_1 \\ x_2 \end{bmatrix} = \begin{bmatrix} 0 \\ 0 \end{bmatrix}$$

which can be written as a system of two linear equations:

$$9x_1 + 6x_2 = 0$$
$$3x_1 + 2x_2 = 0$$

We see that the first equation is three times the second equation. We can therefore simplify the system and obtain

$$3x_1 + 2x_2 = 0$$
$$0x_2 = 0$$

This shows that the system has infinitely many solutions, namely

$$\left\{ (x_1, x_2) : x_2 = t \text{ and } x_1 = -\frac{2}{3}t, \text{ for } t \in \mathbf{R} \right\}$$

In particular, the system has nontrivial solutions; for instance, choosing $t = 3$, we find $x_1 = -2$ and $x_2 = 3$, or choosing $t = -1$, we find $x_1 = 2/3$ and $x_2 = -1$. Any value for t that is different from 0 will yield a nontrivial solution.

If we graphed the two lines corresponding to the two equations when $a = 9$, the lines would be identical. For all other values of a, the two lines would only intersect at the point $(0, 0)$, which corresponds to the trivial solution. (Try it!) ◀

9.2.4 Computing Inverse Matrices (Optional)

We saw in the preceding subsection how to invert 2×2 matrices by solving two systems of linear equations. This method becomes computationally quite intensive when we try to invert larger matrices. Inverting an $n \times n$ matrix would result in n linear systems, each system consisting of n equations in n unknowns. By solving these n systems simultaneously, we can speed up the process quite a bit.

When we look back at Equation (9.9) of Example 11 of Subsection 9.2.3,

$$2b_{11} + 5b_{21} = 1 \qquad \text{and} \qquad 2b_{12} + 5b_{22} = 0$$
$$b_{11} + 3b_{21} = 0 \qquad\qquad\qquad b_{12} + 3b_{22} = 1$$

and rewrite both systems as augmented matrices, namely

$$\left[\begin{array}{cc|c} 2 & 5 & 1 \\ 1 & 3 & 0 \end{array} \right] \qquad \text{and} \qquad \left[\begin{array}{cc|c} 2 & 5 & 0 \\ 1 & 3 & 1 \end{array} \right]$$

we see that each augmented matrix has the same matrix A on its left side. To solve each system, we must perform row transformations until we can read off the solutions. Reading off the solutions is particularly easy when the matrix on the left is transformed into the identity matrix. We do this for the augmented matrix

$$\left[\begin{array}{cc|c} 2 & 5 & 1 \\ 1 & 3 & 0 \end{array} \right] \qquad \begin{array}{c} (R_1) \\ (R_2) \end{array}$$

We divide the first row by 2 in order to get a 1 in the first position. In the second row we want a 0 in the first position; to get this we subtract $2(R_2)$ from (R_1). That is,

$$\begin{array}{c} \frac{1}{2}(R_1) \\ (R_1) - 2(R_2) \end{array} \quad \left[\begin{array}{cc|c} 1 & \frac{5}{2} & \frac{1}{2} \\ 0 & -1 & 1 \end{array} \right] \quad \begin{array}{c} (R_3) \\ (R_4) \end{array}$$

By adding $\frac{5}{2}(R_4)$ to (R_3), we get a 0 in the second position in the first row. Multiplying (R_4) by -1, we get a 1 in the second position in the second row. That is, we obtain the identity matrix on the left side.

$$\begin{array}{c} (R_3) + \dfrac{5}{2}(R_4) \\ (-1)(R_4) \end{array} \quad \left[\begin{array}{cc|c} 1 & 0 & 3 \\ 0 & 1 & -1 \end{array} \right]$$

Rewriting this as a system of linear equations, we see that

$$b_{11} = 3$$
$$b_{21} = -1$$

To find b_{12} and b_{22}, we need to transform the augmented matrix

$$\left[\begin{array}{cc|c} 2 & 5 & 0 \\ 1 & 3 & 1 \end{array} \right]$$

so that the matrix on the left is the identity matrix. This involves the exact same transformations, since the coefficient matrix is the same for both systems of linear equations. Now comes the trick that speeds up the calculation. Instead of doing each augmented matrix separately, we do them simultaneously. Namely, we write

$$\left[\begin{array}{cc|cc} 2 & 5 & 1 & 0 \\ 1 & 3 & 0 & 1 \end{array} \right]$$

and then make row transformations until this is of the form

$$\left[\begin{array}{cc|cc} 1 & 0 & b_{11} & b_{12} \\ 0 & 1 & b_{21} & b_{22} \end{array} \right]$$

It then follows that

$$A^{-1} = \left[\begin{array}{cc} b_{11} & b_{12} \\ b_{21} & b_{22} \end{array} \right]$$

Let's do the calculation:

$$\left[\begin{array}{cc|cc} 2 & 5 & 1 & 0 \\ 1 & 3 & 0 & 1 \end{array}\right] \quad \begin{array}{l} (R_1) \\ (R_2) \end{array}$$

$$\begin{array}{c} \frac{1}{2}(R_1) \\ (R_1) - 2(R_2) \end{array} \left[\begin{array}{cc|cc} 1 & \frac{5}{2} & \frac{1}{2} & 0 \\ 0 & -1 & 1 & -2 \end{array}\right] \quad \begin{array}{l} (R_3) \\ (R_4) \end{array}$$

$$\begin{array}{c} (R_3) + \frac{5}{2}(R_4) \\ (-1)(R_4) \end{array} \left[\begin{array}{cc|cc} 1 & 0 & 3 & -5 \\ 0 & 1 & -1 & 2 \end{array}\right]$$

We recognize the column

$$\left[\begin{array}{c} 3 \\ -1 \end{array}\right]$$

which gives

$$b_{11} = 3 \quad \text{and} \quad b_{21} = -1$$

The column on the right,

$$\left[\begin{array}{c} -5 \\ 2 \end{array}\right]$$

corresponds to the solution

$$b_{12} = -5 \quad \text{and} \quad b_{22} = 2$$

We conclude that

$$A^{-1} = \left[\begin{array}{cc} 3 & -5 \\ -1 & 2 \end{array}\right]$$

as we saw in Example 11.

This method works for larger matrices as well. It is quite efficient, considering that we must solve a large number of linear equations. We illustrate the technique on a 3×3 matrix.

▶ **Example 15**   Find the inverse (if it exists) of

$$A = \left[\begin{array}{ccc} 1 & -1 & -1 \\ 2 & -1 & 1 \\ -1 & 1 & -1 \end{array}\right]$$

Solution

We write

$$\left[\begin{array}{ccc|ccc} 1 & -1 & -1 & 1 & 0 & 0 \\ 2 & -1 & 1 & 0 & 1 & 0 \\ -1 & 1 & -1 & 0 & 0 & 1 \end{array}\right] \quad \begin{array}{l} (R_1) \\ (R_2) \\ (R_3) \end{array}$$

and perform appropriate row transformations in order to get the identity matrix on the left side. The matrix on the right side is then the inverse matrix. The first step is to get 0s below the 1 in the first column.

$$\begin{array}{c} (R_1) \\ 2(R_1) - (R_2) \\ (R_1) + (R_3) \end{array} \left[\begin{array}{ccc|ccc} 1 & -1 & -1 & 1 & 0 & 0 \\ 0 & -1 & -3 & 2 & -1 & 0 \\ 0 & 0 & -2 & 1 & 0 & 1 \end{array}\right] \quad \begin{array}{l} (R_4) \\ (R_5) \\ (R_6) \end{array}$$

Now, we turn to the second column. We want a 1 in the middle and 0s on top and bottom of the second column.

$$
\begin{matrix}
(R_4) - (R_5) \\
(-1)(R_5) \\
(R_6)
\end{matrix}
\left[
\begin{array}{ccc|ccc}
1 & 0 & 2 & -1 & 1 & 0 \\
0 & 1 & 3 & -2 & 1 & 0 \\
0 & 0 & -2 & 1 & 0 & 1
\end{array}
\right]
\begin{matrix}
(R_7) \\
(R_8) \\
(R_9)
\end{matrix}
$$

Next, we turn to the third column. We need 0s in the first two rows and a 1 in the third row. Note that we can only use (R_9) to do the transformations.

$$
\begin{matrix}
(R_7) + (R_9) \\
2(R_8) + 3(R_9) \\
-\dfrac{1}{2}(R_9)
\end{matrix}
\left[
\begin{array}{ccc|ccc}
1 & 0 & 0 & 0 & 1 & 1 \\
0 & 2 & 0 & -1 & 2 & 3 \\
0 & 0 & 1 & -\dfrac{1}{2} & 0 & -\dfrac{1}{2}
\end{array}
\right]
\begin{matrix}
(R_{10}) \\
(R_{11}) \\
(R_{12})
\end{matrix}
$$

Finally, we need to get a 1 in the second row and second column.

$$
\begin{matrix}
(R_{10}) \\
\dfrac{1}{2}(R_{11}) \\
(R_{12})
\end{matrix}
\left[
\begin{array}{ccc|ccc}
1 & 0 & 0 & 0 & 1 & 1 \\
0 & 1 & 0 & -\dfrac{1}{2} & 1 & \dfrac{3}{2} \\
0 & 0 & 1 & -\dfrac{1}{2} & 0 & -\dfrac{1}{2}
\end{array}
\right]
$$

We succeeded in getting the identity matrix on the left side. Therefore,

$$
A^{-1} =
\begin{bmatrix}
0 & 1 & 1 \\
-\dfrac{1}{2} & 1 & \dfrac{3}{2} \\
-\dfrac{1}{2} & 0 & -\dfrac{1}{2}
\end{bmatrix}
$$

◀

This method is reasonably quick as long as the matrices are not too big. If you cannot get the identity matrix on the left side, then the matrix does not have an inverse, as illustrated in the next example.

▶ **Example 16** Find the inverse (if it exists) of

$$
A =
\begin{bmatrix}
2 & 6 \\
1 & 3
\end{bmatrix}
$$

Solution

Since det $A = (2)(3) - (1)(6) = 0$, we already know that A is singular and therefore does not have an inverse. But let's see what happens if we try to find the inverse.

$$
\left[
\begin{array}{cc|cc}
2 & 6 & 1 & 0 \\
1 & 3 & 0 & 1
\end{array}
\right]
\begin{matrix}
(R_1) \\
(R_2)
\end{matrix}
$$

As before, we try to get a 1 in the first row and first column and a 0 in the second row and first column. We find

$$
\begin{matrix}
\dfrac{1}{2}(R_1) \\
(R_1) - 2(R_2)
\end{matrix}
\left[
\begin{array}{cc|cc}
1 & 3 & \dfrac{1}{2} & 0 \\
0 & 0 & 1 & -2
\end{array}
\right]
$$

Since the last row in the left half consists only of 0s, we can never obtain the identity matrix on the left side. That is, our method fails to provide an inverse matrix and we conclude that A does not have an inverse. ◀

9.2.5 An Application: The Leslie Matrix

In this subsection, we discuss age-structured populations with discrete breeding seasons, such as perennial plants in the temperate zone, where reproduction is limited to a particular season of the year. Such populations are described by discrete time models.

We begin with the simplest case. We measure time in units of generations and denote the size of a population at generation t by $N(t)$, where $t = 0, 1, 2, \ldots$. The discrete time analogue of unrestricted growth was discussed in Chapter 2 and is described by

$$N(t + 1) = RN(t) \quad \text{for } t = 0, 1, 2, \ldots \tag{9.14}$$

with $N(0) = N_0$. The constant R can be interpreted as the number of individuals in the next generation per individual in the current generation. We assume that $R \geq 0$. The solution of (9.14) can be found by first computing $N(t)$ for $t = 1, 2$, and 3:

$$N(1) = RN(0)$$

$$N(2) = RN(1) = R[RN(0)] = R^2 N(0)$$

$$N(3) = RN(2) = R[R^2 N(0)] = R^3 N(0)$$

and then recognizing the pattern. We conclude (as in Chapter 2) that

$$N(t) = R^t N(0)$$

Unrestricted growth results in exponential growth of the population. Depending on the value of R, we obtain the following long-term behavior:

$$\lim_{t \to \infty} N(t) = \begin{cases} \infty & \text{if } R > 1 \\ N(0) & \text{if } R = 1 \\ 0 & \text{if } 0 \leq R < 1 \end{cases}$$

(see Figure 9.9).

The population model in (9.14) is the simplest model of discrete time population growth. The quantity R describes the relative change of the population size from generation to generation.

In many species, reproduction is highly age dependent. For instance, periodical cicadas spend 13 to 17 years in the nymphal stage; they only reproduce once in their life. The purple coneflower, a prairie flower, does not reproduce until it is about three years old but then continues to produce seeds throughout its life. To take the life history into account, we will examine discrete time, age-structured models. These models were introduced by Patrick Leslie (1945). They are formulated as matrix models and are based on demographic data. They are used to calculate important demographic quantities, such as the population size in each age class and the growth rate of the population. Leslie's approach is widely used not only in population biology but also in the life insurance industry.

We begin with a numerical example to illustrate this method. Since only females produce offspring, we will only follow the females in the population. We assume that breeding occurs once a year and that we take a census of the population at the end of each breeding season. Individuals that are born during a particular breeding season are of age 0 at the end of this breeding season. If a zero-year-old survives until the end of the following breeding season, it will be of age 1 when we take a census at the end of that breeding season. If a one-year-old survives until the end of the next breeding season, it will be of age 2 at the end of that breeding season, and so on. In our example,

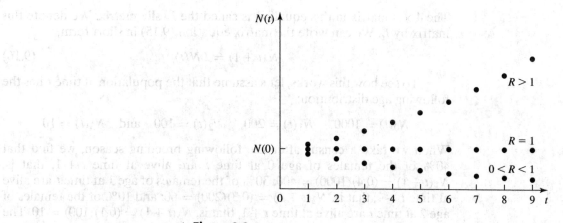

▲ **Figure 9.9**
The population sizes $N(t) = R^t N(0)$ for successive values of t when $R > 1$, $R = 1$, and $0 < R < 1$

we will assume that no individual lives beyond age 3, that is, there are no individuals of age 4 or older in the population. We denote by

$$N_x(t) = \text{number of females of age } x \text{ at time } t$$

where $t = 0, 1, 2, \ldots$. Then

$$N(t) = \begin{bmatrix} N_0(t) \\ N_1(t) \\ N_2(t) \\ N_3(t) \end{bmatrix}$$

is the vector that describes the number of females in each age class at time t.

We assume that 40% of the females of age 0, 30% of the females of age 1, and 10% of the females of age 2 at time t are alive when we take the census at time $t + 1$. That is,

$$N_1(t + 1) = (0.4)N_0(t)$$

$$N_2(t + 1) = (0.3)N_1(t)$$

$$N_3(t + 1) = (0.1)N_2(t)$$

The number of zero-year-old females at time $t + 1$ is equal to the number of female offspring during the breeding season that survive until the end of the breeding season when the census is taken. We assume that

$$N_0(t + 1) = 2N_1(t) + 1.5N_2(t)$$

which means that females reach sexual maturity when they are of age 1. The factor 2 in front of $N_1(t)$ should be interpreted as the average number of surviving female offspring of a one-year-old female. [The factor 1.5 in front of $N_2(t)$ is then the average number of surviving female offspring of a two-year-old female.] There is no contribution of three-year-olds to the newborn class.

We can summarize the dynamics in matrix form:

$$\begin{bmatrix} N_0(t + 1) \\ N_1(t + 1) \\ N_2(t + 1) \\ N_3(t + 1) \end{bmatrix} = \begin{bmatrix} 0 & 2 & 1.5 & 0 \\ 0.4 & 0 & 0 & 0 \\ 0 & 0.3 & 0 & 0 \\ 0 & 0 & 0.1 & 0 \end{bmatrix} \begin{bmatrix} N_0(t) \\ N_1(t) \\ N_2(t) \\ N_3(t) \end{bmatrix} \qquad (9.15)$$

The 4×4 matrix in this equation is called the **Leslie matrix**. We denote this matrix by L. We can write the matrix equation (9.15) in short form,

$$N(t + 1) = LN(t) \tag{9.16}$$

To see how this works, let's assume that the population at time t has the following age distribution:

$$N_0(t) = 1000, \quad N_1(t) = 200, \quad N_2(t) = 100 \quad \text{and} \quad N_3(t) = 10$$

When we take a census after the following breeding season, we find that 40% of the females of age 0 at time t are alive at time $t + 1$, that is, $N_1(t + 1) = (0.4)(1000) = 400$; 30% of the females of age 1 at time t are alive at time $t + 1$, that is, $N_2(t + 1) = (0.3)(200) = 60$; and 10% of the females of age 2 at time t are alive at time $t + 1$, that is, $N_3(t + 1) = (0.1)(100) = 10$. The number of surviving female offspring at time $t + 1$ is

$$N_0(t + 1) = 2N_1(t) + 1.5N_2(t) = (2)(200) + (1.5)(100) = 550$$

We obtain the same result when we use the Leslie matrix and Equation (9.15):

$$
\begin{bmatrix} N_0(t+1) \\ N_1(t+1) \\ N_2(t+1) \\ N_3(t+1) \end{bmatrix} =
\begin{bmatrix} 0 & 2 & 1.5 & 0 \\ 0.4 & 0 & 0 & 0 \\ 0 & 0.3 & 0 & 0 \\ 0 & 0 & 0.1 & 0 \end{bmatrix}
\begin{bmatrix} 1000 \\ 200 \\ 100 \\ 10 \end{bmatrix}
$$

$$
=
\begin{bmatrix} (2)(200) + (1.5)(100) \\ (0.4)(1000) \\ (0.3)(200) \\ (0.1)(100) \end{bmatrix} =
\begin{bmatrix} 550 \\ 400 \\ 60 \\ 10 \end{bmatrix}
$$

To obtain the age distribution at time $t + 2$, we apply the Leslie matrix to the population vector at time $t + 1$, that is,

$$
\begin{bmatrix} N_0(t+2) \\ N_1(t+2) \\ N_2(t+2) \\ N_3(t+2) \end{bmatrix} =
\begin{bmatrix} 0 & 2 & 1.5 & 0 \\ 0.4 & 0 & 0 & 0 \\ 0 & 0.3 & 0 & 0 \\ 0 & 0 & 0.1 & 0 \end{bmatrix}
\begin{bmatrix} 550 \\ 400 \\ 60 \\ 10 \end{bmatrix} =
\begin{bmatrix} 890 \\ 220 \\ 120 \\ 6 \end{bmatrix}
$$

These three breeding seasons are illustrated in Figure 9.10.

From the preceding discussion, we can now find the general form of the Leslie matrix. We assume that the population is divided into $m + 1$ age

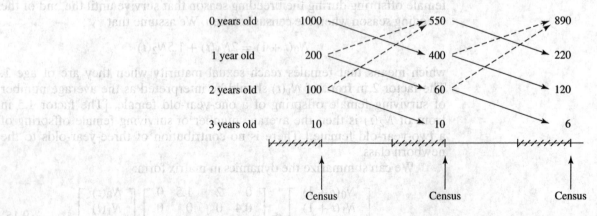

▲ **Figure 9.10**
An illustration of the three breeding seasons

classes. The census is again taken at the end of each breeding season. Then, for the survival of each age class, we find

$$N_1(t + 1) = P_0 N_0(t)$$
$$N_2(t + 1) = P_1 N_1(t)$$
$$\vdots$$
$$N_m(t + 1) = P_{m-1} N_{m-1}(t)$$

where P_i denotes the fraction of females of age i at time t that survive to time $t + 1$. Since P_i denotes a fraction, it follows that $0 \leq P_i \leq 1$ for $i = 0, 1, \ldots, m - 1$. The equation for zero-year-old females is given by

$$N_0(t + 1) = F_0 N_0(t) + F_1 N_1(t) + \cdots + F_m N_m(t)$$

where F_i is the average number of surviving female offspring per female individual of age i. Writing this in matrix form, we find that the Leslie matrix is given by

$$\begin{bmatrix} F_0 & F_1 & F_2 & \cdots & F_{m-1} & F_m \\ P_0 & 0 & \cdots & \cdots & \cdots & 0 \\ 0 & P_1 & 0 & \cdots & \cdots & 0 \\ \vdots & \cdots & \cdots & \cdots & \cdots & \vdots \\ 0 & \cdots & \cdots & 0 & P_{m-1} & 0 \end{bmatrix}$$

This is a matrix in which all elements are 0, except possibly those in the first row and in the first subdiagonal below the diagonal.

▶ **Example 17**   Suppose that the Leslie matrix of a population is of the form

$$L = \begin{bmatrix} 5 & 7 & 1.5 \\ 0.2 & 0 & 0 \\ 0 & 0.4 & 0 \end{bmatrix}$$

Interpret this matrix, and determine what happens if you follow a population for two breeding seasons, starting with 1000 zero-year-old females.

Solution

The population is divided into three age classes, zero-year-olds, one-year-olds, and two-year-olds. The elements in the subdiagonal describe the survival probabilities; that is, 20% of the zero-year-olds survive until the next census and 40% of the one-year-olds survive until the next census. The maximum age is 2 years. Zero-year-olds produce an average of five surviving females per female; one-year-olds produce an average of seven surviving females per female; two-year-olds produce an average of 1.5 surviving females per female.

A population that starts with 1000 zero-year-old females at time 0 has the population vector

$$N(0) = \begin{bmatrix} 1000 \\ 0 \\ 0 \end{bmatrix}$$

Using $N(t + 1) = LN(t)$, we find

$$N(1) = \begin{bmatrix} 5 & 7 & 1.5 \\ 0.2 & 0 & 0 \\ 0 & 0.4 & 0 \end{bmatrix} \begin{bmatrix} 1000 \\ 0 \\ 0 \end{bmatrix} = \begin{bmatrix} 5000 \\ 200 \\ 0 \end{bmatrix}$$

and

$$N(2) = \begin{bmatrix} 5 & 7 & 1.5 \\ 0.2 & 0 & 0 \\ 0 & 0.4 & 0 \end{bmatrix} \begin{bmatrix} 5000 \\ 200 \\ 0 \end{bmatrix} = \begin{bmatrix} 26{,}400 \\ 1000 \\ 80 \end{bmatrix}$$ ◄

Stable Age Distribution We will now investigate what happens when we run such a model for a long time. Let's assume that the Leslie matrix is given by

$$L = \begin{bmatrix} 1.5 & 2 \\ 0.08 & 0 \end{bmatrix}$$

We can compute successive population vectors using the equation

$$N(t+1) = LN(t)$$

That is,

$$N_0(t+1) = 1.5N_0(t) + 2N_1(t)$$

$$N_1(t+1) = 0.08N_0(t)$$

Suppose that we start with $N_0(0) = 100$ and $N_1(0) = 100$. Then

$$\begin{bmatrix} N_0(1) \\ N_1(1) \end{bmatrix} = \begin{bmatrix} 1.5 & 2 \\ 0.08 & 0 \end{bmatrix} \begin{bmatrix} 100 \\ 100 \end{bmatrix}$$

$$= \begin{bmatrix} 350 \\ 8 \end{bmatrix}$$

Continuing in this way, we find the population vectors at successive times, starting at time 0,

$$\begin{bmatrix} 100 \\ 100 \end{bmatrix}, \quad \begin{bmatrix} 350 \\ 8 \end{bmatrix}, \quad \begin{bmatrix} 541 \\ 28 \end{bmatrix}, \quad \begin{bmatrix} 868 \\ 43 \end{bmatrix}, \quad \begin{bmatrix} 1388 \\ 69 \end{bmatrix},$$

$$\begin{bmatrix} 2221 \\ 111 \end{bmatrix}, \quad \begin{bmatrix} 3553 \\ 178 \end{bmatrix}, \quad \begin{bmatrix} 5685 \\ 284 \end{bmatrix}, \cdots$$

(In these calculations, we rounded to the nearest integer.) The first thing that you notice is that the total population is growing [the population size at time t is $N_0(t) + N_1(t)$]. But we can say a lot more. If we look at the successive ratios

$$q_0(t) = \frac{N_0(t)}{N_0(t-1)}$$

we find the following:

t	1	2	3	4	5	6	7
$q_0(t)$	3.5	1.55	1.60	1.5991	1.6001	1.5997	1.6001

That is, $q_0(t)$ seems to approach a limiting value, namely 1.6. The same happens when we look at the ratio

$$q_1(t) = \frac{N_1(t)}{N_1(t-1)}$$

We find

t	1	2	3	4	5	6	7
$q_1(t)$	0.08	3.5	1.536	1.605	1.609	1.604	1.596

Both ratios seem to approach the same value, namely 1.6. In fact, we can show that

$$\lim_{t \to \infty} \frac{N_0(t)}{N_0(t-1)} = \lim_{t \to \infty} \frac{N_1(t)}{N_1(t-1)} = 1.6$$

(We will see in the next section how to get this result analytically from the Leslie matrix.) If we look at the fraction of females of age class 0, that is,

$$p(t) = \frac{N_0(t)}{N_0(t) + N_1(t)}$$

we find

t	0	1	2	3	4	5	6	7
$p(t)$	0.5	0.9777	0.9508	0.9528	0.9526	0.9524	0.9523	0.9524

That is, this fraction also seems to converge. It looks as if about 95.2% of the population are in age class 0 when t is sufficiently large.

Though the population is increasing in size, the fraction of females in age class 0 (and hence also in age class 1) seems to converge. This constant fraction is referred to as the **stable age distribution**.

This method of finding the stable age distribution does not always work, and we will give an example in Problem 68, in which the population does not reach a stable age distribution.

If we start the population in the stable age distribution, the fraction of females in age class 0 will remain the same, namely about 95.2%, and the population will increase by a constant factor each generation, namely by a factor of 1.6. Here is a numerical illustration of these two important properties. A stable age distribution for this population is

$$N(0) = \begin{bmatrix} 2000 \\ 100 \end{bmatrix}$$

(We will learn in the next section how to find this.) If we start with this stable age distribution, then

$$N(1) = LN(0) = \begin{bmatrix} 1.5 & 2 \\ 0.08 & 0 \end{bmatrix} \begin{bmatrix} 2000 \\ 100 \end{bmatrix} = \begin{bmatrix} 3200 \\ 160 \end{bmatrix}$$

The fraction of females in age class 0 remains the same, namely, $3200/3360 = 2000/2100$. We compare this to

$$(1.6) \begin{bmatrix} 2000 \\ 100 \end{bmatrix} = \begin{bmatrix} 3200 \\ 160 \end{bmatrix}$$

which yields the same result. That is, if N denotes a stable age distribution, then

$$LN = \lambda N \tag{9.17}$$

where $\lambda = 1.6$ and $N = [N_0, N_1]'$, with $N_0/(N_0 + N_1) \approx 95.2\%$. [We will see in the next section how to compute λ and the ratio $N_0/(N_0 + N_1)$.] Equation (9.17) is used to determine the stable age distribution. The vector N is called an **eigenvector**; the value λ is called the corresponding **eigenvalue**. A 2×2 matrix has two eigenvalues (which might be identical). In the case of the Leslie matrix, the largest eigenvalue is interpreted as the growth parameter; that is, it determines how the population grows. Its associated eigenvector is a stable age distribution.

In the next section, we will learn how to compute eigenvalues and eigenvectors. Because we will need this in other contexts as well, we will develop the theory in a more general setting. At the end of the next section, we will return to Leslie matrices and see how to compute the stable age distribution and the growth parameter of the population.

9.2.6 Problems

(9.2.1), (9.2.2)

In Problems 1–6, let

$$A = \begin{bmatrix} -1 & 2 \\ 0 & -3 \end{bmatrix},$$

$$B = \begin{bmatrix} 0 & 1 \\ 2 & 4 \end{bmatrix},$$

$$C = \begin{bmatrix} 1 & -2 \\ 1 & -1 \end{bmatrix}$$

1. Find $A - B + 2C$.
2. Find $-2A + 3B$.
3. Determine D so that $A + B = 2A - B + D$.
4. Show that $A + B = B + A$.
5. Show that $(A + B) + C = A + (B + C)$.
6. Show that $2(A + B) = 2A + 2B$.

In Problems 7–12, let

$$A = \begin{bmatrix} 1 & 3 & -1 \\ 2 & 4 & 1 \\ 0 & -2 & 2 \end{bmatrix},$$

$$B = \begin{bmatrix} 5 & -1 & 4 \\ 2 & 0 & 1 \\ 1 & -3 & -3 \end{bmatrix},$$

$$C = \begin{bmatrix} -2 & 0 & 4 \\ 1 & -3 & 1 \\ 0 & 0 & 2 \end{bmatrix}$$

7. Find $2A + 3B - C$.
8. Find $3C - B + \frac{1}{2}A$.
9. Determine D so that $A + B + C + D = \mathbf{0}$.
10. Determine D so that $A + 4B = 2(A + B) + D$.
11. Show that $A + B = B + A$.
12. Show that $(A + B) + C = A + (B + C)$.
13. Find the transpose of

$$A = \begin{bmatrix} -1 & 0 & 3 \\ 2 & 1 & -4 \end{bmatrix}$$

14. Find the transpose of

$$A = \begin{bmatrix} 1 \\ -3 \\ 4 \end{bmatrix}$$

In Problems 15–20, let

$$A = \begin{bmatrix} -1 & 0 \\ 1 & 2 \end{bmatrix},$$

$$B = \begin{bmatrix} 2 & 3 \\ -1 & 1 \end{bmatrix},$$

$$C = \begin{bmatrix} 1 & 2 \\ 0 & -1 \end{bmatrix}$$

15. Compute the following.
 (a) AB
 (b) BA
16. Compute ABC.
17. Show that $AC \neq CA$.
18. Show that $(AB)C = A(BC)$.
19. Show that $(A + B)C = AC + BC$.
20. Show that $A(B + C) = AB + AC$.
21. Suppose that A is a 3×4 matrix and B is a 4×2 matrix. What is the size of the product AB?
22. Suppose that A is a 2×5 matrix, B is a 1×3, matrix, C is a 5×1 matrix, and D is a 2×3 matrix. Which of the following matrix multiplications are defined? Whenever it is defined, state the size of the resulting matrix.
 (a) AB
 (b) AC
 (c) BD
23. Suppose that A is a 4×3 matrix, B is a 1×3 matrix, C is a 3×1 matrix, and D is a 4×3 matrix. Which of the following matrix multiplications are defined? Whenever it is defined, state the size of the resulting matrix.
 (a) BD'
 (b) $D'A$
 (c) ACB
24. Let

$$A = \begin{bmatrix} 1 & 4 & -2 \end{bmatrix}$$

and

$$B = \begin{bmatrix} -1 \\ 2 \\ 3 \end{bmatrix}$$

 (a) Compute AB.
 (b) Compute BA.
25. Let

$$A = \begin{bmatrix} 1 & 3 \\ 0 & -2 \end{bmatrix}$$

and

$$B = \begin{bmatrix} 1 & 2 & 0 & -1 \\ 2 & 1 & 3 & 0 \end{bmatrix}$$

 (a) Compute AB.
 (b) Compute $B'A$.
26. Suppose that

$$A = \begin{bmatrix} 1 & -1 \\ 3 & 0 \\ 5 & 2 \end{bmatrix}$$

and

$$B = \begin{bmatrix} 2 & 4 & 1 \\ 6 & 0 & 0 \end{bmatrix}$$

Show that $(AB)' = B'A'$.

27. Let

$$A = \begin{bmatrix} 2 & 1 \\ -1 & -3 \end{bmatrix}$$

Find A^2, A^3, and A^4.

28. Let

$$I_3 = \begin{bmatrix} 1 & 0 & 0 \\ 0 & 1 & 0 \\ 0 & 0 & 1 \end{bmatrix}$$

Show that $I_3 = I_3^2 = I_3^3$.

29. Let

$$A = \begin{bmatrix} 1 & 3 \\ 0 & -2 \end{bmatrix}$$

and

$$I_2 = \begin{bmatrix} 1 & 0 \\ 0 & 1 \end{bmatrix}$$

Show that $AI_2 = I_2A = A$.

30. Let

$$A = \begin{bmatrix} 1 & 3 & 0 \\ 0 & -1 & 2 \\ -1 & -2 & 1 \end{bmatrix}$$

and

$$I_3 = \begin{bmatrix} 1 & 0 & 0 \\ 0 & 1 & 0 \\ 0 & 0 & 1 \end{bmatrix}$$

Show that $AI_3 = I_3A = A$.

31. Write the following in matrix form.
$$\begin{aligned} 2x_1 + 3x_2 - x_3 &= 0 \\ 2x_2 + x_3 &= 1 \\ x_1 \qquad - 2x_3 &= 2 \end{aligned}$$

32. Write the following in matrix form.
$$\begin{aligned} 2x_2 - x_1 &= x_3 \\ 4x_1 + x_3 &= 7x_2 \\ x_2 - x_1 &= x_3 \end{aligned}$$

33. Write the following in matrix form.
$$\begin{aligned} 2x_1 - 3x_2 &= 4 \\ -x_1 + x_2 &= 3 \\ 3x_1 &= 4 \end{aligned}$$

34. Write the following in matrix form.
$$\begin{aligned} x_1 - 2x_2 + x_3 &= 1 \\ -2x_1 + x_2 - 3x_3 &= 0 \end{aligned}$$

(9.2.3)

35. Show that the inverse of

$$A = \begin{bmatrix} 2 & 1 \\ 1 & 1 \end{bmatrix}$$

is

$$B = \begin{bmatrix} 1 & -1 \\ -1 & 2 \end{bmatrix}$$

36. Show that the inverse of

$$A = \begin{bmatrix} 2 & 3 & 1 \\ 5 & 2 & 3 \\ 1 & 2 & 0 \end{bmatrix}$$

is

$$B = \begin{bmatrix} -\frac{6}{5} & \frac{2}{5} & \frac{7}{5} \\ \frac{3}{5} & -\frac{1}{5} & -\frac{1}{5} \\ \frac{8}{5} & -\frac{1}{5} & -\frac{11}{5} \end{bmatrix}$$

In Problems 37–40, let

$$A = \begin{bmatrix} -1 & 1 \\ 2 & 3 \end{bmatrix},$$

$$B = \begin{bmatrix} 2 & -2 \\ 3 & 2 \end{bmatrix}$$

37. Find the inverse (if it exists) of A.

38. Find the inverse (if it exists) of B.

39. Show that $(A^{-1})^{-1} = A$.

40. Show that $(AB)^{-1} = B^{-1}A^{-1}$.

41. Find the inverse (if it exists) of

$$C = \begin{bmatrix} 2 & 4 \\ 3 & 6 \end{bmatrix}$$

42. Find the inverse (if it exists) of

$$I_3 = \begin{bmatrix} 1 & 0 & 0 \\ 0 & 1 & 0 \\ 0 & 0 & 1 \end{bmatrix}$$

43. Suppose that

$$A = \begin{bmatrix} -1 & 0 \\ 2 & -3 \end{bmatrix}$$

and

$$D = \begin{bmatrix} -2 \\ -5 \end{bmatrix}$$

Find X, so that $AX = D$, by

(a) solving the associated system of linear equations, and

(b) using the inverse of A.

44. (a) Show that if $X = AX + D$, then

$$X = (I - A)^{-1}D$$

provided that $I - A$ is invertible.

(b) Suppose now that

$$A = \begin{bmatrix} 3 & 2 \\ 0 & -1 \end{bmatrix}$$

and

$$D = \begin{bmatrix} -2 \\ 2 \end{bmatrix}$$

Compute $(I - A)^{-1}$, and use your result in (a) to compute X.

45. Use the determinant to determine whether the matrix

$$A = \begin{bmatrix} 2 & -1 \\ 1 & 3 \end{bmatrix}$$

is invertible.

46. Use the determinant to determine whether the matrix

$$A = \begin{bmatrix} -1 & 3 \\ 0 & 3 \end{bmatrix}$$

is invertible.

47. Use the determinant to determine whether the matrix

$$A = \begin{bmatrix} 4 & -1 \\ 8 & -2 \end{bmatrix}$$

is invertible.

48. Use the determinant to determine whether the matrix

$$A = \begin{bmatrix} -1 & 1 \\ -1 & 1 \end{bmatrix}$$

is invertible.

49. Suppose that

$$A = \begin{bmatrix} 2 & 4 \\ 3 & 6 \end{bmatrix}$$

(a) Compute det A. Is A invertible?

(b) Suppose that

$$X = \begin{bmatrix} x \\ y \end{bmatrix}$$

and

$$B = \begin{bmatrix} b_1 \\ b_2 \end{bmatrix}$$

Write $AX = B$ as a system of linear equations.

(c) Show that if

$$B = \begin{bmatrix} 3 \\ 9 \\ 2 \end{bmatrix}$$

then

$$AX = B$$

has infinitely many solutions. Graph the two straight lines associated with the corresponding system of linear equations, and explain why it has infinitely many solutions.

(d) Find a column vector

$$B = \begin{bmatrix} b_1 \\ b_2 \end{bmatrix}$$

so that

$$AX = B$$

has no solutions.

50. Suppose that

$$A = \begin{bmatrix} a & 8 \\ 2 & 4 \end{bmatrix},$$

$$X = \begin{bmatrix} x \\ y \end{bmatrix},$$

and

$$B = \begin{bmatrix} b_1 \\ b_2 \end{bmatrix}$$

(a) Show that when $a \neq 4$, then $AX = B$ has exactly one solution.

(b) Suppose $a = 4$. Find conditions on b_1 and b_2 so that $AX = B$ has (i) infinitely many solutions, and (ii) no solutions.

(c) Explain your results in (a) and (b) graphically.

51. Use the determinant to determine whether

$$A = \begin{bmatrix} 1 & -1 \\ 0 & 2 \end{bmatrix}$$

is invertible. If it is invertible, compute its inverse. In either case, solve $AX = 0$.

52. Use the determinant to determine whether

$$B = \begin{bmatrix} 1 & 1 \\ 2 & 1 \end{bmatrix}$$

is invertible. If it is invertible, compute its inverse. In either case, solve $BX = 0$.

53. Use the determinant to determine whether

$$C = \begin{bmatrix} 1 & 3 \\ 1 & 3 \end{bmatrix}$$

is invertible. If it is invertible, compute its inverse. In either case, solve $CX = 0$.

54. Use the determinant to determine whether

$$D = \begin{bmatrix} -3 & 6 \\ -4 & 8 \end{bmatrix}$$

is invertible. If it is invertible, compute its inverse. In either case, solve $DX = 0$.

(9.2.4)

In Problems 55–58, find the inverse matrix to each given matrix, if it exists.

55.
$$A = \begin{bmatrix} 2 & -1 & -1 \\ 2 & 1 & 1 \\ -1 & 1 & -1 \end{bmatrix}$$

56.
$$A = \begin{bmatrix} -1 & 3 & -1 \\ 2 & -2 & 3 \\ -1 & 1 & 2 \end{bmatrix}$$

57.
$$A = \begin{bmatrix} -1 & 0 & -1 \\ 0 & -2 & 0 \\ -1 & 1 & 2 \end{bmatrix}$$

58.
$$A = \begin{bmatrix} -1 & 0 & 2 \\ -1 & -2 & 3 \\ 0 & 2 & -1 \end{bmatrix}$$

(9.2.5)

In Problems 59–62, assume that breeding occurs once a year and that a census is taken at the end of each breeding season.

59. Assume that a population is divided into three age classes and that 20% of the females of age 0 and 70% of the females of age 1 survive until the end of the following breeding season. Furthermore, assume that females of age 1 have an average of 3.2 female offspring and females of age 2 have an average of 1.7 female offspring. If, at time 0, the population consists of 2000 females of age 0, 800 females of age 1, and 200 females of age 2, find the Leslie matrix and the age distribution at time 2.

60. Assume that a population is divided into three age classes and that 80% of the females of age 0 and 10% of the females of age 1 survive until the end of the following breeding season. Furthermore, assume that females of age 1 have an average of 1.6 female offspring and females of age 2 have an average of 3.9 female offspring. If, at time 0, the population consists of 1000 females of age 0, 100 females of age 1, and 20 females of age 2, find the Leslie matrix and the age distribution at time 3.

61. Assume that a population is divided into four age classes and that 70% of the females of age 0, 50% of the females of age 1, and 10% of the females of age 2 survive until the end of the following breeding season. Furthermore, assume that females of age 2 have an average of 4.6 female offspring and females of age 3 have an average of 3.7 female offspring. If, at time 0, the population consists of 1500 females of age 0, 500 females of age 1, 250 females of age 2, and 50 females of age 3, find the Leslie matrix and the age distribution at time 2.

62. Assume that a population is divided into four age classes and that 65% of the females of age 0, 40% of the females of age 1, and 30% of the females of age 2 survive until the end of the following breeding season. Furthermore, assume that females of age 1 have an average of 2.8 female offspring, females of age 2 have an average of 7.2 female offspring, and females of age 3 have an average of 3.7 female offspring. If, at time 0, the population consists of 1500 females of age 0, 500 females of age 1, 250 females of age 2, and 50 females of age 3, find the Leslie matrix and the age distribution at time 3.

In Problems 63–64, assume the given Leslie matrix L. Determine the number of age classes in the population, the fraction of one-year-olds that survive until the end of the following breeding season, and the average number of female offspring of a two-year-old female.

63.
$$L = \begin{bmatrix} 2 & 3 & 2 & 1 \\ 0.4 & 0 & 0 & 0 \\ 0 & 0.6 & 0 & 0 \\ 0 & 0 & 0.8 & 0 \end{bmatrix}$$

64.
$$L = \begin{bmatrix} 0 & 5 & 0 \\ 0.8 & 0 & 0 \\ 0 & 0.3 & 0 \end{bmatrix}$$

In Problems 65–66, assume the given Leslie matrix L. Determine the number of age classes in the population. What fraction of two-year-olds survive until the end of the following breeding season? Determine the average number of female offspring of a one-year-old female.

65.
$$L = \begin{bmatrix} 1 & 2.5 & 3 & 1.5 \\ 0.9 & 0 & 0 & 0 \\ 0 & 0.3 & 0 & 0 \\ 0 & 0 & 0.2 & 0 \end{bmatrix}$$

66.
$$L = \begin{bmatrix} 0 & 4.2 & 3.7 \\ 0.7 & 0 & 0 \\ 0 & 0.1 & 0 \end{bmatrix}$$

67. Assume that the Leslie matrix is of the form
$$L = \begin{bmatrix} 1.2 & 3.2 \\ 0.8 & 0 \end{bmatrix}$$
Suppose that at time $t = 0$, $N_0(0) = 100$, and $N_1(0) = 0$. Find the population vectors for $t = 0, 1, 2, \ldots, 10$. Compute the successive ratios

$$q_0(t) = \frac{N_0(t)}{N_0(t-1)}$$

and

$$q_1(t) = \frac{N_1(t)}{N_1(t-1)}$$

for $t = 1, 2, \ldots, 10$. What value do $q_0(t)$ and $q_1(t)$ approach as $t \to \infty$? (Take a guess.) Compute the fraction of females of age 0 for $t = 0, 1, \ldots, 10$. Can you find a stable age distribution?

68. Assume that a Leslie matrix is of the form
$$L = \begin{bmatrix} 0 & 5.2 \\ 0.7 & 0 \end{bmatrix}$$
Suppose that at time $t = 0$, $N_0(0) = 100$, and $N_1(0) = 0$. Find the population vectors for $t = 0, 1, 2, \ldots, 10$. Compute the successive ratios

$$q_0(t) = \frac{N_0(t)}{N_0(t-1)}$$

and

$$q_1(t) = \frac{N_1(t)}{N_1(t-1)}$$

for $t = 1, 2, \ldots, 10$. What value do $q_0(t)$ and $q_1(t)$ approach as $t \to \infty$? (Take a guess.) Compute the fraction of females of age 0 for $t = 0, 1, \ldots, 10$. Can you find a stable age distribution?

9.3 LINEAR MAPS, EIGENVECTORS, AND EIGENVALUES

In this section, we will consider maps of the form

$$\mathbf{x} \to A\mathbf{x} \tag{9.18}$$

where A is a 2×2 matrix and $\mathbf{x}$ is a 2×1 column vector (or, simply, vector). Since $A\mathbf{x}$ is a 2×1 vector, this map takes a 2×1 vector and maps it into a 2×1 vector. This allows us to apply A repeatedly: We can compute $A(A\mathbf{x}) = A^2\mathbf{x}$, which is again a 2×1 vector, and so on. We will first look at vectors, then at maps $A\mathbf{x}$, and finally we will investigate iterates of the map A, that is, $A^2\mathbf{x}$, $A^3\mathbf{x}$, and so on.

In the following we will denote vectors by boldface lowercase letters. It follows from the properties of matrix multiplications that the map (9.18) satisfies the following:

1. $A(\mathbf{x} + \mathbf{y}) = A\mathbf{x} + A\mathbf{y}$, and

2. $A(\lambda\mathbf{x}) = \lambda(A\mathbf{x})$, where λ is a scalar.

Because of these two properties, we say that the map $\mathbf{x} \to A\mathbf{x}$ is linear.

We saw an example of such a map in the previous section: If A is a 2×2 Leslie matrix and $\mathbf{x}$ is a population vector at time 0, then $A\mathbf{x}$ represents the population vector at time 1.

Linear maps are important in other contexts as well, and we will encounter them in Chapters 10 and 11. We restrict our discussion to 2×2 matrices but point out that we can generalize the following theory to arbitrary $n \times n$ matrices. (These topics are covered in courses on matrix or linear algebra.)

9.3.1 Graphical Representation

Vectors We begin with a graphical representation of vectors. We assume that $\mathbf{x}$ is a 2×1 matrix. We call $\mathbf{x}$ a column vector or simply a vector. Since a 2×1 matrix has just two components, we can represent a vector in the plane. For instance, to represent the vector

$$\mathbf{x} = \begin{bmatrix} 3 \\ 4 \end{bmatrix}$$

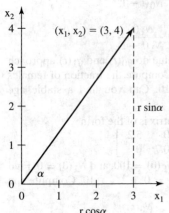

▲ Figure 9.11
The vector $\begin{bmatrix} 3 \\ 4 \end{bmatrix}$ in the $x_1 - x_2$ plane

in the $x_1 - x_2$ plane, we draw an arrow from the origin $(0, 0)$ to the point $(3, 4)$, as illustrated in Figure 9.11. We see from Figure 9.11 that a vector has a length and a direction. The length of the vector $\mathbf{x} = \begin{bmatrix} 3 \\ 4 \end{bmatrix}$, denoted by $|\mathbf{x}|$, is the distance from the origin $(0, 0)$ to the point $(3, 4)$, that is,

$$\text{length of } \mathbf{x} = |\mathbf{x}| = \sqrt{9 + 16} = 5$$

We define the direction of $\mathbf{x}$ as the angle α between the positive x_1-axis and the vector $\mathbf{x}$ (measured counterclockwise), as shown in Figure 9.11. The angle α is in the interval $[0, 2\pi)$. In our example, α satisfies $\tan \alpha = 4/3$.

More generally (see Figure 9.11 again), a vector $\begin{bmatrix} x_1 \\ x_2 \end{bmatrix}$ has length

$$|\mathbf{x}| = \sqrt{x_1^2 + x_2^2}$$

and direction α, where $\alpha \in [0, 2\pi)$ satisfies

$$\tan \alpha = \frac{x_2}{x_1}$$

The angle α is always measured counterclockwise from the positive x_1-axis.

If we denote the length of $\mathbf{x}$ by r, that is, $r = |\mathbf{x}|$, then, as shown in Figure 9.11, since $x_1 = r \cos \alpha$ and $x_2 = r \sin \alpha$, the vector $\mathbf{x}$ can also be written as

$$\mathbf{x} = \begin{bmatrix} r \cos \alpha \\ r \sin \alpha \end{bmatrix}$$

We thus have an alternative way of representing vectors in the plane. We can either use the endpoint (x_1, x_2) or the length and direction (r, α). The first representation leads to our familiar Cartesian coordinate system. The second representation, using the length and direction of the corresponding vector, leads to the **polar coordinate system**. We will use both representations in the following.

▶ **Example 1**   If the length of the vector $\begin{bmatrix} x_1 \\ x_2 \end{bmatrix}$ is 4 and its angle with the positive x_1-axis is 120° (measured clockwise), what is its representation in Cartesian coordinates?

Solution

An angle of 120° measured clockwise from the positive x_1-axis corresponds to an angle 360° − 120° = 240° measured counterclockwise from the positive

x_1-axis (Figure 9.12). Since the length of the vector is 4, we find

$$x_1 = 4\cos(240°) = (4)\left(-\frac{1}{2}\right) = -2$$

$$x_2 = 4\sin(240°) = (4)\left(-\frac{1}{2}\sqrt{3}\right) = -2\sqrt{3}$$

which yields the Cartesian coordinate representation

$$\mathbf{x} = \begin{bmatrix} -2 \\ -2\sqrt{3} \end{bmatrix}$$

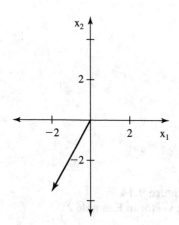

▲ Figure 9.12
The vector in Example 1

Since vectors are matrices, we can use matrix addition to add vectors. For instance,

$$\begin{bmatrix} 1 \\ 2 \end{bmatrix} + \begin{bmatrix} 2 \\ 2 \end{bmatrix} = \begin{bmatrix} 3 \\ 4 \end{bmatrix}$$

This has a graphical interpretation (see Figure 9.13). To add $\begin{bmatrix} 2 \\ 2 \end{bmatrix}$ to $\begin{bmatrix} 1 \\ 2 \end{bmatrix}$, we move the vector $\begin{bmatrix} 2 \\ 2 \end{bmatrix}$ to the tip of the vector $\begin{bmatrix} 1 \\ 2 \end{bmatrix}$ without changing the direction or the length of $\begin{bmatrix} 2 \\ 2 \end{bmatrix}$. The sum of the two vectors then starts at $(0, 0)$ and ends at the point where the tip of the vector that was moved ended. This can also be described in the following way. The sum is the diagonal in the parallelogram that is formed by the two vectors $\begin{bmatrix} 1 \\ 2 \end{bmatrix}$ and $\begin{bmatrix} 2 \\ 2 \end{bmatrix}$. The rules for vector addition are therefore referred to as the **parallelogram law**.

Multiplication of a vector by a scalar is carried out componentwise. For instance, if we multiply $\begin{bmatrix} 1 \\ 2 \end{bmatrix}$ by 2, we find

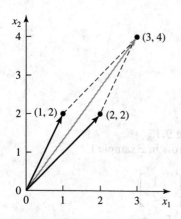

▲ Figure 9.13
Addition of two vectors

$$2\begin{bmatrix} 1 \\ 2 \end{bmatrix} = \begin{bmatrix} 2 \\ 4 \end{bmatrix}$$

This corresponds to changing the length of the vector. The vector $\begin{bmatrix} 1 \\ 2 \end{bmatrix}$ has length $\sqrt{1+4} = \sqrt{5}$; the vector $2\begin{bmatrix} 1 \\ 2 \end{bmatrix}$ has length $\sqrt{4+16} = \sqrt{20} = 2\sqrt{5}$. That is, multiplying the vector $\begin{bmatrix} 1 \\ 2 \end{bmatrix}$ by 2 increases its length by the factor 2. Since 2 is positive, the resulting vector points in the same direction as the vector $\begin{bmatrix} 1 \\ 2 \end{bmatrix}$. If we multiply $\begin{bmatrix} 1 \\ 2 \end{bmatrix}$ by -1, then the resulting vector is $\begin{bmatrix} -1 \\ -2 \end{bmatrix}$, which has the same length as the original vector but points in the opposite direction, as illustrated in Figure 9.14.

▶ **Example 2**   Let

$$\mathbf{u} = \begin{bmatrix} 2 \\ 1 \end{bmatrix}, \qquad \mathbf{v} = \begin{bmatrix} -1 \\ -3 \end{bmatrix}, \qquad \mathbf{w} = \begin{bmatrix} -2 \\ 3 \end{bmatrix}$$

Find $-\frac{1}{2}\mathbf{u}$, $\mathbf{u} + \mathbf{v}$, $\mathbf{v} + \mathbf{w}$, and $-\mathbf{v}$, and illustrate the results graphically.

Solution

$$-\frac{1}{2}\mathbf{u} = -\frac{1}{2}\begin{bmatrix} 2 \\ 1 \end{bmatrix} = \begin{bmatrix} -1 \\ -\frac{1}{2} \end{bmatrix}$$

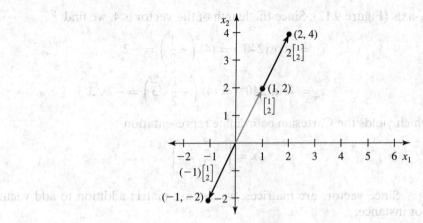

◀ **Figure 9.14**
The vector in Example 2

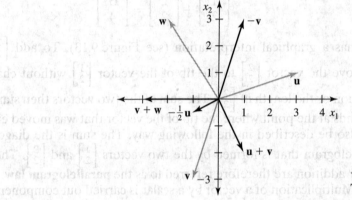

◀ **Figure 9.15**
The vectors in Example 1

$$\mathbf{u} + \mathbf{v} = \left[\begin{array}{c} 2 \\ 1 \end{array} \right] + \left[\begin{array}{c} -1 \\ -3 \end{array} \right] = \left[\begin{array}{c} 1 \\ -2 \end{array} \right]$$

$$\mathbf{v} + \mathbf{w} = \left[\begin{array}{c} -1 \\ -3 \end{array} \right] + \left[\begin{array}{c} -2 \\ 3 \end{array} \right] = \left[\begin{array}{c} -3 \\ 0 \end{array} \right]$$

$$-\mathbf{v} = - \left[\begin{array}{c} -1 \\ -3 \end{array} \right] = \left[\begin{array}{c} 1 \\ 3 \end{array} \right]$$

The results are illustrated in Figure 9.15.

We summarize vector addition, and multiplication of a vector by a scalar, in the following.

Let $\mathbf{x} = \left[\begin{smallmatrix} x_1 \\ x_2 \end{smallmatrix} \right]$ and $\mathbf{y} = \left[\begin{smallmatrix} y_1 \\ y_2 \end{smallmatrix} \right]$. Then

$$\mathbf{x} + \mathbf{y} = \left[\begin{array}{c} x_1 + y_1 \\ x_2 + y_2 \end{array} \right]$$

If a is a scalar, then

$$a\mathbf{x} = \left[\begin{array}{c} ax_1 \\ ax_2 \end{array} \right]$$

If $|\mathbf{x}|$ denotes the length of $\mathbf{x}$, then the length of $a\mathbf{x}$ is the absolute value of a times the length of $\mathbf{x}$, that is, $|a||\mathbf{x}|$.

Linear Maps We will now turn to maps of the form

$$\mathbf{x} \to A\mathbf{x}$$

where A is a 2×2 matrix and $\mathbf{x}$ is a 2×1 vector. Since $A\mathbf{x}$ is a 2×1 vector as well, the map A takes the 2×1 vector $\mathbf{x}$ and maps it into a 2×1 vector. We can thus represent such maps in the plane, and graphically investigate the action of A on the vector $\mathbf{x}$.

The simplest such map is the **identity map**, represented by the identity matrix I_2,

$$\begin{bmatrix} x_1 \\ x_2 \end{bmatrix} \to \begin{bmatrix} 1 & 0 \\ 0 & 1 \end{bmatrix}\begin{bmatrix} x_1 \\ x_2 \end{bmatrix}$$

Since

$$I_2\mathbf{x} = \begin{bmatrix} 1 & 0 \\ 0 & 1 \end{bmatrix}\begin{bmatrix} x_1 \\ x_2 \end{bmatrix} = \begin{bmatrix} x_1 \\ x_2 \end{bmatrix} = \mathbf{x}$$

we conclude that the identity matrix leaves the vector $\mathbf{x}$ unchanged.

Slightly more complicated is the map

$$\begin{bmatrix} x_1 \\ x_2 \end{bmatrix} \to \begin{bmatrix} a & 0 \\ 0 & b \end{bmatrix}\begin{bmatrix} x_1 \\ x_2 \end{bmatrix} = \begin{bmatrix} ax_1 \\ bx_2 \end{bmatrix}$$

This map stretches or contracts each coordinate separately. In Figure 9.16, we show the action of the map on $\begin{bmatrix} 1 \\ 2 \end{bmatrix}$ when $a = 2$ and $b = -\frac{1}{2}$. We find

$$\begin{bmatrix} 2 & 0 \\ 0 & -\frac{1}{2} \end{bmatrix}\begin{bmatrix} 1 \\ 2 \end{bmatrix} = \begin{bmatrix} 2 \\ -1 \end{bmatrix}$$

This map stretches the first coordinate by a factor of 2 and contracts the second coordinate by a factor of 1/2. The minus sign in front of $\frac{1}{2}$ corresponds to reflecting the second coordinate about the x_1-axis.

Another linear map is a **rotation**, which rotates a vector in the $x_1 - x_2$ plane by a fixed angle. The following matrix rotates a vector by an angle θ. If $\theta > 0$, then the rotation is counterclockwise; if $\theta < 0$, the rotation is clockwise by the angle $|\theta|$.

$$R_\theta = \begin{bmatrix} \cos\theta & -\sin\theta \\ \sin\theta & \cos\theta \end{bmatrix}$$

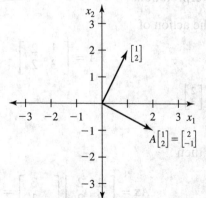

◀ **Figure 9.16**

The action of the matrix A on the vector $\begin{bmatrix} 1 \\ 2 \end{bmatrix}$

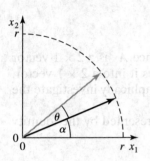

▲ Figure 9.17
The rotation of a vector

To check that this is indeed a rotation, we investigate the action of R_θ on a vector with coordinates (x_1, x_2), as illustrated in Figure 9.17. Using the polar coordinate system, we can write this vector as

$$\begin{bmatrix} x_1 \\ x_2 \end{bmatrix} = \begin{bmatrix} r\cos\alpha \\ r\sin\alpha \end{bmatrix}$$

where r is the length of the vector and α is the angle it forms with the positive x_1-axis.

Applying R_θ, we find

$$\begin{bmatrix} \cos\theta & -\sin\theta \\ \sin\theta & \cos\theta \end{bmatrix}\begin{bmatrix} r\cos\alpha \\ r\sin\alpha \end{bmatrix} = \begin{bmatrix} r(\cos\theta\cos\alpha - \sin\theta\sin\alpha) \\ r(\sin\theta\cos\alpha + \cos\theta\sin\alpha) \end{bmatrix}$$

$$= \begin{bmatrix} r\cos(\theta + \alpha) \\ r\sin(\theta + \alpha) \end{bmatrix}$$

where we used the trigonometric identities

$$\cos(\theta + \alpha) = \cos\theta\cos\alpha - \sin\theta\sin\alpha$$

$$\sin(\theta + \alpha) = \sin\theta\cos\alpha + \cos\theta\sin\alpha$$

We see that the resulting vector still has length r and that the angle with the x_1-axis is $\alpha + \theta$. If $\theta > 0$, as in Figure 9.17, then the vector is rotated counterclockwise by the angle θ.

▶ Example 3 Use a rotation matrix to rotate the vector $\begin{bmatrix} 1 \\ 3 \end{bmatrix}$ counterclockwise by the angle $\pi/3$.

Solution

The corresponding rotation matrix is

$$R_{\pi/3} = \begin{bmatrix} \cos(\pi/3) & -\sin(\pi/3) \\ \sin(\pi/3) & \cos(\pi/3) \end{bmatrix} = \begin{bmatrix} 1/2 & -\sqrt{3}/2 \\ \sqrt{3}/2 & 1/2 \end{bmatrix}$$

Hence, the rotated vector has coordinates

$$\begin{bmatrix} 1/2 & -\sqrt{3}/2 \\ \sqrt{3}/2 & 1/2 \end{bmatrix}\begin{bmatrix} 1 \\ 3 \end{bmatrix} = \begin{bmatrix} \frac{1}{2} - \frac{3}{2}\sqrt{3} \\ \frac{1}{2}\sqrt{3} + \frac{3}{2} \end{bmatrix} = \frac{1}{2}\begin{bmatrix} 1 - 3\sqrt{3} \\ \sqrt{3} + 3 \end{bmatrix} \quad ◀$$

From this brief discussion of linear maps, we see that the map $\mathbf{x} \to A\mathbf{x}$ typically takes the vector $\mathbf{x}$ and rotates, stretches, or contracts it. For an arbitrary matrix A, vectors may be moved in a way that has no simple geometric interpretation.

▶ Example 4 Investigate the action of

$$A = \begin{bmatrix} 1 & 2 \\ 3 & 2 \end{bmatrix}$$

on $\begin{bmatrix} 3 \\ -1 \end{bmatrix}$ and $\begin{bmatrix} 2 \\ 3 \end{bmatrix}$.

Solution

If $\mathbf{x} = \begin{bmatrix} 3 \\ -1 \end{bmatrix}$, then

$$A\mathbf{x} = \begin{bmatrix} 1 & 2 \\ 3 & 2 \end{bmatrix}\begin{bmatrix} 3 \\ -1 \end{bmatrix} = \begin{bmatrix} 1 \\ 7 \end{bmatrix}$$

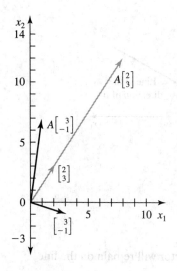

▲ Figure 9.18
The action of the matrix
$A = \begin{bmatrix} 1 & 2 \\ 3 & 2 \end{bmatrix}$ on two vectors

We can compute the outcome of this map, but there does not seem to be a straightforward geometric explanation (see Figure 9.18). On the other hand, if $\mathbf{x} = \begin{bmatrix} 2 \\ 3 \end{bmatrix}$, then

$$A\mathbf{x} = \begin{bmatrix} 1 & 2 \\ 3 & 2 \end{bmatrix} \begin{bmatrix} 2 \\ 3 \end{bmatrix} = \begin{bmatrix} 8 \\ 12 \end{bmatrix} = 4 \begin{bmatrix} 2 \\ 3 \end{bmatrix}$$

The result of $A\mathbf{x}$ is simply a multiple of $\mathbf{x}$ (see Figure 9.18). Such a vector is called an eigenvector, and the stretching or contracting factor is called an eigenvalue. Eigenvectors and eigenvalues are the topic of the next subsection. ◀

9.3.2 Eigenvalues and Eigenvectors

We saw in the preceding subsection that there are matrices and vectors for which the map $\mathbf{x} \to A\mathbf{x}$ takes on a particularly simple form, namely

$$A\mathbf{x} = \lambda\mathbf{x} \tag{9.19}$$

where λ is a scalar. This subsection is devoted to investigating this relationship. We will again restrict our discussion to 2×2 matrices and begin with the following definition.

> **Definition** Assume that A is a square matrix. A nonzero vector $\mathbf{x}$ that satisfies the equation
>
> $$A\mathbf{x} = \lambda\mathbf{x}$$
>
> is an eigenvector of the matrix A and the number λ is an eigenvalue of the matrix A.

Note that we assume that the vector $\mathbf{x}$ is different from the zero vector. (The zero vector $\mathbf{x} = \mathbf{0}$ always satisfies the equation $A\mathbf{x} = \lambda\mathbf{x}$ and thus would not be special.) The eigenvalue λ can be 0, however. We will see that λ can even be a complex number.

The action of A on eigenvectors is of a particularly simple form; if we apply A to an eigenvector $\mathbf{x}$ (that is, compute $A\mathbf{x}$), the result is simply a constant multiple of $\mathbf{x}$. This property of an eigenvector has an important geometric interpretation when the eigenvalue λ is a real number. Namely, if we draw a straight line through the origin in the direction of an eigenvector, then any vector on this straight line will remain on this line after applying the map A (see Figure 9.19).

How can we find eigenvalues and eigenvectors for 2×2 matrices? We will see that a 2×2 matrix has two eigenvalues, which are either distinct or identical. We will only discuss the case in which the two eigenvalues are distinct; the case in which the eigenvalues are identical is more complicated and is covered in courses on linear algebra. For our purposes, it will suffice to look at the case in which the two eigenvalues are distinct.

We will show how to find eigenvalues and eigenvectors by way of example. We will use the same matrix as in Example 4.

▶ **Example 5**   Find all eigenvalues and eigenvectors of

$$A = \begin{bmatrix} 1 & 2 \\ 3 & 2 \end{bmatrix}$$

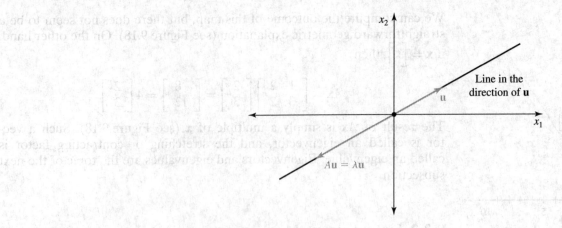

▲ **Figure 9.19**
Any vector on the line in the direction of the eigenvector will remain on the line under the map A

Solution

We are interested in finding $\mathbf{x} \neq \mathbf{0}$ and λ, such that

$$A\mathbf{x} = \lambda\mathbf{x}$$

We can rewrite this as

$$A\mathbf{x} - \lambda\mathbf{x} = \mathbf{0}$$

In order to factor $\mathbf{x}$, we must multiply $\lambda\mathbf{x}$ by the identity matrix $I = I_2$. (Because we will be dealing only with 2×2 matrices, we simply write I instead of I_2.) This yields

$$A\mathbf{x} - \lambda I\mathbf{x} = \mathbf{0}$$

We can now factor $\mathbf{x}$, which results in

$$(A - \lambda I)\mathbf{x} = \mathbf{0}$$

We showed in Section 9.2 that in order to obtain a nontrivial solution ($\mathbf{x} \neq \mathbf{0}$), $A - \lambda I$ must be singular, that is,

$$\det(A - \lambda I) = 0$$

Now,

$$A - \lambda I = \begin{bmatrix} 1 - \lambda & 2 \\ 3 & 2 - \lambda \end{bmatrix}$$

and

$$
\begin{aligned}
\det(A - \lambda I) &= (1 - \lambda)(2 - \lambda) - (2)(3) \\
&= 2 - 3\lambda + \lambda^2 - 6 = \lambda^2 - 3\lambda - 4 \\
&= (\lambda + 1)(\lambda - 4)
\end{aligned}
$$

Thus, we need to solve

$$(\lambda + 1)(\lambda - 4) = 0$$

and find

$$\lambda_1 = -1 \quad \text{and} \quad \lambda_2 = 4$$

These two numbers are the eigenvalues of the matrix A. Each eigenvalue will have its own eigenvector.

To compute the eigenvectors, we carry out the following calculations. If $\lambda_1 = -1$, we must find a nonzero vector $\mathbf{x} = \begin{bmatrix} x_1 \\ x_2 \end{bmatrix}$ such that

$$A \begin{bmatrix} x_1 \\ x_2 \end{bmatrix} = (-1) \begin{bmatrix} x_1 \\ x_2 \end{bmatrix}$$

Writing this as a system of equations, we find

$$x_1 + 2x_2 = -x_1$$
$$3x_1 + 2x_2 = -x_2$$

Simplifying, we obtain

$$2x_1 + 2x_2 = 0$$
$$3x_1 + 3x_2 = 0$$

The two equations are identical and reduce to

$$x_1 + x_2 = 0$$

We are looking for a nonzero vector that satisfies this equation. For instance,

$$\begin{bmatrix} x_1 \\ x_2 \end{bmatrix} = \begin{bmatrix} 1 \\ -1 \end{bmatrix}$$

would be a reasonable choice. To check that $\begin{bmatrix} 1 \\ -1 \end{bmatrix}$ is indeed an eigenvector corresponding to the eigenvalue $\lambda_1 = -1$, we compute

$$\begin{bmatrix} 1 & 2 \\ 3 & 2 \end{bmatrix} \begin{bmatrix} 1 \\ -1 \end{bmatrix} = \begin{bmatrix} -1 \\ 1 \end{bmatrix} = (-1) \begin{bmatrix} 1 \\ -1 \end{bmatrix}$$

that is,

$$A \begin{bmatrix} 1 \\ -1 \end{bmatrix} = (-1) \begin{bmatrix} 1 \\ -1 \end{bmatrix}$$

The vector $\begin{bmatrix} 1 \\ -1 \end{bmatrix}$ is not the only eigenvector associated with $\lambda_1 = -1$. In fact, any vector $a \begin{bmatrix} -1 \\ 1 \end{bmatrix}$, $a \neq 0$, is an eigenvector associated with the eigenvalue -1. For instance, $\begin{bmatrix} -1 \\ 1 \end{bmatrix}$ or $\begin{bmatrix} 2 \\ -2 \end{bmatrix}$ are other choices (see Figure 9.20).

To find an eigenvector associated with $\lambda_2 = 4$, we must solve

$$A \begin{bmatrix} x_1 \\ x_2 \end{bmatrix} = 4 \begin{bmatrix} x_1 \\ x_2 \end{bmatrix}$$

which yields

$$x_1 + 2x_2 = 4x_1$$
$$3x_1 + 2x_2 = 4x_2$$

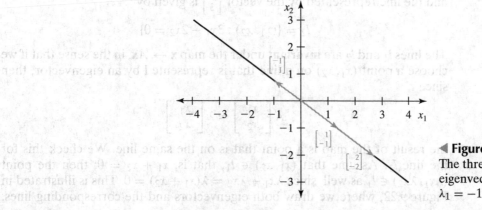

◀ **Figure 9.20**
The three vectors are all eigenvectors to the eigenvalue $\lambda_1 = -1$

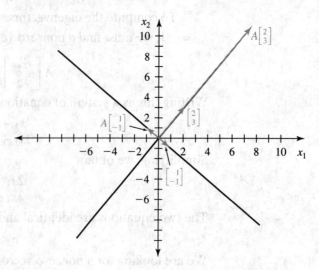

▲ **Figure 9.21**
The two eigenvectors with their corresponding lines. The images of the eigenvectors under the map A remain on their respective lines

Simplifying, we obtain

$$-3x_1 + 2x_2 = 0$$
$$3x_1 - 2x_2 = 0$$

For instance,

$$\begin{bmatrix} x_1 \\ x_2 \end{bmatrix} = \begin{bmatrix} 2 \\ 3 \end{bmatrix}$$

satisfies this. We see that A has two eigenvalues, namely -1 and 4. An eigenvector associated with -1 is $\begin{bmatrix} 1 \\ -1 \end{bmatrix}$, an eigenvector associated with 4 is $\begin{bmatrix} 2 \\ 3 \end{bmatrix}$ (see Figure 9.21). As before, any vector $b\begin{bmatrix} 2 \\ 3 \end{bmatrix}$, $b \neq 0$, is an eigenvector associated with the eigenvalue 4.   ◀

Eigenvectors are not unique; they are determined only up to a multiplicative constant. When the eigenvalues are real, as in our example, all eigenvectors corresponding to a particular eigenvalue lie on the *same* straight line through the origin. The line represented by the vector $\begin{bmatrix} 1 \\ -1 \end{bmatrix}$ is given by

$$l_1 = \{(x_1, x_2) : x_1 + x_2 = 0\}$$

and the line represented by the vector $\begin{bmatrix} 2 \\ 3 \end{bmatrix}$ is given by

$$l_2 = \{(x_1, x_2) : 3x_1 - 2x_2 = 0\}$$

The lines l_1 and l_2 are **invariant** under the map $\mathbf{x} \rightarrow A\mathbf{x}$, in the sense that if we choose a point (x_1, x_2) on a line that is represented by an eigenvector, then since

$$A \begin{bmatrix} x_1 \\ x_2 \end{bmatrix} = \lambda \begin{bmatrix} x_1 \\ x_2 \end{bmatrix}$$

the result of the map is a point that is on the same line. We check this for the line l_1. Assume that $(x_1, x_2) \in l_1$, that is, $x_1 + x_2 = 0$; then the point $(\lambda x_1, \lambda x_2) \in l_1$ as well, since $\lambda x_1 + \lambda x_2 = \lambda(x_1 + x_2) = 0$. This is illustrated in Figure 9.22, where we draw both eigenvectors and the corresponding lines.

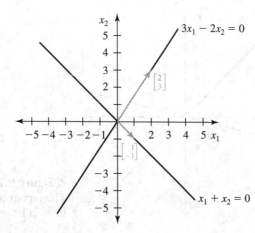

◀ **Figure 9.22**
Eigenvectors and associated lines of the matrix $A = \begin{bmatrix} 1 & 2 \\ 3 & 2 \end{bmatrix}$

We will refer to this as the geometric interpretation of eigenvalues and eigenvectors.

In the following example, we show the procedure for finding eigenvalues and eigenvectors once more, using a different matrix, which will also have real and distinct eigenvalues.

▶ **Example 6**   **(Finding eigenvalues and eigenvectors)** Find the eigenvalues and eigenvectors of the matrix

$$A = \begin{bmatrix} 1 & 4 \\ 1 & -2 \end{bmatrix}$$

Solution

To find the eigenvalues of A, we must solve

$$\det(A - \lambda I) = \det \begin{bmatrix} 1 - \lambda & 4 \\ 1 & -2 - \lambda \end{bmatrix}$$

$$= (1 - \lambda)(-2 - \lambda) - 4 = \lambda^2 + \lambda - 6$$

$$= (\lambda - 2)(\lambda + 3) = 0$$

We find

$$\lambda_1 = 2 \quad \text{and} \quad \lambda_2 = -3$$

To find an eigenvector associated with the eigenvalue $\lambda_1 = 2$, we must determine x_1 and x_2 (not both equal to 0), so that

$$\begin{bmatrix} 1 & 4 \\ 1 & -2 \end{bmatrix} \begin{bmatrix} x_1 \\ x_2 \end{bmatrix} = 2 \begin{bmatrix} x_1 \\ x_2 \end{bmatrix}$$

Writing this as a system of linear equations, we find

$$x_1 + 4x_2 = 2x_1$$
$$x_1 - 2x_2 = 2x_2$$

We see that both equations are the same. Simplifying yields

$$-x_1 + 4x_2 = 0$$

Setting $x_2 = 1$, we find $x_1 = 4$, that is, $\begin{bmatrix} 4 \\ 1 \end{bmatrix}$ is an eigenvector associated with the eigenvalue $\lambda_1 = 2$.

To find an eigenvector associated with the eigenvalue $\lambda_2 = -3$, we must determine x_1 and x_2 (not both equal to 0), so that

$$\begin{bmatrix} 1 & 4 \\ 1 & -2 \end{bmatrix} \begin{bmatrix} x_1 \\ x_2 \end{bmatrix} = -3 \begin{bmatrix} x_1 \\ x_2 \end{bmatrix}$$

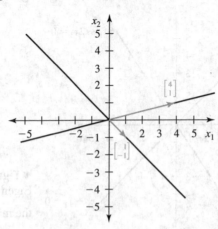

◀ **Figure 9.23**
The eigenvectors and their corresponding lines in Example 6

Writing this as a system of linear equations, we find

$$x_1 + 4x_2 = -3x_1$$
$$x_1 - 2x_2 = -3x_2$$

We see that both equations are the same. Simplifying yields

$$x_1 + x_2 = 0$$

Setting $x_1 = 1$, we find $x_2 = -1$, that is, $\begin{bmatrix} 1 \\ -1 \end{bmatrix}$ is an eigenvector associated ·
with the eigenvalue $\lambda_2 = -3$.

Both eigenvectors and corresponding lines are illustrated in Figure 9.23.

◀

Let's do one last example of finding eigenvalues and eigenvectors. This time, we choose a matrix that will illustrate a case when we can read off eigenvalues from the matrix directly.

▶ **Example 7**   (**Reading off eigenvalues from a matrix**) Find the eigenvalues and eigenvectors of the matrix

$$A = \begin{bmatrix} -2 & 1 \\ 0 & -1 \end{bmatrix}$$

Solution

To find the eigenvalues of A, we must solve

$$\det(A - \lambda I) = \det \begin{bmatrix} -2 - \lambda & 1 \\ 0 & -1 - \lambda \end{bmatrix}$$
$$= (-2 - \lambda)(-1 - \lambda) - (0)(1) = 0$$

We find

$$\lambda_1 = -2 \quad \text{and} \quad \lambda_2 = -1$$

Let's pause for a moment and look back at the matrix A. The eigenvalues we found are identical to the diagonal elements of A! This is because one of the off-diagonal elements of A is equal to 0. Knowing this simplifies finding eigenvalues. If one (or both) off-diagonal elements of the 2×2 matrix A are equal to 0, then the eigenvalues of A are equal to the diagonal elements.

Now, let's find the associated eigenvectors. To find an eigenvector associated with the eigenvalue $\lambda_1 = -2$, we must determine x_1 and x_2 (not both equal to 0), so that

$$\begin{bmatrix} -2 & 1 \\ 0 & -1 \end{bmatrix} \begin{bmatrix} x_1 \\ x_2 \end{bmatrix} = -2 \begin{bmatrix} x_1 \\ x_2 \end{bmatrix}$$

Writing this as a system of linear equations, we find

$$-2x_1 + x_2 = -2x_1$$
$$- x_2 = -2x_2$$

Simplifying either equation yields

$$x_2 = 0$$

This time, we cannot choose a value for x_2 (since $x_2 = 0$). But looking at the first equation $-2x_1 + x_2 = -2x_1$ tells us that any value for x_1 will yield a true identity provided $x_2 = 0$. Since $x_2 = 0$, we cannot choose $x_1 = 0$ (because $\begin{bmatrix} 0 \\ 0 \end{bmatrix}$ is not an eigenvector); any value of $x_1 \neq 0$ will do, however. Let's choose $x_1 = 1$. Then $\begin{bmatrix} 1 \\ 0 \end{bmatrix}$ is an eigenvector associated with $\lambda_1 = -2$.

To find an eigenvector associated with the eigenvalue $\lambda_2 = -1$, we must determine x_1 and x_2 (not both equal to 0), so that

$$\begin{bmatrix} -2 & 1 \\ 0 & -1 \end{bmatrix} \begin{bmatrix} x_1 \\ x_2 \end{bmatrix} = (-1) \begin{bmatrix} x_1 \\ x_2 \end{bmatrix}$$

Writing this as a system of linear equations, we find

$$-2x_1 + x_2 = -x_1$$
$$-x_2 = -x_2$$

This system simplifies to

$$-x_1 + x_2 = 0$$
$$0x_2 = 0$$

The first equation tells us that $x_1 = x_2$; the second equation tells us that we can choose any value for x_2. Choosing $x_2 = 1$, we need to set $x_1 = 1$, and find that $\begin{bmatrix} 1 \\ 1 \end{bmatrix}$ is an eigenvector associated with the eigenvalue $\lambda_2 = -1$. ◀

So far, we have only seen examples in which the eigenvalues were real and distinct. In the next example, we will see that eigenvalues can be complex. When eigenvalues are complex, we will not compute the corresponding eigenvectors, because the eigenvectors will be complex as well, which is beyond the scope of this course.

▶ **Example 8** **(Complex eigenvalues)** Let

$$A = \begin{bmatrix} \cos 30° & -\sin 30° \\ \sin 30° & \cos 30° \end{bmatrix}$$

Describe the action of A on the vector $\begin{bmatrix} 1 \\ 0 \end{bmatrix}$. Compute the eigenvalues of A.

Solution

We recognize that A is a matrix that describes a rotation counterclockwise by 30°. (The matrix A is the matrix R_θ for $\theta = 30°$, as defined in Subsection 9.3.1.) The vector $\begin{bmatrix} 1 \\ 0 \end{bmatrix}$ is thus rotated counterclockwise by 30°, and

$$A \begin{bmatrix} 1 \\ 0 \end{bmatrix} = \begin{bmatrix} \cos 30° \\ \sin 30° \end{bmatrix} = \begin{bmatrix} \sqrt{3}/2 \\ 1/2 \end{bmatrix} \text{ (see Figure 9.24).}$$

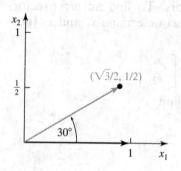

▲ Figure 9.24

The vector $\begin{bmatrix} 1 \\ 0 \end{bmatrix}$ is rotated counterclockwise by 30° in Example 8

We will now compute the eigenvalues associated with this matrix. We set

$$\det (A - \lambda I) = 0$$

Using $\cos 30° = \frac{1}{2}\sqrt{3}$ and $\sin 30° = \frac{1}{2}$, we find

$$\det \begin{bmatrix} \frac{1}{2}\sqrt{3} - \lambda & -\frac{1}{2} \\ \frac{1}{2} & \frac{1}{2}\sqrt{3} - \lambda \end{bmatrix} = 0$$

or

$$\left(\frac{1}{2}\sqrt{3} - \lambda\right)^2 + \frac{1}{4} = 0$$

$$\frac{3}{4} - \lambda\sqrt{3} + \lambda^2 + \frac{1}{4} = 0$$

$$\lambda^2 - \sqrt{3}\lambda + 1 = 0$$

Solving this quadratic equation, we obtain

$$\lambda_{1,2} = \frac{\sqrt{3} \pm \sqrt{3 - 4}}{2}$$

$$= \frac{1}{2}\left(\sqrt{3} \pm i\right)$$

where $i^2 = -1$. This shows that eigenvalues can be complex numbers. ◀

In Chapter 11, we will discuss the stability of equilibria in systems of ordinary differential equations. This will lead us to investigating the eigenvalues of certain linear maps. It will be important to determine whether the real parts of the eigenvalues are positive or negative. In the case where the linear map is given by a 2 × 2 matrix, there is a useful criterion. Consider

$$A = \begin{bmatrix} a & b \\ c & d \end{bmatrix}$$

Then the determinant of A is

$$\det A = ad - bc$$

and the **trace** of A (denoted by tr A) is defined as

$$\boxed{\operatorname{tr} A = a + d}$$

The trace is the sum of the diagonal elements of A. The trace and the determinant of a matrix are related to its eigenvalues, as follows.

The eigenvalues of A satisfy

$$\det (A - \lambda I) = 0$$

or

$$\det \begin{bmatrix} a - \lambda & b \\ c & d - \lambda \end{bmatrix} = 0$$

which yields

$$(a - \lambda)(d - \lambda) - bc = 0$$

$$\lambda^2 - (a + d)\lambda + ad - bc = 0$$

Since $\det A = ad - bc$ and $\operatorname{tr} A = a + d$, we can write this as

$$\lambda^2 - (\operatorname{tr} A)\lambda + \det A = 0 \qquad (9.20)$$

If λ_1 and λ_2 are the two solutions of (9.20), then λ_1 and λ_2 must satisfy

$$(\lambda - \lambda_1)(\lambda - \lambda_2) = 0$$

Multiplying this out, we find that λ_1 and λ_2 satisfy

$$\lambda^2 - (\lambda_1 + \lambda_2)\lambda + \lambda_1\lambda_2 = 0$$

Comparing this with (9.20), we find the following important result.

If A is a 2×2 matrix with eigenvalues λ_1 and λ_2, then

$$\operatorname{tr} A = \lambda_1 + \lambda_2 \quad \text{and} \quad \det A = \lambda_1\lambda_2$$

To prepare for the next result, we make the following observations. Assume that λ_1 and λ_2 are both real and negative. Then the trace of A, which is the sum of the two eigenvalues, is negative, and the determinant of A, which is the product of the two eigenvalues, is positive.

In the case when the eigenvalues are complex conjugates, we have the same result. Assume that λ_1 and λ_2 are complex conjugates and that their real parts are negative. We can then show that the trace of A is negative and the determinant is positive. (We will see an example of this fact later.)

That is, whenever the real parts of the two eigenvalues λ_1 and λ_2 are negative, it follows that $\operatorname{tr} A < 0$ and $\det A > 0$. The converse of this result is also true. That is, if $\operatorname{tr} A < 0$ and $\det A > 0$, then both eigenvalues have negative real parts. This will be a useful criterion in Chapter 11, since it will allow us to determine whether or not both eigenvalues have negative real parts without computing the eigenvalues.

Theorem Let A be a 2×2 matrix with eigenvalues λ_1 and λ_2. Then the real parts of λ_1 and λ_2 are negative if and only if

$$\operatorname{tr} A < 0 \quad \text{and} \quad \det A > 0$$

▶ **Example 9** (Trace and determinant) Use the preceding theorem to show that both of the eigenvalues of

$$A = \begin{bmatrix} -1 & -3 \\ 1 & -2 \end{bmatrix}$$

have negative real parts. Then compute the eigenvalues, and use them to recompute the trace and the determinant of A.

Solution

Since

$$\operatorname{tr} A = -1 - 2 = -3 < 0 \quad \text{and} \quad \det A = (-1)(-2) - (1)(-3) = 5 > 0$$

it follows from the theorem that both eigenvalues have negative real parts. To compute the eigenvalues, we solve

$$\det (A - \lambda I) = 0$$

or

$$\det \begin{bmatrix} -1-\lambda & -3 \\ 1 & -2-\lambda \end{bmatrix} = 0$$

Evaluating this, we find

$$(-1-\lambda)(-2-\lambda) - (1)(-3) = 0$$
$$\lambda^2 + 3\lambda + 5 = 0$$

that is,

$$\lambda_{1,2} = \frac{-3 \pm \sqrt{9-20}}{2}$$
$$= -\frac{3}{2} \pm \frac{1}{2} i\sqrt{11}$$

This also shows that the real parts of both eigenvalues are negative.

We can now use the eigenvalues to recompute the trace and the determinant of A.

$$\operatorname{tr} A = \lambda_1 + \lambda_2 = \left(-\frac{3}{2} + \frac{1}{2} i\sqrt{11}\right) + \left(-\frac{3}{2} - \frac{1}{2} i\sqrt{11}\right) = -3$$

This is the same result that we obtained previously. Note that since λ_1 and λ_2 are complex conjugates, the imaginary parts cancel when we add λ_1 and λ_2. The determinant of A is

$$\det A = \lambda_1 \lambda_2 = \left(-\frac{3}{2} + \frac{1}{2} i\sqrt{11}\right) \left(-\frac{3}{2} - \frac{1}{2} i\sqrt{11}\right)$$
$$= \frac{9}{4} - \frac{1}{4} i^2 (11) = \frac{9}{4} + \frac{11}{4} = \frac{20}{4} = 5$$

which is the same result that we obtained previously. Note that since λ_1 and λ_2 are complex conjugates, we computed a product of the form $(x+iy)(x-iy) = x^2 - (iy)^2 = x^2 + y^2$, which is a real number (and positive). ◀

9.3.3 Iterated Maps (Needed for Section 10.7)

We restrict ourselves to the case in which A is a 2×2 matrix with real eigenvalues. We saw that in this case the eigenvectors define lines through the origin that are invariant under the map A. If the two invariant lines are *not* identical, we say that the two eigenvectors are **linearly independent** (see Figure 9.25). This can be formulated as follows in terms of eigenvectors. If we denote by $\mathbf{u}_1$ and $\mathbf{u}_2$ the two eigenvectors, then $\mathbf{u}_1$ and $\mathbf{u}_2$ are linearly independent if there does not exist a number a such that $\mathbf{u}_1 = a\mathbf{u}_2$. (Linear independence is defined not just for eigenvectors: Any two nonzero vectors $\mathbf{x}_1$ and $\mathbf{x}_2$ are linearly independent if there does not exist a number a such that $\mathbf{x}_1 = a\mathbf{x}_2$.)

The following criterion is useful when we want to determine whether eigenvalues are linearly independent.

Let A be a 2×2 matrix with eigenvalues λ_1 and λ_2. Denote by $\mathbf{u}_1$ the eigenvector associated with λ_1 and by $\mathbf{u}_2$ the eigenvector associated with λ_2. If $\lambda_1 \neq \lambda_2$, then $\mathbf{u}_1$ and $\mathbf{u}_2$ are linearly independent.

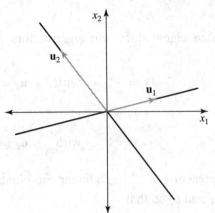

◀ **Figure 9.25**
The two eigenvectors $\mathbf{u}_1$ and $\mathbf{u}_2$ are linearly independent

There are also cases in which $\mathbf{u}_1$ and $\mathbf{u}_2$ are linearly independent even though $\lambda_1 = \lambda_2$. We will, however, be concerned primarily with cases in which $\lambda_1 \neq \lambda_2$. Hence, the preceding criterion will suffice for our purposes. (The other cases are covered in courses on linear algebra.)

A consequence of linear independence is that we can write any vector uniquely as a **linear combination** of the two eigenvectors. Suppose that $\mathbf{u}_1$ and $\mathbf{u}_2$ are linearly independent eigenvectors of a 2×2 matrix; then any 2×1 vector $\mathbf{x}$ can be written as

$$\mathbf{x} = a_1 \mathbf{u}_1 + a_2 \mathbf{u}_2$$

where a_1 and a_2 are uniquely determined. We will not prove this, but will examine what we can do with it.

If we apply A to $\mathbf{x}$ (written as a linear combination of the two eigenvectors of A), then using the linearity of the map A, we find

$$A\mathbf{x} = A(a_1 \mathbf{u}_1 + a_2 \mathbf{u}_2) = a_1 A\mathbf{u}_1 + a_2 A\mathbf{u}_2$$

Now, $\mathbf{u}_1$ and $\mathbf{u}_2$ are both eigenvectors corresponding to A. Hence, $A\mathbf{u}_1 = \lambda_1 \mathbf{u}_1$ and $A\mathbf{u}_2 = \lambda_2 \mathbf{u}_2$. We thus find

$$A\mathbf{x} = a_1 \lambda_1 \mathbf{u}_1 + a_2 \lambda_2 \mathbf{u}_2$$

This representation of $\mathbf{x}$ is particularly useful if we apply A repeatedly to $\mathbf{x}$. Applying A to $A\mathbf{x}$, we find

$$A^2\mathbf{x} = A(A\mathbf{x}) = A(a_1 \lambda_1 \mathbf{u}_1 + a_2 \lambda_2 \mathbf{u}_2) = a_1 \lambda_1 A\mathbf{u}_1 + a_2 \lambda_2 A\mathbf{u}_2$$
$$= a_1 \lambda_1^2 \mathbf{u}_1 + a_2 \lambda_2^2 \mathbf{u}_2$$

where we again used the fact that $\mathbf{u}_1$ and $\mathbf{u}_2$ are eigenvectors of the matrix A. Continuing in this way, we find that

$$A^n\mathbf{x} = a_1 \lambda_1^n \mathbf{u}_1 + a_2 \lambda_2^n \mathbf{u}_2 \tag{9.21}$$

Instead of multiplying A with itself n times (which is rather time consuming), we can use the right-hand side of (9.21), which just amounts to adding two vectors (a much faster task). Are you convinced that this might be useful? If not, the next two examples should convince you.

▶ **Example 10** Let

$$A = \begin{bmatrix} 1 & 2 \\ 3 & 2 \end{bmatrix}$$

Find $A^{10}\mathbf{x}$ for $\mathbf{x} = \begin{bmatrix} 4 \\ 1 \end{bmatrix}$.

Solution

We computed eigenvalues and eigenvectors for the matrix A earlier. We found

$$\lambda_1 = -1 \quad \text{with} \quad \mathbf{u}_1 = \begin{bmatrix} 1 \\ -1 \end{bmatrix}$$

and

$$\lambda_2 = 4 \quad \text{with} \quad \mathbf{u}_2 = \begin{bmatrix} 2 \\ 3 \end{bmatrix}$$

We first represent $\mathbf{x} = \begin{bmatrix} 4 \\ 1 \end{bmatrix}$ as a linear combination of $\mathbf{u}_1$ and $\mathbf{u}_2$. For this, we must find a_1 and a_2 so that

$$\begin{bmatrix} 4 \\ 1 \end{bmatrix} = a_1 \begin{bmatrix} 1 \\ -1 \end{bmatrix} + a_2 \begin{bmatrix} 2 \\ 3 \end{bmatrix}$$

Writing this as a system of linear equations yields

$$a_1 + 2a_2 = 4$$
$$-a_1 + 3a_2 = 1$$

Using the method of elimination, we obtain

$$a_1 + 2a_2 = 4$$
$$5a_2 = 5$$

Hence, $a_2 = 1$ and $a_1 = 4 - 2a_2 = 4 - 2 = 2$. We find

$$\begin{bmatrix} 4 \\ 1 \end{bmatrix} = 2 \begin{bmatrix} 1 \\ -1 \end{bmatrix} + \begin{bmatrix} 2 \\ 3 \end{bmatrix}$$

To compute $A^{10}\mathbf{x}$, we use the right-hand side of (9.21):

$$A^{10}\mathbf{x} = A^{10} \begin{bmatrix} 4 \\ 1 \end{bmatrix} = A^{10} \left(2 \begin{bmatrix} 1 \\ -1 \end{bmatrix} + \begin{bmatrix} 2 \\ 3 \end{bmatrix} \right)$$

$$= 2A^{10} \begin{bmatrix} 1 \\ -1 \end{bmatrix} + A^{10} \begin{bmatrix} 2 \\ 3 \end{bmatrix}$$

$$= 2(-1)^{10} \begin{bmatrix} 1 \\ -1 \end{bmatrix} + 4^{10} \begin{bmatrix} 2 \\ 3 \end{bmatrix}$$

$$= 2 \begin{bmatrix} 1 \\ -1 \end{bmatrix} + 4^{10} \begin{bmatrix} 2 \\ 3 \end{bmatrix} = \begin{bmatrix} 2,097,154 \\ 3,145,726 \end{bmatrix}$$

To compute A^{10} directly would have taken a much longer time.   ◀

The Leslie Matrix Revisited

▶ **Example 11**   In Subsection 9.2.5, we investigated an age-structured population with Leslie matrix

$$L = \begin{bmatrix} 1.5 & 2 \\ 0.08 & 0 \end{bmatrix}$$

Find both eigenvalues and eigenvectors.

Solution

To find the eigenvalues, we must solve

$$\det (L - \lambda I) = 0$$

Since

$$L - \lambda I = \begin{bmatrix} 1.5 - \lambda & 2 \\ 0.08 & -\lambda \end{bmatrix}$$

we find

$$\det(L - \lambda I) = (1.5 - \lambda)(-\lambda) - (2)(0.08)$$
$$= -1.5\lambda + \lambda^2 - 0.16$$

Since

$$\lambda^2 - 1.5\lambda - 0.16 = (\lambda - 1.6)(\lambda + 0.1)$$

the eigenvalues are

$$\lambda_1 = 1.6 \quad \text{and} \quad \lambda_2 = -0.1$$

To compute the corresponding eigenvectors, we start with the larger eigen-value $\lambda_1 = 1.6$. We need to solve

$$\begin{bmatrix} 1.5 & 2 \\ 0.08 & 0 \end{bmatrix} \begin{bmatrix} x_1 \\ x_2 \end{bmatrix} = (1.6) \begin{bmatrix} x_1 \\ x_2 \end{bmatrix}$$

or

$$1.5x_1 + 2x_2 = 1.6x_1$$
$$0.08x_1 \qquad = 1.6x_2$$

Simplifying, we find

$$-0.1x_1 + 2x_2 = 0$$
$$0.08x_1 - 1.6x_2 = 0$$

or

$$x_1 - 20x_2 = 0$$
$$x_1 - 20x_2 = 0$$

From this, it follows that

$$x_1 = 20x_2$$

For instance, $x_2 = 1$ and $x_1 = 20$ satisfies this. Therefore, an eigenvector corresponding to $\lambda_1 = 1.6$ is

$$\mathbf{u}_1 = \begin{bmatrix} 20 \\ 1 \end{bmatrix}$$

The eigenvector corresponding to $\lambda_2 = -0.1$ satisfies

$$\begin{bmatrix} 1.5 & 2 \\ 0.08 & 0 \end{bmatrix} \begin{bmatrix} x_1 \\ x_2 \end{bmatrix} = -0.1 \begin{bmatrix} x_1 \\ x_2 \end{bmatrix}$$

or

$$1.5x_1 + 2x_2 = -0.1x_1$$
$$0.08x_1 \qquad = -0.1x_2$$

Simplifying, we find

$$1.6x_1 + 2x_2 = 0$$
$$0.08x_1 + 0.1x_2 = 0$$

or

$$0.8x_1 + x_2 = 0$$
$$0.8x_1 + x_2 = 0$$

From this, it follows that

$$0.8x_1 = -x_2$$

For instance, $x_1 = 5$ and $x_2 = -4$ satisfies this. Therefore, an eigenvector corresponding to the eigenvalue $\lambda_2 = -0.1$ is

$$\mathbf{u}_2 = \begin{bmatrix} 5 \\ -4 \end{bmatrix}$$

◀

We will now show that the larger eigenvalue determines the growth rate of the population and the eigenvector corresponding to the larger eigenvalue is a stable age distribution. To illustrate this, we assume that the population vector at time 0 is

$$N(0) = \begin{bmatrix} 105 \\ 1 \end{bmatrix}$$

We can compute $N(1)$ by evaluating

$$N(1) = LN(0) = \begin{bmatrix} 1.5 & 2 \\ 0.08 & 0 \end{bmatrix}\begin{bmatrix} 105 \\ 1 \end{bmatrix} = \begin{bmatrix} 159.5 \\ 8.4 \end{bmatrix}$$

If we wanted to compute $N(t)$ for some integer value t, we would need to find

$$N(t) = L^t N(0)$$

This can be computed using (9.21). We need to write $N(0)$ as a linear combination of the two eigenvectors and then apply L^t to this linear combination.

Looking at the eigenvalues, we see that $N(0)$, as a linear combination of the two eigenvectors, is

$$\begin{bmatrix} 105 \\ 1 \end{bmatrix} = 5\begin{bmatrix} 20 \\ 1 \end{bmatrix} + \begin{bmatrix} 5 \\ -4 \end{bmatrix}$$

Now, if we want to compute $N(1)$, we must find

$$N(1) = LN(0) = L\begin{bmatrix} 105 \\ 1 \end{bmatrix} = L\left(5\begin{bmatrix} 20 \\ 1 \end{bmatrix} + \begin{bmatrix} 5 \\ -4 \end{bmatrix}\right)$$

$$= 5L\begin{bmatrix} 20 \\ 1 \end{bmatrix} + L\begin{bmatrix} 5 \\ -4 \end{bmatrix}$$

Since $\begin{bmatrix} 20 \\ 1 \end{bmatrix}$ is an eigenvector corresponding to $\lambda_1 = 1.6$, we find $L\begin{bmatrix} 20 \\ 1 \end{bmatrix} = 1.6\begin{bmatrix} 20 \\ 1 \end{bmatrix}$. Likewise, since $\begin{bmatrix} 5 \\ -4 \end{bmatrix}$ is an eigenvector corresponding to $\lambda_2 = -0.1$, we find $L\begin{bmatrix} 5 \\ -4 \end{bmatrix} = -0.1\begin{bmatrix} 5 \\ -4 \end{bmatrix}$. Hence,

$$N(1) = (5)(1.6)\begin{bmatrix} 20 \\ 1 \end{bmatrix} + (-0.1)\begin{bmatrix} 5 \\ -4 \end{bmatrix}$$

$$= \begin{bmatrix} 159.5 \\ 8.4 \end{bmatrix}$$

which, of course, is the same answer as before. Now, let's find $N(t)$. For this, we need to compute

$$N(t) = L^t N(0) = L^t\begin{bmatrix} 105 \\ 1 \end{bmatrix} = L^t\left(5\begin{bmatrix} 20 \\ 1 \end{bmatrix} + \begin{bmatrix} 5 \\ -4 \end{bmatrix}\right)$$

$$= 5(1.6)^t\begin{bmatrix} 20 \\ 1 \end{bmatrix} + (-0.1)^t\begin{bmatrix} 5 \\ -4 \end{bmatrix}$$

We see from this that $(1.6)^t$ grows much faster than $(-0.1)^t$. In fact, $(-0.1)^t$ tends to 0 as $t \to \infty$. We conclude that the larger eigenvalue determines the growth rate of the population.

It turns out that the larger eigenvalue of a Leslie matrix is always real and positive, provided that a positive fraction of zero-year-old females survive, and either zero-year-old or one-year-old females have offspring. Furthermore, $\lambda_1 \geq |\lambda_2|$. The strict inequality $\lambda_1 > |\lambda_2|$ holds if zero-year-olds can give birth. We thus find the following.

> If L is a 2×2 Leslie matrix with eigenvalues λ_1 and λ_2, then the larger eigenvalue determines the growth parameter of the population.

If λ_1 is the larger eigenvalue and $0 < \lambda_1 < 1$, then the population size decreases over time. If $\lambda_1 > 1$, then the population size increases over time.

We will now show that the eigenvector corresponding to the larger eigenvalue is a stable age distribution. Let λ_1 be the larger eigenvalue and $\mathbf{u}_1$ its corresponding eigenvector; then, if $N(0) = \mathbf{u}_1$, it follows that

$$N(t) = L^t N(0) = L^t \mathbf{u}_1 = \lambda_1^t \mathbf{u}_1$$

To show that $\mathbf{u}_1$ is a stable age distribution, observe that if $\mathbf{u}_1 = \begin{bmatrix} x \\ y \end{bmatrix}$, then the fraction of zero-year-olds at time t is

$$\frac{\lambda^t x}{\lambda^t x + \lambda^t y} = \frac{x}{x + y}$$

which is the same fraction of zero-year-olds at time 0. Furthermore, since $\lambda_1 > 0$, we can choose $\mathbf{u}_1$ so that both entries are positive (a condition that is needed if the entries represent population sizes). We summarize this in the following.

> If L is a 2×2 Leslie matrix with eigenvalues λ_1 and λ_2, then the eigenvector corresponding to the larger eigenvalue is a stable age distribution.

For the matrix

$$L = \begin{bmatrix} 1.5 & 2 \\ 0.08 & 0 \end{bmatrix}$$

the vector

$$\begin{bmatrix} 20 \\ 1 \end{bmatrix}$$

is an eigenvector to the larger eigenvalue, and thus it is a stable age distribution. In Subsection 9.2.5, we claimed that $\begin{bmatrix} 2000 \\ 100 \end{bmatrix}$ was a stable age distribution. In both cases, the fraction of zero-year-olds is the same, namely, $20/21 = 2000/2100$. That is, both vectors represent the same proportion of zero-year-olds in the population. We can check that $\begin{bmatrix} 2000 \\ 100 \end{bmatrix} = 100 \begin{bmatrix} 20 \\ 1 \end{bmatrix}$, which shows that both vectors are eigenvectors. (Recall that if $\mathbf{u}$ is an eigenvector, then any vector $a\mathbf{u}$, $a \neq 0$ is also an eigenvector.) When we list a stable age distribution, we make sure that both entries are positive since they represent numbers of individuals in each age class.

If $\lambda_1 > |\lambda_2|$, then the population vector $N(t)$ will converge to a stable age distribution as $t \to \infty$ provided $N(0) \neq \mathbf{u}_2$. This follows from writing $N(0)$ as a linear combination of the two eigenvectors $\mathbf{u}_1$ and $\mathbf{u}_2$ and applying L^t to the result:

$$L^t N(0) = L^t(a_1 \mathbf{u}_1 + a_2 \mathbf{u}_2) = a_1 \lambda_1^t \mathbf{u}_1 + a_2 \lambda_2^t \mathbf{u}_2$$

$$= a_1 \lambda_1^t \begin{bmatrix} x_1 \\ y_1 \end{bmatrix} + a_2 \lambda_2^t \begin{bmatrix} x_2 \\ y_2 \end{bmatrix}$$

where $\mathbf{u}_1 = \begin{bmatrix} x_1 \\ y_1 \end{bmatrix}$ and $\mathbf{u}_2 = \begin{bmatrix} x_2 \\ y_2 \end{bmatrix}$. Here $a_1 \neq 0$ since $N(0) \neq \mathbf{u}_2$. The fraction of zero-year-olds at time t is

$$\frac{a_1 \lambda_1^t x_1 + a_2 \lambda_2^t x_2}{a_1 \lambda_1^t x_1 + a_2 \lambda_2^t x_2 + a_1 \lambda_1^t y_1 + a_2 \lambda_2^t y_2} \to \frac{x_1}{x_1 + y_1}$$

as $t \to \infty$.

9.3.4 Problems

(9.3.1)

1. Let

$$A = \begin{bmatrix} 2 & 1 \\ 3 & 4 \end{bmatrix},$$

$$\mathbf{x} = \begin{bmatrix} x_1 \\ x_2 \end{bmatrix},$$

and

$$\mathbf{y} = \begin{bmatrix} y_1 \\ y_2 \end{bmatrix}$$

(a) Show by direct calculation that $A(\mathbf{x} + \mathbf{y}) = A\mathbf{x} + A\mathbf{y}$.

(b) Show by direct calculation that $A(\lambda \mathbf{x}) = \lambda(A\mathbf{x})$.

2. Let

$$A = \begin{bmatrix} a_{11} & a_{12} \\ a_{21} & a_{22} \end{bmatrix},$$

$$\mathbf{x} = \begin{bmatrix} x_1 \\ x_2 \end{bmatrix},$$

and

$$\mathbf{y} = \begin{bmatrix} y_1 \\ y_2 \end{bmatrix}$$

(a) Show by direct calculation that $A(\mathbf{x} + \mathbf{y}) = A\mathbf{x} + A\mathbf{y}$.

(b) Show by direct calculation that $A(\lambda \mathbf{x}) = \lambda(A\mathbf{x})$.

In Problems 3–8, represent each given vector $\mathbf{x} = \begin{bmatrix} x_1 \\ x_2 \end{bmatrix}$ in the x_1-x_2 plane, and determine its length and the angle that it forms with the positive x_1-axis (measured counterclockwise).

3. $\mathbf{x} = \begin{bmatrix} 2 \\ 2 \end{bmatrix}$

4. $\mathbf{x} = \begin{bmatrix} -1 \\ 0 \end{bmatrix}$

5. $\mathbf{x} = \begin{bmatrix} 0 \\ 3 \end{bmatrix}$

6. $\mathbf{x} = \begin{bmatrix} -1 \\ -1 \end{bmatrix}$

7. $\mathbf{x} = \begin{bmatrix} -\sqrt{3} \\ 1 \end{bmatrix}$

8. $\mathbf{x} = \begin{bmatrix} 1 \\ -\sqrt{3} \end{bmatrix}$

In Problems 9–12, vectors are given in their polar coordinate representation (length r and angle α measured counterclockwise from the positive x_1-axis). Find the representation of the vector $\begin{bmatrix} x_1 \\ x_2 \end{bmatrix}$ in Cartesian coordinates.

9. $r = 2, \alpha = 30°$

10. $r = 3, \alpha = 150°$

11. $r = 1, \alpha = 70°$

12. $r = 5, \alpha = 215°$

13. Suppose a vector $\begin{bmatrix} x_1 \\ x_2 \end{bmatrix}$ has length 3 and is 15° clockwise from the positive x_1-axis. Find x_1 and x_2.

14. Suppose a vector $\begin{bmatrix} x_1 \\ x_2 \end{bmatrix}$ has length 2 and is 140° clockwise from the positive x_1-axis. Find x_1 and x_2.

15. Suppose a vector $\begin{bmatrix} x_1 \\ x_2 \end{bmatrix}$ has length 5 and is 25° counterclockwise from the positive x_2-axis. Find x_1 and x_2.

16. Suppose a vector $\begin{bmatrix} x_1 \\ x_2 \end{bmatrix}$ has length 4 and is 70° counterclockwise from the negative x_2-axis. Find x_1 and x_2.

In Problems 17–22, find $\mathbf{x} + \mathbf{y}$ for the given vectors $\mathbf{x}$ and $\mathbf{y}$. Represent $\mathbf{x}$, $\mathbf{y}$, and $\mathbf{x} + \mathbf{y}$ in the plane, and explain graphically how you add $\mathbf{x}$ and $\mathbf{y}$.

17. $\mathbf{x} = \begin{bmatrix} 1 \\ 2 \end{bmatrix}$ and $\mathbf{y} = \begin{bmatrix} 3 \\ 2 \end{bmatrix}$

18. $\mathbf{x} = \begin{bmatrix} -1 \\ 0 \end{bmatrix}$ and $\mathbf{y} = \begin{bmatrix} 1 \\ 2 \end{bmatrix}$

19. $\mathbf{x} = \begin{bmatrix} 0 \\ -2 \end{bmatrix}$ and $\mathbf{y} = \begin{bmatrix} 1 \\ -1 \end{bmatrix}$

20. $\mathbf{x} = \begin{bmatrix} -2 \\ -1 \end{bmatrix}$ and $\mathbf{y} = \begin{bmatrix} 1 \\ 2 \end{bmatrix}$

21. $\quad \mathbf{x} = \begin{bmatrix} 1 \\ 0 \end{bmatrix}$ and $\mathbf{y} = \begin{bmatrix} -1 \\ 0 \end{bmatrix}$

22. $\quad \mathbf{x} = \begin{bmatrix} -3 \\ -1 \end{bmatrix}$ and $\mathbf{y} = \begin{bmatrix} -2 \\ 3 \end{bmatrix}$

*In Problems 23–28, compute a***x** *for the given vector* **x** *and scalar a. Represent* **x** *and a***x** *in the plane, and explain graphically how you obtain a***x**.

23. $\quad \mathbf{x} = \begin{bmatrix} -2 \\ 1 \end{bmatrix}$
and $a = 2$

24. $\quad \mathbf{x} = \begin{bmatrix} 3 \\ -1 \end{bmatrix}$
and $a = -1$

25. $\quad \mathbf{x} = \begin{bmatrix} 0 \\ -2 \end{bmatrix}$
and $a = 0.5$

26. $\quad \mathbf{x} = \begin{bmatrix} 3 \\ -9 \end{bmatrix}$
and $a = -1/3$

27. $\quad \mathbf{x} = \begin{bmatrix} -4 \\ 1 \end{bmatrix}$
and $a = 1/4$

28. $\quad \mathbf{x} = \begin{bmatrix} 0.5 \\ 0.25 \end{bmatrix}$
and $a = 4$

In Problems 29–34, let
$$\mathbf{u} = \begin{bmatrix} 3 \\ 4 \end{bmatrix}, \mathbf{v} = \begin{bmatrix} -1 \\ -2 \end{bmatrix},$$
and
$$\mathbf{w} = \begin{bmatrix} 1 \\ -2 \end{bmatrix}$$

29. Compute $\mathbf{u} + \mathbf{v}$ and illustrate the result graphically.
30. Compute $\mathbf{u} - \mathbf{v}$ and illustrate the result graphically.
31. Compute $\mathbf{w} - \mathbf{u}$ and illustrate the result graphically.
32. Compute $\mathbf{v} - \frac{1}{2}\mathbf{u}$ and illustrate the result graphically.
33. Compute $\mathbf{u} + \mathbf{v} + \mathbf{w}$ and illustrate the result graphically.
34. Compute $2\mathbf{v} - \mathbf{w}$ and illustrate the result graphically.

In Problems 35–40, give a geometric interpretation of the map $\mathbf{x} \rightarrow A\mathbf{x}$ *for each given map A.*

35. $\quad A = \begin{bmatrix} 1 & 0 \\ 0 & 1 \end{bmatrix}$

36. $\quad A = \begin{bmatrix} 2 & 0 \\ 0 & -1 \end{bmatrix}$

37. $\quad A = \begin{bmatrix} 0 & -1 \\ 1 & 0 \end{bmatrix}$

38. $\quad A = \begin{bmatrix} 0 & 1 \\ -1 & 0 \end{bmatrix}$

39. $\quad A = \frac{1}{2}\begin{bmatrix} \sqrt{3} & -1 \\ 1 & \sqrt{3} \end{bmatrix}$

40. $\quad A = \frac{1}{2}\begin{bmatrix} \sqrt{2} & -\sqrt{2} \\ \sqrt{2} & \sqrt{2} \end{bmatrix}$

41. Use a rotation matrix to rotate the vector $\begin{bmatrix} -1 \\ 2 \end{bmatrix}$ counterclockwise by the angle $\pi/6$.

42. Use a rotation matrix to rotate the vector $\begin{bmatrix} 4 \\ -1 \end{bmatrix}$ counterclockwise by the angle $\pi/3$.

43. Use a rotation matrix to rotate the vector $\begin{bmatrix} 5 \\ 2 \end{bmatrix}$ counterclockwise by the angle $\pi/12$.

44. Use a rotation matrix to rotate the vector $\begin{bmatrix} -2 \\ -3 \end{bmatrix}$ counterclockwise by the angle $\pi/9$.

45. Use a rotation matrix to rotate the vector $\begin{bmatrix} 2 \\ 1 \end{bmatrix}$ clockwise by the angle $\pi/4$.

46. Use a rotation matrix to rotate the vector $\begin{bmatrix} 1 \\ -2 \end{bmatrix}$ clockwise by the angle $\pi/3$.

47. Use a rotation matrix to rotate the vector $\begin{bmatrix} 5 \\ -3 \end{bmatrix}$ clockwise by the angle $\pi/7$.

48. Use a rotation matrix to rotate the vector $\begin{bmatrix} -2 \\ -3 \end{bmatrix}$ clockwise by the angle $\pi/8$.

(9.3.2)

In Problems 49–56, find the eigenvalues λ_1 *and* λ_2 *and corresponding eigenvectors* $\mathbf{v}_1$ *and* $\mathbf{v}_2$ *for each matrix A. Determine the equations of the lines through the origin in the direction of the eigenvectors* $\mathbf{v}_1$ *and* $\mathbf{v}_2$, *and graph the lines together with the eigenvectors* $\mathbf{v}_1$ *and* $\mathbf{v}_2$, *and the vectors* $A\mathbf{v}_1$ *and* $A\mathbf{v}_2$.

49. $\quad A = \begin{bmatrix} 2 & 3 \\ 0 & -1 \end{bmatrix}$

50. $\quad A = \begin{bmatrix} 0 & 0 \\ 1 & -3 \end{bmatrix}$

51. $\quad A = \begin{bmatrix} 1 & 0 \\ 0 & -1 \end{bmatrix}$

52. $\quad A = \begin{bmatrix} -1 & 0 \\ 0 & 2 \end{bmatrix}$

53. $\quad A = \begin{bmatrix} -4 & 2 \\ -3 & 1 \end{bmatrix}$

54. $\quad A = \begin{bmatrix} 3 & 6 \\ -1 & -4 \end{bmatrix}$

55. $\quad A = \begin{bmatrix} 2 & 1 \\ 2 & 3 \end{bmatrix}$

56. $\quad A = \begin{bmatrix} -3 & -0.5 \\ 7 & 1.5 \end{bmatrix}$

In Problems 57–62, find the eigenvalues λ_1 *and* λ_2 *for each matrix A.*

57. $\quad A = \begin{bmatrix} 4 & 0 \\ 0 & 3 \end{bmatrix}$

58. $\quad A = \begin{bmatrix} -7 & 0 \\ 0 & 6 \end{bmatrix}$

59. $\quad A = \begin{bmatrix} 1 & -3 \\ 0 & 2 \end{bmatrix}$

60. $\quad A = \begin{bmatrix} -1 & 4 \\ 0 & -2 \end{bmatrix}$

61. $\quad A = \begin{bmatrix} -3/2 & 0 \\ -1 & 1/2 \end{bmatrix}$

62. $\quad A = \begin{bmatrix} \sqrt{2} & 0 \\ 2 & -\sqrt{3} \end{bmatrix}$

63. Let

$$A = \begin{bmatrix} 2 & 4 \\ -2 & -3 \end{bmatrix}$$

Without explicitly computing the eigenvalues of A, decide whether the real parts of both eigenvalues are negative.

64. Let

$$A = \begin{bmatrix} -2 & 3 \\ -1 & 1 \end{bmatrix}$$

Without explicitly computing the eigenvalues of A, decide whether the real parts of both eigenvalues are negative.

65. Let

$$A = \begin{bmatrix} 4 & 4 \\ -4 & -3 \end{bmatrix}$$

Without explicitly computing the eigenvalues of A, decide whether the real parts of both eigenvalues are negative.

66. Let

$$A = \begin{bmatrix} 0 & 1 \\ -2 & 1 \end{bmatrix}$$

Without explicitly computing the eigenvalues of A, decide whether the real parts of both eigenvalues are negative.

67. Let

$$A = \begin{bmatrix} 2 & -5 \\ 2 & -3 \end{bmatrix}$$

Without explicitly computing the eigenvalues of A, decide whether the real parts of both eigenvalues are negative.

68. Let

$$A = \begin{bmatrix} 2 & 5 \\ 2 & -3 \end{bmatrix}$$

Without explicitly computing the eigenvalues of A, decide whether the real parts of both eigenvalues are negative.

(9.3.3)

69. Let

$$A = \begin{bmatrix} -1 & 1 \\ 0 & 2 \end{bmatrix}$$

(a) Show that

$$\mathbf{u}_1 = \begin{bmatrix} 1 \\ 0 \end{bmatrix}$$

and

$$\mathbf{u}_2 = \begin{bmatrix} 1 \\ 3 \end{bmatrix}$$

are eigenvectors of A, and that $\mathbf{u}_1$ and $\mathbf{u}_2$ are linearly independent.

(b) Represent

$$\mathbf{x} = \begin{bmatrix} 1 \\ -3 \end{bmatrix}$$

as a linear combination of $\mathbf{u}_1$ and $\mathbf{u}_2$.

(c) Use your results in (a) and (b) to compute $A^{20}\mathbf{x}$.

70. Let

$$A = \begin{bmatrix} -2 & 1 \\ -4 & 3 \end{bmatrix}$$

(a) Show that

$$\mathbf{u}_1 = \begin{bmatrix} 1 \\ 4 \end{bmatrix}$$

and

$$\mathbf{u}_2 = \begin{bmatrix} 1 \\ 1 \end{bmatrix}$$

are eigenvectors of A, and that $\mathbf{u}_1$ and $\mathbf{u}_2$ are linearly independent.

(b) Represent

$$\mathbf{x} = \begin{bmatrix} -1 \\ 2 \end{bmatrix}$$

as a linear combination of $\mathbf{u}_1$ and $\mathbf{u}_2$.

(c) Use your results in (a) and (b) to compute $A^{10}\mathbf{x}$.

71. Let

$$A = \begin{bmatrix} -1 & 0 \\ 3 & 1 \end{bmatrix}$$

Find

$$A^{15} \begin{bmatrix} 2 \\ 0 \end{bmatrix}$$

without using a calculator.

72. Let

$$A = \begin{bmatrix} 4 & -3 \\ 2 & -1 \end{bmatrix}$$

Find

$$A^{30} \begin{bmatrix} -4 \\ -2 \end{bmatrix}$$

without using a calculator.

73. Let

$$A = \begin{bmatrix} 5 & 7 \\ -2 & -4 \end{bmatrix}$$

Find

$$A^{20} \begin{bmatrix} -3 \\ -2 \end{bmatrix}$$

without using a calculator.

74. Let

$$A = \begin{bmatrix} 1 & -1/4 \\ 1/2 & 1/4 \end{bmatrix}$$

Find

$$A^{30} \begin{bmatrix} 1/2 \\ 3/2 \end{bmatrix}$$

without using a calculator.

75. Suppose that

$$L = \begin{bmatrix} 2 & 4 \\ 0.3 & 0 \end{bmatrix}$$

is the Leslie matrix for a population with two age classes.

(a) Determine both eigenvalues.

(b) Give a biological interpretation of the larger eigenvalue.

(c) Find the stable age distribution.

76. Suppose that

$$L = \begin{bmatrix} 1 & 3 \\ 0.7 & 0 \end{bmatrix}$$

is the Leslie matrix for a population with two age classes.

(a) Determine both eigenvalues.

(b) Give a biological interpretation of the larger eigenvalue.

(c) Find the stable age distribution.

77. Suppose that

$$L = \begin{bmatrix} 7 & 3 \\ 0.1 & 0 \end{bmatrix}$$

is the Leslie matrix for a population with two age classes.

(a) Determine both eigenvalues.

(b) Give a biological interpretation of the larger eigenvalue.

(c) Find the stable age distribution.

78. Suppose that

$$L = \begin{bmatrix} 0 & 5 \\ 0.9 & 0 \end{bmatrix}$$

is the Leslie matrix for a population with two age classes.

(a) Determine both eigenvalues.

(b) Give a biological interpretation of the larger eigenvalue.

(c) Find the stable age distribution.

79. Suppose that

$$L = \begin{bmatrix} 0 & 5 \\ 0.09 & 0 \end{bmatrix}$$

is the Leslie matrix for a population with two age classes.

(a) Determine both eigenvalues.

(b) Give a biological interpretation of the larger eigenvalue.

(c) Find the stable age distribution.

9.4 ANALYTIC GEOMETRY

René Descartes (1596–1650) and Pierre de Fermat (1601–1665) are credited with the invention of analytic geometry, which combines techniques from algebra and geometry and provides important tools for multidimensional calculus. In Chapter 10, we will discuss some aspects of multidimensional calculus; we will therefore need a few results from analytic geometry. Our first task will be to generalize points and vectors in the plane to higher dimensions. We will then introduce a product between vectors that will allow us to determine the length of a vector and the angle between two vectors. Finally, we will give a vector representation of lines and planes in three-dimensional space.

9.4.1 Points and Vectors in Higher Dimensions

We are familiar with points and vectors in the plane. To represent points, we use a Cartesian coordinate system in the plane that consists of two axes, the x_1-axis and the x_2-axis, that are perpendicular to each other. Any point in the plane can be represented by an ordered pair (a_1, a_2) of real numbers, where a_1 is the x_1-coordinate and a_2 is the x_2-coordinate. Since we need two numbers to locate a point in the plane, we call the plane "two dimensional."

The plane can thus be thought of as the set of all points (x_1, x_2) with $x_1 \in \mathbf{R}$ and $x_2 \in \mathbf{R}$. We introduce the notation $\mathbf{R}^2$ to denote the set of these points. That is, the two-dimensional plane is

$$\mathbf{R}^2 = \{(x_1, x_2) : x_1 \in \mathbf{R}, x_2 \in \mathbf{R}\}$$

To generalize this to n dimensions, we set

$$\mathbf{R}^n = \{(x_1, x_2, \ldots, x_n) : x_1 \in \mathbf{R}, x_2 \in \mathbf{R}, \ldots, x_n \in \mathbf{R}\}$$

For instance, $\mathbf{R}^3$ is the three-dimensional space. It consists of all points (x_1, x_2, x_3) with $x_i \in \mathbf{R}$ for $i = 1, 2,$ and 3. In three dimensions, we use a "right-handed" Cartesian coordinate system, to represent points and vectors in $\mathbf{R}^3$. This coordinate system consists of three mutually perpendicular axes oriented in a right-handed coordinate system. That is, the axes are perpendicular to each other and oriented so that the thumb of your right hand points along the positive x_1-axis, the index finger of your right hand points along the positive x_2-axis, and the middle finger of your right hand points along the positive x_3-axis. This coordinate system is shown in Figure 9.26.

In four and higher dimensions, we can no longer draw a coordinate system.

We introduced vectors in two dimensions in Section 9.3. A vector is a quantity that has a direction and a magnitude. We saw in Section 9.3 that a vector in two dimensions is an ordered pair that can be represented by a

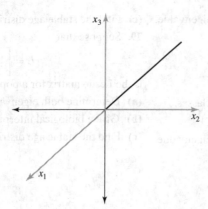

◀ **Figure 9.26**
A right-handed three-dimensional coordinate system. The axes are perpendicular. The x_1-axis points in the direction of the thumb, the x_2-axis in the direction of the index finger, and the x_3-axis in the direction of the middle finger of the right hand

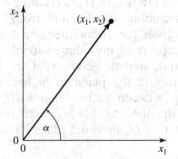

▲ **Figure 9.27**
The vector $\begin{bmatrix} x_1 \\ x_2 \end{bmatrix}$ in the x_1-x_2 plane

directed segment, as illustrated in Figure 9.27. There, the vector $\mathbf{x} = \begin{bmatrix} x_1 \\ x_2 \end{bmatrix}$ is represented by a directed segment with initial point $(0, 0)$ and endpoint (x_1, x_2). The arrow at the tip indicates the direction of the vector. We will generalize this now to n dimensions.

> **Definition** A vector in n-dimensional space is an ordered n-tuple
>
> $$\mathbf{x} = \begin{bmatrix} x_1 \\ x_2 \\ \vdots \\ x_n \end{bmatrix}$$
>
> of real numbers. The numbers $x_1, x_2, \ldots, x_n$ are called the **components** of $\mathbf{x}$.

Vectors in n-dimensional space are also represented by directed segments with initial point $(0, 0, \ldots, 0)$ and endpoint $(x_1, x_2, \ldots, x_n)$. In three dimensions, we can use the three-dimensional right-handed Cartesian coordinate system to visualize vectors. In four and higher dimensions, we can no longer draw the coordinate system. We illustrate them as follows in the plane: To draw a vector $\mathbf{x}$ whose endpoint has coordinates $(x_1, x_2, \ldots, x_n)$, we draw a directed arrow from the origin $(0, 0, \ldots, 0)$ to the point $P = (x_1, x_2, \ldots, x_n)$ (Figure 9.28).

Next, we generalize vector addition to n dimensions. We saw in the preceding section how to add vectors in two-dimensional space. In n dimensions, vector addition is defined in a similar way.

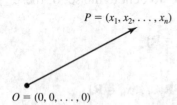

$P = (x_1, x_2, \ldots, x_n)$

$O = (0, 0, \ldots, 0)$

▲ **Figure 9.28**
A vector in n-dimensional space

> **Vector Addition** If
>
> $$\mathbf{x} = \begin{bmatrix} x_1 \\ x_2 \\ \vdots \\ x_n \end{bmatrix} \quad \text{and} \quad \mathbf{y} = \begin{bmatrix} y_1 \\ y_2 \\ \vdots \\ y_n \end{bmatrix}$$
>
> then
>
> $$\mathbf{x} + \mathbf{y} = \begin{bmatrix} x_1 + y_1 \\ x_2 + y_2 \\ \vdots \\ x_n + y_n \end{bmatrix}$$

The geometric interpretation is the same as in two-dimensional space. This law of addition is similarly called the parallelogram law, for reasons that can be seen from Figure 9.29, where the addition of two vectors is illustrated.

We see from Figure 9.29 that vector addition can also be interpreted in the following way: To obtain $\mathbf{x} + \mathbf{y}$, we place the vector $\mathbf{y}$ at the tip of the vector $\mathbf{x}$. The sum $\mathbf{x} + \mathbf{y}$ is then the vector that starts at the same point as $\mathbf{x}$ and ends at the point where the moved vector $\mathbf{y}$ ends, as illustrated in Figure 9.30.

The multiplication of a vector by a scalar also generalizes easily to higher dimensions, and has the same geometric interpretation.

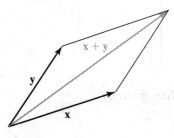

▲ **Figure 9.29**
The parallelogram for vector addition

Multiplication of a Vector by a Scalar If a is a scalar and $\mathbf{x}$ is a vector in n-dimensional space, then

$$a\mathbf{x} = a \begin{bmatrix} x_1 \\ x_2 \\ \vdots \\ x_n \end{bmatrix} = \begin{bmatrix} ax_1 \\ ax_2 \\ \vdots \\ ax_n \end{bmatrix}$$

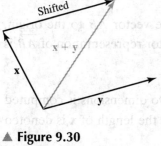

▲ **Figure 9.30**
Adding two vectors

Vector Representation   A vector $\mathbf{x}$ is a directed line segment $\overrightarrow{AB}$ from the initial point A to the terminal point B. A particular representation of $\mathbf{x}$ has the origin at its initial point and the same direction and length as the directed segment $\overrightarrow{AB}$. We can use the terminal point of $\mathbf{x}$ as the representation of $\mathbf{x}$ when its initial point is the origin. We will call this particular representation the **vector representation** of $\overrightarrow{AB}$.

We can use Figure 9.31 to find the vector representation of a directed segment $\overrightarrow{AB}$, from point A with coordinates $(a_1, a_2, \ldots, a_n)$ to point B with coordinates $(b_1, b_2, \ldots, b_n)$. Using the parallelogram law, we see that

$$\overrightarrow{OA} + \overrightarrow{AB} = \overrightarrow{OB}$$

or, solving for $\overrightarrow{AB}$,

$$\overrightarrow{AB} = \overrightarrow{OB} - \overrightarrow{OA}$$

Now, $\overrightarrow{OB}$ has the vector representation $[b_1, b_2, \ldots, b_n]'$, and $\overrightarrow{OA}$ has the vector representation $[a_1, a_2, \ldots, a_n]'$. The difference $\overrightarrow{OB} - \overrightarrow{OA}$ is then simply

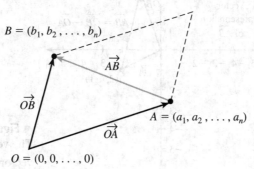

◀ **Figure 9.31**
The vector representation of a directed segment

the difference between the two vectors, that is,

$$\overrightarrow{OB} - \overrightarrow{OA} = \begin{bmatrix} b_1 - a_1 \\ b_2 - a_2 \\ \vdots \\ b_n - a_n \end{bmatrix}$$

The vector representing the line segment $\overrightarrow{AB}$ is thus given by

$$\overrightarrow{AB} = \begin{bmatrix} b_1 - a_1 \\ b_2 - a_2 \\ \vdots \\ b_n - a_n \end{bmatrix}$$

▶ **Example 1**   Find the vector representation of $\overrightarrow{AB}$ when $A = (2, -1)$ and $B = (1, 3)$.

Solution

The vector representation of $\overrightarrow{AB}$ is given by

$$\overrightarrow{AB} = \begin{bmatrix} b_1 - a_1 \\ b_2 - a_2 \end{bmatrix} = \begin{bmatrix} 1 - 2 \\ 3 - (-1) \end{bmatrix} = \begin{bmatrix} -1 \\ 4 \end{bmatrix}$$

We illustrate this graphically in Figure 9.32.

We see from Figure 9.32 that if we shift the vector $\overrightarrow{AB}$ to the origin, its tip ends at $(-1, 4)$ which confirms that the vector representation of $\overrightarrow{AB}$ is $\begin{bmatrix} -1 \\ 4 \end{bmatrix}$. ◀

Length of a Vector   The length of a vector in two dimensions is computed from the Pythagorean theorem. If $\mathbf{x} = \begin{bmatrix} x_1 \\ x_2 \end{bmatrix}$, then the length of $\mathbf{x}$ is denoted by $|\mathbf{x}|$, and we find

$$|\mathbf{x}| = \sqrt{x_1^2 + x_2^2}$$

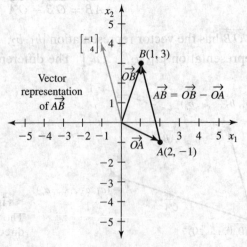

◀ **Figure 9.32**
The vector representation of the vector $\overrightarrow{AB}$

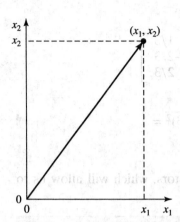

▲ Figure 9.33
The length of the vector $\begin{bmatrix} x_1 \\ x_2 \end{bmatrix}$

as illustrated in Figure 9.33. We can generalize this to n dimensions. Recall that the transpose of a vector is denoted by prime, that is,

$$[x_1, x_2, \ldots, x_n]' = \begin{bmatrix} x_1 \\ x_2 \\ \vdots \\ x_n \end{bmatrix}$$

This allows us to denote column vectors as the transpose of the corresponding row vectors, which is convenient when writing large column vectors.

> The length of a vector $\mathbf{x} = [x_1, x_2, \ldots, x_n]'$ is
>
> $$|\mathbf{x}| = \sqrt{x_1^2 + x_2^2 + \cdots + x_n^2}$$

▶ **Example 2** Find the length of

$$\mathbf{x} = \begin{bmatrix} 1 \\ -3 \\ 4 \end{bmatrix}$$

Solution

The length of $\mathbf{x}$ is given by

$$|\mathbf{x}| = \sqrt{x_1^2 + x_2^2 + x_3^2} = \sqrt{(1)^2 + (-3)^2 + (4)^2} = \sqrt{26} \qquad \blacktriangleleft$$

If we know the length of a vector $\mathbf{x}$, we can **normalize x** to obtain a vector of length 1 in the same direction as $\mathbf{x}$ (see Figure 9.34). We call such a vector a **unit vector** in the direction of $\mathbf{x}$, and denote it by $\hat{\mathbf{x}}$, that is,

$$\hat{\mathbf{x}} = \frac{\mathbf{x}}{|\mathbf{x}|}$$

and, of course, $|\hat{\mathbf{x}}| = 1$. We summarize this in the following.

▲ Figure 9.34
The normalized vector $\hat{x}$ has the same direction as x; its length is 1

> $\dfrac{\mathbf{x}}{|\mathbf{x}|}$ is a vector of length 1 in the direction of $\mathbf{x}$.

▶ **Example 3** Normalize the vector

$$\mathbf{x} = \begin{bmatrix} 3 \\ -6 \\ 6 \end{bmatrix}$$

Solution

We must first find the length of $\mathbf{x}$.

$$|\mathbf{x}| = \sqrt{x_1^2 + x_2^2 + x_3^2} = \sqrt{(3)^2 + (-6)^2 + (6)^2} = \sqrt{81} = 9$$

The unit vector $\hat{\mathbf{x}}$ is then given by

$$\hat{\mathbf{x}} = \frac{\mathbf{x}}{|\mathbf{x}|} = \frac{1}{9} \begin{bmatrix} 3 \\ -6 \\ 6 \end{bmatrix} = \begin{bmatrix} 1/3 \\ -2/3 \\ 2/3 \end{bmatrix}$$

We check that $\hat{\mathbf{x}}$ is indeed a vector of length 1:

$$\hat{\mathbf{x}} = \sqrt{(1/3)^2 + (-2/3)^2 + (2/3)^2} = 1$$

9.4.2 The Dot Product

We will now define the dot product of two vectors, which will allow us to determine the angle between two vectors.

> **Definition** The **scalar product** or **dot product** of two vectors $\mathbf{x} = [x_1, x_2, \ldots, x_n]'$ and $\mathbf{y} = [y_1, y_2, \ldots, y_n]'$ is the number
>
> $$\mathbf{x} \cdot \mathbf{y} = \mathbf{x}'\mathbf{y} = \sum_{i=1}^{n} x_i y_i$$

Note that the result is a scalar (hence the name scalar product). It is also called dot product because the notation uses a dot between $\mathbf{x}$ and $\mathbf{y}$.

▶ **Example 4**   Find the dot product of

$$\mathbf{x} = \begin{bmatrix} 2 \\ 3 \\ 1 \end{bmatrix} \quad \text{and} \quad \mathbf{y} = \begin{bmatrix} -1 \\ 2 \\ 0 \end{bmatrix}$$

Solution

Using the definition of the dot product, we find

$$\mathbf{x} \cdot \mathbf{y} = \begin{bmatrix} 2 & 3 & 1 \end{bmatrix} \begin{bmatrix} -1 \\ 2 \\ 0 \end{bmatrix} = -2 + 6 + 0 = 4$$

The dot product satisfies the following two properties:

1. $\mathbf{x} \cdot \mathbf{y} = \mathbf{y} \cdot \mathbf{x}$
2. $\mathbf{x} \cdot (\mathbf{y} + \mathbf{z}) = \mathbf{x} \cdot \mathbf{y} + \mathbf{x} \cdot \mathbf{z}$

We can use the dot product to express the length of a vector. Recall that in n dimensions the length of a vector $\mathbf{x} = [x_1, x_2, \ldots, x_n]'$ is defined as

$$|\mathbf{x}| = \sqrt{x_1^2 + x_2^2 + \cdots + x_n^2}$$

If we compute the dot product of $\mathbf{x}$ with itself, we find

$$\mathbf{x} \cdot \mathbf{x} = \sum_{i=1}^{n} x_i^2$$

Hence, comparing $|\mathbf{x}|$ and $\mathbf{x} \cdot \mathbf{x}$, we see that the following holds:

$$|\mathbf{x}|^2 = \mathbf{x} \cdot \mathbf{x}$$

The length of a vector is therefore $|\mathbf{x}| = \sqrt{\mathbf{x} \cdot \mathbf{x}}$.

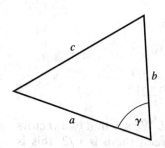

▲ Figure 9.35
The Law of Cosines:
$c^2 = a^2 + b^2 - 2ab\cos\gamma$

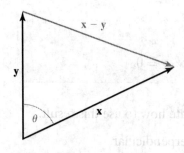

▲ Figure 9.36
The angle between two vectors

The Angle between Two Vectors The other important application of the dot product is that it allows us to find the angle between two vectors. To derive this result, we need one of the trigonometric laws, the law of cosines, as illustrated in Figure 9.35.

Now, let **x** and **y** be two nonzero vectors whose initial points coincide. Then **x** − **y** is a vector that connects the endpoint of **y** to the endpoint of **x**. This follows from the parallelogram law for adding two vectors, namely **y** + **x** − **y** = **x**, as illustrated in Figure 9.36.

Using the law of cosines, we find

$$|\mathbf{x} - \mathbf{y}|^2 = |\mathbf{x}|^2 + |\mathbf{y}|^2 - 2|\mathbf{x}||\mathbf{y}|\cos\theta$$

where θ is the angle between **x** and **y**, as illustrated in Figure 9.36. On the other hand, the length of the vector **x** − **y**, denoted $|\mathbf{x} - \mathbf{y}|$, can be computed using the dot product:

$$|\mathbf{x} - \mathbf{y}|^2 = (\mathbf{x} - \mathbf{y}) \cdot (\mathbf{x} - \mathbf{y})$$
$$= \mathbf{x} \cdot \mathbf{x} - \mathbf{x} \cdot \mathbf{y} - \mathbf{y} \cdot \mathbf{x} + \mathbf{y} \cdot \mathbf{y}$$
$$= |\mathbf{x}|^2 - 2\mathbf{x} \cdot \mathbf{y} + |\mathbf{y}|^2$$

Setting the two expressions for $|\mathbf{x} - \mathbf{y}|^2$ to be equal, we find

$$|\mathbf{x}|^2 + |\mathbf{y}|^2 - 2|\mathbf{x}||\mathbf{y}|\cos\theta = |\mathbf{x}|^2 - 2\mathbf{x} \cdot \mathbf{y} + |\mathbf{y}|^2$$

Solving this for **x** · **y**, we find the following:

$$\mathbf{x} \cdot \mathbf{y} = |\mathbf{x}||\mathbf{y}|\cos\theta \qquad (9.22)$$

The significance of equation (9.22) is that it allows us to find the angle between the two nonzero vectors **x** and **y**, since

$$\theta = \cos^{-1}\left(\frac{\mathbf{x} \cdot \mathbf{y}}{|\mathbf{x}||\mathbf{y}|}\right)$$

Note that $\theta \in [0, \pi)$, since the interval $[0, \pi)$ is used to find the inverse of the cosine function.

▶ Example 5   Find the angle between

$$\mathbf{x} = \begin{bmatrix} 2 \\ 1 \end{bmatrix} \qquad \text{and} \qquad \mathbf{y} = \begin{bmatrix} 1 \\ 1 \end{bmatrix}$$

Solution

To determine the angle θ between **x** and **y**, we use

$$\mathbf{x} \cdot \mathbf{y} = |\mathbf{x}||\mathbf{y}|\cos\theta$$

We find

$$\mathbf{x} \cdot \mathbf{y} = \begin{bmatrix} 2 \\ 1 \end{bmatrix} \cdot \begin{bmatrix} 1 \\ 1 \end{bmatrix} = 2 + 1 = 3$$

$$|\mathbf{x}| = \sqrt{(2)^2 + (1)^2} = \sqrt{5}$$

$$|\mathbf{y}| = \sqrt{(1)^2 + (1)^2} = \sqrt{2}$$

Hence,

$$\cos\theta = \frac{\mathbf{x}\cdot\mathbf{y}}{|\mathbf{x}||\mathbf{y}|} = \frac{3}{\sqrt{5}\sqrt{2}} = \frac{3}{\sqrt{10}}$$

and, therefore,

$$\theta = \cos^{-1}\frac{3}{\sqrt{10}} \approx 18.4° \qquad \text{or} \qquad 0.3218$$

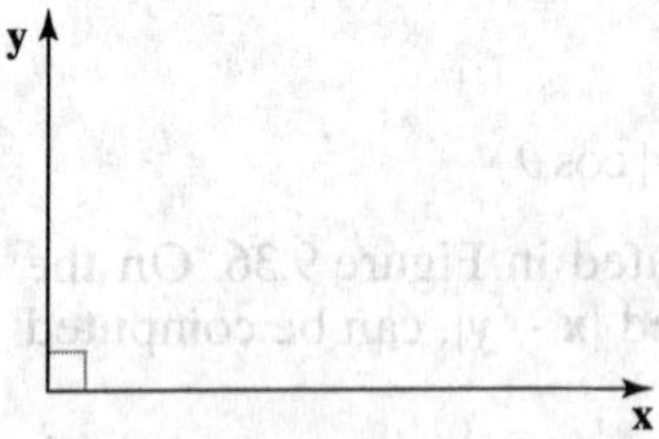

We wish to single out the case in which $\theta = \pi/2$. We say that two vectors are **perpendicular** to each other if the angle between them is $\pi/2$; this is illustrated in Figure 9.37.

An important consequence of (9.22) is that it gives us a criterion with which to determine whether two vectors are perpendicular. Since $\cos(\pi/2) = 0$, we have the following.

▲ **Figure 9.37**
The vectors x and y are perpendicular

Theorem
x and **y** are perpendicular if $\mathbf{x}\cdot\mathbf{y} = 0$.

We will now give two examples that illustrate how to use this result.

▶ **Example 6** Let $\mathbf{x} = \begin{bmatrix} 1 \\ 2 \end{bmatrix}$. Find $\mathbf{y} = \begin{bmatrix} y_1 \\ y_2 \end{bmatrix}$ so that **x** and **y** are perpendicular.

Solution

The vectors **x** and **y** are perpendicular if $\mathbf{x}\cdot\mathbf{y} = 0$. We find

$$\mathbf{x}\cdot\mathbf{y} = \begin{bmatrix} 1 \\ 2 \end{bmatrix}\cdot\begin{bmatrix} y_1 \\ y_2 \end{bmatrix} = y_1 + 2y_2$$

Setting this equal to 0, we obtain the equation

$$y_1 + 2y_2 = 0$$

Any choice of numbers (y_1, y_2) that satisfies this equation would thus give us a vector that is perpendicular to **x**. For instance, if we choose $y_2 = 1$ and $y_1 = -2$, then

$$\mathbf{y} = \begin{bmatrix} -2 \\ 1 \end{bmatrix}$$

is perpendicular to **x**.

▶ **Example 7** Show that the coordinate axes in a two-dimensional Cartesian coordinate system are perpendicular.

Solution

A two-dimensional Cartesian coordinate system is illustrated in Figure 9.38. We see that the x-axis can be represented by the vector $\begin{bmatrix} 1 \\ 0 \end{bmatrix}$ and the y-axis by the vector $\begin{bmatrix} 0 \\ 1 \end{bmatrix}$. Computing the dot product between these two vectors, we find

$$\begin{bmatrix} 1 \\ 0 \end{bmatrix}\cdot\begin{bmatrix} 0 \\ 1 \end{bmatrix} = (1)(0) + (0)(1) = 0$$

and we conclude that the two vectors are perpendicular. Therefore, the x-axis and the y-axis are perpendicular.

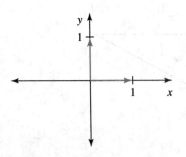

▲ Figure 9.38
The coordinate axes in a two dimensional Cartesian coordinate system are perpendicular

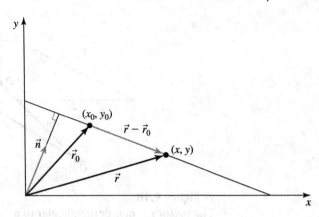

▲ Figure 9.39
The vector $\mathbf{r} - \mathbf{r}_0$ is perpendicular to $\mathbf{n}$

We will now use the dot product, and the result that the dot product between perpendicular vectors is zero, to obtain the equation of a line in two-dimensional space and the equation of a plane in three-dimensional space.

Lines in the Plane We will use the dot product to write equations for lines in the plane. Suppose that we wish to find the solution of the equation of a line through the point (x_0, y_0), which is perpendicular to the vector $\mathbf{n} = \begin{bmatrix} a \\ b \end{bmatrix}$. If (x, y) is another point on the line, then the vector $\mathbf{r} - \mathbf{r}_0$ is perpendicular to $\mathbf{n}$, as illustrated in Figure 9.39.

Therefore,

$$\mathbf{n} \cdot (\mathbf{r} - \mathbf{r}_0) = 0$$

This is called the **vector equation** of a line in the plane. To obtain the **scalar equation** of this line, we set

$$\mathbf{n} = \begin{bmatrix} a \\ b \end{bmatrix}, \quad \mathbf{r} = \begin{bmatrix} x \\ y \end{bmatrix} \quad \text{and} \quad \mathbf{r}_0 = \begin{bmatrix} x_0 \\ y_0 \end{bmatrix}$$

Then

$$\mathbf{n} \cdot (\mathbf{r} - \mathbf{r}_0) = \begin{bmatrix} a \\ b \end{bmatrix} \cdot \begin{bmatrix} x - x_0 \\ y - y_0 \end{bmatrix}$$

$$= a(x - x_0) + b(y - y_0) = 0$$

That is, we find the following.

The line through (x_0, y_0) perpendicular to $\begin{bmatrix} a \\ b \end{bmatrix}$ has the equation

$$a(x - x_0) + b(y - y_0) = 0 \qquad (9.23)$$

 Example 8 Find an equation for the line through $(4, 3)$ perpendicular to $\begin{bmatrix} 1 \\ 2 \end{bmatrix}$.

Solution

Using (9.23), we find

$$(1)(x - 4) + (2)(y - 3) = 0$$

Simplifying yields

$$x + 2y = 10$$

◀

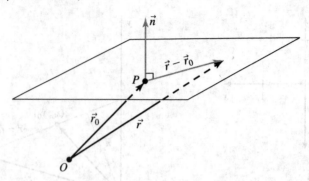

▲ **Figure 9.40**
The vector $\mathbf{r} - \mathbf{r}_0$ is perpendicular to $\mathbf{n}$

Planes in Space We can characterize planes by a point P in the plane, and the vector with initial point P that is perpendicular to all vectors in the plane whose initial points coincide with P, as illustrated in Figure 9.40. If P is the endpoint of $\mathbf{r}_0$, $\mathbf{r} - \mathbf{r}_0$ is a vector in the plane with initial point P, and if $\mathbf{n}$ is a vector that is perpendicular to the plane, then

$$\mathbf{n} \cdot (\mathbf{r} - \mathbf{r}_0) = 0$$

This is called the vector equation of a plane. To obtain a scalar equation of a plane in three-dimensional space, we set

$$\mathbf{n} = \begin{bmatrix} a \\ b \\ c \end{bmatrix}, \qquad \mathbf{r} = \begin{bmatrix} x \\ y \\ z \end{bmatrix} \quad \text{and} \quad \mathbf{r}_0 = \begin{bmatrix} x_0 \\ y_0 \\ z_0 \end{bmatrix}$$

Then $\mathbf{n} \cdot (\mathbf{r} - \mathbf{r}_0) = 0$ results in the scalar equation of the plane through the point (x_0, y_0, z_0) with normal vector $\mathbf{n} = \begin{bmatrix} a, & b, & c \end{bmatrix}'$.

Evaluating the dot product yields

$$a(x - x_0) + b(y - y_0) + c(z - z_0) = 0$$

Summarizing this result, we find the following.

The plane through (x_0, y_0, z_0) perpendicular to $\begin{bmatrix} a, & b, & c \end{bmatrix}'$ has the equation

$$a(x - x_0) + b(y - y_0) + c(z - z_0) = 0 \qquad (9.24)$$

▶ **Example 9** Find the equation of a plane in three-dimensional space through $(2, 0, 3)$ that is perpendicular to $\begin{bmatrix} -1, & 4, & 1 \end{bmatrix}'$.

Solution

Using (9.24), we find

$$(-1)(x - 2) + (4)(y - 0) + (1)(z - 3) = 0$$

Simplifying yields

$$-x + 4y + z = 1 \qquad ◀$$

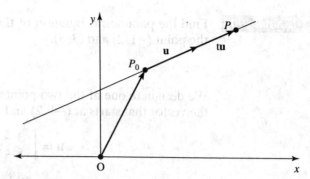

▲ Figure 9.41
The parametric equation of a line

9.4.3 Parametric Equations of Lines

There is another way to write lines in $\mathbf{R}^2$. Looking at Figure 9.41, the line can be described by the vector connecting the points O and P_0, denoted by $\overrightarrow{OP_0}$, and the vector $\mathbf{u}$: Any point P on the line can be thought of as the endpoint of the sum of the vector $\overrightarrow{OP_0}$ and the vector $\overrightarrow{P_0P}$. Now, $\overrightarrow{P_0P}$ is a multiple of the vector $\mathbf{u}$. We find for the equation of the line in vector form

$$\overrightarrow{OP} = \overrightarrow{OP_0} + \overrightarrow{P_0P} = \overrightarrow{OP_0} + t\mathbf{u} \tag{9.25}$$

for some $t \in \mathbf{R}$. If P has coordinates (x, y), P_0 has coordinates (x_0, y_0), and $\mathbf{u} = \begin{bmatrix} u_1 \\ u_2 \end{bmatrix}$, then

$$\begin{bmatrix} x \\ y \end{bmatrix} = \begin{bmatrix} x_0 \\ y_0 \end{bmatrix} + t \begin{bmatrix} u_1 \\ u_2 \end{bmatrix} \tag{9.26}$$

for some $t \in \mathbf{R}$. By varying t, we can reach *any* point on the line. Equation (9.25) [or (9.26)] is called a **parametric equation** in vector form of a line and t is called a **parameter**. We can also write the equation of the line in parametric form for each coordinate separately:

$$x = x_0 + tu_1$$
$$y = y_0 + tu_2$$

for $t \in \mathbf{R}$.

▶ Example 10

Find the parametric equation of the line in the x-y plane that goes through the point $(2, 1)$ in the direction of $\begin{bmatrix} -1 \\ -3 \end{bmatrix}$.

Solution

We find

$$\begin{bmatrix} x \\ y \end{bmatrix} = \begin{bmatrix} 2 \\ 1 \end{bmatrix} + t \begin{bmatrix} -1 \\ -3 \end{bmatrix}, \ t \in \mathbf{R}$$

or $x = 2 - t$ and $y = 1 - 3t$ for $t \in \mathbf{R}$. By eliminating t, we can write the equation in the to us familiar standard form of a line in the x-y plane. Namely, $t = 2 - x$ and, therefore,

$$y = 1 - 3(2 - x) = 1 - 6 + 3x$$

and hence,

$$3x - y - 5 = 0$$

is the standard form of the equation of this line. ◀

▶ **Example 11**   Find the parametric equation of the line in the x-y plane that goes through the point $(-1, 2)$ and $(3, 5)$.

Solution

We designate one of the two points, say $(-1, 2)$, as the point P_0 and let **u** be the vector that starts at $(-1, 2)$ and ends at $(3, 5)$. Then

$$\mathbf{u} = \begin{bmatrix} 3 - (-1) \\ 5 - 2 \end{bmatrix} = \begin{bmatrix} 4 \\ 3 \end{bmatrix}$$

and the parametric equation of this line is

$$\begin{bmatrix} x \\ y \end{bmatrix} = \begin{bmatrix} -1 \\ 2 \end{bmatrix} + t \begin{bmatrix} 4 \\ 3 \end{bmatrix}, \ t \in \mathbf{R} \quad ◀$$

In the first example we saw that by eliminating t, we can obtain the standard form of a linear equation. We can also go from the standard form to the parametric form by introducing a parameter t.

▶ **Example 12**   Find a parametric form of the line in standard form

$$2x - 3y + 1 = 0$$

Solution

The easiest way to parameterize this equation is to set $x = t$, solve the equation for y, and substitute t for x:

$$3y = 2x + 1 \quad \text{is equivalent to} \quad y = \frac{2}{3}x + \frac{1}{3}$$

With $x = t$, we can write the parametric equation as

$$x = t$$
$$y = \frac{2}{3}t + \frac{1}{3}$$

for $t \in \mathbf{R}$.

This is by no means the only way to parameterize the line. If we had chosen $t = \frac{1}{3}(x - 1)$, we would have found

$$x = 3t + 1$$
$$y = \frac{2}{3}(3t + 1) + \frac{1}{3} = 2t + 1 \quad ◀$$

The vector form representation of the line can be used in higher dimensions. For instance, a line in $\mathbf{R}^3$ that goes through a point $P_0 = (x_0, y_0, z_0)$ in the direction of $\mathbf{u} = \begin{bmatrix} u_1 \\ u_2 \\ u_3 \end{bmatrix}$ would have the form

$$\begin{bmatrix} x \\ y \\ z \end{bmatrix} = \begin{bmatrix} x_0 \\ y_0 \\ z_0 \end{bmatrix} + t \begin{bmatrix} u_1 \\ u_2 \\ u_3 \end{bmatrix}, \ t \in \mathbf{R}$$

or, if we write out the coordinates separately,

$$x = x_0 + tu_1$$
$$y = y_0 + tu_2$$
$$z = z_0 + tu_3$$

for $t \in \mathbf{R}$.

▶ Example 13   Find the parametric equation of the line in x-y-z space that goes through the points $(1, -1, 3)$ and $(2, 4, -1)$.

Solution

We designate one point as P_0, say $(2, 4, -1)$, and let **u** be the vector connecting the two points (it does not matter which of the two points we select as the starting point):

$$\mathbf{u} = \begin{bmatrix} 2-1 \\ 4-(-1) \\ -1-3 \end{bmatrix} = \begin{bmatrix} 1 \\ 5 \\ -4 \end{bmatrix}$$

Then the parametric equation of this line is

$$\begin{bmatrix} x \\ y \\ z \end{bmatrix} = \begin{bmatrix} 2 \\ 4 \\ -1 \end{bmatrix} + t \begin{bmatrix} 1 \\ 5 \\ -4 \end{bmatrix}, \quad t \in \mathbf{R}$$

or

$$x = 2 + t$$
$$y = 4 + 5t$$
$$z = -1 - 4t$$

for $t \in \mathbf{R}$.   ◀

Eliminating t in the parametric equation for a line in $\mathbf{R}^3$ is not very useful since it does not yield just one equation as in the case of a line in $\mathbf{R}^2$. We will therefore not do it.

9.4.4 Problems

(9.4.1)

1. Let $\mathbf{x} = [1, 4, -1]'$ and $\mathbf{y} = [-2, 1, 0]'$.

(a) Find $\mathbf{x} + \mathbf{y}$.

(b) Find $2\mathbf{x}$.

(c) Find $-3\mathbf{y}$.

2. Let $\mathbf{x} = [-4, 3, 1]'$ and $\mathbf{y} = [0, -2, 3]'$.

(a) Find $\mathbf{x} - \mathbf{y}$.

(b) Find $2\mathbf{x} + 3\mathbf{y}$.

(c) Find $-\mathbf{x} - 2\mathbf{y}$.

3. Let $A = (2, 3)$ and $B = (4, 1)$. Find the vector representation of $\overrightarrow{AB}$.

4. Let $A = (-1, 0)$ and $B = (2, -4)$. Find the vector representation of $\overrightarrow{AB}$.

5. Let $A = (0, 1, -3)$ and $B = (-1, -1, 2)$. Find the vector representation of $\overrightarrow{AB}$.

6. Let $A = (1, 3, -2)$ and $B = (0, -1, 0)$. Find the vector representation of $\overrightarrow{AB}$.

7. Find the length of $\mathbf{x} = [1, 3]'$.

8. Find the length of $\mathbf{x} = [-1, 4]'$.

9. Find the length of $\mathbf{x} = [0, 1, 5]'$.

10. Find the length of $\mathbf{x} = [-2, 3, -3]'$.

11. Normalize $[1, 3, -1]'$.

12. Normalize $[2, 0, -4]'$.

13. Normalize $[1, 0, 0]'$.

14. Normalize $[0, -3, 1, 3]'$.

(9.4.2)

15. Find the dot product of $\mathbf{x} = [1, 2]'$ and $\mathbf{y} = [3, -1]'$.

16. Find the dot product of $\mathbf{x} = [-1, 2]'$ and $\mathbf{y} = [-2, -4]'$.

17. Find the dot product of $\mathbf{x} = [0, -1, 3]'$ and $\mathbf{y} = [-3, 1, 1]'$.

18. Find the dot product of $\mathbf{x} = [1, -3, 2]'$ and $\mathbf{y} = [3, 1, -4]'$.

19. Use the dot product to compute the length of $[0, -1, 2]'$.

20. Use the dot product to compute the length of $[-1, 4, 3]'$.

21. Use the dot product to compute the length of $[1, 2, 3, 4]'$.

22. Use the dot product to compute the length of $[-1, -2, -3, -4]'$.

23. Find the angle between $\mathbf{x} = [1, 2]'$ and $\mathbf{y} = [3, -1]'$.

24. Find the angle between $\mathbf{x} = [-1, 2]'$ and $\mathbf{y} = [-2, -4]'$.

25. Find the angle between $\mathbf{x} = [0, -1, 3]'$ and $\mathbf{y} = [-3, 1, 1]'$.

26. Find the angle between $\mathbf{x} = [1, -3, 2]'$ and $\mathbf{y} = [3, 1, -4]'$.

27. Let $\mathbf{x} = [1, -1]'$. Find $\mathbf{y}$ so that $\mathbf{x}$ and $\mathbf{y}$ are perpendicular.

28. Let $\mathbf{x} = [-2, 1]'$. Find $\mathbf{y}$ so that $\mathbf{x}$ and $\mathbf{y}$ are perpendicular.

29. Let $\mathbf{x} = [1, -2, 4]'$. Find $\mathbf{y}$ so that $\mathbf{x}$ and $\mathbf{y}$ are perpendicular.

30. Let $\mathbf{x} = [2, 0, -1]'$. Find $\mathbf{y}$ so that $\mathbf{x}$ and $\mathbf{y}$ are perpendicular.

31. A triangle has vertices at coordinates $P = (0, 0)$, $Q = (4, 0)$, and $R = (4, 3)$.
(a) Use basic trigonometry to compute the lengths of all three sides and compute all three angles.
(b) Use the results of this section to repeat (a).

32. A triangle has vertices at coordinates $P = (0, 0)$, $Q = (0, 3)$, and $R = (5, 0)$.
(a) Use basic trigonometry to compute the lengths of all three sides and compute all three angles.
(b) Use the results of this section to repeat (a).

33. A triangle has vertices at coordinates $P = (1, 2, 3)$, $Q = (1, 5, 2)$, and $R = (2, 4, 2)$.
(a) Compute the lengths of all three sides.
(b) Compute all three angles in both radians and degrees.

34. A triangle has vertices at coordinates $P = (2, 1, 5)$, $Q = (-1, -3, 7)$, and $R = (2, -4, 1)$.
(a) Compute the lengths of all three sides.
(b) Compute all three angles in both radians and degrees.

35. Find the equation for the line through $(2, 1)$ that is perpendicular to $[1, 2]'$.

36. Find the equation for the line through $(5, 2)$ that is perpendicular to $[-1, 1]'$.

37. Find the equation for the line through $(1, -2)$ that is perpendicular to $[4, 1]'$.

38. Find the equation for the line through $(0, 1)$ that is perpendicular to $[1, 0]'$.

39. Find the equation for the plane through $(1, 2, 3)$ that is perpendicular to $[0, -1, 1]'$.

40. Find the equation for the plane through $(-1, 0, -3)$ that is perpendicular to $[1, -2, 1]'$.

41. Find the equation for the plane through $(0, 0, 0)$ that is perpendicular to $[1, 0, 0]'$.

42. Find the equation for the plane through $(1, -1, 2)$ that is perpendicular to $[-1, 1, 1]'$.

(9.4.3)

In Problems 43–46, find the parametric equation of the line in the x-y plane that goes through the indicated point in the direction of the indicated vector.

43. $(1, -1)$, $\begin{bmatrix} 2 \\ 1 \end{bmatrix}$

44. $(3, -4)$, $\begin{bmatrix} -1 \\ 2 \end{bmatrix}$

45. $(-1, -2)$, $\begin{bmatrix} 1 \\ -3 \end{bmatrix}$

46. $(-1, 4)$, $\begin{bmatrix} 2 \\ 3 \end{bmatrix}$

In Problems 47–50, find the parametric equation of the line in the x-y plane that goes through the given points. Then eliminate the parameter to find the equation of the line in standard form.

47. $(-1, 2)$ and $(3, 4)$

48. $(2, 1)$ and $(3, 5)$

49. $(1, -3)$ and $(4, 0)$

50. $(2, 3)$ and $(-1, -4)$

In Problems 51–54, parameterize the equation of the line given in standard form.

51. $3x + 4y - 1 = 0$

52. $x - 2y + 5 = 0$

53. $2x + y - 3 = 0$

54. $x - 5y + 7 = 0$

In Problems 55–58, find the parametric equation of the line in x-y-z space that goes through the indicated point in the direction of the indicated vector.

55. $(1, -1, 2)$, $\begin{bmatrix} 1 \\ -2 \\ 1 \end{bmatrix}$

56. $(2, 0, 4)$, $\begin{bmatrix} 1 \\ 2 \\ 3 \end{bmatrix}$

57. $(-1, 3, -2)$, $\begin{bmatrix} -1 \\ -2 \\ 4 \end{bmatrix}$

58. $(2, 1, -3)$, $\begin{bmatrix} 3 \\ -1 \\ 2 \end{bmatrix}$

In Problems 59–62, find the parametric equation of the line in x-y-z space that goes through the given points.

59. $(5, 4, -1)$ and $(2, 0, 3)$

60. $(1, 0, -3)$ and $(4, 1, 0)$

61. $(2, -3, 1)$ and $(-5, 2, 1)$

62. $(1, 0, 4)$ and $(3, 1, 0)$

63. Given is a plane through $(1, -1, 2)$, which is perpendicular to $\begin{bmatrix} 1 \\ 2 \\ 1 \end{bmatrix}$, and a line through the points $(0, -3, 2)$ and $(-1, -2, 3)$. Where do the plane and the line intersect?

64. Given is a plane through $(2, 0, -1)$, which is perpendicular to $\begin{bmatrix} -1 \\ 1 \\ 3 \end{bmatrix}$, and a line through the points $(1, 0, -2)$ and $(-1, -1, 1)$. Where do the plane and the line intersect?

65. Given is a plane through $(0, -2, 1)$, which is perpendicular to $\begin{bmatrix} -1 \\ 1 \\ -1 \end{bmatrix}$. Find a line through $(5, -1, 0)$ that is parallel to the plane.

66. Given is the plane $x + 2y - z + 1 = 0$. Find a line in parametric form that is perpendicular to the plane.

9.5 Key Terms

Chapter 9 Review: Topics *Discuss the following definitions and concepts:*

1. Linear system of equations
2. Solving a linear system of equations
3. Upper triangular form
4. Gaussian elimination
5. Matrix
6. Basic matrix operations: addition, multiplication by a scalar
7. Transposition
8. Matrix multiplication
9. Identity matrix
10. Inverse matrix
11. Determinant
12. Leslie matrix
13. Stable age distribution
14. Vector
15. Parallelogram law
16. Linear map
17. Eigenvalues and eigenvectors
18. Vector addition
19. Multiplication of a vector by a scalar
20. Length of a vector
21. Dot product
22. Angle between two vectors
23. Perpendicular vectors
24. Line in the plane and in space
25. Equation of a plane
26. Parametric equation of a line

9.6 Review Problems

1. Let

$$A = \begin{bmatrix} -1 & 1 \\ 0 & 2 \end{bmatrix}$$

(a) Find Ax when $x = \begin{bmatrix} 1 \\ 1 \end{bmatrix}$. Graph both x and Ax in the same coordinate system.

(b) Find the eigenvalues λ_1 and λ_2 and the corresponding eigenvectors u_1 and u_2 of A.

(c) If u_i is the eigenvector corresponding to λ_i, find Au_i, and explain graphically what happens when you apply A to u_i.

(d) Write x as a linear combination of u_1 and u_2; that is, find a_1 and a_2 so that

$$x = a_1 u_1 + a_2 u_2$$

Show that

$$Ax = a_1 \lambda_1 u_1 + a_2 \lambda_2 u_2$$

and illustrate this graphically.

2. Let

$$A = \begin{bmatrix} 3 & 1/2 \\ -5 & -1/2 \end{bmatrix}$$

(a) Find Ax when $x = \begin{bmatrix} 2 \\ 1 \end{bmatrix}$. Graph both x and Ax in the same coordinate system.

(b) Find the eigenvalues λ_1 and λ_2 and the corresponding eigenvectors u_1 and u_2 of A.

(c) If u_i is the eigenvector corresponding to λ_i, find Au_i, and explain graphically what happens when you apply A to u_i.

(d) Write x as a linear combination of u_1 and u_2; that is, find a_1 and a_2 so that

$$x = a_1 u_1 + a_2 u_2$$

Show that

$$Ax = a_1 \lambda_1 u_1 + a_2 \lambda_2 u_2$$

and illustrate this graphically.

3. Given the Leslie matrix

$$L = \begin{bmatrix} 1.5 & 0.875 \\ 0.5 & 0 \end{bmatrix}$$

find the growth rate of the population, and determine its stable age distribution.

4. Given the Leslie matrix

$$L = \begin{bmatrix} 0.5 & 2.99 \\ 0.25 & 0 \end{bmatrix}$$

find the growth rate of the population, and determine its stable age distribution.

5. Let

$$AB = \begin{bmatrix} 0 & 1 \\ 2 & -1 \end{bmatrix}$$

and

$$A^{-1} = \begin{bmatrix} 4 & -1 \\ 8 & -1 \end{bmatrix}$$

Find B.

6. Let

$$(AB)^{-1} = \begin{bmatrix} -1 & 3 \\ 0 & 2 \end{bmatrix}$$

and

$$B = \begin{bmatrix} 0 & -2 \\ 1 & 3 \end{bmatrix}$$

Find A.

7. Explain two different ways to solve a system of the form

$$a_{11}x_1 + a_{12}x_2 = b_1$$
$$a_{21}x_1 + a_{22}x_2 = b_2$$

when $a_{11}a_{22} - a_{12}a_{21} \neq 0$.

8. Suppose that

$$a_{11}x_1 + a_{12}x_2 = b_1$$
$$a_{21}x_1 + a_{22}x_2 = b_2$$

has infinitely many solutions. If you wrote this system in matrix form $AX = B$, could you find X by computing $A^{-1}B$?

9. Let

$$ax + 3y = 0$$
$$x - y = 0$$

How would you need to choose a so that this system had infinitely many solutions?

10. Let A be a 2×2 matrix and X and B be 2×1 matrices. Assume that $\det A = 0$. Explain how the choice of B affects the number of solutions of $AX = B$.

11. Suppose that

$$L = \begin{bmatrix} 0.5 & 2.3 \\ a & 0 \end{bmatrix}$$

is the Leslie matrix of a population with two age classes. For which values of a does this population grow?

12. Suppose that

$$L = \begin{bmatrix} 0.5 & 2.0 \\ 0.1 & 0 \end{bmatrix}$$

is the Leslie matrix of a population with two age classes.

(a) If you were to manage this population, would you need to be concerned about its long-term survival?

(b) Suppose that you can either improve the fecundity of or the survival of the zero-year-olds. But due to physiological and environmental constraints, fecundity of zero-year-olds will not exceed 1.5, and survival of zero-year-olds will not exceed 0.4. Investigate how the growth rate of the population is affected by changing either the survival or the fecundity of zero-year-olds, or both. What would be the maximum achievable growth rate?

(c) In real situations, what other factors might you need to consider when you decide on management strategies?

MULTIVARIABLE
CALCULUS

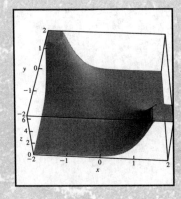

To survive in cold temperatures, humans must maintain a sufficiently high metabolic rate, or regulate heat loss by covering their skin with insulating material. There is a functional relationship that gives the lowest temperature for survival (T_e) as a function of metabolic heat production (M) and whole-body thermal conductance (g_{Hb}). The metabolic heat production depends on the type of activity; some values for humans are summarized in the following table.

Activity	M in Wm^{-2}
Sleeping	50
Working at a desk	95
Level walking at 2.5 mph	180
Level walking at 3.5 mph with a 40-lb pack	350

Data adapted from Landsberg (1969).

The whole-body thermal conductance g_{Hb} describes how quickly heat is lost. The value of g_{Hb} depends on the type of protection; for instance, $g_{Hb} = 0.45 \, mol \, m^{-2} \, s^{-1}$ without clothing, $g_{Hb} = 0.14 \, mol \, m^{-2} \, s^{-1}$ for a wool suit, and $g_{Hb} = 0.04 \, mol \, m^{-2} \, s^{-1}$ for a warm sleeping bag. That is, the smaller g_{Hb}, the better protection from the cold the material provides. The relationship between T_e, M, and g_{Hb} is given by

$$T_e = 36 - \frac{(0.9M - 12)(g_{Hb} + 0.95)}{27.8 g_{Hb}}$$

where M is measured in $W \, m^{-2}$, g_{Hb} is measured in $mol \, m^{-2} \, s^{-1}$, and T_e is measured in degree Celsius (Campbell, 1986). The temperature T_e is a function of two variables, namely M and g_{Hb}; to meet the required temperature, we can change either M (by starting to move when we get cold) or g_{Hb} (by putting on more clothes when we get cold). We can plot T_e as a function of M for different values of g_{Hb}, or plot T_e as a function of g_{Hb} for different values of M, as shown in Figures 10.1 and 10.2.

Looking at Figure 10.1, we see that for a given thermal conductance (say, wearing a wool suit), we need to increase metabolic heat production as it gets colder to stay above the minimum temperature for survival. Looking at Figure 10.2, we see that for a given activity (say level walking at 2.5 mph), we need to decrease

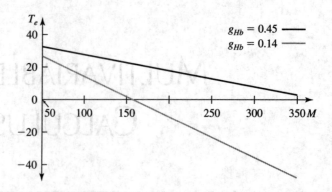

▲ Figure 10.1
The graph of T_e as a function of M for various values of g_{Hb}

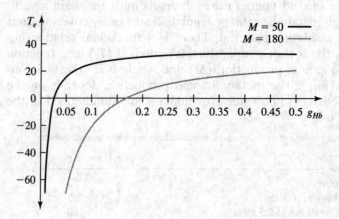

▲ Figure 10.2
The graph of T_e as a function of g_{Hb} for various values of M

thermal conductance (that is, dress warmer) as it gets colder in order to stay above the minimum temperature for survival.

Since we can vary the two variables M and g_{Hb} independently, we can think of T_e as a function of two independent variables. In this chapter, we will discuss functions of two or more independent variables, develop the theory of differential calculus for such functions, and discuss a number of applications.

10.1 FUNCTIONS OF TWO OR MORE INDEPENDENT VARIABLES

To recall the notation and terminology that we used when we considered functions of one variable, let

$$f : [0, 4] \longrightarrow \mathbf{R}$$
$$x \longrightarrow \sqrt{x}$$

The function $y = f(x)$ depends on one variable, namely x. Its domain is the set of numbers that we can use to evaluate $f(x)$, namely, the interval $[0, 4]$. Its range is the set of all possible values $y = f(x)$ for x in the domain of f. We see from Figure 10.3 that the range of $f(x)$ is the interval $[0, 2]$.

We now consider functions for which the domain consists of pairs of real numbers (x, y) with $x, y \in \mathbf{R}$, or, more generally, of n-tuples of real numbers $(x_1, x_2, \ldots, x_n)$ with $x_1, x_2, \ldots, x_n \in \mathbf{R}$. We also call n-tuples points. We use the notation $\mathbf{R}^n$ to denote the set of all n-tuples $(x_1, x_2, \ldots, x_n)$ with

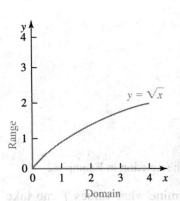

Range

Domain

▲ **Figure 10.3**
The domain and range of a function

$y = \sqrt{x}$

$x_1, x_2, \ldots, x_n \in \mathbf{R},$

$$\mathbf{R}^n = \{(x_1, x_2, \ldots, x_n) : x_1 \in \mathbf{R}, x_2 \in \mathbf{R}, \ldots, x_n \in \mathbf{R}\}$$

For $n = 1$, $\mathbf{R}^1 = \mathbf{R}$, which is the set of all real numbers. For $n = 2$, $\mathbf{R}^2$ is the set of all points in the plane, and so on. Note that n-tuples are *ordered*; for instance, $(2, 3) \neq (3, 2)$. We consider functions whose ranges are subsets of the real numbers; such functions are called real valued.

> **Definition** Suppose $D \subset \mathbf{R}^n$. A **real-valued function** f on D assigns a real number to each element in D and we write
>
> $$f : D \to \mathbf{R}$$
>
> $$(x_1, x_2, \ldots, x_n) \to f(x_1, x_2, \ldots x_n)$$
>
> The set D is the domain of the function f and the set $\{w \in \mathbf{R} : f(x_1, x_2, \ldots, x_n) = w$ for some $(x_1, x_2, \ldots, x_n) \in D\}$ is the range of the function f.

If a function f depends on just two independent variables, we will often denote the independent variables by x and y, and write $f(x, y)$. In the case of three variables, we will often write $f(x, y, z)$. If f is a function of more than three independent variables, it is more convenient to use subscripts to label the variables, for example, $f(x_1, x_2, x_3, x_4)$.

The first example shows how a function of more than one independent variable is evaluated at given points. You should pay attention to the fact that the domain consists of *ordered* n-tuples.

▶ **Example 1**   Evaluate the function

$$f(x, y, z) = \frac{xy}{z^2}$$

at the points $(2, 3, -1)$ and $(-1, 2, 3)$.

Solution

$$f(2, 3, -1) = \frac{(2)(3)}{(-1)^2} = 6$$

$$f(-1, 2, 3) = \frac{(-1)(2)}{(3)^2} = -\frac{2}{9}$$ ◀

The next example shows how to determine the range of a real-valued function for a given domain.

▶ **Example 2**   Let $D = \{(x, y) : 0 \le x \le 1, \ 0 \le y \le 1\}$ and

$$f : \quad D \quad \longrightarrow \mathbf{R}$$

$$(x, y) \longrightarrow x + y$$

Graph the domain of f in the x-y plane and determine the range of f.

Solution

The domain of f is the set D, which consists of all points (x, y) whose x and y coordinates are between 0 and 1. This is the square shown in Figure 10.4.

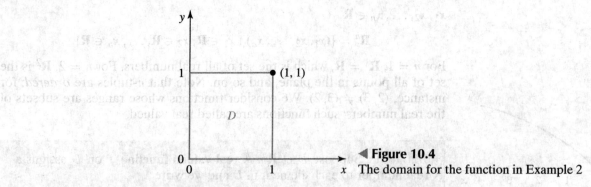

◀ **Figure 10.4**
The domain for the function in Example 2

To find the range of f, we need to determine what values f can take when we plug in points (x, y) from the domain D. The function $z = f(x, y)$ is smallest when $(x, y) = (0, 0)$, namely $f(0, 0) = 0$. The function $z = f(x, y)$ is largest when $(x, y) = (1, 1)$, namely $f(1, 1) = 2$. It takes on all values in between. Hence, the range of f is the set $\{z : 0 \leq z \leq 2\}$. ◀

As in the case of single-variable functions, the domain sometimes needs to be restricted. The next example illustrates how to find the largest possible domain.

▶ **Example 3** Find the largest possible domain for the function

$$f(x, y) = \sqrt{y^2 - x}$$

Solution

The largest possible domain of f is the set of all points (x, y) so that $y^2 - x \geq 0$. We can illustrate this as a set in the x-y plane, where the boundary is given by the graph of $y^2 - x = 0$. The graph of $y^2 - x = 0$ is a parabola that opens to the right. The set of points (x, y) that satisfies $y^2 - x \geq 0$ is then the shaded area in Figure 10.5. It includes the boundary curve.

The easiest way to see this is to use a test point. The boundary curve $y^2 - x = 0$ divides the plane into two regions: the set of points that lie inside the parabola and the set of points that lie outside the parabola. In one of the regions, $y^2 - x > 0$; in the other, $y^2 - x < 0$. If we use a test point from the inside of the parabola, say the point $(1, 0)$, then $y^2 - x = 0 - 1 < 0$; if we use a test point from the outside of the parabola, say the point $(-1, 0)$, then

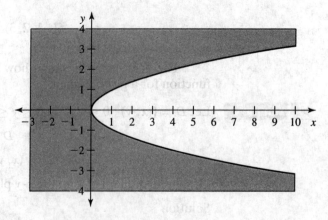

▲ **Figure 10.5**
The domain of $f(x, y)$ in Example 3

$y^2 - x = 0 - (-1) > 0$. We therefore conclude that the set of points for which $y^2 - x \geq 0$ is the outside of the parabola including the boundary curve. ◄

The Graph of a Function of Two Independent Variables It is possible to graph a real-valued function of two variables. Namely, we can write $z = f(x, y)$ and then locate the point (x, y, z) in three-dimensional space. The following definition makes this precise.

> **Definition** If f is a function of two independent variables with domain D, then the graph of f is the set of all points (x, y, z) such that $z = f(x, y)$ for $(x, y) \in D$. That is, the graph of f is the set
>
> $$S = \{(x, y, z) : z = f(x, y),\ (x, y) \in D\}$$

To locate a point (x, y, z) in three-dimensional space, we use the Cartesian coordinate system, which we introduced in Chapter 9. It consists of three mutually perpendicular axes that emanate from a common point, called the origin, which has coordinates $(0, 0, 0)$. The axes are oriented in a right-handed coordinate frame, as explained in Chapter 9. This is shown in Figure 10.6, together with a point P that has coordinates (x_0, y_0, z_0).

Figure 10.6 suggests that we can visualize the graph of $f(x, y), (x, y) \in D$, as lying directly above (or below) the domain D in the x-y plane. The graph of $f(x, y)$ is therefore a **surface** in three-dimensional space, as illustrated in Figure 10.7 for $f(x, y) = 2x^2 - y^2$.

Graphing a surface in three-dimensional space is difficult. Fortunately, good computer software is now available that facilitates this task. We show a few such surfaces, generated by computer software, in Figures 10.7 through 10.10. We will make no effort to learn how to graph such functions but will provide problems in the Problem Section in which surfaces and functions must be matched. To be able to do so, we look at the surfaces in Figures 10.7 through 10.10.

To understand the shape of the graph of $f(x, y) = 2x^2 - y^2$ in Figure 10.7, we can fix a value for x and then "walk" on the surface in the y-direction. Since the function is of the form "constant minus y^2" for fixed x, we expect a curve that has the shape of an upside-down parabola. Looking at the surface

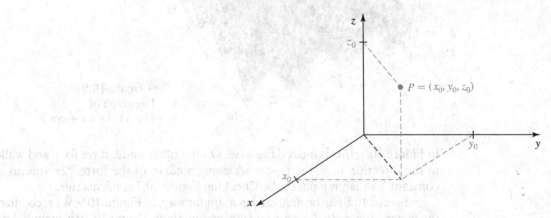

▲ **Figure 10.6**
The right-handed coordinate system with a point

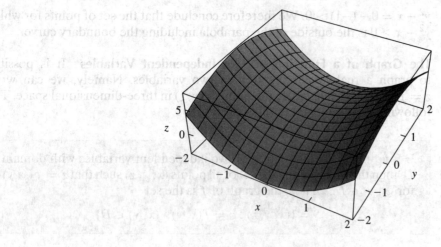

◄ **Figure 10.7**
The graph of $z = f(x, y)$ is a surface in three dimensional space. (Here: $f(x, y) = 2x^2 - y^2$)

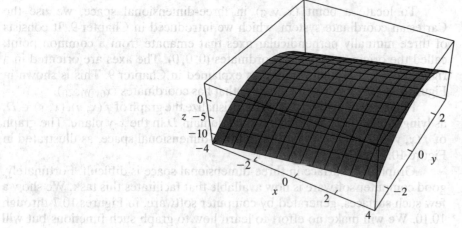

◄ **Figure 10.8**
The graph of $f(x, y) = x - y^2$

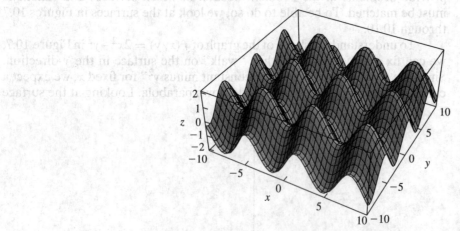

◄ **Figure 10.9**
The graph of $f(x, y) = \sin x + \cos y$

in Figure 10.7, this is indeed the case. On the other hand, if we fix y and walk in the direction of x, then we walk along a curve of the form "$2x^2$ minus a constant," which is a parabola. Checking Figure 10.7 confirms this.

Figure 10.8 can be analyzed in a similar way. In Figure 10.9, we recognize the waves created by the sine and cosine functions. Figure 10.10 is included to show that functions of two variables can result in interesting (and complicated) shapes.

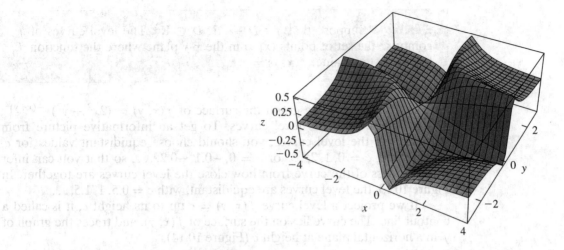

▲ **Figure 10.10**
The graph of $f(x, y) = \dfrac{xy}{1+x^2y^2}$

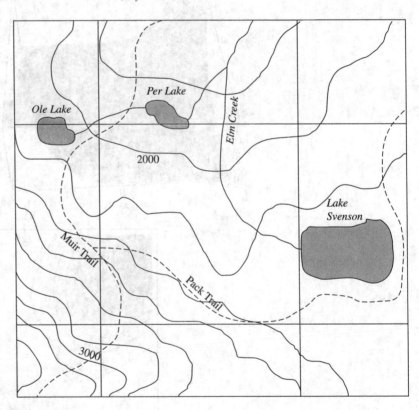

▲ **Figure 10.11**
A topographical map with level curves

Another way to visualize functions is with **level curves** or **contour lines**. This is used, for instance, in topographical maps (see Figure 10.11). There is a subtle distinction between level curves and contour lines, in that level curves are drawn in the function domain whereas contour lines are drawn on the surface. This distinction is not always made, and often the two terms are used interchangeably. In this text, we will almost exclusively use level curves, for which we now give the precise definition.

Definition Suppose that $f : D \to \mathbf{R}$, $D \subset \mathbf{R}^2$. The level curves of f comprise the set of points (x, y) in the x-y plane where the function f has a constant value: $f(x, y) = c$.

In Figure 10.12, we graph the surface of $f(x, y) = (2x^2 + y^2)e^{-(x+y)^2}$, and in Figure 10.13, its level curves. To get an informative picture from the graph of the level curves, you should choose equidistant values for c, for instance, $c = 0, 1, 2, \ldots$ or $c = 0, -0.1, -0.2, \ldots$, so that you can infer the steepness of the curve from how close the level curves are together. In Figure 10.13, the level curves are equidistant, with $c = 0.5, 1, 1.5, \ldots$.

If we project a level curve $f(x, y) = c$ up to its height c, it is called a contour line. The curve lies on the surface of $f(x, y)$, and traces the graph of f in a horizontal plane at height c (Figure 10.14).

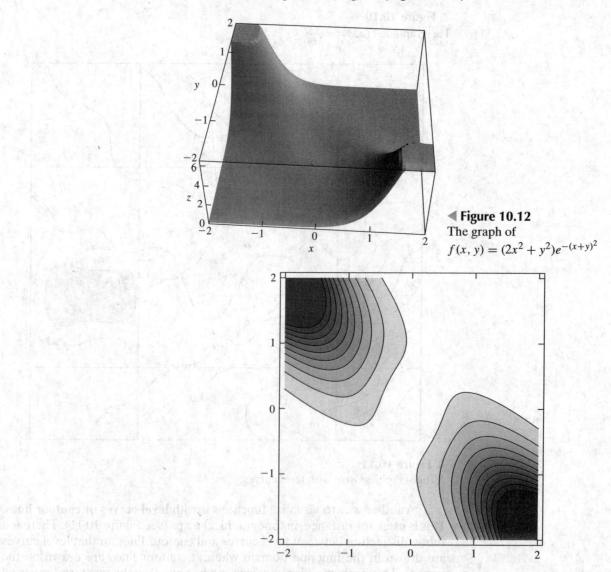

◀ **Figure 10.12**
The graph of
$f(x, y) = (2x^2 + y^2)e^{-(x+y)^2}$

▲ **Figure 10.13**
Level curves for $f(x, y) = (2x^2 + y^2)e^{-(x+y)^2}$

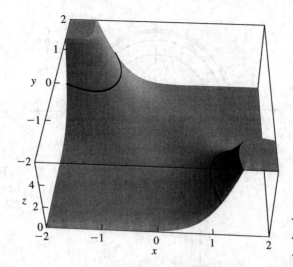

◀ **Figure 10.14**
A contour line for
$f(x, y) = (2x^2 + y^2)e^{-(x+y)^2}$

▶ **Example 4**   Set $D = \{(x, y) : x^2 + y^2 \le 4\}$. Compare the level curves of

$$f(x, y) = 4 - x^2 - y^2 \quad \text{for } (x, y) \in D$$

and

$$g(x, y) = \sqrt{4 - x^2 - y^2} \quad \text{for } (x, y) \in D$$

Solution

The level curve $f(x, y) = c$ is the set of all points (x, y) that satisfy

$$4 - x^2 - y^2 = c \quad \text{or} \quad x^2 + y^2 = 4 - c$$

We recognize this as a circle with center at the origin and radius $\sqrt{4 - c}$, illustrated in Figure 10.15 for $c = 0, 0.5, 1, 1.5,$ and 2.

The level curve $g(x, y) = c$ satisfies

$$\sqrt{4 - x^2 - y^2} = c \quad \text{or} \quad x^2 + y^2 = 4 - c^2$$

This is also a circle with center at the origin but the radius is $\sqrt{4 - c^2}$; it is illustrated in Figure 10.16 for $c = 0, 0.5, 1, 1.5,$ and 2.

To compare the level curves of two functions, we choose $c = 0, 0.5, 1, 1.5,$ and 2 for both $f(x, y)$ and $g(x, y)$. The level curves for $c = 0$ and 1 are the

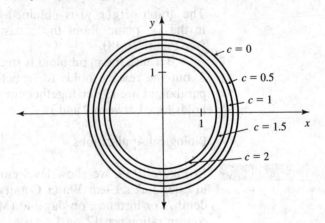

▲ **Figure 10.15**
Level curves for $f(x, y) = 4 - x^2 - y^2$

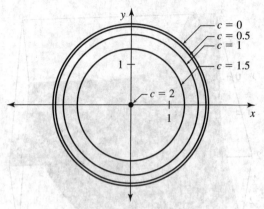

◄ Figure 10.16
Level curves for
$g(x, y) = \sqrt{4 - x^2 - y^2}$

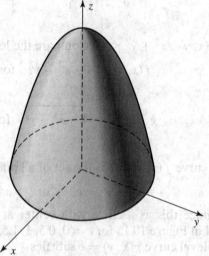

◄ Figure 10.17
The surface of $f(x, y) = 4 - x^2 - y^2$ is a paraboloid

same for the two functions, but the contour lines of $f(x, y)$ for $c = 1.5$ and 2 are a lot closer to the contour line for $c = 1$ than those of $g(x, y)$. We can understand this when we look at the surfaces. The graph of $f(x, y)$ is obtained by rotating the parabola $z = 4 - x^2$ in the x-z plane about the z-axis; its surface is called a paraboloid (see Figure 10.17).

The graph of $g(x, y)$ is obtained by rotating the half circle $z = \sqrt{4 - x^2}$ in the x-z plane about the z-axis; its surface is the top half of a sphere (see Figure 10.18).

Because the paraboloid is steeper than the sphere for c between 1 and 2, but the reverse holds for c between 0 and 1, the contour lines of the paraboloid are closer together for c between 1 and 2, whereas the reverse holds for c between 0 and 1. ◄

Biological Applications

 Example 5 In Figure 10.19, we show level curves for oxygen concentration (in mg/l) in Long Lake, Clear Water County (Minnesota), as a function of date and depth. For instance, on day 140 (May 20, 1998) at 10 m depth, the oxygen concentration was 12 mg/l.

The waterflea *Daphnia* needs a minimum of 3 mg/l oxygen to survive. Suppose that you went out to Long Lake on day 200 (July 19, 1998) and

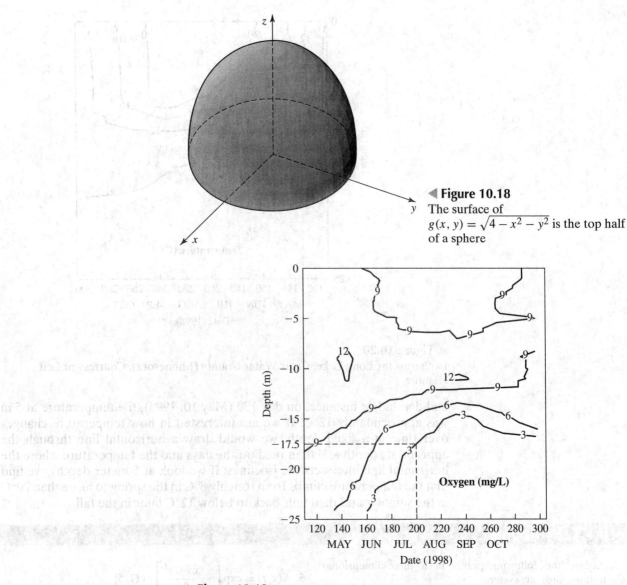

◀ **Figure 10.18**
The surface of
$g(x, y) = \sqrt{4 - x^2 - y^2}$ is the top half
of a sphere

▲ **Figure 10.19**
Level curves for oxygen concentration on Long Lake, Clear Water County
(Minnesota). Courtesy of Leif Hembre

wanted to look for *Daphnia* in the lake. Below what depth could you be fairly
sure not to find any *Daphnia*?

Solution

Because *Daphnia* needs a minimum of 3 mg/l of oxygen, we need to find
the depth on July 19, 1998, below which the oxygen concentration is always
less than 3 mg/l. We see that the 3 mg/l oxygen level curve goes through the
point (200, −17.5). Thus, on July 19, 1998, *Daphnia* needed to stay above
17.5 m; that is, we could have been fairly sure not to have found *Daphnia*
below 17.5 m.   ◀

▶ **Example 6**   In Figure 10.20, we show temperature **isoclines** (called **isotherms**) for Long
Lake, Clear Water County (Minnesota). Temperature isoclines are lines of
equal temperature, that is, level curves. The isoclines are functions of date

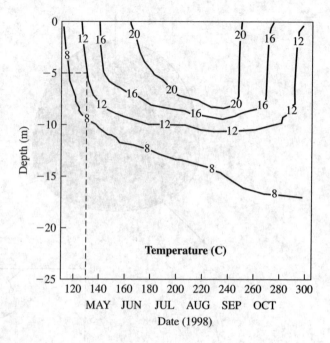

▲ Figure 10.20
Isotherms for Long Lake, Clear Water County (Minnesota). Courtesy of Leif Hembre

and depth. For instance, on day 130 (May 10, 1998), the temperature at 5 m was approximately 12°C. If we are interested in how temperature changes over time at a fixed depth, we would draw a horizontal line through the appropriate depth and then read off the days and the temperature where the horizontal line intersects the isoclines: If we look at 5-meter depths, we find that the temperature climbs from roughly 8°C in the spring to more than 20°C in the summer and then falls back to below 12°C later in the fall. ◀

10.1.1 Problems

1. Locate the following points in a three-dimensional Cartesian coordinate system.

(a) $(1, 3, 2)$

(b) $(-1, -2, 1)$

(c) $(0, 1, 2)$

(d) $(2, 0, 3)$

2. Describe in words the set of all points in $\mathbf{R}^3$ that satisfy the following.

(a) $x = 0$

(b) $y = 0$

(c) $z = 0$

(d) $z \geq 0$

(e) $y \leq 0$

In Problems 3–6, evaluate each function at the given point.

3. $f(x, y) = \dfrac{2x}{x^2 + y^2}$ at $(2, 3)$

4. $f(x, y, z) = \sqrt{x^2 - 3y + z}$ at $(3, -1, 1)$

5. $h(x, t) = \exp\left[-\dfrac{(x - 2)^2}{2t}\right]$ at $(1, 5)$

6. $g(x_1, x_2, x_3, x_4) = x_1 x_4 \sqrt{x_2 x_3}$ at $(1, 8, 2, -1)$

In Problems 7–10, find the largest possible domain and the corresponding range of each function. Determine the equation of the level curves $f(x, y) = c$, together with the possible values of c.

7. $f(x, y) = x^2 + y^2$

8. $f(x, y) = \sqrt{9 - x^2 - y^2}$

9. $f(x, y) = \ln(y - x^2)$

10. $f(x, y) = \exp[-(x^2 + y^2)]$

In Problems 11–14 match each function with the appropriate graph in Figures 10.21–10.24.

11. $f(x, y) = 1 + x^2 + y^2$

12. $f(x, y) = \sin(x) \sin(y)$

13. $f(x, y) = y^2 - x^2$

14. $f(x, y) = 4 - x^2$

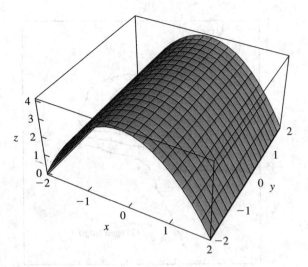

▲ **Figure 10.21**

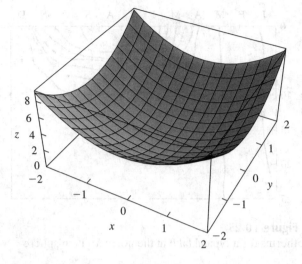

▲ **Figure 10.23**

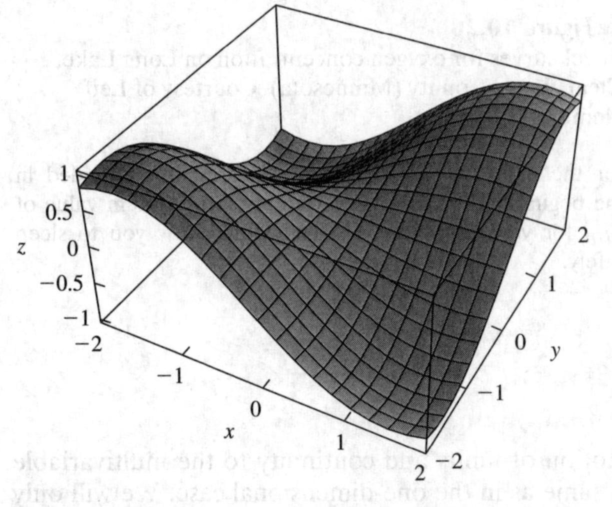

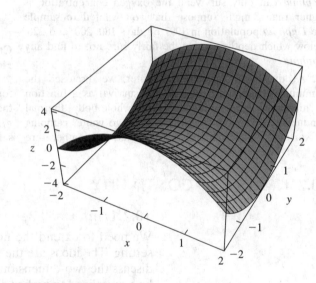

▲ **Figure 10.24**

▲ **Figure 10.22**

15. Let

$$f_a(x, y) = ax^2 + y^2$$

for $(x, y) \in \mathbf{R}$, where a is a positive constant.

(a) Assume that $a = 1$ and describe the level curves. The graph of $f(x, y)$ intersects both the x-z and the y-z plane; show that these two curves of intersection are parabolas.

(b) Assume that $a = 4$. Then

$$f_4(x, y) = 4x^2 + y^2$$

and the level curves satisfy

$$4x^2 + y^2 = c$$

Use a graphing calculator to sketch the level curves for $c = 0, 1, 2, 3$, and 4. These curves are ellipses. Find the curves of intersection of $f_4(x, y)$ with the x-z and the y-z plane.

(c) Repeat (b) for $a = 1/4$.

(d) Explain in words how the surfaces of $f_a(x, y)$ change when a changes.

16. The graph in Figure 10.25 shows isotherms of a lake in the temperate climate of the northern hemisphere.

(a) Use this plot to sketch the temperature profiles in March and June. That is, plot the temperature as a function of depth for a day in March and for a day in June.

(b) Explain how it follows from your temperature plots that the lake is **homeothermic** in March, that is, has the same temperature from the surface to the bottom.

(c) Explain how it follows from your temperature plots that the lake is **stratified** in June, that is, has a warm layer on top (called the **epilimnion**), followed by a region where the temperature changes quickly (called the **metalimnion**), which is then followed by a cold layer deeper down (called the **hypolimnion**).

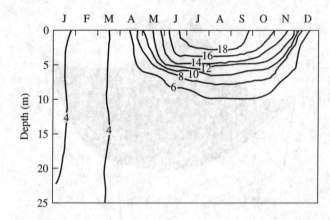

▲ **Figure 10.25**
Isotherms for a typical lake in the northern hemisphere

17. Figure 10.26 shows the oxygen concentration for Long Lake, Clear Water County (Minnesota). The waterflea *Daphnia* can only survive if the oxygen concentration is higher than 3 mg/l. Suppose that you wanted to sample the *Daphnia* population in 1997 on days 180, 200, and 220. Below which depths can you be fairly sure not to find any *Daphnia*?

18. At the beginning of this chapter, we discussed the minimum temperature required for survival as a function of metabolic heat production and whole-body thermal conductance. Suppose that you wish to go winter camping in Northern Minnesota and the predicted low temperature

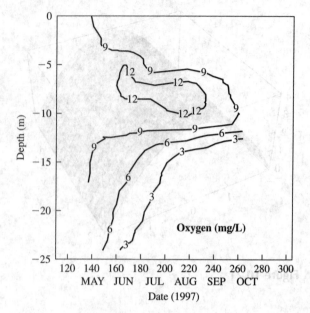

▲ **Figure 10.26**
Level curves for oxygen concentration on Long Lake, Clear Water County (Minnesota). Courtesy of Leif Hembre

for the night is −15°F. Use the information provided in the beginning of the chapter to find the maximum value of g_{Hb} for your sleeping bag that would allow you to sleep safely.

10.2 LIMITS AND CONTINUITY

10.2.1 Informal Definition of Limits

We need to extend the notion of limits and continuity to the multivariable setting. The ideas are the same as in the one-dimensional case. We will only discuss the two-dimensional case but note that everything in this section can be generalized to higher dimensions.

Let's start with an informal definition of limits. We say that the "limit of $f(x, y)$ as (x, y) approaches (x_0, y_0) is equal to L" if $f(x, y)$ can be made arbitrarily close to L whenever the point (x, y) is sufficiently close to the point (x_0, y_0) [but not equal to (x_0, y_0)]. We denote this by

$$\lim_{(x,y)\to(x_0,y_0)} f(x, y) = L$$

The following example shows how to compute limits in simple cases.

 (a) $\lim_{(x,y)\to(0,0)} (x^2 + y^2) = 0^2 + 0^2 = 0$

(b) $\lim_{(x,y)\to(4,-3)} (x^2 + y^2) = 4^2 + (-3)^2 = 25$

(c) $\lim_{(x,y)\to(-1,2)} x^2 y = (-1)^2(2) = 2$ ◀

As in the one-dimensional case, there is a formal definition of limits, which is difficult to use. Fortunately, there are laws similar to those in the one-dimensional case, which allow us to compute limits.

LIMIT LAWS FOR THE TWO-DIMENSIONAL CASE

If a is a constant and if

$$\lim_{(x,y)\to(x_0,y_0)} f(x,y) = L_1 \quad \text{and} \quad \lim_{(x,y)\to(x_0,y_0)} g(x,y) = L_2$$

where L_1 and L_2 are real numbers, then the following hold.

1. **(Addition Rule)**

$$\lim_{(x,y)\to(x_0,y_0)} [f(x,y) + g(x,y)] = \lim_{(x,y)\to(x_0,y_0)} f(x,y)$$
$$+ \lim_{(x,y)\to(x_0,y_0)} g(x,y)$$

2. **(Constant Factor Rule)**

$$\lim_{(x,y)\to(x_0,y_0)} af(x,y) = a \lim_{(x,y)\to(x_0,y_0)} f(x,y)$$

3. **(Multiplication Rule)**

$$\lim_{(x,y)\to(x_0,y_0)} f(x,y)g(x,y) = \left[\lim_{(x,y)\to(x_0,y_0)} f(x,y) \right]$$
$$\times \left[\lim_{(x,y)\to(x_0,y_0)} g(x,y) \right]$$

4. **(Quotient Rule)**

$$\lim_{(x,y)\to(x_0,y_0)} \frac{f(x,y)}{g(x,y)} = \frac{\lim_{(x,y)\to(x_0,y_0)} f(x,y)}{\lim_{(x,y)\to(x_0,y_0)} g(x,y)}$$

provided that $L_2 \neq 0$.

Limits That Exist

▶ **Example 2**

(a)

$$\lim_{(x,y)\to(1,2)} (x^2y + 3x) = \left(\lim_{(x,y)\to(1,2)} x^2y \right) + \left(3 \lim_{(x,y)\to(1,2)} x \right)$$
$$= \left(\lim_{(x,y)\to(1,2)} x^2 \right)\left(\lim_{(x,y)\to(1,2)} y \right)$$
$$+ \left(3 \lim_{(x,y)\to(1,2)} x \right)$$
$$= (1)^2(2) + (3)(1) = 5$$

(b)

$$\lim_{(x,y)\to(-1,3)} \left(2y^2 - \frac{3x}{y} \right) = (2)(3)^2 - \frac{(3)(-1)}{3} = 18 + 1 = 19$$

(c)

$$\lim_{(x,y)\to(2,0)} \frac{4y + 2x}{x^2 + 2xy - 3} = \frac{(4)(0) + (2)(2)}{(2)^2 + (2)(2)(0) - 3}$$
$$= \frac{4}{4-3} = 4$$

◀

We see from Example 2 that limits of polynomials and rational functions can be evaluated simply by evaluating the functions at the respective points, provided that the functions are defined at those points.

Limits That Do Not Exist In the one-dimensional case, there were only two ways in which we could approach a number, namely, from the left or the right. If the two limits were different, we said that the limit did not exist. In two dimensions, there are many more ways that we can approach the point (x_0, y_0), namely, any curve in the x-y plane that ends up at the point (x_0, y_0). We call such curves **paths**.

> If $f(x, y)$ approaches L_1 as $(x, y) \to (x_0, y_0)$ along path C_1, and $f(x, y)$ approaches L_2 as $(x, y) \to (x_0, y_0)$ along path C_2, and if $L_1 \neq L_2$, then $\lim_{(x,y)\to(x_0,y_0)} f(x, y)$ does not exist.

▶ **Example 3** Show that

$$\lim_{(x,y)\to(0,0)} \frac{x^2 - y^2}{x^2 + y^2}$$

does not exist.

Solution

We first let $(x, y) \to (0, 0)$ along the positive x-axis; this is the curve C_1 in Figure 10.27. On C_1, $y = 0$, $x > 0$. Then

$$\lim_{x\to 0+} \frac{x^2}{x^2} = 1$$

Next, we let $(x, y) \to (0, 0)$ along the positive y-axis; this is the curve C_2 in Figure 10.27. On C_2, $x = 0$ and $y > 0$. Then

$$\lim_{y\to 0+} \frac{-y^2}{y^2} = -1$$

Since $1 \neq -1$, we conclude that the limit does not exist. ◀

Unless we have a lot of experience, it is not easy to find paths for which limits differ. Therefore, in the Problem Section, we will always provide the paths along which you should check the limits, in case a limit does not exist.

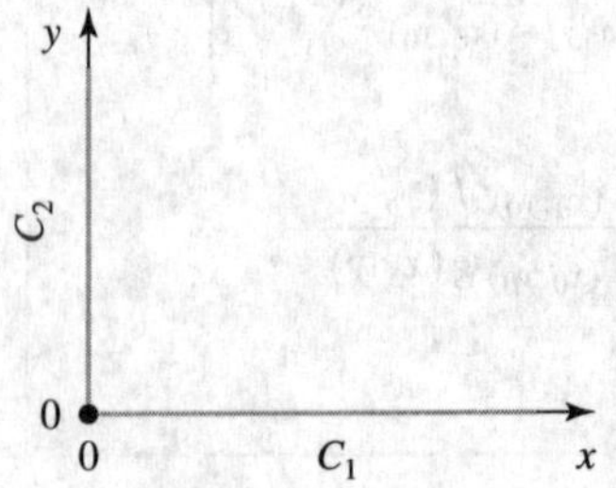

▲ **Figure 10.27**
The paths C_1 and C_2 for Example 3

▶ **Example 4** Show that

$$\lim_{(x,y)\to(0,0)} \frac{4xy}{xy + y^3}$$

does not exist.

Solution

A natural choice for paths to $(0, 0)$ are straight lines of the form $y = mx$. We assume that $m \neq 0$. If we substitute $y = mx$ in the preceding limit, then $(x, y) \to (0, 0)$ reduces to $x \to 0$ and we find

$$\lim_{x\to 0} \frac{4mx^2}{mx^2 + (mx)^3} = \lim_{x\to 0} \frac{4}{1 + m^2 x} = 4$$

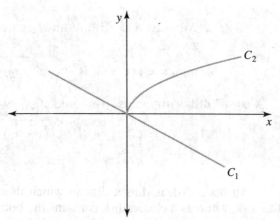

That is, as long as we approach $(0, 0)$ along the straight line $y = mx$, $m \neq 0$, the limit is always 4, irrespective of m. Such a path, labeled C_1, is shown in Figure 10.28.

You might be tempted to say that the limit exists. But let's approach $(0, 0)$ along the parabola $x = y^2$. This is the curve C_2 in Figure 10.28. Substituting y^2 for x then yields

$$\lim_{y \to 0} \frac{4y^3}{y^3 + y^3} = 2 \neq 4$$

Thus we found paths along which the limits differ. Therefore, the limit does not exist. ◀

To show that a limit does not exist, we must identify two paths along which the limits differ. To show that a limit exists, we cannot use paths since the limits along *all* possible paths must be the same; there is no way to check all possible paths. To show that a limit exists, we proceed as in the one-dimensional case; namely, we combine the formal definition of limits and the limit laws to compute limits.

10.2.2 Formal Definition of Limits (Optional)

We recall the formal definition of limits in the one-dimensional case. To show that

$$\lim_{x \to 2} x^2 = 4$$

we must show that whenever x is close to 2 (but not equal to 2), then x^2 is close to 4. Formally, this means that we must show that for every $\epsilon > 0$ there exists $\delta > 0$ so that

$$|x^2 - 4| < \epsilon \qquad \text{whenever} \qquad 0 < |x - 2| < \delta$$

To generalize this idea to higher dimensions, we need a notion of proximity. In the one-dimensional case, we chose an interval centered at a particular point that contained all points within distance δ of the center of the interval, except for the center of the interval. In two dimensions, we replace the interval by a disk. A disk of radius δ centered at the point (x_0, y_0) is the set of all points that are within distance δ of (x_0, y_0).

As with open and closed intervals, there are open and closed disks. We have the following definition.

> **Definition** An **open disk** with radius r centered at $(x_0, y_0) \in \mathbf{R}^2$ is the set
>
> $$B_r(x_0, y_0) = \left\{ (x, y) \in \mathbf{R}^2 : \sqrt{(x - x_0)^2 + (y - y_0)^2} < r \right\}$$
>
> A **closed disk** with radius r centered at $(x_0, y_0) \in \mathbf{R}^2$ is the set
>
> $$\overline{B}_r(x_0, y_0) = \left\{ (x, y) \in \mathbf{R}^2 : \sqrt{(x - x_0)^2 + (y - y_0)^2} \le r \right\}$$

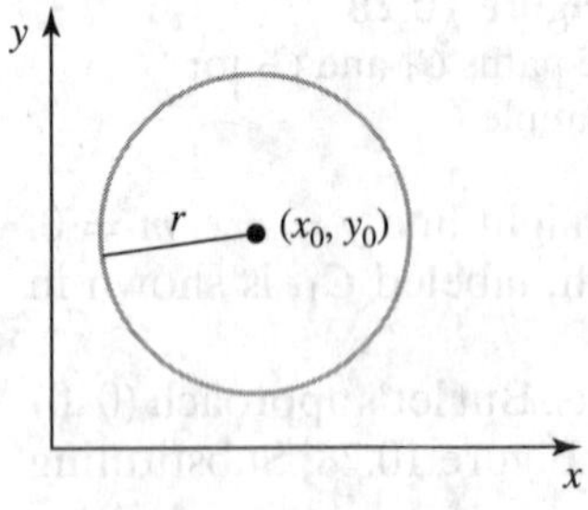

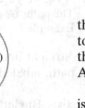

▲ **Figure 10.29**
A closed disk with radius r
centered at (x_0, y_0)

An open disk is thus a disk in which the boundary line is not part of the disk, whereas a closed disk contains the boundary line. This is analogous to intervals: an open interval does not contain the two endpoints (which are the boundary of the interval), whereas a closed interval does contain them. A closed disk is shown in Figure 10.29.

When we defined $\lim_{x \to x_0} f(x)$, we emphasized that the value of f at x_0 is not important. This is expressed in the formal $\epsilon - \delta$ definition by excluding x_0 from the interval $(x_0 - \delta, x_0 + \delta)$, stated as $0 < |x - x_0| < \delta$. We will need to do the same when we generalize the limit to two dimensions. We write $B_\delta(x_0, y_0) - \{(x_0, y_0)\}$ to denote the open disk with radius δ centered at (x_0, y_0), where the center (x_0, y_0) is removed. We can now generalize the notion of limits to two dimensions.

> **Definition** The limit of $f(x, y)$ as (x, y) approaches (x_0, y_0), denoted by
>
> $$\lim_{(x, y) \to (x_0, y_0)} f(x, y)$$
>
> is the number L such that for every $\epsilon > 0$ there exists a $\delta > 0$ so that
>
> $$|f(x, y) - L| < \epsilon \qquad \text{whenever} \qquad (x, y) \in B_\delta(x_0, y_0) - \{(x_0, y_0)\}$$

This is very similar to the definition in one dimension: We require that whenever (x, y) is close (but not equal) to (x_0, y_0), then $f(x, y)$ is close to L.

We provide one example in which we use the formal definition of limits. After this, we will give limit laws, which are extensions of the laws in the one-dimensional case. Using the formal definition is difficult, and we will not need to do so subsequently. But seeing an example might help you to understand the definition.

▶ **Example 5** Let

$$f(x, y) = x^2 + y^2$$

and show that

$$\lim_{(x, y) \to (0, 0)} f(x, y) = 0$$

Solution

We must show that for every $\epsilon > 0$ there exists a $\delta > 0$ such that

$$|x^2 + y^2 - 0| < \epsilon \qquad \text{whenever} \qquad (x, y) \in B_\delta(0, 0) - \{(0, 0)\}$$

Now, $(x, y) \in B_\delta(0, 0)$ if $\sqrt{x^2 + y^2} < \delta$ or $x^2 + y^2 < \delta^2$. We need to show $|x^2 + y^2| < \epsilon$. This suggests that we should choose δ so that $\delta^2 = \epsilon$. Let's try this. We set $\delta = \sqrt{\epsilon}$ for $\epsilon > 0$; then

$$\sqrt{x^2 + y^2} < \delta = \sqrt{\epsilon}$$

implies

$$|x^2 + y^2| < \epsilon$$

But this is what we need to show. In other words, we have shown that for every $\epsilon > 0$, we can find $\delta > 0$ (namely $\delta = \sqrt{\epsilon}$) so that whenever (x, y) is close to $(0, 0)$, then $x^2 + y^2$ is close to 0. ◀

10.2.3 Continuity

The definition of continuity is also analogous to that in the one-dimensional case.

A function $f(x, y)$ is continuous at (x_0, y_0) if the following hold.

1. $f(x, y)$ is defined at (x_0, y_0)
2. $\lim\limits_{(x,y) \to (x_0, y_0)} f(x, y)$ exists
3. $\lim\limits_{(x,y) \to (x_0, y_0)} f(x, y) = f(x_0, y_0)$

▶ **Example 6** Show that

$$f(x, y) = 2 + x^2 + y^2$$

is continuous at $(0, 0)$.

1. $f(x, y)$ is defined at $(0, 0)$, namely

$$f(0, 0) = 2$$

2. To show that the limit exists, we refer to Example 1(a), where we claimed that

$$\lim_{(x,y) \to (0,0)} (x^2 + y^2) = 0$$

which indicates that the limit exists. Using the limit laws, we conclude that

$$\lim_{(x,y) \to (0,0)} f(x, y) = 2 + \lim_{(x,y) \to (0,0)} (x^2 + y^2)$$

exists.

3. Using that

$$\lim_{(x,y) \to (0,0)} (x^2 + y^2) = 0$$

we find

$$\lim_{(x,y) \to (0,0)} (2 + x^2 + y^2) = 2 + \lim_{(x,y) \to (0,0)} (x^2 + y^2) = 2 + 0 = 2$$

Since $f(0, 0) = 2$, this shows that $f(x, y)$ is continuous at $(0, 0)$. ◀

▶ **Example 7**   Show that

$$f(x, y) = \begin{cases} \dfrac{x^2 - y^2}{x^2 + y^2} & \text{for } (x, y) \neq (0, 0) \\ 0 & \text{for } (x, y) = (0, 0) \end{cases}$$

is discontinuous at $(0, 0)$.

Solution

The function $f(x, y)$ is defined at $(0, 0)$. Hence (1) holds. But in Example 3, we showed that

$$\lim_{(x,y)\to(0,0)} f(x, y)$$

does not exist. Therefore, (2) is violated and we conclude that $f(x, y)$ is discontinuous at $(x, y) = (0, 0)$.   ◀

Composition of Functions   Using the definition of continuity and the rules for finding limits, we can show that polynomial functions of two variables (that is, functions that are sums of terms of the form $ax^n y^m$, where a is a constant and n and m are nonnegative integers) are continuous.

We can obtain a much larger class of continuous functions when we allow the composition of functions. For instance, the function $h(x, y) = e^{x^2+y^2}$ can be written as a composition of two functions. Namely, if we set $z = f(x, y) = x^2 + y^2$ and $g(z) = e^z$, then

$$h(x, y) = (g \circ f)(x, y) = g[f(x, y)] = e^{x^2+y^2}$$

Another example is $h(x, y) = \sqrt{y^2 - x}$. Here, $z = f(x, y) = y^2 - x$ and $g(z) = \sqrt{z}$. Then $h(x, y) = (g \circ f)(x, y)$. More generally, if

$$f : D \to \mathbf{R}, \quad D \subset \mathbf{R}^2$$

and

$$g : I \to \mathbf{R}, \quad I \subset \mathbf{R}$$

then the composition $(g \circ f)(x, y)$ is defined as

$$h(x, y) = (g \circ f)(x, y) = g[f(x, y)]$$

We can show that if f is continuous at (x_0, y_0) and g is continuous at $z = f(x_0, y_0)$, then

$$h(x, y) = (g \circ f)(x, y) = g[f(x, y)]$$

is continuous at (x_0, y_0). As an example, let's return to the function

$$h(x, y) = e^{x^2+y^2}$$

With $z = f(x, y) = x^2 + y^2$ and $g(z) = e^z$, $h(x, y) = g[f(x, y)]$ is continuous, since $f(x, y)$ is continuous for all $(x, y) \in \mathbf{R}^2$, and $g(z)$ is continuous for all z in the range of $f(x, y)$. Likewise, $h(x, y) = \sqrt{y^2 - x}$ is continuous for all $(x, y) \in \{(x, y) : y^2 - x \geq 0\}$, since $z = f(x, y) = y^2 - x$ is continuous for all $(x, y) \in \mathbf{R}^2$, and $g(z) = \sqrt{z}$ is continuous for all $z \geq 0$.

10.2.4 Problems

(10.2.1)

In Problems 1–8, use the properties of limits to calculate the following limits.

1. $\lim_{(x,y)\to(1,0)}(x^2 - 3y^2)$

2. $\lim_{(x,y)\to(-1,1)}(2xy + 3x^2)$

3. $\lim_{(x,y)\to(0,2)}\left(4xy^2 - \frac{x+1}{y}\right)$

4. $\lim_{(x,y)\to(1,1)}(x^2 + y^2)$

5. $\lim_{(x,y)\to(0,1)}\frac{2xy-3}{x^2+y^2+1}$

6. $\lim_{(x,y)\to(-1,-2)}\frac{x^2-y^2}{2xy+2}$

7. $\lim_{(x,y)\to(2,0)} \frac{2x+4y^2}{y^2+3x}$

8. $\lim_{(x,y)\to(1,-2)} \frac{2x^2+y}{2xy+3}$

9. Show that

$$\lim_{(x,y)\to(0,0)} \frac{x^2-2y^2}{x^2+y^2}$$

does not exist, by computing the limit along the positive x-axis and along the positive y-axis.

10. Show that

$$\lim_{(x,y)\to(0,0)} \frac{3x^2-y^2}{x^2+y^2}$$

does not exist, by computing the limit along the positive x-axis and along the positive y-axis.

11. Compute

$$\lim_{(x,y)\to(0,0)} \frac{4xy}{x^2+y^2}$$

along the x-axis, the y-axis, and the line $y=x$. What can you conclude?

12. Compute

$$\lim_{(x,y)\to(0,0)} \frac{3xy}{x^2+y^3}$$

along lines of the form $y=mx$, for $m\neq 0$. What can you conclude?

13. Compute

$$\lim_{(x,y)\to(0,0)} \frac{2xy}{x^3+yx}$$

along lines of the form $y=mx$, for $m\neq 0$, and along the parabola $y=x^2$. What can you conclude?

14. Compute

$$\lim_{(x,y)\to(0,0)} \frac{3x^2y^2}{x^3+y^6}$$

along lines of the form $y=mx$, for $m\neq 0$, and along the parabola $x=y^2$. What can you conclude?

(10.2.2)

15. Draw an open disk with radius 2 centered at $(1,-1)$ in the x-y plane, and give a mathematical description of this set.

16. Draw a closed disk with radius 3 centered at $(2,0)$ in the x-y plane, and give a mathematical description of this set.

17. Give a geometric interpretation of the set

$$A = \left\{(x,y)\in \mathbf{R}^2 : \sqrt{x^2+y^2-4y+4} < 3\right\}$$

18. Give a geometric interpretation of the set

$$A = \left\{(x,y)\in \mathbf{R}^2 : \sqrt{x^2+6x+y^2-2y+10} < 2\right\}$$

19. Let

$$f(x,y) = 2x^2+y^2$$

Use the $\epsilon-\delta$ definition of limits to show that

$$\lim_{(x,y)\to(0,0)} f(x,y) = 0$$

20. Let

$$f(x,y) = x^2+3y^2$$

Use the $\epsilon-\delta$ definition of limits to show that

$$\lim_{(x,y)\to(0,0)} f(x,y) = 0$$

(10.2.3)

21. Use the definition of continuity to show that

$$f(x,y) = x^2+y^2$$

is continuous at $(0,0)$.

22. Use the definition of continuity to show that

$$f(x,y) = \sqrt{9+x^2+y^2}$$

is continuous at $(0,0)$.

23. Show that

$$f(x,y) = \begin{cases} \dfrac{4xy}{x^2+y^2} & \text{for } (x,y)\neq(0,0) \\ 0 & \text{for } (x,y)=(0,0) \end{cases}$$

is discontinuous at $(0,0)$. (*Hint:* Use Problem 11.)

24. Show that

$$f(x,y) = \begin{cases} \dfrac{3xy}{x^2+y^3} & \text{for } (x,y)\neq(0,0) \\ 0 & \text{for } (x,y)=(0,0) \end{cases}$$

is discontinuous at $(0,0)$. (*Hint:* Use Problem 12.)

25. Show that

$$f(x,y) = \begin{cases} \dfrac{2xy}{x^3+yx} & \text{for } (x,y)\neq(0,0) \\ 0 & \text{for } (x,y)=(0,0) \end{cases}$$

is discontinuous at $(0,0)$. (*Hint:* Use Problem 13.)

26. Show that

$$f(x,y) = \begin{cases} \dfrac{3x^2y^2}{x^3+y^6} & \text{for } (x,y)\neq(0,0) \\ 0 & \text{for } (x,y)=(0,0) \end{cases}$$

is discontinuous at $(0,0)$. (*Hint:* Use Problem 14.)

27. (a) Write

$$h(x,y) = \sin(x^2+y^2)$$

as a composition of two functions.

(b) For which values of (x,y) is $h(x,y)$ continuous?

28. (a) Write

$$h(x,y) = \sqrt{x+y}$$

as a composition of two functions.

(b) For which values of (x,y) is $h(x,y)$ continuous?

29. (a) Write

$$h(x,y) = e^{xy}$$

as a composition of two functions.

(b) For which values of (x,y) is $h(x,y)$ continuous?

30. (a) Write

$$h(x,y) = \cos(y-x)$$

as a composition of two functions.

(b) For which values of (x,y) is $h(x,y)$ continuous?

10.3 PARTIAL DERIVATIVES

10.3.1 Functions of Two Variables

Suppose that the response of an organism depends on a number of independent variables. To investigate this dependency, a common experimental design is to measure the response when changing one variable while keeping all other variables fixed. As an example, Pisek et al. (1969) measured the net assimiliation of CO_2 of *Ranunculus glacialis*, a member of the buttercup family, as a function of temperature and light intensity. They varied the temperature while keeping the light intensity constant. Repeating this experiment at different light intensities, they were able to determine how net assimiliation of CO_2 changes as a function of both temperature and light intensity.

This experimental design illustrates the idea behind partial derivatives. Suppose that we want to know how the function $f(x, y)$ changes when x and y change. Instead of changing both variables simultaneously, we might get an idea of how $f(x, y)$ depends on x and y when we change one variable while keeping the other variable fixed.

To illustrate this we look at

$$f(x, y) = x^2 y$$

We want to know how $f(x, y)$ changes if we change one variable, say x, and keep the other variable, in this case y, fixed. We fix $y = y_0$, then the change of f with respect to x is simply the derivative of f with respect to x when $y = y_0$. That is,

$$\frac{d}{dx} f(x, y_0) = \frac{d}{dx} x^2 y_0 = 2x y_0$$

Such a derivative is called a partial derivative.

> **Definition** Suppose that f is a function of two independent variables x and y. The **partial derivative** of f with respect to x is defined by
>
> $$\frac{\partial f(x, y)}{\partial x} = \lim_{h \to 0} \frac{f(x + h, y) - f(x, y)}{h}$$
>
> The partial derivative of f with respect to y is defined by
>
> $$\frac{\partial f(x, y)}{\partial y} = \lim_{h \to 0} \frac{f(x, y + h) - f(x, y)}{h}$$

To denote partial derivatives, we use "∂" instead of "d." We will also use the notation

$$f_x(x, y) = \frac{\partial f(x, y)}{\partial x} \qquad \text{and} \qquad f_y(x, y) = \frac{\partial f(x, y)}{\partial y}$$

In the definition of partial derivatives, you should recognize the formal definition of derivatives of Chapter 4. Namely, to compute $\partial f / \partial x$, we look at the ratio of the difference in the f values, $f(x + h, y) - f(x, y)$, and the difference in the x values, $x + h - x$. The other variable, y, is not changed. We then let h tend to 0.

To compute $\partial f(x, y)/\partial x$, we differentiate f with respect to x while treating y as a constant. When we read $\partial f(x, y)/\partial x$, we can say "the partial derivative of f of x and y with respect to x." To read $f_x(x, y)$, we say "f sub x of x and y."

Finding partial derivatives is no different from finding derivatives of functions of one variable, since by keeping all but one variable fixed, computing a partial derivative is reduced to computing a derivative of a function of one variable. You just need to keep straight which of the variables you have fixed and which one you will vary.

▶ Example 1   Find $\partial f/\partial x$ and $\partial f/\partial y$ when

$$f(x, y) = ye^{xy}$$

Solution

To compute $\partial f/\partial x$, we treat y as a constant, and differentiate f with respect to x using the chain rule.

$$\frac{\partial f(x, y)}{\partial x} = \frac{\partial}{\partial x}(ye^{xy}) = ye^{xy}y = y^2 e^{xy}$$

To compute $\partial f/\partial y$, we treat x as a constant, and differentiate f with respect to y using the product rule combined with the chain rule.

$$\frac{\partial f(x, y)}{\partial y} = \frac{\partial}{\partial y}(ye^{xy})$$

$$= 1 \cdot e^{xy} + ye^{xy}x$$

$$= e^{xy}(1 + xy)$$ ◀

▶ Example 2   Find $\partial f/\partial x$ when

$$f(x, y) = \frac{\sin(xy)}{x^2 + \cos y}$$

Solution

We treat y as a constant and use the quotient rule.

$$u = \sin(xy) \qquad\qquad v = x^2 + \cos y$$

$$\frac{\partial u}{\partial x} = y\cos(xy) \qquad\qquad \frac{\partial v}{\partial x} = 2x$$

Hence,

$$\frac{\partial f(x, y)}{\partial x} = \frac{y(x^2 + \cos y)\cos(xy) - 2x\sin(xy)}{(x^2 + \cos y)^2}$$ ◀

Geometric Interpretation   As with ordinary derivatives, partial derivatives are slopes of lines that are tangent to certain curves. We can find these curves on the surface $z = f(x, y)$.

Let's start with the interpretation of $\partial f/\partial x$. We fix $y = y_0$; then $f(x, y_0)$ as a function of x is obtained by intersecting the surface $z = f(x, y)$ with a vertical plane that is parallel to the x-z plane and goes through $y = y_0$. The curve of intersection is the graph of $z = f(x, y_0)$, as illustrated in Figure 10.30.

We can now project this curve onto the x-z plane, which is illustrated in Figure 10.31. The curve is the graph of a function that depends only on x. As such, we can find the slope of the tangent line at any point P with coordinates (x_0, z_0), where $z_0 = f(x_0, y_0)$. The slope of the tangent line is given by the derivative of the function f in the x direction, that is, by $\partial f/\partial x$. This is the geometric meaning of $\partial f/\partial x$. We summarize this in the following.

The partial derivative $\partial f/\partial x$ evaluated at (x_0, y_0) is the slope of the tangent line at the point (x_0, y_0, z_0) with $z_0 = f(x_0, y_0)$ to the curve $z = f(x, y_0)$.

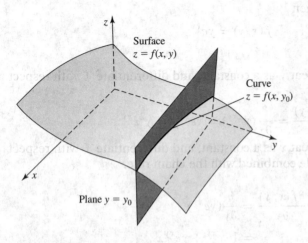

▲ Figure 10.30
The surface of $f(x, y)$ intersected with the plane $y = y_0$

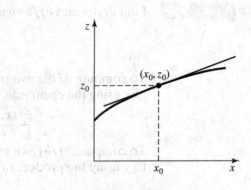

▲ Figure 10.31
The projection of the curve
$z = f(x, y_0)$ onto the x-z
plane

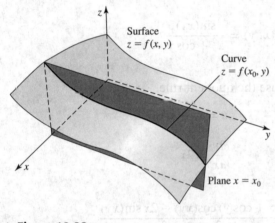

▲ Figure 10.32
The surface of $f(x, y)$ intersected with the plane
$x = x_0$

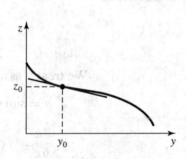

▲ Figure 10.33
The projection of the curve
$z = f(x_0, y)$ onto the y-z
plane

The derivative $\partial f/\partial y$ has a similar meaning. This time we fix $x = x_0$, and intersect the surface $z = f(x, y)$ with a vertical plane that is parallel to the y-z plane and goes through $x = x_0$. The curve of intersection is the graph of $z = f(x_0, y)$, as illustrated in Figure 10.32.

The projection of this curve onto the y-z plane together with the tangent line at (y_0, z_0) is illustrated in Figure 10.33. Analogous to the interpretation of $\partial f/\partial x$, we then have the following geometric meaning of $\partial f/\partial y$.

> The partial derivative $\partial f/\partial y$ evaluated at (x_0, y_0) is the slope of the tangent line at the point (x_0, y_0, z_0) with $z_0 = f(x_0, y_0)$ to the curve $z = f(x_0, y)$.

▶ **Example 3**   Let

$$f(x, y) = 3 - x^3 - y^2$$

Find $f_x(1, 1)$ and $f_y(1, 1)$, and interpret your results geometrically.

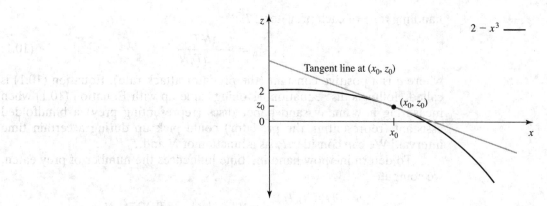

▲ Figure 10.34
The curve of intersection of the graph $z = f(x, y)$ with the plane $y = y_0$. Drawn is the curve in the plane of intersection together with its tangent line at (x_0, z_0)

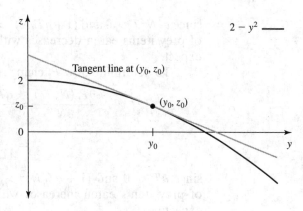

▲ Figure 10.35
The curve of intersection of the graph $z = f(x, y)$ with the plane $x = x_0$. Drawn is the curve in the plane of intersection together with its tangent line at (y_0, z_0)

Solution

We have

$$f_x(x, y) = -3x^2 \quad \text{and} \quad f_y(x, y) = -2y$$

Hence,

$$f_x(1, 1) = -3 \quad \text{and} \quad f_y(1, 1) = -2$$

To interpret $f_x(1, 1)$ geometrically, we fix $y = 1$. The vertical plane $y = 1$ intersects the graph of $f(x, y)$. The curve of intersection has slope -3 when $x = 1$. The projection of the curve of intersection is shown in Figure 10.34.

Similarly, to interpret $f_y(1, 1)$, we intersect the graph of $f(x, y)$ with the vertical plane $x = 1$. The tangent line at the curve of intersection has slope -2 when $y = 1$. The projection of the curve of intersection is shown in Figure 10.35. ◄

A Biological Application

▶ Example 4   Holling (1959) derived an expression for the number of prey items eaten by a predator, P_e, during a time interval T, as a function of prey density, N, and handling time of each prey item, T_h:

$$P_e = \frac{aNT}{1 + aT_h N} \tag{10.1}$$

where a is a positive constant (the predator attack rate). Equation (10.1) is called Holling's disc equation. [Holling came up with Equation (10.1) when measuring how many sand paper discs (representing prey) a blindfolded assistant (representing the predator) could pick up during a certain time interval.] We can consider P_e as a function of N and T_h.

To determine how handling time influences the number of prey eaten, we compute

$$\frac{\partial P_e(N, T_h)}{\partial T_h} = aNT(-1)(1 + aT_h N)^{-2} aN$$

$$= -\frac{a^2 N^2 T}{(1 + aT_h N)^2} < 0$$

since $a^2 N^2 T > 0$ and $(1 + aT_h N)^2 > 0$. Since $\partial P_e / \partial T_h$ is negative, the number of prey items eaten decreases with increasing handling time, as you would expect.

To determine how P_e changes with N, we compute

$$\frac{\partial P_e(N, T_h)}{\partial N} = \frac{aT(1 + aT_h N) - aNTaT_h}{(1 + aT_h N)^2}$$

$$= \frac{aT}{(1 + aT_h N)^2} > 0$$

since $aT > 0$ and $(1 + aT_h N)^2 > 0$. Since $\partial P_e / \partial N$ is positive, the number of prey items eaten increases with increasing prey density, as you would expect. ◀

10.3.2 Functions of More Than Two Variables

The definition of partial derivatives extends in a straightforward way to functions of more than two variables. These are ordinary derivatives with respect to one variable while treating all other variables as constants.

▶ Example 5   Let f be a function of three independent variables x, y, and z:

$$f(x, y, z) = e^{yz}(x^2 + z^3)$$

Find $\partial f / \partial x$, $\partial f / \partial y$, and $\partial f / \partial z$.

Solution

To compute $\partial f / \partial x$, we treat y and z as constants and differentiate f with respect to x.

$$\frac{\partial f}{\partial x} = e^{yz} 2x$$

Likewise,

$$\frac{\partial f}{\partial y} = z e^{yz}(x^2 + z^3)$$

$$\frac{\partial f}{\partial z} = y e^{yz}(x^2 + z^3) + e^{yz} 3z^2$$ ◀

10.3.3 Higher-Order Partial Derivatives

As in the case of functions of one variable, we can define higher-order partial derivatives for functions of more than one variable. For instance, to find the second partial derivative of $f(x, y)$ with respect to x, denoted by $\partial^2 f/\partial x^2$, we compute

$$\frac{\partial^2 f}{\partial x^2} = \frac{\partial}{\partial x}\left(\frac{\partial f}{\partial x}\right)$$

We can write $\partial^2 f/\partial x^2$ also as f_{xx}.

We can also compute **mixed derivatives**. For instance,

$$f_{yx} = \frac{\partial^2 f}{\partial x \partial y} = \frac{\partial}{\partial x}\left(\frac{\partial f}{\partial y}\right)$$

Note the order of "yx" in the subscript of f and the order of $\partial x \, \partial y$ in the denominator: Either notation means that we differentiate with respect to y first.

▶ **Example 6** Set

$$f(x, y) = \sin x + xe^y$$

Find $f_{xx} = \dfrac{\partial^2 f}{\partial x^2}$, $f_{yx} = \dfrac{\partial^2 f}{\partial x \, \partial y}$, and $f_{xy} = \dfrac{\partial^2 f}{\partial y \, \partial x}$.

Solution

$$f_{xx} = \frac{\partial^2 f}{\partial x^2} = \frac{\partial}{\partial x}\left(\frac{\partial f}{\partial x}\right) = \frac{\partial}{\partial x}\left[\frac{\partial}{\partial x}(\sin x + xe^y)\right]$$

$$= \frac{\partial}{\partial x}[\cos x + e^y] = -\sin x$$

$$f_{yx} = \frac{\partial^2 f}{\partial x \, \partial y} = \frac{\partial}{\partial x}\left(\frac{\partial f}{\partial y}\right) = \frac{\partial}{\partial x}\left[\frac{\partial}{\partial y}(\sin x + xe^y)\right]$$

$$= \frac{\partial}{\partial x}[0 + xe^y] = e^y$$

$$f_{xy} = \frac{\partial^2 f}{\partial y \, \partial x} = \frac{\partial}{\partial y}\frac{\partial f}{\partial x} = \frac{\partial}{\partial y}\left[\frac{\partial}{\partial x}(\sin x + xe^y)\right]$$

$$= \frac{\partial}{\partial y}[\cos x + e^y] = e^y$$ ◀

In the preceding example, you see that $f_{xy} = f_{yx}$, implying that the order of differentiation did not matter. Although this is not always the case, there are conditions under which the order of differentiation in mixed partial derivatives does not matter. To state this theorem, we need the notion of an open and a closed disk. We have the following definition.

Definition An **open disk** with radius r centered at $(x_0, y_0) \in \mathbf{R}^2$ is the set
$$B_r(x_0, y_0) = \left\{(x, y) \in \mathbf{R}^2 : \sqrt{(x - x_0)^2 + (y - y_0)^2} < r\right\}$$

A **closed disk** with radius r centered at $(x_0, y_0) \in \mathbf{R}^2$ is the set
$$\overline{B}_r(x_0, y_0) = \left\{(x, y) \in \mathbf{R}^2 : \sqrt{(x - x_0)^2 + (y - y_0)^2} \leq r\right\}$$

An open disk is thus a disk in which the boundary line is not part of the disk, whereas a closed disk contains the boundary line. This is analogous to intervals: An open interval does not contain the two endpoints (which are the boundary of the interval), whereas a closed interval does contain them.

We can now state the mixed derivative theorem.

> **The Mixed Derivative Theorem** If $f(x, y)$ and its partial derivatives f_x, f_y, f_{xy}, and f_{yx} are continuous on an open disk centered at the point (x_0, y_0), then
>
> $$f_{xy}(x_0, y_0) = f_{yx}(x_0, y_0)$$

It is straightforward to define partial derivatives of even higher order. For instance, if f is a function of two independent variables x and y, then

$$\frac{\partial^3 f}{\partial x^2 \, \partial y} = \frac{\partial}{\partial x} \frac{\partial^2 f}{\partial x \, \partial y} = \frac{\partial}{\partial x} \frac{\partial}{\partial x} \frac{\partial f}{\partial y}$$

We illustrate this on

$$f(x, y) = y^2 \sin x$$

for which

$$\frac{\partial^3 f}{\partial x^2 \, \partial y} = \frac{\partial}{\partial x} \frac{\partial}{\partial x} \frac{\partial}{\partial y} (y^2 \sin x)$$

$$= \frac{\partial}{\partial x} \frac{\partial}{\partial x} (2y \sin x)$$

$$= \frac{\partial}{\partial x} (2y \cos x) = -2y \sin x$$

and

$$\frac{\partial^3 f}{\partial x \, \partial y^2} = \frac{\partial}{\partial x} \frac{\partial}{\partial y} \frac{\partial}{\partial y} (y^2 \sin x)$$

$$= \frac{\partial}{\partial x} \frac{\partial}{\partial y} (2y \sin x) = \frac{\partial}{\partial x} (2 \sin x) = 2 \cos x$$

and so on.

The mixed derivative theorem can be extended to higher order derivatives. The order of differentiation does not matter as long as the function and all its derivatives through the order in question are continuous on an open disk centered at the point at which we want to compute the derivative. For instance,

$$\frac{\partial^3 f}{\partial y^2 \, \partial x} = \frac{\partial}{\partial y} \frac{\partial}{\partial y} \frac{\partial}{\partial x} (y^2 \sin x)$$

$$= \frac{\partial}{\partial y} \frac{\partial}{\partial y} (y^2 \cos x)$$

$$= \frac{\partial}{\partial y} (2y \cos x) = 2 \cos x$$

which is the same as $\partial^3 f/(\partial x \, \partial y^2)$.

10.3.4 Problems

(10.3.1)

In Problems 1–16, find $\partial f/\partial x$ and $\partial f/\partial y$ for the given functions.

1. $f(x, y) = x^2 y + xy^2$

2. $f(x, y) = 2x\sqrt{y} - \frac{1}{xy}$

3. $f(x, y) = (xy)^{3/2} - (xy)^{2/3}$

4. $f(x, y) = x^3 y^4 - \frac{1}{x^3 y^4}$

5. $f(x, y) = \sin(x + y)$

6. $f(x, y) = \tan(x - y)$

7. $f(x, y) = \cos^2(x^2 - 2y)$

8. $f(x, y) = \sec(y^2 x - x^2)$

9. $f(x, y) = e^{\sqrt{x+y}}$

10. $f(x, y) = x^2 e^{-xy}$

11. $f(x, y) = e^x \sin(xy)$

12. $f(x, y) = e^{-y^2} \cos(x^2 + y^2)$

13. $f(x, y) = \ln(2x + y)$

14. $f(x, y) = \ln(3x^2 - y)$

15. $f(x, y) = \log_3(y^2 - x^2)$

16. $f(x, y) = \log_5(3xy)$

In Problems 17–24, find the indicated partial derivatives.

17. $f(x, y) = 3x^2 - y - 2y^2; f_x(1, 0)$

18. $f(x, y) = x^{1/3} y - xy^{1/3}; f_y(1, 1)$

19. $g(x, y) = e^{x+3y}; g_y(2, 1)$

20. $h(u, v) = e^u \sin(u + v); h_u(1, -1)$

21. $f(x, z) = \ln(xz); f_z(e, 1)$

22. $g(v, w) = \frac{w^2}{v+w}; g_v(1, 1)$

23. $f(x, y) = \frac{xy}{x^2+2}; f_x(-1, 2)$

24. $f(u, v) = e^{u^2/2} \ln(u + v); f_u(2, 1)$

25. Let
$$f(x, y) = 4 - x^2 - y^2$$
Compute $f_x(1, 1)$ and $f_y(1, 1)$ and interpret these partial derivatives geometrically.

26. Let
$$f(x, y) = \sqrt{4 - x^2 - y^2}$$
Compute $f_x(1, 1)$ and $f_y(1, 1)$ and interpret these partial derivatives geometrically.

27. Let
$$f(x, y) = 1 + x^2 y$$
Compute $f_x(-2, 1)$ and $f_y(-2, 1)$ and interpret these partial derivatives geometrically.

28. Let
$$f(x, y) = 2x^3 - 3yx$$
Compute $f_x(1, 2)$ and $f_y(1, 2)$ and interpret these partial derivatives geometrically.

29. In Example 4, we investigated Holling's disk equation
$$P_e = \frac{aNT}{1 + aT_h N}$$
(see Example 4 for the meaning of this equation). We will now consider P_e as a function of the predator attack rate a and the length of the time interval, T, during which the predator searches for food.

(a) Determine how the predator attack rate, a, influences the number of prey eaten per predator.

(b) Determine how the length of the time interval, T, influences the number of prey eaten per predator.

30. Suppose that the per capita growth rate of some prey at time t depends on both the prey density $H(t)$ at time t and the predator density $P(t)$ at time t. Assume the following relationship:
$$\frac{1}{H}\frac{dH}{dt} = r\left(1 - \frac{H}{K}\right) - aP \qquad (10.2)$$
where r, K, and a are positive constants. The right-hand side of (10.2) is a function of both prey density and predator density. Investigate how an increase in (a) prey density and (b) predator density affects the per capita growth rate of this prey species.

(10.3.2)

In Problems 31–38, find $\partial f/\partial x$, $\partial f/\partial y$, and $\partial f/\partial z$ for the given functions.

31. $f(x, y, z) = x^2 z + yz^2 - xy$

32. $f(x, y, z) = xyz$

33. $f(x, y, z) = x^3 y^2 z + \frac{x}{yz}$

34. $f(x, y, z) = \frac{xyz}{x^2+y^2+z^2}$

35. $f(x, y, z) = e^{x+y+z}$

36. $f(x, y, z) = e^{yz} \sin x$

37. $f(x, y, z) = \ln(x + y + z)$

38. $f(x, y, z) = y \tan(x^2 + z)$

(10.3.3)

In Problems 39–48, find the indicated partial derivatives.

39. $f(x, y) = x^2 y + xy^2; \frac{\partial^2 f}{\partial x^2}$

40. $f(x, y) = y^3(x - y); \frac{\partial^2 f}{\partial y^2}$

41. $f(x, y) = xe^y; \frac{\partial^2 f}{\partial x \partial y}$

42. $f(x, y) = \sin(x + y); \frac{\partial^2 f}{\partial y \partial x}$

43. $f(u, w) = \tan(u + w); \frac{\partial^2 f}{\partial u^2}$

44. $g(s, t) = \ln(s^2 + 3t); \frac{\partial^2 g}{\partial t^2}$

45. $f(x, y) = x^3 \cos y; \frac{\partial^3 f}{\partial x^2 \partial y}$

46. $f(x, y) = e^{x-y}; \frac{\partial^3 f}{\partial y^2 \partial x}$

47. $f(x, y) = \ln(x + y); \frac{\partial^3 f}{\partial x^3}$

48. $f(x, y) = \sin(3xy); \frac{\partial^3 f}{\partial y^2 \partial x}$

49. The functional responses of some predators are sigmoidal; that is, the number of prey attacked per predator as a function of prey density has a sigmoidal shape. If we denote by N the prey density, by P the predator density, by T the time available for searching for prey, and by T_h the handling time of each prey item per predator, then the number of prey encounters per predator as a function of N, T, and T_h can be expressed as

$$f(N, T, T_h) = \frac{b^2 N^2 T}{1 + cN + bT_h N^2}$$

where b and c are positive constants.

(a) Investigate how an increase in the prey density N affects the function $f(N, T, T_h)$.

(b) Investigate how an increase in the time available for search, T, affects the function $f(N, T, T_h)$.

(c) Investigate how an increase in the handling time T_h affects the function $f(N, T, T_h)$.

(d) Graph $f(N, T, T_h)$ as a function of N when $T = 2.4$ hours, $T_h = 0.2$ hours, $b = 0.8$, and $c = 0.5$.

50. In this problem we will investigate how mutual interference of parasitoids affects their searching efficiency for a host. We assume that N is the host density and P is the parasitoid density. A frequently used model for host-parasitoid interactions is the **Nicholson-Bailey** model (Nicholson, 1933; Nicholson and Bailey, 1935), in which it is assumed that the number of parasitized hosts, denoted by N_a, is given by

$$N_a = N[1 - e^{-bP}] \qquad (10.3)$$

where b is the searching efficiency.

(a) Show that

$$b = \frac{1}{P} \ln \frac{N}{N - N_a}$$

by solving (10.3) for b.

(b) Consider

$$b = f(P, N, N_a) = \frac{1}{P} \ln \frac{N}{N - N_a}$$

as a function of P, N, and N_a. How is the searching efficiency b affected when the parasitoid density increases?

(c) Assume now that the fraction of parasitized host depends on the host density; that is, assume that

$$N_a = g(N)$$

where $g(N)$ is a nonnegative, differentiable function. The searching efficiency b can then be written as a function of P and N, namely,

$$b = h(P, N) = \frac{1}{P} \ln \frac{N}{N - g(N)}$$

How does the searching efficiency depend on host density when $g(N)$ is a decreasing function of N? (Use the fact that $g(N) < N$.)

51. Leopold and Kriedemann (1975) measured crop growth rate of sunflowers as a function of leaf area index and percent of full sunlight. (Leaf area index is the ratio of leaf surface area to the ground area the plant covers.) They found that for a fixed level of sunlight, crop growth rate first increases and then decreases as a function of leaf area index. For a given leaf area index, crop growth rate increases with sunlight level. The leaf area index that maximizes the crop growth rate is an increasing function of sunlight. Sketch the crop growth rate as a function of leaf area index for different values of percent of full sunlight.

10.4 TANGENT PLANES, DIFFERENTIABILITY, AND LINEARIZATION

10.4.1 Functions of Two Variables

Tangent Planes Suppose that $z = f(x)$ is differentiable at $x = x_0$. Then the equation of the tangent line of $z = f(x)$ at (x_0, z_0) with $z_0 = f(x_0)$ is given by

$$z - z_0 = f'(x_0)(x - x_0) \qquad (10.4)$$

This is illustrated in Figure 10.36.

We now generalize this to functions of two variables. The analogue of a tangent line is called a **tangent plane**, an example of which is shown in Figure 10.37. Let $z = f(x, y)$ be a function of two variables. We saw in the previous section that the partial derivatives $\partial f/\partial x$ and $\partial f/\partial y$, evaluated at (x_0, y_0), are the slopes of tangent lines at the point (x_0, y_0, z_0) with $z_0 = f(x_0, y_0)$, to certain curves through (x_0, y_0, z_0) on the surface $z = f(x, y)$. These two tangent lines, one in the x-direction, the other in the y-direction, define a unique plane. If, in addition, $f(x, y)$ has partial derivatives that are continuous on an open disk containing (x_0, y_0), then we can show that the tangent line at (x_0, y_0, z_0) to any other smooth curve on the surface $z = f(x, y)$ through (x_0, y_0, z_0) is contained in this plane. This plane is then called the tangent plane.

We will use the two original tangent lines to find the equation of the tangent plane at a point (x_0, y_0, z_0) on the surface $z = f(x, y)$. We take the

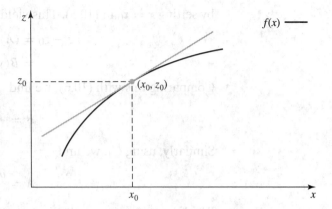

▲ **Figure 10.36**
The curve $z = f(x)$ and its tangent line at the point (x_0, z_0)

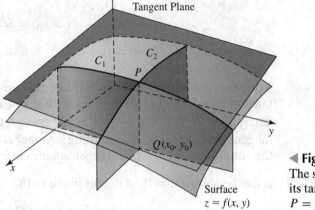

◀ **Figure 10.37**
The surface $z = f(x, y)$ and its tangent plane at $P = (x_0, y_0, z_0)$

curve that is obtained as the intersection of the surface $z = f(x, y)$ with the plane that is parallel to the y-z plane and contains the point (x_0, y_0, z_0)—that is, the plane $x = x_0$—and we denote this curve by C_1. Its tangent line at (x_0, y_0, z_0) is contained in the tangent plane (see Figure 10.37). Likewise, we take the curve of intersection between $z = f(x, y)$ and the plane that is parallel to the x-z plane and contains the point (x_0, y_0, z_0)—that is, the plane $y = y_0$—and we denote this curve by C_2. Its tangent line is contained in the tangent plane (see Figure 10.37).

In Section 9.4, we gave the general equation of a plane, namely

$$z - z_0 = A(x - x_0) + B(y - y_0) \tag{10.5}$$

We will use curves C_1 and C_2 to determine the constants A and B. C_1 satisfies the equation

$$z = f(x_0, y)$$

The tangent line to C_1 at (x_0, y_0, z_0) therefore satisfies

$$z - z_0 = \frac{\partial f(x_0, y_0)}{\partial y}(y - y_0) \tag{10.6}$$

The tangent line at (x_0, y_0, z_0) is contained in the tangent plane at (x_0, y_0, z_0). The equation of the tangent line at (x_0, y_0, z_0) can therefore also be obtained by setting $x = x_0$ in (10.5). This yields

$$z - z_0 = (A)(0) + B(y - y_0)$$
$$= B(y - y_0)$$

Comparing this with (10.6), we find

$$B = \frac{\partial f(x_0, y_0)}{\partial y}$$

Similarly, using C_2, we find

$$A = \frac{\partial f(x_0, y_0)}{\partial x}$$

We thus arrive at the following result.

> If the tangent plane to the surface $z = f(x, y)$ at the point (x_0, y_0, z_0) exists, then that tangent plane has the equation
>
> $$z - z_0 = \frac{\partial f(x_0, y_0)}{\partial x}(x - x_0) + \frac{\partial f(x_0, y_0)}{\partial y}(y - y_0)$$

You should observe the similarity of this equation to the equation of the tangent line in (10.4). We will see that the mere existence of the partial derivatives $\frac{\partial f(x_0, y_0)}{\partial x}$ and $\frac{\partial f(x_0, y_0)}{\partial y}$ is *not* enough to guarantee the existence of a tangent plane at (x_0, y_0); something stronger is needed. But before we discuss the conditions under which tangent planes exist, let's look at an example.

▶ **Example 1**   It can be shown that the tangent plane to the surface

$$z = f(x, y) = 4x^2 + y^2$$

at the point $(1, 2, 8)$ exists. Find its equation.

Solution

First note that the point $(1, 2, 8)$ is contained in the surface $z = f(x, y)$, since $8 = f(1, 2)$. To find the tangent plane, we need to compute the partial derivatives of $f(x, y)$,

$$\frac{\partial f}{\partial x} = 8x \qquad \text{and} \qquad \frac{\partial f}{\partial y} = 2y$$

and evaluate them at $(x_0, y_0) = (1, 2)$,

$$\frac{\partial f(1, 2)}{\partial x} = 8 \qquad \text{and} \qquad \frac{\partial f(1, 2)}{\partial y} = 4$$

Hence, the equation of the tangent plane is

$$z - 8 = 8(x - 1) + 4(y - 2)$$

We can write this in a more compact form, namely

$$z - 8 = 8x - 8 + 4y - 8$$

or

$$8x + 4y - z = 8$$ ◀

Differentiability We will now discuss the conditions under which tangent planes exist. For this, we need to define what it means for a function of two variables to be differentiable. To make the connection with functions of one variable clear, recall that the tangent line is used for linearly approximating $f(x)$ at $x = x_0$. The linear approximation of $f(x)$ at $x = x_0$ is given by

$$L(x) = f(x_0) + f'(x_0)(x - x_0) \tag{10.7}$$

The distance between $f(x)$ and its linear approximation at $x = x_0$ is then

$$|f(x) - L(x)| = |f(x) - f(x_0) - f'(x_0)(x - x_0)| \tag{10.8}$$

From the definition of the derivative, we know that

$$f'(x_0) = \lim_{x \to x_0} \frac{f(x) - f(x_0)}{x - x_0} \tag{10.9}$$

If we divide (10.8) by the distance between x and x_0, $|x - x_0|$, we find

$$\frac{|f(x) - L(x)|}{|x - x_0|} = \left| \frac{f(x) - L(x)}{x - x_0} \right| = \left| \frac{f(x) - f(x_0) - f'(x_0)(x - x_0)}{x - x_0} \right|$$

$$= \left| \frac{f(x) - f(x_0)}{x - x_0} - f'(x_0) \right|$$

Taking the limit $x \to x_0$ and using (10.9), we therefore find that

$$\lim_{x \to x_0} \left| \frac{f(x) - L(x)}{x - x_0} \right| = 0 \tag{10.10}$$

We could say that $f(x)$ is differentiable at $x = x_0$ if (10.10) holds.

For functions of two variables, the definition of differentiability is based on the same idea. Before we can give it, we need to address one more point. In the preceding, we divided by the distance between x and x_0. In two dimensions, the distance between two points (x, y) and (x_0, y_0) is $\sqrt{(x - x_0)^2 + (y - y_0)^2}$.

Definition Suppose that $f(x, y)$ is a function of two independent variables and that both $\partial f/\partial x$ and $\partial f/\partial y$ are defined throughout an open disk containing (x_0, y_0). Set

$$L(x, y) = f(x_0, y_0) + \frac{\partial f(x_0, y_0)}{\partial x}(x - x_0) + \frac{\partial f(x_0, y_0)}{\partial y}(y - y_0)$$

Then $f(x, y)$ is differentiable at (x_0, y_0) if

$$\lim_{(x,y) \to (x_0, y_0)} \left| \frac{f(x, y) - L(x, y)}{\sqrt{(x - x_0)^2 + (y - y_0)^2}} \right| = 0$$

Furthermore, if $f(x, y)$ is differentiable at (x_0, y_0), then $z = L(x, y)$ defines the tangent plane to the graph of f at $(x_0, y_0, f(x_0, y_0))$. We say that $f(x, y)$ is differentiable if it is differentiable at every point of its domain.

The key idea in both the one- and the two-dimensional cases is to approximate functions by linear functions, so that the error in the approximation vanishes as we approach the point at which we approximated the function [x_0 in the one-dimensional case, and (x_0, y_0) in the two-dimensional case].

The following theorem holds, as in the one-dimensional case.

> **Theorem** If $f(x, y)$ is differentiable at (x_0, y_0), then f is continuous at (x_0, y_0).

Differentiability of $f(x, y)$ at (x_0, y_0) means that the function $f(x, y)$ is close to the tangent plane at (x_0, y_0) for all (x, y) close to (x_0, y_0). The mere existence of the partial derivatives $\partial f/\partial x$ and $\partial f/\partial y$ at (x_0, y_0), however, is *not* enough to guarantee differentiability (and consequently the existence of a tangent plane at a certain point). Here is a very simple example that will show what can go wrong with that assumption.

▶ **Example 2** Assume that

$$f(x, y) = \begin{cases} 0 & \text{if } xy \neq 0 \\ 1 & \text{if } xy = 0 \end{cases}$$

Show that $\frac{\partial f(0,0)}{\partial x}$ and $\frac{\partial f(0,0)}{\partial y}$ exist, but $f(x, y)$ is not continuous, and consequently not differentiable, at $(0, 0)$.

Solution

The graph of $f(x, y)$ is shown in Figure 10.38. To compute $\partial f/\partial x$ at $(0, 0)$, we set $y = 0$, then $f(x, 0) = 1$, and therefore

$$\frac{\partial f(0, 0)}{\partial x} = 0$$

Likewise, setting $x = 0$, we find $f(0, y) = 1$ and

$$\frac{\partial f(0, 0)}{\partial y} = 0$$

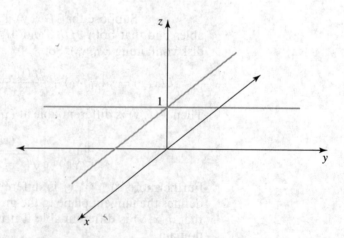

▲ **Figure 10.38**
The graph of the function

$$f(x, y) = \begin{cases} 0 & \text{if } xy \neq 0 \\ 1 & \text{if } xy = 0 \end{cases}$$

Even though $\frac{\partial f(0,0)}{\partial x}$ and $\frac{\partial f(0,0)}{\partial y}$ exist, the function is not continuous at $(0, 0)$

That is, both partial derivatives exist at $(0, 0)$. However, $f(x, y)$ is not continuous at $(0, 0)$. For this it is enough to show that $f(x, y)$ has different limits along two different paths as (x, y) approaches $(0, 0)$. For the first path, denoted by C_1, we choose $y = 0$. Then

$$\lim_{\substack{(x, y) \overset{C_1}{\to} (0,0)}} f(x, y) = 1$$

For the second path, denoted by C_2, we choose $y = x$. Then

$$\lim_{\substack{(x, y) \overset{C_2}{\to} (0,0)}} f(x, y) = 0$$

Since $1 \neq 0$, $f(x, y)$ is not continuous at $(0, 0)$. Since differentiability implies continuity, a function that is not continuous cannot be differentiable. ◀

The definition of differentiability is not easy to use if we actually want to check whether a function is differentiable at a certain point. Fortunately, there is another criterion, which suffices for all practical purposes, that can be used to check whether $f(x, y)$ is differentiable. We saw in Example 2 that the mere existence of the partial derivatives $\partial f / \partial x$ and $\partial f / \partial y$ at a point (x_0, y_0) is not enough to guarantee that $f(x, y)$ is differentiable. However, if the partial derivatives are continuous on a disk centered at (x_0, y_0), then this is enough to guarantee differentiability.

SUFFICIENT CONDITION FOR DIFFERENTIABILITY

> Suppose $f(x, y)$ is defined on an open disk centered at (x_0, y_0), and the partial derivatives $\partial f / \partial x$ and $\partial f / \partial y$ are continuous on an open disk centered at (x_0, y_0). Then $f(x, y)$ is differentiable at (x_0, y_0).

▶ **Example 3** Show that $f(x, y) = 2x^2 y - y^2$ is differentiable for all $(x, y) \in \mathbf{R}^2$.

Solution

We use the sufficient condition for differentiability. First, observe that $f(x, y)$ is defined for all $(x, y) \in \mathbf{R}^2$. The partial derivatives are given by

$$\frac{\partial f}{\partial x} = 4xy \qquad \text{and} \qquad \frac{\partial f}{\partial y} = 2x^2 - 2y$$

Since both $\partial f / \partial x$ and $\partial f / \partial y$ are polynomials, both are continuous for all $(x, y) \in \mathbf{R}^2$ and, hence, $f(x, y)$ is differentiable for all $(x, y) \in \mathbf{R}^2$. ◀

Linearization

> **Definition** Suppose that $f(x, y)$ is differentiable at (x_0, y_0). The linearization of $f(x, y)$ at (x_0, y_0) is the function
>
> $$L(x, y) = f(x_0, y_0) + \frac{\partial f(x_0, y_0)}{\partial x}(x - x_0) + \frac{\partial f(x_0, y_0)}{\partial y}(y - y_0)$$
>
> The approximation
>
> $$f(x, y) \approx L(x, y)$$
>
> is the **standard linear approximation** or the **tangent plane approximation** of $f(x, y)$ at (x_0, y_0).

▶ Example 4   Find the linear approximation of

$$f(x, y) = x^2y + 2xe^y$$

at the point $(2, 0)$.

Solution

The linearization of $f(x, y)$ at $(2, 0)$ is given by

$$L(x, y) = f(2, 0) + \frac{\partial f(2, 0)}{\partial x}(x - 2) + \frac{\partial f(2, 0)}{\partial y}(y - 0)$$

Now, $f(2, 0) = 4$,

$$\frac{\partial f}{\partial x} = 2xy + 2e^y \qquad \frac{\partial f}{\partial y} = x^2 + 2xe^y$$

Hence,

$$\frac{\partial f(2, 0)}{\partial x} = 2 \qquad \frac{\partial f(2, 0)}{\partial y} = 4 + 4 = 8$$

and we find

$$L(x, y) = 4 + 2(x - 2) + 8(y - 0) = 4 + 2x - 4 + 8y$$
$$= 2x + 8y$$

◀

▶ Example 5   Find the linear approximation of

$$f(x, y) = \ln(x - 2y^2)$$

at the point $(3, 1)$ and use it to find an approximation for $f(3.05, 0.95)$. Use a calculator to compute the value of $f(3.05, 0.95)$ and compare it to the approximation.

Solution

The linearization of $f(x, y)$ at $(3, 1)$ is given by

$$L(x, y) = f(3, 1) + \frac{\partial f(3, 1)}{\partial x}(x - 3) + \frac{\partial f(3, 1)}{\partial y}(y - 1)$$

Now, $f(3, 1) = \ln(3 - 2) = \ln 1 = 0$,

$$\frac{\partial f(x, y)}{\partial x} = \frac{1}{x - 2y^2} \quad \text{and} \quad \frac{\partial f(x, y)}{\partial y} = \frac{-4y}{x - 2y^2}$$

Hence,

$$\frac{\partial f(3, 1)}{\partial x} = \frac{1}{3 - 2} = 1 \quad \text{and} \quad \frac{\partial f(3, 1)}{\partial y} = \frac{-4}{3 - 2} = -4$$

and we find

$$L(x, y) = 0 + (1)(x - 3) + (-4)(y - 1)$$
$$= x - 3 - 4y + 4 = x - 4y + 1$$

Using $(3.05, 0.95)$, we get

$$L(3.05, 0.95) = 3.05 - (4)(0.95) + 1 = 0.25$$

Comparing this with $f(3.05, 0.95) \approx 0.2191$, we see that the error of approximation is only about $|0.25 - 0.2191| = 0.031$. ◀

10.4.2 Vector-Valued Functions

So far, we have only considered real-valued functions,

$$f : \mathbf{R}^n \to \mathbf{R}$$

We will now extend our discussion to functions where the range is a subset of $\mathbf{R}^m$, that is,

$$f : \mathbf{R}^n \to \mathbf{R}^m$$

Such functions are **vector-valued functions** since they take on values that are represented by vectors.

$$f : \mathbf{R}^n \to \mathbf{R}^m$$

$$(x_1, x_2, \ldots, x_n) \to \begin{pmatrix} f_1(x_1, x_2, \ldots, x_n) \\ f_2(x_1, x_2, \ldots, x_n) \\ \vdots \\ f_m(x_1, x_2, \ldots, x_n) \end{pmatrix}$$

Here, each function $f_i(x_1, \ldots, x_n)$ is a real-valued function,

$$f_i : \mathbf{R}^n \to \mathbf{R}$$

$$(x_1, x_2, \ldots, x_n) \to f_i(x_1, x_2, \ldots, x_n)$$

We will encounter vector-valued functions where $n = m = 2$ extensively in Chapter 11. As an example, consider a community consisting of two species. Let u and v denote the respective densities of the two species, and $f(u, v)$ and $g(u, v)$ the per capita growth rates of the two species as functions of the densities u and v. We can then write this as a map from $\mathbf{R}^2$ to $\mathbf{R}^2$, where

$$(u, v) \to \begin{pmatrix} f(u, v) \\ g(u, v) \end{pmatrix}$$

Our main task in this subsection will be to define the linearization of vector-valued functions where the domain and the range are $\mathbf{R}^2$. We will motivate this by analogy with the cases of functions $f : \mathbf{R} \to \mathbf{R}$ and $f : \mathbf{R}^2 \to \mathbf{R}$. At the end of this section, we will mention how to generalize this to arbitrary vector valued functions.

We begin with differentiable functions $f : \mathbf{R} \to \mathbf{R}$. The linearization of $f : \mathbf{R} \to \mathbf{R}$ about x_0 is given by

$$L(x) = f(x_0) + f'(x_0)(x - x_0) \tag{10.11}$$

▶ **Example 6** Find the linearization of

$$f(x) = 2 \ln x$$

at $x_0 = 1$.

Solution

The linearization of $f(x)$ at $x = 1$ is given by

$$L(x) = f(1) + f'(1)(x - 1)$$
$$= 0 + (2)(x - 1) = 2x - 2$$

since $f'(x) = 2/x$. ◀

We found the linearization of a real valued function f with domain in $\mathbf{R}^2$, $f : \mathbf{R}^2 \to \mathbf{R}$, in the previous subsection. It is given by

$$L(x, y) = f(x_0, y_0) + \frac{\partial f(x_0, y_0)}{\partial x}(x - x_0) + \frac{\partial f(x_0, y_0)}{\partial y}(y - y_0)$$

We can write this in matrix notation as

$$L(x, y) = f(x_0, y_0) + \left[\begin{array}{cc} \dfrac{\partial f(x_0, y_0)}{\partial x} & \dfrac{\partial f(x_0, y_0)}{\partial y} \end{array} \right] \left[\begin{array}{c} x - x_0 \\ y - y_0 \end{array} \right]$$

▶ **Example 7**   Find the linearization of

$$f(x, y) = \ln x + \ln y$$

at $(1, 1)$.

Solution

We find $f(1, 1) = 0$. Since $f_x(x, y) = \frac{1}{x}$ and $f_y(x, y) = \frac{1}{y}$, it follows that

$$\frac{\partial f(1, 1)}{\partial x} = 1 \quad \text{and} \quad \frac{\partial f(1, 1)}{\partial y} = 1$$

Hence,

$$L(x, y) = 0 + [1, 1] \left[\begin{array}{c} x - 1 \\ y - 1 \end{array} \right]$$
$$= x - 1 + y - 1 = x + y - 2$$ ◀

We will denote vector-valued functions by boldface letters. Suppose that

$$\mathbf{h} : \mathbf{R}^2 \to \mathbf{R}^2$$

$$(x, y) \to \left[\begin{array}{c} f(x, y) \\ g(x, y) \end{array} \right]$$

and assume that all first partial derivatives are continuous on a disk centered at (x_0, y_0). We can linearize each component of the function $\mathbf{h}$. The linearization of f is

$$\alpha(x, y) = f(x_0, y_0) + \frac{\partial f(x_0, y_0)}{\partial x}(x - x_0) + \frac{\partial f(x_0, y_0)}{\partial y}(y - y_0)$$

and the linearization of g is

$$\beta(x, y) = g(x_0, y_0) + \frac{\partial g(x_0, y_0)}{\partial x}(x - x_0) + \frac{\partial g(x_0, y_0)}{\partial y}(y - y_0)$$

We define the vector-valued function $\mathbf{L}(x, y) = \left[\begin{array}{c} \alpha(x,y) \\ \beta(x,y) \end{array} \right]$. The linearization of $\mathbf{h}$ can thus be written in matrix form:

$$\mathbf{L}(x, y) = \left[\begin{array}{c} \alpha(x, y) \\ \beta(x, y) \end{array} \right] = \left[\begin{array}{c} f(x_0, y_0) \\ g(x_0, y_0) \end{array} \right]$$

$$+ \left[\begin{array}{c} \dfrac{\partial f(x_0, y_0)}{\partial x}(x - x_0) + \dfrac{\partial f(x_0, y_0)}{\partial y}(y - y_0) \\ \dfrac{\partial g(x_0, y_0)}{\partial x}(x - x_0) + \dfrac{\partial g(x_0, y_0)}{\partial y}(y - y_0) \end{array} \right]$$

or

$$L(x, y) = \begin{bmatrix} f(x_0, y_0) \\ g(x_0, y_0) \end{bmatrix} + \begin{bmatrix} \dfrac{\partial f(x_0, y_0)}{\partial x} & \dfrac{\partial f(x_0, y_0)}{\partial y} \\ \dfrac{\partial g(x_0, y_0)}{\partial x} & \dfrac{\partial g(x_0, y_0)}{\partial y} \end{bmatrix} \begin{bmatrix} x - x_0 \\ y - y_0 \end{bmatrix}$$

where the matrix

$$\begin{bmatrix} \dfrac{\partial f(x_0, y_0)}{\partial x} & \dfrac{\partial f(x_0, y_0)}{\partial y} \\ \dfrac{\partial g(x_0, y_0)}{\partial x} & \dfrac{\partial g(x_0, y_0)}{\partial y} \end{bmatrix}$$

is a 2×2 matrix that is called the **Jacobi matrix** or the **derivative matrix**. We denote the Jacobi matrix by $(D\mathbf{h})$, that is,

$$(D\mathbf{h})(x_0, y_0) = \begin{bmatrix} \dfrac{\partial f(x_0, y_0)}{\partial x} & \dfrac{\partial f(x_0, y_0)}{\partial y} \\ \dfrac{\partial g(x_0, y_0)}{\partial x} & \dfrac{\partial g(x_0, y_0)}{\partial y} \end{bmatrix}$$

▶ **Example 8** Assume that

$$\mathbf{f} : \mathbf{R}^2 \to \mathbf{R}^2$$

$$(x, y) \to \begin{bmatrix} u \\ v \end{bmatrix}$$

with

$$u(x, y) = x^2 y - y^3 \quad \text{and} \quad v(x, y) = 2x^3 y^2 + y$$

Compute the Jacobi matrix and evaluate it at $(1, 2)$.

Solution

The Jacobi matrix is

$$\begin{bmatrix} \dfrac{\partial u}{\partial x} & \dfrac{\partial u}{\partial y} \\ \dfrac{\partial v}{\partial x} & \dfrac{\partial v}{\partial y} \end{bmatrix} = \begin{bmatrix} 2xy & x^2 - 3y^2 \\ 6x^2 y^2 & 4x^3 y + 1 \end{bmatrix}$$

At $(1, 2)$ we find

$$\begin{bmatrix} 4 & -11 \\ 24 & 9 \end{bmatrix}$$

▶ **Example 9** Let

$$\mathbf{f} : \mathbf{R}^2 \to \mathbf{R}^2$$

$$(x, y) \to \begin{bmatrix} u \\ v \end{bmatrix}$$

with

$$u(x, y) = ye^{-x} \quad \text{and} \quad v(x, y) = \sin x + \cos y$$

Find the linear approximation to $\mathbf{f}(x, y)$ at $(0, 0)$. Compare $\mathbf{f}(0.1, -0.1)$ to its linear approximation.

Solution

We compute the Jacobi matrix first,

$$(D\mathbf{f})(x, y) = \begin{bmatrix} -ye^{-x} & e^{-x} \\ \cos x & -\sin y \end{bmatrix}$$

The linear approximation of $\mathbf{f}(x, y)$ at $(0, 0)$ is

$$\mathbf{L}(x, y) = \begin{bmatrix} \alpha(x, y) \\ \beta(x, y) \end{bmatrix} = \begin{bmatrix} u(0, 0) \\ v(0, 0) \end{bmatrix} + (D\mathbf{f})(0, 0) \begin{bmatrix} x - 0 \\ y - 0 \end{bmatrix}$$

$$= \begin{bmatrix} 0 \\ 1 \end{bmatrix} + \begin{bmatrix} 0 & 1 \\ 1 & 0 \end{bmatrix} \begin{bmatrix} x \\ y \end{bmatrix}$$

$$= \begin{bmatrix} 0 \\ 1 \end{bmatrix} + \begin{bmatrix} y \\ x \end{bmatrix}$$

Using $(x, y) = (0.1, -0.1)$, we find

$$\mathbf{L}(0.1, -0.1) = \begin{bmatrix} 0 \\ 1 \end{bmatrix} + \begin{bmatrix} -0.1 \\ 0.1 \end{bmatrix} = \begin{bmatrix} -0.1 \\ 1.1 \end{bmatrix}$$

and

$$\mathbf{f}(0.1, -0.1) = \begin{bmatrix} -0.1e^{-0.1} \\ \sin 0.1 + \cos(-0.1) \end{bmatrix} = \begin{bmatrix} -0.09 \\ 1.09 \end{bmatrix}$$

We see that the linear approximation is close to the actual value. ◀

We can generalize the Jacobi matrix to functions $\mathbf{f} : \mathbf{R}^n \to \mathbf{R}^m$. If

$$\mathbf{f}(x_1, x_2, \ldots, x_n) = \begin{bmatrix} f_1(x_1, x_2, \ldots, x_n) \\ f_2(x_1, x_2, \ldots, x_n) \\ \vdots \\ f_m(x_1, x_2, \ldots, x_n) \end{bmatrix}$$

where $f_i : \mathbf{R}^n \to \mathbf{R}, i = 1, 2, \ldots, n$, are real-valued functions of n independent variables, then the Jacobi matrix is an $m \times n$ matrix of the form

$$D\mathbf{f} = \begin{bmatrix} \dfrac{\partial f_1}{\partial x_1} & \cdots & \dfrac{\partial f_1}{\partial x_n} \\ \vdots & & \vdots \\ \dfrac{\partial f_m}{\partial x_1} & \cdots & \dfrac{\partial f_m}{\partial x_n} \end{bmatrix}$$

The linearization of $\mathbf{f}$ about the point $(x_1^\star, x_2^\star, \ldots, x_n^\star)$ is then

$$L(x_1^\star, \ldots, x_n^\star) = \begin{bmatrix} f_1(x_1^\star, \ldots, x_n^\star) \\ f_2(x_1^\star, \ldots, x_n^\star) \\ \vdots \\ f_m(x_1^\star, \ldots, x_n^\star) \end{bmatrix} + D\mathbf{f}(x_1^\star, \ldots, x_n^\star) \begin{bmatrix} x_1 - x_1^\star \\ \vdots \\ x_n - x_n^\star \end{bmatrix}$$

10.4.3 Problems

(10.4.1)

In Problems 1–6, the tangent plane at the indicated point (x_0, y_0, z_0) exists. Find its equation.

1. $f(x, y) = 2x^3 + y^2; (1, 2, 6)$
2. $f(x, y) = \sin x + \cos y; (0, 0, 1)$
3. $f(x, y) = e^{x^2 + y^2}; (1, 0, e)$
4. $f(x, y) = \ln(x^2 + y^2); (1, 1, \ln 2)$
5. $f(x, y) = \sin(xy); (1, 0, 0)$
6. $f(x, y) = e^x \cos y; (0, 0, 1)$

In Problems 7–12, show that $f(x, y)$ is differentiable at the indicated point.

7. $f(x, y) = y^2 x + x^2 y; (1, 1)$

8. $f(x, y) = xy - 3x^2; (1, 1)$
9. $f(x, y) = \cos(x + y); (0, 0)$
10. $f(x, y) = e^{x-y}; (0, 0)$
11. $f(x, y) = x + y^2 - 2xy; (-1, 2)$
12. $f(x, y) = \tan(x^2 + y^2); \left(\dfrac{\pi}{4}, -\dfrac{\pi}{4}\right)$

In Problems 13–18, find the linearization of $f(x, y)$ at the indicated point (x_0, y_0).

13. $f(x, y) = \sqrt{x} + 2y; (1, 0)$
14. $f(x, y) = \cos(x^2 y); \left(\dfrac{\pi}{2}, 0\right)$
15. $f(x, y) = \tan(x + y); (0, 0)$

16. $f(x, y) = e^{3x+2y}$; (1, 2)
17. $f(x, y) = \ln(x^2 + y)$; (1, 1)
18. $f(x, y) = x^2 e^y$; (1, 0)
19. Find the linear approximation of
$$f(x, y) = e^{x+y}$$
at (0, 0) and use it to approximate $f(0.1, 0.05)$. Compare the approximation to $f(0.1, 0.05)$ using a calculator.
20. Find the linear approximation of
$$f(x, y) = \sin(x + 2y)$$
at (0, 0) and use it to approximate $f(-0.1, 0.2)$. Compare the approximation to $f(-0.1, 0.2)$ using a calculator.
21. Find the linear approximation of
$$f(x, y) = \ln(x^2 - 3y)$$
at (1, 0) and use it to approximate $f(1.1, 0.1)$. Compare the approximation to $f(1.1, 0.1)$ using a calculator.
22. Find the linear approximation of
$$f(x, y) = \tan(2x - 3y^2)$$
at (0, 0) and use it to approximate $f(0.03, 0.05)$. Compare the approximation to $f(0.03, 0.05)$ using a calculator.

(10.4.2)

In Problems 23–30, find the Jacobi matrix for each given function.

23. $\mathbf{f}(x, y) = \begin{bmatrix} x + y \\ x^2 - y^2 \end{bmatrix}$

24. $\mathbf{f}(x, y) = \begin{bmatrix} 2x - y \\ x^2 \end{bmatrix}$

25. $\mathbf{f}(x, y) = \begin{bmatrix} e^{x-y} \\ e^{x+y} \end{bmatrix}$

26. $\mathbf{f}(x, y) = \begin{bmatrix} (x - y)^2 \\ \sin(x + y) \end{bmatrix}$

27. $\mathbf{f}(x, y) = \begin{bmatrix} \cos(x - y) \\ \cos(x + y) \end{bmatrix}$

28. $\mathbf{f}(x, y) = \begin{bmatrix} \ln(x + y) \\ e^{x+y} \end{bmatrix}$

29. $\mathbf{f}(x, y) = \begin{bmatrix} 2x^2 y - 3y + x \\ e^x \sin y \end{bmatrix}$

30. $\mathbf{f}(x, y) = \begin{bmatrix} \sqrt{x^2 + y^2} \\ e^{-x^2} \end{bmatrix}$

In Problems 31–36, find a linear approximation to each function $f(x, y)$ at the indicated point.

31. $\mathbf{f}(x, y) = \begin{bmatrix} 2x^2 y \\ \frac{1}{xy} \end{bmatrix}$ at (1, 1)

32. $\mathbf{f}(x, y) = \begin{bmatrix} 3x - y^2 \\ 4y \end{bmatrix}$ at (-1, -2)

33. $\mathbf{f}(x, y) = \begin{bmatrix} e^{2x-y} \\ \ln(2x - y) \end{bmatrix}$ at (1, 1)

34. $\mathbf{f}(x, y) = \begin{bmatrix} e^x \sin y \\ e^{-y} \cos x \end{bmatrix}$ at (0, 0)

35. $\mathbf{f}(x, y) = \begin{bmatrix} \frac{x}{y} \\ \frac{y}{x} \end{bmatrix}$ at (1, 1)

36. $\mathbf{f}(x, y) = \begin{bmatrix} (x + y)^2 \\ xy \end{bmatrix}$ at (-1, 1)

37. Find a linear approximation to
$$\mathbf{f}(x, y) = \begin{bmatrix} x^2 - xy \\ 3y^2 - 1 \end{bmatrix}$$
at (1, 2). Use your result to find an approximation for $f(1.1, 1.9)$ and compare the approximation to the value of $f(1.1, 1.9)$ when you use a calculator.

38. Find a linear approximation to
$$\mathbf{f}(x, y) = \begin{bmatrix} x/y \\ 2xy \end{bmatrix}$$
at (-1, 1). Use your result to find an approximation for $f(-0.9, 1.05)$ and compare the approximation to the value of $f(-0.9, 1.05)$ when you use a calculator.

39. Find a linear approximation to
$$\mathbf{f}(x, y) = \begin{bmatrix} (x - y)^2 \\ 2x^2 y \end{bmatrix}$$
at (2, -3). Use your result to find an approximation for $f(1.9, -3.1)$ and compare the approximation to the value of $f(1.9, -3.1)$ when you use a calculator.

40. Find a linear approximation to
$$\mathbf{f}(x, y) = \begin{bmatrix} \sqrt{2x + y} \\ x - y^2 \end{bmatrix}$$
at (1, 2). Use your result to find an approximation for $f(1.05, 2.05)$ and compare the approximation to the value of $f(1.05, 2.05)$ when you use a calculator.

10.5 MORE ABOUT DERIVATIVES (OPTIONAL)

10.5.1 The Chain Rule for Functions of Two Variables

In Section 10.3, we discussed how the net assimilation of CO_2 can change as a function of both temperature and light intensity. If we follow net assimilation of CO_2 over time, we must take into account that both temperature and light intensity depend on time. If we denote by $T(t)$ the temperature at time t, by $I(t)$ the light intensity at time t, and by $N(t)$ the net assimilation of CO_2 at time t, then $N(t)$ is a function of both $T(t)$ and $I(t)$; we can write

$$N(t) = f(T(t), I(t))$$

Net assimilation is thus a composite function.

To differentiate composite functions of one variable, we use the chain rule. Namely, suppose that $w = f(x)$ is a function of one variable and that x depends on t. To differentiate w with respect to t, the chain rule says that

$$\frac{dw}{dt} = \frac{dw}{dx}\frac{dx}{dt} \qquad (10.12)$$

The chain rule can be extended to functions of more than one variable.

CHAIN RULE FOR FUNCTIONS OF TWO INDEPENDENT VARIABLES

> If $w = f(x, y)$ is differentiable and x and y are differentiable functions of t, then w is a differentiable function of t and
> $$\frac{dw}{dt} = \frac{\partial w}{\partial x}\frac{dx}{dt} + \frac{\partial w}{\partial y}\frac{dy}{dt}$$

We will not prove this, but merely outline the steps that lead to this formula. We approximate $w = f(x, y)$ at (x_0, y_0) by its linear approximation

$$L(x, y) = f(x_0, y_0) + \frac{\partial f(x_0, y_0)}{\partial x}(x - x_0) + \frac{\partial f(x_0, y_0)}{\partial y}(y - y_0)$$

If we set $\Delta x = x - x_0$, $\Delta y = y - y_0$, and $\Delta w = f(x, y) - f(x_0, y_0)$, we can approximate $\Delta w = f(x_0+\Delta x, y_0+\Delta y) - f(x_0, y_0)$ by its linear approximation. We find

$$\Delta w \approx \frac{\partial f(x_0, y_0)}{\partial x}\Delta x + \frac{\partial f(x_0, y_0)}{\partial y}\Delta y$$

Dividing both sides by Δt, we find

$$\frac{\Delta w}{\Delta t} \approx \frac{\partial f(x_0, y_0)}{\partial x}\frac{\Delta x}{\Delta t} + \frac{\partial f(x_0, y_0)}{\partial y}\frac{\Delta y}{\Delta t}$$

If we let $\Delta t \to 0$, then

$$\frac{\Delta w}{\Delta t} \to \frac{dw}{dt}, \qquad \frac{\Delta x}{\Delta t} \to \frac{dx}{dt}, \qquad \frac{\Delta y}{\Delta t} \to \frac{dy}{dt}$$

and we find

$$\frac{dw}{dt} = \frac{\partial f(x_0, y_0)}{\partial x}\frac{dx}{dt} + \frac{\partial f(x_0, y_0)}{\partial y}\frac{dy}{dt}$$

or, in short,

$$\frac{dw}{dt} = \frac{\partial w}{\partial x}\frac{dx}{dt} + \frac{\partial w}{\partial y}\frac{dy}{dt}$$

▶ **Example 1** Let

$$w = f(x, y) = x^2 y^3$$

with $x(t) = \sin t$ and $y(t) = e^{-t}$. Find the derivative of $w = f(x, y)$ with respect to t when $t = \pi/2$.

Solution

Using the chain rule, we find

$$\frac{dw}{dt} = \frac{\partial w}{\partial x}\frac{dx}{dt} + \frac{\partial w}{\partial y}\frac{dy}{dt}$$

$$= 2xy^3\frac{dx}{dt} + x^2 3y^2\frac{dy}{dt}$$

$$= 2xy^3\cos t + 3x^2 y^2(-1)e^{-t}$$

Now, $\cos\frac{\pi}{2} = 0$,

$$x\left(\frac{\pi}{2}\right) = \sin\frac{\pi}{2} = 1 \quad\text{and}\quad y\left(\frac{\pi}{2}\right) = e^{-\pi/2}$$

Hence,

$$\left.\frac{dw}{dt}\right|_{t=\pi/2} = (2)(1)(e^{-3\pi/2})(0) - (3)(1)^2 e^{-\pi} e^{-\pi/2} = -3e^{-3\pi/2}$$

We can check the answer directly; namely,

$$w(t) = (\sin^2 t)(e^{-3t})$$

and, therefore,

$$\left.\frac{dw}{dt}\right|_{t=\pi/2} = 2(\sin t)(\cos t)e^{-3t} - 3(\sin^2 t)e^{-3t}\bigg|_{t=\pi/2} = -3e^{-3\pi/2} \quad \blacktriangleleft$$

▶ **Example 2**   Suppose that we wish to predict the abundance of a particular plant species. We suspect that the two major factors influencing the abundance of this plant are nitrogen levels and disturbance levels. Previous studies have shown that an increase in nitrogen in the soil has a negative effect on the abundance of this species; also, an increase in disturbance due to grazing seems to have a negative effect on abundance. If both nitrogen and disturbance due to grazing increase over the next few years, how would the abundance of this species be affected?

Solution

We denote by $B(t)$ the abundance of the plant at time t. We are interested in how $B(t)$ will change over time; that is, we want to find out whether $B(t)$ will increase or decrease over time. For this, we need to compute the derivative of $B(t)$. If we assume that the abundance of the plant is primarily affected by nitrogen and disturbance levels, we can consider B as a function of N, the nitrogen level, and D, the disturbance level. Both N and D change over time and are thus functions of time. The function $B(t)$ is therefore a function of both $N(t)$ and $D(t)$. Using the chain rule for functions of two independent variables, we find

$$\frac{dB}{dt} = \underbrace{\frac{\partial B}{\partial N}}_{<0}\underbrace{\frac{dN}{dt}}_{>0} + \underbrace{\frac{\partial B}{\partial D}}_{<0}\underbrace{\frac{dD}{dt}}_{>0} < 0$$

since abundance B is a decreasing function of both N and D, and both nitrogen and disturbance levels are assumed to increase over the next few years. We thus find that the abundance of the plant will decrease over the next few years. $\quad\blacktriangleleft$

10.5.2 Implicit Differentiation

We discussed implicit differentiation in Section 4.4. This was a useful technique for differentiating a function $y = f(x)$ when y was given implicitly, such as in

$$x^2 y - e^{-y} = 0 \tag{10.13}$$

To find dy/dx, we differentiate both sides with respect to x, keeping in mind that y is a function of x. We obtain

$$2xy + x^2\frac{dy}{dx} - e^{-y}\left(-\frac{dy}{dx}\right) = 0$$

$$2xy + \frac{dy}{dx}(x^2 + e^{-y}) = 0$$

Solving for dy/dx, we find

$$\frac{dy}{dx} = -\frac{2xy}{x^2 + e^{-y}} \tag{10.14}$$

When we look at (10.13), we can define a function $F(x, y)$ as

$$F(x, y) = x^2 y - e^{-y} \quad \text{with } F(x, y) = 0$$

$F(x, y)$ is a function of two variables. We say that the equation $F(x, y) = 0$ defines y implicitly as a function of x.

We will turn to the general case to see why this is a useful way to look at this problem. We think of y as a function of x and define a function $F(x, y)$ such that $F(x, y) = 0$. This defines y implicitly as a function of x. To find the derivative of y with respect to x, we set

$$w = F(u, v) \quad \text{with } u(x) = x \text{ and } v(x) = y$$

This makes w a function of x, that is, $w = w(x)$. We can use the chain rule to differentiate w with respect to x, and we find

$$\frac{dw}{dx} = \frac{\partial F}{\partial u} \frac{du}{dx} + \frac{\partial F}{\partial v} \frac{dv}{dx}$$

Since $\frac{du}{dx} = 1$ and $\frac{dv}{dx} = \frac{dy}{dx}$, we obtain, with $u = x$ and $v = y$,

$$\frac{dw}{dx} = \frac{\partial F}{\partial x} + \frac{\partial F}{\partial y} \frac{dy}{dx} \tag{10.15}$$

Since $F(x, y)$ satisfies $F(x, y) = 0$, we find $w(x) = 0$ for all values of x and, therefore,

$$\frac{dw}{dx} = 0 \tag{10.16}$$

Equating (10.15) and (10.16), we obtain

$$0 = \frac{\partial F}{\partial x} + \frac{\partial F}{\partial y} \frac{dy}{dx}$$

We can solve this for dy/dx,

$$\frac{dy}{dx} = -\frac{\dfrac{\partial F}{\partial x}}{\dfrac{\partial F}{\partial y}} = -\frac{F_x}{F_y}$$

provided that $\partial F/\partial y \neq 0$. We summarize this in the following.

> Suppose that $w = F(x, y)$ is differentiable, and $F(x, y) = 0$ defines y implicitly as a differentiable function of x. Then, at any point where $F_y \neq 0$,
>
> $$\frac{dy}{dx} = -\frac{F_x}{F_y}$$

▶ **Example 3** Find dy/dx if $x^2 y - e^{-y} = 0$.

Solution

We set $F(x, y) = x^2 y - e^{-y}$. We need to find F_x and F_y:

$$F_x = \frac{\partial F}{\partial x} = 2xy$$

$$F_y = \frac{\partial F}{\partial y} = x^2 + e^{-y}$$

Then, since $F(x, y) = 0$,

$$\frac{dy}{dx} = -\frac{F_x}{F_y} = -\frac{2xy}{x^2 + e^{-y}}$$

as in (10.14). Since $x^2 + e^{-y} \neq 0$ for all $(x, y) \in \mathbf{R}^2$, dy/dx is defined for all $(x, y) \in \mathbf{R}^2$ with $F(x, y) = 0$. ◀

The following example shows that we can use this rule to find the derivatives of inverse trigonometric functions.

▶ **Example 4** Find dy/dx for

$$y = \arcsin x$$

Solution

We already know the answer, since we discussed the derivative of $\arcsin x$ earlier, in the context of inverse functions. But let's see how we can use the results of this section to find the answer. Since

$$x = \sin y \quad \text{for} \quad -\frac{\pi}{2} \leq y \leq \frac{\pi}{2}$$

is equivalent to

$$y = \arcsin x \quad \text{for} \quad -1 \leq x \leq 1$$

we can define a function $F(x, y)$ that satisfies $F(x, y) = 0$ and defines y implicitly as a function of x, namely,

$$F(x, y) = x - \sin y$$

Then

$$\frac{dy}{dx} = -\frac{F_x}{F_y} = -\frac{1}{-\cos y} = \frac{1}{\cos y}$$

Since $\sin^2 y + \cos^2 y = 1$, we find

$$\cos y = \sqrt{1 - \sin^2 y} = \sqrt{1 - x^2}$$

Here we must use the fact that $x = \sin y$ is defined for $-\pi/2 \leq y \leq \pi/2$ and $\cos y \geq 0$ for y in this interval. Using this representation for $\cos y$, we find

$$\frac{dy}{dx} = \frac{1}{\sqrt{1 - x^2}}$$

which is defined for $-1 < x < 1$. ◀

10.5.3 Directional Derivatives and Gradient Vectors

Suppose that you are on a sloped surface, like a hillside. Depending on which direction you walk, you must either go uphill, stay at the same level, or go downhill. That is, by choosing a particular direction, you have some control over the steepness of your path. How steep your path is can be described by the slope of the tangent line at your starting point in the direction of your path. This slope is given by the **directional derivative**.

We assume that $z = f(x, y)$ is a differentiable function of two independent variables. We choose a point (x_0, y_0, z_0) with $z_0 = f(x_0, y_0)$ on the surface defined by $z = f(x, y)$. To define the slope of a tangent line at (x_0, y_0, z_0), we must specify a direction in which we wish to go. We know how to deal with this problem when we go in the x- or y-direction. In these

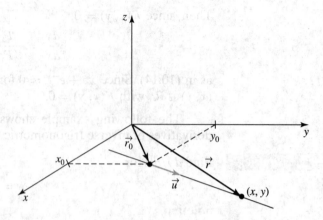

▲ Figure 10.39
Going in the direction of **u** from (x_0, y_0)

cases, the partial derivatives $\partial f/\partial x$ and $\partial f/\partial y$ tell us how $f(x, y)$ changes. We will now explain how we can express the slope when we choose an arbitrary direction.

The first step is to find a way to express what we mean by "going in a certain direction." We start at a point (x_0, y_0) and wish to go in the direction of a unit vector $\mathbf{u} = \begin{bmatrix} u_1 \\ u_2 \end{bmatrix}$ (recall that a unit vector has length 1). This is illustrated in Figure 10.39. From Figure 10.39, we see that

$$\mathbf{r} = \mathbf{r_0} + t\mathbf{u} \tag{10.17}$$

where $\mathbf{r_0} = \begin{bmatrix} x_0 \\ y_0 \end{bmatrix}$, $\mathbf{r} = \begin{bmatrix} x \\ y \end{bmatrix}$, and t is a real number; different values of t get us to different points on the straight line through (x_0, y_0) that points in the direction of **u**. We can write (10.17) also as

$$\begin{bmatrix} x \\ y \end{bmatrix} = \begin{bmatrix} x_0 \\ y_0 \end{bmatrix} + \begin{bmatrix} tu_1 \\ tu_2 \end{bmatrix} \quad \text{for } t \in \mathbf{R} \tag{10.18}$$

Equation (10.18) is called the parametric equation of a line, which we introduced in Section 9.4; t is called the parameter. Since $x = x_0 + tu_1$ and $y = y_0 + tu_2$, we find

$$\frac{dx}{dt} = u_1 \quad \text{and} \quad \frac{dy}{dt} = u_2 \tag{10.19}$$

We can now find out what happens to $f(x, y)$ when we start at (x_0, y_0) and go in the direction of the unit vector **u**, as illustrated in Figure 10.40. For this, we use the parametric equation (10.18) of the line through (x_0, y_0) that is oriented in the direction of **u**:

$$x = x_0 + tu_1 \quad \text{and} \quad y = y_0 + tu_2$$

Since $z(t) = f(x(t), y(t))$, we can use the chain rule to find out how f changes when we vary t (that is, move along the straight line),

$$\frac{dz}{dt} = \frac{\partial f}{\partial x}\frac{dx}{dt} + \frac{\partial f}{\partial y}\frac{dy}{dt} = \frac{\partial f}{\partial x}u_1 + \frac{\partial f}{\partial y}u_2 \tag{10.20}$$

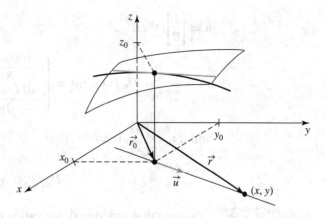

▲ **Figure 10.40**
An illustration of the directional derivative

where we used (10.19) in the last step. Equation (10.20) can be written as a dot product:

$$\frac{dz}{dt} = \begin{bmatrix} \dfrac{\partial f}{\partial x} \\ \dfrac{\partial f}{\partial y} \end{bmatrix} \cdot \begin{bmatrix} u_1 \\ u_2 \end{bmatrix}$$

The first vector in the dot product is called the **gradient** of f.

> **Definition** Assume that $z = f(x, y)$ is a function of two independent variables, and $\partial f/\partial x$ and $\partial f/\partial y$ exist. Then the vector
>
> $$\nabla f(x, y) = \begin{bmatrix} \dfrac{\partial f(x, y)}{\partial x} \\ \dfrac{\partial f(x, y)}{\partial y} \end{bmatrix}$$
>
> is called the gradient of f at (x, y).

The notation ∇f is read "grad f" or "gradient of f." The symbol ∇ is called "del," so you can also say "del f." An alternative notation is grad f.
We can now define the derivative of f in a particular direction.

> **Definition** The directional derivative of $f(x, y)$ at (x_0, y_0) in the direction of the unit vector $\mathbf{u} = \begin{bmatrix} u_1 \\ u_2 \end{bmatrix}$ is
>
> $$D_{\mathbf{u}} f(x_0, y_0) = (\nabla f(x_0, y_0)) \cdot \mathbf{u}$$

Note that in the definition of the directional derivative we assume that $\mathbf{u}$ is a *unit vector*. Choosing a unit vector (as opposed to a vector of some other length) ensures that the directional derivative of $f(x, y)$ agrees with the partial derivatives when we go along the positive x- or y-axis. That is, if

$$\mathbf{u} = \begin{bmatrix} 1 \\ 0 \end{bmatrix}, \text{ then}$$

$$D_{\mathbf{u}} f(x_0, y_0) = \begin{bmatrix} \dfrac{\partial f(x_0, y_0)}{\partial x} \\ \dfrac{\partial f(x_0, y_0)}{\partial y} \end{bmatrix} \cdot \begin{bmatrix} 1 \\ 0 \end{bmatrix} = \dfrac{\partial f(x_0, y_0)}{\partial x}$$

and if $\mathbf{u} = \begin{bmatrix} 0 \\ 1 \end{bmatrix}$, then

$$D_{\mathbf{u}} f(x_0, y_0) = \dfrac{\partial f(x_0, y_0)}{\partial y}$$

▶ **Example 5** Compute the directional derivative of

$$f(x, y) = \sqrt{x^2 + 2y^2}$$

at the point $(-1, 2)$ in the direction of $\begin{bmatrix} -1 \\ 3 \end{bmatrix}$.

Solution

We first compute the gradient vector

$$\nabla f(x, y) = \begin{bmatrix} \dfrac{x}{\sqrt{x^2 + 2y^2}} \\ \dfrac{2y}{\sqrt{x^2 + 2y^2}} \end{bmatrix}$$

Evaluating this at $(-1, 2)$, we find

$$\nabla f(-1, 2) = \begin{bmatrix} \dfrac{-1}{\sqrt{1+8}} \\ \dfrac{4}{\sqrt{1+8}} \end{bmatrix} = \begin{bmatrix} -1/3 \\ 4/3 \end{bmatrix}$$

Since $\begin{bmatrix} -1 \\ 3 \end{bmatrix}$ is not a unit vector, we normalize it first. The vector $\begin{bmatrix} -1 \\ 3 \end{bmatrix}$ has length $\sqrt{1+9} = \sqrt{10}$; hence

$$\mathbf{u} = \dfrac{1}{\sqrt{10}} \begin{bmatrix} -1 \\ 3 \end{bmatrix}$$

and

$$D_{\mathbf{u}} f(-1, 2) = (\nabla f(-1, 2)) \cdot \mathbf{u}$$

$$= \begin{bmatrix} -\dfrac{1}{3} \\ \dfrac{4}{3} \end{bmatrix} \cdot \begin{bmatrix} -\dfrac{1}{\sqrt{10}} \\ \dfrac{3}{\sqrt{10}} \end{bmatrix}$$

$$= \dfrac{1}{3\sqrt{10}} + \dfrac{12}{3\sqrt{10}} = \dfrac{13}{3\sqrt{10}}$$

◀

▶ **Example 6** Compute the directional derivative of

$$f(x, y) = x^2 y - 2y^2$$

at the point $(-3, 2)$ in the direction of $(-1, 1)$.

Solution

We first compute the gradient vector

$$\nabla f(x, y) = \begin{bmatrix} 2xy \\ x^2 - 4y \end{bmatrix}$$

Evaluating this at $(-3, 2)$, we find

$$\nabla f(-3, 2) = \begin{bmatrix} (2)(-3)(2) \\ (-3)^2 - (4)(2) \end{bmatrix} = \begin{bmatrix} -12 \\ 1 \end{bmatrix}$$

The vector that goes from $(-3, 2)$ to $(-1, 1)$ has the form

$$\begin{bmatrix} -1 - (-3) \\ 1 - 2 \end{bmatrix} = \begin{bmatrix} 2 \\ -1 \end{bmatrix}$$

This vector has length $\sqrt{4 + 1} = \sqrt{5}$. Normalizing this vector yields

$$\mathbf{u} = \frac{1}{\sqrt{5}} \begin{bmatrix} 2 \\ -1 \end{bmatrix}$$

and the directional derivative is

$$D_{\mathbf{u}} f(-3, 2) = (\nabla f(-3, 2)) \cdot \mathbf{u}$$

$$= \frac{1}{\sqrt{5}} \begin{bmatrix} -12 \\ 1 \end{bmatrix} \cdot \begin{bmatrix} 2 \\ -1 \end{bmatrix} = \frac{1}{\sqrt{5}}(-24 - 1) = -\frac{25}{\sqrt{5}} = -5\sqrt{5} \ \blacktriangleleft$$

Properties of the Gradient Vector The directional derivative is a dot product. We can therefore write

$$D_{\mathbf{u}} f(x, y) = (\nabla f(x, y)) \cdot \mathbf{u}$$

$$= |\nabla f(x, y)||\mathbf{u}| \cos \theta$$

where θ is the angle between ∇f and $\mathbf{u}$ (see Figure 10.41). Since $|\mathbf{u}| = 1$ ($\mathbf{u}$ is a unit vector), we find

$$D_{\mathbf{u}} f(x, y) = |\nabla f(x, y)| \cos \theta$$

The angle θ is in the interval $[0, 2\pi)$, and $\cos \theta$ is maximal when $\theta = 0$. We therefore find that $D_{\mathbf{u}} f(x, y)$ is maximal when $\mathbf{u}$ is in the direction of ∇f.

We will now show that geometrically, the gradient vector at a point (x_0, y_0) is perpendicular to the level curve $f(x, y) = c$ that passes through this point. The level curve $f(x, y) = c$ is a curve in the x-y plane. It will be useful to think of traveling on this curve starting at time $t = 0$ at a point labeled with $t = 0$. We can then refer to any point on the curve by giving the time t at which we pass through it. We say that we **parameterize** the curve using the parameter t. We write the curve as the vector vector $\mathbf{r}(t) = \begin{bmatrix} x(t) \\ y(t) \end{bmatrix}$, just as in Subsection 9.4.3 where we parameterized lines. If we differentiate $\mathbf{r}(t)$ with respect to t, we obtain

$$\frac{d}{dt} \mathbf{r}(t) = \begin{bmatrix} \dfrac{dx(t)}{dt} \\ \dfrac{dy(t)}{dt} \end{bmatrix}$$

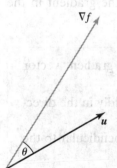

▲ **Figure 10.41**
The angle θ between ∇f and the unit vector $\mathbf{u}$

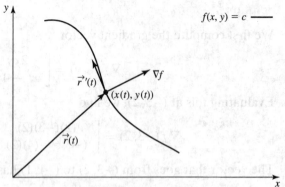

$f(x, y) = c$ ——

◀ Figure 10.42
The gradient is perpendicular
to the level curve

which is the tangent vector at the point $(x(t), y(t))$. Since

$$f(x, y) = c \qquad (10.21)$$

and x and y are both functions of t, we can write $w(t) = f(x(t), y(t)) = c$ and differentiate $w(t)$ with respect to t. We find $dw/dt = 0$, since $w(t)$ is a constant. Using the chain rule to differentiate $f(x(t), y(t))$ yields

$$\frac{d}{dt} f(x(t), y(t)) = \frac{\partial f}{\partial x} \frac{dx}{dt} + \frac{\partial f}{\partial y} \frac{dy}{dt}$$

Hence,

$$\frac{\partial f}{\partial x} \frac{dx}{dt} + \frac{\partial f}{\partial y} \frac{dy}{dt} = 0$$

or

$$\nabla f \cdot \frac{d\mathbf{r}}{dt} = 0$$

This shows that the gradient of f at (x_0, y_0) is perpendicular to the level curve at (x_0, y_0), as illustrated in Figure 10.42.

We summarize these two important properties of the gradient in the following.

Suppose that $f(x, y)$ is a differentiable function. The gradient vector $\nabla f(x, y)$ has the following properties.

1. At each point (x_0, y_0), $f(x, y)$ increases most rapidly in the direction of the gradient vector $\nabla f(x_0, y_0)$.

2. The gradient vector of f at a point (x_0, y_0) is perpendicular to the level curve through (x_0, y_0).

▶ Example 7 Let

$$f(x, y) = x^2 y + y^2$$

In what direction does $f(x, y)$ increase most rapidly at $(1, 1)$?

Solution

The function $f(x, y)$ increases most rapidly at $(1, 1)$ in the direction of $\nabla f(1, 1)$. Since

$$\nabla f(x, y) = \begin{bmatrix} \dfrac{\partial f}{\partial x} \\ \dfrac{\partial f}{\partial y} \end{bmatrix} = \begin{bmatrix} 2xy \\ x^2 + 2y \end{bmatrix}$$

it follows that

$$\nabla f(1, 1) = \begin{bmatrix} 2 \\ 3 \end{bmatrix}$$

That is, $f(x, y)$ increases most rapidly in the direction of $\begin{bmatrix} 2 \\ 3 \end{bmatrix}$ at the point $(1, 1)$.

▶ **Example 8** Find a unit vector that is perpendicular to the level curve of the function

$$f(x, y) = x^2 - y^2$$

at $(1, 2)$.

Solution

The gradient of f at $(1, 2)$ is perpendicular to the level curve at $(1, 2)$. The gradient of f is given as

$$\nabla f(x, y) = \begin{bmatrix} 2x \\ -2y \end{bmatrix}$$

Hence,

$$\nabla f(1, 2) = \begin{bmatrix} 2 \\ -4 \end{bmatrix}$$

To normalize this vector, we divide $\nabla f(1, 2)$ by its length. Since

$$|\nabla f(1, 2)| = \sqrt{(2)^2 + (-4)^2} = \sqrt{4 + 16} = 2\sqrt{5}$$

the unit vector that is perpendicular to the level curve of $f(x, y)$ at $(1, 2)$ is

$$\mathbf{u} = \frac{1}{2\sqrt{5}} \begin{bmatrix} 2 \\ -4 \end{bmatrix} = \frac{1}{\sqrt{5}} \begin{bmatrix} 1 \\ -2 \end{bmatrix} = \begin{bmatrix} \frac{1}{5}\sqrt{5} \\ -\frac{2}{5}\sqrt{5} \end{bmatrix}$$

10.5.4 Problems

(10.5.1)

1. Let $f(x, y) = x^2 + y^2$ with $x(t) = 3t$ and $y(t) = e^t$. Find the derivative of $w = f(x, y)$ with respect to t when $t = \ln 2$.

2. Let $f(x, y) = e^x \sin y$ with $x(t) = t$ and $y(t) = t^3$. Find the derivative of $w = f(x, y)$ with respect to t when $t = 1$.

3. Let $f(x, y) = \sqrt{x^2 + y^2}$ with $x(t) = t$ and $y(t) = \sin t$. Find the derivative of $w = f(x, y)$ with respect to t when $t = \pi/3$.

4. Let $f(x, y) = \ln(xy - x^2)$ with $x(t) = t^2$ and $y(t) = t$. Find the derivative of $w = f(x, y)$ with respect to t when $t = 5$.

5. Let $f(x, y) = \frac{1}{x} + \frac{1}{y}$ with $x(t) = \sin t$ and $y(t) = \cos t$. Find the derivative of $w = f(x, y)$ with respect to t when $t = \pi/4$.

6. Let $f(x, y) = xe^y$ with $x(t) = e^t$ and $y(t) = t^2$. Find the derivative of $w = f(x, y)$ with respect to t when $t = 0$.

7. Find $\frac{dz}{dt}$ for $z = f(x, y)$ with $x = u(t)$ and $y = v(t)$.

8. Find $\frac{dw}{dt}$ for $w = e^{f(x,y)}$ with $x = u(t)$ and $y = v(t)$.

(10.5.2)

9. Find $\frac{dy}{dx}$ if $(x^2 + y^2)e^y = 0$.

10. Find $\frac{dy}{dx}$ if $(\sin x + \cos y)x^2 = 0$.

11. Find $\frac{dy}{dx}$ if $\ln(x^2 + y^2) = 3xy$.

12. Find $\frac{dy}{dx}$ if $\cos(x^2 + y^2) = \sin(x^2 - y^2)$.

13. Find $\frac{dy}{dx}$ if $y = \arccos x$.

14. Find $\frac{dy}{dx}$ if $y = \arctan x$.

15. The growth rate r of a particular organism is affected by both the availability of food and the number of other competitors for the food source. Denote by $F(t)$ the amount of food available at time t and by $N(t)$ the number of competitors at time t. The growth rate r can then be thought of as a function of the two time-dependent variables $F(t)$ and $N(t)$. Assume that the growth rate is an increasing function of food availability and a decreasing function of the number of competitors. How is the growth rate r affected if the availability of food decreases over time while the number of competitors increases?

16. Suppose that you travel along an environmental gradient, along which both temperature and precipitation increase. If the abundance of a particular plant species increases with both temperature and precipitation, would you expect to encounter this species more or less often during your journey? (Use calculus to answer this question.)

(10.5.3)

In Problems 17–24, find the gradient of each function.

17. $f(x, y) = x^3 y^2$

18. $f(x, y) = \dfrac{xy}{x^2 + y^2}$

19. $f(x, y) = \sqrt{x^3 - 3xy}$

20. $f(x, y) = x(x^2 - y^2)^{2/3}$

21. $f(x, y) = \exp\left[\sqrt{x^2 + y^2}\right]$

22. $f(x, y) = \tan\dfrac{x - y}{x + y}$

23. $f(x, y) = \ln\left(\dfrac{x}{y} + \dfrac{y}{x}\right)$

24. $f(x, y) = \cos(3x^2 - 2y^2)$

In Problems 25–30, compute the directional derivative of $f(x, y)$ at the given point in the indicated direction.

25. $$f(x, y) = \sqrt{2x^2 + y^2}$$

at $(1, 2)$ in the direction of $\begin{bmatrix} 1 \\ 1 \end{bmatrix}$.

26. $$f(x, y) = x^2 \sin y$$

at $(-1, 0)$ in the direction of $\begin{bmatrix} 2 \\ -1 \end{bmatrix}$.

27. $$f(x, y) = e^{x+y}$$

at $(0, 0)$ in the direction of $\begin{bmatrix} -1 \\ -1 \end{bmatrix}$.

28. $$f(x, y) = x^3 y^2$$

at $(2, 3)$ in the direction of $\begin{bmatrix} -2 \\ 1 \end{bmatrix}$.

29. $$f(x, y) = 2xy^3 - 3x^2 y$$

at $(1, -1)$ in the direction of $\begin{bmatrix} 3 \\ 1 \end{bmatrix}$.

30. $$f(x, y) = ye^{x^2}$$

at $(0, 2)$ in the direction of $\begin{bmatrix} 4 \\ -1 \end{bmatrix}$.

In Problems 31–34, compute the directional derivative of $f(x, y)$ at the point P in the direction of the point Q.

31. $f(x, y) = 2x^2 y - 3x$, $P = (2, 1)$, $Q = (3, 2)$
32. $f(x, y) = 4xy + y^2$, $P = (-1, 1)$, $Q = (3, 2)$
33. $f(x, y) = \sqrt{xy - 2x^2}$, $P = (1, 6)$, $Q = (3, 1)$

34. $f(x, y) = e^{x-y}$, $P = (2, 2)$, $Q = (1, -1)$
35. In what direction does $f(x, y) = 3xy - x^2$ increase most rapidly at $(-1, 1)$?
36. In what direction does $f(x, y) = e^x \cos y$ increase most rapidly at $(0, \pi/2)$?
37. In what direction does $f(x, y) = \sqrt{x^2 - y^2}$ increase most rapidly at $(5, 3)$?
38. In what direction does $f(x, y) = \ln(x^2 + y^2)$ increase most rapidly at $(1, 1)$?
39. Find a unit vector that is normal to the level curve of the function

$$f(x, y) = 3x + 4y$$

at the point $(-1, 1)$.
40. Find a unit vector that is normal to the level curve of the function

$$f(x, y) = x^2 + \frac{y^2}{9}$$

at the point $(1, 3)$.
41. Find a unit vector that is normal to the level curve of the function

$$f(x, y) = x^2 - y^3$$

at the point $(1, 3)$.
42. Find a unit vector that is normal to the level curve of the function

$$f(x, y) = xy$$

at the point $(2, 3)$.
43. (*Chemotaxis*) Chemotaxis is the chemically directed movement of organisms up a concentration gradient, that is, in the direction in which the concentration increases most rapidly. The slime mold *Dictyostelium discoideum* exhibits this phenomenon. In this case, single-celled amoeba of this species move up the concentration gradient of a chemical called cyclic adenosine monophosphate (cyclic AMP). Suppose the concentration of cyclic AMP at the point (x, y) in the x-y plane is given by

$$f(x, y) = \frac{4}{|x| + |y| + 1}$$

If you place an amoeba at the point $(3, 1)$ in the x-y plane, determine in which direction the amoeba will move if its movement is directed by chemotaxis.
44. Suppose an organism moves down a sloped surface along the steepest line of descent. If the surface is given by

$$f(x, y) = x^2 - y^2$$

find the direction in which the organism will move at the point $(2, 3)$.

10.6 APPLICATIONS (OPTIONAL)

10.6.1 Maxima and Minima

In Section 5.1, we introduced local extrema for functions of one variable. Local extrema can also be defined for functions of more than one independent variable; here, we will restrict our discussion to functions of two variables. Recall that we denoted by $B_\delta(x_0, y_0)$ the open disk with radius δ centered at (x_0, y_0). The following definition, which you should compare to the corresponding definition in Section 5.1, extends the notion of local extrema to functions of two variables.

> **Definition** A function $f(x, y)$ defined on a set $D \subset \mathbf{R}^2$ has a **local (or relative) maximum** at a point (x_0, y_0) if there exists a $\delta > 0$ so that
>
> $$f(x, y) \leq f(x_0, y_0) \quad \text{for all } (x, y) \in B_\delta(x_0, y_0) \cap D$$
>
> A function $f(x, y)$ defined on a set $D \subset \mathbf{R}^2$ has a **local (or relative) minimum** at a point (x_0, y_0) if there exists a $\delta > 0$ so that
>
> $$f(x, y) \geq f(x_0, y_0) \quad \text{for all } (x, y) \in B_\delta(x_0, y_0) \cap D$$

Informally, a local maximum (respectively, a local minimum) is a point that is higher (respectively, lower) than all nearby points. We can define **global (or absolute)** extrema as well. Namely, if the inequalities in the definition hold for all $(x, y) \in D$, then f has an absolute maximum (respectively, minimum) at (x_0, y_0). Figure 10.43 shows an example of a function of two variables with a local maximum at $(0, 0)$.

How can we find local extrema? Recall that in the single-variable case, a horizontal tangent line at a point on the graph of a differentiable function is a necessary condition for the point to be a local extremum (Fermat's theorem). This can be generalized to functions of more than one variable: Looking at Figure 10.43, we see that the tangent plane at the local extremum is horizontal. The equation of a horizontal tangent plane on the graph of a differentiable function $f(x, y)$ at (x_0, y_0) is

$$z = f(x_0, y_0)$$

Comparing this with the general form of a tangent plane (Section 10.4), we see that both $\partial f / \partial x$ and $\partial f / \partial y$ at (x_0, y_0) are equal to 0, or, in other words,

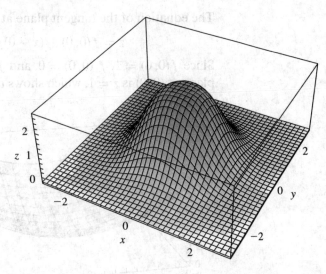

▲ **Figure 10.43**
The graph of a function $f(x, y)$ with a local maximum at $(0, 0)$

$\nabla f(x_0, y_0) = \begin{bmatrix} 0 \\ 0 \end{bmatrix}$. Another way to see this is to argue that if the gradient were nonzero at (x_0, y_0), then the function would increase in the gradient direction and decrease in the opposite direction, so we could not be at a local extremum. We thus have

> If $f(x, y)$ has a local extremum at (x_0, y_0) and if f is differentiable at (x_0, y_0), then
>
> $$\nabla f(x_0, y_0) = \begin{bmatrix} 0 \\ 0 \end{bmatrix} \qquad (10.22)$$

A point (x_0, y_0) that satisfies (10.22) is called a **critical point**; points where $f(x, y)$ is not differentiable are also called critical points. We wish to emphasize that, as in Section 5.1, (10.22) is a *necessary* condition: It only identifies *candidates* for local extrema. Other critical points are also just candidates for local extrema. A further investigation is then needed to determine whether a candidate is indeed a local extremum.

▶ **Example 1**   Figure 10.44 shows the graph of the differentiable function $f(x, y) = x^2 + y^2 + 1$, $(x, y) \in \mathbf{R}^2$. We see that $f(x, y)$ has a local minimum at $(0, 0)$. Show that $\nabla f(0, 0) = \begin{bmatrix} 0 \\ 0 \end{bmatrix}$ and determine the equation of the tangent plane at $(0, 0)$.

Solution

We compute

$$\nabla f(x, y) = \begin{bmatrix} 2x \\ 2y \end{bmatrix}$$

Evaluating $\nabla f(0, 0)$ at $(0, 0)$, we find

$$\nabla f(0, 0) = \begin{bmatrix} 0 \\ 0 \end{bmatrix}$$

The equation of the tangent plane at $(0, 0)$ is given by

$$z = f(0, 0) + (x - 0) f_x(0, 0) + (y - 0) f_y(0, 0)$$

Since $f(0, 0) = 1$, $f_x(0, 0) = 0$, and $f_y(0, 0) = 0$, the equation of the tangent plane at $(0, 0)$ is $z = 1$, which shows that the tangent plane is horizontal.   ◀

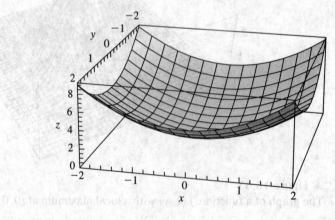

◀ **Figure 10.44**
The graph of the function
$f(x, y) = x^2 + y^2 + 1$

▶ **Example 2** Find all critical points of

$$f(x, y) = x^2 + y^2 + xy, \quad (x, y) \in \mathbf{R}^2$$

Solution

Since the function $f(x, y)$ is differentiable in $\mathbf{R}^2$, the only critical points are points that satisfy $\nabla f(x, y) = \begin{bmatrix} 0 \\ 0 \end{bmatrix}$. Now,

$$\nabla f(x, y) = \begin{bmatrix} 2x + y \\ 2y + x \end{bmatrix} = \begin{bmatrix} 0 \\ 0 \end{bmatrix}$$

We need to solve the system of linear equations

$$2x + y = 0$$
$$x + 2y = 0$$

It follows from the first equation that $y = -2x$. Substituting this into the second equation yields

$$x + 2(-2x) = 0 \quad \text{or} \quad -3x = 0 \quad \text{or} \quad x = 0$$

and, therefore, $y = 0$. The function has one critical point, namely $(0, 0)$. ◀

We will now give a sufficient condition that will allow us to determine whether a candidate for a local extremum is indeed a local extremum and, if so, whether it is a local maximum or a local minimum. The proof of this condition is beyond the scope of this book.

Recall that in the case of a function of one variable, we obtained the following sufficient condition for twice differentiable functions: If $f'(x_0) = 0$ and $f''(x_0) > 0$ [respectively, $f''(x_0) < 0$], then $f(x)$ has a local minimum (respectively, a local maximum) at $x = x_0$. In the multivariable case, there is an analogous condition involving second partial derivatives.

Theorem Suppose the second partial derivatives of f are continuous in a disk centered at (x_0, y_0) and that $\nabla f(x_0, y_0) = \begin{bmatrix} 0 \\ 0 \end{bmatrix}$. Define

$$D = f_{xx}(x_0, y_0) f_{yy}(x_0, y_0) - (f_{xy}(x_0, y_0))^2$$

1. If $D > 0$ and $f_{xx}(x_0, y_0) > 0$, then f has a local minimum at (x_0, y_0).
2. If $D > 0$ and $f_{xx}(x_0, y_0) < 0$, then f has a local maximum at (x_0, y_0).
3. If $D < 0$, then f does not have a local extremum at (x_0, y_0). The point (x_0, y_0) is then called a **saddle**.

In all other cases, the test is inconclusive. We now return to Example 2 and determine whether $(0, 0)$ is a local extremum.

▶ **Example 2** **(continued)** Determine whether the critical point $(0, 0)$ of $f(x, y) = x^2 + y^2 + xy$ in Example 2 is a local maximum or a local minimum (Figure 10.45).

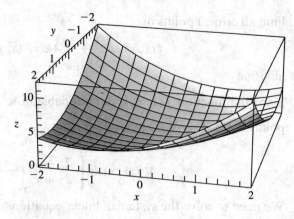

◀ **Figure 10.45**
The graph of the function
$f(x, y) = x^2 + y^2 + xy$

Solution

We need to find all second partial derivatives. Since

$$\frac{\partial f(x, y)}{\partial x} = 2x + y \quad \text{and} \quad \frac{\partial f(x, y)}{\partial y} = x + 2y$$

we have

$$\frac{\partial^2 f}{\partial x^2} = 2, \qquad \frac{\partial^2 f}{\partial y^2} = 2, \qquad \frac{\partial^2 f}{\partial x \partial y} = 1$$

Hence,

$$D = f_{xx} f_{yy} - f_{xy}^2 = (2)(2) - (1)^2 = 3 > 0$$

Since $D > 0$ and $f_{xx} > 0$, we conclude that $(0, 0)$ is a local minimum (Figure 10.45). ◀

▶ **Example 3**   Find all local extrema of

$$f(x, y) = 3xy - x^3 - y^3, \quad (x, y) \in \mathbf{R}^2$$

and classify them according to local maximum, local minimum, or neither.

Solution

The function $f(x, y)$ is differentiable on its domain. The critical points thus satisfy

$$\nabla f(x, y) = \begin{bmatrix} 3y - 3x^2 \\ 3x - 3y^2 \end{bmatrix} = \begin{bmatrix} 0 \\ 0 \end{bmatrix}$$

which yields $y = x^2$ and $x = y^2$. This has the solutions $(0, 0)$ and $(1, 1)$. Now,

$$\frac{\partial^2 f(x, y)}{\partial x^2} = -6x, \qquad \frac{\partial^2 f(x, y)}{\partial y^2} = -6y, \qquad \frac{\partial^2 f(x, y)}{\partial x \partial y} = 3$$

Therefore,

$$D = f_{xx} f_{yy} - (f_{xy})^2 = 36xy - 9$$

At $(1, 1)$, $D = 36 - 9 > 0$. Since $f_{xx}(1, 1) = -6 < 0$, $f(x, y)$ has a local maximum at $(1, 1)$. At $(0, 0)$, $D = -9 < 0$. The critical point $(0, 0)$ is neither a local maximum nor a local minimum since $D < 0$ (Figure 10.46). ◀

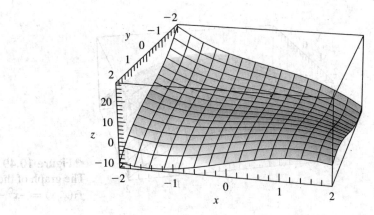

◀ **Figure 10.46**
The graph of the function
$f(x, y) = 3xy - x^3 - y^3$

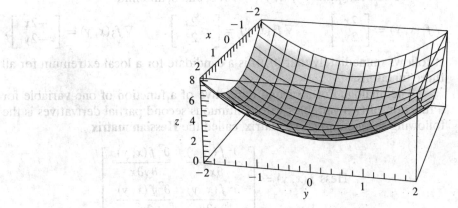

◀ **Figure 10.47**
The graph of the function
$f_1(x, y) = x^2 + y^2$

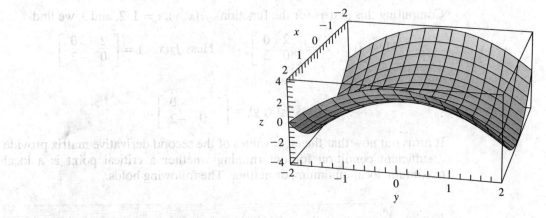

▲ **Figure 10.48**
The graph of the function $f_2(x, y) = x^2 - y^2$

A Sufficient Condition Based on Eigenvalues (Optional)   We will now give
a sufficient condition that is phrased in terms of eigenvalues to determine
whether a candidate for a local extremum is indeed a local extremum, and if
so, what type (that is, local maximum or local minimum). To motivate this
condition, we will look at the following three functions defined for $(x, y) \in \mathbf{R}^2$:

$$f_1(x, y) = x^2 + y^2, \qquad f_2(x, y) = x^2 - y^2, \qquad f_3(x, y) = -x^2 - y^2$$

which are illustrated in Figures 10.47 through 10.49.

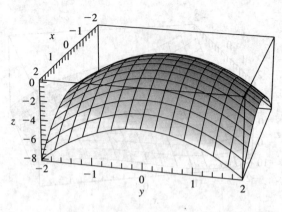

◀ **Figure 10.49**
The graph of the function
$f_3(x, y) = -x^2 - y^2$

We can compute $\nabla f_i(x, y), i = 1, 2,$ and 3, and find

$$\nabla f_1(x, y) = \begin{bmatrix} 2x \\ 2y \end{bmatrix}, \qquad \nabla f_2(x, y) = \begin{bmatrix} 2x \\ -2y \end{bmatrix}, \qquad \nabla f_3(x, y) = \begin{bmatrix} -2x \\ -2y \end{bmatrix}.$$

It follows immediately that $(0, 0)$ is a candidate for a local extremum for all three functions.

The analogue of a second derivative of a function of one variable for a function of two variables with continuous second partial derivatives is the following second derivative matrix, called the **Hessian matrix**,

$$\text{Hess } f(x, y) = \begin{bmatrix} \dfrac{\partial^2 f(x, y)}{\partial x^2} & \dfrac{\partial^2 f(x, y)}{\partial y \partial x} \\ \dfrac{\partial^2 f(x, y)}{\partial x \partial y} & \dfrac{\partial^2 f(x, y)}{\partial y^2} \end{bmatrix}$$

Computing this matrix for the functions $f_i(x, y), i = 1, 2,$ and 3, we find

$$\text{Hess } f_1(x, y) = \begin{bmatrix} 2 & 0 \\ 0 & 2 \end{bmatrix}, \qquad \text{Hess } f_2(x, y) = \begin{bmatrix} 2 & 0 \\ 0 & -2 \end{bmatrix},$$

and

$$\text{Hess } f_3(x, y) = \begin{bmatrix} -2 & 0 \\ 0 & -2 \end{bmatrix}.$$

It turns out now that the eigenvalues of the second derivative matrix provide a sufficient condition for determining whether a critical point is a local maximum, local minimum, or neither. The following holds.

Suppose the second partial derivatives of f are continuous in a disk centered at (x_0, y_0) and that $\nabla f(x_0, y_0) = \begin{bmatrix} 0 \\ 0 \end{bmatrix}$. Then

1. If the two eigenvalues of the second derivative matrix at (x_0, y_0), Hess $f(x_0, y_0)$, are both positive, then f has a local minimum at (x_0, y_0).

2. If the two eigenvalues of Hess $f(x_0, y_0)$ are both negative, then f has a local maximum at (x_0, y_0).

3. If the two eigenvalues of Hess $f(x_0, y_0)$ are of opposite signs, then f does not have a local extremum at (x_0, y_0). The point (x_0, y_0) is then called a **saddle**.

In all other cases, the test is inconclusive. Returning to our example, we need to evaluate Hess $f_i(x, y)$, $i = 1, 2$, and 3, at $(0, 0)$. We see that Hess $f_i(x, y)$, $i = 1, 2$, and 3, do not depend on x or y; thus Hess $f_i(0, 0) =$ Hess $f_i(x, y)$. We can read off the eigenvalues of each of the second derivative matrices evaluated at $(0, 0)$ since they are in diagonal form. We find that the eigenvalues of Hess $f_1(0, 0)$ are both 2; hence $f_1(0, 0)$ is a local minimum, which agrees with the graph in Figure 10.47. The eigenvalues of Hess $f_2(0, 0)$ are 2 and -2, and we conclude that $f_2(0, 0)$ is not a local extremum (see Figure 10.48); the graph of $f(x, y)$ resembles a saddle near $(0, 0)$, hence the name saddle point. The eigenvalues of Hess $f_3(0, 0)$ are both -2, and we conclude that $f_3(0, 0)$ is a local maximum (see Figure 10.49).

We assumed in the second derivative criterion that all second partial derivatives are continuous in a disk centered at (x_0, y_0). This implies that

$$\frac{\partial^2 f(x, y)}{\partial x \, \partial y} = \frac{\partial^2 f(x, y)}{\partial y \, \partial x}$$

That is, the off-diagonal elements of Hess $f(x, y)$ are identical and the Hessian matrix is of the form $\begin{bmatrix} a & c \\ c & b \end{bmatrix}$. Such a matrix is called **symmetric**. We can show that the eigenvalues of a symmetric matrix are always real (see Problem 34). This fact has an important consequence: If the second partial derivatives of f are continuous in a disk centered at (x_0, y_0), then the eigenvalues of Hess $f(x_0, y_0)$ are both real. Provided neither eigenvalue is equal to zero, one of the three cases in our second derivative criterion occurs, which will then allow us to settle the question whether the candidate (x_0, y_0) for which $\nabla f(x_0, y_0) = \begin{bmatrix} 0 \\ 0 \end{bmatrix}$ is a local extremum and of what type. If one or both eigenvalues are equal to zero, we cannot say anything about the nature of the critical point based on the Hessian matrix. This is discussed in Problem 11.

▶ **Example 4** Find the local extrema of

$$f(x, y) = 2x^2 - xy + y^4, \quad (x, y) \in \mathbf{R}^2$$

Solution

We compute

$$\nabla f(x, y) = \begin{bmatrix} 4x - y \\ -x + 4y^3 \end{bmatrix}$$

Setting both partial derivatives equal to 0, we find

$$4x - y = 0 \quad \text{and} \quad -x + 4y^3 = 0$$

It follows from the first equation that $x = y/4$. Substituting this into the second equation, we find

$$-\frac{y}{4} + 4y^3 = 0$$

$$-\frac{y}{4}(1 - 16y^2) = 0$$

yielding

$$y_1 = 0 \quad \text{and} \quad y_2 = \frac{1}{4} \quad \text{and} \quad y_3 = -\frac{1}{4}$$

and, hence, the corresponding x-values are

$$x_1 = 0 \quad \text{and} \quad x_2 = \frac{1}{16} \quad \text{and} \quad x_3 = -\frac{1}{16}$$

Since ∇f is defined for all $(x, y) \in \mathbf{R}^2$, there are no other critical points. The three candidates for local extrema are thus

$$(0, 0) \quad \text{and} \quad \left(\frac{1}{16}, \frac{1}{4}\right) \quad \text{and} \quad \left(-\frac{1}{16}, -\frac{1}{4}\right)$$

The Hessian matrix is of the form

$$\text{Hess } f(x, y) = \begin{bmatrix} 4 & -1 \\ -1 & 12y^2 \end{bmatrix}$$

We evaluate the Hessian matrix at each candidate and compute its eigenvalues.

(i)
$$\text{Hess } f(0, 0) = \begin{bmatrix} 4 & -1 \\ -1 & 0 \end{bmatrix}$$

The eigenvalues satisfy

$$\det \begin{bmatrix} 4 - \lambda & -1 \\ -1 & -\lambda \end{bmatrix} = \lambda(\lambda - 4) - 1 = \lambda^2 - 4\lambda - 1 = 0$$

Thus,

$$\lambda_{1,2} = \frac{4 \pm \sqrt{16 + 4}}{2} = 2 \pm \sqrt{5} \approx \begin{cases} 4.2361 \\ -0.2361 \end{cases}$$

implying that f has a saddle point at $(0, 0)$.

(ii)
$$\text{Hess } f\left(\frac{1}{16}, \frac{1}{4}\right) = \begin{bmatrix} 4 & -1 \\ -1 & \frac{3}{4} \end{bmatrix}$$

The eigenvalues satisfy

$$\det \begin{bmatrix} 4 - \lambda & -1 \\ -1 & \frac{3}{4} - \lambda \end{bmatrix} = (4 - \lambda)\left(\frac{3}{4} - \lambda\right) - 1 = 3 - 4\lambda - \frac{3}{4}\lambda + \lambda^2 - 1$$

$$= \lambda^2 - \frac{19}{4}\lambda + 2 = 0$$

Thus,

$$\lambda_{1,2} = \frac{\frac{19}{4} \pm \sqrt{\frac{361}{16} - 8}}{2} = \frac{19}{8} \pm \frac{1}{8}\sqrt{233} \approx \begin{cases} 4.2830 \\ 0.4670 \end{cases}$$

implying that f has a local minimum at $(\frac{1}{16}, \frac{1}{4})$.

(iii)
$$\text{Hess } f\left(-\frac{1}{16}, -\frac{1}{4}\right) = \begin{bmatrix} 4 & -1 \\ -1 & \frac{3}{4} \end{bmatrix}$$

This is the same matrix as for $(\frac{1}{16}, \frac{1}{4})$ [that is, case (ii)]. We thus conclude that f has a local minimum at $(-\frac{1}{16}, -\frac{1}{4})$ as well.

The graph of $f(x, y)$ is illustrated in Figure 10.50. ◀

We saw in the last example that finding the eigenvalues of the Hessian matrix can be time consuming. There is another criterion that follows from the relationship that expresses the determinant and the trace of a 2×2 matrix in terms of its eigenvalues. Recall that if A is a 2×2 matrix with eigenvalues λ_1 and λ_2, then $\det A = \lambda_1 \lambda_2$ and $\text{tr } A = \lambda_1 + \lambda_2$. If the eigenvalues of A are both real (as is the case for a symmetric matrix), and if $\det A > 0$, then either both λ_1 and λ_2 are positive or both are negative. If, in addition, $\text{tr } A > 0$, then both λ_1 and λ_2 are positive. We thus arrive at the following criterion.

◄ **Figure 10.50**
The graph of the function
$f(x, y) = 2x^2 - xy + y^4$

Suppose the second partial derivatives of f are continuous in a disk centered at (x_0, y_0) and that $\nabla f(x_0, y_0) = \begin{bmatrix} 0 \\ 0 \end{bmatrix}$. Then

1. If det Hess $f(x_0, y_0) > 0$ and tr Hess $f(x_0, y_0) > 0$, then f has a local minimum at (x_0, y_0).
2. If det Hess $f(x_0, y_0) > 0$ and tr Hess $f(x_0, y_0) < 0$, then f has a local maximum at (x_0, y_0).
3. If det Hess $f(x_0, y_0) < 0$, then (x_0, y_0) is not a local extremum; (x_0, y_0) is a saddle point.

Recall that if one of the eigenvalues of Hess $f(x_0, y_0)$ is equal to 0 [or, equivalently, det Hess $f(x_0, y_0) = 0$], then we cannot say anything about the nature of the critical point based on the Hessian matrix. (Such a case is explored in Problem 11.)

► **Example 5** Find and classify the critical points of

$$f(x, y) = x^3 - 4xy + y, \quad (x, y) \in \mathbf{R}^2$$

Solution
We find

$$\nabla f(x, y) = \begin{bmatrix} 3x^2 - 4y \\ -4x + 1 \end{bmatrix} = \begin{bmatrix} 0 \\ 0 \end{bmatrix}$$

when

$$3x^2 - 4y = 0 \quad \text{and} \quad -4x + 1 = 0$$

The second equation yields $x = 1/4$. Substituting this in the first equation, we find

$$\frac{3}{16} - 4y = 0$$

implying $y = 3/64$. Since f is differentiable for all $(x, y) \in \mathbf{R}^2$, there is only one critical point, namely $(\frac{1}{4}, \frac{3}{64})$. To classify the critical point, we compute

$$\text{Hess } f(x, y) = \begin{bmatrix} 6x & -4 \\ -4 & 0 \end{bmatrix}$$

Evaluating this at the critical point, we find

$$\text{Hess } f\left(\frac{1}{4}, \frac{3}{64}\right) = \begin{bmatrix} \frac{3}{2} & -4 \\ -4 & 0 \end{bmatrix}$$

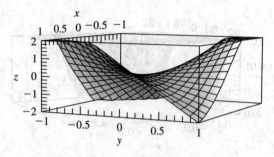

◀ **Figure 10.51**
The graph of the function
$f(x, y) = x^3 - 4xy + y$

Since det Hess $f\left(\frac{1}{4}, \frac{3}{64}\right) = -16 < 0$, we conclude that $f(x, y)$ has a saddle point at $\left(\frac{1}{4}, \frac{3}{64}\right)$ (see Figure 10.51). ◀

▶ **Example 6**   Find and classify the critical points of

$$f(x, y) = \sqrt{x^2 + y^2}, \ (x, y) \in \mathbf{R}^2$$

Solution

We find

$$\nabla f(x, y) = \begin{bmatrix} \dfrac{2x}{2\sqrt{x^2 + y^2}} \\ \dfrac{2y}{2\sqrt{x^2 + y^2}} \end{bmatrix} = \dfrac{1}{\sqrt{x^2 + y^2}} \begin{bmatrix} x \\ y \end{bmatrix}, \ (x, y) \neq (0, 0)$$

Since the gradient of f is undefined at $(0, 0)$, the point $(0, 0)$ is a critical point. There are no other critical points since $\nabla f(x, y) \neq \begin{bmatrix} 0 \\ 0 \end{bmatrix}$ for all $(x, y) \neq (0, 0)$.

Now, $f(x, y) > 0$ for $(x, y) \neq (0, 0)$ and $f(x, y) = 0$ for $(x, y) = (0, 0)$. Therefore, $f(x, y)$ has a local minimum at $(0, 0)$ (see Figure 10.52). Note that we cannot use the Hessian here to decide whether $(0, 0)$ is a local extremum and of what type since the theorem requires that the gradient is zero at the point, but here the gradient is undefined at $(0, 0)$. [The Hessian is also not defined at $(0, 0)$.] ◀

Global Extrema   We will now turn our discussion to global extrema. Recall that for functions of one variable, the extreme value theorem provides us with the existence of global extrema for functions defined on a *closed* interval. The analogue of closed intervals in the two dimensional plane is a **closed** set; similarly, the analogue of an open interval is an **open** set. An example of a closed set is a closed disk; an example of an open set is an open disk (see Section 10.2).

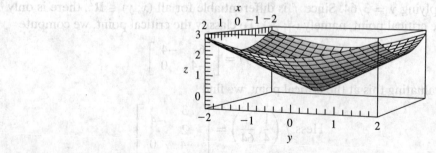

◀ **Figure 10.52**
The graph of the function
$f(x, y) = \sqrt{x^2 + y^2}$

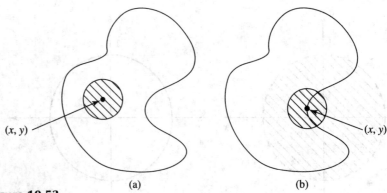

▲ **Figure 10.53**
The point (x, y) on the left is an interior point; the point (x, y) on the right is a boundary point

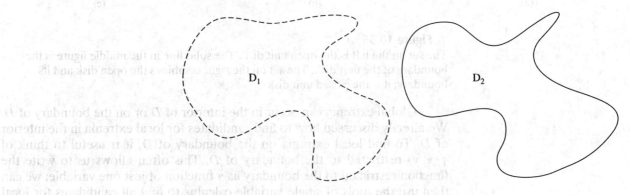

▲ **Figure 10.54**
The set D_1 on the left is open. The set D_2 on the right is closed, the solid line is the boundary

To define these concepts more generally, we start with a set $D \subset \mathbf{R}^2$. A point (x, y) is called an **interior point** of D if there exists a $\delta > 0$ so that the disk centered at (x, y) with radius δ is contained in D, that is, $B_\delta(x, y) \subset D$ (see Figure 10.53a). A point (x, y) is a **boundary point** of D if every disk centered at (x, y) contains both points in D and points not in D; the boundary point (x, y) need not be contained in D (see Figure 10.53b). The **interior** of D consists of all interior points of D; the **boundary** of D consists of all boundary points of D. A set $D \subset \mathbf{R}^2$ is **open** if it consists only of interior points; a set $D \subset \mathbf{R}^2$ is **closed** if it contains all its boundary points (see Figure 10.54).

Most of the time, the domains of our functions will be rectangles or disks. Figure 10.55 illustrates the concepts we just learned on the unit disk. We start with the open unit disk $\{(x, y) : x^2 + y^2 < 1\}$ (Figure 10.55a). Every point in this set is an interior point. The unit circle $\{(x, y) : x^2 + y^2 = 1\}$ is the boundary of the open unit disk (Figure 10.55b). If we combine the open disk and its boundary, we obtain the closed unit disk $\{(x, y) : x^2 + y^2 \leq 1\}$ (Figure 10.55c).

To formulate the extreme value theorem in $\mathbf{R}^2$, we also need the notion of a **bounded** set. A set is bounded if it is contained within some disk.

Extreme Value Theorem in $\mathbf{R}^2$ If f is continuous on a closed and bounded set $D \subset \mathbf{R}^2$, then f has a global maximum and a global minimum on D.

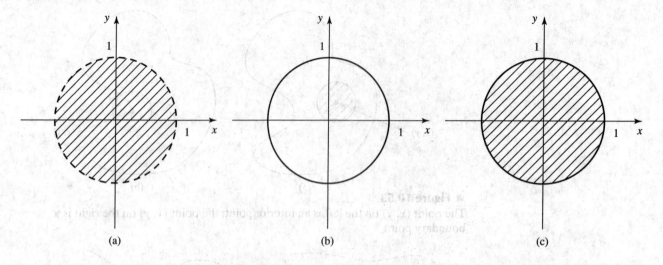

▲ **Figure 10.55**
The set on the left is the open unit disk. The solid line in the middle figure is the boundary of the unit disk. The set on the right combines the open disk and its boundary; it is the closed unit disk

Global extrema can occur in the interior of D or on the boundary of D. We already discussed how to find candidates for local extrema in the interior of D. To find local extrema on the boundary of D, it is useful to think of $f(x, y)$ restricted to the boundary of D. This often allows us to write the function restricted to the boundary as a function of just one variable; we can then use the tools of single variable calculus to find all candidates for local extrema on the boundary of D (see Example 7). To find global extrema for continuous functions defined on a closed and bounded set, we thus proceed as follows:

1. Determine all candidates for local extrema in the interior of D.
2. Determine all candidates for local extrema on the boundary of D.
3. Select the global maximum and the global minimum from the set of points determined in steps 1 and 2.

▶ **Example 7**   Find the global extrema of

$$f(x, y) = x^2 - 3y + y^2, \qquad -1 \le x \le 1, 0 \le y \le 2$$

Solution

The function is defined on a closed and bounded rectangle and is continuous. The extreme value theorem thus guarantees the existence of global extrema. We begin with finding critical points in the interior of the domain,

$$\nabla f(x, y) = \begin{bmatrix} 2x \\ -3 + 2y \end{bmatrix} = \begin{bmatrix} 0 \\ 0 \end{bmatrix}$$

when $x = 0$ and $y = 3/2$. The point $(0, 3/2)$ is in the interior of the domain of f and is thus a critical point with $f(0, 3/2) = -2.25$. There are no other critical points in the interior of the domain of f.

Next, we need to check the boundary values (see Figure 10.56). We start with the line segment C_1, which connects the points $(-1, 0)$ and $(1, 0)$ on the x-axis. On C_1, $y = 0$. Hence, f on C_1 is of the form

$$f(x, 0) = x^2, \qquad -1 \le x \le 1$$

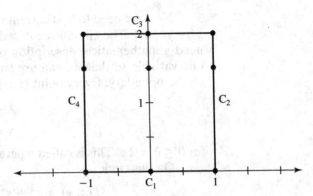

▲ **Figure 10.56**
The domain of $f(x, y)$ in Example 7. The boundary consists of the line segments C_1, C_2, C_3, and C_4

By restricting $f(x, y)$ to the curve $y = 0$, we obtained a function of just one variable. Using single-variable calculus, we find that $f'(x, 0) = 2x = 0$ for $x = 0$. The critical point on C_1 is thus $(0, 0)$ with $f(0, 0) = 0$; in addition, there are the two endpoints $(-1, 0)$ with $f(-1, 0) = 1$ and $(1, 0)$ with $f(1, 0) = 1$.

On C_2, we have $x = 1$, which yields $f(1, y) = 1 - 3y + y^2$, $0 \leq y \leq 2$, which is again a function of just one variable. Now, $f'(1, y) = -3 + 2y = 0$ for $y = 3/2$. Hence, we find a candidate at $(1, 3/2)$ with $f(1, 3/2) = -1.25$; other candidates are the endpoints $(1, 0)$ with $f(1, 0) = 1$ and $(1, 2)$ with $f(1, 2) = -1$.

On C_3, we have $y = 2$ yielding $f(x, 2) = x^2 - 2$. Thus, $f'(x, 2) = 2x = 0$ for $x = 0$, giving the critical point $(0, 2)$ with $f(0, 2) = -2$. Other candidates are the endpoints $(-1, 2)$ with $f(-1, 2) = -1$ and $(1, 2)$ with $f(1, 2) = -1$.

On C_4, $x = -1$ and $f(-1, y) = 1 - 3y + y^2$, which is the same as on C_2. We thus have the additional points $(-1, 3/2)$ with $f(-1, 3/2) = -1.25$, $(-1, 0)$ with $f(-1, 0) = 1$, and $(-1, 2)$ with $f(-1, 2) = -1$.

Comparing all the values of $f(x, y)$ at the candidate points (see Figure 10.56 and the following table), we find that the global minimum is $f(0, 3/2) = -2.25$ and the global maxima are $f(-1, 0) = 1$ and $f(1, 0) = 1$.

(x, y)	$(0, 3/2)$	$(0, 0)$	$(-1, 0)$	$(1, 0)$	$(1, 3/2)$	$(1, 2)$	$(0, 2)$	$(-1, 2)$	$(1, 2)$	$(-1, 3/2)$
$f(x, y)$	-2.25	0	1	1	-1.25	-1	-2	-1	-1	-1.25

◄

▶ **Example 8** Find the absolute maxima and minima of $f(x, y) = x^2 + y^2 - 2x + 4$ on the disk $D = \{(x, y) : x^2 + y^2 \leq 4\}$.

Solution

The function is defined on a closed and bounded disk and is continuous. The extreme value theorem thus guarantees global extrema. We begin with finding critical points in the interior of the domain

$$\nabla f(x, y) = \begin{bmatrix} 2x - 2 \\ 2y \end{bmatrix} = \begin{bmatrix} 0 \\ 0 \end{bmatrix}$$

when $x = 1$ and $y = 0$. Since $x^2 + y^2 = 1 \leq 4$, the point $(1, 0)$ is in the interior of the domain of f and is thus a critical point with $f(1, 0) = 3$. There are no other critical points in the interior of the domain of f.

Next we need to find extrema on the boundary of the domain, the circle $x^2 + y^2 = 4$. The circle is centered at the origin $(0, 0)$ and has radius 2. We need a mathematical description of the circle in terms of a function of just one variable so that we can use single-variable calculus to identify extrema on the boundary. Every point (x, y) on this circle can be written as

$$x = 2 \cos \theta$$

$$y = 2 \sin \theta$$

for $0 \leq \theta < 2\pi$. This is called a **parameterization** of the circle.

On this circle,

$$f(x, y) = x^2 + y^2 - 2x + 4$$
$$= 4 \cos^2 \theta + 4 \sin^2 \theta - 4 \cos \theta + 4$$
$$= 4 - 4 \cos \theta + 4 = 8 - 4 \cos \theta$$

where we used $\sin^2 \theta + \cos^2 \theta = 1$. To find maxima and minima of the single-valued function $g(\theta) = 8 - 4 \cos \theta$, we need to differentiate $g(\theta)$,

$$g'(\theta) = 4 \sin \theta$$

and solve $g'(\theta) = 0$ in $[0, 2\pi)$. We find the two angles $\theta = 0$ and $\theta = \pi$. Now,

$$g(0) = 8 - 4 = 4 \qquad \text{and} \qquad g(\pi) = 8 + 4 = 12$$

The maximum on the boundary is at $\theta = \pi$, which corresponds to the point $(-2, 0)$. The minimum on the boundary is at $\theta = 0$, which corresponds to the point $(2, 0)$.

Comparing the extrema on the boundary to the extremum in the interior of the set D, we find that the global minimum is in the interior at $(1, 0)$ and the global maximum is on the boundary at $(-2, 0)$ (Figure 10.57).

We conclude this section with an application. ◀

▶ **Example 9**   Determine the values of three nonnegative numbers whose sum is 90 and whose product is maximal.

Solution

We denote the three numbers by x, y, and z. Then $x + y + z = 90$; their product is xyz. Since $z = 90 - x - y$, we can write the product as $xyz = xy(90 - x - y)$. Our goal is to maximize this product. We define the function

$$f(x, y) = xy(90 - x - y), \quad x + y \leq 90, \ x \geq 0, \ y \geq 0$$

Since x, y, and z are nonnegative numbers and their sum is equal to 90, the domain is the set $\{(x, y) : x + y \leq 90, x \geq 0, y \geq 0\}$, which is the triangular region bounded by the lines $x = 0$, $y = 0$ and $y = 90 - x$.

We need to find (x, y) so that $f(x, y)$ is maximal. Now,

$$\nabla f(x, y) = \begin{bmatrix} 90y - 2xy - y^2 \\ 90x - x^2 - 2xy \end{bmatrix} = \begin{bmatrix} 0 \\ 0 \end{bmatrix}$$

when

$$y(90 - 2x - y) = 0 \qquad \text{and} \qquad x(90 - 2y - x) = 0$$

Solutions with $x = 0$ or $y = 0$ are points on the boundary. To find solutions in the interior of the domain, we need to solve

$$2x + y = 90$$

$$x + 2y = 90$$

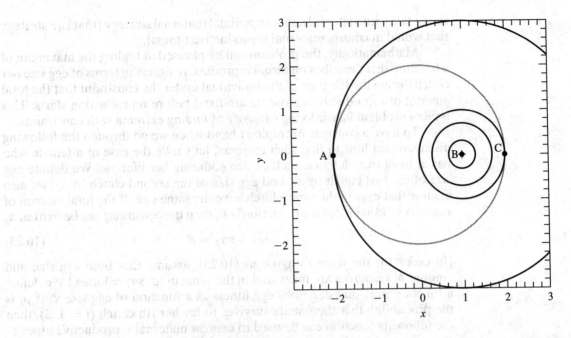

▲ **Figure 10.57**

Shown are the contour lines of the function $f(x, y) = x^2 + y^2 - 2x + 4$ (in black for the values $f(x, y) = c$ with $c = 3.1, 3.5, 4$, and 12) and the boundary of the disk $x^2 + y^2 \leq 4$ (in blue). The point $B = (1, 0)$ is the local minimum in the interior of the disk $x^2 + y^2 \leq 4$, the point $A = (-2, 0)$ is the local maximum on the boundary of the disk, and the point $C = (2, 0)$ is the local minimum on the boundary of the disk

Multiplying the second equation by 2 and subtracting the first equation yields

$$3y = 90$$

or $y = 30$. Using the first equation, $2x = 90 - y = 60$, we find $x = 30$, yielding the candidate $(30, 30)$, which is in the interior of the domain, with $f(30, 30) = 30^3 = 27,000$. There are no other candidates for local extrema in the interior of D.

The function $f(x, y)$ is continuous on a closed and bounded set, namely the triangle with corners $(0, 0)$, $(0, 90)$, and $(90, 0)$. The extreme value theorem guarantees that f has a global maximum on the domain. We see that $f(x, y)$ takes on value 0 on the boundary of the domain. Comparing the values of $f(x, y)$ on the boundary of D to the value at the candidate point $(30, 30)$, we conclude that the function $f(x, y)$ has the global maximum at the interior candidate $(30, 30)$.

The product xyz is therefore maximal when $x = y = z = 30$. ◄

10.6.2 Extrema with Constraints

A number of studies have shown that in butterflies who lay their eggs singly, egg size decreases with maternal age. Begon and Parker (1986) proposed a mathematical model to explain this decline in egg size in terms of a maternal strategy that would optimize reproductive fitness. Basic assumptions of their model are that all resources necessary for egg production are gathered before laying eggs, and that clutch size is fixed (for instance, a single egg per clutch). Under these assumptions, Begon and Parker were able to show that if egg fitness is an increasing and concave down function of egg size, then a decline in egg size with maternal age is an optimal maternal strategy (that is, a strategy that would maximize maternal reproductive fitness).

Mathematically, the problem can be phrased as finding the maximum of a function that describes maternal reproductive fitness in terms of egg size per clutch during the life time of the individual under the constraint that the total amount of reproductive resources are fixed before reproduction starts. This type of problem falls into the category of finding extrema with constraints.

To have a concrete example at hand when we go through the following discussion on how to find such extrema, let's take the case of a female who has at most two clutches, each of size n, during her lifetime. We denote egg size of the first clutch by x_1, and egg size of the second clutch by x_2; we also assume that eggs in the same clutch have the same size. If the total amount of resources available for reproduction is R, then the constraint can be written as

$$nx_1 + nx_2 = R \qquad\qquad (10.23)$$

[In order for the units to agree in (10.23), assume that both egg size and amount of resources are measured in the same units, say calories.] We define a function $\rho(x)$ that describes egg fitness as a function of egg size x. If p_i is the probability that the female survives to lay her ith clutch ($i = 1, 2$), then the following function can be used to express maternal reproductive fitness:

$$f(x_1, x_2) = p_1 n\rho(x_1) + p_2 n\rho(x_2)$$

Our goal is thus to find extrema of the function $f(x_1, x_2)$ under the constraint $nx_1 + nx_2 = R$.

Finding extrema with constraints involves two functions, we describe the constraint, the other the function we wish to maximize. All of the constraints in this section will be of the form

$$g(x, y) = 0$$

For instance, the constraint (10.23) can be written as $g(x_1, x_2) = nx_1 + nx_2 - R = 0$.

We can illustrate the constraint $g(x, y) = 0$ as a set of points in the x-y plane: $\{(x, y) : g(x, y) = 0\}$. These will typically be curves of the sort shown in Figure 10.58. In the following discussion, it will be useful to think of traveling on this curve starting, say, at the point labeled with $t = 0$ and ending at the point with $t = T$. We can then refer to any point on the curve by giving the time t at which we pass through it. We say that we *parameterize* the curve using the parameter t. The curve can thus be written as a vector-valued function

$$\mathbf{r}(t) = \begin{bmatrix} x(t) \\ y(t) \end{bmatrix}$$

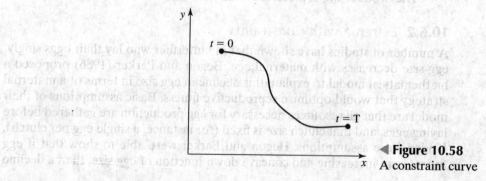

◀ **Figure 10.58**
A constraint curve

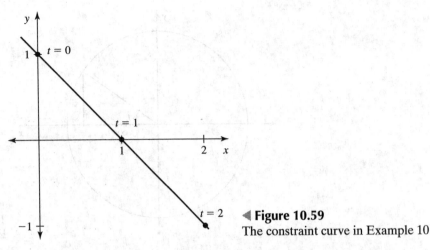

◀ **Figure 10.59**
The constraint curve in Example 10

where at time t, we pass through the point $(x(t), y(t))$, just as in Subsection 9.4.3, where we parameterized lines. The constraint then satisfies the equation $g(\mathbf{r}(t)) = 0$. Let's look at two examples.

▶ **Example 10**   Let

$$g(x, y) = x + y - 1$$

Then the set $\{(x, y) : g(x, y) = 0\}$ describes a curve in the x-y plane, namely the straight line $y = 1 - x$. We can parameterize this curve using

$$x(t) = t \quad \text{and consequently} \quad y(t) = 1 - t$$

This is not the only parameterization, but it is the simplest. At time $t = 0$, for instance, we are at the point $(0, 1)$. At time $t = 1$, we are at $(1, 0)$, at time $t = 2$, we are at $(2, -1)$, and so on, as illustrated in Figure 10.59. This parameterization describes the motion along the line to the right and downward, at the rate of one unit of x per unit of time.   ◀

Note that the parameterization $x = t$ and $y = f(t)$ always works when y is given explicitly as a function of x, that is, $y = f(x)$, as in Example 10. The next example shows a parameterization for the unit circle.

▶ **Example 11**   Let

$$g(x, y) = x^2 + y^2 - 1$$

Then $g(x, y) = 0$ is the circle with radius 1 centered at the origin. A natural parameterization of the unit circle is

$$x(t) = \cos t \quad \text{and} \quad y(t) = \sin t, \ 0 \le t < 2\pi$$

The parameter t describes the angle as illustrated in Figure 10.60. That is, we move counterclockwise around the circle at 1 radian per unit time.   ◀

What is the advantage of such a parameterization? It will allow us to relate the gradient of g at (x, y), $\nabla g(x, y)$, to the tangent on the graph of the constraint curve at (x, y). Let's see how. Let $\mathbf{r}(t)$ be the vector from the origin to the point $(x(t), y(t))$ on the parameterized constraint curve $g(x, y) = 0$. Using the formal definition of the derivative, we define the derivative of $\mathbf{r}(t)$ with respect to t as

$$\mathbf{r}'(t) = \frac{d}{dt}\mathbf{r}(t) = \lim_{\Delta t \to 0} \frac{\mathbf{r}(t + \Delta t) - \mathbf{r}(t)}{\Delta t}$$

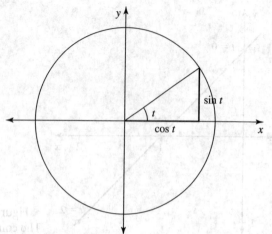

◀ **Figure 10.60**
The constraint curve in
Example 11

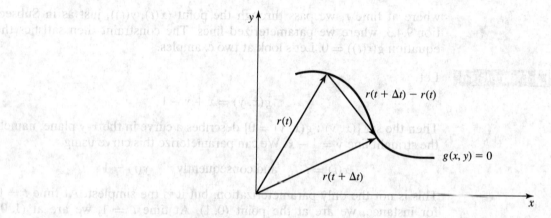

▲ **Figure 10.61**
The vector $\mathbf{r}(t + \Delta t) - \mathbf{r}(t)$

provided the limit exists. The vector $\mathbf{r}(t+\Delta t)-\mathbf{r}(t)$ is illustrated in Figure 10.61. Dividing this vector by Δt only changes its length but not its direction. It thus appears that in the limit $\Delta t \to 0$, the limiting vector $\mathbf{r}'(t)$, if it exists, is tangential to the curve at $(x(t), y(t))$ (see Figure 10.62). We'll try to understand this in the case where the curve $g(x, y) = 0$ can be written as a function that gives y explicitly in terms of x, namely $y = h(x)$ and where $h'(x)$ exists. Recall that the derivative of $\mathbf{r}(t)$ with respect to t is defined componentwise:

$$\frac{d}{dt}\mathbf{r}(t) = \frac{d}{dt}\begin{bmatrix} x(t) \\ y(t) \end{bmatrix} = \begin{bmatrix} \frac{dx(t)}{dt} \\ \frac{dy(t)}{dt} \end{bmatrix}$$

The slope of the tangent line at a point (x, y) on the curve is given by $h'(x)$. Since

$$h'(x) = \frac{dy}{dx} = \frac{dy/dt}{dx/dt}$$

we see that the slope of the tangent line can be expressed as the ratio of the components dx/dt and dy/dt of the vector $\mathbf{r}'(t)$, as illustrated in Figure 10.62, implying that $\mathbf{r}'(t)$ is tangential to the curve at $(x(t), y(t))$.

This has an important implication: Since $g(x, y) = 0$ is a level curve, the gradient of g at (x, y), $\nabla g(x, y)$, is perpendicular to the level curve at (x, y)

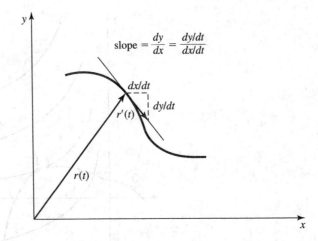

▲ Figure 10.62
The vector $\mathbf{r}'(t)$

and thus to $\mathbf{r}'(t)$ (see Subsection 10.5.3). How can we see this? Since the curve is given by the equation $g(x(t), y(t)) = 0$, it follows that $\frac{d}{dt} g(x(t), y(t)) = 0$. Now, using the chain rule, we find

$$\frac{d}{dt} g(x(t), y(t)) = \frac{\partial g}{\partial x} \frac{dx}{dt} + \frac{\partial g}{\partial y} \frac{dy}{dt} = \nabla g(x, y) \cdot \mathbf{r}'(t)$$

where we used the definition of the dot product in the last step. We thus have

$$\nabla g(x, y) \cdot \mathbf{r}'(t) = 0$$

which implies that the gradient vector $\nabla g(x, y)$ is perpendicular to the tangent vector $\mathbf{r}'(t)$ at $(x(t), y(t))$, as claimed previously. This is an important fact that we will need below when we try to find extrema under constraints.

Now, let's go back to the problem of finding extrema under constraints. We denote the function we wish to optimize by $f(x, y)$ and the constraint by $g(x, y) = 0$. Finding extrema with constraints amounts to restricting the function $f(x, y)$ to the constraint curve and seeking its extrema there. The constraint $g(x, y) = 0$ defines a set of points (x, y) in the x-y plane. The graph of $z = f(x, y)$ is a surface in three-dimensional x-y-z space. Using level curves for $f(x, y)$, we can represent $f(x, y)$ in the x-y plane. We can then graph both the level curves of $f(x, y)$ and the constraint $g(x, y) = 0$ in the same two-dimensional coordinate system (see Figure 10.63). We claim that $f(x, y)$ has a local extremum at the point P in Figure 10.63 under the constraint $g(x, y) = 0$. To see this, imagine traveling on the curve $g(x, y) = 0$ starting at the point Q and traveling in the direction of the arrow. You first intersect the level curve $f(x, y) = c_1$, and later $f(x, y) = c_2$. To make the following discussion more concrete, assume that $c_1 < c_2 < c_3 < c_4$. (Other cases will be discussed in Problems 59 and 60.) The values of f along your travel route then increase until you reach the point P. Once you pass P, the values of f decrease again. Thus, along the curve $g(x, y) = 0$, the function f has a local extremum (in this case, a local maximum) at P.

What characterizes this point P? The level curve through P and the constraint curve touch each other at the point P, that is, they both have the same tangent line. We recall from Subsection 10.5.3 that the gradient of f at the point P is perpendicular to the level curve through P. Combining this with the fact we derived previously—namely, that the gradient of g at P is

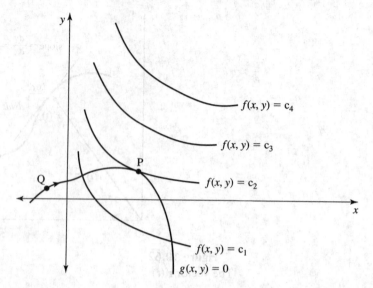

▲ **Figure 10.63**
Level curves and constraints

perpendicular to the tangent line at P on the graph of g—we conclude that if there is an extremum at P, then ∇g and ∇f are parallel.

To state the theorem more generally [and not just in the case where $g(x, y)$ can be written as a function that gives y explicitly in terms of x], we need to require that $\nabla g(x, y) \neq \begin{bmatrix} 0 \\ 0 \end{bmatrix}$ at P. Denoting the coordinates of P by (x_0, y_0), we can then formulate the result.

Lagrange's Theorem Assume that f and g have continuous first partial derivatives and that $f(x, y)$ has an extremum at (x_0, y_0) subject to the constraint $g(x, y) = 0$. If $\nabla g(x_0, y_0) \neq \begin{bmatrix} 0 \\ 0 \end{bmatrix}$, then there exists a number λ so that

$$\nabla f(x_0, y_0) = \lambda \nabla g(x_0, y_0) \qquad (10.24)$$

The number λ is called a **Lagrange multiplier**. Finding candidates for extrema subject to a constraint using Lagrange multipliers is called the **method of Lagrange multipliers**. The condition (10.24) is a necessary condition. We illustrate in the following example how to use Lagrange multipliers to find extrema with constraints.

▶ **Example 12** Find all extrema of

$$f(x, y) = e^{-xy}$$

subject to the constraint $x^2 + 4y^2 = 1$.

Solution

We define $g(x, y) = x^2 + 4y^2 - 1$. Then the constraint is of the form $g(x, y) = 0$. Using the method of Lagrange multipliers, we are looking for (x, y) and λ so that

$$\nabla f(x, y) = \lambda \nabla g(x, y) \qquad \text{and} \qquad g(x, y) = 0$$

Since

$$\nabla f(x, y) = \begin{bmatrix} -ye^{-xy} \\ -xe^{-xy} \end{bmatrix} \quad \text{and} \quad \nabla g(x, y) = \begin{bmatrix} 2x \\ 8y \end{bmatrix}$$

this translates into the set of equations

$$-ye^{-xy} = 2\lambda x \quad \text{and} \quad -xe^{-xy} = 8\lambda y \quad \text{and} \quad x^2 + 4y^2 = 1$$

We can eliminate λ from the first two equations (multiply the first equation by $4y$, the second by x, and take the difference of the two equations). We find

$$-4y^2 e^{-xy} + x^2 e^{-xy} = 0$$

Simplifying yields $e^{-xy}(x^2 - 4y^2) = 0$. Since $e^{-xy} \neq 0$, we obtain $x^2 - 4y^2 = 0$. Combining this with the constraint equation, we get the system

$$x^2 - 4y^2 = 0$$
$$x^2 + 4y^2 = 1$$

We leave the first equation and eliminate y from the second equation by adding the two equations. We find

$$x^2 - 4y^2 = 0$$
$$2x^2 = 1$$

Thus, $x^2 = 1/2$ and $4y^2 = x^2 = 1/2$, implying $y^2 = 1/8$. Solving this, we obtain the following candidates

$$\left(\sqrt{\tfrac{1}{2}}, \tfrac{1}{2}\sqrt{\tfrac{1}{2}} \right), \quad \left(\sqrt{\tfrac{1}{2}}, -\tfrac{1}{2}\sqrt{\tfrac{1}{2}} \right), \quad \left(-\sqrt{\tfrac{1}{2}}, \tfrac{1}{2}\sqrt{\tfrac{1}{2}} \right), \quad \left(-\sqrt{\tfrac{1}{2}}, -\tfrac{1}{2}\sqrt{\tfrac{1}{2}} \right)$$

with

$$f\left(\sqrt{\tfrac{1}{2}}, \tfrac{1}{2}\sqrt{\tfrac{1}{2}} \right) = e^{-1/4}, \quad f\left(\sqrt{\tfrac{1}{2}}, -\tfrac{1}{2}\sqrt{\tfrac{1}{2}} \right) = e^{1/4},$$

$$f\left(-\sqrt{\tfrac{1}{2}}, \tfrac{1}{2}\sqrt{\tfrac{1}{2}} \right) = e^{1/4}, \quad f\left(-\sqrt{\tfrac{1}{2}}, -\tfrac{1}{2}\sqrt{\tfrac{1}{2}} \right) = e^{-1/4}$$

The extreme value theorem applies to the constraint curve because this curve is closed and bounded; we can conclude that maxima and minima exist on this constraint curve and we can select them from our candidates. The maxima are $(-\sqrt{\tfrac{1}{2}}, \tfrac{1}{2}\sqrt{\tfrac{1}{2}})$ and $(\sqrt{\tfrac{1}{2}}, -\tfrac{1}{2}\sqrt{\tfrac{1}{2}})$; the minima are $(\sqrt{\tfrac{1}{2}}, \tfrac{1}{2}\sqrt{\tfrac{1}{2}})$ and $(-\sqrt{\tfrac{1}{2}}, -\tfrac{1}{2}\sqrt{\tfrac{1}{2}})$ (see Figure 10.64). ◄

Looking back at Example 12, we notice that we did not have to compute the actual value of λ to find extrema. This will typically be the case.

In the statement of the result, we mentioned that the condition $\nabla f = \lambda \nabla g$ is a necessary condition. This means that finding (x_0, y_0) so that $\nabla f(x_0, y_0) = \lambda \nabla g(x_0, y_0)$ only identifies candidates for local extrema. Here is an example that illustrates this.

▶ **Example 13**   Use Lagrange multipliers to identify candidates for local extrema of

$$f(x, y) = y$$

subject to the constraint $y - x^3 = 0$, and show that there is one such candidate which is not a local extremum. Furthermore, show that the function $f(x, y)$ subject to the constraint $y - x^3 = 0$ has no global extrema.

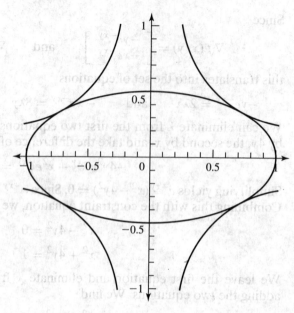

◀ **Figure 10.64**
The level curve of $f(x, y)$
that touches the constraint
curve $g(x, y) = 0$ in
Example 12

Solution

We define $g(x, y) = y - x^3$. Then

$$\nabla f(x, y) = \begin{bmatrix} 0 \\ 1 \end{bmatrix} \quad \text{and} \quad \nabla g(x, y) = \begin{bmatrix} -3x^2 \\ 1 \end{bmatrix}$$

With $y - x^3 = 0$, we obtain

$$0 = -3\lambda x^2 \quad \text{and} \quad 1 = \lambda \quad \text{and} \quad y = x^3$$

Eliminating λ, we find $x = 0$ and thus $y = 0$. We claim that $(0, 0)$ is not a local extremum. The easiest way to see this is to find out what $f(x, y)$ looks like along the constraint curve $y = x^3$. If we use $y = x^3$ to substitute y in the function $f(x, y)$, we obtain a single variable function $h(x) = x^3$, $x \in \mathbf{R}$. We know from single variable calculus that $h(x) = x^3$ has no local extrema on $\mathbf{R}$ even though there is a candidate for a local extremum at $x = 0$ since $h'(x) = 0$ for $x = 0$. Since $\lim_{x \to \infty} h(x) = \infty$ and $\lim_{x \to -\infty} h(x) = -\infty$, the function $f(x, y)$ subject to the constraint $y - x^3 = 0$ has no global extrema. ◀

The method of Lagrange multipliers only identifies candidates for local extrema and, as we saw in the previous example, these candidates may not turn out to be local extrema. Just finding candidates, however, is often good enough if we are interested in global extrema. This is illustrated in the next example.

▶ **Example 14** Suppose you wish to enclose a rectangular plot. You have 1600 ft of fencing. Using that material, what are the dimensions of the study plot that will have the largest area? (See Figure 10.65.)

Solution

We wish to maximize

$$A = xy$$

subject to the constraint $2x + 2y = 1600$. We define

$$f(x, y) = xy \quad \text{and} \quad g(x, y) = 2x + 2y - 1600 = 0$$

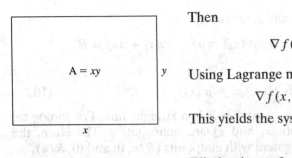

$A = xy$

▲ Figure 10.65
The rectangular plot in
Example 14

Then

$$\nabla f(x, y) = \begin{bmatrix} y \\ x \end{bmatrix} \qquad \text{and} \qquad \nabla g(x, y) = \begin{bmatrix} 2 \\ 2 \end{bmatrix}$$

Using Lagrange multipliers, we need to find (x, y) and λ so that

$$\nabla f(x, y) = \lambda \nabla g(x, y) \qquad \text{and} \qquad 2x + 2y - 1600 = 0$$

This yields the system of equations

$$y = 2\lambda \qquad x = 2\lambda \qquad x + y = 800$$

Eliminating λ from the first two equations, we conclude $x = y$ and thus $2x = 800$ or $x = 400$. For physical reasons, x and y can only take nonnegative values. The constraint thus restricts x and y to the line segment $y = 800 - x$, $0 \le x \le 800$. To see that $f(400, 400)$ gives us a maximum, we compare $f(400, 400)$ to the values at the endpoints of the line segment describing the constraint, $f(0, 800)$ and $f(800, 0)$. Since $f(400, 400) = 160{,}000$ and $f(0, 800) = f(800, 0) = 0$, $f(400, 400)$ is indeed the global maximum. ◀

You probably remember this type of problem from Section 5.4. The method of Lagrange multipliers provides another method for the problems we discussed in Section 5.4. The method of Lagrange multipliers is more general than the method we learned in Section 5.4; it can be used even if we cannot solve the constraint for either x or y to eliminate one of the two variables, as we did in Section 5.4.

As the last example in this section, we return to the motivating example at the beginning of this section where we wished to maximize $f(x_1, x_2) = p_1 n \rho(x_1) + p_2 n \rho(x_2)$ subject to the constraint $nx_1 + nx_2 = R$. We now make the additional assumption that $\rho(x)$ increases at a decelerating rate and satisfies $\rho(0) = 0$ (see Figure 10.66), implying that there is a diminishing return to increasing egg size.

▶ Example 15 Assume that x_1 and x_2 are nonnegative. Maximize

$$f(x_1, x_2) = p_1 n \rho(x_1) + p_2 n \rho(x_2)$$

subject to the constraint $nx_1 + nx_2 = R$, and show that egg size should decline with maternal age.

Solution

We find

$$\nabla f(x_1, x_2) = \begin{bmatrix} p_1 n \rho'(x_1) \\ p_2 n \rho'(x_2) \end{bmatrix} \qquad \text{and} \qquad \nabla g(x_1, x_2) = \begin{bmatrix} n \\ n \end{bmatrix}$$

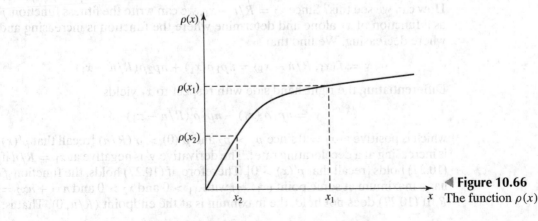

◀ Figure 10.66
The function $\rho(x)$

Thus we need to find (x_1, x_2) and λ so that

$$p_1 n \rho'(x_1) = n\lambda \qquad p_2 n \rho'(x_2) = n\lambda \qquad nx_1 + nx_2 = R$$

Eliminating λ, we obtain

$$p_1 \rho'(x_1) = p_2 \rho'(x_2) \tag{10.25}$$

Now, the constraint curve $nx_1 + nx_2 = R$ is a straight line. For biological reasons, we require that both x_1 and x_2 are nonnegative. Therefore, the constraint curve is the line segment with endpoints $(R/n, 0)$ and $(0, R/n)$.

There might not be a point (x_1, x_2) on this line segment that satisfies (10.25). It is not too difficult to show, however, that there is at most one such point. Namely, solving the constraint curve for x_2, namely $x_2 = R/n - x_1$, and substituting this into (10.25) yields

$$\frac{p_1}{p_2} = \frac{\rho'(R/n - x_1)}{\rho'(x_1)} \tag{10.26}$$

Since $\rho'(x) > 0$ and $\rho''(x) < 0$,

$$\frac{d}{dx_1} \frac{\rho'(R/n - x_1)}{\rho'(x_1)} = \frac{-\rho''(R/n - x_1)\rho'(x_1) - \rho'(R/n - x_1)\rho''(x_1)}{[\rho'(x_1)]^2} > 0$$

it follows that there is at most one value of x_1 so that (10.26) [and thus (10.25)] holds.

We thus have the following situation: If there is a point (x_1, x_2) that satisfies (10.25) and lies on this segment, then there are three candidates for global extrema, namely (x_1, x_2), $(R/n, 0)$, and $(0, R/n)$; otherwise, there are only the two endpoints $(R/n, 0)$ and $(0, R/n)$. We need to select the global maximum from this set of candidates.

Now, $p_1 > p_2$ implies $f(R/n, 0) > f(0, R/n)$. Thus the global maximum cannot occur at the endpoint $(0, R/n)$. We claim that if there is point (x_1, x_2) that satisfies (10.25) and lies on the segment with endpoints $(R/n, 0)$ and $(0, R/n)$, then the global maximum occurs at the point (x_1, x_2); otherwise, it occurs at the endpoint $(R/n, 0)$. Which of the two points yields the global maximum depends on the function ρ and the ratio of the survival probabilities p_1 and p_2. We claim that if

$$\frac{\rho'(R/n)}{\rho'(0)} < \frac{p_2}{p_1} \tag{10.27}$$

then there is a point (x_1, x_2) that satisfies (10.25) and lies on the segment with endpoints $(R/n, 0)$ and $(0, R/n)$. The global maximum then occurs at this point (x_1, x_2); otherwise, the global maximum occurs at the endpoint $(R/n, 0)$. How can we see this? Since $x_2 = R/n - x_1$, we can write the fitness function f as a function of x_1 alone and determine where the function is increasing and where decreasing. We find that

$$y = f(x_1, R/n - x_1) = np_1\rho(x_1) + np_2\rho(R/n - x_1)$$

Differentiating the right-hand side with respect to x_1 yields

$$y' = np_1\rho'(x_1) - np_2\rho'(R/n - x_1)$$

which is positive at $x_1 = 0$ since $p_1 > p_2$ and $\rho'(0) > \rho'(R/n)$ [recall that $\rho'(x)$ is increasing at a decelerating rate]. The derivative y' is negative at $x_1 = R/n$ if (10.27) holds [recall that $\rho'(x) > 0$]. Therefore, if (10.27) holds, the function f has a maximum at some point (x_1, x_2) with $x_1 > 0$ and $x_2 > 0$ and $nx_1 + nx_2 = R$. If (10.27) does not hold, the maximum is at the endpoint $(R/n, 0)$. That is, the strategy $(R/n, 0)$ can be improved by laying eggs in the second clutch if (10.27) holds. If (10.27) does not hold, the optimal strategy is laying all eggs in the first clutch and choosing egg size R/n for each of the n eggs.

To show that egg size should decline, we again use the assumption that $\rho(x)$ is a function with diminishing return, that is, $\rho(x)$ is increasing at a decelerating rate. Since $p_1 > p_2$ (the probability of being alive at a later age is smaller than at an earlier age), $\rho'(x_1) < \rho'(x_2)$ and therefore $x_1 > x_2$ (see Figure 10.66), which implies that egg size should decline. If the optimal strategy is at the endpoint $(R/n, 0)$, then the egg size in the first clutch is R/n and in the second clutch it is 0, implying that egg size should decline in this case as well. ◀

10.6.3 Diffusion

Suppose that we place a sugar cube into a glass of water without stirring the water. The sugar dissolves and the sugar molecules move about randomly in the water. If we wait long enough, the sugar concentration will eventually be uniform throughout the water. This random movement of molecules is called **diffusion** and plays an important role in many processes of life. For instance, gas exchange in unicellular organisms and many small multicellular organisms takes place by diffusion. Diffusion is a slow process, which means that cells have to be close to the surface if they want to exchange gas by diffusion. This limits the size and shape of organisms, unless they evolve different gas exchange mechanisms. (There are large organisms, like kelp, that rely on diffusion for gas exchange, but their blades are extremely thin so that all cells are close to the surface.)

Derivation of the One-Dimensional Diffusion Equation We want to understand what type of microscopic description yields a diffusion equation. We assume that molecules move along the x-axis, and we denote the concentration of these molecules at x at time t by $c(x, t)$. That is, the number of molecules at time t in the interval $[x_1, x_2)$ is given by

$$N_{[x_1, x_2)}(t) = \int_{x_1}^{x_2} c(x, t)\, dx \tag{10.28}$$

Because the molecules move around, the number of molecules in a given interval changes over time. We will express this change as the difference between the net movement of molecules on the left and on the right end of the interval. The quantity that describes this net movement is called the **flux**, and is denoted by $J(x, t)$. We interpret

$J(x, t)\, \Delta t = $ the net number of molecules crossing x from

the left to the right during a time interval of length Δt

That is, if we consider the change in the number of molecules in the interval $[x_0, x_0 + \Delta x)$ during the time interval $[t, t + \Delta)$, then

$$N_{[x_0, x_0+\Delta x)}(t + \Delta t) - N_{[x_0, x_0+\Delta x)}(t)$$
$$= J(x_0, t)\, \Delta t - J(x_0 + \Delta x, t)\, \Delta t \tag{10.29}$$

Dividing both sides of (10.29) by Δt, and letting $\Delta t \to 0$, we find

$$\lim_{\Delta t \to 0} \frac{N_{[x_0, x_0+\Delta x)}(t + \Delta t) - N_{[x_0, x_0+\Delta x)}(t)}{\Delta t} \tag{10.30}$$
$$= J(x_0, t) - J(x_0 + \Delta x, t)$$

The left-hand side of (10.30) is equal to

$$\frac{d}{dt} N_{[x_0, x_0 + \Delta x)}(t) \tag{10.31}$$

Using (10.28), we can write (10.31) as

$$\frac{d}{dt} \int_{x_0}^{x_0 + \Delta x} c(x, t)\, dx$$

When $c(x, t)$ is sufficiently smooth, we can interchange differentiation and integration. (We cannot justify this step here, but it is justified in courses on real analysis.) We find

$$\frac{d}{dt} \int_{x_0}^{x_0 + \Delta x} c(x, t)\, dx = \int_{x_0}^{x_0 + \Delta x} \frac{\partial c(x, t)}{\partial t}\, dx$$

Note that the "d" changed into a "∂" when we moved the derivative inside the integral. Before we moved it inside, we differentiated a function that depended only on t, but once we moved it inside, we differentiated a function that depends on two variables, namely x and t. Summarizing, we arrive at the equation

$$\int_{x_0}^{x_0 + \Delta x} \frac{\partial}{\partial t} c(x, t)\, dx = J(x_0, t) - J(x_0 + \Delta x, t) \tag{10.32}$$

To obtain the diffusion equation, we divide both sides of (10.32) by Δx and take the limit $\Delta x \to 0$. On the left-hand side of (8.7), we find

$$\lim_{\Delta x \to 0} \frac{1}{\Delta x} \int_{x_0}^{x_0 + \Delta x} \frac{\partial}{\partial t} c(x, t)\, dx = \frac{\partial c(x_0, t)}{\partial t}$$

On the right-hand side of (8.7), we find

$$\lim_{\Delta x \to 0} \frac{J(x_0, t) - J(x_0 + \Delta x, t)}{\Delta x} = -\frac{\partial J(x_0, t)}{\partial x}$$

Putting things together, we arrive at

$$\frac{\partial c(x_0, t)}{\partial t} = -\frac{\partial J(x_0, t)}{\partial x} \tag{10.33}$$

A phenomenological law, namely Fick's law, relates the flux to the change in concentration. It holds, for instance, when molecules move around randomly in a solvent. Fick's law says that

$$J = -D \frac{\partial c}{\partial x} \tag{10.34}$$

where D is a positive constant, called the **diffusion constant**. It means that the flux is proportional to the change in concentration; the minus sign in (10.34) means that the net movement of molecules is from regions of high concentration to regions of low concentration. This agrees with our intuition: Going back to our example of sugar dissolving in water, we expect the net movement of sugar molecules to be from regions of high concentration to regions of low concentration so that ultimately the sugar concentration is uniform.

Combining (10.33) and (8.9), we arrive at the **diffusion equation**, namely

$$\frac{\partial c}{\partial t} = D \frac{\partial^2 c}{\partial x^2} \tag{10.35}$$

The diffusion approach is ubiquitous in biology. It is not only used in the description of the random movement of molecules but in a wide array of applications, such as the change in allele frequencies due to random genetic drift, the invasion of alien species into virgin habitat, the directed movement of organisms along gradients of chemicals (chemotaxis), pattern formation, and many more. Equation (10.35) is the simplest form of an equation that incorporates diffusion. In physics, (8.10) is called the heat equation. It describes the diffusion of heat through a solid bar; in this case, $c(x, t)$ represents the temperature at x at time t.

Equation (8.10) is an example of a **partial differential equation**, which is an equation that contains partial derivatives. The theory of partial differential equations is complex and well beyond the scope of this course. We will only be able to discuss some aspects of this equation.

Solving the Diffusion Equation   For most partial differential equations, it is not possible to find an analytical solution; such equations often can only be solved numerically, and even this is typically not an easy task. Fortunately, (8.10) is simple enough that we can find a solution. We claim that

$$c(x, t) = \frac{1}{\sqrt{4\pi Dt}} \exp\left[-\frac{x^2}{4Dt}\right] \tag{10.36}$$

is a solution of (8.10). As in the case of ordinary differential equations, we can check this by computing the appropriate derivatives. For the left-hand side of (8.10), we need the first partial derivative of $c(x, t)$ with respect to t. We find

$$\frac{\partial c(x, t)}{\partial t} = \frac{\partial}{\partial t} \frac{1}{\sqrt{4\pi Dt}} \exp\left[-\frac{x^2}{4Dt}\right]$$

$$= -\frac{1}{2} \frac{4\pi D}{(4\pi Dt)^{3/2}} \exp\left[-\frac{x^2}{4Dt}\right]$$

$$+ \frac{1}{\sqrt{4\pi Dt}} \exp\left[-\frac{x^2}{4Dt}\right] \frac{x^2}{4Dt^2} \tag{10.37}$$

$$= \exp\left[-\frac{x^2}{4Dt}\right] \left\{ \frac{x^2}{4Dt^2\sqrt{4\pi Dt}} - \frac{2\pi D}{4\pi Dt\sqrt{4\pi Dt}} \right\}$$

$$= \frac{1}{2t\sqrt{4\pi Dt}} \exp\left[-\frac{x^2}{4Dt}\right] \left\{ \frac{x^2}{2Dt} - 1 \right\}$$

On the right-hand side of (8.10), we need the second partial derivative of $c(x, t)$ with respect to x. We find

$$\frac{\partial c(x, t)}{\partial x} = \frac{\partial}{\partial x} \frac{1}{\sqrt{4\pi Dt}} \exp\left[-\frac{x^2}{4Dt}\right]$$

$$= \frac{1}{\sqrt{4\pi Dt}} \exp\left[-\frac{x^2}{4Dt}\right] \left(-\frac{2x}{4Dt}\right)$$

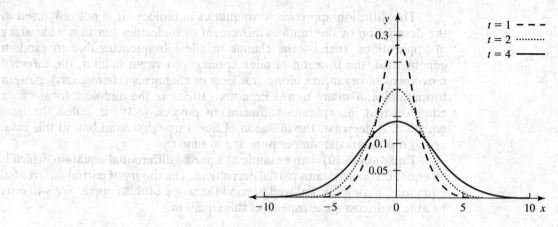

▲ **Figure 10.67**
The solution of the diffusion equation at different times

and

$$\frac{\partial^2 c(x,t)}{\partial x^2} = \frac{-1}{\sqrt{4\pi Dt}} \frac{1}{2Dt} \exp\left[-\frac{x^2}{4Dt}\right]\left\{1 - x\frac{2x}{4Dt}\right\}$$

$$= \frac{1}{2Dt\sqrt{4\pi Dt}} \exp\left[-\frac{x^2}{4Dt}\right]\left\{\frac{x^2}{2Dt} - 1\right\} \tag{10.38}$$

Putting things together, we see that (10.36) satisfies (8.10).

The function in (8.11) is called the **Gaussian density**. Figure 10.67 shows $c(x,t)$ for $t = 1, 2$, and 4; it clearly shows that the concentration $c(x,t)$ becomes more uniform as time goes on.

Diffusion is a very slow process. The diffusion constant D measures how quickly it proceeds. The larger D, the faster the concentration spreads out, and we can show that within t units of time, the bulk of the molecules spread over a region of length of order $\sqrt{t}$.

To give an idea of how slow diffusion is, here are a few examples taken from Yeargers, Shonkwiler, and Herod (1996). Oxygen in blood at 20°C has a diffusion constant of 10^{-5} cm²/s, which means that it takes an oxygen molecule roughly 500 seconds to cross a distance of 1 mm by diffusion alone. Ribonuclease (an enzyme that hydrolyzes RNA [ribonucleic acid]) in water at 20°C has a diffusion constant of 1.1×10^{-6} cm²/s, which means that ribonuclease takes roughly 4672 seconds (or 1 hr 18 min) to cross a distance of 1 mm by diffusion alone. These examples illustrate why organisms frequently rely on other active mechanisms to transport molecules.

The diffusion equation (8.10) can be generalized to higher dimensions. Namely, (8.8) becomes

$$\frac{\partial c}{\partial t} = -\nabla J \tag{10.39}$$

and (8.9) becomes

$$J = -D\nabla c \tag{10.40}$$

Combining (10.39) and (10.40), we find

$$\frac{\partial c}{\partial t} = D\nabla \cdot (\nabla c)$$

where $\nabla \cdot (\nabla c)$ is to be interpreted as a dot product. That is, if $\mathbf{x} = (x_1, x_2, x_3) \in \mathbf{R}^3$, $t \in \mathbf{R}$, then

$$\frac{\partial c}{\partial t} = D \left(\frac{\partial^2 c}{\partial x_1^2} + \frac{\partial^2 c}{\partial x_2^2} + \frac{\partial^2 c}{\partial x_3^2} \right)$$

As a shorthand notation, we define

$$\Delta = \frac{\partial^2}{\partial x_1^2} + \frac{\partial^2}{\partial x_2^2} + \frac{\partial^2}{\partial x_3^2}$$

where Δ is called the **Laplace operator**. We then write

$$\frac{\partial c}{\partial t} = D \, \Delta c$$

More generally, if $\mathbf{x} = (x_1, x_2, \dots, x_n) \in \mathbf{R}^n$, then

$$\Delta = \frac{\partial^2}{\partial x_1^2} + \frac{\partial^2}{\partial x_2^2} + \cdots + \frac{\partial^2}{\partial x_n^2}$$

(Δc is read "Laplacian of c.")

10.6.4 Problems

(10.6.1)

In Problems 1–10, the functions are defined for all $(x, y) \in \mathbf{R}^2$. Find all candidates for local extrema and use the Hessian matrix to determine the type (maximum, minimum, or saddle).

1. $f(x, y) = x^2 + y^2 - 2x$

2. $f(x, y) = -2x^2 - y^2 + 3y$

3. $f(x, y) = x^2 y - 4x^2 - 4y$

4. $f(x, y) = xy - 2y^2$

5. $f(x, y) = -2x^2 + y^2 - 6y$

6. $f(x, y) = x(1 - x + y)$

7. $f(x, y) = e^{-x^2 - y^2}$

8. $f(x, y) = yxe^{-y}$

9. $f(x, y) = x \cos y$

10. $f(x, y) = y \sin x$

11. In this problem we will illustrate that if one of the eigenvalues of the Hessian matrix at a point where the gradient vanishes is equal to 0, then we cannot make any statements about whether the point is a local extremum just based on the Hessian matrix. Consider the following functions:

$$f_1(x, y) = x^2$$
$$f_2(x, y) = x^2 + y^3$$
$$f_3(x, y) = x^2 + y^4$$

Figures 10.68 through 10.70 show graphs of the three functions.

(a) Show that for $i = 1, 2,$ and 3,

$$\nabla f_i(0, 0) = \begin{bmatrix} 0 \\ 0 \end{bmatrix}$$

(b) Show that for $i = 1, 2,$ and 3,

$$\text{Hess } f_i(0, 0) = \begin{bmatrix} 2 & 0 \\ 0 & 0 \end{bmatrix}$$

and determine the eigenvalues of Hess $f_i(0, 0)$.

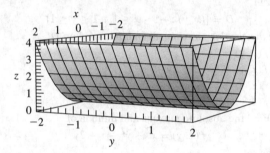

▲ **Figure 10.68**
$f_1(x, y)$ in Problem 11

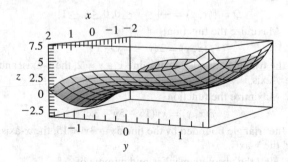

▲ **Figure 10.69**
$f_2(x, y)$ in Problem 11

(c) Since one of the eigenvalues of Hess $f_i(0, 0)$ is equal to 0, we cannot use the criterion stated in the text to determine the behavior of the three functions at $(0, 0)$. Use Figures 10.68 through 10.70 to describe what happens at $(0, 0)$ for each function.

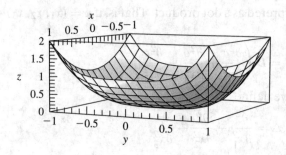

▲ **Figure 10.70**
$f_3(x, y)$ in Problem 11

12. Consider the function
$$f(x, y) = ax^2 + by^2$$

(a) Show that
$$\nabla f(0, 0) = \begin{bmatrix} 0 \\ 0 \end{bmatrix}$$

(b) Find values for a and b so that (i) $(0, 0)$ is a local minimum, and (ii) $(0, 0)$ is a local maximum, and (iii) is a saddle.

In Problems 13–16, the functions are defined on the rectangular domain
$$D = \{(x, y) : -1 \leq x \leq 1, -1 \leq y \leq 1\}$$
Find the absolute maxima and minima of f on D.

13. $f(x, y) = 2x - y$
14. $f(x, y) = 3 - x + 2y$
15. $f(x, y) = x^2 - y^2$
16. $f(x, y) = x^2 + y^2$
17. Find the absolute maxima and minima of
$$f(x, y) = x^2 + y^2 - x + 2y$$
on the set
$$D = \{(x, y) = 0 \leq x \leq 1, -2 \leq y \leq 0\}$$

18. Find the absolute maxima and minima of
$$f(x, y) = x^2 - y^2 + 4x + y$$
on the set
$$D = \{(x, y) = -4 \leq x \leq 0, 0 \leq y \leq 1\}$$

19. Maximize the function
$$f(x, y) = 2xy - x^2y - xy^2$$
on the triangle bounded by the line $x + y = 2$, the x-axis, and the y-axis.

20. Maximize the function
$$f(x, y) = xy(15 - 5y - 3x)$$
on the triangle bounded by the line $5y + 3x = 15$, the x-axis, and the y-axis.

21. Find the absolute maxima and minima of
$$f(x, y) = x^2 + y^2 + 4x - 1$$
on the disk
$$D = \{(x, y) : x^2 + y^2 \leq 9\}$$

22. Find the absolute maxima and minima of
$$f(x, y) = x^2 + y^2 - 6y + 3$$
on the disk
$$D = \{(x, y) : x^2 + y^2 \leq 16\}$$

23. Find the absolute maxima and minima of
$$f(x, y) = x^2 + y^2 + x - y$$
on the disk
$$D = \{(x, y) : x^2 + y^2 \leq 1\}$$

24. Find the absolute maxima and minima of
$$f(x, y) = x^2 + y^2 + x + 2y$$
on the disk
$$D = \{(x, y) : x^2 + y^2 \leq 4\}$$

25. Can a continuous function of two variables have two maxima and no minima? Describe in words what the properties of such a function would be and contrast this behavior with a function of one variable.

26. Suppose $f(x, y)$ has a horizontal tangent plane at $(0, 0)$. Can you conclude that f has a local extremum at $(0, 0)$?

27. Suppose crop yield Y depends on nitrogen (N) and phosphorus (P) concentrations as
$$Y(N, P) = NPe^{-(N+P)}$$
Find the value of (N, P) that maximizes crop yield.

28. Choose three numbers x, y, and z so that their sum is equal to 60 and their product is maximal.

29. Find the maximum volume of a rectangular closed (top, bottom and four sides) box with surface area 48 m².

30. Find the maximum volume of a rectangular open (bottom and four sides, no top) box with surface area 75 m².

31. Find the minimum surface area of a rectangular closed (top, bottom, and four sides) box with volume 216 m³.

32. Find the minimum surface area of a rectangular open (bottom and four sides, no top) box with volume 256 m³.

33. The distance between the origin $(0, 0, 0)$ and the point (x, y, z) is
$$\sqrt{x^2 + y^2 + z^2}$$
Find the minimum distance between the origin and the plane $x + y + z = 1$. (*Hint:* Minimize the squared distance between the origin and the plane.)

34. Given the symmetric matrix
$$A = \begin{bmatrix} a & c \\ c & b \end{bmatrix}$$
where a, b, and c are real numbers, show that the eigenvalues of A are real. (*Hint:* Compute the eigenvalues.)

35. Understanding species richness and diversity is a major concern of ecological studies. A frequently used measure of diversity is the Shannon and Weaver index
$$H = -\sum_{i=1}^{n} p_i \ln p_i$$
where p_i is equal to the proportion of species i, $i = 1, 2, \ldots, n$, and n is the total number of species in the study area. Assume now that a community consists of three species with relative proportions p_1, p_2, and p_3.

(a) Use the fact that $p_1 + p_2 + p_3 = 1$ to show that H is of the form
$$H(p_1, p_2) = -p_1 \ln p_1 - p_2 \ln p_2$$
$$-(1 - p_1 - p_2) \ln(1 - p_1 - p_2)$$

and that the domain of $H(p_1, p_2)$ is the triangular set in the p_1-p_2 plane bounded by the lines $p_1 = 0$, $p_2 = 0$, and $p_1 + p_2 = 1$.

(b) Show that H attains its absolute maximum when $p_1 = p_2 = p_3 = 1/3$.

(10.6.2)

In Problems 36–45, use Lagrange multipliers to find the maxima and minima of the functions under the given constraints.

36. $f(x, y) = 2x - y$; $x^2 + y^2 = 5$

37. $f(x, y) = 3x^2 + y$; $x^2 + y^2 = 1$

38. $f(x, y) = xy$; $x^2 + y^2 = 4$

39. $f(x, y) = xy$; $2x - 4y = 1$

40. $f(x, y) = x^2 - y^2$; $2x + y = 1$

41. $f(x, y) = x^2 + y^2$; $3x - 2y = 4$

42. $f(x, y) = xy^2$; $x^2 - y = 0$

43. $f(x, y) = x^2 y$; $x^2 + 3y = 1$

44. $f(x, y) = x^2 y^2$; $2x - 3y = 4$

45. $f(x, y) = x^2 y^2$; $x^2 - y^2 = 1$

In Problems 46–55, use Lagrange multipliers to find the answers to the indicated problems in Section 5.4.

46. Problem 1

47. Problem 2

48. Problem 3

49. Problem 4

50. Problem 5

51. Problem 6

52. Problem 7

53. Problem 9

54. Problem 12

55. Problem 18

56. Let

$$f(x, y) = x + y, \quad (x, y) \in \mathbf{R}^2$$

with constraint function $xy = 1$.

(a) Use Lagrange multipliers to find all local extrema.

(b) Are there global extrema?

57. Let

$$f(x, y) = x + y$$

with constraint function

$$\frac{1}{x} + \frac{1}{y} = 1, \ x \neq 0, y \neq 0$$

(a) Use Lagrange multipliers to find all local extrema.

(b) Are there global extrema?

58. Let

$$f(x, y) = xy, \quad (x, y) \in \mathbf{R}^2$$

with constraint function $y - x^2 = 0$.

(a) Use Lagrange multipliers to find candidates for local extrema.

(b) Use the constraint $y - x^2 = 0$ to reduce $f(x, y)$ to a single-variable function and use this function to show that $f(x, y)$ has no local extrema on the constraint curve.

59. Explain why $f(x, y)$ has a local extremum at the point P in Figure 10.63 under the constraint $g(x, y) = 0$ if $c_1 > c_2 > c_3 > c_4$.

60. Explain why $f(x, y)$ has a local extremum at the point P in Figure 10.63 under the constraint $g(x, y) = 0$ if $c_1 < c_2$ and $c_2 > c_3 > c_4$.

61. In the introductory example, we discussed how egg size depends on maternal age. Assume now that the total amount of resources available is 10 (in appropriate units), the number of eggs per clutch is 3, the number of clutches is 2, and the egg size in clutch number i is denoted by x_i.

(a) Find the constraint function.

(b) Suppose the fitness function is given by

$$f(x_1, x_2) = \frac{3}{2}\rho(x_1) + \frac{3}{4}\rho(x_2)$$

where $\rho(x) = \frac{2x}{5+x}$. Find the optimal egg sizes for clutch 1 and clutch 2 under the constraint in (a).

62. In the introductory example, we discussed how egg size depends on maternal age. Assume now that the fitness function is given by

$$f(x_1, x_2) = \frac{5}{3}\rho(x_1) + \frac{5}{6}\rho(x_2)$$

with

$$\rho(x) = \frac{3x}{4+x}$$

The constraint function is given by

$$5x_1 + 5x_2 = 7$$

(a) Compare the given functions to the corresponding ones in the text and identify the parameters n, p_1, p_2, and R from the text.

(b) Solve the constraint function for x_2 and substitute your expression for x_2 into the function f. This then yields a function of one variable. Find the domain of this single variable function and use single variable calculus to determine optimal egg sizes for clutch 1 and clutch 2.

(10.6.3)

63. Show that

$$c(x, t) = \frac{1}{\sqrt{8\pi t}} \exp\left[-\frac{x^2}{8t}\right]$$

solves

$$\frac{\partial c(x, t)}{\partial t} = 2\frac{\partial^2 c(x, t)}{\partial x^2}$$

64. Show that

$$c(x, t) = \frac{1}{\sqrt{2\pi t}} \exp\left[-\frac{x^2}{2t}\right]$$

solves

$$\frac{\partial c(x, t)}{\partial t} = \frac{1}{2}\frac{\partial^2 c(x, t)}{\partial x^2}$$

65. A solution of

$$\frac{\partial c(x, t)}{\partial t} = D\frac{\partial^2 c(x, t)}{\partial x^2}$$

is the function

$$c(x, t) = \frac{1}{\sqrt{4\pi Dt}} \exp\left[-\frac{x^2}{4Dt}\right]$$

for $x \in \mathbf{R}$ and $t > 0$.

(a) Show that $c(x, t)$ as a function of x for fixed values of $t > 0$ is (i) positive for all $x \in \mathbf{R}$, (ii) increasing for $x < 0$, decreasing for $x > 0$, (iii) has a local maximum at $x = 0$, and (iv) has inflection points at $x = \pm\sqrt{2Dt}$.

(b) Graph $c(x, t)$ as a function of x when $D = 1$, for $t = 0.01$, $t = 0.1$, and $t = 1$.

66. A solution of

$$\frac{\partial c(x, t)}{\partial t} = D\frac{\partial^2 c(x, t)}{\partial x^2}$$

is the function

$$c(x, t) = \frac{1}{\sqrt{4\pi Dt}}\exp\left[-\frac{x^2}{4Dt}\right]$$

for $x \in \mathbf{R}$ and $t > 0$.

(a) Show that a local maximum of $c(x, t)$ for fixed t occurs at $x = 0$.

(b) Show that $c(0, t)$, $t > 0$, is a decreasing function of t.

(c) Find

$$\lim_{t \to 0^+} c(x, t)$$

when $x = 0$ and when $x \neq 0$.

(d) Use the fact that

$$\int_{-\infty}^{\infty} e^{-u^2/2}\, du = \sqrt{2\pi}$$

to show that for $t > 0$,

$$\int_{-\infty}^{\infty} c(x, t)\, dx = 1$$

(e) The function $c(x, t)$ can be interpreted as the concentration of a substance diffusing in space. Explain the meaning of

$$\int_{-\infty}^{\infty} c(x, t)\, dx = 1$$

and use your results in (c) and (d) to explain why this means that initially (that is, at $t = 0$) the entire amount of the substance was released at the origin.

Mathematically, we can specify such an initial condition (in which the substance is concentrated at the origin at time 0) by the δ-function $\delta(x)$, with the property

$$\delta(x) = 0, \quad \text{for } x \neq 0$$

and

$$\int_{-\infty}^{\infty} \delta(x)\, dx = 1$$

67. The two-dimensional diffusion equation

$$\frac{\partial n(\mathbf{r}, t)}{\partial t} = D\left(\frac{\partial^2 n(\mathbf{r}, t)}{\partial x^2} + \frac{\partial^2 n(\mathbf{r}, t)}{\partial y^2}\right) \quad (10.41)$$

where $n(\mathbf{r}, t)$, $\mathbf{r} = (x, y)$, denotes the population density at the point $\mathbf{r} = (x, y)$ in the plane at time t, can be used to describe the spread of organisms. Assume that a large number of insects are released at time 0 at the point $(0, 0)$. Furthermore, assume that at later times, the density of these insects can be described by the diffusion equation (10.41). Show that

$$n(x, y, t) = \frac{n_0}{4\pi Dt}\exp\left[-\frac{x^2 + y^2}{4Dt}\right]$$

satisfies (10.41).

10.7 SYSTEMS OF DIFFERENCE EQUATIONS (OPTIONAL)

10.7.1 A Biological Example

About 14% of all insect species (and thus about 10% of all species of multicellular animals) are estimated to belong to a group of insects called *parasitoids*. These are insects (mostly in the order Hymenoptera) that lay their eggs on, in, or near the (in most cases, immature) body of another arthropod, which serves as a host for the developing parasitoids. The eggs develop into free living adults while consuming the host.

Parasitoids play an important role in biological control. A successful example is *Trichogramma* wasps that parasitize insect eggs. These wasps are reared in factories for field releases. Every year, millions of hectars of agricultural land are treated with released *Trichogramma* wasps; for instance, to protect sugar cane from the sugar cane borer, *Chilo* spp., in China, or to protect corn fields from the European corn borer, *Ostrinia nubilalis* Hübner, in Western Europe. Another successful example of biological control of an insect pest is the parasitoid wasp *Aphytis melinus* that regulates the red scale (*Aonidiella aurantii*), which damages citrus trees in California.

The importance of parasitoids in pest control stimulated both empirical and theoretical work. Theoretical studies of host-parasitoid interactions go back to Thompson (1924) and Nicholson and Bailey (1935). The work of Nicholson and Bailey was particularly influential. They introduced discrete generation, host-parasitoid models of the form

$$N_{t+1} = bN_t e^{-aP_t}$$

$$P_{t+1} = cN_t[1 - e^{-aP_t}]$$

for $t = 0, 1, 2, \ldots$. Here, N and P denote the population sizes of susceptible hosts and searching adult female parasitoids at times t and $t+1$, respectively. The parameter b is interpreted as the *net growth parameter*. We see from the first equation that hosts grow exponentially in the absence of parasitoids ($P = 0$). The term e^{-aP_t} is the fraction of hosts that are *not* parasitized (and thus $1 - e^{-aP_t}$ is the fraction of hosts that are parasitized) at generation t. Parasitized hosts produce parasitoids. The parameter c is equal to the number of parasitoids produced per parasitized host. Note that only hosts that are not parasitized reproduce.

A numerical simulation of the Nicholson-Bailey equation (Figure 10.71) shows that population sizes oscillate with increasing amplitude until either the parasitoid becomes extinct, followed by an exponential increase of the host, or until the host becomes extinct, followed by the extinction of the parasitoid.

The model behavior disagrees with most empirical studies. (Although some laboratory experiments have produced such unstable behavior.) The model has since been modified in a number of ways to stabilize the dynamics. One such attempt is called the negative binomial model (Griffiths, 1969, May, 1978) in which

$$N_{t+1} = bN_t \left(1 + \frac{aP}{k}\right)^{-k}$$

$$P_{t+1} = cN_t \left[1 - \left(1 + \frac{aP}{k}\right)^{-k}\right]$$

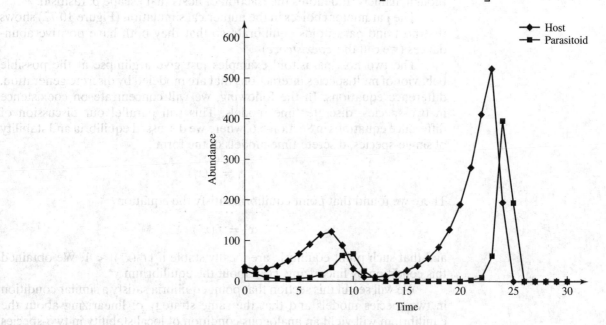

▲ **Figure 10.71**
A numerical simulation of the Nicholson-Bailey equation with $a = 0.023$, $b = 1.5$, and $c = 1$. The population sizes oscillate until extinction occurs

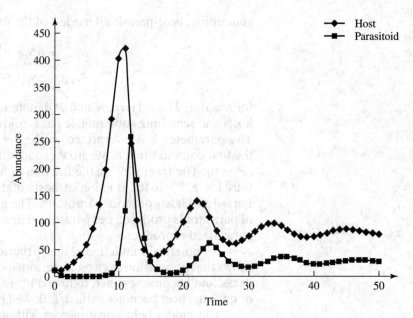

▲ Figure 10.72
A numerical simulation of the negative binomial model with $a = 0.023$, $b = 1.5$,
$c = 1$, and $k = 0.5$. The choices for the parameters a, b, and c are the same as in
Figure 10.71. The choice for k stabilizes the host and parasitoid interactions and
coexistence occurs

The form of this set of equations is quite similar to the Nicholson-Bailey
equation, and the parameters b and c have the same interpretation as before.
The main (and crucial) difference is the term $\left(1 + \frac{aP}{k}\right)^{-k}$ that replaces the
term e^{-aP} in the Nicholson-Bailey model. It has the same interpretation
though, namely, it denotes the fraction of hosts that escape parasitism.

The parameter choices in the numerical simulation (Figure 10.72) shows
that host and parasitoids equilibrate so that they both have positive abundances (we call this *coexistence*).

The two host-parasitoid examples just give a glimpse at the possible
behavior of multispecies interactions that are modeled by discrete generation,
difference equations. In the following, we will concentrate on coexistence
in two-species, discrete time models. This will parallel our discussion of
difference equations in Section 5.6, where we discussed equilibria and stability
of single-species, discrete time models of the form

$$x_{t+1} = f(x_t)$$

There we found that point equilibria satisfy the equation

$$x^\star = f(x^\star)$$

and that such point equilibria are locally stable if $|f'(x^\star)| < 1$. We obtained
this condition by linearizing $f(x)$ about the equilibrium $x^\star$.

We will see in this section that point equilibria satisfy a similar condition
in two-species models and that the same strategy of linearizing about the
equilibrium will yield an analogous condition of local stability in two-species
models. Since investigating nonlinear difference equations will lead us to
the study of linear difference equations, we begin our discussion with linear
difference equations.

10.7.2 Equilibria and Stability in Systems of Linear Difference Equations

Linear difference equations are of the form

$$x_1(t+1) = a_{11}x_1(t) + a_{12}x_2(t) \tag{10.42}$$

$$x_2(t+1) = a_{21}x_1(t) + a_{22}x_2(t) \tag{10.43}$$

where $t = 0, 1, 2, \ldots$. This can be written in matrix form

$$\underbrace{\begin{bmatrix} x_1(t+1) \\ x_2(t+1) \end{bmatrix}}_{\mathbf{x}(t+1)} = \underbrace{\begin{bmatrix} a_{11} & a_{12} \\ a_{21} & a_{22} \end{bmatrix}}_{A} \underbrace{\begin{bmatrix} x_1(t) \\ x_2(t) \end{bmatrix}}_{\mathbf{x}(t)}$$

which shows that linear difference equations are *linear maps*, which we discussed in Section 9.3. First, note that if $\mathbf{x}(0) = \begin{bmatrix} 0 \\ 0 \end{bmatrix}$, then $\mathbf{x}(t) = \begin{bmatrix} 0 \\ 0 \end{bmatrix}$ for all $t = 1, 2, 3, \ldots$. We call $\begin{bmatrix} 0 \\ 0 \end{bmatrix}$ a (**point**) **equilibrium**. More generally, a point equilibrium satisfies the equation

$$\mathbf{x}^{\star} = A\mathbf{x}^{\star}$$

We see that $\begin{bmatrix} 0 \\ 0 \end{bmatrix}$ is a point equilibrium of the linear system $\mathbf{x}(t+1) = A\mathbf{x}(t)$.

We will now investigate what happens when $\mathbf{x}(0) \neq \begin{bmatrix} 0 \\ 0 \end{bmatrix}$.

In Section 9.3, we learned how to compute $\mathbf{x}(t)$ for a given initial condition $\mathbf{x}(0)$ without computing the values of $\mathbf{x}(s)$ for all values of s between 0 and t. We found that if A has two real and distinct eigenvalues λ_1 and λ_2, then we can write any vector $\mathbf{x}(0)$ as a linear combination of its eigenvectors $\mathbf{u}_1$ and $\mathbf{u}_2$ (corresponding to λ_1 and λ_2, respectively),

$$\mathbf{x}(0) = c_1 \mathbf{u}_1 + c_2 \mathbf{u}_2$$

where c_1 and c_2 are real numbers. Using this representation of $\mathbf{x}(0)$, we found that

$$\mathbf{x}(t) = c_1 \lambda_1^t \mathbf{u}_1 + c_2 \lambda_2^t \mathbf{u}_2 \tag{10.44}$$

which we can use to say something about the *long-term behavior* of $\mathbf{x}(t)$, $\lim_{t \to \infty} \mathbf{x}(t)$. Let's return to the question of what happens to $\mathbf{x}(t)$ when $\mathbf{x}(t) \neq \begin{bmatrix} 0 \\ 0 \end{bmatrix}$. We see from the representation of $\mathbf{x}(t)$ in (10.44) that if $|\lambda_1| < 1$ and $|\lambda_2| < 1$, then $\lim_{t \to \infty} \mathbf{x}(t) = \begin{bmatrix} 0 \\ 0 \end{bmatrix}$, regardless of $\mathbf{x}(0)$. We say in this case that $\begin{bmatrix} 0 \\ 0 \end{bmatrix}$ is a **stable equilibrium**. If either $|\lambda_1| > 1$ or $|\lambda_2| > 1$, then the equilibrium $\begin{bmatrix} 0 \\ 0 \end{bmatrix}$ is called **unstable**.

The stability condition for the equilibrium $\begin{bmatrix} 0 \\ 0 \end{bmatrix}$ of $\mathbf{x}(t+1) = A\mathbf{x}(t)$, where A is a 2×2 matrix holds more generally (not just for λ_1 and λ_2 real).

> The point $\begin{bmatrix} 0 \\ 0 \end{bmatrix}$ is a stable equilibrium of
>
> $$\mathbf{x}(t+1) = A\mathbf{x}(t)$$
>
> if both eigenvalues λ_1 and λ_2 of A satisfy
>
> $$|\lambda_1| < 1 \qquad \text{and} \qquad |\lambda_2| < 1$$
>
> If either $|\lambda_1| > 1$ or $|\lambda_2| > 1$, $\begin{bmatrix} 0 \\ 0 \end{bmatrix}$ is an unstable equilibrium.

In the preceding criterion, we no longer require the eigenvalues of A to be real and distinct. However, it is beyond the scope of this book to show the criterion for general λ_1 and λ_2. Here is the first example.

▶ **Example 1** Show that $\begin{bmatrix} 0 \\ 0 \end{bmatrix}$ is an equilibrium of

$$\begin{bmatrix} x_1(t+1) \\ x_2(t+1) \end{bmatrix} = \begin{bmatrix} -0.4 & 0.2 \\ -0.3 & 0.1 \end{bmatrix} \begin{bmatrix} x_1(t) \\ x_2(t) \end{bmatrix}$$

and determine its stability.

Solution

To check that $\begin{bmatrix} 0 \\ 0 \end{bmatrix}$ is an equilibrium, we need to show that $\begin{bmatrix} 0 \\ 0 \end{bmatrix}$ satisfies

$$\begin{bmatrix} 0 \\ 0 \end{bmatrix} = \begin{bmatrix} -0.4 & 0.2 \\ -0.3 & 0.1 \end{bmatrix} \begin{bmatrix} 0 \\ 0 \end{bmatrix}$$

But this is true since $A \begin{bmatrix} 0 \\ 0 \end{bmatrix} = \begin{bmatrix} 0 \\ 0 \end{bmatrix}$ for any matrix A with constant entries. To determine the stability of A, we need to find the eigenvalues of A. We need to solve

$$\det(A - \lambda I) = \det \begin{bmatrix} -0.4 - \lambda & 0.2 \\ -0.3 & 0.1 - \lambda \end{bmatrix}$$

$$= (-0.4 - \lambda)(0.1 - \lambda) + (0.2)(0.3)$$

$$= \lambda^2 + 0.3\lambda + 0.02 = 0$$

The solutions are

$$\lambda_{1,2} = \frac{-0.3 \pm \sqrt{0.09 - 0.08}}{2} = \frac{-0.3 \pm 0.1}{2}$$

and, hence, $\lambda_1 = -0.1$ and $\lambda_2 = -0.2$. Since $|\lambda_1| = |-0.1| < 1$ and $|\lambda_2| = |-0.2| < 1$, it follows that $\begin{bmatrix} 0 \\ 0 \end{bmatrix}$ is stable. ◄

When the eigenvalues λ_1 and λ_2 are complex conjugate, then the criterion for stability can be simplified. Namely, if λ_1 and λ_2 are complex conjugate, then

$$|\lambda_1|^2 = |\lambda_2|^2 = \lambda_1\lambda_2 \tag{10.45}$$

This identity is not obvious at first sight but can be shown when we graph λ_1 and λ_2 and determine their absolute values. Let

$$\lambda_1 = a + ib \qquad \text{and} \qquad \lambda_2 = a - ib$$

be the two complex conjugate eigenvalues of A. In Figure 10.73, we draw λ_1 and λ_2 when both a and b are positive. An application of the Pythagorean theorem shows that (Figure 10.73)

$$|\lambda_1|^2 = a^2 + b^2 \qquad \text{and} \qquad |\lambda_2|^2 = a^2 + b^2 \tag{10.46}$$

Algebraically, we find

$$\lambda_1\lambda_2 = (a + ib)(a - ib) = a^2 - (ib)^2 = a^2 - i^2b^2 = a^2 + b^2$$

since $i^2 = -1$. Combining this with (10.46) shows (10.45). We use (10.45) in the next example.

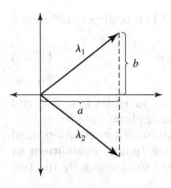

▲ **Figure 10.73**
A graphical illustration of the identity $|\lambda_1|^2 = |\lambda_2|^2 = \lambda_1\lambda_2$

▶ **Example 2** Show that the equilibrium $\begin{bmatrix} 0 \\ 0 \end{bmatrix}$ of

$$\begin{bmatrix} x_1(t+1) \\ x_2(t+1) \end{bmatrix} = \begin{bmatrix} -0.3 & -0.5 \\ 0.7 & 0.15 \end{bmatrix}\begin{bmatrix} x_1(t) \\ x_2(t) \end{bmatrix}$$

is stable.

Solution

To test stability of $\begin{bmatrix} 0 \\ 0 \end{bmatrix}$, we need to solve

$$\det\begin{bmatrix} -0.3 - \lambda & -0.5 \\ 0.7 & 0.15 - \lambda \end{bmatrix} = 0$$

This amounts to solving the quadratic equation

$$(-0.3 - \lambda)(0.15 - \lambda) + (0.5)(0.7) = 0$$

or

$$\lambda^2 + 0.15\lambda + 0.305 = 0 \tag{10.47}$$

Since the discriminant $(0.15)^2 - (4)(1)(0.305) = -1.1975 < 0$, it follows that the two solutions λ_1 and λ_2 of (10.47) are complex conjugate. Without computing λ_1 and λ_2, we can check whether $|\lambda_1|$ and $|\lambda_2|$ are both less than 1 since $|\lambda_1|^2 = |\lambda_2|^2 = \lambda_1\lambda_2 = \det A$. Now,

$$\det A = \begin{bmatrix} -0.3 & -0.5 \\ 0.7 & 0.15 \end{bmatrix}$$

$$= (-0.3)(0.15) - (0.7)(-0.5) = 0.305 < 1$$

Therefore, $|\lambda_1| < 1$ and $|\lambda_2| < 1$ and $\begin{bmatrix} 0 \\ 0 \end{bmatrix}$ is a stable equilibrium. ◀

10.7.3 Equilibria and Stability of Nonlinear Systems of Difference Equations

We saw examples of nonlinear systems of difference equations in the beginning of this section: the Nicholson-Bailey equation and the negative binomial model. The general form of a system of two nonlinear difference equations is

$$x_1(t+1) = F(x_1(t), x_2(t)) \tag{10.48}$$

$$x_2(t+1) = G(x_1(t), x_2(t)) \tag{10.49}$$

where F and G are (nonlinear) functions of the two variables x_1 and x_2 and $t = 0, 1, 2, \ldots$ is the independent variable that denotes time.

As in the previous subsection, we will be interested in equilibria and their stability. We say that the point $(x_1^\star, x_2^\star)$ is a (**point**) **equilibrium** of the system (10.48) and (10.49) if $x_1^\star$ and $x_2^\star$ satisfy simultaneously the two equations

$$x_1^\star = F(x_1^\star, x_2^\star)$$

$$x_2^\star = G(x_1^\star, x_2^\star)$$

▶ **Example 3** Find all equilibria of

$$x_1(t+1) = 2x_1(t)[1 - x_1(t)]$$

$$x_2(t+1) = x_1(t)[1 - x_2(t)]$$

Solution

To find equilibria, we need to solve

$$x_1 = 2x_1(1 - x_1)$$

$$x_2 = x_1(1 - x_2)$$

Multiplying out and rearranging terms yields

$$2x_1^2 - x_1 = 0$$

$$x_2 + x_1 x_2 - x_1 = 0$$

The first equation has solutions $x_1 = 0$ or $\frac{1}{2}$. Solving the second equation for x_2, we find

$$x_2(1 + x_1) = x_1 \qquad \text{or} \qquad x_2 = \frac{x_1}{1 + x_1}$$

If $x_1 = 0$, then $x_2 = 0$; if $x_1 = 1/2$, then $x_2 = (1/2)/(3/2) = 1/3$. Summarizing our results, we found two point equilibria,

$$\begin{bmatrix} 0 \\ 0 \end{bmatrix} \quad \text{and} \quad \begin{bmatrix} 1/2 \\ 1/3 \end{bmatrix} \qquad \blacktriangleleft$$

▶ **Example 4** Find all biologically relevant equilibria of the Nicholson-Bailey model

$$N_{t+1} = bN_t e^{-aP_t}$$

$$P_{t+1} = cN_t[1 - e^{-aP_t}]$$

Solution

To find equilibria, we need to solve

$$N = bNe^{-aP} \tag{10.50}$$

$$P = cN[1 - e^{-aP}] \tag{10.51}$$

The first equation is satisfied for $N = 0$. If we substitute $N = 0$ into the second equation, we find $P = 0$. The system thus has the trivial equilibrium

$(N^\star, P^\star) = (0, 0)$ that corresponds to the state when both host and parasitoid are absent.

If $N \neq 0$, then we can cancel N in equation (10.50) and find

$$e^{aP^\star} = b \qquad \text{or} \qquad P^\star = \frac{1}{a} \ln b$$

We see that in order for $P^\star$ to be positive (this is required to be a biologically reasonable nontrivial equilibrium), we require b to be greater than 1. Now, using $e^{aP^\star} = b$, we find

$$P^\star = cN^\star \left[1 - \frac{1}{b} \right]$$

With $P^\star = \frac{1}{a} \ln b$, this reduces to

$$N^\star = \frac{P^\star}{c[1 - 1/b]} = \frac{\ln b}{ac[1 - 1/b]} = \frac{b}{b-1} \frac{1}{ac} \ln b$$

We see that for $b > 1$, $N^\star > 0$. We conclude that in addition to the trivial equilibrium $(N^\star, P^\star) = (0, 0)$, if $b > 1$, there exists a biologically reasonable, nontrivial equilibrium; that is, an equilibrium in which both the host and the parasitoid densities are positive. This equilibrium is given by

$$N^\star = \frac{b}{b-1} \frac{1}{ac} \ln b \qquad \text{and} \qquad P^\star = \frac{1}{a} \ln b \qquad \blacktriangleleft$$

To determine the stability of point equilibria, we proceed in the same way as in the single species case. We linearize about the equilibria and use the linearized system and what we learned about linear maps in Chapter 9 to derive an analytical condition for local stability. Here is how this works. We start with the general system of difference equations

$$x_1(t + 1) = F(x_1(t), x_2(t)) \tag{10.52}$$

$$x_2(t + 1) = G(x_1(t), x_2(t)) \tag{10.53}$$

and assume it has a point equilibrium $(x_1^\star, x_2^\star)$, which simultaneously satisfies

$$x_1^\star = F(x_1^\star, x_2^\star) \qquad \text{and} \qquad x_2^\star = G(x_1^\star, x_2^\star)$$

To linearize about $(x_1^\star, x_2^\star)$, we write

$$x_1(t) = x_1^\star + z_1(t) \qquad \text{and} \qquad x_2(t) = x_2^\star + z_2(t)$$

where we interpret $z_1(t)$ and $z_2(t)$ as small perturbations, just as in Chapter 5, where we discussed stability of equilibria in difference equations, or in Chapter 8, where we discussed stability of equilibria in differential equations. We now linearize $F(x_1(t), x_2(t))$ and $G(x_1(t), x_2(t))$ about the equilibrium $(x_1^\star, x_2^\star)$. We find

linearization of $F(x_1(t), x_2(t))$ is $F(x_1^\star, x_2^\star) + \left(\frac{\partial F}{\partial x_1} \right)^\star z_1(t) + \left(\frac{\partial F}{\partial x_2} \right)^\star z_2(t)$

linearization of $G(x_1(t), x_2(t))$ is $G(x_1^\star, x_2^\star) + \left(\frac{\partial G}{\partial x_1} \right)^\star z_1(t) + \left(\frac{\partial G}{\partial x_2} \right)^\star z_2(t)$

where $(\cdot)^\star$ means that we evaluate the expression in the parentheses at the equilibrium $(x_1^\star, x_2^\star)$.

With $x_1(t) = x_1^\star + z_1(t)$ and $x_2(t) = x_2^\star + z_2(t)$, we find that approximately

$$x_1^\star + z_1(t+1) \approx \underbrace{F(x_1^\star, x_2^\star)}_{x_1^\star} + \left(\frac{\partial F}{\partial x_1}\right)^\star z_1(t) + \left(\frac{\partial F}{\partial x_2}\right)^\star z_2(t)$$

$$x_2^\star + z_2(t+1) \approx \underbrace{G(x_1^\star, x_2^\star)}_{x_2^\star} + \left(\frac{\partial G}{\partial x_1}\right)^\star z_1(t) + \left(\frac{\partial G}{\partial x_2}\right)^\star z_2(t)$$

Cancelling $x_1^\star$ from the first equation and $x_2^\star$ from the second equation, and writing the resulting approximation in matrix form, we obtain

$$\begin{bmatrix} z_1(t+1) \\ z_2(t+1) \end{bmatrix} \approx \begin{bmatrix} \left(\dfrac{\partial F}{\partial x_1}\right)^\star & \left(\dfrac{\partial F}{\partial x_2}\right)^\star \\ \left(\dfrac{\partial G}{\partial x_1}\right)^\star & \left(\dfrac{\partial G}{\partial x_2}\right)^\star \end{bmatrix} \begin{bmatrix} z_1(t) \\ z_2(t) \end{bmatrix} \tag{10.54}$$

We recognize the 2×2 matrix as the Jacobi matrix of the vector-valued function $\begin{bmatrix} F(x_1, x_2) \\ G(x_1, x_2) \end{bmatrix}$. The right-hand side of (10.54) is a linear map of the form $A\mathbf{x}$ where A is a 2×2 matrix and $\mathbf{x} = \begin{bmatrix} x_1 \\ x_2 \end{bmatrix}$ is a 2×1 vector. In the previous subsection, we found that a linear map

$$\mathbf{x}_{t+1} = A\mathbf{x}_t$$

has the equilibrium $(0, 0)$, which is stable if the absolute values of the two eigenvalues of A are both less than 1. This is the criterion we need to determine the stability of the equilibrium $(x_1^\star, x_2^\star)$ of the system (10.52) and (10.53).

The point equilibrium $\begin{bmatrix} x_1^\star \\ x_2^\star \end{bmatrix}$ of the system (10.52) and (10.53) is locally stable if the two eigenvalues λ_1 and λ_2 of the Jacobi matrix evaluated at $(x_1^\star, x_2^\star)$,

$$\begin{bmatrix} \left(\dfrac{\partial F}{\partial x_1}\right)^\star & \left(\dfrac{\partial F}{\partial x_2}\right)^\star \\ \left(\dfrac{\partial G}{\partial x_1}\right)^\star & \left(\dfrac{\partial G}{\partial x_2}\right)^\star \end{bmatrix}$$

satisfy

$$|\lambda_1| < 1 \quad \text{and} \quad |\lambda_2| < 1$$

If $|\lambda_1| > 1$ or $|\lambda_2| > 1$, the point equilibrium $\begin{bmatrix} x_1^\star \\ x_2^\star \end{bmatrix}$ is an unstable equilibrium.

Note that as in Chapters 5 and 8, the stability analysis is only a local analysis and we must say that an equilibrium is *locally* stable; the local analysis does not reveal anything about global stability.

▶ **Example 5**   Discuss the stability of the equilibria of the system in Example 3.

Solution

In Example 3,

$$F(x_1, x_2) = 2x_1(1 - x_1)$$
$$G(x_1, x_2) = x_1(1 - x_2)$$

The Jacobi matrix is

$$J(x_1, x_2) = \begin{bmatrix} 2 - 4x_1 & 0 \\ 1 - x_2 & -x_1 \end{bmatrix}$$

Evaluating $J(x_1, x_2)$ at the equilibrium $(0, 0)$, we find

$$J(0, 0) = \begin{bmatrix} 2 & 0 \\ 1 & 0 \end{bmatrix}$$

which has eigenvalues $\lambda_1 = 2$ and $\lambda_2 = 0$. Since $|\lambda_1| > 1$, it follows that $(0, 0)$ is an unstable equilibrium.

Evaluating $J(x_1, x_2)$ at the equilibrium $(1/2, 1/3)$, we find

$$J(1/2, 1/3) = \begin{bmatrix} 0 & 0 \\ 2/3 & -1/2 \end{bmatrix}$$

which has eigenvalues $\lambda_1 = 0$ and $\lambda_2 = -1/2$. Since $|\lambda_1| < 1$ and $|\lambda_2| < 1$, it follows that $(1/2, 1/3)$ is a locally stable equilibrium.   ◀

▶ **Example 6**   Show that the nontrivial equilibrium of the Nicholson-Bailey equation is unstable.

Solution

The Nicholson-Bailey equation is of the form

$$F(N, P) = bNe^{-aP}$$
$$G(N, P) = cN[1 - e^{-aP}]$$

To find the Jacobi matrix evaluated at the nontrivial equilibrium, we differentiate F and G and then evaluate the derivatives at the equilibrium $(N^\star, P^\star) = (\frac{b}{b-1} \frac{1}{ac} \ln b, \frac{1}{a} \ln b)$, which we computed in Example 3.

$$\left(\frac{\partial F}{\partial N}\right)^\star = be^{-aP}\Big|_{(N^\star, P^\star)} = 1$$

$$\left(\frac{\partial F}{\partial P}\right)^\star = -abNe^{-aP}\Big|_{(N^\star, P^\star)} = -aN^\star \quad (\text{since } be^{-aP^\star} = 1)$$

$$\left(\frac{\partial G}{\partial N}\right)^\star = c[1 - e^{-aP}]\Big|_{(N^\star, P^\star)} = c\left[1 - \frac{1}{b}\right] \quad \left(\text{since } e^{-aP^\star} = \frac{1}{b}\right)$$

$$\left(\frac{\partial G}{\partial P}\right)^\star = caNe^{-aP}\Big|_{(N^\star, P^\star)} = acN^\star\frac{1}{b}$$

The Jacobi matrix evaluated at $(N^\star, P^\star)$ is then

$$J(N^\star, P^\star) = \begin{bmatrix} 1 & -aN^\star \\ c[1 - \frac{1}{b}] & acN^\star\frac{1}{b} \end{bmatrix}$$

Instead of computing the eigenvalues explicitly, we will first show that the two eigenvalues of the matrix J are complex conjugate if $b > 1$ (this was the condition we found in Example 3 that guaranteed a biologically reasonable, nontrivial equilibrium). The eigenvalues of J satisfy the equation $\det(J - \lambda I) = 0$, that is,

$$(1 - \lambda)\left(acN^\star\frac{1}{b} - \lambda\right) + acN^\star\left(1 - \frac{1}{b}\right) = 0$$

which simplifies to

$$\lambda^2 - \left(1 + \frac{ac}{b}N^\star\right)\lambda + acN^\star = 0$$

The solutions of this equation are complex conjugate if the discriminant

$$\left(1 + \frac{ac}{b}N^\star\right)^2 - 4acN^\star < 0$$

With $N^\star = \frac{b}{b-1}\frac{1}{ac}\ln b$, the discriminant is

$$f(b) = \left(1 + \frac{\ln b}{b - 1}\right)^2 - \frac{4b}{b - 1}\ln b$$

This function depends only on b. Graphing $f(b)$ (see Figure 10.74) shows that $f(b) < 0$ for $b > 1$, thus confirming that the two eigenvalues of J are complex conjugate if $b > 1$.

When we discussed linear systems of difference equations, we derived the identity

$$|\lambda_1|^2 = |\lambda_2|^2 = \lambda_1\lambda_2 = \det J$$

The determinant of J is given by

$$\det J = acN^\star\frac{1}{b} + acN^\star\left(1 - \frac{1}{b}\right) = acN^\star = \frac{b\ln b}{b - 1}$$

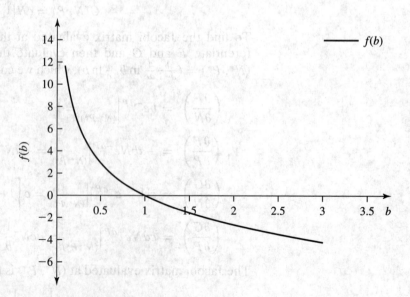

▲ **Figure 10.74**
The graph of $f(b)$ confirms that $f(b) < 0$ for $b > 1$

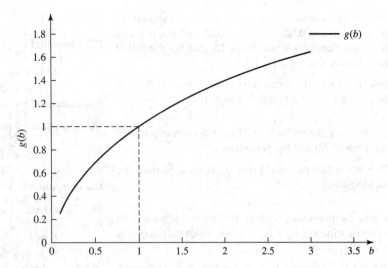

▲ **Figure 10.75**
The graph of $g(b)$ confirms that $g(b) > 1$ for $b > 1$

Graphing $g(b) = \frac{b \ln b}{b-1}$ as a function of b (see Figure 10.75), we see that $g(b) > 1$ for $b > 1$, from which we conclude that for $b > 1$,

$$|\lambda_1|^2 = |\lambda_2|^2 = \lambda_1 \lambda_2 > 1$$

implying that the nontrivial equilibrium is unstable. ◀

10.7.4 Problems

(10.7.1)

Problems 1–6 refer to the Nicholson Bailey host-parasitoid model. Problems 1, 2, 5, and 6 are best done with the help of a spreadsheet but can also be done with a calculator. Nicholson and Bailey introduced the discrete generation, host-parasitoid model of the form

$$N_{t+1} = bN_t e^{-aP_t}$$
$$P_{t+1} = cN_t[1 - e^{-aP_t}]$$

for $t = 0, 1, 2, \ldots$

1. Evaluate the Nicholson-Bailey model for the first 10 generations when $a = 0.02$, $c = 3$, and $b = 1.5$. For the initial host density, choose $N_0 = 5$, and for the initial parasitoid density, choose $P_0 = 0$.

2. Evaluate the Nicholson-Bailey model for the first 10 generations when $a = 0.02$, $c = 3$, and $b = 0.5$. For the initial host density, choose $N_0 = 15$, and for the initial parasitoid density, choose $P_0 = 0$.

3. Show that when the initial parasitoid density is $P_0 = 0$, then the Nicholson-Bailey model reduces to

$$N_{t+1} = bN_t$$

With N_0 denoting the initial host density, find an expression for N_t in terms of N_0 and the parameter b.

4. When the initial parasitoid density is $P_0 = 0$, then the Nicholson-Bailey model reduces to

$$N_{t+1} = bN_t$$

as shown in the previous problem. For which values of b is the host density increasing if $N_0 > 0$; for which values of b is it decreasing? (Assume $b > 0$.)

5. Evaluate the Nicholson-Bailey model for the first 15 generations when $a = 0.02$, $c = 3$, and $b = 1.5$. For the initial host density, choose $N_0 = 5$, and for the initial parasitoid density, choose $P_0 = 5$.

6. Evaluate the Nicholson-Bailey model for the first 25 generations when $a = 0.02$, $c = 3$, and $b = 1.5$. For the initial host density, choose $N_0 = 15$, and for the initial parasitoid density, choose $P_0 = 8$.

Problems 7–12 refer to the negative binomial host-parasitoid model. Problems 7, 8, 11, and 12 are best done with the help of a spreadsheet but can also be done with a calculator. The negative binomial model is a discrete generation, host-parasitoid model of the form

$$N_{t+1} = bN_t \left(1 + \frac{aP_t}{k}\right)^{-k}$$

$$P_{t+1} = cN_t \left[1 - \left(1 + \frac{aP_t}{k}\right)^{-k}\right]$$

for $t = 0, 1, 2, \ldots$

7. Evaluate the negative binomial model for the first 10 generations when $a = 0.02$, $c = 3$, $k = 0.75$, and $b = 1.5$. For the initial host density, choose $N_0 = 5$, and for the initial parasitoid density, choose $P_0 = 0$.

8. Evaluate the negative binomial model for the first 10 generations when $a = 0.02$, $c = 3$, $k = 0.75$, and $b = 0.5$. For the initial host density, choose $N_0 = 15$, and for the initial parasitoid density, choose $P_0 = 0$.

9. Show that when the initial parasitoid density is $P_0 = 0$, then the negative binomial model reduces to

$$N_{t+1} = bN_t$$

With N_0 denoting the initial host density, find an expression for N_t in terms of N_0 and the parameter b.

10. When the initial parasitoid density is $P_0 = 0$, then the negative binomial model reduces to

$$N_{t+1} = bN_t$$

as shown in the previous problem. For which values of b is the host density increasing if $N_0 > 0$; for which values of b is it decreasing? (Assume $b > 0$.)

11. Evaluate the negative binomial model for the first 25 generations when $a = 0.02$, $c = 3$, $k = 0.75$, and $b = 1.5$. For the initial host density, choose $N_0 = 100$, and for the initial parasitoid density, choose $P_0 = 50$.

12. Evaluate the negative binomial model for the first 25 generations when $a = 0.02$, $c = 3$, $k = 0.75$, and $b = 0.5$. For the initial host density, choose $N_0 = 100$, and for the initial parasitoid density, choose $P_0 = 50$.

13. In the Nicholson-Bailey model, the fraction of hosts escaping parasitism is given by

$$f(P) = e^{-aP}$$

(a) Graph $f(P)$ as a function of P for $a = 0.1$ and $a = 0.01$.

(b) For a given value of P, how are the chances of escaping parasitism affected by increasing a?

14. In the negative binomial model, the fraction of hosts escaping parasitism is given by

$$f(P) = \left(1 + \frac{aP}{k}\right)^{-k}$$

(a) Graph $f(P)$ as a function of P for $a = 0.1$ and $a = 0.01$ when $k = 0.75$.

(b) For $k = 0.75$ and a given value of P, how are the chances of escaping parasitism affected by increasing a?

15. In the negative binomial model, the fraction of hosts escaping parasitism is given by

$$f(P) = \left(1 + \frac{aP}{k}\right)^{-k}$$

(a) Graph $f(P)$ as a function of P for $k = 0.75$ and $k = 0.5$ when $a = 0.02$.

(b) For $a = 0.02$ and a given value of P, how are the chances of escaping parasitism affected by increasing k?

16. The negative binomial model can be reduced to the Nicholson-Bailey model by letting the parameter k in the negative binomial model go to infinity. Show that

$$\lim_{k \to \infty} \left(1 + \frac{aP}{k}\right)^{-k} = e^{-aP}$$

(*Hint:* Use L'Hospital's rule.)

(10.7.2)

17. Show that $\begin{bmatrix} 0 \\ 0 \end{bmatrix}$ is an equilibrium of

$$\begin{bmatrix} x_1(t+1) \\ x_2(t+1) \end{bmatrix} = \begin{bmatrix} -0.7 & 0 \\ -0.3 & 0.2 \end{bmatrix} \begin{bmatrix} x_1(t) \\ x_2(t) \end{bmatrix}$$

and determine its stability.

18. Show that $\begin{bmatrix} 0 \\ 0 \end{bmatrix}$ is an equilibrium of

$$\begin{bmatrix} x_1(t+1) \\ x_2(t+1) \end{bmatrix} = \begin{bmatrix} 0.4 & 0.2 \\ 0 & -0.9 \end{bmatrix} \begin{bmatrix} x_1(t) \\ x_2(t) \end{bmatrix}$$

and determine its stability.

19. Show that $\begin{bmatrix} 0 \\ 0 \end{bmatrix}$ is an equilibrium of

$$\begin{bmatrix} x_1(t+1) \\ x_2(t+1) \end{bmatrix} = \begin{bmatrix} -1.4 & 0 \\ -0.5 & 0.1 \end{bmatrix} \begin{bmatrix} x_1(t) \\ x_2(t) \end{bmatrix}$$

and determine its stability.

20. Show that $\begin{bmatrix} 0 \\ 0 \end{bmatrix}$ is an equilibrium of

$$\begin{bmatrix} x_1(t+1) \\ x_2(t+1) \end{bmatrix} = \begin{bmatrix} 0.1 & 0.4 \\ 0.1 & -0.2 \end{bmatrix} \begin{bmatrix} x_1(t) \\ x_2(t) \end{bmatrix}$$

and determine its stability.

21. Show that $\begin{bmatrix} 0 \\ 0 \end{bmatrix}$ is an equilibrium of

$$\begin{bmatrix} x_1(t+1) \\ x_2(t+1) \end{bmatrix} = \begin{bmatrix} 1 & 2 \\ 3 & 2 \end{bmatrix} \begin{bmatrix} x_1(t) \\ x_2(t) \end{bmatrix}$$

and determine its stability.

22. Show that $\begin{bmatrix} 0 \\ 0 \end{bmatrix}$ is an equilibrium of

$$\begin{bmatrix} x_1(t+1) \\ x_2(t+1) \end{bmatrix} = \begin{bmatrix} 1.5 & 02 \\ 0.08 & 0 \end{bmatrix} \begin{bmatrix} x_1(t) \\ x_2(t) \end{bmatrix}$$

and determine its stability.

23. Show that the equilibrium $\begin{bmatrix} 0 \\ 0 \end{bmatrix}$ of

$$\begin{bmatrix} x_1(t+1) \\ x_2(t+1) \end{bmatrix} = \begin{bmatrix} -0.2 & -0.4 \\ 0.6 & 0.1 \end{bmatrix} \begin{bmatrix} x_1(t) \\ x_2(t) \end{bmatrix}$$

is stable.

24. Show that the equilibrium $\begin{bmatrix} 0 \\ 0 \end{bmatrix}$ of

$$\begin{bmatrix} x_1(t+1) \\ x_2(t+1) \end{bmatrix} = \begin{bmatrix} 0.2 & 0.3 \\ -0.5 & -0.4 \end{bmatrix} \begin{bmatrix} x_1(t) \\ x_2(t) \end{bmatrix}$$

is stable.

25. Show that the equilibrium $\begin{bmatrix} 0 \\ 0 \end{bmatrix}$ of

$$\begin{bmatrix} x_1(t+1) \\ x_2(t+1) \end{bmatrix} = \begin{bmatrix} 4.2 & -3.4 \\ 2.4 & -1.1 \end{bmatrix} \begin{bmatrix} x_1(t) \\ x_2(t) \end{bmatrix}$$

is stable.

26. Show that the equilibrium $\begin{bmatrix} 0 \\ 0 \end{bmatrix}$ of

$$\begin{bmatrix} x_1(t+1) \\ x_2(t+1) \end{bmatrix} = \begin{bmatrix} 2 & -4 \\ 5 & -6 \end{bmatrix} \begin{bmatrix} x_1(t) \\ x_2(t) \end{bmatrix}$$

is stable.

(10.7.3)

27. Show that the equilibrium $\begin{bmatrix} 0 \\ 0 \end{bmatrix}$ of

$$x_1(t+1) = \frac{x_2(t)}{4(1+x_1^2(t))}$$

$$x_2(t+1) = \frac{2x_1(t)}{1+x_2^2(t)}$$

is locally stable.

28. Show that the equilibrium $\begin{bmatrix} 0 \\ 0 \end{bmatrix}$ of

$$x_1(t+1) = \frac{3x_2(t)}{1+x_1^2(t)}$$

$$x_2(t+1) = \frac{2x_1(t)}{1+x_2^2(t)}$$

is unstable.

29. Show that the equilibrium $\begin{bmatrix} 0 \\ 0 \end{bmatrix}$ of

$$x_1(t+1) = x_2(t)$$

$$x_2(t+1) = \frac{2x_2(t) - x_1(t)}{2 + x_1(t)}$$

is locally stable.

30. Show that for any $a > 1$, the equilibrium $\begin{bmatrix} 0 \\ 0 \end{bmatrix}$ of

$$x_1(t+1) = x_2(t)$$

$$x_2(t+1) = \frac{ax_2(t) - (a-1)x_1(t)}{a + x_1(t)}$$

is locally stable.

31. Show that $\begin{bmatrix} 0 \\ 0 \end{bmatrix}$ is an equilibrium point of

$$x_1(t+1) = ax_2(t)$$

$$x_2(t+1) = 2x_1(t) - \cos(x_2(t)) + 1$$

Assume $a > 0$. For which values of a is $\begin{bmatrix} 0 \\ 0 \end{bmatrix}$ locally stable?

32. Show that $\begin{bmatrix} 0 \\ 0 \end{bmatrix}$ and $\begin{bmatrix} -\pi \\ \pi \end{bmatrix}$ are equilibria of

$$x_1(t+1) = -x_2(t)$$

$$x_2(t+1) = \sin(x_2(t)) - x_1(t)$$

and analyze their stability.

33. Find all nonnegative equilibria of

$$x_1(t+1) = x_2(t)$$

$$x_2(t+1) = \frac{1}{2}x_1(t) + \frac{2}{3}x_2(t) - x_2^2(t)$$

and analyze their stability.

34. Find all nonnegative equilibria of

$$x_1(t+1) = x_2(t)$$

$$x_2(t+1) = \frac{1}{2}x_1(t) + \frac{1}{3}x_2(t) - x_2^2(t)$$

and analyze their stability.

35. For which values of a is the equilibrium $\begin{bmatrix} 0 \\ 0 \end{bmatrix}$ of

$$x_1(t+1) = \frac{ax_2(t)}{1+x_1^2(t)}$$

$$x_2(t+1) = \frac{x_1(t)}{1+x_2^2(t)}$$

locally stable?

36. For which values of a is the equilibrium $\begin{bmatrix} 0 \\ 0 \end{bmatrix}$ of

$$x_1(t+1) = x_2(t)$$

$$x_2(t+1) = \frac{1}{2}x_1(t) + ax_2(t) - x_2^2(t)$$

locally stable?

37. Denote by $x_1(t)$ the number of juveniles and by $x_2(t)$ the number of adults at time t. Assume that $x_1(t)$ and $x_2(t)$ evolve according to

$$x_1(t+1) = x_2(t)$$

$$x_2(t+1) = \frac{1}{2}x_1(t) + rx_2(t) - x_2^2(t)$$

(a) Show that if $r > 1/2$, there exists an equilibrium $\begin{bmatrix} x_1\star \\ x_2\star \end{bmatrix}$ with $x_1^* > 0$ and $x_2^* > 0$. Find x_1^* and x_2^*.

(b) Determine the stability of the equilibrium found in (a) when $r > 1/2$.

38. Find all biologically relevant equilibria of the Nicholson-Bailey model

$$N_{t+1} = 2N_t e^{-0.2P_t}$$

$$P_{t+1} = N_t \left[1 - e^{-0.2P_t} \right]$$

and analyze their stability.

39. Find all biologically relevant equilibria of the Nicholson-Bailey model

$$N_{t+1} = 4N_t e^{-0.1P_t}$$

$$P_{t+1} = N_t \left[1 - e^{-0.1P_t} \right]$$

and analyze their stability.

40. Find all biologically relevant equilibria of the negative binomial host-parasitoid model

$$N_{t+1} = 4N_t \left(1 + \frac{0.01P_t}{2} \right)^{-2}$$

$$P_{t+1} = N_t \left[1 - \left(1 + \frac{0.01P_t}{2} \right)^{-2} \right]$$

and analyze their stability.

41. Find all biologically relevant equilibria of the negative-binomial host-parasitoid model

$$N_{t+1} = 4N_t \left(1 + \frac{0.01P_t}{0.5} \right)^{-0.5}$$

$$P_{t+1} = N_t \left[1 - \left(1 + \frac{0.01P_t}{0.5} \right)^{-0.5} \right]$$

and analyze their stability.

10.8 Key Terms

Chapter 10 Review: Topics *Discuss the following definitions and concepts:*

1. Real-valued function
2. Function of two variables
3. Surface
4. Level curve
5. Limit
6. Limit laws
7. Continuity
8. Partial derivative
9. Geometric interpretation of a partial derivative
10. Mixed derivative theorem
11. Tangent plane
12. Differentiability
13. Differentiability and continuity
14. Sufficient condition for differentiability
15. Standard linear approximation, tangent plane approximation
16. Vector-valued function
17. Jacobi matrix, derivative matrix
18. Chain rule
19. Implicit differentiation
20. Directional derivative
21. Gradient
22. Local extrema
23. Sufficient condition for finding local extrema
24. Hessian matrix
25. Global extrema
26. The extreme value theorem
27. Diffusion
28. Systems of difference equations
29. Point equilibria and their stability
30. Nicholson-Bailey equation

10.9 Review Problems

1. Suppose that you conduct an experiment to measure germination success of seeds of a certain plant as a function of temperature and humidity. You find that seeds don't germinate at all when the humidity is too low, regardless of temperature; germination success is highest for intermediate values of temperature; and seeds tend to germinate better when you increase humidity levels. Use the preceding information to sketch a graph of germination success as a function of temperature for different levels of humidity. Also sketch the graph of germination success as a function of humidity for different temperature values.

2. Gaastra (1959) measured the effects of atmospheric CO_2 enrichment on CO_2 fixation in sugar beet leaves at various light levels. He found that increasing CO_2 at fixed light levels increases the fixation rate and that increasing light levels at fixed atmospheric CO_2 concentration also increased fixation. If $F(A, I)$ denotes the fixation rate as a function of atmospheric CO_2 concentration (A) and light intensity (I), determine the signs of $\partial F/\partial A$ and $\partial F/\partial I$.

3. In Burke and Grime (1996), a long-term field experiment in a limestone grassland was described.

(a) One of the experiments related total area covered by *indigenous* species to fertility and disturbance gradients. The experiment was designed so that the two variables (fertility and disturbance) could be altered independently. Burke and Grime found that the area covered by indigenous species generally increased with the amount of fertilizer added and decreased with disturbance intensity. If $A_i(F, D)$ denotes the area covered by indigenous species as a function of the amount of fertilizer added (F) and the disturbance intensity (D), determine the signs of $\partial A_i/\partial F$ and $\partial A_i/\partial D$ for their experiment.

(b) In another experiment, Burke and Grime related the total area covered by *introduced* species to fertility and disturbance gradients. Let $A_e(F, D)$ denote the area covered by introduced species as a function of the amount of fertilizer added (F) and the disturbance intensity (D). Burke and Grime found that

$$\frac{\partial A_e}{\partial F} > 0$$

and

$$\frac{\partial A_e}{\partial D} > 0$$

Explain in words what this means.

(c) Compare the responses to fertilization and disturbance to the area covered in the two experiments.

4. Vitousek and Farrington (1997) investigated nutrient limitations in soils of different ages. In the abstract of their paper, they say that

> Walker and Syers (1976) proposed a conceptual model that describes the pattern and regulation of soil nutrient pools and availability during long-term soil and ecosystem development. Their model implies that plant production generally should be limited by N [nitrogen] on young soils and by P [phosphorus] on old soils; N and P supply should be more or less equilibrate on intermediate aged soils.

They tested this hypothesis in fertilizer experiments along a gradient of soil age, where they measured the average diameter increment (in mm/yr) of *Metrosideros polymorpha* trees. Denote by $D(N, P, t)$ the diameter increment (in mm/yr) as a function of the amount of nitrogen added (N), the amount of phosphorus added (P), and the age of the soil (t). Their experiments showed that

$$\frac{\partial D}{\partial t}(N, 0, t) < 0$$

and

$$\frac{\partial D}{\partial t}(0, P, t) > 0$$

for their choices of $N > 0$ and $P > 0$. Explain why their results support the Walker and Syers hypothesis.

5. Find the Jacobi matrix

$$\mathbf{f}(x, y) = \begin{bmatrix} x^2 - y \\ x^3 - y^2 \end{bmatrix}$$

6. Find a linear approximation to $\mathbf{f}(x, y)$ at the indicated point:

$$\mathbf{f}(x, y) = \begin{bmatrix} 2xy^2 \\ \frac{x}{y} \end{bmatrix}$$

at $(1, 1)$.

7. We can compute the average radius of spreading individuals at time t, denoted by r_{avg}. We find

$$r_{avg} = \sqrt{\pi D t} \tag{10.55}$$

(a) Graph r_{avg} as a function of D for $t = 0.1, t = 1$, and $t = 5$. Describe in words how an increase in D affects the average radius of spread.

(b) Show that

$$D = \frac{(r_{avg})^2}{\pi t} \tag{10.56}$$

(c) Equation (10.56) can be used to measure D, the diffusion constant, from field data of mark-recapture experiments, taken from Kareiva (1983), as follows: Marked organisms are released from the release site and then recaptured after a certain amount of time t from the time of release. The distance of the recaptured organisms from the release site is measured. If N denotes the total number of recaptured organisms, d_i denotes the distance of the ith recaptured organism from the release site, and t is the time between release and recapture, use (10.56) to explain why

$$D = \frac{1}{\pi t} \left(\frac{1}{N} \sum_{i=1}^{N} d_i \right)^2$$

can be used to measure D from field data. (Note that the time between release and recapture is the same for each individual in this study.)

SYSTEMS OF DIFFERENTIAL EQUATIONS

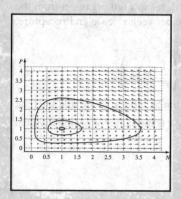

We first encountered systems of differential equations in Section 8.3. There, the emphasis was on understanding where such systems arise and on finding equilibria. (Section 8.3 is not a prerequisite for this chapter.) We will now give a systematic treatment of such systems. Suppose that we are given a set of variables $x_1, x_2, \ldots, x_n$, each depending on an independent variable, say t, so that, $x_1 = x_1(t), x_2 = x_2(t), \ldots, x_n = x_n(t)$, and that the dynamics of the variables are linked by differential equations of the form

$$
\begin{aligned}
\frac{dx_1}{dt} &= g_1(t, x_1, x_2, \ldots, x_n) \\
\frac{dx_2}{dt} &= g_2(t, x_1, x_2, \ldots, x_n) \\
&\vdots \\
\frac{dx_n}{dt} &= g_n(t, x_1, x_2, \ldots, x_n)
\end{aligned}
\tag{11.1}
$$

This set of equations is called a system of differential equations. On the left-hand side of this set of equations are the derivatives of $x_i(t)$ with respect to t. On the right-hand side of each equation, there is a function g_i, which depends on the variables $x_1, x_2, \ldots, x_n$ and on t. We will first look at the case when the functions g_i are linear in the variables $x_1, x_2, \ldots, x_n$; that is, when for $i = 1, 2, \ldots, n$,

$$
g_i(t, x_1, x_2, \ldots, x_n) = a_{i1}(t)x_1 + a_{i2}(t)x_2 + \cdots + a_{in}(t)x_n + f_i(t)
$$

We can write the linear system in matrix form, namely,

$$
\frac{d\mathbf{x}}{dt} = A(t)\mathbf{x}(t) + \mathbf{f}(t)
\tag{11.2}
$$

where

$$
\mathbf{x}(t) = \begin{bmatrix} x_1(t) \\ x_2(t) \\ \vdots \\ x_n(t) \end{bmatrix}
\qquad
A = \begin{bmatrix}
a_{11}(t) & a_{12}(t) & \ldots & a_{1n}(t) \\
a_{21}(t) & a_{22}(t) & \ldots & a_{2n}(t) \\
\vdots & \vdots & \vdots & \vdots \\
a_{n1}(t) & a_{n2}(t) & \ldots & a_{nn}(t)
\end{bmatrix}
$$

and

$$
\mathbf{f}(t) = \begin{bmatrix} f_1(t) \\ f_2(t) \\ \vdots \\ f_n(t) \end{bmatrix}
$$

Equation (11.2) is called a system of **linear first-order equations** (first-order since only first derivatives occur). We will investigate only the case when $\mathbf{f}(t) = \mathbf{0}$ and $A(t)$ does not depend on t. Equation (11.2) then reduces to

$$\frac{d\mathbf{x}}{dt} = A\mathbf{x}(t) \tag{11.3}$$

where

$$A = \begin{bmatrix} a_{11} & a_{12} & \cdots & a_{1n} \\ a_{21} & a_{22} & \cdots & a_{2n} \\ \vdots & \vdots & \vdots & \vdots \\ a_{n1} & a_{n2} & \cdots & a_{nn} \end{bmatrix}$$

Equation (11.3) is called a **homogeneous** linear first-order system with constant coefficients. (*Homogeneous* refers to the fact that $\mathbf{f}(t) = \mathbf{0}$. When $\mathbf{f}(t) \neq \mathbf{0}$, the system is called **inhomogeneous**.) Since the matrix A does not depend on t, all the coefficients are constant; such a system is **autonomous** (we encountered autonomous systems in Section 8.1). Note that A is a square matrix.

In Section 11.1, we will present some of the theory for systems of the form (11.3). In Section 11.2, we will discuss some applications of linear systems. Section 11.3 is devoted to the theory of nonlinear systems, and Section 11.4 to applications of nonlinear systems.

11.1 LINEAR SYSTEMS: THEORY

In this section, we will analyze homogeneous, linear first-order systems with constant coefficients, that is, systems of the form

$$\frac{dx_1}{dt} = a_{11}x_1 + a_{12}x_2 + \cdots + a_{1n}x_n$$

$$\frac{dx_2}{dt} = a_{21}x_1 + a_{22}x_2 + \cdots + a_{2n}x_n$$

$$\vdots \tag{11.4}$$

$$\frac{dx_n}{dt} = a_{n1}x_1 + a_{n2}x_2 + \cdots + a_{nn}x_n$$

where the variables $x_1, x_2, \ldots, x_n$ are functions of t, and the parameters a_{ij}, $1 \leq i, j \leq n$, are constants We can write this in matrix form as

$$\frac{d\mathbf{x}}{dt} = A\mathbf{x}(t)$$

where

$$\mathbf{x}(t) = \begin{bmatrix} x_1(t) \\ x_2(t) \\ \vdots \\ x_n(t) \end{bmatrix} \qquad A = \begin{bmatrix} a_{11} & a_{12} & \cdots & a_{1n} \\ a_{21} & a_{22} & \cdots & a_{2n} \\ \vdots & \vdots & \vdots & \vdots \\ a_{n1} & a_{n2} & \cdots & a_{nn} \end{bmatrix}$$

Most of the time we will restrict ourselves to the case $n = 2$.

▶ **Example 1** Write

$$\frac{dx_1}{dt} = 4x_1 - 2x_2$$

$$\frac{dx_2}{dt} = -3x_1 + x_2$$

in matrix notation.

Solution

We write $x_1 = x_1(t)$ and $x_2 = x_2(t)$. Using the rules for matrix multiplication, we find

$$\begin{bmatrix} \dfrac{dx_1}{dt} \\ \dfrac{dx_2}{dt} \end{bmatrix} = \begin{bmatrix} 4x_1 - 2x_2 \\ -3x_1 + x_2 \end{bmatrix} = \begin{bmatrix} 4 & -2 \\ -3 & 1 \end{bmatrix} \begin{bmatrix} x_1(t) \\ x_2(t) \end{bmatrix}$$

That is,

$$\frac{d\mathbf{x}(t)}{dt} = \begin{bmatrix} 4 & -2 \\ -3 & 1 \end{bmatrix} \mathbf{x}(t) \qquad \blacktriangleleft$$

First, we will be concerned with solutions of (11.4): A solution is an ordered n-tuple of functions $(x_1(t), x_2(t), \ldots, x_n(t))$ that satisfies all n equations in (11.4). Second, we will discuss equilibria: An equilibrium is a point $\hat{\mathbf{x}} = (\hat{x}_1, \hat{x}_2, \ldots, \hat{x}_n)$ so that $A\hat{\mathbf{x}} = \mathbf{0}$. We begin with a graphical approach to visualizing solutions.

11.1.1 The Direction Field

Before we turn to finding solutions for homogeneous, linear first-order systems with constant coefficients, we discuss an important property of solution curves that will allow us to sketch solutions graphically in the x_1-x_2 plane with the help of **direction fields**. Consider

$$\begin{aligned} \frac{dx_1}{dt} &= x_1 - 2x_2 \\ \frac{dx_2}{dt} &= x_2 \end{aligned} \tag{11.5}$$

Imagine now that you are standing at a point (x_1, x_2) in the x_1-x_2 plane and the system (11.5) determines your future location. Where should you go next? The two differential equations tell you how your coordinates will change. To give a concrete example, look at the point $(2, -1)$. We claim that you move along a curve whose tangent line at the point $(2, -1)$ has slope

$$\frac{dx_2}{dx_1} = \frac{dx_2/dt}{dx_1/dt} = \frac{x_2}{x_1 - 2x_2} = \frac{-1}{2 - (2)(-1)} = -\frac{1}{4}$$

Why is this true? A solution of (11.5) that starts at time 0 at the point $(x_1(0), x_2(0))$ is given by points of the form $(x_1(t), x_2(t))$, $t \geq 0$ that satisfy (11.5). An example of a solution curve is shown in Figure 11.1. We see that the solution is a curve in the x_1-x_2 plane, and at each point, we can draw a tangent line whose slope is dx_2/dx_1. The tangent line at $(2, -1)$, for which we computed the slope (namely, $-1/4$) is also drawn in Figure 11.1.

We can draw tangent lines at each point (x_1, x_2) in the x_1-x_2 plane. Knowing all the tangent lines allows us to sketch the corresponding solution curve. This is done by assigning each point (x_1, x_2) in the x_1-x_2 plane a vector $\begin{bmatrix} dx_1/dt \\ dx_2/dt \end{bmatrix}$. This vector has the property that it is tangential to the solution curve that passes through the point (x_1, x_2) and it points in the direction of the solution.

In our example, the vector is of the form $\begin{bmatrix} x_1 - 2x_2 \\ x_2 \end{bmatrix}$, and the slope of the solution curve that goes through (x_1, x_2) is

$$\frac{dx_2}{dx_1} = \frac{dx_2/dt}{dx_1/dt} = \frac{x_2}{x_1 - 2x_2}$$

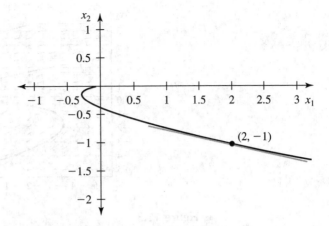

▲ Figure 11.1
Solution curve through $(2, -1)$ with tangent line

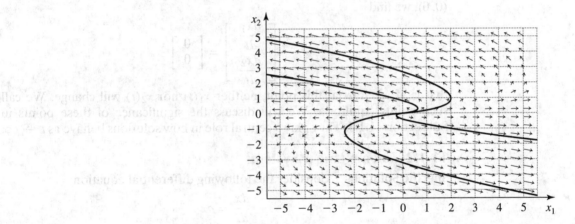

▲ Figure 11.2
The direction field of the system (11.5) together with some solution curves

The collection of these vectors is called a **direction** or **slope field**, and each vector of the direction field is called a **direction vector**. Since (11.5) is an autonomous system, the direction vector only depends on the location of the point (x_1, x_2) but not on t. This implies that the direction field is the same for all times t.

The direction field for (11.5) is shown in Figure 11.2. (Figure 11.2 also contains four solution curves that were computer generated; each curve starts at a different point very close to the origin $(0, 0)$.) The length of the direction vector at a given point tells us how quickly the solution curve passes through the point; it is proportional to $\sqrt{(dx_1/dt)^2 + (dx_2/dt)^2}$. If we are only interested in the direction of the solution curves, we can indicate the direction by small line segments (as shown in Figure 11.3), which often results in a less cluttered picture. Figure 11.3 also contains four solution curves that were computer generated as in Figure 11.2; note that the direction vectors are always tangential to the solution curves. We can therefore use the direction field to sketch solution curves by drawing curves in such a way that the direction vectors are always tangential to the curve.

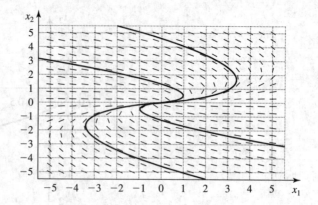

▲ **Figure 11.3**
The direction field of the system (11.5) with some solution curves, where the direction vectors are not drawn to scale

The point $(0, 0)$ is special: When we compute the direction vector at $(0, 0)$, we find

$$
\begin{bmatrix} \dfrac{dx_1}{dt} \\[2mm] \dfrac{dx_2}{dt} \end{bmatrix} = \begin{bmatrix} 0 \\ 0 \end{bmatrix}
$$

that is, if we start at this point, neither $x_1(t)$ nor $x_2(t)$ will change. We call such points equilibria. We will discuss the significance of these points in Subsection 11.1.3. They play a central role in how solutions behave as $t \to \infty$.

11.1.2 Solving Linear Systems

Specific Solutions   Consider the following differential equation:

$$
\frac{dx}{dt} = ax
$$

This is a linear, first-order differential equation with a constant coefficient [that is, it is of the form (11.4) with $n = 1$]. We can find a solution by integrating after separating variables (as we learned in Chapter 8). All solutions are of the form

$$
x(t) = ce^{at}
$$

where c is a constant that depends on the initial condition. An initial condition picks out a specific solution among the set of solutions. For instance, if $x(0) = x_0$, then $c = x_0$.

We will now show that exponential functions are also solutions of systems of linear differential equations. We restrict our discussion to systems of two differential equations. Consider the system

$$
\frac{dx_1}{dt} = a_{11}x_1(t) + a_{12}x_2(t) \tag{11.6}
$$

$$
\frac{dx_2}{dt} = a_{21}x_1(t) + a_{22}x_2(t) \tag{11.7}
$$

which in matrix form is written as

$$
\frac{d\mathbf{x}}{dt} = A\mathbf{x}(t) \tag{11.8}
$$

A solution of (11.8) is a vector-valued function. As in the preceding example $dx/dt = ax$, we will find that (11.8) admits a collection of solutions and if we choose an initial condition, a particular solution will be picked out. We will get to initial conditions later in this subsection. Let's first see what the solutions look like.

We claim that the following vector-valued function

$$\mathbf{x}(t) = \begin{bmatrix} u_1 e^{\lambda t} \\ u_2 e^{\lambda t} \end{bmatrix} = e^{\lambda t} \begin{bmatrix} u_1 \\ u_2 \end{bmatrix} \tag{11.9}$$

where λ, u_1 and u_2 are constants, is a solution of (11.8) for an appropriate choice of values for λ, u_1, and u_2. To see how we must choose λ, u_1, and u_2, we differentiate $\mathbf{x}(t)$ in (11.9). This yields

$$\frac{d\mathbf{x}}{dt} = \begin{bmatrix} u_1 \lambda e^{\lambda t} \\ u_2 \lambda e^{\lambda t} \end{bmatrix} = \lambda e^{\lambda t} \begin{bmatrix} u_1 \\ u_2 \end{bmatrix} \tag{11.10}$$

If $\mathbf{x}(t)$ solves (11.8), then $\mathbf{x}(t)$ must satisfy (11.8). Using (11.10) for the left-hand side and (11.9) for the right-hand side of (11.8), we find

$$\underbrace{\lambda e^{\lambda t} \begin{bmatrix} u_1 \\ u_2 \end{bmatrix}}_{\frac{d\mathbf{x}(t)}{dt}} = \underbrace{A e^{\lambda t} \begin{bmatrix} u_1 \\ u_2 \end{bmatrix}}_{A\mathbf{x}(t)}$$

or, after dividing both sides by $e^{\lambda t}$,

$$\lambda \begin{bmatrix} u_1 \\ u_2 \end{bmatrix} = A \begin{bmatrix} u_1 \\ u_2 \end{bmatrix}$$

This last expression should remind you of eigenvalues and eigenvectors that we encountered in Section 9.3: $\begin{bmatrix} u_1 \\ u_2 \end{bmatrix}$ is an eigenvector corresponding to an eigenvalue λ of A. With this choice, (11.9) is a solution of (11.8).

In this subsection, we will only look at differential equations of the form (11.8) for which the eigenvalues of A are both real and distinct. We will discuss complex eigenvalues in the next subsection (we will not discuss the case when both eigenvalues are identical). We illustrate on the following system how to find specific solutions of a system of the form (11.8) in the case when both eigenvalues of A are real and distinct.

$$\frac{dx_1}{dt} = 2x_1 - 2x_2$$
$$\frac{dx_2}{dt} = 2x_1 - 3x_2 \tag{11.11}$$

Finding eigenvalues **1.** The coefficient matrix of (11.11) is given by

$$A = \begin{bmatrix} 2 & -2 \\ 2 & -3 \end{bmatrix}$$

We first determine the eigenvalues and corresponding eigenvectors of A. To find the eigenvalues of A, we must solve

$$\det(A - \lambda I) = 0$$

That is,

$$\det \begin{bmatrix} 2 - \lambda & -2 \\ 2 & -3 - \lambda \end{bmatrix} = (2 - \lambda)(-3 - \lambda) + 4$$

$$= \lambda^2 + \lambda - 2 = (\lambda - 1)(\lambda + 2) = 0$$

which has solutions

$$\lambda_1 = 1 \quad \text{and} \quad \lambda_2 = -2$$

A solution corresponding to the eigenvalue λ_1

2. To find an eigenvector $\mathbf{u} = \begin{bmatrix} u_1 \\ u_2 \end{bmatrix}$ corresponding to $\lambda_1 = 1$, we solve

$$A \begin{bmatrix} u_1 \\ u_2 \end{bmatrix} = \lambda_1 \begin{bmatrix} u_1 \\ u_2 \end{bmatrix} \quad \text{with } \lambda_1 = 1$$

We find

$$\begin{bmatrix} 2u_1 - 2u_2 \\ 2u_1 - 3u_2 \end{bmatrix} = \begin{bmatrix} u_1 \\ u_2 \end{bmatrix}$$

which can be written as

$$u_1 - 2u_2 = 0$$
$$2u_1 - 4u_2 = 0$$

Both equations reduce to the same equation, namely,

$$u_1 = 2u_2$$

Setting $u_2 = 1$, for instance, we find $u_1 = 2$. An eigenvector corresponding to $\lambda_1 = 1$ is then

$$\mathbf{u} = \begin{bmatrix} 2 \\ 1 \end{bmatrix}$$

(Recall that any nonzero multiple of $\mathbf{u}$ is also an eigenvector corresponding to $\lambda_1 = 1$.)
We claim that

$$\mathbf{x}(t) = e^t \begin{bmatrix} 2 \\ 1 \end{bmatrix}$$

solves (11.11). Let's check. We find

$$\frac{d\mathbf{x}}{dt} = e^t \begin{bmatrix} 2 \\ 1 \end{bmatrix} = \mathbf{x}(t)$$

Since $\begin{bmatrix} 2 \\ 1 \end{bmatrix}$ is an eigenvector corresponding to $\lambda_1 = 1$, $\mathbf{x}(t)$ also satisfies

$$A\mathbf{x}(t) = Ae^t \begin{bmatrix} 2 \\ 1 \end{bmatrix} = e^t A \begin{bmatrix} 2 \\ 1 \end{bmatrix} = e^t \begin{bmatrix} 2 \\ 1 \end{bmatrix} = \mathbf{x}(t)$$

where we used the fact that

$$A \begin{bmatrix} 2 \\ 1 \end{bmatrix} = \lambda_1 \begin{bmatrix} 2 \\ 1 \end{bmatrix} = \begin{bmatrix} 2 \\ 1 \end{bmatrix}$$

since $\begin{bmatrix} 2 \\ 1 \end{bmatrix}$ is an eigenvector to $\lambda_1 = 1$. Therefore,

$$\frac{d\mathbf{x}}{dt} = A\mathbf{x}(t)$$

and we conclude that $\mathbf{x}(t) = e^t \begin{bmatrix} 2 \\ 1 \end{bmatrix}$ is indeed a solution of (11.11).

A solution corresponding to the eigenvalue λ_2

3. We can now repeat the same steps for the eigenvalue $\lambda_2 = -2$. To find
an eigenvector $\mathbf{v} = \begin{bmatrix} v_1 \\ v_2 \end{bmatrix}$ corresponding to the eigenvalue $\lambda_2 = -2$, we
solve

$$\begin{bmatrix} 2v_1 - 2v_2 \\ 2v_1 - 3v_2 \end{bmatrix} = -2 \begin{bmatrix} v_1 \\ v_2 \end{bmatrix}$$

which can be written as

$$4v_1 - 2v_2 = 0$$
$$2v_1 - v_2 = 0$$

The two equations reduce to the same equation, namely,

$$2v_1 = v_2$$

Setting $v_1 = 1$, we find $v_2 = 2$. An eigenvector corresponding to $\lambda_2 = -2$ is then

$$\mathbf{v} = \begin{bmatrix} 1 \\ 2 \end{bmatrix}$$

We set

$$\mathbf{x}(t) = e^{-2t} \begin{bmatrix} 1 \\ 2 \end{bmatrix}$$

and check that

$$\frac{d\mathbf{x}}{dt} = -2e^{-2t} \begin{bmatrix} 1 \\ 2 \end{bmatrix} = -2\mathbf{x}(t)$$

and

$$A\mathbf{x}(t) = Ae^{-2t} \begin{bmatrix} 1 \\ 2 \end{bmatrix} = e^{-2t}A \begin{bmatrix} 1 \\ 2 \end{bmatrix} = -2e^{-2t} \begin{bmatrix} 1 \\ 2 \end{bmatrix} = -2\mathbf{x}(t)$$

where we used the fact that

$$A \begin{bmatrix} 1 \\ 2 \end{bmatrix} = \lambda_2 \begin{bmatrix} 1 \\ 2 \end{bmatrix} = -2 \begin{bmatrix} 1 \\ 2 \end{bmatrix}$$

since $\begin{bmatrix} 1 \\ 2 \end{bmatrix}$ is an eigenvector to $\lambda_2 = -2$. This shows that $\mathbf{x}(t) = e^{-2t} \begin{bmatrix} 1 \\ 2 \end{bmatrix}$ is also a solution of (11.11).

The direction field **4.** We wish to illustrate these two particular solutions in the corresponding direction field. Recall that an eigenvector can be used to construct a straight line through the origin in the direction of the eigenvector. In Figure 11.4, we show the direction field together with the two lines in the direction of the eigenvectors.

If $\mathbf{x}(0)$ is a point on one of the lines defined by the eigenvectors, then the solution $\mathbf{x}(t)$ will remain on that line for *all* later times. The location on the line will change with time. If the corresponding eigenvalue is positive, the solution will move away from the origin; if the eigenvalue is negative, it will move toward the origin, as can be seen from the direction of the direction vectors. The solid line in Figure 11.4 ($2x_1 - x_2 = 0$) corresponds to the eigenvalue $\lambda_2 = -2$ and we see that the direction vectors on this line point towards the origin. The dashed line in Figure 11.4 ($x_1 - 2x_2 = 0$) corresponds to the eigenvalue $\lambda_2 = 1$, and we see that the direction vectors on this line point away from the origin.

The General Solution Suppose that $\mathbf{y}(t)$ and $\mathbf{z}(t)$ are two solutions of

$$\frac{d\mathbf{x}}{dt} = A\mathbf{x}(t) \tag{11.12}$$

that is, but since

$$\frac{d\mathbf{y}}{dt} = A\mathbf{y}(t) \quad \text{and} \quad \frac{d\mathbf{z}}{dt} = A\mathbf{z}(t)$$

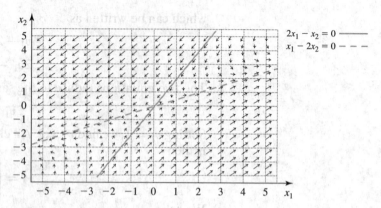

▲ **Figure 11.4**
The direction field of the system (11.11) with the lines in the direction of the eigenvectors

Then the linear combination

$$c_1\mathbf{y}(t) + c_2\mathbf{z}(t) \tag{11.13}$$

also solves (11.12). This can be seen as follows: First, note that

$$\frac{d}{dt}[c_1\mathbf{y}(t) + c_2\mathbf{z}(t)] = c_1\frac{d\mathbf{y}}{dt} + c_2\frac{d\mathbf{z}}{dt}$$

But since $\mathbf{y}(t)$ and $\mathbf{z}(t)$ are both solutions of (11.12),

$$c_1\frac{d\mathbf{y}}{dt} + c_2\frac{d\mathbf{z}}{dt} = c_1 A\mathbf{y}(t) + c_2 A\mathbf{z}(t) \tag{11.14}$$

Using the linearity properties of A, we can write the right-hand side of (11.14) as

$$A[c_1\mathbf{y}(t) + c_2\mathbf{z}(t)]$$

Summarizing this, we find

$$\frac{d}{dt}[\underbrace{c_1\mathbf{y}(t) + c_2\mathbf{z}(t)}_{\mathbf{x}(t)}] = A[\underbrace{c_1\mathbf{y}(t) + c_2\mathbf{z}(t)}_{\mathbf{x}(t)}]$$

which shows that the linear combination $c_1\mathbf{y}(t) + c_2\mathbf{z}(t)$ is also a solution.

Combining solutions, as in (11.13), illustrates the **superposition principle**. Since this is an important principle, we state it in the following.

SUPERPOSITION PRINCIPLE

Suppose that

$$\begin{bmatrix} \dfrac{dx_1}{dt} \\ \dfrac{dx_2}{dt} \end{bmatrix} = \begin{bmatrix} a_{11} & a_{12} \\ a_{21} & a_{22} \end{bmatrix}\begin{bmatrix} x_1(t) \\ x_2(t) \end{bmatrix} \tag{11.15}$$

If

$$\mathbf{y}(t) = \begin{bmatrix} y_1(t) \\ y_2(t) \end{bmatrix} \quad \text{and} \quad \mathbf{z}(t) = \begin{bmatrix} z_1(t) \\ z_2(t) \end{bmatrix}$$

are solutions of (11.15), then

$$\mathbf{x}(t) = c_1\mathbf{y}(t) + c_2\mathbf{z}(t)$$

is also a solution of (11.15).

We saw that $e^{\lambda t} \begin{bmatrix} w_1 \\ w_2 \end{bmatrix}$, where $\begin{bmatrix} w_1 \\ w_2 \end{bmatrix}$ is an eigenvector corresponding to the real eigenvalue λ of A, solves (11.12). If we have two real and distinct eigenvalues λ_1 and λ_2 with corresponding eigenvectors $\mathbf{u}$ and $\mathbf{v}$, then setting $\mathbf{y}(t) = e^{\lambda_1 t} \mathbf{u}$ and $\mathbf{z}(t) = e^{\lambda_2 t} \mathbf{v}$ and using the superposition principle, the following linear combination

$$\mathbf{x}(t) = c_1 e^{\lambda_1 t} \mathbf{u} + c_2 e^{\lambda_2 t} \mathbf{v} \tag{11.16}$$

where c_1 and c_2 are constants, is also a solution of (11.12). The constants c_1 and c_2 depend on the initial conditions. We can show that every solution can be written in the form (11.16); we therefore call a solution of the form (11.16) the general solution. (The situation is more complicated when A has repeated eigenvalues, that is, when $\lambda_1 = \lambda_2$. We do not give the general solution for that case here but will discuss two such examples in Problems 25 and 26.) We summarize our findings in the following.

THE GENERAL SOLUTION

Let

$$\frac{d\mathbf{x}}{dt} = A\mathbf{x}(t) \tag{11.17}$$

where A is a 2×2 matrix with two real and distinct eigenvalues λ_1 and λ_2, and corresponding eigenvectors $\mathbf{u}$ and $\mathbf{v}$. Then

$$\mathbf{x}(t) = c_1 e^{\lambda_1 t} \mathbf{u} + c_2 e^{\lambda_2 t} \mathbf{v} \tag{11.18}$$

is the general solution of (11.17). The constants c_1 and c_2 depend on the initial condition.

We will now check that (11.18) is indeed a solution of (11.17). To do so, we differentiate (11.18) with respect to t, and find

$$\frac{d\mathbf{x}}{dt} = \lambda_1 c_1 e^{\lambda_1 t} \mathbf{u} + \lambda_2 c_2 e^{\lambda_2 t} \mathbf{v}$$

Since $\mathbf{u}$ and $\mathbf{v}$ are eigenvectors corresponding to λ_1 and λ_2, we also find

$$Ax(t) = A(c_1 e^{\lambda_1 t} \mathbf{u} + c_2 e^{\lambda_2 t} \mathbf{v})$$
$$= c_1 e^{\lambda_1 t} A\mathbf{u} + c_2 e^{\lambda_2 t} A\mathbf{v}$$
$$= c_1 e^{\lambda_1 t} \lambda_1 \mathbf{u} + c_2 e^{\lambda_2 t} \lambda_2 \mathbf{v}$$

Hence,

$$\frac{d\mathbf{x}}{dt} = A\mathbf{x}(t)$$

Going back to the example in the beginning of this subsection, the general solution of (11.11)

$$\frac{d\mathbf{x}}{dt} = \begin{bmatrix} 2 & -2 \\ 2 & -3 \end{bmatrix} \mathbf{x}(t)$$

is

$$\mathbf{x}(t) = c_1 e^{t} \begin{bmatrix} 2 \\ 1 \end{bmatrix} + c_2 e^{-2t} \begin{bmatrix} 1 \\ 2 \end{bmatrix} \tag{11.19}$$

An initial condition   **5.** Suppose we know that at time 0,
for (11.11)

$$\mathbf{x}(0) = \begin{bmatrix} -1 \\ 4 \end{bmatrix} \tag{11.20}$$

holds for (11.11). We can then determine the constants c_1 and c_2 in (11.19) so that $\mathbf{x}(t)$ satisfies the initial condition (11.20),

$$\mathbf{x}(0) = c_1 \begin{bmatrix} 2 \\ 1 \end{bmatrix} + c_2 \begin{bmatrix} 1 \\ 2 \end{bmatrix} = \begin{bmatrix} -1 \\ 4 \end{bmatrix}$$

To find c_1 and c_2, we must solve the system of linear equations

$$2c_1 + c_2 = -1$$
$$c_1 + 2c_2 = 4$$

Eliminating c_1 in the second equation yields

$$2c_1 + c_2 = -1$$
$$3c_2 = 9$$

Hence, $c_2 = 3$ and, therefore,

$$2c_1 = -1 - c_2 = -1 - 3 = -4 \qquad \text{or} \qquad c_1 = -2$$

Therefore,

$$\mathbf{x}(t) = -2e^t \begin{bmatrix} 2 \\ 1 \end{bmatrix} + 3e^{-2t} \begin{bmatrix} 1 \\ 2 \end{bmatrix}$$

or, writing it out as separate equations,

$$x_1(t) = -4e^t + 3e^{-2t}$$
$$x_2(t) = -2e^t + 6e^{-2t}$$

To summarize, this solution solves the system

$$\frac{dx_1}{dt} = 2x_1 - 2x_2$$
$$\frac{dx_2}{dt} = 2x_2 - 3x_1$$

which was given in (11.11) with initial condition $x_1(0) = -1$ and $x_2(0) = 4$.

We give one more example of an initial value problem that illustrates all at once the different steps that must be carried out in order to obtain a solution in the case when A has two distinct and real eigenvalues.

▶ **Example 2**   Solve

$$\frac{d\mathbf{x}}{dt} = \begin{bmatrix} 2 & -3 \\ 1 & -2 \end{bmatrix} \mathbf{x}(t) \tag{11.21}$$

with the initial condition

$$\mathbf{x}(0) = \begin{bmatrix} 3 \\ -1 \end{bmatrix} \tag{11.22}$$

Solution

The first step is to find the eigenvalues and the corresponding eigenvectors. To find the eigenvalues, we must solve

$$\det(A - \lambda I) = \det \begin{bmatrix} 2 - \lambda & -3 \\ 1 & -2 - \lambda \end{bmatrix} = (2 - \lambda)(-2 - \lambda) + 3$$

$$= \lambda^2 - 1 = 0$$

which has solutions

$$\lambda_1 = 1 \quad \text{and} \quad \lambda_2 = -1$$

The eigenvector **u** corresponding to the eigenvalue $\lambda_1 = 1$ satisfies

$$\begin{bmatrix} 2 & -3 \\ 1 & -2 \end{bmatrix} \begin{bmatrix} u_1 \\ u_2 \end{bmatrix} = \begin{bmatrix} u_1 \\ u_2 \end{bmatrix}$$

Rewriting this as a system of equations, we find

$$2u_1 - 3u_2 = u_1$$
$$u_1 - 2u_2 = u_2$$

These two equations reduce to the same equation, namely,

$$u_1 - 3u_2 = 0$$

If we set $u_2 = 1$, then $u_1 = 3$ and, therefore,

$$\mathbf{u} = \begin{bmatrix} 3 \\ 1 \end{bmatrix}$$

is an eigenvector corresponding to $\lambda_1 = 1$.

The eigenvector **v** corresponding to the eigenvalue $\lambda_2 = -1$ satisfies

$$\begin{bmatrix} 2 & -3 \\ 1 & -2 \end{bmatrix} \begin{bmatrix} v_1 \\ v_2 \end{bmatrix} = - \begin{bmatrix} v_1 \\ v_2 \end{bmatrix}$$

Rewriting this as a system of equations, we find

$$2v_1 - 3v_2 = -v_1$$
$$v_1 - 2v_2 = -v_2$$

These two equations reduce to the same equation, namely,

$$v_1 - v_2 = 0$$

If we set $v_1 = 1$, then $v_2 = 1$ and, therefore,

$$\mathbf{v} = \begin{bmatrix} 1 \\ 1 \end{bmatrix}$$

is an eigenvector corresponding to $\lambda_2 = -1$. The general solution of (11.21) is therefore

$$\mathbf{x}(t) = c_1 e^t \begin{bmatrix} 3 \\ 1 \end{bmatrix} + c_2 e^{-t} \begin{bmatrix} 1 \\ 1 \end{bmatrix}$$

where c_1 and c_2 are constants. The initial condition (11.22) will allow us to determine the constants c_1 and c_2, namely,

$$\mathbf{x}(0) = c_1 \begin{bmatrix} 3 \\ 1 \end{bmatrix} + c_2 \begin{bmatrix} 1 \\ 1 \end{bmatrix} = \begin{bmatrix} 3 \\ -1 \end{bmatrix}$$

That is, c_1 and c_2 satisfy

$$3c_1 + c_2 = 3$$
$$c_1 + c_2 = -1$$

We solve this system using the standard elimination method,

$$3c_1 + c_2 = 3$$
$$2c_1 = 4$$

Hence, $c_1 = 2$ and $c_2 = 3 - 3c_1 = 3 - 6 = -3$. The solution of (11.21) that satisfies the initial condition (11.22) is therefore

$$\mathbf{x}(t) = 2e^t \begin{bmatrix} 3 \\ 1 \end{bmatrix} - 3e^{-t} \begin{bmatrix} 1 \\ 1 \end{bmatrix}$$

which can also be written in the form

$$x_1(t) = 6e^t - 3e^{-t}$$
$$x_2(t) = 2e^t - 3e^{-t}$$

The direction field, with two lines in the direction of the two eigenvectors and the solution, is shown in Figure 11.5. ◀

You might have noticed that all initial conditions have been formulated at time $t = 0$. This is a natural choice for an initial condition; however, we could have chosen any other time, for instance, $t = 1$. Suppose that in Example 2, the initial condition had been

$$\mathbf{x}(1) = \begin{bmatrix} 2 \\ 1 \end{bmatrix} \tag{11.23}$$

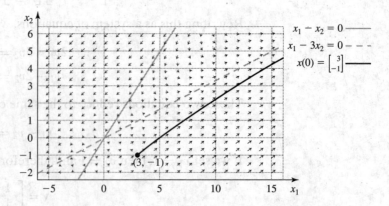

▲ **Figure 11.5**
The direction field of the system (11.21) with the lines in the direction of the eigenvectors and the solution with initial condition $\begin{bmatrix} 3 \\ -1 \end{bmatrix}$

The general solution would still be

$$\mathbf{x}(t) = c_1 e^t \begin{bmatrix} 3 \\ 1 \end{bmatrix} + c_2 e^{-t} \begin{bmatrix} 1 \\ 1 \end{bmatrix}$$

but the constants c_1 and c_2 would satisfy

$$3ec_1 + e^{-1}c_2 = 2$$
$$ec_1 + e^{-1}c_2 = 1$$

We solve this system using the standard elimination method,

$$3ec_1 + e^{-1}c_2 = 2$$
$$2ec_1 = 1$$

Hence, $c_1 = \frac{1}{2e}$ and $c_2 = \frac{e}{2}$. The solution satisfying (11.23) is then

$$\mathbf{x}(t) = \frac{1}{2}e^{t-1}\begin{bmatrix} 3 \\ 1 \end{bmatrix} + \frac{1}{2}e^{1-t}\begin{bmatrix} 1 \\ 1 \end{bmatrix}$$

11.1.3 Equilibria and Stability

In this subsection, we will be concerned with equilibria and stability, two concepts that we encountered in Section 8.2, where we discussed ordinary differential equations. Both concepts can be extended to systems of differential equations. We will restrict ourselves to the case

$$\frac{d\mathbf{x}}{dt} = A\mathbf{x}(t) \tag{11.24}$$

with

$$A = \begin{bmatrix} a_{11} & a_{12} \\ a_{21} & a_{22} \end{bmatrix} \quad \text{and} \quad \mathbf{x}(t) = \begin{bmatrix} x_1(t) \\ x_2(t) \end{bmatrix} \tag{11.25}$$

We say that a point

$$\hat{\mathbf{x}} = \begin{bmatrix} \hat{x}_1 \\ \hat{x}_2 \end{bmatrix}$$

is an equilibrium of (11.24) if

$$A\hat{\mathbf{x}} = \mathbf{0}$$

that is, a point where the direction vector in the corresponding direction field has length 0. If we start a system of differential equations at an equilibrium point, it will remain there for all later times.

To find equilibria of (11.24), we must solve $A\mathbf{x} = \mathbf{0}$. We see immediately that $\hat{\mathbf{x}} = \mathbf{0}$ solves $A\mathbf{x} = \mathbf{0}$. It follows from results in Subsection 9.2.3 that if $\det A \neq 0$, then $(0, 0)$ is the only equilibrium of (11.24). If $\det A = 0$, then there will be other equilibria.

As in Chapter 8, the characterizing property of an equilibrium is that if we start a system in equilibrium, it will stay there for all future times. This does *not* mean that if the system is in equilibrium and is **perturbed** by a small amount (that is, the solution is moved to a nearby point), it will return to the equilibrium (just as in Chapter 8). Whether or not a solution will return to an equilibrium after a small perturbation is addressed by the

stability of the equilibrium. We saw in the previous subsection that, in the case when A has two real and distinct eigenvalues, the solution of (11.24) is given by

$$\mathbf{x}(t) = c_1 e^{\lambda_1 t} \mathbf{u} + c_2 e^{\lambda_2 t} \mathbf{v} \qquad (11.26)$$

where $\mathbf{u}$ and $\mathbf{v}$ are the eigenvectors corresponding to the eigenvalues λ_1 and λ_2 of the matrix A, and the constants c_1 and c_2 depend on the initial condition. Knowing the solution will allow us to study the behavior of the solution (11.26) as $t \to \infty$, and thus address the question of stability, at least when the eigenvalues are real and distinct.

Since A is a 2×2 matrix and all entries of A are real, the eigenvalues of A are either both real or both complex conjugate. We will treat these two cases separately. To simplify our discussion, we again assume that A has two distinct eigenvalues.

The equilibria of (11.24) can be found by solving

$$A\mathbf{x} = \mathbf{0} \qquad (11.27)$$

If $\det A \neq 0$, then (11.27) has only one solution, namely the trivial solution $(0, 0)$ (see Subsection 9.2.3). Since $\det A = \lambda_1 \lambda_2$, where λ_1 and λ_2 are the eigenvalues of A, we see that if λ_1 and λ_2 are both nonzero, then $\det A \neq 0$.

Case 1: A has Two Distinct, Real and Nonzero Eigenvalues   In this case, the equation $A\mathbf{x} = \mathbf{0}$ has only one solution, namely $(0, 0)$, and thus $(0, 0)$ is the only equilibrium. The general solution of (11.24) is given by (11.26), and we can therefore study the behavior of the solution of (11.24) directly by investigating (11.26). We are interested in determining

$$\lim_{t \to \infty} \mathbf{x}(t) = \lim_{t \to \infty} \left[c_1 e^{\lambda_1 t} \mathbf{u} + c_2 e^{\lambda_2 t} \mathbf{v} \right] \qquad (11.28)$$

We see immediately that the behavior of $\mathbf{x}(t)$ is determined by $e^{\lambda_1 t}$ and $e^{\lambda_2 t}$. Recall that

$$\lim_{t \to \infty} e^{\lambda_i t} = \begin{cases} 0 & \text{if } \lambda_i < 0 \\ \infty & \text{if } \lambda_i > 0 \end{cases}$$

We distinguish the following three categories:

1. Both eigenvalues are negative: $\lambda_1 < 0$ and $\lambda_2 < 0$.
2. The eigenvalues are of opposite signs: $\lambda_1 < 0$ and $\lambda_2 > 0$ (or $\lambda_1 > 0$ and $\lambda_2 < 0$).
3. Both eigenvalues are positive: $\lambda_1 > 0$ and $\lambda_2 > 0$.

Representative direction fields for each of the three categories are shown in Figures 11.6 through 11.8. We will discuss each category separately.

Category 1   When both eigenvalues are negative, we conclude from (11.28) that

$$\lim_{t \to \infty} \mathbf{x}(t) = \mathbf{0}$$

regardless of $(x_1(0), x_2(0))$. We say that the equilibrium $(0, 0)$ is **globally stable** since the solution will approach the equilibrium $(0, 0)$ regardless of the starting point (and not just from nearby points). We call $(0, 0)$ a **sink**, or a **stable node**. A direction field for this case is shown in Figure 11.6; the shape of the direction field explains why we call this equilibrium a sink: all solutions "flow" into the origin.

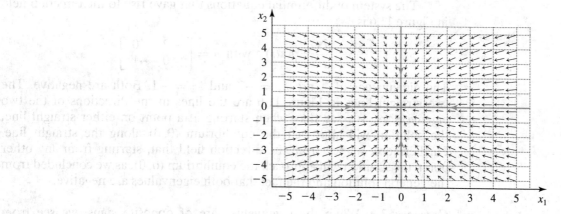

▲ **Figure 11.6**
The direction field of a linear system where both eigenvalues are negative, together with the lines in the direction of the eigenvectors

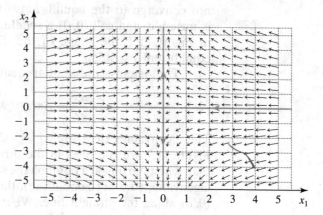

▲ **Figure 11.7**
The direction field of a linear system where both eigenvalues are of opposite sign, together with the lines in the direction of the eigenvectors

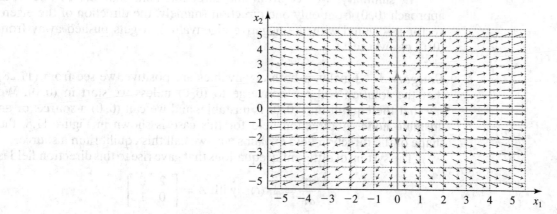

▲ **Figure 11.8**
The direction field of a linear system where both eigenvalues are positive, together with the lines in the direction of the eigenvectors

The system of differential equations that gave rise to the direction field in Figure 11.6 is

$$\frac{d\mathbf{x}}{dt} = A\mathbf{x}(t) \quad \text{with } A = \begin{bmatrix} -2 & 0 \\ 0 & -1 \end{bmatrix}$$

The eigenvalues of A are $\lambda_1 = -2$ and $\lambda_2 = -1$. Both are negative. The two straight lines in Figure 11.6 are the lines in the directions of the two eigenvectors. We see that, when starting at a point on either straight line, the solution will approach the equilibrium $(0, 0)$ along the straight line. Furthermore, we see from the direction field that, starting from any other point, the solution will approach the equilibrium $(0, 0)$, as we concluded from the general solution and the fact that both eigenvalues are negative.

Category 2   When the eigenvalues are of opposite signs, we see from (11.28) that the component of the solution associated with the negative eigenvalue goes to 0 as $t \to \infty$, and the component associated with the positive eigenvalue goes to infinity. That is, unless we start in the direction of the eigenvector associated with the negative eigenvalue, the solution will not converge to the equilibrium $(0, 0)$. We say that the equilibrium $(0, 0)$ is **unstable**, and call $(0, 0)$ a **saddle**. A direction field for this case is shown in Figure 11.7; the shape of the direction field explains why we call this equilibrium a saddle.

The system of differential equations that gave rise to this direction field is

$$\frac{d\mathbf{x}}{dt} = A\mathbf{x}(t) \quad \text{with } A = \begin{bmatrix} -2 & 0 \\ 0 & 1 \end{bmatrix}$$

The eigenvalues of A are $\lambda_1 = -2$ and $\lambda_2 = 1$. The two straight lines in Figure 11.7 are the lines in the directions of the two eigenvectors. We see that, when starting at a point on the straight line corresponding to the negative eigenvalue (the horizontal line), the solution will approach the equilibrium $(0, 0)$ along the straight line. When starting at a point on the straight line corresponding to the positive eigenvalue (the vertical line), the solution will move away from the equilibrium $(0, 0)$ along the straight line. Furthermore, we see from the direction field that, starting from any other point, the solution will eventually move away from the equilibrium $(0, 0)$.

In summary, we see from the direction field that the solution can approach $(0, 0)$ from only one direction (namely, the direction of the eigenvector associated with the negative eigenvalue); it gets pushed away from $(0, 0)$ elsewhere.

Category 3   Finally, if both eigenvalues are positive, we see from (11.28) that the solution will not converge to $(0, 0)$ unless we start in $(0, 0)$. We say that the equilibrium $(0, 0)$ is unstable, and we call $(0, 0)$ a **source**, or an **unstable node**. A direction field for this case is shown in Figure 11.8; the shape of the direction field explains why we call this equilibrium a source.

The system of differential equations that gave rise to this direction field is

$$\frac{d\mathbf{x}}{dt} = A\mathbf{x}(t) \quad \text{with } A = \begin{bmatrix} 2 & 0 \\ 0 & 1 \end{bmatrix}$$

The eigenvalues of A are $\lambda_1 = 2$ and $\lambda_2 = 1$, which are both positive. The two straight lines in Figure 11.8 are the lines in the directions of the two eigenvectors. We see that, when starting at a point on either straight line, the solution will move away from the equilibrium $(0, 0)$ along the straight line.

Furthermore, we see from the direction field that, starting from any other point, the solution will move away from the equilibrium $(0, 0)$.

Case 2: A has Complex Conjugate Eigenvalues   We will not solve the system when A has complex conjugate eigenvalues. Instead, we will look at some examples to see what typical direction fields look like.

Category 1   Let

$$A = \begin{bmatrix} -1 & -1 \\ 1 & 0 \end{bmatrix} \tag{11.29}$$

Then

$$\det(A - \lambda I) = \det \begin{bmatrix} -1 - \lambda & -1 \\ 1 & -\lambda \end{bmatrix} = (-1 - \lambda)(-\lambda) + 1$$

$$= \lambda^2 + \lambda + 1 = 0$$

which gives

$$\lambda_{1,2} = \frac{-1 \pm \sqrt{1 - 4}}{2} = -\frac{1}{2} \pm \frac{i}{2}\sqrt{3}$$

Both eigenvalues are complex and form a conjugate pair. (Note that the real parts of both eigenvalues are negative.) To see what the solutions of

$$\frac{d\mathbf{x}}{dt} = A\mathbf{x}(t)$$

look like, we graph the direction field and solutions $x_1(t)$ and $x_2(t)$ in Figures 11.9 and 11.10, respectively.

We see from the direction field that, starting from any point other than $(0, 0)$, solutions spiral into the equilibrium $(0, 0)$. For this reason, the equilibrium $(0, 0)$ is called a **stable spiral**. When we plot solutions as functions of time, they show oscillations, as illustrated in Figure 11.10. The amplitude of the oscillations decreases over time. We therefore call the oscillations **damped**.

The oscillations are caused by the imaginary part of the eigenvalues; the damping of the oscillations is caused by the negative real part of the eigenvalues. Before we explain this further, we give two more examples, one in which the complex conjugate eigenvalues have positive real parts, the other in which the eigenvalues are purely imaginary.

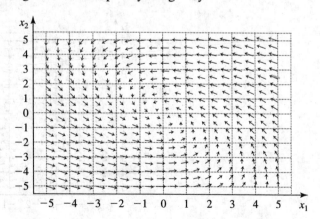

▲ **Figure 11.9**
The direction field for the system with matrix A in (11.29)

Furthermore, we see from the direction field that, starting from any other point, the solution will move away from the equilibrium at $(0, 0)$.

Case 2: A has complex conjugate Eigenvalues. We will not solve the system when A has complex conjugate eigenvalues. Instead, we will look at some specific cases to see what the direction fields look like.

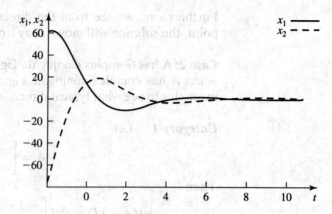

▲ **Figure 11.10**
The solutions for (11.29)

Category 2 Let

$$B = \begin{bmatrix} 1 & -1 \\ 1 & 0 \end{bmatrix}$$

Then

$$\det(B - \lambda I) = \det \begin{bmatrix} 1 - \lambda & -1 \\ 1 & -\lambda \end{bmatrix} = (1 - \lambda)(-\lambda) + 1$$

$$= \lambda^2 - \lambda + 1 = 0$$

which has solutions

$$\lambda_{1,2} = \frac{1 \pm \sqrt{1 - 4}}{2} = \frac{1}{2} \pm \frac{i}{2}\sqrt{3}$$

Both eigenvalues of B are complex and form a complex conjugate pair, but their real parts are positive. To see what the solutions of

$$\frac{d\mathbf{x}}{dt} = B\mathbf{x}(t) \tag{11.30}$$

look like, we graph the direction field and solutions $x_1(t)$ and $x_2(t)$ in Figures 11.11 and 11.12, respectively.

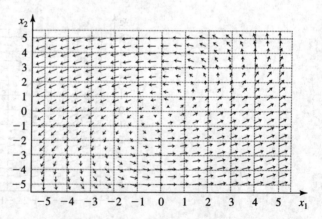

▲ **Figure 11.11**
The direction field for the system with matrix B in (11.30)

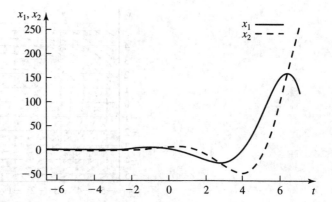

▲ **Figure 11.12**
The solutions for (11.30)

We see from the direction field that, starting from any point other than $(0, 0)$, the solutions spiral out from the equilibrium $(0, 0)$. For this reason, the equilibrium $(0, 0)$ is a **unstable spiral**. When we plot solutions as functions of time, as in Figure 11.12, we see that the solutions show oscillations as before, but this time their amplitudes are increasing. The oscillations are again caused by the imaginary parts of the eigenvalues; the increase in amplitude is caused by the positive real parts.

Category 3 Here is an example where both eigenvalues are purely imaginary. Let

$$C = \begin{bmatrix} 0 & -1 \\ 1 & 0 \end{bmatrix}$$

Then

$$\det(C - \lambda I) = \det \begin{bmatrix} -\lambda & -1 \\ 1 & -\lambda \end{bmatrix}$$
$$= \lambda^2 + 1 = 0$$

which gives

$$\lambda_{1,2} = \pm i$$

Both eigenvalues of C are complex and form a complex conjugate pair, but they are purely imaginary (their real parts are equal to 0). To see what the solutions of

$$\frac{d\mathbf{x}}{dt} = C\mathbf{x}(t) \tag{11.31}$$

look like, we graph both the direction field and solutions $x_1(t)$ and $x_2(t)$ in Figures 11.13 and 11.14.

Looking at solution curves in Figure 11.14, we see that the solutions oscillate as in the previous two examples, but this time the amplitude does not change with time. Looking at the direction field in Figure 11.13, we see that solutions spiral around the equilibrium $(0, 0)$. But since the amplitude of the solutions does not change, the solutions neither approach nor move away from the equilibrium. The equilibrium $(0, 0)$ is called a **neutral spiral** or a **center**. We can show that the solutions form closed curves. (We will analyze this more closely in Example 3, where we will show this direction field with a solution curve.)

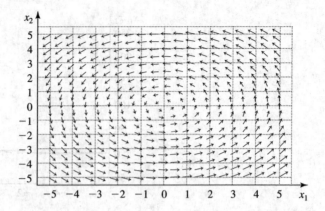

▲ **Figure 11.13**
The direction field for the system with matrix C in (11.31)

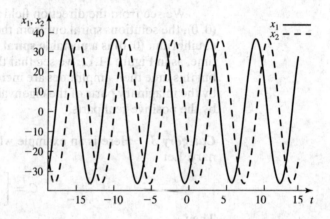

▲ **Figure 11.14**
The solutions for (11.31)

Where do the Oscillations Come from?   We can show that the solutions of (11.24) are given by (11.26), regardless of whether the eigenvalues are real or complex, as long as they are distinct. If the eigenvalues are complex, then the solution contains terms of the form e^z, where z is a complex number.

We can write a complex number z as

$$z = a + ib$$

where a and b are both real. Recall that the number a is called the real part of z, the number b is called the imaginary part of z.

To understand what e^z means when z is complex, we write

$$e^z = e^{a+ib} = e^a e^{ib}$$

The term e^a is a real number; therefore, we know what it means. The term e^{ib} is an exponential with a purely imaginary exponent. We have not encountered this before. The following formula, which we cannot prove here, explains its meaning.

EULER'S FORMULA

$$e^{ib} = \cos b + i \sin b$$

Both the sine and the cosine functions show oscillations; this is the reason why systems of differential equations with complex eigenvalues have solutions that oscillate, as in the following example, where we can guess the answer.

▶ **Example 3**   Show that $x_1(t) = \cos t$ and $x_2(t) = \sin t$ solve

$$\frac{dx_1}{dt} = -x_2$$

$$\frac{dx_2}{dt} = x_1$$

(11.32)

with $x_1(0) = 1$ and $x_2(0) = 0$.

Solution

The associated matrix of coefficients is

$$A = \begin{bmatrix} 0 & -1 \\ 1 & 0 \end{bmatrix}$$

which is the same as matrix C that we saw in (11.31). We therefore know that

$$\lambda_1 = i \qquad \text{and} \qquad \lambda_2 = -i$$

We need to show that

$$x_1(t) = \cos t$$

$$x_2(t) = \sin t$$

solves (11.32). Let's check. We find

$$\frac{dx_1}{dt} = -\sin t = -x_2(t)$$

and

$$\frac{dx_2}{dt} = \cos t = x_1(t)$$

In addition, we must check that our solution satisfies the initial condition. Indeed, $x_1(0) = \cos 0 = 1$ and $x_2(0) = \sin 0 = 0$. The solution for this system is a pair of real-valued functions, as claimed. The solution is shown in Figure 11.15, and the trajectory of $(x_1(t), x_2(t))$ in the x_1-x_2 plane is shown in Figure 11.16.

We see from Figure 11.15 that the solutions in this case are periodic. They show sustained oscillations, that is, the amplitudes of $x_1(t)$ and $x_2(t)$ do not change with time. In Figure 11.16, we see that the solution in the x_1-x_2 plane forms closed curves, indicating that the solutions are periodic.   ◀

Summary   We give a brief summary of the classification of the equilibrium $(0, 0)$ of

$$\frac{d\mathbf{x}}{dt} = A\mathbf{x}(t) \quad \text{with } A = \begin{bmatrix} a_{11} & a_{12} \\ a_{21} & a_{22} \end{bmatrix}$$

Recall that

The trace of A is $a_{11} + a_{22}$.

The determinant of A is $a_{11}a_{22} - a_{12}a_{21}$.

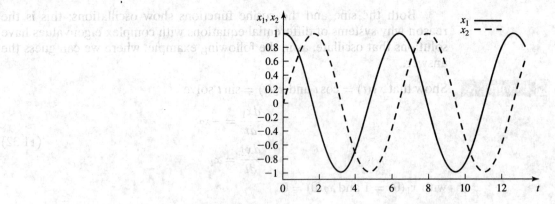

▲ **Figure 11.15**
The solution for Example 3 as functions of t with initial condition $x_1(0) = 1$ and $x_2(0) = 0$

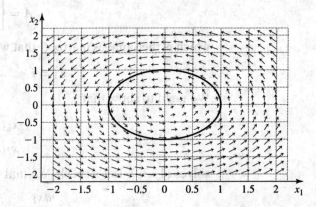

▲ **Figure 11.16**
The trajectory for Example 3 in the direction field with initial condition $x_1(0) = 1$ and $x_2(0) = 0$. The trajectory is a closed curve

We denote the trace of A by τ and the determinant of A by Δ. The eigenvalues of A are found by solving

$$\det(A - \lambda I) = \det \begin{bmatrix} a_{11} - \lambda & a_{12} \\ a_{21} & a_{22} - \lambda \end{bmatrix}$$

$$= (a_{11} - \lambda)(a_{22} - \lambda) - a_{12}a_{21}$$

$$= \lambda^2 - (a_{11} + a_{22})\lambda + a_{11}a_{22} - a_{12}a_{21}$$

$$= \lambda^2 - \tau\lambda + \Delta = 0$$

where we used $\tau = a_{11} + a_{22}$ and $\Delta = a_{11}a_{22} - a_{12}a_{21}$. This gives

$$\lambda_1 = \frac{\tau + \sqrt{\tau^2 - 4\Delta}}{2} \quad \text{and} \quad \lambda_2 = \frac{\tau - \sqrt{\tau^2 - 4\Delta}}{2} \tag{11.33}$$

We see immediately that

If $\tau^2 > 4\Delta$, both eigenvalues are real and distinct.

If $\tau^2 < 4\Delta$, eigenvalues are complex conjugates.

When both eigenvalues are real and distinct—that is, when $\tau^2 > 4\Delta$—we can distinguish the following three cases:

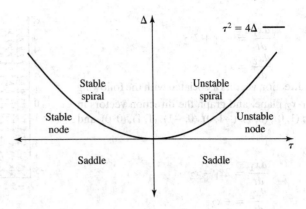

▲ **Figure 11.17**
The stability behavior of a system of two linear, homogeneous differential equations with constant coefficients

1. $\Delta < 0$: $\lambda_1 > 0$, $\lambda_2 < 0$ (saddle)
2. $\Delta > 0$, $\tau < 0$: $\lambda_1 < 0$, $\lambda_2 < 0$ (sink or stable node)
3. $\Delta > 0$, $\tau > 0$: $\lambda_1 > 0$, $\lambda_2 > 0$ (source or unstable node)

When both eigenvalues are complex conjugates—that is, when $\tau^2 < 4\Delta$—we can distinguish the following three cases:

1. $\tau < 0$: both eigenvalues have negative real parts (stable spiral)
2. $\tau > 0$: both eigenvalues have positive real parts (unstable spiral)
3. $\tau = 0$: both eigenvalues are purely imaginary (center)

We can summarize this graphically in the τ-Δ plane as shown in Figure 11.17. The parabola $4\Delta = \tau^2$ is the boundary line between oscillatory and nonoscillatory behavior. The line $\tau = 0$ divides the stable and the unstable regions. The line $\Delta = 0$ separates the saddle and the node regions. The case in which the eigenvalues are identical resides on the boundary line $4\Delta = \tau^2$.

The line $\Delta = 0$ corresponds to the case in which one eigenvalue is equal to 0. As long as the other eigenvalue is not equal to 0, both eigenvalues are again distinct and the solution is of the form (11.26). However, in this case, there are equilibria other than $(0, 0)$. We will discuss two such examples in Problems 57 and 58, and one in Section 11.2.

11.1.4 Problems

(11.1.1)

In Problems 1–4, write each system of differential equations in matrix form.

1.
$$\frac{dx_1}{dt} = 2x_1 + 3x_2$$
$$\frac{dx_2}{dt} = -4x_1 + x_2$$

2.
$$\frac{dx_1}{dt} = x_1 + x_2$$
$$\frac{dx_2}{dt} = -x_1$$

3.
$$\frac{dx_1}{dt} = x_3$$
$$\frac{dx_2}{dt} = -x_1$$
$$\frac{dx_3}{dt} = x_1 + x_2$$

4.
$$\frac{dx_1}{dt} = 2x_2 - x_3$$
$$\frac{dx_2}{dt} = -x_1$$
$$\frac{dx_3}{dt} = 5x_1 + x_3$$

5. Consider

$$\frac{dx_1}{dt} = -x_1 + 2x_2$$
$$\frac{dx_2}{dt} = x_1$$

Determine the direction vectors associated with the following points in the x_1-x_2 plane, and graph the direction vectors in the x_1-x_2 plane: $(1, 0)$, $(0, 1)$, $(-1, 0)$, $(0, -1)$, $(1, 1)$, $(0, 0)$, and $(-2, 1)$.

6. Consider

$$\frac{dx_1}{dt} = 2x_1 - x_2$$
$$\frac{dx_2}{dt} = -x_2$$

Determine the direction vectors associated with the following points in the x_1-x_2 plane, and graph the direction vectors in the x_1-x_2 plane: $(2, 0)$, $(1.5, 1)$, $(1, 0)$, $(0, -1)$, $(1, 1)$, $(0, 0)$, and $(-2, -2)$.

7. Consider

$$\frac{dx_1}{dt} = x_1 + 3x_2$$
$$\frac{dx_2}{dt} = -x_1 + 2x_2$$

Determine the direction vectors associated with the following points in the x_1-x_2 plane, and graph the direction vectors in the x_1-x_2 plane: $(1, 0)$, $(0, 1)$, $(-1, 1)$, $(0, -1)$, $(-3, 1)$, $(0, 0)$, and $(-2, 1)$.

8. Consider

$$\frac{dx_1}{dt} = -x_2$$
$$\frac{dx_2}{dt} = x_1$$

Determine the direction vectors associated with the following points in the x_1-x_2 plane, and graph the direction vectors in the x_1-x_2 plane: $(1, 0)$, $(0, 1)$, $(-1, 0)$, $(0, -1)$, $(1, 1)$, $(0, 0)$, and $(-2, -2)$.

9. In Figures 11.18 through 11.21, direction fields are given. Each of the following systems of differential equations corresponds to exactly one of the direction fields. Match the systems to the appropriate figures.

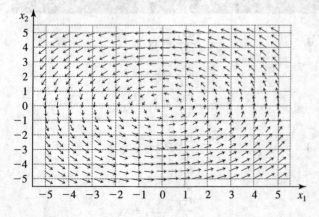

▲ **Figure 11.18**

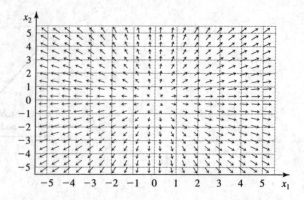

▲ **Figure 11.19**

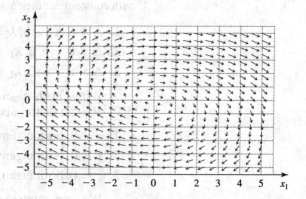

▲ **Figure 11.20**

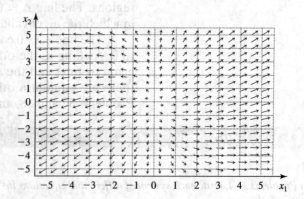

▲ **Figure 11.21**

(a)

$$\frac{dx_1}{dt} = -x_2$$
$$\frac{dx_2}{dt} = x_1$$

(b)

$$\frac{dx_1}{dt} = x_1$$
$$\frac{dx_2}{dt} = x_2$$

(c)
$$\frac{dx_1}{dt} = x_1 + 2x_2$$
$$\frac{dx_2}{dt} = -2x_1$$

(d)
$$\frac{dx_1}{dt} = 2x_1$$
$$\frac{dx_2}{dt} = x_1 + x_2$$

10. The direction field of
$$\frac{dx_1}{dt} = x_1 + 3x_2$$
$$\frac{dx_2}{dt} = 2x_1 + 3x_2$$
is given in Figure 11.22. Sketch the solution curve that goes through the point $(1, 0)$.

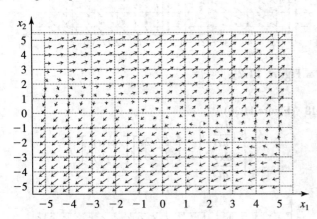

▲ **Figure 11.22**

11. The direction field of
$$\frac{dx_1}{dt} = 2x_1 + 3x_2$$
$$\frac{dx_2}{dt} = -x_1 + x_2$$
is given in Figure 11.23. Sketch the solution curve that goes through the point $(2, -1)$.

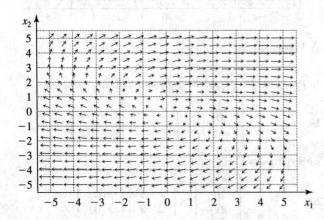

▲ **Figure 11.23**

12. The direction field of
$$\frac{dx_1}{dt} = -x_1 - x_2$$
$$\frac{dx_2}{dt} = -2x_2$$
is given in Figure 11.24. Sketch the solution curve that goes through the point $(-3, -3)$.

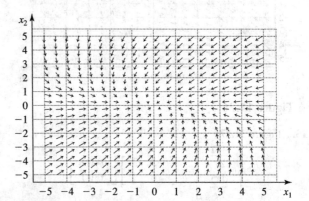

▲ **Figure 11.24**

(11.1.2)

In Problems 13–18, find the general solution to each given system of differential equations, and sketch the lines in the direction of the eigenvectors. Indicate on each line the direction in which the solution would move if starting on that line.

13. (Figure 11.25)
$$\begin{bmatrix} \dfrac{dx_1}{dt} \\ \dfrac{dx_2}{dt} \end{bmatrix} = \begin{bmatrix} 1 & 3 \\ 5 & 3 \end{bmatrix} \begin{bmatrix} x_1(t) \\ x_2(t) \end{bmatrix}$$

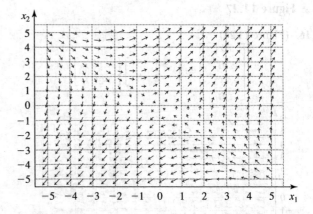

▲ **Figure 11.25**

14. (Figure 11.26)
$$\begin{bmatrix} \dfrac{dx_1}{dt} \\ \dfrac{dx_2}{dt} \end{bmatrix} = \begin{bmatrix} 2 & 1 \\ 4 & -1 \end{bmatrix} \begin{bmatrix} x_1(t) \\ x_2(t) \end{bmatrix}$$

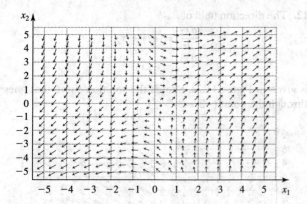

▲ **Figure 11.26**

15. (Figure 11.27)

$$\left[\begin{array}{c} \dfrac{dx_1}{dt} \\ \dfrac{dx_2}{dt} \end{array}\right] = \left[\begin{array}{cc} -3 & 3 \\ 6 & 4 \end{array}\right]\left[\begin{array}{c} x_1(t) \\ x_2(t) \end{array}\right]$$

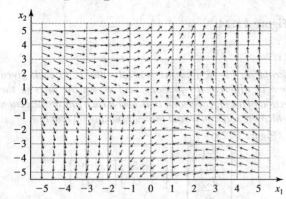

▲ **Figure 11.27**

16. (Figure 11.28)

$$\left[\begin{array}{c} \dfrac{dx_1}{dt} \\ \dfrac{dx_2}{dt} \end{array}\right] = \left[\begin{array}{cc} -5 & 3 \\ -2 & 0 \end{array}\right]\left[\begin{array}{c} x_1(t) \\ x_2(t) \end{array}\right]$$

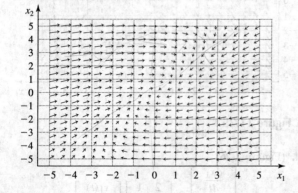

▲ **Figure 11.28**

17. (Figure 11.29)

$$\left[\begin{array}{c} \dfrac{dx_1}{dt} \\ \dfrac{dx_2}{dt} \end{array}\right] = \left[\begin{array}{cc} -2 & 0 \\ -3 & 1 \end{array}\right]\left[\begin{array}{c} x_1(t) \\ x_2(t) \end{array}\right]$$

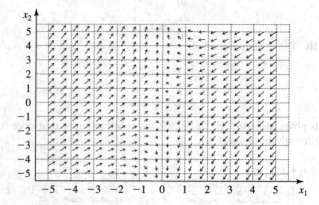

▲ **Figure 11.29**

18. (Figure 11.30)

$$\left[\begin{array}{c} \dfrac{dx_1}{dt} \\ \dfrac{dx_2}{dt} \end{array}\right] = \left[\begin{array}{cc} 5 & 2 \\ 1 & 6 \end{array}\right]\left[\begin{array}{c} x_1(t) \\ x_2(t) \end{array}\right]$$

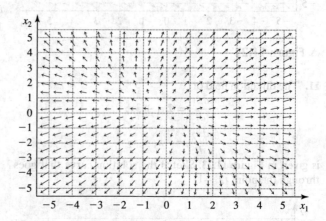

▲ **Figure 11.30**

In Problems 19–24, solve the given initial value problem.

19.
$$\left[\begin{array}{c} \dfrac{dx_1}{dt} \\ \dfrac{dx_2}{dt} \end{array}\right] = \left[\begin{array}{cc} -3 & 0 \\ 4 & 2 \end{array}\right]\left[\begin{array}{c} x_1(t) \\ x_2(t) \end{array}\right]$$
with $x_1(0) = -5$ and $x_2(0) = 5$.

20.
$$\left[\begin{array}{c} \dfrac{dx_1}{dt} \\ \dfrac{dx_2}{dt} \end{array}\right] = \left[\begin{array}{cc} 1 & 3 \\ 2 & 2 \end{array}\right]\left[\begin{array}{c} x_1(t) \\ x_2(t) \end{array}\right]$$
with $x_1(0) = 2$ and $x_2(0) = -1$.

21. $\begin{bmatrix} \dfrac{dx_1}{dt} \\ \dfrac{dx_2}{dt} \end{bmatrix} = \begin{bmatrix} 3 & -2 \\ 0 & 1 \end{bmatrix} \begin{bmatrix} x_1(t) \\ x_2(t) \end{bmatrix}$

with $x_1(0) = 1$ and $x_2(0) = 1$.

22. $\begin{bmatrix} \dfrac{dx_1}{dt} \\ \dfrac{dx_2}{dt} \end{bmatrix} = \begin{bmatrix} -3 & 4 \\ -1 & 2 \end{bmatrix} \begin{bmatrix} x_1(t) \\ x_2(t) \end{bmatrix}$

with $x_1(0) = 1$ and $x_2(0) = 2$.

23. $\begin{bmatrix} \dfrac{dx_1}{dt} \\ \dfrac{dx_2}{dt} \end{bmatrix} = \begin{bmatrix} 4 & 7 \\ 1 & -2 \end{bmatrix} \begin{bmatrix} x_1(t) \\ x_2(t) \end{bmatrix}$

with $x_1(0) = -1$ and $x_2(0) = -2$.

24. $\begin{bmatrix} \dfrac{dx_1}{dt} \\ \dfrac{dx_2}{dt} \end{bmatrix} = \begin{bmatrix} 2 & 6 \\ 1 & 3 \end{bmatrix} \begin{bmatrix} x_1(t) \\ x_2(t) \end{bmatrix}$

with $x_1(0) = -3$ and $x_2(0) = 1$.

In Problems 25 and 26, we discuss the case of repeated eigenvalues.

25. Let

$$\begin{bmatrix} \dfrac{dx_1}{dt} \\ \dfrac{dx_2}{dt} \end{bmatrix} = \begin{bmatrix} 1 & 0 \\ 0 & 1 \end{bmatrix} \begin{bmatrix} x_1(t) \\ x_2(t) \end{bmatrix} \qquad (11.34)$$

(a) Show that

$$A = \begin{bmatrix} 1 & 0 \\ 0 & 1 \end{bmatrix}$$

has repeated eigenvalues, namely $\lambda_1 = \lambda_2 = 1$.

(b) Show that $\begin{bmatrix} 1 \\ 0 \end{bmatrix}$ and $\begin{bmatrix} 0 \\ 1 \end{bmatrix}$ are eigenvectors of A, and

that any vector $\begin{bmatrix} c_1 \\ c_2 \end{bmatrix}$ can be written as

$$\begin{bmatrix} c_1 \\ c_2 \end{bmatrix} = c_1 \begin{bmatrix} 1 \\ 0 \end{bmatrix} + c_2 \begin{bmatrix} 0 \\ 1 \end{bmatrix}$$

(c) Show that

$$\mathbf{x}(t) = c_1 e^t \begin{bmatrix} 1 \\ 0 \end{bmatrix} + c_2 e^t \begin{bmatrix} 0 \\ 1 \end{bmatrix}$$

is a solution of (11.34) that satisfies the initial condition $x_1(0) = c_1$ and $x_2(0) = c_2$.

26. Let

$$\begin{bmatrix} \dfrac{dx_1}{dt} \\ \dfrac{dx_2}{dt} \end{bmatrix} = \begin{bmatrix} 1 & 2 \\ 0 & 1 \end{bmatrix} \begin{bmatrix} x_1(t) \\ x_2(t) \end{bmatrix} \qquad (11.35)$$

(a) Show that

$$A = \begin{bmatrix} 1 & 2 \\ 0 & 1 \end{bmatrix}$$

has repeated eigenvalues, namely $\lambda_1 = \lambda_2 = 1$.

(b) Show that every eigenvector of A is of the form

$$c_1 \begin{bmatrix} 1 \\ 0 \end{bmatrix}$$

where c_1 is a real number different from 0.

(c) Show that

$$\mathbf{x}_1(t) = e^t \begin{bmatrix} 1 \\ 0 \end{bmatrix}$$

is a solution of (11.35).

(d) Show that

$$\mathbf{x}_2(t) = t e^t \begin{bmatrix} 1 \\ 0 \end{bmatrix} + e^t \begin{bmatrix} 0 \\ 0.5 \end{bmatrix}$$

is a solution of (11.35).

(e) Show that

$$c_1 \mathbf{x}_1(t) + c_2 \mathbf{x}_2(t)$$

is a solution of (11.35). (It turns out that this is the general solution.)

(11.1.3)

In Problems 27–36, we consider differential equations of the form

$$\frac{d\mathbf{x}}{dt} = A\mathbf{x}(t)$$

where

$$A = \begin{bmatrix} a_{11} & a_{12} \\ a_{21} & a_{22} \end{bmatrix}$$

The eigenvalues of A will be real, distinct, and nonzero. Analyze the stability of the equilibrium $(0,0)$, and classify the equilibrium according to sink, source, or saddle.

27. $A = \begin{bmatrix} 1 & 0 \\ 1 & -2 \end{bmatrix}$

28. $A = \begin{bmatrix} -2 & 4 \\ 2 & -5 \end{bmatrix}$

29. $A = \begin{bmatrix} 6 & -4 \\ -3 & 5 \end{bmatrix}$

30. $A = \begin{bmatrix} -1 & 3 \\ 2 & -5 \end{bmatrix}$

31. $A = \begin{bmatrix} -3 & -1 \\ 1 & -6 \end{bmatrix}$

32. $A = \begin{bmatrix} -3 & 1 \\ 1 & -2 \end{bmatrix}$

33. $A = \begin{bmatrix} 0 & -2 \\ -1 & 3 \end{bmatrix}$

34. $A = \begin{bmatrix} 1 & 2 \\ 3 & 7 \end{bmatrix}$

35. $A = \begin{bmatrix} -2 & -3 \\ 1 & 3 \end{bmatrix}$

36. $A = \begin{bmatrix} 4 & -1 \\ 5 & -1 \end{bmatrix}$

In Problems 37–46, we consider differential equations of the form

$$\frac{d\mathbf{x}}{dt} = A\mathbf{x}(t)$$

where

$$A = \begin{bmatrix} a_{11} & a_{12} \\ a_{21} & a_{22} \end{bmatrix}$$

The eigenvalues of A will be complex conjugates. Analyze the stability of the equilibrium $(0,0)$, and classify the equilibrium according to stable spiral, unstable spiral, or center.

37. $A = \begin{bmatrix} 1 & 3 \\ -2 & -2 \end{bmatrix}$

38.
$$A = \begin{bmatrix} 2 & -3 \\ 4 & -1 \end{bmatrix}$$

39.
$$A = \begin{bmatrix} 4 & 5 \\ -3 & -3 \end{bmatrix}$$

40.
$$A = \begin{bmatrix} 2 & 2 \\ -6 & -4 \end{bmatrix}$$

41.
$$A = \begin{bmatrix} -1 & 1 \\ -3 & 1 \end{bmatrix}$$

42.
$$A = \begin{bmatrix} 3 & -2 \\ 1 & 2 \end{bmatrix}$$

43.
$$A = \begin{bmatrix} 0 & -1 \\ 1 & 0 \end{bmatrix}$$

44.
$$A = \begin{bmatrix} 0 & -3 \\ 2 & 2 \end{bmatrix}$$

45.
$$A = \begin{bmatrix} 1 & 2 \\ -5 & -3 \end{bmatrix}$$

46.
$$A = \begin{bmatrix} 2 & -3 \\ 3 & -2 \end{bmatrix}$$

In Problems 47–56, we consider differential equations of the form

$$\frac{d\mathbf{x}}{dt} = A\mathbf{x}(t)$$

where

$$A = \begin{bmatrix} a_{11} & a_{12} \\ a_{21} & a_{22} \end{bmatrix}$$

Analyze the stability of the equilibrium $(0, 0)$, *and classify the equilibrium.*

47.
$$A = \begin{bmatrix} -1 & -2 \\ 1 & 3 \end{bmatrix}$$

48.
$$A = \begin{bmatrix} -2 & 2 \\ -4 & 3 \end{bmatrix}$$

49.
$$A = \begin{bmatrix} -1 & -1 \\ 5 & -3 \end{bmatrix}$$

50.
$$A = \begin{bmatrix} 2 & 2 \\ 2 & 3 \end{bmatrix}$$

51.
$$A = \begin{bmatrix} 1 & 3 \\ 2 & 3 \end{bmatrix}$$

52.
$$A = \begin{bmatrix} -1 & 5 \\ -3 & -3 \end{bmatrix}$$

53.
$$A = \begin{bmatrix} -2 & 3 \\ 1 & -4 \end{bmatrix}$$

54.
$$A = \begin{bmatrix} -2 & -7 \\ 1 & 2 \end{bmatrix}$$

55.
$$A = \begin{bmatrix} 3 & -5 \\ 2 & -1 \end{bmatrix}$$

56.
$$A = \begin{bmatrix} 3 & 6 \\ -1 & -4 \end{bmatrix}$$

57. The following system has two distinct real eigenvalues, but one eigenvalue is equal to 0:

$$\frac{d\mathbf{x}}{dt} = \begin{bmatrix} 4 & 8 \\ 1 & 2 \end{bmatrix} \mathbf{x}(t) \qquad (11.36)$$

(a) Find both eigenvalues and associated eigenvectors.

(b) Use the general solution (11.26) to find $x_1(t)$ and $x_2(t)$.

(c) The direction field is shown in Figure 11.31. Sketch the lines corresponding to the eigenvectors. Compute dx_2/dx_1 and conclude that all direction vectors are parallel to the line in the direction of the eigenvector corresponding to the nonzero eigenvalue. Describe in words how solutions starting at different points behave.

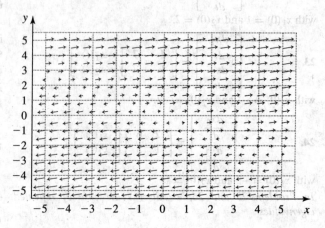

▲ **Figure 11.31**

58. The following system has two distinct real eigenvalues, but one eigenvalue is equal to 0:

$$\frac{d\mathbf{x}}{dt} = \begin{bmatrix} 2 & 4 \\ 3 & 6 \end{bmatrix} \mathbf{x}(t) \qquad (11.37)$$

(a) Find both eigenvalues and associated eigenvectors.

(b) Use the general solution (11.26) to find $x_1(t)$ and $x_2(t)$.

(c) The direction field is shown in Figure 11.32. Sketch the lines corresponding to the eigenvectors. Compute dx_2/dx_1, and conclude that all direction vectors are parallel to the line in the direction of the eigenvector corresponding to the nonzero eigenvalue. Describe in words how solutions starting at different points behave.

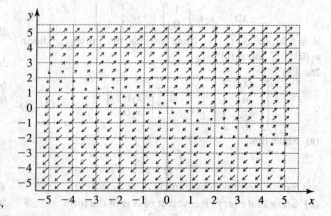

▲ **Figure 11.32**

11.2 LINEAR SYSTEMS: APPLICATIONS

11.2.1 Compartment Models

Compartment models (which we encountered in Chapter 8) describe flow between compartments, such as nutrient flow between lakes or the flow of a radioactive tracer between different parts of an organism. In the simplest situations, the resulting model is a system of linear differential equations.

We will consider a general two-compartment model that can be described by a system of two linear differential equations. A schematic description of the model is given in Figure 11.33.

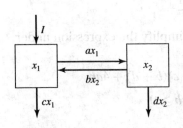

▲ **Figure 11.33**
A schematic description of a general two compartment model

We denote by $x_1(t)$ the amount of matter in compartment 1 at time t, and by $x_2(t)$ the amount of matter in compartment 2 at time t. To have a concrete example in mind, think of $x_1(t)$ and $x_2(t)$ as the amount of water in each of the two compartments. The direction of the flow of matter and the rates at which matter flows are shown in Figure 11.33. We see that matter enters compartment 1 at the constant rate I and moves from compartment 1 to compartment 2 at rate ax_1 if x_1 is the amount of matter in compartment 1. Matter in compartment 1 is lost at rate cx_1. In addition, matter flows from compartment 2 to compartment 1 at rate bx_2 if x_2 is the amount of the matter in compartment 2. Matter in compartment 2 is lost at rate dx_2; there is no external input into compartment 2. The constants I, a, b, c, and d are all nonnegative.

We describe the dynamics of $x_1(t)$ and $x_2(t)$ by the following system of differential equations:

$$\frac{dx_1}{dt} = I - (a+c)x_1 + bx_2$$
$$\frac{dx_2}{dt} = ax_1 - (b+d)x_2 \tag{11.38}$$

If $I > 0$, then (11.38) is a system of inhomogeneous linear differential equations with constant coefficients. Constant input is often important in real situations, such as the flow of nutrients between soil and plants, where nutrients might be added at a constant rate. In the following discussion, however, we will set $I = 0$, since this corresponds to the situation discussed in the previous section (that is, no matter is added over time). It is not difficult to guess how the system behaves when $I = 0$: Either some matter is continually lost so that one or both compartments empty out, or no matter is lost so that at least one compartment will contain matter. We will discuss both cases.

When $I = 0$, then (11.38) reduces to the linear system

$$\frac{d\mathbf{x}}{dt} = A\mathbf{x}(t) \quad \text{with } A = \begin{bmatrix} -(a+c) & b \\ a & -(b+d) \end{bmatrix} \tag{11.39}$$

To avoid trivial situations, we assume that at least one of the parameters a, b, c, or d is positive (otherwise, no material would ever move in the system).

To find the eigenvalues of A, we compute

$$\det \begin{bmatrix} -(a+c) - \lambda & b \\ a & -(b+d) - \lambda \end{bmatrix}$$
$$= (a+c+\lambda)(b+d+\lambda) - ab$$
$$= \lambda^2 + (a+b+c+d)\lambda + (a+c)(b+d) - ab$$
$$= \lambda^2 - \tau\lambda + \Delta = 0$$

where τ is the trace of A and Δ is the determinant of A:

$$\tau = -(a + b + c + d) < 0$$

$$\Delta = (a + c)(b + d) - ab = ad + bc + cd \geq 0$$

The eigenvalues λ_1 and λ_2 are the solutions of $\lambda^2 - \tau\lambda + \Delta = 0$:

$$\lambda_{1,2} = \frac{\tau \pm \sqrt{\tau^2 - 4\Delta}}{2}$$

To check whether the eigenvalues are real, we simplify the expression under the square root,

$$\tau^2 - 4\Delta = (a + b + c + d)^2 - 4(a + c)(b + d) + 4ab$$

$$= [(a + b) - (c + d)]^2 + 4ab \geq 0$$

which implies that both eigenvalues are real.

Since $\Delta \geq 0$ and $\tau^2 - 4\Delta \geq 0$, it follows that $\sqrt{\tau^2 - 4\Delta} \leq |\tau|$. Since in addition, $\tau < 0$, we conclude that both eigenvalues are less than or equal to 0. As long as $\Delta > 0$, both eigenvalues will be strictly negative; then $(0, 0)$ will be the only equilibrium and will be a stable node.

$\Delta > 0$ When $\Delta = ad + bc + cd > 0$, at least one of the terms in $ad + bc + cd$ must be positive; that is, either

$$a, d > 0 \quad \text{or} \quad b, c > 0 \quad \text{or} \quad c, d > 0$$

To see what this means, we return to Figure 11.33. When a and d are both positive, the matter in compartment 1 can move to compartment 2, and matter in compartment 2 can leave the system. Similarly, when both b and c are positive, matter from both compartments can leave the system through compartment 1. If c and d are both positive, then matter can leave both compartments. It follows that in any of these three cases, all matter will eventually leave the system, implying that $(0, 0)$ is a stable equilibrium.

$\Delta = 0$ Then

$$\lambda_1 = \tau = -(a + b + c + d) \quad \text{and} \quad \lambda_2 = 0$$

Since we assume that at least one of the four parameters a, b, c, or d is positive, it follows that $\lambda_1 < 0$; thus, the eigenvalues are distinct. To have $\Delta = 0$, one of the following must hold:

$$c = a = 0 \quad \text{or} \quad d = b = 0 \quad \text{or} \quad c = d = 0$$

If $c = a = 0$, then matter gets stuck in compartment 1; if $d = b = 0$, then matter gets stuck in compartment 2; if $c = d = 0$, no matter will ever leave the system, and the amount of matter present at time t is equal to the amount of matter present at time 0. In other words, the total amount $x_1(t) + x_2(t)$ is constant (it does not depend on t); it is therefore called a **conserved quantity**. This follows from (11.38) as well, since in this case,

$$\frac{dx_1}{dt} = -ax_1 + bx_2$$

$$\frac{dx_2}{dt} = ax_1 - bx_2$$

Adding the two equations, we find

$$\frac{dx_1}{dt} + \frac{dx_2}{dt} = 0$$

Since

$$\frac{dx_1}{dt} + \frac{dx_2}{dt} = \frac{d}{dt}(x_1 + x_2)$$

it follows that $x_1(t) + x_2(t)$ is a constant.

Solving the System When $c = d = 0$ We can write the solution explicitly to find out the equilibrium state. To determine the solution, we must compute the eigenvectors corresponding to the eigenvalues λ_1 and λ_2.

The eigenvector **u** corresponding to $\lambda_1 = \tau$ satisfies

$$\begin{bmatrix} -a & b \\ a & -b \end{bmatrix} \begin{bmatrix} u_1 \\ u_2 \end{bmatrix} = -(a+b) \begin{bmatrix} u_1 \\ u_2 \end{bmatrix}$$

Writing this system out yields

$$-au_1 + bu_2 = -au_1 - bu_1$$
$$au_1 - bu_2 = -au_2 - bu_2$$

This simplifies to one equation, namely,

$$u_1 = -u_2$$

If we set $u_1 = 1$, then $u_2 = -1$, and an eigenvector corresponding to $\lambda_1 = -(a+b)$ is

$$\mathbf{u} = \begin{bmatrix} 1 \\ -1 \end{bmatrix}$$

The eigenvector **v** corresponding to $\lambda_2 = 0$ satisfies

$$\begin{bmatrix} -a & b \\ a & -b \end{bmatrix} \begin{bmatrix} v_1 \\ v_2 \end{bmatrix} = \begin{bmatrix} 0 \\ 0 \end{bmatrix}$$

This simplifies to one equation, namely,

$$-av_1 + bv_2 = 0$$

If we set $v_1 = b$, then $v_2 = a$, and an eigenvector corresponding to $\lambda_2 = 0$ is

$$\mathbf{v} = \begin{bmatrix} b \\ a \end{bmatrix}$$

The general solution when $\Delta = 0$ is, therefore,

$$\mathbf{x}(t) = c_1 e^{-(a+b)t} \begin{bmatrix} 1 \\ -1 \end{bmatrix} + c_2 \begin{bmatrix} b \\ a \end{bmatrix}$$

and

$$\lim_{t \to \infty} \mathbf{x}(t) = c_2 \begin{bmatrix} b \\ a \end{bmatrix}$$

At time 0,

$$\begin{bmatrix} x_1(0) \\ x_2(0) \end{bmatrix} = c_1 \begin{bmatrix} 1 \\ -1 \end{bmatrix} + c_2 \begin{bmatrix} b \\ a \end{bmatrix}$$

that is,

$$c_1 + bc_2 = x_1(0)$$
$$-c_1 + ac_2 = x_2(0)$$

Adding these two equations, we eliminate c_1 and find

$$(a+b)c_2 = x_1(0) + x_2(0)$$

or

$$c_2 = \frac{x_1(0) + x_2(0)}{a+b}$$

Recall that the sum $x_1(t) + x_2(t)$ is a constant for all $t \geq 0$; we set $x_1(0) + x_2(0) = K$. Then

$$\lim_{t\to\infty} x_1(t) = K\frac{b}{a+b} \quad \text{and} \quad \lim_{t\to\infty} x_2(t) = K\frac{a}{a+b}$$

This means that compartment 1 will eventually contain a fraction $b/(a+b)$ of the total amount, and compartment 2 a fraction $a/(a+b)$ of the total amount. These are the relative rates at which matter enters the respective compartments.

▶ **Example 1**

Find the system of differential equations corresponding to the compartment diagram shown in Figure 11.34, and analyze the stability of the equilibrium $(0, 0)$.

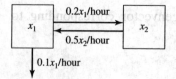

▲ **Figure 11.34**
The compartment diagram for Example 1

Solution

The compartment model is described by the following system of differential equations:

$$\frac{dx_1}{dt} = -(0.1 + 0.2)x_1 + 0.5x_2$$

$$\frac{dx_2}{dt} = 0.2x_1 - 0.5x_2$$

In matrix notation, this is

$$\frac{d\mathbf{x}}{dt} = \begin{bmatrix} -0.3 & 0.5 \\ 0.2 & -0.5 \end{bmatrix} \mathbf{x}(t)$$

To investigate the stability of $(0, 0)$, we find the eigenvalues of the matrix describing the system. That is, we solve

$$\det \begin{bmatrix} -0.3 - \lambda & 0.5 \\ 0.2 & -0.5 - \lambda \end{bmatrix} = (-0.3 - \lambda)(-0.5 - \lambda) - (0.2)(0.5)$$

$$= (0.3)(0.5) + (0.3 + 0.5)\lambda + \lambda^2 - (0.2)(0.5)$$

$$= \lambda^2 + 0.8\lambda + 0.05 = 0$$

We find

$$\lambda_{1,2} = \frac{-0.8 \pm \sqrt{0.64 - 0.2}}{2}$$

$$= \begin{cases} -0.4 + \frac{1}{2}\sqrt{0.44} \approx -0.068 \\ -0.4 - \frac{1}{2}\sqrt{0.44} \approx -0.732 \end{cases}$$

We find that both eigenvalues are negative. Therefore, the equilibrium $(0, 0)$ is a stable node. ◀

▶ **Example 2**   Given the system of differential equations

$$\frac{dx_1}{dt} = -0.7x_1 + 0.2x_2$$

$$\frac{dx_2}{dt} = 0.3x_1 - 0.2x_2$$

determine the parameter for the compartment diagram in Figure 11.33.

Solution

The general compartment diagram describing a linear system with two states is shown in Figure 11.33. Comparing the diagram and the system of equations, we find that $I = 0$ and

$$a + c = 0.7$$
$$b = 0.2$$
$$a = 0.3$$
$$b + d = 0.2$$

Solving this system of equations, we conclude that

$$a = 0.3, \quad b = 0.2, \quad c = 0.4 \quad \text{and} \quad d = 0 \qquad \blacktriangleleft$$

Compartment models are used in pharmacology to study how drug concentrations change within a body. In the simplest of such models, a drug is administered to a person in a single dose; we investigate this case in the next example.

▶ **Example 3**   A drug is administered to a person in a single dose. We assume that the drug does not accumulate in body tissue but is secreted through urine. We denote the amount of drug in the body at time t by $x_1(t)$ and in the urine at time t by $x_2(t)$. Initially,

$$x_1(0) = K \quad \text{and} \quad x_2(0) = 0$$

We describe the movement of the drug between the body and the urine by

$$\frac{dx_1}{dt} = -ax_1(t) \tag{11.40}$$

$$\frac{dx_2}{dt} = ax_1(t) \tag{11.41}$$

That is, the body's amount of the drug decreases at the same rate as the drug accumulates in the urine. We can solve (11.40) directly, and find

$$x_1(t) = c_1 e^{-at}$$

Since $x_1(0) = K = c_1$, we find

$$x_1(t) = K e^{-at}$$

Plugging this into (11.41), we have

$$\frac{dx_2}{dt} = aK e^{-at}$$

or

$$\int dx_2 = \int aK e^{-at} \, dt$$

Hence,

$$x_2(t) = c_2 - Ke^{-at}$$

With $x_2(0) = 0$, we find $0 = c_2 - K$. Hence,

$$x_2(t) = K(1 - e^{-at})$$

The constant a is called the excretion rate in this application. Finding the excretion rate is important for determining how long a drug will remain in the body. We can determine the excretion rate a by plotting $\ln(K - x_2(t))$ versus t. Namely,

$$\ln(K - x_2(t)) = \ln(Ke^{-at})$$
$$= \ln K - at$$

That is, plotting $\ln(K - x_2(t))$ versus t results in a straight line with slope $-a$, which allows us to determine a. ◄

11.2.2 The Harmonic Oscillator (Optional)

We consider a particle that moves along the x-axis. We assume that the acceleration is proportional to the distance to the origin and that the direction of the acceleration always points toward the origin. If $x(t)$ is the location of the particle at time t, then the second derivative of $x(t)$ denotes the acceleration of the particle, and we find

$$\frac{d^2x}{dt^2} = -kx(t) \tag{11.42}$$

for some $k > 0$. This is a **second-order differential equation**, because the derivative of the highest-order derivative in the equation is of order 2. We will use this example to show how a second-order differential equation can be transformed into a system of first order differential equations. To do so, we set

$$\frac{dx}{dt} = v(t)$$

Then

$$\frac{dv}{dt} = \frac{d^2x}{dt^2}$$

and, hence,

$$\frac{dv}{dt} = -kx(t)$$

We thus obtain the following system of first-order differential equations:

$$\frac{dx}{dt} = v$$
$$\frac{dv}{dt} = -kx$$

or

$$\frac{d}{dt}\begin{bmatrix} x \\ v \end{bmatrix} = \begin{bmatrix} 0 & 1 \\ -k & 0 \end{bmatrix}\begin{bmatrix} x \\ v \end{bmatrix}$$

If we denote the matrix by A, then

$$\operatorname{tr} A = 0 \quad \text{and} \quad \det A = k > 0$$

which implies that the eigenvalues of A are complex conjugates with real parts equal to 0. We find

$$\det(A - \lambda I) = \det \begin{bmatrix} -\lambda & 1 \\ -k & -\lambda \end{bmatrix} = \lambda^2 + k = 0$$

Hence,

$$\lambda_1 = i\sqrt{k} \quad \text{and} \quad \lambda_2 = -i\sqrt{k}$$

Thus, we expect the equilibrium $(0, 0)$ to be a neutral spiral (or center) and the solutions to exhibit oscillations whose amplitudes do not change with time. We can solve (11.42) directly: Since

$$\frac{d}{dt} \sin(at) = a \cos(at)$$

$$\frac{d}{dt} \cos(at) = -a \sin(at)$$

we find

$$\frac{d^2}{dt^2} \sin(at) = -a^2 \sin(at)$$

and

$$\frac{d^2}{dt^2} \cos(at) = -a^2 \cos(at)$$

If we set $a = \sqrt{k}$, we see that $\cos(\sqrt{k}t)$ and $\sin(\sqrt{k}t)$ solve (11.42). Using the superposition principle, we therefore obtain the solution of (11.42) as

$$x(t) = c_1 \sin(\sqrt{k}t) + c_2 \cos(\sqrt{k}t)$$

To determine the constants c_1 and c_2, we must fix an initial condition. If we assume, for instance, that

$$x(0) = 0 \quad \text{and} \quad v(0) = v_0 \tag{11.43}$$

then

$$0 = c_2$$

Since $v(t) = dx/dt$, we find

$$v(t) = c_1 \sqrt{k} \cos(\sqrt{k}t) - c_2 \sqrt{k} \sin(\sqrt{k}t)$$

and, therefore,

$$v(0) = c_1 \sqrt{k} = v_0$$

which implies that

$$c_1 = \frac{v_0}{\sqrt{k}}$$

Hence, the solution of (11.42) that satisfies the initial condition (11.43) is given by

$$x(t) = \frac{v_0}{\sqrt{k}} \sin(\sqrt{k}t)$$

The harmonic oscillator is quite important in physics. It describes, for instance, a frictionless pendulum when the displacement from the resting state is not too large.

11.2.3 Problems

(11.2.1)

In Problems 1–6, determine the system of differential equations corresponding to each compartment model and analyze the stability of the equilibrium $(0,0)$. The parameters have the same meaning as in Figure 11.33.

1. $a = 0.5, b = 0.1, c = 0.05, d = 0.02$

2. $a = 0.4, b = 1.2, c = 0.3, d = 0$

3. $a = 2.5, b = 0.7, c = 0, d = 0.1$

4. $a = 1.7, b = 0.6, c = 0.1, d = 0.3$

5. $a = 0, b = 0, c = 0.1, d = 0.3$

6. $a = 0, b = 0.1, c = 0, d = 0$

In Problems 7–12, find the corresponding compartment diagram for each system of differential equations.

7.
$$\frac{dx_1}{dt} = -0.4x_1 + 0.3x_2$$
$$\frac{dx_2}{dt} = 0.1x_1 - 0.5x_2$$

8.
$$\frac{dx_1}{dt} = -0.4x_1 + 3x_2$$
$$\frac{dx_2}{dt} = 0.2x_1 - 3x_2$$

9.
$$\frac{dx_1}{dt} = -1.2x_1$$
$$\frac{dx_2}{dt} = 0.3x_1 - 0.2x_2$$

10.
$$\frac{dx_1}{dt} = -0.2x_1 + 0.4x_2$$
$$\frac{dx_2}{dt} = 0.2x_1 - 0.4x_2$$

11.
$$\frac{dx_1}{dt} = -0.2x_1$$
$$\frac{dx_2}{dt} = -0.3x_2$$

12.
$$\frac{dx_1}{dt} = -x_1$$
$$\frac{dx_2}{dt} = x_1 - 0.5x_2$$

13. Suppose that a drug is administered to a person in a single dose, and assume that the drug does not accumulate in body tissue but is excreted through urine. Denote the amount of drug in the body at time t by $x_1(t)$ and in the urine at time t by $x_2(t)$. If $x_1(0) = 4$ mg and $x_2(0) = 0$, find $x_1(t)$ and $x_2(t)$ if
$$\frac{dx_1}{dt} = -0.3x_1(t)$$

14. Suppose that a drug is administered to a person in a single dose, and assume that the drug does not accumulate in body tissue but is excreted through urine. Denote the amount of drug in the body at time t by $x_1(t)$ and in the urine at time t by $x_2(t)$. If $x_1(0) = 6$ mg and $x_2(0) = 0$, find a system of differential equations for $x_1(t)$ and $x_2(t)$ if it takes twenty minutes for the drug to be at one-half of its initial amount of the drug in the body.

15. A very simple two-compartment model for gap dynamics in a forest assumes that gaps are created by disturbances (wind, fire, and so forth) and that gaps revert back to forest as trees grow in the gaps. We denote by $x_1(t)$ the area occupied by gaps, and by $x_2(t)$ the area occupied by adult trees. We assume that the dynamics are given by
$$\frac{dx_1}{dt} = -0.2x_1 + 0.1x_2 \tag{11.44}$$
$$\frac{dx_2}{dt} = 0.2x_1 - 0.1x_2 \tag{11.45}$$

(a) Find the corresponding compartment diagram.

(b) Show that $x_1(t) + x_2(t)$ is a constant. Denote the constant by A and give its meaning. (*Hint:* Show that $\frac{d}{dt}(x_1 + x_2) = 0$.)

(c) Let $x_1(0) + x_2(0) = 20$. Use your answer in (b) to explain why this implies that $x_1(t) + x_2(t) = 20$ for all $t > 0$.

(d) Use your result in (c) to replace x_2 in (11.44) by $20 - x_1$, and show that this reduces the system (11.44) and (11.45) to
$$\frac{dx_1}{dt} = 2 - 0.3x_1 \tag{11.46}$$
with $x_1(t) + x_2(t) = 20$ for all $t \geq 0$.

(e) Solve the system (11.44) and (11.45), and determine what fraction of the forest is occupied by adult trees at time t when $x_1(0) = 2$ and $x_2(0) = 18$. What happens as $t \to \infty$?

16. A simple model for forest succession can be formulated by a three-compartment model. We assume that gaps in a forest are created by disturbances and are colonized by early successional species, which are then replaced by late successional species. We denote by $x_1(t)$ the total area occupied by gaps at time t, by $x_2(t)$ the total area occupied by early successional species at time t, and by $x_3(t)$ the total area occupied by late successional species at time t. The dynamics are given by
$$\frac{dx_1}{dt} = 0.2x_2 + x_3 - 2x_1$$
$$\frac{dx_2}{dt} = 2x_1 - 0.7x_2$$
$$\frac{dx_3}{dt} = 0.5x_2 - x_3$$

(a) Draw the corresponding compartment diagram.

(b) Show that
$$x_1(t) + x_2(t) + x_3(t) = A$$
where A is a constant, and give the meaning of A.

(11.2.2)

17. Solve
$$\frac{d^2x}{dt^2} = -4x$$
with $x(0) = 0$ and $\frac{dx(0)}{dt} = 6$.

18. Solve
$$\frac{d^2x}{dt^2} = -9x$$
with $x(0) = 0$ and $\frac{dx(0)}{dt} = 12$.

19. Transform the second-order differential equation

$$\frac{d^2x}{dt^2} = 3x$$

into a system of first-order differential equations.

20. Transform the second-order differential equation

$$\frac{d^2x}{dt^2} = -\frac{1}{2}x$$

into a system of first-order differential equations.

21. Transform the second-order differential equation

$$\frac{d^2x}{dt^2} + \frac{dx}{dt} = x$$

into a system of first-order differential equations.

22. Transform the second-order differential equation

$$\frac{d^2x}{dt^2} - 2\frac{dx}{dt} = 3x$$

into a system of first-order differential equations.

11.3 NONLINEAR AUTONOMOUS SYSTEMS: THEORY

In this section, we will develop some of the theory needed to analyze systems of differential equations of the form

$$\frac{dx_1}{dt} = f_1(x_1, x_2, \ldots, x_n)$$

$$\frac{dx_2}{dt} = f_2(x_1, x_2, \ldots, x_n)$$

$$\vdots$$ (11.47)

$$\frac{dx_n}{dt} = f_n(x_1, x_2, \ldots, x_n)$$

where $f_i : \mathbf{R}^n \rightarrow \mathbf{R}$, for $i = 1, 2, \ldots, n$. We assume that the functions f_i, $i = 1, 2, \ldots, n$, do not explicitly depend on t; the system (11.47) is therefore called autonomous. We no longer assume that the functions f_i are linear as in Section 11.1. Using vector notation, the system (11.47) can be written in the form

$$\frac{d\mathbf{x}}{dt} = \mathbf{f}(\mathbf{x})$$

where $\mathbf{x} = [x_1, x_2, \ldots, x_n]'$, and $\mathbf{f}(\mathbf{x})$ is a vector-valued function $\mathbf{f} : \mathbf{R}^n \rightarrow \mathbf{R}^n$ with components $f_i(x_1, x_2, \ldots, x_n) : \mathbf{R}^n \rightarrow \mathbf{R}$, $i = 1, 2, \ldots, n$. The function $\mathbf{f}(\mathbf{x})$ defines a direction field, just as in the linear case.

Unless the functions f_i are linear, it is typically not possible to find explicit solutions of systems of differential equations. If we want to solve such systems, we frequently must use numerical methods. Instead of trying to find solutions, we will focus on point equilibria and their stability, just as in Section 8.2.

The definition of a point equilibrium (as given in Section 8.2) must be extended to systems of the form (11.47). We say that a point

$$\hat{\mathbf{x}} = (\hat{x}_1, \hat{x}_2, \ldots, \hat{x}_n)$$

is a point equilibrium (or simply equilibrium) of (11.47) if

$$\mathbf{f}(\hat{\mathbf{x}}) = \mathbf{0}$$

An equilibrium is also called a **critical point**. As in the linear case, this is a point in the direction field at which the direction vector has length 0, implying that if we start a system of differential equations at an equilibrium point, it will stay there for all later times.

As in the linear case, a solution might not return to an equilibrium after a small perturbation; this is addressed by the stability of the equilibrium. The theory of stability for systems of nonlinear autonomous differential equations is parallel to that in Section 8.2; there is both an analytical and graphical approach that reduces to the theory in Section 8.2 when there is a single differential equation. We will restrict our discussion to systems of two equations in two variables.

11.3.1 Analytical Approach

A Single Autonomous Differential Equation

▶ **Example 1**   Find all equilibria of

$$\frac{dx}{dt} = x(1-x) \tag{11.48}$$

and analyze their stability.

Solution

We developed the theory for single autonomous differential equations in Section 8.2. To find equilibria, we set

$$x(1-x) = 0$$

which yields

$$\hat{x}_1 = 0 \quad \text{and} \quad \hat{x}_2 = 1$$

To analyze their stability, we linearize the differential equation (11.48) about each equilibrium, and compute the corresponding eigenvalue. We set

$$f(x) = x(1-x)$$

Then

$$f'(x) = 1 - 2x$$

The eigenvalue associated with the equilibrium $\hat{x}_1 = 0$ is

$$\lambda_1 = f'(0) = 1 > 0$$

which implies that $\hat{x}_1 = 0$ is unstable.

The eigenvalue associated with the equilibrium $\hat{x}_2 = 1$ is

$$\lambda_2 = f'(1) = -1 < 0$$

which implies that $\hat{x}_2 = 1$ is locally stable.   ◀

The eigenvalue corresponding to an equilibrium of the differential equation

$$\frac{dx}{dt} = f(x) \tag{11.49}$$

is the slope of the function $f(x)$ at the equilibrium value. The reason for this is discussed in detail in Section 8.2; we repeat the basic argument here. Suppose that $\hat{x}$ is an equilibrium of (11.49)—that is, $f(\hat{x}) = 0$. If we perturb $\hat{x}$ slightly (that is, if we look at $\hat{x} + z$ for small $|z|$), we can find out what happens to $\hat{x} + z$ by looking at

$$\frac{dx}{dt} = \frac{d}{dt}(\hat{x} + z) = \frac{dz}{dt}$$

and

$$f(\hat{x} + z) \approx f(\hat{x}) + f'(\hat{x})(\hat{x} + z - \hat{x}) = f'(\hat{x})z$$

In the last step, we used the fact that $f(\hat{x}) = 0$. We find

$$\frac{dz}{dt} \approx f'(\hat{x})z$$

which has approximate solution

$$z(t) \approx z(0)e^{\lambda t} \quad \text{with } \lambda = f'(\hat{x})$$

Therefore, we see that if $f'(\hat{x}) < 0$, then $z(t) \to 0$ as $t \to \infty$ and, hence, $x(t) = \hat{x} + z(t) \to \hat{x}$ as $t \to \infty$; that is, the solution will return to the equilibrium $\hat{x}$ after a small perturbation. In this case, $\hat{x}$ is locally stable. If $f'(\hat{x}) > 0$, then $z(t)$ will not go to 0, which implies that $\hat{x}$ is unstable. The linearization of $f(x)$ thus tells us whether an equilibrium is locally stable or unstable. The exact same method is used for systems of differential equations.

Systems of Two Differential Equations   We consider differential equations of the form

$$\frac{dx_1}{dt} = f_1(x_1, x_2) \tag{11.50}$$

$$\frac{dx_2}{dt} = f_2(x_1, x_2) \tag{11.51}$$

or, using vector notation,

$$\frac{d\mathbf{x}}{dt} = \mathbf{f}(\mathbf{x}) \tag{11.52}$$

where $\mathbf{x} = \mathbf{x}(t) = (x_1(t), x_2(t))$, and $\mathbf{f}(\mathbf{x})$ is a vector-valued function $\mathbf{f}(\mathbf{x}) = (f_1(\mathbf{x}), f_2(\mathbf{x}))$ with $f_i(\mathbf{x}) : \mathbf{R}^2 \to \mathbf{R}$. An equilibrium or critical point, $\hat{\mathbf{x}}$, of (11.52) satisfies

$$\mathbf{f}(\hat{\mathbf{x}}) = \mathbf{0}$$

Suppose that $\hat{\mathbf{x}}$ is a point equilibrium. Then, as in the case of one nonlinear equation, we look at what happens to a small perturbation of $\hat{\mathbf{x}}$. We perturb $\hat{\mathbf{x}}$; that is, we look at how $\hat{\mathbf{x}} + \mathbf{z}$ changes under the dynamics described by (11.52):

$$\frac{d}{dt}(\hat{\mathbf{x}} + \mathbf{z}) = \frac{d}{dt}\mathbf{z} = \mathbf{f}(\hat{\mathbf{x}} + \mathbf{z})$$

The linearization of $\mathbf{f}(\mathbf{x})$ about $\mathbf{x} = \hat{\mathbf{x}}$ is

$$\mathbf{f}(\hat{\mathbf{x}}) + D\mathbf{f}(\hat{\mathbf{x}})\mathbf{z} = D\mathbf{f}(\hat{\mathbf{x}})\mathbf{z}$$

where we used that $\mathbf{f}(\hat{\mathbf{x}}) = \mathbf{0}$. The matrix $D\mathbf{f}(\hat{\mathbf{x}})$ is the Jacobi matrix evaluated at $\hat{\mathbf{x}}$. If we approximate $\mathbf{f}(\hat{\mathbf{x}} + \mathbf{z})$ by its linearization $D\mathbf{f}(\hat{\mathbf{x}})$, then

$$\frac{d\mathbf{z}}{dt} = D\mathbf{f}(\hat{\mathbf{x}})\mathbf{z} \tag{11.53}$$

is the linear approximation of the dynamics of the perturbation $\mathbf{z}$.

We now have a system of linear differential equations that is a good approximation, provided that $\mathbf{z}$ is sufficiently close to $\mathbf{0}$. We learned in Section 11.1 how to analyze linear systems. We saw that eigenvalues of the matrix $D\mathbf{f}(\hat{\mathbf{x}})$ allowed us to determine the nature of the equilibrium. We will use the same approach here, but we emphasize that this is now a *local* analysis, just as in the case of a single differential equation since we only know that the linearization (11.53) is a good approximation as long as we are sufficiently close to the point about which we linearized.

We return to our classification scheme for the linear case,

$$\frac{d\mathbf{x}}{dt} = A\mathbf{x}$$

where $\mathbf{x}(t) = (x_1(t), x_2(t))$ and A is a 2×2 matrix. We let

$$\Delta = \det A \quad \text{and} \quad \tau = \operatorname{tr} A$$

When we linearize a nonlinear system about an equilibrium, the matrix A is the Jacobi matrix evaluated at the equilibrium:

$$A = D\mathbf{f}(\hat{\mathbf{x}})$$

We exclude the following cases: (i) $\Delta = 0$, (ii) $\tau = 0$ and $\Delta > 0$, and (iii) $\tau^2 = 4\Delta$. Namely, when $\Delta = 0$, at least one eigenvalue is equal to 0. When $\tau = 0$ and $\Delta > 0$, the two eigenvalues are purely imaginary. When $\tau^2 = 4\Delta$, the two eigenvalues are identical. Except in these three cases, we can use the same classification scheme as in the linear case (Figure 11.35).

This is possible because the linearized vector field and the original vector field are geometrically very similar close to an equilibrium point (after all, that is the idea behind linearization). This result is known as the Hartman-Grobman theorem, which says that as long as $D\mathbf{f}(\hat{\mathbf{x}})$ has no zero or purely imaginary eigenvalues, then the linearized and the original vector fields are very similar close to the equilibrium. That is,

$$\frac{d\mathbf{x}}{dt} = \mathbf{f}(\mathbf{x}) \quad \text{and} \quad \frac{d\mathbf{z}}{dt} = D\mathbf{f}(\hat{\mathbf{x}})\mathbf{z}$$

behave very similarly for $\mathbf{x} = \hat{\mathbf{x}} + \mathbf{z}$ with $\mathbf{z}$ close to $\mathbf{0}$.

We find the same classification scheme as in the linear case:

- The equilibrium $\hat{\mathbf{x}}$ is a node if both eigenvalues of $D\mathbf{f}(\hat{\mathbf{x}})$ are real, distinct, nonzero, and of the same sign. The node is locally stable if the eigenvalues are negative, and unstable if the eigenvalues are positive.

- The equilibrium $\hat{\mathbf{x}}$ is a saddle if both eigenvalues are real and nonzero but have opposite signs. A saddle is unstable.

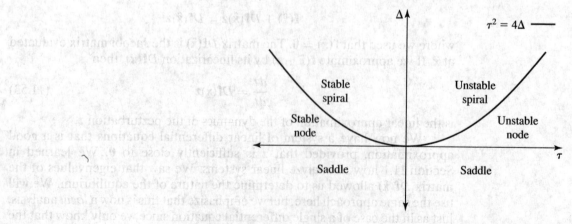

▲ **Figure 11.35**
The stability behavior of a system of two autonomous equations

- The equilibrium $\hat{\mathbf{x}}$ is a spiral if both eigenvalues are complex conjugates with nonzero real parts. The spiral is locally stable if the real parts of the eigenvalues are negative, and unstable if the real parts of the eigenvalues are positive.

When the two eigenvalues are purely imaginary, we cannot determine the stability by linearization.

▶ **Example 2**   Consider

$$\frac{dx_1}{dt} = x_1 - 2x_1^2 - 2x_1x_2$$

$$\frac{dx_2}{dt} = 4x_2 - 5x_2^2 - 7x_1x_2$$

(11.54)

(a) Find all equilibria of (11.54), and   **(b)** analyze their stability.

Solution

(a) To find equilibria, we set the right-hand side of (11.54) equal to 0:

$$x_1 - 2x_1^2 - 2x_1x_2 = 0 \tag{11.55}$$

$$4x_2 - 5x_2^2 - 7x_1x_2 = 0 \tag{11.56}$$

Factoring x_1 in the first equation and x_2 in the second yields

$$x_1(1 - 2x_1 - 2x_2) = 0 \quad \text{and} \quad x_2(4 - 5x_2 - 7x_1) = 0$$

That is,

$$x_1 = 0 \quad \text{or} \quad 2x_1 + 2x_2 = 1$$

and

$$x_2 = 0 \quad \text{or} \quad 7x_1 + 5x_2 = 4$$

Combining the different solutions, we get the four cases

(i) $x_1 = 0$ and $x_2 = 0$

(ii) $x_1 = 0$ and $7x_1 + 5x_2 = 4$

(iii) $x_2 = 0$ and $2x_1 + 2x_2 = 1$

(iv) $2x_1 + 2x_2 = 1$ and $7x_1 + 5x_2 = 4$

First, we will compute the equilibria in these four cases.

Case (i):   There is nothing to compute; the equilibrium is $(\hat{x}_1, \hat{x}_2) = (0, 0)$.

Case (ii):   To find the equilibrium, we must solve the system

$$x_1 = 0$$

$$7x_1 + 5x_2 = 4$$

which has the solutions

$$x_1 = 0 \quad \text{and} \quad x_2 = \frac{4}{5}$$

Hence, the equilibrium is $(\hat{x}_1, \hat{x}_2) = (0, \frac{4}{5})$.

Case (iii): To find the equilibrium, we must solve the system

$$x_2 = 0$$
$$2x_1 + 2x_2 = 1$$

which has the solutions

$$x_2 = 0 \quad \text{and} \quad x_1 = \frac{1}{2}$$

Hence, the equilibrium is $(\hat{x}_1, \hat{x}_2) = (\frac{1}{2}, 0)$.

Case (iv): find the equilibrium, we must solve the system

$$2x_1 + 2x_2 = 1$$
$$7x_1 + 5x_2 = 4$$

We use the standard elimination method: Leaving the first equation, and changing the second by multiplying the first by 7 and the second by 2, and then subtracting the second equation from the first, this system is equivalent to

$$2x_1 + 2x_2 = 1$$
$$4x_2 = -1$$

which has the solutions

$$x_2 = -\frac{1}{4} \quad \text{and} \quad x_1 = \frac{1}{2} - x_2 = \frac{3}{4}$$

Hence, the equilibrium is $(\hat{x}_1, \hat{x}_2) = (\frac{3}{4}, -\frac{1}{4})$.

We can illustrate all equilibria in the direction field of (11.54), which is displayed in Figure 11.36. The equilibria are shown as dots.

(b) To analyze the stability of the equilibria, we compute the Jacobi matrix

$$Df = \begin{bmatrix} \dfrac{\partial f_1}{\partial x_1} & \dfrac{\partial f_1}{\partial x_2} \\ \dfrac{\partial f_2}{\partial x_1} & \dfrac{\partial f_2}{\partial x_2} \end{bmatrix}$$

With

$$f_1(x_1, x_2) = x_1 - 2x_1^2 - 2x_1x_2 \quad \text{and} \quad f_2(x_1, x_2) = 4x_2 - 5x_2^2 - 7x_1x_2$$

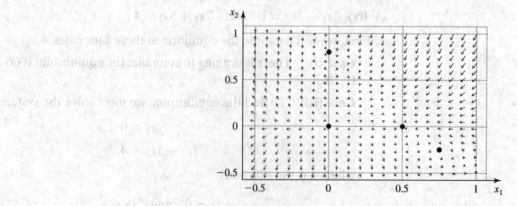

▲ **Figure 11.36**
The direction field of (11.54) together with the equilibria

we find

$$Df(x_1, x_2) = \begin{bmatrix} 1 - 4x_1 - 2x_2 & -2x_1 \\ -7x_2 & 4 - 10x_2 - 7x_1 \end{bmatrix}$$

We will now go through the four cases and analyze each equilibrium.

Case (i): The equilibrium is the point $(0, 0)$. The Jacobi matrix at $(0, 0)$ is

$$Df(0, 0) = \begin{bmatrix} 1 & 0 \\ 0 & 4 \end{bmatrix}$$

Since this matrix is in diagonal form, the eigenvalues are the diagonal elements, and we find $\lambda_1 = 1$ and $\lambda_2 = 4$. Since both eigenvalues are positive, the equilibrium is unstable. Using the same classification as in the linear case, we say that $(0, 0)$ is an unstable node.

The linearization of the direction field about $(0, 0)$ is displayed in Figure 11.37 where we show the direction field of

$$\frac{d\mathbf{x}}{dt} = \begin{bmatrix} 1 & 0 \\ 0 & 4 \end{bmatrix} \mathbf{x}(t)$$

Figure 11.37 confirms that $(0, 0)$ is an unstable node (or source). You should compare Figure 11.37 to the direction field of (11.54) shown in Figure 11.36. You will find that the linearized direction field and the direction field of (11.54) close to the equilibrium $(0, 0)$ are quite similar.

Case (ii): The equilibrium is the point $(0, \frac{4}{5})$. The Jacobi matrix at $(0, \frac{4}{5})$ is

$$Df\left(0, \frac{4}{5}\right) = \begin{bmatrix} 1 - \dfrac{8}{5} & 0 \\ -\dfrac{28}{5} & 4 - 8 \end{bmatrix} = \begin{bmatrix} -\dfrac{3}{5} & 0 \\ -\dfrac{28}{5} & -4 \end{bmatrix}$$

Since this matrix is in lower triangular form, the eigenvalues are simply the diagonal elements, and we find that the eigenvalues of $Df(0, \frac{4}{5})$ are $\lambda_1 = -\frac{3}{5}$ and $\lambda_2 = -4$. Since both eigenvalues are negative, $(0, \frac{4}{5})$ is locally stable. Using the same classification as in the linear case, we say that $(0, \frac{4}{5})$ is a stable node.

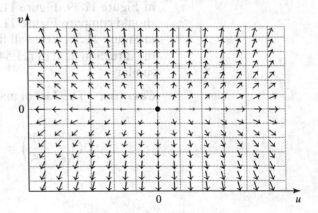

▲ **Figure 11.37**
The linearization of the direction field about $(0, 0)$

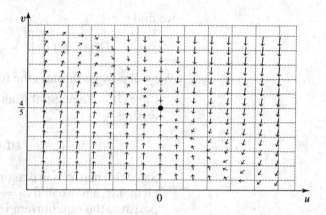

▲ **Figure 11.38**
The linearization of the direction field about $(0, \frac{4}{5})$

The linearization of the direction field about $(0, \frac{4}{5})$ is displayed in Figure 11.38. Figure 11.38 confirms that $(0, \frac{4}{5})$ is a stable node (or sink). You should compare Figure 11.38 to the direction field of (11.54) shown in Figure 11.36. You will find that the linearized direction field and the direction field of (11.54) close to the equilibrium $(0, \frac{4}{5})$ are quite similar.

Case (iii): The equilibrium is the point $(\frac{1}{2}, 0)$. The Jacobi matrix at $(\frac{1}{2}, 0)$ is

$$D\mathbf{f}\left(\frac{1}{2}, 0\right) = \begin{bmatrix} 1-2 & -1 \\ 0 & 4-\frac{7}{2} \end{bmatrix} = \begin{bmatrix} -1 & -1 \\ 0 & \frac{1}{2} \end{bmatrix}$$

Since this matrix is in upper triangular form, the eigenvalues are simply the diagonal elements, and we find that the eigenvalues of $D\mathbf{f}(\frac{1}{2}, 0)$ are $\lambda_1 = -1$ and $\lambda_2 = \frac{1}{2}$. Since one eigenvalue is positive and the other is negative, $(\frac{1}{2}, 0)$ is unstable. Using the same classification as in the linear case, we say that $(\frac{1}{2}, 0)$ is a saddle.
The linearization of the direction field about $(\frac{1}{2}, 0)$ is displayed in Figure 11.39. Figure 11.39 confirms that $(\frac{1}{2}, 0)$ is a saddle. You should compare Figure 11.39 to the direction field of (11.54) shown in Figure 11.36. You will find that the linearized direction field and the direction field of (11.54) close to the equilibrium $(\frac{1}{2}, 0)$ are quite similar.

Case (iv): The Jacobi matrix at the equilibrium $(\frac{3}{4}, -\frac{1}{4})$ is

$$D\mathbf{f}\left(\frac{3}{4}, -\frac{1}{4}\right) = \begin{bmatrix} 1-3+\frac{1}{2} & -\frac{3}{2} \\ \frac{7}{4} & 4+\frac{10}{4}-\frac{21}{4} \end{bmatrix}$$

$$= \begin{bmatrix} -\frac{3}{2} & -\frac{3}{2} \\ \frac{7}{4} & \frac{5}{4} \end{bmatrix}$$

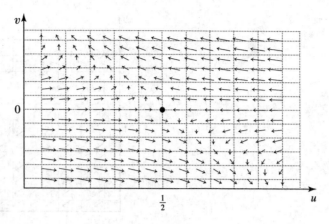

▲ **Figure 11.39**
The linearization of the direction field about $(\frac{1}{2}, 0)$

To find the eigenvalues, we must solve

$$\det \begin{bmatrix} -\dfrac{3}{2} - \lambda & -\dfrac{3}{2} \\[2mm] \dfrac{7}{4} & \dfrac{5}{4} - \lambda \end{bmatrix} = 0$$

Evaluating the determinant on the left-hand side and simplifying yields

$$\left(-\frac{3}{2} - \lambda\right)\left(\frac{5}{4} - \lambda\right) + \left(\frac{3}{2}\right)\left(\frac{7}{4}\right) = 0$$

$$\lambda^2 + \frac{3}{2}\lambda - \frac{5}{4}\lambda - \frac{15}{8} + \frac{21}{8} = 0$$

$$\lambda^2 + \frac{1}{4}\lambda + \frac{3}{4} = 0$$

Solving this quadratic equation, we find

$$\lambda_{1,2} = \frac{-\frac{1}{4} \pm \sqrt{\frac{1}{16} - 3}}{2}$$

$$= -\frac{1}{8} \pm \frac{1}{8}\sqrt{-47} = -\frac{1}{8} \pm \frac{1}{8}i\sqrt{47}$$

That is,

$$\lambda_1 = -\frac{1}{8} + \frac{1}{8}i\sqrt{47} \quad \text{and} \quad \lambda_2 = -\frac{1}{8} - \frac{1}{8}i\sqrt{47}$$

The eigenvalues are complex conjugates, with negative real parts. That is, $(\frac{3}{4}, -\frac{1}{4})$ is locally stable, and we expect the solutions to spiral into the equilibrium when we start close to the equilibrium.

The linearization of the direction field about $(\frac{3}{4}, -\frac{1}{4})$ is displayed in Figure 11.40. Figure 11.40 confirms that $(\frac{3}{4}, -\frac{1}{4})$ is a stable spiral. You should compare Figure 11.40 to the direction field of (11.54) shown in Figure 11.36. You will find that the linearized direction field and the direction field of (11.54) close to the equilibrium $(\frac{3}{4}, -\frac{1}{4})$ are quite similar. ◄

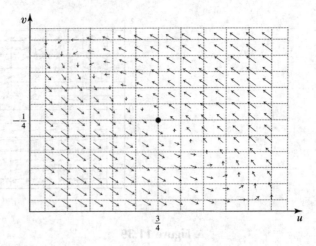

▲ **Figure 11.40**
▲ **Figure 11.40**
The linearization of the direction field about $(\frac{3}{4}, -\frac{1}{4})$

11.3.2 Graphical Approach for 2×2 Systems

In this subsection, we will discuss a graphical approach to systems of two autonomous differential equations. Suppose that

$$\frac{dx_1}{dt} = f_1(x_1, x_2)$$

$$\frac{dx_2}{dt} = f_2(x_1, x_2)$$

which in vector form is

$$\frac{d\mathbf{x}}{dt} = \mathbf{f}(\mathbf{x})$$

The curves

$$f_1(x_1, x_2) = 0$$

$$f_2(x_1, x_2) = 0$$

are called **zero isoclines** or **null clines**, and they represent the points in the x_1-x_2 plane where the growth rates of the respective quantities are equal to zero. This is illustrated in Figure 11.41 for a particular choice of f_1 and f_2. Let's assume that x_1 and x_2 are nonnegative; this restricts the discussion to the first quadrant of the x_1-x_2 plane. The two curves in Figure 11.41 divide the first quadrant into four regions, and we label each region according to whether dx_i/dt is positive or negative. Of course, this depends on the functions $f_1(x_1, x_2)$ and $f_2(x_1, x_2)$. The labeling of the regions in Figure 11.41 should therefore be considered as a choice to be concrete in the following discussion.

The point where both null clines in Figure 11.41 intersect is a point equilibrium or critical point, which we call $\hat{\mathbf{x}}$. We can use the graph to determine the signs of the entries in the Jacobi matrix

$$D\mathbf{f}(\hat{\mathbf{x}}) = \begin{bmatrix} a_{11} & a_{12} \\ a_{21} & a_{22} \end{bmatrix}$$

where $a_{ij} = \frac{\partial f_i}{\partial x_j}(\hat{\mathbf{x}})$. Namely, the entry $a_{11} = \frac{\partial f_1}{\partial x_1}$ is the effect of a change of f_1 in the x_1 direction when we keep x_2 fixed. To determine the sign of a_{11}, follow

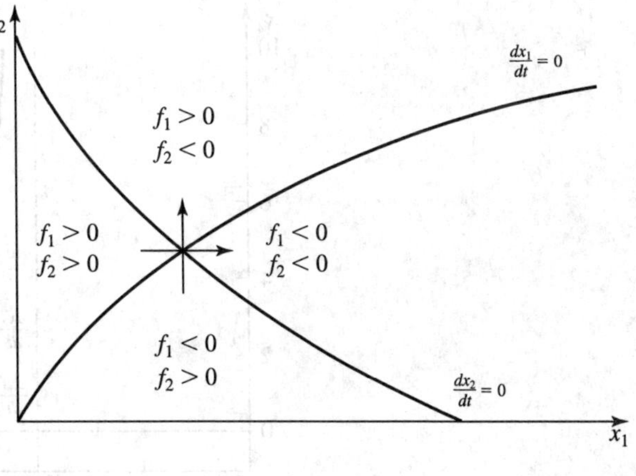

▲ **Figure 11.41**
Graphical approach: Zero isoclines

the horizontal arrow in Figure 11.41: The arrow goes from a region where f_1 is positive to a region where f_1 is negative, implying that f_1 is decreasing and hence $\frac{\partial f_1}{\partial x_1}(\hat{\mathbf{x}}) = a_{11} < 0$. We conclude that the sign of a_{11} in $D\mathbf{f}(\hat{\mathbf{x}})$ is negative, which we indicate in the Jacobi matrix by a minus sign in place of a_{11}:

$$D\mathbf{f}(\hat{\mathbf{x}}) = \begin{bmatrix} - & a_{12} \\ a_{21} & a_{22} \end{bmatrix}$$

Next, we determine the sign of $a_{12} = \frac{\partial f_1}{\partial x_2}(\hat{\mathbf{x}})$. This time, we want to know how f_1 changes at the equilibrium $\hat{\mathbf{x}}$ when we move in the x_2 direction and keep x_1 fixed. This is the direction of the vertical arrow through the equilibrium point. Since the arrow goes from a region where f_1 is negative to a region where f_1 is positive, f_1 increases in the direction of x_2 and, therefore, $a_{12} = \frac{\partial f_1}{\partial x_2}(\hat{\mathbf{x}}) > 0$.

The signs of the other two entries are found similarly, and we obtain

$$D\mathbf{f}(\hat{\mathbf{x}}) = \begin{bmatrix} - & + \\ - & - \end{bmatrix}$$

Thus, the trace of $D\mathbf{f}(\hat{\mathbf{x}})$ is negative and the determinant of $D\mathbf{f}(\hat{\mathbf{x}})$ is positive. Using the criterion stated in Subsection 9.3.2, we conclude that both eigenvalues have negative real parts and, therefore, that the equilibrium is locally stable.

▶ **Example 3**   Use the graphical approach to analyze the equilibrium $(3, 2)$ of

$$\frac{dx_1}{dt} = 5 - x_1 - x_1 x_2 + 2 x_2$$

$$\frac{dx_2}{dt} = x_1 x_2 - 3 x_2$$

Solution

First, note that $(3, 2)$ is indeed an equilibrium of this system. Now, the zero isoclines satisfy

$$\frac{dx_1}{dt} = 0, \quad \text{which holds for } x_2 = \frac{5 - x_1}{x_1 - 2}$$

and

$$\frac{dx_2}{dt} = 0, \quad \text{which holds for } x_2 = 0 \text{ or } x_1 = 3$$

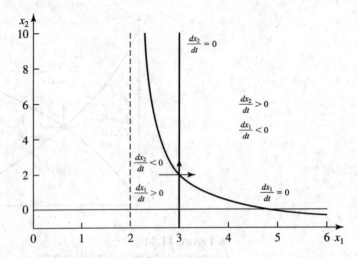

▲ **Figure 11.42**
The zero isoclines in the x_1-x_2 plane

The zero isoclines in the x_1-x_2 plane are drawn in Figure 11.42 The equilibrium $(3, 2)$ is the point of intersection of the zero isoclines $x_1 = 3$ and $x_2 = \frac{5-x_1}{x_1-2}$. The signs of $\frac{dx_1}{dt}$ and $\frac{dx_2}{dt}$ are indicated in Figure 11.42. We claim that

$$D\mathbf{f}(\hat{\mathbf{x}}) = \begin{bmatrix} - & - \\ + & 0 \end{bmatrix}$$

Here is why: To find the sign of $a_{11} = \frac{\partial f}{\partial x_1}$, we need to determine how dx_1/dt changes as we cross the zero isocline of x_1 in the x_1-direction. We see from the graph that dx_1/dt changes from positive to negative when we follow the horizontal arrow while crossing the zero isocline of x_1. Therefore, a_{11} is negative. To see why $a_{22} = 0$, follow the vertical arrow in the x_2 direction. Since the vertical arrow is on the zero isocline of x_2, the sign of dx_2/dt does not change as we cross the equilibrium in the x_2-direction. Therefore, $a_{22} = 0$. The signs of a_{12} and a_{21} follow from observing that if we cross the zero isocline of x_1 in the x_2 direction (the vertical arrow), then dx_1/dt changes from positive to negative, making $a_{12} < 0$. If we cross the zero isocline of x_2 in the direction of x_1 (the horizontal arrow), we see that dx_2/dt changes from negative to positive, making $a_{21} > 0$.

To determine the stability of $\hat{\mathbf{x}}$, we look at the trace and the determinant. Since the trace is negative and the determinant is positive, we conclude that the equilibrium is locally stable. ◀

This simple graphical approach does not always give us the signs of the real parts of the eigenvalues, as illustrated in the following example. Suppose that we arrive at the following Jacobi matrix, where the signs of the entries are

$$\begin{bmatrix} + & - \\ - & - \end{bmatrix}$$

The trace may now be positive or negative (the determinant is negative). Therefore, we cannot conclude anything about the eigenvalues. In this case we would have to compute the eigenvalues or the trace and the determinant.

11.3.3 Problems

(11.3.1)

In Problems 1–4, the point $(0,0)$ is always an equilibrium. Investigate its stability using the analytical approach.

1.
$$\frac{dx_1}{dt} = x_1 - 2x_2 + x_1x_2$$
$$\frac{dx_2}{dt} = -x_1 + x_2$$

2.
$$\frac{dx_1}{dt} = -x_1 - x_2 + x_1^2$$
$$\frac{dx_2}{dt} = x_2 - x_1^2$$

3.
$$\frac{dx_1}{dt} = x_1 + x_1^2 - 2x_1x_2 + x_2$$
$$\frac{dx_2}{dt} = x_1$$

4.
$$\frac{dx_1}{dt} = 3x_1x_2 - x_1 + x_2$$
$$\frac{dx_2}{dt} = x_2^2 - x_1$$

In Problems 5–10, find all equilibria of each system of differential equations, and use the analytical approach to determine the stability of each equilibrium.

5.
$$\frac{dx_1}{dt} = -x_1 + 2x_1(1 - x_1)$$
$$\frac{dx_2}{dt} = -x_2 + 5x_2(1 - x_1 - x_2)$$

6.
$$\frac{dx_1}{dt} = -x_1 + 3x_1(1 - x_1 - x_2)$$
$$\frac{dx_2}{dt} = -x_2 + 5x_2(1 - x_1 - x_2)$$

7.
$$\frac{dx_1}{dt} = 4x_1(1 - x_1) - 2x_1x_2$$
$$\frac{dx_2}{dt} = x_2(2 - x_2) - x_2$$

8.
$$\frac{dx_1}{dt} = 2x_1(5 - x_1 - x_2)$$
$$\frac{dx_2}{dt} = 3x_2(7 - 3x_1 - x_2)$$

9.
$$\frac{dx_1}{dt} = x_1 - x_2$$
$$\frac{dx_2}{dt} = x_1x_2 - x_2$$

10.
$$\frac{dx_1}{dt} = x_1x_2 - x_2$$
$$\frac{dx_2}{dt} = x_1 + x_2$$

(11.3.2)

11. Assume that
$$\frac{dx_1}{dt} = x_1(10 - 2x_1 - x_2)$$
$$\frac{dx_2}{dt} = x_2(10 - x_1 - 2x_2)$$

(a) Graph the zero isoclines.

(b) Show that $(\frac{10}{3}, \frac{10}{3})$ is an equilibrium, and determine its stability using the analytical approach.

12. Assume that
$$\frac{dx_1}{dt} = x_1(10 - x_1 - 2x_2)$$
$$\frac{dx_2}{dt} = x_2(10 - 2x_1 - x_2)$$

(a) Graph the zero isoclines.

(b) Show that $(\frac{10}{3}, \frac{10}{3})$ is an equilibrium, and determine its stability using the analytical approach.

In Problems 13–18, use the graphical approach for 2×2 systems to find the sign structure of the Jacobi matrix at the indicated equilibrium. If possible, determine its stability. Assume that the system of differential equations is given by
$$\frac{dx_1}{dt} = f_1(x_1, x_2)$$
$$\frac{dx_2}{dt} = f_2(x_1, x_2)$$

Furthermore, assume that x_1 and x_2 are both nonnegative. In each problem, the zero isoclines are drawn and the equilibrium we want to investigate is indicated by a dot. Assume that both x_1 and x_2 increase close to the origin and that f_1 and f_2 change sign when crossing their zero isoclines.

13. See Figure 11.43.

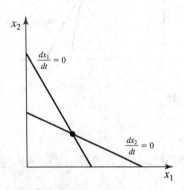

▲ **Figure 11.43**

14. See Figure 11.44.

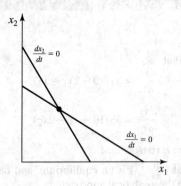

▲ **Figure 11.44**

15. See Figure 11.45.

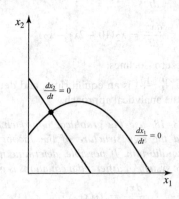

▲ **Figure 11.45**

16. See Figure 11.46.

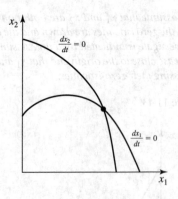

▲ **Figure 11.46**

17. See Figure 11.47.

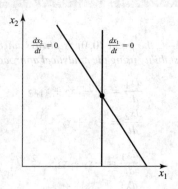

▲ **Figure 11.47**

18. See Figure 11.48.

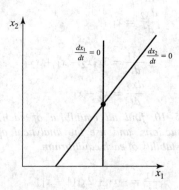

▲ **Figure 11.48**

19. Let

$$\frac{dx_1}{dt} = x_1(2 - x_1) - x_1 x_2$$

$$\frac{dx_2}{dt} = x_1 x_2 - x_2$$

(a) Graph the zero isoclines.

(b) Show that $(1, 1)$ is an equilibrium. Use the graphical approach to determine its stability.

20. Let

$$\frac{dx_1}{dt} = x_1(2 - x_1^2) - x_1 x_2$$

$$\frac{dx_2}{dt} = x_1 x_2 - x_2$$

(a) Graph the zero isoclines.

(b) Show that $(1, 1)$ is an equilibrium. Use the graphical approach to determine its stability.

11.4 NONLINEAR SYSTEMS: APPLICATIONS

11.4.1 The Lotka-Volterra Model of Interspecific Competition

Imagine two species of plants growing together in the same plot. They both use similar resources—namely, light, water, and nutrients. The use of these resources by one individual reduces their availability to other individuals. We call this type of interaction between individuals **competition**. **Intraspecific competition** occurs between individuals of the same species, and **interspecific competition** between individuals of different species. Competition may result in reduced fecundity and/or reduced survivorship. The effects of competition are often more pronounced when the number of competitors is higher.

In this subsection we will discuss the Lotka-Volterra model of interspecific competition, which incorporates density-dependent effects of competition in the manner described previously. It is an extension of the logistic equation to the case of two species. To describe the model, we denote the population size of species 1 at time t by $N_1(t)$ and that of species 2 at time t by $N_2(t)$. Each species grows according to the logistic equation when the other species is absent. We denote their respective carrying capacities by K_1 and K_2, and their respective intrinsic rates of growth by r_1 and r_2. We assume that K_1, K_2, r_1, and r_2 are positive. In addition, the two species may have inhibitory effects on each other. We measure the effect of species 1 on species 2 by the **competition coefficient** α_{21}; the effect of species 2 on species 1 is measured by the competition coefficient α_{12}. The Lotka-Volterra model of interspecific competition is then given by the following system of differential equations:

$$
\begin{aligned}
\frac{dN_1}{dt} &= r_1 N_1 \left(1 - \frac{N_1}{K_1} - \alpha_{12} \frac{N_2}{K_1} \right) \\
\frac{dN_2}{dt} &= r_2 N_2 \left(1 - \frac{N_2}{K_2} - \alpha_{21} \frac{N_1}{K_2} \right)
\end{aligned}
\tag{11.57}
$$

Let's look at the first equation. If $N_2 = 0$, the first equation reduces to the logistic equation $dN_1/dt = r_1 N_1(1 - N_1/K_1)$, as mentioned. To understand the precise meaning of the competition coefficient α_{12}, observe that N_2 individuals of species 2 have the same effect on species 1 as $\alpha_{12} N_2$ individuals of species 1. The term $\alpha_{12} N_2$ thus converts the number of N_2 individuals into "N_1-equivalents." For instance, set $\alpha_{12} = 0.2$ and assume that $N_2 = 20$; then the effect of 20 individuals of species 2 on species 1 is the same as the effect of $(0.2)(20) = 4$ individuals of species 1 on species 1, because both reduce the growth rate by the same amount. A similar interpretation can be attached to the competition coefficient α_{21} in the second equation.

This model takes a very simplistic view of competition; actual competitive interactions are more complicated. Nevertheless, it serves an important purpose. Because of its simple form, it allows us to study the consequences of competition; this can give us valuable insight into more complex situations.

We will analyze the model using both zero isoclines and eigenvalues.

Zero Isoclines The first step is to find the equations of the zero isoclines. To find the zero isocline for species 1, we set

$$
r_1 N_1 \left(1 - \frac{N_1}{K_1} - \alpha_{12} \frac{N_2}{K_1} \right) = 0
$$

The solutions are $N_1 = 0$ or

$$
N_2 = \frac{K_1}{\alpha_{12}} - \frac{1}{\alpha_{12}} N_1
\tag{11.58}
$$

To find the zero isocline for species 2, we set

$$r_2 N_2 \left(1 - \frac{N_2}{K_2} - \alpha_{21} \frac{N_1}{K_2}\right) = 0$$

The solutions are $N_2 = 0$ or

$$N_2 = K_2 - \alpha_{21} N_1 \tag{11.59}$$

The isocline $N_i = 0$, $i = 1, 2$, corresponds to the case in which species i is absent. This is biologically reasonable, because individuals of either species are not created spontaneously—once a species is absent, it remains absent.

The other two isoclines, given by (11.58) and (11.59), are of particular interest because they tell us whether both species can stably coexist. Both isoclines are straight lines in the N_1-N_2 plane, and there are four ways these isoclines can be arranged. This is illustrated in Figures 11.49 through 11.52, together with the corresponding direction fields. The solid dots in each figure are the equilibria. We see that there are always the equilibria $(K_1, 0)$ and $(0, K_2)$. These are the equilibria where only one species is present; they are referred to as **monoculture** equilibria. In addition, there are two cases where an equilibrium exists when both species are present; such an equilibrium is called a nontrivial equilibrium. We will now discuss each case separately.

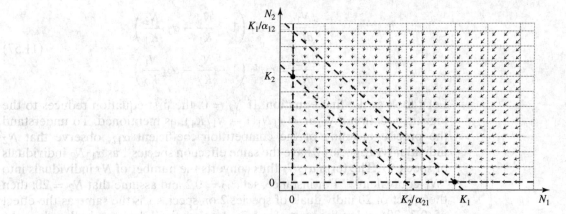

▲ **Figure 11.49**
Case 1: Species 1 outcompetes species 2

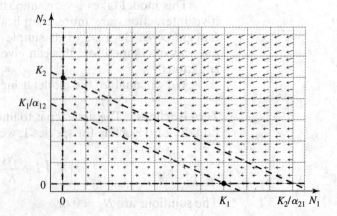

▲ **Figure 11.50**
Case 2: Species 2 outcompetes species 1

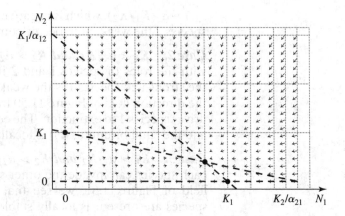

▲ **Figure 11.51**
Case 3: Species 1 and 2 can coexist

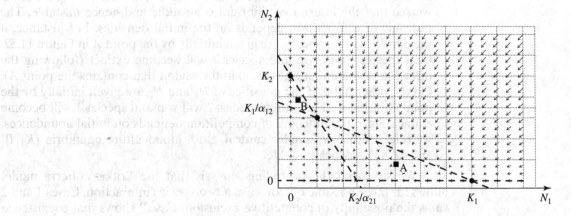

▲ **Figure 11.52**
Case 4: Either species 1 or species 2 wins, depending on the initial condition

Case 1: $K_1 > \alpha_{12}K_2$ and $K_2 < \alpha_{21}K_1$.   When we look at the direction field in Figure 11.49, we see that species 1 drives species 2 to extinction. If both species are present initially, then the abundance of species 2 declines over time, and it will eventually become extinct, whereas species 1 will reach its carrying capacity K_1. We say that species 1 outcompetes species 2, and refer to this as the case of **competitive exclusion**.

We can understand why species 1 is the better competitor by looking at the inequalities $K_1 > \alpha_{12}K_2$ and $K_2 < \alpha_{21}K_1$. The carrying capacity K_1 is the abundance of species 1 that results in a zero per capita growth rate of species 1 in the absence of species 2. Namely, if $N_1 = K_1$ and $N_2 = 0$, then $\frac{1}{N_1}\frac{dN_1}{dt} = 0$. The carrying capacity K_2 has a similar interpretation. To see the effect of species 2 on species 1, we look at the case when species 2 is at its carrying capacity K_2 and the abundance of species 1 is negligible. In this case, the per capita growth rate of species 1 is approximately equal to $r_1(1 - \alpha_{12}K_2/K_1)$, which is positive since $\alpha_{12}K_2/K_1$ is less than 1. This implies that species 2 cannot prevent species 1 from increasing its abundance when rare; we say that species 1 can invade species 2 when rare. On the other hand, since $K_2 < \alpha_{21}K_1$, species 2 cannot invade species 1. Namely, if we set $N_1 = K_1$ and assume that the abundance of species 2 is negligible, then the per capita growth rate of species 2 is approximately

$r_2(1 - \alpha_{21}K_1/K_2)$, which is negative. We say that species 1 is the strong and species 2 the weak interspecific competitor.

Case 2: $K_1 < \alpha_{12}K_2$ and $K_2 > \alpha_{21}K_1$. This is the same as case 1, but with the roles of species 1 and 2 interchanged. That is, species 2 is now the strong and species 1 is the weak interspecific competitor. We see from the direction field in Figure 11.50 that species 2 outcompetes species 1 and drives species 1 to extinction. The equilibrium $(K_1, 0)$ is therefore unstable, and the equilibrium $(0, K_2)$ is locally stable.

Case 3: $K_1 > \alpha_{12}K_2$ and $K_2 > \alpha_{21}K_1$. The two inequalities imply that each species can invade the other species. When we look at the direction field of Figure 11.51, we see that the interior equilibrium where both species are present is locally stable. We say that **coexistence** is possible. The two monoculture equilibria $(K_1, 0)$ and $(0, K_2)$ are unstable.

Case 4: $K_1 < \alpha_{12}K_2$ and $K_2 < \alpha_{21}K_1$. In this case, neither species can invade the other. When we look at the direction field of Figure 11.52, we see that the interior equilibrium is a saddle and hence unstable. The outcome of competition depends on the initial densities. For instance, if the densities of N_1 and N_2 are given initially by the point A in Figure 11.52, then species 1 will win and species 2 will become extinct (following the direction of the direction vectors in the region that contains the point A). On the other hand, if the densities of N_1 and N_2 are given initially by the point B in Figure 11.52, then species 2 will win and species 1 will become extinct. Since the outcome of competition depends on initial abundances, we refer to this as **founder control**. Both monoculture equilibria $(K_1, 0)$ and $(0, K_2)$ are locally stable.

We see from the preceding analysis that the Lotka-Volterra model allows for three possible outcomes in a two-species interaction. Cases 1 and 2 show the possibility of competitive exclusion. Case 3 shows that coexistence is possible. Case 4 shows that, depending on the initial abundances, one or the other species eventually wins; this case is referred to as founder control.

Though we stated that the system (11.57) describes a highly idealized situation of competition, there is a famous example that fits the equations very well. The Russian ecologist G. F. Gause (1934) studied competition between species of the protozoan *Paramecium*. When *Paramecium aurelia* and *P. caudatum* were grown together, *P. aurelia* competitively excluded *P. caudatum*. When *P. caudatum* and *P. bursaria* were grown together, they stably coexisted. Solution curves from (11.57) were fitted to both sets of data by estimating the relevant parameters in (11.57), and an excellent fit was obtained.

We will now turn to using eigenvalues to analyze (11.57).

Eigenvalues We will first determine all possible equilibria. Setting $dN_1/dt = 0$, we find

$$N_1 = 0 \quad \text{or} \quad N_1 + \alpha_{12}N_2 = K_1$$

Setting $dN_2/dt = 0$, we find

$$N_2 = 0 \quad \text{or} \quad \alpha_{21}N_1 + N_2 = K_2$$

There are four possible combinations:

1. The equilibrium $(\hat{N}_1, \hat{N}_2) = (0, 0)$ refers to the case when both species are absent. This is called the trivial equilibrium.

2. The equilibrium $(\hat{N}_1, \hat{N}_2) = (K_1, 0)$ refers to the case when species 2 is absent and species 1 is at its carrying capacity K_1.

3. The equilibrium $(\hat{N}_1, \hat{N}_2) = (0, K_2)$ refers to the case when species 1 is absent and species 2 is at its carrying capacity K_2.

4. The fourth equilibrium is obtained by simultaneously solving

$$N_1 + \alpha_{12} N_2 = K_1$$

$$\alpha_{21} N_1 + N_2 = K_2$$

and requiring that both solutions be positive.

To analyze the stability of these equilibria, we must compute the Jacobi matrix associated with (11.57). We find

$$Df(N_1, N_2) = \begin{bmatrix} r_1 - 2\dfrac{r_1}{K_1}N_1 - \dfrac{r_1\alpha_{12}}{K_1}N_2 & -\dfrac{r_1\alpha_{12}}{K_1}N_1 \\[2mm] -\dfrac{r_2\alpha_{21}}{K_2}N_2 & r_2 - 2\dfrac{r_2}{K_2}N_2 - \dfrac{r_2\alpha_{21}}{K_2}N_1 \end{bmatrix}$$

We look at each equilibrium separately.

1. Evaluating the Jacobi matrix at the trivial equilibrium $(\hat{N}_1, \hat{N}_2) = (0, 0)$, we find

$$Df(0, 0) = \begin{bmatrix} r_1 & 0 \\ 0 & r_2 \end{bmatrix}$$

which is a diagonal matrix; the eigenvalues are therefore

$$\lambda_1 = r_1 \quad \text{and} \quad \lambda_2 = r_2$$

Since $r_1 > 0$ and $r_2 > 0$, the equilibrium $(0, 0)$ is unstable.

2. Evaluating the Jacobi matrix at the equilibrium $(\hat{N}_1, \hat{N}_2) = (K_1, 0)$, we find

$$Df(K_1, 0) = \begin{bmatrix} -r_1 & -r_1\alpha_{12} \\ 0 & r_2\left(1 - \alpha_{21}\dfrac{K_1}{K_2}\right) \end{bmatrix}$$

Since $Df(K_1, 0)$ is an upper triangular matrix, we can immediately read off the eigenvalues. We find

$$\lambda_1 = -r_1 \quad \text{and} \quad \lambda_2 = r_2\left(1 - \alpha_{21}\frac{K_1}{K_2}\right)$$

Since $r_1 > 0$, it follows that $\lambda_1 < 0$. The eigenvalue $\lambda_2 < 0$ when $K_2 < \alpha_{21}K_1$. Therefore, the equilibrium

$$(K_1, 0) \text{ is } \begin{cases} \text{locally stable} & \text{for } K_2 < \alpha_{21}K_1 \\ \text{unstable} & \text{for } K_2 > \alpha_{21}K_1 \end{cases}$$

3. Evaluating the Jacobi matrix at the equilibrium $(\hat{N}_1, \hat{N}_2) = (0, K_2)$, we find

$$Df(0, K_2) = \begin{bmatrix} r_1\left(1 - \alpha_{12}\dfrac{K_2}{K_1}\right) & 0 \\ -r_2\alpha_{21} & -r_2 \end{bmatrix}$$

Since $Df(0, K_2)$ is a lower triangular matrix, we can immediately read off the eigenvalues. We find

$$\lambda_1 = r_1\left(1 - \alpha_{12}\frac{K_2}{K_1}\right) \quad \text{and} \quad \lambda_2 = -r_2$$

Since $r_2 > 0$, it follows that $\lambda_2 < 0$. The eigenvalue $\lambda_1 < 0$ when $K_1 < \alpha_{12}K_2$. Therefore, the equilibrium

$$(0, K_2) \text{ is } \begin{cases} \text{locally stable} & \text{for } K_1 < \alpha_{12}K_2 \\ \text{unstable} & \text{for } K_1 > \alpha_{12}K_2 \end{cases}$$

4. We stated that the fourth equilibrium can be obtained by simultaneously solving

$$N_1 + \alpha_{12}N_2 = K_1$$

$$\alpha_{21}N_1 + N_2 = K_2$$

Using the standard method of elimination, we find

$$N_1 + \alpha_{12}N_2 = K_1$$

$$(\alpha_{21}\alpha_{12} - 1)N_2 = \alpha_{21}K_1 - K_2$$

and, therefore,

$$\hat{N}_2 = \frac{\alpha_{21}K_1 - K_2}{\alpha_{21}\alpha_{12} - 1}$$

and

$$\hat{N}_1 = K_1 - \alpha_{12}\frac{\alpha_{21}K_1 - K_2}{\alpha_{21}\alpha_{12} - 1} = \frac{\alpha_{12}K_2 - K_1}{\alpha_{21}\alpha_{12} - 1}$$

We require that both $\hat{N}_1$ and $\hat{N}_2$ be positive (after all, these are population densities). That is, we require that

$$\frac{\alpha_{21}K_1 - K_2}{\alpha_{21}\alpha_{12} - 1} > 0 \quad \text{and} \quad \frac{\alpha_{12}K_2 - K_1}{\alpha_{21}\alpha_{12} - 1} > 0 \tag{11.60}$$

If $\alpha_{21}\alpha_{12} > 1$, then (11.60) reduces to

$$K_2 < \alpha_{21}K_1 \quad \text{and} \quad K_1 < \alpha_{12}K_2 \tag{11.61}$$

If $\alpha_{21}\alpha_{12} < 1$, then (11.60) reduces to

$$K_2 > \alpha_{21}K_1 \quad \text{and} \quad K_1 > \alpha_{12}K_2 \tag{11.62}$$

It turns out that finding the eigenvalues associated with this nontrivial equilibrium is algebraically rather involved if we use the Jacobi matrix associated with the original system of differential equations (11.57). However, if we investigate the effect of a perturbation directly, then the analysis becomes manageable. To demonstrate what we mean by this, we set

$$z_1 = N_1 - \hat{N}_1 \quad \text{and} \quad z_2 = N_2 - \hat{N}_2$$

Then (z_1, z_2) represents the deviation from the equilibrium. First, observe that

$$\frac{dz_1}{dt} = \frac{dN_1}{dt} \quad \text{and} \quad \frac{dz_2}{dt} = \frac{dN_2}{dt}$$

Substituting $z_1 + \hat{N}_1$ for N_1 and $z_2 + \hat{N}_2$ for N_2 in (11.57) yields

$$\frac{dz_1}{dt} = r_1(z_1 + \hat{N}_1)\left(1 - \frac{z_1 + \hat{N}_1}{K_1} - \alpha_{12}\frac{z_2 + \hat{N}_2}{K_1}\right)$$

$$= r_1(z_1 + \hat{N}_1)\left(\underbrace{1 - \frac{\hat{N}_1}{K_1} - \alpha_{12}\frac{\hat{N}_2}{K_1}}_{=0} - \frac{z_1}{K_1} - \alpha_{12}\frac{z_2}{K_1}\right)$$

and

$$\frac{dz_2}{dt} = r_2(z_2 + \hat{N}_2)\left(1 - \frac{z_2 + \hat{N}_2}{K_2} - \alpha_{21}\frac{z_1 + \hat{N}_1}{K_2}\right)$$

$$= r_2(z_2 + \hat{N}_2)\left(\underbrace{1 - \frac{\hat{N}_2}{K_2} - \alpha_{21}\frac{\hat{N}_1}{K_2}}_{=0} - \frac{z_2}{K_2} - \alpha_{21}\frac{z_1}{K_2}\right)$$

Instead of analyzing the Jacobi matrix associated with (11.57) evaluated at the nontrivial equilibrium $(\hat{N}_1, \hat{N}_1)$, we investigate the equilibrium $(z_1, z_2) = (0, 0)$ of the new system

$$\frac{dz_1}{dt} = -\frac{r_1}{K_1}(z_1 + \hat{N}_1)(z_1 + \alpha_{12}z_2)$$

$$\frac{dz_2}{dt} = -\frac{r_2}{K_2}(z_2 + \hat{N}_2)(z_2 + \alpha_{21}z_1)$$

The Jacobi matrix $J(z_1, z_2)$ is equal to

$$\begin{bmatrix} -\dfrac{r_1}{K_1}(z_1 + \alpha_{12}z_2) - \dfrac{r_1}{K_1}(z_1 + \hat{N}_1) & -\dfrac{r_1\alpha_{12}}{K_1}(z_1 + \hat{N}_1) \\[2ex] -\dfrac{r_2\alpha_{21}}{K_2}(z_2 + \hat{N}_2) & -\dfrac{r_2}{K_2}(z_2 + \alpha_{21}z_2) - \dfrac{r_2}{K_2}(z_2 + \hat{N}_2) \end{bmatrix}$$

Evaluating the matrix $J(z_1, z_2)$ at $(z_1, z_2) = (0, 0)$, we obtain

$$J(0, 0) = \begin{bmatrix} -\dfrac{r_1}{K_1}\hat{N}_1 & -\dfrac{r_1\alpha_{12}}{K_1}\hat{N}_1 \\[2ex] -\dfrac{r_2\alpha_{21}}{K_2}\hat{N}_2 & -\dfrac{r_2}{K_2}\hat{N}_2 \end{bmatrix}$$

Now,

$$\mathrm{tr}(J(0, 0)) = -\frac{r_1\hat{N}_1}{K_1} - \frac{r_2\hat{N}_2}{K_2}$$

$$\det(J(0, 0)) = \frac{r_1r_2}{K_1K_2}\hat{N}_1\hat{N}_2(1 - \alpha_{12}\alpha_{21})$$

The nontrivial equilibrium $(\hat{N}_1, \hat{N}_2)$ satisfies $\hat{N}_1 > 0$ and $\hat{N}_2 > 0$, implying that $\mathrm{tr}(J(0, 0)) < 0$. Since

$$\det(J(0, 0)) > 0 \quad \text{when } 1 - \alpha_{12}\alpha_{21} > 0$$

we find that $(\hat{N}_1, \hat{N}_2)$ is unstable if (11.61) holds, and is locally stable if (11.62) holds.

When we compare the conditions in Case 4 with those in Cases 2 and 3, we see that the interior equilibrium is locally stable when the two boundary (monoculture) equilibria $(K_1, 0)$ and $(0, K_2)$ are unstable, and that the interior equilibrium is unstable when the two boundary equilibria are locally stable. We can also show that both eigenvalues are always real; that is, we do not expect oscillations.

Comparing the two approaches, we find the same results. The graphical approach gives qualitative answers only, albeit rather useful ones. The graphical approach also turns out to be easier. The eigenvalue approach allows us to obtain quantitative answers in terms of eigenvalues, which tell us something about how quickly the system returns to a stable equilibrium after a small perturbation. Such quantitative answers are important if we want to know how the system responds to small perturbations.

11.4.2 A Predator-Prey Model

In this subsection, we investigate a simple model for predation, defined as the consumption of prey by a predator. We assume that the prey is alive when the predator attacks it, and that the predator kills its prey and thus removes it from the population.

There are many patterns of abundance that result from predator-prey interactions; most notably, those situations in nature in which predator and prey abundances appear to be closely linked and show periodic fluctuations. A frequently cited example is the Canada lynx (*Lynx canadensis*) and snowshoe hare (*Lepus americanus*) system. The population abundances show periodic oscillations. A laboratory example of coupled predator-prey oscillations is the azuki bean weevil (*Callosbruchus chinensis*) and its parasitoid wasp (*Heterospilus prosopidis*).

The simplest predator-prey model that exhibits coupled oscillations is the Lotka-Volterra model (Lotka, 1932; Volterra, 1926). We describe this system in Volterra's own words:

> The first case I have considered is that of two associated species, of which one, finding sufficient food in its environment, would multiply indefinitely when left to itself, while the other would perish for lack of nourishment if left alone; but the second feeds upon the first, and so the two species can coexist together.
> The proportional rate of increase of the eaten species diminishes as the number of individuals of the eating species increases, while augmentation of the eating species increases with the increase of the number of individuals of the eaten species.

We will now translate this verbal formulation of the model into a system of differential equations. If we denote by $N(t)$ the abundance of the prey and by $P(t)$ the abundance of the predator, then the model is given by the following system of differential equations:

$$\frac{dN}{dt} = rN(t) - aP(t)N(t)$$
$$\frac{dP}{dt} = baP(t)N(t) - dP(t)$$

(11.63)

where r, a, b, and d are positive constants. Note that the prey increases exponentially in the absence of the predator ($P = 0$). The intrinsic rate of increase of the prey in the absence of the predator is r. The constant a denotes the attack rate and the term aPN is the consumption rate of prey. Predators decline exponentially in the absence of prey; the rate of decline is given by the constant d. The consumption of prey results in an increase of predator abundance. The constant b describes how efficiently predator turn prey into predator offspring.

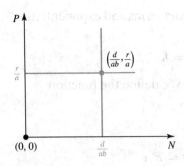

▲ Figure 11.53
Zero isoclines

To find equilibria of (11.63), we set

$$N(r - aP) = 0$$
$$P(abN - d) = 0$$

This yields the zero isoclines $N = 0$ and $P = r/a$ for $dN/dt = 0$, and $P = 0$ and $N = d/ab$ for $dP/dt = 0$. The zero isoclines are shown in Figure 11.53. The points of intersection of the zero isoclines for N and P are equilibria. We find

$$(\hat{N}, \hat{P}) = (0, 0) \quad \text{and} \quad (\hat{N}, \hat{P}) = \left(\frac{d}{ab}, \frac{r}{a} \right)$$

To analyze the stability of these two equilibria, we linearize (11.63). The Jacobi matrix is given by

$$D\mathbf{f}(N, P) = \begin{bmatrix} r - aP & -aN \\ baP & baN - d \end{bmatrix}$$

When $(\hat{N}, \hat{P}) = (0, 0)$, then

$$D\mathbf{f}(0, 0) = \begin{bmatrix} r & 0 \\ 0 & -d \end{bmatrix}$$

This is a diagonal matrix, hence the eigenvalues are given by the diagonal elements

$$\lambda_1 = r > 0 \quad \text{and} \quad \lambda_2 = -d < 0$$

We conclude that $(0, 0)$ is unstable.
When $(\hat{N}, \hat{P}) = (\frac{d}{ab}, \frac{r}{a})$, then

$$D\mathbf{f}\left(\frac{d}{ab}, \frac{r}{a} \right) = \begin{bmatrix} 0 & -\frac{d}{b} \\ rb & 0 \end{bmatrix}$$

To determine the eigenvalues of $D\mathbf{f}(\frac{d}{ab}, \frac{r}{a})$, we must solve

$$\det \begin{bmatrix} -\lambda & -\frac{d}{b} \\ rb & -\lambda \end{bmatrix} = \lambda^2 + rd = 0$$

Solving this, we find

$$\lambda_1 = i\sqrt{rd} \quad \text{and} \quad \lambda_2 = -i\sqrt{rd}$$

That is, both eigenvalues are purely imaginary. This does not allow us to determine the stability of this equilibrium by linearizing about the equilibrium, as pointed out in Section 11.3. Fortunately, however, we can solve (11.63) exactly.

To solve (11.63) exactly, we divide dP/dt by dN/dt. We find

$$\frac{dP/dt}{dN/dt} = \frac{dP}{dN} = \frac{P(abN - d)}{N(r - aP)}$$

Separating variables and integrating, we obtain

$$\int \frac{r - aP}{P} \, dP = \int \frac{abN - d}{N} \, dN$$

Carrying out the integration, we find

$$r \ln P - aP = abN - d \ln N + C$$

where C is the constant of integration. Rearranging terms and exponentiating yields

$$\left(N^d e^{-abN}\right)\left(P^r e^{-aP}\right) = K$$

where $K = e^C$ depends on the initial condition. We define the function

$$f(N, P) = \left(N^d e^{-abN}\right)\left(P^r e^{-aP}\right)$$

and set

$$g(N) = N^d e^{-abN} \quad \text{and} \quad h(P) = P^r e^{-aP}$$

Then we can show that $g(N)$ has its absolute maximum when $N = d/ab$ and $h(P)$ has its absolute maximum when $P = r/a$. The function $f(N, P)$ therefore takes on its absolute maximum at the equilibrium point $(d/ab, r/a)$. We can therefore define level curves

$$f(N, P) = K$$

for $K \leq K_{\max}$, where $K_{\max}$ is the value of f at the equilibrium $(d/ab, r/a)$. We will not be able to demonstrate this here, but these level curves are closed curves. (We show such level curves in Figure 11.54, to convince you that these curves are indeed closed.) These level curves are solution curves.

Solutions of $N(t)$ and $P(t)$ as functions of time corresponding to the closed curves in Figure 11.54 are shown in Figures 11.55 and 11.56. When we plot $N(t)$ and $P(t)$ versus t, we see that the closed trajectories in the N-P plane correspond to periodic solutions for the predator and the prey. The amplitudes of the oscillations depend on the initial condition. Note that the amplitudes in the two figures are different.

These closed trajectories are not stable under perturbations. That is, if a small perturbation changes the value of N or P, the solution will follow a *different* closed trajectory. This property is a major drawback of the model. It implies that if a natural population actually followed this simple model, its abundance would not exhibit regular cycles, because external factors would constantly shift the population to different trajectories. If a natural population exhibits regular cycles, we would expect these cycles to be stable, that is, the population would return to the same cycle after a small perturbation. Such cycles are called **stable limit cycles**. A locally stable equilibrium that is approached by oscillations can be obtained by modifying the original Lotka-Volterra model to include a nonlinear predator response to prey abundance. This is discussed in Problems 19 through 21.

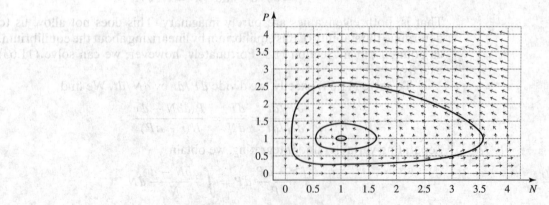

▲ **Figure 11.54**
Solutions for (11.63) in the N-P plane

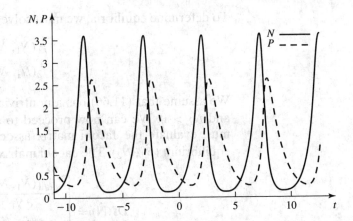

▲ **Figure 11.55**
Solutions for (11.63) as functions of time

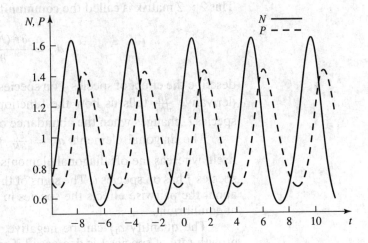

▲ **Figure 11.56**
Solutions for (11.63) as functions of time

11.4.3 The Community Matrix

In this subsection we consider a fairly general multispecies population model, which was initiated by Levins (1970) and further developed by May (1975). The goal is to determine how interactions between pairs of species influence the stability of the equilibria of an assemblage of species.

We assume an assemblage of two species in which the abundance of species i at time t is given by $N_i(t)$. (At the end of this subsection we will generalize this to an assemblage of m species.) Suppose that the following set of differential equations describes the dynamics of this two-species assemblage:

$$\frac{dN_1}{dt} = f_1(N_1(t), N_2(t))$$

$$\frac{dN_2}{dt} = f_2(N_1(t), N_2(t)) \tag{11.64}$$

or, in vector notation,

$$\frac{d\mathbf{N}}{dt} = \mathbf{f}(\mathbf{N}(t)) \tag{11.65}$$

To determine equilibria, we must solve the system of equations

$$f_1(N_1, N_2) = 0$$
$$f_2(N_1, N_2) = 0 \tag{11.66}$$

We assume that (11.66) has a nontrivial solution $\hat{\mathbf{N}} = (\hat{N}_1, \hat{N}_2)$ with $\hat{N}_1 > 0$ and $\hat{N}_2 > 0$. We can now proceed to determine its stability. To do so, we must evaluate the Jacobi matrix associated with the system (11.64) at the equilibrium $(\hat{N}_1, \hat{N}_2)$. The Jacobi matrix at the equilibrium $\hat{\mathbf{N}}$ is given by

$$D\mathbf{f}(\hat{\mathbf{N}}) = \begin{bmatrix} \dfrac{\partial f_1(\hat{N}_1, \hat{N}_2)}{\partial N_1} & \dfrac{\partial f_1(\hat{N}_1, \hat{N}_2)}{\partial N_2} \\[2mm] \dfrac{\partial f_2(\hat{N}_1, \hat{N}_2)}{\partial N_1} & \dfrac{\partial f_2(\hat{N}_1, \hat{N}_2)}{\partial N_2} \end{bmatrix}$$

This 2×2 matrix is called the **community matrix**. Its elements

$$a_{ij} = \frac{\partial f_i(\hat{N}_1, \hat{N}_2)}{\partial N_j}$$

describe the effect of species j on species i at equilibrium, because the partial derivative $\frac{\partial f_i}{\partial N_j}$ tells us how the function f_i, which describes the growth of species i, changes when the abundance of species j changes.

The diagonal elements $a_{ii} = \frac{\partial f_i}{\partial N_i}$ measure the effect species i has on itself, whereas the off-diagonal elements $a_{ij} = \frac{\partial f_i}{\partial N_j}$, $i \neq j$, measure the effect species j has on species i. The signs of the elements a_{ij} thus tell us something about the pairwise effects the species in this assemblage have on each other at equilibrium.

The quantity a_{ij} can be negative, 0, or positive. If $a_{ij} < 0$, then the growth rate of species i is decreased if species j increases its abundance; we therefore say that species j has a negative or inhibitory effect on species i. If $a_{ij} = 0$, then changes in the abundance of species j have no effect on the growth rate of species i. If $a_{ij} > 0$, then the growth rate of species i is increased if species j increases its abundance; we therefore say that species j has a positive or facilitatory effect on species i.

We look at the possible combinations of the pair (a_{21}, a_{12}). This pair describes the interactions *between* the two species in the assemblage. The following table lists all possible combinations.

		a_{12}		
		+	0	−
a_{21}	+	++	+0	+−
	0	0+	00	0−
	−	−+	−0	−−

To interpret this table, take the pair (00), for instance. The pair (00) represents the case in which neither species has an effect on the other species at equilibrium. Let's look at another pair, for instance, (0+). In this case, $a_{21} = 0$ and species 1 has no effect on species 2, but $a_{12} > 0$ and species 2 has a positive effect on species 1.

The case (00) is the simplest, and we will discuss it first. The community matrix in this case is

$$\begin{bmatrix} a_{11} & 0 \\ 0 & a_{22} \end{bmatrix}$$

Since the community matrix is in diagonal form, its eigenvalues are the diagonal elements a_{11} and a_{22}. Hence the equilibrium $(\hat{N}_1, \hat{N}_2)$ is stable, provided that both a_{11} and a_{22} are negative. That is, if neither species has an effect on the other species, a locally stable nontrivial equilibrium, in which both species coexist, only exists if they each have a negative effect on themselves; this means that each species needs to regulate its own population size.

Following May (1975), the remaining eight combinations in the table can be categorized into five biologically distinct types of interactions.

Mutualism or symbiosis (++): *Each species has a positive effect on the other.*

Competition (−−): *Each species has a negative effect on the other.*

Commensalism (+0): *One species benefits from the interaction, whereas the other is unaffected.*

Amensalism (−0): *One species is negatively affected, whereas the other is unaffected.*

Predation (+−): *One species benefits, whereas the other is negatively affected.*

We will now discuss the stability of the nontrivial equilibrium $(\hat{N}_1, \hat{N}_2)$ in all five cases. Recall that we assumed that this equilibrium exists, and that both $\hat{N}_1 > 0$ and $\hat{N}_2 > 0$. The community matrix (the Jacobi matrix of (11.64) at equilibrium) is of the form

$$A = \begin{bmatrix} a_{11} & a_{12} \\ a_{21} & a_{22} \end{bmatrix}$$

Both eigenvalues of A have negative real parts if and only if

$$\text{tr } A = a_{11} + a_{22} < 0 \quad \text{and} \quad \det A = a_{11}a_{22} - a_{12}a_{21} > 0$$

We assume in the following that

$$a_{11} < 0 \quad \text{and} \quad a_{22} < 0 \tag{11.67}$$

so that the first condition $\text{tr} A < 0$ is automatically satisfied. This has the same interpretation as discussed in the case (00), namely both species have a negative effect on themselves or regulate their own population densities.

We will now go through all five cases, and determine under which conditions the nontrivial equilibrium is stable.

Mutualism   We assume (11.67). The sign structure of the community matrix at equilibrium in the case of mutualism is then of the form

$$A = \begin{bmatrix} - & + \\ + & - \end{bmatrix}$$

Since

$$\det A = \underbrace{a_{11}a_{22}}_{>0} - \underbrace{a_{12}a_{21}}_{>0}$$

the determinant of A may be either positive or negative. If $a_{12}a_{21}$ is sufficiently small compared to $a_{11}a_{22}$, then $\det A > 0$, and the equilibrium $(\hat{N}_1, \hat{N}_2)$ is locally stable. In other words, if the positive effects of the species on each other are sufficiently counteracted by their own population control (represented by a_{11} and a_{22}), the equilibrium is locally stable.

Competition   We assume (11.67). The sign structure of the community matrix at equilibrium in the case of competition is then of the form

$$A = \begin{bmatrix} - & - \\ - & - \end{bmatrix}$$

Since

$$\det A = \underbrace{a_{11}a_{22}}_{>0} - \underbrace{a_{12}a_{21}}_{>0}$$

the determinant of A may be either positive or negative. Now, the equilibrium $(\hat{N}_1, \hat{N}_2)$ is locally stable if the negative effects each species has on the other are smaller than the effects each species has on itself. (In this case, $\det A > 0$ and, hence, the equilibrium is locally stable.)

Commensalism and Amensalism   We assume (11.67). The sign structure of the community matrices at equilibrium are then of the form

$$A = \begin{bmatrix} - & 0 \\ + & - \end{bmatrix} \quad \text{or} \quad A = \begin{bmatrix} - & 0 \\ - & - \end{bmatrix}$$

In either case, the determinant is positive; therefore, the equilibrium is locally stable.

Predation   We assume (11.67). The sign structure of the community matrix at equilibrium in the case of predation is then of the form

$$A = \begin{bmatrix} - & - \\ + & - \end{bmatrix}$$

and $\det A$ is always positive. That is, provided a nontrivial equilibrium exists, it is locally stable.

The Multispecies Case   We will now briefly look at the case in which more than two species form an assemblage. Say there are m species. We denote the density of species i at time t by $N_i(t)$. The dynamics are described by the following system of differential equations:

$$\frac{dN_1}{dt} = f_1(N_1(t), N_2(t), \ldots, N_m(t))$$

$$\frac{dN_2}{dt} = f_2(N_1(t), N_2(t), \ldots, N_m(t))$$

$$\vdots$$

$$\frac{dN_m}{dt} = f_m(N_1(t), N_2(t), \ldots, N_m(t))$$

The equilibria are found by solving the following system of equations:

$$f_1(\hat{N}_1, \hat{N}_2, \ldots, \hat{N}_m) = 0$$
$$f_2(\hat{N}_1, \hat{N}_2, \ldots, \hat{N}_m) = 0$$
$$\vdots$$
$$f_m(\hat{N}_1, \hat{N}_2, \ldots, \hat{N}_m) = 0$$

If we assume that $\hat{N} = (\hat{N}_1, \hat{N}_2, \ldots, \hat{N}_m)$ is an equilibrium, then the community matrix at equilibrium is the Jacobi matrix at equilibrium, which is given by

$$D\mathbf{f}(\hat{\mathbf{N}}) = \begin{bmatrix} \dfrac{\partial f_1}{\partial N_1} & \cdots & \dfrac{\partial f_1}{\partial N_m} \\ \vdots & & \vdots \\ \dfrac{\partial f_m}{\partial N_1} & \cdots & \dfrac{\partial f_m}{\partial N_m} \end{bmatrix}$$

As in the case of two species, the element

$$a_{ij} = \frac{\partial f_i(\hat{\mathbf{N}})}{\partial N_j}$$

describes the effect of species j on species i at equilibrium.

11.4.4 A Mathematical Model for Neuron Activity

The nervous system of an organism is a communication network that allows for rapid transmission of information between cells. It consists of nerve cells, called **neurons**. A typical neuron has a cell body that contains the cell nucleus and nerve fibers. Nerve fibers that receive information are called **dendrites**, those that transport information are called **axons**, which provide links to other neurons via **synapses**. A typical vertebrate neuron is shown in Figure 11.57.

Neurons respond to electrical stimuli; this response is exploited when studying neurons. When the cell body of an isolated neuron is stimulated with a very mild electrical shock, the neuron shows no response; increasing the intensity of the electrical shock beyond a certain threshold will trigger a response, namely an impulse that travels along the axon. Increasing the intensity of the electrical shock further does not change the response. This impulse is thus an all-or-none response.

A. L. Hodgkin and A. F. Huxley studied the giant axon of a squid experimentally, and developed a mathematical model for neuron activity. Their work appeared in a series of papers in 1952. It is an excellent example of how experimental and theoretical research can be combined to gain a thorough understanding of a natural system. In 1963, Hodgkin and Huxley were awarded the Nobel prize in physiology/medicine for their work on neurons.

We begin with a brief explanation of how a neuron works. The main players in the functioning of a neuron are sodium (Na^+) and potassium (K^+) ions. The cell membrane of a neuron is impermeable to sodium and potassium ions when the cell is in a resting state. In a typical neuron in its resting state, the concentration of Na^+ in the interior of the cell is about one-tenth of the extracellular concentration of Na^+; the concentration of K^+ in the interior of the cell is about thirty times the extracellular concentration of K^+. When the neuron is in its resting state, the interior of the cell is negatively charged (at $-70\,\text{mV}$) relative to the exterior of the cell.

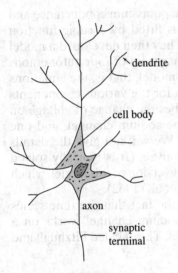

dendrite

cell body

axon

synaptic terminal

▲ **Figure 11.57**
A typical vertebrate neuron

When a nerve cell becomes stimulated, its surface becomes permeable to Na^+ ions, which rush into the cell through sodium channels in the surface. This results in a reversal of polarization at the points where Na^+ ions entered the cell. The surface inside the cell is now positively charged relative to the outside of the cell and becomes permeable to K^+ ions, which rush outside through potassium channels. Because the potassium ions (K^+) are positively charged and move from the inside to the outside of the cell, the polarization at the surface of the cell is again reversed and is now below the polarization of the resting cell. To restore the original concentration of Na^+ and K^+ (and thus the original polarization), energy must be expended to run the so-called sodium and potassium pumps in the surface of the cell to pump the excess Na^+ from the interior to the exterior of the cell and to pump K^+ from the exterior to the interior of the cell.

To trigger such a reaction, the intensity of the stimulus must be above a certain threshold. The described reaction occurs locally on the surface of the cell. This large local change in polarization triggers the same reaction in the neighborhood, which allows the reaction to propagate along the nerve cell—thus creating the observed impulse that travels along the cell. This localized large change in polarization, which is then reversed to the original polarization, is called an **action potential**. An example of an action potential is illustrated in Figure 11.58.

Hodgkin and Huxley measured sodium and potassium conductance and fitted curves to their data. The sodium curve is fitted by a cubic function and the potassium curve by a quartic function. They then developed a model for the action potential. The model consists of a system of four autonomous differential equations. It is a phenomenological model, that is, the equations are based on fitting curves to experimental data for the various components of the model. There is one equation that describes the change of voltage on the cell surface, two equations that describe the sodium channel, and one equation that describes the potassium channel. We will not give the details of this system, as it is far too complicated to analyze. (It is typically solved numerically.) Instead, we will present a simplified version of this model, which was developed by Fitzhugh (1961) and Nagumo et al. (1962).

The Fitzhugh-Nagumo model is based on the fact that the time scales of the two channels are quite different. The sodium channel works on a much faster time scale than the potassium channel. This fact led Fitzhugh and

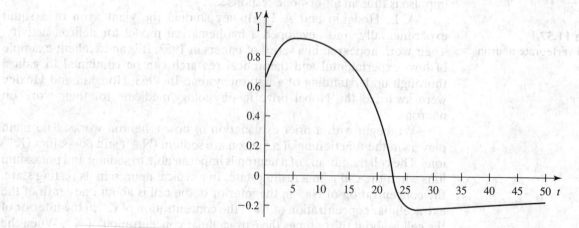

▲ **Figure 11.58**
The action potential

Nagumo to assume that the sodium channel is essentially always in steady state, which allowed them to reduce the four equations of the Hodgkin and Huxley model to two equations. The Fitzhugh-Nagumo model is thus an approximation to the Hodgkin and Huxley model. It retains the essential features of the action potential, but is much easier to analyze.

The Fitzhugh-Nagumo model is described by two variables. One variable, denoted by V, describes the potential of the cell surface. The other variable, denoted by w, models the sodium and the potassium channels. The equations are given by

$$\frac{dV}{dt} = -V(V-a)(V-1) - w$$

$$\frac{dw}{dt} = b(V - cw) \tag{11.68}$$

where a, b, and c are constants that satisfy $0 < a < 1$, $b > 0$, and $c > 0$.

We will analyze the system graphically. The zero isoclines of (11.68) are given by

$$w = -V(V-a)(V-1) \quad \text{and} \quad w = \frac{1}{c}V$$

The important feature of this model is that the zero isocline $dV/dt = 0$ is the graph of a cubic function in the V-w plane. The zero isoclines are illustrated in Figure 11.59.

We see that if c is small, there is just one equilibrium, namely $(0, 0)$; whereas when c is sufficiently large, the line $w = V/c$ intersects the graph of $dV/dt = 0$ three times.

The threshold phenomenon and the action potential are observed when c is small. In this case, there is just one equilibrium, namely $(0, 0)$, which corresponds to the resting state.

We can analyze the stability of $(0, 0)$ by linearizing the system about this equilibrium. We find

$$\mathbf{Df}(V, w) = \begin{bmatrix} -3V^2 + 2V + 2aV - a & -1 \\ b & -bc \end{bmatrix}$$

Hence,

$$\mathbf{Df}(0, 0) = \begin{bmatrix} -a & -1 \\ b & -bc \end{bmatrix}$$

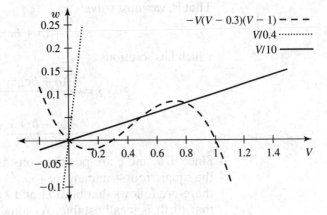

▲ **Figure 11.59**
The zero isoclines of (11.68)

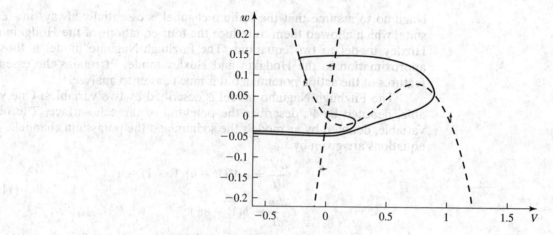

▲ **Figure 11.60**
Solution curves for the Fitzhugh-Nagumo model

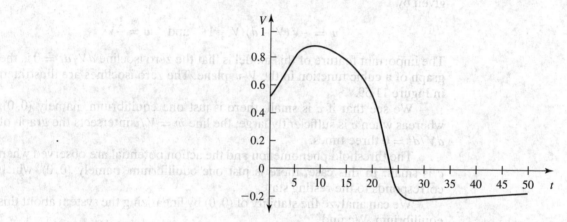

▲ **Figure 11.61**
The action potential when $V(0) > a$

To find the eigenvalues, we compute

$$\det \begin{bmatrix} -a - \lambda & -1 \\ b & -bc - \lambda \end{bmatrix} = (-a - \lambda)(-bc - \lambda) + b = 0$$

That is, we must solve

$$\lambda^2 + (a + bc)\lambda + b(ac + 1) = 0$$

which has solutions

$$\lambda_{1,2} = \frac{-(a + bc) \pm \sqrt{(a + bc)^2 - 4b(ac + 1)}}{2}$$

$$= \frac{-(a + bc) \pm \sqrt{(a - bc)^2 - 4b}}{2}$$

Since a, b, and c are positive constants, it follows that the expression under the square root—namely, $(a + bc)^2 - 4b(ac + 1)$—is smaller than $(a + bc)^2$. It therefore follows that both λ_1 and λ_2 have negative real parts, which implies that $(0, 0)$ is locally stable. As long as $(a - bc)^2 > 4b$, the equilibrium is a stable sink. When $(a - bc)^2 < 4b$, the eigenvalues are complex conjugates and $(0, 0)$ becomes a stable spiral.

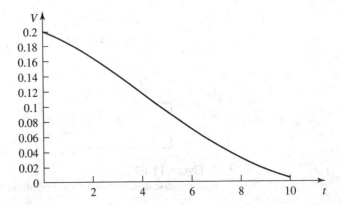

▲ **Figure 11.62**
The initial stimulus dies away quickly when $V(0) < a$

The system mimics the action potential when both eigenvalues are real and negative. If we apply a weak stimulus—that is, increase V to a value less than a—V will quickly return to 0. However, if we apply a strong enough stimulus—that is, $V \in (a, 1)$—the trajectory will move away from the equilibrium point, as shown in Figure 11.60. A plot of voltage versus time reveals that if the stimulus is too weak, $V(t)$ will quickly return to the equilibrium, whereas if the stimulus is large enough, the solution curve of $V(t)$ resembles the action potential. In Figures 11.61 and 11.62, we present two solution curves $V(t)$. In either case, $w(0) = 0$. In Figure 11.61, $V(0) = 0.5 > a$, and we observe an action potential; whereas in Figure 11.62, $V(0) = 0.2 < a$, and the initial stimulus is too small and dies away quickly.

11.4.5 A Mathematical Model for Enzymatic Reactions

The goal of this subsection is threefold. First, we will learn how to model biochemical reactions; second, we will see how mathematical models can be used to understand empirical observations; third, we will introduce an important method, namely by making appropriate assumptions, the number of variables in a model can sometimes be reduced, which typically facilitates the analysis of the model.

The class of biochemical reactions we will study are enzymatic reactions, which are ubiquitous in the living world. Enzymes are proteins that act as catalysts in chemical reactions by reducing the required activation energy for the reaction. Enzymes are not altered by the reaction; they aid in the initial steps of the reaction and control the rate of the reaction by binding the reactants (called substrates) to the active site of the enzyme, thus forming an enzyme-substrate complex that then allows the substrates to react and to form the product. This is illustrated in Figure 11.63.

If we denote the substrate by S, the enzyme by E, and the product of the reaction by P, then an enzymatic reaction can be described by

$$E + S \underset{k_-}{\overset{k_+}{\rightleftharpoons}} ES \overset{k_2}{\longrightarrow} E + P \tag{11.69}$$

where k_+, k_-, and k_2 are the reaction rates in the corresponding reaction steps.

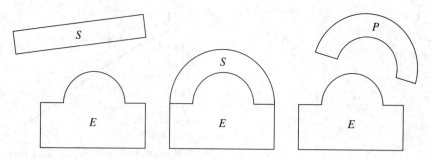

▲ **Figure 11.63**
A schematic description of an enzymatic reaction

We can translate the schematic description of the reaction (11.69) into a system of differential equations. We use the following notation.

$$e(t) = [E] = \text{enzyme concentration at time } t$$
$$s(t) = [S] = \text{substrate concentration at time } t$$
$$c(t) = [ES] = \text{enzyme substrate complex concentration at time } t$$
$$p(t) = [P] = \text{product concentration at time } t$$

Using the mass action law,

$$\frac{ds}{dt} = k_- c - k_+ se$$
$$\frac{de}{dt} = (k_- + k_2)c - k_+ se$$
$$\frac{dc}{dt} = k_+ se - (k_- + k_2)c \qquad (11.70)$$
$$\frac{dp}{dt} = k_2 c$$

Michaelis and Menten (1913) were instrumental in the description of enzyme kinetics, through both experimental and theoretical work. On the experimental side, they developed techniques that allowed them to measure reaction rates under controlled conditions. Their experiments showed a hyperbolic relationship between the rate of the enzymatic reaction dp/dt and the substrate concentration s, which they described as

$$\frac{dp}{dt} = \frac{v_m s}{K_m + s} \qquad (11.71)$$

where v_m is the saturation constant and K_m is the half-saturation constant (that is, if $s = K_m$, then $dp/dt = v_m/2$).

On the theoretical side, they developed a mathematical model for enzyme kinetics that predicted the observed hyperbolic relationship between the substrate concentration and the initial rate at which the product is formed.

In the following, we will analyze (11.70), which is a system of four equations. It is not easy to analyze; we will simplify it, and in the end, we will obtain (11.71). We will also use (11.70) to illustrate that it is sometimes possible to reduce the number of equations in a system.

We claim that the system has a conserved quantity, that is, a quantity that does not depend on time and is therefore constant throughout the reaction.

To find this quantity, note that

$$\frac{de}{dt} + \frac{dc}{dt} = 0$$

That is,

$$\frac{d}{dt}(e + c) = 0$$

which implies that

$$e(t) + c(t) = e_0 \tag{11.72}$$

where e_0 is a constant. Since $e(t) + c(t)$ is constant, we say that the sum $e(t) + c(t)$ is a conserved quantity. The advantage of this is that if we know the initial concentrations $e(0)$ and $c(0)$ and one of the two quantities $e(t)$ or $c(t)$, we immediately know the other, since $e(0) + c(0) = e(t) + c(t)$. This reduces the number of equations from four to three.

To reduce the number of equations even further, we make another assumption. Whereas the existence of a conserved quantity followed from the system of equations, and we could have obtained it without knowing the meaning of these equations, the next assumption requires a thorough understanding of the enzymatic reaction itself, and cannot be deduced from the set of equations (11.70). Briggs and Haldane (1925), who had this understanding, proposed that the rate of formation balances the rate of breakdown of the complex; that is, they assumed that

$$\frac{dc}{dt} = 0$$

This assumption yields the equation

$$0 = k_+ se - (k_- + k_2)c$$

which can be rewritten as

$$\frac{se}{c} = \frac{k_- + k_2}{k_+}$$

We denote this ratio by K_m; that is,

$$K_m = \frac{k_- + k_2}{k_+}$$

and, therefore,

$$\frac{se}{c} = K_m \tag{11.73}$$

Solving (11.72) for e—that is, $e = e_0 - c$—and plugging e into (11.73), we find

$$\frac{s(e_0 - c)}{c} = K_m$$

which we can solve for c. This yields

$$c = \frac{e_0 s}{K_m + s} \tag{11.74}$$

which allows us to rewrite the equation for the rate at which the product is formed. Since $dp/dt = k_2 c$, it follows from (11.74) that

$$\frac{dp}{dt} = \frac{k_2 e_0 s}{K_m + s} \tag{11.75}$$

We can interpret the factor $k_2 e_0$. Namely, if all of the enzyme is complexed with the substrate, then $e = 0$ and therefore $c = e_0$ (since $e + c = e_0$ is a conserved quantity). This implies that the rate at which the product is formed, $dp/dt = k_2 c$, is fastest when $c = e_0$, in which case $dp/dt = k_2 e_0$. We can therefore interpret $k_2 e_0$ as the maximum rate at which this reaction can proceed. We introduce the notation

$$v_m = k_2 e_0$$

and rewrite (11.75) as

$$\frac{dp}{dt} = \frac{v_m s}{K_m + s} \tag{11.76}$$

Equation (11.76) is known as the Michaelis-Menten law and describes the velocity of an enzymatic reaction. We see from Equation (11.76) that the reaction rate dp/dt is limited by the availability of the substrate S.

Equations (11.76) and (11.71) are the same. Whereas (11.71) was derived from fitting a curve to data points that related the measured substrate concentration s to the velocity of the reaction dp/dt, (11.76) is derived from a mathematical model. The mathematical model allows us to interpret the constants v_m and K_m in terms of the enzymatic reaction, and the experiments allow us to measure v_m and K_m.

Microbial Growth in a Chemostat: An Application of Substrate-Limited Growth   A rather simplistic view of microbial growth is that microbes convert substrate through enzymatic reactions into products that are then converted into microbial biomass.

In the following, we will investigate a mathematical model for microbial growth in a chemostat. The growth of microbes will be limited by the availability of the substrate.

A chemostat is a growth chamber in which sterile medium with concentration s_0 of the substrate enters the chamber at a constant rate D. Air is pumped into the growth chamber to mix and aerate the culture. To keep the volume in the growth chamber constant, the content of the growth chamber is removed at the same rate D as new medium enters. A sketch of a chemostat can be seen in Figure 11.64.

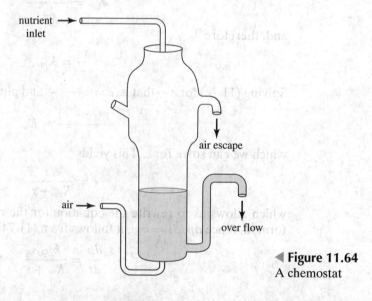

◄ Figure 11.64
A chemostat

We denote the microbial biomass at time t by $x(t)$, and the substrate concentration at time t by $s(t)$. Jacques Lucien Monod was very influential in the development of quantitative microbiology; in (1950) he derived the following system of differential equations to describe the growth of microbes in a chemostat:

$$\frac{ds}{dt} = D(s_0 - s) - q(s)x$$

$$\frac{dx}{dt} = Yq(s)x - Dx \tag{11.77}$$

where $s_0 > 0$ is the substrate concentration of the entering medium, $D > 0$ is the rate at which medium enters or leaves the chemostat, and $Y > 0$ is the yield constant. The function $q(s)$ is the rate at which microbes consume the substrate; this rate depends on the substrate concentration. The yield factor Y can thus be interpreted as a conversion factor of substrate into biomass.

Monod (1942) showed empirically that the uptake rate $q(s)$ fits the hyperbolic relationship

$$q(s) = \frac{v_m s}{K_m + s} \tag{11.78}$$

where v_m is the saturation level and K_m is the half-saturation constant (that is, $q(K_m) = v_m/2$). A graph of $q(s)$ is shown in Figure 11.65. It later occurred to Monod that (11.78) is identical to the Michaelis-Menten law (11.76), which might suggest that microbial growth is governed by enzymatic reactions.

In the following, we will determine possible equilibria of (11.77) and analyze their stability. There is always, of course, the trivial equilibrium, which is obtained when substrate enters a growth chamber void of microbes [that is, when $x(0) = 0$]. In that case, there will be no microbes at later times and, hence, $\frac{dx}{dt} = 0$ for all times $t \geq 0$. The substrate equilibrium is then found by setting $\frac{ds}{dt} = 0$ with $x = 0$:

$$0 = D(s_0 - s)$$

which has solution $s = s_0$. Hence, one equilibrium is

$$(\hat{s}_1, \hat{x}_1) = (s_0, 0) \tag{11.79}$$

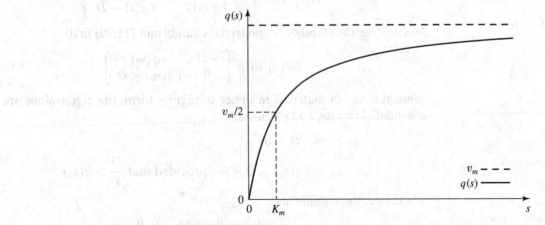

▲ **Figure 11.65**
A graph of $q(s)$

To obtain a nontrivial equilibrium $(\hat{s}_2, \hat{x}_2)$, we will look for an equilibrium with $\hat{x}_2 > 0$. To find this, we solve the simultaneous equations

$$\frac{ds}{dt} = 0 \quad \text{and} \quad \frac{dx}{dt} = 0$$

It follows from $dx/dt = 0$ that

$$q(\hat{s}_2) = \frac{D}{Y} \tag{11.80}$$

We see immediately from Figure 11.65 that (11.80) has a solution $\hat{s}_2 > 0$ if $0 < D/Y < v_m$. Using (11.78) in (11.80), we can compute $\hat{s}_2$, namely,

$$\frac{v_m \hat{s}_2}{K_m + \hat{s}_2} = \frac{D}{Y}$$

which yields

$$\hat{s}_2 = \frac{D K_m}{Y v_m - D} \tag{11.81}$$

which is indeed positive, provided that $0 < D/Y < v_m$. Using (11.80) in $ds/dt = 0$, we find

$$D(s_0 - \hat{s}_2) = \frac{D}{Y} \hat{x}_2$$

or

$$\hat{x}_2 = Y(s_0 - \hat{s}_2) \tag{11.82}$$

from which we can see that $\hat{x}_2 > 0$, provided that $\hat{s}_2 < s_0$. Under the assumption that $0 < D/Y < v_m$ and $\hat{s}_2 < s_0$, we therefore have a nontrivial equilibrium

$$(\hat{s}_2, \hat{x}_2) = \left(\frac{D K_m}{Y v_m - D}, Y(s_0 - \hat{s}_2) \right) \tag{11.83}$$

There are no other equilibria.

To analyze the stability of the two equilibria (11.79) and (11.83), we find the Jacobi matrix $D\mathbf{f}$ associated with the system (11.77),

$$D\mathbf{f}(s, x) = \begin{bmatrix} -D - q'(s)x & -q(s) \\ Y q'(s)x & Y q(s) - D \end{bmatrix}$$

We analyze the stability of the trivial equilibrium (11.79) first:

$$D\mathbf{f}(s_0, 0) = \begin{bmatrix} -D & -q(s_0) \\ 0 & Y q(s_0) - D \end{bmatrix}$$

Since the Jacobi matrix is in upper triangular form, the eigenvalues are the diagonal elements, and we find

$$\lambda_1 = -D < 0$$

$$\lambda_2 = Y q(s_0) - D < 0 \quad \text{provided that} \quad \frac{D}{Y} > q(s_0)$$

Therefore, the equilibrium

$$(s_0, 0) \quad \text{is} \quad \begin{cases} \text{locally stable} & \text{if } \frac{D}{Y} > q(s_0) \\ \\ \text{unstable} & \text{if } \frac{D}{Y} < q(s_0) \end{cases}$$

For the nontrivial equilibrium (11.83), we find

$$A = D\mathbf{f}(\hat{s}_2, \hat{x}_2) = \begin{bmatrix} -D - q'(\hat{s}_2)\hat{x}_2 & -q(\hat{s}_2) \\ Yq'(\hat{s}_2)\hat{x}_2 & Yq(\hat{s}_2) - D \end{bmatrix}$$

Using (11.80), this simplifies to

$$A = \begin{bmatrix} -D - q'(\hat{s}_2)\hat{x}_2 & -\frac{D}{Y} \\ Yq'(\hat{s}_2)\hat{x}_2 & 0 \end{bmatrix}$$

Now,

$$\text{tr } A = -D - q'(\hat{s}_2)\hat{x}_2 < 0$$

and

$$\det A = Dq'(\hat{s}_2)\hat{x}_2 > 0$$

for $\hat{x}_2 > 0$, since $q(s)$ is an increasing function. Therefore, if the nontrivial equilibrium exist—that is, if both $\hat{s}_2 > 0$ and $\hat{x}_2 > 0$—then it is locally stable.

We saw that the nontrivial equilibrium exists, provided that $0 < D/Y < v_m$ and $s_0 > \hat{s}_2$. These two conditions can be summarized as

$$0 < \frac{D}{Y} < v_m \quad \text{and} \quad \frac{DK_m}{Yv_m - D} < s_0$$

If the first inequality holds, then the denominator in the second inequality is positive. Solving the second inequality for D, we then find

$$D < \frac{Ys_0v_m}{s_0 + K_m} = Yq(s_0) \tag{11.84}$$

We can now summarize our results. The chemostat has two equilibria, a trivial one in which microbes are absent, and a nontrivial one that allows stable microbial growth. If $D > Yq(s_0)$, then the trivial equilibrium is the only biologically reasonable equilibrium and it is locally stable. If $D < Yq(s_0)$, both equilibria are biologically reasonable; the trivial one is now unstable and the nontrivial is the locally stable one. Stable microbial growth is therefore possible, provided that the rate at which medium enters and leaves the growth chamber is between 0 and $Yq(s_0)$.

11.4.6 Problems

(11.4.1)

1. Suppose that the densities of two species evolve according to the Lotka-Volterra model of interspecific competition. Assume that species 1 has intrinsic rate of growth $r_1 = 2$ and carrying capacity $K_1 = 20$, and that species 2 has intrinsic rate of growth $r_2 = 3$ and carrying capacity $K_2 = 15$. Furthermore, assume that 20 individuals of species 2 have the same effect on species 1 as 4 individuals of species 1 have on themselves, and 30 individuals of species 1 have the same effect on species 2 as 6 individuals of species 2 have on themselves. Find a system of differential equations that describes this situation.

2. Suppose the densities of two species evolve according to the Lotka-Volterra model of interspecific competition. Assume that species 1 has intrinsic rate of growth $r_1 = 4$ and carrying capacity $K_1 = 17$, and that species 2 has intrinsic rate of growth $r_2 = 1.5$ and carrying capacity $K_2 = 32$. Furthermore, assume that 15 individuals of species 2 have the same

effect on species 1 as 7 individuals of species 1 have on themselves, and 5 individuals of species 1 have the same effect on species 2 as 7 individuals of species 2 have on themselves. Find a system of differential equations that describes this situation.

In Problems 3–6, use the graphical approach to classify the following Lotka-Volterra models of interspecific competition according to "coexistence," "founder control," "species 1 excludes species 2," or "species 2 excludes species 1."

3.
$$\frac{dN_1}{dt} = 2N_1\left(1 - \frac{N_1}{10} - 0.7\frac{N_2}{10}\right)$$
$$\frac{dN_2}{dt} = 5N_2\left(1 - \frac{N_2}{15} - 0.3\frac{N_1}{15}\right)$$

4.
$$\frac{dN_1}{dt} = 3N_1\left(1 - \frac{N_1}{50} - 0.3\frac{N_2}{50}\right)$$
$$\frac{dN_2}{dt} = 4N_2\left(1 - \frac{N_2}{30} - 0.8\frac{N_1}{30}\right)$$

5.
$$\frac{dN_1}{dt} = N_1\left(1 - \frac{N_1}{20} - \frac{N_2}{5}\right)$$
$$\frac{dN_2}{dt} = 2N_2\left(1 - \frac{N_2}{15} - \frac{N_1}{3}\right)$$

6.
$$\frac{dN_1}{dt} = 3N_1\left(1 - \frac{N_1}{25} - 1.2\frac{N_2}{25}\right)$$
$$\frac{dN_2}{dt} = N_2\left(1 - \frac{N_2}{30} - 0.8\frac{N_1}{30}\right)$$

In Problems 7–10, use the eigenvalue approach to analyze all equilibria of the given Lotka-Volterra models of interspecific competition.

7.
$$\frac{dN_1}{dt} = 3N_1\left(1 - \frac{N_1}{18} - 1.3\frac{N_2}{18}\right)$$
$$\frac{dN_2}{dt} = 2N_2\left(1 - \frac{N_2}{20} - 0.6\frac{N_1}{20}\right)$$

8.
$$\frac{dN_1}{dt} = 4N_1\left(1 - \frac{N_1}{12} - 0.3\frac{N_2}{12}\right)$$
$$\frac{dN_2}{dt} = 5N_2\left(1 - \frac{N_2}{15} - 0.2\frac{N_1}{15}\right)$$

9.
$$\frac{dN_1}{dt} = N_1\left(1 - \frac{N_1}{35} - 3\frac{N_2}{35}\right)$$
$$\frac{dN_2}{dt} = 3N_2\left(1 - \frac{N_2}{40} - 4\frac{N_1}{40}\right)$$

10.
$$\frac{dN_1}{dt} = N_1\left(1 - \frac{N_1}{25} - 0.1\frac{N_2}{25}\right)$$
$$\frac{dN_2}{dt} = N_2\left(1 - \frac{N_2}{28} - 1.2\frac{N_1}{28}\right)$$

11. Suppose that two species of beetles are reared together and separately. When species 1 is reared alone, it reaches an equilibrium of about 200. When species 2 is reared alone, it reaches an equilibrium of about 150. When both of them are reared together, they seem to be able to coexist; species 1 reaches an equilibrium of about 180 and species 2 reaches an equilibrium of about 80. If their densities follow the Lotka-Volterra equation of interspecific competition, find α_{12} and α_{21}.

12. Suppose that two species of beetles are reared together. Species 1 wins if there are initially 100 individuals of species 1 and 20 individuals of species 2. But species 2 wins if there are initially 20 individuals of species 1 and 100 individuals of species 2. When the beetles are reared separately, both species seem to reach an equilibrium of about 120. Based on this information and assuming that the densities follow the Lotka-Volterra model of interspecific competition, can you give lower bounds on α_{12} and α_{21}?

(11.4.2)

In Problems 13–14, use a graphing calculator to sketch solution curves of the given Lotka-Volterra predator-prey model in the $N - P$ plane. Also graph $N(t)$ and $P(t)$ as functions of t.

13.
$$\frac{dN}{dt} = 2N - PN$$
$$\frac{dP}{dt} = \frac{1}{2}PN - P$$

with initial conditions

(a) $(N(0), P(0)) = (2, 2)$

(b) $(N(0), P(0)) = (3, 3)$

(c) $(N(0), P(0)) = (4, 4)$

14.
$$\frac{dN}{dt} = 3N - 2PN$$
$$\frac{dP}{dt} = PN - P$$

with initial conditions

(a) $(N(0), P(0)) = (1, 3/2)$

(b) $(N(0), P(0)) = (2, 2)$

(c) $(N(0), P(0)) = (3, 1)$

In Problems 15–16, we investigate Lotka-Volterra predator-prey model.

15. Assume that
$$\frac{dN}{dt} = N - 4PN$$
$$\frac{dP}{dt} = 2PN - 3P$$

(a) Show that this system has two equilibria, the trivial equilibrium $(0, 0)$ and a nontrivial one, in which both species have positive densities.

(b) Use the eigenvalue approach to show that the trivial equilibrium is unstable.

(c) Determine the eigenvalues corresponding to the nontrivial equilibrium. Does your analysis allow you to infer anything about its stability?

(d) Use a graphing calculator to sketch curves in the $N - P$ plane. Also sketch solution curves of the prey and the predator densities as functions of time.

16. Assume that
$$\frac{dN}{dt} = 5N - PN$$
$$\frac{dP}{dt} = PN - P$$

(a) Show that this system has two equilibria, the trivial equilibrium $(0, 0)$ and a nontrivial one, in which both species have positive densities.

(b) Use the eigenvalue approach to show that the trivial equilibrium is unstable.

(c) Determine the eigenvalues corresponding to the nontrivial equilibrium. Does your analysis allow you to infer anything about its stability?

(d) Use a graphing calculator to sketch curves in the $N - P$ plane. Also sketch solution curves of the prey and the predator densities as functions of time.

17. Assume that $N(t)$ denotes the density of an insect species at time t and $P(t)$ denotes the density of its predator at time t. The insect species is an agricultural pest and its predator is used as a biological control agent. Their dynamics are given by the system of differential equations
$$\frac{dN}{dt} = 5N - 3PN$$
$$\frac{dP}{dt} = 2PN - P$$

(a) Explain why

$$\frac{dN}{dt} = 5N \tag{11.85}$$

describes the dynamics of the insect in the absence of the predator. Solve (11.85). Describe what happens to the insect population in the absence of the predator.

(b) Explain why introducing the insect predator into the system can help to control the density of the insect.

(c) Assume that at the beginning of the growing season the insect density is 0.5 and the predator density is 2. You decide to control the insects by using an insecticide in addition to the predator. You are careful, and choose an insecticide that does not harm the predator. After spraying, the insect density drops to 0.01 and the predator density remains at 2. Use a graphing calculator to investigate the long-term implications of your decision to spray the field. In particular, investigate what would have happened to the insect densities if you had decided not to spray the field, and compare your results to the insect density over time that results from your application of the insecticide.

18. Assume that $N(t)$ denotes prey density at time t and $P(t)$ denotes predator density at time t. Their dynamics are given by the system of equations

$$\frac{dN}{dt} = 4N - 2PN$$
$$\frac{dP}{dt} = PN - 3P$$

Assume that initially $N(0) = 3$ and $P(0) = 2$.

(a) If you followed this predator-prey community over time, what would you observe?

(b) Assume that bad weather kills 90% of the prey population and 67% of the predator population. If you continued to observe this predator-prey community, what would you expect to see?

19. An unrealistic feature of the Lotka-Volterra model is that the prey exhibits unlimited growth in the absence of the predator. The model described by the following system remedies this. (Namely, we assume that the prey evolves according to logistic growth in the absence of the predator. The other model features are retained.)

$$\frac{dN}{dt} = 3N\left(1 - \frac{N}{10}\right) - 2PN$$
$$\frac{dP}{dt} = PN - 4P \tag{11.86}$$

(a) Explain why the prey evolves according to

$$\frac{dN}{dt} = 3N\left(1 - \frac{N}{10}\right) \tag{11.87}$$

in the absence of the predator. Investigate the long term behavior of solutions to (11.87).

(b) Find all equilibria of (11.86), and determine their stability using the eigenvalue approach.

(c) Use a graphing calculator to sketch the solution curve of (11.86) in the $N - P$ plane when $N(0) = 2$ and $P(0) = 2$. Also, sketch $N(t)$ and $P(t)$ as functions of time, starting with $N(0) = 2$ and $P(0) = 2$.

20. An unrealistic feature of the Lotka-Volterra model is that the prey exhibits unlimited growth in the absence of the predator. The model described by the following system remedies this. (Namely, we assume that the prey evolves according to logistic growth in the absence of the predator. The other model features are retained.)

$$\frac{dN}{dt} = N\left(1 - \frac{N}{K}\right) - 4PN$$
$$\frac{dP}{dt} = PN - 5P \tag{11.88}$$

where $K > 0$ denotes the carrying capacity of the prey in the absence of the predator. In the following, we will investigate how the carrying capacity affects the outcome of this predator-prey interaction.

(a) Draw the zero isoclines of (11.88) for (i) $K = 10$, and (ii) $K = 3$.

(b) When $K = 10$, the zero isoclines intersect, indicating the existence of a nontrivial equilibrium. Analyze the stability of this nontrivial equilibrium.

(c) Is there a minimum carrying capacity required in order to have a nontrivial equilibrium? If yes, find it and explain what happens when the carrying capacity is below this minimum, and what happens when the carrying capacity is above this minimum.

21. An unrealistic feature of the Lotka-Volterra model is that the prey exhibits unlimited growth in the absence of the predator. The model described by the following system remedies this. (Namely, we assume that the prey evolves according to logistic growth in the absence of the predator. The other model features are retained.)

$$\frac{dN}{dt} = N\left(1 - \frac{N}{20}\right) - 5PN$$
$$\frac{dP}{dt} = 2PN - 8P \tag{11.89}$$

(a) Draw the zero isoclines of (11.89).

(b) Use the graphical approach of Subsection 11.3.2 to determine whether the nontrivial equilibrium is locally stable.

In Problems 22–26, we will analyze how a change in parameters in the modified Lotka-Volterra predator-prey model

$$\frac{dN}{dt} = aN\left(1 - \frac{N}{K}\right) - bPN$$
$$\frac{dP}{dt} = cPN - dP \tag{11.90}$$

affects predator-prey interactions

22. (a) Find the zero-isoclines of (11.90), and determine conditions under which a nontrivial equilibrium exists (that is, an equilibrium in which both prey and predator have positive densities).

(b) Use the graphical approach of Subsection 11.3.2 to show that if a nontrivial equilibrium exists, it is locally stable.

In Problems 23–26, we use the results of Problem 22. Assume that the parameters are chosen so that a nontrivial equilibrium exists.

23. Use the results of Problem 22 to show that an increase in a (the intrinsic rate of growth of the prey) results in an increase of the predator density but leaves the prey density unchanged.

24. Use the results of Problem 22 to show that an increase in b (the searching efficiency) reduces the predator density but has no effect on the equilibrium abundance of the prey.

25. Use the results of Problem 22 to show that an increase in c (the predator growth efficiency) reduces the prey equilibrium abundance and increases the predator equilibrium abundance.

26. Use the results of Problem 22 to show that an increase in K (the prey carrying capacity in the absence of the predator) increases the predator equilibrium abundance but has no effect on the prey equilibrium abundance.

(11.4.3)

In Problems 27–34, classify each community matrix at equilibrium according to the five cases considered in Subsection 11.4.3, and determine whether the equilibrium is stable. (Assume in each case that the equilibrium exists.)

27. $\begin{bmatrix} -1 & -1.3 \\ 0.3 & -2 \end{bmatrix}$

28. $\begin{bmatrix} -3 & -1.2 \\ -1 & -2 \end{bmatrix}$

29. $\begin{bmatrix} -1.5 & 1.6 \\ 2.3 & -5.1 \end{bmatrix}$

30. $\begin{bmatrix} -0.3 & 0 \\ 0.4 & -0.7 \end{bmatrix}$

31. $\begin{bmatrix} -1 & 1.3 \\ 2 & -1.5 \end{bmatrix}$

32. $\begin{bmatrix} -2.7 & 0 \\ -1.3 & -0.6 \end{bmatrix}$

33. $\begin{bmatrix} -5 & -1.7 \\ -2.3 & -0.2 \end{bmatrix}$

34. $\begin{bmatrix} -2.3 & -4.7 \\ 1.2 & -3.2 \end{bmatrix}$

In Problems 35–40, we consider communities composed of two species. The abundance of species 1 at time t is given by $N_1(t)$, the abundance of species 2 at time t is given by $N_2(t)$. Their dynamics are described by

$$\frac{dN_1}{dt} = f_1(N_1, N_2)$$
$$\frac{dN_2}{dt} = f_2(N_1, N_2)$$

Assume that when both species are at low abundances, their abundances increase and that f_1 and f_2 change sign when crossing their zero isoclines. In each problem, determine the sign structure of the community matrix at the nontrivial equilibrium (indicated by a dot), based on the graph of the zero isoclines. Determine the stability of the equilibria if possible.

35. See Figure 11.66.

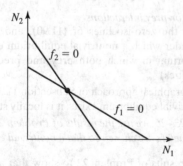

▲ **Figure 11.66**

36. See Figure 11.67.

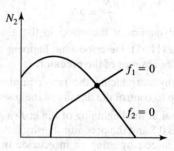

▲ **Figure 11.67**

37. See Figure 11.68.

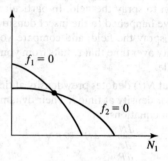

▲ **Figure 11.68**

38. See Figure 11.69.

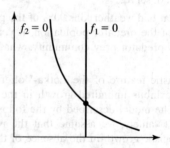

▲ **Figure 11.69**

39. See Figure 11.70.

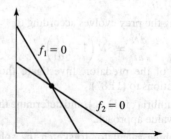

▲ **Figure 11.70**

40. See Figure 11.71.

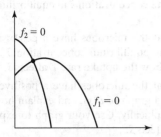

▲ **Figure 11.71**

41. Suppose that the diagonal elements a_{ii} of the community matrix of a species assemblage in equilibrium are negative. Explain why this implies that species i exhibits self-regulation.

42. Consider a community composed of two species. Assume that both species inhibit themselves. Explain why mutualistic and competitive interactions lead to qualitatively similar predictions about the stability of the corresponding equilibria. That is, show that if $A = [a_{ij}]$ is the community matrix at equilibrium for the case of mutualism, and if $B = [b_{ij}]$ is the community matrix at equilibrium for the case of competition, then the following holds: If $|a_{ij}| = |b_{ij}|$ for $1 \leq i, j \leq 2$, then either both equilibria are locally stable or both are unstable.

43. The classical Lotka-Volterra model of predation is given by

$$\frac{dN}{dt} = aN - bNP$$

$$\frac{dP}{dt} = cNP - dP$$

where $N = N(t)$ is the prey density at time t and $P = P(t)$ is the predator density at time t. The constants $a, b, c,$ and d are all positive.

(a) Find the nontrivial equilibrium $(\hat{N}, \hat{P})$ with $\hat{N} > 0$ and $\hat{P} > 0$.

(b) Find the community matrix corresponding to the nontrivial equilibrium.

(c) Explain each entry of the community matrix found in (b) in terms of how individuals in this community affect each other.

44. The modified Lotka-Volterra model of predation is given by

$$\frac{dN}{dt} = aN\left(1 - \frac{N}{K}\right) - bNP$$

$$\frac{dP}{dt} = cNP - dP$$

where $N = N(t)$ is the prey density at time t and $P = P(t)$ is the predator density at time t. The constants $a, b, c, d,$ and K are positive. Furthermore, assume that $d/c < K$.

(a) Find the nontrivial equilibrium $(\hat{N}, \hat{P})$ with $\hat{N} > 0$ and $\hat{P} > 0$.

(b) Find the community matrix corresponding to the nontrivial equilibrium.

(c) Explain each entry of the community matrix found in (b) in terms of how individuals in this community affect each other.

(11.4.4)

45. Use a graphing calculator to study the following example of the Fitzhugh-Nagumo model.

$$\frac{dV}{dt} = -V(V - 0.3)(V - 1) - w$$

$$\frac{dw}{dt} = 0.01(V - 0.4w)$$

Sketch the graph of the solution curve in the V-w plane when (i) $(V(0), w(0)) = (0.4, 0)$, and (ii) $(V(0), w(0)) = (0.2, 0)$.

46. Use a graphing calculator to study the following example of the Fitzhugh-Nagumo model.

$$\frac{dV}{dt} = -V(V - 0.6)(V - 1) - w$$

$$\frac{dw}{dt} = 0.03(V - 0.6w)$$

Sketch the graph of the solution curve in the V-w plane when (i) $(V(0), w(0)) = (0.8, 0)$, and (ii) $(V(0), w(0)) = (0.4, 0)$.

47. Assume the following example of the Fitzhugh-Nagumo model.

$$\frac{dV}{dt} = -V(V - 0.3)(V - 1) - w$$

$$\frac{dw}{dt} = 0.01(V - 0.4w)$$

Assume that $w(0) = 0$. For which initial values of $V(0)$ can you observe an action potential?

48. Assume the following example of the Fitzhugh-Nagumo model.

$$\frac{dV}{dt} = -V(V - 0.6)(V - 1) - w$$

$$\frac{dw}{dt} = 0.03(V - 0.6w)$$

Assume that $w(0) = 0$. For which initial values of $V(0)$ can you observe an action potential?

(11.4.5)

In Problems 49–52, use the mass action law to translate each chemical reaction into a system of differential equations.

49.
$$A + B \xrightarrow{k} C$$

50.
$$A + B \underset{k_-}{\overset{k_+}{\rightleftarrows}} C$$

51.
$$E + S \xrightarrow{k_1} ES \xrightarrow{k_2} E + P$$

52.
$$A + B \xrightarrow{k} A + C$$

53. Show that the following system of differential equations has a conserved quantity, and find it.

$$\frac{dx}{dt} = 2x - 3y$$

$$\frac{dy}{dt} = 3y - 2x$$

54. Show that the following system of differential equations has a conserved quantity, and find it.

$$\frac{dx}{dt} = -4x + 2y$$

$$\frac{dy}{dt} = -y + 2x$$

55. Show that the following system of differential equations has a conserved quantity, and find it.

$$\frac{dx}{dt} = -x + 2xy + z$$

$$\frac{dy}{dt} = -2xy$$

$$\frac{dz}{dt} = x - z$$

56. Suppose that $x(t) + y(t)$ is a conserved quantity. If

$$\frac{dx}{dt} = -3x + 2xy$$

find the differential equation for $y(t)$.

57. The Michaelis-Menten law (equation (11.76)) states that

$$\frac{dp}{dt} = \frac{v_m s}{K_m + s}$$

where $p = p(t)$ is the concentration of the product of the enzymatic reaction at time t, $s = s(t)$ is the concentration of the substrate at time t, and v_m and K_m are positive constants. Set

$$f(s) = \frac{v_m s}{K_m + s}$$

where v_m and K_m are positive constants.

(a) Show that

$$\lim_{s \to \infty} f(s) = v_m$$

(b) Show that

$$f(K_m) = \frac{v_m}{2}$$

(c) Show that for $s \geq 0$, $f(s)$ is (i) nonnegative, (ii) increasing, and (iii) concave down. Sketch a graph of $f(s)$. Label v_m and K_m on your graph.

(d) Explain why we said that the reaction rate dp/dt is limited by the availability of the substrate.

58. The growth of microbes in a chemostat was described by equation (11.77). Using the notation of (11.77) with

$$q(s) = \frac{v_m s}{K_m + s}$$

where v_m and K_m are positive constants, we will investigate how the substrate concentration $\hat{s}$ in equilibrium depends on the uptake rate Y.

(a) Assume that the microbes have a positive equilibrium density. Find the equilibrium concentration $\hat{s}$ algebraically, and investigate how the uptake rate Y affects $\hat{s}$.

(b) Assume that the microbes have a positive equilibrium density. Sketch a graph of $q(s)$, and explain how you would determine $\hat{s}$ graphically. Use your graph to explain how the uptake rate Y affects $\hat{s}$.

59. The growth of microbes in a chemostat was described by equation (11.77). Using the notation of (11.77) with

$$q(s) = \frac{v_m s}{K_m + s}$$

we will investigate how the substrate concentration $\hat{s}$ in equilibrium depends on D, the rate at which the medium enters the chemostat.

(a) Assume that the microbes have a positive equilibrium density. Find the equilibrium concentration $\hat{s}$ algebraically. Investigate how the rate D affects $\hat{s}$.

(b) Assume that the microbes have a positive equilibrium density. Sketch a graph of $q(s)$, and explain how you would determine $\hat{s}$ graphically. Use your graph to explain how the rate D affects $\hat{s}$.

In Problems 60–61, we investigate specific examples of microbial growth described by Equation (11.77). We use the notation of Subsection 11.4.5. In each case, determine all equilibria and their stability.

60.

$$\frac{ds}{dt} = 2(4 - s) - \frac{3s}{2 + s}x$$

$$\frac{dx}{dt} = \frac{sx}{2 + s} - 2x$$

61.

$$\frac{ds}{dt} = 2(4 - s) - \frac{3s}{1 + s}x$$

$$\frac{dx}{dt} = \frac{3sx}{1 + s} - 2x$$

11.5 Key Terms

Chapter 11 Review: Topics *Discuss the following definitions and concepts.*

1. Linear first-order equation
2. Homogeneous
3. Direction field, slope field, direction vector
4. Solution of a system of linear differential equations
5. Eigenvalue, eigenvector
6. Superposition principle
7. General solution
8. Stability
9. Sink or stable node
10. Saddle
11. Source or unstable node
12. Spiral
13. Euler's formula
14. Compartment model
15. Conserved quantity
16. Harmonic oscillator
17. Nonlinear autonomous system of differential equations
18. Critical point
19. Zero isoclines or null clines
20. Graphical approach to stability
21. Lotka-Volterra model of interspecific competition
22. Intraspecific competition, interspecific competition
23. Competitive exclusion, founder control, coexistence
24. Lotka-Volterra predator-prey model

25. Community matrix

26. Fitzhugh-Nagumo model

27. Action potential

28. Michaelis-Menten law

11.6 Review Problems

1. Let $N_1(t)$ and $N_2(t)$ denote the respective sizes of two populations at time t, and assume that their dynamics are given by

$$\frac{dN_1}{dt} = r_1 N_1$$

$$\frac{dN_2}{dt} = r_2 N_2$$

where r_1 and r_2 are positive constants, denoting the intrinsic rate of growth of the two populations. Set $Z(t) = N_1(t)/N_2(t)$, and show that $Z(t)$ satisfies

$$\frac{d}{dt} \ln Z(t) = r_1 - r_2 \qquad (11.91)$$

Solve (11.91), and show that $\lim_{t\to\infty} Z(t) = \infty$ if $r_1 > r_2$. Conclude from this that population 1 becomes numerically dominant when $r_1 > r_2$.

2. Let $N_1(t)$ and $N_2(t)$ denote the respective sizes of two populations at time t, and assume that their dynamics are given by

$$\frac{dN_1}{dt} = r_1 N_1$$

$$\frac{dN_2}{dt} = r_2 N_2$$

where r_1 and r_2 are positive constants, denoting the intrinsic rate of growth of the two populations. Denote the combined population size at time t by $N(t)$; that is, $N(t) = N_1(t) + N_2(t)$. Define the relative proportions

$$p_1 = \frac{N_1}{N} \quad \text{and} \quad p_2 = \frac{N_2}{N}$$

Use the fact that $p_1/p_2 = N_1/N_2$ to show that

$$\frac{dp_1}{dt} = p_1(1 - p_1)(r_1 - r_2)$$

Show that if $r_1 > r_2$ and $0 < p_1(0) < 1$, $p_1(t)$ will increase for $t > 0$, and population 1 will become numerically dominant.

3. An unrealistic feature of the Lotka-Volterra model is that the prey exhibits unlimited growth in the absence of the predator. The model described by the following system remedies this. (Namely, we assume that the prey evolves according to logistic growth in the absence of the predator. The other model features are retained.)

$$\frac{dN}{dt} = 2N \left(1 - \frac{N}{10}\right) - 3PN$$

$$\frac{dP}{dt} = PN - 3P \qquad (11.92)$$

(a) Draw the zero isoclines of (11.92).

(b) Use the graphical approach of Subsection 11.3.2 to determine whether the nontrivial equilibrium is locally stable.

4. Tilman (1982) developed a theoretical framework for studying resource competition in plants. In its simplest form, one species competes for a single resource, for instance, nitrogen. If $B(t)$ denotes the total biomass at time t and

$R(t)$ is the amount of resource available at time t, then the dynamics are described by the following system of differential equations.

$$\frac{dB}{dt} = B[f(R) - m]$$

$$\frac{dR}{dt} = a(S - R) - cBf(R)$$

The first equation describes the rate of biomass change, the function $f(R)$ describes how the species growth rate depends on the resource, and m is the specific loss rate. The second equation describes the resource dynamics; the constant S is the maximal amount of resource in a given habitat. The rate of resource supply (dR/dt) is assumed to be proportional to the difference between the current resource level and the maximal amount of resource; the constant a is the constant of proportionality. The term $cBf(R)$ describes the resource uptake by the plants; the constant c can be considered as a conversion factor.

We assume in the following that $f(R)$ follows the Monod growth function

$$f(R) = \frac{dR}{k + R}$$

where d and k are positive constants.

(a) Find all equilibria. Show that if $d > m$ and $S > mk/(d - m)$, then there exists a nontrivial equilibrium.

(b) Sketch the zero isoclines for the case in which the system admits a nontrivial equilibrium. Use the graphical approach to analyze the stability of the nontrivial equilibrium.

5. The following describes a simple competition model in which two species of plants compete for vacant space. Assume that the entire habitat is divided into a large number of patches. Each patch can be occupied by at most one species. We denote by $p_i(t)$ the fraction of patches occupied by species i. Note that $0 \le p_1(t) + p_2(t) \le 1$. The dynamics are described by

$$\frac{dp_1}{dt} = c_1 p_1(1 - p_1 - p_2) - m_1 p_1$$

$$\frac{dp_1}{dt} = c_2 p_2(1 - p_1 - p_2) - m_2 p_2$$

where c_1, c_2, m_1, and m_2 are positive constants. The first term on the right-hand side of each equation describes the colonization of vacant patches; the second term on the right-hand side of each equation describes how occupied patches become vacant.

(a) Show that the dynamics of species 1 in the absence of species 2 are given by

$$\frac{dp_1}{dt} = c_1 p_1(1 - p_1) - m_1 p_1 \qquad (11.93)$$

and find conditions on c_1 and m_1 so that (11.93) admits a nontrivial equilibrium (an equilibrium in which $0 < p_1 \leq 1$).

(b) Assume now that $c_1 > m_1$ and $c_2 > m_2$. Show that if

$$\frac{c_1}{m_1} > \frac{c_2}{m_2}$$

then species 1 will exclude species 2 if species 1 initially occupies a positive fraction of the patches.

6. (*Paradox of enrichment*) Rosenzweig (1971) analyzed a number of predator-prey models and concluded that enriching the system by increasing the nutrient supply destabilizes the nontrivial equilibrium. We will think of the predator-prey model as a plant-herbivore system, in which plants represent prey and herbivores represent predators. The models analyzed were of the form

$$\frac{dN}{dt} = f(N, P) \qquad (11.94)$$

$$\frac{dP}{dt} = g(N, P) \qquad (11.95)$$

where $N = N(t)$ is the plant abundance at time t and $P = P(t)$ is the herbivore abundance at time t. The models all shared the property that the zero isocline for the herbivore was a vertical line and the zero isocline for the plants was a hump-shaped curve. We will look at one of the models, namely

$$\frac{dN}{dt} = aN\left(1 - \frac{N}{K}\right) - bP\left(1 - e^{-rN}\right)$$

$$\frac{dP}{dt} = cP\left(1 - e^{-rN}\right) - dP \qquad (11.96)$$

(a) Find the zero isoclines for (11.96), and show that (i) the zero isocline of the herbivore ($dP/dt = 0$) is a vertical line in the N-P plane, and (ii) the zero isocline for the plants ($dN/dt = 0$) intersects the N-axis at $N = K$.

(b) Plot the zero isoclines in the N-P plane for $a = b = c = r = 1$ and $d = 0.9$, and three levels of the carrying capacity: (i) $K = 1$, (ii) $K = 4$, and (iii) $K = 10$.

(c) For each of the three carrying capacities, determine whether a nontrivial equilibrium exists.

(d) Use the graphical approach of Subsection 11.3.2 to determine the stability of the existing nontrivial equilibria in (c).

(e) Enriching the community could mean increasing the carrying capacity of the plants. For instance, adding nitrogen or phosphorus to plant communities frequently results in an increase in biomass, which can be interpreted as an increase in the carrying capacity of the plants (the K value). Based on your answers in (d), explain why enriching the community (increasing the carrying capacity of the plants) can result in a destabilization of the nontrivial equilibrium. What are the consequences?

7. The growth of microbes in a chemostat was described by Equation (11.77). We will investigate how the microbial abundance in equilibrium depends on the characteristics of the system.

(a) Assume that $q(s)$ is a nonnegative function. Show that the equilibrium abundance of the microbes, $\hat{x}$, is given by

$$\hat{x} = Y(s_0 - \hat{s})$$

where $\hat{s}$ is the substrate equilibrium abundance. When is $\hat{x} > 0$?

(b) Assume now that

$$q(s) = \frac{v_m s}{K_m + s}$$

Investigate how the uptake rate Y, and the rate D at which new medium enters the chemostat, affect the equilibrium abundance of the microbes.

8. (*Successional niche*) Pacala and Rees (1998) discuss a simple mathematical model of competition to explain successional diversity by means of a successional niche mechanism. In this model, two species, an early successional and a late successional, occupy discrete patches. Each patch experiences disturbances (such as fire) at rate D. After a patch is disturbed, both species are present. Over time, however, the late successional species outcompetes the early successional species, which causes the early successional species to become extinct. This change, from a patch that is occupied by both species to a patch that is occupied by the late successional species only, happens at rate a. We keep track of the number of patches occupied by both species at time t, denoted by $x(t)$, and the number of patches occupied by just the late successional species at time t, denoted by $y(t)$. The dynamics are given by the following system of linear differential equations.

$$\frac{dx}{dt} = -ax + Dy$$

$$\frac{dy}{dt} = ax - Dy \qquad (11.97)$$

where a and D are positive constants.

(a) Show that all equilibria are of the form $(x, ax/D)$.

(b) Find the eigenvalues and eigenvectors corresponding to each equilibrium.

(c) Show that

$$\begin{bmatrix} x(t) \\ y(t) \end{bmatrix} = C_1 \begin{bmatrix} u_1 \\ u_2 \end{bmatrix} + C_2 e^{\lambda_2 t} \begin{bmatrix} v_1 \\ v_2 \end{bmatrix}$$

is a solution of (11.97), where $\begin{bmatrix} u_1 \\ u_2 \end{bmatrix}$ is the eigenvector corresponding to the zero eigenvalue and $\begin{bmatrix} v_1 \\ v_2 \end{bmatrix}$ is the eigenvector corresponding to the nonzero eigenvalue λ_2.

(d) Show that $x(t) + y(t)$ does not depend on t. [*Hint:* Show that $\frac{d}{dt}(x(t) + y(t)) = 0$.] Show that the line $x + y = A$ (where A is a constant) is parallel to the line in the direction of the eigenvector corresponding to the nonzero eigenvalue.

(e) Show that the zero isoclines of (11.97) are given by

$$y = \frac{a}{D}x$$

and that this line is the line in the direction of the eigenvector corresponding to the zero eigenvalue.

(f) Suppose now that $x(t) + y(t) = c$, where c is a positive constant. Show that (11.97) can be reduced to just one equation, namely

$$\frac{dx}{dt} = -(a + D)x + Dc$$

Show that $\hat{x} = c\frac{D}{D+a}$ is the only equilibrium, and determine its stability.

C H A P T E R

PROBABILITY AND STATISTICS

C H A P T E R

12

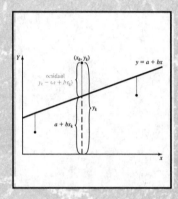

We conclude this book with a chapter on probability and statistics. Though neither field is part of calculus, they both rely on calculus. They provide indispensable tools for life scientists.

Many phenomena in nature are not deterministic. To name just a few examples, the number of eggs laid by a bird, the lifespan of an organism, the inheritance of genes, or the number of people infected during an outbreak of a disease. To deal with the inherent randomness (or stochasticity) of natural phenomena, we need to develop special tools; these tools are supplied by the areas of probability and statistics.

A short description of the role of probability and statistics in the life sciences might be as follows: Probability theory provides tools for modeling randomness and forms the foundation of statistics. Statistics provides tools for analyzing data from scientific experiments.

12.1 COUNTING

It is often necessary to count the ways in which a certain task can be performed. The mathematical field of **combinatorics** deals with such enumeration problems. Frequently, the total number of possible ways is very large, which makes it impractical to write down all possible choices. There are three basic counting principles that will help us to count in a more systematic way; the first is the multiplication principle; the other two follow from it.

12.1.1 The Multiplication Principle

To illustrate the first principle, we look at the following example.

▶ **Example 1**   Imagine that we wish to experimentally manipulate growth conditions for plants, say the grass species big bluestem, *Andropogon gerardi*. We want to grow plants in pots in a greenhouse at two different levels of fertilizer (low and high) and four different temperatures (10°C, 15°C, 20°C, and 25°C). If we want three replicates of each possible combination of fertilizer and temperature treatment, how many pots will we need?

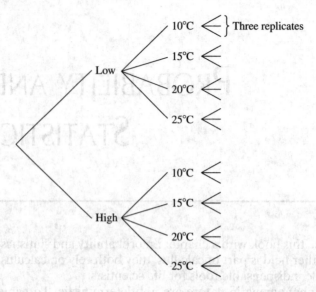

▲ **Figure 12.1**
The tree diagram illustrates how many pots are needed in the experiment described in Example 1

Solution

We can answer this question using a tree diagram, as shown in Figure 12.1. We see from the tree that we will need

$$2 \cdot 4 \cdot 3 = 24$$

pots for our experiment.   ◄

The counting principle that we just used is called the multiplication principle, which we can summarize as follows.

<table>
<tr><td>**MULTIPLICATION PRINCIPLE**</td><td>Suppose that an experiment consists of m ordered tasks. Task 1 has n_1 possible outcomes, task 2 has n_2 possible outcomes, ... , and task m has n_m possible outcomes. The total number of possible outcomes of the experiment is

$$n_1 \cdot n_2 \cdot n_3 \cdots n_m$$</td></tr>
</table>

Looking back at Example 1 of growing big bluestem in pots, we see that the experiment consisted of three tasks: first, to select the fertilizer level; second, to select the temperature; and third, to replicate each combination of fertilizer and temperature three times. The successive tasks are illustrated in the tree diagram, and the total number of pots required for the experiment can be obtained by counting the number of tips of the tree.

We present one more example that illustrates this counting principle.

▶ **Example 2**   Suppose that after a long day in the greenhouse you decide to order pizza. You call a local pizza parlor and learn that there are three choices of crust and five choices of toppings and that you can order the pizza with or without cheese. If you only want one topping, how many different choices do you have for selecting a pizza?

Solution

Your "experiment," which consists of ordering a pizza, involves three tasks. The first task is to choose a crust, the second is to choose the topping, and the third is to decide whether or not you want cheese. Using the multiplication principle, we find that there are

$$3 \cdot 5 \cdot 2 = 30$$

different pizzas that you could order. ◀

12.1.2 Permutations

▶ **Example 3**   Suppose that you grow plants in a greenhouse. To control for spatially varying environmental conditions, you rearrange the pots every other day. If you have six pots arranged in a row on a bench, in how many ways can you arrange the pots?

Solution

To answer this question, imagine that you arrange the pots on the bench from left to right: You have six choices for the leftmost position on the bench, for the next position you can choose any of the remaining five pots, for the third position you can choose any of the remaining four pots, and so on, until there is one pot left that must go to the rightmost position. Using the multiplication principle, we find that there are

$$6 \cdot 5 \cdot 4 \cdot 3 \cdot 2 \cdot 1 = 720$$

ways to arrange the six pots on the bench. ◀

As shorthand notation for the type of descending products of positive integers we encountered in Example 3, we define

$$n! = n(n-1)(n-2)\cdots 3 \cdot 2 \cdot 1$$

and read $n!$ as "n factorial." We can now write $6!$ instead of $6 \cdot 5 \cdot 4 \cdot 3 \cdot 2 \cdot 1$. We define

$$0! = 1$$

Then for $n = 0, 1, 2, \ldots,$

$$(n+1)! = (n+1)n!$$

The quantity $n!$ grows very quickly. Suppose that instead of six pots you have seven; then there are $7! = 5040$ ways to arrange the seven pots. With twelve pots, there are $12! = 479, 001, 600$ possible ways to arrange them.

We will look at another example, and then state a general principle.

▶ **Example 4**   Suppose that a track team has ten sprinters, any four of whom can form a relay team. Assume that each person can run in any position on the team. How many teams can be formed if teams that consist of the same four people in different running orders are considered different teams?

Solution

We select the members of the team in the order in which they run. There are ten available sprinters for the first position. After having chosen a person for the first position, there are nine left, and we can choose any of the nine for the second position. For the third position, we can choose among the eight remaining people and, finally, for the fourth position, we can select a person from the remaining seven. Using the multiplication principle, we find that there are

$$10 \cdot 9 \cdot 8 \cdot 7 = 5040$$

different relay teams.   ◀

In Example 3, we selected $k = 6$ objects from a set of $n = 6$; the order of selection was important. In Example 4, we selected $k = 4$ objects from a set of $n = 10$, where again the order of selection was important. Such selections are called **permutations**. Using the multiplication principle, we can find the number of possible permutations.

PERMUTATIONS

> A **permutation** of n different objects taken k at a time is an *ordered* subset of k out of the n objects. The number of ways that this can be done is denoted by $P(n, k)$, and is given by
>
> $$P(n, k) = n(n - 1)(n - 2) \cdots (n - k + 1)$$

Note that the last term in the product defining $P(n, k)$ is of the form $(n - k + 1)$, because there are k descending factors and the first factor is n.

Returning to Example 3, where we wanted to select six out of six objects in an ordered arrangement, we can now use our permutation rule to compute the number of ways that we can make the selection. Setting $n = 6$ and $k = 6$, we find

$$P(6, 6) = 6 \cdot 5 \cdot 4 \cdot 3 \cdot 2 \cdot 1$$

for the number of different relay teams. Since the product consists of six terms in descending order, starting with 6, the last term is $n - k + 1 = 6 - 6 + 1 = 1$.

Returning to Example 4, where we wanted to select four out of ten objects in an ordered arrangement, we can now use our permutation rule to compute the number of ways that we can make the selection. Setting $n = 10$ and $k = 4$, we find

$$P(10, 4) = 10 \cdot 9 \cdot 8 \cdot 7$$

for the number of different relay teams. Since the product consists of four terms in descending order, starting with 10, the last term is $n - k + 1 = 10 - 4 + 1 = 7$.

Another way to compute $P(n, k)$ follows from the calculation

$$P(n, k) = n(n - 1)(n - 2) \cdot (n - k + 1) \frac{(n - k)(n - k - 1) \cdots 3 \cdot 2 \cdot 1}{(n - k)(n - k - 1) \cdots 3 \cdot 2 \cdot 1}$$

which after simplifying yields

$$P(n, k) = \frac{n!}{(n - k)!} \tag{12.1}$$

Here is one more example before we introduce the third counting principle.

▶ **Example 5**   How many five-letter words with no repeated letters can you form using the 26 letters of the alphabet? (Note that a "word" here need not be in the dictionary.)

Solution

This amounts to choosing 5 letters from 26, where the order is important. Hence there are

$$P(26, 5) = \frac{26!}{21!} = 26 \cdot 25 \cdot 24 \cdot 23 \cdot 22 = 7,893,600$$

different words.   ◄

12.1.3 Combinations

When choosing a permutation, the order of the objects is important. If the order is not important, how can we compute the number of arrangements?

We return to Example 4, where we chose a relay team. The order on the team is important when the members on the team actually run. But if we only wanted to know who was on the team, the order would no longer be important. We saw that there are $10 \cdot 9 \cdot 8 \cdot 7$ different relay teams. But since four people can be arranged in $4! = 4 \cdot 3 \cdot 2 \cdot 1 = 24$ different ways, each choice of four people appears in 24 different teams. If we divide $10 \cdot 9 \cdot 8 \cdot 7$ by $4 \cdot 3 \cdot 2 \cdot 1$, we obtain the number of ways that we can choose four people from a group of ten if the order does not matter. We find that this number is

$$\frac{10 \cdot 9 \cdot 8 \cdot 7}{4 \cdot 3 \cdot 2 \cdot 1} = 210$$

Such unordered selections are called **combinations**. The approach we just used gives us a general formula for the number of combinations, summarized in the following.

COMBINATIONS

A **combination** of n different objects taken k at a time is an *unordered* subset of k out of n objects. The number of ways that this can be done is denoted by $C(n, k)$, and is given by

$$C(n, k) = \frac{n(n-1)(n-2)\cdots(n-k+1)}{k!}$$

Instead of $C(n, k)$, we often write $\binom{n}{k}$, which we read "n choose k." The symbol $\binom{n}{k}$ is called a **binomial coefficient**. Using (12.1), we find

$$C(n, k) = \binom{n}{k} = \frac{P(n, k)}{k!} = \frac{n!}{k!(n-k)!} \tag{12.2}$$

Looking at the rightmost expression, we see that this is equal to $\binom{n}{n-k}$. We therefore find the identity

$$\binom{n}{k} = \binom{n}{n-k} \tag{12.3}$$

The following counting argument also explains this identity. The expression $\binom{n}{k}$ denotes the number of ways that we can select an unordered subset of size k from a set of size n. Instead of choosing the elements that go into the set, we could also choose the elements, that do not go into the set. There are $n - k$ such elements, and we can select them in $\binom{n}{n-k}$ different ways.

Another identity follows from setting $k = 0$ in (12.3), and using (12.2) and $0! = 1$. Namely,

$$\binom{n}{0} = \binom{n}{n} = \frac{n!}{0!n!} = 1$$

This identity can also be understood from the following counting argument. The expression $\binom{n}{0}$ means that we select a subset of size 0 from a set of size n, where the order is not important. There is only one such set, namely, the empty set. Similarly, $\binom{n}{n}$ means that we select a subset of n objects from a set of size n, where the order is not important. There is only one way to do this, namely, we must take the entire set of n objects.

We can use similar reasoning to argue that $\binom{n}{1} = n$; this represents the number of ways that we can choose subsets of size 1, where the order is not important. There are n such subsets, namely, all the singletons. [Actually, the order does not play a role when considering sets with one element, reflected in the fact that $P(n, 1) = n$ as well.]

Here is another example in which we use the rule for counting combinations.

▶ **Example 6** Suppose that you wish to plant five grass species in a plot. You can choose among twelve different species. How many choices do you have?

Solution

Since the order is not important for this selection, there are

$$C(12, 5) = \frac{12!}{5!7!} = \frac{12 \cdot 11 \cdot 10 \cdot 9 \cdot 8}{5 \cdot 4 \cdot 3 \cdot 2 \cdot 1} = 792$$

different ways to make the selection. Alternatively, we could have written

$$C(12, 5) = \binom{12}{5} = \frac{P(12, 5)}{5!}$$

These expressions are equivalent and are evaluated as before. ◀

As a last example in this subsection, we prove the binomial theorem, which we encountered in Chapter 4 when we used the formal definition of derivatives to prove the rule for differentiating the function $f(x) = x^n$, n a positive integer.

▶ **Example 7** Show that if n is a positive integer, then

$$(x + y)^n = \sum_{k=0}^{n} \binom{n}{k} x^k y^{n-k} \tag{12.4}$$

Solution

The term $(x + y)^n$ consists of n factors, each of the form $x + y$. When multiplying out, each factor contributes either an x or a y. Thus the product consists of sums that contain terms of the form $x^k y^{n-k}$ for $k = 0, 1, 2, \ldots, n$. A term of the form $x^k y^{n-k}$ occurs $\binom{n}{k}$ times since there are $\binom{n}{k}$ ways of selecting the factor x k times from among the n factors $(x + y)$. Thus (12.4) follows. ◀

12.1.4 Combining the Counting Principles

Often, the difficult part of counting is to decide which rule to use. To gain experience with this, we discuss several examples in which we combine the three counting principles.

▶ **Example 8** How many different eleven-letter words can be formed using the letters in the word MISSISSIPPI?

Solution

There are four S's, four I's, two P's, and one M. There are 11! ways to arrange the letters, but some of the resulting words will be indistinguishable, since letters that repeat in a word can be swapped without creating a new word. Therefore, we need to divide by the order of the repeated letters. We then find that there are

$$\frac{11!}{4!4!2!1!} = 34,650$$

different words.   ◀

► Example 9 Returning to Example 2, suppose that you want two different toppings on your pizza. How does this change your answer?

Solution

As there are five toppings and the order in which we choose them is not important, we have $\binom{5}{2}$ choices for the toppings. Everything else remains the same, and we find that there are

$$3 \cdot \binom{5}{2} \cdot 2 = 60$$

different pizzas to choose from.   ◀

► Example 10 Suppose that a license plate consists of three letters followed by three digits. How many license plates can there be, if repetition of letters but not of digits is allowed?

Solution

The order is important in this case. For each letter there are 26 choices, since repetition is allowed. There are ten choices for the first digit, nine choices for the second digit, and eight choices for the third digit, since repetition of digits is not allowed. Hence, there are

$$26 \cdot 26 \cdot 26 \cdot 10 \cdot 9 \cdot 8 = 12,654,720$$

different license plates with the above restriction.   ◀

► Example 11 An urn contains six green and four blue balls. You take out three balls at random without replacement. How many different selections contain exactly two green balls and one blue ball? (Assume that the balls are distinguishable.)

Solution

The order in this selection is not important. To obtain two green balls and one blue ball, we select two of the six green and one of the four blue balls and then combine our choices. That is, there are

$$\binom{6}{2}\binom{4}{1} = 60$$

different selections.   ◀

In the preceding example, we explicitly stated that all balls are distinguishable, and we will from now on always assume that the objects we select are distinguishable without explicitly stating this. Our counting principles apply only to distinguishable objects, as assumed in the definitions ("n different objects"). There are cases in physics, where objects are indistinguishable—for instance, electrons—but we will not deal with this here.

▶ **Example 12**   A collection contains seeds for five different annual plants; two produce yellow flowers, and the other three produce blue flowers. You plan a garden bed with three different annual plants from this selection, but you do not want both of the plants with yellow flowers in the bed. How many different selections can you make?

Solution

Possible selections contain either plants with blue flowers only, or one plant with yellow and two with blue flowers. We must choose three different plants out of the available five. There are $\binom{3}{3}\binom{2}{0}$ choices for a flower bed with blue flowers only, and $\binom{3}{2}\binom{2}{1}$ choices for a flower bed with exactly one plant with yellow flowers. Hence, adding up the two cases, there are

$$\binom{3}{3}\binom{2}{0} + \binom{3}{2}\binom{2}{1} = (1)(1) + (3)(2) = 7$$

different selections.

Alternatively, we could have approached this problem in the following way. There are a total of $\binom{5}{3}$ ways to select three plants from the available five. Choices with two plants with yellow flowers and one plant with blue flowers are not acceptable; there are $\binom{3}{1}\binom{2}{2}$ such choices. There are no choices of plants with three yellow and no blue flowers since there are only two plants with yellow flowers. All other choices are acceptable. Hence, there are

$$\binom{5}{3} - \binom{3}{1}\binom{2}{2} = 10 - (3)(1) = 7$$

different selections.   ◀

▶ **Example 13**   A standard deck of cards consists of 52 cards, arranged in four suits, each with 13 different values. In the game of poker, a hand consists of five cards, drawn at random from the deck without replacement.

(a) How many hands are possible?

(b) How many hands consist of exactly one pair (that is, two cards of equal value, the three other cards have different values)?

Solution

(a) There are

$$\binom{52}{5} = 2,598,960$$

different ways of choosing five cards from a deck of 52 cards.

(b) To pick exactly one pair, we first assign the value to the pair (13 ways), then choose their suits [$\binom{4}{2}$ ways]. The three remaining cards all have different values that are different from the pair and from each other [$\binom{12}{3}$ ways]. There are four ways to assign a suit to each card, thus a total of 4^3 ways. Combining the different steps, we find that there are

$$13 \cdot \binom{4}{2}\binom{12}{3} \cdot 4^3 = 1,098,240$$

ways to pick exactly one pair.   ◀

12.1.5 Problems

(12.1.1)

1. Suppose that you want to investigate the influence of light and fertilizer levels on plant performance. You plan to use five fertilizer and two light levels. For each combination of fertilizer and light level, you want four replicates. What is the total number of replicates?

2. Suppose that you want to investigate the effects of leaf damage on the performance of drought stressed plants. You plan to use three levels of leaf damage and four different watering protocols. For each combination of leaf damage and watering protocol, you plan to have three replicates. What is the total number of replicates?

3. *Coleomegilla maculata*, a lady beetle, is an important predator of egg masses of *Ostrinia nubialis*, the European corn borer. *C. maculata* also feeds on aphids and maize pollen. To study its food preferences, you choose two satiation levels for *C. maculata* and combinations of two of the three food sources (that is, either egg masses and aphids, or egg masses and pollen, or aphids and pollen). For each experimental protocol, you want twenty replicates. What is the total number of replicates?

4. To test the effects of a new drug, you plan the following clinical trial. Each patient receives the new drug, an established drug, or a placebo. You enroll 50 patients. In how many ways can you assign them to the three treatments?

5. The Muesli-Mix is a popular breakfast hangout near a campus. A typical breakfast there consists of one beverage, one bowl of cereal, and a piece of fruit. If you can choose among three different beverages, seven different cereals, and four different types of fruit, how many choices for breakfast do you have?

6. To study sex differences in food preferences in rats, you offer one of three choices of food to each rat. You plan to have 12 rats for each food and sex combination. How many rats will you need?

(12.1.2)

7. You plan a trip to Europe, during which you wish to visit London, Paris, Amsterdam, Rome, and Heidelberg. Because you want to buy a railway ticket before you leave, you must decide on the order in which you will visit these five cities. How many different routes are there?

8. Five people line up for a photograph. How many different lineups are there?

9. You have just bought seven different books. In how many ways can they be arranged on your bookshelf?

10. Four cars arrive simultaneously at an intersection. Only one car can go through at a time. In how many different ways can they leave the intersection?

11. How many four-letter words with no repeated letters can you form from the 26 letters of the alphabet?

12. A committee of three people, consisting of a president, a vice president, and a treasurer, must be chosen from a group of ten. How many committees can be selected?

13. Three different awards are to be given to a class of 15 students. Each student can receive at most one award. Count the number of ways these awards can be given out.

14. You have just enough time to play four songs out of ten from your favorite CD. In how many ways can you program your CD player to play the four songs?

(12.1.3)

15. A bag contains 10 different candy bars. You are allowed to choose three. How many choices do you have?

16. During International Movie Week, 60 movies are shown. You have time to see five movies. How many different plans can you make?

17. A committee of three people must be formed from a group of 10. How many committees are there, if no specific tasks are assigned to the members of the committee?

18. A standard deck contains 52 different cards. In how many ways can you select five cards from the deck?

19. An urn contains 15 different balls. In how many ways can you select four balls without replacement?

20. Twelve people wait in front of an elevator that only has room for five. Count the number of ways that the first group of people to take the elevator can be chosen.

21. Four A's and five B's are to be arranged into a nine-letter word. How many different words can you form?

22. Suppose that you want to plant a flower bed with four different plants. You can choose from among eight plants. How may different choices do you have?

(12.1.4)

23. A box contains five red and four blue balls. You choose two balls.

(a) How many possible selections contain exactly two red balls; how many exactly two blue balls; and how many exactly one of each color?

(b) Show that the sum of the number of choices for the three cases in (a) is equal to the number of ways that you can select two balls out of the nine balls in the box.

24. Twelve children are divided up into three groups, of five, four, and three. In how many ways can this be done if the order within each group is not important?

25. Five A's, three B's, and six C's are to be arranged into a 14-letter word. How many different words can you form?

26. A bag contains 45 beans of three different varieties. Each variety is represented 15 times in the bag. You grab nine beans out of the bag.

(a) Count the number of ways that each variety can be represented exactly three times in your sample.

(b) Count the number of ways that only one variety appears in your sample.

27. Let $S = \{a, b, c\}$. List all possible subsets, and argue that the total number of subsets is $2^3 = 8$.

28. Suppose that a set contains n elements. Argue that the total number of subsets of this set is 2^n.

29. In how many ways can Brian, Hilary, Peter, and Melissa sit on a bench if Peter and Melissa want to be next to each other?

30. Paula, Cindy, Gloria, and Jenny have dinner at a round table. In how many ways can they sit around the table if Cindy wants to sit to the left of Paula?

31. In how many ways can you form a committee of three people from a group of seven if two of the people do not want to serve together?

32. In how many ways can you form two committees of three people each from a group of nine if

(a) no person is allowed to serve on more than one committee?

(b) people can serve on both committees simultaneously?

33. A collection contains seeds for four different annual and three different perennial plants. You plan a garden bed with three different plants, and you want to include at least one perennial. How many different selections can you make?

34. In diploid organisms, chromosomes appear in pairs in the nuclei of all cells except gametes (sperm or ovum). Gametes are formed during meiosis, where the number of chromosomes in the nucleus is halved; that is, only one member of each pair of chromosomes ends up in a gamete. Humans have 23 pairs of chromosomes. How may kinds of gametes can a human produce?

35. Expand $(x + y)^4$.

36. Expand $(2x - 3y)^5$.

37. In how many ways can four red and five black cards be selected from a standard deck of cards if cards are drawn without replacement?

38. In how many ways can two aces and three kings be selected from a standard deck of cards if cards are drawn without replacement?

39. In the game of poker, determine the number of ways exactly two pairs can be picked.

40. In the game of poker, determine the number of ways a *flush* (five cards of the same suit) can be picked.

41. In the game of poker, determine the number of ways *four of a kind* (four cards of the same value plus one other cards) can be picked.

42. In the game of poker, determine the number of ways a *straight* (five cards with consecutive values, such as A 2 3 4 5 or 7 8 9 10 J or 10 J Q K A, but not all of the same suit) can be picked.

43. (*Counterpoint*) Counterpoint is a musical term that means the combination of simultaneous voices; it is synonymous to polyphony. In *triple counterpoint*, three voices are arranged such that any voice can take any place of the three possible positions: highest, intermediate, and lowest voice. In how many ways can the three voices be arranged?

44. (*Counterpoint*) Counterpoint is a musical term that means the combination of simultaneous voices; it is synonymous to polyphony. In *quintuple counterpoint*, five voices are arranged such that any voice can take any place of the five possible positions: from highest to lowest voice. In how many ways can the five voices be arranged?

12.2 WHAT IS PROBABILITY?

12.2.1 Basic Definitions

A **random experiment** is an experiment in which the outcome is uncertain but which can be repeated. Tossing a coin and rolling a die are examples of random experiments. The set of all possible outcomes of a random experiment is called the **sample space** and is often denoted by Ω. We look at some examples, where we describe random experiments and give the associated sample space.

 Example 1 — Suppose that we toss a coin labeled heads (H) on one side and tails (T) on the other. If we toss the coin once, the possible outcomes are H and T, and the sample space is therefore

$$\Omega = \{H, T\}$$

If we toss the coin twice in a row, then each outcome is an ordered pair describing the outcome on the first toss followed by the outcome on the second toss, such as HT, which means heads followed by tails. The sample space is

$$\Omega = \{HH, HT, TH, TT\}$$ ◀

Example 2 — Consider a population for which we keep track of the genotype at one locus. We assume that the genes at this locus occur in three different forms, called alleles, denoted by A_1, A_2, and A_3. Furthermore, we assume that the individuals in the population are diploid; that is, the chromosomes occur in pairs. This means that a genotype is described by a pair of genes, such as A_1A_1. Since the order of the chromosomes is not important, the genotype A_1A_2 is the same as A_2A_1. If our random experiment consists of picking one individual out of the population and noting the genotype, then the sample space is given by

$$\Omega = \{A_1A_1, A_1A_2, A_1A_3, A_2A_2, A_2A_3, A_3A_3\} \quad \blacktriangleleft$$

▶ **Example 3**   An urn contains five balls, numbered 1–5. We draw two balls from the urn without replacement and note the numbers drawn.

There is some ambiguity in the formulation of this experiment. We can draw the two balls one after the other (without replacing the first in the urn after having noted its number), or we can draw the two balls simultaneously. In the first case, the sample space consists of ordered pairs (i, j), where the first entry is the number on the first ball and the second entry is the number on the second ball. Because the sampling is done without replacement, the two numbers are different. The sample space can be written as

$$\begin{aligned}\Omega = \{&(1, 2), (1, 3), (1, 4), (1, 5),\\ &(2, 1), (2, 3), (2, 4), (2, 5),\\ &(3, 1), (3, 2), (3, 4), (3, 5),\\ &(4, 1), (4, 2), (4, 3), (4, 5),\\ &(5, 1), (5, 2), (5, 3), (5, 4)\}\end{aligned}$$

or, in short,

$$\Omega = \{(i, j) : 1 \leq i \leq 5, 1 \leq j \leq 5, i \neq j\}$$

In the second case, there is no first or second ball, because we draw the balls simultaneously. We can write the sample space as

$$\begin{aligned}\Omega = \{&(1, 2), (1, 3), (1, 4), (1, 5),\\ &(2, 3), (2, 4), (2, 5),\\ &(3, 4), (3, 5),\\ &(4, 5)\}\end{aligned}$$

or, in short,

$$\Omega = \{(i, j) : 1 \leq i < j \leq 5\}$$

where the first entry of (i, j) represents the smaller of the two numbers on the balls in our sample.

The specifics of the random experiment determine which description of the sample space we prefer. $\quad \blacktriangleleft$

When we perform random experiments, we often consider a particular outcome, or, more generally, a particular subset of the sample space. We call subsets of the sample space **events**. Since an outcome is an element of the sample space, an outcome is an event as well, namely a subset that consists of just one element. We will use the basic set operations to deal with events.

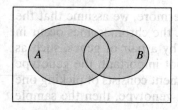

▲ **Figure 12.2**
The union of A and B, $A \cup B$, illustrated in a Venn diagram. The rectangle represents the set Ω

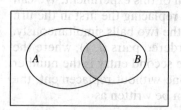

▲ **Figure 12.3**
The intersection of A and B, $A \cap B$, illustrated in a Venn diagram

Basic Set Operations   Suppose that A and B are events of the sample space Ω. The **union** of A and B, denoted by $A \cup B$ (read "A union B") is the set of all outcomes that belong to either A or B (or both). The **intersection** of A and B, denoted by $A \cap B$ (read "A intersected with B") is the set of all outcomes that belong to both A and B. Figures 12.2 and 12.3 show these first two set operations. Figures 12.2 and 12.3, where sets are visualized as "bubbles," are called **Venn diagrams**.

We can generalize the union and intersection of two events to a finite number of events. Let $A_1, A_2, \ldots, A_n$ be a finite number of events. Then

$$\bigcup_{i=1}^{n} A_i = A_1 \cup A_2 \cup \cdots \cup A_n = (A_1 \cup A_2 \cup \cdots \cup A_{n-1}) \cup A_n$$

$$= \left[\begin{array}{l} \text{the set of all outcomes that} \\ \text{belong to at least one set } A_i \end{array} \right]$$

$$\bigcap_{i=1}^{n} A_i = A_1 \cap A_2 \cap \cdots \cap A_n = (A_1 \cap A_2 \cap \cdots \cap A_{n-1}) \cap A_n$$

$$= \left[\begin{array}{l} \text{the set of all outcomes that} \\ \text{belong to all sets } A_i, i = 1, 2, \ldots, n \end{array} \right]$$

The **complement** of A, denoted by A^c, is the set of all outcomes contained in Ω that are not in A (see Figure 12.4). It follows that

$$\Omega^c = \emptyset \quad \text{and} \quad \emptyset^c = \Omega$$

where $\emptyset$ denotes the empty set. Furthermore,

$$(A^c)^c = A$$

When we take complements of unions or intersections, the following two identities are useful (Figure 12.5).

De MORGAN'S LAWS

$$(A \cup B)^c = A^c \cap B^c$$
$$(A \cap B)^c = A^c \cup B^c$$

▶ **Example 4**   Let us consider the coin tossing experiment of Example 1, in which we tossed a coin twice. The sample space is given by

$$\Omega = \{HH, HT, TH, TT\}$$

We denote by A the event that at least one head occurred:

$$A = \{HH, HT, TH\}$$

Let B denote the event that the first toss resulted in tails:

$$B = \{TH, TT\}$$

We see that

$$A \cup B = \{HH, HT, TH, TT\} \quad \text{and} \quad A \cap B = \{TH\}$$

Furthermore,

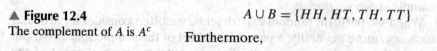

$$A^c = \{TT\} \quad \text{and} \quad B^c = \{HH, HT\}$$

▲ **Figure 12.4**
The complement of A is A^c

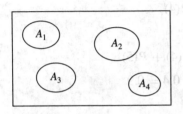

◀ **Figure 12.5**
De Morgan's laws

To see how De Morgan's laws work, we compute both $(A \cup B)^c$ and $A^c \cap B^c$. We find

$$(A \cup B)^c = \emptyset \quad \text{and} \quad A^c \cap B^c = \emptyset$$

which is consistent with De Morgan's first law. The second De Morgan's law claims that $(A \cap B)^c$ and $A^c \cup B^c$ are the same. We find that indeed

$$(A \cap B)^c = \{HH, HT, TT\} = A^c \cup B^c \qquad ◀$$

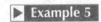

▲ **Figure 12.6**
The four sets A_1, A_2, A_3, and A_4 are disjoint

We say that $A_1, A_2, \dots, A_n$ are **pairwise disjoint** (or, simply, disjoint) if

$$A_i \cap A_j = \emptyset \quad \text{whenever} \quad i \neq j$$

This is illustrated for four sets in the Venn diagram in Figure 12.6.

▶ **Example 5**   Is it true that if $A_1 \cap A_2 \cap A_3 = \emptyset$, then A_1, A_2, and A_3 are pairwise disjoint?

Solution

No. Figure 12.7 shows a counterexample: $A_1 \cap A_2 \cap A_3 = \emptyset$ but $A_2 \cap A_3 \neq \emptyset$, implying that A_1, A_2, and A_3 are not pairwise disjoint.   ◀

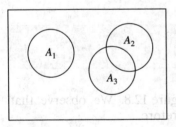

▲ **Figure 12.7**
A counterexample for
Example 5

The Definition of Probability   In the following definition, we assume that the sample space Ω has finitely many elements.

> **Definition** Let Ω be a finite sample space and A and B are events in Ω. A **probability** is a function that assigns values between 0 and 1 to events. The probability of an event A, denoted by $P(A)$, satisfies the following properties:
>
> **1.** For any event $A, 0 \leq P(A) \leq 1$.
> **2.** $P(\emptyset) = 0$ and $P(\Omega) = 1$.
> **3.** For two disjoint events A and B,
>
> $$P(A \cup B) = P(A) + P(B)$$

Note that a probability is a number that is *always* between 0 and 1. If you compute a probability and get either a negative number or a number greater than 1, you know immediately that your answer must be wrong.

▶ **Example 6**   Assume that $\Omega = \{1, 2, 3, 4, 5\}$, and that

$$P(1) = P(2) = 0.2 \qquad P(3) = P(4) = 0.1 \qquad \text{and} \qquad P(5) = 0.4$$

where we wrote $P(i)$ for $P(\{i\})$. Set $A = \{1, 2\}$ and $B = \{4, 5\}$. Find $P(A \cup B)$, and show that $P(\Omega) = 1$.

Solution

Since A and B are disjoint ($A \cap B = \emptyset$), we find

$$P(A \cup B) = P(A) + P(B) = P(\{1, 2\}) + P(\{4, 5\})$$

Since $\{1, 2\} = \{1\} \cup \{2\}$ and $\{4, 5\} = \{4\} \cup \{5\}$, and both are unions of disjoint sets,

$$P(\{1, 2\}) = P(1) + P(2)$$
$$P(\{3, 4\}) = P(3) + P(4)$$

Hence,

$$P(A \cup B) = P(1) + P(2) + P(4) + P(5)$$
$$= 0.2 + 0.2 + 0.1 + 0.4 = 0.9$$

To show that $P(\Omega) = 1$, we observe that

$$\Omega = \{1, 2, 3, 4, 5\} = \{1\} \cup \{2\} \cup \{3\} \cup \{4\} \cup \{5\}$$

Since this is a union of disjoint sets,

$$P(\Omega) = P(1) + P(2) + P(3) + P(4) + P(5)$$
$$= 0.2 + 0.2 + 0.1 + 0.1 + 0.4 = 1$$

◀

In the following, we will derive two additional basic properties of probabilities. The first is

$$P(A^c) = 1 - P(A)$$

To see why this is true, we look at Figure 12.8. We observe that $\Omega = A \cup A^c$, and that A and A^c are disjoint. Therefore,

$$1 = P(\Omega) = P(A \cup A^c) = P(A) + P(A^c)$$

from which the claim follows by rearranging terms.

Note that in property 2 of the definition, we wrote $P(\emptyset) = 0$ *and* $P(\Omega) = 1$. It would have been sufficient to require just one of these two identities. For instance, since $\Omega^c = \emptyset$, we can write $\Omega = \Omega \cup \emptyset$. Now, Ω and $\emptyset$ are disjoint, and $P(\Omega) = 1$. It follows that

$$1 = P(\Omega) = P(\Omega \cup \emptyset) = P(\Omega) + P(\emptyset)$$

and therefore

$$P(\emptyset) = 1 - P(\Omega) = 0$$

The next property allows us to compute probabilities of unions of two sets (which are not necessarily disjoint, as in property 3 of the definition).

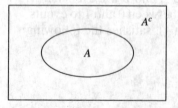

▲ **Figure 12.8**
The set Ω, represented by the rectangle, can be written as a disjoint union of the sets A and A^c

For any sets A and B,

$$P(A \cup B) = P(A) + P(B) - P(A \cap B)$$

This is illustrated in Figure 12.9 and follows from the fact that we count $A \cap B$ twice when we compute $P(A) + P(B)$. The proof of this property is relegated to Problem 15. Here, we give an example where we use both of these additional properties.

 Example 7 Assume that $\Omega = \{1, 2, 3, 4, 5\}$ and that

$$P(1) = P(2) = 0.2, \quad P(3) = P(4) = 0.1 \quad \text{and} \quad P(5) = 0.4$$

Set $A = \{1, 3, 4\}$ and $B = \{4, 5\}$. Find $P(A \cup B)$.

Solution

Observe that A and B are not disjoint. We find

$$A \cap B = \{4\}$$

Using the second of the additional properties, we obtain

$$P(A \cup B) = P(A) + P(B) - P(A \cap B)$$
$$= P(\{1, 3, 4\}) + P(\{4, 5\}) - P(\{4\})$$
$$= (0.2 + 0.1 + 0.1) + (0.1 + 0.4) - (0.1) = 0.8$$

We could have obtained the same result by observing that

$$A \cup B = \{1, 3, 4, 5\} \quad \text{and therefore} \quad A \cup B = \{2\}^c$$

which yields

$$P(A \cup B) = P(\{2\}^c) = 1 - P(\{2\}) = 1 - 0.2 = 0.8 \qquad \blacktriangleleft$$

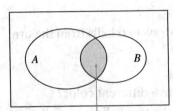

▲ Figure 12.9
To compute $P(A \cup B)$, we add $P(A)$ and $P(B)$, but since we count $A \cap B$ twice, we need to subtract $P(A \cap B)$

12.2.2 Equally Likely Outcomes

An important class of random experiments with finite sample spaces is those in which all outcomes are equally likely. That is, if $\Omega = \{1, 2, \ldots, n\}$, then $P(1) = P(2) = \cdots = P(n)$, where we wrote $P(i)$ for $P(\{i\})$. Then

$$1 = P(\Omega) = \sum_{i=1}^{n} P(i) = nP(1)$$

which implies that

$$P(1) = P(2) = \cdots = P(n) = \frac{1}{n}$$

If we denote by $|A|$ the number of elements in A, and if $A \subset \Omega$ with $|A| = k$, then

$$P(A) = \frac{|A|}{|\Omega|} = \frac{k}{n}$$

In the following two examples, we will discuss random experiments in which all outcomes are equally likely. You should pay particular attention to the sample space, to make sure that you understand that all of its elements are indeed equally likely.

▶ **Example 8**   Toss a fair coin three times and find the probability of the event $A = \{$at least two heads$\}$.

Solution

The sample space in this case is

$$\Omega = \{HHH, HHT, HTH, THH, HTT, THT, TTH, TTT\}$$

All outcomes are equally likely since we assumed that the coin is fair. Since $|\Omega| = 8$, it follows that each possible outcome has probability $1/8$. Thus,

$$P(A) = P(HHH, HHT, HTH, THH)$$
$$= \frac{|A|}{|\Omega|} = \frac{4}{8} = \frac{1}{2}$$

◀

▶ **Example 9**   An urn contains five blue and six green balls. We draw two balls from the urn without replacement.

(a) Determine the sample space Ω and find $|\Omega|$.

(b) What is the probability that the two balls have different color?

(c) What is the probability that at least one of the two balls is green?

Solution

(a) As physical objects, the balls are distinguishable and we can imagine them being numbered from 1 to 11, where we assign the first five numbers to the five blue balls and the remaining six numbers to the six green balls. The sample space for this random experiment then consists of all subsets of size 2 that can be drawn from the set of eleven balls. Each subset of size 2 is then equally likely. The size of the sample space is

$$|\Omega| = \binom{11}{2}$$

since the order in which the balls are removed from the urn is not important, and sampling is done without replacement.

(b) Let A denote the event that the two balls are of different color. To obtain an outcome where one ball is blue and the other is green, we must select one blue ball from the five blue balls, which can be done in $\binom{5}{1}$ different ways, and one green ball from the six green balls, which can be done in $\binom{6}{1}$ ways. Using the multiplication rule from Section 12.1, we find

$$|A| = \binom{5}{1}\binom{6}{1}$$

Hence,

$$P(A) = \frac{|A|}{|\Omega|} = \frac{\binom{5}{1}\binom{6}{1}}{\binom{11}{2}} = \frac{5 \cdot 6}{\frac{11 \cdot 10}{2}} = \frac{6}{11}$$

(c) Let B denote the event that at least one ball is green. It can be written as a union of the two disjoint sets

$$B_1 = \{\text{exactly one ball is green}\}$$
$$B_2 = \{\text{both balls are green}\}$$

Since $B = B_1 \cup B_2$ with $B_1 \cap B_2 = \emptyset$, it follows that

$$P(B) = P(B_1) + P(B_2)$$

Using a similar argument as in (a), we find

$$P(B) = \frac{\binom{5}{1}\binom{6}{1}}{\binom{11}{2}} + \frac{\binom{5}{0}\binom{6}{2}}{\binom{11}{2}} = \frac{5 \cdot 6}{55} + \frac{15}{55} = \frac{9}{11} \quad \blacktriangleleft$$

▶ **Example 10**   Four cards are drawn at random and without replacement from a standard deck of 52 cards. What is the probability of at least two kings?

Solution

We find

$$P(\text{at least two kings}) = 1 - [P(\text{no kings}) + P(\text{one king})]$$

There are $\binom{52}{4}$ ways of selecting four cards. There are four kings in a standard deck of cards. There are $\binom{48}{4}$ ways of selecting a hand of four cards that does not contain any kings and there are $\binom{4}{1}\binom{48}{3}$ ways of selecting a hand of four cards that contains exactly one king. Hence,

$$P(\text{at least two kings}) = 1 - \frac{\binom{48}{4}}{\binom{52}{4}} - \frac{\binom{4}{1}\binom{48}{3}}{\binom{52}{4}}$$

$$= 1 - \frac{48 \cdot 47 \cdot 46 \cdot 45}{52 \cdot 51 \cdot 50 \cdot 49} - \frac{16 \cdot 48 \cdot 47 \cdot 46}{52 \cdot 51 \cdot 50 \cdot 49} \approx 0.0257 \quad \blacktriangleleft$$

An Application from Genetics   Gregor Mendel, an Austrian monk, experimented with peas to study the laws of inheritance. He started these experiments in 1856. His work was fundamental in the understanding of the laws of inheritance. It took over thirty-five more years until Mendel's original work was publicized and his conclusions were confirmed in additional experiments.

We will use the current knowledge about inheritance to determine the likelihood of outcomes of certain crossings. We describe one of Mendel's experiments that studies the inheritance of flower color in peas. Mendel had seeds that produced plants with either red or white flowers. Flower color in Mendel's peas is determined by a single locus on the chromosome. The genes at this locus occur in two forms, called alleles, which we denote by C and c. Since pea plants are diploid organisms, each plant has two genes that determine flower color, one from each parent plant. The following genotypes are thus possible: CC, Cc, and cc. The genotypes CC and Cc have red flowers, whereas the genotype cc has white flowers.

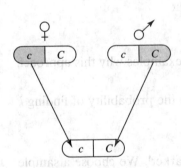

▲ **Figure 12.10**
The cross $Cc \times Cc$ resulted in an offspring of type Cc

▶ **Example 11**   Suppose that you cross two pea plants, both of type Cc. Determine the probabilities for each genotype in the next generation. What is the probability that a randomly chosen seed from this crossing results in a plant with red flowers?

Solution

We denote the offspring of this cross as a pair where the first entry is the maternal and the second entry is the paternal contribution. For instance, an offspring of type (c, C) inherited a c from the mother and a C from the father (see Figure 12.10). We list all possible outcomes of this cross in the sample space

$$\Omega = \{(C, C), (C, c), (c, C), (c, c)\}$$

The laws of inheritance imply that gametes form at random and that, therefore, all outcomes in Ω are equally likely. (This is Mendel's first law.) Since $|\Omega| = 4$, it follows that each outcome has probability 1/4.

Though the sample space has four different outcomes, there are only three different genotypes, since (C, c) and (c, C) denote the same genotype, Cc. In the following, we will often denote the event $\{(C, c), (c, C)\}$ simply by Cc, as is customary in genetics. We therefore find

$$P(CC) = \frac{1}{4} \qquad P(Cc) = \frac{1}{2} \qquad \text{and} \qquad P(cc) = \frac{1}{4}$$

Since both genotypes CC and Cc result in red flowers, it follows that

$$P(\text{red}) = P(\{(C, C), (C, c), (c, C)\}) = \frac{3}{4} \qquad \blacktriangleleft$$

The Mark-Recapture Method The mark-recapture method is commonly used to estimate population sizes. We illustrate this method using a fish population as an example. Suppose that N fish are in a lake, where N is unknown. To get an idea of how big N is, we capture K fish, mark them, and subsequently release them. We wait until the marked fish in the lake have had sufficient time to mix with the other fish. We then capture n fish. Suppose that k of the n fish are marked (assume $k > 0$), then if the fish are well mixed again, the ratio of the marked to unmarked fish in the sample of size n should approximately be equal to the ratio of marked to unmarked fish in the lake,

$$\frac{k}{n} \approx \frac{K}{N}$$

We might therefore conclude that there are about

$$N \approx K\frac{n}{k}$$

fish in the lake. We will explain in the next two examples why this approach makes sense.

▶ **Example 12** Given the mark-recapture experiment, compute the probability of finding k marked fish in a sample of size n.

Solution

There are N fish in the lake, K of which are marked. We choose a sample of size n. Each outcome is therefore a subset of size n, and all outcomes are equally likely. Using the counting techniques from Section 12.1, we find

$$|\Omega| = \binom{N}{n}$$

since the order in the sample is not important.

We denote by A the event that the sample of size n contains exactly k marked fish. To determine how many outcomes contain exactly k marked fish, we argue as follows: We need to select k fish from the K marked ones and $n - k$ fish from the $N - K$ unmarked ones. Selecting the k marked fish can be done in $\binom{K}{k}$ ways; selecting the $n - k$ unmarked fish can be done in $\binom{N-K}{n-k}$ ways. Since each choice of k marked fish can be combined with any choice of the $n - k$ unmarked fish, we use the multiplication principle to find the total number of ways of obtaining a sample of size n with exactly k marked fish, namely,

$$|A| = \binom{K}{k}\binom{N - K}{n - k}$$

Therefore,

$$P(A) = \frac{|A|}{|\Omega|} = \frac{\binom{K}{k}\binom{N-K}{n-k}}{\binom{N}{n}}$$

(This example is of the same basic type as the urn problem in Example 9.) ◄

We will now give an argument that explains why the total number of fish in the lake can be estimated using the formula $N \approx Kn/k$.

▶ Example 13

Assume that there are K marked fish in the lake. We take a sample of size n and observe k marked fish. Show that the value of N that maximizes the probability of finding k marked fish in a sample of size n is the largest integer less than or equal to Kn/k. We use this value as our estimate for the population size N. Since this estimate of N maximizes the probability of what we observe, it is called a **maximum likelihood estimate**.

Solution

We denote by A the event that the sample of size n contains exactly k marked fish. We found in Example 12 that

$$P(A) = \frac{|A|}{|\Omega|} = \frac{\binom{K}{k}\binom{N-K}{n-k}}{\binom{N}{n}}$$

We now consider $P(A)$ as a function of N and denote it by p_N. To find the value of N that maximizes p_N, we look at the ratio p_N/p_{N-1}. (The function p_N is not continuous, since it is only defined for integer values of N. Therefore, we cannot differentiate p_N to find its maximum.) The ratio is given by

$$\frac{p_N}{p_{N-1}} = \frac{\frac{\binom{K}{k}\binom{N-K}{n-k}}{\binom{N}{n}}}{\frac{\binom{K}{k}\binom{N-1-K}{n-k}}{\binom{N-1}{n}}} = \frac{\binom{N-K}{n-k}}{\binom{N-1-K}{n-k}} \frac{\binom{N-1}{n}}{\binom{N}{n}}$$

$$= \frac{(N-K)!(n-k)!(N-1-K-n+k)!}{(n-k)!(N-K-n+k)!(N-1-K)!} \frac{(N-1)!n!(N-n)!}{n!(N-1-n)!N!}$$

When we cancel terms, we find

$$\frac{p_N}{p_{N-1}} = \frac{N-K}{N-K-n+k} \frac{N-n}{N}$$

We will now investigate when this ratio is greater than or equal to 1, since this will allow us to find the values of N where p_N exceeds p_{N-1}. Values of N where p_N exceeds both p_{N-1} and p_{N+1} are local maxima. The ratio p_N/p_{N-1} is greater than or equal to 1 if

$$(N-K)(N-n) \geq N(N-K-n+k)$$

Multiplying this out, we find

$$N^2 - Nn - KN + Kn \geq N^2 - NK - Nn + Nk$$

Simplifying yields

$$Kn \geq kN$$

or

$$N \leq K\frac{n}{k}$$

Thus, $p_N \geq p_{N-1}$ as long as $N \leq Kn/k$. If Kn/k is an integer, then $p_N = p_{N-1}$ for $N = Kn/k$ and both Kn/k and $Kn/k - 1$ maximize the probability of observing k fish in the sample of size n and so both values can be chosen as estimates for the number of fish in the lake. If Kn/k is not an integer, then the largest integer less than Kn/k maximizes the probability p_N. To arrive at just one value, we will always use the largest integer less than or equal to Kn/k to estimate the total number of fish in the lake. ◀

▶ **Example 14**   Assume that there are 15 marked fish in a lake. We take a sample of size 10 and observe 4 marked fish. Find an estimate of the number of fish in the lake based on Example 13.

Solution

It follows from Example 13 that an estimate for the number of fish in the lake, denoted by N, is the largest integer less than or equal to Kn/k where in this example $K = 15, n = 10$, and $k = 4$. Since

$$K\frac{n}{k} = 15 \cdot \frac{10}{4} = 37.5$$

we estimate that there are 37 fish in the lake.

To see that this value indeed maximizes

$$p_N = \frac{\binom{K}{k}\binom{N-K}{n-k}}{\binom{N}{n}}$$

(defined in Example 13), we graph p_N as a function of N for $K = 15, n = 10$, and $k = 4$ (Figure 12.11). Since the sample contains $n = 10$ fish, the number of fish in the lake must be at least 10. Therefore, $p_N = 0$ for $N < 10$. ◀

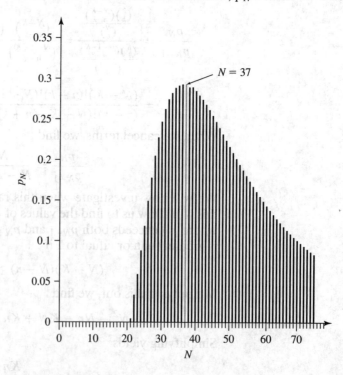

▲ **Figure 12.11**
The function p_N for different values of N when $K = 15, n = 10$, and $k = 4$. The graph shows that p_N is maximal for $N = 37$

12.2.3 Problems

(12.2.1)

In Problems 1–4, determine the sample spaces for each random experiment.

1. The random experiment consists of tossing a coin three times.

2. The random experiment consists of rolling a six-sided die twice.

3. An urn contains five balls numbered 1–5. The random experiment consists of selecting two balls without replacement.

4. An urn contains six balls numbered 1–6. The random experiment consists of selecting five balls without replacement.

In Problems 5–8, assume that
$$\Omega = \{1, 2, 3, 4, 5, 6\}$$
$A = \{1, 3, 5\}$, *and* $B = \{1, 2, 3\}$.

5. Find $A \cup B$ and $A \cap B$.

6. Find A^c and show that $(A^c)^c = A$.

7. Find $(A \cup B)^c$.

8. Are A and B disjoint?

In Problems 9–12, assume that
$$\Omega = \{1, 2, 3, 4, 5\}$$
$P(1) = 0.1$, $P(2) = 0.2$, *and* $P(3) = P(4) = 0.05$. *Furthermore, assume that* $A = \{1, 3, 5\}$ *and* $B = \{2, 3, 4\}$.

9. Find $P(5)$.

10. Find $P(A)$ and $P(B)$.

11. Find $P(A^c)$.

12. Find $P(A \cup B)$.

13. Assume that $P(A \cap B) = 0.1$, $P(A) = 0.4$, and $P(A^c \cap B^c) = 0.2$. Find $P(B)$.

14. Assume that $P(A) = 0.4$, $P(B) = 0.4$, and $P(A \cup B) = 0.7$. Find $P(A \cap B)$ and $P(A^c \cap B^c)$.

15. We will show the second of the additional properties, namely
$$P(A \cup B) = P(A) + P(B) - P(A \cap B) \qquad (12.5)$$
(a) Use a diagram to show that B can be written as a disjoint union of the sets $A \cap B$ and $B \cap A^c$.

(b) Use a diagram to show that $A \cup B$ can be written as a disjoint union of the sets A and $B \cap A^c$.

(c) Use your results in (a) and (b) to show that
$$P(A \cup B) = P(A) + P(B \cap A^c)$$
and
$$P(B \cap A^c) = P(B) - P(A \cap B)$$
Conclude from this that (12.5) holds.

16. If $A \subset B$, we can define the difference between the two sets A and B, denoted by $B - A$ (read "B minus A"). Namely,
$$B - A = B \cap A^c$$
as illustrated in Figure 12.12.

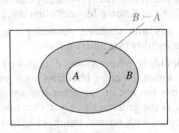

▲ **Figure 12.12**
The set A is contained in B. The shaded area is the difference of A and B, $B - A$

The following difference rule holds:
$$P(B - A) = P(B) - P(A) \qquad (12.6)$$
To show (12.6), we go through the following steps.

(a) Use the diagram in Figure 12.12 to show that B can be written as a disjoint union of A and $B - A$.

(b) Use your result in (a) to conclude that
$$P(B) = P(A) + P(B - A)$$
and show that (12.6) follows from this.

(c) An immediate consequence of (12.6) is the following result. If $A \subset B$, then
$$P(A) \leq P(B)$$
Use (12.6) to show this inequality.

(12.2.2)

17. Toss two fair coins and find the probability of at least one head.

18. Toss three fair coins and find the probability of no heads.

19. Roll a fair die twice and find the probability of at least one 4.

20. Roll two fair dice and find the probability that the sum of the two numbers is even.

21. Roll two fair dice, one after the other, and find the probability that the first number is larger than the second number.

22. Roll two fair dice and find the probability that the minimum of the two numbers will be greater than 4.

23. In Example 11, we considered a cross between two pea plants, each of genotype Cc. Find the probability that a randomly chosen seed from this cross has white flowers.

24. In Example 11, we considered a cross between two pea plants, each of genotype Cc. Now we cross a pea plant of genotype cc with a pea plant of genotype Cc.

(a) What are the possible outcomes of this cross?

(b) Find the probability that a randomly chosen seed from this cross results in red flowers.

25. Suppose that two parents are of genotype Aa. What is the probability that their offspring is of genotype Aa? (Assume Mendel's first law.)

26. Suppose that one parent is of genotype AA and the other is of genotype Aa. What is the probability that their offspring is of genotype AA? (Assume Mendel's first law.)

27. A family has three children. Assuming a $1:1$ sex ratio, what is the probability that all are girls?

28. Color blindness is an X-linked inherited disease. A woman who carries the color blindness gene on one of her X chromosomes but not on the other has normal vision. A man who carries the gene on his only X chromosome is color blind. If a woman with normal vision who carries the color blindness gene on one of her X chromosomes has a son with a man who has normal vision, what is the probability that their son will be color blind?

29. A lake contains an unknown number of fish, denoted by N. You capture 100 fish, mark them, and subsequently release them again. Later, you return and catch ten fish, three of which are marked.

(a) Find the probability that exactly three out of ten fish will be marked. This probability will be a function of N, the unknown number of fish in the lake.

(b) Find the value of N that maximizes the probability you computed in (a), and show that this value agrees with the value we computed in Example 13.

30. An urn contains five blue and three green balls. You remove three balls from the urn without replacement. What is the probability that at least two out of the three balls are green?

31. You select two cards without replacement from a standard deck of 52 cards. What is the probability that both cards are spades?

32. You select five cards without replacement from a standard deck of 52 cards. What is the probability that you get four aces?

33. An urn contains four green, six blue, and two red balls. You take three balls out of the urn without replacement. What is the probability that all three balls are of different colors?

34. An urn contains four green, six blue, and five red balls. You take three balls out of the urn without replacement. What is the probability that all three balls are of the same color?

35. Four cards are drawn at random without replacement from a standard deck of 52 cards. What is the probability of at least one ace?

36. Four cards are drawn at random without replacement from a standard deck of 52 cards. What is the probability of exactly one pair?

37. Thirteen cards are drawn at random without replacement from a standard deck of 52 cards. What is the probability that all are red?

38. Four cards are drawn at random without replacement from a standard deck of 52 cards. What is the probability that all are of different suits?

39. Five cards are drawn at random without replacement from a standard deck of 52 cards. What is the probability of exactly two pairs?

40. Five cards are drawn at random without replacement from a standard deck of 52 cards. What is the probability of three of a kind and a pair (for instance, Q Q Q 3 3)? (This is called a *full house* in poker.)

12.3 CONDITIONAL PROBABILITY AND INDEPENDENCE

Before we define *conditional probability* and *independence*, we will illustrate these concepts using the Mendelian crossing of peas that we considered in the previous section to study flower color inheritance.

Assume that two parent pea plants are of genotype Cc. Suppose you know that the offspring of the cross $Cc \times Cc$ has red flowers; what is the probability that it is of genotype CC? We can find this probability by noting that one of the three equally likely possibilities that produce red flowers [namely, (C, C), (C, c), and (c, C), if we list the types according to maternal and paternal contributions as in Example 11 of the previous section] is of type CC. Hence the probability that the offspring is of genotype CC is 1/3. Such a probability, where we condition on some prior knowledge (as flower color of offspring), is called a **conditional probability**.

Suppose now that the paternally transmitted gene in the offspring of the cross $Cc \times Cc$ is of type C. What is the probability that the maternally transmitted gene in the offspring is of type c? To answer this question, we note that the paternal gene has no impact on the choice of the maternal gene in this case. The probability that the maternal gene is of type c is therefore 1/2. We say that the maternal gene is **independent** of the paternal gene. Knowing which of the paternal genes was chosen does not change the probability of the maternal gene.

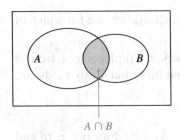

$A \cap B$

▲ **Figure 12.13**
The conditional probability of
A given B is the proportion of
A in the set B, $A \cap B$, relative
to the set B

12.3.1 Conditional Probability

As illustrated in the introduction of this section, conditional probabilities have something to do with prior knowledge. Suppose we know that the event B has occurred and that $P(B) > 0$. Then the conditional probability of the event A given B, denoted by $P(A \mid B)$, is the probability that A will occur given the fact that B has occurred. It is defined as

$$P(A \mid B) = \frac{P(A \cap B)}{P(B)} \tag{12.7}$$

To explain this definition, we look at Figure 12.13. The probability of A given B is the proportion of A in the set B relative to B.

In the following example, we will use the definition (12.7) to repeat the introductory example, and find the probability that an offspring is of genotype CC given that its flower color is red.

▶ **Example 1**   Find the probability that the offspring of a $Cc \times Cc$ cross of pea plants is of type CC, given that its flowers are red.

Solution

Let A denote the event that the offspring is of genotype CC and B be the event that the flower color of the offspring is red. We want to find $P(A \mid B)$. Using (12.7), we have

$$P(A \mid B) = \frac{P(A \cap B)}{P(B)}$$

The *unconditional* probabilities $P(B)$ and $P(A \cap B)$ are computed using the sample space $\Omega = \{(C, C), (C, c), (c, C), (c, c)\}$, whose outcomes all have the same probability. The probability $P(B)$ is the probability that the genotype of the offspring is in the set $\{(C, C), (C, c), (c, C)\}$. Since the sample space has equally likely outcomes, $P(B) = 3/4$. To compute $P(A \cap B)$, we note that $A \cap B$ is the event that the offspring is of genotype CC. Using the sample space Ω, we find $P(A \cap B) = 1/4$. Hence,

$$P(A \mid B) = \frac{P(A \cap B)}{P(B)} = \frac{\frac{1}{4}}{\frac{3}{4}} = \frac{1}{3}$$

which is the same answer that we obtained before.   ◀

Equation (12.7) can be used to compute conditional probabilities, as seen in Example 1. By rearranging terms, however, it can also be used to compute probabilities of the intersection of events. Namely, if we multiply both sides of (12.7) by $P(B)$, we obtain

$$P(A \cap B) = P(A \mid B)P(B) \tag{12.8}$$

In Equation (12.7), we conditioned on the event B. If we condition on A instead, we have the following identity:

$$P(B \mid A) = \frac{P(A \cap B)}{P(A)}$$

Rearranging terms as in (12.8), we find

$$P(A \cap B) = P(B \mid A)P(A) \tag{12.9}$$

Formulas (12.8) and (12.9) are particularly useful for computing probabilities in two-stage experiments. We illustrate the use of these identities in the following example. We will also see that there is a natural choice for which of the two events to condition on.

▶ **Example 2** Suppose that we draw two cards at random without replacement from a standard deck of 52 cards. Compute the probability that both cards are diamonds.

Solution

This can be thought of as a two-stage experiment: We first draw one card and then, without replacing the first one, we draw a second card. We define the two events

$$A = \{\text{the first card is diamond}\}$$
$$B = \{\text{the second card is diamond}\}$$

Then

$$A \cap B = \{\text{both cards are diamonds}\}$$

Should we use (12.8) or (12.9)? Since the first card is drawn first, it will be easier to condition on the outcome of the first draw than on the second draw; that is, we will compute $P(B \mid A)$ rather than $P(A \mid B)$ and then use (12.9). Now,

$$P(A) = \frac{13}{52}$$

since 13 out of the 52 cards are diamonds and each card has the same probability of being drawn. To compute $P(B \mid A)$, we note that if the first card is diamond, then there are 12 diamonds left in the deck of the remaining 51 cards. Therefore,

$$P(B \mid A) = \frac{12}{51}$$

Using (12.9), we find

$$P(A \cap B) = P(B \mid A)P(A) = \frac{12}{51} \cdot \frac{13}{52} = \frac{1}{17}$$ ◀

12.3.2 The Law of Total Probability

We begin this subsection by defining a partition of a sample space. Suppose the sample space Ω is written as a union of n disjoint sets $B_1, B_2, \ldots, B_n$. That is,

(i) $B_i \cap B_j = \emptyset$ whenever $i \neq j$

(ii) $\Omega = \bigcup_{i=1}^{n} B_i$

We then say that the sets $B_1, B_2, \ldots, B_n$ form a **partition** of the sample space Ω (see Figure 12.14).

Let A be an event. We can write A as a union of disjoint sets using this partition of Ω:

$$A = (A \cap B_1) \cup (A \cap B_2) \cup \cdots \cup (A \cap B_n)$$

as illustrated in Figure 12.15.

Since the sets $A \cap B_i$, $i = 1, 2, \ldots, n$, are disjoint, we find

$$P(A) = \sum_{i=1}^{n} P(A \cap B_i)$$

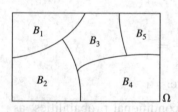

▲ **Figure 12.14**
The sets $B_1, B_2, \ldots, B_5$ form a partition of Ω

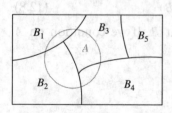

▲ **Figure 12.15**
The set A is a disjoint union of sets of the form $A \cap B_i$

To evaluate $P(A \cap B_i)$, we might find it useful to condition on B_i; that is, $P(A \cap B_i) = P(A \mid B_i)P(B_i)$. Then

$$P(A) = \sum_{i=1}^{n} P(A \mid B_i)P(B_i) \qquad (12.10)$$

Equation (12.10) is known as the **law of total probability**.

▶ Example 3   A test for the HIV virus shows a positive result in 99% of all cases when the virus is actually present and in 5% of all cases when the virus is not present (a *false positive* result). If such a test is administered to a randomly chosen individual, what is the probability that the test result is positive? Assume that the prevalence of the virus in the population is 1/200.

Solution

We set

$$A = \{\text{test result is positive}\}$$

Individuals in this population fall into two groups—those who are infected with the HIV virus and those who are not. These two sets form a partition of the population. If we pick an individual at random from the population, then the person belongs to one of these two groups. We define

$$B_1 = \{\text{person is infected}\}$$

$$B_2 = \{\text{person is not infected}\}$$

Using (12.10), we can write

$$P(A) = P(A \mid B_1)P(B_1) + P(A \mid B_2)P(B_2)$$

Now, $P(B_1) = 1/200$ and $P(B_2) = 199/200$. Furthermore, $P(A \mid B_1) = 0.99$ and $P(A \mid B_2) = 0.05$. Hence

$$P(A) = (0.99)\frac{1}{200} + (0.05)\frac{199}{200} = 0.0547 \qquad ◀$$

We can illustrated the last example using a **tree diagram**, as shown in Figure 12.16.

The numbers on the branches represent the respective probabilities. To find the probability of a positive test result, we multiply the probabilities along the paths that lead to tips labeled "test positive." (The two paths are marked in the tree.) Adding the results along the different paths then yields the desired probability. That is,

$$P(\text{test positive}) = \frac{1}{200}(0.99) + \frac{199}{200}(0.05) = 0.0547$$

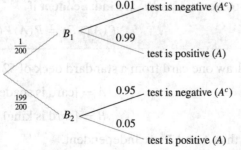

▲ **Figure 12.16**
The blue paths in this tree diagram lead to the event A. The numbers on the branches represent the respective probabilities

as in Example 3. Tree diagrams are quite useful when using the law of total probability.

In the next example, we will return to our pea plants. Recall that red-flowering pea plants are of genotype CC or Cc and that white-flowering pea plants are of genotype cc.

▶ **Example 4** Suppose that you have a batch of red-flowering pea, plants, 20% of which are of genotype Cc and 80% of genotype CC. You pick one of the red-flowering plants at random and cross it with a white-flowering plant. Find the probability that the offspring will produce red flowers.

Solution

A white-flowering plant is of genotype cc. If the red-flowering parent plant is of genotype CC (probability 0.8) is crossed with a white-flowering plant, then all offspring are of genotype Cc and therefore produce red flowers. If the red-flowering parent plants is of genotype Cc (probability 0.2) is crossed with a white-flowering plant, then with probability 0.5 an offspring is of genotype Cc (and therefore red flowering) and with probability 0.5 of genotype cc (and therefore white flowering). We use a tree diagram to illustrate the computation of the probability of a red-flowering offspring (Figure 12.17).

The paths that lead to a red-flowering offspring are marked. Using the tree diagram (or the law of total probability), we find

$$P(\text{red-flowering offspring}) = (0.8)(1) + (0.2)(0.5) = 0.9$$ ◀

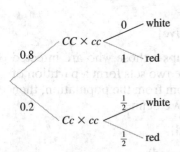

▲ **Figure 12.17**
The tree diagram for Example 4: The blue paths lead to red flowering offspring

12.3.3 Independence

Suppose that you toss a fair coin twice. Let A be the event that the first toss results in heads and B the event that the second toss results in heads. Suppose that A occurs. Does this change the probability that B will occur? The answer is obviously no. The outcome of the first toss does not influence the outcome of the second toss. We can express this mathematically using conditional probabilities. Namely,

$$P(B \mid A) = P(B) \tag{12.11}$$

and we say that A and B are independent. We will not use (12.11) as the definition of independence; rather, we will use the definition of conditional probabilities to rewrite (12.11). Since $P(B \mid A) = P(A \cap B)/P(A)$, Equation (12.11) can be written as

$$\frac{P(A \cap B)}{P(A)} = P(B)$$

Multiplying both sides by $P(A)$, we find $P(A \cap B) = P(A)P(B)$. We use this as our definition.

> Two events A and B are **independent** if
> $$P(A \cap B) = P(A)P(B)$$

▶ **Example 5** You draw one card from a standard deck of 52 cards. Let

$$A = \{\text{card is spade}\}$$

$$B = \{\text{card is king}\}$$

Show that A and B are independent.

Solution

To show that A and B are independent, we compute

$$P(A) = \frac{13}{52} \qquad P(B) = \frac{4}{52}$$

and

$$P(A \cap B) = P(\text{card is king of spade}) = \frac{1}{52}$$

Since

$$P(A)P(B) = \frac{13}{52} \cdot \frac{4}{52} = \frac{1}{52} = P(A \cap B)$$

it follows that A and B are independent.   ◀

We will return to our pea plant example to illustrate how we can compute probabilities of intersections of events when we know that the events are independent.

▶ **Example 6**   What is the probability that the offspring of a $Cc \times Cc$ cross is of genotype cc?

Solution

We computed the answer previously using a sample space with equally likely outcomes. But we can also compute this probability using independence.

In order for the offspring to be of genotype cc, both parents must contribute a c gene. Let

$$A = \{\text{paternal gene is } c\}$$

$$B = \{\text{maternal gene is } c\}$$

Now, it follows from the laws of inheritance that A and B are independent and that $P(A) = P(B) = 1/2$. Hence,

$$P(cc) = P(A \cap B) = P(A)P(B) = \frac{1}{2} \cdot \frac{1}{2} = \frac{1}{4}$$

This is the same result that we obtained when we looked at the sample space $\Omega = \{(C, C), (C, c), (c, C), (c, c)\}$ and observed that all four possible types were equally likely.   ◀

We can extend independence to more than two events. Events $A_1, A_2, \ldots, A_n$ are independent if for any $1 \le i_1 < i_2 < \cdots < i_k \le n$

$$P(A_{i_1} \cap A_{i_2} \cap \cdots \cap A_{i_k}) = P(A_{i_1})P(A_{i_2}) \cdots P(A_{i_k}) \qquad (12.12)$$

To see what this means when we have three events A, B, and C, we write the conditions explicitly. Three events A, B, and C are independent if

$$P(A \cap B) = P(A)P(B) \quad P(A \cap C) = P(A)P(C) \quad P(B \cap C) = P(B)P(C) \qquad (12.13)$$

and

$$P(A \cap B \cap C) = P(A)P(B)P(C) \qquad (12.14)$$

That is, both (12.13) *and* (12.14) must hold.

The number of conditions we must verify increases quickly with the number of events. When there are just two sets, only one condition must be checked—namely, $P(A \cap B) = P(A)P(B)$. When there are three sets, we just saw that there are four conditions—namely, there are $\binom{3}{2}$ conditions involving pairs of sets and $\binom{3}{3}$ conditions involving all three sets. With four sets, there is a total of $\binom{4}{2} + \binom{4}{3} + \binom{4}{4} = 11$ conditions; with five sets, 26 conditions; and with ten sets, 1013 conditions.

We stress that it is *not* enough to check independence between pairs of events to determine whether a collection of sets is independent. However, independence between pairs of sets is an important property itself, and we wish to define it. Namely, events $A_1, A_2, \ldots, A_n$ are **pairwise independent** if

$$P(A_i \cap A_j) = P(A_i)P(A_j) \qquad \text{whenever } i \neq j$$

We provide an example where events are pairwise independent but not independent.

▶ **Example 7** Roll two dice. Let

$$A = \{\text{the first die shows an even number}\}$$
$$B = \{\text{the second die shows an even number}\}$$
$$C = \{\text{the sum is odd}\}$$

Show that A, B, and C are pairwise independent but not independent.

Solution

To show pairwise independence, we compute

$$P(A) = \frac{3}{6} = \frac{1}{2} \qquad P(B) = \frac{3}{6} = \frac{1}{2} \qquad P(C) = \frac{18}{36} = \frac{1}{2}$$

$$P(A \cap B) = \frac{9}{36} = \frac{1}{4} = P(A)P(B)$$

$$P(A \cap C) = \frac{9}{36} = \frac{1}{4} = P(A)P(C)$$

$$P(B \cap C) = \frac{9}{36} = \frac{1}{4} = P(B)P(C)$$

However, the event $A \cap B \cap C = \emptyset$ since, if both dice show even numbers, then the sum of two even numbers cannot be odd. Therefore,

$$P(A \cap B \cap C) = P(\emptyset) = 0 \neq P(A)P(B)P(C)$$

which shows that A, B, and C are not independent. ◀

When events are independent, we can use (12.12) to compute the probability of the intersections of events, as illustrated in the following example.

▶ **Example 8** Assume a $1:1$ sex ratio. A family has five children. Find the probability that at least one of the children is a girl.

Solution

Instead of computing the probability that at least one of the children is a girl, we will look at the complement of this event. This is a particularly useful trick when we look at events that ask for the probability of "at least one." We denote by

$$A_i = \{\text{the } i\text{th child is a boy}\}$$

Then the events $A_1, A_2, \ldots, A_5$ are independent. Let

$$B = \{\text{at least one of the children is a girl}\}$$

Instead of computing the probability of B directly, we look at the complement of B. Now, B^c is the event that all children are boys, which is expressed as

$$B^c = A_1 \cap A_2 \cap \cdots \cap A_5$$

It follows that

$$P(B) = 1 - P(B^c) = 1 - P(A_1 \cap A_2 \cap \cdots \cap A_5)$$

$$= 1 - P(A_1)P(A_2)\cdots P(A_5)$$

$$= 1 - \left(\frac{1}{2}\right)^5 = \frac{31}{32}$$

If we had tried to compute $P(B)$ directly, we would have needed to compute the probability of exactly one girl, exactly two girls, and so on, and then add the probabilities. In this situation, it is quicker and easier to compute the probability of the complement. ◀

12.3.4 The Bayes Formula

In Example 3 of this section, we computed the probability that the result of an HIV test of a randomly chosen individual is positive. For the individual, however, it is much more important to know whether a positive test result actually means that he or she is infected. Recall that we defined

$$A = \{\text{test result is positive}\}$$

$$B_1 = \{\text{person is infected}\}$$

$$B_2 = \{\text{person is not infected}\}$$

We are interested in $P(B_1 \mid A)$; that is, the probability that a person is infected given that the result is positive. We saw in Example 3 that $P(A \mid B_1)$ and $P(A \mid B_2)$ followed immediately from the characteristics of the test. Now, we wish to compute a conditional probability where the roles of A and B_1 are reversed.

Before we compute the probability for this specific example, we look at the general case. We assume that the sets $B_1, B_2, \ldots, B_n$ form a partition of the sample space Ω, A is an event, and the probabilities $P(A \mid B_i)$, $i = 1, 2, \ldots, n$ are known. We are interested in computing $P(B_i \mid A)$. This can be accomplished as follows. Using the definition of conditional probabilities, we find

$$P(B_i \mid A) = \frac{P(A \cap B_i)}{P(A)} \tag{12.15}$$

To compute $P(A \cap B_i)$, we now condition on B_i, that is,

$$P(A \cap B_i) = P(A \mid B_i)P(B_i) \tag{12.16}$$

To evaluate the denominator $P(A)$, we use the law of total probability, that is,

$$P(A) = \sum_{j=1}^{n} P(A \mid B_j)P(B_j) \tag{12.17}$$

Combining (12.15), (12.16), and (12.17), we arrive at the following formula, known as the Bayes formula.

THE BAYES FORMULA

> Let $B_1, B_2, \ldots, B_n$ form a partition of Ω and A be an event. Then
> $$P(B_i \mid A) = \frac{P(A \mid B_i)P(B_i)}{\sum_{j=1}^{n} P(A \mid B_j)P(B_j)}$$

We will now return to our example. We wish to find $P(B_1 \mid A)$, that is, the probability that a person is infected given a positive result. We partition the population into the two sets B_1 and B_2. Then, using the Bayes formula, we find

$$P(B_1 \mid A) = \frac{P(A \mid B_1)P(B_1)}{P(A \mid B_1)P(B_1) + P(A \mid B_2)P(B_2)}$$

$$= \frac{(0.99)\frac{1}{200}}{(0.99)\frac{1}{200} + (0.05)\frac{199}{200}} \approx 0.090$$

The denominator is equal to $P(A)$, which we already computed in Example 3.

This example is worth discussing in more detail. The probability is quite small that a person is infected given a positive result. But you should compare this to the unconditional probability that a person is infected, namely $P(B_1)$. If we look at the ratio of the conditional to the unconditional probability, we find

$$\frac{P(B_1 \mid A)}{P(B_1)} \approx 18.1$$

That is, if a test result is positive, the chance of actually being infected increases by a factor of 18 when compared to a randomly chosen individual in the population who has not been tested. (In practice, if a test result is positive, more than one test is performed to see whether a person is indeed infected or whether the first result was a false positive.)

If a test result is negative, then the probability that the individual is not infected can also be computed using the Bayes formula. We find

$$P(B_2 \mid A^c) = \frac{P(B_2 \cap A^c)}{P(A^c)} = \frac{P(A^c \mid B_2)P(B_2)}{P(A^c)}$$

$$= \frac{(0.95)\frac{199}{200}}{1 - 0.0547} \approx 0.999947$$

where we used $P(A) = 0.0547$, which we computed in Example 3. This result is rather reassuring.

The reason that the probability of being infected given a positive result is so small comes from the fact that the prevalence of the disease is relatively low (1 in 200). To illustrate this, we treat the prevalence of the disease as a variable and compute $P(B_1 \mid A)$ as a function of the prevalence of the disease. That is, we set

$$p = P(\text{a randomly chosen individual is infected})$$

Using the same test characteristics as before, we find (see Figure 12.18)

$$f(p) = P(B_1 \mid A) = \frac{p(0.99)}{p(0.99) + (1-p)(0.05)} = \frac{0.99p}{0.05 + 0.94p}$$

A graph of $f(p)$ is shown in Figure 12.19. We see that, for small p, $f(p)$ (the probability of being infected given a positive result) is quite small. The ratio

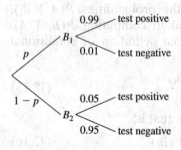

▲ Figure 12.18
The tree diagram for computing the probability that a person is infected given the test came back positive when the prevalence of the disease is p

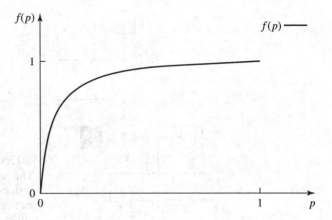

▲ Figure 12.19
The probability of being infected given the test came back positive as a function of
the prevalence of the disease

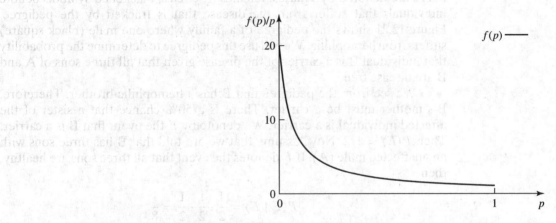

▲ Figure 12.20
The ration $f(p)/p$ as a function of p illustrates by what factor the probability of
being infected increases when the test result is positive compared to the prevalence
of the disease in the population

$f(p)/p$ is shown in Figure 12.20, from which we conclude that though $f(p)$ is
small when p is small, the ratio $f(p)/p$ is quite large for small p.

The Bayes formula is also important in genetic counseling. Hemophilia
is a blood disorder that is characterized by a blood-clotting factor deficiency.
Individuals afflicted with this disease suffer from excessive bleeding. The
disease is caused by an abnormal gene that resides on the X-chromosome. A
female who carries the abnormal gene on one of her X-chromosomes but not
on the other is a carrier of the disease but will not develop symptoms. A male
who carries the abnormal gene on his (only) X-chromosome will develop
symptoms of the disease. Almost all symptomatic individuals are males.

We assume in the following that only one parent carries the abnormal gene. If the father carries the abnormal gene (and thus suffers from
hemophilia), all his daughters will be carriers since they inherit their father's
X-chromosome; but all his sons will be disease free since they inherit their
father's Y-chromosome. If the mother carries the abnormal gene, then her
daughters have a 50% chance of being carriers and her sons have a 50%
chance of suffering from the disease.

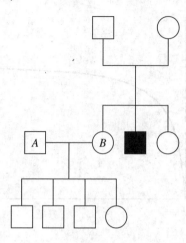

◀ **Figure 12.21**

The pedigree of a family where one member suffers from hemophilia. Squares indicate males, circles females, and the black square shows an afflicted individual

Pedigrees of families show family relationships between individuals. They are indispensable tools to trace diseases of genetic origin. In a pedigree, males are denoted by squares, females by circles; blackened symbols denote individuals that suffer from the disease that is tracked by the pedigree. Figure 12.21 shows the pedigree of a family where one male (black square) suffers from hemophilia. We will use this pedigree to determine the probability that individual B is a carrier of the disease given that all three sons of A and B are disease free.

We see from the pedigree that B has a hemophilic brother. Therefore, B's mother must be a carrier. There is a 50% chance that a sister of the affected individual is a carrier. We denote by E the event that B is a carrier. Then $P(E) = 1/2$. Now, assume that we are told that B has three sons with an unaffected male (A). If F denotes the event that all three sons are healthy, then

$$P(F \mid E) = \frac{1}{2} \cdot \frac{1}{2} \cdot \frac{1}{2} = \frac{1}{8}$$

since if B is a carrier, each son has probability of 1/2 of not inheriting the disease gene and thus being healthy.

We can use Bayes formula to compute the probability that B is a carrier given that none of her three sons has the disease.

$$P(E \mid F) = \frac{P(E \cap F)}{P(F)} = \frac{P(F \mid E)P(E)}{P(F)}$$

To compute the denominator, we must use the law of total probability, as illustrated in the tree diagram in Figure 12.22.

We find

$$P(F) = \frac{1}{2} \cdot \frac{1}{8} + \frac{1}{2} \cdot 1 = \frac{9}{16}$$

Therefore, using the Bayes formula,

$$P(E \mid F) = \frac{\frac{1}{8} \cdot \frac{1}{2}}{\frac{9}{16}} = \frac{1}{9}$$

or, in words, based on the pedigree, the probability that B is a carrier of the gene causing hemophilia given that none of her three sons is symptomatic for the disease is 1/9.

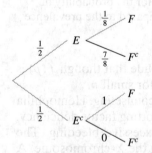

▲ **Figure 12.22**

The sample space is partitioned into two sets, E and E^c, where E is the event that the individual B is a carrier of the hemophilia gene. Based on whether or not B is a carrier, the probability of the event that all three sons are healthy (F) can be computed as shown

12.3.5 Problems

(12.3.1)

1. You draw two cards from a standard deck of 52 cards. Find the probability that the second card is a spade given that the first card is a club.

2. You draw three cards from a standard deck of 52 cards. Find the probability that the third card is a club given that the first two cards were clubs.

3. An urn contains five blue and six green balls. You take two balls out of the urn, one after the other, without replacement. Find the probability that the second ball is green given that the first ball is blue.

4. An urn contains five green, six blue, and four red balls. You take three balls out of the urn, one after the other, without replacement. Find the probability that the third ball is green given that the first two balls were red.

5. A family has two children. The younger one is a girl. Find the probability that the other child is a girl as well.

6. A family has two children. One of their children is a girl. Find the probability that both children are girls.

7. You roll two fair dice. Find the probability that the first die is a 4 given that the sum is 7.

8. You roll two fair dice. Find the probability that the first die is a 5 given that the minimum of the two numbers is a 3.

9. You toss a fair coin three times. Find the probability that the first coin is heads given that at least one head occurred.

10. You toss a fair coin three times. Find the probability that at least two heads occurred given that the second toss resulted in heads.

(12.3.2)

11. A screening test for a disease shows a positive test result in 90% of all cases when the disease is actually present and in 15% of all cases when it is not. Assume that the prevalence of the disease is 1 in 100. If the test is administered to a randomly chosen individual, what is the probability that the result is negative?

12. A screening test for a disease shows a positive test result in 95% of all cases when the disease is actually present and in 20% of all cases when it is not. When the test was administered to a large number of people, 21.5% of the results were positive. What is the prevalence of the disease?

13. A drawer contains three bags numbered 1–3. Bag 1 contains two blue balls, bag 2 contains two green balls, and bag 3 contains one blue and one green ball. You choose one bag at random and take out one ball. Find the probability that the ball is blue.

14. A drawer contains six bags numbered 1–6. Bag i contains i blue balls and 2 green balls. You roll a fair die and then pick a ball out of the bag with the number shown on the die. What is the probability that the ball is blue?

15. You pick two cards from a standard deck of 52 cards. Find the probability that the second card is an ace. Compare this to the probability that the first card is an ace.

16. You pick three cards from a standard deck of 52 cards. Find the probability that the third card is an ace. Compare this to the probability that the first card is an ace.

17. Suppose that you have a batch of red-flowering pea plants of which 40% are of genotype CC and 60% of genotype Cc. You pick one plant at random and cross it with a white-flowering pea plant. Find the probability that the offspring of this cross will have white flowers.

18. Suppose that you have a batch of red- and white-flowering pea plants where all three genotypes $CC, Cc,$ and cc are equally represented. You pick one plant at random and cross it with a white-flowering pea plant. What is the probability that the offspring will have red flowers?

19. A bag contains two coins: One is a fair coin; the other one has two heads. You pick one coin at random and flip it. Find the probability that the outcome is heads.

20. A drug company claims that a new headache drug will bring instant relief in 90% of all cases. If a person is treated with a placebo, there is a 10% chance the person will feel instant relief. In a clinical trial, half the subjects are treated with the new drug; the other half receive the placebo. If an individual from this trial is chosen at random, what is the probability that the person will have experienced instant relief?

(12.3.3)

21. You are dealt one card from a standard deck of 52 cards. If A denotes the event that the card is a spade and if B denotes the event that the card is an ace, determine whether A and B are independent.

22. You are dealt two cards from a standard deck of 52 cards. If A denotes the event that the first card is an ace and B denotes the event that the second card is an ace, determine whether A and B are independent.

23. An urn contains five green and six blue balls. You take two balls out of the urn, one after the other, without replacement. If A denotes the event that the first ball is green and B denotes the event that the second ball is green, determine whether A and B are independent.

24. An urn contains five green and six blue balls. You take one ball out of the urn, note its color, and replace it. You then take a second ball out of the urn, note its color, and replace it. If A denotes the event that the first ball is green and B denotes the event that the second ball is green, determine whether A and B are independent.

25. Assume a 1 : 1 sex ratio. A family has three children. Find the probability of the event A, where

(a) $A = \{$all children are girls$\}$

(b) $B = \{$at least one boy$\}$

(c) $C = \{$at least two girls$\}$

(d) $D = \{$at most two boys$\}$

26. Assume that 20% of a very common insect species in your study area is parasitized. Assume that insects are parasitized independently of each other. If you collect ten specimens of this species, what is the probability that at least one specimen in your sample is parasitized?

27. A multiple choice question has four choices, and a test has a total of ten multiple choice questions. A student only passes the test if he or she answers all questions correctly. If the student guesses the answers to all questions randomly, find the probability that he or she will pass.

28. Assume that A and B are disjoint and that both events have positive probability. Are they independent?

29. Assume that the probability that an insect species lives more than five days is 0.1. Find the probability that in a sample of size 10 of this species at least one insect will still be alive after five days.

30. (a) Use a Venn diagram to show that
$$(A \cup B)^c = A^c \cap B^c$$

(b) Use your result in (a) to show that if A and B are independent, then A^c and B^c are independent.

(c) Use your result in (b) to show that if A and B are independent, then
$$P(A \cup B) = 1 - P(A^c)P(B^c)$$
(12.3.4)

31. A screening test for a disease shows a positive result in 95% of all cases when the disease is actually present and in 10% of all cases when it is not. If the prevalence of the disease is 1 in 50, and an individual tests positive, what is the probability that the individual actually has the disease?

32. A screening test for a disease shows a positive result in 95% of all cases when the disease is actually present and in 10% of all cases when it is not. If a result is positive, the test is repeated. Assume that the second test is independent of the first test. If the prevalence of the disease is 1 in 50, and an individual tests positive twice, what is the probability that the individual actually has the disease?

33. A bag contains two coins; one is a fair coin, and the other one has two heads. You pick one coin at random and flip it. What is the probability that you picked the fair coin given that the outcome of the toss was heads?

34. You pick two cards from a standard deck of 52 cards. Find the probability that the first card was a spade given that the second card was a spade.

35. Suppose a woman has a hemophilic brother and one healthy son. Furthermore, assume that neither her mother nor her father were hemophilic but that her mother was a carrier for hemophilia. Find the probability that she is a carrier of the hemophilia gene.

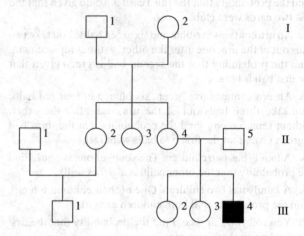

▲ **Figure 12.23**
The pedigree for Problems 36 and 37. The solid black square (individual III-4) represents an afflicted male

The pedigree in Figure 12.23 shows a family in which one member (III-4) is hemophilic. In Problems 36–37, refer to this pedigree.

36. (a) Given the pedigree, find the probability that the individual I-2 is a carrier of the hemophilia gene.

(b) Given the pedigree, find the probability that II-3 is a carrier of the hemophilia gene.

37. (a) Given the pedigree, find the probability that II-3 is a carrier of the hemophilia gene.

(b) Given the pedigree, find the probability that III-2 is a carrier of the hemophilia gene.

(c) Given the pedigree, find the probability that II-2 is a carrier of the hemophilia gene.

12.4 DISCRETE RANDOM VARIABLES AND DISCRETE DISTRIBUTIONS

Outcomes of random experiments frequently are real numbers, such as the number of heads in a coin tossing experiment, the number of seeds produced in a cross between two plants, or the lifespan of an insect. Such numerical outcomes can be described by **random variables**. A random variable is a function from the sample space Ω into the set of real numbers. Random variables are typically denoted by X, Y, or Z, or other capital letters chosen from the end of the alphabet. For instance,

$$X : \Omega \to \mathbf{R}$$

describes the random variable X as a map from the sample space Ω into the set of real numbers.

Random variables are classified according to their range. If the number of values that X can take is either finite or countably infinite,[1] X is called a **discrete random variable**. If X takes on a continuous range of values—for instance, values that range over an interval—X is called a **continuous random variable**. Discrete random variables are the topic of this section; continuous random variables are the topic of the next section.

12.4.1 Discrete Distributions

In the first example, we look at a random variable that takes on a finite number of values.

▶ **Example 1**   Toss a fair coin three times. Let X be a random variable that counts the number of heads in each outcome. The sample space is

$$\Omega = \{HHH, HHT, HTH, THH, HTT, THT, TTH, TTT\}$$

and the random variable

$$X : \Omega \to \mathbf{R}$$

takes on values 0, 1, 2, or 3. For instance,

$$X(HHH) = 3 \quad \text{or} \quad X(TTH) = 1 \quad \text{or} \quad X(TTT) = 0 \quad ◀$$

In the next example, we look at a random variable that takes on a countably infinite number of values.

▶ **Example 2**   Toss a fair coin repeatedly until the first time heads appears. Let Y be a random variable that counts the number of trials until the first time heads shows up. The sample space is

$$\Omega = \{H, TH, TTH, TTTH, \ldots\}$$

and the random variable

$$Y : \Omega \to \mathbf{R}$$

takes on values 1, 2, 3, ... For instance,

$$Y(H) = 1 \quad Y(TH) = 2 \quad Y(TTH) = 3 \quad \ldots \quad ◀$$

We will now turn to the problem of how to assign probabilities to the different values of a random variable X. For the moment, we will restrict the discussion to the case when the range of X is finite.

Let's go back to Example 1. The coin in Example 1 is fair. This means that each outcome in Ω has the same probability, namely 1/8. We can translate this into probabilities for X. For instance,

$$P(X = 1) = P(\{HTT, THT, TTH\})$$
$$= P(HTT) + P(THT) + P(TTH)$$
$$= \frac{1}{8} + \frac{1}{8} + \frac{1}{8} = \frac{3}{8}$$

---

[1]A countably infinite set has infinitely many elements that can be enumerated, like the set of integers. (The set of rational numbers is also countably infinite; the set of real numbers is uncountably infinite.)

We can perform similar computations for all other values of X. The table on the left summarizes the results.

The function $p(x) = P(X = x)$ is called a **probability mass function**. Note that $p(x) \geq 0$ and $\sum_x p(x) = 1$; these are defining properties of a probability mass function.

x	$P(X = x)$
0	1/8
1	3/8
2	3/8
3	1/8

> **Definition** A random variable is called a discrete random variable if it takes on a finite or countably infinite number of values. The probability distribution of X can be described by the probability mass function $p(x) = P(X = x)$, which has the following properties:
>
> **1.** $p(x) \geq 0$
> **2.** $\sum_x p(x) = 1$, where the sum is over all values of X with $P(X = x) > 0$

The probability mass function is one way to describe the probability distribution of a discrete random variable. Another important function that describes the probability distribution of a random variable X is the (cumulative) distribution function $F(x) = P(X \leq x)$. This function is defined for *any* random variable, not just discrete ones.

> **Definition** The **(cumulative) distribution function** $F(x)$ of a random variable X is defined as
>
> $$F(x) = P(X \leq x)$$

Instead of "cumulative distribution function" we will simply say "distribution function."

The probability mass function and the distribution function are equivalent ways of describing the probability distribution of a discrete random variable, and we can obtain one from the other, as illustrated in the following two examples.

▶ **Example 3** Suppose that the probability mass function of a discrete random variable X is given by the table on the left. Find and graph the corresponding distribution function $F(x)$.

Solution

The function $F(x)$ is defined for all values of $x \in \mathbf{R}$. For instance, $F(-2.3) = P(X \leq -2.3) = P(\emptyset) = 0$ or $F(1) = P(X \leq 1) = P(X = -1 \text{ or } 0) = 0.3$. Since $F(x) = P(X \leq x)$, we must be particularly careful when x is in the range of X. To illustrate this, we compute $F(1.4)$ and $F(1.5)$. We find

x	$P(X = x)$
-1	0.1
0	0.2
1.5	0.05
3	0.15
5	0.5

$$F(1.4) = P(X \leq 1.4) = P(X = -1 \text{ or } 0) = 0.1 + 0.2 = 0.3$$

$$F(1.5) = P(X \leq 1.5) = P(X = -1, 0, \text{ or } 1.5) = 0.1 + 0.2 + 0.05 = 0.35$$

The distribution function $F(x)$ is a piecewise defined function. We find

$$F(x) = \begin{cases} 0 & \text{for } x < -1 \\ 0.1 & \text{for } -1 \leq x < 0 \\ 0.3 & \text{for } 0 \leq x < 1.5 \\ 0.35 & \text{for } 1.5 \leq x < 3 \\ 0.5 & \text{for } 3 \leq x < 5 \\ 1 & \text{for } x \geq 5 \end{cases}$$

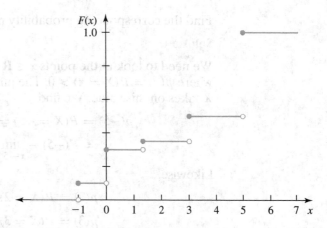

▲ **Figure 12.24**
The distribution function $F(x)$ of Example 3. The solid circles on the left ends of the line segments indicate that the distribution function takes on this value at the points where the function jumps

The graph of $F(x)$ is shown in Figure 12.24. ◀

Looking at Figure 12.24, we see that the graph of $F(x)$ is a nondecreasing and piecewise constant function that takes jumps at those values x where $P(X = x) > 0$. The function $F(x)$ is right continuous; that is, for any $c \in \mathbf{R}$,

$$\lim_{x \to c^+} F(x) = F(c)$$

It is not left continuous everywhere since at values $c \in \mathbf{R}$ where $P(X = c) > 0$,

$$\lim_{x \to c^-} F(x) \neq F(c)$$

For instance, when $c = 3$,

$$\lim_{x \to 3^-} F(x) = 0.35 \neq F(3) = 0.5$$

Furthermore,

$$\lim_{x \to -\infty} F(x) = 0 \quad \text{and} \quad \lim_{x \to \infty} F(x) = 1$$

These properties are characteristic of the distribution function of a discrete random variable.

It is possible to obtain the probability mass function from the distribution function. Let's look at the distribution function of Example 3. The function jumps at $x = 3$ and the jump height is 0.15. Since $F(x) = P(X \leq x)$, it follows that

$$p(3) = P(X = 3) = P(X \leq 3) - P(X < 3)$$

$$= F(3) - \lim_{x \to 3^-} F(x) = 0.5 - 0.35 = 0.15$$

We see that the distribution function jumps at the values of X for which $P(X = x) > 0$. The jump height is then equal to the probability that X takes on this value.

▶ **Example 4** Suppose the distribution function of a discrete random variable X is given by

$$F(x) = \begin{cases} 0 & \text{for} & x < -5 \\ 0.2 & \text{for} & -5 \leq x < 2 \\ 0.6 & \text{for} & 2 \leq x < 3 \\ 0.7 & \text{for} & 3 \leq x < 6.5 \\ 1.0 & \text{for} & x \geq 6.5 \end{cases}$$

Find the corresponding probability mass function.

Solution

We need to look at the points $x \in \mathbf{R}$ where $F(x)$ jumps. Those are the points where $p(x) = P(X = x) > 0$. The jump height is equal to the probability that X takes on this value. We find

$$p(-5) = P(X = -5) = P(X \le -5) - P(X < -5)$$

$$= F(-5) - \lim_{x \to 5^-} F(x) = 0.2 - 0.0 = 0.2$$

Likewise,

$$p(2) = P(X = 2) = 0.6 - 0.2 = 0.4$$

$$p(3) = P(X = 3) = 0.7 - 0.6 = 0.1$$

$$p(6.5) = P(X = 6.5) = 1.0 - 0.7 = 0.3$$

There are no other values of x where $P(X = x) > 0$. ◀

12.4.2 Mean and Variance

Knowing the distribution of a random variable tells us everything about the random variable. In practice, however, it is often impossible or unnecessary to know the full probability distribution of a random variable that describes a particular random experiment. Instead, it might suffice to determine a few characteristic quantities, such as the average value and a measure that describes the spread around the average value.

The Average Value or the Mean of a Discrete Random Variable

▶ **Example 5**   Clutch size can be thought of as a random variable. Let X denote the number of eggs per clutch laid by a certain bird species, and let's assume that the distribution of X is described by the probability mass function in the table on the left. The average number of eggs per clutch is computed as the weighted sum

$$\text{average value} = \sum_x x P(X = x)$$

$$= (1)(0.05) + (2)(0.1) + (3)(0.2)$$

$$+(4)(0.3) + (5)(0.25) + (6)(0.1) = 3.9$$

and we find that the average clutch size is 3.9. ◀

The average value of X is called the **expected value** or **mean** of X and denoted by EX. It is a very important quantity. Here is its definition.

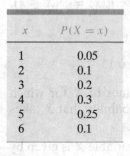

x	$P(X = x)$
1	0.05
2	0.1
3	0.2
4	0.3
5	0.25
6	0.1

Definition If X is a discrete random variable, then the expected value or mean of X is

$$EX = \sum_x x P(X = x)$$

where the sum is over all values of x with $P(X = x) > 0$.

When the range of X is finite, the sum in the definition is always defined. When the range of X is countably infinite, we must sum an infinite number of terms. Such sums can be infinite, depending on the distribution of X. The expected value of X is only defined if both $\sum_{x<0} x P(X = x)$ and $\sum_{x\geq 0} x P(X = x)$ are finite. Determining whether such infinite sums are finite is beyond the scope of this chapter, and we will therefore restrict the discussion to cases where these sums are finite.

To see that the definition of the mean of a discrete random variable coincides with our everyday notion of average values, we look at the following example.

▶ Example 6   On a winter day somewhere in southern Minnesota, the following temperature readings T_k (in Fahrenheit) at hour k were obtained.

k	0	1	2	3	4	5	6	7	8	9	10	11	12
T_k	6	6	6	5	5	5	5	5	8	10	12	12	12

k	13	14	15	16	17	18	19	20	21	22	23
T_k	12	12	12	10	8	8	8	5	5	3	3

The average temperature on that day based on these hourly observations, denoted by $\overline{T}$, is

$$\overline{T} = \frac{1}{24}(6 + 6 + 6 + 5 + 5 + 5 + 5 + 5 + 8 + 10 + 12$$
$$+ 12 + 12 + 12 + 12 + 12 + 10 + 8 + 8 + 8 + 5 + 5 + 3 + 3)$$
$$= \frac{183}{24} = 7.625$$

Rearranging these values according to size, we find

$$\overline{T} = \frac{1}{24}[(3 + 3) + (5 + 5 + 5 + 5 + 5 + 5 + 5) + (6 + 6 + 6)$$
$$+ (8 + 8 + 8 + 8) + (10 + 10) + (12 + 12 + 12 + 12 + 12 + 12)]$$
$$= \frac{1}{24}[(3)(2) + (5)(7) + (6)(3) + (8)(4) + (10)(2) + (12)(6)]$$
$$= 3 \cdot \frac{2}{24} + 5 \cdot \frac{7}{24} + 6 \cdot \frac{3}{24} + 8 \cdot \frac{4}{24} + 10 \cdot \frac{2}{24} + 12 \cdot \frac{6}{24}$$
$$= \sum [\text{temperature}] \times [\text{relative frequency of that temperature}]$$
$$= \frac{183}{24} = 7.625 \qquad \blacktriangleleft$$

In Example 6, we introduced the notion of a **relative frequency**, which tells us how often a value appears in a sample relative to the total sample size. For instance, 3°F appears twice in this sample of 24 measurements, so its relative frequency is 2/24.

If we interpret the relative frequencies as probabilities, we see that $\overline{T}$ in Example 6 is indeed the expected value of the temperature T on that day.

▶ Example 7   The following table contains the number of leaves per basil plant in a sample of 25 basil plants.

16	15	13	16	16
14	16	15	18	17
16	18	16	13	16
16	16	15	15	16
15	18	16	16	15

To find the relative frequency distribution, we must count how often each value occurs and then divide by the sample size, which is 25 in this case. This is summarized in the following table.

No. of leaves	13	14	15	16	17	18
Relative frequency	$\frac{2}{25}$	$\frac{1}{25}$	$\frac{6}{25}$	$\frac{12}{25}$	$\frac{1}{25}$	$\frac{3}{25}$

We interpret relative frequencies as probabilities. If the random variable X denotes the number of leaves per plant with probability distribution given by the relative frequency distribution, then the expected value of the number of leaves per plant is

$$EX = 13 \cdot \frac{2}{25} + 14 \cdot \frac{1}{25} + 15 \cdot \frac{6}{25} + 16 \cdot \frac{12}{25} + 17 \cdot \frac{1}{25} + 18 \cdot \frac{3}{25}$$

$$= 393 \cdot \frac{1}{25} = 15.72$$

Note that though the number of leaves per plant is an integer value, the average number of leaves per plant is not. You would actually lose valuable information if you rounded the average number to the closest integer. ◀

It is important to realize that the expected value of an integer-valued random variable need not be an integer. To emphasize this point, consider the average number of lifetime births expected by women 18 to 34 years old in 1992. (The following data are taken from the U.S. Bureau of the Census 1994.) The number of lifetime births expected by a woman who is not a high school graduate is 2.393; whereas the corresponding number for a woman with a graduate or professional degree is 1.990. If we rounded these numbers to the closest integer, they would be the same, namely 2; we would no longer see the difference between these groups.

We can extend the definition of the expected value of X to the expected value of a function of X. Let $g(x)$ be a function of x. Then

$$Eg(X) = \sum_x g(x) P(X = x) \tag{12.18}$$

▶ Example 8 Compute EX^2 for the random variable X in Example 5.

Solution

Using the probability mass function given in Example 5, we find

$$EX^2 = \sum_x x^2 P(X = x)$$

$$= (1)^2(0.05) + (2)^2(0.1) + (3)^2(0.2) + (4)^2(0.3) + (5)^2(0.25) + (6)^2(0.1)$$

$$= 16.9$$ ◀

The Variance of a Discrete Random Variable   Another important quantity that characterizes the distribution of a random variable is called the **variance**. It describes how spread out the range of the random variable is. To motivate the definition, let's look at the two random variables X and Y, with probability mass functions as follows.

k	$P(X = k)$	$P(Y = k)$
−10	0	0.2
−1	0.2	0
0	0.6	0.6
1	0.2	0
10	0	0.2

We illustrate these two distributions in Figure 12.25. Both random variables have mean 0, but the range of Y is much more spread out than the range of X. To capture this in a single quantity, we will compute a weighted average of the squared distances to the mean, which is called the variance.

> **Definition** For any random variable X with mean μ, the variance of X is defined as
> $$\text{var}(X) = E(X - \mu)^2$$
> If X is a discrete random variable, then
> $$\text{var}(X) = \sum_x (x - \mu)^2 P(X = x)$$

Since the variance is an average value of a squared quantity, it is always nonnegative.

Let's return to the random variables X and Y: Since their means are both equal to 0, their variances are

$$\text{var}(X) = (-1 - 0)^2(0.2) + (0 - 0)^2(0.6) + (1 - 0)^2(0.2) = 0.4$$

$$\text{var}(Y) = (-10 - 0)^2(0.2) + (0 - 0)^2(0.6) + (10 - 0)^2(0.2) = 40$$

We see that the variance of Y is larger than the variance of X, reflecting the fact that the range of Y is more spread out than the range of X.

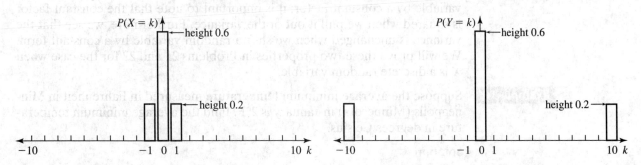

▲ **Figure 12.25**
The probability mass functions of X and Y. The distribution of Y is more spread out than the distribution of X

The variance of X is often denoted by σ^2 (read: "sigma squared"). A quantity that is closely related to the variance is the **standard deviation**, denoted by s.d. or σ. It is defined as the square root of the variance,

$$\text{s.d.} = \sigma = \sqrt{\text{var}(X)}$$

The standard deviation has the advantage that it has the same units as the mean and so can be interpreted more easily than the variance.

▶ **Example 9**   Compute the variance and the standard deviation of the number of leaves per plant in Example 7.

Solution

Denote by X the random variable that counts the number of leaves per plant with probability distribution given in the table of Example 7. In Example 7, we found that $EX = 15.72$. Therefore, the variance of X is

$$\text{var}(X) = (13 - 15.72)^2 \frac{2}{25} + (14 - 15.72)^2 \frac{1}{25} + (15 - 15.72)^2 \frac{6}{25}$$

$$+ (16 - 15.72)^2 \frac{12}{25} + (17 - 15.72)^2 \frac{1}{25} + (18 - 15.72)^2 \frac{3}{25}$$

$$= 1.5616$$

and the standard deviation of X is

$$\text{s.d.}(X) = \sqrt{\text{var}(X)} = \sqrt{1.5616} \approx 1.2496 \qquad \blacktriangleleft$$

We will now collect some important rules regarding expected values and variances. The first rule tells us how to compute the expected value and the variance of a linear transformation of X. This rule holds for any random variable, not just discrete ones.

> Let a and b be constants. Then
>
> $$E(aX + b) = a(EX) + b$$
> $$\text{var}(aX + b) = a^2 \text{var}(X)$$

The first property says that the expected value of a linear function of X is the linear function evaluated at the expected value of X. The second property tells us what happens to the variance when we multiply a random variable by a constant factor; it is important to note that the constant factor is squared when we pull it out of the variance. Furthermore, we see that the variance is unchanged when we shift a random variable by a constant term. We will prove these two properties in Problems 21 and 22 for the case when X is a discrete random variable.

▶ **Example 10**   Suppose the average minimum temperature measured in Fahrenheit in Minneapolis (Minnesota) in January is 2°F. Find the average minimum temperature in degrees Celsius.

Solution

The linear transformation

$$C = \frac{5(F - 32)}{9}$$

converts temperature measured in Fahrenheit (F) into temperature measured in degrees Celsius (C). Hence,

$$EC = E\left[\frac{5(F-32)}{9}\right] = 5\frac{EF-32}{9}$$

$$= 5\frac{2-32}{9} = -\frac{150}{9} \approx -16.67$$

and we find that the average minimum temperature in January in Minneapolis is about $-16.67°$C.

▶ **Example 11** Find a formula that converts the variance of a temperature measured in degrees Celsius into the variance of the temperature measured in degrees Fahrenheit.

Solution

We use the linear transformation of Example 10 that relates a temperature measured in Fahrenheit (F) to the temperature measured in degrees Celsius (C)

$$C = \frac{5(F-32)}{9}$$

Solving this for F, we find

$$F = \frac{9}{5}C + 32$$

Therefore,

$$\mathrm{var}(F) = \mathrm{var}\left(\frac{9}{5}C + 32\right) = \left(\frac{9}{5}\right)^2 \mathrm{var}(C)$$

$$= \frac{81}{25}\mathrm{var}(C) = (3.24)\mathrm{var}(C)$$

It is often necessary to look at sums of random variables. We collect some rules without proof. Let X and Y be two random variables. Then $X + Y$ is also a random variable, and we have

$$E(X + Y) = EX + EY$$

This formula holds for any random variables, not just discrete ones.

▶ **Example 12** Suppose the average number of women who enter a coffee shop during lunch hour is 52.2 and the average number of men is 47.3. Find the average total number of people entering the coffee shop during lunch hour.

Solution

If we denote the number of women by X and the number of men by Y, then we are interested in finding $E(X + Y)$. With $EX = 52.2$ and $EY = 47.3$, we have

$$E(X + Y) = EX + EY = 52.2 + 47.3 = 99.5$$

We can use our rules to find an alternate formula for the variance. We start with

$$(X - \mu)^2 = X^2 - 2X\mu + \mu^2$$

Taking expectations on both sides, we find

$$E(X - \mu)^2 = E(X^2 - 2\mu X + \mu^2)$$

Since the expectation of a sum is the sum of the expectations, the right-hand side simplifies to

$$EX^2 - E(2\mu X) + E\mu^2 = EX^2 - 2\mu EX + \mu^2 = EX^2 - (EX)^2$$

since $\mu = EX$, $E\mu^2 = \mu^2 = (EX)^2$, and $E(2\mu X) = 2\mu EX = 2(EX)^2$. With $E(X - \mu)^2 = \text{var}(X)$, we have

$$\text{var}(X) = EX^2 - (EX)^2$$

This formula is often more convenient to use, since it leads to algebraically simpler expressions. Note that $EX^2 \neq (EX)^2$ unless $\text{var}(X) = 0$ and that $EX^2 \geq (EX)^2$ since $\text{var}(X) \geq 0$. We apply this formula to the random variable X in Example 5.

▶ **Example 13** Use the random variable X in Example 5, the result of Example 8, and the preceding formula to compute the variance of X.

Solution

In Example 5, we found $EX = 3.9$. In Example 8, we computed EX^2 and found

$$EX^2 = 16.9$$

Hence,

$$\text{var}(X) = 16.9 - (3.9)^2 = 1.69 \qquad \blacktriangleleft$$

Joint Distributions   It is often important to investigate the relationship between random variables.

▶ **Example 14** Gout is a type of arthritis in which uric acid is deposited in crystalline form within joints. A medical study might focus on whether the disease prevalence is gender dependent. A survey in 1986 revealed that about 13.6 per 1000 men and 6.4 per 1000 women are affected. We can treat gender as one random variable and the presence of gout as another by defining

$$X = \begin{cases} 1 & \text{if male} \\ 0 & \text{if female} \end{cases}$$

and

$$Y = \begin{cases} 1 & \text{if gout is present} \\ 0 & \text{if gout is not present} \end{cases}$$

The following table lists the number of individuals in each of the four combinations in a study of 10,000 men and 10,000 women.

	$X = 0$	$X = 1$	Total
$Y = 0$	9936	9864	19,800
$Y = 1$	64	136	200
Total	10,000	10,000	20,000

We see that the fraction of individuals in this study that are both male and affected by gout is $136/20{,}000 = 0.0068$. If we interpret this as a probability, we could write

$$P(X = 1, Y = 1) = 0.0068$$

Converting all numbers in the table into relative frequencies produces the **joint probability distribution** of X and Y.

	$X = 0$	$X = 1$	Total
$Y = 0$	0.4968	0.4932	0.99
$Y = 1$	0.0032	0.0068	0.01
Total	0.5	0.5	1.0

In general, when X and Y are discrete random variables, we define the joint probability distribution of X and Y by

$$p(x, y) = P(X = x, Y = y)$$

for all values of x in the range of X and all values of y in the range of Y. We can obtain the distribution of X or of Y, called the **marginal distribution** as follows:

$$p_X(x) = P(X = x) = \sum_y P(X = x, Y = y)$$

$$p_Y(y) = P(Y = y) = \sum_x P(X = x, Y = y)$$

▶ **Example 15** Use the data from Example 14 to determine the probability that a randomly chosen person in this study has gout.

Solution

We want to find the probability that $Y = 1$:

$$P(Y = 1) = P(X = 0, Y = 1) + P(X = 1, Y = 1)$$
$$= 0.0032 + 0.0068 = 0.01$$

We can define conditional probabilities as in Section 12.3:

$$P(X = x \mid Y = y) = \frac{P(X = x, Y = y)}{P(Y = y)} \tag{12.19}$$

provided $P(Y = y) > 0$.

▶ **Example 16** Use the data of Example 14 to determine the probability that a randomly chosen man in the study has gout.

Solution

We want to find $P(Y = 1 \mid X = 1)$. Using (12.19), we find

$$P(Y = 1 \mid X = 1) = \frac{P(X = 1, Y = 1)}{P(X = 1)} = \frac{0.0068}{0.5} = 0.0136$$

or 13.6 per 1000 men. ◄

Whether or not gender influences the prevalence of gout leads us to the concept of **independence** of random variables. The definition of independence of random variables follows from that of the independence of events. Namely, X and Y are **independent** if for any two events A and B,

$$P(X \in A, Y \in B) = P(X \in A)P(Y \in B)$$

When X and Y are discrete random variables, this simplifies to

$$P(X = x, Y = y) = P(X = x)P(Y = y) \tag{12.20}$$

for all values of x in the range of X and for all values of y in the range of Y.

▶ **Example 17**   Use the data of Example 14 to determine whether X and Y are independent.

Solution

We check whether $P(X = 1, Y = 1)$ is equal to $P(X = 1)P(Y = 1)$. We find $P(X = 1, Y = 1) = 0.0068$, $P(X = 1) = 0.5$, and $P(Y = 1) = 0.01$. Consequently, $P(X = 1)P(Y = 1) = (0.5)(0.01) = 0.005$, which is different from $P(X = 1, Y = 1)$. We conclude that X and Y are not independent; they are said to be dependent. ◄

▶ **Example 18**   Suppose that X and Y are two independent discrete random variables with probability mass functions as listed in the table:

k	$P(X = k)$	$P(Y = k)$
-1	0.1	0.3
0	0.0	0.2
1	0.7	0.1
2	0.2	0.4

(a) Find the probability that X takes on value -1 and Y takes on value 2.

(b) Find the probability that X is negative and Y is positive.

Solution

(a) We want to find $P(X = -1, Y = 2)$. Using that X and Y are independent, we have

$$P(X = -1, Y = 2) = P(X = -1)P(Y = 2) = (0.1)(0.4) = 0.04$$

(b) We want to find the probability of the event that X is negative and Y is positive. This is the event $\{X = -1 \text{ and } Y = 1 \text{ or } 2\}$. Since X and Y are independent, we have

$$P(X = -1 \text{ and } Y = 1 \text{ or } 2) = P(X = -1)P(Y = 1 \text{ or } 2)$$
$$= (0.1)(0.1 + 0.4) = (0.1)(0.5) = 0.05 ◄$$

The definition of independence in (12.20) allows us to find the expected value of a product of independent discrete random variables. The following calculation shows the result when X and Y have finite range. Namely, in this case, we have

$$E(XY) = \sum_{x,y} xy P(X = x, Y = y) = \sum_{x,y} xy P(X = x)P(Y = y)$$

$$= \sum_x x P(X = x) \sum_y y P(Y = y) = (EX)(EY)$$

This holds more generally.

> If X and Y are two independent random variables, then
> $$E(XY) = (EX)(EY)$$

▶ **Example 19** For the random variables X and Y in Example 18, find $E(XY)$.

Solution

Since X and Y are independent, we have $E(XY) = (EX)(EY)$. Now,

$$EX = (-1)(0.1) + (0)(0.0) + (1)(0.7) + (2)(0.2) = 1.0$$

and

$$EY = (-1)(0.3) + (0)(0.2) + (1)(0.1) + (2)(0.4) = 0.6$$

Hence,

$$E(XY) = (EX)(EY) = (1.0)(0.6) = 0.6$$ ◀

We can use the rule about the expected value of a product of independent random variables to compute the variance of the sum of two independent random variables. Namely, suppose that X and Y are independent. Then

$$\text{var}(X + Y) = E(X + Y)^2 - [E(X + Y)]^2$$
$$= E(X^2 + 2XY + Y^2) - (EX + EY)^2$$
$$= EX^2 + 2E(XY) + EY^2 - (EX)^2 - 2(EX)(EY) - (EY)^2$$

Since X and Y are independent, $E(XY) = (EX)(EY)$, and the sum simplifies to

$$EX^2 - (EX)^2 + EY^2 - (EY)^2$$

But this we recognize as $\text{var}(X) + \text{var}(Y)$. Hence,

> If X and Y are independent random variables, then
> $$\text{var}(X + Y) = \text{var}(X) + \text{var}(Y)$$

Both formulas, $E(XY) = (EX)(EY)$ and $\text{var}(X + Y) = \text{var}(X) + \text{var}(Y)$, hold for any independent random variables X and Y, not just for discrete ones. However, we will use these identities only in the context of discrete random variables. It is important to keep in mind that these two formulas only hold when X and Y are independent.

In the remaining subsections of this section, we introduce a number of important discrete distributions.

12.4.3 The Binomial Distribution

In this subsection, we will discuss a discrete random variable that models the number of successes among a fixed number of trials. Suppose that you perform a random experiment of repeated trials where each trial has two possible outcomes, success or failure. Each trial is called a **Bernoulli trial**. The trials are independent and the probability of success in each trial is p. We define the random variables $X_k, k = 1, 2, \ldots, n$, as

$$X_k = \begin{cases} 1 & \text{if the } k\text{th trial is successful} \\ 0 & \text{otherwise} \end{cases}$$

Then $P(X_k = 1) = p = 1 - P(X_k = 0)$ for $k = 1, 2, \ldots, n$.

If we repeat these trials n times, we might want to know the total number of successes. We set

$$S_n = \text{number of successes in } n \text{ trials}$$

We can define S_n in terms of the random variables X_k, namely

$$S_n = \sum_{k=1}^{n} X_k \tag{12.21}$$

Since the trials are independent, this representation shows that S_n can be written as a sum of independent random variables, all having the same distribution. We will use this representation subsequently.

The random variable S_n is discrete and takes on values $0, 1, 2, \ldots, n$. To find its probability mass function $p(k) = P(S_n = k)$, we argue as follows. The event $\{S_n = k\}$ can be represented as a string of 0s and 1s of length n, where 0 represents failure and 1 represents success. For instance, if $n = 5$ and $k = 3$, then 01101 could be interpreted as the outcome of five trials, the first resulting in failure, followed by two successes, then a failure, and finally a success. The probability of this particular outcome is easy to compute, since the trials are independent. We find

$$P(01101) = (1 - p)pp(1 - p)p = p^3(1 - p)^2$$

The outcome 01101 is not the only one with three successes in five trials. Any other string of length 5 with exactly three 1s has the same property. To determine the number of different strings with this property, note that there are $\binom{5}{3}$ different ways of placing the three 1s in the five possible positions, and there is exactly one way to place the 0s in the remaining two positions. Hence there are $\binom{5}{3} \cdot 1 = \binom{5}{3}$ different strings of length 5 with exactly three 1s. There is another way to find this. Namely, there are 5! ways of arranging the three 1s and the two 0s if the 0s and the 1s were distinguishable. Since the 0s and the 1s can be rearranged among themselves without changing the outcome, we must divide by the order. We then find that there are

$$\frac{5!}{3!2!} = \binom{5}{3}$$

different outcomes. As all outcomes are equally likely, we find

$$P(S_5 = 3) = \binom{5}{3}p^3(1 - p)^2$$

We can use similar reasoning to derive the general formula. We summarize this in the following.

BINOMIAL DISTRIBUTION

Let S_n be a random variable that counts the number of successes in n independent trials, each having probability p of success. Then S_n is said to be **binomially distributed** with parameters n and p, and

$$P(S_n = k) = \binom{n}{k} p^k (1-p)^{n-k}, \quad k = 0, 1, 2, \ldots, n$$

The random variable S_n is called a **binomial** random variable and its distribution is called the **binomial distribution**.

▶ **Example 20**   Toss a fair coin four times. Find the probability that there are exactly three heads.

Solution

Let S_4 denote the number of heads. If heads denote a success, then the probability of success is $p = 1/2$. S_4 is thus binomially distributed with parameters $n = 4$ and $p = 1/2$. Therefore,

$$P(S_4 = 3) = \binom{4}{3} \left(\frac{1}{2}\right)^3 \left(1 - \frac{1}{2}\right) = 4 \cdot \frac{1}{16} = \frac{1}{4} \quad ◀$$

▶ **Example 21**   In a shipment of ten boxes, each box has probability 0.2 of being damaged. Find the probability of having two or more damaged boxes in the shipment.

Solution

Let S_{10} denote the number of damaged boxes in the shipment. S_{10} is binomially distributed with parameters $n = 10$ and $p = 0.2$. The event of two or more damaged boxes can then be written as $S_{10} \geq 2$. To compute $P(S_{10} \geq 2)$, we use that

$$P(S_{10} \geq 2) = 1 - P(S_{10} < 2) = 1 - [P(S_{10} = 0) + P(S_{10} = 1)]$$

$$= 1 - \left[\binom{10}{0}(0.2)^0(0.8)^{10} + \binom{10}{1}(0.2)(0.8)^9 \right]$$

$$\approx 0.6242 \quad ◀$$

▶ **Example 22**   Down's syndrome or trisomy 21 is a genetic disorder in which three copies of chromosome 21 instead of two copies are present. In the US, the prevalence is about 1 in 700 pregnancies. What is the probability that at least one in 100 pregnancies is affected?

Solution

If S_{100} is the number of pregnancies and $p = 1/700$ is the probability of a pregnancy being affected, then S_{100} is binomially distributed with parameters $n = 100$ and $p = 1/700$. Thus

$$P(S_{100} \geq 1) = 1 - P(S_{100} = 0)$$

$$= 1 - \left(1 - \frac{1}{700}\right)^{100} \approx 0.1332 \quad ◀$$

If we use the representation $S_n = \sum_{k=1}^{n} X_k$ in (12.21) for the binomial random variable S_n, it is straightforward to compute its mean and its variance. We find

$$EX_1 = (1)p + (0)(1-p) = p$$

and, with $EX_1^2 = (1)^2 p + (0)^2(1 - p) = p$,

$$\text{var}(X_1) = EX_1^2 - (EX_1)^2 = p - p^2 = p(1 - p)$$

Since all $X_k, k = 1, 2, \ldots, n$, have the same distribution,

$$ES_n = E \sum_{k=1}^{n} X_k = \sum_{k=1}^{n} EX_k = np \qquad (12.22)$$

In addition, since the X_k are independent,

$$\text{var}(S_n) = \text{var}\left(\sum_{k=1}^{n} X_k\right) = \sum_{k=1}^{n} \text{var}(X_k) = np(1 - p) \qquad (12.23)$$

We give two more applications of the binomial distribution, to become familiar with other situations where this distribution plays an important role.

▶ Example 23   We consider the flowering pea plants again. Suppose that 20 independent offspring result from $Cc \times Cc$ crosses. Find the probability that at most two offspring have white flowers, and compute the expected value and the variance of the number of offspring that have white flowers.

Solution

In a $Cc \times Cc$ cross, the probability of a white-flowering offspring (genotype cc) is 1/4, and the probability of a red-flowering offspring (genotype CC or Cc) is 3/4. The flower colors of different offspring are independent. We can therefore think of this as an experiment with twenty trials, each having success probability 1/4. We want to know the probability of at most two successes. With $n = 20$, $k \le 2$ and $p = 1/4$, we find

$$P(S_{20} \le 2) = P(S_{20} = 0) + P(S_{20} = 1) + P(S_{20} = 2)$$

$$= \binom{20}{0}\left(\frac{1}{4}\right)^0\left(\frac{3}{4}\right)^{20} + \binom{20}{1}\left(\frac{1}{4}\right)^1\left(\frac{3}{4}\right)^{19} + \binom{20}{2}\left(\frac{1}{4}\right)^2\left(\frac{3}{4}\right)^{18}$$

$$\approx 0.0913$$

The expected value of the number of white-flowering offspring is

$$ES_{20} = (20)\left(\frac{1}{4}\right) = 5$$

and the variance is

$$\text{var}(S_{20}) = (20)\left(\frac{1}{4}\right)\left(\frac{3}{4}\right) = \frac{15}{4} = 3.75 \qquad ◀$$

▶ Example 24   Suppose that a woman who is a carrier for hemophilia has four daughters with a man who is not hemophilic. Find the probability that at least one daughter carries the hemophilia gene.

Solution

Each daughter has probability 1/2 of carrying the disease gene, independently of all others. We can think of this as an experiment with four independent trials and success probability 1/2 ("success" in this case is being a carrier). Therefore,

$$P(S_4 \geq 1) = 1 - P(S_4 = 0)$$

$$= 1 - \binom{4}{0} \left(\frac{1}{2}\right)^0 \left(\frac{1}{2}\right)^4$$

$$= 1 - \frac{1}{16} = \frac{15}{16} = 0.9375 \qquad ◄$$

Sampling with and without Replacement Consider an urn with 10 green and 15 blue balls. Sampling balls from this urn can be done with and without replacement. If we sample with replacement, we take out a ball, note its color, and then place the ball back into the urn. The number of balls of a specific color is then binomially distributed.

▶ **Example 25** If we sample five balls with replacement from this urn, what is the probability of three blue balls in the sample?

Solution

Denote the number of blue balls in the sample by S_5. Then S_5 is binomially distributed with n, the number of trials, equal to 5, and success probability $p = 15/(15 + 10) = 3/5$. We find

$$P(S_5 = 3) = \binom{5}{3} \left(\frac{3}{5}\right)^3 \left(\frac{2}{5}\right)^2 = 0.3456 \qquad ◄$$

If we sample without replacement, we take the balls out one after the other without putting them back into the urn, and note their colors. If we take the balls out one after the other without replacing them, then the composition of the urn changes every time we remove a ball, and the number of balls of a certain color is no longer binomially distributed.

▶ **Example 26** If we sample five balls without replacement from the urn, what is the probability of three blue balls in the sample?

Solution

(We encountered a similar problem in Example 9 of Section 12.2.) There are $\binom{25}{5}$ ways of sampling 5 balls from this urn. This is the size of the sample space. In order to have 3 blue balls, we need to select 3 out of the 15 blue balls, which can be done in $\binom{15}{3}$ ways. Since we want a total of 5 balls, we also need to select 2 out of the 10 green balls, which can be done in $\binom{10}{2}$ ways. Combining the blue and the green balls, we find that there are $\binom{15}{3}\binom{10}{2}$ ways of selecting three blue and two green balls from the urn. Therefore, the probability of obtaining three blue balls in a sample of size five when sampling is done without replacement is

$$\frac{\binom{15}{3}\binom{10}{2}}{\binom{25}{5}} = \frac{455 \cdot 45}{53130} \approx 0.3854$$

Note that the answer is different from that in Example 25. ◄

The probability distribution in Example 26 is called the **hypergeometric distribution**. The hypergeometric distribution describes sampling without replacement if two types of objects are in the urn. Suppose the urn has M

green and N blue balls, and a sample of size n is taken from the urn without replacement. If X denotes the number of blue balls in the sample, then

$$P(X = k) = \frac{\binom{N}{k}\binom{M}{n-k}}{\binom{M+N}{n}}, \quad k = 0, 1, 2, \dots, n$$

12.4.4 The Multinomial Distribution

In the previous subsection, we considered experiments where each trial resulted in exactly one of two possible outcomes. We will now extend this to more than two possible outcomes. The distribution is then called the **multinomial distribution**.

▶ **Example 27** To study food preferences in the lady beetle *Coleomegilla maculata*, we present each beetle with three different food choices: maize pollen, egg masses of the European corn borer, and aphids. We suspect that 20% of the time the beetle prefers the aphids, 35% of the time egg masses, and 45% of the time pollen. We carry out this experiment with 30 beetles and find that 8 beetles prefer aphids, 10 egg masses, and 12 pollen. Compute the probability of this event, assuming that the trials are independent.

Solution

We define the random variables

$N_1 = $ number of beetles that prefer aphids

$N_2 = $ number of beetles that prefer egg masses

$N_3 = $ number of beetles that prefer pollen

and the probabilities

$p_1 = P(\text{beetle prefers aphids}) = 0.2$

$p_2 = P(\text{beetle prefers egg masses}) = 0.35$

$p_3 = P(\text{beetle prefers pollen}) = 0.45$

We claim that

$$P(N_1 = 8, N_2 = 10, N_3 = 12) = \frac{30!}{8!10!12!}(0.2)^8(0.35)^{10}(0.45)^{12}$$

The term $\frac{30!}{8!10!12!}$ counts the number of ways that we can arrange 30 objects, 8 of which are of one type, 10 of another, and 12 of a third, which represent the beetles preferring aphids, egg masses, and pollen, respectively. The term $(0.2)^8(0.35)^{10}(0.45)^{12}$ comes from the probability of a particular arrangement, just as in the binomial case. ◀

A more involved example for the multinomial distribution is another of Mendel's experiments, in which he crossed pea plants that had round yellow seeds with plants that had green wrinkled seeds. Roundness and yellow color are dominant traits, and greenness and wrinkled texture are recessive traits. We denote by R the allele for round seeds, by r the allele for wrinkled seeds, by Y the allele for yellow seeds, and by y the allele for green seeds. Then a cross between plants that are homozygous for round yellow seeds (genotype RR/YY) and plants that are homozygous for wrinkled green seeds (genotype rr/yy) is written as

$$RR/YY \times rr/yy$$

This results in offspring of type Rr/Yy. That is, all offspring are heterozygous with round yellow seeds. Crossing plants from this offspring generation then results in all possible combinations, as illustrated in the following table.

	RY	Ry	rY	ry
RY	RR/YY round yellow	RR/Yy round yellow	Rr/YY round yellow	Rr/Yy round yellow
Ry	RR/Yy round yellow	RR/yy round green	Rr/yY round yellow	Rr/yy round green
rY	rR/YY round yellow	rR/Yy round yellow	rr/YY wrinkled yellow	rr/Yy wrinkled yellow
ry	rR/yY round yellow	rR/yy round green	rr/yY wrinkled yellow	rr/yy wrinkled green

The laws of inheritance tell us that each outcome of this cross is equally likely. There are 16 different genotypes, some of which give rise to the same morphological type of seed (for instance, round yellow). The morphological type of the seed is called the **phenotype**. We summarize the phenotypes and their probability distribution in the following table.

	Yellow	Green
Round	9/16	3/16
Wrinkled	3/16	1/16

▶ **Example 28** Suppose that you obtain 50 independent offspring from a cross

$$Rr/Yy \times Rr/Yy$$

where 25 seeds are round yellow, 9 are round green, 12 are wrinkled yellow, and 4 are wrinkled green. Find the probability of this outcome.

Solution

This is another application of the multinomial distribution. Arguing as in the previous example, we find that the probability of this outcome is

$$\frac{50!}{25!9!12!4!} \left(\frac{9}{16}\right)^{25} \left(\frac{3}{16}\right)^9 \left(\frac{3}{16}\right)^{12} \left(\frac{1}{16}\right)^4 \qquad \blacktriangleleft$$

12.4.5 Geometric Distribution

We again consider a sequence of independent Bernoulli trials, namely a random experiment of repeated trials where each trial has two possible outcomes, success or failure, and the trials are independent. As in Subsection 12.4.3, we denote the probability of success by p. This time, however, we define a random variable X that counts the number of trials until the first success. The random variable X takes on values $1, 2, 3, \ldots$ and is therefore a discrete random variable. Its probability distribution is called the **geometric distribution** and is given by

$$P(X = k) = (1 - p)^{k-1} p, \ k = 1, 2, 3, \ldots \tag{12.24}$$

since the event $\{X = k\}$ means that the first $k - 1$ trials resulted in failure (each one has probability $1 - p$ and the trials are independent) followed by a trial that resulted in a success (with probability p and independent of all other trials).

The range of X is the set of all positive integers, and thus X takes on infinitely (though still countably) many values. This is the first time we encounter a random variable of this kind. There are some issues we need to discuss that pertain to this infinite range. For instance, to show that $P(X = k)$ in (12.24) is indeed a probability mass function, we will need to sum up the probabilities from $k = 1$ to $k = \infty$. This means summing up an infinite number of terms and we will need to explain what this means.

To show that (12.24) is a probability mass function, we need to check that $P(X = k) \geq 0$ and that it sums to 1. The first part is straightforward: Since p is a probability, $0 \leq p \leq 1$, which makes $(1 - p)^{k-1} p \geq 0$ for all $k = 1, 2, 3, \ldots$. For the second part, we need to show that

$$\sum_k P(X = k) = 1$$

where the sum ranges over all values of k in the range of X, namely, $k = 1, 2, 3, \ldots$. We write this as

$$\sum_{k=1}^{\infty} P(X = k)$$

and define this **infinite sum** as

$$\sum_{k=1}^{\infty} P(X = k) = \lim_{n \to \infty} \sum_{k=1}^{n} P(X = k)$$

To compute this sum (and later the mean and the variance), we introduce the **geometric series**. The geometric series is the infinite sum

$$\sum_{k=0}^{\infty} q^k = 1 + q + q^2 + q^3 + \ldots$$

The finite sum

$$S_n = \sum_{k=0}^{n} q^k = 1 + q + q^2 + \cdots + q^n$$

can be computed using the following computational trick. Write

$$S_n = 1 + q + q^2 + \cdots + q^n$$
$$q S_n = q + q^2 + q^3 + \cdots + q^n + q^{n+1}$$

and then subtract $q S_n$ from S_n. Most terms cancel and we find

$$S_n - q S_n = 1 - q^{n+1}$$

Factoring S_n on the left-hand side and solving for S_n yields

$$(1 - q) S_n = 1 - q^{n+1}$$

$$S_n = \frac{1 - q^{n+1}}{1 - q}$$

provided $q \neq 1$.

If $|q| < 1$, then $\lim_{n \to \infty} q^{n+1} = 0$, and, therefore,

$$\sum_{k=0}^{\infty} q^k = \lim_{n \to \infty} \sum_{k=0}^{\infty} q^k = \lim_{n \to \infty} \frac{1 - q^{n+1}}{1 - q} = \frac{1}{1 - q} \quad \text{for } |q| < 1$$

These are important results, which we summarize.

For $q \neq 1$,

$$\sum_{k=0}^{n} q^k = \frac{1 - q^{n+1}}{1 - q} \tag{12.25}$$

For $|q| < 1$,

$$\sum_{k=0}^{\infty} q^k = \frac{1}{1 - q} \tag{12.26}$$

We can use these results to check that the probability mass function of the geometric distribution adds up to 1.

$$\sum_{k=1}^{\infty} P(X = k) = \sum_{k=1}^{\infty} (1 - p)^{k-1} p$$

$$= p \sum_{l=0}^{\infty} (1 - p)^l = p \frac{1}{1 - (1 - p)} = p \cdot \frac{1}{p} = 1$$

In the summation, we made the substitution $l = k - 1$. Then the summation range $k = 1, 2, 3, \ldots$ changes to $l = 0, 1, 2, \ldots$, which allows to apply the results for the geometric series we derived in (12.26).

▶ **Example 29** The random experiment consists of rolling a fair die until the first time a six appears. Find the probability that the first six appears at the fifth trial.

Solution

Denote by X the first time a six appears and by p the probability that the die shows a six in a single trial (the success in this experiment). Since the die is fair, all six numbers on the die are equally likely and we find $p = 1/6$. Then

$$P(X = 5) = \left(1 - \frac{1}{6}\right)^4 \frac{1}{6} \approx 0.0804 \qquad \blacktriangleleft$$

▶ **Example 30** Consider a sequence of independent Bernoulli trials with success probability p. Find the probability of no success in the first k trials.

Solution

Denote by X the number of trials until the first success. We want to find the event $\{X > k\}$. Now, the event $\{X > k\}$ can be phrased in terms of a binomial random variable S_k, which counts the number of successes in the first k trials. The event $\{X > k\}$ is equivalent to the event $\{S_k = 0\}$. Therefore,

$$P(X > k) = P(S_k = 0) = (1 - p)^k \qquad \blacktriangleleft$$

▶ **Example 31**

Compare the probability of no success in the first k trials of independent Bernoulli trials to the probability of no success in k trials following n unsuccessful trials.

Solution

If X denotes the number of trials with success probability p, then we want to compare $P(X > k)$ to $P(X > n + k \mid X > n)$. From Example 30, we conclude that

$$P(X > k) = (1 - p)^k$$

To compute the conditional probability $P(X > n + k \mid X > n)$, we use

$$P(X > n + k \mid X > n) = \frac{P(X > n + k, X > n)}{P(X > n)}$$

Since the event $\{X > n + k\}$ is contained in the event $\{X > n\}$,

$$P(X > n + k, X > n) = P(\{X > n + k\} \cap \{X > n\}) = P(X > n + k)$$

and

$$\frac{P(X > n + k, X > n)}{P(X > n)} = \frac{P(X > n + k)}{P(X > n)} = \frac{(1 - p)^{n+k}}{(1 - p)^n} = (1 - p)^k$$

We find that

$$P(X > k) = P(X > n + k \mid X > n)$$

That is, not having had a success in the first n trials does not change the probability of not having successes in the following k trials compared to not having k successes in the first k trials. This is a consequence of the independence of trials. ◀

▶ **Example 32**

If both parents are carriers of a recessive autosomal disease but are not symptomatic for the disease, then there is a 25% chance that a child of theirs will be symptomatic for the diseases. Suppose the parents have three asymptomatic children and plan on having a fourth child. What is the probability that the fourth child will not be symptomatic for the disease?

Solution

Denote by X the waiting time for the first symptomatic child in this family. With "success" probability $p = 1/4$, we find, using Example 31,

$$P(X > 4 \mid X > 3) = P(X > 1) = 1 - P(X = 1) = 1 - \frac{1}{4} = \frac{3}{4}$$

We can also argue as follows: The fact that the first three children are asymptomatic for the disease does not change the probability that their next child will not be symptomatic for the disease since these events are independent. ◀

We will now compute the mean and the variance of the geometric distribution. If X is a geometrically distributed random variable with $P(X = k) = (1 - p)^{k-1} p$, then

$$EX = \sum_{k=1}^{\infty} k(1 - p)^{k-1} p = p \sum_{k=1}^{\infty} k(1 - p)^{k-1}$$

To compute this infinite sum, we need (12.26) and a result that we cannot justify here. Namely, if $|q| < 1$, then

$$\frac{d}{dq} \sum_{k=0}^{\infty} q^k = \sum_{k=0}^{\infty} \frac{d}{dq} q^k$$

In words, the derivative of this infinite sum can be obtained by differentiating each term separately and then taking the sum. Interchanging differentiation and summation when the sum is an infinite sum cannot always be done but can be justified in this case. Using this result, we find

$$\sum_{k=0}^{\infty} \frac{d}{dq} q^k = \sum_{k=0}^{\infty} k q^{k-1} = \sum_{k=1}^{\infty} k q^{k-1}$$

where we used in the last step that the term with $k = 0$ is equal to 0. Since $\sum_{k=0}^{\infty} q^k = \frac{1}{1-q}$ when $|q| < 1$, it follows that

$$\frac{d}{dq} \sum_{k=0}^{\infty} q^k = \frac{d}{dq} \frac{1}{1-q} = \frac{1}{(1-q)^2}$$

and hence,

$$\sum_{k=1}^{\infty} k q^{k-1} = \frac{1}{(1-q)^2}$$

With $q = 1 - p$,

$$EX = p \sum_{k=1}^{\infty} k(1-p)^{k-1} = p \frac{1}{(1-(1-p))^2} = \frac{p}{p^2} = \frac{1}{p}$$

To compute the variance of X, we employ a similar argument. We first derive a result that we will need to carry out the calculation for the variance. Namely, if $|q| < 1$, then

$$\frac{d^2}{dq^2} \sum_{k=0}^{\infty} q^k = \sum_{k=0}^{\infty} \frac{d^2}{dq^2} q^k = \sum_{k=2}^{\infty} k(k-1) q^{k-2}$$

Since

$$\frac{d^2}{dq^2} \frac{1}{1-q} = \frac{2}{(1-q)^3}$$

We have

$$\sum_{k=2}^{\infty} k(k-1) q^{k-2} = \frac{2}{(1-q)^3}$$

To compute the variance, it is useful to first compute

$$E[X(X-1)] = \sum_{k=2}^{\infty} k(k-1) P(X = k) = \sum_{k=2}^{\infty} k(k-1)(1-p)^{k-1} p$$

$$= (1-p)p \sum_{k=2}^{\infty} k(k-1)(1-p)^{k-2} = (1-p)p \frac{2}{(1-(1-p))^3}$$

$$= \frac{2p(1-p)}{p^3} = \frac{2(1-p)}{p^2}$$

Since $E[X(X-1)] = EX^2 - EX$, we find

$$\text{var}(X) = EX^2 - (EX)^2 = E[X(X-1)] + EX - (EX)^2$$

$$= \frac{2(1-p)}{p^2} + \frac{1}{p} - \frac{1}{p^2} = \frac{1-p}{p^2}$$

We summarize our results.

If X is geometrically distributed with $P(X = k) = (1-p)^{k-1}p$, then

$$EX = \frac{1}{p} \quad \text{and} \quad \text{var}(X) = \frac{1-p}{p^2}$$

▶ **Example 33** You roll a fair die until the first time a six appears. How long do you have to wait on average?

Solution

To answer this question, we need to find EX, where X is geometrically distributed with success probability $p = 1/6$. Therefore,

$$EX = \frac{1}{p} = 6$$

In words, on average, you have to roll a die six times until the first time a six appears.　◀

12.4.6 The Poisson Distribution

The Poisson distribution is one of the most important probability distributions. It is used to model, for instance, amino acid substitutions in proteins, or the escape probability of hosts from parasitism, or spatial distributions of plants. It often models "rare events," as we will see.

We say that X is **Poisson** distributed with parameter $\lambda > 0$ if

$$P(X = k) = e^{-\lambda} \frac{\lambda^k}{k!}, \quad k = 0, 1, 2, \ldots$$

The random variable X is a discrete random variable and its range is the set of all nonnegative integers. Thus the range is infinite but still countable.

To show that the probability distribution we defined sums to 1, or to find the mean and the variance of X, we need some additional results. Recall that the Taylor polynomial of order n of $f(x) = e^x$ is

$$P_n(x) = \sum_{k=0}^{n} \frac{x^k}{k!}$$

It turns out (but we cannot show this here) that in the limit as $n \to \infty$,

$$e^x = \lim_{n\to\infty} P_n(x) = \lim_{n\to\infty} \sum_{k=0}^{n} \frac{x^k}{k!}$$

for all $x \in \mathbf{R}$. On the right-hand side, we have an infinite sum and we will use the notation (as in the previous section)

$$\sum_{k=0}^{\infty} \frac{x^k}{k!} = \lim_{n\to\infty} \sum_{k=0}^{n} \frac{x^k}{k!}$$

to denote this infinite sum. Thus, for any $x \in \mathbf{R}$,

$$e^x = \sum_{k=0}^{\infty} \frac{x^k}{k!} \tag{12.27}$$

We can use (12.27) to show that the probability mass function for the Poisson distribution indeed sums up to 1. Namely,

$$\sum_{k=0}^{\infty} P(X = k) = \sum_{k=0}^{\infty} e^{-\lambda} \frac{\lambda^k}{k!} = e^{-\lambda} \sum_{k=0}^{\infty} \frac{\lambda^k}{k!} = e^{-\lambda} e^{\lambda} = 1$$

Furthermore, since $\lambda > 0$, $P(X = k) \geq 0$. Thus the probability mass function for the Poisson distribution satisfies the two conditions of a probability mass function.

▶ **Example 34**   Suppose the number of plants per hectare of a certain species is Poisson distributed with parameter $\lambda = 3$ plants per hectare. Find the probability that there are (a) no plants in a given hectare and (b) at least two plants in a given hectare.

Solution

Denote by X the number of plants in a given hectare. Then X is Poisson distributed with parameter $\lambda = 3$.

(a) The probability that there are no plants in a given hectare is

$$P(X = 0) = e^{-\lambda} = e^{-3} \approx 0.0498$$

(b) The probability that there are at least two plants in a given hectare is

$$P(X \geq 2) = 1 - P(X \leq 1) = 1 - [P(X = 0) + P(X = 1)]$$

$$= 1 - e^{-\lambda}(1 + \lambda) = 1 - e^{-3}(1 + 3) \approx 0.8009 \qquad ◀$$

To find the mean and the variance of a Poisson distributed random variable, we need to use (12.27) repeatedly. Let X be Poisson distributed with parameter λ. Then, formally,

$$EX = \sum_{k=0}^{\infty} k P(X = k) = \sum_{k=0}^{\infty} k e^{-\lambda} \frac{\lambda^k}{k!}$$

$$= e^{-\lambda} \lambda \sum_{k=1}^{\infty} \frac{\lambda^{k-1}}{(k-1)!} = e^{-\lambda} \lambda \sum_{k=0}^{\infty} \frac{\lambda^k}{k!} = \lambda e^{-\lambda} e^{\lambda} = \lambda$$

To compute the variance of X, we begin with computing

$$E[X(X-1)] = \sum_{k=0}^{\infty} k(k-1) P(X = k) = \sum_{k=2}^{\infty} k(k-1) P(X = k)$$

$$= \sum_{k=2}^{\infty} k(k-1) e^{-\lambda} \frac{\lambda^k}{k!} = e^{-\lambda} \lambda^2 \sum_{k=2}^{\infty} \frac{\lambda^{k-2}}{(k-2)!}$$

$$= e^{-\lambda} \lambda^2 \sum_{k=0}^{\infty} \frac{\lambda^k}{k!} = e^{-\lambda} \lambda^2 e^{\lambda} = \lambda^2$$

Therefore,

$$\text{var}(X) = EX^2 - (EX)^2 = E[X(X-1)] + EX - (EX)^2$$
$$= \lambda^2 + \lambda - \lambda^2 = \lambda$$

To summarize our findings,

If X is Poisson distributed with parameter $\lambda > 0$, then
$$EX = \lambda \quad \text{and} \quad \text{var}(X) = \lambda$$

▶ **Example 35** The number of substitutions on a given amino acid sequence during a fixed period of time is modeled by a Poisson distribution. Suppose the number of substitutions on an amino acid sequence of 100 amino acids over a time period of 1,000,000 years is Poisson distributed with average number of substitutions equal to 1. What is the probability that at least one substitution occurred?

Solution

If X denotes the number of substitutions, then X is Poisson distributed with mean 1. Since the mean of a Poisson distribution is equal to its parameter, we find $\lambda = 1$. Hence,

$$P(X \geq 1) = 1 - P(X = 0) = 1 - e^{-\lambda} = 1 - e^{-1} \approx 0.6321 \quad \blacktriangleleft$$

We mentioned previously that the Poisson distribution frequently models rare events. The next result makes this precise. Consider a sequence of independent Bernoulli trials with success probability p. The number of successes among n trials is binomially distributed. We denote by S_n the number of successes in n trials. We consider the case when the number of trials n is very large but the success probability p is very small so that successes are rare. To make this mathematically precise, we will need to take the limit as n tends to infinity such that the product np, which denotes the expected number of successes among n trials, approaches a constant. To achieve this, we need to let p tend to 0 as n tends to infinity. To indicate that the success probability depends on n, we will denote it by p_n. The following result says that the number of successes among a large number of trials is approximately Poisson distributed if the success probability is small.

POISSON APPROXIMATION TO THE BINOMIAL DISTRIBUTION

Suppose S_n is binomially distributed with parameters n and p_n. If $p_n \to 0$ as $n \to \infty$ such that $\lim_{n\to\infty} np_n = \lambda > 0$, then

$$\lim_{n\to\infty} P(S_n = k) = e^{-\lambda} \frac{\lambda^k}{k!}$$

In all the examples for the binomial distribution, the success probability p was fixed and did not depend on n. How should we then interpret the result? It is the precise mathematical formulation. It allows us to use a Poisson distribution as an approximation of the binomial distribution when the number of trials n is large and the success probability p is small. Let's first prove the result before we look at an example.

The key ingredient for proving this result is the following limit. If $\lim_{n\to\infty} x_n = x$, then

$$\lim_{n\to\infty} \left(1 + \frac{x_n}{n}\right)^n = e^x$$

To prove this limit, we show that

$$\lim_{n\to\infty} \ln\left(1 + \frac{x_n}{n}\right)^n = x$$

Now,

$$\lim_{n\to\infty} \ln\left(1 + \frac{x_n}{n}\right)^n = \lim_{n\to\infty} n \ln\left(1 + \frac{x_n}{n}\right)$$

This limit is of the form $\infty \cdot 0$, which suggests that we should rewrite the limit in the form $\frac{0}{0}$ or $\frac{\infty}{\infty}$ and use L'Hospital's rule. We rewrite it in the form

$$\lim_{n\to\infty} n \ln\left(1 + \frac{x_n}{n}\right) = \lim_{n\to\infty} \frac{\ln\left(1 + \frac{x_n}{n}\right)}{\frac{1}{n}} = \lim_{n\to\infty} \frac{\ln\left(1 + \frac{x_n}{n}\right)}{\frac{x_n}{n}} x_n$$

To evaluate this limit, we compute

$$\lim_{y\to 0} \frac{\ln(1 + y)}{y}$$

using L'Hospital's rule. We find

$$\lim_{y\to 0} \frac{\ln(1 + y)}{y} = \lim_{y\to 0} \frac{\frac{1}{1+y}}{1} = 1$$

This result, together with $\lim_{n\to\infty} x_n = x$ and $\lim_{n\to\infty} \frac{x_n}{n} = 0$, yields

$$\lim_{n\to\infty} \frac{\ln\left(1 + \frac{x_n}{n}\right)}{\frac{x_n}{n}} x_n = (1)(x) = x$$

We can now prove the Poisson approximation result. Observe that

$$P(S_n = k) = \binom{n}{k} p_n^k (1 - p_n)^{n-k}$$

We define $\lambda_n = np_n$, and hence, $p_n = \frac{\lambda_n}{n}$. Then

$$P(S_n = k) = \binom{n}{k} \left(\frac{\lambda_n}{n}\right)^k \left(1 - \frac{\lambda_n}{n}\right)^{n-k}$$

$$= \frac{n(n-1)(n-2)\cdots(n-k+1)}{n^k} \frac{\lambda_n^k}{k!} \left(1 - \frac{\lambda_n}{n}\right)^n \left(1 - \frac{\lambda_n}{n}\right)^{-k}$$

Since $\lim_{n\to\infty} np_n = \lim_{n\to\infty} \lambda_n = \lambda$,

$$\lim_{n\to\infty} \frac{n(n-1)(n-2)\cdots(n-k+1)}{n^k} = \lim_{n\to\infty} \frac{n}{n} \frac{n-1}{n} \frac{n-2}{n} \cdots \frac{n-k+1}{n} = 1$$

$$\lim_{n\to\infty} \frac{\lambda_n^k}{k!} = \frac{\lambda^k}{k!}, \qquad \lim_{n\to\infty} \left(1 - \frac{\lambda_n}{n}\right)^{-k} = 1$$

and

$$\lim_{n \to \infty} \left(1 - \frac{\lambda_n}{n}\right)^n = \lim_{n \to \infty} \left(1 + \frac{-\lambda_n}{n}\right)^n = e^{-\lambda}$$

we find

$$\lim_{n \to \infty} P(S_n = k) = (1)\frac{\lambda^k}{k!}e^{-\lambda}(1) = e^{-\lambda}\frac{\lambda^k}{k!}$$

which shows the result.

Let's use the result to compare the Poisson approximation to the exact results of a binomial distribution. This will then illustrate how to use the result we just proved.

▶ **Example 36**   Suppose we toss a biased coin 100 times and denote by S_{100} the number of heads. If the probability of heads is 1/50, compute $P(S_{100} = k)$ for $k = 0, 1$, and 2 exactly and compare to the Poisson approximation.

Solution

Since $n = 100$ and $p = 1/50$, we compare the distribution of S_{100} to a Poisson distribution with parameter $\lambda = np = 100/50 = 2$. We find

$$P(S_{100} = k) = \binom{100}{k}\left(\frac{1}{50}\right)^k\left(\frac{49}{50}\right)^{100-k} \approx e^{-2}\frac{2^k}{k!}$$

For $k = 0$,

$$P(S_{100} = 0) = \left(\frac{49}{50}\right)^{100} \approx 0.1326$$

$$e^{-2} \approx 0.1353$$

For $k = 1$,

$$P(S_{100} = 1) = 100 \cdot \frac{1}{50}\left(\frac{49}{50}\right)^{99} \approx 0.2707$$

$$2e^{-2} \approx 0.2707$$

For $k = 2$,

$$P(S_{100} = 2) = \frac{100 \cdot 99}{2}\left(\frac{1}{50}\right)^2\left(\frac{49}{50}\right)^{98} \approx 0.2734$$

$$e^{-2}\frac{2^2}{2} \approx 0.2707$$

In each case, we see that the approximate value is quite close to the exact value. The advantage is that the Poisson distribution is much easier to calculate than the binomial distribution since the binomial coefficients $\binom{n}{k}$ are computationally intensive.   ◀

In Section 10.7, we discussed host-parasitoid models. Parasitoids are insects that lay their eggs on, in, or near the (in most case, immature) body of another arthropod, which serves as the host for the developing parasitoid. The eggs develop into free living adults while consuming the host. Parasitoids make up about 14% of all insect species. A key component in modeling host-parasitoid interactions is the probability of a host escaping parasitism. The first models were developed by Nicholson and Bailey (1935). They assumed that the probability a host escapes parasitism is given by e^{-aP}, where P is the parasitoid density and a is a positive parameter, called the search efficiency. The next example explains where this functional form of the escape probability comes from.

▶ Example 37   Parasitoid encounters of their hosts are sometimes modeled with a Poisson distribution. Suppose a host is surrounded by P parasitoids. Each parasitoid, independently of all others, has a probability a of encountering the host. We can consider this as a sequence of P Bernoulli trials with success probability a. If no parasitoid encounters the host, the host will escape parasitism. If P is large and a is small, we can use the Poisson approximation. The number of encounters is then approximately Poisson distributed with parameter aP, and we find

$$P(\text{host escapes parasitism}) = e^{-aP}$$

This is the escape probability used in the Nicholson-Bailey host-parasitoid model we discussed in Section 10.7. It is the 0th term of a Poisson distribution and comes about that parasitoids are assumed to search randomly.   ◀

The Poisson approximation plays a crucial role in estimating the time of divergence of species using amino acid sequence data. Sequences of the same protein across different species are compared and the number of pairwise amino acid differences gives an indication of the evolutionary distance between each pair of species. The simplest mathematical model to estimate times of divergence based on amino acid sequences assumes that the probability of a substitution at a given site in the amino acid sequence is the same for all sites and only depends on the time since divergence. Furthermore, all sites are assumed to be independent. The number of amino acid substitutions along a sequence of length n is then binomially distributed with success probability equal to the probability of a substitution at this site, provided that multiple substitutions at this site can be ignored (this holds when the time since divergence is not too long). If the sequence is sufficiently long and the time since divergence is not too long so that the probability of substitution is small, then the number of substitutions is equal to the number of differences between the two sequences and can be approximated by a Poisson distribution.

▶ Example 38   Suppose two vertebrate species diverged about 10 million years ago. Based on other data, we expect that the probability of an amino acid difference between their hemoglobin-α-chains (length 140 amino acids) is about 0.014.

(a) How many amino acid differences would you expect when comparing the two sequences?

(b) What is the probability of finding at least three sites with amino acid differences?

Solution

The number of amino acid differences is approximately Poisson distributed with parameters equal to the product of the length of the sequence and the probability of finding a difference at a given site. We find $\lambda = 140 \cdot 0.014 = 1.96$.

(a) The expected number of amino acid differences is equal to the parameter of the distribution, namely 1.96.

(b) The probability of finding at least three differences is equal to

$$1 - e^{-\lambda}(1 + \lambda + \lambda^2/2) = 1 - e^{-1.96}(1 + 1.96 + 1.96^2/2) \approx 0.3125$$ ◀

12.4.7 Problems

(12.4.1)

1. Toss a fair coin twice. Let X be the random variable that counts the number of tails in each outcome. Find the probability mass function describing the distribution of X.

2. Roll a fair die twice. Let X be the random variable that gives the maximum of the two numbers. Find the probability mass function describing the distribution of X.

3. An urn contains three green and two blue balls. You remove two balls at random without replacement. Let X denote the number of green balls in your sample. Find the probability mass function describing the distribution of X.

4. You draw five cards from a standard deck of 52 cards without replacement. Let X denote the number of aces in your hand. Find the probability mass function describing the distribution of X.

5. Suppose that the probability mass function of a discrete random variable X is given by the following table.

x	$P(X = x)$
-3	0.2
-1	0.3
1.5	0.4
2	0.1

Find and graph the corresponding distribution function $F(x)$.

6. Suppose the probability mass function of a discrete random variable X is given by the following table.

x	$P(X = x)$
-1	0.1
-0.5	0.2
0.1	0.1
0.5	0.25
1	0.35

Find and graph the corresponding distribution function $F(x)$.

7. Let X be a random variable with distribution function

$$F(x) = \begin{cases} 0 & x < -2 \\ 0.2 & -2 \le x < 0 \\ 0.3 & 0 \le x < 1 \\ 0.7 & 1 \le x < 2 \\ 1 & x \ge 2 \end{cases}$$

Determine the probability mass function of X.

8. Let X be a random variable with distribution function

$$F(x) = \begin{cases} 0 & x < 0 \\ 0.1 & 0 \le x < 1.3 \\ 0.3 & 1.3 \le x < 1.7 \\ 0.8 & 1.7 \le x < 1.9 \\ 0.9 & 1.9 \le x < 2 \\ 1 & x \ge 2 \end{cases}$$

Determine the probability mass function of X.

9. Let $S = \{1, 2, 3, \dots, 10\}$, and assume that

$$p(k) = \frac{k}{N}, \quad k \in S$$

where N is a constant.

(a) Determine N so that $p(k)$, $k \in S$, is a probability mass function.

(b) Let X be a discrete random variable with $P(X = k) = p(k)$. Find the probability that X is less than 8.

10. (*Geometric distribution*) In Example 2, we tossed a coin repeatedly until the first heads showed up. Assume that the probability of heads is p with $p \in (0, 1)$. Let Y be a random variable that counts the number of trials until the first heads shows up.

(a) Show that $P(Y = 1) = p$, $P(Y = 2) = (1 - p)p$, and $P(Y = 3) = (1 - p)^2 p$.

(b) Explain why

$$P(Y = j) = (1 - p)^{j-1} p$$

for $j = 1, 2, \dots$. This is called the **geometric distribution**.

(c) Prove that

$$\sum_{j \ge 1} P(Y = j) = 1$$

by showing the following steps.

(i) For $0 \le q < 1$, define

$$S_n = 1 + q + q^2 + \cdots + q^n$$

Show that

$$S_n - q S_n = 1 - q^{n+1}$$

and conclude from this that

$$S_n = \frac{1 - q^{n+1}}{1 - q}$$

(ii) Show that

$$P(Y \le k) = \sum_{j=1}^{k} P(Y = j)$$

$$= p \sum_{j=1}^{k} (1 - p)^{j-1}$$

Use your results in (i) to show that this simplifies to

$$1 - (1 - p)^k$$

and conclude from this that

$$\lim_{k \to \infty} P(Y \le k) = 1$$

which is equivalent to

$$\sum_{j \ge 1} P(Y = j) = 1$$

(12.4.2)

11. The following table contains the number of leaves per basil plant in a sample of size 25.

19	21	20	13	18
14	17	14	17	17
13	15	12	15	17
15	16	18	17	14
14	14	13	20	13

(a) Find the relative frequency distribution.

(b) Compute the average value by (i) averaging the values in the table directly, and (ii) using the relative frequency distribution obtained in (a).

12. The following table contains the number of aphids per plant in a sample of size 30.

15	27	13	2	0	16
26	0	2	1	17	15
21	13	5	0	19	25
12	11	0	16	22	1
28	9	0	0	1	17

(a) Find the relative frequency distribution.

(b) Compute the average value by (i) averaging the values in the table directly, and (ii) using the relative frequency distribution obtained in (a).

13. The following table contains exam scores of 25 students.

7	8	8	3	2
5	6	9	10	6
8	8	7	6	9
10	4	4	8	6
9	10	5	5	8

(a) Find the relative frequency distribution.

(b) Compute the average value by (i) averaging the values in the table directly, and (ii) using the relative frequency distribution obtained in (a).

14. The following table contains the number of flower heads per plant in a sample of size 20.

15	17	19	18	15
17	18	15	14	19
17	15	15	18	19
20	17	14	17	18

(a) Find the relative frequency distribution.

(b) Compute the average value by (i) averaging the values in the table directly, and (ii) using the relative frequency distribution obtained in (a).

15. Suppose that the probability mass function of a discrete random variable X is given by the following table.

x	$P(X = x)$
-2	0.1
-1	0.4
0	0.3
1	0.2

(a) Find EX.

(b) Find EX^2.

(c) Find $E[X(X-1)]$.

16. Suppose that the probability mass function of a discrete random variable X is given by the following table.

x	$P(X = x)$
0	0.3
1	0.3
2	0.1
3	0.1
4	0.2

(a) Find EX.

(b) Find EX^2.

(c) Find $E(2X - 1)$.

17. Suppose that the probability mass function of a discrete random variable X is given by the following table.

x	$P(X = x)$
-3	0.2
-1	0.3
1.5	0.4
2	0.1

Find the mean, the variance, and the standard deviation of X.

18. Suppose that the probability mass function of a discrete random variable X is given by the following table.

x	$P(X = x)$
-1	0.1
-0.5	0.2
0.1	0.1
0.5	0.25
1	0.35

Find the mean, the variance, and the standard deviation of X.

19. Let X be **uniformly distributed** on the set
$$S = \{1, 2, 3, \ldots, 10\}$$
That is,
$$P(X = k) = \frac{1}{10}, \quad k \in S$$

(a) Find EX.

(b) Find var(X).

20. Let X be uniformly distributed on the set
$$S = \{1, 2, 3, \ldots, n\}$$
where n is a positive integer; that is,
$$P(X = k) = \frac{1}{n}, \quad k \in S$$

(a) Find EX.

(b) Find var(X).

Hint: Recall that
$$\sum_{k=1}^{n} k = \frac{n(n+1)}{2}$$
and
$$\sum_{k=1}^{n} k^2 = \frac{n(n+1)(2n+1)}{6}$$

21. Assume that X is a discrete random variable with finite range, and set
$$p(x) = P(X = x)$$

(a) Show that
$$E(aX + b) = \sum_{x}(ax + b)p(x)$$

(b) Use your result in (a) and the rules for finite sums to conclude that
$$E(aX + b) = aEX + b$$

22. Assume that X is a discrete random variable with finite range, and set
$$p(x) = P(X = x)$$

(a) Show that
$$\text{var}(aX + b) = a^2 \sum_{x}(x - EX)^2 p(x)$$

(b) Use your result in (a) and the rules for finite sums to conclude that
$$\text{var}(aX + b) = a^2\text{var}(X)$$

23. Let X and Y be two random variables with joint distribution

	$X = 0$	$X = 1$
$Y = 0$	0.3	0.1
$Y = 1$	0.2	0.4

(a) Find $P(X = 1, Y = 0)$.

(b) Find $P(X = 1)$.

(c) Find $P(Y = 0)$.

(d) Find $P(Y = 0 \mid X = 1)$.

24. Let X and Y be two random variables with joint distribution

	$X = 0$	$X = 1$
$Y = 0$	0.2	0.3
$Y = 1$	0.0	0.5

(a) Find $P(X = 0, Y = 1)$.

(b) Find $P(X = 0)$.

(c) Find $P(Y = 1)$.

(d) Find $P(Y = 1 \mid X = 0)$.

25. Let X and Y be two independent random variables with probability mass function described by the following table.

k	$P(X = k)$	$P(Y = k)$
-2	0.1	0.2
-1	0	0.2
0	0.3	0.1
1	0.4	0.3
2	0.05	0
3	0.15	0.2

(a) Find EX and EY.

(b) Find $E(X + Y)$.

(c) Find var(X) and var(Y).

(d) Find var$(X + Y)$.

26. Let X and Y be two independent random variables with probability mass function described by the following table.

k	$P(X = k)$	$P(Y = k)$
-3	0.1	0.1
-1	0.1	0.2
0	0.2	0.1
0.5	0.3	0.3
2	0.15	0.1
2.5	0.15	0.2

(a) Find EX and EY.

(b) Find $E(X + Y)$.

(c) Find var(X) and var(Y).

(d) Find var$(X + Y)$.

27. We have two formulas for computing the variance of X, namely,
$$\text{var}(X) = E(X - EX)^2$$
and
$$\text{var}(X) = EX^2 - (EX)^2$$

(a) Explain why var$(X) \geq 0$.

(b) Use your results in (a) to explain why
$$EX^2 \geq (EX)^2$$

28. Assume that X is a discrete random variable with finite range. Show that if var$(X) = 0$, then $P(X = EX) = 1$.
(12.4.3)

29. Toss a fair coin ten times. Let X be the number of heads.
(a) Find $P(X = 5)$.
(b) Find $P(X \geq 8)$.
(c) Find $P(X \leq 9)$.

30. Toss a coin with probability 0.3 of heads five time. Let X be the number of tails.
(a) Find $P(X = 2)$.
(b) Find $P(X \geq 1)$.

31. Roll a fair die six times. Let X be the number of times you roll a 6. Find the probability mass function.

32. An urn contains four green and six blue balls. You draw a ball at random, note its color, and replace it. You repeat this four times. Let X denote the total number of green balls you obtain. Find the probability mass function of X.

33. Assume that 20% of all plants in a field are infested with aphids. Suppose that you pick twenty plants at random. What is the probability that none of them carried aphids?

34. To test for a disease that has a prevalence of 1 in 100 in a population, blood samples of 10 individuals are pooled and the pooled blood is then tested. What is the probability that the test result is negative (the disease is not present in the pooled blood sample)?

35. Suppose that a box contains ten apples. The probability that any one apple is spoiled is 0.1. (Assume that the apples are independent.)
(a) Find the expected number of spoiled apples per box.
(b) A shipment contains ten boxes. Find the expected number of boxes that contain no spoiled apples.

36. Toss a fair coin ten times. Let X denote the number of heads. What is the probability that X is within one standard deviation of its mean?

37. A multiple choice exam contains fifty questions. Each question has four choices. Find the expected number of correct answers if a student guesses the answers at random.

38. A TRUE-FALSE exam has 20 questions. Find the expected number of correct answers if a student guesses the answers at random.

39. (*Sampling with and without replacement*) An urn contains 12 green and 24 blue balls.
(a) You take 10 balls out of the urn. Find the probability that 6 of the 10 balls are blue.
(b) You take a ball out of the urn, note its color, and replace it. You repeat this 10 times. Find the probability that 6 of the 10 balls are blue.

40. (*Sampling with and without replacement*) An urn contains K green and $N - K$ blue balls.
(a) You take n balls out of the urn. Find the probability that k of the n balls are green.
(b) You take a ball out of the urn, note its color, and replace it. You repeat this n times. Find the probability that k of the n balls are green.

(12.4.4)

41. Repeat Example 27 when $N_1 = 10$, $N_2 = 14$, and $N_3 = 6$.

42. Repeat Example 27 when $N_1 = 5$, $N_2 = 15$, and $N_3 = 10$.

43. Repeat Example 28 when 20 seeds are round yellow, ten seeds are round green, eight seeds are wrinkled yellow, and two seeds are wrinkled green.

44. Repeat Example 28 when seventeen seeds are round yellow, twenty-two seeds are round green, thirteen seeds are wrinkled yellow, and eight seeds are wrinkled green.

45. An urn contains six green, eight blue, and ten red balls. You take one ball out of the urn, note its color, and replace it. You repeat this six times. What is the probability that you sampled two of each color?

46. An urn contains eight green, four blue, and six red balls. You take one ball out of the urn, note its color, and replace it. You repeat this four times. What is the probability that you sampled two green, one blue, and one red ball?

47. In a $Cc \times Cc$ cross of peas, five offspring are of genotype CC, twelve are of genotype Cc, and six are of genotype cc. What is the probability of this event?

48. Assume a 1 : 1 sex ratio. A woman who is a carrier of hemophilia has four children with a man who is not hemophilic. What is the probability that she has one daughter who is not a carrier, one daughter who is a carrier, one son who is, and one son who is not hemophilic?

(12.4.5)

49. A random experiment consists of flipping a fair coin until the first time heads appear. Find the probability that the first head appears on the kth trial for $k = 1, 2$, and 3.

50. A random experiment consists of rolling a die until the first time a five or a six appears. Find the probability that the first five or six appears on the kth trial for $k = 1, 2, \ldots, 5$.

51. A random experiment consists of flipping a fair coin until the first time heads appear. Find the probability that the first head appears after the third trial.

52. A random experiment consists of rolling a die until the first six appears. Find the probability that the first six appears after the seventh trial.

53. A random experiment consists of flipping a fair coin until the first time heads appear. Find the probability that the first head appears within the first four trials.

54. A random experiment consists of rolling a die until the first time a 1 or a 2 appears. Find the probability that the first 1 or 2 appears within the first five trials.

55. An urn contains 1 black and 14 white balls. Balls are drawn at random, one at a time, until the black ball is selected. Each ball is replaced before the next ball is drawn. Find the probability that at least 20 draws are needed.

56. An urn contains 1 black and $n - 1$ white balls. Balls are drawn at random, one at a time, until the black ball is selected. Each ball is replaced before the next ball is drawn. Find the probability that at least n draws are needed. What happens as $n \to \infty$?

57. An urn contains 5 green and 25 blue balls. Balls are drawn at random, one at a time, until a green ball is selected. Each ball is replaced before the next ball is drawn. Let T denote the first time until a green ball is drawn. Find ET and var(T).

58. An urn contains 10 green and 20 blue balls. Balls are drawn at random, one at a time, until a green ball is selected. Each ball is replaced before the next ball is drawn. Let T denote the first time until a green ball is drawn. Find ET and var(T).

59. An urn contains one black and nine white balls. Balls are drawn at random until the black ball is selected. Find the probability that exactly six balls are needed if (a) each ball is replaced before the next ball is drawn, and (b) balls are not replaced.

60. An urn contains 1 black and $n-1$ white balls. Balls are drawn at random until the black ball is selected. Find the probability that exactly k balls are needed if (a) each ball is replaced before the next ball is drawn, and (b) balls are not replaced.

61. Suppose the waiting time for the first success is geometrically distributed with mean $1/p$.

(a) Find the probability that the first success occurs on the kth trial.

(b) The experiment is repeated after the first success. Assume that the waiting time for the second success has the same distribution as the waiting time for the first success. Find the probability mass function for the distribution of the second success.

62. A Bernoulli experiment with success probability p is repeated until the nth success. Assume that each trial is independent of all others. Find the probability mass function of the distribution of the nth success. (This distribution is called the negative binomial distribution.)

(12.4.6)

63. Suppose X is Poisson distributed with parameter $\lambda = 2$. Find $P(X = k)$ for $k = 0, 1, 2,$ and 3.

64. Suppose X is Poisson distributed with parameter $\lambda = 0.5$. Find $P(X = k)$ for $k = 0, 1, 2,$ and 3.

65. Suppose X is Poisson distributed with parameter $\lambda = 1$.

(a) Find $P(X \geq 2)$.

(b) Find $P(1 \leq X \leq 3)$.

66. Suppose X is Poisson distributed with parameter $\lambda = 0.3$.

(a) Find $P(X \geq 3)$.

(b) Find $P(2 \leq X \leq 4)$.

67. Suppose X is Poisson distributed with parameter $\lambda = 1.5$. Find the probability that X exceeds 3.

68. Suppose X is Poisson distributed with parameter $\lambda = 0.8$. Find the probability that X is at most 2.

69. Suppose X is Poisson distributed with parameter $\lambda = 2$. Find the probability that X is at least 2.

70. Suppose X is Poisson distributed with parameter $\lambda = 0.4$. Find the probability that X is less than 4.

71. Suppose the number of phone calls arriving at a switch board per hour is Poisson distributed with mean 7 calls per hour. Find the probability that no phone calls arrive during a certain hour.

72. Suppose the number of phone calls arriving at a switch board per hour is Poisson distributed with mean 3 calls per hour.

(a) Find the probability that at least one phone call arrives between noon and 1 P.M.

(b) Assuming that phone calls in different hours are independent of each other, find the probability that no phone calls arrive between noon and 2 P.M.

73. Suppose the number of typos on a book page is Poisson distributed with mean 0.5. Find the probability that there is at least one typo on a given page.

74. Suppose the number of typos on a book page is Poisson distributed with mean 0.1.

(a) Find the probability that there are no typos on a page.

(b) How many pages with typos do you expect in a 200-page book?

75. The number of amino acid substitutions on a given amino acid sequence is Poisson distributed with mean 3. What is the probability of at least two substitutions?

76. The number of amino acid substitutions on a given amino acid sequence is Poisson distributed with mean 2. Given that there are substitutions on the sequence, what is the probability that there are at least two substitutions?

77. Suppose the number of customers per hour arriving at the post office is Poisson distributed with a mean of 4 customers per hour.

(a) Find the probability that no customer arrives between 2 and 3 P.M.

(b) Find the probability that exactly two customers arrive between 3 and 4 P.M.

(c) Assuming that the number of customers arriving between 2 and 3 P.M. is independent of the number of customers arriving between 3 and 4 P.M., find the probability that exactly two customers arrive between 2 and 4 P.M.

(d) Assume that the number of customers arriving between 2 and 3 P.M. is independent of the number of customers arriving between 3 and 4 P.M. Given that exactly two customers arrive between 2 and 4 P.M., what is the probability that both arrive between 3 and 4 P.M.?

78. Suppose the number of customers per hour arriving at the post office is Poisson distributed with a mean of 5 customers per hour.

(a) Find the probability that exactly one customer arrives between 2 and 3 P.M.

(b) Find the probability that exactly two customers arrive between 3 and 4 P.M.

(c) Assuming that the number of customers arriving between 2 and 3 P.M. is independent of the number of customers arriving between 3 and 4 P.M., find the probability that exactly three customers arrive between 2 and 4 P.M.

(d) Assume that the number of customers arriving between 2 and 3 P.M. is independent of the number of customers arriving between 3 and 4 P.M. Given that exactly three customers arrive between 2 and 4 P.M., what is the probability that one arrives between 2 and 3 P.M. and two between 3 and 4 P.M.?

79. X and Y are independent Poisson with mean 3.

(a) Find $P(X + Y = 2)$.

(b) Given $X + Y = 2$, find the probability that $X = k$ for $k = 0, 1,$ and 2.

80. X and Y are independent Poisson with mean λ.

(a) Find the distribution of $X + Y$. [*Hint:* $\sum_{k=0}^{n} \binom{n}{k} = 2^n$.]

(b) Given $X + Y = n$, find the probability that $X = k$ for $k = 0, 1, 2, \ldots, n$.

In Problems 81–82, use the Poisson approximation.

81. For a certain vaccine, 1 in 1000 individuals experience some side effects. Find the probability that in a group of 500 people nobody experiences side effects.

82. For a certain vaccine, 1 in 500 individuals experience some side effects. Find the probability that in a group of 200 people at least one person experiences side effects.

83. About 1 in 700 births in the United State are affected by Down's syndrome, a chromosomal disorder. Find the probability that there is at most one case of Down's syndrome among 1000 births by (a) computing the exact probability, and (b) using a Poisson approximation.

84. About 1 in 1000 boys is affected by Fragile X syndrome, a genetic disorder that causes learning difficulties. Find the probability that in a group of 500 boys nobody is affected by this disorder by (a) computing the exact probability, and (b) using a Poisson approximation.

85. (Refer to Example 37.) Suppose a parasitoid has a probability of 0.03 of detecting a given host. If 50 parasitoids are trying to find a particular host, what is the probability that the host will avoid detection?

12.5 CONTINUOUS DISTRIBUTIONS

12.5.1 Density Functions

In the previous section, we discussed random variables that took on a finite (or countably infinite) number of values. In this section, we will discuss random variables that take on a continuum of possible values; they are called **continuous random variables**. Such examples arise, for instance, when we consider the length distribution of an organism. Within an appropriate (species-specific) interval, an individual's length can take on any value.

For illustration, consider the following example that is adapted from de Roos (1996). The waterflea *Daphnia pulex* feeds on the alga *Chlamydomonas rheinhardii*. An important component of the feeding behavior of *Daphnia* is the strong dependence of the amount of food consumed on the size of the individual. To model the feeding behavior of *Daphnia*, we would need to describe the size structure of the population. We view size as a random variable, denoted by X, and determine the fraction of *Daphnia* whose size is less than or equal to x, for all possible values of x. If the population size is large, this fraction can be well approximated by a continuous function, which we denote by $F(x)$. It serves the role of a distribution function so that $F(x) = P(X \leq x)$.

A distribution function completely characterizes the probability distribution of a random variable, as we saw in the previous section. This is not any different for continuous random variables. To describe the probability distribution of a continuous random variable X, we will therefore use its distribution function $F(x)$. The distribution function for a continuous random variable has the same definition as the one for a discrete random variable:

$$F(x) = P(X \leq x)$$

It has the following properties (see Figure 12.26), some of which differ from that of a discrete random variable:

1. $0 \leq F(x) \leq 1$.
2. $\lim_{x \to -\infty} F(x) = 0$ and $\lim_{x \to \infty} F(x) = 1$.
3. $F(x)$ is nondecreasing and continuous.

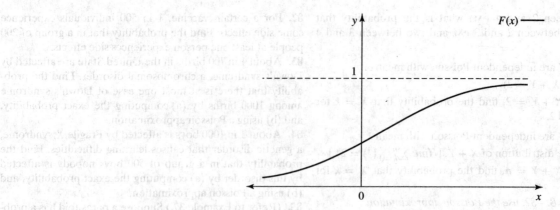

▲ **Figure 12.26**
The distribution function of a continuous random variable

Note that the distribution function of a continuous random variable is continuous. You should contrast this with the distribution function of a discrete random variable, which is piecewise constant and takes jumps at those values of x where $P(X = x) > 0$, though both are nondecreasing and take on values between 0 and 1.

▶ **Example 1**   Show that

$$F(x) = \begin{cases} 1 - e^{-2x} & \text{for } x > 0 \\ 0 & \text{for } x \leq 0 \end{cases}$$

is a distribution function of a continuous random variable.

Solution

[A graph of $F(x)$ is shown in Figure 12.27.] We must check the three properties of distribution functions of continuous random variables.

1. Since $0 \leq 1 - e^{-2x} \leq 1$ for $x > 0$ and $F(x) = 0$ for $x \leq 0$, it follows that $0 \leq F(x) \leq 1$ for all $x \in \mathbf{R}$.

2. Since $F(x) = 0$ for $x \leq 0$ and $\lim_{x \to \infty} e^{-2x} = 0$, it follows that

$$\lim_{x \to -\infty} F(x) = 0 \quad \text{and} \quad \lim_{x \to \infty} F(x) = 1$$

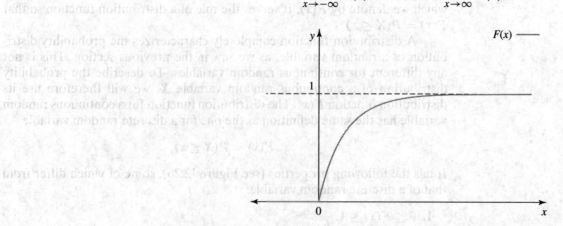

▲ **Figure 12.27**
The distribution function in Example 1

3. To show that $F(x)$ is continuous for all $x \in \mathbf{R}$, note that $F(x)$ is continuous for $x > 0$ and for $x < 0$. To check continuity at $x = 0$, we compute

$$\lim_{x \to 0^+} F(x) = \lim_{x \to 0^+} (1 - e^{-2x}) = 0 = \lim_{x \to 0^-} F(x)$$

which is equal to $F(0) = 0$. Hence, $F(x)$ is continuous at $x = 0$.

To show that $F(x)$ is nondecreasing, we compute $F'(x)$ for $x > 0$.

$$F'(x) = 2e^{-2x} > 0 \text{ for } x > 0$$

which implies that $F(x)$ is increasing for $x > 0$. Since $F(x)$ is continuous for all $x \in \mathbf{R}$ and equal to 0 for $x \le 0$, it follows that $F(x)$ is nondecreasing for all $x \in \mathbf{R}$. ◀

If there is a nonnegative function $f(x)$ so that the distribution function $F(x)$ of a random variable X has the representation

$$F(x) = \int_{-\infty}^{x} f(u) \, du$$

we say that X is a continuous random variable with **(probability) density function** $f(x)$ (see Figure 12.28).

Since $F(x)$ is a distribution function, it follows that

$$\int_{-\infty}^{\infty} f(x) \, dx = 1 \tag{12.28}$$

Any nonnegative function that satisfies (12.28) defines a density function. The function $f(x)$ need not be continuous. In all of our applications, $f(x)$ will be continuous except for possibly a finite number of points. It will frequently be defined as a piecewise continuous function. Using the fundamental theorem of calculus, part I, we can obtain the density function $f(x)$ of a continuous random variable from its distribution function $F(x)$ by differentiating the distribution function. Namely, $f(x) = F'(x)$ at all points x where $F(x)$ is differentiable. At points where $F(x)$ is not differentiable, we set $f(x) = 0$. This ensures that the density function is defined everywhere.

Furthermore, since

$$P(a < X \le b) = P(X \le b) - P(X \le a) = F(b) - F(a)$$

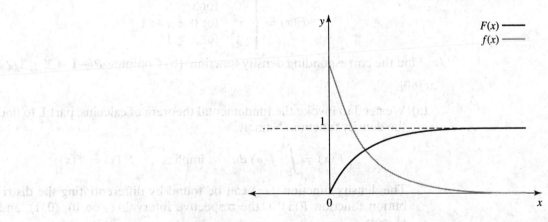

▲ **Figure 12.28**
The distribution function $F(x)$ and corresponding density function $f(x)$ of a continuous random variable

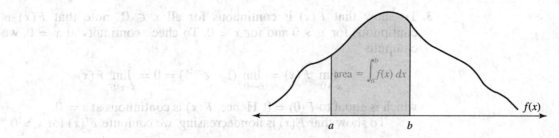

▲ Figure 12.29
The area between the density function $f(x)$ and the x-axis between a and b represents the probability that the random variable X lies between a and b

we find

$$P(a < X \le b) = \int_a^b f(x)\, dx \qquad (12.29)$$

That is, the area under the curve $y = f(x)$ between a and b represents the probability that the random variable takes on values between a and b, as illustrated in Figure 12.29. It is important to realize that in (12.29), probabilities are represented by area; in particular, the *integral* of $f(x)$—not $f(x)$ itself—has this physical interpretation. The density function $f(x)$ does not have an immediate physical interpretation. Later in the section, we will explain how to find $f(x)$ empirically.

In contrast to discrete random variables, for which $P(X \le b)$ and $P(X < b)$ can differ, there is no difference for continuous random variables, since

$$P(X = b) = \int_b^b f(x)\, dx = 0$$

Therefore,

$$\int_a^b f(x)\, dx = P(a < X \le b) = P(a \le X \le b)$$

$$= P(a < X < b) = P(a \le X < b)$$

▶ Example 2 The distribution function of a continuous random variable X is given by

$$F(x) = \begin{cases} 0 & \text{for } x \le 0 \\ x^2 & \text{for } 0 < x < 1 \\ 1 & \text{for } x \ge 1 \end{cases}$$

(a) Find the corresponding density function. (b) Compute $P(-1 \le X \le 1/2)$.

Solution

(a) We need to invoke the fundamental theorem of calculus, part I, to find the density function. Namely,

$$F(x) = \int_a^x f(u)\, du \qquad \text{implies} \qquad F'(x) = f(x)$$

The density function $f(x)$ can be found by differentiating the distribution function $F(x)$ in the respective intervals $(-\infty, 0)$, $(0, 1)$, and $(1, \infty)$:

$$f(x) = F'(x) = \begin{cases} 2x & \text{for } 0 < x < 1 \\ 0 & \text{for } x < 0 \text{ or } x > 1 \end{cases}$$

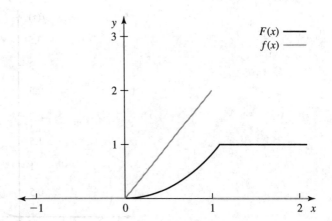

▲ **Figure 12.30**
The distribution function $F(x)$ and corresponding density function $f(x)$ in Example 2

To define the density function everywhere, we set $f(0) = f(1) = 0$. The distribution function together with its density function is shown in Figure 12.30.

(b) Using the distribution function, we immediately find

$$P\left(-1 \le X \le \frac{1}{2}\right) = F\left(\frac{1}{2}\right) - F(-1) = \frac{1}{4} - 0 = \frac{1}{4}$$

If, instead, we use the density function, we must evaluate

$$P\left(-1 \le X \le \frac{1}{2}\right) = \int_{-1}^{1/2} f(x)\, dx = \int_{0}^{1/2} 2x\, dx$$

$$= x^2\Big]_{0}^{1/2} = \frac{1}{4} \qquad \blacktriangleleft$$

The formulas for the mean and the variance of a continuous random variable are analogous to the discrete case. The expectation EX of a continuous random variable with density function $f(x)$ is defined as

$$EX = \int_{-\infty}^{\infty} x f(x)\, dx$$

The expectation of a function of a random variable, $g(X)$, where X has density function $f(x)$, is

$$Eg(X) = \int_{-\infty}^{\infty} g(x) f(x)\, dx$$

The variance of a continuous random variable X with mean μ is defined as

$$\mathrm{var}(X) = E(X - \mu)^2 = \int_{-\infty}^{\infty} (x - \mu)^2 f(x)\, dx$$

The alternate formula that we gave in the previous section holds as well, namely,

$$\mathrm{var}(X) = EX^2 - (EX)^2 = \int_{-\infty}^{\infty} x^2 f(x)\, dx - \left(\int_{-\infty}^{\infty} x f(x)\, dx\right)^2$$

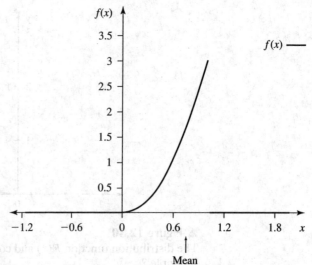

▲ **Figure 12.31**
The density function $f(x)$ in Example 3 together with the location of the mean of X

As a reminder, these integrals are defined on unbounded intervals and $f(x)$ might be discontinuous. To evaluate such integrals, we must use the methods developed in Section 7.4.

▶ **Example 3** The density function of a random variable X is given by (see Figure 12.31)

$$f(x) = \begin{cases} 3x^2 & \text{for } 0 < x < 1 \\ 0 & \text{otherwise} \end{cases}$$

Compute the mean and the variance of X.

Solution

To compute the mean, we must evaluate

$$EX = \int_{-\infty}^{\infty} xf(x)\,dx = \int_0^1 3x^3\,dx = \frac{3}{4}x^4\Big]_0^1 = \frac{3}{4}$$

The mean is indicated in Figure 12.31. To compute the variance, we first evaluate

$$EX^2 = \int_{-\infty}^{\infty} x^2 f(x)\,dx = \int_0^1 3x^4\,dx = \frac{3}{5}x^5\Big]_0^1 = \frac{3}{5}$$

The variance of X is then given by

$$\text{var}(X) = EX^2 - (EX)^2 = \frac{3}{5} - \left(\frac{3}{4}\right)^2 = \frac{3}{80}$$ ◀

▶ **Example 4** The exponential function

$$f(r) = \begin{cases} ae^{-ar} & \text{for } r > 0 \\ 0 & \text{for } r \le 0 \end{cases}$$

where $a > 0$ is a constant, is frequently used to model seed dispersal. The function $f(r)$ is a density function, and $\int_a^b f(r)\,dr$ describes the fraction of seeds dispersed between distances a and b from the source at 0. Find the average dispersal distance.

Solution

We use the preceding formula for the average value,

$$\text{average dispersal distance} = \int_{-\infty}^{\infty} rf(r)\,dr = \int_{0}^{\infty} rae^{-ar}\,dr$$

since $f(r) = 0$ for $r \leq 0$. To evaluate this integral, we must integrate by parts:

$$\text{average dispersal distance} = \int_{0}^{\infty} rae^{-ar}\,dr = \lim_{z\to\infty} \int_{0}^{z} rae^{-ar}\,dr$$

$$= \lim_{z\to\infty} \left[r(-e^{-ar})\right]_{0}^{z} + \lim_{z\to\infty} \int_{0}^{z} e^{-ar}\,dr$$

The first expression is equal to

$$\lim_{z\to\infty}[-ze^{-az} + 0] = -\lim_{z\to\infty} \frac{z}{e^{az}}$$

This is of the form $\frac{\infty}{\infty}$. Using L'Hospital's rule, we find

$$\lim_{z\to\infty} \frac{z}{e^{az}} = \lim_{z\to\infty} \frac{1}{ae^{az}} = 0$$

since $a > 0$. The second expression is

$$\lim_{z\to\infty} \int_{0}^{z} e^{-ar}\,dr = \lim_{z\to\infty} \left[-\frac{1}{a}e^{-ar}\right]_{0}^{z} = \frac{1}{a}$$

Therefore,

$$\text{average dispersal distance} = \frac{1}{a}$$

We will now discuss how to empirically determine $f(x)$. We set

$$F(x) = \int_{-\infty}^{x} f(t)\,dt$$

Then

$$F(x + \Delta x) - F(x) = \int_{-\infty}^{x+\Delta x} f(t)\,dt - \int_{-\infty}^{x} f(t)\,dt$$

$$= \int_{x}^{x+\Delta x} f(t)\,dt$$

If Δx is sufficiently small, then $f(t)$ will not vary much over the interval $[x, x + \Delta x)$, and we approximate $f(t)$ by $f(x)$ over the interval $[x, x + \Delta x)$. Hence,

$$\int_{x}^{x+\Delta x} f(t)\,dt \approx f(x)\Delta x \tag{12.30}$$

for Δx sufficiently small and we can think of $f(x)\Delta x$ as approximately representing the fraction that falls into the interval $[x, x + \Delta x)$. This is an important interpretation which will help us to determine density functions empirically. This should remind you of the Riemann sum approximation that we discussed in Section 6.1. For Δx small, just one rectangle gives a good approximation, as illustrated in Figure 12.32.

To determine the density $f(x)$ empirically, we will use the approximation (12.30). We take a sufficiently large sample from the population and measure the quantity of interest of each individual in the sample. We partition the

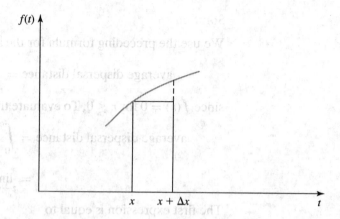

▲ **Figure 12.32**

The area of the rectangle is $f(x)\Delta x$. It is a good approximation of $\int_x^{x+\Delta x} f(t)\, dt$, which is the area under the curve between x and $x + \Delta x$

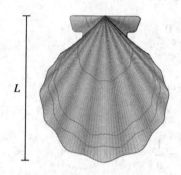

▲ **Figure 12.33**
The saggital length of a shell

interval over which the quantity of interest varies into subintervals of length Δx_i. For each subinterval, we count the number of sample points that fall into the respective subintervals. To display the data graphically, we use a **histogram**, which consists of rectangles whose widths are equal to the lengths of the corresponding subintervals and whose *areas* are equal to the number of sample points that fall into the corresponding subintervals. This is analogous to approximating areas under curves by rectangles. We illustrate this procedure in the following example.

Brachiopods form a marine invertebrate phylum whose soft body parts are enclosed in shells. They were the dominant seabed shelled animals in the Paleozoic, but suffered greatly during the Permian-Triassic mass extinction.[2] They are still present today (with approximately 120 genera) and occupy a diverse range of habitats—but they are no longer the dominant seabed shelled animal, their place having been taken by bivalve mollusks. (The brachiopod story is described in Ward, 1992.)

The brachiopod *Dielasma* is common in Permian reef deposits in the north of England. The following table (adapted from Benton and Harper, 1997) represents measurements of the sagittal length of the shell measured in mm (see Figure 12.33).

Length	5	7	9	11	13	15	17	19	21	23	25
Frequency	3	28	12	2	4	4	6	6	5	3	1

The length measurements are divided into classes $[0, 2)$, $[2, 4)$, $[4, 6)$, ..., $[26, 28)$. The midpoint of each subinterval represents the size class. For instance, length 11 in the table below corresponds to the size class of lengths between 10 mm and 12 mm. The number below each size class, the frequency, represents the number of brachiopods in the sample whose lengths fell into the corresponding size class. For instance, there were two brachiopods in the sample whose lengths fell into the size class $[10, 12)$.

[2]The Permian geological period lasted from 286 to 248 million years ago; the Triassic followed the Permian and lasted from 248 to 213 million years ago. The Permian-Triassic mass extinction is believed to have been the most severe mass extinction of life that has ever occurred.

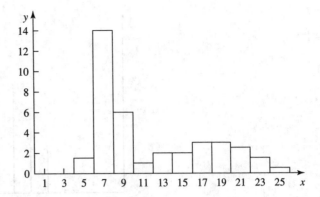

▲ **Figure 12.34**
The histogram for the frequency distribution of the saggital length

To display this data set graphically, we use a histogram, as shown in Figure 12.34. The horizontal axis shows the midpoints of each size class. The graph consists of rectangles whose widths are equal to the length of the corresponding size class and whose *area* is proportional to the number of specimens in the corresponding class. It is very important to note that a histogram represents numbers by area, not by height. For instance, the number of specimens in size class $[8, 10)$ is equal to 12. This size class is represented by 9, the midpoint of the interval $[8, 10)$. Because the width of the size class is 2, the height of the rectangle must be equal to 6 units so that the area of the corresponding rectangle is equal to 12 units. In our example, all size classes are of the same length; of course, this need not be the case in general.

Displaying data graphically has certain advantages. For instance, we see immediately that the size distribution is biased toward smaller sizes, because the rectangles in the histogram that correspond to smaller shell lengths have larger areas. This bias toward smaller sizes might indicate, for instance, that brachiopods suffered a high juvenile mortality.

It is often convenient to scale the vertical axis of the histogram so that the total area of the histogram is equal to 1. One of the advantages of this is that the histogram then does not directly reflect the sample size, as only proportions are represented. This makes it easier to compare histograms from different samples. For instance, if someone else had obtained a different sample of this type of brachiopod in the same location, and if both samples are representative of its length distribution, then both histograms should look similar.

If we scale the total area of the histogram to 1, then the area of each rectangle in the histogram represents the fraction of the sample in the corresponding class. To obtain the fraction of sample points in a certain class, we divide the number of sample points in this class by the total sample size. In our example, the sample size was 74. The fraction in the sample in size class $[8, 10)$, for instance, would then be $12/74 = 0.16$ or 16%.

The choice of the widths of the classes in the histogram is somewhat arbitrary. The goal is to obtain an informative graph. Typically, the larger the sample size, the smaller the widths of the classes can be.

The outline of the normalized histogram (that is, where the total area is equal to 1) can be used as an approximation to the density function $f(x)$ (see Figure 12.35). Typically, the larger the sample size, the smaller the widths of the size classes, and the better the approximation of $f(x)$.

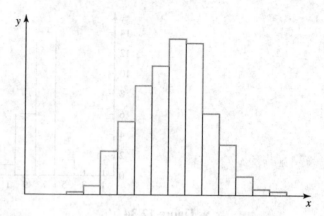

▲ **Figure 12.35**
The histogram as an approximation of the density function

In the following subsections, we will introduce some continuous distributions and their applications.

12.5.2 The Normal Distribution

The normal distribution was first introduced by Abraham De Moivre (1667–1754) in the context of computing probabilities in binomial experiments when the number of trials is large. Gauss showed later that this distribution was important in the error analysis of measurements. It is the most important continuous distribution, and we discuss an application first.

Quantitative genetics is concerned with metric characters, such as plant height, litter size, body weight, and so on. Such characters are called **quantitative characters**. There are many quantitative characters whose frequency distributions follow a bell-shaped curve. For instance, counting the bristles on some particular part of the abdomen (fifth sternite) of a strain of *Drosophila melanogaster*, Mackay (1984) found that the number of bristles varied according to a bell-shaped curve. (This is shown in Figure 12.36, which is adapted from Hartl and Clark, 1989.)

The smooth curve in Figure 12.36 that is fitted to the histogram is proportional to the density function of a **normal distribution** (the curve is not scaled so that the area under the curve is equal to 1). The density function of

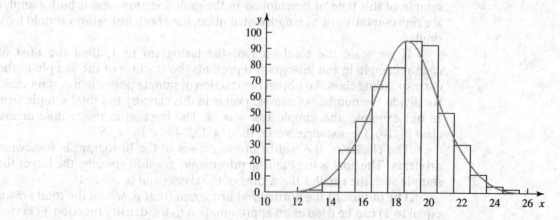

▲ **Figure 12.36**
The number of abdominal bristles

the normal distribution is described by just two parameters, called μ and σ, which can be estimated from data. The parameter μ can be any real number; the parameter σ is a positive real number. The density function is given in the following.

> A continuous random variable X is **normally distributed** with parameters μ and σ if it has density function
>
> $$f(x) = \frac{1}{\sigma\sqrt{2\pi}}e^{-(x-\mu)^2/2\sigma^2}, \qquad -\infty < x < \infty$$

The parameter μ is the mean and the parameter σ is the standard deviation. In Problem 11, we will investigate the shape of the density function of the normal distribution (a graph is shown in Figure 12.37). We collect the properties here.

1. $f(x)$ is symmetric about $x = \mu$.
2. The maximum of $f(x)$ is at $x = \mu$.
3. The inflection points of $f(x)$ are at $x = \mu - \sigma$ and $x = \mu + \sigma$.

Since $f(x)$ is a density function,

$$f(x) \geq 0 \quad \text{and} \quad \int_{-\infty}^{\infty} f(x)\,dx = 1$$

With the tools we have so far, we cannot show that the density function is normalized to 1. In Problem 12, we will show that the mean μ is indeed the expected value of X, namely,

$$\mu = \int_{-\infty}^{\infty} x f(x)\,dx$$

Furthermore, if a quantity X is normally distributed with parameters μ and σ, then

$$P(a \leq X \leq b) = \int_{a}^{b} \frac{1}{\sigma\sqrt{2\pi}}e^{-(x-\mu)^2/2\sigma^2}\,dx$$

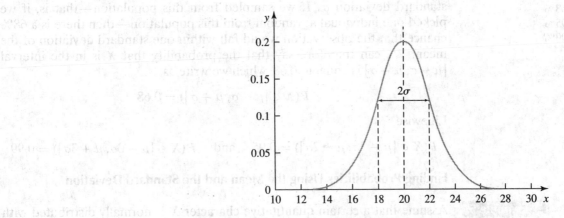

▲ **Figure 12.37**
The graph of the normal density with $\mu = 20$ and $\sigma = 2$. The maximum is at $\mu = 20$; the two inflection points are at 18 and 22, respectively

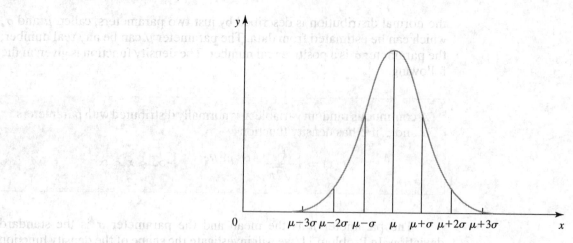

▲ Figure 12.38
The density of the normal distribution with mean μ and standard deviation σ

It is not possible to evaluate this integral using elementary functions; it can only be evaluated numerically. There are tables for the normal distribution with parameters $\mu = 0$ and $\sigma = 1$ that list values for

$$F(x) = P(X \le x) = \int_{-\infty}^{x} \frac{1}{\sqrt{2\pi}} e^{-z^2/2} \, dz$$

[A table for $F(x)$ is reproduced in the Appendix B.] The table for $F(x)$ can be used to obtain probabilities for general μ and σ, and we will see later how to do this.

At the moment, we will only need the following values. The area $A(k)$ under the density function of the normal distribution with mean μ and standard deviation σ between $\mu - k\sigma$ and $\mu + k\sigma$ for $k = 1, 2$, and 3 is given in the table on the left (see Figure 12.38). That is, if a certain quantity X of a population is normally distributed with mean μ and standard deviation σ, then 68% of the population fall within one standard deviation of the mean, 95% fall within two standard deviations of the mean, and 99% fall within three standard deviations of the mean.

These percentages can also be interpreted in the following way. Suppose that a quantity X in a population is normally distributed with mean μ and standard deviation σ. If we sampled from this population—that is, if we picked one individual at random from this population—then there is a 68% chance that the observation would fall within one standard deviation of the mean. We can therefore say that the probability that X is in the interval $[\mu - \sigma, \mu + \sigma]$ is equal to 0.68, which we write as

$$P(X \in [\mu - \sigma, \mu + \sigma]) = 0.68$$

Likewise,

$$P(X \in [\mu - 2\sigma, \mu + 2\sigma]) = 0.95 \quad \text{and} \quad P(X \in [\mu - 3\sigma, \mu + 3\sigma]) = 0.99$$

Finding Probabilities Using the Mean and the Standard Deviation

▶ **Example 5** Assume that a certain quantitative character X is normally distributed with mean $\mu = 4$ and standard deviation $\sigma = 1.5$. Find an interval centered at the mean such that there is a 95% chance (99% chance) that an observation will fall into this interval.

k	$A(k)$
1	68%
2	95%
3	99%

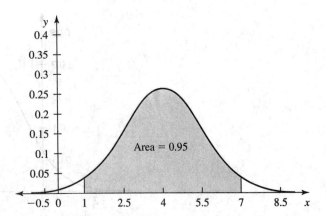

▲ **Figure 12.39**
The normal density with mean 4 and standard deviation 1.5 in Example 5: 95% of
the observations fall into the interval $[1, 7]$

Solution

Since 95% corresponds to a range within two standard deviations of the mean
(see Figure 12.39), the resulting interval is

$$[4 - (2)(1.5), 4 + (2)(1.5)] = [1, 7]$$

We can therefore write

$$P(X \in [1, 7]) = 0.95$$

Similarly, 99% corresponds to a range within three standard deviations of the
mean, resulting in an interval of the form

$$[4 - (3)(1.5), 4 + (3)(1.5)] = [-0.5, 8.5]$$

We can therefore write

$$P(X \in [-0.5, 8.5]) = 0.99 \qquad \blacktriangleleft$$

▶ **Example 6**   Assume that a certain quantitative character X is normally distributed with
mean $\mu = 3$ and standard deviation $\sigma = 2$. We take a sample of size 1. What
is the chance that we observe a value greater than 9?

Solution

Since $9 = 3 + (3)(2)$, we find that we want to know the chance that the
observation is three standard deviations above the mean (see Figure 12.40).
Now, 99% of the population is within three standard deviations of the mean;
therefore, 1% is outside of the interval $[\mu - 3\sigma, \mu + 3\sigma]$. Since the density
function of the normal distribution is symmetric about the mean, it follows
that the area to the left of $\mu - 3\sigma$ and the area to the right of $\mu + 3\sigma$ are the
same. Hence, there is a $(1\%)/2 = 0.5\%$ chance that the observation is above
9. We can therefore write

$$P(X > 9) = 0.005 \qquad \blacktriangleleft$$

▶ **Example 7**   Assume that a certain quantitative character X is normally distributed with
parameters μ and σ. What is the probability that an observation lies below
$\mu + \sigma$?

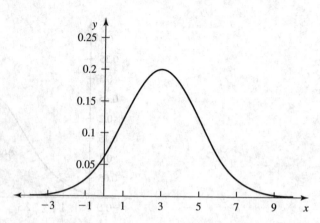

▲ **Figure 12.40**
The normal density with mean 3 and standard deviation 2 in Example 6: 0.5% of the observations are greater than 9

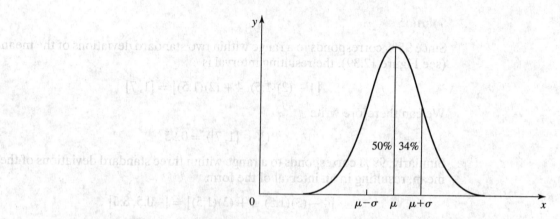

▲ **Figure 12.41**
The area for Example 7

Solution

Since 68% of the population fall within one standard deviation of the mean, and since the density curve is symmetric about the mean, it follows that 34% of the population fall into the interval $[\mu, \mu + \sigma]$. Furthermore, because of the symmetry of the density function, 50% of the population are below the mean. Hence, 50% + 34% = 84% of the population lie below $\mu + \sigma$, as illustrated in Figure 12.41. We can therefore write

$$P(X < \mu + \sigma) = 0.84 \qquad ◄$$

Finding Probabilities Using the Table The table for a normal distribution with mean 0 and standard deviation 1 (see Appendix B) can be used to compute probabilities when the distribution is normal with mean μ and standard deviation σ.

We begin by explaining how to use the table for the normal distribution with mean 0 and standard deviation 1, called the **standard normal distribution**, whose density is given by

$$f(u) = \frac{1}{\sqrt{2\pi}}e^{-u^2/2} \qquad \text{for } -\infty < u < \infty$$

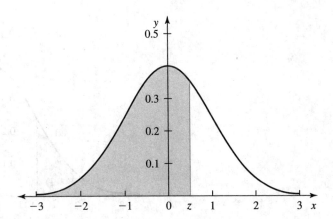

▲ **Figure 12.42**
The area to the left of $x = z$ under the graph of the normal density $f(x)$ is $F(z)$, which is listed in the table for the normal distribution

The table lists values for

$$F(z) = \int_{-\infty}^{z} \frac{1}{\sqrt{2\pi}} e^{-u^2/2} \, du$$

Geometrically, $F(z)$ is the area to the left of the line $x = z$ under the graph of the density function, as illustrated in Figure 12.42.

We interpret $F(z)$ as the probability that an observation is to the left of z. For instance, when $z = 1$, $F(1) = 0.8413$, and we say that the probability that an observation has a value less than or equal to 1, is equal to 0.8413, or that 84.13% of the population has a value less than or equal to 1.

As you can see, the table does not provide entries for negative values of z. To compute such values, we take advantage of the symmetries of the density function. For instance, if we wish to compute $F(-1)$, we see from the graph of the density function that the area to the left of -1 is the same as the area to the right of 1 (see Figure 12.43). We write

$$F(-1) = \int_{-\infty}^{-1} \frac{1}{\sqrt{2\pi}} e^{-u^2/2} \, du = \int_{1}^{\infty} \frac{1}{\sqrt{2\pi}} e^{-u^2/2} \, du$$

$$= \int_{-\infty}^{\infty} \frac{1}{\sqrt{2\pi}} e^{-u^2/2} \, du - \int_{-\infty}^{1} \frac{1}{\sqrt{2\pi}} e^{-u^2/2} \, du$$

$$= 1 - F(1) = 1 - 0.8413 = 0.1587$$

Here we used the fact that the total area is equal to 1.

We can use the table to compute areas under the graph of a normal density with arbitrary mean μ and standard deviation σ. Namely, to find the value of

$$\int_{a}^{b} \frac{1}{\sigma\sqrt{2\pi}} e^{-(x-\mu)^2/2\sigma^2} \, dx$$

where $-\infty < a < b < \infty$, we use the substitution

$$u = \frac{x - \mu}{\sigma} \quad \text{with} \quad \frac{du}{dx} = \frac{1}{\sigma}$$

which yields

$$\int_{a}^{b} \frac{1}{\sigma\sqrt{2\pi}} e^{-(x-\mu)^2/2\sigma^2} \, dx = \int_{(a-\mu)/\sigma}^{(b-\mu)/\sigma} \frac{1}{\sqrt{2\pi}} e^{-u^2/2} \, du$$

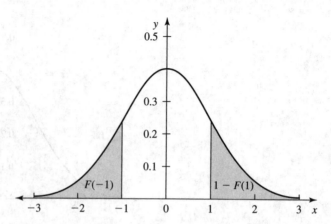

▲ **Figure 12.43**
The area of the shaded region $F(-1)$ is the same as the area of the shaded region $1 - F(1)$.

We recognize the right-hand side as the area under the *standard normal density* between $(a - \mu)/\sigma$ and $(b - \mu)/\sigma$. Therefore, the area under the normal density with mean μ and standard deviation σ between a and b is the same as the area under the standard normal density between $(a - \mu)/\sigma$ and $(b - \mu)/\sigma$ (see Figure 12.44). We illustrate this in the following example.

▶ **Example 8** Suppose that a quantity X is normally distributed with mean 3 and standard deviation 2. Find the fraction of the population that falls into the interval $[2, 5]$, that is, find $P(X \in [2, 5])$.

Solution

To answer this question we must compute

$$\int_2^5 \frac{1}{2\sqrt{2\pi}} e^{(x-3)^2/8} \, dx \tag{12.31}$$

Using the transformation $u = (x - 3)/2$, we find that when

$$x = 2 \quad \text{then } u = \frac{2 - 3}{2} = -\frac{1}{2}$$

and when

$$x = 5 \quad \text{then } u = \frac{5 - 3}{2} = 1$$

Therefore, the area under the normal density with mean 3 and standard deviation 2 between 2 and 5 is the same as the area under the standard normal density between $-1/2$ and 1. Hence, the integral in (12.31) is equal to

$$\int_{-1/2}^1 \frac{1}{\sqrt{2\pi}} e^{-u^2/2} \, du = \int_{-\infty}^1 \frac{1}{\sqrt{2\pi}} e^{-u^2/2} \, du - \int_{-\infty}^{-1/2} \frac{1}{\sqrt{2\pi}} e^{-u^2/2} \, du$$

$$= F(1) - F\left(-\frac{1}{2}\right) = F(1) - \left(1 - F\left(\frac{1}{2}\right)\right)$$

$$= F(1) + F\left(\frac{1}{2}\right) - 1 = 0.8413 + 0.6915 - 1 = 0.5328$$

and $P(X \in [2, 5]) = 0.5328$.

Instead of writing out these integrals, it is easier to determine what we need to compute when we sketch the relevant area under the standard normal

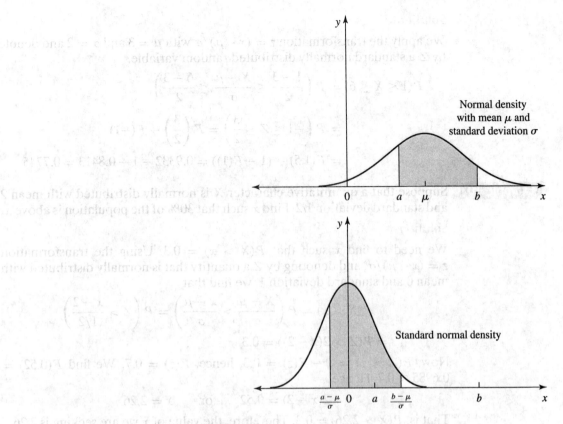

▲ **Figure 12.44**
The area under the normal density with mean μ and standard deviation σ between a and b is the same as the area under the standard normal density between $(a - \mu)/\sigma$ and $(b - \mu)/\sigma$

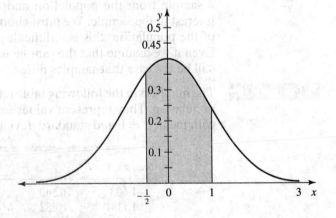

▲ **Figure 12.45**
The area in Example 8

curve. We see from Figure 12.45 that we need to compute $F(1) - F(-\frac{1}{2})$. Since $F(-\frac{1}{2}) = 1 - F(\frac{1}{2})$, it follows that we need to find $F(1) - 1 + F(\frac{1}{2})$, which we computed previously. ◀

▶ **Example 9** Let X be normally distributed with mean 3 and variance 4. Find $P(1 \leq X \leq 6)$.

Solution

We apply the transformation $z = (x - \mu)/\sigma$ with $\mu = 3$ and $\sigma = 2$ and denote by Z a standard normally distributed random variable.

$$P(1 \leq X \leq 6) = P\left(\frac{1-3}{2} \leq \frac{X - \mu}{\sigma} \leq \frac{6-3}{2}\right)$$

$$= P\left(-1 \leq Z \leq \frac{3}{2}\right) = F\left(\frac{3}{2}\right) - F(-1)$$

$$= F(1.5) - (1 - F(1)) = 0.9332 - 1 + 0.8413 = 0.7745 \quad \blacktriangleleft$$

▶ **Example 10** Suppose that a quantitative character X is normally distributed with mean 2 and standard deviation 1/2. Find x such that 30% of the population is above x.

Solution

We need to find x such that $P(X > x) = 0.3$. Using the transformation $z = (x - \mu)/\sigma$, and denoting by Z a quantity that is normally distributed with mean 0 and standard deviation 1, we find that

$$P(X > x) = P\left(\frac{X - \mu}{\sigma} > \frac{x - \mu}{\sigma}\right) = P\left(Z > \frac{x - 2}{1/2}\right)$$

$$= P(Z > 2(x - 2)) = 0.3$$

Now, $P(Z > z) = 1 - F(z) = 0.3$; hence, $F(z) = 0.7$. We find $F(0.52) = 0.6985 \approx 0.7$. Hence,

$$2(x - 2) = 0.52 \quad \text{or} \quad x = 2.26$$

That is, $P(X > 2.26) = 0.3$. Therefore, the value of x we are seeking is 2.26.

$\blacktriangleleft$

A Notes on Samples To obtain information about a quantity, such as size or bristle number, we cannot survey the entire population. Instead, we take a sample from the population and find the distribution of the quantity of interest in the sample. We must choose the sample so that it is representative of the population; this is a difficult problem, which we cannot discuss here. Even if we assume that the sample is representative of the population, it will still be the case that samples differ.

▶ **Example 11** The numbers in the following table represent two samples, each from the same population. They represent values for a quantity that is normally distributed, with mean $\mu = 0$ and standard deviation $\sigma = 1$.

Sample 1				
−1.633	0.542	0.250	−0.166	0.032
1.114	0.882	1.265	−0.202	0.151
1.151	−1.210	−0.927	0.425	0.290
−1.939	0.891	−0.227	0.602	0.873
0.385	−0.649	−0.577	0.237	−0.289

Sample 2				
−0.157	0.693	1.710	0.800	−0.265
1.492	−0.713	0.821	−0.031	−0.780
−0.042	1.615	−1.440	−0.989	−0.580
0.289	−0.904	0.259	−0.600	−1.635
0.721	−1.117	0.635	0.592	−1.362

Both samples are obtained from a table of random numbers that are normally distributed with mean 0 and standard deviation 1 (Beyer, 1991).

(a) Count the number of observations in each sample that fall below the mean $\mu = 0$, and compare the number to what you would expect based on properties of the normal distribution.

(b) Count the number of observations in each sample that fall within one standard deviation of the mean, and compare the number to what you would expect based on properties of the normal distribution.

Solution

(a) Since the mean is equal to 0, to find the number of observations that are below the mean, we simply count the number of observations that are negative. In the first sample, 10 observations are below the mean; in the second sample, 14 observations are below the mean. We expect that half of the sample points are below the mean. Since each sample is of size 25, we expect about 12 or 13 sample points to be below the mean.

(b) Since the standard deviation is 1, we count the number of observations that fall into the interval $[-1, 1]$. In the first sample, there are 19 such observations; in the second sample, there are 18 such observations. To compare this with the theoretical value, note that 68% of the population falls within one standard deviation of the mean. Since the sample size is equal to 25 and $(0.68)(25) = 17$, we expect about 17 observations to fall into the interval $[-1, 1]$.   ◄

The preceding example illustrates an important point. Even if random samples are taken from the same population, they are not identical. For instance, in the preceding example we expect that half of the observations will be below the mean. In the first sample, less than half of the observations will be below the mean; whereas in the second sample, more than half of the observations are below the mean.

As the sample size increases, however, it will reflect the population increasingly better. That is, in order to determine the distribution of a quantitative character, such as the number of bristles in *D. melanogaster*, you would take a sample and find, for instance, the histogram associated with the quantity of interest. If the sample is large enough, the histogram will reflect the population distribution quite well. But if you repeat the experiment, you should not expect the two histograms to be exactly the same. If the sample size is large enough, however, they will be close.

The importance of the normal distribution cannot be overstated. A large part of statistics is based on the assumption that observed quantities are normally distributed. You will probably ask why we can assume the normal distribution in the first place. The reason for this is quite deep, and we will discuss some of it in the next section. At this point, we wish to just give you the gist of it.

Many quantities can be thought of as a sum of a large number of small contributions. We can show that the distribution of any sum of independent random variables, which all have the same distribution with a finite mean and a finite variance, converges to a normally distributed random variable when the number of terms in the sum increases. This result is known as the **central limit theorem** (see Section 12.6).

The central limit theorem is evoked, for instance, in quantitative genetics, namely, many quantitative traits (such as height or birth weight of an organism) are thought of as resulting from numerous genetic and environmental factors, which all act in an additive or multiplicative way. If these factors act in an additive way and are independent, the central limit theorem can be applied directly and the distribution of the trait value will resemble a normal distribution. If the factors act in a multiplicative way, then a logarithmic transformation reduces this to the additive case.

The same reasoning is also used when we consider measurement errors. A measurement error can be thought of as a sum of a large number of independent contributions from different sources that act additively. Measurement errors are thus often assumed to be normally distributed.

12.5.3 The Uniform Distribution

The uniform distribution is in some ways the simplest continuous distribution. We say that a random variable U is **uniformly distributed** over the interval (a, b) if its density function is given by

$$f(x) = \begin{cases} \frac{1}{b-a} & \text{for } x \in (a, b) \\ 0 & \text{otherwise} \end{cases}$$

as illustrated in Figure 12.46.

The reason for the term *uniform* can be seen when we compute the probability that the random variable U falls into the interval $(x_1, x_2) \subset (a, b)$. To compute probabilities of events for a uniformly distributed random variable, we compute the area of a rectangle. We therefore use the simple geometric formula "area is equal to the product of height and length," instead of formally integrating the density function (see Figure 12.47). We find

$$P(U \in (x_1, x_2)) = \begin{bmatrix} \text{area under } f(x) = \frac{1}{b-a} \\ \text{between } x_1 \text{ and } x_2 \end{bmatrix}$$

$$= \frac{x_2 - x_1}{b - a}$$

We see that this probability depends only on the length of the interval (x_1, x_2) relative to the length of the interval (a, b), and *not* on the location of (x_1, x_2) provided that (x_1, x_2) is a subset of (a, b). Therefore, intervals of equal lengths that are contained in (a, b) have equal chances of containing U.

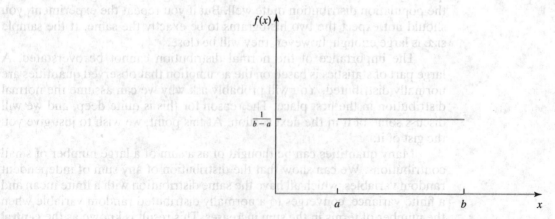

▲ **Figure 12.46**
The density function of a uniformly distributed random variable over the interval (a, b)

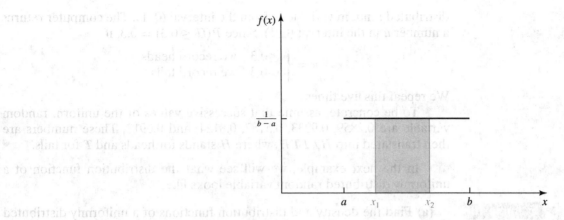

▲ **Figure 12.47**
The probability $P(U \in (x_1, x_2))$ is equal to the area of the shaded region

To find the mean of a uniformly distributed random variable, we evaluate

$$EU = \int_{-\infty}^{\infty} x f(x)\, dx = \int_{a}^{b} x \frac{1}{b-a}\, dx = \frac{1}{b-a} \left[\frac{1}{2} x^2 \right]_{a}^{b}$$

$$= \frac{1}{b-a} \frac{b^2 - a^2}{2} = \frac{a+b}{2}$$

which is the midpoint of the interval (a, b). To find the variance, we first compute

$$EU^2 = \int_{-\infty}^{\infty} x^2 f(x)\, dx = \int_{a}^{b} x^2 \frac{1}{b-a}\, dx = \frac{1}{b-a} \left[\frac{1}{3} x^3 \right]_{a}^{b}$$

$$= \frac{1}{b-a} \frac{b^3 - a^3}{3} = \frac{b^2 + ab + a^2}{3}$$

Then,

$$\mathrm{var}(U) = EU^2 - (EU)^2 = \frac{b^2 + ab + a^2}{3} - \frac{(a+b)^2}{4}$$

$$= \frac{4b^2 + 4ab + 4a^2 - 3a^2 - 6ab - 3b^2}{12} = \frac{b^2 - 2ab + a^2}{12}$$

$$= \frac{(b-a)^2}{12}$$

Uniform distributions are frequently used in computer simulations of random experiments, as in the following example.

▶ **Example 12**   Suppose that you wish to simulate on the computer a random experiment that consists of tossing a coin five times with probability 0.3 of heads. The software that you want to use can generate uniformly distributed random variables in the interval $(0, 1)$. How do you proceed?

Solution

Each trial consists of flipping the coin and then recording the outcome. To simulate a coin flip with probability 0.3 of heads, we draw a uniformly distributed random variable U from the interval $(0, 1)$. The computer returns a number u in the interval $(0, 1)$. Since $P(U \leq 0.3) = 0.3$, if

$$u = \begin{cases} \leq 0.3 & \text{we record heads} \\ > 0.3 & \text{we record tails} \end{cases}$$

We repeat this five times.

To be concrete, assume that successive values of the uniform random variable are 0.2859, 0.9233, 0.5187, 0.8124, and 0.0913. These numbers are then translated into $HTTTH$, where H stands for heads and T for tails. ◀

In the next example, we will see what the distribution function of a uniformly distributed random variable looks like.

▶ **Example 13**

(a) Find the density and distribution functions of a uniformly distributed random variable on the interval $(1, 5)$, and graph both in the same coordinate system.

(b) Suppose that we draw a uniformly distributed random variable from the interval $(1, 5)$. Compute the probability that the first digit after the decimal point is a 2.

Solution

(a) Since the length of the interval $(1, 5)$ is 4, the density function is given by

$$f(x) = \begin{cases} \frac{1}{4} & \text{for } 1 < x < 5 \\ 0 & \text{otherwise} \end{cases}$$

The distribution function $F(x) = P(X \leq x)$ is given by

$$F(x) = \int_{-\infty}^{x} f(u)\, du = \begin{cases} 0 & \text{for } x \leq 1 \\ \int_{1}^{x} \frac{1}{4}\, du = \frac{1}{4}x - \frac{1}{4} & \text{for } 1 < x < 5 \\ 1 & \text{for } x \geq 5 \end{cases}$$

Graphs of $f(x)$ and $F(x)$ are shown in Figure 12.48.

(b) The event that the first digit after the decimal point is a 2 is the event that the random variable U falls into the set

$$A = [1.2, 1.3) \cup [2.2, 2.3) \cup [3.2, 3.3) \cup [4.2, 4.3)$$

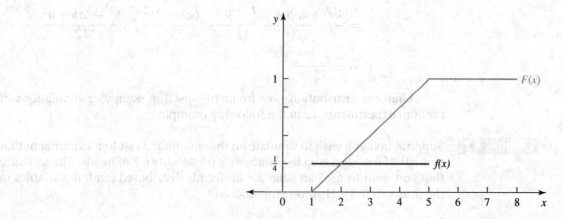

▲ **Figure 12.48**
The density function $f(x)$ and the distribution function $F(x)$ of a uniformly distributed random variable over the interval $(1, 5)$ (see Example 13)

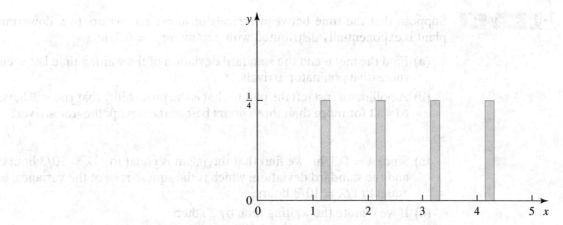

▲ **Figure 12.49**
The probability $P(U \in A)$ in Example 13 is equal to the sum of the areas of the shaded regions

Therefore,

$$P(U \in A) = (4)(0.1)\left(\frac{1}{4}\right) = 0.1$$

as illustrated in Figure 12.49.   ◄

12.5.4 The Exponential Distribution

We give the density function of the exponential distribution first and then explain where this distribution plays a role. We say that a random variable X is **exponentially distributed** with parameter $\lambda > 0$ if its density function is given by

$$f(x) = \begin{cases} \lambda e^{-\lambda x} & \text{for } x > 0 \\ 0 & \text{for } x \le 0 \end{cases}$$

Since

$$\int_0^x \lambda e^{-\lambda u}\, du = -e^{-\lambda u}\Big|_0^x = 1 - e^{-\lambda x}$$

its distribution function $F(x) = P(X \le x)$ is given by

$$F(x) = \begin{cases} 1 - e^{-\lambda x} & \text{for } x > 0 \\ 0 & \text{for } x \le 0 \end{cases}$$

The expected value of X is

$$EX = \int_0^\infty x \lambda e^{-\lambda x}\, dx = \frac{1}{\lambda}$$

and the variance of X is

$$\text{var}(X) = \int_0^\infty \left(x - \frac{1}{\lambda}\right)^2 \lambda e^{-\lambda x}\, dx = \frac{1}{\lambda^2}$$

(The mean and the variance will be calculated in Problems 47 and 48.)

The exponential distribution is frequently used to model waiting times or lifetimes.

▶ **Example 14** Suppose that the time between arrivals of insect pollinators to a flowering plant is exponentially distributed with parameter $\lambda = 0.3$/hr.

(a) Find the mean and the standard deviation of the waiting time between successive pollinator arrivals.

(b) A pollinator just left the plant, what is the probability that you will have to wait for more than three hours before the next pollinator arrives?

Solution

(a) Since $\lambda = 0.3$/hr, we find that the mean is equal to $1/\lambda = 10/3$ hours, and the standard deviation, which is the square root of the variance, is equal to $1/\lambda = 10/3$ hours.

(b) If we denote the waiting time by T, then

$$P(T > 3) = 1 - F(3) = e^{-(0.3)(3)} \approx 0.4066 \qquad \blacktriangleleft$$

▶ **Example 15** Suppose that the life time of an organism is exponentially distributed with parameter $\lambda = (1/200)$ yrs^{-1}.

(a) Find the probability that the organism will live for more than 50 years.

(b) Given that the organism is 100 years old, find the probability that it will live for at least another 50 years.

Solution

We denote the lifetime of the organism by T where T is measured in units of years. Then T is exponentially distributed with parameter $\lambda = (1/200)$ yrs^{-1}.

(a) We want to find the probability that T exceeds 50 years. We compute

$$P(T > 50) = 1 - P(T \leq 50) = 1 - (1 - e^{-50/200})$$
$$= e^{-50/200} = e^{-1/4} \approx 0.7788$$

Note that the units in the exponent canceled out.

(b) We want to find $P(T > 150 \mid T > 100)$. This is a conditional probability. We evaluate it in the following way:

$$P(T > 150 \mid T > 100) = \frac{P(T > 150 \text{ and } T > 100)}{P(T > 100)}$$

Since $\{T > 150\} \subset \{T > 100\}$, it follows that

$$\{T > 150\} \cap \{T > 100\} = \{T > 150\}$$

Therefore,

$$P(T > 150 \text{ and } T > 100) = P(T > 150)$$

We can now continue to evaluate

$$\frac{P(T > 150 \text{ and } T > 100)}{P(T > 100)} = \frac{P(T > 150)}{P(T > 100)} = \frac{e^{-150/200}}{e^{-100/200}}$$
$$= e^{-3/4+1/2} = e^{-1/4} \approx 0.7788$$

This is the same answer as in (a). The fact that the organism has lived for 100 years does not change its probability of it living for another 50 years. We say that this organism does not age. This nonaging property is a characteristic feature of the exponential distribution. Of course, most organisms age. Nevertheless, the exponential distribution is still frequently used to model lifetimes even if the organism ages, in which case it should be considered as an approximation of the real situation. ◀

We discuss this nonaging property in more detail. If T is an exponentially distributed lifetime, then T satisfies the property

$$P(T > t + h \mid T > t) = P(T > h)$$

In words, if the organism is still alive after t units of time, then the probability of living for at least another h units of time is the same as the probability that the organism survived the first h units of time. This implies that death does not become more (or less) likely with age.

This property follows immediately from the following calculation, which is the same as the one that we carried out in Example 15.

$$P(T > t + h \mid T > t) = \frac{P(T > t + h \text{ and } T > t)}{P(T > t)}$$

$$= \frac{P(T > t + h)}{P(T > t)} = \frac{e^{-\lambda(t+h)}}{e^{-\lambda t}}$$

$$= e^{-\lambda h} = P(T > h)$$

For most organisms or objects, the exponential distribution is a poor lifetime model. However, it is the correct distribution to model radioactive decay.

▶ **Example 16** Assume that the lifetime of a radioactive atom is exponentially distributed with parameter $\lambda = 3/$days.

(a) Find the average lifetime of this atom.

(b) Find the time T_h such that the probability that the atom will not have decayed at time t is equal to $1/2$. (The time T_h is called the half-life.)

Solution

(a) The average lifetime is $1/\lambda = (1/3)$ days.

(b) If the random variable T denotes the lifetime of the atom, then T_h satisfies

$$P(T > T_h) = \frac{1}{2}$$

that is,

$$e^{-\lambda T_h} = \frac{1}{2} \quad \text{or} \quad T_h = \frac{\ln 2}{\lambda}$$

With $\lambda = 3/$days, we find

$$T_h = \frac{\ln 2}{3} \text{ days} \approx 0.2310 \text{ days} \qquad ◀$$

▶ **Example 17** The exponential function

$$f(r) = \begin{cases} \lambda e^{-\lambda r} & \text{for } r > 0 \\ 0 & \text{for } r \leq 0 \end{cases}$$

where $\lambda > 0$ is a constant, is frequently used to model seed dispersal (see Example 4 of this section). The function $f(r)$ is a density function, and, for $0 < a < b$, $\int_a^b f(r)\, dr$ describes the fraction of seeds dispersed between distances a and b from the source at 0.

(a) Show that $f(r)$ is a density function.

(b) Show that the fraction of seeds that are dispersed distance R or more declines exponentially with R.

(c) Find R such that 60% of the seeds are dispersed within distance R of the source. How does R depend on λ?

Solution

(a) To show that $f(r)$ is a density function, we need to show that $f(r) \geq 0$ for all $r \in \mathbf{R}$ and that $\int_{-\infty}^{\infty} f(r)\,dr = 1$. Since $\lambda > 0$ and $e^{-\lambda r} > 0$, it follows immediately that $f(r) \geq 0$ for $r > 0$. Combining this with $f(r) = 0$ for $r \leq 0$, we find $f(r) \geq 0$ for all $r \in \mathbf{R}$. To check the second criterion, we need to carry out the integration. Since the function $f(r)$ is a piecewise defined function, we need to split up the integral into two parts:

$$\int_{-\infty}^{\infty} f(r)\,dr = \int_{-\infty}^{0} f(r)\,dr + \int_{0}^{\infty} f(r)\,dr$$

$$= \int_{-\infty}^{0} 0\,dr + \int_{0}^{\infty} \lambda e^{-\lambda r}\,dr$$

The term $\int_{-\infty}^{0} 0\,dr$ is equal to 0. An antiderivative of $\lambda e^{-\lambda r}$ is $-e^{-\lambda r}$. Hence,

$$\int_{0}^{\infty} \lambda e^{-\lambda r}\,dr = \lim_{z \to \infty} \int_{0}^{z} \lambda e^{-\lambda r}\,dr = \lim_{z \to \infty} \left[-e^{-\lambda r}\right]_{0}^{z}$$

$$= \lim_{z \to \infty} \left[-e^{-\lambda z} - (-1)\right] = 1$$

since $\lim_{z \to \infty} e^{-\lambda z} = 0$.

(b) For $R > 0$, let $G(R)$ denote the fraction of seeds that are dispersed distance R or more. Then

$$G(R) = \int_{R}^{\infty} f(r)\,dr = \int_{R}^{\infty} \lambda e^{-\lambda r}\,dr = \lim_{z \to \infty} \int_{R}^{z} \lambda e^{-\lambda r}\,dr$$

$$= \lim_{z \to \infty} \left[-e^{-\lambda r}\right]_{R}^{z} = \lim_{z \to \infty} \left(-e^{-\lambda z} + e^{-\lambda R}\right) = e^{-\lambda R}$$

This shows that $G(R)$ declines exponentially with R.

(c) The number R satisfies

$$0.6 = \int_{0}^{R} \lambda e^{-\lambda r}\,dr$$

Carrying out the integration, we find

$$0.6 = \left[-e^{-\lambda r}\right]_{0}^{R} = 1 - e^{-\lambda R}$$

To find R, we need to solve

$$e^{-\lambda R} = 0.4$$

$$-\lambda R = \ln 0.4$$

$$R = -\frac{\ln 0.4}{\lambda} = \frac{1}{\lambda} \ln \frac{5}{2}$$

where we used in the last step the fact that $-\ln 0.4 = \ln \frac{1}{0.4} = \ln \frac{5}{2}$. We see that $R \propto 1/\lambda$ (that is, the bigger λ, the smaller R), which means that seeds are dispersed more closely to the source for larger values of λ. ◄

▶ **Example 18**   Suppose that you wish to use a computer to generate exponentially distributed random variables but the computer only has software that can generate uniformly distributed random variables in the interval $(0, 1)$. How do you proceed?

Solution

The key ingredient to solving this problem is the following result. If X is a continuous random variable with a strictly increasing distribution function $F(x)$, then $F(X)$ is uniformly distributed in the interval $(0, 1)$. To prove this result, we need to show that

$$P(F(X) \leq u) = u \quad \text{for } 0 < u < 1$$

Now the event $\{F(X) \leq u\}$ is equivalent to the event $\{X \leq F^{-1}(u)\}$, where $F^{-1}(\cdot)$ is the inverse function of $F(\cdot)$. That the inverse function exists follows from the assumption that the distribution function is strictly increasing, which implies that $F(x)$ is one to one. Therefore,

$$P(F(X) \leq u) = P(X \leq F^{-1}(u))$$

Since $F(x) = P(X \leq x)$, we find

$$P(X \leq F^{-1}(u)) = F(F^{-1}(u)) = u$$

where the last step follows from the properties of the inverse function.

How can we use this result to generate exponentially distributed random variables from uniformly distributed random variables? The computer generates a uniformly distributed random variable U, which we interpret as $F(X)$, where X is distributed according to the distribution function $F(x)$. Since

$$U = F(X) \quad \text{is equivalent to} \quad X = F^{-1}(U)$$

we need to find the inverse function of $F(x)$ and compute $F^{-1}(U)$. This is then the random variable X.

In the case of an exponential distribution with parameter λ, the distribution function $F(x)$ is

$$F(x) = 1 - e^{-\lambda x} \quad \text{for } x \geq 0$$

which is strictly increasing on $[0, \infty)$. Set $u = 1 - e^{-\lambda x}$ and solve for x:

$$1 - u = e^{-\lambda x}$$

$$-\frac{1}{\lambda} \ln(1 - u) = x$$

Therefore,

$$F^{-1}(u) = -\frac{1}{\lambda} \ln(1 - u)$$

To be concrete, assume $\lambda = 2$, and that a computer generated the following three uniformly in $(0, 1)$ distributed random variables:

$$u_1 = 0.8890, \quad u_2 = 0.9394, \quad u_3 = 0.3586$$

Then

$$x_1 = F^{-1}(u_1) = -\frac{1}{2}\ln(1 - 0.8890) \approx 1.099$$

$$x_2 = F^{-1}(u_2) = -\frac{1}{2}\ln(1 - 0.9394) \approx 1.402$$

$$x_3 = F^{-1}(u_3) = -\frac{1}{2}\ln(1 - 0.3586) \approx 0.2221$$

are the corresponding realizations of the exponentially distributed random variable. ◄

Aging   Aging is a universal feature of both living organisms and mechanical devices; it is described by a progressive loss of vitality or reliability. A number of mathematical models are used to describe this phenomenon. The starting point for these models is the **survival function** $S(x)$, which is defined as the probability that the individual/device is still alive/functioning at age x. If X is the lifetime, then

$$S(x) = P(X > x) = 1 - P(X \le x) = 1 - F(x)$$

where $F(x)$ is the distribution function of X. We assume now that X is a nonegative, continuous random variable with density function $f(x) > 0$ for $x > 0$. This implies that $F(0) = 0$ and that $F(x)$ is strictly increasing for $x > 0$. The **failure** or **hazard rate** function $\lambda(x)$ is defined as the relative rate of decline of the survival function.

$$\lambda(x) = -\frac{1}{S(x)}\frac{dS}{dx}$$

Note that for $x > 0$, $\lambda(x)$ is positive since $S(x)$ is strictly decreasing for $x > 0$. Since for $x > 0$,

$$S(x) = P(X > x) = \int_x^\infty f(u)\,du$$

we find

$$\lambda(x) = -\frac{1}{S(x)}\frac{dS}{dx} = \frac{1}{P(X > x)}f(x)$$

Consequently, $\lambda(x)\,dx$ can be interpreted as the conditional probability of dying within the age interval $[x, x + dx)$ given the individual is still alive at age $x > 0$.

$$\lambda(x)\,dx = P(X \in [x, x + dx) \mid X > x) \quad \text{for } x > 0$$

Nonaging   Following this interpretation of $\lambda(x)\,dx$, we say that a system does not age if the failure rate $\lambda(x)$ is constant. In this case, for $x > 0$,

$$-\frac{1}{S(x)}\frac{dS}{dx} = \lambda = \text{constant}$$

Separating variables and integrating yields

$$\int \frac{dS}{S} = -\int \lambda\,dx$$

or

$$\ln S(x) = -\lambda x + C_1$$

Thus,

$$S(x) = C_1 e^{-\lambda x}$$

Since X is a nonnegative continuous random variable with $S(0) = P(X > 0) = 1$, we find $C_1 = 1$. Therefore,

$$S(x) = e^{-\lambda x} = 1 - F(x) \quad \text{or} \quad F(x) = 1 - e^{-\lambda x} \quad (12.32)$$

We conclude that X is exponentially distributed with parameter λ. We showed earlier that the exponential distribution has the non-aging property. Equation (12.32) shows that the reverse holds as well. If a device has the nonaging property (that is, constant failure rate), then its lifetime distribution is exponential.

▶ **Example 19**   Suppose the hazard rate function $\lambda(x) = 3/\text{year}$ for $x \geq 0$. Find the probability that an individual will die before age 1 year.

Solution

If $\lambda(x) = 3/\text{year}$, then $S(x) = e^{-3x}$, $x \geq 0$, where x is measured in years. If X denotes the lifetime of the individual, then

$$P(X \leq 1) = 1 - S(1) = 1 - e^{-3} \approx 0.9502$$

There is about a 95% chance that the individual will die before age 1 year. ◀

Aging   When the failure rate function increases with age, then an older device has a higher probability of dying than a younger device, and we say that the system is an **aging system**. Figure 12.50 shows an empirical hazard rate function based on 8,926 males from an inbred line of *Drosophila melanogaster* obtained in the lab of Professor Jim Curtsinger at the University of Minnesota. The smoothed line is a curve fitted to the data. The data are based on daily measurements of survival. The horizontal axis lists the age of individuals. The vertical axis lists $-\ln(N_{x+1}/N_x)$ (on a log scale), where N_x is the number

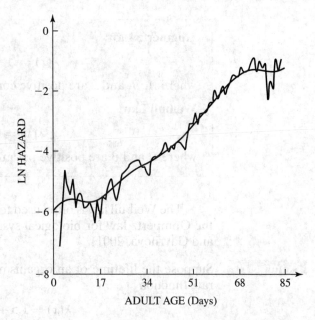

▲ **Figure 12.50**
An empirical hazard rate function (courtesy of Dr. Jim Curtsinger)

of adults alive at age x. This quantity can be interpreted as the hazard rate function, averaged over the interval $[x, x+1]$ (see Problem 65). We see from the graph that the hazard rate function increases with age but seems to level off at very old ages. This is the typical pattern one observes in these kinds of studies.

We again assume that the lifetime X is a nonnegative, continuous random variable with hazard rate function $\lambda(x) > 0$ for $x > 0$ and survival function $S(x)$, $x \geq 0$, with $S(0) = 1$.

Given the hazard rate function $\lambda(x)$, we can find the survival function $S(x)$ by integration. For $x > 0$,

$$-\frac{1}{S(x)} \frac{dS}{dx} = \lambda(x)$$

$$\frac{dS}{S} = -\lambda(x)\,dx$$

$$\ln S(x) = -\int_0^x \lambda(u)\,du + C_1$$

$$S(x) = C \exp\left[-\int_0^x \lambda(u)\,du\right]$$

Since $S(0) = 1$, we find $C = 1$. Therefore,

$$S(x) = \exp\left[-\int_0^x \lambda(u)\,du\right]$$

The two most prominent hazard rate functions that model aging are the **Gompertz law**, in which the hazard rate function increases exponentially with age, and the **Weibull law**, in which the hazard rate function increases according to a power law.

Gompertz Law:

$$\lambda(x) = A + Be^{\alpha x}, \ x \geq 0$$

where A, B, and α are positive constants.

Weibull Law:

$$\lambda(x) = Cx^{\beta}, \ x \geq 0$$

where C and β are positive parameters.

The Weibull law is often used for reliability of technical devices, whereas the Gompertz law for biological systems (see the review article by Gavrilov and Gavrilova, 2001).

▶ **Example 20**   Suppose the lifetime of an organism follows the Gompertz law with hazard rate function

$$\lambda(x) = 1.5 + 0.3e^{0.1x}, \ x \geq 0$$

where x is measured in years. Find the probability that the organism will live for more than one year.

Solution

The survival function is given by

$$S(x) = \exp\left[-\int_0^x (1.5 + 0.3e^{0.1u})\, du\right], \quad x \geq 0$$

We evaluate the integral first. For $x \geq 0$,

$$\int_0^x (1.5 + 0.3e^{0.1u})\, du = 1.5u + \frac{0.3}{0.1}e^{0.1u}\Big|_0^x$$

$$= (1.5x + 3e^{0.1x}) - (0 + 3) = 1.5x + 3e^{0.1x} - 3$$

Therefore,

$$S(x) = \exp[-(1.5x + 3e^{0.1x} - 3)], \quad x \geq 0$$

If X denotes the lifetime of the organism, then the probability that the organism will live for more than one year is

$$P(X > 1) = S(1) = \exp\left[-(1.5 + 3e^{0.1} - 3)\right] \approx 0.1628$$

The organism has about a 16% chance of living for more than one year. ◄

▶ **Example 21** Mortality data from *Drosophila melanogaster* were fitted to a Weibull law. It was found that the following hazard rate function provided the best fit.

$$\lambda(x) = (3 \times 10^{-6})x^{2.5}, \quad x \geq 0$$

where x is measured in days.

(a) Find the probability that an individual will die within the first 20 days.

(b) Find the age at which the probability of still being alive is 0.5.

Solution

(a) The survival function is

$$S(x) = \exp\left[-\int_0^x (3 \times 10^{-6})u^{2.5}\, du\right], \quad x \geq 0$$

We evaluate the integral first. For $x \geq 0$,

$$\int_0^x (3 \times 10^{-6})u^{2.5}\, du = (3 \times 10^{-6})\frac{1}{3.5}u^{3.5}\Big|_0^x$$

$$= (3 \times 10^{-6})\frac{1}{3.5}x^{3.5}$$

Thus the probability that an individual will die within the first 20 days is

$$1 - S(20) = 1 - \exp\left[-(3 \times 10^{-6})\frac{1}{3.5}(20)^{3.5}\right] \approx 0.0302$$

(b) We need to find x so that

$$S(x) = 0.5$$

We solve for x:

$$\exp\left[-(3 \times 10^{-6})\frac{1}{3.5}x^{3.5}\right] = \frac{1}{2}$$

$$-(3 \times 10^{-6})\frac{1}{3.5}x^{3.5} = \ln\frac{1}{2}$$

$$x^{3.5} = \frac{(3.5)(\ln 2)}{3 \times 10^{-6}}$$

$$x = \left(\frac{(3.5)(\ln 2)}{3 \times 10^{-6}}\right)^{1/3.5}$$

$$x \approx 48.746$$

The age where the probability of survival is 0.5 is about 48.7 days. ◀

12.5.5 Problems

(12.5.1)

1. Show that
$$f(x) = \begin{cases} 3e^{-3x} & \text{for } x > 0 \\ 0 & \text{for } x \le 0 \end{cases}$$
is a density function. Find the corresponding distribution function.

2. Show that
$$f(x) = \begin{cases} \frac{1}{2} & 0 < x < 2 \\ 0 & \text{otherwise} \end{cases}$$
is a density function. Find the corresponding distribution function.

3. Determine c so that
$$f(x) = \frac{c}{1 + x^2}, \quad x \in \mathbf{R}$$
is a density function.

4. Determine c so that
$$f(x) = \begin{cases} \frac{c}{x^2} & \text{for } x > 1 \\ 0 & \text{for } x \le 1 \end{cases}$$
is a density function.

5. Let X be a continuous random variable with density function
$$f(x) = \begin{cases} 2e^{-2x} & x > 0 \\ 0 & x \le 0 \end{cases}$$
Find EX and $\text{var}(X)$.

6. Let X be a continuous random variable with density function
$$f(x) = \frac{1}{2}e^{-|x|}$$
for $x \in \mathbf{R}$. Find EX and $\text{var}(X)$.

7. Let X be a continuous random variable with distribution function
$$F(x) = \begin{cases} 1 - \frac{1}{x^3} & \text{for } x > 1 \\ 0 & \text{for } x \le 1 \end{cases}$$
Find EX and $\text{var}(X)$.

8. Let X be a continuous random variable with
$$P(X > x) = e^{-ax}, \quad x \ge 0$$
where a is a positive constant. Find EX and $\text{var}(X)$.

9. Let X be a continuous random variable with density function
$$f(x) = \begin{cases} (a - 1)x^{-a} & \text{for } x > 1 \\ 0 & \text{for } x \le 1 \end{cases}$$
(a) Show that $EX = \infty$ when $a \le 2$.

(b) Compute EX when $a > 2$.

10. Suppose that X is a continuous random variable which takes on only nonnegative values. Set
$$G(x) = P(X > x)$$
(a) Show that
$$G'(x) = -f(x)$$
where $f(x)$ is the corresponding density function.

(b) Assume that
$$\lim_{x \to \infty} xG(x) = 0$$
and use integration by parts and (a) to show
$$EX = \int_0^\infty G(x)\,dx \qquad (12.33)$$

(c) Let X be a continuous random variable with
$$P(X > x) = e^{-ax}, \quad x > 0$$
where a is a positive constant. Use (12.33) to find EX. (If you did Problem 8, compare your answers.)

(12.5.2)

11. Denote by
$$f(x) = \frac{1}{\sigma\sqrt{2\pi}}e^{-(x-\mu)^2/2\sigma^2}$$
for $-\infty < x < \infty$, the density of a normal distribution with mean μ and standard deviation σ.

(a) Show that $f(x)$ is symmetric about $x = \mu$.

(b) Show that the maximum of $f(x)$ is at $x = \mu$.

(c) Show that the inflection points of $f(x)$ are at $x = \mu - \sigma$ and $x = \mu + \sigma$.

(d) Graph $f(x)$ for $\mu = 2$ and $\sigma = 1$.

12. Suppose that $f(x)$ is the density function of a normal distribution with mean μ and standard deviation σ.

Show that

$$\mu = \int_{-\infty}^{\infty} x f(x)\, dx$$

is the mean of this distribution. (*Hint:* Use substitution.)

13. Suppose that a quantitative character is normally distributed with mean $\mu = 12.8$ and standard deviation $\sigma = 2.7$. Find an interval centered at the mean such that 95% (99%) of the population falls into the interval.

14. Suppose a quantitative character is normally distributed with mean $\mu = 15.4$ and standard deviation $\sigma = 3.1$. Find an interval centered at the mean such that 95% (99%) of the population falls into this interval.

In Problems 15–20, assume that a quantitative character is normally distributed with mean μ and standard deviation σ. Determine what fraction of the population falls into the given intervals.

15. $[\mu, \infty)$

16. $[\mu - 2\sigma, \mu + \sigma]$

17. $(-\infty, \mu + 3\sigma]$

18. $[\mu + \sigma, \mu + 2\sigma]$

19. $(-\infty, \mu - 2\sigma]$

20. $[\mu - 3\sigma, \mu]$

21. Suppose that X is normally distributed with mean $\mu = 3$ and standard deviation $\sigma = 2$. Use the table in the appendix to find the following.

(a) $P(X \le 4)$

(b) $P(2 \le X \le 4)$

(c) $P(X > 5)$

(d) $P(X \le 0)$

22. Suppose that X is normally distributed with mean $\mu = -1$ and standard deviation $\sigma = 1$. Use the table in Appendix B to find the following.

(a) $P(X > 0)$

(b) $P(0 < X < 1)$

(c) $P(-1.5 < X < 2.5)$

(d) $P(X > 1.5)$

23. Suppose that X is normally distributed with mean $\mu = 1$ and standard deviation $\sigma = 2$. Use the table in Appendix B to find x such that the following hold.

(a) $P(X \le x) = 0.9$

(b) $P(X > x) = 0.4$

(c) $P(X \le x) = 0.4$

(d) $P(|X - 1| < x) = 0.5$

24. Suppose that X is normally distributed with mean $\mu = -2$ and standard deviation $\sigma = 1$. Use the table in Appendix B to find x such that the following hold.

(a) $P(X \ge x) = 0.8$

(b) $P(X < 2x + 1) = 0.5$

(c) $P(X \le x) = 0.1$

(d) $P(|X - 2| > x) = 0.4$

25. Assume that the mathematics score X on the Scholastic Aptitude Test (SAT) is normally distributed with mean 500 and standard deviation 100.

(a) Find the probability that an individual's score exceeds 700.

(b) Find the math SAT score so that 10% of the students who took the test have that score or greater.

26. In a study of *Drosophila melanogaster* by Mackay (1984), the number of bristles on the fifth abdominal sternite in males was shown to follow a normal distribution with mean 18.7 and standard deviation 2.1.

(a) What percentage of the male population has fewer than 17 abdominal bristles?

(b) Find an interval centered at the mean so that 90% of the male population have bristle numbers that fall into this interval.

27. Suppose the weight of an animal is normally distributed with mean 3720 gr and standard deviation 527 g. What percentage of the population has a weight that exceeds 5000 g?

28. Suppose the height of an adult animal is normally distributed with mean 17.2 in. Find the standard deviation if 10% of the animals have a height that exceeds 19 in.

29. Suppose that X is normally distributed with mean 2 and standard deviation 1. Find $P(0 \le X \le 3)$.

30. Suppose that X is normally distributed with mean -1 and standard deviation 2. Find $P(-3.5 \le X \le 0.5)$.

31. Suppose that X is normally distributed with mean μ and standard deviation σ. Show that $EX = \mu$. [You may use the fact that if Z is standard normally distributed, then $EZ = 0$ and $\text{var}(X) = 1$.]

32. Suppose that X is normally distributed with mean μ and standard deviation σ. Show that $\text{var}(X) = \sigma^2$. [You may use the fact that if Z is standard normally distributed, then $EZ = 0$ and $\text{var}(X) = 1$.]

33. Suppose that X is standard normally distributed. Find $E|X|$.

34. Suppose that the number of seeds a plant produces is normally distributed, with mean 142 and standard deviation 31. Find the probability that in a sample of 5 plants, at least one plant produces more than 200 seeds. Assume that the plants are independent.

35. The total maximum score on a calculus exam was 100 points. The mean score was 74 and the standard deviation was 11. Assume that the scores are normally distributed.

(a) Determine the percentage of students scoring 90 or above.

(b) Determine the percentage of students scoring between 60 and 80 (inclusive).

(c) Determine the minimum score of the highest 10% of the class.

(d) Determine the maximum score of the lowest 5% of the class.

36. The mean weight of female students at a small college is 123 pounds (lb), and the standard deviation is 9 lb. If the weights are normally distributed, determine what percentage of female students weigh (a) between 110 and 130 lb, (b) less than 100 lb, and (c) more than 150 lb.

(12.5.3)

37. Suppose that you pick a number at random from the interval $(0, 4)$. What is the probability that the first digit after the decimal point is a 3?

38. Suppose that you pick a number X at random from the interval $(0, a)$. If $P(X \geq 1) = 0.2$, find a.

39. Suppose that you pick a number X at random from the interval (a, b). If $EX = 4$ and $\text{var}(X) = 3$, find a and b.

40. Suppose that you pick five numbers at random from the interval $(0, 1)$. Assume that the numbers are independent. What is the probability that all numbers are greater than 0.7?

41. Suppose that X_1, X_2, and X_3 are independent and uniformly distributed over $(0, 1)$. Define
$$Y = \max(X_1, X_2, X_3)$$
Find EY. [*Hint:* Compute $P(Y \leq y)$ and deduce from this the density of Y.]

42. Suppose that X_1, X_2, and X_3 are independent and uniformly distributed over $(0, 1)$. Define
$$Y = \min(X_1, X_2, X_3)$$
Find EY. [*Hint:* Compute $P(Y > y)$ and deduce from this the density of Y.]

43. Suppose that you wish to simulate a random experiment that consists of tossing a coin with probability 0.6 of heads ten times. The computer generates the following ten random variables: 0.1905, 0.4285, 0.9963, 0.1666, 0.2223, 0.6885, 0.0489, 0.3567, 0.0719, 0.8661. Find the corresponding sequence of heads and tails.

44. Suppose that you wish to simulate a random experiment that consists of rolling a fair die. The computer generates the following ten random variables: 0.7198, 0.2759, 0.4108, 0.7780, 0.2149, 0.0348, 0.5673, 0.0014, 0.3249, 0.6630. Describe how you would find the corresponding sequence of numbers on the die, and find them.

45. Suppose $X_1, X_2, \ldots, X_n$ are independent random variables with uniform distribution on $(0, 1)$. Define $X = \min(X_1, X_2, \ldots, X_n)$.

(a) Compute $P(X > x)$.

(b) Show that $P(X > x/n) \to e^{-x}$ as $n \to \infty$.

46. Suppose $X_1, X_2, \ldots, X_n$ are independent random variables with uniform distribution on $(0, 1)$. Define $X = \max(X_1, X_2, \ldots, X_n)$.

(a) Find the distribution function of X.

(b) Use Problem 10 to compute EX.

(12.5.4)

47. Let X be exponentially distributed with parameter λ. Find EX.

48. Let X be exponentially distributed with parameter λ. Find $\text{var}(X)$.

49. Suppose that the lifetime of a battery is exponentially distributed, with an average lifespan of three months. What is the probability that the battery will last for more than four months?

50. Suppose that the lifetime of a battery is exponentially distributed, with an average lifespan of two months. You buy six batteries. What is the probability that none of them will last more than two months. (Assume that the batteries are independent.)

51. Suppose that the lifetime of a radioactive atom is exponentially distributed, with an average lifetime of 27 days.

(a) Find the probability that the atom will not decay during the first 20 days after you start to observe it.

(b) Suppose that the atom does not decay during the first 20 days that you observe it. What is the probability that it will not decay during the next 20 days?

52. If X has distribution function $F(x)$, we can show that $F(X)$ is uniformly distributed over the interval $(0, 1)$. Use this fact to generate exponentially distributed random variables with mean 1. [Assume that a computer generated the following four uniformly distributed random variables on the interval $(0, 1)$: 0.0371, 0.5123, 0.1370, 0.9865.]

53. Suppose the lifetime of a technical device is exponentially distributed with mean five years.

(a) Find the probability that the device will have failed after three years.

(b) Given the device has worked for six years, find the probability that it will work for another year.

54. Suppose the lifetime of an organism is exponentially distributed with hazard rate function $\lambda(x) = 2/\text{days}$.

(a) Find the probability that an individual of this species lives for more than three days.

(b) What is the expected lifetime.

55. Suppose the lifetime of a technical device is exponentially distributed with parameter $\lambda = 0.2/\text{year}$.

(a) What is the expected lifetime?

(b) The **median lifetime** is defined as the age x_m at which the probability of not having failed by age x_m is 0.5. Find x_m.

56. The median lifetime is defined as the age x_m at which the probability of not having failed by age x_m is 0.5. If the lifespan of an organism is exponentially distributed and $x_m = 4$ years, what is the hazard rate function?

57. The hazard rate function of an organism is given by
$$\lambda(x) = 0.3 + 0.1e^{0.01x}, \ x \geq 0$$
where x is measured in days.

(a) What is the probability that the organism will live for more than five days?

(b) What is the probability that the organism will live between seven and ten days?

58. The hazard rate function of an organism is given by
$$\lambda(x) = 0.1 + 0.5e^{0.02x}, \ x \geq 0$$
where x is measured in days.

(a) What is the probability that the organism will live less than ten days?

(b) What is the probability that the organism will live for another five days given that it survived the first five days?

59. The median lifetime is defined as the age x_m at which the probability of not having failed by age x_m is 0.5. Use a graphing calculator to numerically approximate the median lifetime if the hazard rate function is
$$\lambda(x) = 1.2 + 0.3e^{0.5x}, \ x \geq 0$$

60. The median lifetime is defined as the age x_m at which the probability of not having failed by age x_m is 0.5. Use a graphing calculator to numerically approximate the median lifetime if the hazard rate function is

$$\lambda(x) = 0.5 + 0.1e^{0.2x}, \ x \geq 0$$

61. The hazard rate function of an organism is given by

$$\lambda(x) = (2 \times 10^{-5})x^{1.5}, \ x \geq 0$$

where x is measured in days.

(a) What is the probability that the organism will live for more than 50 days?

(b) What is the probability that the organism will live between 50 and 70 days?

62. The hazard rate function of an organism is given by

$$\lambda(x) = 0.04x^{3.1}, \ x \geq 0$$

where x is measured in years.

(a) What is the probability that the organism will live for more than three years?

(b) What is the probability that the organism will live for another three years given that it survived the first three years?

63. The median lifetime is defined as the age x_m at which the probability of not having failed by age x_m is 0.5. Find the median lifetime if the hazard rate function is

$$\lambda(x) = (4 \times 10^{-5})x^{2.2}, \ x \geq 0$$

64. The median lifetime is defined as the age x_m at which the probability of not having failed by age x_m is 0.5. Find the median lifetime if the hazard rate function is

$$\lambda(x) = (3.7 \times 10^{-6})x^{2.7}, \ x \geq 0$$

65. Let N_x be the number of individuals that are still alive at age x. Show that

$$-\ln \frac{N_{x+1}}{N_x}$$

can be estimated by

$$\int_x^{x+1} \lambda(u) \, du$$

where $\lambda(x)$ is the hazard rate function at age x.

12.6 LIMIT THEOREMS

12.6.1 The Law of Large Numbers

Prostate-specific antigen (PSA) levels provide a diagnostic tool for prostate cancer detection. They are also used to screen for "biochemical failure" after surgical removal of the prostate. Biochemical failure is defined as a PSA level that exceeds 0.5 ng/ml; it can be an indication that the prostate cancer cells are still in the body of the patient. In a study by Iselin et al. (1999), the PSA levels of 817 men with prostate cancer were followed after surgical removal of the prostate. In 429 of the 817 men in the study, the disease was confined to the prostate. After five years, 8% of the men in the group whose cancer was confined to the prostate had biochemical failure.

Suppose now that you heard of a small study of 30 men whose prostate was surgically removed and in whom the disease was confined to the prostate. After five years, three out of the thirty men, or 10%, experienced biochemical failure, as defined previously. Which of the two figures, 8% or 10% would you deem more reliable? Your answer will likely be 8%. We tend to trust larger studies more than smaller studies. The reason for this is found in a mathematical result, known as the **law of large numbers**, which implies that estimates of proportions become more reliable as the sample size increases.

The following two inequalities will help us to state this result:

Markov's Inequality If X is a nonnegative random variable with $EX < \infty$, then for any $a > 0$,

$$P(X \geq a) \leq \frac{EX}{a}$$

Proof We prove this result when X is a nonnegative, continuous random variable with density function $f(x)$. Let $a > 0$. Then

$$EX = \int_0^\infty xf(x) \, dx = \int_0^a xf(x) \, dx + \int_a^\infty xf(x) \, dx \qquad (12.34)$$

Since $\int_0^a x f(x)\, dx \geq 0$, we find that

$$EX \geq \int_a^\infty x f(x)\, dx \tag{12.35}$$

Using $x \geq a$, the right-hand side of (12.35) can be bounded by

$$\geq \int_a^\infty a f(x)\, dx = a \int_a^\infty f(x)\, dx = a P(X \geq a)$$

Therefore,

$$EX \geq a P(X \geq a) \quad \text{or} \quad P(X \geq a) \leq \frac{EX}{a} \qquad ■$$

The next inequality is a consequence of Markov's inequality.

> **Chebyshev's Inequality** If X is a random variable with finite mean μ and finite variance σ^2, then for $c > 0$,
>
> $$P(|X - \mu| \geq c) \leq \frac{\sigma^2}{c^2}$$

Proof The events $\{|X - \mu| \geq c\}$ and $\{(X - \mu)^2 \geq c^2\}$ are the same. Therefore,

$$P(|X - \mu| \geq c) = P((X - \mu)^2 \geq c^2)$$

The random variable $(X - \mu)^2$ is nonnegative and $E(X - \mu)^2 = \sigma^2 < \infty$ by assumption. We can thus apply Markov's inequality and find

$$P((X - \mu)^2 \geq c^2) \leq \frac{E(X - \mu)^2}{c^2} = \frac{\sigma^2}{c^2} \qquad ■$$

Consider now a sequence of independent random variables $X_1, X_2, \dots, X_n$, all with the same distribution. We say that $X_1, X_2, \dots, X_n$ are **independent and identically distributed** (i.i.d., for short). We assume that $EX_i = \mu$ and $\text{var}(X_i) = \sigma^2$. We can think of this sequence as coming from some random experiment that we repeat n times and X_i is the outcome on the ith trial. For instance, the X_i's could denote the successive outcomes of tossing a coin n times with $X_i = 1$ if the ith toss results in heads and $X_i = 0$ otherwise.

We are interested in the (arithmetic) **average** of $X_1, X_2, \dots, X_n$, denoted by

$$\overline{X}_n = \frac{1}{n} \sum_{i=1}^n X_i$$

In the case of the coin-tossing example, $\overline{X}_n$ would be the fraction of heads in n trials.

Using the rules for expectations, we can find the mean and the variance of $\overline{X}_n$:

$$E\overline{X}_n = E\left(\frac{1}{n} \sum_{i=1}^n X_i\right) = \frac{1}{n} \sum_{i=1}^n EX_i = \frac{1}{n} \sum_{i=1}^n \mu = \frac{1}{n} \cdot n\mu = \mu$$

and, using independence,

$$\text{var}(\overline{X}_n) = \text{var}\left(\frac{1}{n} \sum_{i=1}^n X_i\right) = \frac{1}{n^2} \sum_{i=1}^n \text{var} X_i = \frac{1}{n^2} \sum_{i=1}^n \sigma^2 = \frac{1}{n^2} \cdot n\sigma^2 = \frac{\sigma^2}{n}$$

Applying Chebyshev's inequality with $c > 0$ to $\overline{X}_n$ yields

$$P(|\overline{X}_n - \mu| \geq c) \leq \frac{\text{var}(\overline{X}_n)}{c^2} = \frac{\sigma^2}{nc^2}$$

If we let $n \to \infty$, the right-hand side tends to 0. Since probabilities are (always) nonnegative, we can conclude that

$$\lim_{n\to\infty} P(|\overline{X}_n - \mu| \geq c) = 0 \qquad (12.36)$$

This type of convergence is called **convergence in probability**. More generally, we say that a random variable Z_n converges to a constant γ in probability as $n \to \infty$ if, for any $\epsilon > 0$,

$$\lim_{n\to\infty} P(|Z_n - \gamma| \geq \epsilon) = 0$$

The limit result in (12.36) can thus be expressed as "$\overline{X}_n$ converges to μ in probability." This is the weak law of large numbers.

WEAK LAW OF LARGE NUMBERS	If $X_1, X_2, \ldots, X_n$ are i.i.d. with $E|X_i| < \infty$, then as $n \to \infty$, $\overline{X}_n$ converges to EX_1 in probability.

We proved the result under the additional assumption that $\text{var}(X_i) < \infty$. The weak law of large numbers still holds if $\text{var}(X_i) = \infty$ provided $E|X_i| < \infty$, but the proof of the result would be quite a bit more complicated (and we won't do it).

The weak law of large numbers explains why taking a larger sample improves the reliability of estimates for proportions. In the prostate cancer example at the beginning of this subsection, we were interested in the likelihood of biochemical failure after five years in men who underwent prostate surgery when the cancer was confined to the prostate. We can think of this likelihood as a (to us unknown) probability that we estimate by taking a large sample in which all individuals are considered independent. (Such a sample is called a random sample.) If we set

$$X_i = \begin{cases} 1 & \text{if biochemical failure after five years in individual } i \\ 0 & \text{otherwise} \end{cases}$$

then $\overline{X}_n = \frac{1}{n}\sum_{i=1}^n X_i$ is the fraction of men in the sample that experience biochemical failure after five years. Now, if we set

$$\mu = EX_i = P(X_i = 1)$$

then the law of large numbers tells us that $\overline{X}_n \to \mu$ in probability as $n \to \infty$. Thus, if the sample size is sufficiently large, $\overline{X}_n$ provides a good estimate for the probability of biochemical failure, in the sense that with high probability, $\overline{X}_n$ will be close to μ.

▶ **Example 1** Suppose $X_1, X_2, \ldots, X_n$ are i.i.d. with

$$X_i = \begin{cases} 1 & \text{with probability } p \\ 0 & \text{with probability } 1 - p \end{cases}$$

Set $\overline{X}_n = \frac{1}{n}\sum_{i=1}^n X_i$ and show that $\overline{X}_n$ converges to p in probability as $n \to \infty$.

Solution

Since

$$E|X_i| = |1|(p) + |0|(1-p) = p < \infty \quad \text{and} \quad EX_i = (1)(p) + (0)(1-p) = p$$

we can invoke the law of large numbers and conclude that

$$\overline{X}_n = \frac{1}{n}\sum_{i=1}^{n} X_i \to EX_i = p$$

in probability as $n \to \infty$. ◀

▶ **Example 2**   Suppose $X_1, X_2, \ldots, X_n$ are i.i.d. with

$$X_i = \begin{cases} 1 & \text{with probability } p \\ 0 & \text{with probability } 1-p \end{cases}$$

Use Chebyshev's inequality to find n so that $\overline{X}_n$ will differ from p by less than 0.01 with probability at least 0.95.

Solution

We know from Example 1 that $\overline{X}_n$ converges to p in probability as $n \to \infty$. We want to investigate how fast $\overline{X}_n$ converges to p. More precisely, we want to find n so that

$$P(|\overline{X}_n - p| < 0.01) \geq 0.95$$

or, taking complements,

$$P(|\overline{X}_n - p| \geq 0.01) \leq 0.05$$

Using Chebyshev's inequality, we find

$$P(|\overline{X}_n - p| \geq 0.01) \leq \frac{\text{var}(\overline{X}_n)}{(0.01)^2}$$

Since the X_i's are independent,

$$\text{var}(\overline{X}_n) = \text{var}\left(\frac{1}{n}\sum_{i=1}^{n} X_i\right) = \frac{1}{n^2}\sum_{i=1}^{n} \text{var}(X_i)$$

$$= \frac{1}{n^2}\sum_{i=1}^{n} p(1-p) = \frac{p(1-p)}{n}$$

Therefore,

$$\frac{\text{var}(\overline{X}_n)}{(0.01)^2} = 10,000\frac{p(1-p)}{n}$$

We want this expression less than or equal to 0.05. We don't know p though. The following allows us to find a bound on $p(1-p)$. The function $f(p) = p(1-p)$ is an upside down parabola with roots at $p = 0$ and $p = 1$. Its maximum is at $p = 1/2$, namely, $f(1/2) = 1/4$. Therefore,

$$p(1-p) \leq \frac{1}{4} \quad \text{for all } 0 \leq p \leq 1$$

We thus obtain

$$10,000\frac{p(1-p)}{n} \leq 10,000\frac{1}{4n} \leq 0.05 \quad \text{or} \quad n \geq 50,000$$

We conclude that a sample size of 50,000 would suffice to estimate p within an error of 0.01 with 95% probability. It turns out that Chebyshev's inequality does not give very good estimates and this lower bound on the sample size is much larger than what we would really need. In the next subsection, we will learn a better way to estimate sample sizes. ◀

▶ **Example 3**   (Monte Carlo Integration) Suppose that $f(x)$ is an integrable function on $[0, 1]$ with $f(x) \geq 0$ for $x \in [0, 1]$. Let $U_1, U_2, \ldots, U_n$ be i.i.d. with U_i uniformly distributed on $(0, 1)$. Show that

$$\lim_{n \to \infty} \frac{1}{n} \sum_{i=1}^{n} f(U_i) = \int_0^1 f(x)\, dx \quad \text{in probability}$$

Solution

Define $X_i = f(U_i)$. Then, using that the density function of a uniform distribution on $(0, 1)$ is equal to 1 on $(0, 1)$,

$$E|X_i| = E|f(U_i)| = \int_0^1 |f(x)|\, dx = \int_0^1 f(x)\, dx < \infty$$

and

$$EX_i = Ef(U_i) = \int_0^1 f(x)\, dx$$

The X_i's are i.i.d. We can therefore apply the law of large numbers and conclude that

$$\lim_{n \to \infty} \frac{1}{n} \sum_{i=1}^{n} f(U_i) = \lim_{n \to \infty} \frac{1}{n} \sum_{i=1}^{n} X_i = EX_i = \int_0^1 f(x)\, dx$$

in probability. ◀

12.6.2 The Central Limit Theorem

We now come to a result that is indeed central to probability theory. It says that if we add up a large number of independent and identically distributed random variables with finite mean and variance, then, after suitable scaling, the distribution of the resulting quantity is approximately normally distributed. We will not be able to prove the result here, but we will be able to explore some of its implications.

Central Limit Theorem Suppose $X_1, X_2, \ldots, X_n$ are i.i.d. with mean $EX_i = \mu$ and variance $\text{var}(X_i) = \sigma^2 < \infty$. Define $S_n = \sum_{i=1}^{n} X_i$. Then, as $n \to \infty$,

$$P\left(\frac{S_n - n\mu}{\sqrt{n\sigma^2}} \leq x \right) \to F(x)$$

where $F(x)$ is the distribution function of the standard normal distribution.

▶ **Example 4**   Toss a fair coin 500 times. Use the central limit theorem to find an approximation for the probability of at least 265 heads.

Solution

We define

$$X_i = \begin{cases} 1 & \text{if } i\text{th toss results in heads} \\ 0 & \text{otherwise} \end{cases}$$

Then

$$\mu = EX_i = \frac{1}{2} \quad \text{and} \quad \sigma^2 = \text{var}(X_i) = \frac{1}{2}\left(1 - \frac{1}{2}\right) = \frac{1}{4}$$

Denote by $S_{500} = \sum_{i=1}^{500} X_i$ the number of heads among the first 500 tosses. Then

$$P(S_{500} \geq 265) = P\left(\frac{S_{500} - 500\mu}{\sqrt{500\sigma^2}} \geq \frac{265 - 250}{\sqrt{125}}\right)$$

$$\approx 1 - F\left(\frac{15}{\sqrt{125}}\right) = 1 - F(1.34)$$

$$= 1 - 0.9099 = 0.0901$$

where $F(x)$ is the distribution function of a standard normally distributed random variable and the values of $F(x)$ are obtained from the table in Appendix B. ◀

When the central limit theorem is applied to integer-valued random variables, a correction is typically used. This is motivated by the following example.

▶ **Example 5**   For S_{500} defined in Example 4, use the central limit theorem to find an approximation for $P(S_{500} = 250)$.

Solution

If we applied the central limit theorem without any corrections, we would find

$$P(S_{500} = 250) = P\left(\frac{S_{500} - 500\mu}{\sqrt{500\sigma^2}} = \frac{250 - 250}{\sqrt{125}}\right) \approx P(Z = 0) = 0$$

where Z is a standard normally distributed random variable. We can compare this to the exact value. Namely, S_{500} is binomially distributed with parameters $n = 500$ and $p = 1/2$. We find

$$P(S_{500} = 250) = \binom{500}{250}\left(\frac{1}{2}\right)^{500} \approx 0.036$$

The central limit theorem does not give a good approximation. We can do better by writing the event $\{S_{500} = 250\}$ as $\{249.5 \leq S_{500} \leq 250.5\}$ (see Figure 12.51). Then

$$P(S_{500} = 250) = P(249.5 \leq S_{500} \leq 250.5)$$

$$= P\left(\frac{249.5 - 250}{\sqrt{125}} \leq \frac{S_{500} - 500\mu}{\sqrt{500\sigma^2}} \leq \frac{250.5 - 250}{\sqrt{125}}\right)$$

$$= P(-0.04 \leq Z \leq 0.04)$$

where Z is standard normally distributed. We find

$$P(-0.04 \leq Z \leq 0.04) = 2F(0.04) - 1 = (2)(0.5160) - 1 = 0.032$$

where $F(x)$ is the distribution function of a standard normal distribution. ◀

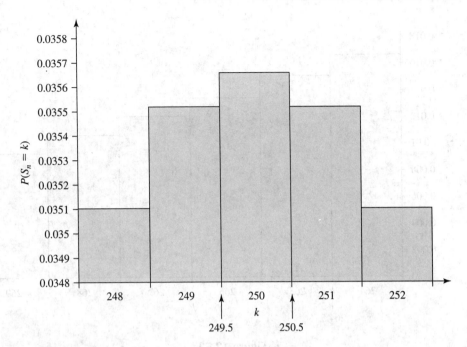

▲ **Figure 12.51**
The histogram correction for Example 5

The correction we employed in Example 5 is called the **histogram correction**.

▶ **Example 6**   Redo Example 4 with the histogram correction.

Solution

We write the event $\{S_{500} \geq 265\}$ as $\{S_{500} \geq 264.5\}$ (see Figure 12.52). Then

$$P(S_{500} \geq 265) = P(S_{500} \geq 264.5) = P\left(\frac{S_{500} - 500\mu}{\sqrt{500\sigma^2}} \geq \frac{264.5 - 250}{\sqrt{125}} \right)$$

$$\approx P(Z \geq 1.30) = 1 - F(1.30) = 1 - 0.9032 = 0.0968$$

where Z is a standard normally distributed random variable with distribution function $F(x)$.   ◀

Quantitative genetics is a field in biology that attempts to explain quantitative differences between individuals that are either of genotypic or environmental origin, such as differences in height, litter size, number of abdominal bristles in *Drosophila*, and so on. Estimates on the number of loci involved in a quantitative trait range from very few, such as five loci for skull length in rabbits (Wright, 1968), to very many, such as 98 loci for abdominal bristles in *Drosophila* (Falconer, 1989).

When many loci are involved in a quantitative trait, the **infinitesimal model** is used to model the genotypic value of that trait. The genotypic value G of a trait is considered as a sum of the contributions of each of the loci involved,

$$G = X_1 + X_2 + \cdots + X_n$$

where in the simplest case, the X_i's are assumed to be independent and identically distributed and represent the contribution of locus i to the genotypic value. If the X_i's have finite mean and variance, and if n is large, the distribution of G can be approximated by a normal distribution.

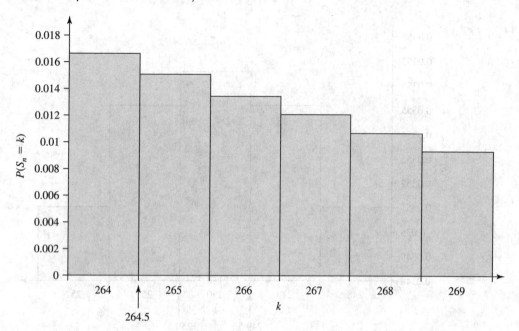

▲ **Figure 12.52**
The histogram correction for Example 6

▶ **Example 7**   Suppose a trait is controlled by 100 loci. Each locus, independently of all others, contributes to the genotypic value of the trait either +1 with probability 0.6 or −0.7 with probability 0.4.

(a) Find the mean value of the trait.

(b) What proportion of the population has a trait value greater than 40?

Solution

(a) The genotypic value of the trait can be written as

$$S_{100} = \sum_{i=1}^{100} X_i$$

with

$$X_i = \begin{cases} 1 & \text{with probability } 0.6 \\ -0.7 & \text{with probability } 0.4 \end{cases}$$

Hence,

$$ES_{100} = \sum_{i=1}^{100} EX_i = \sum_{i=1}^{100}[(1)(0.6) + (-0.7)(0.4)]$$

$$= \sum_{i=1}^{100} 0.32 = 32$$

The mean value of the trait is this 32.

(b) To find the proportion of the population that has a trait value greater than 40, we employ the central limit theorem. We compute the variance of X_i first.

$$EX_i^2 = (1)^2(0.6) + (-0.7)^2(0.4) = 0.796$$

Thus,

$$\text{var}(X_i) = EX^2 - (EX_i)^2 = 0.796 - (0.32)^2 = 0.6936$$

Now,

$$P(S_{100} > 40) = P\left(\frac{S_{100} - 32}{\sqrt{(100)(0.6936)}} > \frac{40 - 32}{\sqrt{69.36}}\right)$$

$$\approx 1 - F(0.96) = 1 - 0.8315 = 0.1685$$

where $F(x)$ is the distribution function of a standard normal distribution. About 17% of the population has trait value greater than 40. ◄

▶ **Example 8** (Estimating sample sizes) Suppose you wish to conduct a medical study to determine the fraction of people in the general population whose total cholesterol level is above 220 g/dl. How large a sample size would we need to estimate the proportion within 0.01 of the true value with probability at least 0.95?

Solution

Define

$$X_i = \begin{cases} 1 & \text{if } i\text{th individual has cholesterol} \geq 220 \text{ mg/dl} \\ 0 & \text{otherwise} \end{cases}$$

and assume that the individuals are selected so that the X_i's are i.i.d. Then

$$\overline{X}_n = \frac{1}{n}\sum_{i=1}^{n} X_i$$

is an estimate of the proportion of individuals whose cholesterol level exceeds 220 mg/dl.

If we set $S_n = \sum_{i=1}^{n} X_i$, then with $p = EX_i$ and $\sigma^2 = \text{var}(X_i)$,

$$\frac{S_n - np}{\sqrt{n\sigma^2}} \quad \text{is approximately standard normally distributed}$$

Dividing both numerator and denominator by n, we find with $\sigma = \sqrt{\text{var}(X_i)} = \sqrt{p(1-p)}$,

$$\sqrt{n}\frac{\overline{X}_n - p}{\sqrt{p(1-p)}} \quad \text{is approximately standard normally distributed}$$

We are interested in finding n in order to estimate the proportion within 0.01 of the true value p with probability at least 0.95.

$$P(|\overline{X}_n - p| \leq 0.01) \geq 0.95$$

We rewrite this as

$$P(-0.01 \leq \overline{X}_n - p \leq 0.01) \geq 0.95$$

or

$$P\left(\sqrt{n}\frac{-0.01}{\sqrt{p(1-p)}} \leq \sqrt{n}\frac{\overline{X}_n - p}{\sqrt{p(1-p)}} \leq \sqrt{n}\frac{0.01}{\sqrt{p(1-p)}}\right) \geq 0.95$$

Since $\sqrt{n}\frac{\overline{X}_n - p}{\sqrt{p(1-p)}}$ is approximately standard normally distributed, we find that the left-hand side is approximately

$$2F\left(\sqrt{n}\frac{0.01}{\sqrt{p(1-p)}}\right) - 1$$

This is ≥ 0.95 if

$$F\left(\sqrt{n}\frac{0.01}{\sqrt{p(1-p)}}\right) \geq 0.975$$

or

$$\sqrt{n}\frac{0.01}{\sqrt{p(1-p)}} \geq 1.96$$

Solving for n, we find

$$n \geq (196)^2 p(1-p)$$

Since we do not know p, we take the worst possible case that maximizes $p(1-p)$. As we saw in Example 2, this occurs for $p = 1/2$. Thus,

$$n \geq (196)^2 \frac{1}{2}\left(1 - \frac{1}{2}\right) = 9604$$

Thus about 9604 individuals would suffice for this study. You should compare this to the result in Example 2, where we solved the same problem (though a different application) using Chebyshev's inequality instead of the central limit theorem. ◀

Remark. Both the normal and the Poisson distribution serve as approximations to the binomial distribution. As a rule of thumb, the approximations are reasonably good when $n \geq 40$. When $np \leq 5$, the Poisson approximation should be used; when $np \geq 5$, the normal approximation should be used.

12.6.3 Problems

(12.6.1)

1. Let X be exponentially distributed with parameter $\lambda = 1/2$. Use Markov's inequality to estimate $P(X \geq 3)$ and compare this to the exact answer.

2. Let X be uniformly distributed over $(1, 4)$.

(a) Use Markov's inequality to estimate $P(X \geq a)$, $1 \leq a \leq 4$, and compare this to the exact answer.

(b) Find the value of $a \in (1, 4)$ that minimizes the difference between the bound and the exact probability computed in (a).

3. Prove Markov's inequality when X is a nonnegative discrete random variable with $EX < \infty$.

4. Let X be a continuous random variable with density $f(x)$ and assume that $X \geq 2$. Why is $EX \geq 2$?

5. Let X be uniformly distributed over $(-2, 2)$. Use Chebyshev's inequality to estimate $P(|X| \geq 1)$ and compare this to the exact answer.

6. Let X be standard normally distributed. Use Chebyshev's inequality to estimate (a) $P(|X| \geq 1)$, (b) $P(|X| \geq 2)$, and (c) $P(|X| \geq 3)$. Compare each case to the exact answer.

7. Suppose X is a random variable with mean 10 and variance 9. What can you say about $P(|X - 10| \geq 5)$?

8. Suppose X is a random variable with mean -5 and variance 2. What can you say about the probability that X deviates from its mean by at least 4?

9. Suppose $X_1, X_2, \ldots, X_n$ are i.i.d. with

$$X_i = \begin{cases} -1 & \text{with probability } 0.2 \\ 1 & \text{with probability } 0.5 \\ 2 & \text{with probability } 0.3 \end{cases}$$

What can you say about $\frac{1}{n}\sum_{i=1}^{n} X_i$ as $n \to \infty$?

10. Suppose $X_1, X_2, \ldots, X_n$ are independent random variables with $P(X_i > x) = e^{-2x}$. What can you say about $\frac{1}{n}\sum_{i=1}^{n} X_i$ as $n \to \infty$?

11. Suppose $X_1, X_2, \ldots, X_n$ are independent random variables with density function

$$f(x) = \frac{1}{\pi(1+x^2)}, \quad x \in \mathbf{R}$$

Can you apply the law of large numbers to $\frac{1}{n}\sum_{i=1}^{n} X_i$? If yes, what can you say about $\frac{1}{n}\sum_{i=1}^{n} X_i$ as $n \to \infty$?

12. How often do you have to toss a coin to determine $P(\text{heads})$ within 0.1 of its true value with probability at least 0.9?

13. A previous study showed that less than 5% of the population suffers from a certain disorder. To get a more accurate estimate on this proportion, you plan to conduct a study. What sample size should you choose if you want to be at least 95% sure that your estimate is within 0.05 of the true value?

14. Assume $Ee^{cX} < \infty$ for $c > 0$. Use Markov's inequality to prove **Bernstein's inequality**.

$$P(X \geq x) \leq e^{-cx} Ee^{cX}$$

for $c > 0$.

(12.6.2)

15. Toss a fair coin 400 times. Use the central limit theorem to find an approximation for the probability of at most 190 heads.

16. Toss a fair coin 150 times. Use the central limit theorem to find an approximation for the probability that the number of heads is at least 70.

17. Toss a fair coin 200 times.

(a) Use the central limit theorem to find an approximation for the probability that the number of heads is at least 120.

(b) Use Markov's inequality to find an estimate for the event in (a) and compare your answer with that in (a).

18. Toss a fair coin 300 times.

(a) Use the central limit theorem to find an approximation for the probability that the number of heads is between 140 and 160.

(b) Use Chebyshev's inequality to find an estimate for the event in (a) and compare your answer with that in (a).

19. Suppose S_n is binomially distributed with parameters $n = 200$ and $p = 0.3$. Use the central limit theorem to find an approximation for $P(99 \leq S_n \leq 101)$ (a) without the histogram correction, and (b) with the histogram correction. (c) Use a graphing calculator to compute the exact probability and compare your answers to those in (a) and (b).

20. Suppose S_n is binomially distributed with parameters $n = 150$ and $p = 0.4$. Use the central limit theorem to find an approximation for $P(S_n = 60)$ (a) without the histogram correction, and (b) with the histogram correction. (c) Use a graphing calculator to compute the exact probability and compare your answers to those in (a) and (b).

21. Suppose a genotypic trait is controlled by 80 loci. Each locus independently of all others contributes to the genotypic value of the trait either $+0.3$ with probability 0.2, -0.1 with probability 0.5, or -0.5 with probability 0.3.

(a) Find the mean value of the trait.

(b) What proportion of the population has a trait value between -12 and -7?

22. Suppose a genotypic trait is controlled by 90 loci. Each locus independently of all others contributes to the genotypic value of the trait either 1.1 with probability 0.7, 0.9 with probability 0.1, or 0.1 with probability 0.2.

(a) Find the mean value of the trait.

(b) What proportion of the population has a trait value less than 72?

23. How often should you toss a coin to be at least 90% certain that your estimate of $P(\text{heads})$ is within 0.1 of its true value?

24. How often should you toss a coin to be at least 90% certain that your estimate of $P(\text{heads})$ is within 0.01 of its true value?

25. To forecast the outcome of a presidential election in which two candidates run for office, a telephone poll is conducted. How many people should be surveyed to be at least 95% sure that the estimate is within 0.05 of the true value? (Assume that there are no undecided people in the survey.)

26. A medical study is conducted to estimate the proportion of people suffering from seasonal affected disorder. How many people should be surveyed to be at least 99% sure that the estimate is within 0.02 of the true value?

In Problems 27–30, S_n is binomially distributed with parameters n and p.

27. For $n = 100$ and $p = 0.01$, compute $P(S_n = 0)$ (a) exactly, (b) using a Poisson approximation, and (c) using a normal approximation.

28. For $n = 100$ and $p = 0.1$, compute $P(S_n = 10)$ (a) exactly, (b) using a Poisson approximation, and (c) using a normal approximation.

29. For $n = 50$ and $p = 0.1$, compute $P(S_n = 5)$ (a) exactly, (b) using a Poisson approximation, and (c) using a normal approximation.

30. For $n = 50$ and $p = 0.5$, compute $P(S_n = 25)$ (a) exactly, (b) using a Poisson approximation, and (c) using a normal approximation.

31. Suppose you want to estimate the proportion of people in the United State who do not believe in evolution. You happen to take a class on evolutionary theory at a U.S. college that is attended by 200 students, all of whom are biology majors. Do you think you would get an accurate estimate if you asked all 200 students in your class? Discuss.

32. A soft drink company introduces a new beverage. One month after its introduction, the company wants to know whether its marketing strategies have reached young adults of ages 18–20. You happen to work part time for the marketing company that is conducting the survey. You are also taking a calculus class at the same time that is attended by 250 students. It would be easy for you to hand out a survey in class. Would you suggest this to your supervisor in the marketing company? Discuss.

33. Clementines are sold in boxes. Each box contains 50 clementines. The probability that a clementine in a box is spoiled is 0.01.

(a) Use an appropriate approximation to determine the probability that a box contains 0, 1, or at least 2 spoiled clementines.

(b) A shipment of clementines with 100 boxes is considered unacceptable if 35% or more of the boxes contain spoiled clementines. What is the probability that a shipment is unacceptable?

34. Turner's syndrome is a rare chromosomal disorder in which girls have only one X chromosome. It affects about 1 in 2000 girls in the United State. About 1 in 10 girls with Turner's syndrome suffer from an abnormal narrowing of the aorta.

(a) In a group of 4000 girls, what is the probability that there are 0, 1, 2, or at least 3 girls affected with Turner's syndrome?

(b) In a group of 170 girls affected with Turner's syndrome, what is the probability that at least 20 of them suffer from an abnormal narrowing of the aorta?

In Problems 35–37, use the following facts: Cystic fibrosis is an inherited disorder that causes abnormally thick body secretions. About 1 in 2500 white babies in the United State have this disorder. About 3 in 100 children with cystic fibrosis develop diabetes mellitus, and about 1 in 5 females with cystic fibrosis are infertile.

35. Find the probability that in a group of 5000 newborn white babies in the United State at least 4 babies suffer from cystic fibrosis.

36. Find the probability that in a group of 1000 children with cystic fibrosis at least 25 will develop diabetes mellitus.

37. Find the probability that in a group of 250 women with cystic fibrosis no more than 60 are infertile.

12.7 STATISTICAL TOOLS

In the preceding sections of this chapter, we learned how to model various random experiments. By assuming the underlying probability distribution of the model, we were able to compute the probabilities of events, such as obtaining white-flowering pea plants.

To understand phenomena in nature, however, we often take the reverse approach, and infer the underlying probability distribution from the observation of events. Based on a collection of observations, called **data**, we will estimate characteristics of the underlying probability distribution. For instance, in the case of the normal distribution, our goal might be to estimate the mean and the variance, which are parameters that describe a normal distribution.

This section will provide some statistical tools to achieve this goal. We assume that all of our observations are presented as numerical data.

12.7.1 Collecting and Describing Data

To learn something about the distribution of a quantity in a population (clutch size, plant height, lifespan, and so forth), we cannot measure the quantity for every individual in the population. Instead, we take a subset of the population, called a **sample**, measure the quantity in the sample, and then try to infer the distribution of the quantity in the population from its distribution in the sample. In order to do so, the sample must be representative of the population. This is a difficult problem, which we cannot address satisfactorily here. It is easy to include a bias in a sample if the observations are chosen from a section of the population with similar characteristics that are not representative of the whole population. For instance, if we want to assess fecundity in a plant that grows in a wide variety of soils but we sample plants on sandy soils only, then our sample might not reflect the fecundity of the entire population.

In the following, we will always assume that our sample is representative of the population; we call such a sample a **random sample**. All observations in a random sample are independent and come from the same distribution (namely, the distribution of the quantity in the entire population). A typical scheme to obtain a random sample of size n is to pick an individual at random from the population, record the quantity of interest, replace the individual, and then select the next individual. This procedure is repeated until a sample of size n is obtained. Replacing the sampled individual after recording the quantity of interest ensures that the population always has the same composition and, hence, all observations have the same distribution. We denote the sample by the vector $(X_1, X_2, \ldots, X_n)$, where X_k is the kth observation. The X_k are independent random variables that all have the same distribution. As in the previous section, we say that the X_k are independent and identically distributed.

The following data set describes the scores of eleven students in two exams. The maximum score on each exam was 10. We will use this data set to explain the definitions in this subsection.

Student	A	B	C	D	E	F	G	H	I	J	K
Exam 1	6	4	9	7	8	7	8	5	7	8	7
Exam 2	8	6	8	7	7	—	9	6	7	8	8

One way to summarize data is to give the number of times a certain category (in this case, exam score) occurs. These numbers are called frequencies. If we divide the frequencies by the total number of observations, we obtain relative frequencies. The list of (relative) frequencies is called a (relative) frequency distribution. (We encountered "frequency distributions" earlier in this chapter.)

To obtain the frequency distribution of exam 1, we count the number of times each score appears. For instance, score 6 appears once, so its frequency is 1; score 7 appears four times, so its frequency is 4; and so on.

The frequencies of the exam scores appear in the following table. (Scores that are not listed in the table have frequency 0.)

Score	4	5	6	7	8	9
Exam 1	1	1	1	4	3	1
Exam 2	0	0	2	3	4	1

To obtain relative frequencies, we divide each frequency of the first exam by 11, because eleven students took the first exam, and each frequency of the second exam by 10, because ten students took the second exam. The relative frequencies of the exam scores appear in the following table. (Scores that are not listed in the table have relative frequency 0.)

Score	4	5	6	7	8	9
Exam 1	1/11	1/11	1/11	4/11	3/11	1/11
Exam 2	0	0	0.2	0.3	0.4	0.1

A quantity that is computed from observations in a sample is called a **statistic**.

The first statistic that we define is the **sample median**. It is the middle of the observations when we order the data according to size. When the number of observations is odd, there is one data point in the middle of the ordered data. If the number of observations is even, we take the average of the two observations in the middle. The list of the ordered exam data follows:

Exam 1: 4,5,6,7,7,**7**,7,8,8,8,9

Exam 2: 6,6,7,7,**7**,**8**,8,8,8,9

Eleven students took the first exam; after ordering the scores, the sixth score is the median, which is 7. Ten students took the second exam; after ordering the scores, the average of the fifth and the sixth score is the median, which is $\frac{7+8}{2} = 7.5$.

The two most important statistics that are used to summarize data are the **sample mean** and the **sample variance**. To define these quantities, we recall that we denoted a sample of size n by the vector $(X_1, X_2, \ldots, X_n)$, where X_k is the kth observation. The sample mean and the sample variance are defined as follows.

$$
\text{Sample mean: } \quad \overline{X}_n = \frac{1}{n} \sum_{k=1}^{n} X_k
$$

$$
\text{Sample variance: } \quad S_n^2 = \frac{1}{n-1} \sum_{k=1}^{n} (X_k - \overline{X}_n)^2
$$

The sample mean is thus the arithmetic average of the observations (we encountered this quantity previously). The sample variance is the sum of the squared deviations from the sample mean, divided by $n - 1$. (We will explain in the next subsection why we divide by $n - 1$ rather than n.)

This definition of the sample variance is not very convenient for computation. There is an alternate form that is typically easier to use. It follows from an algebraic manipulation of the definition of the sample variance:

$$
S_n^2 = \frac{1}{n-1} \sum_{k=1}^{n} (X_k - \overline{X}_n)^2 = \frac{1}{n-1} \sum_{k=1}^{n} (X_k^2 - 2X_k \overline{X}_n + \overline{X}_n^2)
$$

$$
= \frac{1}{n-1} \left[\sum_{k=1}^{n} X_k^2 - 2\overline{X}_n \sum_{k=1}^{n} X_k + n\overline{X}_n^2 \right]
$$

Using the fact that $\sum_{k=1}^{n} X_k = n\overline{X}_n$, this simplifies to

$$
S_n^2 = \frac{1}{n-1} \left[\sum_{k=1}^{n} X_k^2 - n\overline{X}_n^2 \right] = \frac{1}{n-1} \left[\sum_{k=1}^{n} X_k^2 - \frac{1}{n} \left(\sum_{k=1}^{n} X_k \right)^2 \right]
$$

▶ **Example 1** Compute the sample mean and the sample variance for the data of the first exam.

Solution

We find

$$
\sum_{k=1}^{11} X_k = 76 \quad \text{and} \quad \sum_{k=1}^{11} (X_k)^2 = 546
$$

Hence,

$$
\overline{X}_n = \frac{76}{11} \approx 6.9
$$

and

$$
S_n^2 = \frac{1}{n-1} \left[\sum_{k=1}^{n} X_k^2 - \frac{1}{n} \left(\sum_{k=1}^{n} X_k \right)^2 \right]
$$

$$
= \frac{1}{10} \left(546 - \frac{1}{11} (76)^2 \right) \approx 2.09 \qquad \blacktriangleleft
$$

If the sample distribution is summarized in a frequency distribution, the sample mean and the sample variance have the following definitions: Assume that a sample of size n has l distinct values $x_1, x_2, \ldots, x_l$, where x_k occurs f_k times in the sample. Then the sample mean is given by the formula

$$\overline{X}_n = \frac{1}{n} \sum_{k=1}^{l} x_k f_k$$

and the sample variance has the form

$$S_n^2 = \frac{1}{n-1} \left[\sum_{k=1}^{l} x_k^2 f_k - \frac{1}{n} \left(\sum_{k=1}^{l} x_k f_k \right)^2 \right]$$

▶ **Example 2**   Use the frequency distribution of the first exam to calculate the sample mean and the sample variance.

Solution

We find the same answers as in Example 1.

$$\overline{X}_n = \frac{1}{n} \sum_{k=1}^{l} x_k f_k$$

$$= \frac{1}{11} [(4)(1) + (5)(1) + (6)(1) + (7)(4) + (8)(3) + (9)(1)] = \frac{76}{11}$$

and

$$S_n^2 = \frac{1}{n-1} \left[\sum_{k=1}^{l} x_k^2 f_k - \frac{1}{n} \left(\sum_{k=1}^{l} x_k f_k \right)^2 \right]$$

$$= \frac{1}{10} [(16)(1) + (25)(1) + (36)(1) + (49)(4) + (64)(3) + (81)(1)$$

$$- \frac{1}{11}(76)^2]$$

$$= \frac{1}{10} \left[546 - \frac{1}{11}(76^2) \right] \approx 2.09$$ ◀

12.7.2 Estimating Means and Proportions

We saw in Section 12.4 that the mean and the variance are useful parameters for describing the probability distribution of a population. If we wish to learn something about the distribution of some quantity from a random sample of the population, we might want to know what the mean and variance of the population distribution are. We will use the sample mean and the sample variance to estimate the mean and the variance of the population.

It is important to realize that any statistic computed from a sample will vary from sample to sample, since the samples are random subsets of the population. Statistics are therefore random variables with their own probability distributions.

We assume that the population distribution has finite mean μ and finite variance σ^2. These two parameters are unknown to us, and we wish to estimate them by taking a random sample $(X_1, X_2, \ldots, X_n)$ of size n from the population. The X_k are independent and identically distributed according to the distribution of the population with

$$EX_k = \mu \quad \text{and} \quad \text{var}(X_k) = \sigma^2 \quad \text{for } k = 1, 2, \ldots, n \qquad (12.37)$$

To estimate the mean and the variance of the population distribution, we will use the sample mean and the sample variance defined in the previous subsection. The sample mean is the arithmetic average

$$\overline{X}_n = \frac{1}{n} \sum_{k=1}^{n} X_k$$

We stated that statistics are random variables. As such, we can compute their mean and variance. Using (12.37), the mean of the sample mean is

$$E\overline{X}_n = E\left(\frac{1}{n} \sum_{k=1}^{n} X_k\right) = \frac{1}{n} \sum_{k=1}^{n} EX_k = \frac{1}{n}(n\mu) = \mu$$

Using independence of the observations in addition to (12.37), the variance of the sample mean is

$$\mathrm{var}(\overline{X}_n) = \mathrm{var}\left(\frac{1}{n} \sum_{k=1}^{n} X_k\right) = \frac{1}{n^2} \sum_{k=1}^{n} \mathrm{var}(X_k) = \frac{1}{n^2}(n\sigma^2) = \frac{\sigma^2}{n}$$

We see from this that the expected value of the sample mean is equal to the population mean. The spread of the distribution of $\overline{X}_n$ is described by the variance of $\overline{X}_n$. Since the variance of $\overline{X}_n$ becomes smaller as the sample size increases ($\sigma^2/n \to 0$ as $n \to \infty$), we conclude that the sample mean of large samples shows less variation about its mean than the sample mean of small samples. This implies that the larger the sample size, the more accurately the mean of the population can be estimated. In fact, invoking the weak law of large numbers from the previous section, we find that

$$\overline{X}_n \to \mu \quad \text{in probability as } n \to \infty$$

This justifies the use of $\overline{X}_n$ as an estimate for the mean of the distribution. Since $E\overline{X}_n = \mu$, we say that $\overline{X}_n$ is an **unbiased estimator** for μ.

The sample variance is

$$S_n^2 = \frac{1}{n-1} \sum_{k=1}^{n} (X_k - \overline{X}_n)^2$$

To compute its mean, we need the following identity. For any $c \in \mathbf{R}$,

$$\sum_{k=1}^{n} (X_k - c)^2 = \sum_{k=1}^{n} (X_k - \overline{X}_n)^2 + n(\overline{X}_n - c)^2 \tag{12.38}$$

To see why this is true, we expand the left-hand side of (12.38), after adding $0 = \overline{X}_n - \overline{X}_n$ inside the parentheses:

$$\sum_{k=1}^{n} (X_k - c)^2 = \sum_{k=1}^{n} (X_k - \overline{X}_n + \overline{X}_n - c)^2$$

$$= \sum_{k=1}^{n} [(X_k - \overline{X}_n)^2 + 2(X_k - \overline{X}_n)(\overline{X}_n - c) + (\overline{X}_n - c)^2]$$

$$= \sum_{k=1}^{n} (X_k - \overline{X}_n)^2 + 2(\overline{X}_n - c) \sum_{k=1}^{n} (X_k - \overline{X}_n) + n(\overline{X}_n - c)^2$$

Since $\sum_{k=1}^{n} (X_k - \overline{X}_n) = 0$, the middle term is equal to 0, and (12.38) follows.

If we set $c = \mu$ in (12.38), and rearrange the equation, then

$$\sum_{k=1}^{n}(X_k - \overline{X}_n)^2 = \sum_{k=1}^{n}(X_k - \mu)^2 - n(\overline{X}_n - \mu)^2$$

Taking expectations on both sides, and using on the right-hand side the fact that the expectation of a sum is the sum of the expectations, we find

$$E\left(\sum_{k=1}^{n}(X_k - \overline{X}_n)^2\right) = \sum_{k=1}^{n}E(X_k - \mu)^2 - nE(\overline{X}_n - \mu)^2$$

Now, $\sum_{k=1}^{n}(X_k - \overline{X}_n)^2 = (n-1)S_n^2$, $E(X_k - \mu)^2 = \sigma^2$, and $E(\overline{X}_n - \mu)^2 = \text{var}(\overline{X}_n) = \frac{1}{n}\sigma^2$. Hence,

$$(n-1)ES_n^2 = n\sigma^2 - \sigma^2 = (n-1)\sigma^2$$

and, therefore,

$$ES_n^2 = \sigma^2$$

That is, the expected value of the sample variance is equal to the variance of the population. This is the reason that we divided by $n - 1$ instead of n when computing the sample variance. (We will not compute the variance of the sample variance; it is given by a complicated formula. One can show that the variance of the sample variance goes to 0 as the sample size becomes infinite.) Again, invoking the weak law of large numbers, we find that

$$S_n^2 \to \sigma^2 \quad \text{in probability as } n \to \infty$$

Since $ES_n^2 = \sigma^2$, S_n^2 is used as an unbiased estimator for σ^2.

Let's look at an example to illustrate how we would estimate the mean and the variance of a characteristic of a population.

▶ **Example 3**   Suppose that a computer generates the following sample of independent observations from a population.

$$0.0201, 0.8918, 0.9619, 0.1713, 0.0357,$$

$$0.6325, 0.4276, 0.2517, 0.2330, 0.6754$$

Estimate the mean and the variance.

Solution

To estimate the mean, we compute the sample mean $\overline{X}_n$. We sum the ten numbers in the sample and divide the result by 10, which yields

$$\overline{X}_n = 0.4301$$

Thus, our estimate for the mean is 0.4301.

To estimate the variance, we compute the sample variance S_n^2. We square the difference between each sample point and the sample mean and add the results. We then divide this number by 9. We find

$$S_n^2 = 0.1176$$

Thus, our estimate for the variance is 0.1176.   ◀

In the preceding example, we estimated the population mean and the population variance from a sample of size 10. Since we don't know the population parameters, we have no idea how good our estimates are. The following discussion addresses this problem.

Confidence Intervals In scientific publications, we frequently find the sample mean reported under a heading of the form "Mean ± S.E." The expression "Mean" stands for the sample mean $\overline{X}_n$. The expression "S.E." stands for the **standard error**, which serves as an estimate for the standard deviation of the sample mean. It is denoted by $S_{\overline{X}}$ and defined as

$$S_{\overline{X}} = \frac{S_n}{\sqrt{n}} \tag{12.39}$$

where S_n is the square root of the sample variance, called the **sample standard deviation**. This definition is motivated by the following consideration. (Again, we assume that the mean and the variance of the population distribution are finite.) To estimate the population mean μ, we use the sample mean $\overline{X}_n$. The variance of $\overline{X}_n$ gives us an idea how much the distribution of $\overline{X}_n$ varies. Now,

$$\text{var}(\overline{X}_n) = \text{var}\left(\frac{1}{n}\sum_{k=1}^{n} X_k\right) = \frac{1}{n^2}\text{var}\left(\sum_{k=1}^{n} X_k\right)$$

Since the X_k are independent, the variance of the sum is the sum of the variances; in addition, all X_k have the same distribution. Hence

$$\text{var}(\overline{X}_n) = \frac{1}{n^2}n\sigma^2 = \frac{\sigma^2}{n} \tag{12.40}$$

The variance of $\overline{X}_n$ thus depends on another population parameter, namely, the variance σ^2. In problems where we wish to estimate the mean, we typically don't know the variance either. If the sample size is large, however, the sample variance will be close to the population variance. We can therefore approximate the variance of $\overline{X}_n$ by replacing σ^2 in (12.40) by S_n^2, which yields S_n^2/n. The standard error is then the square root of this expression.

Here is an example that illustrates how to determine a sample mean and its standard error.

▶ **Example 4** In a sample of six leaves from a morning glory plant that is infested with aphids, the following numbers of aphids per leaf are found: 12, 27, 17, 35, 14, and 18. Find the sample mean, the sample variance, and the standard error.

Solution

The sample size is $n = 6$. We construct the following table:

$\sum_{k=1}^{n} X_k$	$\sum_{k=1}^{n} X_k^2$	$\overline{X}_n$	S_n^2	$\frac{S_n}{\sqrt{n}}$
123	2907	20.5	77.1	3.58

where we used the formulas

$$\overline{X}_n = \frac{1}{n}\sum_{k=1}^{n} X_k \quad \text{and} \quad S_n^2 = \frac{1}{n-1}\left[\sum_{k=1}^{n} X_k^2 - \frac{1}{n}\left(\sum_{k=1}^{n} X_k\right)^2\right]$$

with $n = 6$. If we wanted to report this result as "Mean ± S.E.," we would then write 20.5 ± 3.58. ◀

What does "Mean ± S.E." mean? When we write "Mean ± S.E.," we specify an interval, namely, [Mean − S.E., Mean + S.E.]. Since we use "Mean" ($= \overline{X}_n$) as an estimate for the population mean μ, we would like this interval to contain μ. Surely, since $\overline{X}_n$ is a random variable, if we took repeated samples and computed such intervals for each sample, not *all* the intervals would contain μ. But maybe we can at least find out what fraction of these intervals (or similar intervals) contain the population mean μ. In other words, we might wish to know before taking the sample what the probability is that the interval [Mean – S.E., Mean + S.E.] or, more generally, [Mean – aS.E., Mean + aS.E.], where a is a positive constant, will contain μ. To be concrete, we will try to determine a so that this probability is equal to 0.95.

If the sample size n is very large, it follows from the central limit theorem that

$$\frac{\overline{X}_n - \mu}{\sigma / \sqrt{n}}$$

is approximately standard normally distributed. If Z is standard normally distributed, then

$$P(-1.96 \le Z \le 1.96) = 0.95$$

Hence, the event

$$-1.96 \le \frac{\overline{X}_n - \mu}{\sigma / \sqrt{n}} \le 1.96 \tag{12.41}$$

has probability approximately 0.95 for n sufficiently large.

Rearranging the terms in (12.41), we find

$$-1.96 \frac{\sigma}{\sqrt{n}} \le \overline{X}_n - \mu \le 1.96 \frac{\sigma}{\sqrt{n}}$$

or

$$\overline{X}_n - 1.96 \frac{\sigma}{\sqrt{n}} \le \mu \le \overline{X}_n + 1.96 \frac{\sigma}{\sqrt{n}}$$

We can thus write

$$P \left(\overline{X}_n - 1.96 \frac{\sigma}{\sqrt{n}} \le \mu \le \overline{X}_n + 1.96 \frac{\sigma}{\sqrt{n}} \right) \approx 0.95 \quad \text{for large } n \tag{12.42}$$

This means that if we repeatedly draw random samples of size n from a population with mean μ and standard deviation σ, then in about 95% of the samples, the interval $[\overline{X}_n - 1.96 \frac{\sigma}{\sqrt{n}}, \overline{X}_n + 1.96 \frac{\sigma}{\sqrt{n}}]$ will contain the true mean μ. Such an interval is referred to as a **95% confidence interval**.

Notice that this interval contains the parameter σ. If we do not know μ, we probably do not know σ either. We might then wish to replace σ by the square root of the sample variance, denoted by S_n. Fortunately, when n is large, S_n will be very close to σ and (12.42) holds approximately when σ is replaced by S_n.

We thus find that for large n,

$$P \left(\overline{X}_n - 1.96 \frac{S_n}{\sqrt{n}} \le \mu \le \overline{X}_n + 1.96 \frac{S_n}{\sqrt{n}} \right) \approx 0.95$$

or, rewriting the event in interval notation,

$$P \left(\mu \in \left[\overline{X}_n - 1.96 \frac{S_n}{\sqrt{n}}, \overline{X}_n + 1.96 \frac{S_n}{\sqrt{n}} \right] \right) \approx 0.95$$

The interval in this expression is of the form

$$[\text{Mean} - (1.96)\text{S.E.}, \text{Mean} + (1.96)\text{S.E.}] \tag{12.43}$$

We thus succeeded in determining what fraction of intervals of the form (12.43) contain the mean. Namely, if n is large, and we take repeated samples each of size n, then approximately 95% of the intervals [Mean − (1.96)S.E., Mean+(1.96)S.E.] would contain the true mean. [Or, equivalently, before we take the sample, with probability 0.95 the interval in (12.43) will contain the actual value of μ.] If we wanted 99% of such intervals to contain the true mean, we would need to replace the factor 1.96 by 2.58, since if Z is standard normally distributed, then $P(-2.58 \le Z \le 2.58) = 0.99$ (95% and 99% are the most frequently used percentages for confidence intervals). Since $P(-1 \le Z \le 1) = 0.68$, we conclude that if the sample size is large, approximately 68% of intervals of the form [Mean − S.E., Mean + S.E.] contain the population mean μ.

We wish to emphasize that the preceding discussion requires that the sample size n is large. Whether or not a particular n can be considered large depends on the population distribution. If the X_k themselves are normally distributed with mean μ and standard deviation σ, then we can show that $\sqrt{n}(\overline{X}_n - \mu)/\sigma$ is standard normally distributed for *all* n. When we replace σ by S_n in (12.41) in this case, then n is considered large for $n \ge 40$. That is, for $n \ge 40$, $\sqrt{n}(\overline{X}_n - \mu)/S_n$ is then approximately standard normally distributed. The more the distribution of the X_k differs from the normal distribution, the larger n needs to be to use the normal distribution as an approximation.

Estimating Proportions Of particular interest is the case of estimating proportions (for instance, the proportion of white-flowering pea plants in a $Cc \times Cc$ cross). If we take a random sample of size n from a population, where a proportion p has a certain characteristic, then the number of observations in the sample with this characteristic is binomially distributed with parameters n and p. If we denote this quantity by B_n, then

$$P(B_n = k) = \binom{n}{k} p^k (1 - p)^{n-k} \quad \text{for } k = 0, 1, 2, \dots, n$$

We consider each single observation in the sample as a success or a failure, depending on whether the observation has the characteristic under investigation. We set

$$X_k = \begin{cases} 1 & \text{if the } k\text{th observation is a success} \\ 0 & \text{if the } k\text{th observation is a failure} \end{cases}$$

Then $B_n = \sum_{k=1}^{n} X_k$ is the total number of successes in the sample, and p is the success probability. The sample mean $\overline{X}_n = \frac{1}{n} \sum_{k=1}^{n} X_k$ is then the fraction of successes in the sample. We find

$$E\overline{X}_n = \frac{1}{n} \sum_{k=1}^{n} EX_k = \frac{1}{n}(np) = p$$

and

$$\text{var}(\overline{X}_n) = \text{var}\left(\frac{1}{n} \sum_{k=1}^{n} X_k\right) = \frac{1}{n^2} n(\text{var}(X_1))$$

$$= \frac{np(1-p)}{n^2} = \frac{p(1-p)}{n}$$

The sample mean $\overline{X}_n$ will serve as an estimate for the success probability p. It is customary to use $\hat{p}$ instead of $\overline{X}_n$; that is, we denote the estimate for p

by $\hat{p}$ (read "p hat"). If we observe k successes in a sample of size n, then

$$\hat{p} = \frac{k}{n}$$

with

$$E\hat{p} = p \quad \text{and} \quad \text{var}(\hat{p}) = \frac{p(1-p)}{n}$$

To find the standard error in the case of k successes in a sample of size n, note that

$$S_n^2 = \frac{1}{n-1}\left[\sum_{j=1}^{n} X_j^2 - \frac{1}{n}\left(\sum_{j=1}^{n} X_j\right)^2\right]$$

$$= \frac{1}{n-1}\left(k - \frac{k^2}{n}\right)$$

$$= \frac{n}{n-1}\frac{k}{n}\left(1 - \frac{k}{n}\right) = \frac{n\hat{p}(1-\hat{p})}{n-1}$$

The standard error of the sample mean p is therefore

$$S_{\hat{p}} = \frac{S_n}{\sqrt{n}} = \sqrt{\frac{\hat{p}(1-\hat{p})}{n-1}} \tag{12.44}$$

In the literature, you will typically find the standard error for proportions as

$$S_{\hat{p}} = \sqrt{\frac{\hat{p}(1-\hat{p})}{n}} \tag{12.45}$$

For n large, the two numbers are very close, so that it does not matter which one you use.

As an example, we consider Mendel's experiment of crossing pea plants.

▶ **Example 5**   To estimate the probability of white-flowering pea plants in a $Cc \times Cc$ cross, Mendel randomly crossed red-flowering pea plants of genotype Cc. He obtained 705 plants with red flowers and 224 plants with white flowers. Estimate the probability of a white-flowering pea plant in a $Cc \times Cc$ cross and give a 95% confidence interval.

Solution

The sample mean is

$$\overline{X}_n = \frac{224}{224 + 705} \approx 0.24$$

The estimate for the probability of a white-flowering pea plant in a $Cc \times Cc$ cross is therefore $\hat{p} = 0.24$. The standard error is

$$\text{S.E.} = \sqrt{\frac{\hat{p}(1-\hat{p})}{n}} = \sqrt{\frac{(0.24)(0.76)}{705 + 224}} \approx 0.014$$

We can thus report the result as 0.24 ± 0.014. Since n is large, we find a 95% confidence interval as

$$[0.24 - (1.96)(0.014), 0.24 + (1.96)(0.014)] = [0.213, 0.267]$$

(We know from the laws of inheritance that the expected value of $\overline{X}_n$ is $p = 0.25$, which is contained in the confidence interval.) ◀

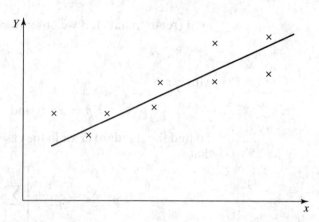

▲ **Figure 12.53**
A straight line fitted to data

12.7.3 Linear Regression

In textbooks, or in scientific literature, you will frequently see plots where a straight line has been fitted to data (as shown in Figure 12.53). The quantities on the horizontal and the vertical axes are linearly related, and a linear model is used to describe the relationship. We denote the quantity on the horizontal axis by x and the quantity on the vertical axis by Y. We think of x as a particular treatment, which is under control of the experimenter, and of Y as the response. When taking measurements of Y, errors are typically present, so that the data points will not lie exactly on the straight line (even if the linear model is correct) but will be scattered around it; that is, Y is not completely determined by x. The degree of scatter is an indication of how much random variation there is. We will see in the following how to separate the random variation from the actual relationship between the two quantities in the case when they are linearly related. We will discuss one particular model in the following.

We assume that x is under the control of the experimenter to the extent that it can be measured without error. The response Y, however, shows random variation. We assume the following linear model

$$Y = a + bx + \epsilon$$

where ϵ is a normal random variable, representing the **error**, with mean 0 and standard deviation σ. The standard deviation of the error does not depend on x, and is thus the same for all values of x.

Our goal is to estimate a and b from data. The data consist of points (x_i, y_i), $i = 1, 2, \ldots, n$. The approach will be to choose a and b so that the sum of the squared deviations

$$h(a, b) = \sum_{k=1}^{n} [y_k - (a + bx_k)]^2$$

is minimized. The deviations $y_k - (a + bx_k)$ are called **residuals**. The procedure of finding a and b is called the **method of least squares** and is illustrated in Figure 12.54. The resulting straight line is called the **least square line** (or **linear regression line**).

▶ Example 6   Given the three points $(0, 2)$, $(1, 0)$, and $(2, 1)$, use the method of least squares to find the least square line.

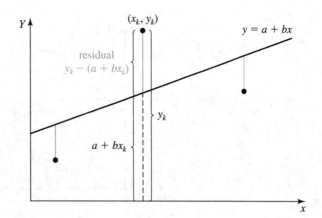

▲ Figure 12.54
The line $Y = a + bx$ is chosen so that the sum of the squared residuals is minimized

Solution

We wish to find a straight line of the form $y = a + bx$. For given values of a and b, the residuals are

$$2 - (a + 0b) \qquad 0 - (a + b) \qquad 1 - (a + 2b)$$

and the sum of their squares is

$$(2 - a)^2 + (a + b)^2 + (1 - a - 2b)^2$$
$$= (4 - 4a + a^2) + (a^2 + 2ab + b^2) + (1 + a^2 + 4b^2 - 2a - 4b + 4ab)$$
$$= 5 - 6a + 3a^2 + 6ab + 5b^2 - 4b$$
$$= (2b^2 + 2b) + (3 + 3b^2 + 3a^2 + 6ab - 6a - 6b) + 2$$
$$= 2(b^2 + b) + 3(1 + b^2 + a^2 + 2ab - 2a - 2b) + 2$$

Grouping the terms in this way allows us to complete the squares; we find that this is then equal to

$$2\left(b + \frac{1}{2}\right)^2 + 3(1 - a - b)^2 + \frac{3}{2} \tag{12.46}$$

(If this looks like magic, don't worry; we will derive a general formula for a and b shortly.) Since (12.46) consists of two squares (plus a constant term), we see that the expression is minimized when the two squares are both equal to 0. We solve

$$b + \frac{1}{2} = 0$$
$$1 - a - b = 0$$

which yields $b = -1/2$ and $a = 3/2$. Therefore, the least square line is of the form

$$y = \frac{3}{2} - \frac{1}{2}x$$

This line together with the three points is shown in Figure 12.55. ◀

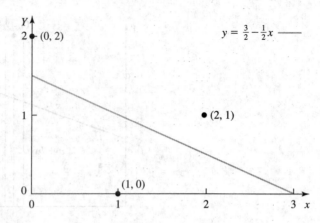

▲ **Figure 12.55**
The least square line together with the three data points of Example 6

We will now derive the general formula for finding a and b. The basic steps will be similar to those in Example 6. We will first rewrite the residuals. We set

$$\overline{x} = \frac{1}{n} \sum_{k=1}^{n} x_k \qquad \text{and} \qquad \overline{y} = \frac{1}{n} \sum_{k=1}^{n} y_k$$

and

$$y_k - (a + bx_k) = (y_k - \overline{y}) + (\overline{y} - a - b\overline{x}) - b(x_k - \overline{x})$$

In the following, we will simply write $\sum$ instead of $\sum_{k=1}^{n}$. If we square the expression and sum over k, we find

$$\sum [y_k - (a + bx_k)]^2 = \sum (y_k - \overline{y})^2 + n(\overline{y} - a - b\overline{x})^2$$
$$+ b^2 \sum (x_k - \overline{x})^2 - 2b \sum (x_k - \overline{x})(y_k - \overline{y})$$
$$+ 2(\overline{y} - a - b\overline{x}) \sum (y_k - \overline{y}) \qquad (12.47)$$
$$- 2b(\overline{y} - a - b\overline{x}) \sum (x_k - \overline{x})$$

The last two terms are equal to 0. We introduce notation to simplify our derivation:

$$SS_{xx} = \sum (x_k - \overline{x})^2 = \sum x_k^2 - \frac{(\sum x_k)^2}{n}$$

$$SS_{yy} = \sum (y_k - \overline{y})^2 = \sum y_k^2 - \frac{(\sum y_k)^2}{n}$$

$$SS_{xy} = \sum (x_k - \overline{x})(y_k - \overline{y}) = \sum x_k y_k - \frac{(\sum x_k)(\sum y_k)}{n}$$

Using this notation, the right-hand side of (12.47) can be written as

$$SS_{yy} + n(\overline{y} - a - b\overline{x})^2 + b^2 SS_{xx} - 2b SS_{xy}$$

The last two terms suggest that we should complete the square:

$$SS_{yy} + n(\overline{y} - a - b\overline{x})^2 + SS_{xx} \left(b^2 - 2b \frac{SS_{xy}}{SS_{xx}} + \left(\frac{SS_{xy}}{SS_{xx}} \right)^2 \right) - \frac{(SS_{xy})^2}{SS_{xx}}$$

$$= n(\overline{y} - a - b\overline{x})^2 + SS_{xx} \left(b - \frac{SS_{xy}}{SS_{xx}} \right)^2 + SS_{yy} - \frac{(SS_{xy})^2}{SS_{xx}}$$

As in Example 6, we succeeded in writing the sum of the squared deviations as a sum of two squares plus an additional term. We can minimize the expression by setting each squared expression equal to 0:

$$\bar{y} - a - b\bar{x} = 0$$

$$b - \frac{SS_{xy}}{SS_{xx}} = 0$$

Solving for a and b yields

$$b = \frac{SS_{xy}}{SS_{xx}}$$

$$a = \bar{y} - b\bar{x}$$

These expressions serve as estimates for a and b, denoted by $\hat{a}$ and $\hat{b}$. Summarizing our results, we have the following.

The least square line (or linear regression line) is given by

$$y = \hat{a} + \hat{b}x$$

with

$$\hat{b} = \frac{\sum_{k=1}^{n}(x_k - \bar{x})(y_k - \bar{y})}{\sum_{k=1}^{n}(x_k - \bar{x})^2} \tag{12.48}$$

$$\hat{a} = \bar{y} - \hat{b}\bar{x} \tag{12.49}$$

We illustrate finding $\hat{a}$ and $\hat{b}$ in the following example.

 Example 7 Fit a linear regression line through the points

$$(1, 1.62), (2, 3.31), (3, 4.57), (4, 5.42), (5, 6.71)$$

Solution

To facilitate the computation, we construct the following table.

x_k	y_k	$x_k - \bar{x}$	$y_k - \bar{y}$	$(x_k - \bar{x})(y_k - \bar{y})$
1	1.62	−2	−2.706	5.412
2	3.31	−1	−1.016	1.016
3	4.57	0	0.244	0
4	5.42	1	1.094	1.094
5	6.71	2	2.384	4.768
$\bar{x}=3$	$\bar{y}=4.326$	$\sum(x_k-\bar{x})^2 =10$	$\sum(y_k-\bar{y})^2 =15.29$	$\sum(x_k-\bar{x})(y_k-\bar{y}) =12.29$

Now,

$$\hat{b} = \frac{12.29}{10} = 1.229$$

$$\hat{a} = 4.326 - (1.229)(3) = 0.639$$

Hence, the linear regression line is given by

$$y = 1.23x + 0.64$$

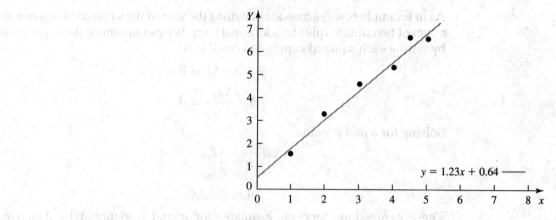

▲ Figure 12.56
The linear regression line and the data points of Example 7

This line and the data points are shown in Figure 12.56. ◄

Now that we know how to fit a straight line to a set of points, we might want to know how good the fit is. To this purpose, we will define a quantity known as the **coefficient of determination**. We motivate its definition as follows. We start with a set of observations (x_k, y_k), $k = 1, 2, \ldots, n$, and assume the linear model $Y = a + bx + \epsilon$. We set

$$\hat{y}_k = \hat{a} + \hat{b}x_k \qquad \text{and} \qquad \overline{y} = \frac{1}{n}\sum_{k=1}^{n} y_k$$

We think of $\hat{y}_k$ as the expected response under the linear model if $x = x_k$. Now, $y_k - \overline{y}$ is the deviation from the observation to the sample mean, $y_k - \hat{y}_k$ is the deviation from the observation to the expected response under the linear model, and $\hat{y}_k - \overline{y}$ is the deviation from the expected response under the linear model to the sample mean. The deviation $\hat{y}_k - \overline{y}$ can be thought of as being explained by the model, and the deviation $y_k - \hat{y}_k$ as the unexplained part due to random variation (the stochastic error). We can write

$$y_k - \overline{y} = (\hat{y}_k - \overline{y}) + (y_k - \hat{y}_k) \tag{12.50}$$

If we look at the total sum of the squared deviations, $\sum(y_k - \overline{y})^2$, and use (12.50), we find

$$\sum(y_k - \overline{y})^2 = \sum [(\hat{y}_k - \overline{y}) + (y_k - \hat{y}_k)]^2$$
$$= \sum(\hat{y}_k - \overline{y})^2 + 2\sum(\hat{y}_k - \overline{y})(y_k - \hat{y}) \tag{12.51}$$
$$+ \sum(y_k - \hat{y}_k)^2$$

We want to show that $\sum(\hat{y}_k - \overline{y})(y_k - \hat{y}) = 0$. To do so, we observe that

$$\hat{y}_k - \overline{y} = (\hat{a} + \hat{b}x_k) - (\hat{a} + \hat{b}\overline{x}) = \hat{b}(x_k - \overline{x})$$
$$y_k - \hat{y}_k = (y_k - \overline{y}) - (\hat{y}_k - \overline{y}) = (y_k - \overline{y}) - \hat{b}(x_k - \overline{x})$$

Therefore,

$$\sum(\hat{y}_k - \overline{y})(y_k - \hat{y}_k) = \sum \hat{b}(x_k - \overline{x})[(y_k - \overline{y}) - \hat{b}(x_k - \overline{x})]$$
$$= \hat{b}\sum(x_k - \overline{x})(y_k - \overline{y}) - \hat{b}^2 \sum(x_k - \overline{x})^2$$

Using (12.48) for one of the $\hat{b}$ in the term $\hat{b}^2$, this becomes

$$\sum(\hat{y}_k - \overline{y})(y_k - \hat{y}_k) = \hat{b}\sum(x_k - \overline{x})(y_k - \overline{y})$$

$$-\hat{b}\frac{\sum(x_k - \overline{x})(y_k - \overline{y})}{\sum(x_k - \overline{x})^2}\sum(x_k - \overline{x})^2 \quad (12.52)$$

$$= 0$$

This allows us to partition the total sum of squares into the explained and the unexplained sums of squares. Namely, continuing with (12.7.3) and using (12.7.3), we find

$$\underbrace{\sum(y_k - \overline{y})^2}_{\text{total}} = \underbrace{\sum(\hat{y}_k - \overline{y})^2}_{\text{explained}} + \underbrace{\sum(y_k - \hat{y}_k)^2}_{\text{unexplained}}$$

The ratio

$$\frac{\text{explained}}{\text{total}} = \frac{\sum(\hat{y}_k - \overline{y})^2}{\sum(y_k - \overline{y})^2}$$

is therefore the proportion of variation that is explained by the model. It is denoted by r^2 and called the coefficient of determination. With $\hat{y}_k - \overline{y} = \hat{b}(x_k - \overline{x})$ and $\hat{b}$ given in (12.48), this can be written as

$$r^2 = (\hat{b})^2\frac{\sum(x_k - \overline{x})^2}{\sum(y_k - \overline{y})^2} = \left(\frac{\sum(x_k - \overline{x})(y_k - \overline{y})}{\sum(x_k - \overline{x})^2}\right)^2\frac{\sum(x_k - \overline{x})^2}{\sum(y_k - \overline{y})^2}$$

$$= \frac{[\sum(x_k - \overline{x})(y_k - \overline{y})]^2}{\sum(x_k - \overline{x})^2\sum(y_k - \overline{y})^2}$$

We summarize this in the following.

The coefficient of determination r^2 is given by

$$r^2 = \frac{[\sum(x_k - \overline{x})(y_k - \overline{y})]^2}{\sum(x_k - \overline{x})^2\sum(y_k - \overline{y})^2}$$

and represents the proportion of variation that is explained by the model.

Returning to Example 7, we find

$$r^2 = \frac{(12.29)^2}{(10)(15.29)} = 0.988$$

That is, 98.8% of the variation is explained by the model.

Since r^2 is the ratio of explained to total variation, it follows that $r^2 \leq 1$. Furthermore, since r^2 is the square of an expression, it is always nonnegative. That is, we have

$$0 \leq r^2 \leq 1$$

The closer r^2 is to 1, the more closely the data points follow the straight line resulting from the linear model. In the extreme case, when $r^2 = 1$, all points lie on the line; there is no random variation.

12.7.4 Problems

(12.7.1)

1. The following data represent the number of aphids per plant found in a sample of ten plants:

$$17, 13, 21, 47, 3, 6, 12, 25, 0, 18$$

Find the median, the sample mean, and the sample variance.

2. The following data represent the number of seeds per flower head in a sample of nine flowering plants:

$$27, 39, 42, 18, 21, 33, 45, 37, 21$$

Find the median, the sample mean, and the sample variance.

3. The following data represent the frequency distribution of seed numbers per flower head in a flowering plant.

Seed Number	Frequency
9	37
10	48
11	53
12	49
13	61
14	42
15	31

Calculate the sample mean and the sample variance.

4. The following data represent the frequency distribution of the numbers of days that it took a certain ointment to clear up a skin rash.

Number of Days	Frequency
1	2
2	7
3	9
4	27
5	11
6	5

Calculate the sample mean and the sample variance.

5. The following data represent the relative frequency distribution of clutch size in a sample of 300 laboratory guinea pigs.

Clutch Size	Relative Frequency
2	0.05
3	0.09
4	0.12
5	0.19
6	0.23
7	0.12
8	0.13
9	0.07

Calculate the sample mean and the sample variance.

6. The following data represent the relative frequency distribution of clutch size in a sample of 42 mallards.

Clutch Size	Relative Frequency
6	0.10
7	0.24
8	0.29
9	0.21
10	0.16

Calculate the sample mean and the sample variance.

7. Let $(X_1, X_2, \ldots, X_n)$ denote a sample of size n. Show that

$$\sum_{k=1}^{n}(X_k - \overline{X}) = 0$$

where $\overline{X}$ is the sample mean.

8. Let $(X_1, X_2, \ldots, X_n)$ denote a sample of size n. Show that

$$n\overline{X}^2 = \frac{1}{n}\left(\sum_{k=1}^{n} X_k\right)^2$$

where $\overline{X}$ is the sample mean.

9. Assume that a sample of size n has l distinct values $x_1, x_2, \ldots, x_l$, where x_k occurs f_k times in the sample. Explain why the sample mean is given by the formula

$$\overline{X} = \frac{1}{n}\sum_{k=1}^{l} x_k f_k$$

10. Assume that a sample of size n has l distinct values $x_1, x_2, \ldots, x_l$, where x_k occurs f_k times in the sample. Explain why the sample variance is given by the formula

$$S^2 = \frac{1}{n-1}\left[\sum_{k=1}^{l} x_k^2 f_k - \frac{1}{n}\left(\sum_{k=1}^{l} x_k f_k\right)^2\right]$$

(12.7.2)

11. Use a graphing calculator to generate three samples of size 10 each from a uniform distribution over the interval $(0, 1)$.

(a) Compute the sample mean and the sample variance of each sample.

(b) Combine all three samples and compute the mean and the sample variance of the combined sample.

(c) Compare your answers in (a) and (b) with the true values of the mean and the variance.

12. Suppose that X is exponentially distributed with mean 1. A computer generates the following sample of independent observations from this population.

$$0.3169, 0.5531, 2.376,$$
$$1.150, 0.6174, 0.1563$$
$$2.936, 1.778, 0.7357$$
$$0.1024$$

Find the sample mean and the sample variance, and compare with the corresponding population parameters.

13. Compute the sample mean and the standard error for the sample in Problem 1.

14. Compute the sample mean and the standard error for the sample in Problem 2.

15. The following data represent a sample from a normal distribution with mean 0 and variance 1.

$$-0.68, 1.22, 1.33, -0.84$$
$$-0.06, 0.50, 0.03, -0.13$$
$$-0.29, -0.47$$

Construct a 95% confidence interval.

16. The following data represent a sample from a normal distribution with mean 0 and variance 1.

$$-1.18, 0.52, 0.36, -0.16$$
$$0.92, 0.68, -0.61, -0.54$$
$$0.15, 1.04$$

Construct a 95% confidence interval.

17. Use a graphing calculator to construct a 95% confidence interval for a sample of size 30 from a uniform distribution over the interval $(0, 1)$. Take a class poll to determine the percentage of confidence intervals that contain the true mean. Discuss the result in class.

18. (a) If X has distribution function $F(x)$, we can show that $F(X)$ is uniformly distributed over the interval $(0, 1)$. Use this fact, a graphing calculator, and the table for the standard normal distribution to generate fifteen standard normally distributed random variables.

(b) Use your data from (a) to construct a 95% percent confidence interval. Take a class poll to determine the percentage of confidence intervals that contain the true mean. Discuss the result in class.

19. To determine the germination success of seeds of a certain plant, you plant 162 seeds. You find that 117 of the seeds germinate. Estimate the probability of germination and give a 95% confidence interval.

20. To test a new drug for lowering cholesterol, 72 people with elevated cholesterol receive this drug; 51 of them show reduced cholesterol levels. Estimate the probability that this drug lowers cholesterol and give a 95% confidence interval.

(12.7.3)

In Problems 21–22, fit a linear regression line through the given points and compute the coefficient of determination.

21. $(-3, -6.3)$, $(-2, -5.6)$, $(-1, -3.3)$, $(0, 0.1)$, $(1, 1.7)$, $(2, 2.1)$

22. $(0, 0.1)$, $(1, -1.3)$, $(2, -3.5)$, $(3, -5.7)$, $(4, -5.8)$

23. Show that the sum of the residuals about the linear regression line is equal to 0.

24. Show that the last two terms in (12.47),

$$2(\bar{y} - a - b\bar{x}) \sum (y_k - \bar{y})$$

and

$$2(\bar{y} - a - b\bar{x}) \sum (x_k - \bar{x})$$

are equal to 0.

25. To determine whether the frequency of chirping crickets depends on temperature, the following data were obtained (Pierce, 1949).

Temperature (F)	Chirps/s
69	15
70	15
72	16
75	16
81	17
82	17
83	16
84	18
89	20
93	20

Fit a linear regression line and compute the coefficient of determination.

26. The initial velocity v of an enzymatic reaction that follows Michaelis-Menten kinetics is given by

$$v = \frac{v_{max} s}{K_m + s} \qquad (12.53)$$

where s is the substrate concentration, and v_{max} and K_m are two parameters that characterize the reaction. The following computer-generated table contains values of the initial velocity v when the substrate concentration s was varied.

s	v
1	4.1
2.5	6.1
5	9.3
10	12.9
20	17.1

(a) Invert (12.53) and show that

$$\frac{1}{v} = \frac{K_m}{v_{max}} \frac{1}{s} + \frac{1}{v_{max}} \qquad (12.54)$$

This is the Lineweaver-Burk equation. If we plot $1/v$ as a function of $1/s$, a straight line with slope K_m/v_{max} and intercept $1/v_{max}$ results. Transform the data using (12.54), and fit a linear regression line to the transformed data. Find the slope and the intercept of the linear regression line, and determine K_m and v_{max}.

(b) Dowd and Riggs (1965) proposed to use the transformation

$$v = v_{max} - K_m \frac{v}{s} \qquad (12.55)$$

and then plot v against v/s. The resulting straight line has slope $-K_m$ and intercept v_{max}. Use (12.55) to transform the data, and fit a linear regression line to the transformed data. Find the slope and the intercept of the linear regression line, and determine K_m and v_{max}.

12.8 Key Terms

Chapter 12 Review: Topics *Discuss the following definitions and concepts:*

1. Multiplication principle
2. Permutation
3. Combination
4. Random experiment
5. Sample space
6. Basic set operations, Venn diagram, De Morgan's laws
7. Definition of probability
8. Equally likely outcomes
9. Mendel's pea experiments
10. Mark-recapture method
11. Maximum likelihood estimate
12. Conditional probability
13. Partition of sample space
14. Law of total probability
15. Independence
16. Bayes formula
17. Random variable
18. Discrete distribution
19. Probability mass function
20. Distribution function of a discrete random variable
21. Mean and variance
22. Joint distributions
23. Binomial distribution
24. Multinomial distribution
25. Geometric distribution
26. Poisson distribution
27. Poisson approximation to the Binomial distribution
28. Continuous random variable
29. Density function
30. Distribution function of a continuous random variable
31. Mean and variance of a continuous random variable
32. Histogram
33. Normal distribution
34. Uniform distribution
35. Exponential distribution
36. Aging
37. Gompertz law
38. Weibull law
39. Law of large numbers
40. Markov's inequality
41. Chebyshev's inequality
42. Central limit theorem
43. Histogram correction
44. Sample
45. Statistic
46. Sample median, sample mean, sample variance, standard error
47. Confidence interval
48. Estimating proportions
49. Linear regression line
50. Coefficient of determination

12.9 Review Problems

1. (a) There are 25 students in a calculus recitation. What is the probability that no two students have the same birthday?

(b) Let p_n denote the probability that in a group of n people, no two people have the same birthday. Show that

$$p_1 = 1 \qquad \text{and} \qquad p_{n+1} = p_n \frac{365 - n}{365}$$

Use this formula to generate a table of p_n for $1 \le n \le 25$.

2. Thirty patients are to be randomly assigned to two different treatment groups. How many ways can this be done?

3. Fifteen different plants are to be equally divided among five plots. How many ways can this be done?

4. Assume that a certain disease is caused by a genetic mutation or appears spontaneously. The disease will appear in 67% of all people with the mutation, and in 23% of all people without the mutation. Assume that 3% of the population carries the disease gene.

(a) What is the probability that a randomly chosen individual will develop the disease?

(b) Given an individual who suffers from the disease, what is the probability that they have the genetic mutation?

5. Suppose that 42% of the seeds of a certain plant germinate.

(a) What is the expected number of germinating seeds in a sample of ten seeds?

(b) You plant ten seeds in one pot. What is the probability that none of the seeds will germinate?

(c) You plant five pots with ten seeds each. What is the expected number of pots with no germinating seeds?

(d) You plant five pots with ten seeds each. What is the probability that at least one pot has no germinating seeds?

6. Suppose that the amount of yearly rainfall in a certain area is normally distributed with mean 27 and standard variation 5.7 (measured in inches).

(a) What is the probability that in a given year the rainfall will exceed 35 inches?

(b) What is the probability that in five consecutive years the rainfall will exceed 35 inches in each year?

(c) What is the probability that in at least one out of ten years the rainfall will exceed 35 inches per year?

7. Suppose that each time a student takes a particular test he or she has a 20% chance of passing. (Assume that consecutive trials are independent.)

(a) What are the chances of passing the test on the second trial?

(b) Given that a student failed the test the first time, what are the chances of passing the test on the second trial?

8. Explain why

$$2^n = \sum_{k=1}^{n} \binom{n}{k}$$

9. A bag contains on average 170 chocolate covered raisins. Production standards require that in 95% of all bags the number of raisins does not deviate from 170 by more than 10. Assume that the number of raisins is normally distributed with mean μ and variance σ^2.

(a) Determine μ and σ.

(b) A shipment contains 100 bags. What is the probability that no bag contains less than 160 raisins?

10. Suppose that two parents are carriers of a recessive gene causing a metabolic disorder. Neither parent has the disease. If they have three children, what is the probability that none of them will be afflicted by the disease? (Note that a recessive gene only causes a disorder if an individual has two copies of the gene.)

11. Suppose that you choose a plant from a large batch of red-flowering pea plants and cross it with a white-flowering pea plant. What percentage of the red-flowering parent plants are of genotype Cc if 90% of the offspring have red flowers?

12. Suppose that a random variable is normally distributed with mean μ and variance σ^2. How would you estimate μ and σ?

13. Let $(X_1, X_2, \ldots, X_n)$ be a sample of size n from a population with mean μ and variance σ^2. Define

$$V = \frac{1}{n} \sum_{k=1}^{n} (X_k - \overline{X})^2$$

where $\overline{X}$ is the sample mean.

(a) If S^2 is the sample variance, show that

$$V = \frac{n-1}{n} S^2$$

(b) Compute EV.

14. Assume that the weight of a certain species is normally distributed with mean μ and variance σ^2. The following data represent the weight (measured in grams) of ten randomly selected individuals from this species.

$$171, 168, 151, 192, 175$$
$$163, 182, 157, 177, 169$$

(a) Find the median, the sample mean, and the sample variance.

(b) Construct a 95% confidence interval for the population mean.

15. (a) Generate five observations (x, y) from a random experiment, where

$$y = 2x + 1 + \epsilon$$

$x = 1, 2, 3, 4, 5$, and ϵ is normally distributed with mean 0 and variance 1.

(b) Use your data from (a) to find the least square line and compare your results to the linear model that describes this experiment.

(c) How much of your data is explained by the model?

16. Suppose X is a continuous random variable with density function

$$f(x) = \begin{cases} 0 & \text{for } x < 1 \\ (r-1)x^{-r} & \text{for } x \geq 1 \end{cases}$$

where r is a constant greater than 1.

(a) For which values of r is $EX = \infty$?

(b) Compute EX for those values of r for which $EX < \infty$.

ANSWERS TO ODD-NUMBERED PROBLEMS

Section 1.1

1. (a) $\{-4, 2\}$ **(b)** $\{-4, 2\}$

3. (a) $\{-1, 5\}$ **(b)** $x = 5$ or $x = 1$

5. (a) $[-\frac{2}{5}, \frac{6}{5}]$,   **(b)** $(-\infty, -\frac{7}{3}) \cup (3, \infty)$,
(c) $(-\infty, -1] \cup [-\frac{1}{7}, \infty)$,   **(d)** $(-\frac{1}{5}, \frac{13}{5})$

7. $x + 3y - 14 = 0$

9. $3x + y + 2 = 0$

11. $7x - 3y + 5 = 0$

13. $4x + 3y - 12 = 0$

15. $2y - 3 = 0$

17. $x + 1 = 0$

19. $3x - y + 2 = 0$

21. $2y - x - 4 = 0$

23. $y + 2x - 2 = 0$

25. $x + 4y - 3 = 0$

27. $x + 2y + 4 = 0$

29. $x + 3y + 4 = 0$

31. $2x + 5y - 22 = 0$

33. $y - x + 6 = 0$

35. $y = 2$

37. $x = -1$

39. $x = -1$

41. $y = 3$

43. (a) $y = kx$, $[x] = $ ft, $[y] = $ cm, $k = \frac{30.5 \text{ cm}}{1 \text{ft}}$ implies $y = (30.5 \frac{\text{cm}}{\text{ft}})x$
(b) **(i)** 183 cm,   **(ii)** $\frac{1159}{12}$ cm,   **(iii)** $\frac{1159}{24}$ cm
(c) **(i)** $\frac{346}{61}$ ft,   **(ii)** $\frac{150}{61}$ ft,   **(iii)** $\frac{96}{61}$ ft

45. $s(t) = (40 \text{ mi/hr})t$, $k = 40$ mi/hr

47. $\frac{1}{(0.305)^2}$ ft^2

49. (a) $[y] =$ liter, $[x] =$ ounces, $y = \left(\frac{1}{33.81} \frac{\text{liter}}{\text{ounces}}\right) x$
(b) $\frac{12}{33.81}$ liters

51. (a) 88 km/hr,   **(b)** 81 mi/hr

53. (a) $C = K - 273.15$,
(b) 77.4 K $= -195.75°$C $= -320.35°$F, 90.2 K $= -182.95°$C $= -297.31°$F; nitrogen gets distilled first since it has the lower boiling point.

55. $(x + 1)^2 + (y - 4)^2 = 9$

57. (a) $(x - 2)^2 + (y - 5)^2 = 9$,
(b) $y = \sqrt{5} + 5$ or $y = -\sqrt{5} - 5$,   **(c)** No

59. center: $(2, 0)$; radius: 4

61. center: $(2, -1)$, radius: 4

63. (a) $\frac{5}{12}\pi$,   **(b)** $255°$

65. (a) $\frac{1}{2}\sqrt{2}$,   **(b)** $-\frac{1}{2}\sqrt{3}$,   **(c)** $\sqrt{3}$

67. (a) $\alpha = -\frac{\pi}{3}$ or $\alpha = \frac{4\pi}{3}$ **(b)** $\alpha = \frac{\pi}{3}$ or $\alpha = \frac{4\pi}{3}$

69. Divide both left and right side by $\cos\theta$.

71. $\{0, \frac{\pi}{3}, \frac{5\pi}{3}\}$

73. (a) $4^{7/3}$,   **(b)** 27,   **(c)** $(625)^k$

75. (a) $x = \frac{1}{16}$,   **(b)** $x = 27$,   **(c)** $x = \frac{1}{100}$

77. (a) $x = -5$,   **(b)** $x = -4$,   **(c)** $x = -3$

79. (a) $\ln 3$,   **(b)** $\log_4(x - 2) + \log_4(x + 2)$,   **(c)** $6x - 2$

81. $x = \frac{1}{3}(\ln 2 + 1)$

83. (a) $x = 3 + e^5$ **(b)** $x = \sqrt{4 + e}$ **(c)** $x = 18$

85. $5 - 7i$

87. $13 + 2i$

89. $15 + 9i$

91. 37

93. $3 + 2i$

95. $6 - 3i$

97. $8 - 2i$

99. $z + \bar{z} = 2a$, $z - \bar{z} = 2bi$

101. $x_1 = \frac{3}{4} + i\frac{\sqrt{7}}{4}$, $x_2 = \frac{3}{4} - i\frac{\sqrt{7}}{4}$

103. $x_1 = -1, x_2 = 2$

105. $x_1 = \frac{3}{8} + i\frac{\sqrt{7}}{8}$, $x_2 = \frac{3}{8} - i\frac{\sqrt{7}}{8}$

107. $x_1 = \frac{7}{3}, x_2 = -1$

109. $x_1 = x_2 = 1$

111. $x_1 = \frac{5 + i\sqrt{47}}{6}$, $x_1 = \frac{5 - i\sqrt{47}}{6}$

Section 1.2

1. range: $y \geq 0$

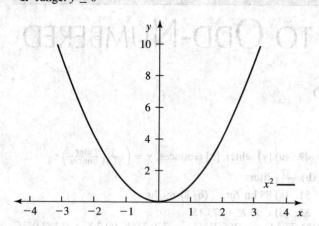

▲ **Figure 1.2.1**

3. range: $y \in [0, 1)$

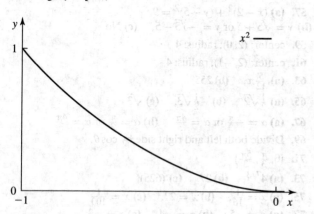

▲ **Figure 1.2.3**

5. **(b)** No, their domains are different.

7. $f(x)$ is odd. $f(-x) = -2x = -f(x)$

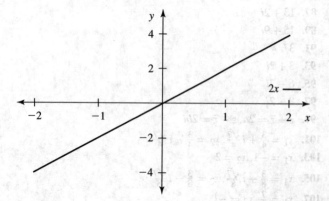

▲ **Figure 1.2.7**

9. $f(x)$ is even. $f(-x) = |3(-x)| = |3x| = f(x)$

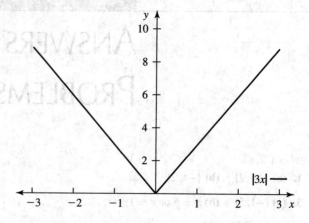

▲ **Figure 1.2.9**

11. $f(x)$ is even. $f(-x) = -|-x| = -|x| = f(x)$

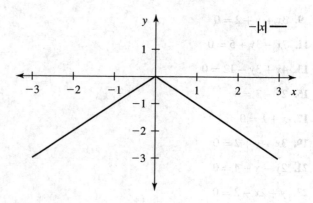

▲ **Figure 1.2.11**

15. **(a)** $(f \circ g)(x) = 1 - 4x^2, x \in \mathbf{R}$,
(b) $(g \circ f)(x) = 2(1 - x^2), x \in \mathbf{R}$

17. $(f \circ g)(x) = 3x, x \geq 9$

19. $(f \circ g)(x) = x, x \geq 0; (g \circ f)(x) = x, x \geq 0$

21. $x^2 > x^4$ for $0 < x < 1$; $x^2 < x^4$ for $x > 1$

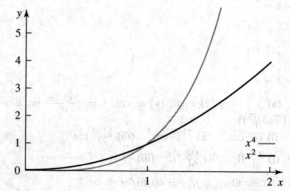

▲ **Figure 1.2.21**

23. They intersect at $x = 0$ or 1.

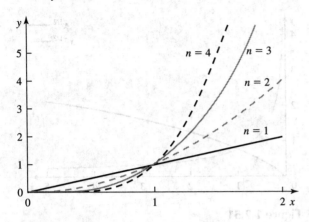

▲ **Figure 1.2.23**

25. (a)

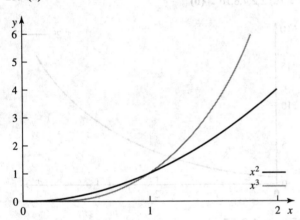

▲ **Figure 1.2.25**

27. (a) $f(-x) = f(x)$ **(b)** $f(-x) = -f(x)$
29. (a) $k = \frac{3}{2}$ **(b)** domain: $0 \le x \le 3$

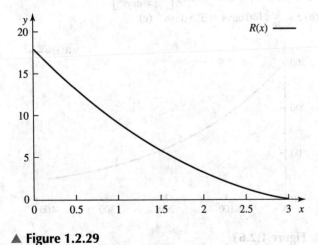

▲ **Figure 1.2.29**

31. $s(t) = t$, polynomial of degree 1
33. domain: $x \neq 1$; range: $y \neq 0$
35. domain: $x \neq -3, 3$; range: **R**
37. $\frac{1}{x} < \frac{1}{x^2}$ for $0 < x < 1$; $\frac{1}{x} > \frac{1}{x^2}$ for $x > 1$; they intersect at $x = 1$.

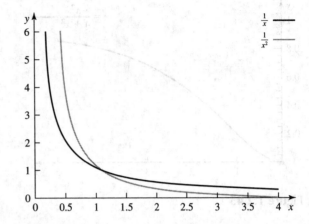

▲ **Figure 1.2.37**

39. (a)

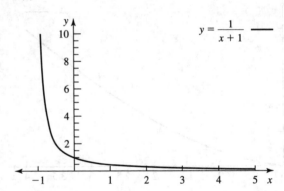

▲ **Figure 1.2.39**

(b) range of $f(x)$ is $(0, \infty)$, **(c)** $x = -\frac{1}{2}$,
(d) exactly one.
41. (a)

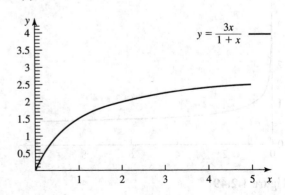

▲ **Figure 1.2.41**

(b) range of $f(x)$ is $[0, 3)$,   **(c)** $x = 2$,
(d) exactly one, $x = \frac{a}{3-a}$.

43. 83.3%, 4.76%

45. (a)

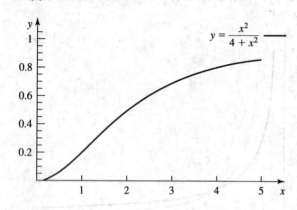

▲ **Figure 1.2.45**

(b) range of $f(x)$ is $[0, 1)$,   **(c)** $f(x)$ approaches 1.

47.

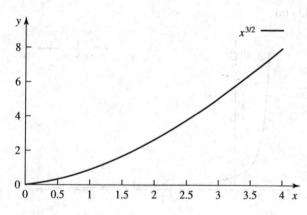

▲ **Figure 1.2.47**

49.

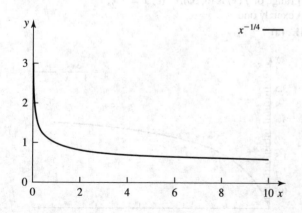

▲ **Figure 1.2.49**

51. (a)

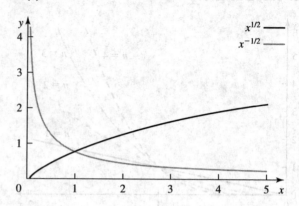

▲ **Figure 1.2.51**

53. increases
55. increases
57. (a) 1,2,4,8,16   **(b)**

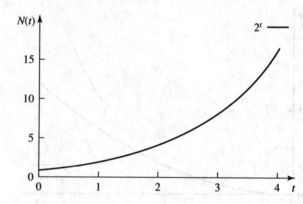

▲ **Figure 1.2.57**

59. $20 \exp\left[-\frac{\ln 2}{5730} 2000\right]$

61. $\lambda = \frac{\ln 2}{7 \, \text{days}}$

63. (a) $W(t) = (300 \, \text{gr}) \exp\left[-\frac{\ln 2}{140 \, \text{days}} t\right]$

(b) $t = \frac{\ln 5}{\ln 2} 140 \, \text{days} \approx 325 \, \text{days}$ **(c)**

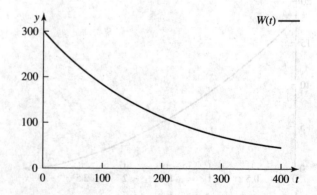

▲ **Figure 1.2.63**

65. $\frac{W(t)}{W(0)} = \exp\left[-\frac{\ln 2}{5730} 15{,}000\right] \approx 16.3\%$

69. (a) yes   **(b)** no   **(c)** yes   **(d)** yes   **(e)** no   **(f)** yes

71. (a) $f^{-1}(x) = \sqrt{x-1}, x \geq 1$ **(b)**

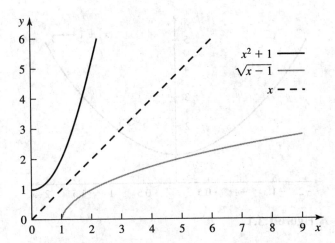

▲ **Figure 1.2.71**

73. $f^{-1}(x) = \sqrt[3]{\frac{1}{x}}, x > 0$

75. $f^{-1}(x) = \log_3 x, x > 0$

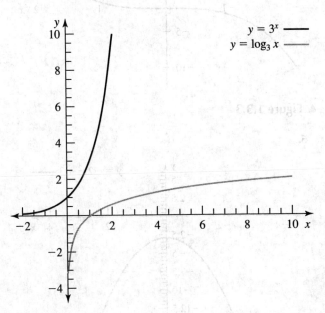

▲ **Figure 1.2.75**

77. $f^{-1}(x) = \log_{1/4} x, x > 0$

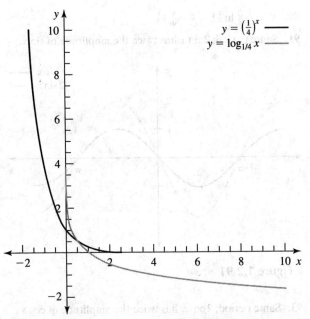

▲ **Figure 1.2.77**

79. $f^{-1}(x) = \log_2(x), x \geq 1$

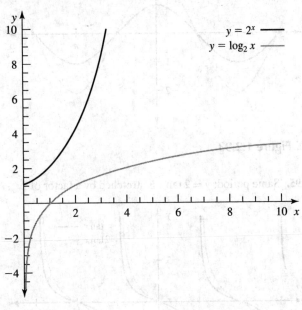

▲ **Figure 1.2.79**

81. (a) x^5,   **(b)** x^4,   **(c)** x^{-5},   **(d)** x^{-4},
(e) x^{-3},   **(f)** x^2

83. (a) $5\ln x$,   **(b)** $6\ln x$,   **(c)** $\ln(x-1)$,   **(d)** $-4\ln x$

85. (a) $e^{x \ln 3}$,   **(b)** $e^{(x^2-1)\ln 4}$,   **(c)** $e^{-(x+1)\ln 2}$,
(d) $e^{(-4x+1)\ln 3}$

87. $\mu = \ln 2$

89. $K = -\frac{3}{4} \ln\left(1 - \frac{4}{3} \cdot \frac{47}{300}\right)$

91. Same period; $2 \sin x$ has twice the amplitude of $\sin x$.

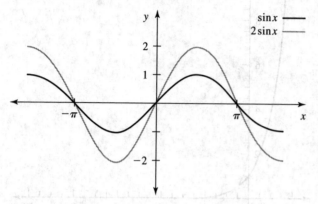

▲ **Figure 1.2.91**

93. Same period; $2 \cos x$ has twice the amplitude of $\cos x$.

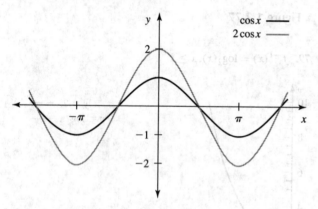

▲ **Figure 1.2.93**

95. Same period; $y = 2 \tan x$ is stretched by a factor of 2.

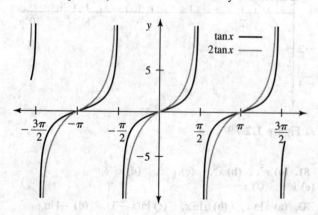

▲ **Figure 1.2.95**

97. amplitude: 3; period: $\frac{\pi}{2}$

99. amplitude: 4; period: 1

101. amplitude: 4; period: 8π

103. amplitude: 3; period: 10

Section 1.3

1.

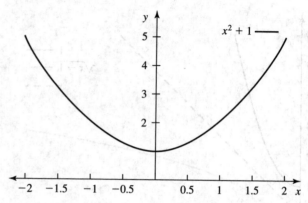

▲ **Figure 1.3.1**

3.

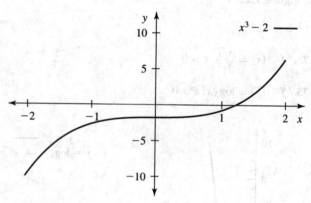

▲ **Figure 1.3.3**

5.

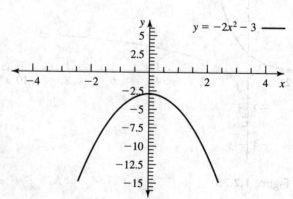

▲ **Figure 1.3.5**

7.

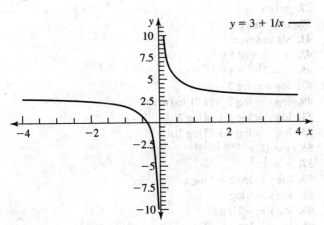

▲ **Figure 1.3.7**

9.

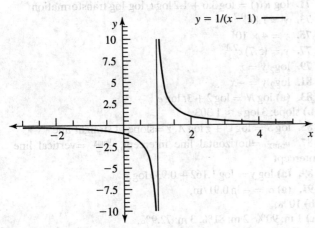

▲ **Figure 1.3.9**

11.

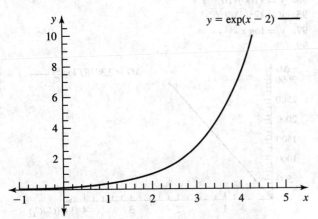

▲ **Figure 1.3.11**

13.

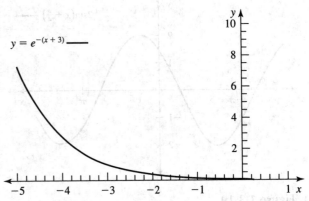

▲ **Figure 1.3.13**

15.

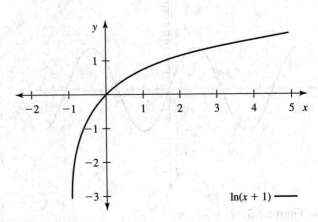

▲ **Figure 1.3.15**

17.

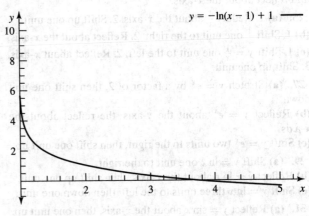

▲ **Figure 1.3.17**

19.

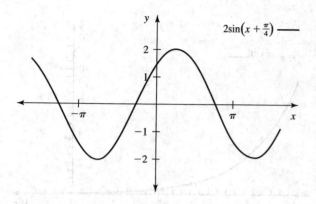

$2\sin\left(x + \frac{\pi}{4}\right)$ ——

▲ **Figure 1.3.19**

21.

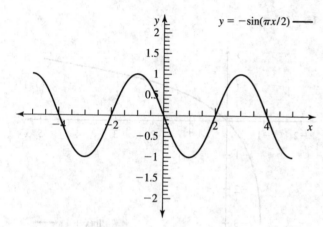

$y = -\sin(\pi x/2)$ ——

▲ **Figure 1.3.21**

23. (a) Shift two units down.
(b) Shift $y = x^2$ one unit to the right and then one unit up.
(c) Shift $y = x^2$ two units to the left, stretch by a factor of 2, and reflect about the x-axis.

25. (a) 1. Reflect $\frac{1}{x}$ about the x-axis. 2. Shift up one unit.
(b) 1. Shift $\frac{1}{x}$ one unit to the right. 2. Reflect about the x-axis.
(c) 1. Shift $y = \frac{1}{x}$ one unit to the left. 2. Reflect about x-axis. 3. Shift up one unit.

27. (a) Stretch $y = e^x$ by a factor of 2, then shift one unit down.
(b) Reflect $y = e^x$ about the y-axis, the reflect about the x-axis.
(c) Shift $y = e^x$ two units to the right, then shift one unit up.

29. (a) Shift $y = \ln x$ one unit to the right.
(b) Reflect $y = \ln x$ about the x-axis, then shift up one unit.
(c) Shift $y = \ln x$ three units to the left, then down one unit.

31. (a) Reflect $y = \sin x$ about the x-axis, then one unit up.
(b) Shift $y = \sin x$ by $\pi/4$ units to the right.
(c) Shift $y = \sin x$ by $\pi/3$ units to the left, then reflect about the x-axis.

35. (b) No, **(c)** No
37. seven
39. one, three
41. six to seven
43. $y = 5 \times (0.58)^x$
45. $y = 3^{1/3} \times (3^{-1/3})^x$
47. $\log y = \log 3 - 2x$
49. $\log y = \log 2 - (1.2)(\log e)x$
51. $\log y = \log 5 + (4 \log 2)x$
53. $\log y = \log 4 + (2 \log 3)x$
55. $y = (2)x^{-(\log 2)/\log 5}$
57. $y = \frac{1}{8}x^2$
59. $\log y = \log 2 + 5 \log x$
61. $\log y = 6 \log x$
63. $\log y = -2 \log x$
65. $\log y = \log 4 - 3 \log x$
67. $\log y = \log 3 + 1.7 \log x$, log-log transformation
69. $\log N(t) = \log 130 + (1.2t) \log 2$, log-linear transformation
71. $\log R(t) = \log 3.6 + 1.2 \log t$, log-log transformation
73. $y = 1.8x^{0.2}$
75. $y = 4 \times 10^x$
77. $y = (5.7)x^{2.1}$
79. $\log_2 y = x$
81. $\log_2 y = -x$
83. (a) $\log N = \log 2 + 3t \log e$
(b) slope: $3 \log e \approx 1.303$
85. $\log S = \log C + z \log A$, z =slope of straight line
87. v_{max} =horizontal line intercept, $\frac{v_{max}}{K_m}$ =vertical line intercept
89. (a) $\log y = \log 1.162 + 0.933 \log x$
91. (a) $\alpha = - \ln 0.91/m$,
(b) 10%,
(c) 1 m: 90%, 2 m: 81%, 3 m: 72.9%,
(e) slope $= \log 0.9 = -\alpha/\ln 10$,
(f) $z = -\frac{1}{\alpha} \ln(0.01) = \frac{\ln(0.01)}{\ln(0.9)}$,
(g) Clear lake: small α; milky lake: large α
93. $y = (100)(10^{1/3})^x$
95. $y = (2^{1/3})(2^{2/3})^x$
97. $y = \log x$
99.

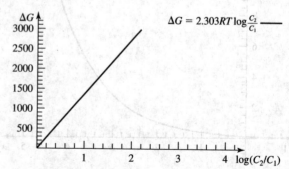

$\Delta G = 2.303 RT \log \frac{C_2}{C_1}$ ——

▲ **Figure 1.3.99**

101.

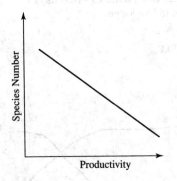

▲ **Figure 1.3.101**

103.

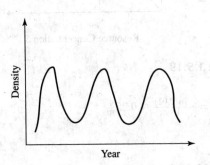

▲ **Figure 1.3.103**

105.

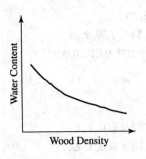

▲ **Figure 1.3.105**

107.

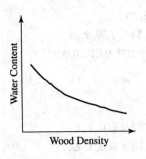

▲ **Figure 1.3.107**

109.

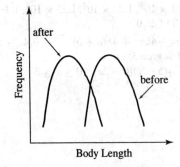

▲ **Figure 1.3.109**

111.

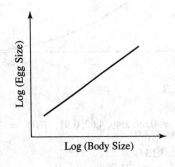

▲ **Figure 1.3.111**

113.

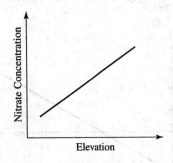

▲ **Figure 1.3.113**

115.

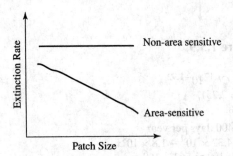

▲ **Figure 1.3.115**

Section 1.5

1. (a) $10^4, 1.1 \times 10^4, 1.22 \times 10^4, 1.35 \times 10^4, 1.49 \times 10^4$
(b) $t = 10 \ln 10 \approx 23.0$

3. (b) $R(x) = -4kx^3 + 4k(a+b)x^2 - kb(4a-b)x + kab^2$, polynomial of degree 3
(c) $R(x) = (0.3)(5-x)(6-2x)^2, 0 \le x \le 3$

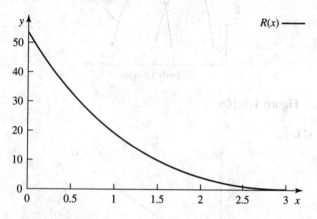

▲ **Figure 1.5.3**

5. (a) $L(t) = 0.69, 2.40, 4.62, 6.91$, $E(t) = 1.72, 2.20 \times 10^4, 2.69 \times 10^{43}, 10^{434}$
(b) 20.93 years, 3.13 ft
(c) $10^{536,000,000}$ years, 21.98 ft
(d) $L = 19.92$ ft, $E = 10^{195 \times 10^6}$ ft

7. $T = \dfrac{\ln 2}{\ln(1+\frac{9}{100})}$, T goes to infinity as q gets closer to 0.

9. (a)

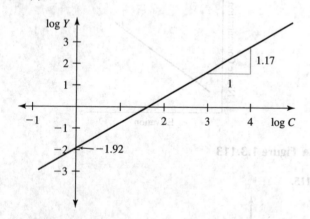

▲ **Figure 1.5.9**

(b) $Y = C^{1.17} 10^{-1.92}$
(c) $Y_p = 2.25 Y_c$
(d) 8.5%

11. (a) 400 days per year
(b) $y = 4.32 \times 10^9 - 1.8 \times 10^8 x$
(c) 376×10^6 to 563×10^6 years ago

13. (a) males: $S(t) = \exp[-(0.019t)^{3.41}]$; females: $S(t) = \exp[-(0.022t)^{3.24}]$

(b) males: 47.27 days; females: 40.59 days
(c) males should live longer

15. (a) $x = k, v = \frac{a}{2}$
(b) $x_{0.9} = 81 x_{0.1}$

17.

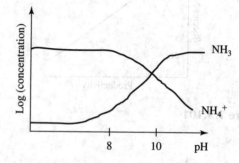

▲ **Figure 1.5.17**

19. $g(s) = \dfrac{v_{\max}}{S_k} S$

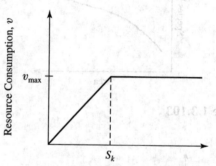

▲ **Figure 1.5.19**

21. (a) $\alpha = -\dfrac{\ln(0.01)}{18 \text{ m}} \approx 0.25 \frac{1}{\text{m}}$
(b) 4.87 m

Section 2.1

1. $0.5, 0.33, 0.25, 0.2, 0.17, 0.14$

3. $1, 0.25, 0.11, 0.063, 0.04, 0.028$

5. $-1, 0, 3, 8, 15, 24$

7. $1, 4, 9, 16, 25, 36$

9. $1, 2.72, 4.11, 5.65, 7.39, 9.36$

11. $1, 0.33, 0.11, 0.037, 0.012, 0.0041$

13. $1, 2, 4, 8, 16, 32$

15. 161 minutes

17. 50 minutes

19. $N(t) = (40)(2^t), t = 0, 1, 2, \ldots$

21. $N(t) = (20)(3^t), t = 0, 1, 2, \ldots$

23. $N(t) = (5)(4^t), t = 0, 1, 2, \ldots$

25. $N(t+1) = 2N(t), N(0) = 20$

27. $N(t + 1) = 3N(t)$, $N(0) = 10$

29. $N(t + 1) = 4N(t)$, $N(0) = 30$

31.

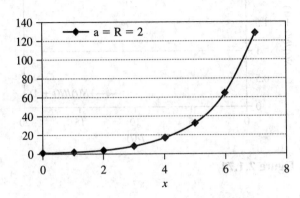

▲ **Figure 2.1.31**

33.

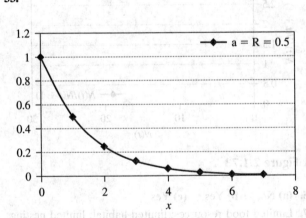

▲ **Figure 2.1.33**

35. 3, 6, 12, 24, 48, 96

37. 2, 6, 18, 54, 162, 486

39. 1, 5, 25, 125, 625, 3125

41. 1024, 512, 256, 128, 64, 32

43. 729, 243, 81, 27, 9, 3

45. 31250, 6250, 1250, 250, 50, 10

47. $N(t) = (15)(2^t)$, $t = 0, 1, 2, \ldots$

49. $N(t) = (12)(3^t)$, $t = 0, 1, 2, \ldots$

51. $N(t) = (24)(4^t)$, $t = 0, 1, 2, \ldots$

53. $N(t) = (5000)\left(\frac{1}{2}\right)^t$, $t = 0, 1, 2, \ldots$

55. $N(t) = (8000)\left(\frac{1}{3}\right)^t$, $t = 0, 1, 2, \ldots$

57. $N(t) = (1200)\left(\frac{1}{5}\right)^t$, $t = 0, 1, 2, \ldots$

59.

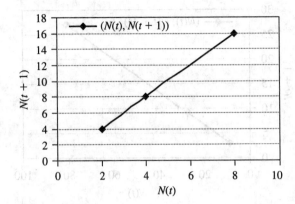

▲ **Figure 2.1.59**

61.

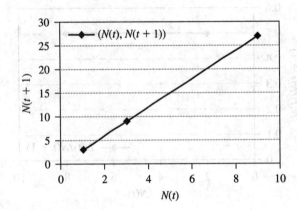

▲ **Figure 2.1.61**

63.

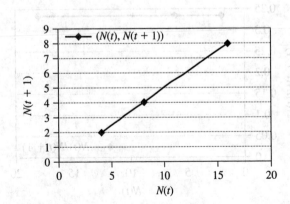

▲ **Figure 2.1.63**

65.

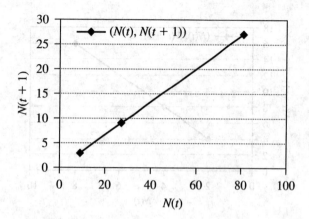

▲ **Figure 2.1.65**

67.

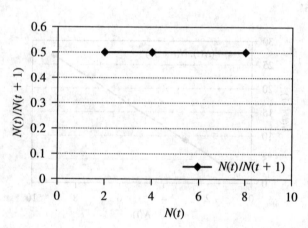

▲ **Figure 2.1.67**

69.

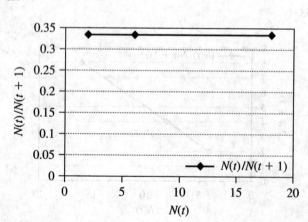

▲ **Figure 2.1.69**

71.

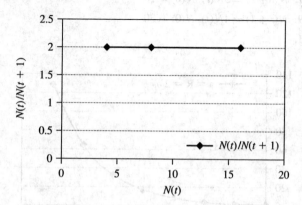

▲ **Figure 2.1.71**

73.

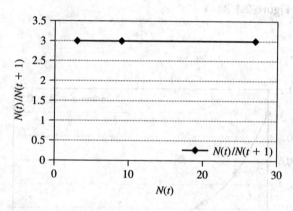

▲ **Figure 2.1.73**

75. (a) No, **(b)** Yes, **(c)** Yes

77. Limited food resources; limited habitat; limited nesting sites.

Section 2.2

1. $0, 1, 2, 3, 4, 5$

3. $\frac{1}{2}, \frac{1}{3}, \frac{1}{4}, \frac{1}{5}, \frac{1}{6}, \frac{1}{7}$

5. $0, -1, 2, -3, 4, -5$

7. $0, \frac{1}{2}, \frac{4}{3}, \frac{9}{4}, \frac{16}{5}, \frac{25}{6}$

9. $6, 7, 8, 9$

11. $\frac{1}{36}, \frac{1}{49}, \frac{1}{64}, \frac{1}{81}$

13. $\frac{6}{7}, \frac{7}{8}, \frac{8}{9}, \frac{9}{10}$

15. $\sqrt{6 + e^6}, \sqrt{7 + e^7}, \sqrt{8 + e^8}, \sqrt{9 + e^9}$

17. $a_n = n, n = 0, 1, 2, \ldots$

19. $a_n = 2^n, n = 0, 1, 2, \ldots$

21. $a_n = \frac{1}{2^n}, n = 0, 1, 2, \ldots$

23. $a_n = (-1)^{n+1}(n + 1), n = 0, 1, 2, \ldots$

25. $a_n = (-1)^{n+1} \frac{1}{(n+2)}, n = 0, 1, 2, \ldots$

27. $a_n = \sin[(n + 1)\pi], n = 0, 1, 2, \ldots$

29. $\frac{1}{2}, \frac{1}{3}, \frac{1}{4}, \frac{1}{5}, \frac{1}{6}; 0$

31. $0, \frac{1}{2}, \frac{2}{3}, \frac{3}{4}, \frac{4}{5}; 1$

33. $1, \frac{1}{2}, \frac{1}{5}, \frac{1}{10}, \frac{1}{17}; 0$

35. $1, -\frac{1}{2}, \frac{1}{3}, -\frac{1}{4}, \frac{1}{5}; 0$

37. $1, \frac{1}{\sqrt{2}}, \frac{1}{\sqrt{3}}, \frac{1}{2}, \frac{1}{\sqrt{5}}; 0$

39. $1, \frac{1}{2}, \frac{1}{3}, \frac{1}{4}, \frac{1}{5}; 0$

41. $0, \frac{1}{2}, \frac{4}{3}, \frac{9}{4}, \frac{16}{5}$; limit does not exist.

43. $0, 1, \sqrt{2}, \sqrt{3}, \sqrt{4}$; limit does not exist.

45. $1, 2, 4, 8, 16$; limit does not exist.

47. $1, 3, 9, 27, 64$; limit does not exist.

49. $a = 0, N = 100$

51. $a = 0, N = 10$

53. $a = 0, N = 100$

55. $a = 0, N = 100$

57. $a = 0, N = 31$

59. $a = 1, N = 99$

61. $a = 1, N = 100$

63. $a = 1, N = 9$

65. N is the largest integer less than or equal to $1/\epsilon$.

67. N is the largest integer less than or equal to $\sqrt{1/\epsilon}$.

69. N is the largest integer less than or equal to $1/\epsilon$.

71. 0

73. 1

75. 1

77. 0

79. 0

81. 1

83. $2, 4, 8, 16, 32$

85. $1, 1, 1, 1, 1$

87. $-6, 16, -28, 60, -116$

89. $\frac{1}{2}, \frac{1}{3}, \frac{1}{4}, \frac{1}{5}, \frac{1}{6}$

91. $2, \frac{5}{2}, \frac{29}{10}, \frac{941}{290}, \frac{969,581}{272,890}$

93. 4

95. -3

97. $2, -2$

99. $-1 + \sqrt{3}, -1 - \sqrt{3}$

101. $0, 5$

103. $5; 5$

105. $0, 2; 2$

107. $0, \frac{1}{2}; \frac{1}{2}$

109. $2, -2; 2$

Section 2.3

1. $N_t = \frac{2N_t}{1 - \frac{1}{15}N_t}$

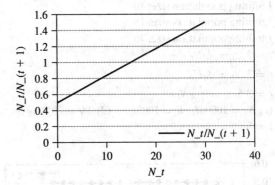

▲ **Figure 2.3.1**

3. $N_t = \frac{1.5N_t}{1 - \frac{0.5}{40}N_t}$

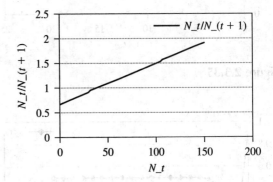

▲ **Figure 2.3.3**

5. $N_t = \frac{2.5N_t}{1 - \frac{1.5}{90}N_t}$

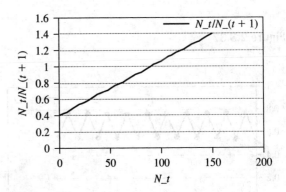

▲ **Figure 2.3.5**

7. $R = 2, K = 20$

9. $R = 1.5, K = 30$

11. $R = 4, K = 450$

13. 0, 90

15. 0, 30

17. 0, 60

19. Limiting population size: 10

21. Limiting population size: 15

23. Limiting population size: 40

25. $r = 2$, $x_t = \frac{1}{20} N_t$

27. $r = 3$, $x_t = \frac{2}{45} N_t$

29. $r = 3.5$, $x_t = \frac{2.5}{105} N_t$

31. (c) $N_t = 1000 M_t$, $K = 1000 L$, **(d)** $M_t = 20$, $L = 5$

33. $z = 6$

35.

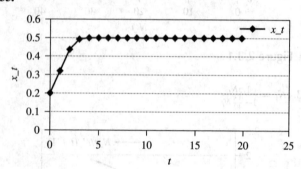

▲ **Figure 2.3.35**

37.

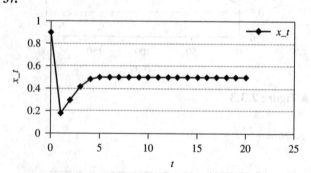

▲ **Figure 2.3.37**

39.

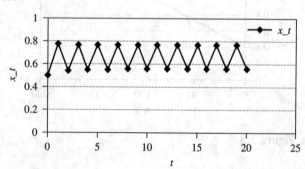

▲ **Figure 2.3.39**

41.

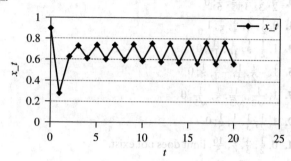

▲ **Figure 2.3.41**

43.

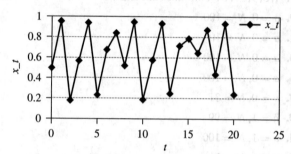

▲ **Figure 2.3.43**

45.

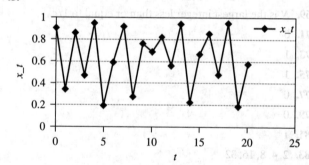

▲ **Figure 2.3.45**

47. Points of intersection: $N = 0$ and $N = 10$

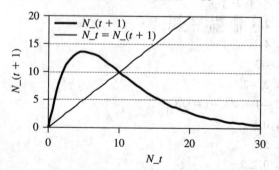

▲ **Figure 2.3.47**

49. Points of intersection: $N = 0$ and $N = 12$

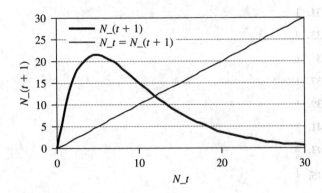

▲ **Figure 2.3.49**

51.

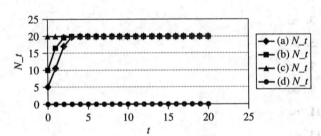

▲ **Figure 2.3.51**

53.

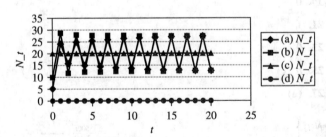

▲ **Figure 2.3.53**

57. One-month-old rabbit pairs produce one pair of rabbits; two-month-old rabbit pairs produce two pairs of rabbits.

Section 2.5

1. 0

3. 40

5. ∞

7. 1

9. 0

11. $a_n = \frac{2n+3}{2n+2}, n = 0, 1, 2, \ldots$

13. $a_n = \frac{n+1}{(n+1)^2+1}, n = 0, 1, 2, \ldots$

15.

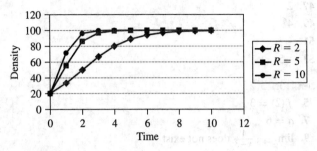

▲ **Figure 2.5.15**

17. $\hat{R}_t \approx 0.8$ for t large; extinction will occur.

19.

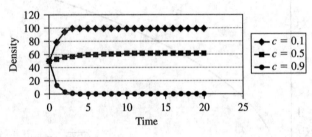

▲ **Figure 2.5.19**

Section 3.1

1. -1

3. $-\frac{1}{2}$

5. $\frac{3}{2}\sqrt{2}$

7. $\frac{4}{3}\sqrt{3}$

9. $e^{-1/2}$

11. 0

13. 7

15. 0

17. 1

19. 0

21. $-\infty$

23. ∞

25. ∞

27. ∞

29. $\frac{1}{6}$

31. $\frac{1}{2}$

33. $\lim_{x \to \infty} f(x) = \lim_{x \to -\infty} f(x) = 0; \lim_{x \to 0} f(x) = \infty$

35. divergence by oscillations

37. -9

39. 54

41. $\frac{53}{3}$

43. $\frac{47}{2}$

45. $\frac{28}{3}$

47. 2

49. 4

51. $-\frac{1}{6}$

53. -5

Section 3.2

5. $f(2) = 3$

7. $a = 6$

9. $\lim_{x \to 3} \frac{1}{x-3}$ does not exist

11. $f(1) \neq \lim_{x \to 1} f(x)$

15. **(b)**

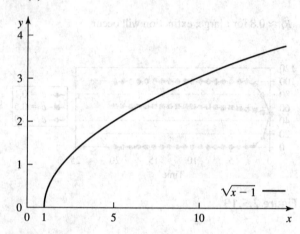

▲ **Figure 3.2.15**

(c) No

17. $x \in \mathbf{R}$

19. $x \neq 1$

21. $x \in \mathbf{R}$

23. $\{x : x < -1 \text{ or } x > 0\}$

25. $\{x \in \mathbf{R} : x \neq \frac{1}{4} + \frac{k}{2}, k \in \mathbf{Z}\}$

27. **(a)** $f(x)$ is not continuous at $x = 0$

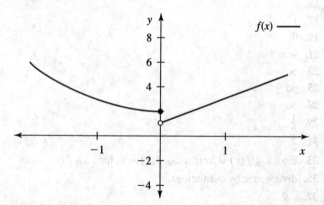

▲ **Figure 3.2.47**

(b) $c = 2$

29. $\frac{1}{2}$

31. 1

33. 3

35. 1

37. 1

39. 1

41. 2

43. $\frac{1}{4}$

45. $\frac{1}{6}$

47. 0

Section 3.3

1. 0

3. ∞

5. 3

7. ∞

9. ∞

11. ∞

13. 4

15. 2

17. 0

19. 0

21. $\frac{3}{2}$

23. 0

25. a

27. **(a)**

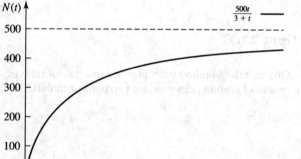

▲ **Figure 3.3.27**

(b) 500   **(c)** 250

29. (a)

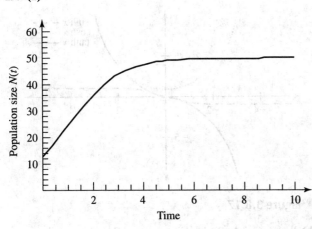

▲ **Figure 3.3.29**

(b) 50

Section 3.4

1. (a)

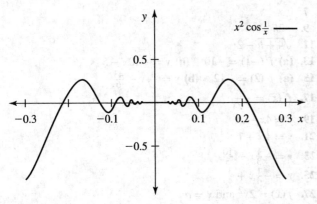

▲ **Figure 3.4.1**

3. (a)

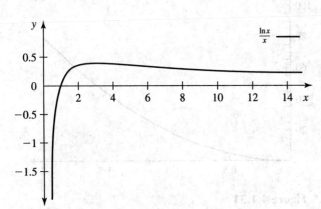

▲ **Figure 3.4.3**

(b) $x \geq e$

5. 1

7. 5

9. π

11. 0

13. 1

15. 0

17. 0

Section 3.5

1. (b) $f(0) = -1$, $f(2) = 3$

3. (b) $f(1) = \sqrt{3} < 2 < f(2) = \sqrt{6}$

5. $f(0) = 1 > 0$, $f(1) = e^{-1} - 1 < 0$

7. $x \approx 0.57$

9. (a) $x \approx -0.67$ **(c)** No

11. (a) $N(10) = 23$ **(b)** $N(10) = 23.5$ (in million)

Section 3.6

1. $0.495 < x < 0.505$

3. $-\sqrt{\frac{11}{30}} < x < \sqrt{\frac{11}{30}}$

5. (b) $1.995 < x < 2.005$

7. (a)

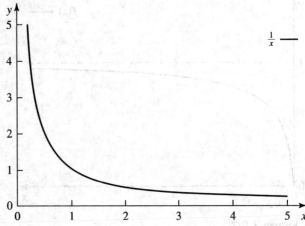

▲ **Figure 3.6.7**

(b) $0 < x < \frac{1}{4}$

9. $\delta = \frac{\epsilon}{2}$

11. $\delta = 6\epsilon - \epsilon^2$ for $0 < \epsilon \leq 3$, $\delta = 9$ for $\epsilon > 3$

13. $\delta = \frac{2}{\sqrt{M}}$

15. $\delta = \frac{1}{\sqrt[4]{M}}$

17. $N = \sqrt{\frac{3}{\epsilon}}$

19. $N = \frac{1}{\epsilon}$

21. $\delta = \frac{\epsilon}{|m|}$

Section 3.8

1. $x \in \mathbf{R}$

3. $x \in \mathbf{R}$

5.

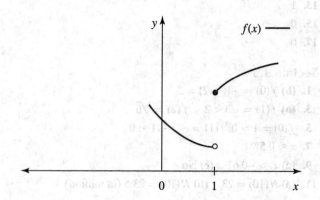

▲ Figure 3.8.5

7.

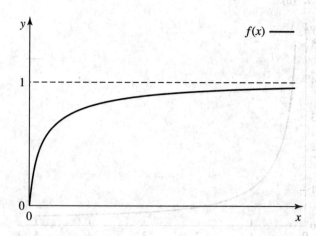

▲ Figure 3.8.7

9. $f(-2) = -2$, $\lim_{x \to -2^+} f(x) = -2$, $\lim_{x \to -2^-} f(x) = -3$

11. $a = 1.24 \times 10^6, k = 5$

13. (a) $g(t) = \begin{cases} 1 & \text{for } \frac{1}{6} + 2k \le x \le \frac{5}{6} + 2k, k = 0, 1, 2, \dots \\ 0 & \text{otherwise} \end{cases}$

(b) $s(t)$ is continuous, $g(t)$ is not continuous

15. (c) $T_h = 0$: $g(N) = aTN = f(N)$

(d) $\lim_{N \to \infty}(aTN) = \infty$; $\lim_{N \to \infty} \frac{aTN}{1 + aT_h N} = \frac{T}{T_h}$

17. (a)

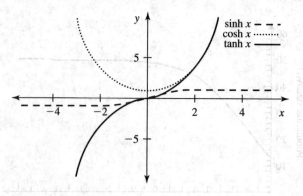

▲ Figure 3.8.17

(b) $\lim_{x \to \infty} \sinh x = \infty$, $\lim_{x \to -\infty} \sinh x = -\infty$, $\lim_{x \to \infty} \cosh x = \infty$, $\lim_{x \to -\infty} \cosh x = \infty$, $\lim_{x \to \infty} \tanh x = 1$, $\lim_{x \to -\infty} \tanh x = -1$

Section 4.1

1. 0

3. 2

5. 0

7. 0

9. $-2h$

11. $\sqrt{4 + h} - 2$

13. (a) $f'(-1) = -10$ (b) $y = -10x - 5$

15. (a) $f'(2) = -12$ (b) $y = \frac{1}{12}x - \frac{43}{6}$

17. $f'(x) = \frac{1}{2\sqrt{x}}$

19. $y = 4x - 2$

21. $y = \frac{1}{4}x + 1$

23. $y = -\frac{1}{3}x - \frac{10}{3}$

25. $y = -\frac{1}{4}x + \frac{3}{4}$

27. $f(x) = 2x^2$ and $x = a$

29. $f(x) = \frac{1}{x^2+1}, a = 2$

31. (a)

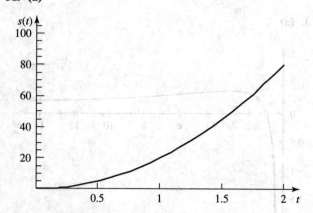

▲ Figure 4.1.31

(b) 40 km/hr,   (c) 40 km/hr

33. (a) $s\left(\frac{3}{4}\right) = 30, s(1) = \frac{160}{3}$
(b) $\frac{280}{3}$ **(c)** $v\left(\frac{3}{4}\right) = 80, |v\left(\frac{3}{4}\right)| = 80$
35. $R^{\star} = 1.25$

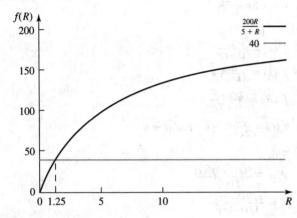

▲ **Figure 4.1.35**

37. $f(N) = 0$ for $N = 0$ and $N = 20$

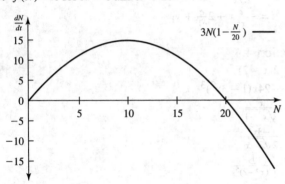

▲ **Figure 4.1.37**

39. $x = 7$ or $x = 4$; reaction ceases when $x = 4$.
41. $\frac{dN}{dt} = 0$ for $N = 0$ or $N = K$
43. B
45.

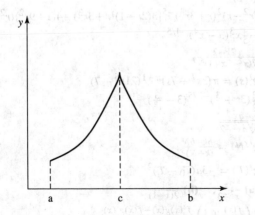

▲ **Figure 4.1.45**

47. $x = -5$
49. $x = -2$
51. $x = 3$
53. $x = -1$
55. $x = \sqrt{1/2}$ and $x = -\sqrt{1/2}$
57. $x = 1$
59. $x = 0$

Section 4.2
1. $6x^2 - 3$
3. $-10x^4 + 7$
5. $-4 - 10x$
7. $35s^6 + 6s - 5$
9. $-2t^3 + 4$
11. $2x \sin \frac{\pi}{3}$
13. $-12x^3 \tan \frac{\pi}{6}$
15. $3t^2 e^{-2} + 1$
17. $3s^2 e^3$
19. $60x^2 - 24x^5 + 72x^7$
21. $24x^2 + \frac{1}{8}$
23. $3ax^2$
25. $2ax$
27. $2rs$
29. $3rs^2x^2 - r$
31. $4(b-1)N^3 - \frac{2N}{b}$
33. $a^3 - 3at^2$
35. $V_0 \gamma$
37. $1 - \frac{2N}{K}$
39. $2rN - 3\frac{r}{K}N^2$
41. $4\frac{2\pi^5}{15}\frac{k^4}{c^2h^3}T^3$
43. $-191x + y - 377 = 0$
45. $8x - \sqrt{2}y - 18 = 0$
47. $y - 21x + 16 = 0$
49. $3y + x - 2 = 0$
51. $x + 24y - 71\sqrt{3} = 0$
53. $\sqrt{3}y - \frac{1}{2}x + \frac{5}{2} = 0$
55. $2ax - y - a = 0$
57. $(a^2 + 2)y - 4ax + 4a = 0$
59. $\frac{1}{3a}x + y + a + \frac{1}{3a} = 0$
61. $2a(a+1)y + \frac{1}{2}(a+1)^2x - 8a^2 - (a+1)^2 = 0$
63. $(0, 0)$
65. $\left(\frac{3}{2}, \frac{9}{4}\right)$
67. $(0, 0)$ and $\left(\frac{2}{9}, -\frac{4}{243}\right)$
69. $(0, 0), \left(-\frac{1}{2}, -\frac{17}{96}\right)$, and $\left(4, -\frac{160}{3}\right)$

71. $(0, 4)$; only point

73. $\left(-\frac{1}{2}, \frac{15}{4}\right)$; only point

75. $\left(\frac{1}{3}\sqrt{3}, \frac{7}{9}\sqrt{3}+2\right), \left(-\frac{1}{3}\sqrt{3}, -\frac{7}{9}\sqrt{3}+2\right)$

77. Tangent line: $y = 2x - 1$

79. $y = 2ax - a^2$

81. $P'(x)$ is a polynomial of degree 3.

Section 4.3

1. $f'(x) = 4 + 2x - 6x^2$

3. $f'(x) = 3x^2(3x - 14x^2) + (x^3 + 17)(3 - 28x)$

5. $f'(x) = x(2x + 3x^2) + (\frac{1}{2}x^2 - 1)(2 + 6x)$

7. $f'(x) = \frac{2x}{5}$

9. $f'(x) = 6(3x - 1)$

11. $f'(x) = -12(1 - 2x)$

13. $g'(s) = 2(4s - 5)(2s^2 - 5s)$

15. $g'(t) = 6(4t - 20t^3)(2t^2 - 5t^4)$

17. $y = x - 1$

19. $y = -56x - 64$

21. $y = -\frac{1}{6}x + \frac{7}{3}$

23. $y = \frac{1}{5}x$

25. $f'(x) = (12x + 5)(1 - x) - (2x - 1)(3x + 4)$

27. $f'(x) = (6x^2 - 12x + 1)(1 - x^2) - 2x(x - 3)(2x^2 + 1)$

29. $f'(x) = a(4x - 3)$

31. $f'(x) = 8ax(x^2 - a)$

33. $g'(t) = 2a(at + 1)$

35. 11

37. $y' = 2f(x) + 2xf'(x)$

39. $y' = -15x^2 f(x) - 5x^3 f'(x) - 2$

41. $y' = 3f'(x)g(x) + 3f(x)g'(x)$

43. $y' = f'(x)g(x) + f(x)g'(x) + 4g(x)g'(x)$

45. $f(B) = Bg(B)$ with $f'(0) = g(0)$

47. $f'(N) = r\left[(a - 2N)\left(1 - \frac{N}{K}\right) - \frac{1}{K}(aN - N^2)\right]$

49. $f'(x) = \frac{1}{(x+1)^2}$

51. $\frac{-4x + 2x^2 + 1}{(1-x)^2}$

53. $f'(x) = \frac{2x^3 - 3x^2 + 3}{(1-x)^2}$

55. $h'(t) = \frac{t^2 + 2t - 4}{(t+1)^2}$

57. $f'(s) = \frac{2s^2 - 4s + 4}{(1-s)^2}$

59. $f'(x) = \frac{1}{2\sqrt{x}}(x - 1) + \sqrt{x}$

61. $f'(x) = \frac{\sqrt{3}}{2\sqrt{x}}(x^2 - 1) + 2x\sqrt{3x}$

63. $f'(x) = 3x^2 + \frac{3}{x^4}$

65. $f'(x) = 4x - \frac{3-6x}{x^4}$

67. $g'(s) = \frac{2s^{-1/3} - s^{-2/3} - 1}{3(s^{2/3} - 1)^2}$

69. $f'(x) = (-2)\left(\sqrt{2x} + \frac{2}{\sqrt{x}}\right) + (1 - 2x)\left(\frac{1}{\sqrt{2x}} - \frac{1}{x^{3/2}}\right)$

71. $y = -8x - \frac{41}{3}$

73. $y = \frac{7}{16}x - 1$

75. $f'(x) = \frac{3a}{(3+x)^2}$

77. $f'(x) = \frac{8ax}{(4+x^2)^2}$

79. $f'(R) = \frac{2Rk + R^2}{(k+R)^2}$

81. $h'(t) = \frac{\sqrt{a}}{2\sqrt{t}}(t - a) + \sqrt{at} + a$

83. $-\frac{5}{18}$

85. $y' = \frac{2xf(x) - x^2 f'(x)}{[f(x)]^2}$

87. $y' = \frac{xf'(x) - f(x)}{[f(x) + x]^2}$

89. $y' = \frac{f'(x)g(x) - 2f(x)g'(x)}{[g(x)]^3}$

91. $y' = \frac{1}{2\sqrt{x}}f(x)g(x) + \sqrt{x}f'(x)g(x) + \sqrt{x}f(x)g'(x)$

93. $y = -\frac{c}{x_1^2}x + 2\frac{c}{x_1} + y_1$

Section 4.4

1. $2(x - 2)$

3. $-24x(1 - 3x^2)^3$

5. $\frac{x}{\sqrt{x^2 + 3}}$

7. $\frac{-3x^2}{2\sqrt{3 - x^3}}$

9. $-\frac{12x^2}{(x^3 - 2)^5}$

11. $\frac{2x - 3}{(2x^2 - 1)^{3/2}}$

13. $\frac{1 - 3x}{\sqrt{2x - 1}(x - 1)^3}$

15. $\frac{2\sqrt{s} + 1}{4\sqrt{s}\sqrt{s + \sqrt{s}}}$

17. $\frac{-9t^2}{(t - 3)^4}$

19. $(r^2 - r)^2(r + 3r^3)^{-5}[3(2r - 1)(r + 3r^3) - 4(1 + 9r^2)(r^2 - r)]$

21. $-\frac{4}{5}x^3(3 - x^4)^{-4/5}$

23. $\frac{2x - 2}{7(x^2 - 2x + 1)^{6/7}}$

25. $g'(s) = \pi(3s^7 - 7s)^{\pi - 1}(21s^6 - 7)$

27. $\frac{2}{5}(3t + \frac{3}{t})^{-3/5}(3 - \frac{3}{t^2})$

29. $\frac{ax}{\sqrt{ax^2 - 2}}$

31. $g'(N) = \frac{bk - bN}{(k+N)^3}$

33. $g'(T) = -3a(T_0 - T)^2$

35. **(a)** $\frac{2x}{x^2 + 3}$, **(b)** $\frac{1}{2(x-1)}$

37. $2\left(\frac{f(x)}{g(x)} + 1\right)\frac{f'(x)g(x) - f(x)g'(x)}{[g(x)]^2}$

39. $\frac{2f(x)f'(x)[g(2x) + 2x] - [f(x)]^2(g'(2x) \cdot 2 + 2)}{[g(2x) + 2x]^2}$

41. $y' = 3(\sqrt{x^3 - 3x} + 3x)^2 \left(\frac{3x^2 - 3}{2\sqrt{x^3 - 3x}} + 3 \right)$

43. $y' = 36x(3x^2 - 1)(1 + (3x^2 - 1)^3)$

45. $y' = 3 \left(\frac{2x+1}{3(x^3-1)^3 - 1} \right)^2 \frac{6(x^3-1)^3 - 2 - 27x^2(2x-1)(x^3-1)}{(3(x^3-1)^3 - 1)^2}$

47. $\frac{dy}{dx} = -\frac{x}{y}$

49. $\frac{dy}{dx} = -\left(\frac{y}{x}\right)^{1/4}$

51. $\frac{dy}{dx} = 4\sqrt{xy} - \frac{y}{x}$

53. $\frac{dy}{dx} = \frac{y}{x}$

55. (a) $y = \frac{4}{3}x - \frac{25}{3}$ **(b)** $y = -\frac{3}{4}x$

57. (a) $y = \frac{3}{4}x - \frac{9}{4}$ **(b)** $y = -\frac{4}{3}x + \frac{136}{9}$

59. (a) $(27)^{1/6}$

(b)

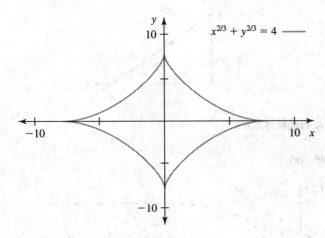

▲ **Figure 4.4.59**

61. $-\frac{2}{3}\sqrt{3}$

63. $-\frac{3}{4}$

65. $\frac{dV}{dt} = 3x^2 \frac{dx}{dt}$

67. $\frac{dS}{dt} = 8\pi r \frac{dr}{dt}$

69. $\frac{dh}{dt} = \frac{1}{100\pi} \frac{m}{min}$

71. $\frac{183}{\sqrt{61}} \frac{mi}{hr}$ for both $t = 20$ min and $t = 40$ min

73. $f'(x) = 3x^2 - 6x$, $f''(x) = 6x - 6$

75. $g'(x) = 2(x+1)^{-2}$, $g''(x) = -\frac{4}{(x+1)^3}$

77. $g'(t) = \frac{9t^2 + 2}{2\sqrt{3t^3 + 2t}}$,

$g''(t) = \frac{27t^4 + 36t^2 - 4}{4(3t^3 + 2t)^{3/2}}$

79. $f'(s) = \frac{3}{4}\left(s^{1/2} - \frac{1}{s}\right)^{-1/2}$,

$f''(s) = -\frac{3}{8} \frac{\frac{1}{2\sqrt{s}} + \frac{1}{s^2}}{(s^{1/2} - \frac{1}{s})^{3/2}}$

81. $g'(t) = -\frac{5}{2}t^{-7/2} - \frac{1}{2}t^{-1/2}$,

$g''(t) = \frac{35}{4}t^{-9/2} + \frac{1}{4}t^{-3/2}$

83. $f(x) = x^5$, $f'(x) = 5x^4$, $f''(x) = 20x^3$, $f'''(x) = 60x^2$,
$f^{(4)}(x) = 120x$, $f^{(5)}(x) = 120$, $f^{(6)} = \cdots = f^{(10)}(x) = 0$

85. $p(x) = 3x^2 + 2x + 3$

87. (a) velocity: $v_0 - gt$, acceleration: $-g$,
(b) $t = \frac{v_0}{g}$

Section 4.5

1. $f'(x) = 2\cos x + \sin x$

3. $f'(x) = 3\cos x - 5\sin x - 2\sec x \tan x$

5. $f'(x) = \sec^2 x + \csc^2 x$

7. $f'(x) = 3\cos(3x)$

9. $f'(x) = 6\cos(3x + 1)$

11. $f'(x) = 4\sec^2(4x)$

13. $f'(x) = 4\sec(1 + 2x)\tan(1 + 2x)$

15. $f'(x) = 6x\cos(x^2)$

17. $f'(x) = 2x\sec(x^2 - 3)\tan(x^2 - 3)$

19. $f'(x) = 6\sin x \cos x$

21. $f'(x) = -8x\sin(x^2)$

23. $f'(x) = -8\sin x \cos x$

25. $f'(x) = -4x\sec^2(1 - x^2)$

27. $f'(x) = -18\tan^2(3x - 1)\sec^2(3x - 1)$

29. $f'(x) = \frac{2x\cos(2x^2 - 1)}{\sqrt{\sin(2x^2 - 1)}}$

31. $g'(s) = \frac{-\sin s}{2\sqrt{\cos s}}$

33. $g'(t) = \frac{2\cos(2t)[\cos(6t) - 1] + 6\sin(6t)[\sin(2t) + 1]}{[\cos(6t) - 1]^2}$

35. $f'(x) = \frac{2x\sec(x^2 - 1)[\tan(x^2 - 1) + \cot(x^2 + 1)]}{\csc(x^2 + 1)}$

37. $f'(x) = 2\cos(2x - 1)\cos(3x + 1) - 3\sin(2x - 1)\sin(3x + 1)$

39. $f'(x) = 6x\sec^2(3x^2 - 1)\cot(3x^2 + 1) - 6x\csc^2(3x^2 + 1)\tan(3x^2 - 1)$

41. $f'(x) = \sec^2 x$

43. $f'(x) = -\frac{1}{1 + \sin x}$

45. $g'(x) = \frac{-6x\cos(3x^2 - 1)}{\sin^2(3x^2 - 1)}$

47. $g'(x) = \frac{-10x\cot(1 - 5x^2)}{\csc^2(1 - 5x^2)}$

49. $h'(x) = -\frac{6\sec^2(2x) - 3}{(\tan(2x) - x)^2}$

51. $h'(s) = 0$

53. $f'(x) = \frac{2(1 + x^2)\cos(2x) - 2x\sin(2x)}{(1 + x^2)^2}$

55. $f'(x) = -\frac{1}{x^2}\sec^2\left(\frac{1}{x}\right)$

57. $f'(x) = \frac{2(\sec x^2)[x(\tan x^2) - \tan x]}{\sec^2 x}$

59. $x = \frac{3}{2} + 3k, k \in \mathbf{Z}$

65. $f'(x) = \frac{x\cos\sqrt{x^2 + 1}}{\sqrt{x^2 + 1}}$

67. $f'(x) = (\cos\sqrt{3x^3 + 3x})\frac{9x^2 + 3}{2\sqrt{3x^3 + 3x}}$

69. $f'(x) = 4x\sin(x^2 - 1)\cos(x^2 - 1)$

71. $f'(x) = 27x^2 \tan^2(3x^3 - 3) \sec^2(3x^3 - 3)$

73. (a) $\frac{dc}{dt} = \frac{\pi}{2}\cos(\frac{\pi}{2}t)$

(b)

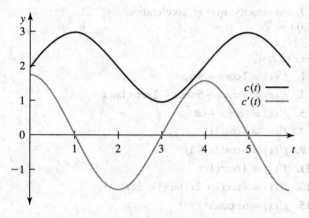

▲ **Figure 4.5.73**

(c) (i) $\frac{dc}{dt} = 0$, (ii) increasing, (iii) $c(t)$ has a horizontal tangent and either a maximum or a minimum.

Section 4.6

1. $f'(x) = 2e^{2x}$

3. $f'(x) = -12e^{1-3x}$

5. $f'(x) = (-4x + 3)e^{-2x^2+3x-1}$

7. $f'(x) = (21x^2 - \frac{1}{2\sqrt{x}})e^{7x^3-\sqrt{x}+3}$

9. $f'(x) = e^x(1 + x)$

11. $f'(x) = xe^{-x}(2 - x)$

13. $f'(x) = \frac{e^x(1+x^2-2x)-2x}{(1+x^2)^2}$

15. $f'(x) = \frac{2e^x-2e^{-x}-2}{(2+e^x)^2}$

17. $f'(x) = (\cos x)e^{\sin x}$

19. $f'(x) = e^{\sin(x^2-1)}\cos(x^2 - 1)2x$

21. $f'(x) = e^x\cos(e^x)$

23. $f'(x) = (2e^{2x} + 1)\cos(e^{2x} + x)$

25. $f'(x) = (1 - \cos x)\exp(x - \sin x)$

27. $g'(s) = \exp[\sec s^2](\sec s^2)(\tan s^2)(2s)$

29. $f'(x) = (\sin x + x\cos x)e^{x\sin x}$

31. $f'(x) = (-3)(2x + \sec^2 x)e^{x^2+\tan x}$

33. $f'(x) = (\ln 2)2^x$

35. $f'(x) = (\ln 2)2^{x+1}$

37. $f'(x) = 2(\ln 5)5^{2x-1}$

39. $f'(x) = 2x(\ln 2)2^{x^2+1}$

41. $h'(t) = (2t)(\ln 2)2^{t^2-1}$

43. $f'(x) = (\ln 2)\frac{1}{2\sqrt{x}}2^{\sqrt{x}}$

45. $f'(x) = (\ln 2)\frac{x}{\sqrt{x^2-1}}2^{\sqrt{x^2-1}}$

47. $h'(t) = \frac{\ln 5}{2\sqrt{t}}5^{\sqrt{t}}$

49. $g'(x) = -2(\ln 2)(\sin x)2^{2\cos x}$

51. $g'(r) = \frac{\ln 3}{5r^{1/5}}3^{r^{1/5}}$

53. 2

55. 0

57. $\frac{1}{\ln 2}$

59. (a) $N(0) = 1$

61. $\frac{dN}{dt} = (\ln 2)N(t)$ which implies that $\frac{dN}{dt}$ is proportional to $N(t)$

63. (a) $\frac{dN}{dt} = \frac{rK(\frac{K}{N(0)}-1)e^{-rt}}{[1+(\frac{K}{N(0)}-1)e^{-rt}]^2}$,

(c)

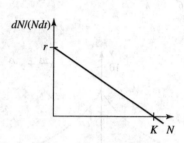

▲ **Figure 4.6.63**

65. (a)

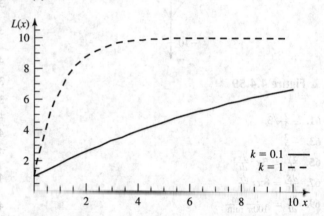

▲ **Figure 4.6.65**

(b) L_∞ is the limiting size and L_0 is the initial size.
(c) The fish with $k = 1$ reaches $L = 5$ more quickly.
(d) With age, the rate of growth decreases.
(e) The larger k, the more quickly the fish grows and reaches its limiting size.

67. $\frac{dW}{dt} = -4W(t)\frac{1}{\text{days}}$

69. $\frac{dW}{dt} = -\frac{1}{5}W(t)\frac{1}{\text{days}}$

71. (a) $W(4) = 6e^{-12}$, (b) half-life: $\frac{\ln 2}{3}$

73. (a) $\frac{dW}{dt} = -(\ln\frac{5}{2})W(t)$, (b) $W(3) = 5(\frac{5}{2})^{-3}$,

(c) half-life: $\frac{\ln 2}{\ln 5 - \ln 2}$

Section 4.7

1. $\frac{d}{dx}f^{-1}(x) = x$

3. $\frac{d}{dx}f^{-1}(x) = \frac{\sqrt{3}}{6\sqrt{x}}$

5. $f^{-1}(x) = \left(\frac{3-x}{2}\right)^{1/3}$, $x \le 3$, $\frac{d}{dx}f^{-1}(x) = -\frac{1}{6}\left(\frac{2}{3-x}\right)^{2/3}$

7. $\frac{d}{dx}f^{-1}(0) = \frac{1}{4}$

9. $\frac{d}{dx}f^{-1}(2) = 4$

11. $\frac{d}{dx}f^{-1}(1) = \frac{1}{2}$

13. $\frac{d}{dx}f^{-1}(\pi) = \frac{1}{2}$

15. $\frac{d}{dx}f^{-1}(0) = 1$

17. $\frac{d}{dx}f^{-1}(\frac{1}{2}) = \frac{2}{3}\sqrt{3}$

19. $\frac{d}{dx}f^{-1}(1) = 1$

21. $\frac{d}{dx}f^{-1}(1) = \frac{1}{2}$

23. $f'(x) = \frac{1}{x+1}$

25. $f'(x) = \frac{-2}{1-2x}$

27. $f'(x) = \frac{2}{x}$

29. $f'(x) = \frac{3}{x}$

31. $f'(x) = 2(\ln x)\frac{1}{x}$

33. $f'(x) = \frac{4(\ln x)^2}{x}$

35. $f'(x) = \frac{x}{x^2+1}$

37. $f'(x) = \frac{1}{x(x+1)}$

39. $f'(x) = \frac{-1}{1-x} - \frac{2}{1+2x}$

41. $f'(x) = (1 - \frac{1}{x})\exp[x - \ln x]$

43. $f'(x) = \cot x$

45. $f'(x) = \frac{2x\sec^2(x^2)}{\tan(x^2)}$

47. $f'(x) = \ln x + 1$

49. $f'(x) = \frac{1-\ln x}{x^2}$

51. $h'(t) = \cos(\ln(3t))\frac{1}{t}$

53. $f'(x) = \frac{2x}{x^2-3}$ for $|x| \ne \sqrt{3}$

55. $f'(x) = \frac{-2x}{(\ln 10)(1-x^2)}$

57. $f'(x) = \frac{3x^2-3}{(\ln 10)(x^3-3x)}$

59. $f'(u) = \frac{4u^3}{(\ln 3)(3+u^4)}$

63. $\frac{dy}{dx} = x^x(\ln x + 1)$

65. $\frac{dy}{dx} = (\ln x)^x[\ln(\ln x) + \frac{1}{\ln x}]$

67. $\frac{dy}{dx} = x^{\ln x}2(\ln x)\frac{1}{x}$

69. $\frac{dy}{dx} = x^{1/x-2}(1 - \ln x)$

71. $\frac{dy}{dx} = [x^x(\ln x + 1)\ln x + x^{x-1}]x^{x^x}$

73. $\frac{dy}{dx} = x^{\cos x}[\frac{\cos x}{x} - (\sin x)(\ln x)]$

75. $\frac{1}{y}\frac{dy}{dx} = 2 + \frac{27}{9x-2} - \frac{x}{2(x^2+1)} - \frac{9x^2}{4(3x^3-7)}$

Section 4.8

1. $\sqrt{65} \approx 8.0625$, error $= 2.42 \times 10^{-4}$

3. $\sqrt[3]{127} \approx \frac{374}{75}$, error $\approx 3.57 \times 10^{-5}$

5. $(0.99)^{25} \approx 0.75$, error ≈ 0.0278

7. $\sin\left(\frac{\pi}{2} + 0.02\right) \approx 1$, error $\approx 2.00 \times 10^{-4}$

9. $\ln(1.01) \approx 0.01$, error $\approx 4.97 \times 10^{-5}$

11. $L(x) = 1 - x$

13. $L(x) = -\frac{1}{4}x + \frac{3}{4}$

15. $L(x) = 1 - 2x$

17. $L(x) = 1 + x$

19. $L(x) = \frac{1}{e}x$

21. $L(x) = 1 + x$

23. $L(x) = 1 - x$

25. $L(x) = x$

27. $L(x) = 1 - nx$

29. $L(x) = x$

31. 100.3

33. $B(1.1) \approx 5.005$

35. $[1.8, 2.2]$

37. $[10.8, 13.2]$

39. $[5.91, 8.87]$

41. $\pm 6\%$

43. $\pm 0.668\%$

45. $\pm 9\%$

47. $\pm 2.4\%$

49. $\pm \frac{(a+b-2x)x}{(a-x)(b-x)}\left(100\frac{\Delta x}{x}\right)$

Section 4.10

1. $f'(x) = -12x^3 - \frac{1}{x^{3/2}}$

3. $h'(t) = \frac{1}{3}\left(\frac{1+t}{1-t}\right)^{2/3}\frac{-2}{(1+t)^2}$

5. $f'(x) = 2e^{2x}\sin\left(\frac{\pi}{2}x\right) + e^{2x}\frac{\pi}{2}\cos\left(\frac{\pi}{2}x\right)$

7. $f'(x) = \frac{\frac{1}{x+1}\ln x - \frac{1}{x}\ln(x+1)}{[\ln x]^2}$

9. $f'(x) = -xe^{-x^2/2}$, $f''(x) = e^{-x^2/2}(x^2 - 1)$

11. $h'(x) = \frac{1}{(x+1)^2}$, $h''(x) = -\frac{2}{(x+1)^3}$

13. $\frac{dy}{dx} = \frac{\cos x + y^2 - 2xy}{x^2 - 2xy}$

15. $\frac{dy}{dx} = 1 - 2(x - y)$

17. $\frac{dy}{dx} = -\frac{x}{y}$, $\frac{d^2y}{dx^2} = -\frac{16}{y^3}$

19. $\frac{dy}{dx} = \frac{1}{x\ln x}$, $\frac{d^2y}{dx^2} = -\frac{\ln x + 1}{(x\ln x)^2}$

21. $5.70\frac{\text{ft}}{\text{sec}}$

23. **(a)** $\frac{dy}{dx} = f'(x)e^{f(x)}$ **(b)** $\frac{dy}{dx} = \frac{f'(x)}{f(x)}$

(c) $\frac{dy}{dx} = 2f(x)f'(x)$

25. (a)

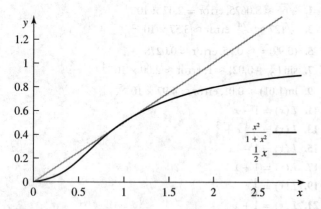

▲ **Figure 4.10.25**

(b) $c = 1$, $y = \frac{1}{2}x$

27. $y - \frac{1}{2}e^{-(\pi/3)^2} = -e^{-(\pi/3)^2}\left(\frac{1}{2}\sqrt{3} + \frac{\pi}{3}\right)\left(x - \frac{\pi}{3}\right)$

29. $y = x$

31. $p(x) = 2x^2 + 4x + 8$

33. (a) $s(5.5) \approx 54.84$ miles

(b) $v(t) = \frac{ds}{dt} = 3\pi + 3\pi \sin(\pi t)$, $a(t) = \frac{dv}{dt} = 3\pi^2 \cos(\pi t)$

(c) $v(t) \geq 0$ **(d)** Three valleys and three hills

35. (b) $N(t) = 100 + 50\cos\left(\frac{\pi}{2}t\right)$

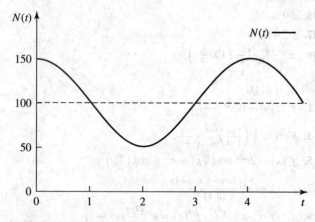

▲ **Figure 4.10.35**

(c) The population size shows oscillations.

37. 9.33%

Section 5.1

1. absolute minimum: $(0, -1)$; absolute maximum: $(1, 1)$

3. absolute minimum: $(\frac{3\pi}{2}, -1)$; absolute maximum: $(\frac{\pi}{2}, 1)$

5. absolute minimum: $(0, 0)$; absolute maximum: $(-1, 1)$ and $(1, 1)$

7. absolute minimum: $(2, e^{-2})$; absolute maximum: $(0, 1)$

9.

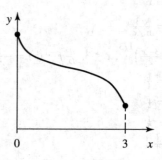

▲ **Figure 5.1.9**

11.

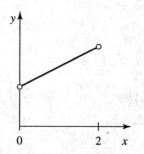

▲ **Figure 5.1.11**

13. local maximum = global maximum = $(-1, 3)$, no local and global minima

15. local minimum = global minimum = $(0, -2)$, local maximum = global maximum = $(-1, -1)$ and $(1, -1)$

17. local minimum = $(-2, -3)$ and $(1, 0)$, global minimum = $(-2, -3)$, local maximum = global maximum = $(0, 1)$

19. $f'(0) = 0$; $f(0) < f(x)$ for all $x \neq 0$

21. $f'(0) = 0$; $f(0) > f(x)$ for all $x \neq 0$

23. $f'(0) = 0$; $f(x) < 0$ for $x < 0$ and $f(x) > 0$ for $x > 0$

25. $f'(-2) = 0$; $f(x) < 0$ for $x < -2$ and $f(x) > 0$ for $x > -2$

27. $f(0) = 0$ and $f(x) > 0$ for $x \neq 0$; $\lim_{x \to 1^-} f'(x) = -1$, $\lim_{x \to 1^+} f'(x) = 1$.

29. $f(1) = f(-1) = 0$ and $f(x) > 0$ for $x \neq 1, -1$; $\lim_{x \to 1^-} f'(x) = -2 \neq \lim_{x \to 1^+} f'(x) = 2$ and $\lim_{x \to -1^-} f'(x) = -2 \neq \lim_{x \to -1^+} f'(x) = 2$.

31. loc min = glob min = $(-1, 0)$ and $(1, 0)$; loc max = glob max = $(0, 1)$ and $(2, 1)$

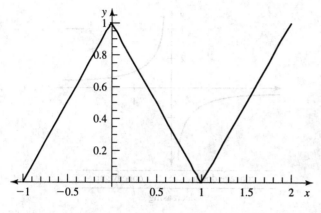

▲ **Figure 5.1.31**

45. (a) Slope: $a + b$, **(b)** $c = \frac{a+b}{2}$

47. Use the mean value theorem and argue that the slope of the secant line connecting $(a, f(a))$ and $(c, f(c))$ and the slope of the secant line connecting $(b, f(b))$ and $(c, f(c))$ have opposite signs, where $c \in (a, b)$ with $f(c) \neq 0$.

49. (a) 0.25 m/s, **(b)** $\frac{3}{100}t^2, 0 < t < 5$, **(c)** $t = \frac{5}{3}\sqrt{3}$ s

51. $0 \le B(3) \le 3$

53. $f(x) = 3, x \in \mathbf{R}$

Section 5.2

1. $y' = 3 - 2x$, $y'' = -2$; $y' > 0$ and y is increasing on $(-\infty, 3/2)$; $y' < 0$ and y is decreasing on $(3/2, \infty)$; $y'' < 0$ and y is concave down.

33. (a)

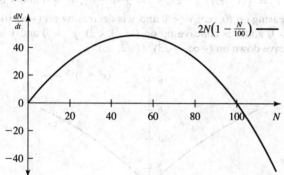

▲ **Figure 5.1.33**

$\frac{dN}{dt}$ is maximal for $N = 50$

(b) $f'(N) = r - \frac{2r}{K}N$

35. (a) Slope: 2,

(b) $x = 1$; guaranteed by the mean value theorem.

37. Slope: 0; $c = 0$

39. $[0, 1]$

41. $f(-1) = 1$, $f(2) = -2$, $\frac{f(2)-f(-1)}{2-(-1)} = -1$

43.

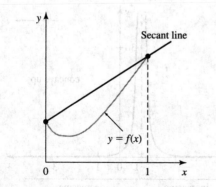

▲ **Figure 5.1.43**

The mean value theorem guarantees such a point.

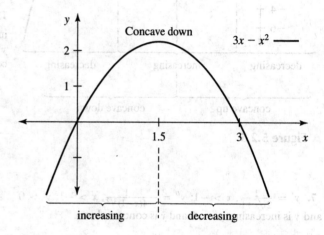

▲ **Figure 5.2.1**

3. $y' = 2x + 1$, $y'' = 2$; $y' > 0$ and y is increasing on $(-1/2, \infty)$; $y' < 0$ and y is decreasing on $(-\infty, -1/2)$; $y'' > 0$ and y is concave up.

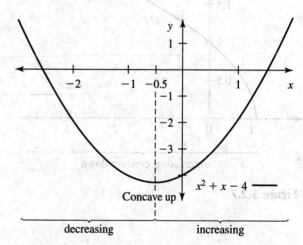

▲ **Figure 5.2.3**

5. $y' = -2x^2+7x-3$, $y'' = -4x+7$; $y' > 0$ and y is increasing on $(1/2, 3)$; $y' < 0$ and y is decreasing on $(-\infty, 1/2) \cup (3, \infty)$; $y'' > 0$ and y is concave up on $(-\infty, 7/4)$; $y'' < 0$ and y is concave down on $(7/4, \infty)$.

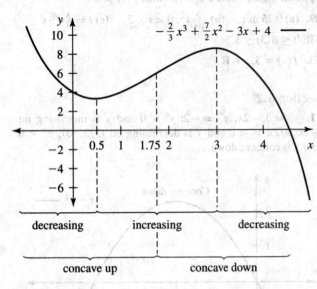

▲ Figure 5.2.5

7. $y' = \frac{1}{2\sqrt{x+1}}$, $x > -1$; $y'' = -\frac{1}{4(x-1)^{3/2}}$, $x > -1$; $y' > 0$ and y is increasing; $y'' < 0$ and y is concave down.

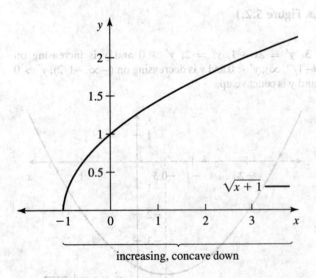

▲ Figure 5.2.7

9. $y' = -\frac{1}{x^2}$, $x \neq 0$; $y'' = \frac{2}{x^3}$, $x \neq 0$; $y' < 0$ and y is decreasing for $x \neq 0$; $y'' < 0$ and y is concave down for $x < 0$; $y'' > 0$ and y is concave up for $x > 0$.

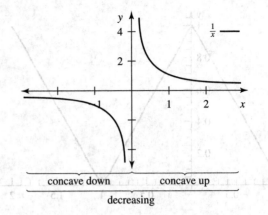

▲ Figure 5.2.9

11. $y' = \frac{1}{3}(x^2 + 1)^{-2/3}2x$, $y'' = \frac{6-2x^2}{9(x^2+1)^{5/3}}$; $y' > 0$ and y is increasing on $(0, \infty)$; $y' < 0$ and y is decreasing on $(-\infty, 0)$; $y'' > 0$ and y is concave up on $(-\sqrt{3}, \sqrt{3})$; $y'' < 0$ and y is concave down on $(-\infty, -\sqrt{3}) \cup (\sqrt{3}, \infty)$.

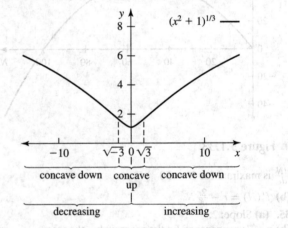

▲ Figure 5.2.11

13. $y' = -\frac{2}{(1+x)^3}$, $y'' = \frac{6}{(1+x)^4}$; $y' > 0$ and y is increasing on $(-\infty, -1)$; $y' < 0$ and y is decreasing on $(-1, \infty)$; $y'' > 0$ and y is concave up for $x \neq -1$.

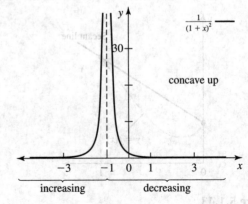

▲ Figure 5.2.13

15. $y' = \cos x$, $y'' = -\sin x$; $y' > 0$ and y is increasing on $(0, \pi/2) \cup (3\pi/2, 2\pi)$; $y' < 0$ and y is decreasing on $(\pi/2, 3\pi/2)$; $y'' > 0$ and y is concave up on $(\pi, 2\pi)$; $y'' < 0$ and y is concave down on $(0, \pi)$.

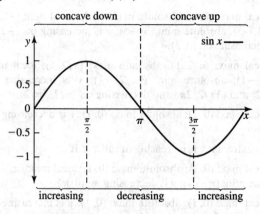

▲ **Figure 5.2.15**

17. $y' = e^x$, $y'' = e^x$; $y' > 0$ and y is increasing for $x \in \mathbf{R}$; $y'' > 0$ and y is concave up for $x \in \mathbf{R}$.

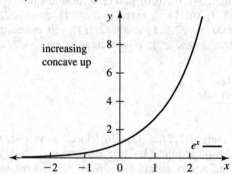

▲ **Figure 5.2.17**

19. $y' = -xe^{-x^2/2}$, $y'' = e^{-x^2/2}(x^2 - 1)$; $y' > 0$ and y is increasing for $x < 0$; $y' < 0$ and y is decreasing for $x > 0$; $y'' > 0$ and y is concave up for $x < -1$ and $x > 1$; $y'' < 0$ and y is concave down for $-1 < x < 1$.

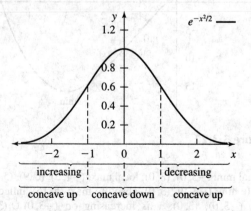

▲ **Figure 5.2.19**

21. (a)

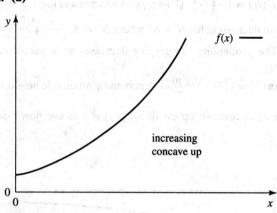

▲ **Figure 5.2.21a**

(b)

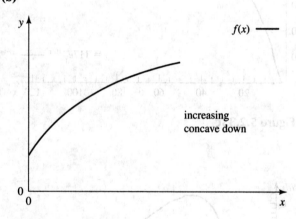

▲ **Figure 5.2.21b**

(c) In (a), $y' > 0$ and $y'' > 0$; in (b), $y' > 0$ and $y'' < 0$.
23. (b) $f(-1) > 0$ and $f(1) < 0$
25. $f'(x)$ is decreasing; use the definition of concave down.
27. (a)

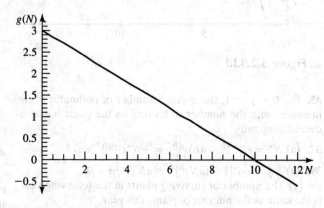

▲ **Figure 5.2.27**

(b) $g'(N) = -\frac{r}{K} < 0$

29. $f'(N) = 1 - \left(\frac{N}{K}\right)^{\theta}(1+\theta)$; $f(N)$ is increasing for $0 < N <$ $N^{\star}$ and decreasing for $N > N^{\star}$ where $N^{\star} = K\left(\frac{1}{1+\theta}\right)^{1/\theta}$.

31. The probability of escaping decreases with parasitoid density.

33. (a) $y' = 117e^{-10/x}\frac{10}{x^2} > 0$; maximum attainable height is 117.

(b) $y(x)$ is concave up on $(0, 5)$; $y(x)$ is concave down on $(5, \infty)$,

(c)

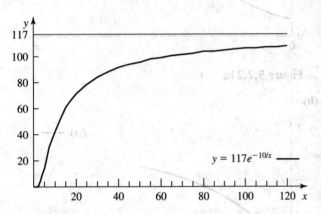

$y = 117e^{-10/x}$

▲ **Figure 5.2.33c**

(d)

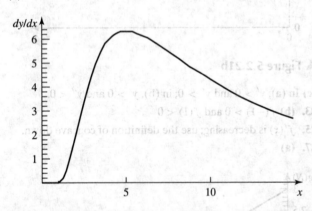

▲ **Figure 5.2.33d**

35. For $0 < \gamma < 1$, the average number of pollinator visits increases with the number of flowers on the plant but at a decelerating rate.

37. (a) $N^{\star} = e^{-aN^{\star}}$, **(b)** $\frac{dN^{\star}}{da} = -N^{\star}e^{-aN^{\star}} < 0$

39. (a) $\frac{dA}{dN} = -R[1 + (aN)^b]^{-2}b(aN)^{b-1}a < 0$,
(b) (v) The number of surviving plants in the following year is the same as the number of plants this year.

41. For $0 < a < 1$, Y is an increasing function of X but $\frac{Y}{X}$ is a decreasing function of X. $Y(X)$ is concave down.,
(b) Juveniles have relatively larger heads than adults.

43. $y = f(x)$ is concave up for $k > 1$ and concave down for $0 < k < 1$.

Section 5.3

1. local max: $(0, 4)$; absolute max: $(0, 4)$; local min: $(-2, 0)$ and $(3, -5)$; absolute min: $(3, -5)$; y is increasing on $(-2, 0)$ and decreasing on $(0, 3)$.

3. local max: $(\pi/2, 1)$; absolute max: $(\pi/2, 1)$; local min: $(3\pi/2, -1)$; absolute min: $(3\pi/2, -1)$; y is increasing on $(0, \pi/2)$ and $(3\pi/2, 2\pi)$ and decreasing on $(\pi/2, 3\pi/2)$.

5. local max: $(0, 1)$; absolute max: $(0, 1)$; y is decreasing on $[0, \infty)$

7. no extrema; y is increasing for all $x \in \mathbf{R}$.

9. local max: $(0, 1)$; absolute max: $(0, 1)$; local min: $(\pi, -1)$; absolute min: $(\pi, -1)$; y is decreasing on $(0, \pi)$.

11. local max: $(0, 1)$; absolute max: $(0, 1)$; y is increasing on $(-\infty, 0)$, and decreasing on $(0, \infty)$.

13. local max: $(-3, 15.5)$; local min: $(2, -16/3)$; y is increasing on $(-\infty, -3) \cup (2, \infty)$; y is decreasing on $(-3, 2)$.

15. local min: $(-1, 0)$, $(0, 0)$, and $(1, 0)$; absolute min: $(-1, 0)$, $(0, 0)$, and $(1, 0)$; local max: $(-\sqrt{2}/2, 1)$ and $(\sqrt{2}/2, 1)$; absolute max: $(-\sqrt{2}/2, 1)$ and $(\sqrt{2}/2, 1)$; y is increasing on $(-1, -\sqrt{2}/2) \cup (0, \sqrt{2}/2)$; y is decreasing on $(-\sqrt{2}/2, 0) \cup (\sqrt{2}/2, 1)$.

19. $(0, -2)$

21. $(\sqrt{2}/2, e^{-1/2})$

23. $(0, 0)$

27. local min: $(-2, 2/3)$, $(3, -16)$; local max: $(-1, 16/3)$, $(5, 16/3)$; absolute min: $(3, -16)$; absolute max: $(-1, 16/3)$, $(5, 16/3)$; inflection point: $(1, -16/3)$ y is increasing on $(-2, -1) \cup (3, 5)$; y is decreasing on $(-1, 3)$; y is concave up on $(1, 5)$; y is concave down on $(-2, 1)$;

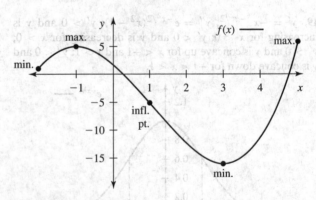

▲ **Figure 5.3.27**

29. local min: $(-3, 0)$, $(3, 0)$; local max: $(-4, 5)$, $(0, 9)$, $(5, 16)$; absolute min: $(-3, 0)$, $(3, 0)$; absolute max: $(5, 16)$; inflection points: $(-3, 0)$, $(3, 0)$; y is increasing on $(-3, 0) \cup (3, 5)$; y is decreasing on $(-4, -3) \cup (0, 3)$; y is concave up on $(-4, -3) \cup (3, 5)$; y is concave down on $(-3, 3)$;

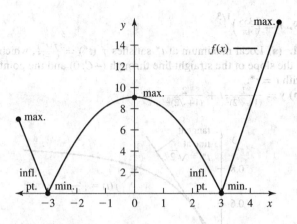

▲ **Figure 5.3.29**

31. no extrema; inflection points: $x = \pi/2 + k\pi$, $k \in \mathbf{Z}$; y is increasing on $\mathbf{R}$; y is concave up on $(\pi/2 + 2k\pi, 3\pi/2 + 2k\pi)$, $k \in \mathbf{Z}$; y is concave down on $(-\pi/2 + 2k\pi, \pi/2 + 2k\pi)$, $k \in \mathbf{Z}$;

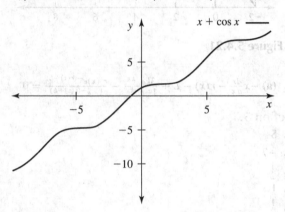

▲ **Figure 5.3.31**

33. local min: $(0, -1)$; absolute min: $(0, -1)$; inflection points at $x = -1/\sqrt{3}$ and $x = 1/\sqrt{3}$; y is increasing on $(0, \infty)$; y is decreasing on $(-\infty, 0)$; y is concave up on $(-1/\sqrt{3}, 1/\sqrt{3})$; y is concave down on $(-\infty, -1/\sqrt{3}) \cup (1/\sqrt{3}, \infty)$;

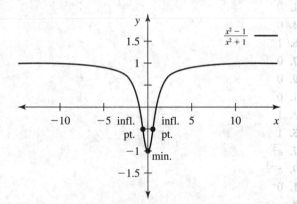

▲ **Figure 5.3.33**

35. **(c)** decreasing for all $x \neq 1$; no extrema;

(d) $f(x)$ is concave down on $(-\infty, 1)$ and concave up on $(1, \infty)$; no inflection points;

(e)

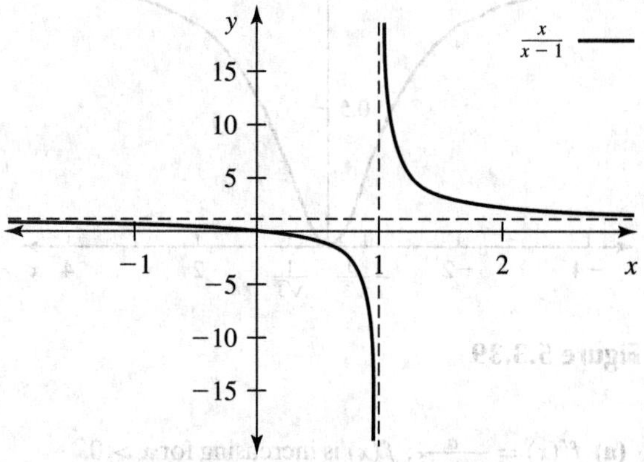

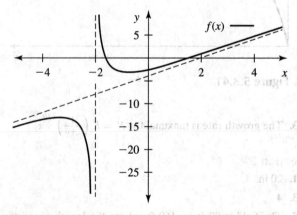

▲ **Figure 5.3.35**

37. **(a)** $\lim_{x \to -2^+} f(x) = \infty$, $\lim_{x \to -2^-} f(x) = -\infty$;

(b) $f(x)$ is increasing on $(-\infty, -2 - \sqrt{6}/2) \cup (-2 + \sqrt{6}/2, \infty)$; $f(x)$ is decreasing on $(-2 - \sqrt{6}/2, -2) \cup (-2, -2 + \sqrt{6}/2)$; local max at $x = -2 - \sqrt{6}/2$; local min at $-2 + \sqrt{6}/2$;

(c) $f(x)$ is concave up on $(-2, \infty)$; $f(x)$ is concave down on $(-\infty, -2)$;

(d) $y = 2x - 4$;

(e)

▲ **Figure 5.3.37**

39. **(a)** $f'(x) = \frac{2x}{(1+x^2)^2}$ and $f(x)$ is increasing for $x > 0$ and decreasing for $x < 0$,

(b) $f(x)$ is concave up for $-1/\sqrt{3} < x < 1/\sqrt{3}$ and concave down for $x < -1/\sqrt{3}$ or $x > 1/\sqrt{3}$. $f(x)$ has two inflection points at $x = \pm 1/\sqrt{3}$,

(c) $\lim_{x \to \infty} = 1$, $\lim_{x \to -\infty} = 1$,

(d)

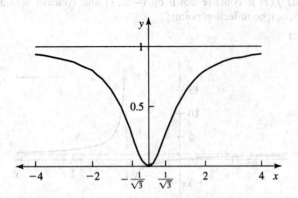

▲ **Figure 5.3.39**

41. (a) $f'(x) = \frac{a}{(a+x)^2}$; $f(x)$ is increasing for $x > 0$,

(b) $f''(x) = -\frac{2a}{(a+x)^3}$; there are no inflection points,

(c) $\lim_{x\to\infty} \frac{x}{a+x} = 1$; there is a horizontal asymptote at $y = 1$,

(d)

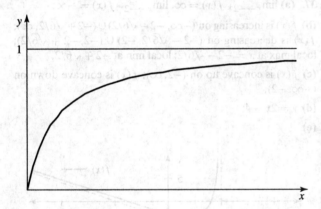

▲ **Figure 5.3.41**

43. The growth rate is maximal for $N = K\left(\frac{1}{1+\theta}\right)^{1/\theta}$.

Section 5.4

1. 20 in.

3. 4

5. The field is 80 ft by 160 ft, where the length along the river is 160 ft.

7. $5(1 + \sqrt{2})$

9. 4

11. (b) $(6/5, 2/5)$ **(c)** local minimum at $x = 6/5$ as in **(b)**.

13. $g'(x) = 2f(x)f'(x)$ has the same sign change as $f'(x)$.

15. The height is equal to the diameter, namely $2\left(\frac{500}{\pi}\right)^{1/3}$.

17. (a) $r = \sqrt{2}, \theta = 2$; **(b)** $r = \sqrt{10}, \theta = 2$

19. $r = \left(\frac{355}{4\pi}\right)^{1/3}$

21. (a) Local maximum at t^* satisfies $f'(t^*) = \frac{f(t^*)}{C+t^*}$, which is the slope of the straight line through $(-C, 0)$ and the point with $t = t^*$.

(b) $y = \frac{1}{(1+\sqrt{2})^2}t + \frac{2}{(1+\sqrt{2})^2}$

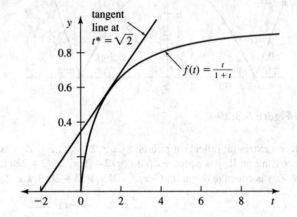

▲ **Figure 5.4.21**

23. (a) $-x\frac{dr}{dx} - r(x) - L + \frac{3ke^{-kx}}{1-e^{-kx}} - \frac{r'(x)e^{-(r(x)+L)}}{1-e^{-(r(x)+L)}} = 0$

Section 5.5

1. 8

3. -7

5. $\frac{1}{2}$

7. 1

9. $\frac{1}{2}$

11. ∞

13. $\frac{\ln 2}{\ln 3}$

15. $\frac{1}{2}$

17. 0

19. ∞

21. 0

23. 0

25. 1

27. 0

29. 0

31. 0

33. 1

35. 1

37. e^3

39. e^{-1}

41. 0

43. ∞

45. 3

47. 0

49. 1

51. $\frac{\ln a}{\ln b}$

53. $\lim_{x\to\infty}\left(1+\frac{c}{x^p}\right)=\begin{cases}\infty & \text{if }0<p<1\\e^c & \text{if }p=1\\1 & \text{if }p>1\end{cases}$

57. Let $y=121e^{-17/x}$. **(a)** $\lim_{x\to0+}\frac{dy}{dx}=0$, $\lim_{x\to\infty}\frac{dy}{dx}=0$
(b) $\frac{17}{2}$ **(c)** Height is increasing at an accelerating rate for $0<x<\frac{17}{2}$ and at a decelerating rate for $x>\frac{17}{2}$.
(d)

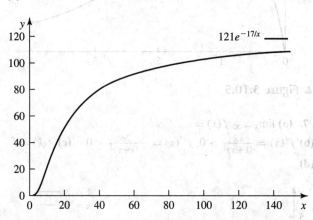

▲ **Figure 5.5.57i**

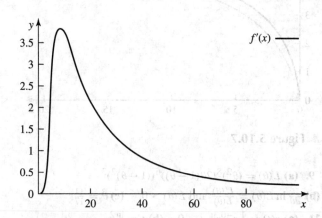

▲ **Figure 5.5.57ii**

Section 5.6

1. (a) $N_5=(10)(1.03)^5$, **(b)** $t=\frac{\ln 2}{\ln 1.03}$

3. (a) $b=1.02$, **(b)** $N_{10}=(20)(1.02)^{10}$, **(c)** $t=\frac{\ln 2}{\ln 1.02}$

5. (a) $b=1+\frac{x}{100}$,
(b) $t=\frac{\ln 2}{\ln(1+x/100)}$; 693.5, 139.0, 70.0, 35.0, 14.2, 7.3

7. (a) 0, **(b)** stable

9. (a) $x=1/2$ and $x=-2$,
(b) $x=1/2$ is locally stable, $x=-2$ is unstable.

11. $x=0$ is unstable, $x=0.5$ is locally stable.

13. (a) $x=0$ is locally stable, $x=1$ is unstable, $x=4$ is locally stable, **(b) (i)** 0, **(ii)** 4

15. (b) 0, **(c)** $1/\beta$, **(d)** $P=2/\beta$ is an inflection point,
(e)

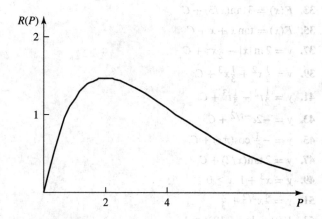

▲ **Figure 5.6.15**

17. (b) To find equilibria, solve $N=10Ne^{-0.01N}$; $N=0$ is another equilibrium.
(c) Oscillations seem to appear and the system does not seem to converge to the nontrivial equilibrium.

19. Equilibria: $N=0$ and $N=50$.
(b) Starting from $N=10$, it appears that the limiting population size is 50. $N=50$ is locally stable.

Section 5.7

1. $\sqrt{7}=2.645751$

3. 0.6529186

5. 1.895494

7. (a) $|x_n|=2^n x_0$, **(b)** ∞

9. (a) $x_0=3$, $x_1=4.166667$, $x_2=4.003333$, $x_3=4.000001$,
(b) $x_0=x_1=x_2=\cdots=4$

Section 5.8

1. $F(x)=\frac{2}{3}x^3+C$

3. $F(x)=\frac{1}{3}x^3+x^2-x+C$

5. $F(x)=\frac{1}{5}x^5-x^3+x+C$

7. $F(x)=x^4-x^2+3x+C$

9. $F(x)=x+\ln|x|+C$

11. $F(x)=x+\frac{1}{x}+C$

13. $F(x)=\ln|1+x|+C$

15. $F(x)=x^5-\frac{5}{3x^3}+C$

17. $F(x)=\ln|2+x|+C$

19. $F(x)=-\frac{1}{3}e^{-3x}+C$

21. $F(x)=e^{2x}+C$

23. $F(x)=-\frac{1}{2}e^{-2x}+C$

25. $F(x)=-\frac{1}{2}\cos(2x)+C$

27. $F(x) = -3\cos(x/3) + C$

29. $F(x) = -\frac{4}{\pi}\cos(\pi x/2) - \frac{6}{\pi}\sin(\pi x/2) + C$

31. $F(x) = \frac{1}{2}\tan(2x) + C$

33. $F(x) = 3\tan(x/3) + C$

35. $F(x) = \tan x + x + C$

37. $y = 2\ln|x| - \frac{1}{2}x^2 + C$

39. $y = \frac{1}{2}x^2 + \frac{1}{3}x^3 + C$

41. $y = \frac{1}{2}t^2 - \frac{1}{3}t^3 + C$

43. $y = -2e^{-t/2} + C$

45. $y = -\frac{1}{\pi}\cos(\pi s) + C$

47. $y = 2\tan(x/2) + C$

49. $y = x^3 + 1,\ x \ge 0$

51. $y = \frac{4}{3}x^{3/2} + \frac{2}{3}$

53. $N(t) = \ln t + 10,\ t \ge 1$

55. $W(t) = e^t,\ t \ge 0$

57. $W(t) = 1 - \frac{1}{3}e^{-3t}$

59. $T(t) = 3 + \frac{1}{\pi} - \frac{1}{\pi}\cos(\pi t)$

61. $y = \frac{e^x - e^{-x}}{2}$

63. $L(x) = 25 - 10e^{-0.1x},\ L(0) = 15$

65. $t = 2.5$ s, $v(2.5) = 80\ \frac{\text{m}}{\text{s}}$

67. (a) The first term on the right-hand side describes evaporation; the second one describes watering.
(b) $a \approx 0.0417$

Section 5.10

1. (b) Absolute maximum at $(1, e^{-1})$;   **(c)** inflection point at $(2, 2e^{-2})$;
(d)

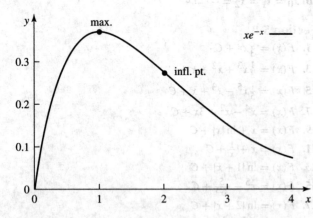

▲ **Figure 5.10.1**

3. (a) $f'(x) = \frac{4}{(e^x + e^{-x})^2} > 0$, hence, $f(x)$ is strictly increasing.

5. (e)

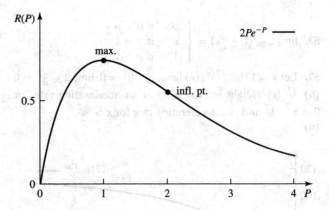

▲ **Figure 5.10.5**

7. (a) $\lim_{x\to\infty} f(x) = c$

(b) $f'(x) = \frac{ck}{(k+x)^2} > 0,\ f''(x) = -\frac{2ck}{(k+x)^3} < 0$ **(c)** $f(k) = \frac{c}{2}$

(d)

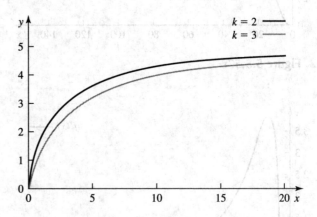

▲ **Figure 5.10.7**

9. (a) $L(\theta) = (\theta^2)^8(2\theta(1-\theta))^6((1-\theta)^2)^3$

(b) $\frac{d}{d\theta}\ln L(\theta) = \frac{L'(\theta)}{L(\theta)}$ and $L(\theta) > 0$.   **(c)** $\hat{\theta} = \frac{11}{17}$

11. (a) $c(t) = \frac{1}{3}e^{-0.3t},\ t \ge 0$ **(b)** $t = \frac{\ln 2}{0.3}$

13. (a) $t = v_0/g,$ **(b)** $\frac{v_0^2}{2g},$ **(c)** 0,   **(d)** $t = \frac{2v_0}{g}$

Section 6.1

1. 0.21875

3. 0.46875

5. $\sqrt{1} + \sqrt{2} + \sqrt{3} + \sqrt{4}$

7. $2^0 + 2^1 + 2^2 + 2^3 + 2^4 + 2^5$

9. $x^0 + x^1 + x^2 + x^3 + x^4$

11. $(-1)^0 + (-1)^1 + (-1)^2 + (-1)^3$

13. $\left(\frac{1}{n}\right)^2\frac{1}{n} + \left(\frac{2}{n}\right)^2\frac{1}{n} + \left(\frac{3}{n}\right)^2\frac{1}{n} + \cdots + \left(\frac{n}{n}\right)^2\frac{1}{n}$

15. $\sum_{k=1}^{6} k$

17. $\sum_{k=2}^{5} \ln k$

19. $\sum_{k=1}^{n} 2k$

21. $\sum_{k=1}^{n} q^{k-1}$

23. 670

25. 112

27. $\frac{n(n+1)(2n+1)}{3}$

29. 0

33. 1.36

35. 14

37. 0

39. $\int_{a}^{b} x \, dx = \frac{1}{2}b^2 - \frac{1}{2}a^2$

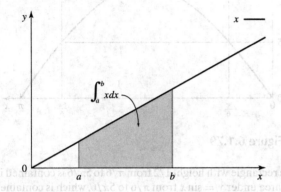

▲ **Figure 6.1.39**

41. $\int_{0}^{1} x^2 \, dx$

43. $\int_{-3}^{2} (2x - 1) \, dx$

45. $\int_{-3}^{3} (x - 1)(x + 2) \, dx$

47. $\int_{-5}^{2} e^x \, dx$

49. $\lim_{\|P\|\to 0} \sum_{k=1}^{n} \sqrt{c_k + 1} \Delta x_k$ where $\|P\|$ is a partition of $[2, 6]$, $c_k \in [x_{k-1}, x_k]$ and $\Delta x_k = x_k - x_{k-1}$

51. $\lim_{\|P\|\to 0} \sum_{k=1}^{n} \ln c_k \Delta x_k$ where $\|P\|$ is a partition of $[1, e]$, $c_k \in [x_{k-1}, x_k]$ and $\Delta x_k = x_k - x_{k-1}$

53. $\lim_{\|P\|\to 0} \sum_{k=1}^{n} g(c_k) \Delta x_k$ where $\|P\|$ is a partition of $[0, 5]$, $c_k \in [x_{k-1}, x_k]$ and $\Delta x_k = x_k - x_{k-1}$

55.

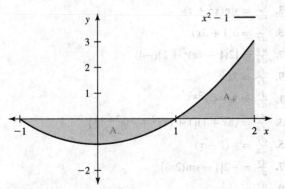

▲ **Figure 6.1.55**

$\int_{-1}^{2} (x^2 - 1) \, dx = -A_- + A_+$

57.

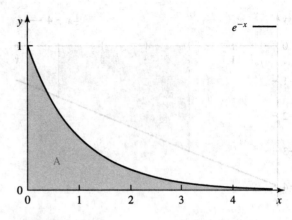

▲ **Figure 6.1.57**

$\int_{0}^{5} e^{-x} \, dx = A$

59.

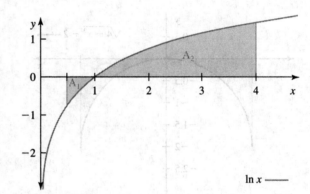

▲ **Figure 6.1.59**

$\int_{1/2}^{4} \ln x \, dx = A_2 - A_1$

61.

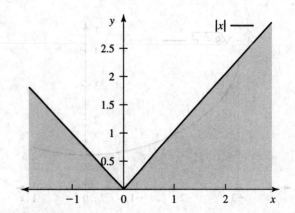

▲ **Figure 6.1.61**

$\int_{-2}^{3} |x| \, dx = \frac{13}{2}$

63.

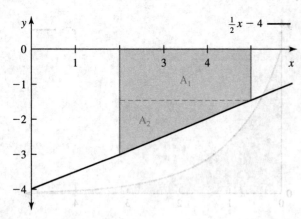

▲ Figure 6.1.63

$\int_2^5 (\frac{1}{2}x - 4)\, dx = -\frac{27}{4}$

65.

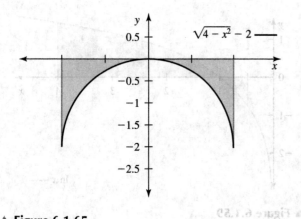

▲ Figure 6.1.65

$\int_{-2}^2 \left(\sqrt{4 - x^2} - 2\right) dx = 2\pi - 8$

67.

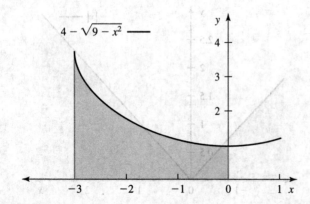

▲ Figure 6.1.67

$\int_{-3}^0 \left(4 - \sqrt{9 - x^2}\right) dx = 12 - \frac{9}{4}\pi$

69. 0

71. 4

73. 0

75. $x \geq x^2$ for $0 \leq x \leq 1$

77. $\sqrt{x} \geq 0$ for $x \geq 0$ and $\sqrt{x} \leq 2$ for $0 \leq x \leq 4$

79.

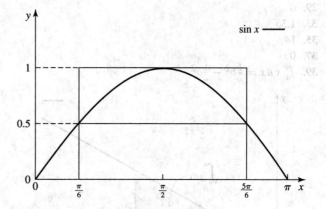

▲ Figure 6.1.79

The rectangle with height $1/2$ from $\pi/6$ to $5\pi/6$ is contained in the area under $y = \sin x$ from $\pi/6$ to $5\pi/6$, which is contained in the rectangle with height 1 from $\pi/6$ to $5\pi/6$.

81. $a = \pi/2$

83. $a = 3$

Section 6.2

1. $\frac{dy}{dx} = x^2$

3. $\frac{dy}{dx} = 4x - 3$

5. $\frac{dy}{dx} = \sqrt{2 + x}$

7. $\frac{dy}{dx} = \sqrt{1 + \sin^2 x}$

9. $\frac{dy}{dx} = xe^{4x}$

11. $\frac{dy}{dx} = \frac{1}{x+3}$

13. $\frac{dy}{dx} = \sin(x^2 + 1)$

15. $\frac{dy}{dx} = 3(1 + 3x)$

17. $\frac{dy}{dx} = [2(1 - 4x)^2 + 1](-4)$

19. $\frac{dy}{dx} = 2x^2$

21. $\frac{dy}{dx} = 3(1 + e^{3x})$

23. $\frac{dy}{dx} = (6x + 1)[1 + (3x^2 + x)e^{3x^2+x}]$

25. $\frac{dy}{dx} = -(1 + x)$

27. $\frac{dy}{dx} = -2[1 - \sin(2x)]$

29. $\frac{dy}{dx} = -\frac{1}{x^2}$

31. $\frac{dy}{dx} = -2x \sec(x^2)$

33. $\frac{dy}{dx} = 2[1 + (2x)^2] - (1 + x^2)$

35. $\frac{dy}{dx} = 9x^2 \ln x - 4x \ln x$

37. $\frac{dy}{dx} = (1 + 3x^2) \sin(x + x^3) + 4x \sin(2 - x^2)$

39. $x - \frac{1}{3}x^3 + C$

41. $x^3 - x^2 + C$

43. $\frac{1}{6}x^3 + \frac{3}{2}x^2 - \frac{1}{3}x + C$

45. $x^{3/2} \left(\frac{4}{5}x - \frac{2}{3} \right) + C$

47. $\frac{2}{7}x^{7/2} + C$

49. $\frac{2}{9}x^{9/2} + \frac{7}{9}x^{9/7} + C$

51. $\frac{2}{3}x^{3/2} + 2\sqrt{x} + C$

53. $\frac{1}{3}x^3 - x + C$

55. $-\frac{1}{3}x^3 + \frac{5}{2}x^2 - 6x + C$

57. $\frac{1}{2}e^{2x} + C$

59. $-3e^{-x} + C$

61. $-e^{-x^2/2} + C$

63. $-\frac{1}{2}\cos(2x) + C$

65. $\frac{1}{3}\sin(3x) + C$

67. $\frac{1}{3}\tan(3x) + C$

69. $\sec x + C$

71. $\frac{1}{2}\ln|\sec(2x)| + C$

73. $\tan x + \ln|\sec x| + C$

75. $4\tan^{-1} x + C$

77. $\sin^{-1} x + C$

79. $\ln|x + 1| + C$

81. $x - \ln|x| + C$

83. $\ln|x - 3| + C$

85. $-\ln|x + 3| + C$

87. $5(x - \tan^{-1} x) + C$

89. $\frac{3^x}{\ln 3} + C$

91. $-\frac{2^{-x}}{\ln 2} + C$

93. $\frac{1}{3}x^3 + \frac{2^x}{\ln 2} + C$

95. $\frac{2}{3}x^{3/2} + 2e^{x/2} + C$

97. -6

99. $-\frac{1}{3}$

101. 3

103. $\frac{28}{3}$

105. $\frac{1}{2}$

107. $\frac{1}{2}$

109. $\frac{\pi}{4}$

111. $\frac{\pi}{6}$

113. $\frac{1}{2}\ln 2$

115. $\frac{1}{3}(1 - e^{-3})$

117. 1

119. 1

121. $\ln \frac{3}{2}$

123. $\frac{1}{2}$

125. $f(x) = 4x$

Section 6.3

1. $e^2 + 1$

3. $\frac{9}{2}$

5. $\frac{4}{3}$

7. $\frac{14}{3} - \ln 4$

9. $2\ln 2 - \ln 3$

11. $\frac{3}{2}$

13. $\frac{2}{3}$

15. $\frac{16}{3}$

17. (a) $N(t) = 101 - e^{-t}$ (b) $1 - e^{-5}$
(c) $N(5) - N(0) = \int_0^5 e^{-t}\, dt$; shaded area.

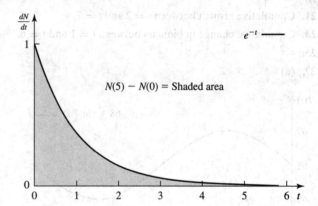

▲ **Figure 6.3.17**

19. (a)

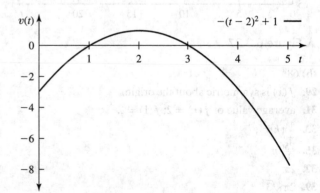

▲ **Figure 6.3.19(a)**

(b) Particle moves to the left for $0 \le t \le 1$ and $3 \le t \le 5$, it moves to the right for $1 \le t \le 3$.
(c) $s(t) = 2t^2 - \frac{1}{3}t^3 - 3t$; signed area between $v(u)$ and the horizontal axis from 0 to t

(d)

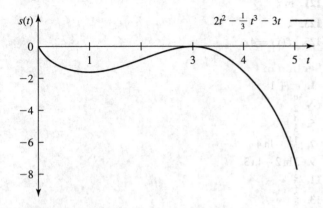

▲ **Figure 6.3.19(d)**

The rightmost position: $s(0) = s(3) = 0$; the leftmost position: $s(5) = -\frac{20}{3}$.

21. Cumulative growth between $t = 2$ and $t = 7$.

23. Cumulative change in biomass between $t = 1$ and $t = 6$.

25. $-\frac{2}{3}$

27. (a)

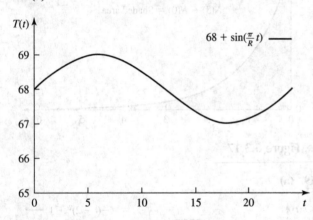

▲ **Figure 6.3.27**

(b) 68

29. $f(x)$ is symmetric about the origin.

31. average value of $f(x) = 2$; $f(1) = 2$

33. $\frac{1}{3}\pi h r^2$

35. $\frac{256}{15}\pi$

37. 2π

39. $2\pi\sqrt{3}$

41. $\frac{2}{15}\pi$

43. $\frac{\pi}{2}\left(e^4 + e^{-4} - 2\right)$

45. $\pi\left(\frac{\pi}{2} - 1\right)$

47. $\frac{32}{5}\pi$

49. $\pi \ln 3$

51. $\frac{3}{10}\pi$

53. (a) $2\sqrt{5}$ **(b)** $2\sqrt{5}$

55. $\frac{8}{27}\left[(10)^{3/2} - \left(\frac{13}{4}\right)^{3/2}\right]$

57. $\frac{14}{3}$

59. $\int_{-1}^{1} \sqrt{1 + 4x^2}\, dx$

61. $\int_{0}^{1} \sqrt{1 + e^{-2x}}\, dx$

63. (a) $\frac{\pi}{2}$ **(b)** $\frac{\pi}{2}$

65. $f'(a) = \frac{1}{2}(e^a - e^{-a})$

Section 6.5

1. $3.38\ \frac{m^3}{s}$

3. (a)

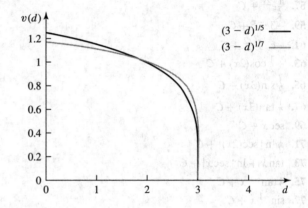

▲ **Figure 6.5.3**

(b) $v(D) = 0$, $v'(d) < 0$ for $d \in [0, D)$

5. (a) $\bar{v} = v(d_1)$

Section 7.1

1. $\frac{2}{3}(x^2 + 3)^{3/2} + C$

3. $-\frac{1}{3}(1 - x^2)^{3/2} + C$

5. $\frac{5}{3}\sin(3x) + C$

7. $-\frac{7}{12}\cos(4x^3) + C$

9. $\frac{1}{2}e^{2x+3} + C$

11. $-e^{-x^2/2} + C$

13. $\frac{1}{2}\ln|x^2 + 4x| + C$

15. $3\ln|x + 4| + C$

17. $\frac{2}{3}(x + 3)^{3/2} + C$

19. $\frac{2}{3}(2x^2 - 3x + 2)^{3/2} + C$

21. $-\ln|1 + x - 2x^2| + C$

23. $\frac{1}{4}\ln|1 + 2x^2| + C$

25. $e^{x^2} + C$

27. $e^{1+\ln x} + C$

29. $-\frac{2}{3\pi}\cos\left(\frac{3\pi}{2}x + \frac{\pi}{4}\right) + C$

31. $\frac{1}{2}\tan^2 x + C$

33. $\frac{1}{3}(\ln x)^3 + C$

35. $\frac{2}{5}(5+x)^{5/2} - \frac{10}{3}(5+x)^{3/2} + C$

37. $\ln|ax^2 + bx + c| + C$

39. $\frac{1}{n+1}[g(x)]^{n+1} + C$

41. $-e^{-g(x)} + C$

43. $\frac{1}{3}(10^{3/2} - 1)$

45. $\frac{7}{2025}$

47. $-e^{-9/2} + 1$

49. $\frac{3}{8}$

51. $\frac{1}{2}$

53. $4 + 3\ln 3$

55. $\frac{1}{2}$

57. $2(e^{-1} - e^{-3})$

59. $\ln|\sin x| + C$

Section 7.2

1. $x\sin x + \cos x + C$

3. $\frac{2}{3}x\sin(3x) + \frac{2}{9}\cos(3x) + C$

5. $-2x\cos x + 2\sin x + C$

7. $xe^x - e^x + C$

9. $x^2 e^x - 2xe^x + 2e^x + C$

11. $\frac{1}{2}x^2 \ln x - \frac{1}{4}x^2 + C$

13. $\frac{1}{2}x^2 \ln(3x) - \frac{1}{4}x^2 + C$

15. $x\tan x + \ln|\cos x| + C$

17. $\frac{1}{2}\left(\sqrt{3} - \frac{\pi}{3}\right)$

19. $2\ln 2 - 1$

21. $2\ln 4 - \frac{3}{2}$

23. $1 - 2e^{-1}$

25. $\frac{1}{2} + \frac{1}{4}(\sqrt{3} - 1)e^{\pi/3}$

27. $\frac{2e^{-3x}}{36+\pi^2}\left[\pi\sin\left(\frac{\pi}{2}x\right) - 6\cos\left(\frac{\pi}{2}x\right)\right] + C$

29. $\frac{1}{2}x[\sin(\ln x) - \cos(\ln x)] + C$

31. $\int \cos^2 x\, dx = \frac{1}{2}\sin x\cos x + \frac{1}{2}x + C$

33. (b) $\int \arcsin x\, dx = x\arcsin x + \sqrt{1 - x^2} + C$

35. (b) $\frac{1}{2}(\ln x)^2 + C$

37. (b) $-\frac{1}{3}x^2 e^{-3x} - \frac{2}{9}e^{-3x} - \frac{2}{27}e^{-3x} + C$

39. $2\sqrt{x}\sin(\sqrt{x}) + 2\cos(\sqrt{x}) + C$

41. $-e^{-x^2/2}(2 + x^2) + C$

43. $\sin x\, e^{\sin x} - e^{\sin x} + C$

45. $2\sqrt{x}e^{\sqrt{x}} - 2e^{\sqrt{x}} + C$

47. $(\sqrt{x} + 1)^2 \ln(\sqrt{x} + 1) - \frac{1}{2}(\sqrt{x} + 1)^2 - 2(\sqrt{x} + 1)$
$\ln(\sqrt{x} + 1) + 2(\sqrt{x} + 1) + C$

Section 7.3

1. $-\frac{1}{2}xe^{-2x} - \frac{1}{4}e^{-2x} + C$

3. $\ln|\sin x| + C$

5. $-\cos(x^2) + C$

7. $\frac{1}{4}\tan^{-1}\left(\frac{x}{4}\right) + C$

9. $x - 3\ln|x + 3| + C$

11. $\frac{1}{2}\ln|x^2 + 3| + C$

13. $\int \ln x\, dx = x\ln x - x + C$

15. $\frac{4}{9}(x - 2)^{9/4} + \frac{8}{5}(x - 2)^{5/4} + C$

17. $2e^2$

19. $\frac{\pi}{2}$

21. $\frac{1}{2}$

23. (a) $\frac{1}{x(x-2)} = \frac{1}{2}\left[\frac{1}{x-2} - \frac{1}{x}\right]$

(b) $\frac{1}{2}\ln\left|\frac{x-2}{x}\right| + C$

25. $\frac{1}{3}\tan^{-1}\left(\frac{x-2}{3}\right) + C$

27. $\frac{1}{5}\ln\left|\frac{x-3}{x+2}\right| + C$

29. $\frac{1}{6}\ln\left|\frac{x-3}{x+3}\right| + C$

31. $\frac{1}{3}\ln\left|\frac{x-2}{x+1}\right| + C$

33. $x - 5\ln|x + 2| + 2\ln|x + 1| + C$

35. $x + 2\ln\left|\frac{x-2}{x+2}\right| + C$

37. $2 - \ln 5 + \ln 3$

39. $\frac{1}{2}\ln 2$

41. $-\ln 2$

43. $\frac{\pi}{4} - \frac{1}{2}\ln 2$

45. $\frac{1}{1+x} + \ln\left|\frac{x}{x+1}\right|$

47. $-\frac{2}{x+1} + \ln\left|\frac{x+1}{x-1}\right|$

49. $\frac{1}{108}\ln\left|\frac{x+3}{x-3}\right| - \frac{1}{36}\left(\frac{1}{x+3} + \frac{1}{x-3}\right) + C$

51. $-\frac{1}{x} - \tan^{-1} x$

Section 7.4

1. infinite interval; 1

3. infinite interval; π

5. infinite interval; 2

7. infinite interval; 2

9. infinite interval; 0

11. integrand discontinuous; 6

13. integrand discontinuous; 2

15. integrand discontinuous at $x = 0$; -2

17. infinite interval; $\frac{1}{2}$

19. integrand discontinuous at $x = 0$; integral divergent

21. integrand discontinuous at $x = 1$; 0

23. infinite interval; integral divergent

25. infinite interval; integral divergent

27. integrand discontinuous at $x = \pm 1$; 0

29. integrand discontinuous at $x = \pm 1$; integral divergent

31. $c = 3$

35. **(b)** $0 \le \int_1^\infty e^{-x^2} \, dx \le \lim_{z \to \infty} \int_1^z e^{-x} \, dx = e < \infty$

37. **(b)** $\int_1^\infty \frac{1}{\sqrt{1+x^2}} \, dx \ge \lim_{z \to \infty} \frac{1}{2} \int_1^z \frac{1}{x} \, dx = \infty$; divergent

39. For $x \ge 1$: $0 \le e^{-x^2/2} \le e^{-x}$; convergent

41. For $x \ge 1$: $\frac{1}{\sqrt{x+1}} \ge \frac{1}{2\sqrt{x}}$; divergent

43. **(a)** Use L'Hospital's rule
(b) Show that $\lim_{x \to \infty} \frac{\ln x}{\sqrt{x}} = 0$ and use a graphing calculator to show that for $x > 74.2$, $2 \ln x \le \sqrt{x}$
(c) Use for $x > 74.2$ that $e^{-\sqrt{x}} \le e^{-2\ln x} = \frac{1}{x^2}$ and conclude that the integral is convergent

Section 7.5

1. 2.328

3. 0.6292

5. $M_4 \approx 0.6912$; error ≈ 0.0019

7. $M_4 \approx 5.3838$; error ≈ 0.0505

9. $T_4 \approx 2.3438$

11. $T_3 \approx 0.6380$

13. $T_5 = 20.32$; error ≈ 0.32

15. $T_4 \approx 1.8195$; error ≈ 0.0661

17. $n = 82$

19. $n = 58$

21. $n = 92$

23. $n = 50$

25. **(a)** $M_5 = 0.245$; $T_5 = 0.26$; $|\int_0^1 x^3 \, dx - M_5| = 0.005$; $|\int_0^1 x^3 \, dx - T_5| = 0.01$
(c) $0.6433 \le \int_0^1 \sqrt{x} \, dx \le 0.6730$

Section 7.6

1. $\frac{x}{2} + \frac{3}{4} \ln |2x - 3| + C$

3. $\frac{1}{2} \left(x\sqrt{x^2 - 16} - 16 \ln |x + \sqrt{x^2 - 16}| \right) + C$

5. $6 - 16e^{-1}$

7. $\frac{2}{9} e^3 + \frac{1}{9}$

9. $\frac{1}{2} e^{\pi/6} - \frac{1}{4}(\sqrt{3} - 1)$

11. $-2e^{-x/2}(x^2 + 4x + 7) + C$

13. $\frac{1}{20} [10x - 6 + \sin(10x - 6)] + C$

15. $x\sqrt{\frac{9}{4} + x^2} + \frac{9}{4} \left(\ln |x + \sqrt{\frac{9}{4} + x^2}| \right) + C$

17. $\frac{4e^{2x+1}}{16+\pi^2} \left[2\sin \left(\frac{\pi}{2} x \right) - \frac{\pi}{2} \cos \left(\frac{\pi}{2} x \right) \right] + C$

19. $2 \ln 2$

21. $\frac{x}{2} (\sin(\ln(3x)) + \cos(\ln(3x))) + C$

Section 7.7

1. $L(x) = 1 + 2x$

3. $L(x) = 1 + x$

5. $L(x) = \ln 2$

7. $P_5(x) = 1 - \frac{x^2}{2!} + \frac{x^4}{4!}$

9. $P_6(x) = \frac{120}{5!} x^5 = x^5$

11. $P_3(x) = \sqrt{2} + \frac{1}{2\sqrt{2}} x - \frac{1}{16\sqrt{2}} x^2 + \frac{1}{64\sqrt{2}} x^3$; $P_3(0.1) \approx 1.4491$; $f(0.1) = \sqrt{2.1}$; $|f(0.1) - P_3(0.1)| \approx 3.34 \times 10^{-7}$

13. $P_5(x) = x - \frac{x^3}{3!} + \frac{x^5}{5!}$; $P_5(1) \approx 0.8417$; $f(1) \approx 0.8415$; $|P_5(1) - f(1)| \approx 1.96 \times 10^{-4}$

15. $P_2(x) = x$; $P_2(0.1) = 0.1$; $f(0.1) \approx 0.10033$; $|P_2(0.1) - f(0.1)| \approx 3.35 \times 10^{-4}$

17. **(a)** $P_3(x) = x - \frac{x^3}{3!}$
(b) $\lim_{x \to 0} \frac{P_3(x)}{x} = 0$ and $P_3(x)$ approximates $f(x) = \sin x$ at $x = 0$

19. $P_3(x) = 1 + \frac{1}{2}(x - 1) - \frac{1}{8}(x - 1)^2 + \frac{1}{16}(x - 1)^3$; $P_3(2) \approx 1.4375$; $f(2) \approx 1.4142$; $|P_3(2) - f(2)| \approx 0.023$

21. $P_3(x) = \frac{1}{2}\sqrt{3} - \frac{1}{2}(x - \frac{\pi}{6}) - \frac{1}{4}\sqrt{3}(x - \frac{\pi}{6})^2 + \frac{1}{12}(x - \frac{\pi}{6})^3$; $P_3(\frac{\pi}{7}) \approx 0.9010$; $f(\frac{\pi}{7}) \approx 0.9010$; $|P_3(\frac{\pi}{7}) - f(\frac{\pi}{7})| \approx 6.861 \times 10^{-5}$

23. $P_3(x) = e^2 + e^2(x - 2) + \frac{e^2}{2}(x - 2)^2 + \frac{e^2}{6}(x - 2)^3$; $P_3(2.1) \approx 8.1661$; $f(2.1) \approx 8.1662$; $|P_3(2.1) - f(2.1)| \approx 3.14 \times 10^{-5}$

27. $n = 10$

29. $n = 2$

31. $P_2(x) = 0$; error term: $f(x) - P_2(x) = f(x)$

33. **(b)** Use $x = 1$: $\tan^{-1} = \frac{\pi}{4}$

Section 7.9

1. $-\frac{1}{9}(1 - x^3)^3 + C$

3. $-2e^{-x^2} + C$

5. $\frac{6}{7}(1 + \sqrt{x})^{7/3} - \frac{3}{2}(1 + \sqrt{x})^{4/3} + C$

7. $\frac{1}{6} \tan(3x^2) + C$

9. $\frac{1}{2} x^2 \ln x - \frac{1}{4} x^2 + C$

11. $\tan x \ln(\tan x) - \tan x + C$

13. $\frac{1}{2} \tan^{-1} \left(\frac{x}{2} \right) + C$

15. $- \ln | \cos x| + C$

17. $\frac{e^{2x}}{5} (2 \sin x - \cos x) + C$

19. $2e^{x/2} + C$

21. $-\frac{1}{2} \sin x \cos x + \frac{x}{2} + C$

23. $\ln |\frac{x-1}{x}| + C$

25. $x - 5 \ln |x + 5| + C$

27. $\ln |x + 5| + C$

29. $\frac{1}{2} x^2 + 3x + 4 \ln |x - 1| + C$

31. $4 + \ln 3$

33. $1 - e^{-1/2}$

35. $\frac{\pi}{8}$

37. 4

39. $\frac{\pi}{6}$

41. divergent

43. divergent

45. 2

47. $-\frac{1}{4}$

49. $e - e^{1/\sqrt{2}}$

51. **(a)** $M_4 = 0.625$ **(b)** $T_4 = 0.75$

53. **(a)** $M_5 \approx 0.6311$ **(b)** $T_5 \approx 0.6342$

55. $P_3(x) = 2x - \frac{4}{3}x^3$

57. $P_3(x) = (x-1) - \frac{1}{2}(x-1)^2 + \frac{1}{3}(x-1)^3$

59.

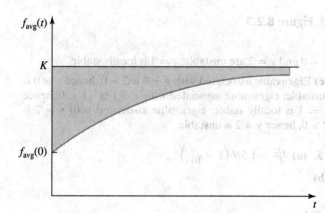

▲ **Figure 7.9.59**

Section 8.1

1. $y = \frac{1}{2}x^2 - \cos x + 1$

3. $y = \ln x$ for $x > 0$

5. $x(t) = 2 - \ln(1-t)$ for $t < 1$

7. $s(t) = \frac{2}{9}(3t+1)^{3/2} + \frac{7}{9}$ for $t \geq -\frac{1}{3}$

9. $V(t) = \sin t + 5$

11. $y = 2e^{3x}$

13. $x(t) = 5e^{2-2t}$

15. $h(s) = \frac{1}{2}(9e^{2s} - 1)$

17. $N(t) = 20e^{0.3t}$, $N(5) = 20e^{1.5} \approx 90$

19. **(a)** $N(t) = Ce^{rt}$

(b) $\log N(t) = \log C + (r \log e)t$; to determine r, graph $N(t)$ on a semi-log graph - the slope is then $r \log e$.

(c) 1. Obtain data at various points in time. 2. Plot on semi-log paper. 3. Determine slope of resulting straight line. 4. Divide slope by $\log e$, this yields r.

21. **(a)** $N(t) = \frac{100}{10-t}$

(b)

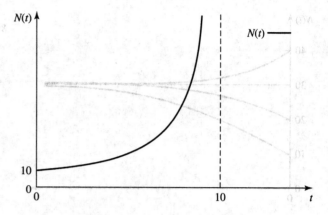

▲ **Figure 8.1.21**

$\lim_{t \to 10^-} N(t) = \infty$

23. **(a)** $L_\infty = 123$; $k = \frac{1}{27} \ln \frac{244}{123} \approx 0.0254$

(b) $L(10) = 123 - 122e^{-0.254} \approx 28.37$ in.

(c) $t = \frac{1}{0.0254} \ln \frac{122}{12.3} \approx 90.33$ months

25. $y = \frac{2}{3e^{-x}-2}$

27. $y = \frac{5}{1+4e^{5x}}$

29. $y = \frac{3}{1-\frac{2}{5}e^{-6(x-1)}}$

31. $y = \frac{1}{\frac{1}{C}e^{-x}-1}$

33. $y = 1 \pm (-2(x+C))^{-1/2}$

35. **(b)** **(i)** $y = \frac{2-2e^{4x}}{1+e^{4x}}$ **(ii)** $y = 2$ **(iii)** $y = \frac{\frac{2}{3}e^{4x}+2}{1-\frac{1}{3}e^{4x}}$

37. $N(t) = \frac{200}{1+3e^{-0.34t}}$, $\lim_{t \to \infty} N(t) = 200$

39. **(a)** $N(t) = \frac{50}{1+4e^{-1.5t}}$ **(b)** $N(t) = \frac{50}{1-\frac{4}{9}e^{-1.5t}}$

(c)

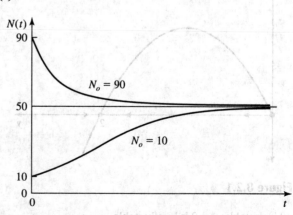

▲ **Figure 8.1.39**

(d) $\lim_{t \to \infty} N(t) = 50$ in both   **(a)** and   **(b)**

41. (a) $\frac{dN}{dt} = 5N\left(1 - \frac{N}{30}\right)$

(b)

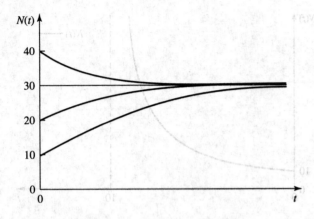

▲ **Figure 8.1.41**

43. (a) $p(t) = \frac{1}{\frac{1-p_0}{p_0}e^{-st/2}+1}, t \geq 0$

(b) $t = \frac{2}{0.01} \ln 9 \approx 439.4$

(c) $\lim_{t \to \infty} p(t) = 1$, which means that eventually the population will consist only of A_1A_1 types.

45. $y = \sqrt{x^2 + 2x + 4}$

47. $y = -1 + 3\exp\left[1 - e^{-x}\right]$

49. $y = 6x - 7$

51. $r(t) = \exp\left[1 - e^{-t}\right]$

53. $\frac{dc}{c} = k\frac{dm}{m}$

55. $\frac{dy}{dx} = \frac{1}{7.7}\frac{y}{x}$

Section 8.2

1. (a) $y = 0, 2$

(b)

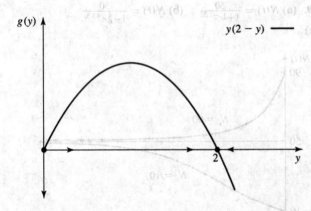

▲ **Figure 8.2.1**

$y = 0$ is unstable; $y = 2$ is locally stable.
(c) Eigenvalue associated with $y = 0$ is $2 > 0$, hence $y = 0$ is unstable; eigenvalue associated with $y = 2$ is $-2 < 0$, hence $y = 2$ is locally stable.

3. (a) $y = 0, 1, 2$

(b)

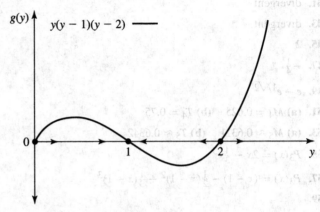

▲ **Figure 8.2.3**

$y = 0$ and $y = 2$ are unstable; $y = 1$ is locally stable.
(c) Eigenvalue associated with $y = 0$ is $2 > 0$, hence $y = 0$ is unstable; eigenvalue associated with $y = 1$ is $-1 < 0$, hence $y = 1$ is locally stable; eigenvalue associated with $y = 2$ is $2 > 0$, hence $y = 2$ is unstable.

5. (a) $\frac{dN}{dt} = 1.5N\left(1 - \frac{N}{100}\right)$

(b)

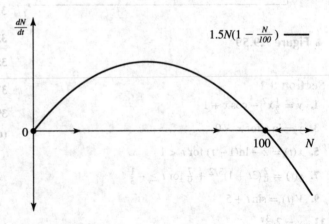

▲ **Figure 8.2.5**

$N = 0$ is unstable; $N = 100$ is stable
(c) Eigenvalue associated with $N = 0$ is $1.5 > 0$, hence $N = 0$ is unstable; eigenvalue associated with $N = 100$ is $-1.5 < 0$, hence $N = 100$ is locally stable. Same results as in **(b)**.

7. (a) $K = 2000$ **(b)** $t = \frac{1}{2}\ln 199 \approx 2.65$ **(c)** 2000

9. (a) $N \approx 52.79$ is unstable; $N \approx 947.21$ is locally stable
(b) The maximal harvesting rate is $rK/4$.

11.

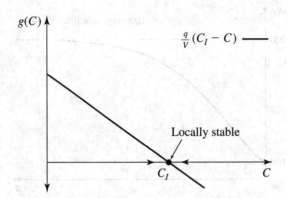

▲ **Figure 8.2.11**

The equilibrium C_I is locally stable.

13. (a) $\frac{dC}{dt} = \frac{0.2}{400}(3 - C)$

(b) $C(t) = 3 - 3e^{-t/2000}, t \geq 0; \lim_{t\to\infty} C(t) = 3$

(c) $C = 3$ is locally stable.

15. (a) Equilibrium concentration: $C_I = 254$

(b) $T_R = \frac{1}{0.37} \approx 2.703$

(c) $T_R = \frac{1}{0.37} \approx 2.703$

(d) They are the same.

19. $T_R = \frac{12.3\times10^9}{220}$ seconds ≈ 647.1 days; $C(T_R) \approx 0.806\frac{\text{mg}}{\text{l}}$

21. (a)

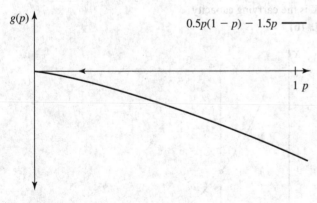

▲ **Figure 8.2.21**

(b) $p = 0$ is locally stable

(c) $g'(0) = -1 < 0$, which implies that 0 is locally stable.

23. (a) $\frac{dp}{dt}$ describes the rate of change of $p(t)$; $cp(1 - p - D)$ describes colonization of vacant undestroyed patches; $-mp$ describes extinction.

(b)

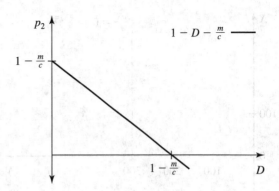

▲ **Figure 8.2.23**

(c) $D < 1 - \frac{m}{c}$; $p_1 = 0$ is unstable; $p_2 = 1 - D - \frac{m}{c}$ is locally stable

25. (a) $N = 0$, $N = 17$, and $N = 200$

(b) $N = 0$ is locally stable; $N = 17$ is unstable; $N = 200$ is locally stable

(c)

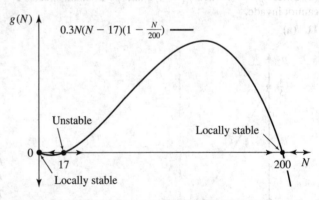

▲ **Figure 8.2.25**

Section 8.3

1. (a) $R_0 = 1.5 > 1$, the disease will spread

(b) $R_0 = \frac{1}{2} < 1$, the disease will not spread

3. $R_0 = 0.9999 < 1$, the disease will not spread

5. (a)
$$\frac{dN}{dt} = N_I - 5N - 0.02NX + X$$
$$\frac{dX}{dt} = 0.02NX - 2X$$

(b) equilibrium: $(\hat{N}, \hat{X}) = (100, N_I - 500)$, this is a nontrivial equilibrium provided $N_I > 500$

7. (a)
$$\frac{dN}{dt} = 200 - N - 0.01NX + 2X$$
$$\frac{dX}{dt} = 0.01NX - 3X$$

(b)

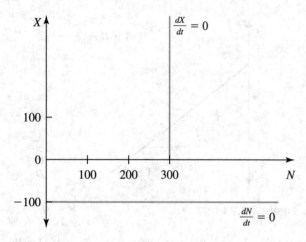

▲ **Figure 8.3.7**

(c) No nontrivial equilibria

9. (a) Equilibria: $(0, 0)$, $(0, 2/3)$, $(1/2, 0)$

(b) Since $\frac{dp_2}{dt} < 0$ when $p_1 = 1/2$ and p_2 is small, species 2 cannot invade.

11. (a)

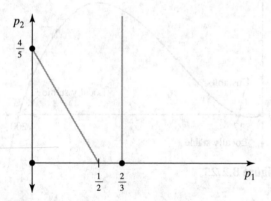

▲ **Figure 8.3.11**

(b) equilibria: $(0, 0)$, $(2/3, 0)$, $(0, 4/5)$

Section 8.5

1. (a) $\frac{dT}{dt}$ is proportional to the difference between the temperature of the object and the temperature of the surrounding medium.

(b) $t = \frac{1}{0.013} \ln \frac{9}{4} \approx 62.38$ minutes

3. (a) $N(t) = N(0)e^{r_e t}$

(b) $N(t) = \dfrac{K}{1 + \left(\frac{K}{N_0} - 1\right)e^{-r_l t}}$

(c) $r_e \approx 0.691$; $K = 1001$: $r_l \approx 1.382$; $K = 10{,}000$: $r_l \approx 0.701$

5. (a)

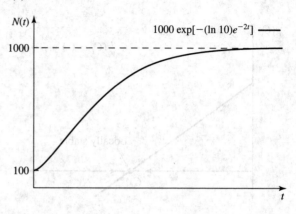

▲ **Figure 8.5.5(a)**

(c)

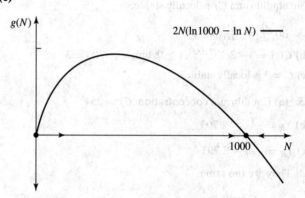

▲ **Figure 8.5.5(c)**

$N = 0$ is unstable; $N = 1000$ is locally stable
K is the carrying capacity

7. (b)

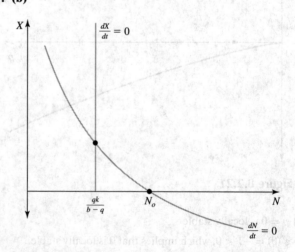

▲ **Figure 8.5.7**

(c) $\hat{N} = \frac{qk}{b-q} > 0$ if $b > q$, $\hat{X} = \left(k + \frac{qk}{b-q}\right)\frac{q}{b}\left(\frac{N_0(b-q)}{qk} - 1\right)$

Section 9.1

1. $\{(4, 3)\}$

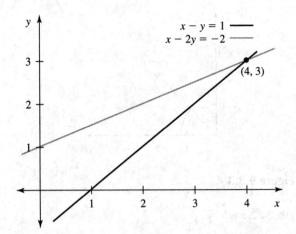

▲ **Figure 9.1.1**

3. No solution; the lines are parallel.

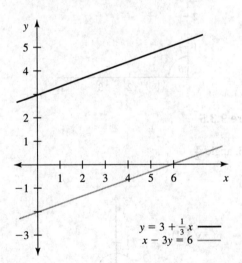

▲ **Figure 9.1.3**

5. (a) $c = 10$ (b) $c \neq 10$ (c) No

7. Zach bought five fish and six plants.

11. $x = 1, y = -1, z = 0$

13. $x = 2, y = 0, z = -3$

15. $x = 1, y = 1, z = -2$

17. $\{(x, y, z) : x = 2 - t, y = 1 + t, z = t, t \in \mathbf{R}\}$

19. underdetermined; $\{(x, y, z) : x = 7 + t, y = t + 2, z = t, t \in \mathbf{R}\}$

21. overdetermined; no solution

23. 750 gr of SL 24-4-8; 1000 gr of SL 21-7-12; $\frac{11000}{17}$ gr of SL 17-0-0.

Section 9.2

1. $\begin{bmatrix} 1 & -3 \\ 0 & -9 \end{bmatrix}$

3. $D = \begin{bmatrix} 1 & 0 \\ 4 & 11 \end{bmatrix}$

7. $\begin{bmatrix} 19 & 3 & 6 \\ 9 & 11 & 4 \\ 3 & -13 & -7 \end{bmatrix}$

9. $D = \begin{bmatrix} -4 & -2 & -7 \\ -5 & -1 & -3 \\ -1 & 5 & -1 \end{bmatrix}$

13. $A' = \begin{bmatrix} -1 & 2 \\ 0 & 1 \\ 3 & -4 \end{bmatrix}$

15. (a) $\begin{bmatrix} -2 & -3 \\ 0 & 5 \end{bmatrix}$

(b) $\begin{bmatrix} 1 & 6 \\ 2 & 2 \end{bmatrix}$

17. $AC = \begin{bmatrix} -1 & -2 \\ 1 & 0 \end{bmatrix}$

$CA = \begin{bmatrix} 1 & 4 \\ -1 & -2 \end{bmatrix}$

$AC \neq CA$

19. $(A + B)C = AC + BC = \begin{bmatrix} 1 & -1 \\ 0 & -3 \end{bmatrix}$

21. 3×2

23. (a) 1×4 (b) 3×3 (c) 4×3

25. (a) $\begin{bmatrix} 7 & 5 & 9 & -1 \\ -4 & -2 & -6 & 0 \end{bmatrix}$

(b) $\begin{bmatrix} 1 & -1 \\ 2 & 4 \\ 0 & -6 \\ -1 & -3 \end{bmatrix}$

27. $A^2 = \begin{bmatrix} 3 & -1 \\ 1 & 8 \end{bmatrix}$

$A^3 = \begin{bmatrix} 7 & 6 \\ -6 & -23 \end{bmatrix}$

$A^4 = \begin{bmatrix} 8 & -11 \\ 11 & 63 \end{bmatrix}$

31. $\begin{bmatrix} 2 & 3 & -1 \\ 0 & 2 & 1 \\ 1 & 0 & -2 \end{bmatrix} \begin{bmatrix} x_1 \\ x_2 \\ x_3 \end{bmatrix} = \begin{bmatrix} 0 \\ 1 \\ 2 \end{bmatrix}$

33. $\begin{bmatrix} 2 & -3 \\ -1 & 1 \\ 3 & 0 \end{bmatrix} \begin{bmatrix} x_1 \\ x_2 \end{bmatrix} = \begin{bmatrix} 4 \\ 3 \\ 4 \end{bmatrix}$

35. $AB = \begin{bmatrix} 1 & 0 \\ 0 & 1 \end{bmatrix}$, $BA = \begin{bmatrix} 1 & 0 \\ 0 & 1 \end{bmatrix}$

37. $A^{-1} = \begin{bmatrix} -3/5 & 1/5 \\ 2/5 & 1/5 \end{bmatrix}$

39. $A^{-1} = \begin{bmatrix} -3/5 & 1/5 \\ 2/5 & 1/5 \end{bmatrix}$ and show that $(A^{-1})^{-1} = A$

41. C does not have an inverse.

43. (a) $x_1 = 2, x_2 = 3$ **(b)** $A^{-1} = \begin{bmatrix} -1 & 0 \\ -2/3 & -1/3 \end{bmatrix}$,
$x_1 = 2, x_2 = 3$

45. $\det A = 7$, A is invertible

47. $\det A = 0$, A is not invertible

49. (a) $\det A = 0$, A is not invertible
(b)
$$2x + 4y = b_1$$
$$3x + 6y = b_2$$
(c) solution: $\{(x, y) : x = \frac{3}{2} - 2t, y = t, t \in \mathbf{R}\}$
(d) The system has no solutions when $\frac{b_1}{2} \neq \frac{b_2}{3}$

51. $\det A = 2$, A is invertible
$$A^{-1} = \begin{bmatrix} 1 & 1/2 \\ 0 & 1/2 \end{bmatrix}$$
$$X = \begin{bmatrix} 0 \\ 0 \end{bmatrix}$$

53. C^{-1} does not exist. $\{(x, y) : x = -3t, y = t, t \in \mathbf{R}\}$

55. $\begin{bmatrix} 1/4 & 1/4 & 0 \\ -1/8 & 3/8 & 1/2 \\ -3/8 & 1/8 & -1/2 \end{bmatrix}$

57. $\begin{bmatrix} -2/3 & -1/6 & -1/3 \\ 0 & -1/2 & 0 \\ -1/3 & 1/6 & 1/3 \end{bmatrix}$

59. $L = \begin{bmatrix} 0 & 3.2 & 1.7 \\ 0.2 & 0 & 0 \\ 0 & 0.7 & 0 \end{bmatrix}$
$$N(2) = \begin{bmatrix} 2232 \\ 580 \\ 280 \end{bmatrix}$$

61. $L = \begin{bmatrix} 0 & 0 & 4.6 & 3.7 \\ 0.7 & 0 & 0 & 0 \\ 0 & 0.5 & 0 & 0 \\ 0 & 0 & 0.1 & 0 \end{bmatrix}$
$$N(2) = \begin{bmatrix} 1242 \\ 934 \\ 525 \\ 25 \end{bmatrix}$$

63. four age classes; 60% of one-year old survive until the end of the following breeding season; 2 is the average number of female offspring of a two-year-old.

65. four age classes; 20% of two-year old survive until the end of the following breeding season; 2.5 is the average number of female offspring of a one-year-old.

67. $q_0(t)$ and $q_1(t)$ seem to converge to 2.3; it appears that 74% of females will be of age 0 in the stable age distribution.

Section 9.3

3.

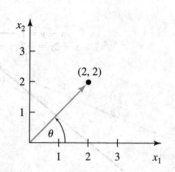

▲ **Figure 9.3.3**

length: $2\sqrt{2}$, angle $\frac{\pi}{4}$

5.

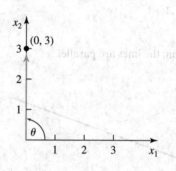

▲ **Figure 9.3.5**

length: 3, angle $\frac{\pi}{2}$

7.

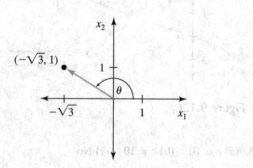

▲ **Figure 9.3.7**

length: 2, angle $\frac{5\pi}{6}$

9. $\begin{bmatrix} \sqrt{2} \\ 1 \end{bmatrix}$

11. $\begin{bmatrix} \cos 70° \\ \sin 70° \end{bmatrix}$

13. $x_1 = 3\cos 15°, x_2 = -3\sin 15°$

15. $x_1 = 5\cos 115°, x_2 = 5\sin 115°$

17.

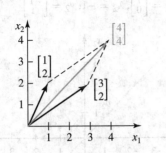

▲ **Figure 9.3.17**

$$\begin{bmatrix} 4 \\ 4 \end{bmatrix}$$

19.

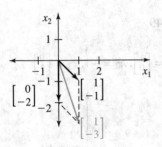

▲ **Figure 9.3.19**

$$\begin{bmatrix} 1 \\ -3 \end{bmatrix}$$

21.

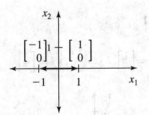

▲ **Figure 9.3.21**

$$\begin{bmatrix} 0 \\ 0 \end{bmatrix}$$

23.

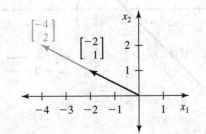

▲ **Figure 9.3.23**

$$\begin{bmatrix} -4 \\ 2 \end{bmatrix}$$

25.

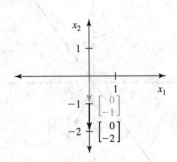

▲ **Figure 9.3.25**

$$\begin{bmatrix} 0 \\ -1 \end{bmatrix}$$

27.

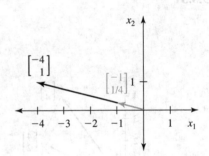

▲ **Figure 9.3.27**

$$\begin{bmatrix} -1 \\ 1/4 \end{bmatrix}$$

29.

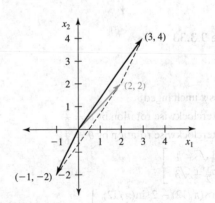

▲ **Figure 9.3.29**

$$\begin{bmatrix} 2 \\ 2 \end{bmatrix}$$

31.

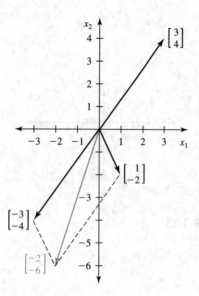

▲ Figure 9.3.31

$$\begin{bmatrix} -2 \\ -6 \end{bmatrix}$$

33.

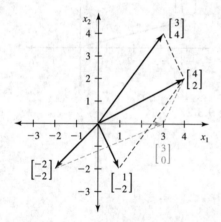

▲ Figure 9.3.33

$$\begin{bmatrix} 3 \\ 0 \end{bmatrix}$$

35. leaves **x** unchanged

37. counterclockwise rotation by $\theta = \frac{\pi}{2}$

39. counterclockwise rotation by $\theta = \frac{\pi}{6}$

41. $\begin{bmatrix} -\frac{1}{2}\sqrt{3} - 1 \\ -\frac{1}{2} + \sqrt{3} \end{bmatrix}$

43. $\begin{bmatrix} 5\cos(\pi/12) - 2\sin(\pi/12) \\ 5\sin(\pi/12) + 2\cos(\pi/12) \end{bmatrix}$

45. $\begin{bmatrix} \sqrt{2} + \sqrt{2}/2 \\ -\sqrt{2} + \sqrt{2}/2 \end{bmatrix}$

47. $\begin{bmatrix} 5\cos(-\pi/7) + 3\sin(-\pi/7) \\ 5\sin(-\pi/7) - 3\cos(-\pi/7) \end{bmatrix}$

49. $\lambda_1 = 2$, $\mathbf{v}_1 = \begin{bmatrix} 1 \\ 0 \end{bmatrix}$, $\lambda_2 = -1$, $\mathbf{v}_2 = \begin{bmatrix} -1 \\ 1 \end{bmatrix}$

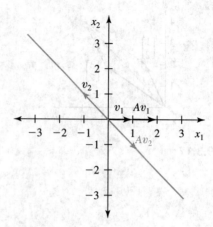

▲ Figure 9.3.49

51. $\lambda_1 = 1$, $\mathbf{v}_1 = \begin{bmatrix} 1 \\ 0 \end{bmatrix}$, $\lambda_2 = -1$, $\mathbf{v}_2 = \begin{bmatrix} 0 \\ 1 \end{bmatrix}$

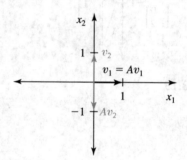

▲ Figure 9.3.51

53. $\lambda_1 = -2$, $\mathbf{v}_1 = \begin{bmatrix} 1 \\ 1 \end{bmatrix}$, $\lambda_2 = -1$, $\mathbf{v}_2 = \begin{bmatrix} 2 \\ 3 \end{bmatrix}$

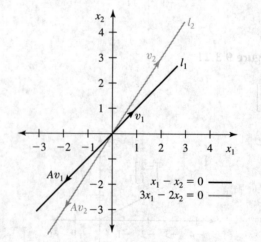

▲ Figure 9.3.53

55. $\lambda_1 = 1, \mathbf{v}_1 = \begin{bmatrix} 1 \\ -1 \end{bmatrix}, \lambda_2 = 4, \mathbf{v}_2 = \begin{bmatrix} 1 \\ 2 \end{bmatrix}$

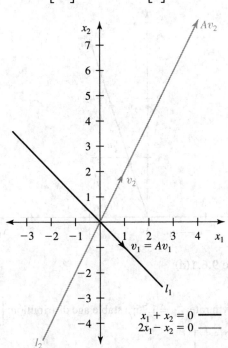

▲ **Figure 9.3.55**

57. $\lambda_1 = 4, \lambda_2 = 3$

59. $\lambda_1 = 1, \lambda_2 = 2$

61. $\lambda_1 = -\frac{3}{2}, \lambda_2 = \frac{1}{2}$

63. The real parts of both eigenvalues are negative.

65. The real parts of both eigenvalues are not negative.

67. The real parts of both eigenvalues are negative.

69. **(a)** $l_1: x_2 = 0; l_2: -3x_1 + x_2 = 0$; since l_1 and l_2 are not identical, $\mathbf{u}_1$ and $\mathbf{u}_2$ are linearly independent.
(b) $\mathbf{x} = 2\mathbf{u}_1 - \mathbf{u}_2$ **(c)** $A^{20}\mathbf{x} = \begin{bmatrix} -1048574 \\ -3145728 \end{bmatrix}$

71. $\begin{bmatrix} -2 \\ 6 \end{bmatrix}$

73. $\begin{bmatrix} (-7)3^{20} + 4(-2)^{20} \\ (2)3^{20} - 4(-2)^{20} \end{bmatrix}$

75. **(a)** $\lambda_1 = 1 + \sqrt{2.2}, \lambda_2 = 1 - \sqrt{2.2}$
(b) The larger eigenvalue corresponds to the growth rate.
(c) 89.2% are in age class 0, 10.8% are in age class 1 in the stable age distribution.

77. **(a)** $\lambda_1 = \frac{1}{2}(7 + \sqrt{50.2}), \lambda_2 = \frac{1}{2}(7 - \sqrt{50.2})$
(b) The larger eigenvalue corresponds to the growth rate.
(c) 98.6% are in age class 0, 1.4% are in age class 1 in the stable age distribution.

79. **(a)** $\lambda_1 = \sqrt{0.45}, \lambda_2 = -\sqrt{0.45}$
(b) The larger eigenvalue corresponds to the growth rate.
(c) 88.17% are in age class 0, 11.83% are in age class 1 in the stable age distribution.

Section 9.4

1. **(a)** $\begin{bmatrix} -1 \\ 5 \\ -1 \end{bmatrix}$

(b) $\begin{bmatrix} 2 \\ 8 \\ -2 \end{bmatrix}$

(c) $\begin{bmatrix} 6 \\ -3 \\ 0 \end{bmatrix}$

3. $\begin{bmatrix} 2 \\ -2 \end{bmatrix}$

5. $\begin{bmatrix} -1 \\ -2 \\ 5 \end{bmatrix}$

7. $\sqrt{10}$

9. $\sqrt{26}$

11. $\begin{bmatrix} 1/\sqrt{11} \\ 3/\sqrt{11} \\ -1/\sqrt{11} \end{bmatrix}$

13. $\begin{bmatrix} 1 \\ 0 \\ 0 \end{bmatrix}$

15. 1

17. 2

19. $\sqrt{5}$

21. $\sqrt{30}$

23. $\cos\theta = 1/\sqrt{50}, \theta \approx 1.429$

25. $\cos\theta = 2/\sqrt{110}, \theta \approx 1.379$

27. $\begin{bmatrix} 1 \\ 1 \end{bmatrix}$

29. $\begin{bmatrix} -2 \\ 1 \\ 1 \end{bmatrix}$

31. **(a,b)** $\overline{PQ} = 4, \overline{QR} = 3, \overline{PR} = 5$, angle $QPR = \tan^{-1}(3/4) \approx 36.9°$, angle $PRQ = 90° - \tan^{-1}(3/4) \approx 53.1°$, angle $RQP = 90°$.

33. **(a)** $\overline{PQ} = \sqrt{10}, \overline{QR} = \sqrt{2}, \overline{PR} = \sqrt{6}$
(b) angle $QPR = \cos^{-1}(7/\sqrt{60}) \approx 25.4° = 0.442$, angle $PRQ = \cos^{-1}(-1/\sqrt{12}) \approx 106.8° = 1.864$, angle $RQP = \cos^{-1}(3/\sqrt{20}) \approx 47.9° = 0.835$.

35. $x + 2y = 4$

37. $4x + y = 2$

39. $-y + z = 1$

41. $x = 0$

43. $x = 1 + 2t$ and $y = -1 + t$ for $t \in \mathbf{R}$

45. $x = -1 + t$ and $y = -2 - 3t$ for $t \in \mathbf{R}$

47. $2y - x - 8 = 0$

49. $y - x + 4 = 0$

51. $\begin{bmatrix} x \\ y \end{bmatrix} = \begin{bmatrix} 0 \\ 1/4 \end{bmatrix} + t\begin{bmatrix} 1 \\ -3/4 \end{bmatrix}, t \in \mathbf{R}$

53. $\begin{bmatrix} x \\ y \end{bmatrix} = \begin{bmatrix} 0 \\ 3 \end{bmatrix} + t \begin{bmatrix} 1 \\ -2 \end{bmatrix}, t \in \mathbf{R}$

55. $\begin{bmatrix} x \\ y \\ z \end{bmatrix} = \begin{bmatrix} 1 \\ -1 \\ 2 \end{bmatrix} + t \begin{bmatrix} 1 \\ -2 \\ 1 \end{bmatrix}, t \in \mathbf{R}$

57. $\begin{bmatrix} x \\ y \\ z \end{bmatrix} = \begin{bmatrix} -1 \\ 3 \\ -2 \end{bmatrix} + t \begin{bmatrix} -1 \\ -2 \\ 4 \end{bmatrix}, t \in \mathbf{R}$

59. $\begin{bmatrix} x \\ y \\ z \end{bmatrix} = \begin{bmatrix} 5 \\ 4 \\ -1 \end{bmatrix} + t \begin{bmatrix} 3 \\ 4 \\ -4 \end{bmatrix}, t \in \mathbf{R}$

61. $\begin{bmatrix} x \\ y \\ z \end{bmatrix} = \begin{bmatrix} 2 \\ -3 \\ 1 \end{bmatrix} + t \begin{bmatrix} 7 \\ -5 \\ 0 \end{bmatrix}, t \in \mathbf{R}$

63. $(-5/2, -1/2, 9/2)$

65. $\begin{bmatrix} x \\ y \\ z \end{bmatrix} = \begin{bmatrix} 5 \\ -1 \\ 0 \end{bmatrix} + t \begin{bmatrix} 1 \\ 1 \\ 0 \end{bmatrix}, t \in \mathbf{R}$

Section 9.6

1. (a) $A\mathbf{x} = \begin{bmatrix} 0 \\ 2 \end{bmatrix}$

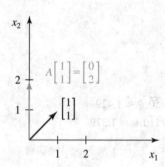

▲ **Figure 9.6.1(a)**

(b) $\lambda_1 = -1, \mathbf{u}_1 = \begin{bmatrix} 1 \\ 0 \end{bmatrix}, \lambda_2 = 2, \mathbf{u}_1 = \begin{bmatrix} 1 \\ 3 \end{bmatrix}$
(c)

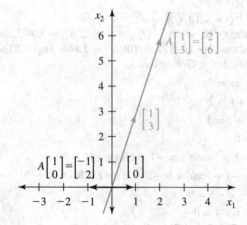

▲ **Figure 9.6.1(c)**

(d) $a_1 = 2/3, a_2 = 1/3$

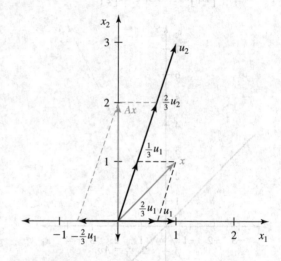

▲ **Figure 9.6.1(d)**

3. Growth rate: $\lambda_1 = 1.75$; a stable age distribution: $\begin{bmatrix} 35 \\ 10 \end{bmatrix}$

5. $\begin{bmatrix} -2 & 5 \\ -2 & 9 \end{bmatrix}$

7. 1. Gaussian elimination; 2. Write in matrix form $AX = B$ and find the inverse of A, then compute $X = A^{-1}B$

9. $a = -3$

11. For $\frac{5}{23} < a \le 1$, the population will grow.

Section 10.1

1.

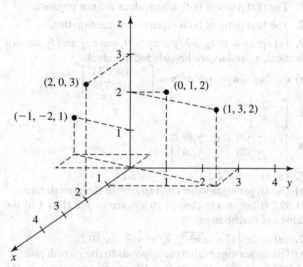

▲ **Figure 10.1.1**

3. $\frac{4}{13}$

5. $e^{-1/10}$

7. domain: $\{(x, y) : x \in \mathbf{R}, y \in \mathbf{R}\}$; range: $\{z : z \geq 0\}$; level curves: $x^2 + y^2 = c$, circle with radius $\sqrt{c}$ centered at $(0, 0)$

9. domain: $\{(x, y) : y > x^2, x \in \mathbf{R}\}$; range: $\{z : z \in \mathbf{R}\}$; level curves: $y = e^c + x^2$, parabolas shifted in $y > 0$ direction by e^c

11. Figure 10.23

13. Figure 10.24

15. level curve: $x^2 + y^2 = c$, circle centered at origin with radius $\sqrt{c}$; intersection with x-z plane: $z = x^2$, intersection with y-z plane: $z = y^2$

(b)

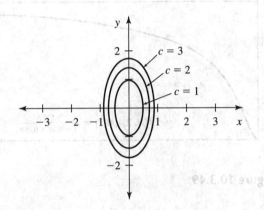

▲ **Figure 10.1.15(b)**

intersection with x-z plane: $z = 4x^2$, intersection with y-z plane: $z = y^2$

(c)

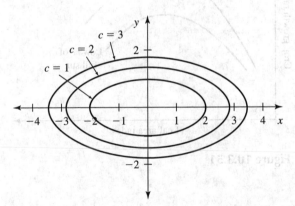

▲ **Figure 10.1.15(c)**

intersection with x-z plane: $z = \frac{1}{4}x^2$, intersection with y-z plane: $z = y^2$

(d) The intersection with the y-z plane is always $z = y^2$. When $a = 1$, the resulting surface could be obtained by rotating the curve $z = y^2$ about the z-axis. When $0 < a < 1$, then the resulting surface is still a paraboloid but longer along the x-axis than along the y-axis. When $a > 1$, the paraboloid is longer along the y-axis than along the x-axis.

17. Day 180: 22 m; day 200: 18 m; day 220: 14 m

Section 10.2

1. 1

3. $-\frac{1}{2}$

5. $-\frac{3}{2}$

7. $\frac{2}{3}$

9. Along positive x-axis: 1; along positive y-axis: -2

11. Along x-axis: 0; along y-axis: 0; along $y = x$: 2

13. Along $y = mx, m \neq 0$: 2; along $y = x^2$: 1; the limit does not exist.

15.

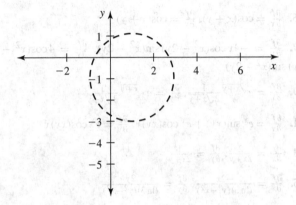

▲ **Figure 10.2.15**

$\{(x, y) : (x - 1)^2 + (y + 1)^2 < 4\}$

17. The boundary is a circle with radius 3, centered at $(0, 2)$. The boundary is not included.

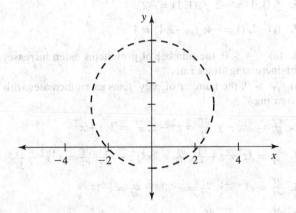

▲ **Figure 10.2.17**

19. Choose $2\delta^2 = \epsilon$.

21. 1. $f(x, y)$ is defined at $(0, 0)$. 2. $\lim_{(x,y)\to(0,0)}(x^2 + y^2)$ exists. 3. $f(0, 0) = 0 = \lim_{(x,y)\to(0,0)}(x^2 + y^2)$

23. In Problem 11, we showed that $\lim_{(x,y)\to(0,0)} \frac{4xy}{x^2+y^2}$ does not exist. Hence $f(x, y)$ is discontinuous at $(0, 0)$

25. In Problem 13, we showed that $\lim_{(x,y)\to(0,0)} \frac{2xy}{x^3+yx}$ does not exist. Hence $f(x, y)$ is discontinuous at $(0, 0)$

27. (a) $h(x, y) = g[f(x, y)]$ with $f(x, y) = x^2 + y^2$ and $g(z) = \sin z$; **(b)** the function is continuous for all $(x, y) \in \mathbf{R}^2$

29. (a) $h(x, y) = g[f(x, y)]$ with $f(x, y) = xy$ and $g(z) = e^z$; **(b)** the function is continuous for all $(x, y) \in \mathbf{R}^2$

Section 10.3

1. $\frac{\partial f}{\partial x} = 2xy + y^2$, $\frac{\partial f}{\partial y} = x^2 + 2xy$

3. $\frac{\partial f}{\partial x} = \frac{3}{2}y\sqrt{xy} - \frac{2y}{3(xy)^{1/3}}$, $\frac{\partial f}{\partial y} = \frac{3}{2}x\sqrt{xy} - \frac{2x}{3(xy)^{1/3}}$

5. $\frac{\partial f}{\partial x} = \cos(x + y)$, $\frac{\partial f}{\partial y} = \cos(x + y)$

7. $\frac{\partial f}{\partial x} = -4x\cos(x^2 - 2y)\sin(x^2 - 2y)$, $\frac{\partial f}{\partial y} = 4\cos(x^2 - 2y)\sin(x^2 - 2y)$

9. $\frac{\partial f}{\partial x} = e^{\sqrt{x+y}}\frac{1}{2\sqrt{x+y}}$, $\frac{\partial f}{\partial y} = e^{\sqrt{x+y}}\frac{1}{2\sqrt{x+y}}$

11. $\frac{\partial f}{\partial x} = e^x\sin(xy) + e^x\cos(xy)y$, $\frac{\partial f}{\partial y} = e^x\cos(xy)x$

13. $\frac{\partial f}{\partial x} = \frac{2}{2x+y}$, $\frac{\partial f}{\partial y} = \frac{1}{2x+y}$

15. $\frac{\partial f}{\partial x} = \frac{-2x}{(\ln 3)(y^2-x^2)}$, $\frac{\partial f}{\partial y} = \frac{2y}{(\ln 3)(y^2-x^2)}$

17. 6

19. $3e^5$

21. 1

23. $\frac{2}{9}$

25. $f_x(1, 1) = -2$, $f_y(1, 1) = -2$

27. $f_x(-2, 1) = -4$, $f_y(-2, 1) = 4$

29. (a) $\frac{\partial P_e}{\partial a} > 0$: the number of prey items eaten increases with increasing attack rate.
(b) $\frac{\partial P_e}{\partial T} > 0$: the number of prey items eaten increases with increasing T.

31. $\frac{\partial f}{\partial x} = 2xz - y$, $\frac{\partial f}{\partial y} = z^2 - x$, $\frac{\partial f}{\partial z} = 2yz + x^2$

33. $\frac{\partial f}{\partial x} = 3x^2y^2z + \frac{1}{yz}$, $\frac{\partial f}{\partial y} = 2x^3yz - \frac{x}{y^2z}$, $\frac{\partial f}{\partial z} = x^3y^2 - \frac{x}{yz^2}$

35. $\frac{\partial f}{\partial x} = e^{x+y+z}$, $\frac{\partial f}{\partial y} = e^{x+y+z}$, $\frac{\partial f}{\partial z} = e^{x+y+z}$

37. $\frac{\partial f}{\partial x} = \frac{1}{x+y+z}$, $\frac{\partial f}{\partial y} = \frac{1}{x+y+z}$, $\frac{\partial f}{\partial z} = \frac{1}{x+y+z}$

39. $2y$

41. e^y

43. $2\sec^2(u + w)\tan(u + w)$

45. $-6x\sin y$

47. $\frac{2}{(x+y)^3}$

49. (a) $\frac{\partial f}{\partial N} > 0$: the number of prey encounters per predator increases as the prey density increases.
(b) $\frac{\partial f}{\partial T} > 0$: the function increases as the time for search increases.
(c) $\frac{\partial f}{\partial T_h} < 0$: the function decreases as the handling time T_h increases.
(d)

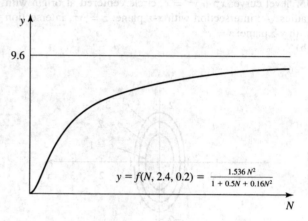

$$y = f(N, 2.4, 0.2) = \frac{1.536\,N^2}{1 + 0.5N + 0.16N^2}$$

▲ **Figure 10.3.49**

51.

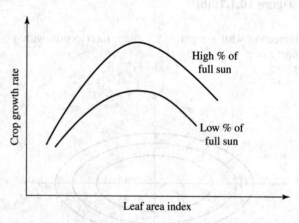

▲ **Figure 10.3.51**

Section 10.4

1. $7 = 6x + 4y - z$

3. $z - 2ex = -e$

5. $z - y = 0$

7. $f(x, y)$ is defined in an open disk centered at $(1, 1)$ and is continuous at $(1, 1)$.

9. $f(x, y)$ is defined in an open disk centered at $(0, 0)$ and is continuous at $(0, 0)$.

11. $f(x, y)$ is defined in an open disk centered at $(-1, 2)$ and is continuous at $(-1, 2)$.

13. $L(x, y) = \frac{1}{2}x + 2y + \frac{1}{2}$

15. $L(x, y) = x + y$

17. $L(x, y) = x + \frac{1}{2}y - \frac{3}{2} + \ln 2$

19. $L(x, y) = 1 + x + y$, $L(0.1, 0.05) = 1.15$, $f(0.1, 0.05) \approx 1.1618$

21. $L(x, y) = 2x - 3y - 2$, $L(1.1, 0.1) = -0.1$, $f(1.1, 0.1) \approx -0.0943$

23. $Df(x, y) = \begin{bmatrix} 1 & 1 \\ 2x & -2y \end{bmatrix}$

25. $Df(x, y) = \begin{bmatrix} e^{x-y} & -e^{x-y} \\ e^{x+y} & e^{x+y} \end{bmatrix}$

27. $Df(x, y) = \begin{bmatrix} -\sin(x-y) & \sin(x-y) \\ -\sin(x+y) & -\sin(x+y) \end{bmatrix}$

29. $Df(x, y) = \begin{bmatrix} 4xy+1 & 2x^2-3 \\ e^x \sin y & e^x \cos y \end{bmatrix}$

31. $L(x, y) = \begin{bmatrix} 4x+2y-4 \\ -x-y+3 \end{bmatrix}$

33. $L(x, y) = \begin{bmatrix} e(2x-y) \\ 2x-y-1 \end{bmatrix}$

35. $L(x, y) = \begin{bmatrix} x-y+1 \\ y-x+1 \end{bmatrix}$

37. $L(1.1, 1.9) = \begin{bmatrix} -0.9 \\ 9.8 \end{bmatrix}$, $f(1.1, 1.9) \approx \begin{bmatrix} -0.88 \\ 9.83 \end{bmatrix}$

39. $L(1.9, -3.1) = \begin{bmatrix} 25 \\ -22.4 \end{bmatrix}$, $f(1.9, -3.1) \approx \begin{bmatrix} 25 \\ -22.382 \end{bmatrix}$

Section 10.5

1. $18 \ln 2 + 8$

3. $\dfrac{\frac{\pi}{3} + \frac{\sqrt{3}}{4}}{\sqrt{\frac{\pi^2}{9} + \frac{3}{4}}}$

5. 0

7. $\dfrac{dz}{dt} = \dfrac{\partial f}{\partial x} u'(t) + \dfrac{\partial f}{\partial y} v'(t)$

9. $-\dfrac{2x}{2y + x^2 + y^2}$

11. $-\dfrac{2x - 3y(x^2+y^2)}{2y - 3x(x^2+y^2)}$

13. $\dfrac{dy}{dx} = -\dfrac{1}{\sqrt{1-x^2}}$ for $0 \le y \le \pi$ and $-1 \le x \le 1$

15. The growth rate decreases over time.

17. $\operatorname{grad} f = \begin{bmatrix} 3x^2 y^2 \\ 2x^3 y \end{bmatrix}$

19. $\operatorname{grad} f = \dfrac{1}{2\sqrt{x^3 - 3xy}} \begin{bmatrix} 3x^2 - 3y \\ -3x \end{bmatrix}$

21. $\operatorname{grad} f = \dfrac{\exp[\sqrt{x^2+y^2}]}{\sqrt{x^2+y^2}} \begin{bmatrix} x \\ y \end{bmatrix}$

23. $\operatorname{grad} f = \dfrac{x^2 - y^2}{x^2 + y^2} \begin{bmatrix} \frac{1}{x} \\ -\frac{1}{y} \end{bmatrix}$

25. $\frac{2}{3}\sqrt{3}$

27. $-\sqrt{2}$

29. $\frac{3}{2}\sqrt{10}$

31. $D_{\mathbf{u}} f(2, 1) = \dfrac{13}{\sqrt{2}}$

33. $D_{\mathbf{u}} f(1, 6) = -\dfrac{1}{4\sqrt{29}}$

35. $f(x, y)$ increases most rapidly in the direction of $\begin{bmatrix} 5 \\ -3 \end{bmatrix}$ at the point $(-1, 1)$.

37. $f(x, y)$ increases most rapidly in the direction of $\begin{bmatrix} 5/4 \\ -3/4 \end{bmatrix}$ at the point $(5, 3)$.

39. $\begin{bmatrix} 3/5 \\ 4/5 \end{bmatrix}$

41. $\dfrac{1}{\sqrt{733}} \begin{bmatrix} 2 \\ -27 \end{bmatrix}$

43. The amoeba will move in the direction of $\begin{bmatrix} -4/25 \\ -4/25 \end{bmatrix}$

Section 10.6

1. $f(x, y)$ has a local minimum at $(1, 0)$

3. $f(x, y)$ has saddle points at $(2, 4)$ and $(-2, 4)$

5. $f(x, y)$ has a saddle at $(0, 3)$

7. $f(x, y)$ has a local maximum at $(0, 0)$.

9. $f(x, y)$ has saddle points at $(0, \pi/2 + k\pi)$ for $k \in \mathbf{Z}$

11. (c) Figure 10.65: $f(x, y)$ stays constant for fixed x; there is neither a maximum nor a minimum at $(0, 0)$. Figure 10.66: saddle point at $(0, 0)$, Figure 10.67: local minimum at $(0, 0)$.

13. Absolute maximum: $(1, -1)$; absolute minimum: $(-1, 1)$

15. Absolute maxima: $(1, 0)$ and $(-1, 0)$; absolute minima: $(0, 1)$ and $(0, -1)$

17. Absolute maxima: $(0, 0)$, $(1, 0)$, $(1, -2)$, and $(0, -2)$; absolute minimum: $(1/2, -1)$

19. Absolute maximum at $(2/3, 2/3)$; absolute minima occur at all points along the boundary of the domain.

21. Absolute minimum: $(-2, 0)$; absolute maximum: $(3, 0)$

23. Absolute minimum: $(-1/2, 1/2)$; absolute maximum: $(1/\sqrt{2}, -1/\sqrt{2})$

25. Yes.

27. Absolute maximum at $(N, P) = (1, 1)$.

29. Maximum volume is 64 m^3.

31. The minimum surface area is 216 m^2.

33. The minimum distance is $1/\sqrt{3}$.

37. Absolute maxima: $(-\sqrt{35}/6, 1/6)$, $(\sqrt{35}/6, 1/6)$; absolute minimum: $(0, -1)$

39. Absolute minimum: $(1/4, -1/8)$; no maxima

41. Absolute minimum: $(12/13, -8/13)$; no maxima.

43. Local minimum: $(0, 1/3)$; no absolute minima; absolute maxima: $(1/\sqrt{2}, 1/6)$, $(-1/\sqrt{2}, 1/6)$.

45. Absolute minima: $(1, 0)$ and $(-1, 0)$; no maxima.

49. The total length of the fence is 96 ft.

51. Largest possible area is 4.

53. Smallest perimeter is 4.

55. $r = \sqrt{A}$, $\theta = 2$, perimeter is $4\sqrt{A}$

57. Local minimum at $(2, 2)$; no absolute extrema.

61. (a) $3x_1 + 3x_2 = 10$,
(b) absolute maximum at $\left(\frac{65-40\sqrt{2}}{3}, \frac{40\sqrt{2}-55}{3} \right)$

Section 10.7

1. $N_t = 5, 7.5, 11.25, 16.875, 25.31, 37.97, 56.95, 85.43,$
$128.14, 192.22, 288.33; P_t = 0$ for $t = 0, 1, 2, \ldots, 10$

3. $N_t = b^t N_0$

5. $N_t = 5, 6.79, 9.89, 14.67, 21.86, 32.59, 48.51, 71.67, 102.92,$
$128.47, 68.40, 0.71, 0.02, 0.03, 0.04, 0.06; P_t = 5, 1.43, 0.57, 0.34,$
$0.30, 0.39, 0.76, 2.18, 9.18, 51.81, 248.67, 203.69, 2.09, 0.0022,$
$0, 0$

7. $N_t = 5, 7.5, 11.25, 16.88, 25.31, 37.97, 56.95, 85.43, 128.14,$
$192.22, 288.33; P_t = 0$ for $t = 0, 1, 2, \ldots 10$

9. $N_t = b^t N_0$

11. (rounded to the closest integer) $N_t = 100, 79, 37, 16, 10,$
$10, 13, 17, 24, 34, 47, 61, 67, 54, 32, 18, 13, 13, 16, 20, 27, 36, 45,$
$51, 48, 36; P_t = 50, 141, 164, 80, 27, 10, 5, 3, 3, 4, 8, 19, 49, 94,$
$99, 59, 27, 13, 8, 6, 7, 10, 17, 33, 58, 72$

13. (a)

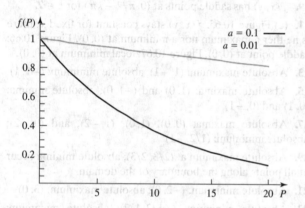

▲ **Figure 10.7.13**

(b) Increasing a, increases the chances of escaping parasitism.

15. (a)

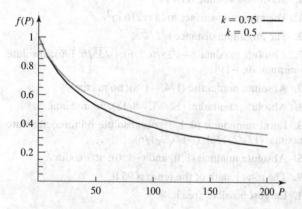

▲ **Figure 10.7.15**

(b) Increasing k, increases the chances of escaping parasitism.

17. Stable
19. Unstable
21. Unstable
23. Stable
25. Unstable
27. Eigenvalues: $1/\sqrt{2}, -1/\sqrt{2}$
29. Eigenvalues: $0.5 + i0.5, 0.5 - i0.5$
31. For $0 < a < 1/2$, $(0, 0)$ is locally stable.
33. $(0, 0)$ is unstable; $(1/6, 1/6)$ is locally stable.
35. $(0, 0)$ is locally stable if $-1 < a < 1$.
37. (a) If $r > 1/2$, then $(r - 1/2, r - 1/2)$ is an equilibrium.
(b) For $1/2 < r < 3/2$, the equilibrium $(r - 1/2, r - 1/2)$ is locally stable.
39. $(0, 0)$ is unstable; $((40 \ln 4)/3, 10 \ln 4)$ is unstable.
41. $(0, 0)$ is unstable; $(1000, 750)$ is locally stable.

Section 10.9

1.

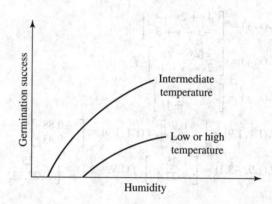

▲ **Figure 10.9.1(i)**

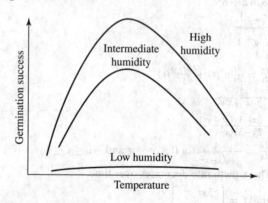

▲ **Figure 10.9.1(ii)**

3. (a) $\frac{\partial A_i}{\partial F} > 0$, $\frac{\partial A_i}{\partial D} < 0$
(b) $\frac{\partial A_e}{\partial F} > 0$: area covered by introduced species increased with the amount of fertilizer added.
$\frac{\partial A_e}{\partial D} > 0$: area covered by introduced species increased with disturbance intensity (note that this is the opposite to (a)).
(c) Fertilization had a positive effect in both cases. That is, fertilization increased the total area covered of both introduced and indigenous species. Disturbance intensity had a negative effect on indigenous species and a positive effect on introduced species.

5. $D\mathbf{f}(x, y) = \begin{bmatrix} 2x & -1 \\ 3x^2 & -2y \end{bmatrix}$

7. (a)

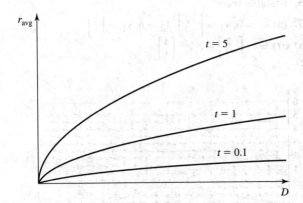

▲ **Figure 10.9.7**

(b) Use **(a)**: $r_{avg}^2 = \pi Dt$ and solve for D

(c) r_{avg} = arithmetic average = $\frac{1}{N} \sum_{i=1}^{N} d_i$ and use formula for D in **(b)**

Section 11.1

1. $\frac{d\mathbf{x}}{dt} = \begin{bmatrix} 2 & 3 \\ -4 & 1 \end{bmatrix} \mathbf{x}(t)$

3. $\frac{d\mathbf{x}}{dt} = \begin{bmatrix} 0 & 0 & 1 \\ -1 & 0 & 0 \\ 1 & 1 & 0 \end{bmatrix} \mathbf{x}(t)$

5. $(1, 0)$: $\begin{bmatrix} -1 \\ 1 \end{bmatrix}$; $(0, 1)$: $\begin{bmatrix} 2 \\ 0 \end{bmatrix}$; $(-1, 0)$: $\begin{bmatrix} 1 \\ -1 \end{bmatrix}$; $(0, -1)$: $\begin{bmatrix} -2 \\ 0 \end{bmatrix}$;
$(1, 1)$: $\begin{bmatrix} 1 \\ 1 \end{bmatrix}$; $(0, 0)$: $\begin{bmatrix} 0 \\ 0 \end{bmatrix}$; $(-2, 1)$: $\begin{bmatrix} 4 \\ -2 \end{bmatrix}$;

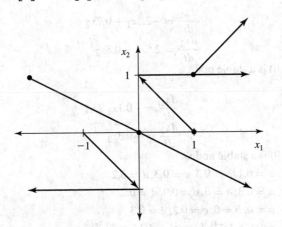

▲ **Figure 11.1.5**

7. $(1, 0)$: $\begin{bmatrix} 1 \\ -1 \end{bmatrix}$; $(0, 1)$: $\begin{bmatrix} 3 \\ 2 \end{bmatrix}$; $(-1, 1)$: $\begin{bmatrix} 2 \\ 3 \end{bmatrix}$; $(0, -1)$: $\begin{bmatrix} -3 \\ -2 \end{bmatrix}$;
$(-3, 1)$: $\begin{bmatrix} 0 \\ 5 \end{bmatrix}$; $(0, 0)$: $\begin{bmatrix} 0 \\ 0 \end{bmatrix}$; $(-2, 1)$: $\begin{bmatrix} 1 \\ 4 \end{bmatrix}$;

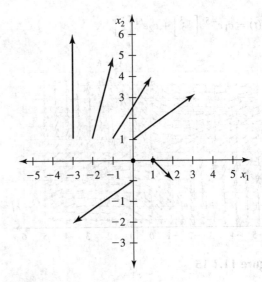

▲ **Figure 11.1.7**

9. Figure 11.18: (a); Figure 11.19: (b); Figure 11.20: (c); Figure 11.21: (d)

11.

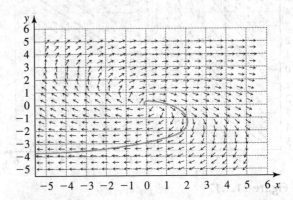

▲ **Figure 11.1.11**

13. $\mathbf{x}(t) = c_1 e^{6t} \begin{bmatrix} 3 \\ 5 \end{bmatrix} + c_2 e^{-2t} \begin{bmatrix} 1 \\ -1 \end{bmatrix}$

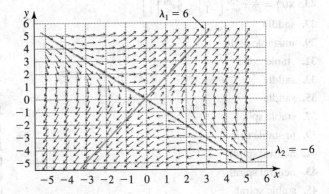

▲ **Figure 11.1.13**

15. $\mathbf{x}(t) = c_1 e^{-5t} \begin{bmatrix} -3 \\ 2 \end{bmatrix} + c_2 e^{6t} \begin{bmatrix} 1 \\ 3 \end{bmatrix}$

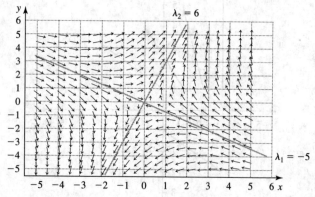

▲ **Figure 11.1.15**

17. $\mathbf{x}(t) = c_1 e^{-2t} \begin{bmatrix} 1 \\ 1 \end{bmatrix} + c_2 e^{t} \begin{bmatrix} 0 \\ 1 \end{bmatrix}$

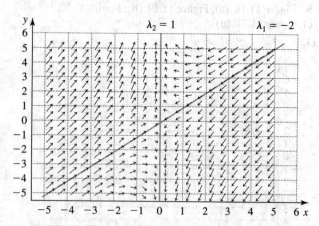

▲ **Figure 11.1.17**

19. $\mathbf{x}(t) = e^{-3t} \begin{bmatrix} -5 \\ 4 \end{bmatrix} + e^{2t} \begin{bmatrix} 0 \\ 1 \end{bmatrix}$

21. $\mathbf{x}(t) = e^{t} \begin{bmatrix} 1 \\ 1 \end{bmatrix}$

23. $\mathbf{x}(t) = \frac{13}{8} e^{-3t} \begin{bmatrix} 1 \\ -1 \end{bmatrix} - \frac{3}{8} e^{5t} \begin{bmatrix} 7 \\ 1 \end{bmatrix}$

27. saddle

29. unstable node

31. stable node

33. saddle

35. saddle

37. stable spiral

39. unstable spiral

41. neutral spiral

43. neutral spiral

45. stable spiral

47. saddle

49. stable spiral

51. saddle

53. stable node

55. unstable spiral

57. **(a)** $\lambda_1 = 0$, $\mathbf{v}_1 = \begin{bmatrix} 2 \\ -1 \end{bmatrix}$; $\lambda_2 = 6$, $\mathbf{v}_1 = \begin{bmatrix} 4 \\ 1 \end{bmatrix}$

(b) $\mathbf{x}(t) = c_1 \begin{bmatrix} 2 \\ -1 \end{bmatrix} + c_2 e^{6t} \begin{bmatrix} 4 \\ 1 \end{bmatrix}$

(c)

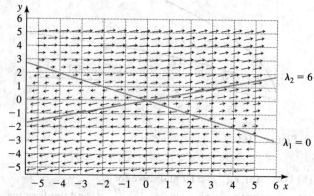

▲ **Figure 11.1.57**

All direction vectors are parallel. The solutions move parallel to the eigenvector corresponding $\lambda_2 = 6$.

Section 11.2

1.
$$\frac{dx_1}{dt} = -0.55x_1 + 0.1x_2$$
$$\frac{dx_2}{dt} = 0.5x_1 - 0.12x_2$$

$(0, 0)$ is a stable node.

3.
$$\frac{dx_1}{dt} = -2.5x_1 + 0.7x_2$$
$$\frac{dx_2}{dt} = 2.5x_1 - 0.8x_2$$

$(0, 0)$ is a stable node.

5.
$$\frac{dx_1}{dt} = -0.1x_1$$
$$\frac{dx_2}{dt} = -0.3x_2$$

$(0, 0)$ is a stable node.

7. $a = 0.1, b = 0.3, c = 0.3, d = 0.2$

9. $a = 0.3, b = 0, c = 0.9, d = 0.2$

11. $a = 0, b = 0, c = 0.2, d = 0.3$

13. $x_1(t) = 4e^{-0.3t}, x_2(t) = 4(1 - e^{-0.3t})$

15. **(a)** $a = 0.2, b = 0.1, c = 0, d = 0$

(b) The constant is the total area.

(d) $x_2(t) = 20 - x_1(t) = \frac{40}{3} + \frac{14}{3} e^{-0.3t}$, $\lim_{t \to \infty} x_1(t) = \frac{20}{3}$,
$\lim_{t \to \infty} x_2(t) = \frac{40}{3}$

17. $x(t) = 3\sin(2t)$

19. $\frac{dx}{dt} = v, \frac{dv}{dt} = 3x$

21. $\frac{dx}{dt} = v, \frac{dv}{dt} = x - v$

Section 11.3

1. saddle

3. saddle

5. $(0, 0)$: unstable node; $(0, 4/5)$: saddle; $(1/2, 0)$: saddle; $(0.5, 0.3)$: stable node

7. $(0, 0)$: unstable node; $(0, 1)$: saddle; $(1, 0)$: unstable node; $(1/2, 1)$: saddle

9. $(0, 0)$: saddle; $(1, 1)$: unstable spiral

11. (a)

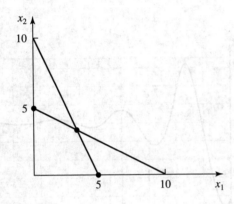

▲ **Figure 11.3.11**

(b) $(10/3, 10/3)$ is a stable node.

13. $\text{tr} < 0, \det =?$

15. $\text{tr} =?, \det < 0$

17. $\text{tr} < 0, \det > 0$; equilibrium is locally stable

19. (a)

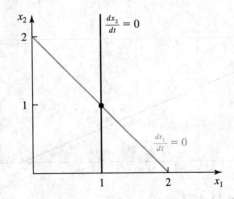

▲ **Figure 11.3.19**

(b) $\text{tr} < 0, \det > 0$: $(1, 1)$ is locally stable

Section 11.4

1.

$$\frac{dN_1}{dt} = 2N_1 \left(1 - \frac{N_1}{20} - \frac{N_2}{100} \right)$$
$$\frac{dN_2}{dt} = 3N_2 \left(1 - \frac{N_2}{15} - \frac{N_1}{75} \right)$$

3. Species 2 outcompetes species 1.

5. Founder control

7. $(0, 0)$: unstable (source); $(18, 0)$: unstable (saddle); $(0, 20)$: stable (sink)

9. $(0, 0)$: unstable (source); $(35, 0)$: stable (sink); $(0, 40)$: stable (sink); $(85/11, 100/11)$: unstable (center)

11. $(\alpha_{12}, \alpha_{21}) = (1/4, 7/18)$

13.

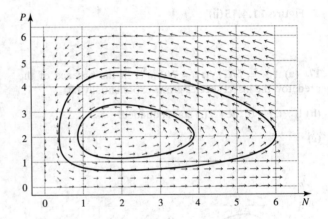

▲ **Figure 11.4.13**

15. (a) trivial equilibrium: $(0, 0)$; nontrivial equilibrium: $(3/2, 1/4)$

(b) $\lambda_1 = 1, \lambda_2 = -3$: $(0, 0)$ is an unstable saddle

(c) $\lambda_1 = i\sqrt{3}, \lambda_2 = -i\sqrt{3}$: purely imaginary eigenvalues, linear stability analysis cannot be used to infer stability of equilibrium.

(d)

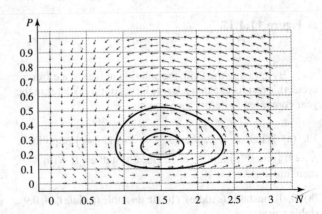

▲ **Figure 11.4.15(i)**

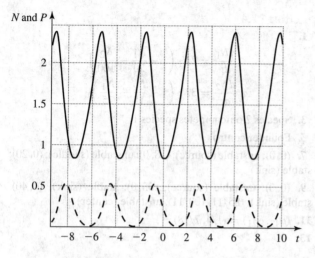

▲ Figure 11.4.15(ii)

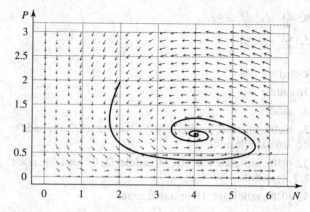

▲ Figure 11.4.19(i)

17. (a) $\frac{dN}{dt} = 5N$, $N(t) = N(0)e^{5t}$; in the absence of the predator, the insect species grows exponentially fast.

(b) If $P(t) > 0$, then $N(t)$ stays bounded.

(c)

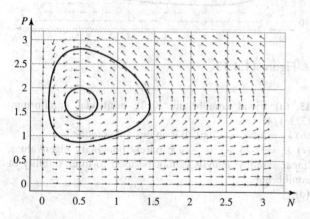

▲ Figure 11.4.17

By spraying the field, the solution moves to a different cycle; this results in a much larger insect outbreak later in the year compared to before the spraying.

19. (a) When $P = 0$, then $\frac{dN}{dt} = 3N\left(1 - \frac{N}{10}\right)$; equilibria: $\hat{N} = 0$ (unstable) and $\hat{N} = 10$ (locally stable). If $N(0) > 0$, then $\lim_{t \to \infty} N(t) = 10$

(b) $(0, 0)$: unstable (saddle); $(10, 0)$: unstable (saddle); $(4, 0.9)$: stable spiral

(c)

▲ Figure 11.4.19(ii)

21. (a)

▲ Figure 11.4.21

(b) $\operatorname{tr}D\mathbf{f}(\hat{N}, \hat{P}) < 0$ and $\det D\mathbf{f}(\hat{N}, \hat{P}) > 0$: the nontrivial equilibrium is locally stable.

23. $\hat{N} = \frac{d}{c}$ does not depend on a; hence, it remains unchanged if a changes. $\hat{P} = \frac{a}{b}(1 - \frac{d/c}{K})$ is an increasing function of a; hence, the predator equilibrium increases when a increases.

25. $\hat{N} = \frac{d}{c}$ is a decreasing function of c; hence, the prey abundance decreases as c increases. $\hat{P} = \frac{a}{b}(1 - \frac{d/c}{K})$ is an increasing function of c; hence, the predator abundance increases as c increases.

27. predation; locally stable

29. mutualism; locally stable

31. mutualism; unstable

33. competition; unstable

35. $\begin{bmatrix} - & - \\ - & - \end{bmatrix}$; trace negative; determinant undetermined

37. $\begin{bmatrix} - & - \\ - & - \end{bmatrix}$; trace negative; determinant undetermined

39. $\begin{bmatrix} - & - \\ - & - \end{bmatrix}$; trace negative; determinant undetermined

41. If $a_{ii} < 0$, then the growth rate of species i is negatively affected by an increase in the density of species i. This is referred to as self regulation.

43. (a) $(\hat{N}, \hat{P}) = (\frac{d}{c}, \frac{a}{b})$

(b) $\begin{bmatrix} 0 & -b\frac{d}{c} \\ c\frac{a}{b} & 0 \end{bmatrix}$

(c) $a_{11} = a_{22} = 0$: neither species has an effect on itself; $a_{12} = -b\frac{d}{c} < 0$: prey is affected negatively by predators; $a_{21} = c\frac{a}{b} > 0$: predators are affected positively by prey.

45.

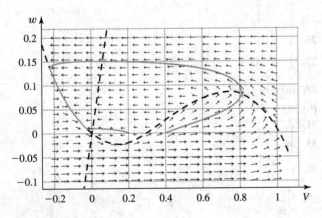

▲ **Figure 11.4.45**

47. $V(0) > 0.3$

49. $\frac{dc}{dt} = kab$

51. $e = [E], s = [S], c = [ES], p = [P], \frac{de}{dt} = -k_1 es + k_2 c,$ $\frac{ds}{dt} = -k_1 es, \frac{dc}{dt} = k_1 es - k_2 c, \frac{dp}{dt} = k_2 c$

53. $\frac{dx}{dt} + \frac{dy}{dt} = 0, x(t) + y(t)$ is constant.

55. $\frac{dx}{dt} + \frac{dy}{dt} + \frac{dz}{dt} = 0, x(t) + y(t) + z(t)$ is constant.

57. (c)

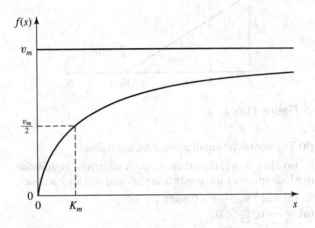

▲ **Figure 11.4.57**

(d) Since $\frac{dp}{dt} = f(s)$, the reaction rate is a function of s and hence, the availability of s determines the reaction rate.

59. (a) $\hat{s} = \frac{DK_m}{Yv_m - D}$; $\hat{s}$ is an increasing function of D.

(b)

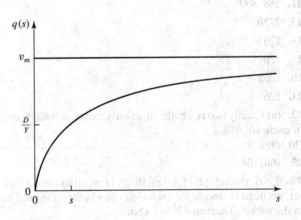

▲ **Figure 11.4.59**

The s coordinate of the point of intersection of the graph of $q(s)$ and the horizontal line $f(s) = D/Y$ is the equilibrium.

61. $(4, 0)$: unstable; $(2, 2)$: stable

Section 11.6

1. $Z(t) = Z(0)e^{(r_1 - r_2)t}$

3. (a)

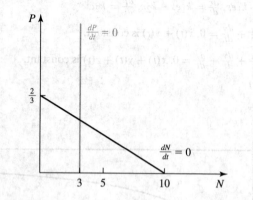

▲ **Figure 11.6.3**

(b) The nontrivial equilibrium is locally stable.

5. (a) If $c_1 > m_1$, then there exists a nontrivial equilibrium in which species 1 has positive density and species 2 is absent.

7. (a) $\hat{x} = Y(s_0 - \hat{s}) > 0$ when $\hat{s} < s_0$

(b) $\frac{\partial \hat{s}}{\partial D} > 0$; $\frac{\partial \hat{s}}{\partial Y} < 0$

Section 12.1

1. 40

3. 120

5. 84

7. 120

9. 5040

11. 358, 800

13. 2730

15. 120

17. 120

19. 1365

21. 126

23. (a) exactly two red balls: 10; exactly two blue balls: 6; one of each: 20

(b) total: 36

25. 168, 168

27. $\emptyset, \{a\}, \{b\}, \{c\}, \{a, b\}, \{a, c\}, \{b, c\}, \{a, b, c\}$; to form a subset, we need to decide for each element whether it should be in the subset. There are $2^3 = 8$ choices.

29. 12

31. 30

33. 31

35. $x^4 + 4x^3 y + 6x^2 y^2 + 4xy^3 + y^4$

37. $\binom{26}{4}\binom{26}{5}$

39. $4\binom{13}{2}\binom{4}{2}\binom{4}{2}\binom{11}{1}$

41. $\binom{13}{1}\binom{4}{4}\binom{12}{1}\binom{4}{1}$

43. 3!

Section 12.2

1. $\Omega = \{HHH, HHT, HTH, THH, HTT, THT, TTH, TTT\}$

3. $\Omega = \{(i, j) : 1 \le i < j \le 5\}$

5. $A \cup B = \{1, 2, 3, 5\}$, $A \cap B = \{1, 3\}$

7. $\{4, 6\}$

9. 0.6

11. 0.25

13. 0.5

15. (a)

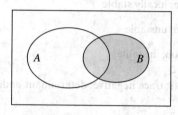

▲ **Figure 12.2.15(a)**

(b)

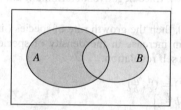

▲ **Figure 12.2.15(b)**

17. $\frac{3}{4}$

19. $\frac{5}{36}$

21. $\frac{5}{12}$

23. $\frac{1}{4}$

25. $\frac{1}{2}$

27. $\frac{1}{8}$

29. (a) $\frac{\binom{N-100}{7}\binom{100}{3}}{\binom{N}{10}}$

(b) 333

31. $\frac{1}{17}$

33. $\frac{12}{55}$

35. $1 - \frac{\binom{48}{4}}{\binom{52}{4}}$

37. $\frac{\binom{26}{13}}{\binom{52}{13}}$

39. $\frac{\binom{13}{2}\binom{4}{2}\binom{4}{2}\binom{11}{1}\binom{4}{1}}{\binom{52}{5}}$

Section 12.3

1. $\frac{12}{51}$

3. $\frac{3}{5}$

5. $\frac{1}{2}$

7. $\frac{1}{6}$

9. $\frac{4}{7}$

11. 0.8425

13. $\frac{1}{2}$

15. $P(\text{first card is an ace}) = P(\text{second card is an ace}) = \frac{1}{13}$

17. 0.3

19. $\frac{3}{4}$

21. A and B are independent.

23. A and B are not independent.

25. (a) $\frac{1}{8}$ (b) $\frac{7}{8}$ (c) $\frac{1}{2}$ (d) $\frac{7}{8}$

27. $\left(\frac{1}{4}\right)^{10}$

29. $1 - (0.9)^{10}$

31. 0.1624

33. $\frac{1}{3}$

35. $\frac{1}{3}$

37. (a) $\frac{1}{2}$ (b) $\frac{1}{2}$ (c) $\frac{1}{3}$

Section 12.4

1. $P(X = 0) = \frac{1}{4}, P(X = 1) = \frac{1}{2}, P(X = 2) = \frac{1}{4}$

3. $P(X = 0) = \frac{\binom{3}{0}\binom{2}{2}}{\binom{5}{2}}, P(X = 1) = \frac{\binom{3}{1}\binom{2}{1}}{\binom{5}{2}},$
$P(X = 2) = \frac{\binom{3}{2}\binom{2}{0}}{\binom{5}{2}}$

5.
$$F(x) = \begin{cases} 0, & x < -3 \\ 0.2, & -3 \le x < -1 \\ 0.5, & -1 \le x < 1.5 \\ 0.9, & 1.5 \le x < 2 \\ 1, & x \ge 2 \end{cases}$$

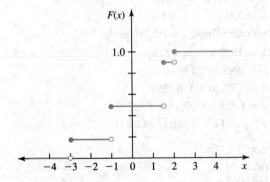

▲ **Figure 12.4.5**

7. $P(X = -1) = 0.2, P(X = 0) = 0.1, P(X = 1) = 0.4,$
$P(X = 2) = 0.3$

9. (a) $N = 55$ (b) $\frac{28}{55}$

11. (a)

k	12	13	14	15	16	17	18	19	20	21
p_k	1/25	4/25	5/25	3/25	1/25	5/25	2/25	1/25	2/25	1/25

(b) average value: $\frac{396}{25} = 15.84$

13. (a)

k	2	3	4	5	6	7	8	9	10
p_k	1/25	1/25	2/25	3/25	4/25	2/25	6/25	3/25	3/25

(b) average value $\frac{171}{25} = 6.84$

15. (a) -0.4 (b) 1.0 (c) 1.4

17. $EX = -0.1, \text{var}(X) = 3.39, \text{s.d.} = \sqrt{3.39}$

19. (a) $EX = \frac{55}{10}$ (b) $\text{var}(X) = 8.25$

23. (a) 0.1 (b) 0.5 (c) 0.4 (d) 0.2

25. (a) $EX = 0.75, EY = 0.3$ (b) $E(X + Y) = 1.05$
(c) $\text{var}(X) = 1.7875, \text{var}(Y) = 3.01$
(d) $\text{var}(X + Y) = \text{var}(X) + \text{var}(Y) = 4.7975$

27. (a) Since $(X - EX)^2 \ge 0$, it follows that $E(X - EX)^2 \ge 0$;
therefore, $\text{var}(X) \ge 0$
(b) Since $\text{var}(X) = EX^2 - (EX)^2 \ge 0$, it follows that
$EX^2 \ge (EX)^2$

29. (a) $\binom{10}{5}(0.5)^{10}$ (b) $(0.5)^{10}\left[\binom{10}{8} + \binom{10}{9} + \binom{10}{10}\right]$
(c) $1 - (0.5)^{10}$

31. $P(X = k) = \binom{6}{k}\left(\frac{1}{6}\right)^k\left(\frac{5}{6}\right)^{6-k}, k = 0, 1, 2, \ldots, 6$

33. $(0.8)^{20}$

35. (a) 1 (b) $(10)(0.9)^{10}$

37. 12.5

39. (a) $\frac{\binom{24}{6}\binom{12}{4}}{\binom{36}{10}}$ (b) $\binom{10}{6}\left(\frac{2}{3}\right)^6\left(\frac{1}{3}\right)^4$

41. $\frac{30!}{10!14!6!}(0.2)^{10}(0.35)^{14}(0.45)^6$

43. $\frac{40!}{20!10!8!2!}\left(\frac{9}{16}\right)^{20}\left(\frac{3}{16}\right)^{10}\left(\frac{3}{16}\right)^8\left(\frac{1}{16}\right)^2$

45. $\frac{6!}{2!2!2!}\left(\frac{6}{24}\right)^2\left(\frac{8}{24}\right)^2\left(\frac{10}{24}\right)^2$

47. $\frac{23!}{5!12!6!}\left(\frac{1}{4}\right)^5\left(\frac{1}{2}\right)^{12}\left(\frac{1}{4}\right)^6$

49. $1/2, 1/4, 1/8$

51. $1/8$

53. $15/16$

55. $\left(\frac{14}{15}\right)^{19}$

57. $ET = 6, \text{var}(T) = 30$

59. (a) $\left(\frac{9}{10}\right)^5 \frac{1}{10}$ (b) $\frac{9}{10}\frac{8}{9}\frac{7}{8}\frac{6}{7}\frac{5}{6}\frac{1}{5}$

61. (a) $(1 - p)^{k-1}p$ (b) $\binom{k-1}{1}p^2(1 - p)^{k-2}$

63.

k	0	1	2	3
$P(X = k)$	e^{-2}	$2e^{-2}$	$2e^{-2}$	$\frac{4}{3}e^{-2}$

65. (a) $1 - 2e^{-1}$ (b) $e^{-1}\left(1 + \frac{1}{2} + \frac{1}{6}\right)$

67. $1 - e^{-1.5}\left[1 + 1.5 + \frac{(1.5)^2}{2} + \frac{(1.5)^3}{6}\right]$

69. $1 - 3e^{-2}$

71. e^{-7}

73. $1 - e^{-0.5}$

75. $1 - 4e^{-3}$

77. (a) e^{-4} **(b)** $13e^{-4}$ **(c)** $32e^{-8}$ **(d)** $\frac{1}{4}$

79. (a) $18e^{-6}$ **(b)** $1/4, 1/2, 1/4$

81. $P(X = 0) \approx e^{-0.5}$

83. (a) $1 - \left(\frac{699}{700}\right)^{1000}$ **(b)** $1 - e^{-1000/700}$

85. $e^{-1.5}$

Section 12.5

1. distribution function: $F(x) = 1 - e^{-3x}$ for $x \geq 0$, $F(x) = 0$ for $x \leq 0$

3. $c = \frac{1}{\pi}$

5. $EX = \frac{1}{2}$, $\text{var}(X) = \frac{1}{4}$

7. $EX = \frac{3}{2}$, $\text{var}(X) = \frac{3}{4}$

9. (b) $EX = \frac{a-1}{a-2}$

11. (d)

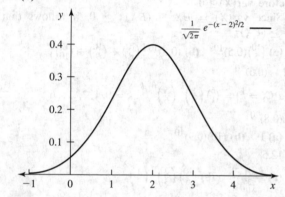

▲ **Figure 12.5.11**

13. 95%: $(7.4, 18.2)$, 99%: $(4.7, 20.9)$

15. 50%

17. 99.5%

19. 2.5%

21. (a) 0.6915 **(b)** 0.383 **(c)** 0.1587 **(d)** 0.0668

23. (a) $x = 3.56$ **(b)** $x = 1.5$ **(c)** $x = 0.5$ **(d)** $x = 1.34$

25. (a) 0.0228 **(b)** $x = 628$

27. 0.76%

29. 0.8185

33. $E|X| = 2$

35. (a) 0.0735 **(b)** 0.6068 **(c)** 88 **(d)** 56

37. 0.1

39. $a = 1, b = 7$

41. $EY = \frac{3}{4}$

43. $HHTHHTHHHT$

45. (a) $(1-x)^n$ **(b)** Use L'Hospital's rule on $\lim_{n \to \infty} \ln(1 - \frac{x}{n})^n$

47. $EX = \frac{1}{\lambda}$

49. $e^{-4/3}$

51. (a) $e^{-20/27}$ **(b)** $e^{-20/27}$

53. (a) $1 - e^{-3/5}$ **(b)** $e^{-1/5}$

55. (a) 5 years **(b)** $\ln 2/0.2$ years

57. (a) $\exp\left[-(1.5 + 10e^{0.05} - 10)\right]$

(b) $\exp\left[-(2.1 + 10e^{0.07} - 10)\right] - \exp\left[-(3 + 10e^{0.1} - 10)\right]$

59. Solution of $1.2x + (0.6)e^{0.5x} - 0.6 - \ln 2 = 0$ is approximately 0.451.

61. (a) $\exp\left[-(2 \times 10^{-5})\frac{(50)^{2.5}}{2.5}\right]$ **(b)** 0.1477

63. $x_m \approx 30.4$

Section 12.6

1. Exact probability: $e^{-3/2}$; Markov's inequality: $P(X \geq 3) \leq \frac{2}{3}$

5. Exact: $P(|X| \geq 1) = 0$; Chebyshev's inequality: $P(|X| \geq 1) \leq \frac{1}{3}$

7. $\frac{9}{25}$

9. $\frac{1}{n}\sum_{i=1}^{n} X_i$ converges to $3/2$ as $n \to \infty$

11. Since $E|X_i| = \infty$, we cannot apply the law of large numbers as stated in Section 12.6.

13. The sample size should be at least 380.

15. 0.1587

17. (a) 0.0023 **(b)** 0.83

19. (a) 0.1192 **(b)** 0.182 **(c)** 0.1521

21. (a) -11.2 **(b)** 0.579

23. 69

25. 385

27. (a) 0.3660 **(b)** 0.3679 **(c)** 0.0582

29. (a) 0.1849 **(b)** 0.1755 **(c)** 0.1896

31. Likely not.

33. (a) 0.6065, 0.3033, 0.0758 **(b)** 0.8461

35. 0.1429

37. 0.0485

Section 12.7

1. median: 15; sample mean: 16.2; sample variance: 180.2

3. $\overline{X} = 11.93$; $S^2 = 3.389$

5. $\overline{X} = 5.69$; $S^2 = 3.465$

11. (c) true values: $\mu = 0.5$, $\sigma^2 = \frac{1}{12}$

13. $\overline{X} = 16.2$; S.E. $= 4.245$

15. $[-0.3993, 0.5213]$

19. $\hat{p} = 0.72$; $[0.651, 0.789]$

21. $y = 1.92x - 0.92$; $r^2 = 0.9521$

25. $y = 0.201x + 0.481$; $r^2 = 0.841$

Section 12.9

1. (a) 0.431

3. 168, 168, 000

5. (a) $EX = 4.2$ **(b)** 0.0043 **(c)** 0.0215 **(d)** $1 - (0.0043)^5$

7. (a) 0.16 **(b)** 0.2

9. (a) $\mu = 170$, $\sigma = 6.098$ **(b)** 0.4013

11. 20%

13. (b) $EV = \frac{n-1}{n}\sigma^2$

REFERENCES

W.C. Allee (1931) *Animal Aggregations. A Study in General Sociology*. University of Chicago Press, Chicago.

H.G. Andrewartha and L.C. Birch (1954) *The Distribution and Abundance of Animals*. University of Chicago Press, Chicago.

M. Begon and G.A. Parker (1986) Should egg size and clutch size decrease with age? *Oikos* **47**: 293–302.

T.S. Jr. Bellows (1981) The descriptive properties of some models for density dependence. *Journal of Animal Ecology* **50**: 139–156.

M.J. Benton (1997) Models for the Diversification of Life. *TREE* **12**: 490–495.

M. Benton and D. Harper (1997) *Basic Paleontology*. Addison Wesley and Longman.

P.J. Bohlen, P.M. Groffman, C.T. Driscoll, T.J. Fahey, and T.G. Siccama (2001) Plant-soil-microbial interactions in a northern hardwood forest. *Ecology* **82**: 965–78.

R. Borchert (1994) Soil and stem water storage determine phenology and distribution of tropical dry forest trees. *Ecology* **75**: 1437–49.

T. Boulinier, J.D. Nichols, and J.E. Hines (2001) Forest fragmentation and bird community dynamics: inference at regional scales. *Ecology* **82**: 1159–69.

G.E. Briggs and J.B.S. Haldane (1925) A note on the kinematics of enzyme actions. *Biochemical Journal* **19**: 338–339.

J.L. Brooks and S.I. Dodson (1965) Predation, body-size and composition of plankton. *Science* **150**: 28–35.

M.J.W. Burke and J.P. Grime (1996) An experimental study of plant community invasibility. *Ecology* **77**: 776–790.

CRC Standard Probability and Statistics Tables and Formulae (1991). Editor, W.H. Beyer. CRC Press.

M.L. Cain, S.W. Pacala, J.A. Silander, Jr., and M.-J. Fortin (1995) Neighborhood models of clonal growth in the white clover *Trifolium repens*. *The American Naturalist* **145**: 888–917.

G.S. Campbell (1986) *An Introduction to Environmental Biophysics*. Springer, New York.

M.J. Crawley (1997) Plant-Herbivore Dynamics. In *Plant Ecology*, ed. M.J. Crawley, pp.401–474. Blackwell Science.

J.W. Dalling, K. Winter, and J.D. Nason (2001) The unusual life history of *Alseis blackiana*: a shade-persistent pioneer tree? *Ecology* **82**: 933–45.

C. Darwin (1859) *On the Origin of Species by Means of Natural Selection*. John Murray, London.

C. Darwin (1959) *The Origin of Species*. Reprinted in Penguin Classics 1985. Penguin Books, Ltd., England.

D.L. DeAngelis (1992) *Dynamics of Nutrient Cycling and Food Webs*. Chapman and Hall.

W.R. DeMott, R.D. Gulati, and K. Siewertsen (1998) Effects of phosphorus-deficient diets on the carbon and phosphorus balance of *Daphnia magna*. *Limnol. Oceanogr.* **43**: 1147–1161.

A.M. de Roos (1996) A gentle introduction to physiologically structured population models. In *Structured-Population Models in Marine, Terrestrial, and Freshwater Systems*, eds. S. Tuljapurkar and H. Caswell. Chapman & Hall.

R.H. Dott, Jr., and R.L. Batten (1976) *Evolution of the Earth*. Second edition. McGraw-Hill Book Company.

J.E. Dowd and D.S. Riggs (1965) A comparison of estimates of Michaelis-Menten kinetic constants from various linear transformations. *The Journal of Biological Chemistry* **240**: 863–869.

C.M. Duarte and S. Agustí (1998) The CO_2 balance of unproductive aquatic ecosystems. *Science* **281**: 234–236.

R. Fitzhugh (1961) Impulses and physiological states in theoretical models of nerve membrane. *Biophysics Journal* **1**: 445–466.

D.S. Falconer (1989) *Introduction to Quantitative Genetics*. Logmans, New York.

D.S. Falconer and T.F.C. Mackay (1996) *Introduction to Quantitative Genetics*. Fourth Edition, Longman.

D.J. Futuyama (1995) *Science on Trial*. Second edition. Sinauer.

P. Gaastra (1959) Mededelinger van de Landbouwhogeschool Te Wageningen, The Netherlands. In F.B. Salisbury and C.W. Ross (1978) *Plant Physiology*. Wadsworth.

L.A. Gavrilov and N.S. Gavrilova (2001) The reliability theory of aging and longevity. *Journal of Theoretical Biology* **213**: 527–545.

G.F. Gause (1934) *The Struggle for Existence*. Williams & Wilkins, Baltimore (reprinted 1964 by Hafner, New York).

Z.M. Gliwicz (1990) Food treshold and body size in caldocerans. *Nature* **343**: 683–40.

P.R. Grant (1982) Variation in the size and shape of Darwin's finch eggs. *Auk* **99**: 15–23.

P.R. Grant, I. Abbott, D. Schluter, R.L. Curry, and L.K. Abbott (1985) Variation in the size and shape of Darwin's finches. *Biological Journal of the Linnean Society* **25**: 1–39.

K.J. Griffiths (1969) The importance of coincidence in the functional and numerical responses of two parasites of the European pine sawfly, *Neodiprion sertifer*. *Canadian Entomologist*, **101**: 673-713.

J.B.S. Haldane (1957) The cost of natural selection. *J. Genet.* **55**: 511–524.

D.J. Hall (1964) An experimental approach to the dynamics of a natural population of *Daphnia galeata mendota*. *Ecology* **45**: 94–111.

D.L. Hartl and A.G. Clark (1989) *Principles of Population Genetics*. Second Edition. Sinauer.

R.W. Herschy (1995) *Streamflow Measurements*. 2nd edition. E&FN Spon.

L. Holdridge, W. Grenke, W. Hatheway, T. Liang, and J. Tosi, Jr. (1971) *Forest environments in tropical life zones: A pilot study*. Oxford: Pergamon Press.

C.S. Holling (1959) Some characteristics of simple types of predation and parasitism. *Can. Ent.* **91**: 385–398.

H.S. Horn (1971) *The adaptive geometry of trees*. Monographs in Population Biology 3. Princeton University Press. Princeton, NJ.

C.E. Iselin, J.E. Robertson, and D.F. Paulson (1999) Radical perineal prostatectomy: oncological outcome during a 20-year period. *The Journal of Urology* **161**: 163–168.

Y. Iwasa, T.J. de Jong and P.G.L. Klinkhamer (1995). Why pollinators visit only a fraction of the open flowers on a plant and its consequence for fitness curves of plants. *Journal of Evolutionary Biology* **8**: 439–453.

T.H. Jukes and C.R. Cantor (1969) Evolution of protein molecules. pp. 21-132. *In* H.N. Munro (ed.), *Mammalian Protein Metabolism III*. Academic Press, New York.

P.M. Kareiva (1983) Local movement in herbivorous insects: applying a passive diffusion model to mark-recapture field experiments. *Oecologia* **57**: 322–327.

W.O. Kermack and A.G. McKendrick: Contributions to the mathematical theory of epidemics. *Proc. Roy. Soc.* **A 115**: 700–721 (1927); **138**: 55–83 (1932); **141**: 94–122 (1933).

K.B. Krauskopf and D.K. Bird (1995) *Introduction to Geochemistry*. Page 268. McGraw-Hill, Inc., New York.

H.E. Landsberg (1969) *Weather and Health, an Introduction to Biometerology*. Garden City, New Jersey. Doubleday.

A.C. Leopold and P.E. Kriedemann (1975) *Plant growth and development*. McGraw Hill, New York.

P. Leslie (1945) On the use of matrices in certain population mathematics. *Biometrika* **33**: 183–212.

R. Levins (1969). Some demographic and genetic consequences of environmental heterogeneity for biological control. *Bull. Entomol. Soc. Am.* **15**: 237–240.

R. Levins (1970) Community equilibria and stability, and an extension of the competitive exclusion principle. *Am. Nat.* **104**: 413–423.

D.C. Lloyd (1987) Selection of offspring size at independence and other size-versus-number strategies. *American Naturalist* **129**: 800–817.

A.J. Lotka (1932) The growth of mixed populations: two species competing for a common food supply. *Journal of the Washington Academy of Sciences* **22**: 461–469.

R.H. MacArthur and E.O. Wilson (1963). An equilibrium theory of insular zoogeography. *Evolution* **17**: 373–387.

T.F.C. Mackay (1984) Jumping genes meet abdominal bristles: Hybrid dysgenesis-induced quantitative variation in *Drosophila melanogaster*. *Genet. Res.* **44**: 231–237.

R.M. May (1975) *Stability and Complexity in Model Ecosystems* Princeton, Princeton University Press.

R.M. May (1976) Simple mathematical models with very complicated dynamics. *Nature* **261**: 459-461.

R.M. May (1978) Host-parasitoid systems in patchy environments: a phenomenological model. *Journal of Animal Ecology*, **47**: 833–843.

L. Michaelis and M.I. Menten (1913) Die Kinetik der Invertinwirkung. *Biochem. Z.* **49**: 333–369

J.L. Monod (1942) *Recherches sur la Croissance des Cultures Bacteriennes.* Hermann, Paris.

J.L. Monod (1950) La technique de culture continué: theorie et applications. *Ann. Instit. Pasteur* **79**: 390–410.

B. Moss (1980) *Ecology of Freshwaters.* Blackwell Scientific Publication.

J.S Nagumo, S. Arimoto and S. Yoshizawa (1962) An active pulse transmission line simulating nerve axon. *Proc. IRE.* **50**: 2061–2071.

S. Nee and R.M. May (1992) Dynamics of metapopulation: habitat destruction and competitive coexistence. *Journal of Animal Ecology* **61**: 37–40.

A.J. Nicholson (1933) The balance of animal populations. *Journal of Animal Ecology* **2**: 131–178

A.J. Nicholson and V.A. Bailey (1935) The balance of animal populations. *Proceedings of the Zoological Society of London* **3**: 551–598.

K.J. Niklas (1994) *Plant Allometry. The Scaling of Form and Process.* The University of Chicago Press. Chicago.

R.T. Oglesby (1977) Relationships of fish yield to lake phytoplankton, standing crop, production, and morphoedaphic factors. *J. Fish. Res. Bd. Can.* **34**: 2271–2279.

J.G. Owen (1988) On productivity as a predictor of rodent and carnivore density. *Ecology* **69**: 1161–65.

S.W. Pacala and M. Rees (1998) Models suggesting field experiments to test two hypotheses explaining successional diversity. *American Naturalist* **152**: 729–737.

R. Pearl and L.J. Reed (1920) On the rate of growth of the population of the United States since 1790 and its mathematical representation. *Proceedings of the National Academy of Sciences* **6**: 275–288.

G.W. Pierce (1949) *The Songs of Insects.* Cambridge, Mass. Harvard University Press.

A. Pisek, W. Larcher, W. Moser and I. Pack (1969) Kardinale Temperaturbereiche der Photosynthese und Grenztemperaturen des Lebens der Blätter verschiedener Spermatophyten. III. Temperaturabhängigkeit und optimaler Temperaturbereich der Netto-Photosynthese. *Flora, Jena* **158**: 608–630.

S.D. Pletcher (1998) Mutation and the evolution of age-specific mortality rates : experimental results and statistical developments. Thesis (Ph. D.)–University of Minnesota.

F.W. Preston (1962) The canonical distribution of commonness and rarity. *Ecology* **43**: 185–215, 410–431.

M.J. Reiss (1989) *The Allometry of Growth and Reproduction.* Cambridge University Press.

D.A. Roff (1992) *The Evolution of Life Histories.* Chapman & Hall.

M.L. Rosenzweig (1971) Paradox of Enrichment: Destabilization of Exploitation Ecosystems in Ecological Time. *Science* **171**: 385–387.

M.L. Rosenzweig and Z. Abramsky (1993) How are diversity and productivity related? pp. 52-65. *In* R.E. Ricklefs and D. Schluter (eds.) *Species Diversity in Ecological Communities.* University of Chicago Press, Chicago.

J. Roughgarden (1996) *Theory of Population Genetics and Evolutionary Ecology.* 2nd edition. Prentice Hall.

S.M. Stanley (1979) Macroevolution: Pattern and Process. San Francisco: W.H. Freeman & Co, 332.

Statistical Abstract of the United States 1994, U.S. Bureau of the Census, 14th edition. Washington, DC, 1994.

R.W. Sterner (1997) Modeling interactions between food quality and quantity in homeostatic consumers. *Freshwater Biology* **38**: 473–482.

R.W. Sterner and J.J. Elser (2002) *Ecological Stoichiometry: The Biology of Elements from Molecules to the Biosphere*. Princeton University Press. Princeton, NJ.

C.R. Taylor, G.M.O. Maloiy, E.R. Weibel, U.A. Langman, J.M.Z. Kaman, H.J. Seeherman, and B N.C. Heglund (1980). Design of the mammalian respiratory system. III. Scaling maximum aerobic capacity to body mass: wild and domestic mammals. *Respir. Physiol* **44**: 25–37.

W.R. Thompson (1924) La théorie Mathématique de l'action des parasites entomophages et le facteur dy hasard. *Annales de la Faculté des Sciences de Marseille* **2**: 69–89.

D. Tilman (1982) *Resource Competition and Community Structure*. Princeton University Press, Princeton, NJ.

D. Tilman (1994) Competition and biodiversity in spatially structured habitats. *Ecology* **75**: 2–16.

M.M. Tilzer, W. Geller, U. Sommer, and H.H. Stable (1982) Kohlenstoffkreislauf und Nahrungsketten in der Freiwasserzone des Bodensees. *Konstanzer Blätter für Hochschulfragen* **73**: 51.

J.W. Valentine (1985) Biotic diversity and clade diversity. In *Phanerozoic Diversity Patterns: Profiles in Macroevolution*, Valentine, J.W., ed., pp. 419–424. Princeton University.

P.M. Vitousek and H. Farrington (1997) Nutrient limitation and soil development: Experimental test of a biogeochemical theory. *Biogeochemistry* **37**: 63–75.

V. Volterra (1926) Fluctuations in the abundance of species considered mathematically. *Nature* **118**: 558–560.

T.D. Walker (1985) Diversification functions and the rate of taxonomic evolution. In *Phanerozoic Diversity Patterns: Profiles in Macroevolution*, Valentine, J.W., ed., pp. 311–334. Princeton University Press.

T.D. Walker and J.W. Valentine (1984) Equilibrium models of evolutionary species diversity and the number of empty niches. *American Naturalist* **124**: 887–899.

P.D. Ward (1992) *On Methuselah's Trail-Living Fossils and the Great Extinctions* by P.D. Ward, Freeman.

S. Wright (1968) *Evolution of Populations, Vol 1. Genetics and Biometrik Foundation*. University of Chicago Press, Chicago.

E.K. Yeargers, R.W. Shonkwiler and J.V. Herod (1996) *An Introduction to the Mathematics of Biology: with Computer Algebra Models*. Birkhäuser, Boston.

INDEX

Algebra

Quadratic Formula

The solutions of the quadratic equation $ax^2 + bx + c = 0$ are given by

$$x = \frac{-b \pm \sqrt{b^2 - 4ac}}{2a}.$$

Factorial notation

For each positive integer n,

$$n! = n(n-1)(n-2)\cdots 3 \cdot 2 \cdot 1\,;$$

by definition, $0! = 1$.

Radicals

$$\sqrt[n]{x^m} = \left(\sqrt[n]{x}\right)^m = x^{m/n}$$

Exponents

$$(ab)^r = a^r b^r \qquad a^r a^s = a^{r+s} \qquad x^{-n} = \frac{1}{x^n}$$

$$(a^r)^s = a^{rs} \qquad \frac{a^r}{a^s} = a^{r-s}$$

Binomial Formula

$$(x + y)^2 = x^2 + 2xy + y^2$$
$$(x + y)^3 = x^3 + 3x^2 y + 3xy^2 + y^3$$
$$(x + y)^4 = x^4 + 4x^3 y + 6x^2 y^2 + 4xy^3 + y^4$$

In general, $(x + y)^n = x^n + \binom{n}{1}x^{n-1}y + \binom{n}{2}x^{n-2}y^2$

$$+ \cdots + \binom{n}{k}x^{n-k}y^k$$

$$+ \cdots + \binom{n}{n-1}xy^{n-1} + y^n,$$

where the binomial coefficient $\binom{n}{m}$ is the integer $\frac{n!}{m!(n-m)!}$

Special Factors

$$x^2 - a^2 = (x - a)(x + a)$$
$$x^3 + a^3 = (x + a)(x^2 - ax + a^2)$$
$$x^3 - a^3 = (x - a)(x^2 + ax + a^2)$$
$$x^4 - a^4 = (x^2 - a^2)(x^2 + a^2)$$

Geometry

Distance Formulas

Distance on the real number line:

$$d = |a - b|$$

Distance in the coordinate plane:

$$d = \sqrt{(x_1 - x_2)^2 + (y_1 - y_2)^2}$$

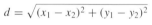

Equations of Lines and Circles

Slope-intercept equation:
$$y = mx + b$$

Point-slope equation:
$$y - y_1 = m(x - x_1)$$

Circle with center (h, k) and radius r:

$$(x - h)^2 + (y - k)^2 = r^2$$

Triangle area:
$$A = \frac{1}{2}bh$$

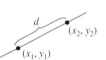

Trapezoid area:
$$A = \frac{b_1 + b_2}{2}h$$

Rectangle area:
$$A = bh$$

Circle area:
$$A = \pi r^2$$

Circumference:
$$C = 2\pi r$$

Sphere volume:
$$V = \frac{4}{3}\pi r^3$$
Surface area:
$$A = 4\pi r^2$$

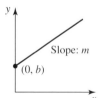

Sector of Circle
(θ in radians)

$$\text{Area} = \frac{\theta r^2}{2}$$
$$s = r\theta$$

Cone volume:
$$V = \frac{1}{3}\pi r^2 h$$
Curved surface area:
$$A = \pi r \sqrt{r^2 + h^2}$$

Cylinder volume:
$$V = \pi r^2 h$$
Curved surface area:
$$A = 2\pi rh$$

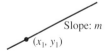

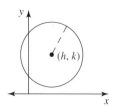